B.M. Johri · K.B. Ambegaokar · P.S. Srivastava

Comparative Embryology of Angiosperms Vol. 1

With 362 Figures
in two Volumes

Springer-Verlag Berlin Heidelberg GmbH

Prof. Dr. Brij M. Johri
Dr. Kunda B. Ambegaokar
Department of Botany
University of Delhi
Delhi 110007
India

Dr. Prem S. Srivastava
Department of Botany
Hamdard University
Hamdard Nagar
New Delhi 110062
India

ISBN 978-3-642-76397-7 ISBN 978-3-642-76395-3 (eBook)
DOI 10.1007/978-3-642-76395-3

Originally published by Springer-Verlag Berlin Heidelberg in 1992
Softcover reprint of the hardcover 1st edition 1992

Dedicated to

Professor Karl Schnarf
(1879–1947)

Professor Panchanan Maheshwari
(1904–1966)

Preface

The concept for the present book dates back to the early 1960s when Professor P. Maheshwari FRS (deceased May 1966) invited scientists to contribute to a multiauthored book – *Comparative Embryology of Angiosperms*. Soon thereafter, Professor Gwenda L. Davis (University of New England, Armidale, Australia) joined the University of Delhi as a Visiting Professor and announced that her book, *Systematic Embryology of Angiosperms*, would soon be published (it appeared in 1966). Therefore, Professor Maheshwari postponed his project for the time being.

The only book on embryology available until then (in english) was by Prof. Maheshwari, published in 1950 (McGraw-Hill, New York) although G. L. Davis' book (without illustrations) was also useful, but to a limited extent. The rapid increase in knowledge in the years that followed, with the advent of new techniques such as ultramicroscopy (TEM and SEM), histochemistry, and the study of organic and inorganic materials, provided impetus to the study of embryology, transgressing its traditional confinements to 'compound' microscopy. This resulted in the emergence of new ideas and methodologies, which helped in a better interpretation of data previously restricted to only the description of developmental aspects.

As new information on embryology accumulated, the need for a comprehensive, illustrated book on comparative embryology that could serve researchers as a reference work became imperative.

In 1979, when Dr. Klaus Peters of Springer-Verlag visited BMJ with an invitation to prepare a treatise on *Embryology of Angiosperms*, the book, *Experimental Embryology of Vascular Plants* (ed. B.M. Johri), was already in its final stage and was published by Springer-Verlag in 1982. The multiauthored book *Embryology of Angiosperms* (ed. B.M. Johri) was published by Springer-Verlag in 1984, and the outline of the book, *Comparative Embryology of Angiosperms*, was prepared simultaneously and given to Klaus Peters. Although a very large collection of relevant literature was available to us at the Department of Botany, University of Delhi, it took almost 10 years to survey the exhaustive literature and to complete the manuscript on Comparative Embryology.

This book endeavours to consolidate the scattered information on angiosperm embryology, to present a coordinated and coherent interpretation of significant data accumulated thus far, and to elucidate the interrelationship of taxa based on comparative embryological findings.

Comparative Embryology of Angiosperms is a review of the developmental processes leading to sexual reproduction in flowering plants, and provides illustrated source material on sexual reproduction, primitive and advanced embryological characters, and phylogenetic relationships. On the basis of embryological data and certain evidence from other areas of study, special emphasis has been placed on the relationship among and within the families and orders of angiosperms. For our work Engler's classification system for angiosperms proposed in *Syllabus der Pflanzenfamilien*, revised by Melchior (1964), has been adopted.

This treatise summarizes nearly 5000 publications (up to 1991) on angiosperm embryology; its resulting ultimate size prompted us to divide it into two volumes. For the convenience of the reader, the preface, acknowledgements and the table of contents are presented in both volumes. For economic reasons, the references for the entire work and the plant index are included only in Volume 2.

Each order begins with general features, followed by the embryology of its constituent families, culminating in critical taxonomic considerations on the basis of comparative embryologic data, and the placement of each family in six chosen systems of classification is discussed.

A comprehensive bibliography is given at the end of each family and includes the citations referred to in the text. Additional citations (not mentioned in the text) are provided for those who wish to obtain further information.

When appropriate, inaccuracies in observation and interpretation are pointed out, alternative interpretations offered, gaps in our knowledge highlighted, and prospects outlined.

Comparative Embryology of Angiosperms provides useful information to students and research workers involved in angiosperm embryology, taxonomists, plant breeders, agriculturists and horticulturists, and plant scientists interested in the evolutionary tendencies in flowering plants.

University of Delhi, July 1992

B.M. Johri
K.B. Ambegaokar
P.S. Srivastava

Acknowledgements

The preparation of *Comparative Embryology of Angiosperms* took almost a decade. We could not have accomplished this undertaking without the academic support and encouragement from our colleagues at the University of Delhi, and plant scientists in India and abroad. Over the last 55 years, B.M. Johri has had the privilege of discussions with embryologists on the reproductive biology of angiosperms, especially at national and international conferences. He is grateful to all of them.

We are indebted to fellow embryologists who generously provided relevant literature and advance information about ongoing research projects. To name a few: Prof. Martin Bopp (Heidelberg), Dr. F. Bouman and Dr. W. Verkerke (Amsterdam), the late Dr. R. Dahlgren (Copenhagen), Dr. M.I. Hakki (Berlin), Prof. U. Hamann (Bochum), Dr. R. Wunderlich (Vienna), Prof. M. Yakovlev and Prof. T. Batygina (Leningrad), Dr. H. Tobe (Chiba, Japan), Dr. P.H. Raven (St. Louis, USA), Prof. A.E. Cocucci (Cordoba, Argentina), the late Dr. Thanikoimani (Pondichery, India), Prof. K.K. Lakshmanan (Coimbatore, India), Dr. M.P. Nayar and Dr. B.D. Sharma (Botanical Survey of India, Calcutta), Prof. S. Chanda (Bose Institute, Calcutta) and our colleagues (at the University of Delhi) Prof. H.Y. Mohan Ram, Prof. S.C. Maheshwari, Prof. R.N. Kapil, Prof. N.N. Bhandari and Dr. Ashok Bhatnagar.

For advance information regarding the recently published (1989) Dedicatory Volume in honour of Prof. A. Lebégue, *Some Aspects and Orientations in Plant Embryology*, we wish to thank the editors, Prof. J. Paré and Prof. M. Bugnicourt (University of Picardie, Amiens, France).

For the translation of Russian literature we are obliged to Dr. Balram Sharma (Indian Agric. Res. Inst., New Delhi) and Dr. (Mrs.) Akhilesh Tiwari (University of Delhi); and to Dr. (Mrs.) Sheela Srivastava (University of Delhi) for German literature. Dr. M.A. Rau (Mysore) very kindly prepared a number of schematic figures on the development of endosperm in the Sympetalae. Dr. F. Bouman provided scanning electron micrographs of seeds which appear in this volume. Thanks are also extended to Mr. Dev Raj Khosla (Indian Natl. Sci. Acad., New Delhi) who prepared photocopies of literature and illustrations, so essential for checking the manuscript. Mr. S.B. Ambegaokar (Delhi) was kind enough to compare over 1800 pages of the final manuscript with the earlier draft. Dr. (Mrs.) Kapoor ably assisted us in the collection of literature and indexing the Contents and References. Dr. Kapoor and Dr. Manoj Kumar (University of Delhi) also helped in the preparation of plates.

We express our appreciation to Mr. G.A. Shastri and Ms. Sushil Kaur for easy access to the University Library, and to Dr. (Mrs.) Vatsala (Publ. Direc., CSIR, New Delhi), Dr. (Mrs.) Nina Dhiman and Ms. Binny Mathur (Editorial Assistants, Phytomorphology), and our research scholars Mr. N. Syed Abbas Ali and Ms. Babeeta Chrungu for their help in checking the periodicals. Mr. Krishan Lal, artist (University of Delhi), very competently designed and prepared several figures, and labelled all the illustrations. Mr. D. Daniel and Mr. S.K. Das (University of Delhi) attended to photographic work with professional competence.

We are indebted to Dr. Dieter Czeschlik, Life Science Editor, Springer-Verlag, Heidelberg, for his immense patience in waiting to see the completion of this work. Moreover, we would like to thank Mrs. Jean von dem Bussche for copyediting "Sympetalae" (pp. 615–829) in 1987; Dr. G. Forbath, Copy Editor, Ms. Carola Wissmeier, Secretary to Dr. D. Czeschlik, and Ms. Renate Münzenmayer, Production Editor, Springer-Verlag, for the meticulous handling of this publication. For the high quality of printing by MacMillan India, Bangalore, we have great appreciation.

We are obliged to *Janab* Hakim Abdul Hameed *Sahab* (Chancellor), Prof. M. Amin (Vice Chancellor) and Prof. Muhammed Iqbal (Dean, Faculty of Science) of the Hamdard University, New Delhi, for their interest and encouragement in this project.

Thanks are also due to Ms. Madhumati, Mr. Aniruddha, Ms. Deepshikha, and Mr. Deepak, research scholars at the Department of Botany, Hamdard University for cross-checking the text and references in the final proofs, and for preparing plant indices.

We would like to make special mention of Mr. R.K. Gupta (University of Delhi) for expert secretarial assistance, and typing twice the manuscript of about 2000 pages, and Mrs. Krishna Thakur (University of Delhi) for periodical secretarial assistance. Mr. Sukhbir Singh and Mr. Chandra Pal (University of Delhi) have been very helpful in attending to the day-to-day routine work during the preparation of this book.

The Indian National Science Academy, New Delhi, and Springer-Verlag, Heidelberg, provided partial financial support for this book.

We also wish to record gratitude to Mrs. Raj Johri (wife of B.M. Johri) and Dr. (Mrs.) Sheela Srivastava (wife of P.S. Srivastava) for their support, forbearance and interest during the long periods when *Comparative Embryology of Angiosperms* was in preparation. Dr. Lalit M.

Johri (son) and Mrs. Meera Johri (daughter-in-law) keenly followed the progress of this book and encouraged an expeditious completion.
Once again, many thanks to all those who helped us in various ways.

May 28, 1992

B.M. Johri
K.B. Ambegaokar
P.S. Srivastava

Contents Volume 1

Contents Volume 2

II Dicotyledons: Sympetalae

1 The Background

The history of "embryology" dates back to the age-old tradition of Arabs and Assyrians of dusting the male inflorescence of the date palm on to the female palm to obtain a good supply of dates. This religious ceremony indicated some sort of sex-blending. But, until the experiments of Camerarius (1694), no attention was paid to sexuality in plants. The only efforts till then were of Grew (1682) who treated the stamens as the male organs (of flowers) which induced the ovary to produce the fruit.

Camerarius (1694) observed that in mulberry, *Mercurialis annua*, *Ricinus* and *Zea mays*, the anthers are essential for preparing a young plant in seeds contained in the ovary. He was not aware of the exact role played by the anthers. His work was later confirmed by Kölreuter (1761) who gave a detailed account of the role of insects in pollination. While working on the interaction of pollen grains with the stigma of flowers, he was able to ascertain the specificity of pollen grains. He produced hybrids in *Dianthus*, *Hyoscyamus*, *Matthiola* and *Nicotiana*.

Meanwhile, the discovery of the microscope provided an insight into the then unknown world of microorganisms. The excited microscopists focussed their attention on whatever object they came across. The Italian mathematician, astronomer and good microscopist, Amici (1824), observed the splitting of a pollen grain of *Portulaca*, emergence of pollen tube, and entry of the pollen tube into the stigma. But he could not ascertain the fate of the pollen tube. Brogniart (1827) also examined several pollinated pistils. He observed the pollen tubes (spermatic tubules) but imagined that they burst and discharged the "spermatic granules". He also visualized the passing of spermatic granules through the style, and their reaching the placenta, as spermatozoids do in animals. Amici (1830), still doubtful about the process, continued his studies on *Portulaca oleracea*, *Hibiscus syriacus*, etc., and discovered that the pollen tube itself elongates and reaches the ovule.

The false notion of Schleiden (1837) – the propounder of the "cell theory" with Schwann (1838, 1839; see EJ Gardner 1978) – caused great confusion. According to him, the extreme tip of the pollen tube grows into a vesicle and then develops into an embryo. The embryo sac acts simply as an "incubator". However, Amici (1844, 1847), in spite of strong opposition, established the fact that it is not the tip of the pollen tube that develops into an embryo, but a "germinal vesicle" already present in the embryo sac in the ovule. The germinal vesicle becomes "fertilized" by the fluid from the pollen tube and develops into an embryo.

The well-illustrated works of Hofmeister (1849), finally confirmed Amici's observations and settled the relationship between pollen tube and the egg. He (Hofmeister 1847–1861) also provided detailed information on the formation of tetrads, organization of embryo sac, and development of the embryo in several genera. Hanstein (1870) traced the sequence of early cell divisions in the embryogeny of *Capsella* and *Alisma.*

Following Hofmeister, Strasburger (1879) presented a comprehensive account of development of the Polygonum type of embryo sac, its organization, and formation of secondary (fused polar nuclei) nucleus in *Polygonum divaricatum.* His most significant contribution is the discovery of syngamy (Strasburger 1884). He could illustrate the two male gametes in a pollen tube, their discharge in the embryo sac, and fusion of one of the male gametes with the egg cell (syngamy) in *Monotropa* and other plants. The fate of the other male gamete, its fusion with the two polar nuclei (triple fusion), was discovered by Nawaschin (1898). Polyembryony was also reported (Ernst 1918, Schnarf 1929), and the concept of "apomixis" had been put forward by Juel (1898, 1900) and Murbeck (1901).

By the end of nineteenth century, the major facts about the development of male and female gametophytes, endosperm and embryo became known (see Coulter and Chamberlain 1903).

2 Diagnostic Characters

A number of diagnostic features facilitate the study of comparative embryology, and are also significant for taxonomic considerations. These characters are mentioned below (see Tobe 1989).

Anther. Number of sporangia, developmental type of anther wall, endothecium fibrous or non-fibrous, tapetum secretory (glandular) or amoeboid (intrusive, invasive, plasmodial), cytokinesis simultaneous or successive, quadripartition by cell-plate formation or furrowing, tetrads, polyads, microspores, pollen grains shed singly, or clumped in dyads, tetrads, pollinia or massulae, at two- or three-celled stage, shape, ornamentation of exine, germinal furrow, pores and allied structures, special type of pollen as in Cyperaceae, Epacridaceae and Loranthaceae.

Ovule. Vascularized or non-vascularized, ortho-, hemiana-, amphi-, campylo-, ana-, circino-tropous or incompletely differentiated (reduced), ategmic, uni- or bitegmic, micropyle formed by one or both integuments, endothelium present or absent, tenui- or crassinucellate, formation of nucellar cap, operculum, epistase, hypostase, and ovular appendages.

Megasporogenesis. Reduction divisions, tetrad, dyad or coenomegaspore. If coenomegaspore, distribution of nuclei 2 + 2, 1 + 1 + 1 + 1, 1 + 3, and 1 + 2 + 1 (see Sects. 3.5, 3.6).

Female Gametophyte. Monosporic: Polygonum or Oenothera type; bisporic: Allium or Endymion type; tetrasporic: Adoxa, Plumbago, Penaea, Peperomia, Drusa, Plumbagella or Fritillaria type, variations in embryo sac development in the same taxon, micropylar, lateral or chalazal extension, embryo sac haustoria, synergid/s ephemeral or persistent, synergid haustoria, number of antipodals, ephemeral or persistent, uni- or multinucleate, antipodal haustoria, special features.

Pollination. Mode of pollination: anemophily, hydrophily, entomophily; structural adaptation of pollen and stigma, in two-celled pollen grain division of generative cell in pollen tube, on stigma or in the style, time lag between pollination and double fertilization, special mode of pollination as in *Ficus* and *Yucca*.

Fertilization. Entry of pollen tube in embryo sac: porogamous, mesogamous or chalazogamous; discharge of gametes, syngamy, triple fusion or fusion of sperm with secondary nucleus, persistent remnants of pollen tube, entry of more than one pollen tube and polyspermy, and fertilization of one or both synergids.

Endosperm. Nuclear, Helobial or Cellular type; stage of proembryo at wall formation in Nuclear (and Helobial) type, micropylar and chalazal haustoria, one or both aggressive, extension into circumjacent tissue, number of nuclei and nuclear behaviour, persistence of haustoria in developing seed, wall formation in haustoria, development of endosperm proper, reserve food, rumination due to integument/s or pericarp, special behaviour of peripheral layers of endosperm in consuming seed-coat and even pericarp (partially or completely).

Embryo. Stage of endosperm at division of zygote, quadrant, octant, derivation of embryonal organs, Crucifer, Asterad, Solanad, Caryophyllad, Chenopodiad and Piperad type of embryogeny, fully differentiated and organless (undifferentiated) embryo, delayed differentiation, suspensor absent, short or massive, uni- or multinucleate, ephemeral or persistent, suspensor haustoria, special feature/s.

Polyembryony. Cleavage and/or budding of zygotic embryonal mass, proliferation of suspensor, accessory embryo/s from haploid or diploid (fertilized) synergid, somatic (diploid) embryo from nucellar or integumentary cells, developmental pattern of accessory embryos.

Apomixis. Diplospory, apospory, auto- or pseudogamous development of endosperm, parthenogenetic development of egg cell in reduced and unreduced embryo sac, obligate and facultative apomict.

Seed-coat. Unspecialized, exo-, meso- or endotestal, exo- or endotegmic, testal-tegmic, "naked" seeds (without seed-coat).

Seed. Exalbuminous or albuminous, food reserve in endosperm, perisperm, and/or embryo, nature of food reserve, intraseminal leaf primordia, cotyledons equal or unequal, fleshy or thin, foliaceous, folded or convolute, embryo chloro- or non-chlorophyllous; appendages on seed; hairs, wing, aril, elaiosome.

When more than one type of development occurs in the same taxon, it is necessary to calculate their relative percentage. For example, on the basis of counts of several microscopic fields, mention the percentage of tetrahedral, isobilateral, decussate, linear and T-shaped microspore tetrads, types of embryo sac in ovules, and types of endosperm and embryogeny in developing seed (see Schmid 1982, Tobe and Raven 1987c).

The first comprehensive account of comparative embryology of angiosperms is by Schnarf (1931), followed by Davis (1966), B.M. Johri (1970a), Yakovlev (1981, 1983), Poddubnaya-Arnoldi (1982), and Batygina and Yakovlev (1985, 1987). The embryogeny has been discussed by Johansen (1950), and seed anatomy by Netolitzky (1926), Corner (1976) and Takhtajan (1985, 1988). Several examples of the role of embryological characters in taxonomic considerations have been cited by P. Maheshwari (1950, 1956, 1963b, 1964), B.M. Johri (1963b, 1967), Rau (1970), Palser (1975), Herr (1984), and Tobe and Raven (1983a, b, 1984b, c, d, e, f, 1987a, b, c).

R. Dahlgren (1975b, 1977a, 1983a) and R. Dahlgren and Clifford (1982) have designed a series of diagrams for comparative representation of several embryological characters in various families.

3 Developmental Aspects

3.1 Anther

A typical anther is tetrasporangiate with two microsporangia in each lobe (see Bhandari 1984). When transverse septa develop in each microsporangium (Sect. 18.6), it becomes polysporangiate, as in the Annonaceae. In the Onagraceae (Tobe and Raven 1986a, b), transverse as well as longitudinal septa develop in each microsporangium (Sect. 39.13). At dehiscence, the septa break down in each lobe and the mature anther becomes bilocular (see Engler 1876). In *Ficus ottoniifolia* (Verkerke 1986a), each of the four sporangia dehisces independently (Sect. 10.4).

In some taxa, as in Malvaceae (Sect. 35.5), the anthers are bisporangiate. *Arceuthobium minutissimum* (Bhandari and Nanda 1968a) shows a horseshoe-shaped archesporium around a central columella (Sect. 12.8).

Wall. The young anther is a homogeneous mass of parenchymatous cells covered by a distinct epidermis. It becomes four-lobed, and a single or a group of two or more hypodermal archesporial cells differentiates in each lobe. The archesporial cell/s divides periclinally separating the primary parietal layer (below the epidermis) and the primary sporogenous tissue. The latter may undergo mitotic divisions resulting in secondary sporogenous tissue. The primary parietal layer, by further periclinal divisions, gives rise to the anther wall (see Periasamy and Swamy 1964).

According to Bhandari and Khosla (1982), in *Triticale* the primary parietal layer divides periclinally and produces the endothecium, middle layer, and secretory tapetum. These three layers are concentric and surround the sporogenous tissue. The connective tissue does not contribute to the development of inner tapetum.

The middle layers are generally rich in reserve food. Their number varies, and may increase up to four or five layers, as in *Daphniphyllum himalayense* (A.K. Bhatnagar and Garg 1977) and *Nelumbo nucifera* (S.C. Gupta and Ahluwalia 1979). They are usually resorbed during development, but some of them may persist or develop fibrous thickenings as in members of the Liliaceae and Zingiberaceae.

On the basis of the formation of middle layers, Davis (1966) classifies the development of anther wall into four types: Basic, Dicotyledonous, Monocotyledonous and Reduced.

Basic Type. The primary parietal layer divides periclinally and produces two secondary parietal layers. Both the layers undergo further periclinal divisions, and the one towards the epidermis differentiates into endothecium followed by a middle layer. The other secondary parietal layer produces another middle layer, and the tapetum. The middle layers may undergo further divisions. Examples: Anacardiaceae, Lecythidaceae, Rhamnaceae, Vitaceae.

Dicotyledonous Type. Comparable to the "Basic" type, the outer secondary parietal layer produces the endothecium and middle layer; the inner layer functions directly as tapetum. Examples: Acanthaceae, Aristolochiaceae, Bignoniaceae, Convolvulaceae.

Monocotyledonous Type. The outer secondary parietal layer develops directly into endothecium, while the inner gives rise to the middle layer and tapetum. Examples: Alismataceae, Amaranthaceae, Aponogetonaceae, Cactaceae, Chenopodiaceae, Cruciferae.

Reduced Type. The primary parietal layer, by a periclinal division, produces the endothecium and tapetum. The middle layer is absent. "Reduced" type of development is reported in *Antigonon leptopus* (Agoram and Krishnamurthy 1980) and *Gaultheria* spp. (Chou 1952). Davis (1966) refers this type to the members of Lemnaceae and Najadaceae. B.L. Gupta (1935) confirms the absence of a middle layer in *Wolffia arrhiza*, S.C. Maheshwari (1954) in *W. microscopica*, Lawalrée (1952) in *Lemna minor*, and S.C. Maheshwari and Kapil (1963a) in *L. paucicostata* (Fig. 61.3.1G–L, N). Lodkina's (1977) report of a middle layer in *Lemna gibba, L. paucicostata, L. trisulca, Najas flexillus, N. graminea, N. marina* and *Wolffia microscopica* requires confirmation.

According to Davis (1966), the Basic type occurs in nine families, while the Dicotyledonous type in the majority of dicotyledonous families and one monocotyledonous family, Taccaceae. The Monocotyledonous type includes the majority of monocotyledonous families as well as several dicotyledonous families. Two different types of wall formation occur in only a few families: Combretaceae, Euphorbiaceae, Sterculiaceae and Thymelaeaceae. Davis (1966) mentions that variation occurs in development in the same species (*Stellaria media*, Caryophyllaceae). A.K. Bhatnagar and Kapil (1979) report Basic, Dicotyledonous and Monocotyledonous types in *Bischofia javanica.*

Varghese and Chowdhury (1972) question the validity of the terms suggested by Davis. They point out that these terms have only a descriptive value, and serve a limited purpose. In our opinion, judging from the variations in the development of anther wall, it does not serve any useful purpose to stick to the terminology of Davis.

The epidermal layer undergoes only anticlinal divisions, covers the inner growing tissue, and its function is protective. The epidermal cells may develop a cuticle on the outer tangential wall, or may contain tannin as in *Hibbertia* (Sastri 1958b). They develop fibrous thickenings in some members of Dilleniaceae and Zingiberaceae. The tapetum-like behaviour of the epidermal cells (Sect. 65.2) in *Zeuxine longilabris* (Karanth et al. 1979) is unique, and needs further study.

Endothecium. Closer inwards to the epidermis lies the endothecium, which is responsible for the dehiscence of the anther (see Chauhan 1979, Gerendy and French 1988). Its cells enlarge and develop callose thickenings on their inner tangential walls. These radiate outwards at the time of maturation of anther. The thickenings are usually absent in the region of dehiscence of the anther in cleistogamous flowers, and in some members of Dipterocarpaceae, Ericaceae and Hydrocharitaceae. Rarely, the endothecial thickenings extend on the dorsal side into the connective as in *Bupleurum tenue* (S.C. Gupta and M. Gupta 1964), and they extend and unite (in the region of the connective) surrounding the staminal bundle, as in *Eucommia ulmoides* (Eckardt 1963). Occasionally, the endothecium becomes biseriate. It is five or six layers thick in *Nicotiana tabacum* (Jose and Singh 1968) and *Voacanga foetida* (Maheswari Devi 1971).

From cleared anthers, sections and macerations, French (1985a, b) examined endothecial thickenings in Araceae, 106 representative species of 20 genera in Pathoideae and Monsteroideae, and representative species of 37 genera of Calloideae, Lasioideae and Philodendroideae. The wall thickenings in the endothecium are based on an annular–helical pattern. Variations in the thickenings are related to differences in cell shape, cell orientation, intergradation, presence of branched and reticulate thickenings, and various types of discontinuous thickenings. The pattern of wall thickenings can be used as a basis of classification of the Araceae.

The absence of endothecial thickenings is a derived condition often correlated with porate dehiscence (Davis 1966). However, Araceae, Philesiaceae and Rapateaceae have porate dehiscence and also endothecial thickenings (R. Dahlgren and Clifford 1982, see also Greenday and French 1988). Families with nonporate dehiscence and which lack endothecial thickenings include Hydrocharitaceae (Ernst-Schwarzenbach 1945), *Musa errans* (Juliano and Alcala 1933), *Oryza* (Juliano and Aldama 1937), and *Xyris* (see French 1985a). Untawale and Bhasin (1973) report annular and helical thickenings in *Musa*. Endothecial thickenings are also absent in some cleistogamous flowers as in *Amphicarpum* (Ritzrow 1908); in *Colocasia* (Pothoideae), *Cercestis* and *Rhektophyllum* (Lasioideae), and *Philodendron* (Philodendroideae) of the Araceae (French 1985b).

French (1986a) made a further study of endothecial thickenings in representative species from 43 genera of Colocasioideae, Aroideae and Pistioideae. All endothecial cells are anticlinally elongated with bars or reticulate thickenings on anticlinal walls and variably discontinuous thickenings on periclinal walls. Several genera show unusual and distinctive patterns. In *Stylochiton* (Aroideae), the endothecial cells are periclinally elongated and exhibit continuous helices and rings. Other genera with a similar pattern occur in Monsteroideae and Lasioideae. *Acorus* (Araceae) exhibits a unique pattern of stellate thickenings.

Endothecial thickenings in mature anther (ready to dehisce) should be studied from suitably stained dissected wholemounts and sections in various planes.

Tapetum. The tapetum is the innermost layer of the anther wall (see Warmke and Overman 1972) and originates from the derivatives of the primary parietal layer. Presumably, it differentiates from the sterilized sporogenous tissue in *Potamogeton crispus* (B.L. Gupta 1934). Lakshmanan and Rajeshwari (1979) report the origin of tapetum from peripheral sporogenous layers in *Syringodium isoetifolium*. Its dual origin has been observed in *Tarenna asiatica* (Periasamy and Parmeswaran 1965). Towards the outer side the tapetum is derived from the primary parietal layer, whereas on the inner side it develops from the cells of the connective. According to Periasamy and Swamy (1966), such a development occurs in all the angiosperms.

In *Annona squamosa* (Periasamy and Kandaswamy 1981), the origin of the tapetum is trimorphic. The archesporial tissue is uniseriate and alternate archesporial cells develop into sterile septa so that the microsporangium appears polysporangiate. The outer tapetum is derived from the primary parietal layer, the inner from the connective, and both are secretory. The septa in the anther locule are formed due to the sterility of the archesporial cells and develop into plasmodial tapetum (Fig. 18.6.1C–J).

Pacini et al. (1985) discuss the "form, function and possible phylogeny" of the tapetum in the Embryophyta.

The tapetum is mainly a nourishing tissue. In a young anther, the tapetal cells remain in continuity with the microspore mother cells through plasmodesmata through which the nutrients are provided to the latter, till the formation of meiocytes. These connections are plugged with callose thickenings, as soon as meiosis is completed.

The tapetal cells are uninucleate in the beginning, but in the majority of plants mitotic divisions of their nuclei are not accompanied by cytokinesis (see Cooper 1933). Thus, they become two- to four-nucleate or, at times, the number goes up to 16. Very often, due to fusion the nuclei become polyploid. Polyploidy may also be attained due to endomitosis, polyteny, or by formation of restitution nuclei (Figs 3.1.1, 19.2.1).

The tapetum may become irregularly biseriate, and in *Calotropis* it is multilayered. The tapetum is designated as amoeboid (invasive, intrusive, plasmodial) or secretory (glandular), according to its behaviour. The nature of tapetum is a diagnostic feature in the majority of families. However, both types may occur in the same family, in different genera. Rarely, in *Anemonopsis macrophylla* (Tobe 1974b), both types occur in different anthers of the same flower. The appearance of perforations in the tapetal cells of the anther is reported in *Acalypha cilliata* (P.K. Mukherjee 1964).

Amoeboid Type. The walls of tapetal cells break down, protoplasts enlarge and enter the anther sac, and by their fusion periplasmodium is formed around the microspore mother cells. The microspore mother cells, or microspores, or two-celled pollen grains remain embedded in the periplasmodium (Fig. 52.3.1A–E). Periplasmodial formation occurs at different developmental stages of microspores in different species. At the tetrad stage, the tapetal cells produce an enzyme callase which disorganizes the callose walls around the microspore tetrads and the microspores are released in the periplasmodium. The plasmodium is gradually absorbed by the developing pollen grains. In *Rhoeo spathacea* (Nanda and Gupta 1977), an extra-tapetal membrane develops on the outside of the periplasmodium, next to endothecium, and encloses the microspores and periplasmodium. Nanda and Gupta also report raphides in the periplasmodium. Later on the membrane breaks down, and the mature pollen grains are released. The development of an "extra-tapetal" membrane has been previously reported in some members of Compositae (J. Heslop-Harrison 1969, see also Gupta and Nanda 1972).

Pacini and Juniper (1983) studied ultrastructural details of the development of amoeboid tapetum in *Arum italicum.* The anther wall comprises the epidermis, fibrous endothecium, three middle layers, and two-layered tapetum. At the premeiotic phase of microspore mother cells, the tapetal cells are connected with each other by plasmodesmata, but these are absent in the tapetal walls facing the middle layer and the microspore mother cell (Fig. 3.1.2A). During the meiotic divisions of mother cells, the tapetum shows conspicuous changes. At leptotene (Fig. 3.1.2B), the plasmodesmata in the tapetal cells widen to form cytomictic channels; the tangential walls disappear at pachytene (Fig. 3.1.2C). During meiosis I, anaphase (Fig. 3.1.2D), the radial walls of tapetal cells also disappear and microtubules lie parallel to the inner radial walls of mother cells. The tapetal cytoplasm starts to intrude between the meiocytes; microtubules appear at that region (Fig. 3.1.2E).

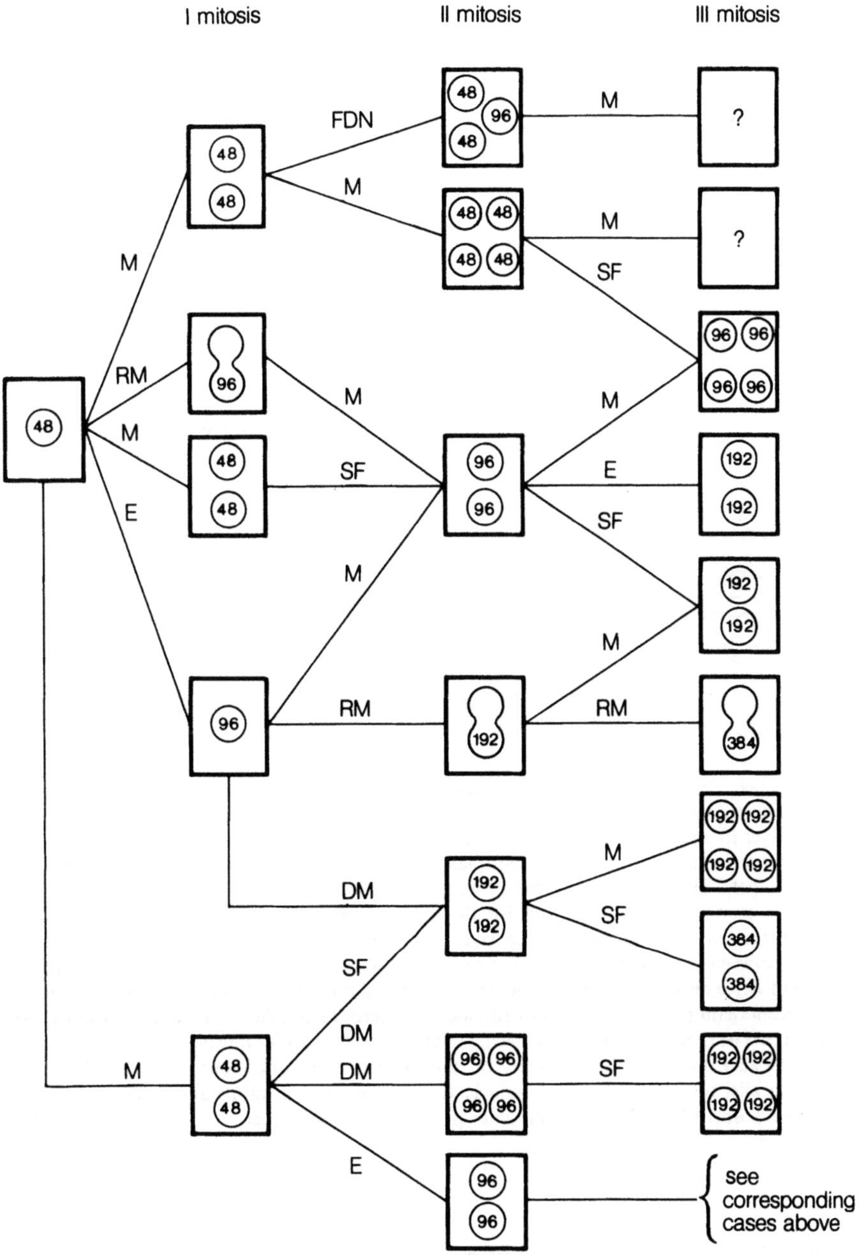

Fig. 3.1.1. Schematic representation of the methods of polyploidization in the cells of the internal anther tapetum of *Solanum tuberosum* (2n = 48). *DM* diplochromosome mitosis; *E* endomitosis; *FDN* fusion of daughter nuclei; *M* mitosis; *RM* restitutional mitosis; *SF* spindle fusion; *SF/DM* Spindle fusion in diplochromosome mitosis. (Adapted from Avanzi 1950, see D'Amato 1984)

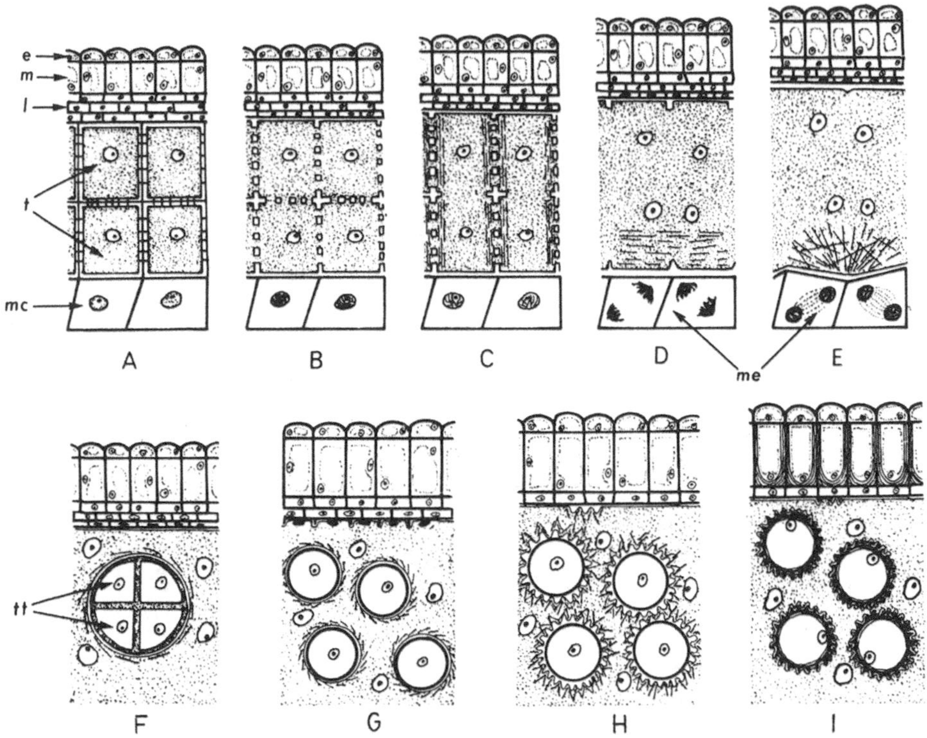

Fig. 3.1.2A–I. *Arum italicum*, amoeboid tapetum (semi-diagrammatic). **A** Premeiotic phase; the tapetal cells (*t*) are connected by plasmodesmata which are absent in the tapetal walls facing the middle layer (*l*) and microspore mother cells (*mc*); *e* epidermis; *m* mechanical layer/endothecium, **B** Leptotene (microspore mother cells), plasmodesmata widen. **C** Pachytene; tangential walls between tapetal cells have disappeared. Note microtubules (*thin lines*) along radial walls of tapetal cells. **D** Meiosis (*me*) I, anaphase; radial walls of tapetal cells have also disappeared. Microtubules lie parallel to the inner radial wall of meiocytes. **E** Meiosis I, telophase. Tapetal cytoplasm starts to intrude between meiocytes and at this site many microtubules appear in random orientation. **F** Tetrad (*tt*) embedded in multinucleate periplasmodium. Microtubules surround the tetrad. **G** Microspore (in periplasmodium) surrounded by microtubules; *dark line* exine. **H** Microspores, later stage. The tapetal plasmalemma facing microspores retracts leaving cone-shaped spaces; this retraction also occurs at the periphery of tapetum but is discontinuous. Microtubules are no longer parallel to exine surface. **I** Uninucleate pollen grain with intine and exine. Two middle layers are absorbed in the periplasmodium (**F–H**), thickenings appear in endothecium (**I**). (After Pacini and Juniper 1983)

Figure 3.1.2D and E show the formation of multinucleate periplasmodium in which the tetrads, microspores and uninucleate pollen grains are embedded (Fig. 3.1.2F–I). Microtubules surround the tetrad facing the callose wall and the microspore. The tapetal plasmalemma facing the microspores retracts from the exine surface leaving roughly cone-shaped spaces (Fig. 3.1.2H). Microtubules no longer lie parallel to the exine surface. By the localized retraction of the plasmamembrane, the spaces left on the microspore surface and tapetal periphery become filled with a material forming "spines" (Fig. 3.1.2I); the intine also develops.

The amoeboid type mostly occurs in the monocotyledons.

Secretory Type. The secretory or glandular tapetum is characteristic of most of the dicotyledons. The tapetal cells remain intact in situ till the maturation of pollen grains, and provide nutrition. A tapetal membrane develops on the inner tangential surface of tapetal cells (S.C. Gupta and Ahluwalia 1979, see also Shivanna et al. 1979). The tapetal cells develop pro-orbicules. The sporopollenin, a substance resistant to chemical decomposition, is produced in the mitochondria of tapetal cells (J. Heslop-Harrison 1963). The sporopollenin is deposited on the pro-orbicules. As the microspores are released in the anther locule, the inner periclinal and radial walls of the tapetal cells dissolve. According to Bhandari and Ram Kishori (1973), at the tetrad stage the protoplasts of the tapetal cells develop a membrane in *Nigella damascena*, and the orbicules become studded onto this membrane. Later, they are deposited on the tectum to form the exine-sculpture of pollen grains. Orel et al. (1977) describe the development of a sporopollenin film within the tapetal cells during the formation of microspore tetrads in five species of the Gramineae, and beet, potato and tomato.

The tapetum synthesizes certain substances which are released into the anther locule during maturation of pollen grains. These substances contain fibro-granular proteins and carotinoid-containing lipid droplets. The fibro-granular proteins enter the exine of pollen grain, and are stored in the region of bacula. These wall-bound proteins are extra-cellular and capable of recognizing the compatibility of stigma. The carotinoid-containing lipid granules spread over the exine and help in pollination. They are described as "pollen-kitt". The tapetum is a physiologically active layer of the anther wall (see Hesse 1981, Knox 1984, Shivanna and Johri 1985, Pacini et al. 1985).

The malfunctioning of tapetal cells has an adverse effect on the development of the endothecium and pollen fertility (Nanda and Gupta 1974, Vijayaraghavan and Ratnaparkhi 1979, see also Warmke and Overman 1972). The tapetal cells cause sterility of pollen grains when they become overstretched and form a persistent band around the pollen grains as in *Allium cepa* (Nanda and Gupta 1974) and *Belamcanda chinensis* (S.C. Gupta and Nanda 1973). The endothecial cells also remain small, and do not develop fibrous thickenings. The accumulation of a large quantity of micronutrients in the tapetum also affects the development of pollen grains. The malfunctioning and hypertrophied condition of tapetal cells exhibits callase activity which leads to the failure of formation of callose walls around the tetrads.

3.2 Microsporogenesis

Sporogenous Tissue. The sporogenous tissue may consist of a single vertical row of hypodermal cells as in some members of Cactaceae, Compositae, Cucurbitaceae and Malvaceae, or may undergo mitotic divisions to produce several rows. Each cell of the archesporium is potentially capable of developing into spores, but sometimes several cells take up the nourishing function and become sterile (see Dickinson 1982). The development of trabuculate partitions dividing the sporogenous tissue into groups occurs in *Hydrocera triflora* (Venkateswarlu and Laxminarayan 1957). In some members of Annonaceae, and in *Myristica malabarica* (N.C. Nair 1972), the microsporangium is septate. In *Cananga* (only in a few anthers), the partitions develop due to sterilization of sporogenous tissue, whereas in *Annona* only one microspore mother cell occurs in each compartment and the septa develop tapetum. *Goniothalamus* shows multilayered transverse septa, and each microspore mother cell has its own tapetum (Parulekar 1970). Several sterile transverse septa develop between sporogenous cells of *Parkia biglandulosa* (Dnyansagar 1954b). Each compartment contains several microspore mother cells nourished by its own tapetum.

Septate microsporangia also occur in Balsaminaceae, Convolvulaceae, Gentianaceae, Plumbaginaceae, Rhizophoraceae (Lersten 1971), Flacourtiaceae, Myrsinaceae and Sterculiaceae (Endress and Vosser 1975), and Onagraceae (Tobe and Raven 1986a).

An abnormal behaviour of sporogenous cells (in an anther) is reported in *Leptomeria billardierii* (Ram 1959c). Some of the cells enlarge, the nucleus divides, the daughter nuclei divide again, and again for the third time. Thus, eight nuclei are produced, and these organize more or less like an embryo sac. This feature has not been reported so far in any other angiosperm (see B.M. Johri and Ambegaokar 1984).

Microspore Mother Cells. The microspore mother cells are usually uninucleate. However, occasionally, polynucleate condition has been observed in *Santalum obtusifolium* (S.P. Bhatnagar 1966), triticales (Kempanna et al. 1971) and in *Secale-Agropyrum* hybrids (Poddubnaya-Arnoldi and Makhaline 1969). Perhaps, this condition is due to the absence of cytokinesis. The microspore mother cells are interconnected, and are also connected with the tapetal cells through plasmodesmata. As they enter the phase of meiotic divisions, they lose contact with the tapetal cells, become enclosed in thick callose walls, but remain interconnected through bigger channels.

Meiosis. The microspore mother cells undergo meiotic divisions. Meiosis I is a heterotypic division, and reduction occurs in the number of chromosomes. Meiosis II is a homotypic division, only mitosis occurs. Cytokinesis follows meiotic divisions of the microspore mother cells. The interconnections between the microspore mother cells become obliterated, and the tetrads become independent.

Cytomixis. The migration of nuclear material from cell to cell is designated as cytomixis (Gates 1911). It takes place at all stages of meiotic divisions in pollen mother cells. The presence of cytoplasmic links in between the meiocytes provides evidence for migration of nuclear materials. Anucleate, polyploid and aneuploid pollen mother cells are the consequence of cytomixis. In case of complete migration, the donor cell becomes empty and the recipient polyploid (see Whelan 1974).

Cytomixis in pollen mother cells has been observed throughout the year, both in fixed and fresh flower buds of *Ervatamia divaricata*. It occurs to the maximal degree from March to June; the temperature fluctuation affects cytomixis. Sometimes, most of the pollen mother cells of an anther are involved in cytomixis while the development in the adjacent anthers is normal; direct fusion of two pollen mother cells also occurs. The adjacent pollen mother cells may be linked by two or more cytoplasmic channels of varying breadth. Cytomixis is more common at prophase I but occurs almost at every stage up to tetrad formation.

The extrusion of chromatin and cytomixis in pollen mother cells affect the viability of gametes. It is associated with high pollen sterility, and usually there is no seed formation.

In *Nigella sativa* (A.K. Datta and Biswas 1984), cytomixis plays a significant role in the production of aneuploid gametes, and as a result thereof trisomic and other aneuploids may appear in the subsequent generations.

Sinha (1985) observed the migration of chromatic material in 11%–18% of cells at metaphase I in *Bacopa monnieri*, *Centranthera tranquobaria*, *Limnophila indica*, *Lindernia ciliata*, *L. parviflora*, *L. verbaenifolia*, *Mazus pumilus* and *Sutera dissecta*. There is 26%–40% cytomixis in *Dopatrium junceum*, *Lindenbergia indica* and *L. crustacea* at metaphase I, 30% at anaphase I and II in *Lindenbergia indica*, 15% in *L. crustacea* and 20% in *D. junceum*.

The flow or migration of chromatin material from one pollen mother cell into another is unidirectional. A pollen mother cell donates its chromatin material to one cell and might receive chromatin simultaneously from another cell, forming a chain.

In *Rauwolfia serpentina* (N. Banerjee and Sharma 1988), the frequency of cytomixis showed significant increase during monsoon months (9.62%). It was predominant in prophase I to early metaphase I. The regular occurrence of cytomixis in all the individuals of a particular population of *R. serpentina* in successive years suggests a genetic control, and its manifestation depends on the physiological state of the tissue.

Cytokinesis. Cytokinesis is either simultaneous or successive. The former is characteristic of the dicotyledons, and the latter of the monocotyledons. The occurrence of both types in different genera of the same family is also reported. However, the occurrence of both types of cytokinesis, as in *Catharanthus pusillus* (Bhasin 1971), *Conyza stricta* (H.P. Sharma and Murty 1978) and *Vernonia cinerea* (Dakshini and Dadlani 1978), is rare.

In the simultaneous type, the first meiotic division of microspore mother cell is not accompanied by wall formation. The two-nucleate condition is immediately followed by meiosis II. The callose wall intrudes through centripetal furrows, unites in the centre, and separates the microspores. Quadripartition is also brought about by the laying down of cell plates after meiosis II. In contrast, in the successive type, the first meiotic division separates two dyads, a cell plate develops between the two nuclei and extends to the sides centrifugally. Each dyad undergoes meiosis II, resulting in a tetrad. The cell plate formation takes place as after meiosis I (see Longly and Waterkeyn 1979).

Tetrads. The arrangement of microspores in a tetrad exhibits five patterns. Generally, they are tetrahedral and isobilateral. In addition, they may be linear, T-shaped and decussate. All five types are reported in *Aristolochia elegans* (Fig. 21.2.1G–M), and *Sparganium erectum* (Fig. 62.3.1H). Linear tetrads also occur in *Halophila ovata* (Kausik and Rao 1942). The microspores form uniseriate chains and the individual tetrads are separated by sterile cells (Fig. 52.4.7C). The variation in tetrad formation depends more or less on the shape of microspore mother cells (Chennaveeraiah and Mahabale 1962), and the orientation of spindles during meiosis II. In *Typha*, Sahasi and Ikuse (1974) attribute the formation of various types to rotation and sliding of the microspores.

The microspores (in a tetrad) are interconnected and surrounded by a thick callose wall. The microspores in a massula of Orchidaceae are also interconnected. The massulae of *Zeuxine strateumatica* (Vijayaraghavan and Shukla 1980) are held together by elastic, refractive viscin threads. These threads are PAS-positive and probably contain elastin, but are not autofluorescent.

In *Allium tuberosum* and *Cyclamen persicum* (Bhandari et al. 1981), the callose is deposited within the primary wall around each microspore mother cell. The primary wall persists until the late tetrad stage. In *A. tuberosum*, the primary wall dissolves simultaneously with the callose and the microspores are released. Bhandari et al. (1981) suggest that the primary wall assists the callose in preventing the entry of macromolecules, and isolating the microspores.

Polyspory. Occasionally, the microspores in a tetrad undergo further divisions and polyspory occurs producing polyads. The number of microspores goes up to eight in *Hyphaene* (Mahabale and Cheenaveeraiah 1957), 11 in *Cuscuta reflexa* (Fig. 48.4.1A–E) and 22 in *Thunbergia mysorensis* (Khaleel and Boraiah 1973). Compound pollen grains occur in several families (Fig. 48.4.1L–S). In *Brachycome ciliaris* (Davis 1964a), the compound pollen grains become four-nucleate, due to failure of cytokinesis (see B.M. Johri and Tiagi 1952).

In Podostemaceae (Sect. 28.2), both simultaneous (*Indotristicha* and *Terniola*) and successive (*Dicraea*) cytokinesis follows meiotic divisions in microspore mother cells. The microspores separate in pairs, and form "double pollen grains" in *Dicraea* (Fig. 28.2.1A), *Griffithella* (Razi 1949), *Hydrobryum*, *Podostemon* (Magnus 1913), *Angolaea*, *Mniopsis* and *Lophogyne* (Went 1910, 1912). Double pollen grains are also reported in *Bulbostylis* and *Cyperus* (C.K. Shah 1962), *Cuscuta* (Fedortshuk 1931, B.M. Johri and Tiagi 1952), *Marianthus* (Sundri and Narayana 1977) and *Strychnos* (Mohrbutter 1936). The pollen grains adhere together to form a pollinium as in some members of Asclepiadaceae (Fig. 47.7.1A, D), Mimosaceae (Fig. 26.19.1A) and Orchidaceae. The microspores change their shape and size as the pollinium rounds off.

In Orchidaceae, the microspores of a tetrad may become free from one another, or adhere together to form a compound pollen grain. In tribes

Neottieae and Ophrydeae, the compound grains adhere together to form a massula (Fig. 65.2.2L).

Due to callase activity, the callose wall around the tetrad is digested, and the microspores are released in the anther locule except in some families where they remain in permanent tetrads as in Ericaceae, Juncaceae (Fig. 55.2.1A–D), Monimiaceae, Phylidraceae (Fig. 54.18.1A–G), Winteraceae, etc.

3.3 Male Gametophyte

Two- and Three-Celled Pollen Grains. The development of male gametophyte starts soon after the uninucleate pollen grains are produced. However, overwintering may occur in the species of colder regions. Vacuolation occurs in the pollen grain and its nucleus takes up a wallward position. The mitotic division is always unequal separating a larger vegetative cell from the smaller generative cell (see Steffen 1963a, Hageman 1981). The mitotic spindle is either symmetrical or asymmetrical. It may be conical or flattened. A curved cell-plate separates the two nuclei, and extends up to the intine. The curved wall grows between the intine and plasmalemma of the generative cell. The generative cell gets separated from the wall of the pollen grain, and comes to lie in the cytoplasm of vegetative cell.

The generative cell may divide before, or after the dehiscence of anther. Meanwhile, the vegetative cell continues to grow, and accumulates reserve food in the form of starch and fat. It becomes rich in proteins, and its nucleus may undergo endoduplication and become highly polyploid. The generative cell also elongates. This is the two-celled stage of pollen grain at which shedding occurs in the majority of angiosperms.

When the first mitotic division occurs in a microspore within a tetrad, the position of the generative cell is usually predetermined due to certain genetic factors, and this feature is of taxonomic significance. The generative cell may be cut off towards the centre (proximal pole) (Fig. 54.18.1B, C), or the outer wallward side (distal pole) (Fig. 55.2.1D). The polarity can be ascertained only when the pollen grains are attached in a tetrad. As the microspores in a tetrad, or a massula, are interconnected, the mitotic divisions usually occur simultaneously.

In several species, the generative cell divides when the pollen grains are still within the anther locule, and the two male gametes are formed. The division is mitotic, and cytokinesis occurs by cell-plate formation. The male gametes are oval or spindle-shaped and invested by a sheath of cytoplasm. Then the pollen grains are liberated at the three-celled stage.

If the pollen grains are dispersed at the two-celled stage, the division of generative cell may occur after the deposition of pollen on the stigma, or in the pollen tube (Fig. 3.3.1A–C) while traversing the style or, rarely, after reaching the embryo sac.

The formation of supernumerary sperm nuclei, through the division of sperm cells, is an abnormality and is known as polyspermy.

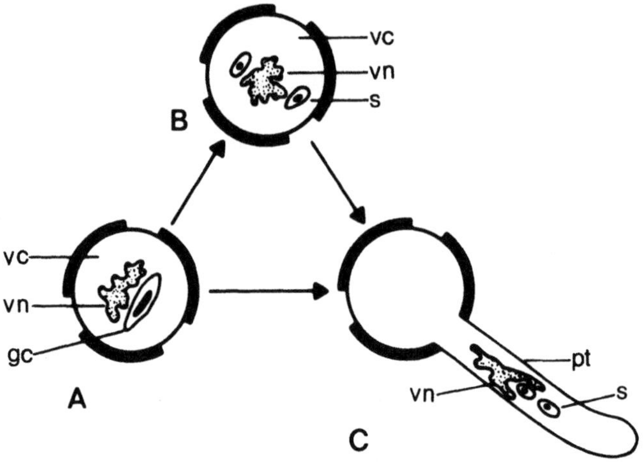

Fig. 3.3.1A,B. Bi- and tri-celled pollen grains; *vc* vegetative cell; *vn* vegetative nucleus; *gc* generative cell. *s* sperm cell. **C** Germinated pollen grain with vegetative nucleus and two sperm cells in pollen tubes (*pt*). (Adapted from Dexheimer (1970)

Usually, the pollen grains are disseminated singly, but they may remain attached in dyads (Fig. 28.2.1A), tetrads (Fig. 55.2.1A–D), pollinia (Fig. 26.19.1A), or massula (Fig. 65.2.1L).

The occurrence of supernumerary nuclei in pollen grains is not very common, e.g., in Pittosporaceae (Sundri and Narayana 1977, L.L. Narayana and Sundri 1981, 1983).

The cytoplasm of two- or three-celled pollen grain is rich in carbohydrates, lipids, proteins and amino acids (see H.G. and I. Baker 1979). It also contains enzymes for pollen germination and tube elongation. The enzymes are mostly hydrolytic; their nature depends on the type of reserve food in the pollen grains and the tissue through which the tube has to grow. The two- and three-celled pollen grains exhibit physiological differences in their behaviour (B.M. Johri and Shivanna 1977). The two-celled pollen grains remain viable for a longer time, and show a high percentage of germination in vitro, as compared to the three-celled grains.

Pollen Grains in Cyperaceae and Epacridaceae. Cyperaceae (Fig. 63.2.1A–H) present a peculiar type of pollen development. The meiotic divisions of microspore mother cells are followed by simultaneous cytokinesis in *Carex*, *Cyperus*, *Kyllinga*, *Lipocarpha* and *Scirpus*. Of the four nuclei thus formed (Fig. 63.2.1A–D), three occupy wallward position while the fourth lies in the centre. Walls develop in between the three nuclei, and the large functional microspore is separated from the rest (Fig. 63.2.1E). The three nuclei may undergo a further division, but they soon degenerate. However, the functional microspore remains enclosed in the wall of the tetrad, and is often described as pseudomonad.

Epacridaceae (Fig. 42.6.1A–J) exhibits another noteworthy example with a variety of pollen types (Smith-White 1948, 1955a, Venkata Rao 1961b). Venkata Rao (1961b) presumes that there is evidence of parallelism between the development of male and female gametophytes of angiosperms in various patterns of pollen development in the Epacridaceae. He compares the full tetrad type with all four functional microspore nuclei to the tetrasporic embryo sac, the Pentachondra type with two functional microspore nuclei to the bisporic type, and the Styphellia type to the monosporic type of embryo sac. This comparison is of no consequence.

Pollen Embryo Sacs. An abnormal behaviour of microspore nuclei is also exhibited in the formation of pollen embryo sacs. Stow (1930, 1934) observed binucleate, four-nucleate and well-developed eight-nucleate "embryo sacs" in the anthers of *Hyacinthus orientalis*. Vijayaraghavan and Ratnaparkhi (1977) report that 10% of anthers of *Heuchera micrantha* show a tendency towards the formation of pollen-embryo sacs. The elongated microspores with well-organized eight-nucleate embryo sac-like structures simulate the Polygonum, Allium or Adoxa type of female gametophyte. In addition, enlarged microspores show nuclear arrangements resembling the Penaea or Plumbago type. An increase in the number of nuclei in a microspore also occurs in *Capsicum annuum* var. *acuminatum* but they do not get organized (Dharamdhaj and Prakash 1978).

Pollen Grains in Aquatic Plants. Filiform (thread-like) pollen grains (Fig. 52.9.1D–F) occur in marine plants (sea grasses) of Hydrocharitales – *Halophila, Thalassia*, and Zosterales – *Amphibolis, Cymodocea, Diplanthera, Halodule, Phyllospadix, Posidonia, Syringodium* and *Zostera* (Fig. 52.8.1A–I). The microspores are arranged linearly parallel to the locule in *Diplanthera* (Lakshmanan 1968), *Halodule* (Yamashita 1976b) and *Syringodium* (Lakshmanan and Rajeshwari 1979). The microspores elongate after they are released from the tetrad, become filamentous and 1–5 mm long. The pollen grains lack a well-defined exine, and there is no specialization for shedding. These long rope-like masses are carried in the water currents, and may adhere to the bibrachiate styles of the submerged stigmas of female flowers (Fig. 52.9.1A–E).

The pollen grains of sea-grasses exhibit three different adaptation to the environment. The spherical grains of Hydrocharitaceae are embedded in slime (Fig. 52.4.2A, B), they may be ellipsoidal and are released in tetrads within a slime tube as in *Halophila*, or may be filiform.

Pollen Wall. The microspores synthesize part of their walls from the substances produced by their own cytoplasm while in tetrad condition (Heslop-Harrison 1968a, b, c, 1971, Banerjee and Barghoorn 1971b). The initial wall termed primexine is of cellulose, and is discontinuous opposite the endoplasmic reticulum. The gaps oriented parallel to plasmalemma mark the position of germ-pores. The cytoplasm of the microspore also synthesizes precursors of sporopollenin. They are polymerized and deposited on the lamellae in such a way that various patterns of exine of mature pollen grains become predetermined. The wall is made up of radially-elongate rod-like structures, the baculae, their lower ends spread out forming a continuous foot layer. The upper ends of baculae may also expand, and give rise to a roof, the tectum; such pollen grains are tectate.

The uninucleate pollen grain is densely cytoplasmic with a prominent nucleus. It enlarges due to increase in cytoplasm and vacuolation. In a free state, the pollen grain initiates formation of intine, and the innermost layer of exine (see Godwin 1968). In the formation of intine, the proteins synthesized by the cytoplasm of microspores are incorporated. The plasmalemma of the pollen cytoplasm develops radially oriented cytoplasmic tubules containing proteins, in the developing intine. As the intine grows, these tubules with proteins are cut off from the plasmalemma by a layer of intine which is free from tubules. Thus, the intine proteins are the products of the pollen grain cytoplasm (gametophytic). The intine proteins are concentrated more in the region of germ-pores in aperturate pollen grains, whereas they are distributed throughout in the non-aperturate pollen grains of monocotyledons (Knox 1971, 1984, Shivanna 1978, Chanda et al. 1979, Shivanna and Johri 1985, see also Stanley and Linskens 1974, Walker 1974a, b, Walker and Doyle 1975).

The proteins synthesized by the tapetum (sporophytic) penetrate through the micropores of the tectum and are stored in the spaces between the baculae, whereas the lipid substances spread over the tectum layer.

The exine is differentiated into two layers, the outer sculptured layer, the sexine, and inner non-sculptured layer, the nexine. The sexine comprises the baculae with tectum in tectate grains; in non-tectate grains the baculae may remain separate or join to form various patterns. The exine is variously described according to its ornamentation, such as psilate (smooth), scabrate (very fine projections), verrucate (warty), echinate (spiny), reticulate (network-like), etc.

The morphology of pollen grain is usually considered on the basis of its shape, size, thickness and ornamentation of exine. The most important feature is the form of aperture, their number, distribution and position (designated NPC: N = number, P = position, C = character). The aperture is designated as "treme" (Erdtman 1969). When the apertures are regularly distributed over the surface, the arrangement is called "nomotreme", irregular distribution is "anomotreme", and absence of apertures is referred to as "atreme" or inaperturate.

The number of apertures varies from zero to eight (N_0–N_8) and, accordingly, mono-, bi-, tri-, tetra- and so on is prefixed to the character of pollen grain.

The position of aperture may be distal (outwards in a tetrad), i.e. "anatreme", or the apertures may be distributed on the equatorial plane at equal distance as in "zonotreme".

The two basic forms of apertures (morphoform) are the porate and colpate. The apertures have a circular outline in porate form; in colpate grains there are elongated furrows (colpi) along the axis. The compound forms are designated colporate or pororate. In a colporate pollen grain the outer region bears a furrow with a pore in the centre.

In a very useful and informative discussion, Blackmore et al. (1987) emphasize ontogeny and phylogeny of all aspects of microsporogenesis, male gametophyte and pollen grains on the basis of 26 character states.

3.4 Ovule

The ovule, or the megasporangium, is attached to the placenta of the ovary by a stalk, the funicle. It is a mass of potentially fertile tissue called nucellus, enveloped by one or two integuments (Fig. 3.4.1A–F; see also Bouman 1974).

Types of Ovules. According to the position of the ovule, in relation to hilum, the mature ovules are classified as ortho-, ana-, hemiana-, campylo- and amphitropous (Figs. 3.4.1A–F, 3.4.2A–H). In orthotropous form, the micropyle lies in the line of hilum, above it, whereas it is reversed in the anatropous type. The curvature is much less than the anatropous form in campylotropous type (see Bocquet 1959). The ovule is at right angles to the funiculus in hemianatropous type. The amphitropous form occurs when the ovule, as well as the embryo sac, curves like a horseshoe. In Aristolochiaceae (*Bragantia*- N.C. Nair and Narayanan 1961), Cactaceae and Plumbaginaceae the ovule becomes anatropous due to unequal growth of the axis of funiculus, then curves further till the micropyle faces upward and develops into circinotropous type.

The ortho-, ana-, hemiana- campylo- and amphitropous (Fig. 3.4.2A–E) forms of ovules are considered as basic forms. The rest of the forms have been derived by various degrees of curvature as in circinotropous form (Fig. 3.4.2F–H).

The position of the ovule is also described according to its orientation. The raphe may be dorsal (abaxial: away from the axis) or ventral (adaxial: towards the axis). The micropyle may point upward (epitropous), downward (hypotropous) or to a side (lateral: pleurotropus). The position of raphe is a diagnostic feature in the families Buxaceae, Euphorbiaceae and Theacae. Variations have been observed in orientation of ovules in the same ovary. In *Pittosporum floribundum* (Sheela and Narayana 1966), the ovules are hypotropous in the upper region of ovary and epitropous in the lower region; the raphe is dorsal.

According to Bor (1978), the ovule consists of two separate developmental units: the funicular or basiovular unit which gives rise to raphe, chalaza and outer integument (together forming testa), and the nucellar unit from which the inner integument is derived.

Integument/s. The ovules generally have one or two integuments and are designated uni- and bitegmic. The nucellus may be massive or scanty and is designated crassi- and tenui-nucellate.

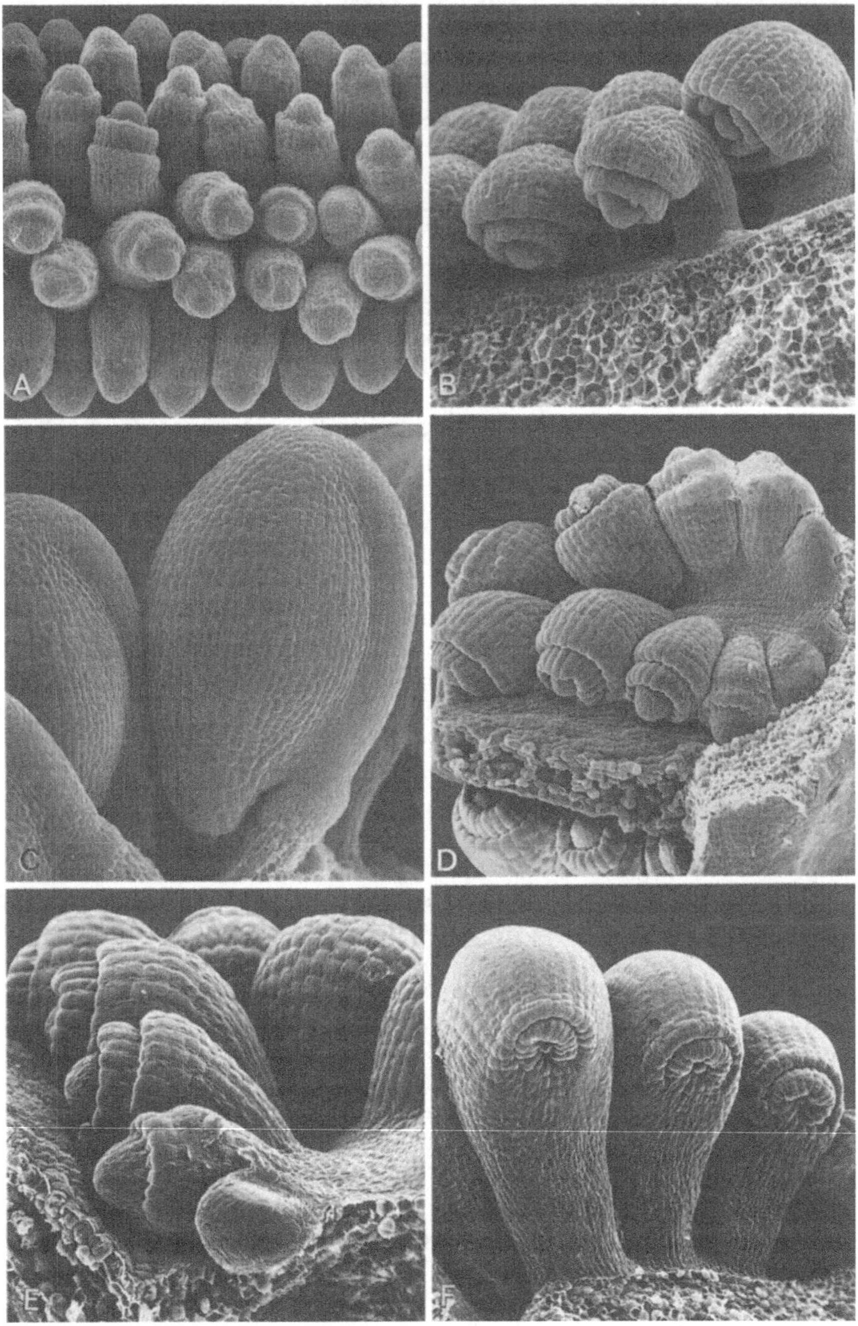
A
B
C
D
E
F

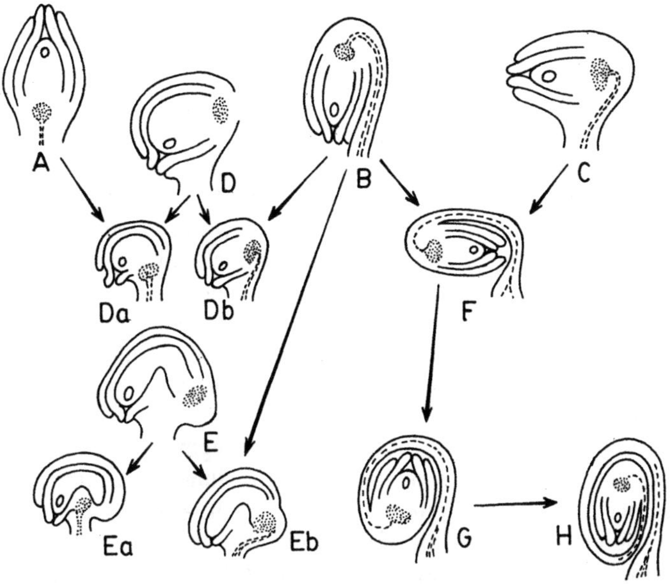

Fig. 3.4.2A–H. Relationship between different types of ovule. **A** Orthotropous. **B** Anatropous. **C** Hemianatropous. **D** Campylotropous. **Da** Ortho-campylotropous. **Db** Ana-campylotropous. **E** Amphitropous. **Ea** Ortho-amphitropous. **Eb** Ana-amphitropous. **F** Hypertropous. **G** Ortho-circinotropous. **H** Ana-circinotropous. (After Savchenko 1973, see Poddubnaja-Arnoldi 1976)

W.R. Philipson (1977) classifies the ovules into four types: Bitegmic crassinucellate (Type A), Bitegmic tenuinucellate (Type B), Unitegmic crassinucellate (Type C), and Unitegmic tenuinucellate (Type D).

Ovules of Type A have been observed in Magnoliidae, Hamamelidae, Caryophyllidae, and most members of Dilleniidae and Rosidae (subclasses of Cronquist 1968). In a number of families of Dilleniidae, the ovules are of Type B, and in a few families of Type D. A large number of families of the Rosidae have ovules of Type C and D. In the subclass Asteridae, most members have ovules of Type D.

The families which consistently have ovules of Type B form a subclass, the Thealean-Primulalean complex. For the families with ovules of Type C and D, W.R. Philipson suggests the term "Unitegminae".

The transition from ovules of Type A to Type D probably took place in two steps, since a few families under the Unitegminae have crassinucellate ovules: Araliaceae and Umbelliferae. A gradation from Type A to C, with some examples of D, occurs in the Celastrales.

Fig. 3.4.1A–F. SEM micrographs of developing ovules. **A** *Passiflora racemosa*, ovular primordia with initiation of anatropous curvature, and inner and outer integuments. **B, C** *P. verspitilio*, overgrowth of nucellus beyond integuments (**B**), and mature anatropous (**C**) ovule. **D–F** *Agrostemma gracile*, ovular primordia with initiation and development of integuments (**D, E**), and mature (**F**) ovules. (After Bouman 1984)

Type D has remained very stable and is known in about half of the dicotyledons. This type is of much significance as a diagnostic feature in systematics.

In angiosperms, the inner integument is always dermal except in *Codiaeum variegatum* and *Euphorbia milii* (Bor and Bouman 1974), *E. geniculata* (Bor and Kapil 1975), and *Chrozophora obliqua* (Bor and Kapil 1976). The outer integument is generally subdermal, but its dermal origin has been observed in Dichapetalaceae – *Dichapetalum mombattense* (Boesewinkel and Bouman 1980), Erythroxylaceae – *Erythroxylum coca* (Boesewinkel and Geenen 1980), Geraniaceae – *Erodium, Geranium, Monsonia, Pelargonium* and *Sarcocaulon* (Boesewinkel and Been 1979), Linaceae – *Linum usitatissimum* (Boesewinkel 1980a), and Rutaceae – *Ruta graveolens* and *Zanthoxylum simulans* (Boesewinkel 1977). Both the integuments are of dermal origin in Centrospermae, Parietales and Capparales.

Bouman and Calis (1977) suggest the origin of unitegmic condition due to integumentary shifting. In *Eranthis hiemalis*, the inner integument of the pseudo-crassinucellate ovule originates by the periclinal divisions of the dermal cells. The outer integument develops subdermally, proximal to the inner one. Due to activity in the subdermal initials, the inner integument is shifted upward; its growth is arrested. The incision between the outer and inner integument ends at or above the level of nucellar apex. Thus, a single massive integument surrounds the nucellus. In *Delphinium nudicaule*, the joint activity of the initials of outer and inner integument gives rise to a single, mainly subdermal, massive integument. The activity of the dermal initials of the inner integument is arrested.

The formation of unitegmic ovule by integumentary shifting has also been observed in *Helleborus niger* (Bouman and Calis 1977). Here the integument appears bifid in transmedian longitudinal sections of older ovules. In the same way, unitegmic ovules are also formed in *Gentiana cruciata* (Bouman and Schrier 1979).

In Polygonaceae, Trapaceae and some members of Euphorbiaceae, the nucellus produces a long beak which projects through and beyond the micropyle (Fig. 3.4.3A). In Commelinaceae (Chikkannaih 1964), the apical region of the nucellus is broad, and there is no well-organized micropyle.

In some taxa, an additional integument develops between the outer and the inner integument but usually it remains arrested (Fig. 3.4.3B).

When the integument/s do not develop, the ovule is designated ategmic (Fig. 3.4.3C).

In bitegmic ovules the two integuments arise independently at the base of the ovule. They overtop the nucellus leaving a narrow opening called the micropyle.

The opening of the inner integument is called endostome, and that of the outer integument exostome. The exostome and the endostome may not lie in the same alignment producing a zigzag micropyle. In the majority of families with bitegmic ovules, the inner integument overgrows the outer and forms the micropyle. The unitegmic ovules with a massive integument have a long, narrow micropyle. The unitegmic ovules of Sabiaceae (*Meliosma* spp. – M.V.S.

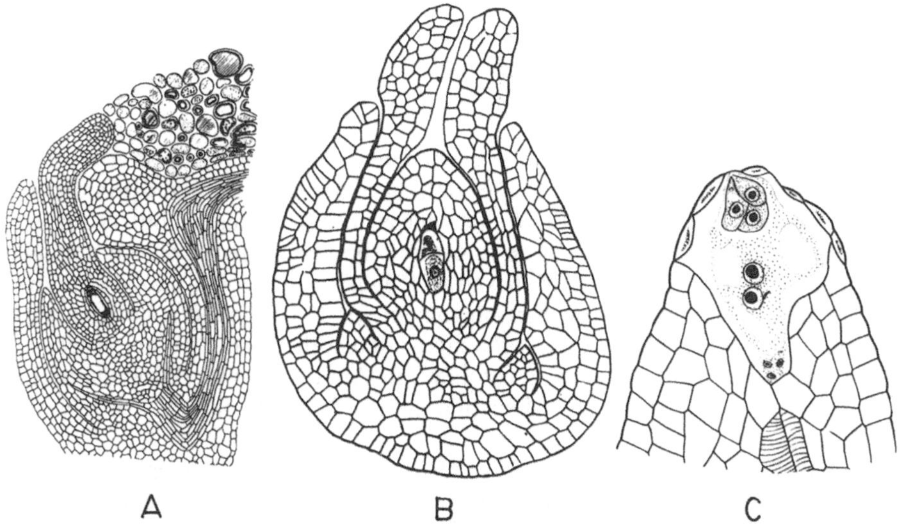

Fig. 3.4.3A. *Baliospermum montanum*, longisection of ovule showing nucellar beak, obturator, and vascular strand in inner integument. (After P.N. Rao 1970). **B** *Cananga odorata*, longisection of ovule with poorly developed middle integument. (After Corner 1949). **C** *Exocarpus bidwillii*, longisection of ategmic "ovule". (S.P. Bhatnagar and Joshi 1965)

Raju 1952a) and Orchidaceae (*Gastrodia elata* – Abe 1976) also do not develop a micropyle.

Sunder Rajan (1976) considers the micropyle formed by both the integuments as primitive. He compares this condition with six other primitive characters occurring in different families: (1) trees or shrubs, (2) actinomorphic flowers, (3) free petals, (4) axile placentation, (5) crassinucellate ovule, and (6) Nuclear type of endosperm. The author does not cite the examples of families, but relies on the data provided by Davis (1966). This information is not at all helpful in judging the primitiveness of a family.

An apparently poorly developed middle integument occurs in *Canangium*, *Mezzettia*, *Xylopia* (Corner 1949), *Cananga* (Fig. 3.4.3B), *Myristica* (N.C. Nair 1972), *Nelumbo nucifera* (S.C. Gupta and Ahluwalia 1977) and *Sarcandra* (Vijayaraghavan 1964).

The aril is often considered as the third integument which, in our opinion, is not justified at all. The aril arises from the funicle, at the base of the integuments, and from the funicle and exostome in *Myristica* (Fig. 18.8.2A–C). It grows during post-fertilization stages and surrounds the ovule as in Aizoaceae, Crossosomataceae, or merely remains as a vestige at the base as in Bixaceae. It is thin-walled and contains oil and starch in Passifloraceae. It is fimbriate in Crossosomataceae, vascularized in Myristicaceae and Paeoniaceae, and with coenocytic cells in Dilleniaceae (*Acrotrema* – Fig. 22.2.2A–I).

The bitegmic ovules of annonaceous genera and *Myristica* (N.C. Nair 1972), where both the middle integument and aril are present, have four integuments.

The number of integuments indicates the evolutionary status of the family. The bitegmic condition occurs usually in dicotyledonous Polypetalous families, whereas the unitegmic condition is associated with most of the Sympetalous families. Exceptions occur in both cases (see de Boer and Bouman 1972).

In the ovules of Anacardiaceae, the bitegmic condition occurs in *Rhus* and *Lannea* (Kelkar 1958a, b) due to the splitting of the integumentary primordia. In *Pistacia* spp. (Grundwag 1976), the ovule is initially unitegmic but, at a later stage, the integument splits close to the micropyle forming a partial second integument. According to P. Maheshwari (1950), the unitegmic condition may arise due to elimination of one integument, or the fusion of two separate primordia. The Santalales and Rafflesiaceae depict the former feature, whereas the accretion of integuments is exhibited in Anacardiaceae, Balsaminaceae, Betulaceae, Haloragaceae, Lecythidaceae, Marcgraviaceae, Rosaceae (*Sanguisorba*), and others.

Van Heel (1970a, b, 1971a, b, c, 1973, 1976) attributes an evolutionary significance to the formation of distally lobed integuments. The bitegmic ovule of *Exochorda racemosa* shows a three- to four-lobed outer integument and a bilobed inner integument; while the single integument in *Juglans regia* may exhibit four to seven, entire or fringed, lobes in the distal region (see also Leroy 1955).

In the ovules of Centrospermales and Cactales, there is a prominent air-space between the outer and inner integument at the base, and it is a characteristic feature of these orders.

The occurrence of stomata on the integuments is a common feature in several families such as Malvaceae, Sapotaceae and Umbelliferae. In Amaryllidaceae and Moringaceae, the integuments become chlorophyllous in the young seed.

Generally, the vascular supply in an ovule terminates at the chalaza and branches there. The occurrence of integumentary vascular system is treated as a primitive feature. The bitegmic ovules of Magnoliaceae (*Magnolia*) and Euphorbiaceae (*Dalechampia*) have an independent vascular system in each integument. The outer integument is vascularized in several families: Berberidaceae, Betulaceae, Boraginaceae, Compositae, Cyrillaceae, Leguminosae, Moringaceae, Ranunculaceae, Sterculiaceae. In *Hyphaene indica* (Mahabale and Chennaveeraiah 1957), 16–17 vascular bundles extend in the outer integument. Rarely, the bundles occur only in the inner integument (*Baliospermum* – Fig. 3.4.3A; *Croton, Jatropha, Ricinus*), whereas in Myristicaceae both outer and inner integument as well as the aril are vascularized.

The development of operculum occurs in the members of Lemnaceae (S.C. Maheshwari 1956b), and Araceae (S.C. Maheshwari and Khanna 1956). The inner integument proliferates at its micropylar tip, and forms a cap-like structure. Its cells are pigmented, and their walls are thickened.

Endothelium. In the ovule, where the nucellus is ephemeral, the innermost layer of the integument in contact with the embryo sac develops into integumentary tapetum or endothelium (Fig. 3.4.4A–E). The cells elongate radially, become richly cytoplasmic, and perform the function of transport of nutrients to the embryo sac.

The flow of nutrients from the integumentary cells to the embryo sac depends on the chemical nature of the intervening cell walls of endothelial cells, and the embryo sac. The flow may be restricted to the micropylar, submicropylar, or chalazal region (see Kapil and Tiwari 1978a).

In some members of Scrophulariaceae, the unequal radial elongation of the endothelial cells causes rumination in the endosperm. In *Celsia coromandeliana* (Kapoor et al. 1975), some cells develop thickenings on the inner tangential and radial walls, and protect endosperm and embryo.

The cells of the endothelium may become multinucleate (Fig. 3.4.4B), or the ploidy level of the nuclei increases due to endomitosis. In *Pedicularis palustris* (Steffen 1955), the nuclei become 32-ploid. The endothelium is multilayered (Fig. 3.4.4C) in the chalazal region of the embryo sac in *Tithonia rotundifolia* (Pullaiah 1978a), and is supposed to be a characteristic feature of the tribe Heliantheae. In *Melampyrum pratense* (D. Raju and Arekal 1977), the inner epidermis of the integument becomes obliterated in the micropylar region, during maturation of the embryo sac. Finally, the integumentary tapetum develops from the hypodermal layer. The endothelium is a common feature in unitegmic, tenuinucellate ovules of the Sympetalae. However, in Acanthaceae and Loganiaceae the endothelium develops only in some members. Its absence in Diapensiaceae is a taxonomic feature.

In *Flaveria repanda* (S. Misra 1964), *Melampodium divaricatum* (Maheswari Devi and Pullaiah 1976) and sunflower (Fig. 3.4.4D), the endothelial cells divide in all planes and produce a multilayered, massive structure which destroys the embryo sac. Perhaps, this may be due to the accumulation of nutritive substances (see Savchenko 1960).

In *Utricularia flexuosa* (R. Khan 1963), the tangential cells separating the endothelial cells disorganize, and their nuclei accumulate in a nodule which bulges into the degenerating embryo sac (Fig. 3.4.4E). Rarely, the biseriate endothelium of *Chelone glabra* (Arekal 1963a) also develops such nodules.

Nucellus. Enclosed within the integuments lies the reproductive homogeneous tissue, the nucellus. An archesporial cell, or a group of cells, differentiates in the hypodermal region of the nucellus in tenuinucellate ovules. The archesporial cell/s cut/s off a parietal cell in crassinucellar ovules. The parietal cell may remain undivided as in *Flagellaria indica* (Subramanyam and Narayana 1972), or may undergo divisions to produce a parietal tissue. The cells of the nucellar epidermis divide periclinally, and add to the parietal tissue. The megaspore mother cell becomes deep-seated either due to the growth of parietal tissue and derivatives of nucellar epidermis, or due to the tissue produced by the nucellar epidermis alone. In the latter case, the ovule is termed as pseudo-crassinucellar (see Davis 1966) as in *Cordia* (Tasneem 1966) and *Foeniculum vulgare* (S.C. Gupta 1964, 1970).

The occurrence of twin nucelli within a common integument is not very uncommon.

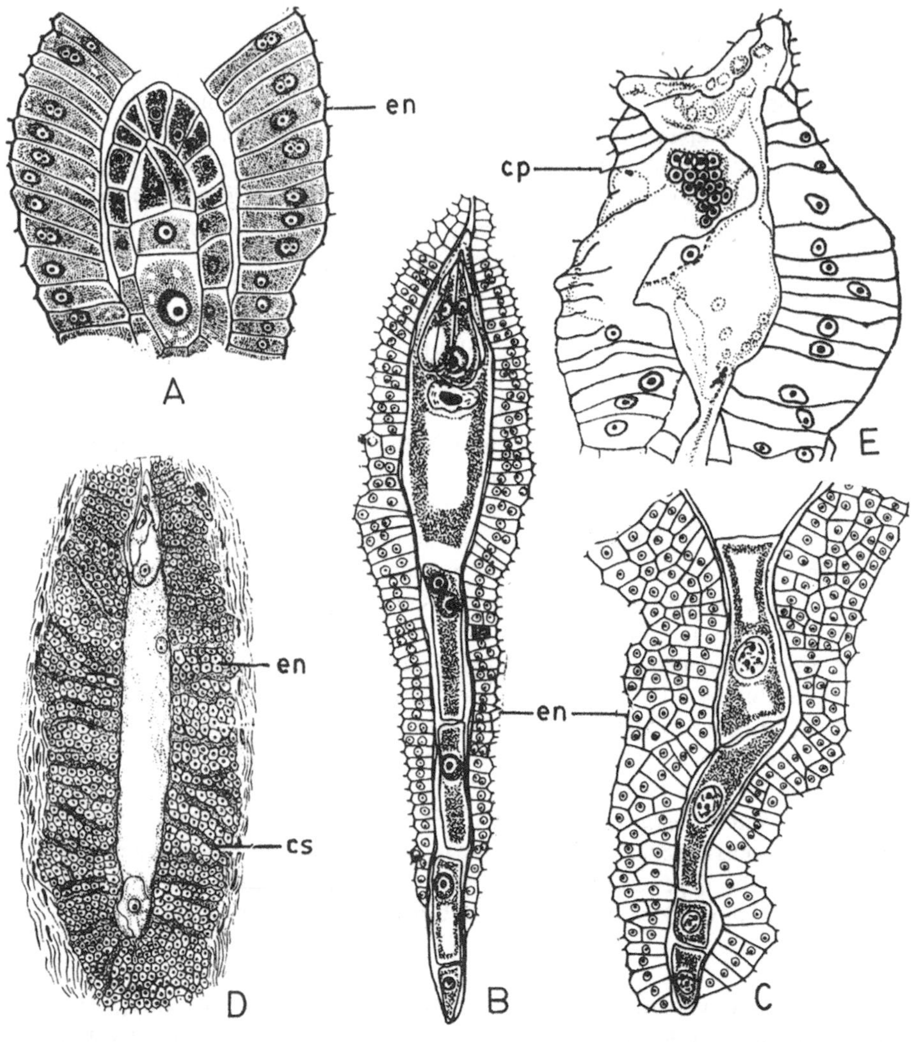

Fig. 3.4.4A–E. Endothelium. **A** *Volutarella ramosa*, T-shaped tetrad of megaspores, endothelium (*en*) with prominent uninucleate radially elongate cells. (After P.K. Deshpande 1964a), **B, C** *Tithonia rotundifolia*. **B** Mature embryo sac with multinucleate endothelial cells. Of the four antipodals one is binucleate; **C** Chalazal portion of embryo sac invested by multilayered endothelium, antipodal cells polyploid. (After Pullaiah 1978a). **D** *Helianthus annuus*, part of ovule at mature embryo sac, endothelium highly proliferated and interrupted by cuticular septae (*cs*). (After Savchenko 1969). **E** *Utricularia flexuosa*, portion of ovule with degenerated embryo sac. Due to break down of endothelial cells a coenocytic pouch (*cp*) has been formed. (After Khan 1963)

Sometimes, the derivatives of nucellar epidermis form a nucellar cap as in Lemnaceae (S.C. Maheshwari 1954). In *Elettaria* (Panchaksharappa 1962b) and *Hedychium* (Sachar and Arora 1963), the cells of the nucellar epidermis below the micropyle elongate radially and form a nucellar pad. Such a radial elongation of nucellar cells also occurs in Aizoaceae and Cactaceae. In Leitneriaceae, Polygonaceae, Trapaceae, and some members of Euphorbiaceae, the nucellus develops a beak which projects through the micropyle. The nucellar cap, nucellar pad, and nucellar beak are features of taxonomic importance in several families.

Vascular elements have been observed in the nucellus of *Acalypha* (Sect. 29.9), *Casuarina* (Sect. 4.2), *Duabanga* (Sect. 39.7) and several other taxa.

The nucellus is generally consumed during the growth of the embryo sac, but in Centrospermales (Fig. 16.6.1A) and some members of Proteaceae, Piperaceae and Zingiberaceae it stores food and persists in the seed as perisperm. In Podostemaceae (Figs 28.2.1D, E), and *Cenarrhenes nitida* (Proteaceae, Venkata Rao 1967), the nucellar cells below the embryo sac break down forming a pseudo-embryo sac which provides nourishment to the developing gametophyte and embryo.

In the ovules of *Cabomba caroliniana* (Ramji and Padmanabhan 1965), *Halophila ovata* (Lakshmanan 1963a), and *Vateria indica* (Fig. 22.11.1G–I), the nucellar growth in height is due to the presence of a basal meristem.

The nucellar tissue, at the base of the embryo sac, in *Quercus gambelii* (Brown and Mogensen 1972), grows upward and forms a column in the embryo sac. This structure is termed postament.

A rare feature in *Halophila ovata* (Fig. 52.4.3C), *Tamarix* (Fig. 37.15.2A–C) and *Pandanus parvus* (Cheah and Stone 1975; see also Fagerlind 1940b) is that the walls of nucellar cells at the base of the embryo sac break down and their nuclei migrate into the embryo sac.

The nucellar cells become multinucleate in the chalazal region in *Samadera indica* (N.C. Nair and Joseph 1957); the nuclei fuse and exhibit polyploidy. They reach up to 256n in *Trapa natans* (Nägl 1962, Trela-Sawicka 1978).

Hypostase and Epistase. Anatomical differentiation occurs during the formation of hypostase and epistase. Some of the nucellar cells, at the base of the embryo sac, become specialized and delimit the downward growth of the embryo sac. This structure is called hypostase. The cells are usually sclerenchymatous, or suberized (*Cannabis*), or they may remain thin-walled and become secretory. According to Venkata Rao (1963), the hypostase acts as a connecting link between the vascular strand and the embryo sac, and helps in the transportation of nutrients. Autoradiographic studies have shown that the hypostase is not concerned in the conduction of nutrients (see Kapil 1974). The hypostase probably serves to maintain the water balance in seed during dormancy (Johansen 1928a, b; see also van Tieghem 1901). It persists in seed in several plants. Some of the cells at the nucellar tip become thick-walled, and differentiate as epistase. In *Rhamnus wightii* and *Zizyphus glabrata* (see N.C. Nair 1970b), these cells are filled with dark contents.

In *Turnera fournieri* (Tiwari 1983), a thick-walled cup-shaped hypostase differentiates at the megaspore mother-cell stage; apparently devoid of cytoplasm. At the periphery the cells are partially thick-walled but completely thick-walled in the centre. The funicular provascular strand of the ovule runs almost up to the level of the hypostase. The cell walls of the hypostase predominantly consist of callose with a considerable amount of pectin; lignin is scanty. Presumably, the callose serves as a "molecular filter" which regulates the flow of nutrients to the embryo sac, and prevents the inflow of the unwanted large molecules. The occurrence of callose in the hypostase has also been detected in *Antirrhinum* (Rodkiewicz 1967), *Pentstemon* and *Digitalis* (Kuran 1972).

Obturator. An obturator is an outgrowth of the placenta, funicle, integument or style which bridges the nucellus or the micropyle to the adjoining integuments. It is presumed that it guides the course of pollen tube from the style or the ovary wall to the micropyle. It is either in the form of thin-walled compact tissue, or of loose hairy outgrowths. The cells may be glandular. Usually, in the majority of families, it develops from the funicle. The obturator is of placental origin in *Nelumbo*, and stylar in Thymeleaceae. Occasionally, both placental and funicular obturator occur in Aizoaceae. In Urticaceae, the inner integument develops hairy outgrowths which penetrate into the style as in *Leucosyke* and *Myriocarpa* (Fagerlind 1944b).

Incompletely Differentiated (Reduced) Ovules. The reports on "naked" ovules in the Apocynaceae, Olacaceae and Rubiaceae are inaccurate (see Eames 1961). In these families, there is a tendency for the reduction of nucellus.

Guignard (1917a, b) suggested a series of reduction of nucellus in unitegmic, tenuinucellate ovules of Apocynaceae (Sect. 47.6). In *Cerbera*, the nucellus is a single layer of cells which invests the megaspore tetrad, and is distinct from the integument. The nucellus collapses on the sides of the megaspore mother cell in *Vinca*, whereas in *Apocynum* it is represented by only three or four cells above the megaspore mother cell, and is continuous with the epidermis of the integument.

A similar series of reduction of nucellus can be traced in Rubiaceae (Fig. 47.8.2A–F). In *Phyllis*, the nucellus is a uniseriate layer, the lateral nucellar cells have disappeared in *Bouvardia* and *Vaillantia*, and the nucellus is represented by only a couple of cells below the micropyle. A few radially elongate cells in the micropylar region above the archesporial tissue represent the nucellus in *Rubia*, and only one or two cells in *Oldenlandia*. In *Houstonia*, the nucellar cells are not distinguishable from the integument.

Maheswari Devi (1971) reports reduction of nucellus in four genera of the tribe Plumiereae: *Catharanthus*, *Holarrhena*, *Rauwolfia*, and *Voacanga*.

Ovary-Ovule Complex. There is an ovary-ovule complex in the Loranthaceae (Sect. 12.8), Santalaceae (Sect. 12.6) [except *Comandra* (M. Ram 1957), *Iodina* (S.P. Bhatnagar and Sabharwal 1968) and *Thesium* (S.P. Bhatnagar and Agarwal 1961) where the ovules are unitegmic)], and Opiliaceae (Sect. 12.4), and a distinct tendency of reduction in the number of ovules and placenta. *Exocarpus* (M. Ram 1959a) reaches an extreme, and there is no distinction between the placenta and the ovule. Reduction also occurs in the integuments, and nucellus. In *Leptomeria* (M. Ram 1959b), the integument is more or less suppressed, and the nucellus is represented by a single cell.

In the Loranthaceae (Fig. 12.8.1$A_{1,2}$– $F_{1,2}$), the "mamelon" represents the placenta, the ovules of normal structure have not been observed. In *Helixanthera ligustrina* (P. Maheshwari and Johri 1950a), the placenta is absent. In *Amyema* (Dixit 1958a), the placenta may be absent or a rudimentary placenta may develop due to the elongation of sporogenous cells. *Helicanthes elastica* (B.M. Johri et al. 1957) shows a conspicuous conical unlobed placenta. The mamelon is basally three-lobed in an unilocular ovary in *Macrosolen cochinsinensis* (P. Maheshwari and B. Singh 1952) and *Peraxilla tetrapetala* (S. Prakash 1960a). The ovary is three-chambered with a basally three-lobed placenta in *Nuytsia floribunda* (R. Narayana 1958b), and four-chambered with a completely four-lobed placenta in *Lysiana exocarpi* (R. Narayana 1958a).

The maximum reduction is reported in *Balanophora* (Fagerlind 1945a) where the entire gynoecium is affected (Sect. 13.2). The ovular locule and stylar canal do not differentiate, and the archesporial cell differentiates in the centre of the basal region of a simplified carpel (Eames 1961). Ontogenetically, this simplified carpel is considered as the nucellar tissue (Arekal and Shivamurthy 1979).

3.5 Megasporogenesis

Archesporium and Sporogenous Tissue. The hypodermal archesporium is usually one- or two-celled, but multicellular archesporium develops in some families like Casuarinaceae, Rosaceae and Sterculiaceae. In *Casuarina equisetifolia* (Swamy 1948a), vascular elements differentiate in the centre of sporogenous tissue (from the procambial strand). A few tracheids have been observed among the sporogenous cells in the mamelon of *Amyema miquelii* (Dixit 1954).

In crassinucellar ovules, the archesporial cell cuts off a parietal cell which is not the case in tenuinucellar and pseudo-crassinucellar ovules. Generally, a single archesporial /sporogenous/ megaspore mother cell is functional.

Dyad, Tetrad and Coenomegaspore. The megaspore mother cell (Fig. 3.5.1A) has a cellulosic wall and is connected with the circumjacent cells through plasmodesmata. During megasporogenesis, callose (Rodkiewicz 1970) is deposited on the megaspore mother-cell wall, plasmodesmatal connections become obliterated, and the megaspore mother cell becomes isolated from the adjoining tissue (see Kapil and Bhatnager 1981). It undergoes meiotic divisions. The heterotypic division results in the formation of a dyad (Fig. 3.5.1B). The homotypic division occurs in each dyad cell and a linear or T-shaped tetrad is produced (Fig. 3.5.1C, E). Suppression of homotypic division in one of the dyads results in the formation of a triad (Fig. 3.5.1G) as in Cactaceae,

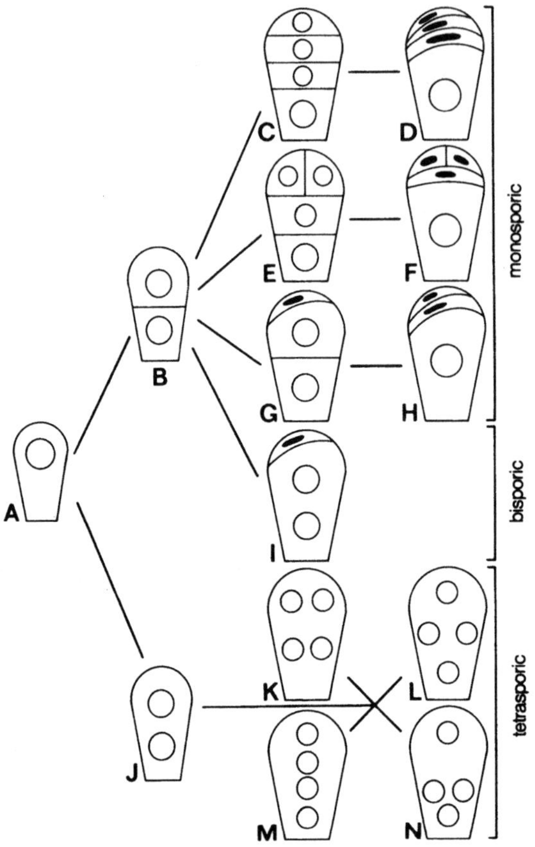

Fig. 3.5.1A–N. **A** Megaspore mother cell. **B** Dyad following heterotypic division (meiosis I). **C** Tetrad following homotypic division (meiosis II). **D** Three megaspores degenerate, chalazal functional. **E** T-shaped tetrad. **F** Chalazal megaspore functional, three degenerated megaspores. **G** Two megaspores with degenerated dyad. **H** Functional megaspore with degenerated dyad and one megaspore: **I** Lower dyad binucleate, upper degenerated. **J** Meiosis I without wall formation. **K–L** Meiosis II without wall formation. 2 + 2 (**K**), 1 + 1 + 1 + 1 (**L**), 1 + 2 + 1 (**M**) and 1 + 3 (**N**) arrangement of nuclei in coenomegaspore. **A–H** Monosporic, **I** bisporic, and **J–N** tetrasporic type of development of embryo sac. (After Bouman 1984)

Connaraceae and Stachyuraceae. In Crassulaceae, isobilateral tetrads have also been observed; this appears doubtful. Rembert (1971) observed ten patterns of megaspore tetrads in Leguminales, seven in monosporic type (including triad) and three in bisporic type of the embryo sac.

The absence of wall formation in megaspore mother cell, following meiosis I and II, leads to the development of a four-nucleate coenomegaspore (Fig. 3.5.1I–N; see also Schnarf 1936, Smith 1973b, Swamy 1974b, Willemse 1981).

During megasporogenesis, a notable feature is the development of callose thickenings around the non-functional megaspores. In the Polygonum type of embryo sac, the callose appears first in the chalazal region of megaspore mother cell, spreads over its entire surface, and isolates it from the circumjacent maternal tissue. Then the callose is retained only around the micropylar dyad, or around three upper megaspores so that the functional chalazal megaspore does not have any callose. The reverse condition has been noted in the Oenothera type. Its appearance first in the micropylar region, isolation of the megaspore mother cell for a short period, and later on disappearance from the micropylar dyad and the functional micropylar megaspore denote that the absence of callose is necessary for further growth of a megaspore. The distribution of callose only on transverse walls has been observed in homeopolar tetrad where any one of the megaspores develops. In the bisporic Endymion type of embryo sac of *Oenothera lamarckiana*, callose disappears from the micropylar dyad.

The functional activity of the megaspore and dyad depends on its larger size, viability for a longer period, and the occurrence of maximal organelles. In *Epipactis*, the chalazal megaspore has a porous wall which allows the inflow of metabolites, whereas the callosic walls in three chalazal megaspores, and chalazal dyad result in the loss of activity of megaspore dyad.

The accumulation of plastids containing starch in a megaspore mother cell, dyad, or a megaspore denotes its functional activity (see K.V.O. Dahlgren 1927a).

The chalazal megaspore is functional in the majority of taxa (Fig. 3.5.1C–F), and the development of embryo sac is of the Polygonum type. The micropylar megaspore functions in Balanophoraceae, Onagraceae, and Theaceae (*Camellia*). Occasionally, the epichalazal megaspore (one above the chalazal) is functional in *Cannabis sativa* (Mohan Ram and Nath 1964). At times, all the four megaspores of a tetrad develop concurrently and four binucleate megaspores (two-nucleate embryo sacs) are formed as in *Casuarina equisetifolia* (Swamy 1948a), and *Iphigenia indica* (Sulbha 1954a).

The formation of megaspore haustoria in *Sedum chrysanthum* (Fig. 25.5.2A–D) and *S. sempervivoides* (Mauritzon 1933a) is an unusual feature.

3.6 Female Gametophyte

Types of Embryo Sac

Depending on the number of megaspore nuclei taking part in its formation, the embryo sac is designated as Mono-, Bi- and Tetrasporic (Fig. 3.5.1A–N). P. Maheshwari (1937, 1941, 1946, 1947, 1948), B.M. Johri (1963a, b), and Willemse and van Went (1984) have reviewed different types of embryo sacs in angiosperms.

Monosporic Type. The monosporic type is represented by the Polygonum and Oenothera types. The chalazal megaspore of a tetrad develops into an embryo sac in the former, while the micropylar develops in the latter. The other three megaspores degenerate (Fig. 3.6.1).

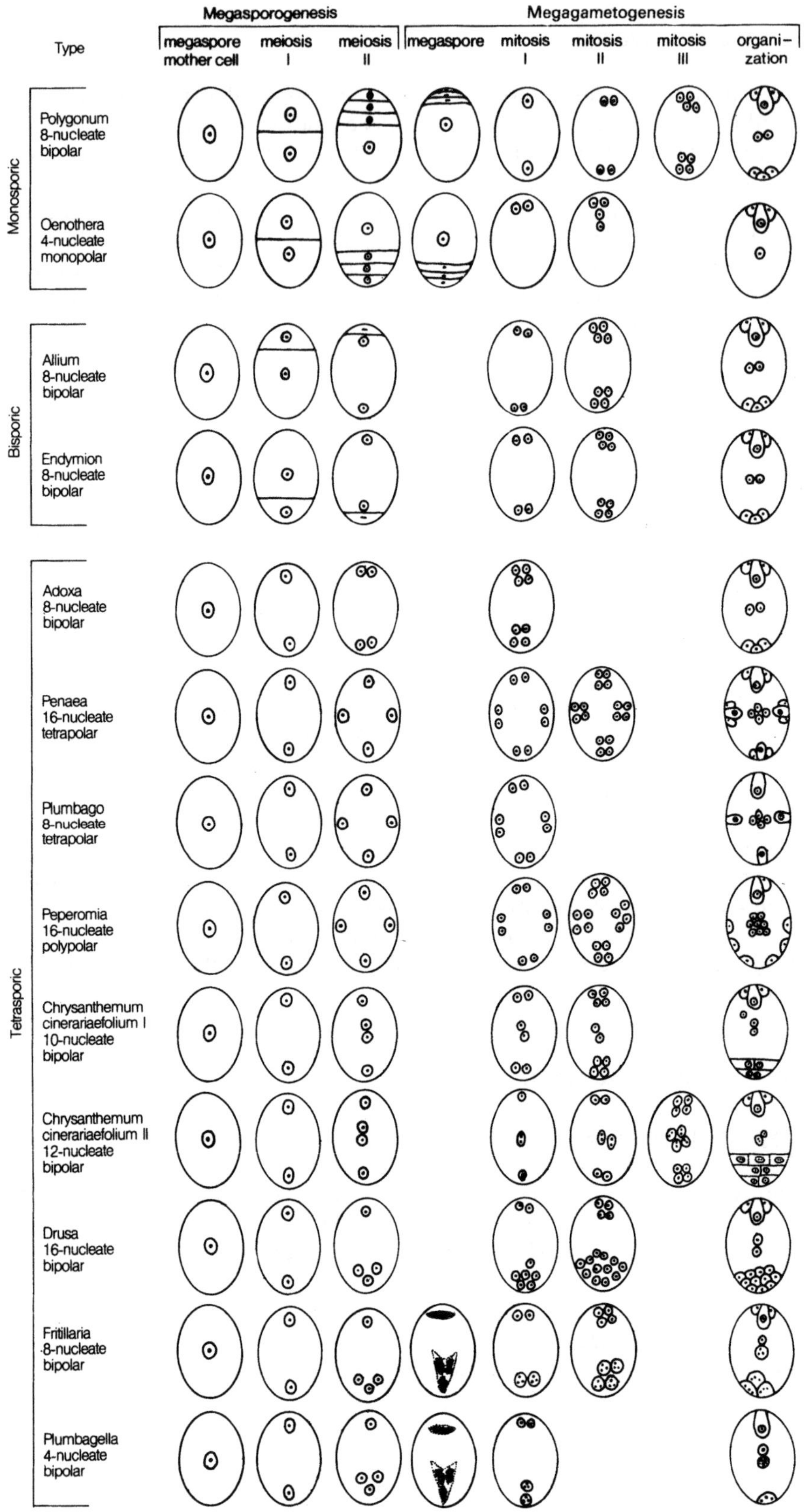

Megasporogenesis
Megagametogenesis
Type
megaspore mother cell
meiosis I
meiosis II
megaspore
mitosis I
mitosis II
mitosis III
organization
Monosporic
Polygonum 8-nucleate bipolar
Oenothera 4-nucleate monopolar
Bisporic
Allium 8-nucleate bipolar
Endymion 8-nucleate bipolar
Tetrasporic
Adoxa 8-nucleate bipolar
Penaea 16-nucleate tetrapolar
Plumbago 8-nucleate tetrapolar
Peperomia 16-nucleate polypolar
Chrysanthemum cinerariaefolium I 10-nucleate bipolar
Chrysanthemum cinerariaefolium II 12-nucleate bipolar
Drusa 16-nucleate bipolar
Fritillaria 8-nucleate bipolar
Plumbagella 4-nucleate bipolar

Polygonum Type. In the Polygonum type of embryo sac, the nucleus of the chalazal megaspore undergoes three consecutive mitotic divisions. Concurrently, the embryo sac grows along the micropylar-chalazal axis, the nuclei are separated to the poles by a central vacuole, and polarity is established at the binucleate stage. During maturation, the three nuclei from the micropylar quartet differentiate as egg apparatus consisting of the egg cell and two synergids. The fourth one is the upper polar nucleus. The three nuclei of the chalazal quartet organize into antipodal cells except in some taxa like *Cassytha*, *Scilla*, *Thismia*, *Xylopia* and *Xyris* where they degenerate before organization into cells. The fourth nucleus functions as the lower polar nucleus which usually moves upward and fuses with the upper polar nucleus to form a secondary nucleus. However, variations occur in the behaviour of the polar nuclei. They may fuse in the micropylar region of the embryo sac, in the centre, or chalazal region. Likewise, they may fuse before fertilization, or may remain appressed till fertilization.

Sometimes, the chalazal nuclei of the four-nucleate embryo sac divide earlier than the micropylar resulting in a six-nucleate embryo sac as in *Macrosolen cochinchinensis* (P. Maheshwari and Singh 1952); subsequently the micropylar nuclei also divide. Due to the suppression of division of chalazal nuclei, the embryo sac may be six- or seven-nucleate as in Alismataceae, Butomaceae and Lythraceae. A rare condition of the absence of central vacuole, during development and maturation of the embryo sac, has been noted in *Aneilema paniculatum* (Chikkannaiah 1969). This needs confirmation.

In *Balanophora* (Fagerlind 1945a, Arekal and Shivamurthy 1978, Engell 1979), the egg apparatus is always organized in the lower pole of the gametophyte, while the nuclei in the upper pole fuse successively (Sect. 13.2).

Oenothera Type. In the Oenothera type, the lower three megaspores degenerate (Fig. 39.13.2B–E), and the nucleus of the functional micropylar megaspore undergoes mitotic division. A notable feature is that at the two-nucleate stage of embryo sac the nuclei do not migrate to the two poles, but stay at the micropylar pole and undergo one more mitotic division. These four nuclei organize into the egg apparatus and upper polar nucleus (Fig. 3.6.1). This feature is characteristic of the Onagraceae. In the mono- and bisporic type of embryo sac in *Schisandra chinensis*, Swamy (1964a) describes four-nucleate embryo sacs with a similar organization. The primary chalazal nucleus is absent in both cases.

Bisporic Type – Allium and Endymion type. Bisporic development occurs when either the chalazal or micropylar dyad undergoes the second meiotic division. The former condition occurs in the Allium type (Fig. 3.6.1), the latter in the Endymion type. The non-functional dyad, in both types, degenerates. The functional dyad undergoes two mitotic divisions forming an eight-nucleate embryo sac. The organization is similar to that in the Polygonum type.

◄

Fig. 3.6.1. Schematic representation to show the development of different types of embryo sacs. (See Willemse and van Went 1984)

Tetrasporic Type. In the tetrasporic type, cytokinesis does not follow meiotic divisions in the megaspore mother cell, and a coenomegaspore with four haploid nuclei is formed (Fig. 3.5.1A–N).

The variations in the arrangement of the four megaspore nuclei in a coenomegaspore lead to the formation of different types of embryo sacs (Fig. 3.6.1): (1) Two megaspore nuclei lie at each pole (Adoxa type), (2) four megaspore nuclei lie in a crosswise manner – one at each pole and two placed laterally (Penaea, Plumbago and Peperomia type), (3) one megaspore nucleus lies at each pole, and the other two remain in the centre (Chrysanthemum cineraraefolium type), (4) the arrangement of the nuclei is 1 + 3, i.e. one at the micropylar pole and three at the chalazal pole (Drusa, Fritillaria and Plumbagella type).

Adoxa Type. The nuclei (2 + 2) of the coenomegaspore undergo one mitotic division to form an eight-nucleate embryo sac. The organization is similar to that of the Polygonum type (Fig. 3.6.1).

Penaea Type. The nuclei (1 + 1 + 1 + 1) of the coenomegaspore undergo two mitotic divisions forming 16 nuclei, in four quartets. A single nucleus from each quartet migrates to the centre, and these four nuclei behave as polars. The three nuclei at the micropylar pole organize into the egg apparatus; the three nuclei in the other groups also organize into cells but do not function as egg-apparatus (Fig. 3.6.1).

Plumbago Type. The arrangement of nuclei in the coenomegaspore is crosswise (1 + 1 + 1 + 1). The nuclei undergo only one mitotic division forming eight nuclei in four groups. One nucleus from each group migrates to the centre and functions as polar nucleus. The nucleus at the micropylar pole organizes as the egg cell; a synergid is absent. The remaining three nuclei also organize into cells but soon degenerate (Fig. 3.6.1).

Peperomia Type. The arrangement of the four nuclei in the coenomegaspore is again crosswise (1 + 1 + 1 + 1) or irregular. Each nucleus undergoes two mitotic divisions forming four quartets, or the nuclei are irregularly distributed in the embryo sac. The two nuclei at the micropylar pole differentiate as the egg cell and a synergid. Eight nuclei migrate to the centre and unite to form an octoploid secondary nucleus. The remaining six nuclei organize as peripheral cells at the chalazal end, and function as antipodals (Fig. 3.6.1). This behaviour is characteristic of *Peperomia pellucida* (Fagerlind 1939a). Occasionally, the egg and one synergid organize at the micropylar pole, and 14 nuclei fuse in the centre to form the secondary nucleus. This happens in *P. hispidula* (D.S. Johnson 1914).

Chrysanthemum cinerariaefolium Type. The nuclei of the coenomegaspore show 1 + 2 + 1 arrangement. Two variations occur in this type, due to different behaviour of the two centrally located megaspore nuclei. In the first case, these two nuclei do not undergo any division, while the nuclei at the poles divide twice and the embryo sac becomes ten-nucleate. The three nuclei of the micropylar quartet organize into the egg apparatus, the fourth one, along with two central nuclei, behave as polar nuclei. The four nuclei at the chalazal pole are cut off as antipodal cells (Fig. 3.6.1).

In the second case, the two centrally-situated megaspore nuclei fuse. The latter and two nuclei at the poles undergo two mitotic divisions resulting in three quartets, two with haploid nuclei, one at each pole, and the third with diploid nuclei in the centre. The three nuclei of the micropylar quartet differentiate as egg apparatus. The fourth one and one diploid nucleus from the central quartet function as polar nuclei. The remaining seven nuclei (three diploid and four haploid) organize as antipodals (Fig. 3.6.1).

Drusa Type. The arrangement of nuclei in a coenomegaspore is 1 + 3. Each nucleus undergoes two mitotic divisions to form a 16-nucleate embryo sac. The three nuclei at the micropylar pole differentiate as egg apparatus, the remaining one migrates to the centre as the polar nucleus. Eleven antipodal cells are cut off and the remaining one behaves as the lower polar nucleus (Fig. 3.6.1).

Fritillaria Type. The arrangement of the nuclei in a coenomegaspore is again 1 + 3. These nuclei undergo two mitotic divisions. During the first division, the three chalazal spindles fuse to produce two triploid daughter nuclei. A haploid egg apparatus differentiates from the micropylar quartet. The remaining haploid nucleus recedes to the centre as the upper polar nucleus. One of the triploid nuclei from the chalazal quartet behaves as lower polar nucleus; the other three nuclei are cut off as triploid antipodal cells (Fig. 3.6.1).

Bambacioni (1928a), for the first time, discovered that in *Fritillaria* the spindles of the three chalazal nuclei fuse during the mitotic division resulting in two triploid nuclei. This phenomenon is referred to as the "Bambacioni effect".

The Fritillaria type in *Clintonia* is noteworthy. Pahuja and Kumar (1970) reinvestigated the development in *C. undensis* var. *alpina.* The arrangement of four megaspore nuclei in a coenomegaspore is 1 + 3. The micropylar nucleus undergoes two successive mitotic divisions to produce a micropylar quartet which later organizes into the egg apparatus and upper polar nucleus. The three chalazal megaspore nuclei fuse and degenerate. Thus, the antipodals and the lower polar nucleus are absent.

Pahuja and Kumar (1970) designate this reduced Fritillaria type as biphasic (two mitotic divisions after the meiotic divisions of megaspore mother cell), and pseudomonosporic (single megaspore nucleus functional).

Plumbagella Type. The arrangement of megaspore nuclei in a coenomegaspore is 1 + 3, and a single mitotic division leads to the formation of a mature four-nucleate embryo sac. The three chalazal spindles fuse during division resulting in two triploid daughter nuclei. A single nucleus at the micropylar pole differentiates as the egg cell, the other behaves as the upper polar nucleus. The lower polar nucleus from the chalazal pole is triploid. A single triploid antipodal cell is cut off (Fig. 3.6.1).

Variations in the Types of Embryo Sacs
The development of more than one type of embryo sac in a family, or genus, or even in a species, is not uncommon. Kapil and Prakash (1966) recorded five different types (Polygonum, Endymion, Drusa, Penaea and Adoxa) of development in *Delosperma cooperi* (Aizoaceae). Dharamadhaj and Prakash (1978) observed twin embryo sacs, one monosporic and the other bisporic, in the same nucellus of an ovule of *Capsicum annuum* var. *acuminatum*. However, according to them, the formation of the monosporic type is more frequent in summer, and that of the bisporic type in winter. The temperature may have a definite role in determining the type of embryo sac. These observations support the results obtained by Hjelmqvist and Grazi (1965).

Fagerlind (1944a) classified tetrasporic embryo sacs into ten categories. The criteria used are: (1) number of mitotic divisions (phases) intervening between meiosis in megaspore mother cell and organization of mature female gametophyte, (2) number of poles (or groups) at which the nuclei are arranged in a coenomegaspore, and (3) presence or absence of fusion of nuclei during mitotic divisions.

Thus, the development in the Adoxa and Plumbago type is monophasic (only one mitotic division occurs), but the arrangement of nuclei is bipolar (two groups) in Adoxa and tetrapolar (four groups) in the Plumbago type. Similarly, it is biphasic (two mitotic divisions) in the Penaea type, with tetrapolar arrangement of nuclei.

Swamy and Krishnamurthy (1975) discuss a new approach to the classification of embryo sacs. According to them, in the previous system, the number of megaspore nuclei involved in the constitution of the female gametophyte, and the number of divisions intervening between the megaspore mother cell and the mature embryo sac are emphasized. They consider the development of egg apparatus from a single primary micropylar nucleus as more "stable and stereotyped". The development at the chalazal pole of the embryo sac is more variable, and has less functional significance.

In their proposed system, the successive steps involved in the ontogeny of embryo sac are: (1) attainment of haploidy, (2) establishment of polarity leading to the formation of egg apparatus mother nucleus, (3) elaboration phase including steps leading to the formation of egg apparatus, and (4) mature organization. They recognize four basic types of embryo sac development, acknowledging the time of polarity as the guiding factor. The proposed system consists of "Supra-Homotypic" and "Homotypic" categories (Fig. 3.6.2).

Supra-Homotypic Category. Polarity is established after one more division following the homotypic division, i.e. at the two-nucleate stage of embryo sac, and a single primary antipodal nucleus is present – Polygonum type. The Oenothera type is similar but the primary antipodal nucleus is absent.

Homotypic Categories. *Category I.* Characterized by the establishment of polarity immediately following homotypic division. A single primary antipodal nucleus is present – Allium, also Endymion type. *Category II.* Polarity is established following homotypic division, and three primary antipodal nuclei are present – Penaea type; it also gives rise to the Drusa (even Peperomia),

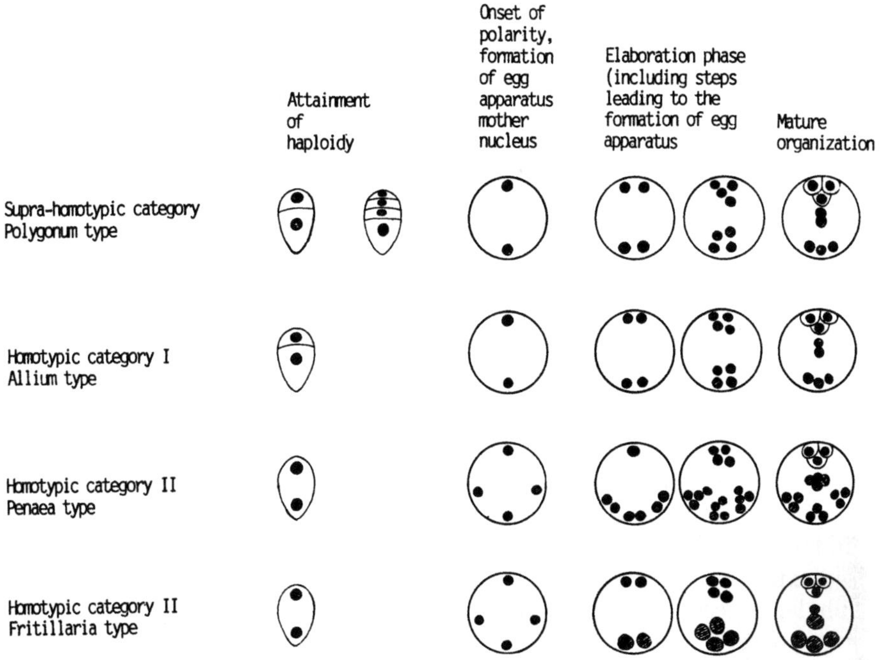

Fig. 3.6.2. Schematic representation of ontogenetic sequence of embryo sac types, based on the onset of polarity (micropylar nuclei *black*, chalazal *shaded*) after attaining haploidy. (After Swamy and Krishnamurthy 1975)

Plumbago and Adoxa types. The Fritillaria type, also the Plumbagella type, belong to Category II; the chief difference is that there is nuclear fusion and triploid nuclei are formed in the chalazal region.

While discussing the "Supra-Homotypic" and "Homotypic" categories, Swamy and Krishnamurthy (1975) state that at the two-nucleate stage of embryo sac the micropylar nucleus should be designated "egg apparatus mother nucleus", and the chalazal one as "primary antipodal nucleus". Since the micropylar and chalazal nuclei also give rise to upper and lower polar nuclei, Nagendran and Arekal (1976) emphasize that the upper nucleus be designated as "micropylar quartet mother nucleus", and the lower as "chalazal quartet mother nucleus".

The classification of embryo sacs proposed by Swamy and Krishnamurthy, and the nomenclature suggested by them and Nagendran and Arekal for the micropylar and chalazal nuclei of two-nucleate embryo sacs are of no special significance, and confusing. The conventional classification and nomenclature are much simpler, and convincing.

3.7 Mature Embryo Sac

In the majority of angiosperms, the embryo sac remains confined to the nucellus (within the limits of integuments) during its growth and maturation. But in *Pentaphragma horsfieldii* (Kapil and Vijayaraghavan 1962) and *Torenia violacea* (Yamazaki 1954), it becomes extra-micropylar at the four-nucleate stage. In *Lepidagathis trinervis* (Ambegaokar 1959) and *Nelsonia campestris* (Mohan Ram and Masand 1963), the apical end grows beyond the micropyle in the schizogenous cavity in the funicular region, and the egg apparatus differentiates in this cavity. An extreme condition is observed in *Exocarpus* spp. (Ram 1959a), where the embryo sac grows in the placenta (in the absence of a true ovule) and beyond it so that the egg apparatus becomes exposed (Fig. 3.4.3C).

Loranthaceae show a different pattern. Several embryo sacs grow concurrently up to varying lengths in the stylar and stigmatic region (Fig. 12.8.2F). The embryo sac exhibits maximum length in *Moquiniella rubra* (B.M. Johri and Raj 1969) where it grows up to the tip of stigma as in *Helixanthera lingustrina* (Fig. 12.8.2C), but (in *Moquiniella*) it curves backward and becomes inverted J-shaped (Fig. 12.8.2D).

The embryo sac develops micropylar and chalazal extensions, as in Acanthaceae, Santalaceae, Stylidiaceae and Opiliaceae. A singular condition is reported in *Exocarpus*, and finger-like outgrowths are produced in the apical and middle region of the embryo sac (Fig. 12.6.3A–C). The apex of the embryo sac extends in the stylar canal and broadens laterally. It develops several tubular processes which grow downward from the micropylar region of the embryo sac along the side of the placenta (Fig. 12.6.3A). Some of these haustorial processes extend upward beyond the level of the egg apparatus (Fig. 12.6.3B), and may reach the outer epidermis of the style. The development of a branched micropylar caecum in *Nuytsia floribunda* (R. Narayana 1958b) is also rare (Fig. 12.8.2E).

The mature embryo sac usually shows three-celled egg apparatus, three antipodal nuclei or cells, and two polar nuclei or secondary nucleus (Masand and Kapil 1966, B.M. Johri and Bhatnagar 1973). The synergids are absent in the Plumbago and Plumbagella types, and antipodals in the Oenothera type (Fig. 3.6.1). The number, as well as the ploidy-level, of the antipodal cells varies.

The mature embryo sac in Podostemaceae exhibits a different pattern (Fig. 28.2.1D, E), and a detailed description is given along with the family (Sect. 28).

The egg apparatus consists of the egg cell and two synergids (see Favre Duchartre 1975). It remains within the limits of the embryo sac wall except in Compositae, Ericaceae and Stylidiaceae where the synergids project into the micropyle. The egg cell shows a distinct polarity with its nucleus embedded in the dense cytoplasm at the lower end; the upper portion is occupied by a vacuole. In *Plumbago capensis* (Cass 1972), at the micropylar end the egg cell shows conspicuous wall ingrowths, the filiform apparatus. In the absence of synergids, the egg cell has taken over the function of the synergid, in addition to its gametic function.

The egg cell of the aposporous (nucellar) embryo sac of *Cortaderia jubata* (M.N. Philipson 1978b), rarely exhibits an extension (haustorium) which enters the micropyle, adjoining the synergid haustorium.

The accumulation of starch grains in the embryo sac is a common feature in several families, such as Acanthaceae, Aizoaceae, Bruniaceae, Cactaceae,

Crassulaceae, Frankeniaceae, Portulacaceae and Tiliaceae. In *Blepharis maderaspatensis* (Phatak and Ambegaokar 1963), the starch grains in the mature embryo sac are in such abundance that they mask the nuclei. The starch is gradually consumed, and disappears during post-fertilization stages. Compound starch grains have been reported in *Cansjera rheedii* (Swamy 1960). Fat globules occur in Sonneratiaceae.

Multiple Embryo Sacs. The concurrent development of two megaspore mother cells leads to the development of twin tetrads as in *Turnera ulmifolia* (Vijayaraghavan and Kaur 1966), and twin embryo sacs as in *Linum perenne* (D. Rao 1966). Rarely, mono- and bisporic gametophytes may develop side by side as in *Capsicum annuum* var. *acuminatum* (Dharmadhaj and Prakash 1978).

Six embryo sacs develop concurrently in *Cassytha filiformis* (Sastri 1962), ten or more in *Pterospermum suberifolium* (Venkata Rao 1952b), and 20–25 in *Casuarina equisetifolia* (Swamy 1948a).

Synergids. The synergids are pyriform and "hooked", or the tip is prolonged into a "beak". The cell wall is present only in the micropylar region. Similar to the egg, they also exhibit polarity. The nuclei lie in the upper narrow region, while the lower region is occupied by a vacuole. They show a characteristic structure, the filiform apparatus, at the tip. The apical "caps" have been reported in *Helianthemum vulgare* (Kapil and R. Maheshwari 1964).

The synergids are haustorial in Compositae – *Calendula* and *Ursinea* (K.V.O. Dahlgren 1924), *C. fulgida*, *C. maritima*, *C. persica*, *C. tunetana* (Ahlstrand 1979d) and *Osteospermum sinuatum* (Fig. 3.7.1A), *Cotula* (Davis 1961c), *Elephantopus* (Pullaiah 1979a) and *Ursinia anthemoides* (see Poddubnaya-Arnoldi 1976); Crassulaceae – *Sedum sempervivoides* (Mauritzon 1933a); Santalaceae – *Quinchamalium chilense* (Fig. 3.7.1B–D); Gramineae – *Cortaderia jubata* (Fig. 3.13.2D) and *C. selloana* (M.N. Philipson 1977, 1978b).

The role of synergids in the process of fertilization has attracted much attention. They secrete chemotropic substances and guide the growth of pollen tube; one of the synergids serves as a landing site for the pollen tube and provides nutrition (see Vijayaraghavan and Bhat 1983).

Following two successful pollinations, the sections of 75 ovules of *Hordeum vulgare* (cv Bonus) were examined to determine the pattern of early synergid degeneration (Mogensen 1984). The degeneration of both synergids is equally as likely as the degeneration of only one synergid. When only one synergid degenerates, it is mostly located towards the placenta (proximal).

Antipodals. The antipodals show much variation in number, size, ploidy-level, and may be ephemeral or persistent. Normally, there are three antipodal cells, but in a large number of plants they undergo secondary multiplication; the largest number is reported in Gramineae. The antipodals do not organize into cells in *Cassytha*, *Hydnora*, *Scilla*, *Thismia* and *Xyris*. The antipodal cells become polyploid, due to endomitosis or polyteny. They may be multinucleate and multinucleolate.

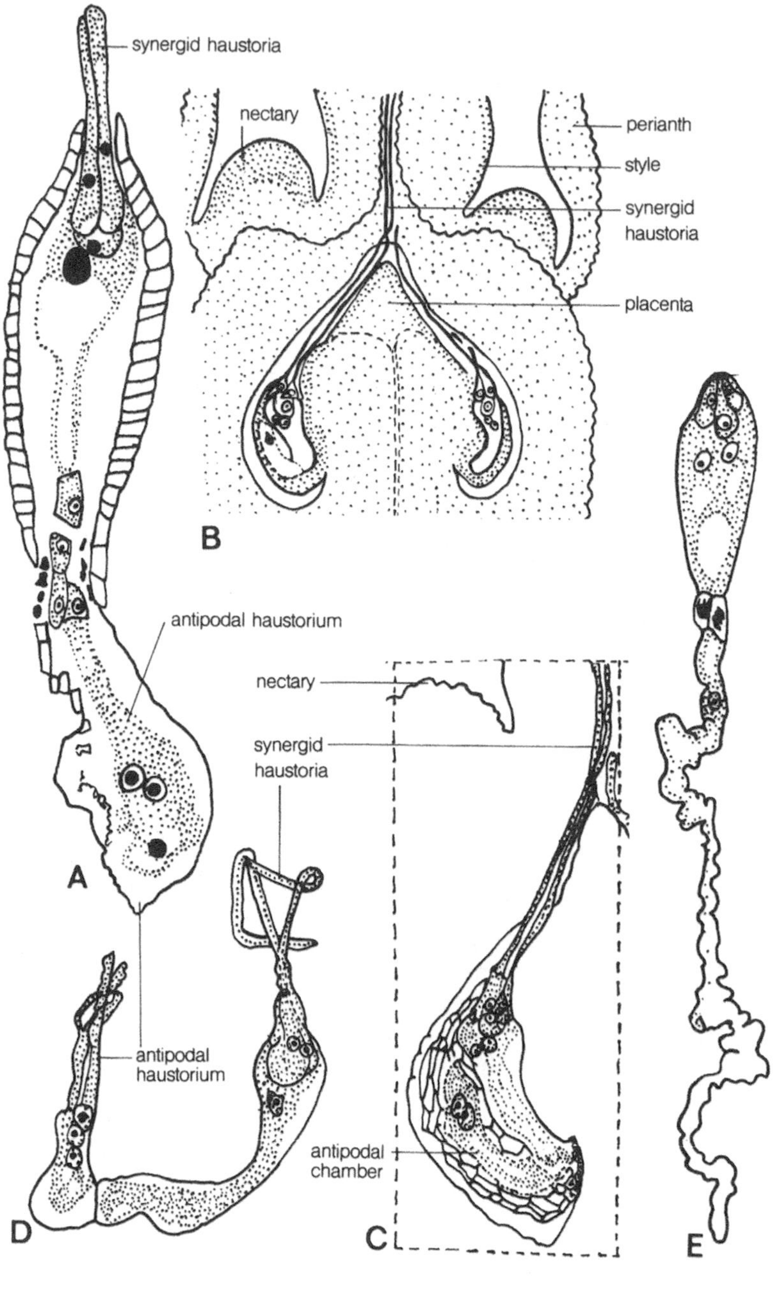

Fig. 3.7.1A. *Osteospermum sinuatum*, mature embryo sac with synergid haustoria, and an antipodal haustorium. (After Ahlstrand 1979d). **B–D** *Quinchamalium chilense*. **B** Longisection of lower portion of gynoecium; **C** Embryo sac from **B** (*left*). **D** Mature embryo sac (from dissection). Note extensive synergid haustoria in **B–D**, and branched antipodal haustorium in **D**. (After Agarwal 1962). **E** *Galium mollugo*, mature embryo sac with much-elongated basal antipodal cell. (After Fagerlind 1937)

At times, the antipodal cells enlarge considerably as in *Aconitum napellus* (Osterwalder 1898), *Argemone* (Sachar 1955) and *Ranunculus tripartitus* (T.K. Patel and Cook 1972), and become haustorial. In *Zea mays* (Diboll and Larson 1966), the antipodals are separated from one another by incomplete walls, and those adjacent to chalaza show conspicuous wall ingrowths. In *Grindelia* and *Prionopsis*, Howe (1975) observed that one or both antipodal cells grow laterally into the integuments. These outgrowths invade the endothelial cells, and occasionally approach the surface of the ovule. Howe's observations need confirmation. Their role in the nutrition of the embryo sac is undisputable.

A unique condition has been observed in *Quinchamalium chilense* (B.M. Johri and Agarwal 1965). The three nuclei at the chalazal end of the embryo sac are contained in an antipodal chamber which develops into a prominent haustorium (Fig. 3.7.1C, D). The haustorium grows through the funiculus, and branches profusely as it reaches the tip of placenta.

Osteospermum sinuatum (Fig. 3.7.1A) shows a conspicuous antipodal haustorium. In *Gallium mollugo* (Fig. 3.7.1E), the basal antipodal cell forms a long tubular haustorium.

The antipodals exhibit an unusual feature in the hybrids and polyploid varieties of Gramineae. According to Beaudry (1951), the antipodals alter the nutrient supply causing the break down of endosperm in a cross between *Elymus virginicus* L. × *Agropyron repens* (L.) Beauv. In an octoploid variety of *Hierochloë odorata*, Weimarck (1967a) noticed a degenerated mature embryo sac with five multinucleate antipodals. He considered it as an abnormality due to degeneration. B.M. Johri and Ambegaokar (1975) made a comparative study of the development of antipodals in relation to development of endosperm in three varieties of hexaploid triticales (a cross between *Triticum* and *Secale cereale*). The variety T 6437 shows maximum shrinkage of seeds due to failure of the development of endosperm. In this variety, the antipodals exhibit maximal size with hypertrophied nuclei, and they persist for a much longer time as compared to the varieties Triticale-1 and Bronco 90, which show normal development of endosperm.

Central Cell. The central cell is separated from the egg cell and synergids by plasma membranes, and from the antipodals by thin walls often traversed by plasmodesmata. The outer wall of the central cell may show transfer cell-like wall ingrowths. Thus, it is well-located for absorption, conduction and secretion of nutrients. It also serves as a repository for starch, lipids and proteins (Kapil and Bhatnagar 1983). During post-fertilization stages, the wall labyrinth of the central cell increases and new organelles appear.

Normally, there are two polar nuclei in the central cell, except in the Oenothera, Penaea, Plumbago, Peperomia and Chrysanthemum cinerariaefolium types (Fig. 3.6.1). When there are more than two, all of them come together and fuse simultaneously to form a secondary nucleus. Sometimes, the two polar nuclei remain distinct till triple fusion. The latter is said to be responsible for the stimulation of metabolic activity within the central cell.

3.8 Pollination

During pollination, the pollen grains are brought over to the stigmatic surface by various agencies. The role of style as a "sense-organ" is discussed by Gilissen (1976). Pollination in land plants has been studied for a much longer period than in aquatic plants. This aspect has received some attention in the description of the families (see also E.G. Williams et al. 1986).

The primitiveness of wind-pollination has been a matter of much discussion. There is no doubt, however, that the aquatic habit of higher plants and hydrophily are derived conditions. Most aquatic plants are still pollinated in the air, like other terrestrial plants: *Nymphaea*, *Alisma* and *Hottonia* are good examples of entomophily, *Potamogeton* or *Myriophyllum* for anemophily, and *Lobelia dortmanna* for autogamy. Some aquatics have taken advantage also of the water for pollination (Faegri and van der Pijl 1971).

Water-pollination may take place either on the surface of water (ephydrophily), a further step in the development from entomophily or anemophily. Not all pollinations taking place in water are hydrophilous. Many small, autogamous terrestrial plants may flower even when submerged, and the autogamous mechanism functions even under these circumstances, generally in an air bubble confined within a flower (Faegri and van der Pijl 1971).

Ephydrophily is unique among abiotic pollinations in that it takes place in a two-dimensional medium. Compared with a three-dimensional medium of anemophily or hyphydrophily, this makes possible a great saving of pollen.

In the classical example of *Vallisneria*, the entire male flower is released instead of individual pollen grains; the pollen itself therefore does not touch the water surface. Small depressions in the surface around the floating female flowers cause the male flower to slide down and the anthers attach themselves to the stigma. This effective mode of pollination results in the reduction of the number of pollen grains (ca 72) in a male flower. A similar mechanism is also reported in *Hydrilla* (explosive anthers), and *Lemna trisulca* where the whole plant rises up to the surface.

Water may come into the picture as a pollinating agent also in terrestrial plants, e.g. in rain-pollination. During rain the flowers fill with water up to a certain level. Pollen grains float on the surface of this water, and eventually reach the stigmas that are exposed at the level of water. Primarily, rain-pollination is autogamous, but a mechanism of allogamy is also conceivable with rain-splash, even if it is not very effective. So far, no flower is known which is dependent exclusively on rain-pollination.

It is essential for all forms of hydrophily that the pollen is not killed by immersion in water. Ephydrophily and rain-pollination presume water-repellent pollen. In hyphydrophilous plants pollen grains endure wetting.

Cox and Knox (1986, see also Knox and Ducker 1976) have given an excellent account for four taxa. The marine sea-grasses *Halodule pinifolia* and *Halophila ovalis*, and fresh-water aquatics *Lepilaena cylindrocarpa* and *Ruppia polycarpa* are surface-pollinated. The pollen floats on the surface upon release from the anthers. The individual pollen grains join together in large floating rafts that function as search vehicles. Initially, each search vehicle represents the products of a single anther but subsequently through collisions even larger search vehicles of mixed parentage are formed. Pollen and stigma differ significantly between the marine and freshwater types (Tables 3.8.1). In the freshwater genera *Lepilaena* and *Ruppia*, the pollen is ellipsoidal or spherical while the floating stigmas present a small, round target (the indusiate stigma plus the radius of surface depression in the surface plane). In the marine genera *Halophila* and *Halodule*, however, the pollen is filiform as are floating stigmas which present a linear target of decreasing length as the tide rises. In all types they are either long and filamentous or borne on long filamentous peduncles which (after fertilization) retract the developing fruit beneath the surface.

In *Lepilaena* and *Ruppia*, anthesis occurs under water or at the surface. If the *Ruppia* inflorescence is under water, a gas bubble propels the mass of pollen to the water surface. There the pollen grains line up side by side forming search vehicles. Surface and water currents propel these vehicles, and on encountering the stigma, they adhere. In *Lepilaena*, the boat-like anthers are released under water, float to the surface, and dehisce. These search vehicles drift through the influence of wind and waves, some are trapped in the slight depressions in the water surface formed by the floating tripartite stigmas of the apocarpous female inflorescences. In both these freshwater systems, search vehicles which fail to encounter stigmas fuse into a yellow hydrophobic surface film that accumulates towards the edge of the pond.

In *Halodule*, the tiny stamens become erect and dehisce when exposed to the air at low tide. The anthers express the filiform pollen as a flocculent mass. As the tide comes in, the pollen grains form search vehicles of large diameter which float on the surface. The long linear stigmas also float and pollination occurs through the collision of a search vehicle with the floating stigmas. In *Halophila*, the staminate flowers produce three anthers which are adnate until just before dehiscence at low tide. At dehiscence, the pollen tetrads (which remain held in their mucilaginous tubes) are released as flocculent masses that rapidly expand into search vehicles. The long floating filaments of the three branched stigmas form linear targets for the search vehicles.

After a detailed consideration of four systems of aquatic pollination, pollination in a few land plants may be taken up. In grasses (Gramineae) and sedges (Cyperaceae), pollination is typically anemophilous. Pant et al. (1982a, b) observed entomophilous pollination in *Dichanthium annulatum*, *Cyperus rotundus* and *Scirpus lacustris*. The inflorescences of these plants are regularly visited for foraging by two types of insects. Those which collect pollen grains on their hind-legs – *Apis indica*, *Micrapis florea*, *Megapis dorsata* and *Melipona laeviceps*, and those which collect pollen grains on their hind-legs, thorax and abdomen – *Nomia andrenina*, *N. westwoodii*, *Lasioglossum* (*Ctenonomia*) *albescens*, *Halictus* (*Seladonia*) *lucidipennis* and *Evylaeus* sp.

Some of the insects are pollen specific, while others visit different flowers for pollen and nectar.

Armbruster (1984) has drawn attention to the fact that instead of the usual nectar, some flowers produce resin, e.g. *Dalechampia* (Euphorbiaceae) and *Clusia* (Guttiferae). Bees that pollinate these flowers collect and use the resin in their nest construction. Floral resins are slow in hardening as compared to

Table 3.8.1. Characteristics of surface-pollinated genera (Adapted after Cox and Knox 1986)

	Cymodoceaceae *Halodule* n = 114	Hydrocharitaceae *Halophila* n = 5	Zannichelliaceae *Lepilaena* n = 172	Ruppiaceae *Ruppia* n = 300
Habitat	Marine	Marine	Freshwater	Freshwater
Pollen Morphology				
Shape	Filiform	Ellipsoidal[a]	Spherical	Boomerang
Length (μm)	1000	80	50	140
Search Vehicle Diameters				
Median (μm)	5000	4690	338	1361
Range	1000–22000	1320–7520	118–2020	220–1817
Transport	Floats	Floats	Floats	Floats
Stigma Characteristics				
Shape	Filiform	Filiform	Indusiate	Indusiate
Length (μm)	22000	13000	776	1222
Placement	Floats	Floats	Floats	Floats

[a] Embedded in long mucilaginous tube.

many other plant resins. This facilitates collection and storage by bees, and probably enhances the attractiveness of the flowers. Floral resins are presumably of great utility and dependability and may be an especially important resource to certain bees, which are major pollinators of many species of tropical plants. Bees appear to forage floral resin using a strategy of energy efficiency: large bees collect resin only from sources with a large amount of resin; small bees collect resin from sources with a small as well as a large amount of resin.

Resin secretion in flowers may have primarily arisen as a defence against herbivores, and secondarily as an "award" to the pollinator. These bees may also collect some pollen, besides resin. This aspect of floral resin attracting pollinators requires further study.

Ant pollination systems are remarkably rare. Beattie et al. (1984) point out that the pollen exposed to ants for brief periods exhibits reduced viability, reduced per cent of germination, and shorter pollen tubes relative to control pollen. Pollination with ant-borne pollen on *Lycopersicon peruvianum* resulted in lower seed-set than with control pollen. This disruption of pollination processes could exert a powerful selection pressure against the evolution of ant-pollination systems. It is suggested that the nest-building and brood-rearing habit of ants requires that they secrete large amounts of antibiotics to combat pathogenic microorganisms. It is these secretions that disrupt pollen function; bees and wasps exhibit very different nesting behaviour. The eggs, larvae and pupae lie exposed in the nest, unprotected by either a wax cell as in bees, or a paper one as in wasps. Consequently, there are no chemical barriers to their co-evolving with flowers as pollinators.

Intracarpellary Pollen Grains. In *Butomopsis lanceolata* (B.M. Johri 1936b, see also B.M. Johri and Ambegaokar 1984), occasionally the pollen grains were observed in the stylar canal. In one instance a pollen grain had found its way into the ovary and germinated on the surface of an ovule. The occurrence of intrastylar and intraovarian pollen grains is a rare feature, and has been observed in *Amianthium muscaetoxicum* (Eunus 1951a), *Annona* (Adatia 1946), *Bootia cordata*, *Butomus umbellatus*, *Hydrocleis nymphoides*, *Limnocharis emarginata* and *Trillium* (B.M. Johri and Bhatnagar 1957), *Erythronium americanum* (Haque 1951), *Fritillaria roylei* and *Lilium tigrinum* (Vasil and M.M. Johri 1964), and *Ottelia alismoides* (Islam 1950).

Pollen Germination. When the pollen germinates, mostly it produces a single tube. Polysiphonous condition refers to the emergence of several pollen tubes (Fig. 3.8.1A). Sometimes, three to five tubes emerge and one or two tubes may branch (Fig. 3.8.1B, C). The generative cell and tube nucleus migrate into separate branches (Fig. 3.8.1C). In *Lilium regale* (Fig. 3.8.1D), the pollen tube becomes coiled and produces one or two side branches. The main tube continues to elongate even after the generative cell and tube nucleus migrate into a branch. Branched pollen tubes are common in the Casuarinaceae (Fig. 4.2.1B) and Betulaceae (Figs. 9.2.1, 9.2.2A).

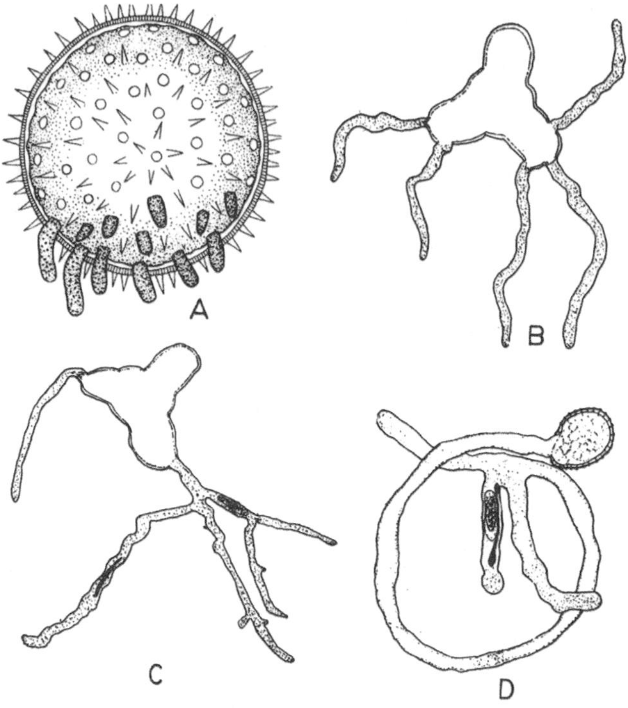

Fig. 3.8.1A–D. Polysiphonous germination of pollen grains. **A** *Lavathera* sp., emergence of many tubes. (After Poddubnaya-Arnoldi 1976). **B, C** *Clarkia elegans*. **B** Pollen with five unbranched tubes; **C** Pollen with one un-branched tube, and second tube branched with generative cell and tube nucleus in different branches. **D** *Lilium regale*, the pollen tube has coiled; the generative cell and tube nucleus are in the same branch. (After Johnston 1959)

Usually, the pollen grains germinate singly (Fig. 65.2.1A). In taxa in which pollen grains are shed in tetrads (Fig. 65.2.1H–K), or pollinia (Fig. 26.19.1B), or massulae (Fig. 65.2.1M), pollen germination occurs in composite units. All the four microspores in a tetrad may germinate synchronously; the outermost microspores in a massulae germinate first and then the others.

Post-Pollination Events. Using a semi-vivo pollination technique, G.B. and D.L. Mulcahy (1985) draw attention to the influence of the ovary on the growth of pollen tube.

Pollination on cut styles of *Nicotiana elata* and *Petunia hybrida* showed poorer growth of pollen tubes as compared to growth on styles intact with the ovary. The presence of other floral parts has no effect on the growth of pollen tubes.

Following a detailed study of six species of the Umbelliferae, eight of Cruciferae, six of *Solanum* and ten unrelated species, Cruden and Lyon (1985) emphasize that there is a positive correlation between stigma depth – approximate distance a pollen tube has to grow to reach the exogenous source of nutrients in the transmitting tissue – and pollen grain size, and *not* between pollen grain size and stylar length. All these pollen grains have a starchy food reserve. Absence of a correlation between pollen grain size and stylar length is inconsistent with the hypothesis that pollen grains must contain sufficient nutrients to sustain the growth of the tube up to the ovule. In fact, as early as 1830, Amici had suggested that the pollen tubes, during their growth, obtain nutrients from the transmitting tissue. In recent years, many experimental studies have demonstrated the utilization of stylar nutrients by the growing pollen tubes.

Pollen-Pistil Interaction. The surface of the stigma is the landing and germinating platform for pollen grains, and it also participates in the recognition and acceptance/rejection (compatibility/incompatibility) reactions (for stigma characteristics, see J. Heslop-Harrison 1981).

Pollen grains have a sticky or oily surface in many plants. They adhere to the exudate of wet stigmas, or to the papillae of dry stigmas. The stigmatic exudate protects the stigma against excessive desiccation, and provides a suitable medium for pollen germination. The papillae of dry stigmas are covered with a pellicle of external proteins which are hydrophylous (Shivanna 1978). Following pollen adhesion to the papillae, they become hydrated, and the wall-bound proteins of pollen grain are released. The cuticle of stigma is dissolved, and pollen tubes enter the transmitting tissue of the stigma. The transmitting tissue in the style provides mechanical and chemical guidance for growing pollen tubes as well as the nutrients (see Kroh and Helsper 1974). In the ovary, transmitting tissue (the obturator) provides additional guidance for the growth of pollen tubes.

Based on physiological studies, the entire mechanism from the germination of pollen grain up to the discharge of male gametes in the embryo sac has been divided into several subsystems. These proceed effectively only when the interaction of pollen tube with the tissues of pistil is healthy. Otherwise, the process may terminate at several subsystems. The following flow chart (adapted after E.G. Williams et al. 1982, Knox 1982, 1984, Shivanna 1982,

Compatible Pollination	Incompatible Pollination
Pollen lands on Stigma Pollen hydration	
1. Pollen grains germinate and produces tubes	Do not germinate
2. Tubes grow through the transmitting tissue of the stigma and enter style	Tubes abort on stigmatic surface
3. Tubes grow through stylar canal/transmitting tissue	Tubes abort in the upper part of style
4. Tubes ėnter ovary	Tubes fail to enter ovary
5. Tubes enter micropyle of ovules	Tubes fail to enter micropyles
6. Tubes enter embryo sacs	Tubes do not enter embryo sacs
7. Tubes discharge male gametes in the embryo sac	Tubes fail to release male gametes
8. Fertilization occurs	Fertilization fails

Shivanna and Johri 1985, Knox et al. 1986) indicates the arrest points (*right*) in incompatible crosses compared to normal pollination (*left*).

In a pollen tube the cytoplasm and most of the organelles are located in the tip-region. As the tube grows and elongates, callose plugs develop on the inner side of the tube in a centripetal manner in the region away from the tip so that the backward flow of cytoplasm in the tube is checked. The growth of the tube is unidirectional through the transmitting tissue, guided by the architecture of the transmitting tissue.

In compatible pollinations, the pollen tubes effectively utilize the nutrients of the transmitting tissue (see Sassen 1974).

More than one tube can reach the embryo sac, but only one tube penetrates the degenerated synergid via the filiform apparatus. Sperm cells, vegetative nucleus and tube cytoplasm with many amyloplasts are discharged into this synergid.

Usually, the pollen tube passes through the micropyle to the embryo sac and exhibits porogamy. But in Anacardiaceae (*Pistacia*), Betulaceae (Figs. 9.2.1, 9.2.2A), Casuarinaceae, and Juglandaceae the pollen tube penetrates the chalaza, often branches there, and one of the branches passes upward through the nucellus and penetrates through the micropyle (extravesicular chalazogamy). In *Corylus* (see Swamy 1948a), and rarely in *Eucommia ulmoides* (Eckardt 1963), the pollen tube enters the embryo sac through the chalazal end, traverses through it and then reaches the egg cell (intra-vesicular chalazogamy). In mesogamous mode, the pollen tube penetrates through the integuments and reaches the micropyle. Reayat Khan (1970) coined the term "exogamy" for fertilization in Lentibulariaceae. The apex of the embryo sac is extraovular and fertilization occurs in the ovarian cavity. In Loranthaceae, true ovules are absent. In many taxa the tip of embryo sacs grow into the stylar canal and stigma, and the pollen tubes meet the embryo sacs in this region.

3.9 Double Fertilization

After pollination, one of the synergids shows signs of degeneration. Synergid degeneration far ahead of the pollen tube penetration into the ovary (Jensen et al. 1983) is another evidence of this interdependence, or reciprocal influence of the two gametophytes well before they come into contact with each other. The pollen tube penetrates the embryo sac, and enters the degenerated synergid through the filiform apparatus. It bursts in the chalazal region of the degenerated synergid. A coiled persistent pollen tube inside the embryo sac has been observed in *Passiflora calcarata* (M.V.S. Raju 1956b). The contents of the pollen tube – two male gametes, sometimes vegetative nucleus, and cytoplasm – are liberated through a pore which is either terminal (*Petunia hybrida* – van Went 1970, *Plumbago zeylanica* – Russell 1980a, b), or subterminal (*Gossypium hirsutum* – Jensen and Fisher 1968).

Double fertilization in angiosperms (see Steffen 1963b, Korobova 1974, Battaglia 1981, Belyaeva 1981, van Went and Willemse 1984) follows pollination. In this process one of the male gametes (sperm) fuses with the egg cell (syngamy), while the other male gamete fuses with the secondary nucleus (Fig. 3.9.1) or the two polar nuclei (triple fusion) (Figs. 3.9.2A–D, 3.9.3A).

Rarely, more than one pollen tube may enter the embryo sac (for pollen-ovule ratio see Cruden 1977, Cruden and Jensen 1979). As many as three tubes and sperm cells have been reported in *Cuscuta reflexa* (Fig. 3.9.3B).

Frequently, two darkly staining structures have been observed in the discharged pollen tube, or in the degenerated synergid, or lying close to the egg apparatus. Both structures are of nuclear material as revealed by ultrastructural studies and radioautographs. These so-called "X"-bodies may represent the degenerated vegetative nucleus of the pollen tube and degenerated nucleus of the synergid.

The additional pollen tubes, on entering the embryo sac, may fertilize the intact synergid leading to the formation of a proembryo (Fig. 3.12.1G).

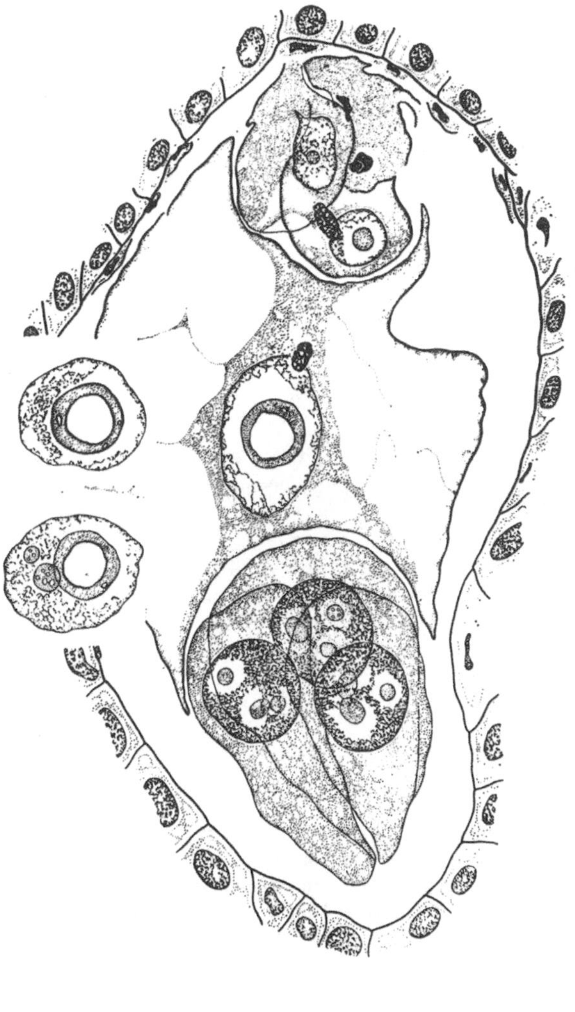

Fig. 3.9.1. *Anemone nemorasa*, double fertilization. *Inset*, zygote nucleus and primary endosperm nucleus. (After Yoffe 1969)

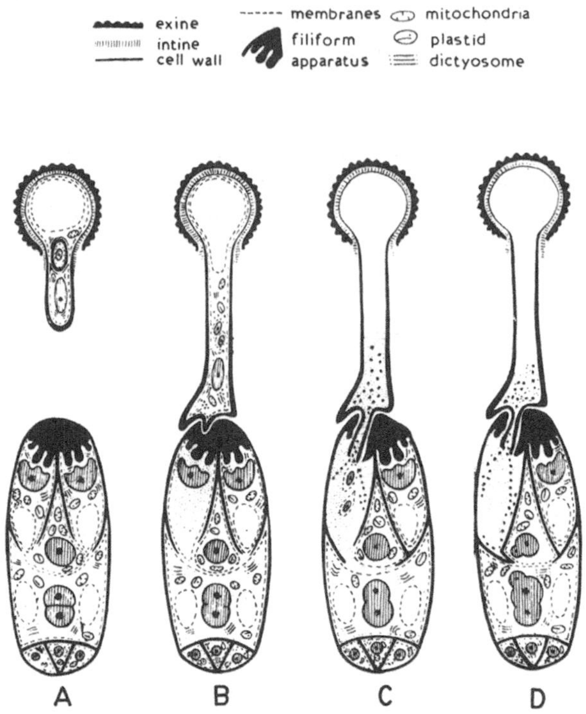

Fig. 3.9.2. Diagrammatic representation of male and female gametophyte. **A** Pollen tube and embryo sac. **B** Entry of pollen tube. **C** Discharge of male gametes inside the degenerated synergid. **D** Syngamy and triple fusion. There is no wall in the chalazal region of egg cell and synergids. Note the distribution of mitochondria, plastids and dictyosomes in the male and female gametophyte. (After Cocucci 1969)

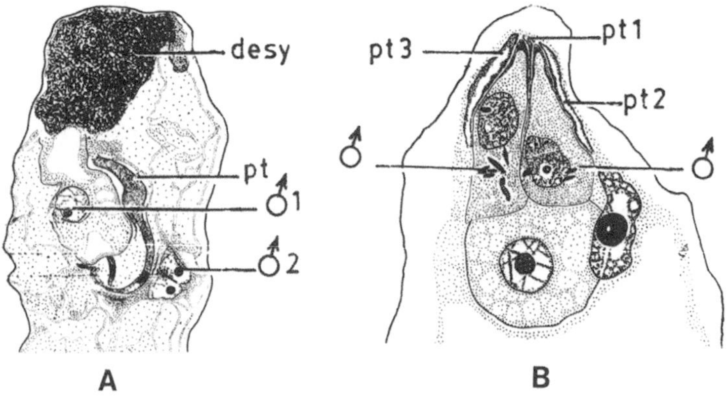

Fig. 3.9.3A, B. Fertilization. **A** *Camellia sinensis*, micropylar portion of embryo sac to show syngamy and triple fusion; *desy* degenerated synergid; *pt* pollen tube; (After Sethi 1965). **B** *Cuscuta reflexa*, upper part of fertilized embryo sac with zygote and primary endosperm nucleus; there are remnants of three pollen tubes, seven (or eight?) sperms are lying in one synergid and three in the other. (After B.M. Johri and Tiagi 1952)

The synergid bursts and the male gametes enter the egg cell, or the central cell through the bridges fromed due to the fusion of plasma membranes (see Kapil and Bhatnagar 1975).

The egg cell is said to have an attraction zone (Belyaeva 1977). As the sperms emerge from the synergid, one after the other, the first one is attracted towards the egg cell. As soon as it comes in contact with the wall of the egg cell, the attraction ceases. The second sperm is carried to the secondary nucleus.

Our information on the movement of male gametes to the egg and polars or secondary nucleus is very meagre. There is no certainty that the amoeboid movement of the male gametes, or the streaming of cytoplasm of the central cell, brings them in contact with the egg and polars/secondary nucleus.

At the time of fertilization, the male gamete is at the late telophase of mitosis (i.e. in the form of a spireme) initiated at spermatogenesis, while the egg nucleus is in the "aging" or "maturing" phase (Steffen 1951, Gerassimova-Navashina 1960, 1982).

Attention is invited to the detailed studies of Russell (1980a, b, 1982, 1983, 1984) and Russell and Cass (1981, 1983, 1988) on fertilization in *Plumbago* and *Plumbagella.*

Persistent Pollen Tubes. After fertilization, the remnants of the pollen tube persist for some time in the micropyle as well as in the embryo sac, or only in the embryo sac. There are several instances of the remnants persisting in the mature seed (see P. Maheshwari 1950, B.M. Johri and Ambegaokar 1984). From time to time it has been argued that after the discharge of gametes the persistent pollen tubes have a haustorial role, absorb nutrients from the circumjacent tissue, and pass on to the embryo until it is able to establish its own source of nutrition (see Longo 1903, Cook 1909, Venkata Rao 1952a, D. Singh 1963). Werner (1915) and Täckholm (1915) mention that in the Onagraceae the pollen tube becomes greatly broadened in the micropyle, and often sends out branches into the outer integument and nucellus. At the same time, the pollen tube grows towards the embryo sac destroying the cells in its path. The tube is recognizable until the embryo sac has enlarged considerably. Werner and Täckholm presume that the pollen tube absorbs nourishment from the adjoining tissue and transmits it to the embryo. In *Oxybaphus* (Cooper 1949), the pollen tube remains recognizable throughout the development of seed and probably transports nutritive materials from the secretory cells of the funiculus to the embryo.

Tilquin et al. (1983) refer to haustorial pollen tubes in *Fuchsia boliviana.* The pollen tubes do not collapse (disappear) after fertilization, but become even more conspicuous and persist for a considerable time, and remain recognizable in the mature seed. Tilquin et al. assume that after fertilization the pollen tube absorbs food materials still present in the style and partly also from the central region of the ovary and conveys these to the embryo until it can obtain nourishment from the endosperm. They strengthen their argument from the well-known fact that the pollen grain does not contain adequate reserve food to sustain the total growth and elongation of the tube which absorbs nourishment from the transmitting tissue of the style.

Indra and Krishnamurthy (1985) emphasize that in *Ottelia alismoides* the pollen tube forms a bulla (balloon-like swelling) as it enters the micropyle. It grows into a synergid and discharges the two male gametes and some cytoplasm. Soon after fertilizations, the micropyle in the region of outer integument becomes extremely wide; now the actual micropyle is said to be formed only by the inner integument. The "bulla" formation occurs at two or more sites between the obturator (glandular tissue) and the synergid.

Indra and Krishnamurthy further state: "There is a substantial increase in cytoplasmic content of the pollen tube in the post-fertilization stages especially in the 'bulla' regions as indicated by all staining procedures". The persistent pollen tube finally bursts at the bulla above the inner integument. The pollen tube is enucleate, and the increase in cytoplasm is due to absorption of reserve food material from outside the ovule (stylar glandular transmitting tissue) and transmission to the embryo sac. Indra and Krishnamurthy, using histochemical techniques, found only indirect evidence of the absorption of nutrients from outside the ovule and their transmission to the embryo sac.

The widening of the micropyle (exostome) is very doubtful and unconvincing. P. Maheshwari and Johri (1950b) also examined post-fertilization development in *O. alismoides* but did not observe either the widening of the micropyle or the formation of the bulla.

Briggs (1989) studied early post-fertilization development in *Swainsona greyana*. He reported that the pollen tube and the penetrated synergid persist throughout proembryonic development until the heart-shaped stage. After fertilization, the pollen tube swells, forms a bulla within the nucellar cap cells and stains for insoluble carbohydrates and proteins. Two days after anthesis, wall ingrowths develop along the suspensor cell wall in contact with the penetrated synergid before developing along all of the outer walls of the basal suspensor cells. A pollen tube-synergid-proembryo nutrient pathway is postulated.

Herscovitch and Martin (1989, 1990) made a histochemical and ultrastructural study of pollen-pistil interaction in *Grevillea banksii*; Martin and Herscovitch (1989) reported the behaviour of the pollen tube in *G. banksii* and *G. sericea*. In *G. banksii*, after fertilization of the ovules, the pollen tube branches near or at the micropyle. The enucleate branches invade the ovary wall where further branching occurs and the branches ramify intercellularly in the ovarian tissue. Branching was not observed in the styles of excised flowers, or in the ovular tissue. Martin and Herscovitch's conclusion on the haustorial function of pollen tube branches, which invade the ovary wall, is based on circumstantial evidence. They used fluorescence microscopy after decolourizing aniline blue staining of thick sections or entire ovaries macerated in 8 N KOH for 8 h or more. This is the first and only report that the pollen tubes branch and "parasitize" the ovary wall.

P. Maheshwari (1950) and B.M. Johri and Ambegaokar (1984) discussed the role of persistent pollen tubes and emphasized that these are merely the remnants of a "dead" structure which are not at all concerned in absorbing and transmitting the nutrients.

This controversy could be resolved by examining live dissected material with a phase contrast microscope. Zhou and Yang (1982, 1985) have developed a technique for enzymatic isolation of live pre- and post-fertilized embryo sacs. Success would depend on the proper selection of the material at the right stage.

Dissections, tran- and longisections, fluorescence, electron and interference microscopy, and histochemical studies to follow up the food reserve in the ovular and ovarian tissue from the time of entry and branching of the pollen tube up to the cessation of growth of the pollen tube would be very rewarding.

Another approach is to label the metabolites in emasculated flowers with $^{14}CO_2$, pollinating the pistil with normal (unlabelled) pollen, and then to demonstrate the ^{14}C-labelled metabolites in the pollen tube branches in the ovular/ovarian tissue. This would be a convincing demonstration that (during post-fertilization development) the persistent tubes absorb labelled metabolites from the adjoining tissue, and actually have a haustorial function.

The labelling can be achieved (see Coe 1954) by feeding a leafy branch with emasculated flowers (ovules at the mature embryo sac stage) with $^{14}CO_2$ for 36 h or more. Due to photosynthetic activity, the metabolites in the stigma, style, ovary and ovules will become labelled with ^{14}C. By autoradiography of sections of frozen/fixed material (following pollination and fertilization), labelled compounds in the bulla, unbranched/branched pollen tubes in the ovular-/ovarian tissue can be detected.

So far, none of the investigators have used these procedures to demonstrate the haustorial function of the pollen tube (see B.M. Johri 1992). Assumptions are essentially based on indirect and circumstantial evidence.

3.10 Endosperm

Endosperm, a product of double fertilization, develops only in angiosperms. A male gamete fuses with the polar nuclei, or the secondary nucleus, and the primary endosperm nucleus is produced. The division of the primary endosperm nucleus, and repeated divisions of the daughter nuclei lead to the formation of endosperm tissue (Swamy and Ganapathy 1957a, Bhatnagar and Sawhney 1981, Vijayaraghavan and Prabhakar 1984, Bhojwani and Bhatnagar 1986).

The primary endosperm nucleus is not formed in Podostemaceae. In Trapaceae (Sect. 39.3) and Orchidaceae (Sect. 65.2), either the division of the primary endosperm nucleus is suppressed, or only a few divisions occur but all the daughter nuclei degenerate.

The nutritive role of endosperm has long been recognized. It nurses the embryo from the proembryo stage till it becomes self-sufficient, and regulates the pattern of development (Krishnamurthy 1988). The endosperm contains growth regulators, and coconut milk, sometimes the extract of corn endosperm at milk stage, is added to the nutrient medium for growing tissues in cultures. In *Salix* (Håkansson 1954), the endosperm is ephemeral and provides only growth regulators.

The peripheral cells of endosperm develop phellogen in the "naked" seeds of *Crinum defixum* (B.S.M. Dutt 1962), while the inner cells are chlorophyllous and store starch. The mature endosperm stores carbohydrates, proteins or fatty food reserve. The chemical composition of the storage products forms the basis for taxonomic classification of monocotyledons (see Huber 1977).

The developing endosperm derives nutrients from the food reserves stored in the nucellus and integuments. The development of chalazal, micropylar or secondary endosperm haustoria leads to partial or entire absorption of integuments in several families. The peripheral layers of the endosperm become cutinized, and serve as protective layers as in *Andrographis* spp. (Mohan Ram 1960b, Mohan Ram and Masand 1963), *Elytraria acaulis* (B.M. Johri and Singh 1959) and *Haplanthus tentaculatus* (Phatak and Ambegaokar 1961a).

Depending on its mode of development, the endosperm is said to be of the Nuclear (Figs. 3.10.1A–D, 3.10.2A–F), Helobial (Fig. 3.10.1E–L) and Cellular (Fig. 3.10.1M–O) type.

Nuclear Type. The primary endosperm nucleus divides followed by free-nuclear divisions (Fig. 3.10.1A–D). The free nuclei either aggregate in the micropylar or chalazal region, or in both, or may become distributed in a film of cytoplasm around the central vacuole. At times, some of the nuclei may fuse to form polyploid nuclei as in *Stemona tuberosa* (Swamy 1964b).

The endosperm remains free-nuclear throughout in *Floerkea proserpinacoides* (P. Maheshwari and Johri 1956) and *Oxyspora* (Subramanyam 1951a). In the majority of plants, however, it does become cellular. Wall formation occurs at different stages of development in different species, as early as the four-nucleate endosperm in *Coffea arabica* (Mendes 1941), or even after the formation of 1000 free nuclei as in *Anemone* (Bhandari 1968). Wall formation is centripetal or it is initiated at the micropylar end and proceeds to the chalaza (Fig. 3.10.3D). It commences from the chalazal end and proceeds upward in *Carya* (McKay 1947), *Elatine* (Frisendahl 1927), *Frankenia* (Walia and Kapil 1965) and *Pistacia* (Grundwag 1976).

The single layer of endosperm in *Begonia crenata* (Swamy and Parmeswaran 1960) undergoes periclinal divisions; the outer layer develops into an aleurone layer while the inner produces the endosperm tissue. The peripheral layer of the cellular endosperm becomes meristematic in Gramineae, and contributes cell layers to the inner endosperm. Later, the peripheral layer develops into an aleurone layer.

Fig. 3.10.1A–O. Types of endosperm. **A–D** *Scoliopus bigelvii*, Nuclear type. (After Berg 1959). **E–L** Helobial type. **E–J** *Philydrum lanuginosum*. (After Hamann 1966b). **K–L** *Sparganium erectum*. **K** Lower part of embryo sac; *ME* micropylar endosperm chamber; *CE* chalazal endosperm chamber; *Ant* antipodals; **L** Chalazal chamber and remnants of antipodals (*AntR*). (After M. Muller-Doblies 1969a). **M–O** Cellular type. **M** *Arisaema wallichianum*, note the large chalazal haustorium corresponding to the chalazal chamber in Helobial type. (After S.C. Maheshwari and Khanna 1956); **N** *Spirodela polyrhiza*. (After S.C. and N. Maheshwari 1963); **O** *Lemna paucicostata*, the entire endosperm is cellular and a chalazal haustorium is absent. (After S.C. Maheshwari and Kapil 1963b)

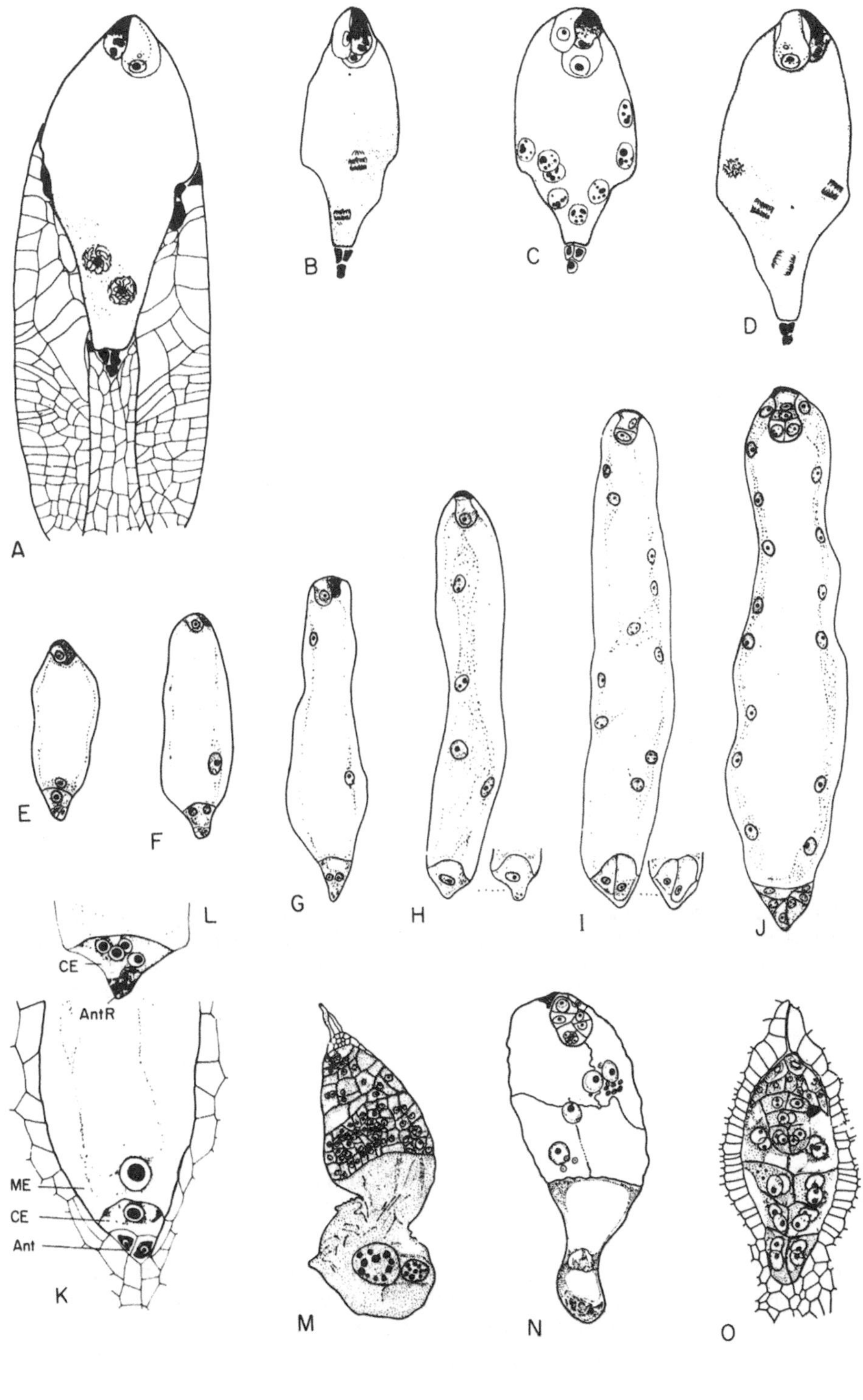
A
B
C
D
E
F
G
H
I
J
L
CE
AntR
ME
CE
Ant
K
M
N
O

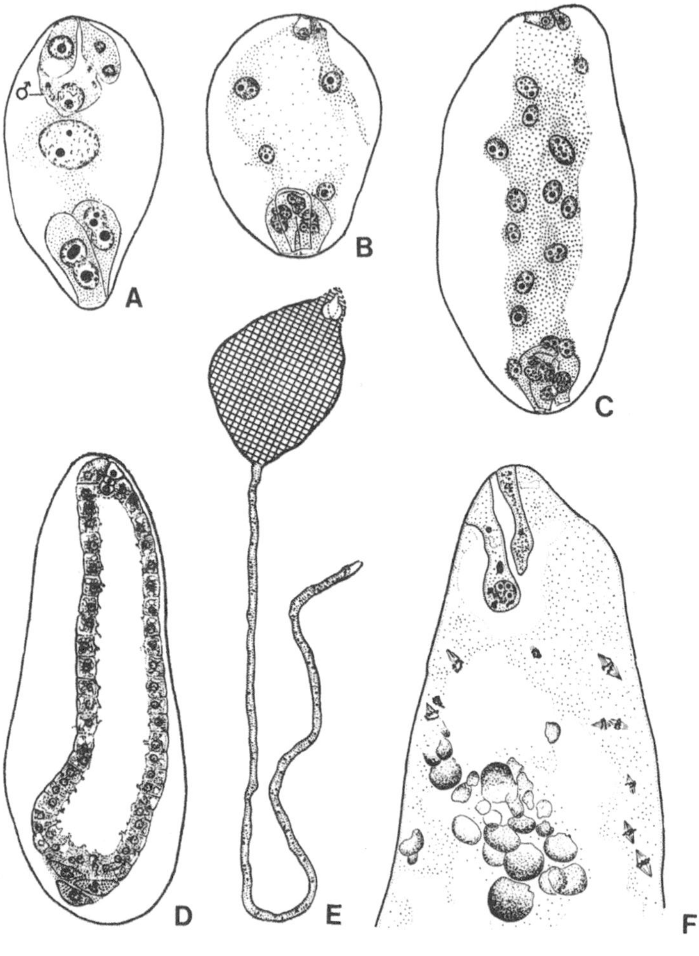

Fig. 3.10.2A–F. Nuclear endosperm. **A–D** *Caltha palustris.* **A** Fertilized embryo sac, syngamy has not yet been completed; **B** Four-nucleate endosperm, the antipodals have become polyploid; **C** 16-nucleate endosperm, the zygote has not yet divided; **D** Centripetal wall formation at three-celled proembryo. (After Kapil and Jalan 1962). **E** *Echinocystis lobata*, the endosperm proper has become cellular and there is a long coenocytic chalazal haustorium, embryo dicotyledonous (dissected wholemount). (After Chopra and Seth 1977). **F** *Foeniculum vulgare*, micropylar portion of embryo sac with zygote, and endosperm nuclei in division. Note enucleate endosperm vesicles. (After S.C. Gupta 1964)

The development of micropylar haustorium, as in *Frankenia hirsuta* (Walia and Kapil 1965), is a rare feature. It becomes cellular at a very late stage.

The formation of chalazal haustorium has been noted in Anacardiaceae (*Pistacia*), Cucurbitaceae (Fig. 3.10.2E), Euphorbiaceae, Leguminoseae and Proteaceae. Streaming movement of the protoplasm has been observed in the vermiform appendage of *Grevillea robusta* (Kausik 1941a), and coenocytic haustorium of *Trichosanthes anguina* (Chopra 1955). The length of the chalazal haustorium is maximal (19,000 μm) in *Sechium edule* (D. Singh 1964a). The development of unicellular finger-like appendages all over the surface of the endosperm in *Lomatia polymorpha* (Venkata Rao 1962) is unique. It increases the absorptive surface.

Another noteworthy feature in the Nuclear type of endosperm is the formation of enucleate (Fig. 3.10.2F) and nucleate cytoplasmic nodules. Enucleate nodules have been reported in *Passiflora* (M.V.S. Raju 1956b) and *Foeniculum vulgare* (S.C. Gupta 1964). According to Al-Attar (1977), the formation of such nodules distributes the endosperm nuclei to the periphery and prevents their accumulation around the zygote. Otherwise, it causes hindrance to the growth of the embryo. Nucleate nodules have been reported in *Capsella bursa-pastoris* (P. Maheshwari and Sachar 1954b). Both enucleate and nucleate nodules occur in *Carica papaya* (Dathan and Singh 1976). D. Singh (1964b) states that the nodules are much better developed in plants at higher altitudes, and is a possible method of filling up the vacuoles of the endosperm. However, R.C. Misra (1966) and Prasad (1969, 1974) attribute nutritional significance to them since the nodules persist for only a short duration in the developing endosperm.

Helobial Type. While reviewing the Helobial type of endosperm, Swamy and Parmeswaran (1963) suggest certain norms: (1) The primary endosperm nucleus lies in the chalazal region of the embryo sac, (2) it divides in situ, (3) the micropylar chamber is larger than the chalazal and separated either by a wall or a membrane, (4) the nucleus of the micropylar chamber moves away from the wall, undergoes free-nuclear divisions, and later the tissue becomes cellular, (5) the nuclear division in the chalazal chamber is strictly free-nuclear and it becomes haustorial, (6) if the chalazal chamber becomes coenocytic, the few initial divisions are synchronous; however, the chalazal chamber lags behind in the number of nuclei and their rate of formation, and (7) if wall-formation occurs in the chalazal chamber, it is significantly belated not only in its own ontogeny but also as compared with the time of initiation of cells in the micropylar chamber (see Fig. 3.10.1E–M).

After the formation of micropylar and chalazal endosperm chambers, usually the nucleus of the micropylar chamber divides first. However, there are exceptions, and the nucleus of the chalazal chamber may divide earlier than that of the micropylar as in *Phylidrum* (Kapil and Walia 1965). The free-nuclear divisions in the micropylar and chalazal chamber may or may not be synchronous.

Asynchronous free-nuclear divisions in the chalazal chamber have been noted in *Sparganium* spp. and *Typha* sp. (Asplund 1972, 1973). Wall formation in the micropylar as well as the chalazal chamber is not a rare feature, e.g. *Trillium undulatum* (Swamy 1948c), *Lindmania pendulifolia* (Lakshmanan 1967), *Phylidrum lanuginosum* (Kapil and Walia 1965), *Sparganium* spp. and *Typha* (Asplund 1972, 1973). In the latter four genera, the chalazal chamber becomes coenocytic by free-nuclear divisions, or cellularization occurs. In addition, in Philydraceae, Bromeliaceae, Sparaganiaceae and Typhaceae, wall formation is initiated first in the chalazal and then in the micropylar chamber.

Swamy and Parmeswaran (1963), Swamy and Krishnamurthy (1973) and Swamy (1974a) discuss several instances of Helobial type of endosperm in dicotyledonous families, and draw attention to their either Nuclear or Cellular ontogeny. According to Swamy and Parmeswaran, the formation of true Helobial endosperm is confined only to the monocotyledonous families; if it occurs in dicotyledonous families, it is only a modification of the Cellular or Nuclear type of endosperm. However, typical Helobial type of endosperm has been observed in dicotyledonous families, such as Santalaceae and Saxifragaceae.

Cellular Type. The division of the primary endosperm nucleus is followed by wall formation. The subsequent divisions may be followed by transverse walls resulting in a linear row of cells as in *Cercidiphyllum* (Swamy and Bailey 1949) and *Sarracenia* (Shreve 1906), or the walls may be crosswise as in *Parrotiopsis* (Kapil and Kaul 1972).

The majority of sympetalous families are characterized by the formation of either micropylar, or chalazal, or both types of endosperm haustoria. The sequence of planes of wall formation, following the first few divisions, exhibits several subtypes of development. These are discussed along with the respective families. Rau has drawn up a comparative schematic representation of endosperm types in sympetalous families: Labiatae, Verbenaceae (Fig. 48.10.2A–F), Scrophulariaceae (Fig. 48.15.6A–K_4), Gesneriaceae (Fig. 48.22.1A–K), and Campanulaceae (Fig. 51.2.2A–F_1).

Usually, for obtaining nutrients and transporting them to the developing endosperm, the haustoria grow in the direction of the tissue rich in food reserves. Their hypertrophied, often polyploid nuclei indicate high metabolic activity. The order Santalales presents extreme cases of development of the haustoria. In *Opilia*, the long tubular chalazal haustorium penetrates the pedicel of the flower, whereas in *Cansjera rheedii* (Swamy 1960) it branches profusely and grows towards the base of the ovule. In addition, secondary branches also develop (Fig. 12.4.2A, B). In *Iodina rhombifolia* (Bhatnagar and Sabharwal 1968), the aggressive chalazal haustorium branches extensively at the free end (Fig. 12.6.7A, B).

Both, the inter- and intracellular growth of the haustorium have been observed in *Nothapodytes foetida* (Swamy and Ganapathy 1957b). The basal endosperm cell develops chain-like connections with the chalazal tissue (through the perforations in the intervening walls), and absorbs nutrients (Fig. 33.13.1A–C). The outline of the nucleus appears most unusual, and exhibits a high-ploidy level.

The development of micropylar and chalazal haustoria is noteworthy in Loasaceae (Fig. 37.19.1A–G), Scrophulariaceae (*Pedicularis sylvatica* – Fig. 48.15.3A, B), Gesneriaceae (*Klugia notoniana* – Fig. 48.22.3A–C), and Orobanchaceae (*Aeginetia indica* – Fig. 48.24.1H, I, J, L, M).

In *Rhinanthus* spp. (B. Tiagi 1972), secondary haustoria develop from the central cellular endosperm after the inactivation of aggressive micropylar and chalazal haustorium. The formation of cambium-like strips in the central cellular endosperm is unique (for further details, refer to Sect. 48.15).

The asymmetrical development of the central cellular endosperm, formation of "basal apparatus", ruminations, and the development of primary and secondary haustoria characterize the endosperm in Acanthaceae (Sect. 48.19).

The formation of a "composite" endosperm in Loranthaceae is unknown elsewhere in angiosperms. As several embryo sacs grow concurrently in an ovary, each develops endosperm which fuse to form a "composite" structure (Fig. 12.8.3A, B).

Which of the types – Nuclear, Helobial, Cellular – is primitive? One can derive the Helobial from the Nuclear, and Cellular from the Helobial type. These three types are randomly distributed in the primitive and advanced angiosperms, as is also true of several other embryological characters. Generally, the Nuclear type is common in the Polypetalae, Cellular in the Sympetalae, and Helobial in the Monocotyledons. In Saxifragaceae (Sect. 26.7), all the three types occur in different taxa, but both Cellular and Helobial types may occur in *Boykinia*, *Mitella*, *Ribes* and *Tiarella* (Rau 1970).

Mitrasacme alsinoides (Fig. 47.2.2A–G) shows a Cellular–Nuclear–Cellular type of development. In a uniseriate four-celled endosperm, nuclear divisions occur in the micropylar and chalazal cell. Finally, the entire endosperm becomes cellular.

As the Nuclear type of endosperm has been observed in a larger number of taxa than the other two types, it may be considered to be primitive.

What is the morphological nature of the endosperm? It is 2n in *Butomopsis* (Fig. 53.3.2F, G), 3n in most plants, 5n in *Plumbago* (Sect. 44.2) and *Fritillaria* (Sect. 54.2), and up to 15n in *Peperomia* (Sect. 20.3). Whatever the ploidy-level of the endosperm, the development always conforms to one of the three types, and its main function is to nourish the embryo during development and maturation, and during germination (in albuminous seeds).

When the primary endosperm nucleus is 2n (upper polar + sperm cell), it is truly sporophytic and genetically identical with the zygote but the two behave quite differently – one produces the endosperm and the other the embryo. This also disproves Sargant's (1900) view that the participation of the lower polar causes imbalance and the primary endosperm nucleus (2n) is unable to produce the embryo. Now we know that it is not two polars but up to 14 which are concerned in the formation of the primary endosperm nucleus. Therefore, the only interpretation could be that the endosperm is a sui generis tissue.

In the absence of endosperm, the embryo usually aborts, unless nutrition can be obtained otherwise as in Trapaceae (Fig. 39.3.2N, O). Krishnamurthy (1988) emphasizes that the endosperm exercises a hormonal control on the growth and differentiation of embryo.

Balanophora (Fig. 13.2.3A, B) is so far the only example where, at the time of germination of the fruit on the host rootlet, some of the endosperm cells produce tubular slender sticky extensions for anchorage to the young rootlet.

The endosperm has long been presumed to be a "dead" tissue. B.M. Johri and Bhojwani (1965) clearly demonstrated, for the first time, that in (in vitro) cultures the endosperm of *Exocarpus cupressiformis* (Santalaceae) generates shoot buds.

Ruminate Endosperm. The ruminate condition of endosperm occurs during the maturation of seed so that the surface of endosperm becomes uneven. The ruminations are caused due to localized activity of the seed-coat, or endosperm. On morphological basis, Periasamy (1962a) distinguishes seven types: Annona (Fig. 3.10.3A–C), Passiflora (Fig. 3.10.3D), Myristica (Fig. 3.10.3E), Spigelia (Fig. 3.10.3F), Verbascum (Fig. 3.10.3G), Coccoloba (Fig. 3.10.3H, I) and Elytraria (Fig. 3.10.3K). In all these types, except Elytraria, irregularities occur in the growth of integument/s which bring about the rumination of endosperm. In the Elytraria type, during the development of seed, localized regions in the peripheral layers of cellular endosperm show active growth so that it becomes ruminate (Fig. 3.10.3K). In this process the seed-coat is consumed by the endosperm, and only the remnants persist.

An excęssive elongation and enlargement of cells of the seed-coat has been observed in the Annona type (Fig. 3.10.3A–C). In Annonaceae, the cells of the outer, inner, or middle (Fig. 3.10.3C) integument elongate and cause ruminations. They are not vascularized. In some members of Annonaceae, the ruminations are plate-like and lie one above the other (Fig. 3.10.3A).

In the Passiflora type, the ovules are bitegmic, and the inner epidermis of testa and outer epidermis of tegmen show unequal radial elongation of cells which cause ruminations (Fig. 3.10.3D). In the Verbascum type, the ovules are unitegmic, and the cells of the persistent endothelium elongate unequally (Fig. 3.10.3G).

In *Degeneria* (Swamy 1949a), *Tiliacora* (Sastri 1954a), and some members of the Myristicaceae (Fig. 3.10.3E), Leeaceae and Vitaceae, the ingrowths develop due to the meristematic activity of localized areas of tegmen. The position of the meristematic zones is not constant except in Vitaceae and Leeaceae. In Myristicaceae, the ingrowths develop due to localized meristematic activity in testa and the massive chalaza (Periasamy 1962a, see also N.C. Nair 1972). According to Mohana Rao (1974b), the inner layers of the inner integument produce the ingrowths which become vascularized.

The ovules are bitegmic in *Coccoloba*, and unitegmic in *Spigelia*. In the Coccoloba (Fig. 3.10.3H, I) and Spigelia (Fig. F) type, the mature seed-coat comprises only the epidermis of testa. The epidermal cells of testa and of the massive chalaza, during the growth of seed, develop irregular infoldings due to cell enlargement. The endosperm tissue grows in between the infoldings, and ruminations are produced.

The Spigelia type of ruminate endosperm is common in Rubiaceae (Fig. J), but the condition in *Borreria hispida* (Periasamy and Vaidyanathan 1980) is rather rare. In the pachychalazal seed, the epidermal layer (of seed-coat) along the boundary between the strophiole and ovular part forms two inward extensions into the mature endosperm making it three-lobed and ruminate along the ventral face.

The endosperm also consumes the strophiolar tissue so that the strophiole is reduced to only a few cell layers at the region immediately adjoining the septum. The growth of the endosperm into the strophiole also makes it ruminate.

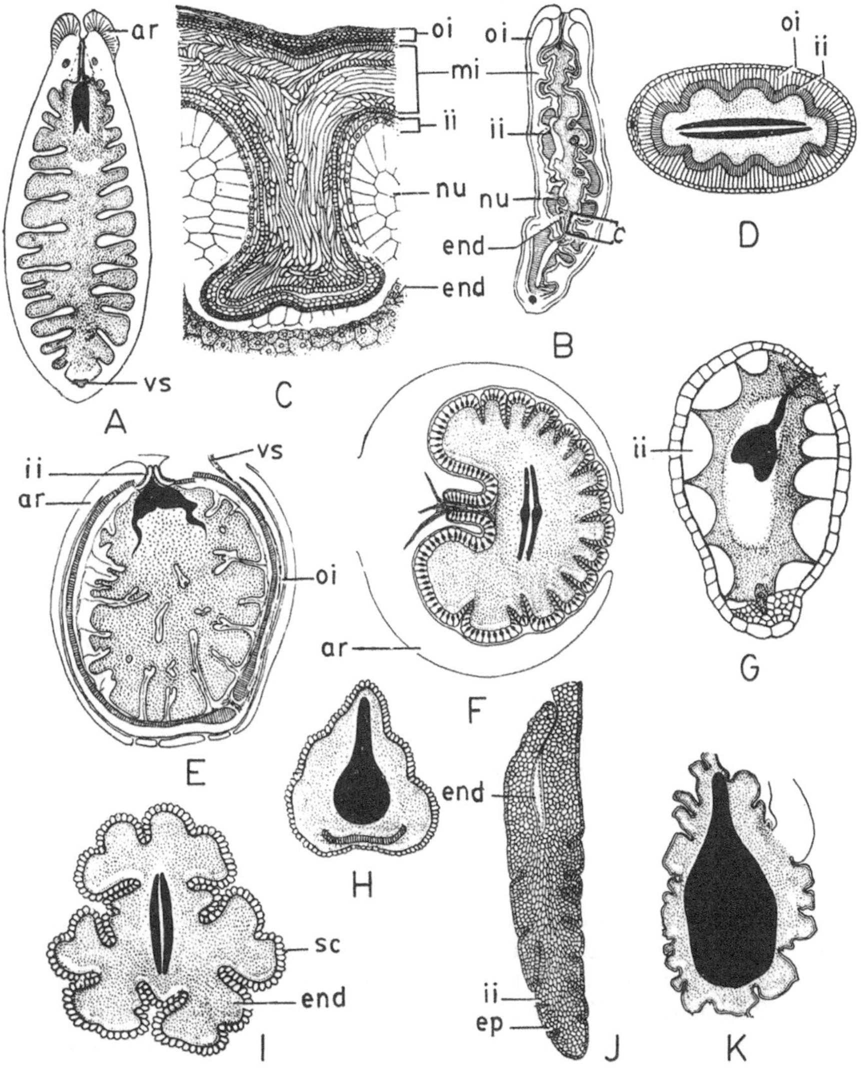

Fig. 3.10.3A–K. Ruminate endosperm. **A** *Annona squamosa*, longisection of mature seed with aril (*ar*) and vascular strand (*vs*); embryo *black*. **B, C** *Cananga odorata*. **B** Trans-median longisection of immature seed with outer (*oi*), middle (*mi*) and inner integument (*ii*), endosperm (*end*) and nucellus (*nu*); **C** *Inset C* from **B**. (After Periasamy 1962a). **D** *Passiflora calcarata*, transection of mature seed with inner (*ii*) and outer integument (*oi*). (After M.V.S. Raju 1956b). **E** *Myristica fragrans*, longisection of mature seed with aril (*ar*), inner integument (*ii*), and vascular strand (*vs*). (After Voigt 1888). **F** *Ervatamia heyneana*, transection of mature seed with aril. (After Periasamy 1962a). **G** *Tetranema mexicana*, longisection of mature seed with inner integument (*ii*). (After Krishna Iyengar 1942a). **H, I** *Coccoloba uvifera*. **H** Longisection of mature seed; **I** Transection of seed *end* endosperm; *sc* seed-coat (exotesta). **J** *Psychotria congesta*, longisection of immature seed. (**H–J** after Periasamy 1962a). **K** *Andrographis serpyllifolia*, longisection of mature seed. (After Mohan Ram 1960b)

In *Bentinckia codapanna* (Padmanabhan and Regupathy 1981), the endocarp is closely appressed to the thin seed-coat. The endocarp, the seed-coat and the outer layers of endosperm, as a unit, produce sinuous inpushings which impart a semblance to ruminate endosperm. The undulations originate in the fruit-wall even before the endosperm becomes cellular.

According to van Heel (1970a, 1971a, b), the labyrinth seed is an extreme form of rumination; the seed cut in any plane exhibits ruminations. In *Kingiodendron pinnatum*, the testa first encroaches on the endosperm and then on cotyledons. It penetrates in the lobes and crevices of the cotyledons and makes the seed ruminate giving it a labyrinth appearance. Other examples of labyrinth seeds are met with in Convolvulaceae, Hernandiaceae and Anacardiaceae (see Boesewinkel and Bouman 1984).

3.11 Embryo

The process of syngamy results in the formation of zygote. The zygote undergoes, apparently, a resting period. In some species it shrinks as in *Syringodium* (Lakshmanan and Rajeshwari 1979). The polarity becomes more pronounced, and differentiation occurs in the apical and basal region of the zygote. The apical region is an active centre of protein synthesis and cell division, while the basal part accumulates osmotically active substances. In sympetalous families like Acanthaceae, Gesneriaceae, and several others the micropylar part of the embryo sac elongates considerably. The apical region of the zygote elongates into a long tube-like structure so that the tip with the nucleus comes to lie in the central part of cellular endosperm. The zygote, in *Acrotrema arnottianum* (Swamy and Periasamy 1955), exhibits the unique development of a mantle (Fig. 22.2.1A, B). As the growth proceeds, the zygote emerges out of the mantle.

Early Embryogeny. Usually, a transverse division occurs in the zygote, and separates a spherical terminal cell *ca* from the basal cell *cb* (see Sivaramkrishna 1978). The division of zygote may rarely be vertical as in Loranthaceae (Sect. 12.8), or vertical/oblique as in *Quercus gambelii* (Brown and Mogensen 1972). The next division may be transverse in both the terminal (*ca*) and basal (*cb*) cell, and a linear tetrad is produced (Fig. 3.11.5). The daughter cells of *ca* are designated *l* and *l′*. The derivatives of *cb* are designated *m* and *ci* (see Fig. 3.11.5). The term proembryo is applied from the two-celled condition till the initiation of organs in the embryonal mass.

In the formation of a T-shaped type of tetrad, a vertical division occurs in *ca*, and the two juxtaposed hemispherical cells form the tier *q* (Fig. 3.11.1A). *q* gives rise to a quadrant by vertical divisions in *l* and *l′*, or by vertical division in each of the two juxtaposed cells, perpendicular to the previous one (Fig. 3.11.1B). The vertical walls do not intersect, but a small intercept is left between them.

Subsequent to the formation of a quadrant, the proembryo develops into an octant. The formation of octant is of two types (Fig. 3.11.2A–F). In the first

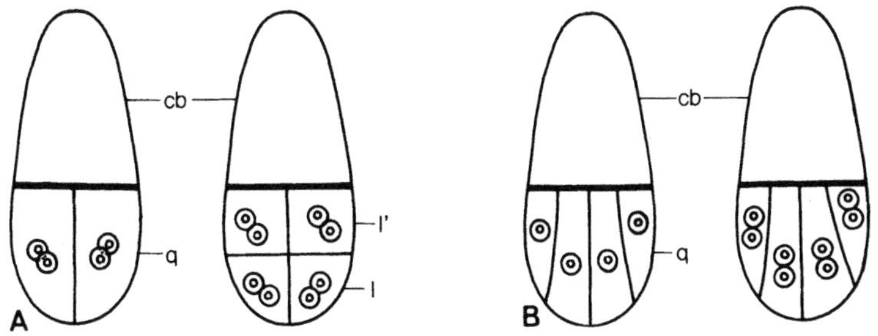

Fig. 3.11.1A, B. Quadrants (four cells at distal end of proembryo) produce octants (eight cells) either in two tiers (**A**), or one tier (**B**). When in two tiers, apical tier is designated *l* and the basal tier *l'*. (After Natesh and Rau 1984)

instance, in a linear type of tetrad, one more vertical division occurs in each cell of the quadrant, at right angles to the previous one, forming two superposed tiers (*l*, *l'*) of four cells each (Fig. 3.11.2A, B). In a T-shaped tetrad, similarly, the two tiers *l* and *l'* develop by a transverse division in each cell of *q* (Fig. 3.11.2C, D). In the second instance, instead of undergoing a transverse division, the four cells of tier *q* divide vertically in such a way that each quadrant is divided into unequal cells – one adjoining the axis and the other towards the periphery (Fig. 3.11.2E, F). They form together a single tier of eight cells.

The variations in the octants occur due to the plane of formation of wall, following nuclear division in each cell of a quadrant. The walls are laid down in three planes: transverse, vertical or longitudinal, and oblique (periclinal to curved wall). According to Takao (1976), theoretically, the octant can show 15 patterns, but in *Impatiens textorii* he observed only five: (1) One of the cells of the quadrant divides in vertical plane, and the other three in transverse plane. Then the octant is composed of one pair of juxtaposed cells, and three pairs of superposed cells. (2) If two cells of the quadrant divide vertically, and the other two transversely, the octant shows two pairs of juxtaposed and two pairs of superposed cells. (3) An octant consists of three pairs of juxtaposed cells, and one pair of superposed cells. (4) All the four pairs of cells in an octant are juxtaposed. (5) In an octant two pairs of cells are juxtaposed, the other two cells of the quadrant divide by oblique walls so that the daughter cells are arranged circumaxially.

The development of proembryo up to an octant stage and the formation of two types of octant are similar in dicotyledons and monocotyledons (Swamy 1962, see also Guignard and Mestre 1969a, Guignard 1975a). The morphogenetic potential of the cells is described later (see also D'Amato 1977).

Embryogenic Classification. The system of classification by Souèges (1937a–f) and Mestre (1967) is mainly based on the origin of four-celled proembryo, and organ differentiation from each cell. The rate at which these cells divide varies in different species, but the planes of their divisions and the disposition of cells formed are specific and predetermined. The four-celled proembryo has a

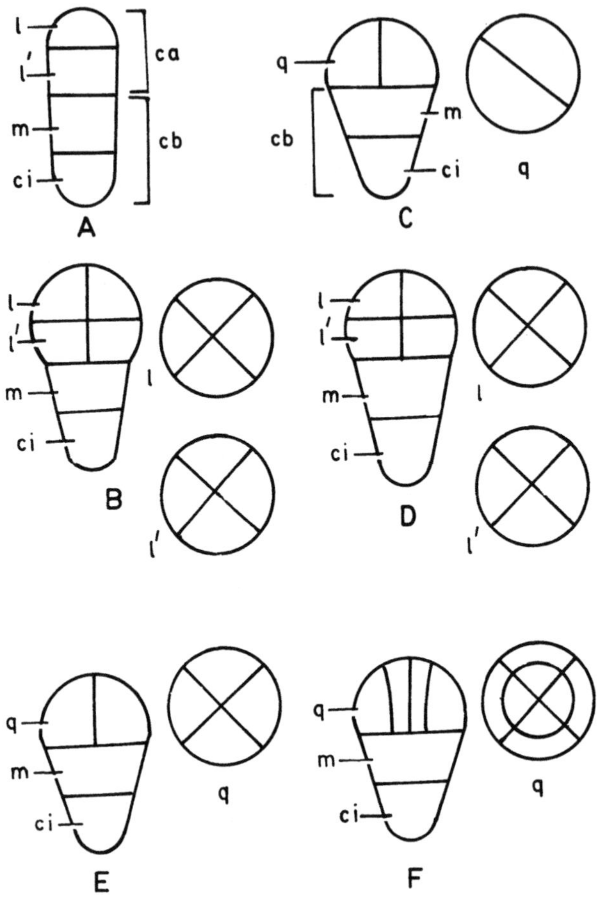

Fig. 3.11.2 A–F. Types of octant formation during early embryogeny. (After Swamy 1962)

specific, predetermined potential blue-print of the mature embryo (see Brogniart 1827, Crété 1963a, Guignard 1975b, Lebégue 1989).

A transverse division of the zygote gives rise to the terminal cell (*ca*) and the basal cell (*cb*). The destiny of these two cells is represented by Megarchetypes – *pvt* stem tip, *phy* hypocotyledonary part, *icc* root cylinder, *iec* hypophysis (root epidermis), *co* root cap, *s* suspensor, *pco* cotyledonary region proper:

Megarchetype I:	cb = pvt + phy + icc + iec + co + s
	ca = pco
Megarchetype II:	cb = phy + icc + iec + co + s
	ca = pco + pvt
Megarchetype III:	cb = $\frac{1}{2}$ phy + icc + iec + co + s
	ca = pco + pvt + $\frac{1}{2}$ phy
Megarchetype IV:	cb = iec + co + s
	ca = pco + pvt + phy + icc

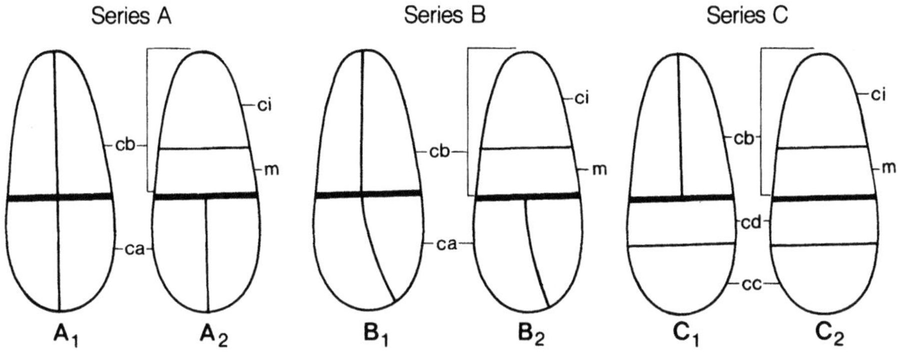

Fig. 3.11.3. Proembryonal tetrads, classification based primarily on the plane of division in apical cell; vertical in *Series A*, oblique in *Series B*, and transverse in *Series C*. Further subdivision takes into account the mode of division of basal cell which may be either vertical. (A_1, B_1, C_1) or transverse (A_2, B_2, C_2). (After Natesh and Rau 1984)

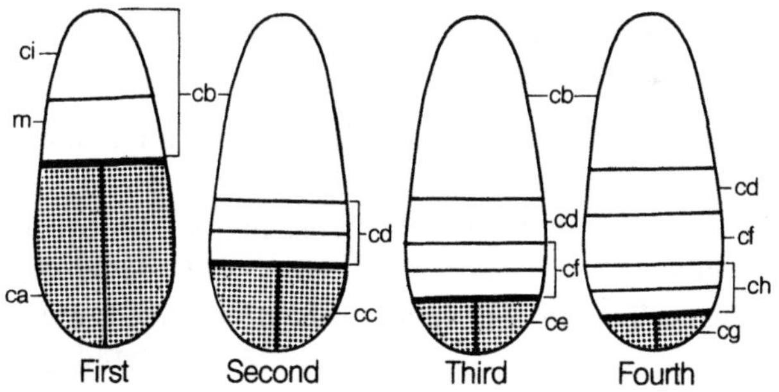

Fig. 3.11.4. Proembryonal tetrads of Series C to show tetrads of first, second, third and fourth order, respectively. The potential embryonal pole (*shaded*) shifts progressively to the extreme chalazal end. (After Natesh and Rau 1984)

Megarchetype V: cb = co + s
ca = pco + pvt + phy + icc + iec

Megarchetype VI: cb = s
ca = pco + pvt + phy + icc + iec + co

The Megarchetypes are divided into Series A, B and C, depending on the variation in the plane of division of the terminal cell (*ca*) of the two-celled proembryo. When the division is in vertical plane, the tetrad is referred to Series A, when oblique, to Series B, and a division in transverse plane to Series C (Fig. 3.11.3).

When the terminal cell (*ca*) and a part of the basal cell (*cb*) give rise to embryo proper, it is referred to period I. This potentiality shifts from the basal cell (*cb*) to the terminal cell (*ca*) in Period II, and then to the apical derivative of the terminal cell (*cc*) in Period III, and so on (Fig. 3.11.4). Thus, there is a progressive series of shift in potentiality to the apical derivatives.

In the Series C, the laws governing the embryogeny are applicable to the apical derivatives (see Crété 1963a).

Simplified System of Mestre (1967). In view of the fact that, very often, many irregularities occur in the sequence of divisions during early stages, Mestre formulated a simplified sytem. According to him, only the potentialities of the first two cells, apical and basal, are of relevance and he recognizes only three Megarchetypes:

Megarchetype A: ca = cotyledons only
cb = stem + root

Megarchetype B: ca = stem, cb = root (in general)

Sub-type B_1: ca = stem (major part of hypocotyl)
cb = minor part of hypocotyl and root

Sub-type B_2: ca = entire hypocotyl (stem)
cb = root

Megarchetype C: ca = stem and root
cb = only suspensor, i.e. the basal cell has no contribution in the development of new plant

Johansen's System. Schnarf (1929) and Johansen (1945) recognize six principal types of embryos among dicotyledons (Fig. 3.11.5). P. Maheshwari (1950) distinguished them as follows:

I. The terminal cell of the two-celled proembryo divides by a longitudinal wall:
  i) The basal cell has only a minor role, or none, in the subsequent development of the embryo — Crucifer or *Onagrad* type
  ii) The basal and terminal cells both contribute to the development of the embryo — Asterad type

II. The terminal cell of the two-celled proembryo divides by a transverse wall.
  1. The basal cell plays only a minor role, or none, in the subsequent development of the embryo:
  i) The basal cell usually forms a suspensor of two or more cells — Solanad type
  ii) The basal cell does not divide further, and the suspensor, if present, is always derived from the terminal cell — Caryophyllad type
  2. The basal and terminal cells both contribute to the development of the embryo — Chenopodiad type

III. The zygote divides by a vertical wall — Piperad type

Embryogenic types, as suggested by Souèges, Johansen, and P. Maheshwari are compared in the following Table:

The derivatives of the two-celled proembryo, *ca* (apical cell) and *cb* (basal cell), lead to six types of embryogeny (Fig. 3.11.5).

Table 3.11.1. Types of Embryogeny in Angiosperms

R. Souèges (1939–1951)	D.A. Johansen (1950)	P. Maheshwari (1950)
1) Period I: Series 'A' and 'B'– Group I. Megarchetypes IV–VI	Onagrad (includes Myosurus, Juncus, Capsella, Veronica, Mentha, Trifolium, Ruta and Lilium variations)	Crucifer/Onagrad
2) Period I: Series 'A' and 'B'– Groups I and II. Megarchetypes I–III	Asterad (includes Muscari, Geum, Polygonum, Urtica, Lamium, Oxalis, Senecio and Poa variations)	Asterad
3) Period I: Series 'C'– Group III (mostly). Megarchetypes IV – V.	Solanad (includes Hyoscyamus, Nicotiana, Papaver, Sherardia and Linum variations)	Solanad
4) Period I: Series 'C'– Groups III–VIII. Megarchetypes III, V.	Chenopodiad (includes Chenopodiad, Polemonium and Myosotis variations)	Chenopodiad
5) Period II (also a few species in period III and IV). Series 'A', 'B', 'C'– Groups IX–XVI. Megarchetypes I–VI	Caryophyllad (includes Sagina, Vaccaria, Hypecoum, Corydalis, Sedum, Medicago, Drosera, Fumaria, Myriophyllum, Ruppia, Sagittaria, Zannichellia and Aglaonema variations)	Caryophyllad
6) —	Piperad	Piperad

1) Crucifer (Onagrad) Type. The general plan of development is given below:

The apical cell *ca* (Fig. 3.11.6A) divides vertically; the two daughter cells undergo another vertical division, at right angles to the previous one, to form a quadrant *q* (Fig. B). A transverse division gives rise to two superposed tiers of an octant designated *l* and *l′* (Fig. C, D). Periclinal division in each cell of an octant demarcates the protoderm from the axial zone (Fig. E, F). Further divisions in the cells of protoderm are always anticlinal, and they keep pace with the growing axial zone. The cells of the axial zone divide vertically, separating the central cells, the plerome initials (*pl*) from the outer periblem (*pe*) initials (Fig. F–I). The axial tissue divides in both the planes. Vertical divisions are more frequent in the plerome, and the tiers *l* and *l′* give rise to the cotyledons and stem tip.

The basal cell *cb* divides transversely to produce *m* and *ci* (Fig. B). A transverse division in *m* produces tiers *d* and *f*, and *ci* gives rise to *n* and *n′* (Fig. E–G). The tiers *m* and *ci* contribute to the suspensor. The lowermost cell of the suspensor functions as hypophysis (Fig. H, I). It divides transversely, and both the cells undergo crosswise vertical divisions so that eight cells are produced in two tiers (Fig. I, J). The upper four cells produce root cap and root epidermis, while the lower four give rise to root cortex.

2) Asterad Type. The development is according to the following plan.

The terminal cell *ca* of the two-celled proembryo (Fig. 3.11.7A) undergoes a vertical division (Fig. B). The two sister cells undergo one more vertical

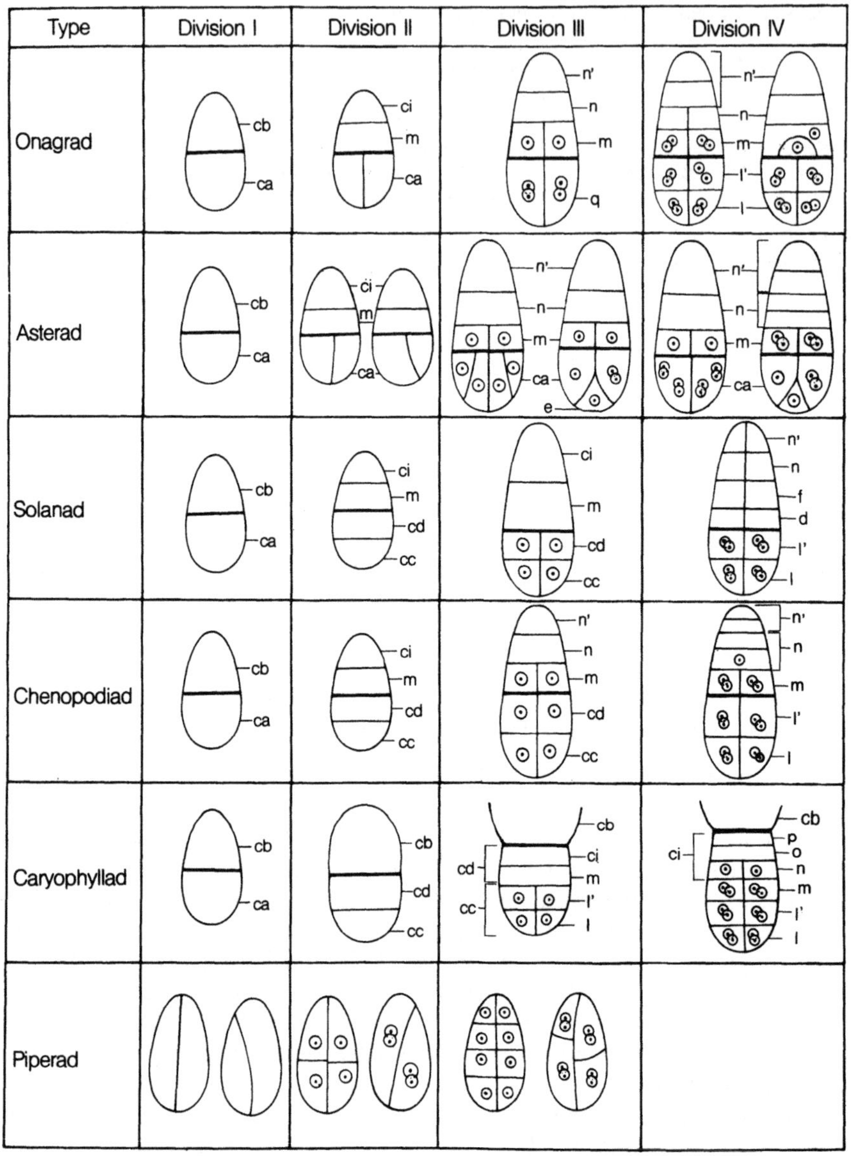

Fig. 3.11.5. Schematic representation of main types of embryogeny; based on Schnarf (1929) and Johansen's (1950) system of classification. (After Natesh and Rau 1984)

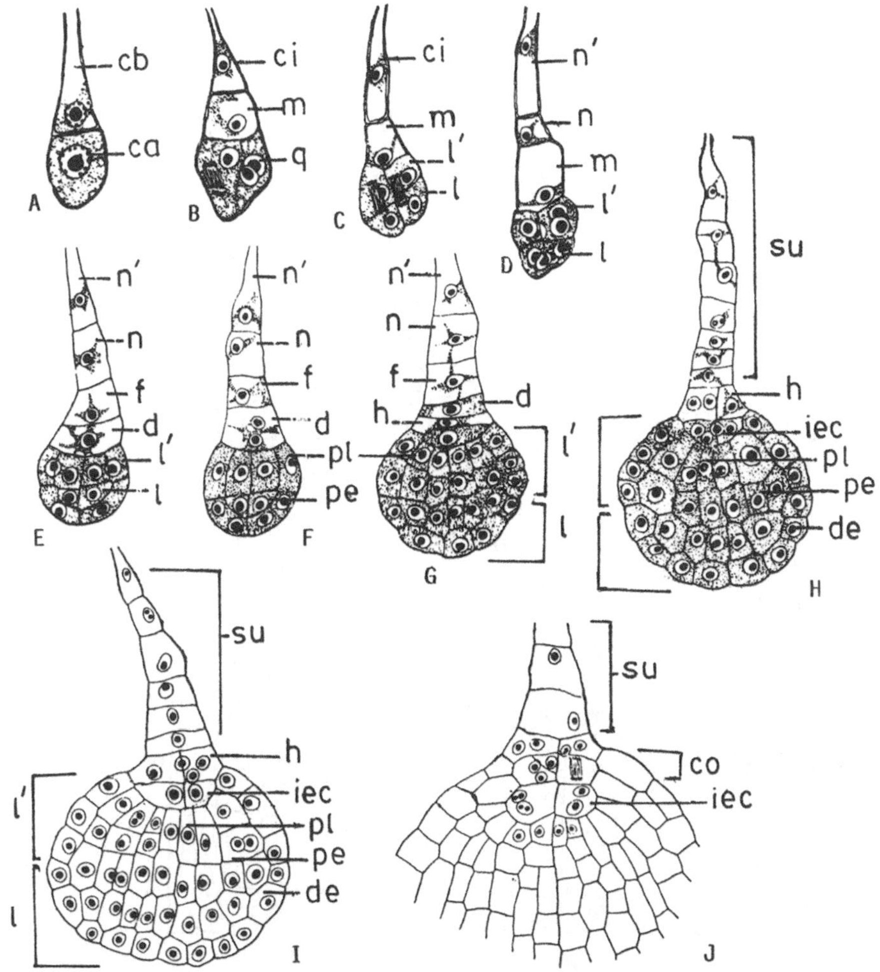

Fig. 3.11.6A–J. *Lophanthus chinensis*, embryogeny of Onagrad type; *su* suspensor. (After Crété 1942)

division, in a plane at right angles to the previous one forming a quadrant *q* (Fig. C–F). Each cell of a quadrant divides by an obliquely oriented vertical wall producing two unequal cells, one towards the central axis and the other towards the periphery. Thus, in a single tier of an octant, an axial group of four cells is separated from the peripheral four cells (Fig. G, H). Further divisions in *q* give rise to the initials of cotyledons and stem tip (Fig. G–K).

The basal cell *cb* divides transversely and produces tiers *m* and *ci*. The tier *ci* divides transversely, and produces *n* and *n′* (Fig. C). The tier *n* divides vertically (Fig. G–I), while *n′* divides transversely to give rise to *o* and *p* (Fig. G, H). Of the derivatives of *cb*, *m* gives rise to hypocotyl and radicle, *n* to the remaining

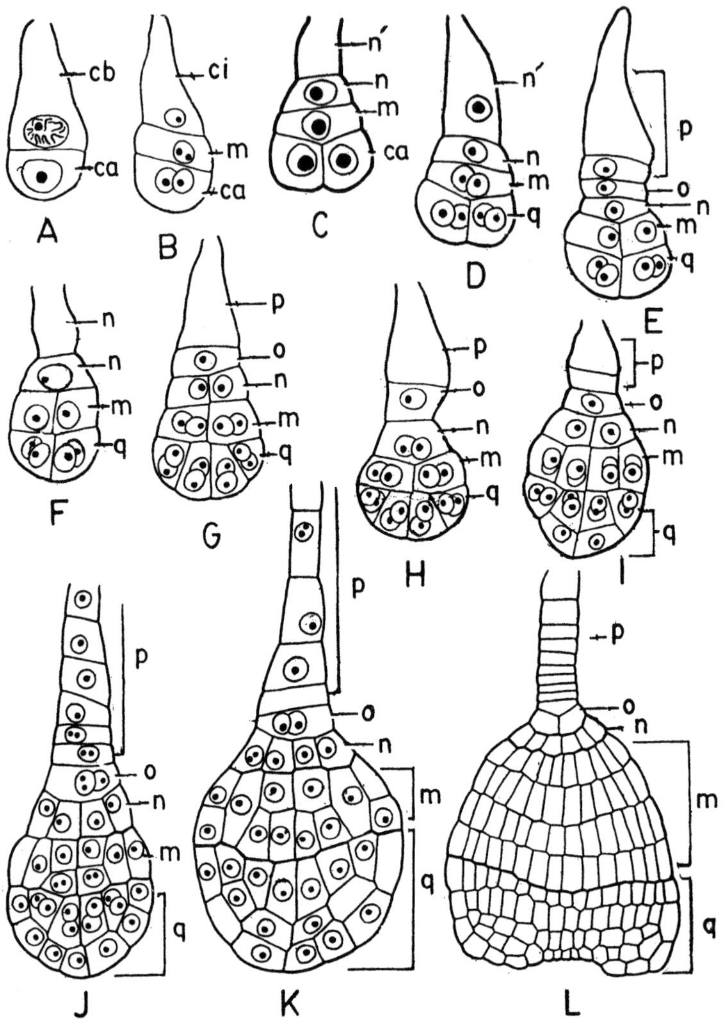

Fig. 3.11.7A–L. *Bidens biternata*, embryogeny of Asterad type. (After P.K. Deshpande 1964b)

part of root tip, and *o* and *p* are the initials of root cap and suspensor, respectively (Fig. G–L).

3) Solanad Type. The development is according to the following scheme:

The cells *ca* and *cb* of the two-celled proembryo (Fig. 3.11.8A) undergo transverse divisions to form a four-celled linear proembryo – these cells are designated *l*, *l′*, *m* and *ci* (Fig. B). The tiers *l* and *l′* divide vertically to form a quadrant (Fig. C). Another vertical division in each cell produces an octant (Fig. D). The tier *l* gives rise to the cotyledons and shoot apex, and *l′* to

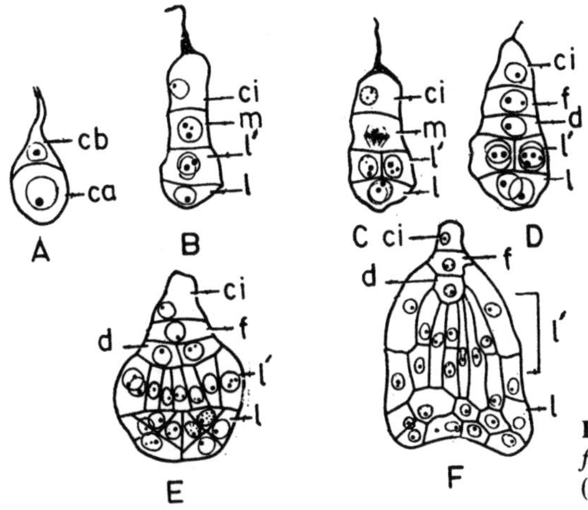

Fig. 3.11.8A–F. *Pentaphragma horsfieldii*, embryogeny of Solanad type. (After Kapil and Vijayaraghavan 1965)

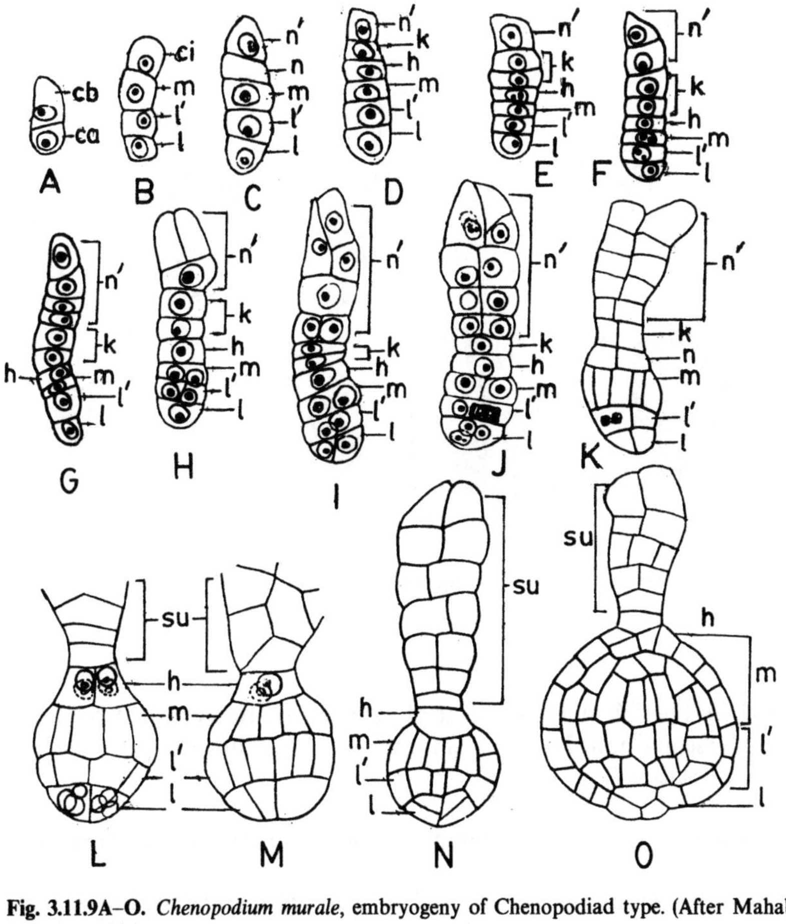

Fig. 3.11.9A–O. *Chenopodium murale*, embryogeny of Chenopodiad type. (After Mahabale and Solanky 1954c)

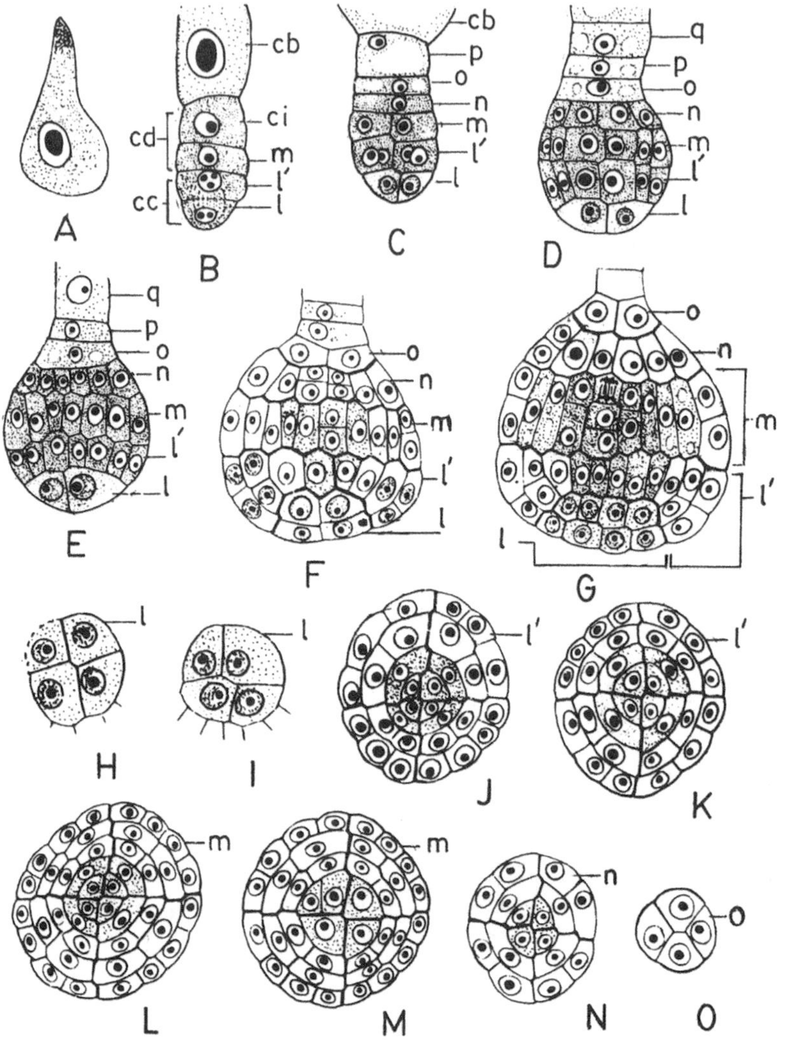

Fig. 3.11.10A–Q. *Stellaria media*, embryogeny of Caryophyllad type. (After Ramji 1975)

hypocotyl, and periblem and plerome of the root. *m* produces the tiers *d* and *f* by a transverse division. The tier *d* is the initial of root tip, and *f* and *ci* are the initials of suspensor (Fig. D–F).

4) Chenopodiad Type. The development is according to the following plan:

Transverse divisions in tiers *ca* and *cb* produce a four-celled proembryo: *l*, *l'*, *m* and *ci* (Fig. 3.11.9A, B). *ci* divides transversely to produce *n* and *n'*, and

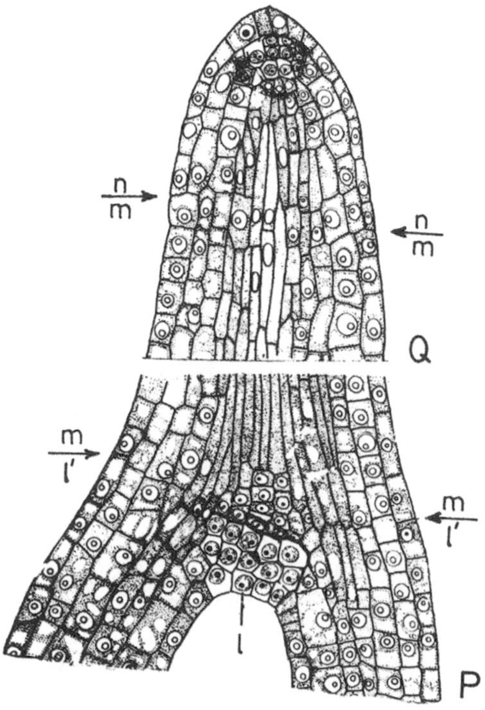

Fig. 3.11.10P, Q (*Contd.*)

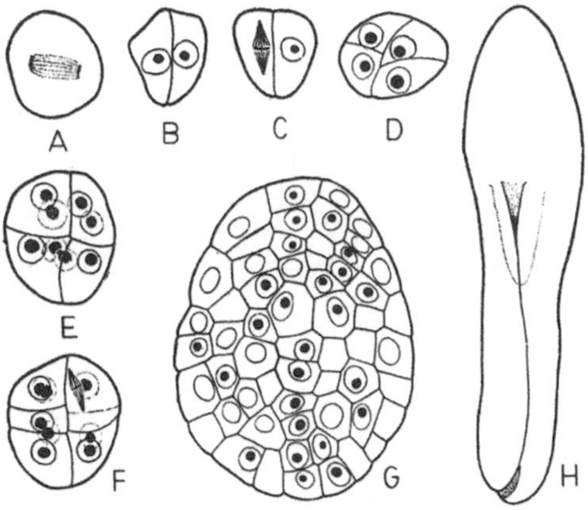

Fig. 3.11.11A–H. *Comandra umbellata*, embryogeny of Piperad type. (After M. Ram 1957)

by another transverse division *n* gives rise to *h* and *k* (Fig. C, D). Further divisions in *n'* produce a multicellular suspensor (Fig. E–O). The embryo proper develops from the tiers *l*, *l'*, *m* and *h* (Fig. E–O). In *Chenopodium murale* (Mahabale and Solanky 1954c), a vertical division occurs in *l'* followed by a vertical wall in *l* forming a quadrant (Fig. H, I); *m* also divides vertically (Fig. H–J).

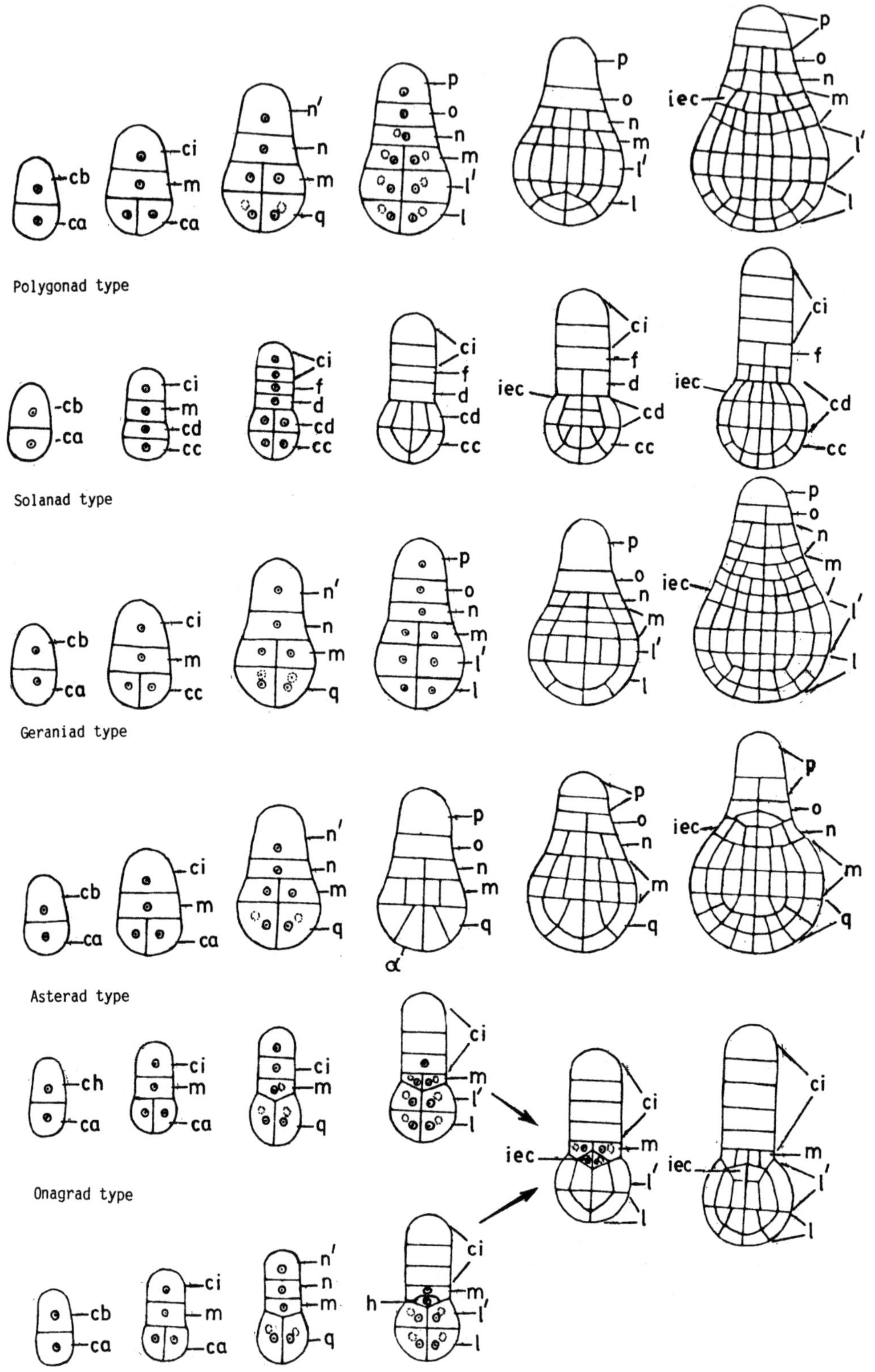
Polygonad type
Solanad type
Geraniad type
Asterad type
Onagrad type

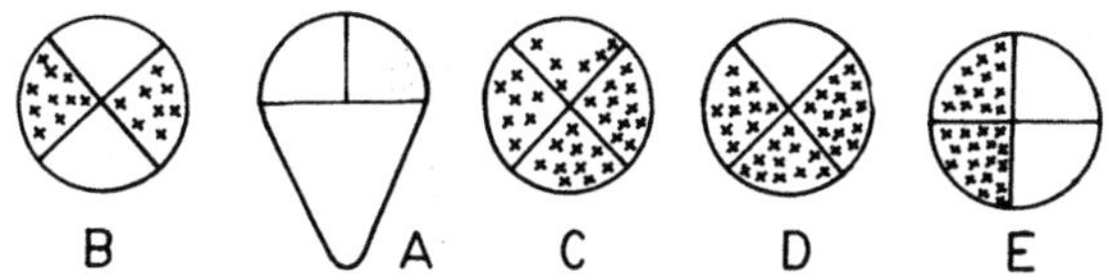

Fig. 3.11.13A–E. Proembryo at quadrant stage, initials of cotyledon *cross-marked*. **A** Vertical section of proembryo. **B** Transection of terminal tier of dicotyledonous proembryo; two opposite sectors denote loci of cotyledons. **C–E** Transections of terminal tier of monocotyledonous proembryo; there is a single locus of cotyledon. **C** Condition in Philydraceae; **D** Pontederiaceae, Sparganiaceae and Iridaceae; **E** Hydrocharitaceae, Potamogetonaceae and Amaryllidaceae. (After Lakshmanan 1972)

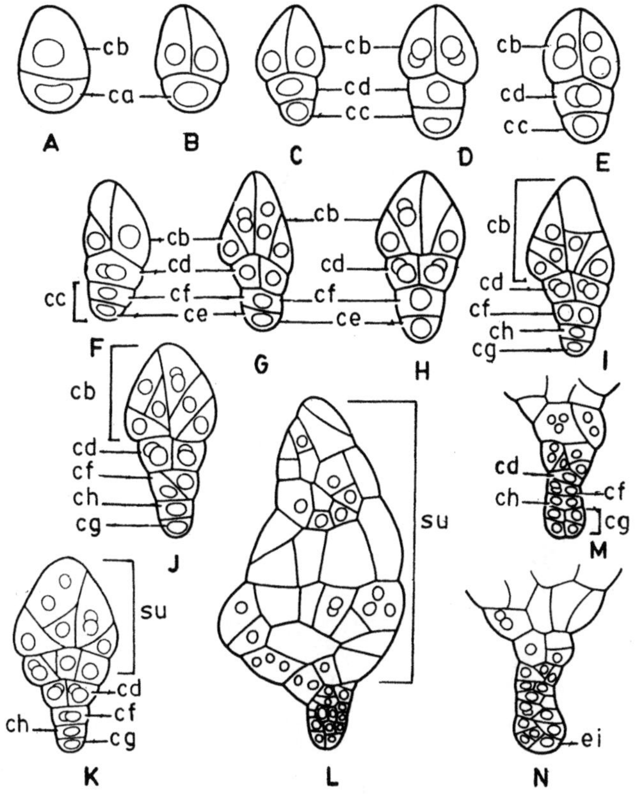

Fig. 3.11.14A–N. *Tulipa kolpakowskiana*, early embryogeny. **A** Two-celled proembryo. **B** Vertical division in *cb*. **C** Tetrad. **D** Second vertical division in *cb*. **E** Later stage, vertical division in *cb* and *cd*. **F** Transverse division in *cc* to form *ce* and *cf*. **G, H** Later stages. **I–K** Transverse division in *ce* to form *cg* and *ch*, vertical divisions in *cf* and *cd*. **L** Proembryo with a well-developed suspensor (*su*). **M** Octant (*cg*). **N** Proembryo with epiphysis initial (*ei*). (After Haccius and Hausner 1972)

Fig. 3.11.12. Developmental types of embryo: Polygonad, Solanad, Geraniad, Asterad and Onagrad. (After Yamazaki 1982)

After the formation of quadrant and octant, the protoderm is first demarcated in tier *m*, then in *l'*, and finally in *l* (Fig. K–O).

The derivatives of tier *l* give rise to the two cotyledons and shoot tip, and of *l'* to the lower and *m* to the upper part of hypocotyl. *h* functions as hypophysis, and gives rise to root-tip. Thus, the embryo proper develops from the derivatives of both *ca* and *cb*.

5) Caryophyllad Type. The general plan of development is given below:

The zygote (Fig. 3.11.10A) divides transversely into the apical cell *ca* and basal cell *cb*. A transverse division in *ca* produces tiers *cc* and *cd*; by one more

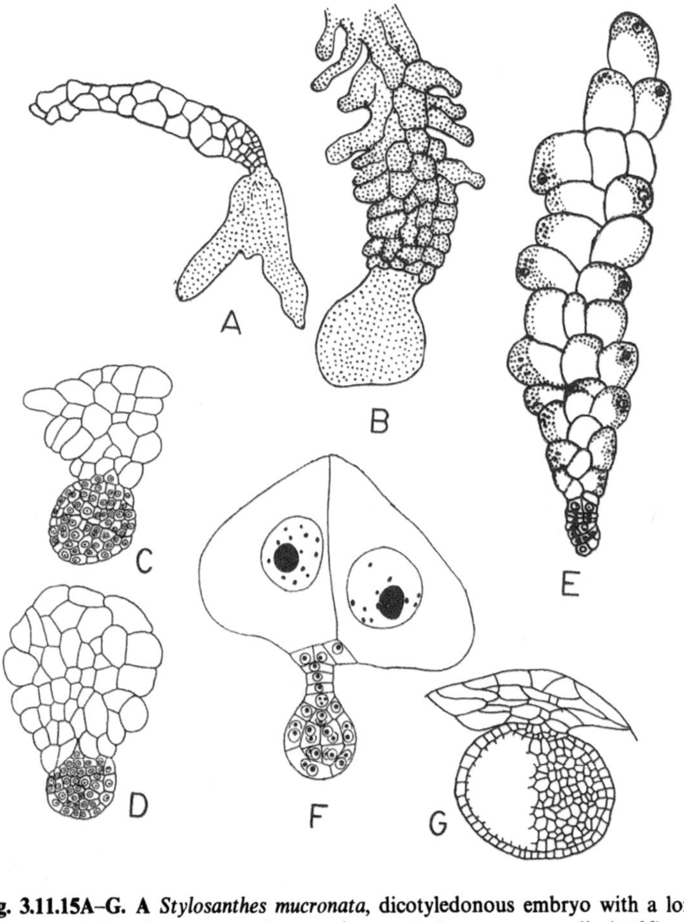

Fig. 3.11.15A–G. A *Stylosanthes mucronata*, dicotyledonous embryo with a long, multicellular suspensor. **B** *Crotalaria stricta*, the prominent tubular suspensor cells (*middle region*) remain in intimate contact with endosperm cells. (After Anantaswamy Rau 1953). **C, D** *Sesbania grandiflora*, multicelled conspicuous suspensor. (After Anantaswamy Rau 1951d). **E** *Rubia cordifolia*, proembryo with vesicular suspensor cells. (After Venkateswarlu and Rajeswara Rao 1958). **F** *Haloragis micrantha*, proembryo with two juxtaposed hypertrophied suspensor cells. (After Nijalingappa 1975). **G** *Glottidium vesicarium*, globular proembryo with fan-shaped suspensor. (After N. Prakash and Herr 1979)

division in the same plane *cc* gives rise to *l*, *l′*, and *cd* to *m* and *ci* (Fig. B). The tier *ci*, by two successive transverse divisions, produces tiers *n*, *o* and *p* (Fig. C).

The tier *cb* remains undivided and becomes vesicular (Fig. C) in *Stellaria media* (Ramji 1975).

The tier *l* gives rise to epicotyl (shoot apex), *l′* to cotyledons, *m* to hypocotyl, and *n* to the tip of hypocotyl. The derivatives of the tier *o* form the root-cap, and of *p* and *cb* give rise to the suspensor (Fig. B–Q). Further details are given under "Morphogenetic Concept".

Thus, the embryo proper is derived from the daughter cells of *ca* only; in addition, the tier *ca* also contributes to the formation of suspensor. It may be noted that the tiers *m*, *n*, *o* and *p* are the derivatives of *cb* in the Crucifer, Asterad, Solanad and Chenopodiad type. However, in the Caryophyllad type they are the derivatives of *cd*, a daughter cell of *ca*.

6) Piperad Type. The division of the zygote is invariably vertical as in Loranthaceae, Piperaceae, and occasionally in some other taxa. The division is obliquely vertical in *Scabiosa succia* (Souèges 1937d). *Comandra umbellata* (Ram 1957) is an exception in Santalaceae. The zygote divides vertically (Fig. 3.11.11A), and the two daughter cells (Fig. B, C) undergo transverse divisions to form a quadrant (Fig. D) and an octant (Fig. E). Subsequent divisions give rise to a globular proembryo (Fig. F, G) which differentiates into a dicotyledonous embryo (Fig. H). The suspensor is absent.

In contrast, in Loranthaceae the vertical division of the zygote is followed by a series of transverse divisions producing a long, biseriate suspensor (Sect. 12.8). In *Balanophora*, the first two or three divisions in the zygote are vertical (Zweifel 1939).

Yamazaki's Classification. Yamazaki (1974, 1982) proposes a system of classification of angiosperms on the basis of patterns of differentiation of initials of the root cortex (*iec*). He emphasized that it is a constant feature not only in the families, but even in the orders. He recognised five types of embryogeny (Fig. 3.11.12): Polygonad type, Solanad type, Geraniad type, Asterad type, and Onagrad type – Ruta variation and Onagrad variation.

1) Polygonad Type. The cell *ca* of the two-celled proembryo gives rise to shoot apex and cotyledons. The cell *m*, daughter cell of *cb*, produces two tiers. The central cells of the upper tier produce hypophysis (*iec*), while the tier *m* gives rise to hypocotyl.

2) Solanad Type. The terminal cell *ca* divides transversely to form tiers *cc* and *cd*. The derivatives of *cc* form the initials of shoot apex and cotyledons. *cd* forms two tiers. The central cells of the upper tier produce *iec*, whereas the lower tier produces the entire hypocotyl.

3) Geraniad Type. The cell *ca* gives rise to shoot apex, cotyledons and hypocotyl. The tier *m*, daughter cell of *cb*, produces three or more cell layers. *iec* is derived from the central cells of the lowermost tier of *m*. The upper tiers of *m* give rise to the root-cap, and a part of the suspensor.

4) Asterad Type. The terminal cell *ca* directly gives rise to tier *q* which develops into shoot apex and cotyledons. *cb* divides transversely to produce *m* and *ci*; *ci* gives rise to *n* and *n'*, and *n'* to *o* and *p*. The tier *n* (or *m*) consists of one layer of cells, and the central cells give rise to the hypophysis initial (*iec*). Sometimes, the tier *m* is two- or three-layered, and the hypophysis initial is derived from the lower tier, while the upper one produces the root-cap. The embryogeny then conforms to the Calendula variation.

5) Onagrad Type. The cell *ca* and its derivatives form the initials of shoot apex, cotyledons, and hypocotyl. In the Ruta variation the cell *m*, lower daughter cell of *cb*, undergoes two longitudinal divisions, followed by a transverse wall. The lower four cells form the initials of the root cortex.

In the Onagrad variation, the cell *m* cuts off a lenticular cell *h*, which divides twice longitudinally to produce the four-celled hypophysis.

Lastly, Yamazaki mentions the "Irregular" type where the segmentation and differentiation of the root cortex is irregular, and difficult to follow. The zygote may divide transversely or longitudinally.

In our opinion, there is no merit in Yamazaki's concept of the differentiation of initials of root cortex (*iec*) as a basis of classification of angiosperm families and orders. The embryogenic classification, as discussed by Natesh and Rau (1984), is certainly the most valid system and has been widely used for over four decades since Schnarf (1939) and Johansen (1945) proposed the six basic types.

The embryogenic classification, according to these five types (except Piperad), is inadequate to classify various types of variations which have been described as "unclassified or abnormal" (P. Maheshwari 1950) or "irregular" (Crété 1963a). Swamy and Padmanabhan (1962) point out that variations occur even in the laying down of walls at different stages within the same species.

These embryogenic classifications consist of a hierarchy of types and subtypes which have been widely criticized. First, most investigators have not studied sufficient material to emphasize the range of variation. Second, when more than one type of embryogeny has been observed, their relative percentage has never been studied. Third, the sequence of early cell divisions on which the categories are based vary a good deal in the proembryos of the same species; therefore, the categories are no more valid (see Wardlaw 1955, Swamy and Padmanabhan 1962, Ramji 1975, Periasamy 1977, Lersten 1983).

Padmanabhan (1967, 1968), Ramji (1975), and Swamy and Krishnamurthy (1980) emphasize the importance of organization of meristems and organogenesis. The entire developmental sequence is possibly controlled by physical, biochemical and genetic factors. The zygote exhibits distinct polarity, and the terminal pole is far more active. According to Swamy and Krishnamurthy (1980), the formation of a partition wall in the zygote depends on the "principle of minimal area and distribution of dictyosomes. During embryo development, there occurs a gradient of decreasing growth activity from the shoot pole to the suspensor/root pole". The systems based on cell lineages fail to explain these features.

Morphogenetic Concept. Padmanabhan (1967, 1968) and Ramji (1975) interpret embryogeny on the basis of morphogenic phases. Ramji (1975) distinguishes four phases: (1) filamentous phase following transverse divisions in zygote and its derivatives (Fig. 3.11.10A, B), (2) onset of vertical divisions (three-dimensional phase) (Fig. C), (3) periclinal phase (Fig. D, F), and (4) block-meristem phase (Fig. F–O).

1) Filamentous Phase. In *Stellaria media* (Ramji 1975), the zygote (Fig. 3.11.10A) divides transversely forming *ca* and *cb*. *ca*, by three successive divisions in the same plane, gives rise to *l*, *l'*, *m* and *ci* (Fig. B); *cb* remains undivided. Thus, there is much more activity (cell divisions) at the terminal pole. Ramji (1975) also draws attention to the differential staining reaction of the protoplast. This is much more intense in *ca* than *cb*.

2) Onset of Vertical Divisions or Three-Dimensional Phase. Vertical walls are laid in tiers *l'*, *m* and then in *l* (Fig. C). The vertical walls in *l'* and *m* do not coincide, or are perpendicular to each other so that a small sector of the original wall (intercept) is left at the point of contact. During later stages, this intercept helps in determining the bilateral symmetry of the proembryo.

3) Periclinal Phase. Periclinal divisions parallel to the surface of the embryo occur in *m* and *l'* separating a group of axial cells in the centre (Fig. D). The periclinal division in *n* occurs when the tiers *m* and *l'* become six-celled (Fig. E). The activity of the cells of tier *l* is very slow as they are the potentials of the shoot-apex, which grows vigorously during the germination of seed. The four cells of tier *l* (Fig. F) divide by oblique transverse walls, separating the outer cover of four protoderm cells, and inner group of four cells. This results in separation of a continuous protodermal layer around the central axial zone. The increase in the number of cells in the axial zone is due to vertical divisions. The central zone of four axial cells along with the encircling layer gives rise to procambium; the cells lying between this zone and the protoderm give rise to cortex (Fig. G; for further details of each tier, apex upward, see Fig. H–O).

4) Block Meristem Phase or Organ Indication Phase. The outward globular appearance (Fig. F, G) of the proembryo is retained with the increase in the number of cells. The cells of the central axial zone divide in a tangential plane, whereas those of outer layers undergo divisions in a radial plane.

The cells of the subprotodermal layer of *l'*, on either side of the apex of epicotyl, divide anticlinally and differentiate as initials of cotyledons, whereas those in the centre elongate to form the rib meristem, later producing pith of the cotyledonary node. Tier *m* gives rise exclusively to the hypocotyl. The three zones – protoderm, initials of cortex, and axial region – are well-demarcated due to differential staining of the protoplasts. Frequent divisions in transverse plane (Fig. G) take place in central cells, which develop into the procambium. Tier *n* forms the tip of the hypocotyl, which is the potential root.

Figures P and Q show longisections of apices of epicotyl and hypocotyl. (The author includes radicle in hypocotyl.) The cortical cells developed from the tier *l′* are in continuation with those produced by tier *m* (Fig. P). In the region *n* the subcortical layer is developed by the periclinal division of protoderm cells, and the remaining two layers are in continuation with *m*. Thus, the differentiation of cortical tissue in the tiers *m* and *n* differs (Fig. P, Q).

Periasamy (1977) proposes a new system of classification of embryos based on morphogenetic significance. Accordingly, he classifies 45 species of dicotyledonous and monocotyledonous embryos into five groups. Periasamy emphasizes that the external surface of the embryo and its derivatives towards the inner side are subject to two different environments. The criteria used for classification are: (1) number of cell tiers that comprise the proembryo at the time when the first series of internal cells are segmented, (2) relative location of the tiers with varying cell number, and (3) terminal or subterminal location of the tier in which the first internal cells are segmented.

The embryogeny in *Paeonia* shows a totally different type of development (see Sect. 22.3).

Dicot and Monocot Embryos. The similarity of development between the dicotyledonous and monocotyledonous type of embryos has been traced by Lakshmanan (1972). Apparently, the external appearance of the proembryo up to the globular stage is similar in both. Swamy and his co-workers (see Lakshmanan 1972) have demonstrated that the epicotyl and cotyledon arise from one and the same terminal tier (*q*) in monocotyledons. The epicotyl shows belated development, while the cotyledon overgrows in early stages. Consequently, the epicotyl assumes a lateral position. The subterminal tier and its derivatives contribute to the hypocotyledonary region.

Lakshmanan (1972) also discusses the differences between the ontogeny of the two types of embryos (Fig. 3.11.13A–E). The main difference lies in the number of sectors of a quadrant, and their position in a quadrant, which contribute to the formation of cotyledon(s). In a dicotyledonous embryo, the two opposite cells of a terminal quadrant give rise to two cotyledons (Fig. B). In monocotyledons, the number of sectors of a quadrant involved in the formation of a single cotyledon varies (Fig. C–E). "It is practically almost all the four (except a few cells derived from one of the quadrant cells) in the Philydraceae, three cells of the quadrant in the Pontederiaceae, Sparganiaceae, and Iridaceae, and adjacent two cells in the Hydrocharitaceae, Potamogetonaceae, and Amaryllidaceae that constitute the cotyledonary locus" (Lakshmanan 1972).

Guignard and Ly-Thi-Ba (1989) have reviewed the embryogeny of the Helobiae.

There is a single cotyledonary locus in monocotyledons, and two in dicotyledons. Likewise, the morphogenetic potential of a shoot apex differentiates at the end of the first vertical division in the terminal cell in monocotyledons, and at the end of the second vertical division in dicotyledons (see also Swamy and Krishnamurthy 1980).

It is customary to classify monocotyledonous proembryos, like those of dicotyledons, according to the embryogenic systems of classification by Souèges (see Crété, 1963a), and Johansen (1950). However, Figure 3.11.13A–E clearly indicates the difference between the cotyledonary loci of the two.

In taxa such as Hydrocharitales, Aponogetonales, Potamogetonales, Najadales, Commelinales, Xyridales, Zingiberales, Liliales, Amaryllidales, Orchidales, Graminales, etc., equal volumes of the shoot apex are involved in the organization of the cotyledon (single) and epicotyl, whereas in Alismataceae, Juncaceae, Pontederiaceae, Typhales, etc., a larger volume of the shoot apex takes part in the development of the cotyledon (Swamy 1979).

The embryogeny in *Triticum* (Batygina 1969) exhibits the orientation of oblique walls at the onset of division of the zygote (Fig. 58.2.3A, B). They are disposed in such a way that a continuous dorsiventral symmetry is maintained during the entire development of the embryo (Fig. C–T). The vegetative point (shoot apex) originates from the apical cell *ca* and is terminal (Fig. N), the scutellum initial originates from the cell *ca* as well as from *cb* (*m* and *n*) and is apical–lateral in origin (Fig. N). The region of intensive colour and high mitotic activity of cells denotes the origin of meristems during organogenesis. The dorsiventral symmetry of the embryo (Fig. C–T), inception of oblique walls (Fig. A–H, B′–F′), displacement of tiers *m* and *ci*, and early differentiation of scutellum led Batygina (1969) to propose a new type, the Graminad type.

Juguet (1989) discusses the problems of total number of cells in a proembryo when the symmetry becomes established in the monocots and dicots. According to him, the dicot embryo has an "axial" and the monocot embryo a "bilateral" symmetry. The cotyledon and the future stem-tip both develop in lateral–apical position but become respectively terminal and lateral. Juguet reemphasizes "that the quarrel between "terminalists" and "lateralists" is a futile one". We support this view, but consider the cotyledon/s as lateral and the shoot apex as terminal structures, both in monocots and dicots.

Delayed Embryo Differentiation. Haccius (1971) and co-workers (Haccius and Hausner 1972, Haccius and Bhandari 1975, Haccius and Philip 1979) draw attention to "embryogenesis with delayed embryo-differentiation". The embryogeny in *Tulipa* (Haccius and Hausner 1972) is a good example, and shows the formation of a vertical wall in the apical cell at the end of the fifth or sixth cell generation, initiating the globular stage of the proembryo. The major portion of the derivatives of *ca* and all of *cb* do not contribute to the formation of the embryo proper.

After the first cell generation, the zygote produces two superposed cells *ca* and *cb* (Period I, see Fig. 3.11.4). In the second generation (Period II), *ca* produces tiers *cc* and *cd*, in the third (Period III) *ca* gives rise to *ce* and *cf*, in the fourth (Period IV) *ce* to *cg* and *ch*, and the fifth and sixth generations continue similarly.

In *Tulipa kolpakowskiana* (Haccius and Hausner 1972), a vertical division occurs first in the basal cell *cb* of the two-celled proembryo (Fig. 3.11.14A–C). Successive vertical divisions occur in each of these two sister cells up to the end of the third cell generation (Fig. D–H). The derivatives of *cb* undergo divisions in all planes and the basal portion becomes massive (Fig. I–L).

The tier *ca* undergoes a transverse division forming *cc* and *cd* (Fig. C–E). The tier *cc* divides transversely to form *ce* and *cf*, while *cd* undergoes a vertical division (Fig. F, G). At the end of the fourth cell generation (Fig. H–J), *ce* produces two superposed daughter cells, *cg* and *ch*, whereas *cf* and the derivatives of *cd* divide vertically. The fifth cell generation begins with the transverse division of *cg* (Fig. M), *ch* divides vertically. The vertical wall is laid down in the terminal cell mostly at the end of the fifth or sixth cell generation (Fig. L); M and N are later stages. The embryogeny of *Tulipa* can be classified to Period V or VI of Souèges system. It cannot be referred to Johansen's system, as in this classification the provision is only up to Period II (Caryophyllad type). It may be noted that, during early embryogeny, vigorous growth is restricted to the basal region only and the differentiation of the embryo is delayed. The derivatives of *cb* and a considerable portion of the cells derived from *ca* do not contribute to the formation of the embryo proper.

Suspensor. The suspensor is usually uniseriate with uninucleate cells in young proembryos. However, the occurrence of massive suspensor with multinucleate cells is not uncommon (Fig. 48.4.4J–O; see also Raghavan 1986a, b). The highly coiled suspensor of multinucleate cells (Fig. 24.3.1A–F) in *Isomeris arborea* was incorrectly interpreted as endosperm nodules by Billings (1937). During the earlier stages of development the suspensor cells exhibit endomitotic polyploidization in *Trapa natans* (Nägl 1962). The highest ploidy-level noted is 256×. The polyploid level corresponds to 1024× in *Tropaeolum majus*, and 409× in *Phaseolus coccineus* (Nägl 1962). The occurrence of giant chromosomes in *Phaseolus* (Nägl 1962) and *Gagea lutea* (Hasitschka-Jenschke 1959, 1962) indicates a high metabolic activity of suspensor cells.

In *Stylosanthes mucronata* (Fig. 3.11.15A), at the dicotyledonous stage of the embryo the suspensor is long and multicellular. In *Crotalaria stricta* (Fig. 3.11.15B), the peripheral suspensor cells in the middle region of a long multicellular suspensor develop tubular projections which penetrate into the endosperm tissue. The suspensor is fairly broad, multicellular and conspicuous in *Sesbania grandiflora* (Fig. 3.11.15C, D). The suspensor cells in *Rubia cordifolia* (Fig. 3.11.15E) become prominently vesicular. The two juxtaposed cells (daughter cells of *cb*) enlarge and their nuclei become hypertrophied in *Haloragis micrantha* (Fig. 3.11.15F). The suspensor is fan-shaped in *Glottidium vesicarium* (Fig. 3.11.15G).

A four-celled suspensor with coenocytic cells is reported in *Lathyrus latifolius* (Nägl 1962), and a broad and conspicuous suspensor in *Phaseolus coccineus* (Yeung and Clutter 1978).

The extensive suspensor haustorium in Podostemaceae (Figs. 28.2.2K–N, 28.2.3A–C), Trapaceae (Fig. 39.3.1N, O) and Orchidaceae (Fig. 65.2.3E–H) serves the function of absorption and transport of nutrients in the absence of endosperm tissue. In *Dicraea stylosa* (Fig. 28.2.3A–C), the binucleate basal cell of the two-celled proembryo develops conspicuous tubular haustorial branches. During later stages, the suspensor pushes the embryo into the pseudoembryo sac which provides nutrients in the absence of endosperm.

The development of haustoria (Sect. 26.5) from the suspensor in Crassulaceae (Mauritzon 1933a, Subramanyam 1963, 1967) is well known. The hypha-like branches of the haustorial micropylar cell of the suspensor grow into the integument and the raphe, and in some members become extraovular and penetrate the funicle. The Orchidaceae (Sect. 65.2) show much variation in the organization of the suspensor (Swamy 1942a). A well-developed branched suspensor persists in the globular incompletely differentiated embryo in the mature seed (Fig. 65.2.3H) of *Dendrobium nobile* (Poddubnaya-Arnoldi 1967).

The role of suspensor has been revealed by ultra-structural, histochemical and in vitro cultural studies. P. Schulz and Jensen (1969) observed that in *Capsella* the octant proembryo has a six-celled suspensor. The wall of the micropylar terminal vesicular cell bears projections on its inner surface. The wall ingrowths are more concentrated at the micropylar tip and are similar to "transfer cells" (see Gunning and Pate 1969). The latter perform the function of transport of nutrients from the adjoining tissue to the developing embryo.

The histochemical studies in *Brassica campestris* (Malik and Vermani 1975) also support these observations. Malik and Vermani further suggest that the suspensor has a secretory function. Bhalla et al. (1979) experimented with enzymatic reactions occurring in the suspensor haustorium (in the carpellary region) and the suspensor of *Tropaeolum majus*. Their data indicate that the suspensor haustorium absorbs nutritive material from the adjoining carpellary tissue and the suspensor transports it to the young proembryo.

The suspensor has an important role in embryogenesis by acting as a site for the synthesis of growth regulators required by the embryo (Alpi et al. 1975). While experimenting with the major role of gibberellins in characterizing the physiological role of the suspensor, Cionini et al. (1976) demonstrated that GA_3 (10^{-8} to 10^{-6} M) replaces the function of the suspensor (in *Phaseolus coccineus*) when suspensor-deprived embryos of heart-shaped and early cotyledonary stages (0.5–1.5 mm long) are cultured in vitro.

Chlorophyllous Embryos. Yakovlev and Zhukova (1980) review the occurrence of chlorophyll in the embryos of 438 species, belonging to 224 genera of 72 families. The seeds of 1094 species of 666 genera of 182 families have been investigated so far. The greening of embryos is a genetic feature (Yakovlev and Zhukova 1975), and in several species the chlorophyll disappears in the mature embryo.

Chlorophyllous embryos invariably occur in Brassicaceae (Cruciferae), Convolvulaceae, Dipsacaceae, Papilionaceae, Plumbaginaceae, Rhamnaceae and Tiliaceae, whereas the families Betulaceae, Bignoniaceae, Caprifoliaceae, Fagaceae, Oleaceae, and Saxifragaceae have totally non-chlorophyllous embryos. The rest of the 58 families, such as Anacardiaceae, Apocynaceae, Malvaceae, Myrtaceae, Scrophulariaceae and others, have both chlorophyllous and non-chlorophyllous embryos. The monocotyledonous families rarely show green embryos.

The chlorophyllous embryo may be considered as a diagnostic feature, e.g. closely related families Capparaceae and Cruciferae have chlorophyllous embryos. On this basis, Celastraceae (green embryo) can be distinguished from Aquifoliaceae (non-green embryo), Dipsacaceae from Morinaceae, and Plumbaginaceae from Primulaceae.

3.12 Polyembryony

Polyembryony is of common occurrence in angiosperms. In most cases, the additional embryos do not develop up to maturity and degenerate early. Polyembryony has been classified by Schnarf (1929), P. Maheshwari and Sachar (1963), Yakovlev (1967), and Bouman and Boesewinkel (1969). The classification suggested by P. Maheshwari and Sachar (1963) is followed in the text (see also Lakshmanan and Ambegaokar 1984).

True Polyembryony. 1) Embryos arising within an embryo sac, either by cleavage of the zygotic proembryo, or from the synergid/s. 2) Embryos arising from the tissues lying outside the embryo sac, i.e. nucellus and integument/s (adventive embryony).

False Polyembryony. This includes twin nucelli formed due to cleavage of the nucellar primordium (*Pterocarya fraxinifolia* – Bouman and Boesewinkel 1969), or the development of two or more embryo sacs within the same nucellus either due to multicellular archesporium, or development of more than one megaspore from a single tetrad, or concurrent development of sporic and aposporic embryo sacs.

Cleavage Polyembryony. Polyembryony may be due to proliferation, or budding, or splitting of the zygotic proembryo. All these have been observed in *Eulophia epidendraea* (Fig. 65.2.3E, F). Narayanan (1968) reported cleavage polyembryony in 86 percent of ovules of *Streblus taxoides.* Two to four embryos are produced due to cleavage of the zygotic proembryo, and the twin embryos develop up to the dicotyledonous stage (Fig. 3.12.1A). Mathew and Chaphekar (1977) also report cleavage of the zygotic proembryo in *Stachyurus chinensis* (Fig. 3.12.1B, C).

The proliferation of proembryonal and suspensor cells giving rise to additional embryos is a common feature in the members of Acanthaceae (Fig. 3.12.1D–F), Actinidiaceae, Santalaceae, Zygophyllaceae and others.

Synergid Polyembryony. In several families the persistent synergid develops into an embryo, e.g. Amaryllidaceae, Apocynaceae (Plumiereae), Burmanniaceae, Cucurbitaceae, Stachyuraceae, Tamaricaceae and some others. In most cases the synergid does not get fertilized. B.M. Johri and Kak (1954) reported the presence of two pollen tubes in the embryo sac of *Tamarix ericoides* (Fig. 3.12.1G). The synergid embryo appears to have developed after fertilization. Similar fertilized synergids have also been reported in Alismataceae (B.M. Johri 1936a).

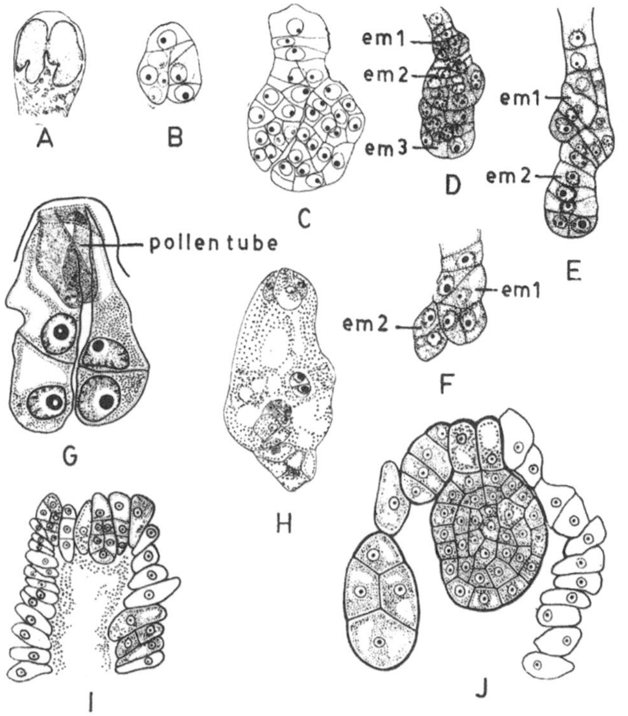

Fig. 3.12.1A–J. Polyembryony. **A** *Streblus taxoides*, longisection of seed, twin dicotyledonous embryos result from cleavage of zygotic embryo. (After Narayanan 1968). **B, C** *Stachyurus chinensis*, budding/cleavage of proembryo. (After Mathew and Chaphekar 1977). **D** *Barleria prionites*, additional proembryo *em 1* formed by proliferation of a suspensor cell (?), and *em 3* by proliferation of proembryo *em 2*. (After Phatak and Ambegaokar 1956). **E** *Sericocalyx scaber*, additional proembryo *em 1* formed by budding of a suspensor cell. (After Phatak and Ambegaokar 1961b). **F** *Lepidagathis trinervis*, additional proembryo *em 2* formed by proliferation of proembryo *em 1*. (After Ambegaokar unpublished). **G** *Tamarix ericoides*, portion of embryo sac showing two-celled zygotic and two-celled synergid proembryo; note remnants of two pollen tubes. (After B.M. Johri and Kak 1954). **H** *Paspalum scrobiculatum*, embryo sac showing antipodal (nucellar?) proembryo; note the adjoining antipodal cells. (After Narayanaswami 1954). **I, J** *Melampodium divaricatum*, endothelium with some cells developing into proembryonal structures. (After Maheswari Devi and Pullaiah 1976)

Sachar (1955) reported synergid polyembryony in *Argemone mexicana*. The synergids are usually ephemeral, but one of them may persist and develop into a proembryo. This proembryo is smaller, has an independent suspensor, and lies juxtaposed to the zygotic proembryo.

In *Dioscorea composita*, the zygotic and synergid proembryos develop until a late stage. They follow a uniform pattern of development up to organogenesis in *Nechamandra alternifolia* (Lakshmanan 1963b).

The development of haploid proembryo from the synergid, along with the diploid zygotic proembryo, is reported in several taxa: *Lilium martagon*, *Arabis lyallii*. The suspensor of the zygotic as well as of the synergid proembryo of

Table 3.12.1. *Capsicum frutescens*, multiple seedlings in reciprocal crosses of two varieties (After Morgan and Rappleye 1954)

Varieties	Twins				Triplet
	Unattached			Conjoined	Unattached
	Sexual–sexual 2n 2n	Maternal–sexual n 2n	Maternal–maternal n n	Sexual–sexual 2n 2n	Sexual–sexual–sexual 2n 2n 2n
Non-pungent, non-elongated, dark-green immature fruit × pungent, elongated yellow immature fruit	72	11	1	7	1
Pungent, elongated, yellow, immature fruit × non-pungent, non-elongated, dark-green immature fruit	10	4	0	3	0
Total 109	82	15	1	10	1

A. lyallii elongates into similar filamentous structures with a small embryonal mass at the apical end.

In *Cuscuta*, polyembryony is rare, but in *C. reflexa* (B.M. Johri and Tiagi 1952) a two-celled synergid proembryo has been observed along with the bicelled zygotic proembryo. Accessory pollen tubes and supernumerary male gametes have also been observed in the embryo sac (Fig. 3.9.3B). However, it could not be determined whether the synergid proembryo is haploid or diploid.

In *Pennisetum squamulatum* (Sindhe et al. 1980), the diploid synergid proembryo develops along with the zygotic proembryo. The suspensor of the zygotic proembryo is typically multicellular as in the Gramineae, and that of the synergid is partly uniseriate and partly biseriate.

In *Azadirachta indica* (N.C. Nair and Kanta 1961), at times one or both synergids develop into haploid proembryos. In *Carthamus tinctorius* (Maheswari Devi and Pullaiah 1977), the triplets develop due to fertilization of the egg and both synergids.

Multiple seedlings arise in *Capsicum frutescens* (2n = 24) from individual polyembryonic seeds, following reciprocal crosses of two varieties (Table 3.12.1).

The haploid members of twin seedlings arose from reduced cells of the female gametophyte, presumably the synergid (the antipodals degenerate prior to fertilization). All diploid members of the multiple seedlings were of sexual origin (adventive embryos do not develop in this sexual species). Diploid–diploid twins could arise from twin embryo sacs. The conjoined diploid–diploid twins presumably originated from the cleavage of zygotic embryo. Haploid–haploid twins could originate from the synergids, if the zygotic embryo degenerates. Most of the unattached 2n-2n twins developed within a single embryo sac by fertilization of a synergid and the egg. It is likely that the pollen tube had supernumerary sperm cells, or more than one pollen tube entered the embryo sac. The origin of the unattached triplet could not be traced. The possibilities are: (a) polyspermy can lead to fertilization of the egg and both the synergids, and (b) fertilization of the egg in three embryo sacs in the same nucellus could also lead to a 2n-2n-2n triplet – rather rare. The study of polyembryony in hybrids and their n/2n constitution needs better attention.

Antipodal Polyembryony. The formation of antipodal embryos is not so frequent. These have been observed in *Alangium lamarkii* (Gopinath 1943), *Paspalum scrobiculatum* (Fig. 3.12.1H), *Rudbeckia sullivanti* (Battaglia 1955), *Ulmus americana* (Shattuck 1905), *U. capensis* (Guignard and Mestre 1966) and *U. glabra* (Ekdahl 1941). In *Solidago canadensis*, Pullaiah (1978b) observed two proembryos developing at the antipodal end but these did not mature to full size.

In our opinion, the development of embryos from antipodals needs confirmation. When the antipodal cells divide, the configuration of cells may simulate embryogenic patterns. Since advanced stages have never been recorded, it is not at all appropriate to interpret such cell assemblages as antipodal embryos. We would rather interpret them as internal divisions in antipodal cells – a feature already known in Dipsacaceae (Fig. 50.5.7C–G). This is probably also true of endothelial (Fig. 3.12.1I, J), and integumentary proembryos in *Spiranthes cernua* (Swamy 1948d).

Adventive Polyembryony. The development of proembryo/s from the integument/s is less frequent: *Spiranthes cernua* (Swamy 1948d). Either a single cell, or a group of two or three cells of the inner epidermis (endothelium) of the inner integument proliferate to produce the additional embryo/s (Fig. 3.12.1I, J). Integumentary embryos are known in *Carthamus tinctorius* (Maheswari Devi and Pullaiah 1977), *Datura* (Satina et al. 1950), and *Spiranthes* (Swamy 1948e).

The formation of embryos from the nucellar cells is a common feature. The cells of the nucellus, in the micropylar region, lying on the periphery of the embryo sac proliferate and nucellar embryos are initiated. As they grow, they come to lie in the embryo sac and attain various stages of development. In *Opuntia dillenii* (P. Maheshwari and Chopra 1955), the egg degenerates, several nucellar embryos develop, and one of them reaches up to the mature stage. The nucellar embryos are frequent in Anacardiaceae (mango), Cactaceae (*Mammillaria* – Fig. 17.2.1A–D), Combretaceae, Gramineae, Malpighiaceae, Rutaceae, Urticaceae and others. Polyembryonate seeds have been reported in Myrtaceae.

False polyembryony is reported in Loranthaceae (Sect. 12.8), Gramineae (Sect. 58.9) and some other taxa.

3.13 Gametophytic Apomixis

Apomixis, believed earlier to be a rare phenomenon, has now been observed frequently in families like Compositae, Gramineae, Rosaceae and many others. As described by Winkler (1908), apomixis is a process in contrast to amphimixis (sexual reproduction) where the sporophyte is produced from the gametophyte without sexual fusion.

Classification. The classification of apomixis is rather diverse. P. Maheshwari (1950) divides it into four classes: (1) non-recurrent apomixis, (2) recurrent apomixis, (3) adventive embryony, and (4) formation of vegetative propagules. The formation of vegetative propagules is merely a form of vegetative reproduction.

Battaglia (1963) adopted several terminologies, and defined apomixis as the production of sporophyte from the gametophyte without sexual fusion. He classifies apomixis into two types: recurrent and non-recurrent. Unlike Fagerlind (1940a), Gustafsson (1946, 1947a, b), P. Maheshwari (1950) and Battaglia do not include (and rightly so) adventive embryony and vegetative reproduction in "apomixis".

Solntzeva (1969) accepts Battaglia's classification, but includes adventive embryony in apomixis. The vegetative reproduction is totally excluded. While defining apomixis, Solntzeva emphasizes: (1) apomixis is a special and peculiar mode of seed reproduction; (2) this mode of reproduction implies the use of morphological structures inherent in seed reproduction – flower, ovule, embryo sac, embryo, seed and fruit; (3) in this mode of reproduction disturbances may occur both in sporogenesis and in sexual process, and (4) production of a new organism takes place through the formation of an embryo. In her scheme of classification of apomixis in angiosperms, she describes the cytological disturbances which occur during sporogenesis and sexual process.

Davis (1966, 1972) also follows Battaglia's classification. J. Heslop-Harrison (1972) does not recognize the non-recurrent type as a part of apomixis as it includes the process of meiosis in the formation of reduced embryo sacs.

Types of Apomixis. In a "non-recurrent" type, or in the formation of a reduced embryo sac, the megaspore mother cell undergoes regular meiotic divisions. The megasporogenesis and gametogenesis are normal, and the components of the embryo sac carry a haploid set of chromosomes. Syngamy does not occur, and the haploid egg cell gives rise to the embryo by parthenogenesis, pseudogamy, or androgenesis.

In parthenogenesis, the egg cell divides without pollination. Depending on the reduced and unreduced embryo sac, parthenogenesis is referred to as haploid or diploid, respectively. Parthenogenesis can also be induced experimentally as an effect of temperature.

In pseudogamy, pollination occurs but the male gamete disintegrates either inside or outside the egg cell. It provides only a stimulus as in Compositae and Rubiaceae. The development of haploid embryo from the sperm nucleus, i.e. androgenesis, is reported in *Crepis* (Gerassimova 1936). Here the egg nucleus degenerates, and the sperm nucleus functions in the cytoplasm of the egg cell.

The embryo may develop from some other cell of the reduced embryo sac, such as the synergid. It is termed "haploid apogamy".

The second type of apomixis, "recurrent", occurs in a non-reduced (diploid) embryo sac. It is effected by (1) aneuspory, (2) gonial apospory, (3) somatic apospory, and (4) synkaryogenesis.

Nogler (1984) critically analyzes the phenomenon of gametophytic apomixis, and suggests appropriate terminologies. The following account is based on his review.

Apomixis or asexual (agamic) reproduction by seeds, i.e. agamospermy (seed apomixis) covers two aspects.

1) Adventive Embryony. One or more embryos develop from the nucellar cells (Sect. 3.12). Only the endosperm is produced in the embryo sac.

2) Gametophytic Apomixis. The embryo sac develops (diplospory) from an unreduced generative cell (sporogenous or megaspore mother cell) either directly by mitosis (Antennaria type), or indirectly by modified meiosis resulting in an unreduced restitution nucleus (Taraxacum and Ixeris type). Unreduced embryo sac can also result from somatic cells of the nucellus (apospory).

In diplospory and apospory, there is no meiosis and fertilization of the egg does not occur. The unreduced egg develops parthenogenetically. The formation of endosperm depends on the fertilization of polar nuclei/secondary nucleus.

When both egg cell and polar nuclei are not fertilized, the development of embryo and endosperm is autonomous.

Diplospory (Generative Apospory)

Antennaria Type. The megaspore mother cell becomes vacuolated (as in functional megaspore), undergoes three successive mitoses producing an unreduced eight-nucleate embryo sac: *Antennaria, Erigeron, Eupatorium, Hieraceum, Poa alpina, P. nervosa* and others.

Taraxacum Type. The megaspore mother cell does enter meiosis but forms a meiotic restitution nucleus. The second meiotic division leads to an unreduced dyad (instead of a reduced tetrad). Following three successive mitoses, the lower dyad gives rise to an unreduced eight-nucleate embryo sac: *Taraxacum, Erigeron, Chondrilla* and others.

Ixeris Type. As in Taraxacum type, the meiotic restitution nucleus undergoes a division but is not followed by a wall. The coenomegaspore contains two unreduced (instead of four reduced) nuclei. Two successive mitotic divisions produce an unreduced eight-nucleate embryo sac: *Ixeris dentata, Rudbeckia, Erigeron.*

Female Meiosis in Diplosporous Apomicts

Various possibilities of meiosis in megaspore mother cell in diplosporous apomicts are shown in Fig. 3.13.1A–E. During normal meiosis, a linear tetrad develops, the upper three megaspores degenerate, and the lower is functional (Fig. 3.13.1A). Disturbed meiosis results in irregular distribution of chromosomes leading to aneuploidy (Fig. 3.13.1B) and the embryo sac is always diploid. The meiotic restitution nucleus undergoes meiosis II followed by a wall, and the lower unreduced dyad forms a Taraxacum type of embryo sac (Fig. 3.13.1C). If the wall is not formed after meiosis II, the embryo sac is of the Ixeris type (Fig. 3.13.1D). If meiosis does not occur at all, the embryo sac is of the Antennaria type (Fig. 3.13.1E).

Apospory (Somatic Apospory)

The unreduced embryo sacs develop from the centrally and/or laterally placed nucellar cells, or cells in the chalazal region (Fig. 3.13.2A–D), or even from the cells of the inner integument. The aposporous embryo sac initials enlarge, become vacuolate, compress and resorb the circumjacent tissue, and develop into one or more embryo sacs.

Cortaderia jubata (M.N. Philipson 1978b) is an excellent example of somatic apospory (Fig. 3.13.2D) and non-pseudogamy. In this species pollen grains are not produced; the embryos develop in the aposporous embryo sacs,

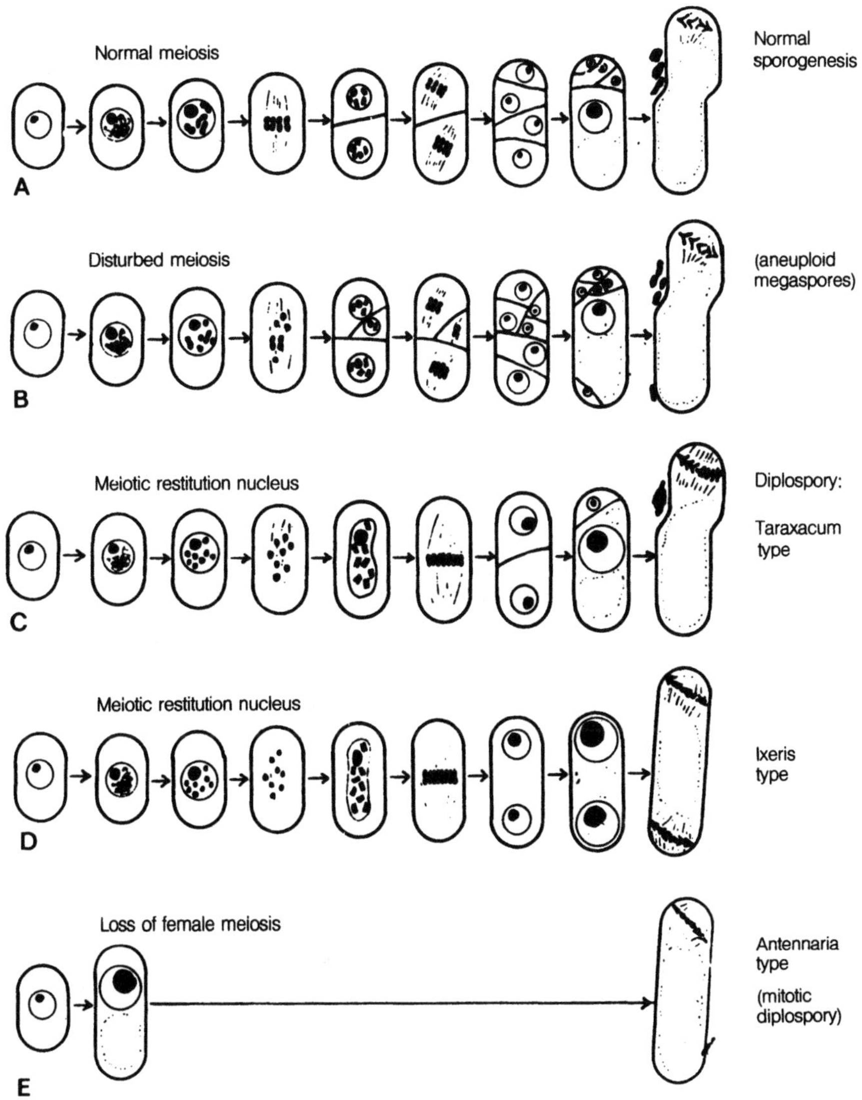

Fig. 3.13.1A–E. Female meiosis in diplosporous apomicts. (After Nogler 1984, modified from Fagerlind 1944a)

produced from the nucellus; the embryos are well developed by the time the inflorescence emerges.

Hieraceum Type. The aposporous embryo sac is bipolar, and is widely distributed: some Compositae, *Hypericum perforatum*, *Poa pratensis*, *P. granitica*, *Elatostema acuminatum* and others.

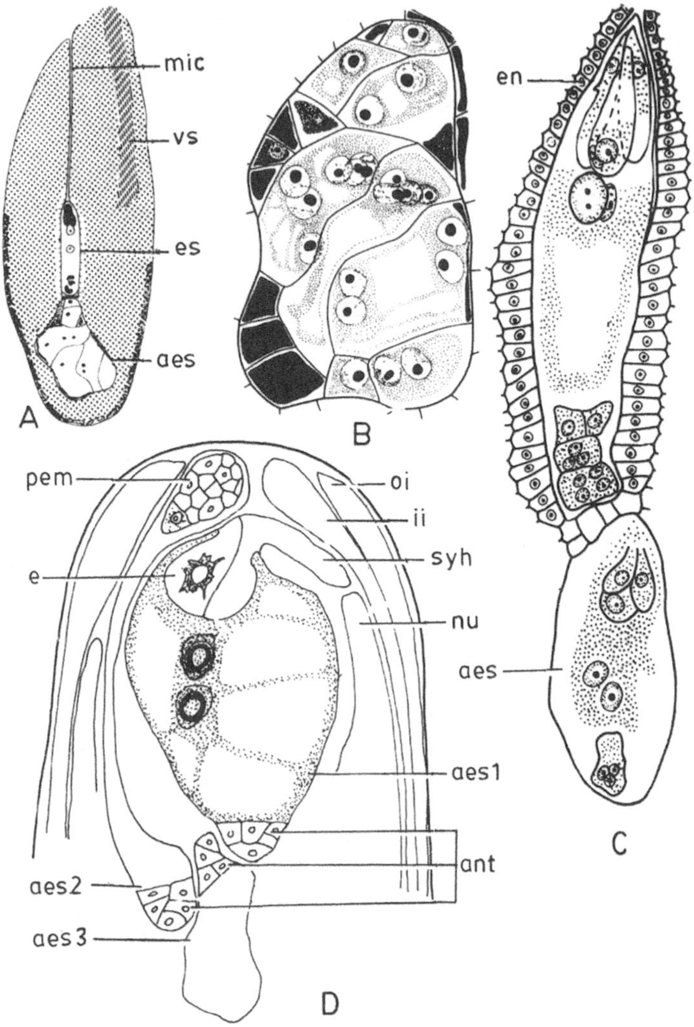

Fig. 3.13.2A–D. Somatic apospory. **A, B** *Cliftonia monophylla*. **A** Longisection of ovule with normal embryo sac (*es*); several aposporous embryo sacs (*aes*) develop in the chalazal region. *mic* micropyle. *vs* vascular supply; **B** Aposporous embryo sacs in the chalazal region (of another ovule). (After Vijayaraghavan and Dhar 1978). **C** *Melampodium divaricatum*, normal embryo sac with endothelium (*en*), and an aposporous embryo sac at chalazal end. (After Maheswari Devi and Pullaiah 1976). **D** *Cortaderia jubata*, longisection of young bitegmic seed to show three somatic embryo sacs: *aes 1*, *aes 2*, *aes 3*. *aes 1* contains the undivided egg (*e*), a synergid with its haustorium (*syn*) deflected from the micropyle, two polar nuclei, and five antipodals (*ant*); *aes 2* contains an inverted (?) globular proembryo (*pem*) and five antipodals; there were also a few endosperm cells which are not shown; *aes 3* has four antipodals at the micropylar end (reversed polarity), the rest of the embryo sac is empty; *oi*, *ii* outer and inner integument; *nu* nucellus. (After M.N. Philipson 1978b)

Panicum Type. The unreduced embryo sac is monopolar. The nucleus of the aposporous initial undergoes two successive mitotic divisions, and of these four nuclei three organize into egg apparatus and the fourth is the single polar. The lower polar nucleus and the antipodals are absent: common in Panicoideae and Andropogoneae.

Female Meiosis in Aposporous Apomicts

In aposporous apomicts, the sexual and aposporous embryo sacs often develop concurrently (Fig. 3.13.2A–C). However, the megaspore mother cell before undergoing normal meiosis, or later, or even the early embryo sacs, may remain arrested and degenerate. Thus, only the aposporous embryo sacs develop. When the aposporous initials take the lead, the chances of development of sexual embryo sacs decline since the megaspore mother cell must undergo meiosis and in the meantime the aposporous embryo sac is already well advanced.

The diplosporous and aposporous embryo sacs usually have the same organization as sexual eight-nucleate embryo sacs but various abnormalities may occur.

Endosperm and Embryo in Apomicts

When pollination does not occur, the endosperm develops from the polar nuclei without fertilization. This mode is referred to as "autonomous" and the egg develops parthenogenetically: many Compositae, *Burmannia coelestis*, *Elatostema*, *Cortaderia jubata* and others.

When fertilization of polar nuclei is necessary, the development of endosperm is said to be "pseudogamous": most apomictic Rosaceae (except *Alchemilla*, *Aphanes*), apomictic Gramineae (except *Calamagrostis*, *Cortaderia jubata*), *Zephyranthes texana*, *Ranunculus auricomus*.

The apomicts in the same genus show either autogamy or pseudogamy; exceptions do occur. The ploidy-level of endosperm is more variable in pseudogamous than in autogamous apomicts.

In many pseudogamous apomicts, the development of embryo begins only after the formation of endosperm as in sexual plants: *Rubus*, *Hypericum perforatum*, *Ranunculus auricomus*, *Panicum maximum*, *Pennisetum ciliare*.

In *Poa*, *Parthenium* and *Tripsacum*, also pseudogamous apomicts, precocious development of embryo occurs in most ovules while still in the bud; the endosperm develops only later after the fertilization of polar nuclei.

In autonomous apomicts both types of development of embryo occurs, but precocious embryony prevails: *Hieraceum*, *Taraxacum*. The endosperm may be initiated before anthesis, and in *Crepis* and *Cortaderia jubata* the egg cell divides only after the endosperm formation has begun.

Hemigamy

In certain pseudogamous apomicts, while one male gamete fertilizes the polar nuclei, the other male gamete enters the egg cell, does not fuse with the nucleus, and degenerates as in *Atamosco* (*Zephyranthes*). In *Rudbeckia*, the male gamete may even divide once (Fig. 51.9.4A–G) or several times before the daughter nuclei degenerate, or even participate in the formation of the embryo (especially the suspensor) which can become a chimera (Solntzeva 1974, P.N. Rao and Narayanan 1980). Hemigamy is not a prerequisite for embryo differentiation, but only an "additional feature in certain pseudogamous apomicts" (Coe 1953).

In "androgenesis" (male parthenogenesis), after the entry of the male gamete in the egg cell, the egg nucleus degenerates and the embryo originates from the male gamete.

Savidan (1983) experimented to determine how far the sexuality in facultative apomicts can be modified following hybridization. He observed that, in *Panicum maximum*, apomixis is associated with sexual reproduction in biotypes from South Africa and the Antilles. Three of the hybrids exhibited high percentage of sexuality, while the others appeared to be obligate or nearly so.

The average percentage of sexual reproduction in apomictic hybrids was 5.6, as compared to 8.1 in apomictic forms. He concluded that there is no increase of sexuality in apomicts due to hybridization.

Further information on apomixis is given along with the families where it has been reported.

3.14 Seed

The act of fertilization brings about several changes in the structure of ovule, and a seed is produced (see Linskens 1973). The outer and inner integument undergo structural modifications, and are termed testa and tegmen, respectively. They together constitute the seed-coat. The aril, if present, may grow and envelope the seed partially, or completely. The nucellus is usually consumed during development. But, if it persists as in Centrospermales, it is called perisperm (Fig. 16.6.1A). As the seed gets detached from the funicle, a scar is left on its surface, known as hilum. In anatropous ovules it leaves a longitudinal ridge, termed raphe. The endosperm and embryo occupy the major portion of seed. The endosperm is a storage tissue and, according to its presence or absence, the seeds are called albuminous (endospermous) or exalbuminous (non-endospermous). In a ripe seed, in the majority of families, the embryo is fully differentiated into the radicle, plumule, and cotyledons, and maintains its dormancy. In some members of Ranunculaceae, Burmanniaceae (Fig. 54.16.1J), Orchidaceae (Fig. 65.2.3H), and some parasitic plants such as *Orobanche* (see Sect. 48.24) the embryo is apparently undifferentiated.

Embryo. Martin (1946) studied the size of embryo in relation to the endosperm, and differences in size, shape and position of the embryo in the seeds of 1287 genera of angiosperms. The information is very revealing. He used as measurement quarter units of a circle. When the embryo is smaller than a quarter, it is designated "small", a quarter as "quarter", more than two quarters as "half", and more than three units as "dominant". The embryo is called "total" when it occupies the entire seed, as in exalbuminous seeds.

The embryo in a mature seed is also classified according to its position and shape into three divisions and a number of subdivisions (Martin 1946, see also Radford et al. 1974). R. Dahlgren and Clifford (1982) give a good account of two divisions and subdivisions, especially with reference to the monocotyledons (see Fig. 3.14.1A–Y).

1) Basal division. Embryo usually relatively small and restricted to lower part of seed (except certain grasses), seeds medium to large.

1a) Rudimentary. Embryo small, globular to ovate-oblong, (not truly rudimentary), relatively undifferentiated, massive starchy or non-starchy endosperm: most Magnoliales, Laurales (increasing embryo volume, reduction in endosperm), Piperales (perisperm partly takes over the function of endosperm), *Erythronium, Ptychosperma.*

1b) Broad. Embryo as wide as or wider than long, basal, globular or lenticular, in copious endosperm, peripheral or nearly so: *Aneilema, Mayaca,* Eriocaulaceae, Rapateaceae, Xyridaceae, *Nymphaea, Houttuynia.*

1c) Capitate. Embryo more or less basal, distally expanded, in copious endosperm: *Dioscorea, Strelitzia, Commelina, Tradescantia,* most Cyperaceae.

1d) Lateral. In relation to endosperm and ovule symmetry, "basal or basolateral", discoid or lenticular, surrounded by copious endosperm: Poaceae.

2) Axile Division. Embryo small to total, central (axile), straight or variously curved, seeds minute to large.

2a) Linear. Axial embryo much larger than broad, straight, curved or coiled, cotyledons not expanded, endosperm present or absent: several genera of Hydrocharitaceae, Alismataceae, *Zosteraceae, Najas, Uvularia, Calochortus, Erythronium, Lilium, Sparganium* and many others.

2b) Minute. Axial embryo in minute seed, less than 0.2 mm long, minute and undifferentiated to almost total size of seed: Burmanniaceae, Orchidaceae.

The cotyledons show much variation: a) foliate embryo with expanded and thick cotyledons (in an axile position) bent upon the hypocotyl, b) foliate embryo with cotyledons usually thin and extensively expanded and folded in various ways, c) axial embryo usually erect with thick cotyledons which overlap and encase the somewhat short hypocotyl, the endosperm is scanty or absent, d) cotyledons narrow or expanded, perisperm central or lateral; such a peripheral embryo is large and elongate, arcuate, annular, spirolobal (incumbent cotyledons folded once) or straight, and e) foliate erect embryo with cotyledons thin to thick and slightly expanded to broad (Martin 1946, Radford et al. 1974).

Embryos with a swollen hypocotyl or radicle are referred to as "macropodous" and are associated with exalbuminous seeds: Najadales, Zos-

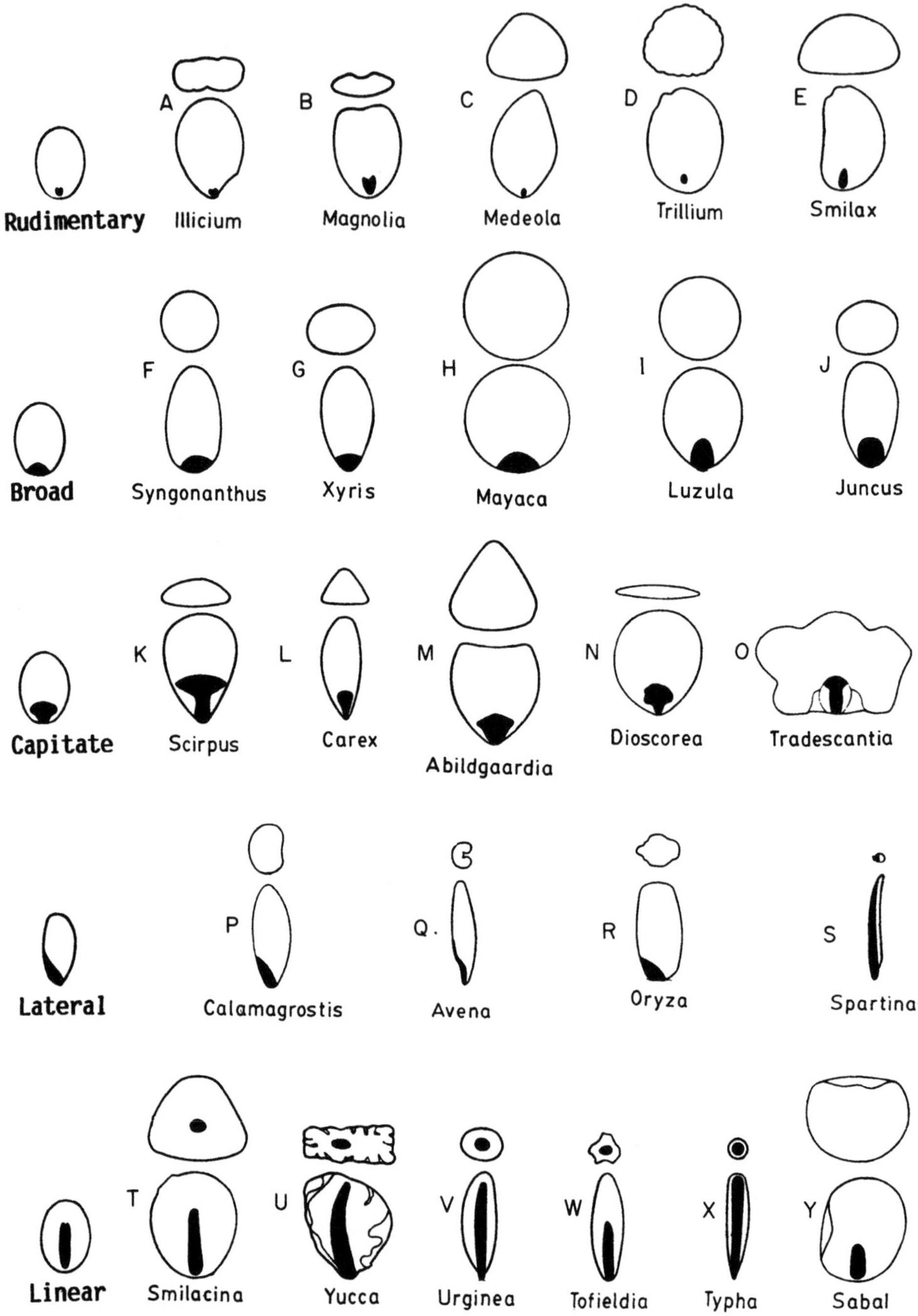

Fig. 3.14.1A–Y. Embryo types, tran- and longisections; an example is cited under each figure. **A–E** Rudimentary. **F–J** Broad. **K–O** Capitate. **P–S** Lateral. **T–Y** Linear. (After Martin 1946, see R. Dahlgren and Clifford 1982)

terales (except Juncaginaceae and Schaeuchzeriaceae) and Hydrocharitales (Thallassioideae and Halophiloideae of Hydrocharitaceae) – all submerged aquatics; not known in dicots (R. Dahlgren and Clifford 1982).

Classification of Seed. During the post-fertilization growth of ovules, sometimes the chalazal region shows conspicuous changes, especially in the ovules with late development of the endosperm. The cells, between the integument and nucellus, undergo intercalary growth pushing the integument in the micropylar region. The chalaza becomes massive (Fig. 3.14.2A–D), and the seed is said to be pachychalazal. In a perichalazal seed the growth is unilateral, and the chalaza exhibits bilateral symmetry (Fig. 3.14.2E–J). A peculiar feature, the formation of chalazosperm, has been noted in the seed of Cynastraceae. The chalazosperm (Fig. 3.14.5E) serves as a storage tissue, in the absence of nucellus and endosperm in the seed.

The development of ingrowths from the chalaza, inner and outer integument, and unequal growth of the peripheral endosperm leads to a ruminate condition of the endosperm (Fig. 3.10.3A–K). This is a diagnostic feature for considerations of systematic assignment.

Morpho-anatomical study of seed is often the basis of classification. The shape, size, colour, nature of surface, and position of hilum (in legumes) is taken into consideration. R.C. Misra (1970) classified the seeds of *Plantago* on the basis of external morphology. Clifford and Smith (1969), Chowdhury and Buth (1970), Ambegaokar (1976), and Behl and Tiagi (1976) used the above criteria, along with internal morphology, to classify legumes. On a similar basis the

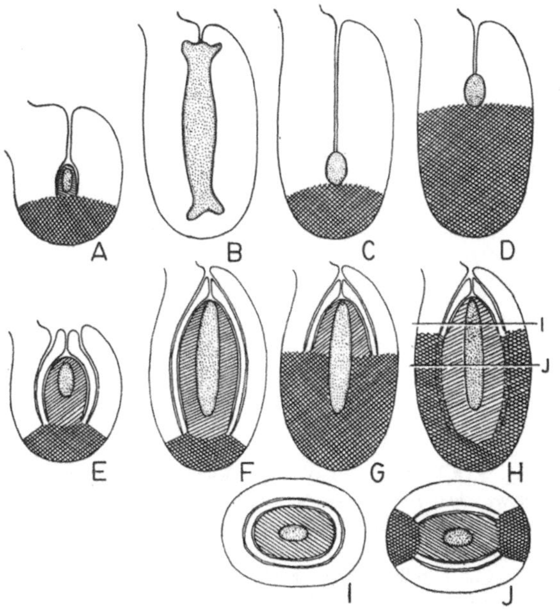

Fig. 3.14.2A–J. Variations in the growth of chalaza during seed development (diagrammatic). All ovules anatropous (chalaza-*cross hatched*). Endosperm and embryo sac *stippled*. Nucellus *striped*. **A–D** Unitegmic ovule, median longisections. **A** Ovule at fertilization; **B** Immature seed with endosperm; **C, D** Immature seed with normal (**C**) and massive (**D**) chalaza. **E–J** Bitegmic ovules; **E–H** median longisections; **I, J** cross-sections of **H** at respective levels; **E** Ovule at fertilization; **F–H** Immature seeds with normal (**F**), massive (**G**) and perichalaza (**H**). (After Periasamy 1962a)

seeds of Cruciferae have been classified by Martin (1946) and Vaughan and Whitehouse (1971), and of Convolvulaceae by Sampathkumar and K.R. Ayyangar (1982).

The sequential study of the developmental changes in the structure of ovule, after fertilization till the formation of mature seed, provides several important features for phylogenetic classification.

Vascular Supply. The vascular supply of the seed is a diagnostic feature. Usually, a well-differentiated raphal bundle terminates at the chalaza (Fig. 3.14.3A). In Annonaceae, Cucurbitaceae, Vitaceae, Polygalaceae, the single bundle may extend in the median plane into the outer integument (post-chalazal bundle/s) to the micropyle (Fig. 3.14.3B). In Meliaceae, Connaraceae,

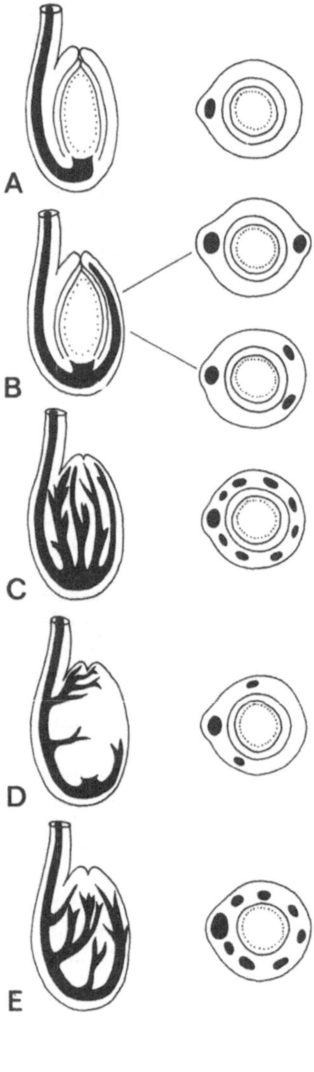

Fig. 3.14.3A–E. Types of seed vasculature: *left row* side views, *right row* transverse sections. **A** Seed with well developed raphal bundle. **B** One postchalazal bundle in the median plane, or two postchalazal bundles. **C** Many post-chalazal bundles. **D** *Pre-raphe bundles.* **E** Pachychalazal bundles. (After Boesewinkel and Bouman 1984)

Dichapetalaceae, a number of post-chalazal bundles run up to the micropyle in the outer integument (Fig. 3.14.3C). In Meliaceae (*Dysoxylon*, *Trichilia grandifolia*) and Connoraceae, in addition to the raphal, chalazal and testal bundles, pre-raphal bundles run from the beginning of raphe in the direction of the micropyle (Fig. 3.14.3D). In Meliaceae, Tropaeolaceae, *Glycosmis*, a network of vascular bundles originates from the chalaza and (by intercalary growth) replaces the seed-coat partly or wholly (Fig. 3.14.3E). This condition is termed pachychalaza.

The shape of the seed mostly depends on the shape of the ovule. In the majority of angiosperms, the ovule is anatropous and gives rise to a normal seed (Fig. 3.14.4E). All the other types of seeds can be derived from the anatropous seed as shown in Figure 3.14.4A–N.

The "hilar" seed in some Meliaceae, *Erythrina* and *Aesculus* has an enlarged hilum (Fig. 3.14.4C) and a greater/smaller part of the surface of seed-coat consists of the extended hilum.

In Meliaceae, Sapindaceae, Lauraceae and Tropaeolaceae, which have pachychalazal seeds, the place of attachment of the integument/s is extended (Fig. 3.14.4F, I, K). Below the place of attachment, there is intercalary growth so that the endosperm and embryo are contained in the pachychalaza.

In *Ricinus* and *Myristica*, the zone below the nucellus and above the attachment of outer integument proliferates and results in a short inner and a long outer integument (Fig. 3.14.4F, G). In these endopachychalazal seeds, the seed-coat may remain either endotestal or exotegmic.

In Annonaceae and Meliaceae, in the perichalazal seeds, the extension of the chalaza takes place only in the median plane of the seed (Fig. 3.14.4J) and one of the vascular bundles extends around the periphery from funicle to the micropyle.

Campylotropous seeds are reported in Papaveraceae, Capparidaceae, Cactaceae, Leguminosae, Geraniaceae, Malvaceae and other taxa. The antiraphal side is strongly extended (Fig. 3.14.4L–N). Mostly, from the chalazal or raphal direction, or inner integument at the raphal side, pressure is exerted on the embryo sac so that it becomes U- or V-shaped. The U- or V-shaped embryo becomes twice as large as the seed.

The obcampylotropous seed is the converse of the campylotropous seed. The raphal side extends, and the antiraphe remains small (Fig. 3.14.4B) as in Leguminosae and Vitaceae.

The orthotropous seed in one-seeded fruits in Piperaceae, Polygonaceae, Urticaceae and Juglandaceae lacks the raphe, and the hilum is situated opposite the micropyle (Fig. 3.14.4H).

Corner (1976), in his well-illustrated book *Seed Anatomy of Dicotyledons*, elaborately discusses the bearing of seed anatomy on the phylogeny of the orders, families, and genera. He defines several terminologies which have been followed in the text.

Seed-Coat. According to Corner (1976, see also Rauh et al. 1975), during development the cells of the testa and tegmen may undergo periclinal divisions so that the number of cell layers increases, i.e. they become multiplicative.

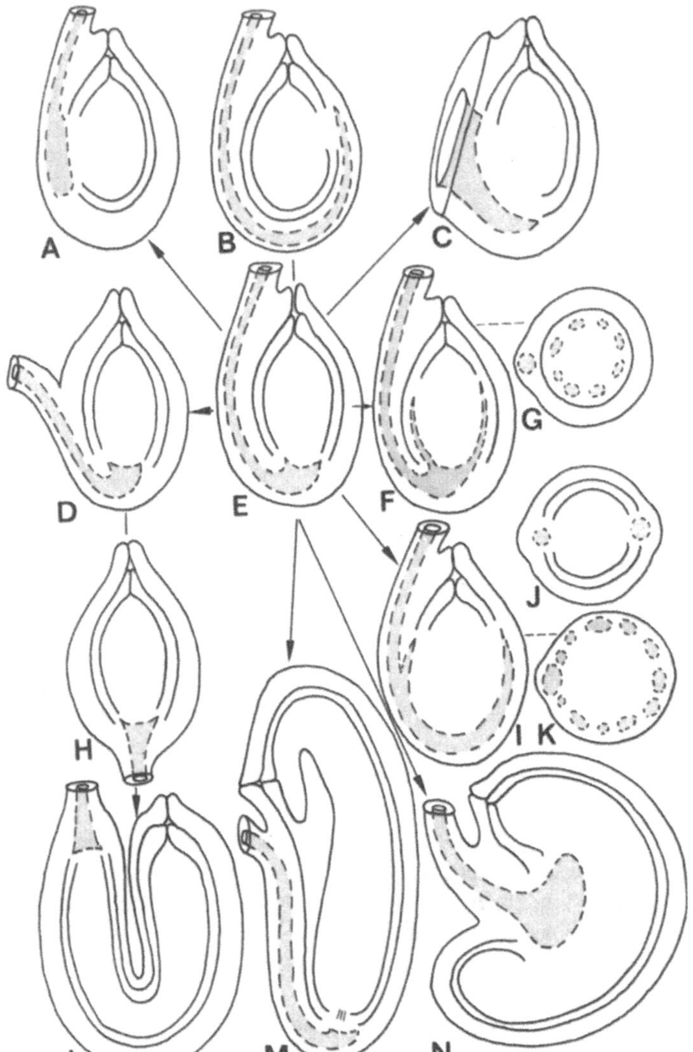

Fig. 3.14.4A–N. Possible derivation of seed types from normal anatropous seed; mostly longisections, **G**, **J** and **K** transections. **A** Chalaza shifted in raphal direction. **B** Obcampylotropous seed, chalaza shifted in antiraphal direction. **C** Hilar seed. **D** Hemianatropous seed. **E** Normal anatropous seed. **F** Pachychalazal seed with short inner integument. **G** Transection. **H** Orthotropous seed. **I** Pachychalazal seed with short inner and outer integuments. **J** Perichalazal seed in transection. **K** Transection of **I**. **L–N** Different types of campylotropous seeds. (After Boesewinkel and Bouman 1984)

Sometimes, some of the cell layers collapse during maturation, or one of the integuments may collapse and degenerate. The seed is then designated as "testal" (Annonaceae – Sect. 18.6, Cucurbitaceae – Sect. 38.2), or "tegmic" (Podostemaceae – Sect. 28.2) as the case may be. In unitegmic ovules, the seed is described as testal (sympetalous families). The integument, i.e. testa (or tegmen), differentiates into three regions: outer epidermis – exotesta (or exotegmen); middle region – mesophyll; inner epidermis – endotesta (or endotegmen).

In the seed-coat, a protective covering, has the mechanical tissue either in the outer epidermis, the mesophyll, or inner epidermis of the testa. Accordingly, the seed is designated "exotestal" (Leguminosae – Sect. 26.19), "mesotestal" (Theaceae – Sect. 22.12), or "endotestal" (Magnoliaceae – Sect. 18.2, Geraniaceae – Sect. 29.4). Likewise, the disposition of mechanical elements in the outer and inner epidermis of the tegmen forms "exotegmic" (Malvaceae Sect. 35.5, Sterculiaceae – Sect. 35.7), and "endotegmic" (Zygophyllaceae – Sect. 19.6) seeds. The mechanical tissue comprises lignified cells, compact palisade-like cells (Leguminosae), fibres, or tracheidal thickenings (see R.P. Singh 1985).

The seed of pomegranate (*Punica granatum*) is mesotestal. The exotesta is a thin-walled, palisade layer, and is pulpy. It is called sarcotesta. In *Magnolia*, the outer part of testa becomes fleshy, and develops into "sarcotesta" (Sect. 18.2). The sarcotestal seeds are dispersed by animals.

Generally, the seed contains endosperm of uniform surface. However, localized peripheral growth of endosperm, ingrowths of the inner and/or outer integument, chalaza, and even pericarp leads to a condition when the endo sperm is called ruminate (Fig. 3.10.3).

Takhtajan (1985, 1988) has further added to our knowledge of the seeds of monocots and dicots.

Appendages. The modifications of the seed which help in dispersal are epidermal hairs (coma), aril, arillode, caruncle, strophiole, elaiosome, wing, and jaculator (Figs 3.14.5 A–F, 3.14.6 A–F, 3.14.7 A–J). According to Pijl (1955), these appendages are classified on the basis of site of origin. Bor (1978) states that "in an anatropous ovule, hilum, raphe, chalaza and exostome are merely topographical zones of the testa. The accessory seed appendages arising from the surface have a common origin".

The epidermal cells of the testa often develop into long, unicellular, hair-like, thin or thick-walled structures, either all over the surface, or in patches, or at localized sites as in Acanthaceae, Asclepiadaceae, Malvaceae, and Salicaceae. These seeds are called "comose".

In euphorbiaceous seed the free-end of the testa proliferates, and develops a tissue on the side of or all round the funicle. It differentiates into a thick, whitish caruncle (Fig. 3.14.5A).

In Linaceae, the tip of the testa develops an outgrowth which grows downward and covers the seed partially. It is termed an arilloid or false aril (Eames 1961). Kapil and Vasil (1963) regard it as an excessively developed caruncle.

The aril may be thin- or thick-walled and, usually, arises from the funicle, outer integument, or both (Fig. 3.14.5B). In a nutmeg seed it develops from the funicle, and exostome (Mohana Rao 1974b). The orthotropous ovule of *Hitchenia caulina* (Panchaksharappa 1962b) shows a similar feature. Accord-

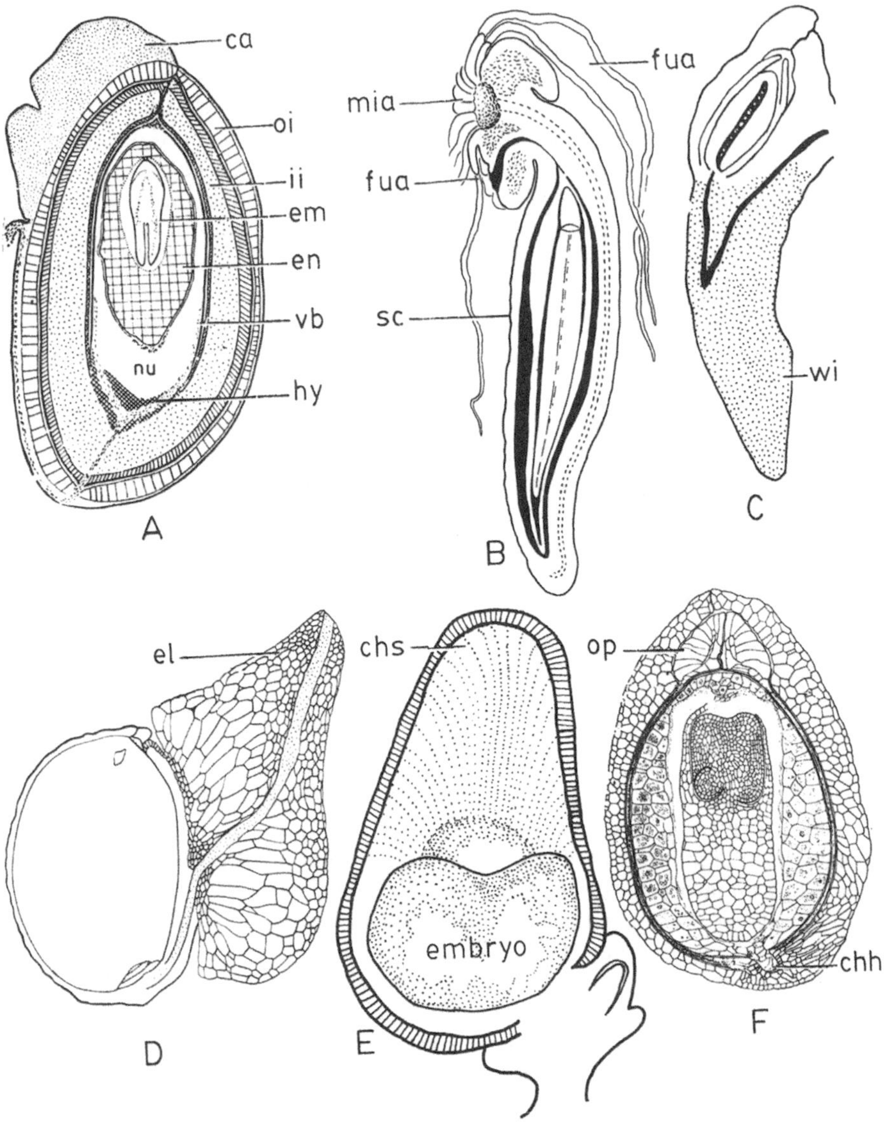

Fig. 3.14.5. **A** *Acalypha indica.* longisection of seed with caruncle (*ca*); outer integument (*oi*); inner integument (*ii*); embryo (*em*); endosperm (*en*); vascular bundle (*vb*); nucellus (*nu*) and hypostase (*hy*). (After B.M. Johri and Kapil 1953). **B** *Sarawakodendron* sp., longisection of seed in side view; *fua* funicular aril; *mia* micropylar aril; *sc* seed. (After Corner 1976). **C** *Cercidiphyllum japonicum*, longisection of ovule, the triangular extension of raphe and chalaza later develop into a thin membraneous wing (*wi*). (After Swamy and Bailey 1949). **D** *Trillium ovatum*, seed with elaiosome (*el*). (After Berg 1958). **E** *Cynastrum* sp., mature seed to show chalazosperm (*chs*). (After Melchior 1964). **F** *Lemna pausicostata*, longisection of seed to show operculum (*op*), endosperm with chalazal haustorium (*chh*), and embryo (*centre*). (After S.C. Maheshwari and Kapil 1964)

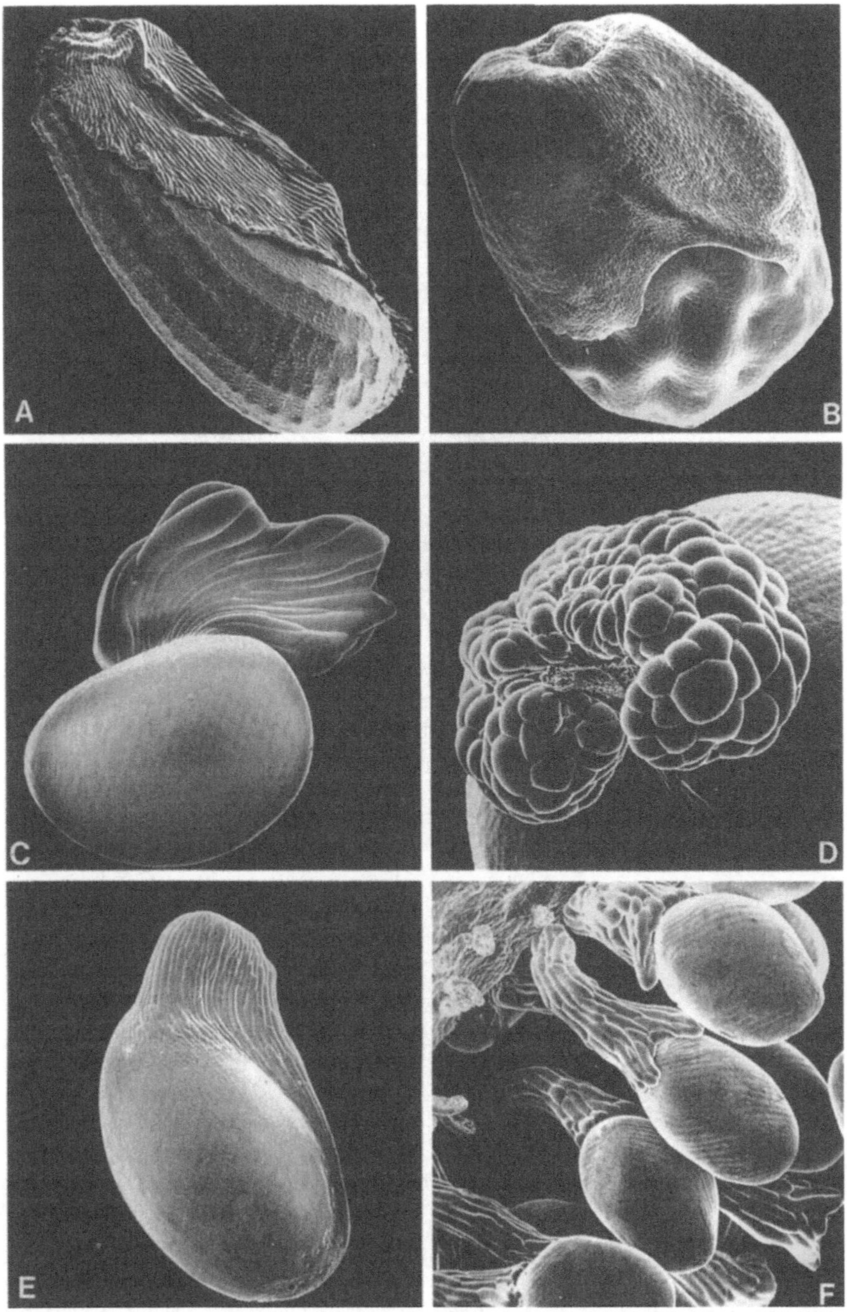

Fig. 3.14.6A–F. Types of seed arils, SEM micrographs. **A** *Turnera ulmifolia*, raphal axis. **B** *Passiflora suberosa*, funicular outgrowth. (After Kloos and Bouman 1980). **C** *Chelidonium majus*, strophiole. **D** *Chionodoxa luciliae*, caruncle. **E** *Viola odorata*, Exotestal elaiosome. **F** *Episcia melittifolia*, funicular aril. (After Boesewinkel and Bouman 1984)

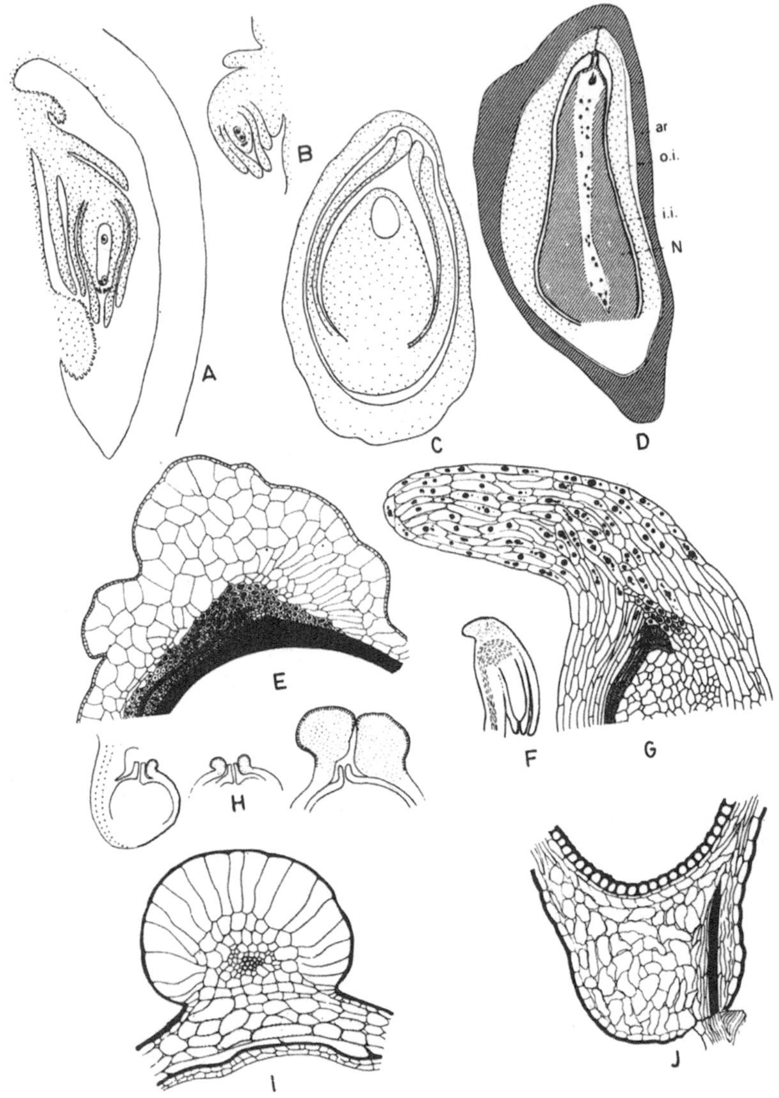

Fig. 3.14.7A–J. Aril and strophiole. **A** *Asphodelus fistulosus*, longisection through ovary showing an ovule with aril and obturator. **B** *Bulbine annua*, longisection of bitegmic ovule with aril. **C** *Eremurus himalaicus*, ovule surrounded by aril. (After Stenar 1928). **D** *Kniphofia praecox*, longisection of ovule at young proembryo with nuclear endosperm. Note the aril (*ar*), outer (*oi*) and inner integument (*ii*), and nucellus (*N*). (After Schnarf and Wunderlich 1939). **E** *Iris ruthenica*, upper part of seed with strophiole. **F, G** *Galanthus nivalis*. **F** Ovule; **G** Strophiole. **H** *Scilla* sp. to show development of strophiole on the outer integument at the micropyle. (After Bresinsky 1963). **I** *Scoliopus bigelovii*, lower part of seed with strophiole. (After Berg 1959). **J** *Luzula campestris*, strophiole on seed. (After Bresinsky 1963)

ing to V. Puri (1970), the morphological nature of arils is variable. Mohana Rao (1974b) states: "the single anatomically and functionally similar structure arises from two morphologically different sites". Mohana Rao (1974b) also mentions that it is superfluous to classify various outgrowths from the functional, and ecological point of view.

Various types of aril are shown in Figure 3.14.6A–F, and aril and strophiole in Figure 3.14.7A–J.

In the seeds of *Cercidiphyllum*, the chalaza and raphe together produce a triangular outgrowth which develops into a membranous wing in mature seed (Fig. 3.14.5C). A wing also develops in Hamamelidaceae, but due to the elongation of the tip of the outer integument.

The elaiosome arises as an outgrowth of raphe, or hilum. It contains food material, and attracts ants which help in seed dispersal (Amaryllidaceae, Aristolochiaceae, Liliaceae – *Trillium*; Fig. 3.14.5D). Excessive chalazal growth in the seeds of *Cynastrum* produces a tissue known as chalazosperm (Fig. 3.14.5E) which stores food reserves.

Several acanthaceous genera are characterized by the development of a jaculator or retinaculum. It is a hook-like outgrowth of the funiculus, and ejects the seed out of the capsule.

Sometimes, during post-fertilization, the cells of tegmen in the micropylar region undergo excessive expansion and develop into an operculum (Fig. 3.14.5F) as in *Lemna paucicostata* (S.C. Maheshwari and Kapil 1964). In the mature seed, rest of the tegmen collapses and is represented by two refractive cuticular thickenings which enclose the endosperm and embryo. In a dormant seed the operculum fits at the top as a stopper and, during germination, it is thrown off so that the embryo pushes out.

3.15 Nutrition of the Anther, Ovule and Seed

Anther. Initially, the floral vascular supply transports nutrients to the various parts. As soon as the tapetum differentiates in the microsporangium, it provides nutrition, or the nutrition passes through it (Pacini et al. 1985), for the development of microspore mother cells, tetrads/microspores/pollen grains – including "clumped" pollen grains: dyads (Fig. 28.2.1A), tetrads (Fig. 18.5.1A, C, D), pollinia (Fig. 47.7.1A–D) and massulae (Fig. 65.2.1L, M).

The absorbing external surface of each grain is progressively reduced. In the massulae, the pollen has no direct contact with the tapetum. The inner exine walls, or sometimes the intine itself, or joined grains are generally reduced or even discontinuous so that the nutrients may pass easily through and ensure the nutrition of the inner spores/pollen. This also ensures a synchronous pollen development.

According to Pacini et al. (1985), all the substances produced by the secretory tapetum – starch and/or proteins in plastids, soluble polysaccharides in the vacuoles – reach the microspores/pollen grains via the locular fluid. In the periplasmodial tapetum, the cytoplasm adheres closely to the microspores. In the secretory tapetum, the cells are polarized both in structure and function;

this is not so in the amoeboid tapetum. The inner and outer tangential faces of secretory tapetum differ in their capacity to release substances.

The locular cavity increases in size during anther development, the secretory tapetum continuously supplies locular fluid until it starts to degenerate. The enzyme callase, produced by the tapetum, depolymerizes the callosic walls of tetrads and the microspores are released. The tapetum then becomes very active and has a multiple function. From the second meiotic division onwards, PAS-positive material present in tapetal vacuoles passes into the locule.

The tapetum also produces: (a) sporophytic proteins and enzymes which are deposited next to the aperture regions or in the interaperturate areas; (b) tryphine which, in contrast to pollenkitt, comprises a mixture of hydrophobic and hydrophilic substances often containing cytoplasmic elements (degenerated organelles); (c) pollenkitt, a hydrophobic oily layer containing mainly lipids and carotenoids.

The tapetal cell organelles (mostly plastids, perhaps other organelles too) form lipid masses which are deposited outside the mature pollen grains just before they are released.

In polysporangiate microsporangia of Annonaceae (Fig. 18.6.1C–J), Onagraceae (Fig. 39.13.1A–I), and some other families, each sublocule has its independent tapetum which enhances the nutritive function (Tobe and Raven 1986a). It is well known that when the tapetum disfunctions, fertile pollen grains are not produced.

Ovule. The funicular vascular supply may not extend beyond the chalaza, branch at the base of nucellus, extend into one or both integuments and may even branch there. The euphorbiaceous ovule is a good example (Landes 1946). With the development of the ovule into a seed, the vascular supply becomes very pronounced (Figs 3.14.4A–E, 3.14.5A–N). Besides the vascular supply, there are several other strategies to provide the nutrition, e.g. by various types of haustoria.

Megaspore Haustoria. In Crassulaceae, each megaspore of a linear tetrad produces a tubular haustorium, apically from the micropylar megaspore and laterally from the other megaspores (Fig. 26.5.2A–D). These haustoria intertwine, extend up to the nucellar epidermis, and contain starch.

Embryo Sac Haustoria. The apical extension of the embryo sac, and even the chalazal one are common in Acanthaceae (Sect. 48.19). In *Exocarpus* (Fig. 12.6.3A–C), several tubular haustoria directed upward develop from the micropylar end; some haustoria grow towards the chalazal region. Sometimes, flap-like haustoria, directed towards the chalaza, grow all round the middle region of the embryo sac.

In Podostemaceae (Fig. 28.2.1D, E), the central cells of the nucellus break down, a plasmodium (pseudo-embryo sac) is formed, and after fertilization the embryo grows in this pseudo-embryo sac (Fig. 28.2.3A–C).

The numerous embryo sacs in *Utricularia* derive nutrition from the placental and chalazal nutritive tissue. The endothelium is also very conspicuous and the nutrients pass through it (Figs. 48.25.3, 48.25.4).

The endothelium (see Kapil and Tiwari 1978) is a characteristic feature of the Sympetalae (Fig. 3.4.4A–E), and is actively concerned in the transport of nutrients.

In *Helianthus annuus*, the endothelium is many-layered and interrupted by transverse cuticular septa (Fig. 3.4.7D). Its role in the efficient transport of nutrients should be carefully examined.

The occasional reports of the differentiation of endothelium in some monocots appear doubtful (see Swamy and Krishnamurthy 1970), and deserve to be examined histochemically.

Synergid and Antipodal Haustoria. Synergid haustoria are reported in Compositae (Sect. 51.9). The most conspicuous synergid haustoria develop in *Quinchamalium chilense* (Figs. 12.6.5A–C, 12.6.6A–C), and the tip extends up to one-third of the basal region of the style. *Cortaderia* (Fig. 3.13.2D; see also Fig. 58.2.1), an apomict, shows lateral synergid haustoria which contain starch.

In *Q. chilense* (Fig. 12.6.5A–E), the antipodal nuclei do not organize into cells. The tubular antipodal chamber functions as a haustorium.

The basal antipodal cell in Rubiaceae (Fig. 47.8.3F, G) elongates a great deal and becomes tortuous.

The frequent occurrence of polyploid nuclei and multi-nucleate antipodal cells indicates a high metabolic activity (see D'Amato 1984). As early as 1898, in *Aconitum napellus*, Osterwalder observed giant antipodal cells with the tip reaching up to the egg apparatus.

Haustorial Pollen Tube. For almost nine decades, the controversy about the haustorial role of the persistent remnants of the pollen tube has been going on (Longo 1903, Venkata Rao 1952a, Tilquin et al. 1983, Indra and Krishnamurthy 1985, Briggs 1989).

The latest report on haustorial pollen tubes, which parasitize the ovary wall during post-fertilization development in *Grevillea banksii*, was given by Martin and Herscovitch (1989).

P. Maheshwari and Johri (1950b) and B.M. Johri and Ambegaokar (1984) have critically examined this assumption and conclude that the "dead" remains of the pollen tube do not have any haustorial role (see also Sect. 3.9).

Endosperm Haustoria. In most plants, the endosperm is the main nutritive tissue and supports the development of the embryo. The haustoria are the most spectacular structures and augment the nutritive function.

A prominent chalazal haustorium, which invades the adjoining tissue, develops in Leguminosae and Cucurbitaceae (Fig. 38.2.2A–I). The endosperm is of the Nuclear type; the haustorium may remain free-nuclear or become partially or wholly cellular. It becomes coiled in Proteaceae (Figs 11.2.2D–H, K, L, 11.2.3A–C); in Caryophyllaceae (Fig. 16.10.2B, C) superficial hairy growths occur. In Icacinaceae (Fig. 33.13.1A, B), the chalazal haustorium shows many interesting features.

In the Helobial type, the haustoria are not very common. In *Monochoria* (Ono 1928), the bilateral "saccate" nuclear micropylar haustoria are very conspicuous (see Sect. 54.13).

In the Cellular type, only a chalazal haustorium may develop: Olacaceae (Fig. 12.2.1A, B), Opiliaceae (Figs 12.4.1A–F, 12.4.2A, B), Santalaceae (Figs 12.6.4D, E, 12.6.7A, B), Myzodendraceae (Fig. 12.7.1G, H, J) and others. In Olacaceae, Opiliaceae and Myzodendraceae, the haustorium invades the placenta and extends as far as the pedicel.

Both micropylar and chalazal haustoria develop in Balsaminaceae (Fig. 31.11.2D, E, G–I), Cyrillaceae (Fig. 33.2.1C–F), Loasaceae (Fig. 37.19.1A–G), Hydrophyllaceae (Fig. 48.5.1E, F), Verbenaceae (Fig. 48.8.3A–C), Labiatae (Fig. 48.10.3), Scrophulariaceae (Figs. 48.15.3A, B, 48.15.4O–T), Acanthaceae (Fig. 48.19.3H, I), Gesneriaceae (Fig. 48.22.3A–C), Lentibulariaceae and others.

Both micropylar and chalazal haustoria may or may not branch. In many cases, the chalazal haustorium is not at all aggressive and collapses. The micropylar haustoria invade the exotesta as in *Aeginetia* (Fig. 48.24.1E, H, I, J, L–N).

Suspensor Haustoria. The suspensor haustorium is very aggressive and invades the integument (Figs. 26.5.1G, 26.5.3A–E). The uppermost cell of the suspensor enlarges considerably, and it develops many haustorial extensions (Figs. 28.2.1F, 28.2.2N, 28.2.3A–C). The suspensor may comprise biseriate elongated multinucleate cells as in Leguminosae (Sect. 26.19), or a uniseriate row of multinucleate bulbous cells (Figs. 48.4.4H–P, 48.4.7A–D, N, O). The suspensor cells may invade the endosperm (Figs. 47.8.4A, B). In *Plumbago*, the group of suspensor cells is much larger than the proembryo (Fig. 44.2.6F–K). There are many other variations. The most unusual filamentous highly coiled suspensor of uniseriate multinucleate cells is reported in *Isomeris* (Fig. 24.3.1A–F′).

Haustorial Cotyledons. In Hippocastinaceae (Fig. 31.7.1B–E), one cotyledon remains enclosed in the liquid endosperm to absorb nutrients for the growth and maturation of the other cotyledon. In *Yucca*, the cotyledon remains permanently embedded in the perisperm to absorb nutrients for germination and establishment of the seedling.

Transfer Cells. The cells actively concerned in short-distance transport of nutrients develop wall ingrowths to increase the absorptive surface (see Pate and Gunning 1972, B.M. Johri and Ambegaokar 1984). The wall ingrowths are simple and papillate, or reticulate or labyrinth when branched. An important feature is that in transfer cells the plasma membrane follows the contour of the wall ingrowths, and the middle lamella is absent.

The embryo sac wall, synergids, antipodals, endosperm, perisperm, embryo suspensor, and even pericarp have transfer cell ingrowths.

After fertilization, wall ingrowths are initiated at the micropylar region of the embryo sac. They increase in size and persist up to the heart-shaped embryo. These ingrowths are especially concentrated around the embryo. At the chalazal end, the wall ingrowths are usually short-lived.

The central cell is the largest cell of the embryo sac, and its wall usually has tuberculate wall ingrowths adjacent to the nutrient-rich nucellar cells. In *Helianthus annuus*, the micropylar part of the central cell has wall ingrowths which project towards the synergids. In this taxon, the synergids are reported to lack the filiform apparatus; which requires confirmation (see Berger and Erdelská 1973, Vijayaraghavan et al. 1988).

Bhandari and Mal (1989) report a prominent wall ingrowth, in *Scilla siberica*, close to the chalazal end of embryo sac; it separates a small chalazal and a large micropylar chamber. This ingrowth is interrupted by one or more gaps through which the metabolites, and even nuclei, can pass from one chamber to the other. The authors compare it to a transfer cell.

A filiform apparatus in the synergids has been known since long. The ingrowths of the wall of the synergid are synonymous with the filiform apparatus (see Willemse and van Went 1984). It is said to be actively concerned in a secretion into the micropyle which possibly directs the tip of the pollen tube into the synergid. Probably, the filiform apparatus in the two synergids have different and specific functions, and the one which receives the pollen tube is in a degenerate condition as compared to the other healthy synergid.

The tetrasporic eight-nucleate embryo sac of *Plumbago* lacks synergids, and the filiform apparatus has been demonstrated in the apical region of the egg cell (Cass 1972, see also Russell 1989). This is likely to be the case in *Plumbagella.*

The persistent antipodal cells may show massive and highly branched wall ingrowths, and tuberculate wall projections. The wall ingrowths proliferate after fertilization in *Nigella damascena, Papaver somniferum, Scilla sibirica* and *Zea mays.* During the free-nuclear endosperm, wall ingrowths occur all over the antipodal walls except in a small area at the chalazal region, in *Aconitum vulparia* (see Vijayaraghavan et al. 1988). The major pathway for the flow of nutrients into the embryo sac is the chalazal region, and the wall ingrowths are involved in translocation.

The inner and outer epidermis of the endosperm show cell wall ingrowths, and are actively concerned in nutritive processes. Brisson and Peterson (1975) demonstrated wall ingrowths on the inner wall of the endosperm haustoria of *Vaccinium macrocarpum.* This may be a common feature in other endosperm haustoria, and needs to be investigated.

The suspensor anchors the proembryo/embryo to the embryo sac, and the circumjacent tissue (endosperm) is nutritionally rich. The cells of the suspensor show prominent ingrowths, especially on the side walls; there are no ingrowths on the periclinal walls. The uppermost basal cell, especially large in some taxa as in *Capsella bursa-pastoris*, shows conspicuous wall ingrowths and is actively concerned in transporting the nutrients to the embryo (Gunning and Pate 1974, Vijayaraghavan et al. 1988).

Wall ingrowths also occur on the epidermis of cotyledons, epidermis of endosperm facing the perisperm as in *Mesembryanthemum.* Aleurone transfer cells have been observed at the placenta of grass seeds.

In *Vicia*, transfer cells occur in endothelial cells, epidermis of endosperm, suspensor cells, and cotyledonary epidermis facing the endosperm (Gunning and Pate 1974).

In *Pappophorum subbulbosum* (Poaceae; Rost et al. 1984), transfer cells occur in the outer pericarp epidermis (continuous across the placental pad), and discontinuous layer of the inner pericarp epidermis (absent across the placental pad). From the inside, several partially collapsed cell layers of the nucellar projection occur which have wall-ingrowths. The aleurone is a continuous uniseriate layer around the entire caryopsis (except the placental pad where it is crushed and contains remnants of wall-ingrowths). At maturity, the caryopsis-coat comprises the layers of pericarp, integument and nucellus. Numerous other examples of transfer cells are cited and illustrated in various families.

According to Buell (1952a), there are two periods of accumulation of food reserve (fats, starch, possibly proteins) in the ovary wall, placenta and ovules (chiefly nucellus) about the time of fertilization: (a) during the development of the embryo sac, and (b) from the globular proembryo up to the maturation of seed.

At first, the food reserve in the nucellus is mobilized, followed by the food reserve in the integuments. During its entry into the ovule, the cells adjoining the micropyle nourish the pollen tube, and then the filiform apparatus in the synergid. To begin with, the proembryo is nourished by the suspensor. The globular proembryo and, up to the maturation of the embryo, the endosperm provides the nutrients. Additional supply of nutrients becomes available from the circumjacent tissue through the development of various types of haustoria, their disposition, and duration of aggressive activity. The aggressive activity, and the duration thereof, vary from plant to plant, and also depend on the metabolic status of the parent tissue. The branching, or labyrinth condition, of wall-ingrowths is directly correlated with the absorbtive activity.

Nutritive Role of Seed-Coat. With elaborate physiological and biochemical studies, Murray (1987) has demonstrated the dynamic nutritive role of testa and tegmen in legume seeds. In *Pisum sativum*, the post-fertilized embryo sac enlarges, the central vacuole fills with a solute-enriched fluid. This fluid is considered to arise as a secretion from the tegmen, and is distinct from the endosperm. The secretion of sucrose by tegmen and testa is essentially irreversible in vivo. The flow of nutrients through the maternal tissues, and the relationship of seed-coat metabolism to the nutrition of the developing embryo, have been reported. Essentially, all the metabolites acquired by the developing embryo must first pass through a variable number of seed-coat cells. This involves the "unloading" of solutes from the phloem sieve elements of seed-coat veins, and the processing of certain compounds remained in seed-coat cells prior to secretion. The composition of the embryo sac reservoir of nutrients alters predictably as the seed develops. Further investigations on the nutritive role of the seed-coat for the developing embryo, especially in ex-albuminous seeds, would be very significant. The accumulation of reserve food in the seed-coat has been observed in many taxa, e.g. in *Butomopsis lanceolata* (B.M. Johri 1936b) and in *Indotristicha ramosissima* (Fig. 28.2.1F, G).

Finally, the food reserves accumulate in the endosperm in albuminous seeds, sometimes in the perisperm, and in the embryo in exalbuminous seeds. Jacobsen (1984) gives a good account of the reserve food in seeds, and their mobilization during germination.

3.16 Comparative Studies

The classificatory systems of angiosperms by Bentham and Hooker (1965a, b, c), Hutchinson (1973), Takhtajan (1980), Cronquist (1981), and R. Dahlgren (1980b) are frequently used for comparative "Taxonomic Considerations". Attention is also invited to the updated classification of angiosperms by G. Dahlgren (1989).

Other classifications/relationships mentioned in the following pages are those of Baillon (1874, 1892), Benson (1957, 1962), Bessey (1897, 1915), Corner (1981), Cronquist (1957, 1965, 1968, 1981, 1983), Dumortier (1829), Ehrendorfer (1973), Endlicher (1836–1840), Good (1957), Gunderson (1943, 1950), Hallier (1905, 1912), de Jussieu (1789), Kuijt (1969), Lawrence (1951), Netolitzky (1926), Radford et al. (1974), Rendle (1959, 1971), Schürhoff (1926), A.C. Smith (1972), Stebbins (1974a, b), Thorne (1963, 1968, 1976, 1977, 1983) and Wettstein (1935).

Robyns (1972) proposes a new system of orders and families of Sympetalae; Yamazaki (1974) does so on the basis of embryology. W.R. Philipson (1975) draws attention to evolutionary lines within the dicotyledons. The systematics, phylogeny and evolution of the monocotyledons have been highlighted by Ankerman (1927), Deyl (1955), Doyle (1973), Fritsh (1932) and Huber (1977).

Brewbaker (1967) points out the phylogenetic significance of two- and three-celled pollen grains. Erdtman (1945, 1952) refers to the systematic significance of dyads and tetrads of microspores, and the importance of pollen morphology in taxonomy. Canright (1963) uses pollen morphology as a basis of phylogeny of some Ranalian families. Walker and Skvarla (1975) propose a new concept in the evolutionary morphology of angiosperms on the basis of primitively columella-less pollen.

Sporne (1969) considers the ovule as an indicator of the evolutionary status of angiosperms. W.R. Philipson (1974, 1977) uses ovular morphology as a basis of classification of dicotyledons. Bouman (1971b) points out the usefulness of developmental and structural studies of integuments for consideration of taxonomic and phylogenetic problems. Sunder Rajan (1976) considers the ovular micropyle as an indicator of primitiveness in angiosperms. Bocquet and Bersier (1960) report the systematic value of teratological ovular development.

Tobe (1989, see also Cocucci 1973a, 1983, Hamann 1977, di Fulvio 1981) emphasizes the role of embryology for a systematic and evolutionary study. Schnarf (1933), Kapil (1962, see also Kapil and Bhatnagar 1980), P. Maheshwari (1964), B.M. Johri (1967), Yakovlev (1974), Palser (1975), Poddubnaya-Arnoldi (1982) and Herr (1984) also draw attention to the importance of embryology as a basis for taxonomic considerations. Fagerlind (1944a) refers to the phylogeny of female gametophyte on the basis of tetrasporic embryo sacs (see also Battaglia 1989). A.C. Joshi (1938d) states that the distribution of Fritillaria type of embryo sac indicates the mono- or polyphyletic origin of angiosperms; a wholly incorrect assumption. Other embryological characters used for phylogenetic relationships and classification are "events leading up to fertilization" (Gerasimova-Navashina 1961), embryo development (Yakovlev 1969, Periasamy 1977, Yamazaki 1982), and seed structure (Wunderlich 1967, Vaughan 1968, Corner 1976, Barthlott 1981).

Single embryological characters are not as useful for phylogenetic and evolutionary considerations as a group of characters. Anatomical (Solereder 1908, Eames 1951, Metcalfe 1961, Metcalfe and Chalk 1950), and chaemotaxonomic data (Hegnauer 1963, 1964, 1969, 1973, Frohne and Jensen 1973, Gibbs 1974, Jensen et al. 1975, R. Dahlgren 1983a, b) provide significant support to embryological data.

Most of the above characteristics, and several others, are appropriately discussed in respect of taxonomic considerations. There is no doubt that a comparative analysis of cumulative data is useful in the delimitation of species, genera, families and orders.

I Dicotyledons: Archichlamydeae

4 Casuarinales

The order Casuarinales comprises the monotypic family Casuarinaceae.

4.1 General Features

The order Casuarinales exhibits the following embryological features: (1) acolpate, three-porate pollen grains, two-celled at dispersal stage; (2) orthotropous, bitegmic and crassinucellar ovule with multicellular archesporium, and micropyle formed by inner integument; (3) sporogenous tissue contributes to the development of conducting strand, and formation of a massive parietal tissue; (4) concurrent development of several Polygonum type of embryo sacs with chalazal caecum; (5) persistence of pollen tube and its profuse branching at the chalaza; (6) extravesicular chalazogamy; (7) Nuclear type of endosperm; (8) Onagrad type of embryogeny; (9) exalbuminous seeds with protective crystalliferous, sclerotic pericarp.

4.2 Casuarinaceae

The anther wall comprises an epidermis, fibrous endothecium, one or two middle layers, and tapetum with two-nucleate cells in *Casuarina equisetifolia* (Swamy 1944a).

The acolpate, three-porate pollen grains with thick exine are two-celled at the dispersal stage (Moseley 1948).

The bicarpellary ovary contains one fertile orthotropous, bitegmic and crassinucellate ovule. The inner integument is crushed except in the apical region where it organizes the micropyle. The nucellus projects through the micropyle in *Casuarina montana* (Swamy 1948a). The presence of nucellar tracheids has been noted.

A procambial strand differentiates in the chalaza and extends upward as vascular strand, up to the sporogenous mass (Swamy 1944a). Some of the centrally situated sporogenous cells seem to elongate and contribute to the extension of the strand into the nucellus. In spite of such diversion of a considerable part of the archesporium, as many as 20–25 embryo sacs develop in the same nucellus.

The hypodermal archesporium is multicellular. Each cell cuts off a parietal cell. The parietal tissue is five to six cells thick. Meiosis occurs in the megaspore mother cells, producing linear tetrads. Due to concurrent development of megaspores up to two- to four- or eight-nucleate stage, nearly 20–25 embryo sacs are organized. The chalazal megaspore develops into a Polygonum type of

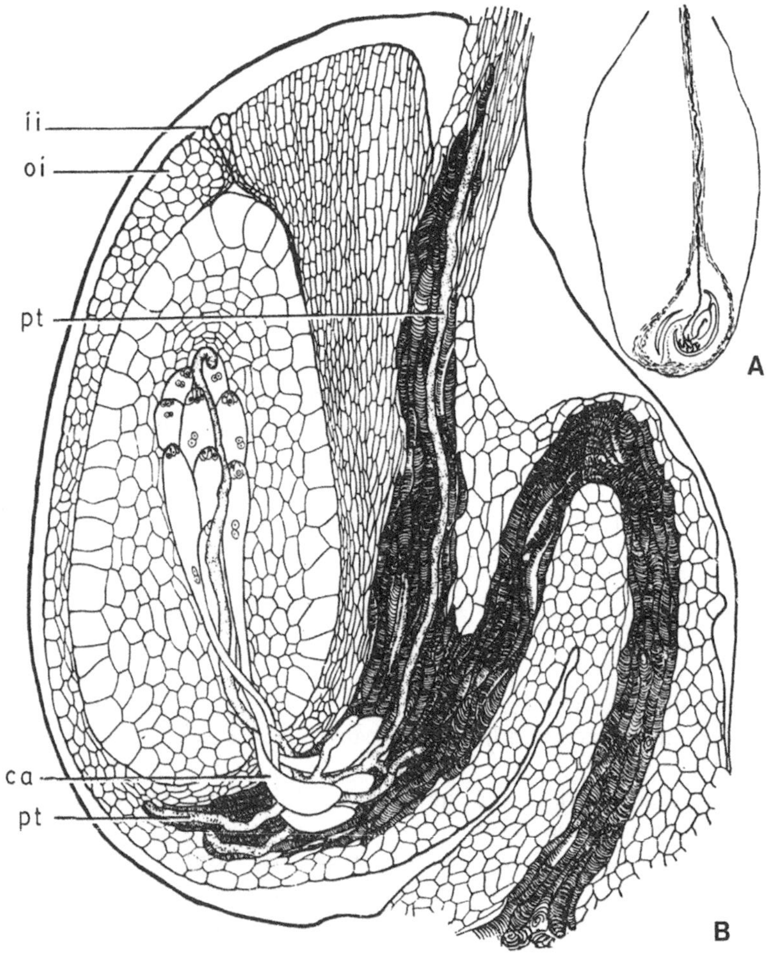

Fig. 4.2.1A, B. *Casuarina equisetifolia.* **A** Longisection of ovary, *central black line* denotes pollen tube. **B** Longisection of ovule to show mature embryo sacs (nine), and path of pollen tube (*pt*). The pollen tube branches in the chalazal region; *ii* inner integument; *oi* outer integument; *ca* embryo sac caecum. (After Swamy 1948a)

embryo sac. The chalazal end of gametophytes form caeca with terminal pouches extending up to the funiculus (Fig. 4.2.1B). The synergids are hooked, and the antipodals lie in caeca. In *C. montana* (Swamy 1948a), rarely the persistent synergid becomes binucleate. This is doubtful.

The growth of the pollen tube continues intercellularly, for about 45–60 days. It grows downward, along the vascular strand, to the chalaza and branches there profusely. Generally, the branches end blindly and some of them may even reach up to the outer epidermis of outer integument, in the basal region. Ultimately, the branch containing the male gametes makes its way upward, between the chalazal caeca of the embryo sacs, passes through the nucellus, and penetrates the embryo sac at a point adjacent to the egg apparatus (Fig. 4.2.1A, B). The mode of fertilization is chalazogamous, and such a penetration of the pollen tube into the embryo sac at the egg apparatus is described as extravesicular (Swamy 1948a).

The development of endosperm is of the Nuclear type, wall formation is initiated at the micropylar end and proceeds centripetally as far as the chalaza. The tissue in the micropylar region is compact. During its growth the embryo consumes the entire endosperm.

The zygote undergoes a transverse division when the micropylar half of the embryo sac contains cellular endosperm. A further transverse division produces a linear row of three cells. The terminal cell of the three-celled proembryo develops into the embryo proper, the middle and basal cell into the suspensor. As the basal cell enlarges, its nucleus becomes hypertrophied. The embryogeny appears to conform to the Onagrad type. Polyembryony due to fertilization of multiple embryo sacs is common. Parthenocarpy has also been reported (Swamy 1948a).

The exalbuminous seed contains an erect, spathulate, large embryo with a well-differentiated radicle and two cotyledons. The testa shrivels, or its outer epidermis may develop thick outer walls. The sclerotic pericarp with crystalliferous tissue gives protection to the dry fruit.

TC Frye (1903), MF Moseley Jr (1948), BGL Swamy (1944a, 1948a), M Treub (1891)

4.3 Taxonomic Considerations

Casuarinaceae. Casuarinaceae show gymnospermous morphological characters. The flowers are extremely reduced, naked and monoecious. The male flower is represented by a single stamen, and the female flower by a tricarpellate pistil.

Several contradictory opinions have been expressed concerning the systematic position of Casuarinaceae. Engler and Prantl (1931), Wettstein (1935), Hjelmqvist (1948) and Rendle (1959) treat it as the most primitive taxon of dicotyledons. Wettstein and Rendle presume its origin from Ephedraceae, and Wettstein retains it in Amentiferae. Hallier (1912), Bessey (1915), Hutchinson (1967) and Tippo (1938) recognize it to be an advanced member. Hutchinson (1973) places it with amentiferous orders, and derives the entire group from the

Hamamelidales. Tippo (1938) and Cronquist (1968, 1981) support this view. Moseley (1948) considers it as "moderately specialized", and on the basis of its anatomy and floral morphology suggests its derivation from Hamamelidaceae as a result of reduction of floral organs during the course of evolution. R. Dahlgren (1977a, 1980b, 1981) places the order Casuarinales in the superorder Hamamelidanae.

The embryological features, such as three-four-five-colpate pollen grains, persistent endosperm, Solanad type of embryogeny, and winged seed with massive sclerotic testa of Hamamelidaceae, differ from those of Casuarinaceae. The wood structure is similar in both families.

Eames (1961) assigns Casuarinaceae to Betulaceae, on the basis of pollen structure, two-trace stamen, and other floral characters. Both the families exhibit true chalazogamy, Polygonum type of embryo sac, and Nuclear endosperm. However, unitegmic ovule with single archesporium, persistent endosperm, and Asterad type of embryogeny in Betulaceae do not occur in Casuarinaceae.

Casuarinaceae resemble Betulaceae and Hamamelidaceae in the belated development of ovule, massive sporogenous tissue giving rise to multiple embryo sacs, and delay in syngamy in relation to anthesis.

Melchior (1964) places Casuarinales near Juglandales. Both the orders show several common embryological features such as three-porate pollen, orthotropous crassinucellate ovule, parietal tissue, Polygonum type of embryo sac, chalazogamy and delayed fertilization, Nuclear endosperm, and a single-layered seed-coat.

Casuarinaceae are undoubtedly angiosperms; the simplicity of flower is due to reduction. The bitegmic ovule with multicellular archesporium, and Polygonum type of embryo sac are markedly angiospermous. The double-trace stamen is a primitive character, and occurs in Proteaceae and Betulaceae. The pollen grains show affinity to Betulaceae and Myricaceae. Chalazogamy has little or no phylogenetic significance. Eames (1961) considers Casuarinaceae as a highly specialized family, derived from primitive stock, and which has remained isolated.

5 Juglandales

The order Juglandales comprises two families: Myricaceae and Juglandaceae.

5.1 General Features

This order is characterised by (1) three-porate pollen grains, (2) basal, orthotropous, unitegmic ovules with massive nucellus, (3) single archesporial cell and well-developed parietal tissue, (4) Polygonum type of embryo sac; occasionally tetrasporic in Juglandaceae, (5) delayed fertilization and chalazogamy in Juglandaceae, (6) Nuclear endosperm, (7) Asterad type of embryogeny, and (8) single-layered seed-coat.

5.2 Myricaceae

The embryology has not been fully investigated.

The ovule is basal, orthotropous, unitegmic and crassinucellate in *Myrica*, and pendulous and anatropous in *Canacomyrica*. The nucellus is elongated, and the embryo sac is located in its upper region. The growth of the integument is arrested, occasionally, and then the micropyle does not organize in *Myrica gale* (Håkansson 1955). The vascular supply branches at the chalaza and passes into the integument.

The archesporial cell cuts off a parietal cell, the latter produces six to eight layers of parietal tissue. Meiosis in the megaspore mother cell results in a linear tetrad. The chalazal megaspore develops into a Polygonum type of embryo sac. The polar nuclei fuse before fertilization, and antipodals are ephemeral. A hypostase of lignified cells develops. Twin embryo sacs may occur.

The growth of pollen tube is slow, and fertilization is delayed. The entry of pollen tube is porogamous.

The development of endosperm precedes the division of zygote by about 4 weeks. Yen (1950) reported the period as 6 to 7 weeks. The development of endosperm is of the Nuclear type. A noteworthy feature is the formation of nucleated cytoplasmic nodules in *Myrica gale* (Håkansson 1955). Wall formation is initiated at the 32-nucleate stage of endosperm; at first cells develop in the micropylar region and proceed towards the chalaza. The entire endosperm becomes cellular. It is generally ephemeral, or may persist as a single layer with oil and aleurone grains as storage products. Its main role is to produce growth substances (Håkansson 1955).

The exalbuminous seed is small, and contains a straight embryo. The seed-coat is represented by the single outer epidermal layer of thick-walled cells.

Å Håkansson (1955), TK Yen (1950)

5.3 Juglandaceae

The anther wall comprises the radially elongate epidermis, an endothecium, two to four middle layers, and glandular tapetum with multinucleate cells. The microspore mother cells undergo meiotic divisions, and tetrahedral and decussate tetrads are produced. Cytokinesis is simultaneous.

The three- to seven-porate pollen grains are two-celled at the dispersal stage. The formation of male gametes occurs in pollen grains before germination, in *Carya illinoinensis* (Solntzeva and Ogebuadze 1974). The male nuclei are spherical in the beginning, become curved afterwards, and remain surrounded by a thin film of cytoplasm (Finn 1925).

The ovule is orthotropous, unitegmic and crassinucellate with lobulate exostome. The single, deeply situated archesporial cell cuts off a parietal cell, and the parietal tissue is well developed. The cells of the nucellar epidermis divide and add to the parietal tissue. In *Pterocarya fraxinifolia*, Bouman and Boesewinkel (1969) report twin nucelli in an ovule due to cleavage of nucellar primordium. The megaspore mother cell undergoes meiotic divisions, producing a linear or T-shaped tetrad. Twin tetrads develop in *Juglans* (Langdon 1934). The chalazal megaspore produces a Polygonum type of embryo sac (Fig. 5.3.1A–D). Twin and triple embryo sacs (per ovule) are formed in *Carya illinoinensis* (Ogebuadze 1977). There are three small antipodals. Karsten (1902) reported tetrasporic type of development in *J. cordiformis*, which needs confirmation.

Extravesicular chalazogamy (see Swamy 1948a) occurs in *Juglans cinerea*, *J. nigra* (Nawaschin and Finn 1913) and *J. regia* (Nast 1941). However, in

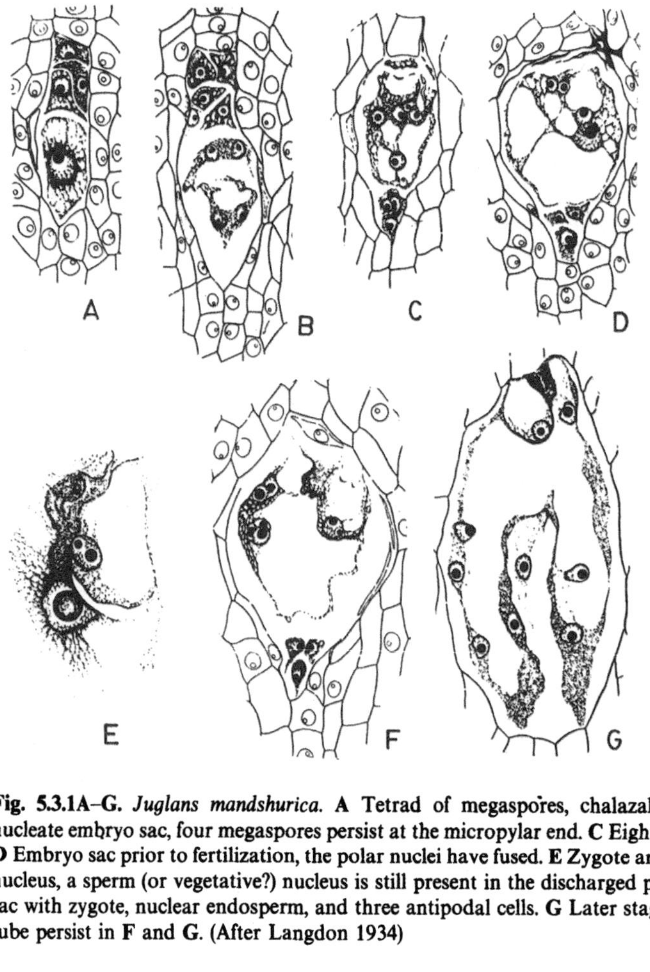

Fig. 5.3.1A–G. *Juglans mandshurica.* **A** Tetrad of megaspores, chalazal functional. **B** Four-nucleate embryo sac, four megaspores persist at the micropylar end. **C** Eight-nucleate embryo sac. **D** Embryo sac prior to fertilization, the polar nuclei have fused. **E** Zygote and primary endosperm nucleus, a sperm (or vegetative?) nucleus is still present in the discharged pollen tube. **F** Embryo sac with zygote, nuclear endosperm, and three antipodal cells. **G** Later stage; remnants of pollen tube persist in **F** and **G**. (After Langdon 1934)

Hicoria (J.G. and N.C. Woodroof 1927) porogamy is reported. As many as two (Langdon 1934) or five (Nawaschin and Finn 1913) pollen tubes enter the embryo sac in *Juglans*. Fertilization is delayed. Figure 5.3.1E shows the zygote and primary endosperm nucleus.

The development of endosperm is of the Nuclear type, and a large number of free nuclei are produced around the central vacuole (Fig. 5.3.1F, G). Wall formation begins from the micropylar end in *Juglans* (Nast 1941), and from the chalazal (needs reinvestigation) end in *Carya* (McKay 1947). The endosperm is reduced to a single layer of cells, and stores fatty substances.

According to Langdon (1934), the division of the zygote is transverse. The apical cell produces cotyledons and stem tip, whereas the basal cell gives rise to hypocotyl, root axis, root-cap, and a multicellular suspensor. Ogebuadze (1977) reports Asterad or Piperad type of embryogeny in *Carya*.

In *Carya* (Solntzeva and Ogebuadze 1974), the zygote undergoes a resting period of 1.5–2 months. A period of 2–2.5 months elapses between the division of zygote and initiation of differentiation of organs. Complete ripening of fruit begins 5 to 6 months after fertilization.

Twin nucelli are reported in *Pterocarya fraxinifolia* (Bouman and Boesewinkel 1969). This is due to the cleavage of nucellar primordium after the initiation of integument. Such nucelli contain normal embryo sacs.

The seed is large and the massive embryo has contorted cotyledons. The seed-coat is highly vascularized, and reduced to the outer epidermal layer of large, radially elongate, thin-walled cells. Stomata occur in the epidermal layer.

F Bouman and FD Boesewinkel (1969), WW Finn (1925), WA van Heel (1976), G Karsten (1902), LM Langdon (1934), JF Leroy (1955), JG Luza and VS Polito (1988), JW McKay (1947), CG Nast (1941), SG Nawaschin (1895a), SG Nawaschin and WW Finn (1913), NV Ogebuadze (1977), MP Solntzeva and NV Ogebuadze (1974), DE Stone and CR Broome (1971), BGL Swamy (1948a), DR Whitehead (1965), JG Woodroof (1924), JG and NC Woodroof (1927), NC Woodroof (1928)

5.4 Taxonomic Considerations

Myricaceae, Juglandaceae. Bentham and Hooker (1965c) place Myricaceae and Juglandaceae in Unisexuales, Engler and Diels (1936) place Myricaceae in Myricales, and Juglandaceae in Juglandales, and Hutchinson (1973), R. Dahlgren (1977a, 1980b) and Cronquist (1968, 1981) agree with this arrangement. Hjelmqvist (1948) considers Juglandaceae to be advanced over Myricaceae, and includes both the families in the order Juglandales. Hutchinson (1973) maintains that the family Myricaceae is derived from Rosalean stocks ancestral to the Hamamelidales. Bessey (1915) considers the origin of Juglandaceae through Sapindales. R. Dahlgren (1977a) places both the orders, Juglandales and Myricales, near Casuarinales in his superorder Rosiflorae. Juglandaceae bear affinity with Betulaceae and Fagaceae in floral and gross anatomical characters.

In addition to the general features of the order, Juglandaceae is characterized by the occurrence of chalazogamy, and Asterad or Piperad type of embryogeny. In Myricaceae the mode of fertilization is porogamous. Nucleate cytoplasmic nodules occur in the endosperm, and the endosperm is ephemeral.

Both the families show close affinities, and the order is homogeneous.

The features of Sapindales, comparable to Juglandales are: two-celled pollen grains, crassinucellar ovules, Polygonum type of embryo sac, chalazogamy in *Anacardium* and *Pistacia* (Anacardiaceae), and Nuclear endosperm with cytoplasmic nodules in Anacardiaceae. Cronquist (1968, 1981) negates the relationship, as the wood of Juglandaceae is primitive.

For the relationships of Juglandales with Fagales, refer to Section 9.4.

6 Balanopales

6.1 Balanopaceae

The monotypic family Balanopaceae represents the order Balanopales. The embryology has not been investigated.

7 Leitneriales

The order Leitneriales comprises two families: Leitneriaceae and Didymelaceae.

7.1 General Features

Leitneriales show: (1) secretory anther tapetum, (2) three- to six-colporate pollen grains with reticulate exine, (3) amphitropous, bitegmic crassinucellar ovule with a slender nucellar beak, (4) unicellular archesporium with a massive parietal tissue, (5) Polygonum type of embryo sac with ephemeral antipodal cells, (6) Nuclear endosperm, (7) persistent massive endosperm, and (8) perisperm in seed.

7.2 Leitneriaceae

Pfeiffer (1912) gives a good account of the embryology.

The anther wall comprises persistent epidermis, fibrous endothecium, an ephemeral middle layer, and glandular tapetum. The microspore mother cells undergo meiotic divisions. The tetrads are tetrahedral, isobilateral and decussate.

The pollen grains are three- to six-colporate with reticulate surface.

The ovule is amphitropous, bitegmic and crassinucellate in *Leitneria floridana*. Both the integuments take part in the formation of micropyle. A single archesporial cell differentiates, and cuts off a parietal cell which produces a massive parietal tissue. The nucellar epidermal cells, by periclinal divisions, produce a slender nucellar beak. The nucellus is persistent. The megaspore mother cell, by meiotic divisions, gives rise to a linear tetrad. The chalazal megaspore develops into a Polygonum type of embryo sac. The synergids are non-vacuolate (requires confirmation), and the antipodals are evanescent.

The development of the endosperm is of the Nuclear type, the free nuclei lie peripherally. After 18 days (of its formation) the endosperm starts becoming cellular. In the chalazal region the endosperm is composed of loosely packed large cells with irregular nuclei. The endosperm occupies the entire seed, and is persistent and starchy.

The zygote divides at the 35-nucleate stage of the endosperm. The divisions are oblique, and cell formation is irregular. The proembryo, before the differentiation of cotyledons, is pear-shaped with a massive suspensor. The mature embryo is large and straight, with two thin and broad cotyledons.

The seed is endospermic and perispermous. The outer epidermis of the testa is thin-walled. The mesophyll cells are thick-walled with reticulate striations. The inner epidermis of the tegmen comprises thick-walled cells with reticulate markings.

WM Pfeiffer (1912)

7.3 Didymelaceae

The embryology has not been studied in detail.

There is a brief account of *Didymeles medagascariensis* by Leandri (1937, see also Yakovlev 1981). The anther is tetralocular (bilocular according to Hutchinson 1973) and extrose. The ovule is hemianatropous and bitegmic. The exostome forms a tube which is broad in the apical region. The seed is exalbuminous (copious endosperm according to Hutchinson 1973) with a large embryo which has flat-concave cotyledons and a short radicle.

H Baillon (1892), J Hutchinson (1973), J Leandri (1937), MS Yakovlev (1981)

7.4 Taxonomic Considerations

Leitneriaceae. Bentham and Hooker (1965c) place Leitneriaceae in Unisexuales, and Bessey (1915) in Ranales. Engler and Prantl (1931), Hutchinson (1973), and R. Dahlgren (1977a) consider it as a monotypic family of Leitenerиales. Hutchinson traces its ancestry in Rosales. Hjelmqvist (1948) includes this family in the Amentiferae, in the line of Myricaceae. Cronquist (1968, 1981) and others derive it, by reduction, from Hamamelidales.

The embryological features of Leitneriaceae resemble those of Myricaceae except in the structure of ovule and seed-coat, and the occurrence of endosperm and perisperm in seed. The presence of starchy endosperm, and reticulate thickenings on the walls of testa and endotegmen suggest that this family may not be related directly with other Amentiferae (see also Thorne 1973). In *Leitneria* the division of zygote is followed by a transverse or longitudinal wall. Further information about its embryogeny may be helpful in determining the systematic position.

Leitneriaceae show affinities with Geraniales in having tricolporate pollen grains, bitegmic crassinucellar ovule, Polygonum type of embryo sac with ephemeral antipodals, porogamous entry of pollen tube, Nuclear endosperm, and albuminous seed with perisperm.

The mature pollen grains, ovule, development of embryo sac and endosperm, and mesotestal seed-coat are comparable to that in Hamamelidales.

8 Salicales

8.1 General Features

The representative monotypic family Salicaceae shows:

(1) secretory anther tapetum, (2) two-celled pollen grains, (3) uni- or bitegmic crassinucellate ovule, formation of parietal tissue, (4) Polygonum type of embryo sac, (5) porogamous entry of pollen tube, (6) Nuclear endosperm, and (7) Asterad or Onagrad type of embryogeny. A central cavity is present in the endosperm of *Salix*, and the exalbuminous small seed is adapted to wind dispersal.

8.2 Salicaceae

The anther wall comprises the epidermis, fibrous endothecium, ephemeral middle layers, and glandular tapetum of binucleate cells. There are one to four middle layers in *Salix* (Chamberlain 1897), and two in *Populus* (Nagaraj 1952).

The pollen grains are inaperturate and granulose in *Populus*, and three-colpate with reticulate exine in *Salix*. They are two-celled at the dispersal stage.

The ovule is anatropous and crassinucellar. In *Populus deltoides* (Nagaraj 1952), it is initially bitegmic but, in later stages, only the outer integument can be distinguished. The ovules are unitegmic in *P. tremuloides* (Nagaraj 1952) and *Salix*. The archesporium is single-celled in *P. tremuloides* and *Salix* (P. Maheshwari and Roy 1951), but five- to six-celled in *P. deltoides*. The functional archesporial cell cuts off a parietal cell. The megaspore mother cell undergoes meiotic divisions, tetrads are T-shaped, and the chalazal megaspore functions.

In a short paper, P. Maheshwari and Roy (1951) contradict Chamberlain's observations (1897) on *Salix*, and point out that the embryo sac is of the Polygonum and not Allium or Adoxa type. The micropylar tip of embryo sac comes in contact with the integument. The synergids and antipodals persist up to fertilization.

Fertilization is porogamous, and is not "delayed" as in other amentiferous families.

The endosperm is of the Nuclear type. Multinucleate cytoplasmic nodules are formed in *Salix cinerea* (Håkansson 1954). Wall formation in endosperm occurs in the micropylar and chalazal region at the heart-shaped embryo in *Salix*. In *Populus* the endosperm completely fills the central cavity, whereas a distinct cavity persists in *Salix*. According to Håkansson (1954): "the endosperm of *Salix* has no storage function, being of ephemeral nature. It has certain physiological functions during this period. The formation of growth substances, inciting the growth of the fertilized ovules is the function. When no endosperm is formed, the ovules remain small. A second function of *Salix* endosperm is probably a certain influence on the development of the embryo".

The embryogeny is of the Asterad type Polygonum-variation in *Salix* (Souèges 1923f), and Onagrad type in *P. deltoides*. Synergid embryos are reported in *Salix* (Chamberlain 1897), but this needs confirmation.

The seed is small with a straight embryo, bears a plumage of aseptate hairs at the base (from the placenta), and is adapted to wind dispersal. The testa comprises only the outer epidermal layer of the integument, with the outer and radial walls thickened.

CJ Chamberlain (1897), Å Håkansson (1954), P Maheshwari and SK Roy (1951), M Nagaraj (1952), R Souèges (1923f), H Takeda (1936)

8.3 Taxonomic Considerations

Salicaceae. Bentham and Hooker (1965c) place Salicaceae among anomalous families of Monochlamydeae. Hallier (1912) and Bessey (1915) include it in Caryophyllales. Melchior (1964), Hutchinson (1967, 1973), R. Dahlgren (1977a) and Cronquist (1968, 1981) treat it as a lone member of Salicales. Hutchinson considers it as a descendant of hamamelidaceous ancestors, Cronquist (1981) of violaceous ancestors, whereas Hjelmqvist (1948) regards it as an advanced order within the Amentiferae.

Salicaceae share several embryological features with Hamamelidaceae, such as anatropous crassinucellate ovule, Polygonum type of embryo sac, Nuclear endosperm, and seeds adapted to wind dispersal (comose in the former and, occasionally, winged in the latter).

Ephemeral endosperm occurs in Myricaceae, and unitegmic crassinucellar ovule and Nuclear type of endosperm are similar to those in Juglandaceae. However, the primitive features of Juglandaceae, such as presence of integumentary vascular bundle, massive nucellus and delayed fertilization, are absent in Salicaceae. Moreover, reduced flower, scanty nucellus, seed adapted to wind dispersal, and rapid germination of seed indicate its advanced nature (in the Amentiferae). The small comose seed with a well developed exotesta is comparable to that in Tamaricaceae (Corner 1976).

9 Fagales

Fagales comprises two families: Betulaceae and Fagaceae.

9.1 General Features

As members of a common order Fagales, Betulaceae and Fagaceae show some similarities: (1) inferior ovary with one to many ovules, anatropous crassinucellar ovule, (2) Polygonum type of embryo sac with ephemeral antipodal cells, (3) delayed fertilization, and Nuclear type of endosperm. The differences are more pronounced: (1) The Betulaceae have (1) three- to four-porate pollen grains, (2) unitegmic ovules, (3) single-celled archesporium, (4) chalazogamy, (5) massive persistent endosperm, and (6) Asterad type of embryogeny. *Alnus* exhibits unreduced parthenogenesis, and antipodal, synergid and nucellar polyembryony. On the other hand, the Fagaceae show: (1) three- to six-colporate pollen grains, (2) bitegmic ovule with multicellular archesporium, (3) postament and chalazal caecum of embryo sac, (4) porogamous fertilization, (5) endosperm with a large central vacuole, (6) embryogeny of either Piperad or Asterad type, and (7) exalbuminous seed.

9.2 Betulaceae

The embryology has not been fully studied.

The pollen grains are three-, four- or five-porate.

The ovule is anatropous, unitegmic and crassinucellar with a short straight micropyle. The hypodermal archesporial cell probably cuts off a parietal cell, producing parietal tissue. Nawaschin (1894) traced the origin of this tissue to nucellar epidermis which Håkansson (1957) doubted. The megaspore mother cell undergoes meiotic divisions resulting in a linear tetrad. The chalazal megaspore develops into a Polygonum type of embryo sac. The polar nuclei fuse before fertilization; the antipodals are ephemeral.

The sporogenous tissue is massive in *Carpinus betulus*, and several young embryo sacs occur in Coryloideae (Endress 1977a). The embryo sac develops a long chalazal caecum in *Corylus*. The fusion of the polar nuclei occurs in the caecum, whereas the antipodals degenerate in situ.

Extravesicular chalazogamy (see Swamy 1948a) is reported in *Alnus glutinosa*, *A. viridis*, *Betula alba* and *B. ermani* (Fig. 9.2.1). Intravesicular chalazogamy (Benson 1894) occurs in *Carpinus betulus* and *Corylus avellana*.

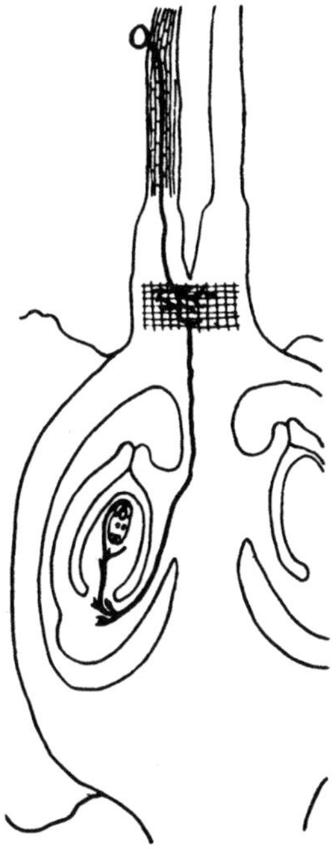

Fig. 9.2.1. *Betula ermani*, longisection of gynoecium with unitegmic ovule at mature eight-nucleate embryo sac. The pollen grain germinates on the style, the tube passes through the ovary, enters the ovule at the chalaza, and then moves upward in the nucellus up to the egg apparatus. The pollen tube gives out branches on its way. (See Yakovlev 1981)

The pollen tube penetrates the embryo sac at the chalazal end, and then reaches the egg apparatus (Swamy 1948a). This needs confirmation. Branching of the pollen tube occurs in *Betula alba* (Benson 1894), *Betula* sp., *Carpinus* and *Corylus* (Nawaschin 1895b). In *B. costata* (Fig. 9.2.2A), the pollen tube enters at the chalaza and puts out branches which traverse through the nucellus. There is again profuse branching at the micropylar end, and further branching occurs inside the embryo sac. The entry of pollen tube at the chalazal end of the embryo sac is reported in *Alnus incana* (Fig. 9.2.2D). A period of 3 to 4 months elapses between pollination and fertilization in *A. glutinosa* (Benson 1894).

The development of the endosperm is of the Nuclear type. Later the endosperm becomes cellular, crushes the nucellus, and fills up the chamber completely (Håkansson 1957). The development is similar in *Alnus rugosa* (R.H. Woodworth 1930a) and *Betula verrucosa* (Håkansson 1957). The massive endosperm fills the seed, and contains aleurone grains and oil as reserve food.

The embryogeny conforms to the Asterad type.

As the viability of pollen grains is very low, and as meiotic divisions do not occur during the formation of the embryo sac, 50% of the seeds contain diploid egg-embryos. Thus, *Alnus* exhibits unreduced parthenogenesis. In addition, polyembryony due to the activity of antipodals, synergids, nucellus, and endosperm is reported in *Alnus rugosa* (R.H. Woodworth 1930a). The concurrent development of multiple embryo sacs also gives rise to additional embryos. Nucellar, synergid and zygotic polyembryony also occurs in *A. incana*, *B. cycoviensis*, *B. pendula* and *Duschekia kamschatica* (Fig. 9.2.2B–G). The presence of four to seven cotyledonary buds on certain embryos suggests

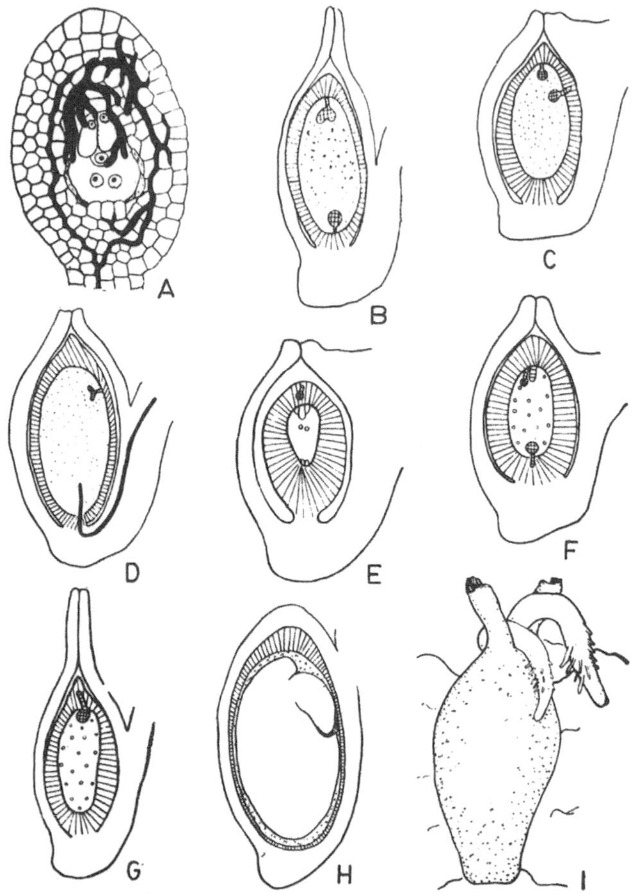

Fig. 9.2.2A–I. **A** *Betula costata*, **B**, **G** *Duschekia kamschatica*, **C** *B. pendula*, **D**, **F** *Alnus incana*, **E** *B. cycoviensis*, **H** *B. davurica*, **I** *D. fruticosa*. Nucellus striped in **B–H**. **A** Longisection of nucellus with branched pollen tube. **B**, **C** Longisection of young seeds with proembryo from the zygote and antipodal (**B**), and zygote and nucellus (**C**). Endosperm *dotted in* **B–D**, **H**. **D** Chalazogamous entry of pollen tube, note a nucellar embryo. **E** Mature embryo sac with a nucellar proembryo above the egg apparatus. **F** Proembryos from zygote, synergid, and nucellus (chalazal end). *Empty circles* denote endosperm nuclei. **G** Nucellar proembryo and endosperm. **H** Mature embryo in seed (longisection). **I** Germinated fruit with two embryos. (See Yakovlev 1981)

fusion of embryos forming a composite mass. P. Maheshwari (1950) suggested that some of the conclusions need confirmation, particularly the formation of embryos from endosperm.

The seed is small and may be pachychalazal. The unitegmic condition must be the result of connation of two integuments (Corner 1976).

The seed of *Betula davurica* (Fig. 9.2.2H) contains a large embryo with scanty endosperm. Two embryos may mature in *Duschekia fruticosa* (Fig. 9.2.2I).

The passage of the pollen tube through the style, its branching in the stylar region, entry through the funiculus, entry into the embryo sac at the chalazal end, branching at the chalazal end of nucellus, branching in the nucellus, at its micropylar end, inside the embryo sac, process of fertilization, parthenogenesis, polyembryony, and structure of the seed (persistence of nucellus as perisperm, if any) require thorough reinvestigation.

M Benson (1894), EJH Corner (1976), PK Endress (1977a), Å Håkansson (1957), AD Macdonald and DH Mothersill (1987), P Maheshwari (1950), JC Mestre (1964), SG Nawaschin (1894, 1895b), BGL Swamy (1948a), A Szwabowicz (1976), RH Woodworth (1929, 1930a, b), MS Yakovlev (1981)

9.3 Fagaceae

The anther wall comprises four layers; the innermost is the tapetum with two- to multinucleate cells. Simultaneous cytokinesis results in the formation of tetrahedral tetrads in *Quercus* (Stairs 1964).

The pollen grains are three- to six- or seven-aperturate (colpate), and two-celled at the dispersal stage.

The ovule is anatropous, bitegmic, unitegmic in *Nothofagus* (Poole 1952), and crassinucellar. The outer integument is extremely long in Fagaceae (*Fagus sylvatica*), and perhaps partially serves as an obturator (see Endress 1977a). In *Fagus* (Hjelmqvist 1957a), the inner integument is reduced, thinner and irregularly folded with papillate elevations on the surface. A 20- to 60-celled archesporium differentiates in the hypodermal region, only one cell functions. True parietal cells are not produced in *Quercus robur* (Hjelmqvist 1953). In *Fagus* a parietal cell is produced giving rise to several cell layers. Meiotic divisions in megaspore mother cell result in a linear or T-shaped tetrad. The chalazal megaspore develops into a Polygonum type of embryo sac. Adoxa type of development is reported in *Quercus alba* (Turkel et al. 1955).

In *Quercus* the embryo sac develops a caecum reaching up to the base of nucellus. Another interesting feature is the development of a central resistant column of cells below the antipodals, in continuity with the vascular trace. It protrudes into the embryo sac and persists until late stages of the embryo. It is referred to as "postament" (Poole 1952, Brown and Mogenson 1972).

The polar nuclei fuse prior to fertilization. The antipodals are inconspicuous and ephemeral.

Fertilization is porogamous. The formation of zygote is delayed, as compared to the primary endosperm nucleus. Two months elapse after pollination to complete the nuclear fusion of gametes (Hjelmqvist 1953).

After fertilization, the primary endosperm nucleus migrates into the caecum while the antipodals remain in situ at the apex of postament, where they degenerate. In *Q. alba* the antipodals move to the chalazal end of the caecum.

The development of endosperm is of the Nuclear type. The aggregation of nuclei in the chalazal region of the embryo sac and their fusion is very common. In *Quercus robur* (Conrad 1900), wall formation is initiated in the micropylar region after the formation of hundreds of nuclei and differentiation of dermatogen in the globular proembryo. The cellular endosperm occupies the upper one-third portion of the embryo sac, while a large vacuole occupies the lower part. The endosperm is consumed by the developing embryo.

In *Q. coccinea* (Stairs 1964), the embryo requires more than two months for its complete development. Johansen (1950) and Davis (1966) refer the embryogeny to the Onagrad type. Hjelmqvist (1957a) suggests that due to oblique wall formation and variability in cleavage patterns, the embryogeny (as a whole) in Fagales cannot be assigned to any type so far described.

In *Q. gambelii* (Brown and Mogensen 1972), the division of zygote is followed by an oblique or nearly vertical wall, producing a small apical and a large basal cell. As the division of zygote is followed by almost vertical partition, the embryogeny can be assigned to the Piperad type. If it is considered that this division is followed by an oblique wall, and as apical and basal cell both contribute to the formation of embryo proper, it can be assigned to the Asterad type. Brown and Mogensen (1972) assigned the embryogeny to a variation of Piperad or Asterad type rather than the Onagrad type. Hjelmqvist (1953) presumed the varying positions of additional embryos to be due to fertilization of multiple embryo sacs.

The seed is exalbuminous, has a large straight embryo with thick or plicate cotyledons, and is enclosed in a nut. The testa is thin-walled, and contains branches of vascular bundle. The tegmen gets crushed.

RC Brown and HL Mogensen (1972), AH Conrad (1900), GL Davis (1966), PK Endress (1977a), H Hjelmqvist (1953, 1957a), DA Johansen (1950), HL Mogensen (1970a), M Okamoto (1989), AL Poole (1952), AP Singh and HL Mogensen (1975, 1976), GR Stairs (1964), HS Turkel, AL Rebuck and AR Grove Jr (1955)

9.4 Taxonomic Considerations

Betulaceae, Fagaceae. Bentham and Hooker (1965c) combine Betulaceae and Fagaceae in Cupiliferae, and Hallier (1912) in Amentiferae. Wettstein (1935), Engler and Diels (1936), Rendle (1959) and Hutchinson (1973) include them in Fagales. Hutchinson assigns the genus *Corylus* to the family Corylaceae. Hjelmqvist (1948) subdivides the Fagales into Fagales and Betulales, within Amentiferae. Bessey (1915) assigns the two families with Juglandaceae, and treats the three as derived from sapindaceous ancestors. Tippo (1938) derives them from hamamelidaceous stocks, and considers Betulaceae as more primitive. Cronquist (1968, 1981) postulates their derivation from ancestors similar to Hamamelidales.

Betulaceae and Juglandaceae share several similar features, such as unitegmic ovule with single archesporium, chalazogamy, delayed fertilization, Nuclear endosperm, and Asterad type of embryogeny. The occurrence of unitegmic ovules in Betulaceae, Juglandaceae and *Nothofagus* (Fagaceae) is a derived condition due to reduction of bitegmic ovules.

The multicellular archesporium and delayed syngamy in Fagaceae indicate primitiveness. Exalbuminous seeds with a large embryo and contorted cotyledons, and vascularized testa, occur in Juglandaceae and Fagaceae. All the three families exhibit primitive embryological features.

While dealing with the general similarities of inflorescence, flower, pollen, fruit and seed in the Hamamelidales-Fagales group, Endress (1977a) emphasizes embryological features, such as belated development of ovule, and delayed syngamy in relation to anthesis. Coryloideae in Betulaceae, and *Nothofagus* in Fagaceae show close affinity to Hamamelidales during earlier stages of development (up to fertilization). The similarities are: number of integuments (except *Nothofagus*), outer integument partially functioning as obturator, massive sporogenous tissue in the ovule (Coryloideae), development of multiple embryo sacs (Coryloideae), and Polygonum type of embryo sac.

10 Urticales

The Urticales comprise five families: Rhoipteleaceae, Ulmaceae, Eucommiaceae, Moraceae and Urticaceae.

10.1 General Features

The Urticales show (1) secretory anther tapetum, (2) simultaneous cytokinesis following meiotic divisions of pollen mother cells, (3) two- to six-porate pollen grains with simple exine, two-celled at the dispersal stage, (4) anatropous (orthotropous in Urticaceae), bitegmic, crassi-nucellar ovule, weakly-crassinucellar and unitegmic in Eucommiaceae, both crassi- and tenuinucellar in Moraceae, (5) Polygonum type of embryo sac in all members except Ulmaceae where it is of the Adoxa type or an intermediate type between Adoxa and Drusa, (6) distinct hypostase, (7) antipodals small and ephemeral, they proliferate in Ulmaceae, Moraceae and Urticaceae, (8) Nuclear type of endosperm except in Eucommiaceae where it is Cellular; chalazal haustorium develops in Moraceae and Urticaceae, (9) embryogeny of the Asterad type in Moraceae and Urticaceae, and of the Onagrad or Solanad type in Ulmaceae, (10) polyembryony frequent, (11) apospory in Urticaceae, (12) seeds exalbuminous except in Ulmaceae, and (13) seed-coat unspecialized and either the testa, or tegmen, or both, become crushed; remain thin and membraneous if they persist.

10.2 Rhoipteleaceae

The embryology has not been investigated. The anther has four lobes which are connected into two thecae. The pollen grains are triangular.

The ovule is hemianatropous, bitegmic, attached to the septum by a short ascending funicle.

The seed is exalbuminous with a straight embryo which has a slender radicle and thick cotyledons.

J Hutchinson (1973), MS Yakovlev (1981)

10.3 Ulmaceae

The anther is tetrasporangiate. The anther wall comprises an epidermis, fibrous endothecium, two ephemeral middle layers in *Ulmus* (Shattuck 1905) and three or four in *Holoptelea* (Capoor 1937b), and a glandular tapetum with binucleate cells. The tapetal cells are multinucleate in *Zelkova serrata* (Yamada and Yoshida 1979). Partial sterilization of sporogenous tissue, which probably provides nourishment to the remaining cells, occurs in *Holoptelea*. The microspore mother cells undergo meiotic divisions giving rise to tetrahedral and decussate tetrads. Cytokinesis is simultaneous.

The pollen grains are two-celled at the dispersal stage. In *Holoptelea*, division of the generative cell occurs on the stigma.

The ovule is anatropous, bitegmic and crassinucellar with a single hypodermal archesporial cell. Both the integuments form the micropyle in *Celtis* and *Ulmus*, while in *Holoptelea* and *Zelkova* only the inner one. The nucellus develops a beak, and the hypostase is well developed.

The hypodermal archesporial cell cuts off a parietal cell (Shattuk 1905). On the other hand, R.I. Walker (1938, 1950) reported that the single hypodermal archesporial cell functions directly as megaspore mother cell in *Ulmus americana*, *U. fulva*, *U. glabra* var. *camperdownii*, *U. pumila* and *U. racemosa*. The megaspore mother cell undergoes meiotic divisions and a coenomegaspore of four linearly arranged nuclei is produced. The nuclei undergo one further division to produce eight-nucleate, seven-celled gametophyte. This happens in 40%–60% of ovules of *Ulmus*, while in the remaining ovules the four megaspore nuclei assume a 1 + 3 position. The micropylar nucleus undergoes two more mitotic divisions, while three chalazal nuclei undergo one or more divisions. Mature gametophytes with four to 12 antipodals have been observed. According to R.I. Walker (1950), the development of female gametophyte in *Ulmus* is transitional between the Adoxa type and a modified Drusa type.

The Adoxa type is also reported by Leliveld (1935) and D'Amato (1940).

Zelkova serrata (Yamada and Yoshida 1979) has a single ovule in the ovary. The megaspore mother cell undergoes meiosis producing a coenomegaspore which shows 1 + 1 + 1 + 1, 1 + 2 + 1, 2 + 2 arrangement of nuclei, or even a 1 + 3 arrangement. The two lower nuclei may divide earlier than the upper two or all the four nuclei may divide simultaneously and an eight-nucleate embryo sac is formed. If some of the nuclei divide again, the embryo sac has more than eight nuclei and one to four nuclei in the chalazal end degenerate. The organized embryo sac has three-celled egg apparatus, two polars, and a variable number of antipodal cells (some nuclei do not organize into cells). The mature embryo sacs are of the Adoxa type, sometimes Drusa type with a variable number of antipodal nuclei (always less than 11). In our opinion a reinvestigation may be very useful.

In *Celtis occidentalis* (Jönsson 1879/80) and *Holoptelea integrifolia* (Capoor 1937b), the gametophyte is of the Polygonum type. Supernumerary antipodals occur in *Holoptelea.*

Capoor (1937b) reported some abnormalities in *Holoptelea.* In several cases, due to the elongation of funiculi, the ovules grow through the style reaching as far as the stigma. Therefore, the stylar canal remains completely open.

In *Ulmus* the pollen tubes branch, anastomose, and wander profusely through the funiculus, integuments and, occasionally, in the nucellus. Fertilization is porogamous, and the pollen tubes persist up to the two-celled proembryo (Shattuck 1905). Several features reported in *Ulmus* require reinvestigation.

The development of endosperm is of the Nuclear type, and the nuclear divisions are mostly synchronous. Later, it becomes cellular.

The development of embryo is of the Onagrad type in *Holoptelea*, and Solanad in *Ulmus.*

In *U. americana*, the presence of antipodal embryos, extra embryos at the micropylar region, and pseudopolyembryony due to the presence of two embryo sacs in an ovule are common (Shattuck 1905). Ekdahl (1941), and Guignard and Mestre (1966) reported antipodal embryos in *U. campestris*, and *U. glabra*, respectively.

The seed is small and exalbuminous with straight embryo; the cotyledons are fleshy and the radicle short. The testa is crushed in *Ulmus.* It persists in *Holoptelea*; the thick-walled outer epidermis contains brown pigment. The tegmen is crushed.

SP Capoor (1937b), F D'Amato (1940), WC Dickison and EM Swetrer (1970), I Ekdahl (1941), JL Guignard and JC Mestre (1966), H Hjelmqvist and F Grazi (1965), B Jönsson (1879/80), JA Leliveld (1935), CH Shattuck (1905), T Takaso (1987), T Takaso and H Tobe (1990), H Tobe (1987), RI Walker (1938, 1950), H Yamada and O Yoshida (1979)

10.4 Eucommiaceae

The anther wall comprises persistent epidermis, endothecium, one or two ephemeral middle layers, and secretory tapetum with two- to six-nucleate cells in *Eucommia ulmoides* (Tang 1962). During early development of the anther, the epidermal cells between the theca become detached. The endothecium is extensively developed. In *E. ulmoides* it is two- to three-layered, extends to the connective, and surrounds the staminal vascular bundle on either side (Eckardt 1963). The microspore mother cells undergo meiotic divisions, and tetrahedral tetrads are produced. Cytokinesis is simultaneous.

The pollen grains are tricolporoidate, and shed at two-celled stage.

The ovule is anatropous, weakly crassinucellar and unitegmic with a long micropylar canal. The cells of the nucellar epidermis divide periclinally to form a small nucellar cap. The multicellular archesporium cuts off parietal cells. The development of the female gametophyte is of the Polygonum type. One of the embryo sacs develops to maturity, while the others may develop up to four-nucleate stage and collapse.

Fertilization is porogamous. In two cases intravesicular chalazogamy has been observed in *Eucommia* (Eckardt 1963). Branching of the pollen tube is also reported.

The development of endosperm is of the Cellular type.

In *Eucommia ulmoides* the zygote remains in resting condition even up to 6 weeks; the embryogeny has not been studied. Tang (1962) reports suspensor polyembryony.

The seed is albuminous with a straight embryo. The seed-coat is membraneous and thin-walled.

T Eckardt (1957, 1963), H Harms (1930), SH Tang (1962), O Tippo (1940)

10.5 Moraceae

The reproductive biology of *Ficus* is highly specialized and is dealt with separately. The following account refers to other members of the Moraceae.

The wall of the tetrasporangiate anther comprises heavily cutinized epidermis, fibrous endothecium, three to four ephemeral middle layers, and secretory tapetum. According to S. Kumar (1978), in *Morus* spp. the development of anther wall conforms to the "Basic type" (see Davis 1966). Resin occurs in the endothecial cells, and connective of *Cannabis sativa* (Mohan Ram and Nath 1964). The tapetal cells are two- to multinucleate, and show nuclear fusions. It becomes irregularly biseriate. The microspore mother cells undergo meiotic divisions, simultaneous cytokinesis, and produce tetrahedral and decussate, rarely isobilateral, tetrads.

The pollen grains are biporate with a smooth exine, and two-celled at the dispersal stage. S.P. Singh (1954) observed pollen grains in the ovary wall of *Morus*.

The ovule is hemianatropous, anatropous, bitegmic and crassinucellate with a micropyle formed by the inner integument. The cells of the nucellar epidermis produce a nucellar cap. The corky hypostase is well developed.

The archesporium is unicellular, except in *Morus* where it is multicellular but only one cell functions. It cuts off a parietal cell, the parietal tissue is well developed. The megaspore mother cell undergoes meiotic divisions, gives rise to a linear or T-shaped tetrad, and the chalazal megaspore is functional. Frequently, the sub-chalazal megaspore develops in *Cannabis sativa* (Mohan Ram and Nath 1964). The development of the female gametophyte is of the Polygonum type. The synergids are usually small except in *Artocarpus* (Subba Rao 1940a). The antipodals are ephemeral in *Dorstenia contrayerva* and *D. drakena*. Modilewski (1908b) reports proliferation of antipodals, and parthenogenesis in *D. contrayerva* and *D. drakena*.

In *Morus alba* (S.P. Singh 1954), very few embryo sacs reach maturity, and generally parthenocarpic fruits are produced.

The endosperm is of the Nuclear type, but later becomes cellular. In *Artocarpus*, free-nuclear condition persists until a late stage of embryogeny (Davis 1966). In *Cannabis*, enucleate cytoplasmic vesicles develop in the nuclear endosperm which ultimately becomes cellular (Mohan Ram and Nath 1964). M. Ram (1960) had pointed out earlier that a chalazal free-nuclear haustorium also develops. Mohan Ram and Nath state that enucleate vesicles also occur in the chalazal haustorium later which becomes cellular with binucleate vacuolate cells (see also M. Ram 1960).

The embryogeny conforms to the Asterad type in *Streblus* (Anantaswamy Rau 1942) and *Dorstenia* (Padmanabhan 1961b). Tricotyly is reported in *Morus multicaulis* (R.M. Datta 1941). Cleavage and nucellar polyembryony occur in *S. taxoides* (Narayanan 1968), and the supernumerary embryos are always confined to the micropylar part of the embryo sac.

Srivastava and Tandon (1985) point out that in *Morus indica* and *M. laevigata* adventive polyembryony originates from cells of the ovary and nucellus. The adventive embryo initially has a well-defined nucleus and dense cytoplasm. These embryos grow only up to the four- to eight-celled stage and then degenerate. Of the 65 adventive embryos studied, 30 were observed in open-pollinated flowers, 23 in intra- and interspecific crosses, and 12 in unpollinated (isolated) female flowers. Possibly, the pollen grains stimulate the formation of adventive embryos.

Gonial apospory is reported in several species of *Dorstenia* (see Davis 1966).

The seeds are small to large, albuminous or exalbuminous, enclosed in the woody endocarp. The embryo is straight or curved, cotyledons flat and thin or thick or convolute (Corner 1976).

The outer epidermis of outer integument and the inner epidermis of inner integument form the seed-coat in *Broussonetia* and *Morus*. In *Artocarpus* (Subba Rao 1940a), the outer integument alone forms the seed-coat.

Ficus

For a proper growth and development of the syconium (Fig. 10.5.1), *Ficus* has a mutualistic symbiotic dependence on a sycophilous wasp. *F. religiosa* (B.M. Johri 1989, B. M. Johri and Konar 1955, 1956, Galil and Eisikowitch 1968a) and *F. ottoniifolia* (Verkerke 1986a) have been studied in detail.

At the apex of the syconium is an ostiole with inwards directed bracts on the inside (Fig. 10.5.2A–C).

F. religiosa and *F. ottoniifolia* are monoecious, and each syconium contains male and female flowers (Figs. 10.5.3A–D, 10.5.4A–D, 10.5.5A–D). On the inner fleshy wall of the syconium, *F. religiosa* (B.M. Johri and Konar 1956) has 11–19 pedicellate male flowers around the closed ostiole, each flower has three hooded perianth lobes and a stamen; 105–113 long-styled or seed flowers and 80–81 short-styled or gall flowers, each pedicellate flower has five perianth lobes, and an uniovulate ovary with a laterally attached style with papillate

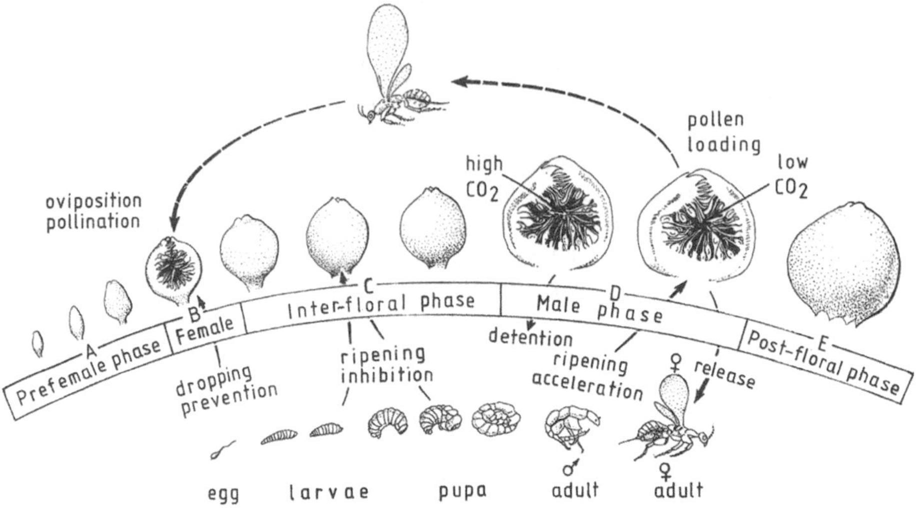

Fig. 10.5.1. Concurrent symbiotic developmental cycle of fig syconium and wasp, phases **A–E**. The female wasp loaded with pollen (phase **D**) leaves the old syconium, and enters a young syconium (phase **B**). Following oviposition, pollination occurs. The development of the wasp is also shown. (Adapted after B.M. Johri and Konar 1956, Galil 1977)

stigma. In *F. ottoniifolia* (Verkerke 1986a), there are 60 pedicellate male flowers, each with two cucullate perianth segments and one stamen (Figs. 10.5.3A, 10.5.5D). There are 1200–1400 female flowers, each flower has three to four membraneous perianth segments, a uniovulate ovary and a single terminal style (Fig. 10.5.3A, D). The seed flowers may be sessile, shortly pedicellate and sometimes slightly stipitate, and have a style up to 1800 μm long (Fig. 10.5.3A). During phase A, the stigma of a gall flower develops two unequal lobes (Fig. 10.5.4C, D). The transmitting tissue bordering the stylar canal consists of two layers of cytoplasm-rich cells.

The dithecous anther (Fig. 10.5.5A–D) undergoes normal development. Abnormal pollen grains develop due to arrested cytokinesis after meiosis. The pollen grains are shed at the two-celled stage, and the generative cell divides in the pollen tube.

The ovule is hemianatropous, bitegmic and crassinucellate (Figs. 10.5.8A, C, G, 10.5.9A, B, 10.5.10A–C). The outer integument remains short and the inner alone forms the micropyle. The apical cells of the nucellar epidermis divide and produce a conical cap. The development of the embryo sac is of the Polygonum type, and shows usual organization. The polar nuclei fuse before fertilization.

Grover (1969) reports that both Polygonum and Allium type of embryo sacs develop in *F. carica*. The Allium type of development needs confirmation.

The most interesting part of the study of figs is the pollination, concurrent development of the sycophilous chalcid wasp, emergence of the pollen-bearing wasp, and its entrance into a second generation young syconium (Fig. 10.5.1).

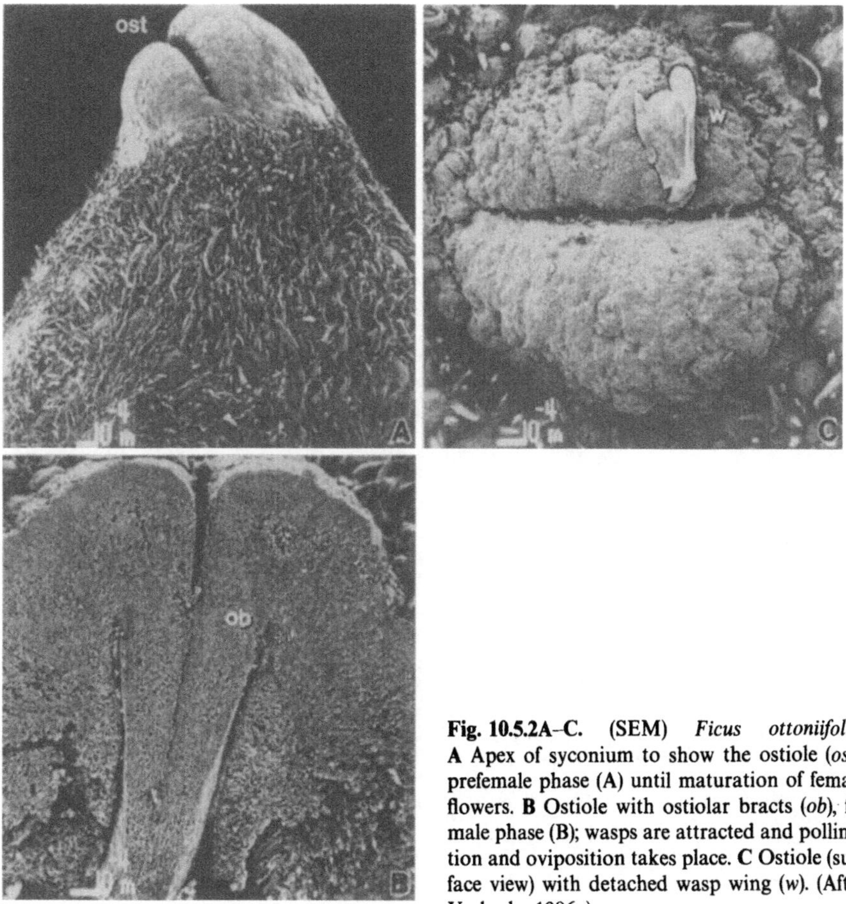

Fig. 10.5.2A–C. (SEM) *Ficus ottoniifolia.* **A** Apex of syconium to show the ostiole (*ost*), prefemale phase (A) until maturation of female flowers. **B** Ostiole with ostiolar bracts (*ob*), female phase (B); wasps are attracted and pollination and oviposition takes place. **C** Ostiole (surface view) with detached wasp wing (*w*). (After Verkerke 1986a)

B.M. Johri and Konar (1956), Galil and Eisikowitch (1968a, b), Chopra and Kaur (1969), Grover (1969), Grover (née Kaur) and Chopra (1971), Galil (1977) and Verkerke (1986a, 1987a, b) have studied these aspects (see also B.M. Johri 1989). Their findings are summarized.

It was long believed that the pollen vector (agaonid wasp *Blastophaga psenes* in *Ficus carica*) effected pollination passively by rubbing against the open anthers on its way out of the old syconium and touching the receptive stigmas on entering the new one. It has now been discovered that pollination is not always as simple as that. Several fig pollinators have specific organs which serve to transport pollen from old to young syconia. The loading and unloading through these organs entails very characteristic behavioural patterns not encountered in other pollination syndromes.

In *Ficus sycomorus* and *F. religiosa* (Galil 1977), and many other species, the number of male flowers is comparatively small and the wasps do not

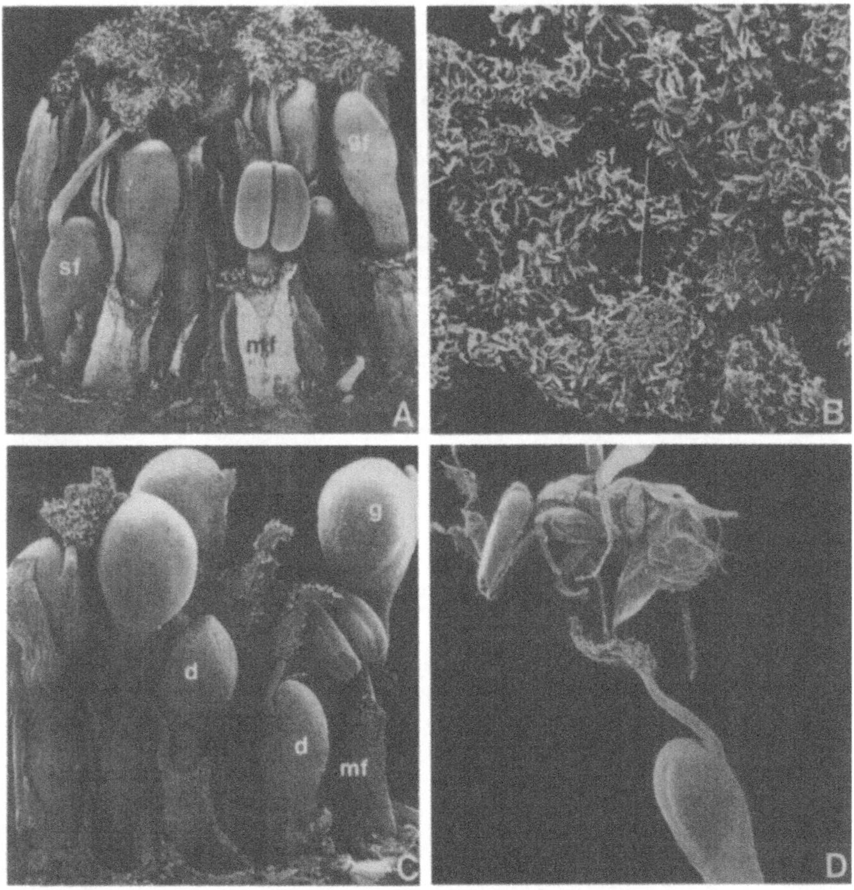

Fig. 10.5.3A–D. (SEM) *Ficus ottoniifolia*, flowers at the time of fertilization, phase B; perianth partly removed in **A**, **C**, **D**. **A** Syconium wall lined with seed (*sf*), gall (*gf*) and male flowers (*mf*). Note the synstigma. **B** Synstigma (surface view), note crushed stigmatic papillae of gall flower (*arrow*); *sf* seed flower stigma. **C** Syconial wall lined with male and gall (*g*) flowers, and drupes (*d*). **D** Gall flower, stigma penetrated by ovipositor of the wasp (*Courtella gabonensis*). (After Verkerke 1986a)

become dusted with pollen on their way out. Sporadically, many pollen grains have been observed on their bodies as they leave the fig. Many of the fig wasps [*Ceratosolen arabicus* (Figs. 10.5.6A, 10.5.7C), pollinator of *F. sycomorus*, and *Blastophaga quadraticeps* (Figs. 10.5.6B, C, 10.5.7A, B) of *F. religiosa*] have thoracic pollen pockets closed from the outside by a movable lid. Many other wasp species of the *Urostigma* section have, in addition to open pockets, coxal corbiculae in the form of longitudinal depressions along the coxae (proximal joints) of the forelegs, which also serve to transport pollen. The corbiculae are fringed on one side by a row of long and stiff hairs (comb) which aid in the loading and unloading of pockets.

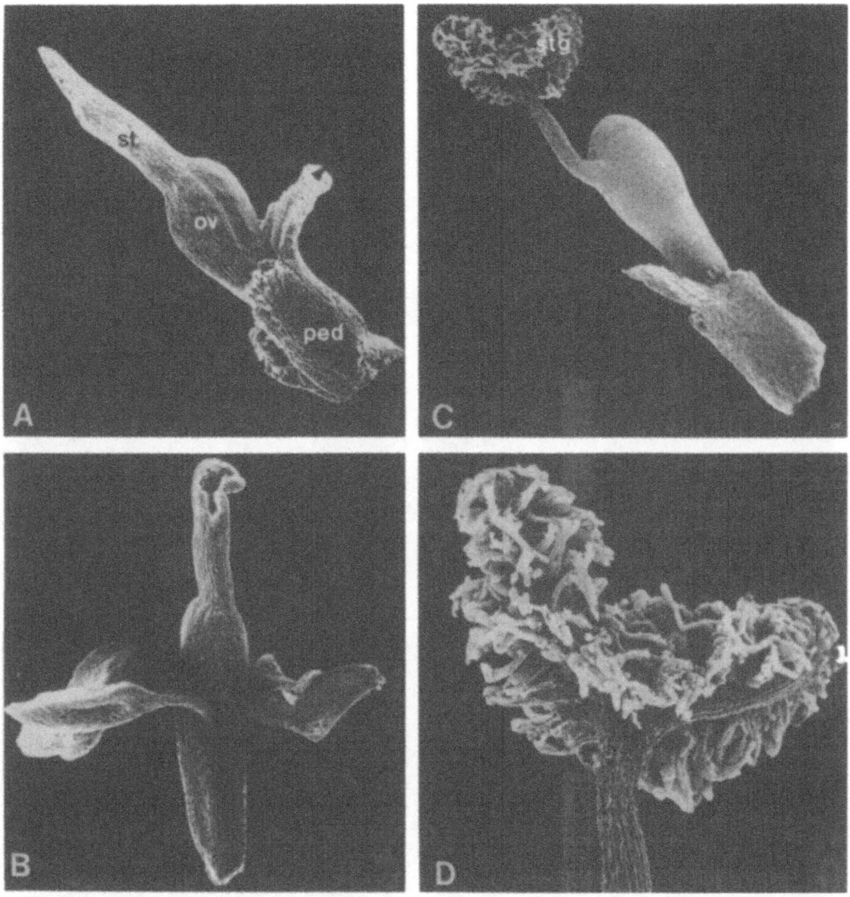

Fig. 10.5.4A–D. (SEM) *Ficus ottoniifolia*, prefertilization development of gall flower, perianth partly removed. **A, B** Prefemale phase until maturation of female flower. Note the pedicel (*ped*), ovary (*ov*) and style (*st*). **C** Female phase, wasps are attracted and pollination and oviposition takes place; *stg* stigma, **D** Stigma from **C**. (After Verkerke 1986a)

Before leaving the fig, the female wasp approaches the anthers and with the aid of the arolia (lobes or pads) of the forelegs, raises pollen into its pockets which open in response to touch. After entering the receptive fig (Fig. 10.5.6D), the above movements are repeated in reverse order at the end of each act of oviposition. Pollination proper is effected by striking the tarsi of forelegs one against the other, or by delicate touching of the stigmas.

Not all agaonid wasps have pollen pockets or corbiculae. The pollen vector of the common fig, like those of a whole group of species belonging to the subgenus *Ficus*, has no specific pollen containers. By way of compensation, however, the plants concerned produce an abundance of pollen which ensures passive coating of the wasps with pollen as they pass between the numerous stamens.

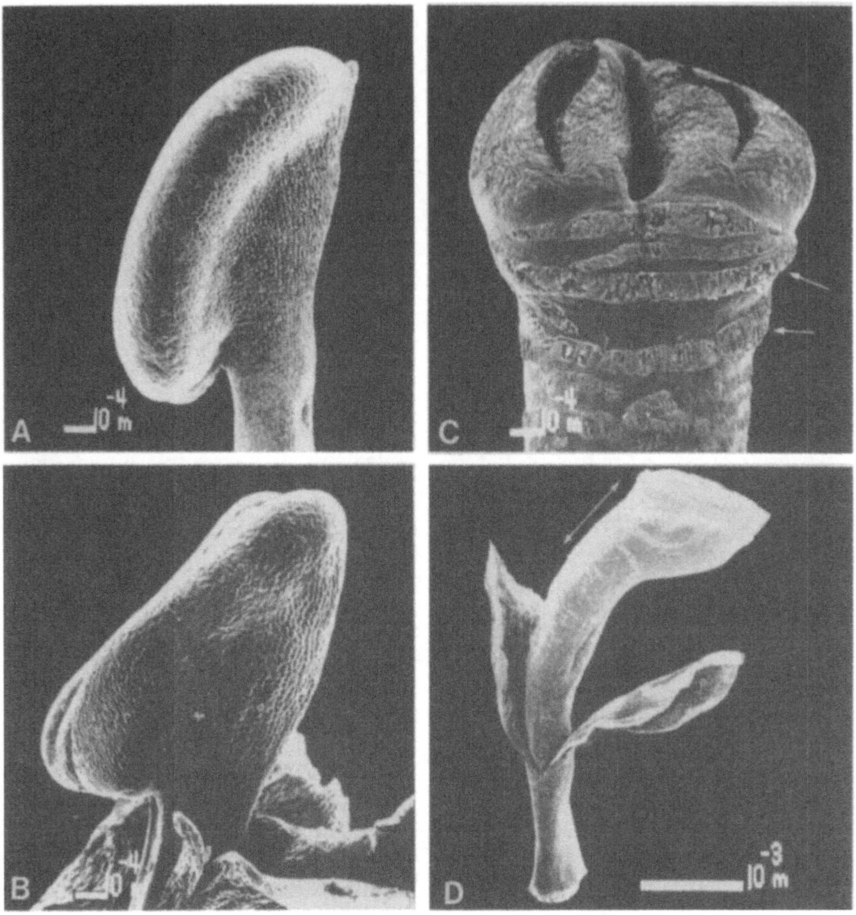

Fig. 10.5.5A–D. (SEM) *Ficus ottoniifolia*, development of male flower, perianth partly removed; **A, B, D** lateral view, **C** ventral view. **A** Anther oblique, phase A–C. **B** Apical thickening of filament initiates tilting of anther, early phase D. **C, D** Ventral filament elongation (*double arrow*) causes both tilting of anther and opening of stomium (**C**), and tanniferous epidermal cells become disrupted (*arrow*). (After Verkerke 1986a)

Blastophaga psenes, pollinator or *F. carica*, is an exception since the pollen is stored in the intersegmental and pleural invaginations on the body of the wasp.

In *F. religiosa*, the syconia are inhabited exclusively by *Blastophaga quadraticeps* (Agaonidea, Agaoninae). Galil and Eisikowitch (1969) mention five developmental phases of the syconia (Fig. 10.5.1A–E).

The female flowers mature at phase B and vary in style length – some sessile, or almost sessile with a comparatively long style, or the flowers are pedicelled with short styles. The long curved stigmas are situated approximately at the same distance from the receptacular wall. They are interlaced and

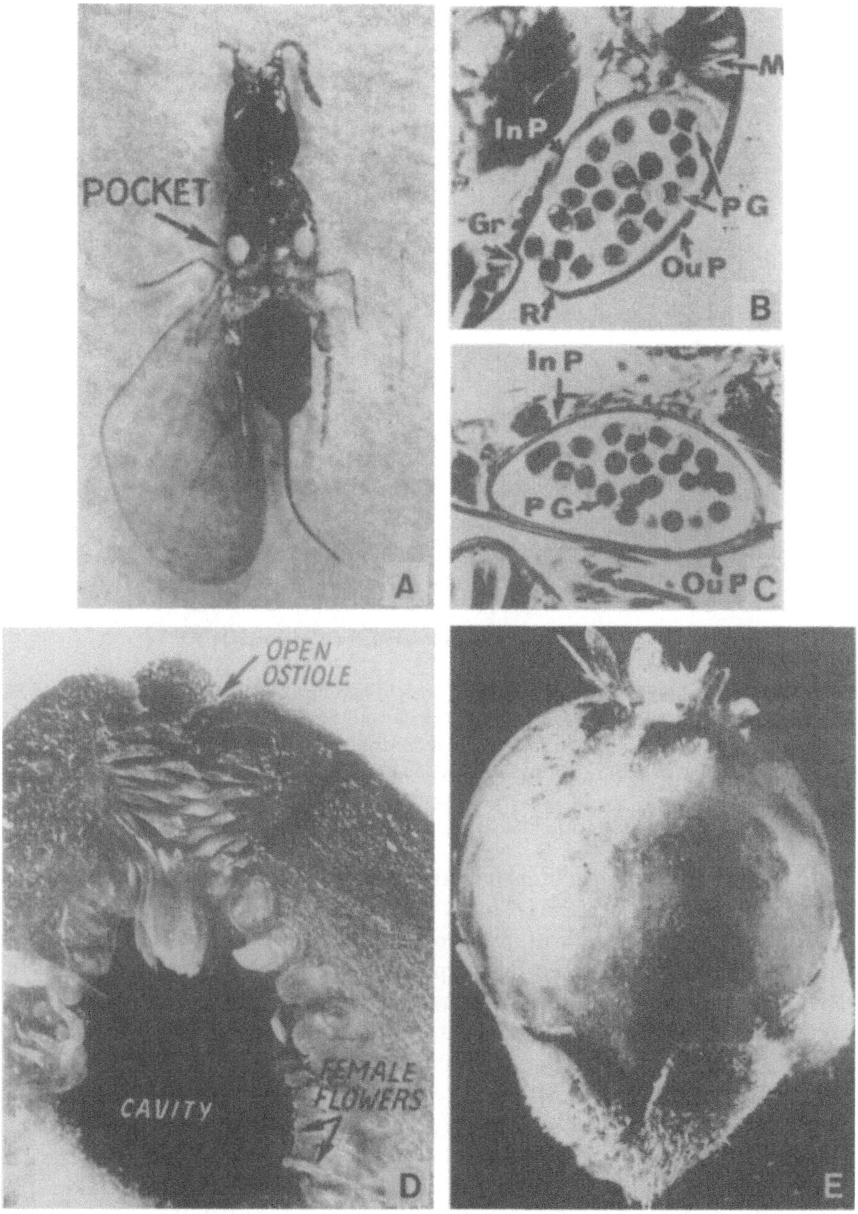

Fig. 10.5.6. **A** *Ceratosolen arabicus*, ventral view of female to show thoracic pollen-pocket (*arrow*). **B, C** *Blastophaga quadraticeps*. **B** Transection through slightly open pollen-pocket close to metathorax muscles (*M*) of insect. It contains pollen grains (*PG*), and has an opening (*R*) near groove (*Gr*) bound by inner (*In P*) and outer plate (*Ou P*); **C** Longisection of closed pollen-pocket containing pollen grains (*PG*), and bound by inner (*In P*) and outer plate (*Ou P*). **D** *Ficus* sp., longisection through receptive syconium, female-phase with open ostiole and female flowers lining the central cavity. **E** *F. religiosa*, syconium with a tuft of detached wings (at the ostiole) of *B. quadraticeps*. (After Galil 1977)

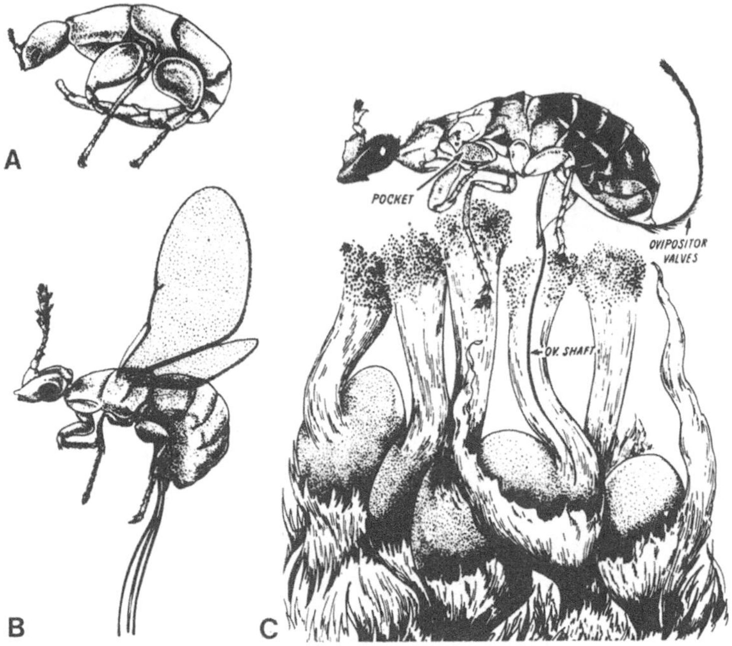

Fig. 10.5.7A, B. *Blastophaga quadraticeps*, male (**A**) and female (**B**), specific pollinators of *Ficus religiosa*. **C** *Ceratosolen arabicus*, wingless female with broken antennae, ovipositing the pistils of female flowers during pollination (pollen is contained in a pocket), by inserting ovarian shaft (*ov. SHAFT*). The ovipositor valves facilitate the process. (After Galil 1977)

stuck together forming a cup-shaped chamber connected by the styles with the ovaries below (Fig. 10.5.3A). This is the compound "synstigma" (Fig. 10.5.3B). It prevents the wasps within from slipping down between the styles and ovaries. Consequently, the fig-wasps trapped in the stigmatic chamber can oviposit only through the synstigma and styles.

At phase D, the synstigma shrinks and withers so that the way from the ovaries into the cavity of the syconium opens. The male flowers mature and the stamens (one per flower) project into the cavity. Most anthers dehisce (Fig. 10.5.5C) by narrow slits (some do not open at all) but the pollen does not get released.

The male wasps now emerge from the galls. Copulation takes place through a hole in the gall (while the female wasp is still within). On emerging from the gall, the female does not leave the syconium immediately. After stretching her body and wings, she approaches male flowers pushing her head into the open anther, or even into a closed anther, thereby causing it to open. Thus, the female becomes coated with pollen. The wasp even eats some pollen since pollen grains have been detected in its alimentary canal.

Now the wasp leaves the syconium through the ostiole. Its scales loosen and the wasp crawls through the slits. If the scales remain adpressed, one to four exit holes are made by the males, in the zone of the ostiolar scales. The syconium wall is thin and the male wasps bite their way through the wall. Only afterwards do the females appear in the outlet and fly away. The wasp-galls are now perforated and easily distinguishable from the fertile ovaries containing seeds.

The ripe anthers "attract" the young female wasps immediately after they emerge from the gall. After once visiting the anther, the female wasp would not be attracted again. The attractant is not known. The male wasps have no role in pollination and are not attracted to the anthers.

The pollen-loaded females, on escape from the older syconia, are attracted to the young second generation syconia (Fig. 10.5.1). At this time the ostiolar scales somewhat loosen, and the female wasp forces its way into the syconium (Fig. 10.5.6D, E). During this process, it looses the wings and flagellae of the antennae (Fig. 10.5.7C). Generally, only one wasp enters the syconium; more than one wasp may also enter. In some figs there is a one-to-one relationship, i.e. only the wasp of a particular species is associated with each fig species. In other fig species, more than one species of wasps may be concerned.

Once having succeeded in entering the syconial cavity, she wanders about on the synstigma and pollination occurs. The wasp stumbles and falls down between attempts to oviposit (Fig. 10.5.3D). The hypogonium guides the ovipositor into the stylar canal of the gall flower (Fig. 10.5.7C). When the ovipositor reaches the curved part of the stylar canal, it forces its way between the raphal strand and nucellus. The ovipositor punctures the tissues. Following "recognition", an egg is deposited between the cytoplasm-rich cells at the junction of the integuments and the nucellus (Figs. 10.5.8A, 10.5.9A). Near the wasp egg there are some drops of a secreted fluid. On withdrawal of the ovipositor, the injured tissues become restored to their original condition, due to growth.

The pollen germinates on the stigmatic papillae (Fig. 10.5.8C, D). The tube grows through the style, the generative cell divides to form two male gametes, and double fertilization is brought about as usual (see Chopra and Kaur 1969, Verkerke 1987b). The ovule may not have been fertilized at the time of oviposition. After oviposition the synstigma and style turn brown, and the stigma falls apart opening up the central syconial cavity.

Usually, gall and some seed-flowers are oviposited but both receive pollen tubes and double fertilization occurs. Oviposition and fertilization are followed by rapid growth of the ovaries and the syconium.

Galil and Eisikowitch (1968a) have experimentally demonstrated that the pollen-loaded wasps with amputated ovipositor do bring about pollination, but the wasp-egg cannot be deposited. Therefore, purely seed syconia were obtained. It has been concluded that the growth of syconium is dependent on pollination and not on the oviposition of the wasp-egg.

In both gall and seed flowers, the endosperm and embryo develop, and in gall flowers the wasp larva grows concurrently (Fig. 10.5.8A–M). The larva migrates from the nucellus into the embryo sac (Figs. 10.5.8G–M, 10.5.9B) causing abortion of the young embryo. The endosperm continues to develop and the free-nuclear endosperm becomes cellular (Figs. 10.5.8E–M, 10.5.9E,

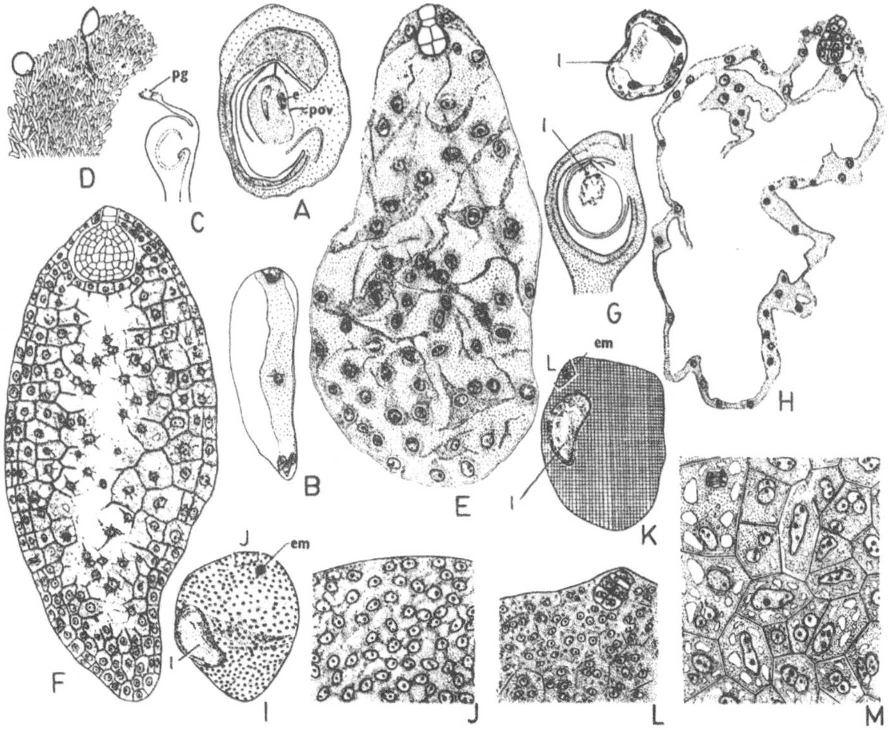

Fig. 10.5.8A–M. *Ficus religiosa.* **C–E** dissected wholemounts, rest sections; **E, F** from seed flowers, others from gall flowers. **A** Longisection ovary of gall flower after oviposition (*pov* path of oviposition) and deposition of egg (*e*) of *Blastophaga quadraticeps*, between nucellus and inner integument (diagrammatic). **B** Embryo sac (from **A**) with egg (synergids not seen), secondary nucleus and antipodals. **C** Short-styled pollinated (*pg*, pollen grains) female flower. **D** Germinated pollen grain on stigma, from **C**. **E** Free-nuclear endosperm and young proembryo, from long-styled flower. **F** Later stage, endosperm partly cellular and late globular proembryo. **G** Gall flower, longisection of ovary with larva (*l*) of wasp. **H** Larva, free-nuclear endosperm and globular proembryo from **G**. **I** Free-nuclear endosperm, proembryo (*black*) and larva. **J** *Inset* from **I**, free-nuclear endosperm. **K** Advanced stage, cellular endosperm, proembryo (*em black*) and larva (diagrammatic). **L** *Inset* from **K**, note wall formation in endosperm and proembryo. **M** Later stage, several cells of the endosperm are multinucleate and others have polyploid nuclei of irregular shape. (After B.M. Johri and Konar 1956)

10.5.10B, G, H). The larva consumes the endosperm gradually and almost none of it is left during metamorphosis of the insect. Most of the ovular tissue is resorbed; remnants of nucellus and outer epidermis of the outer integument persist. Ultimately, the wasp fills the gall almost completely.

In *F. religiosa* (B.M. Johri and Konar 1956), the endosperm nuclei do not divide synchronously in seed flowers. About 60 free nuclei are present at the quadrant proembryo. Centripetal walls are laid down at the early globular proembryo. The endosperm cells are usually uninucleate, occasionally binucleate and the nuclei may fuse. The peripheral cells elongate tangentially and divide periclinally recalling cambial activity (Fig. 10.5.8F). At the heart-shaped embryo, the cellular endosperm fills the embryo sac.

In gall flowers, the earlier stages of endosperm development are comparable to those in seed flowers. In Figure 10.5.8G, H, the wasp larva has digested the nucellus and lies close to the embryo sac containing the free-nuclear endosperm and quadrant proembryo. Later stages are shown in Figure 10.5.8I–L. At the octant proembryo, cleavage furrows followed by wall formation produce multinucleate vacuolated cells; the nuclei fuse forming irregular polyploid masses (Fig. 10.5.8L, M). Further nuclear divisions may or may not be accompanied by wall formation. As the larva grows, it encroaches on the endosperm. Finally, the proembryo aborts, and the entire endosperm is consumed so that the drupelet contains only the wasp.

The development of embryo in a seed flower is quite normal, including tricotyly which is also known in other plants, such as *Magnolia grandiflora*, and *Opuntia dillenii* (P. Maheshwari and Chopra 1955). This is the normal condition in *Degeneria vitiensis* (Swamy 1949a).

In gall flowers, the growth of the proembryo beyond the octant stage is arrested, but the endosperm continues to develop. Obviously, it is not for lack of nutrition that the embryo fails to develop, but the *Blastophaga* may secrete some hormone/enzyme which inhibits the growth of the embryo.

Verkerke (1986a) has studied the fig–fig wasp symbiosis in *Ficus ottoniifolia* for which the pollinator is *Courtella gabonensis* (Fig. 10.5.3D) or *C. camerunensis*. The pollen reaches maturity only 4 days after the organization of the Polygonum type of embryo sac. It remains enclosed in the anther for 6 weeks. Although not mentioned, the large number of male flowers indicates that the female wasp gets dusted with pollen while forcing its way out through the ostiole. Prefertilization gall flowers are shown in Figure 10.5.4A–D. In galls the enlarged cells of the outer epidermis of the pericarp and the collapsed layers of the underlying tissues persist at maturity (Fig. 10.5.9A–F). In "seeds" the pericarp differentiates into epicarp, mesocarp and sclerotic endocarp; a few layers of outer integument persist, while the inner integument collapses (Fig. 10.5.10A–H). In main features, the development of endosperm, embryo, seeds/fruit and syconium is similar to that in *F. religiosa* (B.M. Johri and Konar 1956).

Finally, the wasp exits from the gall (Fig. 10.5.1).

The gynodioecious *F. aspergifolia* (Verkerke 1987a) has (a) figs which contain both male and female gall flowers, and (b) figs with female flowers which produce only seeds. The ovules in the two types of female flowers differ in relation to the different functions, and the author uses the term "ovule dimorphism".

In *F. religiosa* (B.M. Johri and Konar 1956), in the drupelet the seed shows an exotesta (outer epidermis of outer integument), a few collapsed layers of inner integument and a conspicuous endotegmen, and cellular endosperm with reserve food. The curved embryo has only a few layers of endosperm around it except for a broad zone on the funicular side. The pericarp can be distinguished into a few layers of epicarp, a single-layered thick-walled mesocarp, and

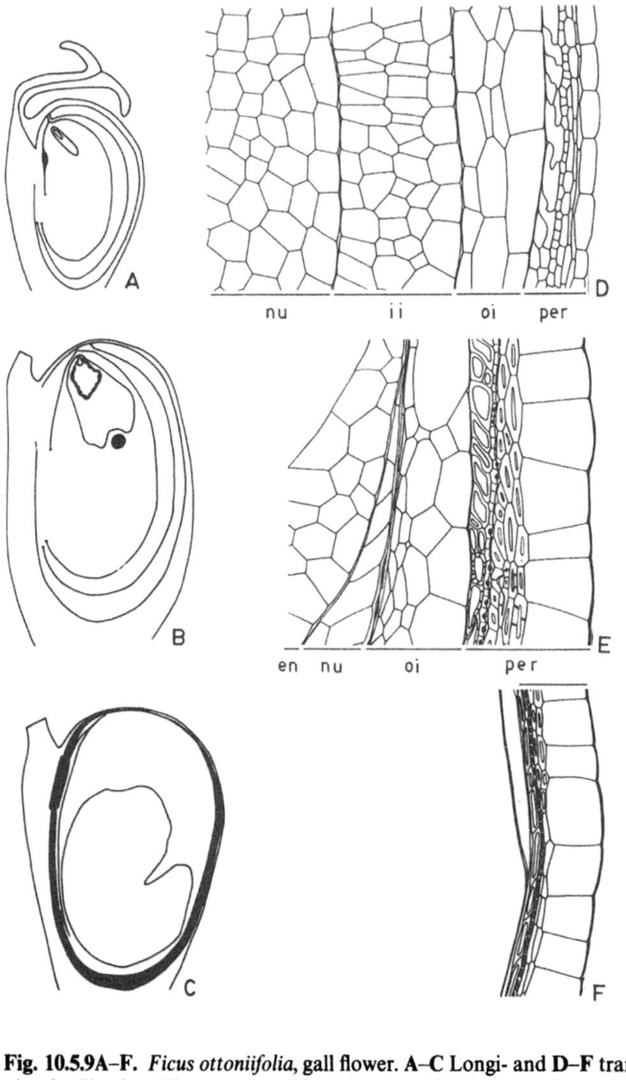

Fig. 10.5.9A–F. *Ficus ottoniifolia*, gall flower. **A–C** Longi- and **D–F** transections. **A** Carpel, 3 days after fertilization. The wasp egg (*black*) has been laid between the cytoplasm-rich raphal cells and the nucellus. Note asymmetric growth of nucellus. **B** Ovary, 9 days after fertilization. The embryo sac contains a globular proembryo and free-nuclear endosperm. The larva (*black*) lies below the embryo sac. **C** Ovary, 30 days after fertilization. **D** Pericarp (*per*), outer (*oi*) and inner (*ii*) integuments, and nucellus (*nuc*) from **A**. **E** The inner layers of cells of pericarp have become sclerified (*black* region in **C**), the inner integument has collapsed, and nucellus is followed by cellular endosperm (*end*), from lateral region of **C**. **F** Except for a part of the pericarp, all the other tissues have collapsed, from apical region of **C**. (After Verkerke 1986a)

a sclerotic endocarp. The wall of the syconium is parenchymatous with a few sclereids in the outer region. It becomes pulpy in a ripe fig.

The ripe syconia with seeds (protected by the stony endocarp) are eaten by birds. The seeds are not damaged at all while passing through the alimentary

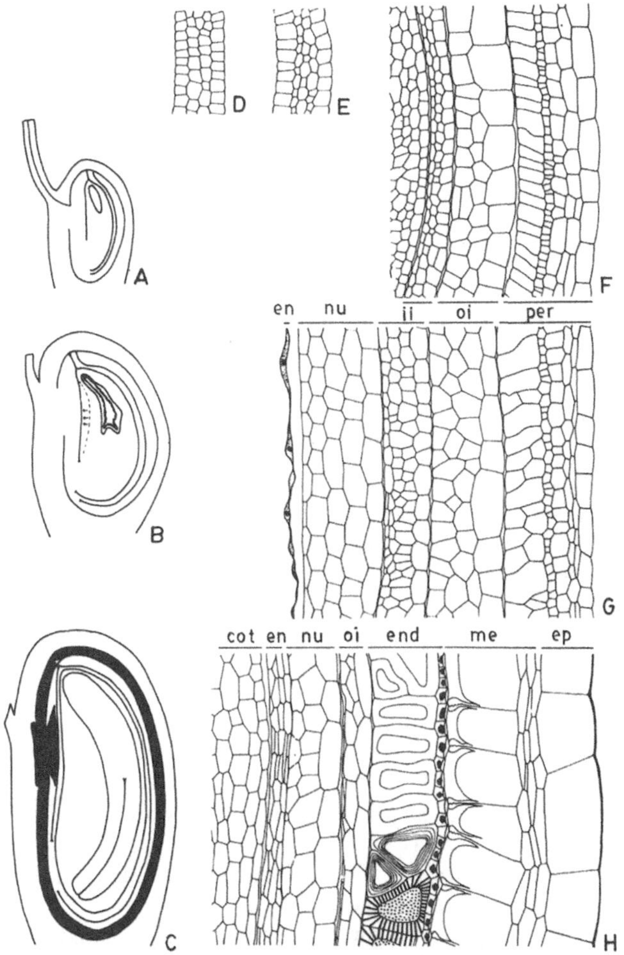

Fig. 10.5.10A–H. *Ficus ottoniifolia*, seed flower. **A, B** Longisection of seed carpels. **A** At mature embryo sac when pollination occurs (phase B). **B** 3 days after fertilization when the embryo sac contains globular proembryo and free-nuclear endosperm; the nucellar epidermal cells elongate (*arrows*). **C** Longisection of drupe with dicotyledonous embryo (end of phase C). **D–H** Cross-sections. **D, E** Pre-fertilization ovary wall. **F** Pericarp (*per*), outer (*oi*) and inner (*ii*) integuments and part of nucellus from **A**. **G** *Per, oi, ii*, nucellus (*nu*) and free-nuclear endosperm (*en*), from **B**. **H** Epicarp (*ep*), mesocarp (*me*) and endocarp (*end*) followed by *oi, nu, en* and cotyledonary (cot) tissue, from **C**. (After Verkerke 1986a)

canal, and retain 100% viability. This also accounts for the widespread distribution of *Ficus*.

In *F. sycamorus* (see Galil 1977), which does not set seed in the Middle East, the developing larvae of *Sycophaga sycomori* exert a direct influence on the development of the syconium. The occupied syconia do not drop and also resist premature ripening, which enables the contained wasp brood to complete its developmental cycle.

After impregnation of the females in the gall, the male wasps tunnel through the syconial wall. As carbon dioxide (10% in *F. religiosa*) escapes, and the internal atmosphere equilibrates with the external, the females widen the fertilization holes and emerge from the galls. Generally, high carbon dioxide content (Fig. 10.5.1, phase D) in the internal atmosphere of fruits inhibits ripening, whereas its depletion promotes ripening (see Galil 1977).

The relationship between the sycophilic wasp, developing in the ovary, and the syconium is not merely a dietary one, for the insect becomes involved in certain vital processes of the fig. This is an instance of coevolution where both – the fig and the wasp – have become highly specialized. This mutualistic symbiosis must have become established in ancient times (see Murray 1985).

The fig–fig-wasp biology has been fully investigated but there are still many unanswered questions which deserve attention.

M Anantaswamy Rau (1942), I Banerji and A Hakim (1954), RN Chopra and H Kaur (1969), EJH Corner (1976), BC Das (1961), RM Datta (1941), GL Davis (1966), NK Dwivedi (1990), J Galil (1977), J Galil and D Esikowitch (1968a, b, 1969), J Galil and Y Snitzer-Pasternak (1970), H Grover (1969), H Grover (née Kaur) and RN Chopra (1971), BM Johri (1989), BM Johri and RN Konar (1955, 1956), H (Harjinder) Kaur (1970a), S Kumar (1978), P Maheshwari and RN Chopra (1955), J Modilewski (1908a, b), HY Mohan Ram and R Nath (1964), MG Murray (1985), CR Narayanan (1968), G Neeman and J Galil (1978), D Padmanabhan (1961b), M Ram (1960), SP Singh (1954), BL Srivastava and S Tandon (1985), AM Subba Rao (1940a), BGL Swamy (1949a), O Tippo (1938), W Verkerke (1986a, 1987a, b, 1990), RC Verma, A Sarkar and BC Das (1986)

10.6 Urticaceae

Simultaneous cytokinesis accompanies meiosis in microspore mother cells. The tetrads are tetrahedral and decussate.

The male gametes are rod-shaped in *Urtica* (Strasburger 1910). The pollen grains are two-celled at the dispersal stage.

The ovule is basal, orthotropous, bitegmic and crassinucellar with the inner integument forming the micropyle. By periclinal divisions the cells of the nucellar epidermis add to the parietal tissue. The archesporial cell cuts off a parietal cell. The parietal tissue is well developed. The megaspore mother cell undergoes meiotic divisions. The chalazal megaspore of a linear tetrad develops into a Polygonum type of embryo sac. The polar nuclei fuse before fertilization. The antipodal cells multiply in *Urtica.* A hypostase is conspicuous. An obturator develops from the outer integument in *Elatostema machaerophyllum*, and from the inner epidermis of the ovary wall in *E. pendunculosum* (Fagerlind 1944b).

The development of endosperm is of the Nuclear type, which later becomes cellular. The chalazal region of endosperm develops into a short haustorium in *Urtica cannabina* and *U. pilulifera.* It is richly cytoplasmic with hypertrophied nuclei (see Schnarf 1931).

The embryogeny conforms to the Asterad type. Supernumerary embryos are produced from synergids, antipodal cells and nucellus. Sometimes, multiple embryo sacs fuse followed by polyembryony. If the developing embryos lie close together, they may fuse. Unreduced parthenogenesis occurs in apomictic species of *Elatostema* (Fagerlind 1944b).

Somatic apospory of Hieracium type occurs in *E. acuminatum* and *E. penninerve*; aneuspory of Taraxacum type in *E. eurynchum* (2% embryo sacs), *E. machaerophyllum* (88% embryo sacs) and *E. peltifolium* (Fagerlind 1944b). Gonial apospory has also been noted in *E. eurynchum* (98% embryo sacs), *E. machaerophyllum*, *E. rostratum* and *E. sinuatum* (Fagerlind 1944b).

The testa and tegmen become crushed, and there is no definite seed-coat.

G Bentham (1880), F Fagerlind (1944b), GG Franchi and E Pacini (1980), FM Jarrett (1976), N Prakash, LR Bohm and J Maze (1979), K Schnarf (1931), M Sharma (1964), R Souèges (1920c, 1921), E Strasburger (1910)

10.7 Taxonomic Considerations

The order Urticales is considered as phylogenetically advanced, and as the terminal taxon of an evolutionary line from which Fagales have also been derived (Lawrence 1951). Engler and Diels (1936) include Ulmaceae, Rhoiptel-eaceae, Moraceae and Urticaceae in Urticales. Eucommiaceae has been placed in Rosales by Harms (1930), near Hamamelidaceae. Tippo (1940) and Wettstein (1935) assign it to Urticales. Hutchinson (1973) includes all the families in Urticales, except Rhoipteleaceae which has been placed in Juglan-dales. He assigned the genus *Cannabis* to a separate family, Cannabaceae (Cannabinaceae). Bentham and Hooker (1965c) place it in Urticaceae, whereas Engler and Prantl (1931) and Melchior (1964) assign it to Moraceae.

Ulmaceae. The development of embryo sac of the tetrasporic Adoxa type or of an Intermediate type between Adoxa and Drusa marks out the family Ulma-ceae from the rest. The persistent remnants of the pollen tube, Onagrad or Solanad type of embryogeny, and exalbuminous seed with curved embryo and folded cotyledons are additional features in which the family Ulmaceae differs from the other members. Otherwise, secondary multiplication of anti-podals, Nuclear endosperm, and polyembryony occur in Moraceae and Urticaceae. Rest of the features are in general conformity with those of the order.

Eucommiaceae. Ulmaceae and Eucommiaceae have unilocular, bicarpellary ovary with axile placentation, a single anatropous, pendulous ovule with nucellar cap or beak, and samaroid fruit, but the seed is exalbuminous in Ulmaceae. In addition, Eucommiaceae show several noteworthy features. The fibrous thickenings in the endothecium develop extensively. The thickenings are also present in the region of the connective, and around the staminal vascular bundle. The unitegmic ovule, a rare occurrence of intravesicular chalazogamy, branched pollen tubes, Cellular endosperm, and a long resting period of zygote for about 6 weeks distinguish the family and makes its assignment to Urticales anomalous. Unitegmic ovules with multicellular archesporium and intravesicular chalazogamy occur in Fagales (Betulaceae –

Corylus), and there is a long resting period of zygote in Juglandales (Juglandaceae – *Carya illinensis*).

Eucommiaceae are considered to be advanced derivatives of Hamamelidales, or of stocks ancestral to them. The reduced flowers, presence of secretory ducts, Cellular endosperm, and samaroid fruit of *Eucommia* adapted to wind pollination are some of the features allied to Hamamelidaceae.

C.C. Berg (1977) considers the position of Eucommiaceae in Urticales doubtful. Cronquist (1968, 1981) and R. Dahlgren (1975a, 1981) erect a separate order, Eucommiales. Due to the presence of iridoid substances, R. Dahlgren (1975a) suggests its position near Cornales (see also Jensen et al. 1975).

Moraceae, Urticaceae. Moraceae and Urticaceae show a close alliance. The two-celled pollen grains, hemianatropous bitegmic ovules with prominent nucellar cap, Polygonum type of embryo sac, secondary multiplication of antipodals, Nuclear endosperm with chalazal haustorium, Asterad type of embryogeny, nucellar polyembryony, parthenocarpy, and albuminous seeds are the similarities. But the occasional formation of an Allium type of embryo sac, unusual type of pollination and seed development in *Ficus*, tricotyly, cleavage polyembryony, and seed with curved embryo and a definite seed-coat distinguish Moraceae. In Urticaceae the embryo is straight, and a definite seed-coat is absent.

The presence of unreduced parthenogenesis, antipodal, synergid and nucellar polyembryony, polyembryony due to multiple embryo sacs, gonial and somatic apospory in *Elatostemma* spp. are outstanding features comparable to those of *Alnus* (Betulaceae – Fagales).

Mohan Ram and Nath (1964) made a detailed study of *Cannabis sativa*, and support its assignment to an independent family Cannabinaceae, erected by Rendle (1959) and Hutchinson (Cannabaceae) (1973). The family segregates itself from Moraceae in the (1) absence of milky latex, (2) resin in endothecial cells and connective of anther, (3) corky hypostase, (4) frequent development of Polygonum type of embryo sac from the epichalazal megaspore, (5) tubular chalazal endosperm haustorium, and (6) enucleate vesicles in the nuclear endosperm. The position of Cannabinaceae is rather isolated. Anatomical and cytological data (Tippo 1938, Raven 1975) also support this view.

Cronquist (1968, 1981) and Takhtajan (1969, 1980) place the order Urticales near Hamamelidales, whereas Thorne (1968), Stebbins (1974a), R. Dahlgren (1975a, 1981) and C.C. Berg (1977) place it close to Malvales. The phytochemical data (Meeuse 1975) as well as the morphology of inflorescence (Endress 1977a), fruit and seed do not support any relationship of Urticales with Hamamelidales.

11 Proteales

The order Proteales comprise a single family, Proteaceae.

11.1 General Features

The characteristic features of Proteales are: (1) secretory anther tapetum, (2) simultaneous cytokinesis of pollen mother cells, (3) bitegmic, crassinucellate ovules with well-developed parietal tissue, a nucellar cap, hypostase, and epistase, (4) Polygonum type of embryo sac with ephemeral antipodals, (5) porogamous mode of fertilization, (6) Nuclear endosperm, and (7) Asterad type of embryogeny.

The occurrence of varying degrees of development of the endosperm haustoria in Grevilloideae, and their total absence in Proteoideae is noteworthy. The nucellus becomes the source of nourishment in exceptional cases, as in *Cenarrhenes nitida*. In this species pseudo-embryo sacs also develop. The nucellus persists as perisperm (endosperm scanty) in *Helicia*, *Lambertia*, etc. The seed structure is elaborately developed. The seeds are winged in some members. Perichalazal growth occurs in *Helicia*, *Macadamia* and some other taxa.

11.2 Proteaceae

The anther is tetralocular. The hypodermal archesporial cells cut off primary sporogenous cells, and primary parietal cells. The anther wall (Fig. 11.2.1A) comprises an epidermis, fibrous endothecium, two or three ephemeral middle layers, and secretory tapetum. Uninucleate vacuolate tapetal cells occur in *Macadamia ternifolia* (Kausik 1938b), and binucleate in *Hakea* (Venkata Rao 1969a) and *Lambertia* (Venkata Rao 1968). The tapetum is two-layered (Fig. 11.2.1A) on the side of the connective in *Lomatia tinctoria* (Venkata Rao 1962). Following meiosis in microspore mother cells (Fig. 11.2.1B), simultaneous cytokinesis occurs by furrowing, and pollen tetrads are tetrahedral and decussate.

The pollen grains are two-celled when shed, and have two to three or seven apertures (poroid) (Fig. 11.2.1C–E).

The ovule is anatropous, hemianatropous, or orthotropous, bitegmic and crassinucellar. The inner integument grows before the outer one covering the nucellus and leaving a wide micropyle (Fig. 11.2.1F–P). The archesporium is multicellular except in *Hakea* (Venkata Rao 1969a) where it is one-celled. Only one cell is functional, and a parietal cell is cut off. The apex of the nucellus is glandular, and a nucellar cap develops in *Lomatia*. In the chalazal region of the embryo sac a zone of meristematic tissue develops above the pad of tannin-filled cells. Epistase and hypostase are present. Meiotic divisions occur in megaspore mother cell, and a linear, T-shaped or, rarely, inverted T-shaped tetrad is produced. The chalazal megaspore is functional. The female gametophyte (Fig. 11.2.2A) is of the Polygonum type in *Protea lepidocarpon* (Ballantine 1909). The three ephemeral antipodals are inconspicuous. The synergids are broad at the base, and hooked.

Venkata Rao (1967) reported two types of mature ovules. Those of *Cenarrhenes*, *Conospermum*, *Protea*, *Stirlingia* and *Synaphea* have a massive nucellus comprising large, thin-walled, highly vacuolate parenchyma, and

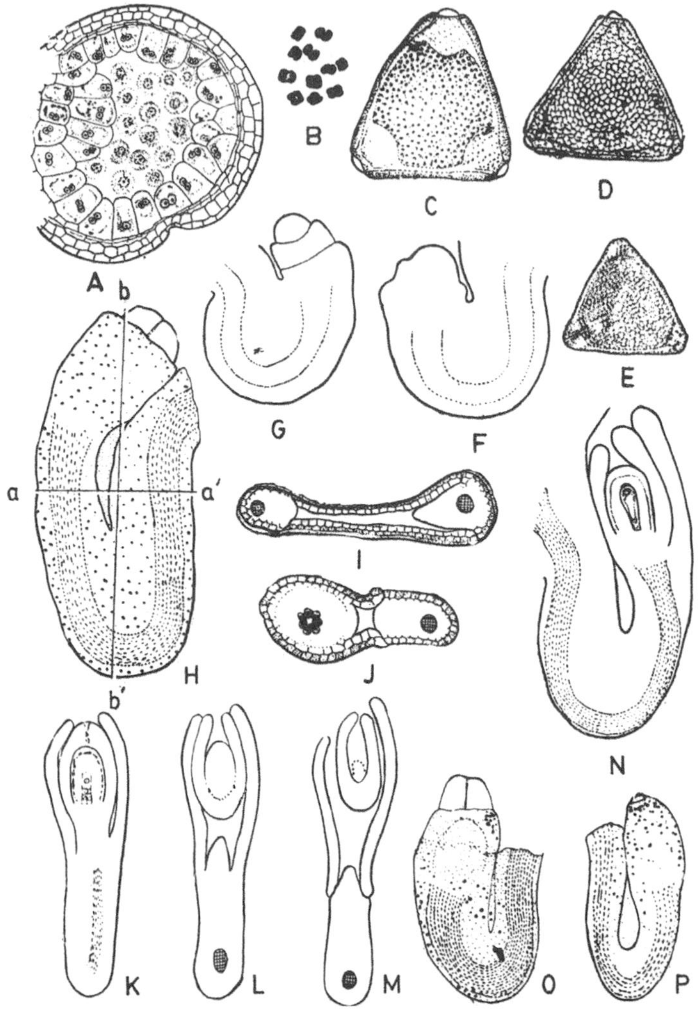

Fig. 11.2.1.A *Lomatia polymorpha*, transection of young anther lobe at microspore mother cell, the binucleate secretory tapetal cells are two-layered at places. **B** *L. tinctoria*, diakinesis showing 11 bivalents. **C–E** Poroid pollen grains of *L. salaifolia* (**C**), *L. polymorpha* (**D**) and *L. hirsuta* (**E**). The exine is ornamented. **F–M** *L. polymorpha*; **F–H** Stages in the development of ovule. **I, J** Cross-sections of ovule along the plane *a-a'* marked in **H**; **K–M** Longisections of ovule along plane *b-b'* marked in **H**. **N** *L. tinctoria*, median longisection of ovule cut parallel to the flat surface, the outer integument is absent on the funicular side. Note the conspicuous horseshoe-shaped growth of funicle in **H**, **K–N**. It develops into the seed-wing. **O, P** Ovule of *L. hirsuta* and *L. dentata* (**P**). (After Venkata Rao 1962)

small embryo sac. Starch grains are present in *Protea* and *Stirlingia*. In *Agastachys*, *Bellendena*, *Franklandia* (Venkata Rao 1965), *Persoonia*, and most of the members of Grevilleoideae, the ovules have scanty nucellus confined to the sides of the embryo sac which elongates and extends up to the zone of tannin-bearing cells in the chalaza. *Isopogon trilobus* shows a prominent parietal tissue.

A unique feature is described in the ovule of *Cenarrhenes nitida*, belonging to Persoonieae. Before fertilization, the central region of nucellus disorganizes resulting in a cavity which becomes filled with nutritive fluid. The chalazal end of the embryo sac comes in contact with the lower part of the cavity. This is somewhat akin to the pseudo-embryo sac of Podostemaceae, and the nucellus serves as a source of nutrition. The cells of the hypostase are absorbed.

Fertilization is porogamous. The cells of the glandular epistase help in directing and nourishing the pollen tube. In *Franklandia* the remnants of pollen tube persist up to the globular proembryo.

The development of endosperm is of the Nuclear type. Generally, free-nuclear divisions occur in the endosperm, and wall formation commences from the micropylar end (Fig. 11.2.2B–P). Cytokinesis occurs in varying degrees, in different genera. The endosperm in *C. nitida* is an exception. The embryo sac does not enlarge after fertilization; the free-nuclear endosperm is scanty, and probably does not become cellular. The nucellus persists as perisperm, and is the chief source of nutrition.

In *Bellendena* and *Persoonia* the elongation of the embryo sac (by digesting the nucellus) is slow. The entire endosperm becomes cellular, and haustorial activity has not been noticed.

Grevilleoideae show a rapid elongation of the embryo sac. The nucellus is absorbed.

In *Knightia* and *Orites revoluta*, there is no sharp demarcation between the upper cellular and lower free-nuclear region of endosperm. The entire endosperm becomes cellular, and is afterwards absorbed. The endosperm of *Brabejum* (Jordaan 1946), *Macadamia* (Kausik 1938b) and Branksieae is similar to that of *Knightia*, except that the nuclear region is haustorial.

In *Orites accularia* and *O. revoluta*, the nuclear region is narrow and swollen at the distal end. Secondary haustoria arise from the lower part of the cellular region. In *O. diversifolia*, the nuclear region is thrown into worm-like folds. It becomes partly or wholly cellular in mature seed.

The chalazal ñuclear region is haustorial in *Lambertia* (Venkata Rao 1968), whereas some secondary haustorial outgrowths are present on the sac-like distal end in *Stenocarpus*. Finger-shaped hypha-like outgrowths, all over the surface of endosperm in *Lomatia* (Venkata Rao 1962), are noteworthy (Fig. 11.2.3A–C). The chalazal nuclear part becomes coiled and appears like an appendage; ultimately it is absorbed.

In *Buchinghamia* several haustorial lobes arise from the cellular region, and increase the absorptive surface.

The endosperm is highly evolved in *Grevillea* (Kausik 1938c, 1939a, 1941a, 1942) and *Hakea* (Venkata Rao 1969a). In *Grevillea* the upper cellular region shows haustorial activity in some species, otherwise it is even in outline. The lower region develops into a coenocytic, haustorial, "vermiform" appendage (Fig. 11.2.2C, E, G, H, K, L). Streaming movement of protoplasm has been noticed in the coenocytic region (Kausik 1941a). According to Venkata Rao (1967), the nuclear region of the endosperm in *G. pinnatifida* forms multinucle-

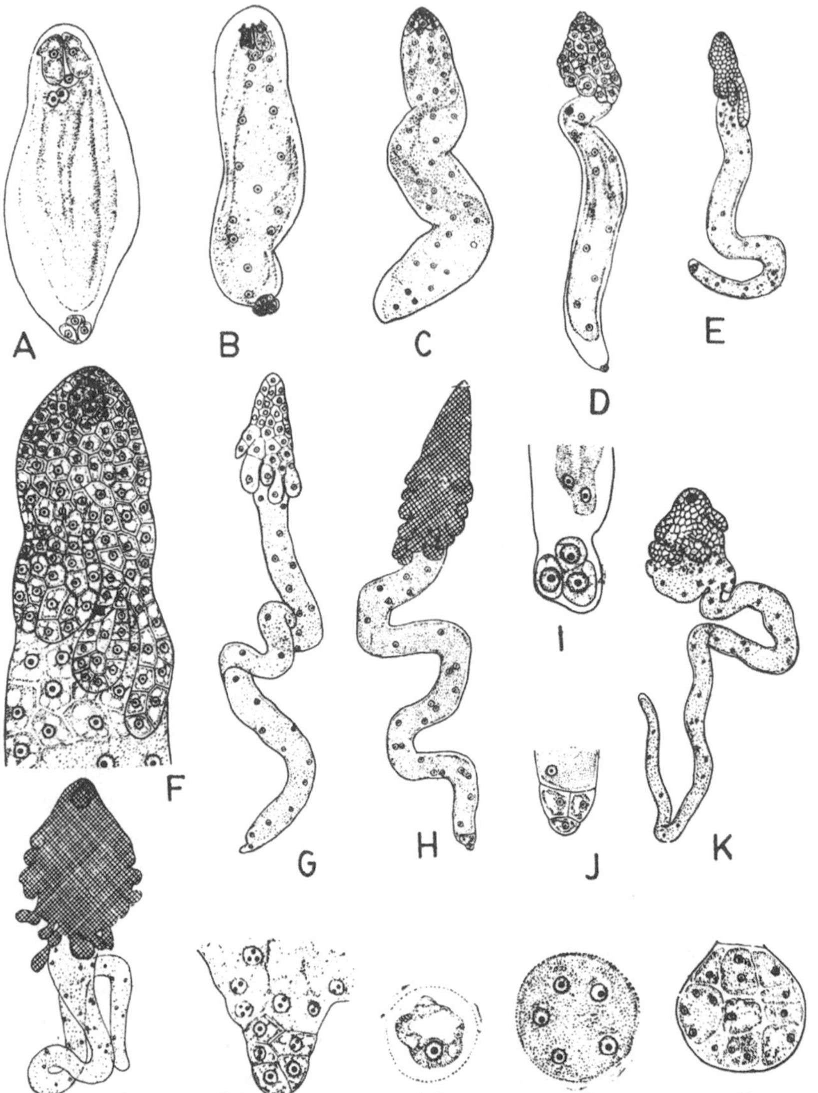

Fig. 11.2.2A–P. *Grevillea banksii.* **A** Mature embryo sac (from dissection). **B** Zygote with remnants of synergids and antipodals, and nuclear endosperm. **C, D** Endosperm showing cytokinesis from the micropylar end. **E–H** Cellular region of endosperm with secondary haustorial lobes at globular proembryo. Chalazal free-nuclear endosperm forms a "vermiform appendage". **I, J** Persistent antipodals. **K, L** Embryo sac with heart-shaped embryo, lobed cellular endosperm, and coiled free-nuclear chalazal appendage. **M** Secondary haustorial lobe. **N–P** Development of secondary haustorial lobe in cellular region. (After Venkata Rao 1966)

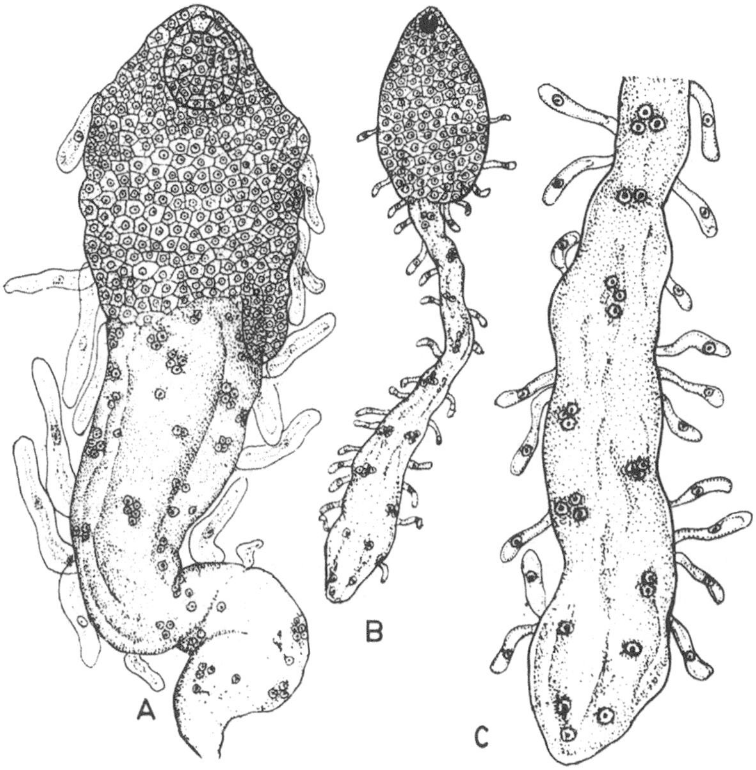

Fig. 11.2.3.A *Lomatia polymorpha*, endosperm (dissected wholemount) with coenocytic chalazal haustorium, uninucleate hairy projections occur on the surface of cellular and haustorial region. **B, C** *L. tinctoria.* **B** Same as **A**; **C** Coenocytic chalazal haustorium with one-celled projections on the surface. (After Venkata Rao 1962)

ate cells. In *G. alpina* and *G. bankskii*, finger-like and sac-like secondary haustoria grow at the base of the cellular endosperm (Fig. 11.2.2E–H, K, L), and later on the secondary haustoria coalesce. The secondary haustoria in *G. banksii* become coenocytic, but do not coalesce (Fig. 11.2.2M–P). A wide sac-like haustorium with secondary haustoria develops in *Hakea*. Later, this part becomes cellular and stores food.

A progressive evolution in the development of haustorium has been observed.

The embryogeny conforms to the Asterad type-Penaea variation (Johansen 1950) in *Franklandia* (Venkata Rao 1965) and *Lomatia* (Venkata Rao 1962). The mature embryo shows foliaceous cotyledons with prominent basal lobes.

The seed is small with a thin seed-coat, or is large with a massive seed-coat as in *Macadamia*, and is exalbuminous except in *Bellendena*. The structure of the seed-coat varies in different genera. The outer epidermis of the testa may consist of thin-walled, papillose cells, or cells with thickened, pitted outer walls as in *Grevillea*, or longitudinally elongate fibres as in *Hakea*, or it may get crushed. The mesophyll is sclerotic in *Macadamia*, but perishes in other genera. The wing, when present, develops from the testa. The tegmen consists of lignified fibers, or fibrous cells. In indehiscent fruits, it is undifferentiated and gets crushed.

In *Macadamia* (Venkata Rao 1970), the growth of integuments is arrested after fertilization, but the growth of chalaza is very conspicuous. The sclerotic pachychalazal zone forms the seed-coat.

A winged seed of *Lomatia tinctoria* is shown in Figure 11.2.4A. The wing is formed by a horseshoe-shaped vascularized extension of the funicle (Fig. 11.2.1H, K–N). In surface view the immature wing shows polygonal richly cytoplasmic cells, but at maturity the walls become thick and the cells lose contents (Fig. 11.2.4B, C). The exotesta is tanniniferous, and the endotesta becomes thick-walled (Fig. 11.2.4D, E). Only the two inner layers of tegmen and a few layers of nucellar tissue persist in the seed.

Banksia ornata is a common dominant in heaths in South-eastern Australia, which are frequently swept by fires which kill plants. Regeneration is only from seed, suggesting that the future of the species is almost completely fire-dependent. Other fire-dependent species are *B. serrata* and *B. serratifolia*.

The multiple fruit consists of a number of follicles attached to a central rachis. The junction between the valves of the follicle has a layer of inter-digitating cells.

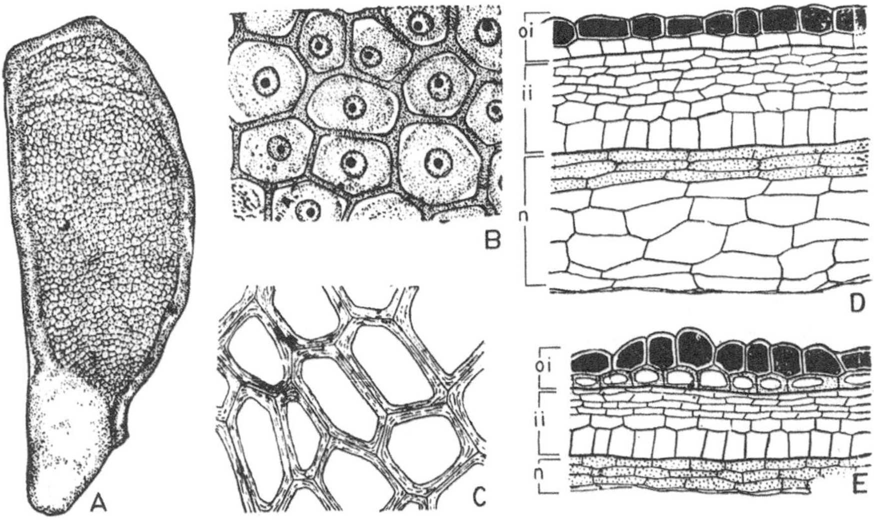

Fig. 11.2.4A–C. *Lomatia tinctoria.* **A** Seed with a conspicuous funicular wing (cf. Fig. 11.2.1H, K–N); **B, C** Surface view of young (**B**) and mature (**C**) seed-wing. **D–E** *L. polymorpha*, development of seed-coat. **D** Immature seed, the outer (*oi*) integument is two-layered and the inner (*ii*) seven-layered, followed by several layers of nucellus (*n*); the outer layers of nucellus are richly cytoplasmic; **E** Mature seed, the exotesta is tanniniferous, and the endotesta is thick-walled. Only the endotegmen, hypodermis, and a few layers of nucellar tissue persist at maturity, the rest of the layers collapse. (After Venkata Rao 1962)

AJ Ballantine (1909), P Brough (1933), DA Johansen (1950), PG Jordaan (1946), SB Kausik (1938a, b, c, 1939a, 1940a, 1941a, 1942), ARH Martin and JC Herscovitch (1989), A Messeri (1928), FG Smith (1968), C Venkata Rao (1960, 1961a, 1962, 1963, 1964, 1965, 1966, 1967, 1968, 1969a, b, c, 1970), AB Wardrop (1983)

11.3 Taxonomic Considerations

Bentham and Hooker (1965c) include the family Proteaceae in the Series Daphnales along with Thymelaeaceae and Elaeagnaceae. Engler and Prantl (1931), Rendle (1971) and Melchior (1964) place the Proteales between Urticales and Santalales. Hutchinson (1973) sought its phyletic relationship with Thymelaeaceae. Takhtajan (1969, 1980) and Cronquist (1981) relate it to Elaeagnaceae. R. Dahlgren (1981) erects the superorder Proteiflorae with the order Proteales, and family Proteaceae, and places it between the superorders Fabiflorae and Myrtiflorae.

Proteaceae seem to be a primitive monochlamydous family. It shows primitive features such as a double-trace stamen as in Casuarinaceae and Betulaceae.

The family Proteaceae resembles Thymelaeaceae and Elaeagnaceae. All the three families have several common features: secretory anther tapetum, simultaneous cytokinesis of pollen mother cells, anatropous, bitegmic, crassinucellar ovule with Polygonum type of embryo sac, Nuclear endosperm, and Asterad type of embryogeny. Moreover, Elaeagnaceae show ephemeral antipodals, and chalazal endosperm haustorium, as in Proteaceae.

In Elaeagnaceae the seed is exotestal with outer epidermis forming a lignified palisade layer. In Proteaceae and Thymelaeaceae it is the outer epidermis of tegmen that comprises the lignified cells. A good deal of variation occurs in the structure of seed in Proteaceae. By evaluating the seed structure, Corner (1976) rejects its alliance with Thymelaeaceae and Elaeagnaceae.

Venkata Rao (1967) wrongly emphasizes the resemblance of Proteaceae to Santalaceae and Loranthaceae. To some extent, the development of chalazal endosperm haustorium may be compared with that in the Santalaceae, but the embryological differences are more marked.

12 Santalales

The Santalales comprise two suborders, Santalineae with six families: Olacaceae, Dipentodontaceae, Opiliaceae, Grubbiaceae, Santalaceae and Myzodendraceae; and Loranthineae with a single family: Loranthaceae and subfamilies Loranthoideae and Viscoideae.

12.1 General Features

The Santalales show the following general features: (1) secretory anther tapetum, (2) two- or three-celled pollen grains, (3) anatropous, uni- or bitegmic tenuinucellate ovule; very often the distinction between the nucellus and integument is absent, (4) development of caecum at micropylar and chalazal end of the embryo sac; in Loranthoideae the micropylar tip of the embryo sac extends to varying length in the stylar canal and stigma, (5) Polygonum type of embryo sac with ephemeral antipodal cells; Allium type in Olacaceae and Viscoideae, (6) occurrence of synergid and antipodal haustoria, (7) Cellular type of endosperm; Helobial type in Olacaceae and Santalaceae, (8) development of micropylar and chalazal haustoria, (9) composite endosperm in Loranthoideae, (10) the division of zygote is transverse; it may be vertical in Santalaceae and Viscoideae; in Loranthoideae it is always vertical, and (11) the albuminous seed lacks seed-coat.

In this order the placental-ovular complex presents a rather confusing picture. The ovules are referred to as bitegmic, unitegmic, and ategmic (undifferentiated).

Ategmic ovules are reported in Olacaceae (*Aptandra, Liriosma, Olax imbricata, Ptychopetalum* and *Ximenia*), Opiliaceae (*Agonandra, Opilia* and *Schoepfia*), Santalaceae (*Buckleya, Chloretum, Leptomeria, Mida, Osyris, Quinchamalium, Santalum* and *Scleropyrum*), and Myzodendraceae (*Myzodendron*).

12.2 Olacaceae

The anther wall comprises an epidermis, fibrous endothecium, two or three ephemeral middle layers, and a secretory tapetum. The endothecium becomes irregularly biseriate in *Strombosia* (Agarwal 1961b). The tapetal cells are two- or four-nucleate. They are uninucleate in *Ximenia americana* (Sankara Rao and Shivaramaiah 1977). The microspore mother cells undergo meiotic divisions resulting in tetrahedral, isobilateral and decussate tetrads. Cytokinesis is simultaneous.

The three-colporate pollen grains are two- or three-celled at the dispersal stage in *Olax stricta* and *Strombosia*, and three-celled in *O. scandens* and *O. wightiana* (Agarwal 1963a). They are triporate in *S. ceylanica* (Shamanna 1961).

In *Olax* the ovary is trilocular below and unilocular above. The ovules are pendulous on the placenta and directed downward (Agarwal 1963a).

The ovule is anatropous and tenuinucellar. A single cell above the archesporial cell represents the nucellus in *Anacolosa frutescens* (Fagerlind 1947a). There is a tendency towards reduction of the nucellus which C.F. Reed (1955) has tried to correlate with the degree of parasitism. The ovules are bitegmic in *Coula, Heisteria, Macrotheca* and *Minquartia.* A single massive integument with a long micropyle develops in *Anacolosa* (Fagerlind 1948), *Olax scandens, O. stricta* and *O. wightiana* (Shamanna 1954, S. Agarwal 1963a). The ovules are undifferentiated (ategmic) in *Aptandra, Liriosma, Olax imbricata, Ptychopetalum petiolatum* and *Ximenia americana* (Sankara Rao and Shivaramaiah 1977).

The archesporial cell functions directly as megaspore mother cell. In *X. americana* there are two to three archesporial cells, but only one functions and produces a parietal cell. The parietal tissue is two to three layers thick. The megaspore mother cell undergoes meiosis and produces a linear or T-shaped tetrad. The chalazal megaspore develops into a Polygonum type of embryo sac. The development is of the Allium type in *Olax wightiana* (Shamanna 1954, Agarwal 1963a). The tip of the embryo sac extends up to the base of stigma in *Aptandra*, and only up to the base of style in *Olax scandens* and *O. wightiana*, and *Chaunochiton*, *Ptychopetalum* and *Schoepfia* (Fagerlind 1948). The embryo sac remains confined to the ovule in *Anacolosa*, *Olax stricta* and *Strombosia* (Fig. 12.2.1A). A lateral caecum of the embryo sac ramifies through the funiculus, and reaches up to the apex of the placental column in *Olax*. A similar lateral haustorium develops in some ovules of *Ximenia*.

The synergids are hooked. The antipodals are ephemeral. They organize into cells in *Strombosia* and *Ximenia*, and are haustorial in *Ximenia* (Sankara Rao and Shivaramaiah, 1977). In *Olax* all the three ephemeral nuclei remain in one cell, or there are two cells one of which is binucleate. The embryo sac contains starch grains in *Strombosia* and *Ximenia*.

The endosperm is of the Cellular type in *Strombosia* (Agarwal 1963b), the chalazal tip develops into a uninucleate haustorium extending up to the apex of placenta. The development is of the Helobial type in *Olax* (Agarwal 1963a). The chalazal chamber becomes tetranucleate, haustorial, and elongates downward through the ovule, grows in the ovarian cavity, penetrates the base of the

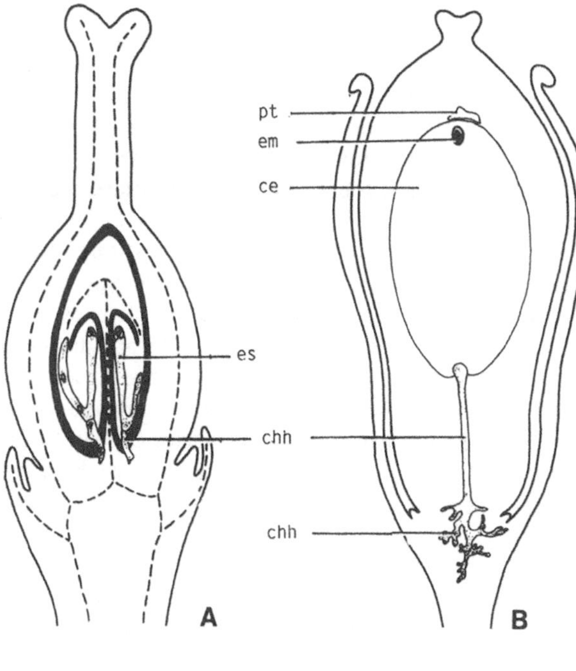

Fig. 12.2.1A,B. *Olax stricta*. **A** Longisection of ovary to show two embryo sacs (*es*) with chalazal haustorium (*chh*) extending up to the receptacle. **B** Longisection of ovary at globular proembryo (*em*), remnants of pollen tube (*pt*), cellular endosperm (*ce*), and chalazal endosperm haustorium extending from the base of endosperm up to receptacle of young fruit. (After Agarwal 1963a)

ovary, and branches profusely in the pedicel (Fig. 12.2.1B). Repeated nuclear divisions occur in the micropylar chamber, and most of the nuclei migrate into the lateral caecum. Wall formation takes place in the micropylar chamber and lateral caecum, producing endosperm proper. The endosperm is persistent and copious, and contains oil or starch (in *Coula*) as storage product.

The zygote divides transversely, and shows a conspicuous multiseriate suspensor in *Strombosia.*

The developing endosperm consumes the integument, and therefore the seed-coat is absent. The fruit is a pseudo-drupe. The drupe-like or nut-like fruits in some members are the largest in this alliance of families.

S Agarwal (1961b, 1963a, b), F Fagerlind (1947a, 1948), BM Johri (1962a, b), BM Johri and SP Bhatnagar (1960), M Ram (1970a), CF Reed (1955), K Sankara Rao and G Shivaramaiah (1977), S Shamanna (1954, 1961)

12.3 Dipentodontaceae

The embryology has not been investigated.

12.4 Opiliaceae

The anther wall comprises an epidermis, fibrous endothecium, one or two ephemeral middle layers, and secretory tapetum. The tapetal cells are binucleate in *Opilia* (Shamanna 1955) and multinucleate in *Cansjera* (Swamy 1960). Microspore mother cells undergo meiotic divisions, followed by simultaneous cytokinesis resulting in tetrahedral tetrads.

The three-colporate pollen grains are two-celled at the dispersal stage.

In Opiliaceae there is a tendency for the reduction of the placental-ovular complex. The ovule (directed downward) is borne on a central placental column in *Rhopalopilia* (Fagerlind 1948) and *Opilia* (Swamy and Dayanand Rao 1963). The placental column is rather small in *Cansjera* (Swamy 1960). It shows further reduction, and in *Agonandra* there is an undifferentiated central mound bearing laterally placed embryo sacs (Fagerlind 1948).

The ovule is anatropous, unitegmic and tenuinucellar with a long micropylar canal. The nucellus is extremely reduced. The nucellar epidermis, integument and micropyle are short-lived in *Cansjera* and *Opilia.* The archesporial cell functions as megaspore mother cell. The latter undergoes meiotic divisions and produces a linear tetrad. The chalazal megaspore gives rise to a Polygonum type of embryo sac. The synergids are hooked, and the antipodals are ephemeral. Compound starch grains occur in the embryo sac of *Cansjera.* The mature embryo sac is characteristically bent like the letter U, in which the chalazal arm is much longer. With the initiation of caecum, the antipodal nuclei migrate into the "pouch" and degenerate. The enlargement of the micropylar part of embryo sac crushes the integument and the egg apparatus becomes exposed. In *Opilia* the chalazal caecum develops into a long, tubular structure reaching up to the level of thalamus, leaving the antipodals in situ.

The development of endosperm is of the Cellular type. In *Opilia* the division of the primary endosperm nucleus occurs at the junction of a cyst-like micropylar chamber and a long tubular chalazal caecum (Fig. 12.4.1A, B). It is followed by wall formation. The micropylar chamber develops into the endosperm proper, while the uninucleate, tubular chalazal chamber develops

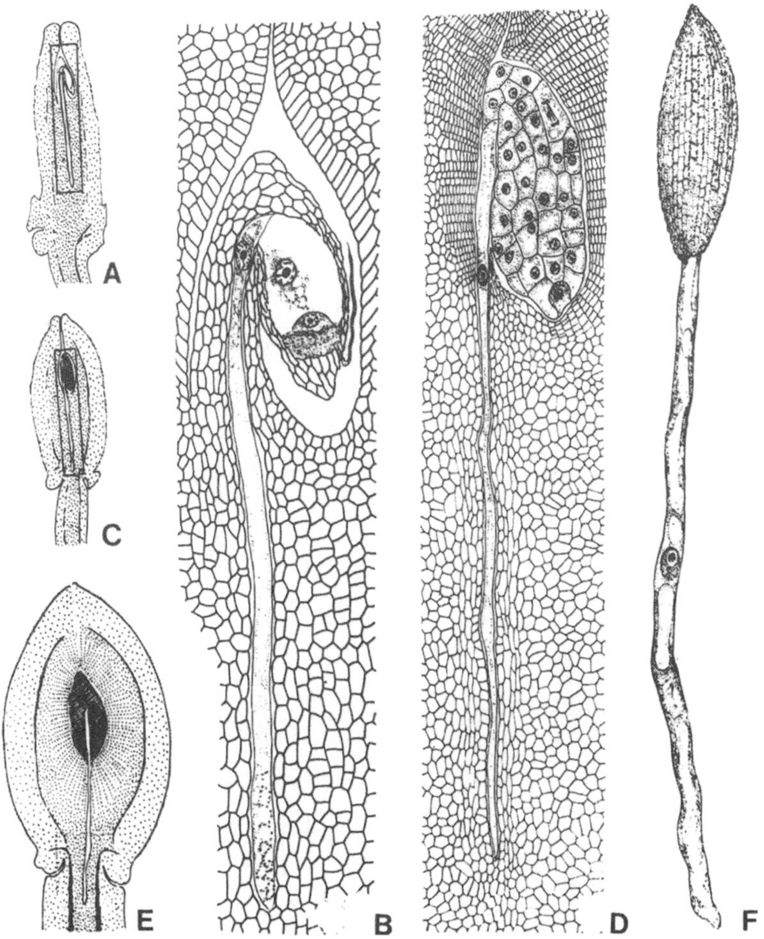

Fig. 12.4.1A–F. *Opilia amentacea.* **A** Median longisection of ovary, outline diagram for **B**. **B** *Inset* from **A**, the embryo sac contains zygote, uninucleate micropylar endosperm chamber, and uninucleate tubular chalazal haustorium. **C** *Outline* for **D**, endosperm *black*. **D** *Inset* from **C**, embryo sac with zygote, cellular endosperm, and uninucleate chalazal haustorium which reaches up to the base of ovary. **E** Longisection of fruit. The chalazal tubular haustorium penetrates as far as the pedicel, endosperm *black*. The ovary wall shows parenchymatous and sclerotic zones. **F** Cellular endosperm and associated uninucleate chalazal haustorium, dissected from young fruit. (After Swamy and Dayanand Rao 1963)

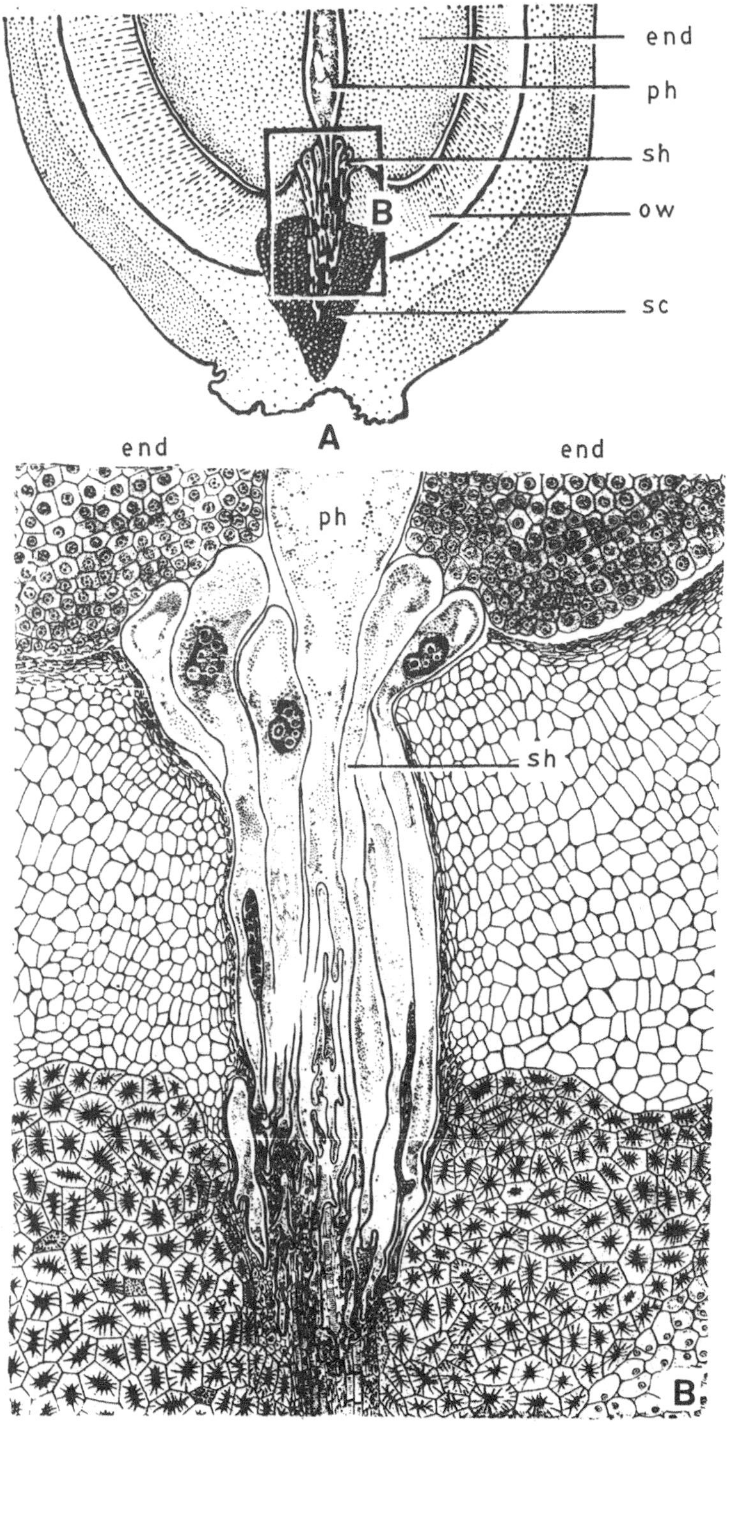
end
ph
sh
B
ow
sc
A
end
end
ph
sh
B

into a haustorium. It elongates parallel to the axis, and penetrates the pedicel (Fig. 12.4.1C–F).

In *Cansjera*, during the maturation of the embryo sac, the solid basal region of the ovary exhibits considerable growth due to meristematic activity. This region also shows a peripheral and axial vascular system. The division of primary endosperm nucleus is followed by a wall resulting in a micropylar and a chalazal chamber (caecum). As the Cellular endosperm consumes the wall of the ovule, the endosperm comes to lie in contact with the inner wall of the ovary. The unicellular chalazal chamber grows intercellularly, and its lower end extends downward up to the axial vascular system. It looks spindle-shaped, and its nucleus becomes hypertrophied. When the chalazal chamber extends up to the point of separation of the peripheral and axial vascular supply, it develops three or four branches which again branch and grow towards the pedicel (Fig. 12.4.2A). The endosperm cells adjoining the primary endosperm haustorium (chalazal chamber) also develop into tubular, branched secondary haustoria. The ultimate branches of both primary and secondary haustoria penetrate and ramify in the axial vasculature in the basal part of the ovary. These haustoria destroy the primary walls of the adjoining cells.

The lateral growth of the haustoria is limited by the development of a zone of sclereids in the basal region of the ovary (Fig. 12.4.2B).

The zygote divides transversely. A suspensor is present.

The seed is albuminous. The seed-coat is thin-walled and gets crushed.

F Fagerlind (1948), M Ram (1970b), S Shamanna (1955), BGL Swamy (1960), BGL Swamy and J Dayanand Rao (1963)

12.5 Grubbiaceae

The embryology has not been fully worked out.

The pollen grains are three-colpate and two- or three-celled at the shedding stage.

Fagerlind (1947d) studied *Grubbia tomentosa*. The ovule is anatropous, unitegmic and tenuinucellate with a long micropylar canal. The sporogenous cell functions as megaspore mother cell. The latter undergoes meiotic divisions, and produces a linear tetrad. The chalazal megaspore develops into a Polygonum type of embryo sac which elongates and comes to lie in the micropylar canal. The presence of an endothelium is reported.

The massive cellular endosperm, at the cotyledonary stage of the embryo, shows chalazal and micropylar haustoria. The epidermis of the endosperm is cuticularized. The reserve food in endosperm and embryo is proteinacous. The embryo occupies the entire length of seed.

Fig. 12.4.2A, B. *Cansjera rheedii*. **A** Longisection of lower portion of fruit to show endosperm (*end*), primary (*ph*) and secondary haustoria (*sh*); *ow* ovary wall; *sc* sclerenchymatous pad. **B** *Inset* from **A**. (After Swamy 1960)

S Carlquist (1977, 1978), F Fagerlind (1947d), BM Johri and SP Bhatnagar (1960), M Ram (1970b, c), Ph van Tieghem (1897)

12.6 Santalaceae

The anther wall comprises an epidermis, fibrous endothecium, two to three middle layers, and multinucleate tapetum. A tapetum with uninucleate cells is reported in *Osyris* (P.C. Joshi 1960). The epidermis of the anther wall is persistent and cutinized in *Santalum*. In *Comandra* the endothecial thickenings are feebly developed, whereas in *Thesium* they show a stellate pattern. The tapetum is multilayered in *Osyris* and *Santalum*. Meiotic divisions in the microspore mother cells followed by simultaneous cytokinesis due to furrowing result in tetrahedral, isobilateral, linear and T-shaped tetrads.

Pollen grains are two-celled at the shedding stage in *Comandra, Exocarpus* (M. Ram 1957, 1959a), *Leptomeria* (M. Ram 1959b), *Osyris* (P.C. Joshi 1960), *Thesium* (S.P. Bhatnagar and Agarwal 1961) and *Quinchamalium* (B.M. Johri and Agarwal 1965). They are three-celled in *Santalum* (Paliwal 1956), *Leptomeria* (M. Ram 1959b) and *Mida* (S.P. Bhatnagar 1960).

Cytomixis was observed between the microspore mother cells. Polynucleate pollen mother cells are reported in *Santalum obtusifolium* (S.P. Bhatnagar 1966). Sometimes, the number of microspores in a tetrad may be fewer or more than four.

An interesting feature in *Leptomeria billardierii* (M. Ram 1959c) is the formation of eight-nucleate embryo sac-like structures, originating from the sporogenous cells in the anther.

The ovule may be pendulous, straight, hemianatropous, anatropous or orthotropous. There is no distinction between the nucellus and integument, except in *Comandra, Iodina* and *Thesium* where the ovule is unitegmic. In *Exocarpus* there is no organized placenta, and the ovule does not differentiate (see Fagerlind 1959, S.P. Bhatnagar and Joshi 1965).

The archesporial cell functions directly as megaspore mother cell. The latter undergoes meiotic divisions and produces a linear tetrad. Usually, the chalazal megaspore functions and gives rise to a Polygonum type of embryo sac, except in *Buckleya* (B. Raj 1964) and *Iodina* (S.P. Bhatnagar and Sabharwal 1966, 1968) where it is of the Allium type. Endymion type of development is reported in *Buckleya* (B. Raj 1964).

The embryo sac remains confined to the ovule in *Choretrum* (see S.P. Bhatnagar 1970), *Iodina* (S.P. Bhatnagar and Sabharwal 1966), *Osyris* (Paliwal 1956, P.C. Joshi 1960), *Scleropyrum* (L.N. Rao 1942) and *Thesium* (Paliwal 1956, S.P. Bhatnagar and Agarwal 1961).

The formation of extra-ovular prolongations of the embryo sac is a noteworthy feature of the family. In *Comandra umbellata* (M. Ram 1957), a lateral caecum originates in the micropylar region of the embryo sac, on the funicular side of the ovule, and grows into the long and twisted placenta (Fig. 12.6.1A, B). Even at the four-nucleate stage the chalazal end of the embryo sac elongates, and grows beyond the ovule. Finally, the caecum comes

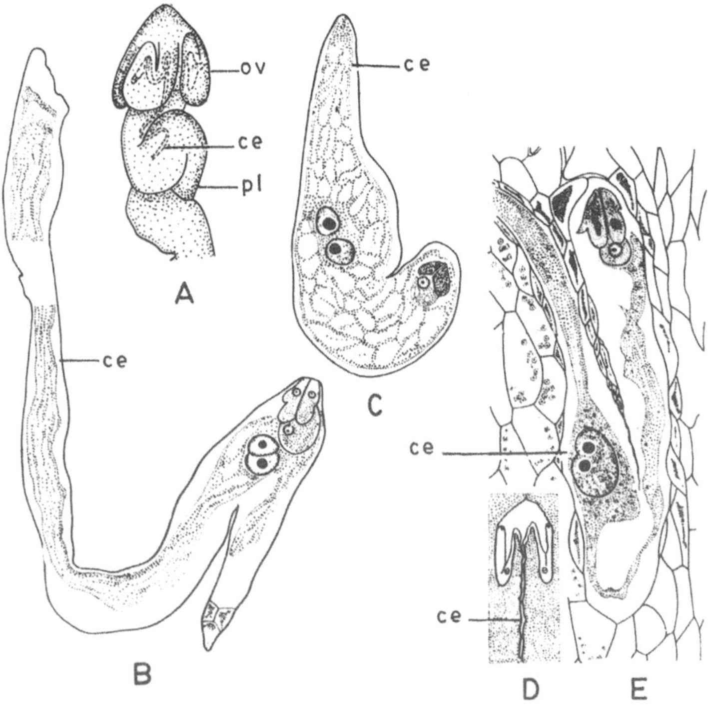

Fig. 12.6.1A, B. *Comandra umbellata* (wholemount). **A** Placenta (*pl*), *dotted lines* indicate the extension of the embryo sac; *ce* caecum; *ov* ovule. **B** Mature embryo sac with a well-developed lateral caecum. (After M. Ram 1957). **C–E** *Leptomeria cunninghamii*. **C** Embryo sac with caecum (wholemount). **D** Portion of ovary and placenta to show extension of embryo sac (*ce*). **E** Portion of embryo sac from **D**, with egg apparatus and polar fusion nucleus. (After M. Ram 1957)

to lie in the ovarian cavity (Fig. 12.6.1C–E) in *Leptomeria* (Ram 1959b), *Mida* (S.P. Bhatnagar 1960), and *Quinchamalium* (Agarwal 1961a, B.M. Johri and Agarwal 1965).

The placental column is straight and elongated with pendulous ovules (Fig. 12.6.2A) in *Santalum album* (Sedgley 1982), and the embryo sac develops a chalazal caecum which grows into the placenta leaving the antipodals in situ. The development of endosperm is of the Helobial type. Free-nuclear divisions occur in the micropylar chamber followed by cellularization. The chalazal chamber remains uninucleate throughout (Fig. 12.6.2B–D).

In *Exocarpus*, the development of the embryo sac is of the Polygonum type with an eight-nucleate gametophyte. The embryo sac develops lateral finger-like haustorial extensions in *E. menziesii* (Fig. 12.6.3A). In a mature embryo sac the growth in the apical region is rather extensive (Fig. 12.6.3B). In the fertilized embryo sac of *E. strictus* (Ram 1959a), the haustoria are produced in the middle region (Fig. 12.6.3C).

Due to the apical growth of the gametophyte, the epidermal cells of the "placenta" are crushed and the embryo sac becomes exposed (Fig. 12.6.4A–C).

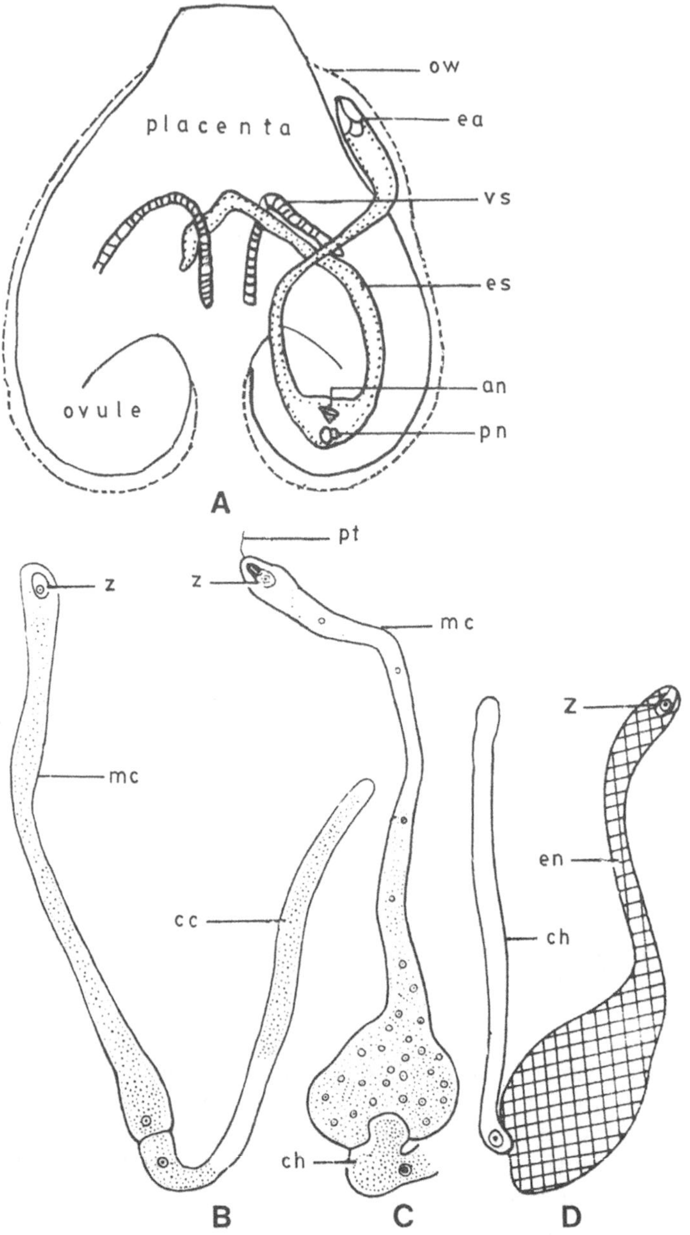

Fig. 12.6.2A. *Santalum acuminatum*, longisection (semi-diagrammatic) of placenta with one embryo sac; *ow* inner ovary wall; *ea* egg apparatus; *vs* vascular supply; *es* embryo sac; *an* antipodals; *pn* polar nuclei. (After Sedgley 1982). **B–D** *S. album*. **B** Embryo sac with zygote (*z*) and two-celled endosperm-micropylar (*mc*) and chalazal chamber (*cc*). **C** Upper part of embryo sac with zygote (*z*), and coenocytic micropylar chamber (*mc*) and portion of uninucleate chalazal haustorium (*ch*); *pt* remnants of pollen tube. (After S.P. Bhatnagar 1959). **D** Later stage, cellular endosperm (*en*) and uninucleate chalazal haustorium. (After B.M. Johri and Bhatnagar 1969)

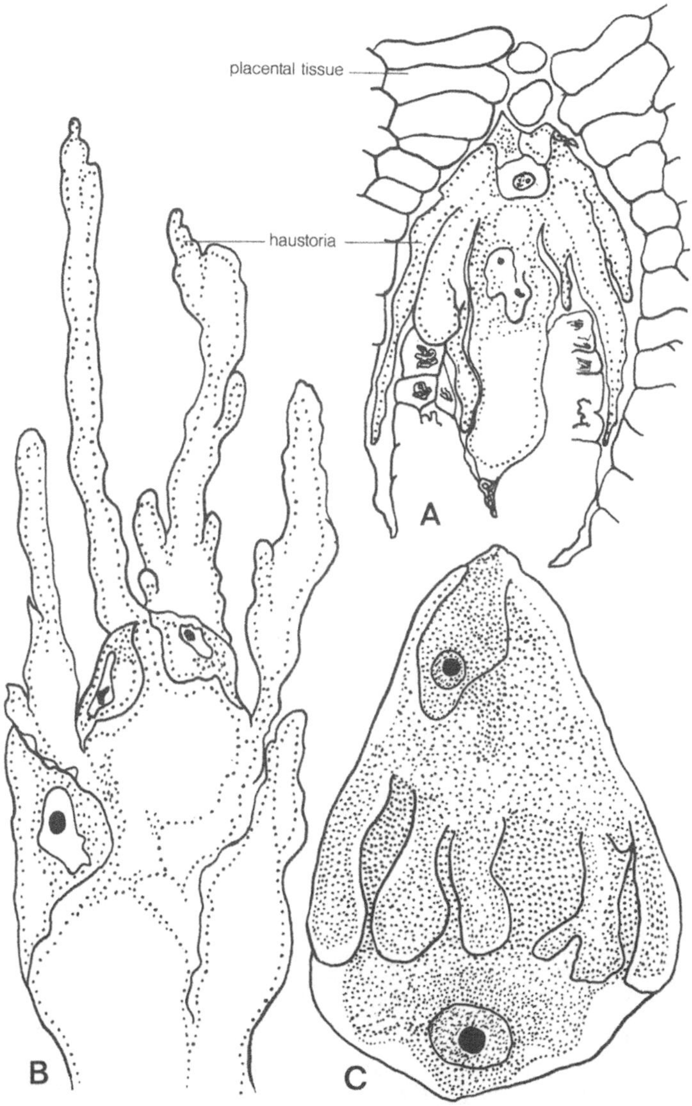

Fig. 12.6.3A, B. *Exocarpus menziesii.* **A** Organized embryo sac with finger-like haustoria growing downward; **B** Later stage. There is profuse development of haustoria growing upward. (After Fagerlind 1959). **C** *E. sparteus*, embryo sac with haustoria from the middle region. (After M. Ram 1959a)

The development of endosperm and young embryo is shown in Figure 12.6.4B–H.

The synergids are generally hooked, except in *Quinchamalium* where the tips elongate and protrude beyond the embryo sac, even before fertilization. They form extensive haustoria (Figs. 12.6.5A–D, 12.6.6A–C) penetrating up to

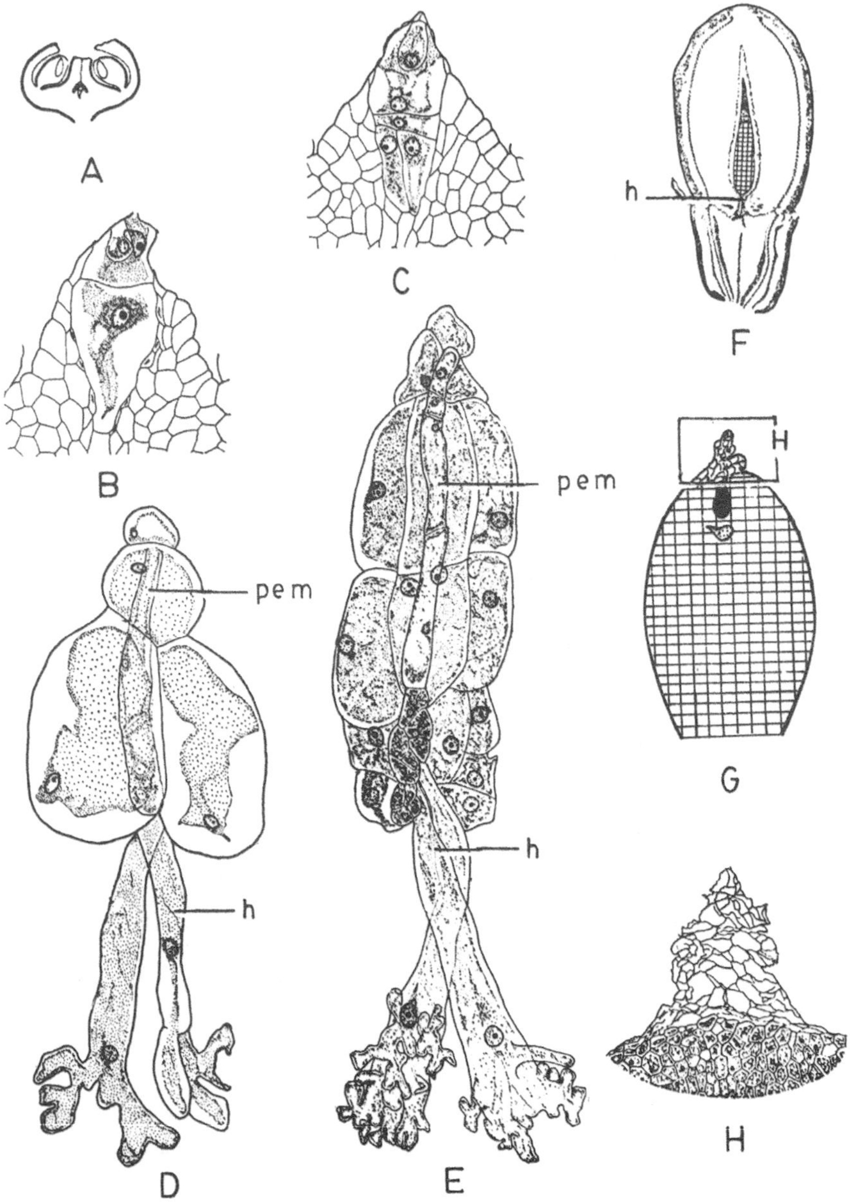

Fig. 12.6.4A–H. *Exocarpus sparteus.* **A** Longisection of flower. **B** Zygote and two-celled endosperm. **C** Zygote and four-celled endosperm. **D** Elongated tubular proembryo (*pem*) and endosperm with haustoria (*h*), note the finger-like tips. **E** Same, later stage. **F** Longisection of fruit to show endosperm and embryo (diagrammatic). **G** Globular proembryo (*black*), degenerated suspensor, and endosperm (*cross-hatched*). **H** *Inset* from **G**. (After M. Ram 1959a)

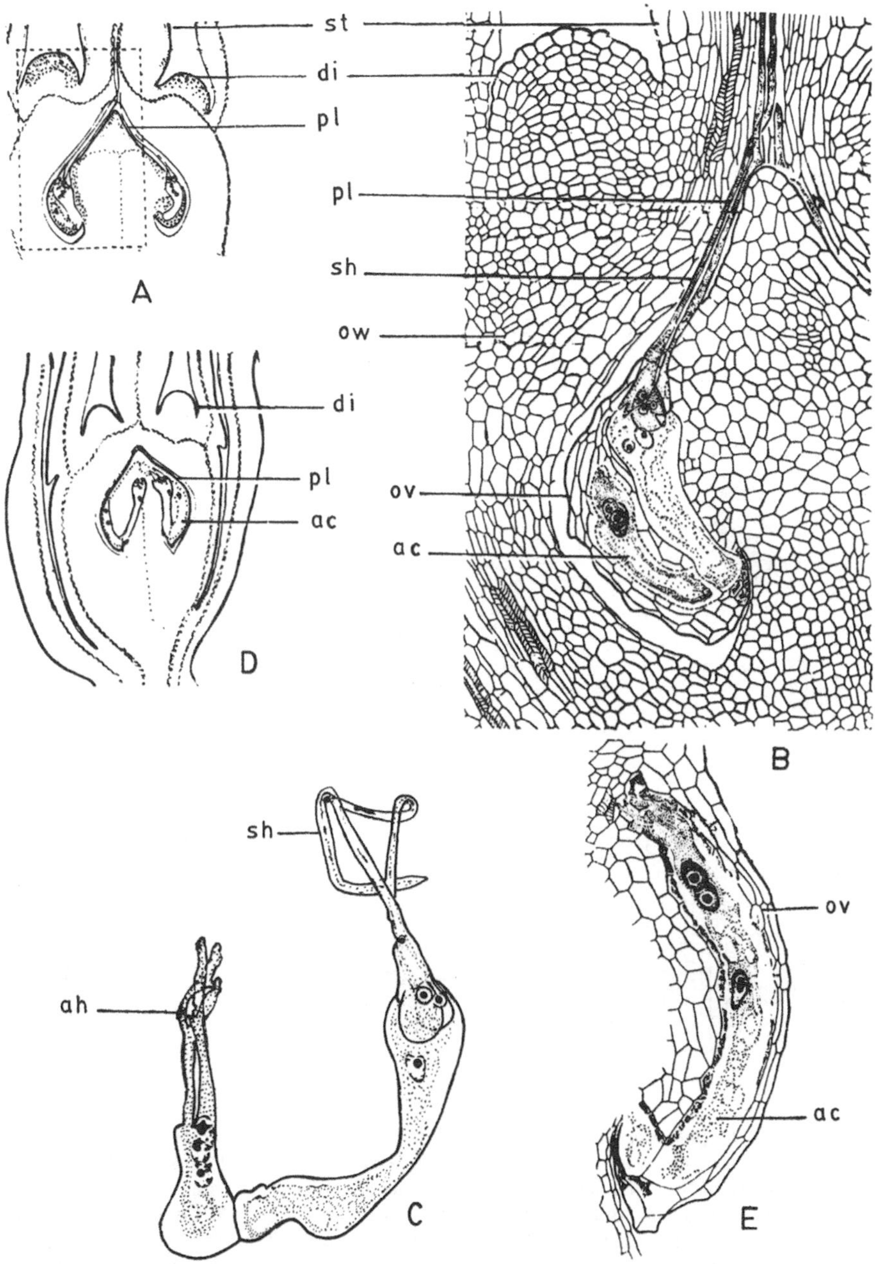

Fig. 12.6.5A–E. *Quinchamalium chilense.* **A** Portion of ovary (*l.s.*) to show conical placenta (*pl*) bearing two ovules; *st* style; *di* disc. **B** *Inset* from **A**. The tip of embryo sac has become extra-ovular and grown along the outer margin of the placenta (*pl*). The synergid haustoria (*sh*) reach up to the base of style (*st*), and the antipodal chamber (*ac*) extends into the funiculus; *ow* ovary wall; *ov* ovule. **C** Mature embryo sac to show extensive synergid (*sh*) and antipodal (*ah*) haustoria (wholemount). **D** Longisection of ovary showing two embryo sacs. **E** Antipodal chamber (*ac*) from **D**. (After Agarwal 1962)

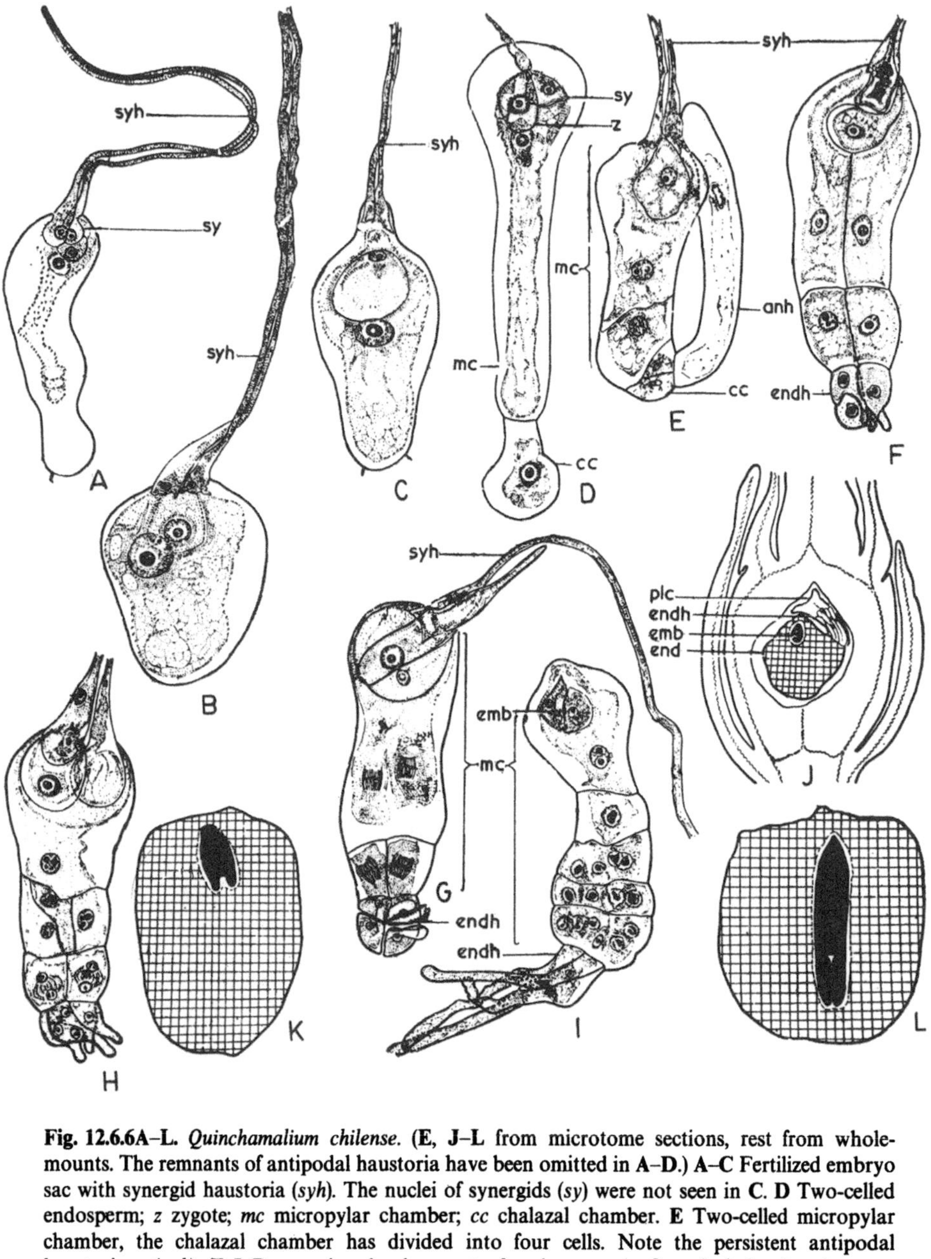

Fig. 12.6.6A–L. *Quinchamalium chilense*. (**E**, **J–L** from microtome sections, rest from wholemounts. The remnants of antipodal haustoria have been omitted in **A–D**.) **A–C** Fertilized embryo sac with synergid haustoria (*syh*). The nuclei of synergids (*sy*) were not seen in **C**. **D** Two-celled endosperm; *z* zygote; *mc* micropylar chamber; *cc* chalazal chamber. **E** Two-celled micropylar chamber, the chalazal chamber has divided into four cells. Note the persistent antipodal haustorium (*anh*). **F–I** Progressive development of endosperm (*end*) and chalazal endosperm haustoria (*endh*). **J** Longisection of ovary to show endosperm at globular proembryo (*emb*). The endosperm haustoria extend through the funicle; *plc* placenta. **K**, **L** Endosperm at early and late dicotyledonous embryo. (After Agarwal 1962)

one-third the length of style (Agarwal 1961a, 1962). In *Comandra* the synergids are ephemeral.

The polar nuclei fuse before fertilization. The entry of pollen tube is porogamous. Persistent pollen tubes have been observed in *Santalum* (S.P. Bhatnagar 1959).

The formation of a chalazal caecum, leaving the antipodals in situ, is also common. The three antipodals are generally ephemeral, but persist in *Mida* and *Santalum*. In *Quinchamalium* a three-nucleate antipodal chamber elongates and develops into a haustorium (Fig. 12.6.5E) which invades the placental tissue. In *Scleropyrum*, L.N. Rao (1942) reported supernumerary free antipodal nuclei and their migration into the chalazal chamber. A lateral caecum, extending along the vascular supply, is reported in *Comandra*.

The development of endosperm is of the Cellular type, except in *Mida* where it is Helobial. Generally, the division of primary endosperm nucleus results in the formation of micropylar and chalazal chamber. The former

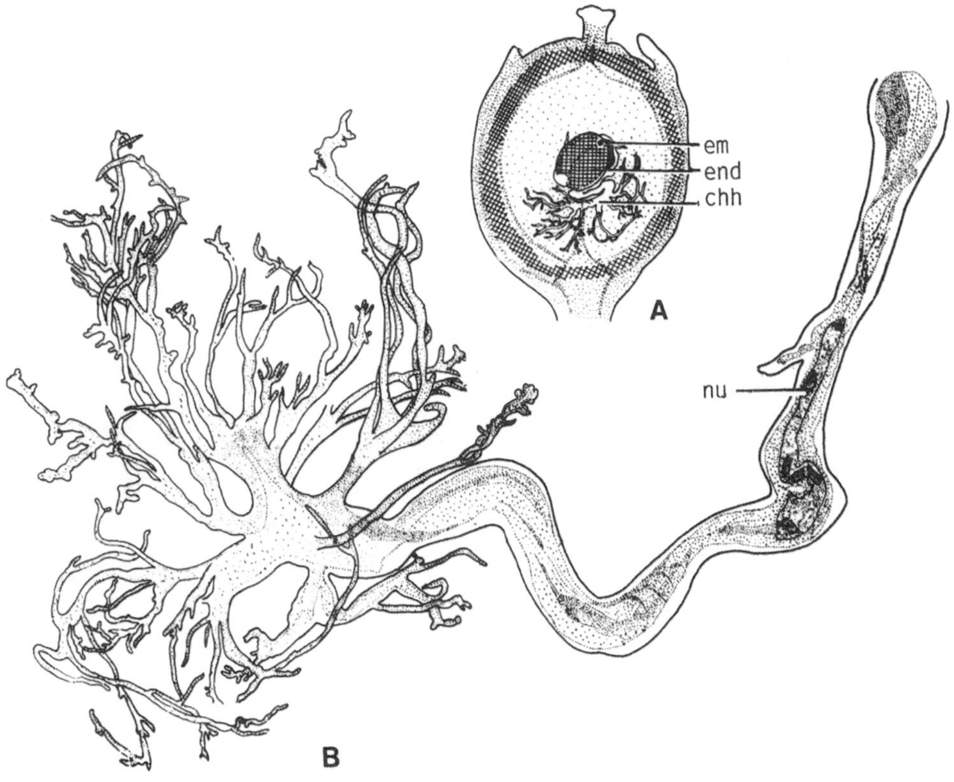

Fig. 12.6.7A, B. *Iodina rhombifolia.* **A** Longisection (diagrammatic) of fruit at globular proembryo (*em*), the endosperm (*end*) is well developed; note the aggressive chalazal haustorium (*chh*). **B** Haustorium with many-branched lower end and hypertrophied nucleus (*nu*). (After S.P. Bhatnagar and Sabharwal 1966)

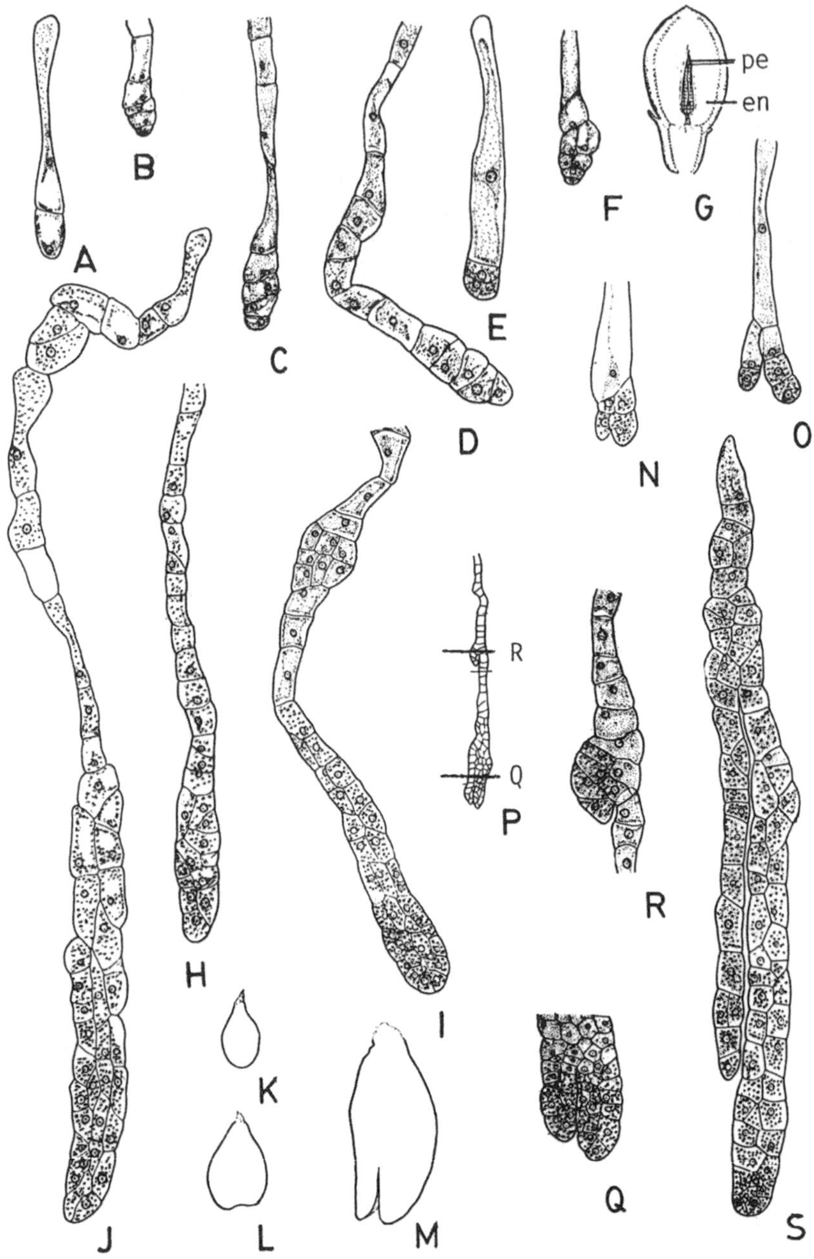

Fig. 12.6.8A–S. *Exocarpus sparteus*, embryogeny. **A** Two-celled proembryo. **B–D** Uniseriate proembryos. **E** Transverse division in terminal cell. **F** Multicelled apical region. **G** Longisection (diagrammatic) of fruit showing endosperm (*en*) and proembryo (*pe*). **H** Proembryo from **G**. **I, J** Progressive stages of proembryo, terminal region shows small richly cytoplasmic cells, and suspensor comprises elongated vacuolate cells. **K–M** Globular (**K**), heart-shaped (**L**) and dicotyledonous (**M**) embryo (diagrammatic). **N–S** Polyembryony. **Q, R** Portions marked *Q* and *R* in **P**. **S** Both proembryos are long and filamentous. (After Ram 1959a)

develops into the endosperm proper, and the latter forms a haustorium (Fig. 12.6.6D). In *Mida* the chalazal chamber forms a uninucleate haustorium. Free-nuclear divisions in the micropylar chamber are followed by wall formation.

The lateral caecum acts as a primary haustorium in *Comandra*. It cuts off a small cell and its derivatives produce secondary haustoria. Three to four cells, formed in the same manner in *Mida*, elongate and function as secondary haustoria. In addition, the micropylar chamber of endosperm produces numerous finger-like processes towards the base.

The chalazal chamber divides by a vertical wall in *Exocarpus* (Fig.12.6.4C). In *E. sparteus* the daughter cells produce numerous lobed outgrowths at the base (Fig. 12.6.4D, E). In *E. cupressiformis* the haustorium becomes multicelled.

In *Quinchamalium* four cells at the base of endosperm develop into independent haustoria (Fig. 12.6.6E–I). They elongate and extend along the course of antipodal haustorium of the embryo sac, reaching up to the apex of placenta (Agarwal 1962). The endosperm proper is well developed (Fig. 12.6.6J–L).

A profusely branched, aggressive, coralloid haustorium with a highly hypertrophied nucleus is reported in *Iodina* (Fig. 12.6.7A, B). The nucleus increases to almost 20-times its original size. In *Thesium* (Erbrich 1965) the nucleus becomes highly polyploid (12 n-384) due to endomitosis.

The zygote divides transversely, except in *Comandra* and *Quinchamalium* where the division is vertical. A suspensor may or may not develop. Further divisions are variable. Figure 12.6.8 shows the development of embryo (A–M), and cleavage (N–S) and suspensor polyembryony (P, R) in *E. sparteus*.

The additional embryo in *Mida*, *Santalum* and *Quinchamalium* arises from the zygote of twin embryo sacs.

The pericarp directly surrounds the endosperm which consumes the endocarp, and the stony mesocarp gives protection. In *Quinchamalium*, as the ovary wall is consumed, the persistent perianth functions as "pericarp". In *Iodina* the persistent perianth becomes an additional protective layer.

S Agarwal (1961a, 1962), SP Bhatnagar (1959, 1960, 1965a, 1966, 1970), SP Bhatnagar and S Agarwal (1961), SP Bhatnagar and PC Joshi (1965), SP Bhatnagar and G Sabharwal (1966, 1968), P Erbrich (1965), F Fagerlind (1959), F Gagnepain and E Boureau (1947), GS Iyengar (1937), BM Johri (1962a, b), BM Johri and S Agarwal (1965), BM Johri and SP Bhatnagar (1960, 1969, 1970), PC Joshi (1960), RL Paliwal (1956), R Pilger (1935), B Raj (1964, 1969), M Ram (1957, 1959a, b, c), LN Rao (1942), M Sedgley (1982), FH and EC Smith (1942), HU Stauffer (1957), Ph van Tieghem (1896)

12.7 Myzodendraceae

The unigeneric monotypic family, Myzodendraceae, has fascinated botanists for their curious structure and reproductive biology. *Myzodendron* is essentially a dioecious genus; Skottsberg (1913) reports bisexual flowers. The staminate flowers, without a calyx and corolla (Hutchinson 1973), have two or three stamens. The bisporangiate (monothecous) anther (Fig. 12.7.1A, B) dehisces by terminal slits.

The pistillate flower has a three-lobed stigma on the top of the superior (not inferior as generally reported) ovary. The calyx is adnate to the ovary and free at the top. There are no petals and staminodes. Within the ovary, there are three reduced ovules at the apex of a short placental column (Fig. 12.7.1A, G–I), each hanging down into a pocket of the ovarian cavity. The upper part of each pocket becomes confluent with the others so that the ovary is trilocular below and unilocular above (Kuijt 1969).

The embryology has not been studied in sufficient detail.

The pollen grains are spheroidal, spinuliferous, and polyporate. The 3 to 19 apertures are scattered randomly over the entire surface. The ektexine is represented only by spines and a thin granular layer between them. Endexine is thick in the region bordering the pore, and extends into the pore (Feuer 1981).

The tenuinucellate ovule is highly reduced and undifferentiated (ategmic), and is represented only by the nucellus (Fig. 12.7.1C–F). With the growth of the endosperm, the scanty nucellus is completely crushed (Fig. 12.7.1G–I). This organization is not very different from that in Santalaceae and Olacaceae.

The archesporial cell functions directly as megaspore mother cell. The latter undergoes meiotic divisions, and the chalazal megaspore (Fig. 12.7.1C) of a linear or T-shaped tetrad develops into a Polygonum type of embryo sac (Fig. 12.7.1D, E). The polar nuclei fuse before fertilization; the three antipodals are ephemeral.

In *Myzodendron punctulatum* (Skottsberg 1913), the chalazal end of the embryo sac forms a prominent tubular caecum which extends into the placental column (Fig. 12.7.1F).

The development of the endosperm is of the Cellular type. The first cross-wall in the caecum is formed following the division of the primary endosperm nucleus. With the development of the endosperm proper, uniseriate cells are also formed in the haustorium (caecum; Fig. 12.7.1H, I). The haustorium branches, extends into the funicle and reaches the pedicel of the flower through the placenta (Fig. 12.7.1H, J). The development of endosperm is comparable to that in Olacaceae.

According to Johnson (1889), at first the chalazal haustorium consists of a uniseriate row of cells. Due to the dissolution of the walls, the haustorium becomes coenocytic. Skottsberg (1913) reports only a uninucleate chalazal haustorium which branches at the tip (see M. Ram 1970d). In view of the contradictory reports, a detailed reinvestigation is necessary.

The development of embryo has not been studied. It is reported to lack a suspensor. Figure 12.7.2 shows a globular proembryo (A), mature embryo (B), and disposition of embryo in the endosperm with a prominent haustorium (C).

Only one of the three ovules in an ovary matures into a seed, and it is lifted up by the elongation of the placental column.

The fruit is a three-angled or winged small achene. The naked (without testa) seed has a fleshy chlorophyllous endosperm, and a fleshy central embryo (Fig. 12.7.2C).

The seedling has a curious structure. It shows a small thick-walled cup

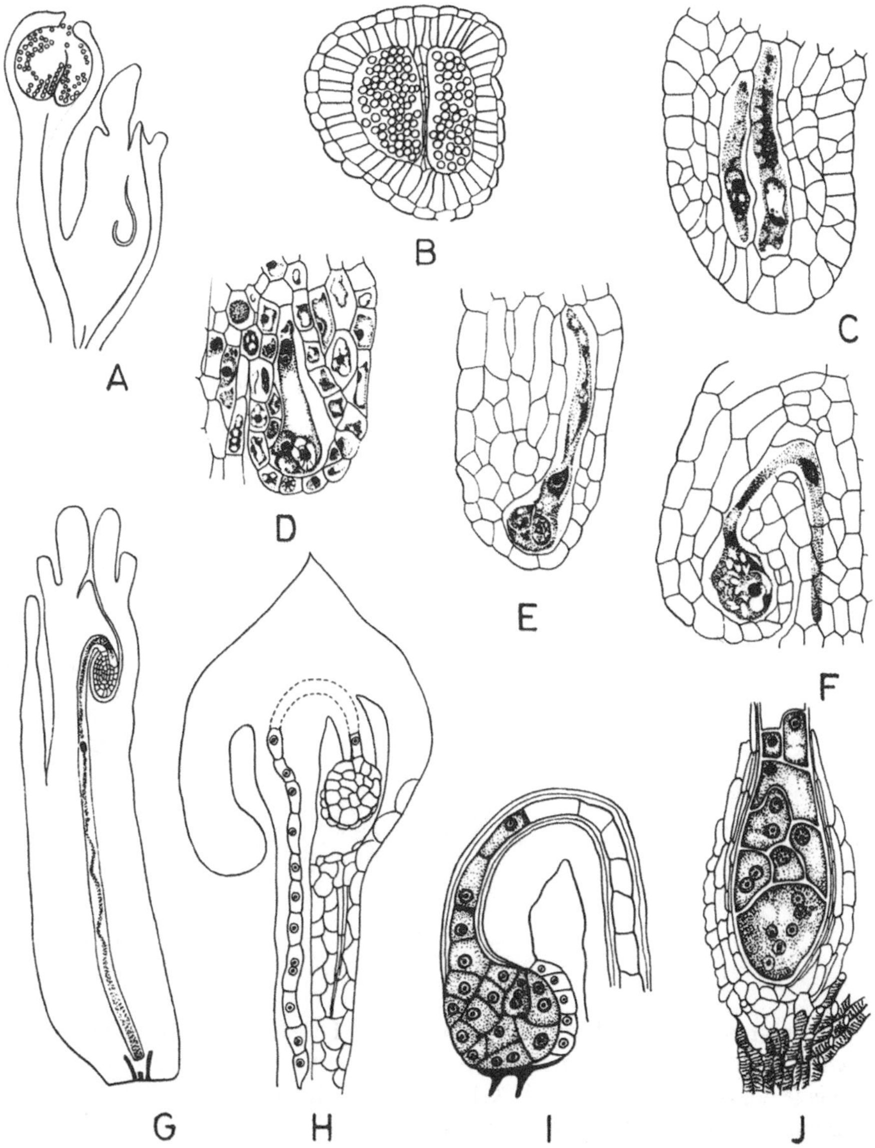

Fig. 12.7.1.A *Myzodendron linearifolium*, longisection of flower; stamen with bisporangiate anther, and gynoecium with superior ovary and undifferentiated (ategmic) pendulous ovule (nucellus only). **B** *M. recurvum.* Anther with pollen, the wall shows the epidermis and endothecium. **C** *M. quadriflorum.* Nucellus with two functional chalazal megaspores. **D–J** *M. punctulatum.* **D** Embryo sac with egg apparatus and polar nuclei, antipodals have degenerated; **E** Later stage, secondary nucleus lying close to egg apparatus; **F** Fertilized embryo sac with zygote, primary endosperm nucleus in chalazal caecum; **G** Zygote, endosperm proper, and unseriate endosperm cells in caecum/haustorium; **H** Gynoecium, the haustorium extends through the placenta up to the base of the pedicel; **I** Placenta, later stage of endosperm than in **G**; **J** Tip of chalazal haustorium close to the base of vascularized pedicel, the endosperm cells are uni-, bi- and multinucleate. (**A–F**, **I** After Skottsberg 1913, **G**, **H** and **J** after Johnson 1889)

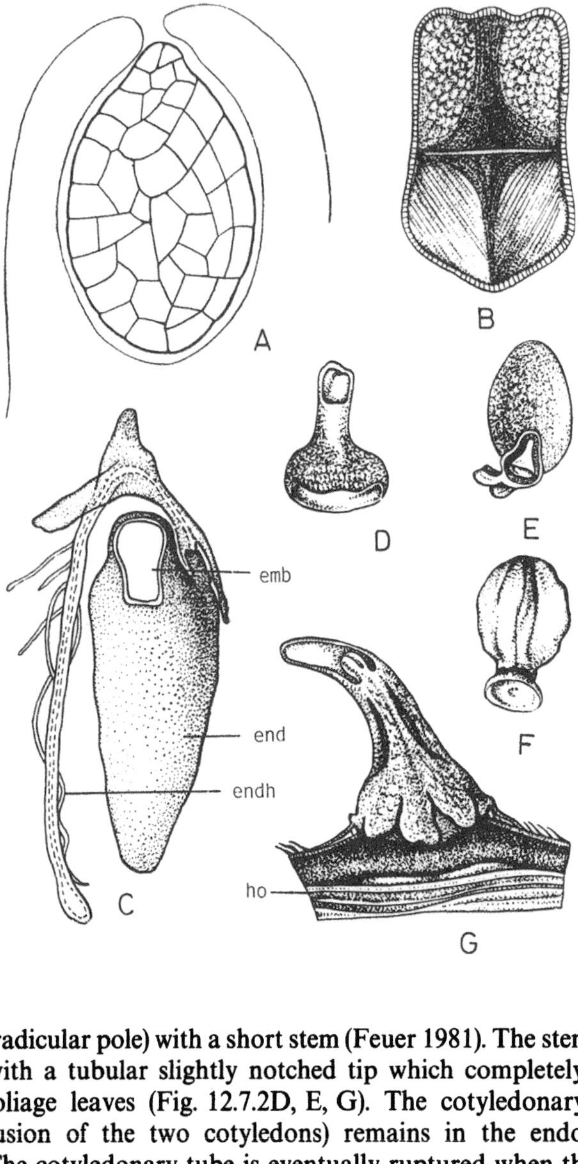

Fig. 12.7.2A–C. *Myzodendron punctulatum*, **A, B** Globular proembryo (**A**) and differentiated (mature?) embryo (**B**); **C** Embryo (*emb*) and endosperm (*end*) with conspicuous haustorium (*endh*). **D–G** *M. brachystachyum*. **D** Germinated embryo with perforated trumpet-shaped radicular end, and cotyledonary tube (hypocotyl) with first pair of leaves at the apex of notched tube; **E** Fruit, embryo as in **D**; **F** Fruit, embryo at later stage than in **E**. Note the cup-shaped adhesive membrane; **G** Seedling attached to host (*ho*) stem. The shoot has not yet emerged from the cotyledonary tube. (**A–C** After Johnson 1889, **D–G** after Hooker 1846)

(radicular pole) with a short stem (Feuer 1981). The stem is really the hypocotyl with a tubular slightly notched tip which completely encloses the first two foliage leaves (Fig. 12.7.2D, E, G). The cotyledonary tube (formed by the fusion of the two cotyledons) remains in the endosperm for some time. The cotyledonary tube is eventually ruptured when the foliage leaves emerge (Kuijt 1969).

On germination on the host surface, the radicle emerges from the achene, quickly expands, a thin adhesive membrane extends radially, and penetration follows from the axial part of the haustorial disc (Fig. 12.7.2E, F). An early stage of the host and semi-parasite is shown in Figure 12.7.2G.

TB Batygina and MS Yakovlev (1987), S Feuer (1981), JD Hooker (1846), J Hutchinson (1973), T Johnson (1889), BM Johri and SP Bhatnagar (1960), J Kuijt (1969), M Ram (1970d), C Skottsberg (1913), JC Willis (1973)

12.8 Loranthaceae: Loranthoideae

The representatives of Loranthoideae have been studied at the University of Delhi for over 25 years. The chief events in their life history are shown in Figure 12.8.7. B.M. Johri (1987a, see also S.P. Bhatnagar and Johri 1983) has reviewed the reproductive biology of mistletoes (Loranthaceae and Viscaceae).

The anther wall comprises the epidermis, fibrous endothecium, one or two ephemeral middle layers, and glandular tapetum of binucleate cells. The fibrous thickenings in the endothecium are absent in *Amylotheca dictyophleba* (B. Raj 1970a), *Lepeostegeres gemmiflorus* (Dixit 1958b) and *Macrosolen cochinchinensis* (P. Maheshwari and Singh 1952). In *Loranthus* (*Hyphear*) *europaeus* (Pisek 1924), and at times in *Scurrula*, the tapetum is amoeboid. The microspore mother cells undergo meiotic divisions, simultaneous cytokinesis, and produce tetrahedral and decussate tetrads. Because the microspores are triradiate, they are so arranged in a tetrad that all the three tips of each microspore touch the tips of the other microspores.

Polyspory with five or six spores has been observed in *Lysiana exocarpi* (R. Narayana 1958a).

Sterile transverse partitions in pollen sacs have been reported in some members: *Amyema gravis*, *Elytranthe* (Schaeppi and Steindl 1942), *Helixanthera ligustrina* (B.M. Johri and Bhatnagar 1972), *Taxillus cuneatus* (R. Narayana 1955) and *Tapinostemma acaciae* (B.M. Johri and S. Prakash 1965). Finally, the septae break down and the microsporangium becomes continuous.

The pollen grains are triangular with three slightly curved concavo-convex arms. Sometimes, the pollen may be four-armed as in *Barathranthus* (S. Prakash 1963), *Helicanthes* (B.M. Johri et al. 1957), *Lepeostegeres* (Dixit 1958b), *Lysiana* (R. Narayana 1958a) and *Moquiniella* (B.M. Johri and Raj 1969). Spherical pollen grains have been observed in *Atkinsonia* (S. Prakash 1961), *Tupeia* (van Tieghem 1895, Danser 1933, Smart 1952) and *Ixocactus* (Kuijt 1967, 1969). Usually, the pollen grains are tricolpate, but in *Ixocactus* they are tetracolpate (see B.M. Johri and Bhatnagar 1972). They are two-celled at the dispersal stage.

B.M. Johri (1956) reports that in *Helixanthera ligustrina* some of the microspore mother cells fail to undergo meiosis, and form large restitution nuclei. Then an exine develops (within the original wall of the mother cell) simulating a pollen grain. In a few cases such pollen grains also showed a pollen tube. Other microspore mother cells in these microsporangia developed normally.

Ovary-Ovule Complex

B.M. Johri and Bhatnagar (1972) discuss the morphology, structure and disposition of placenta (mamelon), from four-lobed condition in four-chambered ovary (*Lysiana*) to complete absence of the placenta (*Helixanthera*). B.M. Johri and Ambegaokar (1984) and B.M. Johri (1987b) reconsider the status of the ovary-ovule complex under six categories (Fig. 12.8.1A_1–F_2).

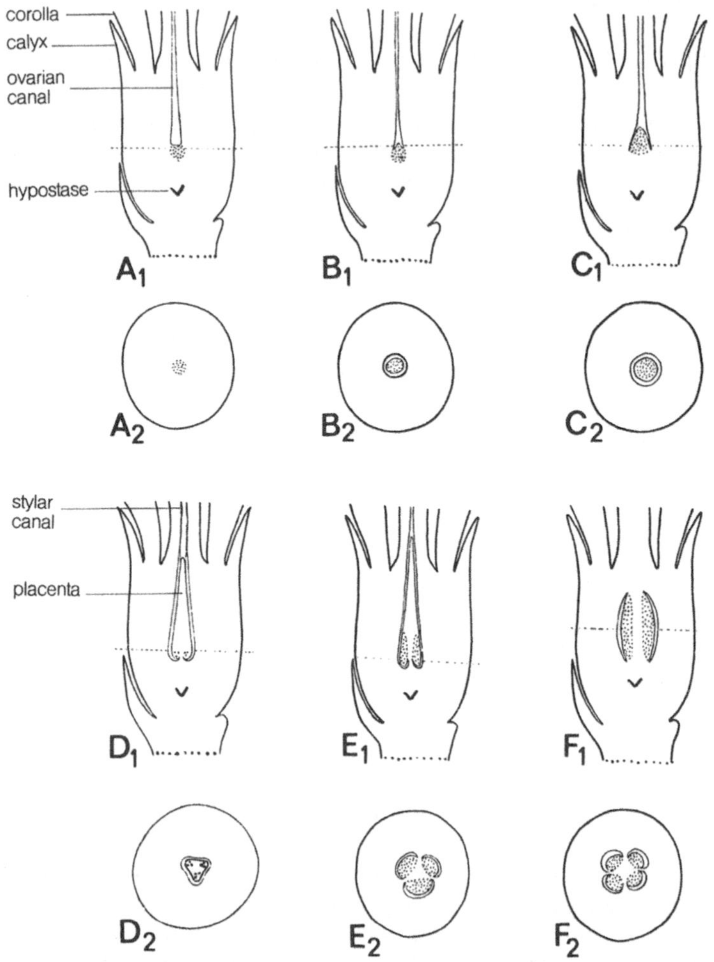

Fig. 12.8.1A$_1$–F$_2$. Loranthoideae, longi- and transections (diagrammatic) of ovaries, level of transections indicated by a *dotted line* in longisections. **A$_1$, A$_2$** *Helixanthera ligustrina*, sporogenous tissue (*stipped*) at the base of stylar canal. **B$_1$, B$_2$** *Amyema preissii*, sporogenous tissue lifts the base of the ovarian canal. **C$_1$, C$_2$** *Helicanthes elastica*, sporogenous tissue in an unlobed mamelon. **D$_1$, D$_2$** *Macrosolen cochinchinensis*, sporogenous cell/s in each of the three basal lobes of mamelon in an unilocular ovary. **E$_1$, E$_2$** *Nuytsia floribunda*, sporogenous tissue in each lobe of the mamelon in a trilocular ovary. **F$_1$, F$_2$** *Lysiana exocarpi*, sporogenous tissue in each lobe in a tetralócular ovary. (Modified after R. Narayana 1958a, Kapil and Vasil 1963)

1) Placenta absent, archesporium differentiates in the subepidermal region below the base of the stylar canal: *Helixanthera ligustrina* (Fig. 12.8.1A$_1$, A$_2$), *Dendrophthoe falcata* (B. Singh 1952a), *Barathranthus axanthus* (S. Prakash 1963), *Moquiniella rubra* (B.M. Johri and Raj 1969) and *Struthanthus vulgaris* (Venturelli 1981a).

2) Placenta transitory, archesporium differentiates as in category 1, but the base of stylar canal is pushed up as a result of elongation of sporogenous tissue: *Dendrophthoe pentandra*, *Scurrula montana* (Schaeppi and Steindl 1942), *D. neelgherrensis* (R. Narayana 1956), *Amyema miquelii* (Fig. 12.8.1B_1, B_2), *Tolypanthus involucratus* (Dixit 1961) and *Struthanthus vulgaris* (Venturelli 1981a). In *A. miquelii* (Dixit 1954), the mamelon shows tracheidal thickenings in some cells. In *A. miquelii* (Dixit 1954), the mamelon shows tracheidal thickenings in some cells.

3) Placenta prominent and unlobed, free from the carpellary wall, ovary completely unilocular, archesporium differentiates in the entire subepidermal region of the placenta: *Amyema gravis* (Schaeppi and Steindl 1942), *Helicanthes elastica* (Fig. 12.8.1C_1, C_2), *Tapinostemma acaciae* (B.M. Johri and Prakash 1965).

4) Placenta three-lobed at the base only, free from the ovary wall, ovary completely unilocular, archesporium differentiates in each lobe of placenta: *Macrosolen cochinchinensis* (Fig. 12.8.1D_1,D_2).

5) Placenta three-lobed at the base, fused with the ovary wall in between the lobes, free at the tip, ovary trilocular below and unilocular above, archesporium differentiates in each lobe: *Nuytsia floribunda* (Fig. 12.8.1E_1, E_2), *Lepeostegeres gemmiflorus* (Dixit 1958b), *Amylotheca dictyophleba* (Raj 1970a), *Tripodanthus acutifolius* and *T. flagellaris*.

6) Placenta four-lobed throughout, fused with the ovary wall in between the lobes, except at the tip where it is free, ovary tetralocular, archesporium differentiates in each lobe: *Lysiana exocarpi* (Fig. 12.8.1F_1, F_2), *Peraxilla tetrapetala*.

In contrast, *Phthirusa pyrifolia* (Kuijt and Weberling 1972) lacks the ovarian cavity which is characteristic of the family. The archesporium differentiates just above the collenchymatous pad (hypostase). Such a condition is not known in any other member of Loranthaceae, and *Phthirusa* needs to be reinvestigated.

Cocucci (1983, see also Cocucci and Venturelli 1982) does not accept the interpretation of mamelon as placenta. He refers to the ovules as "discrete" and "collective". Discrete ovules are equated to the lobes of mamelon, and unlobed mamelon with a massive sporogenous tissue to a collective ovule. In our opinion the equivalence of mamelon to a placenta is fully justified.

Distinct ategmic ovules comprise *only* the nucellus (Fig. 12.8.1$D_{1,2}$, $E_{1,2}$, $F_{1,2}$). The concept of a collective ovule is erroneous, since in the Loranthaceae there are no indications as if a number of ategmic ovules had fused to form a common structure, the so-called collective ovule.

The archesporium is uni- or multicellular (*Amyema*, *Lysiana*, *Peraxilla*), function/s directly as megaspore mother cells which undergo meiotic divisions and produce linear tetrads. Sometimes ⊥-shaped tetrads occur in *Amyema preissii* (Dixit 1958b). The chalazal megaspore develops into a Polygonum type of embryo sac (Figs. 12.8.2B, C, E, 12.8.7J).

Sometimes, the subchalazal megaspore develops (*A. preissii*). An ovary may show archesporial and megaspore mother cells, tetrads (*A. preissii*), and two- and four-nucleate embryo sacs (*L. exocarpi* – R. Narayana 1958a).

The polar nuclei fuse below the egg apparatus as in *Tapinostemma acaciae* (B.M. Johri and S. Prakash 1965). The antipodals may be ephemeral or persistent (Figs. 12.8.4F, 12.8.5D). Occasionally, only two cells organize, one binucleate.

The formation of thick-walled saucer-shaped (collenchymatous pad) or tubular (collenchymatous tube) hypostase is a characteristic feature of Loranthoideae (Figs. 12.8.1 longisections, 12.8.2A, F, 12.8.5B, G, 12.8.6A, F, 12.8.7K′–P).

The elongation of embryo sacs is noteworthy (Fig. 12.8.2F). The lower end of the embryo sac extends up to the base (caecum) of hypostase, leaving the antipodals in situ (Fig. 12.8.4F). In *Nuytsia floribunda* (R. Narayana 1958b), the lower end of the embryo sac pierces the hypostase and grows beyond it. The upper end extends up to the base of style in *Amylotheca*, *Atkinsonia* (Fig. 12.8.2A), *Elytranthe*, *Macrosolen* (P. Maheshwari and Singh 1952) and *Peraxilla*, and slightly above the base of style in *Lepeostegeres* (Dixit 1958b). It grows to almost half- or two-thirds the height of style in *Amyema* (Dixit 1958a), *Dendrophthoe* (B. Singh 1952a), *Helicanthes* (B.M. Johri and Agrawal 1954), *Lysiana* (R. Narayana 1958a), *Nuytsia* (R. Narayana 1958b), *Struthanthus* (Venturelli 1984c), *Tolypanthus* (Dixit 1961) and *Tripodanthus* (Venturelli 1983). It reaches the stigma in *Barathranthus* (S. Prakash 1963), *Helixanthera hookeriana* (Schaeppi and Steindl 1942), *Scurrula* (Agrawal 1954), *Tapinostemma* (B.M. Johri and S. Prakash 1965) and *Tupeia* (Smart 1952). In *Helixanthera ligustrina* (Fig. 12.8.2C), it comes to lie just below the stigmatic epidermis (Fig. 12.8.2F). After reaching the apex of the stigma in *Moquiniella* (B.M. Johri and Raj 1965, 1969), the tips of embryo sacs curve backward into the style (Fig. 12.8.2D). Some of the embryo sacs, however, remain arrested in the style (Fig. 12.8.2F). In different genera of Loranthoideae, the tips of embryo sacs grow to varying heights in the style and stigma (Fig. 12.8.2F). There is no correlation in the disposition and structure of the mamelon and

Fig. 12.8.2. **A** *Atkinsonia ligustrina*, longisection of ovary, mature embryo sacs. **B** *Amyema miquelii*, tip of Polygonum type of embryo sac with egg apparatus and upper polar nucleus (at two-thirds the height of style). The tip of embryo sac in *Tripodanthus* also reaches up to two-thirds the height of the style. **C** *Helixanthera ligustrina*, longisection of stigma with tips of five unfertilized and fertilized embryo sacs, one reaches as far as the stigmatic epidermis. One of the embryo sacs shows two-celled proembryo. **D** *Moquiniella rubra*, longisection of stigma with tips of four embryo sacs (es_1–es_4), three reach up to the stigma and curve backward so that the egg apparatus/proembryo acquires reverse polarity. Two embryo sacs show zygote and one four-celled proembryo. **E** *Nuytsia floribunda*, tip of embryo sac in the stylar tissue. Note the lateral caecum. **F** Longisection of carpel. The names of different taxa indicate the height to which the tip of the embryo sac reaches in the style and stigma. (**A** After S. Prakash 1961, **B**, **F** after B.M. Johri and Bhatnagar 1972, **C** after P. Maheshwari and Johri 1950a, **D** after B.M. Johri and Raj 1965, **E** after R. Narayana 1958b)

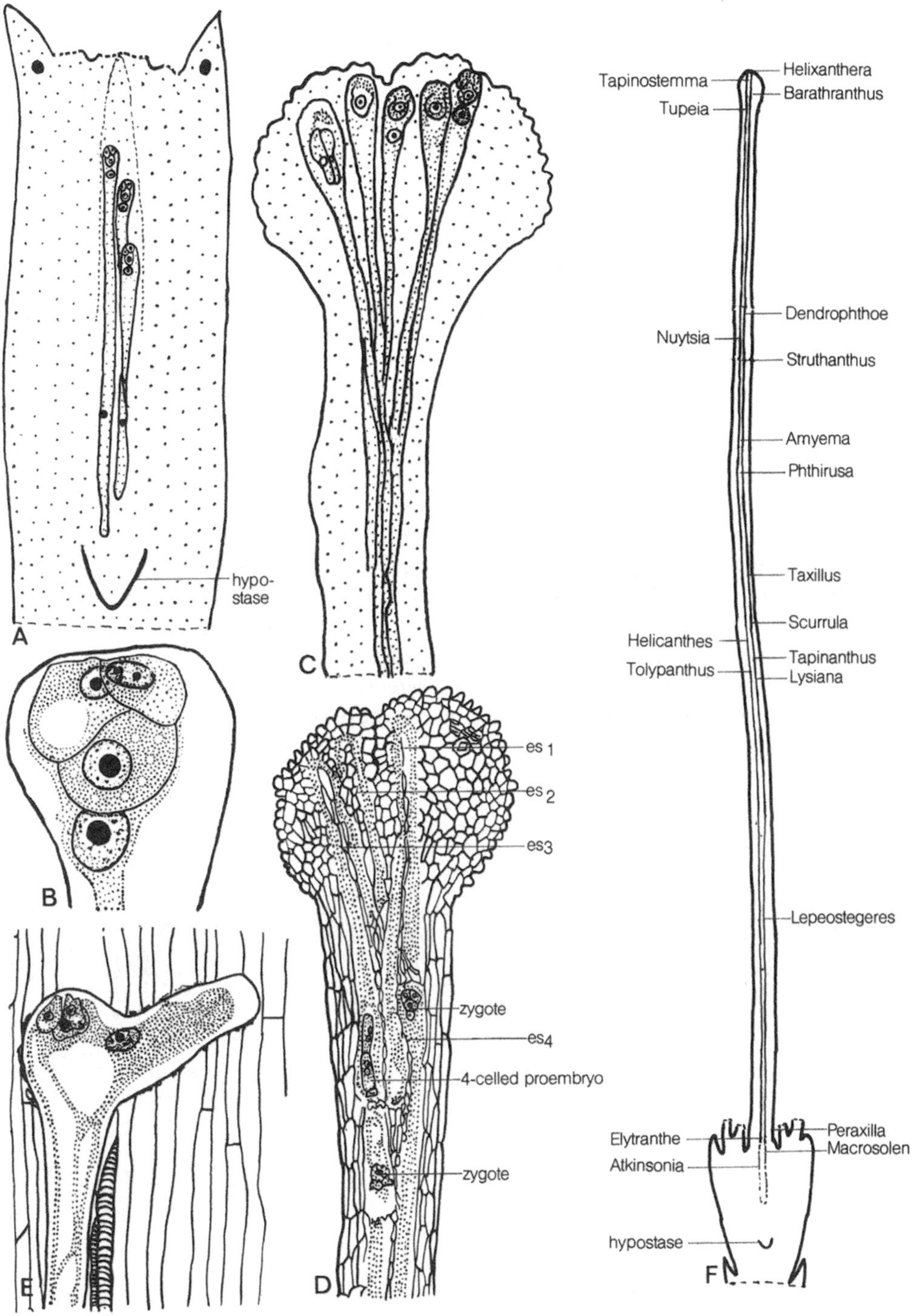
A
hypo-
stase
B
C
D
es 1
es 2
es3
zygote
es4
4-celled proembryo
zygote
E
F
Tapinostemma
Tupeia
Helixanthera
Barathranthus
Dendrophthoe
Nuytsia
Struthanthus
Amyema
Phthirusa
Taxillus
Scurrula
Helicanthes
Tolypanthus
Tapinanthus
Lysiana
Lepeostegeres
Elytranthe
Atkinsonia
Peraxilla
Macrosolen
hypostase

length of the embryo sac. This feature is not known in any other angiosperm. The significance, evolutionary or functional, is not clear.

The formation of six-nucleate embryo sacs, due to earlier division of the two chalazal nuclei of a four-nucleate embryo sac, is quite common. Several embryo sacs develop simultaneously in the same ovary. They digest the contents of the adjoining tissue rich in starch. In *Nuytsia* (Fig. 12.8.2E) and *Atkinsonia* (S. Prakash 1961), a lateral caecum develops at the tip of embryo sac, invading the circumjacent tissue.

The primary endosperm nucleus travels down to the lower part of the embryo sac. Its division is followed by a wall (P. Maheshwari and Singh 1952). The development of endosperm is of the Cellular type. The endosperm formation occurs in several embryo sacs in the same ovary (Fig. 12.8.3A); due to the growth of individual endosperm the ovarian tissue is destroyed. The endosperms from different embryo sacs fuse to form a composite structure (Fig. 12.8.3B).

The collenchymatous pad (hypostase) is often shifted to a lateral position due to the downward growth of the endosperm, as in *Tolypanthus lagenifer* (Dixit 1961).

In *Tapinostemma* (B.M. Johri and Prakash 1965), the basal end of the endosperm produces finger-like haustorial processes (Fig. 12.8.4F). These processes invade the adjoining tissue, and sometimes penetrate even through the thick-walled cells of the hypostase (Fig. 12.8.4I). Such a development has not been observed in any other member of Loranthoideae.

The only exception is *Struthanthus vulgaris* (Fig. 12.8.5A, C–G, I), where a composite structure is not formed and the endosperm develops from only one

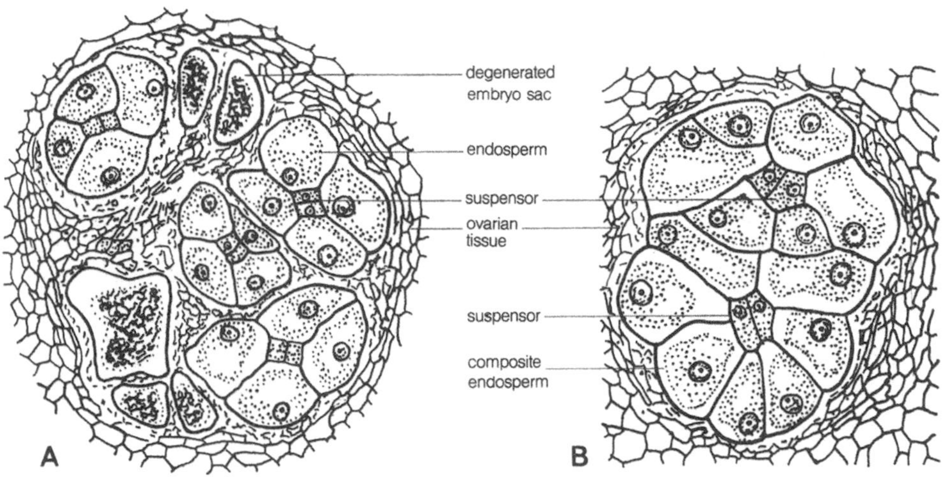

Fig. 12.8.3A, B. *Tolypanthus involucratus.* **A** Transection of ovary with four embryo sacs each containing four-seriate endosperm and biseriate suspensor (*t.s.*) of proembryos. Five embryo sacs have degenerated. **B** Later stage, composite endosperm with two biseriate suspensors. (After Dixit 1961)

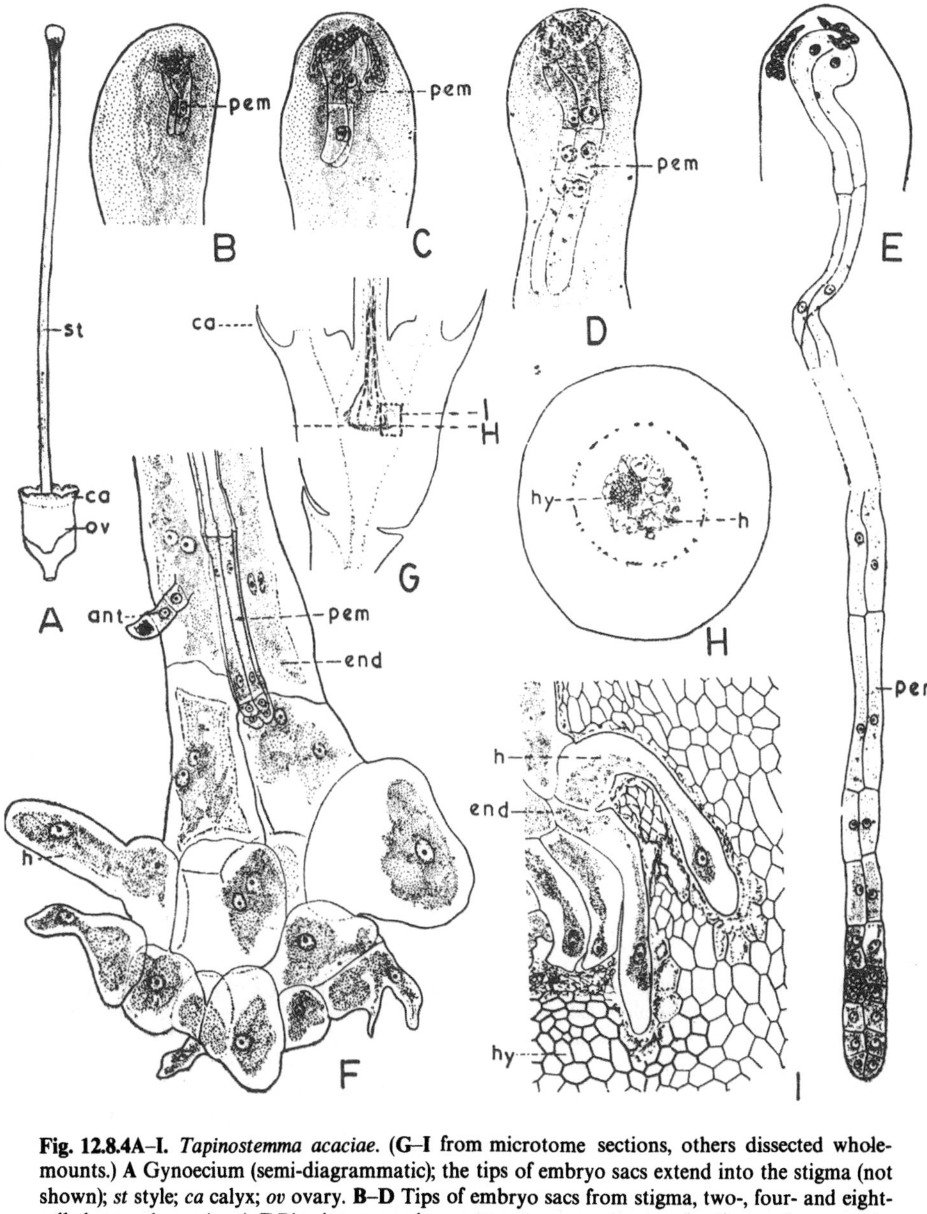

Fig. 12.8.4A–I. *Tapinostemma acaciae*. (**G–I** from microtome sections, others dissected whole-mounts.) **A** Gynoecium (semi-diagrammatic); the tips of embryo sacs extend into the stigma (not shown); *st* style; *ca* calyx; *ov* ovary. **B–D** Tips of embryo sacs from stigma, two-, four- and eight-celled proembryos (*pem*). **E** Biseriate proembryo with a long suspensor and embryonal mass at the tip. **F** Lower part of endosperm (*end*), cells with haustorial (*h*) processes. The tip of the proembryo is in the upper middle part of the endosperm; *ant* persistent antipodal cells. **G** Longisection of ovary to show lower end of several embryo sacs. **H** Cross-section of ovary at level *H* marked in **G**. **I** *Inset I* from **G**, haustorial (*h*) extensions of endosperm cells; *hy* hypostase. (After B.M. Johri and Prakash 1965)

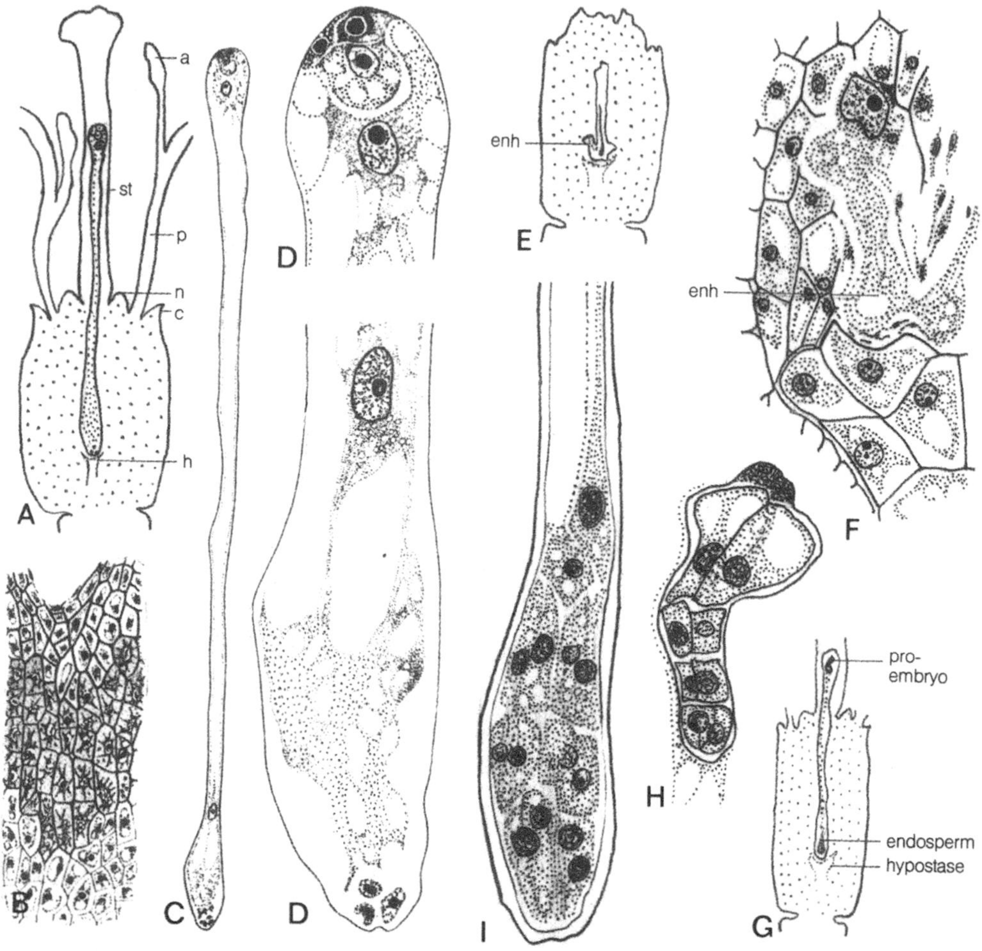

Fig. 12.8.5A–I. *Struthanthus vulgaris*. **A** Longisection of flower to show elongation of embryo sacs in style; *a* anther of epipetalous stamen; *st* style; *p* petal; *n* nectary; *c* calyculus; *h* hypostase. **B** Hypostase with thick-walled cells, from **A**. **C** Eight-nucleate embryo sac from **A**. **D**, **D′** Micropylar (**D**) and chalazal (**D′**) region from **C**. **E** Ovary with an embryo sac containing endosperm (formed from a *single* embryo sac) with lateral haustorium (*enh*). **F** Endosperm haustorium from **E**. **G** Portion of ovary with style to show an embryo sac with biseriate proembryo, and cellular endosperm. **H** Proembryo from **G**. **I** Endosperm from **G**. (After Venturelli 1981a)

embryo sac as in other angiosperms. This feature needs to be re-examined even though the author confirms that her findings are correct (pers. comm. 1987).

The division of zygote is vertical (Fig. 12.8.4B) (Griffith 1843), and transverse divisions follow to form a long biseriate proembryo (Figs. 12.8.4B–E, 12.8.5G, H). The terminal tier of the proembryo undergoes repeated divisions forming an embryonal mass. Several proembryos develop simultaneously in the style, but only a few grow down to the ovary. Normally, a single embryo reaches maturity.

With continued growth of the suspensor, it elongates, grows through the endosperm, and reaches the ovary. Figure 12.8.3A shows four suspensors (t.s.), one in each endosperm, B shows two suspensors in a composite endosperms. In Figure 12.8.4F, a proembryo has entered the composite endosperms (l.s.). The endosperm and the proembryo/s grow concurrently.

When the embryonal mass reaches the base of the endosperm, it gradually develops into a globular, heart-shaped and dicotyledonous embryo (Figs. 12.8.6A, D, 12.8.7M–P). With the progressive growth and upward shift of the embryo, the suspensor/s become coiled.

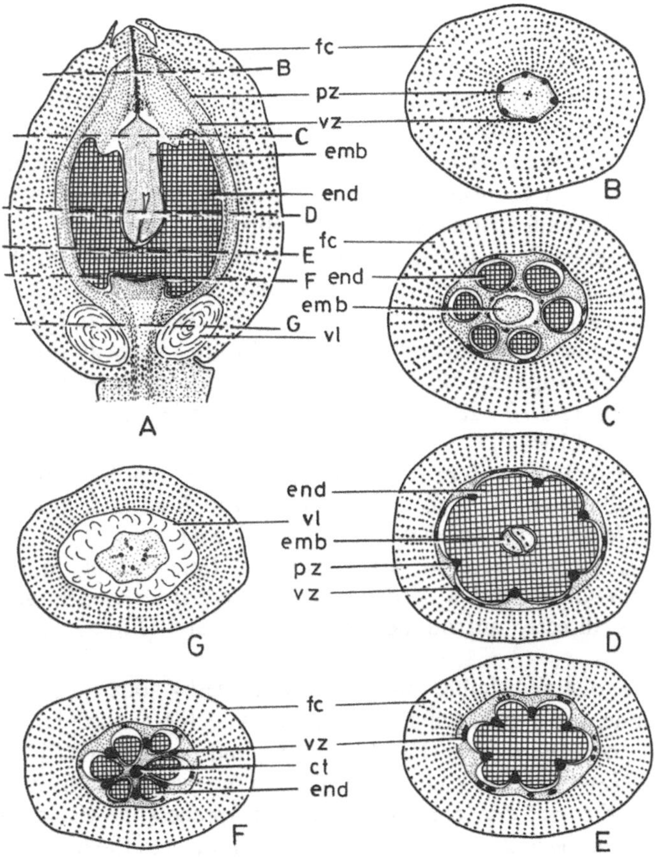

Fig. 12.8.6A–G. *Lepeostegeres gemmiflorus.* **A** Longisection of fruit, the pericarp shows fleshy coat (*fc*), parenchymatous zone (*pz*), vascular zone (*vs*), and viscid layer (*vl*) restricted only to the base. The endosperm (*end*) encloses the embryo (*emb*) except at the radicular end. The *black thick line* in the middle and at the base of endosperm represents the collenchymatous tube (*ct*, *black* central circle in **F**). **B–G** Transections of fruit at respective levels marked in **A**. **B** Pericarp; **C** Six apical prongs (*cross-hatched*) of endosperm, and radicular end of embryo (*emb*); **D** Six-lobed endosperm with vascular supply in between each lobe. Note the two cotyledons in the middle. **E**, **F** Cross-sections at levels below the embryo; **F** shows six separate lobes of endosperm; **G** Note the circular viscid layer (*vl*). (After Dixit 1958b)

In a mature embryo, the radicular end (= hypocotyl; a true radicle is absent) projects beyond the endosperm (Figs. 12.8.6A, C, 12.8.7P). The radicular end is generally studded with papillate projections and hairs (*Lysiana exocarpi* – R. Narayana 1958a). *Amylotheca dictyophleba, Amyema pendula, Dendrophthoe neelgherrensis* and *Helicanthes elastica* also show similar outgrowths (see B.M. Johri and Bhatnagar 1972).

The mature embryo may show two separate cotyledons, or the cotyledons may become closely appressed to each other (except in the region of plumule). Such a condition has been designated as "pseudomonocotyledonous", and is reported in *Amyema, Barathranthus, Dendrophthoe, Helicanthes, Helixanthera, Lysiana, Scurrula, Tapinostemma* and *Tolypanthus* (see B.M. Johri and Bhatnagar 1972).

Maturity of the endosperm is accompanied by several conspicuous features. At the base it grows peripherally around the hypostase. The lateral growth brings about a lobed condition due to the growth of endosperm around the vascular strands. At the apex, the endosperm grows in between the vascular strands. Thus, the mature endosperm looks like a flower vase with a number of lobes and apical prongs corresponding to the number of vascular strands (Fig. 12.8.6A–G). The endosperm cells contain tannin and starch.

Venturelli (1981a) has formulated a reproductive calendar (as in Brazil) for *Struthanthus vulgaris* (Fig. 12.8.7A–P). In an elegant study, she emphasizes the correlation between the size of bud, flower and ovary, and temporal development of microsporogenesis (Fig. 12.8.7A–E), megasporogenesis (Fig. 12.8.7G–I), male (Fig. 12.8.7E, F) and female gametophyte (Fig. 12.8.7J, K), pollination (Fig. 12.8.7E′, K′), fertilization (Fig. 12.8.7K, L), endosperm and embryo (Fig. 12.8.7M–P).

The male and female buds appear in January, bloom in April, followed by entomophilous pollination in April and fertilization in May and June. The endosperm, embryo and fruit develop from July to November.

The fruit is a pseudodrupe in *Atkinsonia* and *Gaiadendron*, in other members it is a pseudoberry. The seed is "naked" since it lacks a testa, and the endosperm is directly surrounded by the pericarp (Figs. 12.8.6A–E, 12.8.7M–P).

The pericarp is distinguishable into (a) an outer fleshy zone followed by (b) the viscid zone, (c) a parenchymatous zone, and (d) a vascular zone (Fig. 12.8.6A–G). In *Lepeostegeres* (Fig. 12.8.6A, G) and *Peraxilla* (B.M. Johri and Bhatnagar 1972), the viscid zone is restricted to the base of the fruit only.

The birds pick the fruit, try to crush it and in this process the viscin (from viscid layer enclosing the endosperm and the embryo) sticks to the beak. In its efforts to get rid of the tenaciously sticking viscin, the bird causes wounds on the bark and the endosperm sticks to the bark. On germination, the radicular end produces a discoid hold-fast and establishes itself on the host. The plumular leaves also develop, and finally the seedling.

Further studies of the uninvestigated genera, especially from Africa, Australia and South America, would be very rewarding.

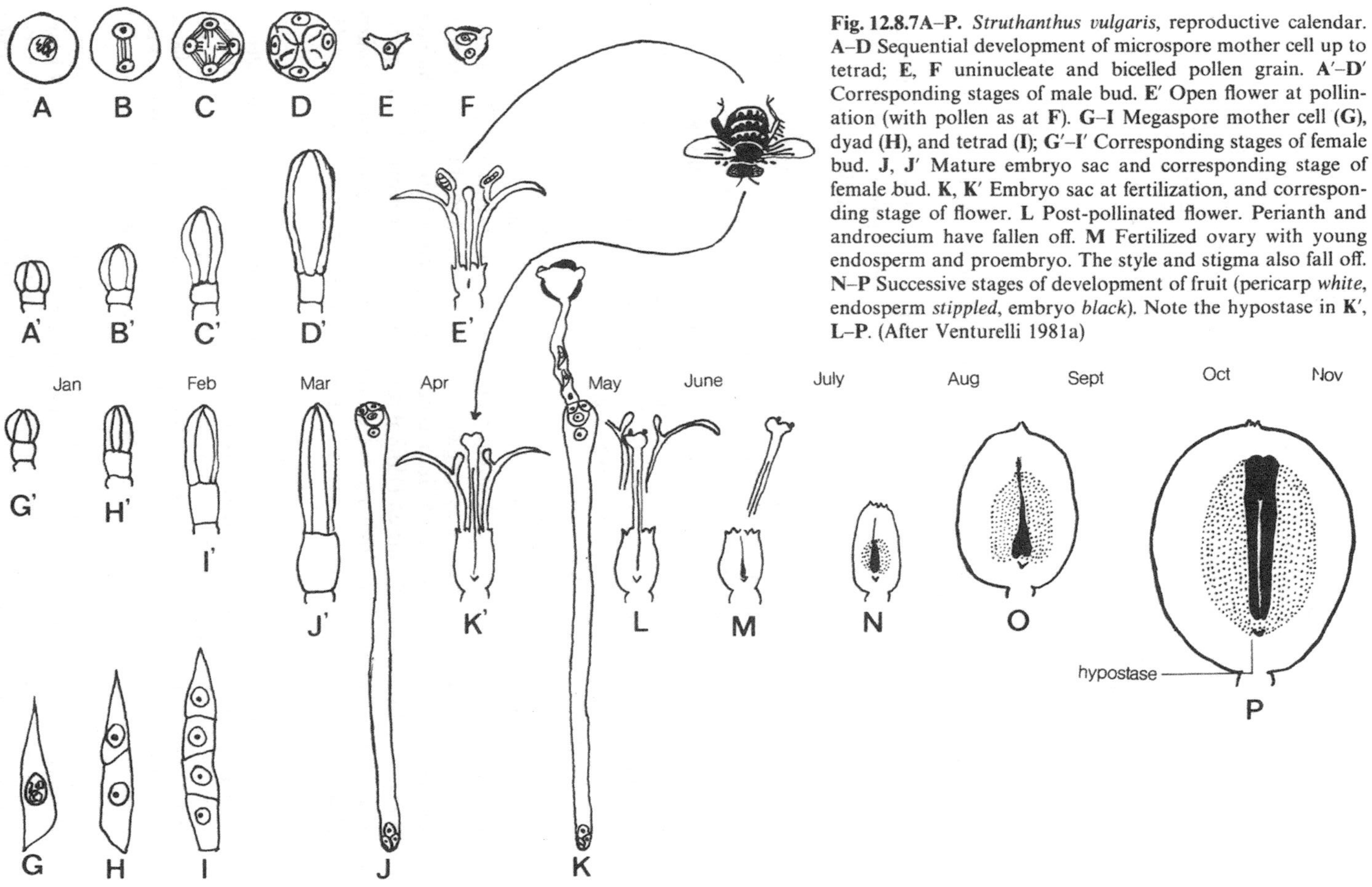

Fig. 12.8.7A–P. *Struthanthus vulgaris*, reproductive calendar. **A–D** Sequential development of microspore mother cell up to tetrad; **E**, **F** uninucleate and bicelled pollen grain. **A′–D′** Corresponding stages of male bud. **E′** Open flower at pollination (with pollen as at **F**). **G–I** Megaspore mother cell (**G**), dyad (**H**), and tetrad (**I**); **G′–I′** Corresponding stages of female bud. **J**, **J′** Mature embryo sac and corresponding stage of female bud. **K**, **K′** Embryo sac at fertilization, and corresponding stage of flower. **L** Post-pollinated flower. Perianth and androecium have fallen off. **M** Fertilized ovary with young endosperm and proembryo. The style and stigma also fall off. **N–P** Successive stages of development of fruit (pericarp *white*, endosperm *stippled*, embryo *black*). Note the hypostase in **K′**, **L–P**. (After Venturelli 1981a)

JS Agrawal (1954), BA Barlow (1983), SP Bhatnagar and S Chandra (1968), SP Bhatnagar and R Garg (1984), SP Bhatnagar and BM Johri (1983), AE Cocucci and M Venturelli (1982, 1984), BH Danser (1933), SN Dixit (1954, 1958a, b, 1961), W Griffith (1843), BM Johri (1956, 1987a, b), BM Johri and JS Agrawal (1954), BM Johri and KB Ambegaokar (1984), BM Johri and SP Bhatnagar (1972), BM Johri and S Prakash (1965), BM Johri and B Raj (1965, 1969), BM Johri, JS Agrawal and S Garg (1957), RN Kapil and IK Vasil (1963), J Kuijt (1967, 1969), J Kuijt and F Weberling (1972), P Maheshwari and BM Johri (1950a), P Maheshwari and B Singh (1952), P Maheshwari, BM Johri and SN Dixit (1957), R Narayana (1954, 1955, 1956, 1958a, b), A Pisek (1924), S Prakash (1960b, 1961, 1963), B Raj (1970a, b), H Schaeppi and F Steindl (1942, 1945), B Singh (1952a), C Smart (1952), Ph van Tieghem (1895), M Treub (1883), M Venturelli (1980, 1981a, b, 1983, 1984a, b, c)

Loranthaceae: Viscoideae

Bhandari and Vohra (1983) and B.M. Johri (1987b) have reviewed the embryology of Viscoideae.

In *Arceuthobium minutissimum* (Bhandari and Nanda 1968a), the structure of the anther is rather unique. There is a single annular pollen sac forming a continuous ring around a central column of sterile cells called "columella" (Fig. 12.8.8A–F).

The anther wall comprises an epidermis, fibrous endothecium, one or two ephemeral middle layers, and glandular tapetum. In *A. minutissimum* the endothecium is absent, and fibrous thickenings develop in the epidermis which is designated epithecium (Fig. 12.8.8G–L). Simultaneous cytokinesis follows meiotic divisions in the microspore mother cell, and the tetrads are tetrahedral and decussate in *Korthalsella opuntia* (Bhandari and Chaube 1969).

The spherical pollen grains are deeply trilobed, and two-celled at the dispersal stage (Fig. 12.8.8M–P).

The central ovarian papilla (mamelon = placenta) develops from the base of the ovary (Fig. 12.8.9A), and contains one or more archesporial cells. The true ovules, with nucellus and integuments, do not differentiate. The archesporial cells function directly as megaspore mother cells, and the development of the embryo sac is of the Allium type (Fig. 12.8.9B–L). In *Arceuthobium douglasii*, Bhandari and Nanda (1968b) refute the earlier report of tetrasporic type (Jones and Gordon 1965). Bisporic Endymion type is reported in *Ginalloa* (Rutishauser 1937), *Korthalsella* (Rutishauser 1937, Schaeppi and Steindl 1945,

Fig. 12.8.8A–P. *Arceuthobium minutissimum.* **A** Transection of perianth lobe bearing stamen primordium at archesporial stage. **B** Transection of perianth lobe bearing sessile stamen with horseshoe-shaped archesporium. **C–E** Stamen has differentiated into columella (*col*) and horseshoe-shaped archesporium surrounding it. **F** Transection of stamen to show horseshoe-shaped archesporium and sterile central columella. **G, I, K, M** Longisection of male flowers to show stamens at various stages of development. **H** *Inset H* from **G**. Some cells of the outermost archesporial layer have formed parietal layer. The tapetum has already differentiated on the inner side. **J, L** *Inset J* and *L* in **I** and **K**. The anther wall shows epidermis, two- to three-layered glandular tapetum and sporogenous cells. Note the disintegration of wall layer in **L**. **N** Longisection of dehisced anther from **M**. The epidermis has a cuticle and fibrous bands (exothecium). Note also the remnants of degenerated columella. **O, P** Uni- and bi-celled pollen grain. (After Bhandari and Nanda 1968a)

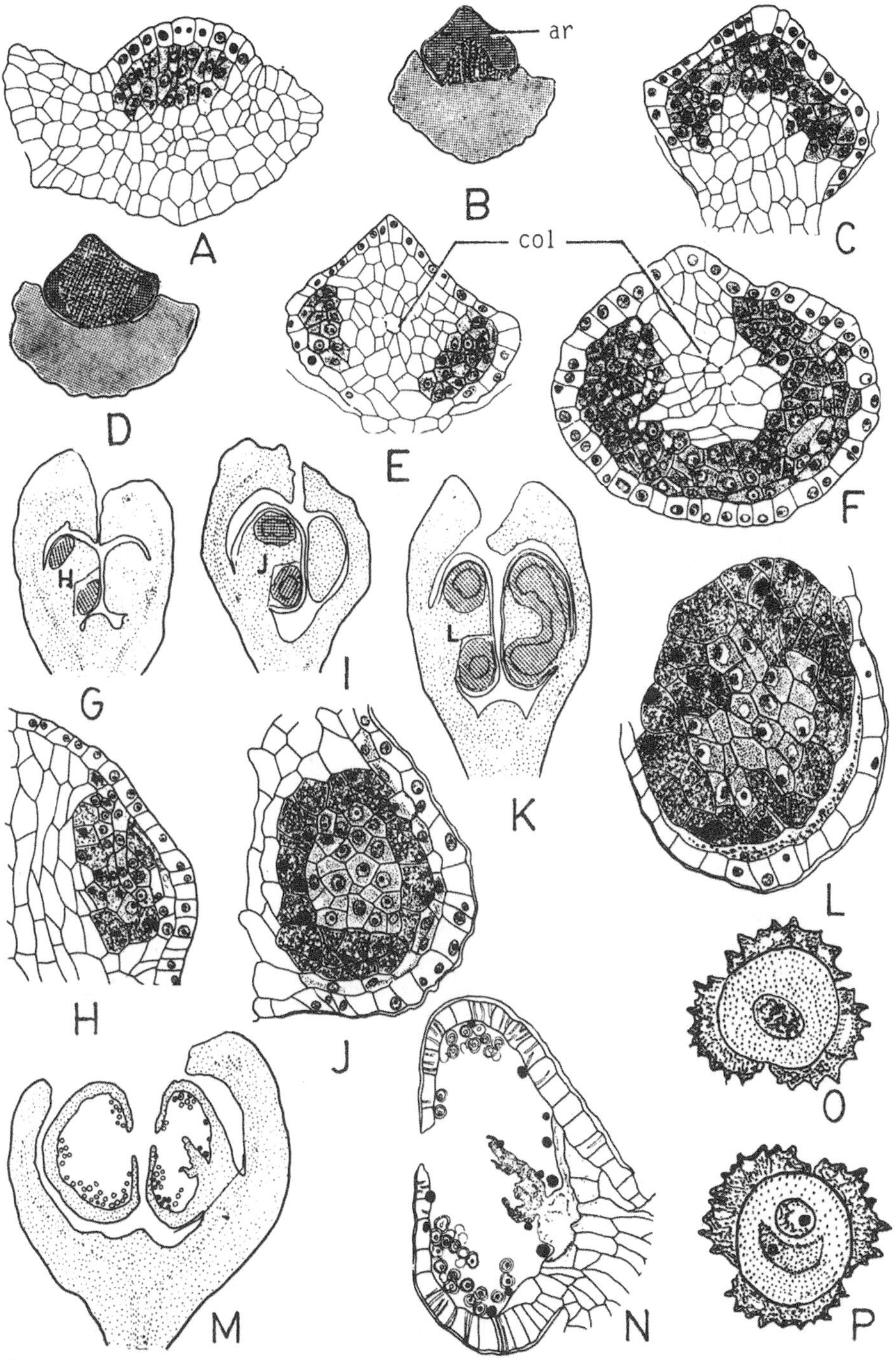
ar
col
A
B
C
D
E
F
G
H
I
J
K
L
H
J
L
M
N
O
P

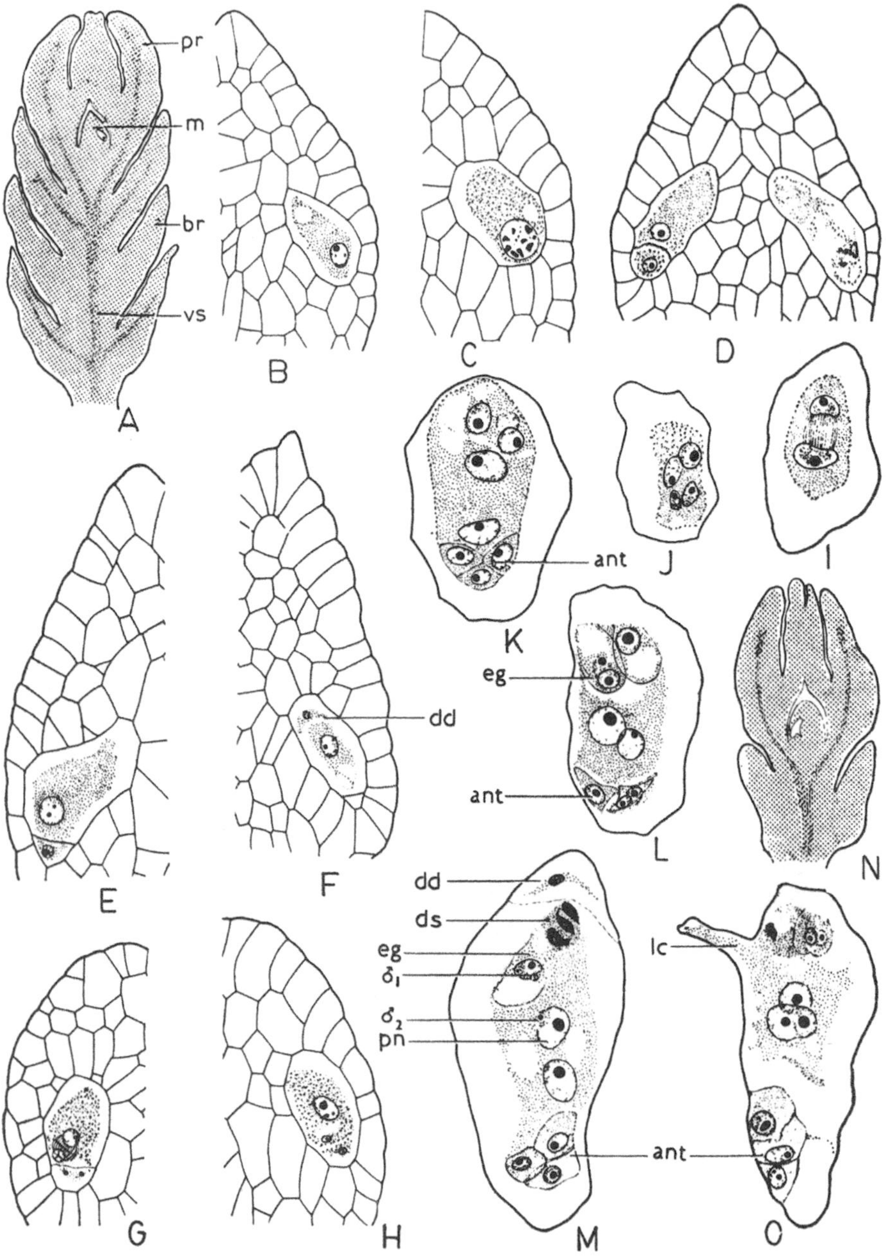

Fig. 12.8.9A–O. *Arceuthobium minutissimum.* **A** Longisection of terminal pistillate flower shows bracts (*br*), mamelon (*m*) and perianth (*pr*); *vs* vascular supply. **B, C** Longisection of mamelon with megaspore mother cell. **D** Mamelon with megaspore mother cell (*right*) and dyad (*left*). **E, F** Dyad. In **E** the chalazal dyad cell is functional, and in **F** micropylar; *dd* degenerated dyad. **G** Nucleus of lower dyad cell in division; the upper dyad shows two degenerated nuclei. **H** Lower dyad cell still uninucleate; the upper dyad cell has divided transversely but both the nuclei have

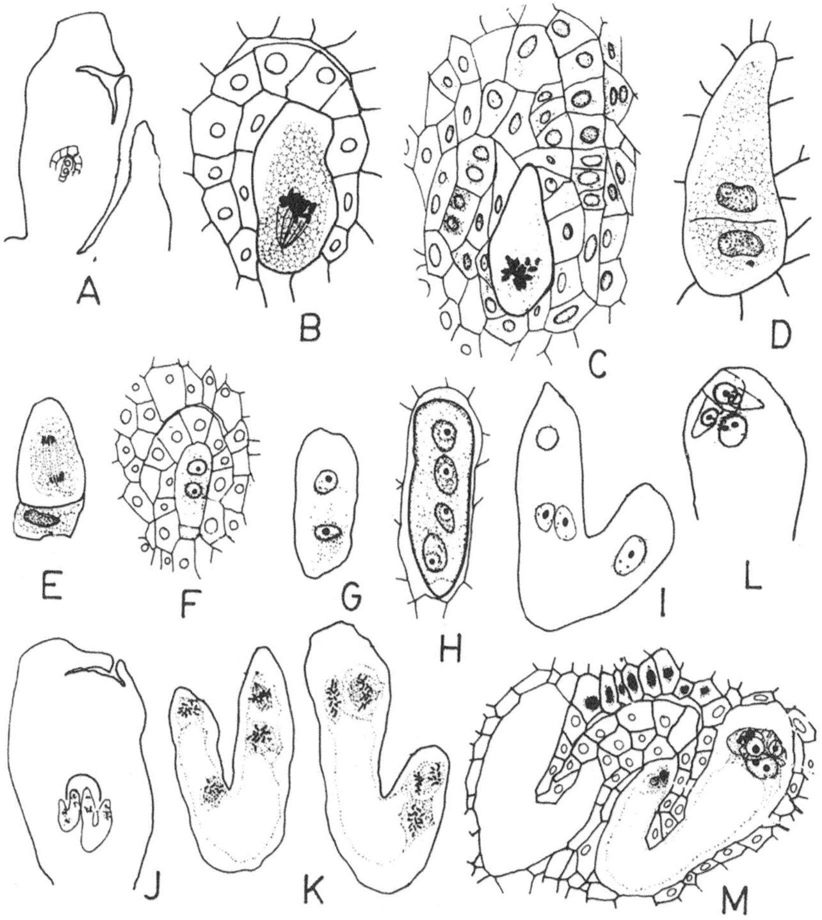

Fig. 12.8.10A–M. *Korthalsella japonica.* **A** Longisection of female flower at binucleate embryo sac. **B, C** Megaspore mother cell in metaphase I. **D** Dyad. **E** Micropylar dyad at anaphase. **F** Binucleate embryo sac with "ovular epidermis", from **A. G, H** Two- and four-nucleate embryo sac. **I** Four-nucleate embryo sac with the chalazal region curving laterally and upward. **J** Female flower with two curved four-nucleate embryo sac, nuclei in division. **K** Embryo sacs from **J. L** Micropylar portion of embryo sac with egg apparatus and secondary nucleus. **M** Mature embryo sac (*right*). (From Yoshida and Kawaguchi 1970)

Correa 1958, Yoshida and Kawaguchi 1970), and *Viscum* (Schaeppi and Steindl 1945).

In *Arceuthobium* and *Viscum* the embryo sac is straight, or slightly curved, and in *A. minutissimum* (Bhandari and Nanda 1968a) the fertilized embryo sac develops a lateral caecum (Fig. 12.8.9N–O).

degenerated. **I–K** Two-, four- and eight-nucleate bisporic embryo sac; *ant* antipodals. **L** Organized embryo sac; *eg* egg cell. **M** Embryo sac at fertilization; *ds* degenerated synergid; *pn* polar nuclei. **N** Longisection of pistillate flower at fertilization. **O** Embryo sac from **N** to show syngamy and triple fusion, and lateral caecum (*lc*). (After Bhandari and Nanda 1968a)

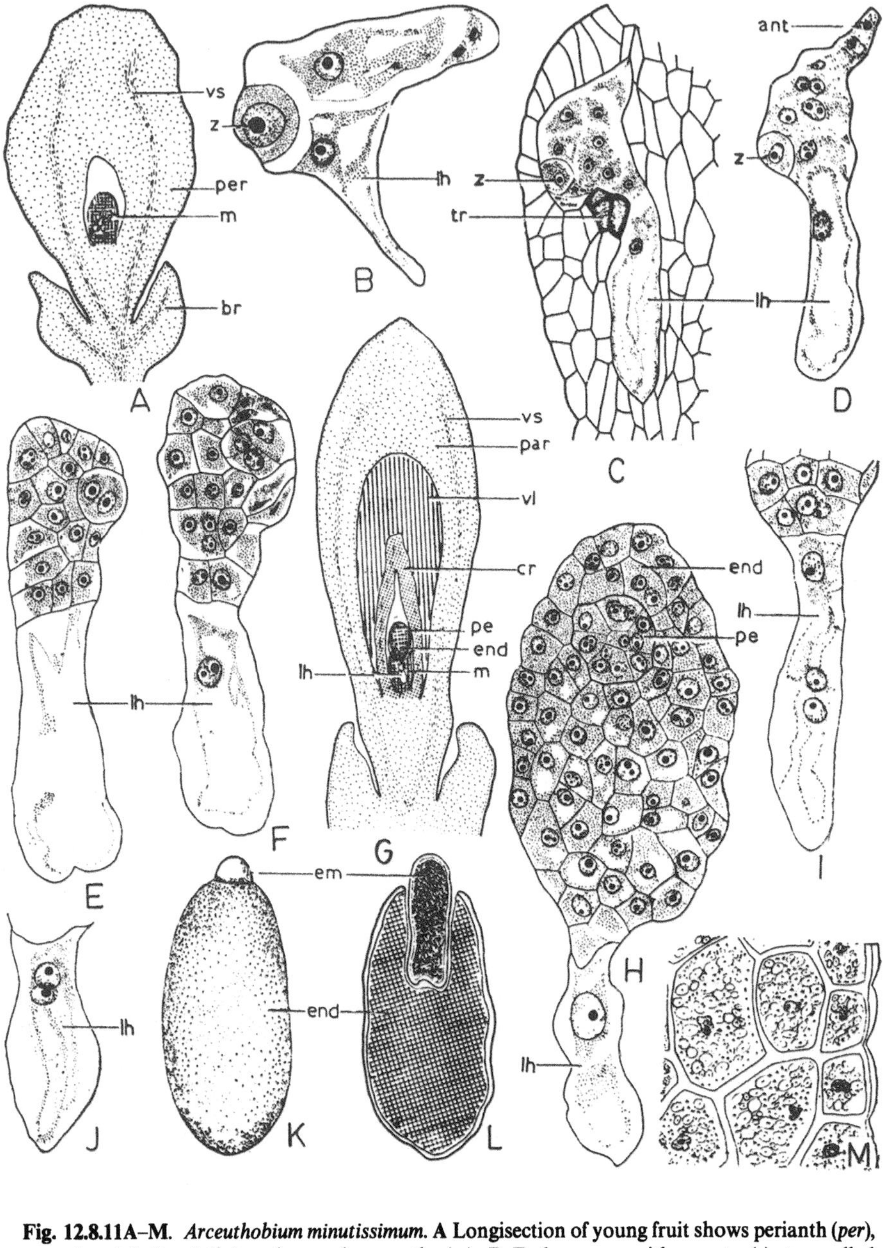

Fig. 12.8.11A–M. *Arceuthobium minutissimum.* **A** Longisection of young fruit shows perianth (*per*), mamelon (*m*), bract (*br*) and vascular supply (*vs*). **B** Embryo sac with zygote (*z*), two-celled endosperm, and lateral haustorium (*lh*); the degenerated antipodals persist. **C–F** Progressive development of endosperm. Note tracheids (*tr*) in the mamelon in **C**. The proembryo is two-celled in **E** and three-celled in **F**. **G** Longisection of fruit at preglobular proembryo (*pe*) with endosperm (*end*) and lateral haustorium (*lh*). The pericarp shows exocarp, vascular supply (*vs*), parenchymatous zone (*par*), viscid layer (*vl*) and crest (*cr*). **H** Endosperm, proembryo, and lateral haustorium from **G**. **I, J** Endosperm haustorium, three-celled in **I** and binucleate in **J**. **K, L** Mature endosperm encloses the embryo; **K** wholemount, **L** longisection almost at the same stage. Note the protruding radicular portion of embryo and absence of haustorium. **M** Mature endosperm tissue with starch-packed cells, the epidermis is covered with a cuticle. (After Bhandari and Nanda 1968a)

The bisporic embryo sac is U-shaped in *Dendrophthora, Ginalloa, Korthalsella* (Fig. 12.8.10A–M) and *Phoradendron*. At the four-nucleate stage it elongates and the basal end curves upward (carrying the nuclei) and penetrates the carpellary tissue (Fig. 12.8.10I–M). The chalazal end, thus, comes to lie at a higher level than the original micropylar end, and the egg apparatus differentiates at the upper pole. Generally, only one embryo sac develops in each papilla.

The development of endosperm is of the Cellular type (Fig. 12.8.11A, B). The division of the primary endosperm nucleus results in an unequal micropylar and chalazal chamber. The endosperm proper develops either from the micropylar chamber, or from both the chambers (Fig. 12.8.11C–H). In *Arceuthobium* (Bhandari and Nanda 1968a), the caecum contains one to three nuclei and behaves as a haustorium (Fig. 12.8.11H–J). The endosperm

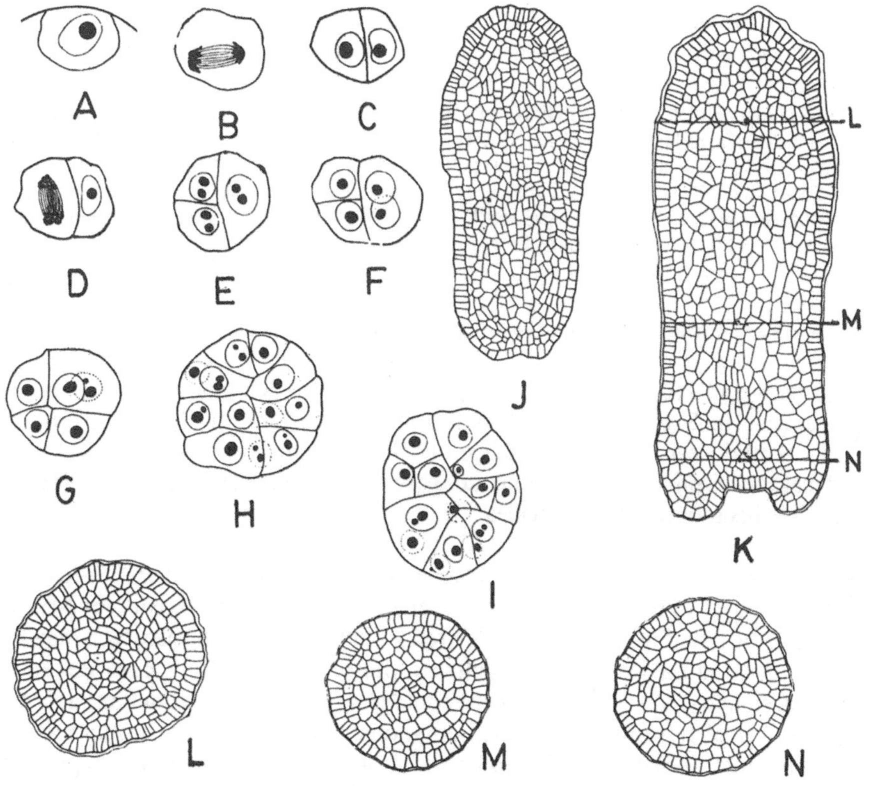

Fig. 12.8.12A–N. *Arceuthobium minutissimum.* **A** Zygote. **B, C** Vertical division in zygote. **D–I** Stages in the development of globular proembryo. **J** Longisection of bipolar torpedo-shaped proembryo; there is a continuous protoderm, and cotyledonary primordia have been initiated. **K** Longisection of mature embryo with thick cuticle, continuous protoderm, well-developed radicle, small cotyledons and shoot-apex, and absence of suspensor. **L–N** Cross-sections at levels *L, M* and *N* marked in *K*. (After Bhandari and Nanda 1968a)

is massive, and stores reserve food (Fig. 12.8.11K–M). The cells contain chlorophyll.

The division of zygote is followed by a transverse wall, except in *Arceuthobium* (Bhandari and Nanda 1968a, Correa 1958) and *Korthalsella* (Correa 1958) where it is obliquely vertical (Fig. 12.8.12A–C). Bhandari and Chaube (1969), and Yoshida and Kawaguchi (1970) describe the formation of a transverse wall in the zygote of *K. japonica* and *K. opuntia*. The suspensor does not differentiate. According to Cohen (1963), the embryogeny conforms to the Penaea variation of Asterad type in *A. americanum*, *A. campylopodum* and *A. douglasii*. Bhandari and Nanda report Dendrophthora variation of Piperad type in *A. minutissimum* (Fig. 12.8.12A–N).

The mature embryo is reduced. Only traces of vestigial cotyledons, as slight depressions on the cotyledonary pole, are reported in *A. pusillum* (Tainter 1968). In *A. minutissimum* (Cohen 1963, Bhandari and Nanda 1968a), *K. opuntia* (Rutishauser 1937) and *Viscum* the cotyledons are rudimentary.

The fruit is baccate, and the pericarp can be distinguished into four or five zones. In *Arceuthobium* it shows a leathery exocarp followed by parenchymatous zone, vascular zone, crest of viscid layer, and remnants of endocarp.

NN Bhandari and RK Chaube (1969), NN Bhandari and K Indra (1969), NN Bhandari and K Nanda (1968a, b), NN Bhandari and SCA Vohra (1983), SP Bhatnagar and S Chandra (1968), DM Calder (1983), LI Cohen (1963), JP Correa (1958), JD Hooker (1856), BM Johri (1962a, 1987a, b), BL Jones and CC Gordon (1965), J Kuijt (1955, 1969), HJ Lam (1948), R Narayana (1956), A Rutishauser (1935, 1937), H Schaeppi and F Steindl (1945), H Schulle (1933), FH Tainter (1968), HH York (1913), O Yoshida and H Kawaguchi (1970)

12.9 Taxonomic Considerations

Bentham and Hooker (1965c) group Loranthaceae, Santalaceae and Balanophoraceae in Santalales. Hutchinson (1973) placed Loranthaceae, Grubbiaceae, Santalaceae, Myzodendraceae and Balanophoraceae in Santalales, and Olacaceae, Opiliaceae, Octoknemataceae, Aptandraceae, Dipentodontaceae and Medusandraceae in Olacales.

Olacaceae. Kuijt (1969) regards Olacaceae – or some part thereof – as the starting point of four separate evolutionary lines: (a) Santalaceae, (b) Myzodendraceae could have only been derived from Olacaceae, (c) Viscaceae may be traceable to santalacean plants similar to present-day epiphytic parasites such as *Phacellaria*, and (d) Loranthaceae could not have evolved from Santalaceae. The flower of Loranthaceae is dichlamydous, and can scarcely have a monochlamydous ancestor in Santalaceae. Loranthaceae, therefore, represent the fourth evolutionary line originating in the Olacaceae, possibly through a *Chaunochiton*-like species.

Van Tieghem (1896), C.F. Reed (1955) and Hutchinson (1959, 1973) raise the family Olacaceae to an ordinal rank, Olacales. B.M. Johri and Bhatnagar (1960) compare the embryological features of Olacaceae and Santalaceae (Table 12.9.1), and conclude that both the families show marked similarities

Table 12.9.1. Embryology of Olacaceae and Santalaceae (After B.M. Johri and Bhatnagar 1960)

	Olacaceae	Santalaceae
Habit	Autotrophic or root parasite	Root parasite
Flower	Unisexual or bisexual, mono- or dichlamydeous, disc present	Unisexual or bisexual, monochlamydeous, disc present
Anther tapetum	Glandular, cells 2- or 4-nucleate	Glandular, cells uni- or multinucleate
Pollen	Tricolpate, 2- or 3-celled	Tricolpate, 2- or 3-celled
Ovary	Superior, inferior or semi-inferior, unilocular, sometimes chambered at base and unilocular above	Superior, semi-inferior or inferior, unilocular or 2–5-chambered at base and unilocular above
Ovule	1–5, borne on central placenta, anatropous, differentiation into nucellus and integument lacking (*Olax imbricata*), unitegmic (*Olax wightiana*, *Strombosia*), or bitegmic (*Coula*, *Ximenia*)	1–5, pendulous, borne on central placenta, hemianatropous or anatropous, differentiation into nucellus and integument lacking (*Santalum*), or unitegmic (*Thesium*); ovule absent in *Exocarpus*.
Embryo sac	Polygonum or Allium type, micropylar end may or may not extend beyond the ovule, chalazal caecum is formed	Polygonum, tip of embryo sac may or may not extend beyond the ovule, chalazal caecum is formed
Endosperm	Cellular and chalazal haustorium unicellular in *Strombosia*; Helobial in *Olax* and chalazal chamber haustorial	Cellular, or Helobial (*Santalum*, *Mida*), chalazal haustorium uni- or multicellular, secondary haustoria develop in *Osyris* and *Iodina*
Division of zygote	Zygote divides transversely; suspensor multiseriate	Zygote divides transversely or longitudinally (*Comandra*), suspensor present or absent

except for the presence of calyx and bitegmic ovules in some cases in Olacaceae. Even the comparison of the placental-ovular complex of Olacaceae and Santalaceae (Fagerlind 1947a, 1948) and floral anatomy (F.H. and E.C. Smith 1942) indicate their close alliance. B.M. Johri and Bhatnagar emphasize that there is no justification to raise Olacaceae to ordinal rank as suggested by Van Tieghem, C.F. Reed and Hutchinson.

The genus *Octoknema* has been given a family status: Octoknemataceae. Only scanty data are available about their embryology. According to Fagerlind (1948), the structure of wood, pollen grains, ovary, placenta, and development of the embryo sac resemble the members of Olacaceae. Fagerlind (1948), Stauffer (1957) and B.M. Johri and Bhatnagar (1960) do not feel the necessity of assigning it to a separate family. Its merger with Olacaceae appears to be more appropriate.

Opiliaceae. The spherical pollen grains, Polygonum type of embryo sac, and unitegmic ovule in Opiliaceae resemble those of *Thesium*, whereas the undifferentiated ovule of *Agonandra* shows similarity with that of *Exocarpus*. The chalazal caecum penetrating the thalamus in *Opilia*, and the much-branched chalazal haustorium with its secondary haustoria in *Cansjera* (Swamy 1960), are comparable to the condition in Santalaceae. Fagerlind (1948) suggests its inclusion as the tribe Opilieae under Santalaceae (see also B.M. Johri and Bhatnagar 1960).

Grubbiaceae. Endlicher (1840) and de Candolle (1857) established Grubbiaceae as a distinct family; Bentham and Hooker (1965c) do not recognize the family status and reduce it to the tribe Grubbieae under Santalaceae.

Van Tieghem (1897) studied the structure of flower and fruit of *Grubbia*. The bicarpellate ovary with axile placentation, anatropous, unitegmic ovule, and integument developing into seed-coat (as in Grubbiaceae) has no parallel in the families of Santalineae. Consequently, he refers the family to Hamamelidales, with which we do not agree.

In Grubbiaceae the ovule has a thick integument with a long micropyle; the embryo sac lies in the micropyle. The endosperm develops micropylar and chalazal haustoria in *Grubbia* (Fagerlind 1947b); the peripheral layer of endosperm is thickly cuticularized and the food reserve is proteinaceous. The embryo occupies the entire length of seed. In view of these features, Fagerlind (1948) relates Grubbiaceae to Ericaceae. B.M. Johri and Bhatnagar (1960) point out that similar features also occur in Olacaceae and Santalaceae (see Schulle 1933, Pilger 1935, M. Ram 1957, 1970a, S.P. Bhatnagar 1959). Therefore, there is no justification for relating Grubbiaceae to Ericaceae.

The pollen grains are different from those of Santalaceae. The occurrence of an endothelium, integument developing into seed-coat, and micropylar endosperm haustorium in *Grubbia* warrant the erection of a separate family, the Gubbiaceae, in the Santalineae.

Santalaceae. Apart from the general features of the order Santalales, Santalaceae exhibit several interesting features: (1) polynucleate pollen in *Santalum obtusifolium* (S.P. Bhatnagar 1966), (2) embryo sac-like structures from male archesporial/sporogenous cells in *Leptomeria billardierii* (M. Ram 1959c), (3) unitegmic (*Comandra*, *Iodina* and *Thesium*), or undifferentiated ovule (*Exocarpus*), (4) Polygonum, Allium and Endymion type of embryo sac; formation of extra-ovular extensions of embryo sac, apical and lateral caeca in *E. menziesii* (Fagerlind 1959), (5) antipodal and synergid haustoria in *Quinchamalium*, (6) Cellular, and Helobial (*Mida* and *Santalum*) type of endosperm, (7) profusely branched haustoria of *Iodina* with hypertrophied nucleus, (8) transverse division of zygote; vertical in *Comandra* and *Quinchamalium*, (9) cutinized epidermis of endosperm with proteinaceous food reserve, (10) absence of seed-coat, and (11) embryo occupying major part of the seed.

The position of *Exocarpus* has been controversial. Gagnepain and Boureau (1947) suggest that *Exocarpus* should be assigned to a separate family, Exocarpaceae, and placed near the Taxaceae under gymnosperms. Lam (1948) includes *Exocarpus* in Protangiospermae along with *Casuarina* and *Salix*. M. Ram (1959a) investigated the embryology of *Exocarpus cupressiformis*, *E. sparteus* and *E. strictus* and emphasizes their affinities. The flower has a typical angiospermous structure. The anther wall develops normally, pollen grains are two-celled, there is no differentiation of ovule, the archesporium differentiates at the apex of the "placenta", and development of the embryo sac is of the Polygonum type. The Cellular type of endosperm develops a chalazal haustorium. The division of zygote is transverse, and the embryo shows a suspensor and two cotyledons. The seed is enclosed in pericarp. These features indicate its alliance with Santalaceae. It differs from the latter in having a straight, instead of a U- or N-shaped embryo sac, and a long, proliferated suspensor (see B.M. Johri 1963b, P. Maheshwari and Kapil 1963). The embryological features are not comparable to those of *Casuarina* or *Salix*. Therefore, *Exocarpus* should be assigned to Santalaceae.

Myzodendraceae. The genus *Myzodendron* has been assigned to the tribe Osyrideae in Santalaceae. Rendle (1959) confirmed its inclusion in Santalaceae. The common features between *Myzodendron* and Santalaceae are: (1) receptacular ovary, (2) three pendulous and undifferentiated ovules on a free central placenta, (3) Polygonum type of embryo sac, and (4) Cellular type of endosperm with a conspicuous chalazal haustorium. *Myzodendron* differs from Santalaceae in the structure of flower, pollen grains, and presence of rudimentary testa. Embryologically, *Myzodendron* (Kuijt 1969) shows clear affinities with Olacaceae. The ovules are undifferentiated, a parenchyma layer covers the embryo sac, and the antipodal end of embryo sac forms a long tubular haustorium which invades the placental column. The first cross-wall, isolating the haustorium, is probably formed after the division of the primary endosperm nucleus [cf. *Cansjera* (Opiliaceae), Swamy 1960]. The haustorium reaches up to the base of flower, and ramifies in that region. The features mentioned here are scarcely different from that of the Olacaceae. Kuijt emphasizes that *Myzodendron* could have been derived directly from Olacaceae. The erection of a separate family Myzodendraceae appears justified and is supported by studies on pollen morphology (Feuer 1981). A more detailed account would be helpful to determine its systematic assignment.

After considering the comparative embryology of the order Santalales,

B.M. Johri and Bhatnagar (1960) state that the suborder Santalineae should include four families – Olacaceae, Grubbiaceae, Santalaceae and Myzodendraceae – and the family Santalaceae should comprise six tribes: Comandreae, Thesieae, Osyrideae, Santaleae, Opilieae and Anthoboleae.

Loranthaceae. The suborder Loranthinae includes a single family, Loranthaceae. P. Maheshwari et al. (1957), B.M. Johri and Bhatnagar (1960), S.P. Bhatnagar and Johri (1983), Bhandari and Vohra (1983) and B.M. Johri (1987b, see also 1963b) discuss the affinities of the two subfamilies, Loranthoideae and Viscoideae. The Loranthoideae are characterized by the presence of calyculus, triradiate pollen grains, except in *Atkinsonia* and *Tupeia*, Polygonum type of embryo sac, in some members extension of female gametophyte into the style and stigma, formation of composite endosperm, and division of zygote followed by a vertical wall. The embryo has a long, tortuous suspensor. The viscid zone lies outside the vascular zone in the pericarp. In contrast, the Viscoideae usually do not have a calyculus (except in male flowers of *Viscum articulatum* and *V. orientale*), the pollen grains are spherical, and the U-shaped embryo sac is bisporic (Allium type). The endosperm develops from a single embryo sac. The zygote divides transversely, except in *Arceuthobium minutissimum* (Correa 1958) and *Korthalsella* (Rutishauser 1935, Correa 1958), and the embryo usually lacks a suspensor. The viscid zone lies inward to the vascular zone (see B.M. Johri 1963b, P. Maheshwari and Kapil 1966).

In view of the foregoing account, the elevation of Loranthoideae and Viscoideae to the family rank as Loranthaceae and Viscaceae (Table 12.9.2) is justified (B.M. Johri and Bhatnagar 1960, B.M. Johri 1963b, see also Kuijt 1969, Bhandari and Vohra 1983). Van Tieghem (1896) erected two independent orders, Loranthales and Viscales. B.M. Johri and Bhatnagar (1960) do not consider it necessary.

Cocucci (1983, see also Cocucci and Venturelli 1982) emphasizes the importance of gynoecium, mamelon, ovules (lobes of mamelon) and embryo in the taxonomy of Loranthaceae (Fig. 12.9.1). The primitive condition of gynoecium occurs in *Tripodanthus* where the ovary contains discrete ovules, i.e. they are ategmic but with well-differentiated nucellus (megasporangium) (Fig. 12.9.1). The two diverging tendencies concern the orientation of ovules met with in the tribes Lorantheae and Lysianeae.

In the first category (Elytrantheae and Lysianeae), the nucellar axis (micropylar-chalazal axis) is so oriented that the chalaza points towards the apex, becoming more parallel to the longitudinal axis of the ovary. The nucellus (lobe of mamelon) fuses with the carpellary tissue. The "compitum" (mamelon) enlarges and develops into amyliferous tissue in which the embryo sacs lie: *Peraxilla* → *Lepeostegeres* → *Lysiana* (Fig. 12.9.1).

In the second category (Tapinostemmeae and Lorantheae), the nucellar axis is erect and the chalaza points downward, i.e. towards the inferior position (Fig. 12.9.1). The ovary becomes unilocular, and the ovules unite to form a "collective" ovule (mamelon without lobes). The next step is the reduction and sinking of the collective ovule till the disappearance of the mamelon: *Tapinostemma* → *Helicanthes* → *Amyema* → *Moquiniella* (Fig. 12.9.1).

Table 12.9.2. Embryology of Santalaceae, Loranthoideae (Loranthaceae) and Viscoideae (Viscaceae) (After Bhandari and Vohra 1983)

	Santalaceae	Loranthaceae	Viscaceae
Pollen grains	Spherical or oblong, 2- or 3-celled	Triradiate, spherical in *Atkinsonia*, *Ixocactus*, *Oryctanthus* and *Tupeia*; 2-celled	Spherical, 2-celled
Ovary	Superior, semi-inferior, or inferior, unilocular or 2–5-chambered at base and unilocular above	Inferior, unilocular, often 3- or 4-chambered	Inferior, unilocular
Ovule	1–5, borne on central placenta, hemiana- or anatropous, uni-, bi- or ategmic (undifferentiated), ovule does not at all differentiate in *Exocarpus*	Ovule does not develop, placenta (mamelon) unlobed and conical (*Helicanthes*), 3- or 4-lobed (*Lysiana*), free or attached to ovary wall (*Lepeostegeres*)	Placenta (mamelon) conical (basal attachment), bilobed in Phoradendreae, absent in *Viscum*
Embryo sac	Monosporic, Polygonum type, may remain confined to the ovule or the tip may grow out towards the stylar canal	Monosporic, Polygonum type, multiple embryo sacs except in *Phthirusa pyrifolia* and *Struthanthus vulgaris*, tip remains limited to the ovary, or reaches up to different heights in the style and stigma. The base elongates up to the hypostase, in *Nuytsia* beyond the hypostase	Bisporic, Allium type (*Arceuthobium*, *Eubrachion*, *Viscum*), or Endymion (*Korthalsella*, *Phoradendron*), embryo sac usually straight but U-shaped in *Arceuthobium* and *Korthalsella*, lower end bends, extends beyond the papilla, and proceeds upward towards the carpellary tissue
Caecum	Chalazal caecum common, micropylar caecum reported in *Comandra*	Chalazal caecum common (antipodals left in situ), micropylar caecum in *Atkinsonia* and lateral in *Nuytsia*	Lateral caecum in *Arceuthobium* and *Korthalsella*
Endosperm	Cellular or Helobial type (*Mida*, *Santalum*), chalazal haustorium develops, secondary haustoria in *Comandra*, *Iodina*, *Mida*; non-chlorophyllous	Cellular, composite structure except in *Phthirusa* and *Struthanthus*, chalazal haustorium reported in *Tapinostemma*; non-chlorophyllous	Cellular endosperm develops individually in each embryo sac, non-aggressive chalazal haustorium in Phoradendreae; chlorophyllous
Embryo	Division of zygote transverse, vertical in *Comandra*, massive suspensor in *Mida*, small in *Leptomeria*, and absent in *Osyris* and *Thesium*	Division of zygote vertical, suspensor biseriate, long and tortuous	Division of zygote transverse, vertical in *Arceuthobium* and *Korthalsella*, suspensor absent in *Eubrachion ambiguum*
Testa	Integuments become crushed during endosperm development, testa absent	Absent; ovules lack differentiation of integument	Absent; ovules lack integument
Pericarp	Parenchymatous epicarp and stony mesocarp persist, parenchymatous endocarp consumed by endosperm	Leathery epicarp followed by viscid zone (mesocarp) which is external to vasculature of ovary, endocarp parenchymatous	Leathery or parenchymatous epicarp, viscid zone (mesocarp) internal to vasculature of ovary, endocarp parenchymatous

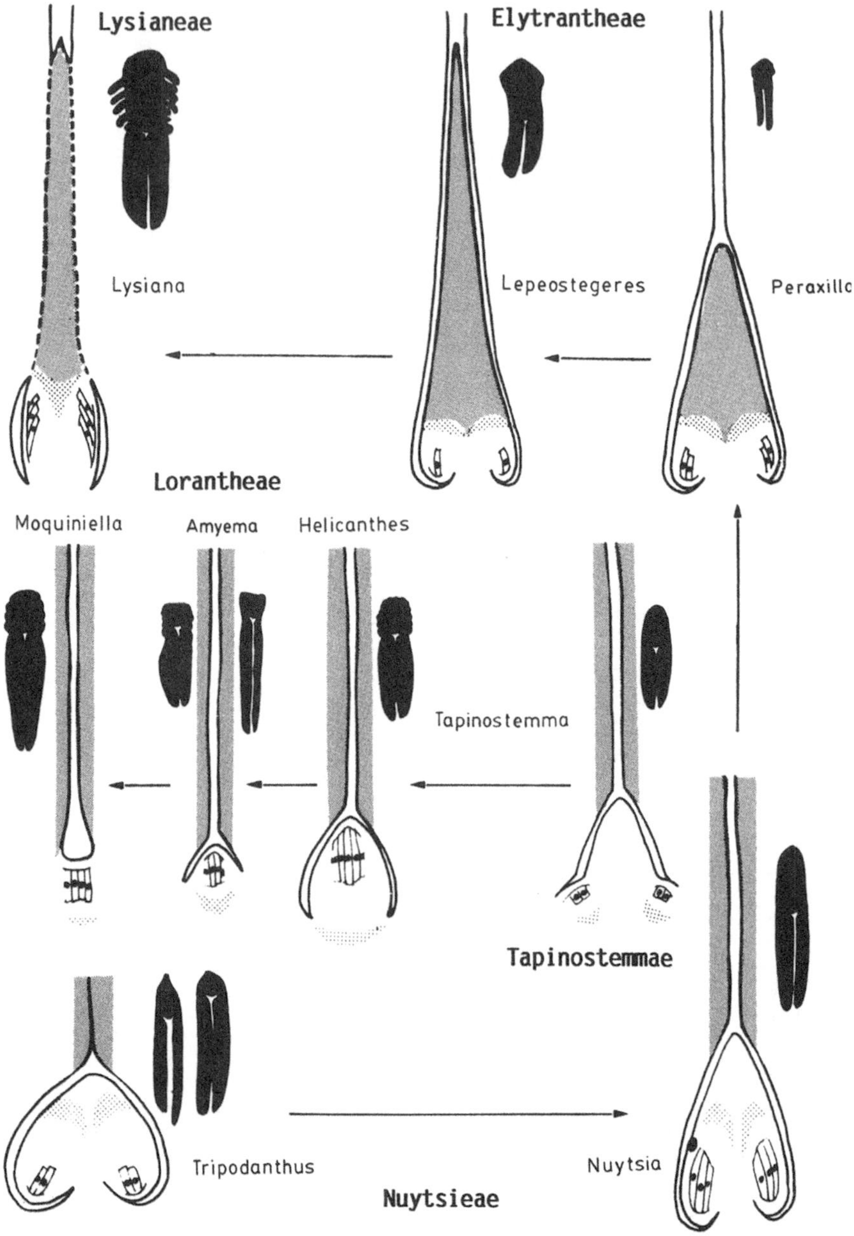

Fig. 12.9.1. Loranthaceae (Loranthoideae), evolutionary tendencies based on "gynoecium organization and embryo specialization". Embryo *black*, amyliferous tissue in style *shaded*, placenta *dotted*. (After Cocucci and Venturelli 1982)

With the above consideration Cocucci (1983) also correlates the condition of cotyledons in the mature embryo, i.e. free or fused. *Peraxilla* to *Lysiana* the cotyledons are free (in *Lysiana* the hypocotyl shows projections), and *Tapinostemma* to *Moquiniella* the cotyledons are partially fused. The primitive genus *Tripodanthus* shows both types of embryos – free as well as fused cotyledons (Fig. 12.9.1).

According to Danser (1933), who revised Engler and Prantl's (1897) classification of Loranthaceae, there are only three tribes: Elytrantheae, Nuytsieae and Lorantheae (see also Melchior 1964, B.M. Johri and Bhatnagar 1972).

1) Elytrantheae: Ovary multilocular, placenta with three to six basal lobes, each lobe represents only the nucellus (ategmic ovule). The placenta is amyliferous and provides nutrition to the developing megagametophyte. The style lacks amyliferous tissue. The tip of embryo sacs extends only up to the basal part of style, e.g. *Lepeostegeres, Macrosolen, Peraxilla.*

2) Nuytsieae: Ovary trilocular at the base, placenta with three basal lobes, each lobe represents only the nucellus. The style has amyliferous tissue and provides nourishment to the developing megagametophyte. The tips of embryo sacs reach up to two-thirds the height of styles, e.g. *Nuytsia.*

3) Lorantheae: Ovary unilocular, placenta unlobed when present, the tip of embryo sacs extends up to the middle of style or as far as the stigma, e.g. *Helicanthes, Helixanthera.*

On the basis of (a) uni-/multilocular ovary, (b) superior/inferior placenta, free/fused mamelon with tissue of stylar canal, (c) discrete/"collective" ovule, Cocucci (1983) suggests two additional tribes: Lysianeae and Tapinostemmae (Fig. 12.9.1).

4) Lysianeae: Ovary multilocular with "collective" ovule. The mamelon is fused with the tissue of stylar canal. The tip of embryo sacs reaches up to the middle of style, e.g. *Lysiana.*

5) Tapinostemmae: Ovary unilocular with discrete ovules. Mamelon is present. The ovules are free from one another, but remain imbedded in carpellary tissue. The tip of embryo sacs extends up to the stigma, e.g. *Tapinostemma.*

We do not see any justification for the erection of the additional tribes, Lysianeae and Tapinostemmae, since they do not show distinctive features. *Lysiana* is well placed in the tribe Elytrantheae, and *Tapinostemma* in Lorantheae.

Taxonomic significance cannot be attached to the presence or absence and structure of the placenta (B.M. Johri and Ambegaokar 1984, see also B.M. Johri 1987a). In the tribe Elytrantheae, the placenta is four-lobed in a tetralocular ovary (*Lysiana, Peraxilla*), or three-lobed in a unilocular ovary (*Macrosolen*). In Nuytsieae the placenta is three-lobed in a trilocular ovary (*Nuytsia*). In Lorantheae the placenta may be absent, rudimentary (*Amyema,*

Dendrophthoe), or well developed, unlobed, and free from the carpellary wall (*Helicanthes, Lepeostegeres*). Therefore, Cocucci's (1983) emphasis on mamelon as a taxonomic character is not justified at all.

The differences between the two suborders Santalineae and Loranthineae are well-marked. In Santalineae the ovules are distinct, except in *Agonandra* (Opiliaceae) and *Exocarpus* (Santalaceae). The development of the embryo sac is of the Polygonum type, and the embryo sac may develop micropylar and chalazal haustoria. The development of endosperm is either Cellular or Helobial. The zygote divides transversely except in *Comandra*. During seed development, the integuments are consumed, and ultimately there is no testa. The pericarp does not show the presence of a viscid layer. On the other hand, in Loranthineae true ovules are absent; in an inferior ovary a mamelon may develop in some members. Development of the embryo sac is of the Polygonum type in Loranthaceae, and Allium type in Viscaceae. In Loranthaceae the embryo sacs extend to varying lengths in the stylar canal and stigma, and micropylar and chalazal caeca may be formed in some members. The development of endosperm is always Cellular, and the endosperm is composite in Loranthaceae. The zygote divides vertically in Loranthaceae, and in *Arceuthobium* and *Korthalsella* of Viscaceae, otherwise a transverse division occurs in the members of Viscaceae. As true ovules do not differentiate, the testa is absent. A viscid layer develops in the pericarp (see B.M. Johri and Bhatnagar 1960).

In view of such marked differences, in our opinion Santalinae and Loranthineae should both be raised to ordinal rank – Santalales and Loranthales (Table 12.9.2). Santalales would comprise the families Olacaceae, Opiliaceae, Grubbiaceae, Santalaceae and Myzodendraceae; Loranthales would comprise Loranthaceae and Viscaceae.

13 Balanophorales

Balanophorales include only one family, the Balanophoraceae.

13.1 General Features

Balanophorales show: (1) absence of fibrous thickenings in endothecium, (2) three-celled pollen grains, (3) highly reduced ovule, (4) micropylar megaspore of an inverted T-shaped tetrad, or a triad, produces a Polygonum type of embryo sac; Allium in *Corynaea*, (5) organization of egg apparatus and upper polar nucleus in the chalazal region of the embryo sac, and successive fusion of four nuclei in the micropylar region in *Balanophora*, (6) Cellular type of endosperm, (7) Piperad type of embryogeny; absence of suspensor, (8) undifferentiated, minute embryo in the so-called "seed", and (9) unusual mode of germination of seed.

13.2 Balanophoraceae

The family comprises herbaceous fleshy root parasites devoid of chlorophyll and stomata. The flowers are mostly male and female, rarely hermaphrodite. The anther wall comprises the persistent epidermis, a fibrous endothecium does not differentiate, the parietal layers are ephemeral, and the epidermis abuts directly on the tapetum in *Balanophora* (Fagerlind 1945a).

In *Corynaea crassa* var. *sprucei* (Engell 1979), the male flower has three perianth lobes. The three stamens form a synandrium which encloses a sterile lobe in the centre. In the synandrium (Engell 1979), there are three, four, five or six compartments. Beneath the elongated epidermal cells there are one to three middle layers, and a secretory tapetum of uni- or binucleate cells; fibrous endothecium and stomium do not develop. A multilayered epidermis covers the compartments as a cap. Cytokinesis is successive, and the tetrads are tetrahedral. In *Helosis* the tapetum is plasmodial (Umikar 1920), and cytokinesis simultaneous.

The family is highly multipalynous with colpate or porate forms. Sometimes, even inaperturate forms occur. The pollen grains are three-celled at the dispersal stage. The male flower, perhaps, opens by throwing off the epidermal cap. In *Helosis* (Umikar 1920), the cap does not develop; this genus is considered as an apomict.

The female flowers of *Balanophora* and *Corynaea* have a rather different structure, and have been treated separately.

In *Balanophora abbreviata* (Shivamurthy and Arekal 1980), the female floral primordia develop on the inflorescence axis above the region of male flowers, while in *B. fungosa* spp. *indica* (Shivamurthy and Arekal 1980) they develop on the surface of the female inflorescence. They appear as finger-like protuberances amidst the spadicles. The mature flower is extremely reduced in size and structure, archegonium-like without vasculature and perianth. Floral development exactly simulates ontogenetic events normally noted in the nucellus of an angiosperm ovule. It is concluded that this female flower is ontogenetically a nucellus but, functionally, a gynoecium.

Concerning *Corynaea crassa*, Engell (1979) states: "A nucellus primordium develops in the bottom of the ovary and in this nucellus probably two archesporial cells will be differentiated. From these two cells a crassinucellate ovule will develop with two deep-lying sporogenous cells and without development of an integument". He calls this condition of ovule as ategmic and crassinucellate. This is not a correct interpretation. The ovule is indeed undifferentiated.

The archesporial cell is enclosed in a long, beaked "placenta". The megaspore mother cell undergoes meiotic divisions; the chalazal dyad is much smaller than the micropylar. Both the dyads undergo meiosis II and an inverted T-shaped tetrad develops (Fig. 13.2.1A–E) in *Balanophora abbreviata*, *B. dioica*, *B. fungosa* subsp. *indica* var. *indica* (Arekal and Shivamurthy 1978), *B. elongata* (Fagerlind 1945b), and *B. globosa* and *B. thwaitesii* (Teryokhin and Yakovlev 1967). A triad has been observed in *B. abbreviata*. The uppermost megaspore develops further, and produces an embryo sac of the Polygonum

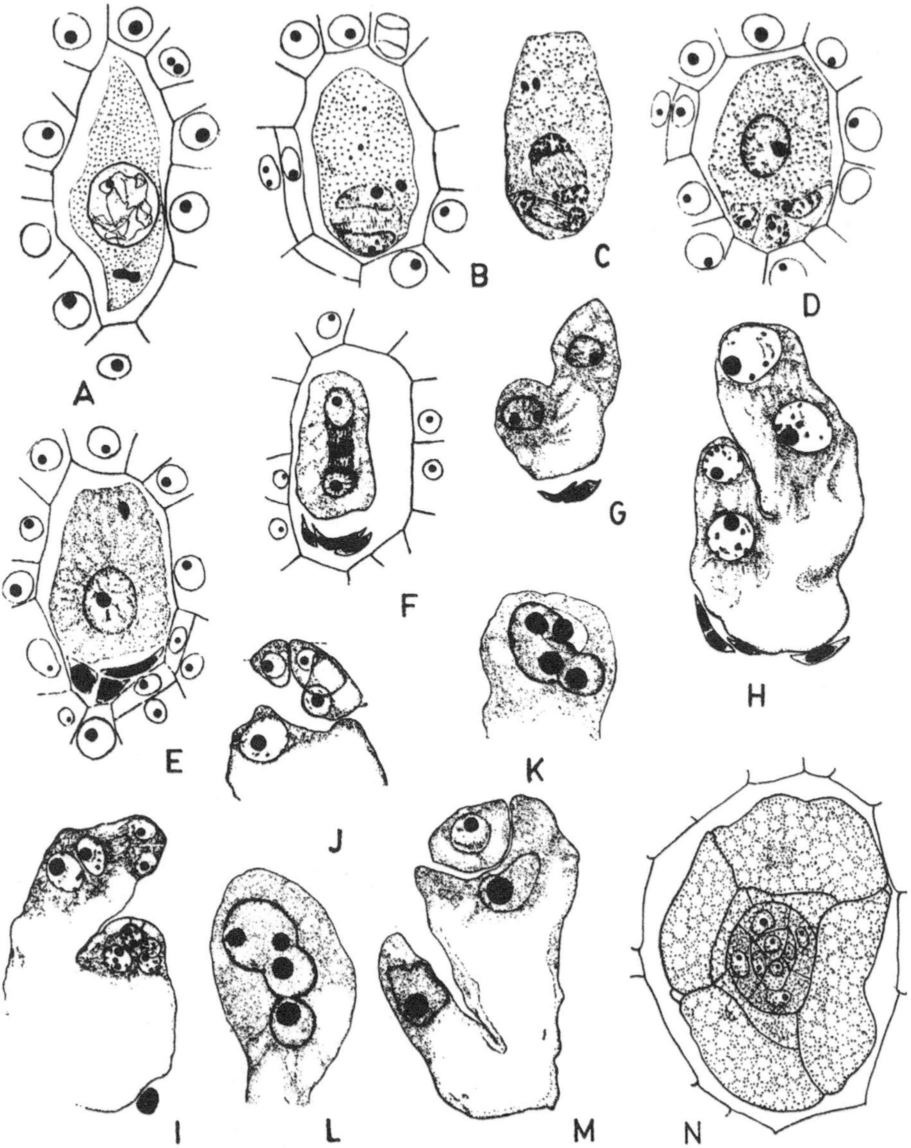

Fig. 13.2.1A–N. *Balanophora dioica.* **A** Megaspore mother cell in prophase. **B** Telophase, meiosis I. **C** Telophase, meiosis II. **D, E** Megaspore tetrad, upper megaspore functional. **F** Functional megaspore nucleus at late telophase. **G** Binucleate embryo sac, chalazal end curves upward. **H** Four-nucleate embryo sac. The chalazal end is larger with prominent nuclei. Degenerated megaspore nuclei are at the base. **I** Eight-nucleate, unorganized embryo sac. **J** Upper portion of embryo sac with egg cell, two synergids, and upper polar nucleus. **K, L** Fusion of antipodal nuclei. **M** Embryo sac with zygote, primary endosperm nucleus (?), and fused antipodal. **N** Transection of embryo sac to show cellular endosperm enclosing proembryo. (After Ekambaram and Panje 1935)

type (Fig. 13.2.1F–L). Similar observations have been reported in *B. dioica* (Ekambaram and Panje 1935) and *Langsdorffia hypogea* (Fagerlind 1945c).

During its development, the embryo sac shows "reversed polarity". At the two-nucleate stage a tubular process originates at the basal end in *Balanophora* (Fig. 13.2.1G), and at the apical end in *Langsdorffia*. This caecum grows rapidly, and comes to lie at a higher level than the original upper micropylar end of the embryo sac. The chalazal nucleus passes into the caecum. Figure 13.2.1H shows a four-nucleate embryo sac. During the differentiation of eight nuclei (Fig. 13.2.1I), four nuclei in the caecum develop into the egg apparatus and upper polar nucleus (Fig. 13.2.1J), those in the morphologically micropylar end fuse successively to form an irregularly lobed nucleus (Fig. 13.2.1K–M). In *B. dioica* (Ekambaram and Panje 1935), this nucleus assumes a more or less irregular outline. During endosperm formation it degenerates with the collapse of the antipodal arm. Thus, the lower polar nucleus is absent, and also the antipodal cells (Arekal and Shivamurthy 1978). However, the fusion of two polar nuclei in the embryo sac has been reported in *B. abbreviata*, *B. indica* (Zweifel 1939) and *B. thwaitesii* (Teryokhin and Yakovlev 1967). The mature embryo sac is U- or J-shaped (Fig. 13.2.1I, M). Bisporic Allium type is reported in *Ditepalanthus afzelli* and *Helosis cavennensis* (Fagerlind 1938a). However, Chodat and Bernard (1900) described Endymion type in *Helosis*. This needs reinvestigation.

In *Corynaea crassa* var. *sprucei* (Engell 1979), the flowers as well as the inflorescences are unisexual (Fig. 13.2.2A); sometimes both male and female flowers develop on the same branch (Fig. 13.2.2B). The young flowers are embedded in paraphyses or trichomes which function as nectaries.

The insects, perhaps, perforate the anthers and pollinate the flower.

The female flowers of *Corynaea* are pedicillate and devoid of perianth lobes (Fig. 13.2.2B). The gynoecium is bicarpellate and forms a compressed ovary with two styles. One, sometimes two, archesporial cells develop in the "placenta" but only one functions. The micropylar dyad degenerates, and the development of the embryo sac is of the Allium type. The mature embryo sac consists of the egg apparatus, a central cell with secondary nucleus, and three antipodal cells (Fig. 13.2.2C, D1–D3). The placenta degenerates, and the embryo sac becomes "naked".

The development of endosperm is of the Cellular type. The division of primary endosperm nucleus is followed by a wall separating a small micropylar from a large chalazal chamber. The micropylar chamber gives rise to the endosperm proper, while the chalazal develops into a large, uninucleate haustorium. The latter is crushed by the developing endosperm tissue.

The embryogeny conforms to the Piperad type. The first, as well as the second division is vertical in *Balanophora* (Zweifel 1939). In *B. dioica* (Ekambaram and Panje 1935) the third division is periclinal in all the four quadrants (Fig. 13.2.1N). The embryo consists of four central and four peripheral cells. The nuclei are all arranged in the equatorial region of the embryo. When the embryo lies deep in the endosperm, the next division in the embryo is transverse, but takes place only in the peripheral cells. The fully formed embryo consists of 12 cells. Treub (1898) reported five-celled embryo in

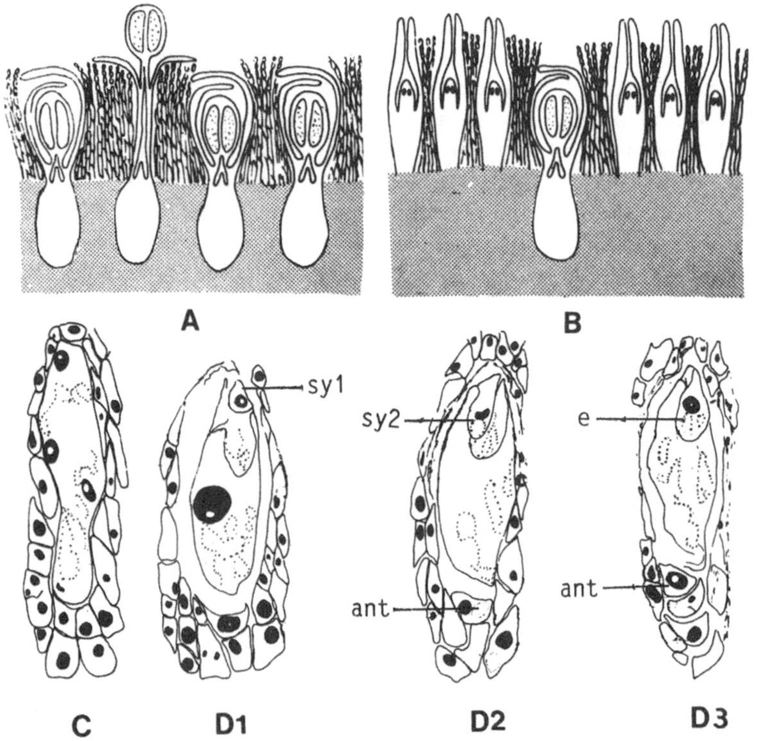

Fig. 13.2.2A–D. *Corynaea crassa* var. *sprucei*. **A** Schematic drawing of one tall and three short male flowers and paraphyses. Tall flower with unfolded perianth on the top of paraphyses. Short flowers with folded perianth. Three stamens with fused anthers and filaments; there is a small sterile cone above the receptacle in each flower. **B** Monoecious side-shoot with a single male flower in the centre, and some six pedicellate, naked (without perianth) flowers with bicarpellate gynoecium surrounded by paraphyses. **C** Four-nucleate embryo sac. **D1–D3** Successive serial longisections of a mature embryo sac; **D1** Egg cell, one synergid (*sy1*), and secondary nucleus; **D2** Egg cell, another synergid (*sy2*), central cell, and one antipodal (*ant*); **D3** Egg cell (*e*), central cell, and three antipodals. The nucellus has degenerated. (After Engell 1979)

B. elongata, while Ernst (1914) reported six-celled condition in the same species. The embryo in the mature seed is unorganized. A suspensor is absent. Teryokhin and Yakovlev (1967) describe embryogenesis as of "non-oriented" zygote cleavage type. The development of embryo requires further investigation. Polyembryony has been reported in *B. globosa* and *Helosis guianensis*.

Fig. 13.2.3A–D. *Balanophora abbreviata*, germination of "seed" and establishment of parasite on host rootlet; longisection, semidiagrammatic. **A** Germination of "fruit" on host rootlet; note the anchoring endosperm tubules (*et*), embryo (*em*), endosperm (*end*), fruit wall (*fw*), host-root cortex (*hrc*), and host-root central cylinder (*hrcl*). **B** Germination of "fruit" shows entry of primary

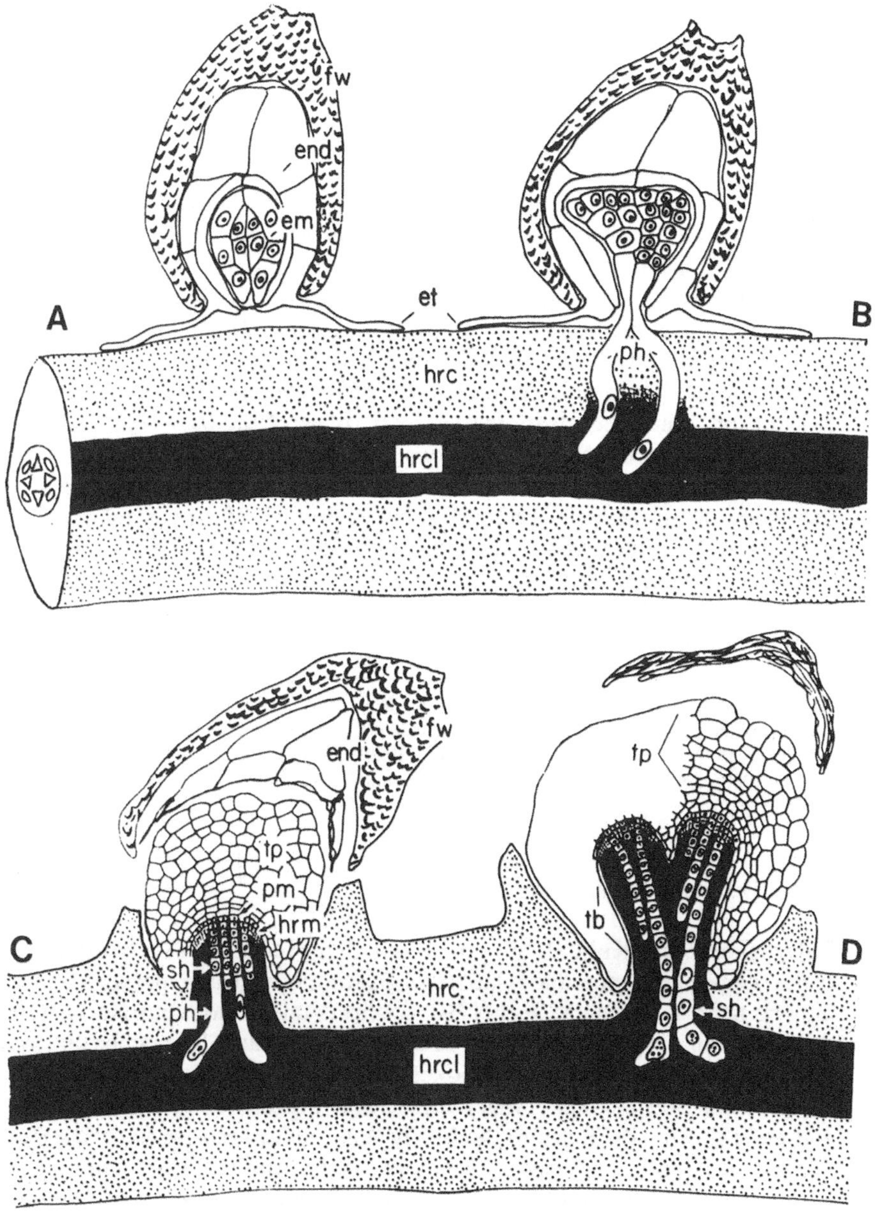

haustoria (*ph*) into host root; the fruit contains a multicellular organless embryo. **C** Tuber primordium on host rootlet showing organization of secondary haustorial cells and tuber bundle; *hrm* host-root meristem; *pm* meristem of parasite; *sh* secondary haustorium; *tp* tuber parenchyma. **D** Tuber primordium, slightly later stage, to show forked tuber bundle (*tb*). (After Arekal and Shivamurthy 1976)

In aposporous embryo sacs of *Balanophora globosa* (Ernst 1913), the lower polar nucleus of the embryo sac degenerates, and the upper polar nucleus alone gives rise to endosperm. *B. globosa* and *B. japonica* (Kuwada 1928, Zweifel 1939, Fagerlind 1945b) exhibit Taraxacum type of apospory. The first meiotic division of megaspore mother cell ends in the formation of a restitution nucleus. Meiosis II results in a dyad of unreduced cells, and the lower one develops into an eight-nucleate embryo sac.

The germination of "seed' in *Balanophora*, as described by Arekal and Shivamurthy (1976), is unique and needs special mention. Typical seeds are absent, and the ripe fruit comprises an undifferentiated embryo, few-celled endosperm, and sclerotic pericarp. These fruits, in the vicinity of host rootlets, break open. The endosperm cells produce tubular, slender, sticky extensions for anchorage of the fruit to the young host-rootlets (Fig. 13.2.3A, B). The undifferentiated embryo develops four to five tubular processes (as primary haustorium) which penetrate the vascular tissue of the host (Fig. 13.2.3B). It also forms a tuber by repeated divisions, and emerges out of the pericarp above the ground. The cells of the central cylinder of the host-rootlet become active in the region of contact (Fig. 13.2.3C). A lateral branch of the central cylinder of host enters the tuber, and branches dichotomously (Fig. 13.2.3D). The tuber develops secondary haustorium in the form of chains of cells which penetrate the central vascular cylinder of the host, and absorb food material. The tuber grows, and develops flowering scapes.

GD Arekal and GR Shivamurthy (1976, 1978), R Chodat and Ch Bernard (1900), AJ Eames (1961), T Ekambaram and RR Panje (1935), K Engell (1979), A Ernst (1913, 1914), F Fagerlind (1938a, 1945a, b, c, 1959), JD Hooker (1856), Y Kuwada (1928), LN Rao (1932), GR Shivamurthy and GD Arekal (1980), ES Teryokhin and MS Yakovlev (1967), M Treub (1898), O Umikar (1920), R Zweifel (1939)

13.3 Taxonomic Considerations

Balanophoraceae. Kuijt (1969) points out that Hooker's (1856) emphasis on Halorrhagaceae as possible relatives of Balanophoraceae, especially *Gunnera*, may be nearest the mark. Juel (1910) confirms this notion, but Harms (1935) does not refer to it in his treatment of Balanophoraceae. This may be due to the fact that in *Genera Plantarum* Hooker himself abandoned the idea.

Bentham and Hooker (1965c), Engler and Prantl (1931), Wettstein (1935), and Hutchinson (1973) group the family Balanophoraceae in Santalales, with Santalaceae and Loranthaceae. Balanophoraceae show distinct affinities with both the families. The common features with Santalaceae are: parasitic mode, extreme reduction of ovule, monosporic eight-nucleate or, sometimes, Allium type of female gametophyte, and Cellular type of endosperm with chalazal haustorium. The Balanophoraceae resemble the Loranthaceae-Viscoideae (*Korthalsella*) in reversed polarity of the embryo sac, vertical division of zygote, and absence of suspensor. The testa is absent in Santalales, and an undifferentiated embryo in the mature "seed" generally occurs in parasites.

For further discussion, see Section 39.19, Cynomoriaceae.

The presence of synandrium, extreme reduction of gynoecium without any histological differentiation, and the peculiar mode of germination of "seed" may be some of the reasons for erection of a separate order Balanophorales (see Melchior 1964). This order is placed between Santalales and Medusandrales; the affinities with Santalales are referred to above but we do not have any embryological data for Medusandrales.

With our present knowledge it does not seem possible to discuss the taxonomic assignment any further. We must await more detailed information on taxa which can be considered as relatives of this family (see also Kuijt 1969).

14 Medusandrales

This order is represented by a single family Medusandraceae; the embryology has not been investigated.

15 Polygonales

The Polygonales includes a single family, Polygonaceae.

15.1 General Features

The Polygonaceae are: (1) highly multipalynous with colpate, colporate or porate pollen grains, the number and distribution of pores vary widely, the exine is elaborately reticulate, granulose or spinuloid, (2) the pollen grains are three-celled at dissemination, (3) the ovules are orthotropous, bitegmic, and crassinucellate with a prominent nucellar beak, (4) multicellular archesporium, (5) Polygonum type of embryo sac (antipodals ephemeral), (6) persistent remnants of pollen tube, (7) Nuclear type of endosperm, ruminate in *Coccoloba*, (8) Asterad type of embryogeny, and (9) straight embryo in an albuminous seed with unspecialized seed-coat.

15.2 Polygonaceae

The anther is tetrasporangiate, and the anther wall comprises an epidermis, fibrous endothecium, ephemeral middle layer, and secretory tapetum. The development of anther wall is of the Reduced type (Davis 1966) in *Antigonon leptopus* (Agoram and Krishnamurthy 1980), otherwise the development is of the Monocotyledonous type (Maheswari Devi and Manorama 1985). The tapetal cells become two- to four-nucleate. Simultaneous cytokinesis in the microspore mother cells follows meiosis, and the tetrads are tetrahedral and isobilateral.

The family is highly multipalynous, with colpate, colporate or porate pollen grains. The bronchi of exine may be well organized and small. Pollen polymorphism has been observed in *Polygonum* spp. (Maheswari Devi and Manorama 1984). The largest pollen grain is three-nucleate and embryo sac-like. The pollen grains are two-celled at the dispersal stage in *Rumex uesicarius*. In *Polygonum alatum*, *P. capitatum*, *P. chinense* and *P. molle* the pollen grains are three-celled, and triaperturate with reticulate exine (Maheswari Devi and Manorama 1985).

The ovule is orthotropous, bitegmic (or unitegmic through connation), and crassinucellar with a hypostase. The inner integument forms the micropyle. A nucellar break is prominent, and projects beyond the micropyle in *Polygonum persicaria* (Souèges 1919). The hypodermal archesporium is multicellular (Dudgeon 1918), but only one cell functions. It cuts off a parietal cell. The parietal tissue is five to six layers thick. The megaspore mother cell undergoes meiotic divisions giving rise to a linear or T-shaped tetrad. The chalazal megaspore is functional.

Development of the female gametophyte is of the Polygonum type. The synergids are hooked in *Rumex*. The three antipodals become multinucleate, but are ephemeral. The first clear account of the Polygonum type of embryo sac development was reported in *Polygonum diverícatum* (Strasburger 1879).

Fertilization is porogamous, and the pollen tube is persistent and unbranched. The synergids remain intact during the entry of the pollen tube (Mahony 1935). This needs confirmation.

Development of the endosperm is of the Nuclear type. Wall formation is initiated at the micropylar end at the globular proembryo in *Polygonum pensylvanicum* (Neubauer 1971). A meristematic layer differentiates at the periphery in the cellular endosperm. The cells in the central region become multinucleate. After 10 days of its formation the meristematic layer develops into an aleurone layer. The nucellus persists as a thin covering. The chalazal region of endosperm remains free-nuclear in *Fagopyrum esculentum* (see Davis 1966). This aspect requires further study.

The embryogeny conforms to the Asterad type (Neubauer 1971, B.A. Rao and Mukherjee 1974) in *Polygonum*. According to Guignard (1966), the embryogeny in *Rumex aquaticus* resembles that of *P. persicaria*.

The mature seed is pachychalazal with ruminate endosperm in *Antigonon leptopus* (Agoram and Krishnamurthy 1980). The ruminations conform to the "Coccoloba" type (Periasamy 1962a). The invaginations of seed-coat are restricted to the portion of seed derived from the chalazal region. They develop first at the chalazal end, and proceed upward.

The seed contains a small and straight embryo with starchy endosperm. The seed-coat is unspecialized with the outer epidermis of testa showing reticulate thickenings in *Rumex*.

In *Coccoloba uvifera* (Periasamy 1964a), during post-fertilization development, the young seed becomes tightly enclosed by the ovary wall. In a bitegmic young seed the endosperm is surrounded by a well-developed nucellus

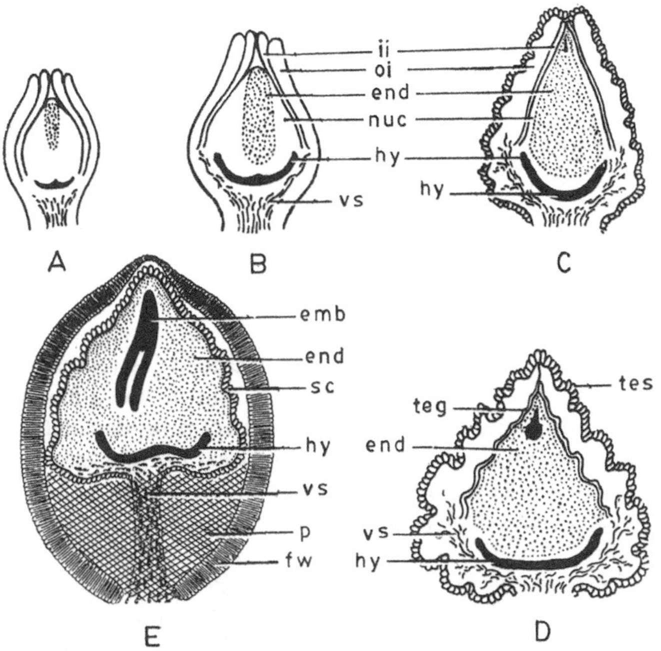

Fig. 15.2.1A–E. *Coccoloba uvifera*, development of seed (semidiagrammatic). **A** Longisection of ovule, endosperm *dotted*. **B** Later stage; *ii* inner integument; *oi* outer integument; *end* endosperm; *nuc* nucellus; *hy* hypostase; *vs* vascular supply. **C** Early globular proembryo (*black*), nucellus mostly consumed by endosperm, outer integument overarches the inner integument, and epidermis of the testa has become wavy. **D** Globular proembryo, nucellus completely consumed, tegmen (*teg*) wavy, and epidermis of testa (*tes*) thrown into folds. **E** Longisection of mature fruit to show dicotyledonous embryo (*emb*), ruminate endosperm (*end*), and excessive growth of chalaza; *sc* seed-coat; *p* pericarp; *fw* fruit wall. (After Periasamy 1964a)

(Fig. 15.2.1A); the chalazal tissue becomes meristematic, increases in breadth, and assumes a shallow cup-like appearance. A hypostase also develops (Fig. 15.2.1B). The vascular supply branches below the chalaza, and the branches reach up to the base of integuments. During its growth the endosperm consumes the nucellus (Fig. 15.2.1C, D).

The outer epidermis of the outer integument divides only anticlinally, the cells enlarge, become papillate, and accumulate dark-brown contents. The inner layers of the outer integument become meristematic and develop into irregular (mostly three) lobes. The chalazal growth is also massive and irregular. The two-layered inner integument shows undulations (Fig. 15.2.1D).

At first the epidermis of the outer integument becomes irregular, then this irregularity proceeds inward and the entire seed-coat assumes undulated appearance (Fig. 15.2.1D).

As the endosperm becomes cellular and grows, it consumes the nucellus and integument except for the lobed outer epidermis of outer integument which alone persists and renders the endosperm ruminate (Fig. 15.2.1E). The epidermis of the fruit wall develops into a palisade-like layer of radially elongate sclereids with heavily lignified walls (Fig. 15.2.1E). During the maturation of fruit, tissues of the fruit wall disorganize except for the epidermis.

S Agoram and KV Krishnamurthy (1980), HR Bhargava and RH Sawhney (1958), J Brandbyge (1989), GL Davis (1966), Y Doida (1960), W Dudgeon (1918), K Engell (1973), JL Guignard (1966), E Kirsten (1973), H Maheswari Devi and K Manorama (1984, 1985), KL Mahony (1935), PK Mukherjee (1972), BF Neubauer (1971), K Periasamy (1962a, 1964a), BA Rao and PK Mukherjee (1974), VS Rao (1936b), BS Siva Rao (1952), R Souèges (1919), E Strasburger (1879)

15.3 Taxonomic Considerations

Bentham and Hooker (1965c) place the family Polygonaceae in Curvembryae, and Bessey (1915) in Caryophyllales. Hallier (1912) recognizes the affinities between Caryophyllales and Polygonales, and includes them in his Centrospermae. Engler and Diels (1936) and Hutchinson (1973) treat Polygonaceae as a separate order, Polygonales. Hutchinson (1973) considers it as a reduced degraded type of Caryophyllales. Polygonales have their ancestry in Caryophyllales (Lawrence 1951).

Cronquist (1968, 1981) places Polygonales prior to Plumbaginales, and states that both are related to Caryophyllales. All the three orders contain anthocynins, and have food storage tissue in seed. R. Dahlgren (1981) places Plumbaginales (position uncertain) and Polygonales in superorder Polygoniflorae. The relationship of Polygonaceae with Plumbaginaceae is discussed in Section 44.3.

Polygonaceae. Polygonaceae show closest affinity with Caryophyllaceae, and in both the families the pollen grains are colpate, colporate or porate. The bi-, tri- or tetracarpellary unilocular ovary has a basal ovule, occasionally also in Caryophyllaceae. Further resemblance is a bitegmic, crassinucellar ovule with multicellular archesporium, Polygonum type of embryo sac with ephemeral antipodal cells, Nuclear type of endosperm, and albuminous seed. However, the Asterad type of embryogeny, small straight embryo in mature seed, absence of perisperm, and unspecialized seed-coat differ from those of Caryophyllaceae. The latter family has campylotropous ovule, Caryophyllad type of embryogeny, and curved embryo (around the perisperm) in mature seed. The seed is exotestal with reticulate markings on the exotesta. The mesophyll is unspecialized, and the tegmen is crushed in Caryophyllaceae.

On the whole, the embryological data support the phyletic position assigned to the order.

16 Centrospermae

The order Centrospermae comprises 12 families, distributed over four suborders: Phytolaccineae – Phytolaccaceae, Gyrostemonaceae, Achatocarpaceae, Nyctaginaceae, Molluginaceae, Aizoaceae; Portulacineae – Portulacaceae, Basellaceae; Caryophyllineae – Caryophyllaceae; Chenopodiineae – Dysphaniaceae, Chenopodiaceae, Amaranthaceae. The family Didieraceae is appended to it.

16.1 General Features

The Centrospermae show: (1) secretory anther tapetum, (2) simultaneous cytokinesis in pollen mother cells, (3) three-celled pollen grains, two- to three-celled in Nyctaginaceae, (4) ana-campylotropous, bitegmic, crassinucellar ovule with a characteristic air-space (between the integuments) at the base of ovule, (5) curved nucellus producing a nucellar cap, persists as mealy perisperm, (6) uni- or multicelled archesporium, (7) massive parietal tissue, (8) Polygonum type of embryo sac, mono-, bi- and tetrasporic embryo sacs occur in Nyctaginaceae and Aizoaceae, (9) endosperm of the Nuclear type, (10) Caryophyllad type of embryogeny in Phytolaccaceae, Molluginaceae, Portulacaceae, Caryophyllaceae, Asterad in Nyctaginaceae, Solanad in Aizoaceae, Chenopodiad in Basellaceae, Chenopodiaceae, Amaranthaceae, Didieraceae, and (11) generally exotestal arillate seed with a large curved embryo enclosing a massive perisperm.

16.2 Phytolaccaceae

The anther wall comprises the epidermis, fibrous endothecium, two or three ephemeral middle layers, and secretory tapetum of binucleate cells. The cells become multinucleate in *Phytolacca decandra* (Lewis 1905). Simultaneous cytokinesis follows meiotic divisions of microspore mother cells, and results in tetrahedral and isobilateral tetrads.

The pollen grains are tricolporate and three-celled at the dispersal stage in *Rivina humilis* (A.C. Joshi 1936a), and two-celled in *Phytolacca.*

The ovule is basal, ana-campylotropous, bitegmic and crassinucellate. The inner integument organizes the micropyle. An air-space is present between the two integuments, at the base of ovule. The cells of the nucellar epidermis, by periclinal divisions, produce a massive nucellar tissue. It is strongly curved, and the epidermal cells below the micropyle are stretched radially. This tissue becomes 12–14 cells thick in *Rivina* (A.C. Joshi 1936a). A funicular obturator and hypostase also differentiate.

The archesporial cell cuts off a parietal cell. The parietal tissue is massive. The megaspore mother cell undergoes meiotic divisions giving rise to a linear tetrad. The chalazal megaspore produces a Polygonum type of embryo sac. The synergids are elongate with a pointed apex, the polar nuclei fuse before fertilization, and the antipodals persist up to fertilization. In *Rivina humilis* (Kajalė 1954a), the synergids are persistent, and the antipodal cells undergo secondary multiplication after fertilization.

The development of endosperm is of the Nuclear type. Wall formation commences at the late globular proembryo, and finally it becomes completely cellular. The enodsperm stores abundant starch.

The development of embryo follows Chenopodiad type in *R. humilis*, and Caryophyllad type in *Phytolacca dioica* (Kajale 1944a, 1954a). The suspensor is massive.

The seed is small and exotestal, and contains a large curved embryo enclosing the starchy perisperm. The scanty endosperm persists as one or two layers in the micropylar half of the seed. The seed is arillate in capsular genera, but the aril is absent in drupaceous genera. A vestigial aril occurs in *Rivina*. The outer epidermis of testa is composed of radially elongate, thick-walled cells with brown contents. The mesophyll and the inner epidermis become crushed. The tegmen is represented only by its inner epidermis since the outer collapses. The cells develop bar-like thickenings on the radial walls, and have brown contents.

AC Joshi (1936a), LB Kajale (1944a, 1954a), JF Lewis (1905), T Rocén (1927), O Rohweber (1970), P Satyanarayana and LL Narayana (1985), D Singh and CP Taneja (1985), MS Yakovlev (1983)

16.3 Gyrostemonaceae

The embryology has not been investigated.

16.4 Achatocarpaceae

The embryology has not been studied.

The anthers are bilocular. The ovule is basal and campylotropous. Seed exarillate, solitary, erect, lenticular, with strongly bent embryo surrounding the mealy endosperm, cotyledons linear. Fruit one-seeded berry.

J Hutchinson (1973), JC Willis (1973)

16.5 Nyctaginaceae

The anther wall comprises the epidermis, fibrous endothecium, two ephemeral middle layers, and a secretory tapetum of binucleate cells. The microspore mother cells undergo meiotic divisions, cytokinesis is simultaneous, and the tetrads tetrahedral and isobilateral.

The pollen grains are three- to four-colporate and two-celled at the dispersal stage. They are three-celled in *Oxybaphus micranthus* (Rocén 1927). Sterility of pollen grains occurs in *Boerhaavia repanda* (Bhargava 1932) and *Bougainvillea glabra* (D.C. Cooper 1931a).

The ovule is anatropous or campylotropous, and crassinucellar. It is bitegmic in *Bougainvillea, Mirabilis, Oxybaphus* and *Pisonia* with the inner integuments forming the micropyle, and unitegmic in *Ambronia* (Rocén 1927) and *Boerhaavia* (P. Maheshwari 1929, Bhargava 1932). The integumentary vascular bundle extends up to the micropyle in *Mirabilis*. Funicular obturator develops in *Pisonia aculeata* (Venkateswarlu 1947b). The single archesporial cell cuts off a parietal cell. The parietal tissue is massive. The nucellar epidermis produces a prominent nucellar beak in *Boerhaavia diffusa* (P. Maheshwari 1929) and *B. repanda* (Bhargava 1932). The megaspore mother cell, following meiotic divisions, produces a linear tetrad. In *Bougainvillea* (Das and Ramkrishnan 1974), a number of megaspore mother cells undergo meiotic divisions, and several linear tetrads are produced; most of them degenerate. The female gametophyte is of the Polygonum type. The synergids are ephemeral, and the three persistent antipodals multiply producing six to seven cells.

Several abnormalities in the formation of tetrads and female gametophyte have been reported. In *Oxybaphus nyctagineus* (Rocén 1927), a megaspore tetrad may contain a binucleate cell in the middle, and one uninucleate cell on either side. The binucleate cell never develops further. The suppression of one or more transverse walls in the linear tetrad in *Mirabilis jalapa* results in the formation of bi-, tri- or tetrasporic types of embryo sac (Hjelmqvist and Grazi 1965). Normally, monosporic development occurs. Sometimes, the partition walls are incomplete and the adjoining mature embryo sacs may fuse. Hjelmqvist and Grazi (1965) attributed these variations to the effect of temperature.

The entry of the pollen tube is porogamous. It finds its way between the two synergids leaving them intact (D.C. Cooper 1949). The pollen tube persists throughout the development of seed in *Oxybaphus*, and its haustorial role is suggested (D.C. Cooper 1949). An accessory pollen tube is reported in *Boerhaavia diffusa* (P. Maheshwari 1929). The persistence of both the synergids, and the haustorial role of the pollen tube are questionable.

Development of the endosperm is of the Nuclear type. After the formation of 64 free nuclei in *Oxybaphus* (D.C. Cooper 1949), the embryo becomes aggressive and the nuclear divisions in endosperm slow down. The endosperm becomes cellular in the micropylar region, but a larger part is absorbed in the free-nuclear state in *Boerhaavia* (Kajale 1938) and *Mirabilis jalapa* (Woodcock 1929).

The zygote divides transversely producing the apical and basal cells. The basal cell undergoes transverse divisions resulting in a filamentous six-celled proembryo. Simultaneously, a longitudinal division occurs in the apical cell followed by a vertical wall. Then the vertical divisions proceed in the second, third and fourth cell from the apex in *Boerhaavia* (Kajale 1936, 1938). Such a sequence of wall formation in the family Nyctaginaceae is totally different from that of the other related families of the order Centrospermae. The micropylar cells develop into a long and slender suspensor in *B. diffusa*, and a short and massive suspensor in *B. repanda* (Kajale 1938). Another peculiar feature is the formation of periblem from the dermatogen. It is not a sister layer of pleurome as in other angiosperms. The embryogeny conforms to the Asterad type. In *Oxybaphus* (see Crété 1963a), it conforms to Period I, Series A, Megarchetype III (Asterad type). The embryogeny in *Pisonia aculeata* (Venkateswarlu 1947b) is also of the Asterad type.

Synergid embryos have been observed in *B. repanda* (Kajale 1938).

The seed is small (except in *Pisonia*), with or without perisperm. The outer persistent epidermis of testa is composed of thick-walled cells with brown contents. These cells are not lignified. The mesophyll and inner epidermis of testa become crushed. The inner epidermis is thick-walled in *Mirabilis*. The inner epidermis of tegmen persists only in *Mirabilis* and *Pisonia*. Its cells show fine striations on the tangential walls. Otherwise, the tegmen becomes crushed along with the outer epidermis.

HR Bhargava (1932), DC Cooper (1931a, 1949), P Crété (1963a), VLD Das and V Ramkrishnan (1974), H Hjelmqvist and F Grazi (1965), LB Kajale (1936, 1937a, 1938), P Maheshwari (1929), P Maheshwari and BM Johri (1950b), T Rocén (1927), R Souèges (1938c), J Venkateswarlu (1947b), EF Woodcock (1929)

16.6 Molluginaceae

The anther wall comprises the epidermis, fibrous endothecium, an ephemeral middle layer, and secretory tapetum with bi- and trinucleate cells in *Limeum indicum*. Übisch granules occur in *Orygia decumbens* (H.S. Narayana and Lodha 1963), and nuclear fusions take place in the tapetal cells. Following meiotic divisions, cytokinesis is simultaneous, and the microspore mother cells produce tetrahedral, decussate and isobilateral tetrads.

The pollen grains are tricolporate; three-celled at the dispersal stage. Occasionally, they germinate in situ in *Mollugo nudicaulis* (Bhargava 1934).

The ovule is anatropous, bitegmic and crassinucellar with inner integument forming the micropyle. An air-space is present between the integuments, at the base of ovule, on one side. The nucellus is strongly curved in *Limeum indicum* (H.S. Narayana and Jain 1962). The cells of the nucellar epidermis divide to form a massive nucellar cap. A single hypodermal archesporial cell cuts off a parietal cell. Multicellular archesporium occurs in *Orygia*. The parietal tissue is two or three layers thick. The megaspore mother cell undergoes meiotic divisions resulting in a linear or T-shaped tetrad. The chalazal megaspore develops into a Polygonum type of embryo sac. The synergids are beaked, the polar nuclei fuse before fertilization, and the antipodals may be ephemeral. The development of a hypostase is a common feature. A funicular obturator develops in *Limeum*, whereas in *Orygia* the epidermis of the placenta develops multicellular filamentous outgrowths in the vicinity of micropyle.

Development of the endosperm is of the Nuclear type. Wall formation is initiated around the proembryo at the globular stage in *Limeum* and *Orygia decumbens* (H.S. Narayana and Lodha 1963), and finally the endosperm becomes completely cellular. The cells are uni- or binucleate, and store starch. In a mature seed the endosperm persists as a cap around the radicle.

The embryogeny conforms to the Caryophyllad type-Sagina variation in *Gisekia pharnaceoides* (A.C. Joshi and Rama Rao 1936), but in *Limeum indicum* H.S. Narayana and Jain (1962) report Caryophyllad type-Drosera variation. H.S. Narayana and Lodha (1963), however, suggest Solanad type-Linum variation in *Orygia decumbens*.

The reniform seed of *Orygia* contains a large curved embryo on the hilar side, with a massive mealy perisperm. The endosperm persists as one or two layers in the micropylar half of seed. Both the integuments contribute to the formation of seed-coat. The outer epidermis of testa is tanniniferous, and the inner is thin-walled. The outer epidermis of tegmen collapses due to the enlargement of the inner epidermis. The cells of the latter have a banded appearance due to bar-like thickenings on the radial walls (Fig. 16.6.1A, B).

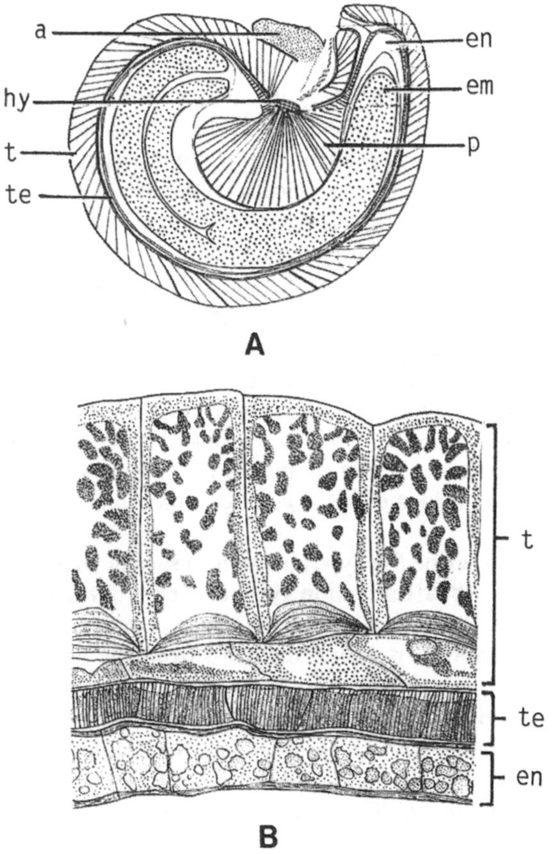

Fig. 16.6.1A, B. *Orygia decumbens.* **A** Longisection of mature camplyotropous seed. There is copious perisperm (*p*), scanty endosperm (*en*), curved embryo (*em*), and a short funicular aril (*a*); *hy* hypostase; *t* testa; *te* tegmen. **B** Transection of seed-coat to show two-layered testa and fibrous endotegmen. Outer epidermal cells of the testa are radially elongate, and lamellate thickenings occur on inner walls. Exotegmen has collapsed. (After H.S. Narayana 1962a)

The funicular aril surrounds the seed in *Mollugo oppositifolia*; it persists as a vestige in *Gisekia*, *M. nudicaulis* and *Orygia*. It is absent in *Limeum*.

HR Bhargava (1934), AC Joshi and V Rama Rao (1936), HS Narayana and K Jain (1962), HS Narayana and BC Lodha (1963), BS Siva Rao (1975)

16.7 Aizoaceae

The anther wall comprises an epidermis, fibrous endothecium, single ephemeral middle layer, and glandular tapetum with two- to four-nucleate cells in *Glinus lotoides* (H.S. Narayana and Lodha 1972). Simultaneous cytokinesis accompanies meiotic divisions in microspore mother cells. The tetrads are tetrahedral and decussate.

The pollen grains are tricolpate, and three-celled at the dispersal stage.

The ovule is ana-campylotropous, bitegmic and crassinucellar. The inner integument forms the micropyle. Both the integuments are two-layered. The apical cells form a collar-like structure in *Aptenia cordifolia* (N. Prakash 1967a) and *Delosperma cooperi* (Kapil and Prakash 1966). An air-space occurs between the integuments at the base. The nucellar epidermis is radially stretched in *Mesembryanthemum*, *Tetragonia* and *Trianthema*. The archesporium is uni- or multicellular. The archesporial cell cuts off a parietal cell, and the parietal tissue is well developed. The megaspore mother cell undergoes meiotic divisions to produce a linear tetrad. The chalazal megaspore develops into a Polygonum type of embryo sac. The synergids are hooked, polar nuclei fuse before fertilization, and antipodals are ephemeral. The latter may or may not organize into cells in *Delosperma*. Funicular obturator occurs in *Glinus lotoides* (H.S. Narayana and Lodha 1972), *Tetragonia tetragonioides* (N. Prakash 1967b), and *Trianthema argentina* (Cocucci 1961), and placental in *Aptenia cordifolia*. An operculum is formed in *Hypertelis bowkeriana* (Bregman and Bouman 1983).

Kapil and Prakash (1966) describe five types of embryo sacs in *Delosperma cooperi*: 14% Polygonum type, 18% Endymion type, and 68% tetrasporic – 47% Drusa type, 35% Penaea type, and 18% Adoxa type. Occasionally, an aposporic embryo sac (with reversed polarity) develops from the additional archesporial cell.

Development of the endosperm is of the Nuclear type. A weak chalazal haustorial caecum is reported in *Hereroa hesperantha* (N. Prakash 1967c). Wall formation is initiated at the globular proembryo. It is restricted to the micropylar region in *Aptenia cordifolia* and *Tetragonia tetragonioides* (N. Prakash 1967a, b).

The embryogeny conforms to Solanad type-Linum variation in *Glinus*. Solanad type also occurs in *Sesuvium portulacastrum*, *Trianthema decandra* and *T. monogyna*. Suspensor polyembryony occurs in *A. cordifolia*.

The exotestal and endotegmic seed contains a large curved embryo, well-developed perisperm, and scanty endosperm (H.S. Narayana 1962a). The testa is multiplicative in *Trianthema*. The outer epidermis of testa is thick-walled with brown contents, the inner one is unspecialized. The outer epidermis of tegmen is thick and persistent in *Mesembryanthemum*. The inner epidermis is uniformly thick. The aril surrounds the seed in *Trianthema* and *Sesuvium*. It is absent in *Aptenia cordifolia*, *Hereroa hesperantha* and *Tetragonia tetragonioides* (N. Prakash 1967a, b, c).

HR Bhargava (1935a), R Bregman and F Bouman (1983), AE Cocucci (1961), VR Dnyansagar and RS Malkhede (1963), RN Kapil and N Prakash (1966), HS Narayana (1962a), HS Narayana and BC Lodha (1972), F Pax (1889), N Prakash (1967a, b, c), W Schmid (1925)

16.8 Portulacaceae

The anther wall comprises the epidermis, fibrous endothecium, a single ephemeral middle layer, and secretory tapetum of binucleate cells. The tapetal cells become multinucleate, and the nuclei fuse in *Portulaca tuberosa* (Raghavan and Srinivasan 1941c). Granular cutinization occurs on the inner wall of tapetal cells. The archesporium is a single row of cells which function as microspore mother cells (Kajale 1942). They undergo meiotic divisions giving rise to tetrahedral and isobilateral tetrads. Cytokinesis is simultaneous.

The pollen grains are three- to four-colporate, and three-celled at the shedding stage.

The ovule is ana-campylotropous, bitegmic and crassinucellar. The inner integument forms the micropyle. At the base of the ovule an air-space occurs between the two integuments in *Portulaca*.

The hypodermal archesporial cell cuts off a parietal cell, the parietal tissue is three or four layers thick. The apical cells of the nucellus elongate radially. Following meiotic divisions, the megaspore mother cell produces a linear or T-shaped tetrad. The chalazal megaspore gives rise to the Polygonum type of embryo sac. The synergids are hooked, and the antipodals are small and ephemeral. Starch grains appear at the binucleate stage of the embryo sac, these have also been observed in the egg cell (D.C. Cooper 1940).

In *Portulaca* three to four hours elapse between pollination and fertilization. Through the micropyle the pollen tube enters the embryo sac between the two synergids, without damaging any of them. The tip of the pollen tube bifurcates, one branch passes to the egg and the other to the secondary nucleus (D.C. Cooper 1940). The fate of synergids, and branching of pollen tube need confirmation.

The development of endosperm is of the Nuclear type. It becomes cellular at the globular proembryo in *Portulaca tuberosa*.

The embryogeny conforms to Period I, Series A, Megarchetype III (Onagrad type) in *Claytonia*, and Period II, Series A, Megarchetype III (Caryophyllad type) in *Portulaca* (Souèges 1938b). The occurrence of three embryos in an embryo sac is reported in *Portulaca oleracea* (Kajale 1942). The largest of the three is zygotic, the origin of the other two is disputable. The mature embryo is curved.

The seed is small, reniform, albuminous, with a strophiole in *Claytonia* and *Talinum*, and has starchy perisperm. The seed is exotestal with the outer epidermis of cuboid cells with thickened and lignified outer wall. They may be finely pitted, papilliform or with short hair-like processes. The inner epidermis is a crystal-layer in *Portulaca*; otherwise it is crushed. The cells of the inner epidermis of tegmen have bar-like thickenings on the radial walls in *Portulaca*. The rest of the tissue is crushed (Corner 1976).

The seeds of *Portulaca crytopetala* and *Talinum patens* develop an operculum. In the seeds of *Anacampseros rufescens*, adapted to wind dispersal, the outer layer of the outer integument is papery and can be separated from the inner layer of the outer integument, forming a balloon (Bregman and Bouman 1983).

R Bregman and F Bouman (1983), R Carolin (1987), MT Cook (1903), DC Cooper (1935a, 1940), EJH Corner (1976), B Haccius (1954), LB Kajale (1942), AM Khan (1973, 1975), G Klebs (1881–1885), WH Murdy and MEB Carter (1987), TS Raghavan and AR Srinivasan (1941c), T Rocén (1927), R Souèges (1938b, 1945c)

16.9 Basellaceae

The anther wall comprises persistent epidermis, fibrous multiseriate endothecium, ephemeral middle layer, and glandular tapetum with multinucleate cells. The microspore mother cells undergo meiotic divisions, cytokinesis is simultaneous, and the tetrads tetrahedral and isobilateral.

The family is highly multipalynous, and the pollen grains are three-celled at the dispersal stage in *Basella rubra* (Maheswari Devi and Pullaiah 1975).

The ovule is ana- to campylotropous, bitegmic and crassinucellar with inner integument forming the micropyle. The inner integument overgrows the outer one. The cells of the nucellar epidermis divide and form a nucellar beak. A prominent air-space develops between the integuments in the basal region. The archesporial cell cuts off a parietal cell, the parietal tissue is well developed. The megaspore mother cell undergoes meiotic divisions forming a linear tetrad. The chalazal megaspore develops into Polygonum type of embryo sac, polar nuclei fuse before fertilization, antipodals are small and ephemeral, and starch grains are present in the mature embryo sac.

The development of endosperm is of the Nuclear type. Cytokinesis occurs at the globular proembryo in *Basella*.

The embryogeny conforms to Chenopodiad type in *Basella rubra* (Maheswari Devi and Pullaiah 1975).

The seed is small, exarillate, and contains perisperm. The outer epidermis of testa comprises short, radially elongate cells with thick outer walls. The tegmen is represented only by the inner epidermis. The cells have streak-like thickenings on the radial walls.

GL Hindmarsh (1965), H Maheswari Devi and T Pullaiah (1975)

16.10 Caryophyllaceae

The anther is bilocular, and the wall comprises the epidermis, fibrous endothecium, one or two ephemeral middle layers, and secretory tapetum of binucleate cells. Simultaneous cytokinesis accompanies meiosis in microspore mother cells, and tetrahedral and isobilateral tetrads are produced in *Stellaria media* (P.C. Joshi 1936, see also P.S. and L.L. Narayana 1986).

The pollen grains are three- or four-colporate, pancolpate, and pantoporate. They are three-celled at the dispersal stage. In *Polycarpea nebulosa* (Long 1965), some pollen grains precociously produce abortive pollen tubes within the anther. Normal pollen tubes develop on the stigma. A similar condition occurs in *Thylacospermum* (P.C. Joshi 1936).

The ovule is campylotropous, bitegmic and crassinucellar (Fig. 16.10.1A) with the micropyle formed by inner integument (S. Pal and Murty 1972). The cells of the nucellar epidermis undergo periclinal divisions. The hypodermal archesporium is uni- or multi-celled. A single cell cuts off a parietal cell. The

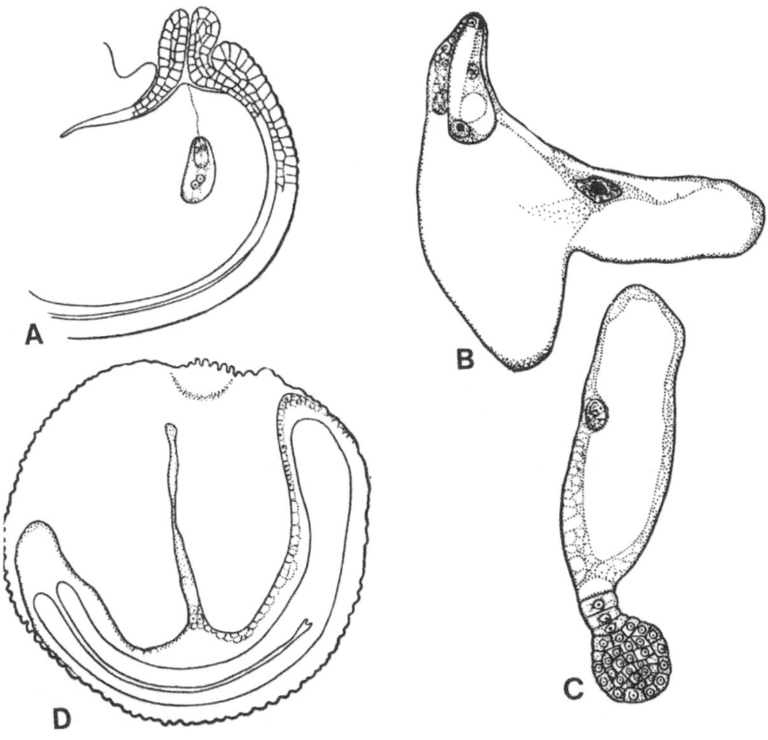

Fig. 16.10.1A–D. *Agrostemma githago*. **A** Longisection of ovule at eight-nucleate embryo sac. **B** Embryo sac with lateral extension containing secondary nucleus. **C** Proembryo with a large uninucleate suspensor. **D** Ovule at dicotyledonous embryo. Note the persistent lateral extension of the embryo sac. (After Rocén 1926)

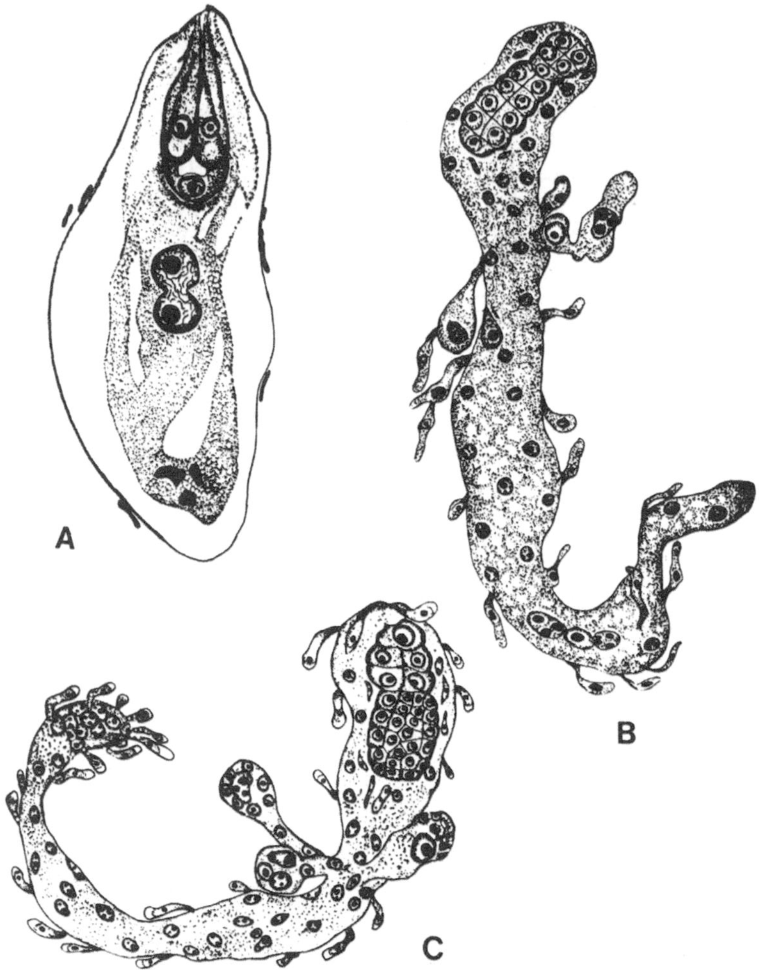

Fig. 16.10.2A, B. *Acanthophyllum krascheninnikovii*, **B** *A. albidium*. **A** Mature embryo sac. **B, C** Nuclear endosperm with finger-like appendages at globular proembryo. (See Yakovlev 1983)

parietal tissue is single-layered in *Stellaria media* (P.C. Joshi 1936), and two- to three-layered in *S. paniculata* (S. Pal and Murty 1974). The megaspore mother cell undergoes meiotic divisions. The chalazal megaspore of a linear or T-shaped tetrad develops into a Polygonum type of embryo sac (Fig. 16.10.2A). The synergids are hooked, polar nuclei fuse before fertilization, and antipodals are ephemeral. The embryo sac may develop a lateral caecum (Fig. 16.10.1B, see also D).

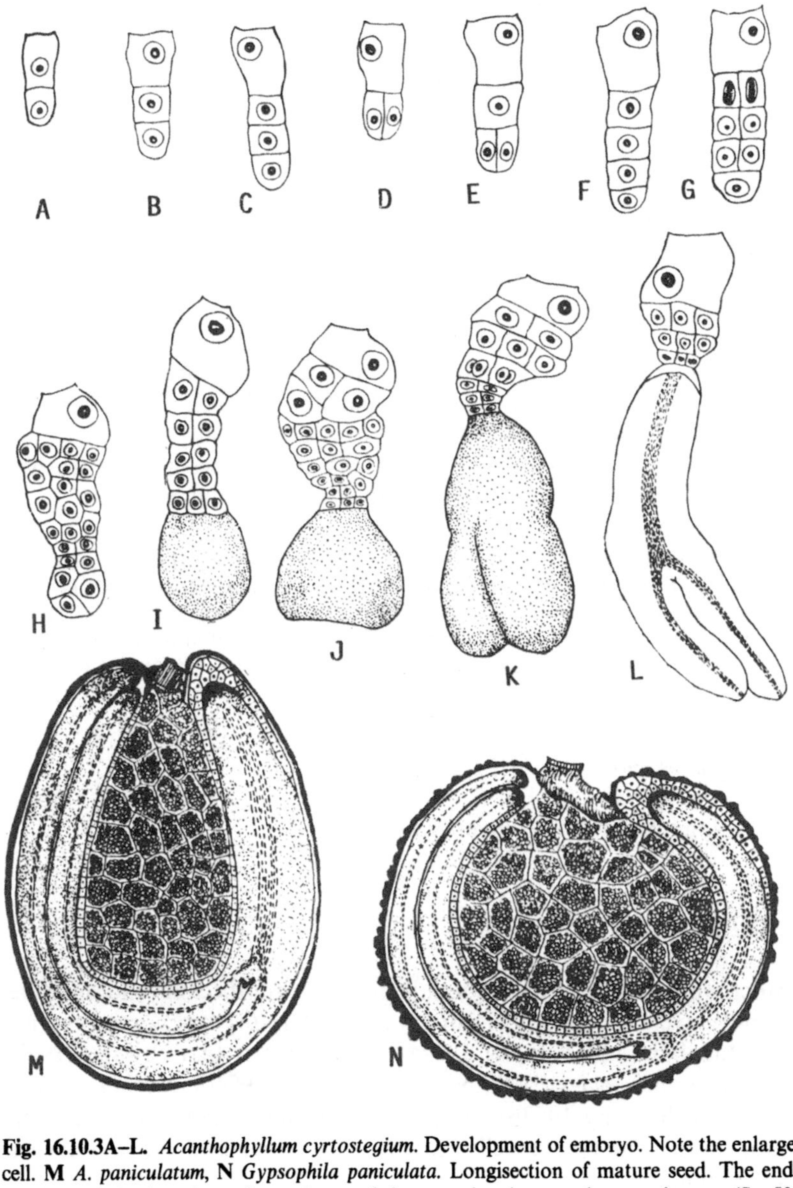

Fig. 16.10.3A–L. *Acanthophyllum cyrtostegium*. Development of embryo. Note the enlarged basal cell. **M** *A. paniculatum*, **N** *Gypsophila paniculata*. Longisection of mature seed. The endosperm persists only in the micropylar region, and the curved embryo encloses perisperm. (See Yakovlev 1983)

The development of endosperm is of the Nuclear type. In some taxa, the endosperm develops superficial extensions (Fig. 16.10.2B, C). Rocén (1927) further classified the development of endosperm into three types, according to the extent of cellular endosperm present in the seed: (1) Silene type – the endosperm occupies the entire seed, (2) Melandrium type – the endosperm occupies only the micropylar half of seed, and (3) Heliosperma type – a single layer of endosperm is present around the hypocotyl of embryo.

The embryogeny corresponds to the Caryophyllad type, and the basal cell of the conspicuous suspensor is vesicular and haustorial (Figs. 16.10.1C, 16.10.3A–L). The mature embryo is strongly curved around the copious mealy perisperm (Fig. 16.10.3M, N).

The seed is small, reniform, usually exarillate, and perispermous. In *Stegnosperma halimifolium* and *S. watsonii*, an aril develops from the chalazal region and completely surrounds the seeds (P.S. and L.L. Narayana 1986). The outer epidermis of testa comprises tabular cells with thick outer walls, the inner epidermis has tabular cells with thick inner walls, and the mesophyll is crushed. The tegmen is generally crushed, but the inner epidermis may persist.

JC Audran and M Batcho (1981), KM Buell (1952a, b), IP Chaudhury and HS Narayana (1978), RE Crang and MA Millay (1972), WD Creasy (1961), JL Guignard and JC Mestre (1967), PC Joshi (1936), K Kudlicka, M Kuras and B Rodkiewicz (1981), RW Long (1965), M Longevialle (1970, 1971), PS and LL Narayana (1986), W Newcomb and LC Fowke (1973, 1974), N Pal (1951d, 1952), S Pal (1974, 1975), S Pal and YS Murty (1972, 1974), MV Ramji (1975), T Rocén (1926, 1927)

16.11 Dysphaniaceae

The embryology has not been investigated.

16.12 Chenopodiaceae

The anther wall comprises an epidermis, fibrous endothecium and an ephemeral middle layer. The tapetum is mostly of the secretory type: *Arthrocnemum indicum* (Mahabale and Solanky 1954a), *Beta vulgaris* (Artschwager 1927), *Chenopodium ambrosioides* (Mahabale and Solanky 1954b), *Kochia scoparia* (Mahabale and Solanky 1953b) and *Suaeda maritima* (Maheswari Devi and Rama Rao 1976). The tapetal cells are uni-, bi- or four-nucleate and nuclear fusions are common.

In *A. indicum*, *C. ambrosioides*, *K. scoparia* and *S. fruticosa* (see Lakshmanan and Dulcy 1982), the cell walls break down and the protoplasts form a mass at the periphery.

In *Atriplex repens*, Lakshmanan and Dulcy (1982) report that in some loculi of a few anthers the tapetum is "amoeboid". The tapetal cells enlarge abnormally and, at the microspore stage, their inner walls break down and the protoplasts protrude into the loculi. The protoplast is densely cytoplasmic and forms tubular extensions around the microspores. According to Lakshmanan and Dulcy, the protoplasts are aggressive "haustorial" structures. If such extensions occur in the oppositely placed tapetal cells, these processes ramify in the locule around the microspores. They degenerate along with, or later than, the secretory tapetal cells.

The tapetum, in fact, is not of the amoeboid type, and such behaviour of tapetal protoplasts is of the intermediate type between the amoeboid and secretory. There is no question of the tapetal protoplasts being aggressive and haustorial. These protoplasts provide nutrition to the developing pollen grains.

Amoeboid type of tapetum is reported in *Chenopodium murale* (Mahabale and Solanky 1954c), and both secretory and amoeboid in male sterile *Beta vulgaris* (Artschwager 1947). The report of amoeboid tapetum in *Arthrocnemum indicum* by Mahabale and Solanky (1954a) seems to be erroneous.

All cases of aberrant behaviour of tapetum should be reexamined.

Sterile pollen grains occur in some loculi where the endothecium lacks fibrous thickenings, and the tapetum does not organize.

The microspore mother cells undergo meiotic divisions, and produce isobilateral and tetrahedral tetrads. Simultaneous cytokinesis occurs by furrowing.

The pollen grains are three-celled at the shedding stage. The occurrence of two vegetative nuclei in the pollen grains of double size is reported in *Chenopodium album* (Bhargava 1936), *Kochia scoparia* and *Suaeda fruticosa* (Mahabale and Solanky 1953a, b). Four-nucleate pollen grains with two vegetative and two male nuclei have been observed in *Arthrocnemum indicum* (Mahabale and Solanky 1954a), *Atriplex hymenelytra* (Billings 1934), *C. album* (Bhargava 1936) and *C. ambrosioides* (Mahabale and Solanky 1954b). The development of male gametophytes needs further study.

The ovule is campylotropous, bitegmic and crassinucellar. The inner integument forms the micropyle. A prominent air-space occurs between the two integuments in the chalazal region. The archesporium is one-, three- or four-celled in *Salicornia brachiata* and *S. fruticosa* (Maheswari Devi and Rama Rao 1982). The archesporial cell cuts off a parietal cell, the parietal tissue is not very conspicuous except in *Chenopodium ambrosioides* and *Salicornia brachiata* (Maheswari Devi and Rama Rao 1982). The nucellar epidermis divides, and produces a nucellar beak. Its cells elongate and become richly protoplasmic in the micropylar region. The megaspore mother cell undergoes meiotic divisions and produces a linear tetrad. The chalazal megaspore develops into a Polygonum type of embryo sac. The synergids have prominent hooks, and the three antipodal cells may be ephemeral or persist up to fertilization. In *Beta vulgaris* supernumerary antipodals are produced. Their number goes up to 15 in *C. amaranticolor*. The polar nuclei fuse before fertilization. Allium type of development of the embryo sac is reported in *Suaeda maritima* (Maheswari Devi and Rama Rao 1976). Twin embryo sacs occur in *Arthrocnemum*. A chalazal caecum develops in *Beta* and *Kochia*.

The pollen grains germinate in two to three hours on the stigma in *B. vulgaris*. Entry of the pollen tube is porogamous, and the richly cytoplasmic nucellar beak guides the growth of the pollen tubes (Artschwager and Starrett

1933). Double fertilization occurs, and the pollen tube persists up to six-celled proembryo.

The development of endosperm is of the Nuclear type. The endosperm becomes completely cellular at a young stage of the proembryo in *Arthrocnemum indicum* (Mahabale and Solanky 1954a). One or two layers of endosperm persist in the micropylar half of the seed.

The embryogeny conforms to the Chenopodiad type. The suspensor is massive except in *Arthrocnemum* and *Kochia* where it is uniseriate with a large basal haustorial cell. The mature embryo is curved or, generally, spirally-coiled. Polyembryony is reported in *Beta* and *Suaeda maritima* (Maheswari Devi and Rama Rao 1976).

The seed is small with starchy perisperm, and spirally coiled embryo. The integuments are reduced to exotesta and endotegmen. The cells of outer epidermis of the testa are somewhat radially elongated with their outer walls strongly thickened, and contain tannin. The inner epidermis of tegmen is persistent, and shows thickenings on the radial and inner walls.

E Artschwager (1927, 1947), E Artschwager and RC Starrett (1933), HR Bhargava (1936), FH Billings (1934), L Bruun and P Olesen (1989), GO Cooper (1935), MI Hakki (1971, 1973), B Jassem (1980), RL Jones (1974), KK Lakshmanan and EJ Dulcy (1982), TS Mahabale and IN Solanky (1953a, b, 1954a, b, c), H Maheswari Devi and MS Rama Rao (1976, 1982), DD Nautiyal and SC Gupta (1985), G Psaras (1984), R Souèges (1920a), ME Williams (1932), HJ Wilms (1980a, b, 1981a, b)

16.13 Amaranthaceae

The anther wall comprises an epidermis, fibrous endothecium, single ephemeral middle layer, and secretory tapetum of binucleate cells. The fibrous thickenings of endothecium are absent in *Achyranthes aspera* (Kajale 1937b). Übisch granules are studded on the inner walls of tapetal cells. In *Alternanthera paronychioides*, after the degeneration of tapetum, the Übisch granules come to lie appressed to the fibrous endothecium. The microspore mother cells undergo meiotic divisions and produce tetrahedral and isobilateral tetrads. Decussate tetrads are reported in *Psilostachys* (Bakshi 1952). Simultaneous cytokinesis occurs by furrowing.

The pantoporate pollen grains are three-celled at the shedding stage.

The basal ovule is orthotropous in *Psilostachys* (Bakshi 1952), campylotropous in *Celosia cristata* (Pal 1968) and amphitropous in *Amaranthus*. It is bitegmic and crassinucellar. The micropyle is formed by the inner integument, except in *Gomphrena* where both integuments are concerned. An air-space is present between the integuments in the basal region. A single cell of the multicellular archesporium cuts off a parietal cell producing parietal tissue. A nucellar cap is present. Following meiosis the megaspore mother cell produces a linear or T-shaped tetrad. The chalazal megaspore develops into a Polygonum type of embryo sac. The synergids are hooked, polar nuclei fuse before fertilization, and antipodals are ephemeral except in *Pupalia lappacea* (Kajale 1940, 1954b) where they increase up to 30–40 cells. In *Psilostachys* the embryo sac develops a chalazal caecum.

Double fertilization occurs as usual. The entry of accessory pollen tubes into the embryo sac is common in *Digera arvensis* (A.C. Joshi and Kajale 1937). This aspect needs further study.

The development of endosperm is of the Nuclear type. Wall formation commences in the micropylar part, and proceeds to the chalazal region. It becomes cellular at the globular proembryo. In *Alternanthera sessilis* the endosperm remains free-nuclear in the chalazal region.

The development of embryo follows the Chenopodiad type. In *Gomphrena celosioides* (Padhye 1962a), the embryogeny is of the Chenopodiad type, or Period I, Series C, Megarchetype V of Embryonic Group 3 of Souèges' system of classification.

The seeds are exotestal and endotegmic with funicular aril, and starchy perisperm. Both the integuments contribute to the formation of the seed-coat. The outer epidermis of testa is a mechanical layer of radially elongate cells with strongly thickened outer walls, and contains tannin. The inner epidermis is thin-walled. The inner epidermis of tegmen alone persists, and shows thickenings on the radial and inner walls.

In *Aerva tomentosa* (Sachar and Murgai 1958, 1959), the megaspore mother cell directly functions as uninucleate diploid embryo sac; meiosis does not occur (diplospory). Sometimes, mitotic division in the megaspore mother cell is followed by a wall and a "dyad" is formed. The micropylar dyad degenerates and the chalazal develops into a two-, four- and eight-nucleate embryo sac. The synergids, or even the egg apparatus, may degenerate. Due to the elongation of the embryo sac, the antipodals are shifted to a lateral position. Occasionally, twin embryo sacs develop, probably from two megaspore mother cells. The polar nuclei fuse to form the secondary nucleus which assumes the function of primary endosperm nucleus.

There is extreme rarity of male plants, and an almost total absence of pollen grains. This prevents fertilization, and pollen tubes have not been observed in the ovules. The Nuclear type of endosperm develops autonomously. The multinucleolate nuclei have an irregular outline. Cytokinesis occurs first at the micropylar end, and then proceeds to the chalaza.

The unfertilized unreduced egg develops parthenogenetically, and embryogeny is comparable to the Chenopodiad type. The suspensor may proliferate. The mature embryo is dicotyledonous.

Sachar and Murgai conclude that *Aerva tomentosa* is an apomict.

TS Bakshi (1952, 1954), AC Joshi (1936b), AC Joshi and LB Kajale (1937), LB Kajale (1935, 1937b, 1940, 1954b), R Mendoza (1973), MD Padhye (1962a), A Pal (1968), V Puri and B Singh (1935), RC Sachar and P Murgai (1958, 1959)

16.14 Didiereaceae

The embryology [see Schölch (1963) and Rauh and Schölch (1965)] has not been studied in sufficient detail, the following account is based on *Alluaudia dumosa*, *A. humbertii*, *A. montagnacii*, *A. procera* and *Didierea magascariensis*.

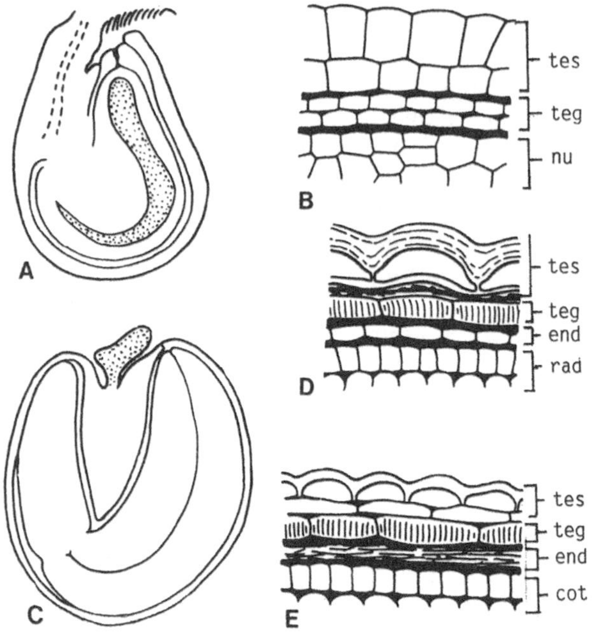

Fig. 16.14.1A–E. Didiereaceae, structure of seed. **A, B** *Alluaudia dumosa*; **A** Longisection of immature seed, micropyle is formed by inner integument; **B** Portion of seed-coat, there is a two-layered testa (*tes*) and two-layered tegmen (*teg*); *nu* nucellus. **C, D** *A. ascendens*; **C** Longisection of seed with curved embryo, and vestigial aril (*stippled*); **D** Portion of seed-coat from radicular region of **C** (endotesta and exotegmen crushed), exotesta with thick outer walls, endotegmen has thickenings, and there is a persistent layer of endosperm (*end*); *rad* radicular region of embryo. **E** *Didierea mirabilis*, portion of seed-coat in cotyledonary region, testa is two-layered, endotegmen shows thickenings, endosperm has degenerated, followed by a layer of cotyledon (*cot*). (After Perrot and Guérin 1903, see Corner 1976)

The embryological features of Didiereaceae are: secretory anther tapetum, simultaneous cytokinesis in pollen mother cells, bi- and tri-celled pollen, crassinucellate, ana- to campylotropous ovules with two-layered integuments, micropyle formed by the inner integument only, occasional presence of a nucellar cap formed by the division of nucellar epidermis, Polygonum type of embryo sac, Nuclear type of endosperm, Chenopodiad type of development of the embryo, exalbuminous (?), arillate seed with curved embryo (Fig. 16.14.1C), perisperm absent, and the structure of seed-coat is simple (Fig. 16.14.1A–E). The outer epidermal cells of testa develop thickenings, the inner may collapse (Fig. 16.14.1D) or persist (Fig. 16.14.1E). The outer epidermis of tegmen also collapses, and the inner develops bar-like thickenings (Fig. 16.14.1D, E). The endosperm is consumed; the presence of a single layer above the radicle is doubtful (see Corner 1976).

EJH Corner (1976), BM Johri (1967), E Perrot and P Guérin (1903), W Rauh and HF Schölch (1965), HF Schölch (1963)

16.15 Taxonomic Considerations

Engler and Prantl (1931) place all the 12 families (Melchior 1964) in Centrospermae. Hallier (1912) derives them from his Ranalian Podophylleae. Hutchinson (1973) distributes them over four orders: Nyctaginaceae in Thymelaeales; Didieraceae in Sapindales; Molluginaceae, Caryophyllaceae, Aizoaceae and Portulacaceae in Caryophyllales; Phytolaccaceae, Gyrostemonaceae, Chenopodiaceae, Amaranthaceae and Basellaceae in Chenopodiales.

There is a general uniformity of characters in the order.

The suborder Phytolaccineae comprises six families: Phytolaccaceae, Gyrostemonaceae, Achatocarpaceae, Nyctaginaceae, Molluginaceae and Aizoaceae.

Phytolaccaceae. Apart from the characters of the order, Phytolaccaceae have one basal ovule in each loculus of two-celled ovary, funicular obturator, persistent antipodals, Chenopodiad or Caryophyllad type of embryogeny, and schizocarpic fruit.

Nyctaginaceae. The position of Nyctaginaceae is rather controversial. The antipodals undergo secondary multiplication. The formation of mono-, bi- and tetrasporic embryo sacs in *Mirabilis jalapa* (Hjelmqvist and Grazi 1965) is similar to that in Aizoaceae. Kajale (1938) discusses the development of embryo in Nyctaginaceae (*Boerhaavia*) in relation to that in Caryophyllaceae, Amaranthaceae, Chenopodiaceae and Molluginaceae. According to him, there is a close similarity in the development of embryo in Centrospermales except in Nyctaginaceae. The proembryo is multicellular and uniseriate; four or five cells take part in the development of embryo proper. All the cells that form the embryo differentiate before any longitudinal division takes place in the proembryonal cells. The first longitudinal division occurs in the second, third or fourth cell from the apex, and then extends towards the apex. Contrary to this, in Nyctaginaceae the first vertical division occurs in the apical cell. Thus, the sequence in which longitudinal divisions appear in the proembryo is quite different from that in other families. Secondly, the periblem originates from the dermatogen, and is not a sister layer of plerome as in other angiosperms. All these developmental sequences need further investigation.

Venkateswarlu (1947c) supports the inclusion of Nyctaginaceae in Thymelaeales. The affinities between the two are: (1) secretory anther tapetum, (2) simultaneous cytokinesis by furrowing in pollen mother cells, (3) three-celled pollen grains, (4) presence of nucellar cap, (5) micropyle formed by inner integument, (6) supernumerary antipodals, and (7) Nuclear type of endosperm becoming cellular afterwards. Moreover, Nyctaginaceae differ from the rest of the families of Centrospermales in Asterad type of embryogeny in *Oxybaphus viscosus* (Souèges 1938c), *Boerhaavia diffusa*, and *B. repanda* (Kajale 1940). In Thymelaeales, Thymelaeaceae, Geissolomataceae and Penaeaceae have Asterad type of embryogeny. If embryogeny is the basis, Venkateswarlu (1947c) favours the placement of Nyctaginaceae with the members of Thymelaeales as also suggested by Hutchinson (1973). The characters cited by Venkateswarlu (1947c), except the embryogeny, also occur in Centrospermales.

Corner (1976) states that the seed structure of Nyctaginaceae is typically Centrospermalean, and rejects its placement in Thymelaeales. Takhtajan (1969, 1980) considers it as an advanced family with entomophilous mode of pollination and exarillate seed, and places it in Caryophyllales.

Ficoideae. Bentham and Hooker (1965c) place Ficoideae (Ficoidaceae) along with Cactaceae in Ficoidales, and split it into three tribes: Mesembryeae, Aizoideae and Mollugineae. Hutchinson (1973) erects two independent families, Molluginaceae and Aizoaceae (see also Melchior 1964).

Molluginaceae and Aizoaceae. Molluginaceae and Aizoaceae show marked similarity in the structure and development of anther, ovule, embryo sac, and endosperm, embryogeny of the Caryophyllad or Solanad type, and perispermous seed. A.C. Joshi and Rama Rao (1936) segregate Aizoaceae from Molluginaceae, on the basis of funicular aril, number of cell layers in outer integument, occurrence of periclinal divisions in nucellar epidermis, and presence of starch grains in the embryo sac. In addition, funicular aril, airspace between and at the base of integuments, and nucellar cap are poorly developed in Aizoaceae. Sometimes, the seed is exalbuminous (*Aptenia*). Aizoaceae also exhibit several types of development of the embryo sac. Schmid (1925) reports Adoxa type in *Mesembryanthemum pseudotruncatella*, but this needs confirmation. Kapil and Prakash (1966) report five types of embryo sacs (Sect. 16.7) in *Delosperma*.

N. Prakash (1967a) investigated the embryology of several genera of Aizoaceae, and concluded that there is no marked distinction between Molluginaceae and Aizoaceae, and Molluginaceae should be merged with Aizoaceae. This needs further study.

Portulacaceae and Basellaceae. The suborder Portulacineae comprises Portulacaceae and Basellaceae. Bentham and Hooker (1965c) treat Basellaceae as a subfamily Baselleae under Chenopodiaceae. Eichler (1878), Engler and Prantl (1931), Rendle (1959) and Lawrence (1951) raise it to a family rank under Centrospermales. Hutchinson (1973) recognizes Basellaceae as an independent family in Chenopodiales – an order evolved from Caryophyllales. Bessey (1915), Gunderson (1950), Takhtajan (1980) and Cronquist (1981) place it in Caryophyllales. Melchior (1964) includes Portulacaceae, Basellaceae, Chenopodiaceae and Caryophyllaceae in Centrospermales (Table 16.9.1). According to Takhtajan (1969, 1980), Basellaceae is related to Portulacaceae and differs from Chenopodiaceae. Portulacaceae and Basellaceae are in close alliance in most of the essential features which are in conformity with those of the order. However, Basellaceae do show some differences. The unilocular ovary with a single ovule on basal placenta, nucellar beak, Chenopodiad type of embryogeny, and drupaceous fruit do not match with those of Portulacaceae.

Chenopodiad type of embryogeny occurs in Chenopodiaceae. Chenopodiaceae show occasional presence of amoeboid tapetum, campylotropous ovule with an air space between the integuments, Polygonum as well as Allium type of embryo sac, and secondary multiplication of antipodals. These features clearly indicate the differences between Basellaceae and Chenopodiaceae. Maheswari Devi and Pullaiah (1975) favour the retention of Basellaceae in Centrospermales, as an independent family (Table 16.9.1).

Caryophyllaceae. Caryophyllaceae represent the suborder Caryophyllineae. Caryophyllaceae are characterized by free-central placentation. Eichler (1878), Wettstein (1935) and Rendle (1959) seek their ancestry in Phytolaccaceae (Table 16.9.1). On the other hand, Bessey (1915) supports their Ranalian ancestry. Phytolaccaceae and Caryophyllaceae both show glandular anther tapetum, three-celled pollen grains, ana-campylotropous ovule with a nucellar beak, Nuclear type of endosperm, and Caryophyllad type of embryogeny. According to Corner (1976), Caryophyllaceae has a simplified seed of Centrospermae. Embryologically, Caryophyllaceae are much closer to Phytolaccaceae than to Ranales. Ranalian families show variations in the number of integuments, and development of the embryo sac. Again, the campylotropous ovule with nucellar beak, ephemeral antipodals, Caryophyllad type of embryogeny and Centrospermalean seed do not occur in Ranales.

Stegnosperma. P.S. and L.L. Narayana (1986), who have studied the embryology of *S. halimifolium* and *S. watsonii*, suggest (without much justification) that this taxon should be raised to family rank, Stegnospermataceae, in the vicinity of Phytolaccaceae as a possible connecting link between the Phytolaccaceae and Caryophyllaceae. *Stegnosperma* is well-placed in the Caryophyllaceae, and we do not agree with P.S. and L.L. Narayana.

The suborder Chenopodiineae consists of Dysphaniaceae, Chenopodiaceae, Amaranthaceae and Didiereaceae. Chenopodiineae is supposed to have evolved from Caryophyllineae.

Chenopodiaceae, Amaranthaceae. Chenopodiaceae are characterized by fleshy habit, whereas Amaranthaceae are usually distinguished by the presence of scarious bracts. Embryologically, both Chenopodiaceae and Amaranthaceae have close affinities: secretory anther tapetum, bitegmic crassinucellar ovules with endostome forming the micropyle, presence of an air-space, massive nucellus (with nucellar beak or cap) persisting as perisperm, Polygonum type of embryo sac with persistent antipodals, Nuclear endosperm, and Chenopodiad embryogeny. The structure of seed is typically Centrospermalean. Their inclusion in the same order is justified. They also show affinities with Caryophyllaceae.

Didiereaceae. Didiereaceae have been included in the orders Sapindales, Centrospermales and Euphorbiales. The embryological features of Didiereaceae and Centrospermae are very similar: (1) secretory anther tapetum, (2) simultaneous cytokinesis in pollen mother cells, (3) trinucleate pollen, (4) ana- to campylotropous ovule, crassinucellate, (5) micropyle formed by inner integument, (6) occasional presence of nucellar cap formed from the divisions of nucellar epidermis, and (7) Chenopodiad type of embryogeny (Schölch

Table 16.9.1. Embryology of Phytolaccaceae, Aizoaceae, Portulacaceae, Basellaceae, Caryophyllaceae and Chenopodiaceae (Adapted after Maheswari Devi and Pullaiah 1975)

	Phytolaccaceae	Aizoaceae	Portulacaceae	Basellaceae	Caryophyllaceae	Chenopodiaceae
Anther	Tetrasporangiate	Tetrasporangiate	Tetrasporangiate	Tetrasporangiate	Bisporangiate	Tetrasporangiate
Wall	–	Basic type	Monocotyledonous	Monocotyledonous	Mono- or dicotyledonous	Monocotyledonous
Endothecium	Fibrous	Fibrous	Fibrous	Fibrous, multiseriate	Fibrous	Fibrous
Tapetum	Glandular, cells bi- or multinucleate	Glandular, cells bi- or four-nucleate	Glandular, cells bi- or multinucleate	Glandular, cells multinucleate	Glandular, cells binucleate	Glandular, cells uni- or binucleate; occasionally amoeboid
Tetrads	Tetrahedral, isobilateral	Tetrahedral, isobilateral, decussate	Tetrahedral, isobilateral	Tetrahedral, isobilateral	Tetrahedral, decussate	Tetrahedral, isobilateral, decussate
Pollen grain	Two- (*Phytolacca*) or three-celled (*Rivina*)	3-celled	3-celled	3-celled	3-celled	3-celled
Ovule	Anacampylotropous, bitegmic, crassinucellate	Anacampylotropous, bitegmic, crassinucellate, funicular obturator develops	Anacampylotropous, bitegmic, crassinucellate	Ana- or campylotropous, bitegmic, crassinucellate	Campylotropous, bitegmic, crassinucellate, nucellar beak conspicuous	Campylotropous, bitegmic, crassinucellate, nucellar beak conspicuous

Micropyle	Formed by inner integument	Inner integument	Inner integument	Inner integument	Inner integument	Inner integument
Air-space between the integuments at their base	Present	Present	Present	Present	Present	Present
Archesporium	2- or 3-celled	Multicelled	2- or 3-celled	2- or 3-celled	Uni- or multicelled	2- or 3-celled
Embryo sac	Polygonum type	Polygonum, Endymion, Drusa, Adoxa or Penaea	Polygonum	Polygonum	Polygonum	Polygonum; Allium type in *Sueda maritima*
Antipodal cells	Persistent, secondary multiplication in *Rivina*	Ephemeral	Ephemeral	Ephemeral	Ephemeral	Ephemeral, sometimes supernumerary due to divisions
Endosperm	Nuclear type	Nuclear, chalazal haustorium in *Hereroa hesparantha*	Nuclear	Nuclear	Nuclear, in some taxa superficial extensions develop	Nuclear
Embryogeny	Caryophyllad type, massive suspensor	Solanad, occasional suspensor polyembryony	Onagrad or Caryophyllad, occasional polyembryony	Chenopodiad	Caryophyllad, basal cell of suspensor haustorial	Chenopodiad, suspensor massive
Seed	Small, exotestal, perispermous	Exotestal, and endotegmic, perispermous	Small, exotestal, albuminous, perispermous	Small, exotestal, perispermous	Small, exotestal, perispermous	Small, exotestal and endotegmic, perispermous

1963). The structure of vegetative organs, pericarp, testa, and development of the embryo sac are also comparable (see B.M. Johri 1967). Schölch (1963) concludes that Didiereaceae should be placed in Centrospermales between Cactaceae and Portulacaceae, side by side with Nyctaginaceae. Wunderlich (1967) considers embryological, palynological and biochemical data of Didiereaceae, and favours their separation from Sapindaceae and placement in Centrospermae. Takhtajan (1980) places Didiereaceae in the Caryophyllales which includes a number of families assigned to Centrospermae by Melchior (1964).

17 Cactales

The family Cactaceae alone represents the order Cactales.

17.1 General Features

The embryological features are: (1) secretory anther tapetum with multinucleate cells, (2) simultaneous cytokinesis in pollen mother cells, (3) three-celled pollen grains, (4) bitegmic, crassinucellate ovule with an air-space between the integuments in the chalazal region, (5) micropyle formed by the swollen tip of the inner integument, (6) curved nucellus forming nucellar cap, (7) multicellular archesporium, (8) Polygonum type of embryo sac with ephemeral antipodals, (9) Nuclear type of endosperm, (10) occasional degeneration of zygotic embryo, (11) nucellar and suspensor polyembryony, and (12) perispermous, albuminous, arillate seed.

17.2 Cactaceae

The anther wall comprises the epidermis, fibrous endothecium, single middle layer, and tapetum in *Mammillaria carnea* (Y.D. Tiagi 1961). The epidermis is lignified and papillate in *Opuntia dillenii* (Y.D. Tiagi 1954). The tapetum is secretory with multinucleate cells; nuclear fusions are common. Meiotic divisions in microspore mother cells are followed by simultaneous cytokinesis. The tetrads are tetrahedral and decussate.

Pollen grains are three-colpate, pantoporate, and three-celled at the shedding stage.

The funicle is branched, and two or more ovules are borne on the same stalk (P. Maheshwari and Chopra 1955). The inner surface of the funicle is covered with papillate cells. The ovule is anatropous, hemianatropous, campylotropous or circinotropous. It is bitegmic and crassinucellate. The micropyle is formed by the swollen apex of the inner integument which protrudes and approaches the funicle. The cells of the nucellar epidermis produce a nucellar cap. The cells in the region of tip are radially elongate. In the chalazal region of ovule a prominent air-space occurs between the two integuments.

Y.D. Tiagi (1970a) has studied *M. tenuis*, *O. dillenii*, *O. ficusindica* and *Zygocactus truncatus*. A single cell from the hypodermal multicellular archesporium cuts off a parietal cell which gives rise to a parietal tissue. The megaspore mother cell undergoes meiotic divisions producing a triad, or linear or T-shaped tetrad. The chalazal megaspore develops into a Polygonum type of embryo sac. The synergids are hooked, and exhibit filiform apparatus in *M. tenuis* (Y.D. Tiagi 1957) and *O. dillenii* (P. Maheshwari and Chopra 1955). The antipodals are ephemeral; they may become binucleate. The embryo sac contains starch grains.

Double fertilization occurs, and persistent pollen tubes have been observed both in the micropyle and nucellus.

The development of endosperm is of the Nuclear type. It becomes cellular around the embryo in *Astrophytum*, *Thelocactus* and *Toumeya*, and persists as a cap in the mature seed (Engleman 1960). In *M. tenuis* (Y.D. Tiagi 1957) and *Opuntia* (P. Maheshwari and Chopra 1955), the endosperm remains free-nuclear throughout. In *Cereus* and *Ferocactus* (Kapil and Prakash 1969), the chalazal region of the nuclear endosperm contains dense cytoplasm, and the nuclei become hypertrophied. It forms a haustorium. Cytokinesis occurs at the dicotyledonous stage of embryo, and only a few cells are formed around the radicle.

The earlier divisions of zygote and its derivatives are transverse, resulting in a filamentous proembryo in *Astrophytum*, *Thelocactus*, *Toumeya* (Engleman 1960), and *Ripsalis* (Mauritzon 1934a). Kapil and Prakash (1969) report an oblique division of zygote in *Cereus jamacaru*, presence of a massive suspensor, and formation of additional embryos due to proliferation of suspensor cells. The mature embryo has well-developed cotyledons in primitive genera, but remains undifferentiated in advanced members.

The zygotic embryo generally degenerates, and additional embryos develop from the nucellus in several species of *Mammillaria* (Fig. 17.2.1A–D), *Opuntia* (P. Maheshwari and Chopra 1955, Naumova 1978), and *Pereskia* (Y.D. Tiagi 1970a). Generally, only one embryo reaches maturity. Sometimes, in *O. dillenii* (P. Maheshwari and Chopra 1955), due to fusion of embryonal masses, embryos with three and five cotyledons are formed. The seeds are polyembryonate in *O. elata* (Naumova 1978); the embryos are of nucellar origin. The embryogenesis in nucellar and sexual embryos is comparable, but the sexual embryos take 30 days to mature, while the nucellar ones only 15–20 days (Naumova 1978).

The seed is perispermous (perisperm absent in *Astrophytum myriostigma* – Engleman 1960), albuminous, and prominently exotestal.

The seeds of cactaceous taxa vary in size (approx 0.5 mm to 10 mm), shape, colour, epidermal structure, and position of funicle and micropyle (Barthlott and Voit 1979). The micropyle and hilum are sometimes separated by a zone of sclerotic cells of testa.

The hilum-micropylar zone, together with raphe and chalaza, form only a small area of seed structure. The embryo is oblong or globular, straight or

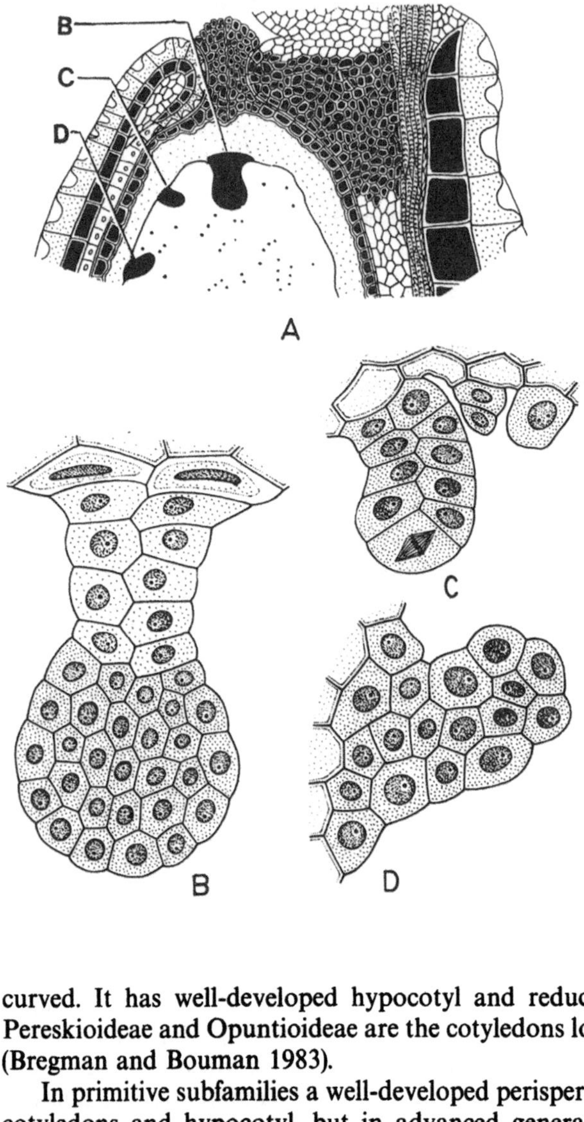

Fig. 17.2.1A–D. *Mammillaria tenuis*, nucellar polyembryo. **A** Longisection of micropylar portion of ovule to show embryo sac with three adventive embryos which have developed from the nucellus. **B–D** Nucellar proembryo marked *B*, *C* and *D* in **A**, note the suspensor in **B**. (After Y. D. Tiagi 1956, 1970a)

curved. It has well-developed hypocotyl and reduced cotyledons. Only in Pereskioideae and Opuntioideae are the cotyledons longer than the hypocotyl (Bregman and Bouman 1983).

In primitive subfamilies a well-developed perisperm is located between the cotyledons and hypocotyl, but in advanced genera the perisperm has disappeared and the function of storage has been taken over by the hypocotyl. A "hilum cup" is formed due to shrinkage of funicular cells, and aril or strophiole is formed due to the growth of funicular tissue. These cells form a collar around the hilum–micropylar zone, or develop into an amorphous mass. The strophiole aids in seed dispersal, and in water uptake prior to germination. Opercula are restricted to the subfamilies Cactoideae (Cereoideae) and Pereskioideae, and are absent in Opuntioideae.

The outer epidermis of testa is a mechanical layer with strongly thickened cells. The inner epidermis is thin-walled, and may persist. The inner epidermis of tegmen alone persists, and shows bar-like thickenings on radial walls. In

Opuntioideae the funicle, with its papillate outgrowths, becomes pulpy. The seeds of many cacti are strophiolate (Buxbaum 1955), and the swelling of funicle may contribute to its formation (Engleman 1960).

I Banerji and S Sen (1954), W Barthlott and G Voit (1979), R Bregman and F Bouman (1983), P Brough (1923), F Buxbaum (1944, 1955, 1958), RN Chopra (1957), EM Engleman (1960), RN Kapil and N Prakash (1969), P Maheshwari and RN Chopra (1954, 1955), J Mauritzon (1934a), TN Naumova (1978), M Neumann (1935), F Ritter (1980), YD Tiagi (1954, 1956, 1957, 1959, 1961, 1970a)

17.3 Taxonomic Considerations

Cactaceae. Bentham and Hooker (1965a) place Cactaceae in the order Ficoidales, between Passiflorales and Umbellales. Engler and Diels (1936) and Hutchinson (1973) group them with parietalean families. Y.D. Tiagi (1970a) negates the relationship with parietalean families, except for parietal type of placentation. Wettstein (1935) and Thorne (1963) associate them with Aizoaceae and Portulacaceae. Cronquist (1981) considers them to be allied to Caryophyllales; R. Dahlgren (1980b, 1981) places them in Caryophylliflorae (Caryophyllales). According to Hutchinson (1973), the position of Cactaceae (Cactales) is rather problematic.

P. Maheshwari (1950, 1964) emphasizes the relationship of Cactaceae with Aizoaceae and Portulacaceae. The embryological similarities are: (1) glandular anther tapetum with two- to four-nucleate cells, (2) simultaneous cytokinesis, (3) three-celled pollen grains, (4) bitegmic and crassinucellate ovules with curved nucellus, (5) occurrence of air-space between the two integuments in the chalazal region, (6) micropyle formed by the swollen tip of the inner integument, (7) presence of nucellar cap, (8) Polygonum type of embryo sac, (9) Nuclear type of endosperm, and (10) perisperm as storage tissue in the seed (Mauritzon 1934a, Neumann 1935, P. Maheshwari 1950, 1964, Banerji and Sen 1954, Kapil and Prakash 1969).

According to Martin (1946) and Kapil and Prakash (1969), the seed structure of Cactaceae is similar to that of Centrospermales. The radial elongation of nucellar epidermal cells is comparable to that in Aizoaceae. Engleman (1960) compares the funicular hump of Centrospermales with the strophiole of cacti. Wunderlich (1967) does not recognize the ordinal rank of Cactaceae, and she suggests their inclusion in Centrospermae.

P. Maheshwari (1950) cites several similarities between Portulacaceae and Cactaceae: (1) microspore mother cells in one or two (rarely more than two) rows in the pollen sac, (2) micropyle in close proximity to the funiculus, (3) T-shaped tetrad of megaspores, (4) long synergids, and (5) embryo sac constricted in the middle region (Neumann 1935).

P. Maheshwari (1950) rejects any affinity between Cactaceae and Passifloraceae. In the latter, periclinal divisions of nucellar epidermis are absent, the ovule is completely anatropous with a perfectly symmetrical nucellus, and the outer integument overgrows the inner one, and both form the micropyle. There is no parallel to these features in Cactaceae.

On the basis of primitive characters in the floral construction of Cactaceae, Y.D. Tiagi (1970a) mentions its relationship with Ranales or Ranalian ancestors. The floral anatomy, nodal type, and basic chromosome number are similar to those of Calycanthaceae. The similarities between Cactaceae and Calycanthaceae are: (1) secretory anther tapetum, (2) simultaneous cytokinesis of pollen mother cells, (3) multicellular female archesporium, (4) Polygonum type of embryo sac with ephemeral antipodals, and (5) nucellar polyembryony (S.L. Mathur 1968). On the other hand, the differences between the two are also well marked. Calycanthaceae show (1) two-celled, two- or three-sulcate pollen grains, (2) anatropous ovule with micropyle formed by both the integuments, (3) absence of parietal cell, (4) absence of double fertilization in some members, and (5) autonomous development of endosperm according to Cellular type (S.L. Mathur 1968). The seed is exalbuminous and endotestal, and does not show any similarity with that of Cactaceae.

Cactaceae form a bridge between Aizoaceae and Portulacaceae. Embryologically, the similarities are more pronounced with Aizoaceae. Buxbaum (1944) derives Cactaceae from Phytolaccaceae, and considers its parallel evolution with Aizoaceae.

18 Magnoliales

The Magnoliales comprise 22 families in six groups: (1) Magnoliaceae, Degeneriaceae, Himantandraceae, Winteraceae, Annonaceae, Eupomatiaceae, Myristicaceae, Canellaceae; (2) Schisandraceae, Illiciaceae; (3) Austrobaileyaceae, Trimeniaceae, Amborellaceae, Monimiaceae, Calycanthaceae, Gomortegaceae, Lauraceae, Hernandiaceae; (4) Tetracentraceae, Trochodendraceae; (5) Eupteleaceae; (6) Cercidiphyllaceae.

18.1 General Features

The Magnoliales are a large order, and exhibit structural variations: (1) secretory anther tapetum, occasionally amoeboid in Winteraceae and Monimiaceae, (2) simultaneous cytokinesis in pollen mother cells, successive in Myristicaceae, and both types in Annonaceae, (3) two-celled pollen grains, sterilization of sporogenous tissue in Annonaceae and Myristicaceae, (4) pollen grains shed in tetrads in Winteraceae, Monimiaceae, and rarely in Annonaceae, (5) anatropous, bitegmic and crassinucellate ovule, sometimes unitegmic in Monimiaceae (*Siparuna*), a middle integument also occurs in Annonaceae and Myristicaceae, (6) single-celled archesporium, multicellular in Annonaceae, Calycanthaceae, uni- or multicellular in Magnoliaceae, Monimiaceae and Lauraceae, (7) Polygonum type of embryo sac, Polygonum and Allium type in Schisandraceae and Monimiaceae, an embryo sac haustorium (chalazal) develops in Lauraceae and Trochodendraceae, (8) antipodals ephemeral, secondary multiplication in Monimiaceae and Lauraceae, (9) Cellular type of endosperm except in Myristicaceae and Canellaceae, Nuclear as well as Cellular in Lauraceae, ruminate condition in Degeneriaceae, Annonaceae, Myristicaceae and Canellaceae, (10) development of embryo irregular in Magnoliaceae and Winteraceae, Onagrad type in Magnoliaceae, Annonaceae, Schisandraceae and Lauraceae, Asterad in Schisandraceae, Illiciaceae, Monimiaceae and Lauraceae, Piperad in Lauraceae, Chenopodiad in Trochodendraceae, and Solanad type in Cercidiphyllaceae and Eupteleaceae, and (11) seed generally endotestal. Sarcotesta occurs in Magnoliaceae and Degeneriaceae, sclerotic exotesta in Winteraceae, Canellaceae and Schisandraceae, and woody mesotesta in Annonaceae.

18.2 Magnoliaceae

The anther wall comprises the epidermis, endothecium with reticulate thickenings, three to four ephemeral middle layers, and glandular tapetum with multinucleate cells. The endothecium is persistent, and irregularly biseriate. The tapetum is two- to three-layered in *Liriodendron* (Maneval 1914) and *Michelia* (Hayashi 1965). Übisch granules are studded on the inner wall of the tapetum. After meiotic divisions, the microspore mother cells produce tetrahedral, isobilateral, and T-shaped tetrads. Simultaneous cytokinesis takes place by furrowing.

The pollen grains are monosulcate (Hayashi 1960), and two-celled at the dispersal stage; three-celled in *Liriodendron tulipifera* (Kaeiser and Boyce 1962). Sometimes, they remain united in tetrads. During germination the pollen tube is not restricted to the sulcus (Hayashi 1965).

In *Talauma candollei*, the pollen grains are abnormal and contain a variable number of nuclei. In *Magnolia ovata* (Kapil and Bhandari 1964), the abnormal pollen grain has two generative cells and two vegetative nuclei.

The ovule is anatropous, bitegmic, and crassinucellar; both integuments form the micropyle. The outer integument is vascularized, and has stomata on its surface (Mohana Rao 1975a). Funicular obturator has been noted in *Liriodendron* (Kaeiser and Boyce 1962). A hypostase also differentiates. A single cell from the hypodermal multicellular archesporium cuts off a parietal cell; the parietal tissue is massive. The megaspore mother cell undergoes meiotic divisions, and produces a linear or T-shaped tetrad. The chalazal megaspore develops into a Polygonum type of embryo sac. The synergids and antipodals are ephemeral. The polar nuclei fuse before fertilization.

The endosperm is of the Cellular type. In *Magnolia obovata* (Kapil and Bhandari 1964), the chalazal chamber develops into a two- or three-celled haustorium.

The zygote divides by transverse, obliquely transverse, or vertical wall. Development of the embryo is either irregular, or conforms to the Onagrad type. The basal cell of the embryo divides to form a group of cells. In *Magnolia grandiflora* (Ly-Thi-Ba et al. 1970), the embryogeny corresponds to Period I, Megarchetype VI of the Souèges system of classification (Onagrad type).

The seed of *Magnolia* projects outside the fruit due to the elongation of funicule. The seed is massive with a small embryo, and oily endosperm. Both the cotyledons are well developed in a ripe fruit (Hayashi 1965). The seed is sarcotestal, and the testa is multiplicative with woody endotesta. The outer part of testa is fleshy, oily, and vascularized. The inner part, the endotesta, is made up of two to ten layers of radially elongate lignified cells with a single crystal in each cell.

NN Bhandari (1971), JE Canright (1952, 1953), PK Endress and LD Hufford (1989), A Guard (1943), Y Hayashi (1960, 1963b, 1964, 1965, 1966, 1984), M Kaeiser and SG Boyce (1962), RN Kapil and NN Bhandari (1964), Ly-Thi-Ba, JL Guignard and JC Mestre (1970), WE Maneval (1914), PR Mohana Rao (1975a), D Padmanabhan (1960a), HN Stoudt (1960), LB Thien (1974), K Yasui (1937)

18.3 Degeneriaceae

The anther wall comprises the epidermis, fibrous endothecium, three or four ephemeral middle layers, and glandular tapetum with binucleate cells. The fibrous thickenings extend to the connective region. The microspore mother cells undergo meiotic divisions, and tetragonal tetrads are produced. Cytokinesis is simultaneous. At the time of separation of microspores, a narrow groove develops on the distal-fold, and differentiates as a germinal furrow.

The pollen grains are monocolpate and two-celled, and shed singly. The generative cell is cut off towards the distal end in *Degeneria vitiensis* (Swamy 1949a).

The ovule is anatropous, bitegmic and crassinucellar, and the micropyle is formed by the inner integument. The funicular obturator of elongated cells is conspicuous. The raphal bundle extends beyond the chalaza (Fig. 18.3.1F, G). The single hypodermal archesporial cell divides into a parietal and a sporogenous cell. The parietal tissue is massive. The megaspore mother cell undergoes meiotic divisions, and produces a linear tetrad. The chalazal megaspore develops into a Polygonum type of embryo sac. The antipodals are ephemeral.

The endosperm is of the Cellular type. The mature endosperm is ruminate in *Degeneria vitiensis* (Periasamy 1962a). In later stages the endosperm grows vigorously, and pushes against the ingrowths of seed-coat (Fig. 18.3.1A–H). It contains oil as storage product.

The zygote divides transversely; vertical walls divide both the apical and basal cells producing two superposed tiers of two cells each (Fig. 18.3.2A–F). Further divisions are irregular (Fig. 18.3.2G–R). The mature embryo shows three or four cotyledons, a short suspensor, and a bulbous hypocotyl (Fig. 18.3.2S, T).

The albuminous, pachychalazal seed contains a small embryo.

The testa is multiplicative, and lacks fibres. The outer epidermis consists of thin-walled palisade-like cells with a thick cuticle. The mesophyll is thin-

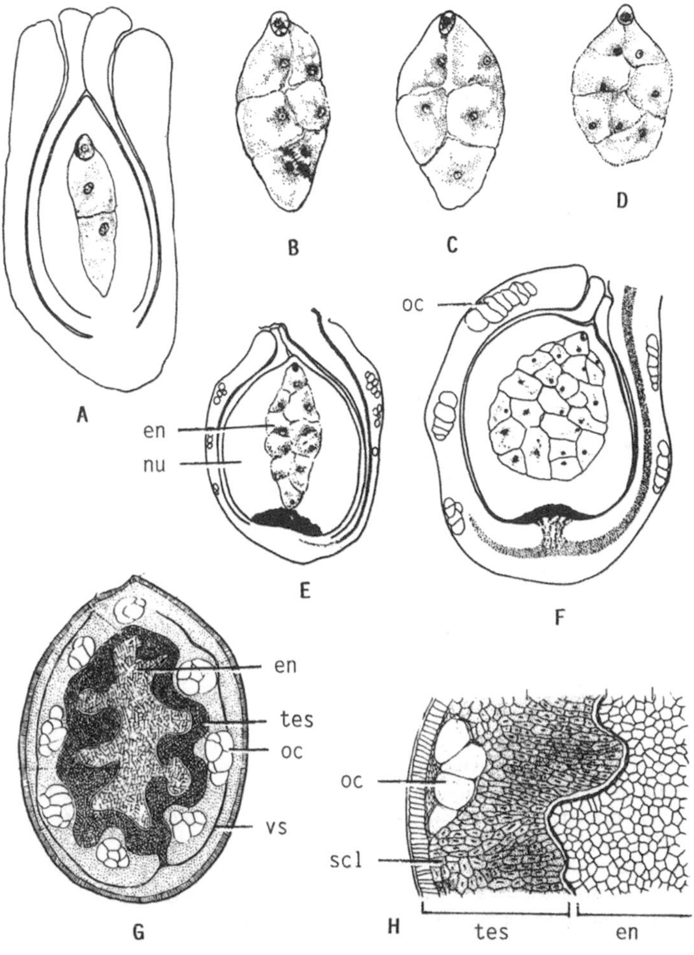

Fig. 18.3.1A–H. *Degeneria vitiensis*. **A** Bitegmic ovule, the embryo sac contains zygote and two-celled endosperm. **B–D** Early stages in the development of endosperm. **E–F** Later stages, oil-bearing cells (*oc*) differentiate in testa, the nucellar cells (*nu*) at chalaza degenerate (*black*); *en* endosperm. **G** Median longisection (diagrammatic) of mature seed; *tes* testa; *vs* vascular supply. **H** Portion from **G**; *scl* sclerenchyma. (After Swamy 1949a)

walled, and contains groups of oil cells. The endotesta is woody composed of thick-walled lignified fibrils. The inner cells undergo irregular divisions, and develop ingrowths. The tegmen gets crushed (Fig. 18.3.1G, H).

IW Bailey and AC Smith (1942), NN Bhandari (1970a, 1971), OA Dahl and JR Rowley (1965), BGL Swamy (1949a), K Periasamy (1962a)

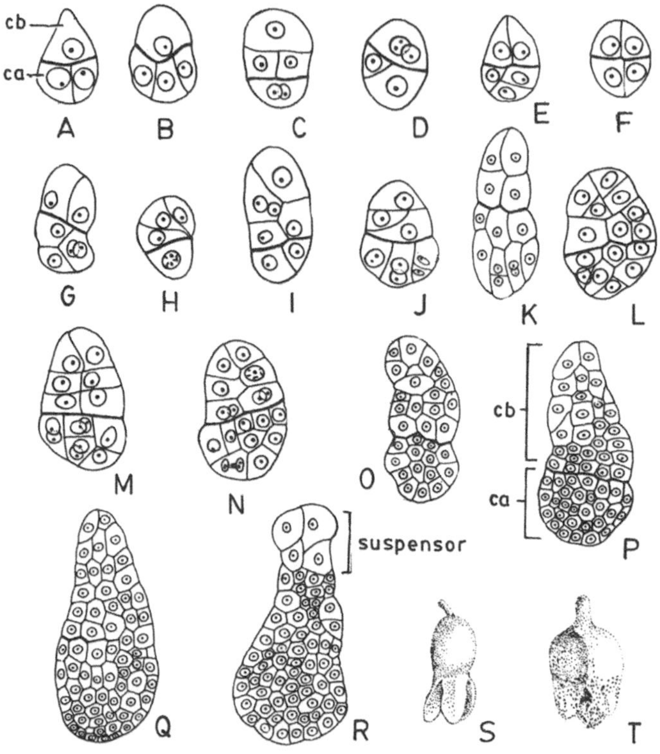

Fig. 18.3.2A–T. *Degeneria vitiensis*, embryogeny (*heavy line* indicates boundary between the daughter cells derived from basal (*cb*) and terminal (*ca*) cell of two-celled proembryo). **A** Three-celled proembryo, vertical division in *ca*. **B–K** Stages in the development of proembryo; there is no definite pattern. **L–Q** The derivatives of *ca* and *cb* form an ovoid embryo. **R** Differentiation of suspensor at the micropylar end. The demarcation between the derivatives of *ca* and *cb* could not be made out. **S, T** Embryos dissected from mature seeds. There are three cotyledons in each, the hypocotyl is swollen, and the suspensor is stalk-like. (After Swamy 1949a)

18.4 Himantandraceae

The development of endosperm, embryo and seed has not been investigated.

The four elongated microsporangia lie embedded in the tissue of laminar stamens. The anther wall is six layers thick below the epidermis, endothecium shows fibrous thickenings, tapetum is secretory with binucleate cells, and the anther dehisces by longitudinal slits. Successive cytokinesis follows meiotic divisions in pollen mother cells, the tetrads are tetrahedral, decussate, isobilateral, linear and T-shaped in *Galbulimima belgraveana* (N. Prakash et al. 1984).

The mature pollen grains are single, monosulcate, scabrate, atectate, spheroidal and two-celled. They do not contain starch.

The ovule has a funicular obturator, and is anatropous, bitegmic and crassinucellate with the inner integument forming the micropyle. The raphal

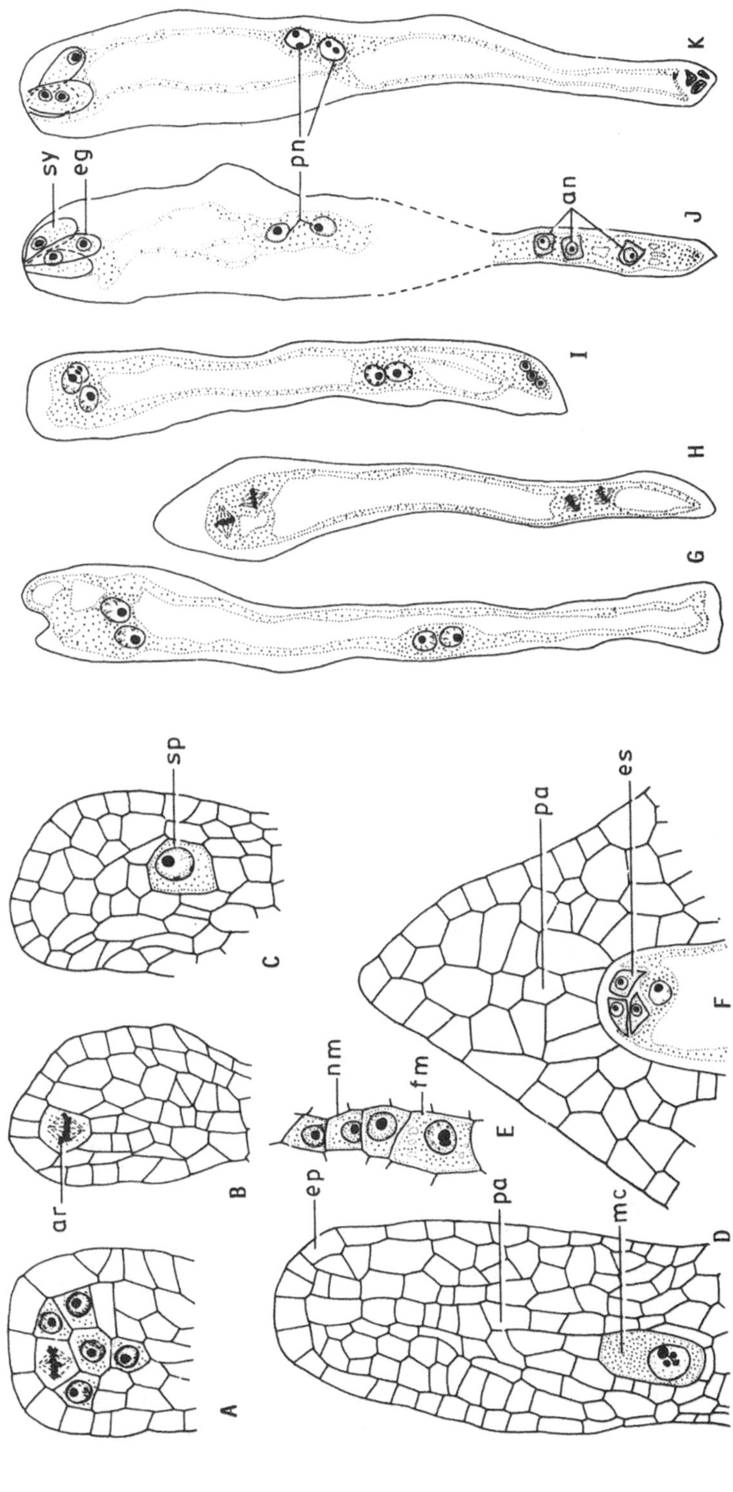

Fig. 18.4.1A–K. *Galbulimima belgraveana.* **A** Longisection of nucellus with a group of archesporial cells. **B** Archesporial cell (*ar*) in division. **C** Sporogenous cell (*sp*). **D** Megaspore mother cell (*mc*); *pa* parietal tissue; *ep* nucellar epidermis. **E** Linear tetrad; *fm* functional and *nm* nonfunctional megaspore. **F** Upper part of embryo sac (*es*) with egg apparatus and upper polar nucleus. **G** Four-nucleate embryo sac. **H** Nuclei in division. **I** Eight-nucleate embryo sac. **J** Mature embryo sac; *sy* synergid; *eg* egg; *pn* polar nuclei; *an* antipodal cells. **K** Later stage, the antipodals have degenerated. (After N. Prakash et al. 1984)

bundle terminates at the chalaza. The nucellus and the parietal tissue are massive so that the megaspore mother cell becomes deep-seated (Fig. 18.4.1A–D). Meiotic divisions in megaspore mother cell result in a linear tetrad (Fig. 18.4.1E), and the chalazal megaspore develops into a Polygonum type of embryo sac (Fig. 18.4.1F–K). Twin embryo sacs develop from two megaspore mother cells (N. Prakash et al. 1984).

The seeds have a small embryo with oily endosperm (Hutchinson 1973).

IW Bailey, CG Nast and AC Smith (1943), PK Endress (1977b), J Hutchinson (1973), N Prakash, DB Foreman and SJ Griffith (1984)

18.5 Winteraceae

The anther wall comprises the epidermis, fibrous endothecium, three or four middle layers, and tapetum. The epidermal cells contain phenol-like substances in *Zygogynum* (Swamy 1952). The tapetum is glandular in *Drimys winteri* (Bhandari and Venkataraman 1968), *Pseudowintera axillaris, Zygogynum* (Sampson 1963), and amoeboid in *Pseudowintera colorata* (Bhandari 1963a). The tapetal cells become polyploid. The fibrous endothecial thickenings extend to two or three layers of connective tissue (Fig. 18.5.1A–C). The meiotic divisions in microspore mother cells result in the formation of tetrahedral and decussate tetrads. Cytokinesis is simultaneous. In *Zygogynum* the divisions in microspores are synchronous; in other taxa the microspores in tetrads are at different stages of development.

The (modified) monocolpate (Hayashi 1965) pollen grains are shed as tetrads at the two-celled stage. The generative cell is cut off towards the proximal pole (Fig. 18.5.1A, C, D).

The ovule is laminar, anatropous, bitegmic, crassinucellar, and the micropyle is formed by the inner integument. In *Zygogynum* both the integuments are concerned. A hypodermal archesporial cell cuts off a parietal cell, the parietal tissue is massive. The cells of nucellar epidermis also divide and add to it. The megaspore mother cell undergoes meiotic divisions, and a linear or T-shaped tetrad is produced. The chalazal megaspore develops into a Polygonum type of embryo sac. The synergids exhibit filiform apparatus; the antipodals degenerate after fertilization.

Double fertilization occurs, and triple fusion precedes syngamy.

Development of the endosperm is of the Cellular type (Bhandari and Venkataraman, 1968). The cells are thin-walled, and store oil. It persists in the seed. According to Strasburger (1905), the development of the endosperm is of the Nuclear type in *Drimys winteri*, which needs confirmation.

The zygote divides transversely, and further divisions are irregular. The embryo is minute and organless (undifferentiated) in *Drimys*, and dicotyledonous in *Pseudowintera.*

The seed is exotestal with outer epidermis composed of lignified cells. The mesophyll is thin-walled. The tegmen is somewhat thick-walled in *Zygogynum* and *Belliolum*. In other members it does not persist. The cells of the seed-coat contain tannin.

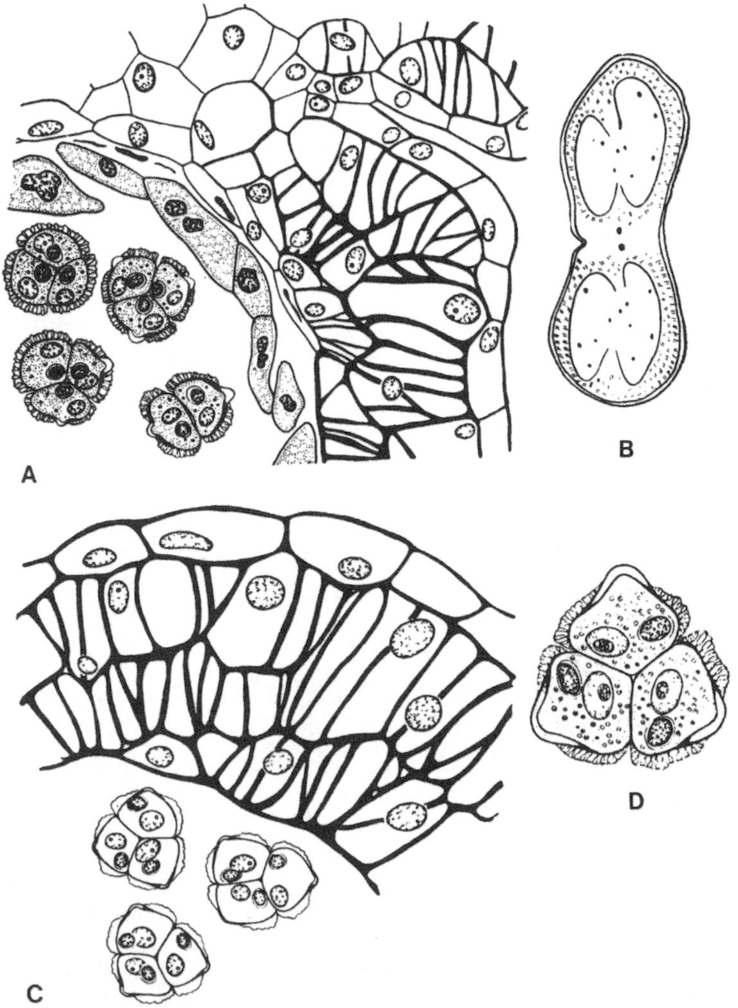

Fig. 18.5.1A–D. *Bubbia howeana.* **A** Portion of cross-section of anther. The epidermis is followed by two (or three) layers of fibrous endothecium and glandular tapetum. Two-celled pollen in tetrads. **B** Mature anther (cross-section), *stippled* portion represents fibrous endothecium. **C** Portion of mature anther with epidermis, fibrous endothecium, two-celled pollen in tetrads. **D** Mature bicelled pollen tetrad. (See Yakovlev 1981)

PR Bhagawathi (1938), NN Bhandari (1963a, 1971), NN Bhandari and R Venkataraman (1968), G Gottsberger, I Silberbauer-Gottsberger and F Ehrendorfer (1980), Y Hayashi (1965), FB Sampson (1963, 1970, 1974), E Strasburger (1905), BGL Swamy (1952), W Vink (1970, 1977), MS Yakovlev (1981)

18.6 Annonaceae

The anther wall comprises the epidermis, fibrous endothecium, three or four middle layers, and tapetum (Fig. 18.6.1A). The fibrous thickenings are absent

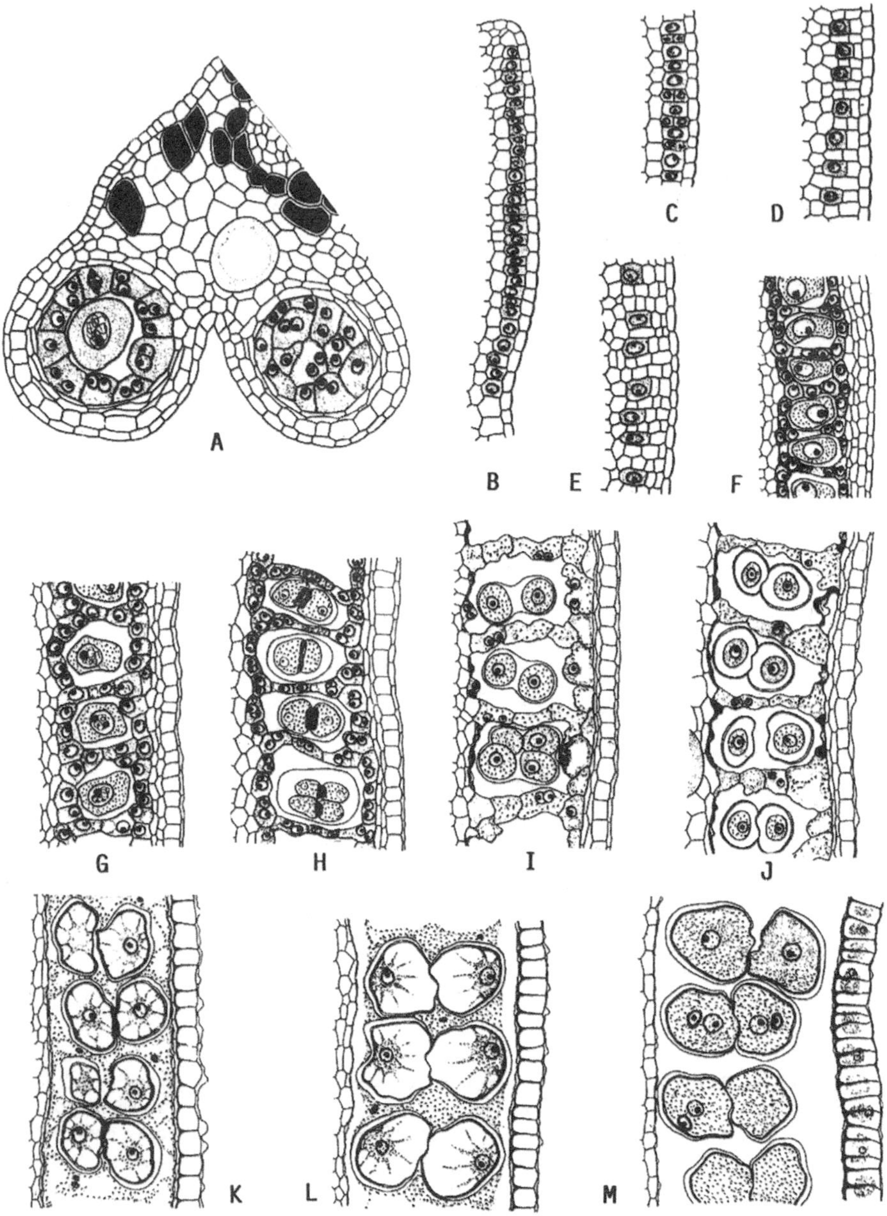

Fig. 18.6.1A–M. *Annona squamosa.* **A** Transection of half-anther through septum; one locule contains a sporogenous cell and the other shows only tapetal cells. **B–M** Longisection of portions of anther; **B** Single row of archesporial cells; **C** Alternate archesporial cells divide and differentiate as sterile septal initials; **D** Formation of primary parietal layer; **E** Three wall layers, and one- or two-layered sterile septa alternate with sporogenous cells; **F, G** The innermost parietal layer and septal layer form tapetum; **H, I** Formation of dyads and tetrads; **J** Microspores, the tapetum is still intact; **K, L** Uninucleate pollen grains embedded in tapetal periplasmodium; **M** Bicelled pollen grains, periplasmodium has been consumed. The epidermis becomes compressed (cf. **G–M**); endothecium shows fibrous thickenings in **M.** (After Periasamy and Kandasamy 1981)

in the endothecial cells of *Annona* (Juliano 1935a), which appears doubtful. The tapetum is glandular in *Artabotrys*, *Cananga* and *Miliusa*, and amoeboid in *Annona*, *Goniothalamus* and *Uvaria*. The tapetal cells are two- to four-nucleate, and due to fusion the nuclei become polyploid (Fig. 18.6.1A).

The male archesporium is multicellular in *Artabotrys* (Asana and Adatia 1947) and *Miliusa* (Periasamy and Swamy 1959), and it forms a row (Fig. 18.6.1B) in *Annona* (Juliano 1935a), *Asimina* (Locke 1936), *Cananga* (Periasamy and Swamy 1959), and *Mandora* and *Xylopia* (Lecompte 1896). Due to sterilization of sporogenous tissue (Fig. 18.6.1C–E), septa are produced so that individual or even groups of microspore mother cells become separated (Juliano 1935a, Periasamy and Swamy 1959). In *Annona* (Fig. 18.6.1F–J), *Asimina* and *Cananga*, the sterile septa behave as tapetum. The transverse septa are multilayered in *Goniothalamus*, and every spore mother cell has its own tapetum on all the sides (Parulekar 1970).

Periasamy and Thangavel (1988) point out that in *Xylopia nigricans*, in each lobe of the anther, up to 13 groups of single microspore mother cells differentiate separated by sterile septa. The cells adjoining the microspore mother cell organize into a single-layered glandular tapetum, and its cells remain uninucleate throughout. There may be 50–60 such microsporangia in each anther.

The tapetum is secretory and usually of parietal origin. Some of the sporogenous cells may also be involved. Periasamy and Kandasamy (1981) point out that in *Annona squamosa* the peripheral tapetum is of the secretory type but collapses. The cells of the transverse septa form a periplasmodium (Fig. 18.6.1K, L) at the tetrad stage. Probably, the peripheral tapetum also contributes to the periplasmodium. This aspect deserves further study.

The microspore mother cells undergo meiotic divisions, and cytokinesis is successive in *Annona* (Juliano 1935a, Sastri 1957), *Uvaria* (Sastri 1957), and *Xylopia* (Periasamy and Thangavel 1988), and simultaneous in *Annona* (Samuelsson 1914), *Asimina*, *Cananga*, *Polyalthia* and *Saccopetalum* (Sastri 1957). In *Cananga* and *Miliusa*, furrowing starts at the first meiotic division, but is completed only at the end of the second division. The tetrads are tetrahedral, isobilateral and decussate.

In *Annona* the pollen grains remain loosely attached in tetrads, in pairs in *Cananga*, and are free in *Artabotrys* and *Miliusa*. They are monocolpate and two-celled (Fig. 18.6.1M) at the shedding stage. The generative cell is cut off at the distal end.

Periasamy and Thangavel (1988) have tabulated the main features of the anther in Annonaceae – four, four to eight or numerous sporangia per anther; sporogenous tissue uniseriate with sterile partitions, biseriate or massive; cytokinesis successive or simultaneous; tapetum secretory, partly secretory and partly plasmodial, uni-, bi- or multinucleate cells; pollen grains single, dyads, loosely held tetrads, or tetrads enclosed in a common wall.

The ovule is anatropous, bitegmic or tritegmic, and crassinucellar. The micropyle is formed by both the integuments. The funicular vascular supply reaches up to the tip of the outer integument. In *Cananga*, *Mezzettia* and *Xylopia* a middle integument is present. In *C. odorata* (Periasamy and Swamy

1961), by its meristematic centripetal growth, the nucellus increases during post-fertilization development. An aril is common (Corner 1949), and a hypostase in *Artabotrys* and *Asimina*. One to multicelled hypodermal archesporium cuts off a parietal cell; the parietal tissue is massive. After meiotic divisions, the megaspore mother cell produces a linear tetrad. The chalazal megaspore develops into a Polygonum type of embryo sac. The antipodals are uninucleate. The inner region of the carpel wall, due to its meristematic activity, fills the locule with spongy tissue (Periasamy and Swamy 1961).

The endosperm is of the Cellular type. During early stages of development, the endosperm forms a uniseriate row of superposed cells, 20 in *Cananga odorata* and 30 in *Miliusa wightiana* (Periasamy and Swamy 1961). In later stages it becomes ruminated due to the folds of the seed-coat (Periasamy 1962a). In a mature seed it stores oil, and the cell walls become thick and pitted.

The embryogeny corresponds to the Onagrad type-Trifolium variation in *Polyalthia longifolia* (Sastri 1955a), and Onagrad type-Lotus variation in *Cananga odorata* (Periasamy and Swamy 1961). Rarely, two embryos develop concurrently in *Annona squamosa*.

The seed is massive, albuminous, arillate, perichalazal, and mesotestal. The testa is multiplicative. The outer epidermis of the testa is composed of radially elongate cells with stomata. The mesophyll differentiates into outer longitudinal fibres, and inner oblique or transverse fibres. The innermost layer of cells, together with fibres, in some parts, grows at a rapid rate and forms ingrowths. Ruminations in *Artabotrys odoratissimus* (Mohana Rao 1975b), *Miliusa wightiana* (Periasamy and Swamy 1961), and *Polyalthia longifolia* (Mohana Rao 1979) are plate-like and give a storied appearance in trans-longitudinal plane. The plates show irregular contour in *Cananga odorata*. The ruminations bifurcate and anastomose with the adjoining ones (Periasamy and Swamy 1961). The ruminations are conspicuously absent along the median and transmedian planes in *C. odorata*, and in transmedian plane in *Polyalthia*. Accordingly, the endosperm appears four- or two-lobed. The ruminations are formed by the outer, middle, or inner integument. In the majority of members, the outer integument develops ingrowths. In *Cananga odorata* (Periasamy and Swamy 1961), the middle and inner integument contribute to the formation of ruminations. However, in *Anaxagorea javanica* (Corner 1949) the inner integument alone forms the ruminations.

RD Adatia (1946), JJ Asana and RD Adatia (1947), JE Canright (1962), EJH Corner (1949), G Gottsberger (1989a, b), WB Herms (1907), JB Juliano (1935a), RK Lampton (1957), H Lecompte (1896), JF Locke (1936), PR Mohana Rao (1975b, 1979, 1982), EM Norman and D Clayton (1986), NK Parulekar (1970), K Periasamy (1962a), K Periasamy and MK Kandasamy (1981), K Periasamy and BGL Swamy (1959, 1961), K Periasamy and S Thangavel (1988), G Samuelsson (1914), RLN Sastri (1955a, 1957), A Thomas and B Lugardon (1972), H van der and F Bouman (1988), JW Walker (1971)

18.7 Eupomatiaceae

The embryology has not been fully investigated.

The anther is tetralocular in *Eupomatia laurina* (see Yakovlev 1981). The wall shows epidermis, fibrous endothecium, one middle layer, and collapsed tapetum at the binucleate pollen grain. The meiotic divisions in the microspore mother cells are followed by simultaneous cytokinesis. The tetrads are tetrahedral, isobilateral, decussate, occasionally linear. When the pollen grains are fully formed, the epidermal cells show conical projections. The endothecium

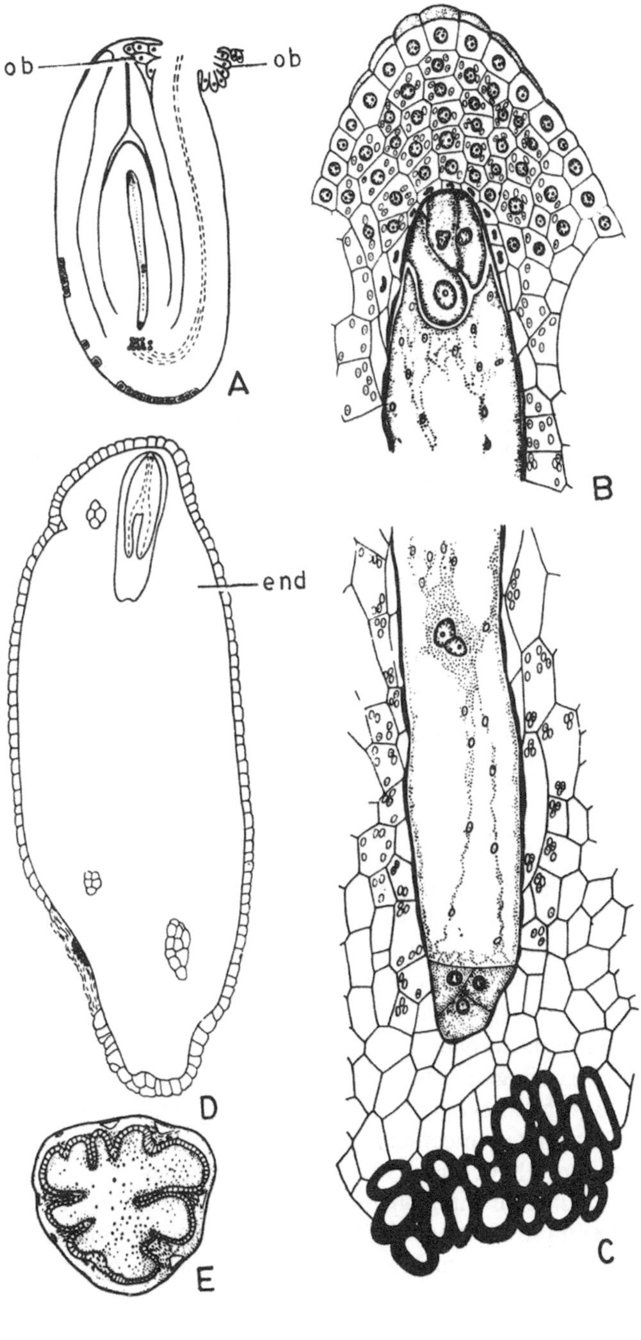

Fig. 18.7.1A–E. *Eupomatia laurina.* **A** Longisection of ovule at mature embryo sac, note funicular obturator (*ob*). **B** Micropylar part of ovule with embryo sac containing egg apparatus. **C** Chalazal portion of ovule; the embryo sac shows two polar nuclei, three antipodal cells, and starch. Note the well-developed hypostase. **D** Longisection of seed with embryo and invagination of testa into endosperm (*end*). (See Yakovlev 1981). **E** Cross-section of seed with ruminations in endosperm. (After Corner 1976)

becomes very prominent, and it alone persists at dehiscence. The two-celled pollen grains may germinate in situ.

There are two or three ovules in a carpel in lateral position. They are anatropous, bitegmic and crassinucellate with micropyle formed by the inner integument in *Eupomatia laurina* (Mohana Rao 1983). The outer integument is three or four cells thick, and the inner integument two or three cells thick. The hypostase and obturator are well developed at mature embryo sac (Fig. 18.7.1A).

The mature embryo sac is eight-nucleate (Fig. 18.7.1B, C). At the heart-shaped embryo the cellular endosperm is well developed. It is ruminated by the invasions of testa (Fig. 18.7.1D, E).

The embryo is small, and embedded in abundant oily endosperm.

The mesotestal seeds are pale-brownish, uneven, subangular, and coarsely (but shallowly) pitted. The raphe bundle remains unbranched and ends at the chalaza.

The young testa comprises six layers of cells. The outer epidermis consists of large flattened, tabular cells with thick cuticle, followed by four layers of fibres. The epidermis is crushed. The fibres elongate in longitudinal axis. In mature seed the outer epidermis of testa divides periclinally and produces a subepidermal layer of thin-walled parenchyma. At the point of rumination, the epidermal cells and fibres elongate transversely. The ruminations extend into the tegmen.

Of the three layers of tegmen, the outer epidermal cells elongate along longitudinal axis. The inner epidermal cells enlarge, and the middle layer becomes compressed between the two. In the mature seed only the inner epidermis persists.

The fruit is an urceolate, turbinate berry with one to three pale-brownish seeds in each locule.

EJH Corner (1976), AT Hotchkiss (1958), PR Mohana Rao (1983), MS Yakovlev (1981)

18.8 Myristicaceae

The anther wall comprises the epidermis, fibrous endothecium, two ephemeral middle layers, and glandular tapetum with uninucleate cells. Sterile septa divide the microsporangium into two or three compartments in *Myristica malabarica* (N.C. Nair 1972). The microspore mother cells undergo meiotic divisions, and produce isobilateral tetrads. Cytokinesis is successive.

The pollen grains are monocolpate, and two-celled at the dispersal stage.

The ovule is anatropous, bitegmic and crassinucellar. The inner integument forms the micropyle. Both the integuments are massive (Fig. 18.8.1A–E), and the outer epidermis of the outer integument contains stomata in *Myristica fragrans* and *M. malabarica* (N.C. Nair 1972). In both species, N.C. Nair reports a primordium of middle integument, between the inner and outer integuments, at mature embryo sac stage. It forms a collar only on antiraphe-side; Mohana Rao (1974b) does not mention its occurrence in *M. fragrans*.

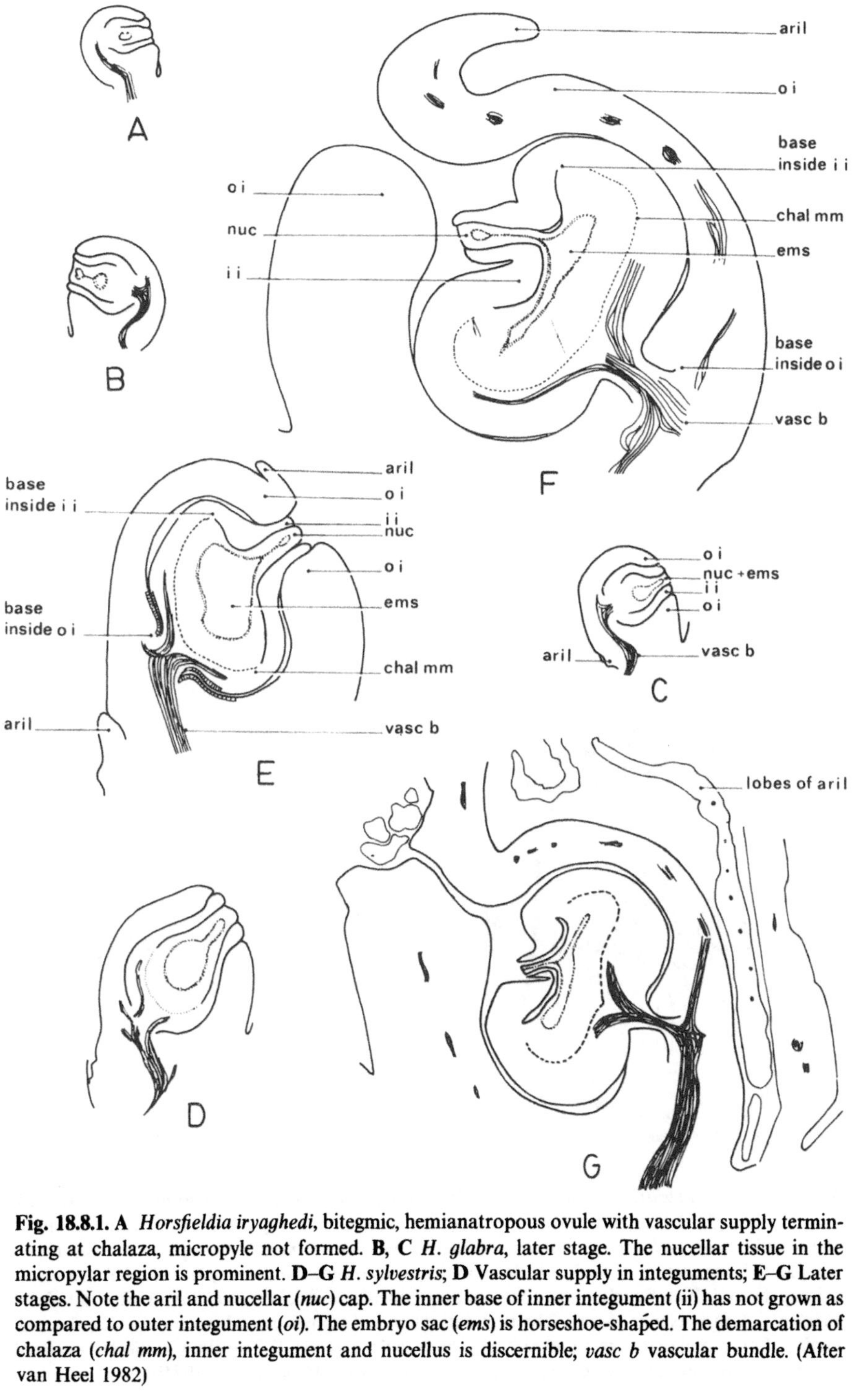

Fig. 18.8.1. **A** *Horsfieldia iryaghedi*, bitegmic, hemianatropous ovule with vascular supply terminating at chalaza, micropyle not formed. **B, C** *H. glabra*, later stage. The nucellar tissue in the micropylar region is prominent. **D–G** *H. sylvestris*; **D** Vascular supply in integuments; **E–G** Later stages. Note the aril and nucellar (*nuc*) cap. The inner base of inner integument (ii) has not grown as compared to outer integument (*oi*). The embryo sac (*ems*) is horseshoe-shaped. The demarcation of chalaza (*chal mm*), inner integument and nucellus is discernible; *vasc b* vascular bundle. (After van Heel 1982)

In *Myristica fragrans* (Voigt 1888, Periasamy 1961, N.C. Nair 1972), and in *Horsfieldia sylvestris* and *Knema laurina* (van Heel 1982), there is meristematic activity below the insertion of the inner integument. The newly formed tissue adds to the chalaza; the tip of the inner integument persists only at the

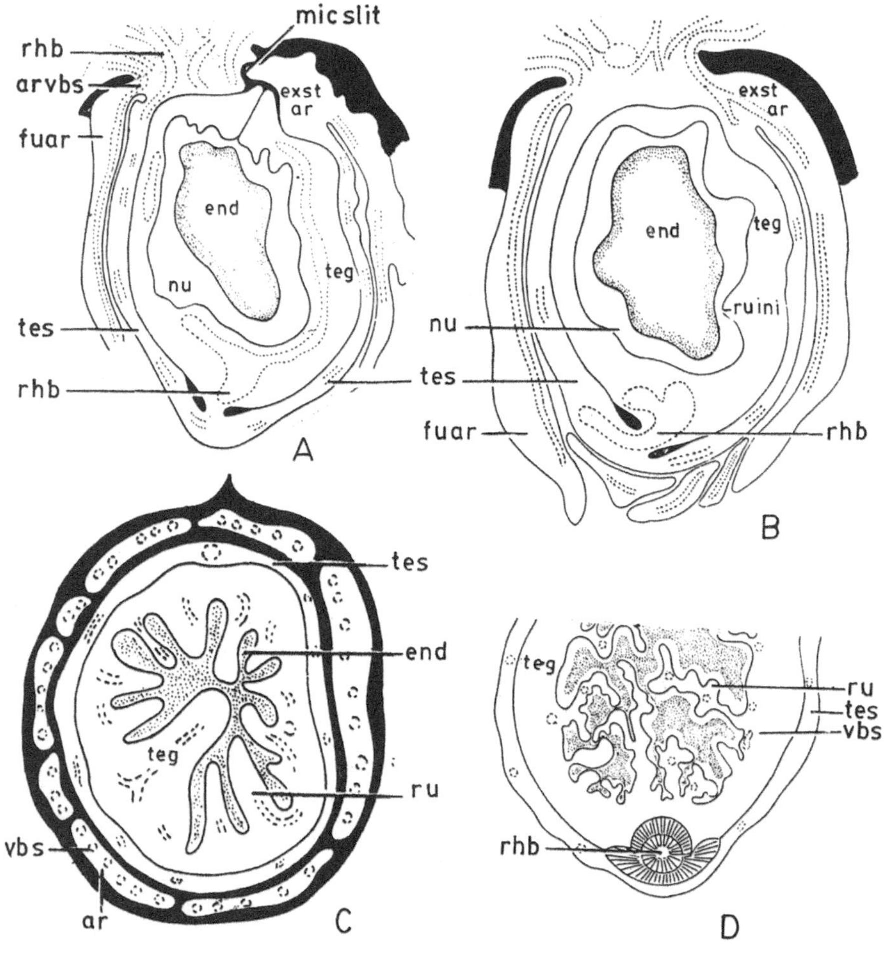

Fig. 18.8.2A–D. *Myristica fragrans.* **A** Median longisection of seed. Exostomal aril (*exst ar*), funicular aril (*fuar*), testa (*tes*) and tegmen (*teg*) are vascularized. Tegmen develops ruminations in the nucellus (*nu*); *arvbs* aril vascular bundles; *end* endosperm; *micslit* micropylar slit; *rhb* raphe bundle. **B** Longisection of seed, a little away from micropyle. Note entry of vascular strands at micropyle, and from funicle into exostome and funicular aril lobes; *ruini* rumination initials of tegmen. **C** Cross-section of fruit. The ruminations (*ru*) of tegmen in endosperm are vascularized, encircled by vascularized testa and aril (*ar*); *vbs* vascular strands. **D** Translongisection of chalazal region of seed. Palisade layers encircle raphe bundle. (After Mohana Rao 1974b)

micropylar region (Fig. 18.8.1D–G). The chalaza becomes massive and the embryo sac lies in it.

A single archesporial cell cuts off a parietal cell; the parietal tissue is massive. The megaspore mother cell undergoes meiotic divisions and produces a linear tetrad. The chalazal megaspore develops into a Polygonum type of embryo sac. According to van Heel (1982), the embryo sac undergoes enormous expansion. A postament also develops.

The entry of the pollen tube is porogamous. The pollen tube persists up to the young proembryo stage.

The development of endosperm is of the Nuclear type. Finally, it becomes cellular during later stages. It is thin-walled, ruminate, and stores starch and oil.

The mature embryo is small with divergent cotyledons, and a short radicle. Embryogeny has not been studied.

During the maturation of seed the chalazal tissue becomes massive and tanniferous. Tegmic ruminations occur at the chalaza and the vascular strand terminates at the base of the inner integument. According to N.C. Nair (1972), the boundary between nucellus and the inner integument remains vague in older ovules, and the ruminations are chalazal as well as lateral and originate from the inner integument. The inner basal portion of the inner integument shifts upward, the outer portion of base appears to remain close to the base of the outer integument. In *Horsfieldia* spp., the chalaza may be shortly columnar and lateral. The ruminations may be chalazal as well as lateral. In *Knema laurina* (van Heel 1982), the chalaza is tubular and ruminations develop from it as nine ridges around the endosperm.

Mohana Rao (1974b) and Roth and Lindorf (1974b) state that in *M. fragrans* the original condition of the two integuments does not change (Fig. 18.8.2A, B). Moreover, there is no pachychalazal growth and the inner seed-coat is formed by the inner integument. The cells of tegmen, on the inner side of the vascular bundle, develop ruminations which contain oil cells (Fig. 18.8.2C, D). Corner (1976) agrees with the interpretation of Mohana Rao (1974b) about the endotestal seed, normal chalaza, and tegmic ruminations.

The seed of Myristicaceae has exostomal and funicular aril (Fig. 18.8.2B). The cells of the outer epidermis of testa are thickly cuticularized. The mesophyll is composed of thin-walled cells. The inner epidermis consists of much-elongated tubular palisade cells with lignified pitted walls, and contain calcium oxalate crystals. The inner subepidermal layer is also a palisade layer, but is unlignified. Corner (1976) reports that the outer epidermis of tegmen is a single layer of longitudinally lignified fibres, compact or incomplete in places with thin-walled cells (*Horsfieldia irya*, *Myristica lowiana*), or it is a single layer of short sclerotic or tracheidal cells (*Knema*), or unspecialized (*M. fragrans*) at the chalaza with a counter-palisade of lignified cells recurved from the endotesta. The inner zone contains oil cells.

NN Bhandari (1971), EJH Corner (1976), WA van Heel (1982), AC Joshi (1943, 1946), PR Mohana Rao (1974b), NC Nair (1972), K Periasamy (1961), I Roth and H Lindorf (1974b), RLN Sastri (1955b), A Voigt (1888)

18.9 Canellaceae

The embryology of Canellaceae came to our attention through Parmeswaran's investigations (1961, 1962) on *Canella alba*, *Warburgia stuhlmannii* and *W. ugadensis*.

The anther wall comprises the epidermis, fibrous endothecium, two ephemeral middle layers, and glandular tapetum with two- to four-nucleate cells (Fig. 18.9.1A, B, F, G). Nuclear fusions may occur, and the cells become tetraploid. Übisch granules are present on the inner wall of tapetal cells. The microspore mother cells undergo meiotic divisions to produce tetrahedral tetrads (Fig. 18.9.1C, D). Cytokinesis is simultaneous.

The generative cell is produced towards the proximal pole. The pollen grains are monocolpate, and two-celled at the shedding stage (Fig. 18.9.1E).

The ovule is anatropous, bitegmic and crassinucellate with a zigzag micropyle (Fig. 18.9.1M). A single hypodermal archesporial cell cuts off a parietal cell. The parietal tissue is massive. The megaspore mother cell, by meiotic divisions, produces a linear tetrad. The chalazal megaspore develops into a Polygonum type of embryo sac (Fig. 18.9.1H–M). The polar nuclei fuse before fertilization. The antipodal cells persist up to fertilization.

The endosperm is oily, and ruminate in *Cinnamosma*. The mature embryo is small. The development of endosperm and embryo has not been studied.

Both the integuments form the seed-coat. The outer epidermis of the testa is composed of sclerosed cells. The mesophyll is thin-walled, and contains oil-cells. The testa is infiltrated with black contents. The tegmen is crushed.

NN Bhandari (1971), N Parmeswaran (1961, 1962), TK Wilson (1966)

18.10 Schisandraceae

The anther wall comprises an epidermis, fibrous endothecium, one or two ephemeral middle layers, and a glandular tapetum with multinucleate cells. Nuclear fusions occur in the tapetal cells. The epidermal cells become papillate, and contain tannin. Etherial oil cells are present in the connective. The microspore mother cells undergo meiotic divisions, and the tetrads are tetrahedral and decussate. Cytokinesis is simultaneous.

The pollen grains are heteropolar, hexacolpate, and shed at the two-celled stage (Hayashi 1960, Vijayaraghavan and Dhar 1975).

The ovule is ana- to campylotropous, bitegmic and crassinucellate (Hayashi 1963b). The micropyle is organized by both the integuments. The archesporium is multicelled in *Schisandra* and *Kadsura* (Hayashi 1963b). In *S. michaux* (Kapil and Jalan 1964), an archesporial cell cuts off a parietal cell; the parietal tissue is four layers thick. The megaspore mother cell undergoes meiotic divisions to produce dyads. The upper dyad fails to divide or degenerates, while the lower one divides in *S. michaux* (Kapil and Jalan 1964). The development of female gametophyte is of the Polygonum type; the antipodals are ephemeral. In *S. chinensis*, Swamy (1964a) describes monosporic Polygonum type, and bisporic Allium type embryo sacs. Polygonum type is

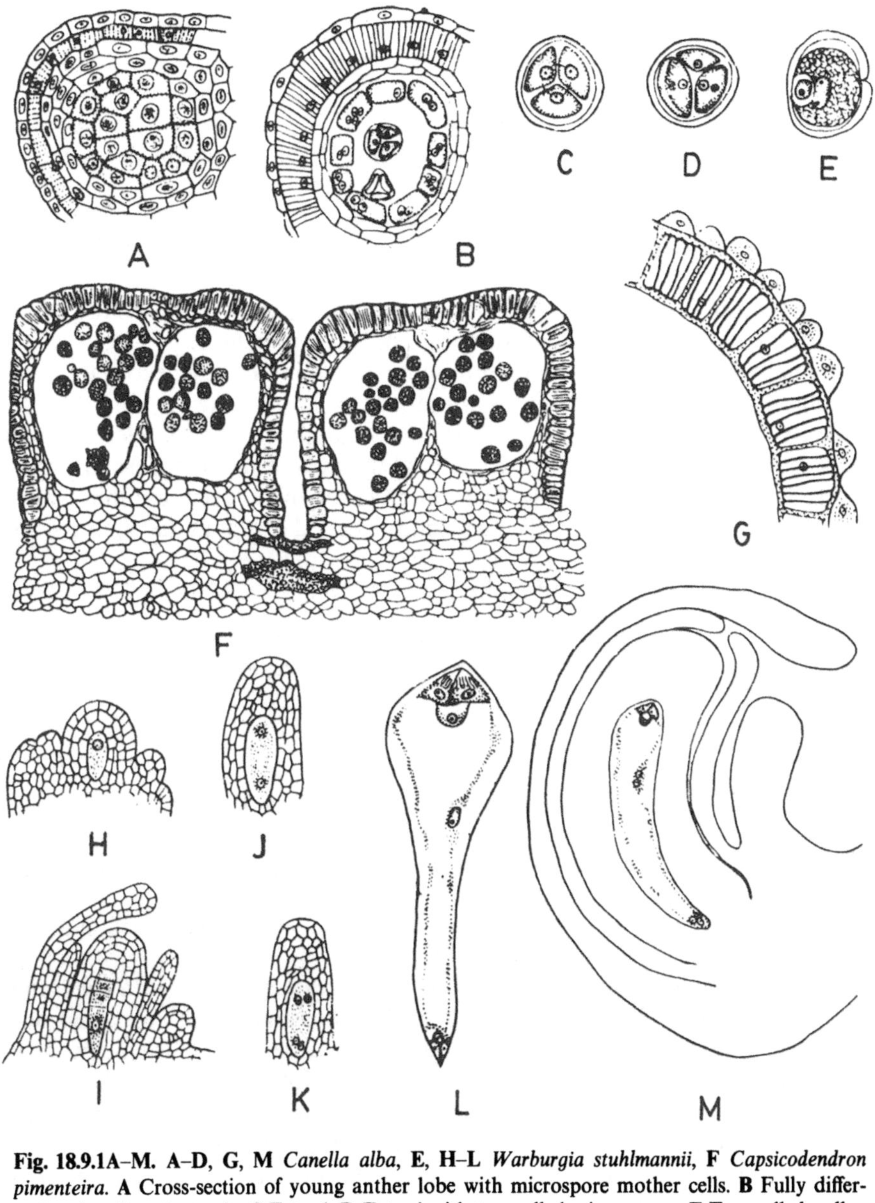

Fig. 18.9.1A–M. **A–D, G, M** *Canella alba*, **E, H–L** *Warburgia stuhlmannii*, **F** *Capsicodendron pimenteira*. **A** Cross-section of young anther lobe with microspore mother cells. **B** Fully differentiated wall, and tetrads. **C** Tetrad. **D** Tetrad with two-celled microspores. **E** Two-celled pollen grain. **F** Cross-section of staminal column with a pair of microsporangia in each anther lobe. **G** Portion of wall from mature anther. **H** Upper part of ovule with megaspore mother cell. **I** Linear tetrad. **J–L** Two-, four-nucleate and mature embryo sac. **M** Longisection of ovule at mature embryo sac. Note endostome and exostome. (**A–E** and **G–M** After Parmeswaran 1962, **F** after Wilson 1966)

exhibited by 50% ovules. The nucleus of the chalazal megaspore divides twice to form four nuclei lying at the micropylar pole of the embryo sac. They organize into an egg apparatus, and upper polar nucleus. In the Allium type, the two nuclei of chalazal dyad divide once, and the four nuclei again organize as the egg apparatus and upper polar nucleus. The antipodals and the lower polar nucleus are missing. Yoshida (1962) also describes similar development in *S. chinensis*.

The development of endosperm is of the Cellular type (Hayashi 1963b). The mature endosperm stores starch and oil.

The zygote divides by a transverse wall, and the apical cell plays a major role in the development of the embryo. The embryogeny corresponds to the Asterad or Onagrad type, and there is a massive suspensor.

The seed is small, campylotropous, albuminous and exotestal with a minute embryo. The cotyledons appear as two knobs of tissue until the time of seed dispersal (Hayashi 1963b). The outer epidermis of the vascularized testa consists of macrosclereids with hexagonal facets, and the hypodermal two to three layers of brachysclereids. The inner two or three layers are thin-walled with scattered sclerotic cells. The tegmen is crushed except at the micropyle. It is thick-walled.

The pericarp is thick and fleshy.

NN Bhandari (1970b, 1971), Y Hayashi (1960, 1963b), K-L Huynh (1976), S Jalan and RN Kapil (1964), RN Kapil and S Jalan (1964), BGL Swamy (1964a), MR Vijayaraghavan and U Dhar (1975, 1977), O Yoshida (1962)

18.11 Illiciaceae

The anther wall comprises an epidermis, fibrous endothecium, two or three ephemeral middle layers, and biseriate glandular tapetum of binucleate cells. The microspore mother cells undergo meiotic divisions producing tetrahedral tetrads; cytokinesis is simultaneous.

The pollen grains are tricolporate, and shed at the two-celled stage in *Illicium religiosum* (Hayashi 1960).

The ovule is anatropous, bitegmic and crassinucellate in *I. anisetum* (Hayashi 1963a). The micropyle is formed by exostome. It is thickened and lignified. The archesporial cell cuts off a parietal cell; the parietal tissue is massive. The megaspore mother cell undergoes meiotic divisions and produces a linear, or rarely T-shaped tetrad. The chalazal megaspore produces a Polygonum type of embryo sac. The three antipodals are ephemeral.

Development of the endosperm is of the Cellular type. The endosperm is copious and stores oil.

The embryogeny conforms to the Asterad type in *I. anisetum* (Hayashi 1963a). In a ripe seed the embryo shows two knob-like cotyledons.

The seed is albuminous and exotestal. The testa is multiplicative. The outer epidermis comprises microsclereids followed by thin-walled mesophyll. The inner two to three layers consist of sclerotic cells. The tegmen generally disappears.

NN Bhandari (1971), Y Hayashi (1960, 1963a)

18.12 Austrobaileyaceae

The stamens are laminar and massive with thecae borne on the adaxial surface (Fig. 18.12.1A). The anther wall (Fig. 18.12.1B) comprises an epidermis, well-differentiated fibrous endothecium, two to five middle layers, and irregularly biseriate secretory tapetum of parietal origin. The tapetal cells are binucleate in *Austrobaileya scandens* (Endress 1980a). During meiosis I of the microspore mother cells, a rudimentary cell plate is formed (Fig. 18.12.1C, D). Figure 18.12.1E shows meiosis II. Cytokinesis is simultaneous, and the tetrads are tetrahedral (Fig. 18.12.1F).

In *A. scandens* (Endress and Honegger 1980), the pollen grains are globose, anasulcate, tectate-perforate, regulose (reticulate) and columellate. The margin of sulcus is well marked; the endexine is considerably laminated at the border of aperture, and slightly laminated outside the aperture.

The pollen grain is two-celled at the dispersal stage (Fig. 18.12.1G).

The carpels are described as conduplicate in *A. scandens* by Bailey and Swamy (1949). However, according to Endress (1980a), they are ascidiate up to the stigmatic region, with a short plicate transition at the tip of style.

The ovules are anatropous, bitegmic and crassinucellate (Fig. 18.12.1H–K) with the inner or both integuments forming the micropyle. The raphe vascular strand extends into the outer integument along the antiraphe side. Tanniniferous cells occur on the ovular surface, the inner epidermis of the inner integument, chalazal region of vascular bundle, and base of nucellus. The nucellar epidermis undergoes periclinal divisions, and forms a "cap". The deeply seated megaspore mother cell (Fig. 18.12.1H), in *A. scandens* (Endress 1980a), undergoes meiotic divisions and a triad (possibly the upper dyad remains undivided) is formed (Fig. 18.12.1I). The development of the embryo sac is presumably of the Polygonum type; the antipodals are ephemeral (Fig. 18.12.1K).

The perichalazal, sarcotestal, albuminous seed is ruminate (Fig. 18.12.2A). The ruminations (Fig. 18.12.2B, C) are of the Annona type (Periasamy 1963). The testa is differentiated into outer thin-walled, and inner lignified tissue. The tegmen is crushed.

IW Bailey and BGL Swamy (1949), NN Bhandari (1971), PK Endress (1980a), PK Endress and R Honegger (1980), K Periasamy (1963)

18.13 Trimeniaceae

The embryology has not been studied in sufficient detail.

In *Trimenia papuana* the pollen sac is rudimentary and attached to the innermost tepal (Fig. 18.13.1A). The connective may extend beyond the anthers (Fig. 18.13.1B, C). In *T. neocaladonica* and *Piptocalyx moorei*, the anthers are much broader (Fig. 18.13.1D, E).

The anthers are tetrasporangiate (Fig. 18.13.1F–H), and the wall comprises persistent epidermis, fibrous endothecium (restricted only to the external part,

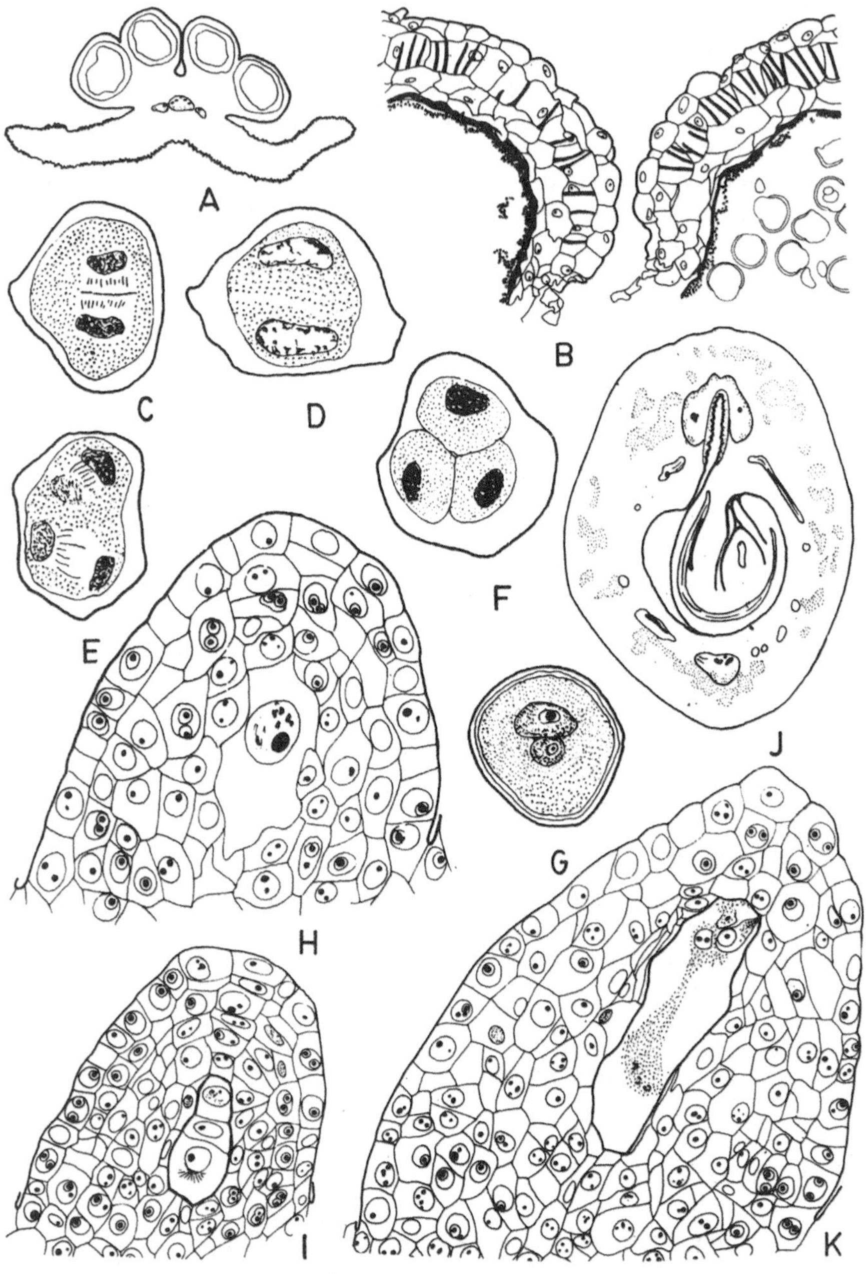

Fig. 18.12.1A–K. *Austrobaileya scandens.* **A** Transection, laminar stamen with two bisporangiate thecae on adaxial surface. **B** Part of anther at anthesis. **C–F** Pollen mother cell in meiosis I and II. **G** Two-celled pollen grain. **H** Longisection of nucellus at megaspore mother cell. **I** Linear triad. **J** Transection of ovary, ovule at mature embryo sac. **K** Nucellus with mature embryo sac. (After Endress 1980a)

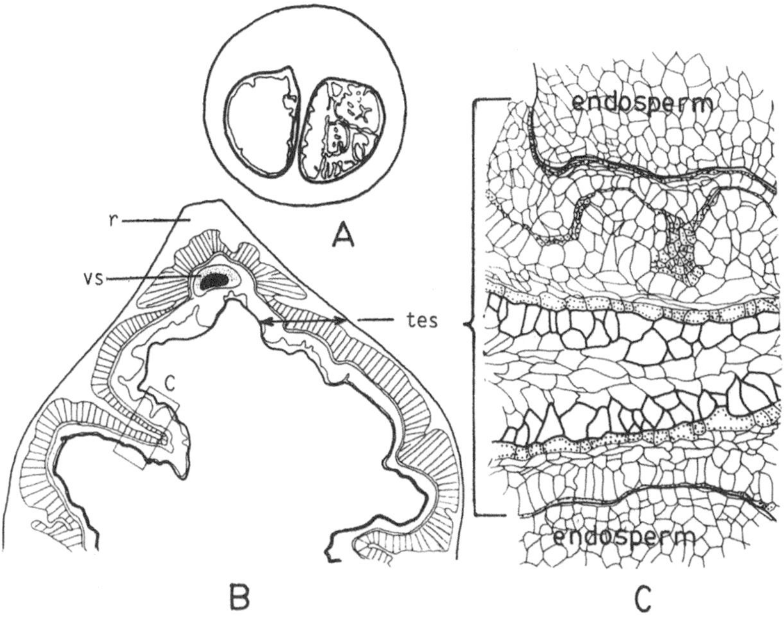

Fig. 18.12.2A–C. *Austrobaileya scandens.* **A** Transection of fruitlet with two ruminate seeds, the right seed cut superficially. **B** Part of transection of seed to show ruminations; lignified tissue *hatched*, followed by tanniniferous tissue marked with a *thin line*. **C** *Inset* C from **B** (lignified cells with *thick walls*, tanniniferous cells *dotted*). (After Endress 1980a)

absent in the region of the connective), and one or two ephemeral middle layers. The secretory tapetum shows binucleate cells in *Piptocalyx moorei* and *Trimenia papuana* (Endress and Sampson 1983). The microspore mother cells undergo meiosis, cytokinesis is successive, and the tetrads isobilateral and decussate.

The pollen is biaperturate with the centres of apertures located at the two pointed ends of each grain. In *T. papuana* the pollen is inaperturate, globose, spherical and polyporate (Sampson and Endress 1984). Both types have rugulate structure. The exine is tectate-columellate. In *T. papuana* and *T. weinmanniifolia*, the innermost tectum, columella and foot-layer have a partly granular form. The endexine is lamellate in non-apertural regions. The pollen is shed at the two-celled stage.

The surface of ovary is marked by longitudinal shallow furrows indicating the position of anthers in the bud. The stigma is hairy. A strand from the ventral bundle supplies the ovule, and extends up to the micropyle. A hypostase differentiates.

The ovule is anatropous and bitegmic (Fig. 18.13.2A–G). The outer integument has a short and discrete tip which forms the micropyle (Fig. 18.13.2A). The outer integument is six to seven layers thick, and the inner three to four layers. The outer epidermis of outer integument has radially elongate cells, and it develops into a stony layer in the seed. The inner epidermis of the inner

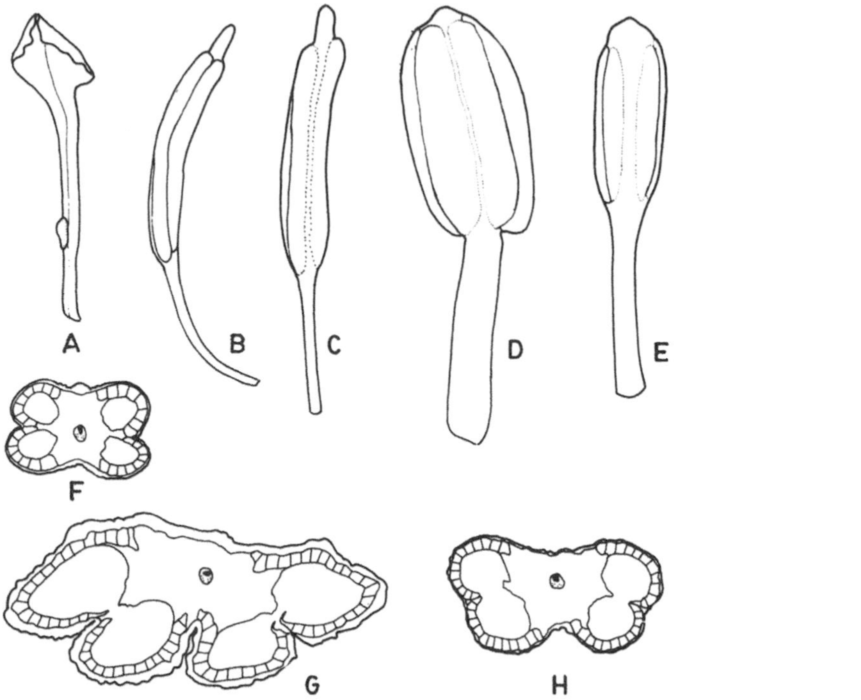

Fig. 18.13.1A–H. Stamens. **A–C, F** *Trimenia papuana* **D, G** *T. neocaledonica*, **E, H** *Piptocalyx moorei*. **A** Innermost tepal with rudimentary pollen sac (adaxial view) from mature bud. **B** Stamen at anthesis, lateral view (right side – adaxial). **C–E** Stamens at anthesis (abaxial view). Note the prolonged connective in **B** and **C**. In **D** and **E** the anthers are much broader and extrosely dehiscent. **F–H** Transections of tetralocular anthers shortly before dehiscence (adaxial side above), endothecium *hatched*. (After Endress and Sampson 1983)

integument is tanniniferous, and it forms a massive cap over the micropyle in *T. neocaladonica* and *P. moorei* (Endress and Sampson 1983).

At the time of initiation of integument in the ovule, the archesporium is not distinct. But, later, multicellular archesporium differentiates (Fig. 18.13.2E, F). There is no information about the formation of a parietal cell. After meiosis several embryo sacs grow simultaneously but only one reaches maturity. Megasporogenesis has not been described.

The embryo sac is long and narrow, and occupies the entire length of nucellus (Fig. 18.13.2G). The egg apparatus and polar nuclei lie at the apex, and the antipodals probably at the base of the embryo sac.

The development of endosperm and embryo has not been worked out.

PK Endress and FB Sampson (1983), FB Sampson and PK Endress (1984)

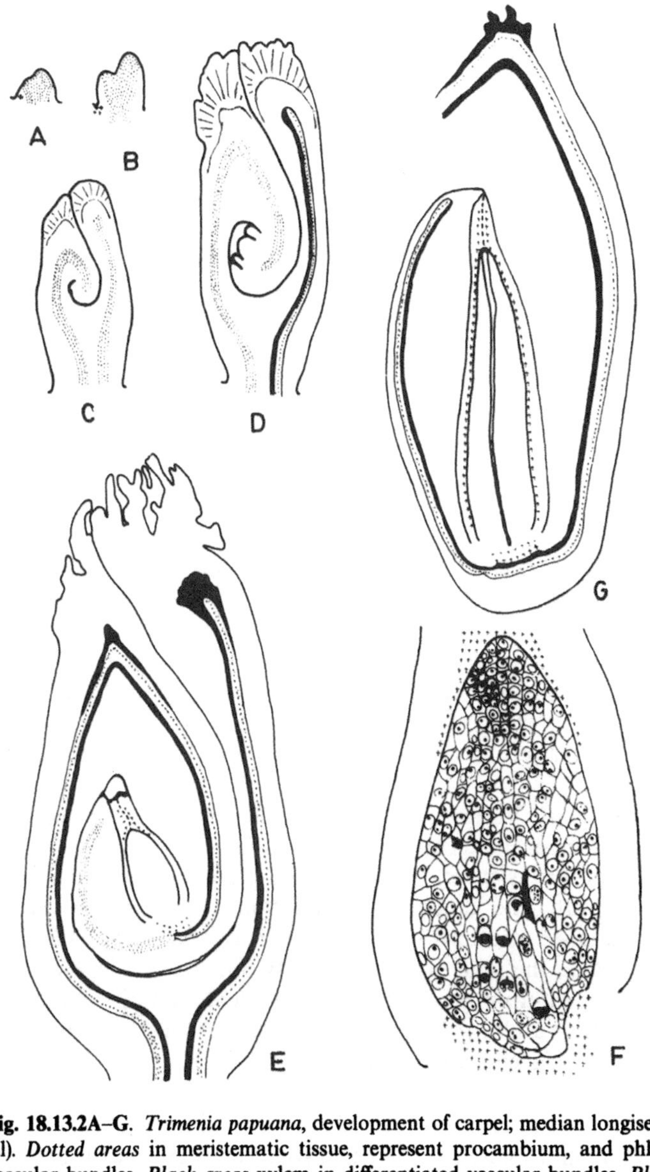

Fig. 18.13.2A–G. *Trimenia papuana*, development of carpel; median longisections (left side adaxial). *Dotted areas* in meristematic tissue, represent procambium, and phloem in differentiated vascular bundles. *Black areas* xylem in differentiated vascular bundles. *Plus signs* tanniniferous tissue. *Black lines in stigmatic region* cell rows. **A** Origin of ventral "cross-zone". **B** "Chair-like" stage. **C** Early epidermal periclinal divisions in future stigmatic region (cell rows of epidermal origin indicated); inception of inner integument. **D** Rotation of ovule almost completed, ventral vascular bundle still procambial. **E** Ovule with meiocytes, integumentary vascular bundle still procambial. **F** Nucellus with numerous meiocytes (from **E**). **G** Ovule at anthesis, elongated nucellus, and long tubular embryo sac. (After Endress and Sampson 1983)

18.14 Amborellaceae

Information on embryology is very scanty. The family includes a single genus, *Amborella* (Willis 1973).

The anthers are bilocular and dehiscence occurs by slits. The microspore tetrads are tetrahedral, and the pollen grains are large.

Each carpel contains a single anatropous ovule (orthotropous according to Willis 1973) which is bitegmic and crassinucellate. The inner integument is three layers thick, and the outer one five to seven layers.

The seed has a small and straight embryo with copious endosperm. The mature embryo displays only initial stages of cotyledonary primordia.

IW Bailey and BGL Swamy (1948), J Hutchinson (1973), J Perkins and E Gilg (1911), JC Willis (1973)

18.15 Monimiaceae

In some genera the flowers are highly specialized (Endress 1980b). An extreme case is that of *Tambourissa*. The male and female flowers, on the same inflorescence, are globular and fig-like (Fig. 18.15.1A, B). The carpels are situated around the floral cup (Fig. 18.15.1C, D). The secretory transmitting tissue covering the carpels is continued into the short channel at the mouth of the floral cup (Fig. 18.15.1D, E). This constitutes the "hyperstigma". All the genera with hyperstigma are shown in Fig. 18.15.1F–J. Other taxa lack hyperstigma.

In *Hortonia angustifolia* (Endress 1980c), the fertile stamens have two basal appendages, and their inner epidermis is nectariferous.

The anther wall comprises an epidermis, fibrous endothecium, two or three ephemeral middle layers, and a glandular tapetum of binucleate cells in *Laurelia novae-zelandiae* (Sampson 1969c) and *Tambourissa religiosa* (Endress 1980b). In *H. angustifolia* the tapetum is irregularly one- or two-layered. The tapetum is periplasmodial in *Antherosperma* (Sastri 1963). Meiotic divisions in microspore mother cells result in the formation of tetrahedral tetrads. Cytokinesis is simultaneous. The centripetal plates do not develop until the end of meiosis II.

The pollen grains are in permanent tetrads in *Hedycarya* sp. (Money et al. 1950, Sampson 1969a) and *Laurelia*. The generative cell is cut off towards the distal face (Sampson 1969a). They are inaperturate, two- to three-sulcate and shed at the two- or three-celled stage.

The development of carpel in *Hortonia* is shown in Fig. 18.15.2A–D, and a longisection of ovule at mature embryo sac in E.

The ovule is anatropous and crassinucellate, unitegmic in *Siparuna* (Heilborn 1931), bitegmic in *Hortonia* (Fig. 18.15.2F, G), *Laurellia* (Sampson 1969c), and *Tambourissa*. The micropyle is organized by both the integuments. The cells of the nucellar epidermis divide periclinally to produce a nucellar cap. The female archesporium is uni- to multicelled. An archesporial cell cuts off

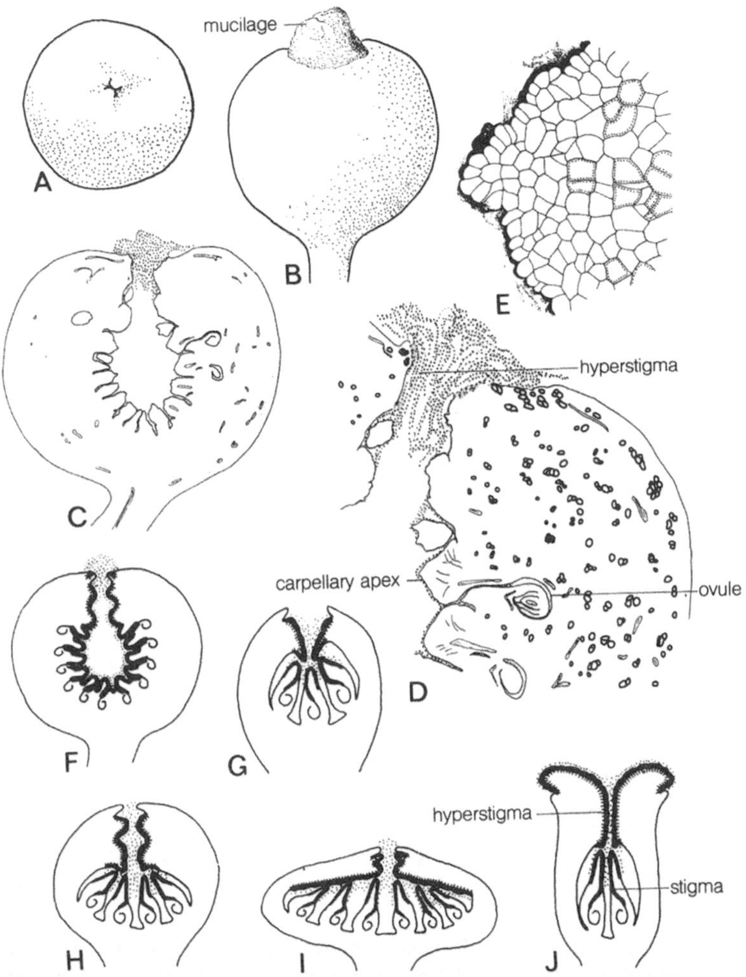

Fig. 18.15.1A–J. **A–F** *Tambourissa religiosa*, **G** *Kibara*, **H** *Wilkiea macrophylla*, **I** *W. huegeliana*, **J** *Hennecartia*. **A, B** Female flower, before (**A**) and at anthesis (**B**). **A** As seen from above, mucilage removed from pore. **B** Lateral view, with mucilage. **C** Entire female flower in longisection. **D** Upper portion (*right*) of **C**. Floral channel with hyperstigma (mucilage *dotted*), and upper carpels. Note groups of stone cells. **E** Carpellary apex with secretory epidermis from **D**, mucilage *dotted*. **F–J** Schematic longisections of female flowers. *Dark lines studded with short stubs* represent hyperstigma (secretory), *plain dark lines* represent stigma and ventral slit, mucilage *dotted*. (After Endress 1980c)

a parietal cell; the parietal tissue is massive. The megaspore mother cell (Fig. 18.15.2F) undergoes meiotic divisions and produces a linear tetrad. The chalazal megaspore (Fig. 18.15.2G) gives rise to a Polygonum type of embryo sac in *Hortonia*, *Mollinedia* (Peter 1920) and *Laurelia*. A peculiar feature has been observed in *Siparuna*. The chalazal megaspore forms a long, tortuous

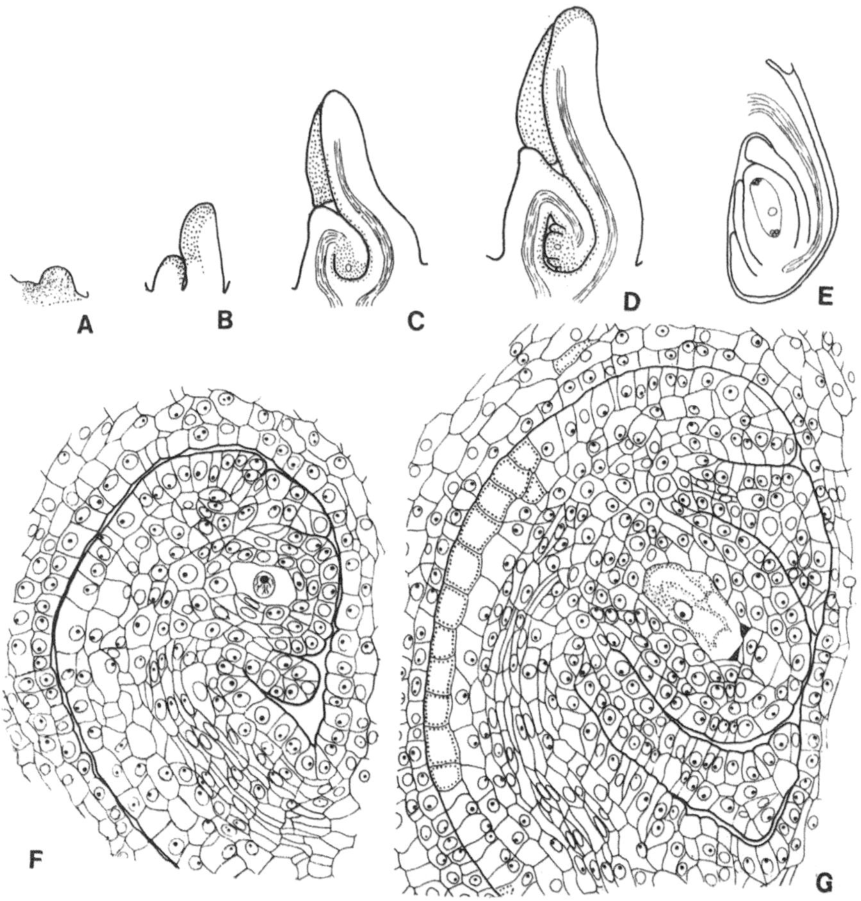

Fig. 18.15.2A–G. *Hortonia angustifolia*. **A, B** Initiation of ovular primordium. **C, D** Longisection of carpels to show the curvature of ovule, the stigma has ventral slits. **E** Longisection of bitegmic, crassinucellate ovule at mature embryo sac (seven-nucleate, polars fused). **F, G** Longisection of ovule at megaspore mother cell (**F**) and functional megaspore (**G**). Tannin-bearing cells marked with *bold dots* in **G**. (After Endress 1980b)

tube with vesicular tip. The tip bursts in another cell, and discharges its contents. This cell forms a four- to five-nucleate unorganized female gametophyte. These observations require confirmation, and a further investigation is called for. In *Peumus* the upper of the two dyad cells degenerates, and the lower is functional giving rise to an Allium type of female gametophyte. An increase in the number of antipodal cells (5–20) has been reported in *Peumus boldus* (see Davis 1966). A hypostase is present.

At anthesis, in *Tambourissa religiosa*, the hyperstigma secretes mucilage (Fig. 18.15.1B–D) so that there is a continuous transmitting medium for the pollen tube from the mouth of the floral cup (where pollen is deposited) up to the micropyle. Syrphids and drosophilids have been observed as pollinators of *T. peltata* (see Endress 1980b). The pollinators deposit their eggs in the male and female flowers by piercing the wall of the floral cup. The adult pollinator flies out of the male flower carrying pollen which is deposited in the mucilage at the mouth of hyperstigma. It would be interesting to know if there are symbiotic relationships with insect parasites serving as pollinators.

The development of endosperm is of the Cellular type. It is massive, and stores oil.

The embryogeny corresponds to the Asterad type.

The seed is albuminous, endotestal with a small straight embryo; the cotyledons may be appressed or divergent. The outer two or three layers of the testa comprise thin-walled cells, and oil cells. The inner epidermis shows lignified tracheidal cells. The tegmen is thin-walled, and may not persist.

NN Bhandari (1971), GL Davis (1966), PK Endress (1979, 1980b, c), O Heilborn (1931), LL Money, IW Bailey and BGL Swamy (1950), J Peter (1920), FB Sampson (1969a, b, c, 1975, 1977), RLN Sastri (1963)

18.16 Calycanthaceae

The anther wall comprises an epidermis, fibrous endothecium, one or two ephemeral middle layers, and secretory tapetum in *Calycanthus fertilis* (S.L. Mathur 1968). Following meiosis, the microspore mother cells produce tetrahedral and decussate tetrads. Cytokinesis is simultaneous.

The pollen grains are two- to three-sulcate, and shed at the three-celled stage. Occasionally, they remain united in tetrads. Their germination in situ has also been observed (S.L. Mathur 1968).

The ovule is anatropous, bitegmic and crassinucellate. The micropyle is organized by both the integuments. The exostome and endostome are lined with microscopic hairs. The nucellar cells divide to produce a nucellar cap. The archesporium is multicellular. These cells function as megaspore mother cells, undergo meiotic divisions, and produce linear tetrads. The chalazal megaspore develops into a Polygonum type of embryo sac (Peter 1920). Supernumerary embryo sacs and a hypostase develop. The three antipodals are ephemeral; one of the polar nuclei also degenerates. Syngamy and triple fusion (the polars may fuse and form secondary nucleus) do not occur in *Calycanthus fertilis* (Davis 1966, S.L. Mathur 1968). According to Peter (1920), Schürhoff (1923) and Brofferia (1930), *Calycanthus* and *Chimonanthus* are apomicts. While Peter and Brofferia suggest nucellar polyembryony as the source of supernumerary embryos, Schürhoff (1923) considers the possibility of somatic parthenogenesis (see Schnarf 1931).

The development of endosperm is autonomous and of the Cellular type.

The development of a sexual embryo is shown in Figure 18.16.1A–E and its structure in H–K; nucellar embryos are common (Fig. 18.16.1F, G).

The seed is endotestal, exalbuminous and enclosed in a hard pericarp. The outer epidermis of testa comprises thin-walled cells with brown contents. It is

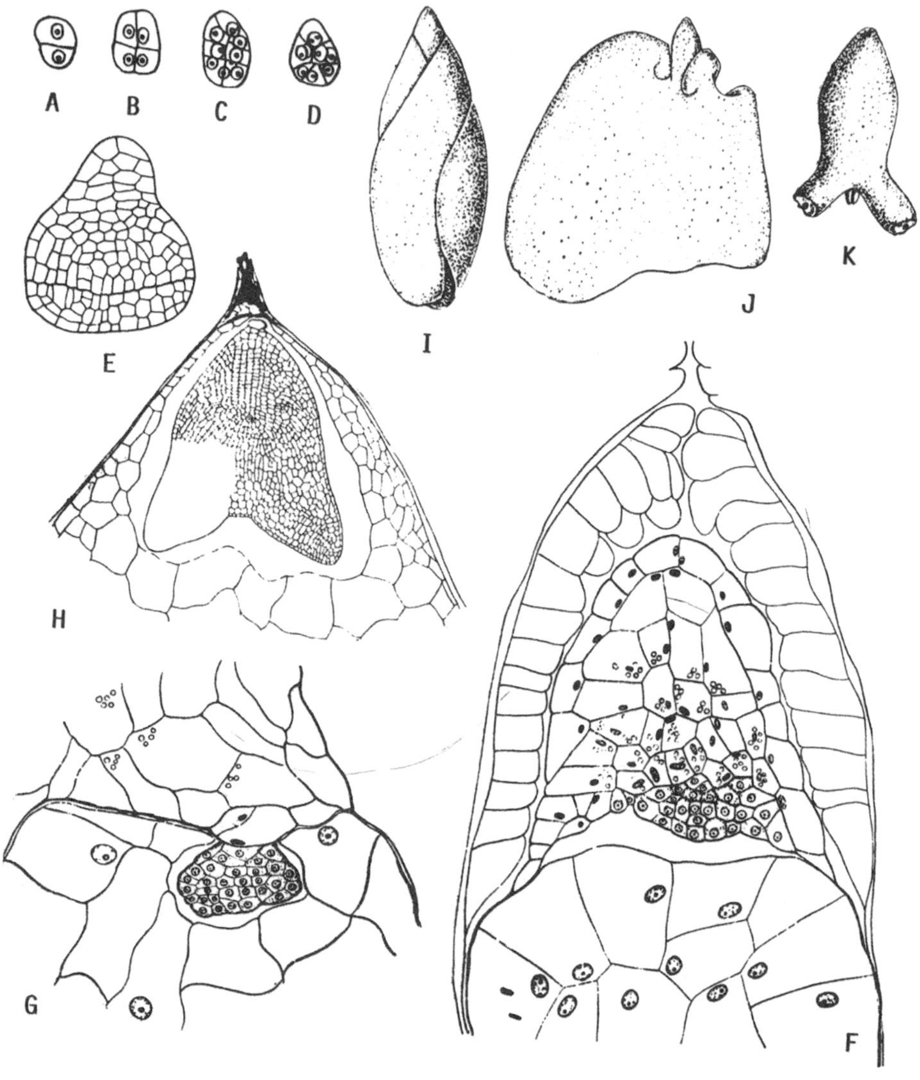

Fig. 18.16.1A–E. *Chimonanthus fragrans*, Embryogeny. **F–H** *C. occidentalis*; **F** Nucellar cap with meristematic zone adjacent to cellular endosperm; **G** Portion of nucellus with multicellular proembryo surrounded by cellular endosperm; **H** Heart-shaped sexual embryo and cellular endosperm, note remnants of pollen tube. **I–K** *C. praecox*; **I** Mature convolute embryo; **J** Embryo with expanded cotyledons; **K** Hypocotyl and shoot-apex, cotyledons removed. (**A**–**E** After Ly-Thi-Ba 1962, for **F–K** see Yakovlev 1981)

slightly lignified in *Chimonanthus*. The inner epidermis is formed of tangentially elongate, lignified cells with reticulate thickenings. The tegmen is crushed.

NN Bhandari (1971), J Brofferia (1930), GL Davis (1966), Ly-Thi-Ba (1962), SL Mathur (1968), J Peter (1920), K Schnarf (1931), PN Schürhoff (1923), YD Tiagi (1963)

18.17 Gomortegaceae

The embryology has not been investigated.

18.18 Lauraceae

The anther wall comprises a persistent epidermis, fibrous endothecium, two middle layers, and amoeboid tapetum. The tapetum is glandular in *Cassytha* (Sastri 1962). The tapetal cells may be bi-, tetra- or multi-nucleate. The microspore mother cells undergo meiotic divisions to produce tetrahedral, isobilateral, T-shaped and linear tetrads. Cytokinesis is successive.

The pollen grains are acolpate and two-celled at the shedding stage.

The ovule is pendulous, anatropous, bitegmic and crassinucellate. Generally, the micropyle is produced by both the integuments. Only the outer integument is concerned in *Cinnamomum* and *Litsea* (Sastri 1958a), and the inner in *Sassafras* (Coy 1928). In *Cassytha* the integuments do not grow beyond the nucellus so that the micropyle is broad, and the overarching funiculus comes in direct contact with the nucellus. The archesporium is one-celled in *Laurus* and *Umbellularia*, and multicelled in *Cassytha* and *Cinnamomum*. The archesporial cells cut off parietal cells, and a massive parietal tissue is produced. The epidermal cells of the nucellus also divide and add to it. Several megaspore mother cells function. They undergo meiotic divisions, and linear tetrads are produced. The chalazal megaspore gives rise to a Polygonum type of embryo sac. In *Cassytha* (Sastri 1956), in a single ovule as many as six embryo sacs develop simultaneously. They show tubular haustorial processes in the direction of micropyle, protruding beyond the nucellus and entering the overarching funicule. The antipodals are ephemeral, but in *Laurus nobilis* (Bambacioni-Mezzetti 1935) they undergo multiplication. In *Cassytha* they do not organize into cells.

In *Persea americana* (Schroeder 1952, Tomer and Gottreich 1978), the ovules are frequently sterile either due to the absence of formation of embryo sacs, or their early degeneration. Twin embryo sacs may originate from two archesporial cells.

The endosperm is of the Nuclear type. Later, it becomes cellular at the globular proembryo in *Cinnamomum* (Sastri 1958a); in *Cassytha* (Sastri 1962) it is ab initio Cellular.

The embryogeny corresponds to the Piperad type in *Sassafras* (Coy 1928), Asterad type in *Persea americana* (Schroeder 1952), and Onagrad type-Trifolium variation in *Cinnamomum iners* and *Litsea sebifera* (Sastri 1958a). The division of zygote is transverse in *Cassytha* (Sastri 1962) and *Laurus nobilis*; further development is irregular.

The seed is exalbuminous, and endotestal. The embryo is large with thick, massive cotyledons, and a very short radicle. The testa alone forms the seedcoat. The outer zone of testa comprises thin-walled cells, and the outer epidermis contains tannin. The inner epidermis consists of longitudinally and transversely elongate narrow tracheidal cells with annular or spiral thickenings. The endocarp is stony.

V Bambacioni-Mezzetti (1935, 1938), E Battaglia (1947a), NN Bhandari (1971), JK Choudhury and JN Mitra (1953), GV Coy (1928), PK Endress (1972), RO Gardner (1974), PR Mohana Rao (1980, 1986), F Pax (1891), RLN Sastri (1952, 1956, 1958a, 1962, 1963), CA Schroeder (1952), M Sedgley (1976, 1979), G Täckholm and E Soderberg (1918), E Tomer and S Gazit (1979), E Tomer and M Gottreich (1975, 1978), MS Yakovlev (1981)

18.19 Hernandiaceae

The embryology has not been studied in sufficient detail (Table 18.24.2).

The flowers are unisexual. The anthers are bilocular with glandular appendages. The tapetum is of the amoeboid type. The pollen mother cells

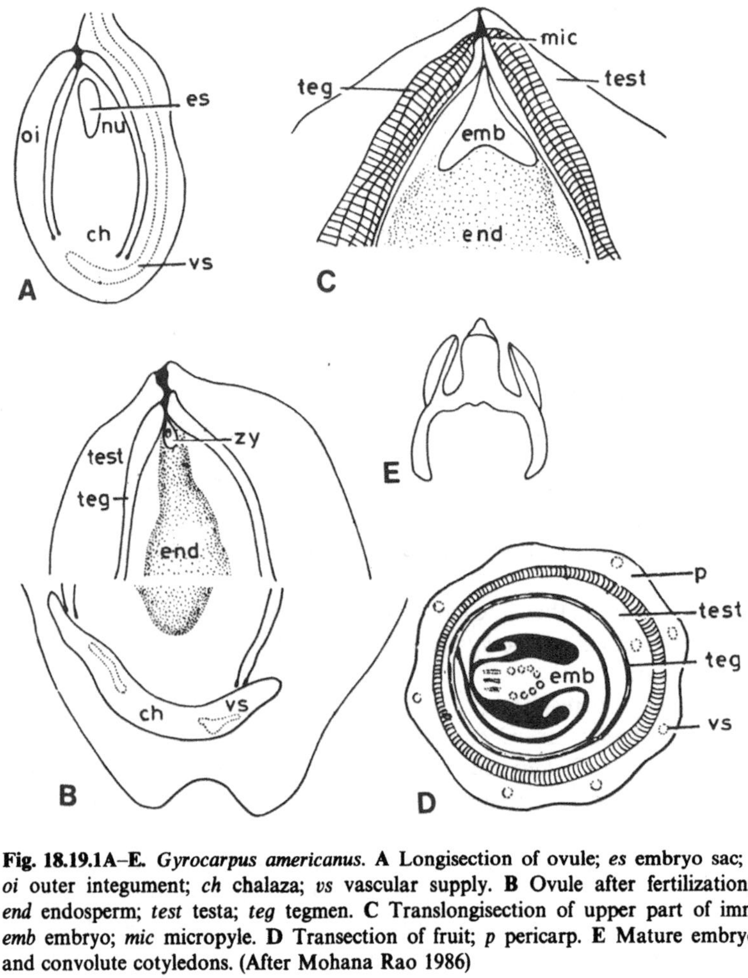

Fig. 18.19.1A–E. *Gyrocarpus americanus.* **A** Longisection of ovule; *es* embryo sac; *nu* nucellus; *oi* outer integument; *ch* chalaza; *vs* vascular supply. **B** Ovule after fertilization; *zy* zygote; *end* endosperm; *test* testa; *teg* tegmen. **C** Translongisection of upper part of immature seed; *emb* embryo; *mic* micropyle. **D** Transection of fruit; *p* pericarp. **E** Mature embryo with leafy and convolute cotyledons. (After Mohana Rao 1986)

undergo meiosis and successive cytokinesis. The pollen grains are two-celled at shedding, and the exine shows spinescent outgrowths (see Yakovlev 1981).

The carpel is two-trace as in Lauraceae. The vascular bundle enters the funiculus, reaches up to the base of nucellus and branches dichotomously in the chalaza (Fig. 18.19.3 C, E). The raphal bundle is reported to enter the outer integument in *Hernandia peltata.*

The single pendulous ovule in the inferior ovary is at first apotropous and then becomes anatropous. It is bitegmic and crassinucellar with micropyle formed by both integuments (Fig. 18.19.1A–C). The inner integument is three- to five-layered at the base, and more massive in the micropylar region. The apical region contains large starch grains. The outer integument is massive, and multiplicative (Fig. 18.19.3A, B).

The distally lobed integuments in *Hernandia peltata* (van Heel 1971b) are absent in *Gyrocarpus americanus* (Mohana Rao 1986).

The embryo sac of *Gyrocarpus* is eight-nucleate. The endosperm is of the Cellular type. The embryogeny (Fig. 18.19.2A–J) is of the Onagrad type [Period I, Series A, Megarchetype VI (B.S.M. Dutt 1978b)].

In *Hernandia*, the exalbuminous testal pachychalazal seed, enclosed in a woody endocarp (Corner 1976, Hutchinson 1973), has a straight embryo with large cotyledons which are plano-convex, folded or lobate and more or less connate in the ruminations of the perisperm (Fig. 18.19.2K, L cf. Fig. 18.19.1D, E). The folded perisperm is massive. This is an example of labyrinth seed (van

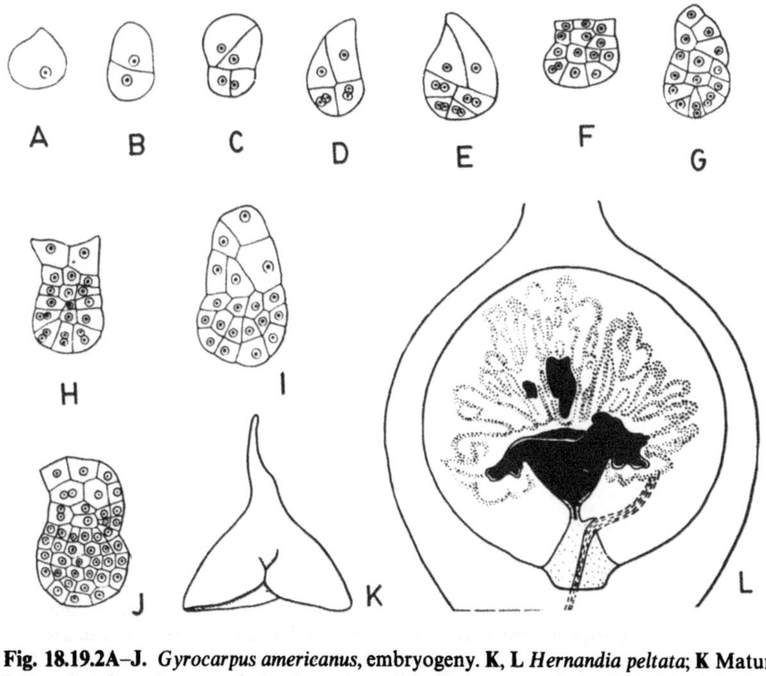

Fig. 18.19.2A–J. *Gyrocarpus americanus*, embryogeny. **K, L** *Hernandia peltata*; **K** Mature embryo; **L** Longisection of mature fruit, the embryo (*black*) shows two large folded (or lobed) cotyledons and ruminate perisperm. (A–J After B.S.M. Dutt 1978b, **K, L** After van Heel 1971a)

Heel 1971a). The chalazal region of the ovule develops enormously to form the body of the seed. The wall of this region becomes ruminate from profuse lobings of the enlarging nucellus and the cotyledons eventually bulge into these lobings to substitute the nucellus (Corner 1976).

In *Gyrocarpus* the testa is thick and undifferentiated, the tegmen becomes crushed (Fig. 18.19.3A, B). The pachychalaza and ruminate outgrowths are absent (Mohana Rao 1986).

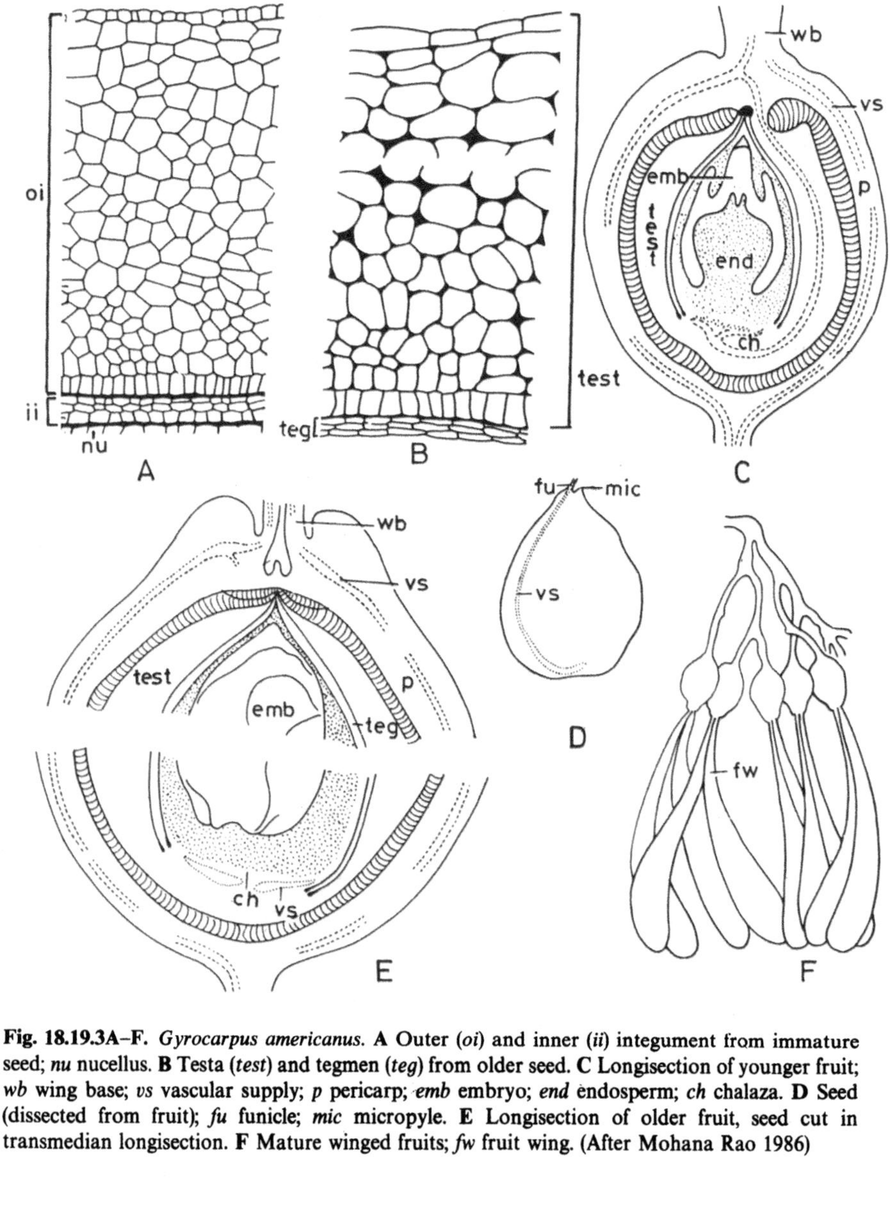

Fig. 18.19.3A–F. *Gyrocarpus americanus.* **A** Outer (*oi*) and inner (*ii*) integument from immature seed; *nu* nucellus. **B** Testa (*test*) and tegmen (*teg*) from older seed. **C** Longisection of younger fruit; *wb* wing base; *vs* vascular supply; *p* pericarp; *emb* embryo; *end* endosperm; *ch* chalaza. **D** Seed (dissected from fruit); *fu* funicle; *mic* micropyle. **E** Longisection of older fruit, seed cut in transmedian longisection. **F** Mature winged fruits; *fw* fruit wing. (After Mohana Rao 1986)

The seed closely adheres to the inner epidermis of the pericarp, and can be easily dissected out from a ripe and undried fruit (Figs. 18.19.1D, E, 18.19.3C, D).

In *Gyrocarpus americanus* (Mohana Rao 1986), the ovoid fruit has eight longitudinal ridges and two spatulate wings which extend from the distal end (Fig. 18.19.3F). The wing epidermis has unicellular hairs and stomata. The wings help in the dispersal of dry fruit.

BSM Dutt (1978b, 1979a, 1980), EJH Corner (1976), PK Endress (1972), WA van Heel (1970a, 1971a, b, c), J Hutchinson (1973), K Kubitzki (1969), PR Mohana Rao (1986), RLN Sastri (1958a, 1963), CF Shutts (1960), AL Takhtajan (1988), MS Yakovlev (1981)

18.20 Tetracentraceae

The information on the embryology is very scanty.

The anther is tetralocular with lateral microsporangia which open by slits (Hutchinson 1973). The endothecium with fibrous thickenings differentiates only on the outer side of anther lobe. The mature pollen grain is bicelled.

The ovule is anatropous, bitegmic and crassinucellate in *Tetracentron* sp.

The albuminous seed has a minute embryo with primordial cotyledons. The endosperm is oily.

JL Brewbaker (1967), J Hutchinson (1973), CG Nast and IW Bailey (1945)

18.21 Trochodendraceae

The anther wall comprises an epidermis, fibrous endothecium, two or three ephemeral middle layers, and biseriate glandular tapetum with two-nucleate cells. The epidermal cells develop protuberances and cuticular fibrillar thickenings along their outer surface in *Trochodendron aralioides* (Yoffe 1962). The endothecium shows broad bands of fibrous thickenings. The microspore mother cells undergo meiotic divisions and produce tetrahedral tetrads. Cytokinesis is simultaneous.

The pollen grains are tricolpate, and shed at the two-celled stage.

The ovule is anatropous, bitegmic, and crassinucellate. The micropyle is produced by the inner integument. The chalazal end of the ovule develops into a long tapering projection. The single hypodermal archesporial cell cuts off a parietal cell. The parietal tissue is three- to four-layered. Following meiotic divisions, the megaspore mother cell produces a linear or T-shaped tetrad. The chalazal megaspore develops into a Polygonum type of embryo sac. The antipodals are uninucleate, and persistent.

The development of endosperm (Yoffe 1965) is of the Cellular type. The endosperm is thin-walled, and oily.

The zygote divides transversely followed by another transverse division in both the cells producing a four-celled linear proembryo. The terminal cell forms the octant, and the remaining cells produce the suspensor.

The embryogeny appears to be of the Chenopodiad type.

The seed is small, endotestal and exotegmic with a very small embryo. The outer epidermis of the testa is thin-walled, whereas the inner comprises cuticularized cells with thick inner and radial walls. The outer epidermis of tegmen shows small tangentially elongate thick-walled cells; the inner one is thin-walled.

The seeds of *Trochodendron aralioides* (Mohana Rao 1981) are linear and ellipsoid with micropylar and chalazal projections. The testa is two layers thick and the inner layer forms the characteristic crystal layer. The tegmen is three layers thick, and the outer epidermis is a conspicuous fibrous layer. The raphe bundle terminates below the hypostase. The fruit is a follicle. The mature pericarp is 30 layers thick, and the ground tissue is parenchymatous.

NN Bhandari (1971), H Harms (1897a), PR Mohana Rao (1981), CG Nast and IW Bailey (1945), MD Yoffe (1962, 1965)

18.22 Eupteleaceae

The anther wall comprises the epidermis, fibrous endothecium, two or three ephemeral middle layers, and secretory tapetum with one- to four-nucleate cells in *Euptelea polyandra* (Endress 1969a). Simultaneous cytokinesis accompanies meiotic divisions in microspore mother cells, and the tetrads are tetrahedral.

The pollen grains are two-celled at the dispersal stage.

The ovule is pendulous, anatropous, bitegmic and crassinucellate with micropyle formed by the inner integument. The archesporial cell cuts off a parietal cell. Due to repeated divisions of the latter and of nucellar epidermal cells, the parietal tissue is very conspicuous. The megaspore mother cell undergoes meiotic divisions producing a linear tetrad. The chalazal megaspore develops into a Polygonum type of embryo sac. Two months elapse between pollination and fertilization.

The development of endosperm is of the Cellular type.

The embryogeny is irregular. The hypocotyl is very conspicuous in the mature dicotyledonous embryo.

The endotestal seed is small with a minute embryo. The outer epidermal cells of the testa are thin-walled with reddish-brown contents. The mesophyll and inner epidermal cells are sclerotic. The tegmen comprises thin-walled cells.

NN Bhandari (1971), PK Endress (1969a, b), R Lemesle (1946), CG Nast and IW Bailey (1946)

18.23 Cercidiphyllaceae

Swamy and Bailey (1949) have made a detailed study of *Cercidiphyllum japonicum*.

The anther tapetum is secretory with binucleate cells. Simultaneous cytokinesis follows meiotic divisions in microspore mother cells forming tetrahedral and decussate tetrads.

The pollen grains are tricolpate. Their furrows are very broad and ill-defined, and are generally described as immersed. They are shed at the two-celled stage.

The ovule is anatropous, bitegmic and crassinucellate. Both the integuments form the micropyle. The hypodermal archesporial cell cuts off a parietal cell producing five or six layers of parietal tissue. The nucellar epidermis undergoes periclinal divisions. The megaspore mother cell undergoes meiotic divisions and produces a linear tetrad. The chalazal megaspore develops into a Polygonum type of embryo sac. The egg apparatus is prominent, and the antipodals are small. The fertilized embryo sac elongates conspicuously.

The development of endosperm is of the Cellular type (Fig. 18.23.1A–E). A large micropylar and a small chalazal chamber are produced following the

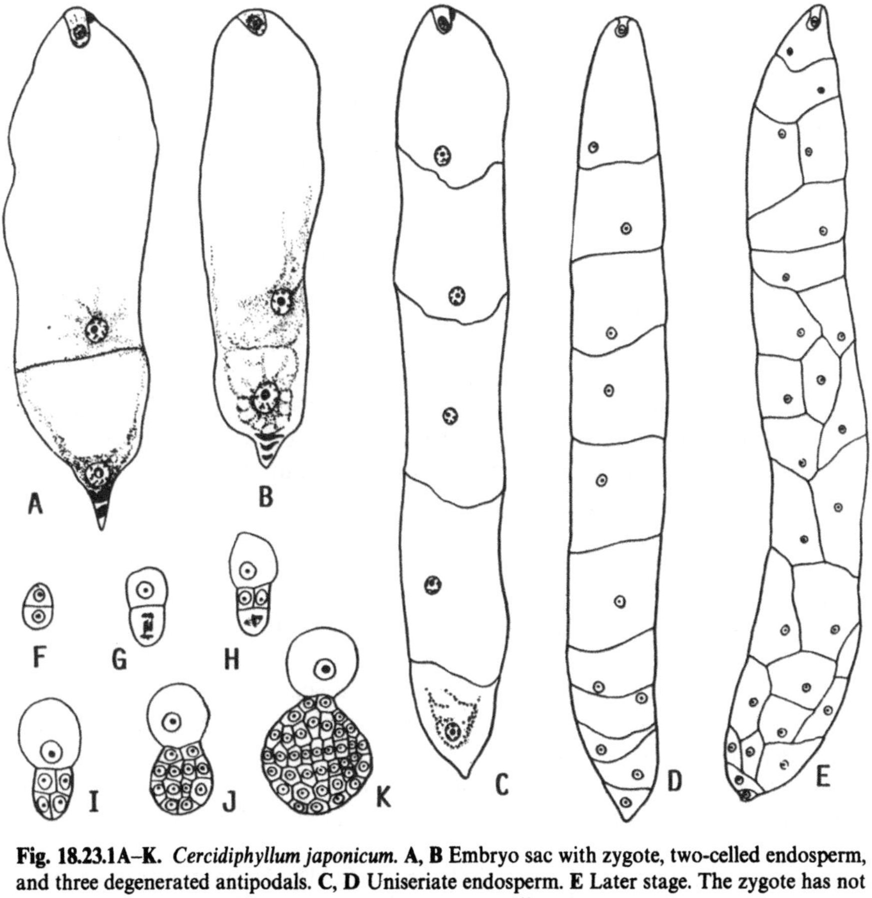

Fig. 18.23.1A–K. *Cercidiphyllum japonicum*. **A, B** Embryo sac with zygote, two-celled endosperm, and three degenerated antipodals. **C, D** Uniseriate endosperm. **E** Later stage. The zygote has not divided in **A–E**. **F–K** Embryogeny. (After Swamy and Bailey 1949)

transverse division of the primary endosperm nucleus. Both the chambers undergo transverse divisions, and the endosperm gives a uniseriate appearance; vertical and oblique walls develop afterwards.

The zygote divides transversely forming apical and basal cell. The apical cell develops into a quadrant; the basal cell does not divide further and becomes enlarged (Fig. 18.23.1 F–I). J and K are proembryos. The embryogeny corresponds to the Caryophyllad type. The mature embryo is large with a well-developed hypocotyl.

The seed is small, winged and albuminous. The chalazal wing has a curved vascular bundle. The testa is thin-walled, and the tegmen is crushed.

NN Bhandari (1971), L Diels (1936), H Harms (1918), K Subramanyam (1962), BGL Swamy and IW Bailey (1949)

18.24 Taxonomic Considerations

Bhandari (1970a, b, 1971) has comprehensively reviewed the embryology of Magnoliales, and discusses the relationships (Table 18.24.1).

Melchior (1964) includes 22 families (of Magnoliales) in six groups, while Cronquist (1981) reckons only ten families. Takhtajan (1980) distributes 20 families in three orders: Magnoliales (8), Laurales (10) and Illiciales (2); R. Dahlgren (1980b, 1981) in four orders: Magnoliales (6), Laurales (8) and Illiciales (2) in Magnoliiflorae, and Trochodendrales (2) in Rosiflorae (see also Cronquist 1975a).

Previously, several genera of doubtful affinity had been included in Magnoliaceae. Now they have been assigned to separate families: *Drimys* to Winteraceae, *Illicium* to Illiciaceae, *Euptelea* to Eupteleaceae, *Cercidiphyllum* to Cercidiphyllaceae, and *Schisandra* and *Kadsura* to Schisandraceae (see also Hutchinson 1973). According to Bailey and Smith (1942), and Swamy (1949a), Magnoliaceae are closely allied to Degeneriaceae and Himantandraceae (Table 18.24.1). Hutchinson places *Degeneria* in Winteraceae. Now it has been removed from Winteraceae, and Bailey and Smith (1942) recognize it as an independent family, Degeneriaceae (see also Melchior 1964).

Degeneriaceae. Magnoliaceae and Degeneriaceae (Table 18.24.1) have several common features indicating their alliance: secretory anther tapetum, simultaneous cytokinesis in pollen mother cells, monocolpate two-celled pollen grains with generative cell produced at distal end, anatropous, bitegmic, crassinucellate ovules, Polygonum type of embryo sac with ephemeral antipodals, and funicular obturator. The Cellular type of endosperm develops a two-celled chalazal haustorium in Magnoliaceae; it is ruminate in Degeneriaceae. The seed structure is more or less alike. A sarcotesta develops in Magnoliaceae and Degeneriaceae. The follicular fruit opens by a dorsal suture in Magnoliaceae, and a ventral suture in Degeneriaceae.

Himantandraceae. According to Bailey et al. (1943), Himantandraceae, Magnoliaceae and Degeneriaceae are closely related. *Eupomatia*, *Galbulimima*,

Magnoliaceae and Annonaceae form a coherent group (Endress 1980b). *Galbulimima* has not been fully investigated.

Himantandraceae, Magnoliaceae, Degeneriaceae and Annonaceae resemble each other in several features (Table 18.24.1): tetrasporangiate, abaxial anthers with a massive wall, secretory tapetum of binucleate cells, two-celled monosulcate pollen, anatropous, bitegmic, crassinucellate ovule with massive parietal tissue, and Polygonum type of embryo sac.

The differences are less conspicuous; cytokinesis in pollen mother cells is simultaneous in Annonaceae, Degeneriaceae and Magnoliaceae, and successive in Himantandraceae and Annonaceae. The organization of microspore tetrads shows much variation (N. Prakash et al. 1984).

The embryological characters confirm a close relationship between Himantandraceae, Annonaceae and Degeneriaceae (Table 18.24.1).

Winteraceae. Bentham and Hooker (1965a) include *Drimys* in the tribe Winterae of Magnoliaceae. Hutchinson (1973) recognizes Winteraceae as a separate family which has some distinctive characters: extensive fibrous endothecium in anther wall, pollen grains in permanent tetrads, monoporate with conspicuous to minute reticulations, generative cell cut off towards the proximal pole, phenolic compounds in the outer epidermis of outer integument, embryo sac of the Polygonum type, endosperm of the Cellular type, embryogeny irregular, and exotesta sclerotic. The family shows Magnolian (Table 18.24.1) affinity, and is related to Illiciaceae and Degeneriaceae (Willis 1973).

Annonaceae. Bentham and Hooker (1965a) place Annonaceae with Magnoliaceae in Ranales and Hutchinson (1973) with Eupomatiaceae in Annonales.

Annonaceae show certain affinities with Magnoliaceae and Degeneriaceae: transversely septate anthers in some genera (similar to Myristicaceae), simultaneous or successive cytokinesis in pollen mother cells, tetragonal tetrads, monocolpate pollen grain, anatropous, bitegmic, crassinucellar ovule with vascularized outer integument. The embryo sac is of the Polygonum type with ephemeral antipodals, as also in Magnoliaceae. Cellular type of endosperm with ruminations also occurs in Degeneriaceae. Onagrad type of embryogeny, and follicular fruit is reported in Magnoliaceae. Its woody mesotesta shows affinity to Illiciaceae. Annonaceae definitely belongs to the Magnolian complex (Table 18.24.1).

Eupomatiaceae. Eupomatiaceae have been previously included in Annonaceae, but now they are treated as a separate family (Hutchinson 1973). In both families the ovule is anatropous, bitegmic and crassinucellate, with oily ruminate endosperm. The fibrous testa forming ruminations is common to both families. The seeds are mesotestal, but in Annonaceae there is a well-developed perichalaza, which is absent in Eupomatiaceae. Eupomatiaceae and Annonaceae probably arose from the same ancestral stock but, in the course of evolution, they deviated considerably (Mohana Rao 1983).

Table 18.24.1. Embryology of Magnoliaceae (Sect. 18.2), Degeneriaceae (18.3), Himantandraceae (18.4), Winteraceae (18.5), Annonaceae (18.6), Schisandraceae (18.10) and Illiciaceae (18.11). (See Hayashi 1960, Padmanabhan 1960a, Kapil 1962, Bhandari 1963a, Kapil and Bhandari 1964, Kapil and Jalan 1964, Bhandari and Venkataraman 1968, N. Prakash et al. 1984)

	Magnoliaceae	Degeneriaceae	Himantandraceae	Winteraceae
Anther tapetum	1–3-layered, secretory, cells multinucleate	1-layered, secretory, cells 2-nucleate	1- or 2-layered, secretory, cells 2-nucleate	1-layered, secretory, cells become polyploid; amoeboid in some taxa
Cytokinesis	Simultaneous	Simultaneous	Successive	Simultaneous
Tetrad	Isobilateral, decussate	Tetragonal	Tetrahedral, isobilateral, decussate, linear, T-shaped	Tetrahedral, decussate
Pollen	Monosulcate, 2-celled	Monocolpate, 2-celled, generative cell cut off towards distal end	Monosulcate, 2-celled	Monocolpate, shed in tetrads, 2-celled, generative cell cut off towards proximal pole
Ovule	Anatropous, bitegmic, crassinucellate, micropyle formed by both integuments	Funicular obturator, anatropous, bitegmic, crassinucellate, micropyle formed by inner integument	Funicular obturator, anatropous, bitegmic, crassinucellate, micropyle formed by inner integument	Laminar, anatropous, bitegmic, crassinucellate, micropyle formed by inner or both integuments
Embryo sac	Polygonum type, synergids and antipodals ephemeral	Polygonum, synergids and antipodals ephemeral	Polygonum	Polygonum, antipodals degenerate after fertilization
Endosperm	Cellular type, in *Magnolia obovata* the chalazal chamber develops into a 2- or 3-celled haustorium	Cellular, ruminate in *Degeneria vitiensis*	–	Cellular
Embryo	Onagrad type or irregular, suspensor massive	–	–	Dicotyledonous with small suspensor in *Pseudowintera*, undifferentiated and organless in *Drimys*

Annonaceae	Schisandraceae	Illiciaceae
1-layered, secretory, cells 2–4-nucleate; amoeboid in some taxa	1-layered, secretory cells multinucleate	2-layered, secretory, cells 2-nucleate
Simultaneous or successive	Simultaneous	Simultaneous
Tetrahedral, isobilateral, decussate	Tetrahedral, decussate	Tetrahedral
Anasulcate, tectate, atectate, tetrads in a common wall, loose tetrads, in pairs, or free, 2-celled, generative cell cut off at distal end	Heteropolar, hexacolpate, 2-celled	Tricolporate, 2-celled
Anatropous, bi- or tritegmic, crassinucellate, micropyle formed by inner integument	Ana- to campylotropous, bitegmic, crassinucellate, micropyle formed by both integuments	Anatropous, bitegmic, crassinucellate, micropyle formed by outer integument
Polygonum	Polygonum, antipodals ephemeral; in *Schisandra chinensis* 50% embryo sacs of Polygonum type and others Allium type, the lower polar and antipodals are not formed	Polygonum, antipodal cells ephemeral
Cellular, lobed due to rumination	Cellular	Cellular
Onagrad	Asterad or Onagrad, suspensor short and inconspicuous, mature embryo shows only primordia of cotyledons	Asterad, mature embryo shows only primordia of cotyledons

Table 18.24.1. (*Contd.*)

	Magnoliaceae	Degeneriaceae	Himantandraceae	Winteraceae
Seed	Sarcotestal, testa multiplicative woody endotesta, albuminous, embryo small	Pachychalazal, albuminous, embryo small, suspensor short, hypocotyl bulbous, 3 or 4 cotyledons	Albuminous, embryo small	Exotestal, albuminous, embryo small

Myristicaceae. Hutchinson (1973) places Myristicaceae in Laurales, and Canellaceae in Magnoliales. Parmeswaran (1962) records their close affinities, and states that they have diverged from a common ancestor. Myristicaceae and Canellaceae have simultaneous cytokinesis of pollen mother cells, two-celled monocolpate pollen grains, generative cell cut off towards the proximal pole, anatropous, bitegmic, crassinucellar ovule, Polygonum type of embryo sac, ruminate endosperm, and a small embryo. The presence of oil cells and trilacunar node are additional features. A.C. Joshi (1946) relates Myristicaceae to Annonaceae. Both have trilacunar nodes, trimerous flowers, five-layered anther wall with secretory tapetum, fibrous endothecium, single row of microspore mother cells, sterile septa in anther, single-furrowed two-celled pollen grains with reticulate exine, anatropous, bitegmic, crassinucellate ovule with post-fertilization chalazal growth, Polygonum type of embryo sac with ephemeral antipodal cells, Onagrad type of embryogeny, ruminate endosperm, fleshy aril, and follicular fruit. The middle integument in Myristicaceae is rudimentary (N.C. Nair 1972). Both have complicated seed structure; the mesotesta is woody. The exotegmen is sclerotic in Myristicaceae, and in Annonaceae it forms a sclerotic endostome.

Myristicaceae differ from Lauraceae (Table 18.24.2), and need not be transferred to Laurales. The occasional occurrence of amoeboid tapetum, successive cytokinesis in pollen mother cells, acolpate pollen grains, embryo sac haustoria, secondary multiplication of antipodals, occasional Cellular type of endosperm, and exalbuminous seed of Lauraceae have no parallel in Myristicaceae. The embryogeny may also reflect their relationships. The seed structure differs in both. N.C. Nair (1972) negates the relationship of Myristicaceae with Magnoliaceae. He states that as the family has several distinctive features, it should be raised to an independent order, Myristicales.

Canellaceae. The taxonomic position of Canellaceae is rather debatable. Hallier (1912), Wettstein (1935), Melchior (1964) and Hutchinson (1973) associate them with Magnoliales, whereas Engler and Diels (1936) and Rendle (1959) place them in Parietales. Embryologically, they closely resemble Myristicaceae. Their exotestal seed shows alliance with Winteraceae. But the syncarpous ovary with parietal placentation does not go well with Magnolian families.

Annonaceae	Schisandraceae	Illiciaceae
Mesotestal, arillate, pachychal-azal, albuminous	Exotestal, albuminous, embryo small	Exotestal, albuminous, embryo small

Schisandraceae and Illiciaceae. Schisandraceae and Illiciaceae are grouped together in the Magnoliales (Melchior 1964), and Illiciales (R. Dahlgren 1980a, b, Takhtajan 1980). Bentham and Hooker (1965a), Engler and Prantl (1931) and Rendle (1952) include the tribe Schisandrae in Magnoliaceae. *Illicium* has been placed in the tribe Winterae of Magnoliaceae by Bentham and Hooker (1965a). Hutchinson (1973) treats both as separate families, Schisandraceae and Illiciaceae, in Magnoliales. Kapil and Jalan (1964) have evaluated their embryological features, and conclude that they are not closely related. Schisandraceae have hexacolpate pollen grains, Polygonum, sometimes modified bisporic type of embryo sac, Asterad or Onagrad type of embryogeny, and a berry type of fruit with succulent pericarp, in contrast to tricolpate pollen grains, Polygonum type of embryo sac, Asterad type of embryogeny, and follicetum with sclerotic pericarp of Illiciaceae. According to Corner (1976), *Schisandra* is primitive to *Illicium* due to a thick testa, palisade-like endotesta, and absence of lignified endocarp, whereas it is advanced due to campylotropous small-sized seed. The exotestal structure of the seed allies *Illicium* with Schisandraceae and Winteraceae, not with Magnoliaceae. (see Table 18.24.1)

The third group of the order Magnoliales comprises eight families: Austrobaileyaceae, Trimeniaceae, Amborellaceae, Monimiaceae, Calycanthaceae, Gomortegaceae, Lauraceae and Hernandiaceae. All these families, except Amborellaceae and Calycanthaceae, belong to Laurales, according to Hutchinson (1973). Calycanthaceae belong to Rosales.

Formerly, several genera were included in Monimiaceae. Now they have been assigned to separate families: *Austrobaileya* to Austrobaileyaceae, *Trimenia* and *Piptocalyx* to Trimeniaceae, *Amborella* to Amborellaceae, *Calycanthus* to Calycanthaceae, and *Gomortega* to Gomortegaceae.

Sufficient embryological data are not available to determine the systematic position of the above families, except Austrobaileyaceae, Monimiaceae, Calycanthaceae and Lauraceae. These four families show close affinities.

Austrobaileyaceae. Bentham and Hooker (1965a) associate Austrobaileyaceae with Monimiaceae in series Microembryae, whereas Engler and Prantl (1889–1897) place them in Ranales along with Magnoliaceae, Myristicaceae and Monimiaceae. Bailey and Swamy (1949) also relate them to Monimiaceae. Cronquist (1968, 1981) treats them as an isolated taxon of Magnoliales.

Table 18.24.2. Embryology of Hernandiaceae (Sect. 18.19), Lauraceae (18.18), *Cassytha* (18.18), Myristicaceae (18.8), Monimiaceae (18.15), Calycanthaceae (18.16) and Lactoridaceae (20.5). (See Bhandari 1971, Bouman 1971a, Dutt 1978b, Endress 1979, 1980a, b, Mathur 1968, Mohana Rao 1980, 1986, Sastri 1956)

	Hernandiaceae	Lauraceae	*Cassytha*	Myristicaceae
Anther	Bisporangiate with glandular appendages, tapetum amoeboid, cytokinesis successive, pollen 2-celled	Bi-tetrasporangiate, wall 5-layered, tapetum amoeboid, cytokinesis successive, tetrads tetrahedral, isobilateral, T-shaped, linear, pollen 2- celled	Bisporangiate, tapetum secretory, cells 2-, 4- or multinucleate	Tetrasporangiate, in *Myristica* each microsporangium becomes divided into 2 or 3 compartments by sterile septa, wall 5-layered, tapetum secretory, cells uninucleate, cytokinesis successive, tetrads isobilateral, pollen 2-celled
Ovule	Anatropous, bitegmic, outer vascularized, crassinucellate; exo- and endostome; normal or pachychalaza; chalaza vascularized; nucellus causes ruminations in chalaza	Anatropous, bitegmic, crassinucellate; endostome; peri- or pachychalaza; periphery of chalaza vascularized; nucellus absorbed or persists in ruminations	Anatropous, bitegmic, crassinucellate; micropyle occluded by nucellus, funiculus. in direct contact with nucellus	Anatropous, funicular aril conspicuous and vascularized, bitegmic, crassinucellate, postament, endostome, pachychalaza, postchalazal branches extend into tegmen and ruminations
Embryo sac	Not studied	Polygonum type, antipodals usually ephemeral but multiply in *Laurus nobilis*	Polygonum, supernumerary embryo sacs common, antipodal nuclei do not organize into cells, tips of embryo sacs produce haustorial tubular extensions in the direction of "micropyle", protrude beyond the nucellus and enter the funicle	Polygonum, in *Horsfieldia* the embryo sac undergoes much expansion
Endosperm	Cellular type	Nuclear, becomes cellular at globular proembryo	Cellular, abundant, reserve food starchy	Nuclear, becomes cellular, ruminate, oily or starchy

Monimiaceae	Calycanthaceae	Lactoridaceae
Tetrasporangiate, wall 4- to 6-layered, tapetum secretory or amoeboid, cells binucleate, cytokinesis simultaneous, tetrads tetrahedral, pollen free or remain attached in tetrads, 2- or 3-celled	Tetrasporangiate, wall 4–5-layered, tapetum secretory, cytokinesis simultaneous, tetrads tetrahedral, decussate, pollen 3-celled, occasionally remain attached in tetrads	Pollen grains united in tetrads
Anatropous, uni- or bitegmic, crassinucellate; hypostase; exostome, nucellar beak persists at endostome; nucellar cap; perichalaza; raphal bundle extends up to two-thirds of length of seed	Anatropous, bitegmic crassinucellate, nucellar cap; hypostase; both exo- and endostome lined with microscopic hairs; raphal bundle expands into shallow cupular chalaza	Anatropous, bitegmic, tenuinucellate (?), endostome
Polygonum and Allium type, antipodals 5–20	Polygonum, supernumerary embryo sacs develop, one polar may degenerate, antipodals ephemeral	Not studied
Cellular, massive, stores oil	Autonomous, Cellular type	Nuclear, later becomes cellular, copious and oily food reserve

Table 18.24.2. (*Contd.*)

	Hernandiaceae	Lauraceae	*Cassytha*	Myristicaceae
Embryo	Onagrad type, straight with large cotyledons, more or less connate in the ruminations of perisperm	Piperad, Asterad or Onagrad, large, thick massive cotyledons, radicle short	Zygote divides transversely, further development irregular, small with well-developed cotyledons	Small, cotyledons divergent, radicle short
Seed	Exalbuminous, testal, pachychalazal, large subglobose	Exalbuminous, endotestal	Moderately large, albuminous, exo- and endotesta and two outer layers of mesophyll persist	Albuminous, arillate (exostomal and funicular), seed-coat formed by both testa and tegmen
Fruit	Basically achene, sometimes 2–4-winged samara	Drupe or berry, persistent perianth surrounds base, pericarp with oil cells, endocarp stony	Drupe, enclosed by inflated calyx tube, pericarp with outer and inner epidermis and inner hypodermis of palisade cells	Fleshy to subwoody, dehiscent drupe

Hutchinson (1973), Takhtajan (1973, 1980), Thorne (1976) and R. Dahlgren (1977a, 1980b, 1981) associate Austrobaileyaceae with Monimiaceae in Laurales.

Austrobaileyaceae are characterized by spiral flowers, laminar, massive stamens with thecae on adaxial surface (similar to Magnoliales), anasulcate, distally aperturate, pollen grains with irregularly reticulate exine, two-celled pollen grains at the dispersal stage, several ascidiate closed carpels with transversely, strongly bilobed stigma (weakly bilobed in Annonaceae–*Goniothalamus*).

The resemblances with the Annonaceae are: floral arrangement, staminal structure, and sarcotestal, albuminous, reticulate, perichalazal seed. The ruminations in the endosperm are of the Annona type. Austrobaileyaceae can be segregated from the Annonaceae on the basis of spiral perianth, imbricate tepals, anasulcate pollen with reticulate-regulate exine, and ascidiate carpels (Endress 1980a).

On the contrary, infrequent presence of amoeboid tapetum, successive cytokinesis, monocolpate pollen grains, several ovules per carpel, embryo sac with haustoria, and secondary multiplication of antipodals of Lauraceae are not met with in Austrobaileyaceae.

The similarity of features with Monimiaceae is more pronounced. Secretory anther tapetum, simultaneous cytokinesis in pollen mother cells, structure of ovule and seed, and the type of embryo sac exhibit their close affinity.

Monimiaceae	Calycanthaceae	Lactoridaceae
Asterad	Sexual embryo uncommon, nucellar polyembryony common, cotyledons convolute	Minute, organless embryo
Albuminous, endotestal, embryo small and straight, cotyledons adpressed or divergent, oil cells in outer layers of testa	Exalbuminous, endotestal	Albuminous, testa and tegmen collapse, meso- and endocarp provide protection
Achenes or drupe, enclosed by hypanthium	Pericarp hard, encloses seed	Follicle, beaked

However, occasional occurrence of amoeboid anther tapetum, unitegmic ovule, Allium type of embryo sac, and supernumerary antipodals segregate Monimiaceae from Austrobaileyaceae. The most important feature, which separates these two families, is the type of pollen grains. They are monocolpate, aperturate and globose, and are shed singly in Austrobaileyaceae, whereas the pollens are inaperturate, two- to three-sulcate in Monimiaceae, and are often dispersed in tetrads.

Endress (1980a) has discussed the resemblances with Magnoliales, and differences with Annonaceae. He concludes that Austrobaileyaceae point to a closer relationship with Magnolian, and not Lauralian families, and they should not be included in the Annonaceae. He emphasizes that "the pollen resembles, more than in almost all other Ranales (and angiosperms), *Clavatipollenites*, the earliest known angiosperm pollen fossils of the lower Cretaceous (Endress and Honegger 1980)". Therefore, Austrobaileyaceae are one of the most prominent "living fossils among the angiosperms".

Further information about the development of endosperm and embryo of *Austrobaileya* may prove helpful in taxonomic assignment.

Trimeniaceae. Endress (1987b) points out that the Trimeniaceae (Sect. 18.13) seem to be the closest relatives of Chloranthaceae (Sects. 20.4, 20.6). The stamens are the visually attractive organs in the entomophilous members of both families. The pollen is polyporate in some members of both families

(*Trimenia* and *Sarcandra*), and reticulate. The drupe-like fruits are berries (except *Hedyosmum*), since the hard layer is produced by testa and not by the ovary wall. Another relatively closely related relict family is the Amborellaceae (Sect. 18.14). As in Chloranthaceae, the ovule is pendant and orthotropous. The pollen is inaperturate as in some Trimeniaceae. The Chloranthaceae also share some features with the Piperaceae (Sect. 20.3) and, therefore, have frequently been placed in the Piperales (Cronquist 1981). A combination of unusual characters points to a possible phylogenetic relationship with the Hamamelidaceae (Sect. 26.3) via Trochodendraceae (Sect. 18.21). The Chloranthaceae seem to have a key position and probably represent a link between Laurales (Sect. 18.18) and Piperales within the Magnoliidae, as well as between Magnoliidae and Hamamelididae in the context of the primitive angiosperms. This issue can be further considered when we have adequate embryological data on the related families.

Lauraceae, Monimiaceae. Lauraceae are related to Monimiaceae (Table 18.24.2): secretory or amoeboid anther tapetum, simultaneous cytokinesis of pollen mother cells, bitegmic, crassinucellar ovule with a massive parietal tissue, Polygonum type of embryo sac with secondary increase in antipodals, Cellular type of endosperm, Asterad type of embryogeny, and tracheidal endotesta. But, occasionally, successive cytokinesis, Nuclear or Cellular type of endosperm, and Piperad or Onagrad type of embryogeny occur in Lauraceae. The embryo sac haustorium in Lauraceae resembles that of Trochodendraceae. Monimiaceae are segregated due to the occurrence of pollen grains in permanent tetrads, nucellar cap, unitegmic ovule in some members, and occasional occurrence of Allium type of embryo sac.

Cassytha. The systematic position of the genus *Cassytha* (Sastri 1962) has been much debated (Bhandari 1971). The structure and development of anther, pollen, ovule, and seed-coat with endocarp closely resemble that of Lauraceae (Table 18.24.2). However, Bhandari erects a separate family, Cassythaceae, due to several dissimilarities reported in *Cassytha*: (1) broad nucellar tip, wide micropyle formed by both integuments, (2) multiple "haustorial" embryo sacs which become extramicropylar and invade the funiculus, (3) ephemeral (three) antipodal nuclei which do not organize into cells, and (4) Cellular type of endosperm.

Hortonia, endemic to Ceylon (Sri Lanka), is considered as especially primitive in Monimiaceae. It has been raised to the rank of a subfamily, Hortonioideae (Money et al. 1950), or even of a family, Hortoniaceae (Smith 1972). Endress (1980b) states that *Hortonia* has an intermediate position between Atherospermataceae and Monimiaceae, showing greater affinity with the latter. Further consideration should await a detailed study of reproductive aspects.

Endress (1980b) concludes that "the evolutionary step from the floral type of *Hortonia* to the type of *Tambourissa* represents an extraordinary morphological grade comparable to the step from gymnospermy to angiospermy".

Calycanthaceae. Morphologically, Calycanthaceae resemble Magnoliaceae and Annonaceae. Due to opposite arrangement of leaves, cup-like receptacle, and exalbuminous seed with a large embryo, Hutchinson (1973) relates them to Rosaceae. Their floral anatomy, nodal type, and basic chromosome number are similar to Cactaceae (Y.D. Tiagi 1970a). The relationship with Cactaceae has already been discussed (see Cactales, Sect. 17.3). Embryologically, Calycanthaceae are allied to Monimiaceae (Table 18.24.2) in the development of anther, ovule, embryo sac, and seed-coat. The formation of more than one embryo sac per ovule, and the Cellular type of endosperm also occur in Lauraceae [*Cassytha filiformis* (Sastri 1963)]. But, the absence of amoeboid anther tapetum, ephemeral antipodals, lack of fertilization due to male sterility in some members and, consequently, autonomous development of endosperm, and nucellar polyembryony are distinct features of Calycanthaceae. They have no relationship with the Rosales, or Myrtales, but are allied to Magnoliaceae, Annonaceae and Eupomatiaceae.

Hernandiaceae. Hernandiaceae are closely allied to Lauraceae (Table 18.24.2). The presence of two-trace carpel, amoeboid anther tapetum, multicellular female archesporium, Cellular type of endosperm (as in *Cassytha*), and the presence of helical thickenings in the seed are some of the similarities. The embryogeny resembles that of *Cinnamomum* and *Litsea* (Sastri 1958a). The study of wood anatomy (Shutts 1960) also supports their alliance.

Pax (1891) established the family Hernandiaceae and included the genera *Gyrocarpus*, *Sparattanthelium*, *Illigera* and *Hernandia*. Mohana Rao (1986) has discussed whether all the four genera should be included in the Hernandiaceae, or *Hernandia* and *Illigera* in Hernandiaceae, and *Gyrocarpus* and *Sparattanthelium* in Gyrocarpaceae (see Takhtajan 1969). In his later classification, Takhtajan (1980) does not recognize the family Gyrocarpaceae. The relationships of the Hernandiaceae have been pointed out with Lauraceae, Gomortegaceae, Myristicaceae, Calycanthaceae, Lactoridaceae, Monimiaceae and Cassythaceae; except Myristicaceae, all the others are included in the order Laurales by Takhtajan (1980). The chief embryological characters are listed in Table 18.24.2 (Gomortegaceae have been omitted because the embryology has not been studied).

This table clearly brings out that the Hernandiaceae are quite distinct from Myristicaceae, Calycanthaceae, Lactoridaceae and Monimiaceae, but are close to Lauraceae (*Cassytha* is included in Lauraceae by Melchior 1964). Further resemblances with Lauraceae are: non-aperturate, spheroidal and tenuexinous pollen grains, presence of unicellular hairs, rubiaceous type stomata, needle-type crystals in leaf and cortex, medium-length vessel members with simple perforations, alternate intervascular pitting, three-cell wide homogeneous rays, paratracheal parenchyma, septate fibres of medium length with simple pits, p-type of sieve element plastids, and unilacunar nodes (see Mohana Rao 1986).

Of the four families – Tetracentraceae, Trochodendraceae, Eupteleaceae and Cercidiphyllaceae – of Magnoliales (Melchior 1964), Trochodendraceae and Cercidiphyllaceae are retained in Magnoliales by Hutchinson (1973). Previously, *Euptelea* had been included in the Trochodendraceae (Hutchinson

1959, 1973); it has been separated as an independent family, Eupteleaceae (see Willis 1973).

Eupteleaceae. On embryological grounds, Endress (1969a, b) relates Eupteleaceae to *Cercidiphyllum*, and retains them in Magnoliales. In conformity with the order Magnoliales, Eupteleaceae show: (1) secretory anther tapetum, (2) simultaneous cytokinesis of pollen mother cells, (3) tricolpate, two-celled pollen grains, (4) anatropous, bitegmic, crassinucellar ovule, (5) Polygonum type of embryo sac, and (6) Cellular type of endosperm. The endotestal seed corresponds to that of Trochodendraceae.

Cercidiphyllaceae. The family Cercidiphyllaceae has been included in Hamamelidaceae by Baillon (1871), in Magnoliaceae by Maximowicz (1872), Bessey (1915) and Lemsle (1946). It is treated as an independent family, Cercidiphyllaceae, by Harms (1918), Takhtajan (1966, 1980), Cronquist (1968, 1981) and Thorne (1968). Swamy and Bailey (1949) include it in Ranales, as it is intimately related to Trochodendraceae and Eupteleaceae.

The pollen grains of *Tetracentron*, *Trochodendron*, some members of Hamamelidaceae, and *Cercidiphyllum* are tricolpate, but, in the last-named genus, the furrows are considerably broader and poorly defined, and the exine is also different. In *Cercidiphyllum* the nucellar epidermal cells divide periclinally, and the embryo sac elongates conspicuously after fertilization. The early development of endosperm is rather typical, and reminds one of *Sarracenia purpurea* (Shreve 1906). The seed is small with a large embryo, and the chalazal wing has a vascular bundle. But, this simplified seed is not primitive (Corner 1976).

According to Bhandari (1971), Trochodendraceae and Cercidiphyllaceae are closely related. Both have anatropous, bitegmic, crassinucellate ovule with a tapering chalazal projection, a single-celled hypodermal female archesporium, massive parietal tissue, Polygonum type of embryo sac, Cellular type of endosperm, and tapering wing of the seed.

19 Ranunculales

The order Ranunculales comprises two suborders: Ranunculineae and Nymphaeineae with seven families: Ranunculineae – Ranunculaceae, Berberidaceae, Sargentodoxaceae, Lardizabalaceae, Menispermaceae; Nymphaeineae – Nymphaeaceae and Ceratophyllaceae.

19.1 General Features

The order Ranunculales exhibits the following features: (1) secretory anther tapetum, amoeboid in Ceratophyllaceae, and of both types in Ranunculaceae and Nymphaeaceae, (2) simultaneous cytokinesis in pollen mother cells, successive in Nymphaeineae, and of both types in Berberidaceae, (3) variations in the number of integuments in Ranunculaceae and Menispermaceae, ovules undifferentiated (ategmic), unitegmic, or bitegmic; usually bitegmic, only unitegmic in Ceratophyllaceae; anatropous, crassinucellate, crassi- or tenuinucellate in Ranunculaceae and Nymphaeaceae, orthotropous in Lardizabalaceae, and hemi- to amphitropous in Menispermaceae, (4) Polygonum type of embryo sac; mono-, bi- or tetrasporic in Ranunculaceae, and mono- or tetrasporic in Berberidaceae, (5) antipodals quite prominent, except in Lardizabalaceae, (6) Nuclear type of endosperm, Cellular in Lardizabalaceae and Ceratophyllaceae, and Nuclear, Cellular and Helobial in Nymphaeaceae, (7) embryogeny of Onagrad type in Menispermaceae, Onagrad, Solanad and irregular in Ranunculaceae, Onagrad and Solanad in Berberidaceae, and Asterad type in suborder Nymphaeineae, and (8) seed mostly exotestal.

19.2 Ranunculaceae

The anther wall comprises the epidermis, and fibrous endothecium; *Anemone* and *Aquilegia* lack the thickenings (Bhandari 1968). There are two or three ephemeral middle layers, and secretory tapetum except in *Actaea spicata* (Jalan 1963) where it is amoeboid. Both types of tapetum are reported in different anthers of the same flower in *Anemonopsis macrophylla* (Tobe 1974b). The development of wall layers is of the Basic type (Davis 1966) in *Nigella damascena* (Vijayaraghavan and Marwah 1969). Both the middle layers persist in *Aquilegia*, and only one in *Ceratocephalus falcatus* (Bhandari and Asnani 1968). Übisch granules are present on the tangential and radial walls of tapetal cells in *Helleborus foetidus* and *Nigella* (Bhandari et al. 1976). The tapetal cells are bi-, tri- or multinucleate. The nuclei fuse and become polyploid.

Turala-Szybowska (1984) analyzed the mitotic behaviour of the tapetal cell in *Ranunculus fluitans* during microsporogenesis. Mitotic divisions and vacuolation in tapetal cells start in the central part of the internal tapetum, and the tapetal layer situated in the apical part of the anther locule. In these regions the tapetum may be more than one layer thick. During the premeiotic phase of microspore mother cells, the tapetal cells remain uninucleate (2n) (Fig. 19.2.1A). While the mother cells undergo meiotic divisions, the tapetal cells become binucleate (2n) either by mitosis or by disturbed mitosis (Fig. 19.2.1B). Due to inhibited prophase the nuclei of the binucleate tapetal cells become tetraploid, or due to mitosis the cells become four-nucleate (2n), or due to disturbed mitosis the cells become three-nucleate (two diploid nuclei and one tetraploid nucleus). A further inhibited prophase in tetraploid nuclei results in eight- or 16-ploid nuclei. In addition, some of the four-nucleate diploid cells undergo inhibited prophase to form four-nucleate tetraploid cells. At uninucleate pollen grain stage, 7% of tapetal cells were 2n and binucleate, 19% 4n binucleate, 51% 8n binucleate, and 19% 16n binucleate. Other tapetal cells were 1% 2n four-nucleate, 2% 2n four-nucleate, and 1% with two nuclei 2n and one 4n (Fig. 19.2.1C). Thus, polyploidization occurs due to endomitosis, disturbed mitosis, or inhibited prophase.

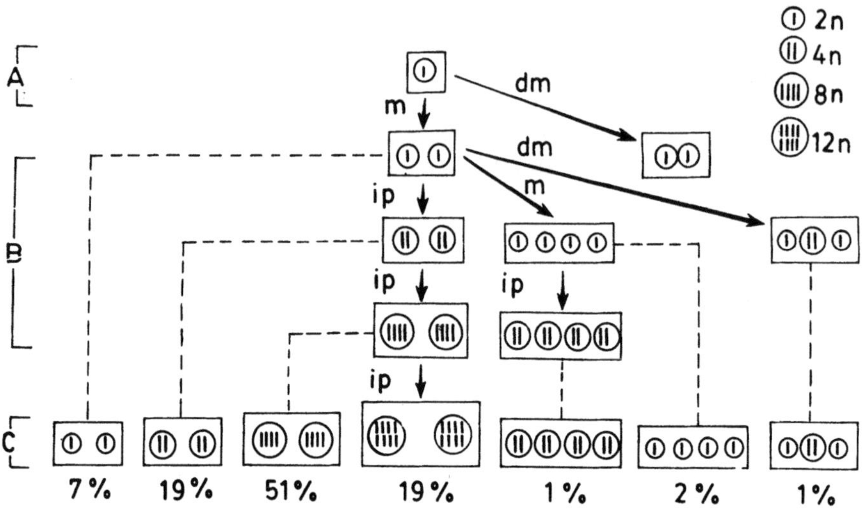

Fig. 19.2.1A–C. *Ranunculus fluitans* Lam. (2n = 16), main pathways in the development of tapetal cells. **A** At pre-meiotic stage of microspore mother cells. Mitosis (*m*) and disturbed mitosis (*dm*) occur in diploid nuclei of tapetal cells. **B** At meiotic divisions I and II in microspore mother cells. Mitosis in tapetal cells occurs without cytokinesis (2n), or mitosis may get inhibited (*ip*) at prophase. Then the nuclei of these cells remain 4n or 8n. When disturbed mitosis occurs, the cell contains two diploid nuclei and one tetraploid nucleus. **C** At primary pollen grain stage. The mitosis in tapetal cells may get inhibited in octaploid nuclei so that the cell becomes 12-ploid. The percentage of nuclei with different ploidy-level is also shown. (After Turala-Syzbowska 1984)

Simultaneous cytokinesis follows meiotic divisions of microspore mother cells, and the tetrads are tetrahedral and decussate.

The tricolporate pollen grains have smooth exine, and are two-celled at the dispersal stage.

The ovules are anatropous, unitegmic in *Anemone*, *Ceratocephalus*, *Clematis*, *Helleborus* and *Ranunculus*, bitegmic in *Actaea*, *Adonis*, *Caltha* and *Thalictrum*, and undifferentiated sterile ovules in *Adonis*, *Anemone* and *Clematis*; tenuinucellate in *Clematis gouriana* and *R. sceleratus*, crassinucellate in *Adonis*, *Aquilegia*, *Caltha*, *Ceratocephalus*, *Ranunculus*, *Thalictrum* and *Trollius*, both tenui- and crassinucellate in *Nigella damascena* and *Adonis annua*, and occasionally pseudocrassinucellate in *A. annua*, *H. foetidus* and *N. damascena* (see Vijayaraghavan 1970a). They are unitegmic and pseudocrassinucellar in *Hepatica nobilis* var. *japonica* (Tobe et al. 1975), and *Trautvetteria japonica* (Tobe 1973). Bitegmic crassinucellar ovules are reported in *Pulsatilla aurea*, and unitegmic tenuinucellar in *P. chinense* (Sun and Wang 1984).

The archesporium is uni- or multicelled. The archesporial cell functions directly as megaspore mother cell, or cuts off a parietal cell as in *Caltha palustris* (Kapil and Jalan 1962). The parietal tissue is two or three layers thick in *Actaea*, *Ceratocephalus* and *Trollius*. The cells of the nucellar epidermis also contribute to the formation of parietal tissue. A nucellar "cap" is prominent in *Ceratocephalus falcatus.*

The megaspore mother cell undergoes meiotic divisions, and produces a linear or T-shaped tetrad. The chalazal megaspore develops into a Polygonum type of embryo sac in *Actaea* (Jalan 1963), *Anemone* (Bhandari 1968), *Anemonella, Aquilegia, Batrachium, Caltha, Ceratocephalus, Delphinium, Helleborus, Myosurus, Ranunculus* and *Thalictrum* (see Schnarf 1929), *Ceratocephalus orthoceros* (Kordyum 1959), *Cimicifuga acerina* (Tobe 1974a), *Clematis gouriana* (*Vijayaraghavan* 1962), *Hepatica* (Tobe et al. 1975), *Nigella* (Vijayaraghavan and Marwah 1969), *Pulsatilla chinense* (Sun and Wang 1984), *Thalictrum javanicum* (Vijayaraghavan 1970a) and *Trollius europaeus* (Bhandari and Kapil 1964). The embryo sac is of the Allium type in fertile and sterile ovules of *Adonis* (Bhandari 1962, 1966), *Ceratocephalus falcatus* (Bhandari and Asnani 1968) and *Trollius* (Bhandari and Kapil 1964). The occurrence of mono- and bisporic types of embryo sac development, in different species of the same genus, has been noted in *Ceratocephalus* and *Trollius.* Picci (1964) reported tetrasporic-Pyrethrum type in *Thalictrum angustifolium* var. *flavum.* Tetrasporic type occurs in the sterile ovules of *Anemone obtusifolia* (Bhandari 1968), and *Clematis gouriana* (Vijayaraghavan 1962). Such a tendency is also reported in *Adonis annua* (Bhandari 1962). The variation in the development of embryo sacs is to such an extent that, in the same species, mono-, bi- and tetrasporic types occur, e.g. in *Anemone vitifolia* (Bhandari, 1965).

In *Pulsatilla chinense* (Sun and Wang 1984), sometimes the embryo sac shows two egg apparatus and six antipodals; presumably due to the dissolution of intervening wall of the twin embryo sacs.

Fish (1970) reports several variations in development of the embryo sac of *Clematis* spp. In fertile ovules, in all species, the development is generally of the Polygonum type, and in sterile ovules of tetrasporic type. In the sterile ovules of *C. orientalis* he observed that the four megaspore nuclei lying in the centre of coenomegaspore fuse in pairs. These two divide once mitotically. The secondary four-nucleate embryo sacs show the egg, two polar nuclei, and one antipodal, or the egg, one polar nucleus, and two antipodals. If another division occurs, a secondary eight-nucleate embryo sac develops. In one case he noticed the egg apparatus at the chalazal end of the embryo sac. The embryo sacs in the sterile ovules finally degenerate soon after the fertilization of fertile ovules. The sterile ovules of *C. vesicolor* exhibit normal eight-nucleate embryo sacs. However, in sterile ovules of *Hepatica nobilis* var. *japonica* (Tobe et al. 1975) only the archesporial cell differentiates and further development is arrested.

The embryo sac develops a caecum at its chalazal end in *Nigella damascena,* and the antipodals are left behind in situ (Vijayaraghavan and Marwah 1969). In Ranunculaceae the antipodals are noteworthy. They are ephemeral in *Adonis aestivalis* (Bhandari 1963b). They enlarge, undergo endomitosis, and become polyploid in *Caltha palustris* (Kapil and Jalan 1962), *Hepatica* (Tobe et al. 1975), and *Thalictrum* (Picci 1969). They are octaploid in *Helleborus niger,* 32-ploid in *Anemone hepatica,* and 64- or 128-ploid in *Aconitum* (Tschermak-Woess 1956).

Osterwalder (1898) reported large antipodal cells with polyploid nuclei in the post-fertilized embryo sac (containing young globular proembryo and free-nuclear endosperm) of *Aconitum napellus*. Zhukova and Sokolovskaya (1977) also studied the antipodals (in the same species), before fertilization. Bohdanowicz and Turala-Szybowska (1985, 1987) examined the antipodal cells in *A. vulparia*. In mature embryo sac the antipodals are very rich in organelles, have giant endopolyploid nuclei with lobed outline especially at the chalazal side. The antipodals resemble transfer cells with marked wall ingrowths which are particularly well-developed in the chalazal part. During the development of free-nuclear endosperm, the antipodal cells enlarge considerably, the wall ingrowths increase in size and number, the cytoplasm becomes dense with weak vacuolation, and they are at their most active metabolic state for transfer and secretory activity.

Syngamy precedes triple fusion in *Helleborus*; in *Adonis* one sperm fuses either with the upper or lower polar nucleus prior to polar fusion. Whether this is the normal condition, it requires further study.

The development of endosperm is of the Nuclear type. Enucleate cytoplasmic vesicles are present in *Thalictrum japonicum* (Vijayaraghavan and Bhandari 1970). Wall formation is initiated either from the micropylar or chalazal region, and is centripetal. The initiation of walls occurs at 256-nucleate stage in *Adonis* (Bhandari 1966), and ca. 1000-nucleate in *Anemone* (Bhandari 1968) at the globular stage of proembryo. Kordyum (1957) reported a chalazal endosperm haustorium in *Nigella sativa*.

The embryogeny also exhibits variation. It is irregular in *Adonis* (Bhandari 1966), Onagrad type in *Actaea spicata* (Jalan 1963), *Anemone* (Bhandari 1968), *Caltha palustris* (Kapil and Jalan 1962), *Ceratocephalus calcaratus* (Fig. 19.2.2A–Q), *Thalictrum japonicum* (Vijayaraghavan and Bhandari 1970), *Ranunculus sceleratus* (Souèges 1914) and *R. tripartitus* (Patel and Cook 1972), and Solanad type in *Helleborus foetidus* (Vijayaraghavan 1970a) and *Nigella damascena* (Vijayaraghavan and Marwah 1969). Tobe et al. (1975) report Caryophyllad or Solanad type in *Hepatica nobilis* var. *japonica*.

Somatic apospory of Hieracium type has been reported in *Ranunculus* spp. (Håfliger 1943, Rousi 1956). In *R. auricomus* (Rozanova 1932, Håfliger 1943, Rutishauser 1954), the egg develops pseudogamously, but triple fusion is normal. Antipodal embryos occur in *Nigella arvensis* (Derschau 1918); this needs confirmation.

In *Eranthis hiemalis* (Fig. 19.2.3), at the shedding stage the mature seed has a 0.1-mm long pear-shaped undifferentiated embryo with a conspicuous suspensor. The cotyledonary primordia and the radicle develop in the soil. The process of intraseminal growth and differentiation continues for about eight months, and the mature embryo attains a length of 2 mm. It comprises 90% cotyledons, and 10% hypocotyl and radicle. In contrast to other angiosperms, the cotyledons are closely adpressed to each other in the upper half region (called "cotyledonary sheath" by the author); the lower half of the cotyledons is free.

The seed is small, albuminous, ex-arillate, with a minute embryo. The chalaza is swollen in *Caltha*, and a swollen raphe-elaiosome is present in

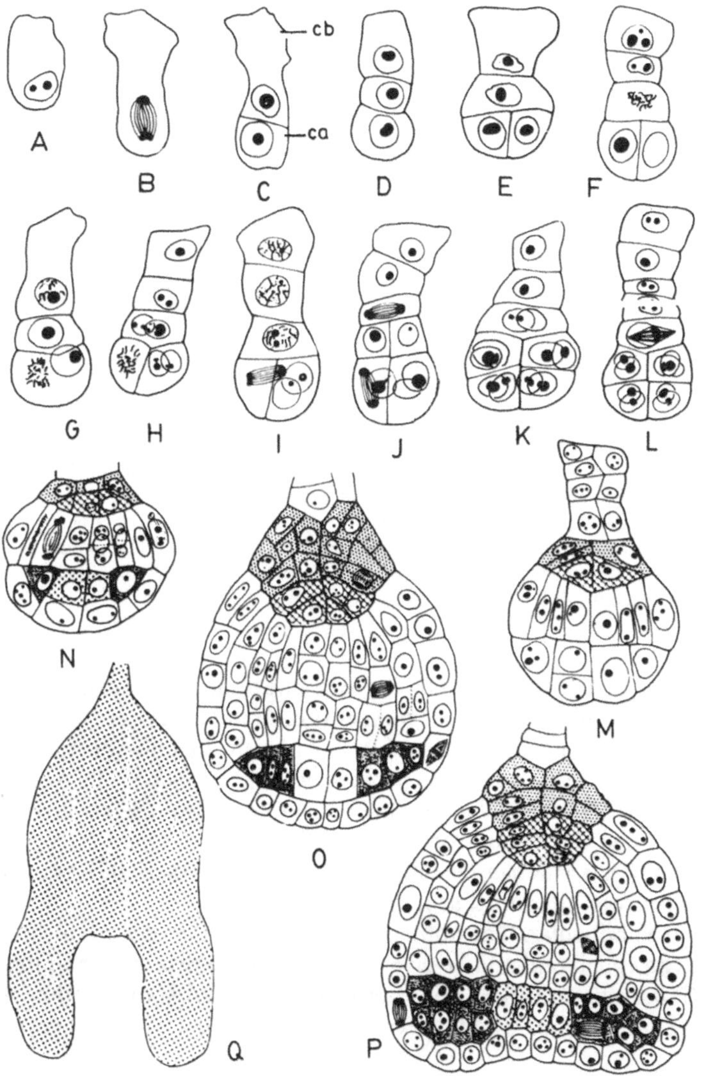

Fig. 19.2.2A–Q. *Ceratocephalus calcaratus*, development of embryo. **A–C** Transverse division in zygote to form the cells *ca* and *cb*. **D, E** Three- and four-celled proembryo. **F–J** Development of quadrant. **K–L** Proembryo at octant stage. **M–O** Globular proembryo. Note the delimitation of dermatogen, cotyledonary and shoot apex initials. **P** Longisection of early heart-shaped embryo. **Q** Longisection of mature dicotyledonous embryo (diagrammatic). (After Bhandari and Asnani 1968)

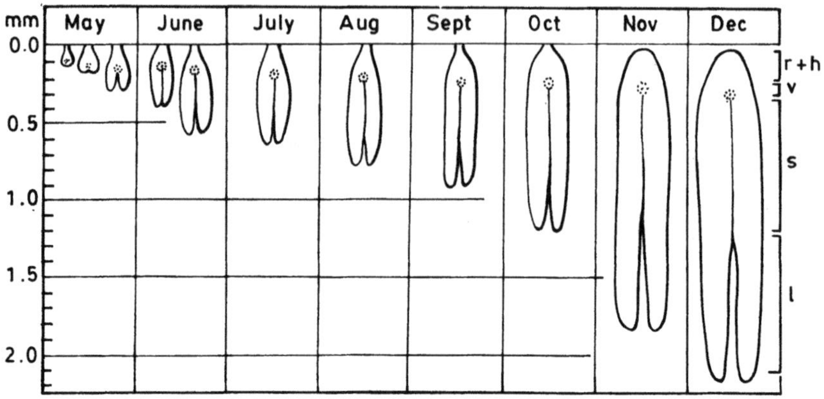

Fig. 19.2.3. *Eranthis hiemalis*, diagrammatic representation of the maturation of embryo; *r* + *h* radicle + hypocotyl; *v* shoot apex; *s* sheath; *l* lamina of cotyledons. In mature seed (0.1 mm long – stage A) the embryo is organless (undifferentiated). The formation of cotyledonary primordia and radicle takes place in the soil (stages B, C). Then the embryo grows in the region of cotyledonary sheath (*s*), and not in the hypocotyl (*h*) region. Until December there is no further change in the outer form, but the embryo grows larger and procambial tissue develops. The mature embryo represents 90% of cotyledons, and only 10% of hypocotyl with radicle. Germination begins in December or early January. (After Haccius 1963)

Helleborus. The seed-coat is formed by both the integuments in bitegmic ovules. The outer epidermis of the outer integument is hard in *Caltha*, and sclerotic in *Aquilegia*. The inner integument has fine bar-like thickenings on the radial walls in *Trollius*. The testa is crushed in *Anemone* and *Clematis*. The tegmen is more or less crushed, but in *Helleborus* the outer epidermis is thick-walled, and the inner shows fine thickenings on radial walls.

In several respects *Circaeaster* differs from other members of Ranunculaceae and is, therefore, described separately.

The anther is bilocular. The anther wall comprises the epidermis, fibrous endothecium, an ephemeral middle layer and secretory tapetum with uninucleate cells in *C. agrestis* (Junell 1930). After meiosis in microspore mother cells, cytokinesis is simultaneous, and the tetrads are tetrahedral. The mature pollen is two-celled.

The ovule is orthotropous, unitegmic and tenuinucellate. The integument is two to three cell layers thick. The archesporial cell functions as megaspore mother cell. The megaspore tetrad is linear or T-shaped, and the chalazal megaspore develops into a Polygonum type of embryo sac. The polars fuse before fertilization. The antipodals are ephemeral, or may persist up to a few divisions in the endosperm.

The path of the pollen tube is chalazogamous (Junell 1930) or mesogamous. The pollen tube penetrates the integument near the antipodals, enters the embryo sac, elongates up to the synergids, ruptures and pours its contents into one of them. This synergid degenerates, whereas the other synergid remains healthy for some time. The remnants of pollen tube persist. The entry of pollen tube and fertilization should be reinvestigated.

The development of the endosperm is of the Cellular type. Division of the primary endosperm nucleus is followed by a vertical wall and then there is an asymmetric transverse division separating two small cells at the chalazal end. The two upper larger cells undergo another transverse division.

The zygote divides by a transverse wall and the linear proembryo shows five cells. The development of embryo probably conforms to the Chenopodiad type.

The greater part of seed is occupied by oily and proteinaceous endosperm. The large, terate, straight embryo shows only primordial cotyledons. Its differentiation takes place during germination. The seed-coat is crushed so that the endosperm remains in contact with the pericarp (Corner 1976).

H Babis (1966, 1973, 1974, 1976a, b, 1978, 1979a, b, c), NN Bhandari (1962, 1963b, 1965, 1966, 1968), NN Bhandari and S Asnani (1968), NN Bhandari and P Chitralekha (1989), NN Bhandari and RN Kapil (1964), NN Bhandari and Ram Kishori (1973), NN Bhandari, Ram Kishori and S Natesh (1976), J Bohdanowicz and K Turala-Szybowska (1985, 1987), F Bouman (1978), F Bouman and JIM Calis (1977), M Brouland (1975), M Brouland and C Guerven (1978), M Brouland, C Guerven and C Le Cog (1976), M Charzynska (1973), EJH Corner (1976), JM Coulter (1898), GL Davis (1966), MV Derschau (1918), RK Fish (1970), M Fougêre-Rifot (1975, 1981), Å Gustafsson (1946, 1947a, 1947b), B Haccius (1963), E Håfliger (1943), M Hesse (1978), T Hiroshi (1973, 1974), T Hiroshi, A Toshiko and I Osamu (1975), R Izmailow (1965, 1973), S Jalan (1963), S Jalan and NN Bhandari (1963), S Junell (1930), RN Kapil and S Jalan (1962), KY Kavathekar and A Pillai (1981), EL Kordyum (1957, 1959), M Kumazawa (1972), O Langlet (1928), C Le Cog, M Brouland and C Guerven (1978), Ly-Thi-Ba (1971, 1974), CR Metcalfe (1936), Y Miyaji (1927), Mu-Xi-jin (1983, 1984), Mu-Xi-jin and Wang Fu-Xiong (Wang Fu-hsiung) (1985b), A Osterwalder (1898), E Pacini, M Cresti and G Sarfatti (1972), TK Patel and CDK Cook (1972), V Picci (1964, 1969), E Pogan and H Wcislo (1981), M Rifot (1973), PN Robeva-Davidova and MA Fileva-Mikhajlova (1979), A Rousi (1956), MA Rozanova (1932), A Rutishauser (1954), RLN Sastri (1969b), J Sâto (1972), K Schnarf (1929), B(Bahadur) Singh (1936), R Souèges (1974), AC Sun and FH Wang (1984), M Tamura (1972), M Tamura and I Mizumoto (1972, 1974), H Tobe (1973, 1974a, 1974b, 1981), H Tobe, T Adachi and O Yoshida (1975), H Tobe and RC Keating (1985), Z Trela-Sawicka (1974, 1977), K Turala-Szybowska (1981, 1984), K Turala-Szybowska and J Wolańska (1990), E Tschermak-Woess (1956), M Ünal (1978), MR Vijayaraghavan (1962, 1970a), MR Vijayaraghavan and NN Bhandari (1970), MR Vijayaraghavan and KN Marwah (1968, 1969), CLF Woodcock and PR Bell (1968), M Yoffe (1969), GYA Zhukova and RB Sokolovskaya (1977)

19.3 Berberidaceae

The anther wall comprises the epidermis, fibrous endothecium, two or three ephemeral middle layers, and secretory tapetum. The thickenings of the endothecium are U-shaped in *Berberis umbellata* (Sastri 1969a). The nuclei of the tapetal cells divide mitotically, and as many as eight nuclei occur in each cell. Nuclear fusions take place and they become polyploid (B.M. Johri 1935c). The microspore mother cells undergo meiotic divisions, and give rise to tetrahedral and isobilateral tetrads. Cytokinesis takes place by furrowing.

The mature pollen grain is three-colpate, and shed at the two-celled stage. The generative cell is ovoid.

The ovule is anatropous, bitegmic and crassinucellate. Integumentary vascular bundles occur in *Caulophyllum thalictroides*, and funicular obturator and hypostase in *Berberis aristana*. The outer integument takes part in the formation of micropyle; both integuments in *Nandina domestica* (Ehdaie and

Russell 1984), and the micropylar passage is zig-zag. The archesporium is multicellular, but only one cell functions. It cuts off a parietal cell in some species of *Mahonia* (see Schnarf 1931). The parietal cell is not cut off in *Jeffersonia* (Andrews 1895), and *Podophyllum* (Lubliner 1925). In *Berberis nepalensis* it may or may not be cut off (B.M. Johri 1935c). The nucellar epidermal cells divide to produce a massive tissue. The megaspore mother cell undergoes meiotic divisions, and produces a linear tetrad. The chalazal megaspore develops into a Polygonum type of embryo sac. Modified Peperomia type of embryo sac is reported in *Caulophyllum*, but this needs confirmation. The synergids are ephemeral; Mauritzon (1936a) mentions haustorial synergids in *Berberis* and *Mahonia* which appears to be incorrect. The antipodals are very conspicuous and persistent. The polar nuclei fuse before fertilization. The nucellus, hypostase and mature embryo sac contain starch grains in *N. domestica.*

The development of endosperm is of the Nuclear type. The embryogeny conforms to the Onagrad type–Myosurus variation; Solanad type in *Berberis aristata.* The embryo is tricotyledonous in *Podophyllum peltatum.* Polyembryony occurs in *B. aristata.*

The seed is albuminous and exotestal. The embryo is well-developed with a long radicle. The testa alone forms the seed-coat. The outer epidermis comprises thick-walled, lignified, pitted, longitudinally elongate fibres, whereas the inner epidermis consists of small, thick-walled cells. The outer layers of mesophyll are slightly thickened and aerenchymatous, the inner one is crushed. A vestigial aril occurs rarely (Corner 1976).

FM Andrews (1895), EJH Corner (1976), M Ehdaie and SD Russell (1984), BM Johri (1935c), OK Lubliner (1925), Ly-Thi-Ba (1976), J Mauritzon (1936a), JW Nowicke and JJ Skvarla (1981), RLN Sastri (1969a), K Schnarf (1931), YF Shen (1954), K Subramanyam (1962), SD Swanson and SH Sohmer (1976)

19.4 Sargentodoxaceae

The embryology of the family Sargentodoxaceae has not been investigated.

19.5 Lardizabalaceae

The anther wall comprises the epidermis, fibrous endothecium, two or three ephemeral middle layers, and secretory tapetum. It is of the amoeboid type in *Stauntonia hexaphylla* (Yoshida and Nakajima 1978). The tapetal cells are two- to four-nucleate; the nuclei fuse to form polyploid masses. The microspore mother cells undergo meiotic divisions, and form tetrahedral (Fig. 19.5.1A) and decussate tetrads. Cytokinesis is simultaneous.

Pollen grains are three-colporate, and shed at the two-celled stage.

The young ovule is orthotropous, bitegmic and crassinucellate. It becomes campylotropous in *Akebia* spp. (Yoshida and Michikawa 1973), and anatropous in *Decaisnea* (Swamy 1953c) and *Stauntonia hexaphylla* (Yoshida and

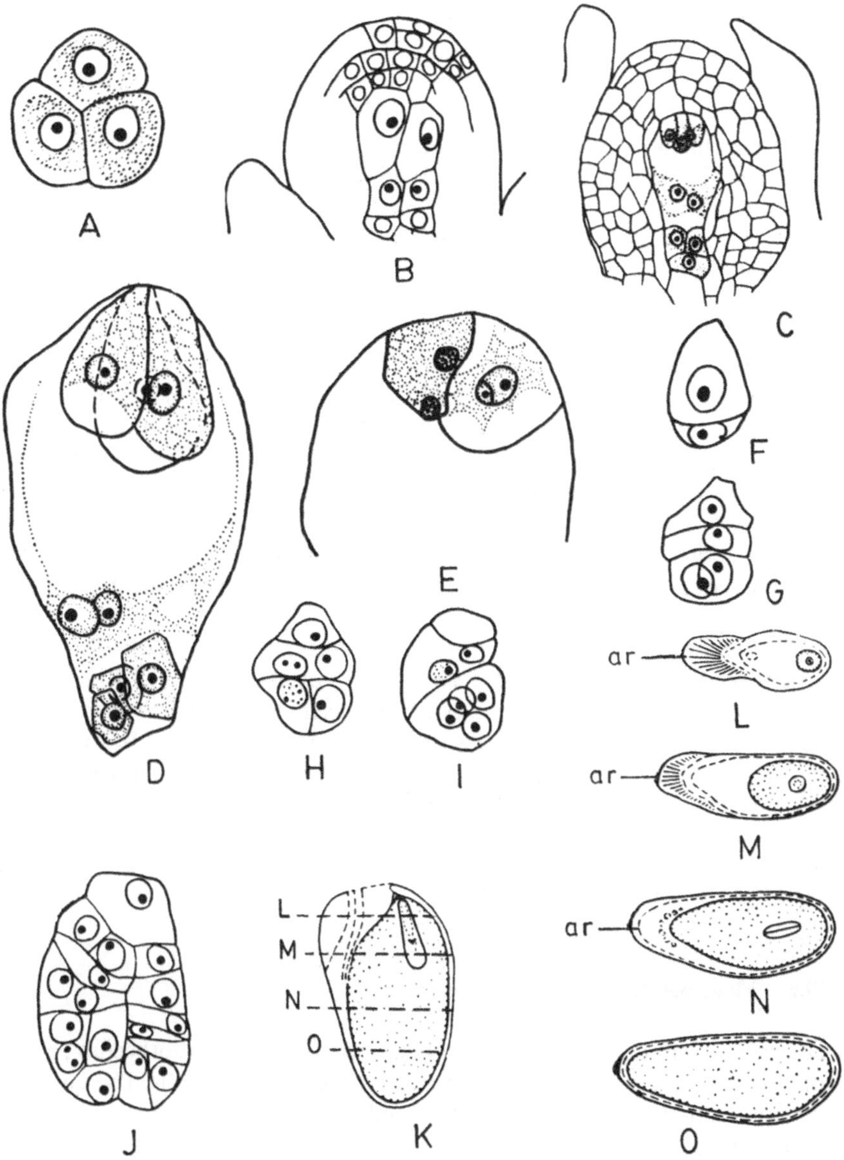

Fig. 19.5.1A–O. A, B, F–J *Stauntonia hexaphylla,* **C** *Akebia quinata,* **D, E** *A. pentaphylla,* **K–O** *A. lobata.* **A** Tetrahedral tetrad of microspores. **B** Sporogenous cells in nucellus. **C, D** Mature embryo sac. **E** Micropylar portion of embryo sac to show syngamy. **F–J** Embryogenesis. **K** Longisection of seed. **L–O** Transections of **K** at respective levels; *ar* aril. (**A, B, F–J** After Yoshida and Nakajima 1978, **C–E** after Yoshida and Michikawa 1973, **K–O** after Corner 1976)

Nakajima 1978). The micropyle is formed by the inner integument only. The hypodermal archesporial cell cuts off a parietal cell. Multiple archesporium occurs in *S. hexaphylla*, but only one cell functions. The parietal cell produces a well-developed parietal tissue. The cells of the nucellar epidermis also divide forming two to three layers (Fig. 19.5.1B, C). The megaspore mother cell undergoes meiotic divisions, produces a T-shaped, or, rarely, linear tetrad in *Holboellia latifolia* (S.P. Bhatnagar 1965b). In *Stauntonia hexaphylla*, during the formation of a linear tetrad, the micropylar dyad divides earlier than the chalazal, but in all cases the chalazal megaspore develops into a Polygonum type of embryo sac (Fig. 19.5.1C, D). The antipodals are small and ephemeral in *Akebia pentaphylla*, *A. trifoliata* (Yoshida and Michikawa 1973), *Holboellia latifolia* (S.P. Bhatnagar 1965b), and *S. hexaphylla* (Yoshida and Nakajima 1978), but persist in *Decaisnea insignis* (Swamy 1953c).

The polar nuclei fuse before fertilization, and the secondary nucleus lies near the antipodals in *Decaisnea*. During double fertilization in *Akebia trifoliata* (Yoshida and Michikawa 1973) one sperm nucleus fuses first with one polar nucleus and then triple fusion occurs. Syngamy in *A. pentaphylla* is shown in Fig. 19.5.1E.

The endosperm is of the Cellular type. Both micropylar and chalazal chamber contribute to the formation of endosperm. It contains oil as a storage product.

The embryogeny conforms to the Asterad type (Fig. 19.5.1F–J).

The seed is albuminous with a vestigial aril in *Akebia* (Fig. 19.5.1K–O). The outer epidermis of testa comprises narrow cells with thick inner and radial walls. The mesophyll cells show brown, unlignified walls. In *Akebia* the mesophyll is composed of longitudinal fibres. The tegmen and inner epidermis of the testa do not show any remarkable feature.

SP Bhatnagar (1965b), EJH Corner (1976), BGL Swamy (1953c), O Yoshida and A Michikawa (1973), O Yoshida and Y Nakajima (1978)

19.6 Menispermaceae

The anther wall comprises persistent epidermis, fibrous endothecium, ephemeral middle layer, and secretory tapetum with uninucleate cells. The tapetum is irregularly biseriate. Cytokinesis is simultaneous, and the tetrads are tetrahedral in *Cocculus triloba* (Yoshida and Yamada 1973).

The tricolpate pollen grains are two-celled at the dispersal stage.

The ovule is hemianatropous or amphi- to campylotropous, bitegmic (*Cocculus* and *Tiliacora*), or unitegmic (*Cissampelos*, *Cocculus*, *Stephania* and *Tinospora*), and crassinucellar. The micropyle is formed by the inner integument. The outer integument is smaller in *Cocculus villosus* (A.C. Joshi 1937a). The single hypodermal archesporial cell cuts off a parietal cell, the parietal tissue is well-developed. The cells of the nucellar epidermis also divide, and form a cap two to four cells thick. The megaspore mother cell undergoes meiotic divisions, and produces a linear or T-shaped tetrad. The chalazal megaspore develops into a Polygonum type of embryo sac. The synergids are egg-like in *Tinospora*. The polar nuclei fuse before fertilization. The three antipodals are large, and they persist up to or even after fertilization in *Cocculus* and *Tinospora* (A.C. Joshi 1937a, 1939a). The number of antipodals increases in *Tinospora*.

The development of the endosperm is of the Nuclear type. Wall formation is initiated (at the globular proembryo) in the micropylar region, and gradually progresses up to the chalaza. In *Tiliacora racemosa* and *Tinospora cordifolia* (Sastri 1964), the infoldings of the outer integument lead to ruminations (Annona type – Periasamy 1962a), whereas in *Cissampelos* and *Cocculus* the infoldings of fibrous endocarp cause a wavy surface of the endosperm (weakly ruminate) (Sastri 1964). The endosperm is oily.

The embryogeny conforms to the Onagrad type. The zygote divides six to eight weeks after fertilization in *Tinospora cordifolia* (Sastri 1964). In *Cocculus triloba* (Yoshida and Yamada 1973), the growth of embryo is said to be vertical, horizontal, globular, and irregular. This interpretation is not quite correct even though the earlier stages of embryogeny show considerable variation.

The exarillate seed is albuminous with a large, curved embryo. It is characterized by an epiphysis. The seed-coat is unspecialized, and generally gets crushed. The outer epidermis of testa persists in some species. It comprises tabular cells with thin, lignified walls. The thick-walled endocarp gives protection in *Cocculus triloba* (Yoshida and Yamada 1973).

AC Joshi (1937a, 1939a), AC Joshi and BVR Rao (1934, 1935), PM Mathew (1958), K Periasamy (1962a), RLN Sastri (1954a, b, 1964), B Singh (1934), NK Tiwary (1934), O Yoshida and H Yamada (1973)

19.7 Nymphaeaceae

The anther wall comprises the epidermis, fibrous endothecium, two to five middle layers, and tapetum. The endothecial thickenings are well-developed in *Brasenia*, *Nelumbo*, *Nymphaea* and *Victoria* (P. Khanna 1965a, 1967), and are absent in *Cabomba* (Padmanabhan and Ramji 1966). They are masked by tannin in *Nelumbo nucifera* (S.C. Gupta and Ahluwalia 1979). The two middle layers persist in *Nelumbo*. The tapetum is reported to be amoeboid in *Castalia* and *Nuphar* (see Schnarf 1931), and secretory with uni- to multinucleate cells in other members. This aspect needs further study. In *Nelumbo* and *Euryale* (P. Khanna 1964a), the cells are multi-nucleate. "A large number of Übisch granules are studded on a prominent tapetal membrane on the inner tangential surface of tapetal cells at the uninucleate microspore stage" (S.C. Gupta and Ahluwalia 1979). The microspore mother cells undergo meiotic divisions and form tetrahedral and isobilateral tetrads. Cytokinesis is successive in *Cabomba*, and simultaneous in *Castalia*, *Nelumbo* and *Nuphar* (see Schnarf 1931, S.C. Gupta and Ahluwalia 1979), and *Brasenia* (P. Khanna 1965a). Compound pollen grains are reported in *Nymphaea* and *Victoria* (P. Khanna 1967).

Pollen grains are shed at the three-celled stage in *Euryale*, *Nymphaea* and *Victoria* (P. Khanna 1964a, 1967) and at the two-celled stage in *Cabomba*

(Padmanabhan and Ramji 1966). In *Nelumbo*, the pollen grains are three-colpate, although in all other members they are monocolpate (see P.K.K. Nair 1970).

The pollen grains in *Barclaya longifolia* (Schneider 1978) are spherical, bean-shaped or psilate to scabrate with a perforate tectum. They are inaperturate.

Sometimes, in *Nelumbo* the pollen tubes enter an embryo sac in which the synergids have degenerated before its entry. This needs confirmation.

The ovule is anatropous (initially orthotropous in *Nelumbo*), bitegmic, and crassinucellate. It is orthotropous (permanently) in *B. longifolia*. The integument forms a beak, or an operculum develops. The nucellar tissue lies below the operculum. The micropyle is formed by the inner integument (Fig. 19.7.1A). In *Cabomba caroliniana* (Ramji and Padmanabhan 1965), the nucellar tissue grows in height due to the presence of a basal meristem. A glandular, aerenchymatous, arillate outgrowth from the funicle has been observed in *Euryale*, *Nymphaea* and *Victoria*. S.C. Gupta and Ahluwalia (1977, 1979) report occasional occurrence of a middle integument in *Nelumbo nucifera*. The funicle is massive with air-spaces in this species, and produces secretory hairs which function as an obturator. A single hypodermal archesporial cell cuts off a parietal cell. The parietal tissue is three to four layers in *Victoria*, two to six layers in *Nelumbo* (S.C. Gupta and Ahluwalia 1979), and only a single layer in *Nymphaea* (P. Khanna 1967). The megaspore mother cell undergoes meiotic divisions producing a linear or T-shaped tetrad. The chalazal megaspore develops into a Polygonum type of embryo sac (Fig. 19.7.1B, C). The antipodals are ephemeral.

In *Barclaya longifolia* (Schneider 1978), the embryo sac has an elongated tubular extension which has been interpreted as "haustorium-like". This assumption is incorrect and the extension should be considered as an embryo sac caecum. Similar extension of the embryo sac is also reported in *Nuphar* (Chifflot 1902), *Nymphaea* (Conrad 1905) and *Euryale* (P. Khanna 1964a). The caecum enlarges in the basal nucellar region, and in a mature embryo sac contains three antipodals. The basal nucellar tissue grows in files of radiating rows, and after fertilization forms perisperm. It is packed with packets of starch grains.

In *Ondinea purpurea* (Schneider 1983), anthesis occurs over a three-day period. First-day flowers are protogynous and characterized by colourful reflexed perianth and staminal parts. The pollen-receptive stigmas of first-day flowers secrete a fluid that fills the small stigmatic cup. As potential pollinators are attracted and attempt to land on the distally exposed gynoecium, they contact the stigmatic fluid which loosens pollen from the insect's body and pollination is achieved. Second- and third-day flowers are functionally staminate.

During double fertilization, syngamy and triple fusion occur simultaneously.

According to P. Khanna (1965a), the antipodals multiply and become multinucleate (Fig. 19.7.1C), and haustorial in *Nelumbo nucifera* during post-fertilization stages. However, according to S.C. Gupta and Ahluwalia (1979),

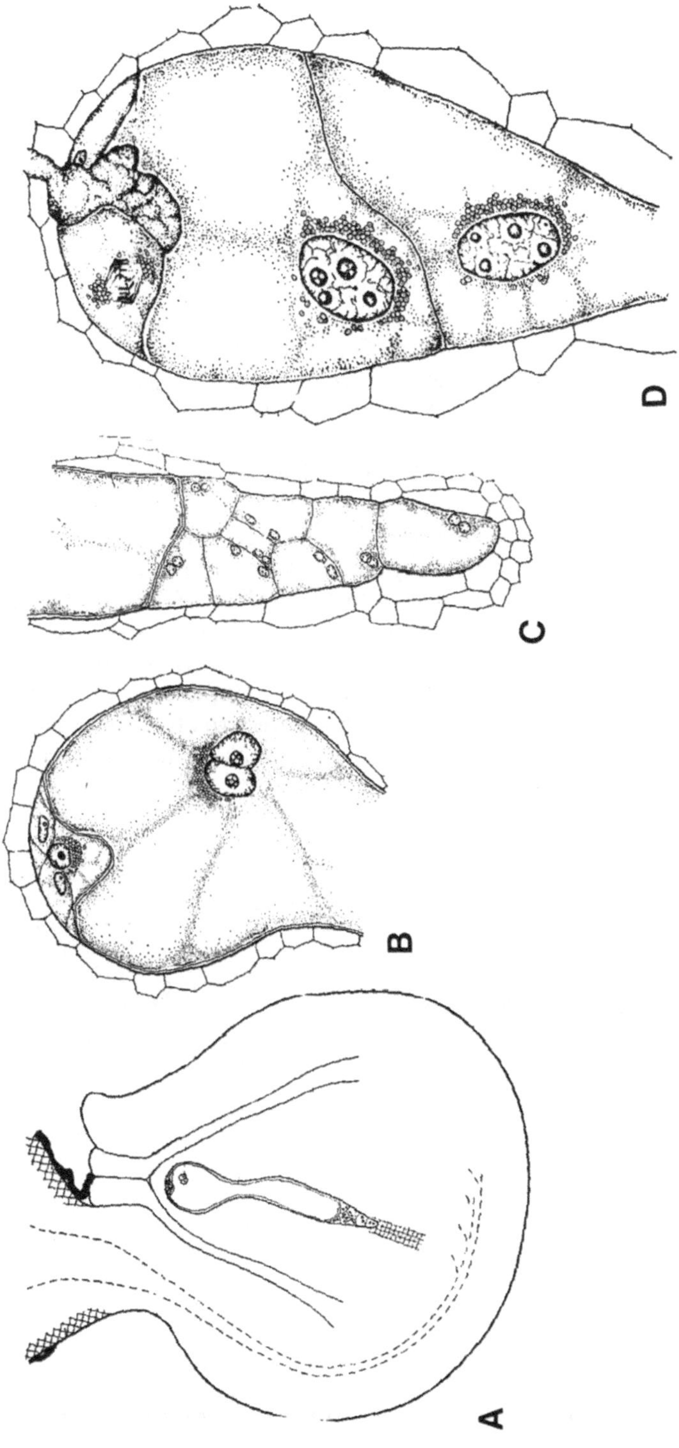

Fig. 19.7.1A–D. *Nelumbo nucifera.* **A** Longisection of ovule at pollination. **B, C** Micropylar (**B**) and chalazal (**C**) part of embryo sac, the antipodals are binucleate. **D** Upper part of embryo sac at two-celled endosperm, the zygote nucleus is in division, note the synergid and remnants of pollen tube. (After Batygina et al. 1983)

there is no multiplication of antipodals in this species. This feature needs further investigation.

The endosperm is of the Cellular type in *Castalia ampla*, *C. pubescens*, *Nelumbo nucifera* (Fig. 19.7.1D), *Nuphar luteuma* (see Schnarf 1929), *Ondinea purpurea* (Schneider and Ford 1978) and *Victoria cruziana* (P. Khanna 1967), and Helobial type in *Nymphaea stellata* (P. Khanna 1967). A tube-like structure develops from the primary chalazal chamber in *Castalia* and *Nymphaea* (M.T. Cook 1906).

In *Euryale* and *Nelumbo* (P. Khanna 1965a), the endosperm is Nuclear. Wall formation commences from the micropylar end.

The embryogeny corresponds to the Asterad type in *Castalia odorata*, *Euryale ferox* and *Nymphaea advena*. S.C. Gupta and Ahuja (1967) report that in *Nelumbo* two distinct meristematic zones representing cotyledonary initials differentiate in the apical region of the proembryo, on either side of the plumule. The suspensor is absent. The cotyledons show meristematic activity at the tips, become massive, and usually one of them grows rapidly. In mature seed it covers the arrested cotyledon at the distal end.

The embryo is large in *Nelumbo* (Fig. 19.7.2A–M), as compared to other members of the family. The growth of plumule is precocious, and in mature seed the shoot apex develops four leaves with primordium of the fifth. The first two leaves do not bear sheaths, but they are present in the latter two leaves and the primordium of the fifth. The growth of the radicle is suppressed, and adventitious roots develop at the nodal region (S.C. Gupta and Ahuja 1967).

The seed is exotestal with abundant starch in perisperm. It is exalbuminous in *Nelumbo* (Fig. 19.7.3) and arillate in *Euryale*, *Nymphaea* and *Victoria*. The

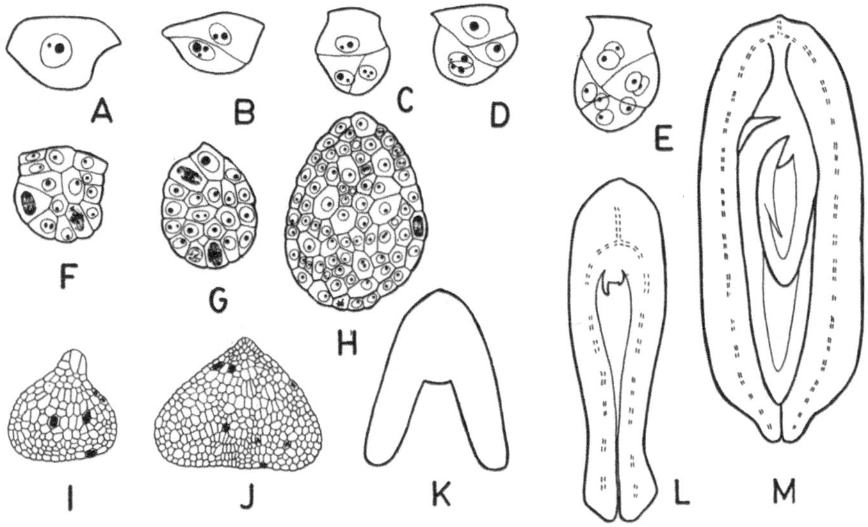

Fig. 19.7.2A–M. *Nelumbo nucifera.* Development of a dicotyledonous embryo. (After Batygina et al. 1983)

testa alone forms the seed-coat. The outer epidermis comprises thick-walled, lignified, columnar cells. The mesophyll and the inner epidermis are unspecialized.

The seed of *Ondinea purpurea* (Schneider and Ford 1978) is small, hard, brown, arillate and ellipsoidal with slightly protruding and flattened micropyle-hilum end. The surface is marked by longitudinal parallel ridges, converging at the micropylar and chalazal end (Fig. 19.7.4A). The raphe is a conspicuous ridge (Fig. 19.7.4B). The sclereid coat is covered by a thin cuticle and each cell has highly thickened radial and outer tangential walls. At the micropylar region the cells of tegmen differentiate into macrosclereids to form an operculum (Fig. 19.7.4C) which helps in rupturing the testa during germination. The endosperm is one or two layers thick, but the rest of the seed is filled with perisperm gorged with starch grains. The operculate seed has a minute embryo, and the plumular leaves are enclosed by two cotyledons (Fig. 19.7.4C). The embryo bears two large separate cotyledons, concave to conform to the shape of epicotyl. The first and second plumular leaves lie decussate to the cotyledons (Fig. 19.7.5A–C).

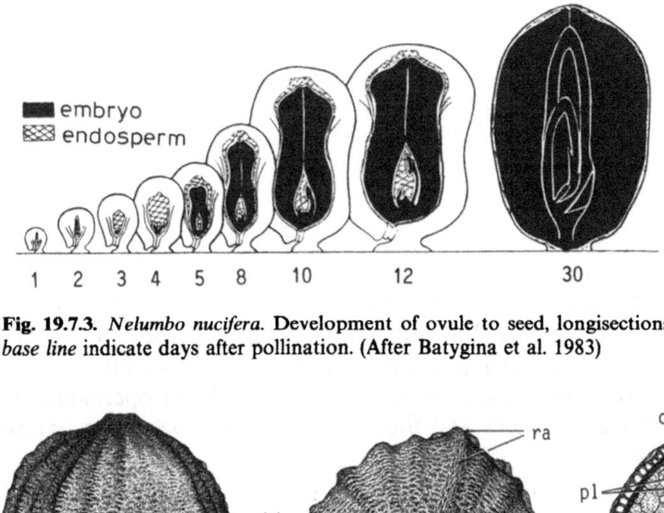

Fig. 19.7.3. *Nelumbo nucifera.* Development of ovule to seed, longisections. The *numbers at the base line* indicate days after pollination. (After Batygina et al. 1983)

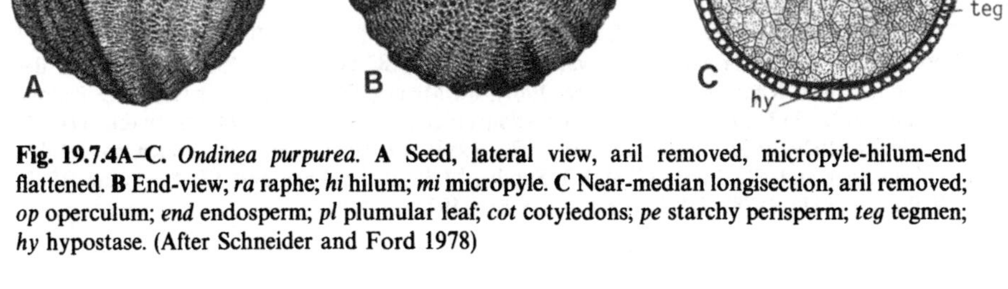

Fig. 19.7.4A–C. *Ondinea purpurea.* **A** Seed, lateral view, aril removed, micropyle-hilum-end flattened. **B** End-view; *ra* raphe; *hi* hilum; *mi* micropyle. **C** Near-median longisection, aril removed; *op* operculum; *end* endosperm; *pl* plumular leaf; *cot* cotyledons; *pe* starchy perisperm; *teg* tegmen; *hy* hypostase. (After Schneider and Ford 1978)

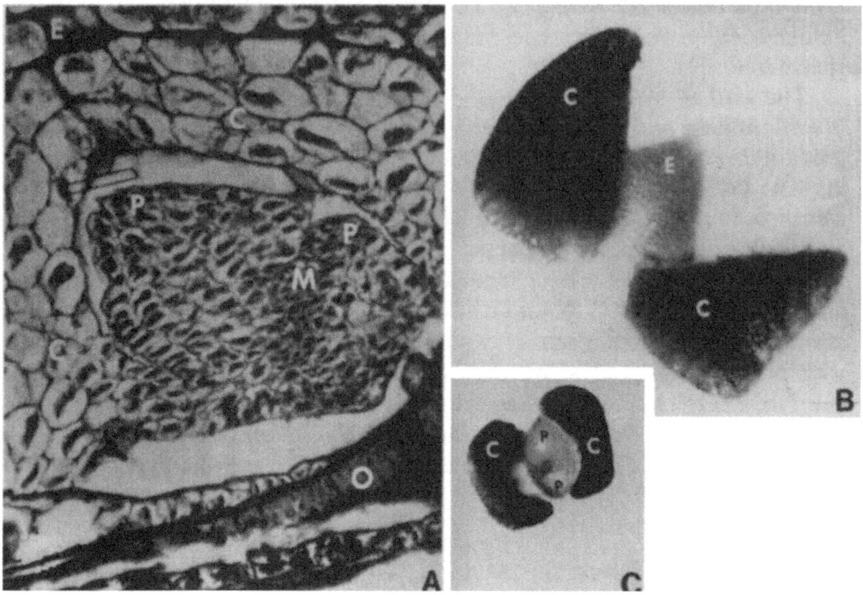

Fig. 19.7.5A–C. *Ondinea purpurea.* **A** Near-median longisection through embryo; *E* epicotyl; *C* cotyledons; *P* plumular leaves; *M* apical meristem; *O* operculum. **B** Dissected mature embryo with two separate cotyledons (*C*) pulled away from epicotyl (*E*). **C** Excised embryo (wholemount), distal view to show concave cotyledons (*C*) which conform to the shape of epicotyl; *P* first and second plumular leaves decussate to cotyledons. (After Schneider and Ford 1978)

The exotestal seed of *Barclaya longifolia* (Schneider 1978) is exarillate and bears large radiating spines except in the micropylar region (spiny testa). The spine is covered with cuticular papillae. The tip of the spine bends into a hook-like structure and becomes a means of dispersal. The cells of operculum, in *B. longifolia*, form macrosclereids and those in the chalaza become sclerified forming a hypostase.

Upon germination, the cotyledons always remain in the seed forcing the operculum. The operculum ruptures the surrounding testa and the hypocotyl-radicle axis and epicotyl emerge from the seed (Conrad 1905). The cotyledons provide reserve food from the perisperm to the developing sporophyte.

TB Batygina, GE Kolesova and VE Vasiljeva (1983), JBJ Chifflot (1902), HS Conrad (1905), MT Cook (1906), SC Gupta and R Ahluwalia (1977, 1979), SC Gupta and R Ahuja (1967), C den Hartog (1970), P Khanna (1964a, 1965a, 1967), KI Meier (1960), PKK Nair (1970), D Padmanabhan (1970a), D Padmanabhan and MV Ramji (1966), MV Ramji and D Padmanabhan (1965), K Schnarf (1929, 1931), EL Schneider (1978, 1983), EL Schneider and J Buchanan (1980), EL Schneider and EG Ford (1978), KT Sundri, M Radhakrishnan and LL Narayana (1982), JW Walker (1976, 1978), R Wheeler and KA Lye (1975)

19.8 Ceratophyllaceae

The embryology has not been studied in detail. The innermost layer of anther wall develops into amoeboid tapetum. Due to the absence of endothecial thickenings, the dehiscence of anther is irregular. The microspore mother cells undergo meiotic divisions, and cytokinesis is successive (Sastri 1955c).

The pollen grains are two-celled at the dispersal stage, in *Ceratophyllum demersum* (Sastri 1955c).

The ovule is anatropous, unitegmic, and crassinucellar. The archesporial cell cuts off a parietal cell. The parietal tissue is well developed. The megaspore mother cell undergoes meiotic divisions giving rise to a linear or T-shaped tetrad. The chalazal megaspore develops into a Polygonum type of embryo sac. The polar nuclei fuse before fertilization. The antipodals are small.

The development of endosperm is of the Cellular type. The division of the primary endosperm nucleus results in the formation of a micropylar and chalazal chamber. The micropylar chamber undergoes repeated divisions and comprises small endosperm cells. The chalazal chamber divides only once.

The embryogeny conforms to the Asterad type-Penaea variation. The mature embryo is well developed, and green with a conspicuous plumule.

The exalbuminous seed has a seed-coat of one to two layers of thin-walled cells. The endocarp gives protection.

RLN Sastri (1955c), K Schnarf (1929, 1931), E Strasburger (1902), K Subramanyam (1962)

19.9 Taxonomic Considerations

Bentham and Hooker (1965a) place Ranunculaceae, Berberidaceae, Menispermaceae and Nymphaeaceae under Ranales; Lardizabalaceae are treated as a tribe under Berberidaceae. Hutchinson's Ranales (1973) comprises Paeoniaceae, Helleboraceae, Ranunculaceae, Nymphaeaceae, Podophyllaceae, Ceratophyllaceae and Cabombaceae. Sargentodoxaceae, Lardizabalaceae, Menispermaceae, Nandinaceae, Circaeasteraceae and Berberidaceae are included in Berberidales.

Paeonia, once included in Ranunculaceae, has been raised to an independent family, Paeoniaceae (see Y.D. Tiagi 1970b). Embryological features (Sect. 19.2) also support its independent status (Sect. 22.3). Hutchinson (1973) raised the tribe Helleboraceae to family rank. Vijayaraghavan (1970a) investigated *Helleborus foetidus* and other multiovular genera. He is not in favour of the separation of Helleboraceae from Ranunculaceae on embryological grounds.

Ranunculaceae. Ranunculaceae show a wide range of variation in floral structure, as well as embryology. Variations occur in the type of anther tapetum, number of integuments, extent of development of nucellus, types of embryo sac, and embryogeny. The anther tapetum is secretory or amoeboid;

the tricolporate pollen grains are two- or three-celled. The anatropous ovule may be bitegmic, unitegmic or undifferentiated (ategmic), and crassi- or tenuinucellar. The development of embryo sac is mono-, bi- or tetrasporic. Even in one and the same species, *Anemone vitifolia* (Bhandari 1965), all the three types occur. The antipodals are exceptionally large, and persistent with polyploid nuclei. The ploidy level in some species is very high. The embryogeny is either of the Onagrad or Solanad type, or irregular. The albuminous, exarillate seed has a straight, minute embryo.

Glaucidium. A monotypic genus, *Glaucidium*, is usually included in Ranunculaceae (Engler and Gilg 1924, Melchior 1964, Cronquist 1981). Melchior places it in the subfamily Hydrastoideae of Ranunculaceae. Hutchinson (1973) includes *Glaucidium* in Helleboraceae – a family separated from Ranunculaceae. Langlet (1928) and Miyaji (1927) suggest that this genus should be assigned to Berberidaceae. The relationship with *Paeonia* has also been pointed out (see Tobe 1981). Tamura (1972) emphasizes that *Glaucidium* should be raised to a unigeneric family. R. Dahlgren (1975b) and Takhtajan (1980) accept the family rank of Glaucidiaceae.

Several workers have pointed out that on the basis of morphology, anatomy, cytotaxonomy and chemotaxonomy, *Glaucidium* should be removed from Ranunculaceae (see Tobe 1981).

We do not have enough information on the embryology of Hydrastidaceae, and the family rank of Paeoniaceae is already well-established. The development of embryo in *Paeonia* is so unusual (Sect. 22.3) and different from that of *Glaucidium* that any relationship with *Paeonia* is out of question.

Tobe (1981) has studied the embryology of *G. palmatum*. The embryology of *Glaucidium*, Ranunculaceae and Berberidaceae is compared in Table 19.9.1 to show the similarities and differences between them. Tobe (1981) points out that the embryology of *Glaucidium* is so distinctive that it confirms the family rank of Glaucidiaceae. The similarities between *Glaucidium* and Ranunculaceae are far more significant than the differences. Therefore, we do not agree with Tobe's conclusions, and wish to emphasize that *Glaucidium* is correctly placed in Ranunculaceae.

Circaeaster. The genus *Circaeaster* (Melchior 1964) is included in Ranunculaceae. However, embryologically it shows significant differences. In *Circaeaster* the ovule is orthotropous and unitegmic. The embryo sac is monosporic, Polygonum type. Fertilization is chalazo-, or mesogamous, and endosperm is of the Cellular type. The mature embryo has only primordial cotyledons and differentiation occurs during germination. The seed-coat is crushed and pericarp covers the endosperm.

In *Eranthis hiemalis, Ranunculus ficaria* and some other taxa of Ranunculaceae the embryo lacks organs at the shedding of seeds. The growth is intraseminal (Natesh and Rau 1984).

According to Hutchinson (1973), the placement of *Circaeaster* with *Kingdonia* in Ranunculaceae is not justified. He considers it as a very reduced

Table 19.9.1. Embryology of *Glaucidium*, Ranunculaceae and Berberidaceae (Adapted after Tobe 1981)

	Glaucidium	Ranunculaceae	Berberidaceae
Ovule	Numerous, anatropous, bitegmic	Numerous, anatropous, bitegmic, sometimes unitegmic	Numerous, anatropous, hemianatropous in *Podophyllum* and *Ranzania*, bitegmic
Nucellus	Nucellar cap 15–20 layers thick, pseudo-crassinucellar	2 or 3 layers, tenui-, crassi- or pseudocrassinucellar	6 to 7 layers, pseudo- or crassinucellar
Integument	Outer integument vascularized up to the tip, inner integument of dermal origin, outer subdermal, inner integument 3–5 layers thick, outer 7–10 layers	Unvascularized, inner integument of dermal origin, outer subdermal, inner integument 2–4 layers thick, outer 4–6 layers	Unvascularized, inner integument 2–3 layers thick, outer 4–8 layers
Micropyle	Formed by inner integument	Inner integument	Both integuments
Archesporium	Multicelled, 3–4 cells function, parietal cell not formed	Uni- or multicelled, only one cell functions, parietal cell may or may not be formed	Multicelled, only one cell functions, parietal cell may or may not be formed
Embryo sac	Polygonum type, one, rarely up to 3 embryo sacs	Polygonum type, only one	Polygonum type, Peperomia type in *Caulophyllum*; only one
Antipodal cells	Small, ephemeral or persistent	Large, commonly multinucleate, persistent	Large with hypertrophied nuclei, persistent
Endosperm	Nuclear type	Nuclear	Nuclear
Embryo	Unclassified, comparable to Onagrad	Onagrad or Solanad	Onagrad

relative of Podophyllaceae or Berberidaceae. There is substantial embryological evidence to raise *Circaeaster* to a family rank, Circaeasteraceae, near Ranunculaceae and Berberidaceae.

Berberidaceae. Berberidaceae show close alliance with Ranunculaceae (Table 19.9.1). Both have secretory anther tapetum, tricolporate two-celled pollen grains, anatropous, bitegmic and crassinucellar ovule, mono- or tetrasporic embryo sac with conspicuous antipodals, Nuclear endosperm, and Onagrad and Solanad type of embryogeny. The seed structure is also comparable to some extent. The similarities between the two are very pronounced, and B.M. Johri (1935c) states that on the basis of embryology Berberidaceae can be included in Ranunculaceae. However, there are certain features, such as occasional occurrence of successive cytokinesis in the pollen mother cells, presence of integumentary vascular bundle, massive nucellus, absence of bisporic embryo sac, and well-developed embryo with a long radicle, which separate it from Ranunculaceae.

Nandina. *Nandina* has been traditionally classified within the Berberidaceae (Cronquist 1981). However, *Nandina* shows aberrant characters in vegetative structure, floral anatomy and cytology which justify its segregation from Berberidaceae and its raising to family rank, Nandinaceae (see Hutchinson 1973, Metcalfe and Chalk 1950, Nowicke and Skvarla 1981, Shen 1954). Embryology does not support such a segregation. In *Nandina domestica* (Ehdaie and Russell 1984), the hypostase is well developed. The micropyle is zigzag. In *Berberis* and *Mahonia* the synergids persist and are haustorial (Mauritzon 1936a), in *B. nepalensis* ephemeral (B.M. Johri 1935c). In *Nandina* the synergids degenerate about the time of fertilization. *Nandina* does not differ significantly from *Berberis* and *Mahonia*, and there is no justification to raise it to an independent family, Nandinaceae.

Lardizabalaceae. Lardizabalaceae have been treated as a tribe of Berberidaceae by Bentham and Hooker (1965a). They can be distinguished from Berberidaceae on the basis of ana- or orthotropous ovule with a nucellar cap. The micropyle is formed by the inner integument alone. The archesporium is single-celled, parietal tissue is well developed, and the antipodals are ephemeral. Only the monosporic type of embryo sac development occurs. Lardizabalaceae alone have Cellular type of endosperm in the suborder Ranunculineae. The family deserves an independent status (Table 19.9.2).

Menispermaceae. Menispermaceae are a multipalynous family with hemianatropous to amphitropous, uni- or bitegmic, crassinucellar ovule. Their embryological features, such as secretory anther tapetum, simultaneous divisions in pollen mother cells, Polygonum type of embryo sac, Nuclear endosperm, and Onagrad type of embryogeny, also occur in Ranunculaceae. The ruminate nature of endosperm is their special feature.

On the whole, the four families Ranunculaceae, Berberidaceae, Lardizabalaceae and Menispermaceae (Table 19.9.2) of the suborder Ranunculineae

Table 19.9.2. Embryology of Berberidaceae, Lardizabalaceae and Menispermaceae (Adapted after Sastri 1969a)

	Berberidaceae	Lardizabalaceae	Menispermaceae
Anther wall	5- or 6-layered	5–6-layered	4–5-layered
Tapetum	Secretory, cells 4–8-nucleate	Secretory, cells 2–4-nucleate; amoeboid in *Stauntonia hexaphylla*	Secretory, cells 1- or 2-nucleate
Cytokinesis	Simultaneous	Simultaneous	Simultaneous
Pollen	2-celled, 3 colpate	2-celled, 3-colporate	2-celled, 3-colpate
Ovule	Anatropous, bitegmic, crassinucellate	Ortho-, campylo- or anatropous, bitegmic, crassinucellate	Hemiana-, amphi- or campylotropous, uni- or bitegmic, crassinucellate
Parietal tissue	Massive, develops from nucellar epidermis	Well developed	Well developed
Archesporium	Uni- or multicellular, one cell functions	Uni- or multicellular, one cell functions	Uni- or multicellular, one cell functions
Megaspore tetrad	Linear	T-shaped or linear	T-shaped or linear
Embryo sac	Polygonum type	Polygonum	Polygonum
Polar nuclei	Fuse before fertilization	Fuse before fertilization	Fuse before fertilization
Antipodal cells	3, large, persistent	3, small, ephemeral, sometimes persistent	3, small, ephemeral
Endosperm	Nuclear type	Cellular	Nuclear, ruminate
Embryo	Onagrad type	Asterad	Onagrad
Seed	Exotestal, albuminous, rarely arillate	Exotestal, albuminous, vestigial aril in *Akebia*	Seed-coat generally crushed, or endotegmic, exarillate, albuminous

exhibit close affinities. Cronquist (1968, 1981) and R. Dahlgren (1975a, 1980b, 1981) include them in the order Ranunculales.

Nymphaeaceae. Nymphaeaceae and Ceratophyllaceae are vessel-less aquatics, and constitute the suborder Nymphaeineae. The family Nymphaeaceae comprises three subfamilies: Nymphaeoideae, Cabomboideae and Nelumbonoideae (Melchior 1964).

Cabomba. *Cabomba* differs from other members of the family in the (1) absence of endothecial thickenings in anther wall, (2) two-nucleate tapetal cells, (3) successive cytokinesis in pollen mother cells, (4) two-celled pollen grains, and (5) nucellar growth due to a basal meristem. Its inclusion in the subfamily Cabomboideae is justified. R. Dahlgren (1977a, 1980b) treats Cabomboideae as an independent family, Cabombaceae, in the superorder Nymphaeiflorae.

Nelumbo. The systematic position of *Nelumbo* is controversial. Assuming that *Nelumbo* shows monocotyly, Cook (1906) and Seaton (1908) place it in Monocotyledons, while Arber (1920), Lawrence (1951) and Hutchinson (1973) retain it in Dicotyledons (reticulate venation of leaf; tricolpate pollen).

Li (1955) and Takhtajan (1959, 1980) propose the erection of the order Nelumbonales, which has been supported by Snigirevskaya (1964) and P. Khanna (1965a). Takhtajan emphasizes features such as tricolpate pollen grains, morphology of the gynoecium, absence of suspensor, absence of idioblasts, and the basic chromosome number. P. Khanna (1965a) draws attention to additional features: persistence of middle layers of the anther wall, secondary multiplication of antipodals and their mutlinucleate condition (proliferation of antipodals is refuted by S.C. Gupta and Ahluwalia née Ahuja – 1979), Nuclear endosperm, exalbuminous seed with a large embryo, and absence of suspensor, perisperm, and hypostase.

On the basis of phytochemical data (absence of ellagic acid and presence of benzylisoquinoline alkaloids), and tricolpate pollen grains, R. Dahlgren (1975a, 1977a, 1980b) erects the order Nelumbonales with the lone family Nelumbonaceae, in superorder Magnoliiflorae.

S.C. Gupta and Ahuja (1967) and S.C. Gupta and Ahluwalia née Ahuja (1977) emphasize the dicotyledonous nature of the embryo in *Nelumbo* (see also Batygina et al. 1983). They also refer to the occurrence of a middle integument, as in *Cananga odorata* (Periasamy and Swamy 1961), and to the open carpel with papillate ovarian canal comparable to that in *Degeneria* (Swamy 1949a). The floral structure strongly supports Ranalian affinity. The ovarian canal, stigmatic hairs, obturator, middle integument, and ovarian hump distinguish the genus *Nelumbo* (S.C. Gupta and Ahluwalia née Ahuja 1977).

The precocious development of embryo forming the shoot apex with four to five leaves and sheaths, two massive cotyledons, and suppression of radicle in mature seed are noteworthy (S.C. Gupta and Ahuja 1977). These investigators favour the position of *Nelumbo* in Nymphaeaceae, in the Ranalian complex, and refute its segregation until other members are fully investigated.

Ondinea. Structural similarities between *Ondinea* and *Nymphaea* together with similar pollination and seed dispersal syndromes lend support to den Hartog's (1970) placement of *Ondinea* in Nymphaeaceae. Further support for the inclusion of *Ondinea* in the Nymphaeaceae s. str. comes from comparative studies of stem vascularization (see Schneider 1983).

Huber (1977), in his proposed system of Monocotyledons, erects a superorder Nymphaeiflorae with Nymphaeales – an order with some dicotyledonous characters – and transfers Nymphaeaceae from Dicotyledons to Monocotyledons.

Cronquist (1968, 1981) and Takhtajan (1969, 1980) consider the Monocotyledons to be derived from Dicotyledons. The aquatic or semiaquatic plants like proto-Nymphaeales must be a transitional series to proto-Alismatales. R. Dahlgren et al. (1985) emphasize that Nymphaeales do show some features of alismatiflorous monocotyledons as laticifers, stipules, extrose anthers, laminar-dispersed placentation and Helobial type of endosperm (rare in Nymphaeaceae). *Cabomba* has liliiflorean floral diagram. The presence of two cotyledons, S-type sieve-element plastids, and perispermous seed do suggest that Nymphaeales are not the starting point of Alismatales. The seeds of Nymphaeales have scanty endosperm and a well-developed perisperm as in Piperales. The absence of perisperm and endosperm in the seeds of Alismatiflorae distinguish them from Nymphaeales.

The reports that Nymphaeaceae s.l. are monocotyledonous, or ancestors to monocotyledons, based on the presence of a single bilobed cotyledon, have also been debated. The studies of Schneider (1983) and Batygina et al. (1983) on the development of embryo have clearly demonstrated that *the embryo has two distinct cotyledons, and not a single bilobed cotyledon.*

Therefore, the assignment of Nymphaeaceae to moncots is not justified at all. Embryological evidences put forward by S.C. Gupta and Ahluwalia née Ahuja (1977) and Batygina et al. (1983) support the dicotyledonous affinities of Nymphaeaceae. Melchior's inclusion of this family in the Ranalian group is in order.

Ceratophyllaceae. Ceratophyllaceae are closely allied to Nymphaeaceae. The similarities are: (1) amoeboid anther tapetum, (2) successive cytokinesis in pollen mother cells, (3) anatropous, crassinucellate ovule with single-celled archesporium, (4) Polygonum type of embryo sac, (5) Cellular type of endosperm, and (6) Asterad type of embryogeny with a straight mature embryo (Sastri 1955c, Subramanyam 1962).

According to Sundri et al. (1982), the Ceratophyllaceae are closely related to *Cabomba* in abaxial microsporangia (Eames 1961), absence of benzylisoquinoline alkaloids, presence of tannins, S-type plastids in sieve tubes, voluminous starchy perisperm (R. Dahlgren 1980b, Gibbs 1974), and the same number of chromosomes, i.e. $2n = 24$ (Darlington and Wylie 1955). Ceratophyllaceae are, therefore, related to Nymphaeaceae in general, and *Cabomba* in particular, rather than to *Nelumbo*.

The presence of unitegmic ovules, exalbuminous seed, and thin-walled testa segregate it from Nymphaeaceae. The former two features indicate its more advanced character. Ceratophyllaceae have been derived from a nymphaeaceous ancestry (Cronquist 1968, 1981, R. Dahlgren 1981).

20 Piperales

The order Piperales comprises Saururaceae, Piperaceae, Chloranthaceae and Lactoridaceae.

20.1 General Features

The general features of the order are: (1) secretory anther tapetum, (2) simultaneous cytokinesis of pollen mother cells, (3) monosulcate pollen grains, two-celled at the dispersal stage, (4) orthotropous, bitegmic, crassinucellate ovule; unitegmic in *Peperomia*, tenuinucellate in *Houttuynia*, (5) single-celled archesporium with parietal tissue, (6) monosporic type of embryo sac; tetrasporic Fritillaria type in Piperaceae, and Peperomia type in *Peperomia*, (7) a large number of antipodals in Piperaceae and *Chloranthus*; ephemeral in *Sarcandra* and Saururaceae, (8) Cellular endosperm; Nuclear type in *Piperaceae*, (9) embryogeny of Piperad type in Piperaceae, and Onagrad in Chloranthaceae (*Sarcandra*), and (10) tegmic seed with a minute embryo and perisperm; perisperm is totally absent in Chloranthaceae.

20.2 Saururaceae

The anther wall comprises the epidermis, fibrous endothecium, single middle layer, and secretory tapetum with binucleate cells. The microspore mother cells undergo meiotic divisions, and the tetrads are tetrahedral, isobilateral and decussate. Cytokinesis is simultaneous.

The pollen grains are monosulcate with a smooth exine. They are two-celled at the dispersal stage.

Houttuynia cordata (Takahashi 1986) is a monotypic genus represented by only one species. Failure of cytokinesis after meiosis results in the formation of coenocytic microspores which vary widely in shape and size. The tapetal cells do not show any demonstrable abnormality, but the microspores degenerate after callose dissolution. Later, the loculus contains numerous coenocytic microspores which consist only of ektexine. This species is completely, or nearly completely, male sterile.

The ovule is orthotropous, bitegmic and crassinucellate; tenuinucellate in *Houttuynia cordata* (Y.S. Murty 1960). The outer integument is two-layered and gets crushed in mature seed. The inner integument is three-layered in *Houttuynia* and *Saururus* (Y.S. Murty 1960), and four-layered in *Anemopsis* (Quibell 1941). A zigzag micropyle, formed by both the integuments, occurs in

Houttuynia. The single hypodermal archesporial cell functions directly in *Houttuynia* (Y.S. Murty 1960); a parietal cell is produced in *Anemopsis* and *Saururus* and the parietal tissue is two layers thick. The megaspore mother cell undergoes meiotic divisions. The chalazal megaspore of a linear or T-shaped tetrad, or a triad, develops into a Polygonum type of embryo sac. The antipodals are ephemeral, and the polar nuclei fuse before fertilization (Fig. 20.2.1A–D).

The development of endosperm is of the Cellular type (Fig. 20.2.1E–G) in *Saururus* (Johnson 1900). The chalazal chamber develops into a haustorium, and the micropylar produces the endosperm proper. Yoshida (1961) reports Helobial type of development in *S. loureiri*. The development of endosperm requires further study.

The embryogeny is of the Asterad type (Fig. 20.2.1H-Q).

The endospermic seed (Fig. 20.2.1G) is small, and has abundant perisperm (Y.S. Murty 1959a). The testa gets crushed. The tegmen forms the seed-coat. Its outer and inner epidermis becomes thick-walled.

P Crété (1957c), DS Johnson (1900), YS Murty (1959a, 1960), M Takahashi (1986), CH Quibell (1941), MVS Raju (1961), O Yoshida (1961)

20.3 Piperaceae

Of the four microsporangia in the anther in *Piper* (Kanta 1962), the development of one, two or three sporangia may be suppressed. Even when both thecae appear normal externally, one may be completely sterile.

The anther wall comprises persistent epidermis, fibrous endothecium, one or two ephemeral middle layers, and secretory tapetum of binucleate cells. The microspore mother cells undergo meiotic divisions; cytokinesis is simultaneous.

The tetrads are isobilateral, decussate and tetrahedral, and usually the microspores separate. Occasionally, in *Peperomia pellucida* and *P. sandersii* (Y.S. Murty 1959a), for some time the four microspores remain attached together in a linear row, or in some other configuration by small connections. After separation such microspores show one or two small protuberances. [This arrangement recalls the condition in *Halophila* (Kausik and Rao 1942)]. Pollen grains of about double the normal size have been observed occasionally in *P. fraseri*.

Pollen grains are monocolpate, and shed at the two-celled stage. In *Peperomia* they are non-sulcate, and show warty protuberances.

The ovule is orthotropous and crassinucellate. It is bitegmic in *Heckeria* (see Schnarf 1931) and *Piper* (Kanta 1962), and unitegmic in *Peperomia* (Y.S. Murty 1959a). The micropyle is produced by both the integuments in *Heckeria*, but only the inner integument in *Piper*.

The archesporial cell cuts off a parietal cell, and four- to five-layered parietal tissue is formed in *Heckeria* (P. Maheshwari and Gangulee 1942). The megaspore mother cell undergoes meiotic divisions and a coenomegaspore

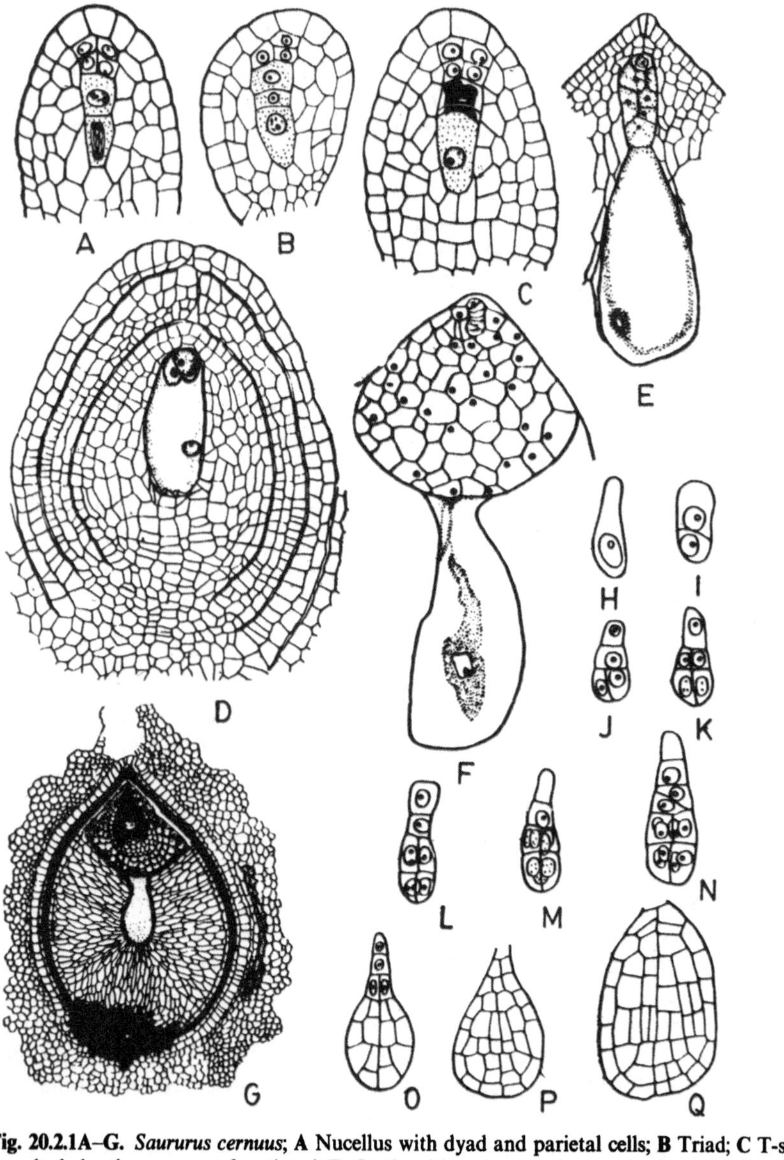

Fig. 20.2.1A–G. *Saururus cernuus*; **A** Nucellus with dyad and parietal cells; **B** Triad; **C** T-shaped tetrad, chalazal megaspore functional; **D** Ovule with mature embryo sac, micropyle zigzag; **E, F** Cellular endosperm with chalazal haustorium, zygote in **E** and proembryo in **F**; **G** Longisection of mature seed, embryo (black) surrounded by endosperm (remnants of the chalazal haustorium persist). Note the perisperm and hypostase. **H–Q** *Houttuynia cordata*, embryogenesis. (**A–C**, **F** After M.V.S. Raju 1961, **D**, **E**, **G** after Johnson 1900, **H–Q** after Crété 1957c)

with four nuclei is produced. Cytokinesis does not occur, and the arrangement of nuclei is 1 + 3.

Further development is of the Fritillaria type in *Heckeria* (P. Maheshwari and Gangulee 1942, Swamy 1945b), *Piper* (Swamy 1944b, Kanta 1962), *P. medium* var. *ceanothifolium* (Maugini 1953), and *P. gaudichandianum* (Y.S. Murty 1959a). In *Piper nigrum* (Kanta 1962), the triploid antipodal cells multiply to form a massive tissue of a hundred or more cells. In *Peperomia* (Fisher 1914), the megaspore mother cell undergoes meiotic divisions and produces a four-nucleate coenomegaspore, the arrangement of the nuclei is tetrahedral. Rudimentary walls appear in between the nuclei, but later they disappear. Further development is of 16-nucleate Peperomia type (Fisher 1914, Johnson 1914, Y.S. Murty 1959b). In several species of *Peperomia* the egg is associated with a single synergid (Fig. 20.3.1A).

A high degree of sterility of ovules, and embryolessness is reported in *Piper nigrum* (Kanta 1961, 1962). It may be due to the failure of fertilization, or excess of deposition of calcium oxalate crystals in embryo sacs.

A forked pollen tube is reported in *Peperomia* (Fagerlind 1939a). One of its branches passes to the egg, the other to the secondary nucleus. This needs confirmation. Figure 20.3.1A–C shows syngamy.

The development of endosperm is of the Cellular type in *Heckeria* and *Peperomia*, (Fig. 20.3.1C–F), and Nuclear type in *Piper*. Later, it becomes cellular.

The embryogeny generally conforms to the Piperad type (Fig. 20.3.1D, E). The division of the zygote is followed by a vertical, sometimes transverse or oblique wall (Kanta 1962). The embryo of *Peperomia* (Johnson 1914) is an undifferentiated globular mass, showing 8–10 cells in *P. hispidula*. The histogens, except for epidermal initials or epidermal cells, are usually unrecognizable in such embryos. The embryo in *Macropiper* (Balfour 1957) shows a high degree of differentiation since not only epidermis and ground tissue are present, but also a provascular cone and cotyledonary lobes. In *Piper nigrum* (Kanta 1961, 1962), older embryos show two well-differentiated cotyledons. The embryo has prominent ridges and furrows and appears hexagonal when seen from the top.

The seed is endotegmic, and contains a small, starchy endosperm, and copious perisperm with starch and oil as storage products (Fig. 20.3.1G). The embryo is microscopic. The testa gets almost crushed. The tegmen is distinct only in the micropylar region; it is crushed into a brown layer around the seed.

In *Peperomia* (Y.S. Murty 1959a), the outer layer of integument is crushed, the inner layer becomes brown and thickened, and its inner wall develops numerous infoldings (few or absent at the base of seed) which change the contour of the outermost layer of nucellus. The remnants of the outer layer and the thick brown inner layer form the protective covering of the mature seed.

In *Piper nigrum* (Kanta 1962), the innermost layer of the inner integument has densely cytoplasmic radially elongate cells. Near the micropyle they give out finger-like projections which meet similar processes, directed outward from the nucellar epidermis so that the two fit tightly against each other.

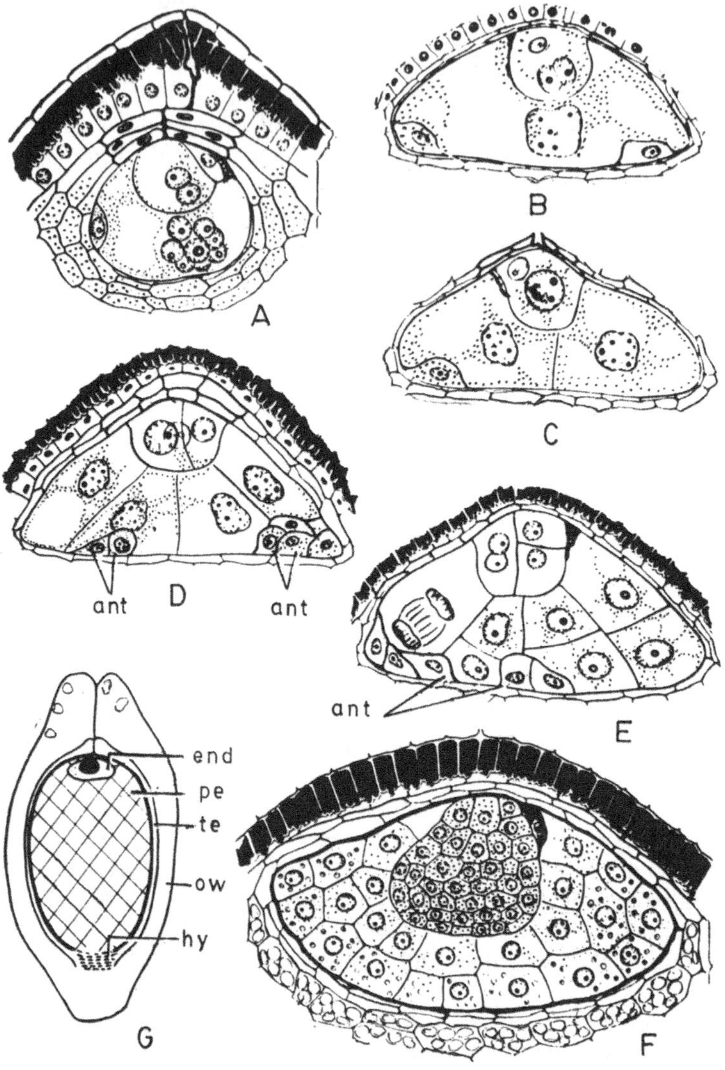

Fig. 20.3.1A. *Peperomia pellucida*, fertilized embryo sac, male and female nuclei in egg cell, one synergid, central cell with a group of polar nuclei, and one lateral antipodal. **B–E** *P. blanda*; **B** Syngamy, primary endosperm nucleus, and two antipodals; **C–E** Progressive stages of development of proembryo and cellular endosperm; *ant* antipodals. **F, G** *P. eburnea*; **F** Immature seed with embryo and endosperm; **G** Mature fruit. Embryo *black*; *end* endosperm; *pe* perisperm; *te* testa; *ow* ovary wall; *hy* hypostase. (See Yakovlev 1981)

During the maturation of seed, the outer integument and outer layers of the inner integument collapse. The inner epidermis of tegmen becomes thick-walled and filled with tannin. A few layers of cells just below the perisperm, at the base of ovule, also become thick-walled and tanniniferous. The epidermis of pericarp and several layers below it become thick-walled. Thus, the embryo and endosperm are well protected.

E Balfour (1957), F Fagerlind (1939a), GC Fisher (1914), DS Johnson (1914), AC Joshi (1944), K Kanta (1961, 1962), SB Kausik and PVK Rao (1942), P Maheshwari and H Gangulee (1942), E Maugini (1953), YS Murty (1959a, b), ZI Nikiticheva (1981), K Schnarf (1931), KS Semple (1974), BGL Swamy (1944b, 1945b)

20.4 Chloranthaceae

The anther wall comprises the epidermis, fibrous endothecium, two or three middle layers, and secretory tapetum of binucleate cells. The epidermal cells contain tannin, and a "spiny" cuticle covers the surface in *Sarcandra irvingbaileyi* (Vijayaraghavan 1964). The microspore mother cells undergo meiotic divisions producing isobilateral and tetrahedral tetrads. Cytokinesis is simultaneous.

The pollen grains are inaperturate, and two-celled at the shedding stage.

The ovule is orthotropous, bitegmic and crassinucellar. The micropyle is formed by the inner integument initially, but it is overgrown by the outer integument. A rudimentary third integument develops in *Sarcandra* (Vijayaraghavan 1964).

The hypodermal archesporial cell cuts off a parietal cell. The parietal tissue is well developed. The megaspore mother cell undergoes meiotic divisions and gives rise to a linear tetrad. The chalazal megaspore develops into a Polygonum type of embryo sac. One of the synergids degenerates before fertilization. The antipodals are small; supernumerary antipodals are formed (by their multiplication) in *Chloranthus japonicus* (Yoshida 1957). They are ephemeral in *Sarcandra* (Yoshida 1957, 1959, 1960). A hypostase differentiates.

Development of the endosperm is of the Cellular type (Fig. 20.4.1A, B). It is copious, persistent, and stores oil.

The embryogeny conforms to the Onagrad type in *Chloranthus* and *Sarcandra* (Fig. 20.4.1C–M). In the ripe seed the embryo is at the globular stage in *Chloranthus* and *Hedyosmum nutans* (Edwards 1920), but dicotyledonous in *Sarcandra* (Fig. 20.4.1N, O).

The seed is albuminous. The outer epidermis and mesophyll of testa comprise thin-walled cells, and scattered sclerotic cells. The inner epidermis comprises radially elongate cells with a fine cellulosic reticulum and crystals on their walls. The outer epidermis of tegmen is thin-walled; the inner is a compact layer of longitudinally elongate cells with firm brown walls. The tegmen may get crushed. In *Hedyosmum* the seed-coat is unspecialized.

J Edwards (1920), PK Endress (1971, 1987a, b), BGL Swamy (1953b, d), MR Vijayaraghavan (1964), O Yoshida (1957, 1959, 1960)

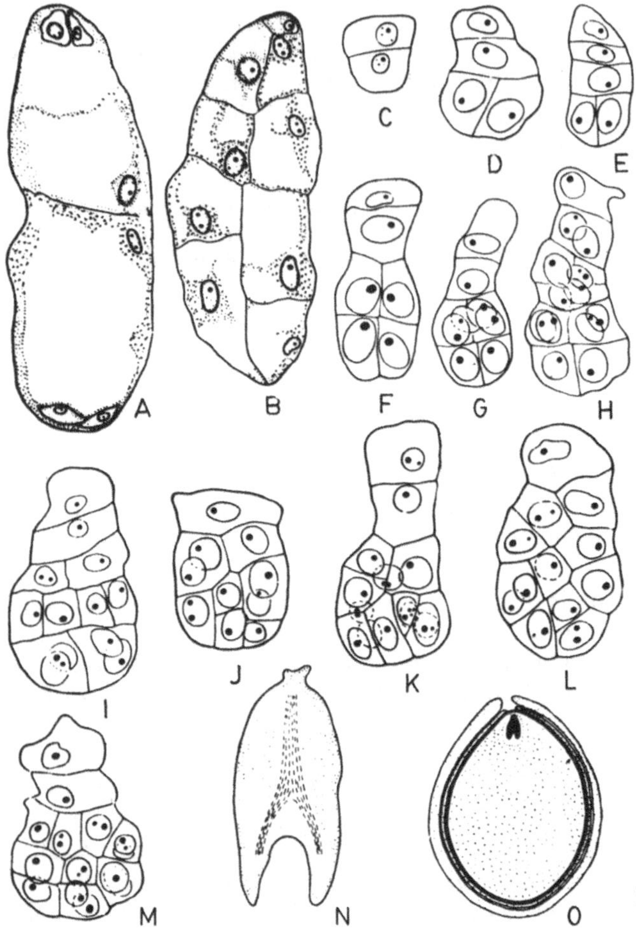

Fig. 20.4.1A–O. *Sarcandra irvingbaileyi.* **A** Embryo sac with zygote, persistent synergid, two-celled endosperm, and two antipodal cells. **B** Zygote and eight-celled endosperm. **C–M** Development of proembryo. **N** Dicotyledonous embryo. **O** Longisection of seed with dicotyledonous embryo and endosperm (diagrammatic). (After Vijayaraghavan 1964)

20.5 Lactoridaceae

The embryological data are very meagre (see Table 18.24.2). The pollen grains remain attached in tetrads. The anatropous bitegmic ovule is tenuinucellar with a cap-like nucellus. The inner integument forms the micropyle. The development of endosperm is of the Nuclear type (Cronquist 1981). The mature seed has copious oily endosperm, and a minute undifferentiated embryo.

F Bouman (1971a), S Carlquist (1964), EJH Corner (1976), A Cronquist (1981), MS Yakovlev (1981)

20.6 Taxonomic Considerations

Bentham and Hooker (1965c) include Piperaceae and Chloranthaceae in Piperales. Engler and Diels (1936) and Hutchinson (1973) include Saururaceae also.

Saururaceae, Piperaceae. Saururaceae and Piperaceae have trilacunar or multilacunar nodes, secretory anther tapetum, simultaneous cytokinesis in pollen mother cells, monosulcate, two-celled pollen grains, orthotropous, bitegmic, crassinucellate ovules, and seed with a minute embryo and abundant perisperm. It appears that the two families are closely allied. However, in Saururaceae the ovule may be tenuinucellate as in *Houttuynia*, both the integuments organize the micropyle, and the archesporium is single-celled. The development of embryo sac is of the Polygonum type with ephemeral antipodals. The Cellular endosperm has a chalazal haustorium. The division of zygote is transverse. In contrast, in Piperaceae the endostome alone forms the micropyle. The female archesporium is multicellular. The development of the embryo sac is of the Fritillaria type, and the number of antipodals is more than 100. The endosperm is of the Nuclear type. The division of zygote is vertical, and the embryogeny is of the Piperad type.

In the family Piperaceae, the position of the genus *Peperomia* is rather debatable. In contrast to the other members of the family, *Peperomia* shows non-sulcate pollen grains with warty protuberances, unitegmic ovules with single-celled archesporium, Peperomia type of embryo sac, Cellular endosperm, and perisperm with starch and oil. These are rather marked differences. Its assignment to a separate family is appropriate.

Chloranthaceae. Chloranthaceae differs from the other two families in having unilacunar nodes (Money et al. 1950). Brongniart (see Swamy 1953b) assigns Chloranthaceae to Santalineae, and Endlicher to Ceratophyllaceae (see Vijayaraghavan 1964). Bessey (1915) places them in Ranales. Vijayaraghavan (1964) investigated the embryology of *Sarcandra irvingbaileyi*.

Chloranthaceae show much more affinity with Saururaceae than Piperaceae in the structure and development of anther, ovule, embryo sac, and endosperm. The division of zygote is also transverse as in Saururaceae. But it can be distinguished due to monosulcate pollen grains with reticulate exine, Cellular type of endosperm without any haustorium, total absence of perisperm, well-developed persistent endosperm, and Onagrad type of embryogeny. Moreover, in Saururaceae and Piperaceae the seeds are tegmic, whereas in Chloranthaceae the seeds have a well-developed seed-coat with exotestal and exotegmic fibres. The inner epidermis of the testa is a crystal layer, and that of the tegmen a compact palisade layer.

Vijayaraghavan (1964) compares the embryological features of Chloranthaceae with those of Santalineae and Ceratophyllaceae. He negates any relationship with both taxa, and points out some affinities with Ranales. The shoot apex of *Sarcandra* is similar to the other vesselless Ranalian genera, especially *Trochodendron* (Ramji 1961). The occurrence of the third rudimentary integument in *Sarcandra* also suggests Ranalian affinity.

The ancestry of Piperales is sought in Magnoliales. The presence of etherial oil cells and uniaperturate pollen shows Magnolian affinities. The position of the families Piperaceae, Chloranthaceae and Saururaceae in the order Piperales, assigned by Hutchinson (1973) and Cronquist (1981) is appropriate. Takhtajan (1980) includes Chloranthaceae in Laurales, and R. Dahlgren (1981) in Magnoliales; the other two families in Piperales.

21 Aristolochiales

The order Aristolochiales comprises three families: Aristolochiaceae, Rafflesiaceae and Hydnoraceae.

21.1 General Features

The order Aristolochiales is characterized by (1) secretory anther tapetum, (2) simultaneous or successive cytokinesis in pollen mother cells, (3) two-celled pollen grains, (4) anatropous, bitegmic, and crassinucellate ovule, (5) single-celled archesporium, massive parietal tissue, (6) Polygonum type of embryo sac, (7) prominent hypostase, (8) Cellular type of endosperm; ruminate in Aristolochiaceae, (9) albuminous seed with a minute embryo, and (10) exo- or endotegmic seeds.

21.2 Aristolochiaceae

The anther wall comprises persistent epidermis, fibrous endothecium, two or three middle layers, and a secretory tapetum of binucleate cells. In *Aristolochia* (B.M. Johri and Bhatnagar 1955a), at places the endothecium and tapetum become biseriate. Übisch granules are present on the inner wall of tapetal cells. The microspore mother cells undergo meiotic divisions. Cytokinesis is simultaneous in *A. europaeum*, *A. sipho* (Täckholm and Söderberg 1918), *A. bracteata* (B.M. Johri and Bhatnagar 1955a) and *Bragantia wallichi* (N.C. Nair and Narayanan 1961), and successive in *Aristolochia clematitis* (Samuelsson 1914), *A. elegans* (B.M. Johri and Bhatnagar 1955a) and *A. fimbriata* (Täckholm and Söderberg 1918). The tetrads are linear, T-shaped, isobilateral, tetrahedral and decussate. Polyspory occurs.

Pollen grains are dispersed at the two-celled stage. They are inaperturate, monosulcate, or three-colpoidate.

The ovule is anatropous, bitegmic, and crassinucellate. It is circinotropous in *Bragantia* (N.C. Nair and Narayanan 1961). The micropyle is formed by endostome. A nucellar cap is produced. A single hypodermal archesporial cell

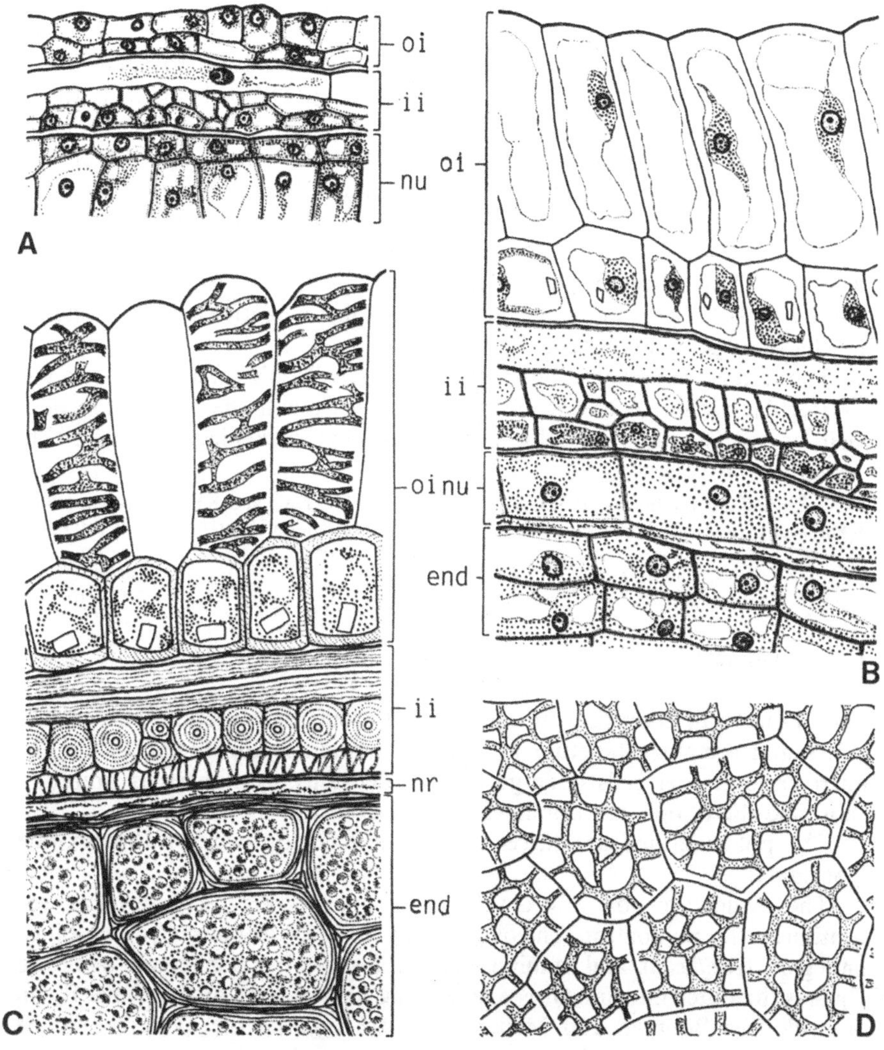

Fig. 21.2.1A–D. *Aristolochia bracteata.* Development of seed-coat, transections. **A** Outer (*oi*) and inner (*ii*) integument, and part of nucellus (*nu*). **B** Later stage, the cells of the exotesta have elongated radially. Cellular endosperm (*end*) is also shown. **C** Reticulate thickenings in exotesta, and thick-walled endotesta with crystals. The exo- and mesotegmen become fibrous, the sclerenchymatous fibres are at right angles to each other. The endotegmen is also fibrous. The endosperm is thick-walled because of deposition of hemicellulose. *nr* nucellus. **D** Funicular cells with reticulate thickenings. (After B.M. Johri and Bhatnagar 1955a)

cuts off a parietal cell. The latter gives rise to a massive parietal tissue. The megaspore mother cell undergoes meiotic divisions, and a linear or T-shaped tetrad is produced. The chalazal megaspore develops into a Polygonum type of embryo sac. A hypostase is prominent in *Bragantia*. The antipodals persist after fertilization and may exhibit haustorial activity. The synergids are ephemeral.

Development of the endosperm is of the Cellular type; ruminations of Annona type occur in *Apama* (Periasamy 1962a). The endosperm is thick-walled, and stores oil and starch.

The embryogeny is probably of the Solanad type. Synergid polyembryony occurs in *A. bracteata* (B.M. Johri and Bhatnagar 1955a).

The development of the seed-coat is represented in Figure 21.2.1A–C. The outer epidermis of testa comprises thick-walled, lignified, radially elongate cells. It is followed by thick-walled cells with crystals. The two outer layers of tegmen consist of crossed lignified fibres. The inner epidermis develops reticulate thickenings, and contains tannin.

The seed is triangular, albuminous with a small embryo, and the testa has an irregular surface. Both the integuments are fused with the raphe for most of their length. The funicular cells develop reticulate thickenings (Fig. 21.2.1D).

BM Johri and SP Bhatnagar (1955a), A Leeman (1927), NC Nair and KR Narayanan (1961), K Periasamy (1962a), G Samuelsson (1914), G Täckholm and E Söderberg (1918), RL Wyatt (1955)

21.3 Rafflesiaceae

The anther wall comprises an epidermis, fibrous endothecium, two middle layers and secretory tapetum in *Cytinus hypocistis* (Guzowska 1964). The microspore mother cells undergo meiotic divisions, cytokinesis is simultaneous, and the tetrads are isobilateral and decussate.

The pollen grains are three- or four-colpate, or two- or three-porate (poroidate).

The ovules are not fully developed in open flowers, and development continues after pollination. The ovule is ana- or orthotropous (*Cytinus*), bi- or unitegmic (*Rafflesia*), and tenuinucellate. *Pilostyles* has a bitegmic ovule. In *Cytinus* the outer integument is a small ring-like swelling at the chalazal end of ovule (see Bouman and Meijer 1987). In *Rafflesia* the ovule should be interpreted as unitegmic because the chalazal swelling (outer integument) is not very prominent (Fig. 21.3.1A–C). The nucellus is single-layered. The archesporial cell functions directly as megaspore mother cell. Development of the embryo sac is of the Polygonum type. The antipodals are well developed.

The development of the endosperm is of the Nuclear type. Later, it becomes cellular and stores fatty food reserve.

The division of zygote and its daughter cells is transverse, and the embryogeny probably conforms to the Caryophyllad or Solanad type (Fig. 21.3.1D–Z). In a mature seed, the unorganized (Fig. 21.3.2) embryo is only four-celled in *Mitrastemon*, and has ten cells in six tiers in *Cytinus* (see Davis 1966). In *Pilostyles*, the embryo consists of eight cells; two cells may be comparable to suspensor, and the embryo proper comprises three tiers of two cells each (see Kuijt 1969).

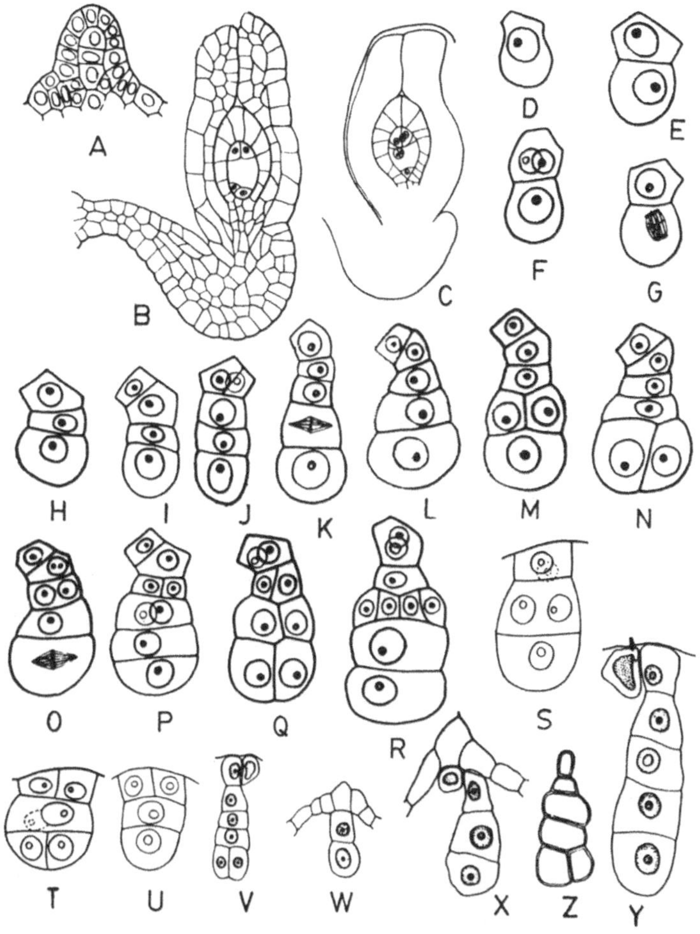

Fig. 21.3.1. **A–C** *Rafflesia patma*, longisection of ovules, mature embryo sac in **C**. **D–R** *Cytinus ruber*. **S–U** *R. patma*, **V** *R. hasseltii*, **W–Z** *Pilostyles ingae*, early embryogeny. (**A–C**, **S–V** After Ernst and Schmid 1913, for **D–R** see Yakovlev 1981, **W–Z** after Endriss 1902)

The minute seed is albuminous with only a few-celled embryo (Fig. 21.3.2). The seed-coat (outer epidermis of tegmen) is sclerotic with pitted cells. In *Rafflesia* the sclerenchymatous outermost layer of tegmen forms the seed-coat. The chalazal swelling also becomes sclerenchymatous. In *Cytinus*, the chalazal swelling is fleshy and may serve as an elaisome. In *Pilostyles*, the seeds have a single compact layer of sclereids surrounding the endosperm. The sclereids have very thick walls with a complex system of pits (see Kuijt 1969, Corner 1976).

The four tribes – Apodantheae, Rafflesieae, Mitrastemoneae, Cytineae – of Rafflesiaceae show striking differences in ovular structure, reflected in the

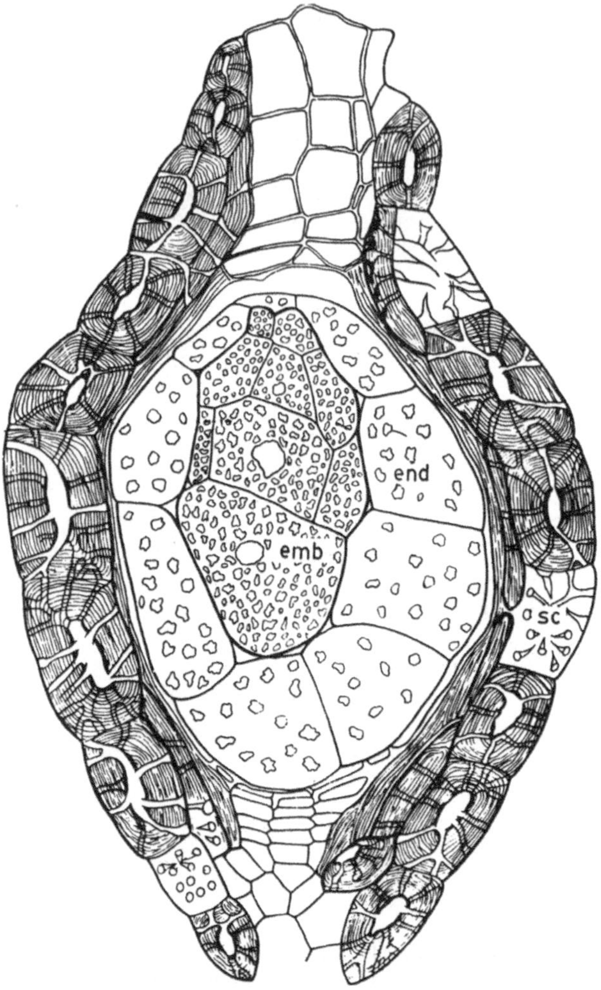

Fig. 21.3.2. *Cytinus ruber*. Longisection of seed with endosperm (*end*) and undifferentiated embryo (*emb*). The testa comprises sclerotic cells (*sc*) with strongly pitted walls. (See Yakovlev 1981)

mature seeds. These are always exotegmic but show differences in the type of cell wall thickenings (Bouman and Meijer 1987).

F Bouman and W Meijer (1987), EJH Corner (1976), GL Davis (1966), W Endriss (1902), A Ernst and E Schmid (1913), I Guzowska (1964), J Kuijt (1969), K Schnarf (1931), K Watanabe (1936), MS Yakovlev (1981)

21.4 Hydnoraceae

The development of anther, pollen grains, and endosperm has not been investigated.

The ovule is orthotropous, unitegmic, and tenuinucellate. The archesporial cell functions as megaspore mother cell. Development of the embryo sac is of the Allium type in *Prosopanche bertoniensis* (Chodat 1916), and tetrasporic Adoxa type in *Hydnora africana* (Dastur 1922). The polar nuclei fuse before fertilization, and the antipodals are ephemeral.

The embryogeny in *Hydnora africana* conforms to the Solanad type, but this needs confirmation (see Davis 1966).

The seed is minute with an inconspicuous embryo, copious endosperm (Hutchinson 1973), and a single layer of perisperm. The outer epidermis of seed is thin-walled, and the inner epidermis has thickened inner walls, except in *Hydnora*.

R Chodat (1916), AE Ćocucci (1975, 1976), RH Dastur (1922), GL Davis (1966), J Hutchinson (1973), K Schnarf (1931)

21.5 Taxonomic Considerations

Hallier (1912) considers the origin of Aristolochiales from Lardizabalaceae (Ranales), and Wettstein (1935) from myrtaceous and annonaceous ancestors. Hutchinson (1973) derives it from herbaceous members of Berberidales (Ranales), and includes Aristolochiaceae, Hydnoraceae, Cytinaceae and Nepenthaceae in Aristolochiales. Bessey (1915) assigns Aristolochiaceae to Myrtales.

Cronquist (1968, 1981) places the order Aristolochiales in the subclass Magnoliidae, and R. Dahlgren (1980b, 1981) treats it in the superorder Magnoliiflorae. The presence of etherial oil cells, the uniaperturate pollen, and seeds with copious endosperm in Aristolochiaceae clearly indicate their Magnolian ancestry.

Aristolochiaceae. Aristolochiaceae show close affinities with Annonaceae: (1) secretory anther tapetum, (2) simultaneous or successive cytokinesis, (3) anatropous, bitegmic, and crassinucellate ovule with a massive parietal tissue, (4) Polygonum type of embryo sac, (5) Cellular endosperm which becomes ruminate due to infoldings of the seed-coat, Annona type in *Apama siliquosa* (Periasamy 1962a), and (6) albuminous seed.

Wettstein's view that they have annonaceous ancestors is generally accepted.

Rafflesiaceae, Hydnoraceae. Rafflesiaceae are allied to Hydnoraceae. Both have ana- or orthotropous (*Cytinus*), unitegmic (*Rafflesia*) and tenuinucellate ovules; the single archesporial cell functions directly as megaspore mother cell. The minute, albuminous seed has an undifferentiated embryo, and thin-walled or pulpy exotesta (absent in *Rafflesia*). Both families differ from Aristolochiaceae in triaperturate pollen grains, tenuinucellate ovule, bi- or tetrasporic embryo sac in Hydnoraceae, Nuclear endosperm in Rafflesiaceae, and absence of ruminations. The endotegmic seed of *Rafflesia* negates its affinities with that

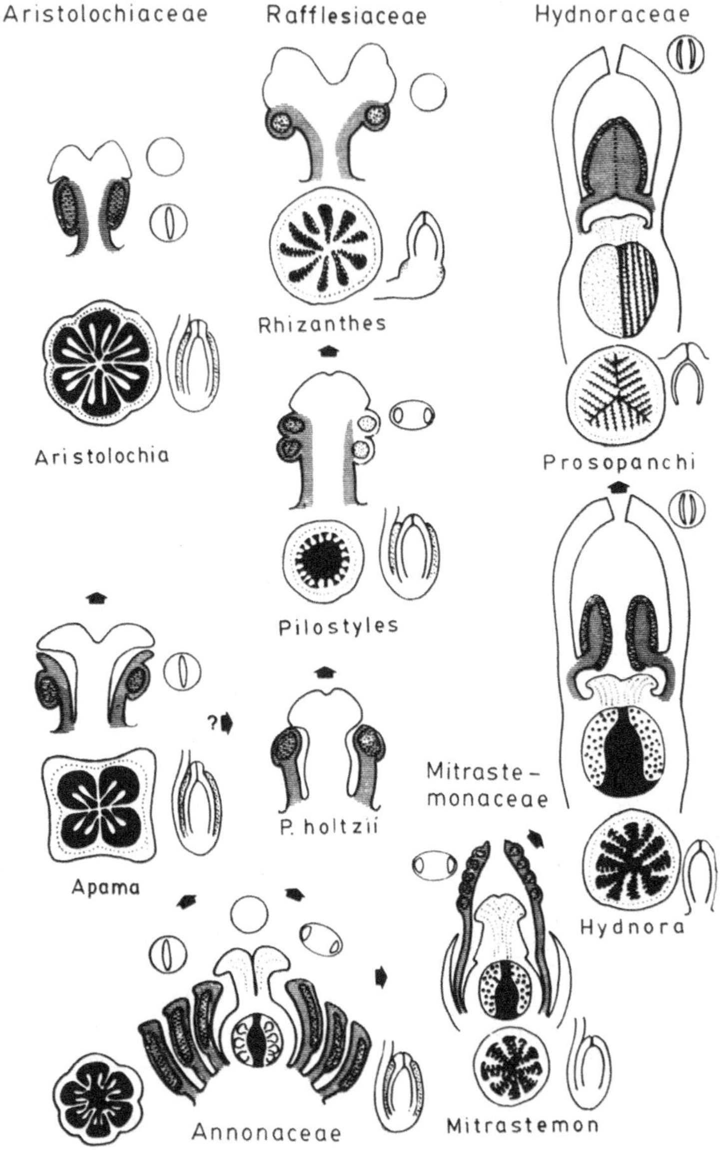

Fig. 21.5.1. Phylogenetic relationships between Annonaceae (embryo sac Polygonum type, endosperm Cellular type), Mitrastemonaceae (embryo sac Polygonum type, endosperm Cellular type), Aristolochiaceae (embryo sac Polygonum type, endosperm Cellular type), Rafflesiaceae (embryo sac Polygonum type, endosperm Cellular type), and Hydnoraceae (embryo sac Allium type in *Prosopanche* and Adoxa type in *Hydnora*, endosperm Cellular type). Each family is represented by a longisection of flower, or only the androecium (*shaded*), with style and stigma, transection of ovary, longisection of ovule, and the pollen type. (After Cocucci 1976, see also Cocucci 1983)

of Aristolochiaceae, while the exotegmic seed of *Apodanthus* shows relationship with Piperaceae and Saururaceae (Corner 1976).

Hydnoraceae can be singled out from the other families due to bisporic Allium, tetrasporic Adoxa type of embryo sac development, and perispermous seed.

R. Dahlgren (1980b, 1981) and Cronquist (1981) include Rafflesiaceae and Hydnoraceae in Rafflesiales (see also Kuijt 1969).

Hydnoraceae (Sect. 21.4) are related to Rafflesiaceae (21.3) and Aristolochiaceae (21.2). The floral and embryological characters indicate that the annonaceous stock may be considered to be the basis for three different evolutionary lines. One leads to Aristolochiaceae, the second to Rafflesiaceae, and the third to Hydnoraceae with Mitrastemonaceae (21.3) in an intermediate position (Fig. 21.5.1). We agree with Cocucci's (1983) interpretation.

22 Guttiferales

The order Guttiferales comprises sixteen families, distributed over four suborders: suborder Dilleniineae – Dilleniaceae, Paeoniaceae, Crossosomataceae, Eucryphiaceae, Medusagynaceae and Actinidiaceae; Ochnineae – Ochnaceae, Dioncophyllaceae, Strasburgeriaceae and Dipterocarpaceae; Theineae – Theaceae, Caryocaraceae, Marcgraviaceae, Quiinaceae and Guttiferae (Clusiaceae); Ancistrocladineae—Ancistrocladaceae.

The embryology of the families Eucryphiaceae and Medusagynaceae of Dilleniineae, Strasburgeriaceae of Ochnineae, Quiinaceae of Theineae, and Ancistrocladaceae of Ancistrocladineae has not been investigated. The development of the embryo of Ochnaceae and Marcgraviaceae either has not been fully worked out.

22.1 General Features

The order is characterized by (1) secretory anther tapetum, (2) simultaneous cytokinesis in pollen mother cells, (3) two-celled pollen grains at the dispersal stage; in Marcgraviaceae they germinate in situ, (4) anatropous, bitegmic ovules, unitegmic in Actinidiaceae, fusion of integuments at places in Marcgraviaceae, and appressed integuments in Dilleniaceae, (5) Polygonum type of embryo sac, Allium type in Theaceae, (6) Nuclear endosperm, Cellular type in Actinidiaceae and Marcgraviaceae, (7) zygotic mantle, and coenocytic aril in a member of Dilleniaceae, peculiar embryogeny in Paeoniaceae, Onagrad type in Dilleniaceae and Crossosomataceae, Solanad type in Actinidiaceae and Theaceae, and Asterad type in Dipterocarpaceae, and (8) pachychalazal seed in Ochnaceae, mesotestal in Theaceae, endotestal in Dilleniaceae, and exotestal in other members.

22.2 Dilleniaceae

The anther is tetrasporangiate. The anther wall comprises four layers in *Acrotrema* (Swamy and Periasamy 1955) and *Dillenia indica* (Sastri 1958b), and five layers in *Hibbertia* (Sastri 1958b). The epidermis is tannin-filled in *H. stricta* and *D. indica* (Sastri 1958b), and is a persistent and dominant layer in *Acrotrema*. The persistent epidermis in *Wormia burbidgei* (Nagaraja Rao 1961) shows fibrous thickenings. The endothecium develops fibrous thickenings in *H. stricta* (Sastri 1958b), but in *Acrotrema* and *H. billardieri* only a few cells acquire thickenings. They are totally absent in *Dillenia*. The one or two middle layers are ephemeral. The tapetum is secretory, and of binucleate cells in *Acrotrema*, *Dillenia* and *Hibbertia*. Amoeboid tapetum is reported in *Wormia suffruticosa* (Paetow 1931), which needs confirmation. Simultaneous cytokinesis by furrowing follows meiotic divisions in the microspore mother cells. The tetrads are tetrahedral, isobilateral, linear and rhomboidal.

Pollen grains are two-celled at the shedding stage; they are three colporate. A large protein crystal occurs in the pollen of *Wormia suffruticosa* (Paetow 1931), and disappears during its maturation.

The ovule is anatropous, bitegmic and crassinucellate. A small space is present between the closely appressed integuments in the micropylar region. The micropyle is zigzag, or closed by the exostome.

In *D. indica* the inner integument gets crushed soon after fertilization, while in *H. stricta* the outer epidermis of the inner integument becomes radially elongate and develops spiral thickenings in *Acrotrema*, *Hibbertia* and *Wormia*. Due to further curvature, after fertilization, the ovule becomes amphitropous (Sastri 1958b). A hypodermal archesporial cell cuts off a parietal cell which gives rise to a two- to three-layered parietal tissue. A nucellar cap has been noted. The megaspore mother cell produces a linear or T-shaped tetrad, and the chalazal megaspore gives rise to a Polygonum type of embryo sac. The synergids are elongated and hooked in *Hibbertia*, and pear-shaped in *Dillenia* (Nagaraja Rao 1955a). The antipodals are ephemeral.

The endosperm is of the Nuclear type, and becomes completely cellular due to centripetal wall formation at the globular stage of proembryo in *Hibbertia* (Sastri 1958b).

A unique feature (see Vijayaraghavan et al. 1990) is the presence of a zygotic mantle in the embryo sac (Fig. 22.2.1A). It is pronounced in *Acrotrema* and *Wormia*, but absent in *Dillenia*. The inner surface of the zygotic membrane becomes gelatinous and lamellated, and the outer surface swells and grows out in the form of a coralloid protuberance between the adjoining endosperm cells (Fig. 22.2.1B). In the mature seed most of the space is occupied by the endosperm, and the embryo is very small.

The embryogeny conforms to the Onagrad type.

The seed of *Acrotrema* is arillate and the aril develops from the funicle (Fig. 22.2.2A, B). During post-fertilization stages, the aril covers the seed partially (Fig. 22.2.2C, D). The embryo sac becomes curved due to the growth of raphe (Fig. 22.2.2C). To begin with, the cells of the aril are uninucleate

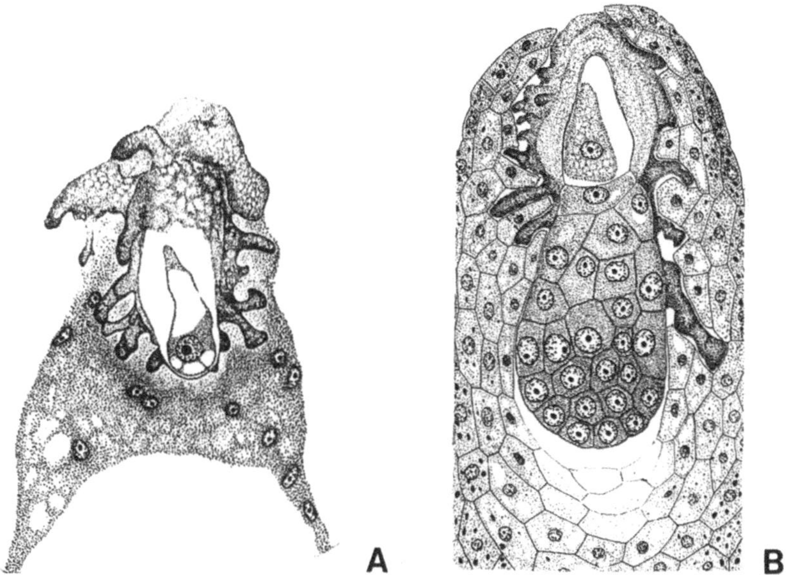

Fig. 22.2.1A–B. *Acrotrema arnottianum.* **A** Zygote, mantle with appendages, and nuclear endosperm. **B** Proembryo with one-celled suspensor, mantle, and cellular endosperm. (After Swamy and Periasamy 1955)

and tubular, then they elongate and become coenocytic (Fig. 22.2.2E–I). In *Hibbertia* the aril is rudimentary.

The albuminous seed is endotestal. At fertilization, the outer integument is two-layered, and the inner one three-layered. The outer epidermis of testa is tanniniferous, and becomes compressed in the mature seed. The cells of the inner epidermis of testa undergo sclerification, and show branched pits. The cells of the outer epidermis of tegmen elongate and become spirally thickened, but collapse in the mature seed. The inner epidermis of tegmen alone persists and becomes tanniniferous. The middle layer of tegmen, and nucellus also collapse.

JD Elizabeth and KK Lakshmanan (1984), JE Dulcy and KK Lakshmanan (1981, 1984), A Nagaraja Rao (1955a, 1957a, 1961), W Paetow (1931), RLN Sastri (1958b), BGL Swamy and K Periasamy (1955), MR Vijayaraghavan, Sudesh and Nidhi Gautham (1990)

22.3 Paeoniaceae

The anther wall comprises an epidermis, two to three layers of endothecium, one or two ephemeral middle layers, and secretory tapetum of binucleate cells. The tapetum becomes biseriate. The microspore mother cells undergo meiotic

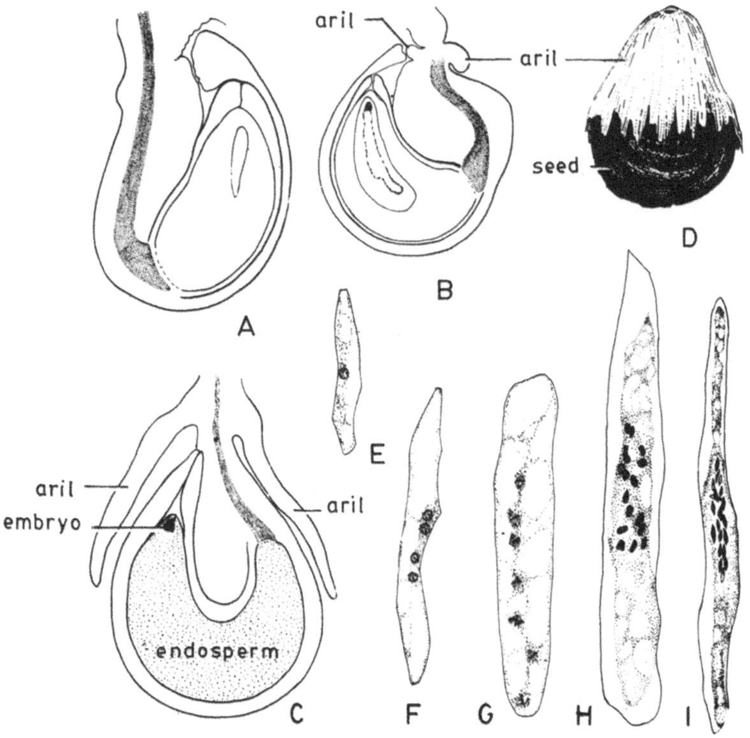

Fig. 22.2.2A–I. *Acrotrema arnottianum.* **A** Longisection of anatropous, bitegmic, arillate ovule. **B** At globular proembryo. The aril is funicular, and the endosperm of the Nuclear type. **C** Immature seed at heart-shaped embryo. **D** The aril covers the upper half of mature seed. **E** Uninucleate cell of aril. **F–I** Elongated coenocytic cells of aril. (After Swamy and Periasamy 1955)

divisions, cytokinesis is simultaneous, and the tetrads are tetrahedral and decussate.

The pollen grains are two-celled at the shedding stage. They are por(oid)ate, prolate and spheroidal. The sexine is thinner than nexine, and is reticulate.

The ovule is anatropous, bitegmic and crassinucellate. The outer integument is massive, and forms the micropyle. A multicellular archesporium differentiates in the hypodermal region in *Paeonia anomala* (Murgai 1962). The formation of a parietal layer is doubtful. The megaspore mother cell undergoes meiotic divisions and linear or T-shaped tetrad is produced. The chalazal megaspore produces a Polygonum type of embryo sac. The polar nuclei fuse in the centre of the embryo sac. The antipodals are large, persistent and multinucleolate in *P. delavayi* (Murgai 1962). They may undergo divisions and produce accessory cells. Formation of twin embryo sacs is reported in *P. tenuifolia* (Murgai 1962). Yakovlev and Yoffe (see Y.D. Tiagi 1970b) observed the concurrent formation of several embryo sacs, and their fusion. A prominent hypostase is present.

The nucellus is absorbed by the embryo sac even before pollination.

During double fertilization one of the synergids perishes (Y.D. Tiagi 1970b). According to Yakovlev and Yoffe (1957), both the synergids remain intact, which is doubtful.

Development of the endosperm is of the Nuclear type. The free nuclei in the chalazal region fuse, and form irregular polyploid masses. Wall formation is centripetal, and commences from the micropylar end at an early stage of proembryo. In *P. officinalis* (Murgai 1962), the free-nuclear chalazal part of the endosperm develops into a haustorium, when the proembryo has a coenocytic suspensor. The haustorium attains its maximum activity and becomes com-

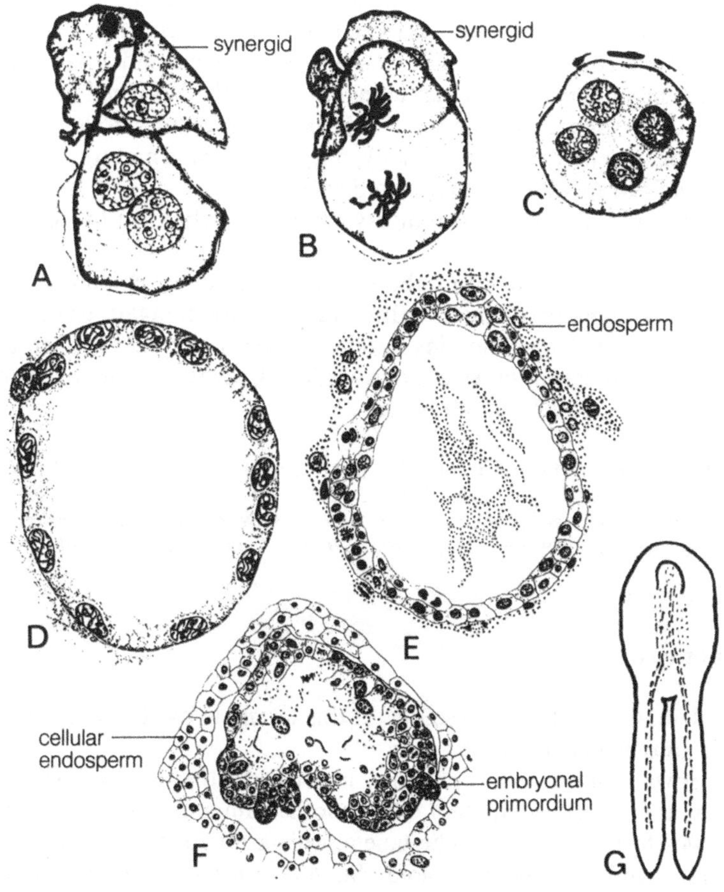

Fig. 22.3.1A–G. *Paeonia lactiflora*. **A** Two-nucleate, proembryo. **B** Both nuclei in division. **C** Four-nucleate proembryo. **D** Multinucleate proembryo with peripheral arrangement of nuclei. The dark nucleus in the upper region (*left*) is an endosperm (free-nuclear) nucleus. **E** Wall formation has taken place in the proembryo, the endosperm is still free-nuclear. **F** The proembryo undergoes active cell division, and embryonal primordia differentiate. Note the cellular endosperm. **G** Dicotyledonous embryo. (After Yakovlev 1969)

pletely cellular at the early globular stage of the proembryo. It degenerates at an advanced stage of globular proembryo. The endosperm is persistent, and stores fat and aleurone grains.

The embryogeny of *Paeonia* has become a controversial issue. The findings of Yakovlev and Yoffe (1957, 1961) have been supported by Cave et al. (1961), Matthiessen (1962) and Moscov (1964).

The nucleus of the zygote undergoes repeated free-nuclear divisions to form a coenocyte and the nuclei become distributed around a large central vacuole (Fig. 22.3.1A–D). Centripetal wall formation occurs producing a proembryonal mass (Fig. 22.3.1E). Several peripheral cells develop meristematic centres, and produce embryonal primordia (Fig. 22.3.1F). Only one of these primordia matures into a dicotyledonous embryo (Fig. 22.3.1G). Such a development is unparalleled in angiosperms, and reminds one of gymnosperms.

Contrary to the above mode, Murgai (1959, 1962) reported that the division of zygote is followed by a wall resulting in an apical and a basal cell. The apical cell may undergo one longitudinal division; the basal cell becomes coenocytic with a vacuole in the centre. It becomes cellular followed by proembryonal meristematic centres. Finally, only one of these matures into an embryo. Murgai's observations require confirmation.

The mature embryo is small and dicotyledonous surrounded by copious endosperm.

The testa is massive, and forms the seed-coat. Its outer epidermis and hypodermis are composed of thick-walled, tangentially elongate cells. The mesophyll is thin-walled, and contains starch. Its inner layers and tegmen get crushed. The inner epidermis of testa is thin-walled. A slightly vascularized aril develops at the base (Camp and Hubbard 1963).

WH Camp and MM Hubbard (1963), K Carniel (1967), MS Cave, HJ Arnott and SA Cook (1961), M Kumazawa (1935), A Matthiessen (1962), IV Moscov (1964), Mu-Xi-Jin and Wang Fu (1985a), P Murgai (1959, 1962), YD Tiagi (1970b), JL Walters (1962), WC Worsdell (1908), R Wunderlich (1966), MS Yakovlev (1951, 1959, 1961, 1965, 1967, 1969), MS Yakovlev and MD Yoffe (1957, 1961)

22.4 Crossosomataceae

A major contribution to the embryology of this family is by Kapil and Vani (1963) on *Crossosoma californicum.*

The anther wall comprises an epidermis, fibrous endothecium, two or three middle layers, and secretory tapetum. The endothecium and tapetum become irregularly two- to three-layered. The tapetal cells are two- to four-nucleate; the nuclei fuse and become polyploid. Simultaneous cytokinesis follows meiotic divisions of microspore mother cells, and tetrahedral and decussate tetrads are produced.

The pollen grains are tricolpate, and two-celled at the shedding stage.

The ovule is amphitropous, bitegmic and crassinucellar. The micropyle is organized by both the integuments. A hypostase differentiates in the chalazal region.

The hypodermal archesporium is multicellular. A massive parietal tissue is produced. The functional megaspore mother cell undergoes meiotic divisions and produces a linear tetrad. The chalazal megaspore develops into a Polygonum type of embryo sac. The antipodals divide and form five or six cells.

The entry of pollen tube is porogamous, and triple fusion precedes syngamy.

The development of the endosperm is of the Nuclear type. Enucleate cytoplasmic vesicles, and aggregation of nuclei in the chalazal region has been noted. Wall formation commences from the micropylar region at the globular proembryo, and proceeds to the chalazal region.

The embryogeny corresponds to the Onagrad type. The mature embryo is large and curved.

The seed is arillate, and thinly albuminous. The testa is highly sclerified. The inner epidermis of the inner integument develops fibrous thickenings. A fimbriate aril, surrounding the seed, develops from the funicle.

RN Kapil (1970a), RN Kapil and RS Vani (1963)

22.5 Eucryphiaceae

The embryology has not been studied.

22.6 Medusagynaceae

The embryological data are very meagre.

The anther opens longitudinally. The microspore tetrads are tetrahedral. The pollen grains are two-, three- or four-porate.

The anatropous ovule has a long funicle in *Medusagyne oppositifolia* (Engler and Melchior 1925).

The seeds are winged (Hutchinson 1973).

A Engler and K Melchior (1925), J Hutchinson (1973)

22.7 Actinidiaceae

The anther is tetrasporangiate, and its wall consists of the epidermis, fibrous endothecium, two or three ephemeral middle layers, and secretory tapetum of multinucleate cells. Due to fusions the tapetal nuclei become polyploid. Raphides are present in the anther wall and connective. Simultaneous cytokinesis, following meiotic divisions of microspore mother cells, results in tetrahedral tetrads.

The pollen grains are two-celled at the dispersal stage.

The ovule is anatropous, unitegmic and tenuinucellar. The nucellus is thin and ephemeral, and calcium oxalate crystals occur in the ovule. A persistent conspicuous endothelium and a well-developed hypostase also develop. Of the two or three hypodermal archesporial cells, one functions directly as megaspore mother cell. It undergoes meiotic divisions and gives rise to a linear tetrad. The chalazal megaspore develops into a Polygonum type of embryo sac in *Actinidia* (Vijayaraghavan 1965). The synergids and antipodals are ephemeral. Nagaraja Rao (1953a) reported inverted polarity in the embryo sac of *Saurauia nepalensis.*

Development of the endosperm is of the Cellular type.

The embryogeny follows the Solanad type. Additional embryos are produced due to the proliferation of suspensor cells in *A. chinensis* (Crété 1944).

The seed is arillate, exalbuminous, and exotestal. The inner walls of the cells are highly thickened and lignified; the upper part of the cell contains tannin and a prominent vacuole. The endothelium appears as a compressed layer.

DI Brundell (1975), P Crété (1944), WC Dickison (1972), CF Harvey and LG Fraser (1988), ME Hopping (1976), ME Hopping and EM Jerram (1979), HL Li (1952), A Nagaraja Rao (1953a), R Schmid (1978a, b), R Souèges (1943b), Ph van Tieghem (1899), MR Vijayaraghavan (1965, 1970b), J White (1990)

22.8 Ochnaceae

The embryogeny has not been investigated.

The anther wall comprises the epidermis, fibrous endothecium, an ephemeral middle layer, and secretory tapetum of binucleate cells in *Ochna atropurpurea* and *O. kirkii* (L.L. Narayana 1975). The microspore mother cells undergo meiotic divisions, and produce tetrahedral tetrads. Cytokinesis is simultaneous.

The pollen grains are three-colpate and two-celled at the shedding stage.

The ovule is anatropous, bitegmic, and tenuinucellate. The integuments are free only at the micropyle. A thin-walled prominent hypostase develops. The hypodermal archesporial cell develops directly into the megaspore mother cell. An additional archesporial cell has been observed in the inner integument in *Ochna atropurpurea* (R.B. Ghosh 1964). Following meiotic divisions, the megaspore mother cell produces a linear or T-shaped tetrad. The chalazal megaspore develops into a Polygonum type of embryo sac (Fig. 22.8.1A–C, E). The formation of a "false" endothelium from the nucellar epidermis has been reported in *O. squarrosa* (Chikkannaiah 1954). The antipodals are large with hypertrophied nuclei.

Somatic apospory of the Hieracium type occurs in *O. serrulata* (Fig. 22.8.1 D, F–L); F, G, K, L show one, I two and J three aposporous "embryo sacs". Generally, somatic cells develop into an eight-nucleate embryo sac, but in *Ochna* the embryo sac shows fewer nuclei.

The development of the endosperm is of the Nuclear type; wall formation commences from the micropylar end. The endosperm is soon absorbed by the developing embryo.

The formation of a synergid embryo has been observed in *O. serrulata* (Chiarugi and Francini 1930). In some species integumentary embryos may also develop (Fig. 22.8.2A, B).

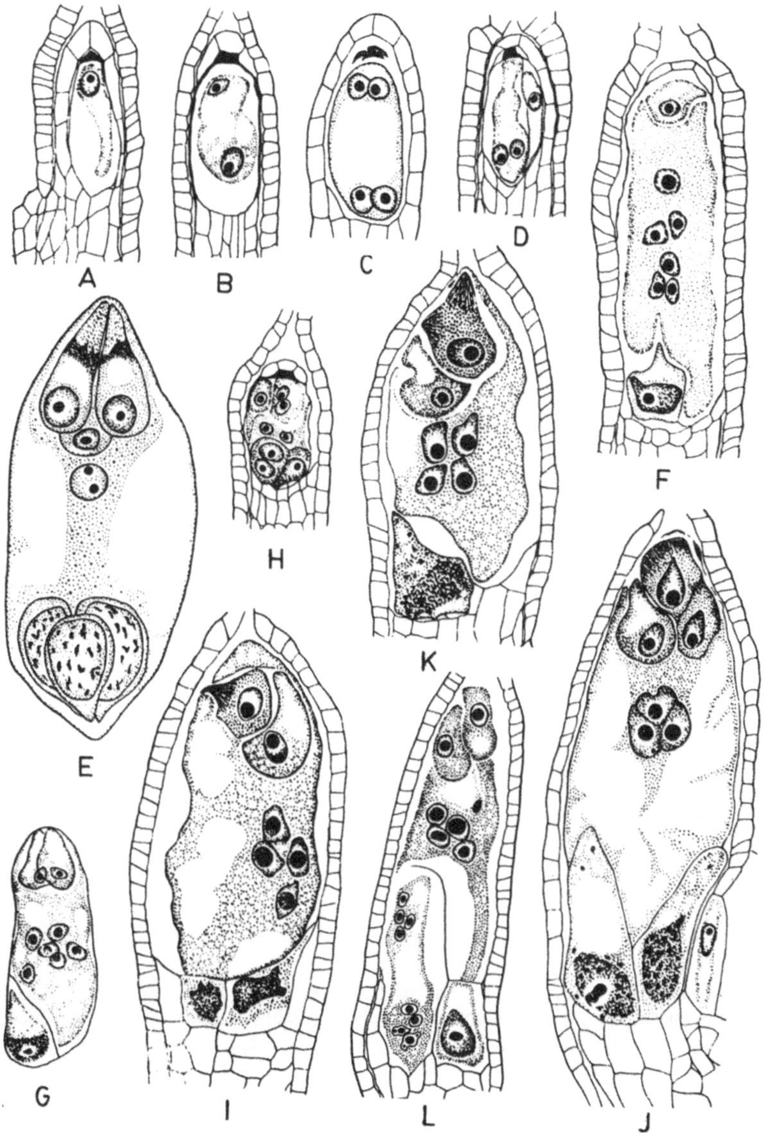

Fig. 22.8.1A–L. A, B, D, F–L *Ochna serrulata*, **C, E** *O. squarrosa*. **A** Chalazal functional megaspore. **B, C** Bi- and tetranucleate embryo sac. **D** Binucleate embryo sac and an aposporous embryo sac initial. **E** Mature embryo sac with hypertrophied antipodal nuclei. **F–H** Abnormal embryo sacs. **I** Zygote, persistent healthy synergid, and four-nucleate endosperm. Note one or two aposporous embryo sacs in **F, G** and **I–L**. **J** Embryo sac with zygote and two egg-like synergids, and primary endosperm nucleus (?). One cell of "endothelium" has differentiated like a functional megaspore. **K** Zygote with hypertrophied synergid, and four-nucleate endosperm. **L** Three embryo sacs enclosed within a common endothelium. One uninucleate, other unorganized eight-nucleate (two lower nuclei tri-nucleolate) and one embryo sac with zygote, synergid and four-nucleate endosperm. (**A, B, D, F–L** After Chiarugi and Francini 1930, **C, E** after Chikkanaiah 1954)

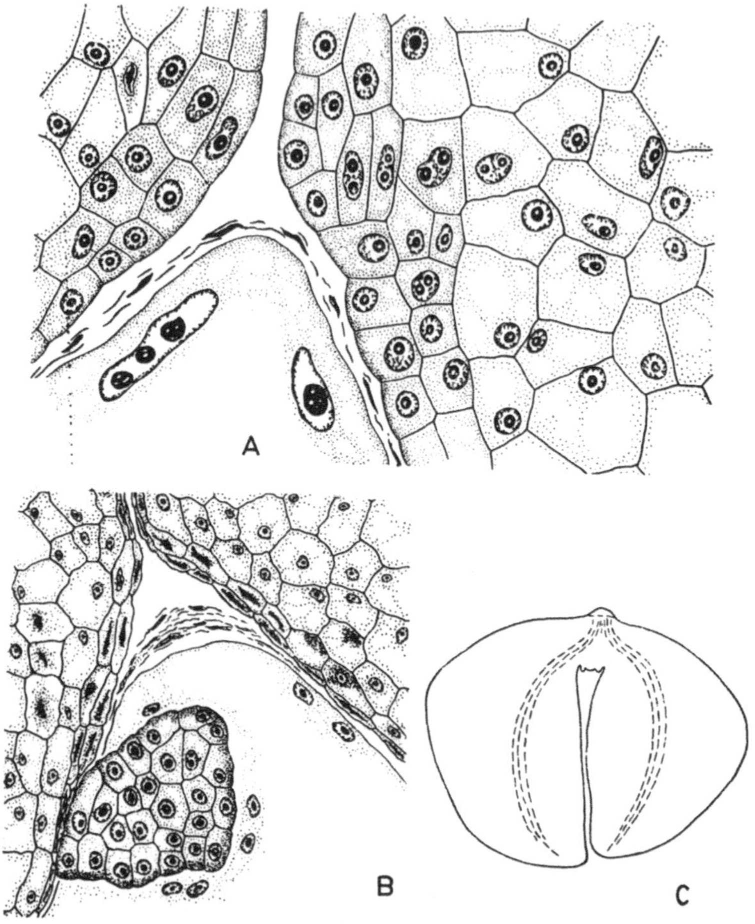

Fig. 22.8.2A–C. **A, C** *Ochna multiflora*, **B** *O. atropurpurea*. **A** Micropylar part of embryo sac with free endosperm nuclei; the inner integumentary cells are richly cytoplasmic. **B** Development of integumentary embryo, lying inside the embryo sac surrounded by free endosperm nuclei **C** Mature embryo. (See Yakovlev 1983)

The mature embryo is straight with thick cotyledons, and a short radicle (Fig. 22.8.2C).

An exalbuminous pachychalazal seed is enclosed in the sclerotic endocarp of a drupe. The massive seed-coat of 14–18 cell-layers is highly aerenchymatous, thin-walled and vascularized. It gets crushed during further development, except for the outer epidermis which is brownish and membraneous.

A Chiarugi and E Francini (1930), PS Chikkannaiah (1954), EJH Corner (1976), RB Ghosh (1964), LL Narayana (1975)

22.9 Dioncophyllaceae

The embryology has not been studied in sufficient detail.

The anthers are tetralocular, introse and open longitudinally. The pollen grains are three- to four-colporate.

The ovule is anatropous, bitegmic, crassinucellate, with the inner integument longer than the outer (Fig. 22.9.1A–C). The micropyle is formed by the inner integument, which has a beak-shaped process at the apex. The outer integument forms a cover (Fig. 22.9.1D, E).

The archesporial cell cuts off a parietal cell, the parietal tissue is two-layered. In *Triphyophyllum peltatum* (Schmid 1964), the megaspore mother cell undergoes meiotic divisions, and the chalazal megaspore of a linear tetrad develops into Polygonum type of embryo sac (Fig. 22.9.1F–H).

The large circular winged seed has a large discoid-obconic dicotyledonous embryo (Fig. 22.9.1I–K) and copious starchy endosperm in *Triphyophyllum* (Cronquist 1981).

A Cronquist (1981), R Schmid (1964)

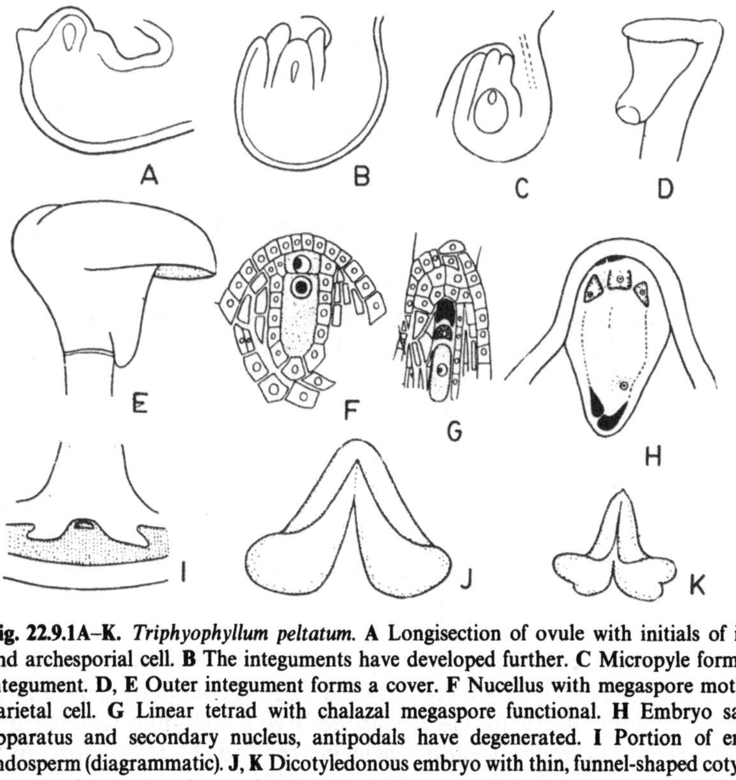

Fig. 22.9.1A–K. *Triphyophyllum peltatum.* **A** Longisection of ovule with initials of integuments and archesporial cell. **B** The integuments have developed further. **C** Micropyle formed by inner integument. **D, E** Outer integument forms a cover. **F** Nucellus with megaspore mother cell and parietal cell. **G** Linear tetrad with chalazal megaspore functional. **H** Embryo sac with egg apparatus and secondary nucleus, antipodals have degenerated. **I** Portion of embryo with endosperm (diagrammatic). **J, K** Dicotyledonous embryo with thin, funnel-shaped cotyledons. The margins of cotyledons have bifurcated in **K**. (After Schmid 1964)

22.10 Strasburgeriaceae

The embryology has not been studied.

22.11 Dipterocarpaceae

The anther comprises cutinized epidermis, endothecium, three ephemeral middle layers, and glandular tapetum of binucleate cells. The endothecium is devoid of fibrous thickenings in *Shorea talura*, *S. robusta*, *Vateria indica* and *Hopea racophloea* (Nagaraja Rao 1953c, 1955b, c, 1956, 1962). The dehiscence of anther takes place due to disorganization of cells between the two loculi. The inner walls of the tapetal cells are studded with globular bodies. The microspore mother cells undergo meiotic divisions and produce tetrahedral, isobilateral and decussate tetrads. Quadripartition of the tetrads takes place by centripetal furrowing.

The pollen grains are three-colpate, and shed at the two-celled stage (A. Kaur et al. 1986).

The ovule is anatropous, bitegmic, and crassinucellate. The inner integument forms the micropyle (Fig. 22.11.1A), and is fairly broad at the micropylar end; the outer integument contains a vascular strand.

A hypodermal archesporial cell cuts off a parietal cell. The parietal tissue is well developed. The megaspore mother cell undergoes meiotic divisions, and produces a linear or T-shaped tetrad. The chalazal megaspore develops into a Polygonum type of embryo sac. The antipodals persist up to fertilization. The polar nuclei fuse in the centre of the embryo sac.

The entry of pollen tube is porogamous, and one of the synergids is destroyed.

The development of endosperm is of the Nuclear type (Fig. 22.11.1A). The free-nuclear endosperm, in the lower half of the embryo sac, is aggressive and absorbs nourishment from the adjoining nucellar and integumentary tissue. During later stages, the endosperm consumes the tissue of seed-coat, irregularly leading to a lobed appearance (Fig. 22.11.1B–I). Some of the endosperm nuclei lie in these lobes (Fig. 22.11.1D–H). The chalazal region of ovule shows meristematic activity (Fig. 22.11.1G–I). Further studies may reveal some more interesting features.

Wall formation is initiated in the micropylar region at an early stage of proembryo (Fig. 22.11.1A), and proceeds towards the chalazal region. In *Vateria* nucleated endosperm nodules are formed. The lower part of the embryo sac remains free-nuclear for a long time. Finally, it becomes completely cellular. The mature endosperm is fatty.

The zygote divides transversely forming an apical and a basal cell. Vertical divisions occur in both cells resulting in a globular tetrad. Further divisions are irregular. The embryogeny is of the Solanad type, Chenopodiad in *Shorea macroptera* (A. Kaur et al. 1986). The mature embryo is well-differentiated, and broad with unequal fleshy cotyledons enclosing the radicle. Polyembryony is reported in *S. robusta* (R.B. Ghosh and Shahi 1957).

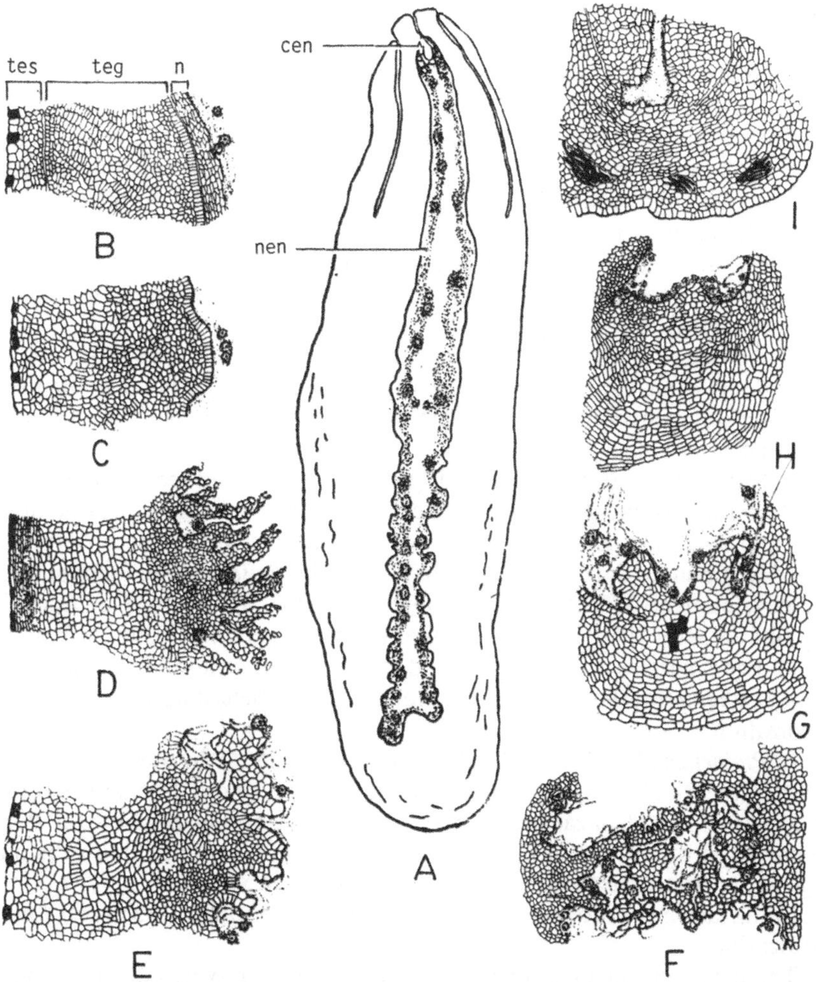

Fig. 22.11.1A–I. *Vateria indica.* **A** Longisection of bitegmic ovule, the two integuments are separate only in the upper region. The embryo sac has globular proembryo and cellular endosperm (*cen*) around it, nuclear endosperm (*nen*) in the rest of the embryo sac. The nuclear endosperm invaginates into the integument. **B** Portion of seed-coat to show testa (*tes*), tegmen (*teg*), nucellus (*n*), and free-nuclear endosperm. **C** Portion of seed-coat in the fused region of testa and tegmen, the nucellus has collapsed, endosperm nuclear. **D**, **E** Invaginations of nuclear endosperm into seed-coat, endosperm nuclei lie in the lobes of seed-coat. **F** Surface view of seed to show the invaginations of endosperm in seed-coat. **G–I** Basal region of ovule; stages of invagination of endosperm in the chalazal tissue. (After Nagaraja Rao 1955d)

The seed is exalbuminous, and exotegmic. Otherwise, the seed-coat is unspecialized and appears thin and papery. The exotegmen is sclerotic. The woody capsule protects the seeds.

RB Ghosh and R Shahi (1957), A Kaur, K Jong, VE Sands and E Soepadmo (1986), G Maury (1980), A Nagaraja Rao (1953c, 1955b, c, d, 1956, 1962), K Swarupanandan (1986)

22.12 Theaceae

The anther wall comprises an epidermis, fibrous endothecium, two ephemeral middle layers, and secretory tapetum of binucleate cells. Crystals (druses) are present in the connective region in *Ploiarium alternifolium* (N. Prakash and Lau 1976). Simultaneous cytokinesis follows meiotic divisions of microspore mother cells, and results in the formation of tetrahedral, isobilateral and decussate tetrads.

The pollen grains are three-colpate, and shed at the two-celled stage. They are tricolporate, subporolate, reticulate and heterobrochate in *Camellia sasanqua* (Mathew 1978).

The ovule is anatropous, bitegmic, and tenuinucellate. The inner integument alone forms the micropyle. The vascular supply reaches up to the tip of the outer integument in *Camellia sinensis* (Kapil and Sethi 1963). A hypostase has been observed.

A hypodermal archesporial cell (or two- to three-celled archesporium) develops into the megaspore mother cell. The latter undergoes meiotic divisions, and dyads are produced. In *Thea* (Mikatadze 1975), the micropylar dyad may develop up to the binucleate embryo sac. The chalazal dyad functions and the development of female gametophyte is of the Allium type in *Camellia* (Kapil and Sethi 1963, Mathew 1978) and *Thea* (Fagerlind 1939b, Mikatadze 1975). The synergids are ephemeral and exhibit hooks or hood-like expansions in *C. sasanqua* (Mathew 1978). The antipodals may persist up to or even after fertilization.

In *Ploiarium alternifolium* (N. Prakash and Lau 1976) the megaspore tetrad is either linear or T-shaped. Sometimes, the upper dyad cell degenerates before its division. The chalazal megaspore is functional and gives rise to a Polygonum type of embryo sac. The antipodals do not organize into cells and degenerate before the fusion of polar nuclei or soon after it.

Fertilization is porogamous, and the remnants of pollen tube persist after fertilization.

The development of endosperm is of the Nuclear type, later it becomes cellular. The mature seed is exalbuminous.

Development of the embryo corresponds to the Solanad type. The cotyledons show backward growth on either side of the radicle (Sethi 1965). Adventitious embryos originate from the placental cells, as well as the inner integument (Cavara 1899). Cavara also mentions the formation of twin embryos from the same egg apparatus. Supernumerary embryos and twin seedlings have been observed by Venkataramani (1950) and Wu (1960). Placental origin of embryos is doubtful; development of integumentary embryos need further study.

In a mature seed of *Ploiarium* the embryo is straight, narrow, cylindrical and dicotyledonous with a short uniseriate suspensor.

The seed is mesotestal. The testa alone forms the seed-coat. The cells of the outer epidermis are thin-walled with lignified inner walls. The cells of outer zone, except the outer epidermis, undergo sclerification and develop into a hard shell. The inner zone is slightly thickened, and contains tannin. The tegmen gets crushed.

In *P. alternifolium* (Corner 1976), the cells of the second layer of seed-coat are lignified, undulate and oblong, but whether they belong to endotesta or exotegmen is uncertain.

SB Bawa (1970a), F Cavara (1899), EJH Corner (1976), F Fagerlind (1939b), CE Kobuski (1948), RN Kapil and SB Sethi (1963), CJ Mathew (1978), TA Mikatadze (1975), N Prakash and YY Lau (1976), SB Sethi (1965), KS Venkataramani (1950), HK Wu (1960)

22.13 Caryocaraceae

The embryology has not been fully studied.
The anther is tetralocular and opens longitudinally.
The ovule is anatropous and bitegmic.
The endosperm, in mature seed, is scanty with two or three layers of thin-walled cells near the chalaza. The cells contain fats.

The embryo of *Caryocar glabrum* (Pilger 1925) has a strongly thickened hypocotyl which stores proteins and fats. The embryo of *Anthodiscus* is thin, long and spirally coiled. The cotyledons are small and hooked.

The seed-coat is formed by thin-walled aerenchymatous testa in *Caryocar*. The mesotesta and tegmen collapse; the mesocarp and endocarp give protection.

EJH Corner (1976), R Pilger (1925)

22.14 Marcgraviaceae

The embryogeny has not been investigated.

The major contribution to the embryology of Marcgraviaceae is that of Swamy (1948b) on *Marcgravia cuyuniensis* and *M. purpurea*.

The epidermis of the anther wall is persistent, and tanniniferous. The endothecium shows fibrous thickenings, and is followed by two or three ephemeral middle layers. The tapetum is secretory, and consists of uninucleate cells. Cytokinesis is simultaneous.

The pollen grains are three-colpate, and germinate in situ.

The ovule is anatropous, bitegmic, and tenuinucellar. The integuments are two layers thick, and the inner integument forms the micropyle. The outer epidermis of the outer integument contains phenol-like substances. The integuments show signs of fusion at varying levels from the base. The nucellus is highly reduced (Fig. 22.14.1A–C, G).

The hypodermal archesporial cell functions directly as megaspore mother cell. Following meiotic divisions it produces a linear or T-shaped tetrad. The chalazal megaspore develops into a Polygonum type of embryo sac. The

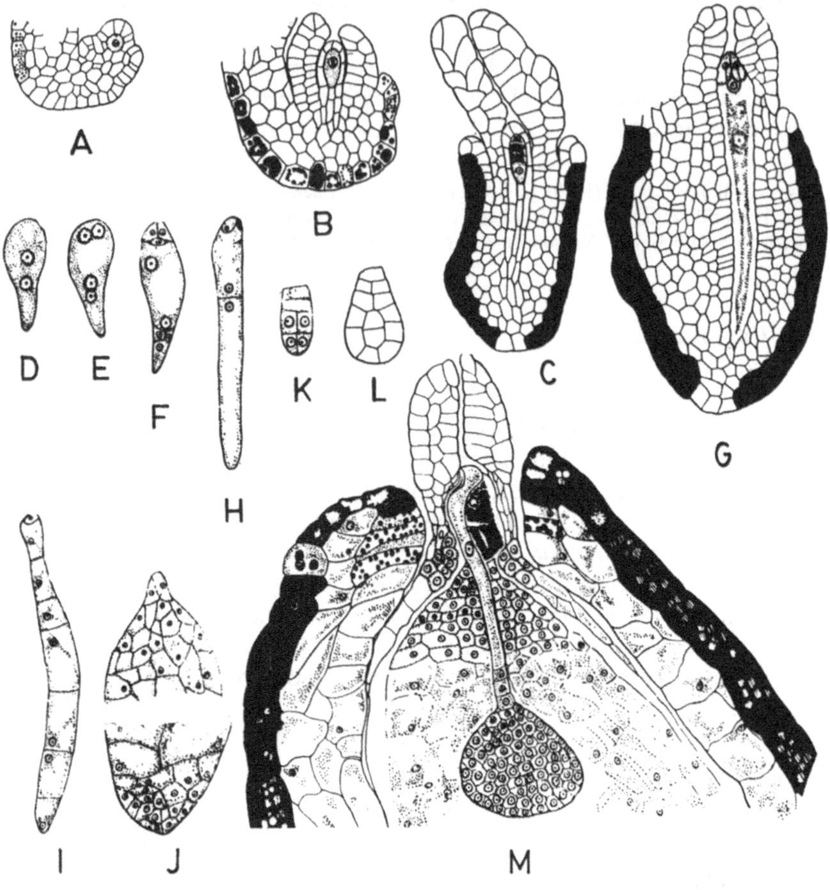

Fig. 22.14.1A–M. **A–C, G, M** *Marcgravia purpurea*, **D–F, H–L** *M. umbellata*. **A, B** Longisections of ovule at hypodermal archesporial and megaspore mother cell. **C** Longisection of bitegmic ovule at linear tetrad. Note the endostome. **D–F** Two-, four- and eight-nucleate embryo sac. **G** Longisection of ovule at mature embryo sac, the antipodals have degenerated. **H** Zygote and two-celled endosperm. **I** Zygote and uniseriate endosperm. **J** Cellular endosperm. **K–L** Proembryo. **M** Longisection of micropylar portion of bitegmic ovule at globular proembryo with a long suspensor, endosperm cellular. (**A–C, G, M** After Swamy 1948b, **D–F, H–L** after Mauritzon 1939a)

antipodals are ephemeral. The antipodal end of the gametophyte extends in the chalaza. An incipient endothelium develops in this region (Fig. 22.14.1A–G).

The endosperm is of the Cellular type. The first few divisions are transverse resulting in an eight-celled uniseriate tissue. Longitudinal divisions occur later, and the cell at the extreme tip of cellular endosperm, towards the micropyle, develops into a non-aggressive haustorium (Fig. 22.14.1H, I, J, M).

Figure 22.14.1K, L shows early proembryos. The terminal cell of the suspensor develops into a tube, and the tip of the tube comes in contact with the endosperm haustorium (Fig. 22.14.1M).

The seed is arillate, and slightly albuminous. The outer epidermis contains orange-brown deposits, and the rest of the seed-coat collapses.

J Mauritzon (1939a), BGL Swamy (1948b)

22.15 Quiinaceae

The embryology has not been investigated.

22.16 Guttiferae (Clusiaceae)

The anther wall comprises the epidermis containing darkly stained bodies, fibrous endothecium, two ephemeral middle layers, and glandular tapetum of binucleate cells. The microspore mother cells undergo meiotic divisions giving rise to tetrahedral, isobilateral, and decussate tetrads. Cytokinesis takes place by centripetal cleavage furrows.

The pollen grains are tricolpate, and shed at the two-celled stage in *Hypericum* (Nagaraja Rao 1957b).

The ovule is anatropous, bitegmic, and tenuinucellate. The micropyle is organized by both the integuments. The hypodermal archesporial cell functions directly as megaspore mother cell (Fig. 22.16.1A–C). Following meiotic divisions, it gives rise to a linear tetrad. The chalazal megaspore develops into a Polygonum type of embryo sac (Fig. 22.16.1D–F) in *Garcinia mangostina* (Sprecher 1919), *G. livistonia* (V. Puri 1939b) and *Hypericum* (Schnarf 1914, Nagaraja Rao 1957b).

Sometimes, two embryo sacs develop concurrently (Fig. 22.16.1E), due to the presence of two megaspore mother cells (V. Puri 1939b, Nagaraja Rao 1957b).

The antipodals are ephemeral; their multiplication is reported in *Hypericum gentianoides* and *H. punctatum* (see Davis 1966).

The development of endosperm (Fig. 22.16.1G, R) is of the Nuclear type (Stenar 1938a, Swamy 1946a), and not Helobial type as reported by Palm (1922). A chalazal "cyst" is formed at eight- to 16-nucleate stage of endosperm (Fig. 22.16.1H), but it gradually merges with the general endosperm. Wall formation is initiated only when the cotyledons differentiate in the embryo.

Autonomous development of endosperm has been observed in *H. virginicum* (Myers 1964). A partition wall develops following division of the primary endosperm nucleus. It suggests Helobial type. The development of endosperm requires further study.

The embryogeny conforms to the Solanad type. In *Hypericum* (Nagaraja Rao 1957b), it conforms to Megarchetype IV and Period II of Souèges system

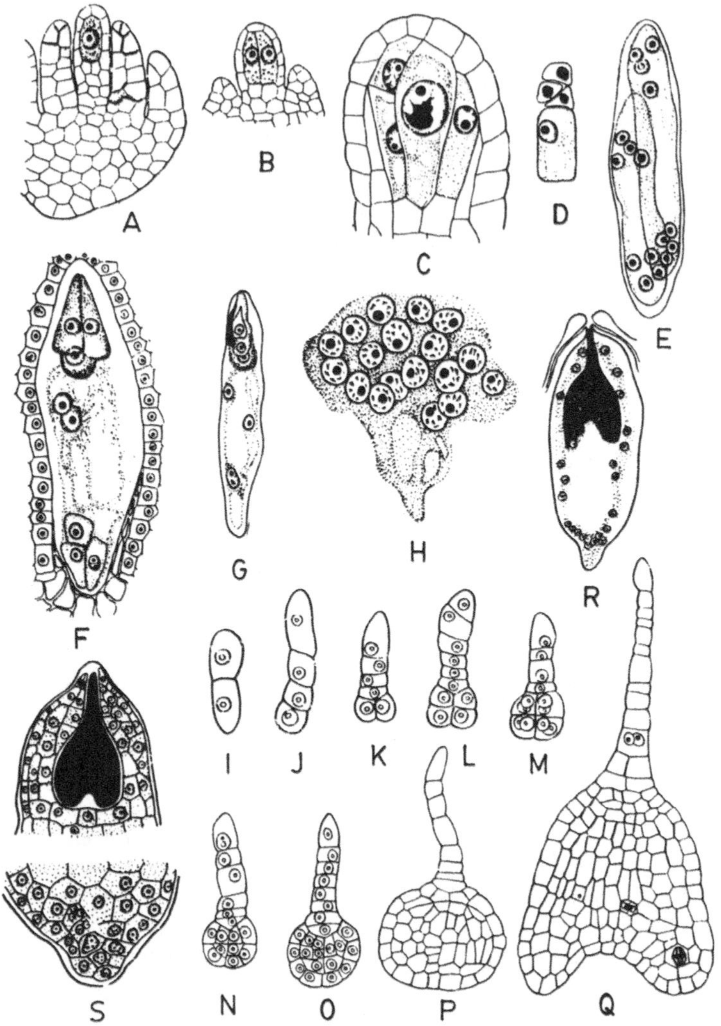

Fig. 22.16.1A–S. A, B, D–G, I–O, R, S *Hypericum patulum*, **C** *Garcinia livingstonii*, **H, P, Q** *H. mysorense*. **A** Bitegmic ovule with megaspore mother cell. **B** Two megaspore mother cells. **C** Two megaspore mother cells and a dyad. **D** Tetrad of megaspores. **E** Twin embryo sacs, each with eight nuclei. **F** Mature embryo sac, and endothelium. **G** Embryo sac with two-celled proembryo, remnants of pollen tube, and nuclear endosperm. **H** Aggregation of endosperm nuclei in the chalazal region of the embryo sac. **I–Q** Development of embryo. **R** Early dicotyledonous embryo (*black*) and free-nuclear endosperm. **S** Endosperm cellular. (**A, B, D–S** After Nagaraja Rao 1957b, **C** after V. Puri 1939b)

of classification (Fig. 22.16.1I–P). Later stages of embryo are shown in Figure 22.16.1Q–S.

Polyembryony, either from synergid by apogamy, or by the proliferation of suspensor cells, is reported in *H. tetrapterum* (Bugnicourt 1969). Myers (1964) described apomictic embryo sacs from the aposporic nucellar cells in *H. virginicum*. Parthenocarpic fruits have been observed in *Garcinia mangostina* (Horn 1940). The fruit contains seed-like structures which may or may not contain the adventive embryo.

The seed is exalbuminous, and arillate with a large embryo in Clusieae. It is exotestal in Clusieae, mesotestal in Callophylleae, and the testa is unspecialized in Garcinieae. In *Hypericum* the seed is exarillate, thinly albuminous, and exotegmic. The testa is represented by the outer epidermis alone. The cells have slightly thickened radial and inner walls, and contain tannin. The exotegmen is thick-walled with stellate-undulate facets. The mesophyll, and inner epidermis are unspecialized.

M Bugnicourt (1969, 1971a, b), KVO Dahlgren (1923), GL Davis (1966), DA Govindappa (1956a), CL Horn (1940), O Myers Jr (1964), A Nagaraja Rao (1957b), B Palm (1922), V Puri (1939b), K Schnarf (1914), YN Seetharam (1989), R Souèges (1925a), A Sprecher (1919), H Stenar (1938a), BGL Swamy (1946a)

22.17 Ancistrocladaceae

The embryology has not been studied in detail.

The anther is tetralocular, introse and opens longitudinally in *Ancistrocladus tectorius* (Keng 1967). The anther wall comprises the epidermis, fibrous endothecium with very large cells, and two ephemeral middle layers (Keng 1967).

The solitary ovule has a short funicle. It is hemianatropous and bitegmic. The outer integument is fairly thick, and the inner integument has only two to three layers.

The development of the endosperm is of the Cellular type. It is starchy and ruminate (Netolitzky 1926).

The mature embryo is relatively small and straight. The testa intrudes between the folds of the cotyledons which enclose the radicle (Hutchinson 1973).

EJH Corner (1976), J Hutchinson (1973), H Keng (1967), F Netolitzky (1926)

22.18 Taxonomic Considerations

The suborder Dilleniineae includes the families Dilleniaceae, Paeoniaceae, Crossosomataceae, Eucryphiaceae, Medusagynaceae and Actinidiaceae.

Bentham and Hooker (1965a), Hallier (1912) and Bessey (1915) include Dilleniaceae in Ranales. The genus *Paeonia* has been placed in Ranunculaceae. Hutchinson (1973) places Dilleniaceae in Dilleniales, and Paeoniaceae in Ranales. Bentham and Hooker treat Crossosomataceae as a member of Dilleniaceae. Engler and Prantl (1897) assign the genus *Crossosoma* to a separate family, Crossosomataceae, in Rosales. Bessey (1915), Wettstein (1935),

Gundersen (1950), Lawrence (1951), Benson (1962), Hutchinson (1973) and Takhtajan (1980) support this view. Lemesle (1948, 1955) proposed the transfer of Crossosomataceae to an intermediate position between Paeoniaceae and Ranunculaceae. Cronquist (1981) places Dilleniaceae and Paeoniaceae in the order Dilleniales, and Crossosomataceae in Rosales.

All the three families – Dilleniaceae, Paeoniaceae and Crossosomataceae – have several common features, such as development of pollen grains, bitegmic, crassinucellate ovule with the micropyle formed by both the integuments, Polygonum type of embryo sac, Nuclear endosperm, and albuminous arillate seeds (Table 22.18.1).

Dilleniaceae. Dilleniaceae can be distinguished from the others due to the absence of fibrous thickenings in endothecium (anther wall), and differentiation of hypostase and endothelium. In the lower region the integuments are closely appressed to each other, but the micropyle is quite wide (see Sastri 1958b). They also have a nucellar cap, a single-celled archesporium, and ephemeral antipodals. The occurrence of the zygotic mantle and coenocytic aril in *Acrotrema* are noteworthy features. The embryogeny is probably of the Onagrad type, and the mature embryo is small.

Paeoniaceae. Paeoniaceae and Dilleniaceae have centrifugal stamens, scalariform vessels, abundant wood rays, a massive bitegmic ovule with the outer integument longer than the inner one and exotestal seed. Corner (1976) favours its assignment close to Dilleniaceae. On the basis of wood anatomy, Keefe and Moseley (1978) support its inclusion in Dilleniales.

On the basis of vascular anatomy (Worsdell 1908), wood anatomy (Kumazawa 1935), palynology (Wodehouse 1936), cytology (Gregory 1941), and floral anatomy (Eames 1951) it has been suggested that *Paeonia* should be assigned to a separate family, Paeoniaceae.

R. Dahlgren (1981) includes Paeoniaceae in Paeoniales, close to Dilleniaceae in the Dilleniales. Takhtajan (1980) agrees with both these orders except that he also includes Crossosomataceae in the Dilleniales. Cronquist (1981) includes both Dilleniaceae and Paeoniaceae in Dilleniales. Y.D. Tiagi (1970b) emphasizes that while *Paeonia* differs from Ranunculaceae, it resembles Dilleniaceae and Crossosomataceae (Table 22.18.1). Willis (1973) states that *Paeonia* is related to Ranunculaceae. Hutchinson (1973) refutes any relationship of *Paeonia* with the Dilleniaceae. He considers it to be much more related to Helleboraceae (separated from Ranunculaceae) than to the Magnoliaceae.

The justification for an independent family, the Paeoniaceae, is provided by Subramanyam (1960) and B.M. Johri (1963b). The embryological features in which *Paeonia* differs from other members of the Ranunculaceae (Table 22.3.1) are: bitegmic ovule, massive outer integument longer than the inner, presence of hypostase, long embryo sacs (sometimes multiple gametophytes), persistent antipodals and their multiplication, chalazal endosperm haustorium, a very unusual mode of development of the embryo, and seed-coat derived from the testa alone.

A comparison of the embryology of *Paeonia*, Ranunculaceae, Dilleniaceae and Crossosomataceae (Table 22.18.1) indicates that Paeoniaceae are much more closely related to Ranunculaceae, than to the other two families. Moreover, Paeoniaceae are correctly placed under the Ranales (Hutchinson 1973), and there is no need to raise them to an ordinal rank, Paeoniales. Also, there is no justification to include the Paeoniaceae in the order Dilleniales.

Crossosomataceae. Crossosomataceae can be distinguished from Dilleniaceae (Table 22.18.1) on the basis of campylotropous ovule with multicellular archesporium, massive parietal tissue, hypostase, secondary multiplication of antipodals, and a large curved embryo (Kapil and Vani 1963). Nuclear endosperm shows enucleate vesicles, and aggregation of nuclei in the chalazal region. The seed has a fimbriate aril in *Crossosoma*. The embryogeny and seed structure are comparable to that in Dilleniaceae. The separation of *Crossosoma* from Dilleniaceae to a separate family Crossosomataceae is justified. Kapil and Vani (1963) recognize its separate family rank, and negate its relationship to Rosales or Ranales. The members of the order Ranales exhibit variations in the number of integuments, nature of ovule, types of embryo sacs, large, persistent polyploid antipodals, and occasionally irregular embryogeny. None of the above-mentioned three families show marked similarities with Ranales (Ranunculales of Cronquist 1981).

Actinidiaceae. Actinidiaceae have been treated previously as a component family of Dilleniaceae (see Lawrence 1951). Hutchinson (1973) considers them as a member of Theales. The genus *Saurauia* is treated as a separate unigeneric family (Saurauiaceae) in Theales by Hutchinson. Vijayaraghavan (1965, 1970b) points out that Actinidiaceae show marked differences from Dilleniaceae (Tables 22.18.1, 22.18.2). They have unitegmic, tenuinucellate ovules on axile placenta, raphides in ovary wall, persistent endothelium, hypostase, Cellular type of endosperm, absence of zygotic mantle, and Solanad type of embryogeny. These features support their separation from Dilleniaceae. He further considers the affinities with Marcgraviaceae, and suggests transfer to Theales. The ovules of *Marcgravia* show fusion of integuments at varying levels from base upward. Mauritzon (1939a) considers it as an approach to unitegmic condition in *Actinidia*. Both Actinidiaceae and Marcgraviaceae (Table 22.18.2) have secretory anther tapetum, ephemeral nucellus, Polygonum type of embryo sac, ephemeral antipodals, and Cellular type of endosperm. Centrifugal stamens, hypostase, and Solanad type of embryogeny occur in Theales. Therefore, the assignment of Actinidiaceae to Theales, as proposed by Vijayaraghavan (1970b), is appropriate.

Takhtajan (1980) and R. Dahlgren (1975a, 1981) include Actinidiaceae in Ericales. Poricidal dehiscence of anthers, unitegmic, tenuinucellate ovule with endothelium, Cellular type of endosperm, Solanad type of embryogeny, and presence of iridoids are some of the features allied to Ericales. But the remaining features enumerated by P. Maheshwari (1950) and Palser (1961) for Ericaceae are absent in Actinidiaceae.

Table 22.18.1. Embryology of *Paeonia*, Ranunculaceae, Dilleniaceae and Crossosomataceae

	Paeonia	Ranunculaceae	Dilleniaceae	Crossosomataceae
Stamen	Spirally arranged, centrifugal	Spirally arranged, centripetal	Numerous, not spirally arranged	Numerous, not spirally arranged
Anther tapetum	Irregularly 1- or 2-layered, glandular, cells binucleate	1-layered, glandular, cells uni- to multinucleate; sometimes amoeboid	1-layered, glandular, cells binucleate	Irregularly 2–3-layered, glandular, cells 2–4-nucleate
Pollen	2-celled	2-celled	2-celled	2-celled
Ovule	Anatropous, bitegmic, outer integument vascularized, crassinucellate, parietal tissue formed only by divisions in nucellar epidermis, nucellus absorbed by embryo sac even before pollination, hypostase present	Anatropous, uni-, bi- or ategmic, integuments non-vascularized, tenui- or crassinucellate, rarely nucellar epidermis divides and contributes to parietal tissue, hypostase rarely present	Ana- to amphitropous, bitegmic, integuments non-vascularized, crassinucellate, 2–3-layered parietal tissue, nucellar cap is formed	Amphitropous, bitegmic, integuments non-vascularized, crassinucellate, parietal tissue massive, hypostase present
Archesporium	Multicelled, several megaspore mother cells function	Uni- or multicelled, one cell functions	Unicelled	Multicelled, only one megaspore mother cell functions
Embryo sac	Polygonum type, twin to multiple embryo sacs may develop	Polygonum, rarely Allium type; tetrasporic Pyrethrum type in *Thalictrum angustifolium*	Polygonum type	Polygonum

Antipodal cells	Large and persistent, multinucleate, secondary multiplication occurs	Persistent, ephemeral in *Adonis*, each with one or more nuclei, become polyploid	Ephemeral	Divide and forms 5 or 6 cells
Endosperm	Nuclear type, chalazal haustorium develops	Nuclear, chalazal haustorium present (*Nigella*)	Nuclear, chalazal haustorium absent	Nuclear, enucleate cytoplasmic vesicles occur, chalazal haustorium absent
Embryogeny	The zygote undergoes free-nuclear divisions followed by centripetal wall formation and differentiation of several peripheral embryonal primordia, only one primordium develops into a dicotyledonous embryo	Onagrad type, rarely Solanad or irregular	Onagrad; a "mantle" develops around the zygote	Onagrad
Seed	Arillate, albuminous, testa massive and epidermis and hypodermis are thick-walled, mesophyll, inner epidermis and tegmen become crushed	Ex-arillate, albuminous in bitegmic ovules, testa formed chiefly by outer integument, in achenes seed-coat is crushed	Arillate, albuminous, exotestal	Arillate, thinly albuminous, exotestal
Fruit	Follicle	Achene	Dehiscent or baccate	Bivalved capsule

Table 22.18.2. Embryology of Actinidiaceae, Theaceae and Marcgraviaceae (See Mauritzon 1939a, Vijayaraghavan 1965)

	Actinidiaceae	Theaceae	Marcgraviaceae
Anther tapetum	Secretory, cells binucleate	Secretory, cells binucleate	Secretory, cells binucleate
Ovules	Numerous, anatropous, unitegmic, tenuinucellar, hypostase present	4 or 5, anatropous, bitegmic, tenuinucellar, micropyle formed by inner integument; hypostase present	Numerous, anatropous, bitegmic[a], tenuinucellar, micropyle formed by inner integument, hypostase absent
Embryo sac	Polygonum type	Allium	Polygonum
Antipodals	Degenerate before fertilization	Degenerate after fertilization	Antipodal nuclei (cells not formed) degenerate before fertilization
Endosperm	Cellular type	Nuclear	Cellular, the uppermost cell becomes conspicuous, penetrates into the micropyle and behaves as a haustorium
Embryogeny	Solanad type	Solanad	—
Polyembryony	Occurs	Occurs	—
Seed-coat	Exotestal, outer epidermis persists, cells filled with tannin, endothelium compressed	Mesotestal, outer integument persists, cells filled with tannin	Exotestal, outer epidermis of outer integument persists, cells filled with tannin

[a] According to Mauritzon (1939a), the two integuments of *Marcgravia* fuse to varying length from the base upward. He points out that this condition is an intermediate step in the derivation of unitegmic condition in *Actinidia*. Although the integuments arise distinctly and in succession, during later stages their identity at the chalazal half of the ovule becomes obliterated. This is partly due to addition of cells which increase their thickness, and partly to mutual pressure. The cells of the outermost layers of the outer integument become filled with orange-brownish contents (perhaps some phenolic compounds) at the time the embryo sac is ready for fertilization. The endothelium is restricted only to the chalazal half of the embryo sac.

The presence of unitegmic and tenuinucellate ovules, endothelium towards the chalazal end of the embryo sac, Polygonum type of embryo sac, small and ephemeral antipodal cells, and Cellular type of endosperm in *Saurauia* indicate its close affinity to Actinidiaceae. Its removal from Actinidiaceae to a separate family (Saurauiaceae) by Hutchinson (1973) does not seem proper on embryological grounds.

Of the ten families belonging to the last three suborders, embryological data are available for only five families: Ochnaceae, Dipterocarpaceae, Theaceae, Marcgraviaceae and Guttiferae. Bentham and Hooker (1965a) place Ochnaceae in Geraniales, and Dipterocarpaceae in Guttiferales along with Guttiferae and Hypericaceae. Hutchinson (1973) groups Ochnaceae and Dipterocarpaceae in Ochnales, Theaceae, Actinidiaceae, Saurauiaceae and Marcgraviaceae in Theales, and Hypericaceae and Clusiaceae in Guttiferales.

Schnarf (1931) states that the Ochnaceae, Marcgraviaceae, Theaceae and Guttiferae are embryologically similar. Cronquist (1981) places the first three families in his Theales. The secretory anther tapetum, simultaneous cytokinesis in pollen mother cells, two-celled tricolpate pollen grains, anatropous, bitegmic and tenuinucellate ovules, absence of parietal cell, Polygonum type of embryo sac, Nuclear type of endosperm, and exalbuminous seed are some of the common features. The endosperm is of the Cellular type, and the seed is albuminous in Marcgraviaceae. Apospory and synergid polyembryony occur in Ochnaceae and Guttiferae.

Ochnaceae. The distinctive features of Ochnaceae are: prominent hypostase, false endothelium (from nucellus), hypertrophied antipodals, and membraneous seed-coat. It also differs from Geraniales in embryological features, such as presence of a well-developed parietal tissue, hypertrophied antipodals, and exalbuminous seed. Corner (1976) prefers to treat the similarity of seed structure of Ochnaceae to that of Simarubaceae (Rutales). The study of its embryogeny would be helpful in determining the phyletic position. At present, its assignment to Guttiferales, rather than to Geraniales, appears to be more satisfactory.

Dipterocarpaceae. Dipterocarpaceae show several notable features. The fibrous thickenings of endothecium are absent in several members, outer integument is vascularized, a meristematic zone is present at the base of nucellus, and the embryo sac is highly aggressive in the chalazal region (in *Vateria*). The Nuclear endosperm contains uni- or multinucleate vesicles. The embryogeny is probably of the Asterad type, or irregular, and the seeds are exotegmic.

Theaceae and Marcgraviaceae. The affinities of Theaceae and Marcgraviaceae (Table 22.18.2) have already been discussed (see paragraph preceding Ochnaceae). The presence of a vascular strand in the outer integument of ovule, Allium type of embryo sac, Solanad type of embryogeny, and sclerotic mesotesta in seed-coat segregate Theaceae. The signs of fusion of integuments, Cellular type of endosperm, and arillate albuminous seed distinguish the family Marcgraviaceae.

Ploiarium. The embryology of *P. alternifolium* has been studied by N. Prakash and Lau (1976). This taxon shares several features with *Camellia* (Kapil and Sethi 1963, Sethi 1965), but there are some differences too. In *Ploiarium* the embryo sac is monosporic, Polygonum type, the antipodals do not organize into cells, hypostase is absent, the shape of ovule and structure of embryo are markedly different, and whether the seed is endotestal or exotegmic is not certain. In *Camellia* the embryo sac is bisporic, Allium type, antipodals organize as cells, hypostase is present, and seed is mesotestal. We do not have any information on the embryology of *Bonnetia.* According to Hutchinson (1973), *Bonnetia* and *Ploiarium* should both be included in Bonnetiaceae; and N. Prakash and Lau support the transfer of *Ploiarium* from Theaceae to Bonnetiaceae. Melchior (1964) assigns *Bonnetia* to Theaceae. The differences in the embryology of *Camellia* and *Ploiarium* do warrant their separation and transfer of *Ploiarium* to Bonnetiaceae. An embryological study of *Bonnetia* may provide further support.

Hypericaceae. Wettstein (1935), Engler and Diels (1936), and Melchior (1964) include Hypericaceae in Guttiferae.

There is a controversy about the development of endosperm in *Hypericum.* Schnarf (1914) describes it as of the Nuclear type, whereas Palm (1922) mentions Helobial type. According to Swamy (1946a), the endosperm is of the Nuclear type in *H. mysorense*; a coenocytic cyst develops in the chalazal region at the 16-nucleate stage of endosperm. It persists for a long time, and then degenerates. Nagaraja Rao (1957b) states that the coenocytic cyst in the Nuclear type of endosperm merges with the general cytoplasm in *H. patulum* and *H. mysorense.* Govindappa (1956a) made similar observations in *H. japonicum.* Myers (1964) reports the fusion of polar nuclei to form a secondary nucleus, prior to opening of the flower. Fertilization may not occur, and the endosperm develops autonomously. The division of secondary nucleus is followed by wall formation in *H. virginicum.* He also mentions the occurrence of aposporic embryo sacs from nucellar cells. The development of endosperm in *Hypericum* needs further investigation.

Guttiferae. Vestal (1937) and Hutchinson (1973) consider that the family Guttiferae seems to have been derived from Theaceae, or from stocks ancestral to them. The exotegmic seed-coat segregates Guttiferae from Theaceae where the seed-coat is mesotestal. In Guttiferae the exotegmen has characteristic large, lignified, tabular, stellate cells. Mesotestal condition occurs only in the subfamily Calophylloideae (Guttiferae).

23 Sarraceniales

The Sarraceniales, an order of insectivorous plants, are considered by most taxonomists to comprise a natural taxon. It has three families: Sarraceniaceae, Nepenthaceae and Droseraceae.

The development of anther, pollen, endosperm, embryo and seed structure has not been investigated in Nepenthaceae. The embryogeny in Sarraceniaceae needs detailed study.

23.1 General Features

The characteristic features of the order Sarraceniales are: (1) secretory anther tapetum of binucleate cells, (2) simultaneous divisions in pollen mother cells, (3) colporate, two-celled pollen grains at the dispersal stage; united in tetrads in Nepenthaceae and Droseraceae, (4) anatropous, bitegmic, tennuinucellar ovule; unitegmic in Sarraceniaceae, (5) Polygonum type of embryo sac, conspicuous antipodals, and a nucellar jacket in Droseraceae, (6) Nuclear type of endosperm with chalazal haustorium; Cellular type in Sarraceniaceae, and (7) endospermic seed, winged in Sarraceniaceae.

23.2 Sarraceniaceae

Shreve (1906) has given a detailed account of embryology of *Sarracenia purpurea* (see also Nichols 1908, Livingston 1950).

The anther wall comprises persistent epidermis, fibrous endothecium, two or three ephemeral middle layers, and secretory tapetum of binucleate cells in *Sarracenia purpurea* (Shreve 1906). The microspore mother cells undergo meiotic divisions. Cytokinesis is simultaneous, and the tetrads are tetrahedral.

The pollen grains have eight meridional grooves in *Sarracenia*, and are two-celled at the dispersal stage.

The ovule is anatropous, unitegmic, and tenuinucellar. The archesporial cell functions directly as megaspore mother cell. It undergoes meiotic divisions, and a linear or T-shaped tetrad is produced. The chalazal megaspore develops into a Polygonum type of embryo sac (Fig. 23.2.1A–F). Sometimes twin gametophytes may develop (Fig. 23.2.1G). A hypostase differentiates. An endothelium develops at the two-nucleate embryo sac. The polar nuclei fuse at the time of fertilization, and the three antipodals are pushed to one side in a pouch during downward elongation of the embryo sac.

The development of the endosperm is of the Cellular type (Fig. 23.2.1H–J). The primary endosperm nucleus undergoes a transverse division. Due to repeated transverse divisions, a row of eight cells is produced. Longitudinal divisions follow later on. The endosperm stores aleurone grains.

A transverse division of zygote occurs after the formation of about 150 endosperm cells. The terminal cell of a uniseriate row of five to seven cells of the proembryo develops into the embryo proper. The suspensor is strongly curved (Fig. 23.2.1J–L).

The seed is endospermic. The outer epidermis of the testa has thick outer walls with conspicuous pores in lateral walls. The raphe develops a wing on one side.

M Livingston (1950), M Nichols (1908), F Shreve (1906)

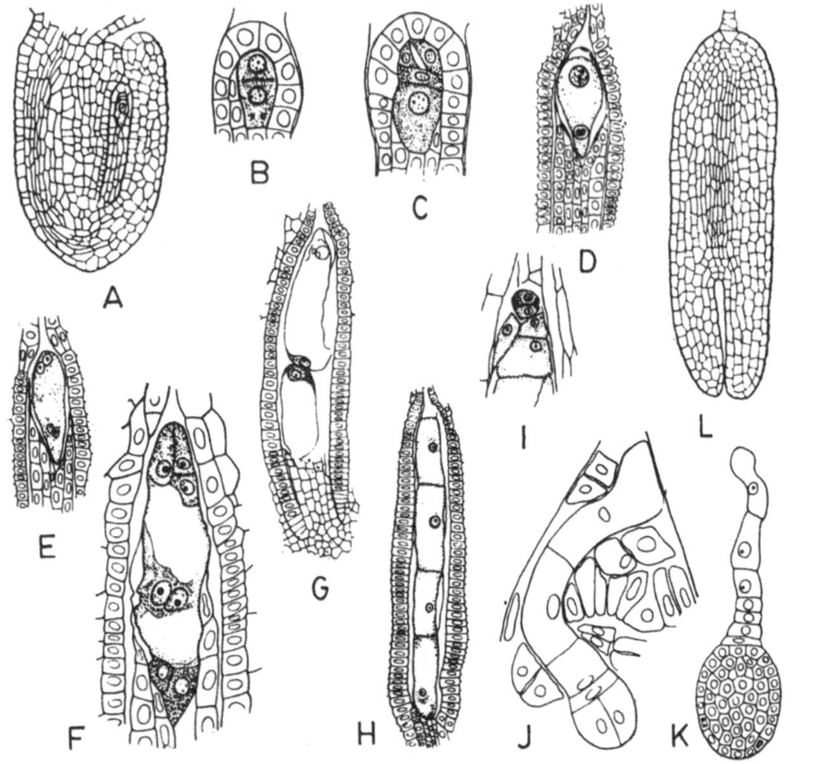

Fig. 23.2.1A–L. *Sarracenia purpurea*. **A** Longisection of young ovule. **B** Megaspore mother cell at telophase I. **C** T-shaped tetrad. **D, E** Two- and four-nucleate embryo sac. **F** Mature embryo sac and endothelium. **G** Twin embryo sacs lying one upon the other. **H** Four-celled endosperm, and prominent endothelium. **I** Micropylar part of embryo sac with two-celled proembryo and a few endosperm cells. **J** Early globular proembryo with suspensor. **K** Later stage. **L** Mature embryo. (After Shreve 1906)

23.3 Nepenthaceae

The development of anther, pollen, endosperm and embryo has not been investigated.

The pollen grains remain attached in permanent tetrads.

The ovule is anatropous, bitegmic, and crassinucellate in *Nepenthes* (Stern 1917). The archesporial cell cuts off a parietal cell; the parietal tissue is single-layered. The megaspore mother cell undergoes meiotic divisions. The chalazal megaspore of a linear tetrad develops into a Polygonum type of embryo sac (see Davis 1966).

GL Davis (1966), R Kühl (1933), K Stern (1917), K Subramanyam, LL Narayana and KT Sundari (1985)

23.4 Droseraceae

The anther wall comprises the epidermis, fibrous endothecium, an ephemeral middle layer, and secretory tapetum. The epidermis is persistent and irregular. The tapetal cells are binucleate; nuclear fusions may occur producing tetraploid and octaploid nuclear masses. Venkatasubban (1950) described occasional intrusion of tapetal cells into the pollen sac dividing it into chambers, and the formation of false periplasmodium in *Drosera indica* and *D. peltata*. The microspore mother cells undergo meiotic divisions resulting in tetrahedral and isobilateral tetrads. During simultaneous cytokinesis, definite walls are laid down to organize the microspores. Compound pollen grains are formed due to the failure of their separation in *D. burmanni* (Narasimhachar 1949).

The generative cell develops in the distal region. Occasionally, in *D. longifolia, D. rotundifolia* and their hybrids (Rosenberg 1909), the vegetative and generative cells are quite similar to each other. The pollen grains are three-colpate, pantoporate (e.g. *Drosophyllum*), and spinulate. They are two-celled at the shedding stage. Sometimes, they germinate in situ. The male gametophyte deserves further study.

The ovule is anatropous, bitegmic, and crassinucellate. The inner integument forms the micropyle (Fig. 23.4.1A, E, H). The hypodermal archesporial cell cuts off a parietal cell in *D. burmanni* (Narasimhachar 1949, Patankar 1956). There are one to three archesporial cells. A single layer of parietal cells develops. In *Drosera rotundifolia* (Pace 1912), *D. indica* (Narasimhachar 1951b) and *D. peltata* (Venkatasubban 1950), the archesporial cell directly functions as the megaspore mother cell. The latter divides meiotically producing a T-shaped tetrad in *Aldrovanda vesiculosa*. The chalazal megaspore develops into a Polygonum type of embryo sac (Fig. 23.4.1A–F). The antipodals are conspicuous in *D. peltata* and *D. burmanni*. The cells of the nucellus are radially elongated, and form a jacket around the embryo sac (false endothelium) in *D. indica* (Narasimhachar 1951b).

During double fertilization triple fusion precedes syngamy.

Development of the endosperm is of the Nuclear type (Fig. 23.4.1G, H), and free nuclei aggregate in the chalazal region. Centripetal wall formation is initiated in the micropylar region, after the formation of 32–64 free nuclei. The chalazal part remains free-nuclear for a fairly long time, and acts as a haustorium. Ultimately, it becomes cellular. The persistent endosperm contains plenty of starchy food reserve.

A transverse division in the zygote results in an apical, and a basal cell. The apical cell develops into the embryo proper, and also contributes to the formation of the suspensor. The basal cell may or may not divide, and add to the suspensor in *D. burmanni* and *D. indica*. The embryogeny (Fig. 23.4.2A–I) may be of the Caryophyllad type. In *Dionaea muscipula* the terminal cell of the two-celled proembryo divides vertically, indicating the Asterad or Onagrad type of development. The embryogeny of *Drosera* (see Crété 1963a) conforms to Period II, Megarchetype IV, Group XI (Caryophyllad type). The mature embryo is microscopic.

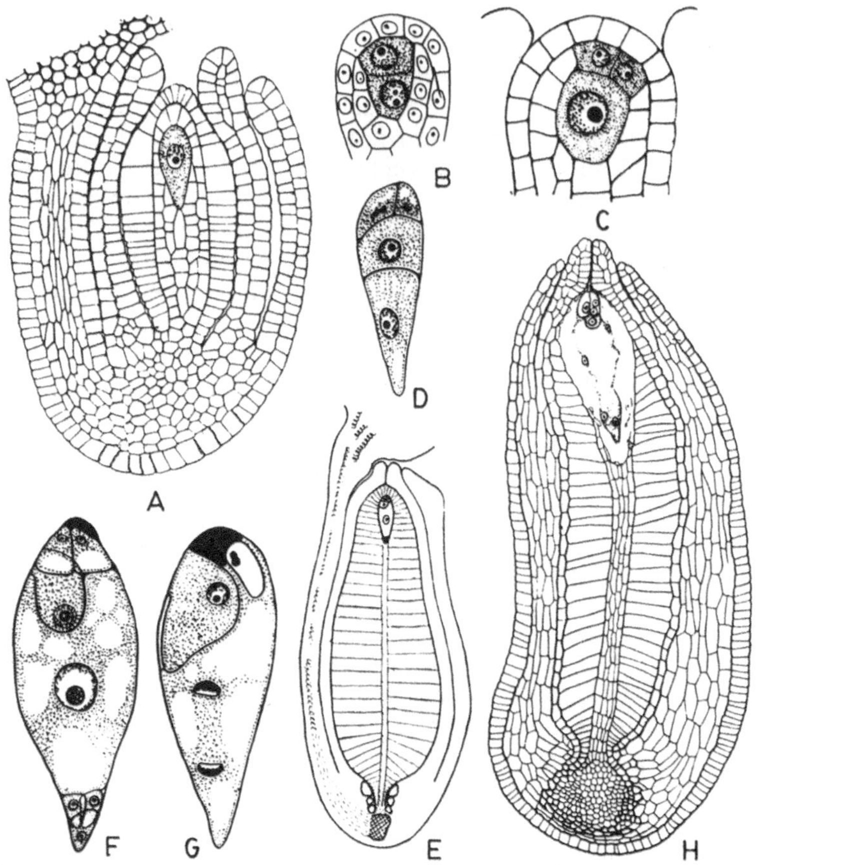

Fig. 23.4.1A–G. *Aldrovanda vesiculosa.* **A** Longisection of anatropous, bitegmic, crassinucellate ovule at megaspore mother cell; **B, C** Megaspore mother cell with parietal cell/s; **D** T-shaped tetrad; **E** Longisection of ovule at mature embryo sac (diagrammatic). Note the conspicuous nucellus in **E** and **H**; **F** Embryo sac from **E**; **G** Embryo sac with zygote, remnants of synergid and pollen tube, and division of primary endosperm nucleus. **H** *Dionaea muscipula,* longisection of ovule with well-developed hypostase, also in **E**. The embryo sac shows zygote, synergids, and free-nuclear endosperm. One synergid has two nuclei. There are no remnants of pollen tube. (See Batygina and Yakovlev 1985)

The outer epidermis of testa is thin-walled with radially elongate cells (see Fig. 23.4.2J). The inner epidermis of tegmen is persistent only in *Drosera intermedia* and *D. rotundifolia*. The inner and radial walls of cells are lignified.

FD Boesewinkel (1989), P Crété (1963a), SG Narasimhachar (1949, 1951b), L Pace (1912), TB Patankar (1956), JE Planchon (1848), O Rosenberg (1909), R Souèges (1936b), H Takahashi (1988), K Venkatasubban (1950)

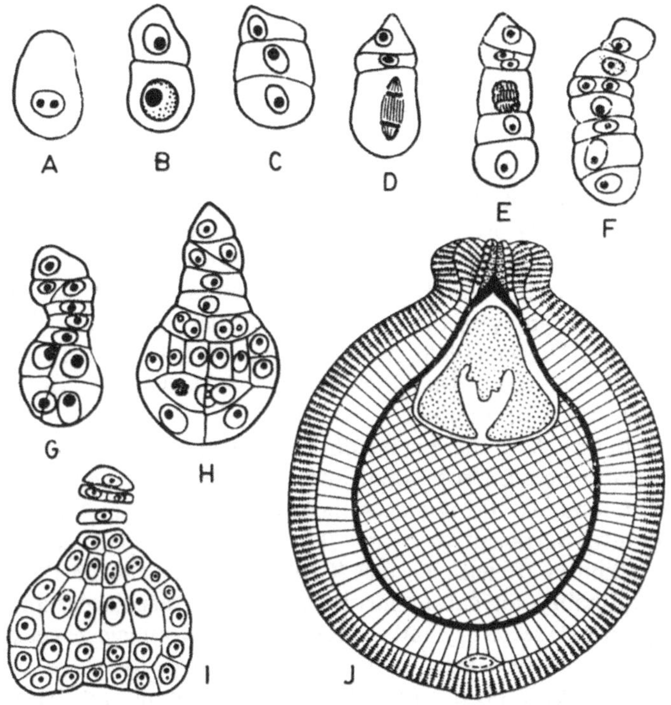

Fig. 23.4.2A–J. **A–I** *Drosera burmanni*; embryogenesis of Caryophyllad type. **J** *Aldrovanda vesiculosa*, longisection of endospermic seed (diagrammatic). Outer epidermis of testa composed of palisade-like cells with transverse thickenings on radial walls followed by mesophyll; tegmen thickened (*bold black line*). Embryo *dotted*, nuclear endosperm *cross-hatched*. (**A–I** After Patankar 1956, for **J** see Batygina and Yakovlev 1985)

23.5 Taxonomic Considerations

Wettstein (1935) places Droseraceae in Parietales due to parietal placentation, while other taxonomists assign all the three families to Saxifragaceae on morphological basis. Engler and Diels (1936) and Melchior (1964) group these families – Sarraceniaceae, Nepenthaceae and Droseraceae – in Sarraceniales. Hutchinson (1973) transfers Nepenthaceae to Aristolochiales, and retains Sarraceniaceae and Droseraceae in Sarraceniales, near Saxifragaceae. R. Dahlgren (1975a, 1981) erects three orders: Sarraceniales (Sarraceniaceae) in superorder Corniflorae; Theales (Nepenthaceae) and Droserales (Droseraceae) in superorder Theiflorae.

Nepenthaceae, Droseraceae and Sarraceniaceae. Scanty data are available on the embryology of Nepenthaceae. It has a bitegmic, crassinucellate ovule with a single layer of parietal cells, and Polygonum type of embryo sac. Nepenthaceae and Droseraceae both show pollen grains in permanent tetrads. The formation of a parietal layer is mentioned in *Drosera burmanni* (Patankar 1956), otherwise the ovules are tenuinucellate. Its affinities to Aristolochiales cannot be determined due to lack of sufficient information.

The Droseraceae show occasional development of false periplasmodium, compound pollen grains, bitegmic ovules with single-layered parietal tissue and a nucellar jacket, Nuclear type of endosperm with chalazal haustorium, a minute embryo, and sometimes fibrous endotegmen.

The Sarraceniaceae and Droseraceae are always grouped together. But Bentham and Hooker (1965a) treat the family Sarraceniaceae separately, and place it between Nymphaeaceae and Papaveraceae. Takhtajan (1980) treats Sarraceniaceae as the sole family in the order Sarraceniales, and places this order near Ranunculales. Thorne (1968) places it in Sarraceniiflorae between Theiflorae and Annoniflorae, and R. Dahlgren (1977a, 1981) includes the order Sarraceniales in superorder Corniflorae.

According to de Buhr (1975), there is insufficient evidence to support the alliance between Sarraceniaceae, Droseraceae and Nepenthaceae except their insectivorous habit. Cronquist (1981) places these three families in Nepenthales. Morphologically, Sarraceniaceae are akin to Theales, particularly Theineae, and their affinities are with Theales (Cronquist 1968, Thorne 1968, see de Buhr 1975). The embryological similarities between Sarraceniaceae and members of Theineae are: (1) secretory anther tapetum with two-nucleate cells, (2) simultaneous cytokinesis, two-celled pollen grains, (3) anatropous, bi- or unitegmic tenuinucellate ovule, (4) differentiation of endothelium and hypostase, (5) archesporial cell functions directly as megaspore mother cell, (6) Polygonum type of embryo sac, and (7) tanniniferous seed as in Actinidiaceae, Marcgraviaceae, Theaceae and Aquifoliaceae (see de Buhr 1975).

Following Thorne's (1968) classification, de Buhr (1975) proposes that the family Sarraceniaceae should be treated as a suborder Sarraceniineae near Theineae in Theales.

Pace (1912) transfers the genus *Parnassia* of Saxifragaceae to Droseraceae, on account of the structure of ovule and embryo sac. V.K. Sharma (1968b) compares the embryology of *Parnassia* with Saxifragaceae and Droseraceae (see Rosales, Sect. 26). The bitegmic, tenuinucellar ovule has a single-celled archesporium which functions as megaspore mother cell, and T-shaped tetrads, Polygonum type of embryo sac, persistent antipodals, and Nuclear type of endosperm are common features between *Parnassia* (V.K. Sharma 1968b) and Droseraceae. However, the embryogeny of *P. palustris* (Lebègue 1953) conforms to Period I, Megarchetype III, Group I (Asterad type). It is very similar to that in *Claytonia perfoliata* (Portulacaceae) (see Rau 1962). The embryogeny of *Drosera* conforms to Period II, Megarchetype IV, Group XI (Caryophyllad type), and is far removed from that of *Parnassia*. V.K. Sharma (1968b) suggests that *Parnassia* forms a link between Saxifragaceae and Droseraceae (see Table 26.21.1).

Nepenthaceae and Sarraceniaceae show exotestal seeds similar to those of Saxifragaceae. Droseraceae (*Drosophyllum*) has fibrous exotegmen (Corner 1976).

24 Papaverales

The order Papaverales comprises four suborders: Papavarineae represented by Papaveraceae; Capparineae by Capparaceae, Cruciferae and Tovariaceae; Resedineae by Resedaceae; Morigineae by Moringaceae.

24.1 General Features

The order Papaverales shows: (1) secretory anther tapetum, (2) simultaneous cytokinesis in pollen mother cells, (3) two-celled pollen grains, (4) bitegmic, crassinucellate ovule, (5) Polygonum type of embryo sac with ephemeral antipodals except in Papaveraceae and *Maerua* (Capparaceae), (6) Nuclear type of endosperm, and (7) Chenopodiad and Caryophyllad type of embryogeny in Papaveraceae; Onagrad in Capparaceae, Cruciferae and Resedaceae; Onagrad, Caryophyllad and Solanad in Capparaceae; Asterad in Moringaceae. In all the families, except Moringaceae, the testa is thin-walled and the inner integument (tegmen) serves the function of protection. In Moringaceae the mesotesta is fibrous, and tegmen is membraneous. On the whole, the order appears to be homogeneous except for embryogeny.

24.2 Papaveraceae

The anther wall comprises the epidermis, fibrous endothecium, three ephemeral middle layers, and secretory tapetum with multinucleate cells. The number of middle layers varies: two in *Eschscholzia californica* (Sachar and Mohan Ram 1958), only one in *Fumaria* (H.B. Saksena 1954). The nuclei of tapetal cells fuse to form irregular polyploid masses. The tapetum, at times, becomes two- to three-layered (Sachar 1953). The microspore mother cells undergo meiotic divisions resulting in tetrahedral and isobilateral tetrads. Cytokinesis is simultaneous.

The family is highly multipalynous. The pollen grains are three-colpate, colporate, pantoporate, and pantocolpate. They are two-celled at the dispersal stage.

The transformation of anthers into carpellary tissue bearing naked ovule and stigmatic surfaces, and differentiation of pollen sacs in the ovary wall in *Argemone mexicana* are occasional features (Sachar 1953).

The ovule is anatropous, bitegmic, and crassinucellar. Both integuments take part in the formation of the micropyle. The ovule is campylotropous in *Fumaria*. Stomata occur on the integuments in *Argemone mexicana* (Sachar 1955). In *Eschscholzia* (Sachar and Mohan Ram 1958), the inner integument projects beyond the outer, and forms the micropyle. A single cell from uni- or multicellular archesporium cuts off a parietal cell. The megaspore mother cell undergoes meiotic divisions, and produces a linear or T-shaped tetrad. The chalazal megaspore develops into a Polygonum type of embryo sac. The antipodals become polyploid and hypertrophied, and are persistent (A.C. Joshi 1933). In *Papaver rhoeas*, they become 64- or 128-ploid (Hasitschka-Jenschke 1956). They undergo secondary multiplication in *Dendromecon* (Berg 1967).

In *P. somniferum* (Bhandari and Bhargava 1983), at the mature embryo sac stage, the antipodals show transfer cell wall ingrowths from the basal wall into the cytoplasm. The ingrowths become conspicuous after fertilization. The degenerated antipodals show scanty cytoplasm dispersed between the labyrinth of wall ingrowths. A similar condition is reported in *Chelidonium majus* and *E. californica* (Fougère-Rifot 1979).

Double fertilization occurs, and branching of pollen tubes is common in *Papaver* and *Argemone* (Sachar 1953).

The development of the endosperm is of the Nuclear type. The nuclei fuse in the chalazal region to form irregular masses in *Eschscholzia*. The rest of the endosperm becomes cellular at the young globular proembryo. It stores food in the form of oil. In *Argemone*, the entire endosperm becomes cellular due to the activity of the peripheral layer. Cytokinesis occurs in *Fumaria* after the formation of several hundred endosperm nuclei, at the globular proembryo.

The division of zygote is transverse, rarely oblique. Further development is irregular in *Argemone*. The embryogeny conforms to Period I, Series C, Megarchetype V (Chenopodiad type) in *Papaver rhoeas* (Souèges 1926b, c, 1928). It conforms to Period II, Series C, Megarchetype V (Caryophyllad type) in *Eschscholtzia* (Souèges 1949a), and to Period II, Series B, Megarchetype VI, Group XIII or XIV (Caryophyllad type) in *Fumaria* (Souèges 1941e, see Crété 1963a).

Solanad type is reported in *Platystemon*, and Caryophyllad in *Hypecoum*. The basal cell/s become hypertrophied and polyploid (see Yakovlev 1981).

The origin of an additional embryo in *Argemone mexicana* is controversial. Sachar (1953, 1956a) states that it must be due to the development of a persistent synergid. Souèges (1949b) and Crété (1956a), however, consider that it must be due to the proliferation of suspensor cells. This aspect requires further study.

A notable feature in *Eschscholtzia californica* (Sachar and Mohan Ram 1958) is the occurrence of four, rarely five, cotyledons in the embryo.

The seed is albuminous with a small chlorophyllous embryo. The exotesta is composed of elongate cells with their outer walls thickened. The inner layer consists of radially elongate thin-walled cells with fibrillar cellulose reticulum, and contain crystals or brown pigment. The exotegmen comprises narrow longitudinal fibres, and the endotegmen is a tracheidal layer.

A rudimentary aril develops in *Sanguinaria canadensis* (Nagaraja Rao and Shamanna 1963), and an exostomal-funicular aril in *Dendromecon*. Some cells in the hilum-region develop into an elaiosome in *Dicentra* (Berg 1969). The elaiosome is very conspicuous in *Corydalis* and *Chelidonium* (Fig. 24.2.1A–C); the seed surface of *Argemone* is shown in Fig. 24.2.1D.

RY Berg (1967, 1969), NN Bhandari and M Bhargava (1983), NN Bhandari, M Bhargava and A Sachdeva (1985), PK Bose and I Banerji (1933), P Crété (1956a, 1963a), ÅE Dahl (1989), M

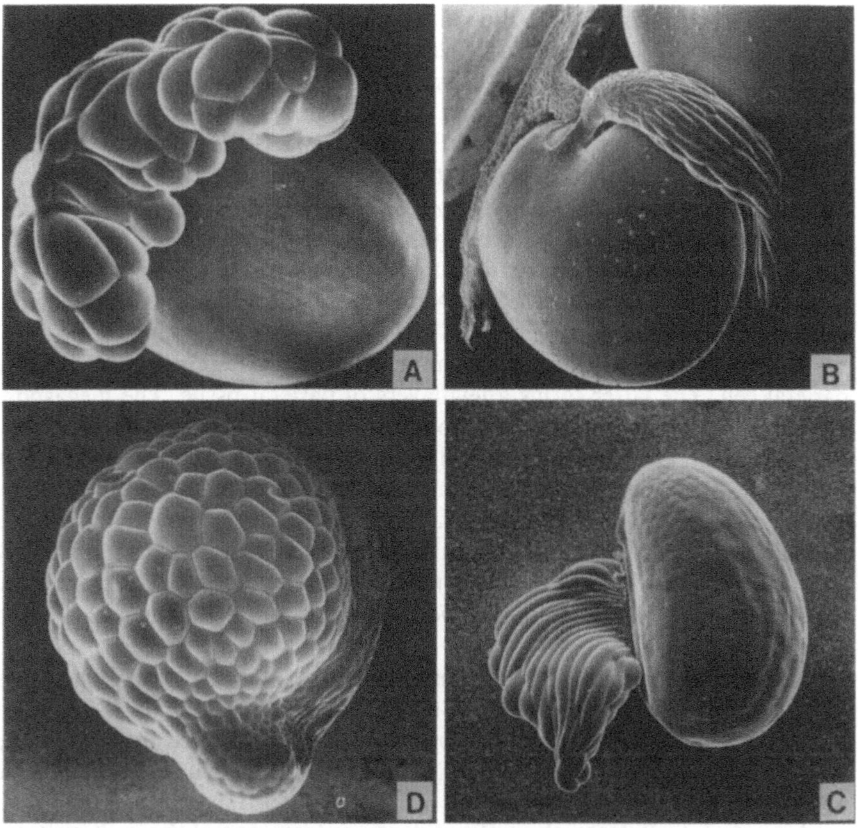

Fig. 24.2.1A–D. (SEM micrographs) **A**, **B** *Corydalis cava*, **C** *Chelidonium majus*, **D** *Argemone platiceras*. **A–C** Seeds with elaiosome. **D** Seed. (Courtesy Dr F. Bouman, Amsterdam)

Fougère-Rifot (1979), V Gupta and LC Lamba (1985), G Hasitschka-Jenschke (1956), JA Jernstedt and C Clark (1979), AC Joshi (1933), RN Kapil and A Nanchahal (1989), A Nagaraja Rao and S Shamanna (1963), RC Sachar (1953, 1955, 1956a), RC Sachar and HY Mohan Ram (1958), HB Saksena (1954), EL Schneider and DM Nichols (1984), RP Singh and AM Khan (1974), R Souèges (1926b, c, e, 1928, 1941e, f, 1949a, b), MS Yakovlev (1981)

24.3 Capparaceae

The anther wall comprises an epidermis, fibrous endothecium, two or three ephemeral middle layers, and secretory tapetum of multinucleate cells. In *Cadaba* (H.S. Narayana 1965) even the epidermal cells develop fibrous thickenings. The tapetum becomes biseriate in *Capparis*. Simultaneous cytokinesis follows meiotic divisions in microspore mother cells, resulting in tetrahedral, decussate and isobilateral (*Cleome monophylla* – A.V.N. Rao 1967) tetrads.

The pollen grains are two- to three-colporate. They are two-celled at the dispersal stage in *Capparis* (H.S. Narayana 1962c, Hanumantha Rao and Prakasa Rao 1977), *Cadaba* and *Crataeva* (H.S. Narayana 1965), and three-celled in *Capparis* and *Cleome* (Tasneem and Kusuma Kumari 1970). Binucleate microspore mother cells and germination of pollen grains in situ have been observed in *Capparis decidua* (H.S. Narayana 1962c).

The ovule is campylotropous, bitegmic, and crassinucellar with a zigzag micropyle. It is anatropous in *Crataeva.* The outermost layer of the inner integument develops into a stony layer in *Isomeris arborea* (P. Maheshwari and Khan 1953). Rudimentary stomata develop along the funicular side of the ovules of *I. arborea* (Sachar 1956b). A hypostase differentiates, and tracheidal cells occur in the nucellus.

A hypodermal cell of uni- or multicellular archesporium produces a parietal cell. The latter forms a few layers of parietal tissue. The megaspore mother cell undergoes meiotic divisions, produces a linear or T-shaped tetrad (T.S. Raghavan and Venkatasubban 1941a), or a linear triad – the upper one is a dyad cell (P. Maheshwari and Khan 1953). The chalazal megaspore develops into a Polygonum type of embryo sac. The synergids are ephemeral in *Isomeris* (P. Maheshwari and Sachar 1954a). The antipodals are usually ephemeral but in *Maerua arenaria* (V.S. Rao 1936a) they are large and occupy nearly half the space of the embryo sac. Twin embryo sacs occur in *Capparis aphylla* (Vaidya 1961a).

The development of the endosperm is of the Nuclear type. It becomes cellular at the late globular or heart-shaped stage of proembryo. P. Maheshwari and Khan (1953) report, in *Isomeris arborea,* nucleated nodules in the middle of the embryo sac, and cambium-like strips in cellular endosperm. The nodules also become cellular.

According to Billings, the embryo develops from one of the multinucleate endosperm nodules. P. Maheshwari and Sachar (1954a), and Sachar (1956b), refute Billings's assumption (1937) of endospermic origin of the embryo in *I. arborea.* The abnormal position of the embryo is partly due to the extension of the micropylar end of the embryo sac leaving the embryo behind. It *does not* arise from the endosperm. The zygote produces a long, uniseriate, filamentous proembryo (Sachar 1956b). The suspensor becomes massive, haustorial, and coiled (Fig. 24.3.1A–F′). Its cells are vesicular and multinucleate (P. Maheshwari and Khan 1953, P. Maheshwari and Sachar 1954a). The embryo proper develops from the two or three uninucleate cells at the tip of suspensor (Fig. 24.3.1D, F).

The embryogeny corresponds to the Caryophyllad type in *Capparis* and *Cadaba,* Solanad in *Crataeva* (H.S. Narayana 1962c, 1965), and Onagrad in *Cleome chelidonii* (T.S. Raghavan 1937) and *Gynandropsis pentaphylla* (Phouphas 1952). Nucellar polyembryony is reported in *Capparis frondosa* (Mauritzon 1934b).

The seed is campylotropous, and exalbuminous with a folded embryo. The testa comprises thin-walled cells with or without stomata. The tegmen develops into a protective tissue. The exotegmen comprises palisade cells with thickened walls or longitudinal fibres followed by lignified endotegmen.

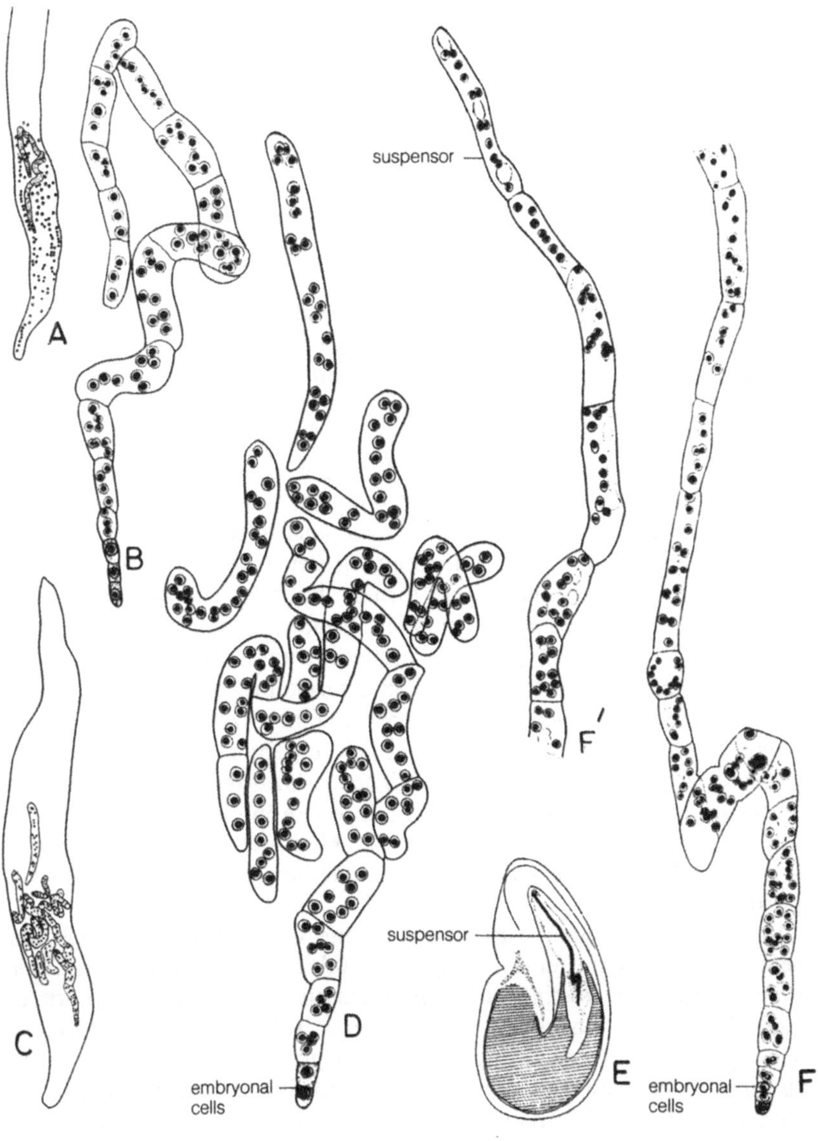

Fig. 24.3.1A–F. *Isomeris arborea.* **A** Lower part of embryo sac (wholemount) with proembryo and free-nuclear endosperm. **B** Proembryo from **A**. **C** Embryo sac (wholemount). Note the proembryo with a long suspensor; endosperm omitted. **D** Proembryo with a highly coiled, uniseriate, suspensor with coenocytic cells, from **C**, **E** Older ovule, the proembryo has a long suspensor. **F**, **F'** Proembryo with suspensor from **E**. The uninucleate cells at the tip (**B**, **D**, **F'**) are progenitors of the embryo proper. (After Sachar 1956b)

V Arunalakshmi (1989), FH Billings (1937), B Hanumantha Rao and PS Prakasa Rao (1977), P. Maheshwari and R Khan (1953), P. Maheshwari and RC Sachar (1954a, 1955), J Mauritzon (1934b), HS Narayana (1962c, 1965, 1970a), MY Orr (1921), C Phouphas (1952), TS Raghavan (1937), TS Raghavan and KR Venkatasubban (1941a), AVN Rao (1967), PS Rao and B Hanumantha Rao (1975, 1976, 1977), VS Rao (1936a, c, 1938), RC Sachar (1956b), F Tasneem and P Kusuma Kumari (1970), NK Tiwary (1936), PB Vaidya (1961a)

24.4 Cruciferae

The anther wall comprises the epidermis, fibrous endothecium, one to three middle layers, and glandular tapetum of binucleate cells. Following meiotic divisions, the microspore mother cells produce tetrahedral, isobilateral and decussate tetrads. Cytokinesis is simultaneous.

The pollen grains are three-colpate and shed at the three-celled stage, except in *Cardamine* and *Chenopodifolia* where they are two-celled. Degeneration of anther, arrested growth of pollen grains, and their improper differentiation have been noted in *Erysimum cheiranthoides* (Belyayeva 1968).

The ovule is ana- to campylotropous (*Brassica juncea* – Sulbha 1957), and bitegmic with a zigzag micropyle formed by both the integuments. It is crassinucellar in *B. campestris* (Ahuja and Bhaduri 1956), and tenuinucellar in *Malcolmia africana*. The hypodermal archesporium is one- to many-celled, but only one cell develops further. It cuts off a parietal cell in *B. juncea*, whereas it functions directly as megaspore mother cell in *Arabidopsis thalianum* and *Erophila verna*. The megaspore mother cell undergoes meiotic divisions and produces a linear or T-shaped tetrad. The chalazal megaspore develops into a Polygonum type of embryo sac. The three antipodals are ephemeral, and may become multinucleate (Sulbha 1957). Hypostase develops in *Matthiola tricuspidata* (Renzoni-Cela 1969). An endothelium, and twin embryo sacs are reported in *B. juncea*.

In *Brassica campestris* (S.P. Bhatnagar and Kaur 1984), the style is hollow, whereas in other Cruciferae the style is solid: *Brassica rapa*, *Capsella bursa-pastoris*, *Diplotaxis tenuifolia* and *Raphanus sativus*. It contains a uniform canal which is lined by a single row of conspicuous thick-walled cells (canal cells). They constitute the transmitting tissue for the growth of pollen tube. The canal cells are larger than the cortical cells, and bulge into the canal. These vacuolated cells are rich in metabolites, and their cytoplasm contains proteins and starch.

During pollination and pollen tube growth, the canal cells shrivel, lose contents, walls break down, and the collapsed cells form a loose mass of cells around the canal.

Development of the endosperm is of the Nuclear type. The endosperm forms prominent nodular, nuclear aggregations in *Capsella bursa-pastoris* (P. Maheshwari and Sachar 1954b). At the initial stage each nodule contains only two to three nuclei, and they are confined to the chalazal region. In later stages they become five- to six-nucleate, and occur in the lower half of the embryo sac. Enucleate cytoplasmic nodules are reported in *Brassica campestris*

(Ahuja and Bhaduri 1956, Rathore and Singh 1968). Nuclear fusions occur in the endosperm of *B. juncea* (Il'ina 1962). The tissue becomes cellular.

The embryogeny corresponds to the Onagrad type-Capsella variation (R.C. Misra 1962, Rathore and Singh 1968). In *Capsella* it conforms to Period I, Series A, Megarchetype IV. The suspensor is thin and filamentous, and the uppermost cell is vesicular. The embryogeny in *Lunaria annua* (see Rodionova 1983) differs from the Capsella variation in the occurrence of a linear tetrad of embryo in addition to the T-shaped, four-celled proembryo. There is no distinct hypophysis cell, and the differentiation of histogens is delayed.

The development of embryo of *Brassica napus* (Tykarska 1976) conforms to Onagrad type, or Megarchetype IV of Souèges:

$$\mathrm{ca} = \mathrm{pvt} + \mathrm{pco} + \mathrm{phy} + \mathrm{icc}$$

$$\mathrm{cb} = \mathrm{iec} + \mathrm{co} + \mathrm{s}$$

Tykarska (1979) describes the development in three phases. Phase I includes growth of the apical cell (*ca*) up to its division into two layers of octants (*l* and *l'*). Phase II comprises the increase of spherical proembryo and change of its symmetry from radial to bilateral. Phase III concerns the growth of embryo from heart-shaped up to the end of embryogenesis. In each phase the growth is very vigorous in the beginning, but later declines.

According to Ramji (1975), the bilateral symmetry is determined much earlier, at the formation of an intercept at the quadrant and octant stage during phase I, and not during phase II as claimed by Tykarska.

Polyembryony is reported in several genera. In *Arabis halleri* and *A. hirsuta* (Mianny 1972), the supernumerary embryo develops from synergid, either by apogamy or by fertilization. In *A. lyalli* (Lebègue 1948b), it develops from an unfertilized synergid. Reduced pseudogamy, i.e. the formation of an embryo without fertilization from a reduced egg, has been observed in *Brassica campestris* × *B. oleracea* var. *gummifera* (Noguchi 1928). *A. holboellii* (Böcher 1947, 1951) exhibits aneuspory of the Taraxacum type: aneuspory results due to irregularities in meiosis during megaspore formation. In this taxon the lower cell of an unreduced dyad develops into an eight-nucleate embryo sac. The same species also shows unreduced pseudogamy. The sperm degenerates inside or outside the embryo sac, but may provide stimulus to the egg to develop pseudogamously.

The seed is endotestal, and exalbuminous. The outer epidermis of testa comprises cells with reticulate thickenings on their radial walls. The mesophyll either collapses or is collenchymatous. The inner epidermis consists of cells with U-shaped thickenings on the inner and radial walls. The outer epidermis of tegmen is pigmented; the inner one consists of palisade cells.

Spooner (1984) studied the sexual population of *Dentaria laciniata*. Following meiosis, the microspore mother cells undergo simultaneous cytokinesis. The tetrads are tetrahedral, and the pollen grains three-celled. The archesporial cell functions directly as megaspore mother cell, a triad is formed, and the embryo sac is of the Polygonum type. The endosperm is of the Nuclear type (later becomes cellular), and the embryogeny is of the Onagrad type-Capsella variation. The seeds germinate readily in nature. Similar development occurs in *D. diphylla* up to embryo sac formation but it does not organize and degenerates. The reproductive events and other data indicate that the eastern North American species of *Dentaria* may form a sexual polyploid complex with some sexual populations and some sterile ones.

YR Ahuja and PN Bhaduri (1956), LY Belyayeva (1968), R Bergfeld, T Kühnl and P Schopfer (1980), SP Bhatnagar and I Kaur (1984), TW Böcher (1947, 1951), F Bouman (1975), R Czapik (1974), J Eigner (1973), S Ellerström and L Zagorcheva (1977), D Gerlach-Cryse (1970), I Grant, WD Bevers-dorf and RL Peterson (1986), MW Green (1978), Å Håkansson (1956), MI Hakki (1974), GM Il'ina (1962), J Kallarackal and SP Bhatnagar (1978), M Kroh and A Munting (1967), M Kuras (1978), LC Lamba (1975), LC Lamba and BB Arora (1981a, b), Lebègue (1948b), T Mackiewicz (1973), P Maheshwari and RC Sachar (1954b), SJ Mayer, AA Snow and ML Stanton (1986), N Mianny (1972), RC Misra (1962, 1966), Y Noguchi (1928), G Norton and JF Harris (1975), PM Pechan (1988), K Prasad (1977a, b, 1979), MV Ramji (1975), RKS Rathore and RP Singh (1968), GC Renzoni-Cela (1969), GB Rodionova (1972, 1983), P Schulz and WA Jensen (1986), C Simoncioli (1974), LD Smith and DH Scarisbrick (1990), W Soerodikoesomo (1970), DM Spooner (1984), AD Stead, IN Roberts and HG Dickinson (1979), K Sulbha (1957), MJ Sumner and L van Caeseele (1989), T Tykarska (1976, 1979), L van Caeseele, JT Mills, M Sumner and R Gillespie (1981), JG Vaughan (1968), JG Vaughan and JM Whitehouse (1971), MS Yakovlev and GK Alimova (1976), MD Yoffe (1952)

24.5 Tovariaceae

The development of anther and embryo has not been investigated.

The anther tapetum is of the secretory type. The microspore mother cells undergo meiotic divisions, and cytokinesis is simultaneous.

The pollen grains are two-celled at the dispersal stage.

In *Tovaria pendula* (Mauritzon 1934b), the ovule is campylotropous, bitegmic, and crassinucellar with a zigzag micropyle. The archesporial cell cuts off a parietal cell which subsequently divides repeatedly. The megaspore mother cell undergoes meiotic divisions. The chalazal megaspore of a linear tetrad develops into a Polygonum type of embryo sac. The antipodals are ephemeral.

The development of endosperm is of the Nuclear type. Later it becomes cellular.

The seed is small, and exarillate. The embryo is bent, contains oil, and is surrounded by copious endosperm. The testa is thin-walled. The outer epidermis of tegmen, in the region of endostome and raphe, comprises palisade cells with reticulate thickenings on the walls.

J Hutchinson (1973), J Mauritzon (1934b), P Schürhoff (1926)

24.6 Resedaceae

The development of the anther wall has not been investigated.

The microspore mother cells undergo meiotic divisions, and tetrahedral and isobilateral tetrads are produced. Cytokinesis is simultaneous.

The pollen grains are two-celled at the dispersal stage.

The ovule is ana- to campylotropous, bitegmic, crassi- or tenuinucellar with a single vascular strand in the integument. The archesporium is multicellular, and each cell cuts off a parietal cell. Several megaspore mother cells undergo meiotic divisions in *Reseda odorata* (Gori 1957). The chalazal megaspore of linear tetrads develops into Polygonum type of embryo sac. The polar nuclei fuse before fertilization, and antipodals are ephemeral. A hypostase develops.

The development of the endosperm is of the Nuclear type. Later it becomes cellular.

The embryogeny corresponds to the Onagrad type. The suspensor is uniseriate.

The seed is reniform, exotegmic, and albuminous with a curved embryo. The outer epidermis of testa consists of papillate cells with thick outer walls, and the inner epidermis is a crystal layer in *Oligomeris linifolia* (D. Singh and Gupta 1968). In *Reseda odorata* (D. Singh and Gupta 1968), in the mature seed the testa becomes thin and papery. The outer epidermis of tegmen forms the mechanical layer of thick-walled, lignified, palisade-like cells. In *Reseda* it forms ridges and furrows. The inner epidermis has dark contents.

P Crété (1955a), C Gori (1957), D Singh and S Gupta (1968), MS Yakovlev (1983)

24.7 Moringaceae

The anther wall comprises an epidermis, fibrous endothecium, a single ephemeral middle layer, and a glandular tapetum of binucleate cells (H.S. Narayana 1962b). The development of the anther wall follows the "Basic" type (Davis 1966) in *Moringa concanensis* (B.S.M. Dutt 1978a). The endothecial thickenings extend to four or five layers in the sterile theca. Übisch granules are studded on the inner walls of tapetal cells. The microspore mother cells undergo meiotic divisions, and produce tetrahedral, isobilateral and decussate tetrads. Cytokinesis is simultaneous, and quadripartitioning occurs by furrowing.

The pollen grains are three-colpate, and shed at the two-celled stage. B.S.M. Dutt and Prakasa Rao (1977) report the presence of pollen grains in the stylar canal of *M. concanensis*.

The ovule is anatropous, bitegmic, and crassinucellate. The integuments are vascularized; the hypostase is persistent. A single hypodermal archesporial cell cuts off a parietal cell which gives rise to three to four layers of parietal tissue. Following meiotic divisions, the megaspore mother cell produces a T-shaped or linear tetrad, sometimes a triad (V. Puri 1941). The development of the female gametophyte is of the Polygonum type. The antipodals are ephemeral.

The occurrence of twin nucelli, twin embryo sacs, egg-like synergids, and supernumerary nuclei in the embryo sac has been reported in *Moringa oleifera* (V. Puri 1934).

The endosperm is of the Nuclear type, and the nuclei are crowded at the micropylar end. Wall formation commences from the micropylar end at the late globular stage of proembryo, and the entire endosperm becomes cellular.

The development of the embryo corresponds to the Asterad type in *M. oleifera* (H.S. Narayana 1962b). The suspensor is absent. However, B.S.M. Dutt (1978c) states that the suspensor is persistent and disintegrates only in mature seed. The embryogeny conforms to Myosurus-variation, Onagrad type in *M. concanensis* (Period I, Megarchetype IV). The embryo consists of thick cotyledons, and the radicle is short. The development of the embryo requires further study.

The seed is mesotestal, exalbuminous, and the testa alone forms the seed-coat. It is multiplicative. The outer and inner epidermis of testa are thin-walled. The thin-walled mesophyll, containing chlorophyll and starch, develops into a thick-walled, pitted, strongly lignified compact tissue. At an early stage, the meristematic activity of hypodermal cells, at three corners, gives rise to wings which become vascularized. The tegmen persists as a thin membraneous layer.

RM Datta and JN Mitra (1947), GL Davis (1966), BSM Dutt (1978a, c, 1979b), BSM Dutt and PS Prakasa Rao (1977), HS Narayana (1962b, 1970b), K Periasamy and Indira (1986), V Puri (1934, 1941), FL Rutgers (1923)

24.8 Taxonomic Considerations

Bentham and Hooker (1965a) place Cruciferae, Capparidaceae, Resedaceae and Papaveraceae in Parietales. Hutchinson (1973) includes Papaveraceae and Fumariaceae in Rhoeodales; Capparaceae and Moringaceae in Capparales; Cruciferae (Brassicaceae) in Cruciferales (Brassicales), and Resedaceae in Resedales. Engler and Diels (1936) include all the families in Rhoeodales. Cronquist (1968, 1981) treats the orders Papaverales and Capparales separately as they differ in the nature of secretory cells, chemical substances, and orientation of stamens. Palynological and serological data also support this separation. He assigns Papaveraceae under Magnoliidae, and Capparales under Dilleniidae.

Papaveraceae. Papaveraceae are characterized by hypertrophied persistent antipodals which undergo secondary multiplication, Chenopodiad and Caryophyllad type of development of embryo, thick-walled exotesta, and endotestal crystal layer. The exotegmen is fibrous with tracheidal endotegmen.

Fumaria. Melchior (1964) places the genus *Fumaria* in Papaveraceae. Hutchinson (1973) and Cronquist (1981) recognize it as a separate family, Fumariaceae, and place it in Rhoeodales along with Papaveraceae. *Fumaria* differs from Papaveraceae in the absence of latex; zygomorphic flowers with spurs, two lateral tripartite stamens, anther wall with a single middle layer, pantoporate pollen grains, campylotropous ovule, cytokinesis in the nuclear endosperm after the formation of several hundred free nuclei (H.B. Saksena

1954), and embryogeny conforms to Period II, Megarchetype VI, Group XIII and, at times, Group XIV (Souèges 1941d, e) – Caryophyllad type. H.B. Saksena (1954) justifies the erection of the family Fumariaceae.

Capparaceae. H.S. Narayana (1970a) states that Capparales are different from Papaverales. Capparales (Capparaceae, Tovariaceae, Moringaceae and Resedaceae) have three-colpate spheroidal pollen with sexine as thick as nexine, while Papaverales (Papaveraceae and Fumariaceae) are characterized by two- to nine-colpate, rupate, and rugate pollen, or pollen provided with irregular apertures. Also, Capparales are different from Papaverales in ephemeral antipodals, and large polar nuclei. According to H.S. Narayana, Tovariaceae and Capparaceae have probably originated from a common ancestor. He suggests the erection of the orders Papaverales and Capparales.

Capparaceae resemble most to Resedaceae. The campylotropous ovule with parietal tissue and hypostase, Polygonum type of embryo sac with ephemeral antipodals (rarely hypertrophied as in *Maeruva*), Nuclear type of endosperm, Onagrad type of embryogeny, curved embryo, albuminous seed, and thin-walled testa and fibrous tegmen are common features. In addition, Capparaceae show cytoplasmic endosperm nodules as in Cruciferae, and Solanad type of embryogeny. The long, massive suspensor in *Isomeris* is an unusual feature.

Resedaceae. According to D. Singh and Gupta (1967), the endotegmen in the seed of Resedaceae has been misinterpreted as the third integument. The seed-structure of Resedaceae is comparable to violaceous seed. Papaveraceae, Capparaceae and Resedaceae have exotegmic seed. As in the seeds of Capparaceae, the exotegmen is one- to many-layered. D. Singh and Gupta consider Capparaceae as the "Basic" type from which seed-structure of some other families can be derived (see also Netolitzky 1926).

Cruciferae show close alliance with Capparaceae. The occurrence of chlorophyllous embryos in both families indicate a close affinity (Yakovlev and Zhukova 1975). The exalbuminous seed with characteristic endotesta, and fruit of Cruciferae indicate greater specialization.

Tovariaceae. Scanty information is available on Tovariaceae, but the development of ovule and endosperm is in general conformity with that of the order.

Moringaceae. De Jussieu (1789) and Hallier (1912) place *Moringa* in Leguminoseae. Lindley recognizes it as a separate family, Moringaceae, and assigns to the order Violales. Hutchinson (1973) associates Moringaceae with Capparaceae in the order Capparales. V. Puri (1941) considers it to be related to Cruciferae and Capparaceae. R.M. Datta and Mitra (1947) support Lindley, and relate Moringaceae to Violaceae and Tovariaceae, on the basis of zygomorphy of flower, perigyny, and parietal placentation. H.S. Narayana (1970b) compares the embryology of Moringaceae with that of Violaceae, Leguminoseae and Polygalaceae, and negates any relationship with these families. The presence of myrosin cells in Moringaceae is similar to that in Capparaceae and

Resedaceae. The structure of anther, pollen, ovule, embryo sac, and endosperm is similar to that in Capparaceae and Cruciferae. A hypostase, and exalbuminous seeds also occur in Cruciferae. However, the occurrence of a short gynophore, anatropous ovule, Asterad type of embryogeny, thin-walled testa containing chlorophyll, lignified mesotesta, and membraneous tegmen are their characteristic features. Consequently, Moringaceae may have diverged from the primitive Capparaceae. The mesotestal seed-coat indicates ancestry to Theales.

25 Batales

The order Batales is represented by the family Bataceae. Its embryology has not been studied.

26 Rosales

The order Rosales comprises 19 families, distributed over four suborders: Hamamelidineae – Platanaceae, Hamamelidaceae, Myrothamnaceae; Saxifragineae – Crassulaceae, Cephalotaceae, Saxifragaceae, Brunelliaceae, Cunoniaceae, Davidsoniaceae, Pittosporaceae, Byblidaceae, Roridulaceae, Bruniaceae; Rosineae – Rosaceae, Neuradaceae, Chrysobalanaceae; Leguminosineae – Connaraceae, Leguminosae, Krameriaceae.

26.1 General Features

The order exhibits a great deal of variation, but some common features are: (1) secretory anther tapetum, both secretory and amoeboid in Saxifragaceae, (2) simultaneous cytokinesis in pollen mother cells, (3) two-celled pollen grains, two- to three-celled in Papilionaceae, and three-celled in Pittosporaceae, (4) sterile transverse septa delimit groups of archesporial cells (in the pollen sac), each group is encircled by its own tapetum, and pollinia are produced in Mimosaceae; pollen embryo sacs observed in Saxifragaceae, (5) ovule ortho-, ana-, hemiana- or campylotropous, crassinucellar, tenuinucellar in Pittosporaceae and Byblidaceae, and both types in Saxifragaceae and Caesalpiniaceae, usually bitegmic, unitegmic in Brunelliaceae, and uni- and bitegmic in Saxifragaceae and Rosaceae; integumentary vascular strand present in Mimosaceae and Caesalpiniaceae, (6) unicellular archesporium, multicellular in Crassulaceae and Cunoniaceae, and uni- and multicellular in Papilionaceae, (7) Polygonum type of embryo sac, Allium type in Crassulaceae, Rosaceae, Saxifragaceae and Papilionaceae, and Adoxa type in Saxifragaceae, (8) Nuclear type of endosperm in Cunoniaceae, Pittosporaceae, Rosaceae, Connaraceae and Leguminosae, Cellular in Crassulaceae, Nuclear and Cellular in Hamamelidaceae, and Nuclear, Cellular and Helobial types in Saxifragaceae; endosperm haustoria develop in Crassulaceae, Rosaceae, Byblidaceae and Leguminoseae, (9) embryogeny follows Solanad type in Hamamelidaceae, Caryophyllad in Crassulaceae, Solanad or Caryophyllad in Saxifragaceae, Asterad in Pittosporaceae, Caryophyllad and Asterad in Rosaceae, Onagrad in Mimosaceae, Asterad in Caesalpiniaceae, and Asterad, Onagrad, Solanad and Caryophyllad in Papilionaceae, (10) exotestal seed, mesotestal in Rosaceae, and (11) apospory in Rosaceae.

26.2 Platanaceae

The embryology has not been fully investigated.

The anther tapetum is glandular with two- to four-nucleate cells in *Platanus acerifolia*. Following meiosis in microspore mother cells, cytokinesis is simultaneous (Bretzler 1924). The pollen grains are three-colpate and two-celled.

The ovule in *Platanus* is orthotropous, bitegmic and crassinucellate. The micropyle is formed by both the integuments, and the vascular bundle terminates at the chalaza. The archesporial cell cuts off a parietal cell; the parietal tissue is well developed. The development of the embryo sac is of the Polygonum type (Bretzler 1924).

The development of the endosperm is of the Cellular type.

The seeds are very small with a thin endosperm, and linear embryo with flat cotyledons. The seed-coat is almost crushed.

E Bretzler (1924), J Brouwer (1924), PK Endress (1977a), PR Mohana Rao (1974a), T Nicoloff (1911), K Schnarf (1931), MS Yakovlev (1981)

26.3 Hamamelidaceae

The anther wall comprises an epidermis, fibrous endothecium, two or three middle layers, and secretory tapetum. The tapetum becomes two- to three-layered in *Exbucklandia populnea* (Kaul and Kapil 1974). The tapetal cells are multinucleate, their nuclei fuse and become polyploid. Orbicules are studded on the inner wall of tapetal cells. Simultaneous cytokinesis follows meiotic divisions in microspore mother cells. The tetrads are tetrahedral, isobilateral and decussate.

The pollen grains are tri-, tetra- and pentacolpate. They are two-celled at the dispersal stage.

The ovule is anatropous, bitegmic and crassinucellate; both integuments form the micropyle. The outer integument elongates, and serves as a partial obturator. The cells of the nucellar epidermis undergo periclinal divisions. The hypostase is thick-walled. The sporogenous tissue is massive in *Corylopsis willmotiae* (Endress 1977a). The archesporial cell cuts off a parietal cell. The parietal tissue is eight to ten layers thick in *Parrotiopsis jaquemontiana* (Kapil and Kaul 1972). The megaspore mother cell undergoes meiotic divisions, and produces a linear or T-shaped tetrad. The second megaspore (from the micropylar end) develops into a binucleate embryo sac, or the lower two develop up to the two-nucleate stage. However, the chalazal megaspore develops into a Polygonum type of embryo sac. One of the synergids may persist up to the octant stage of proembryo. The polar nuclei fuse before fertilization. The antipodals are generally persistent.

During double fertilization, triple fusion precedes syngamy. The pollen tube undergoes overwintering in *Hamamelis* (Endress 1977a).

The development of the endosperm is of the Cellular type in *Parrotiopsis jaquemontiana* (Kaul 1969, Kapil and Kaul 1972), and Nuclear type in *Exbucklandia populnea* (Kaul and Kapil 1974) and *Hamamelis virginiana* (Mathew 1980). In *Exbucklandia*, the endosperm becomes cellular after a few nuclear divisions. It is persistent and stores fats and proteins.

The development of the embryo conforms to the Solanad type in *Parrotiopsis*, and Chenopodiad type-Hamamelis variation in *H. virginiana*. The mature embryo has two large leafy cotyledons in *Exbucklandia*.

The seeds are winged; the wing develops from the free-end of the outer integument. Its cells contain tannin. The testa is massive and sclerotic. The tegmen is thin-walled and tanniniferous (Mohana Rao 1974a).

PK Endress (1977a), FF Flint (1957a, b, 1959), LD Hufford and PK Endress (1989), RN Kapil and U Kaul (1972), U Kaul (1969), U Kaul and RN Kapil (1974), CJ Mathew (1980), AP Melikian (1973), PR Mohana Rao (1974a), F Niedenzu (1891), E Shaw and RD Gibbs (1961), KY Tong (1930), RF Thorne (1973)

26.4 Myrothamnaceae

The development of the embryo has not been fully investigated.

The anther wall comprises the epidermis, endothecium, three or four middle layers, and secretory tapetum of binucleate cells. Cytokinesis is simultaneous, and the tetrads are tetrahedral (Jager-Zürn 1966).

The binucleate pollen grains remain in permanent tetrads in *Myrothamnus flabellifolia* and *M. moschata* (Jager-Zürn 1966).

The ovule is anatropous, bitegmic, and crassinucellate. The inner integument forms the micropyle in *M. flabellifolia*. The outer integument overgrows the inner one in *M. moschata* (Jager-Zürn 1966). The archesporial cell cuts off a parietal cell. The parietal tissue is three to four layers thick. The megaspore mother cell undergoes meiotic divisions and the chalazal dyad develops into a bisporic Allium type of embryo sac. The antipodals are ephemeral.

The development of the endosperm is of the Nuclear type. The embryogeny (Jager-Zürn 1966) appears to be of the Crucifer type.

I Jäger-Zürn (1966), E Shaw and RD Gibbs (1961)

26.5 Crassulaceae

The anther wall comprises the epidermis, fibrous endothecium, two middle layers, and secretory tapetum with uninucleate cells. The microspore mother cells undergo meiotic divisions, and produce tetrahedral and isobilateral tetrads. Cytokinesis is simultaneous, and furrows develop centripetally.

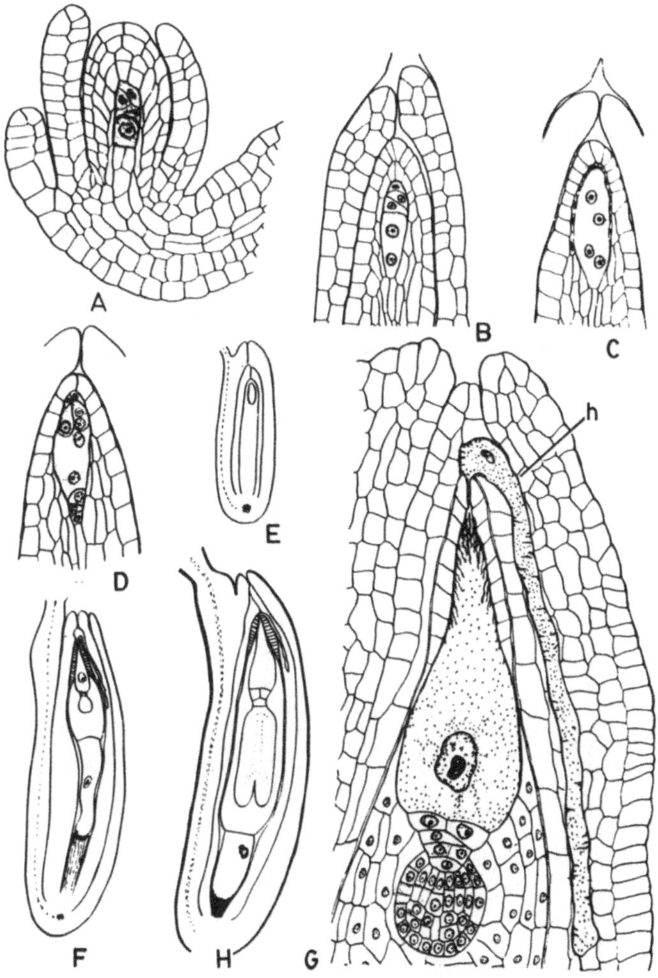

Fig. 26.5.1A. *Kalanchoe faustii*, bitegmic ovule with functional chalazal megaspore, the upper three megaspores have degenerated. **B–H** *Sedum hispanicum* (**E**, **G** and **H** diagrammatic); **B–D** Two- and four-nucleate, and mature embryo sac; **E** Anatropous, bitegmic ovule; **F** Outline figure for **G**; **G** Micropylar region of ovule. The globular proembryo has a highly expanded haustorial uppermost cell of suspensor. Note the wall-ingrowths (transfer cell) in the upper portion. The suspensor haustorium (*h*) invades the inner integument. Cellular endosperm surrounds the proembryo; **H** Mature seed. The lower cell in **F** and **H** is the chalazal endosperm haustorium. (See Batygina and Yakovlev 1985)

The pollen grains are three- or four-colporate and two-celled at the shedding stage.

The ovule is anatropous, bitegmic, and crassinucellate (Fig. 26.5.1A, E–H). The archesporial cell cuts off a parietal cell. Following meiotic divisions, the megaspore mother cell produces linear, isobilateral, T-shaped or even oblique T-shaped tetrad (Fig. 26.5.1A, B). A peculiar feature is the development of megaspore haustoria in *Sedum sempervivoides* (Mauritzon 1933a), and *S. chrysanthum* (Fig. 26.5.2A–D). The multiple archesporium produces several megaspore mother cells, each gives rise to a tetrad. Each megaspore develops a tubular haustorial process which grows upward towards the micropyle. Several such megaspore haustoria intertwine, and form a tangle in the upper part of nucellus. Some of them become extranucellar, and lie at the base of the micropyle. Development of the embryo sac is of the Polygonum type (Fig. 26.5.1A–D). Allium type of development is reported in *Sedum fabaria*, *S. populifolium* and *S. populifolium* var. *notarjanni* (Mauritzon 1933a). The synergids show a cap-like structure at the apex, or from their apex arise three or four slender tubular processes in *Sedum*. The synergids develop haustoria in *S. sempervivoides* (Mauritzon 1933a). The polars fuse before fertilization. There are three small antipodals which develop haustoria in *S. fabaria* (Mauritzon 1933a). The embryo sac grows considerably towards the chalaza, becomes

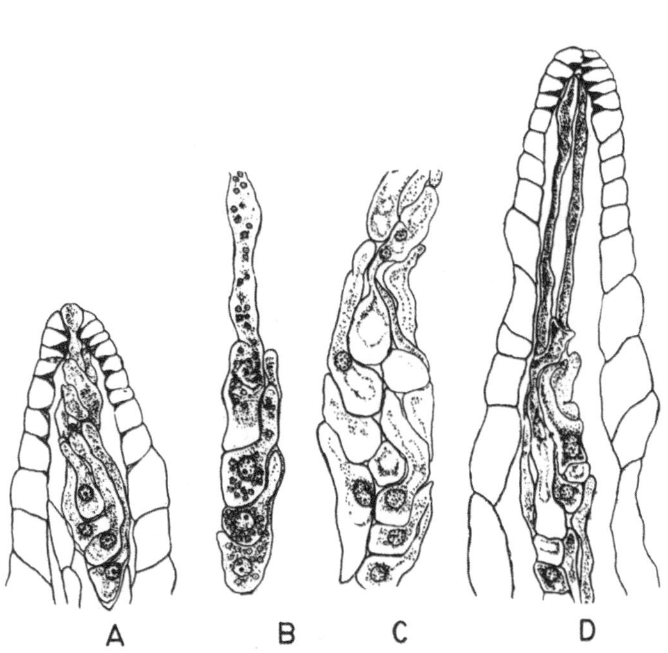

Fig. 26.5.2A–D. *Sedum chrysanthum.* **A** Upper part of nucellus to show a haustorium from each megaspore of tetrad. There are additional haustoria (at the apex) from other megaspores in the same nucellus. **B** Megaspores and haustoria contain starch grains. **C, D** Groups of megaspore tetrads with haustoria. A binucleate embryo sac has developed in **C**. (After Subramanyam 1967)

haustorial, and accumulates a large quantity of starch. These observations need confirmation.

The entry of pollen tube is porogamous, and double fertilization occurs.

The development of the endosperm is of the Cellular type (Fig. 26.5.1F–H). The division of primary endosperm nucleus is followed by a transverse wall producing a larger micropylar and a smaller chalazal chamber. The chalazal

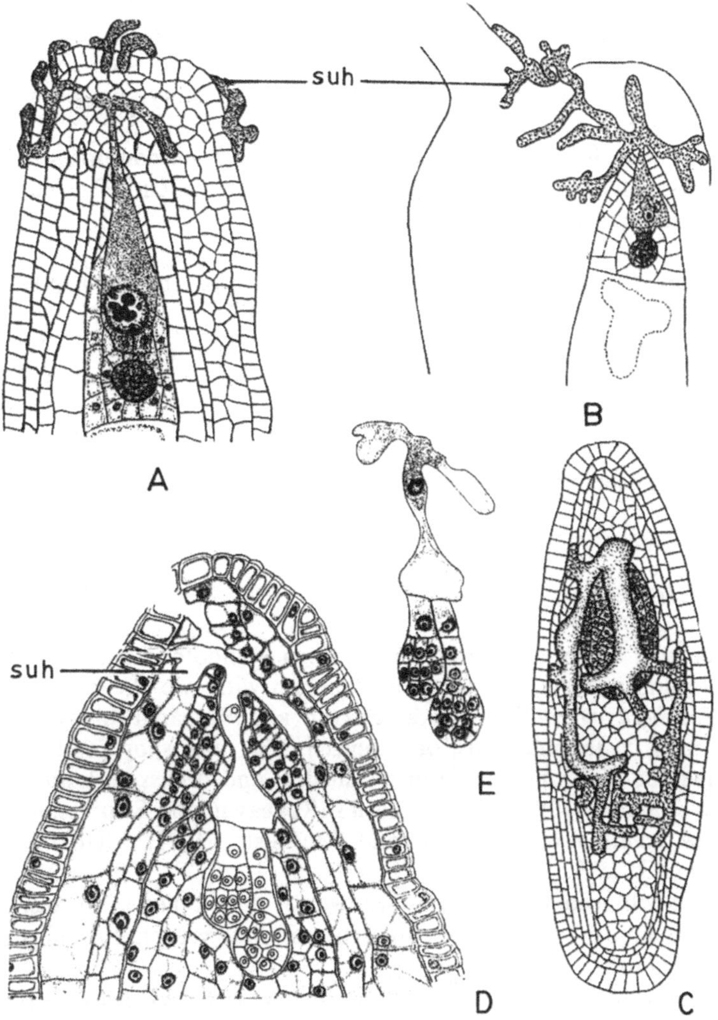

Fig. 26.5.3. A *Pistorinia hispanica*, micropylar portion of ovule at globular proembryo, the suspensor haustorium becomes extra-ovular. **B** *Sedum acre*, longisection of upper portion of seed, the suspensor haustorium (*suh*) branches and ramifies in the seed. **C** *Aeonium bentejui*, longisection (paradermal) of ovule at globular proembryo, the branched suspensor haustorium invades the integument. (After Mauritzon 1933a). **D, E** *S. spectabile*; **D** Micropylar part of seed showing twin embryos and cellular endosperm, suspensor haustorium (*suh*) invades the seed-coat; **E** Twin embryos with basal cell as suspensor haustorium, from **D**. (After Vignon-Fétré 1966)

chamber develops into a uninucleate haustorium. Five subtypes of development are recognized, according to the mode of wall formation in the micropylar chamber: Acre, Spathulata, Rupestre, Crassula, and Tatra subtype (Mauritzon 1933a).

The embryogeny conforms to the Caryophyllad type-Sedum variation, or Period II, Series A, Megarchetype IV (Souèges 1936d, Crètè 1946c, Subramanyam 1963). Sometimes, twin proembryos with a common basal cell may develop (Fig. 26.5.3D, E). An interesting feature in embryogeny is the haustorial behaviour of the basal cell. It becomes hypertrophied and vesicular, and develops fine haustorial processes penetrating the seed-coat (Figs. 26.5.1F–H, 26.5.3A–E). The presence of an antipodal embryo has been noted in *Sedum fabaria* (Mauritzon 1933a); this needs confirmation.

The testa alone forms the seed-coat. Its outer epidermis comprises radially elongate cells with thickened outer walls and brownish contents. The outer two layers are parenchymatous.

P Crètè (1946c), J Mauritzon (1930, 1933a, 1939b), R Souèges (1925c, 1927, 1936d, e), K Subramanyam (1962, 1963, 1967, 1970a), J Vignon-Fétré (1966)

26.6 Cephalotaceae

The embryology has not been investigated.

26.7 Saxifragaceae

The anther wall comprises the epidermis, fibrous endothecium, one or two ephemeral middle layers, and secretory tapetum with two- to four-nucleate cells in *Heuchera micrantha* (Vijayaraghavan and Ratnaparkhi 1977). The epidermis may be tanniniferous, or cutinized. The microspore mother cells undergo meiotic divisions and form tetrahedral and decussate tetrads, isobilateral in *Vahlia* (*Bistella*) *oldenlandioides* (Raghavan and Srinivasan 1942) and *Bauera capitata* (N. Prakash and McAllister 1977). Cytokinesis is simultaneous.

The three-colporate pollen grains are two- or three-celled at the dispersal stage, and are dimorphic. They are two-colpate in *Bauera capitata*, and gorged with starch grains.

Some noteworthy features have been observed in *Heuchera micrantha* (Vijayaraghavan and Ratnaparkhi 1977). Idioblasts and clustered crystals occur in the connective tissue. It also shows fibrous thickenings. The tapetal cells become four-nucleate. In addition, occurrence of "pollen embryo sacs" is unique. The uninucleate microspores, at times, enlarge, undergo three successive divisions, and give rise to eight-nucleate embryo sac-like structures. The organization simulates Polygonum, Allium or Adoxa type of female gametophyte. Further divisions of nuclei occur in some cases. The nuclear arrangements simulate Penaea and Plumbago variations, but the cells are not organized. The formation of pollen embryo sacs is associated with the degeneration of tapetal cells.

The ovules are anatropous, bitegmic, and crassinucellate in most species of the subfamilies Penthoroideae, Saxifragoideae (*Bistella* is tenuinucellate; *Peltiphyllum* and *Saxifraga pallida* are unitegmic), and Baueroideae; bitegmic and tenuinucellate in Brexioideae and Parnassioideae; unitegmic and tenuinucellate in Kirengshomoideae and Hydrangeoideae (Rau 1970). The micropyle is organized by both the integuments.

In *Ribes* spp. (Mauritzon 1936a, see also Batygina and Yakovlev 1985), the archesporium consists of one or two cells. It cuts off a parietal cell which develops a small parietal tissue. The sporogenous cell functions directly as megaspore mother cell. The megaspore tetrads are linear, and occasionally T-shaped. The chalazal megaspore develops into a Polygonum type of embryo sac. The polar nuclei fuse before or at the time of fertilization, and the antipodals are ephemeral in *Saxifraga* and *Bistella*. However, the antipodals are prominent in *Parnassia nubicola* (V.K. Sharma 1968b), and their number increases up to five in *Parnassia* and *S. diversifolia* (Saxena 1964a, b), and seven in *Bauera capitata* (N. Prakash and McAllister 1977). At times, the development of the embryo sac is bisporic in *S. pallida* (Rau and Sharma 1967). In *Bistella oldenlandioides*, the embryo sac grows laterally, and an endothelium with binucleate cells develops. Twin embryo sacs are common in *Bergenia ciliata* (Saxena 1969b), *S. diversifolia* (Saxena 1963, 1964a), and *S. verginensis* (Chapman 1933).

The position of the secondary nucleus is variable. In the species with Helobial type of endosperm development it lies near the antipodals, otherwise in the upper half of the embryo sac.

Rodgersia asculifolia, *Tellima grandiflora* and *Tolmiea menziesii* (Ratnaparkhi 1972, Vijayaraghavan and Ratnaparkhi 1975) show Polygonum as well as Allium type of development. The embryo sac becomes extramicropylar in *Philadelphus tomentosus* (Vijayaraghavan and Ratnaparkhi 1975).

A good deal of variation occurs in the mode of endosperm formation. It may be Cellular, Helobial, or Nuclear. Both Cellular and Helobial types occur in *Boykinia*, *Mitella*, *Ribes* and *Tiarella* (Rau 1970), Helobial type in *Chrysosplenium* and *Saxifraga* (Rau and Sharma 1967, Kaplan 1976), and Cellular type in *Astilbe*, *Bergenia* (Saxena 1969a, b), *Kirengeshoma*, *Penthorum* and *Bistella*. Nuclear type is recorded in *Bauera*, *Brexia*, *Fendlera*, *Francoa* and *Parnassia*. Wall formation is initiated at the globular proembryo in *Parnassia nubicula*, and a single layer persists in seed. In *Bistella*, four uninucleate cells of the chalazal endosperm become richly cytoplasmic, suggesting their haustorial role. In *Mitella pentandra*, a non-aggressive cellular haustorium is formed from the chalazal chamber.

The embryogeny corresponds to the Caryophyllad type or Period II, Megarchetype IV (Rau and Sharma 1967). *Parnassia palustris* belongs to the Asterad type, or Period I, Megarchetype III (Lebègue 1953). Synergid polyembryony occurs in *Bergenia delavayi* (Lebègue 1949). *Penthorum* and *Tellima* also show the presence of additional embryos.

The seed is small, and albuminous with a minute embryo. The testa is represented by radially elongate or thick-walled papillate cells. The tegmen gets crushed, and persists only in the micropylar region. The inner epidermis of tegmen is thick-walled and persistent in *Heuchera*. The SEM studies of seed coats of four species of *Saxifraga* (Kaplan 1976) show differences in the number and distribution of humps and papillae, and the fine structure of cuticle.

TB Batygina and MS Yakovlev (1985), CR Bensel and BF Palser (1975), M Chapman (1933), A Engler (1894), K Kaplan (1976), K Kaplan and M Strohschneider (1984), MS Hils, WC Dickison, TW Lucansky and WL Stern (1954), A Lebègue (1949, 1953), C Le Cog (1967), Vya Mandryk (1972), Vya Mandryk and MA Andrusyak (1969), J Mauritzon (1933a, 1936a), L Pace (1912), BP Pandey and NP Saxena (1977), N Prakash and EJ McAllister (1977), TS Raghavan and VK Srinivasan (1942), S Ratnaparkhi (1972), MA Rau (1970), MA Rau and VK Sharma (1967), NP Saxena (1963, 1964a, b, 1969a, b, 1970, 1971, 1973, 1976), VK Sharma (1968b), R Souèges (1936a), K Subramanyam (1962), MR Vijayaraghavan and S Ratnaparkhi (1975, 1977, 1979), IL Wiggins (1959)

26.8 Brunelliaceae

The embryology has not been studied.

26.9 Cunoniaceae

The embryogeny has not been investigated.

The anther wall comprises the epidermis containing tannin, fibrous endothecium, a single middle layer, and secretory tapetum with binucleate cells in *Weinmannia fraxinea* (Govil and Saxena 1976). Simultaneous cytokinesis follows meiosis in microspore mother cells, and the tetrads are tetrahedral.

The pollen grains are two-celled at the dispersal stage.

The ovule is anatropous, bitegmic, and crassinucellar with a zigzag micropyle. The archesporium is multicellular, and each cell cuts off a parietal cell. The parietal tissue is well developed. The cells of the nucellar epidermis also divide and increase the parietal tissue. Kennedy and Prakash (1981) report a nucellar cap and hypostase in *Callicoma serratifolia*.

A single megaspore mother cell functions, and undergoes meiotic divisions. The chalazal megaspore of a linear or T-shaped tetrad develops into a Polygonum type of embryo sac. Epichalazal megaspore functions in *Schizomeria* and, occasionally, in *Weinmannia*. The polar nuclei fuse before fertilization, and the secondary nucleus lies at the chalazal end of embryo sac. The antipodals are ephemeral. A hypostase differentiates in *Weinmannia*.

The development of the endosperm is of the Nuclear type. Later, it becomes cellular. It is persistent.

The seed is exotegmic. The outer epidermis of the testa comprises tangentially elongate cells; the cells of the inner epidermis are vacuolate. The outer epidermis of tegmen forms the sclerotic layer with radially elongate cells; the radial walls are strongly lignified. The inner epidermal cells are tangentially elongate, and contain tannin.

KVO Dahlgren (1939a), CM Govil and NP Saxena (1976), MJ Kennedy and N Prakash (1981), J Mauritzon (1939b)

26.10 Davidsoniaceae

The embryology has not been worked out.

26.11 Pittosporaceae

The anther wall comprises the epidermis, fibrous endothecium, one or two ephemeral middle layers, and secretory tapetum (Fig. 26.11.1A). The tapetum is irregularly bi- and tri-seriate in *Billardiera cymosa, B. lehmamiana, B. latifolia* (L.L. Narayana and Sundri 1983) and *Pittosporum heterophyllum* (L.L. Narayana and Sundri 1978). The tapetal cells are binucleate in *Hymenosporum flavum* and *Pittosporum floribundum*, and vesicular and multinucleate (Fig. 26.11.1B) in *Bursaria spinosa* (Sheela and Narayana 1966, L.L. Narayana and Sundri 1978). In *Marianthus bignonaceus* (Sundri and Narayana 1977), they are multinucleate and become polyploid due to nuclear fusions. Simultaneous cytokinesis in microspore mother cells is followed by furrowing, and tetrahedral and decussate (*Sollya heterophylla* – L.L. Narayana and Sundri 1979) tetrads are produced; rhomboidal and isobilateral tetrads are recorded in *P. heterophyllum*.

The three-colporate pollen grains are three-celled at the dispersal stage (Fig. 26.11.1C); two-celled in *Billardiera verifolia* (L.L. Narayana and Sundri 1983). At times, the pollen grains germinate in situ in *Marianthus*. Several abnormalities, such as double pollen grains, occurrence of two to six nuclei in a pollen grain, and both the nuclei of pollen grain of similar size, are reported in *Marianthus* (Sundri and Narayana 1977).

The orientation of ovules in the ovary of *Pittosporum floribundum* shows variation. These are hypotropous in the micropylar half and epitropous in the chalazal half.

The ovule is anacampylotropous, unitegmic, and tenuinucellar. The hypodermal archesporial cell functions directly as megaspore mother cell, undergoes meiotic divisions, and produces a triad, or linear, or T-shaped tetrad. The chalazal megaspore develops into a Polygonum type of embryo sac. The polar nuclei fuse prior to fertilization, and the antipodals are ephemeral (Fig. 26.11.1D–I).

The development of the endosperm is of the Nuclear type (Fig. 26.11.1J). Five- to six-nucleate nodules, invested by a cytoplasmic membrane, occur in *Bursaria spinosa* (L.L. Narayana and Sundri 1978). Later, the endosperm becomes cellular (Fig. 26.11.1K) and packed with food reserve.

The embryogeny (Fig. 26.11.1L–R) conforms to the Geum-variation of Asterad type (L.L. Narayana and Sundri 1978, 1979). The mature embryo is very small with two to five cotyledons. Suspensor polyembryony has been noted in *Bursaria spinosa* (Fig. 26.11.1S) and *Pittosporum* spp. (L.L. Narayana and Sundri 1978, 1979).

The seed-coat comprises three to 12 layers of aerenchymatous cells. The outer epidermis consists of compact and thick-walled cells.

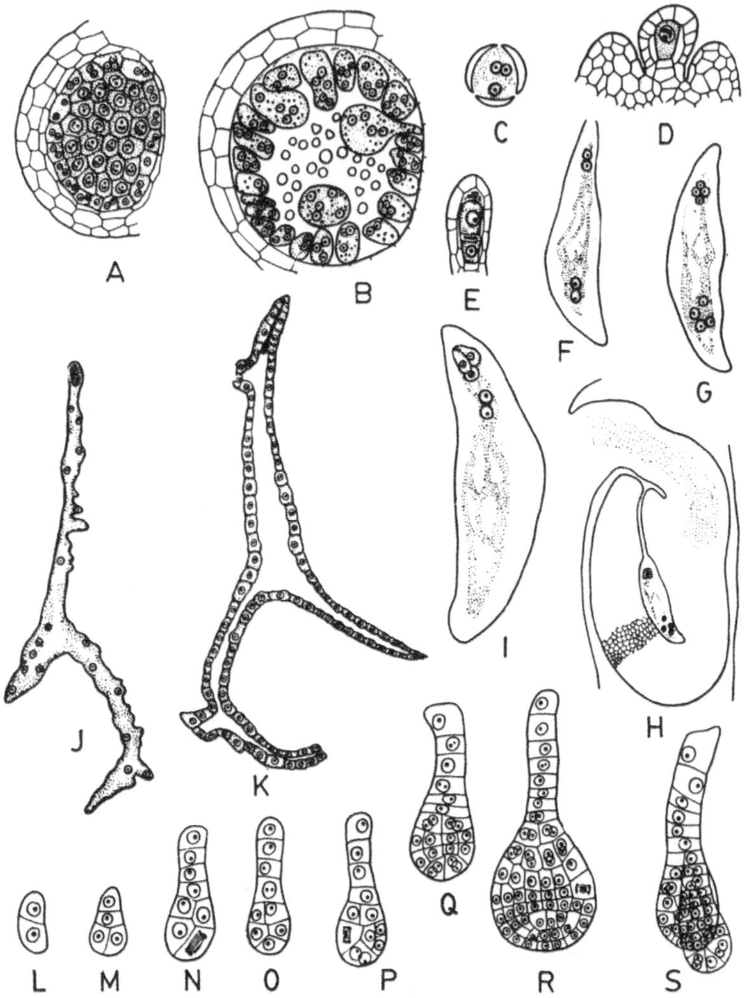

Fig. 26.11.1A–S. *Bursaria spinosa.* **A** Cross-section of anther lobe at microspore mother cell. **B** Later stage, tapetal cells multinucleate and protoplasts project inward. **C** Three-celled pollen. **D** Hypodermal megaspore mother cell. **E** Linear tetrad. **F, G** Four- and eight-nucleate embryo sac. **H** Longisection of unitegmic ovule at eight-nucleate embryo sac. **I** Embryo sac with egg-apparatus and two polar nuclei. **J** Nuclear endosperm at globular proembryo. **K** Cellular endosperm at globular proembryo. **L–R** Embryogeny. **S** Twin proembryos, additional embryo from proliferation of suspensor. (After L.L. Narayana and Sundri 1978)

The fruit is a loculicidal capsule with seven or eight layers thick wall. The outer epidermis is followed by two or three layers of thin-walled cells and four to six layers of lignified cells.

G Bremer (1915), LL Narayana and KT Sundri (1977, 1978, 1979, 1981, 1983, 1984), R Sheela and LL Narayana (1966), KT Sundri and LL Narayana (1977)

26.12 Byblidaceae

The development of anther, pollen, female gametophyte, embryo, and seedcoat has not been investigated.

The differentiation of endothecium only at the apex of anther has been noted. The outer walls of cells are thickened.

The ovule is anatropous, bitegmic, and tenuinucellate. The inner integument forms the micropyle. The archesporial cell functions directly as megaspore mother cell. An endothelium differentiates.

The development of endosperm is probably of the Cellular type, and multicellular terminal haustoria are formed in *Byblis gigantea* (Lang 1901).

The seed is small, and albuminous. The cells of the outer epidermis of testa are tangentially elongate, and have thick and pitted walls.

FX Lang (1901)

26.13 Roridulaceae

The anther is tetrasporangiate. The development of anther wall is of the Basic type in *Roridula gorgonias* (Vani-Hardev 1972). Fibrous thickenings are absent in the endothecium. The secretory tapetum is irregularly biseriate, and nuclear divisions and fusions occur (Fig. 26.13.1A). Simultaneous cytokinesis follows meiotic divisions in microspore mother cells. The tetrads are tetrahedral.

The pollen grains are tricolporoidate, and tenuinexinous with reticulate sexine and striate nexine. They are two-nucleate at the dispersal stage (Fig. 26.13.1B, C).

The ovule is anatropous, unitegmic and tenuinucellate (Fig. 26.13.1J). The archesporial cell develops directly into the megaspore mother cell. The latter undergoes meiotic divisions, and the chalazal megaspore of the linear tetrad develops into a Polygonum type of embryo sac (Fig. 26.13.1D–K). The synergids are elongated with tapering ends. The polar nuclei fuse before fertilization. The three antipodals persist up to fertilization. A hypostase differentiates.

Fertilization is porogamous, and the pollen tube persists for some time (Fig. 26.13.1K, L).

The development of endosperm is of the Cellular type. The division of primary endosperm nucleus results in a micropylar and a chalazal chamber (Fig. 26.13.1L). A vertical division occurs in the micropylar chamber followed by further divisions so that it becomes completely cellular (Fig. 26.13.1M, N). Vani-Hardev (1972) states that the derivatives of the micropylar chamber "encroach into the micropyle and behave like a haustorium".

The first and a few subsequent divisions in the chalazal chamber are transverse. Except the lowermost cell, all the other derivatives of the chalazal chamber divide vertically. The penultimate zone of chalazal endosperm undergoes active cell divisions forming a distinct tissue of small, elongated cells with dense cytoplasm and prominent nuclei.

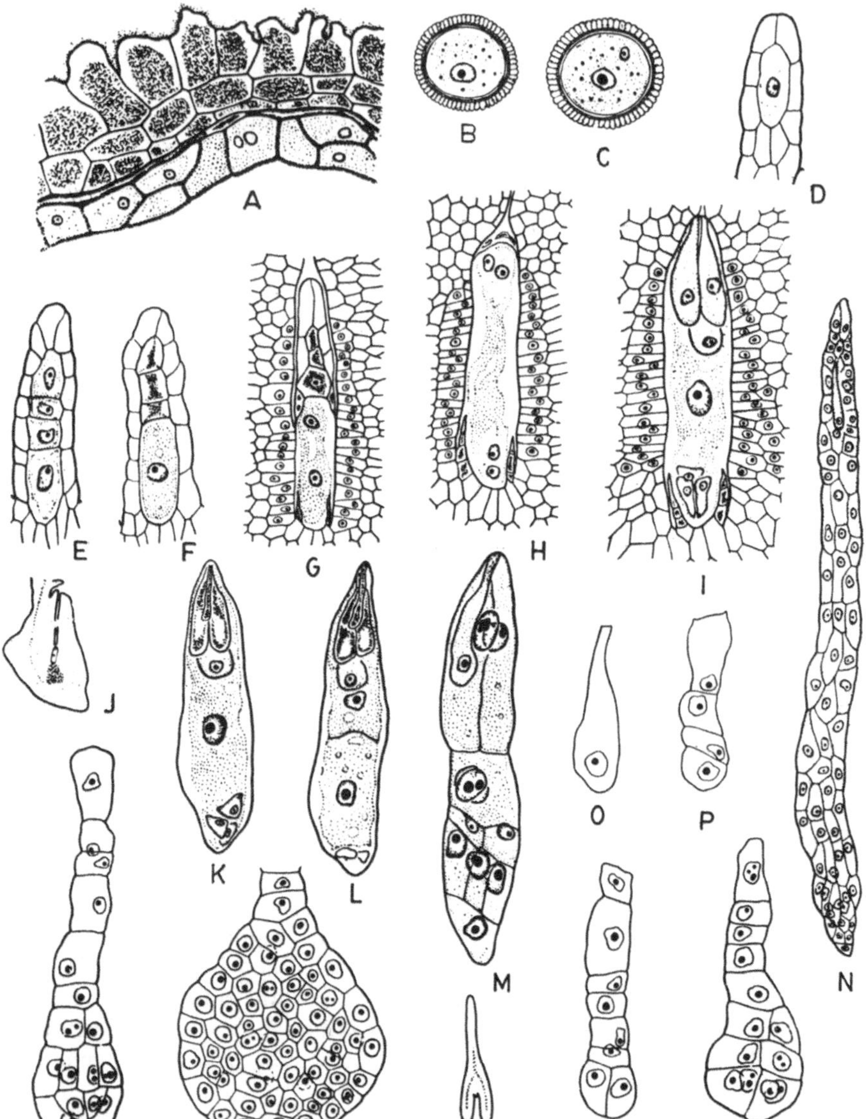

Fig. 26.13.1A–U. *Roridula gorgonias.* **A** Portion of anther wall at uninucleate pollen grain, tapetum irregularly biseriate. **B, C** Uninucleate and bi-celled pollen grain. **D** Hypodermal megaspore mother cell. **E** Linear tetrad. **F** Chalazal megaspore functional. **G, H** Bi- and four-nucleate embryo sac, note endothelium in **G–I. I, K** Mature embryo sac. **J** Longisection of ovule, note the hypostase. **L** Embryo sac with zygote, remnants of pollen tube and antipodals, and two-celled endosperm. **M** Zygote with multicellular endosperm. **N** Later stage. **O–T** Embryogeny. **U** Early dicotyledonous embryo. (**A–T** After Vani-Hardev 1972, **U** after Solereder 1908 – see Batygina and Yakovlev 1985)

The embryogeny conforms to the Solanad type (Fig. 26.13.1O–T). The seeds are small and albuminous with a small embryo (Fig. 26.13.1U), the endosperm is tanniniferous. The cells of the testa become thick-walled. The epidermal cells enlarge, and their tangential and radial walls become thick.

EJH Corner (1976), J Hutchinson (1973), R Marloth (1903), Vani-Hardev (1972)

26.14 Bruniaceae

The development of anther, pollen grains, endosperm, embryo, and seed-coat has not been investigated.

The ovule is anatropous, unitegmic (due to fusion of two primordia), and crassinucellar with a long micropyle. The mode of formation of parietal tissue is uncertain; it is represented by a single layer (Saxton 1910). The megaspore mother cell undergoes meiotic divisions, and produces a linear tetrad. The chalazal megaspore develops into a Polygonum type of embryo sac. The polar nuclei fuse before fertilization. The antipodal cells are ephemeral. Starch grains are present in the mature embryo sac.

EJH Corner (1976), J Mauritzon (1939b), WI Saxton (1910)

26.15 Rosaceae

The anther wall comprises the epidermis, fibrous endothecium, two ephemeral middle layers, and secretory tapetum of multinucleate cells in *Exochorda racemosa* (Johnston 1961). Simultaneous cytokinesis follows meiotic divisions in microspore mother cells, resulting in tetrahedral and, rarely, decussate tetrads. Johnston reported stickiness of chromosomes at anaphase I in *Exochorda*.

The pollen grains are tricolporate, and two-celled at the dispersal stage.

The ovule is ana- to hemianatropous, uni- or bitegmic, and crassinucellate. Unitegmic condition is reported in *Adenostoma*, *Alchemilla*, *Aphanes*, *Aruncus*, *Geum*, *Holodiscus*, *Neviusia*, *Potentilla*, *Rhodotypos*, *Rubus*, *Sibiraea* and *Ulmaria* (see Davis 1966). In bitegmic ovules both the integuments, or only the inner integument, take part in the formation of micropyle. The micropyle is not organized in *Prinsepia utilis* (M. Sharma and Narayana 1971). In *Sanguisorba minor* (Nordberg 1967), a single integument is formed by the accretion of the two. The two sides fuse, and surround the nucellus. In such cases the micropyle is absent. The development of postament and hypostase is reported in *Rubus idaeus* (Topham 1970). An obturator develops in *Eriobotrya japonica* (H.S. Narayana and Sharma 1976), *Malus sieboldii* (Hjelmqvist 1957b) and *Prinsepia utilis* (M. Sharma and Narayana 1971).

The hypodermal archesporium is multicellular. The archesporial cells cut off parietal cells, and the parietal tissue is well developed. The cells of the nucellar epidermis divide, and add to the parietal tissue. The megaspore mother cells undergo meiotic divisions. Generally, only one megaspore mother cell develops into a linear or T-shaped tetrad, but dyads, triads or tetrads, functional megaspores, two-, four- and eight-nucleate embryo sac are reported in the same ovule (H.S. Narayana and Sharma 1976). Similar features have been noted in *Rubus spectabilis* (Virdi and Eaton 1969) and *Sanguisorba* (Nordberg 1967). The chalazal megaspore develops into a Polygonum type of embryo sac. At times, the second megaspore from the micropylar end develops in *Prunus* and *Cerasus* (Baciu 1972). The antipodals are ephemeral, but in *Alchemilla* spp. they are persistent, and multiply in *A. orbiculata*. The polar nuclei fuse prior to fertilization.

In *Alchemilla, Aphanes, Cotoneaster, Rubus* and *Sorbus* the megaspore mother cell degenerates. One or more secondary megaspore mother cells differentiate, undergo meiosis, and produce embryo sacs in the normal way.

Allium type of development is reported by H.S. Narayana and Sharma (1976). They also point out several abnormalities in the pattern of embryo sac, arrangement of nuclei, presence of plural embryo sacs, and multiple embryo sac containing variable nuclei in "odd-ploid" members of *Malus* and *Rubus.*

The carpelloid stamens of flowers of *Rubus trivials* (Shealy and Herr 1973) show anatropous, unitegmic, and crassinucellar ovules. The development of embryo sacs in such ovules is of the Polygonum type.

The development of endosperm is of the Nuclear type. In *Prunus* and *Rubus* the chalazal part of the embryo sac develops a long haustorial tube and remains free-nuclear. In *Sanguisorba* a single layer of cells is produced, the rest of the endosperm remains free-nuclear.

The embryogeny conforms to the Asterad type in *Rubus idaeus* (Topham 1970), and Caryophyllad with some deviation from Sedum variation in *R. caesium* and *R. idaeus* (Mandryk and Petrus 1971). Synergid polyembryony, either by fertilization or by apogamy, is reported in *Geum rivale* (Juguet 1966a). Adventive polyembryony (apogamous embryos) occurs in *Potentilla* (Popova and Kolesnikov 1972).

A tendency to apospory is predominant in the family. Somatic apospory of Hieracium type occurs in *Malus hupehensis* (Dermen 1936), *M. sieboldii* (Hjelmqvist 1957b) and *Sanguisorba minor* (Nordberg 1967). In *M. sieboldii* the development of the original megaspore mother cell is suppressed, and some of the cells in the vicinity develop into mature embryo sacs. The pollen tube enters the embryo sac, but the secondary nucleus alone is fertilized; pseudogamy occurs.

Normal fertilization of apomictic eggs is reported in *Malus* and *Rubus* spp. (see Battaglia 1963). H.S. Narayana and Sharma (1976) refer to the tendency of somatic apospory in diploid and polyploid species of *M. hupahensis* and *M. sieboldii.*

Diplospory is strongly exhibited in *Potentilla arenaria* (Czapik 1962), *Rubus caesius* and *R. idaeus* (Mandryk and Petrus 1971). *R. caesius* is an obligate apomict. The unreduced eggs develop due to pseudogamy.

In *Alchemilla arvensis* (Böös 1924), the megaspore mother cell enters meiosis but the nucleus soon degenerates. The adjoining cells, potentially sporogenous, undergo mitotic divisions producing a few parietal cells and two to six axial cells which may divide meiotically or mitotically. The reduced embryo sac must be fertilized before the development of embryo. The unreduced embryo sacs function without fertilization.

In some species of *Potentilla* (Rutishauser 1943), there is a multicelled archesporium. Embryo sacs may develop from these cells by mitotic divisions as in *P. vera* (generative apospory), or from the chalazal cells of the nucellus (somatic apospory) as in *P. canescens*, *P. praecox* and *P. argentea* (see also P. Maheshwari 1950).

Parthenogenetic development of reduced or unreduced eggs occurs in *Alchemilla*, *Crataegus*, *Malus* (Gustafsson 1946), and *Aphanes* (Hjelmqvist 1959).

In *Malus* (Einset 1945, Olden 1953) and *Rubus* spp. (Crane 1940a, b, Einset 1951), the gametophytes develop as usual. Pollination is essential for embryogenesis. Fertilization does not occur even if the sperms enter the embryo sac, and the development of reduced egg is pseudogamous (reduced pseudogamy).

Nordberg (1967) states that in *Sanguisorba* both tetraploid and hexaploid hybrids show a tendency to form unreduced embryo sac. If the egg cell differentiates, it does not develop into an embryo. Then the apomictic embryo formation follows. The unreduced egg cells develop parthenogenetically, or after fertilization. Only the hybrids produce embryos by facultative apospory.

Campbell et al. (1985) studied ovule and embryo sac development in tetraploid ($n = 34$) individuals of *Amelanchier laevis*. Nomarski differential interference contrast microscopy of cleared whole-ovules, and microtome-stained sections indicated failure of meiosis in ovules. The megasporocyte or its derivatives degenerate, and one to six nearby cells develop into aposporous initials. One of these produces a Polygonum type of embryo sac. The egg forms a proembryo parthenogenetically, but seed maturation requires pollination. This apospory and pseudogamy conforms to the general pattern common in other apomictic genera of Maloideae (Rosaceae).

Unreduced embryo sac may arise from unreduced megaspore mother cell which undergoes mitotic divisions (diplospory). Apospory involves a nucellar cell close to archesporium. Only in *Malus* and *Crataegus* do apomicts produce endosperm autonomously without pollination (Olden 1953). In spite of autonomous development of embryo from egg, seeds may not complete development if endosperm formation is not stimulated by fertilization of polar nuclei.

Apomixis is rare in diploids. In Maloideae all apomictic taxa are either triploid or tetraploid.

Generally, the seed has a sclerotic mesotesta in woody members of the family. In herbaceous members the testa is thin-walled. The outer epidermis is a thin-walled palisade layer with reticulate thickenings. In several members the outer wall of epidermis is thick, and becomes mucilaginous. The inner epidermis of the outer integument, and that of the inner integument, are thin-walled. In unitegmic members the seed-coat is unspecialized.

S Asker (1970, 1977), E Baciu (1971, 1972), I Banerji and KL Chaudhari (1944), E Battaglia (1963), NP Berezenko (1971), G Böös (1924), SC Campbell, CW Greene, BF Neubauer and JM Higgins

(1985), TJ Costa and KAD Mackenzie (1990), MB Crane (1940a, b), R Czapik (1962, 1981, 1985), GM Darrow and GF Waldo (1933), GL Davis (1966), H Dermen (1936, 1949), GJ Dowrick (1961), J Einset (1945, 1948, 1951), Å Gustafsson (1930, 1946), H Hjelmqvist (1957b, 1959, 1962), GW Johnston (1961), M Juguet (1966a), M Labrecque and D Barabe (1984), A Lijefors (1953), P Maheshwari (1950), Vyu Mandryk and YY Petrus (1971), HS Narayana and MK Sharma (1976, 1979), G Nordberg (1967), EJ Olden (1947, 1953), DF Petrov (1939), E Pimienta and VS Polito (1983), LK Popova and SM Kolesnikov (1972), K Ptak (1990), KM Purohit and G Panigrahi (1983), H Rutishauser (1943), M Sharma and HS Narayana (1971), HE Shealy Jr and JM Herr Jr (1973), UP Singh (1959), GL Smith (1963), PB Topham (1970), BV Virdi and GW Eaton (1969)

26.16 Neuradaceae

The embryology has not been investigated.

26.17 Chrysobalanaceae

The embryogeny has not been studied in detail.

The anthers are tetrasporangiate. The anther wall comprises five layers: persistent epidermis, fibrous endothecium, two middle layers, and glandular tapetum with binucleate cells in *Chrysobalanus icaco* (Tobe and Raven 1984b). The anther dehisces by longitudinal slits. Meiosis in microspore mother cells is followed by simultaneous cytokinesis. The tetrads are tetrahedral, occasionally decussate.

The pollen grains are two-celled at the shedding time.

The ovary bears two ovules ascending from the base. The ovule is anatropous, bitegmic and tenuinucellar. The vascular supply shows post-chalazal branches (Netolitzky 1926). The outer integument is more than five layers thick, the inner three to five layers. The outer integument becomes highly tanniniferous and develops into a thick protective structure. The micropyle is formed by both the integuments, and is zigzag in *Licania michauxii* (Tobe and Raven 1984b). A hypostase develops in *C. icaco*, but is absent in *Licania*. The endothelium is a common feature. The archesporium comprises three to five cells. The megaspore mother cells undergo meiotic divisions. Usually there are two to three functional megaspores per nucellus, but only one develops into a modified Polygonum type of embryo sac.

The nucleus of the chalazal megaspore divides mitotically twice to form a four-nucleate embryo sac. The two chalazal nuclei soon unite and move towards the central region of the embryo sac. A mitotic division occurs in the two micropylar nuclei to form four nuclei, three organize into an egg apparatus, while the fourth functions as the upper polar nucleus. The latter unites with the nucleus formed by the fusion of two chalazal nuclei. There are no nuclei at the chalazal end, and, therefore, the antipodal cells are not formed. The mature embryo sac usually contains the egg apparatus and one central nucleus.

Fertilization is porogamous and the development of endosperm is of the Nuclear type. The endosperm remains scanty throughout seed development.

The fusion of two chalazal nuclei, and then with the upper polar nucleus requires confirmation, and development of the endosperm deserves detailed study.

The zygote develops into a globular proembryo with a short suspensor. It later differentiates into a dicotyledonous embryo. The cotyledons are rather large, unfolded and fleshy (Willis 1973).

The seed is ovoid in *Chrysobalanus icaco*, and slightly curved-terete in *Licania apetala*. The seed-coat is smooth. It is mesotestal. Thirty to 50 cells thick tissue (mesotesta) intervenes between the endotesta and exotesta. It is composed principally of undulated "vessels". In *L. michauxii* the mesotesta is much thinner and three to four cells thick. In *L. rigida* Vaughan (1970) reports single-layered endosperm, and five- to six-layered testa that contains vascular bundles. The tegmen collapses.

GT Prance (1972), F Netolitzky (1926), H Tobe and P Raven (1984b), JG Vaughan (1970), JC Willis (1973)

26.18 Connaraceae

The development of anther, pollen, and embryo has not been worked out.

The ovule is hemianatropous with pronounced preraphe (with free integuments between the hilum and the micropyle on the raphe side), bitegmic, and crassinucellar. The micropyle is formed by the outer integument, and funicle is absent (Corner 1976). The single archesporial cell cuts off a parietal cell, and the parietal tissue is two cells thick. The megaspore mother cell undergoes meiotic divisions; the micropylar dyad may not divide, resulting in a triad. The chalazal megaspore of a linear triad develops into a Polygonum type of embryo sac in *Connarus lucçus* (Mauritzon 1939b). The polar nuclei fuse before fertilization, and the secondary nucleus lies near the antipodals.

Development of the endosperm is of the Nuclear type, and it becomes cellular after the formation of numerous free nuclei. The embryo has thick, straight cotyledons, but thin in albuminous seeds (Corner 1976).

Corner (1976) summarizes the main features of connaraceous seed: preraphe form which separates the micropyle widely from the hilum, black testa with firm but not lignified exotestal palisade, fibrous exotegmen, preraphe vascular bundle, chalazal aril, large embryo, and often bean-like smell of the crushed tissue.

EJH Corner (1976), J Mauritzon (1939b)

26.19 Leguminosae (Fabaceae)

The family Leguminosae comprises three subfamilies: Mimosoideae, Caesalpinioideae and Faboideae. The subfamilies are described independently.

Mimosoideae (Mimosaceae)

The anther wall comprises the epidermis, fibrous endothecium, two ephemeral middle layers, and secretory tapetum of uninucleate cells. An intermediate form between the secretory and amoeboid tapetum is reported in *Adenanthera pavonia* (Dnyansagar 1954a). During meiosis in microspore mother cells, the tapetal cells lose their contact with each other and the protoplasts wander between the microspore mother cells and get absorbed. This is doubtful and needs careful reinvestigation.

The microspore mother cells undergo meiotic divisions, and cytokinesis is simultaneous. Simple pollen grains are formed only in a few plants, e.g. *Neptunia* (B. Singh and Shivapuri 1935), *Leucaena* (Dnyansagar 1949) *Prosopis* and *Desmanthus* (Dnyansagar 1957).

In *Dichrostachys cinerea* (Dnyansagar 1954b), occasionally, a transverse septum of tapetal cells is formed in each anther lobe so that its sporogenous mass becomes partitioned into two compartments. In *Parkia biglandulosa* (Dnyansagar 1954b), several sterile transverse septa appear and, eventually, each compartment contains only one pollinium nourished by its own tapetum. In *Mimosa pudica* (Narasimhachar 1951a), the tetrads form pollinia, but in *M. hamata* (Dnyansagar 1951a) single tetrads as well as twin tetrads form pollinia. In *Albizzia lebbek* (P. Maheshwari 1931), *Acacia baileyana* (Newman 1934), and *Calliandra* (Dnyansagar 1958) all the microspores (in a sporangium) together form a single pollinium.

The pollen grains in a polyad (twin tetrad) are unicolpate, trizonocolpate, or trizonocolporate with a smooth surface. They are shed together in masses of eight or 16 pollen grains (Fig. 26.19.1A) in *Acacia dealbata* (Poddubnaya-Arnoldi 1976), *A. farnasiana* (Narasimhachar 1948) and *A. leucophloea* (Ugemuge 1982). Compound pollen grains also occur in *Mimosa hamata*. The

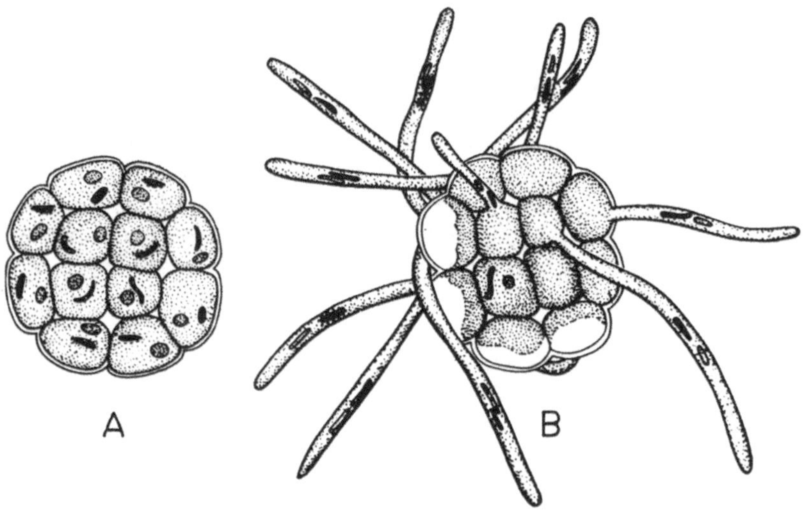

Fig. 26.19.1A, B. *Acacia dealbata*. **A** Pollinium with two-celled pollen grains. **B** Germinated pollen grains. (After Poddubnaya-Arnoldi 1954, see 1964)

pollen grains are two-celled at the shedding stage; three-celled in *Prosopis juliflora* (Deshpande and Gomkale 1982).

The peripheral pollen grains in a pollinium in *A. dealbata* (Fig. 26.19.1B) germinate earlier than the central ones.

In *Albizzia lebbek* (P. Maheshwari 1931), degeneration occurs in anthers only after the formation of microspores. The sterility is partly due to the failure in the mechanism of pollination.

The ovules are ana-, campylo-, or amphitropous, bitegmic, and crassinucellate. The micropyle is usually formed by the outer integument. In *Acacia baileyana* (Newman 1934) it is produced by both the integuments, and in *Parkia biglandulosa* only the endostome is formed. The vascular strand develops in the outer integument up to the tip of micropyle in *Leucaena glauca* (Dnyansagar 1954e). A nucellar cap is produced in *Acacia auriculaeformis* (Dnyansagar 1958).

The archesporium is uni- or multicelled. An archesporial cell functions directly as megaspore mother cell in *Leucaena glauca* (Dnyansagar 1949), or it cuts off a parietal cell. The parietal tissue is well developed in *Desmanthus virgatus* (Gaekwad and Phatak 1952). The megaspore mother cell undergoes meiotic divisions. The chalazal megaspore of a linear or T-shaped tetrad develops into a Polygonum type of embryo sac. The polar nuclei fuse before or at the time of fertilization. The antipodals are persistent.

The development of endosperm is of the Nuclear type. Wall formation is initiated at the micropylar end, and proceeds to the chalaza. In *Mimosa hamata* (Dnyansagar 1951a), wall formation occurs at the four-celled proembryo, whereas the endosperm remains free-nuclear up to the dicotyledonous embryo in *Neptunia triquetra* and *Prosopis spicigera* (Dnyansagar 1954c). The chalazal part remains free-nuclear for a fairly long time, assumes a vesicular form, and becomes haustorial. In *Prosopis juliflora* (B.M. Johri and Garg 1959), the haustorium shows finger-like processes at the chalazal tip. In older seeds it shrinks and collapses, and only traces persist.

The embryogeny conforms to the Onagrad type-Trifolium variation, or Period I, Series B, Megarchetype VI (Dnyansagar 1954e). In several members the embryo lacks a suspensor (Lersten 1983): *Acacia baileyana* (Newman 1934), *Adenanthera pavonia* (Dnyansagar 1958), *Albizzia lebbek* (Pillai et al. 1974), *Mimosa hamata* (Dnyansagar 1951a, 1954c), and *Prosopis spicigera* (Dnyansagar 1954a, 1957).

The seed contains a large, straight embryo with a short radicle and thick cotyledons. A funicular aril is present in some members, such as *Pithecolobium.* The seed-coat is similar to that in Papilionaceae.

PA Buss Jr and NR Lersten (1975), MA Caccavari (1986), B Dell (1980), PK Deshpande and KD Gomkale (1982), VR Dnyansagar (1949, 1951a, b, 1952, 1954a, b, c, d, e, 1955, 1955–56, 1957, 1958, 1970), S Feuer, CJ Niezgoda and LI Nevling (1985), LK Gaekwad and VG Phatak (1952), BM Johri and S Garg (1959), J Kenrick and RB Knox (1979, 1982), DA Larston and CW Lewis (1962), NR Lersten (1971, 1982, 1983), P Maheshwari (1931), NC Nair and S Kahate (1961), SG Narasimhachar (1948, 1951a), IV Newman (1934), SK Pillai, K Ramasita and D Ishwar (1974), VA Poddubnaya-Arnoldi (1976), B Roy (1933), VG Serrato, P Modenesi, G Roti-Michelozzi and L Bevilacqua (1986), B Singh and TN Shivapuri (1935), K Tybirk (1989), NR Ugemuge (1982)

Caesalpinioideae (Caesalpiniaceae)

The anther wall comprises the epidermis, fibrous endothecium, three ephemeral middle layers, and glandular tapetum of two- to multinucleate cells (Venkatesh 1956a). The tapetal cells are tetranucleate in *Cassia didymobotrya* (M.L. Sethi 1930), and *C. tora* (R.M. Datta 1934). The tapetum is periplasmodial in *Cassia absus*, *C. kleinii* and *C. mimosoides* (Venkatesh 1956a). The microspore mother cells undergo meiosis, cytokinesis is simultaneous, and tetrahedral, decussate, isobilateral and T-shaped tetrads are produced.

The pollen grains are three-(four)-zonoaperturate and smooth. They are two-celled at the dispersal stage.

The ovule is anatropous, bitegmic, and crassinucellate with a zigzag micropyle. The outer integument is thicker than the inner one, and the integumentary vascular strand extends up to the tip of micropyle (Rau 1950a, d).

The archesporium is uni- or multicelled, parietal cell is cut off, and the parietal tissue is well developed. A nucellar cap is also a common feature (see Dnyansagar 1970).

The megaspore mother cell undergoes meiotic divisions. The chalazal or epichalazal megaspore of a linear or T-shaped tetrad develops into a Polygonum type of embryo sac. In *Cassia nictitans* (Rembert 1969a), the micropylar dyad fails to undergo meiosis II, and the chalazal megaspore of the lower dyad functions. The polar nuclei fuse prior to fertilization, and the three antipodals persist, except in *Cassia mimosoides*, and *Parkinsonia*. Occasionally, double linear tetrads occur in *Gleditsia triacanthos* (Rembert 1969a).

The development of the endosperm is of the Nuclear type. Wall formation is initiated at the micropylar end, and extends to the chalaza. The chalazal part remains free-nuclear, and develops into a vesicular haustorium. In some species of *Cassia* (Anantaswamy Rau 1951a, Pantulu 1951), the haustorium is tubular with a vesicular tip. In *C. mimosoides* and *C. occidentalis* (Rau 1951a), one more free-nuclear bulbous portion lies below the cellular endosperm. In *C. sophera* (B.M. Johri and Garg 1959), the haustorial nuclei become spindle-shaped. The haustorium collapses in older seeds.

The embryogeny conforms to the Asterad type in *Bauhinia acuminata* (Gomkale and Deshpande 1979) and *Cassia occidentalis* (Pantulu 1951). The suspensor is poorly developed or is absent (see Lersten 1983). Guignard (1881) refers to a small suspensor in *Caesalpinia mimosoides*, *Cassia corymbosa* and *Cercis siliquastrum*, and its absence in *Ceratonia siliqua*, *Gymnocladus dioica*, *Parkinsonia aculeata* and *Poinciana gilliesii*. A short suspensor of cells larger than those of the proembryo, but not sharply delimited from it, develops in *Caesalpinia mimosoides*, *Cassia corymbosa* and *Cercis siliquastrum* (Guignard 1881).

In *Cassia occidentalis*, the suspensor is uniseriate, filamentous, of about five short cells (Pantulu 1951). Deshpande and Thakkar (see Suri and Deshpande 1981) misinterpret the suspensor in *C. mimosoides* as endosperm haustoria.

Nucellar polyembryony is of common occurrence in *B. acuminata* (Gomkale and Deshpande 1979).

The seed-coat is more or less similar to that of Mimosaceae. The linea lucida is absent, and the inner epidermis of testa develops into hour-glass cells. The funicle is thickened, and arilloid.

L Bevilacqua, P Profumo, P Gastaldo and P Barella (1988), PA Buss Jr and NR Lersten (1975), EJH Corner (1951, 1976), RM Datta (1934), PK Deshpande and SJ Thakkar (1982), VR Dnyansagar (1955–56, 1970), KD Gomkale (1980), KD Gomkale and PK Deshpande (1979), L Guignard (1881), BM Johri and S Garg (1959), NR Lersten (1971, 1983), SK Mukherjee (1952), PK Mukherjee and OS Thakur (1959), JV Pantulu (1945, 1951), AK Paul (1937), M Anantaswamy Rau (1950a, d, 1951a), DH Rembert Jr (1969a), ML Sethi (1930), RK Suri and PK Deshpande (1981), M Takahashi (1989), NR Ugemuge (1982), CS Venkatesh (1956a), J Venkateswarlu and V Seshavataram (1971)

Faboideae (Papilionaceae)

The anther wall comprises the epidermis, fibrous endothecium, two ephemeral middle layers, and secretory tapetum of uninucleate cells. The sterile sporogenous tissue divides the loculus into several chambers in *Desmodium glutinosum*, *D. illinense* (Buss et al. 1969), and *Phaseolus aconitifolius* (Deshpande and Bhasin 1974). Shukla (1954) reported intermediate behaviour of tapetal cells in *Lens esculenta.* By the time uninucleate pollen grains are formed, the walls of the tapetal cells collapse and the protoplasts form a peripheral layer around the microspores. The periplasmodium disappears at the binucleate stage of pollen. Calcium oxalate crystals have been observed in *Desmodium* (Buss et al. 1969). The microspore mother cells undergo meiotic divisions followed by simultaneous cytokinesis by furrowing. The tetrads are tetrahedral, decussate (Misra and Sahu 1970), and isobilateral (Mercy et al. 1974).

The pollen grains are spheroidal, and trizonocolpate in *Indigofera enneaphylla* (Untawale and Deshpande 1968). They are two- or three-celled at the dispersal stage. Occasionally, they germinate in situ in *Cicer arietinum* and *C. soonagaricum* (Mercy et al. 1974). They are polysiphonous in *Lupinus.*

The ovule is anatropous, hemianatropous, amphitropous or campylotropous, bitegmic, and crassinucellate with a zigzag micropyle.

An obturator develops in *Amorpha glabra* (Kostrikova 1972), and an endothelium differentiates in *Alfalfa* (Reeves 1930b), *Erythrina indica* (Chandravadana 1963), *Tephrosia* (Narang and Chandra 1977), and *Vicia faba* (Mitchell 1975).

The hypodermal archesporium is uni- or multicelled. A single archesporial cell cuts off a parietal cell, and the parietal tissue is one- to two-layered. The megaspore mother cell undergoes meiotic divisions, and the tetrads are linear, T-shaped, oblique-linear and ⊥-shaped (Rembert 1969b). The chalazal megaspore develops into a Polygonum type of embryo sac. Sometimes, epichalazal megaspore functions as in *Trifolium repens* (Martin 1914) and *Vicia faba* (Mitchell 1975). The polar nuclei fuse before fertilization, and the antipodals are ephemeral. Allium type of development is reported in *Laburnum anagyroides* (Rembert 1966) and *Pongamia*, and Oenothera type in *Vigna unguiculata* (Ojeaga and Samyaolu 1970).

Several abnormal megagametophytes have been reported in *Sesbania aculeata* and *C. aegyptica* (Salgare 1974a, b): three-nucleate, six-nucleate, nine-nucleate, and 15-nucleate with eight nuclei at the micropylar end and seven at the chalazal end. The origin of abnormality, and the frequency of occurrence, have not received attention.

A nucellar cap is present in *Milletia ovalifolia* (N. Pal 1960). In *Indigofera enneaphylla* (Deshpande and Untawale 1971) also a cap-like structure develops; at the late globular proembryo the nucellar epidermis becomes glandular. A reinvestigation is suggested. The nucellar epidermal layer persists in the micropylar region of seed in *Desmodium gangeticum* (Pantulu 1942), *Phaseolus vulgaris* (Weinstein 1926) and *P. aconitifolius* (Deshpande and Bhasin 1974). In *Zornia diphylla* (Deshpande and Bhasin 1976), the tip of embryo sac extends into the micropyle, and comes in direct contact with the outer integument. The nucellar cells in contact with the inner integument elongate and give endothelium-like appearance. The chalazal end of the nucellus is extensively developed in *Phaseolus* spp. (Seshavatharam 1978). The chalazal tissue comprises polygonal, thick-walled, refractive cells in the seeds of *Cicer arietinum* (Behl and Tiagi 1977a).

In *Vigna catjang* the organization of a tapetum-like layer, successively by the nucellus, inner integument, endosperm and, finally, even by the epidermal layer of the hypophysis region of embryo is noteworthy (Anantaswamy Rau, 1951b). These features require confirmation.

Trifolium pratense (Hindmarsh 1964) gives an extremely low yield. Nearly nine per cent of the microspore tetrads are abnormal and uninucleate; sterile pollen grains are produced. The sterility of ovule, formation of twin gametophytes, arrested growth of embryo sacs, and degeneration of egg and synergid are some of the causes of low yield. In addition, the nucellus forms a plug and prevents the entry of pollen tube through the micropyle.

In *T. repens* (Rembert 1977), the young gametophyte grows vigorously, and its expansion is checked by the adjoining nucellus. This causes damage to the gametophyte which results in reduced seed-set. Further information is necessary to confirm this observation.

In *Indigofera enneaphylla* (Deshpande and Untawale 1971), the pollen tube enters the embryo sac between the egg and one synergid without destroying the latter, and remnants of the pollen tube persist. It is rather unlikely that the pollen tube does not destroy one of the synergids. Cook (1924b) states that in *Crotalaria sagittalis* the pollen tube is "parasitic" and destroys the contents of the embryo sac. Perhaps Cook based his observations on fertilized embryo sacs which had degenerated in the normal course. There is no record of parasitic behaviour of pollen tube in angiosperms.

Occasionally, single fertilization occurs in *Arachis hypogea* (B.W. Smith 1956), which needs confirmation.

Sterility in *Crotalaria juncea* (Samal 1936) is due to the failure of growth of pollen tube, so that fertilization does not occur.

The development of the endosperm is of the Nuclear type. Endosperm nodules develop in *Lupinus albus*, *L. luteus* and *L. augustifolius* (Jaranowski 1962). Wall formation is totally absent in *Stylosanthes mucronata* (Anantaswamy Rau 1953). Only one or two layers of cells are formed around the embryo in *Clitoria ternata* (Rau 1951e) and *Lathyrus* spp. (Terzijski and Dumanova 1974).

In *Arachis hypogea* (Prakash 1960a) and *Indigofera enneaphylla* (Deshpande and Untawale 1971), wall formation in endosperm commences from the periphery to the centre, while in most members it commences from the micropylar region. Wall formation is initiated at the globular proembryo but, in some cases, it may be initiated only after the formation of cotyledonary primordia in the embryo.

In most taxa the endosperm becomes cellular in the micropylar region, and in the chalazal region it remains free nuclear for a long time. The latter functions as a haustorium and encroaches upon the chalazal tissue (Anantaswamy Rau 1951f, 1953, B.M. Johri and Garg 1959, Sehgal and Gandhi 1985). The free nuclei in the chalazal region become hypertrophied, and undergo amitotic and endomitotic divisions and nuclear fusions (Terzijski and Dumanova 1974). Amitotic divisions in most cases have turned out to be aberrant mitotic figures. The haustorium may be broad, or narrow and tubular. The haustorium becomes cellular in *Cyamopsis psoralioides* and *Desmodium pulchellum* (Anantaswamy Rau 1953). A pad of thick-walled cells, similar to postament, checks the encroachment of haustorium in the chalazal region.

The development of the endosperm in *Trifolium repens* × *T. uniflorum* (Chen and Gibson 1971), and *T. semipilosum* × *T. repens* (White and Williams 1976) is abnormal. In the former the endosperm collapses during early stages of development, and the growth of hybrid embryo is arrested. According to Chen and Gibson, this must be due to genetic imbalance of the endosperm itself, or unfavourable interaction of the endosperm with the maternal tissue.

Deshpande and Thakkar (see Suri and Deshpande 1981) state: "... in *Cassia mimosoides* ... the cells of the endosperm situated above the suspensor not only elongate but show a tendency to divide mitotically". Such elongated multinucleate cells, according to them, take up haustorial function; so far this is the only record of micropylar endosperm haustorium for the entire Leguminosae. Further, "The endosperm cells, in the micropylar region above the suspensor, in *C. absus*, elongate to form a cap-like structure." It is obvious that in both cases the authors misinterpret suspensor cells as "endosperm haustoria" and "endosperm cap". They must have examined non-median sections which did not have the suspensor connected to the embryo sac wall. Such an error could have been avoided if they had also examined dissections and transections.

Variations occur in the development of embryo (see Rau 1954, Seshavatharam 1969). The embryogeny conforms to the Onagrad, Asterad, Caryophyllad and Solanad type.

In Papilionaceae (Lersten 1983), the suspensor shows interesting features and there is much diversity in the form and structure. It may be massive, pear-shaped, haustorial, extremely elongated, thread-like, or rudimentary. It is large with inflated cells toward the base in *Sophora flavescens* (Nägl 1962). It has bulging cells in *Cyamopsis psoralioides* (Rau 1954), and *Glottidium vescarium*

(N. Prakash and Herr 1979). The suspensor is biseriate and filamentous in *Atylosia barbata* (Crété 1958c), and massive in *Hedysarum coronarium* (Guignard 1881). In *Vicia* the haustorial suspensor protrudes into the micropyle.

In the tribe Loteae the suspensor is a short, biseriate filament but the ploidy levels increase in the polytenic suspensor nuclei (Freed and Grant 1976). Extremely inflated suspensor cells develop in *Anthyllis tetraphylla* (Guignard 1881) and *A. vulneraria* (Souèges 1951a). In the tribe Viscicae the suspensor is most distinctive and uniform, and consists of four cells in each tier, each containing several nuclei at various ploidy-levels (Nägl 1962): *Lathyrus* (Guignard 1881), *Lens* (Crété 1948a) and *Pisum* (Cooper 1938, Yoffe 1957). A long, slender, biseriate filamentous suspensor develops in *Cicer arietinum* (Hegelmaier 1880).

In *Crotalaria juncea* (Samal 1936), the suspensor is much larger and massive and like a carrot tap-root with epidermal cells protruding from the proximal half. The suspensor cells contain chloroplasts and the tubular peripheral cells have a haustorial role (Anantaswamy Rau 1950b).

In *Phaseolus coccineus* (Yeung and Clutter 1978), at the globular proembryo the suspensor is broad, massive and multiseriate and shows conspicuous wall-ingrowths. The nuclei in the basal region of the cells are hypertrophied.

According to Terzijski and Dumanova (1974), "the deleterious effects of the quantitative reduction of the endosperm are compensated by the differentiation of a well-developed suspensor, composed of four hypertrophied coenocytic structures in *Lathyrus*".

Cleavage polyembryony is reported in *Crotalaria sagittalis* (Cook 1924b), and nucellar polyembryony in *Millettia ovalifolia* (N. Pal 1960). Dhawan and Bhojwani (1985) observed twin embryos in the mature seeds of *Leucaena leucocephala*, 0.7% in CVK-28 and 0.2% in CVK-8. Both the embryos were viable; their origin was not determined.

In *Melilotus alba* (Hubam) × *M. messanensis* (Jaranowski 1961), the diploid zygotic embryo, 6 days after pollination, showed budding resulting in an additional embryo from the massive globular proembryo. When a cross with changed partners was made, that is *M. messanensis* × *M. alba* (Hubam), sometimes haploid-diploid twins developed. It is assumed that the diploid zygotic embryo and the haploid synergid embryo would have started developing at the same time. But, at this stage, 140 h after pollination, the zygotic embryo takes the lead and grows faster than the synergid embryo. Two dissimilar embryos resulted from a cross between *M. messanensis* and *M. polonica* × *M. alba*; keeping the male parent *M. polonica* and using *M. alba* (Hubam) as female, two haploid and one diploid proembryo developed.

The embryos produced with *M. polonica* × *M. alba* as male parent had an elongated suspensor, and a higher potential for the development of additional embryos. In Figure 26.19.5B, C, there are haploid-diploid twins but with a different structure of the suspensor.

While the embryology of *Arachis hypogea* is normal, in *A. hagenbeckii* (Bharathi and Murty 1984) the embryo aborts at the octant stage due to absence of endosperm and hyperplastic growth of endothelium. In *A. glabrata* (Bharathi and Murty 1984), some of the embryos reach the heart-shaped stage, but viable seeds are not formed. The endosperm is of the Nuclear type with a tubular coenocytic chalazal haustorium. The embryological differences between wild and cultivated species account for the differences in their fruit characters only to a limited extent. The failure of seed-set in *A. hagenbeckii* is due to malfunction of endosperm.

The testa is generally multiplicative. The outer epidermis of testa comprises a palisade layer of thick-walled malpighian cells with linea lucida and hexagonal facets. The outer wall may become mucilaginous. The outer hypodermis is differentiated as hour-glass cells, or osteosclereids. It is followed by the aerenchymatous mesophyll. The rest of the tissue becomes compressed. The tegmen is unspecialized, and it gets crushed (Corner 1951). In *Canavalia* spp. the osteosclereids form two or three hypodermal layers (Ambegaokar 1976).

The structure of epidermal macrosclereids and hypodermal osteosclereids form the basis for classification of seeds (see Chowdhury and Buth 1970, Ambegaokar 1976, Behl and Tiagi 1976, 1977a, b).

S (Shikha) Agarwal (1984), KB Ambegaokar (1976), M Anantaswamy Rau (1950b, c, 1951b, c, d, e, f, 1953, 1954, 1955), AI Atabekova and CH Ling (1962), HM Behl and B Tiagi (1976, 1977a, b, 1978), NL Benikova (1952), RS Benavente, H Skorupska, RG Palmer and RC Shoemaker (1989), LR Bevilacqua, F Fossati and G Dondero (1987), M Bharathi, PB Kirti, VR Murty and NGP Rao (1982), M Bharathi and VR Murty (1984), JA Biddle (1979), CL Briggs (1989), RA Brink and DC Cooper (1936), MM Brown (1917), PA Buss Jr and NR Lersten (1975), PA Buss Jr, FD Gallen and NR Lersten (1969), BG Cameron and N Prakash (1990), J Cebrat (1967), P Chandravadana (1963, 1965), C Chen and PB Gibson (1971), S Chopra, S Agarwal and RP Singh (1984), KA Chowdhury and GM Buth (1970), MT Cook (1924b), DC Cooper (1938), EJH Corner (1951), P Crété (1948a, 1958c), B Dell (1980), PK Deshpande and RK Bhasin (1974, 1976), PK Deshpande and AG Untawale (1971), V Dhawan and SS Bhojwani (1985), MW Folsom and DD Cass (1989), A Ganesan, SS Shah and MS Swaminathan (1957), VR Dnyansagar (1970), RR Dute, CM Peterson and AE Kushing (1989), HM Farley and AH Hutchinson (1941), HJ Freed and WF Grant (1976), GP and RA George and JM Herr Jr (1979), L Guignard (1881), TM Halward and HT Stalker (1987), WA van Heel (1974), F Hegelmaier (1880), GJ Hindmarsh (1964), M Hossaert and M Valevo (1988), J Jaranowski (1961, 1962), R Jayalakshmi and KK Lakshmanan (1990), SS Jha (1985), SS Jha and AK Pandey (1989), BM Johri and S Garg (1959), AC Joshi (1938b), MK Kandasamy and M Vivekanandan (1984), AT Kapuskar (1964), JC Kennel and HT Horner (1985), JR King (1947), LN Kostrikova (1972), S Krupko (1973), NR Lersten (1982, 1983), KH Makde (1971), JN Martin (1914), ST Mercy, SN Kakkar and TM Varghese (1974), RC Misra and RC Sahu (1970), JP Mitchell (1975), DR Murray (1987), UR Murty, NGP Rao, PB Kirit and M Bharathi (1981), W Nägl (1962, 1974), AK Narang and N Chandra (1977), OO Ojeaga and MO Samyaolu (1970), N Pal (1960), KK Pandey (1955), AK Pandey, SS Jha and A Jha (1990), JV Pantulu (1942), HE Pattee and SC Mohapatra (1987), N Prakash and YY Chan (1976), N Prakash and JM Herr Jr (1979), S Prakash (1960a), KA Pramanik and K Thothathri (1986), NS Rangaswamy and L Nandkumar (1985), HL Reed (1924), RG Reeves (1930a, b), DH Rembert Jr (1966, 1969b, 1971, 1977), SA Salgare (1974a, b), KK Samal (1936), N Sangduen, GL Kreitner and EL Sorensen (1983a, b), JH Schaffner (1901), E Schnepf and W Nägl (1970), CB Sehgal and V Gandhi (1985), V Seshavatharam (1969, 1978, 1982), N Seth and MR Vijayaraghavan (1990), TC Shukla (1954), BW Smith (1956), DL Smith (1971), R Souèges (1949c, 1951a), R Souèges and P Crété (1952), D Subramanian (1983), D Terziyski (1971), VR Tilton, LW Wilcox and RC Palmer (1984), D Terzijski and A Dumanova (1974), NR Ugemuge and MD Padhye (1989), AG Untawale and PK Deshpande (1968), AJ Weinstein (1926), DWR White and E Williams (1976), B Wojciechowska (1971), EC Yeung (1980), EC Yeung and MJ Cavey (1988), EC Yeung and ME Clutter (1978), MD Yoffe (1957)

26.20 Krameriaceae

The embryology has not been investigated in sufficient detail. Simpson and Skvarla (1981) studied the pollen morphology and ultrastructure, and Verkerke (1986b) ovule ontogeny and seed-coat development of *Krameria*.

The anthers are tetrasporangiate, and dehiscence occurs by terminal pores. The isopolar, spheroidal, three-(four)-colporate to triporate pollen grains do not contain an appreciable amount of starch, and are rarely collected by bees (Simpson and Skvarla 1981).

The ovular primordia are trizonate, the inner and outer integuments are of dermal origin. The ovule curves anatropically, at the same time it rotates

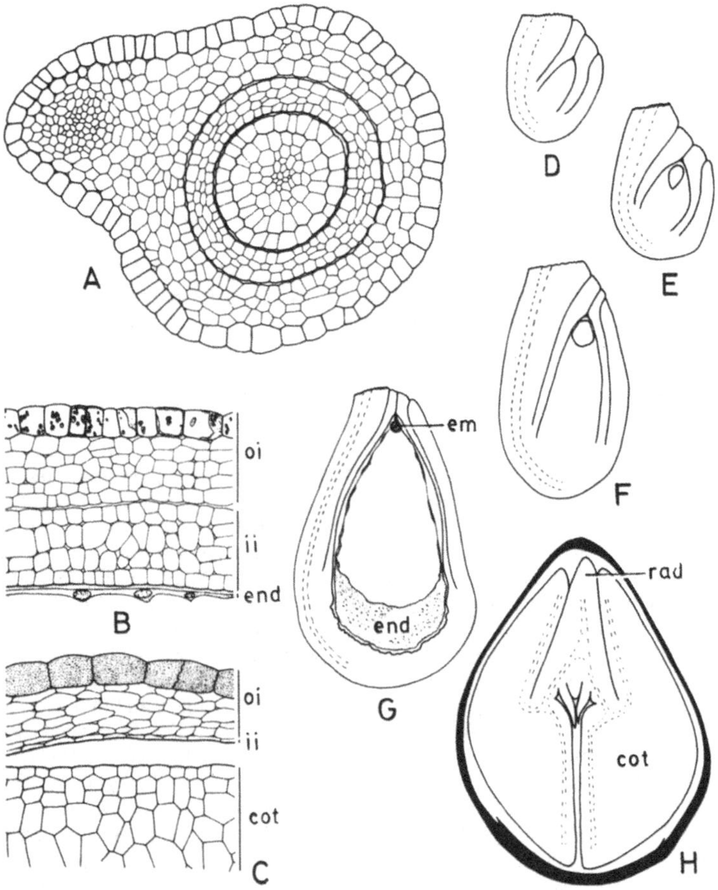

Fig. 26.20.1A–H. *Krameria lanceolata.* **A–C** Transections; **A** Mature ovule, note outer and inner integument, and nucellus; **B, C** Seed-coat, immature (**B**) and mature seed; *oi, ii* outer and inner integument; *end* endosperm. The nucellus has been resorbed in **B**, and endosperm in **C**. **D–G** Longisections of fertilized ovules. Embryo sac in **E** and **F**, proembryo (*em*) and endosperm in **G**. **H** Longisection of seed; *rad* radicle; *cot* cotyledon. (After Verkerke 1986b).

outward over 90 degrees. Eventually, the superimposed curvature and rotation lead to an anatropous, epitropous ovule with a curved funicle. The curved funicle and the raphe have anomocytic stomata.

The pyriform ovule has a narrow and elongated nucellus. The inner integument is three to four cells thick, and the outer three to five cells (Fig. 26.20.1A). The micropyle is formed entirely by the inner integument and the endostome is very narrow. The raphal vascular strand extends into the outer integument. A hypostase-like inner layer of flattened strongly tanniniferous cells extends over the whole nucellus-chalaza connection.

The hypodermal archesporial cell cuts off a parietal cell which undergoes a few divisions. Following meiosis, the megaspore mother cell produces a linear, sometimes T-shaped, tetrad and the chalazal megaspore produces the embryo sac. Further details have not been worked out.

The structure of the testa and tegmen is shown in Figure 26.20.1B, C, and an enlargement of post-fertilized ovule in D–G. The embryo sac is at first situated apically in the nucellus, then (at the globular proembryo) it enlarges a great deal and consumes the nucellus as well as the vast chalazal region (Fig. 26.20.1 cf. E–G).

The endosperm is of the Nuclear type, and most of the nuclei aggregate in the chalazal region of the embryo sac which is lined by hypertrophied cells. Wall formation does not occur in the endosperm, and it is consumed by the growing embryo.

The embryogeny has not been investigated. The mature seed (Fig. 26.20.1H) contains a large embryo with two plano-convex cotyledons. At the upper end each cotyledon extends backward and the two auricles enclose the radicle leaving only its apex exposed. In a cross section there would be four auricles (?), a condition reported earlier in *Floerkea* by P. Maheshwari and Johri (1956).

During seed development, the inner integument thickens by periclinal divisions (Fig. 26.20.1B). The tegmen is mostly absorbed in the ripe seed, and only the innermost layer and cuticle persist. The outer integument also thickens, and the tanniniferous exotesta with slightly thickened walls and a few layers of parenchymatous cells persist (Fig. 26.20.1C). The mesophyll is traversed by numerous vascular bundles (Corner 1976).

The exalbuminous seed, without any especial development of seed-coat, is enclosed in a barbed fruit.

TB Batygina and MS Yakovlev (1985), EJH Corner (1976), J Hutchinson (1973), P Maheshwari and Johri (1956), BB Simpson and JJ Skvarla (1981), W Verkerke (1986b)

26.21 Taxonomic Considerations

The suborder Hamamelidineae comprises Platanaceae, Hamamelidaceae and Myrothamnaceae. Wettstein (1935), Takhtajan (1966, 1980) and Cronquist (1968, 1981) treat these families in Hamamelidales. Hutchinson (1973) groups them in his Hamamelidales alongwith Bruniaceae and Tetracentraceae. According to Hallier (1903, 1912), Tippo (1938), Tang (1943), Takhtajan (1966, 1969, 1980) and Cronquist (1968, 1981), Hamamelidaceae (with inferior or semi-inferior ovary) are derived from Magnoliales. Bentham and Hooker (1965a), Baillon (1871, 1874), Bessey (1915), Harms (1930), Melchior (1964), Endress (1967), Jay (1968) and Hutchinson (1973) presume that Hamamelidaceae arose from a Protorosalian stock with affinities with the living Rosaceae, Saxifragaceae and Cunoniaceae. According to Tippo (1938), Hamamelidaceae gave rise to Casuarinaceae, Fagales and Urticales.

Platanaceae, Myrothamnaceae. The embryology of Platanaceae and Myrothamnaceae has not been adequately worked out. Platanaceae show alliance with Hamamelidaceae: three-colpate pollen grains, bitegmic, crassinucellate ovule, raphe bundle ending at chalaza, formation of micropyle by both integuments, oil and aleurone as reserve food in endosperm, and a narrow embryo with flat cotyledons. The simplified, much-reduced seed of Platanaceae is probably derived from hamamelidaceous seed (Mohana Rao 1974a). Anatomical and cytological data support their close relationship, as also does a chemotaxonomic study (Jay 1968, 1969).

Hamamelidaceae. Hamamelidaceae have secretory anther tapetum, simultaneous cytokinesis in pollen mother cells, anatropous, bitegmic, crassinucellar ovule with a massive parietal tissue, Polygonum type of embryo sac, and Cellular type of endosperm in some members as in Magnoliaceae. The development of an endosperm haustorium is absent in the family, and Solanad type of embryogeny occurs in Eupteleaceae of the magnoliaceous families. Embryologically, Myrothamnaceae differ from Platanaceae and Hamamelidaceae, to some extent, but show similarity with Monimiaceae, another magnoliaceous family: two-celled pollen grains in permanent tetrads, anatropous, crassinucellate ovule, and bisporic Allium type of embryo sac. Hamamelidaceae, Myrothamnaceae and Platanaceae, the phytochemically allied families forming a natural group (Shaw and Gibbs 1961), are related to Magnoliales. Hamamelidaceae also show alliance with Saxifragaceae. Both have massive parietal tissue, Polygonum type of embryo sac, Nuclear type of endosperm, and Solanad type of embryogeny.

The suborder Saxifragineae includes Crassulaceae, Cephalotaceae, Saxifragaceae, Bruneliaceae, Cunoniaceae, Davidsoniaceae, Pittosporaceae, Byblidaceae, Roridulaceae and Bruniaceae.

Crassulaceae. The family Crassulaceae has been placed in the order Rosales, adjacent to Saxifragaceae, by Bentham and Hooker (1965a), Bessey (1915), Wettstein (1935), Engler and Diels (1936), and Rendle (1952). Hutchinson (1973) includes the families Crassulaceae and Saxifragaceae in Saxifragales.

Subramanyam (1962) compares the embryological features of Crassulaceae and Saxifragaceae, and points out their close similarities: multilayered anther wall, simultaneous cytokinesis in pollen mother cells, three to polycolpate, two-celled pollen grains, indefinite number of crassinucellate, usually bitegmic ovules, uni- or multicelled archesporium, Polygonum type of embryo sac, Cellular type of endosperm with chalazal haustorium, and Caryophyllad type of embryogeny. But secretory anther tapetum, apocarpous superior ovary with crassinucellate, bitegmic ovule, total absence of endothelium, development of haustoria from megaspores, synergids and antipodals, and haustorial behaviour of the basal cell of two-celled proembryo are features characteristic of Crassulaceae.

Crassula aquatica has a mode of life somewhat similar to Podostemaceae. According to Subramanyam (1962), its reduced endosperm is a transitional stage to complete suppression of endosperm formation in Podostemaceae. The Crassulaceae have affinities with Hydrostachyaceae (Sect. 27.2) and Podostemaceae (Sect. 28.2).

Saxifragaceae. Saxifragaceae show diversity of features as regards anther tapetum, integuments, parietal tissue, and uni- or multi-celled archesporium. The occasional presence of Allium type of embryo sac and endothelium, varying position of primary endosperm nucleus, Nuclear, Cellular or Helobial type of endosperm development, and Solanad or Caryophyllad type of embryogeny segregate it from the other members of Rosales.

Occurrence of pollen embryo sacs in *Heuchera micrantha* is rather interesting.

Pace (1912) relates *Parnassia* to Droseraceae due to structural similarity of ovule and embryo sac (see Sarraceniales, Sect. 23). Hallier (1901) includes *Parnassia* in Sarraceniales; in 1908 he proposed to place the genus in a separate family near Ranunculaceae, Nymphaeaceae, Droseraceae and Sarraceniaceae. Hutchinson (1959, 1973) also assigns *Parnassia* to a separate family, Parnassiaceae. V.K. Sharma (1968b) compares the embryological features of *Parnassia* with Saxifragaceae, Droseraceae and Hypericaceae (Table 26.21.1). He proposes that *Parnassia* be assigned to a separate family, Parnassiaceae, which forms a link between Saxifragaceae and Droseraceae, and shows similarities with Hypericaceae. *Parnassia* differs from Saxifragaceae: tenuinucellate, bitegmic ovule, large persistent antipodals, Nuclear type of endosperm, and Asterad type of embryogeny. The embryogeny in *Parnassia* belongs to Period I (Asterad type) in Souèges system of classification, whereas in the other three families it corresponds to Period II (Caryophyllad type).

Similarly, *Vahlia* (syn. *Bistella*, subfamily Vahlioideae) also differs in certain respects from the family Saxifragaceae: tenuinucellate ovule, presence of endothelium with binucleate cells, chalazal growth of embryo sac, and richly cytoplasmic chalazal endosperm cells which probably have a haustorial function (Saxena 1973). Anatomical data also support its segregation from Saxifragaceae to an independent family, Vahliaceae.

Hutchinson (1973) assigns the genera *Parnassia* and *Bistella* to separate families, Parnassiaceae and Vahliaceae. In fact, Saxifragaceae are not a uniform family from a systematic and embryological point of view, as stated by Melchior (1964).

Table 26.21.1. Embryology of Guttiferales (Sect. 22.16), Droseraceae (23.4), Saxifragaceae (26.7), *Parnassia*, *Bauera* and Cunoniaceae (26.9) (See V.K. Sharma 1968b, N. Prakash and McAllister 1977)

	Guttiferales	Droseraceae	Saxifragaceae	*Parnassia*	*Bauera*	Cunoniaceae
Anther tapetum	Secretory, cells 2-nucleate	Secretory, cells 2-nucleate; sometimes pseudoplasmodial	Secretory, cells 2–4-nucleate	Secretory, cells 1- or 2-nucleate	Secretory, cells uninucleate	Secretory, cells binucleate
Cytokinesis	Simultaneous	Simultaneous	Simultaneous	Simultaneous	Simultaneous	Simultaneous
Tetrads	Tetrahedral, isobilateral, decussate	Tetrahedral, isobilateral	Tetrahedral, isobilateral, decussate	Tetrahedral	Tetrahedral, decussate	Tetrahedral
Pollen	Free, tricolpate, 2-celled	Free or united in tetrads, generative cell cut off distally, tricolpate, 2-celled	Free, tricolporate or bicolpate, 2- or 3-celled, pollen embryo sacs reported in *Heuchera micrantha*	Free, tricolporate, 2-celled	Free, bicolpate, 2-celled	Free, 2-celled
Ovule	Anatropous, bitegmic, tenuinucellate, micropyle formed by both integuments	Anatropous, bitegmic, tenui- or crassinucellate, micropyle formed by inner integument	Anatropous, uni- or bitegmic, tenui- or crassinucellate, micropyle formed by both integuments in bitegmic ovules	Anatropous, bitegmic, tenuinucellate, micropyle formed by inner integument	Anatropous, bitegmic, crassinucellate, micropyle zigzag and formed by both integuments	Ana- to hemianatropous, bitegmic crassinucellate, micropyle zigzag and formed by both integuments, hypostase present
Tetrad	Linear	T-shaped	Linear or T-shaped	Linear or T-shaped, triads (upper dyad and two megaspores) common	Linear, twin tetrads develop	Linear or T-shaped

Embryo sac	Polygonum type, antipodals, ephemeral, supernumerary in *Hypericum*	Polygonum, antipodals conspicuous, nucellar cells elongate radially around the embryo sac	Polygonum, sometimes bisporic, antipodals ephemeral or persistent, supernumerary in *Saxifraga*	Polygonum, antipodals persistent, supernumerary	Polygonum, rarely supernumerary antipodals occur, ephemeral	Polygonum antipodals ephemeral
Endosperm	Nuclear type, a chalazal cyst is formed, becomes cellular	Nuclear, nuclei aggregate in chalazal region which functions as a haustorium, entire endosperm becomes cellular	Cellular, Helobial or Nuclear, the entire endosperm becomes cellular in Helobial or Nuclear type	Nuclear, becomes cellular	Nuclear, later stages not examined	Nuclear, becomes cellular
Embryo	Solanad or Caryophyllad type	Caryophyllad, Asterad (?) or Onagrad (?)	Caryophyllad or Asterad	Asterad	—	—
Polyembryony	Apogamous from synergid, proliferation of suspensor cells, aposporous from nucellar cells	Absent	Synergid polyembryony	Absent	Absent	Absent
Seed-coat	Exotegmic, mesotestal or unspecialized	Not multiplicative, in some taxa testa larger than tegmen with an air-space between them	Not multiplicative, testa of radially-elongate or thick-walled cells, in *Heuchera* thick-walled endotegmen persists	Multiplicative, exotestal (?)	Multiplicative	Exotegmic
Seed	Arillate, exalbuminous, embryo large in Clusieae	Small, exarillate, albuminous, embryo microscopic, cotyledons short	Small, exarillate, albuminous, embryo minute	Exarillate, exalbuminous, embryo large	Exarillate, albuminous, embryo large	Exarillate, albuminous, embryo small

Bauera. N. Prakash and McAlister (1977) investigated the embryology of *Bauera capitata*. Melchior (1964) includes the genus *Bauera* in Saxifragaceae, but Cronquist (1968), Thorne (1968) and Takhtajan (1969) place it in Cunoniaceae. Airy Shaw (1973) and Hutchinson (1973) raise it to a monogeneric family, Baueraceae. N. Prakash and McAlister (1977) compare the embryological features of *Bauera*, Cunoniaceae and Saxifragaceae (Table 26.21.1). Even though there are several similarities within the three, Saxifragaceae differ in (1) occasional amoeboid tapetum, (2) uni- or bitegmic ovules, (3) single-celled archesporium, (4) sometimes persistent antipodals, (5) absence of starch in the embryo sac, and (6) absence of tannin in endosperm. Dickison (1975a) supports the placement of *Bauera* in or near the Saxifragaceae, on the basis of floral morphology and anatomy.

Bauera shows close resemblance to Cunoniaceae in the type of pollen grains, endosperm, and seed structure.

After assessing the available embryological features, N. Prakash and McAlister (1977) propose an independent family, Baueraceae, between Saxifragaceae and Cunoniaceae, as also suggested by Airy Shaw (1973) and Hutchinson (1973). We do not consider that there is enough justification to assign *Bauera* to an independent family, Baueraceae. Its inclusion in the family Cunoniaceae is fully justified (see also Takhtajan 1980, Batygina and Yakovlev 1985).

Cunoniaceae. Cunoniaceae are assigned to the order Cunoniales (Hutchinson 1973). They have been treated in Saxifragaceae under different status as series, tribe, or sub-order (see Govil and Saxena 1976). They show close alliance with Saxifragaceae in embryological features (Table 26.21.1): anatropous, bitegmic and crassinucellar ovule with multicellular archesporium, well-developed parietal tissue, Polygonum type of embryo sac, ephemeral antipodals, and Nuclear type of endosperm. The knowledge about their embryogeny may add some more facts. The study of floral anatomy (Dickison 1975b) supports their position near Saxifragaceae.

Pittosporaceae. Pittosporaceae and Byblidaceae are placed in Pittosporales by Hutchinson (1973). Pittosporaceae show secretory anther tapetum, simultaneous cytokinesis in pollen mother cells, unitegmic tenuinucellate ovule, single-celled female archesporium, Polygonum type of embryo sac with ephemeral antipodals, and Nuclear type of endosperm. All these features are met with in Saxifragaceae as well as Umbelliferae. According to L.L. Narayana and Sundri (1978), the Pittosporaceae show a close alliance with Saxifragaceae in floral anatomy, vegetative anatomy, palynology, and chemotaxonomy. R. Dahlgren (1977b) places the order Pittosporales in his non-iridoid sympetalous superorder Araliiflorae (Umbelliflorae). Pittosporaceae have sympetalous corolla. The embryological features such as pollen grains, ovule, embryo sac and endosperm are similar to those of Umbelliferae.

The tenuinucellar, unitegmic ovule occurs, occasionally, in Saxifragaceae, but the three-celled pollen grains at the dispersal stage, and two to five cotyledons of mature embryo, have no similarity in Rosales.

The bitegmic, tenuinucellar ovules with endothelium, and the probably Cellular type of endosperm with haustoria in Byblidaceae, are resemblances with *Bistella* (Saxifragaceae), and differences with Pittosporaceae.

Byblidaceae, Roridulaceae. Planchon (1848) indicated the affinities of *Roridula* with Ochnaceae, and Engler (1907) places it in Ochnaceae. Bentham and Hooker (1965a), Eichler (1875), Engler and Prantl (1894), Netolitzky (1926), Solereder (1908) and Wettstein (1911) treat it under Droseraceae along with *Byblis*. Diels (1930a), however, excludes *Roridula* from Droseraceae. Hutchinson (1973), Riley (1963) and Metcalfe and Chalk (1950) treat it under Byblidaceae. Hutchinson does not recognize their independent family status and considers Roridulaceae as a synonym of Byblidaceae (see also Willis 1973). Due to its non-insectivorous habit, Marloth (1925) transfers *Roridula* from the family Byblidaceae to Roridulaceae. Diels (1930b), Erdtman (1952), Good (1957), Melchior (1964), Takhtajan (1980) and Wettstein (1935) agree with this assignment.

Bitegmic ovule with false endothelium developing from the nucellus, occurrence of gonial or somatic apospory, and single-layered testa in Ochnaceae are differences with Roridulaceae. The floral structure and wood anatomy also differ, and the two families do not show any alliance.

Though the "tentacles" of Roridulaceae are always compared with those of Droseraceae, they do not trap insects. They secrete a resinous balsam-like substance, and not digestive enzymes as in Droseraceae. In addition, some of the features, such as pollen in tetrads, bitegmic crassi- or tenuinucellar ovule with false endothelium, Caryophyllad type of embryogeny, and hard seed-coat formed by both the integuments in Droseraceae, do not match with those of Roridulaceae. The floral structure, anatomy, acid physiology of tentacles, and stem anatomy are also different in both (Vani-Hardev 1972).

Vani-Hardev (1972) compares the morphological and embryological features of *Roridula* and *Byblis* (Table 26.21.2). *Byblis* is a perennial carnivorous herb with morphologically and physiologically different tentacles from those of *Roridula*. The wood anatomy is also different. *Roridula* has reflexed anthers with poricidal dehiscence (introrse anthers with apical slits in *Byblis*), tricarpellary, syncarpous, trilocular ovary with pendulous ovules, ovary wall without tentacles (bicarpellary, syncarpous, bilocular ovary with erect ovule, and ovary wall studded with tentacles in *Byblis*), chalazal megaspore develops into a Polygonum type of embryo sac (micropylar megaspore functional in *Byblis*), and Cellular type of endosperm without prominent haustoria (the haustoria are terminal and prominent in *Byblis*).

The segregation of *Roridula* from Byblidaceae (a non-iridoid family) and its elevation to an independent family, Roridulaceae (see Marloth 1925), are justified on embryological basis.

Chemotaxonomic data show the presence of small quantities of iridoid in the leaves of *Roridula* (Jensen et al. 1975) which supports its placement in or near Ericales. Hallier (1912) suggests its relationship with Clethraceae. The embryological similarity between Roridulaceae and Ericales (particularly Clethraceae) are: (1) absence of fibrous endothecium in anther wall, (2)

Table 26.21.2. Floral morphology and embryology of Roridulaceae and Byblidaceae (See Lang 1901, Vani-Hardev 1972)

	Roridulaceae (*Roridula*)	Byblidaceae (*Byblis*)
Inflorescence	Long, slender, racemes	Solitary, axillary
Stamens	5, free, reflexed in bud, anthers more or less cylindrical, connectives form massive nectaries, dehiscence poricidal, endothecium lacks fibrous thickenings	5, slightly fused with petals at the base, anthers conical, connective non-nectariferous, anthers introse dehiscing by small apical pores, endothecium weakly fibrous
Pollen	2-celled	2-celled
Gynoecium	Tricarpellary syncarpous, trilocular ovary, wall devoid of tentacles, ovules pendulous	Bicarpellary syncarpous, bilocular ovary, wall studded with tentacles, ovules erect
Ovule	Anatropous, unitegmic, tenuinucellate	Anatropous, bitegmic, tenuinucellate, inner integument forms the micropyle, endothelium differentiates
Embryo sac	Chalazal megaspore produces Polygonum type of embryo sac	Micropylar megaspore produces Polygonum type of embryo sac
Endosperm	Cellular type, without prominent haustoria	Cellular (?), prominent multicelled micropylar and chalazal haustoria
Embryo	Solanad type	—
Seed	Endospermic, exotestal	Endospermic, exotestal

glandular tapetum with binucleate cells, (3) two-celled pollen shed singly as in Clethraceae, (4) tricarpellary gynoecium as in Clethraceae, (5) unitegmic, tenuinucellate ovule with endothelium, (6) single-celled archesporium which develops into megaspore mother cell, (7) Polygonum type of embryo sac, and synergids with long tapering tips, (8) fusion of polar nuclei before fertilization, (9) Cellular type of endosperm which becomes haustorial (haustoria weakly developed in Clethraceae), (10) Solanad type of embryogeny, and (11) loculicidal capsule.

The placement of Roridulaceae near or in the Ericales is supported by embryological and chemotaxonomic data.

Bruniaceae. The unitegmic, crassinucellate ovule, and Polygonum type of embryo sac with ephemeral antipodals in Bruniaceae can be compared with those of *Saxifraga pallida* (Rau and Sharma 1967).

Rosaceae. Rosaceae, Neuradaceae and Chrysobalanaceae comprise the suborder Rosineae. No information is available on the Neuradaceae. Bentham and Hooker (1965a), Engler and Diels (1936), and Hutchinson (1973) place Rosaceae in Rosales, near Leguminoseae. The occurrence of carpelloid stamens, and spirally arranged free carpels on elongated torus in some genera indicate their direct Ranalian origin. The accretion of integuments in *Sanguisorba minor* leading to unitegmic condition suggests reduction.

The nature of anther, pollen, crassinucellar ovule with multicellular archesporium, Polygonum and Allium type of embryo sac, Nuclear type of endosperm, and Caryophyllad type of embryogeny in Rosaceae are resemblances with Saxifragaceae. The development of chalazal haustorium from the Nuclear type of endosperm, and its absence in seed shows affinity with Leguminoseae. Asterad type of development of the embryo occurs in *Parnassia* and Caesalpiniaceae. Apospory is prevalent in Rosaceae. Mesotestal seed occurs in woody members, and is comparable to that of Hamamelidaceae, Theaceae and Myrtaceae.

Chrysobalanaceae. Traditionally, the family Chrysobalanaceae is assigned to the order Rosales (Melchior 1964, Hutchinson 1973, Takhtajan 1980, Cronquist 1981). However, wood anatomy (Metcalfe and Chalk 1950) differs significantly from that of Rosales. Palynologically also they are different (Erdtman 1952).

Tobe and Raven (1984a) point out that Chrysobalanaceae show affinity with Theaceae, Ochnaceae and even Clusiaceae (Guttiferae).

Chrysobalanaceae and Theaceae resemble each other in several embryological features: (1) anatropous, bitegmic, tenuinucellate ovule with hypostase, (2) multicelled female archesporium, (3) Nuclear type of endosperm, (4) mesotestal exalbuminous seed, and (5) the tegmen becomes crushed. The Theaceae have Allium type of embryo sac with persistent antipodals, and polyembryony occurs.

With Ochnaceae also Chrysobalanaceae show some resemblances: (1) anatropous, bitegmic, tenuinucellar ovule with hypostase, (2) Nuclear type of endosperm which later becomes cellular but is absorbed by the developing embryo, and (3) exalbuminous seed. But the dissimilarities are more significant. In Ochnaceae there is a (1) false endothelium which differentiates from the nucellus, (2) Polygonum type of embryo sac with large antipodals containing hypertrophied nuclei, (3) apospory of Hieracium type, and (4) Polyembryony.

Like Chrysobalanaceae, Clusiaceae also have anatropous, bitegmic, tenuinucellate ovule, Nuclear type of endosperm, and mesotestal exalbuminous seed (Clusieae). However, Clusiaceae have one-celled female archesporium, and Polygonum type of embryo sac with well-developed antipodals. Polyembryony and apomictic embryo sacs have also been reported.

R. Dahlgren (1983a, b) and R. Dahlgren and Thorne (1984) suggest a relationship between Chrysobalanaceae and Myrtales. The resemblances are: (1) glandular anther tapetum, (2) pollen of Chrysobalanaceae resembles that of Aizateaceae, (3) micropyle formed by both integuments, endothelium differentiation in *Axinandra* (Crypteroniaceae), (4) antipodals ephemeral or absent, (5) modified Polygonum type of embryo sac, (6) endosperm of Nuclear type, and (7) exalbuminous seed. But the thicker inner integument and tenuinucellate ovule in Chrysobalanaceae do not match with Myrtales.

According to Wettstein (1933) and Prance (1972), Chrysobalanaceae stand between Rosaceae and Fabaceae (Papilionaceae). The differences with Fabaceae are more prominent: The anther tapetal cells are binucleate in Chrysobalanaceae, multinucleate in Fabaceae. The ovule is tenuinucellate with short-lived nucellus, crassinucellate with persistent nucellar cap in Fabaceae. The mesotestal seed of Chrysobalanaceae differs from the exotestal seed of Fabaceae.

Chrysobalanaceae share many embryological features with Rosaceae (Table 26.21.3): (1) persistent anther epidermis, (2) fibrous endothecium, (3) two-celled pollen grains, (4) multicelled female archesporium, (5) thick outer and inner integument, (6) Polygonum type of embryo sac, and (7) mesotestal seed with fibrous testa.

There are also important differences: Chrysobalanaceae have tenuinucellar ovule with endothelium, and scanty nucellus which disintegrates at an early stage. In Rosaceae the nucellus is massive, endothelium is absent, and apospory is common.

The embryology of Chrysobalanaceae, Rosaceae, Theaceae and Mrytales is compared in Table 26.21.3.

There is no doubt that there are some similarities in the embryological features of Chrysobalanaceae, and Ochnaceae, Clusiaceae, Fabaceae and Myrtales but these similarities do not warrant any close relationship. The affinities with Theaceae can be discussed more meaningfully when further embryological data on this taxon become available, as also pointed out by Tobe and Raven (1984a). Wettstein's (1933) suggestion that Chrysobalanaceae stand between Fabaceae and Rosaceae is more appropriate, and the assignment of Chrysobalanaceae to Rosales (Melchior 1964) is justified.

The suborder Leguminosineae comprises Connaraceae, Leguminosae and Krameriaceae.

Connaraceae. The occurrence of bitegmic, crassinucellar ovule, Polygonum type of embryo sac, and Nuclear type of endosperm is common to Cunoniaceae and Leguminoseae. Exotegmic seeds occur in Cunoniaceae and Connaraceae.

Bentham and Hooker (1965a), Engler and Prantl (1931), Rendle (1959) and Lawrence (1951) consider Leguminoseae as a family, while Hutchinson (1959, 1973) assigns to them an ordinal rank. The family comprises three subfamilies: Mimosoideae (Mimosaceae), Caesalpinioideae (Caesalpiniaceae) and Faboideae (Papilionaceae).

These three subfamilies show similar features among themselves and also in conformity with those of the order: secretory anther tapetum, two-celled pollen grains, bitegmic, crassinucellate ovules, Polygonum type of embryo sac, Nuclear type of endosperm with chalazal haustorium, and characteristic seed-coat. However, they differ from one another in several respects, and their separation to independent families has been proposed by several taxonomists.

Mimosaceae. Mimosaceae are in closer alliance with Caesalpiniaceae than with Papilionaceae. They are considered as primitive. The formation of sterile septa in anther locule, pollinia of single tetrads, twin tetrads, and all the microspores in a sporangium, shedding of pollen grains singly or in masses, vascular strand reaching up to the micropyle in the integument, bitegmic tenui- or crassinucellar ovule, persistent antipodals, Onagrad type of embryogeny, and arillate seeds in some members segregate Mimosaceae from the others. Dnyansagar (1970) proposed the division of Mimosaceae into two tribes, Prosopoideae and Mimoseae, according to single and compound pollen grains, respectively.

Caesalpiniaceae. Caesalpiniaceae show occasional occurrence of anther tapetum forming periplasmodium, bitegmic crassinucellar ovule with an integumentary vascular strand reaching up to the micropyle, Polygonum type of embryo sac with persistent antipodals, and Asterad type of embryogeny. Both families, Mimosaceae and Caesalpiniaceae, are homogeneous.

Papilionaceae. In Papilionaceae, occasional occurrence of division of the sporogenous tissue by sterile septa as in Mimosaceae, and the anther tapetum forming periplasmodium (as in some members of Caesalpiniaceae) are noteworthy. The glandular nucellar epidermis in *Indigofera*, and the organization of a tapetum-like layer successively by the nucellus, inner integument, endosperm and, finally, even the epidermal layer of the hypophysis region of embryo in *Vigna* (Anantaswamy Rau 1951b) are worth mentioning.

Development of the endothelium, ephemeral antipodals, abnormal megagametophytes, occasional occurrence of Allium type of embryo sac, single fertilization in some members (rather doubtful), and absence of integumentary vascular strand differentiate Papilionaceae from the Mimosaceae and Caesalpiniaceae.

Table 26.21.3. Embryology of Chrysobalanaceae, Rosaceae, Fabaceae, Theales and Myrtales (Adapted after Tobe and Raven 1984b)

	Chrysobalanaceae	Rosaceae	Fabaceae (Papilionaceae)	Theales	Myrtales
Anther	Tetrasporangiate, wall 4–5 layers thick, tapetum secretory of binucleate cells, cytokinesis simultaneous	Bisporangiate, wall 4 layers, tapetum secretory of multinucleate cells, cytokinesis simultaneous	Bisporangiate, wall 4–5 layers, tapetum secretory of uninucleate cells, cytokinesis simultaneous	Bisporangiate, wall 4–5 layers, tapetum secretory of binucleate cells, cytokinesis simultaneous	Tetrasporangiate, wall 4–5 layers, tapetum secretory of binucleate cells, cytokinesis simultaneous
Ovule	Anatropous, bitegmic, tenuinucellar, outer integument 5 layers thick, inner 3–5 layers, micropyle formed by both integuments, endothelium and hypostase present, archesporium multicellular	Ana- or hemianatropous, uni- or bitegmic, crassinucellate, outer integument 3–14 layers, inner 2–6 layers, micropyle formed by both integuments, endothelium absent, hypostase present, archesporium multicellular	Ana-, hemiana- or amphitropous, bitegmic, crassinucellate, outer integument 2–10 layers, inner 2–3 layers, micropyle formed by both integuments, endothelium rare, hypostase absent, archesporium uni- or multicellular	Anatropous, bitegmic, tenuinucellar, outer integument 3–10 layers, inner 3–11 layers, micropyle formed by inner integument, endothelium absent, hypostase present, archesporium 1–3-celled	Ana- or campylotropous, bitegmic (unitegmic in *Eugenia*), crassinucellar, outer integument 2–4 or 6–8 layers, inner 2–4 layers, micropyle formed by both integuments, endothelium absent, hypostase present, archesporium 1-celled

Embryo sac	Polygonum type, at the 4-nucleate stage the two chalazal nuclei fuse and the diploid nucleus does not divide	Polygonum or Allium	Polygonum, Oenothera or Allium	Allium	Polygonum
Antipodals	Not formed	3, persistent and multiply	3, ephemeral	3, persist after fertilization	3, ephemeral
Endosperm	Nuclear type	Nuclear	Nuclear	Nuclear	Nuclear
Embryo	—	Asterad or Caryophyllad type	Onagrad, Asterad, Caryophyllad or Solanad	Solanad	Solanad or Onagrad
Seed	Exalbuminous, mesotestal, tegmen crushed	Albuminous, mesotestal, tegmen unspecialized	Exalbuminous, exotestal, tegmen crushed	Exalbuminous or albuminous, mesotestal, tegmen crushed	Exalbuminous, exotestal, tegmen crushed
Polyembryony	—	Synergid polyembryony	Cleavage and nucellar	Placental and inner integumental	Nucellar
Apospory	—	Hieracium type, pseudogamy, diplospory	—	—	Nucellar

The embryogeny varies a great deal: Asterad, Onagrad, Caryophyllad and Solanad type. Similarly, variations occur in the structure of suspensor.

Lersten (1983) has reviewed the structure of suspensor in the Leguminosae. The Mimosoideae and Caesalpinioideae have a rudimentary or a small suspensor in a few taxa; mostly the embryo is without a suspensor. The Papilionoideae have many suspensorless taxa, others are with small suspensor, and still others with very conspicuous suspensor. The scheme of tribal delimitation and relationships for the Papilionoideae is fully supported by the distribution and structure of suspensor. The enormous spherical, many-celled suspensor is unique, as is the variable biseriate filamentous to columnar suspensor in *Lupinus*. There seems to be no correlation between the amount of endosperm and final suspensor size.

Dnyansagar (1970) proposes to split the family Papilionaceae, as suggested also by Souèges and Crété (1952), and Souèges (1956).

In contrast to the variability in the development of embryo, Papilionaceae show marked similarity in the structure of the seed-coat. In an exotestal, arillate seed, the epidermal palisade layer with linea lucida is followed by a layer of hypodermal hour-glass cells in the testa. A similar structure of seed, with some variation, occurs in the other two families.

Krameriaceae. Hutchinson (1973), Takhtajan (1980) and Cronquist (1981) include Krameriaceae in the order Polygalales along with several common families.

Simpson and Skvarla (1981) compare the pollen morphology of Krameriaceae, Leguminosae, Malpighiaceae, Trigoniaceae, Vochysiaceae, Tremandraceae and Polygalaceae, and conclude that the pollen studies do not provide convincing evidence for affinities with any family.

Verkerke (1986b, see also Corner 1976) points out that ovule ontogeny and seed-coat development of *Krameria* are very different from those of Leguminosae; there are though some exceptions. The thickened inner integument with an inner cuticle, and the absence of a vestigial endotesta separate Krameriaceae also from Polygalaceae. The endotegmic seeds of Krameriaceae suggest affinity with Malpighiaceae. The ovule ontogeny and seed-coat development of Krameriaceae differ much less from those in Malpighiaceae. Some members of the Malpighiaceae have pyriform, anatropous, bitegmic ovule with a curved funicle. There are other resemblances too. For the present, it may be emphasized that the Krameriaceae have no affinity with either Leguminosae or Polygalaceae, and the affinity with Malpighiaceae may be further confirmed following detailed studies of malpighiaceous seed and the reduction trends they exhibit.

Rosales. According to Bessey's system of classification (1915), Rosales are placed near Ranales. Hutchinson (1973) considers that the Saxifragales, a herbaceous group, are closely connected with Ranales and slightly advanced over them. Subramanyam (1962) compares the embryological features of both orders. The structure of anther, pollen, embryo sac, and small antipodals (except Ranunculaceae and Berberidaceae) are similar in both. Cellular type of endosperm occurs, except in some Magnoliaceae (sensu stricto), Lauraceae, Ranunculaceae and Berberidaceae. The peculiar development of megaspore tube (haustorium) in two species of Crassulaceae is also reported in one member of Lauraceae (Bambacioni-Mezzetti 1935). Subramanyam also considers such features as multiple archesporium, presence of starch grains in mature embryo sac, and enlargement of basal cell of two-celled embryo in Crassulaceae. He concludes that there are several similar features in Crassulaceae, Saxifragaceae, and Ranalian families (Magnoliales and Ranales of Melchior 1964), and suggests the possible origin of Rosales from Ranales, as assumed also by Bessey and Hutchinson.

27 Hydrostachyales

The order Hydrostachyales comprises a single family, Hydrostachyaceae.

The development of anther wall and seed-coat has not been investigated.

27.1 General Features

The order Hydrostachyales exhibits the following embryological features: (1) secretory anther tapetum, (2) simultaneous cytokinesis, (3) non-aperturate pollen grains formed in permanent tetrads, (4) anatropous, unitegmic, tenuinucellar ovule on parietal placenta, (5) a narrow elongated nucellus with a tiny gametophyte located at its apex, (6) Polygonum type of embryo sac, (7) starch grains in mature embryo sac, (8) Cellular type endosperm (Stachys type), and (9) embryogeny of Onagrad type-Capsella variation.

27.2 Hydrostachyaceae

The anther wall shows secretory anther tapetum with binucleate and diploid cells (Jäger-Zürn 1965). The cytokinesis in microspore mother cells is simultaneous.

The non-aperturate mature pollen grains are shed in permanent tetrads.

The ovule is anatropous, unitegmic and tenuinucellar. The archesporial cell functions directly as megaspore mother cell. The latter undergoes meiotic divisions producing a linear tetrad. The chalazal megaspore develops into a Polygonum type of embryo sac; polar nuclei fuse before fertilization. The mature embryo sac contains starch grains. An endothelium also differentiates.

The development of endosperm is of the Cellular type, Stachys type of Schnarf (1917). The division of primary endosperm nucleus is followed by a transverse wall. The micropylar chamber divides transversely forming a linear row of three cells. The uppermost cell develops into an aggressive micropylar haustorium, the central cell gives rise to cellular endosperm, and the chalazal cell does not form a haustorium.

The embryogeny conforms to the Onagrad type-Capsella variation, and there is no distinct suspensor haustorium.

G Erdtman (1945, 1952), I Jäger-Zürn (1965), J Mauritzon (1933b, 1939b), W Rauh and I Jäger-Zürn (1966), K Schnarf (1917, 1931), K Subramanyam (1962)

27.3 Taxonomic Considerations

Hydrostachyaceae. Hallier (1901, 1903, 1905, 1908) assigns the genus *Hydrostachys* to Podostemaceae; Engler and Prantl (1931) places Hydrostachyaceae in Rosales along with Podostemaceae, Crassulaceae, Saxifragaceae, etc. Hutchinson (1973) assigns it to the order Podostemales along with Podostemaceae. According to Melchior (1964), Hydrostachyales occupy an intermediate position between Rosales and Podostemales. Cronquist (1968) and Takhtajan (1969, 1980) place Hydrostachyaceae in Scrophulariales. Subsequently, Cronquist (1981) includes them in Callitrichales. R. Dahlgren (1977a, 1980b, 1981) erected the order Hydrostachyales near Scrophulariales.

Subramanyam (1962) compares the embryological features of Crassulaceae with those of Podostemaceae and Hydrostachyaceae. They all have a large number of small ovules, and a narrow elongated nucellus with tiny embryo sac located at the apex. Hydrostachyaceae and Crassulaceae are closely allied and have Polygonum type of embryo sac which contains starch grains, and the endosperm is of the Cellular type.

Jäger-Zürn (1965) discusses the morphology and embryology of Hydrostachyaceae and comments on their systematic position. In contrast to Podostemaceae, Hydrostachyaceae exhibit simultaneous type of cytokinesis in pollen mother cells (both types in the former), pollen in permanent tetrads (single grains or dyads in the former), unitegmic ovules on parietal placenta (bitegmic ovules on central placenta in the former), long chalaza, absence of pseudo-embryo sac, Polygonum type of embryo sac with endothelium and starch grains in mature gametophyte, Cellular (Stachys type – transitory) type of endosperm with micropylar haustorium, Onagrad type of embryogeny, and absence of suspensor haustorium.

Jäger-Zürn (1965) suggests the inclusion of Hydrostachyaceae in tubiflorous families, Scrophulariales, on the basis of similarity in the development of gametophytes, endosperm, and embryo.

28 Podostemales

28.1 General Features

The order Podostemales is represented by the family Podostemaceae which shows several important features: (1) binucleate glandular tapetum, (2) pollen grains in pairs in some members, (3) anatropous, bitegmic, tenuinucellate ovules, (4) formation of pseudo-embryo sac, (5) Apinagia, Dicraea, and Podostemum type (bisporic) of development of female gametophyte, (6) absence of triple fusion and, consequently, absence of endosperm, (7) Solanad type of embryogeny, (8) development of haustoria from the binucleate, basal cell of suspensor, and (9) tegmic seed.

28.2 Podostemaceae

The anther wall comprises the epidermis, fibrous endothecium, an ephemeral middle layer, and glandular tapetum with binucleate cells. They are uninucleate in *Griffithella* (Razi 1949). The microspore mother cells undergo meiotic divisions, and produce tetrahedral and isobilateral tetrads. Cytokinesis is simultaneous in *Indotristicha* (Chopra and Mukkada 1966) and *Terniola* (Mukkada 1969), and successive in *Dicraea stylosa* (Mukkada 1962). In *Dicraea* the microspores separate in pairs to form double pollen grains.

Pollen grains are three-colpate, and shed at two-celled stage individually or in pairs in *Angolaea*, *Lophogyne*, *Myopsis* (Went 1910, 1912), *Dicraea* (Fig. 28.2.1A) *Griffithella* and *Lawia* (Razi 1949), and *Podostemon* (Magnus 1913).

In *Indotristicha ramosissima* (Mukkada and Chopra 1973), the ovary contains abundant starch grains; the cells of the inner epidermis of the ovary wall are binucleate and simulate tapetum (Fig. 28.2.1B, C).

The ovule is anatropous, bitegmic and tenuinucellate. The outer integument develops before the inner one, and organizes the micropyle. The pseudo-embryo sacs (in the nucellar region) have a nutritive role in the absence of endosperm. These develop before fertilization in all the genera of Podostemoideae, and after fertilization in Tristichoideae (Arekal and Nagendran 1977b). The nucellar cells located below the embryo sac undergo longitudinal stretching, their walls break down, the protoplasts unite, and a pseudo-embryo sac is formed (Fig. 28.2.1D, E). The inner integument encloses this unique structure. Arekal and Nagendran (1975a) designate it as "nucellar plasmodium".

The hypodermal archesporial cell functions directly as megaspore mother cell. The latter undergoes the first meiotic division giving rise to micropylar and chalazal dyads.

The development of the female gametophyte (Fig. 28.2.1D, E) is rather peculiar. Battaglia (1971) reviewed the different types and, according to him, only two types occur: Apinagia type and Dicraea type, and the third, Podostemum type, is doubtful. However, Arekal and Nagendran (1975b) confirm the occurrence of Podostemum type in 30% ovules of *Hydrobryopsis sessilis*.

1. Apinagia type: Battaglia (1971) subdivided this type into A, B, C and D forms. The C and D forms are mere variations.

Apinagia A form:	five-nucleate, four-celled,
Apinagia B form:	four-nucleate, four-celled,
Apinagia C form:	five-nucleate, five-celled, and
Apinagia D form:	six-nucleate, four-, five- or six-celled.

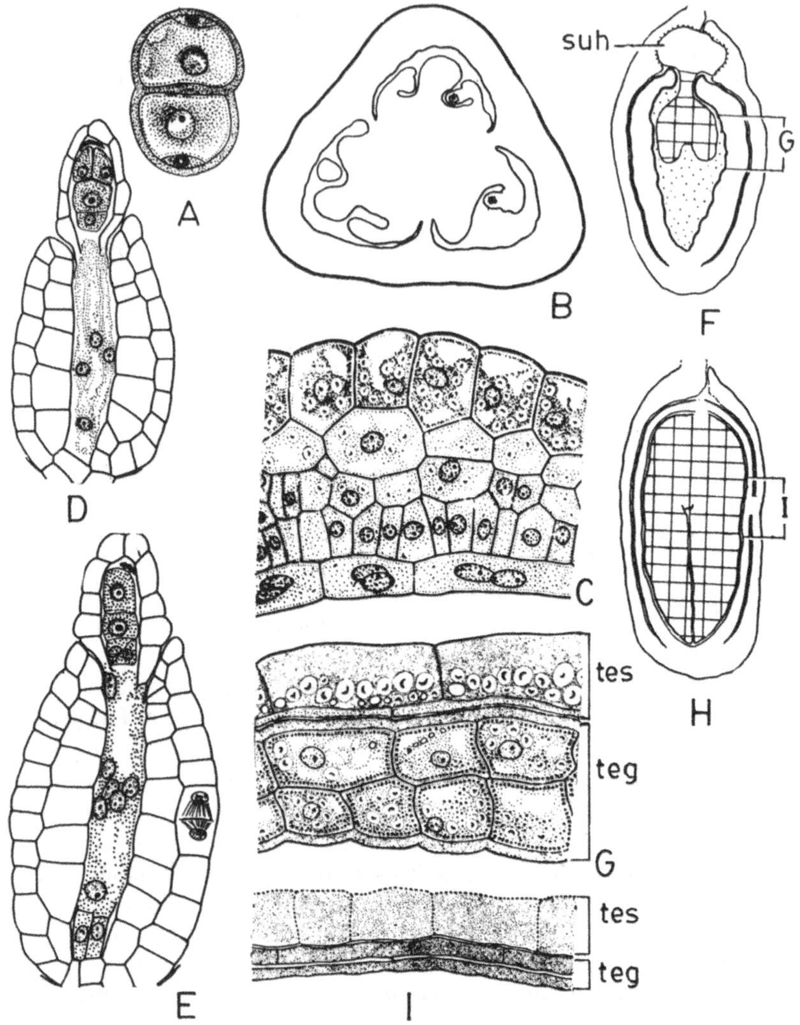

Fig. 28.2.1A. *Dicraea stylosa*, double pollen grain, each with generative and vegetative cell. (After Mukkada 1962). **B, C** *Indotristicha ramosissima*; **B** Transection of ovary at archesporial cell; **C** Portion of ovary wall from **B**, the cells of inner epidermis are binucleate. (After Mukkada and Chopra 1973). **D** *Griffithella hookeriana*, organized embryo sac with juxtaposed synergids, egg, and a polar nucleus (Apinagia type). The micropylar dyad has degenerated, the four-nucleate pseudo-embryo sac is conspicuous. (After Nagendran and Arekal 1976). **E** *Zeylanidium johnsonii*, organized embryo sac with egg, upper polar nucleus, and two synergids in chalazal position (Dicraea type). The outer integument is not shown in **D** and **E**. (After Arekal and Nagendran 1977a). **F–I.** *I. ramosissima*; **F** Longisection of immature seed at heart-shaped embryo (*hatched*) located in pseudo-embryo sac, the suspensor (*suh*) is haustorial; **G** Portion of seed-coat, *inset G* in **F**, note the starch grains in testa (*tes*) and tegmen (*teg*), and thickenings in tegmen; **H** Longisection of mature seed, the embryo (*hatched*) occupies the entire pseudo-embryo sac; **I** Portion of seed-coat, *inset I* in **H**, the testa (*tes*) is mucilagenous, and tegmen (*teg*) thick-walled. (After Mukkada and Chopra 1973)

In Apinagia A form the nucleus of the micropylar dyad may divide, but it degenerates. The chalazal dyad undergoes second meiotic division. Of the two nuclei, the chalazal megaspore nucleus degenerates. The upper spore nucleus divides twice, and all the four nuclei organize into two juxtaposed synergids followed by a large egg cell, and a small lower cell (Fig. 28.2.1D). The last-named cell is very often termed as an antipodal, or upper polar nucleus, or polar cell. As it degenerates after fertilization, and as there is no nuclear fusion with the male cell, Battaglia (1971) prefers to call it an antipodal.

Apinagia A form occurs in *Anastrophea abyssinica* (Jäger-Zürn 1967), *A. divertens*, *A. imthurnii* (Went 1909, 1910), *A. versteegiana* (Went 1910, 1926), *Cladopus nymani* (Went 1926), *Indotristicha ramosissima* (Chopra and Mukkada 1966), *Inversodicraea minutiflora* (Jäger-Zürn 1967), *Mourera fluviatilis* (Went 1909, 1910), *Oserya perpusilla* (Went 1910, 1926), *Terniola zeylanica* (Magnus 1913, Razi 1949, 1955, Mukkada 1969), *Tristicha trifaria* (Went 1910, 1926, Jäger-Zürn 1967), and *Zeylanidium olivaceum* (Arekal and Nagendran 1977a) (cf. Battaglia 1971).

In Apinagia B form there is a precocious degeneration and disappearance of the chalazal megaspore nucleus. The embryo sac becomes four-nucleate, four-celled; the arrangement of cells is similar to that in A form. This type has been observed in *Dalzellia zeylanica*, *Indotristicha ramosissima* (Arekal and Nagendran 1977b), *Farmeria indica* (Arekal and Nagendran 1975a), *Hydrobryum griffithii* (Nagendran et al. 1976), *Griffithella hookeriana* (Razi 1949, 1955), *Podostemon ceratophyllum* (Hammond 1937), *Polypleurum filifolium* (Nagendran et al. 1977), and *Zeylanidium olivaceum* (Razi 1955) (cf. Battaglia 1971). Nagendran and Arekal (1976) reinvestigated the development of female gametophyte in *G. hookeriana* and confirmed its classification. Nagendran (1974) considered the Apinagia type as monosporic, as all the nuclei are derived from only one megaspore nucleus.

In Apinagia C form, noted by Razi (1949, 1955) in *Lawia* (*Terniola*) *zeylanica*, the chalazal megaspore nucleus is cut off as a small antipodal cell. The embryo sac is five-nucleate, five-celled.

The Apinagia D form occurs as an abnormal feature (Razi 1949). The chalazal megaspore nucleus undergoes second somatogenic division producing a six-nucleate and four-, five- or six-celled stage. The lower chalazal cell, with derivatives of the chalazal megaspore nucleus, forms a three-nucleate antipodal cell (Razi 1949).

After examining seven species of Podostemaceae conforming to Apinagia type of embryo sac, Arekal and Nagendran (1977b) confirm that the chalazal megaspore nucleus never persists in the organized embryo sac, as reported by Chopra and Mukkada (1966), Jäger-Zürn (1967) and Mukkada (1969). Moreover, it does not divide to produce a five-, or six-celled embryo sac, as reported by Razi (1949, 1955). According to Arekal and Nagendran, the C and D forms of the Apinagia type (Battaglia 1971) do not occur in Podostemaceae.

2. Dicraea type: This was proposed by Magnus (1913) in *Dicraea elongata*. The first meiotic division in the megaspore mother cell results in the formation of two dyads. The micropylar dyad generally degenerates. The nucleus of the chalazal dyad divides. According to Magnus (1913), this division is followed by a wall. However, Mukkada (1962) confirmed that no such wall is laid down in *D. stylosa*. The development is of the bisporic four-nucleate type. The two nuclei of the chalazal dyad undergo simultaneous mitotic division. The four nuclei are arranged as an egg cell in the centre, superposed by a synergid at the micropylar end, and two juxtaposed antipodal cells (Fig. 28.2.1E). The synergid and egg are sister cells, and the polar nucleus is absent. Dicraea type occurs in *D. elongata* (Magnus 1913, Mukkada 1964), *D. stylosa* (Mukkada 1962, 1964) and *Hybrobryopsis sessilis* (Arekal and Nagendran 1975b).

Arekal and Nagendran (1977a) noted Dicraea type in *Zeylanidium johnsonii*. Nagendran (1974) designated it Polypleurum type due to the change in the name of the genus *Dicraea* (Hall 1971). The development of the embryo sac in *Polypleurum dichotomum* and *P. munnarense* (Nagendran et al. 1977) is according to the Dicraea type, as described by Mukkada (1962, 1964) and Razi (1966). However, Nagendran et al. differ with the interpretation of Magnus (1913), Mukkada (1962), Razi (1966) and Battaglia (1971). According to these authors, Dicraea type of embryo sac consists of one synergid, one egg – both considered sister cells, and two antipodal cells. In Podostemaceae, there is a trend in the elimination of nuclei at the chalazal end.

Arekal and Nagendran (1975b) refute the formation of antipodal cell/s in this family. This is essentially true except when the degenerated chalazal nucleus of the two-nucleate embryo sac persists up to the stage of organized gametophyte.

Arekal and Nagendran give a new interpretation to the Dicraea type of embryo sac, and consider the organized gametophyte "an inverted image of the Podostemum type". The mature gametophyte invariably comprises four cells – two synergids, egg cell, and one polar cell. If the Dicraea type is regarded as a case of permanent reverted polarity (inverted image), the position of the egg cell remains the same in all types of embryo sacs of Podostemaceae. The interpretation that a polar cell tops the egg and the two synergids are below the egg, completely changes the conventional structure of the egg apparatus and disposition of the upper polar nucleus. For the present, the interpretation of Arekal and Nagendran should be accepted with reservation.

3. Podostemum type: The micropylar dyad degenerates as in the Dicraea type. The development is of the bisporic, four-nucleate type. The nucleus of the chalazal dyad undergoes two consecutive divisions, and the resulting four nuclei organize as two juxtaposed synergids at the micropylar pole followed by an egg cell and a polar cell at the chalazal end. Arekal and Nagendran (1975b) report this type in 30% of the ovules of *Hydrobryopsis sessilis*.

4. Willisia type: Arekal and Nagendran (1976) investigated *Willisia selaginoides*, and suggested an additional type, "Willisia". The development of the embryo sac is bisporic and four-nucleate; the chalazal dyad is functional. The components of the embryo sac are disposed in decussate or isobilateral manner. The organized embryo sac shows two large juxtaposed synergids confined to the micropylar half of the ovoid gametophyte, while the egg and polar cell, also juxtaposed, occupy the chalazal half of gametophyte. Thus, according to these investigators, four types of embryo sacs occur in Podostemaceae – Apinagia (monosporic ?), Podostemum, Polypleurum and Willisia (all bisporic). We do not see enough justification for adding the Willisia type, although there is some variation in the organization of the embryo sac.

Nagendran (1974) has raised a fundamental issue. "Is the embryo sac of Podostemaceae bisporic?". Of the three main types of gametophytes in the family, Polypleurum and Podostemum are "truly bisporic" since both the

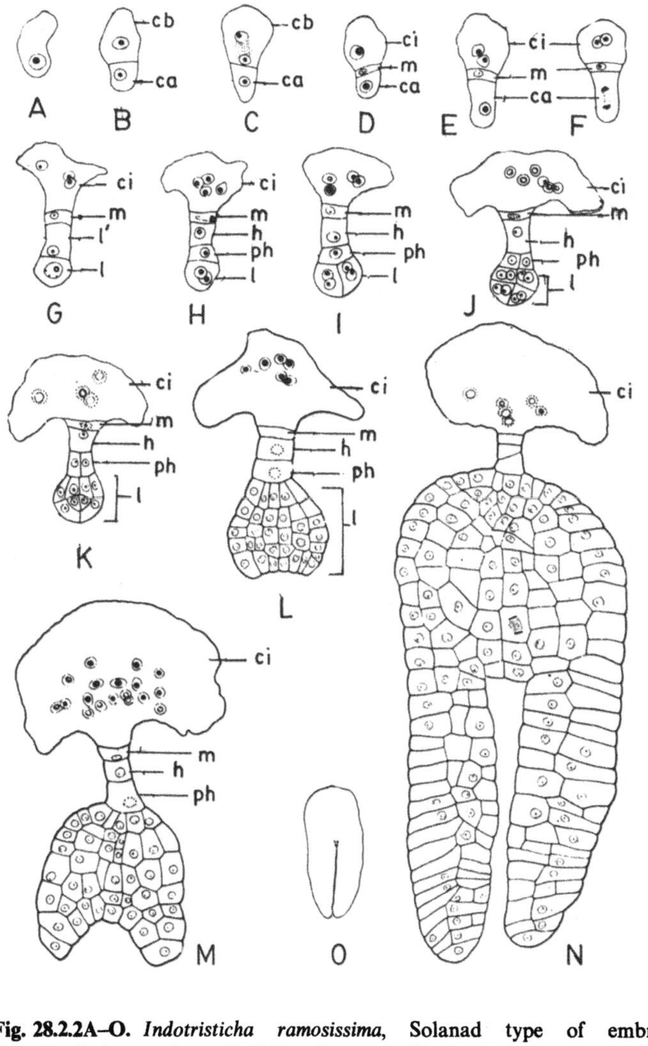

Fig. 28.2.2A–O. *Indotristicha ramosissima*, Solanad type of embryogeny. **A** Zygote. **B** Two-celled proembryo. **C–G** Development of four-celled linear proembryo. **H–J** Formation of octant. **K, L** Later stages. **M** Heart-shaped embryo. **N** Early dicot embryo. The upper-most cell (*ci*) of suspensor (**G–N**) becomes coenocytic and haustorial. **O** Mature embryo, the suspensor (including the haustorial cell) has disorganized. (After Mukkada and Chopra 1973)

megaspore nuclei of the lower dyad cell produce the mature four-nucleate embryo sac. The third type – Apinagia – is interpreted as monosporic because only the micropylar nucleus of the lower dyad cell produces all the four nuclei present in the organized embryo sac. "Whether the chalazal megaspore nucleus persists in the mature embryo sac or not, it is logical and reasonable to conclude that the Apinagia type of embryo sac development be treated as monosporic and tetranucleate". The fact that the chalazal megaspore nucleus persists (may even be ephemeral) in the embryo sac is adequate justification to designate the development as bisporic and not monosporic. Another similar example is *Butomopsis lanceolata* (B.M. Johri 1935e), and Roper (1952) gave a more or less similar reasoning for considering the embryo sac in Alismaceae as monosporic. This has already been discussed at great length by B.M. Johri (1963a).

The entry of pollen tube is porogamous, and syngamy is normal. The second male gamete may not discharge from the pollen tube (Chopra and Mukkada 1966) or, if it is discharged, may not reach and fuse with the polar nucleus. The polar nucleus degenerates, consequently there is no triple fusion and endosperm formation. In the absence of endosperm, the pseudo-embryo

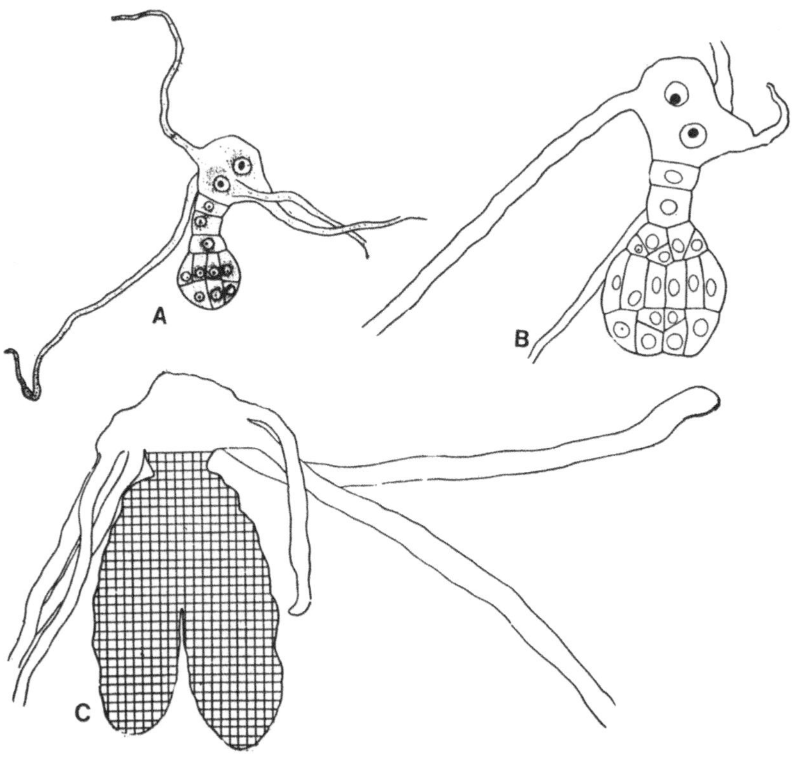

Fig. 28.2.3A–C. *Dicraea stylosa* (wholemount). **A** Globular proembryo with haustorial branches of suspensor. **B, C** Older stages of embryo with suspensor haustoria. (After Mukkada 1962)

sac plays the nutritive role and the developing embryo comes to lie in it (Fig. 28.2.1F, H; cf D, E).

The embryogeny conforms to the Solanad type (Fig. 28.2.2A–N). The uppermost cell of the suspensor enlarges, becomes coenocytic and haustorial (Fig. 28.2.2B–N). It develops thin-walled haustorial branches which penetrate the integuments (Fig. 28.2.3A–C).

The seeds are minute, and show a tegmic seed-coat (Fig. 28.2.1G, I). The cells of the outer epidermis of testa have thick pectic walls and a narrow lumen. They become mucilaginous. The cells of the tegmen are thickened and lignified, and only the thickenings of the walls persist in mature seed (Fig. 28.2.1I).

GD Arekal and CR Nagendran (1975a, b, 1976, 1977a, b), E Battaglia (1971), RN Chopra and AJ Mukkada (1966), JB Hall (1971), BL Hammond (1937), I Jäger-Zürn (1967), BM Johri (1935e, 1963a), RN Kapil (1970b), W Magnus (1913), J Mauritzon (1933b), AJ Mukkada (1962, 1964, 1969), AJ Mukkada and RN Chopra (1973), CR Nagendran (1974), CR Nagendran and GD Arekal (1976, 1979), CR Nagendran, VV Anand and GD Arekal (1980), CR Nagendran, GD Arekal and K Subramanyam (1977), CR Nagendran, K Subramanyam and GD Arekal (1976), BA Razi (1949, 1955, 1966), RB Roper (1952), K Subramanyam (1962), FAFC Went (1909, 1910, 1912, 1926)

28.3 Taxonomic Considerations

Podostemaceae. Podostemaceae are placed with Crassulaceae, Saxifragaceae, and Hydrostachyaceae in Rosales by Engler and Prantl (1931). Hutchinson (1926, 1973) groups them with Hydrostachyaceae in the order Podostemales. P. Maheshwari (1945), on the basis of embryological findings, concludes that it is "almost certain that Podostemaceae are much reduced apetalous derivatives of the Crassulaceae". Subramanyam (1962) compares the embryological features of Crassulaceae with those of Podostemaceae and Hydrostachyaceae, and emphasizes that the differences between these families are probably due to their special mode of life.

Podostemaceae show alliance with Crassulaceae in secretory anther tapetum, two-celled pollen grains, bitegmic ovules, embryo sac located in the upper part of nucellus, presence of starch in pollen and embryo sac, haustorial basal cell of suspensor, and "dwarf" type of seed with a capsular fruit. However, Podostemaceae show several unique features. The pollen grains are shed in pairs in some members. The Apinagia, Dicraea, and Podostemum types of embryo sac exhibit reduction in the development of gametophyte. The pseudo-embryo sac, or "nucellar plasmodium", serves a nutritive role, and is not met with elsewhere. The absence of triple fusion and degeneration of the polar nucleus lead to a total absence of endosperm. The embryogeny corresponds to the Solanad type. The haustorial basal cell of suspensor is non-septate. The thin haustorial branches penetrate the integuments which are rich in starch. The seeds are distinctly tegmic, and remind one of Piperaceae and Saururaceae (Corner 1976). In Crassulaceae, the above-mentioned features are absent. The seed of Crassulaceae is testal, the tegmen gets crushed.

Podostemaceae, likewise, also differ from Hydrostachyaceae (see Hydrostachyales, Sect. 27).

The embryological features justify the erection of an independent order, Podostemales, with a single family, Podostemaceae.

29 Geraniales

The order Geraniales comprises three suborders and nine families: Limnanthineae – Limnanthaceae; Geraniineae – Oxalidaceae, Geraniaceae, Tropaeolaceae, Zygophyllaceae, Linaceae and Erythroxylaceae; Euphorbiineae – Euphorbiaceae and Daphniphyllaceae.

29.1 General Features

The order Geraniales is characterized by the (1) occurrence of secretory anther tapetum (amoeboid in *Erodium* and *Geranium*), (2) simultaneous cytokinesis in pollen mother cells, (3) three-colporate pollen grains, (4) anatropous, bitegmic ovules, integumentary vascular bundle in Euphorbiaceae, (5) development of the embryo sac mostly of the Polygonum type, Allium type in Oxalidaceae and Euphorbiaceae, and tetrasporic in Limnanthaceae and Euphorbiaceae, (6) development of the endosperm of the Nuclear type, chalazal haustorium present in Linaceae, the endosperm is ruminate in Oxalidaceae and *Guaicum* (Zygophyllaceae), (7) embryogeny conforms to the Asterad type in Oxalidaceae and Geraniaceae, and Solanad type in Zygophyllaceae, Linaceae and Erythroxylaceae, and (8) outer epidermis of the inner integument is generally fibrous, or the seeds are exotegmic.

29.2 Limnanthaceae

The anther wall comprises the epidermis, fibrous endothecium, two ephemeral middle layers, and secretory tapetum with binucleate cells. Following meiotic divisions in microspore mother cells, tetrahedral and decussate tetrads are produced. Cytokinesis occurs by furrowing (P. Maheshwari and B.M. Johri 1956, N. Mathur 1956).

The pollen grains are tetracolporate, and two-celled at the dispersal stage in *Floerkea proserpinacoides* and *Limnanthes douglasii.*

The ovule is anatropous (Fig. 29.2.1A, B), bitegmic, and tenuinucellar with integumentary vascular supply reaching up to the tip of integument. The archesporium is multicellular, only one cell functions as megaspore mother cell. It undergoes meiotic division resulting in a larger nucleus which remains in the centre, and a smaller nucleus which moves to the chalazal end. The latter usually degenerates as in *Limnanthes*, or may rarely divide as in *Floerkea.* The central nucleus undergoes second meiotic division and again produces two unequal nuclei. The larger one remains in the centre of the embryo sac, while the other migrates downward and often divides.

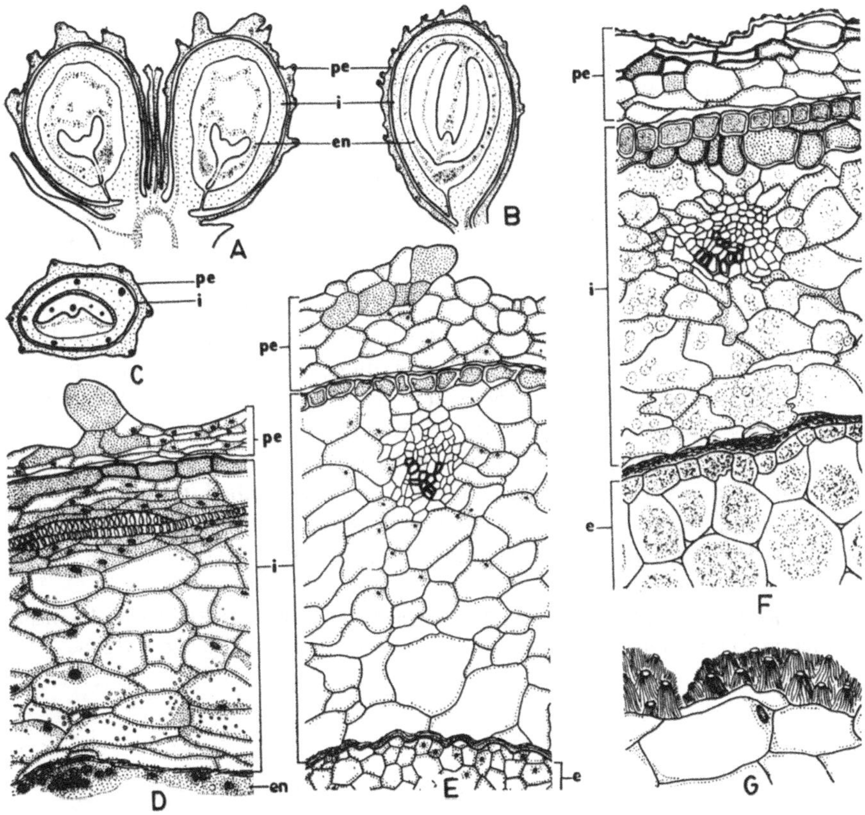

Fig. 29.2.1A–G. *Floerkea proserpinacoides.* **A** Longisection of fruits at early dicotyledonous embryo to show pericarp (*pe*), testa (*i*) and nuclear endosperm (*en*). Note the persistent gynobasic style. **B** Longisection of older fruit. **C** Transection of fruit. **D** Portion of ovary wall, vascularized testa (*i*) and nuclear endosperm from seed approximately at the same age as in **B**. The cells of the testa contain starch. **E** Pericarp and testa from **C**, and peripheral cells of embryo (*e*). **F** Pericarp, testa (*i*), degenerated remnants of endosperm, and portion of a cotyledon (*e*), from **B**. **G** Thickenings on the outer tangential walls of epidermal cells of pericarp. (After P. Maheshwari and Johri 1956)

The upper nucleus divides mitotically, but the daughter nuclei do not become separated by a vacuole. These nuclei undergo one more mitotic division, and the four daughter nuclei organize into egg apparatus and upper polar nucleus. Sometimes, the chalazal end has two nuclei which may be the daughter nuclei of the chalazal nucleus formed after Meiosis I or, if it degenerates, these nuclei may have originated from the lower nucleus formed after Meiosis II. One of these functions as lower polar nucleus, while the other may degenerate.

The mature embryo sac contains only the egg apparatus and two polar nuclei, the antipodal cells are absent. The synergids lack a basal vacuole in *Floerkea.*

The development of the embryo sac in *Floerkea* and *Limnanthes* is tetrasporic and biphasic. Occasionally, six- or seven-nucleate gametophytes have been observed in both taxa.

Double fertilization occurs, and one of the synergids persists up to the globular proembryo.

Development of the endosperm is of the Nuclear type, and it remains nuclear throughout (Fig. 29.2.1A–D). During its development the endosperm digests the cells of the inner integument and forms a haustorial pouch at the micropylar end, towards the funicular side. Nearly 30 to 35 nuclei aggregate in it in *Floerkea*.

The zygote divides transversely. The terminal cell develops into the embryo proper, while the basal cell produces a long suspensor. The mature embryo comprises two large cotyledons, a small radicle, and stem tip with a pair of leaves. The cotyledons show backward extensions (auricles) with vascular supply; they bifurcate on either side of the radicle.

The seed is exalbuminous. The vascularized testa consists of several layers of thin-walled cells (Fig. 29.2.1D–F) which become lignified in the micropylar and chalazal region. The outer tangential walls of the epidermal cells of pericarp develop pyramidal thickenings (Fig. 29.2.1G).

F Fagerlind (1939b), BM Johri (1970a, b), BM Johri and P Maheshwari (1951), P Maheshwari and BM Johri (1956), CT Mason (1950), N Mathur (1956), A Russell (1919), H Stenar (1925a)

29.3 Oxalidaceae

The anther wall comprises the epidermis, fibrous endothecium, one or two middle layers, three or four in *Averrhoa* and *Biophytum* (Thathachar 1942a), and secretory tapetum. The tapetal cells are binucleate, except in *Averrhoa bilimbi*, *A. carambola* and *Biophytum sensitivum* (Thathachar 1942a). The microspore mother cells undergo meiotic divisions and form tetrahedral tetrads. Simultaneous cytokinesis occurs by furrowing.

The pollen grains are three-colporate, and shed at the two-celled stage.

The ovule is anatropous, bitegmic, and tenuinucellate. It is crassinucellate in *A. bilimbi* and *A. carambola* (Thathachar 1942a, Govil 1978, Boesewinkel 1985b). The micropyle is organized by both the integuments. An endothelium differentiates in the inner epidermis of the inner integument. The archesporium is uni- or multicellular. A single cell develops directly into megaspore mother cell. The latter undergoes reduction divisions, and produces a linear or T-shaped tetrad. The chalazal megaspore develops into a Polygonum type of embryo sac in *Averrhoa* (Thathachar 1942a, Govil 1978), *Biophytum* (Govindappa and Boriah 1956) and *Oxalis latifolia* (L.L. Narayana 1959). According to Herr and Maryhruce (1968), in *Oxalis corniculata* the development is mostly of the Allium type, Polygonum type occurs less frequently. The synergids are pyriform, and antipodals are ephemeral.

The entry of pollen tube is porogamous, and double fertilization occurs.

Development of the endosperm is of the Nuclear type, it becomes cellular after the formation of about 32 free nuclei. The mature seed is albuminous with oil as storage product. It is subruminate in *Biophytum* and *Oxalis*.

The embryogeny conforms to the Asterad type (L.L. Narayana 1962). The development of the embryo of *Oxalis* is assigned to Period I, Series A, Megarchetype II (Asterad type).

The outer epidermis of testa comprises a short palisade layer, while the inner one contains crystals (Boesewinkel 1985b). The mesophyll is thin-walled. The outer epidermis of tegmen is composed of longitudinal fibres, and the hypodermis of tracheidal fibres. The tegmen develops folds causing rumination in the endosperm in *Biophytum* and *Oxalis*. The rest of the tissue is crushed.

FD Boesewinkel (1985b), CM Govil (1978), CM Govil and Maninder Kaur (1989), DA Govindappa and G Boriah (1956), CJ Guth and SG Weller (1986), JM Herr Jr (1972), JM Herr Jr and D Maryhruce (1968), VL Kumari and LL Narayana (1980), LL Narayana (1959, 1962, 1970a), W Noll (1935), Souèges (1939d), T Thathachar (1942a), J Venkateswarlu (1936b)

29.4 Geraniaceae

The anther wall comprises the epidermis, fibrous endothecium, a middle layer, and secretory tapetum with multinucleate cells in *Monsonia* (H.S. Narayana and Arora 1963), *Erodium stephanianum*, *Geranium* and *Pelargonium hortatum* (Kumar 1976). Amoeboid tapetum is reported in *Erodium corsicum* (Neogodi 1937), *Geranium robertianum* and *G. pyrenaicum* (Schürhoff 1924). The formation of polyploid nuclei due to nuclear fusions occurs in *Geranium grevellianum*, *G. nepalenses* and *G. ocellatum*. Meiotic divisions in microspore mother cells result in tetrahedral and decussate tetrads. Simultaneous cytokinesis occurs by furrowing.

The pollen grains are tricolpate and shed at the two-celled stage in *M. senegalensis* (H.S. Narayana and Arora 1963), *Erodium* and *Monsonia* (Kumar 1976), and at the three-celled stage in *E. corsicum* (Neogodi 1937), *E. cicutarium*, *P. zonale* (see Davis 1966), and *Geranium* and *P. hortatum* (Kumar 1976).

The ovule is ana- to campylotropous, bitegmic, and crassinucellate. The outer integument organizes the micropyle. An endothelium is absent. The archesporial cell cuts off a parietal cell, and the latter gives rise to a four- to six-layered parietal tissue. A nucellar cap, obturator and hypostase are reported in *Monsonia senegalensis* (H.S. Narayana and Arora 1963). The hypostase is suberized. Following meiotic divisions, the megaspore mother cell produces a linear tetrad. The chalazal megaspore develops into a Polygonum type of embryo sac. The antipodals are ephemeral. Multicellular archesporium, twin megaspore mother cells, and T-shaped tetrads are reported in *G. nepalense* and *G. wallichianum* (Kumar 1976). There is a high degree of abortion in the developmental stages of the embryo sac in *Pelargonium hortatum* (Kumar 1976).

Double fertilization occurs, and the entry of pollen tube is porogamous. The remnants of the pollen tube persist for some time after fertilization.

The development of the endosperm is of the Nuclear type, later it becomes cellular and is mostly consumed by the developing embryo.

The embryogeny conforms to the Asterad type in *Erodium circutarium* (Souèges 1923b), *E. stephanianum* and *G. wallichianum* (Kumar 1976), and Onagrad type in *G. molle* (Souèges 1923c).

The seed is mainly exotegmic and campylotropous. The outer epidermis of testa comprises round, thin-walled cells. The inner epidermis shows thickenings on the outer walls, exhibits a light line, and contains crystals. The outer epidermis of tegmen is a compact layer of thick-walled, lignified cells. The inner one shows thin-walled, compressed cells (Boesewinkel and Been 1979).

FD Boesewinkel (1988), FD Boesewinkel and W Been (1979), GL Davis (1966), A Kumar (1976), HS Narayana and PK Arora (1963), LL Narayana (1970b), G Neogodi (1937), PN Schürhoff (1924), R Souèges (1923b, c), AH Tsai, PM Harney and RL Peterson (1973)

29.5 Tropaeolaceae

The anther wall comprises the epidermis, endothecium, one or two ephemeral middle layers, and secretory tapetum (L.L. Narayana 1970c). Meiotic divisions in microspore mother cells result in tetrahedral tetrads. Simultaneous cytokinesis is followed by ephemeral cell-plate formation (Bolenbach 1928).

The pollen grains are tricolpate, and shed at the three-celled stage (LL. Narayana 1970c).

The ovule is anatropous, bitegmic, and tenuinucellar with the micropyle formed by the inner integument. The archesporial cell directly functions as megaspore mother cell. The latter undergoes meiotic divisions, and produces a linear or T-shaped tetrad. The chalazal megaspore develops into a Polygonum type of embryo sac. The synergids are pyriform, and the antipodals ephemeral. Polar nuclei fuse before fertilization.

The entry of pollen tube is porogamous. During its entry it does not destroy any of the synergids (Walker 1947). This is doubtful.

The development of the endosperm is of the Nuclear type, and it remains nuclear throughout (Walker 1947). However, L.L. Narayana (1970c) mentions that it becomes cellular.

In *Tropaeolum majus* (Walker 1947), the zygote divides transversely, followed by another transverse division in each cell. The apical cell of the four-celled linear proembryo gives rise to the embryo proper, the subapical cell forms the suspensor, and the two remaining cells give rise to suspensor haustoria. At first vertical and then transverse divisions in the four-celled proembryo results in a 14–16-celled stage.

The behaviour of the derivatives of the two uppermost cells of the proembryo is very unusual. They divide repeatedly to produce the basal mass of cells which give rise to an extensive suspensor haustorial system.

The peripheral cells of the basal cell mass, close to the suspensor, on the opposite side of the funiculus, become meristematic and a root-like haustorium

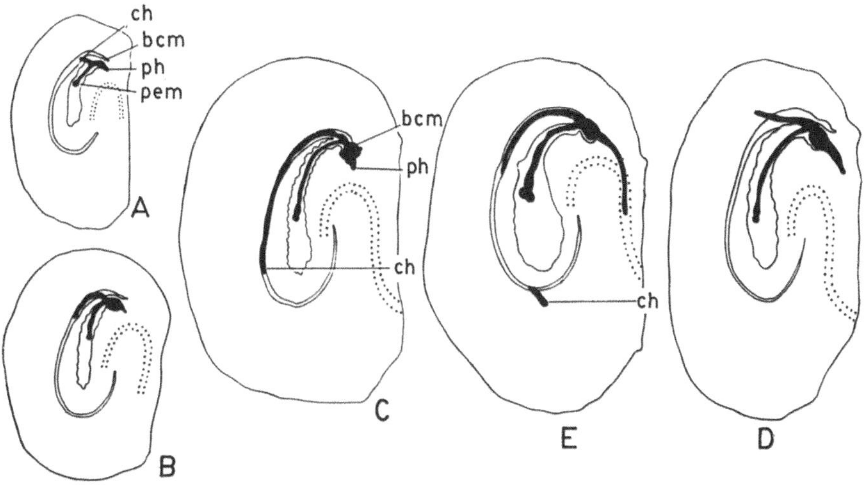

Fig. 29.5.1A–E. *Tropaeolum majus.* **A** Longisection of ovary to show proembryo (*pem*) with suspensor, and origin of carpellary (*ch*) and placental (*ph*) haustorial branches, and basal cell mass (*bcm*). **B** Later stage. **C** Longisection of ovary to show globular proembryo (*black*) with a long suspensor. Note the carpellary (*ch*) and placental (*ph*) haustorial branch. **D** Advanced stage. **E** Longisection of ovary to show heart-shaped embryo with collar and suspensor, the carpellary haustorial branch (*ch*) extends around the ovule, opposite to the funiculus, and enters the carpel wall. The placental branch enters the nucellus and grows close to the vascular bundle. The basal cell mass shows peripheral projections. (After R.I. Walker 1947)

is initiated which penetrates the integument near the micropyle, grows around the ovule in the carpellary cavity, and into the carpel wall. The haustorium penetrates the carpel wall opposite to the micropyle (Fig. 29.5.1A–E).

Shortly after the formation of carpellary suspensor haustorium, another protuberance arises from the peripheral cells of the basal cell mass, close to the funiculus (Fig. 29.5.1A–C). The apical cells are smaller and appear to be meristematic, while the remaining portion consists of large vacuolate cells. This is the placental suspensor haustorium which grows through the integuments and funiculus, reaches up to the vascular bundle of placenta, and grows along its curvature (Fig. 29.5.1D, E).

Besides the carpellary and placental suspensor haustoria, the peripheral cells of the basal mass produce finger-like haustorial cells. The nuclei of these cells are located at the base, whereas the upper portion is vacuolate.

The proembryo and the suspensor grow concurrently with the suspensor haustoria and, in due course, the globular proembryo develops into a heart-shaped embryo with epicotyl surrounded by nuclear endosperm. As the heart-shaped embryo grows to dicotyledonous form, the cells located at the junction of embryo and suspensor proliferate, and elongate to form a collar. These cells are also haustorial in function.

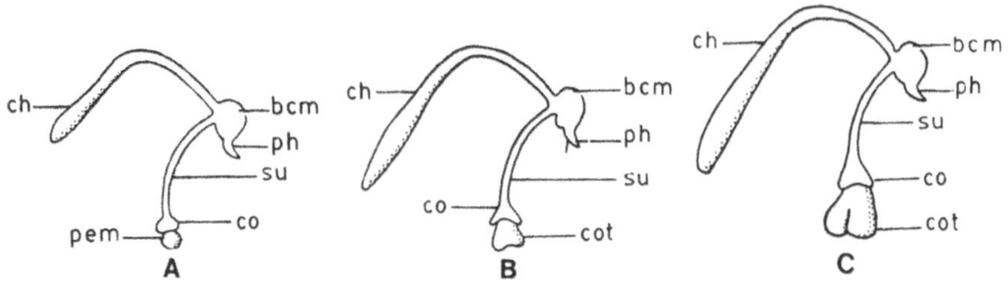

Fig. 29.5.2A–C. *Tropaeolum majus*, semi-diagrammatic drawings of dissected globular (**A**), heart-shaped (**B**), and cotyledonary embryo (**C**); *bcm* basal cell mass; *ch* carpellary haustorium; *co* collar; *cot* cotyledon; *pem* proembryo; *ph* placental haustorium; *su* suspensor. (After Bhalla et al. 1979)

According to Bhalla et al. (1982), embryo development is inhibited until the suspensor has fully grown, and ramifications become established in the placentum and fruit wall. Figure 25.5.2A–C represents the disposition of the haustorial system at globular, heart-shaped and dicotyledonous embryo.

The mature embryo is dicotyledonous with an epicotyl, hypocotyl, and remnants of suspensor haustorium in the integument.

Bhalla et al. (1979) point out that the basal cells of the suspensor in the majority of species display wall labyrinths and resemble transfer cells. The wall-ingrowths in the micropylar region of basal cells indicate that their main function is absorption of nutrients from integuments and other tissues.

After 5 days of pollination the embryo is very small but the suspensor is well developed with two distinct haustoria (Bhalla et al. 1982). At the globular proembryo the suspensor is 2 mm long, while the carpellary haustorium traverses about 3 mm around the ovule, protrudes outside and its tip becomes embedded in the inner fruit wall. At the heart-shaped embryo the suspensor proper is 4 mm long, the rosette of apical cells grows in width, and the carpellary haustorium is 6 mm long.

The seed is pachychalazal and exalbuminous with a straight embryo. Tiwari et al. (1977) point out that at the mature embryo sac the chalaza shows massive growth. A meristematic inwardly curved cup-shaped strip develops in the chalaza above the vascular strand. Anti- and periclinal divisions lead to diffuse growth. The cells derived from the meristematic tissue enlarge and grow along the micropylar-chalazal axis. The endosperm develops all along the periphery of the embryo sac. The embryo sac with the embryo and its well-developed suspensor haustorial system, and the endosperm consume the adjoining nucellar, integumentary, and chalazal tissue. The embryo proper is deep-seated in the chalaza. The suspensor haustoria and the embryo consume the remaining integumentary covering. Finally, the integumentary tissue shrivels completely.

In a mature seed, the chalazal tissue gives protection to the embryo, and can be differentiated into somewhat periclinally elongated cells of the epidermis which do not show any thickening, eight to ten layers of parenchymatous tissue with vascular strands, and three to four layers of tanniniferous tissue. The meristematic zone and the innermost parenchymatous zone are resorbed (Tiwari et al. 1977).

PL Bhalla and CP Malik (1982), PL Bhalla, MB Singh and CP Malik (1979, 1982), A Bolenbach (1928), M Chadefaud (1974), EJH Corner (1976), W Nägl and S Kuhner (1976), LL Narayana (1970c), SC Tiwari, F Bouman and RN Kapil (1977), RI Walker (1947)

29.6 Zygophyllaceae

The anther wall comprises the epidermis, fibrous endothecium, two or three middle layers, and secretory tapetum with binucleate cells. In *Peganum harmala* (Kapil and Ahluwalia 1963), the epidermis is heavily cutinized. The middle layers persist in *Balanites roxburghii* (N.C. Nair and Jain 1956), *Guaiacum officinale* (Masand 1970) and *Seetzenia orientalis* (H.S. Narayana and Prakasa Rao 1963). The outer middle layer may develop fibrous thickenings. The tapetum becomes multinucleate, and irregularly biseriate in *Peganum* and *Seetzenia*. The nuclei become polyploid due to nuclear fusions. Simultaneous cytokinesis follows meiotic divisions in microspore mother cells, and the tetrads are tetrahedral and decussate.

The pollen grains are three-colporate, and shed at the two-celled stage in *Peganum*, *Seetzenia* and *Zygophyllum* (Masand 1963b), and three-celled in *Fagonia* (N.C. Nair and Gupta 1961). In *Peganum*, some of the pollen grains germinate in situ.

The ovule is anatropous, hemianatropous or campylotropous, bitegmic and crassinucellate. The micropyle is organized by the inner integument in *Fagonia*, *Tribulus* and *Zygophyllum*, and by both the integuments in *Guaiacum*, *Peganum* and *Seetzenia*. An endothelium is very prominent, except in *Peganum* where it is absent. The nucellus is well developed in *Peganum*, in contrast to the other members. A hypostase develops.

The archesporium is multicellular, and each cell cuts off a parietal cell. Supernumerary megaspore mother cells, dyads, triads and tetrads have been observed in *Peganum*. Only one embryo sac of the Polygonum type develops. The synergids are pyriform, and the antipodals ephemeral.

The development of the endosperm is of the Nuclear type, and wall formation commences from the micropylar end at the young proembryo stage. In *Fagonia* and *Seetzenia* the chalazal endosperm nuclei become amoeboid and multinucleolate. In *S. orientalis* (H.S. Narayana and Prakasa Rao 1963), the caecum-like lowermost cell of the endosperm at the chalazal end is large and multinucleate. Testal ingrowths in *Guaiacum* (Masand 1970) bring about ruminations in endosperm.

The embryogeny conforms to the Solanad type. Synergid polyembryony occurs in *Peganum* (Kapil and Ahluwalia 1963); proliferation of suspensor cells is reported in *Zygophyllum* (Masand 1963b). The embryogeny of *Tribulus* and *Zygophyllum* corresponds to Period II, Series A, Megarchetype II (Caryophyllad type), and that of *Peganum* to Period I, Series C, Megarchetype IV (Solanad type).

The outer epidermis of testa is slightly thick-walled and lignified in *Peganum* (Corner 1976). It gets crushed in other members. The mesophyll is thin-walled. The inner epidermis comprises small, lignified cells with crystals in *Tribulus*; the cells are radially elongate and large and do not contain crystals in *Peganum*. The inner epidermis of tegmen alone persists, and serves as the mechanical layer. It is composed of narrow, longitudinal, lignified rectangular cells.

EJH Corner (1976), BM Johri (1963b), RN Kapil and K Ahluwalia (1963), P Masand (1963b, 1970), HS Narayana and CG Prakasa Rao (1963), NC Nair and I Gupta (1961), NC Nair and RK Jain (1956), MA Rau (1962), R Souèges (1953)

29.7 Linaceae

The anther wall comprises the epidermis, fibrous endothecium, one or two middle layers, and secretory tapetum with two- to multinucleate cells. A stomium has been observed in *Ochthocosmus africanus* (D. Rao and Narayana 1965). Polyploid nuclei produced by nuclear fusions occur in the tapetum of *Linum flavum*. In *O. africanus*, *Ixonanthes grandiflora* and *I. icosandra* (D. Rao and Narayana 1965), granular thickenings occur on the inner surface of tapetum. The microspore mother cells undergo meiotic divisions, and produce tetrahedral tetrads. Simultaneous cytokinesis occurs by furrowing.

The three-colpate pollen grains are shed at the two-celled stage in *O. africanus* and *Roucheria griffithiana* (D. Rao and Narayana 1965), and three-celled in *Durandea pentagyna*, *Hugonia mystax*, *Linum grandiflorum*, *L. rubrum* and *Reinwardtia trigyna* (L.L. Narayana 1964a).

The ovule is anatropous, bitegmic and crassinucellate; tenuinucellate in *L. mysorens* (Dorasami and Gopinath 1945). A tendency towards tenuinucellate condition has been observed in *Radiola linoides* (Mauritzon 1934d), and *L. perenne* (L.L. Narayana 1970d). Both conditions occur in *Anisadenia*. In *Octhocosmus* and *Ixonanthes* the nucellus is massive. An endothelium, obturator and hypostase are present.

In *Humiria balsamifera* (Boesewinkel 1985a), the ovule is trizonate and both the integuments are of dermal derivation. They are initially two- to three-layered. The micropyle is organized by both integuments.

A hypodermal archesporial cell cuts off a parietal cell. The parietal tissue is one to four layers thick. The megaspore mother cell undergoes meiotic divisions, and produces a linear tetrad, linear or T-shaped in *H. balsamifera*. The chalazal megaspore develops into a Polygonum type of embryo sac. The antipodals are ephemeral. Twin embryo sacs are present in *L. perenne* (see L.L. Narayana 1970d).

The development of endosperm is of the Nuclear type; Helobial in *Linum flavum*, *L. perenne* and *L. usitatissimum* var. *hungaricum* (Schürhoff 1924) which is doubtful. Erdelská (1967) reports the formation of a pectic wall at the time of division of the primary endosperm nucleus. It has no connection with the nuclear divisions. The nuclei pass through the perforation of the wall from

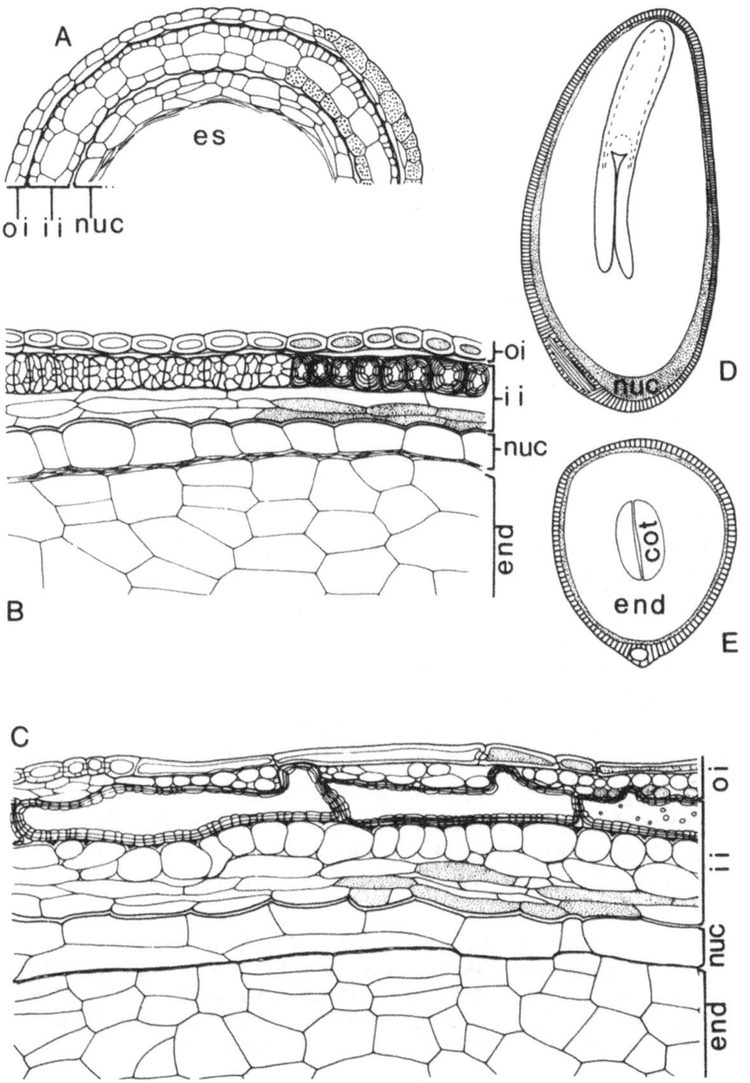

Fig. 29.7.1A–E. *Humiria balsamifera.* **A** Cross-section of a portion of mature ovule to show the disposition of outer (*oi*) and inner integument (*ii*), and nucellus (*nuc*) surrounding the embryo sac (*es*). The cells of the middle layer of inner integument enlarge. Its inner epidermis and the outer epidermis of the outer integument contain tannin. **B** Cross-section of a portion of mature seed-coat, testa two-layered and tegmen and nucellus multiplicative. The cells of the middle layer of tegmen do not divide but become stretched; *end* endosperm. **C** Longisection of a portion of mature seed-coat. Exotegmen thick-walled, the original middle layer of tegmen is longitudinally stretched. **D, E** Longi- and transection of mature seed; *cot* cotyledon. (After Boesewinkel 1985a)

micropylar region to chalazal region and, hence, the development is of the Nuclear and not Helobial type (in *Linum austriacum*).

In *Humiria balsamifera* the development of endosperm is of the Nuclear type. Later, it becomes cellular and the cells contain oil.

A chalazal haustorium is present in *Anisadenia* and *Linum* (L.L. Narayana 1970d), and *Reinwardtia* (L.L. Narayana 1964a).

The embryogeny conforms to the Solanad type. The development in *Linum* corresponds to Period I, Series C, Megarchetype III (Asterad type), and in *Radiola linoides* to Period I, Series C, Megarchetype IV (Chenopodiad type) (see Crété 1963a). Haploid-diploid twins are reported in *Linum usitatissimum* (Kappert 1933).

The cells of the outer epidermis of testa are thick-walled, often mucilaginous in *Linum*. The mesophyll is sclerotic in *Hugonia*. In *L. usitatissimum* the outer epidermis of tegmen comprises longitudinal, thick-walled, pitted fibres. The inner epidermis, endothelium, develops into rectangular cells with pitted walls and contains tannin (Boesewinkel 1980a).

The seed is exotestal and exotegmic. The outer integument remains two- to three-layered throughout, while the inner is multiplicative and contains tannin (Fig. 29.7.1A–C). The outer epidermis of the outer integument (exotesta) consists of thick-walled cells with lamellate and lignified pitted walls. These cells contain tannin and are stretched in longitudinal direction of the seed (Fig. 29.7.1B, C). The cells of the original middle layer of the inner integument do not divide, remain thin-walled, and gradually become stretched perpendicular to the longitudinal axis of seed. The cells of the outer epidermis of the inner integument develop into fibrous exotegmen of thick-walled, lignified pitted cells, stretched in longitudinal direction of the seed (Fig. 29.7.1A–C).

The mature seed of *Humiria* contains a straight embryo with rather short plano-convex cotyledons (Fig. 29.7.1D, E). One to three layers of nucellus persist in the mature seed and are covered by a thick cuticle.

G Bentham (1862b), FD Boesewinkel (1980a, 1984a, 1985a), P Crété (1963a), U Dhar and MR Vijayaraghavan (1979), LS Dorasami and DM Gopinath (1945), R Dulberger (1989), O Erdelská (1967), S Ghosh and KR Shivanna (1984), DM Gopinath (1942), H Kappert (1933), J Mauritzon (1934d), LL Narayana (1964a, 1970d), LL Narayana and D Rao (1978), D Rao and LL Narayana (1965), PN Schürhoff (1924), DL Secor and SD Russell (1988), R Souèges (1937a)

29.8 Erythroxylaceae

The anther wall comprises the epidermis, fibrous endothecium, three ephemeral middle layers, and secretory tapetum with binucleate cells. The microspore mother cells divide meiotically, and produce tetrahedral and decussate tetrads. Simultaneous cytokinesis occurs by furrowing.

The pollen grains are tricolporate, and shed at the three-celled stage (L.L. Narayana 1964b). The pollen grains have a uniform pattern of ornamentation of exine (stenopalynous).

The conspicuously elongated ovule is anatropous, bitegmic and crassinucellar in *Erythroxylum coca* (Boesewinkel and Geenen 1980). Both the integuments are of dermal origin. The inner integument is multiplicative. The outer integument is three layers thick, and remains shorter than the inner one. The inner integument alone forms the micropyle (L.L. Narayana 1970e). The endothelium extends up to the base of nucellus. The archesporial cell cuts off a parietal cell, the parietal tissue is two- or three-layered. The megaspore mother cell undergoes meiotic divisions, and produces a linear tetrad. The chalazal megaspore develops into a Polygonum type of embryo sac. The synergids are beaked, and the antipodals ephemeral. The mature embryo sac lies in direct contact with the endothelium (L.L. Narayana 1960b).

The endosperm is of the Nuclear type. Wall formation is initiated at the globular proembryo. Eventually, it becomes cellular and stores starch.

The embryogeny conforms to the Solanad type. The mature embryo is straight with thick, flat cotyledons.

The outer epidermis of testa is thin-walled. The outer epidermis of tegmen comprises thick-walled, lignified, pitted fibres with sclerotic cells at the endostome.

FD Boesewinkel and J Geenen (1980), J Mauritzon (1934c), LL Narayana (1960b, 1964b, 1970e), D Rao (1968a), PN Schürhoff (1924)

29.9 Euphorbiaceae

The anther wall comprises the epidermis, fibrous endothecium, one or two ephemeral middle layers, and secretory tapetum. The fibrous thickenings are reported to be absent in the endothecium in *Chrozophora rottleri* (G. Sharma 1956), but Kapil (1956b) reports the thickenings in *C. obliqua* and *C. parviflora.* The tapetal cells are binucleate; multinucleate in *Sebastiania chamaelea* (N.C. Nair and Maitreyi 1962). The tapetum is amoeboid in *Acalypha indica* (B.M. Johri and Kapil 1953). The microspore mother cells undergo meiotic divisions. Cytokinesis is simultaneous, and the tetrads are tetrahedral, isobilateral and decussate. Polyspory is reported in *Euphorbia corollata* (Lyon 1898). Degeneration of tapetal cells and pollen mother cells occurs in *Casava* (P.N. Rao and Sarveswara Rao 1976), and the pollen grains are sterile in *Euphorbia dulcis* (Kapil 1961).

Pollen grains are generally triporate. They are non-aperturate, quadriporate, hexaporate, or 10- to 12-porate. The exine is smooth or reticulate. They are two- or three-celled at the dispersal stage.

The ovule is anatropous, hemianatropous, or orthotropous (*Melanthesa*), bitegmic and crassinucellate. Both the integuments are vascularized in *Dalechampia*, only the inner in *Baliospermum*, *Croton*, *Jatropha*, *Ricinus* and others. The vascular supply terminates at the base of the inner integument in *Euphorbia geniculata* (S.P. Singh 1959), whereas it terminates at the chalaza in *Manihot esculentum* (P.N. Rao and Sarveswara Rao 1976). The vascular supply enters the nucellus in *Acalypha indica* (B.M. Johri and Kapil 1953). The obturator is generally placental in origin; of stylar origin in *Breynia* (Venkateswarlu et al. 1974). It is absent in *Cicca* and *Phyllanthus.* In many taxa the nucellar beak is long, slender and prominent (Fig. 3.4.6A). The beak protrudes beyond the micropyle, and a proper micropyle is not formed. Otherwise, the micropyle is produced by both the integuments, or the outer integument only. A hypostase develops during post-fertilization stages.

The archesporium is uni- or multicelled. The archesporial cell cuts off a parietal cell, and the parietal tissue is massive. The nucellar cap is very prominent in *Tragia involucrata* var. *angustifolia* (N.C. Nair and Abraham 1980). The megaspore mother cell undergoes meiosis. The chalazal megaspore of a linear or T-shaped tetrad develops into a Polygonum type of embryo sac, in the majority of the species. Monosporic five-nucleate gametophyte occurs in *Codiaeum veriegatum* (Galimberti 1963). Three nuclei at the micropylar end form the egg apparatus, and the two polar nuclei form a secondary nucleus.

Sanchez's (1938) report of Adoxa type of embryo sac in *Euphorbia heterophylla* has been corrected by P. Maheshwari (1942), and the gametophyte is of the Polygonum type. Dessiatoff (1911) refers to 16-nucleate gametophyte in *E. virgata* which is again of the Polygonum type, as pointed out by Modilewski (1911). In *E. pulcherrima*, Donati (1913) mentions a 16-nucleate embryo sac but K.D. Sharma (1955) considers the development to be of the Adoxa or Polygonum type.

The antipodals are ephemeral in *Hevea* (Nagaraja Rao 1964a), *Phyllanthus niruri* and *E. thymifolia*, and persistent in *Putranjiva roxburghii* and *Treuia nudiflora* (Banerji and Dutt 1944).

There is much variation in the development of the embryo sac in Euphorbiaceae. Bisporic, Allium type occurs in *E. amygdaloides*, *E. characias* (D'Amato 1939), *E. lagascae* (D'Amato 1946a) and *E. mauritanica* (Ventura 1933). Penaea type is reported in *E. procera*, *E. palustris* (Modilewski 1909a, 1910, 1911), *Acalypha australis* (Tateishi 1927), *A. tricolour* (Swamy and Balakrishnan 1946), and *A. malabarica* (Mukherjee 1958). "Acalypha indica" type – a modified Penaea type (two polar and two lateral groups of two cells each, and eight nuclei fuse in the centre) – occurs in *A. indica* (P. Maheshwari and Johri 1940, 1941, B.M. Johri and Kapil 1953), *A. fallax* (Banerji 1949) and *A. ciliata* (Kajale and Murthy 1954). Peperomia type is reported in *A. lanceolata* (Thathachar 1952). *Euphorbia dulcis* (Kapil 1961) shows Fritillaria type of development. Drusa type has been observed in *Mallotus japonicus* (Ventura 1933), and Adoxa type in *E. pulcherrima*. Chrysanthemum parthenium type (16-nucleate) and C. cineraraefolium type (12-nucleate) are reported in *E. epithymoides* (Cesca 1969). *E. dulcis* is an apomict.

The development of the endosperm is of the Nuclear type. It is autonomous in *E. dulcis*. The primary endosperm nucleus is three- to ten-ploid; during early stages nuclear divisions are synchronous. Enucleate cytoplasmic nodules are present in the micropylar region of the embryo sac in *Sebastiana chamaelea* (N.C. Nair and Maitreyi 1962) and *E. helioscoria* (R.P. Singh 1969). A coenocytic chalazal haustorium develops in *Croton klotzschianus*, *C. sparsiflorus* (Venkateswarlu and P.N. Rao 1963), *C. bonplandianum* (R.P. Singh and Chopra 1970) and *C. oblongifolius* (A. Pal 1974). The endosperm finally becomes cellular. Wall formation is initiated at the 32-nucleate stage in

Acalypha lanceolata (Thathachar 1952), 64-nucleate in *A. indica* (B.M. Johri and Kapil 1953) and 256-nucleate in *A. australis* (Tateishi 1927). The laying down of walls commences at the micropylar region, and proceeds to the chalaza, or is centripetal. The endosperm stores starch, fat globules and proteins.

The embryogeny generally conforms to the Onagrad type (Mukherjee 1961a, 1964, 1965, 1968, N.C. Nair and Abraham 1963, 1980, N.C. Nair and Maitreyi 1962, Mukherjee and Padhye 1964, Dnyansagar and Tiwari 1966, Samant 1973, Prakasa Rao and Rajeswari Devi 1974, Sarveswara Rao 1978). In addition, four other major types of development of embryo have been noted: Asterad, Chenopodiad, Solanad and Piperad. The occurrence of Piperad type in *Euphorbia rothiana* is refuted by Venkateswarlu and Rao (1967) who consider the development to be of Onagrad type. Other reports of the Piperad type should be reinvestigated.

Nucellar polyembryony occurs in *E. dulcis* (Kapil 1961).

The seed is albuminous with a large, almost straight embryo with a short radicle, and thin flat cotyledons. A caruncle is characteristic of Crotonoideae. In *Acalypha indica* (B.M. Johri and Kapil 1953), lateral flowers develop into carunculate seeds, whereas the terminal flower lacks caruncle. It develops from the outer integument, at the micropylar region, after fertilization. The caruncle is absent in Phyllanthoideae.

The seed is exotegmic. The testa is thin-walled, pulpy or dry. Its inner epidermis is a thin-walled layer of palisade cells. The outer epidermis of tegmen is a mechanical palisade layer of malpighian cells in Crotonoideae, and sclerotic in Phyllanthoideae. The testa is corky in *Sapium* spp.

PN Bajpai (1965), I Banerji (1949), I Banerji and MK Dutt (1944), J Bor and F Bouman (1974), J Bor and RN Kapil (1975, 1976), A Bruni and B Tosí (1978), G Cesca (1969), S Chopra (1970), S Chopra and RP Singh (1969), F D'Amato (1939, 1946a), N Dessiatoff (1911), VR Dnyansagar and HN Tiwari (1966), G Donati (1913), MK Dutt (1943), ML Frean (1979), Z Galimberti (1963), P Gori (1977), JS Greenwood and JD Bewley (1982), MI Hakki (1985), I Hurusawa (1954), BM Johri and RN Kapil (1953), LB Kajale (1954c), LB Kajale and KSN Murthy (1954), LB Kajale and GV Rao (1943), RN Kapil (1956a, b, 1960, 1961), RN Kapil and AK Bhatnagar (1972), SS Kelkar (1960), M Landes (1946), FN Lyon (1898), P Maheshwari (1942, 1945), P Maheshwari and OR Chowdhury (1937), P Maheshwari and BM Johri (1940, 1941), J Modilewski (1909a, 1910, 1911), LA Morrow (1979), PK Mukherjee (1957, 1958, 1961a, c, 1962, 1964, 1965, 1968), PK Mukherjee and MD Padhye (1964), PK Mukherjee and KN Santhianathan (1985), M Murgia, HJ Wilms, M Cresti and G Cesca (1986), A Nagaraja Rao (1964a), NC Nair and V Abraham (1963, 1980), NC Nair and M Maitreyi (1962), JW Nowicke and JJ Skvarla (1984), A Pal (1974), F Pax (1896a), VG Phatak and SD Savant (1955), PS Prakasa Rao and KR Rajeswari Devi (1974), PN Rao (1962, 1970, 1976), KR Rajeswari Devi and PS Prakasa Rao (1975), MVS Raju and A Nagaraja Rao (1953), SS Raman (1954), PN Rao and D Sarveswara Rao (1976), PS Rao, DK Prakasa and KR Rajeswari Devi (1975, 1977), EUB and CS Reddi (1985), L and M Rossignol and R Haicour (1987), K Samant (1973), ST Sanchez (1938), D Sarveswara Rao (1978), AM Schwarzenbach (1972), G Sharma (1956), KD Sharma (1955), RP Singh (1954, 1956, 1962, 1965, 1968, 1969, 1970a, b), RP Singh and S Chopra (1970), RP Singh and JL Jain (1965), SP Singh (1959), RK Srivastava and GP Agarwal (1953), BGL Swamy and BP Balakrishna (1946), S Tateishi (1927), T Thathachar (1952, 1953), T Thathachar and A Nagaraja Rao (1956), J Venkateswarlu and PN Rao (1963, 1967, 1975b), J Venkateswarlu, PN Rao and D Sarveswara Rao (1974), M Ventura (1933, 1934)

29.10 Daphniphyllaceae

The embryogeny has not been investigated.

In *Daphniphyllum himalayense* (A.K. Bhatnagar and Garg 1977), the anthers are fused and the connective slightly projects beyond the anthers (Fig. 29.10.1A–D). In transections the anthers show four to eight microsporangia (Fig. 29.10.1E–H). The anther wall comprises the epidermis, endothecium with thickened radial and inner tangential walls, four or five middle layers, and two- to three-layered glandular tapetum of multinucleate cells (Fig. 29.10.1I, J). Following meiosis I and II, there is simultaneous cytokinesis

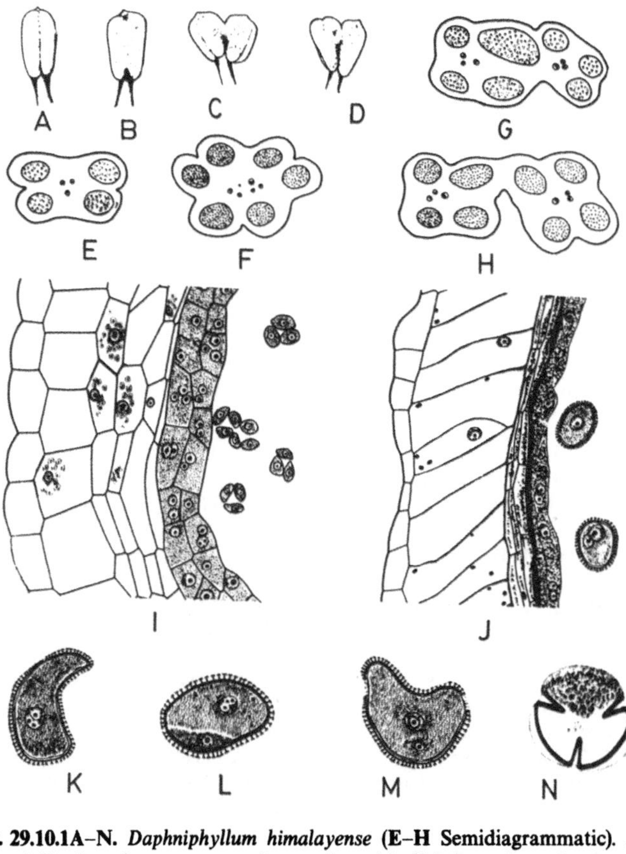

Fig. 29.10.1A–N. *Daphniphyllum himalayense* (**E–H** Semidiagrammatic). **A, B** Dorsal (**A**) and ventral (**B**) view of a tetrasporangiate anther with slightly prolonged connective. **C, D** Dorsal and ventral views of fused stamens. **E** Cross-section of **A** with four microsporangia, the connective has three vascular traces. **F–H** Transections of fused anthers with six, seven and eight microsporangia. The connective has two separate groups of vascular traces. **I** Part of transection of young anther at uninucleate pollen, there are four middle layers, and biseriate secretory tapetum with two- or three-nucleate cells. **J** Portion of mature anther at two-celled pollen grains, endothecial cells radially elongated, middle layers and tapetum have collapsed. **K** Uninucleate, intectate microspore, section passes through one aperture. **L, M** Two-celled pollen grain. **N** Palynogram shows three colpi, apocolpium, and reticulate pattern of exine. (After A.K. Bhatnagar and Garg 1977)

and mostly tetrahedral, occasionally isobilateral and decussate tetrads are formed. The mature pollen grains are two-celled (Fig. 29.10.1K–M). They are three-zonocolporoidate with a thick reticulate exine (Fig. 29.10.1N).

The ovule is anatropous, bitegmic and crassinucellate with a small obturator. The micropyle is closed by the exostome (Corner 1976). The archesporium is single-celled; formation of a parietal cell is uncertain. The development of the embryo sac follows the Polygonum type in *Daphniphyllum macropodum* var. *humile* (Sâto 1972). The antipodals are ephemeral.

The development of the endosperm is of the Cellular type (see Schnarf 1931).

The seed is albuminous and exarillate. The endosperm is thin-walled and oily, and the mature embryo is minute and straight. The testa and tegmen are thin-walled, and get crushed (Mohana Rao 1974a).

The fruit is a drupe, and endocarp is lignified with two layers of cross fibres; mesocarp has scattered stone-cells (Corner 1976).

AK Bhatnagar and M Garg (1977), EJH Corner (1976), PR Mohana Rao (1974a), Y Sâto (1972), K Schnarf (1931), M Ventura (1930)

29.11 Taxonomic Considerations

Limnanthaceae. Limnanthaceae are a monotypic family of the suborder Limnanthineae. Engler and Prantl (1897) place it in the order Sapindales, and Hutchinson (1973) in Geraniales. Lindley (1833, quoted in Takhtajan 1966) created the order Floerkeales. P. Maheshwari and Johri (1956) evaluate the embryological features of Limnanthaceae, and conclude that Limnanthaceae differ both from Sapindales and Geraniales (Table 29.11.1).

The Limnanthaceae comprise only herbaceous members with perfect flowers, and stamens in two whorls. They show: (1) persistent middle layers in anther wall, (2) two- to four-colporate pollen grains, (3) gynobasic style, (4) unitegmic, tenuinucellate, basal-parietal ovules, (5) an unusual type of tetrasporic embryo sac, (6) straight embryo with large, upwardly extended cotyledons (bifurcate in *Floerkea*), (7) pachychalazal, unitegmic seed without a characteristic exotegmen, and (8) schizocarpic fruit.

The Sapindales are shrubs or trees with compound leaves, and unisexual flowers with a single whorl of stamens. The seed is exo- or mesotestal, and the embryo is generally crumpled with curved cotyledons.

The occasional occurrence of amoeboid anther tapetum, three-colporate pollen grains, bitegmic crassinucellate ovules, mostly monosporic embryo sac, and nucellar polyembryony in some members of Sapindales and Geraniales differentiate them from Limnanthaceae.

P. Maheshwari and Johri (1956) propose that Limnanthaceae should be raised to an ordinal rank, Limnanthales, with *Limnanthes* as the type genus.

Oxalidaceae, Geraniaceae, Tropaeolaceae. Oxalidaceae, Geraniaceae and Tropaeolaceae have been treated as sub-families of Geraniaceae by Bentham

Table 29.11.1. Embryology of Limnanthaceae, Geraniales and Sapindales (see P. Maheshwari and Johri 1956)

	Limnanthaceae	Geraniales	Sapindales
Anther tapetum	Glandular, cells binucleate	Glandular, amoeboid in some taxa, cells bi- or multinucleate	Glandular, amoeboid in some taxa, cells 1–4-nucleate
Pollen	2–4-colporate, 2-celled	Usually 3-colporate, 2- or 3-celled	Usually 3-colporate, 2-celled
Ovule	Anatropous, unitegmic tenuinucellate, integument vascularized	Anatropous, bitegmic, crassinucellate (Linaceae, Oxalidaceae), integument vascularized in Euphorbiaceae	Anatropous, bitegmic, crassinucellate, tenuinucellate in Balsaminaceae, integument rarely vascularized
Embryo sac	Tetrasporic, but development differs from that in tetrasporic members of Euphorbiaceae and Malpighiaceae	Mostly monosporic, tetrasporic in Malpighiaceae, mono-, bi- or tetrasporic in Euphorbiaceae	Mostly monosporic, Allium type in *Hydrocera triflora* (Balsaminaceae)
Endosperm	Nuclear type, in *Floerkea* a coenocytic pouch is formed adjacent to the funicular-side	Nuclear	Nuclear, a large multinucleate micropylar haustorium is formed in *Hydrocera triflora*
Embryo	Straight, cotyledons large, there is an upward extension (auricle) and bifurcation of cotyledons in *Floerkea*, nucellar polyembryony absent	Straight, nucellar polyembryony occurs in Euphorbiaceae	Curved, cotyledons crumpled, nucellar polyembryony occurs in Anacardiaceae
Seed	Exalbuminous	Exalbuminous in Geraniaceae, Tropaeolaceae, Balsaminaceae, albuminous in Oxalidaceae	Exalbuminous in Sapindaceae, Sabiaceae and Anacardiaceae, albuminous in Melianthaceae
Fruit	Schizocarpic, 1-seeded achene (nutlet), indehiscent, style persists until a late stage	Capsular, dehiscent, style often persists in fruit	Berry, capsule, or drupe

and Hooker (1965a). Engler and Prantl (1931), Hutchinson (1973), Takhtajan (1966, 1980) and Cronquist (1968, 1980b, 1981) recognize their family rank and include them in Geraniales.

All the three families have similar embryological features indicating their close affinities (L.L. Narayana 1970a, b, c). However, each also has some distinctive features.

In Oxalidaceae the ovules are tenuinucellate with endothelium (crassinucellate in *Averrhoa*), occasional Allium type of embryo sac development, subruminate endosperm in *Biophytum* and *Oxalis*, and endospermic seeds. The seed structure is comparable with that of Caricaceae.

Geraniaceae show occasional presence of amoeboid anther tapetum, crassinucellate ovules, absence of endothelium, and presence of nucellar cap, parietal tissue, obturator, suberized hypostase, and exalbuminous seed.

Oxalidaceae, Geraniaceae and Tropaeolaceae exhibit extreme multipalyny (L.L. Narayana 1970a, b, c). Oltmann (1971) concluded that palynologically they are not similar. Chemotaxonomic studies (Hegnauer 1973) also support this view. Chadefaud (1974), on the basis of floral morphology of Tropaeolaceae, suggests their placement in Capparales. R. Dahlgren (1980b, 1981) includes them in a separate order, Tropaeolales (position uncertain), next to Geraniales. Tropaeolaceae have tenuinucellar ovule, and aggressive suspensor haustoria. Tiwari et al. (1977) point out that Tropaeolaceae differ from the families of Geraniales in ovular characters, and their inclusion in Geraniales is not justified. The outer integument is of subdermal origin (dermal in others), and the seeds are pachychalazal. The massive chalaza forms the seed-coat; crystalliferous and sclerotic cells are absent in the inner epidermis of the outer integument, whereas the characteristic exotegmen does not develop into a fibrous layer. In view of these facts, Dahlgren's separation of Tropaeolaceae from Geraniales deserves attention.

Embryologically, these three families (L.L. Narayana 1970a, b, c) resemble Balsaminaceae (Sapindales, Sect. 31).

Zygophyllaceae. Most phylogenists retain the family Zygophyllaceae in Geraniales. The genus *Balanites* is placed in Simaroubaceae by Bentham and Hooker (1965a), as a subfamily of Zygophyllaceae by Engler and Prantl (1931), as a subfamily Balanitoideae by Melchior (1964), and in a monogeneric family Balanitaceae in Malpighiales by Hutchinson (1973). The other genus *Peganum* is included in Rutaceae by Hooker (1885), and in Zygophyllaceae by Hutchinson (1973). Souèges (1953) and Rau (1962) suggest the creation of a separate family Peganaceae, to be placed near Linaceae.

Peganum differs from Rutaceae: three-colporate pollen grains, occurrence of supernumerary dyads and tetrads of megaspores, and absence of nucellar cap and nucellar polyembryony. In contrast, it resembles Zygophyllaceae in essential features (Table 29.11.2). The occurrence of biseriate tapetum, massive nucellus, and the absence of endothelium in *Peganum* are features similar to those of *Seetzenia* (Zygophyllaceae). However, (1) heavily cutinized epidermis of anther wall, (2) occasional germination of pollen in situ, (3) a persistent synergid and synergid polyembryony, and (4) presence of sclereids in pericarp

Table 29.11.2. Embryology of *Peganum*, Zygophyllaceae (Sect. 29.6), *Balanites* and Simaroubaceae (30.4). (See N.C. Nair and Jain 1956, Kapil and Ahluwalia 1963, Masand 1970, Corner 1976)

	Peganum	Zygophyllaceae	*Balanites*	Simaroubaceae
Anther tapetum	Irregularly biseriate, secretory, cells multinucleate	Secretory, cells binucleate	Secretory, cells multinucleate	Secretory, cells bi- or multinucleate
Cytokinesis	Simultaneous	Simultaneous	Simultaneous	Simultaneous
Tetrads	Tetrahedral, isobilateral, decussate	Tetrahedral, decussate	Tetrahedral, decussate	Tetrahedral, isobilateral, decussate
Pollen	Tricolporate, 2-celled	Tricolporate, 3-celled	Tricolpate, 2-celled	Triporate, 2-celled
Ovule	Funicular small ligulate aril, anatropous, bitegmic, crassinucellate, micropyle formed by both integuments, zigzag, feebly developed hypostase	Ana-, hemiana- or campylotropous, bitegmic, crassinucellate, micropyle formed by both or inner integument, endothelium differentiates, hypostase develops	Campylotropous, bitegmic, crassinucellate, endothelium conspicuous, cells 1–3-nucleate, micropyle formed by both integuments, zigzag, there is a space in the chalazal region between the nucellus and inner integument, and the inner and outer integument	Ana-, hemiana- or campylotropous, uni- or bitegmic, crassinucellate, integuments separated by a space in chalazal region, micropyle formed by both integuments, zigzag, obturator, nucellar cap and hypostase develop
Tetrad	Multiple dyad, triad and tetrad	Linear	Linear, T-shaped	Linear, T-shaped

Embryo sac	Polygonum type, antipodals uninucleate, persist up to 2-nucleate endosperm	Polygonum, antipodal ephemeral	Polygonum	Polygonum, antipodals ephemeral, twin embryo sacs may develop
Endosperm	Nuclear type, becomes cellular, thick-walled, reserve food oil and starch	Nuclear, becomes cellular	Nuclear	Nuclear, becomes cellular; chalazal haustroium in *Suriana*
Embryo	Solanad type, suspensor conspicuous	Solanad or Caryophyllad	—	Onagrad
Polyembryony	Synergid polyembryony	Suspensor polyembryony in *Zygophyllum*	Absent	In *Brucea* an additional embryo may develop
Seed coat	Not multiplicative, exotestal palisade, less distinct thin-walled endotestal palisade, endotegmen lignified fibriform, cells of seed-coat contain oil globules and starch	Not multiplicative	Testa thick, some cells contain crystals, others sclerotic and pitted, vascularized	Testa multiplicative, testa and tegmen become more or less crushed
Seed	Small, subconic, angled, brown, arillate (?), albuminous, a few nucellar cells persist at the micropylar end	Small to medium size, exarillate, albuminous or exalbuminous	Regulose, yellow, exalbuminous, embryo with thick cotyledons	Small to fairly large, exarillate, enclosed in woody endocarp, albuminous or exalbuminous, perisperm reported in *Brucea*, embryo straight or curved, cotyledons thick

support its assignment to a separate tribe, Peganoideae under Zygophyllaceae. The embryogeny of *Peganum harmala* (Souèges 1953, Kapil and Ahluwalia 1963) is similar to that of *Radiola* (Linaceae – see Rau 1962), and differs from that of Rutaceae and other members of Zygophyllaceae.

B.M. Johri (1963b, see also Nair and Jain 1956) summarizes the embryological features of *Balanites*. It shows secretory anther tapetum, Polygonum type of embryo sac, and Nuclear type of endosperm similar to that in Zygophyllaceae and Simaroubaceae (Table 29.11.2). The syncarpous gynoecium, two- or three-celled pollen grains at the dispersal stage, bitegmic, crassinucellate, campylotropous ovule – anatropous in Zygophyllaceae – and conspicuous endothelium occur in both *Balanites* and Zygophyllaceae. In Simaroubaceae the gynoecium is apo- or syncarpous, mature pollen is two-celled, the ovule may be ana-, hemiana-, campylo- or amphitropous, and uni- or bitegmic with a massive nucellus and a nucellar cap. The endothelium is absent, and embryogeny is of the Onagrad type. Therefore, *Balanites* is more closely related to Zygophyllaceae than to Simaroubaceae. The knowledge about its embryogeny may be helpful in determining its phyletic relationship. Masand (1970) discusses the systematic position of *Peganum* and *Balanites*, and suggests their assignment to separate subfamilies, Balanitoideae and Peganoideae, in Zygophyllaceae, under Geraniales. We agree with this arrangement.

Linaceae, Erythroxylaceae. Bentham and Hooker (1965a), Bessey (1915), Engler and Prantl (1931), Melchior (1964), and Takhtajan (1966, 1980) place Linaceae under Geraniales. Bentham and Hooker (1965a) treat Erythroxylaceae as a tribe Erythroxylae under Linaceae, while Engler and Prantl (1931), Melchior (1964), and Takhtajan (1966, 1980) treat them as an independent family under Geraniales. Hutchinson (1973) places both Linaceae and Erythroxylaceae in Malpighiales.

Both the families have essentially a similar type of development of anther, pollen, ovule, embryo sac, endosperm and embryo (L.L. Narayana 1970d, e). Linaceae are multipalynous, whereas Erythroxylaceae are uni-palynous. There is an occasional occurrence of Helobial endosperm, chalazal endosperm haustorium, haploid-diploid twin embryos; obturator and hypostase are absent in Erythroxylaceae. This may justify its separation from Linaceae. Otherwise, in both families, the testa is thin-walled as in Oxalidaceae, but lacks the crystal layer. The characteristic exotegmen occurs in both; the endothelium contains tannin.

Humiria. Melchior (1964) places *Humiria* in Linaceae, subfamily Humirioideae (Humiriaceae). Takhtajan (1980) and R. Dahlgren (1980b, 1981) treat Humiriaceae as a separate family in the order Geraniales which also includes Linaceae, Hutchinson (1973) in Malpighiales and Cronquist (1981) in Linales.

Comparable to Linaceae, in *Humiria* (Boesewinkel 1985a) the ovule is anatropous, bitegmic and crassinucellate, the embryo sac of the Polygonum type, and endosperm of the Nuclear type. In the seed-coat the middle layer of the inner integument remains thin-walled and gradually becomes stretched perpendicular to the longitudinal axis of seed and resembles the "cross cells" of *Linum*. In some respects, *H. balsamifera* differs from other members of Linaceae: there is no endothelium, obturator, or hypostase. The seed is exotestal and exotegmic. A few layers of nucellus persist in the mature seed. Mucilaginous wall formation of exotesta, as in *Linum*, is absent.

The embryology of *H. balsamifera* resembles that of other members of Linaceae, and the differences are not so significant as to warrant the erection of a separate family, Humiriaceae. The placement of *Humiria* in a subfamily Humirioideae, of Linaceae, is quite appropriate.

The suborder Euphorbiineae comprises Euphorbiaceae and Daphniphyllaceae.

Euphorbiaceae. Bentham and Hooker (1965c) place Euphorbiaceae in Unisexuales, Hallier (1912) in Passionales, Wettstein (1935) in Tricoccae, and Hutchinson (1973) and Cronquist (1968, 1981) in Euphorbiales. Bentham and Hooker (1965c) treat Buxaceae as a tribe Buxeae under Euphorbiaceae. Embryologically, it is different from Euphorbiaceae (Celastrales, Sect. 33).

Baillon (1874) includes Dichapetalaceae as a tribe Dichapetalae in Euphorbiaceae. Engler and Diels (1936) place them in Geraniales, and Melchior (1964) in Thymelaeaceae. As embryological information is not available, their relationships cannot be determined.

Wunderlich (1967) calls attention to the similarities between Euphorbiaceae and Thymaeflorae (carotenoid pollen, epitropous ovule with obturator, supernumerary antipodals, and straight embryo).

In fact, embryologically, Euphorbiaceae are a very heterogeneous family. It shows secretory or amoeboid tapetum, two- or three-celled pollen grains, anatropous, hemianatropous or orthotropous, bitegmic and crassinucellate ovules, placental obturator, nucellar beak, hypostase, massive parietal tissue, several types of embryo sac development, Nuclear type of endosperm, different types of development of embryo, and exotegmic seeds.

Thatachar (1953) proposes the relationship of Euphorbiaceae with Malpighiaceae. P.N. Rao (1970) observed apparent similarity between the two, but points out that Polygonum type of embryo sac occurs in the majority of genera, whereas bi- and tetrasporic types occur only in *Acalypha*, *Euphorbia* and *Mallotus*. Failure of fertilization, and frequent polyembryony occur commonly in Malpighiaceae.

Radcliffe–Smith (1987) emphasizes that the Euphorbiaceae differ from (a) Buxaceae which have apotropous ovules with a dorsal raphe, (b) Aextoxicaceae with apotropous, anatropous ovules borne in only one of the two loculi, and seeds with ruminate endosperm, (c) Daphniphyllaceae with a minute apical embryo in copious endosperm, (d) Pandaceae with ortho- or anatropous ovules which lack obturator, (e) Stilaginaceae with small drupaceous fruits, often flattened and somewhat obliquely oriented endocarp, (f) Icacinaceae where the ovules have a dorsal raphe, (g) Hymenocardiaceae with large extrose anthers with a terminal gland which folds back after dehiscence, and bi- or triporate pollen, (h) Bischofiaceae with spherical pollen grains, and fruit a small globose fleshy drupe, and (i) Androstachydaceae with inaperturate pollen grains. The author does not consider phylogenetic relationships.

Daphniphyllaceae. The genus *Daphniphyllum* was once placed under the tribe Phyllantheae of Euphorbiaceae (Hooker 1885), under subfamily Phyllanthoideae of Euphorbiaceae (Wettstein 1935), and as a tribe Daphniphylleae in the subfamily Phyllanthoideae (Rendle 1959). Engler and Prantl (1931), Rosenthal (1931) and Scholz (1964) assign it to a separate family, Daphniphyllaceae, close to Euphorbiaceae. Croziat (1941) finds its alliance with Eucommiaceae, and Thorne (1968) classifies it in Pittosporales. Hallier (1912), Hutchinson (1973), R. Dahlgren (1977a) and Takhtajan (1980) include it in the order Hamamelidales. Airy Shaw (1973) agrees with this assignment, but relates it also with Magnoliales. Hurusawa (1954) and Cronquist (1981) erect the order Daphniphyllales for the monogeneric family Daphniphyllaceae.

The heteropolar, three-colporoidate and intectate pollen grains of *Daphniphyllum himalayense* (A.K. Bhatnagar and Garg 1977) have no resemblance with those of Euphorbiaceae. The absence of an obturator and a Cellular type of endosperm in the family do not justify its alliance with Euphorbiaceae. The occurrence of apomixis (see Schnarf 1931) and precocious degeneration of antipodals (Sâto 1972) seem to be similar features between the two. The wood anatomy is also distinct from that in Euphorbiaceae (Metcalfe and Chalk 1950).

In contrast, it shows resemblance with Hamamelidaceae in floral structure, endothecial thickenings on the radial and inner tangential walls, multinucleate tapetal cells, similar pattern of two-celled pollen grains (A.K. Bhatnagar and Garg 1977), and the structure of wood (Metcalfe and Chalk 1950).

Both Daphniphyllaceae and Hamamelidaceae have anatropous, bitegmic, and crassinucellate ovules, Polygonum type of embryo sac (Sâto 1972), Cellular type of endosperm, large seeds, and straight embryo (Mohana Rao 1974a). In Daphniphyllaceae, however, the entire seed-coat collapses during development, and iridoid substances are present.

30 Rutales

The order Rutales includes 12 families distributed in three suborders. Rutineae comprises seven families – Rutaceae, Cneoraceae, Simaroubaceae, Picrodendraceae, Burseraceae, Meliaceae and Akaniaceae; Malpighiineae three families – Malpighiaceae, Trigoniaceae and Vochysiaceae; and Polygalineae two families – Tremandraceae and Polygalaceae.

30.1 General Features

The order Rutales exhibits: (1) secretory anther tapetum, (2) simultaneous cytokinesis in pollen mother cells, (3) bitegmic, crassinucellate ovule, massive parietal tissue and nucellar cap, (4) hypostase, (5) Polygonum type of embryo sac with ephemeral antipodals, Polygonum and Allium type in Rutaceae, and Allium and Penaea type in Malpighiaceae, (6) Nuclear type of endosperm, (7) Asterad type of embryogeny in Rutaceae (sometimes Onagrad or Solanad), Cneoraceae and Polygalaceae, and Onagrad type in Simaroubaceae and Meliaceae, (8) nucellar polyembryony in Rutaceae, Meliaceae and Malpighiaceae, and (9) endospermic seeds, except in Malpighiaceae.

30.2 Rutaceae

The anther wall comprises persistent epidermis, fibrous endothecium, two or three ephemeral middle layers, and glandular tapetum with multinucleate cells. Nuclear fusions occur in the tapetal cells in *Aegle marmelos*. The microspore mother cells undergo meiotic divisions, simultaneous cytokinesis occurs by furrowing, and the tetrads are tetrahedral, decussate and isobilateral.

Pollen grains are shed at the two-celled stage. They are two- to eight-colporate with reticulate exine.

The ovule is anatropous, bitegmic and crassinucellate. The zigzag micropyle is formed by both the integuments. A third integument develops in *Coleonema album* (Desai 1962a). A nucellar cap develops in *Chloroxylon* and *Glycosmis* (L.L. Narayana 1963a). The hypodermal archesporium is multicellular. A single cell cuts off a parietal cell, the parietal tissue is well developed. The megaspore mother cell undergoes meiotic divisions, and produces a linear or T-shaped tetrad. The chalazal megaspore develops into a Polygonum type of embryo sac. The antipodals are ephemeral. A hypostase differentiates. Large-scale degeneration of gametophytes occurs in *Aegle marmelos* (B.M. Johri and Ahuja 1956). Sâto (1971) reports Polygonum type of embryo sac in *Aucuba japonica*. But, in some cases, the development of female gametophyte is peculiar. The two megaspores of a tetrad, either micropylar or chalazal, develop into four-nucleate embryo sacs. The separating wall, between the two embryo sacs, disintegrates and an eight-nucleate embryo sac is produced. This suggests bisporic development and needs confirmation.

In *Aegle marmelos* (B.M. Johri and Ahuja 1956); triple fusion occurs normally. The egg apparatus degenerates, syngamy has not been observed.

The development of endosperm is of the Nuclear type. Centripetal wall formation is initiated at the micropylar end, after the formation of about 256 free nuclei. The chalazal end of the embryo sac grows into a coiled aggressive haustorium (B.M. Johri and Ahuja 1957), or several tubular projections develop in the chalazal region in *Coleonema album*, *Empleurum serrulatum*, *Ptelea trifoliata* and *Triphasia aurantiola* (Mauritzon 1935a).

The embryogeny conforms to the Onagrad type in *Ruta graveolens* (Johansen 1950), Solanad in *Citrus trifoliata*, and Asterad in *Ravenia spectabilis* (R.B. Ghosh 1969, 1975).

Nucellar polyembryony occurs in the majority of genera. As many as six embryos have been noted in *Xanthoxylum* (Desai 1962b). However, only one matures. In *Citrus* 13 viable embryos occur in a seed; fertilization is essential

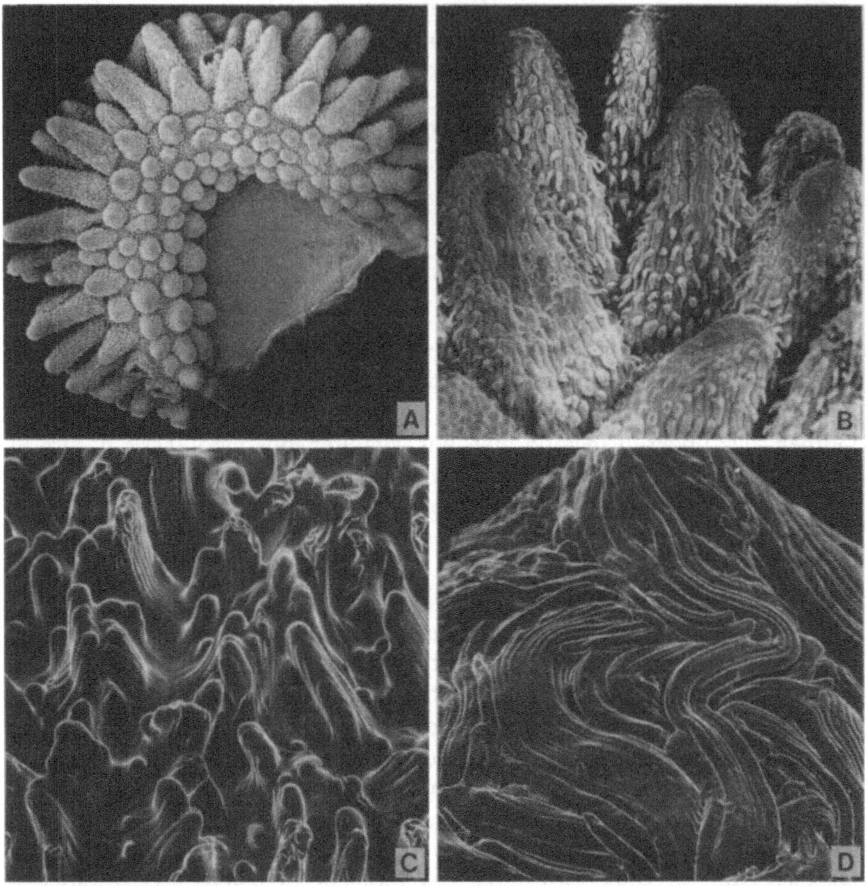

Fig. 30.2.1A–D (SEM) **A**, **B** *Thamnosma africanum*, **C**, **D** *Murraya paniculata*. **A** Seed; **B** Portion of seed-coat. **C** Young seed-coat with trichomata; **D** Same, with adpressed trichomata. (Courtesy Dr F. Bouman, Amsterdam)

for the development of these embryos (H.J. Webber and Batchelor 1943). At times, more than two embryo sacs develop in the same ovule, resulting in two zygotic embryos (Bacchi 1943).

The outer epidermis of testa is fibrous. The mesophyll is either sclerotic, or with scattered stone cells. The cells of tegmen are tracheidal with spiral and annular thickenings. The seed and seed-coat are shown in Figure 30.2.1A–D.

O Bacchi (1943), I Banerji (1954), FD Boesewinkel (1977, 1978), DK Bruck and DB Walker (1985), RS Chakravarthy (1935, 1936), C Chandra and G Shankar (1964), S Desai (1961, 1962a, b), HB Frost (1938), K Furusato (1951, 1953), K Furusato, Y Ohta and K Ishibashi (1957), RB Ghosh (1969, 1975, 1983a, b), HT Horner Jr and NR Lersten (1971), L Iglesias, H Lima and JP Simon (1974), M Iwamasa, V Isamu and N Masao (1970), DA Johansen (1950), BM Johri and MR Ahuja (1956, 1957), B Longo (1908), J Mauritzon (1935a), LL Narayana (1963a), V Narmatha Bai and KK Lakshmanan (1982), AP Pieringer and GJ Edwards (1967), Y Sâto (1971), R Souèges (1926d), DR Thakur and BS Bajwa (1971), MD Vijaya (1986), A Wakana and S Uemoto (1988), JM Webber (1940), HJ Webber and LD Batchelor (1943)

30.3 Cneoraceae

The development of anther and pollen has not been studied. We have some information on *Cneorum tricoccum* (Schürhoff 1924, Souèges 1955, Rau 1962). Boesewinkel (1984b) has made a detailed study of the ovule and seed-coat of this taxon.

The pollen grain is two-celled at dispersal.

The anatropous ovule is bitegmic and crassinucellate (Boesewinkel 1984b). The ovular primordium is trizonate. Both the integuments are two- to three-layered (Fig. 30.3.1A–C), the inner one is longer than the outer and forms the micropyle. The primary parietal cell produces a massive parietal tissue, and there is a distinct nucellar cap of about four cells high.

An obturator is present, but is hardly distinguishable in a mature seed from the outer integument on the funicular side. Gradually, the ovule changes from anatropy to campylotropy due to the intercalary growth of the raphal region, appreciable growth in the length of the nucellus and the integuments on the antiraphal side, and the relatively small ovarian locule and adpression of the seed against the endocarp.

The chalazal megaspore develops into a Polygonum type of embryo sac. The polar nuclei fuse before fertilization, and the secondary nucleus comes to lie near the three antipodals. The endosperm is of the Nuclear type but later becomes cellular due to centripetal wall formation (Fig. 30.3.1C–F). The embryogeny conforms to the Asterad type, or Period I, Group II, Megarchetype II (Souèges 1955, Rau 1962). The mature embryo is strongly curved due to the change in the ovule from anatropy to campylotropy.

During the maturation of seed (Fig. 30.3.1C–F), the nucellus and inner integument are crushed and resorbed. The inner layer of the outer integument undergoes division so that it becomes four- to five-layered. The outer tangential walls of the exotesta become thick, the cells of the meso- and endotesta show conspicuous intercellular spaces. A distinct nucellar cuticle and nucellar remains are present close to the seed-coat. The pericarp has taken over the protective function of the seed-coat. The inner layers of the pericarp become multilayered and form the sclerotic endocarp. The mesocarp and the endocarp become lignified. The seeds are uncinately conduplicate with fleshy endosperm and uncinate embryo (Hutchinson 1973). The cotyledons are situated in the transmedian plane. The endosperm and embryo are rich in lipids, and have scanty starch.

FD Boesewinkel (1984b), J Hutchinson (1973), MA Rau (1962), PN Schürhoff (1924), R Souèges (1955)

30.4 Simaroubaceae

The anther wall comprises the epidermis, fibrous endothecium, two or three ephemeral middle layers, and secretory tapetum with bi- or multinucleate cells.

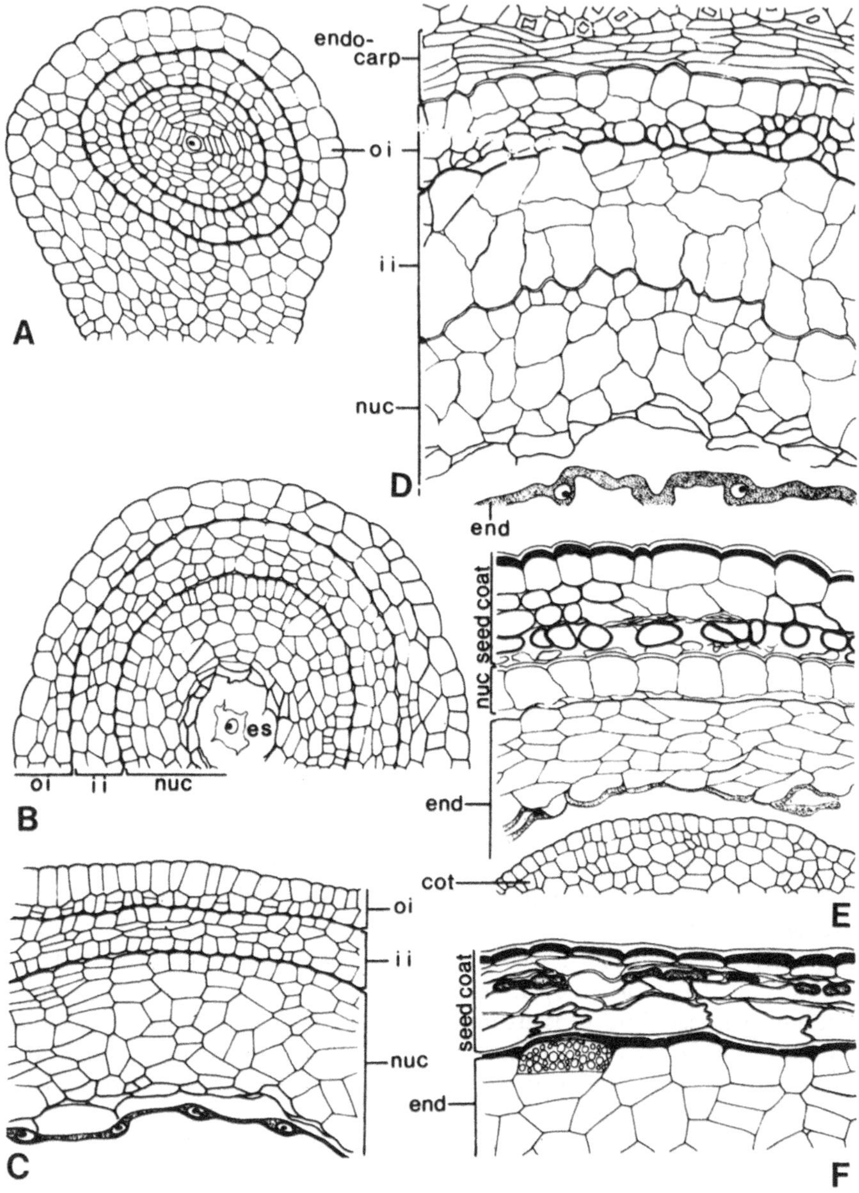

Fig. 30.3.1A–F. *Cneorum tricoccum.* **A, B** Transection of ovules; *oi, ii* outer and inner integument; *nuc* nucellus; *es* embryo sac. **C–F** Transection of seed-coat (endocarp is shown in **D**), nucellus prominent in **C** and **D**, mostly crushed in **E**, and completely in **F**. Endosperm (*end*) nuclear in **C, D**, and cellular in **E, F**; *cot* cotyledon. (After Boesewinkel 1984b)

Due to nuclear fusions the tapetal cells become polyploid in *Samadera indica* (N.C. Nair and Joseph 1957) and *Brucea amarissima* (N.C. Nair and Sukumaran 1960). The microspore mother cells undergo meiotic divisions, cytokinesis is simultaneous, and the tetrads are tetrahedral, decussate and isobilateral.

The triporate pollen grains have smooth exine, and are two-celled at the dispersal stage.

The ovules are hemianatropous in *Ailanthus excelsa* (L.L. Narayana 1957) and *Brucea*, anatropous in *Samadera* (N.C. Nair and Joseph 1957), and campylotropous in *Balanites roxburghii* (L.L. Narayana 1957) and *Suriana maritima* (Rau 1940). They are bitegmic, unitegmic in *Suriana*, and crassinucellate. The vascular strand extends beyond the chalaza in *Ailanthus excelsa* (L.L. Narayana 1957). The integuments are separated by a space in the chalazal region. The micropyle is zigzag. A hypostase is present in *Brucea*, *Samadera* and *Suriana*. A nucellar cap develops except in *Balanites*. An obturator occurs in *Brucea*, *Irvingia* and *Picrodendron* (Wiger 1935). In *Samadera* the nucellar cells become two- to 14-nucleate in the chalazal region, and the nuclei fuse and exhibit polyploid condition (N.C. Nair and Joseph 1957).

The archesporium is uni- or multicelled, only one cell is functional. It cuts off a parietal cell, the parietal tissue is massive except in *Balanites*. An endothelium develops in *Balanites* (L.L. Narayana 1957). The megaspore mother cell undergoes meiotic divisions and produces a linear tetrad, T-shaped tetrads in *Ailanthus* and *Brucea*. The chalazal megaspore develops into a Polygonum type of embryo sac. The antipodals are ephemeral. Twin embryo sacs occur in *Suriana*.

Development of the endosperm is of the Nuclear type. At the chalazal end of the embryo sac, an unbranched haustorium develops in *Suriana maritima* (Rau 1940). Later, the endosperm becomes cellular.

The embryogeny conforms to the Onagrad type. An additional embryo is produced in *Brucea*.

The mature seed contains a straight or curved embryo with a short radicle, and thick cotyledons. The outer epidermis of testa is thin-walled. Its mesophyll contains idioblasts in *Simarouba*, and the inner epidermis is slightly lignified with fine spirals in *Brucea*, *Picrasma* and *Simarouba*. The tegmen becomes crushed. The nucellus persists for a long time. The seed is enclosed in a woody endocarp.

In *Suriana* the outer epidermis of seed-coat is thick-walled, and the inner few layers are thin-walled.

NC Nair and RK Jain (1956), NC Nair and TC Joseph (1957), NC Nair and NP Sukumaran (1960), LL Narayana (1957), D Rao (1970a), MA Rau (1940), J Wiger (1935)

30.5 Picrodendraceae

The embryology has not been investigated.

30.6 Burseraceae

The embryogeny has not been investigated.

The anther wall comprises the epidermis, fibrous endothecium, two ephemeral middle layers, and secretory tapetum of binucleate cells. The epidermal cells are papillose in *Boswellia* (L.L. Narayana 1959, 1960c). Degeneration of anthers is reported in *Santiria*, *Balsamodendron* (Shukla 1954) and *Bursera* (L.L. Narayana 1960b). After meiotic divisions, the microspore mother cells produce tetrahedral tetrads. Cytokinesis is simultaneous, and quadripartition occurs by furrowing.

The three-colporate pollen grains are shed at the two-celled stage. In *Boswellia* and *Garuga* (L.L. Narayana 1960c), they are three-celled at the shedding stage.

The ovule is hemianatropous, bitegmic and crassinucellate. The inner integument alone organizes the micropyle. Unitegmic condition occurs in *Commiphora*, *Canarium asperum*, *C. oleosum* and *Santiria rubiginosa* (Wiger 1935). Shukla (1954) has reported two integuments in *Commiphora* (*Balsamodendron*).

The hypodermal archesporial cell cuts off a parietal cell. A massive parietal tissue is produced. A nucellar cap develops due to periclinal divisions of the nucellar epidermal cells. Following meiotic divisions, the megaspore mother cell gives rise to a linear tetrad. The functional chalazal megaspore develops into a Polygonum type of embryo sac. More than one megaspore mother cell develops up to the tetrad stage, but only one survives in *Balsamodendron* (Shukla 1954). The antipodals are ephemeral.

The pollen tube is monosiphonous in *Garuga* (R.B. Ghosh 1970). Its entry is porogamous, and double fertilization occurs.

The development of endosperm is of the Nuclear type, later it becomes cellular. Rarely, the cells become multinucleate in *Garuga* (R.B. Ghosh 1970). In *Bursera* (Srivastava 1964) a few free nuclei are distributed around the vacuole. The endosperm stores oil.

The mature embryo is large with folded or palmately lobed cotyledons, and a short radicle.

The outer epidermis of testa consists of thin-walled tabular cells. The inner epidermis persists as a pellicle of small, lignified, tracheidal cells with fine spiral and annular thickenings. The tegmen becomes crushed.

RB Ghosh (1970), LL Narayana (1959, 1960c), D Rao (1970b), RD Shukla (1954), GN Srivastava (1964), J Wiger (1935)

30.7 Meliaceae

The anther wall comprises the epidermis, fibrous endothecium, two or three ephemeral middle layers, and secretory tapetum with multinucleate cells. The epidermis is persistent and papillate in *Azadirachta indica* (N.C. Nair and Kanta 1961), *Cipadessa baccifera* (L.L. Narayana 1958), *Melia azedarach*

(N.C. Nair 1959b), *Naregamia alata* (N.C. Nair 1959a) and *Walsura piscidia* (R.B. Ghosh 1966b). The nuclei of tapetal cells fuse and polyploidy occurs. In *Cipadessa* star-shaped crystals are present in the middle layers. The microspore mother cells undergo meiotic divisions, and produce tetrahedral, decussate and isobilateral tetrads; cytokinesis is simultaneous. Bernardo and Ramirez (1959) reported several irregularities in microsporogenesis in "lanzones" (a variety of *Lansium domesticum*) leading to sterility, and formation of polyspores. The "duku" and "langsat" varieties of *L. domesticum* (N. Prakash et al. 1977) show extensive degeneration of microspore mother cells, and formation of viable pollen is totally absent. However, the endothecium develops characteristic fibrous thickenings.

The tetraporate pollen grains are two- or three-celled at the shedding stage. Polyspory has been observed in *Azadirachta indica*, *Cipadessa baccifera* (L.L. Narayana 1958), *Melia azedarach* and *Naregamia alata* (N.C. Nair 1959a, b, N.C. Nair and Kanta 1961).

The ovule is anatropous, bitegmic and crassinucellate. The inner integument produces the micropyle, except in *Dysoxylum* (Paetow 1931) where the outer integument is also concerned. The nucellar epidermal cells divide to produce a prominent nucellar cap. Formation of an obturator is reported in *Cipadessa*, *Lansium domesticum* (N. Prakash et al. 1977), *Melia*, *Narembergia* and *Sandoricum* (Juliano 1934).

The archesporium is uni- to multi-celled, only one cell functions. It cuts off a parietal cell. The parietal tissue is four to seven cells thick. The nucellar tissue comprises eight to ten cell layers in *Aphanamixis polystachya* (R.B. Ghosh 1966a), it is even more extensive in *Walsura piscidia* (R.B. Ghosh 1966b). The megaspore mother cell divides meiotically, and produces a linear, or sometimes T-shaped tetrad. The chalazal megaspore develops into a Polygonum type of embryo sac. The synergids may simulate the egg. The antipodals are ephemeral. In *Melia azedarach*, at times, they become multinucleate. The polar nuclei fuse before fertilization. The degeneration of gametophytes is common in *Melia azedarach* (N.C. Nair 1959b). The development of multiple embryo sacs has been noted in *Aphanamixis* (R.B. Ghosh 1966a), *Azadirachta* (N.C. Nair and Kanta 1961), *Dysoxylum* (Wiger 1935), *Lansium* (Fig. 30.7.1A, B) and *Melia* (N.C. Nair 1959b).

Double fertilization occurs, and the pollen tube persists for some time after fertilization. All varieties of *L. domesticum* (Bernardo et al. 1961, N. Prakash et al. 1977) develop apomictically.

The development of the endosperm is of the Nuclear type. The nuclei aggregate in the chalazal region of the embryo sac. Multinucleate endosperm nodules have been reported in *Azadirachta indica* (N.C. Nair and Kanta 1961). The endosperm becomes cellular at the heart-shaped stage of proembryo, except in *Dysoxylum* (Wiger 1935) where it remains free-nuclear in the chalazal region. It stores oil.

The development of endosperm is autonomous in the "duku" and "langsat" varieties (N. Prakash et al. 1977).

The embryogeny conforms to the Crucifer type in *Naregamia alata* (N.C. Nair 1959a). It is assigned to Period I, Megarchetype III, Group V

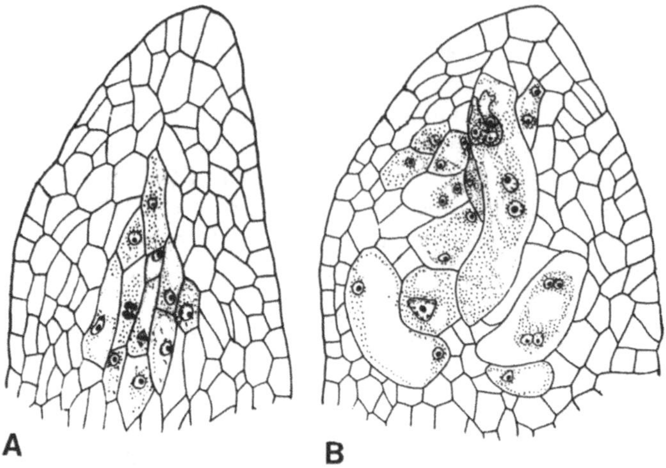

Fig. 30.7.1A, B. *Lansium domesticum*, longisections of nucelli showing a number of sporogenous cells (**A**), and a sexual and five uni-, five bi- and one four-nucleate aposporous embryo sac (**B**). (After N. Prakash et al. 1977)

(Chenopodiad) in *Azadirachta indica* (Garudamma 1956), and Period I, Megarchetype III (Asterad) in *Cipadessa baccifera* (L.L. Narayana 1958). N.C. Nair and Kanta (1961), however, report that the sequence of development is variable in *A. indica*. Synergid polyembryony occurs in *A. indica* (Garudamma 1956, N.C. Nair and Kanta 1961) and *Melia azedarach* (N.C. Nair 1959b). In *Aphanamixis polystachya* (R.B. Ghosh 1962a), nucellar polyembryony has been noted in 25–30 per cent of ovules. It is common in *L. domesticum* (N. Prakash et al. 1977). The egg cell may develop parthenogenetically up to two-celled proembryo, but degenerates. The nucellar embryo reaches maturity. The seeds are polyembryonate in the "langsat" variety, whereas in "duku" a single embryo matures in 70% of the seeds. An unusual feature is the formation of additional embryos by budding of the cotyledons of nucellar embryos.

The seed is albuminous and exotegmic, and shows much variation in different genera. The massive seeds are arillate, the aril arises from the funicle and exostome as in *Aglaia*, *Aphanamixis*, *Dysoxylon*, *Lansium* and *Naregamia*. In *Lansium* the succulent aril covers the entire seed in spite of the degeneration of seed within. The seeds are winged in *Cedrella* and *Cedrelopsis*.

In many genera the seed is pachychalazal. The outer epidermis of testa comprises thick-walled, cuticularized cells with stomata. They contain tannin in *Melia*. The mesophyll is thin-walled, and the inner epidermis contains crystals. The outer epidermis of tegmen consists of longitudinal, lignified, pitted fibres, and the inner contains tannin.

In *Trichilia grandifolia* (Boesewinkel 1981), the seed is exalbuminous, exarillate, and pachychalazal with a massive raphe bundle. The chalaza and outer integument form the sarcotesta with slightly thickened cells and intercellular spaces. The inner epidermis of testa contains crystals. The tegmen gets crushed, except at the micropyle where it becomes lignified.

According to Corner (1976), the expansion of the chalaza is associated with an extensive vascular supply. These vascular bundles are concerned in the transport of nutrients to the large embryo.

"The pachychalaza, the lack of endosperm, the strongly multiplicative integuments, the very large seed, the sarcotesta, the pre-raphe bundles, and perhaps also the loss of an aril, indicate a derived condition of the seed of *Trichilia grandiflora*" (Boesewinkel 1981).

FA Bernardo, CC Jessena Jr and DA Ramirez (1961), FA Bernardo and DA Ramirez (1959), FD Boesewinkel (1981), EJH Corner (1976), GK Garudamma (1956, 1957), SG Gavade (1963), RB Ghosh (1962a, 1966a, b), JB Juliano (1934), NC Nair (1958, 1959a, b, 1970a), NC Nair and K Kanta (1961), LL Narayana (1958), W Paetow (1931), N Prakash, AL Lim and R Manurung (1977), J Wiger (1935)

30.8 Akaniaceae

The embryology has not been investigated.

30.9 Malpighiaceae

The anther wall (Fig. 30.9.1A, B) comprises persistent epidermis, fibrous endothecium, two ephemeral middle layers, and secretory biseriate tapetum with two- to four-nucleate cells. The tapetum forms periplasmodium in *Stigmaphyllon periplocaefolium* (Munshi and Farooq 1969). The microspore mother cells undergo meiotic divisions, cytokinesis occurs by furrowing, and the tetrads (Fig. 30.9.1B–D) are rhomboidal, decussate, isobilateral and tetrahedral in *Malpighia coccigera* (Siddiqui 1968).

The pollen grains are two-celled at the dispersal stage (Fig. 30.9.1E). They degenerate prior to dehiscence of anthers in *S. periplocaefolium* (Munshi and Farooq 1969).

The ovule is semiana- or anatropous, bitegmic and crassinucellar with micropyle showing endostome, or exostome, or both (Fig. 30.9.1L). The nucellus projects beyond the micropyle in *Hiptage* and *Stigmatophyllum*. The archesporium is multicellular (Fig. 30.9.1F). Each hypodermal archesporial cell cuts off a parietal cell, and the parietal tissue is massive. A single megaspore mother cell functions, undergoes meiotic divisions, cytokinesis is absent, and a coenomegaspore is produced (Fig. 30.9.1G, H). The development of the embryo sac is of the Penaea type in *M. urens*, *M. glabra* (Fig. 30.9.1I–K) and *S. periplocaefolium*. The secondary nucleus (4n) is produced by the union of four nuclei, one from each quartet. Cortini (1958) reported *Allium* type, Penaea type, and several modifications of Penaea type in *Malpighia fulcata*, and also "Malpighia" type in this species. Allium type of development (Fig. 30.9.1M–S) occurs in *Galphimia gracilis* (Stenar 1937a), *M. glauca* (Subba Rao 1939) and *Thryallis glauca* (B. Singh 1959). The antipodals are ephemeral and polars fuse before fertilization.

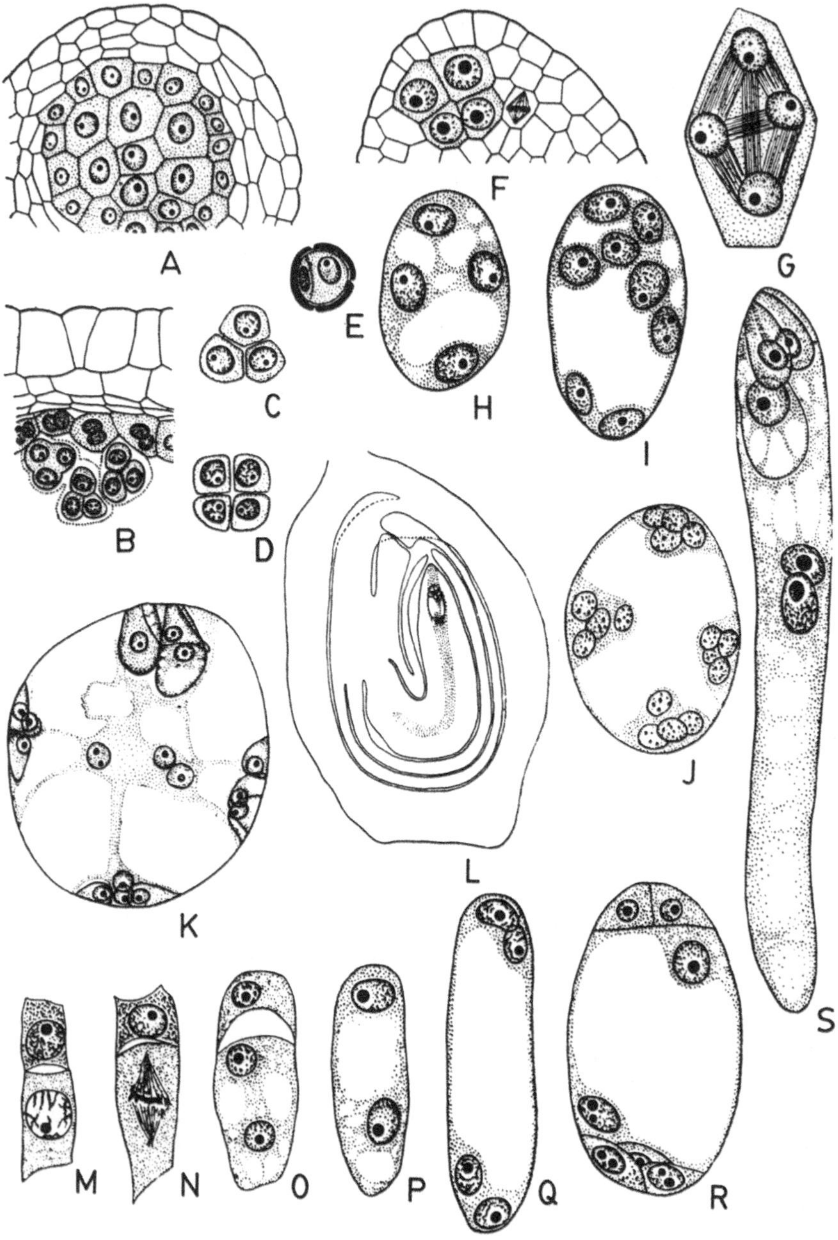

Fig. 30.9.1A–S. A, F, G, K *Malpighia glabra*, **B–E, L** *Thryallis glauca*, **H–J** *M. urens*, **M–S** *Galphimia gracilis*. **A** Cross-section of anther lobe at microspore mother cells. **B** Tetrads, tapetal cells binucleate. **C, D** Tetrahedral and isobilateral tetrad. **E** Two-celled pollen grain. **F** Nucellus with hypodermal archesporium. **G** Meiosis II. **H** Embryo sac with cross-wise arrangement of megaspore nuclei. **I** Embryo sac with eight nuclei. **J** 16 nuclei arranged in four quartets. **K** Embryo sac, Penaea type. **L** Longisection anatropous, bitegmic ovule at mature embryo sac. **M** Dyad. **N** Chalazal dyad functional. **O, P** Binucleate embryo sac. Micropylar dyad persists in **O. Q** Four-nucleate embryo sac. **R** Embryo sac, Allium type. **S** Mature embryo sac. (**A, F, G, K** After B. Singh 1961a, **B–E, L** after B. Singh 1959, **H–J, M–S** after Stenar 1937a)

Pollination and fertilization do not take place as the pollen degenerates, stigma is nonreceptive, and embryo sacs abort before the opening of flowers. Subba Rao (1940b) describes failure of fertilization in *Banisteria laurifolia*, *Hiptage madablota* and *Stigmatophyllum aristatum*.

The development of the endosperm is of the Nuclear type. Small, nucleated nodules develop in the endosperm during early stages in *Thryallis glauca* (B. Singh 1959). The nuclei aggregate in the chalazal region to form a haustorium in *Thryallis glauca* (Fig. 30.9.2A) and *M. glabra* (B. Singh 1961a, b). The chalazal haustorium remains free-nuclear in *M. glabra*, whereas it becomes

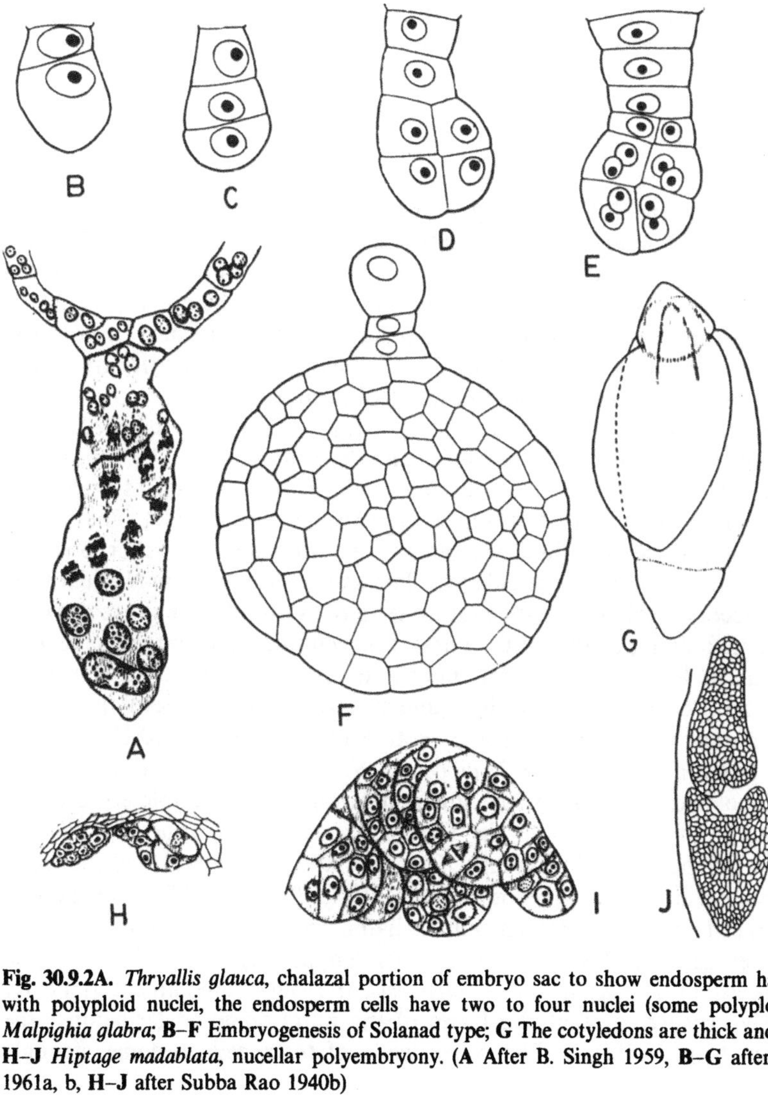

Fig. 30.9.2A. *Thryallis glauca*, chalazal portion of embryo sac to show endosperm haustorium with polyploid nuclei, the endosperm cells have two to four nuclei (some polyploid). **B–G** *Malpighia glabra*; **B–F** Embryogenesis of Solanad type; **G** The cotyledons are thick and unequal. **H–J** *Hiptage madablata*, nucellar polyembryony. (**A** After B. Singh 1959, **B–G** after B. Singh 1961a, b, **H–J** after Subba Rao 1940b)

cellular in *Thryallis*. Due to failure of fertilization in several members, the secondary nucleus forms endosperm autonomously.

The failure of fertilization is a common feature in Malpighiaceae; *M. glabra* is an exception and the embryo develops from the zygote. The embryogeny (Fig. 30.9.2B–F) conforms to the Solanad type (see Davis 1966). Nucellar polyembryony is a common feature (Fig. 30.9.2H–J).

The seed is exalbuminous with a straight or curved embryo with thick cotyledons (Fig. 30.9.2G). The testa is subaerenchymatous; the tegmen forms the main protective part. The outer epidermis of tegmen is fibrous, and the inner sclerotic. The inner epidermis is firmly adherent to the nucellus.

C Cortini (1958), GL Davis (1966), AN Fotidar (1939), M Munshi and M Farooq (1969), SG Narasimhachar (1938), PN Schürhoff (1924), SA Siddiqui (1968), B Singh (1959, 1961a, b), H Stenar (1937a), AM Subba Rao (1937, 1939, 1940b, 1941), W Verkerke (1984), W Verkerke and F Bouman (1980), H Wirz (1910)

30.10 Trigoniaceae

The embryology has not been fully studied. Boesewinkel (1987) gives a good account of the ovule and seed of Trigoniaceae.

The ovular primordium is trizonate, and gradually an anatropous, bitegmic, tenuinucellar ovule is formed in *Trigonia cipoensis*. The outer integument is suppressed on the funicular side. The zigzag micropyle is formed by both the integuments. The dermal cells of the funicle, facing the micropyle, are somewhat papilliferous and may function as an obturator.

The hypodermal sporogenous cell functions directly as the megaspore mother cell. The report of a parietal layer in *Trigonia parviflora* by Mauritzon (1936b) is erroneous. The eight-nucleate embryo sac contains starch, and probably conforms to the Polygonum type (Boesewinkel 1987). The scanty nucellus collapses so that the inner integument surrounds the embryo sac.

After fertilization the ovule enlarges. The endosperm is of the Nuclear type and later becomes cellular. In the mature seed, the walls of the endosperm cells are thickened and pectinaceous. These cells contain one large or several globules as in Vochysiaceae. The endosperm has more fats than starch, as reserve food. Between the inner integument and endosperm develops a conspicuous cuticle which reacts strongly with Sudan IV (Fig. 30.10.1B).

The seed-coat anatomy of *T. cipoensis*, *T. rugosa* and *T. villosa* is quite similar. The middle layers of the inner integument are completely crushed (Fig. 30.10.1A, B). The cells of the fibrous exotegmen enlarge and develop tracheidal lignified wall thickenings. The inner and middle layers of the outer integument divide so that it becomes thicker (Fig. 30.10.1B). All or some of the epidermal cells grow out into unicellular hairs, up to 5 mm long, and become thick-walled and lignified over their entire length (Fig. 30.10.1A–C). In the mature seeds of *T. rugosa* and *T. cipoensis*, the outer integument is nearly completely crushed so that the bases of the hairs come to lie against the exotegmen (Fig. 30.10.1C). The outer walls of testa cells become strongly thickened.

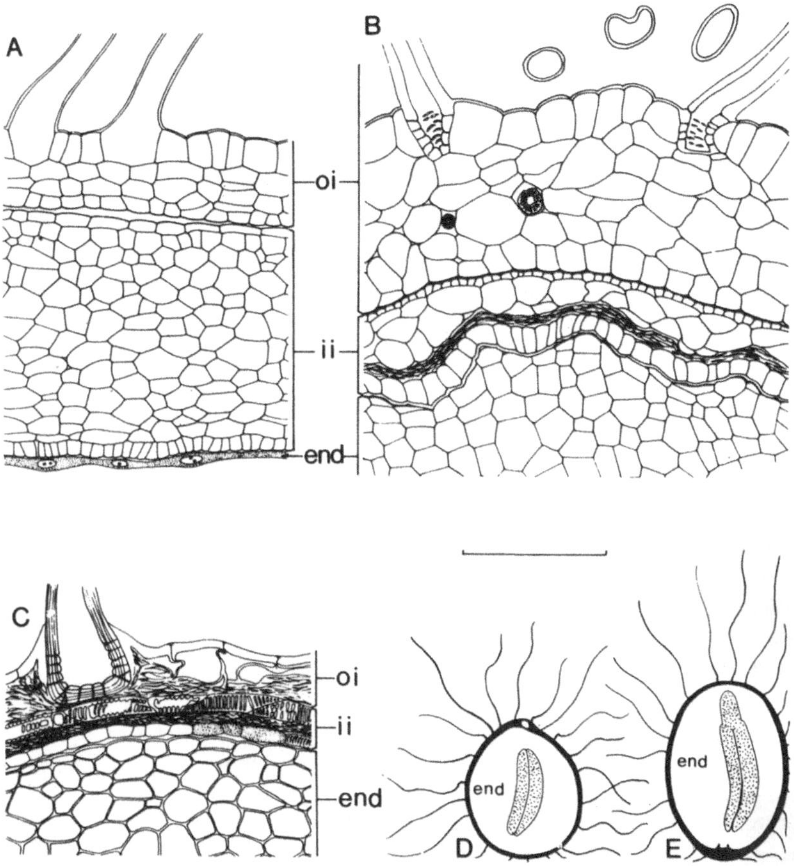

Fig. 30.10.1A–E. A, B, D, E *Trigonia cipoensis*, **C** *T. rugosa*. **A** Longisection of seed-coat, note free-nuclear endosperm; *oi*, *ii* outer and inner integument; *end* endosperm. **B** Transection, the epidermal cells of testa elongate into unicellular thick-walled hairs with strongly pitted wall in basal region. The mesophyll of the tegmen has collapsed. **C** Longisection, mature seed-coat. **D** Transection, hairy seed. Cotyledons *stippled*. **E** Longisection. (After Boesewinkel 1987)

The mature seeds of *T. cipoensis*, *T. rugosa* and *T. villosa* are covered with white cottony hairs. The embryo is linear with flat cotyledons (Fig. 30.10.1D, E). The albuminous seeds are adapted to anemo- or hydrochorous dispersal.

FD Boesewinkel (1980a, b, c, 1985a, 1987), EJH Corner (1976), J Mauritzon (1936b), W Verkerke (1985)

30.11 Vochysiaceae

The embryology of Vochysiaceae has not been fully investigated. The ovule and seed structure have been described by Boesewinkel and Venturelli (1987, see also Mauritzon 1936b, Davis 1966).

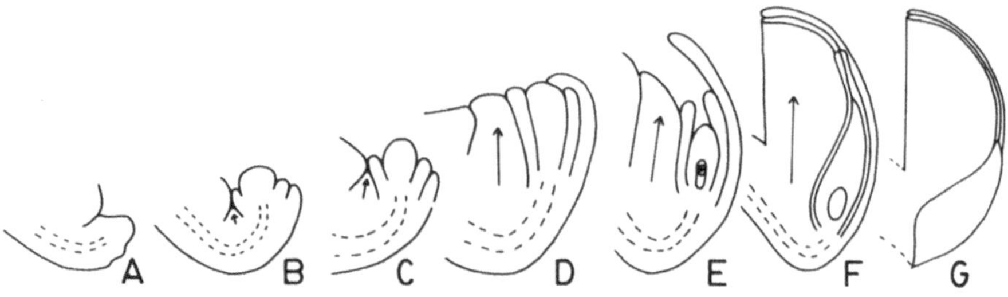

Fig. 30.11.1A–G. *Qualea cordata.* **A–F** Development of ovule. **G** Seed. Outline diagrams, *arrow* points to wing. (After Boesewinkel and Venturelli 1987)

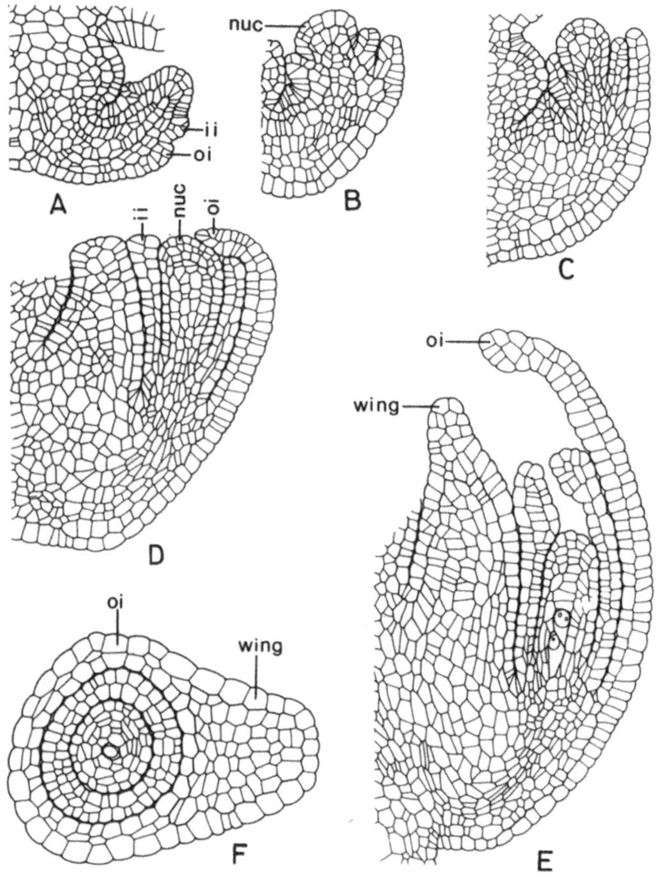

Fig. 30.11.2A–F. *Qualea cordata*, development of ovule, Longi- **(A–E)** and transection **(F)**; *oi, ii* outer and inner integument; *nuc* nucellus. (After Boesewinkel and Venturelli 1987)

The development of the ovule is shown in Fig. 30.11.1A–F. The ovular primordium is trizonate. The differentiation and structure of inner and outer integuments is represented in Fig. 30.11.2A–F. The ovule is bitegmic, hemi-anatropous and crassinucellate with a massive and elongated nucellus (Fig. 30.11.1F). The nucellus is completely resorbed but the basal part (postament) persists in the mature seed in *Qualea cordata*. The inner integument is three to four layers thick, and the outer six to seven layers. The micropyle is formed by both the integuments, and the exostome extends beyond the endostome (Fig. 30.11.1E–G). The funicle is short and elongated. Between the tip of the outer integument and the funicle develops a flat tissue which gives rise to the wing in the seed (Fig. 30.11.1B–G).

The archesporial cell has not been observed. The megaspore mother cell (sometimes two cells) is situated deep in the nucellus (Fig. 30.11.2E). The development of the embryo sac, endosperm and embryo has not been studied.

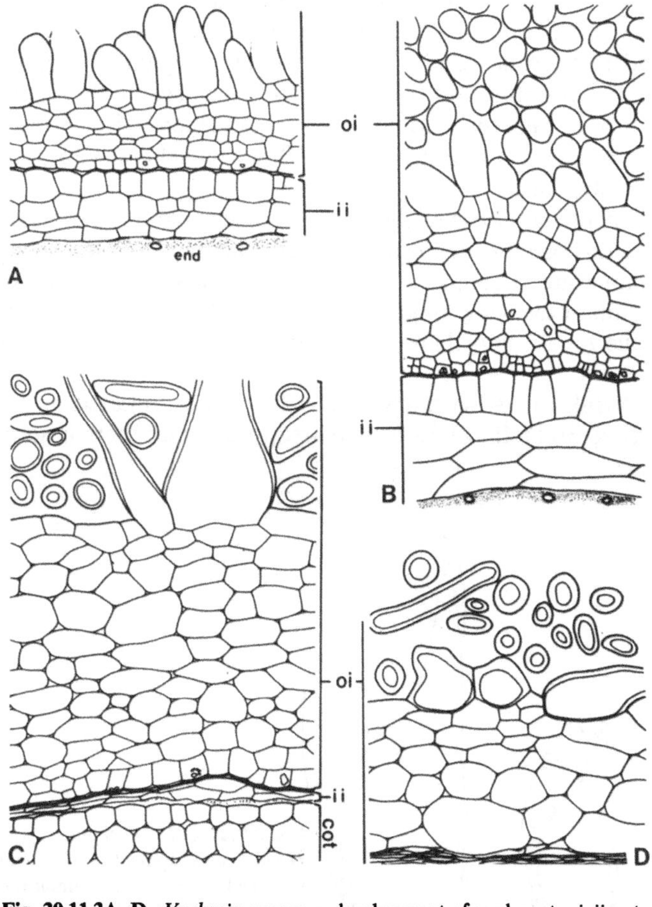

Fig. 30.11.3A–D. *Vochysia pygmaea*, development of seed-coat; *oi*, *ii* outer and inner integument; *end* endosperm; *cot* cotyledon. (After Boesewinkel and Venturelli 1987)

The endosperm is of the Nuclear type (Fig. 30.11.3B), wall formation was not observed, the seed is exalbuminous. The mature embryo completely fills the seed cavity and the cotyledons are convolute in *Callisthene fasiculata* and *C. major* (Boesewinkel and Venturelli 1987), often plicate or flat (Corner 1976). In *Vochysia pygmaea* (Boesewinkel and Venturelli 1987), the straight embryo has two large spirally folded cotyledons which are rich in lipids.

The development of the seed-coat is variable. In *V. pygmaea*, the parenchymatous cells of the inner integument enlarge, become vacuolate, and are gradually consumed by the endosperm (Fig. 30.11.3A–D). The outer integument increases in thickness. Shortly after fertilization, some of the radially elongate epidermal cells grow out into unicellular hairs (Fig. 30.11.3A). Subsequently, the hairs elongate and become thick-walled (Fig. 30.11.3C, D). The cells of the middle layers enlarge and intercellular air-spaces appear. The inner layers get crushed and the mature seed-coat comprises only four to seven layers of the outer integument (Fig. 30.11.3D).

In *Qualea cordata* and *Q. densiflora* (Boesewinkel and Venturelli 1987), the inner integument remains two- to three-layered. The endotegmen is tanniniferous and forms the inner layer of the seed-coat. The outer layers elongate, become fibrous, and thick-walled and pitted. The outer integument remains two-layered. The radial and inner walls of the inner layer become thickened, and small spaces appear between endotesta and exotegmen. The exotesta remains thin-walled. The seed-coat in *Callisthene* (Boesewinkel and Venturelli 1987) resembles that of *Qualea.*

Among the anemochorous seeds of the Vochysiaceae three types of wings can be distinguished. In *Vochysia,* the seed is covered with hairs and the dorsal unilateral (antiraphal) wing is compounded of matted hairs. The hairs partaking in the formation of the wing remain thin-walled. These hairs grow into the empty part of the locule, and therefore the wing simulates a cast of the locule (Fig. 30.11.4A–D).

In *Qualea* (Boesewinkel and Venturelli 1987), the wing is initiated on the funicle-raphal side (Fig. 30.11.2E, F). In hemianatropous ovules there is some

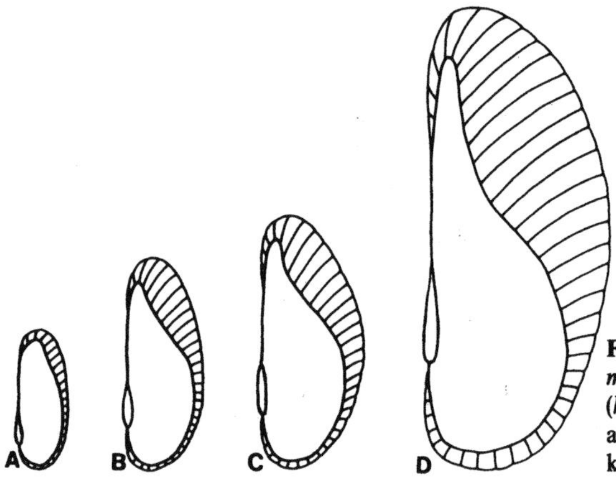

Fig. 30.11.4A–D. *Vochysia pygmaea,* development of wing (*hatched*), longisection of immature and mature seed. (After Boesewinkel and Venturelli 1987)

space for the development of outer integument on the funicular side. The seed wing is a lateral outgrowth of the outer and middle layer of the outer integument, and is inserted median-ventrally. This type of wing formation is not known in any other plant family. In *Callisthene* also the wing originates from the outer integument, but it is inserted transmedian-circumferentially.

FD Boesewinkel and M Venturelli (1987), EJH Corner (1976), GL Davis (1966), J Mauritzon (1936b), FA Stafleu (1952)

30.12 Tremandraceae

The development and structure of anther, pollen grain, embryo sac, endosperm, and embryo has not been worked out.

The ovule is anatropous, bitegmic and crassinucellate. In *Tetratheca* the outer integument is four- and the inner 15-layered, the inner epidermis forms an endothelium. Both the integuments form the micropyle. A parietal cell is cut off by the archesporial cell, and probably it produces a parietal layer in *Tetratheca* sp. The embryo sac shows three antipodals (Mauritzon 1936b).

Seeds small, glabrous or hairy with a conspicuous chalazal appendage (aril?), copious endosperm, and microscopic embryo. Seed-coat thin-walled, outer layers tangentially elongate, and inner layer strongly sclerotic.

JA Biddle and DC Christophel (1978), EJH Corner (1976), GL Davis (1966), J Mauritzon (1936b), JC Willis (1973)

30.13 Polygalaceae

The anther wall comprises the epidermis, fibrous endothecium, one or two ephemeral middle layers, and glandular tapetum with binucleate cells. At maturity only the epidermis and endothecium persist. The microspore mother cells divide meiotically, cytokinesis is simultaneous, and the tetrads are tetrahedral.

The three-colporate pollen grains are three-celled in *Polygala* and *Salomonia*, otherwise two-celled. In *Salomonia* (Nagaraja Rao 1964b), they mostly germinate in situ.

The ovule is trizonate, anatropous, bitegmic and crassinucellate. The outer integument is two- to multilayered and the inner two-layered. Only the outer integument forms the micropyle in *Salomonia*, otherwise both the integuments are concerned and the micropyle is zigzag. In *Salomonia* the nucellus extends up to the tip of the inner integument.

Periclinal divisions occur in the dermal layer of the nucellus giving rise to a nucellar cap. Much of the parental tissue is resorbed by the growing embryo sac.

A hypodermal archesporial cell cuts off a parietal cell which produces eight to ten layers of parietal tissue. The megaspore mother cell undergoes meiotic divisions, produces mostly a linear, sometimes a T-shaped tetrad, and the chalazal megaspore develops into a Polygonum type of embryo sac. A pseudomonosporic, tetrasporic type of development (only one nucleus of the four-nucleate coenomegaspore functions) is reported in *Polygala polygama* (Reeves 1957). This needs confirmation. The antipodals are usually ephemeral, but they are persistent in *P. erioptera* (Mukherjee 1961b), and multiply up to 12 in *P. sibrica* (Srinivasachar 1942). A hypostase is present.

The entry of pollen tube is porogamous, and double fertilization occurs.

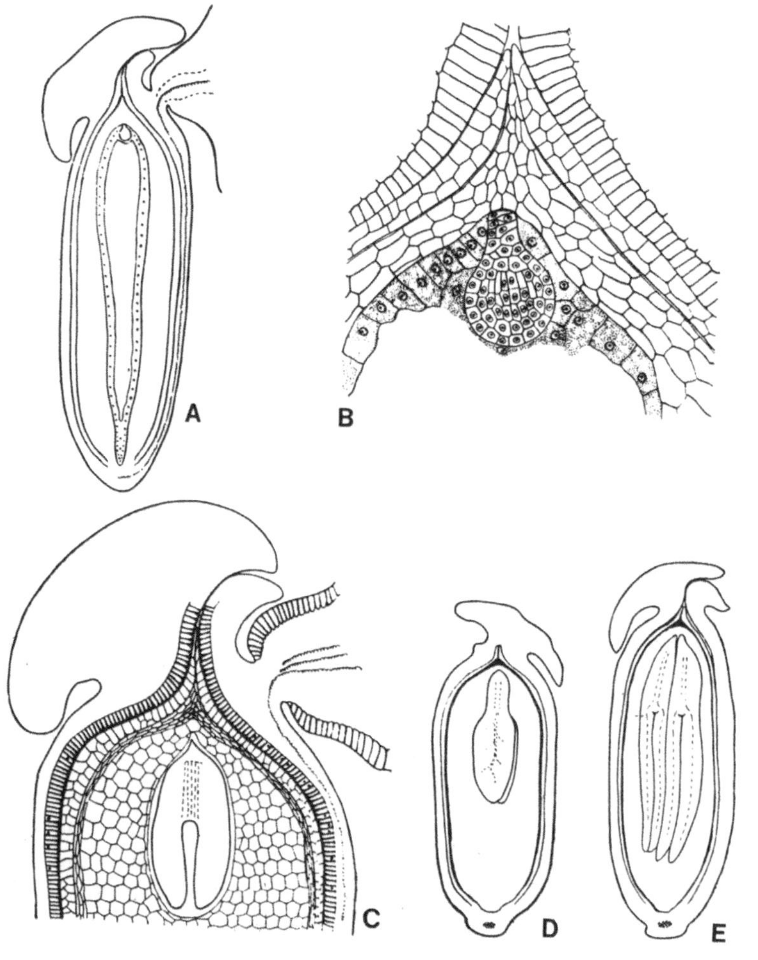

Fig. 30.13.1A–E. *Polygala major*. **A** Longisection of seed at globular proembryo, endosperm nuclear. Note the aril (strophiole) in **A**, **C–E**. **B** Later stage, endosperm partly cellular. **C** Longisection of micropylar region at dicotyledonous embryo. **D** Longisection of seed at dicotyledonous embryo; cellular endosperm not shown in **D** and **E**. **E** Twin embryos. (See Batygina and Yakovlev 1985)

The endosperm is of the Nuclear type. Centripetal wall formation (Fig. 30.13.1A–C) is initiated from the micropylar end at the late globular stage of the proembryo. The mature seed contains only a few layers of endosperm. In *Xanthophyllum fragrans* (Verkerke 1984), copious endosperm envelops foliaceous cotyledons. Both endosperm and cotyledons have fatty food reserve. In contrast, in *X. arnottianum* (Verkerke 1984) the endosperm is scanty and free-nuclear, and does not become cellular. The embryo is rich in starch grains, but not in fatty substances.

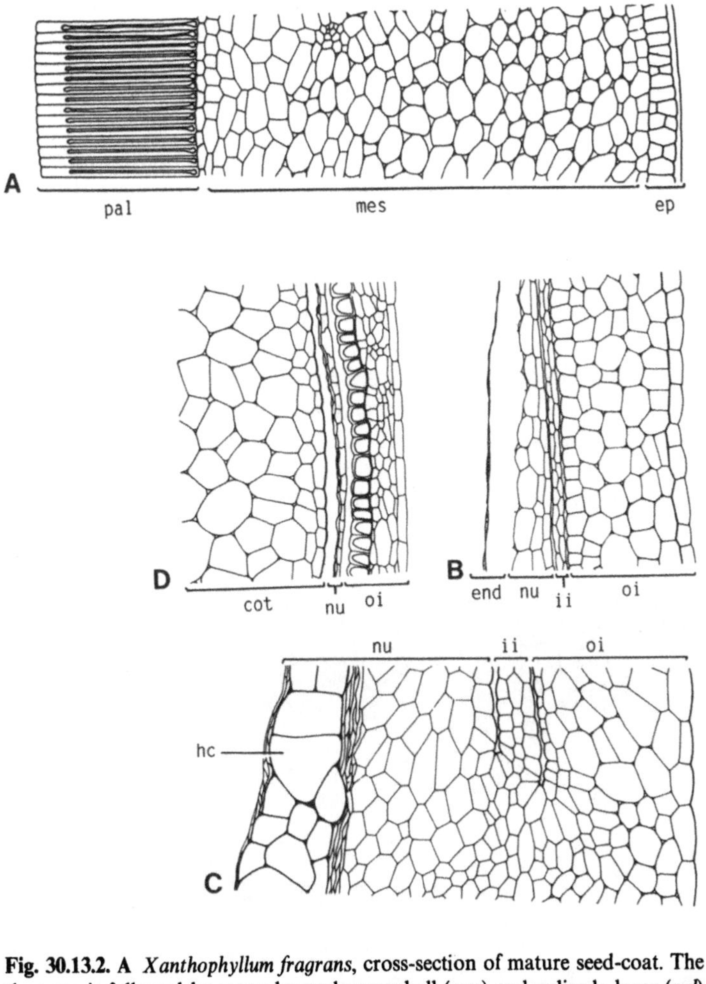

Fig. 30.13.2. A *Xanthophyllum fragrans*, cross-section of mature seed-coat. The epidermis (*ep*) of the testa is followed by many-layered mesophyll (*mes*) and palisade layer (*pal*) of the endotesta, tegmen collapses. **B–D** *X. arnottianum*, longisections; **B** Seed-coat at free-nuclear endosperm (*end*), many-layered outer integument (*oi*) and two-layered inner integument (*ii*); *nu* nucellus; **C** Section at lower end of seed through junction of integument and nucellus; *hc* hypertrophied cells; **D** Mature seed-coat, endotesta shows U-shaped thickenings, remnants of nucellus persist, tegmen collapses, and endosperm is consumed; *cot* cotyledon. (After Verkerke 1984)

The embryogeny corresponds to the Asterad type, and sometimes twin embryos may develop (Fig. 30.13.1E).

The seed is endotestal, and arillate in *Polygala* (Figs. 30.13.1A, C–E, 30.13.3A). The aril covers the seed, and may be pulpy or horny. The testa is multiplicative and forms the seed-coat. The outer epidermis comprises subsclerotic cells. The mesophyll is parenchymatous with thick or thin walls. The inner epidermis consists of a compact palisade of thick-walled cells (Fig. 30.13.2A) with crystals. The thick wall is faintly laminated and prominently pitted.

The structure of seed-coat in *Xanthophyllum arnottianum* is somewhat different (Fig. 30.13.2B–D). The testa is four- to five-layered and thin-walled. The cells of the inner epidermis do not elongate to form a true palisade layer. The tegmen collapses. The remnants (two- to three-layers) of nucellus persist. In the chalazal region some of the perisperm cells become hypertrophied (Fig. 30.13.2C). The seed is exalbuminous.

Two features of the testa are very prominent and of great taxonomic value: length of the palisade cell, and the presence of a parenchymatous middle layer (see Verkerke and Bouman 1980).

In *Polygala chamaebuxus*, *P. myrtifolia*, *P. vayredae* and *P. vulgaris* (Fig. 30.13.3A, B), the seed is covered with many one-celled striate acicular hairs (Fig. 30.13.3B) which are up to 500 μm long and 8 μm wide. The hairs are typically surrounded by their radially arranged neighbouring cells (Verkerke and Bouman 1980). In *P. klotzschii* the acicular hair is curled and thick-walled. In *Bredemeyera lucida*, in addition to the acicular thin-walled hairs, from the apex of the thickened unlobed hook-shaped yellow exotome aril arises a tuft of white hairs. Upon drying the thick-walled hairs stand erect and the thin-walled hairs tend to spread forming an umbrella-like structure (Verkerke 1985).

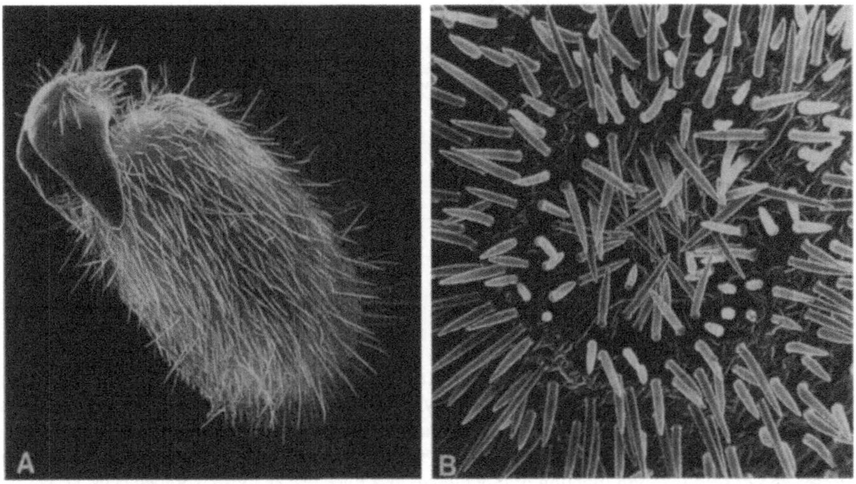

Fig. 30.13.3A, B. (SEM) *Polygala vulgaris*, **A** Seed with elaiosome and hairs. **B** Portion of chalazal region. (Courtesy Dr F. Bouman, Amsterdam)

In the mature seed the white to pale-yellow exostome aril is very prominent. It is typically three-lobed (Fig. 30.15.3A), and the lobes grow downward along the seed to one-fourth of its length. There is a loose tuft of hairs at the top, but the lobes of the aril are globose (Verkerke and Bouman 1980).

EJH Corner (1976), PK Mukherjee (1961b), A Nagaraja Rao (1964b), JH Reeves (1957), PN Schürhoff (1924), R Souèges (1941d), D Srinivasachar (1942), W Verkerke (1984, 1985), W Verkerke and F Bouman (1980), H Wirz (1910)

30.14 Taxonomic Considerations

Bentham and Hooker (1965a) include Polygalaceae, Tremendraceae and Vochysiaceae in Polygalinae, and the rest of the families in Geraniales, Engler and Diels (1936) place all the families in Geraniales, and distribute them over three suborders. Hutchinson (1973) distributes them in five suborders.

Rutaceae and Meliaceae. Rutaceae and Meliaceae are placed in Geraniales by Bentham and Hooker (1965a). Wettstein (1935) includes them together in Terebinthales, whereas Hutchinson (1973) segregated Meliaceae from Rutaceae due to the absence of gland-dotted leaves, and occurrence of a staminal tube. Embryologically, both the families are closely related. The multicellular archesporium, occasional development of Allium type of embryo sac, chalazal endosperm haustorium in *Aegle marmelos*, and the Solanad and Asterad type of embryogeny distinguish Rutaceae from Meliaceae. The seed in several members of Meliaceae is arillate, and pachychalazal. In both families the exotesta is thick-walled; it is sclerotic in Rutaceae. The tegmen is tracheidal in Rutaceae, and fibrous in Meliaceae.

Cneoraceae. Cneoraceae is represented by the genus *Cneorum*. Rau (1962) and B.M. Johri (1963b) have discussed the systematic position of Cneoraceae. Endlicher (1836–1840) includes them in Rutales near Burseraceae, while Bentham and Hooker (1965a) place *Cneorum* in Simaroubaceae. Engler and Prantl (1897) accept the family status of Cneoraceae but transfer them to Geraniales, and add that they are closely related to Zygophyllaceae. Hutchinson (1973) places them near Celastraceae in Celastrales. On the basis of embryogeny (Period I, Megarchetype II, Group II–Asterad type), Souèges (1955) relates them to Rhamnaceae, and suggests its grouping in Rhamnales. The embryogeny in the investigated taxa of Zygophyllaceae (*Tribulus terrestris* and *Zygophyllum fabago*) conforms to Period II, Series B, Megarchetype II, Group IX (Rau 1962) – Caryophyllad type.

The tribe Ruteae of Rutaceae resembles Cneoraceae in ovule and seed characters. The crassinucellate, bitegmic, primitive ovule is rather large and comparable to rutalean ovule in the massive parietal tissue, nucellar cap, and deep-seated embryo sac. The dermal initiation of integuments, initially two-layered, resembles the ovule in Rutoideae. The slightly multiplicative nature of the inner integument and the enlargement of its cells also show rutalean affinity. The seed-coat anatomy is derived and simplified (Boesewinkel 1984b). The campylotropous ovule, partially crushed exotesta, thick-walled endo- and meso-testa, and the tracheidal nature of the inner layer of inner integument of *Ruta graveolens* (Boesewinkel 1977) show affinity with Cneoraceae. Melchior (1964) has correctly placed Cneoraceae next to Rutaceae in his order Rutales.

Simaroubaceae and Burseraceae. Bentham and Hooker (1965a), Bessey (1915), and Engler and Prantl (1931) place the two embryologically allied families, Simaroubaceae and Burseraceae, together in Geraniales. Hallier (1912) includes them in Terebinthales, and Hutchinson (1973) and Takhtajan (1980) in Rutales. These two families resemble Zygophyllaceae (Sect. 29.6), Rutaceae, and Meliaceae in several respects, such as secretory anther tapetum, simultaneous cytokinesis, two-celled pollen, anatropous, bitegmic, crassinucellate ovule, Polygonum type of embryo sac, and Nuclear type of endosperm. The seeds are endospermic in both, and tracheidal exotesta gives protection; the tegmen is crushed.

Simaroubaceae can be distinguished because they are eurypalynous, and show hypostase, obturator, chalazal endosperm haustorium (*Suriana*), and perisperm as in Rutaceae. Perhaps the knowledge about the embryogeny of Burseraceae may add further to their alliance.

Malpighiaceae, Trigoniaceae and Vochysiaceae together comprise the suborder Malpighiineae.

Malpighiaceae. Malpighiaceae are included in Geraniales by Bentham and Hooker (1965a) and Hutchinson (1973), Takhtajan (1980) gives them an ordinal rank, while Cronquist (1981) treats Malpighiaceae as a primitive family in Polygalales.

Malpighiaceae are definitely distinct from the rest of the families of Rutales. Biseriate anther tapetum, nucellar beak, bisporic Allium and tetrasporic Penaea type of embryo sac development, failure of fertilization, autonomous development of the endosperm in some members, and exalbuminous seed have no parallel in Rutales. The Nuclear type of endosperm with nucleated cytoplasmic vesicles, free-nuclear chalazal haustorium in *Malpighia*, and haustorium becoming cellular in *Thryllis* are worth mentioning. Nucellar polyembryony is a common feature. The seed with thin-walled testa, and fibrous and sclerotic exo- and endotegmen resembles sapindaceous seed.

Trigoniaceae. The ovule and seed anatomy of *Trigonia cipoensis* (Boesewinkel 1987) is similar to that of *Linum* and related taxa. They share several other characters: trizonate ovule, dermal integuments, inner integument becomes thicker by repeated divisions in middle layers. The seed-coat structure of *Trigonia* also corresponds with that of *Linum*, *Erythroxylum* and *Humiria* (Boesewinkel 1980a, b, 1985a). The fibrous exotegmen in *Trigonia* is strongly lignified and forms the principal mechanical layer of the seed-coat. The outer layer of the outer integument differentiates in various ways in Trigoniaceae and families of the Linales. These similarities indicate a relationship between Trigoniaceae and Linales, also with other families, such as Vochysiaceae,

Malpighiaceae and Tremandraceae, but not with Polygalaceae. The family Polygalaceae (including Xanthophyllaceae) differs from the majority of the "Polygalales" in its ovule and seed structure (Verkerke 1985). According to Corner (1976), Trigoniaceae cannot be classified either near the Polygalaceae if Trigoniaceae have exotegmic fibres (see also under Polygalaceae).

Vochysiaceae. The Vochysiaceae (Boesewinkel and Venturelli 1987) are generally considered to be related to Polygalaceae, Trigoniaceae, Malpighiaceae, Tremandraceae and Krameriaceae. Some or all of these families have often been placed in the order Polygalales. The ovule and the endotestal seed type with an undifferentiated compressed or resorbed inner integument of Polygalaceae (Verkerke 1985) differ from those of the other families of the conventional Polygalales, which are mostly characterized by a persistent inner integument with an inner pigment layer and some form of an exotegmen. The ovules of Trigoniaceae (Boesewinkel 1987) and Tremandraceae (Mauritzon 1936b) have a thick inner integument. The ovule of *Trigonia* is similar to *Linum*. The seeds of Trigoniaceae, Tremandraceae and Malpighiaceae have a fibrous exotegmen. On the basis of seed structure, these families correspond fairly well with those of the Linaceae complex (Boesewinkel 1980a, b, 1987). The taxonomic relationships must await detailed embryological studies (see also under Polygalaceae).

Polygalinae include Tremendraceae and Polygalaceae. Wettstein (1935) placed Polygalaceae and Tremendraceae in Terebinthales. Hutchinson (1973) assigned Tremendraceae to Pittosporales, and Polygalaceae to Polygalales. As the information on Tremendraceae is scanty, their phyletic relationship cannot be discussed.

Polygalaceae. Boesewinkel (1987) points out that, except for the family Polygalaceae, he agrees with Heimsch (1942) who considered that Polygalaceae, Trigoniaceae, Tremandraceae, Malpighiaceae, Vochysiaceae and also Zygophyllaceae are interrelated and, although more specialized in general habit, floral structure and wood anatomy, these families are probably related to the Linaceae, Erythroxylaceae and Humiriaceae.

The more primitive representatives of the Polygalaceae have crassinucellar ovules with a subdermal vascularized outer integument and a two-layered inner integument. The endotesta is the most protective layer of the seed-coat.

The main embryological features of Polygalaceae are: simultaneous cytokinesis in pollen mother cell, two- or three-celled pollen, trizonate, anatropous, bitegmic, crassinucellate ovule, zigzag micropyle formed by both integuments, hypostase differentiates, Polygonum type of embryo sac, usually ephemeral antipodals, Nuclear type of endosperm which later becomes cellular, Asterad type of embryogeny, endotestal, arillate, hairy albuminous seed, and conspicuous embryo.

On morphological and embryological grounds a relationship of Polygalaceae has been suggested with Malpighiaceae, Trigoniaceae, Vochysiaceae, Tremandraceae and Zygophyllaceae. The relationships of Polygalaceae with Trigoniaceae, Vochysiaceae, Tremendraceae and Krameriaceae (all these families – except Tremendraceae – included in Polygalales by Hutchinson 1973) can be discussed with finality only when the embryology of these four families has been studied in sufficient detail.

On the basis of ovule and seed structure, Boesewinkel (1987) negates relationship of Polygalaceae with Trigoniaceae (see also Corner 1976), and Boesewinkel and Venturelli (1987) with Vochysiaceae. It has been suggested by them that the Order Polygalales (sensu Heimsch 1942) should comprise the family Polygalaceae only.

31 Sapindales

The order Sapindales comprises four suborders and ten subfamilies: Coriariineae – Coriariaceae; Anacardiineae – Anacardiaceae; Sapindineae – Aceraceae, Bretschneideraceae, Sapindaceae, Hippocastanaceae, Sabiaceae, Melianthaceae and Aextoxicaceae, and Balsamineae – Balsaminaceae.

31.1 General Features

The order Sapindales is characterized by (1) secretory anther tapetum, simultaneous cytokinesis of pollen mother cells, two-celled pollen grains, (2) anatropous, bitegmic, crassinucellate ovule, unitegmic in Sabiaceae and tenuinucellate in Balsaminaceae, parietal cell or tissue absent in Balsaminaceae, (3) Polygonum type of embryo sac, Polygonum and Allium type in Balsaminaceae, (4) Nuclear type of endosperm, Helobial in Sabiaceae, Cellular with haustorium in Balsaminaceae, (5) embryogeny of Onagrad type in Coriariaceae, Anacardiaceae, and Aceraceae, and Asterad type in Sapindaceae, Melianthaceae and Balsaminaceae, and (6) extensive simplification of seed; generally testal, tegmen gets crushed, testa or tegmen thin-walled, marked differentiation of mechanical elements absent; exalbuminous except in Melianthaceae.

On the whole, the development in the families is more or less uniform with few exceptions.

31.2 Coriariaceae

The anther is bisporangiate, and the wall comprises papillate epidermis, fibrous endothecium, two ephemeral middle layers, and secretory tapetum of one- to four-nucleate cells in *Coriaria nepalensis* (V.K. Sharma 1968a). The endothecial thickenings appear just before dehiscence, which has been contradicted by M. Garg (1980). The microspore mother cells undergo meiotic divisions, cytokinesis is simultaneous, and the tetrads are decussate and tetrahedral.

The pollen grains are three zonoporate (sometimes two- or four-zonoporate), and three-zonocolporate with perforate tectum and small spines. They are two- or three-celled at the dispersal stage; M. Garg's report of one-celled condition is incorrect. All reports of one-celled mature pollen grains require reinvestigation.

The ovule is anatropous, bitegmic and crassinucellate. The inner integument forms the micropyle. The outer integument forms a flap-like covering on either side of the funiculus. The single archesporial cell cuts off a parietal cell, the parietal tissue is only few-layered. Following meiotic divisions, the megaspore mother cell produces a linear or T-shaped tetrad. The chalazal megaspore develops into a Polygonum type of embryo sac. The antipodals are ephemeral.

Development of the endosperm is of the Nuclear type.

The embryogeny conforms to the Onagrad type, and is assigned to Period I, Megarchetype IV (V.K. Sharma 1968a). The mature embryo is straight.

The seed is exotestal, and mostly exalbuminous. The testa alone forms the seed-coat. The outer epidermis comprises thick-walled cells with tannin followed by thin-walled mesophyll and inner epidermis.

M Garg (1980), J Grimm (1912), J Mauritzon (1936b), MA Rau and VK Sharma (1970a), VK Sharma (1968a)

31.3 Anacardiaceae

The anther wall comprises the epidermis, fibrous endothecium, three ephemeral middle layers, and secretory tapetum of binucleate cells in *Pistacia chinensis* (Copeland 1955). A single middle layer develops in *Rhus mysorensis* (Kelkar 1958a). Simultaneous cytokinesis follows meiotic divisions in microspore mother cells. The tetrads are tetrahedral.

The pollen grains are tricolporate, spherical with pitted wall, and are two-celled at the dispersal stage.

The ovule is anatropous – apotropous in *Blepharocarya depauperata* (Wannan et al. 1987) – uni- or bitegmic and crassinucellar. In *P. vera* (Grundwag and Fahn 1969), the micropyle faces an outgrowth of the funiculus which may be regarded as an obturator. Both the integuments fuse to form a single structure in *P. vera* and *Anacardium occidentale* (Copeland 1955, 1961). In this connection, Grundwag and Fahn point out that in *P. vera* the outer integument is developed to various extent in different ovules. An outer integument, wholly or partially developed, is also known in other species of Anacardiaceae, e.g. in *Rhus toxicodendron* (Grimm 1912), *R. mysorensis* (Kelkar 1958a), *Lannea coromandelina* (Kelkar 1958b), and *Schinus mole* (Copeland 1959). The cells of the nucellar epidermis divide, and add to the parietal tissue. The archesporial cell cuts off a parietal cell, the latter produces a massive tissue. The megaspore mother cell undergoes meiotic divisions, and a linear or T-shaped tetrad is produced. The chalazal megaspore develops into a Polygonum type of embryo sac. The three remaining megaspores do not degenerate at the same time, and twin embryo sacs develop in *Nothopegia* (Kelkar 1974). The polars fuse before fertilization, and antipodals persist until then.

Kelkar (1958a), Copeland (1959) and Grundwag (1976) state that in some species of the tribe Rhoideae the chalazal megaspore of the linear tetrad gives rise to an eight-nucleate embryo sac of the Polygonum type. But in *Rhus lancea* (von Teichman and Robbertse 1986b) the micropylar megaspore produces an eight-nucleate embryo sac which should be considered to be of the Polygonum and not the Oenothera type. In the latter, the embryo sac is always four-nucleate – the egg apparatus and upper polar nucleus.

In *Pistacia vera* (Grundwag and Fahn 1969), the seed-set is low due to sterility of ovules, degeneration of megaspore mother cells, megaspores, and even embryo sacs.

The entry and course of pollen tube is variable in Anacardiaceae. It is strictly porogamous in *Spondias* and *Semecarpus* (Srinivasachar 1940a). According to Grundwag and Fahn (1969), in *P. vera* "germinating pollen tubes penetrate the short style, pass the locule and enter the funiculus of the ovule. In the funiculus, the pollen tubes continue to grow through the vascular bundle up to the chalaza. From the chalaza the tube enters the nucellus through the hypostase, surrounds the embryo sac and penetrates it at its micropylar end" (Fig. 31.3.1A). After pollination the pollen tube takes 24 h to reach the egg apparatus. Grundwag and Fahn observed one instance of direct penetration of a pollen tube from the hypostase to the chalazal end of the embryo sac. Copeland (1955, 1961) also reported chalazogamy in *P. chinensis* and *Anacardium occidentale*, and Grundwag (1976) in *P. atlantica*. All instances of chalazogamy require careful reinvestigation, and cleared wholemounts of ovules, longi- and transections, and dissections should be examined.

The development of the endosperm is of the Nuclear type. In *Mangifera indica* (Sachar and Chopra 1957), nuclear fusions occur (in the chalazal region) producing polyploid nuclei with a variable number of nucleoli. Nucleated cytoplasmic nodules develop in *M. indica* var. Paho (Sachar and Chopra 1957). Wall formation commences in the micropylar region along the periphery, and extends towards the centre. In *Mangifera* the lower part remains free-nuclear for a considerable time but, finally, becomes cellular. Such a free-nuclear region is absent in *Anacardium occidentale* (Copeland 1961). The endosperm is consumed by the developing embryo.

In *Pistacia vera* (Grundwag and Fahn 1969), the zygote divides 10–12 weeks after fertilization. This interval is at least partly due to delay in syngamy. If, however, there is a case of aneuspory, there is a possibility that the male gamete activates the division of the egg nucleus without fusion.

The embryogeny conforms to the Onagrad type-Euphorbia variation in *Rhus* and *Lannea* (Kelkar 1958a, 1961). The development is irregular in *P. vera* (Fig. 31.3.1B–F).

The zygotic embryo degenerates at an early stage in *Mangifera*. The nucellar cells in the micropylar region produce supernumerary embryos. Nucellar polyembryony is recorded in four varieties of mango: *Olour*, *Higgins*,

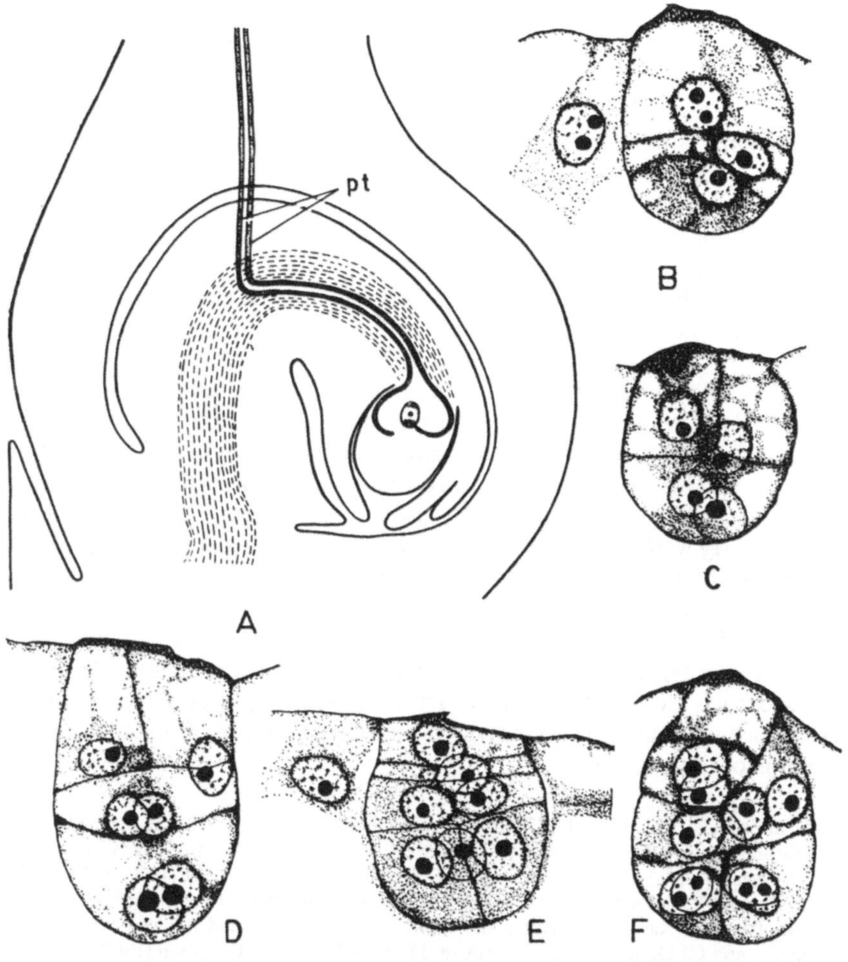

Fig. 31.3.1A–F. *Pistacia vera.* **A** Longisection of ovary. The pollen tubes (*pt*) traverse through the chalazal vascular strand to the tip of mature embryo sac. **B–F** Early embryogeny. There is an endosperm nucleus each in **B** and **E**. (See Batygina and Yakovlev 1985)

Paho and *Combodia* (Sachar and Chopra 1957). As the adventive embryos lie very close to each other, they exhibit various degrees of fusion in early stages of development. A synergid-embryo develops in *Lannea coromandelica* (Kelkar 1961).

The integuments are unspecialized, and generally get crushed except for the inner epidermis of tegmen, which is persistent, and shows bar-like thickenings on radial walls. The exotesta may become the protective layer. The endocarp is also protective.

In *Pistacia* (Copeland 1955, Grundwag and Fahn 1969), the seeds are endotegmic with abundantly pitted radial cell walls of endotegmen. The seeds of *Rhus* (von Teichman and Robbertse 1986b) are endotestal, and the radial walls of the persistent endotesta show lignified thickenings with pits. The remnants of endosperm and cotyledons contain fats and abundant proteinaceous reserve materials.

Von Teichman and Robbertse (1986a) studied the development and structure of the exo-, meso- and endocarp of the drupe of *Sclerocarya birrea* subsp. *caffra*. The mature exocarp comprises the outer epidermis with stomata and lenticels, subepidermal collenchyma, and parenchymatous layers with secretory cells. The fleshy parenchymatous mesocarp (sarcocarp) also contains secretory tissue, and develops only after the differentiation and lignification of endocarp. The endocarp (sclerocarp) consists of brachysclereids, fibres and vascular elements, and develops from the inner epidermis and adjacent tissue of the ovary wall. An "operculum" is a well-defined part of the endocarp, and it develops from the inner epidermis and some cells of the subepidermal cells of the ovary wall.

In *Rhus lancea* (von Teichman and Robbertse 1986b), the mature loose and papery exocarp consists of outer epidermis and a sclerified hypodermis. The mesocarp is not a typical sarcocarp since the ridges and inner layers are sclerenchymatous. The four-layered endocarp develops from the inner epidermis of the ovary wall.

HF Copeland (1955, 1959, 1961), EJH Corner (1976), J Grimm (1912), M Grundwag (1976), M Grundwag and A Fahn (1969), JB Juliano (1937), SS Kelkar (1954, 1958a, b, 1961, 1974), P Maheshwari (1934a), P Maheshwari and RC Sachar (1955), L Radlkofer (1879), RC Sachar and RN Chopra (1957), PK Sen and PC Malik (1940), D Srinivasachar (1940a), I von Teichman (1988), I von Teichman and PJ Robbertse (1986a, b), I von Teichman and AE van Wyk (1988), BS Wannan and CJ Quinn (1990), BS Wannan, JT Waterhouse and CJ Quinn (1987)

31.4 Aceraceae

The anther wall comprises the epidermis, fibrous endothecium, two ephemeral middle layers, and secretory tapetum of two- to three-nucleate cells; nuclear fusions occur in the tapetal cells. The microspore mother cells undergo meiosis followed by centripetal furrows and simultaneous cytokinesis. The tetrads are tetrahedral and isobilateral.

The three-colporate pollen grains are shed at the two-celled stage.

The ovule is anatropous, bitegmic and crassinucellate; the inner integument forms the micropyle. The nucellar epidermal cells divide to produce an eight to ten layers thick cap. The hypodermal archesporial cell cuts off a parietal cell, and the parietal tissue is several layers thick. The megaspore mother cell undergoes meiotic divisions, and produces a linear tetrad. The chalazal megaspore develops into a Polygonum type of embryo sac. The polar nuclei fuse before fertilization. The antipodals are ephemeral. At times, they multiply in *Acer oblongum* (Khushalani 1963b). The mature embryo sac contains starch.

The development of the endosperm is of the Nuclear type.

The zygote divides transversely. The development of the embryo appears to be of the Onagrad type in *Acer saccharinum* (Haskell and Postlethwait 1971). Suspensor polyembryony is reported in *A. platanoides*.

The seed is campylotropous with a curved embryo and flat, thick and curved, or thin and plicate, cotyledons. The thin testa alone forms the seed-coat. The outer epidermis comprises short columnar cells with thick or thin outer walls, followed by thin-walled mesophyll. The inner epidermis contains crystals.

DA Haskell and SN Postlethwait (1971), I Khushalani (1963b), CD Pigott and SJ Warr (1989)

31.5 Bretschneideraceae

The embryology has not been investigated.

H Tobe and C-I Peng (1990)

31.6 Sapindaceae

The anther wall comprises the epidermis, fibrous endothecium, two ephemeral middle layers, and secretory tapetum of uninucleate cells. Following meiotic divisions, the microspore mother cells produce tetrahedral and isobilateral tetrads. Simultaneous cytokinesis is due to centripetal furrowing.

The three-colporate pollen grains are shed at the two-celled stage. They are three-celled in *Allophylus cobbe* (S. Mathur 1980).

The ovule is ana- to campylotropous, bitegmic and crassinucellate. The inner integument forms the micropyle in *Cardiospermum halicacabum* (N.C. Nair and Joseph 1960), otherwise both the integuments are concerned. The hypodermal archesporial cell cuts off a parietal cell, and the parietal tissue is well developed. The megaspore mother cell divides by meiotic divisions, and produces a linear tetrad. The chalazal megaspore develops into a Polygonum type of embryo sac. The synergids are hooked, and the antipodals ephemeral in *Litchi chinensis* (Banerji and Chaudhuri 1944). In *Cardiospermum* (N.C. Nair and Joseph 1960), they multiply up to 14 cells and become multinucleate. A hypostase develops, and an obturator is present in *L. chinensis*.

The entry of the pollen tube is porogamous. It passes between the wall of the embryo sac and the synergid in *Cardiospermum*. Kadry (1946) describes the difference between the two male nuclei. The one which fertilizes the egg is swollen, and rounded in front and tapering behind. It is much longer than the other spherical male nucleus. Double fertilization occurs.

Development of the endosperm is of the Nuclear type, and it remains free-nuclear throughout in *Cardiospermum* (Kadry 1946).

The embryogeny conforms to the Asterad type. The mature embryo is either straight or curved with thin and spirally inrolled cotyledons, or the cotyledons may be thick and curved or straight. Parthenocarpy is reported in *Dodonea viscosa* (A.C. Joshi 1938a).

The seed is arillate, sarcotestal and exalbuminous. The aril develops, after fertilization, from the funicle and is rudimentary and vestigial in *Cardiospermum* and *Dodonea*. The testa is multiplicative. The outer epidermis comprises a palisade of columnar cells with brown walls. The mesophyll is differentiated into an outer compact thick or pulpy tissue, and an inner aerenchymatous tissue. The inner epidermis is thin-walled. The outer epidermis of tegmen alone persists, and consists of fibres and sclerotic endostome.

I Banerji and KL Chaudhuri (1944), KL Chaudhuri (1940), E David (1938), N Gulati and S Mathur (1977), AC Joshi (1938a), AER Kadry (1946), S Mathur (1980), J Muller and PW Leenhouts (1976), NC Nair and TC Joseph (1960), A Pfeiffer (1891), JE Planchon (1845b)

31.7 Hippocastanaceae

The structure and development of anther wall, embryo, and seed-coat have not been investigated.

The anther tapetum is glandular. Simultaneous cytokinesis follows meiotic divisions in microspore mother cells.

The pollen grains are two-celled at the dispersal stage.

The ovule is amphitropous (Fig. 31.7.1A), bitegmic and crassinucellar in *Aesculus woerlitzensis* (List and Steward 1965). The nucellus is massive with a nucellar cap, and the epistase consists of thick-walled refractive cells. A hypostase also develops.

The multicellular archesporium cuts off parietal cells. Development of the embryo sac is of the Polygonum type with a dilated (vesicle) chalazal region (Fig. 31.7.1B, C).

During fertilization and post-fertilization development the ovule, the nucellus, and the embryo sac grow conspicuously in the chalazal region. The nucellus can be demarcated into four regions: (1) vacuolate cells with small nuclei in the micropylar region, (2) large cells around the middle of the embryo sac, (3) smaller cells which undergo rapid divisions in the chalazal region, and (4) hypostase of radiating files of cells which are in contact with the vascular bundle at the base of the ovule.

The vesicular end of the embryo sac grows and encroaches upon the nucellus, and occupies the major portion of the ovule (Fig. 31.7.1B–D). The vesicle is filled with a fluid. The solutes enter the vesicle through the converging files of cells of the hypostase, which remain in contact with the chalazal end of the embryo sac.

Development of the endosperm is of the Nuclear type. The vesicle contains gelatinous (liquid) endosperm with numerous nuclei. The endosperm (Fig. 31.7.1E) is consumed by the developing embryo (Corner 1976).

In *A. woerlitzensis* the zygote divides transversely, vertically or obliquely. The mature embryo is dicotyledonous (Fig. 31.7.1D–J). During earlier stages

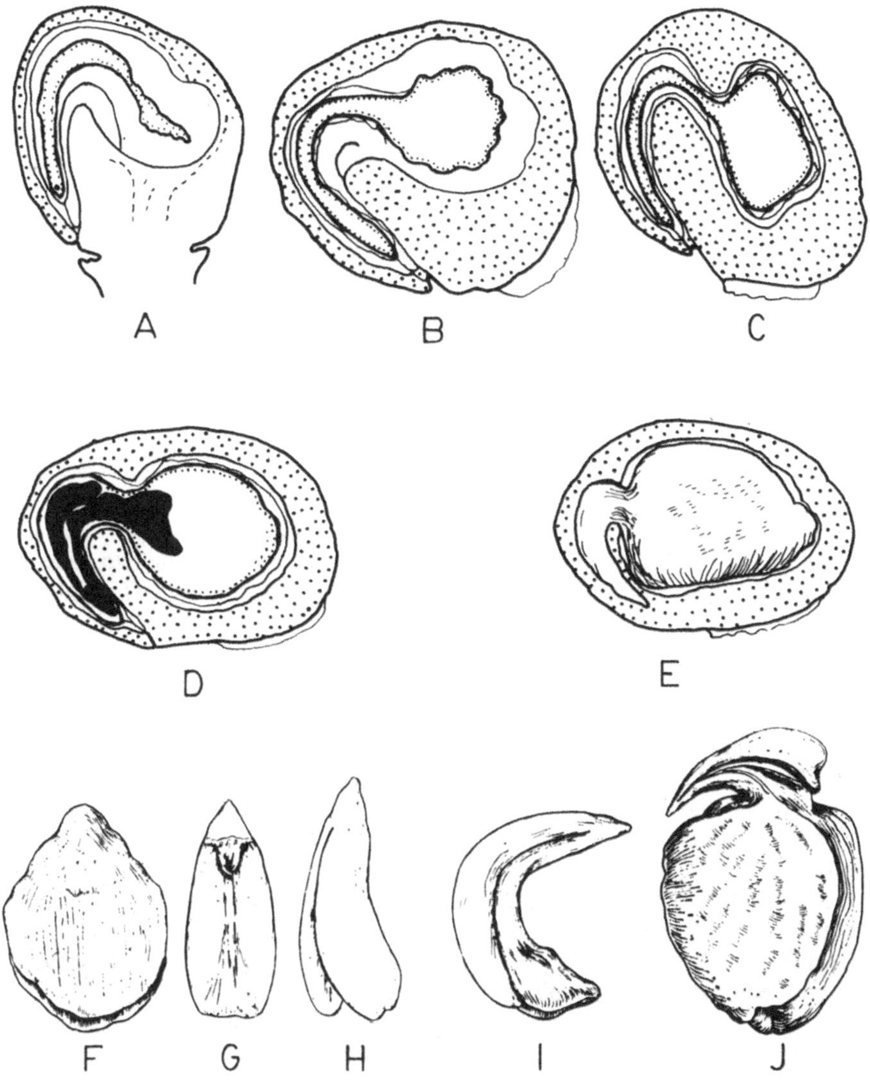

Fig. 31.7.1A–J. *Aesculus woerlitzensis.* **A** Longisection of amphitropous ovule with testa (*sparsely dotted*), tegmen, and well-developed nucellus. The embryo sac (*thinly dotted*) is at the zygote stage. **B**, **C** Later stage (endosperm and embryo not shown). The chalazal end of the embryo sac expands to form a vesicle (crushing the nucellus) which becomes filled with a fluid (liquid endosperm). **D** Longisection of ovule at dicotyledonous embryo. The larger cotyledon is suctorial, extends into the vesicle and absorbs nutritive fluid. **E** Longisection of seed with mature embryo occupying the entire vesicle; nucellus and endosperm have been absorbed. **F**, **G** Young embryo to show bilateral symmetry. **H–J** Embryos with unequal cotyledons. (After List and Steward 1965)

the growth of the two cotyledons is symmetrical (Fig. 31.7.1F, G). As the cotyledons grow up to the curved region of the embryo sac, one of them pushes into the vesicle, grows inside it and absorbs the vesicular fluid (Fig. 31.7.1D). The endosperm, and eventually the nucellus disappear. In mature seed the larger cotyledon folds in the vesicle and fills it.

The seed is pachychalazal (?) and exotestal. It is overgrown with a highly vascular wall, and tough seed-coat. The outermost layer of testa comprises palisade cells with outer and radial walls thickened. The mesophyll is aerenchymatous, and the inner epidermis is unspecialized. The thin-walled tegmen persists around the radicle.

EJH Corner (1976), GL Davis (1966), A List and FC Steward (1965), F Pax (1896b), K Schnarf (1931)

31.8 Sabiaceae

The development of pollen, embryo, and seed-coat has not been investigated.

The anther wall comprises the epidermis, fibrous endothecium, two ephemeral middle layers, and glandular tapetum with binucleate cells. Their nuclei fuse and divide.

The pollen grains are tricolporate, and two-celled when shed. The generative cell is elongated in *Meliosma arnottiana*, and lenticular in *M. wighttii* (M.V.S. Raju 1952a).

The ovule is hemianatropous, unitegmic and crassinucellate. The integument fails to organize the micropyle. The archesporial cell cuts off a parietal cell, and the parietal tissue is well developed. The megaspore mother cell divides meiotically, and produces a linear tetrad. The chalazal megaspore develops into a Polygonum type of embryo sac. The antipodals persist in *M. arnottiana* (M.V.S. Raju 1952a). The chalazal end of the embryo sac develops a diverticulum which grows towards the funiculus. The polar nuclei fuse before fertilization, and the secondary nucleus migrates to the diverticulum leaving the antipodals in situ.

Double fertilization occurs; supernumerary pollen tubes enter the embryo sac. Triple fusion takes place in the diverticulum.

The development of the endosperm is of the Helobial type. The endosperm proper develops from the micropylar chamber. The chalazal chamber becomes binucleate, and develops haustorial processes.

J Mauritzon (1936b), MVS Raju (1952a)

31.9 Melianthaceae

The anther wall comprises the persistent cutinized epidermis, endothecium, three to five middle layers, and secretory tapetum in *Melianthus major* (S. Mathur 1977). The endothecial cells are small and cutinized, and lack fibrous thickenings in the region of dehiscence; the rest of the cells develop fibrous thickenings. One or two middle layers persist. The tapetum is multiseriate towards the anterior and posterior end. Its cells are binucleate, and minute globules (Übisch) are studded on the inner tangential wall. The microspore mother cells undergo meiotic divisions. Cytokinesis occurs by furrowing, as well as cell-plate formation. When simultaneous, the tetrads are tetrahedral, isobilateral and decussate; when successive, they are linear, T-shaped and isobilateral.

The elliptical pollen grains are monad and tricolporate. They are two- or three-celled at the dispersal stage.

Sterility is very high in *M. major* (Khushalani 1963a). Due to cytological disturbances, the differentiation of generative and vegetative nuclei does not occur properly. Many pollen grains contain two to five nuclei due to supernumerary divisions.

The ovule is anatropous, bitegmic and crassinucellate. In *M. major* (S. Mathur and Gulati 1979), the inner integument organizes the micropyle and forms an operculum. In *Bersama* both the integuments form the micropyle. The epistase and hypostase are well developed. The hypodermal archesporial cell cuts off a parietal cell, and the parietal tissue is six to seven layers thick. A nucellar cap and a nucellar beak develop due to periclinal divisions of nucellar epidermal cells. After meiotic divisions, the megaspore mother cell produces a linear, T- or L-shaped, or decussate tetrad. Mostly, either the third megaspore from the micropylar end, or the chalazal megaspore functions. The mature embryo sac is of the Polygonum type. The antipodals are ephemeral. The polar nuclei fuse before or at pollination.

The entry of the pollen tube is porogamous, and it destroys one of the synergids. Double fertilization occurs, and the remnants of the pollen tube with darkly stained bodies persist for a long time.

The development of the endosperm is of the Nuclear type. It remains nuclear throughout which needs confirmation.

The embryogeny conforms to the Asterad type-Penaea variation in *M. major* (Steyn 1975).

The seed is albuminous with a short funicular aril in *Bersama*. The outer epidermis of testa is a palisade of thick-walled cells only in the region of the micropyle. It is followed by thin-walled mesophyll with styloids. The inner epidermis is thin-walled, and the tegmen gets crushed.

M Gürke (1896), I Khushalani (1963a), S Mathur (1977), S Mathur and N Gulati (1979), J Mauritzon (1936b), E Steyn (1975)

31.10 Aextoxicaceae

The development of anther, pollen, embryo sac, endosperm and embryo has not been investigated.

The ovule is anatropous, bitegmic and crassinucellar with the inner integument forming the micropyle. The nucellar beak is massive in *Aextoxicon punctatum* (Mauritzon 1936b).

In a mature seed the endosperm is ruminate, and the cotyledons orbicular-cordate (Hutchinson 1973).

J Hutchinson (1973), J Mauritzon (1936b)

31.11 Balsaminaceae

The anther wall comprises the epidermis, fibrous endothecium, two middle layers, and secretory tapetum. A peculiar feature in *Hydrocera triflora* (Venkateswarlu and Lakshminarayana 1957) and *Impatiens leschenaultii* (L.L. Narayana and Sayeeduddin 1959) is the development of trabeculate partitions of tapetal (doubtful?) cells dividing the sporogenous tissue into patches. The tapetal cells and those of trabeculae become two- to four-nucleate. In *Impatiens* (L.L. Narayana and Sayeeduddin 1959, L.L. Narayana 1963b, 1965), however, such partitions are produced due to sterilization of sporogenous tissue. The cells of septa and tapetum become binucleate. The microspore mother cells undergo meiotic divisions, and produce tetrahedral or isobilateral tetrads. Cytokinesis is simultaneous.

The pollen grains are three- to five-colporate, and shed at the two-celled stage. The generative cell is elongated. Its division occurs by a constriction in *Impatiens balsamina* (Raghavan et al. 1939). This indicates amitotic division which is, obviously, incorrect.

The ovule is anatropous, bitegmic and tenuinucellate; the inner integument forms the micropyle. The integuments are fused all along their length, except at the micropyle. They are free in *Impatiens* (L.L. Narayana 1965). The endothelium is well developed (Fig. 31.11.2F–H).

The hypodermal archesporial cell functions directly as megaspore mother cell. The embryo sac develops according to the Polygonum and Allium type. If of the Polygonum type, the tetrads may be linear or T-shaped. Polygonum type has been observed in *Impatiens inconspicua, I. levengi, I. maculata, I. parviflora, I. radiata, I. tenella, I. tripetala* (L.L. Narayana 1963b, 1965), and *I. balsamina* and *I. textori* (Takao 1966, 1968). Allium type occurs in *Hydrocera triflora* (Venkateswarlu and Lakshminarayana 1957), *I. arguta* (L.L. Narayana 1963b), *I. leschenaultii* (L.L. Narayana and Sayeeduddin 1959), *I. oliveri* (L.L. Narayana 1965) and *I. roylei* (K.V.O. Dahlgren 1934a). The polar nuclei fuse before fertilization, and the antipodals are ephemeral.

The development of the endosperm is of the Cellular type (Figs. 31.11.1A, B, E, F, 31.11.2A, B). Due to transverse partitions the embryo sac becomes four-chambered in *Hydrocera triflora* (Venkateswarlu and Lakshminarayana 1957). The micropylar chamber, with the zygote, develops into an aggressive haustorium (Fig. 31.11.2D, E, G–I). It extends beyond the micropyle, becomes vesicular, and produces branches which ramify in the funicular and placental tissue. Its nucleus becomes hypertrophied. The micropylar haustorium is multinucleate in *H. triflora* (Venkateswarlu and Lakshminarayana 1957). In *I. radiata* (L.L. Narayana 1965), the haustorium, instead of passing into the placenta, ramifies through the fused integuments along the

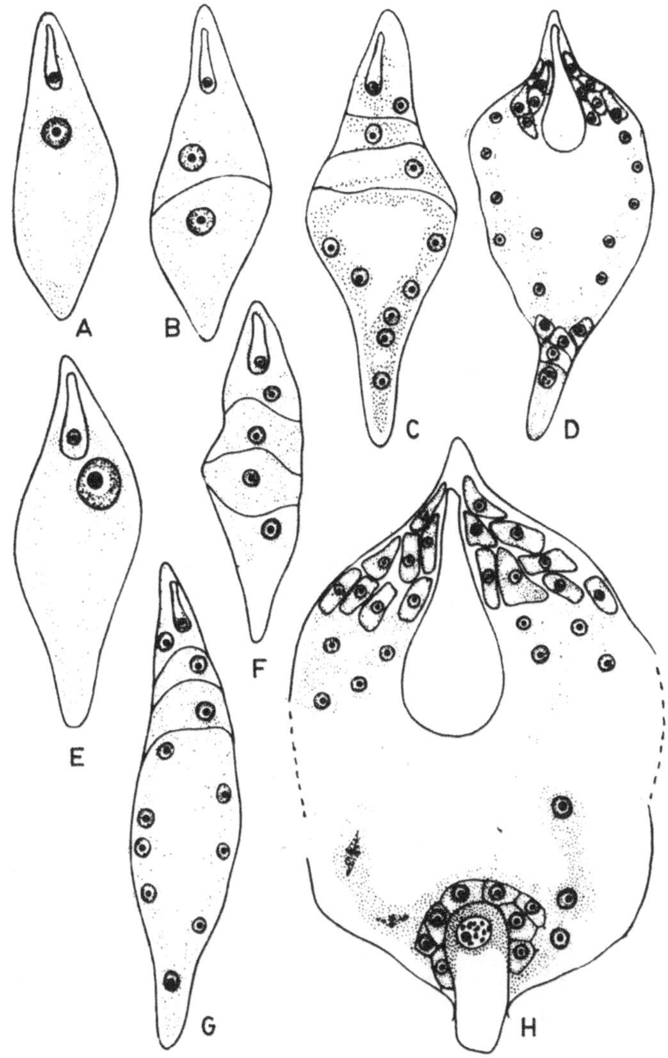

Fig. 31.11.1A–D. *Impatiens arguta*; **A** Embryo sac with zygote and primary endosperm nucleus; **B** Formation of micropylar and chalazal endosperm chamber; **C** Zygote, three-celled micropylar chamber and multinucleate chalazal chamber; **D** Globular proembryo, cellular endosperm at both poles, and free-nuclear in centre. Micropylar endosperm haustorium is absent, but a chalazal haustorium is cut off. **E–H** *I. levengi*; **E** Embryo sac with zygote and primary endosperm nucleus; **F** zygote, three-celled micropylar chamber, and chalazal chamber; **G** Zygote, three-celled micropylar chamber and multinucleate chalazal chamber; **H** Later stage, cellular endosperm at both poles, free-nuclear endosperm in centre, uninucleate chalazal haustorium, micropylar haustorium absent. (After L.L. Narayana 1965)

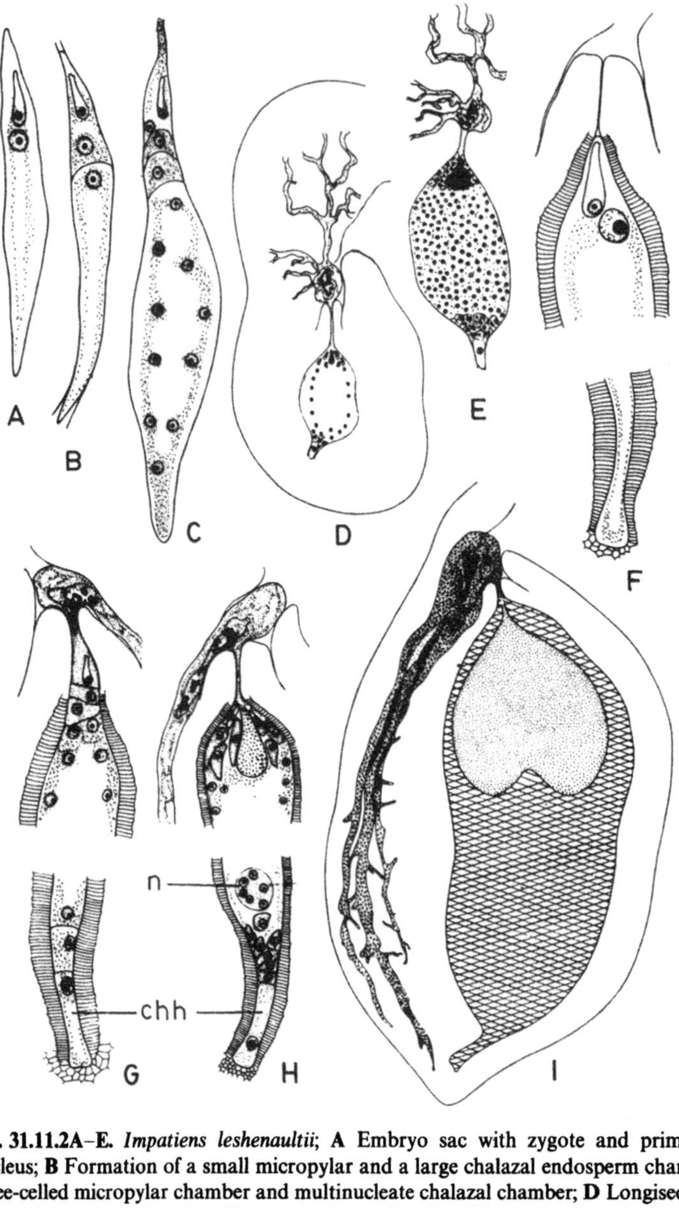

Fig. 31.11.2A–E. *Impatiens leshenaultii*; **A** Embryo sac with zygote and primary endosperm nucleus; **B** Formation of a small micropylar and a large chalazal endosperm chamber; **C** Zygote, three-celled micropylar chamber and multinucleate chalazal chamber; **D** Longisection of ovule to show micropylar and chalazal haustoria, beginning of wall formation in endosperm at micropylar and chalazal end, middle region free-nuclear. Micropylar haustorium ramifies in the placenta; **E** Same as **D**, later stage (dissected whole-mount). **F–I** *I. radiata*; **F** Embryo sac with endothelium, zygote and primary endosperm nucleus. Note the endothelium in **F–H**; **G, H** Micropylar and chalazal haustorium (*chh*), beginning of wall formation in endosperm at micropylar and chalazal end, and free-nuclear endosperm in centre. Note nucleate nodules (*n*) in free-nuclear endosperm in **H**; **I** Immature seed (cleared in chloral hydrate) to show the much-branched micropylar endosperm haustorium reaching as far as the chalaza. Cellular endosperm *hatched*, embryo *stippled*. (After L.L. Narayana 1965)

raphe. The development of haustorium remains arrested in *I. arguta* and *I. levengi* (L.L. Narayana 1965). The middle tiers develop into the endosperm proper. Free-nuclear divisions occur in the chalazal chamber, and a few cells are formed; the basal cell develops into a small haustorium (Figs. 31.11.1C, D, G, H, 31.11.2C–E, G, H). In the free-nuclear part multinucleate cytoplasmic endosperm nodules (Fig. 31.11.2H) develop in *Impatiens.*

The embryogeny conforms to the Asterad type-Erodium variation in *Hydrocera triflora. Impatiens textorii* (Takao 1976) exhibits polymorphism in octant formation.

The seed is exalbuminous, and mostly pachychalazal. The seed-coat is formed of two to five layers of fused integuments. In *Hydrocera* the outer region is composed of thick-walled sclerotic cells. The endothelium gets crushed. The tegmen does not persist.

KVO Dahlgren (1934a), G Heimsch (1942), J Hutchinson (1973), LL Narayana (1963b, 1965, 1970f), LL Narayana and M Sayeeduddin (1959), D Ramadevi and LL Narayana (1990), TS Raghavan, KR Venkatasubban and HD Wülff (1939), R Souèges (1945a), K Steffen (1951), S Takao (1966, 1968, 1976), JL van Went (1989), J Venkateswarlu and L Lakshminarayana (1957)

31.12 Taxonomic Considerations

Bentham and Hooker (1965a), and Engler and Diels (1936) place all the families in Sapindales. Hutchinson (1973) includes Balsaminaceae in Geraniales, and all others in Sapindales.

Coriariaceae. Coriariaceae represent the suborder Coriariineae. They are considered as an isolated member, and are in the nature of a relict family (Melchior 1964). Mauritzon (1936b) relates them to Celastrales. Takhtajan (1980) includes them in Rutales, along with Anacardiaceae. V.K. Sharma (1968a) compares the characters of Coriariaceae with Empetraceae, Anacardiaceae and Salvadoraceae, and points out that they are embryologically similar to Anacardiaceae (Table 31.12.1). It does not show any affinity with Empetraceae. The flap-like covering on either side of the funiculus develops from the outer integument, and the persistent fleshy perianth in fruit is noteworthy. The Onagrad type of embryogeny does not match with the Asterad type in Salvadoraceae. V.K. Sharma (1968a) suggests that Coriariaceae may be considered to represent an isolated family distantly related to the arborescent families included in Sapindales of Engler.

The study of seed anatomy reveals certain similarities between Coriariaceae and Anacardiaceae; the stony endocarp provides protection in Anacardiaceae. The outer epidermis of testa is protective in both. The tegmen is thin. The nucellus and endosperm are absent in the ripe seed (Mohana Rao 1974a).

Anacardiaceae. The suborder Anacardiineae includes a single family, Anacardiaceae. Anacardiaceae differ both from Coriariaceae and Aceraceae:

Table 31.12.1. Embryology of Coriariaceae (Sect. 31.2), Anacardiaceae (31.3), Salvadoraceae (33.11) and Empetraceae (42.5). (Adapted after V.K. Sharma 1968a, Maheswari Devi 1972, Corner 1976)

	Coriariaceae	Anacardiaceae	Salvadoraceae	Empetraceae
Anther	Bisporangiate	Tetrasporangiate	Tetrasporangiate	Tetrasporangiate
Wall	5-layered, endothecium fibrous	6-layered, endothecium fibrous	5-layered, endothecium fibrous	4–5-layered, endothecium non-fibrous
Tapetum	Secretory, cells 1–4-nucleate	Secretory, cells binucleate	Secretory	Secretory, cells multinucleate
Cytokinesis	Simultaneous	Simultaneous	Simultaneous	Simultaneous
Pollen	2–4-zonoporate, 3-zonocolporate, 2- or 3-celled	3-colporate, 2-celled	3-porate, 2-celled	Remain attached in pairs, individual grain 3-colpate or colporate, 2-celled
Ovule	Anatropous, bitegmic, crassinucellate, inner integument forms micropyle, outer integument forms a flap-like covering on either side of funiculus	Anatropous, sometimes apotropous, bitegmic, various degrees of fusion leading to unitegmy, crassinucellate, funicular obturator develops	Anatropous, bitegmic, crassinucellate, micropyle formed by inner integument	Anatropous, unitegmic, tenuinucellate, endothelium differentiates, funicular obturator is reported
Tetrad	Linear, T-shaped	Linear, T-shaped	Linear	Linear
Embryo sac	Polygonum type, antipodals ephemeral	Polygonum, antipodals ephemeral, twin embryo sacs may occur	Polygonum, antipodals simulate egg apparatus	Polygonum, antipodals persist up to fertilization
Endosperm	Nuclear type, becomes cellular	Nuclear, becomes cellular	Nuclear, becomes cellular	Cellular, micropylar haustorium 2- and chalazal 1-nucleate, both bulge at the extremities
Embryogeny	Onagrad type	Onagrad	Asterad	Caryophyllad
Polyembryony	Absent	Synergid, and nucellar polyembryony	Absent	Cleavage polyembryony
Seed coat	Not multiplicative, exotesta thick-walled	Unspecialized	Exotesta of thinly-cuticularized cells with mucilaginous radial and inner walls	Exotesta forms seed-coat
Seed	Exotestal, mostly exalbuminous, embryo straight	More or less exalbuminous, embryo large, stony endocarp protective	Exotestal, exalbuminous, embryo straight, cotyledons thick	Exotestal, albuminous, embryo straight

(1) bitegmic condition of ovule due to splitting of integumentary primordium, (2) hypostase with phenolic compounds, (3) occasional occurrence of chalazogamy, (4) considerable interval between fertilization and division of zygote, (5) nucellar polyembryony, and (6) protective endocarp.

Anacardiaceae also show marked similarities with Coriariaceae (Table 31.12.1) and Aceraceae: (1) secretory anther tapetum, (2) two-celled pollen grains, (3) anatropous, bitegmic, crassinucellate ovule, (4) Polygonum type of embryo sac with ephemeral antipodals, and (5) Nuclear type of endosperm. All the three families, Coriariaceae, Anacardiaceae and Aceraceae, have Onagrad type of development of embryo, and exalbuminous seed. Anacardiaceae and Coriariaceae also have exotestal seed-coat. There is no doubt that these three families are closely related.

In Anacardiaceae the unspecialized testa may degenerate to give unitegmic seeds. In some taxa the inner epidermis of the tegmen becomes thick-walled. None of these features are common to the Rutales or Sapindales and, therefore, any relationship should be ruled out (Corner 1976, see also von Teichman and Robbertse 1986b). On the basis of seed-coat and pericarp structure, there is possibly a relationship between Anacardiaceae and Julianaceae. Von Teichman and Robbertse emphasize that the seed-coat structure supports the generally accepted view that there is a strong phylogenetic affinity between Anacardiaceae and Burseraceae.

Blepharocarya. Wannan et al. (1987) discuss the affinities of *Blepharocarya*, which has been placed either in Sapindaceae or Anacardiaceae. Radlkofer (1879) had included it in the Anacardiaceae. Anacardiaceae, Burseraceae and Julianiaceae are closely related. Burseraceae have two epitropous ovules per locule, as compared to one ovule in the other two families. The tricolpate and prolate pollen grain in *B. involucrigera* is of the Rhus type (Anacardiaceae). The ovule is clearly apotropous with a dorsal raphe and a superior micropyle. These pollen and ovule characters are also common to Anacardiaceae. The secretory canals in the stem and wood clearly point to the anacardian affinity. On the basis of a detailed study of floral morphology, stem anatomy, pollen and ovule structure, Wannan et al. conclude that *Blepharocarya* should be "readmitted" into the Anacardiaceae. Final confirmation must await fuller embryological study of this taxon.

The suborder Sapindineae comprises Aceraceae, Bretschneideraceae, Sapindaceae, Hippocastanaceae, Sabiaceae, Melianthaceae, and Aextoxicaceae.

Aceraceae. The relationships of Aceraceae have already been discussed along with Anacardiaceae (Sect. 31.12).

Bretschneideraceae. *Bretschneidera*, formerly included in Hippocastanaceae (Pax 1896b), is treated as a separate family Bretschneideraceae, related to Moringaceae (Engler and Diels 1936). The embryology has not been investigated, and its taxonomic assignment cannot be discussed.

Sapindaceae. Sapindaceae can be distinguished from the other families of the suborder on the basis of: (1) anacampylotropous ovule with hypostase and obturator, (2) persistent supernumerary antipodals, and (3) arillate, sarcotestal, exotegmic seeds. The embryogeny is of the Asterad type similar to that in Melianthaceae and Balsaminaceae. In spite of these differences, Sapindaceae show embryological features in conformity with the order Sapindales.

Hippocastanaceae. The characteristic features of Hippocastanaceae are: (1) amphitropous ovule, massive nucellus, thick-walled epistase, and hypostase, (2) embryo sac with a vesicle at the chalazal end, (3) asymmetric development of cotyledons (functioning as an organ of absorption of nutrients from the vesicle), (4) insignificant development of suspensor, and (5) variability in the plane of division of the zygote (see List and Steward 1965).

Even though the above features mark out Hippocastanaceae, the other embryological features like development of anther, pollen, ovule, nucellus, and embryo sac show alliance with the other families of Sapindales. The exalbuminous, pachychalazal seed resembles that of Balsaminaceae. It follows that Hippocastanaceae are correctly placed in Sapindales.

Sabiaceae. Sabiaceae is said to be closely allied to Sapindaceae. Its unitegmic ovule with dome-shaped nucellus, formation of a diverticulum at the chalazal end of the embryo sac, and Helobial type of endosperm with haustorial processes from the chalazal chamber single it out from the rest of the families. Its position in Sapindales is inappropriate.

Melianthaceae. Melianthaceae were considered by Bentham and Hooker (1965a) and Hallier (1912) as constituents of Sapindaceae. Gürke (1896, see Lawrence 1951) separated them on morphological grounds. Embryologically also, they differ from Sapindaceae in several respects. Melianthaceae have persistent epidermis, three to five middle layers, and biseriate secretory anther tapetum. Cytokinesis in pollen mother cells is by furrowing, as well as by cell-plate formation. The pollen grains are three-nucleate at the dispersal stage. The adjacent sporangia do not coalesce, and each anther lobe dehisces independently (S. Mathur 1977).

The bitegmic, crassinucellate ovule, Polygonum type of embryo sac, and Nuclear type of endosperm occur in Melianthaceae as well as in Sapindaceae. The endosperm never becomes cellular in Melianthaceae as also in Hippocastanaceae (Khushalani 1963b, List and Steward 1965).

The seed is arillate. The endosperm is massive, horny and ameloid in *Bersama* (see Corner 1976). No other family shows endospermic seeds in Sapindales. The seed structure also differs to some extent from that of Sapindaceae. The seeds of Sapindaceae have palisade-like exotesta, and sclerotic endostome. The exotegmen is composed of longitudinally elongate lignified, or non-lignified, fibres. In Melianthaceae the exotesta is palisade-like only in the region of the micropyle, and the tegmen is almost crushed. The innermost layer of the inner integument is radially elongate at the micropyle (Khushalani 1963b). On embryological ground, Melianthaceae should be treated as an independent and specialized family.

Balsaminaceae. Balsaminaceae are treated as a subfamily of Geraniaceae by Bentham and Hooker (1965a). Engler and Prantl (1931), Hutchinson (1973) and Takhtajan (1980) considered them as an independent family in Geraniales. This family is usually placed in Geraniales, except by Engler and Prantl (1931) and Melchior (1964) who transferred Balsaminaceae to the order Sapindales, especially due to the nature of ovular attachment (pendulous). R. Dahlgren (1977a) erected a separate order Balsaminales in superorder Rutanae.

Balsaminaceae differ from Sapindales in the formation of trabeculate partitions in the pollen sac, fusion of integuments up to the tip (except in *Impatiens maculata* and *I. parviflora*), tenuinucellate ovule with endothelium, Polygonum as well as Allium type of embryo sac, Cellular type of endosperm with micropylar haustorium, and nucleated nodules in the free-nuclear part of the endosperm. Seeds are small- to medium-size, exarillate, exalbuminous, and mostly pachychalazal. The seed-coat is formed by expanded chalaza. Corner (1976) mentions that in *Impatiens* there may be a complete series of species to show the origin of the unitegmic pachychalazal seed from the bitegmic.

According to L.L. Narayana (1970f), Balsaminaceae are closely related to Oxalidaceae, Geraniaceae and Tropeaolaceae. The features of similarities are: (1) secretory anther tapetum, (2) cytokinesis by furrowing, (3) two-celled pollen grains, (4) anatropous, bitegmic ovules, (5) early fusion of polar nuclei, and ephemeral antipodals, (6) porogamous entry of pollen tube, and (7) Asterad type of embryogeny. Anatomically (Heimsch 1942) and cytologically (Warburg 1938) also Balsamineae resemble these three families. L.L. Narayana (1970f) negates the inclusion of Balsaminaceae under Sapindales, on the basis of the nature of ovular attachment as proposed by Engler and Prantl (1931).

The embryological features cited by L.L. Narayana (1970f) as affinities of these three families occur in most of the sapindaceous families. Asterad type of embryogeny occurs in Sapindaceae and Melianthaceae. These affinities do not warrant the separation of Balsaminaceae from Sapindales. In our opinion there is adequate justification to erect the order Balsaminales, as already suggested by R. Dahlgren (1977a).

32 Julianiales

The order Julianiales comprises a single family, Julianiaceae. The embryology of this family has not been investigated.

33 Celastrales

The order Celastrales comprises three suborders with 13 families: Celastrineae – Cyrillaceae, Pentaphylacaceae, Aquifoliaceae, Corynocarpaceae, Pandaceae,

Celastraceae, Staphyleaceae, Hippocrateaceae, Stackhousiaceae and Salvadoraceae; Buxineae – Buxaceae; Icacineae – Icacinaceae and Cardiopteridaceae.

33.1 General Features

The order Celastrales is characterized by: (1) secretory anther tapetum, (2) simultaneous cytokinesis in pollen mother cells, (3) two-celled pollen grains except in Stackhousiaceae, (4) anatropous ovules, orthotropous in Celastraceae; variable number of integument, and nature of nucellus, (5) Polygonum type of embryo sac, (6) Nuclear type of endosperm, Cellular in Cyrillaceae and Aquifoliaceae, and both types in Buxaceae, endosperm haustoria in Cyrillaceae and Icacinaceae, and (7) probably Solanad type of embryogeny in Cyrillaceae, Solanad in Celastraceae, probably Caryophyllad type in Aquifoliaceae and Corynocarpaceae, Asterad in Stackhousiaceae and Salvadoraceae, and Onagrad in Buxaceae.

33.2 Cyrillaceae

The development of the embryo has not been investigated.

The anther wall comprises the epidermis, fibrous endothecium, two ephemeral middle layers, and secretory tapetum of binucleate cells. The epidermis, endothecium, and middle layers contain tannin in *Cyrilla racemiflora* and *Cliftonia monophylla* (Vijayaraghavan and Dhar 1978). A few cells of the connective also develop fibrous thickenings, and contain tannin and druses. Simultaneous cytokinesis follows meiotic divisions in microspore mother cells. The tetrads are tetrahedral.

The pollen grains are tricolporate, and two-celled at the dispersal stage.

The ovule is anatropous, unitegmic and tenuinucellar with a long and narrow micropyle. Druses, tannin-filled cells, and stomata occur in the funicle in *Cyrilla* (Vijayaraghavan and Dhar 1978). The inner epidermis of the integument develops into an endothelium. A single hypodermal cell functions directly as megaspore mother cell. The latter, by meiotic divisions, produces a linear tetrad. The chalazal megaspore develops into a Polygonum type of embryo sac. The polar nuclei fuse before fertilization, the secondary nucleus lies in the centre of the embryo sac. The antipodals persist up to fertilization.

The development of the endosperm is of the Cellular type (Fig. 33.2.1A–G). Due to two consecutive transverse divisions of the primary endosperm nucleus, a row of four cells is formed in *Cliftonia* (Vijayaraghavan and Dhar 1978). The two middle cells develop into the endosperm proper (Fig. 33.2.1B–F). The two terminal cells develop into haustoria. The micropylar haustorium (Fig. 33.2.1B–F) is one-celled, binucleate and aggressive, and penetrates the funicular cells. Rarely, it becomes three- to four-nucleate in *Cyrilla*. The chalazal haustorium is one-celled, binucleate (Fig. 33.2.1C–F), and persists up to the dicotyledonous stage of the embryo.

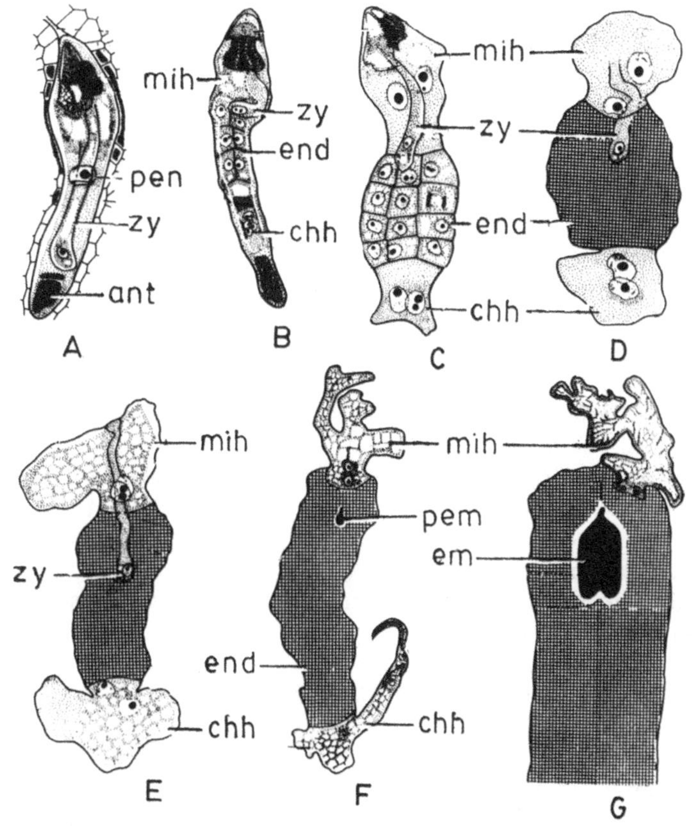

Fig. 33.2.1A–G. *Cyrilla racemiflora.* **A** Embryo sac with zygote (*zy*), degenerated synergid, remnants of pollen tube, primary endosperm nucleus (*pen*), and degenerated antipodal cells (*ant*). **B** Embryo sac with zygote, degenerated synergid, binucleate micropylar haustorium (*mih*), endosperm proper (*end*), four-nucleate chalazal haustorium (*chh*), and degenerated mass of antipodal cells. **C, D** Embryo sac with elongated tubular zygote, binucleate micropylar haustorium, cellular endosperm proper, and binucleate chalazal haustorium. **E** Bulbous endosperm haustoria. **F** Embryo sac at globular proembryo. The well-developed branched micropylar haustorium is binucleate, and the chalazal haustorium four-nucleate. **G** Upper portion of embryo sac at dicotyledonous embryo. (After Vijayaraghavan and Dhar 1978)

The degeneration of egg and zygote is frequent. The zygote elongates (Fig. 33.2.1A–E) before division, and the mature embryo is large, straight and dicotyledonous.

A noteworthy feature of the family is the occurrence of aposporic embryo sacs of Hieracium type (Battaglia 1963) in *Cliftonia monophylla* (Vijayaraghavan 1970c, Vijayaraghavan and Dhar 1978). In the chalazal region, some of the nucellar cells differentiate as megaspore mother cells and behave as embryo sac mother cells. They enlarge, undergo mitotic divisions, and develop into two- and four-nucleate embryo sacs. An eight-nucleate unorganized embryo sac was also observed. These are non-functional, and degenerate during the development of endosperm.

The integumentary cells become vacuolate and degenerate at the dicotyledonous stage of the embryo. The mature seed does not have any seed-coat (Vijayaraghavan and Dhar 1978). The multilayered pericarp is tanniniferous, and contains druses.

E Battaglia (1963), HF Copeland (1953), JL Thomas (1960, 1961), MR Vijayaraghavan (1969, 1970c), MR Vijayaraghavan and U Dhar (1978)

33.3 Pentaphylacaceae

The development of anther, pollen, embryo sac, endosperm and embryo has not been investigated.

The ovule is ana- or campylotropous, bitegmic and crassinucellar in *Pentaphylax euryoides* (Mauritzon 1936b). A long micropyle is organized by the inner integument. The parietal tissue is scanty. The embryo sac is eight-nucleate and, later, becomes curved.

Seeds are more or less winged, with scanty endosperm and horseshoe-shaped embryo (Hutchinson 1973).

A Cronquist (1981), J Hutchinson (1973), J Mauritzon (1936b)

33.4 Aquifoliaceae

The development of anther, embryo and seed-coat has not been studied in detail.

In *Ilex cornuti* (Copeland 1963), the endothecium is ribbed, the tapetum with binucleate cells is not strictly a single layer, and the three-grooved pollen grains have a papillate surface. They are two-celled at the shedding stage.

The ovule is apotropous, unitegmic and crassinucellar. One to three cells in the hypodermal region cut off parietal cells. The parietal tissue is one or two layers thick in *Ilex aquifolium* and *I. cornuti* (Copeland 1963). At times, two megaspore mother cells function. The megaspore mother cell undergoes meiotic divisions, and produces a linear tetrad. The chalazal megaspore develops into a Polygonum type of embryo sac. An endothelium and hypostase differentiate. The polars fuse before fertilization, and the secondary nucleus is prominent. The antipodals are inconspicuous.

Development of the endosperm is of the Cellular type. A transverse division of the primary endosperm nucleus results in a micropylar and a chalazal chamber. Both chambers undergo longitudinal divisions; linear cell formation (Schürhoff 1921) occurs rarely.

The embryogeny is probably of the Caryophyllad type.

According to Hu and Ferreira (1989), the ripe fruit in *Ilex* contains seeds at the early heart-shaped embryo; further development to maturity occurs later.

The seed is endospermic with a minute embryo. The cells of the integument become lignified at the micropyle. The endothelial layer accumulates tannin. The remnants of nucellus persist in mature seed.

HF Copeland (1963), JM Herr Jr (1959a, b, c, 1961), CY Hu and AG Ferreira (1989), PN Schürhoff (1921)

33.5 Corynocarpaceae

The development and structure of anther, pollen grain, embryo sac, and embryo have not been investigated.

The ovule is anatropous, bitegmic and crassinucellate in *Corynocarpus laevigata* (Pigott 1927). The inner integument forms the micropyle. The archesporial cell cuts off a parietal cell. The cells of the nucellar epidermis divide and produce three to four layers.

In a mature embryo sac, the three antipodal cells multiply.

The development of the endosperm is of the Nuclear type.

The embryogeny probably corresponds to the Caryophyllad type (see Davis 1966).

The seed is exalbuminous, and contains a large embryo with minute radicle. The massive (20–40 layers) testa is thin-walled, and gets crushed. The tegmen also gets crushed.

Possibly the seed is pachychalazal as in Anacardiaceae (Corner 1976).

EJH Corner (1976), GL Davis (1966), EM Pigott (1927)

33.6 Pandaceae

The embryology has not been investigated.

The anther is bilocular, and introse and opens longitudinally. The ovules are ortho- or anatropous, without obturator. The seeds are usually flattened, rarely ovoid, with copious fatty endosperm, and a straight embryo which has thin, flat, cordate cotyledons.

A Cronquist (1981), J Hutchinson (1973), J Vaughan and JA Rest (1969), JC Willis (1973)

33.7 Celastraceae

The anther wall comprises the epidermis, fibrous endothecium, two or three ephemeral middle layers, and secretory tapetum. The epidermal cells are papillate, and contain tannin. In *Gymnosporia* (Adatia and Gavde 1962), the endothecium becomes irregularly biseriate. The tapetal cells are multinucleate. In *Celastrus paniculata* (Adatia and Gavde 1962), they are eight-nucleate; the nuclei fuse and become polyploid. The microspore mother cells divide meiotically and produce tetrahedral and decussate tetrads. Simultaneous cytokinesis occurs by furrowing.

The pollen grains are two-celled at the shedding stage. In *Gymnosporia rothiana* they germinate in situ. Pollen grains fail to develop in *Euonymus japonica* (Copeland 1966); this species is an apomict.

The ovule is orthotropous, bitegmic and tenuinucellate in *Gymnosporia*, and weakly crassinucellate in *Celastrus* and *Elaeodendron* (Adatia and Gavde 1962). Both the integuments form the micropyle. The hypodermal archesporium is uni- to multicellular. It cuts off a parietal cell in *Celastrus* and *Elaeodendron*, otherwise it functions directly as megaspore mother cell. At times, two mother cells may be present. The megaspore mother cell undergoes meiotic divisions and produces a linear or T-shaped tetrad. The chalazal megaspore develops into a Polygonum type of embryo sac. The antipodals are

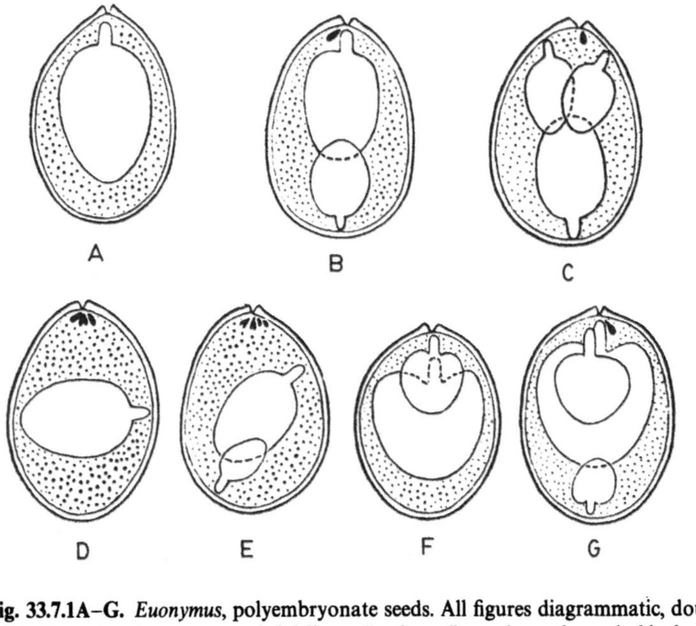

Fig. 33.7.1A–G. *Euonymus*, polyembryonate seeds. All figures diagrammatic, double seed-coat is not indicated, endosperm *stippled*. Minute "embryos" at micropylar pole *black*. **A–E** *E. americanus*; **A** Monoembryonate seed with full-sized embryo (3.0–3.5 mm); **B** Three-embryonate seed, a globular proembryo (0.25 mm) and a large embryo (2–3 mm) at micropylar pole, and a smaller embryo (1–2 mm) at chalazal pole; **C** Four-embryonate seed, a globular proembryo (0.25 mm) and two small embryos (1–2 mm) at micropylar pole, and a larger embryo (2–3 mm) at chalazal pole (mostly integumentary); **D** Two globular proembryos (0.25 mm) and a heart-shaped (0.25–0.5 mm) embryo at micropylar end, and a larger (probably integumentary) embryo (3.0–3.5 mm) lies horizontally in the middle of the embryo sac; **E** Six-embryonate seed. Two globular proembryos and two heart-shaped embryos (0.05–0.25 mm) at the micropylar pole, one smaller (1 mm) and a larger embryo (2–3 mm) lie obliquely opposite to each other. **F, G** *E. alatus* var. *apterus*; **F** Two-embryonate seed with cotyledons of smaller embryo (1–2 mm) clasping the radicle of larger one (3.0–3.5 mm); **G** Four-embryonate seed with one globular proembryo (0.25–0.5 mm) and two embryos (2–3 mm, 1–2 mm) at micropylar end, and a smaller one (under 0.25 mm) at chalazal pole. (Adapted after Brizicky 1964)

ephemeral. The polar nuclei fuse in the centre of the embryo sac. An endothelium differentiates by the time of tetrad formation. In *Euonymus japonica*, half the ovules abort, and the egg apparatus generally degenerates.

The development of endosperm is of the Nuclear type. Centripetal wall formation occurs after the formation of about 200 nuclei.

The embryogeny corresponds to the Solanad type (Adatia and Gavde 1962). It is reported to be of Caryophyllad type-Sagina variation in *Celastrus scandens* and *Euonymus europaeus* (Anderson 1931).

The endosperm in *E. japonica* (Copeland 1966) develops autonomously, and the embryo originates from a cell of the inner integument close to the micropyle.

Polyembryony occurs in *Celastrus paniculata* (Adatia and Gavde 1962). Brizicky (1964) reports integumentary (?), and nucellar embryos (Fig. 33.7.1A–G), and parthenocarpic development of fruit in *Euonymus* spp. Naumova (1970a, b) observed different patterns of embryogenesis in the integumentary and zygotic embryos.

The seed is arillate. The aril may develop from the outer integument, or funicle, or both. The outer epidermal cells of testa are thickly cuticularized, and are followed by aerenchymatous mesophyll with scattered sclerotic cells. The inner epidermis is thin-walled. The outer epidermis of tegmen comprises thick-walled, lignified, pitted, longitudinal fibres. The inner epidermis is represented by the endothelium.

RD Adatia and SG Gavde (1962), A Anderson (1931), GK Brizicky (1964), HF Copeland (1966), H Ding (1967), T Loesener (1942), J Mauritzon (1936b), TN Naumova (1970a, b), R Pilger (1915)

33.8 Staphyleaceae

The embryogeny has not been investigated.

The anther wall comprises the epidermis, fibrous endothecium, a single ephemeral middle layer, and secretory tapetum of binucleate cells. The microspore mother cells undergo meiotic divisions, and produce tetrahedral tetrads. Simultaneous cytokinesis occurs by furrowing.

The pollen grains are shed at the two-nucleate stage.

The ovule is anatropous, bitegmic and crassinucellate. The micropyle is formed by both the integuments. The hypodermal archesporial cell cuts off a parietal cell. The parietal tissue is well developed, and a nucellar cap also develops. The megaspore mother cell divides meiotically, and produces a linear tetrad. The chalazal megaspore gives rise to a Polygonum type of embryo sac. The synergids are hooked, and exhibit filiform apparatus in *Turpinia* (L.L. Narayana 1960a). The three antipodals are small.

The development of the endosperm is of the Nuclear type in *Staphylea pinnata* (Guèrin 1901) and *S. trifoliata* (Riddle 1905). The endosperm stores fatty food reserve.

The mature embryo is straight with thick flat cotyledons, and a short radicle.

The seed is large and albuminous. The testa is multiplicative. The outer epidermis comprises a palisade of thick-walled cells followed by a sclerotic mesophyll. The inner epidermis is thin-walled. The tegmen gets crushed.

P Guèrin (1901), LL Narayana (1960a), LC Riddle (1905)

33.9 Hippocrateaceae

The embryogeny and structure of seed-coat have not been studied.

The anther wall comprises the papillate persistent epidermis, fibrous endothecium, two ephemeral middle layers, and secretory tapetum of multinucleate cells. Simultaneous cytokinesis occurs by furrowing, and the tetrads are tetrahedral and decussate.

The tricolpate, rarely tetracolpate, pollen grains with reticulate exine are two-celled at the dispersal stage.

The ovule is anatropous, bitegmic and tenuinucellate with the inner integument forming the micropyle in *Hippocratea grahamii* (Adatia and Gavde 1962). The archesporial cell functions directly as megaspore mother cell. In *Salacia oblonga* the archesporial cell cuts off a parietal cell which subsequently produces a parietal tissue. The megaspore mother cell undergoes meiotic divisions. The chalazal megaspore of the linear tetrad develops into a Polygonum type of embryo sac. The antipodals are ephemeral, and polar nuclei fuse before fertilization. An endothelium develops in *Hippocratea.*

The development of the endosperm is of the Nuclear type.

The fruit is schizocarpic, and the seeds are exalbuminous, winged and exarillate.

RD Adatia and SG Gavde (1962), J Mauritzon (1936a, c), J Miers (1872)

33.10 Stackhousiaceae

The anther wall comprises the epidermis, fibrous endothecium, a single middle layer, and secretory tapetum of binucleate cells. The microspore mother cells divide meiotically, and produce tetrahedral, isobilateral and decussate tetrads. Simultaneous cytokinesis is accompanied by furrowing.

The pollen grains are shed at the three-celled stage.

The ovule is anatropous, bitegmic, tenuinucellate and develops a hypostase. The micropyle is formed by the outer integument, the inner one collapses. The hypodermal archesporial cell functions directly as megaspore mother cell. The latter, following meiotic divisions, produces linear or T-shaped tetrad. The chalazal megaspore develops into a Polygonum type of embryo sac. The embryo sac comes to lie in direct contact with the integument after the degeneration of nucellus. The three antipodals are multinucleate in *Stackhousia linariaefolia* (Narang 1953), and their number increases up to 15 in *S. monogyna* and some other species (see Mauritzon 1936c).

The development of endosperm is of the Nuclear type. The peripheral layer of cytoplasm around the central vacuole contains 60–70 free nuclei. Cytoplasmic nodules develop in the endosperm. Due to centripetal wall formation, the endosperm becomes cellular at the globular proembryo.

The embryogeny conforms to the Asterad type. The mature embryo is large and straight, with rather thick cotyledons.

The seed is albuminous. The outer epidermis of testa comprises tangentially elongate cells containing tannin. The mesophyll and inner epidermis is thin-walled. The inner integument is absorbed by the developing embryo sac, before anthesis. The funicular vascular bundle extends in the antiraphe up to the micropyle.

J Mauritzon (1936c), N Narang (1953)

33.11 Salvadoraceae

The anther wall comprises the epidermis, fibrous endothecium, two ephemeral middle layers, and secretory tapetum. The microspore mother cells undergo meiotic divisions resulting in tetrahedral and isobilateral tetrads. Simultaneous cytokinesis occurs by furrowing.

The triporate pollen grains are shed at the two-celled stage.

The ovule is anatropous, bitegmic and crassinucellar. The micropyle is organized by the inner integument in *Salvadora persica*, and by the outer integument and inner lip of the inner integument in *Azima tetracantha* (Maheswari Devi 1972). The archesporium is uni- to multicellular; the parietal tissue is well developed. The megaspore mother cell undergoes meiotic divisions, and produces a linear tetrad. The chalazal megaspore develops into a Polygonum type of embryo sac. The antipodal cells simulate egg apparatus. Sometimes twin embryo sacs are formed.

The entry of pollen tube is porogamous, and syngamy and triple fusion occur more or less simultaneously.

The development of the endosperm is of the Nuclear type. The nuclei with cytoplasm accumulate in the micropylar region around the embryo. Later, a similar accumulation of nuclei also occurs in the chalazal region where they fuse and become polyploid. Finally, the endosperm becomes cellular, and is consumed by the developing embryo.

Embryogeny corresponds to the Asterad type-Oxalis variation (Maheswari Devi 1972). The mature embryo is straight with thick flat cotyledons, and a well-developed radicle.

The seed is exotestal in *Azima*. The outer epidermis of testa comprises columnar, thinly cuticularized cells with mucilaginous radial and inner walls, and contains crystals. The mesophyll is thin-walled, except at the hilum and micropyle where it is lignified.

G Bentham (1876b), H Maheswari Devi (1972), J Mauritzon (1936b)

33.12 Buxaceae

The anther wall comprises the persistent epidermis, fibrous endothecium, two or three ephemeral middle layers, and secretory tapetum of binucleate cells. The microspore mother cells undergo meiotic divisions, cytokinesis is simultaneous, and tetrahedral and isobilateral tetrads are produced.

The pollen grains are two- or three-celled at the dispersal stage.

The ovule is anatropous, bitegmic and crassinucellate with a false endostome in *Buxus*. The archesporial cell cuts off a parietal cell which produces a thick parietal tissue. The megaspore mother cell undergoes meiotic divisions. The chalazal megaspore of a linear tetrad develops into a Polygonum type of

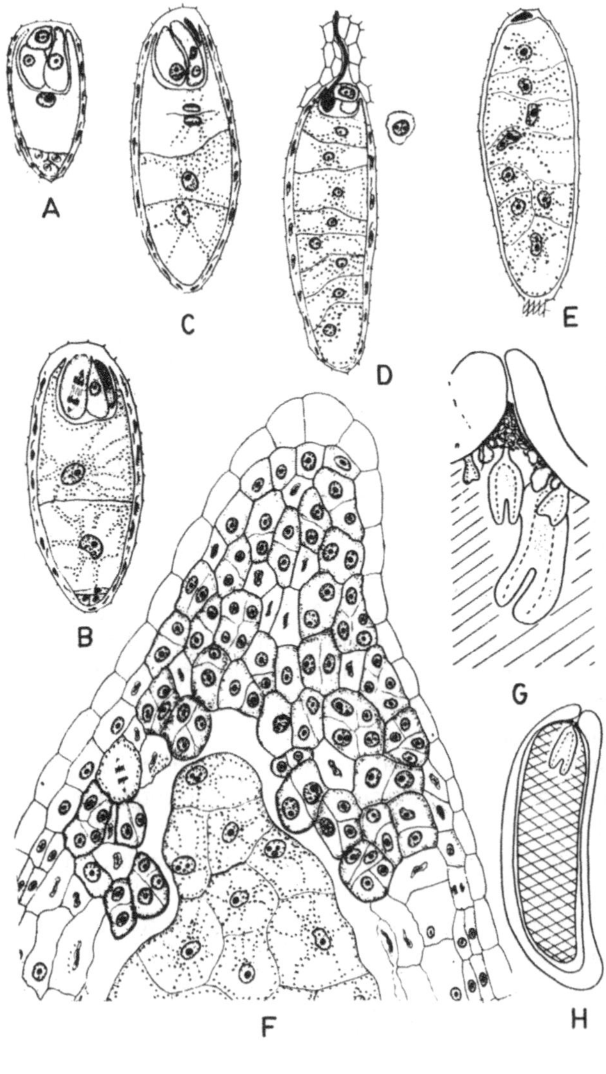

Fig. 33.12.1A–H. A–C *Notobuxus acuminata*, **D, H** *Buxus sempervirens*, **E–G** *Sarcococca humilis*. **A** Mature embryo sac; **B** Fertilized embryo sac, zygote in division, persistent synergid, remnants of pollen tube, and two-celled endosperm; **C** Three-celled endosperm, zygote undivided. **D** Embryo sac with zygote (*inset*), remnants of pollen tube, and multicellular endosperm. **E** Embryo sac with remnants of degenerated egg apparatus, and cellular endosperm; **F** Micropylar region of nucellus. Most of the cells function as initials of apomictic embryos; note a number of proembryos especially around the endosperm; **G** Micropylar region of ovule with a number of nucellar embryos. **H** Ovule with sexual, young dicotyledonous embryo surrounded by endosperm (*cross-hatched*). (After Naumova 1979, 1980, see Yakovlev 1981)

embryo sac (Fig. 33.12.1A). The polar nuclei fuse prior to fertilization. The antipodals are persistent, and multiply (Wiger 1935).

Simmondsia represents a young taxon which appears to have made a comparatively recent transition from insect-mediated pollination (entomophily) to anemophily. Three major alterations have occurred in this transition: (1) loss of the androecium, (2) reduction in flower number per inflorescence to one, and (3) extension of the peduncle. These are selectively advantageous for anemophily.

The development of endosperm is of the Nuclear type in *Buxus* and *Pachysandra*, and Cellular (Fig. 33.12.1B–E) in *Sarcococca* and *Simmondsia* (see Schnarf 1931, Davis 1966).

Development of the embryo is of the Onagrad type. Nucellar polyembryony (see Schnarf 1931, Davis 1966) occurs (Fig. 33.12.1 F, G) in several species of *Sarcococca* (Orr 1923), and in *S. prunifolia* at least one nucellar embryo develops autonomously while the egg degenerates (Wiger 1935).

The albuminous seed is small and exo-mesotestal with a straight or curved embryo with flat cotyledons (Fig. 33.12.1H). The outer epidermis of testa is lignified, and the mesophyll is crushed except in *Sarcococca* where one to two layers are sclerotic. The inner epidermis is palisade-like, and lignified at the micropylar end in *Buxus*. The inner epidermis of tegmen is lignified in *Sarcococca*, otherwise the tegmen is crushed. The seed of *Buxus* shows two small waxy-horny white lobes, on either side of the micropyle, as rudimentary funicular aril or caruncle.

EJH Corner (1976), GL Davis (1966), TN Naumova (1979, 1980), KJ Niklas and SL Buchman (1985), MY Orr (1923), TL Rost, AD Simper, P Schell and S Allen (1977), K Schnarf (1931), J Wiger (1935)

33.13 Icacinaceae

The embryogeny has not been studied.

The anther wall comprises the epidermis, radially elongate thick-walled endothecium, one or two ephemeral middle layers, and secretory tapetum of five- to six-nucleate cells. The microspore mother cells undergo meiotic divisions, cytokinesis is simultaneous, and the tetrads are tetrahedral.

The family is multipalynous with three-colpate and inaperturate elements (P.K.K. Nair 1970). The pollen grains are two-celled at the dispersal stage in *Gomphandra polymorpha* (Padmanabhan 1961c).

The ovule is anatropous, unitegmic and crassinucellate with a well-developed vascular supply in the integument. The integument splits (in young condition) in *Phytocrene* (Fagerlind 1945d), and the splits remain free at the micropyle. A weak obturator develops from the placenta. The archesporium is multicellular. The archesporial cells cut off parietal cells, and the parietal tissue is five to six layers thick. The megaspore mother cell undergoes meiotic divisions giving rise to a linear tetrad. The chalazal megaspore develops into a Polygonum type of embryo sac. At the four-nucleate stage, the embryo sac absorbs the nucellus, extends towards the micropyle and comes to lie between the integument and raphe of the ovule. The synergids lack vacuoles. The polar nuclei fuse before fertilization. The antipodals are ephemeral.

The development of the endosperm is of the Nuclear type. In *Gomphandra*, the zygote divides almost at the same time as the walls are laid down in the free-nuclear endosperm. The inner cells of the endosperm are large and highly vacuolate.

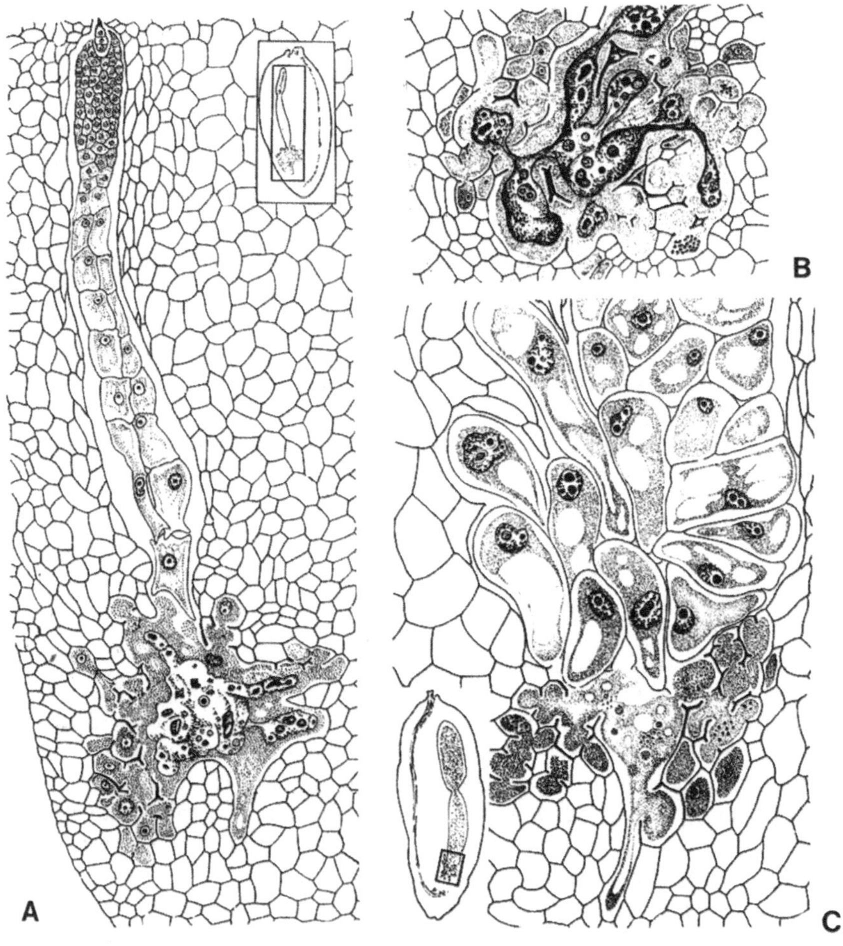

Fig. 33.13.1A–C. *Nothapodytes foetida.* **A** Longisection of part of ovule (*inset* outline diagram) to show three-celled proembryo. Cellular endosperm proper, and intracellular growth of chalazal haustorium with polyploid nucleus. **B** Nucleus of chalazal haustorium with adjoining tissue. **C** Longisection of part of ovule (*inset* outline diagram), later stage than **A**. The endosperm cells in the vicinity of disintegrated chalazal haustorium have become hypertrophied. The chalazal cells have been incorporated into the haustorium in **A–C**. (After Swamy and Ganapathy 1957b)

In *Nothapodytes foetida*, Swamy and Ganapathy (1957b) mention a very unusual type of endosperm haustorium (Fig. 33.13.1A–C). The basal cell of the endosperm proper becomes haustorial and develops a narrow, sac-like intercellular extension in the chalazal region. As the cytoplasm of the haustorial cell comes in contact with the intervening walls of adjacent cells of the chalazal tissue, an aperture develops in the wall at the point of contact, and the haustorium establishes protoplasmic connection with the affected cell. The nucleus of the affected cell becomes amoeboid, develops several nucleoli-like bodies and loses its membrane. Its chromatin becomes coarsely granular, and mixes with the cytoplasm. Finally, the cytoplasm and nucleus of the affected cell are incorporated into the haustorium. Such chain-like cytoplasmic connections develop in the chalazal tissue, and the haustorium continues to enlarge. During its aggressive growth, the haustorium affects groups of cells, and the fragments of their walls become included in the cytoplasm (Fig. 33.13.1A).

The nucleus of the haustorium becomes highly amoeboid and develops numerous nucleoli-like bodies. As the groups of protoplasts are absorbed by the haustorium, the nucleus of the haustorium extends into the affected cells, becomes lobed and dilates at the tip. The several lobes of the nucleus remain connected with the main body. The entire structure gives a very distorted appearance (Fig. 33.13.1B).

The cells of the endosperm proper exhibit distinct zonation, the upper region of smaller compact cells and the lower of larger loosely arranged cells.

When the activity of the haustorium declines, at the initiation of cotyledonary primordia in the embryo, the fragments of the cell walls of the absorbed protoplasts swell and become gelatinous. Two to three tiers of endosperm cells adjoining the haustorium become vacuolate, and develop tubular prolongations which grow intercellularly in the surrounding chalazal tissue. Their nuclei also enlarge, and give a distorted appearance (Fig. 33.13.1C). However, their haustorial behaviour is uncertain.

The seed has a small embryo. The outer layers of integument comprise thick-walled, elongate cells followed by thin-walled hexagonal cells.

F Fagerlind (1945d), J Mauritzon (1936c), PKK Nair (1970), D Padmanabhan (1961c), BGL Swamy and PM Ganapathy (1957b)

33.14 Cardiopteridaceae

The embryology has not been investigated.

33.15 Taxonomic Considerations

Engler and Diels (1936) place all the families in Sapindales. Hutchinson (1973) assigns Buxaceae to Hamamelidales, Staphyleaceae to Sapindales, and Aquifoliaceae, Salvadoraceae, Cyrillaceae, Icacinaceae, Celastraceae, Stackhousiaceae and Hippocrateaceae to Celastrales (see Table 33.15.1).

Table 33.15.1. Embryology of Cyrillaceae, Aquifoliaceae, Celastraceae, Hippocrateaceae, Stackhousiaceae, Salvadoraceae and Icacinaceae (See Vijayaraghavan and Dhar 1978)

	Cyrillaceae	Aquifoliaceae	Celastraceae	Hippocrateaceae
Tapetum	Secretory, cells binucleate	Secretory, cells binucleate	Secretory, cells multinucleate	Secretory, cells multinucleate
Tetrad	Tetrahedral	—	Tetrahedral and decussate	Tetrahedral and decussate
Pollen grains	2-celled	2-celled	2-celled	2-celled
Ovule	Unitegmic, tenuinucellate	Unitegmic, crassinucellate	Bitegmic, weakly crassinucellate, or tenuinucellate	Bitegmic, tenui- or crassinucellate
Embryo sac	Polygonum type	Polygonum	Polygonum	Polygonum
Antipodal cells	Uninucleate, persist up to fertilization	Uninucleate, inconspicuous	Uninucleate, ephemeral	Uninucleate, ephemeral
Endosperm	Cellular type	Cellular	Nuclear	Nuclear
Endosperm haustoria	Micropylar and chalazal haustoria formed	Absent	Absent	Absent
Seed-coat	Absent	Massive	Massive	Massive

Cyrillaceae. It has been suggested that the Cyrillaceae are related to Ericaceae and Celastraceae (see Thomas 1961). R. Dahlgren (1980b) places them in Ericales. Traditionally, Cyrillaceae have remained associated with Aquifoliaceae. Vijayaraghavan (1969, 1970c) compares the embryological features of Aquifoliaceae, Celastraceae, Cyrillaceae and Ericaceae. Cyrillaceae show close affinities with Ericaceae in unitegmic, tenuinucellate ovules, presence of endothelium, Polygonum type of embryo sac with a broad micropylar and narrow chalazal end, uninucleate persistent antipodals, Cellular type of endosperm with chalazal and micropylar haustoria, and probably Solanad type of embryogeny. Cyrillaceae resemble Clethraceae in (1) presence of fibrous endothecium, (2) druses in the cells of connective, and (3) pollen grains shed singly (Vijayaraghavan and Dhar 1978). Occurrence of apospory (Vijayaraghavan 1969) is a noteworthy feature in the family. Cyrillaceae differ from Celastraceae in the nature of ovule, absence of parietal tissue, type of endosperm, and seed structure. Nucellar polyembryony is absent in Cyrillaceae. Thus, embryological characters support its close association with Ericales (Clethraceae), rather than with Celastrales.

Aquifoliaceae. Several embryological features of Aquifoliaceae, such as two-celled pollen grains, unitegmic ovule, Polygonum type of embryo sac, endothelium, and Cellular type of endosperm (Copeland 1963), correspond with those of Cyrillaceae (Table 33.15.1). However, apotropous and crassinucellate condition of the ovule, inconspicuous antipodals, probably Caryophyllad type

Stackhousiaceae	Salvadoraceae	Icacinaceae
Secretory, cells binucleate	Secretory, cells binucleate	Secretory, cells multinucleate
Tetrahedral, decussate and isobilateral	Tetrahedral and isobilateral	Tetrahedral
3-celled	2-celled	2-celled
Bitegmic, tenuinucellate	Bitegmic, crassinucellate	Unitegmic, crassinucellate
Polygonum	Polygonum	Polygonum
Each cell becomes multinucleate	Uninucleate, ephemeral	Uninucleate, ephemeral
Nuclear	Nuclear	Nuclear
Absent	Absent	Absent; conspicuous chalazal endosperm haustorium reported in *Nothapodytes*
Massive	Massive	Massive

of embryogeny, and minute embryo differ from those of Cyrillaceae. Caryophyllad type of embryogeny occurs in Celastraceae, and probably in Corynocarpaceae.

Corynocarpaceae. Corynocarpaceae show affinities with Celastraceae. The common features are: bitegmic, crassinucellate ovule, parietal tissue, Nuclear type of endosperm, and Caryophyllad type of embryogeny. But anatropous ovule, absence of endothelium, secondary multiplication of antipodals, and pachychalazal exalbuminous seed with a large embryo distinguish Corynocarpaceae.

Celastraceae. Celastraceae have several special features, such as superior ovary with two ovules in each locule, orthotropous, bitegmic, tenui- or crassinucellar ovule, endothelium, ephemeral antipodals, Solanad and Caryophyllad (*Celastrus* and *Euonymus*) type of embryogeny, nucellar polyembryony, and arillate albuminous seed with sclerotic mesotesta, and fibrous exotegmen. The wings are absent. The fruit is drupaceous, capsular, or follicular.

Hippocrateaceae. Hallier (1912) combines the genus *Hippocratea* with Celastraceae. *Hippocratea* resembles Celastraceae (Table 33.15.1): (1) imbricate perianth in two whorls, each of five members, (2) glandular anther tapetum, (3) tricolpate, rarely tetracolpate, pollen grains with reticulate exine, (4) tricarpellary syncarpous ovary, (5) anatropous, bitegmic ovules, (6) Polygonum type of embryo sac, presence of an endothelium, and (7) Nuclear type of endosperm (Adatia and Gavde 1962). Adatia and Gavde suggest separation of the genus *Hippocratea*, and its placement in Hippocrateaceae. *Hippocratea* has tenuinucellate ovules with micropyle formed by the inner integument, exalbuminous, exarillate, winged seed, and schizocarpic fruit.

Staphyleaceae. Bentham and Hooker (1965a) united Staphyleaceae with Sapindaceae. Bessey (1915), Wettstein (1935) and Rendle (1959) include them in Celastrales, but Hutchinson (1973) placed them in Sapindales.

Staphyleaceae differ from Sapindaceae in having numerous ovules, copious endosperm, and straight embryo. On the other hand, Staphyleaceae have several common features with Celastraceae. Two-celled pollen grains, crassinucellate, bitegmic ovule with exo- and endostome, parietal cell, Polygonum type of embryo sac, Nuclear type of endosperm, sometimes arillate, albuminous seeds with sclerotic mesophyll, and capsular, infrequently berry-like fruit. The fibrous exotegmen of seed has been lost in this family (Corner 1976). The proper position of Staphyleaceae is in Celastrales rather than in Sapindales.

Stackhousiaceae. Narang (1953) discusses the affinities of Stackhousiaceae in relation to Rhamnaceae, Scrophulariaceae, Lobeliaceae and Celastrales. She concludes that the assignment of Stackhousiaceae to Celastrinae appears to be justified (see also Mauritzon 1936c). Stackhousiaceae closely resemble Celastraceae (Table 32.15.1) and the common features are: bitegmic, tenuinucellate ovule with a single archesporial cell, Polygonum type of embryo sac, Nuclear type of endosperm, and albuminous seeds. However, it can be distinguished due to three-celled pollen grains, supernumerary antipodals, hypostase, cytoplasmic nodules in the endosperm, Asterad type of embryogeny, and a large straight embryo (Narang 1953). The seeds are exarillate, and the inner integument is absorbed before anthesis. Supernumerary antipodals occur in Corynocarpaceae, and Asterad type of embryogeny in Salvadoraceae (Maheswari Devi 1972).

Salvadoraceae. Salvadoraceae show several features in conformity with those of the order Celastrales, and family Celastraceae (Table 32.15.1). They have anatropous bitegmic ovules, Polygonum type of embryo sac, and exarillate seeds as in Stackhousiaceae, with which it is placed. The seed is exotestal and exalbuminous, and lacks celastraceous fibrous exotegmen.

Buxaceae. Buxaceae are lone members of the suborder Buxineae. This family was included as a tribe of Euphorbiaceae by Bentham and Hooker (1965c). It has been treated as a distinct family in the order Euphorbiales by Wettstein (1935), Rendle (1959) and Cronquist (1981); Hutchinson (1973) placed Buxaceae in Hamamelidales, and Bessey (1915) and Melchior (1964) in Celastrales.

Buxaceae have three-loculed superior ovary with two pendulous collateral ovules with a dorsal raphe in each locule, and glossy black carunculate seeds as in Euphorbiaceae. Buxaceae differ in the absence of latex, anatropous ovule, fleshy endosperm, straight embryo, and loculicidal capsule, or berry-like fruit. In Euphorbiaceae latex is present, the ovule has a ventral raphe, and the fruit is schizocarpic. The development of endosperm is always of the Nuclear type, while in Buxaceae both Cellular and Nuclear types occur.

According to Dang-Van-Liem (1962), Buxaceae show affinities with Celastraceae. Bitegmic crassinucellar ovule with parietal tissue, Polygonum type of embryo sac, and Nuclear type of endosperm are common to both. Cellular endosperm occurs in Cyrillaceae and Aquifoliaceae in Celastrales. Asterad type of embryogeny is also present in Stackhousiaceae and Salvadoraceae. But Buxaceae do not resemble at all, either Euphorbiaceae or Celastraceae, in the anatomical structure of seed (see Mohana Rao 1974b).

Buxaceae resemble Hamamelidaceae in the type of pollen grain, ovule, and the development of embryo sac and endosperm. However, the embryogeny and seed structure differ.

Embryologically, Buxaceae are related more to the families of the suborder Celastrineae than to Hamamelidaceae and Euphorbiaceae.

Icacinaceae. The suborder Icacinae comprises Icacinaceae and Cardiopteridaceae. The embryology of the latter family has not been studied. In conformity with the order Celastrales, Icacinaceae have anatropous ovule, Polygonum type of embryo sac, and Nuclear endosperm. The ovules of *Gomphandra* and *Gonocarpus* are unitegmic, but in *Phytocrene* (Fagerlind 1945d) two primordia of integuments are present and the integuments are free at the micropyle. The development of endosperm is of the Nuclear type. The formation of protoplasmic connections between the chalazal nucellar cells, due to the activity of endosperm haustorium in *Nothapodytes*, is unique.

34 Rhamnales

34.1 General Features

The order Rhamnales comprises Rhamnaceae, Vitaceae and Leeaceae. These families show close resemblance in several major respects: secretory multinucleate tapetum, simultaneous cytokinesis, three-colporate pollen grains, presence of hypostase, massive parietal tissue, nucellar beak in Rhamnaceae and Vitaceae, Polygonum type of embryo sac, ephemeral antipodals, Nuclear type of endosperm, Asterad type of embryogeny, ruminate endosperm, and perichalazal seed in Vitaceae and Leeaceae.

34.2 Rhamnaceae

The anther wall comprises the epidermis, fibrous endothecium, two ephemeral middle layers, and secretory tapetum of two- to four-nucleate cells. Granular markings on the inner surface of tapetal cells are reported in *Zizyphus mauritiana* (Kajale 1944b). Only the fibrous endothecium remains intact in the mature anther. The microspore mother cells divide meiotically, and produce tetrahedral and isobilateral tetrads. Simultaneous cytokinesis occurs by centripetal furrowing.

The pollen grains are two-celled at the dispersal stage. They are three-colporate with reticulate exine.

The ovule is anatropous, bitegmic and crassinucellate. The micropyle is organized by the outer integument. The hypostase in *Gouania microcarpa* (see N.C. Nair 1970b) is massive, and projects as a cone into the nucellus. An epistase is present in *Zizyphus glabrata* (see N.C. Nair 1970b). The funicular vascular strand reaches almost up to the tip of the outer integument, except in *Berchemia polyphylla* var. *leioclada* (Kajale 1944b, see also N.C. Nair 1970b) where it terminates at the chalaza. The archesporium consists of one, two or more cells. In *Z. glabrata* there are 12–16 cells. These cells cut off parietal cells which produce a massive parietal tissue. The cells of the nucellar epidermis also divide, and add to the parietal tissue. In *Zizyphus rotundifolia* (Arora 1953), the nucellus forms a beak protruding through the micropyle. The megaspore mother cell undergoes meiotic divisions, and produces a linear or T-shaped tetrad. The chalazal megaspore develops into a Polygonum type of embryo sac in *Colletia cruciata, C. spinosa, Hovenia dulcis, Paliurus aculeatus* and *Rhamnus alaternus* (Dolcher 1947). Allium type of development is reported in *Zizyphus cenoplia* (see N.C. Nair 1970b), *Z. mauritiana* (Kajale 1944b), *Z. rotundifolia* (Arora 1953), and *Z. sativa* (Chiarugi 1930). The antipodals are generally ephemeral, but they persist in *Z. rotundifolia* and are coenocytic and four-nucleate in *Z. sativa* (Srinivasachar 1940b). Supernumerary antipodals are present in *Paliurus* (Dolcher 1947). The polars fuse at the time of fertilization. Multiple embryo sacs have been observed in *Colletia ferox, Gouania tillaefolia, Noltea africana, Ventilago bombaiensis* and *V. caluculata* (see N.C. Nair 1970b), *Hovenia dulcis* (Dolcher 1947), *Zizyphus mauritiana* (Kajale 1944b) and *Z. rotundifolia* (Arora 1953).

The remnants of pollen tubes persist in *Z. rotundifolia.*

The endosperm is of the Nuclear type. Wall formation commences from the micropylar region at an early stage of proembryo in *Zizyphus*, and the entire endosperm becomes cellular except in *Rhamnus wightii* where the nuclei in the chalazal region fuse into one or more irregularly shaped polyploid nuclei.

The embryogeny conforms to the Asterad type in *Ceanothus azureus* (Souèges 1939c) and *Rhamnus frangula* (Souèges 1941c). It is Asterad, rarely Solanad, type in *Zizyphus* (see N.C. Nair 1970b).

The embryogeny (Asterad type) of *Rhamnus* corresponds to Period I, Series B, Megarchetype II (Souèges 1941c).

Synergid polyembryony is reported in *Z. rotundifolia* (Arora 1953).

The seed is exotestal with a straight embryo. The testa is firm, smooth and multiplicative.The outer epidermis is a palisade of thick-walled cells with linea lucida separating the upper lignified part of the cell from the lower thin-walled region. The mesophyll is aerenchymatous with scattered sclerotic cells as in *Ventilago*, or is crushed. The tegmen also gets crushed, except for the inner epidermis. The rectangular cells of the inner epidermis have scalariform thickenings on radial walls.

N Arora (1953), A Chiarugi (1930), T Dolcher (1947), LB Kajale (1944b), NC Nair (1970b), EC Prichard (1955), R Souèges (1939c, 1941c), D Srinivasachar (1940b)

34.3 Vitaceae

The anther wall comprises the epidermis, fibrous endothecium, two ephemeral middle layers, and secretory tapetum. The tapetal cells are multinucleate, their nuclei fuse and become polyploid. The tapetum of binucleate cells produces a periplasmodium in *Cayratia carnosa* (Adatia et al. 1950). The microspore mother cells undergo meiotic divisions resulting in tetrahedral and, rarely, isobilateral and decussate tetrads. The cytokinesis is simultaneous.

The pollen grains are tricolporate with striate exine, and are dispersed at the three-celled stage, except in *Cayratia carnosa* and *Cyphostemma setosum* (N.C. Nair and Bajaj 1966) where they are dispersed at the two-celled stage.

The ovule is anatropous, bitegmic and crassinucellate, the inner integument forms the micropyle. In *Cissus trilobata* (N.C. Nair 1965), the much-elongated inner integument is thrown into folds. The hypostase is prominent, and well developed. In *Ampelocissus latifolia* and *Cayratia carnosa* (Kashyap 1958), the cells at the basal region of the ovary elongate and function as obturator. It develops by proliferation of the placental cells in *Cissus quadrangularis* (Goyal 1959, N.C. Nair and Mani 1960). The cells lining the ovarian cavity become richly cytoplasmic in *Cissus pallida* (N.C. Nair and Parasuraman 1962). The archesporial cell cuts off a parietal cell, the parietal tissue is well developed. The cells of the nucellar epidermis divide periclinally and produce a beak. The nucellus persists for a fairly long time. The megaspore mother cell undergoes meiotic divisions, and produces a linear or T-shaped tetrad. The chalazal megaspore develops into a Polygonum type of embryo sac. The synergids are hooked in *Ampelocissus latifolia* (Kashyap 1958) and *Cyphostemma stetosum* (N.C. Nair and Bajaj 1966), and larger than the egg. The antipodals are ephemeral. The polar nuclei fuse before fertilization. Supernumerary nuclei have been noted at the antipodal end in *Cayratia carnosa* (Adatia et al. 1950, Kashyap 1955); this may be due to the migration of nucellar cells or nuclei into the embryo sac (Kashyap 1955). Of special interest is the elongation of the tip of the embryo sac in *Cissus pallida* (N.C. Nair and Parasuraman 1954, 1962) and *C. vitiginea* (N.C. Nair unpublished), through the micropyle, until it occupies the narrow space between the ovule and ovary wall. Finally, it becomes closely appressed to the richly cytoplasmic inner epidermis of ovary wall and, probably, draws nourishment from it. In *Cissus quadrangularis* (Goyal 1959), it becomes extramicropylar and lodges itself in the obturator. Twin embryo sacs, derived from two adjacent megaspore mother cells, have been observed in *Vitis pallida* (N.C. Nair and Parasuraman 1954).

One synergid is destroyed during the passage of the pollen tube. Accessory pollen tubes are present in *C. pallida*.

Development of the endosperm is of the Nuclear type, wall formation is initiated at the chalazal end, and then at the micropylar end in *Vitis* (Kashyap 1958). The central region remains free-nuclear for a considerable period. Finally, it becomes completely cellular and stores fatty food reserve. Each cell contains a druse.

The endosperm becomes ruminate due to localized ingrowths of the integument. The formation of ruminate endosperm is of the Annona type (Periasamy 1962b). The seed becomes perichalazal after fertilization. In *Cayratia* the ingrowths are dome-shaped with two small longitudinal ridges on the inner side. In *Ampelocissus, Cissus, Tetrastigma* and *Vitis*, two ingrowths develop away from the chalaza. The ingrowths are plate-like and "hanging" in *Cissus*, ridge-like in *Ampelocissus*, narrow at the base and widening out at the free-end in *Tetrastigma*, and plate-like but not hanging in *Vitis*. In *Cyphostemma* (N.C. Nair and Bajaj 1966), the two plate-like ingrowths have lateral branches.

The embryogeny conforms to the Asterad type-Geum variation. There is natural parthenocarpy in *Vitis*. The fruit lacks the embryo and endosperm, but attains normal size due to the growth of the ovary wall.

The perichalazal (except *Cayratia*) seed is endospermic. The small embryo has flat thin cotyledons and a short radicle. The testa is multiplicative. The outer portion is thin-walled, and the mesophyll contains bundles of raphides. The inner epidermis divides periclinally, becomes two to six cells thick, and differentiates as woody endotesta (N.C. Nair 1970c). It comprises thick-walled, pitted, lignified cells with crystals. The tegmen has an outer tracheidal part, and the inner epidermis of oblong cells with thick brownish walls.

RD Adatia, BN Mulay and GR Hingorani (1950, 1953), F Gagnepain (1910), E Gilg (1896), S Goyal (1959), G Kashyap (1955, 1958), BN Mulay, NC Nair and MSRK Sastry (1953), NC Nair (1965, 1970c), NC Nair and YPS Bajaj (1966), NC Nair and KV Mani (1960), NC Nair and V Parasuraman (1954, 1962), NC Nair and PL Suri (1957), JE Planchon (1887), K Periasamy (1962a, b), K Süssenguth (1953)

34.4 Leeaceae

The anther wall comprises the epidermis, fibrous endothecium, two ephemeral middle layers, and glandular tapetum of multinucleate cells; the nuclei fuse and become polyploid. The endothecium alone persists. The microspore mother cells undergo meiotic divisions, cytokinesis is simultaneous, and tetrads are tetrahedral, isobilateral and decussate.

The pollen grains are three-colporate with smooth exine, and three-celled at the dispersal stage in *Leea sambusina* (N.C. Nair and Nambisan 1957).

The ovule is anatropous, bitegmic and crassinucellate. Both the integuments form the micropyle. The hypodermal archesporial cell cuts off a parietal cell, the parietal tissue is six to eight layers thick. The nucellar epidermis divides periclinally and produces two or three layers. A hypostase develops. The megaspore mother cell undergoes meiotic divisions, produces a linear tetrad, and the chalazal megaspore develops into a Polygonum type of embryo sac. The antipodals are ephemeral. The polar nuclei fuse before fertilization.

The development of the endosperm is of the Nuclear type. It is quiescent in the beginning but, later, becomes aggressive and destroys the nucellus. The cells become thick-walled, and store fatty substances. Each cell contains a druse. The seed shows extensive perichalazal growth. The perichalaza is pushed into the seed as a broad, forked structure. Four more ingrowths are produced due to the meristematic activity of the middle layer of the outer integument, two in the region of raphe and two laterally. They cause ruminations in the endosperm.

The development of the embryo follows the Asterad type. The embryo is dicotyledonous and small.

The seed is endotestal, and the testa is multiplicative. The outer epidermis comprises slightly thick-walled cells. The mesophyll contains tannin. The inner epidermis divides and produces four layers of cells which are lignified, and enclose a single simple crystal in each cell. The outer epidermis of tegmen is spirally thickened, while the inner contains tannin.

EJH Corner (1976), NC Nair (1968, 1970d), NC Nair and PN Nambisan (1957)

34.5 Taxonomic Considerations

Rhamnaceae. Bentham and Hooker (1965a) placed Rhamnaceae and Vitaceae (*Leea* is a genus in Vitaceae) along with Celastrineae and Stackhousiaceae in the cohort Celastrales. Wettstein (1935), Engler and Diels (1936), Rendle (1959), Lawrence (1951) and Melchior (1964) placed Rhamnaceae with Vitaceae in a distinct order, Rhamnales. Hutchinson (1973) included Heteropyxidaceae and Elaeagnaceae, along with Rhamnaceae and Vitaceae, in Rhamnales, and treated *Leea* as a genus in Vitaceae. Bessey (1915), Prichard (1955) and Ganapathy and Subramanyam (1962, see N.C. Nair 1970b) consider that Rhamnaceae are related to Vitaceae, and show many features common to Celastraceae.

Rhamnaceae are characterized by two-celled pollen grains with reticulate exine, a vascular strand traversing the outer integument up to its tip, epistase, uni- or multi-celled archesporium, and Allium type of embryo sac in the genus *Zizyphus*. The antipodal cells multiply and are coenocytic. Synergid polyembryony and, rarely, Solanad type of embryogeny occur in the family. The seed is exotestal, and has a palisade with linea lucida in the outer epidermis of the testa. None of these features are present in Vitaceae or Leeaceae.

The allied families Vitaceae (Sect. 34.3) and Leeaceae (Sect. 34.4) differ from Rhamnaceae (Sect. 34.2) in several respects. R. Dahlgren (1977a) points out: "Rhamnales should be divided. Vitaceae and Leeaceae being placed in a separate order." This is quite justifiable on embryological grounds.

How far are Rhamnaceae related to Stackhousiaceae and Celastraceae, as treated by Bentham and Hooker (1965a)? Stackhousiaceae differ from Rhamnaceae in a single middle layer in anther wall, tenuinucellate ovules on basal placenta, absence of nucellar beak, and single-celled female archesporium, the embryo sac consumes inner integument, and seed-coat is formed by the outer integument alone. Similarities between the two are extension of funicular vascular strand beyond the chalaza, Polygonum type of embryo sac, Nuclear type of endosperm, and Asterad type of embryogeny (Mauritzon 1936c, Narang 1953).

Rhamnaceae stand out from Celastraceae in crassinucellate ovules with a vascular strand passing through the integument, a prominent hypostase, Allium type of embryo sac in *Zizyphus*, fusion of polars at the time of fertilization, and in the structure of seed. Therefore, Rhamnaceae need not be placed in Celastrales along with Celastraceae and Stackhousiaceae.

Palynologically, Rhamnaceae and Elaeagnaceae are similar. Corner (1976) reports similarity between their seed-structure. Some embryological characters like the presence of a multi-nucleate secretory tapetum, tricolpate pollen grains, anatropous, bitegmic and crassinucellate ovules, and of a well-developed hypostase are features common to the two families (Rau and Sharma 1970b). Yet, the presence of a basal ovule, absence of integumentary vascular strand, parietal tissue in the ovule, polars fusing prior to fertilization, and chalazal embryo sac haustorium indicate that Elaeagnaceae are different from Rhamnaceae.

Vitaceae. Vitaceae, on the other hand, have two or three-celled pollen grains, and rarely amoeboid tapetum, the inner integument is thrown into folds, and extra-nucellar embryo sacs occur in *Cissus*. The obturator, proliferation of nucellar tissue which collapses in the seed, and parthenocarpy are common features in Vitaceae. The micropyle is organized by the inner integument in Vitaceae and Rhamnaceae. The endosperm is quiescent in the beginning in Vitaceae and Leeaceae, stores fatty substances with a druse in each cell, and is ruminate. The seed shows perichalazal growth, sarcotesta with raphides, woody endotesta, and tracheidal exotegmen.

Planchon (1887, see N.C. Nair 1970c) divided *Vitis* into a number of genera. *Cissus* was divided into three sections: *Eucissus*, *Cayratia* and *Cyphostemma*. Gagnepain (1910) removed *Cayratia* to a generic rank, and retained *Cyphostemma* under *Cissus*. In *Cissus* the embryo sac is haustorial and extends beyond the ovule by absorbing the nucellus (N.C. Nair and Parasuraman 1962). The seed structure differs from that of *Cyphostemma*. In *Cyphostemma setosum* (N.C. Nair and Bajaj 1966), eight to ten layers of nucellar cells persist at the micropyle. The ruminations take place by two median and two lateral folds. The inward projections extend throughout the major length of the dorsal side. In *Cissus* there are two median folds which are oblique, the lateral fold is absent. The perichalazal growth is along the median plane nearly up to the micropylar region on the dorsal side, whereas in *Cyphostemma* it extends throughout the major length on the dorsal side and reaches only up to the ventral side (N.C. Nair and Bajaj 1966). The seed-coat has brownish contents in *Cissus*, but not in *Cyphostemma*. In *Cayratia* the ingrowth in the seed is a single dome-shaped structure with two small longitudinal ridges on the inner side.

Planchon (1887) removed *Leea* from Vitaceae; Dumortier (see Süssenguth 1953) and Bartling (1830) gave *Leea* a family rank, and placed Leeaceae between Vitaceae and Meliaceae (see N.C. Nair 1970d).

N.C. Nair (1970c) points out that in the organization of perichalaza, basal growth of seed after fertilization, formation of ingrowths from free integumentary portion, enormous increase of nucellus after fertilization, late development of endosperm, and in the rumination of seed, Vitaceae resemble Annonaceae (Corner 1949, Periasamy 1962a, Periasamy and Swamy 1961).

Leeaceae. The three-celled pollen grains at the dispersal stage, a small nucellar cap, long persistence of nucellus, micropyle formed by both the integuments, an extensive perichalazal growth with four ingrowths in addition to two chalazal ingrowths, and absence of sarcotesta with raphides in seed distinguish Leeaceae from Vitaceae (N.C. Nair and Nambisan 1957). Concerning rumination of the endosperm, Periasamy (1962b) concludes that *Leea* differs from the other genera of Vitaceae by the large number of ingrowths and a pronounced activity of nutritive layers of the outer integument in the mature seed-coat. According to Periasamy, *Leea* need not be removed from Vitaceae and should be considered as a genus lowest in the phylogenetic series. N.C. Nair (1968), on the other hand, discusses several morphological and embryological similarities and dissimilarities, and places it close to Vitaceae under a monogeneric family Leeaceae.

35 Malvales

The Malvales comprise four suborders: Elaeocarpineae – Elaeocarpaceae; Sarcolaenineae – Sarcolaenaceae; Malvineae – Tiliaceae, Malvaceae, Bombacaceae and Sterculiaceae; and Scytopetalineae – Scytopetalaceae.

35.1 General Features

The order Malvales is characterized by: (1) secretory anther tapetum, amoeboid in Malvaceae, (2) simultaneous divisions in pollen mother cells, (3) two-celled pollen, (4) anatropous, bitegmic, crassinucellate ovule with well-developed parietal tissue, tenuinucellar in Scytopetalaceae, (5) nucellar cap, absent in Elaeocarpaceae and Scytopetalaceae, (6) Polygonum type of embryo sac with ephemeral antipodals, supernumerary in Malvaceae, (7) hypostase, (8) fusion of polar nuclei at the time of fertilization, (9) Nuclear type of endosperm, and cytokinesis at a very late stage of embryo development, (10) Asterad type of embryogeny, Onagrad in *Muntingia* (Elaeocarpaceae), (11) endospermic seed with a straight embryo, curved in Malvaceae, and (12) seed with outer epidermis of tegmen comprising lignified, longitudinal fibres.

The order is homogeneous as suggested by Venkata Rao and Sambasiva Rao (1952), and Venkata Rao (1954a, b).

35.2 Elaeocarpaceae

The anther wall comprises the epidermis, endothecium, two or three ephemeral middle layers, and secretory tapetum of binucleate cells. The epidermal cells have spinescent structures, and contain tannin in *Elaeocarpus* (Venkata Rao 1953b). In this taxon the endothecial cells develop into stone cells with thick walls and prominent pits (Venkata Rao 1953b). The endothecium has fibrous thickenings in *Muntingia* (Venkata Rao 1951a). The microspore mother cells divide meiotically, and produce tetrahedral tetrads. Cytokinesis is simultaneous.

Pollen grains are three-colporate and two-celled at the shedding stage.

In *Aristotelia*, *Elaeocarpus* and *Muntingia*, the ovule is anatropous, bitegmic and crassinucellate with a zigzag micropyle formed by both integuments. The archesporial cell cuts off a parietal cell, and six- to seven-layered parietal tissue develops in *Muntingia* (Venkata Rao 1952a). The megaspore mother cell undergoes meiotic divisions and produces a linear tetrad. The chalazal megaspore develops into Polygonum type of embryo sac. In *Aristotelia manqui* and *A. raceskosa* (Mauritzon 1934c), usually the subchalazal megaspore functions. The polars fuse before fertilization. The antipodals are ephemeral. An endothelium and hypostase develop in *Elaeocarpus*. A socket of thick-walled cells is present around the chalazal end of the embryo sac in *Muntingia*.

The development of the endosperm is of the Nuclear type. The nuclei accumulate in the chalazal region. The endosperm becomes cellular at the quadrant stage of proembryo, and the mature seed is endospermic.

The embryogeny conforms to the Onagrad type. The mature embryo is straight with thick cotyledons.

The seed is albuminous, and the testa and tegmen are multiplicative. The exotesta is mostly thick-walled. The outer epidermis is hairy, and the inner one contains crystals. The outer epidermis of the tegmen comprises longitudinal lignified fibres. The mesophyll is aerenchymatous (densely sclerotic in *Aristotelia* – Corner 1976), and the inner epidermis has thick brownish walls.

EJH Corner (1976), J Mauritzon (1934c), C Venkata Rao (1951a, 1952a, 1953b)

35.3 Sarcolaenaceae

The embryology has not been studied.

The anthers are bilocular with longitudinal dehiscence. The ovule is anatropous, erect to more often pendulous (Hutchinson 1973, Cronquist 1981). The seed is hairy in *Xerochlamys* (Hutchinson 1973). It contains a fleshy or horny endosperm, and the embryo is straight with flat and plaited cotyledons.

A Cronquist (1981), J Hutchinson (1973)

35.4 Tiliaceae

The anther wall comprises the epidermis, fibrous endothecium, ephemeral middle layer, and secretory tapetum of two- to four-nucleate cells. The epidermal cells contain rosette crystals in *Corchorus* and *Triumfetta* (Venkata Rao and Sambasiva Rao 1952). The microspore mother cells undergo meiotic divisions, and tetrahedral and isobilateral tetrads are produced. Simultaneous cytokinesis occurs by furrowing (Banerji 1933).

The pollen grains are two-celled at the dispersal stage. The exine is reticulate.

The ovule is anatropous, bitegmic and crassinucellate. The micropyle is organized by the outer integument, the growth of the inner integument is arrested. The nucellar epidermal cells divide periclinally and produce three layers. The archesporium is multicellular in *Corchorus capsularis* (Banerji 1932), only one cell functions. In *Triumfetta rhomboidea* four or five cells are functional. The hypodermal archesporial cell cuts off a parietal cell and the parietal tissue is five layers thick. The megaspore mother cell undergoes meiotic divisions, tetrads are linear or T-shaped, and chalazal megaspore develops into a Polygonum type of embryo sac. The synergids occupy half the length of the embryo sac in *Corchorus* (Banerji 1932). The polar nuclei do not fuse till fertilization. The antipodals are ephemeral.

In *T. rhomboidea* the obturator is hairy.

The development of endosperm is of the Nuclear type. Wall formation commences from the micropylar end at the initiation of cotyledonary primordia in the embryo, and the endosperm becomes cellular throughout.

The development of embryo is of the Asterad type-Urtica variation. Polyembryony is reported in *C. olitorius*.

Ganesan et al. (1957) state that in the cross *Corchorus olitorius* × *C. capsularis*, a hybrid embryo and Nuclear endosperm develop after normal fertilization. However, the embryo aborts at an early globular stage. They suggested that probably somatoplastic sterility is involved. Iyer (1962) has demonstrated that there is no somatoplastic sterility.

The development of endosperm and embryo has been studied in *Corchorus olitorius*, *C. capsularis*, *C. olitorius* × *C. capsularis*, and *C. capsularis* × *C. olitorius* (Iyer et al. 1961, Iyer 1962). The developmental stages in selfed *C. capsularis* are essentially similar to those of *C. olitorius*. In the latter the endosperm remains free-nuclear up to 11 days after pollination, when the proembryo is globular. Thirteen days after pollination the embryo becomes heart-shaped and the upper half of the embryo sac contains cellular endosperm; fifteen days later the entire endosperm becomes cellular and the embryo has well-differentiated cotyledons. Twentyone days after pollination the cotyledons enlarge, and 30 days after pollination the cotyledons expand and become leafy. Much of the endosperm is now consumed.

In the selfed *C. capsularis*, the embryo matures earlier by almost a week or 10 days, and is somewhat bigger.

In the cross *C. olitorius* × *C. capsularis*, fertilization and early embryogeny are comparable to the reciprocal cross, and the two parents. In *C. olitorius* × *C. capsularis*, the endosperm remains free-nuclear from seven to 30 days after cross-pollination, and the proembryo does not grow beyond the eight- to ten-celled stage. Vasculation in the proembryo occurs 11 days after cross-pollination, and the endosperm shrinks and collapses on the fifteenth day. The seeds are shrivelled with aborted embryo and remnants of endosperm.

In the reciprocal cross *C. capsularis* × *C. olitorius*, the flowers abscised and dried up. The ovules showed that the pollen tubes reached the micropyle but fertilization did not take place.

In *C. capsularis* grafted on *C. olitorius* root-stock, there was an increase in the percentage of fruit-set and marked development of hybrid embryo and endosperm. The globular proembryos were much larger than in straight crosses, but they failed to differentiate and aborted. The endosperm also remained free-nuclear until degeneration after 3 weeks.

When *C. olitorius* was pollinated with *C. capsularis* grafted on the former, occasionally late stages of heart-shaped embryos with cellular endosperm were observed.

The seed is endospermic with a straight embryo. The testa is membraneous, the inner epidermis contains crystals. The outer epidermis of the tegmen comprises a palisade of thick-walled lignified cells. The mesophyll is aerenchymatous and gets crushed. The inner epidermal cells are cuboidal with brownish walls.

I Banerji (1932, 1933), MR Chaudhury (1961), RM Datta (1954, 1955), AP Ganesan, SS Shah and MS Swaminathan (1957), RD Iyer (1962), RD Iyer, K Sulbha and MS Swaminathan (1961), J Mauritzon (1934d), C Venkata Rao and KV Sambasiva Rao (1952), R Souèges (1941b)

35.5 Malvaceae

Venkata Rao (1954a, b, 1955a) has reviewed the embryology of Malvaceae, and P. Maheshwari and Rangaswamy (1960) of *Gossypium*.

The anther is bisporangiate, and its wall comprises the epidermis, fibrous endothecium, an ephemeral middle layer, and amoeboid tapetum which forms the periplasmodium. The microspore mother cells divide meiotically, and produce isobilateral and tetrahedral tetrads. Simultaneous cytokinesis occurs by furrowing.

The mature pollen grains are two-celled at the dispersal stage, and packed with starch. They are three-colporate with reticulate or spinous markings. They contain fatty food reserve and, in later stages, starch, as in *Althaea* and *Malva* (Luxemburg 1927). Occasionally, some dyads persist up to the mature stage in *Malvastrum tricuspidatum*, and tetrads in *Hibiscus labatus* (Gaur 1979). This needs confirmation.

The ovule is campylotropous, bitegmic and crassinucellate with a zigzag micropyle formed by both the integuments. A hairy obturator of stylar origin develops in *Abutilon indicum*, and from the placenta in *Hibiscus micranthus* (Venkata Rao 1955a). A hypostase also differentiates. Numerous stomata occur on the surface of the ovary. The archesporium is uni- to multicellular in *Gossypium* (P.C. Joshi et al. 1967). The archesporial cell cuts off a parietal cell; the parietal tissue is massive, and a nucellar cap develops. The nucellus persists as perisperm. Accessory sporogenous cells differentiate from the parietal tissue, only one cell functions. The megaspore mother cell divides meiotically, and produces a linear or T-shaped tetrad. The chalazal megaspore develops into a Polygonum type of embryo sac. The antipodals undergo supernumerary divisions forming as many as 19–24 cells in *G. herbaceum* (P.C. Joshi et al. 1967) and these persist up to fertilization.

Double fertilization occurs, and is porogamous. Venkata Rao (1952a) points out that the pollen tube is persistent, and serves as a haustorial organ in *Hibiscus hirsutus*, *H. micranthus*, *H. solandra*, *Malachra* (= *Malache*) *capitata*, *Pavonia zeylanica* and *Sida veronicaefolia* (*S. humilis*). The haustorial function of the pollen tube has been questioned (Sect. 3.9), and it is only the remnants of the discharged tube that persist during post-fertilization stages, sometimes even in a mature seed (see P. Maheshwari and Johri 1950b).

The pollen tube branches in several species (Venkata Rao 1952a). Polysiphonous germination of pollen has been observed in *Hibiscus esculentus* (Purewal and Randhawa 1947) and *Malva neglecta* (Stenar 1925b), as many as 14 pollen tubes have been noted in the latter species. In *Abelmoschus esculentus* (S. Chandra and Bhatnagar 1975a, b), syngamy and triple fusion take place simultaneously.

The time-relations concerning pollination, growth of pollen tube, and fertilization have been studied in *Gossypium arboreum* and *G. hirsutum*, and their reciprocal crosses (Pundir 1972). The stigma is papillate, and the pollen tube grows through the single transmitting strand, along the placenta, and enters the zigzag micropyle. Generally, only a single pollen tube emerges from a germinating pollen grain, but polysiphonous conditions often occur in *G. arboreum*. The bulging of tubes is a common feature in *G. hirsutum*, and coiling of tube has been observed in a cross *G. hirsutum* × *G. arboreum*. The interval between pollination and fertilization is 24–48 h in *G. arboreum*, *G. hirsutum* and *G. arboreum* × *G. hirsutum*, but 72–120 h in the cross *G. hirsutum* × *G. arboreum*.

The endosperm is of the Nuclear type and several hundred nuclei are formed before wall formation is initiated at the late globular stage of proembryo in *Hibiscus micranthus*, and at the dicotyledonous stage in *Pavonia zeylanica* (Venkata Rao 1955a). A few layers of endosperm persist in *Abutilon indicum*, *Pavonia zeylanica*, *Sida cordifolia* and other taxa.

In *Gossypium arboreum*, *G. hirsutum* and *G. arboreum* × *G. hirsutum*, the primary endosperm nucleus divides 24 h after pollination but after 72–120 h in *G. hirsutum* × *G. arboreum* (Pundir 1972). In the parent species about 2000 free nuclei are produced on the seventh or eighth day after pollination, whereas in the hybrid the endosperm is 100- to 200-nucleate on the fifth day. A chalazal diverticulum develops in *G. arboreum.* The nuclei become polyploid due to endomitosis and nuclear fusions, and centripetal wall formation is initiated at the micropylar pole on the seventh or eighth day (after pollination) in the parent species. In the hybrid the endosperm nuclei become abnormally large and dumb-bell-shaped and form clumps. The cytoplasm becomes vacuolate, and the endosperm degenerates 10 to 14 days after pollination; wall formation never occurs in the hybrid endosperm. In *Gossypium* (P.C. Joshi et al. 1967), only five or six layers of endosperm persist in the seed.

In *G. arboreum* a young dicotyledonous embryo develops 15 or 16 days after pollination, and 17 or 18 days in *G. hirsutum.* The embryo aborts 12–14 days after pollination in the cross *G. arboreum* × *G. hirsutum,* and after 7 days in the reciprocal cross. The hybrid embryo remains small and shows many abnormalities (Pundir 1972).

S. Chandra and Bhatnagar (1975a, b) have made a detailed time-relation study of post-fertilization development of the ovule in *Abelmoschus esculentus.* The primary endosperm nucleus divides 10–12 h after pollination, while the zygote divides after 16–18 h. Free-nuclear divisions continue in the endosperm until about 2000 nuclei are formed, 5 days after pollination. Synchronous divisions occur only up to the 64- or 128-nucleate stage. At this time the proembryo is globular.

Five days after pollination, centripetal wall formation commences from the micropylar region, proceeds to chalaza, and continues up to 8 days after pollination. Concurrently, cotyledons are initiated, and 7 days after pollination the embryo becomes dicotyledonous, and the root cap also differentiates. The suspensor comprises seven to ten cells.

Eight to 11 days after pollination, the embryo is fully differentiated. The cotyledons become leafy. During the growth of the embryo, most of the starch-rich endosperm is consumed. Finally, in a mature seed, the embryo occupies most of the space, the cotyledons curl, and remnants of endosperm persist in the funicular region and a layer or two between the folds of the cotyledons.

The embryogeny conforms to the Asterad type. Mortier (1969) assigned the development of embryo in *Althaea officinalis* to Period I, Megarchetype II (Asterad type). The mature embryo is dicotyledonous, and the cotyledons become foliar with an inwardly curved margin in *Gossypium.* During elongation, they curl around each other and around the hypocotyl (Fig. 35.5.2A).

Polyembryony has been observed in *G. barbadense* (P.C. Joshi et al. 1967, B.M. Johri and Pundir 1976); haploid-diploid twins are also reported in this taxon (Harland 1936).

The outer epidermis of testa comprises cubical cells with thick, lignified outer walls, and contains tannin (Fig. 35.5.2B). In *Gossypium* (P. C. Joshi et al. 1967), some of the cells develop into hairs (fibres) except in the region of the micropyle. The development of hairs is in three successive crops: at anthesis, 4 to 6 days after pollination, and 8 to 12 days after pollination (Fig. 35.5.1A–I). At maturity the fibres are distinguishable into lint and fuzz. The lint fibres are longer with thinner walls, and show convolutions on the surface. The convolutions are absent on fuzz fibres. The mesophyll contains tannin cells, and is followed by a starch-containing inner epidermal layer. The tegmen is multipli-

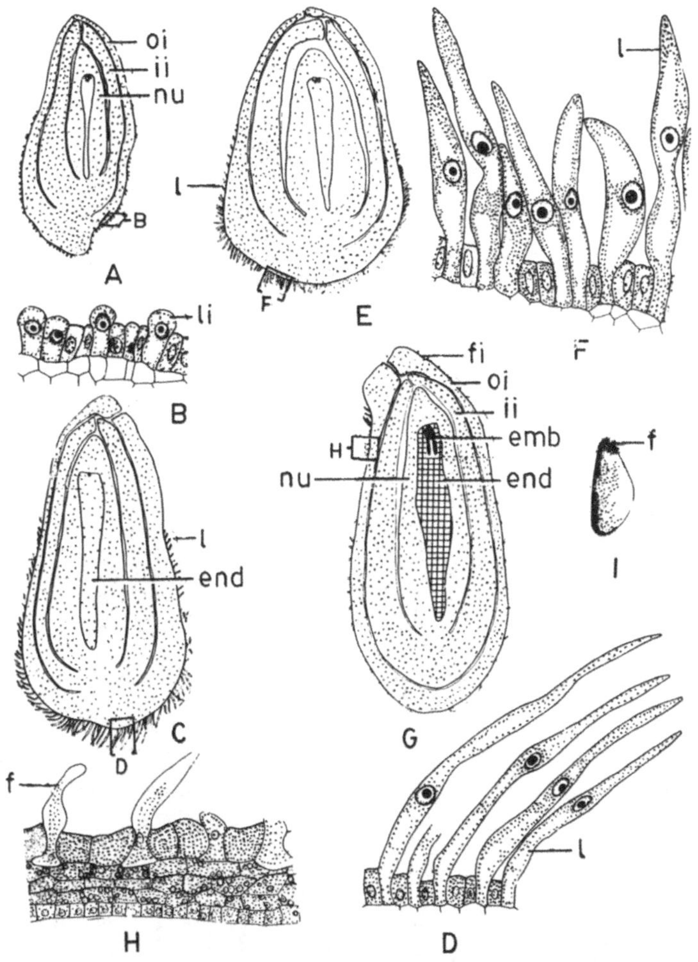

Fig. 35.5.1A–I. *Gossypium barbadense* var. Andrews, development of fibres; *emb* embryo; *end* endosperm; *f* fuzz hairs; *fi* fuzz initial; *ii* inner integument; *l* lint hair; *li* lint initial; *nu* nucellus; *oi* outer integument. **A, C, E, G** Outline diagrams for **B**; **D, F, H**, respectively. The ovules represent stages at anthesis (**A**), and two (**C**), five (**E**) and 10–12 (**G**) days after pollination. **B, D, F** *Inset* from **A, C** and **E**, development of lint hairs. **H** *Inset* from **G**, differentiation of fuzz. **I** Mature seed after the removal of lint; fuzz is mostly limited to the micropylar end. (After P.C. Joshi et al. 1967)

cative (Fig. 35.5.2B). The outer epidermis is a palisade layer of lignified pitted cells (Fig. 35.5.2B–E). In surface view the cells show radiating thickenings in the lumen. The mesophyll shows an outer tanniniferous, starchy zone of four to five layers, and an inner colourless starchy zone of nine or ten layers in *Gossypium*. The inner epidermis forms the fringe layer (Fig. 35.5.2F). Its cells do not contain nuclei and cytoplasm, and the walls develop thickenings.

In a cross *G. arboreum* × *G. hirsutum*, and in the reciprocal cross, the development of seed-coat and fibers is arrested 7 to 10 days after pollination (Pundir 1972).

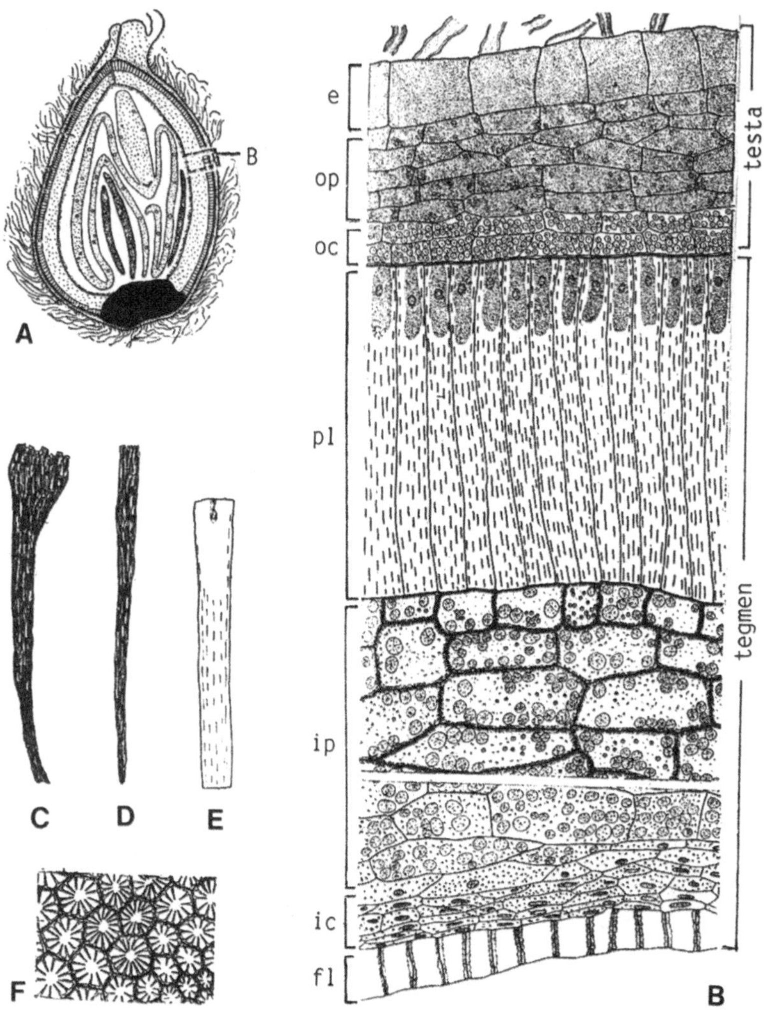

Fig. 35.5.2A–F. A, B, E, F *Gossypium herbaceum* var. *Jaydhar*, **C, D** *G. arboreum* var. *Bhoj*. **A** Longisection of mature seed studded with hairs, 40 days after pollination. Note well-developed hypostase, remnants of endosperm, and embryo with highly coiled cotyledons. **B** *Inset* B from **A**, to show pigmented hairy epidermis (*e*) of testa, outer pigmented zone (*op*), and outer colourless zone (*oc*). The palisade layer (*pl*) comprises exotegmen followed by inner pigmented zone (*ip*), inner colourless zone (*ic*); the inner epidermis of tegmen forms a fringe layer (*fl*) with conspicuous thickenings. The outer and inner colourless zone contain compound starch grains. (After Ramachandani et al. 1966). **C–E** Palisade cells of exotegmen (from macerated wholemount). **F** cells of fringe layer (transection) with radiating thickenings on inner wall. (After P.C. Joshi et al. 1967)

BS Ahuja (1964), T Ashley (1975a, b, c), R Balasubramanium (1931), S Chandra and SP Bhatnagar (1975a, b, 1976), HK Chaudhari (1978), PB Christensen (1986), PC Datta and A Chowdhury (1981), EJ Duncan (1970), HL Edlin (1935), GH Egley and RN Paul (1981), PA Fryxell (1968, 1969), RD Gaur (1979), UR Gore (1932), SC Harland (1936), WA Jensen and DB Fischer (1968), WA Jensen, ME Ashton and CA Beasley (1983), WA Jensen, P Schulz and ME Ashton (1977), BM Johri and NS Pundir (1976), PC Joshi, AM Wadhwani and BM Johri (1967), P Kumar and D Singh (1990), U Lantis (1912), A Luxemburg (1927), P Maheshwari and BM Johri (1950b), P Maheshwari and NS Rangaswamy (1960), J Mortier (1969), NS Pundir (1972), SS Purewal and GS Randhawa (1947), S Ramchandani, PC Joshi and NS Pundir (1966), B Singh (1967), RP Singh and MPS Chauhan (1984), R Souèges (1922b), H Stenar (1925b), M Takahashi and J Kouchi (1988), WA van Heel (1966), C Venkata Rao (1952a, 1954a, b, 1955a), DM Winter (1960), CJ Webb (1984)

35.6 Bombacaceae

The anther wall comprises the epidermis, fibrous endothecium, two or three ephemeral middle layers, and secretory tapetum of multinucleate cells, Nuclear fusions occur in the tapetal cells. Übisch granules are studded on the inner walls of tapetum in *Ochroma lagopus* (Venkata Rao 1953a). The microspore mother cells increase secondarily. They undergo meiotic divisions, simultaneous cytokinesis occurs by furrowing, and tetrahedral tetrads are produced.

Pollen grains are two-celled at the shedding stage. They are triporate, and smooth-walled.

The ovule is anatropous, bitegmic and crassinucellar with a zigzag micropyle formed by both the integuments. The hypodermal archesporial cell cuts off a parietal cell. The parietal tissue is massive, and a nucellar cap is formed. A hypostase also develops. Following meiotic divisions, the megaspore mother cell produces a T-shaped tetrad. The chalazal megaspore develops into a Polygonum type of embryo sac (Banerji 1942a), and it contains starch grains. The antipodals are ephemeral. The embryo sac becomes tubular at the chalazal end; a postament develops in *Ochroma* (Venkata Rao 1953a). In *Eriodendron anfractuosum* (Thirumalachar and Razi 1941), sometimes the mature embryo sac shows reversed polarity.

Fertilization is porogamous. The pollen tube branches and persists up to the globular proembryo. Perhaps, it may serve as a means of transport of nutrients from the integuments to the embryo (Venkata Rao 1954a, b). There is no evidence that a discharged pollen tube is concerned in short transport of metabolites. In *Eriodendron* (Thirumalachar and Razi 1941), the chalazal end of the embryo sac becomes haustorial, and penetrates into the underlying tissue.

Development of the endosperm is of the Nuclear type. It becomes cellular at the globular stage of pro-embryo.

The embryogeny conforms to the Asterad type in *Ochroma*, and Caryophyllad in *Bombacopsis glabra* (Duncan 1970). The mature embryo is straight with thin and flat, or thick and folded, cotyledons. Adventive polyembryony is reported in *Bombacopsis* (Duncan 1970).

The seeds are large with a massive seed-coat. Both testa and tegmen are multiplicative. The outer epidermis of testa comprises tangentially flattened

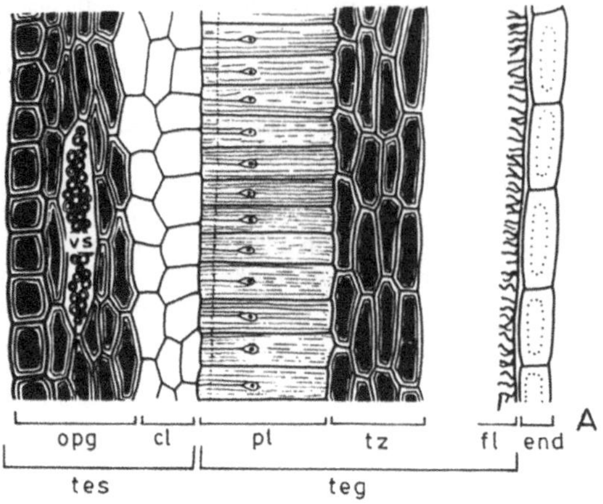

Fig. 35.6.1A–D. *Hampea nutricia*. **A** Part of seed-coat, cross-section; *cl* colourless zone; *end* endosperm; *fl* fringe layer; *opg* outer pigmented layer; *pl* palisade layer; *tes* testa; *teg* tegmen; *tz* tanniniferous zone. **B** Outer epidermis of testa with stomata, surface view. **C** Macrosclereids, cross-section. **D** Fringe layer, surface view. (After R.P. Singh and Chauhan 1984)

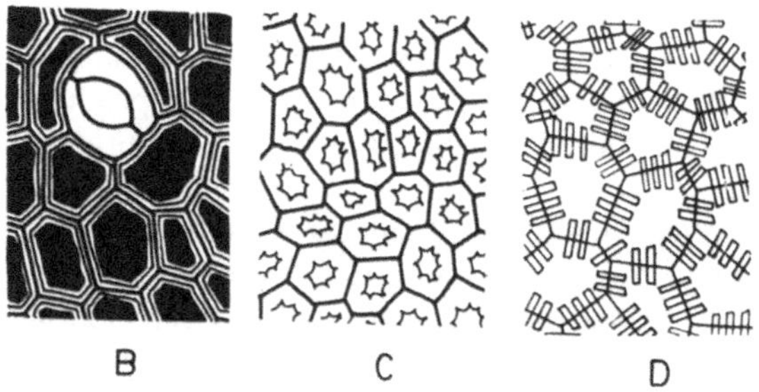

cells with tannin, followed by aerenchymatous mesophyll. The inner epidermis contains crystals. Hairy outgrowths develop from the outer epidermis in many genera. The tegmen consists of a palisade of radially elongate, lignified cells. The mesophyll is aerenchymatous with stellate cells. The inner epidermis is a fringe layer with irregular thickenings on the anticlinal walls, and may contain tannin (Fig. 35.6.1A–D). The chalaza is massive. The aril, when present, is either funicular or micropylar.

I Banerji (1942a), EJ Duncan (1970), BM Johri (1963b), MJ Thirumalachar and BA Razi (1941), C Venkata Rao (1953a, 1954a, b)

35.7 Sterculiaceae

The major contributions to the embryology of Sterculiaceae are by Venkata Rao (1948, 1949, 1950, 1951b, 1952b, 1953c).

The anther wall comprises the persistent epidermis, fibrous endothecium, two or three ephemeral middle layers, and secretory tapetum with multinucleate cells. The tapetum is biseriate in *Dombeya* and *Pterospermum* (Venkata Rao 1949). False periplasmodium has been noted in *Guazuma tomentosa*, *Melochia corcorifolia* (Venkata Rao 1951b) and *Waltheria indica* (Venkata Rao 1949). This needs reinvestigation. The sporogenous tissue increases secondarily. The microspore mother cells undergo meiotic divisions and produce tetrahedral tetrads. Cytokinesis is simultaneous.

Pollen grains are shed at the two-celled stage.

The ovule is anatropous, bitegmic and crassinucellate with a zigzag micropyle in *Dombeya klienhovia, Pentapetes* and *Sterculia* (Venkata Rao 1953c), *Melochia* (Venkata Rao 1951b), *Pterospermum* (Venkata Rao 1952b) and *Waltheria* (Venkata Rao 1950). The micropyle is produced by the outer integument alone in *Abroma, Buettneria* and *Guazuma* (Venkata Rao 1953c). In *Helicteres* (Venkata Rao 1953c), the growth of integuments is arrested, and the nucellus protrudes and forms a cap. The occurrence of integumentary vascular bundles is a common feature. In *Sterculia* (Venkata Rao 1953c), six to nine strands extend up to the micropylar end. The vascular supply has numerous anastomosing branches in *Pterospermum*. A hypostase develops in *Helicteres* and *Pterospermum*, and postament in *Abroma*. An obturator develops from the funicule in *Abroma angusta* and *Sterculia colarata* (Venkata Rao 1953c).

The archesporium is multicellular. As many as 100 archesporial cells have been observed in *Pterospermum suberifolium* (Venkata Rao 1952b), and ten or more embryo sacs may develop concurrently, but only two to three mature. Usually, a single archesporial cell cuts off a parietal cell which produces a parietal tissue. The cells of the nucellar epidermis also divide to form a nucellar cap. The megaspore mother cell divides meiotically and produces a linear tetrad. The chalazal megaspore develops into a Polygonum type of embryo sac. In *Pterospermum* (Venkata Rao 1949, 1952b), a fairly large number of archesporial cells develop directly into megaspore mother cells. Several tetrads are produced. Double and triple eight-nucleate embryo sacs are common. The antipodals are ephemeral. Reversed polarity in the embryo sac has been observed in *Helicteres isora.*

In *Buettneria herbacea* the obturator is hairy. The radially elongate cells of the funicule become richly cytoplasmic in *Abroma, Helicteres* and *Sterculia.*

Double fertilization occurs, and the entry of pollen tube is porogamous.

The endosperm is of the Nuclear type. In *Abroma* and *Pentapetes*, the nuclei in the chalazal region become large and polyploid. Later, the endosperm becomes cellular at the initiation of cotyledonary primordia. The endosperm stores starch and fatty substances.

The embryogeny conforms to the Asterad type. The mature embryo is straight with flat cotyledons. Cleavage polyembryony has been recorded in *Theobroma cacao* (Martinson 1972a). The cleavage of the zygote results in the formation of unequal diploid seedlings. Haploid seedlings are produced after spontaneous development of the egg cell following triple fusion. Martinson (1972b) observed two mature embryo sacs within the same nucellus. A case of two embryo sacs in separate nucelli, but enclosed within common integuments, has also been reported. Twin and multiple embryo sacs have been observed by Venkata Rao (1949, 1952b).

The testa and tegmen are multiplicative. The outer epidermis is of short, palisade cells, or of longitudinally elongate cells with stomata and peltate scales. The mesophyll is pulpy with mucilage sacs, or with lignified stellate cells. The inner epidermis is thin-walled. In winged seeds the wing is formed by the exotesta and mesophyll. The outer epidermis of tegmen comprises longitudinal or short, lignified malpighian cells. The mesophyll is similar to that of the testa, and gets more or less crushed. The inner epidermis may comprise thin-walled or thick-walled cells with bar-like thickenings.

EE Cheesman (1927), EJ Duncan and AW Todd (1972), VA Martinson (1972a, b), PR Mohana Rao (1976), YML Sharma (1938), C Venkata Rao (1948, 1949, 1950, 1951b, 1952b, 1953c)

35.8 Scytopetalaceae

Vijayaraghavan and Dhar (1976) have contributed to our knowledge of embryology of *Scytopetalum tieghemii*, and have studied the male and female gametophytes, seed-coat and pericarp. We have no information on the development of endosperm and embryo.

The anther wall comprises the epidermis, fibrous endothecium, one or two ephemeral middle layers, and secretory tapetum of two- or three-nucleate cells in *Scytopetalum tieghemii* (Vijayaraghavan and Dhar 1976). Druses occur in the connective. The microspore mother cells undergo meiotic divisions, cytokinesis is simultaneous, and the tetrads are tetrahedral. A thick callose wall surrounds the meiocytes, and persists up to the tetrad stage.

The triporate pollen grains are two-celled at the dispersal stage.

The ovule is anatropous, bitegmic and tenuinucellate. The inner integument is longer than the outer one, and forms the micropyle. An endothelium differentiates at the megaspore mother cell stage. It is irregularly biseriate with uninucleate cells.

The archesporial cell functions as megaspore mother cell. The latter divides meiotically and produces a linear tetrad. The chalazal megaspore develops into a Polygonum type of embryo sac. The antipodals are ephemeral.

The seeds are albuminous and exarillate. The endosperm is sometimes ruminate as in *Rhaptopetalum* and *Scytopetalum*. The seed-coat is about ten cells thick, the outer epidermis comprises large cells and the outer wall is almost hair-like in *Oubanguia*. Hairs develop along the raphe in *Brazzeia* and *Rhaptopetalum*, and the seed is completely hairy in *Pierrina*. There is post-chalazal extension of the vascular bundle almost upto the micropyle with branches to the seed-coat entering the ruminations in *Rhaptopetalum* (Corner 1976)

EJH Corner (1976), MR Vijayaraghavan and U Dhar (1976)

35.9 Taxonomic Considerations

Bentham and Hooker (1965a) include Malvaceae, Tiliaceae and Sterculiaceae in Malvales. Engler and Prantl (1931) group all the families – Elaeocarpaceae,

Tiliaceae, Malvaceae, Bombacaceae, Sterculiaceae and Scytopetalaceae – in Malvales. Hutchinson (1973) places Malvaceae in Malvales, and the rest of the families in Tiliales.

Elaeocarpaceae. The suborder Elaeocarpineae comprises a single family, Elaeocarpaceae. Cronquist (1968, 1981) treats Elaeocarpaceae as a primitive family in Malvales, with probably Thealean ancestry; R. Dahlgren (1977a, 1981) considers its position in Malvales as doubtful. This family shows close alliance with Tiliaceae and, to some extent, with Scytopetalaceae. It resembles Tiliaceae in the development of anther, pollen, embryo sac, and endosperm. Even the seed structure is more or less similar in both. Elaeocarpaceae and Scytopetalaceae develop an endothelium; absent in other families of Malvales. However, a nucellar cap, and mucilage sacs and canals in the ovary wall and ovule – common features in Malvales – are absent in both families. Also, the Onagrad type of embryogeny of Elaeocarpaceae does not occur in any other family of Malvales.

The suborder Malvineae comprises Tiliaceae, Malvaceae, Bombacaceae and Sterculiaceae.

Tiliaceae. The deep-seated archesporial cell develops directly as megaspore mother cell in *Corchorus acutangulus* (Venkata Rao and Sambasiva Rao 1952). In such cases, the ovules in the Tiliaceae may be considered as tenuinucellar, as in Scytopetalaceae. Even though a parietal cell is not cut off, there is conspicuous "parietal" tissue of nucellar origin. Therefore, it is incorrect to call it tenuinucellar; at best this condition should be considered "pseudo-crassinucellar". The synergids are exceptionally long in *Corchorus*. The pollen tube is unbranched. Cytokinesis in the Nuclear endosperm is very late as in Sterculiaceae. It occurs at the time of initiation of cotyledonary primordia in the embryo. The outer epidermis of the testa is membraneous, while the inner one forms a characteristic crystal layer. The seed is exotegmic, the inner epidermis has brown walls. Tiliaceae are closer to Elaeocarpaceae and Sterculiaceae than to Malvaceae.

Malvaceae. The Malvaceae show several features distinct from the rest of the members of Malvales. One-celled anthers, amoeboid anther tapetum, multiporate spinescent pollen grains, campylotropous ovule, multiplication of antipodals in some members, branching pollen tube, curved mature embryo with folded cotyledons, comose, albuminous, perispermous seeds, and outer epidermis of the testa of columnar cells with lignified outer walls.

Bombacaceae. Bombacaceae are treated as a tribe of Malvaceae. They differ from Malvaceae in: secondary increase of microspore mother cells, secretory anther tapetum, smooth-walled triporate pollen grains, ephemeral antipodals, presence of starch in embryo sac, massive chalaza, and arillate seeds with scanty endosperm. The embryo is straight. This family should not be treated as a tribe of Malvaceae (Venkata Rao 1954a), and erection of a separate family Bombacaceae is justified.

Hampea. Engler and Prantl (1894), Edlin (1935) and Hutchinson (1973) place the genus *Hampea* in the family Bombacaceae. However, on the basis of the structure of embryo, Fryxell (1968, 1969) places it in the tribe Gossypieae, family Malvaceae.

The pollen grains are multiporate and echinate as in Gossypieae. In Bombacaceae they are triporate and smooth (Erdtman 1952).

The single-layered endosperm fills the cotyledonary folds. The cells are thin-walled and contain food reserve.

The embryo of *Hampea nutricia* (R.P. Singh and Chauhan 1984) is nearly straight and gland-dotted. It has conduplicate, punctate, asymmetrical cotyledons and gossypol ducts.

The seed is exotegmic. The testa (eight to ten layers) is differentiated into three zones. Stomata occur in the outer epidermis, the cells show slightly thickened outer walls, and have dark-brown contents. The mesophyll (three to four layers) is pigmented and compact. The inner zone (two to three layers) is colourless.

The outer epidermis of the tegmen forms a sclerified, fibrous palisade layer which has a distinct light-line. In a cross-section these cells show a star-shaped lumen. Inner to it lie four thick-walled tanniniferous layers. The inner epidermis of the tegmen, in contact with the endosperm, is a fringe layer, as in Malvaceae. In surface view the cells of fringe tissue are polygonal and thickened along anticlinal walls with simple pits.

The seed is arillate as in *Gossypium cunninghamii* and *G. populifolium* (Fryxell 1969).

The structure of the embryo, a testa differentiated into three zones, the exotegmen as palisade layer, and the inner epidermis of tegmen as fringe layer, clearly suggest the position of *Hampea* in/near Malvaceae. Palynological data support these findings (Fryxell 1969, R.P. Singh and Chauhan 1984).

Sterculiaceae. The occurrence of integumentary vascular bundles is a common feature in Sterculiaceae. Nearly six to nine strands extend up to the micropylar tip (*Sterculia*), or anastomosing branches (*Pterospermum*) may develop. Ten or more embryo sacs develop concurrently from about 100 archesporial cells in *Pterospermum*, but only two or three mature. The winged seed contains copious endosperm, but perisperm is absent. The seed is exotegmic; the inner epidermis of the testa does not contain crystals. They are present in all the other families.

Scytopetalaceae. Edlin (1935) considers Scytopetalaceae as the most primitive family, due to the disposition of their stamens; many stamens are arranged in several rows. He compares this feature with multiseriate stamens of Magnoliaceae. The affinities of the family with Elaeocarpaceae (see earlier under Sect. 35.9) have already been dealt with. In addition, the development of anther, structure and type of pollen, and development of the embryo sac are in conformity with the other families of the order. But Scytopetalaceae show tenuinucellar ovule with the inner integument longer than the outer, and the inner integument forms the micropyle. Again, a hypostase-like strand is prominent, but a socket of thick-walled cells around the chalaza is absent (Vijayaraghavan and Dhar 1976). According to Willis (1973), Scytopetalaceae show affinities probably with Olacaceae rather than Malvales. Vijayaraghavan and Dhar (1976) negate any relationship with Olacaceae. Knowledge about the development of endosperm, embryo, and structure of seed may indicate the affinities with other members of Malvales.

36 Thymelaeales

The order Thymelaeales comprises the families Geissolomataceae, Penaeaceae, Dichapetalaceae, Thymelaeaceae and Elaeagnaceae.

36.1 General Features

The order Thymelaeales is characterized by anatropous, bitegmic, crassinucellate ovule, micropyle formed by both integuments, presence of parietal tissue, Polygonum type of embryo sac, tetrasporic Penaea type in Penaeaceae, Nuclear endosperm, Cellular in Geissolomataceae, and Asterad type of embryogeny.

36.2 Geissolomataceae

The development and the structure of anther, pollen grain, embryo sac, and seed-coat have not been studied.

The ovule is anatropous, bitegmic and crassinucellate in *Geissoloma marginatum* (Stephens 1910). Both the integuments form the micropyle. Earlier stages of development are not available, but the embryo sac is probably of the Polygonum type (Stephens 1910). The antipodals are ephemeral. The polar nuclei fuse in the chalazal region of the embryo sac prior to fertilization. Abundant starch grains occur in the mature embryo sac.

Development of the endosperm is of the Cellular type.

The embryogeny conforms to the Asterad type-Penaea variation. The suspensor is said to be absent, which requires confirmation.

Seeds have scanty endosperm. The embryo has long and linear cotyledons (Hutchinson 1973).

J Hutchinson (1973), EL Stephens (1910)

36.3 Penaeaceae

The anther is tetrasporangiate, and the anther wall comprises a persistent and hairy epidermis, ephemeral endothecium, two middle layers, and glandular tapetum with two-nucleate cells in *Penaea mucronata* and *Saltera sarcocolla* (Tobe and Raven 1984f). The development of the anther wall is of the "Basic" type (Davis 1966). The microspore mother cells undergo meiotic divisions, cytokinesis is simultaneous, and tetrads are tetrahedral and decussate, rarely isobilateral.

The pollen grain is two-celled at the shedding stage.

The ovule is anatropous, bitegmic and crassinucellate. Both the integuments are two-layered, and form the micropyle. In younger ovules, at the base the integuments are close to each other. As the chalazal tissue grows, the base of integuments become separated. The vascular supply extends in the base of integuments.

The archesporium is one-celled. It cuts off a parietal cell, the parietal tissue is three to four layers thick. In *Brachysiphon* and *Penaea* (Stephens 1909) and *Saltera* (Tobe and Raven 1984f), the megaspore mother cell develops into a tetrasporic 16-nucleate embryo sac of Penaea type. The 16 nuclei are arranged in four crosswise quartets. A nucleus from each quartet moves into the centre and the four nuclei fuse to form a tetraploid secondary nucleus. The three cells of each triad simulate egg apparatus. The micropylar triad alone is functional. The synergids are not well organized, and true antipodal cells are absent.

Fertilization is porogamous. The development of the endosperm is of the Nuclear type, and it becomes cellular at the globular proembryo (Stephens 1909).

The embryogeny is of the Onagrad type in *P. mucronata* and *Saltera sarcocolla* (Tobe and Raven 1984f). At an early stage of development the hypocotyl stores food reserve as oil.

The proembryo is massive and lacks suspensor. The two cotyledons are equal and small.

The mature seed is exalbuminous and ellipsoidal. In *P. mucronata* the seed is exotestal, endotesta is represented by crystals, whereas in *Saltera sarcocolla* the seed is endotestal and contains numerous large crystals. Exo- and endotegmen are tracheidal and fibrous.

GL Davis (1966), EL Stephens (1908, 1909), H Tobe and PH Raven (1984f)

36.4 Dichapetalaceae

Microsporogenesis, male gametophyte, and embryogeny have not been studied.

The ovule is anatropous, bitegmic and tenuinucellate in *Dichapetalum mombattense* (Boesewinkel and Bouman 1980). The initiation of the inner integument takes place in a zone about three dermal cells broad. The outer integument is completely of dermal origin, and three to five layers thick. It is shorter than the inner one. The inner integument forms the micropyle. In *Tapura africana* (Boesewinkel and Bouman 1980), the cells of the inner layer of the inner integument differentiate into an endothelium. A funicular obturator develops in *D. mombattense.*

The hypodermal archesporial cell functions directly as megaspore mother cell. The chalazal megaspore of a linear tetrad develops into a Polygonum type of embryo sac.

Development of the endosperm is of the Nuclear type. Later it becomes cellular and only one layer is formed. The seed is exalbuminous (Hutchinson 1973).

The large straight embryo is rich in starch, and poor in lipids. In *D. cymulosum* the peripheral layers of cotyledons contain chlorophyll.

The mature seed is ellipsoid in *D. mombattense.* The seed-coat is covered with fine ramifications of the often anastomozing vascular bundles. The exotesta consists of tanniniferous cells with walls thickened with pectic substances. The cuticle is relatively smooth but spirally striate. The exotesta is followed by a layer of compressed cells with scattered groups of xylem elements of the vascular bundle. These elements show helical thickenings.

The structure of the seed-coat of *D. cymulosum, D. madagascariense, D. pallidum* and *Tapura africana* is similar to that in *D. mombattense* (Boesewinkel and Bouman 1980). In *D. bangii* the epidermis becomes multilayered. The inner epidermis of the testa consists of a fairly thick layer of flattened cells with vestiges of vascular bundles. The testa contains scattered lignified and pitted sclereids.

FD Boesewinkel and F Bouman (1980), EJH Corner (1976), J Hutchinson (1973), W Punt (1975)

36.5 Thymelaeaceae

The anther wall comprises the epidermis, fibrous endothecium, two ephemeral middle layers, and secretory tapetum. The tapetal cells are binucleate in *Daphne,* and four-nucleate in *Wikstroemia* (Venkateswarlu 1947a). The microspore mother cells undergo meiotic divisions, simultaneous cytokinesis is followed by furrowing, and the tetrads are tetrahedral and isobilateral. Polyspory (see Venkateswarlu 1947a) has been reported in *Daphne odorata* (tetrad with 5–7 cells) and *Wikstroemia indica* (tetrad with five cells). This is usually due to (a) laggard chromosomes which organize into micronuclei, or (b) supernumerary division of one or more nuclei of the tetrad.

The pollen grains are pantoporate, and three-celled at the dispersal stage. Several variations have been observed (Venkateswarlu 1945, 1947a). In *Thymelaea arvensis* one pollen grain showed two vegetative and two sperm nuclei (due to division of generative cell and vegetative nucleus), and another showed one vegetative nucleus and four sperms (due to supernumerary division of two original sperms). In *Wikstroemia canescens,* in an anther lobe with degenerated pollen, one pollen grain contained two vegetative and two generative cells.

The ovule is anatropous, bitegmic and crassinucellate with an endostome. A hypostase develops. An obturator is present in *Thymelaea.* The epidermal cells of nucellus contribute to the parietal tissue. The archesporial cell cuts off a parietal cell; the parietal tissue is nine to ten cells thick in *Daphne*, and four to five in *Wikstroemia*. The megaspore mother cell undergoes meiotic divisions, and produces a linear or T-shaped tetrad. The chalazal megaspore develops into a Polygonum type of embryo sac. The polar nuclei fuse before fertilization. There are six antipodals in *Wikstroemia*, and *Daphne* has an even larger number.

The development of the endosperm is of the Nuclear type. Wall formation is initiated at the differentiation of cotyledons in the embryo. Only a few layers of endosperm are formed so that the central region (called central vacuole by the author) is never filled up with endosperm (Venkateswarlu 1947a).

The embryogeny conforms to the Asterad type-Erodium variation. In *Daphne mezereum* it is of the Asterad type-Penaea variation (Souèges 1942a).

Nucellar polyembryony is reported in *Wikstroemia indica* (Winkler 1906), and synergid or nucellar polyembryony in *Daphne cannabina* (Venkateswarlu 1946, 1947a).

In a full-grown but immature seed of *Craterosiphon*, *Dicranolepis* and *Synaptolepis*, the nucellus shows a peripheral network of tracheids (Corner 1976).

The seed is exotegmic. The outer and the inner epidermis of the testa are thick-walled. The outer epidermis of the tegmen comprises a palisade of thick-walled, lignified, pitted columnar cells. The thin-walled mesophyll gets crushed. The inner epidermal cells are thinly lignified with reticulate thickenings (Corner 1976).

EJH Corner (1976), F Fagerlind (1940b), A Fuchs (1938), C Garagaty-Feissly (1969–70), SB Kausik (1940b), J Osawa (1913), R Souèges (1942a), K Tan (1980), J Venkateswarlu (1945, 1946, 1947a, c), H Winkler (1906)

36.6 Elaeagnaceae

The anther wall comprises the epidermis, fibrous endothecium, an ephemeral middle layer, and secretory tapetum of two- to four-nucleate cells. The microspore mother cells undergo meiotic divisions, cytokinesis is simultaneous, and the tetrads are tetrahedral and decussate.

The three-colporate pollen grains are shed at the three-celled stage.

The ovule is anatropous, bitegmic and crassinucellate. A funicular obturator of glandular cells has been observed in the micropylar region in *Elaeagnus conferta* (V.K. Sharma 1966). A prominent hypostase develops after fertilization. The archesporium consists of one to three cells, cuts off a parietal cell, and the parietal tissue is five to seven layers thick. The megaspore mother cell undergoes meiotic divisions, and the development of the embryo sac is of the Polygonum type. The antipodals are ephemeral.

The development of the endosperm is of the Nuclear type. The free-nuclear endosperm in the chalazal region exhibits aggressive haustorial activity. Wall formation is initiated in the micropylar region and proceeds towards the chalaza in *E. conferta*; however, the chalazal endosperm remains free-nuclear till its absorption (V.K. Sharma 1966).

The embryogeny corresponds to the Asterad type, and is assigned to Period I, Megarchetype II, Group I (Rau and Sharma 1970b). The mature embryo is straight with large, thick cotyledons, and a short radicle.

The seed is exalbuminous, and exotestal. The testa is highly multiplicative, and forms the seed-coat. The outer epidermis comprises radially elongate, thick-walled lignified cells with stellate lumina, and is followed by thick-walled aerenchymatous mesophyll. The tegmen is unspecialized, and gets crushed.

A Fuchs (1936), MA Rau and VK Sharma (1970b), VK Sharma (1966), B Singh (1964), R Wunderlich (1967)

36.7 Taxonomic Considerations

Engler and Diels (1936) place all the families, except Dichapetalaceae, in Myrtiflorae, under suborder Thymelaeneae. Hutchinson (1973) includes Geissolomataceae, Penaeaceae and Thymelaeaceae along with Nyctaginaceae in Thymelaeales, and Elaeagnaceae in Rhamnales.

Geissolomataceae. Geissolomataceae and Penaeaceae have plurilocular ovary, and embryogeny of the Asterad type-Penaea variation. In Geissolomataceae the embryo sac is of the Polygonum type, polar nuclei fuse in the chalazal region of the embryo sac before fertilization, and the endosperm is of the Cellular type. In Penaeaceae the development of the embryo sac is of the tetrasporic Penaea type, and endosperm of the Nuclear type. Even though Geissolomataceae are inadequately investigated, the two characters, i.e. Polygonum type of embryo sac with ephemeral antipodals and Cellular type of endosperm, justify the creation of a separate family Geissolomataceae as suggested by Stephens (1910).

Penaeaceae. Tobe and Raven (1984f) enumerate seven embryological features of Penaeaceae comparable to the order Myrtales: (1) glandular anther tapetum, (2) crassinucellate ovule, (3) two-layered inner integument, (4) micropyle formed by both integuments, (5) antipodals s. str. absent, (6) endosperm of the Nuclear type, and (7) mature seed exalbuminous. They mention that Penaeaceae is correctly placed in Myrtales sensu R. Dahlgren and Thorne (1984).

Penaeaceae have some distinctive features: (1) hairy anther epidermis, (2) ephemeral endothecium, (3) unusual development of chalaza to separate the base of the two integuments, and (4) Penaea type of tetrasporic 16-nucleate embryo sac.

On the basis of ephemeral endothecium, Tobe and Raven (1984f) compare Penaeaceae with the families Oliniaceae, Aizateaceae, and Rhyncocalycaceae.

Melchior (1964) and Hutchinson (1973) assign Penaeaceae to Thymelaeales, next to Geissolomataceae. According to recent data, Penaeaceae should be assigned to Myrtales and excluded from Thymelaeales as also mentioned by Thorne (1968), Carlquist and Debuhr (1970), Takhtajan (1969, 1980), R. Dahlgren and Thorne (1984), and Johnson and Briggs (1984).

Dichapetalaceae. Baillon (1874) includes Dichapetalaceae as a tribe Dichapetalae in Euphorbiaceae. Engler and Diels (1936) place it in Geraniales, Melchior (1964) in Thymalaeales, and Hutchinson (1973) in Rosales.

The embryological data for Dichapetalaceae are rather meagre. The ovule is tenuinucellar with Polygonum type of embryo sac, and the seed is simple, exotestal, and exalbuminous.

The ovule is crassinucellate in Euphorbiaceae, and the seed is exotegmic. The embryological data do not support relationship with Euphorbiaceae.

In Geraniales the ovules are tenuinucellate with Polygonum type of embryo sac, and the endosperm is of the Nuclear type as in Dichapetalaceae. But the seeds are exotegmic in Geraniales. The assignment of Dichapetalaceae to Geraniales cannot be confirmed unless further data are available.

The ovule, in Thymelaeales, is crassinucellate and the endosperm is of the Nuclear type except in Geissolomataceae where it is of the Cellular type. In Penaeaceae the embryo sac is of the Penaea type. The seed is exotegmic in Thymelaeales, whereas it is exotestal in Dichapetalaceae. According to Corner (1976), it is difficult to assess the significance of exotestal seed due to its simplicity.

Embryological data do not support the relationship with Rosales with crassinucellate ovule and mesotestal seed.

According to Hutchinson (1973), it is rather difficult to place Dichapetalaceae in any of these orders as they do not show any affinity with either Euphorbiaceae or Thymelaeaceae. We must await further information to discuss the taxonomic assignment of Dichapetalaceae.

Thymelaeaceae. Thymelaeaceae have secretory anther tapetum, simultaneous cytokinesis, three-nucleate pollen grains, tracheidal network in the periphery of nucellus, Polygonum type of embryo sac, secondary multiplication of antipodal cells, obturator, Nuclear endosperm, embryogeny of the Asterad type-Penaea and -Erodium variation, and nucellar or synergid polyembryony. Both the integuments take part in the formation of the seed-coat. The seed is exalbuminous, and exotegmic with the outer epidermis of the inner integument forming the palisade of sclerotic cells.

Venkateswarlu (1947a, c), while considering the systematic position of Thymelaeaceae, states that embryogeny plays a very important role in its classification. Thymelaeaceae, Geissolomataceae and Penaeaceae have Asterad type of development of embryo. The latter two have Penaea variation, as in Thymelaeaceae; *Poivrea coccinea* (Combretaceae) of Myrtiflorae also shows Erodium variation of the Asterad type, as in Thymelaeaceae. Thymelaeaceae are related to Geissolomataceae and Penaeaceae on the one hand, and Combretaceae on the other.

Thymelaeaceae (Sect. 36.5) and Nyctaginaceae (Sect. 16.5) have unilocular ovary, and indehiscent fruit. The embryogeny of Nyctaginaceae is of the

Asterad type-Polygonum variation (Kajale 1940), and it differs from the rest of the members of Centrospermales (Sect. 16). Venkateswarlu (1947b) enumerates several embryological similarities between these two families, and favours the inclusion of Nyctaginaceae with Thymelaeaceae, Geissolomataceae and Penaeaceae, as also suggested by Hutchinson (1973).

Elaeagnaceae. Elaeagnaceae show several features similar to Thymelaeaceae: secretory anther tapetum, simultaneous cytokinesis, three-celled pollen, anatropous, bitegmic, crassinucellate ovule, presence of obturator, Polygonum type of embryo sac, haustorial activity of endosperm at the chalazal end, and Asterad type of embryogeny. According to V.K. Sharma (1966), Elaeagnaceae are a highly specialized family related to Thymelaeaceae. Wunderlich (1967) refutes their alliance with Thymelaeaceae, on the basis of seed structure which is typically rhamnaceous (see Corner 1976). The relationship of Elaeagnaceae with Rhamnaceae has already been discussed (see Rhamnales, Sect. 34.5). Takhtajan (1980) treats them in a separate order Elaeagnales, close to the Proteales.

37 Violales

The order Violales comprises six suborders and 20 families: suborder Flacourtiineae – Flacourtiaceae, Peridiscaceae, Violaceae, Stachyuraceae, Scyphostegiaceae, Turneraceae, Malesherbiaceae, Passifloraceae and Achariaceae; Cistineae – Cistaceae, Bixaceae, Sphaerocepalaceae, and Cochlospermaceae; Tamaricineae – Tamaricaceae, Frankeniaceae and Elatinaceae; Caricineae – Caricaceae; Loasineae – Loasaceae; Begoniineae – Datiscaceae and Bignoniaceae.

37.1 General Features

Variations occur in this fairly large order; the following general features can be enumerated:

(1) Secretory anther tapetum, (2) simultaneous cytokinesis in pollen mother cells, (3) two-celled pollen grains, except in Frankeniaceae, (4) anatropous bitegmic ovules, unitegmic in Loasaceae and Begoniaceae, and orthotropous in Flacourtiaceae and Cistaceae, (5) presence of parietal tissue except in Bixaceae, Frankeniaceae and Loasaceae, (6) Polygonum type of embryo sac, tetrasporic in Tamaricaceae and Caricaceae, and biosporic in Stachyuraceae and Datiscaceae, (7) Nuclear endosperm, Cellular in Loasaceae, (8) ruminate endosperm in Passifloraceae, (9) micropylar haustoria develop in Frankeniaceae, and both micropylar and chalazal in Loasaceae, (10) embryogeny of Asterad type in Violaceae, Onagrad in Turneraceae, Passifloraceae, Car-

icaceae, Datiscaceae and Begoniaceae, and Solanad in Stachyuraceae, Loasaceae, Cistaceae, Tamaricaceae, Frankeniaceae and Elatinaceae, occasionally Piperad in Passifloraceae, and (11) structure of seed-coat reflects the affinities of the families of this order.

37.2 Flacourtiaceae

The staminal primordia arise irregularly on a flat, ring-like mound in *Caloncoba echinata* (Endress and Vosser 1975). The pollen sacs are transversely septate, and the sterile cells are present ab initio in the single-layered archesporium. Later, the archesporium becomes massive.

The anther wall comprises persistent epidermis, fibrous endothecium, one or two ephemeral middle layers, and secretory tapetum with two- to four-nucleate cells. The microspore mother cells undergo meiosis, cytokinesis is simultaneous, and the tetrads are tetrahedral and isobilateral.

The pollen grains are two-celled at the dispersal stage.

The ovule is orthotropous (*Casearia tomentosa* – Narayanaswami and Sawhney 1959), hemiana- (*Idesia polycarpa* – Corti 1948) or anatropous (*Kiggelaria* – see Davis 1966, *Arechavaletaia uruguayensis* – Fig. 37.2.1A), bitegmic and crassinucellate. The inner integument forms the micropyle; but

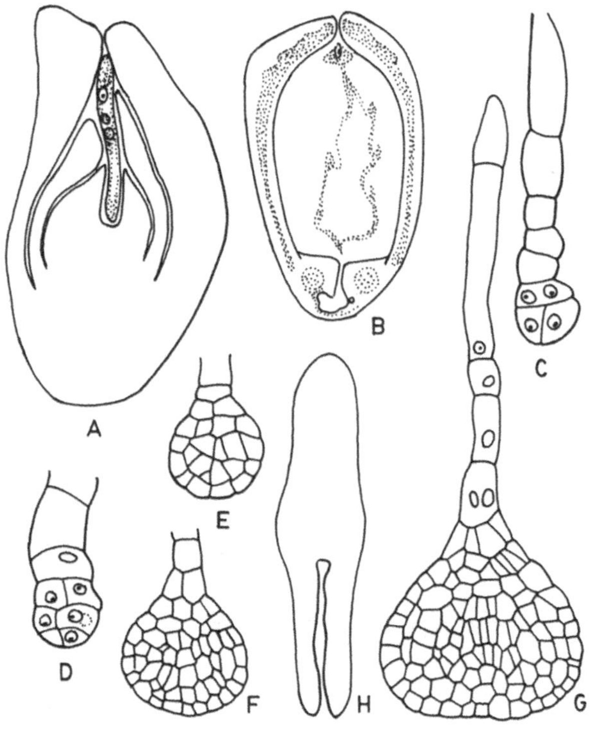

Fig. 37.2.1A. *Arechavaletaia uruguayensis*, longisection of bitegmic ovule, the tip of the mature embryo sac extends into the style (After Ventura 1938). **B** *Caloncoba glauca*, tranlongisection of young seed. Note the sclerified seed-coat, proembryo, nuclear endosperm, and vesicular chalazal caecum of embryo sac. (After van Heel 1977). **C–H** *Idesia polycarpa*, development of embryo. (After Tohda 1971)

both integuments in *A. uruguayensis*. The vascular strands reach up to the tip of the outer integument in *Hydnocarpus laurifolia* (Dathan and Singh 1979). The tip of the ovule bends like a hook after fertilization. A sclerenchymatous hypostase differentiates. The archesporial cell cuts off a parietal cell. The parietal tissue is three to seven layers thick. The megaspore mother cell undergoes meiotic divisions, and produces a linear tetrad. Variations occur in the functional behaviour of megaspores but, usually, the chalazal one develops into a Polygonum type of embryo sac in *Casearia tomentosa* (Narayanaswami and Sawhney 1959), *Flacourtia cataphracta*, *F. ramontchi* (Gopinath 1946), *F. indica* (Dathan and Singh 1973a), *Hydnocarpus laurifolia* (Dathan and Singh 1979), *Idesia polycarpa* (Corti 1948), and *Kiggelaria africana*. The embryo sac breaks through the nucellus and reaches the exostome of the micropyle in *K. africana*, and *Arechavaletaia uruguayensis* (Fig. 37.2.1A). In *F. indica* the parietal layers are crushed and the tip of the embryo sac lies below the nucellar epidermis. The antipodals are ephemeral. The polar nuclei fuse before or at the time of fertilization.

Development of the embryo is shown in Figure 37.2.1C–H. It has a conspicuous suspensor.

The development of the endosperm is of the Nuclear type (Fig. 37.2.1B). Enucleate cytoplasmic nodules occur in *Hydnocarpus*. Through centripetal wall formation, the endosperm becomes cellular at the young globular stage of proembryo in *F. indica* (Dathan and Singh 1973a). In *Hydnocarpus* the mature endosperm contains oil and calcium oxalate crystals.

In *H. laurifolia* the seeds are large, sculptured, endospermous and have a large embryo with thin cotyledons which show prominent bulgings in the position of veins. The integuments are multiplicative in *Hydnocarpus*, *Gynocardia* and *Pangium*, and non-multiplicative in *Erythrospermum*, *Flacourtia*, and *Scaphocalyx*. In the latter two genera the seeds are enclosed in a woody pseudotesta formed by the inner layers of the ovary wall (see Dathan and Singh 1979).

The seed is exotegmic, and contains one or two layers of perisperm in *F. indica* (Dathan and Singh 1973a). The outer layer of testa contains a red pigment. The cells of the other layers are lignified with numerous simple pits. The outer epidermis of the tegmen forms the main sclerenchymatous layer, the inner is tanniniferous. The middle layer may persist. The hypostase connects the main sclerenchymatous layer.

In *Hydnocarpus laurifolia* (Dathan and Singh 1979), the mature seed-coat consists of four zones: (1) outer zone of 50–90 layers of thin-walled compressed and deformed cells with patches of tanniniferous cells mostly in the peripheral region, (2) a broad sclerenchymatous zone, (3) inner parenchyma with scattered tanniniferous cells, and (4) inner zone of tanniniferous layers. The sclerenchymatous zone is of dual origin, and can be distinguished into three regions: (1) three to fourteen outer layers of brachysclereids, (2) middle region of a single layer of macrosclereids, and (3) inner region of brachysclereids with patches of thick-walled tanniniferous cells.

Corner (1976) mentions that in Flacourtiaceae the structure of seed falls in two groups:

1) Flacourtia group (Flacourtia proper) has a fibrous exotegmen as in Celastraceae. The funicular aril is vascularized in *Casearia*.

2) Hydnocarpus group has a very thick pachychalazal seed-coat without any trace of the fibrous exotegmen, even round the micropyle where the two vestigial integuments persist.

The recent discovery of a substance mangostin in the bark of *Hydnocarpus*, in contrast with mangeferin in *Flacourtia*, is significant and a pointer to the unnatural assemblage of these genera in one family (Gunasekera and Sultanbawa 1973, Corner 1976).

R Corti (1948), EJH Corner (1976), ASR Dathan and D Singh (1972b, 1973a, 1979), GL Davis (1966), PK Endress and P Vosser (1975), DM Gopinath (1946), SP Gunasekera and NU Sultanbawa (1973), S Narayanaswami and S Sawhney (1959), H Tohda (1971), WA van Heel (1973, 1974, 1977, 1979), M Ventura (1938)

37.3 Peridiscaceae

The embryology has not been investigated.

The anther is bilocular opening lengthwise, introse. Ovary unilocular with several ovules hanging from the top. Fruit unilocular and single-seeded. The seeds lack endosperm and have a curved embryo.

J Hutchinson (1973), AL Takhtajan (1966), JC Willis (1973)

37.4 Violaceae

The anther wall comprises the epidermis, fibrous endothecium, two ephemeral middle layers, and secretory tapetum of binucleate cells. There are three or four middle layers in *Viola odorata* (D. Singh and Gupta 1963, D. Singh 1966). Übisch granules are studded on the inner walls of tapetal cells in *Ionidium suffruticosum* (Mukherjee and Venugopalan 1957). The microspore mother cells undergo meiotic divisions, cytokinesis is simultaneous, and the tetrads are tetrahedral, isobilateral and decussate. Polyspory occurs in *V. odorata* (D. Singh 1966).

The pollen grains are three- to five-colporate, prolate to spheroidal, and shed at the two-celled stage. In *V. riviniana* (West 1930), the generative cell may divide in the pollen grain, or in the pollen tube. Madge (1929) reports that the vegetative nucleus increases in length. The germination of pollen takes place in situ in cleistogamous flowers of *Viola biflora*, *V. odorata* var. *paraecox* and *V. sylvatica* (Madge 1929).

The ovules are anatropous, bitegmic and crassinucellate. The outer and inner integuments form a zigzag micropyle in *Viola* and *Ionidium*, and the inner integument alone forms the micropyle in *Hybanthus concolor* (Andrews 1910). The nucellar epidermis undergoes periclinal divisions, and produces a two- to three-layered nucellar cap. The archesporium is uni- or multi-celled.

The archesporial cell cuts off a parietal cell. The megaspore mother cell divides meiotically, and forms a linear or T-shaped tetrad. The chalazal megaspore develops into a Polygonum type of embryo sac. The antipodals are small and ephemeral. In *Hybanthus* (Andrews 1910), *Viola cucullata*, *V. pedata* (Bliss 1912) and *V. odorata* (D. Singh and Gupta 1968) the antipodals are fairly large. A double hypostase is present in *Viola cinerea* var. *stocksii*, *V. tricolor* and *Ionidium suffruticosum* (D. Singh 1963, D. Singh and Gupta 1968).

The pollen tube enters the synergid and forces its way through its base (Madge 1929). This needs confirmation.

The development of endosperm is of the Nuclear type. Wall formation is centripetal, and occurs after the formation of numerous nuclei in *Ionidium* and *Viola* (D. Singh 1963). The endosperm is massive and persistent, and stores fatty food reserve.

The embryogeny corresponds to the Asterad type, and the mature embryo is erect and spatulate.

Both the integuments are three-layers thick, and contribute to the formation of testa and tegmen in *Viola*. The seed is albuminous and exotegmic with a patch of chalazal sclereids and a hypostase. It is arillate in some taxa. The cells of the outer epidermis of the testa show lamellar thickenings. They are radially-elongate in *Viola*. The middle layer contains tannin in *Viola*, and is sclerenchymatous in *Ionidium*. Each cell of the inner epidermis contains a single crystal. The exotegmen is excessively sclerified and pitted. The middle layer may persist or collapse, while the inner epidermis persists and becomes thick-walled.

FM Andrews (1910), MC Bliss (1912), ASR Dathan and D Singh (1973c), E Kuta (1990), M Madge (1929), PK Mukherjee and P Venugopalan (1957), MVS Raju (1958), D Singh (1963, 1966, 1970a), D Singh and S Gupta (1963, 1968), R Souèges (1937e), G West (1930)

37.5 Stachyuraceae

The development of anther and pollen grains has not been investigated.

The ovule is anatropous, bitegmic and crassinucellate. The zigzag micropyle is formed by both the integuments. An aril develops from the apical region of the funiculus in *Stachyurus praecox* (Sâto 1976b) and *S. chinensis* (Mathew and Chaphekar 1977), and grows towards the chalazal end of the ovule. Corner (1976) mentions that the exostome is also involved in the formation of aril. The apical cells of the nucellar epidermis divide periclinally to form a two- to three-cell thick cap. The archesporial cell cuts off a parietal cell, and the parietal tissue is three layers thick. The megaspore mother cell undergoes meiotic divisions, and produces a triad, a linear or T-shaped tetrad. A triad may develop in *S. chinensis*, the tetrad is T-shaped in *S. matsuzakii* (Sâto 1976b). The chalazal megaspore develops into a Polygonum type of embryo sac. The base of the mature embryo sac extends into the chalaza and curves convexly. The antipodals are ephemeral, and the polar nuclei fuse before fertilization.

An early stage of bisporic Endymion type of development is reported in *S. chinensis* (Mathew and Chaphekar 1977). This needs confirmation.

The development of the endosperm is of the Nuclear type. Later, it becomes cellular. It is thin-walled and stores fatty food reserve.

The embryogeny conforms to the Solanad type in *S. chinensis* (Mathew and Chaphekar 1977). The mature embryo is small and straight with flat cotyledons and a short radicle. Cleavage (Fig. 3.12.1B, C) and synergid polyembryony have been noted in *S. chinensis*.

The seed is small, arillate, and albuminous. The testa is wholly sclerotic, and the tegmen gets crushed.

EJH Corner (1976), CJ Mathew and M Chaphekar (1977), Y Sâto (1976b)

37.6 Scyphostegiaceae

The embryology has not been fully investigated.

The anther is tetralocular. Simultaneous cytokinesis occurs in the microspore mother cells.

The ovule is anatropous, bitegmic and crassinucellate. The micropyle is formed by both the integuments, the endostome is long and the exostome short. The integuments are three to four layers thick. The nucellus divides periclinally to form a nucellar cap which persists in the mature seed.

The archesporial cell cuts off a parietal cell. The megaspore mother cell undergoes meiotic divisions forming a linear tetrad, and the chalazal megaspore develops into an eight-nucleate Polygonum type of embryo sac.

Development of the endosperm is of the Nuclear type. It becomes cellular, contains starch, and persists in the mature seed.

The exotegmic seed is albuminous and perispermous. The straight embryo has oval cotyledons and occupies more than half the length of the arillate seed. The aril develops from the funiculus and outer integument, and surrounds the basal part of seed. There is an intercalary extension of the endostome and exostome to form a pseudo-funicle, in conjunction with the raphe (Corner 1976).

EJH Corner (1976), WA van Heel (1967), BGL Swamy (1953a), MS Yakovlev (1983)

37.7 Turneraceae

The anther wall comprises the epidermis, fibrous endothecium, two ephemeral middle layers, and secretory tapetum of multinucleate cells. The nuclei fuse and become polyploid. The stomata of haplocheilic type, and fibrous thickenings occur in the region of the connective. The microspore mother cells undergo meiotic divisions, cytokinesis is simultaneous, and the tetrads are tetrahedral and decussate. Polyspory is quite common.

The pollen grains are three-colporate with simple exine ornamentations. They are two-celled at the dispersal stage.

The ovule is anatropous, bitegmic and crassinucellate; the micropyle is produced by both the integuments. The hypodermal archesporium is uni- or multicelled. The archesporial cell cuts off a parietal cell. The parietal tissue is only a few cells thick. The nucellar epidermis also adds to it. The megaspore mother cell undergoes reduction divisions, gives rise to a linear or T-shaped tetrad, and the chalazal megaspore develops into a Polygonum type of embryo sac. The presence of supernumerary megaspore mother cells, dyads, triads and tetrads is a common feature. Twin embryo sacs are also present in *Turnera ulmifolia* (Vijayaraghavan and Kaur 1966). The antipodals are ephemeral. The hypostase is traversed by the funicular vascular trace.

The entry of the pollen tube is porogamous, and double fertilization occurs. The remnants of the pollen tube persist up to a late stage of the embryo.

Development of the endosperm is of the Nuclear type. The nuclei aggregate around the embryo, and in the chalazal region. Wall formation occurs at the

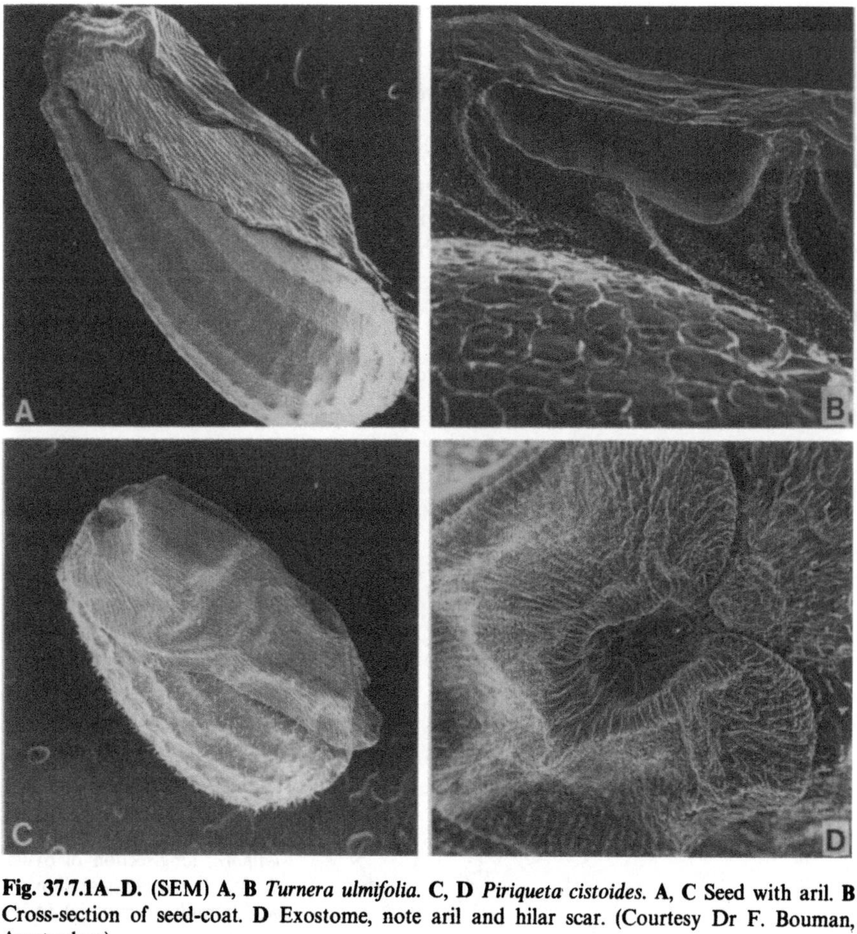

Fig. 37.7.1A–D. (SEM) **A, B** *Turnera ulmifolia.* **C, D** *Piriqueta cistoides.* **A, C** Seed with aril. **B** Cross-section of seed-coat. **D** Exostome, note aril and hilar scar. (Courtesy Dr F. Bouman, Amsterdam)

globular proembryo, and is centripetal. The cells are isodiametric, and contain starch.

The embryogeny conforms to the Onagrad type. In *Turnera ulmifolia* var. *angustifolia* (M.V.S. Raju 1956a), the embryogeny is assigned to Period I, Megarchetype I, Group I (Crucifer type). The mature embryo is dicotyledonous, and curved with a short suspensor.

The seeds are small, exotegmic and arillate (Fig. 37.7.1A, C, D). The outer epidermis of the testa contains tannin. The cells of the inner epidermis enlarge. The outer epidermis of the tegmen comprises longitudinally elongate, pitted, sclerotic cells with their ends raised to form ridges (Fig. 37.7.1B). The inner epidermis contains tannin. The aril is funicular, and arises around the apex of funicle. It is white, sickle-shaped and non-vascularized (M.V.S. Raju 1956a).

D Kaur (1970), MS Mahalingappa (1975), MVS Raju (1956a), N Ramaswamy and B Bir (1984), MR Vijayaraghavan and D Kaur (1966)

37.8 Malesherbiaceae

The embryology has not been fully investigated.

The anther is tetralocular.

The ovule is anatropous, pleurotropous, bitegmic and crassinucellate. The micropyle is formed by both the integuments. The outer integument is two layers thick, three in the micropylar region; the inner one is three-layered. The massive chalaza forms a process which resembles pecten (Fig. 37.8.1A). The seed is albuminous with aleurone grains and fat globules. The large embryo is dicotyledonous (Fig. 37.8.1B).

H Harms (1925), J Hutchinson (1973), SM Ricardi (1967)

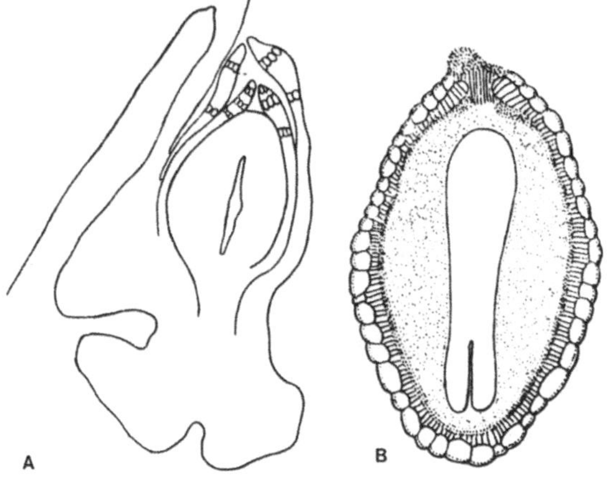

Fig. 37.8.1. A *Malesherbia scarlatiflora*, longisection of ovule. **B** *M. lirana*, longisection of seed (semidiagrammatic). (After Ricardi 1967)

37.9 Passifloraceae

The anther wall comprises the epidermis, fibrous endothecium, two or three ephemeral middle layers, and secretory tapetum of multinucleate cells. The tapetum becomes biseriate in *Passiflora calcarata* and *P. foetida* (M.V.S. Raju 1952b, 1956b). The epidermal cells are fairly large at the dehiscence slit in *P. calcarata.* The microspore mother cells undergo meiotic divisions resulting in tetrahedral and isobilateral tetrads. Cytokinesis is simultaneous.

The pollen grains are three- to twelve-colporate in *Passiflora* and the colpi are arranged in three groups; the colpi of each group coalesce at the pole. The exine is reticulate. They are two-celled at the dispersal stage.

The ovule is anatropous except in *P. calcarata* (M.V.S. Raju 1956b) where it is orthotropous. It is bitegmic and crassinucellate. Both the integuments contribute to the formation of the micropyle. The archesporium is unicelled, except in *P. foetida* (Padhye and Deshpande 1960). It cuts off a parietal cell. The parietal tissue is massive. The nucellar epidermal cells also divide and contribute to it. Following meiotic divisions, the megaspore mother cell forms a linear tetrad. The chalazal megaspore develops into a Polygonum type of embryo sac, the antipodals are ephemeral.

The entry of the pollen tube is porogamous, and double fertilization occurs. The pollen tube grows deep in the embryo sac, gets twisted, and persists for a long time. In *P. adnophylla*, Cook (1909) pointed out that the tube becomes haustorial, and absorbs all the contents. He considers that this is an instance of parasitic condition of the male gametophyte over the female one. Such a presumption is entirely incorrect since the discharged pollen tube does not show any activity.

Persistent and twisted pollen tubes have also been observed in *P. calcarata* (M.V.S. Raju 1956b), *P. foetida* (D. Singh 1962) and *P. leschenaultii* (M.V.S. Raju 1952b).

The development of the endosperm is of the Nuclear type. The nuclei are distributed around a central vacuole, and aggregate at both the poles. The chalazal tubular region behaves as a haustorium in *Passiflora* and *Tacsonia* (Dathan and Singh 1973b). Centripetal wall formation commences after the formation of a large number of nuclei. The peripheral cells simulate meristematic cells; some cells in the chalazal region are plurinucleate in *P. calcarata* (M.V.S. Raju 1956b). Enucleated cytoplasmic nodules have been observed in the chalazal region (M.V.S. Raju 1956b), and finally they merge with the general cytoplasm. The mature endosperm is ruminate.

The embryogeny in *P. foetida* (D. Singh 1962) corresponds to the Onagrad type-Myosurus variation. Padhye (1963), however, reports that in 70% of ovules the division of the zygote is vertical and the development conforms to the Piperad type. This aspect needs confirmation.

The seeds are endotegmic and arillate with a large spatulate embryo. The aril is succulent and coloured. The testa is multiplicative and pulpy. The outer epidermis is covered with a smooth cuticle in *P. foetida* (D. Singh 1962), *P. hirsuta* and *P. holoserecea* (Netolitzky 1926), and is echinulate in *P. calcarata* (M.V.S. Raju 1956b). The middle layer is parenchymatous. The thin-walled inner epidermal cells are unequally radially elongate, more in the region of ingrowths. The inner epidermis of the tegmen is a sclerotic layer with radially elongate thick-walled, lignified, pitted cells. In *Adenia* the tegmen does not become sclerotic. The inner tangential walls of the middle layer become thickened in *P. foetida* (D. Singh 1962), and the inner epidermis contains granules. The endosperm becomes ruminate due to unequal corresponding radial elongation of testal and tegmal cells (Fig. 3.10.4D). The aril arises as a collar from the funicle. It comprises thin-walled cells containing oil and starch. It is brownish-pink in *P. calcarata*, and colourless in *P. foetida*.

MT Cook (1909), ASR Dathan and D Singh (1973b), A Kloos and F Bouman (1980), F Netolitzky (1926), MD Padhye (1963), MD Padhye and BG Deshpande (1960), MVS Raju (1952b, 1956b), D Singh (1962, 1970b)

37.10 Achariaceae

The embryology has not been studied.

The anther is bilocular, opens lengthwise, and has a broad connective (Hutchinson 1973). The ovule is anatropous. The arillate seed has copious endosperm (Willis 1973), and a small straight embryo.

H Harms (1925), J Hutchinson (1973), JC Willis (1973)

37.11 Cistaceae

The anther wall comprises the epidermis, fibrous endothecium, single ephemeral middle layer, and secretory tapetum of two- to four-nucleate cells. The nuclei divide, and the daughter nuclei fuse. The microspore mother cells undergo meiotic divisions, cytokinesis is simultaneous through furrowing, and the tetrads are tetrahedral, isobilateral and decussate. Polyspory is also reported.

The pollen grains with three germ pores are two-celled at the dispersal stage; the generative cell divides in the pollen tube.

The ovule is orthotropous, bitegmic and crassinucellate. Both the integuments form the micropyle. The tip of the integuments are swollen in the micropylar region. The cells of the nucellar epidermis undergo divisions, and a nucellar cap is formed. The archesporium is multicellular. The cells cut off parietal cells; the parietal tissue is well developed. The megaspore mother cell undergoes meiotic divisions, and produces a linear tetrad. The chalazal megaspore usually develops into a Polygonum type of embryo sac; micropylar megaspore in *Helianthemum alpestre*. The third megaspore from the micropyle is reported to be functional in *H. halimifolium* (Chiarugi 1925). The synergids have a cap-like structure at the tip, and the antipodals are ephemeral. Multiple embryo sacs develop in *H. vulgare* (Kapil and Maheshwari 1964). A hypostase is present.

The development of the endosperm is of the Nuclear type. The basal prolongation is common. The endosperm is 64-nucleate at the early globular proembryo, and wall formation is initiated at the early heart-shaped stage in *H. vulgare*. The endosperm is parenchymatous, and stores starch.

The embryogeny conforms to the Solanad type, and the embryo has a uniseriate suspensor. The development of embryo in *H. guttatum* (Souèges 1937b) is assigned to Period I, Series C, Megarchetype IV (Solanad type). The mature embryo is curved.

The seed is small, albuminous and exotegmic (Kapil and R. Maheshwari 1964). The testa persists as a membraneous covering, is rich in starch and shows thick-walled, branched, septate fungal mycelium (fungus could not be identified). The outer epidermis of tegmen comprises dark-brown, thick-walled, lignified, columnar cells forming a bulge in the chalazal region. The inner epidermis contains tannin.

JG Boursnell (1950), A Chiarugi (1925), R Corpal, F Pevez-Gargia and JM Pita (1989), RN Kapil and R Maheshwari (1964), R Souèges (1937b)

37.12 Bixaceae

The epidermis of the mature anther is papillate. The endothecial thickenings are confined around the slits at the end of thecae. The two or three middle layers are ephemeral, and the secretory tapetum is irregularly two-layered, and has binucleate cells. Übisch granules are present on the inner tangential walls of the tapetum. Meiotic divisions in microspore mother cells are followed by simultaneous cytokinesis, and the tetrads are tetrahedral and decussate.

The pollen grains are tricolporate, and two- or three-celled at the dispersal stage. Pollen grains with five nuclei have been observed in *Bixa orellana* (H. Kaur 1970b).

The ovule is anatropous, bitegmic and pseudo-crassinucellate. The outer integument overgrows the inner one, and both form a zigzag micropyle. The cells of the nucellar epidermis divide and form a nucellar cap. A single cell of the uni- or bicelled archesporium functions directly as the megaspore mother cell. The latter undergoes meiotic divisions, and produces a linear or T-shaped tetrad. Twin tetrads are also formed. The chalazal megaspore develops into a Polygonum type of embryo sac. Sometimes, an epichalazal megaspore becomes binucleate. The antipodals are ephemeral, and polar nuclei lie in the middle of the embryo sac. Mature embryo sacs with supernumerary polar nuclei and egg apparatus have been observed in *Bixa orellana* (Chopra and Kaur 1965a). Twin embryo sacs also develop concurrently (Dathan and Singh 1972a).

The entry of the pollen tube is porogamous, and double fertilization occurs. The remnants of the pollen tube persist. One of the synergids persists up to an early stage of proembryo.

Development of the endosperm is of the Nuclear type. The nuclei aggregate at the micropylar end. After the formation of a large number of nuclei, wall formation is initiated at the micropylar region and proceeds to the chalaza. Except for the outer one or two layers, the endosperm stores starch.

The division of zygote is transverse; vertical walls are formed in the apical and basal cell to produce a globular proembryo. Further development is irregular (H. Kaur 1969). The mature embryo has conspicuous elongated cotyledons.

The seed is albuminous with fibrous exotegmen. The outer epidermis of the testa shows hairs and stomata. In *Bixa* the epidermal cells in the chalazal region are sclerotic. The mesophyll contains multinucleate dye cells and oil cells, followed by the inner epidermis. The outer epidermis of the tegmen consists of lignified pitted cells. The mesophyll is parenchymatous. The inner hypodermal layer comprises hour-glass-shaped cells with thickenings, followed by thin-walled cells of inner epidermis. The cells of the latter are pitted in the anticlinal walls.

RN Chopra and H Kaur (1965a), ASR Dathan and D Singh (1972a), H Kaur (1969, 1970b), J Mauritzon (1936a)

37.13 Sphaerosepalaceae

The embryology has not been studied in sufficient detail.

The anther has a broad, glandular filament with two thecae opening introsely through longitudinal slits (Hutchinson 1973).

The anatropous ovule has a massive tanniniferous chalaza. In most species of *Rhopalocarpus* (*Sphaerosepalum*), the endotegmen has a smooth surface but in *R. lucidus* (Fig. 37.13.1A, C), *R. similis* and *R. triplinervius*, the endotegmen forms conicle or cylindrical outgrowths which ruminate the endosperm.

The mature embryo (Fig. 37.13.1B, D, E) has the hypocotyl, primordial root, and leafy cotyledons which are cordate at the base and lobed along the margin (Fig. 37.13.1D, E). The cotyledons are branched in some species of *Rhopalocarpus* and *Dialyceras*. They are bilobed with fimbriate margin in *R. thouarsianus*.

R Capuron (1962), J Hutchinson (1973)

37.14 Cochlospermaceae

The development of anther, pollen, embryo sac, and embryo has not been worked out.

The ovule is anacampylotropous, bitegmic and crassinucellate with a zigzag micropyle in *Cochlospermum gossypium* (Dathan and Singh 1972a). The nucellar epidermis contributes to the massive parietal tissue.

Development of the endosperm is of the Nuclear type in *C. orinoccense* (see Schnarf 1931). Wall formation is centripetal.

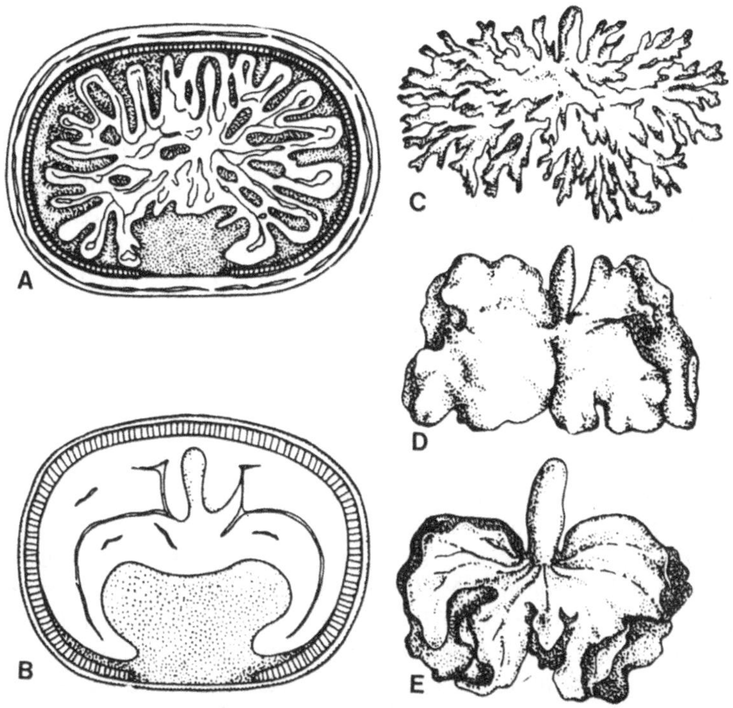

Fig. 37.13.1A–E. **A**, **C** *Rhopalocarpus lucidus*, **B**, **D** *R. thouarsianus*, **E** *Dialyceras parvifolium*. **A** Longisection of young seed with ruminate endosperm, ruminations are due to ingrowths of endotegmen. The chalaza is massive in **A** and **B**. **B** Mature seed (diagrammatic) with two bilobed cotyledons and hypocotyl. **C** Ruminate endosperm. **D**, **E** Embryo with radicle and two-lobed cotyledons. (After Capuron 1962)

The seed is reniform, exarillate and hairy along the dorsal side. It has copious endosperm, and a large curved embryo with broad cotyledons and a short radicle. The outer epidermis of the testa comprises subcuboid or transversely elongate cells. The mesophyll is aerenchymatous, the inner epidermis is of thin-walled subtabular cells. The outer epidermis of tegmen is a palisade layer. The outer hypodermis consists of large, radially elongate cells with brown and oily contents.

EJH Corner (1976), ASR Dathan and D Singh (1972a), J Mauritzon (1936a), K Schnarf (1931)

37.15 Tamaricaceae

The anther wall comprises a prominent, persistent thickly cuticularized epidermis, fibrous endothecium, two ephemeral middle layers, and secretory tapetum. The tapetal cells are large with amoeboid polyploid nuclei. Simultaneous cytokinesis follows meiotic divisions of microspore mother cells. Furrowing is centripetal, and the tetrads are isobilateral, tetrahedral and decussate.

The pollen grains are (two)-three-(four)-colpate, and two-celled at the dispersal stage.

The ovule is anatropous, bitegmic and crassinucellate (Fig. 37.15.2A). The inner integument grows faster than the outer one, and forms the micropyle. At the chalazal end, the epidermal cells elongate and grow into nucleated hairs. The hypodermal archesporial cell cuts off a parietal cell (Fig. 37.15.1A, B). The parietal tissue is single-layered. The megaspore mother cell undergoes meiotic divisions, and the development of the embryo sac is of the tetrasporic type.

Variations occur in the formation of the female gametophyte. It is of the Fritillaria type in *Myricaria germanica*, *Tamarix africana* (Battaglia 1941), *T. chinensis* (V. Puri 1939a), *T. dioica* (A.C. Joshi and Kajale 1936), *T. ericoides* (Y.M.L. Sharma 1939, 1940), *T. gallica* (Battaglia 1941), *T. odessana*, *T. parviflora* (Hjelmqvist and Grazi 1964, see also Battaglia 1988), *T. pentandra* and *T. troupii* (B.M. Johri and Kak 1954). Adoxa type has been observed in *T. aestivalis*, *T. africana*, *T. gallica* and *T. odessana* (Mauritzon 1936a), *T. parviflora* (Hjelmqvist and Grazi 1964), *T. pentandra* and *T. tetrandra* (Mauritzon 1936a) and *T. troupii* (B.M. Johri and Kak 1954). Chrysanthemum cinerariaefolium type occurs in *T. africana* and *T. gallica* (Battaglia 1941). Fritillaria, Adoxa and Chrysanthemum cinerariaefolium types occur in *T. troupii*, whereas, in addition to these three, Drusa type also occurs in *T. pentandra* (B.M. Johri and Kak 1954). Five types, i.e. Fritillaria, Chrysanthemum cinerariaefolium, Drusa, Adoxa and Plumbagella, have been recorded in *T. odessana* and *T. parviflora* (Hjelmqvist and Grazi 1964). Hjelmqvist and Grazi (1964) state that simpler reduced types are more frequent at low temperatures, whereas complicated types are frequent at high temperatures. Abnormal embryo sacs also occur in *T. pentandra.*

Since the Fritillaria type occurs more commonly in *Myricaria* and also in *Tamarix*, it may be considered to be characteristic of this family.

The pollen tube destroys one of the synergids. Sometimes, supernumerary pollen tubes have been observed in the embryo sac; one or both the synergids may be fertilized and develop into additional embryos (Figs. 37.15.2K, L, 37.15.3F).

Table 37.15.1. Distribution (%) of various types of embryo sacs in *Tamarix*

Type of embryo sac	*T. gallica*[a]	*T. africana*[a]	*T. pentandra*[b]	*T. troupii*[b]
Fritillaria	90	43	38	73
Drusa	3	10	8	0
Adoxa	2	0	48	14
Chrysanthemum *cinerariaefolium*	5	47	6	13

[a] After Battaglia (1941); [b] after B.M. Johri and Kak (1954)

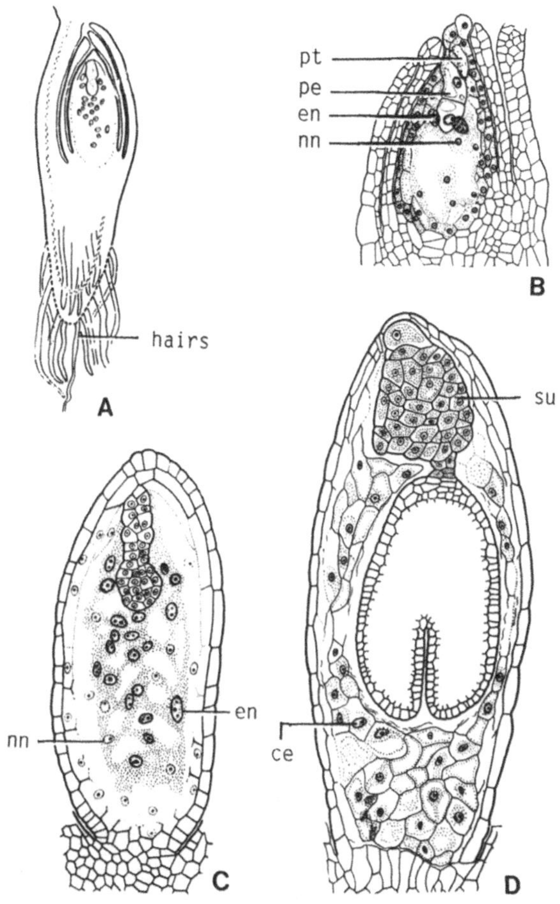

Fig. 37.15.1A–D. *Tamarix ericoides*. **A** Longisection of ovule at early globular proembryo. At the chalazal end the epidermal cells elongate and form a tuft of hairs (plume). **B** Upper part of ovule at bicelled proembryo (*pe*) and free-nuclear endosperm (*en*). The tangential walls of the innermost layer of nucellar cells break down and the nuclei (*nn*) migrate into the embryo sac. The tip of the nucellus has elongated and blocks the micropyle; *pt* pollen tube. **C** Nucellus with embryo sac from **A**; early globular proembryo and nucellar nuclei (*nn* smaller, lighter) mixed with endosperm nuclei (*en* larger, darker). **D** Dicotyledonous embryo with conspicuous suspensor (*su*), and cellular endosperm (*ce*). (After B.M. Johri and Kak 1954)

Development of the endosperm is of the Nuclear type (Fig. 37.15.1A–C). Besides the Nuclear type, Mauritzon (1936a) reports the Cellular type also in *T. tetrandra*, which is erroneous.

Wall formation in the endosperm is initiated in *T. troupii* at the globular proembryo, but in *T. ericoides* at the heart-shaped embryo. As the cotyledons differentiate, cellular endosperm fills the embryo sac (Fig. 37.15.1D). The endosperm is consumed during the maturation of seed (Fig. 37.15.3E, F); the remnants of a layer of nucellus persist (Fig. 37.15.3G).

After fertilization, the nucellar cells in contact with the gametophyte break down and their nuclei migrate into the embryo sac (Fig. 37.15.1A–C). The nucellar nuclei can be easily identified from the endosperm nuclei by their smaller size. Gradually, with the increase in the number of endosperm nuclei, the nucellar nuclei degenerate and are no more recognizable at the advanced globular proembryo. This feature has not been reported in any other species of *Tamarix*.

The development of the embryo in *T. troupii* and *T. ericoides* is of the Solanad type (Fig. 37.15.2A–J). A transverse division of the zygote generates the cells *ca* and *cb*. The division of *ca* produces the tiers *l* and *l′*, and *cb* forms *m* and *ci*. *m* divides transversely into *d* and *f* which undergo transverse and longitudinal divisions. *ci* gives rise to *n* and *n′*, and later divisions are irregular.

The derivatives of *ci* and *f* produce a massive suspensor of more or less vesicular cells which are richly cytoplasmic and contain abundant starch (Figs. 37.15.1D, 37.15.2I, J). The suspensor has a haustorial function and draws upon the reserve food from the adjacent cells of the nucellus and the endosperm. Their remnants persist in the mature seed (Fig. 37.15.3E, G).

The peripheral cells of the suspensor in *T. troupii* bulge out prominently.

The tiers *l* and *l′* divide longitudinally, twice, at right angles to each other, generating the quadrant and octant. Further divisions are followed by the delimitation of the dermatogen, and progressively the globular, heart-shaped and dicotyledonous embryo are formed.

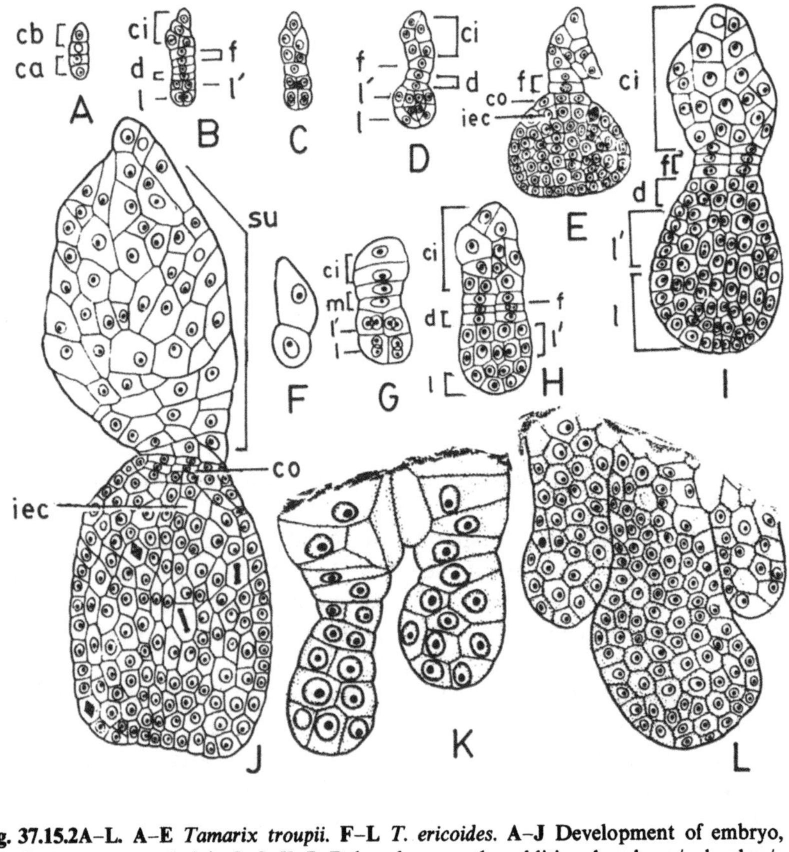

Fig. 37.15.2A–L. **A–E** *Tamarix troupii*. **F–L** *T. ericoides*. **A–J** Development of embryo, note prominent suspensor (*su*) in **I, J, K, L** Polyembryony, the additional embryo/s develop/s from fertilized synergid/s. (**A–J** After B.M. Johri and Kak 1954, **K, L** after Y.M.L. Sharma 1939)

The derivatives of *d* produce the tiers *co* and *iec* (Fig. 37.15.2J); *co* gives rise to root cap and *iec* to initials of the central cylinder of root.

Synergid polyembryony has been reported in *Tamarix ericoides* (Fig. 37.15.2K, L). One of the seeds showed a fully developed large embryo and an equally well developed somewhat smaller embryo (presumably from a fertilized synergid) in lateral position (Fig. 37.15.3F, G).

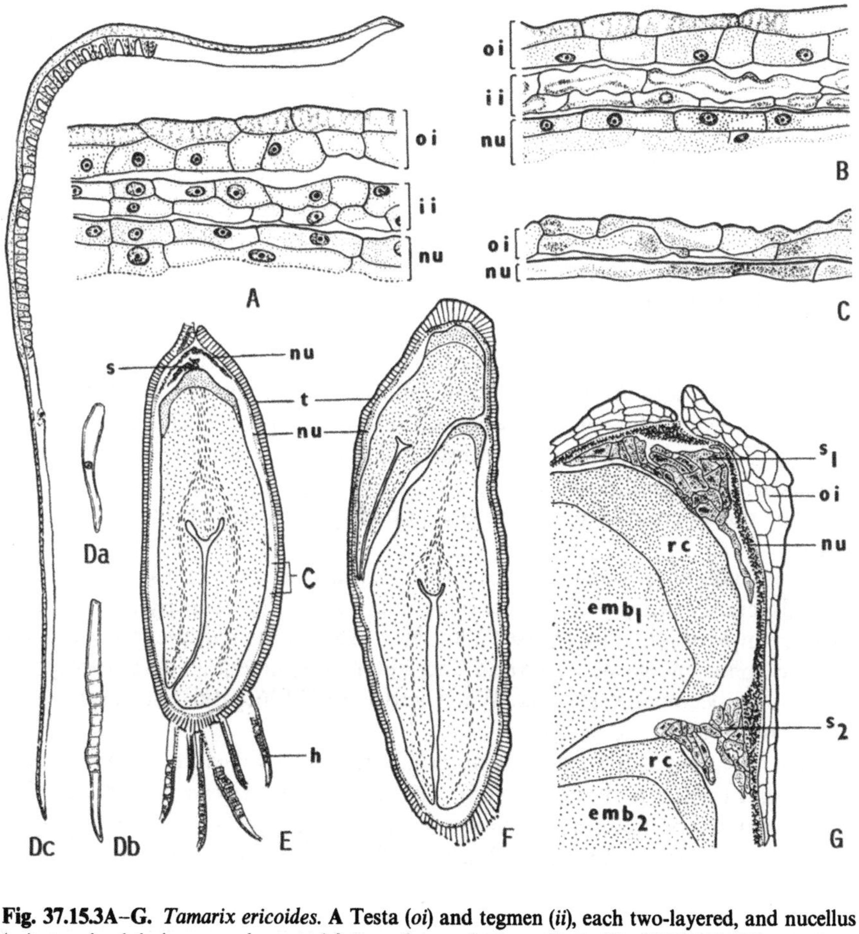

Fig. 37.15.3A–G. *Tamarix ericoides*. **A** Testa (*oi*) and tegmen (*ii*), each two-layered, and nucellus (*nu*) at early globular proembryo and free-nuclear endosperm (from **Fig. 37.15.2A**). **B** Later stage, the inner layer of nucellus has collapsed (from immature seed at the stage shown in **Fig. 35.15.2D**). **C** Testa and remnants of nucellus. **Da–Dc** Development of hair, from chalazal tuft. **E** Longisection of mature seed with chalazal thick-walled hairs (*h*), note the testa (*t*), remnants of nucellus (*nu*) and suspensor (*s*). **F** Seed with fully formed twin embryos. **G** Portion of seed (from a different successive section) shown in **F**. Note the region of micropyle, and remnants of the suspensor (S_1, S_2) of both the embryos (emb_1, emb_2) and nucellus; *oi* outer integument/testa; *rc* root cap. (After B.M. Johri and Kak 1954)

Traub (1939) observed 16 seedlings from one seed in *Myricaria cauliflora*; he did not trace the origin of polyembryony or the seedlings.

Each of the inner and outer integument is two-layered. The seed-coat comprises the longitudinally elongate thick-walled and pitted exotesta and thin-walled endotesta; the tegmen becomes crushed (Fig. 37.15.3A–C). The exotestal, exalbuminous seed has a large straight embryo, a layer of nucellus (perisperm), and a chalazal tuft of hairs (Figs. 37.15.3C–G).

Soon after fertilization, epidermal hairs develop at the elongated chalazal end of the ovule (Fig. 37.15.1A). They continue to enlarge during the growth and maturation of the seed. At first the hairs are one-celled and uninucleate. As the cotyledons differentiate, these cells repeatedly divide transversely to form a uniseriate filament. The thin-walled basal and apical cells remain uninucleate and considerably longer than the other cells which develop characteristic lignified thickenings (Fig. 37.15.3Da–Dc, E).

E Battaglia (1941, 1943, 1988), A Frisendahl (1912), H Hjelmqvist and F Grazi (1964), BM Johri and D Kak (1954), AC Joshi and LB Kajale (1936), J Mauritzon (1936a), V Puri (1939a), YML Sharma (1939, 1940), R Souèges (1924d, 1937b), HF Traub (1939)

37.16 Frankeniaceae

The anther wall comprises cutinized epidermis, endothecium, two ephemeral middle layers, and secretory tapetum of binucleate cells. The endothecium is fibrous, except in the region of stomium. The microspore mother cells undergo meiotic divisions, and produce tetrahedral, decussate and isobilateral tetrads. Cytokinesis is simultaneous.

The pollen grains are tricolporate, and two- or three-celled at the shedding stage.

The ovules are borne on a long funicle, and are anatropous, bitegmic and pseudo-crassinucellate. The inner integument forms the micropyle. An unusual feature in *Frankenia hirsuta* (see Yakovlev 1983) is the conspicuous extension of the inner integument beyond the outer integument. Consequently, the micropyle becomes very long. Walia and Kapil (1965) did not observe such a feature in the same species.

An incipient third integument originates from the outer integument in *F. hirsuta* (Walia and Kapil 1965). The nucellar epidermis may divide. The archesporial cell functions directly as megaspore mother cell. It undergoes meiotic divisions, produces a linear or T-shaped tetrad, and the chalazal

Fig. 37.16.1A–L. *Frankenia pulverulenta.* **A** Fertilized embryo sac. **B, C** Embryo sac with zygote and free-nuclear endosperm. **D** Later stage with two-celled proembryo. **E, F** Embryo sac at globular proembryo, endosperm nuclei aggregate in the micropylar region. **G, I, K** Outline diagrams for H, J and L. **H** Micropylar portion of embryo sac with multicellular micropylar haustorium, cellular endosperm, and part of suspensor. **J** Two-celled micropylar haustorium (each cell coenocytic), and cellular endosperm. **L** Micropylar region of embryo sac with coenocytic micropylar haustorium, from mature seed. (After Walia and Kapil 1965)

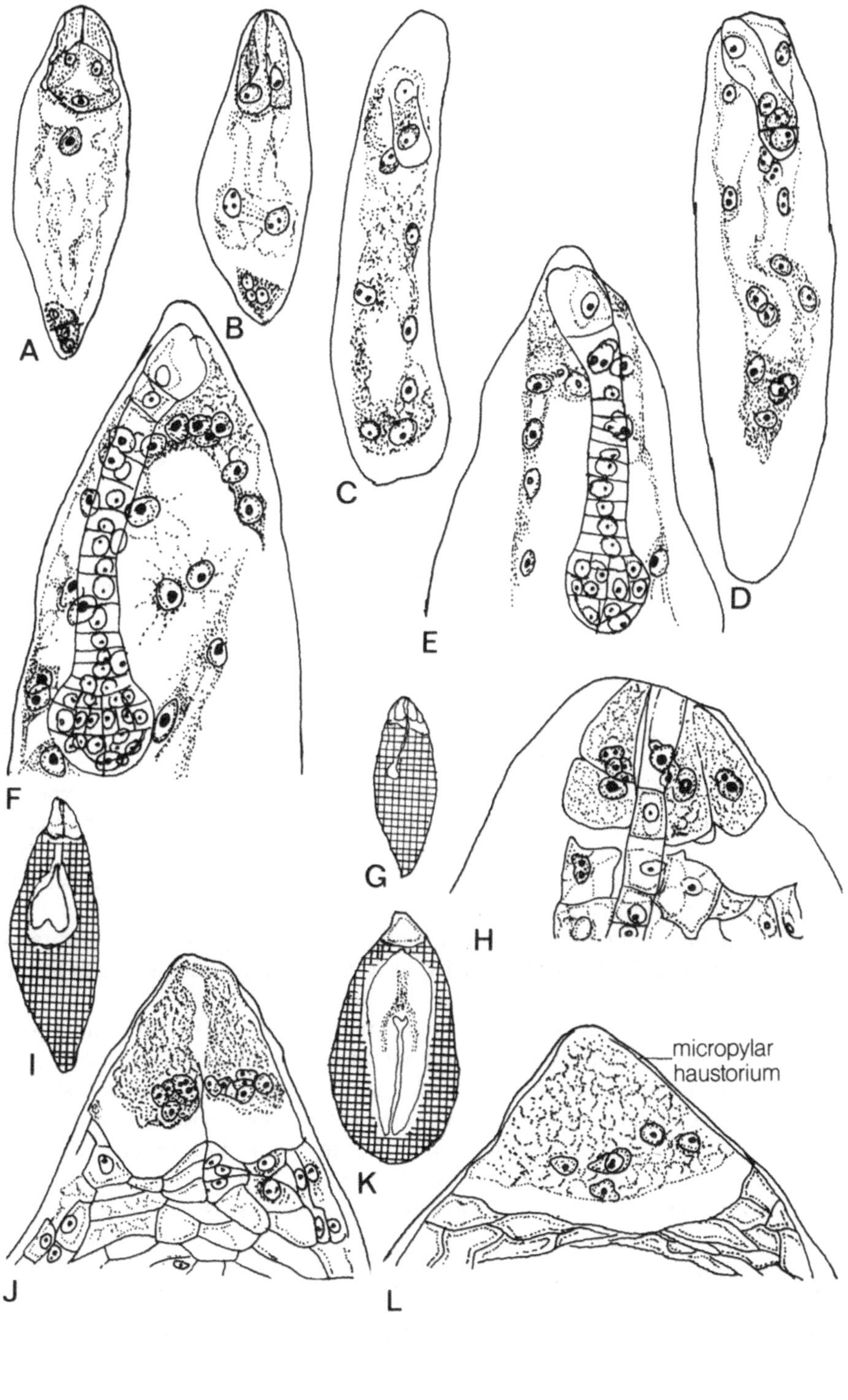
A
B
C
D
E
F
G
H
I
J
K
L
micropylar haustorium

megaspore develops into a Polygonum type of embryo sac. The antipodals persist up to fertilization. The polar nuclei fuse near the egg.

The entry of the pollen tube is porogamous.

The development of the endosperm is of the Nuclear type (Fig. 37.16.1A–F). The nuclei aggregate in the micropylar and chalazal regions; nuclear fusions occur in the micropylar region. Wall formation starts in the chalazal region, and proceeds upward. The coenocytic micropylar part behaves as a haustorium which persists in the mature seed (Fig. 37.16.1G–L). Occasionally, one or more vertical walls are laid down in it (Fig. 37.16.1G–J). The chalazal cells are large and multinucleate. The endosperm cells contain starch and crystals.

The embryogeny conforms to the Solanad type. The mature embryo is dicotyledonous, and straight.

The outer thin-walled epidermis of the testa comprises large cells, and each cell develops a thick-walled papilla; the cells of the inner epidermis are tangentially elongate. The outer epidermis of the tegmen collapses, whereas the inner epidermal cells elongate radially, get filled with tannin, and the inner and tangential walls develop cuticle.

G Bentham (1962a), A Gundersen (1927), K Walia and RN Kapil (1965), MS Yakovlev (1983)

37.17 Elatinaceae

The anther wall comprises the epidermis, fibrous endothecium, two ephemeral middle layers, and secretory tapetum of binucleate cells. The epidermal cells contain tannin in *Bergia ammonioides* (Kajale 1939). The microspore mother cells undergo meiotic divisions. Cytokinesis is simultaneous, and the tetrads are tetrahedral and isobilateral.

The pollen grains are oblate, triangular with three germ pores, and two-celled in *Bergia.* They are three-celled in *Elatine*, and germinate in situ in the cleistogamous flowers of *E. triandra.*

The ovule is anatropous, bitegmic and crassinucellar with micropyle formed by both the integuments. The archesporium is one- to three-celled, only one is functional. It cuts off a parietal cell; the parietal tissue is massive in *B. campensis* (Raghavan and Srinivasan 1940). The divisions in nucellar epidermis also add to the parietal tissue. The megaspore mother cell undergoes meiotic divisions, and T-shaped or linear tetrads are produced. The chalazal megaspore develops into a Polygonum type of embryo sac. The antipodals persist up to fertilization, and the polar nuclei fuse prior to fertilization.

Development of the endosperm is of the Nuclear type. Centripetal wall formation commences from the micropylar pole in *Bergia* at the globular proembryo (Kajale 1939, Dathan and Singh 1971a), and from the chalazal pole upward in *Elatine* (Frisendahl 1927). Three to five layers persist in mature seed.

The embryogeny corresponds to the Solanad type. The suspensor is uniseriate, three-celled.

The seed is exotegmic, albuminous with a large embryo. Both the integuments contribute to the formation of seed-coat. The outer epidermis of the testa is thin-walled, unspecialized, and contains tannin. The inner epidermis gets crushed, or shows annular thickenings on radial walls. The outer epidermis of the tegmen comprises thick-walled, pitted, stellate cells. They may be lignified. The inner thin-walled cells contain tannin in *Bergia*.

ASR Dathan and D Singh (1971a), A Frisendahl (1927), LB Kajale (1939), R Lemesle (1929), TS Raghavan and VK Srinivasan (1940)

37.18 Caricaceae

The anther wall comprises the epidermis, fibrous endothecium, an ephemeral middle layer, and glandular tapetum. The microspore mother cells undergo meiotic divisions, and produce tetrahedral tetrads. Cytokinesis is simultaneous.

The pollen grains are three-colporate with reticulate exine; they are two-celled at the dispersal stage.

The ovule is anatropous, bitegmic and crassinucellate with micropyle formed by both the integuments. The archesporial cell cuts off a parietal cell; the parietal tissue is massive. The development of the female gametophyte is of the Polygonum type (D. Singh 1970c). Tetrasporic development is reported in *Carica candamarensis*, *C. chryopetata*, *C. papaya* and *C. pentagora* (Heilborn 1928). The antipodals are ephemeral. A hypostase is present.

The entry of the pollen tube is porogamous, and double fertilization occurs. The pollen tube is broad and its remnants persist. The pollen tube takes 10 days to reach the nucellus (under greenhouse conditions in *Carica papaya* – Foster 1943).

Development of the endosperm is of the Nuclear type. Cytoplasmic nodules are nucleate as well as enucleate in *Carica papaya* (D. Singh 1960); only enucleate nodules in *C. candamarensis* (Dathan and Singh 1976). Wall formation is centripetal in the endosperm which stores fatty food reserve.

The embryogeny conforms to the Onagrad type-Trifolium variation in *Carica* (Johansen 1950). Dathan and Singh (1976) describe it as of irregular type. The division of zygote is transverse, and a vertical division occurs in the apical cell. The basal cell divides transversely, obliquely or vertically. The mature embryo is large and spatulate. The development of the embryo requires reinvestigation.

The seed is albuminous, and exotegmic with multiplicative pulpy testa. The outer region of the testa is thin-walled (sarcotesta). The inner region comprises compact brown-walled cells. The inner epidermis contains calcium oxalate crystals. The outer epidermis of the tegmen comprises thick-walled lignified longitudinal fibres, the mesophyll may be sclerotic, and the inner epidermal cells are slightly thickened and contain tannin.

SP Agharkar and I Banerji (1930), ASR Dathan and D Singh (1970, 1976), JB Fischer (1980), LT Foster (1943), O Heilborn (1928), DA Johansen (1950), D Singh (1960, 1970c)

37.19 Loasaceae

The development of anther and pollen has not been studied.

The ovule is anatropous, tenuinucellate and unitegmic with a fairly long micropyle. The archesporial cell functions directly as megaspore mother cell. The latter undergoes meiotic divisions, and produces a linear tetrad. According to Kratzer (1918), the micropylar megaspore is functional which appears doubtful. In *Loasa bergii* (Garcia 1962a), the chalazal megaspore forms a Polygonum type of embryo sac. The antipodals persist. The embryo sac shows a constriction in the middle region separating the micropylar and chalazal regions. The endothelium is well developed, and in *Blumenbachia hieronymi* (Garcia 1962b) it surrounds only the middle region of the embryo sac.

Development of the endosperm is of the Cellular type. In *Blumenbachia hieronymi* (Garcia 1962b), the micropylar and chalazal chambers are formed following the division of primary endosperm nucleus. The micropylar chamber develops into a haustorium of two binucleate cells with hypertrophied nuclei (Fig. 37.19.1A, B). The chalazal chamber undergoes two successive transverse divisions and four cells are formed in a row. The basal cell develops into the chalazal haustorium and its branches encroach upon integumentary cells. The upper two cells divide and give rise to the cellular endosperm. The uppermost cell also contributes to the endosperm proper, but some of its daughter cells become haustorial.

The uninucleate chalazal haustorium of *Loasa burgii* is balloon-shaped (Fig. 37.19.1C–E). It remains healthy up to the initiation of cotyledons in the proembryo. The micropylar haustorium is branched, coenocytic, with hypha-like projections (Fig. 37.19.1C). The branches ramify inside the ovule and funiculus and reach up to the placenta.

The chalazal haustorium of *Cajophora silvestris* (Garcia 1962a) is elongated with a much-branched terminal portion containing a large hypertrophied nucleus (Fig. 37.19.1F, G). In *C. lateritia* (Kratzer 1918) the haustorium remains unbranched. The micropylar haustorium is short and broad with many nuclei.

In *Mentzelia laevicaulis* (Garcia 1962a) also the multicellular chalazal haustorium (with conspicuous nuclei) is unbranched. The micropylar haustorium is short and broad.

The embryogeny conforms to the Solanad type (Garcia 1962a), and it is assigned to Period I, Megarchetype V in *Loasa lateritia* (see Crètè 1946a, b). The mature embryo is straight with short cotyledons.

The exotestal seed is albuminous with thin-walled oily endosperm. The outer epidermis of the testa is variously thickened and of characteristic cells;

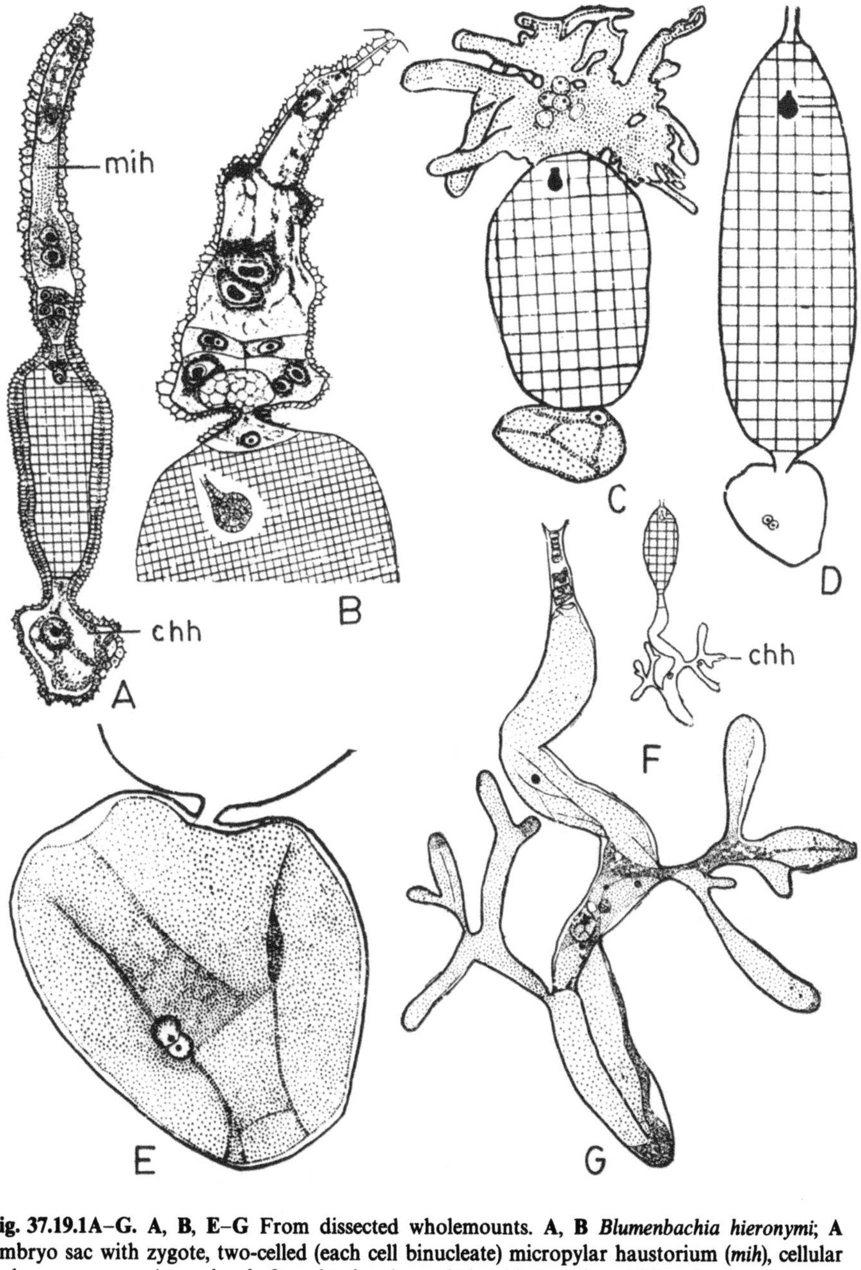

Fig. 37.19.1A–G. A, B, E–G From dissected wholemounts. **A, B** *Blumenbachia hieronymi*; **A** Embryo sac with zygote, two-celled (each cell binucleate) micropylar haustorium (*mih*), cellular endosperm proper (*cross-hatched*), and uninucleate chalazal haustorium (*chh*); **B** Upper portion of embryo sac to show globular proembryo, cellular endosperm and two-celled micropylar haustorium (After Garcia 1962b). **C–E** *Loasa bergii*; **C** Proembryo (*black* in **C, D**), cellular endosperm, and much-branched micropylar and vesicular chalazal haustorium; **D** Outline figure for **E**; **E** Chalazal haustorium from **D**. **F, G** *Cajophora silvestris*; **F** Proembryo, cellular endosperm and chalazal haustorium; **G** Chalazal haustorium from **F**. (After Garcia 1962a)

mesophyll and endothelium generally get crushed. In *Blumenbachia* the outer epidermis is composed of longitudinal fibre-like cells with strong yellowish-brown thickenings on radial walls. In *Cajophora* the outer epidermis consists of large radially elongate cells with stout brown walls, and the radial walls have reticulate thickenings (Corner 1976).

EJH Corner (1976), P Crètè (1946a, b), V Garcia (1962a, b), P Goldblatt, P Bernhardt and J Manning (1989), P Kooiman (1974), J Kratzer (1918)

37.20 Datiscaceae

The embryology has not been fully studied.

The ovule is dizonate in *Datisca cannabina* (Boesewinkel 1984c), and both the integuments are of dermal origin. They are two cell layers thick, and only

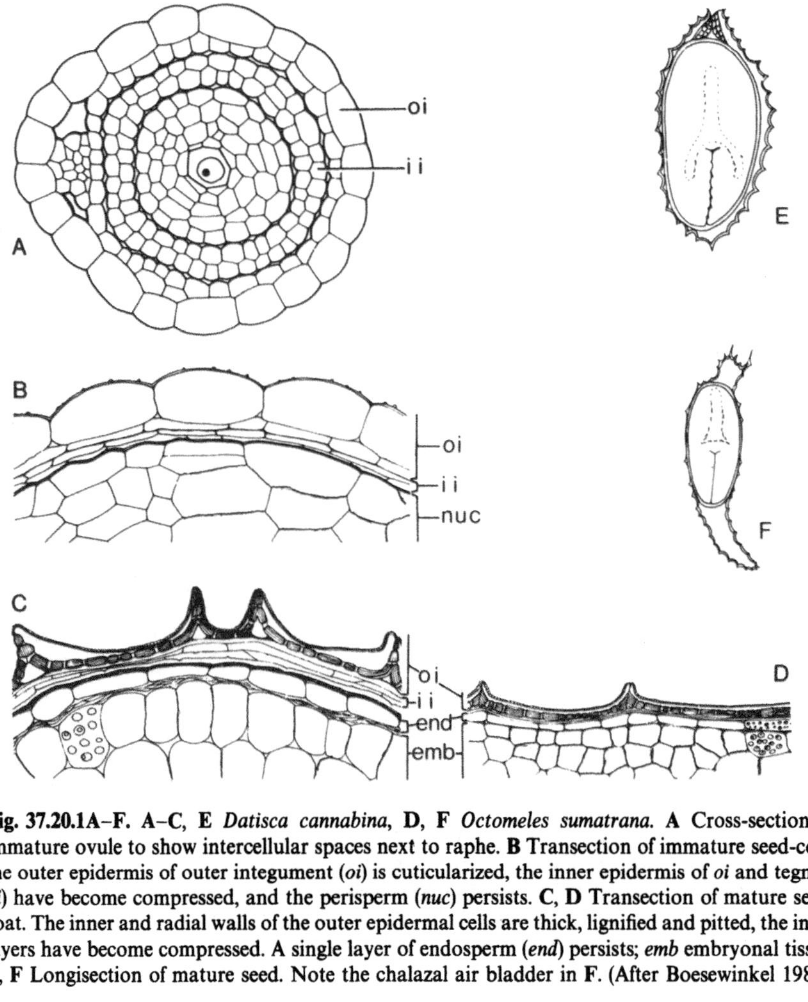

Fig. 37.20.1A–F. A–C, E *Datisca cannabina*, **D, F** *Octomeles sumatrana*. **A** Cross-section of immature ovule to show intercellular spaces next to raphe. **B** Transection of immature seed-coat, the outer epidermis of outer integument (*oi*) is cuticularized, the inner epidermis of *oi* and tegmen (*ii*) have become compressed, and the perisperm (*nuc*) persists. **C, D** Transection of mature seed-coat. The inner and radial walls of the outer epidermal cells are thick, lignified and pitted, the inner layers have become compressed. A single layer of endosperm (*end*) persists; *emb* embryonal tissue. **E, F** Longisection of mature seed. Note the chalazal air bladder in **F**. (After Boesewinkel 1984c)

the tip of the outer integument becomes multilayered. Both the integuments organize the micropyle. Before the maturation of the ovule, longitudinally oriented intercellular spaces appear on the side of the raphal bundle (Fig. 37.20.1A) and in the chalazal region.

The archesporial cell cuts off a parietal cell, the parietal tissue is well developed. The apical cells of the nucellar epidermis undergo a periclinal division. The development of the embryo sac conforms to the Allium type in *D. cannabina* (Himmelbaur 1909). The antipodals are persistent. The polar nuclei fuse prior to fertilization.

Development of the endosperm is of the Nuclear type, later it becomes cellular.

The embryogeny conforms to the Onagrad type.

The operculate (Fig. 37.20.2B) and exotestal small seed has a straight and large embryo (Fig. 37.20.1E) rich in fats and aleurone, but lacks starch. The

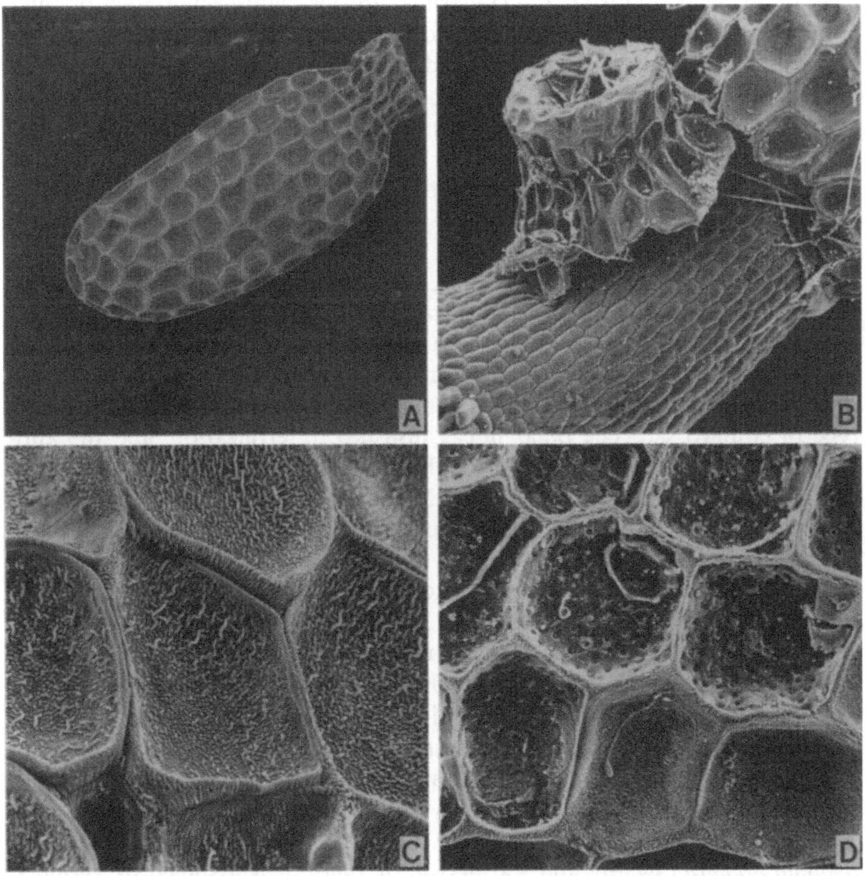

Fig. 37.20.2A–D. (SEM) *Datisca cannabina*. **A** Seed. **B** Germinated seed with seed lid. **C** Portion of seed-coat. **D** Same, outer walls partly removed to show inner pitted walls. (Courtesy Dr F. Bouman, Amsterdam)

outer epidermal cells of the testa enlarge and become cuticularized (Figs. 37.20.1B, E, F, 37.20.2A–C). The inner and the radial walls become thick, lignified and pitted (Figs. 37.20.1C, D, 37.20.2D). The tegmen becomes crushed. The perisperm persists for a while but is finally absorbed. A layer of endosperm persists in the mature seed. The intercellular spaces in the chalazal region become so prominent that an air bladder develops at the tapering chalazal region in mature seed (Fig. 37.20.1F). There is no special rupture layer to lift the operculum (Fig. 37.20.2B).

FD Boesewinkel (1984c), P Crété (1952a, b), W Himmelbaur (1909), P Leins and R Bonnery-Brachten (1977), MA Rau (1962)

37.21 Begoniaceae

The anther wall comprises the epidermis, fibrous endothecium, two or three ephemeral middle layers, and secretory tapetum of binucleate cells. The epidermal cells are protuberant and papillose. The microspore mother cells undergo meiotic divisions and produce tetrahedral tetrads. Simultaneous cytokinesis is reported in *Begonia malabarica* (Maheswari Devi and Naidu 1978).

The pollen grains are two-celled at the dispersal stage. Failure of formation of pollen grains, due to degeneration of pollen mother cells, is reported in *B. anamalayana* (Maheswari Devi and Naidu 1978).

The ovule is anatropous, bitegmic, and crassinucellate (Fig. 37.21.1A, B); weakly crassinucellate in *Begonia picta* (Madhulata 1956). Both the integuments form the zigzag micropyle. The hypodermal archesporial cell cuts off a parietal cell. The megaspore mother cell undergoes meiotic divisions and produces a linear tetrad. The chalazal megaspore develops into a Polygonum type of embryo sac. The polar nuclei fuse before fertilization, and the antipodals are ephemeral. A hypostase differentiates. The pseudo-endothelium is well developed (Maheswari Devi and Naidu 1978, see also Boesewinkel and de Lange 1983) but, at times, it accumulates reserve food which causes degeneration of the embryo sac.

Development of the endosperm is of the Nuclear type (Fig. 37.21.1C, D). The nuclei aggregate in the chalazal region with denser cytoplasm. When 30–40 free-nuclei have been formed, they arrange peripherally around the central vacuole. Wall formation results in a single layer which undergoes periclinal divisions. The inner layer divides repeatedly and gives rise to the rest of endosperm (Swamy and Parmeswaran 1960). The outer layer does not undergo any further divisions, stores aleurone grains and oil, and persists in the mature seed (Fig. 37.21.1E).

The embryogeny conforms to the Onagrad type-Lythrum variation in *Begonia crenata* (Swamy and Parmeswaran 1960). It is assigned to Period I, Series A, Megarchetype IV (Crucifer type).

The seed-coat (Boesewinkel and de Lange 1983) comprises an operculum, collar cells and the remaining testa cells (Fig. 37.21.2A, B). The entire micro-

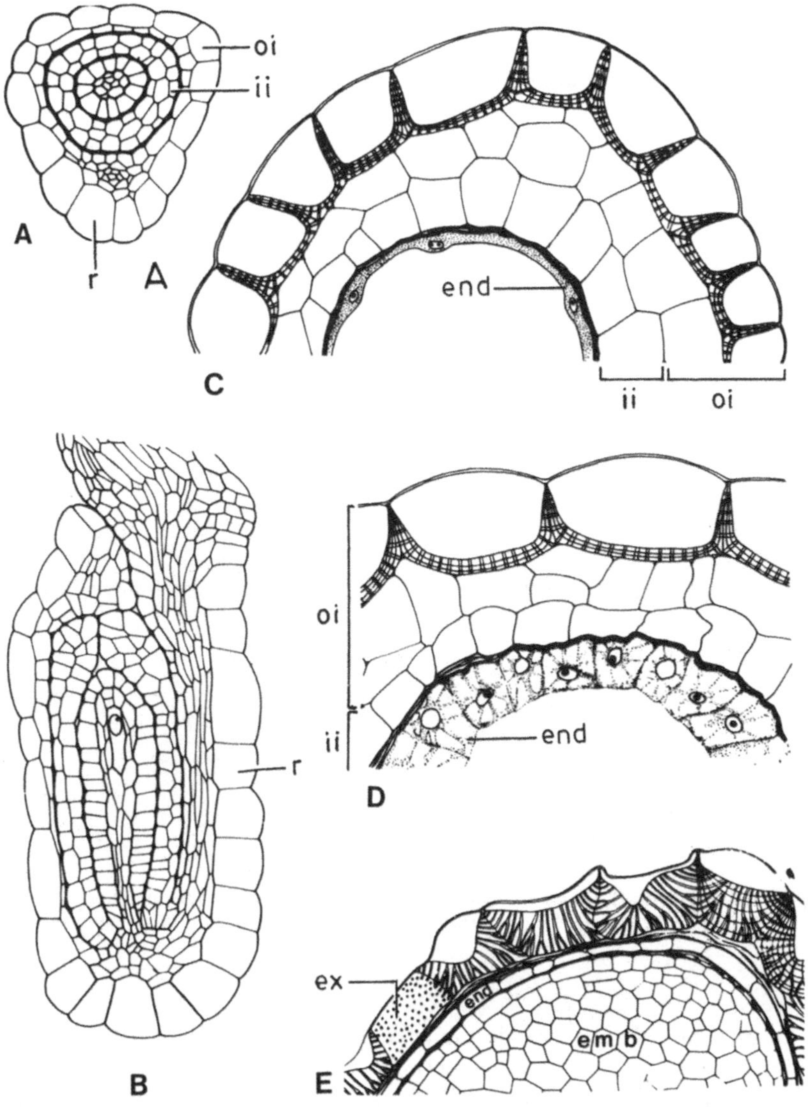

Fig. 37.21.1A–E. *Begonia squamulosa*. **A** Transection of ovule; *oi* outer integument; *ii* inner integument; *r* raphe. **B** Longisection at megaspore mother cell. **C–E** Transections, progressive development of seed-coat. Note the thickenings of exotesta (*ex*); *end* endosperm; *emb* embryo. (After Boesewinkel and Lange 1983)

pylar-hilar region of seed forms the operculum (see lid) which separates along preformed lines of rupture; the walls between the operculum and collar cells split along the middle lamella. Next to the operculum lies a ring of collar cells (Fig. 37.21.2A, B), an exclusive feature of begonian seed. The collar cells are specialized testa cells and the ring occupies nearly one-third of the seed. Sometimes, the collar cells, along with adjacent cells, become twisted.

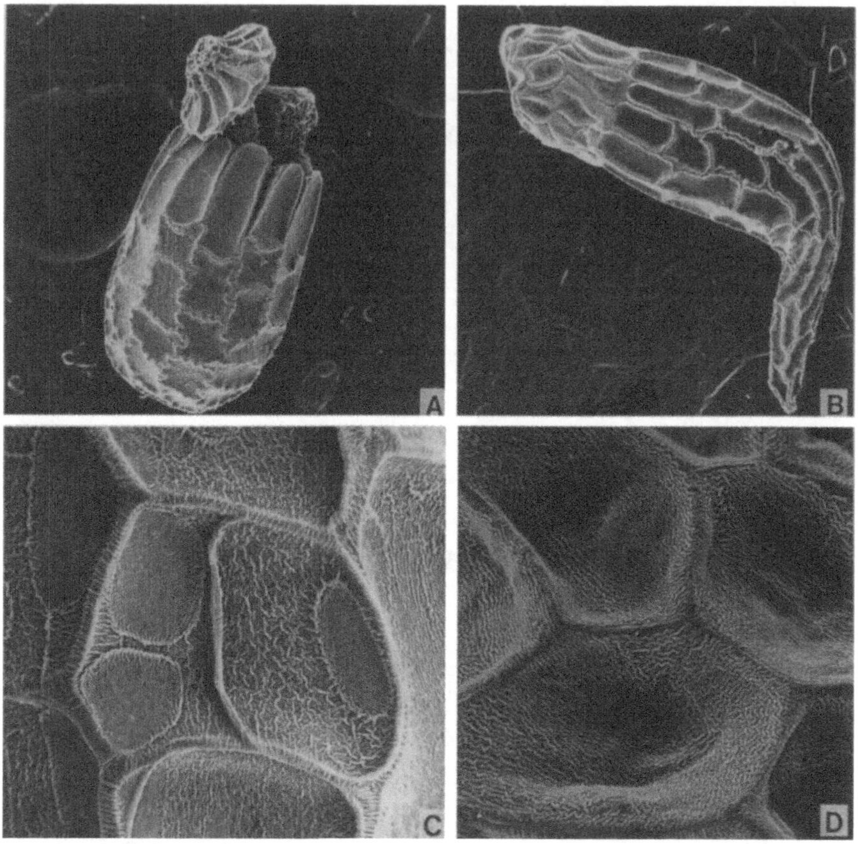

Fig. 37.21.2A–D. (SEM) **A** *Begonia leptetrida*, seed with seed-lid, after germination. **B** *B. patula*, seed. **C** *B. epipsila*, **D** *B. seychellensis*, portion of seed-coat. (Courtesy Dr F. Bouman, Amsterdam)

In *Begonia solananthera* (de Lange and Bouman 1986), the collar cells are elongated; the anticlinal walls between the cells are straight but finely undulated near their chalazal end. The cells bordering the collar are longer and form a ring. The operculum is funnel-like and sunken with hilum in *B. integerrima*, *B. procumbens* and *B. solananthera*. In *B. thomeana* the seeds show a marked adaptation to wind dispersal, and both ends of the seed consist of strongly swollen cells which function as balloons.

During germination the operculum is lifted at the top of the radicle, thus clearing the way for the emergence of seedling.

The ellipsoid seed is small, and only the testa forms the seed-coat represented by the outer epidermis (Figs. 37.21.1E, 37.21.2A–D). The cells expand in surface area. Their inner periclinal walls are strongly lignified with scattered simple pits, whereas the outer walls remain thin. They are filled with amber-coloured contents.

FD Boesewinkel and A de Lange (1983), F Bouman and A de Lange (1983), A de Lange and F Bouman (1986), Madhulata (1956), H. Maheswari Devi and KC Naidu (1978, 1982), W Preil and D Keyser (1976), R Souèges (1939a), BGL Swamy and N Parmeswaran (1960)

37.22 Taxonomic Considerations

Bentham and Hooker (1965a) distribute the families belonging to Violales in four-cohorts: Parietales, Caryophyllales, Guttiferales and Passiflorales. Engler and Diels (1936) include all the families of Violales (Melchior 1964) in Parietales under seven suborders, whereas Hutchinson (1973) places them in seven orders. According to Melchior (1964), if Guttiferales are removed from Parietales, the ·remaining taxon, i.e. Violales, forms a homogeneous and natural group.

The suborder Flacourtiineae comprises the families Flacourtiaceae, Peridiscaceae, Violaceae, Stachyuraceae, Scyphostegiaceae, Turneraceae, Malesherbiaceae, Passifloraceae and Achariaceae.

Flacourtiaceae. Flacourtiaceae closely resemble Violaceae in embryological features, yet there are important variations: a characteristic hook-like structure at the tip of the embryo sac, extra-nucellar embryo sac, variation in the functional behaviour of megaspores, and elongation of the embryo up to half the length of the embryo sac. The structure of seed-coat in Flacourtiaceae varies to such an extent that it forms two groups, Flacourtia group with fibrous exotegmen, and Hydnocarpus group with pachychalaza and without a trace of fibrous exotegmen (Corner 1976). Fibrous exotegmen occurs in Violaceae. Its embryogeny may reveal some more facts.

Hutchinson (1973) includes Flacourtiaceae in Bixales along with Bixaceae, Cistaceae, Cochlospermaceae (see Tables 37.22.1 and 37.22.2) and Peridiscaceae. Bixaceae differ from Flacourtiaceae in endothecial thickenings confined to dehiscence slits, micropyle formed by both integuments, one or two archesporial cells, absence of a parietal cell, and irregular embryogeny. The embryo does not elongate in Bixaceae and Cistaceae, and dye cells are present in the outer integument in Bixaceae. Cistaceae show a cap-like structure at the tip of synergid, Solanad type of embryogeny, and a bulge at the base of integuments in the region of attachment. Moreover, the seeds are exarillate. Therefore, Flacourtiaceae are rightly placed near Violaceae.

Violaceae. Bentham and Hooker (1965a) place Violaceae in Parietales along with Papaveraceae, Resedaceae, etc., Bessey (1915) in Guttiferales between Turneraceae and Bixaceae, and Hutchinson (1973) and R. Dahlgren (1980b, 1981) in Violales.

D. Singh (1963, 1970a) states that the structure of seed in Violaceae resembles that of Cistaceae and Bixaceae of suborder Cistineae, Caricaceae of Caricineae, and Resedaceae of the order Papaverales (Melchoir 1964). The seed is endospermic with an erect and spatulate embryo in the first three taxa. In

Resedaceae the embryo is curved, and the seed is exotegmic with tanniniferous hypostase. Both the integuments take part in the formation of seed-coat. The inner epidermis of the outer integument contains calcium oxalate crystals in Violaceae, Caricaceae and Resedaceae. The outer epidermis of the inner integument is the main mechanical layer in Violaceae, Cistaceae, Bixaceae and Resedaceae. In Violaceae and Resedaceae sclereids in the chalazal region plug the gap in sclerenchymatous layer. The sclereids are absent in the other three families. Thus, the development and structure of violaceous seed show greater similarities to those of Resedaceae. But the anatropous ovule, Asterad type of embryogeny, and massive endosperm in Violaceae differ from campylotropous ovule, Onagrad type of embryogeny, curved embryo, and scanty endosperm in Resedaceae (D. Singh 1963). While considering the structure of seeds of Capparidaceae, Resedaceae and Violaceae, D. Singh and Gupta (1968) suggest that these families are closely related, and have probably been derived from a common ancestor similar to the existing Capparidaceae, but with anatropous ovules, through parallel evolution.

Stachyuraceae. Stachyuraceae, a unigeneric family, have been included in Theaceae (Ternstroemiaceae – Bentham and Hooker 1965a), but have been separated on morphological grounds. Hutchinson (1973) retains Stachyuraceae in Hamamelidales near Buxaceae. Both Stachyuraceae and Buxaceae have superior ovary. Takhtajan (1969, 1980) places them in Violales near Flacourtiaceae. Stachyuraceae show several violaceous features: anatropous, bitegmic, crassinucellate ovule with micropyle formed by both integuments, presence of funicular aril, Polygonum type of embryo sac (Sâto 1976b), Nuclear endosperm, and Solanad type of embryogeny.

Mathew and Chaphekar (1977) and Mathew (1978) observed several important embryological similarities between Stachyuraceae and Theaceae: absence of integumentary tapetum, occasional presence of bisporic embryo sac, fusion of polar nuclei before fertilization, Solanad type of embryogeny, similar structure of suspensor, and presence of polyembryony. According to them, Stachyuraceae have affinities with Flacourtiaceae and Hamamelidaceae, the families with monosporic (Polygonum) type of embryo sac, and Theaceae, the family with bisporic type. They support the inclusion of Stachyuraceae near Theaceae in the order Theales, according to the classification of Cronquist (1968; in 1981, Stachyuraceae have been included in Violales), and negate its relationship with Hamamelidaceae. This alliance is not convincing.

In contrast to the exotegmic fibrous seed-coat of Flacourtiaceae (of Flacourtia group) and Violaceae, the tegmen in Stachyuraceae gets crushed, whereas the outer epidermis of the testa becomes sclerotic and forms the seed-coat (Mohana Rao 1974a). The absence of fibrous exotegmen negates its alliance with Flacourtiaceae (see Takhtajan 1969, 1980). Stachyuraceae show resemblances to Hamamelidaceae, where the entire testa is sclerotic and the tegmen gets crushed. The theaceous testa is rather of a different nature (Mohana Rao 1974a). It has a thin-walled epidermis followed by many layers of pitted and lignified sclerenchyma cells (mesotesta).

The foregoing discussion is clearly indicative of the relationship between Stachyuraceae and Hamamelidaceae.

Turneraceae. Turneraceae and Passifloraceae (Table 37.22.2) of suborder Flaucortiineae are embryologically allied. Turneraceae is placed in Violales by Takhtajan (1966, 1980), and in Loasales by Hutchinson (1973). Vijayaraghavan and Kaur (1966) and D. Kaur (1970) have compared the embryological features of Turneraceae, Passifloraceae, Flacourtiaceae and Loasaceae. Turneraceae show marked dissimilarities with Loasaceae in two-celled pollen grains, anatropous, bitegmic, crassinucellate ovules, ephemeral antipodal cells, Nuclear endosperm without haustoria, Onagrad type of embryogeny, and presence of aril (D. Kaur 1970). In most of these respects, it closely resembles Passifloraceae. The seed-coat is formed by a single integument in Loasaceae, whereas in Turneraceae and Passifloraceae it is formed by both integuments. The cells of the outer epidermis of the inner integument are obliquely oriented in Turneraceae and appear transitional to fibres, whereas the exotegmen is sclerotic with radially oriented cells in Passifloraceae.

Passifloraceae. Bentham and Hooker (1965a) place Passifloraceae in Passiflorales with Turneraceae (Table 37.22.2) and Cucurbitaceae. Hutchinson (1973) includes them in Passiflorales, and places the order close to Cucurbitales. Takhtajan (1980) assigns them close to Turneraceae in Violales. D. Singh (1970b) has summarized the embryological features of Cucurbitaceae in which they differ from Passiflorales: flask-shaped nucellus, equal thickness of integuments, vascular supply extending up to the tip of the outer integument, endosperm with chalazal haustorium, seed-coat from outer integument alone, formation of main sclerenchymatous layer from the outer epidermis of the outer integument, insignificant endosperm in seed, and absence of aril. The presence of ruminate endosperm, and succulent aril enclosing the seed are characteristic of Passifloraceae, otherwise this family shows close affinities with Turneraceae (Vijayaraghavan and Kaur 1966).

Cistaceae. The suborder Cistineae of Violales (Melchior 1964) comprises Cistaceae, Bixaceae, Sphaerosepalaceae and Cochlospermaceae. The embryology of Sphaerosepalaceae has not been investigated, and meagre data are available about Cochlospermaceae (see Schnarf 1931, Dathan and Singh 1972a). Bentham and Hooker (1965a) place Cistaceae and Bixaceae in Parietales, whereas Hutchinson (1973) includes them in Bixales. The three families – Cistaceae, Bixaceae and Cochlospermaceae – resemble one another (Table 37.22.1) in secretory tapetum, bitegmic, crassinucellate ovules with nucellar cap, single vascular strand terminating at the chalaza, Polygonum type of embryo sac, frequent occurrence of twin embryo sacs, Nuclear endosperm, spatulate embryo (curved in Cochlospermaceae), both integuments form seed-coat, and outer epidermis of the inner integument forms the main sclerenchymatous layer in seed.

Orthotropous ovule with a characteristic bulge of the basal cells of the outer integument in the region of attachment with the nucellus, multicellular archesporium with a massive parietal tissue, synergids with cap-like structure at the base, Solanad type of embryogeny, and association of testa with fungus (Kapil and Maheshwari 1964) are the distinguishing features of Cistaceae.

Bixaceae. Bixaceae show the presence of Übisch granules, endothecial thickenings confined to dehiscence slits, single archesporial cell, absence of parietal tissue, irregular embryogeny, presence of dye cells and oil cells in testa, and exotegmen develops into thick palisade-like cells and hypodermal layer (from inside) into hour-glass cells (H. Kaur 1969). Cochlospermaceae are closely allied to Bixaceae. The distinction between the seed structure of the two families is so meagre that Corner (1976) does not segregate them (Table 37.22.1).

The suborder Tamaricineae comprises Tamaricaceae, Frankeniaceae and Elatinaceae. Bentham and Hooker (1965a) place Frankeniaceae and Tamaricaceae along with Caryophyllaceae in Caryophyllineae, and Elatineae in Guttiferales with Hypercineae. Bessey (1915) assigns all the three families to Caryophyllales. Hutchinson (1973) includes Frankeniaceae and Tamaricaceae with Fouquieriaceae in Tamaricales, and Elatinaceae in Caryophyllales. R. Dahlgren (1980b, 1981) erected the order Tamaricales with Tamaricaceae and Frankeniaceae in superorder Violiflorae.

Tamaricaceae. The Tamaricaceae have several important features: two-celled pollen grains, nucellar parietal cells, tetrasporic (Fritillaria, Adoxa, Chrysanthemum cineraraefolium, Drusa and Plumbagella) embryo sac, Nuclear endosperm without haustoria, Solanad type of embryogeny with a massive suspensor, frequent occurrence of polyembryony, exalbuminous, exarillate, simple exotestal seed, and seed-coat derived from the outer integument only (the outer epidermis is a mechanical layer, the inner one contains crystals).

The caryophyllaceous characters, such as free-central placentation, three- to six- or 12-porate pollen grains, multicelled archesporium, Caryophyllad type of embryogeny, curved embryo, and perisperm (A.C. Joshi 1936a, Pal 1952) are absent in all the three families; Tamaricaceae, Frankeniaceae and Elatinaceae. The presence of parietal cells in the nucellus, absence of endosperm haustoria, exalbuminous seed, and seed-coat formed only by the outer integument (A.C. Joshi 1936a, Pal 1952) in Caryophyllaceae are features similar in Tamaricaceae. As several types of tetrasporic embryo sacs do not occur in any other violaceous family, B.M. Johri and Kak (1954) conclude that Hutchinson's erection of a separate order Tamaricales is justified.

Frankeniaceae. Walia and Kapil (1965) studied the embryology of *Frankenia hirsuta* and *F. pulverulenta*. In contrast to Tamaricaceae, Frankeniaceae have three-celled pollen grains, single female archesporium directly functions as megaspore mother cell, Polygonum type of embryo sac, Nuclear endosperm with two- to four-celled micropylar haustorium, albuminous seed with a straight embryo without suspensor (in mature embryo), seed-coat derived from both integuments, and finger-like projections from the outer epidermis of the testa which gives a warty·appearance.

Table 37.22.1. Embryology of Cistaceae, Bixaceae and Cochlospermaceae (see Dathan and Singh 1972a)

	Cistaceae	Bixaceae	Cochlospermaceae
Pollen grains	Usually 3-colporate, suboblate-prolate, 2-celled	3-porate, or 3-colporate, oblate, 2- or 3-celled	3-colporate, prolate, spheroidal, subprolate
Gynoecium	Tricarpellary or 5–10 carpels	2-, rarely 3-carpellary	3–5-carpellary
Ovule	Orthotropous, bitegmic, crassinucellate with nucellar cap	Anatropous, bitegmic, crassinucellate with nucellar cap	Ana- or anacampylotropous, bitegmic, crassinucellate with nucellar cap
Integuments	Outer integument 2-layered, inner 4-layered	Outer integument 4-layered, inner 6–7-layered	Outer integument 4-layered, inner 6–7-layered
Vascular supply	Single, terminating at chalaza	Single, terminating at chalaza	Single, terminating at chalaza
Embryo sac	Polygonum type, twin embryo sacs occur, antipodals ephemeral	Polygonum, twin embryo sacs frequent, antipodals ephemeral	Polygonum type, twin embryo sacs frequent, antipodals ephemeral
Endosperm	Nuclear type	Nuclear	Nuclear
Embryo	Spatulate, bent, coiled or folded	Spatulate, large, cotyledons broad, incurved at the apex	Curved, radicle rather short, cotyledons ligulate
Mature seed	Ovate, endospermous	Ovate, endospermous	Reniform, endospermous
Chalazal modifications	Chalazal region distinguishable into six zones, podium and tanniniferous postament conspicuous	8 zones, podium and tanniniferous postament conspicuous	7 zones, podium and tanniniferous postament conspicuous
Seed-coat	Both integuments form seed-coat, distinguishable into 3 zones, outer epidermis of inner integument forms the sclerenchymatous layer	Both integuments, distinguishable into 8 zones, outer epidermis of inner integument forms the main sclerenchymatous layer, dye cells in outer parenchyma, inner hypodermis forms hour-glass-shaped cells	Both integuments, distinguishable into 6 zones, outer epidermis of inner integument forms mechanical layer, gum cavities in inner parenchyma

Frankeniaceae resemble Elatinaceae in secretory two-layered tapetum, simultaneous cytokinesis, Polygonum type of embryo sac, Nuclear endosperm with initiation of wall formation from the chalazal end, and Solanad type of embryo.

Walia and Kapil (1965) emphasize that the assignment of Frankeniaceae to Tamaricales of Hutchinson and Takhtajan, and to Caryophyllinae of Bentham and Hooker, is unjustifiable. There is also a remote possibility of its relationship with Fouquieriaceae or Plumbaginaceae. The proper position of Frankeniaceae is in the order Parietales, near Elatinaceae, as suggested by Wettstein (1935) and Rendle (1952).

Elatinaceae. In addition to the features mentioned along with Frankeniaceae, Elatinaceae have two-celled pollen, a zigzag micropyle, massive parietal tissue, ephemeral antipodals, and exotegmic seeds with massive perisperm and scanty endosperm. The presence of parietal cells in nucellus, and exalbuminous seed with perisperm and absence of endosperm haustoria show the similarity with Caryophyllaceae. Takhtajan (1980) places Elatinaceae in Theales. Corner (1976), on the basis of seed structure, relates it to Hypericaceae. Dathan and Singh (1971a) also confirm that the cells of the outer epidermis of the inner integument are similar to those of *Hypericum*. The massive nucellus in Elatinaceae is exceptional.

Caricaceae. The suborder Caricineae comprises the lone taxon Caricaceae. Hutchinson (1973) places Caricaceae in Cucurbitales next to Begoniaceae and Datiscaceae. D. Singh (1963) refutes their affinities with Cucurbitaceae on embryological grounds. Most embryological features are in conformity with those of the order Violales. Tetrasporic embryo sac in some species, persistent pollen tube, cytoplasmic endosperm nodules, and sarcotestal seed are its special features. The Nuclear endosperm, and Onagrad type of embryogeny resemble those of Begoniaceae and Datiscaceae. The structure of the seed is comparable to that of Violaceae.

Loasaceae. The suborder Loasineae comprises Loasaceae. Bentham and Hooker (1965a) include it in Passiflorales, Engler and Diels (1936) in Parietales, and Hutchinson (1973) in Loasales along with Turneraceae. Cronquist (1981) places it in Violales, Takhtajan (1980) in Loasales, whereas R. Dahlgren (1981) assigns it to superorder Loasiflorae, order Loasales, in sympetalous group with iridoid compounds. Vijayaraghavan and Kaur (1966) point out marked dissimilarities between Turneraceae and Loasaceae (Table 37.22.2).

The hemianatropous, unitegmic, tenuinucellate ovule, archesporial cell directly functioning as megaspore mother cell, presence of endothelium, persistent (rarely ephemeral) antipodals, and Cellular endosperm with aggressive micropylar and chalazal haustoria (Garcia 1962a) show the affinities with sympetalous families. Garcia (1962a) compares Solanad type of embryogeny with that in Solanaceae (Polemoniales), wherein iridoid compounds are absent. As the androecium is extremely variable in Loasaceae, Wagenitz (1977a, b) did not assign them to any of the existing orders in Asteridae. According to

Table 37.22.2. Embryology of Flacourtiaceae, Turneraceae, Passifloraceae and Loasaceae (See Vijayaraghavan and Kaur 1966, D. Kaur 1970)

	Flacourtiaceae	Turneraceae	Passifloraceae	Loasaceae
Anther tapetum	Secretory, cells 2–4-nucleate	Secretory, cells multinucleate	Secretory, cells multinucleate	—
Pollen grains	2-celled, colporate	2-celled, colporate	2-celled, colporate	—
Ovule	Orthotropous, bitegmic, crassinucellate, hypostase present	Anatropous, bitegmic, crassinucellate, hypostase present	Anatropous, rarely orthotropous, bitegmic, crassinucellate, hypostase absent	Anatropous, unitegmic, tenuinucellate; hypostase absent
Endothelium	Absent	Absent	Absent	Present
Embryo sac	Polygonum type, antipodal cells ephemeral	Polygonum type, antipodals ephemeral	Polygonum type, antipodals ephemeral	Polygonum type, antipodals persistent, rarely ephemeral
Endosperm	Nuclear type, haustoria absent	Nuclear, haustoria absent	Nuclear, haustoria generally absent, ruminate	Cellular, both micropylar and chalazal haustoria develop
Embryogeny	—	Onagrad type	Onagrad	Solanad
Seed	Arillate, exotegmic, seed-coat formed by both integuments	Arillate, exotegmic, seed-coat formed by both integuments	Arillate, endotegmic, seed-coat formed by both integuments	Exarillate, exotestal

R. Dahlgren (1977b), iridoids and seco-iridoids occur in Loasaceae, and they belong to "Iridoid Sympetalae" allied to Dipsacales. In Dipsacales Solanad type of embryogeny occurs in *Morina longifolia* (Vijayaraghavan and Sarveshwari 1968).

Datiscaceae and Begoniaceae. The families Datiscaceae and Begoniaceae form the suborder Begoniineae. Bentham and Hooker (1965a) place them in Passiflorales along with Loasaceae. Hutchinson (1973) includes them in Cucurbitales. The similarities between these two families are: anatropous, bitegmic ovules with micropyle formed by both integuments, single archesporium producing a parietal cell, Nuclear endosperm, Onagrad type of embryogeny, and exarillate exotestal seed. The Allium type of embryo sac and the exalbuminous seed of Datiscaceae differ from Polygonum type of embryo sac with endothelium, and persistent aleurone layer of endosperm in Begoniaceae. The collar cells of the testa, typical of a begonian seed, are absent in the seed of Datiscaceae. Both families differ from Cucurbitaceae (Sect. 38.2, 3) in several respects. Their position in Violales is quite satisfactory.

38 Cucurbitales

The family Cucurbitaceae represents the order Cucurbitales.

38.1 General Features

The order Cucurbitales has the following embryological features: (1) glandular anther tapetum, (2) two- or three-celled pollen grains, (3) anatropous, bitegmic, crassinucellate ovule with vascular supply reaching up to the tip of integument, micropyle formed by both integuments, (4) Polygonum type of embryo sac with ephemeral, rarely haustorial, antipodals, occasional presence of Allium type of embryo sac, (5) polysiphonous persistent pollen tube, (6) Nuclear type of endosperm with tubular haustorium, occurrence of enucleate cytoplasmic nodules, and coenocytic "mounds", (7) Onagrad or Solanad type of embryogeny, (8) nucellar or synergid polyembryony, and apomictic embryo sacs, (9) straight embryo with flat cotyledons, and (10) exalbuminous seed with characteristic exotestal seed-coat.

38.2 Cucurbitaceae

The development of the anther wall is of the "Dicotyledonous type" (Maheswari Devi and Naidu 1984a). It comprises the epidermis, fibrous endothecium, two ephemeral middle layers, and secretory tapetum of binucleate cells in *Bryonopsis lacinosa* (Maheswari Devi and Naidu 1984a). The tapetum forms a periplasmodium in *Luffa cylindrica*. Simultaneous cytokinesis follows meiotic divisions of microspore mother cells, and the tetrads are tetrahedral, isobilateral and decussate (Maheswari Devi and Naidu 1984a).

The family is multipalynous with various apertural forms. The pollen grains are usually shed at the two-celled stage, as in *Luffa tuberosa* (Sarada and Pullaiah 1985), but three-celled condition is reported in *B. lacinosa*.

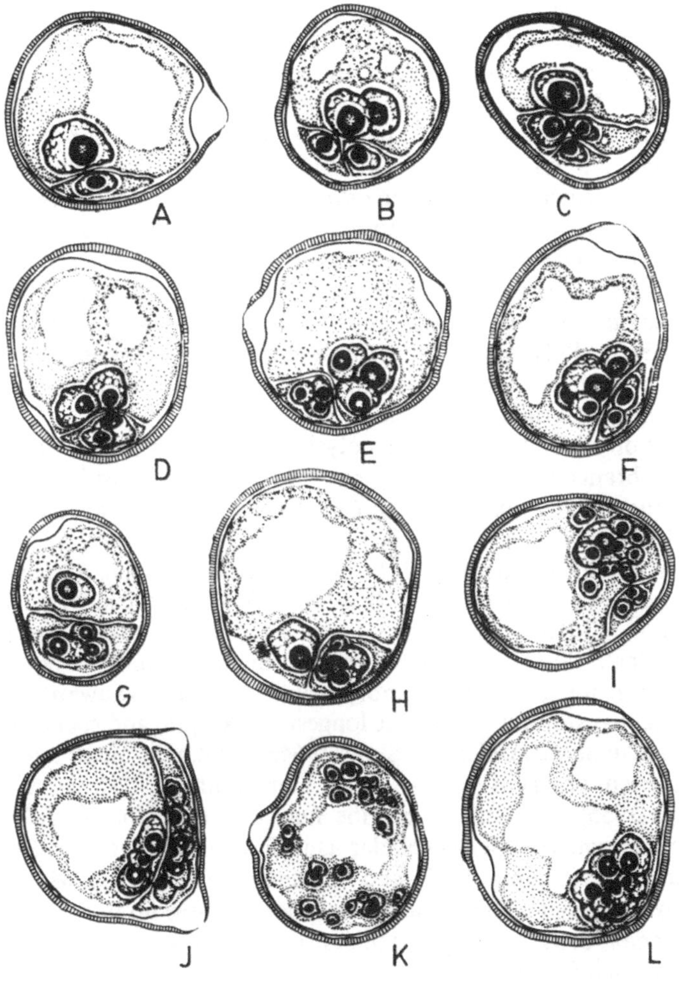

Fig. 38.2.1A–L. Variations in the development of the male gametophyte in tetraploid muskmelon (*Cucumis melo*). **A** Two-celled normal pollen grain. **B, C** Pollen grains with two (**B**) and three (**C**) generative cells, and bi- (**B**) and uninucleate (**C**) vegetative cell. **D–F** Two, three and four nuclei in vegetative cell, and two- (**D–F**) and three-nucleate generative cell. **G–I** Uni- (**G, H**) and multinucleate vegetative cell, and three- (**G, I**) and four-nucleate generative cell. **J** Three-nucleate vegetative and six-nucleate generative cell. **K, L** Pollen grains with 23 and nine free nuclei. In several pollen grains, the nuclei in vegetative and generative cell are of unequal size. (After Dzevaltosky and Litvinenko 1973, see Poddubnaya–Arnoldi 1976)

In a tetraploid variety of *Cucumis melo* (see Poddubnaya-Arnoldi 1976), the normal pollen grain is two-celled (Fig. 38.2.1A), but variations occur. The vegetative cell may be uni- (C, G, H), bi- (B, D) or multinucleate (E, F, I, J). The generative cell also shows variations. It may be bi- (B, D), tri- (C, E, G, H, I) or multinucleate (J). The pollen grain is multinucleate in K and L and does not show the delimitation into the vegetative and generative cell.

The ovules are anatropous, bitegmic and crassinucellate. The funicular bundle continues unbranched up to the tip of the outer integument. The micropyle has both exostome and endostome, or only the endostome, as in *Melothria maderaspatana* (B.M. Johri and Roy Chowdhury 1957) and *Luffa tuberosa*. The cells of the nucellar epidermis divide and produce a nucellar beak as in *M. leucocarpa* (Dathan 1974a) and *L. tuberosa*. The hypodermal archesporial cell cuts off a parietal cell. The parietal tissue is well developed. The megaspore mother cell undergoes meiotic divisions producing a linear or T-shaped tetrad. The chalazal megaspore develops into a Polygonum type of embryo sac. The development of the embryo sac from the epichalazal megaspore is reported in *Lagenaria* (Dzevaltovsky 1963). The synergids are fairly large, and the three antipodals are ephemeral. Allium type of embryo sac occurs in *Benincasa cerifera* (Chopra and Agarwal 1960) and *B. hispida* (Dzevaltovsky 1963); Polygonum type of development occurs rarely. In *B. cerifera* twin megaspore mother cells and twin embryo sacs are frequent. A hypostase develops in *Citrullus* and *Melothria*.

The germination of pollen grains is polysiphonous. The pollen tube is persistent, it may branch and dilate at the base of nucellar beak, as in *Marah macrocarpa* (Dathan and Singh 1971b). In *Citrullus vulgaris* (Gulyayev 1961), two to four pollen tubes enter the micropyle.

Development of the endosperm is of the Nuclear type. The chalazal region extends into a long tubular haustorium (Fig. 38.2.2A–I), except in *Blastania garcini* (Chopra 1953), *Ctenolepis garcinii* and *Corallocarpus conocarpus* (D. Singh 1964a). Centripetal wall formation is initiated at the micropylar region at the globular proembryo, and proceeds downward. The size of the chalazal haustorium varies. The longest (19,000 μm) and coenocytic haustorium occurs in *Sechium edule* (D. Singh 1964a). Protoplasmic streaming movement has been observed in the living haustorium of *Trichosanthes anguina* (Chopra 1955). In many genera the nuclei become polyploid. The haustorium may become completely cellular as in *Bryonia* (Chopra and Basu 1965), *Coccinia indica* (Chopra 1955) and *C. lanatus* (D. Singh 1964a), or remains partially coenocytic. The coenocytic portion is active up to the heart-shaped stage of embryo, and then disorganizes. The nuclei become amoeboid and lobulate, and then fragment (Chopra and Sachar 1963). The seed is non-endospermic.

Enucleate cytoplasmic nodules occur in several members (D. Singh 1955a, b, 1964b, Dathan and Singh 1971b). D. Singh (1964b) suggests that their formation is an active method to fill the vacuole. The peripheral cells in the chalazal region of the endosperm proper develop secondary haustoria in *Luffa* spp. (D. Singh 1964a). In *Melothria maderaspatana* coenocytic "mounds" (in the chalazal region) serve as efficient absorbing organs.

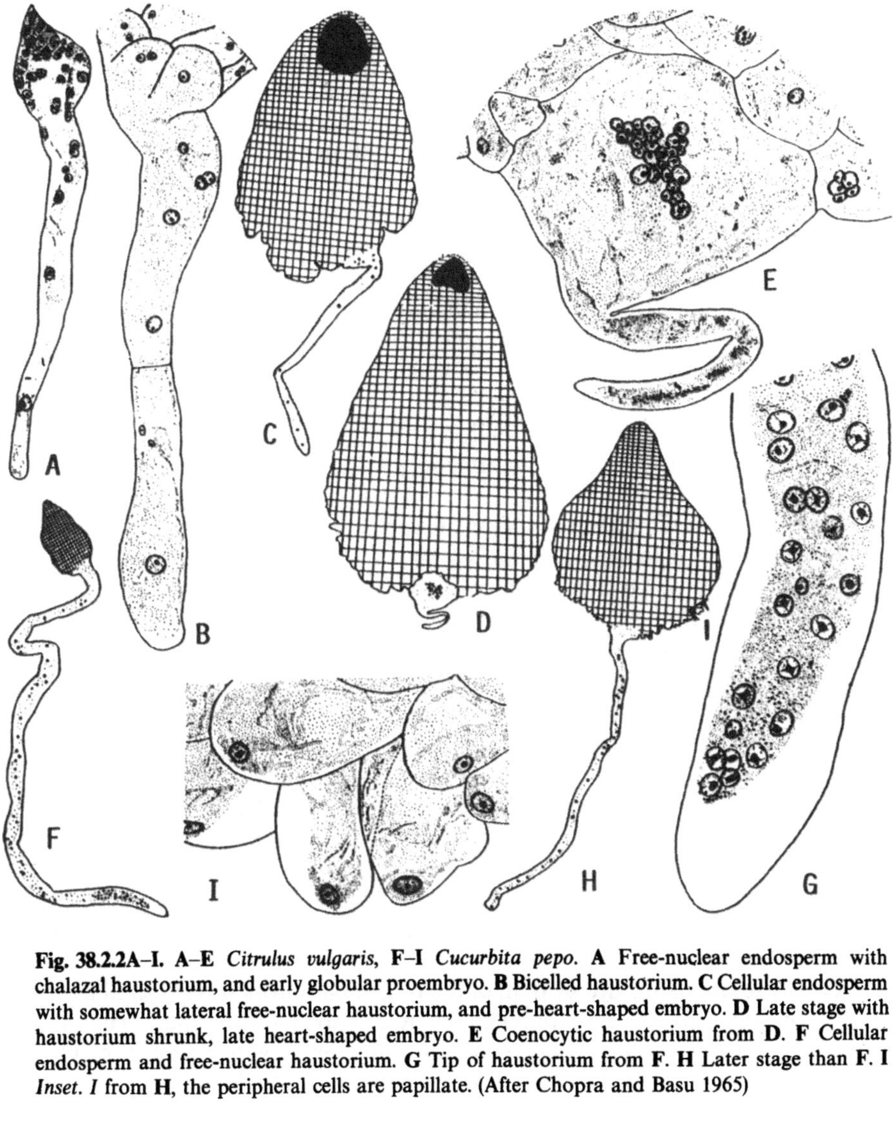

Fig. 38.2.2A–I. **A–E** *Citrulus vulgaris*, **F–I** *Cucurbita pepo*. **A** Free-nuclear endosperm with chalazal haustorium, and early globular proembryo. **B** Bicelled haustorium. **C** Cellular endosperm with somewhat lateral free-nuclear haustorium, and pre-heart-shaped embryo. **D** Late stage with haustorium shrunk, late heart-shaped embryo. **E** Coenocytic haustorium from **D**. **F** Cellular endosperm and free-nuclear haustorium. **G** Tip of haustorium from **F**. **H** Later stage than **F**. **I** *Inset. I* from **H**, the peripheral cells are papillate. (After Chopra and Basu 1965)

The embryogeny corresponds to Myosurus variation of the Onagrad type in *Citrulus colocynthis*, *Dicaelospermum ritchiei*, *Melothria maderaspatana* (D. Singh 1961b), *Cucumis anguria* (Zhalatov and Dzevaltovsky 1975), *Cyclanthera explodens* (Crété 1962), *Ecballium elaterium* (Crété 1960), *Sicyos angulata* (Crété 1958a), *Marah macrocarpa* (Dathan and Singh 1971b) and *Melothria leucocarpa* (Dathan 1974a); Asterad type in *Dactyliandra welwitschii* (Dathan 1974b); Nicotiana variation of the Solanad type in *Cucumis melo* var. *pubescens*, *Fevillea cordifolia* (D. Singh 1955a) and *Luffa tuberosa* (Sarada and Pullaiah 1985).

Nucellar polyembryony has been reported in *Cucumis melo* var. *pubescens* (D. Singh 1955a) and *Momordica charantia* (J.S. Agarwal and Singh 1957), synergid embryo in *Sicyos* (Crètè 1958a), and polyembryonate seeds in *Hodgsonia* (see D. Singh 1970d). In *Marah macrocarpa* (Dathan and Singh 1971b), besides the dicotyledonous condition, some embryos with one or two additional imperfectly developed cotyledons are referred to by the authors as hemitri- or hemitetra-cotyledonary.

Mihov and Zagorcheva (1966) report apomictic embryo sacs, and synergid polyembryony in *M. macrocarpa.*

The development and structure of seed-coat has been studied in several genera (B. Singh 1952b, 1953, D. Singh 1961a, 1965a, 1967, 1970d, D. Singh and Dathan 1972a, b, 1974). The tegmen collapses and testa alone forms the seed-coat. Corner (1976) regroups the genera on the basis of the anatomical structure of the seed-coat.

Following the 'formation of endosperm, the nucellus increases a good deal, but most of it disintegrates and only a few layers are left around the endosperm.

Degeneration of the inner integument varies in different genera. In *Bryonia dioica*, *Ecballium elaterium* and *Sicyos angulata*, there is complete degeneration by the time the mature embryo sac is formed (Fig. 38.2.3A, C, E, F). In *Luffa cylindrica* (Fig. 38.2.3D), the inner integument persists in young seeds at the globular proembryo.

Soon after the radial elongation of exotestal cells, they divide tangentially (Fig. 38.2.3A, B, *ex*, ex_1, ex_2, ex_3). The inner layer (ex_2) does not divide further (Fig. 38.2.4A–H). The second division gives rise to the layers ex_1 and ex_3 (Figs. 38.2.3A–C, 38.2.4A–H). In *Echinocystis wrightii*, both ex_1 and ex_3 contribute to the formation of seed hypodermis (Fig. 38.2.4G, H). The epidermal cells at the extreme tip and base of the seeds of *B. dioica*, *Cyclanthera explodens*, *Lagenaria vulgaris* and *L. cylindrica* are always very small. The subepidermal tissue, or seed hypodermis (*hy*) lies between the epidermis and ex_2. The cells of ex_2 become sclerenchymatous, and may elongate to form a palisade-like tissue.

The structure of mature seed, in different genera, shows several similarities. The exotestal cells develop lignified or rod-shaped thickenings. The seed hypodermis is one- to many-layered and usually remains thin-walled. Next to the hypodermis is a single-layered protective covering which is heavily lignified in mature seeds. This is characteristic of the seeds of Cucurbitaceae.

The seed is exalbuminous, and has a large erect and spatulate embryo. The cotyledons contain fatty food reserve.

JS Agrawal and SP Singh (1957), JJ Asana and RD Adatia (1932), G Balasubramanium and DD Sunderaraj (1957), AK Chakravorti (1947), SVS Chauhan (1970), DP Cholakhyan (1965), RN Chopra (1953, 1955), RN Chopra and S Agarwal (1958, 1960), RN Chopra and B Basu (1965), RN Chopra and RC Sachar (1963), RN Chopra and PN Seth (1977), EJH Corner (1976), P Crété (1958a, 1960, 1962), ASR Dathan (1974a, b), ASR Dathan and D Singh (1971b), PK Deshpande, SM Bhuskute and KH Makde (1984), AK Dzevaltovsky (1963), VA Gulyayev (1961), SC Gupta and VM Rajeswari (1977), BM Johri and C Roy Chowdhury (1957), B Longo (1901, 1903), H Maheswari Devi and KC Naidu (1983, 1984a, b), A Mihov and L Zagorcheva (1966), ER Nikam

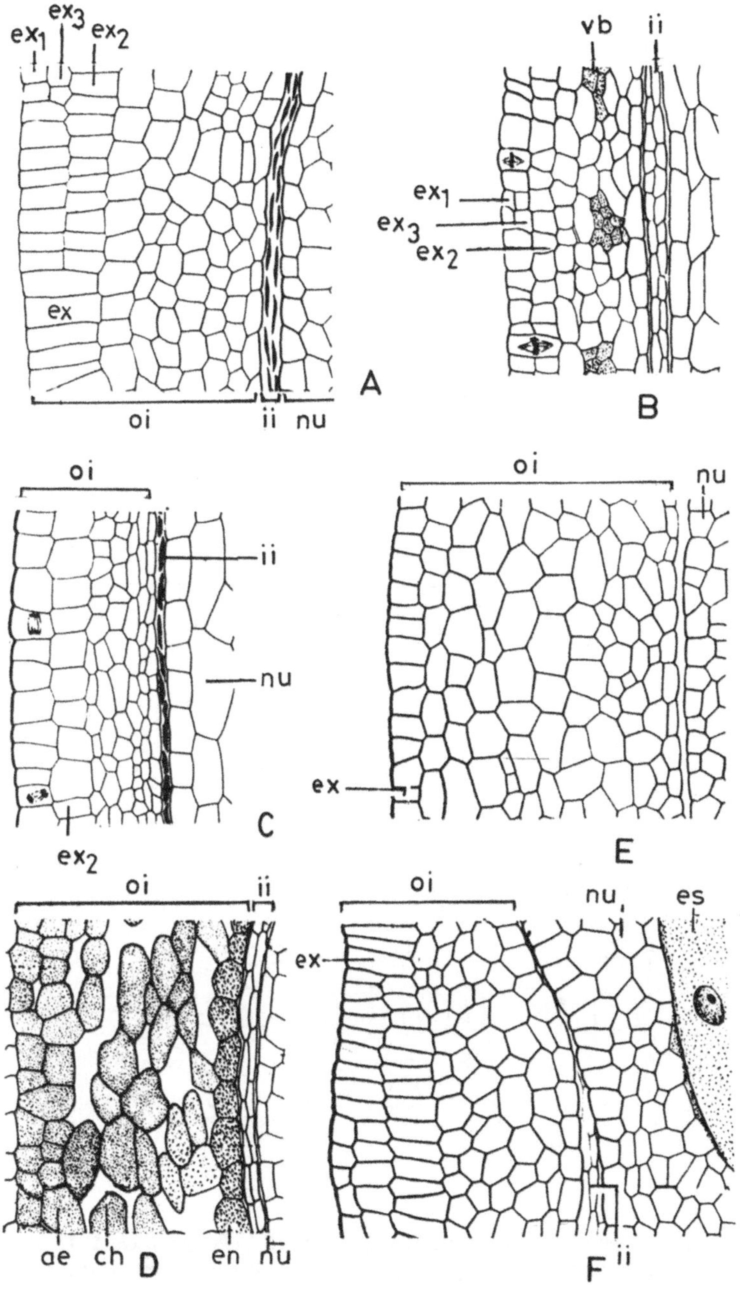

Fig. 38.2.3A–F. **A** *Bryonia dioica*, **B** *Cyclanthera explodens*, **C** *Ecballium elaterium*, **D** *Luffa cylindrica*, **E, F** *Sicyos angulata* – **E** from side and **F** near the base of nucellar beak. Longisections parts of integuments; *ae* aerenchyma; *ch* chlorenchyma; *en* endotesta; *es* part of embryo sac; *ex*, ex_1; ex_2, ex_3, exotesta and its derivatives; *nu* nucellus; *oi* outer integument; *vb* vascular bundle. (After B. Singh 1953)

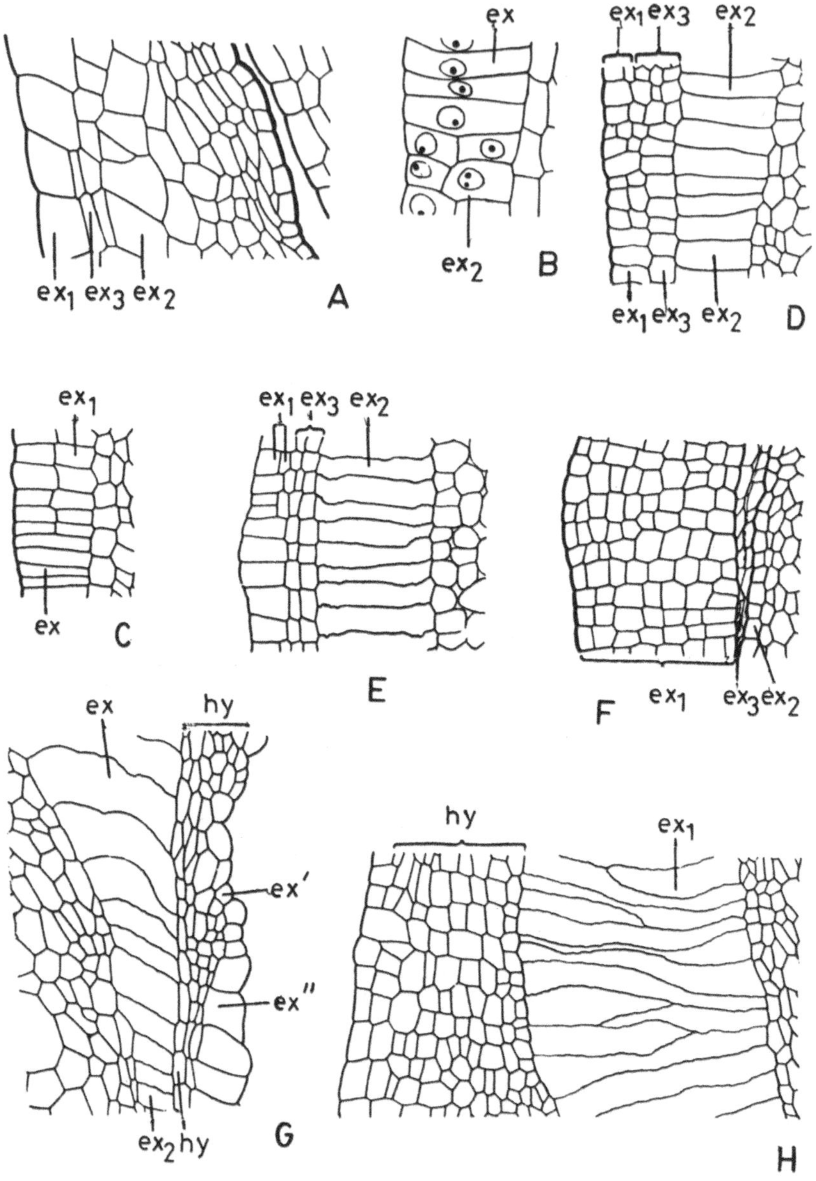

Fig. 38.2.4A–H. **A** *Cucurbita pepo*, **B** *Lagenaria vulgaris*, **C–E** *Luffa cylindrica*, **F–H** *Sicyos angulata*. Longisections parts of outer integument to show divisions of exotesta (*ex*) and their derivatives (ex_1, ex_2, ex_3); *ex′* small-celled exotesta; *ex″* large-celled exotesta; *hy* hypodermis. (After B. Singh 1953)

(1976), VA Poddubnaya-Arnoldi (1976), D Sarada and T Pullaiah (1985), M Sedgley and MA Blesing (1985), B Singh (1952b, 1953), D Singh (1955a, b, 1961a, b, 1964a, b, 1965a, b, 1967, 1970d), D Singh and ASR Dathan (1971, 1972a, b, 1974), Sridhar and D Singh (1986), K Turala (1969, 1971), K Turala-Szybowska (1974, 1975), IK Vasil (1960), G Zhalatov and AK Dzevaltovsky (1975)

38.3 Taxonomic Considerations

Cucurbitaceae. Bentham and Hooker (1965a) place Cucurbitaceae in Passiflorales; Wettstein (1935) and Engler and Diels (1936) allied them with Campanulaceae. Hutchinson (1973) includes them in Cucurbitales along with Begoniaceae, Datiscaceae and Caricaceae. He considered them to be derived from Passiflorales. Bessey (1915) assigns them to Loasales with begonias. Melchior (1964) and Cronquist (1968, 1981) relate and derive them from Violales. R. Dahlgren (1977a, 1981) also assigns them to Violales.

The embryological features of Campanulaceae (Sect. 51.2), such as unitegmic, tenuinucellate ovules without integumentary vascular supply, Cellular endosperm with micropylar and chalazal haustoria, endospermic seeds, and seed-coat formed by the epidermis alone or with some of the hypodermal layers of the integument, do not resemble those of Cucurbitaceae (D. Singh 1970d). D. Singh also negates their affinities with some families of Parietales. They also differ palynologically (Erdtman 1952).

Cucurbitaceae have always been allied to Passifloraceae, but the embryological features of both differ in several respects. Cucurbitaceae show similarities with Datiscaceae and Begoniaceae of Violales: anatropous, bitegmic, crassinucellate ovule with micropyle formed by both integuments, Polygonum type of embryo sac, Nuclear endosperm, Onagrad type of embryogeny, and exotestal seed with straight embryo. The occasional occurrence of Allium type of embryo sac, and exalbuminous seed are reported in Datiscaceae. But, the development of their seed-coat, which forms the basis of generic classification, has no parallel elsewhere. Their ordinal rank is fully justified, and Cucurbitales form an advanced and terminal taxon of Polypetalae (D. Singh 1970d).

39 Myrtiflorae

The order Myrtiflorae includes three suborders – Myrtineae, Hippuridineae and Cynomoriineae. Myrtineae comprise 15 families: Lythraceae, Trapaceae, Crypteroniaceae, Myrtaceae, Dialypetalanthaceae, Sonneratiaceae, Punicaceae, Lecythidaceae, Melastomaceae, Rhizophoraceae, Combretaceae, Onagraceae, Oliniaceae, Haloragaceae and Theligonaceae; Hippuridineae of Hippuridaceae; Cynomoriineae of Cynomoriaceae.

R. Dahlgren and Thorne (1984) discuss the circumscription, variation and relationships of the Myrtales, and Tobe and Raven (1983a) have attempted an analysis of the embryological characters of the various families of this order.

On this basis, Tobe and Raven (1984d) suggest a separate family Rhynchocalycaceae for *Rhynchocalyx*, and Alzateaceae (Tobe and Raven 1984e) for *Alzatea*. Tobe and Raven (1987c) consider that *Heteropyxis* should be placed in a subfamily of Myrtaceae, or in a separate but related family. Is the grouping of *Axinandra* (Tobe and Raven 1983b) with *Crypteronia* and *Dactylocladus* in Crypteroniaceae justified, or should it be grouped with *Rhyncocalyx* and *Alzatea?*

39.1 General Features

The order Myrtiflorae is characterized by (1) secretory anther tapetum, amoeboid in Lecythidaceae, (2) simultaneous cytokinesis in microspore mother cells, (3) anatropous, bitegmic and crassinucellate ovule, campylotropous in Theligonaceae, and unitegmic and tenuinucellar in Theligonaceae and Hippuridaceae, (4) marked fusion of integuments in Haloragaceae; integumentary vascular strand occurs in Myrtaceae and Rhizophoraceae, (5) development of embryo sac of the Polygonum type, Penaea type in Combretaceae, Oenothera in Onagraceae, and Peperomia in *Gunnera* (Haloragaceae), (6) development of endosperm of the Nuclear type, Cellular type in *Gunnera*, Hippuridaceae and Cynomoriaceae, and endosperm absent in Trapaceae and *Lopezia coronata* (Onagraceae), (7) embryogeny of the Onagrad type in Lythraceae, Sonneratiaceae, Melastomaceae, Onagraceae and Hippuridaceae, Solanad type in Trapaceae and Myrtaceae, Asterad type in Combretaceae, Caryophyllad type in Haloragidaceae, and Chenopodiad type in Theligonaceae, and (8) tracheidal exotegmen common in most families, except where the testa is sclerotic, or the stony endocarp gives protection to seed.

39.2 Lythraceae

The anther wall comprises the epidermis, fibrous endothecium, two ephemeral middle layers, and secretory tapetum of two- to six-nucleate cells. The microspore mother cells undergo meiosis, cytokinesis is simultaneous, and tetrads tetrahedral. The sporogenous tissue frequently degenerates in *Ammannia baccifera*, *A. pentandra*, *Lawsonia inermis* and *Nesaea myrtifolia*.

The pollen grains are three-colpate and two-celled at the dispersal stage (A.C. Joshi and Venkateswarlu 1936a).

The ovule is anatropous, bitegmic and crassinucellate; amphitropous in *Nesaea* (A.C. Joshi and Venkateswarlu 1936a). The zigzag micropyle is organized by both the integuments. A nucellar cap develops in *Pemphis acidula* (Venkateswarlu 1937c). A hypostase is also organized in most members. Frequently, two nucelli develop in an ovule and have common integuments or, at least, a common outer integument.

The archesporium is multicellular. A single archesporial cell cuts off a parietal cell, and the parietal tissue is four to five layers thick. The megaspore mother cell undergoes meiotic divisions and forms a linear tetrad, occasionally

T-shaped in *Lawsonia* (A.C. Joshi and Venkateswarlu 1935a), *Nesaea* and *Woodfordia* (A.C. Joshi and Venkateswarlu 1936a). The chalazal megaspore develops into a Polygonum type of embryo sac. Sometimes, in *Lawsonia* the epichalazal megaspore develops. The synergids develop a small vacuole at the apex, in addition to the large chalazal vacuole. The polar nuclei fuse before fertilization, and the antipodals degenerate at a very early stage. A reduction in the number of nuclei in the embryo sac, due to suppression of nuclear divisions, occurs in *Lawsonia* (A.C. Joshi and Venkateswarlu 1935a).

The chalazal end of the embryo sac is pouch-like, and penetrates the vascular strand in *Ammannia* and *Pemphis* (A.C. Joshi and Venkateswarlu 1936a).

Multiple embryo sacs occur in *Lawsonia inermis*, and inverted polarity of the embryo sac is reported in *Woodfordia floribunda* (A.C. Joshi and Venkateswarlu 1935c).

The development of the endosperm is of the Nuclear type, and the nuclei aggregate in polar regions. Wall formation is initiated as the three-histogenic layers differentiate in the embryo.

Embryogeny corresponds to the Onagrad type-Capsella variation. A hypophysis develops from the superior daughter cell of the penultimate cell of the proembryo. The suspensor is uni- or multiseriate.

The seed is exotegmic, and exalbuminous with a straight embryo and flat or convolute cotyledons. Both the integuments contribute to the formation of the seed-coat. The outer epidermis of the testa is a palisade of cuboid cells. The mesophyll persists only in *Pemphis* and is sclerotic. The inner epidermis is a crystal layer with lignified pitted cells. The outer epidermis of the tegmen comprises longitudinally elongate tracheids with different types of thickenings. The inner epidermis is thin-walled.

AC Joshi (1939b), AC Joshi and J Venkateswarlu (1935a, b, c, 1936a, b), J Mauritzon (1934e), D Oliver (1894), BB Smith and JM Herr Jr (1971), R Souèges (1925b), G Tischler (1917), H Tobe and PH Raven (1984e), J Venkateswarlu (1937c)

39.3 Trapaceae

The anther wall comprises persistent epidermis, fibrous endothecium, two or three ephemeral middle layers, and secretory tapetum of multinucleate cells. The tapetum is irregularly biseriate, and its inner wall is studded with Übisch granules (M. Ghosh 1954). The microspore mother cells undergo meiotic divisions, and produce tetrahedral and decussate tetrads. The cytokinesis is simultaneous.

The pollen grains are pyramidal with three much-folded meridional crests. They are two-celled at the shedding stage; three-celled in *Trapa astrachanica* (see Batygina and Yakovlev 1985).

The ovule is anatropous, bitegmic and crassinucellate, with a prominent hypostase. The massive nucellar beak protrudes through the micropyle. The archesporium is one- to two-celled, cuts off a parietal cell, and the parietal

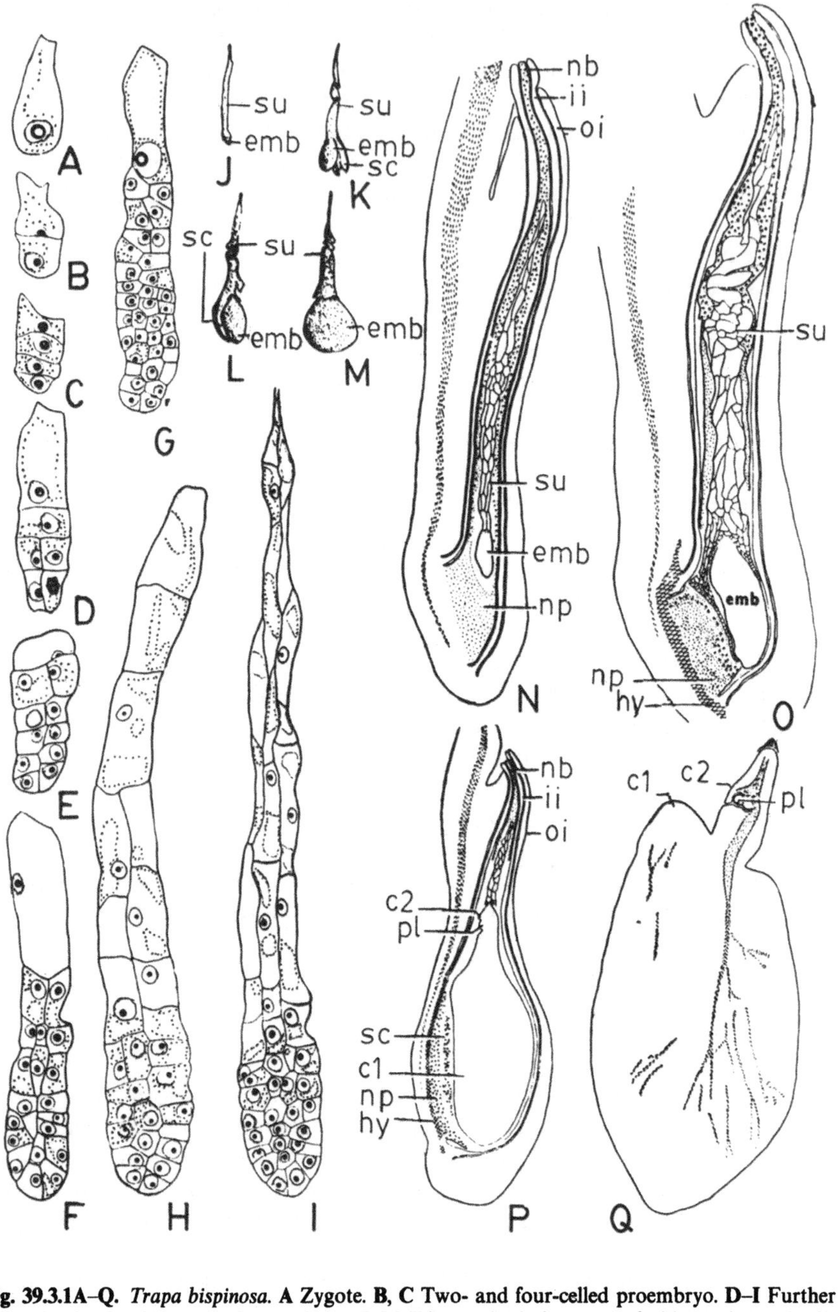

Fig. 39.3.1A–Q. *Trapa bispinosa.* **A** Zygote. **B, C** Two- and four-celled proembryo. **D–I** Further stages in the development of proembryo. **J–M** Dissected wholemount of older embryos (*emb*). Note the massive suspensor (*su*) and also the suspensor collar (*sc*) in **K** and **L**. **N, O** Longisection of ovules showing disposition of embryo. Note the nucellar beak (*nb*), inner (*ii*) and outer (*oi*) integument, nucellar pad (*np*), and hypostase (*hy*). The suspensor (*su*) is massive. **P** Longisection of seed at dicotyledonous (*c*1, *c*2) embryo; *pl* plumule. **Q** Longisection of embryo. One of the cotyledons (*c*2) is highly reduced. (After M. Ram 1956)

tissue is massive in *Trapa bispinosa* (M. Ghosh 1954, Ram 1956). A parietal cell is not cut off in *T. natans* (Trela-Sawicka 1978), which needs confirmation. Periclinal divisions occur in the nucellar epidermis. The nucellar cells undergo endomitosis and become polyploid in *T. natans*. The highest ploidy-level is 256 n (Nägl 1962). The megaspore mother cell produces a linear, sometimes isobilateral or T-shaped tetrad. The chalazal megaspore develops into a Polygonum type of embryo sac. The antipodals have not been observed in *T. bispinosa* (M. Ghosh 1954, Ram 1956). In *T. natans* (Ishikawa 1918), the antipodals are uninucleate, and become hyperthophised and probably polyploid.

The primary endosperm nucleus moves towards the chalazal region, and persists up to the formation of a several-celled proembryo. It undergoes only one or two free-nuclear divisions in *T. natans*. These nuclei finally degenerate, and the endosperm tissue is not formed. The nutritive function (for the growth of the embryo) is taken over by the nucellus and suspensor.

The division of zygote is transverse. Both the cells undergo transverse divisions and produce a linear tetrad (Fig. 39.3.1A–C). The three apical cells produce the embryo. The embryogeny corresponds to the Solanad type (Fig. 39.3.1C–I). The suspensor is long, coiled and multiseriate, and acts as a haustorium (Fig. 39.3.1I–P). Its cells exhibit endomitotic polyploidization. The suspensor forms a collar around the embryo proper (Fig. 39.3.1O). The embryo has two unequal cotyledons, one extremely reduced (Fig. 39.3.1P, Q). The embryo develops at the expense of the nucellus.

The testa is multiplicative, and can be distinguished into three zones in a young seed. The outer and inner zones are parenchymatous, and the inner zone shows reticulate thickenings on the walls. The middle zone comprises suberized cells (Corner 1976). In mature seed the testa is reduced to a thick-walled outer epidermis, and irregular tannin-containing cells of the inner epidermis. The outer epidermis of tegmen consists of narrow tracheidal cells. The inner epidermis is crushed. The fruit is a large one-seeded drupe (Ishikawa 1918).

TB Batygina and MS Yakovlev (1985), EJH Corner (1976), M Ghosh (1954), RM Ishikawa (1918), BM Johri (1963b), P Maheshwari (1950, 1964), W Nägl (1962), M Ram (1956), Z Trela-Sawicka (1978)

39.4 Crypteroniaceae

Tobe and Raven (1983b, 1987a, b) have made a critical study of the embryology and relationships of *Axinandra zeylanica, Crypteronia paniculata, C. cumingii* and *Dactylocladus stenostachys*. All these taxa share the embryological features of the Order Myrtales.

The anther is tetrasporangiate, and the wall comprises the epidermis, endothecium, one (two in *Axinandra*) middle layer, and secretory tapetum of one- or two-nucleate cells. The cells of the connective adjacent to the pollen sacs develop weak fibrous thickenings. At maturity the anther wall consists only of the persistent epidermis; the next three layers collapse.

The microspore mother cells undergo meiotic divisions, followed by simultaneous cytokinesis. The tetrads are tetrahedral (88%), decussate (10%) and isobilateral (2%) in *Axinandra* (Tobe and Raven 1983b). *Crypteronia* shows tetrahedral (66%) and often decussate (34%) tetrads (Tobe and Raven 1987a). *Dactylocladus*, on the basis of 50 selected tetrads, has tetrahedral (84%) and occasionally decussate (16%) tetrads (Tobe and Raven 1987b). These are the first comprehensive data on the quantification of various types of microspore tetrads, and this procedure should invariably be followed by all investigators (see also Schmid 1982).

The pollen grains are two-celled at the shedding stage.

The ovule is anatropous, bitegmic and crassinucellate; a hypostase does not develop. Due to degeneration of the nucellus at the four-nucleate stage of the embryo sac, an endothelium differentiates in *Axinandra*. In *Crypteronia* the chalazal tissue below the embryo sac increases but is disrupted, during post-fertilization development, due to the elongation of the embryo sac.

The hypodermal archesporial cell cuts off a parietal cell and the parietal tissue is two-layered. The megaspore mother cell undergoes meiosis, and the chalazal megaspore of the linear tetrad gives rise to a Polygonum type of embryo sac. The antipodals are ephemeral. In *Axinandra* the walls are not laid down between the nuclei but in *Crypteronia* and *Dactylocladus* antipodal cells are formed.

The development of the endosperm is of the Nuclear type. In *Axinandra* wall formation occurs at the proembryo stage when several thousand free endosperm nuclei have been formed. Cell formation in *Crypteronia* occurs at the young dicotyledonous embryo. The young seed is full of cellular endosperm but all of it is consumed by the developing embryo.

The embryogenic type is not mentioned. In *Axinandra* the zygote divides after the formation of several thousand endosperm nuclei. The suspensor is short and massive, and the mature embryo is straight and flattened; the cotyledons are not folded. In *Crypteronia* and *Dactylocladus*, the suspensor is short (but not massive). Polyembryony does not occur in any of the three taxa.

The seed is endotestal and endotegmic. In *Axinandra*, *Crypteronia* and *Dactylocladus* the exotesta is persistent and has longitudinally elongate cells, in the first taxon it collapses at places. The endotesta is persistent and cryptalliferous in the first two taxa, but degenerates in *Dactylocladus*. The exotegmen collapses. The endotegmen is persistent and tanniniferous. The seeds are winged. In *Axinandra* the membraneous wing on the chalazal side develops from the funiculus. In *Crypteronia* the flat micropylar wing is formed by the funicular tissue and apical tissue of the outer integument. The chalazal wing is narrow and long, and is probably formed by the derivatives of the epidermal cells of the funiculus. *Dactylocladus* has a flat membraneous wing along the entire margin of the seed. The wing is formed by the twofold outer epidermis and is mostly two layers thick.

J Muller (1975), JE Planchon (1845a), R Schmid (1982), H Tobe and PH Raven (1983a, b, 1987a, b), GJCM van Vliet (1975)

39.5 Myrtaceae

The anther wall comprises four to five layers. The endothecium develops fibrous thickenings; sometimes absent in *Angophora floribunda* (N. Prakash 1969d). The middle layers are ephemeral, and the tapetum is secretory with binucleate cells. It is two- to three-layered in the connective region in *Darwinia* (N. Prakash 1969c). The meiotic divisions in microspore mother cells are followed by simultaneous cytokinesis. The tetrads are tetrahedral and decussate, 84% and 16%, respectively, in *Heteropyxis dehniae* (Tobe and Raven 1987c).

The pollen grains are two-celled at the dispersal stage. Sterile one-celled pollen grains occur in *Angophora* and *Eucalyptus stellata* (see Davis 1969).

In the tetrasporangiate anther of *Heteropyxis* (Tobe and Raven 1987c), the two microsporangia in each lobe are separated by a large amount of connective tissue. At maturity, each microsporangium dehisces independently by a longitudinal slit. In other Myrtaceae, the sporangia coaleasce and dehisce by a common slit.

The ovules are ana- to campylotropous, bitegmic and crassinucellate with an endothelium and hypostase. The zigzag micropyle is formed by both the integuments. The ovules are unitegmic in *Eugenia* (S.K. Roy 1960, 1962a) and *Syzygium* (Narayanswami and Roy 1960b, S.K. Roy and Sahai 1962). A vascular bundle traverses the integument. The nucellus is massive in *Myrtus* (S.K. Roy 1962b), and a nucellar cap develops in *E. myrtifolia* (S.K. Roy 1960). The archesporial cell is subhypodermal (this is doubtful and, perhaps, earlier stages were not examined), and cuts off a parietal cell. The parietal tissue is well-developed in *Eugenia bracteata* (S.K. Roy 1955) and *Syzygium caryophyllifolium* (S.K. Roy and Sahai 1962). However, Narayanswami and Roy (1960a, b) report the absence of parietal cell in *S. cumini* and *Psidium*. The deep-seated megaspore mother cell undergoes meiotic divisions, a linear tetrad is produced, and the chalazal megaspore develops into a Polygonum type of embryo sac. The antiopodals are ephemeral. The embryo sacs are normally five-nucleate in *S. caryophyllifolium* (S.K. Roy and Sahai 1962). The chalazal end of the embryo sac develops into a haustorium, and invades the underlying tissue in *E. jambos* (S.K. Roy 1953).

S.K. Roy (1953) reports more than eight nuclei in the embryo sac, which may be due to the fusion of adjacent embryo sacs. A hypostase and postament are formed in *Psidium*.

In *Heteropyxis canescens* and *H. dehniae* (Tobe and Raven 1987c), the embryo sac is bisporic of the Allium type, the antipodals are ephemeral. This condition is not reported in any other Myrtaceae.

Several abnormalities have been observed: unorganized embryo sacs, incompletely organized embryo sacs, twin embryo sacs, and multiple embryo sacs. Sterile ovules without embryo sacs occur in *Eugenia malaccensis* (S.K. Roy 1960) and *Melaleuca* (N. Prakash 1973). Twin nucelli with embryo sacs, enveloped in a common integument, develop in *E. malaccensis*. Narayanswami and Roy (1960a) observed a case of reversed polarity, and presence of supernumerary polar nuclei in *Psidium*. Somatic nucellar aposporous embryo sacs are formed in *E. malaccensis*, *Syzygium cumini* (Narayanswami and Roy 1960b) and *S. caryophyllifolium* (S.K. Roy and Sahai 1962).

The entry of the pollen tube is porogamous, and double fertilization occurs.

Development of the endosperm is of the Nuclear type. The nuclei aggregate in the chalazal region, they fuse and form giant nuclei in *Psidium*. The endosperm remains free-nuclear up to the globular stage of proembryo in *E. bracteata* (S.K. Roy 1955). Wall formation is initiated at the micropylar end, and finally the entire endosperm becomes cellular.

In *Heteropyxis* (Tobe and Raven 1987c), wall formation probably occurs only in a small area around the proembryo/embryo, and does not extend throughout the embryo sac. The seed is exalbuminous.

The embryogeny conforms to the Solanad type in *Callistemon lanceolatus*; *Eucalyptus cinerea, E. cordata, E. deanei, E. gigentea, E. macarthuri, E. occidentalis, E. pauciflora, E. stellulata, E. myrtifolia* and *Feijoa sellowiana* (Polunina 1963) and *Heteropyxis* (Tobe and Raven 1987c), and Onagrad type in *Kunzea capitata* (N. Prakash 1969a) and *Myrtus communis* (S.K. Roy 1962b). The development of the embryo is irregular in *Angophora, Darwinia* and *Psidium*. In *Angophora* di- or tricotyledony, and a collar-like structure at the junction of hypocotyl and radicle has been observed.

The egg apparatus, or the zygote, degenerates in *Eugenia jambos, Syzygium caryophyllifolium* and *S. cumini*. The nucellar cells (in the micropylar region) proliferate, and several embryos are produced. However, the seed contains only one embryo. Polyembryonate seeds occur in *S. caryophyllifolium* (S.K. Roy and Sahai 1962). Johnson (1936) observed two to 21 embryos in the seeds of *E. hookeri*. According to van der Pijl (1934), the formation of adventive embryos begins before pollination, but they attain full size only after pollination and formation of the endosperm. The cells of the inner integument, in the vicinity of the micropyle, also give rise to adventive embryos. Haploid embryo, developing from unfertilized synergid, is reported in *Eugenia delavayi*. Polyembryony is absent in *Eugenia bracteata* (S.K. Roy 1955) and *Psidium*.

The testa is thick and extremely sclerotic. The exotesta comprises thickened lignified cells. The mesophyll is thin-walled. The inner epidermis either contains crystals or becomes strongly thickened. The tegmen generally gets crushed, or persists and contains tannin (Corner 1976, Tobe and Raven 1987c).

The crescent-shaped seed in *Heteropyxis* (Tobe and Raven 1987c) is laterally depressed with a long tapering wing on the micropylar side, and a flat membraneous wing at the chalazal end. The upper wing is formed from the derivatives of the inner and outer integument, and the lower from the derivatives of the outer integument. The wings are not known in other Myrtaceae.

Polunina (1957, 1959) studied the influence of genetic factors as against the physical factor, temperature, in *Eucalyptus*. In *E. cinerea* flowering occurs in October to December (spring in Australia, and autumn in Russia). In Italy, in *E. camaldulensis* and *E. cinerea*, sporogenesis and early stages of the embryo sac develop in summer, but in winter in Australia. The flowering cycle of *E. cinerea* does not show any alteration in adjustment to the conditions of the northern hemisphere; anthesis follows maturation of the embryo sac. But in

E. camadulensis there is an interval between differentiation of the embryo sac and anthesis. The flowering is postponed until the return of summer.

According to Davis (1968b), the embryological processes are rather slow, but regular in *Eucalyptus* spp. In *E. melliodora* in Australia, the meiosis in the megaspore mother cell occurs during winter (May–August), while in Armidale (Australia) it occurs during heavy frost. The enlargement of the egg apparatus follows the differentiation of the embryo sac, but there is a delay (period of cytological stasis) between anthesis and post-fertilization development.

Further studies of this nature and a complete follow-up of the development of seed would be very useful, especially in plants of economic importance separated by space and time.

DV Beardsell, EG William and RB Knox (1989), EJH Corner (1976), R Dahlgren and RF Thorne (1984), AN Drinnan and PY Ladiges (1989), GL Davis (1968b, 1969), E Gauba and LD Pryor (1958, 1961), AM Johnson (1936), LAS Johnson and BG Briggs (1984), S Narayanswami and SK Roy (1960a, b), L van der Pijl (1934), NN Polunina (1957, 1958a, b, 1959, 1963), N Prakash (1969a, b, c, d, 1973), SK Roy (1953, 1955, 1960, 1962a, b), SK Roy and R Sahai (1962), R Schmid (1980), NK Tiwary (1926), H Tobe and PH Raven (1983a, b, 1984e, 1987c), M Sedgley (1989), M Sedgley, FC Hand and AR Griffin (1989)

39.6 Dialypetalanthaceae

The embryology has not been investigated.

39.7 Sonneratiaceae

The anther wall comprises the epidermis, fibrous endothecium, two or three ephemeral middle layers, and secretory tapetum of binucleate cells. Cytokinesis is simultaneous.

The pollen grains are three-porate and two-celled at the dispersal stage. In *Sonneratia apetala* they contain starch (Venkateswarlu 1937a).

The ovule is anatropous, bitegmic and crassinucellate with a zigzag micropyle. The occurrence of vascular strands in the nucellus is reported in *Duabanga sonneratioides* (Venkateswarlu 1936a). Twin nucelli, within common integuments, occur in *Duabanga*.

The archesporium is uni- or multicelled. The archesporial cell cuts off a parietal cell, the parietal tissue is four to six cells thick. The megaspore mother cell undergoes meiotic divisions, and the chalazal megaspore of a linear or T-shaped tetrad develops into a Polygonum type of embryo sac. The antipodals degenerate early, and polar nuclei fuse before fertilization. The mature embryo sac is, thus, four-nucleate (egg apparatus and secondary nucleus) in *Duabanga* and *Sonneratia* (Venkateswarlu 1936a, c).

The development of the endosperm is of the Nuclear type (Venkateswarlu 1937a). The nuclei accumulate in the polar regions, and the endosperm becomes cellular at an advanced stage of proembryo. The endosperm does not fill the entire seed, and the central region remains hollow (the author prefers to call it a central vacuole in the endosperm).

The embryogeny corresponds to the Onagrad type. The suspensor is uni- or biseriate.

Seeds are filiform with an appendage at each end in *Duabanga*, and hard, thick and ellipsoid in *Sonneratia* (Corner 1976).

The exalbuminous seed shows a minute and straight embryo with flat cotyledons in *Duabanga*, and a large curved embryo with contorted cotyledons in *Sonneratia*. The outer epidermis of testa is thin-walled, mesophyll densely sclerotic, and the inner epidermis forms a sclerotic layer in *Sonneratia*, or a crystal layer in *Duabanga*. The outer epidermis of tegmen comprises narrow tracheidal cells with spiral thickenings in *Sonneratia*, or lignified fibres in *Duabanga*. The inner epidermis is thin-walled.

EJH Corner (1976), P Crété (1956g), AC Joshi (1939b), J Mauritzon (1939b), J Venkateswarlu (1936a, c, 1937a)

39.8 Punicaceae

The development and structure of anther, pollen and embryogeny has not been studied.

The ovule is anatropous, bitegmic and crassinucellate with a long funicle in *Punica granatum* (King 1947). The outer integument forms the micropyle.

The archesporial cell produces a parietal cell, and the parietal tissue is three-layered. The megaspore mother cell undergoes meiotic divisions, a linear tetrad is produced, and the chalazal megaspore develops into a Polygonum type of embryo sac. The polar nuclei fuse in the centre of the embryo sac, and the antipodals are ephemeral.

Development of the endosperm is of the Nuclear type.

The seeds are exalbuminous, and embryo straight with contorted cotyledons. The outer epidermis of testa comprises a palisade of very long, thin-walled, columnar pulpy cells (sarcotesta), mesophyll of many layers of sclerotic cells and inner layers of radially elongate cells (mesotesta), and inner epidermis of longitudinal tracheids with annular-spiral lignified thickenings (Corner 1976).

EJH Corner (1976), JR King (1947), J Mauritzon (1939b)

39.9 Lecythidaceae

The embryogeny has not been worked out.

The anther wall comprises the epidermis, fibrous endothecium, two ephemeral middle layers, and an amoeboid tapetum of binucleate cells. The microspore mother cells undergo meiotic divisions, cytokinesis is simultaneous through furrows, and tetrads are tetrahedral. A periplasmodium is formed at the tetrad stage.

The pollen grains are three-celled at the dispersal stage.

The ovule is anatropous, bitegmic and tenuinucellate. The micropyle is formed by the inner integument. The integuments are thick and the outer integument contains a single vascular trace in *Nepoleona imperialis* (Venkateswarlu 1952a, c). An endothelium develops in *N. imperialis.* In *Baringtonia acutangula* the inner integument collapses (Venkateswarlu 1952a). The archesporial cell functions directly as megaspore mother cell, undergoes meiotic divisions, the tetrad is linear or T-shaped, and the chalazal megaspore develops into a Polygonum type of embryo sac. The polar nuclei fuse at the time of fertilization, and antipodals are ephemeral. Twin embryo sacs from two megaspore mother cells, and reversed polarity of an eight-nucleate embryo sac, have been observed in *N. imperialis* (Venkateswarlu 1952c).

The development of endosperm is of the Nuclear type.

The outer epidermis of testa comprises thick-walled lignified palisade cells, followed by aerenchymatous mesophyll with scattered sclerotic cells. The inner epidermis is thin-walled. The tegmen gets crushed.

J Mauritzon (1939b), J Venkateswarlu (1952a, c)

39.10 Melastomaceae

The anther is tetralocular, and its wall comprises the epidermis, endothecium, middle layers, and secretory tapetum of uninucleate cells. The endothecium is heavily thickened in *Mouriria* (Ruys 1925), fibrous thickenings are poorly developed in *Sonerila* (Subramanyam 1944). Generally, there is a single middle layer; five to seven layers in *Melastoma* (Subramanyam 1948a) and two to three layers persist and develop thickenings. The tapetum is biseriate in *Heyneanum* and *Memecylon* (Subramanyam 1942). The microspore mother cells undergo meiotic divisions, cytokinesis is simultaneous, and tetrads tetrahedral.

The three-colporate pollen grains are three-celled at the shedding stage.

The ovule is anatropous, bitegmic and crassinucellate with a zigzag micropyle formed by both the integuments. The hypodermal archesporial cell cuts off a parietal cell, the parietal tissue is two to six cells thick. The cells of the nucellar epidermis divide and produce a nucellar cap, or the cells become conspicuously enlarged and radially arranged in *Melastoma.* The megaspore mother cell undergoes meiotic divisions, a linear or T-shaped tetrad is formed, and the chalazal megaspore gives rise to a Polygonum type of embryo sac. The antipodals rarely organize into cells except in *Oxyspora paniculata* (Subramanyam 1951a). A hypostase develops.

Etheridge and Herr (1968) report that in *Rhexia mariana,* along with the formation of a nucellar cap, a 'pore' appears at the apex of the nucellus. This pore probably results from the suppression of anticlinal divisions in the nucellar epidermis which is subsequently pulled apart with nucellar enlargement. This feature is not known in any other taxon.

In *Osbeckia rosea* (Subramanyam 1946), the micropylar megaspore of a tetrad develops up to four-nucleate embryo sac. Subramanyam (1942, 1944, 1951a) observed double tetrads in *Sonerila*, and twin embryo sacs in *Leandra* and *Oxyspora*.

Development of the endosperm is of the Nuclear type, and the nuclei accumulate in the micropylar and chalazal region. It remains free-nuclear throughout, and is absorbed by the developing embryo.

Development of the embryo is of the Onagrad type in *Leandra cordifolia*, *Memecylon heyneanum*, *Osbeckia capsularis*, *O. hispidissima*, *O. wightiana* (Subramanyam 1942), *Sonerila wallichii* (Subramanyam 1944) and *Oxyspora paniculata* (Subramanyam 1951a). The suspensor is short, multicellular and massive in *Oxyspora*.

The structure of the mature embryo is a taxonomic feature. In *Memecylon umbellatum*, the cotyledons are foliaceous and wrinkled, with the outer cotyledon crumpled over the inner one; the hypocotyl is long. The mature embryo is curved with foliaceous and rolled cotyledons in *M. arnottianum* and *M. gardneri*; the hypocotyl is short. The cotyledons are thick and fleshy in *M. rivulare*; the hypocotyl is short.

Nucellar polyembryony is reported in *Osbeckia hispidissima* (Subramanyam 1942). Due to the proliferation of suspensor cells, additional embryos are formed in *Sonerila wallichii* (Subramanyam 1944).

The seed is exalbuminous. The outer epidermis of testa comprises radially elongate thick-walled cells followed by thin-walled or sclerotic cells. The inner epidermis contains crystals, or gets crushed. The tegmen persists only at the endostome.

P Crété (1957a), AL Etheridge and JM Herr Jr (1968), T Morley (1953), JD Ruys (1925), K Subramanyam (1942, 1944, 1946, 1948a, b, 1951a), H Tobe and PH Raven (1984a), A Ziegler (1925)

39.11 Rhizophoraceae

The development of anther and pollen has not been investigated.

The ovule is anatropous, bitegmic and crassinucellate with a zigzag micropyle. The outer integument contains a vascular strand in *Rhizophora mucronata* (Carey 1934).

In *Cassipourea elliptica* (Juncosa 1984a), a non-viviparous member of Rhizophoraceae, two anatropous ovules [material from La Selva (Costa Rica) and Panama] are present in each carpel of a tricarpellary superior ovary. The material from Llorona (Costa Rica) rarely showed viable seeds. In the basal region the nucellus degenerates. The endothelium contains tannin. The micropyle is formed by the inner integument.

The archesporial cell cuts off a parietal cell, the parietal tissue is two-layered. The megaspore mother cell undergoes meiotic divisions, and chalazal megaspore of a linear tetrad develops into a Polygonum type of embryo sac. The polar nuclei fuse before fertilization; antipodals are ephemeral.

In *Polygonanthus amazonicus* (Tobe and Raven 1988a), the megaspore mother cell produces a dyad as usual. The upper smaller one fails to divide. The lower one divides, and the chalazal megaspore of the triad gives rise to a Polygonum type of embryo sac.

In *C. elliptica*, megasporogenesis is completed by the time the flower opens. The chalazal megaspore develops into a Polygonum type of embryo sac only after pollination.

The development of the endosperm is of the Nuclear type, later it becomes cellular. The endosperm protrudes through the micropyle in *Ceriops candolleana* and *Rhizophora mucronata*, and in *Ceriops* it also develops a chalazal process.

In *C. elliptica*, the nuclei aggregate at the micropylar and chalazal pole, and the central region is occupied by dense cytoplasm. Centripetal wall formation occurs at the globular proembryo. Subsequently, the central region gets filled with radial files of endosperm cells. In *Bruguiera exaristata* (Juncosa 1984b), by the time the embryo fills the seed, much of the cellular endosperm is consumed. The outer layer of the endosperm shows "transfer cells". Then there is a layer or two of large vacuolate integumentary cells encircled by the integumentary compact transfer tissue. The integumentary parenchyma is tannin-filled.

The zygote divides by a transverse (Figs. 39.11.1A, C), longitudinal (Fig. 39.11.1B), or an oblique wall (Figs. 39.11.1D, E). The orientation of the second cell generation is also variable. The proembryo derived from a transverse division of the zygote is followed by longitudinal division of the terminal

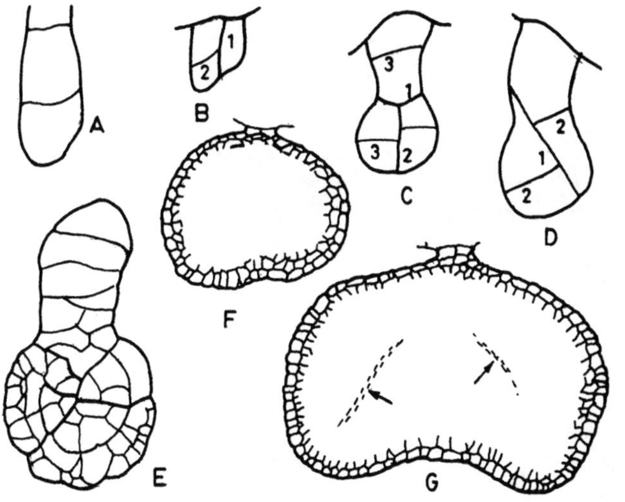

Fig. 39.11.1A–G. *Cassipourea elliptica.* **A–E** Proembryos derived from transverse (**A, C**), longitudinal (**B**) and oblique wall (**D, E**) in zygote; sequence of walls shown in **B, C, D**; protoderm demarcated in **E**, suspensor uniseriate (typical of Le Selva specimens). **F** Initiation of cotyledons. **G** Early heart-shaped embryo, provascular elements have differentiated (*arrows*), suspensor short and multiseriate (common in Llorona specimens). (Adapted after Juncosa 1984a)

cell. The protoderm differentiates in a globular proembryo (Fig. 39.11.1F). Some derivatives of the apical cell contribute to the multiseriate upper part of the suspensor. There is "delayed embryo differentiation" (see Haccius 1971). The specimens from La Selva (Costa Rica) showed a uniseriate suspensor (Fig. 39.11.1E), whereas those from Osa Peninsula (Costa Rica) had shorter multiseriate suspensor. The heart-shaped embryo shows lateral expansion (Fig. 39.11.1G). The cotyledons are initiated on the slightly longitudinally elongate embryo, and each contains a single vascular strand. *Cassipourea elliptica* has a phased pattern of growth like that of *Rhizophora*. The period of elongation of the hypocotyl overlaps the beginning of the resumption of growth in cotyledons. The growth of the hypocotyl is more rapid between *a* and *c*, and of cotyledons between *c* and *e* (Fig. 39.11.2A). Initially, there are two strands, one extending from the median bundle of each cotyledon down to the lower hypocotyl (Fig. 39.11.2B). Figure 39.11.2C shows cleared cotyledons with vasculature.

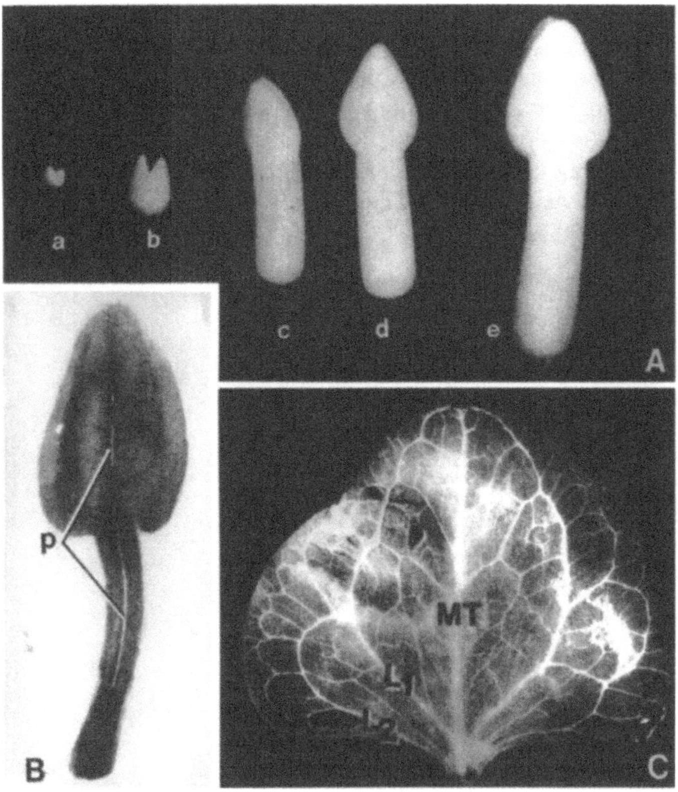

Fig. 39.11.2A–C. *Cassipourea elliptica.* **A** Successive differentiation of embryo (*a–e*). The hypocotyl is very prominent. **B** Embryo from mature seed; *p* protoxylem. **C** Cleared cotyledons; *MT* metaxylem; L_1, L_2 lateral veins. (After Juncosa 1984a)

In *Rhizophora* (Carey 1934), the mature embryo contains two almost wholly fused cotyledons.

According to Corner (1976, see also Tobe and Raven 1988b), the medium-sized seeds are albuminous or exalbuminous (*Cassipourea*, *Crossostylis*, *Weihea*), or arillate, winged in *Macarisia*. The testa is thick with a distinct exotesta (*Carallia*), or large unspecialized pellicular testa (*Anisophyllea*, *Rhizophora*, etc.).

In contrast to other Rhizophoraceae, in *Cassipourea elliptica* (Juncosa 1984a) the germination is epigeal with foliaceous cotyledons. In viviparous members, the cotyledons are fleshy and non-photosynthetic. During germination, the growth is diffuse and intercalary, rather than apical. Apical growth of radicle appears later, and initial vascular differentiation is alternate in the tissue produced.

In *Bruguiera exaristata* (Juncosa 1984b), from Queensland (Australia), germination occurs by the growth of hypocotyl as in non-viviparous plants. The provascular development follows cotyledonary initiation. In the embryo the radicular apex differentiates subterminally. Later, this apex becomes terminal and develops meristematic tissue protected by coleorhiza (formed by the lobes of the outer cortex).

The first two pairs of epicotyledonary leaves are decussate, but the third pair of primorida is initiated at an angle of about 70° to the previous part of the vegetative apices.

Rhizophora mangle (Juncosa 1982), tribe Rhizophorae, exhibits viviparous germination. The initiation of cotyledon and vasculature, apical anatomy of the radicle, and vascular differentiation in *R. mangle*, are highly specialized as compared to non-viviparous *Bruguiera exaristata* (Juncosa 1984b) and *Cassipourea elliptica* (Juncosa 1984a).

Development of the endosperm is of the Nuclear type, as in other members. The inner integument collapses during endosperm development. The multiseriate suspensor gets separated from the testa due to the breakdown of the basal cell followed by the intrusive growth of the endosperm (requires confirmation). Thus, the embryo moves deep into the endosperm before it begins to extrude from the micropyle. The embryo is usually extruded by expansion of the endosperm, and "germination" is effected by the growth of the endosperm rather than of the embryo. Due to the pressure of development of the endosperm, the micropyle remains wide open. The cotyledons arise as a single toroidal primordium. The apex differentiates into two or three cotyledons, i.e. only the apical one-third portion is divided. Later, these lobes fuse marginally to form a narrow tube in the centre.

The axis of the embryo connects the shoot and radicular apical meristems (Fig. 39.11.3A, B). Basipetal development of cotyledonary strands soon connect with the hypocotylary provascular cylinder. The radicular meristem arises somewhat more immersed in the embryonic tissue (Fig. 39.11.3A, B).

The diffuse intercalary meristem below the node causes the elongation of the axis. Growth takes place for several months and the mature radicle is 15–30 cm.

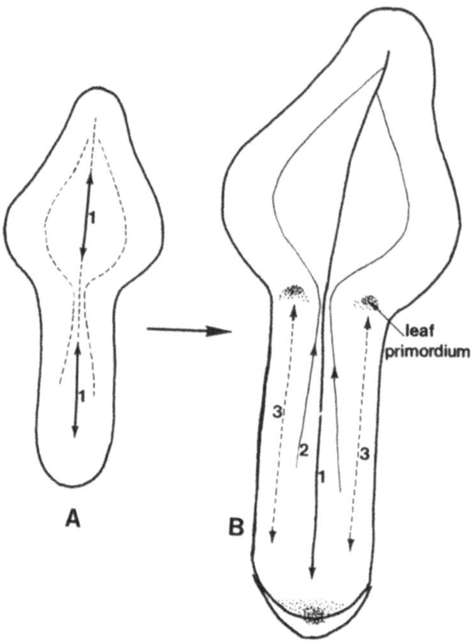

Fig. 39.11.3 A, B. *Cassipourea elliptica*, vascular differentiation in young (**A**) and old (**B**) embryo; *expanded* portion cotyledons, *thick stalk* hypocotyl. *Dotted lines* are major provascular strands. Each cotyledon is vascularized by a single median trace associated with a single provascular lacuna. *Numbers* indicate order of xylem differentiation in various separate loci. (After Juncosa 1984a)

When the apical region of cotyledons is restricted by the integument, growth ceases. Most of the endosperm is crushed or digested by this time; the outer layer of endosperm develops wall ingrowths (transfer cells).

G Carey (1934), EJH Corner (1976), B Haccius (1971), AM Juncosa (1982, 1984a, b), LSS Kumar 1942), H Tobe and PH Raven (1988a, b)

39.12 Combretaceae

The major contribution to the embryology of the family Combretaceae is by Venkateswarlu and Prakasa Rao (1972), on about 18 species and 9 genera representing all the tribes.

The anther wall comprises the epidermis, endothecium with fibrous thickenings, except in the region of stomium, one to six middle layers, and tapetum. The endothecial cells contain tannin and phenolic substances in *Lumnitzera racemosa* and *Terminalia arjuna.* The number of middle layers varies in different genera: single layer in *Guiera senegalensis*; two in *Poivrea coccinea* (Venkateswarlu 1952b), *Combretum decandrum, C. extensum, C. ovalifolium, Lumnitzera racemosa* and *Quisqualis indica*; three in *Anogeissus acuminata, A. latifolia, Bucida buceros, Calycopteris floribunda, Combretum grandiflorum, Terminalia arjuna, T. bellerica, T. catappa* and *T. paniculata* (Venkateswarlu and Prakasa Rao 1972). The cells of the connective, adjacent to middle layers, develop thickenings in *Anogeissus acuminata*, *A. latifolia*, *Lumnitzera racemosa* and *Terminalia arjuna*. The tapetum is secretory and one-, two- or multi-layered, and the cells are uni- or binucleate. It is multi-layered and three- to four-nucleate in *Combretum grandiflorum*. The cells become polyploid due to nuclear fusion. Following meiotic divisions in microspore mother cells, cytokinesis takes place by furrowing, and the tetrads are tetrahedral and isobilateral.

The pollen grains are ridged, triporate with a thick and smooth exine. They are two-celled at the dispersal stage. Sterility of sporogenous tissue and pollen is very common.

The ovule is anatropous, bitegmic and crassinucellate with a zigzag micropyle. In *Guiera senegalensis* the inner integument alone forms the micropyle. The nucellus undergoes repeated divisions to form a cap in *Bucida*, *Combretum*, *Lumnitzera*, *Terminalia* and *Quisqualis*. A funicular obturator is present in *Combretum*, *Poivrea* and *Quisqualis*. A hypostase also develops.

The archesporium is uni- to multicellular. A single archesporial cell cuts off a parietal cell, the parietal tissue is massive. The megaspore mother cell undergoes meiotic divisions, and the tetrad is linear or T-shaped. The chalazal megaspore develops into a Polygonum type of embryo sac. In *Guiera senegalensis*, due to aggressive growth at the tip, the embryo sac absorbs the parietal tissue and nucellar cap, and extends into the micropyle. The antipodals are usually ephemeral (Venkateswarlu and Prakasa Rao 1972).

Penaea type of embryo sac is reported in *Combretum paniculatum* and *C. pincianum* (Mauritzon 1939b).

In *G. senegalensis* the antipodals persist and increase up to six. Twin embryo sacs are produced in *Terminalia belerica* (Nagaraj 1955b) and *T. catappa* (M. Nagaraj 1954c), and triple gametophytes in *Calycopteris* (Venkateswarlu and Prakasa Rao 1972). The degeneration of tetrad and embryo sac has been observed.

The endosperm is of the Nuclear type. The nuclei accumulate in the micropylar and chalazal region. The number of nuclei is generally not very large. Wall formation is initiated in the micropylar region at the globular stage of proembryo. The endosperm is consumed during development of the embryo.

The embryogeny conforms to the Asterad type-Erodium variation. Nucellar polyembryony occurs in *Poivrea coccinea* (Venkateswarlu 1952b), and synergid polyembryony in *Terminalia chebula* (M. Nagaraj 1954a), *Combretum decandrum*, *C. extensum*, *C. grandiflorum*, *Calycopteris floribunda*, *Guiera senegalensis*, *Lumnitzera racemosa* and *Terminalia catappa*.

The outer epidermis of testa is thin-walled. The mesophyll is either sclerotic or of tracheidal cells. The inner epidermis comprises short, tangentially elongate tracheidal cells. The outer epidermis of tegmen consists of longitudinally elongate, lignified fibres with fine, close, spiral thickenings. The mesophyll and inner epidermis get crushed.

F Fagerlind (1941a), SM Jorapur and HD Noronha (1963), J Mauritzon (1939b), M Nagaraj (1954a, b, c, 1955a, b), N Pal (1951a), PS Prakasa Rao (1963), J Venkateswarlu (1947d, 1952b), J Venkateswarlu and PS Prakasa Rao (1972)

39.13 Onagraceae

The anther wall comprises the epidermis, fibrous endothecium, one or two middle layers, and secretory tapetum of binucleate cells.

Tobe and Raven (1986a) describe some unusual features. The sporogenous tissue in some species of 17 genera becomes divided by transverse septa composed only of the tapetum (Fig. 39.13.1A–C), or of both the tapetum and the parenchyma. In *Calylophus, Gaura, Clarkia, Heterogaura, Ludwigia latifolia* and *L. linearis*, the septa are transverse and mostly parenchymatous, or of tapetum only. The "packets" of sporogenous tissue appear spheroidal to ellipsoidal. In *L. linearis* each packet has its own tapetum; the pollen mother cells undergo meiosis synchronously. Finally, all the packets in an anther lobe fuse.

In *Hauya elegans*, the sporogenous tissue becomes divided by transverse as well as vertical septa (Fig. 39.13.1D–I). As many as ten packets have been observed in a transection. In a radial longisection, a large number of packets are arranged in a staggered pattern. The frequency of septa dividing the sporogenous tissue is 1.63 in *Gongylocarpus fruticulosus.*

In *Camissonia* spp. and *Gayophytum humile*, the transverse septa originate from the tapetum.

In *Gaura coccinea, G. longiflora* and *G. mutabilis*, by transverse parenchymatous septa, the sporogenous tissue becomes divided into cylindrical packets of varying length.

According to Tobe and Raven (1986a), the evolution of the anther has been accompanied by an increase in tapetum relative to the sporogenous tissue. The increased ratio of tapetum to sporogenous tissue relates to the distinctive surface structure of pollen grains, such as viscin threads.

The maximum number of septa (88 from four flowers with 16 anthers) occurs in *Gongylocarpus fruticulosus.* Of the 88 septa, 20 were formed by parenchymatous tissue and the rest by tapetum.

Following meiotic divisions in microspore mother cells, simultaneous cytokinesis gives rise to tetrahedral tetrads.

The three-colporate pollen grains are two-celled at the dispersal stage.

The ovule is anatropous, bitegmic and crassinucellate with micropyle formed by both the integuments (Fig. 39.13.2F). A hypostase is common. An epistase is present in *Clarkia elegans* and *Zauschneria californica* (Johansen 1928a, b), and a funicular obturator in *Stenosiphon linifolium* (Johansen 1930b). The archesporial cell cuts off a parietal cell which produces parietal tissue. The megaspore mother cell undergoes meiotic divisions, forms a linear tetrad, and the micropylar megaspore mostly develops into a four-nucleate Oenothera type of embryo sac (Fig. 39.13.2A–H).

Two megaspore mother cells occur in *Jussieua repens, Ludwigia octovalvis, Oenothera odorata* and *Taraxia ovata*, and several in *Clarkia* and *Godetia* (Täckholm 1915). T-shaped tetrads are reported in *Anogra pallida, L. parviflora, Zauschneria latifolia*, and *Fuchsia* var. *Diadem.* Twin linear tetrads, and four binucleate embryo sacs due to concurrent development of four megaspores have been observed in *O. odorata* (Subramanyam and Govindu 1948).

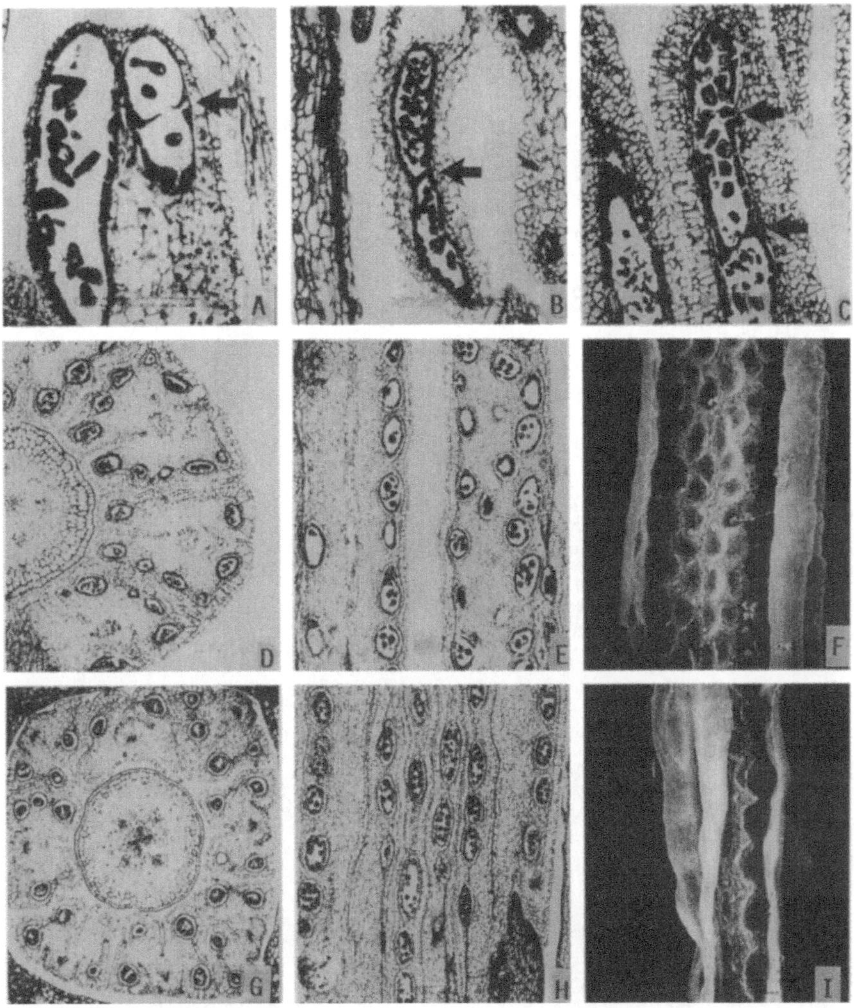

Fig. 39.13.1A–I. Polysporangiate anthers, **A–C**, **E**, **H** longisections, **D**, **G** transections, **F**, **I** SEM preparations. **A** *Lopezia racemosa*, **B** *Circea cordata*, **C** *Fuchsia jimenenzii*, **D–F** *Hauya elegans*, **G–I** *H. heydeana*. **A–C** Sporogenous tissue divided by tapetal septum (*arrow*). **D** Sporogenous tissue divided into packets (groups) by vertical parenchymatous septa (as many as ten packets in t.s.). **E** Tangential longisection of anther (*left*) with many spheroidal or ellipsoidal packets arranged linearly; in radial longisection (*right*) the packets show a scattered arrangement. **F** Remnants of parenchymatous septa on the inner surface of anther wall after dehiscence. **G** Anthers in some of which vertical septa have developed in the sporogenous tissue (not as conspicuous as in **D**). **H** Many spheroidal or elliptical packets of sporogenous cells. **I** Remnants of septa show that the packets are not as scattered as in **F**. (After Tobe and Raven 1986a)

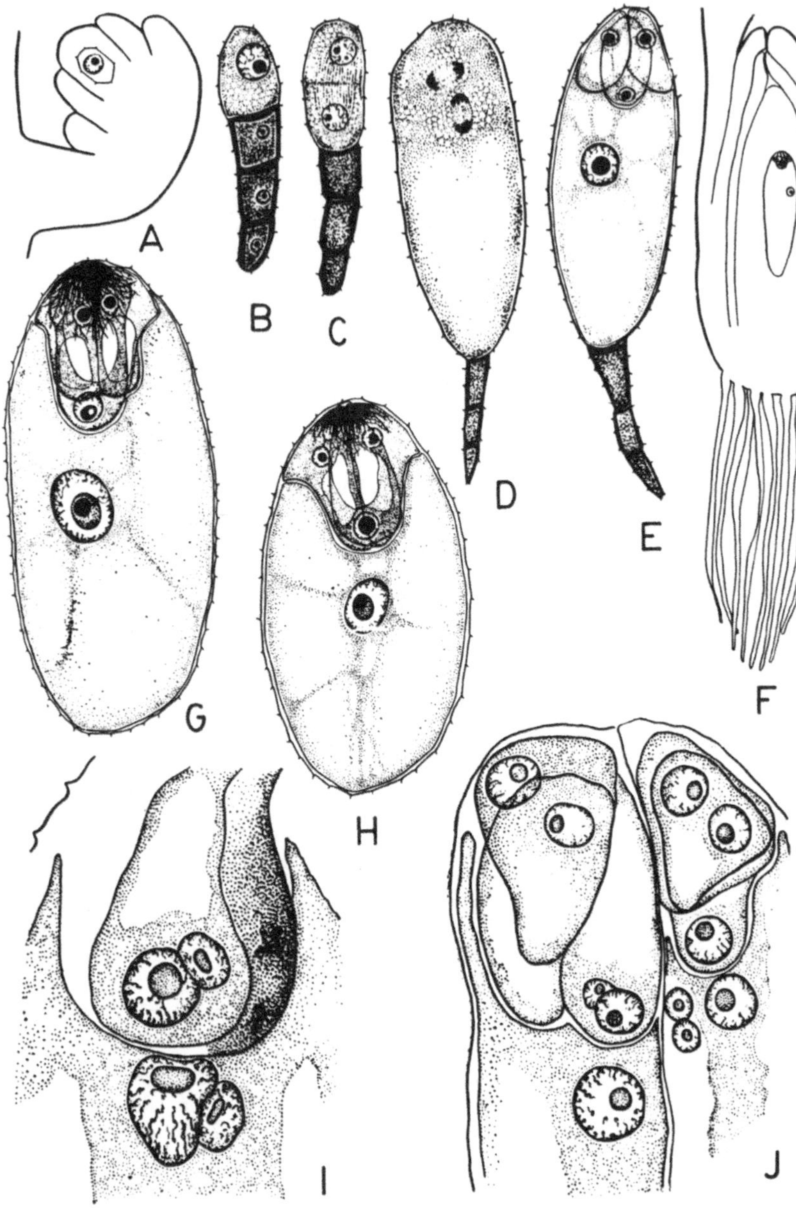

Fig. 39.13.2A–J. **A–G** *Chamerion dodonaei.* **A** Bitegmic ovule at megaspore mother cell; **B, C** Micropylar functional megaspore of linear tetrad; **D** Embryo sac with nuclei at anaphase; **E** Embryo sac with egg apparatus and single polar nucleus: Three degenerated megaspores in **B–E**; **F** Longisection of ovule at mature embryo sac, note hairy outgrowths from the base of ovule; **G** Mature gametophyte. **H** *Ludwigia palustris*, mature gametophyte. **I** *Chamerion angustifolium*, double fertilization, persistent pollen tube shows "x" bodies. **J** *Oenothera biennis*, micropylar portion of twin embryo sacs; syngamy and primary endosperm nucleus in one, in the other the two sperm nuclei have not yet effected fertilization. (See Batygina and Yakovlev 1985)

Ovular sterility is common in *Gayophytum ramosissium* (Johansen 1933). In *O. muricata* (Rodkiewicz et al. 1971), either micropylar or chalazal megaspore may develop. Abnormal three- to five- and seven-nucleate embryo sacs, and twin embryo sacs (Fig. 39.13.2J) are reported in *Jussieua repens* (Khan 1942). Normally, in *Epilobium* sp. the gametophyte is of the Oenothera type (Gachechiladze 1974a, b). However, rarely Polygonum type of development may occur in *E. hirsutum* and *E. nervosum.*

Abnormal embryo sacs also develop in *Zauschneria latifolia* (Johansen 1931b): (1) the micropylar megaspore of the tetrad does not develop further and remains uninucleate, (2) the micropylar nuclei do not organize into egg apparatus, and (3) the polar nucleus fuses with the egg nucleus. All these reports should be re-examined on a quantitative (%) basis.

Figure 39.13.2I shows double fertilization, and two X-bodies in the pollen tube. Of the twin embryo sacs in Figure 39.13.2J, one shows syngamy and primary endosperm nucleus, and in the other double fertilization has not yet taken place.

The development of the endosperm is of the Nuclear type. Khan (1942) noted ca. 900 free nuclei, but their final fate was not ascertained. In *Ludwigia octovalvis*, the endosperm becomes completely cellular; in *Zauschneria latifolia* (Johansen 1931b) the endosperm nuclei divide amitotically. Both reports require confirmation. Total absence of cell wall formation in endosperm has been observed in *Lopezia coronata.* (Täckholm 1914).

The embryogeny conforms to the Onagrad type. The suspensor is short and inconspicuous (Souèges 1920b).

Polyembryony is reported in *Jussieua repens* (Khan 1942). In one ovule two embryos were lying side by side and (in the absence of embryo sac/s) are interpreted as of nucellar origin. In *Taraxia ovata* (Johansen 1931a), "apogamous" embryos were frequently observed, probably from the continued growth of all the four megaspores of a tetrad. The embryo comprised a linear row of ten to 14 cells with a head-like cellular growth at the micropylar end. Apogamous nature is presumed due to the absence of endosperm, and filamentous growth is attributed to the activity of the tetrad, ruling out the possibility of failure of meiosis in the megaspore mother cell. The embryo is, therefore, considered to be haploid. In *Zauschneria latifolia* (Johansen 1931b), besides the perfectly normal zygotic embryos, there were embryos from fertilized synergid, embryos without any regularity in cell organization and hypophysis, and apomictic embryos. A more detailed investigation is necessary before the origin of non-zygotic embryos can be confirmed.

The seed of *Epilobium* has a conspicuous tuft of hairs at the micropylar pole (Fig. 39.13.3A, B).

The exalbuminous seed has a straight embryo with two equal cotyledons. The outer epidermis of the testa is thin-walled, papillate, or hairy. The inner epidermis is a crystal layer with inner and radial walls strongly thickened and lignified. The outer epidermis of the inner integument comprises longitudinal lignified fibres, whereas the inner one is thin-walled and gets crushed, or it may be represented by longitudinally elongate cells with subreticulate thickenings on the inner walls. The fruit is usually a coiled capsule.

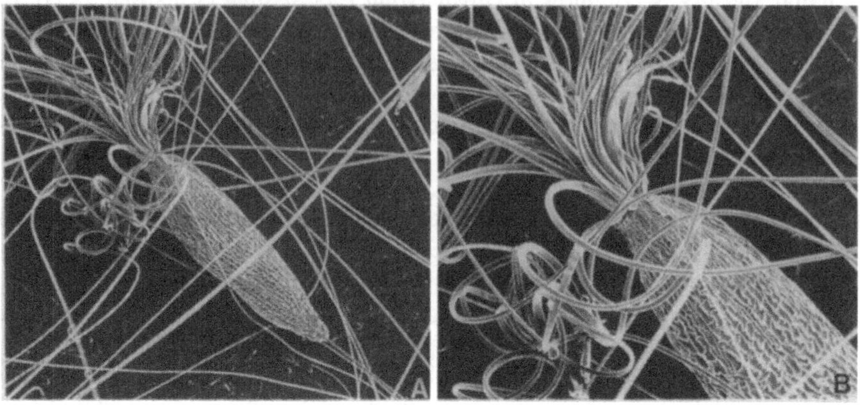

Fig. 39.13.3A, B. (SEM) *Epilobium angustifolium.* **A** Seed. **B** Micropylar portion. (Courtesy Dr F. Bouman, Amsterdam)

F Bartels (1956), HD Dickinson and J Lawson (1975), MI Gachechiladze (1974a, b), RR Gates (1911), Y Heslop-Harrison (1990), RL Hulbary and A Nagaraja Rao (1959), M Ishikawa (1918), DA Johansen (1928a, b, 1930a, b, 1931a, b, 1933, 1934), R Khan (1942), A Lebègue (1948a, c), P Maheshwari (1941, 1950), P Maheshwari and BL Gupta (1934), J Modilewski (1909b), A Nagaraja Rao (1959), CE O'Neal (1923), B Rodkiewicz, J Bednara and H Pora (1971), V Seshavatharam (1970), R Souèges (1920b, 1935a), K Subramanyam and HC Govindu (1948), G Täckholm (1914, 1915), JP Tilquin, K de Brouwer, A Mathieu and M Calier (1983), H Tobe and PH Raven (1985, 1986a, b, 1988a), J Venkateswarlu (1952c)

39.14 Oliniaceae

Mauritzon (1939b) gave a brief account of the embryology of *Olinia ventosa.* The detailed studies of Tobe and Raven (1984c) are based on *O. emarginata* and *O. ventosa.*

In the tetrasporangiate anther the sporogenous tissue in a pollen sac may be divided into two groups by a tapetal partition. The Dicotyledonous type of wall (Davis 1966) comprises the epidermis, an endothecium, a middle layer, and a glandular tapetum of binucleate cells. Since all the other layers degenerate, the wall of the mature anther consists only of the persistent papillate epidermal cells. Dehiscence occurs by longitudinal slits. The cells of the connective tissue adjacent to pollen sacs are radially elongated and may develop fibrous thickenings.

The microspore mother cells undergo meiosis followed by simultaneous cytokinesis, and in *O. emarginata* the tetrads (counted on the basis of 50 selected tetrads, see Schmid 1982) are usually (84%) tetrahedral, occasionally (14%) decussate and rarely (2%) isobilateral. The heteropolar (one pole colpate and the other heterocolpate) pollen grains are two-celled at shedding.

Each locule of the five-locular ovary has two or three pendulous, campylotropous, bitegmic, crassinucellar ovules. It is neither apotropous (anatropous, Cronquist 1981) nor hemitropous (hemianatropous, Cronquist 1981). The integuments are of dermal origin, the outer one is three- and the inner two-layered. Both the integuments form the micropyle. The integuments are unvascularized, and Mauritzon's report (1939b) of a small vascular branch extending into the outer integument in *O. ventosa* is erroneous.

The archesporial cell cuts off a parietal cell, and the parietal tissue is massive. The megaspore mother cell undergoes meiosis and the chalazal megaspore of the T-shaped tetrad gives rise to the Polygonum type of embryo sac. The three antipodals are ephemeral and degenerate before fertilization. Thus, the mature embryo sac is five nucleate – egg apparatus and two polars.

The entry of the pollen tube is porogamous. A hypostase differentiates after fertilization. The development of the endosperm is of the Nuclear type; whether wall formation occurs could not be ascertained. Embryogenesis has not been studied. An embryo from a mature seed showed two equally developed unfolded cotyledons; it lacked a conspicuous suspensor.

The seed-coat appears to comprise three- or four-layered testa. The cells of the exo- and mesotesta enlarge and contain pigments. The cells of the endotesta are radially elongated. The cells of the exotegmen are extremely tangentially elongated and fibrous (but not tracheidal), those of endotegmen collapse. Remnants of nucellar epidermis persist close to the seed-coat.

The exotegmic mature seed is ovoid and ellipsoidal, and is not winged. It is exalbuminous. The embryo has a short radicle and irregularly convolute cotyledons (Hutchinson 1973, Willis 1973).

A Cronquist (1981), GL Davis (1966), J Hutchinson (1973), J Mauritzon (1939b), V Patel, JJ Skvarla and PH Raven (1983), VS Rao and R Dahlgren (1969), R Schmid (1982), H Tobe and PH Raven (1984c), JC Willis (1973)

39.15 Haloragaceae

The anther wall comprises the epidermis, fibrous endothecium, single ephemeral middle layer and glandular tapetum of uni- or binucleate cells in *Myriophyllum intermedium* (Bawa 1969a). The tapetal cells are binucleate in *Haloragis colensoi* (Kapil and Bawa 1968), and two- to five-nucleate in *Laurembergia hirsuta* (Nagaraj and Nijalingappa 1974). Due to nuclear fusion polyploidy may occur in tapetal cells. Simultaneous cytokinesis follows meiosis in microspore mother cells, and tetrahedral and decussate tetrads are produced in *G. macrophylla* (Samuels 1912).

The pollen grains are four- to five-porate and three-celled at the dispersal stage. They are two-nucleate (two-celled) in *Gunnera macrophylla* and *G. chilensis* (Modilewski 1908a).

The ovule is anatropous, bitegmic and crassinucellar. Both integuments form the micropyle. In *M. intermedium*, the integuments are closely appressed to each other, and are separate only in the micropylar region. A weak funicular obturator is present, and a hypostase differentiates in *Haloragis colensoi* (Kapil and Bawa 1968) and *H. micrantha* (Nijalingappa 1975).

The archesporial cell cuts off a parietal cell. The parietal tissue is four- to five-layered. The megaspore mother cell undergoes meiotic divisions, the tetrad is linear or, rarely, T-shaped, and the chalazal megaspore develops into a Polygonum type of embryo sac. The polar nuclei fuse before fertilization, and the secondary nucleus lies either near the antipodals, or in the middle of embryo sac. The antipodals are persistent. The development of embryo sac is of the Peperomia type in *Gunnera chilensis* and *G. macrophylla* (Samuels 1912, Modilewski 1908a).

The remnants of the pollen tube persist up to ten- to 12-celled proembryo.

Development of the endosperm is of the Nuclear as well as the Cellular type. Nuclear type occurs in *Laurembergia brevipes* (Bawa 1969b), *L. hirsuta* (M. Nagaraj and Nijalingappa 1974), *L. javanica* (Bley 1925) and *Myriophyllum alterniflorum* (Stolt 1928). Cytokinesis occurs at eight- to 16-nucleate stage. The endosperm is of the Cellular type in *Haloragis asperrima*, *H. colensoi* (Kapil and Bawa 1968), *H. micrantha* (Nijalingappa 1975) and *M. intermedium* (M. Nagaraj and Nijalingappa 1967). In *M. intermedium* (Bawa 1969a), the division of the primary endosperm nucleus and subsequent divisions are followed by transverse walls so that a linear row of eight cells is formed. In other members the division of the primary endosperm nucleus is followed either by a vertical or a transverse wall.

The embryogeny conforms to the Caryophyllad type. The basal cell of the two-celled proembryo undergoes vertical division. These two cells enlarge, become haustorial, and their nuclei become hypertrophied (Fig. 3.11.16F). The suspensor haustorium is absent in *Laurembergia hirsuta* and *Gunnera*. In *Gunnera* the mature embryo is minute and obcordate. In other members the embryo is large and cylindrical.

The seeds are small, exarillate, endosperm copious with starchy food reserve, and embryo straight (Corner 1976).

The seed-coat is simplified within the small nut or drupe. The outer epidermis and hypodermis of the outer integument form the testa. The inner integument and nucellus disorganize except for a few cells at the tip.

SB Bawa (1969a, b, 1970b), F Bley (1925), EJH Corner (1976), BM Johri (1963b), RN Kapil and SB Bawa (1968), J Modilewski (1908a), M Nagaraj and BHM Nijalingappa (1967, 1974), BHM Nijalingappa (1975), JA Samuels (1912), H Schnegg (1902), R Souèges (1940c), KAH Stolt (1928), N Virkki (1962)

39.16 Theligonaceae (Cynocrambaceae)

Kapil and Mohana Rao (1966a) have made a detailed study of *Theligonum cynocrambe*.

The anther wall comprises the papillate cutinized epidermis, fibrous endothecium, an ephemeral middle layer, and glandular tapetum of uninucleate cells. The microspore mother cells undergo meiotic divisions, cytokinesis is simultaneous, and the tetrads are tetrahedral and isobilateral.

The pollen grains are six-porate, suboblate, three-celled at the dispersal stage and contain abundant starch.

The ovule is campylotropous, unitegmic and tenuinucellate. At the four-nucleate stage of gametophyte, the nucellus begins to disorganize, and the mature embryo sac comes in direct contact with the integument. A hypodermal archesporial cell functions directly as megaspore mother cell, undergoes meiotic divisions, produces a linear tetrad, and the chalazal megaspore develops into a Polygonum type of embryo sac. The antipodals undergo mitosis and become two- to four-nucleate in *Theligonum cynocrambe* (Kapil and Mohana Rao 1966a). The polar nuclei fuse in the central region of the embryo sac.

The development of the endosperm is of the Nuclear type. The primary endosperm nucleus divides and two, four and eight free nuclei are produced. At the quadrant proembryo, the entire endosperm becomes cellular in *T. cynocrambe* (Kapil and Mohana Rao 1966a). At maturity the endosperm becomes curved, is thin-walled with highly vacuolate cells, and contains starch. Only a few cells around the embryo disintegrate, and most of the endosperm persists in the seed.

The embryogeny conforms to the Chenopodiad type. The mature embryo is dicotyledonous and horseshoe-shaped.

The single integument is multiplicative, becomes thick, and the vacuolate cells contain starch. The thin-walled testa gets crushed, and the hard shell of the endocarp gives protection to the seed.

RN Kapil and PR Mohana Rao (1966a), R Wunderlich (1971)

39.17 Hippuridaceae

The wall of the tetrasporangiate anther shows the epidermis, a conspicuous fibrous endothecium which is two-seriate close to the stomium, ephemeral middle layer, and glandular tapetum of binucleate cells. The cytokinesis in pollen mother cells is simultaneous. The four- to six-colpate pollen grains are three-celled at shedding.

The ovule is anatropous, unitegmic, tenuinucellate, or reduced (ategmic). A nucellar cap, and a feebly developed funicular obturator are present. The archesporial cell functions directly as megaspore mother cell, undergoes meiosis, and produces a linear tetrad in *Hippuris vulgaris* (Juel 1911). The chalazal megaspore develops into a Polygonum type of embryo sac. The three small antipodals are ephemeral.

The pollen tube penetrates the embryo sac laterally (see Kapil 1962). This aspect needs further study.

The endosperm is of the Cellular type, and inconspicuous micropylar and chalazal haustoria develop.

The embryogeny is of the Onagrad type. The embryo is large, and a suspensor haustorium is absent (see B.M. Johri 1963b). Most of the endosperm is consumed during the maturation of seed.

TB Batygina and MS Yakovlev (1987), SB Bawa (1970b), BM Johri (1963b), HO Juel (1910, 1911), RN Kapil (1962), R Souèges (1922a)

39.18 Cynomoriaceae

The anther wall comprises the epidermis, fibrous endothecium, ephemeral middle layer, and glandular tapetum of binucleate cells. Following reduction divisions in pollen mother cells, simultaneous cytokinesis produces tetrahedral tetrads. The mature pollen grains are two-celled in *Cynomorium songaricum* (see Batygina and Yakovlev 1987).

The pendulous ovule in an unilocular ovary is unitegmic and pseudo-crassinucellar. The integumentary cells are appressed to each other so that the micropylar canal becomes extremely narrow in *C. coccineum* (Steindl 1945) and *C. songaricum*. The cells of the nucellar epidermis undergo periclinal divisions.

The archesporial cell functions directly as megaspore mother cell, undergoes meiotic divisions, the tetrad is T-shaped, and the development of the embryo sac is of the Polygonum type in *C. songaricum*. The antipodals are not ephemeral (Fig. 39.18.1A–G).

At the time of fertilization, the cells lining the micropylar canal become mucilaginous, or glandular, and perhaps provide nutrition to the pollen tube.

The development of the endosperm is of the Cellular type, and considerable endosperm is produced by the time the proembryo becomes globular (Fig. 39.18.1A–G). The endosperm does not develop any haustoria. During the growth of the embryo, the endosperm is consumed.

When the endosperm is four-celled, the zygote divides by an oblique transverse wall (Fig. 39.18.1C). The derivatives of both apical and basal cell produce an undifferentiated suspensorless embryo (Fig. 39.18.1D–J). The development can perhaps be referred to the Asterad type. It would be interesting to study the germination of such an organless embryo (see also Kuijt 1969). The fruit is a nut (Willis 1973).

TB Batygina and MS Yakovlev (1987), HO Juel (1903, 1910), RN Kapil (1962), J Kuiji (1969), F Steindl (1945), JC Willis (1973)

39.19 Taxonomic Considerations

The suborder Myrtineae of Engler and Diels corresponds to that of Melchior (1964), except that they place the family Oliniaceae in Thymelaeineae, and Theligonaceae in Hippuridineae. Melchior (1964) assigns Oliniaceae and Theligonaceae to his suborder Myrtineae, and places Hippuridaceae in the

Fig. 39.18.1A–J. *Cynomorium songaricum.* **A** Fertilized embryo sac with zygote and primary endosperm nucleus. Note the remnants of synergid in **A–D**, and persistent antipodal cells in **A–G**. **B** Zygote and two-celled endosperm. **C** Zygote in division, endosperm four-celled. **D–F** Two-, four- and eight-celled proembryo, and progressive development of endosperm. **G** Delimitation of protoderm in proembryo, endosperm conspicuous. **H–J** Later stages of proembryo (**H, I**), and undifferentiated mature embryo (**J**). A few endosperm nuclei and cell walls are shown in **H** and **I**, the endosperm has been consumed in **J**. (See Batygina and Yakovlev 1987)

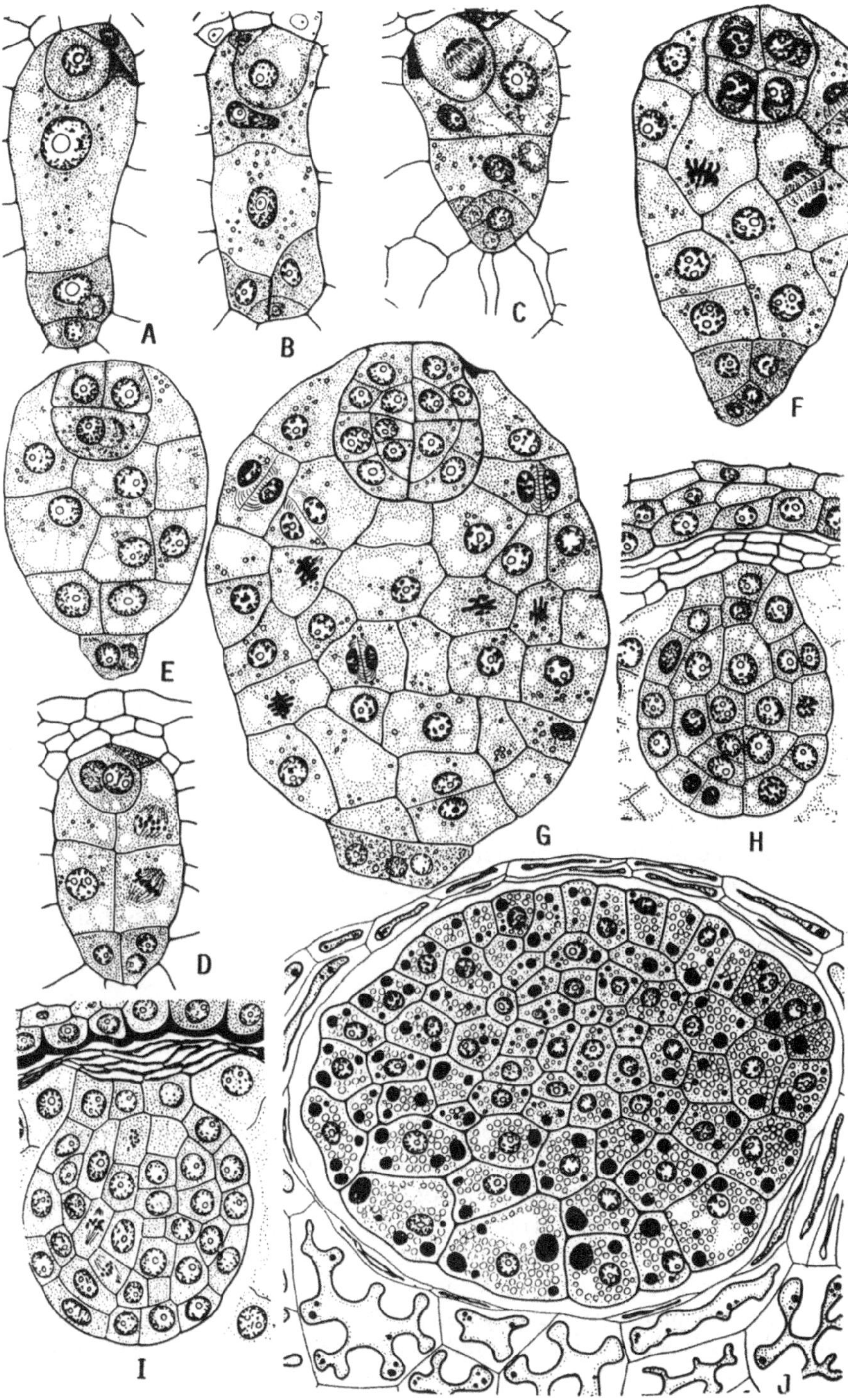
A
B
C
D
E
F
G
H
I
J

suborder Hippuridineae, and Cynomoriaceae in Cynomoriineae. Hutchinson (1973) includes Cunoniaceae and Oliniaceae in Cunoniales, *Hippuris* in Haloragidaceae, and assigns *Callitriche* to a separate family, Callitrichaceae.

Bentham and Hooker (1965a), Bessey (1915), Engler and Gilg (1924) and Wettstein (1935) group Lythraceae, Sonneratiaceae and Onagraceae along with Myrtaceae, Rhizophoraceae and Combretaceae in the order Myrtales (see also Hutchinson 1973). Hutchinson (1926) considers that Lythraceae and allied families show affinities to Caryophyllaceae, and Lythrales are related to Caryophyllales. A.C. Joshi (1939b) considered the relationship of Lythraceae and allied families with Caryophyllaceae, and negates their alliance on embryological grounds. According to him, the right place for these families is in Myrtales only.

Hutchinson (1973) transfers Lythraceae to the order Myrtales, in association with Myrtaceae, Sonneratiaceae, Punicaceae, Melastomataceae and others. According to him, the Myrtales are related to and derived from the same stock as the Guttiferales. Cronquist (1981) treats Lythraceae, Sonneratiaceae, Punicaceae and Melastomataceae, along with eight other families, in Myrtales.

Lythraceae, Sonneratiaceae, Punicaceae and Melastomataceae. Venkateswarlu (1937a, c) treats Lythraceae, Sonneratiaceae, Punicaceae and Melastomataceae as embryologically closely related. These families show similarity in the type of ovule, well-developed parietal tissue, micropyle formed by both integuments, Polygonum type of embryo sac, ephemeral antipodals (they do not even organize in Melastomataceae), Nuclear type of endosperm, Onagrad type of embryogeny, and exalbuminous seed. The endosperm is consumed in the free-nuclear state in Melastomataceae. The seed structure of Sonneratiaceae and Punicaceae is more or less comparable to that of Lythraceae. In Melastomataceae the outer epidermis of testa is sclerotic, but the inner one forms a crystal layer as in the other three families. The tegmen is present only at the endostome. Judging from these facts, Hutchinson's (1973) assignment of Lythraceae to Myrtales is fully justified.

Trapaceae. The genus *Trapa* has been treated under Onagraceae, as an appendage to Onagraceae, as an isolated member of Haloragaceae, and as a separate family Trapaceae. On an embryological basis, P. Maheshwari (1950, 1964), M. Ghosh (1954) and M. Ram (1956) favour the erection of a monogeneric family, Trapaceae (Table 39.19.1). The inferior tetralocular ovary with axile placentae bearing numerous ovules, and a loculicidal capsule in Onagraceae, differ from the semi-inferior bilocular ovary with a single ovule in each loculus, and a large one-seeded drupe of *Trapa* (P. Maheshwari 1950, 1964). In contrast to Onagraceae, *Trapa* shows irregularly biseriate anther tapetum, massive nucellar beak, Polygonum type of embryo sac, absence of endosperm, Solanad type of embryogeny, embryo with a long coiled suspensor, arrested growth of one cotyledon, and suberized mesotesta. The erection of the family Trapaceae is fully justified.

Table 39.19.1. Embryology of Trapaceae (Sect. 39.3) and Onagraceae (39.13) (See Ghosh 1954, Ram 1956, P. Maheshwari 1964, Corner 1976)

	Trapaceae	Onagraceae
Anther tapetum	Irregularly biseriate, secretory, cells multinucleate	Secretory, cells binucleate
Cytokinesis	Simultaneous	Simultaneous
Tetrads	Tetrahedral, decussate	Tetrahedral
Pollen	Pyramidal, 3 much-folded meriodonal crests, usually 2-celled, 3-celled in *Trapa astrachanica*	Tricolporate, 2-celled
Ovule	Anatropous, bitegmic, crassinucellate, true micropyle absent due to long nucellar beak, nucellar cells undergo endomitosis and become polyploid, prominent hypostase	Anatropous, bitegmic, crassinucellate, micropyle formed by both integuments, funicular obturator and epistase reported in some taxa; hypostase common
Embryo sac	Polygonum type, antipodals absent in *Trapa bispinosa*, uninucleate becoming hypertrophied and probably polyploid in *T. natans*	Oenothera, multiple megaspore mother cells, tetrads and embryo sacs develop, lower polar and antipodals are not formed
Endosperm	Primary endosperm nucleus persists up to 10–12-celled proembryo in *T. bispinosa*, 2 to 4 endosperm nuclei in *T. natans* which degenerate	Nuclear type, only a few nuclei are formed which degenerate
Embryo	Solanad type, suspensor long, coiled, multiseriate, haustorial, cells undergo endomitosis and become polyploid, one cotyledon well developed, the other highly reduced	Onagrad, suspensor short and inconspicuous, polyembryony reported in *Jussieua repens* and *Zauschneria latifolia*
Seed-coat	Testa multiplicative, except exotesta all the testal and tegmic tissue becomes crushed	Not multiplicative, inner and radial walls of endotesta strongly thickened and lignified, exotegmen comprises longitudinal lignified fibres
Seed	Exarillate, exalbuminous, enclosed in a 1-seeded drupe	Small, angular, corky, exarillate, exalbuminous, straight embryo with 2 equal cotyledons; fruit a coiled capsule with many seeds

Myrtaceae. The embryological features of Myrtaceae are in general conformity with those of the order. The integument contains a vascular strand as in Rhizophoraceae. However, the occurrence of a nucellar cap, hypostase, postament, endothelium, nucellar polyembryony, apospory, and Solanad and Onagrad type of embryogeny mark its individuality. The sclerotic testa with a crystal layer and crushed tegmen show resemblance to those of Melastomataceae, and to some extent to the mesotestal seed of Lecythidaceae.

Crypteroniaceae. Crypteroniaceae (Tobe and Raven 1987b) have been considered to be closely related to Melastomataceae (R. Dahlgren and Thorne 1984). Embryologically, these families do not differ very much from each other. The anther tapetal cells are usually two-nucleate, the endothecium degenerates much earlier, the nucellus also degenerates early, and the endotegmen persists and contributes to the seed-coat in Crypteroniaceae. In Melastomataceae the tapetal cells are uninucleate, the endothecium persists almost up to the dehiscence of the anther, the nucellus persists at least up to early post-fertilization stages, and the endotegmen is mostly crushed except near the micropyle. The available embryological data do support a relationship between Crypteroniaceae and Melastomataceae (Table 39.19.2).

Tobe and Raven (1984e) suggest that the Crypteroniaceae may also be related to Oliniaceae, Penaeaceae, Rhynchocalycaceae, and Alzateaceae because all the five families have an ephemeral endothecium. However, Crypteroniaceae differ from the other four families in the ephemeral nature of the nucellus. Crypteroniaceae do not have campylotropous ovule as in Oliniaceae; tetrasporic 16-nucleate Penaea type of embryo sac as in Penaeaceae; multicelled female archesporium as in Rhynchocalycaceae; nor is the micropyle formed by the inner integument only, and do not have multicelled female archesporium and bisporic Allium type of embryo sac as in Alzateaceae (Tobe and Raven 1984d, e). The embryological features of Crypteroniaceae are distinct from those of the related families Oliniaceae, Penaeaceae, Rhynchocalycaceae and Alzateaceae. These five families may form a monophyletic group within the order Myrtales (Tobe and Raven 1984f).

The embryological data (Tobe and Raven 1987b) indicate that *Crypteronia* and *Dactylocladus* are more closely related to one another than either is to *Axinandra* (see Sect. 39.4). *Crypteronia* has several ordinal features that characterize Myrtales: glandular anther tapetum, crassinucellate ovule, micropyle formed by both integuments, ephemeral antipodals, endosperm of the Nuclear type, and exalbuminous mature seed (Tobe and Raven 1987a). The placement of Crypteroniaceae in Myrtales is appropriate. Melchior (1964) and Cronquist (1981) also include Crypteroniaceae in Myrtiflorae near Lythraceae, Trapaceae and Myrtaceae.

Rhynchocalyx. The assignment of *Rhynchocalyx* is rather controversial. It has been related to the family Crypteroniaceae (Table 39.19.2) which includes *Rhynchocalyx* together with *Crypteronia, Dactylocladus, Axinandra* and *Alzatea.* The pollen morphology differs in each member. *Rhynchocalyx* and *Alzatea* differ from the other three genera of Crypteroniaceae sensu lato in their wood anatomy, and share characteristics with Lythraceae and Melastomataceae (van Vliet 1975). *Rhynchocalyx* is closer to Lythraceae, Oliniaceae and Melastomataceae because of nodal anatomy. Airy Shaw (1973) assigns it to Lythraceae, whereas R. Dahlgren and Thorne (1984) point out that it differs from Lythraceae in having petals and stamens on the hypanthial rim and sclereids in its leaf petioles.

Embryologically, *Rhynchocalyx* shows several Myrtalean characters: (a) glandular anther tapetum, (b) crassinucellate ovule, (c) two-layered inner integument, (d) ephemeral antipodals, (e) Nuclear type of endosperm, and (f) exalbuminous mature seed (Tobe and Raven 1984d). It differs only in that the micropyle is formed by the inner integument alone, indicating a derived position (Table 39.19.2).

The presence of ephemeral endothecium and micropyle formed by the inner integument segregates *Rhynchocalyx* from Lythraceae. In *Rhynchocalyx* (a) the septum between microsporangia on each side of the anther is persistent (septum of anther collapses in Lythraceae), nucellar subdermal cells elongate radially forming a jacket around the megaspores (radially elongated nucellar subdermal cells are absent in Lythraceae); (b) the homotypic meiotic division occurs earlier in the micropylar cell of the dyad than in the chalazal cell, starch grains are absent in the nucellus (homotypic division occurs later in the micropylar cell of dyad than in the chalazal cell, starch grains are present in the nucellus in Lythraceae); (c) the endosperm is scanty (many free nuclei aggregate in the micropylar/chalazal region in Lythraceae).

In view of the similarities of embryological characters with other Myrtales, *Rhynchocalyx* belongs to the order Myrtales. Due to several features of difference with Lythraceae, Tobe and Raven (1984d) propose a separate family Rhynchocalycaceae.

Alzatea. The assignment of the genus *Alzatea* of Myrtales is also controversial. *Alzatea* resembles Myrtales in: (a) glandular anther tapetum, (b) crassinucellate ovule, (c) two-layered inner integument, (d) ephemeral antipodals, (e) Nuclear type of endosperm, and (f) exalbuminous mature seed. *Alzatea*, like *Rhynchocalyx*, has a micropyle formed by the inner integument alone. In addition, *Alzatea* shows bisporic Allium type of embryo sac. Thus, it is an exceptional member of Myrtales (see Table 39.19.2).

Airy Shaw (1973) assigns it to Lythraceae. But the endothecium is ephemeral in *Alzatea*, the starch grains are absent in nucellus, micropyle is formed by the inner integument, the endosperm is scanty and, moreover, it has Allium type of embryo sac. Thus, *Alzatea* can be isolated from the Lythraceae. The only similarity common to *Alzatea*, *Rhynchocalyx* and Lythraceae is the multicelled ovular archesporium (Table 39.19.2). Tobe and Raven (1984e) propose erection of an independent family Alzateaceae in the order Myrtales.

Heteropyxis. Melchior (1964) assigns *Heteropyxis* to Myrtaceae. In view of its morphology, wood anatomy, floral anatomy, and palynology, Schmid (1980) also includes *Heteropyxis* in Myrtaceae. Embryologically, this taxon is closely allied to Myrtaceae except that (a) the two microsporangia of a theca are

Table 39.19.2. Embryology of Crypteroniaceae, Myrtaceae, *Alzatea*, *Rhynchocalyx*, Lythraceae and Melastomataceae (See Tobe and Raven 1983a, b, 1984a, d, e, 1987a, b, c)

	Crypteroniaceae	*Alzatea*	*Rhynchocalyx*	Myrtaceae	Lythraceae	Melastomataceae
Anther	Tetrasporangiate	Tetrasporangiate	Tetrasporangiate	Tetrasporangiate	Tetrasporangiate	Tetrasporangiate
Anther wall	—	Dicot type	Basic	Basic	Dicot	—
Epidermis	Persistent	Persistent	Persistent	Persistent	Not persistent (?)	Persistent
Endothecium	Ephemeral	Ephemeral	Ephemeral	Fibrous	Fibrous	Fibrous, except in *Melastoma* and *Oxyspora*
Middle layer/s	Ephemeral	Ephemeral	Ephemeral	Ephemeral	Ephemeral	Ephemeral
Tapetum	Glandular, cells 2-nucleate	Glandular, cells 2-nucleate	Glandular, cells 2-nucleate	Glandular, cells 2-nucleate	Glandular, cells 2–6-nucleate	Glandular, cells uninucleate
Cytokinesis	Simultaneous	—	Simultaneous	Simultaneous	Simultaneous	Simultaneous
Tetrad	Tetrahedral and decussate	Tetrahedral and decussate	Tetrahedral and decussate	Tetrahedral and decussate	Tetrahedral and isobilateral	Tetrahedral
Pollen	2-celled	Pseudocolpi absent, 2-celled	Heterocolpate, 2-celled	Tricolporate, pseudocolpi absent, 2-celled	Tricolporate, pseudocolpi may be present, 2-celled	Tricolporate, 2-celled
Ovule	Anatropous	Anatropous	Anatropous	Anatropous	Anatropous	Anatropous
Integument/s	Bitegmic, both two-layered	Bitegmic, inner longer and 2-layered, outer 2–4-layered	Bitegmic, inner longer than outer, both 2-layered	Bi- or unitegmic (*Syzygium*), inner 2-layered	Bitegmic, inner 2-layered	Bitegmic, inner 2-layered
Nucellus	Crassinucellate	Crassinucellate	Crassinucellate	Crassinucellate	Crassinucellate	Crassinucellate
Micropyle	Both integuments form micropyle	Inner integument	Inner integument	Both integuments (few exceptions)	Both integuments	Both integuments
Hypostase	Absent	Present	Absent	Present in some genera	Present	Absent

Female archesporium	1-celled, rarely 2-celled	4–8-celled, only one functions	3–5-celled, only one functions	1-celled	Multicelled, only one functions	1-celled
Tetrad	Linear	Dyad	Linear	Linear	Linear	Linear
Embryo sac	Polygonum type	Allium	Polygonum	Polygonum	Polygonum	Polygonum
Antipodals	Ephemeral	Ephemeral	Ephemeral	Ephemeral	Ephemeral	Ephemeral
Fertilization	Porogamous	Porogamous (?)	Porogamous	Porogamous	Porogamous	Porogamous
Endosperm	Nuclear type, becomes cellular	Nuclear, probably wall formation does not occur	Nuclear, few free nuclei, wall formation does not occur	Nuclear, becomes cellular	Nuclear, becomes cellular	Nuclear, wall formation does not occur
Embryogeny	—	Onagrad type	Onagrad	Solanad and Onagrad	Onagrad	Onagrad
Suspensor	Short	Short	Short	Short, or absent	Short	Short, massive
Embryo	Straight, cotyledons not folded	Cotyledons same size	Cotyledons folded inside	Cotyledons same size	Cotyledons flat or convolute	Cotyledons foliaceous, and crumpled or rolled, or thick and fleshy (*Memecylon*)
Polyembryony	Absent	Absent	Absent	Usual in *Syzygium*	Absent	Occasional
Seed	Seed-coat contributed by both integuments, wing formed on the micropylar and chalazal end, chalazal wing extremely long and narrow; exalbuminous	Exotestal, membraneous wing originates from the funicle and outer integument all round the seed; exalbuminous	Seed-coat contributed by both testa and tegmen, wing of funicular origin on the micropylar side only; exalbuminous	Exotestal, exalbuminous	Exotegmic, both testa and tegmen contribute to seed-coat, exalbuminous	Exotestal, exalbuminous

remotely separate from each other and dehisce by their respective longitudinal slits, instead of a common slit, (b) the ovule and seed have wings on both ends, and (c) the embryo sac is bisporic (Allium type).

Tobe and Raven (1987c) point out that *Heteropyxis* could be considered either as a subfamily of Myrtaceae distinct from all other members of the family, or as a separate but related family.

Lecythidaceae. Bentham and Hooker (1965a) include some genera of Lecythidaceae in Myrtaceae. Lecythidaceae are a distinct family showing amoeboid tapetum, three-celled pollen, bitegmic tenuinucellar ovule with endothelium, vascularized integument and, occasionally, reversed polarity of the embryo sac (Venkateswarlu 1952a). They show alliance with Myrtaceae due to simultaneous cytokinesis in pollen mother cells, vascularized integument, endothelium (rarely), Polygonum type of embryo sac, and Nuclear type of endosperm. The structure of seed, showing a sclerotic mesotesta and crushed tegmen, also indicates their alliance. Otherwise, they do not show similarity with the rest of the families included in Myrtiflorae. The evidence on embryogeny may help in determining their systematic position.

Rhizophoraceae. Rhizophoraceae closely resemble myrtaceous families in the type of ovule, vascularized integument, parietal tissue, Polygonum type of embryo sac with ephemeral antipodals, and Nuclear type of endosperm. The protrusion of endosperm through the micropyle in *Ceriops* and *Rhizophora*, formation of chalazal endosperm processes in *Ceriops*, and closely appressed cotyledons in the mature embryo are noteworthy features.

According to Tobe and Raven (1988b), the Rhizophoraceae are monophyletic. The vegetative, reproductive (including embryological) and chemical data suggest a close relationship to Elaeocarpaceae and Celastraceae, and possibly to Erythroxylaceae.

Combretaceae. Combretaceae are assigned a position between Melastomataceae and Rhizophoraceae, and between Onagraceae and Oliniaceae. The embryology of Combretaceae is comparable to the general pattern of development in Melastomataceae, Lythraceae, Punicaceae and Sonneratiaceae.

The genus *Guiera* differs from the rest of the members of Combretaceae in a single middle layer in the anther wall, micropyle formed by the inner integument, extension of the embryo sac at the micropylar end by absorbing parietal tissue and nucellar cap, and persistent antipodals multiplying up to six (Venkateswarlu and Prakasa Rao 1972). The separation of the genus *Guiera* from the tribe Combretae to a monogeneric tribe Guieraeae is warranted. The absorption of nucellus by the embryo sac is considered as a transitional stage in the evolution of nucellar tissue. Venkateswarlu and Prakasa Rao (1972) suggest that it may perhaps lead to tenuinucellate condition in Lecythidaceae.

The development of tetrasporic Penaea type of embryo sac has been observed only in *Combretum paniculatum* and *C. pincianum* (Mauritzon 1939b). Otherwise, the gametophyte is of the Polygonum type. To some extent, the structure of seed resembles that of Punicaceae. The presence of funicular obturator (*Poivrea*), tetrasporic embryo sac in some members, and Asterad type of embryogeny are special features of Combretaceae.

Onagraceae. Hutchinson (1973) includes Onagraceae in Onagrales. The genus *Trapa* is now removed from Onagraceae and raised to an independent family, Trapaceae. There is a close similarity between Lythraceae and Onagraceae in the development of anther, pollen, ovule, endosperm, embryo, and structure of seed. But the development of Oenothera type of embryo sac singles it out from the rest. The degeneration of antipodals occurs at a very early stage in Lythraceae. In *Lawsonia* (Lythraceae), the embryo sac shows reduction in the number of nuclei to such an extent that the Oenothera type of embryo sac can be derived from that of the Lythraceae (A.C. Joshi and Venkateswarlu 1936a).

In both families, Lythraceae and Onagraceae, the embryogeny is of the Onagrad type. A hypophysis differentiates much earlier and develops from the superior daughter cell of the cell *m* in Lythraceae, whereas in Onagraceae it develops directly from the cell *m* (A.C. Joshi and Venkateswarlu 1936a). In Lythraceae the suspensor is uni- or multiseriate, it is short and inconspicuous in Onagraceae.

Oliniaceae. Engler and Diels (1936) place Oliniaceae in Thymelaineae, whereas Hutchinson (1973) includes them in Cunnoniales. The available embryological data show their alliance with the members of Myrtineae. The Oliniaceae are presumed to be related to Thymelaeaceae (see Cronquist 1981). The similarities in ovule morphology suggest a closer relationship of Oliniaceae with the Lythraceae and Myrtaceae than with Thymelaeaceae. The relationship has been emphasized between Oliniaceae and both Crypteroniaceae and Melastomataceae (Takhtajan 1980, R. Dahlgren and Thorne 1984), or with Penaeaceae (Johnson and Briggs 1984).

Melchior (1964) places Oliniaceae in the suborder Myrtineae of Myrtiflorae. The Oliniaceae show a number of embryological characters of the Myrtales (Sect. 39.1). The ephemeral endothecium is also known in *Axinandra*, *Rhynchocalyx*, *Alzatea*, *Crypteronia*, *Dactylocladus*, as well as Penaeaceae (Tobe and Raven 1983a, b, 1984c, d, e, f, 1987a, b). The campylotropous ovule in Oliniaceae is unknown in any other Myrtales, except in *Mouiri* subg. *Taphroxylon* (Melastomataceae – Morley 1953). Oliniaceae are also unique in their pollen which is heteropolar, one pole colpate and the other heterocolpate. The pollen is said to have half-pseudocolpi (Patel et al. 1983), a condition not known in any other flowering plant.

Tobe and Raven (1984c) point out that the totality of similarities and dissimilarities in embryological characters does not suggest a close relationship of the Oliniaceae to any other group of Myrtales; rather, these features strongly support their isolated position within the order. The Oliniaceae are more closely related to Crypteroniaceae s. st., Rhynchocalycaceae, Alzateaceae and Penaeaceae than to any other. The exclusive characters of the ovule and pollen fully justify the erection of a new family Oliniaceae.

Haloragaceae. Bentham and Hooker (1965a) place the genera *Haloragis*, *Laurembergia*, *Myriophyllum*, *Gunnera* and *Callitriche* in Haloragaceae. Engler and Prantl, and Melchior (1964) include Haloragaceae in Myrtiflorae. Wettstein (1935) and Rendle (1952) recognize Haloragaceae as distinct from Hippuridaceae. In addition, Wettstein separates *Gunnera*, elevates it to the rank of a family – Gunneraceae, and recognizes it as an advanced taxon. Hutchinson (1973) assigns *Gunnera* and *Hippuris* to Haloragaceae (Haloragidaceae), and *Callitriche* to Callitrichaceae. Cronquist (1981) treats Haloragaceae and Gunneraceae as separate families in the order Haloragales. (This order is considered to be a reduced derivative of Myrtales.) R. Dahlgren (1981) erected an independent order Gunnerales for the family Gunneraceae, in superorder Rosiflorae. In fact, the family Haloragaceae (Bentham and Hooker 1965a) is heterogeneous, and the genera *Callitriche*, *Gunnera* and *Hippuris* have now been removed. According to Corner (1976), the seed structure in Haloragaceae bears no resemblance either with Lythraceae or Myrtaceae.

The embryological data support the assignment of Haloragaceae to Myrtiflorae. There is no need to erect the order Haloragales.

B.M. Johri (1963b) and Bawa (1970b) evaluate the embryological features of *Haloragis*, *Gunnera*, *Hippuris* and Callitrichaceae. The occurrence of three- to five-colpate two-celled pollen grains, bicarpellary unilocular ovary with a single ovule, micropyle formed by the inner integument alone, Peperomia type of embryo sac, and absence of suspensor haustorium distinguish *Gunnera* from other members of Haloragaceae (Table 39.19.3). Cytological data also confirm the separation of *Gunnera* (Virkki 1962). Its assignment to a family or order is a matter of choice.

Hippuridaceae. The position of *Hippuris* is debatable. Bentham and Hooker (1965a) place it in Ranales, and consider it to be derived from Nymphaeaceae. Melchior (1964) raises it to the family Hippuridaceae under suborder Hippuridineae. Thorne (1968) includes it in Cornales. R. Dahlgren (1977a, b, 1981) erects the order Hippuridales (Hippuridaceae) in superorder Lamiiflorae, close to the orders Scrophulariales and Hydrostachyales.

The aquatic habit, bisexual reduced flowers, four- to six-colpate pollen grains, monocarpellary unilocular ovary with single pendulous, unitegmic, tenuinucellate ovule (with nucellar cap), Onagrad type of embryogeny, and absence of suspensor haustorium definitely segregate Hippuridaceae from Haloragaceae, and justify their present position. Anatomical data also support this arrangement (Metcalfe and Chalk 1950).

In recent systems of classification, Hippuridaceae is placed in or near Scrophulariales, an iridoid sympetalous group (R. Dahlgren 1975a, Wagenitz 1977a). The nature of ovule (except nucellar cap), endosperm (haustoria absent), and embryogeny resembles that in Scrophulariales. The apetalous reduced flower with inferior monocarpellary ovary, single ovule, and absence of endosperm haustoria may justify their assignment to an independent order Hippuridales.

Cynomoriaceae. The suborder Cynomoriineae is represented by a single family Cynomoriaceae, and Melchior (1964) places the latter next to Hippurida-

Table 39.19.3. Floral morphology and embryology of *Haloragis*, *Hippuris* and *Gunnera* (Adapted from Kapil 1962, Hutchinson 1973, Batygina and Yakovlev 1987)

	Haloragis	*Hippuris*	*Gunnera*
Flower	Bracteate, bisexual, tetramerous, arranged in glomerules in leaf axis	Bracteate, bisexual, rarely unisexual, solitary, perianth forms a rim around the tip of ovary	Ebracteate, bisexual, bimerous, crowded in large spikes or panicles
Stamen/s	4 + 4, obdiplostamonous	Single stamen placed medianly	2, antipetalous
Anther wall	4-layered, tapetum secretory, cells 2-nucleate	4-layered, tapetum secretory, cells 2-nucleate	4-layered, tapetum secretory, cells 2-nucleate
Cytokinesis	Simultaneous	Simultaneous	Simultaneous
Pollen	4–5-porate, 3-celled	4–6-colpate, 3-celled	4–5-porate, 2-celled
Gynoecium	Tetracarpellary, tetralocular, single pendulous ovule in each locule	Monocarpellary, unilocular, single pendulous ovule	Bicarpellary, unilocular, solitary ovule
Ovule	Anatropous, bitegmic, crassinucellar, feeble funicular obturator	Anatropous, unitegmic, or ategmic, tenuinucellar, feeble funicular obturator	Anatropous, bitegmic, crassinucellar, obturator absent
Nucellus	Parietal tissue formed, nucellar cap absent	Parietal tissue not formed, nucellar cap present	Parietal tissue formed, nucellar cap absent
Embryo sac	Polygonum type, antipodals 3, persistent	Polygonum, antipodals 3, ephemeral	Peperomia, secondary nucleus formed by fusion of 6 nuclei, antipodals 6, ephemeral
Fertilization	Porogamous	Pollen tube penetrates the embryo sac laterally	Porogamous
Endosperm	Cellular type	Cellular	Cellular
Embryo	Caryophyllad type, large, cylindrical, suspensor haustorium present	Onagrad type, large, cylindrical, suspensor haustorium absent	Caryophyllad (?), minute, obcordate, without suspensor haustorium
Seed coat	Outer epidermis and hypodermis of outer integument form testa	—	—
Seed	Small, exarillate, albuminous	Exalbuminous	Albuminous

ceae (Hippuridineae). Wettstein (1935) assigns Cynomoriaceae to Santalales. As the members are root parasites, Cronquist (1981) retains *Cynomorium* in Santalales, as a genus in Balanophoraceae. Hutchinson (1973) also includes *Cynomorium* in Balanophoraceae. Kuijt (1969) states: "Although my acquaintance with Balanophoraceae is a shallow one, I am unable to support this separation" of *Cynomorium*, and assignment to an independent family Cynomoriaceae. (Initially van Tieghem split Balanophoraceae into a series of separate families including Cynomoriaceae – see Kuijt.) While *Cynomorium* and *Balanophora* are both root parasites, there are other resemblances between the two.

The similarities and dissimilarities in the embryology of Balanophoraceae (Sect. 13.2), Hippuridaceae (Sect. 39.17) and Cynomoriaceae (Sect. 39.18) are: (a) non-fibrous endothecium in Balanophoraceae and fibrous in the other two, (b) glandular tapetum of binucleate cells, (c) successive cytokinesis in Balanophoraceae, simultaneous in the other two families, (d) pollen grain three-celled in Balanophoraceae, and two-celled in the other two, (e) undifferentiated ovule in Balanophoraceae, anatropous, unitegmic, tenuinucellate ovule in Hippuridaceae and Cynomoriaceae, (f) Polygonum type of embryo sac, which becomes U- or J-shaped with reversed polarity in Balanophoraceae, (g) antipodals absent in Balanophoraceae, ephemeral in Hippuridaceae, and persistent in Cynomoriaceae, (h) endosperm of the Cellular type; a large chalazal haustorial cell in Balanophoraceae, inconspicuous micropylar and chalazal haustoria in Hippuridaceae, and none in Cynomoriaceae, (i) embryogeny of the Piperad type in Balanophoraceae, Onagrad in Hippuridaceae, and Asterad (?) in Cyanomoriaceae, (j) embryo without suspensor and undifferentiated in Balanophoraceae and Cynomoriaceae, with suspensor and typically dicotyledonous and large in Hippuridaceae, (k) seed "naked" and albuminous in Balanophoraceae, seed with testa but more or less exalbuminous in the other two families, and (1) during germination of "fruit" in Balanophoraceae, the "endosperm cells also germinate" and form tubes – *this feature is not known in any other angiosperm.*

In view of the above facts, *Cynomorium* is closer to *Balanophora* (see also Kuijt 1969) than to *Hippuris*. The families Cynomoriaceae and Balanophoraceae should be included in the Order Balanophorales, and neither in Santalales nor in Myrtiflorae.

Theligonaceae. The genus *Theligonum* is included in Urticaceae, Phytolaccaceae and Euphorbiaceae. Dumortier (1829) and Takhtajan (1980) raise it to a family rank, Theligonaceae, in the order Gentianales near Rubiaceae. Engler and Diels (1936) place Theligonaceae with Hippuridaceae in the suborder Hippuridineae of Myrtiflorae. Cronquist (1981) includes them in Rubiales along with the Rubiaceae. R. Dahlgren (1975a) and Wagenitz (1977a) include *Theligonum* in Rubiaceae. Kapil and Mohana Rao (1966a) discuss the morphology and embryology of *Theligonum cynocrambe*, and compare it with Phytolaccaceae, Urticaceae and Euphorbiaceae. They conclude that erection of the family Theligonaceae is fully justified. It does not show any relationship with either Haloragaceae (Sect. 39.15) or Hippuridaceae (Sect. 39.17). The proper place for Theligonaceae is in the order Centrospermales (Sect. 16) near the families Phytolaccaceae and Amaranthaceae. The embryological features of similarity are: (1) monocarpellary ovary with a solitary ovule, (2) secretory anther tapetum, (3) three-celled pollen grains, (4) Polygonum type of embryo sac, (5) Nuclear type of endosperm, (6) presence of raphides, and (7) unicellular, non-glandular hairs, and rubiaceous stomata (Kapil and Mohana Rao 1966a). However, the perisperm, a typical centrospermalean character, is absent in Theligonaceae.

In a very detailed and critical review, Wunderlich (1971) discusses the systematic position of *Theligonum* which has been rather problematic. Several morphological, anatomical, cytological and embryological features of *Theligonum* have been compared with those of Rubiaceae. The common embryological features are: uninucleate anther tapetum, absence of endothelium, Nuclear type of endosperm, albuminous seeds, and chromosomes n = 10 or 11.

The common features with the Anthospermeae, Spermacoceae and Gulieae (tribes of Rubiaceae) are: three-celled pollen grains, single ovule in the ovary, micropyle turned towards the base and outwards, embryogeny probably of the Solanad type, comparable development of seed, and unspecialized testa.

Theligonum has some modified rubiaceous characters, like campylotropous ovule, and horseshoe-shaped curved embryo.

Wunderlich remarks: "the relations to the Rubiaceae are largely hidden so that they had not been recognized before". She concludes that *Theligonum* certainly belongs to Sympetalae (tubular perianth of female flower, ovule tenuinucellate and unitegmic). It shows a close relationship with Rubiaceae. On account of the peculiar combination of features, it should be assigned to a separate tribe "Theligoneae", closely related to Anthospermeae, and also to Spermacoceae and Galieae.

We have some reservations in accepting the assignment proposed by Wunderlich, and would leave the question open for a future consideration.

40 Umbelliflorae

The order Umbelliflorae comprises seven families: Alangiaceae, Nyssaceae, Davidiaceae, Cornaceae, Garryaceae, Araliaceae and Umbelliferae.

40.1 General Features

The order Umbelliflorae is characterized by: (1) secretory anther tapetum, (2) simultaneous cytokinesis, (3) three-colporate two-celled pollen grains; three-celled in Umbelliferae, (4) anatropous, unitegmic ovule (5) Polygonum type of embryo sac, (6) persistent antipodals in Araliaceae and Cornaceae (rarely), and their secondary multiplication in Alangiaceae and Garryaceae, (7) Nuclear type of endosperm; Cellular in Davidiaceae and Cornaceae; both types occur in Alangiaceae, (8) Solanad type of embryogeny in Garryaceae and Umbelliferae, Onagrad or Chenopodiad in Araliaceae, and Asterad in Cornaceae, and (9) testal, albuminous seed with thin-walled seed-coat; outer epidermis lignified in Nyssaceae.

40.2 Alangiaceae

The development of pollen and embryo has not been investigated.

The anther wall comprises the epidermis, fibrous endothecium, single ephemeral middle layer, and secretory tapetum of binucleate cells. The tetrads are isobilateral.

The ovule is anatropous, unitegmic and crassinucellate, and the funicular vascular strand reaches up to the apex of integument. The nucellar epidermis divides periclinally and produces three layers. There is a hypostase of four to five layers. An endothelium develops after fertilization.

The single archesporial cell cuts off a parietal cell which produces a layer of cells. The megaspore mother cell forms a linear tetrad, and the chalazal megaspore gives rise to a Polygonum type of embryo sac. The antipodals increase up to 12 or more, and persist up to fertilization in *Alangium lamarkii* (Gopinath 1945); they are ephemeral in *A. handelii.*

The endosperm is of the Nuclear type in *A. lamarkii* (Gopinath 1945), but Cellular in *A. chinense* (Mitra and Datta 1949). Further studies on endosperm development are necessary.

Apogamous antipodal embryos are described in *A. lamarkii* (Gopinath 1943), but this is doubtful and needs reinvestigation.

The seed contains copious starchy endosperm, the embryo is straight and conspicuous, and the seed-coat is thin-walled.

DM Gopinath (1943, 1945), BM Johri (1963b), JN Mitra and RM Datta (1947, 1949)

40.3 Nyssaceae

The embryogeny has not been investigated.

The anther wall comprises the epidermis, fibrous endothecium, three ephemeral middle layers, and secretory tapetum in *Nyssa sylvatica* (Mohana Rao 1972a). Cytokinesis is simultaneous, and the tetrads are tetrahedral and decussate.

The pollen grains are tricolporate and two-celled at the shedding stage.

The ovule is anatropous, unitegmic and pseudocrassinucellar. The nucellar epidermal cells divide periclinally producing a single layer at the apex, and three layers laterally. A hypostase develops in *Nyssa biflora.* The hypodermal archesporial cell functions directly as megaspore mother cell, and forms a linear or T-shaped tetrad. The chalazal megaspore develops into a Polygonum type of embryo sac. The antipodals are ephemeral. The polar nuclei fuse before fertilization. Tandon and Herr (1971) reported Allium type of embryo sac in 20% of the ovules of *N. aquatica* and *N. biflora*.

The endosperm is of the Nuclear type, wall formation is initiated after the formation of eight free nuclei, and before the division of zygote in *N. sylvatica* (Mohana Rao 1972a).

The seed contains a large embryo, and abundant endosperm. The integument consists of four to five layers at the megaspore mother cell stage. Due to periclinal divisions the number of layers increases up to 48–50. The seed-coat is smooth and, while the outer epidermis is thick-walled and lignified, the inner layers are thin-walled.

The ovary wall is 25–28 layers thick at the megaspore mother cell stage. The thickness of wall increases to a large extent, and the pericarp differentiates into three zones: the outer zone with large cells, prominent nuclei and calcium oxalate crystals, middle zone of tangentially elongate cells with dense cytoplasm, and inner zone of 18–20 layers comprising thick-walled, elongated, tubular fibres with a narrow lumen.

RH Eyde (1966), PR Mohana Rao (1972a), SR Tandon and JM Herr Jr (1971), W Wangerin (1910a)

40.4 Davidiaceae

The structure of anther, pollen grain, embryo and seed-coat has not been fully studied (see Schnarf 1931).

The pollen grains are tricolpoidorate.

The ovule is unitegmic and crassinucellate in *Davidia involucrata* (Horne 1909). Whether bitegmic condition occurs, requires confirmation (see Corner 1976). The nucellar epidermal cells divide periclinally to produce a cap. An endothelium differentiates, but further details are not known. The archesporium is multicelled; the formation of parietal cell/s requires confirmation.

The development of the embryo sac conforms to the Polygonum type. The antipodals are conspicuous and, sometimes, their number increases and they show haustorial activity during post-fertilization stages. This aspect needs further investigation. Multiple and abnormal embryo sacs are reported.

The development of endosperm is of the Cellular type.

The seeds have fleshy endosperm, and a large embryo (Hutchinson 1973).

EJH Corner (1976), AS Horne (1909), J Hutchinson (1973), K Schnarf (1931)

40.5 Cornaceae

The anther wall comprises the epidermis, fibrous endothecium, two or three ephemeral middle layers, and secretory tapetum. The microspore mother cells produce tetrahedral tetrads, and cytokinesis is simultaneous.

The tricolporate pollen grains have a smooth exine, are oblate and spheroidal, and two-celled at the dispersal stage.

The ovule is anatropous, unitegmic and crassinucellar. An endothelium develops at the tetrad stage in *Helwingia japonica* (Sâto 1976a). The cells of the nucellar epidermis undergo divisions. The archesporium is uni- or multi-celled, cuts off a parietal cell, and the parietal tissue is three to four layers thick in *Cornus* (Chopra and Kaur 1965b). The megaspore mother cell forms a linear tetrad, and the chalazal megaspore develops into a Polygonum type of embryo sac in *Aucuba japonica* (Sâto 1971), *A. japonica* var. *borealis*, *Cornus controversa*, *Helwingia japonica* (Sâto 1976a), *C. svecica* (Håkansson 1923), *C. alba* (Fagerlind 1939c), *C. alternifolia*, *C. ammomum*, *C. macrophylla*, *C. oblonga* and *C. stolonifera* (Chopra and Kaur 1965b). In *A. japonica*, sometimes, the embryo sac develops from the lower two, or rarely the upper two, megaspores of the tetrad. In either case each megaspore becomes binucleate, the intervening wall disorganizes, and the four nuclei divide once so that an eight-nucleate embryo sac of usual organization is produced. This mode of development is essentially bisporic. Twin and multiple embryo sacs occur in *C. oblonga*. The antipodals are ephemeral. The polar nuclei fuse before fertilization. The development of the embryo sac is pseudomonosporic in *C. florida* (Morse 1907). Håkansson (1923) reported four free megaspore nuclei in *C. mas*, and Fagerlind (1939c) and D'Amato (1946b) confirmed Fritillaria type of development. D'Amato also reported five- and seven-nucleate (Pyrethrum parthenifolium type) embryo sac in *C. mas*.

B.B. Smith (1973a, 1975) studied the qualitative and quantitative development of five species of *Cornus*: *C. alternifolia*, *C. ammomum*, *C. asperifolia*, *C. florida* and *C. stricta* (Table 40.5.1). *C. florida* differs from the rest in several features, such as multicellular archesporium functioning directly as megaspore mother cells, many linear tetrads, and persistent synergid and antipodals. The maximal increase in the size of female gametophyte occurs at the four-nucleate stage in *C. florida*, in others at the eight-nucleate stage. B.B. Smith proposed the removal of *C. florida* from the genus *Cornus* to the monotypic taxon *Cynoxylon floridum* (L) Ref. (see also Small 1933).

The endosperm is of the Cellular type (Håkansson 1923). The cells are thin-walled and store starch.

The embryogeny corresponds to Period I, Megarchetype II in *Aucuba japonica* (Mestre and Guignard 1970).

The seed-coat is thin-walled, and gets more or less crushed.

RN Chopra and H Kaur (1965b), F D'Amato (1946b), H Eyde (1967), F Fagerlind (1939c), A Håkansson (1923), H Harms (1897b), H Kaur (1970c), JC Mestre and JL Guignard (1970), WC Morse (1907), Y Sâto (1971, 1976a), BB Small (1933), BB Smith (1973a, 1975), W Wangerin (1910b)

40.6 Garryaceae

The anther wall comprises the epidermis, fibrous endothecium, one or two ephemeral middle layers, and glandular tapetum of binucleate cells. The nuclei

Table 40.5.1. Embryology of five species of *Cornus* (See B.B. Smith 1973a)

	C. alternifolia	*C. ammomum*	*C. asperifolia*	*C. florida*	*C. stricta*
Nucellus	Crassinucellate	Crassinucellate	Crassinucellate	Tenuinucellate	Crassinucellate
Archesporial cells	2 or more	1	2 or more	Many	1
Parietal cell/s	1	1	2	None	1
Sporogenous cells	1	1	2	Many	1
Tetrad/s	Linear or T-shaped	Linear or T-shaped	Linear, 2	Linear, many	Linear
Degeneration of megaspore	Variable	Variable	First middle 2, then uppermost	Variable	Variable
Synergids	Degenerate rapidly	Degenerate	Persist until polar nuclei fuse	Persist until polar nuclei fuse	Persist until polar nuclei fuse
Antipodals	Persist until polar nuclei fuse	Degenerate	Persist until polar nuclei fuse	Persist until polar nuclei fuse	Persist until polar nuclei fuse

divide and nuclear fusion results in polyploid condition. The microspore mother cells undergo meiotic divisions, cytokinesis is simultaneous, and the tetrads are tetrahedral and decussate.

The tricolporate pollen grains are shed at the two-celled stage in *Garrya douglasi* (Kapil and Mohana Rao 1966b).

The ovule is anatropous, unitegmic and crassinucellate. The cells of the funiculus, adjacent to placenta, divide and form a collar around the micropylar region of the ovule. A hypostase differentiates after fertilization. The hypodermal cell cuts off a parietal cell. The parietal tissue is four to five layers thick. The megaspore mother cell produces a linear tetrad, and the chalazal megaspore develops into a Polygonum type of embryo sac. The antipodal cells divide into five to seven cells which persist till the endosperm becomes cellular (Kapil and Mohana Rao 1966b).

Development of the endosperm is of the Nuclear type, and wall formation is initiated at the micropylar and chalazal region at the 16-nucleate stage.

The embryogeny conforms to the Solanad type. Suspensor polyembryony is reported in *Garrya veatchii* (Mohana Rao 1963), and the additional embryo develops up to the globular stage.

The seed is small, and sarcotestal with a minute embryo. The endosperm stores abundant starch. The seed-coat is 12–15 layers thick, and the epidermis comprises thin-walled columnar cells. A few outer thin-walled layers persist, while the inner ones get crushed.

RH Eyde (1964), RN Kapil and PR Mohana Rao (1966b), PR Mohana Rao (1963), MF Moseley Jr and RM Beeks (1955), W Wangerin (1910c)

40.7 Araliaceae

The epidermis of the anther shows bulges at maturity, and the fibrous endothecium is biseriate in the region of the connective in *Tieghemopanax sambucifolius* (Mohana Rao 1972b). Two ephemeral middle layers are reported in *Brassia actinophylla* (Gopinath 1944), *Hedera australiana, Panax fruticosa* (D. Singh 1954) and *Tieghemopanax*. The tapetum is secretory with uninucleate cells in *H. australiana*, binucleate in *P. fruticosa* and multinucleate in *B. actinophylla* and *Tieghemopanax*. In *Tieghemopanax* nuclear fusions result in polyploid condition. The microspore mother cells undergo meiotic divisions, and produce tetrahedral and decussate tetrads. The cytokinesis is simultaneous.

The pollen grains are oblate, spheroidal, subporolate, and three-colporate, and shed at the two-celled stage.

The ovule is anatropous, unitegmic and tenui- or crassinucellar with a narrow micropyle. The nucellus disintegrates rapidly. A funicular obturator occurs in *Hedera australiana, Panax fruticosa, Polyscias pinnata* (Gopinath 1944) and *Tieghemopanax*. An endothelium and hypostase also differentiate. The archesporium is one-celled, except in *P. fruticosa* where it may be two-celled. The archesporial cell functions directly as megaspore mother cell, or may produce a parietal cell which does not divide further (Gopinath 1944). The megaspore mother cell gives rise to a linear tetrad. The chalazal megaspore develops into a Polygonum type of embryo sac. The antipodals are ephemeral, and the polars fuse before fertilization. The micropylar megaspore develops into embryo sac in *Hedera nepalensis* (D. Singh 1954). The development of more than one megaspore up to two- or four-nucleate stage is reported in *H. australiana* and *Polyscias pinnata.* Occasionally, a bisporic embryo sac occurs in *H. australiana* and *P. pinnata*; a tendency to Oenothera type is reported in *Brassia actinophylla.* The antipodals persist in *B. actinophylla* and *H. venulosum* (Gopinath 1944).

An embryo sac with reversed polarity is reported in *H. venulosum* (Gopinath 1944).

The development of the endosperm conforms to the Nuclear type, wall formation occurs at the eight-nucleate stage in *Tieghemopanax.* The endosperm is persistent, and stores abundant reserve food. It is ruminate in *Arthrophyllum diversifolium* and *Hedera helix,* due to the infoldings of the outer epidermal layer of the seed-coat (Periasamy 1962a).

The embryogeny conforms to the Onagrad type in *Aralia cachemiria, A. cordata, A. racemosa, Fatsia japonica* and *Hedera helix*; Chenopodiad type in *Hedera australiana* and *Panax fruticosa.*

The seed-coat consists of only the outer epidermis. Its cells are tangentially elongate, slightly thick-walled and contain tannin. It is thin and membraneous.

L Ducamp (1902), DM Gopinath (1944, 1945), PR Mohana Rao (1972b), K Periasamy (1962a), EM Pigott (1915), D Singh (1954)

40.8 Umbelliferae

The development of the anther wall is of the dicotyledonous type. It comprises the epidermis, fibrous endothecium, an ephemeral middle layer, and secretory tapetum of binucleate cells. The endothecium develops fibrous thickenings on the dorsal side in *Bupleurum tenue* (S.C. and M. Gupta 1964). The tapetum becomes biseriate near the septum, and the cells are multinucleate in some taxa (S.C. Gupta 1970). The microspore mother cells undergo meiotic divisions, cytokinesis is simultaneous, and tetrads are tetrahedral, decussate and isobilateral.

The pollen grains are shed at the three-celled stage. The male cells are located at each pole of the ellipsoidal pollen showing fixed polarity (Paliwal 1951, S.C. Gupta 1964). This aspect needs further study.

The ovule is anatropous, unitegmic and tenuinucellar. A pseudocrassinucellar ovule occurs in *Foeniculum vulgare* (S.C. Gupta 1964). A hypostase differentiates as early as the archesporial stage, and persists up to the maturation of the embryo sac. The endothelium differentiates at the megaspore tetrad stage in *Bupleurum,* and extends beyond the nucellus around the micropylar canal. An obturator also develops.

The archesporium is uni- or bi-celled. The hypodermal archesporial cell functions directly as megaspore mother cell, the chalazal megaspore of a linear tetrad develops into a Polygonum type of embryo sac. In *Pimpinella bracteata* and *P. candolleana* (Maheswari Devi and Santa Ram 1985), sometimes the tetrad is T-shaped. The occurrence of isobilateral tetrad is reported in *Peucedanum grande* (G.L. Shah 1953); this needs confirmation. Allium type of development has been observed in *Bupleurum aureum*, Drusa type in *Bowlesia tenera* and *Drusa oppositifolia*, and Penaea type in *Azorella trifurcata* (Håkansson 1923, 1927, 1952).

In most taxa, double fertilization occurs as usual. However, in *Pimpinella candolleana* and *P. monoica* (Maheswari Devi and Santa Ram 1985), syngamy is said to occur only after the formation of a large number of free endosperm nuclei. This aspect needs confirmation, also whether this is an abnormal feature limited to only a few ovules, or common one.

The endosperm is of the Nuclear type. The nuclei aggregate at the micropylar end, and enucleate nodules occur in the micropylar region in *Anethum graveolens*, *Coriandrum sativum*, *Daucus carota*, *Foeniculum vulgare* (Fig. 3.10.2F), *Heracleum sphondylium*, *Psammogeton biternatum* (S.C. Gupta 1970), *Cuminum cyminum* (Sehgal 1965), *Pseudorlaya pumilla* and *Torilis nodosa* (Al-Attar 1977) and *Pimpinella* spp. (Maheswari Devi and Santa Ram 1985). Wall formation is initiated ca. at the 64-nucleate stage in *Anethum* (Dhakre 1968), and ca. at the 256-nucleate stage in *Coriandrum* and *Foeniculum* (S.C. Gupta 1964).

The embryogeny conforms to the Solanad type. The development of a synergid proembryo has been observed in *Pimpinella candolleana* (Maheswari Devi and Santa Ram 1985). The synergid embryo is much smaller than the zygotic. Budding of suspensor is reported in *Coriandrum sativum* and *Foeniculum vulgare* (S.C. Gupta 1964). *Cuminum cyminum* (Sehgal 1965) and *P. candolleana* (Maheswari Devi and Santa Ram 1985). The origin of an additional embryo has not been traced in *Trachyspermum ammi* (Sehgal 1965).

The developing seed of *Foeniculum* is often attacked by a phytophagus chalcid fly *Systole albipennis* which causes embryolessness (S.C. Gupta 1962a, b). It lays its eggs in the mericarp with globular proembryo and partially, or completely, cellular endosperm (Fig. 40.8.1A). The eggs develop at the cost of the embryo and endosperm (Fig. 40.8.1B). Finally, in mature fruits the insect escapes by boring a hole in the fruit wall (Fig. 40.8.1C–E). A curious feature is that the larva of *Systole* may itself get parasitized by a hyperparasite, either *Liodontomerus* or *Tetrastichus* (Fig. 40.8.1F). In this case the larva of *Systole* is destroyed, and no appreciable damage is caused to the embryo or endosperm.

The seed is endospermic with a small embryo. The endosperm stores oil as reserve food. The seed-coat is almost crushed except for the outer epidermis which remains thin-walled.

The fruit is a schizocarp. In each mericarp six vittae alternate with six vascular bundles. The vittae are located in furrows. The outer epidermis of the mericarp comprises anticlinally elongate, thickly cuticularized cells. Most of the cells of the pericarp become sclerotic. The inner epidermal cells have prominent fibrous thickenings.

In *Bupleurum tenue*, stomata occur on stalks of the compound umbels and umbellets, bracts of involucre and involucel, pedicels of flowers, both surfaces

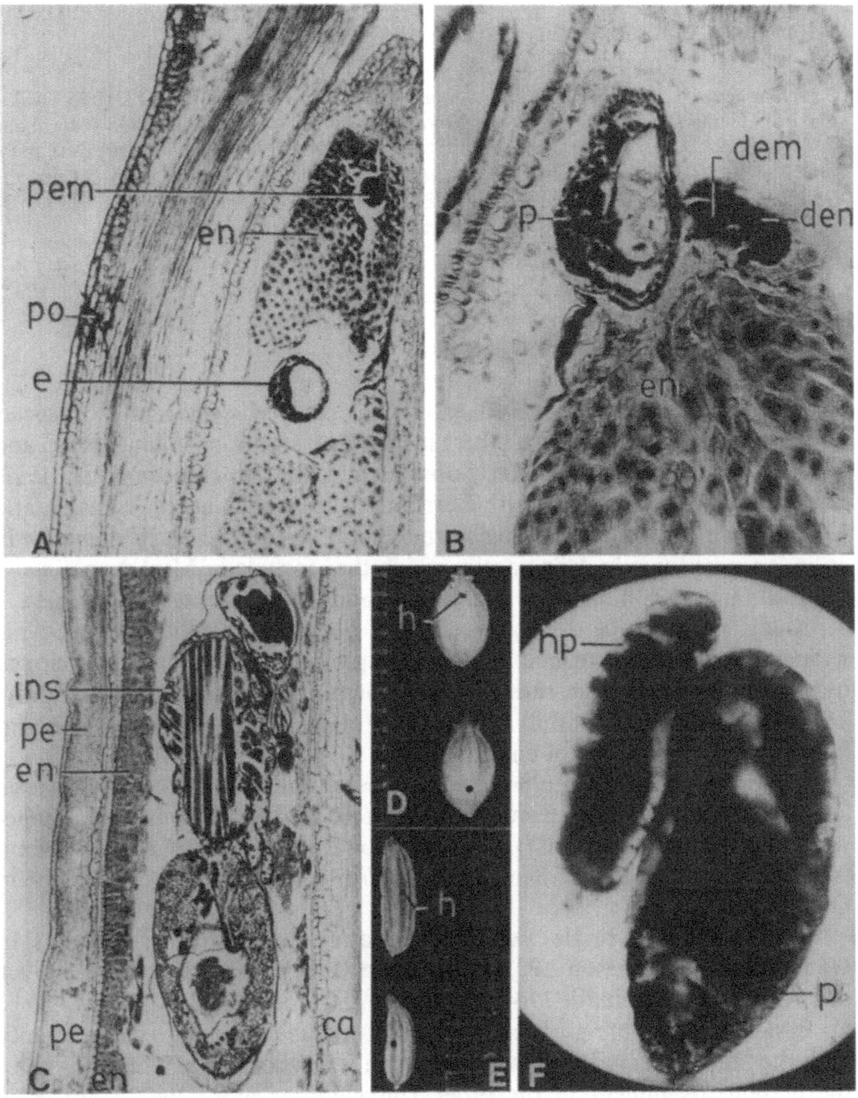

Fig. 40.8.1A–F. *Foeniculum vulgare.* **A** Longisection of immature mericarp with proembryo (*pem*), cellular healthy endosperm (*en*), and egg (*e*) of parasite (*Systole albipennis* Walk.); *po* path of ovipositor. **B** Later stage, the developing parasite (*p*) has caused degeneration of proembryo (*dem*) and adjoining endosperm (*den*). **C** Longisection of mature pericarp with adult insect (*ins*) in seed; *ca* carpophore; *pe* pericarp. **D, E** Fruits of *Coriandrum sativum* (**D**) and *Foeniculum* (**E**) with the exit holes (*h*) of parasites. **F** Hyperparasite (*hp*) larva lying adpressed to the host parasite (*p*, *S. albipennis*). (After S.C. Gupta 1970)

of petals, outer epidermis of pericarp, and stylopodium (S.C. Gupta et al. 1965). Their development is of the syndetochellic type, and are anomocytic in stylopodium, and anisocytic in rest of the organs.

RD Adatia and GL Shah (1952), AAA Al-Attar (1977, 1979), HA Borthwick (1931), JS Dhakre (1968), SC Gupta (1962a, b, 1964, 1970), SC and M Gupta (1964), SC Gupta, GS Paliwal and M Gupta (1965), Å Håkansson (1923, 1927, 1952), R Hegnauer (1971), H Maheswari Devi and A Santa Ram (1985), JC Mestre and JL Guignard (1971), RL Paliwal (1951), AM Pigni, G Corsi and EM Capelletti (1986), RL Rodriguez (1971), CB Sehgal (1965), GL Shah (1953), D Singh and JS Gupta (1956), R Souèges (1926a), J Szujkó-Lacza (1976, 1978)

40.9 Taxonomic Considerations

Alangiaceae. Bentham and Hooker (1965a) place *Alangium* in Cornaceae. Hutchinson (1926) recognizes the independent status of Alangiaceae, and includes them in Umbelliflorae. Subsequently, he (Hutchinson 1959, 1973) transfers Alangiaceae, along with Cornaceae and Araliaceae, to Araliales, and Umbelliferae to Umbellales. Engler and Prantl (1931) also place *Alangium* in an independent family Alangiaceae, which they include in Myrtiflorae.

Takhtajan (1980) includes Alangiaceae and Cornaceae in the Cornales.

Alangiaceae can be distinguished from Cornaceae on the basis of ovule with a vascular strand reaching up to the tip of integument, a well-developed hypostase, persistent and, rarely, supernumerary antipodals, antipodal polyembryony (which is doubtful), absence of Fritillaria type of embryo sac, and Nuclear or Cellular type of endosperm.

The embryological features justify the independent status of Alangiaceae.

The Myrtiflorae differ from Alangiaceae in an inferior ovary with numerous bitegmic ovules, several types of embryo sac – generally of the Polygonum type, tetrasporic 16-nucleate in species of *Combretum*, and Oenothera type in Onagraceae.

Alangium has a single, pendulous, anatropous, unitegmic ovule as in Umbelliferae (Håkansson 1923), Cornaceae (Morse 1907, Fagerlind 1939c), and Araliaceae (Gopinath 1944, D. Singh 1954).

B.M. Johri (1963b) rightly concludes that Alangiaceae have a closer affinity with the families included in Araliales, than with those in Myrtiflorae. Thus, the present assignment in the order Umbelliflorae along with Cornaceae, Araliaceae and Umbelliferae is in order.

Nyssaceae and Davidiaceae. Bentham and Hooker (1965a), Harms (1897b) and Hallier (1912) include *Nyssa* and *Davidia* in Cornaceae. Harms places *Camptotheca* and *Nyssa* in the subfamily Nyssoideae, and *Davidia* in Davidioideae, under Cornaceae. De Jussieu (see Takhtajan 1959) created the family Nyssaceae. Takhtajan (1959) and Melchior (1964) include Nyssaceae in Umbelliflorae along with Cornaceae, and raise the genus *Davidia* to family status Davidiaceae, next to Nyssaceae. In his later classification, Takhtajan (1980) includes the families Davidiaceae, Nyssaceae, Alangiaceae and Cornaceae, along with some other families, in Cornales. Hutchinson (1959, 1969, 1973) assigns Nyssaceae to Araliales near Cornaceae.

Some of the similarities between *Nyssa* and Cornaceae are: three-porate pollen grains, unitegmic, pendulous, anatropous ovule with a single hypodermal archesporium, unequal dyad cells, Polygonum type of embryo sac, and fusion of polar nuclei prior to fertilization (Tandon and Herr 1971). The differences with Cornaceae are: pseudo-crassinucellar ovules, absence of parietal cells, degeneration of non-functional megaspores from chalazal to micropylar, degeneration of apical region of nucellus during maturation of megagametophyte, presence of hypostase, and occasional development of bisporic embryo sac (Tandon and Herr 1971). Mohana Rao (1972a) reports Nuclear type of endosperm, and thick-walled lignified epidermis of the seed-coat. The latter feature does not occur in any other member of Umbelliflorae. Tandon and Herr relate *Nyssa* to Cornaceae, while Mohana Rao (1972a) supports its Umbellalian alliance.

The embryology of the genus *Camptotheca* has not been investigated, and its affinity cannot be discussed.

Anatropous unitegmic ovule, Polygonum type of embryo sac, albuminous seed with straight embryo, and drupaceous fruit are known in Nyssaceae and Davidiaceae. However, crassinucellar ovule, multicellular female archesporium, occasional development of supernumerary antipodals with haustorial activity, and Cellular endosperm (Horne 1909) distinguish Davidiaceae from Nyssaceae. As the embryological features show pronounced differences, both *Nyssa* and *Davidia* deserve independent family status.

Cornaceae. Engler and Diels (1936) place Cornaceae in Umbelliflorae, Hutchinson (1973) in Araliales, and Takhtajan (1959, 1980) assigns them to the order Cornales. Several genera, such as *Alangium*, *Garrya* and *Nyssa*, have been removed from Cornaceae, and given independent family status, which is fully supported by embryological features. No doubt, Cornaceae closely resemble Umbelallian families but have certain distinctive features: presence of uni- or multicelled female archesporium, parietal tissue, obturator, and endothelium; absence of hypostase, occasional occurrence of tetrasporic (Fritillaria) embryo sac in *Cornus mas* and pseudomonosporic in *C. florida*, Cellular type of endosperm, and nucellar apospory (Table 40.9.1). *Mastixia* of the subfamily Mastixiodeae is supposed to form a link between Cornaceae and Araliaceae, on anatomical grounds. The embryology of *Mastixia* has not been investigated. Mohana Rao (1972a) favours the segregation of Cornaceae into a separate order Cornales, on embryological grounds.

Garryaceae. Wangerin (1910b), Hallier (1912) and Bessey (1915) include the genus *Garrya* in Cornaceae. Hjelmqvist (1948) suggests ordinal rank, and points out the affinities of Garryales with the Umbelliflorae. Hutchinson (1973) includes Garryaceae in Araliales. Takhtajan (1959, 1980) and Melchior (1964) group them with Umbelliflorae (Cornales). Kapil and Mohana Rao (1966b) discuss the embryological features, and emphasize that Garryaceae differ from Salicaceae, Chloranthaceae or Piperaceae, and are allied to Alangiaceae and

Table 40.9.1. Embryological features of Cornaceae, Araliaceae and Umbelliferae (See Mohana Rao 1972a)

	Cornaceae	Araliaceae	Umbelliferae
Anther	Bilocular, anther-wall 5–6-layered, endothecium fibrous, tapetum secretory, cytokinesis simultaneous	Bilocular, anther wall 4–5-layered, endothecium fibrous, tapetum secretory, cytokinesis simultaneous	Bilocular, anther wall 4-layered, endothecium fibrous, tapetum secretory, cytokinesis simultaneous
Tetrad	Tetrahedral	Tetrahedral and decussate	Tetrahedral, decussate and isobilateral
Pollen	2-celled, oblate, spheroidal, 3-colpate	2-celled, oblate, spheroidal, subporolate, 3-colporate	3-celled, bilateral, subporolate, 3-colporate
Ovule	Anatropous, unitegmic, crassinucellate, do not abort	Anatropous, unitegmic, tenui- or crassinucellate, abortive ovules common	Anatropous, unitegmic, tenuinucellate, abortive ovules common
Embryo sac	Mono-, bi- or tetrasporic, antipodals ephemeral	Mono- or bisporic, antipodals ephemeral or persistent	Mono-, bi- or tetrasporic, antipodals persist until early stages of endosperm
Endosperm	Cellular type	Nuclear	Nuclear
Embryo	Asterad type	Onagrad or Chenopodiad	Solanad
Polyembryony	Nucellar apospory	—	Suspensor polyembryony
Seed-coat	Thin-walled, collapses	Thin and membraneous	Outer epidermis of testa remains thin-walled
Fruit	Drupe	Berry or drupe	Schizocarp

Cornaceae of the Umbelallian complex. The occurrence of oblate spheroidal three-colporate pollen grains, anatropous unitegmic crassinucellate ovule, Polygonum type of embryo sac, and berry-like fruits are similarities with Cornaceae. But a well-developed hypostase, persistent antipodals, Nuclear type of endosperm, and suspensor polyembryony distinguish *Garrya* from Cornaceae (Kapil and Mohana Rao 1966b). Solanad type of embryogeny, and suspensor polyembryony occur in Umbelliferae. Therefore, the removal of *Garrya* from Cornaceae, erection of the family Garryaceae and its inclusion in Umbelliflorae are fully justified.

Araliaceae. Bentham and Hooker (1965a), Bessey (1915), Wettstein (1935) and Engler (1964) assign Araliaceae to the order Umbelliflorae along with Umbelliferae and Cornaceae. Takhtajan (1966, 1969, 1980) groups Araliaceae and Umbelliferae (Apiaceae) under Araliales. Due to their woody habit, Hutchinson (1973) treats Araliaceae, along with Cornaceae, in a separate order Araliales. Araliaceae are embryologically allied to Umbelliferae, rather than Cornaceae (see Mohana Rao 1972b). The presence of abortive ovules, single, rarely double, female archesporium, hypostase, obturator, mono- or bisporic embryo sac, Nuclear type of endosperm, and structure of seed are similar to those of Umbelliferae. Berry (rarely drupe) type of fruit, eight-carpellate ovary with one to 15 locules, tenui- or crassinucellar ovules, and Onagrad or Chenopodiad type of embryogeny differentiate Araliaceae from Umbelliferae (Mohana Rao 1972b).

Umbelliferae. The family Umbelliferae is a natural taxon. It is characterized by schizocarpic fruit, and bilateral, subporolate, three-colporate, three-celled pollen grains at the dispersal stage. A parietal cell is absent, the ovule is pseudocrassinucellar in *Foeniculum vulgare*. A hypostase and endothelium differentiate at the megaspore mother cell stage. The endothelium grows beyond the nucellus in the micropylar region. The occasional presence of bisporic (Allium) and tetrasporic (Penaea and Drusa) types of embryo sac is also reported. The formation of enucleate endosperm vesicles is restricted to Umbelliferae. The association with parasites causes embryolessness (S.C. Gupta 1962a, b).

Cronquist (1968, 1981) maintains two distinct orders, Cornales and Umbellales (Apiales), and segregates them on morphological, anatomical, and phytochemical basis. He mentions the anatomical findings (of a Russian botanist Tamanschian) which reveal close affinities between Umbellales and Cornales.

In his new system of classification, R. Dahlgren (1977b, 1980b, 1981) places Araliaceae and Apiaceae (Umbelliferae) in the non-iridoid Sympetalous order Araliales (superorder Araliflorae), and the remaining families in the iridoid-Sympetalous order Cornales (superorder Corniflorae). A consideration under the Sympetalae is wholly unwarranted.

While discussing the new aspects of systematics of Asteridae (Sympetalae), Wagenitz (1977b) refers to two connecting lines, one between Cornaceae and Caprifoliaceae, confirmed by serological data (Hildebrand and Fairbrothers

1970a, b), and the other between Araliales and Asterales (Asteriflorae) based on chemotaxonomical investigations of Hegnauer (1971, 1973), Bohlmann et al. (1973), and Wagenitz (1977b). The polyacetylenes and sesquiterpene lactones, which are rare in other angiosperms, occur frequently in these two orders.

Cornales and Araliales have been treated as independent orders by Takhtajan (1980).

The order Umbelliflorae is a more or less natural grouping with the families Alangiaceae, Nyssaceae, Davidiaceae, Cornaceae, Garryaceae, Araliaceae and Umbelliferae. There is no justification to erect the orders Cornales and Araliales.

GPSR Compliance

The European Union's (EU) General Product Safety Regulation (GPSR) is a set of rules that requires consumer products to be safe and our obligations to ensure this.

If you have any concerns about our products, you can contact us on ProductSafety@springernature.com

In case Publisher is established outside the EU, the EU authorized representative is:

Springer Nature Customer Service Center GmbH
Europaplatz 3
69115 Heidelberg, Germany

Batch number: 10755036

Printed by Printforce, the Netherlands

Comparative Embryology of Angiosperms summarizes in two volumes nearly 5,000 publications (up to 1991) and thus gives a comprehensive account of reproductive biology of flowering plants. Each order begins with general features, followed by the embryology of its constituent families and culminates in critical taxonomic considerations on the basis of comparative embryologic data. Further, the placement of each family in six chosen classification systems is discussed.

Comparative Embryology of Angiosperms provides useful information to students and research workers, taxonomists, plant breeders, agriculturists and horticulturists, and those interested in the evolutionary tendencies in flowering plants.

ISBN 978-3-642-76397-7
9 783642 763977 01
ISBN 978-3-642-76397-7

B.M. Johri · K.B. Ambegaokar · P.S. Srivastava

Comparative Embryology of Angiosperms Vol. 2

With 362 Figures
in two Volumes

Springer-Verlag Berlin Heidelberg GmbH

Prof. Dr. Brij M. Johri
Dr. Kunda B. Ambegaokar
Department of Botany
University of Delhi
Delhi 110007
India

Dr. Prem S. Srivastava
Department of Botany
Hamdard University
Hamdard Nagar
New Delhi 110062
India

ISBN 978-3-642-76397-7 ISBN 978-3-642-76395-3 (eBook)
DOI 10.1007/978-3-642-76395-3

Originally published by Springer-Verlag Berlin Heidelberg in 1992
Softcover reprint of the hardcover 1st edition 1992

Dedicated to

Professor Karl Schnarf
(1879–1947)

Professor Panchanan Maheshwari
(1904–1966)

Preface

The concept for the present book dates back to the early 1960s when Professor P. Maheshwari FRS (deceased May 1966) invited scientists to contribute to a multiauthored book – *Comparative Embryology of Angiosperms*. Soon thereafter, Professor Gwenda L. Davis (University of New England, Armidale, Australia) joined the University of Delhi as a Visiting Professor and announced that her book, *Systematic Embryology of Angiosperms*, would soon be published (it appeared in 1966). Therefore, Professor Maheshwari postponed his project for the time being.

The only book on embryology available until then (in english) was by Prof. Maheshwari, published in 1950 (McGraw-Hill, New York) although G.L. Davis' book (without illustrations) was also useful, but to a limited extent. The rapid increase in knowledge in the years that followed, with the advent of new techniques such as ultramicroscopy (TEM and SEM), histochemistry, and the study of organic and inorganic materials, provided impetus to the study of embryology, transgressing its traditional confinements to 'compound' microscopy. This resulted in the emergence of new ideas and methodologies, which helped in a better interpretation of data previously restricted to only the description of developmental aspects.

As new information on embryology accumulated, the need for a comprehensive, illustrated book on comparative embryology that could serve researchers as a reference work became imperative.

In 1979, when Dr. Klaus Peters of Springer-Verlag visited BMJ with an invitation to prepare a treatise on *Embryology of Angiosperms*, the book, *Experimental Embryology of Vascular Plants* (ed. B.M. Johri), was already in its final stage and was published by Springer-Verlag in 1982. The multiauthored book *Embryology of Angiosperms* (ed. B.M. Johri) was published by Springer-Verlag in 1984, and the outline of the book, *Comparative Embryology of Angiosperms*, was prepared simultaneously and given to Klaus Peters. Although a very large collection of relevant literature was available to us at the Department of Botany, University of Delhi, it took almost 10 years to survey the exhaustive literature and to complete the manuscript on Comparative Embryology.

This book endeavours to consolidate the scattered information on angiosperm embryology, to present a coordinated and coherent interpretation of significant data accumulated thus far, and to elucidate the interrelationship of taxa based on comparative embryological findings.
Comparative Embryology of Angiosperms is a review of the developmental processes leading to sexual reproduction in flowering plants, and provides illustrated source material on sexual reproduction, primitive and advanced embryological characters, and phylogenetic relationships. On the basis of embryological data and certain evidence from other areas of study, special emphasis has been placed on the relationship among and within the families and orders of angiosperms.
For our work Engler's classification system for angiosperms proposed in *Syllabus der Pflanzenfamilien*, revised by Melchior (1964), has been adopted.
This treatise summarizes nearly 5000 publications (up to 1991) on angiosperm embryology; its resulting ultimate size prompted us to divide it into two volumes. For the convenience of the reader, the preface, acknowledgements and the table of contents are presented in both volumes. For economic reasons, the references for the entire work and the plant index are included only in Volume 2.
Each order begins with general features, followed by the embryology of its constituent families, culminating in critical taxonomic considerations on the basis of comparative embryologic data, and the placement of each family in six chosen systems of classification is discussed.
A comprehensive bibliography is given at the end of each family and includes the citations referred to in the text. Additional citations (not mentioned in the text) are provided for those who wish to obtain further information.
When appropriate, inaccuracies in observation and interpretation are pointed out, alternative interpretations offered, gaps in our knowledge highlighted, and prospects outlined.
Comparative Embryology of Angiosperms provides useful information to students and research workers involved in angiosperm embryology, taxonomists, plant breeders, agriculturists and horticulturists, and plant scientists interested in the evolutionary tendencies in flowering plants.

University of Delhi, July 1992

B.M. Johri
K.B. Ambegaokar
P.S. Srivastava

Acknowledgements

The preparation of *Comparative Embryology of Angiosperms* took almost a decade. We could not have accomplished this undertaking without the academic support and encouragement from our colleagues at the University of Delhi, and plant scientists in India and abroad. Over the last 55 years, B.M. Johri has had the privilege of discussions with embryologists on the reproductive biology of angiosperms, especially at national and international conferences. He is grateful to all of them.

We are indebted to fellow embryologists who generously provided relevant literature and advance information about ongoing research projects. To name a few: Prof. Martin Bopp (Heidelberg), Dr. F. Bouman and Dr. W. Verkerke (Amsterdam), the late Dr. R. Dahlgren (Copenhagen), Dr. M.I. Hakki (Berlin), Prof. U. Hamann (Bochum), Dr. R. Wunderlich (Vienna), Prof. M. Yakovlev and Prof. T. Batygina (Leningrad), Dr. H. Tobe (Chiba, Japan), Dr. P.H. Raven (St. Louis, USA), Prof. A.E. Cocucci (Cordoba, Argentina), the late Dr. Thanikoimani (Pondichery, India), Prof. K.K. Lakshmanan (Coimbatore, India), Dr. M.P. Nayar and Dr. B.D. Sharma (Botanical Survey of India, Calcutta), Prof. S. Chanda (Bose Institute, Calcutta) and our colleagues (at the University of Delhi) Prof. H.Y. Mohan Ram, Prof. S.C. Maheshwari, Prof. R.N. Kapil, Prof. N.N. Bhandari and Dr. Ashok Bhatnagar.

For advance information regarding the recently published (1989) Dedicatory Volume in honour of Prof. A. Lebègue, *Some Aspects and Orientations in Plant Embryology*, we wish to thank the editors, Prof. J. Paré and Prof. M. Bugnicourt (University of Picardie, Amiens, France).

For the translation of Russian literature we are obliged to Dr. Balram Sharma (Indian Agric. Res. Inst., New Delhi) and Dr. (Mrs.) Akhilesh Tiwari (University of Delhi); and to Dr. (Mrs.) Sheela Srivastava (University of Delhi) for German literature. Dr. M.A. Rau (Mysore) very kindly prepared a number of schematic figures on the development of endosperm in the Sympetalae. Dr. F. Bouman provided scanning electron micrographs of seeds which appear in this volume. Thanks are also extended to Mr. Dev Raj Khosla (Indian Natl. Sci. Acad., New Delhi) who prepared photocopies of literature and illustrations, so essential for checking the manuscript. Mr. S.B. Ambegaokar (Delhi) was kind enough to compare over 1800 pages of the final manuscript with the earlier draft. Dr. (Mrs.) Kapoor ably assisted us in the collection of literature and indexing the Contents and References. Dr. Kapoor and Dr. Manoj Kumar (University of Delhi) also helped in the preparation of plates.

We express our appreciation to Mr. G.A. Shastri and Ms. Sushil Kaur for easy access to the University Library, and to Dr. (Mrs.) Vatsala (Publ. Direc., CSIR, New Delhi), Dr. (Mrs.) Nina Dhiman and Ms. Binny Mathur (Editorial Assistants, Phytomorphology), and our research scholars Mr. N. Syed Abbas Ali and Ms. Babeeta Chrungu for their help in checking the periodicals. Mr. Krishan Lal, artist (University of Delhi), very competently designed and prepared several figures, and labelled all the illustrations. Mr. D. Daniel and Mr. S.K. Das (University of Delhi) attended to photographic work with professional competence.

We are indebted to Dr. Dieter Czeschlik, Life Science Editor, Springer-Verlag, Heidelberg, for his immense patience in waiting to see the completion of this work. Moreover, we would like to thank Mrs. Jean von dem Bussche for copyediting "Sympetalae" (pp. 615–829) in 1987; Dr. G. Forbath, Copy Editor, Ms. Carola Wissmeier, Secretary to Dr. D. Czeschlik, and Ms. Renate Münzenmayer, Production Editor, Springer-Verlag, for the meticulous handling of this publication. For the high quality of printing by MacMillan India, Bangalore, we have great appreciation.

We are obliged to *Janab* Hakim Abdul Hameed *Sahab* (Chancellor), Prof. M. Amin (Vice Chancellor) and Prof. Muhammed Iqbal (Dean, Faculty of Science) of the Hamdard University, New Delhi, for their interest and encouragement in this project.

Thanks are also due to Ms. Madhumati, Mr. Aniruddha, Ms. Deepshikha, and Mr. Deepak, research scholars at the Department of Botany, Hamdard University for cross-checking the text and references in the final proofs, and for preparing plant indices.

We would like to make special mention of Mr. R.K. Gupta (University of Delhi) for expert secretarial assistance, and typing twice the manuscript of about 2000 pages, and Mrs. Krishna Thakur (University of Delhi) for periodical secretarial assistance. Mr. Sukhbir Singh and Mr. Chandra Pal (University of Delhi) have been very helpful in attending to the day-to-day routine work during the preparation of this book.

The Indian National Science Academy, New Delhi, and Springer-Verlag, Heidelberg, provided partial financial support for this book.

We also wish to record gratitude to Mrs. Raj Johri (wife of B.M. Johri) and Dr. (Mrs.) Sheela Srivastava (wife of P.S. Srivastava) for their support, forbearance and interest during the long periods when *Comparative Embryology of Angiosperms* was in preparation. Dr. Lalit M.

Johri (son) and Mrs. Meera Johri (daughter-in-law) keenly followed the progress of this book and encouraged an expeditious completion. Once again, many thanks to all those who helped us in various ways.

May 28, 1992

B.M. Johri
K.B. Ambegaokar
P.S. Srivastava

Contents Volume 2

II Dicotyledons: Sympetalae

III Monocotyledons

Contents Volume 1

References see Vol. 2 p. 1013
Plant Index see Vol. 2 p. 1174

II Dicotyledons: Sympetalae

41 Diapensiales

Diapensiales are represented by a single family, Diapensiaceae.

41.1 General Features

The family Diapensiaceae shows: (1) a well-developed endothecium, secretory anther tapetum with binucleate cells, (2) tricolpate pollen grains liberated individually at the two- or three-celled stage, (3) unitegmic, hemianatropous to campylotropous, tenuinucellate ovule without any endothelium, (4) Polygonum type of embryo sac with persistent antipodals; supernumerary antipodals occur in *Shortia*, (5) Cellular type of endosperm without any haustoria, (6) Solanad (?) type of embryogeny, and (7) seed-coat of one to three layers of cells.

41.2 Diapensiaceae

Palser (1963) investigated megasporogenesis, female gametophyte, and the development of endosperm in eight species: *Diapensia lapponica, D. lapponica* subsp. *ovata, Galax aphylla, Pyxidanthera barbulata, P. brevifolia, Schizocodon* sp., *Shortia galacifolia* and *S. uniflora.*

The anther is bisporangiate in *G. aphylla* and *D. lapponica*, otherwise tetrasporangiate in other taxa. The anther wall comprises persistent epidermis, fibrous endothecium, two to three ephemeral middle layers, and secretory tapetum with binucleate cells. Fibrous thickenings of the endothecium are much pronounced along the line of dehiscence and on abaxial walls in *Galax* and *Pyxidanthera* (Palser 1963). The dehiscence of anther is by an apical or basal pore, or by a slit.

The pollen grains are tricolporate, and two- or three-celled at the dispersal stage. They are liberated singly.

The ovule is hemianatropous to campylotropous, unitegmic, and tenuinucellate. The archesporium is single-celled in *Galax aphylla* and *Shortia galacifolia*, and one- to three-celled in *Diapensia lapponica* (Palser 1963). The single archesporial cell develops into the megaspore mother cell. The latter undergoes meiotic divisions, and the chalazal megaspore of the linear or T-shaped tetrad develops into a Polygonum type of embryo sac. The polar nuclei lie near the egg apparatus, and fuse before fertilization. In *S. galacifolia* the antipodals undergo secondary multiplication, and their number goes up to 25–40. They persist till the early stages of seed development. The mature embryo sac contains starch. One of the synergids is destroyed during fertilization.

After the breakdown of the nucellus, the embryo sac comes in contact with the integument, but an endothelium is not formed. Palser (1963) distinguishes Diapensiaceae from the Ericales on the basis of absence of endothelium. However, Yamazaki (1966) reports endothelium in *S. uniflora*. This aspect needs further attention.

Development of the endosperm is of the Cellular type. The division of the primary endosperm nucleus is followed by a transverse wall. Each cell undergoes a vertical division, and all the four cells give rise to the endosperm proper. In *Shortia galacifolia* (Palser 1963) and *S. uniflora* (Yamazaki 1966), the division of the primary endosperm nucleus may be followed by a transverse or longitudinal wall. Endosperm haustoria are absent in Diapensiaceae.

In *Diapensia lapponica* (see Crété 1964a), the embryogeny conforms to Period I, Megarchetype V, Group VIII, wherein $cb = co + s$, probably Solanad type. In *Shortia uniflora* (Yamazaki 1966), the embryogeny does not agree with that of *Diapensia*. The zygote divides transversely to form two daughter cells, *ca* and *cb*, and each divides transversely, resulting in a linear tetrad, *l*, *l'*, *m* and *ci*. *ci* gives rise to tiers *n* and *n'*.

In *D. lapponica*, *ca* contributes to the formation of root cylinder (*icc*) and hypophysis (*iec*), whereas in *S. uniflora m*, a derivative of *cb*, gives rise to these regions.

Yamazaki (1966), while discussing the affinity of *Shortia* with Ericales, wrongly mentions that Ericalean embryogeny conforms to Megarchetype V. In fact, according to Veillet-Bartoszewska (1957a, 1959a, 1960a, b, 1961a), the embryogeny conforms to Period II, Megarchetype III (Caryophyllad type).

The testa is composed of two to three layers. The cells of the outer epidermis have annular thickenings on their radial walls. The mesophyll and inner epidermis do not persist in *S. uniflora* (Yamazaki 1966). The seeds are minute with copious fleshy endosperm, and a cylindrical central embryo (Hutchinson 1973).

EJH Corner (1976), P Crété (1964a), PE Graef (1958), O Hagerup (1928), J Hutchinson (1973), B Palser (1963), JD Reynolds (1968), G Samuelsson (1913), M Veillet-Bartoszewska (1957a, 1959a, 1960a, b, 1961a), T Yamazaki (1966)

41.3 Taxonomic Considerations

Bentham and Hooker (1965b), R. Dahlgren (1980b, 1981), and Takhtajan (1980) assign Diapensiaceae to Ericales. Hallier (1912) and Wettstein (1935) consider them as an advanced taxon of Bicornes. Engler and Gilg (1924), and Cronquist (1981) raise them to a distinct order, Diapensiales.

Some of the embryological features of Diapensiaceae are not comparable with those of the order Ericales (see also Sect. 42): well-developed endothecium in anther wall, single pollen grains, absence of endothelium, supernumerary persistent antipodals, and absence of endosperm haustoria. Palser (1963) suggests the exclusion of the family Diapensiaceae from Ericales. The embryological data favour its present position as an independent order related to Ericales.

42 Ericales

The order Ericales comprises five families: Clethraceae, Pyrolaceae, Ericaceae, Epacridaceae and Empetraceae.

42.1 General Features

P. Maheshwari (1950) and Palser (1961) enumerate several embryological features of Ericales: (1) absence of fibrous layer in anther wall except in Clethraceae and *Enkianthus* (Ericaceae), (2) secretory anther tapetum, (3) compound pollen grains, pollen grains shed either in tetrads or singly, (4) two-celled at the shedding stage, (5) a fluted hollow style (which connects the lumen of the ovary) along which the pollen tubes make their way into the ovary, (6) anatropous, unitegmic, tenuinucellar ovule with endothelium, single-celled archesporium, (7) Polygonum type of embryo sac, (8) mature embryo sac broader at the micropylar end and narrower at the chalazal end, (9) Cellular type of endosperm with micropylar and chalazal haustoria; Nuclear in *Rhododendron* and some species of *Vaccinium*, (10) considerable elongation of zygote, (11) Asterad type of embryogeny in Clethraceae, Caryophyllad in Pyrolaceae, Caryophyllad and Solanad in Ericaceae and Epacridaceae, and Caryophyllad in Empetraceae, (12) albuminous seed with fleshy endosperm and straight embryo, and (13) single-layered seed-coat formed from the outermost layer of the integument.

42.2 Clethraceae

In *Clethra alnifolia* (Kavaljian 1952, Veillet-Bartoszewska 1960d), the anther wall comprises the epidermis, fibrous endothecium, two ephemeral middle layers, and secretory tapetum with binucleate cells. The microspore mother cells undergo meiotic divisions, cytokinesis is simultaneous, and tetrads are tetrahedral.

The tricolporate pollen grains (Hu 1960, see P.S. Ganapathy 1970) are two-celled at the dispersal stage in *C. alnifolia* (Kavaljian 1952), and are shed singly.

The ovule is anatropous, unitegmic, and tenuinucellar. The archesporial cell functions as megaspore mother cell. The latter undergoes meiotic divisions and develops into a linear tetrad. The chalazal megaspore gives rise to a Polygonum type of embryo sac. An endothelium differentiates in the lower two-thirds region of the embryo sac. The polar nuclei fuse before fertilization; the antipodals are arranged in a linear row, largest at the base.

The development of endosperm is of the Cellular type. The next division, following the division of primary endosperm nucleus, is transverse so that a row of four cells is produced. The micropylar and chalazal haustoria differentiate terminally, they are weakly developed (Samuelsson 1913, P.S. Ganapathy 1970).

The embryogeny conforms to Period I, Megarchetype IV in *C. alnifolia* (Veillet-Bartozewska 1960d). The division in *ca* is longitudinal. The suspensor is uniseriate (P.S. Ganapathy 1970) of at least ten cells (see Davis 1966).

The seed-coat is single-layered. The seeds are small and ovoid, albuminous (food reserve oily and proteinaceous), and with a straight cylindrical embryo. In *Clethra barbinervis*, *C. bodineri* and *C. cavaleriei*, the seeds are irregular and extremely reticulate. The North American species have highly compressed and winged seeds (see Hu 1960, P.S. Ganapathy 1970, Corner 1976). The fruit is a loculicidal capsule.

HF Copeland (1933), EJH Corner (1976), GL Davis (1966), PS Ganapathy (1970), Shiu-Ying Hu (1960), LG Kavaljian (1952), G. Samuelsson (1913), M Veillet-Bartoszewska (1960d)

42.3 Pyrolaceae

The development of anther has not been fully investigated.

The anther tapetum is glandular. The tetrads are isobilateral and decussate.

The pollen grains are two-celled at the dispersal stage. They are shed singly (monad) in *Monotropa*, otherwise they remain in permanent tetrads (Wallace 1975). Wallace's reference to single pollen grains as monads is improper. This term is applied in Cyperaceae and Epacridaceae where only one microspore, out of four, functions. The other three degenerate but remain associated with the functional pollen grain.

The ovule is anatropous, unitegmic, and tenuinucellar. The archesporial cell functions directly as megaspore mother cell. The latter undergoes meiotic divisions, develops a linear or T-shaped tetrad, and the chalazal megaspore gives rise to a Polygonum type of embryo sac. The polar nuclei fuse before fertilization. The antipodals may persist up to the first few divisions of the endosperm. An endothelium differentiates at the four-nucleate stage of the embryo sac.

The degeneration of megaspore mother cells, dyads, tetrads, and four-nucleate and mature embryo sacs is frequent in *Pterospora andromedea* (Bakshi 1959). Occasionally, Allium type of development occurs.

The development of endosperm is of the Cellular type. A transverse division of primary endosperm nucleus is followed by another transverse division so that a row of four cells is produced. The haustoria differentiate terminally. They are short and one-celled.

The embryogeny conforms to the Caryophyllad type (see Davis 1966). According to Teryokhin (1962), embryogenesis in Pyroleae conforms to Period I, Series C, Megarchetype V (Solanad type?). The embryo is reduced.

The minute, winged seeds of *Pterospora* have fleshy endosperm, and a minute embryo. The seed-coat comprises two layers. The epidermal cells are empty. The adjacent walls of these two layers become thickened, and have bordered pits (Bakshi 1959). The fruit is a loculicidal capsule.

In *Monotropa uniflora* (Olson 1980), the ripe seed is unitegmic, tenuinucellate and albuminous, with wing-like integumentary extensions. The central core of the seed contains embryo and endosperm (Fig. 42.3.1A, B). The seed is exotestal with two to three layers of longitudinally elongate cells, the outer epidermis is a mechanical layer. The inner tangential and radial walls of the

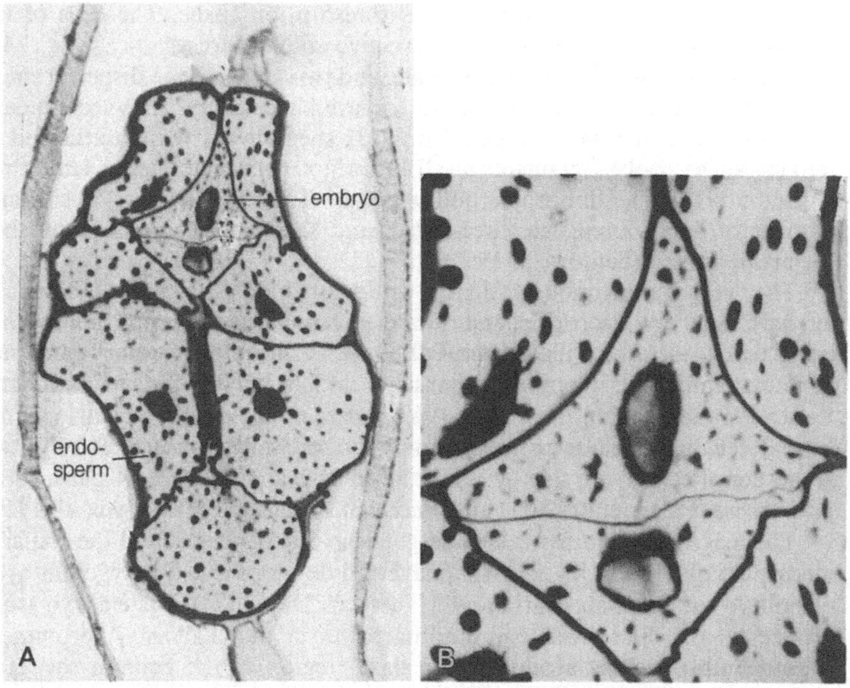

Fig. 42.3.1A, B. *Monotropa uniflora.* **A** Longisection of seed at shedding stage; two-celled proembryo is surrounded by 10–14 endosperm cells. The seed-coat is thin. **B** Two-celled proembryo from **A**. (After Olson 1980)

outer epidermal cells are thickened, while the outer tangential walls remain thin. The inner epidermis collapses. At maturity the desiccation of seed-coat causes the thin outer tangential walls of the mechanical layer to collapse and form ridges.

TS Bakshi (1959), HF Copeland (1941), EJH Corner (1976), GL Davis (1966), PS Ganapathy (1967a), AR Olson (1980), M Pyykkō (1968), JK Small (1914), ES Teryokhin (1962, 1963), GD Wallace (1975)

42.4 Ericaceae

The anther wall comprises the epidermis, two or three middle layers, and glandular tapetum with multinucleate cells. An exothecium develops at the extreme apical region of anther in *Kalmia*, *Phyllodoce* and *Rhodothamnus* (P.S. Ganapathy and Palser 1964). The tapetum is of parietal origin, but it is said to differentiate from the sporogenous tissue in three species of *Vaccinium* (Batygina et al. 1963). This aspect needs confirmation. The development of a distinct endothecium in *Enkianthus* (Safijowska 1960) is an unusual feature; the tapetal cells are binucleate in *Vaccinium* (Venkateswarlu and Maheswari Devi 1973). The microspore mother cells undergo meiotic divisions, simultaneous cytokinesis is followed by furrowing, and the tetrads are tetrahedral, occasionally decussate.

The dehiscence of anthers occurs by resorption tissue. The cells of the resorption tissue and those of the connective contain druses.

The pollen grains are three-colporate, and two-celled at the dispersal stage. They either remain united in tetrads, or are shed singly. In Vaccinioideae (Venkateswarlu and Maheswari Devi 1973), they always remain attached in tetrads, all the pollen grains of a tetrad do not germinate at the same time (Wallace 1975). The dimorphic pollen grains of *Enkianthus* vary in size and number of germ-pores, are three-celled, and shed singly; occasionally they germinate in situ (Safijowska 1960).

The ovule is anatropous or slightly campylotropous, unitegmic, tenuinucellar, with a narrow micropyle, and devoid of vasculature, except in *Enkianthus* (see Palser et al. 1971, Villamil and Palser 1980). The integumentary epidermis contains tannin; the hypostase is densely "cytoplasmic" (P.S. Ganapathy and Palser 1964, Stushnoff and Palser 1969, Palser et al. 1971). The single archesporial cell functions as megaspore mother cell in *V. retusum*. Rarely, two archesporial cells, lying side by side, may enlarge and divide in *Rhododendron yunnanense* (Palser et al. 1971) and *Vaccinium leschnaultii* (Reddy et al. 1966).

The megaspore mother cell undergoes meiotic divisions, and the chalazal megaspore of a linear, T- or ⊥-shaped tetrad develops into a polygonum type of embryo sac (Venkateswarlu and Maheswari Devi 1973). The embryo sac is broader at the micropylar end, and narrower at the chalazal. A prominent endothelium develops around the chalazal region of the embryo sac. It is biseriate in *Vaccinium retusum* and *V. serratum* (Venkateswarlu and Maheswari Devi 1973). The synergids are hooked and extend into the micropyle. The large egg projects beyond the synergids. The two polar nuclei fuse prior to or at the time of fertilization (*Rhododendron* and *Vaccinium*). The antipodals are T-shaped, and usually ephemeral. Supernumerary antipodals occur in *Leucothoë racemosa* and *L. recurva*. The mature embryo sac contains starch. Allium type of development is reported in *Cassiope mertensiana* (Palser 1952).

The development of endosperm is either of the Nuclear or Cellular type (Fig. 42.4.1A–F). Nuclear type occurs in *Rhododendron japonicum* × *R. mucronatum* (Creech 1955), *Vaccinium augustifolium* (Bell 1957), *V. leschenaultii* (Reddy et al. 1966), *V. nummularia*, *V. retusum* and *V. serratum* (Venkateswarlu and Maheswari Devi 1973). According to Venkateswarlu and Maheswari Devi (1973), the division of the primary endosperm nucleus is followed by two free-nuclear divisions, and the eight daughter nuclei arrange in three groups, two at each pole and four in the centre. Walls are laid down separating the two terminal cells, each two-nucleate, which develop into haustoria. The central four-nucleate cell gives rise to the cellular endosperm proper.

Cellular type of endosperm occurs in *Daboecia*, *Kalmia*, *Keiophyllum*, *Loiseleuria*, *Phyllodoce* and *Rhodothamnus* (P.S. Ganapathy and Palser 1964), *Rhododendron ferrugineum*, *R. yunnanense* (Veillet-Bartoszewska 1959a, Palser et al. 1971), and *Vaccinium* spp. (Stushnoff and Palser 1969).

The division of the primary endosperm nucleus separates a micropylar chamber from the chalazal chamber (Fig. 42.4.1A, C, D). A transverse division in each chamber results in a row of four cells (Fig. 42.4.1B, E); a vertical division in each cell forms four superposed tiers, each of two cells (Fig. 42.4.1F shows an earlier stage). The terminal tiers develop into haustoria, while the two central tiers produce the cellular endosperm proper (P.S. Ganapathy and Palser 1964, Palser et al. 1971).

The micropylar haustorium is larger than the chalazal, and shows extensive growth except in *Daboecia* (P.S. Ganapathy and Palser 1964). The intervening wall of the chalazal cells disintegrates and the haustorium becomes two- to four-nucleate.

In *Tripetaleia bracteata* and *T. paniculata* (Yamazaki 1975), the primary endosperm nucleus divides transversely; another transverse division occurs in each cell and a row of four cells is formed. A vertical division occurs in each of the upper two cells, while the lower two undergo transverse division. The endosperm is composed of eight cells in six tiers. The terminal cells develop into haustoria. The wall between the two micropylar cells disintegrates, and the micropylar haustorium becomes binucleate. The micropylar haustorium develops into a large balloon-shaped cell with vacuolate cytoplasm, and eight or ten nuclei in *T. bracteata* and 10–16 in *T. paniculata*. The chalazal haustorium is two-nucleate, and ellipsoidal with vacuolate cytoplasm. The number of nuclei may increase up to four to six.

Stevans (1919) observed both Nuclear and Cellular type of endosperm in *Vaccinium corymbosum*. Otherwise, Stushnoff and Palser (1969) describe Cellular type of development in five taxa of *Vaccinium* – diploid, tetraploid and hexaploid species. Stevan's observations need confirmation.

The embryogeny conforms to the Solanad type (P.S. Ganapathy and Palser 1964, Stushnoff and Palser 1969). The suspensor is linear and four- to five-celled. According to Veillet-Bartoszewska, the embryogeny conforms to Period

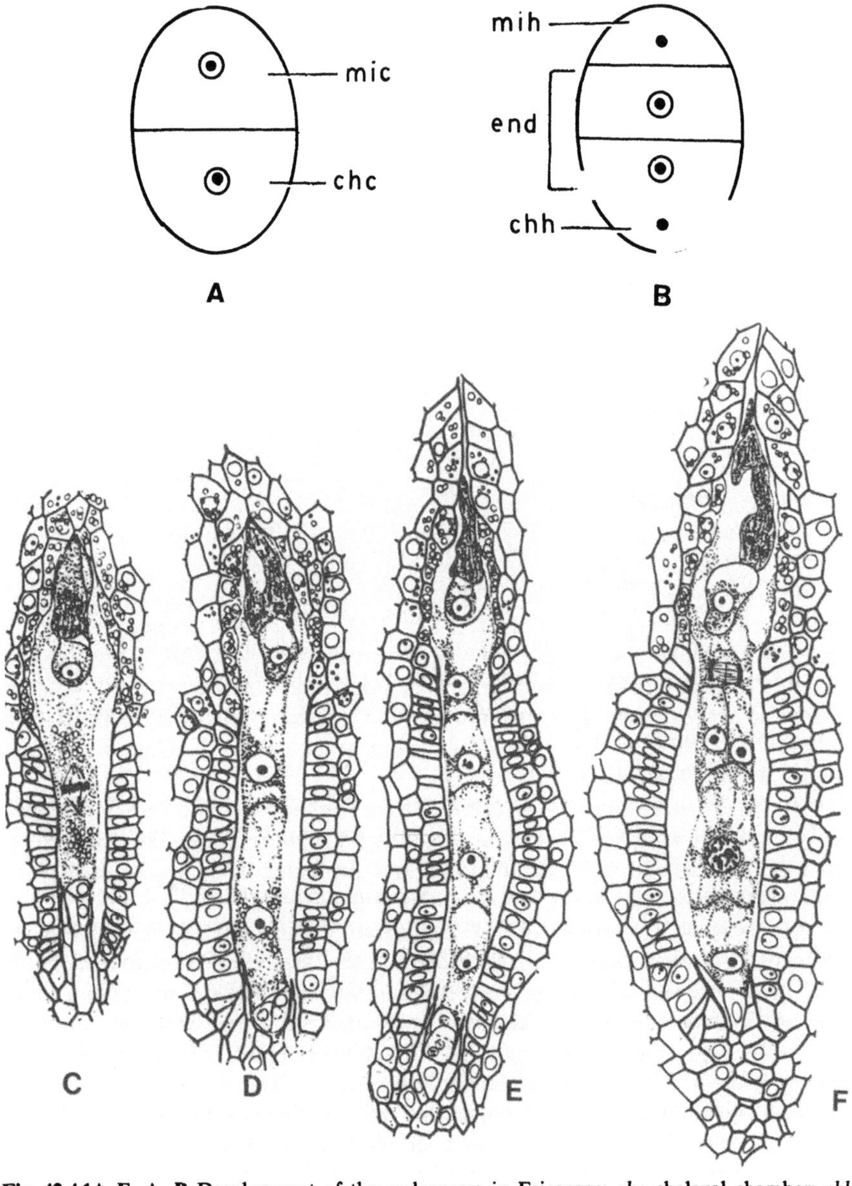

Fig. 42.4.1A–F. A, B Development of the endosperm in Ericaceae; *chc* chalazal chamber; *chh* chalazal haustorium; *end* endosperm initials; *mic* micropylar chamber; *mih* micropylar haustorium (Courtesy Dr M.A. Rau, Mysore, 1980). **C–F** *Rhododendron yunnanense*; **C** Embryo sac with zygote, primary endosperm nucleus in metaphase; **C–F** show remnants of antipodals, starch grains and conspicuous endothelium; **D** Zygote, and micropylar and chalazal endosperm chamber; **E** Zygote and four-celled endosperm, both the chambers have divided transversely; **F** Zygote, endosperm proper (two middle cells, one vertically divided), and micropylar (nucleus in telophase) and chalazal haustorium. The tips of synergids extend into the micropyle in **E, F**. (After Palser et al. 1971)

II, Series A, Megarchetype III, wherein $cc = pco + pvt + \frac{1}{2}phy$, $cd = \frac{1}{2}phy + iec + icc + co + s$ (partly) and $cb = s$. This corresponds to the Caryophyllad type in *Calluna vulgaris* (Veillet-Bartoszewska 1961a), *Erica tetralix* (Veillet-Bartoszewska 1960a), *Gaultheria shallon* (Veillet-Bartoszewska 1959a), *Ledum palustre* (Veillet-Bartoszewska 1960b) and *Rhododendron ferrugineum* (Veillet-Bartoszewska 1957a). A similar destination of initials occurs in the embryogeny of *Tripetaleia bracteata* and *T. paniculata* (Yamazaki 1975). According to Yamazaki (1975), the embryogeny is of the Polygonad type-Sagina variation.

The small seed is albuminous, with a wing in Rhododendrioideae. The seed-coat is exotestal. The outer epidermis generally shows thick pitted and lignified inner and radial walls; short hair-like outgrowths occur in *Daboecia*. The mesophyll and inner epidermis get crushed, but in *Daboecia* the outer hypodermis is persistent and slightly thick-walled (Corner 1976).

TB Batygina, ES Teryokhin, GK Alimova and MS Yakovlev (1963), HP Bell (1957), YL Chou (1952), HF Copeland (1943), EJH Corner (1976), JL Creech (1955), P Crété (1964a), O Drude (1889), PS Ganapathy (1967b), PS Ganapathy and BF Palser (1964), V Kaul, JL Rouse and EG Williams (1986), S Mirick and J Guinn (1981), AR Olson (1980), BF Palser (1951, 1952, 1961), BF Palser, WR and MN Philipson (1971, 1985, 1989), BF Palser, JL Rouse and EG Williams (1987, 1989), KB Reddy, S Murthy and R Narayana (1966), LD Safijowska (1960), G Samuelsson (1913), NE Stevans (1919), PF Stevens (1971), C Stushnoff and BF Palser (1969), M Veillet-Bartoszewska (1957a, 1958a, 1959a, 1960a, b, 1961a), J Venkateswarlu and H Maheswari Devi (1973), P Villamil and BF Palser (1980), GD Wallace (1975), L Watson (1965), EG Williams, V Kaul, JL Rouse and BF Palser (1986), T Yamazaki (1966, 1974, 1975)

42.5 Empetraceae

The embryology has not been investigated in sufficient detail.

The anther wall comprises an epidermis, non-fibrous endothecium, one or two middle layers, and glandular tapetum with multinucleate cells. Fibrous thickenings appear in the region of dehiscence, as well as in the epidermal cells. The microspore mother cells undergo meiotic divisions, cytokinesis is simultaneous, and the tetrads are tetrahedral and decussate (P.S. Ganapathy 1967b).

The pollen grains remain attached in tetrads. The individual pollen grain is three-colpate or colporate, and two-celled at the dispersal stage.

The ovule is anatropous, unitegmic, and tenuinucellar. A funicular obturator is reported. The archesporial cell functions as megaspore mother cell. Meiotic divisions result in a linear tetrad, and the chalazal megaspore develops into a Polygonum type of embryo sac. The embryo sac is broader at the micropylar end and narrower at the chalazal end in *Empetrum nigrum* (Samuelsson 1913). The polar nuclei fuse before fertilization, and the antipodals persist up to fertilization. An endothelium differentiates around the embryo sac.

The development of endosperm is of the Cellular type. The first few divisions, following the division of primary endosperm nucleus, produce a row of four cells in *Empetrum nigrum* (Samuelsson 1913). The terminal cells develop into micropylar and chalazal haustoria. The remaining two cells give rise to the endosperm proper. The micropylar haustorium is binucleate and the chalazal uninucleate, and the haustoria bulge at the extremities. They persist up to the dicotyledonous stage of embryo (see P. Maheshwari 1950).

The embryogeny conforms to Period II, Megarchetype III, Group II (Caryophyllad) in *Corema album* (Veillet-Bartoszewska 1959b). Cleavage polyembryony is reported in *Empetrum nigrum* (see Davis 1966).

The albuminous exotestal seed with a straight embryo has a single-layered seed-coat formed by the outermost layer of the testa.

GL Davis (1966), D Don (1827), PS Ganapathy (1967b), O Hagerup (1927), P Maheshwari (1950), G Samuelsson (1913), M Veillet-Bartoszewska (1959b)

42.6 Epacridaceae

The anther is bisporangiate. The epidermis of the anther wall develops into the exothecium; a fibrous endothecium is absent. The exothecium is ribbed in *Sprengelia sprengelioides* (Copeland 1954). The tapetum is glandular, and its cells are uninucleate in *Styphelia longiflora*. The microspore mother cells undergo meiotic divisions, cytokinesis is simultaneous, and tetrads and polyads (Fig. 42.6.1A–F) develop.

A variety of pollen types is recorded (Erdtman 1952, Smith-White 1955a, 1959, Venkata Rao 1961b). Erdtman (1952) classifies pollen grains in the family into two groups: (1) isodynamosporous (all spores of the same size, i.e. full tetrad), and (2) heterodynamosporous (all spores not of the same size, i.e. monads, dyads, and triads).

The reasons are not known for one nucleus becoming functional and the other three microspore nuclei migrating downward and, ultimately, degenerating. This may be due to unequal distribution of cytoplasm and genetic constitution of nuclei (Venkata Rao 1961b).

Venkata Rao (1961b) describes six types of pollen grains:

1) **Single grains.** The microspores of a tetrad separate into spherical, tricolporate pollen grains, and are dispersed at the two-celled stage – *Richea milligani* and *R. procera*. In *R. sprengelioides*, polyads occur (Fig. 42.6.1A). The remaining four species of the genus *Richea* show pollen grains in tetrads.

2) **Full tetrad type or T-type.** All the microspores of the tetrad are functional and remain united in tetrads. The individual grain is tricolporate – four spp. of *Richea*, all genera of Epacrideae, and some species of *Acrotriche*, *Cyathodes*, *Lissanthe*, *Pentachondra* and *Trochocarpa*.

3) **A-type or Astroloma type.** All the microspores of a tetrad are similar to begin with. Later, one to three develop into viable pollen grains and remain united (Fig. 42.6.1B–D). Rarely, all four may be fertile or sterile (nullads).

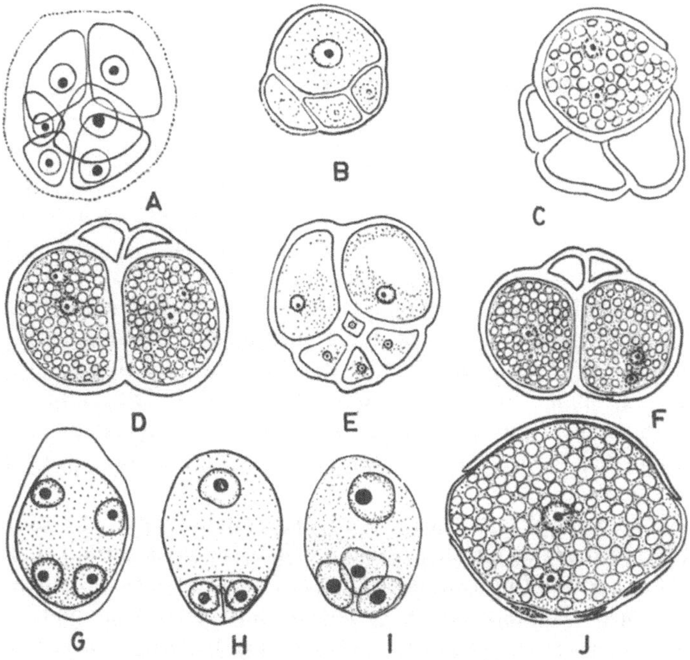

Fig. 42.6.1A–J. Development of pollen. **A** *Richea sprengelioides*, polyad. **B–E** *Cyathodes dealbata*, A type or Astroloma type; **B, C** Tetrad with single fertile pollen grain; **D** Tetrad with two fertile pollen grains; **E** Polyad with two fertile pollen grains. **F** *Pentachondra involucrata*, P type or Pentachondra type, twin fertile pollen grains. **G–J** *Astroloma humifusum*, S type or Styphelia type; **G** All the four microspore nuclei in the tetrad are of the same size; **H** One microspore nucleus in the upper part, and three in the lower; **I** Later stage, one functional microspore nucleus separated from others; the lower three nuclei are still healthy; **J** Monad with two-celled pollen, and remnants of three degenerated microspores. (After Venkata Rao 1961b)

Polyads are of common occurrence: *Astroloma conostephioides*, *A. pinifolius*, *Cyathodes dealbata* (Fig. 42.6.1E) and *Monotoca scoparia*.

4) P-type or Pentachondra type. Two microspores in a tetrad degenerate regularly while two remain functional – *Pentachondra involucrata* (Fig. 42.6.1F).

5) S′-type or modified Styphelia type. Three microspores of each tetrad invariably degenerate while only one remains functional. The mature pollen grain consists of a single large grain (monad) – *Leucopogon* spp. (Smith-White 1955a, b).

6) S-type or Styphelia type. Three microspore nuclei formed during meiotic divisions of the pollen mother cell group together, and the fourth lies on the opposite side in an isolated position (Fig. 42.6.1G–I). After wall formation, the three nuclei form small non-functional cells and degenerate. The isolated nucleus develops into a large tricolporate, functional pollen grain (monad).

The latter accumulates starch and becomes two-celled – *Astroloma humifusum* (Fig. 42.6.1J), and *Styphelia adscendens.*

Smith-White (1955a) recognises four types, and suggests their evolutionary sequence. Venkata Rao (1961b) added two more types and pointed out parallelism in the homology of development of male and female gametophytes. The type in which all the four microspore nuclei (produced after meiotic divisions of pollen mother cells) are functional resembles the tetrasporic development of female gametophyte, P-type of pollen grains in which two microspores are functional to bisporic type, and monad type (S′-type and S-type) to monosporic type. This parallelism is not at all convincing.

The ovule is anatropous, unitegmic and tenuinucellar. The archesporial cell functions as megaspore mother cell, undergoes meiotic divisions, and forms a linear tetrad. The chalazal megaspore develops into a Polygonum type of embryo sac. The polar nuclei fuse before fertilization. The antipodals are ephemeral. In *Leucopogon ericoides* (Copeland 1954), supernumerary antipodals are reported. An endothelium develops around the lower two-thirds of the embryo sac.

The development of endosperm is of the Cellular type. The transverse division of the primary endosperm nucleus is followed by a further transverse division so that a row of four cells is produced. The two middle cells give rise to endosperm proper. The terminal cells develop into micropylar and chalazal haustoria. The micropylar haustorium is two-celled. In *Brachyloma dephnoides* (Copeland 1954), one of the cells is large, balloon-shaped, and three-nucleate; the smaller cell is seven-nucleate. The chalazal haustorium is rudimentary in this taxon. In *Styphelia longifolia* (Brough 1924), only the micropylar haustorium is formed.

The embryogeny conforms to the Caryophyllad type, but in *Dracophyllum secundum* there are certain resemblances to the Solanad type (see Davis 1966).

Seeds exotestal with straight embryo in the middle of fleshy endosperm (Hutchinson 1973, Corner 1976). The seed-coat is one- to many-layered.

P Brough (1923, 1924), HF Copeland (1954), EJH Corner (1976), GL Davis (1966), G Erdtman (1952), O Hagerup (1927), J Hutchinson (1973), MS McGlone (1978), BR Paterson (1961), S Smith-White (1948, 1955a, b, 1959), C Venkata Rao (1961b), M Veillet-Bartoszewska (1961b)

42.7 Taxonomic Considerations

Clethraceae. Bentham and Hooker (1965b) treat Clethraceae as a tribe of Ericaceae. Hallier (1912) and Wettstein (1935) consider them as a primitive taxon of Bicornes, and derive them from Ochnaceae. Bessey (1915) treats Clethraceae as a primitive taxon in Ericales; Hutchinson (1973) and Cronquist (1981) recognize them as a family in the order Ericales.

Clethraceae show close similarity with Ericaceae (Table 42.7.1): two-celled pollen grains, unitegmic, tenuinucellate ovules, presence of endothelium, Polygonum type of embryo sac, Cellular type of endosperm with micropylar and chalazal haustoria, straight embryo, and one-layered seed-coat

(P.S. Ganapathy 1970). However, Clethraceae can be distinguished from Ericaceae and treated as an independent family owing to: (1) extensive fibrous endothecium, (2) single pollen grains, (3) trilocular ovary, and (4) weakly developed endosperm haustoria.

Pyrolaceae. Hallier (1912), Bessey (1915), and Engler and Prantl (1931) include Pyrolaceae in Ericales (Table 42.7.1). Pyrolaceae comprise the subfamilies Pyroloideae and Monotropoideae (Melchior 1964). Small (1914, see Hutchinson 1973) treats Monotropoideae as a separate family, Monotropaceae, on the basis of their saprophytic habit and single pollen grains. Hutchinson (1973) recognizes both Pyrolaceae and Monotropaceae as separate families. However, Melchior (1964) does not recognize the family status of Monotropaceae.

Teryokhin (1963) considers that Pyroleae and Monotropae are appropriately placed in Ericaceae, but are of secondary origin due to their transition to a parasitic mode of life. He distinguishes three tribes of Monotropoideae: (1) Pterosporae – embryos with well differentiated shoot and root zones, endosperm of 30–40 cells, and three layers thick integument: (2) Pyroleae – endosperm and embryo similar to Pterosporae, but the integument is only two layers thick; and (3) Monotropeae – embryo of three to four cells, endosperm of nine to ten cells, and integument two layers thick.

The embryological features, such as absence of endothecium, united pollen grains, similar type of embryo sac, endosperm and embryo, and albuminous seed with a thin seed-coat are resemblances with Ericaceae. Thus, there is every justification for the placement of Pyrolaceae in Ericales. However, the pentacarpellary pentalocular ovary, minute (undifferentiated) organless mature embryo, and loculicidal fruit are distinctive features of Pyrolaceae.

Ericaceae. The family Ericaceae exhibits the characteristic embryological features of the order (Table 42.7.1). The occurrence of resorption tissue in the anther, pollen grains always in tetrads, four-locular ovary, rarely Allium type of embryo sac, Nuclear type of endosperm, ***Rhododendron*** and ***Vaccinium***, and Solanad and Caryophyllad type of embryogeny are additional features of Ericaceae.

Vaccinium. The position of *Vaccinium* is debatable. Bentham and Hooker (1965b) and Hutchinson (1973) treat this genus as a separate family, Vacciniaceae. Drude (1889, see Venkateswarlu and Maheswari Devi 1973) assigns it to a lower taxonomic status as a subfamily under Ericaceae. This position is accepted by Engler and Diels (1936), Rendle (1938) and Gunderson (1950). Venkateswarlu and Maheswari Devi (1973) evaluate the embryological features of *Vaccinium* and other subfamilies of Ericaceae. The uniformity in embryological data supports its retention as a subfamily, Vaccinioideae, of Ericaceae, and negates its elevation to family rank. Anatomical investigations also support this status (Watson 1965).

Empetraceae. Don (1827, see P. Maheshwari 1950) first recognized the family Empetraceae, and relates it to Euphorbiaceae and Celastraceae. Bentham and

Table 42.7.1. Embryology of Clethraceae, Pyrolaceae, Ericaceae and Epacridaceae (Adapted after Vijayaraghavan and Dhar 1978)

	Clethraceae	Pyrolaceae	Ericaceae	Epacridaceae
Tapetum	Secretory, cells binucleate	Secretory, cells binucleate	Secretory, cells multinucleate	Secretory, cells uninucleate
Microspore tetrad	Tetrahedral	Isobilateral and decussate	Tetrahedral and decussate	Monads, dyads, triads, tetrads, polyads
Pollen	2-celled, shed singly	2-celled, mostly remain attached in tetrads	2-celled, remain attached in tetrads or shed singly	3-celled
Ovule	Anatropous, unitegmic, tenuinucellate	Anatropous, unitegmic, tenuinucellate	Anatropous, unitegmic tenuinucellate	Anatropous, unitegmic, tenuinucellate
Embryo sac	Polygonum type with broad micropylar and narrow chalazal end	Polygonum, broad micropylar and narrow chalazal end	Polygonum, broad micropylar and narrow chalazal end	Polygonum, broad micropylar and narrow chalazal end
Antipodal cells	In a linear row, largest at the base	Uninucleate, persist after fertilization	Uninucleate, ephemeral	Uninucleate, ephemeral
Endosperm	Cellular type, first two divisions transverse resulting in a row of 4 cells	Cellular, first two divisions transverse resulting in a row of 4 cells	Cellular; Nuclear in some taxa	Cellular, first two divisions transverse resulting in a row of 4 cells
Endosperm haustoria	Micropylar and chalazal haustoria weakly developed	Micropylar and chalazal haustoria formed	Micropylar and chalazal haustoria formed	Micropylar and chalazal haustoria formed
Embryo	Asterad type	Caryophyllad	Solanad	Caryophyllad
Seed-coat	Exotestal	Exotestal	Exotestal	Exotestal
Seed	Small, without wings, albuminous	Winged, albuminous	Winged in Rhododendroideae, albuminous	Not winged, albuminous

Hooker (1965b) treat them as an anomalous family. P. Maheswari (1950) refers to the views of Agardh (1858), Gray (1858), Baillon (1892) and Hallier (1912), who consider Empetraceae to be related to Ericaceae (Table 42.7.1). However, Bessey (1915) relates them to Sapindales, and Rendle (1952) and Hutchinson (1973) to Celastrales. P. Maheshwari (1950) confirms their position in Ericales by evaluating the embryological data put forward by Samuelsson (1913).

Epacridaceae. Engler and Gilg include Epacridaceae in Ericales; they erect two suborders – Epacridineae with bilocular anthers, and Ericineae with tetralocular anthers. The occurrence of pollen tetrads and single pollen grains distinguish the family Epacridaceae which shows some comparable characters with Ericaceae (Table 42.7.1). In both taxa n = 13 or multiples thereof (Venkata Rao 1961b). In addition, Epacridaceae exhibit Ericalean features such as ana- to campylotropous, unitegmic, tenuinucellar ovule with a prominent endothelium, Polygonum type of embryo sac, Cellular type of endosperm with haustoria, Solanad or Caryophyllad type of embryogeny, and albuminous seed. According to Venkata Rao (1961b), Epacridaceae and Ericaceae have probably evolved from common ancestors.

Cronquist (1981) includes Cyrillaceae in Ericales. Takhtajan (1980) groups Actinidiaceae, Cyrillaceae and Grubbiaceae in Ericales. R. Dahlgren (1980b, 1981) places Actinidiaceae (Saurauiaceae), Cyrillaceae, Roridulaceae and Diapensiaceae in Ericales, and treats Monotropaceae and Pyrolaceae also in Ericales. The relationships of Grubbiaceae (Sect. 12.9), Actinidiaceae (Sect. 22.18), Roridulaceae (Sect. 26.21), Cyrillaceae (Sect. 33.15), Diapensiaceae (Sect. 41.3) and Pyrolaceae (Sect. 42.7) are discussed under the relevant sections.

Cronquist (1981) traces the ancestry of Ericales in Theales. R. Dahlgren (1977a) considers the families Actinidaceae, Cyrillaceae and Clethraceae as transitional between Ericales and Theales.

43 Primulales

The order Primulales comprises Theophrastaceae, Myrsinaceae, and Primulaceae.

43.1 General Features

The Primulales show the following embryological features: (1) fibrous endothecium, secretory anther tapetum with uninucleate cells, (2) tricolporate, two-celled pollen grains, (3) anatropous, bitegmic, and tenuinucellar ovule with micropyle formed by both integuments, (4) multicellular archesporium, (5) Polygonum type of embryo sac with ephemeral antipodals, and differentiation of endothelium, (6) Nuclear type of endosperm, (7) Onagrad type of embryogeny in Myrsinaceae, and Caryophyllad in Primulaceae, and (8) albuminous seed.

43.2 Theophrastaceae

The embryology has not been fully investigated.

A multilayered tapetum in the anther wall is mentioned by Schnarf (1931). A reinvestigation is desirable.

There are numerous ovules immersed in mucilage (Hutchinson 1973). The ovule is ana- to campylotropous, bitegmic, and tenuinucellar. Both the integuments organize the micropyle. The archesporial cell functions as megaspore mother cell in *Claviga japonica* (K.V.O. Dahlgren 1916). The meiotic divisions are followed by wall formation resulting in a linear tetrad. The chalazal megaspore develops into a Polygonum type of embryo sac. The synergids are hooked in *Theophrasta macrophylla* (see Davis 1966), polar nuclei fuse before fertilization, and the antipodals are ephemeral.

Development of the endosperm is of the Nuclear type.

Fruit baccate or drupaceous, indehiscent, one to many seeds, endosperm copious, not ruminate, embryo erect with conspicuous cotyledons (Hutchinson 1973).

KVO Dahlgren (1916), GL Davis (1966), IC Gardner, IM Miller and A Scott (1981), J Hutchinson (1973), J Mauritzon (1935b), K Schnarf (1931)

43.3 Myrsinaceae

The anther is tetralocular, and the wall comprises the epidermis, fibrous endothecium, single middle layer, and secretory tapetum with uninucleate cells. Fibrous thickenings also occur in the middle layer in *Maesa dubia* (Sankara Rao 1972). In each pollen sac the archesporium becomes separated by transverse sterile partitions in *Aegiceras corniculatum* (Periasamy and Sivaramakrishnan 1980) and *A. majus* (Lersten 1971). Each group develops into a microsporangium, and the pollen sac contains 12 to 16 microsporangia in a row. The anther remains polysporangiate until dehiscence. Each "sporangium" has its own secretory tapetum with binucleate cells. The microspore mother cells undergo meiotic divisions, cytokinesis is simultaneous, and the tetrads are tetrahedral. Sterility occurs in the development of anther and male gametophyte in *Maesa dubia* (Sankara Rao 1972). The anther loculi fail to differentiate, and most of the microspore mother cells do not undergo meiosis. However, a few small shrunken non-viable pollen grains develop in unhealthy abnormal anthers.

The triporate pollen grains are two-celled at the dispersal stage.

The ovule is anatropous, bitegmic and tenuinucellar. In *Aegiceras*, Karsten (1891) reported a unitegmic, tenuinucellate ovule, which has been confirmed by Carey and Fraser (1932) and W.R. Philipson (1974).

Both the integuments organize the micropyle. In *Maesa* (Sankara Rao 1972), the archesporium is unicellular. The archesporial cell functions as megaspore mother cell. Following meiotic divisions, a linear tetrad is produced and the chalazal megaspore develops into a Polygonum type of embryo sac. A prominent endothelium develops at the tetrad stage. Its cells are usually uninucleate, rarely binucleate.

In *Maesa* (Sankara Rao 1972), at the four-nucleate stage of embryo sac the chalazal end elongates downward, crushing the adjoining cells. The antipodal cells lie in this so-called caécum which Sankara Rao mistakenly considers as a haustorium. At the eight-nucleate stage (when the ovule becomes anatropous), the polar nuclei lie close to the antipodals. The embryo sac now develops two lateral caeca, and "one of these becomes highly aggressive and grows up into the integumentary tissue". . . . "The secondary nucleus lies midway between the chalazal caecum and the lateral caeca". This plant needs to be reinvestigated.

The development of endosperm is of the Nuclear type. Later, it becomes cellular. De Jongh's (1938) account of endosperm in *Ardisia crispa* is erroneous.

The embryogeny is probably of the Onagrad type, and a long suspensor is formed. Polyembryony is common (see Davis 1966). De Jongh (1938) reports that in *A. crispa* the pollen grains rarely germinate, and the embryo arises from the integumentary tissue.

The seed is small and albuminous with fatty food reserve. The embryo is straight or slightly curved. In *Aegiceras*, the cylindrical embryo germinates within the pericarp. The embryo has two connate cotyledons which form a tube and enclose the plumule (Hutchinson 1973).

The testa is multiplicative. The outer epidermis is thin-walled, and bears stomata on the surface in *Embelia*. In *Aegiceras*, the outer two to three layers of testa are thin-walled, and the inner three layers comprise cells with brown walls. In *Jacquinia*, the thick-walled outer epidermal cells are flattened. The cells of the hypodermal layer have thickenings on anticlinial walls. The thin-walled mesophyll contains rhomboic crystals, or is usually crushed. The tegmen is multiplicative, and is also crushed (Corner 1976). In *Embelia* (Netolitzky 1926), the tegmen consists of two to four layers of thick-walled cells with brown contents.

Myrsinaceae show an interesting symbiotic association of bacteria. De Jongh (1938) studied *Ardisia crispa*, and the symbiont is *Bacterium foliicola* (?). The bacteria adhere to the radicular pole of the embryo, where they remain in a resting but reversible state. At germination, the bacteria are pushed over the cotyledons where they infect the axillary buds; a bacterial mass also finds its way to the terminal bud (Fig. 43.3.1A).

Colourless mobile aerobes were isolated from the seeds. The bacteria occur on the plant on active and dormant meristems (except root and anther meristem), and in the foliar nodules. They are non-acid fast, gram-negative rods with a tendency to become non-motile with a mucilaginous membrane, and to form bacteroids. They are unable to fix nitrogen. The exact function needs to be further investigated.

The bacteria are active in active stem meristems; inactive but reversible to the active state in dormant buds where they may renew activity after several years of rest. The bacteria are enclosed within floral parts (Fig. 43.3.1A–C) at flower formation. They are locked within the carpels and at the top of placenta

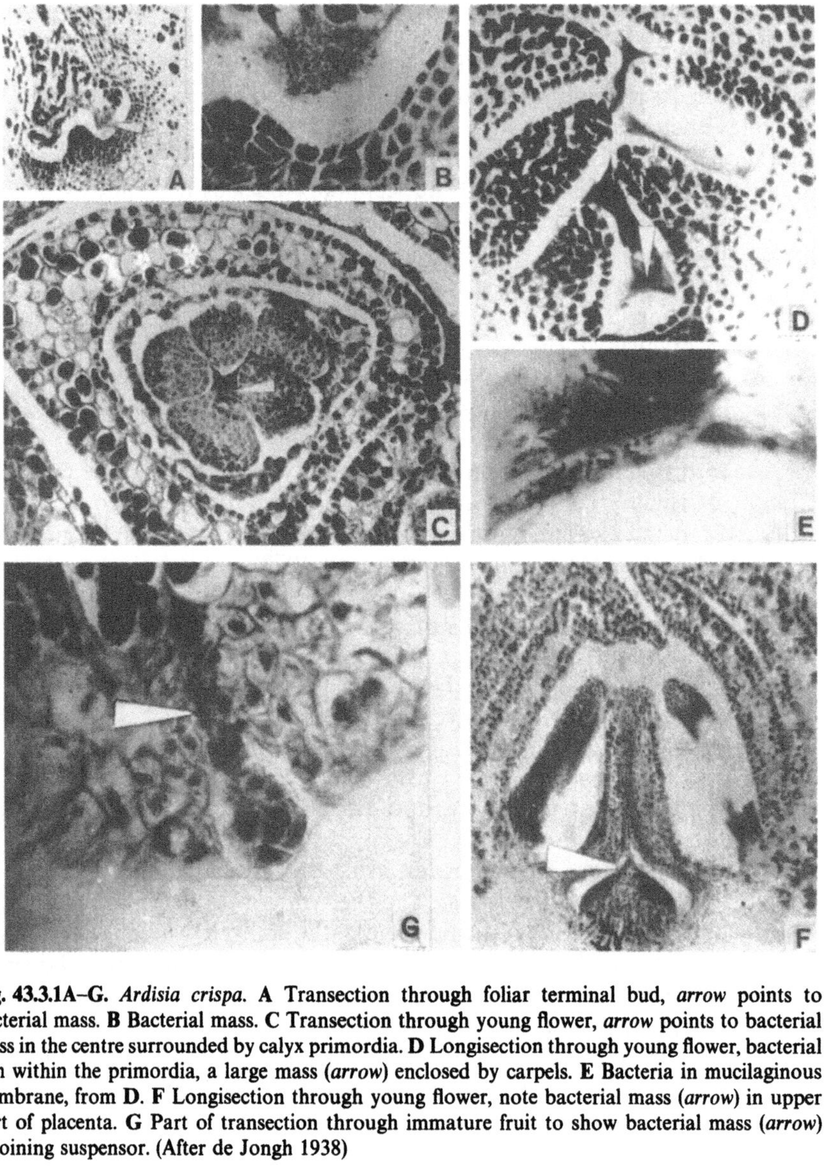

Fig. 43.3.1A–G. *Ardisia crispa.* **A** Transection through foliar terminal bud, *arrow* points to bacterial mass. **B** Bacterial mass. **C** Transection through young flower, *arrow* points to bacterial mass in the centre surrounded by calyx primordia. **D** Longisection through young flower, bacterial film within the primordia, a large mass (*arrow*) enclosed by carpels. **E** Bacteria in mucilaginous membrane, from **D**. **F** Longisection through young flower, note bacterial mass (*arrow*) in upper part of placenta. **G** Part of transection through immature fruit to show bacterial mass (*arrow*) adjoining suspensor. (After de Jongh 1938)

(Fig. 43.3.1D–F), finally surrounded by the ovules, and are entrapped by the growing integuments at the micropylar region.

In an immature fruit, the bacterial mass lies adjoining the suspensor of the developing embryo (Fig. 43.3.1G).

The bacteria grow in the inflorescence, young flower buds, older buds, and immature and mature fruit. The active or inactive state of bacteria and the amount of bacterial film depends on the meristematic or dormant condition of the floral and fruiting organs.

In meristematic tissues of young floral buds and immature fruit (Fig. 43.3.1G), the bacteria are in abundance and in an active state.

In well differentiated buds and young fruit, which remain dormant for a shorter or longer period, the bacteria are fewer and remain in a resting state. They resume their activity as the organs start developing. In a mature fruit, they are immotile.

In a cross-section of an inflorescence, the five meristematic floral buds are enclosed within the bracts. The bacterial film finds its way inside and surrounds each floral bud. It not only remains there, but also penetrates within the carpels. The carpels surround the rod-shaped bacteria. The role of bacteria in flowers and fruits is unknown, but this is the only method for the bacteria to reach the mature seed where they persist between the embryo and endosperm.

The bacteria-free plant, the "cripple", has a juvenile appearance. It is, therefore, obvious that the bacteria are concerned in the formation of normal plants as compared to "cripples".

The embryology and bacterial symbiosis in *Ardisia crispa* require a thorough reinvestigation.

G Carey and L Fraser (1932), EJH Corner (1976), GL Davis (1966), IC Gardner, IM Miller and A Scott (1981), J Hutchinson (1973), De Jongh Ph (1938), G Karsten (1891), NR Lersten (1971), F Netolitzky (1926), K Periasamy and R Shivaramakrishnan (1980), WR Philipson (1974), K Sankara Rao (1972)

43.4 Primulaceae

The anther wall comprises the epidermis, fibrous endothecium, a single ephemeral middle layer, and secretory tapetum with uninucleate cells. The epidermal cells contain tannin in *Primula floribunda* (Subramanyam and Narayana 1968). The tapetum is biseriate in *P. stuartii* (Subramanyam and Narayana 1976). The microspore mother cells undergo meiotic divisions, and the tetrads are tetrahedral and decussate. Simultaneous cytokinesis takes place by furrowing.

The three-colporate pollen grains are two-celled at the dispersal stage. Sometimes, they germinate in situ in *Anagallis pumila* (M.V.S. Raju 1952c).

The ovule is anatropous, bitegmic, and tenuinucellar. The narrow micropyle is formed by both the integuments. The hypodermal archesporium is multicellular. A single archesporial cell develops into the megaspore mother cell. The latter undergoes meiotic divisions, and the chalazal megaspore of the linear tetrad develops into a Polygonum type of embryo sac. The synergids are elongated, polar nuclei fuse before fertilization, and antipodals are ephemeral. In *Lysimachia vulgaris* (K.V.O. Dahlgren 1916), the antipodals enlarge and simulate the synergids. An endothelium develops around the embryo sac, and contains tannin after fertilization in *Primula floribunda*. The synergids may remain intact after fertilization in *Anagallis pumila* (M.V.S. Raju 1952c). This is doubtful, and appears to be based on a wrong interpretation.

Development of the endosperm is of the Nuclear type. Nuclear divisions are non-synchronous and centripetal wall formation is initiated at the globular proembryo in *P. floribunda*. Cytokinesis occurs at eight- to 16- or 32-nucleate endosperm in *Anagallis pumila* (M.V.S. Raju 1952c), whereas walls develop after the formation of ca. 1000 free nuclei in *Primula officinalis* (K.V.O. Dahlgren 1916). The endosperm contains amyloids.

The embryogeny conforms to the Caryophyllad type, or Period II, Group IX (Souèges 1937c, M.V.S. Raju 1952c). An additional embryo is formed due to the budding of suspensor in *P. auricula* (Veillet-Bartoszewska 1957c).

The seed contains horny endosperm, and an erect elliptic embryo. Both the integuments form the seed-coat. The outer epidermis of testa and tegmen contain tannin. The endothelial cells contain tannin and calcium oxalate crystals in *Anagallis*.

KVO Dahlgren (1916), MVS Raju (1952c), AJ Richards and HB Ibrahim (1982), R Souèges (1937c), K Subramanyam and LL Narayana (1968, 1976), M Veillet-Bartoszewska (1957b, c)

43.5 Taxonomic Considerations

Bentham and Hooker (1965b) group Plumbaginaceae, Primulaceae and Myrsineae (Myrsinaceae) in Primulales. Engler and Prantl (1931), Melchior (1964), and Cronquist (1981) include Theophrastaceae, Myrsinaceae and Primulaceae in Primulales. Wettstein (1935) considers Myrsinaceae as an advanced family in Primulales. Hutchinson (1973) places Primulaceae and Plumbaginaceae in the order Primulales, and relates Primulales to Caryophyllaceae. He erects the order Myrsinales to include Myrsinaceae, Theophrastaceae and Aegicerataceae. R. Dahlgren (1980b, 1981) includes the order Primulales in the superorder Primuliflorae, and places it next to Plumbaginiflorae.

Theophrastaceae, Myrsinaceae, Primulaceae. The embryological features common to these families are: anatropous, bitegmic, tenuinucellar ovule, Polygonum type of embryo sac, Nuclear type of endosperm, and albuminous seed with erect embryo.

Myrsinaceae. Myrsinaceae and Primulaceae show secretory anther tapetum, two-celled pollen grains, micropyle formed by both the integuments, multicellular female archesporium, absence of parietal tissue, Polygonum type of embryo sac, presence of endothelium, Nuclear endosperm, and albuminous seed with erect embryo.

Myrsinaceae have some distinctive features: polysporangiate (transversely septate) anthers, unitegmic, tenuinucellate ovule in *Aegiceras*, symbiotic association of bacteria in *Ardisia crispa*, and embryo sac caeca in *Maesa*.

The multilayered anther tapetum (which is doubtful), and unicellular female archesporium of Theophrastaceae are in contrast to the unilayered anther tapetum, and multicellular female archesporium of Myrsinaceae and Primulaceae.

The embryogeny in Myrsinaceae is of the Onagrad type, and that in Primulaceae of the Caryophyllad type. In Myrsinaceae, the testa alone forms the seed-coat, while in Primulaceae both testa and tegmen contribute to the seed-coat.

Primulaceae. Hutchinson (1973, see also Corner 1976) points out that Primulaceae are related to Caryophyllaceae; Primulaceae show herbaceous habit. The embryological similarities between these families are: tricolporate pollen grains, anatropous bitegmic ovules, Polygonum type of embryo sac, Nuclear endosperm, Caryophyllad type of embryogeny, and small seeds with funicular aril. However, in Caryophyllaceae the ovule is crassinucellate with a nucellar cap, the seed is perispermous with a curved embryo, and the endosperm persists only around the radicle. In contrast, in Primulaceae the ovule is tenuinucellar, the seed albuminous (perisperm absent), and the embryo erect. There is thus some relationship between Primulaceae and Caryophyllaceae.

The embryological features of Primulaceae differ considerably from those of Plumbaginaceae (see Sect. 44.2), and the relationships of Primulales and Plumbaginales are discussed under Sect. 44.3.

Cronquist (1981) relates Primulales to Ebenales, and assumes that both may have a common ancestry in Theales. The orders Primulales, Ebenales and Theales have several common features: secretory anther tapetum, simultaneous cytokinesis, two-celled pollen grains, anatropous bitegmic (unitegmic in Ebenales) ovule, Polygonum type of embryo sac (Allium in Theaceae), and Nuclear type of endosperm. Iridoid and seco-iridoid compounds are absent in all the three orders – Primulales, Ebenales and Theales (R. Dahlgren 1977b).

In view of the several common embryological features in Theophrastaceae, Myrsinaceae and Primulaceae, the families are appropriately grouped in the order Primulales. It may be added that we should await further detailed studies on Theophrastaceae.

44 Plumbaginales

The family Plumbaginaceae represents the order Plumbaginales.

44.1 General Features

Plumbaginaceae are characterized by the following features: (1) occurrence of fibrous endothecium, two- to multinucleate tapetal cells, (2) three-, five- or six-colpate two- or three-celled pollen grains, (3) ana- or circinotropous, bitegmic, crassinucellar ovule with parietal tissue, (4) micropyle formed by inner integument, (5) predominently Plumbago and Plumbagella type of embryo sac, (6) absence of synergids, and occurrence of filiform apparatus in egg cell, (7) tetraploid secondary nucleus, (8) egg-like antipodal cell, (9) Nuclear type of endosperm, (10) Caryophyllad type of embryogeny, straight embryo with multicellular suspensor, and (11) exotestal small seed with firm crystalline-granular endosperm.

44.2 Plumbaginaceae

The anther wall comprises the epidermis, fibrous endothecium, single middle layer, and tapetum with two- to multinucleate cells. The microspore mother cells undergo meiotic divisions, cytokinesis is simultaneous, and tetrads tetrahedral or isobilateral.

The pollen grains are three-, five- or six-colpate with spinulous exine, and two- to three-celled at the dispersal stage.

Russell (1984) has made a detailed study of the structure of sperm cells in *Plumbago zeylanica.* Figure 44.2.1 (reconstructed) shows two sperms S_{vn} and S_{ua}, and vegetative nucleus *VN*. S_{vn} has majority of mitochondria (*m*) with very

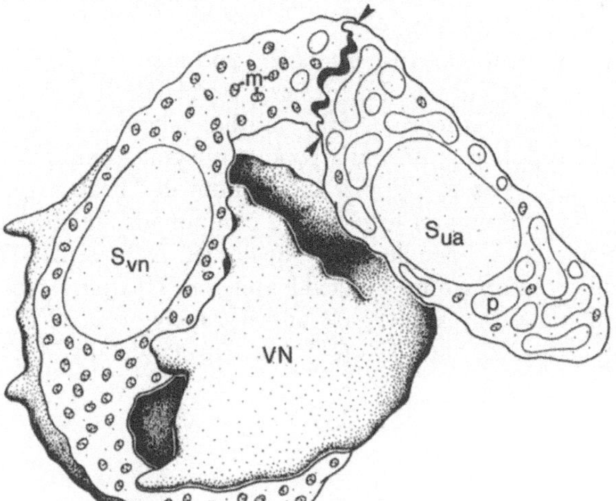

Fig. 44.2.1. *Plumbago zeylanica,* vegetative nucleus (*VN*) and two sperm cells – superimposed profiles of mitochondria (*m*) and plastids (*p*), in pollen grain. One sperm cell (S_{vn}) contains the majority of mitochondria and two plastids near the sperm cross wall (*arrowheads*). The other sperm cell S_{ua} contains most of the plastids and significantly fewer mitochondria. Note the gradation between the main body of S_{vn} and its cellular projection. (After Russell 1984)

few plastids, while S_{ua} contains plastids (*p*) and only a few mitochondria (Russell 1984).

The ovule is anatropous or circinotropous (*Armeria* and *Vogelia*), bitegmic and crassinucellar with the micropyle formed by the inner integument. The hypodermal archesporial cell cuts off a parietal cell; the parietal tissue is three layers thick in *Vogelia indica* (K.L. Mathur and Khan 1941). The megaspore mother cell undergoes meiotic divisions, and forms a four-nucleate coeno-megaspore (Fig. 44.2.2A–F). The nuclei undergo one mitotic division and eight nuclei (in four pairs) are produced (Fig. 44.2.2G). One nucleus at the micropylar pole organizes as egg cell, one nucleus from each pair migrates to the centre, and these four nuclei fuse to form a tetraploid secondary nucleus (Fig. 44.2.2H–J). The remaining three nuclei degenerate. The development of a Plumbago type of embryo sac also occurs in *Plumbago capensis* (Haupt 1934), *Ceratostigma plumbaginoides*, *P. zeylanica* (K.V.O. Dahlgren 1937), *Vogelia*

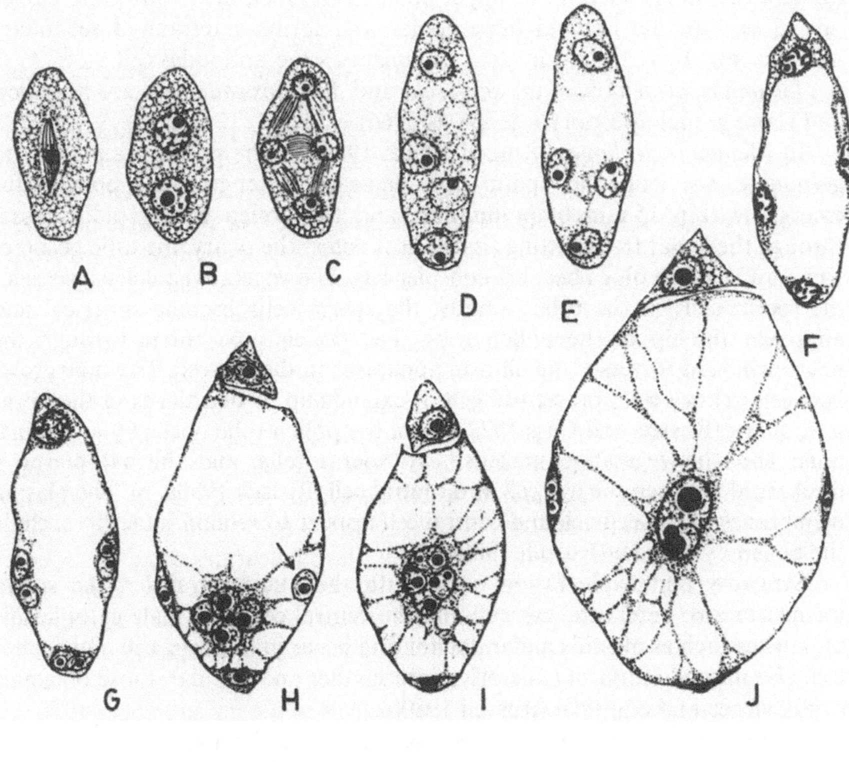

Fig. 44.2.2A–J. *Plumbago scandens.* **A** Megaspore mother cell at metaphase I. **B** Two-nucleate stage. **C** Late telophase II, megaspore nuclei connected by secondary fibres. **D, E** Coeno-megaspore with four nuclei. **F** Later stage. **G** Eight-nucleate stage. **H** Gametophyte with egg, one lateral and one chalazal cell, four polar nuclei and a lateral supernumerary polar nucleus (*arrow*). **I** Mature embryo sac with secondary nucleus (six nucleoli), and one cell at the chalazal end. The two lateral nuclei *may* have fused along with polars. **J** Mature gametophyte with remnants of chalazal and two lateral cells. The egg has been delimited in **H–J**. (After Boyes and Battaglia 1951a)

indica (K.L. Mathur and Khan 1941), and *C. willmottianum*, *P. coccinea* and *P. scandens* (Boyes and Battaglia 1951b).

The synergids are absent. The wall of the egg cell develops a replica of the wall of synergid (filiform apparatus) in *Plumbago capensis* (Cass 1972). It produces inward projections, and at anthesis the projections develop extensively and some even branch.

Penaea type of embryo sac development is reported in *Statice* (*Limonium*), while *Armeria bupleuroides* exhibits Fritillaria type (Fagerlind 1938b). K.L. Mathur and Khan (1941) suggested a reinvestigation of K.V.O. Dahlgren's work on the occurrence of Adoxa type in the family.

Fagerlind (1938b) and Boyes (1939a, b) describe Plumbagella type in *Plumbagella micrantha*. The three nuclei of the four-nucleate coenomegaspore pass to the chalazal end (1 + 3). During mitotic division the three chalazal spindles fuse so that the embryo sac has two haploid nuclei at the micropylar pole and two triploid at the chalazal. One of the haploid nuclei organizes as egg and the other functions as upper polar nucleus. Of the two triploid nuclei, one fuses with the haploid upper polar and forms a tetraploid secondary nucleus; the other triploid nucleus organizes as the antipodal cell.

Limonium oleaefolium var. *confusum* and *L. transwallianum* are apomicts and show gonial apospory (see Davis 1966).

In *Plumbago zeylanica* (Russell 1980a, 1982), as the pollen grains start to germinate, the lenticular sperm cells decrease in length. The pollen tube emerges within 15 min of pollination, and the pollen tube rapidly passes through the stylar transmitting tissue. As it enters the ovary, the tube becomes wider and callose plugs become conspicuous. The vegetative nucleus precedes the sperm cells in the tube. Finally, the sperm cells become spherical and approach the tip of the pollen tube. The pollen tube enters through the micropyle, and traverses the filiform apparatus in the egg cell. The tube grows between the egg cell and central cell. It extends up to two-thirds of the lower part of egg (Russell and Cass 1978). Then the pollen tube opens by a terminal pore, and the vegetative nucleus, two sperm cells, and the cytoplasm is discharged between the egg cell and central cell (Russell 1980a, b). The plasma membranes of the egg cell and central cell appear to remain intact to exclude the pollen cytoplasm (Russell 1980b).

Syngamy and triple fusion occur with the nuclear transfer; the sperm cytoplasm also enters the egg cell and the central cell. The male cytoplasmic organelles such as plastids and mitochondria occur in both egg cell and central cell. The mitochondria of male origin are smaller and circular, those of female origin larger and elliptical (Russell 1980b).

Russell and Cass (1981) observed that, at least in the mature pollen tube of *Plumbago zeylanica*, the organelles of generative cell are polarized so that the first sperm cell (the leading one), which fertilizes the marginal egg cell, inherits the plastids, and the second sperm cell, ready to fertilize the polar nuclei, inherits the mitochondria. The mode of fertilization has been very critically examined in *Plumbago*.

After fertilization the chalazal cell of the embryo sac degenerates first, followed by the lateral cells.

Russell (1980a, b) concludes that there is a possibility of male cytoplasmic inheritance in the endosperm and embryo of *P. zeylanica.*

Development of the endosperm is of the Nuclear type. In *P. zeylanica* (Russell and Cass 1981), cytokinesis occurs approximately at the 16-nucleate stage, 2 days after anthesis. The endosperm becomes cellular, and stores starch.

The embryogeny corresponds to Period II, Series A, Megarchetype V (Crété 1963a), i.e. Caryophyllad type. The multicellular suspensor is massive (Fig. 44.2.3A–K). The mature embryo is straight.

The small seed is albuminous or exalbuminous; the exotesta persists.

The development of seed requires further attention.

JW Boyes (1939a, b), JW Boyes and E Battaglia (1951a, b), DD Cass (1972), P Crété (1963a), KVO Dahlgren (1937), F D'Amato (1943, 1949), GL Davis (1966), F Fagerlind (1938b, c), AW Haupt (1934), H Hjelmqvist (1967), B-Q Hung and SD Russell (1989), KL Mathur (1940), KL Mathur and

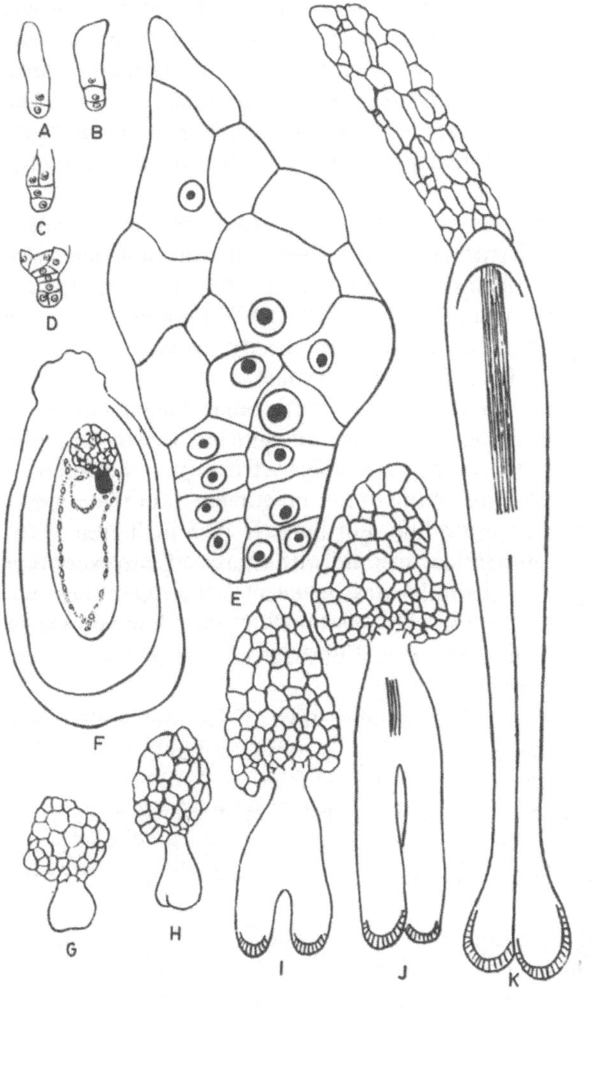

Fig. 44.2.3A–K. *Plumbago europaeum.* **A–E** Early embryogenesis. **F** Longisection of ovule with embryo sac showing proembryo with suspensor and nuclear endosperm. **G, H** Heart-shaped embryo with globular suspensor. **I–K** Dicotyledonous embryo with massive suspensor. (**A–D** After Vaeillet-Bartoszewska 1958b, **E–K** after Yakovlev 1983)

R Khan (1941), SD Russell (1980a, b, 1982, 1983, 1984, 1989), SD Russell and DD Cass (1978, 1981, 1983, 1988), M Veillet-Bartoszewska (1958b), R Wunderlich (1959)

44.3 Taxonomic Considerations

Plumbaginaceae. Bentham and Hooker (1965a) and Hutchinson (1973) include Plumbaginaceae in Primulales along with Primulaceae. Hallier (1912) groups it with Centrospermae. Engler and Prantl (1931), Wettstein (1935), Rendle (1971) and Melchior (1964) recognize their ordinal rank, and their affinity with Centrospermae. R. Dahlgren (1980b) erects the superorder Malviflorae to include the order Plumbaginales.

Primulales (Sect. 43) and Plumbaginales (Sect. 44) show some similarity in embryological features: fibrous endothecium in anther wall, colporate pollen grains, bitegmic ovules, Nuclear endosperm, and albuminous exotestal seed. However, the differences between the two are much more pronounced: Plumbaginales (Plumbaginaceae) have circinotropous crassinucellar ovule, characteristic Plumbago or Plumbagella type of embryo sac, absence of synergids, tetraploid secondary nucleus, filiform apparatus in egg cell, and Caryophyllad type of embryogeny.

Wunderlich (1967) emphasizes that Plumbaginaceae resemble Centrospermales in many respects. Takhtajan (1980) assigns his Caryophyllales and the Plumbaginales under his Subclass Caryophyllidae. Melchior (1964) places Plumbaginales within the Sympetalae. Schnarf (1931, 1933) points out that in respect of embryology the Plumbaginaceae are strictly isolated within the Sympetalae, and correspond with the Centrospermae.

In Plumbaginaceae the micropyle is formed by the inner integument alone, even during post-fertilization stages. The outer integument is two- or three-layered, and the inner integument is always two-layered (except in the region of the micropyle). The outer epidermis of the outer integument, and the inner of the inner integument contain a tannin-like substance (K.V.O. Dahlgren 1916).

These features of the integument correspond with those of Centrospermae. However, there are important dissimilarities between Centrospermales and Plumbaginaceae which do not support this relationship. The Plumbaginaceae show circinotropous ovule, Plumbago and Plumbagella type of embryo sac, absence of perisperm, and straight embryo.

Likewise, Plumbaginales and Polygonales have trinucleate pollen, and a single, basal, bitegmic, crassinucellar ovule in a compound unilocular ovary. The Plumbaginales are placed next to Polygonales in Cronquist's (1981) system. R. Dahlgren (1977b) considers the similarities between these orders, as they have Nuclear type of endosperm, and starch-rich well-developed endosperm in mature seed, but rejects their affinity with Caryophyllales on the basis of chemical data.

R. Dahlgren (1980b) mentions that the position of Plumbaginales is uncertain in the Malviflorae. The occurrence of solitary anatropous ovule, instability in embryo sac formation, and the presence of an obturator suggest their euphorbiaceous affinity, and the structure of its seed-coat has greater similarity with that in the Rhamnaceae or Elaeagnaceae.

The embryological data clearly favour assignment of the family Plumbaginaceae to a separate order, Plumbaginales, or to a superorder Plumbaginiflorae, and indicate their advanced condition over Primulales.

45 Ebenales

The order Ebenales comprises two suborders: Sapotineae with two families – Sapotaceae and Sarcospermataceae, and Ebenineae with five families – Ebenaceae, Styracaceae, Lissocarpaceae, Symplocaceae and Hoplestigmataceae.

45.1 General Features

The Ebenales exhibit the following features: (1) secretory anther tapetum, (2) tricolporate, two-celled pollen grains, (3) anatropous, unitegmic, tenuinucellar ovule, bitegmic in Ebenaceae and *Styrax*, (4) Polygonum type of embryo sac, early degeneration of embryo sac in Ebenaceae, (5) Nuclear type of endosperm, Cellular in Styracaceae and Symplocaceae, and ruminate in Ebenaceae, (b) Caryophyllad type of embryogeny in Styracaceae; polyembryony in Ebenaceae and Symplocaceae, and (7) albuminous testal seed; the testa becomes crushed in Styracaceae and Symplocaceae.

45.2 Sapotaceae

Stomata occur on the anthers of *Mimusops elengi* (S.P. Bhatnagar and Gupta 1969). The stomata are of the anomocytic type, and undergo haplocheilic development. The anther wall comprises the epidermis, fibrous endothecium, two or three ephemeral middle layers, and secretory tapetum with multinucleate cells. The tapetal cells become polyploid due to nuclear fusions. According to Hindmarsh (1965), the tapetum is secretory and irregularly biseriate in *Bassia brachyptera*, *B. divaricata* and *B. patenticuspis*. In *B. paradoxa*, the tapetum is either secretory or "amoeboid" in the sporangia of anthers of the same flower. The protoplasts of the tapetal cells protrude into the locule but do not fuse and form periplasmodium. Such a behaviour can only be considered as intermediate between secretory and amoeboid types.

The microspore mother cells undergo meiotic divisions, cytokinesis is simultaneous, and tetrads tetrahedral, decussate and isobilateral.

The pollen grains are two-, three- or four-colporate with reticulate exine. They are two-celled at the dispersal stage in *M. elengi* (S.P. Bhatnagar and Gupta 1970).

The ovule is anatropous, unitegmic, and tenuinucellate. There is a distinct hypostase in *Madhuca indica* and *Mimusops elengi* (S.N. Murthy 1941a).

Bhatnagar and Gupta contradict Murthy's observations concerning *M. elengi*. The archesporial cell functions as megaspore mother cell, undergoes meiotic divisions, and the tetrad is linear or T-shaped. The embryo sac is of the Polygonum type in *M. elengi* (S.P. Bhatnagar and Gupta 1970). The polar nuclei fuse before fertilization, and antipodals are ephemeral. The chalazal end of the embryo sac extends into a finger-like projection.

Development of the endosperm is of the Nuclear type. In *M. elengi* (S.P. Bhatnagar and Gupta 1970), with the formation of ca. 1000 nuclei, cytoplasm aggregates around the peripheral nuclei at the micropylar end. Wall formation is centripetal and progresses from the micropylar to the chalazal end. The mature endosperm contains a fatty food reserve. While the mature seed of *M. elengi* has a conspicuous endosperm, in other taxa it is scanty or absent (Vaughan 1970, Corner 1976).

In *M. elengi*, even the fully differentiated dicotyledonous embryo has a massive, multicellular suspensor. The mature embryo has a short radicle, and thin, foliaceous, vascularized cotyledons (S.P. Bhatnagar and Gupta 1970).

Yamazaki (1971) describes embryogeny in *Pouteria obovata* var. *dubia*. The zygote elongates, and by irregular divisions forms a few-celled proembryo. The cell *ca*, by a transverse division, forms the tiers *cc* and *cd*. *cc* produces *l* and *l'*, *cd* gives rise to *m* and *n*. The basal cell *cb* forms a multiseriate elongated suspensor.

The mature embryo has two large cotyledons, and a short thick hypocotyl.

Frequently, the young embryonal mass splits at the tip and each branch develops into a mature embryo.

The seeds, often with shiny testa, are medium-sized to large, exarillate, endosperm scanty with oily food reserve, and hilum pale or white, sometimes linear and as long as the seed as in *Achras*.

According to S.P. Bhatnagar and Gupta (1970), in *Mimusops elengi*, of the 25- to 30-layered integument about six outer layers accumulate tannin and transform into sclereids with thick and lignified walls, branched pit canals, and narrow lumen. Meanwhile, tannin continues to accumulate in the inner layers followed by their sclerification. Finally, the entire integument becomes sclerified in the mature seed.

The seed-coat is multiplicative in most layers, including the outer epidermis. The outer part (8–25 or more cells thick) is composed of heavily lignified, sclerotic, pitted cells; the inner part (as thick as the outer), thin-walled and aerenchymatous gets crushed. The outer epidermis comprises small, lignified cells; the inner epidermis gets crushed (Corner 1976).

SP Bhatnagar and M Gupta (1969, 1970), EJH Corner (1976), GJ Hindmarsh (1965), SN Murthy (1941a), JG Vaughan (1970), T Yamazaki (1971)

45.3 Sarcospermataceae

The embryology has not been investigated.

45.4 Ebenaceae

The anther wall comprises the epidermis, endothecium, two middle layers, and multinucleate tapetal cells. The endothecium lacks fibrous thickenings. The microspore mother cells undergo meiotic divisions, cytokinesis is simultaneous, tetrads are tetrahedral, isobilateral and decussate.

The pollen grains are three-colporate with reticulate exine. They are two-celled in *Diospyros virginiana*, but one-celled in *D. kaki* at the dispersal stage. A one-celled condition of mature pollen grain (Haque 1911) is doubtful, and needs confirmation.

The ovule is anatropous, bitegmic, and tenuinucellar; the inner integument forms the micropyle (Fig. 45.4.1B, C). The archesporial cell functions as megaspore mother cell, undergoes meiotic divisions, produces a linear tetrad, and the chalazal megaspore develops into a Polygonum type of embryo sac (Haque 1911). The three antipodals are ephemeral. The endothelium is well developed in *D. kaki* (Yamazaki 1972).

Development of the endosperm is of the Cellular type in *Diospyros virginiana* (Haque 1911) and *D. glaucifolia* (Fig. 45.4.1A). In *D. kaki* (Yamazaki 1972), the division of the primary endosperm nucleus is followed by a

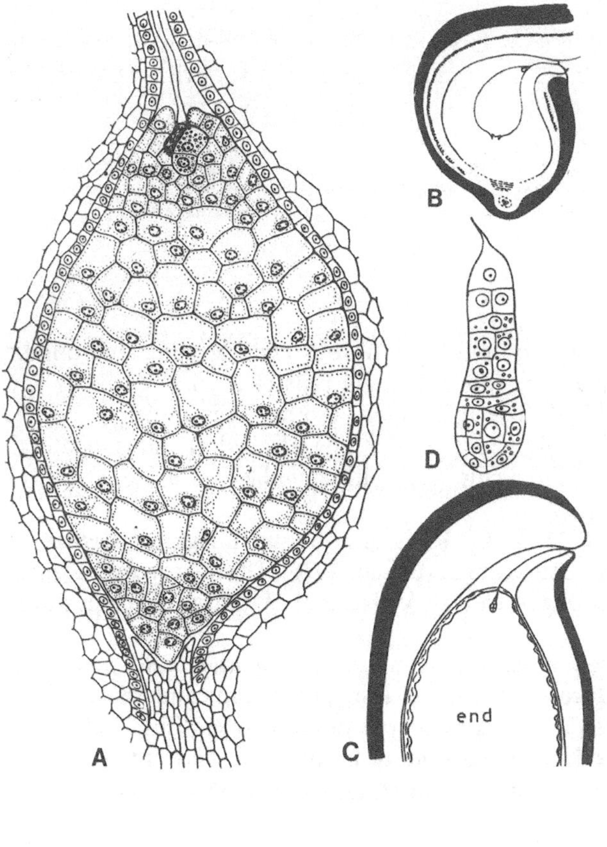

Fig. 45.4.1. **A** *Diospyros glaucifolia*, longisection of ovule with two-celled proembryo, remnants of pollen tube, cellular endosperm, and endothelium. Portion of tegmen surrounds the endosperm. **B** *D. lotus*, longisection of ovule at the stage shown in **A**, **C**, **D** *D. kaki*; **C** Micropylar portion of bitegmic ovule, micropyle formed by the inner integument, globular proembryo, and ruminate endosperm (*end*); **D** Globular proembryo. (After Yakovlev 1983)

transverse wall, another transverse wall in each cell is followed by a longitudinal wall. Further divisions are irregular. The young endosperm presents a ruminate appearance (Fig. 45.4.1C). The mature endosperm is elliptical and stores oil. The irregular development of the seed-coat affects the surface of the endosperm, and makes it ruminate – Annona type (Periasamy 1962a). In *D. chloroxylon*, slight ridges develop from the chalazal part of the seed-coat, and in *D. tomentosa* longitudinal flanges arise from the chalazal end of the seed-coat (Periasamy 1966).

In *Diospyros kaki* (Yamazaki 1972), the zygote divides transversely to form the cells *ca* and *cb*. *ca* undergoes vertical division, followed by transverse division, and gives rise to *l* and *l'*, *cb* divides transversely and forms *cc* and *ci*; *cc* gives rise to *m* and *n*. Figure 45.4.1D shows an early globular proembryo. *ci*, by repeated divisions in both planes, produces a multiseriate massive suspensor. The embryogeny and organogenesis probably correspond to the Polygonad type-Chenopodiad variation or Chenopodiad type.

Polyembryony and freak embryos are reported in *D. virginiana* (Haque 1911). Further details are not available.

In parthenocarpic fruits of *D. virginiana*, Woodburn (1911) mentions seed-like structures which develop "integuments" and "some sort" of "endosperm". Further details have not been provided by the author, and his observations seem to be erroneous.

The albuminous, exarillate, pachychalazal seed contains an embryo with small, thin, foliaceous cotyledons. The testa is multiplicative. The outer epidermis comprises palisade or large radially elongate cells with hexagonal facets, and finely pitted inner and radial walls. The mesophyll is thin-walled. The tegmen becomes mucilaginous and is obliterated (Corner 1976).

EJH Corner (1976), SM Haque (1911), K Periasamy (1962a, 1966), YM Sobajima, M Ishida and A Inaba (1975), J Venkateswarlu (1948), WL Woodburn (1911), T Yamazaki (1972)

45.5 Styracaceae

The anther wall comprises an epidermis, fibrous endothecium, two middle layers, and secretory tapetum with binucleate cells in *Styrax officinalis* var. *californica* (Copeland 1938). The microspore mother cells undergo meiotic divisions, cytokinesis is simultaneous, and tetrads tetrahedral and isobilateral.

The pollen grains are three-colporate with spinuliferous exine, and two-celled at the dispersal stage.

The ovule (Fig. 45.5.1D, F, G, K) is anatropous, bitegmic (*Styrax*) or unitegmic (*Halesia*), and tenuinucellar (Corner 1976). The micropyle is formed by the inner integument (Fig. 45.5.1G). A placental obturator develops at the megaspore mother cell.

The archesporial cell functions as megaspore mother cell, undergoes meiotic divisions, produces a linear or T-shaped tetrad, and the chalazal megaspore develops into a Polygonum type of embryo sac (Fig. 45.5.1A–C, E, F). The polar nuclei fuse at the time of fertilization, and antipodals are ephemeral. The embryo sac contains starch grains (Fig. 45.5.1F, H, I).

Development of the endosperm (Fig. 45.5.1J–L) is of the Cellular type.

The zygote divides about a month after fertilization (Copeland 1938). The development of the embryo conforms to the Polygonad type-Solanad variation, or Solanad type. According to Veillet-Bartoszewska (1960c), the embryogeny in *S. officinalis* refers to Period II, Series C, Megarchetype V (Caryophyllad type). In *Styrax japonicus* (Yamazaki 1970a), the cell *ca* (of the two-celled proembryo) contributes to *pvt* + *pco* + *phy* + *icc* + *iec* + *co* and partly to *s*, and *cb* to suspensor. The suspensor is broad and multiseriate in *S. officinalis.*

The albuminous (oily food reserve) seed has a slightly curved embryo, and a thick vascularized testa which becomes crushed.

HF Copeland (1938), EJH Corner (1976), E Manshard (1936), F Netolitzky (1926), M Veillet-Bartoszewska (1960c), T Yamazaki (1970a)

45.6 Lissocarpaceae

The embryology has not been studied.

45.7 Symplocaceae

In *Symplocos racemosa* (Gulati and Ravishankar 1980, Ravishankar and Gulati 1980), the anther wall comprises the cutinized epidermis, fibrous endothecium, an ephemeral middle layer, and two layers of secretory tapetum with multinucleate cells. The epidermal cells contain crystals and tannin. The microspore mother cells undergo asynchronous meiotic divisions, cytokinesis is successive as well as simultaneous, and the tetrads are linear, T-shaped, isobilateral, decussate or tetrahedral.

The microspores exhibit dimorphism in size, and remain attached in tetrads up to dehiscence. The tri-, or tetraporate pollen grains have spinuliferous exine, and are shed at the two-celled stage. Germination may occur in situ.

The ovule is anatropous, unitegmic and tenuinucellate in *S. chinensis* (Yamazaki 1970b). The hypodermal archesporial cell functions as megaspore mother cell, undergoes meiotic divisions, the tetrad is linear, and the chalazal megaspore develops into a Polygonum type of embryo sac. The antipodals are ephemeral. The endothelium organizes at the eight-nucleate stage of gametophyte.

Development of the endosperm is of the Cellular type. The division of the primary endosperm nucleus, and the subsequent division of daughter nuclei are transverse, followed by irregular divisions. The mature endosperm is ovoidal and contains oil, but no starch.

The embryogeny conforms to the Solanad type in *S. chinensis* (Yamazaki 1970b). The cell *ca* of the two-celled proembryo gives rise to *pvt* + *pco* + *phy* + *icc* + *iec*, whereas *cb* gives rise to *co* and *s*.

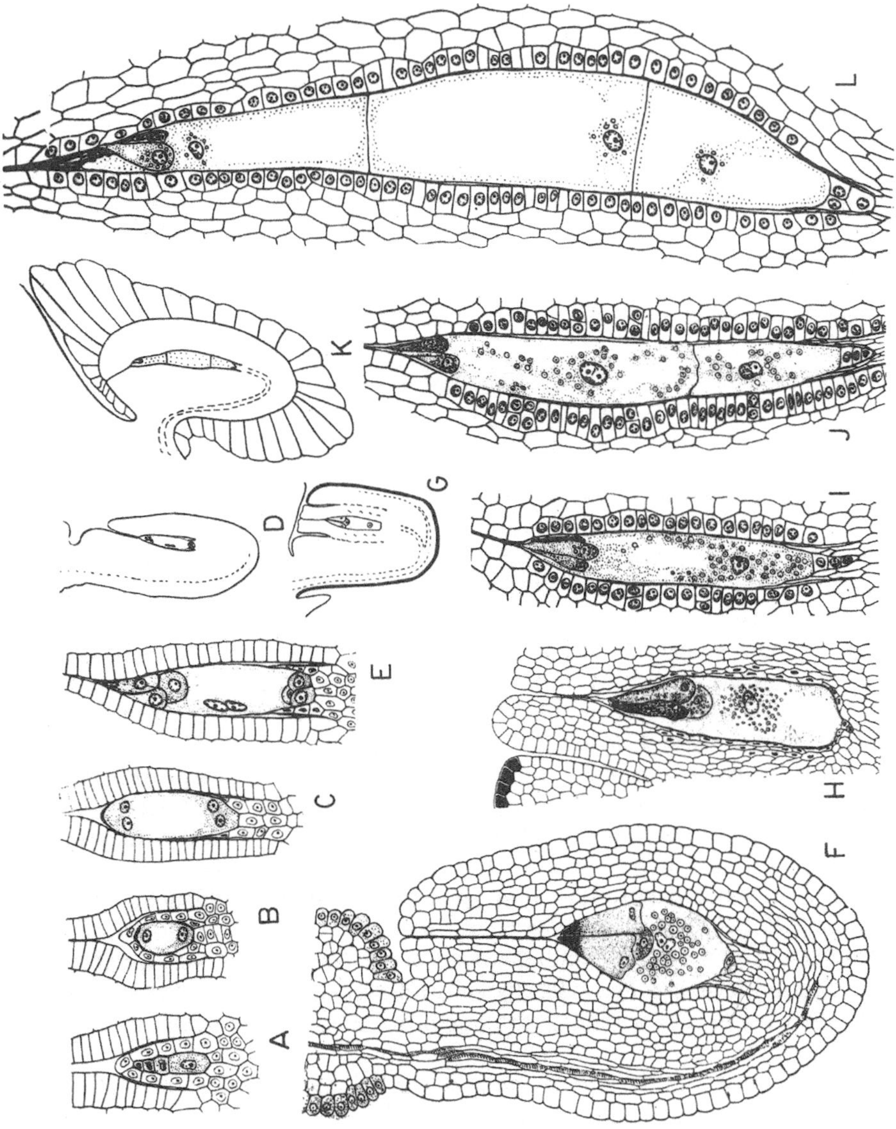
A
B
C
D
E
F
G
H
I
J
K
L

The suspensor consists of five cells; the cell adjoining the embryo proper undergoes vertical division.

The albuminous, exarillate seed has a straight or curved embryo with very short cotyledons. The testa is thin. The seed is enclosed in the stony endocarp (Hutchinson 1973, Corner 1976).

EJH Corner (1976), M Gulati and S Ravishankar (1980), M Gürke (1897), J Hutchinson (1973), H Nagamasu (1989a, b), S Ravishankar and N Gulati (1980), T Yamazaki (1970b)

45.8 Hoplestigmataceae

The embryology has not been investigated.

45.9 Taxonomic Considerations

Bentham and Hooker (1965b), Engler and Prantl (1897), Melchior (1964) and Cronquist (1981) place all the families in one order, Ebenales. Hutchinson (1973) distributes them in two orders: Ebenales – Ebenaceae, Sapotaceae and Sarcospermataceae, and Styracales – Lissocarpaceae, Styracaceae and Symplocaceae. Cronquist (1981) traces the ancestry of Ebenales in Theales. The embryological characters of Sapotaceae, Ebenaceae, Styracaceae and Symplocaceae are compared in Table 45.9.1.

Sapotaceae. The suborder Sapotineae is represented by Sapotaceae. Apart from the characters of the order, this family shows well-developed fibrous endothecium in the anther wall, four-, five- or six-colporate pollen grains with reticulate exine, superior multilocular ovary with a single ascending tenuinucellate ovule in each locule, chalazal extension of the embryo sac (*Mimusops*), absence of endothelium, occasional presence of hypostase, and mesotestal seed. The fruit, a one-seeded berry, contains a pulpy middle zone with a number of

Fig. 45.5.1A–L. A–E *Halesia carolina*, F *Pterostyrax hispidus*, G, H *Styrax obassia*, I–L *Alniphyllum fortunei*. **A** Longisection of nucellus with linear tetrad, upper three degenerate. Note the endothelium in A–C, E. **B**, **C** Bi- and four-nucleate embryo sac. **D** Longisection of anatropous, unitegmic ovule at mature embryo sac. **E** Eight-nucleate embryo sac. **F** Longisection of anatropous, unitegmic ovule with obturator and an eight-nucleate organized embryo sac. **G** Longisection of bitegmic ovule, embryo sac with egg apparatus and secondary nucleus. **H** Portion of ovule to show embryo sac with zygote, remnants of pollen tube, one synergid (the other degenerated), and primary endosperm nucleus. **I** Embryo sac with zygote, remnants of pollen tube and synergid, primary endosperm nucleus, and three antipodal cells. Note the endothelium in I–L. **J** Zygote with degenerated synergid, two-celled endosperm, and three antipodals. **K** Longisection of ovule, embryo sac with zygote and three-celled endosperm. **L** Portion of ovule, embryo sac contains zygote, remnants of pollen tube, degenerated synergid, three-celled endosperm, and three antipodal cells. (See Yakovlev 1983)

Table 45.9.1. Embryology of Sapotaceae, Ebenaceae, Styracaceae and Symplocaceae (Adapted from Yamazaki 1970a, b, 1974)

	Sapotaceae	Ebenaceae	Styracaceae	Symplocaceae
Ovule	Anatropous, unitegmic, tenuinucellar	Anatropous, bitegmic, tenuinucellar	Anatropous, bitegmic, tenuinucellar	Anatropous, unitegmic, tenuinucellar
Endothelium	Absent	Present	Absent	Present
Embryo sac	Polygonum type, antipodals ephemeral	Polygonum, antipodals ephemeral	Polygonum, antipodals ephemeral	Polygonum
Endosperm	Nuclear type	Cellular	Cellular	Cellular
Embryogeny	Polygonad type-Chenopodiad variation	Polygonad type-Chenopodiad variation	Polygonad type-Solanad variation	Polygonad type-Solanad variation
Mature embryo	2 large cotyledons, and a short thick hypocotyl	2 large cotyledons, and a massive straight hypocotyl	2 large cotyledons, and a massive straight hypocotyl	2 small cotyledons, and a curved elongated hypocotyl
Seed-coat	Multiplicative, massive, differentiated into 3 zones	Lignified cells derived from outer integument	Testa massive, later gets crushed, tegmen thin and membraneous	Thin and membraneous, seed enclosed in stony endocarp
Seed	Endosperm scanty, embryo large	Massive ruminate endosperm, embryo moderate size	Endosperm not very massive, embryo large	Endosperm not very massive, embryo large

laticiferous ducts. Corner (1976) compares the massive sclerotic outer part of seed-coat with the sclerotic mesotesta of bitegmic theacous seed where the tegmen becomes obliterated. He considers its affinity with ancestral Actinidiaceae; in both families (Sapotaceae and Actinidiaceae) the ovules are unitegmic.

Ebenaceae. Sapotaceae and Ebenaceae are usually classified together (Table 45.9.1). The Ebenaceae exhibit some features similar to Sapotaceae: reticulate exine of pollen, Polygonum type of embryo sac, and sometimes Nuclear type of endosperm. However, in Ebenaceae the laticiferous tissue is absent, and the flowers are generally unisexual. The multilocular superior ovary contains two (one) pendulous, anatropous, bitegmic ovules in each locule. The endothecium lacks fibrous thickenings. Early degeneration of embryo sac, well-differentiated endothelium, ruminate endosperm, polyembryony, parthenocarpy, and exotestal seed segregate them from the other members of Ebenales.

Styracaceae. Styracaceae show three-colporate, spinuliferous, two-celled pollen grains, and early degeneration of testa similar to that in Symplocaceae. The bitegmic, tenuinucellar condition of ovule (*Styrax*) resembles that in Ebenaceae. However, the superior, half-inferior or inferior, imperfectly septate ovary with parietal placentation in the upper region and axile in the lower region, Cellular type of endosperm, and Solanad type of embryogeny are quite distinct.

Symplocaceae. The position of Symplocaceae is rather controversial. Even though Hutchinson (1973) associates them with Styracaceae in Styracales, R. Dahlgren (1980b) shifts them to Cornales. The assignment of Symplocaceae to Cornales is not at all justified, as only a few embryological characters are common: unitegmic tenuinucellate ovule, and Cellular type of endosperm. Airy Shaw (1973) emphasizes their close relation to Theaceae, and negates their affinity to Styracaceae. Cronquist (1981) confirms the association of Symplocaceae and Styracaceae (Table 45.9.1). Both families have several more or less identical embryological features: (1) tricolporate, spinuliferous, two-celled pollen grains, (2) anatropous, unitegmic (*Halesia* and *Symplocos*), tenuinucellar ovule, (3) Polygonum type of embryo sac, (4) Cellular type of endosperm, (5) probably Solanad type of embryogeny, (6) broad, multiseriate suspensor, (7) albuminous seed; the branches of vascular strands reach up to the micropyle in testa (Corner 1976), and (8) degeneration of testa in mature seed.

The main difference between Symplocaceae and Styracaceae is: in Styracaceae the ovule may be bitegmic (*Styrax*), and an endothelium is absent; whereas in Symplocaceae the ovule is unitegmic, tenuinucellar with a prominent endothelium. This feature indicates its similarity to the ovules of Ericales or Tubiflorae. On the other hand, Theaceae have bitegmic ovules, Allium type of embryo sac, Nuclear type of endosperm, and exalbuminous mesotestal seed. An evaluation of embryological features indicates that Symplocaceae are related to Styracaceae rather than to Theaceae.

46 Oleales

The family Oleaceae represents the order Oleales.

46.1 General Features

Oleaceae exhibit the following features: (1) secretory anther tapetum, (2) simultaneous cytokinesis in microspore mother cells, (3) two- or three-celled, tricolporate pollen grains, (4) anatropous, unitegmic, tenuinucellar ovule with persistent endothelium around the lower half of embryo sac, (5) mono- and bisporic (Allium type – *Olea*) embryo sac with hooked synergids, and small ephemeral antipodals, (6) Nuclear and Cellular type of endosperm, absence of endosperm haustoria, (7) Caryophyllad and Solanad type of embryogeny, and (8) albuminous exotestal seed, and straight embryo with a long uniseriate suspensor.

46.2 Oleaceae

The anther wall comprises the epidermis, fibrous endothecium, one middle layer, and glandular tapetum in *Jasminum calophyllum*. The endothecium is biseriate in *J. angustifolium* and *J. sambac* (Maheswari Devi 1975). The tapetal cells are multinucleate, nuclear fusions occur and the nuclei become polyploid. The tapetum is irregularly biseriate in *Fraxinus velutina* (Copeland 1960) and *Jasminum pubescens* (Y.R. Bhargava 1980). The microspore mother cells undergo meiotic divisions, cytokinesis is simultaneous and tetrads tetrahedral, isobilateral, occasionally T-shaped and linear.

The pollen grains are three- or four-colpate to colporoidate, and two-celled at the dispersal stage. They are three-celled in *Jasminum calophyllum*. In some species of *Jasminum*, the pollen grains are multinucleate, polymorphic or occur in chains. In *Olea europaea* (Pacini and Franchi 1982), the pollen grains sometimes germinate in situ.

The ovule is anatropous, orthotropous in *Linociera intermedia*, unitegmic and tenuinucellar. The archesporium is multicellular in *Fraxinus*, and one- to multicelled in *Jasminum*. A single cell functions as megaspore mother cell, undergoes meiotic divisions, and the tetrad is linear or T-shaped. The chalazal megaspore develops into a Polygonum type of embryo sac in *Syringa villosa* (Ivanov and Sharenkova 1970). Multiple embryo sacs develop in *Jasminum* (Maheswari Devi 1975). The polar nuclei fuse before fertilization, and antipodals are ephemeral. An endothelium differentiates, and at times it covers only the lower half of the embryo sac. Bisporic (Allium) type of embryo sacs occur in *Olea chrysophylla* (Anderson 1931), and *O. europaea* (Anderson 1931, King 1938, Messeri 1952).

Development of the endosperm is of the Cellular type in *Syringa vulgaris* (Stolt 1921, see Schnarf 1929). The first two divisions are transverse.

In *Jasminum* spp. (Maheswari Devi 1975), the development of the endosperm is of the Nuclear type. Cytokinesis occurs at the 32- to 64-nucleate stage, and proceeds from the micropylar to the chalazal region. Finally, the entire tissue becomes cellular. In *J. callophylum* the intervening walls of the multiple embryo sacs break down and the adjoining endosperms fuse. This aspect needs further study.

Due to irregular growth of the inner epidermis of seed-coat, the endosperm becomes ruminate in *Linociera intermedia.*

The embryogeny corresponds to Period II, Megarchetype III, Group 11 (i.e. Caryophyllad type) in *Ligustrum ovalifolium* (Souèges 1942b). Maheswari Devi (1975) reports Solanad type-Nicotiana variation in *Jasminum calophyllum.* Polyembryony occurs in *Fraxinus velutina* (Copeland 1960), and *J. callophylum.* In the latter, the additional embryo may originate from the fertilized synergid.

The seeds are small, exarillate, albuminous (endosperm cells thick-walled), with a straight embryo. In *Olea europaea* (Vaughan 1970), the reserve food in the endosperm and embryo is oil and aleurone grains. The seed-coat is exotestal, the outer epidermis is a palisade of cells with thick outer walls, and a thick-walled inner epidermis (endothelium) persists. Sometimes, the enlarged endothelial cells project into the endosperm (Corner 1976).

A Anderson (1931), G Bentham (1876a), YR Bhargava (1980), HF Copeland (1960), EJH Corner (1976), VSR Das and KN Rao (1966), MC Fernandez and MI Rodríguez-Garcia (1989), K George and S Geethamma (1984), MA Ivanov and EA Sharenkova (1970), JR King (1938), PM Kuriachen and YS Dave (1989), K Lakshminarayana (1978a), H Maheswari Devi (1975), A Messeri (1952), E Pacini and GG Franchi (1982), NK Patel (1960), K Schnarf (1929), R Souèges (1942b, c), GP Steinbauer (1943), KAH Stolt (1921), JG Vaughan (1970), A Vijayalakshmi and KK Lakshmanan (1985)

46.3 Taxonomic Considerations

Oleaceae. Bentham and Hooker (1965b) group Oleaceae in Gentianales, Engler and Diels (1936) in Contortae – suborder Olineae, R. Dahlgren (1980b) in Oleales – superorder Gentianiflorae, and Takhtajan (1980) in Oleales. Hutchinson (1973) places Oleaceae in Loganiales along with Loganiaceae. Cronquist (1981) assigns it to Scrophulariales.

Oleaceae show their alliance with Menyanthaceae. Both have glandular anther tapetum, simultaneous cytokinesis in pollen mother cells, anatropous, unitegmic, tenuinucellar ovule, endothelium, Polygonum type of embryo sac with ephemeral antipodals, and Cellular type of endosperm. From Menyanthaceae (Sect. 47.5) Oleaceae differ in various types of pollen grains (three- to four-colporate or colporoidate), occasional bisporic type of embryo sac, and Caryophyllad type of embryogeny (Period II, Megarchetype III). According to Crété (1963a), the embryogeny of Oleaceae is quite distinct, and has no parallel in Sympetalae.

Oleaceae resemble Loganiaceae in glandular anther tapetum, three-celled pollen grains, unitegmic tenuinucellar ovule, Polygonum type of embryo sac,

Nuclear as well as Cellular type of endosperm, and exotestal seed. As mentioned earlier, Oleaceae have three- to four-colpate, colporoidate pollen grains in contrast to tri- or penta-aperturate pollens of Loganiaceae, multicellular female archesporium, occasional bisporic type of embryo sac, and unique type of embryogeny. Loganiaceae show both secretory and amoeboid anther tapetum, and Nuclear as well as Cellular type of endosperm.

While Oleaceae resemble Menyanthaceae as well as Loganiaceae, the Oleaceae show some characteristic features which justify raising them to an ordinal rank – Oleales. In addition, the occurrence of seco-iridoids (Jensen et al. 1975) with the oleuropein group, mostly restricted to this family, strengthens the position in a separate order near the Gentianales.

De Jussieu (1789) and Takhtajan (1980) assign the genus *Nyctanthes* to Oleaceae; Agardh (1858, see Hutchinson 1973) erects the family Nyctanthaceae. Airy Shaw (1951) proposes the transfer of *Nyctanthes* from Oleaceae to Verbenaceae, and Melchior (1964) agrees with this assignment. Kapil and Vani (1966) evaluate the morphological, anatomical and embryological features of *Nyctanthes arbortristis* with Oleaceae and Verbenaceae, and conclude that *Nyctanthes* should be retained in Oleaceae. George and Geethamma (1984) investigated the cytology, floral morphology and stomatal type (anomocytic) of *N. arbortristis* and confirm its assignment to Oleaceae. The taxonomic assignment of this genus is further considered under Verbenaceae (Sect. 48.8).

47 Gentianales

Gentianales comprise the families Loganiaceae, Desfontainiaceae, Gentianaceae, Menyanthaceae, Apocynaceae, Asclepiadaceae and Rubiaceae.

47.1 General Features

Gentianales exhibit the following features: (1) secretory anther tapetum; secretory and amoeboid types occur in Loganiaceae and Gentianaceae, (2) simultaneous cytokinesis in pollen mother cells, simultaneous as well as successive in Apocynaceae and Asclepiadaceae, (3) two-celled pollen grains; pollinia in Asclepiadaceae, (4) unitegmic tenuinucellar ovule, (5) Polygonum type of embryo sac, mono-, bi- and tetrasporic in Rubiaceae, and mono- and bisporic in Asclepiadaceae, (6) ephemeral antipodals, secondary multiplication in Gentianaceae, Asclepiadaceae and Rubiaceae, and haustorial in Rubiaceae, (7) Nuclear endosperm in Gentianaceae and Apocynaceae, Nuclear and Cellular type in Loganiaceae, Asclepiadaceae and Rubiaceae, and Cellular type in Menyanthaceae, (8) endosperm haustoria in Loganiaceae, Apocynaceae and Rubiaceae; ruminate condition in Loganiaceae, Apocynaceae and Rubiaceae, (9) embryogeny of the Solanad type in Gentianaceae, Asclepiadaceae and

Rubiaceae, Asterad in Menyanthaceae, and Asterad, Onagrad, Caryophyllad and Chenopodiad in Apocynaceae, and (10) albuminous and exotestal seed.

47.2 Loganiaceae

The anther is bisporangiate. The wall comprises the epidermis, fibrous endothecium, one or two middle layers, and secretory tapetum with uni- or binucleate cells in *Strychnos innocua* (Bendre 1975). Due to nuclear fusions the tapetal cells become uninucleate with polyploid nuclei in *S. nux-vomica* (Maheswari Devi and Lakshminarayana 1980). The tapetum is biseriate in the connective region. An amoeboid condition is reported in *Fagraea fragrans, Strychnos laurina, S. sansibariensis* (Mohrbutter 1936), *Mitrasacme alsinoides* (Yamazaki 1963), and *Spigelia splendens* (Juel 1915). Stellate crystals occur in the hypodermal cells of the stomial groove in *S. innocua* (Bendre 1975). Simultaneous cytokinesis follows meiotic divisions in the microspore mother cells, and the tetrads are tetrahedral, isobilateral and decussate.

The pollen grains are tri- to penta-aperturate with a smooth exine. They are shed at the three-celled stage in *Strychnos potatorum* (Maheswari Devi and Lakshminarayana 1980).

The ovule is anatropous, unitegmic, and tenuinucellar (Fig. 47.2.1A). An endothelium differentiates in *Polypremum procumbens* (Moore 1948) and *Strychnos potatorum* (Maheswari Devi and Lakshminarayana 1980). The archesporial cell functions as megaspore mother cell, undergoes meiotic divisions, and the chalazal megaspore of the linear tetrad (Fig. 47.2.1B) develops into a Polygonum type of embryo sac (Fig. 47.2.1C, D). In *P. procumbens*, the tip of the embryo sac extends into the micropyle and reaches up to the surface of the ovule. The antipodals are ephemeral.

Development of the endosperm is of the Nuclear type in *Fagraea fragrans* (Mohrbutter 1936), *Spigelia splendens* (K.V.O. Dahlgren 1922) and *Strychnos nux-vomica* (Maheswari Devi and Lakshminarayana 1980). Synchronous free-nuclear divisions follow the division of the primary endosperm nucleus. The endosperm nuclei are distributed on the periphery of the central vacuole of the embryo sac. Cytokinesis occurs at the 16- or 32-nucleate stage of endosperm.

In *P. procumbens* (Moore 1948), the endosperm is of the Cellular type. Division of the primary endosperm nucleus is followed by a transverse wall (Fig. 47.2.1E). Each cell undergoes a vertical division to produce two tiers of two cells each (Fig. 47.2.1F). Subsequent divisions are irregular. The micropylar and chalazal portion of the endosperm is cylindrical, and are often described as "necks". These comprise small and densely cytoplasmic cells. The cells in the central region are large and vacuolate, and form the main body of the endosperm (Fig. 47.2.1G, H). The endosperm becomes ruminate in *Spigelia splendens* (Periasamy 1962a), due to the ingrowths of the seed-coat.

According to Yamazaki (1963), in *Mitrasacme alsinoides* var. *indica* the division of the primary endosperm nucleus is followed by a transverse wall (Fig. 47.2.2A, B). Both the cells again divide transversely, and a row of four cells is formed (Fig. 47.2.2C). In each of the two middle cells a small lateral cell

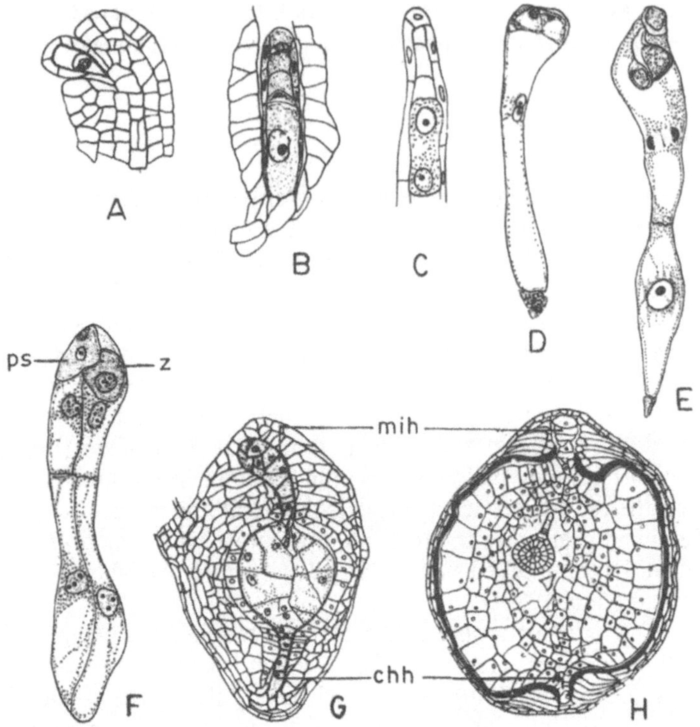

Fig. 47.2.1A–H *Polypremum procumbens.* **A** Megaspore mother cell. **B** Tetrad, chalazal megaspore functional. **C** Binucleate embryo sac. **D** Mature embryo sac. **E** Zygote, two-celled endosperm, nucleus of micropylar chamber in division. **F** Zygote (*z*), and four-celled endosperm; vertical division in micropylar and chalazal chamber; *ps* persistent synergid. **G** Immature seed (l.s.) to show endosperm proper with multicellular micropylar (*mih*) and multicellular chalazal (*chh*) haustorium. **H** Immature seed with endosperm at globular proembryo. Remnants of micropylar and chalazal haustorium persist. The inner wall of the endothelium has thickened. (After Moore 1948)

is cut off on the funicular side. Free-nuclear divisions occur in the terminal cells (Fig. 47.2.2D, E) which become 16-nucleate; the remaining four cells (two large, and two small lateral cells) become binucleate. Cytokinesis occurs in four large cells (Fig. 47.2.2F, G). Yamazaki considers the small cells as haustorial, without any justification. He erroneously interprets this mode of development as of the Helobial type; Davis (1966) considers it "modified Helobial" type, and points out that the lateral cells comprise a non-aggressive haustorium. The development of the endosperm is undoubtedly of the Cellular type (see also Swamy 1974a), and the two lateral binucleate cells are of no significance. Further detailed investigations would be rewarding.

In *Mitrasacme alsinoides* var. *indica* (Yamazaki 1963), the embryogeny corresponds to the Solanad type-Nicotiana variation. The suspensor is two- or three-celled. Yamazaki contradicts this account since it conforms to the Caryophyllad type (P. Maheswari 1950). The development of the embryo should be reinvestigated.

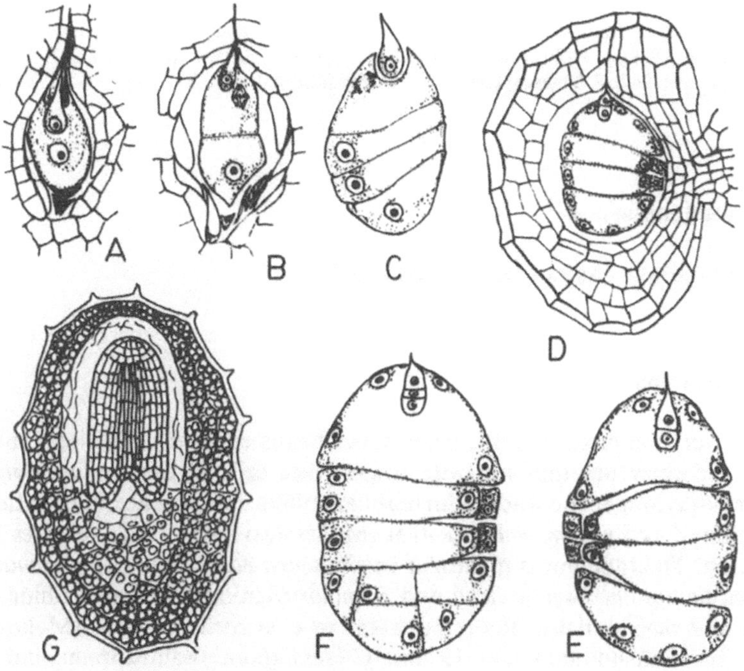

Fig. 47.2.2A–G. *Mitrasacme alsinoides*. **A** Longisection of nucellus, embryo sac contains zygote and primary endosperm nucleus. **B** Zygote and two-celled endosperm. **C** Zygote and uniseriate four-celled endosperm. **D** Longisection of ovule with two-celled proembryo and six-celled endosperm. Free-nuclear divisions have occurred in the micropylar and chalazal chamber. Each of the two middle cells has cut off an uninucleate lateral cell. **E** Embryo sac from **D**. **F** Three-celled proembryo, multicelled endosperm, micropylar chamber coenocytic, chalazal chamber cellular, two middle cells two- and three-nucleate, and lateral cells binucleate. **G** Longisection of seed with dicotyledonous embryo and cellular endosperm. *Thick* inner epidermis (testa) has prominent projections on lateral walls. (After Yamazaki 1963)

The small to rather large, albuminous, exotestal seed contains a minute embryo. The testa comprises the outer epidermis of lignified cells, the inner layers collapse. In *M. alsinoides* var. *indica* (Yamazaki 1963), the outer tangential walls of the epidermis are thin-walled and collapse. The inner tangential and lateral walls are thick, and the lateral walls appear as projections on the seed surface. The mesophyll is crushed. In *Fagraea* the outer wall of the outer epidermis is mucilaginous; in *Polypremum* (Moore 1948) the inner epidermis persists as thickened inner wall.

In *Strychnos* the seed is orbicular lenticular with a tuft of unbranched hairs. The outer epidermal cells are radially elongate with pitted radial inner walls; their outer walls elongate into lignified hairs. The hairs may branch dichotomously in *S. nux-vomica*. The endosperm is thick-walled and oily. The embryo is small with thin flat cotyledons. The cotyledons are thick in *S. potatorum* (Corner 1976).

AM Bendre (1975), EJH Corner (1976), KVO Dahlgren (1922), GL Davis (1966), P Dop (1913), HO Juel (1915), P Maheshwari (1950), H Maheswari Devi and K Lakshminarayana (1980), S Maldonado (1989), C Mohrbutter (1936), RJ Moore (1948), K Periasamy (1962a), W Punt and PW Leenhouts (1967), R Souèges (1940b), HG Svensson (1925), BGL Swamy (1974a), T Yamazaki (1963)

47.3 Desfontainiaceae

The embryology has not been investigated.

47.4 Gentianaceae

The anther wall comprises the epidermis, fibrous endothecium, single middle layer, and secretory tapetum with uninucleate cells in *Centaurium ramosissimum* (Vijayaraghavan and Padmanabhan 1969). The absence of endothecial thickenings in *Exacum pedunculatum* (Maheswari Devi 1962) requires confirmation. The tapetum is glandular in *Conscora pauciflora, E. pedunculatum, Hoppea fastigiata, Swertia lawii* and *S. minor* (Sankara Rao and Chinnappa 1983), and plasmodial at the tetrad stage in *Conscora decurrens* (Maheswari Devi and Lakshminarayana 1977a), *C. sessiliflora* (Balasubramanian and Padmanabhan 1985), *Gentiana pedicellata* and *G. quadriferia*. The remnants of plasmodium persist till the maturation of pollen grains in *G. pedicellata* and *G. quadriferia* (Sankara Rao and Chinnappa 1983), whereas they are absorbed much earlier in *C. diffusa* (Maheswari Devi 1962).

In *C. pauciflora* (Sankara Rao and Chinnappa 1983), the tapetum is derived from the parietal layer and it invests the massive sporogenous tissue. Some of the sporogenous cells are functional, whereas others form sterile patches which divide the sporogenous tissue. The sterile cells elongate and develop more conspicuously towards the placentoids. These trabeculate cells are uninucleate and devoid of cytoplasm.

Following reduction divisions in microspore mother cells, cytokinesis is simultaneous. The pollen grains remain united in tetrads in species of *Adenolisianthus, Calolisianthus, Helia, Irlbachia, Macrocarpaea, Pagaea, Prepusa* and *Symbolanthus*, or in polyads in species of *Calolisianthus, Chelonanthus, Irlbachia, Lisianthus* and *Pagaea* (Sankara Rao and Chinnappa 1983). In other genera the pollen grains are single. The pollen grains are tri-, tetra-, or penta-colporate or pericolpate, or one- to six-porate with smooth, scabrous, striate, striato-reticulate or reticulate surface patterns. They are two- or three-celled at the shedding stage in *Conscora diffusa*, and both conditions occur in the same anther. The two-celled condition at maturity needs confirmation.

The ovule is anatropous, unitegmic, and tenuinucellar. The archesporium is uni- to multicellular. A single archesporial cell functions as megaspore mother cell, undergoes meiotic divisions, and the chalazal megaspore of the linear tetrad develops into a Polygonum type of embryo sac. An endothelium differentiates in *Exacum pumilum*. The synergids are beaked with long tapering ends; the polar nuclei fuse before fertilization. The antipodals persist in *Centaurium*, are weakly developed in *Swertia carolinensis* (McCoy 1949), and ephemeral in *S. corymbosa* (Sankara Rao 1975). According to Maheswari Devi and Lakshminarayana (1977a), there are only two antipodals (one binucleate) in *Swertia*. Secondary multiplication of antipodals occurs in *S. carolinensis* and *S. minor* (Sankara Rao and Nagaraj 1982). The latter species has six to eight antipodals, their tapering chalazal end penetrates the integument. The nuclei divide and fuse to form polyploid masses. They persist till the initiation of cytokinesis in endosperm, and four-celled proembryo.

Development of the endosperm is of the Nuclear type. Centripetal wall formation sets in at the 16- or 32-nucleate stage in *Centaurium*; a large number of free nuclei develop in *Canscora* and *Swertia* before the initiation of walls. After the formation of many nuclei, simultaneous wall formation occurs in *Hoppea dichotoma* (Arekal 1961a), throughout the endosperm at four- to six-celled proembryo.

Development of the embryo conforms to Period II, Series C, Megarchetype V (Caryophyllad type) in *H. dichotoma* (Arekal 1961a) and *Swertia minor* (Sankara Rao and Nagaraj 1982). Maheswari Devi (1962) and Maheswari Devi and Lakshminarayana (1977a) consider the embryogeny to be of the Solanad type-Physalis II variation. Synergid (unfertilized) polyembryony has been reported in *Swertia angustifolia* (Maheswari Devi and Lakshminarayana 1977a).

The small seed contains fatty food reserve in endosperm, and a minute embryo. The outer epidermis of the seed-coat has thickened radial and inner walls. The mesophyll and inner epidermis are thin-walled and often become crushed.

GD Arekal (1961a), V Balasubramanian and D Padmanabhan (1985), F Bouman and A Louis (1989), F Bouman and S Schrier (1979), P Crété (1949a, b, c, 1955d, 1959), GL Davis (1966), U Drexler and MI Hakki (1979), K Lakshminarayana and H Maheswari Devi (1985), H Maheswari Devi (1962), H Maheswari Devi and K Lakshminarayana (1977a, 1978), G Maiti and ML Banerji (1976), RW McCoy (1949), K Sankara Rao (1975, 1978), K Sankara Rao and CC Chinappa (1983), K Sankara Rao and M Nagaraj (1982), MR Vijayaraghavan and U Padmanabhan (1969)

47.5 Menyanthaceae

The anther wall comprises the epidermis, fibrous endothecium, one or two middle layers, and secretory tapetum with multinucleate cells. Nuclear fusions occur in tapetal cells. The microspore mother cells undergo meiotic divisions, cytokinesis is simultaneous, and tetrads tetrahedral.

The triporate, spiraperturate pollen grains are three-celled at the dispersal stage.

The ovule is anatropous, unitegmic, and tenuinucellar. The funicular vascular strand extends up to the micropyle. An archesporial cell functions as megaspore mother cell, undergoes meiotic divisions, produces a linear tetrad, and the chalazal megaspore develops into a Polygonum type of embryo sac. The polar nuclei fuse before fertilization. The antipodals are ephemeral in

Limnanthemum cristatum (Srinivasan 1941) and *Nymphoides peltatum* (Stover 1932), whereas they persist in *L. indicum* and *Menyanthes trifoliata.* An endothelium differentiates; a hypostase is reported in *L. cristatum.*

Development of the endosperm is of the Cellular type; further details are not available (see Maheswari Devi 1962). According to Davis (1966), in *Villarsia reinformis* a row of eight cells is formed before longitudinal divisions occur.

The development of the embryo conforms to Period I, Series A, Megarchetype II (Asterad type) in *Menyanthes trifoliata* (Souèges 1943c). Maheswari Devi (1962) also reports Asterad type in *L. cristatum* and *L. indicum*, and synergid polyembryony in *L. indicum.*

The seed is small and albuminous with a minute embryo; the endosperm contains starch and "albumin crystals" (Maheswari Devi 1962). The cells of the outer epidermis of the seed-coat have thickened and pitted outer walls. The mesophyll and inner epidermis contain several rows of sclerotic cells in *Menyanthes* (Corner 1976). The seed is hairy in *Nymphoides peltatum.*

EJH Corner (1976), P Crété (1956b), GL Davis (1966), BM Johri (1963b), AA Lindsey (1938), H Maheswari Devi (1962), R Souèges (1943c), AR Srinivasan (1941), KAH Stolt (1921), E Stover (1932)

47.6 Apocynaceae

The anther wall comprises the epidermis, fibrous endothecium, two middle layers, and secretory tapetum with uninucleate cells in *Anodendron paniculatum* (Balasubramanian 1983). The endothecium is weakly developed in *Cerbera odollam* and *Vallaris heyneii* (Anantaswamy Rau 1940b), whereas it is five- to six-layered in *Voacanga foetida* (Maheswari Devi 1971). There are ten to 14 wall layers in *V. foetida.* The microspore mother cells undergo reduction divisions, and simultaneous cytokinesis occurs by furrowing. Successive cytokinesis is reported in *Apocynum androsaefolium, Holarrhena antidysenterica* (Latto 1974), and *Plumeria diffusa* and *P. rubra* (Chauhan 1979). Successive as well as simultaneous cytokinesis is reported in *Catharanthus pusillus* (Bhasin 1971) and *Rauvolfia canescens* (Meyer 1938). The tetrads are tetrahedral and isobilateral. Decussate, T-shaped and linear tetrads have been observed in *H. antidysenterica* (Latto 1974).

The pollen grains are triporate and three-celled at the dispersal stage (Maheswari Devi 1970). They are two-celled in *Catharanthus, Holarrhena* and *Plumeria.* Frequently, they germinate in situ in *Plumeria* (Chauhan 1979).

The ovule is hemianatropous or anatropous, orthotropous in *Rauvolfia sumatrana* (Maheswari Devi and Lakshminarayana 1980), unitegmic and tenuinucellar. The ovules show a series of reductions (Guignard 1917a) so that from a one-layered nucellus the lateral cells get crushed but the apical cells persist, and finally only a few apical nucellar cells persist (see also Sect. 47.8) but are no more separately distinguishable from the integument. Eames (1961) considers this feature as a "vestigial" character.

The archesporium is one-celled, but rarely two- to three-celled in *Mascarenhasia micrantha* (Maheswari Devi and Lakshminarayana 1980). The archesporial cell functions as megaspore mother cell, undergoes meiotic divisions, the tetrad is linear or T-shaped, and the chalazal megaspore develops into a Polygonum type of embryo sac. The synergids are elongated in *Carissa carandas, Vallaris heyneii* (Anantaswamy Rau 1940b), and *Holarrhena.* The polar nuclei fuse before fertilization; the antipodals are ephemeral. Starch grains accumulate in the embryo sac. An endothelium also differentiates. Occasionally, twin and abnormal embryo sacs with fewer or more than eight nuclei occur, and degeneration of the embryo sacs is common.

The development of the endosperm is of the Nuclear type. It becomes cellular by centripetal wall formation during the globular stage of the proembryo. A chalazal haustorium develops in *Catharanthus pusillus, C. roseus* (Maheswari Devi 1971) and *Lochnera pusilla* (Murty and Chauhan 1966). The endosperm is ruminate in *Ervatamia heyneana* and *Voacanga grandifolia* (Periasamy 1962a, 1963).

The development of the embryo conforms to the Asterad type in *Voacanga foetida*; Onagrad type in *Catharanthus roseus, C. pusilus* and *Plumeria* (Chauhan 1979); Chenopodiad type in *Rauvolfia tetraphylla* (Maheswari Devi 1971); Caryophyllad type in *C. roseus* (Bhasin 1971), *Holarrhena antidysenterica* (Latoo 1974), *Lochnera rosea* (Dnyansagar and Sudhakaran 1977), and *Nerium oleander*. Maheswari Devi (1971) reported synergid polyembryony in Plumiereae; whether n or 2n is not mentioned.

The seed shows ruminate endosperm in *Ervatamia* and *Voacanga* (Periasamy 1963). During post-fertilization stages a massive chalazal growth leads to the elongation of seed. A rapid meristematic activity also occurs in the hypodermal layers of chalazal and integumentary regions. The epidermis divides only anticlinally, and inwardly directed infoldings appear. The endosperm replaces the integumentary portion, except its epidermis, and acquires rumination (Fig. 3.10.4F) as in *Spigelia* (K.V.O. Dahlgren 1922).

The endosperm is thick-walled and ruminate in *Neokeithia* (rumination due to activity of integument as in *Voacanga*). The straight embryo has flat or folded cotyledons.

The seed is small or large, albuminous, elate or arillate, or with silky hairs. The epidermal cells of seed-coat develop into long hairs as in *Strophanthus*, or groups of cells show projections as in *Catharanthus*.

In *Chilocarpus* the funicular non-vascularized aril is smooth or lobulate, and comprises large cells with lignified walls. The seed-coat comprises the outer epidermis of the testa composed of cuboid cells with thickened, lignified brown walls (Corner 1976).

Tabernaemontana shows a waxy, vascularized aril. It develops from the funicle and raphe round the base of the ovule. The cells contain orange-coloured oil globules. Later, it becomes mucilaginous. The seed-coat comprises the outer epidermis of clavate, thick-walled cells with brown contents. Due to excessive anticlinal divisions, the outer epidermis is thrown into folds, and the thick-walled and oily endosperm becomes ruminate (Periasamy 1963). The embryo is straight with subcordate cotyledons (Corner 1976).

In *Lepinia* the seed is exarillate and albuminous, and the epidermal cells have brown walls. The endosperm shows thick pitted walls. The embryo is straight with flat cotyledons. The integument extrudes between the vascular bundles of the placenta; the endosperm follows its contour and gives a false appearance of rumination (Corner 1976).

M Anantaswamy Rau (1940a, b), A Anderson (1931), V Balasubramanian (1980, 1982, 1983), N Banerji and AK Sharma (1988), RK Bhasin (1971), F Bouman and S Schier (1979), TS Chauhan (1979), EJH Corner (1976), P Crété (1956b), KVO Dahlgren (1922), M De and AK Sharma (1983), VR Dnyansagar and IV Sudhakaran (1977), AJ Eames (1961), ME Fallen (1985), TC Frye and EB Blodgett (1905), GL Guignard (1917a), PSH Khan (1970), K Lakshminarayana (1988), LC Lamba (1974), CS Latto (1974), H Maheswari Devi (1962, 1970, 1971), H Maheswari Devi and K Lakshminarayana (1980), S Meyer (1938), YS Murty and TS Chauhan (1966), K Periasamy (1962a, 1963), GS Srivastava and V Chandra (1958), KC Sud (1984a, b)

47.7 Asclepiadaceae

The anther is usually bisporangiate, tetrasporangiate in *Secamone* (see Davis 1966). The wall comprises the epidermis, four or five middle layers, and secretory tapetum. Fibrous endothecium is reported in *Ceropegia elegans, Vincetoxicum nigrum* (see Davis 1966), *Caralluma attenuata, Cryptostegia grandiflora* (Venkata Rao and Rama Rao 1954), *Ceropegia juncea, Pergularia daemia* and *Tylophora indica* (Maheswari Devi 1964). Fibrous endothecium is absent in *Calotropis gigantea* (Maheswari Devi 1964), *Daemia extensa* (Biswas 1957) and *Oxystelma esculentum* (Maheswari Devi and Lakshminarayana 1977b). The tapetum is uniseriate with multinucleate cells in *Leptadenia reticulata, Pentatropis spiralis* and *Tylophora cordifolia* (Mulay et al. 1965). It is multiseriate with uninucleate cells in *Cynanchum callialata, Pergularia daemia* and *T. indica* (Maheswari Devi 1964), and multiseriate with multinucleate cells in *Oxystelma esculentum.*

The primary sporogenous tissue functions as microspore mother cells as in *Oxystelma*, or may undergo secondary increase as in *Cryptostegia*. The mother cells undergo meiotic divisions, cytokinesis is simultaneous in *Cryptostegia* (Venkata Rao and Rama Rao 1954) and *Hemidesmus* (Maheswari Devi 1964), and successive in *Caralluma, Daemia*, Cynanchoideae (Maheswari Devi 1964) and *Oxystelma*. The tetrads are tetrahedral and linear. Nirula and Richharia (1945) reported linear, T-shaped, tetrahedral and isobilateral tetrads in *Hemidesmus indicus.*

The pollen grains are three-celled, and remain aggregated in pollinia. In *Pergularia daemia* (Vijayaraghavan and Shukla 1976), a pair of pollinia is attached to the translator (Fig. 47.7.1A). The translator comprises (a) main-body (B, C), (b) two lateral blades on either side (B, C), and (c) a pair of retinaculum (B). The retinaculum helps in the attachment of the lateral blade to the acellular portion of the pollinium (A). The pollinium is covered by a persistent tapetal membrane of sporopollenin. It shows stronger autofluorescence (Fig. 47.7.1D) than the pollen grains (Vijayaraghavan and Shukla

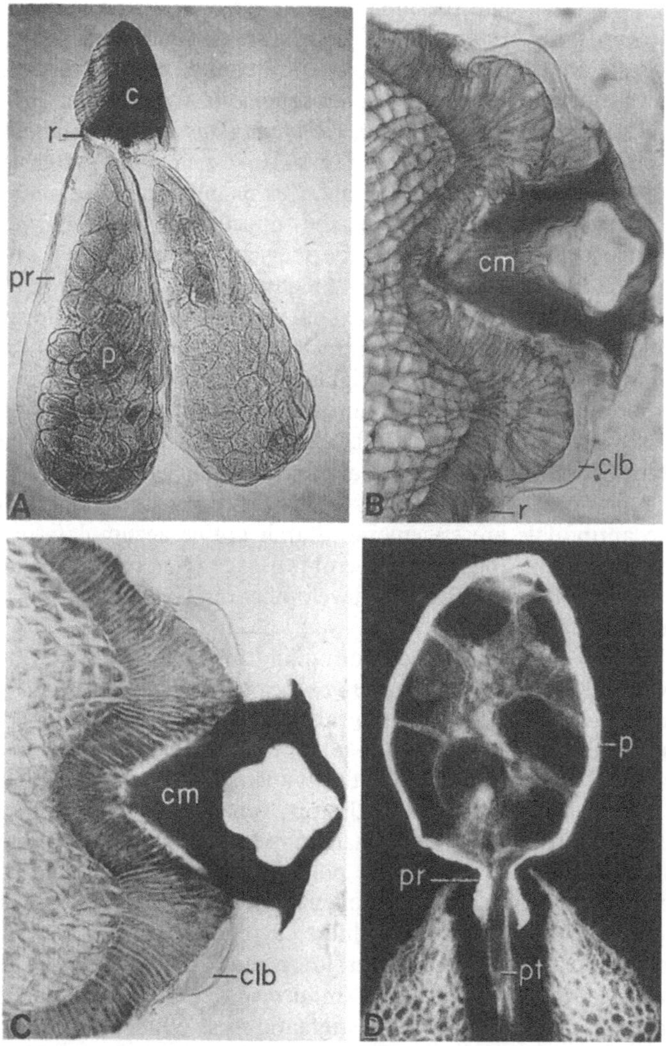

Fig. 47.7.1A–D. *Pergularia daemia.* **A** Wholemount of a pair of pollinia (*p*) with translator which has a corpusculum comprising the main body (*c*), a lateral blade on either side, and a pair of retinaculum (*r*) which spans the corpusculum to the pollinia. The ridge (*pr*) and the beak are two acellular extensions of the pollinium. **B, C** Transections of corpuscula to show the main body (*cm*), lateral blades (*clb*) and retinacula (*r*). (After Vijayaraghavan and Shukla 1976). **D** Transection of germinated pollinium, viewed under a dark-field UV-fluorescence microscope. Autofluorescent tapetal membrane around aggregate pollen (*p*) is discontinuous only at one place to form the germ pore. The unilateral extensions of the tapetal membrane orient the growing pollen tubes (*pt*) towards the receptive site of stigma. The two extensions of the tapetal membrane form the ridge of the "pollinium". The anther flaps clasp at this region. (After Vijayaraghavan and Shukla 1978a)

1977). The tapetal membrane is discontinuous at one place which functions as germ-pore. The pollinium gets attached to the stigma in such a way that the germ-pore faces the receptive site (of stigma). The tapetal membrane develops acellular extensions which help the unidirectional growth of the pollen tube (Fig. 47.7.1D) towards the receptive site (Vijayaraghavan and Shukla 1978a, b).

The ovules are anatropous, unitegmic, or ategmic, and tenuinucellar. Unitegmic ovules are reported in *Caralluma attenuata, Cryptostegia grandiflora* (Venkata Rao and Rama Rao 1954), *Caralluma adscendens* (Maheswari Devi and Lakshminarayana 1978) and *Oxystelma esculentum* (Maheswari Devi and Lakshminarayana 1977b). The nucellus usually collapses early. In *Asclepias crussavica* (see Eames 1961), the nucellus is "vestigial" and represented by two cells at the apex of the megaspore mother cell, and is continuous with the epidermal cells of the integument.

Undifferentiated (ategmic) ovules occur in *Araujia, Asclepias, Gomphocarpus* and *Marsdenia* (Dop 1903), *Cynanchum vincetoxicum* (Seefeldner 1912). The ategmic condition should be reinvestigated.

The archesporial cell functions as megaspore mother cell, undergoes meiotic divisions, the tetrad is linear, and the chalazal megaspore develops into a Polygonum type of embryo sac. The egg apparatus is much elongated, and in *Ceropegia* it occupies nearly half the embryo sac. The polar nuclei fuse before fertilization; the antipodals are ephemeral. Starch grains accumulate in the mature embryo sac. Twin and multiple embryo sacs (*Calotropis procera* – Sabet 1931) have been observed. Bisporic development in *Cynanchum fuscatum* (Pardi 1933) requires confirmation.

Pant et al. (1982a) studied 17 species of pollinators, including a few ants, beetles and nocturnal moths, on the flowers of *Calotropis procera, Leptadenia reticulata, Sarcostemma secamone* and *Wattakaka volubilis*. The pentamerous flowers of these asclepiads show a corona lobe attached to each petal. In the centre, in contact with the corona lobes, is the gynostegium with a pentangular stigmatic disc with receptive surfaces at the tip. Ten pollinia (five pairs) are attached to the stigmatic notch by the retinaculum. In each pair, one pollinium is more conspicuous than the other. The pollinia are so situated that their convex side (on which germination of pollen grain takes place) is situated downward (in the stigmatic notch) to facilitate easy pollination.

The flowers are visited by pollinators for nectar. The pollinators, like *Micrapis florea* and *Apis indica*, which visit *Calotropis procera* are so small that they manage to enter the flowers through stigmatic slits. While escaping from the flower, pollinia become attached to the hairs of the hindlegs. The larger insects, such as *Chlorion lobatum, Megapis dorsata, Schollia* sp. and *Xylocopa fenestrata*, carry pollinia on the forelegs. In *S. secamone* the insects carry pollinia on the fine hairs on mouth parts, or forelegs and hindlegs. Several other insects visit the flowers only for nectar, but do not carry the pollinia. In *C. procera* and *S. secamone*, the pollinia germinate unilaterally.

The flowers of *Leptadenia reticulata* also show the same construction. The method of germination of pollinia is apical in *L. reticulata*, and basal in *W. volubilis*.

The development of the endosperm is of the Nuclear type. The nuclei become distributed on the periphery of the embryo sac around a large central vacuole. Centripetal wall formation is initiated at 16- or 32-nucleate stage of the endosperm (Venkata Rao and Rama Rao 1954). The endosperm stores fatty food reserve.

The embryogeny conforms to Solanad type-Linum variation in *Calotropis gigantea* (Maheswari Devi 1964) and *Daemia extensa* (Biswas 1957). The derivatives of the apical (tier *l*) as well as the basal cell (tiers *m* and *ci*) contribute to the formation of a uniseriate elongated suspensor.

In *Oxystelma esculentum* (Maheswari Devi and Lakshminarayana 1977b), the zygote divides transversely; the daughter cells (*ca* and *cb*) again divide transversely to form a row of four cells (*l*, *l'*, *m*, *ci*). Transverse divisions occur in each of these four cells. The daughter cells of *l* function as the initials of cotyledons, stem tip, and hypocotyledonary region; those of *l'* produce root tip and root cap. The derivatives of *m* and *ci* give rise to uniseriate suspensor. The embryogeny keys out to Solanad type-Physalis II variation.

The seed is compressed and albuminous (fatty food reserve); embryo straight with flat cotyledons (Hutchinson 1973).

The seed-coat is generally reduced to the outer epidermis, or with a few outer layers of mesophyll as flattened crystal cells as in *Calotropis* and *Morrenia*. The outer epidermis may have thick or thin (*Periploca*) walls (Corner 1976).

A tuft of silky hairs (coma) develops at the micropylar region of the seed. In *Asclepias syriaca* (Pearson 1948), the elongation of coma cells occurs after the endosperm becomes eight-nucleate, six to eight days after pollination. The elongation of cells occurs from centrally located cells outward, and continues for 26 days after pollination. The hairs are conical, and the maximal basal width is attained 34 days after pollination.

I Biswas (1957), EJH Corner (1976), GL Davis (1966), BD Deshpande and P Joneja (1962), P Dop (1903), AJ Eames (1961), WW Finn (1925), TC Frye (1901), GL Guignard (1917a, b), V Gupta and MR Vijayaraghavan (1985), J Hutchinson (1973), P Khanna (1964c), K Lakshminarayana (1978b), K Lakshminarayana and M Suresh Babu (1988), H Maheswari Devi (1962, 1964), H Maheswari Devi and K Lakshminarayana (1977b, 1978, 1989), RJ Moore (1946), BN Mulay, BD Deshpande and U Tolani (1965), RL Nirula (1945), RL Nirula and R Richharia (1945), DD Pant, DD Nautiyal and SK Chaturvedi (1982a, b), P Pardi (1933), U Patwardhan (1953), NL Pearson (1948), YS Sabet (1931), G Seefeldner (1912), TH Sylla and F Albers (1989), C Venkata Rao and S Rama Rao (1954), MR Vijayaraghavan and AK Shukla (1976, 1977, 1978a, b), K Viswanathan and KK Lakshmanan (1984), RE Woodson (1941)

47.8 Rubiaceae

The anther wall comprises the epidermis, fibrous endothecium, one or two middle layers, and secretory tapetum with uninucleate cells in *Knoxia corymbosa* (Shivaramaiah and Ganapathy 1961). A dual origin of the tapetum is described in *Tarenna asiatica* (Periasamy and Parmeswaran 1965). It is biseriate in the connective region. Two cell layers in the connective region also develop fibrous thickenings. The microspore mother cells undergo meiotic divisions, cytokinesis is simultaneous, and the tetrads tetrahedral, isobilateral and decussate.

The pollen grains are two- or three-colporate or three-pororate with smooth exine (see Johanson 1987). They are two- or three-celled at the dispersal stage. Occasionally, the pollen grains germinate in situ in *Hymenodictyon excelsum* (S.B. and S.A. Siddiqui 1979), *Oldenlandia corymbosa* (Farooq 1958a) and *O. dichotoma* (S.A. and S.B. Siddiqui 1968a).

Chennaveeraiah and Sivakumar (1983) report an unusual feature of the formation of buds on the pollen grains of *Ophiorrhiza*. Up to the formation of microspore tetrads the development is quite normal (Fig. 47.8.1A), and uninucleate pollen grains are formed (Fig. 47.8.1B). The pollen mitosis shows 11 chromosomes (Fig. 47.8.1C). In the uninucleate pollen grain, the intine bulges out of the three germinal pores, and forms buds (Fig. 47.8.1D). In Figure 47.8.1E, the pollen grain shows three pollen buds, one at each germinal pore. Both pollen and pollen buds contain nuclear fragments. These fragments in the pollen buds are of the vegetative nucleus (Fig. 47.8.1F, G). Sometimes, secondary pollen buds are produced and they also contain chromosomal fragments (Fig. 47.8.1H). Finally, the pollen buds become separated from the pollen grain (Fig. 47.8.1I), and the mature pollen does not have any buds. It contains only the generative cell (Fig. 47.8.1J). These authors observed a similar feature in *Canthium angustifolium* and *C. parviflorum* (pers. comm. 1986).

In *Mussaenda flava* and *M. glabrata* (Priyadarshan and Ramachandran 1984) also buds have been observed at the uninucleate microspore stage. There are three buds in *M. flava*, usually four, sometimes five or even six in *M. glabrata*. There is fragmentation of vegetative nucleus and the fragments move into the pollen buds. In both species, the buds become detached 40 to 50 h before anthesis. The germ-pores heal and the mature pollen does not show any bud.

The formation of buds and the elimination of the vegetative nucleus clearly show that, at least in these systems, the latter has no special role.

The ovule is anatropous, unitegmic and tenuinucellate. According to Fagerlind (1936a, b), the obturator in *Putoria* and *Phyllis* is homologous with the strophiole, but is reduced. He (1937) assumes that in Rubiaceae the strophiole is a second integument. In *Borreria hispida*, Farooq (1952) mentions an obturator as strophiole, and it covers that portion of the ovule which faces the ovarian septum. In fact, according to Kapil et al. (1980), the "strophiole can be defined as a glandular or spongy proliferation, only limited to the raphal region". Fagerlind's (1937) assumption is erroneous.

Wunderlich (1971) has illustrated the structure of the ovule in *Diodia*. In *D. teres*, the ovules develop an integumentary obturator, and in *D. virginiana* the obturator develops at the micropylar as well as the chalazal end. An aril is reported at the base of the ovule in *Pavetta* sp. The nucellus is highly reduced.

Fagerlind (1937) traces the reduction of ovule in this family (see also Eames 1961). The nucellus is represented by a few lateral and apical cells

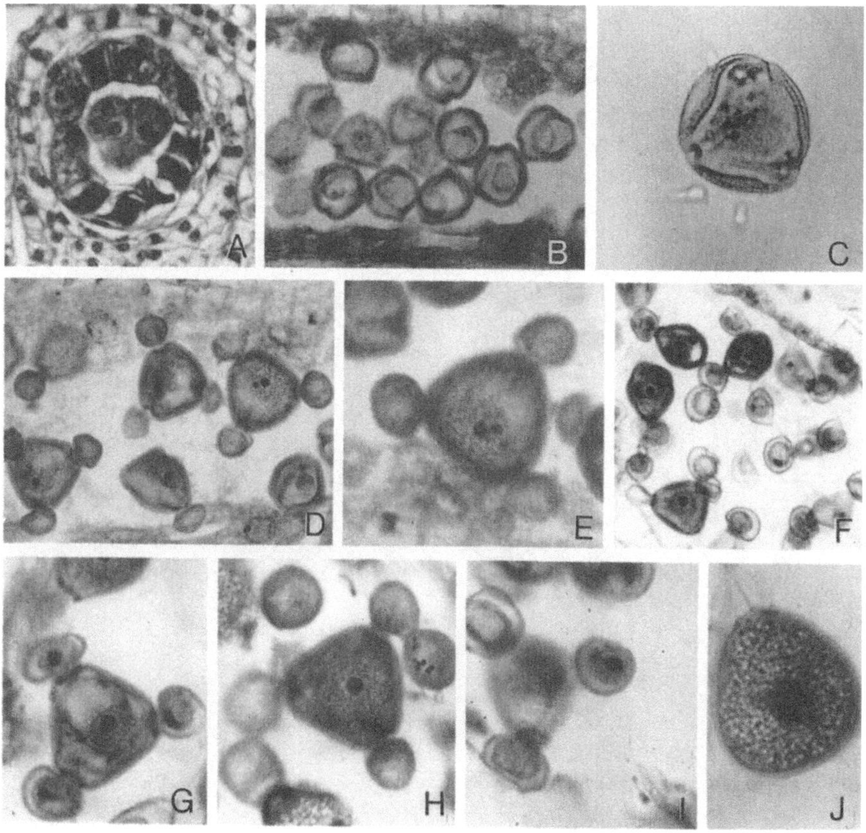

Fig. 47.8.1A–J. **A–C, H** *Ophiorrhiza harrisiana,* **D–F, G, I, J** *O. hirsutula.* **A** Transection of anther locule at tetrad stage. **B** Longisection of anther locule with uninucleate pollen grains. **C** Pollen nucleus mitosis shows 11 chromosomes. **D** Longisection of anther locule, pollen grains with buds at pores. **E** Two-celled pollen grain with pollen buds. **F** Portion of anther locule to show pollen, and pollen buds containing nuclear fragments. **G** Pollen grain with fragments of vegetative nucleus in three pollen buds. **H** Pollen grain with fragments of vegetative nucleus in secondary pollen bud. **I** Separated pollen buds contain nuclear fragments. **J** Mature pollen grain at dehiscence with generative cell, and devoid of pollen buds. (After Chennaveeraiah and Shivkumar 1983. Courtesy Professor M.S. Chennaveeriah, Dharwar, India)

(Fig. 47.8.2A), then only the apical cells persist (Fig. 47.8.2B, C), the nucellar cells elongate (Fig. 47.8.2D), only one cell persists (Fig. 47.8.2E), and finally there is no nucellus at all (Fig. 47.8.2F) (see also Sect. 47.6).

The nucellus is one- or two-celled in *Anotis wightiana* (Shivaramaiah 1971) and *Borreria stricta* (S.A. and S.B. Siddiqui 1968b), and only one-celled in *Dentella repens* (Maheswari Devi and Raju 1980) and *Oldenlandia* spp. (Raghavan and Rangaswami 1941, Farooq and Inamuddin 1969). The nucellus has more or less merged with the integument in *Coffea* sp. (Roth and Lindorf

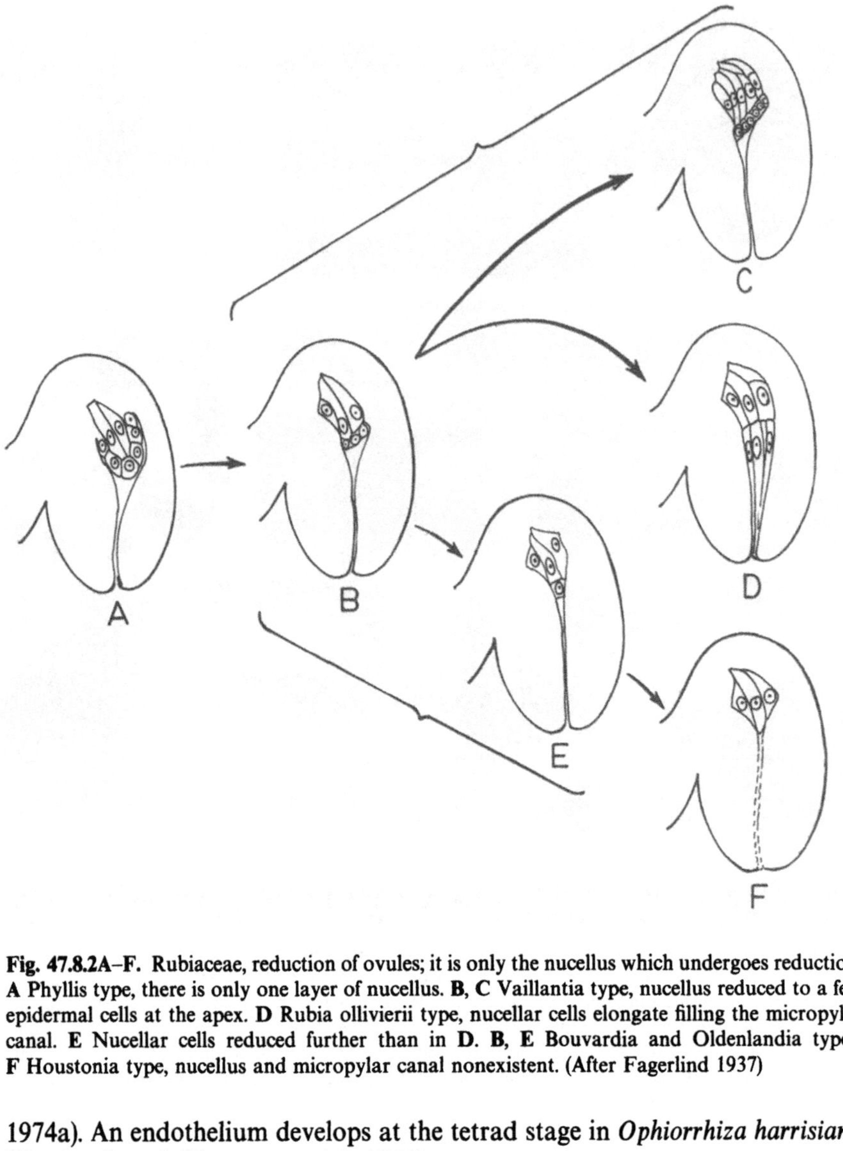

Fig. 47.8.2A–F. Rubiaceae, reduction of ovules; it is only the nucellus which undergoes reduction. **A** Phyllis type, there is only one layer of nucellus. **B, C** Vaillantia type, nucellus reduced to a few epidermal cells at the apex. **D** Rubia ollivierii type, nucellar cells elongate filling the micropylar canal. **E** Nucellar cells reduced further than in **D. B, E** Bouvardia and Oldenlandia types. **F** Houstonia type, nucellus and micropylar canal nonexistent. (After Fagerlind 1937)

1974a). An endothelium develops at the tetrad stage in *Ophiorrhiza harrisiana* (Gopinath and Chennaveeraiah 1961).

The archesporium is one- to multicelled (Fig. 47.8.3A, B). One or more cells function as megaspore mother cell/s (Fig. 47.8.3C), undergo meiotic divisions, and linear (Fig. 47.8.3C–E), T- or ⊥-shaped tetrads are produced. According to Davis (1966), in *Galium lucidum* and *Putoria calabrica* the megaspores develop haustorial processes which invade the integument and reach the funicle. In recent studies none of the investigators have observed extensive development of megaspore haustoria, and this aspect needs detailed study.

Farooq (1960, see also P. Maheshwari 1950) reports that in *Galium asperifolium* "the megaspores germinate by tube formation" (mentioned only in the

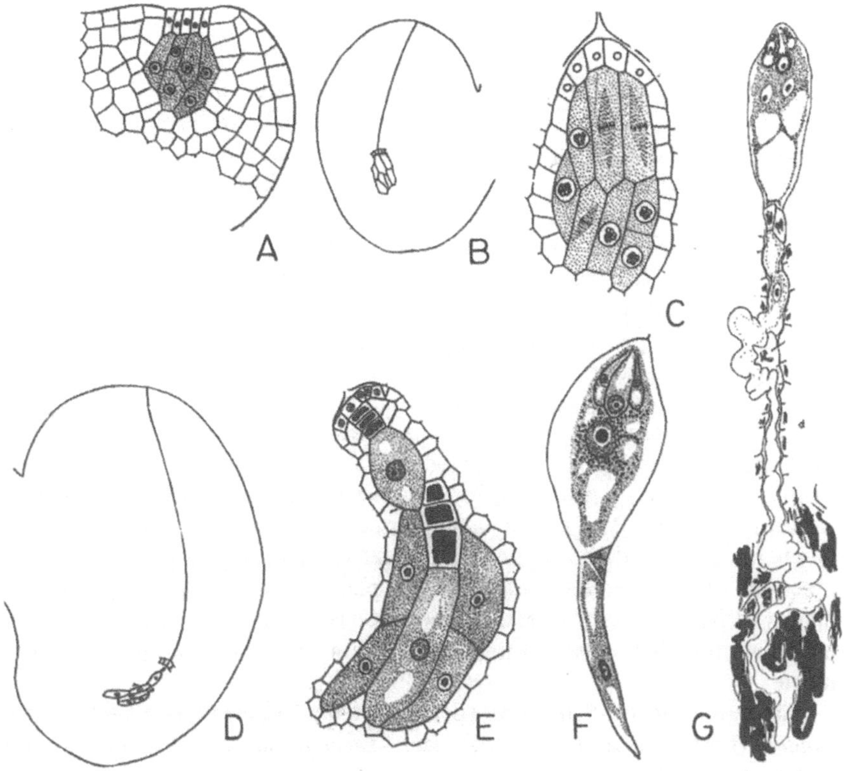

Fig. 47.8.3A–F. *Rubia cordifolia.* **A** Longisection of portion of ovule with five nucellar cells, multicellular archesporium, and integumentary tissue; **B** Longisection of ovule, outline figure for **C**; **C** Nucellar epidermis and megaspore mother cells; **D** Longisection of ovule shows a long micropyle, massive integument, and a few-celled nucellar epidermis; **E** Two megaspore tetrads with functional megaspore, and four megaspore mother cells, from **D**; **F** Mature embryo sac with egg apparatus, secondary nucleus, two small antipodals, and basal elongated antipodal haustorium. (After Venkateswarlu and Rajeswara Rao 1958). **G** *Galium mollugo*, embryo sac with basal elongated antipodal haustorium. (After Fagerlind 1937)

"Summary", and not in the text). He does not give any details, nor illustrate this point.

The chalazal megaspore develops into a Polygonum type of embryo sac. The "entire embryo sac is organized in the micropylar canal" in *G. asperifolium* (Farooq 1960). This appears to be unusual, and needs reinvestigation. Twin embryo sacs develop in *Borreria hispida* (Farooq 1959), *B. stricta* (S.A. and S.B. Siddiqui 1968b) and *G. asperifolium* (Farooq 1960).

The synergids are hooked in *O. nudicaulis* (Farooq and Inamuddin 1969). They lack hooks and have tapering ends in *Rubia cordifolia* (Venkateswarlu and Rajeswara Rao 1958) and *Tarenna asiatica* (Periasamy and Parmeswaran 1965). The antipodals are ephemeral in *Dentella repens*, *Oldenlandia alata*

(Raghavan and Rangaswami 1941) and *O. corymbosa* (Farooq 1958a). They persist up to the 16-nucleate stage of endosperm in *O. nudicaulis* (Farooq and Inamuddin 1969). Supernumerary antipodals have been reported in *Asperula montana*, *Coffea arabica* (see Davis 1966) and *Diodia virginiana* (see Schnarf 1929). Their number goes up to ten. The antipodals are prominent and the lowermost (of the three) undergoes an unusual tubular elongation in *Borreria stricta* (S.A. and S.B. Siddiqui 1968b), *Putoria calabrica* (Fagerlind 1936a), *Rubia cordifolia* (Fig. 47.8.3F) and *Galium mollugo* (Fig. 47.8.3G). There is a similar earlier report in *Asperula*, *Callipeltis cucularia*, *Crucianella* and *Diodia* (see Davis 1966). Whether the elongated antipodal has any haustorial function has not been studied.

The Allium type of embryo sac develops in *Scyphiphora hydrophyllacea*, and Peperomia type in *Crucianella* spp. (see Davis 1966) and *Rubia oliverii* (Fagerlind 1937). It may be noted that the development is of the Polygonum type in *R. cordifolia* (Venkateswarlu and Rajeswara Rao 1958). These and other species should be reinvestigated.

In *Oldenlandia corymbosa* (Farooq 1958a), the solid style is devoid of transmitting tissue. The pollen tubes pass through the intercellular spaces of the stigma, style and the septum, and enter the placenta. They diverge in all directions, wind over the surface of the placenta, and ultimately reach the micropyle. Thus, the course of the pollen tube is partially exotropic. In one case one pollen tube had entered the micropyle while the other had penetrated the funicle.

The development of the endosperm is of the Cellular type in *Ophiorrhiza mungos* (P.M. Ganapathy 1956) and *Tarenna asiatica* (Periasamy and Parmeswaran 1965).

The division of the primary endosperm nucleus is followed by a transverse wall, and another transverse division in each daughter cell forms a row of four cells. Further divisions are irregular. Besides the Cellular type of endosperm, the Nuclear type also occurs in *O. dichotoma* (S.A. and S.B. Siddiqui 1968a). This is doubtful and requires reinvestigation.

The development is of the Nuclear type, and cellularization occurs at the four-celled proembryo in *R. cordifolia* (Venkateswarlu and Rajeswara Rao 1958), and the octant stage in *Dentella repens* (Raghavan and Rangaswami 1941) and *Mussaenda hirsutissima* (Narmatha Bai and Lakshmanan 1983). In *Borreria stricta* (S.A. and S.B. Siddiqui 1968a), cytokinesis occurs after the formation of about 200 nuclei, at the four-celled proembryo.

Gradually, the entire endosperm becomes cellular. It consumes the integument except the epidermis, and the tissue of strophiole. Enucleate cytoplasmic nodules develop in *Oldenlandia corymbosa* (Farooq 1958a) and *O. nudicaulis* (Farooq and Inamuddin 1969). Their fate has not been traced.

The embryogeny conforms to the Solanad type in *Mussaenda hirsutissima* (Narmatha Bai and Lakshmanan 1983). The suspensor is uniseriate and non-haustorial in *Borreria hispida* (Farooq 1959), *Hymenodictyon excelsum* (S.B. and S.A. Siddiqui 1979), *Richardsonia pilosa* (Lloyd 1902), *Oldenlandia dichotoma* (S.A. and S.B. Siddiqui 1968a) and *O. verticillaris* (Narmatha Bai and Lakshmanan 1984).

The uppermost cell of the suspensor becomes large and vesicular in *B. stricta* (S.A. and S.B. Siddiqui 1968b). In *Sherardia arvensis* (Souèges 1925d), the derivatives of the basal cell (tiers *m* and *ci*) give rise to a massive suspensor which is filamentous towards the apical end, and the cells enlarge and become vesicular at the basal end. In *Rubia cordifolia* (Venkateswarlu and Rajeswara Rao 1958), the cells of the suspensor enlarge, become vesicular, and simulate a bunch of grapes (Fig. 47.8.4A). In *Asperula* (see P. Maheshwari 1950), six to seven apical cells of the suspensor are filamentous, while the basal cells form uninucleate balloon-shaped structures which bulge out into the endosperm tissue. In *Galium asperifolium* (Farooq 1960), the basal cells of the suspensor bulge out into the endosperm but are not as well developed as in *Asperula*. These cells undergo transverse and longitudinal divisions producing filamentous branches. In *G. saccharatum* (Fagerlind 1937) also the suspensor cells are bulbous and haustorial, and invade the cellular endosperm (Fig. 47.8.4B).

The outer epidermis of the integument undergoes anticlinal divisions, and also develops ingrowths (ruminations) into the mesophyll. Subsequently, the endosperm grows in between the ingrowths, and becomes ruminate. Such a condition is reported in *B. hispida* (Farooq 1952), *Psychotria*, *Randia* (Corner 1976) and *Tarenna asiatica* (Periasamy and Parmeswaran 1965).

According to Houk (1938), in the seed of *Coffea arabica*, the endosperm is poorly developed, and is absorbed by the growing embryo. Houk points out that only the perisperm stores food reserve. Vaughan (1970) clearly mentions that in coffee seeds the endosperm cells contain oil drops, and small irregularly shaped aleurone grains. Since the ovules are unitegmic and tenuinucellate, and the scanty nucellus is crushed, there is no possibility of the perisperm storing the food reserve. The testa comprises longitudinally elongate sclerotic cells.

In *Borreria hispida* (= *Spermacoce hispida*), Farooq (1952) mentions that the endosperm cells are thick-walled and its cells contain starch. The growing embryo encroaches upon the obturator (strophiole) and consumes its central

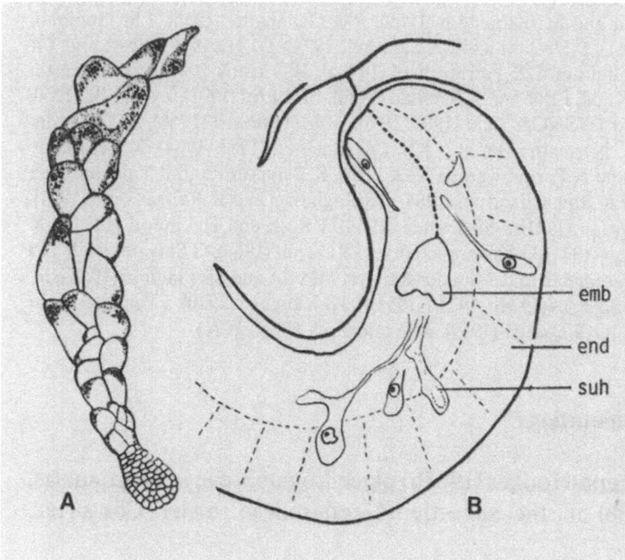

Fig. 47.8.4. A *Rubia cordifolia*, late globular proembryo with haustorial suspensor. (After Venkateswarlu and Rajeswara Rao 1958). **B** *Galium saccharatum*, longisection of portion of seed at heart-shaped embryo (*emb*). The inflated suspensor cells (*suh*) push inside the cellular endosperm (*end*). (After Fagerlind 1937)

part. The same feature is fully described by Periasamy and Vaidyanathan (1980).

In the strophiolate seed of *B. hispida* (Periasamy and Vaidyanathan 1980), during earlier stages of development of the endosperm, the massive chalaza pushes the free-nuclear endosperm towards the anti-raphe side. Later, the seed-coat (at the ovule-strophiole junction) forms two ingrowths (ruminations) into the cellular endosperm. Concurrently, the endosperm consumes most of the strophiolar tissue, and becomes trilobed. *B. hispida* is a rare example (in Rubiaceae) in which the endosperm lies partly within the ovule, and partly in the strophiolar region.

In Rubiaceae, the seed-coat comprises only the outer integumentary epidermis with a few layers of mesophyll. The vascular bundle extends up to the micropyle in pachychalazal seeds. The outer epidermis is thin-walled or thickened with lignin. The thickenings may be all round the cells; radial and inner walls are thickened in *Gardenia* and *Randia*, bar-like thickenings occur in *Guettarda* and *Randia*, and reticulate in *Cinchona*. The mesophyll is thick-walled in *Hydrophylax* (Corner 1976).

The seed is pachychalazal, minute to large, strophiolate, and albuminous with a very small to large embryo (Corner 1976).

In some members, the seed is covered by placental outgrowths. According to Periasamy and Parmeswaran (1962), these outgrowths arise during the earlier stages of development of the ovule in *Randia malabarica* and *Tarenna asiatica*. Their growth ceases after fertilization in *Tarenna*. In *Randia*, at the time of fertilization, the ovules become embedded in the placental outgrowths. The outgrowths continue to grow concurrently with the development of seed, and invest it, except for a small pore at the micropylar end. In a ripe fruit, the outgrowths become pulpy and contain tannin. Due to ruminations in the seed, the pulpy portion shows an undulated surface.

MS Chennaveeraiah and PM Sivakumar (1983), EJH Corner (1976), MB Crane (1940a, b), P Crété (1956e), GL Davis (1966), AJ Eames (1961), F Fagerlind (1936a, b, 1937), M Farooq (1952, 1953, 1958a, 1959, 1960), M Farooq and M Inamuddin (1969), PM Ganapathy (1956), PM Gopinath and MS Chennaveeraiah (1961), S Hashmi and SA Siddiqui (1974), DJ Hicks, R Wyatt and TR Meagher (1985), GR Hildebrand and DE Fairbrothers (1970a), WG Houk (1938), JT Johanson (1987), AC Joshi (1938c), RN Kapil, J Bor and F Bouman (1980), FE Lloyd (1902), Ly-Thi-Ba (1960, 1961), H Maheswari Devi and PVSNGK Raju (1980, 1990), P Maheshwari (1950), AJT Mendes (1941), BG Murray (1990), V Narmatha Bai and KK Lakshmanan (1983, 1984), K Periasamy (1962a, 1964b), K Periasamy and N Parmeswaran (1962, 1965), K Periasamy and R Vaidyanathan (1980), PM Priyadarshan and K Ramachandran (1984), TS Raghavan and K Rangaswami (1941), TS Raghavan and AR Srinivasan (1941b), SS Raman (1954), I Roth and H Lindorf (1974a), K Schnarf (1929), G Shivaramaiah (1971), G Shivaramaiah and RC Dutt (1964), G Shivaramaiah and PS Ganapathy (1961), G Shivaramaiah and S Sunder Rajan (1973), SA and SB Siddiqui (1968a, b), SB and SA Siddiqui (1979), R Souèges (1924b, 1925d, 1951b), JG Vaughan (1970), J Venkateswarlu and G Rajeswara Rao (1958), G Wagenitz (1959), R Wunderlich (1967, 1971)

47.9 Taxonomic Considerations

Loganiaceae. Bentham and Hooker (1965b) place Loganiaceae in Gentianales, Engler and Diels (1936) in the suborder Gentianineae, order Contortae.

Loganiaceae have been split into two subfamilies – Loganioideae and Buddleioideae (see Maheswari Devi and Lakshminarayana 1980). Wettstein (1935) raises their status to two distinct families – Loganiaceae and Buddleiaceae. Melchior (1964) retains Loganiaceae in Gentianales, but transfers Buddleiaceae to suborder Solanineae, order Tubiflorae. Hutchinson (1973) assigns Loganiaceae to Loganiales, along with Potaliaceae, Buddleiaceae, Antoniaceae, Spigeliaceae, Strychnaceae and Oleaceae.

On embryological basis, K.V.O. Dahlgren (1922) proposes the division of Loganiaceae into two subfamilies – Loganioideae with Nuclear endosperm and absence of endothelium, and Buddleioideae with Cellular endosperm and presence of endothelium. Loganioideae show both Nuclear (*Strychnos*) and Cellular (*Polypremum*) type of endosperm, and endothelium occurs in *Strychnos potatorum* (Maheswari Devi and Lakshminarayana 1980) and *Polypremum procumbens* (Moore 1948). Maheswari Devi and Lakshminarayana (1980) do not support the splitting of the family into two subfamilies, Loganioideae and Buddleioideae. Because of the many resemblances in the embryology of *Spigelia* and *Strychnos* with those of the Loganiaceae, Spigeliaceae and Strychnaceae need not be raised to the status of separate families; these genera are well placed in Loganiaceae.

Loganiaceae (see Hutchinson 1973) differ from Buddleiaceae in tri- to penta-aperturate pollen, three-celled at the dispersal stage. Nuclear or Cellular type of endosperm without haustoria, and Solanad type of embryogeny. There is no valid ground to consider Buddleiaceae as a subfamily of Loganiaceae. Also, its assignment to Tubiflorae is justified on the basis of embryology (see Sects. 48.14 and 28).

Buddleiaceae. It may be emphasized that, as compared to Loganiaceae, Buddleiaceae (see Melchior 1964) is closer to Scrophulariaceae. Both have tricolporate pollen, two-celled at the dispersal stage, Polygonum type of embryo sac, Cellular endosperm with well-developed micropylar and chalazal haustoria (*Buddleia*), Onagrad type of embryogeny, and exotestal seed (Sects. 48.14 and 15).

Embryological data are meagre for commenting on the relationships of Potaliaceae and Antoniaceae.

Oleaceae cannot be included in Loganiales (Hutchinson 1973), and the erection of an independent order Oleales is justified (Sect. 46.3).

Gentianaceae. Bentham and Hooker (1965b) and Rendle (1952) split the family Gentianaceae into two subfamilies: Menyanthoideae including marshy plants, and Gentianoideae including all the others. Lindsey (1938), Wettstein (1935) and Hutchinson (1973) raise them to two independent families, Menyanthaceae and Gentianaceae. Menyanthaceae and Gentianaceae have similar chemical compounds (Hegnauer 1969), but have different anatomical features (Lindsey 1938). R. Dahlgren (1980b) includes these families in the order Gentianales, and Cronquist (1981) places Gentianaceae in Gentianales and Menyanthaceae in Solanales.

Arekal (1961a), Maheswari Devi (1962), and Vijayaraghavan and Padmanabhan (1969) point out striking differences in their embryology and support the erection of separate families (Table 47.9.1).

B.M. Johri (1963a) evaluates their morphological and embryological features and some of the dissimilarities are: Menyanthoideae are aquatic or marshy plants with alternate leaves, and valvate or imbricate aestivation of the corolla. The anther tapetum is of the secretory type with bi- or multinucleate cells, nuclear fusions are common. The ovules are borne on suppressed parietal placenta; an endothelium is present. The development of the endosperm is of the Cellular type. The embryogeny conforms to the Asterad type. In contrast, Gentianoideae are terrestrial plants with opposite, decussate leaves, and twisted corolla. The anther tapetum may be amoeboid (*Canscora*), or secretory with uninucleate cells. The ovules are borne on well-developed parietal placentae. The placentae fuse to form a bilocular ovary. An endothelium is absent. The development of the endosperm is of the Nuclear type. The embryogeny corresponds to the Solanad type. The embryological features (Table 47.9.1) fully support the erection of the families Menyanthaceae and Gentianaceae (see also Melchior 1964, Hutchinson 1973).

Apocynaceae and Asclepiadaceae. Apocynaceae and Asclepiadaceae have been placed in Gentianales by Bentham and Hooker (1965b), in the suborder Gentianoideae of order Contortae by Engler and Diels (1936), and in Apocynales by Hutchinson (1973). Both the families show similarities in simultaneous or successive cytokinesis in pollen mother cells, three-celled pollen grains, unitegmic ovules with single archesporium, Polygonum type of embryo sac, and Nuclear type of endosperm. Both appear to have originated from a common stock. However, each family also has distinctive features.

Apocynaceae show hemiana- to anatropous tenuinucellar ovules, ephemeral antipodals, chalazal endosperm haustorium in *Catharanthus* and *Lochnera*, and ruminate endosperm in *Ervatamia* and *Voacanga*. The embryogeny is extremely variable: Onagrad, Asterad, Caryophyllad or Chenopodiad type.

Asclepiadaceae are characterized by the occasional absence of fibrous endothecium, pollen grains in tetrads or pollinia, hemianatropous ovules, occasionally Cellular type of endosperm (*Calotropis procera* – Sabet 1931), Solanad type of embryogeny, comose seeds, and absence of iridoid compounds. In the sub-family Periplocoideae, the anthers are tetralocular with stomium median to the anther loculi, cytokinesis in pollen mother cells is simultaneous, and pollen grains remain attached in tetrads. In the more specialized subfamily Cynanchoideae, the bilocular anthers have stomium in terminal position, cytokinesis is successive, and pollinia are formed (Maheswari Devi 1964).

In view of the above similarities and some differences, the relationship of Apocynaceae and Asclepiadaceae is justified on the basis of embryology, as well as their inclusion in the order Gentianales. The characters are not sufficiently distinctive to assign them to a separate order, Apocynales.

Table 47.9.1. Embryology of Gentianoideae and Menyanthoideae (see Arekal 1961a, Maheswari Devi 1962, Vijayaraghavan and Padmanabhan 1969)

	Gentianoideae	Menyanthoideae
Anther tapetum	Amoeboid or secretory, 1-layered, cells uninucleate	Secretory, 2-layered on proximal side, cells 2- or multinucleate
Ovule	Anatropous, unitegmic, tenuinucellar	Anatropous, unitegmic, tenuinucellar
Endothelium	Generally absent	Present
Female archesporium	1- to many-celled	1-celled
Embryo sac	Polygonum type, 3 uni- or multinucleate antipodal cells, secondary multiplication in *Swertia*	Polygonum type, 3 uninucleate ephemeral antipodal cells
Endosperm	Nuclear type	Cellular type
Embryogeny	Solanad or Caryophyllad type, suspensor of 2 or 3 cells	Asterad type, suspensor of 2 or 3 cells

Rubiaceae. Bentham and Hooker (1965b) refer Rubiaceae, along with Caprifoliaceae, to Rubiales. Engler and Diels (1936) include it in Rubiales, along with Adoxaceae, Dipsacaceae, Caprifoliaceae and Valerianaceae (i.e., Dipsacales of Melchior).

According to Cronquist (1981), Rubiales comprise two families – Rubiaceae and Theligonaceae, while Hutchinson (1973) adds the family Dialypetalanthaceae. Hutchinson derives Rubiales from the Loganiales stock.

R. Dahlgren (1980b) places Rubiaceae (including Theligonaceae) in the order Gentianales along with Loganiaceae. Rubiaceae have complex secoiridoids (Jensen et al. 1975).

The embryology of Rubiaceae is very unusual. The family is multipalynous, there is extreme reduction of nucellus, the development of the embryo sac is variable (Polygonum, Allium and Peperomia type), the antipodals are ephemeral or persistent, they may undergo secondary multiplication, or the lowest antipodal cell may elongate a good deal (haustorial ?), the development of the endosperm is generally Nuclear, sometimes Cellular (the endosperm may be ruminate), the suspensor is haustorial and may be branched, the seed is strophiolate, pachychalazal, and placental outgrowths develop in some members.

Rubiaceae differ from Loganiaceae in the different types of embryo sacs, variability in the structure of antipodals, suspensor haustoria and their branching, and a complex structure of seed. The embryogeny in Rubiaceae is of the Solanad type, Solanad and Caryophyllad in Gentianaceae, and Asterad in Menyanthaceae. Apocynaceae exhibits Onagrad, Asterad, Chenopodiad and Caryophyllad type. The formation of translator and pollinia occurs exclusively in Asclepiadaceae in Gentianales; the strophiolate, pachychalazal seed of Rubiaceae differs from the comose seed of Apocynaceae and Asclepiadaceae. The above data indicate that the embryology of Rubiaceae differs from that of other families of Gentianales.

The assignment of Rubiaceae and Caprifoliaceae in the order Rubiales (Bentham and Hooker 1965b) is also debatable. Embryologically, these two families do not have many common features. The reduction of the nucellar tissue, sometimes development of Peperomia type of embryo sac, significant tapering elongation of basal (of the three) antipodal cell (haustorial ?), occasional occurrence of supernumerary antipodals, Nuclear type of endosperm, Solanad type of embryogeny with suspensor haustoria, and strophiolate, pachychalazal seed of Rubiaceae, all these features are quite distinct from those of the Caprifoliaceae (Sect. 50.2).

The embryology of the family Dialypetalanthaceae, of Rubiales (Hutchinson 1973), has not been investigated.

Rubiaceae differ not only from the families of Gentianales, but also from Caprifoliaceae. Therefore, on the basis of embryology, its assignment to a separate order Rubiales is justified (see Cronquist 1981).

In view of certain similarities with Gentianales, Rubiales should be placed close to Gentianales.

48 Tubiflorae

Tubiflorae are a large order comprising 26 families, distributed over five suborders: Convolvulineae, Boragininineae, Verbenineae, Solanineae, Myoporineae and Phrymineae.

48.1 General Features

Tubiflorae, even though a very large order, exhibit considerable uniformity: (1) secretory anther tapetum, (2) unitegmic, tenuinucellar ovule mostly with endothelium, (3) Polygonum type of embryo sac, (4) Cellular type of endosperm with haustoria; Nuclear in Polemoniaceae, Convolvulaceae, Hydrophyllaceae, Boraginaceae and Solanaceae, and (5) embryogeny that varies a great deal; Onagrad type occurs in the majority of the families.

48.2 Polemoniaceae

The anther wall comprises the epidermis, fibrous endothecium, an ephemeral middle layer, and secretory tapetum in *Gilia capitata*, *Polemonium caeruleum* and *P. pulcherrimum* (Kapil et al. 1968). Occasionally, the endothecium becomes biseriate. The tapetal cells are binucleate in *G. capitata* (S. Kumar and Jain 1977). The nuclei fuse and become tetraploid in *P. caeruleum* (Sunder Rao 1940a). Amoeboid type of tapetum forming periplasmodium is reported by Govil (1970a) in *Phlox paniculata.* A false periplasmodium occurs in *P. caeruleum* (Sunder Rao 1940a). The microspore mother cells undergo meiotic divisions, cytokinesis is simultaneous, and tetrads tetrahedral and decussate.

The pollen grains are polyporate with a smooth exine. They are two-celled at the dispersal stage.

The ovule is hemianatropous, unitegmic, and tenuinucellate. It has a prominent endothelium, and a hypostase (Kapil et al. 1968). The archesporial cell functions as megaspore mother cell, undergoes meiotic divisions, and a linear tetrad is formed. The chalazal megaspore develops into a Polygonum type of embryo sac. The synergids are hooked, polar nuclei fuse just before fertilization, and the antipodals persist up to fertilization.

The development of the endosperm is of the Nuclear type. The first few divisions, following the division of the primary endosperm nucleus, are synchronous. The initiation of walls occurs at the 64-nucleate stage in *Gilia*, and 128- or 256-nucleate in *Polemonium* (Kapil et al. 1968). The cellular endosperm stores oil.

In *Polemonium caeruleum* (Souèges 1939b), the proembryo consists of a tetrad of four superposed cells, *l*, *l'*, *m* and *ci*; *m* and *ci* develop from the basal cell. The tier *l* divides and produces an epiphysis initial. It becomes indistinguishable by the time the cotyledons have formed. The suspensor consists of four cells. The embryogeny conforms to the Chenopodiad type-Myosotis variation. In *Gilia capitata* (Kapil et al. 1968), the apical cell divides vertically and the basal transversely. The suspensor comprises five or six cells.

The seed is small and albuminous with a straight or slightly curved embryo. Usually, the seed-coat comprises the outer epidermis and the cells contain mucilage. The cells are cuboid (*Gilia*), or flattened (*Phlox*), or thickened with spiral thickenings (*Polemonium caeruleum*) on the inner and radial walls. The cells of the mesophyll either show bar-like thickenings, or get crushed. The inner epidermis is pigmented, and the radial walls are thickened (Corner 1976). In *Gilia capitata* (Kapil et al. 1968), the cavities formed by the broken mesophyll become filled with mucilage; occasionally starch grains occur.

EJH Corner (1976), ML Dawson (1936), CM Govil (1970a), V Grant and DH Wilken (1986), HO Juel (1915), RN Kapil, PN Rustagi and R Venkataraman (1968), S Kumar and DK Jain (1977), HA Miller and RH Wetmore (1945), R Souèges (1939b, 1945b), Y Sunder Rao (1940a)

48.3 Fouquieriaceae

The embryology has not been fully investigated.

The tetrasporangiate anther dehisces introsely by longitudinal slits, and lacks appendages. The anther tapetum is glandular.

The pollen grains are tricolporate, elliptical (Henrickson 1973), and shed at the two-celled stage (Brewbaker 1967).

The ovule is anatropous, bitegmic and tenuinucellate with the micropyle formed by a long endostome. An endothelium differentiates; the epistase is cap-like and distinct. The hypodermal archesporial cell functions as megaspore mother cell, and undergoes meiotic divisions resulting in a linear tetrad. The chalazal megaspore develops into a Polygonum type of embryo sac (R. Khan 1943a). The mature embryo sac develops a lateral L-shaped haustorial arm. The polar nuclei fuse before fertilization. The three antipodals persist up to fertilization.

The development of endosperm is of the Cellular type. The haustorial lateral arm of the embryo sac contains the nucleus of the chalazal chamber (Govil 1970b). The mature endosperm is oily.

The embryogeny conforms to the Asterad type.

The seed is broad and albuminous with a small straight embryo. It has a marginal wing and exotestal hairs. The testa and tegmen are multiplicative.

JL Brewbaker (1967), R Dahlgren, SR Jensen and BJ Nielson (1976), CM Govil (1970b), J Henrickson (1973), DA Johansen (1936), R Khan (1943a), J Mauritzon (1936a)

48.4 Convolvulaceae

The anther wall comprises the epidermis, fibrous endothecium, an ephemeral middle layer, and secretory tapetum with multinucleate cells. The microspore mother cells undergo meiotic divisions, cytokinesis is simultaneous, and tetrads tetrahedral, isobilateral and decussate.

Rarely, in *Cuscuta reflexa* (B.M. Johri and Tiagi 1952), after meiosis, microspore nuclei may undergo supernumerary divisions. Thus, the polyad contains as many as ten or 11 microspores of different size (Fig. 48.4.1A–E). The nucleus of one or more microspores in a polyad may divide again so that it becomes two- to four-nucleate (Fig. 48.4.1A, C).

The pollen grains are spherical in *C. reflexa*, and usually with five (sometimes four or six) germ pores (Fig. 48.4.1F, G). The nucleus undergoes a mitotic division so that a generative and a tube cell is produced (Fig. 48.4.1H). The generative cell may also divide, forming two spherical sperm cells which later elongate. Thus, the mature pollen may be two- or three-celled (Fig. 48.4.1I). Often, multinucleate microspores may also develop, and each may contain two (Fig. 48.4.1J) or five (Fig. 48.4.1K) nuclei.

Sometimes, the microspore mother cells undergo meiotic divisions, but cytokinesis is completely or partially arrested. Consequently, several variations arise. (1) Cytokinesis is irregular, the microspores may not separate, and each pollen grain may contain one, two or three nuclei (Fig. 48.4.1L, M). (2) Cytokinesis is incomplete, the furrows are arrested, and the pollen grain with exine shows three or four nuclei (Fig. 48.4.1N–P). (3) Cytokinesis is totally absent and the pollen grain with exine contains two to five nuclei (Fig. 48.4.1Q–S).

The pollen grains in Figure 48.4.1Q–S are larger in size, and have larger nuclei. It appears that they have developed directly from the microspore mother cells without meiotic divisions, but mitotic divisions have occurred. Such pollen grains develop exine, and the nuclei lie in a common cytoplasm.

Rarely, pollen sacs (Fig. 48.4.2A, C) differentiate in *C. reflexa* in the ovary wall (close to the stigmatic region). The microspore mother cells are surrounded by a glandular tapetum of uni- or binucleate cells, followed by three or four layers of starchy cells (Fig. 48.4.2B). Figure 48.4.2D shows uninucleate pollen grains with other characteristic features of a mature anther. Earlier or later stages of development were not observed.

The pollen grains are polyporate or colpate, with spinulate or smooth exine. At the dispersal stage, they are two-celled in *Jacquemontia tenuifolia* (H. Kaur 1981b), and three-celled in *Evolvolus alsinoides* (B. Tiagi and Gupta 1963).

In *Cuscuta epithymum*, Fedortschuk (1931) reports compound pollen grains with: (1) two vegetative nuclei and the generative cell in division, (2) one vegetative nucleus and three sperm cells, (3) one vegetative nucleus and two sperm cells, (4) one vegetative nucleus, two sperm cells, and a prothalial cell. A pollen grain showed two protoplasts – one with a vegetative nucleus, and the other with a vegetative as well as generative cell.

The ovule is anatropous and unitegmic with vascularized integument. In *Evolvulus alsinoides* (B. Tiagi and Gupta 1963), the vascular strand terminates at the chalaza. The ovule is tenui- or crassinucellar. A placental obturator is a common feature. A parietal cell is reported in *Convolvulus arvensis* (K.L. Mathur 1934), *Ipomoea hispida* (H. Kaur 1981a), *I. obscura* (H. Kaur 1969), *I. reptans* (Chan and Hillson 1968), *I. trifida* (Kenyan1928), and *Operculina turpethum* (Bobde 1963). The parietal cell is absent in *Cuscuta* spp.

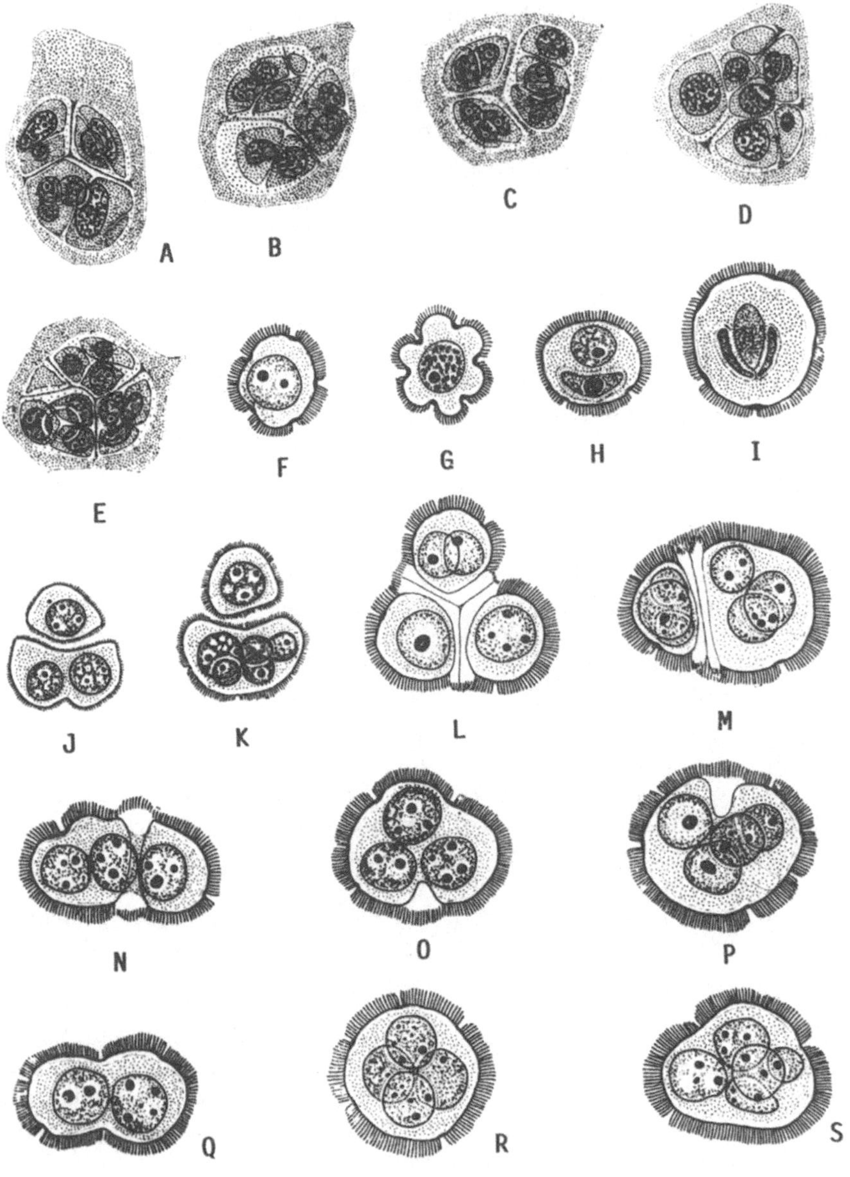

Fig. 48.4.1A–S. *Cuscuta reflexa.* **A–E** Polyads. **F** Uninucleate pollen with five germinal furrows. **G** Six germinal furrows. **H** Two-celled pollen. **I** Three-celled pollen. **J** Uni- and binucleate microspore with incomplete cytokinesis. **K** Uni-, and five-nucleate microspore. **L, M** Three and two microspores in a common exine, in **L** one is binucleate, and in **M** both are trinucleate. **N–P** Multinucleate pollen formed due to arrested cytokinesis. **Q–S** Multinucleate pollen formed either by failure of cytokinesis following meiosis in microspore mother cell, or by division/s of the nucleus of uninucleate pollen. (Adapted from B.M. Johri and Tiagi 1952)

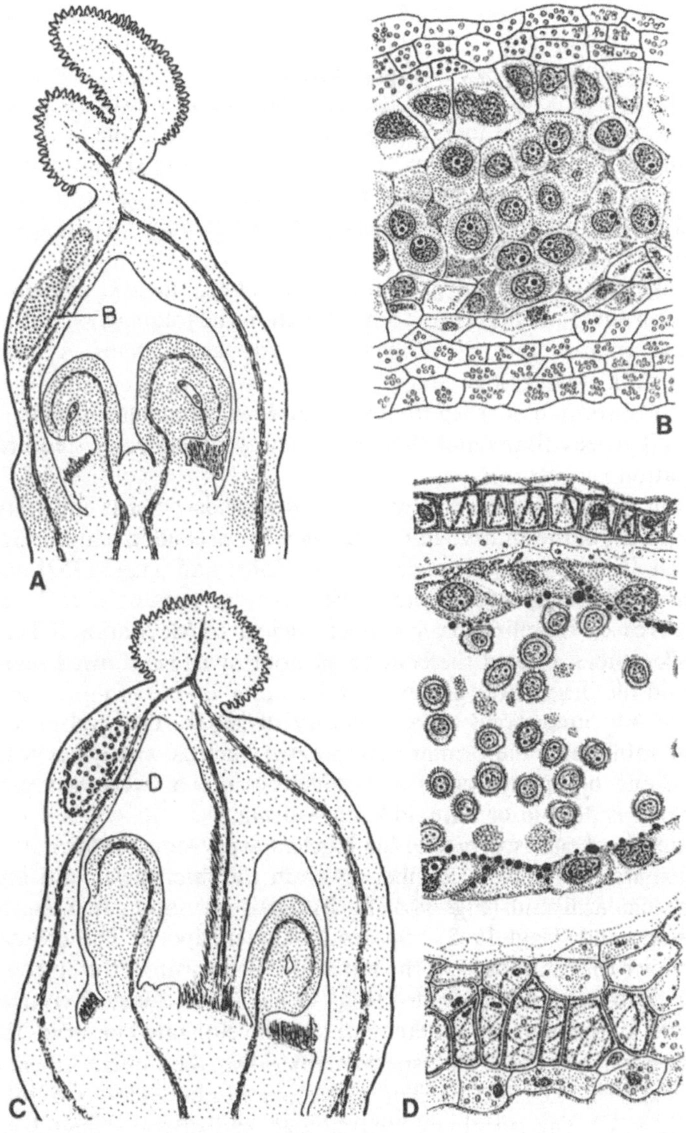

Fig. 48.4.2A–D. *Cuscuta reflexa.* **A, C** Longisection of abnormal ovaries, outline diagrams for **B** and **D. B** Part of "microsporangium" with microspore mother cells. **D** Uninucleate pollen grains with exine (probably formed directly from microspore mother cells without meiosis), tapetum with polyploid nuclei and orbicules, two ephemeral middle layers, fibrous endothecium, and conspicuous epidermis. (After B.M. Johri and Tiagi 1952)

(B.M. Johri and Nand 1934, B. Tiagi 1951a), and *Evolvulus alsinoides* (B. Tiagi and Gupta 1963). The archesporium is generally unicellular, a multicellular condition occurs in *Jacquemontia*. The chalazal megaspore of a linear tetrad develops into a Polygonum type of embryo sac. The synergids are hooked, polar nuclei fuse at the time of fertilization, and antipodals are ephemeral.

The Allium type of embryo sac occurs in *Cuscuta reflexa* (B.M. Johri and Nand 1934, B.M. Johri and B. Tiagi 1952). Twin embryo sacs have also been reported in this species. If the intervening wall breaks down, the common embryo sac contains two egg apparatus, four polar nuclei, and six antipodals. Another embryo sac showed two egg cells, two synergids, an irregularly lobed nucleus formed by the fusion of six (?) nuclei, and six free nuclei in the lower region of the embryo sac.

Parvathi et al. (1978) report Polygonum type of embryo sac in *C. chinensis*, and Allium type in some ovules. They do not mention the frequency of either type, although it appears that the Polygonum type is dominant. A more detailed investigation is suggested.

Whether the variation in microsporogenesis, and male and female gametophyte are a result of environmental influence, or any other factor, requires further investigation.

Double fertilization is reported. One of the synergids persists until late stages of development of the embryo in *Cuscuta monogyna* (Fedortschuk 1931), *C. lupuliformis* (B. Tiagi 1965a), *C. reflexa* (B.M. Johri and Tiagi 1952) and *Evolvulus alsinoides* (B. Tiagi and Gupta 1963).

In an unusual case of fertilization in *Cuscuta reflexa* (B.M. Johri and Tiagi 1952), three pollen tubes entered the embryo sac; one synergid showed seven male gametes and the other three gametes. Presumably, in one or more tubes the male gamete had undergone supernumerary divisions. The zygote had enlarged considerably, and the primary endosperm nucleus was lying in its vicinity. Rarely, one of the synergids also develops into a two-celled proembryo; whether it is diploid or haploid is not certain.

The development of endosperm is of the Nuclear type. After the formation of several nuclei, wall formation commences from the micropylar end and proceeds to the chalazal end (Fig. 48.4.3A, B). Wall formation in *Cuscuta reflexa* (B.M. Johri and Tiagi 1952) appears to be centripetal. The cellular endosperm occupies the entire seed at the initiation of stem tip of the embryo.

The cellular endosperm is thin-walled (Fig. 48.4.3C), and does not contain any food reserve. As the stem tip differentiates in the elongated embryo, the walls of the peripheral cells of the endosperm break down, their nuclei fuse and become polyploid (Fig. 48.4.3D, E). The polyploid nuclei undergo mitotic divisions (Fig. 48.4.3F). The peripheral "plasmodial" endosperm is aggressive and consumes the inner layers of integument. The central cellular endosperm gelatinizes (Fig. 48.4.3G, H) and is consumed by the developing embryo.

The embryogeny conforms to the Caryophyllad type-Myosotis variation in *Argyreia* and *Ipomoea* (Johansen 1950), Caryophyllad type-Fumaria variation in *Ipomoea pes-tigridis* (Jose 1963), Solanad type in *I. reptans* (Chan and Hillson 1968) and Onagrad type-Lotus variation in *Convolvulus arvensis* (Souèges 1937f). The suspensor is well developed, and haustorial in *Cuscuta*

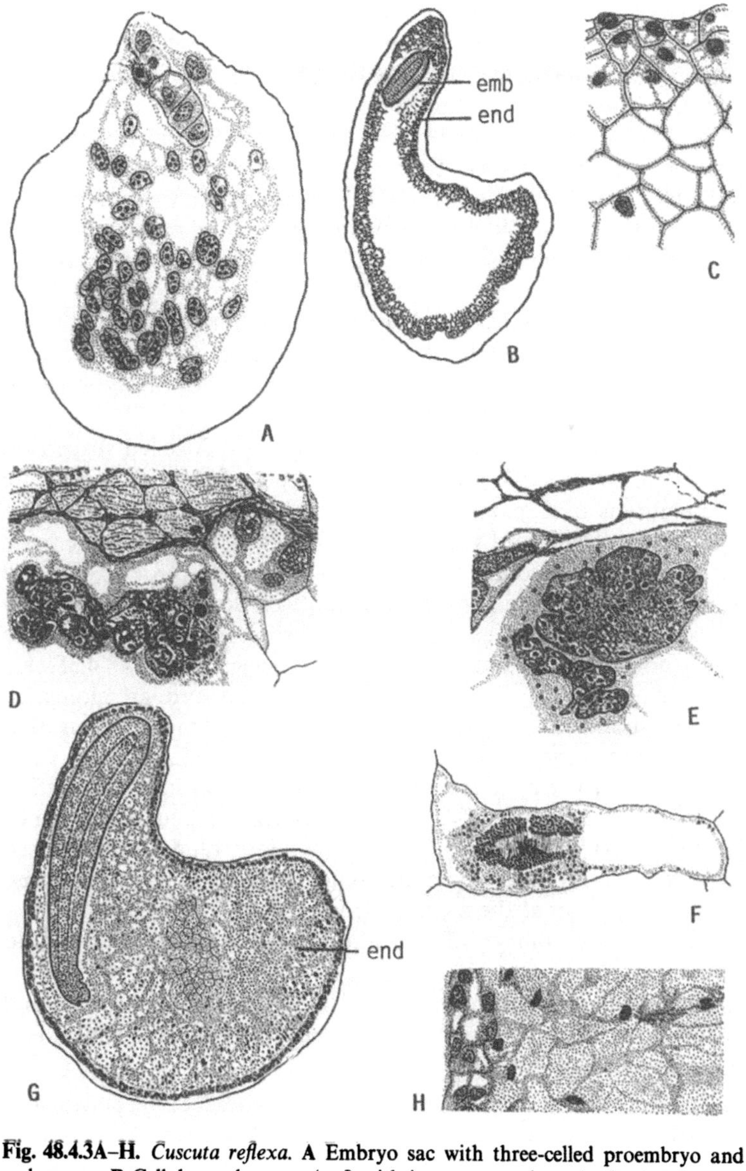

Fig. 48.4.3A–H. *Cuscuta reflexa.* **A** Embryo sac with three-celled proembryo and free-nuclear endosperm. **B** Cellular endosperm (*end*) with immature embryo (*emb*). **C** Portion of endosperm from **B**. **D, E** Aggressive activity of endosperm adjacent to integument, the nuclei are polyploid and irregular. **F** Mitotic division of polyploid endosperm nucleus. **G** Longisection of seed shows gelatinized endosperm (*end*). **H** Portion of mucilaginous matrix, from **G**. (After B.M. Johri and Tiagi 1952)

reflexa (B.M. Johri and Tiagi 1952) and *I. obscura* (H. Kaur 1969). The embryo is leafy and green, except in *Cuscuta* where it is undifferentiated and coiled.

The embryo in *Cuscuta reflexa* remains undifferentiated, and does not conform to any of the embryogenic types. B.M. Johri and Tiagi (1952) erroneously assigned it to the Solanad type. The zygote divides transversely into a terminal cell *a*, and a basal cell *b* (Fig. 48.4.4A). The terminal cell and its derivatives divide transversely so that a row of about six to eight cells is produced (Fig. 48.4.4B–K). The lower five tiers divide repeatedly to form a globular proembryo (Fig. 48.4.4L–P).

Usually the basal cell *b* does not divide. However, sometimes it may divide vertically (Figs. 48.4.4G, K, L, 48.4.5A) or transversely (Fig. 48.4.4N, O). The tier *c* (derivative of *a*) also divides vertically (Fig. 48.4.4L, M), or transversely (48.4.4O).

The tier *d* (derivative of *a*) divides transversely and longitudinally so that a number of cells are produced between the suspensor and globular proembryo. These cells always remain uninucleate, and protrude into the adjacent coenocytic suspensor cells (Fig. 48.4.4P).

The derivatives of basal and apical cells produce a multicelled suspensor. According to Fedortschuk (1931), two different types of suspensor cells are formed in the embryos of *C. monogyna* and *C. epithymum*. In *C. reflexa* the suspensor may be coenocytic and vesicular (Fig. 48.4.4D–P). Rarely, the basal cell may not divide, but becomes vesicular (Figs. 48.4.4M, 48.4.5B–D) and contains as many as 70 nuclei. Sometimes, the well-developed proembryo shows highly reduced, vesicular, coenocytic suspensor cells. In the other type the suspensor is non-vesicular and the cells are uninucleate (Fig. 48.4.5E–K).

The divisions in the globular proembryo continue and an elongated embryo shows shoot tip at its apex (Fig. 48.4.5L, M, P–U). During maturation of the undifferentiated embryo, the suspensor cells disorganize (Fig. 48.4.5N, O). Further growth results in a mature coiled embryo without any differentiation of cotyledons or radicle. The entire vascularized coiled region can be considered as hypocotyl (Fig. 48.4.5Q–U). The vascular strands of the embryo do not show differentiation into xylem and phloem. Spirally arranged scales, perhaps protective, develop at the stem tip (Fig. 48.4.5S, U).

The suspensor in the embryo of *Cuscuta reflexa* is unique and has no parallel in the angiosperms.

The exotestal seed is smooth, rough or hairy (*Ipomoea carnea*), and albuminous with a curved embryo. The cotyledons are folded, the endosperm is starchy; *Cuscuta* is an exception.

Fig. 48.4.4A–P. *Cuscuta reflexa*, embryogeny. **A–K** Sequence of divisions in the terminal cell of bicelled proembryo to establish the tiers *d*, *e*, *f*, *g*, *h* and a_1–a_5. Note the formation and disposition of vesicular coenocytic suspensor cells b_1–b_2 and c_1–c_2; also c_{2a} and c_{2b} in **P**. **L–N** Formation of embryonal mass from the cells *e*, *f*, *g*, *h* and *a*; the cell *d* gives rise to two tiers of cells. **O, P** Advanced stages. Note the haustorial synergid (*s*) in **A–D, F, H–J, L, O**, and remnants of pollen tube (*pt*) in **A, D, I.** (After B.M. Johri and Tiagi 1952)

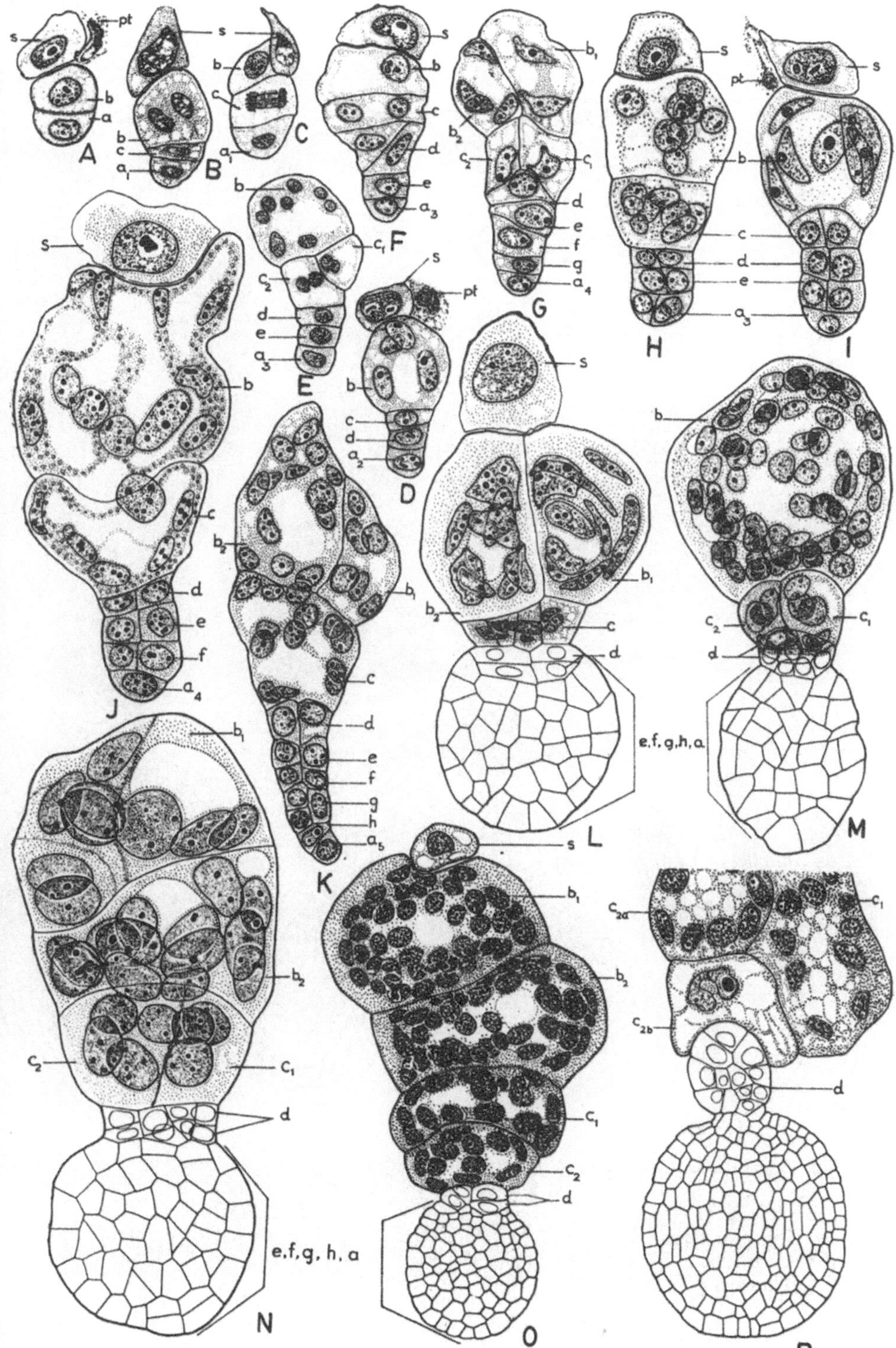

s
pt
b
a
A
B
C
a1
c
F
c1
c2
d
e
a3
E
D
a2
G
b1
b2
f
g
a4
H
I
J
K
h
a5
L
M
N
O
P
e, f, g, h, a
c2a
c2b

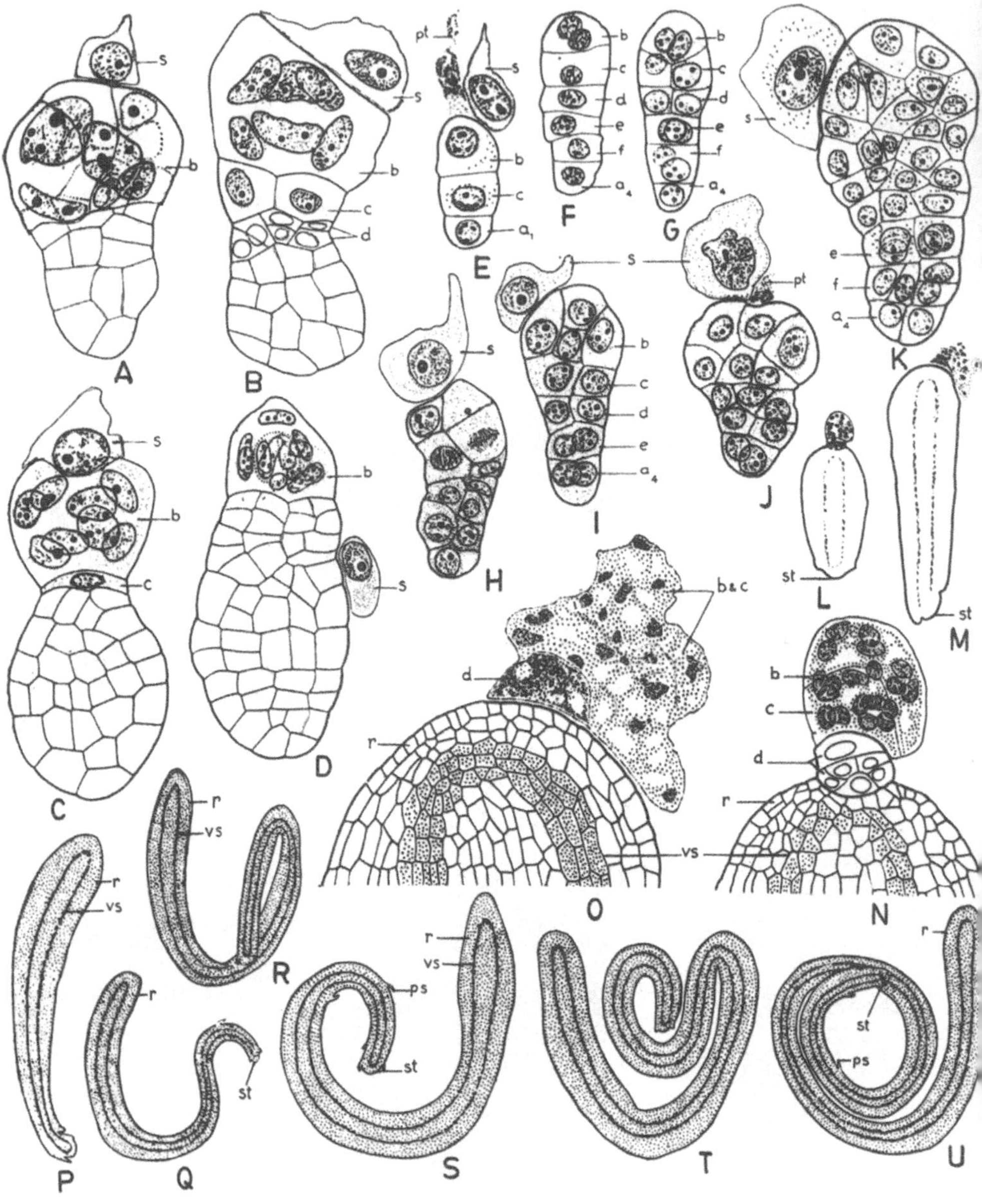

Fig. 48.4.5A–U. *Cuscuta reflexa.* **A–D** Proembryos with reduced vesicular coenocytic suspensor cells; *s* persistent haustorial synergid. **E–K** Proembryos without vesicular suspensor; *pt* remnant of pollen tube. **L, M** Advanced stages of embryos with stem tip (*st*). **N, O** Enlargement of radicular end of embryos shown in **L, M**; note degeneration of haustorial suspensor, and absence of root cap. **P–U** Maturation of coiled embryo; *ps* plumular scale; *r* radicular end; *vs* vascular strand. (After B.M. Johri and Tiagi 1952)

In *Cuscuta reflexa* (B.M. Johri and Tiagi 1952), only a few layers of endosperm persist around the coiled embryo. The outer epidermal layer of testa consists of thin-walled, tabular cells. The subepidermal layer is also thin-walled, followed by one or two layers of lignified palisade tissue, and about three layers of parenchymatous cells. In the funicular region, the epidermal and hypodermal cells elongate and become lignified.

In *Ipomoea carica*, *I. obscura* and *I. sindica* (Govil 1971), the outer epidermal cells of testa are tangentially elongated. The subepidermal layer consists of radially elongated cells followed by a single lignified palisade layer (*I. purpurea*). The palisade tissue consists of two or three layers in *I. obscura*, *I. sindica*, *Merremia emargina*, *Rivea hypocrateriformis*, and five layers in *I. carnea* and *I. vitifolia* (Govil 1971). The mesophyll and the inner epidermis become crushed.

According to Govil (1971), the seeds of Convolvulaceae have a characteristic "pad" between the funiculus and micropyle. This pad comprises three or four layers of compact parenchymatous cells, followed by a radially elongated sclerenchymatous palisade layer. The function of the pad is not explained.

In *Cuscuta pedicellata* (Lyshade 1984), the seed-coat consists of epidermis, two palisade layers, and mesophyll of parenchymatous cells. The epidermis contains starch and mucilage, the parenchymatous cells also contain starch but become compressed. The endosperm is starchy with a peripheral aleurone layer. The cell walls of the endosperm gelatinize.

In the dry seeds of *C. pedicellata*, the surface is deeply pitted due to invaginated epidermal cell walls. As the seeds become wet, these walls bulge outward and the seed surface becomes papillose. This structure of seed-coat epidermis is believed to function in wind dispersal and rapid water uptake. Seed dormancy, common to the genus, does not occur in this species.

PL Bobde (1963), SCK Chan and CJ Hillson (1968, 1971), W Fedortschuk (1931), CM Govil (1970c, 1971), DA Johansen (1950), BM Johri (1951, 1987c), BM Johri and S Nand (1934), BM Johri and B Tiagi (1952), JS Jose (1963), H Kaur (1969, 1981a, b), H Kaur and RP Singh (1970, 1981), FMG Kenyan (1928), OB Lyshade (1984), P Maheshwari (1944c), KL Mathur (1934), LJ Musselman (1984), A Parvathi, LL Narayana and BSM Dutt (1978), KVR Rao (1940), VS Rao (1944), R Sampathkumar and KR Iyyangar (1978, 1982), R Souèges (1937f), ES Teryokhin and VA Kotov (1988), B Tiagi (1951a, 1965a), B Tiagi and PK Gupta (1963)

48.5 Hydrophyllaceae

Di Fulvio (1987, 1989a, b) has made a detailed study of the embryology of this family.

The anther wall comprises the epidermis, fibrous endothecium, one middle layer, and glandular tapetum with binucleate cells in *Nama jamaicense* (di Fulvio 1989a) and uninucleate in *Hydrolea* (di Fulvio 1989b). Simultaneous cytokinesis follows meiotic divisions of microspore mother cells. The tetrads are tetrahedral. The pollen is shed at the two-celled stage.

The ovule is anatropous, unitegmic, and tenuinucellar. An endothelium differentiates at the tetrad stage in *Nama* and *Hydrolea*, and extends from the micropylar region almost up to the basal region of the embryo sac. The archesporial cell functions as megaspore mother cell, undergoes meiotic divisions, and forms a linear tetrad. The chalazal megaspore develops into a Polygonum type of embryo sac, polar nuclei fuse before fertilization, and the antipodals are ephemeral. The antipodal nuclei do not organize into cells in *Hydrolea zeylanica* (Mitra 1947a); this needs confirmation. Di Fulvio (1989b) reports ephemeral antipodal cells in several species of *Hydrolea*.

According to Svensson (1925), the development of the endosperm is of four types:

1) Tanacetifolia form. Nuclear type and also basal endosperm haustoria as in *Phacelia malvifolia* and *P. tanacetifolia*.

2) Parryi form. Cellular-Nuclear-Cellular as in *Phacelia parryi*. The primary endosperm nucleus undergoes a transverse division followed by another transverse division so that a row of three or four cells is produced. Each cell undergoes free-nuclear divisions for a while, and wall formation is initiated in the third cell (from the micropylar cell), and then in all the other cells. Finally, the endosperm becomes cellular. Thus, a nuclear phase is interpolated between two cellular phases. This mode of development is considered "Intermediate" between Cellular and Nuclear type. On the basis of ontogeny, Swamy (1974a) considers that the development is of the Cellular type and, therefore, the question of an Intermediate type does not arise.

3) Congesta form. Cellular type, haustoria mostly absent; rarely, a basal haustorium may develop as in *Phacelia divaricata*, *P. viscida*, *Hydrolea spinosa* (Kainradl 1927), *H. zeylanica* (Mitra 1947a), *Phacelia congesta*, *P. fimbriata*, *P. grandiflora*, *P. linearis*, *P. whitlavia* and *Romanzoffia sitchensis* (Svensson 1925).

Kainradl (1927) reports a massive terminal, branched, unicellular haustorium in *H. spinosa*. In *H. zeylanica* (Mitra 1947a), the division of primary endosperm nucleus is followed by a transverse wall, and by further transverse divisions a row of four to eight cells is formed. Longitudinal divisions take place in the two middle layers. Further divisions are in both planes. The formation of micropylar or chalazal endosperm haustorium has not been observed.

4) Nemophila form. Cellular type with chalazal or micropylar haustorium as in *Nemophila aurita*, *N. automoria*, *N. insignis* and *N. maculata* (Svensson 1925). The primary endosperm nucleus divides transversely (Fig. 48.5.1A), the upper cell again divides transversely and gives rise to a row of three cells. The central cell produces the endosperm proper; the upper terminal cell develops into an uninucleate micropylar haustorium, and the lowest cell into a uninucleate chalazal haustorium (Fig. 48.5.1B–D). In *N. insignis*, the chalazal haustorium develops a lateral branch which penetrates into the funiculus, and comes in contact with the starchy cells of the placenta (Fig. 48.5.1E). In *N. aurita*,

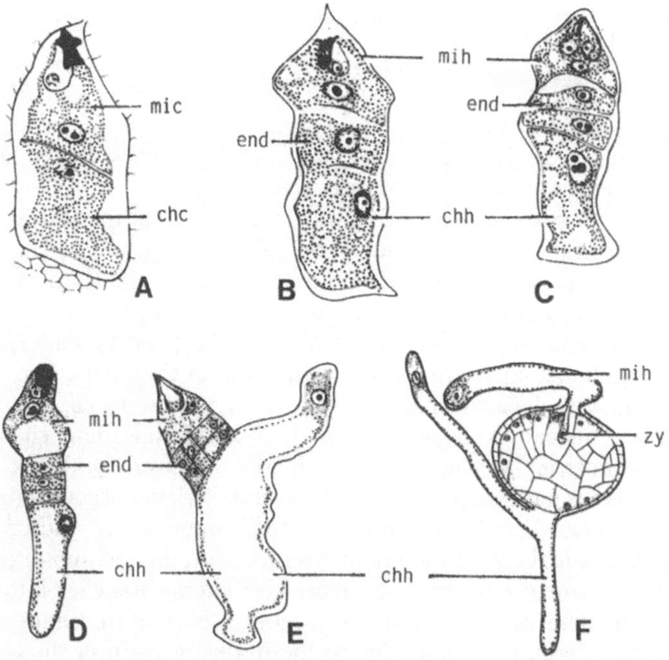

Fig. 48.5.1A–E. *Nemophila insignis*, **A** Embryo sac with zygote, micropylar (*mic*) and chalazal (*chc*) endosperm chamber. Remnants of pollen tube persist in **A**, **B** and **D**; **B** Three-celled endosperm, micropylar (*mih*) and chalazal (*chh*) haustorium, and endosperm (*end*) proper initial; **C** Binucleate micropylar haustorium, two-celled endosperm proper, and uninucleate chalazal haustorium; **D** Micropylar and chalazal haustorium, three-celled endosperm proper; **E** Lateral growth of chalazal haustorium. **F** *N. aurita*. Cellular endosperm, lateral growth of uninucleate micropylar haustorium, and branched uninucleate chalazal haustorium. The zygote (*zy*) has not divided in **A–F**. (After Svensson 1925)

the micropylar as well as the chalazal haustorium becomes aggressive (Fig. 48.5.1F).

Di Fulvio (1987, 1989a, b) has made a detailed study of endosperm development in *Hydrophyllaceae*.

The endosperm is of the Cellular type and can be referred to the Nemophila type of Svensson (1925, see also di Fulvio 1987). The first and the second division are transverse resulting in a row of four cells. (a) In *Eucrypta*, a vertical division produces four tiers of biseriate cells. The upper tier forms a two-celled micropylar haustorium without any lateral branch. The other three tiers give rise to the endosperm proper, the chalazal haustorium is absent. (b) In *Ellisia*, *Nemophila*, *Hydrophyllum* and *Pholistoma*, the two middle tiers of the monoseriate row of four cells produce the Cellular endosperm proper. The micropylar and the chalazal tier produce unicellular well-developed haustoria with hypertrophied nucleus. In *Ellisia* and *Nemophila*, the micropylar haustorium does not develop a lateral branch. In *Pholistoma* and *Hydrophyllum*, a lateral branch is formed and grows towards the placenta. The nucleus migrates into the lateral branch. At the junction of the endosperm proper, the chalazal haustorium produces a single lateral branch which grows towards the funiculus and sometimes penetrates the placenta in *Ellisia*, *Nemophila* and *Pholistoma*. The nucleus migrates into the lateral branch. In *Hydrophyllum* spp., the chalazal haustorium may develop one or two lateral branches. In the latter case, the nucleus does not migrate into any of the branches.

In *Nama jamaicense* (di Fulvio 1989a), the development of the endosperm conforms to Svensson's (1925) "Congesta Form". The micropylar and chalazal haustoria are poorly developed and differentiate only after the third mitotic cycle in endosperm. The reserve food material is lipids and proteins.

In *Hydrolea* (di Fulvio 1989b), the division of the primary endosperm nucleus results in almost equal micropylar and chalazal chambers. The next division occurs in the chalazal chamber with a larger lower cell. Another transverse division in each cell generates a four-celled uniseriate chalazal haustorium. The micropylar chamber also divides transversely. While the lower cell produces the endosperm proper, the terminal cell undergoes three or four transverse divisions and the daughter cells comprise the micropylar haustorium. The terminal cell of the micropylar haustorium and subterminal cell of the chalazal haustorium send out aggressive lateral branches into the integument at the time of division of the zygote. The mature endosperm becomes ruminate (Verbascum type) due to the inward growth of the endothecial cells.

The embryogeny conforms to Period I, Series C, Megarchetype V (Solanad type) in *Phacelia tanacetifolia* (Crété 1946d). Solanad type is also reported in *Hydrolea zeylanica* (Mitra 1947a). The proembryo usually consists of a row of five cells before vertical division occurs in the apical cell. The two apical cells produce the embryo proper, the two middle cells the hypophysis, and the two basal cells the suspensor. In *Nama* and several other species of *Hydrolea* also the embryogeny is of the Solanad type (di Fulvio 1987).

In *Nama* (di Fulvio 1989a), the seed-coat consists of the outer epidermis – the cells develop an inward arching thickening along the inner tangential walls and most of the radial walls, and the inner epidermis (endothelium) with thickened inner tangential and partly thickened radial walls. The mesophyll collapses. In *Hydrolea* (di Fulvio 1989b) also the seed-coat comprises the exotesta and endotesta. The cells of exotesta develop small pectocellulosic spicules on their inner tangential walls. The endotesta (endothelium) is the main protective layer. It is thick, brownish and tanniniferous with a well developed cuticle. The outer tangential walls of the endospermal epidermis are thick.

The seeds are small with copious fleshy endosperm and a straight embryo. In *Hydrolea* they show eight to ten longitudinal ridges corresponding to the lobes of the ruminate endosperm.

RY Berg (1985), GD Chance and D Bacon (1984), P Crété (1946d), GL Davis (1966), TE di Fulvio (1987, 1989a, b), E Kainradl (1927), JN Mitra (1947a), K Schnarf (1931), HG Svensson (1925), BGL Swamy (1974a), HD Wulf and TS Raghavan (1937)

48.6 Boraginaceae

The anther wall comprises persistent cutinized epidermis, fibrous endothecium, two ephemeral middle layers, and glandular tapetum of multinucleate cells in *Trichodesma zeylanicum* (Tasneem 1967). A fibrous endothecium is absent in *Cordia obliqua* (Khaleel 1975); this requires confirmation. The tapetum may become biseriate in *Ehretia laevis* (B.M. Johri and Vasil 1956), and the nuclei become polyploid in *Mertensia platyphylla* (P. Khanna 1964b). Simultaneous cytokinesis follows meiotic divisions in microspore mother cells, and the tetrads are tetrahedral, decussate and isobilateral. In *Cynoglossum amabile* (Millsaps 1940), the tetrads are only of the decussate type. Polyspory occurs in *Ehretia laevis*. Degeneration of microspores is common in *C. obliqua*.

The family is multipalynous. The pollen grains are triporate in *Coldenia procumbens* (Venkateswarlu and Atchutarammurthi 1955b), *Cordia alba* (Khaleel 1975), *Rotula aquatica* (Nagaraj and Tasneem 1967) and *Trichodesma zeylanicum* (Tasneem 1967). The pollen grains exhibit eight prominent ridges in *Ehretia laevis* (B.M. Johri and Vasil 1956). They are tricolpate in *Heliotropium indicum* (P.K. Pal 1963), multicolpate in *H. ovalifolium* (P.K. Pal 1963), and dumbbell-shaped in *Cynoglossum denticulatum* (Khaleel 1974). The pollen grains are two- or three-celled at the dispersal stage.

In *Cordia obliqua* (Khaleel 1975), the sporogenous tissue and microsporocytes degenerate so that the pollen grains do not develop. The pollen grains in *C. alba* (Khaleel 1975) are incompatible, and there is no fruit-set.

The ovule is ortho- (*Cordia sebestena*), ana- , hemiana- or ana-campylotropous. It is unitegmic and tenuinucellate; crassinucellate in *Coldenia* (Venkateswarlu and Atchutarammurthi 1955a, b), *Ehretia*, *Ixorhea tschudiana* (di Fulvio 1978), and *Rotula*; pseudocrassinucellar in *Cordia alba* (Tasneem 1966).

An obturator occurs in *Trichodesma* (Tasneem 1967). A hypostase differentiates in *Cortesia cuneifolia* (di Fulvio 1965) and *Cynoglossum denticulatum* (Khaleel 1974). The endothelium develops usually at the functional megaspore stage. It is said to be absent in *Cordia obliqua* (Khaleel 1975).

A single vascular strand enters the ovule through the funicle, and terminates at the base of the integument in *C. obliqua* (Khaleel 1975), *Ehretia laevis* (B.M. Johri and Vasil 1956), *Heliotropium* spp. (P.K. Pal 1963) and *Rotula aquatica* (Nagaraj and Tasneem 1967). It branches in the integument in *Cynoglossum*, *Echium*, *Lithospermum* and *Pulmonaria*. In *Cynoglossum amible* (Millsaps 1940), *C. denticulatum* (Khaleel 1974) and *Trichodesma amplexicaule* (P. Khanna 1964b), the branches run close to the outer epidermis of the integument. In *Cordia alba* (Khaleel 1975), the vascular strand branches at the base of the integument, one branch traverses towards the embryo sac, while the other enters the integument. Three to seven post-chalazal branches have been observed in *Anchusa*, *Borago*, *Nonnea* and *Symphytum*; one of the branches reaches almost up to the micropyle (Corner 1976).

The hypodermal archesporial cell functions as megaspore mother cell. A parietal cell and parietal tissue is reported in *Ehretia laevis* (B.M. Johri and Vasil 1956), *Heliotropium scabrum*, *H. strigosum* (Khaleel 1978), and *Rotula aquatica* (Nagaraj and Tasneem 1967). The megaspore mother cell undergoes meiotic divisions, produces a linear or T-shaped tetrad, and the chalazal megaspore develops into a Polygonum type of embryo sac. The synergids are hooked. The polar nuclei fuse just before or at fertilization. The antipodals are ephemeral; in *Heliotropium scabrum* (Khaleel 1978) and *I. tschudiana* they persist up to four-celled endosperm. A lateral diverticulum of the embryo sac has been observed in Anchuseae (Svensson 1925).

Bisporic embryo sacs are reported in *Anchusa officinalis*, *Cortesia cuneifolia* (di Fulvio 1965), *Lycopsis arvensis* (Svensson 1925), occasionally in *Cynoglossum amabile* (Millsaps 1940), *Cordia* spp. (Khaleel 1975) and *Ehretia laevis* (B.M. Johri and Vasil 1956).

The occurrence of twin embryo sacs in some members is due to the activity of second and third megaspores (Nagaraj and Tasneem 1968).

The development of the endosperm is Nuclear to Cellular with several intermediate types. Svensson (1925) suggested five types: Borago, Lycopsis, Echium, Lappula and Myosotis. Khaleel (1974) added the Cynoglossum type (see Fig. 48.6.1A, B).

1) Borago type. This is the normal Nuclear type of development, and later the endosperm becomes cellular: *Borago officinalis*, *Onosma echioides* (Svensson 1925), *Trichodesma amplexicaule*, *T. indicum* and *T. zeylanicum* (Khaleel 1977).

2) Lycopsis type. The primary endosperm nucleus undergoes a free-nuclear division followed by another similar division so that four free nuclei are produced. A lateral wall separates two nuclei from a binucleate larger chamber. Free-nuclear divisions occur in both the chambers, cytokinesis takes place only in the larger chamber, and the nuclei in the lateral chamber degenerate. The development is of the Nuclear type: *Anchusa*, *Lycopsis arvensis*, *Nonnea* spp., *Pulmonaria* and *Symphytum* of Anchuseae, and *Cerinthe* and *Lithospermum* of Lithospermae (Svensson 1925).

3) Echium type. The division of the primary endosperm nucleus is followed by an oblique wall separating a small lateral chamber from the large chamber. The lateral chamber divides first and produces two cells. Free-nuclear divisions occur in all the three cells. The larger chamber is more vacuolate, and wall formation occurs very late. Schnarf (1929) and P. Maheshwari (1950) consider this sequence as modified Helobial type. However, Swamy and Parmeswaran (1963, see also Swamy 1974a) do not recognize its affinity to the Helobial type, as the first division in the lateral chamber is followed by wall formation: *Echium plantagineum* of Echieae (Svensson 1925).

4) Lappula type. A vertical wall follows the division of the primary endosperm nucleus, and ends blindly in the chalazal region. The second division is followed by a vertical wall at right angles to the first, it also ends blindly. Free-nuclear divisions are followed by wall formation in the micropylar region, while in the chalazal region cytokinesis is absent: *Asperugo procumbens*, *Krynitzkia barbigera* and *Lappula echinata* of the tribe Erithricheae, and

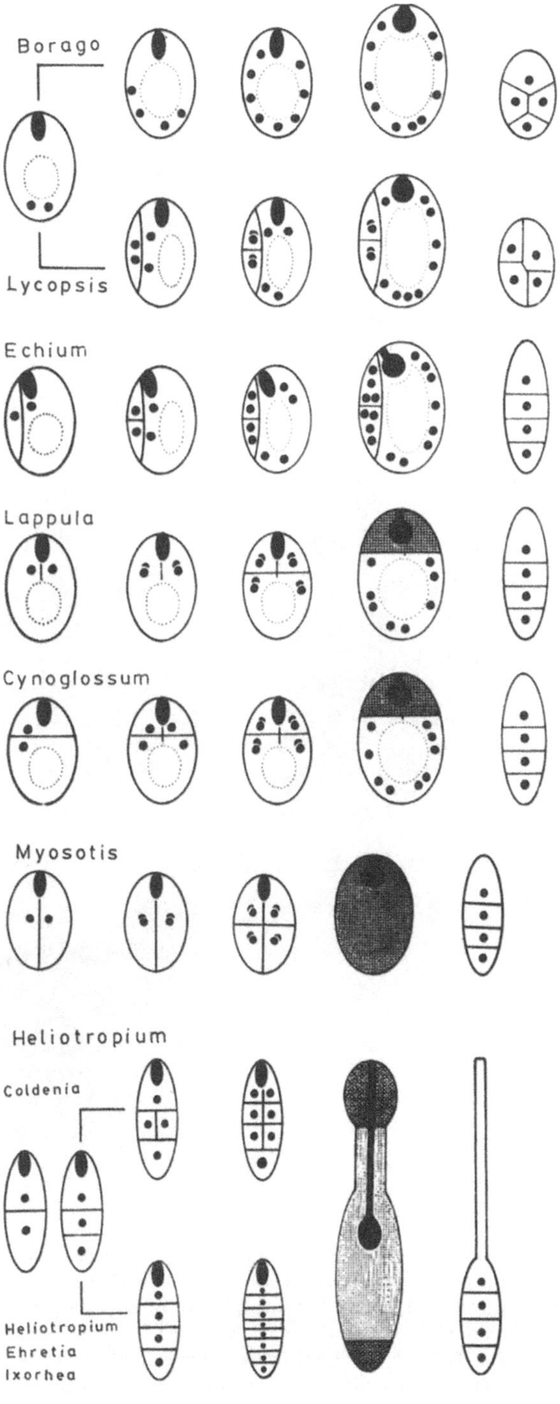

Fig. 48.6.1. Boraginaceae, progressive development of different types of endosperm. Cellular endosperm *cross-hatched*, embryo *black*. The *first four figures* in each row refer to endosperm; the *last row* represents four-celled proembryo.

The subfamily Boraginoideae is characterized by Borago, Lycopsis, Echium, Lappula, Cynoglossum and Myosotis types.

Helitropioideae, Ehretioideae and Cordioideae show Heliotropium type. In *Coldenia* there is a vertical division in the central chamber of the three-celled endosperm, whereas a transverse division occurs in *Heliotropium*, *Ehretia* and *Ixorhea*. The haustoria develop terminally. The four-celled proembryo (with a long suspensor) is similar in all the four taxa. (After di Fulvio 1981)

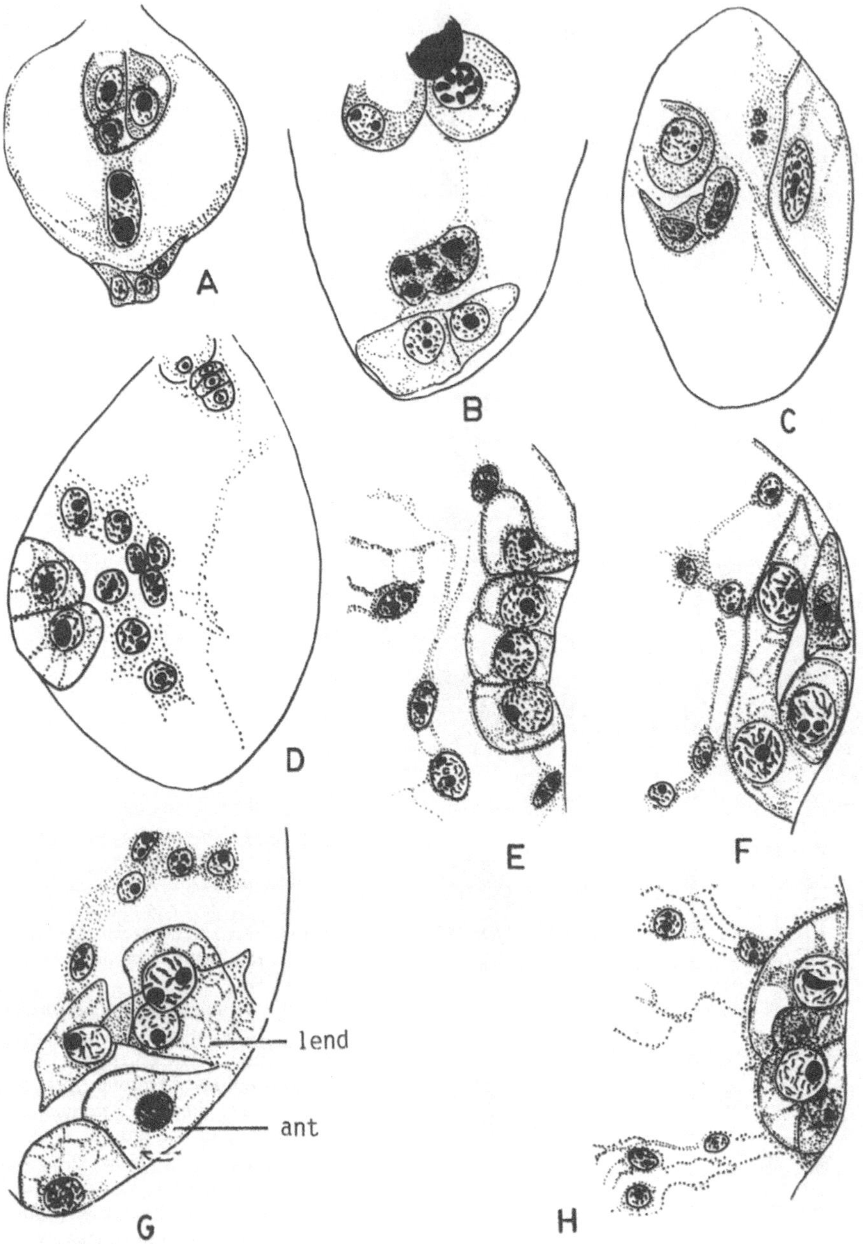

Fig. 48.6.2A–H. *Cerinthe minor.* **A** Mature embryo sac. **B** Embryo sac with remnants of pollen tube, persistent synergid, zygote, and a large primary endosperm nucleus (presumably formed by fusion of several polar nuclei?), and only two antipodal cells. **C** Zygote, and two-celled endosperm with a lateral cell cut off by vertical division of primary endosperm nucleus, the other endosperm nucleus has undergone free-nuclear division. The two cells lying below the zygote appear to be displaced antipodal cells (?). **D** Proembryo with free-nuclear endosperm proper (central chamber),

Cynoglossum officinale, Lindelofia longiflora, Omphalodes linifolia, O. verna and *Solenanthus apernninus* of Cynoglosseae (Svensson 1925), and *Plagiobothrys tenellus* (Fig. 48.6.3A–C).

5) **Myosotis type.** The division of the primary endosperm nucleus is vertical, and the next division is transverse. Further divisions are followed by walls and the entire endosperm becomes completely cellular: *Myosotis* sp. (Svensson 1925).

6) **Cynoglossum type.** A wall follows the division of the primary endosperm nucleus separating a small micropylar and a large chalazal chamber. Further development in the micropylar chamber is cellular. An incomplete vertical wall divides the chalazal chamber, the wall soon disintegrates, free-nuclear divisions take place till the dicot stage of embryo, and finally wall formation occurs: *Cynoglossum denticulatum* (Khaleel 1974).

7) **Heliotropium type.** Normal Cellular type of development is the rule but the subsequent behaviour does not resemble any of the types, and hence this additional type. Successive transverse divisions following the division of primary endosperm nucleus give rise to a linear row of six to eight cells in *Ehretia, Ixorhea* and *Heliotropium.* The terminal cells develop into haustoria, whereas the remaining cells give rise to endosperm proper. In *Coldenia,* the transverse division of the primary endosperm nucleus is followed by another transverse division in the chalazal region. The central cell develops into the endosperm proper, while the terminal cells develop into haustoria. The micropylar haustorium is four-celled in *Ehretia laevis* (B.M. Johri and Vasil 1956), *Coldenia procumbens* (Venkateswarlu and Atchutarammurthi 1955b) and *Heliotropium* spp. (P.K. Pal 1963), and two-celled in *H. scabrum* and *H. strigosum* (Khaleel 1978). The chalazal haustorium is four-celled in *Ehretia laevis* and *Heliotropium* spp. (P.K. Pal 1963), two-nucleate in *Coldenia procumbens* (Venkateswarlu and Atchutarammurthi 1955a, b), and uninucleate in *Heliotropium ovalifolium* (P.K. Pal 1963), *H. scabrum* and *H. strigosum* (Khaleel 1978).

The development of the endosperm in *Cerinthe major* is assigned to the Lycopsis type (Svensson 1925), but that of *C. minor* (Malecka 1981) differs from the Lycopsis type. The mature embryo sac is seven-nucleate with the egg apparatus, secondary nucleus and three antipodals (Fig. 48.6.2A). The synergids are large and prominent with diploid nuclei; diploidy is attained due to endomitosis. During fertilization one of the synergids degenerates, while the nucleus of the other becomes tetraploid. In Figure 48.6.2B the primary endosperm nucleus has an irregular outline and appears to be the product of

---

and two-celled lateral endosperm. **E** Portion of embryo sac to show free-nuclear endosperm proper, and four-celled lateral uniseriate endosperm. **F** Free-nuclear endosperm proper, and three celled lateral endosperm, with one cell binucleate. **G** Free-nuclear endosperm proper, and three-celled lateral endosperm (*lend*), two antipodal cells (*ant*) also persist. **H** Free-nuclear endosperm proper, and two healthy and two degenerated (?) cells of lateral endosperm. (After Malecka 1981) fusion of more than two polar nuclei. Under abnormal conditions the duplication of chromosomes takes place in the primary endosperm nucleus, due to endomitosis or inhibited prophase.

The division of the primary endosperm nucleus is followed by an oblique wall which separates a small lateral cell from the large cell (Fig. 48.6.2C). The nucleus of the large cell undergoes free-nuclear divisions (Fig. 48.6.2D). The mitotic divisions are regular, and the nuclei remain triploid. The zygote divides when the micropylar cell has 10–15 free endosperm nuclei.

The lateral cell divides by a wall perpendicular to the long axis of the embryo sac. These cells may undergo one more mitotic division so that four lateral cells are formed (Fig. 48.6.2E). Rarely, in one cell, the division may not be followed by cytokinesis so that it remains binucleate (Fig. 48.6.2F). The lateral cells, along with two persistent antipodals, form a 'lateral complex' (Fig. 48.6.2G). The degree of ploidy of the nuclei of the lateral cells corresponds to 3n, 6n and 12n. In rare cases, one or two lateral cells may collapse (Fig. 48.6.2H).

Sometimes, the endosperm nuclei in the vicinity of lateral cells may become surrounded by walls and form a multicelled 'lateral complex' together with lateral cells.

The embryogeny is variable. Onagrad type occurs in *Ehretia laevis* (B.M. Johri and Vasil 1956) and *Heliotropium scabrum* (Khaleel 1978); Asterad type in *Cynoglossum officinale* (Crété 1955b), *Trichodesma amplexicaule* (P. Khanna 1964b), *T. indicum* and *T. zeylanicum* (Tasneem 1967), and *Omphalodes linifera* (Souèges 1958c); Solanad type in *Heliotropium curassavicum, H. ovalifolium* (P.K. Pal 1963), *H. strigosum* (Khaleel 1978); Chenopodiad type in *Coldenia procumbens* (Venkateswarlu and Atchutarammurthi 1955b), *Cynoglossum denticulatum* (Khaleel 1974), *Eritrichium strictum* (Crété 1953a), *Heliotropium indicum* (P.K. Pal 1963), *H. peruvianum* (Souèges 1943d), *Plagiobothrys tenellus* (Fig. 48.6.3A–T) and *Rotula aquatica* (Nagaraj and Tasneem 1967).

The seed is albuminous with a straight or curved embryo. The outer epidermis of the testa comprises palisade cells which have thick lignified outer walls with pits. The mesophyll is crushed. In *Cynoglossum denticulatum* (Khaleel 1974), the seed-coat comprises four or five layers of thin-walled cells. The seeds remain enclosed in the nutlets; the nutlets are released. The seed-coat in *Rotula aquatica* (Nagaraj and Tasneem 1971) has slight thickenings on the radial and inner tangential walls. The hard stony pericarp gives protection to the seeds.

Apospory occurs frequently in Boraginaceae. The aposporous embryo sacs develop from the nucellar tissue in the chalazal region. In *Cynoglossum denticulatum* (Khaleel 1974), of the four or five embryo sacs only one reaches up to the eight-nucleate stage. There are six to eight embryo sacs in *Cordia alba*, and six to twelve in *C. obliqua* (Khaleel 1975). In *C. alba*, the development is usually arrested at the binucleate stage, sometimes the embryo sac reaches maturity. The normal embryo sacs do not develop endothelium and degenerate. Perhaps the embryo develops from the aposporous embryo sac. In *Cordia sebestena* (Hanumantha Rao and Prakasa Rao 1984), nucellar apospory occurs in 20% of ovules.

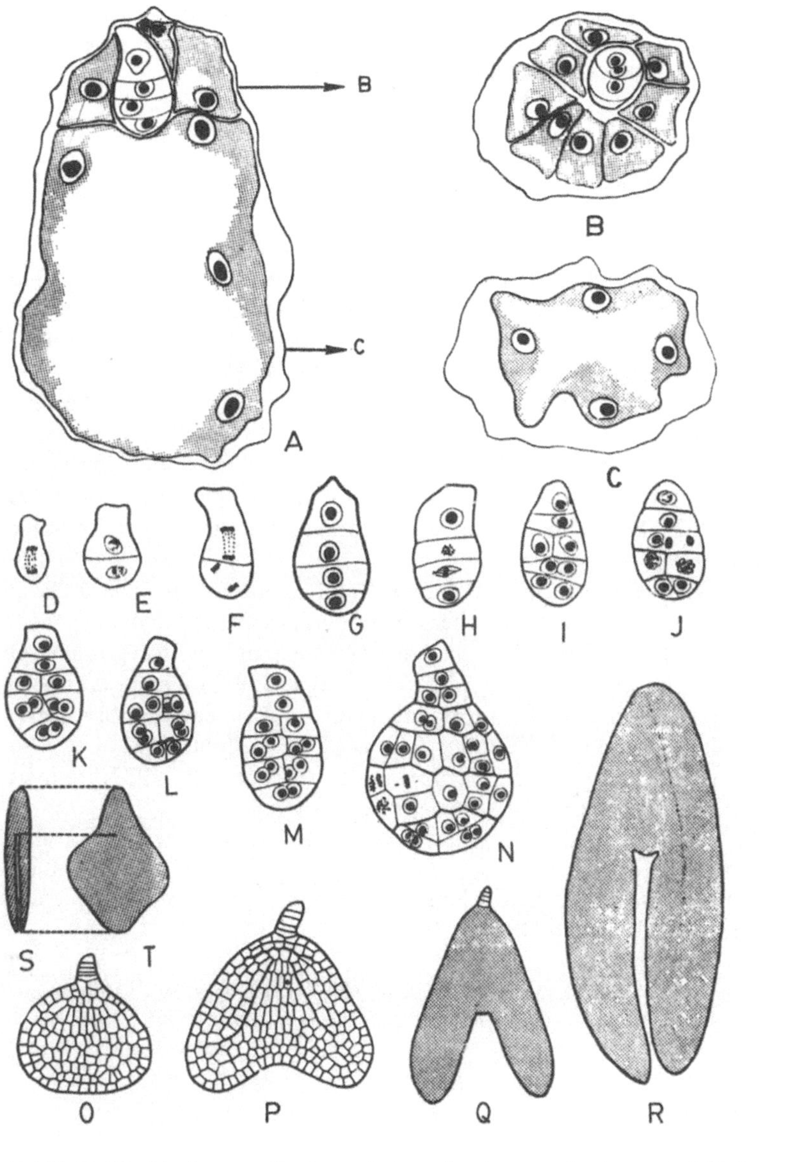

Fig. 48.6.3A–T. *Plagiobothrys tenellus.* Lappula type of endosperm development. **A** Embryo sac with four-celled proembryo, cellular micropylar endosperm chamber, and coenocytic chalazal chamber. **B, C** Cross-sections at levels marked in **A**. **D–R** Progressive development of dicotyledonous embryo, Chenopodiad type. **S** Longisection of embryo. **T** Face view, the levels of radicle and stem tip are indicated. (After di Fulvio 1966)

M Anantaswamy Rau (1951g), SS Bhattacharya and JN Mitra (1952), EJH Corner (1976), P Crété (1953a, 1955b), TE di Fulvio (1965, 1966, 1978, 1979, 1980), B Hanumantha Rao and PS Prakasa Rao (1980, 1984), HH Hilger (1989), BM Johri and IK Vasil (1956), BM Joshi and NK Dwivedi (1980a), S Junnel (1938), TF Khaleel (1974, 1975, 1977, 1978, 1982), P Khanna (1964b), P Maheshwari (1950), J Malecka (1981), V Millsaps (1940), M Nagaraj and F Tasneem (1967, 1968, 1971),

PK Pal (1963), BA Ramamurthi (1958), K Schnarf (1929, 1931), R Souèges (1943d, 1958c), HG Svensson (1922, 1925), BGL Swamy (1974a), BGL Swamy and N Parmeswaran (1963), F Tasneem (1966, 1967), J Venkateswarlu and B Atchutarammurthi (1955a, b)

48.7 Lennoaceae

The embryology has not been investigated in sufficient detail.

The anther wall comprises persistent epidermis, well-developed endothecium, an ephemeral middle layer, and glandular tapetum with one- or two-nucleate cells. Copeland's (1935) figure of *Pholisma arenarium* does not show fibrous thickenings in the endothecium (see Davis 1966).

Ammobroma and *Lennoa* have three-colporate grains, and *Lennoa* may occasionally show a four-colporate condition. In *Pholisma* four-colporate grains predominate, but three- and five-colporate pollen grains also occur. The pollen structure is useful in considerations of the systematics of the family (Drugg 1962). The pollen grains are shed singly at the two-celled stage (see Schnarf 1931).

The ovule is anatropous, unitegmic, and tenuinucellar. The archesporial cell functions as megaspore mother cell, undergoes meiotic divisions, the tetrad is linear, and the chalazal megaspore develops into a Polygonum type of embryo sac in *Lennoa madreporoides* (Süssenguth 1927). The polar nuclei fuse before fertilization, and the antipodals persist during early embryogeny.

The development of the endosperm is of the Cellular type.

The embryogeny in *L. madreporoides* (Süssenguth 1927) follows the Caryophyllad type, and the suspensor is much reduced.

The seeds are small with copious endosperm, and a spherical undifferentiated embryo (Willis 1973, see also Hutchinson 1973). The cells of the outer epidermis show reticulate thickenings. The mesophyll and inner epidermis become crushed (Corner 1976).

HF Copeland (1935), EJH Corner (1976), GL Davis (1966), WS Drugg (1962), J Hutchinson (1973), JN Mitra (1947a), K Schnarf (1931), K Süssenguth (1927), JC Willis (1973)

48.8 Verbenaceae

The anther wall comprises the epidermis, fibrous endothecium, a single middle layer, and secretory tapetum with multinucleate cells in *Lippia nodiflora* (J.K. Maheshwari 1954). Fibrous thickenings also occur in the connective region. The tapetal protoplasts wander into the locule, but do not form a true periplasmodium in *Lantana* (see Padmanabhan 1970b). The microspore mother cells undergo meiotic divisions, cytokinesis is simultaneous, and the tetrads are tetrahedral, decussate and isobilateral.

The family is multipalynous. The tricolpate pollen grains are two- or three-celled at the dispersal stage.

The ovule is anatropous, unitegmic, and tenuinucellar. The nucellus collapses early. At the megaspore mother cell stage, an endothelium develops in *Lippia* but is absent in the micropylar region.

The archesporial cell functions as megaspore mother cell, undergoes meiotic divisions, produces a linear tetrad, and the chalazal megaspore develops into a Polygonum type of embryo sac. The synergids are beaked. The polar nuclei fuse before fertilization. In *Lippia nodiflora* (J.K. Maheshwari 1954), the polars fuse after the entry of the pollen tube into the embryo sac. The antipodals are small and ephemeral in *Lippia*, *Tectona* and *Vitex* (N. Pal 1951b). They are multinucleate in *Lantana* (Thathachar 1940, Padmanabhan 1959), and undergo secondary multiplication in *Clerodendron aculeatum* and *Holmskioldia sanguinea*, and *C. phlomidis* (K.C. Misra 1939).

In *Avicennia officinalis* (Padmanabhan 1960b), due to the arrested growth of the integument, the micropyle is not organized. Consequently, the embryo sac lies exposed in the locule. The nucellus is ephemeral and the endothelium does not differentiate. The chalazal megaspore of the tetrad develops into an eight-nucleate embryo sac. Rarely, two micropylar megaspores remain healthy up to two-nucleate embryo sac, or a single megaspore persists up to the four-nucleate stage. Sometimes, a megaspore or a two-nucleate embryo sac persists during post-fertilization stages (Fig. 48.8.2A, B). A similar feature has been observed in *Bouchea incrassata* (Junell 1934). A non-functional megaspore persists in *Citharexylum ilicifolium* and *Petraea volubilis* (Junell 1934).

Kapil and Vani (1966) report that in *Nyctanthes arbor-tristis* 50% of embryo sacs develop according to the Polygonum type; the tetrad may be linear or T-shaped. The micropylar megaspores degenerate, and mitotic divisions in the chalazal megaspore produce two-, four- and eight-nucleate embryo sacs. The antipodals are small and ephemeral. Twin embryo sacs may develop (with a common endothelium), one bi- and the other four-nucleate.

There is a tendency for bisporic development of the embryo sac, Endymion or Allium type. The dyad cells may become binucleate, or only the upper dyad becomes binucleate, or a binucleate embryo sac develops from the chalazal dyad.

G. Mathur and Mohan Ram (1986) have studied pollination in *Lantana camara*. The freshly opened flowers are yellow (rich in β-carotene) and undergo post-anthesis chromatic alteration to orange, scarlet and magenta due to synthesis of delfinidin monoglucoside. Pollination is brought about by *Thrips hawaiiensis* and *Haplothrips tenuipennis* (Thysanoptera). The flowers are bisexual with a trilobed papillate stigma. The viscous stigmatic exudate is rich in lipids and reducing sugars. The floral structure does not permit the thrips to reach the nectar; they feed on stigmatic exudate and pollen. The pollen production and pollen viability are higher in the lower pair of anthers than in the upper. The flowers are self-fertile but require a biotic agent for pollination. Between thrips and butterflies the former are regular and more efficient pollinators and lead to a higher fruit-set. The butterflies appear in certain months, and even if they are excluded there is no marked decrease in fruit-set. The thrips selectively visit yellow flowers, and avoid flowers of other colours because the yellow colour is associated with the availability of food rewards.

The flower colour changes after pollination and preference of yellow flowers by thrips significantly enhances the efficiency of pollination and seed production.

In *Nyctanthes arbor-tristis*, the development of endosperm is of the Cellular type. The primary endosperm nucleus is located in the centre of the embryo sac. Its division is followed by a transverse wall. Further divisions give rise to a multicellular, biseriate endosperm. The haustoria do not develop.

In other members, the development conforms to the Stachys and Brunella type.

The Stachys type has been observed in *Clerodendron phallax* (Junell 1934) and *Verbena officinalis* (Crété 1942). The division of primary endosperm nucleus is followed by a transverse wall forming a micropylar and a chalazal chamber. A transverse division occurs in the micropylar chamber so that a linear row of three cells is formed (Fig. 48.8.1A, B). Figure 48.8.1C shows

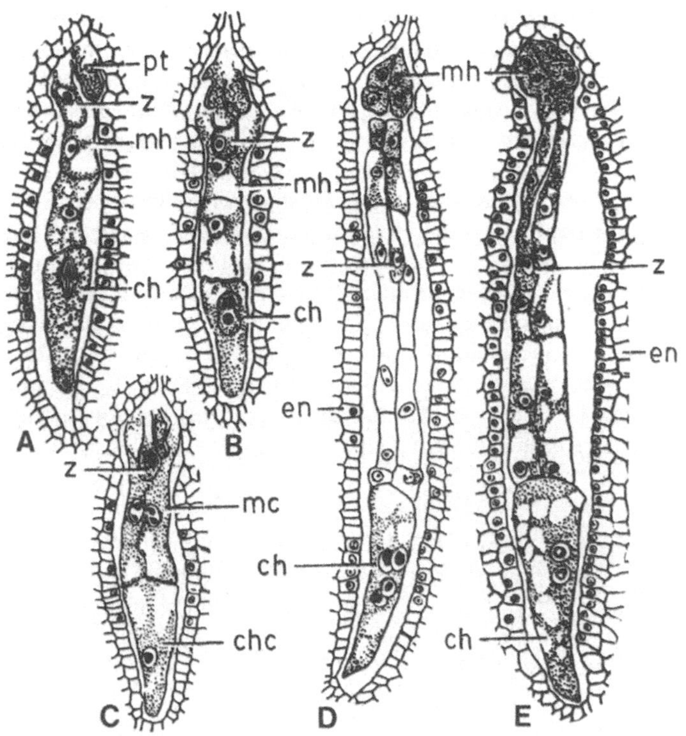

Fig. 48.8.1A–E. *Verbena officinalis*, development of endosperm. **A–C** Stachys type. **A** Three-celled endosperm, the two upper cells are derived from micropylar chamber. The zygote (*z*) is undivided in **A–E**; *pt* remnants of pollen tube; *mh* micropylar haustorium; *ch* chalazal haustorium; **B** Embryo sac with uninucleate micropylar haustorium (*mh*), endosperm proper initial (middle cell), and binucleate chalazal haustorium (*ch*); **C** Embryo sac shows Brunella type of endosperm development, the micropylar chamber (*mc*) is vertically divided; *chc* chalazal chamber. **D, E** Embryo sac with zygote (*z*), endosperm proper, four-celled micropylar haustorium (*mh*), and four-nucleate chalazal haustorium (*ch*) in **D** and binucleate in **E**. The endothelium (*en*) is very prominent in **A–E**. (After Crété 1942).

a vertical division in the micropylar chamber. The terminal cells develop into haustoria, whereas the middle cell produces the cellular endosperm proper.

In the Brunella type, the division of primary endosperm nucleus results in a micropylar and a chalazal chamber. A vertical wall followed by transverse walls in the micropylar chamber gives rise to two tiers of two superposed cells each. The upper tier develops into the micropylar haustorium, the middle tier produces the endosperm proper, and the chalazal chamber develops into a haustorium (Fig. 48.8.1C–E).

The micropylar haustorium (Fig. 48.8.1D, E) is two-celled in *Avicennia marina* and *Clerodendron phallax* (Junell 1934), and four-celled in *Lantana camara* (Crété 1942, Padmanabhan 1959), *Priva cordifolia* (Thirumaran and Lakshmanan 1984), *Stachytarpheta indica* (Thathachar 1940), *S. trichotoma* (Junell 1934) and *Verbena officinalis* (Crété 1942).

In the multicelled micropylar haustorium of *Lippia nodiflora* (J.K. Maheshwari 1954), the uppermost tier differentiates first. It functions for a short time only, followed by the cells below it, and then the next one functions similarly. This type of successive tiers of cells functioning as haustorium is also reported in *Priva cordifolia* (Thirumaran and Lakshmanan 1981).

The chalazal haustorium is unicellular and binucleate in *Lantana camara* (Crété 1942), *Lippia nodiflora* (N. Pal 1951b, J.K. Maheshwari 1954), *Stachytarpheta* (Thathachar 1940), *Tectona grandis* (N. Pal 1951b), *Verbena hispida* (Junell 1934), and two- to four-nucleate (Fig. 48. 8.1D, E) in *Verbena officinalis* (Crété 1942). A four-celled chalazal haustorium is reported in *Callicarpa macrophylla* (Junell 1934).

The development of the endosperm in *Avicennia officinalis* (Padmanabhan 1964) is very interesting. The division of primary endosperm nucleus is followed by a transverse wall forming a micropylar and a chalazal chamber (Fig. 48.8.2A, B). The micropylar chamber also divides by a transverse wall (Fig. 48.8.2C, D). The upper cell undergoes a vertical division and both the cells function as micropylar haustorium (Fig. 48.8.2E, F). The central cell divides repeatedly and produces the endosperm proper (Fig. 48.8.2D–G). The chalazal chamber develops into a uninucleate haustorium (Fig. 48.8.2E–G).

The two-celled micropylar haustorium exhibits a peculiar mode of development. One cell is eight-nucleate and the other four-nucleate, and the intervening wall has plasmodesmata (Fig. 48.8.3A). The haustorium develops papillate projections on its surface. The lower pole of the eight-nucleate cell puts out an outgrowth which extends as far as the chalaza and develops four to five branches (Fig. 48.8.3B). During further growth of these branches, the ramifications penetrate the ovule and reach up to the placenta. The nuclei lie scattered in the cytoplasm, and one or two nuclei may enter the lateral branches (Fig. 48.8.3B). The older as well as the newly formed branches grow wider. In the placenta the ramifications appear like a tracheal system. The growth of these branches is intercellular, and they dissolve the middle lamella of adjoining cells (Fig. 48.8.3C). The wall of the micropylar haustorium is thick and resistant to macerating agents. The nuclei of the micropylar haustorium undergo one or two divisions so that it becomes 16- to 32-nucleate. The four-nucleate sister cell does not grow, but remains in contact with the endosperm cells through plasmodesmata.

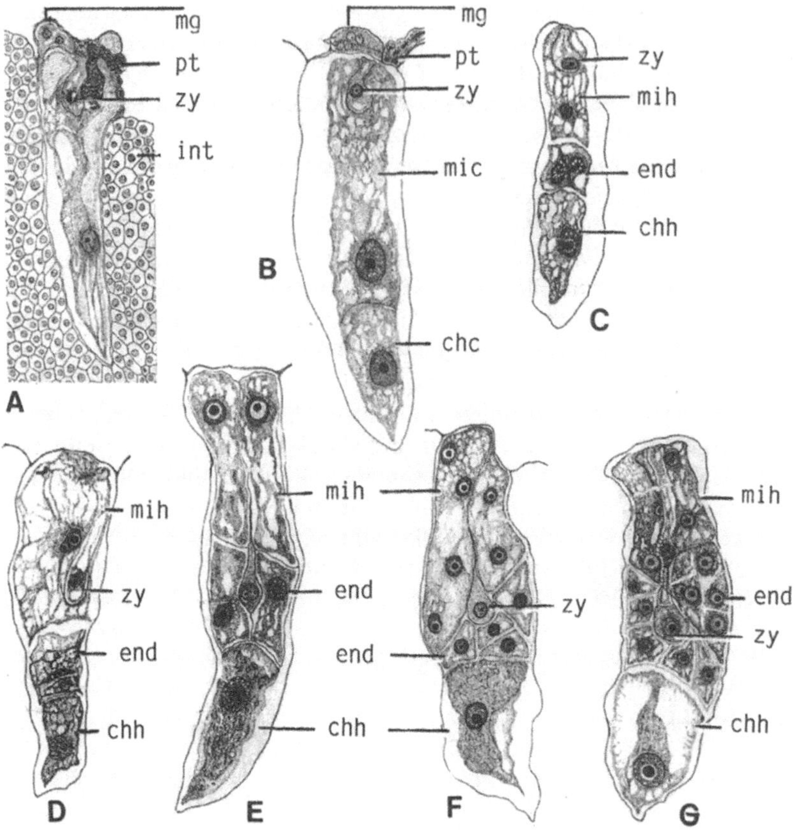

Fig. 48.8.2A–G. *Avicennia officinalis.* **A** Embryo sac with zygote (*zy*) and primary endosperm nucleus. Note binucleate persistent megaspore (*mg*) and remnants of pollen tube (*pt*). The micropyle has not been organized; *int* integument. **B** Two-celled endosperm – micropylar (*mic*) and chalazal chamber (*chc*). Two binucleate persistent megaspores (*mg*) and remnants of pollen tube (*pt*). **C, D** Differentiation of micropylar (*mih*) and chalazal haustorium (*chh*), and endosperm proper (*end*). **E** Later stage, micropylar haustorium two-celled. **F** Of the two-celled micropylar haustorium (*mih*), one cell is four-nucleate and the other binucleate; endosperm proper (*end*) and uninucleate chalazal haustorium (*chh*). **G** Later stage. The zygote (*zy*) has not divided in **B–G**. (After Padmanabhan 1964)

Due to extensive growth of the cellular endosperm, part of it, along with the embryo, becomes extra-ovular and comes to lie in the ovary. Thus, the micropylar haustorium in contact with the cellular endosperm also becomes exposed.

The chalazal haustorium is unicellular and uninucleate, and active only during the early stages. The nucleus becomes hypertrophied and its wall adjoining the chalaza is thick, but the rest of the nuclear wall remains thin.

The cellular endosperm surrounds the developing embryo. According to Swamy and Padmanabhan (1961a), the cellular endosperm does not contain

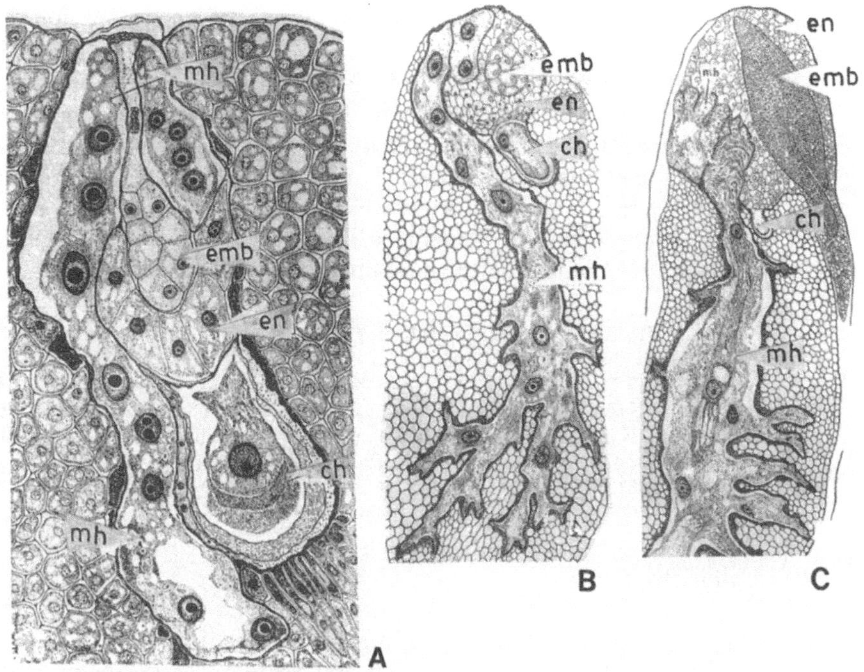

Fig. 48.8.3A–C. *Avicennia officinalis.* **A** The embryo sac shows early globular proembryo (*emb*), two-celled micropylar haustorium (*mh*), endosperm proper (*en*), and uninucleate chalazal haustorium (*ch*). The *right* arm of micropylar haustorium has four nuclei, and the *left* arm eight nuclei. The latter extends up to the chalaza. **B, C** Longisections through ovules, successive stages in the development of the micropylar haustorium. In both cases the haustorium invades and ramifies in the placenta. (After Padmanabhan 1966)

any food reserve. It transports the nutrients from the aggressive micropylar haustorium to the embryo. The suspensor is massive, remains in contact with the endosperm, and becomes a means of transport of food from the endosperm to the embryo.

In *Priva cordifolia* (Thirumaran and Lakshmanan 1981), the chalazal haustorium is one-celled and four- to six-nucleate. It develops tubular extensions into the nucellar tissue. The narrow tips of these extensions branch further. The cells of the endosperm proper in the vicinity of the chalazal haustorium are densely cytoplasmic. During the maturation of seed, the chalazal haustorium becomes inactive and gradually degenerates.

The embryogeny is of the Onagrad type in *Caryopteris mastacanthus* (Jaitly et al. 1968b), *Lantana camara* (Crété 1942) and *Verbena officinalis* (Souèges 1935b). Solanad type is reported in *Lippia nodiflora* (N. Pal 1951b) but, according to J.K. Maheshwari (1954), the embryogeny is of the Crucifer type (Onagrad). This aspect needs further study.

In *Nyctanthes arbor-tristis* (Kapil and Vani 1966), the embryogeny conforms to Period II, Megarchetype III (Caryophyllad type). The suspensor is long and uniseriate. The cotyledons are large and leafy, and occupy the entire seed cavity.

The seed is small, exalbuminous or with scanty endosperm, and a straight embryo. In *N. arbor-tristis* (Kapil and Vani 1966), the seed-coat is vascularized. In *Lippia nodiflora* (J.K. Maheshwari 1954), the outer epidermis and hypodermis of the testa are tangentially stretched and have cutinized walls. The outer two or three layers of mesophyll show well-developed reticulate thickenings. One to three layers of mesophyll comprise lignified cells in *Lippia*, *Tectona* and *Vitex* (Corner 1976); otherwise the mesophyll becomes crushed.

HK Airy Shaw (1951), EJH Corner (1976), P Crété (1942), K George and S Geethamma (1984), S Jaitly, JL Guignard and JC Mestre (1968b), S Junell (1934), RN Kapil and RS Vani (1966), JK Maheshwari (1954), G Mathur and HY Mohan Ram (1978, 1986), KC Misra (1937, 1939), HN Moldenke (1946), V Narmatha Bai and KK Lakshmanan (1982), D Padmanabhan (1959, 1960b, 1962a, 1964, 1970b), N Pal (1951b, c), MR Saxena (1975, 1981), R Souèges (1935b), MY Stant (1952), BGL Swamy and D Padmanabhan (1961a, 1962), T Thathachar (1940), K Thirumaran and KK Lakshmanan (1981, 1984)

48.9 Callitrichaceae

The embryology has not been studied in sufficient detail.

The anther tapetum is secretory with binucleate or, occasionally, multinucleate cells. Simultaneous cytokinesis follows meiotic divisions of microspore mother cells (see Schnarf 1931).

The pollen grains are inaperturate, or provided with thin irregular rifts. They are three-celled at the dispersal stage.

The ovule is anatropous and unitegmic with a massive integument (Figs. 48.9.3A, 48.9.4A). The innermost layer of integument differentiates into the endothelium. The archesporial cell functions as megaspore mother cell, undergoes meiotic divisions, and the chalazal megaspore develops into a Polygonum type of embryo sac (Jørgensen 1923). The polar nuclei fuse before fertilization, and the antipodals are small and ephemeral.

In *Callitriche heterophylla* and *C. palustris* (Philbrick 1984), the flowers are unisexual (Fig. 48.9.1A, B), and hydrophily has been reported. The pollen germinates within the intact unopened anthers, and pollen tubes grow through the vegetative tissues to effect self-fertilization. In both species the staminate and pistillate flowers often occur in the same leaf axil, or in opposite axils at the same node (Fig. 48.9.1A, B). The internodes are short, and flowering takes place in the upper four to six nodes of a floating rosette. After flowering, the internodes elongate and additional leaves and flowers are produced at the apex. The floating rosette becomes submerged after flowering, and fruit development occurs in water.

The pollen (in the underdeveloped intact anther) germinates, pollen tubes grow from the unopened anther sac downward through the stamenal filament (Fig. 48.9.2A–F), emerge from the base of the filament, grow across the vegetative tissue and upward into the ovary of the pistillate flower through its base. In the same plant, pollen tubes also grow across the width of the stem to

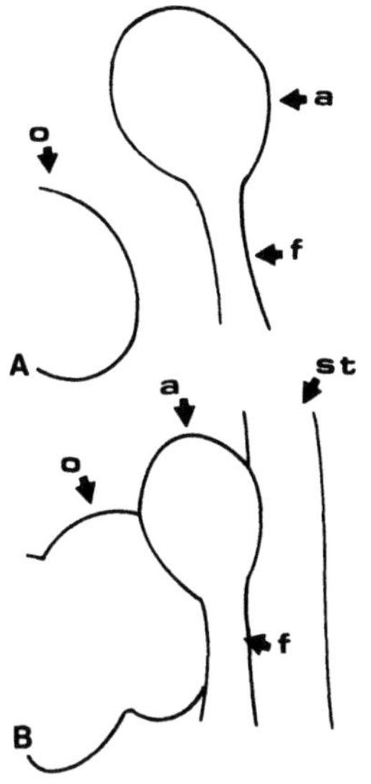

Fig. 48.9.1A, B. *Callitriche heterophylla.* **A** Idealized silhouette of Fig. 48.9.2**C** and **D** to show the location of staminate flower (*a* anther, *f* filament) and ovary (*o*). **B** Idealized silhouette of Fig. 48.9.2**E** and **F** to show location of staminate flower and ovary and their orientation to stem (*st*). (After Philbrick 1984)

the pistillate flower in the opposite leaf axil, or the tubes grow upward or downward the stem to flowers located at the next node.

The styles are linear in pistillate flowers. The pollen grains that land on stigma germinate, and the pollen tubes pass through the style to the ovary and reach the micropyle. In mature floating rosette flowers, pollen tubes were observed growing through the stamenal filaments in 85% of the sample. Only 10% of the typical stylar pattern of pollen tube growth was exhibited. It is possibly an adaptation to outcrossing via hydrophily, to ensure sexual reproduction in a changing environment. It is reported, for the first time, that pollen tubes traverse large segments of vegetative tissue, within a plant, to effect self-fertilization.

The mechanism of pollination and growth of pollen tube through the internode is a most unusual feature. Philbrick has based his observations merely on fluorescence microscopy. In our opinion it would have been very profitable to examine microtome sections as well as dissections without which his observations cannot be substantiated.

The development of the endosperm is of the Cellular type (Fig. 48.9.3A–I). The division of the primary endosperm nucleus is followed by a transverse wall, separating a large micropylar and a small chalazal cell (Fig. 48.9.3A, B). A vertical division occurs in each cell (Fig. 48.9.3C, D). Then the two upper cells divide transversely (Fig. 48.9.3E), and the uppermost tier develops into a micropylar haustorium. The middle tier gives rise to the endosperm proper, and the lowest tier develops into a chalazal haustorium (Fig. 48.9.3F–I) (Souèges 1952a). The 12-celled stage of endosperm in *Callitriche* (see P. Maheshwari 1950) can be distinguished into terminal cells and central cells (Fig. 48.9.4B). The micropylar haustorium of two to four long twisted cells is aggressive (Fig. 48.9.4C, D). The chalazal haustorium is eight-celled

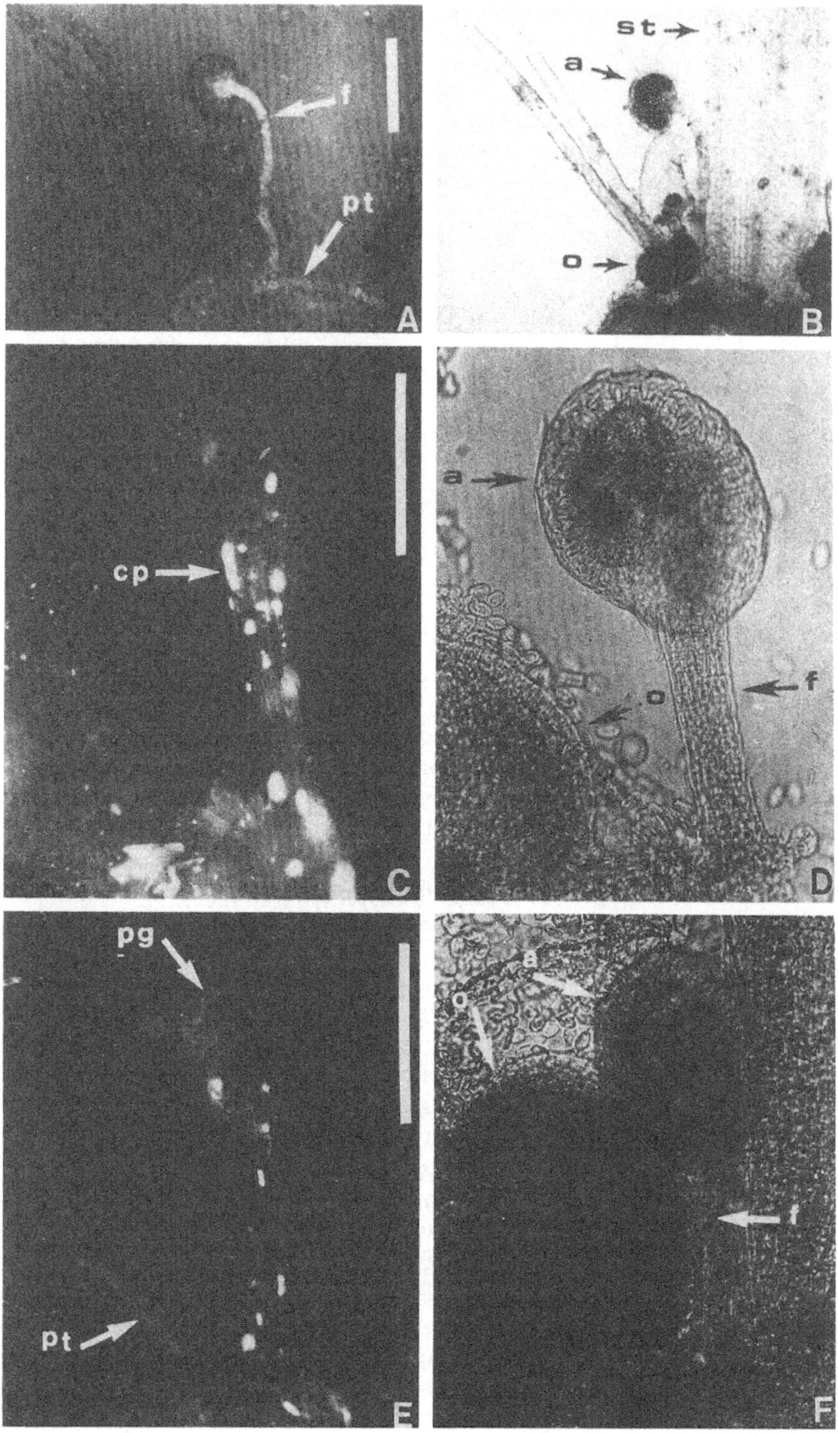
f
pt
A
st
a
o
B
cp
C
a
o
f
D
pg
pt
E
a
o
f
F

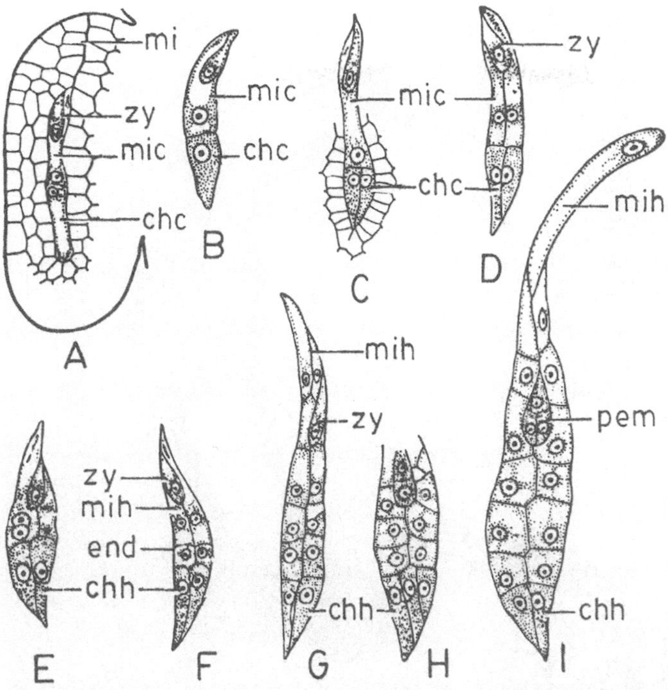

Fig. 48.9.3A–I. *Callitriche vernalis.* **A** Longisection of unitegmic (*mi* micropyle), tenuinucellar ovule. The embryo sac shows zygote (*zy*) and two-celled endosperm – micropylar (*mic*) and chalazal (*chc*) chamber; antipodals have degenerated. **B** Embryo sac with zygote, and two-celled endosperm. **C** Three-celled endosperm, vertical division in chalazal chamber. **D** Vertical division in both micropylar and chalazal chamber. **E** Nucleus of one cell of micropylar chamber in division, chalazal haustorium (*chh*) two-celled. **F** Terminal cells develop into micropylar (*mih*) and chalazal (*chh*) haustorium, central tier (*end*) produces endosperm proper. **G** Later stage. **H** Embryo sac at two-celled proembryo, endosperm, and chalazal (*chh*) haustorium. **I** Later stage. (After Souèges 1952a)

(Fig. 48.9.4E), located in a hook-like structure of the embryo sac (Jørgensen 1923, see also Davis 1966).

The embryogeny (see Rau 1962) in C. verna (Souèges 1952a) conforms to Period I, Group I, and Megarchetype IV with a tetrad in the A_2 category (Onagrad type). The suspensor is long and filamentous.

Fig. 48.9.2A–F. *Callitriche heterophylla.* [**A**, **C**, **E** Fluorescence (scale bar 100 µm) and **B**, **D**, **F** bright-field microscopy] **A**, **B** Staminate and pistillate flowers are at the same node. **A** The pollen tube (*pt*) grows in the filament (*f*) and traverses the node, cf. **B**; **B** Location of stem (*st*), anther (*a*) and ovary (*o*), cf. **A**. **C**, **D** Same, underdeveloped staminate flower and a portion of pistillate flower. Note fluorescing callose plugs (*cp*) that indicate pollen tubes within the filament in **C**, cf. **D** and Fig. 48.9.1**A**; **D** Location of anther (*a*), filament (*f*) of staminate flower and portion of ovary (*o*) of a pistillate flower, cf. **C** and Fig. 48.9.1**A**. **E** Pollen grains (*pg*) in the underdeveloped anther and pollen tubes (*pt*) that have grown through the filament, a pollen tube (*pt*) has entered the ovary through its base, cf. Fig. 48.9.2**F** and Fig. 48.9.1**B**. **F** Location of ovary (*o*) of pistillate flower and anther (*a*) and filament (*f*) of staminate flower. (After Philbrick 1984)

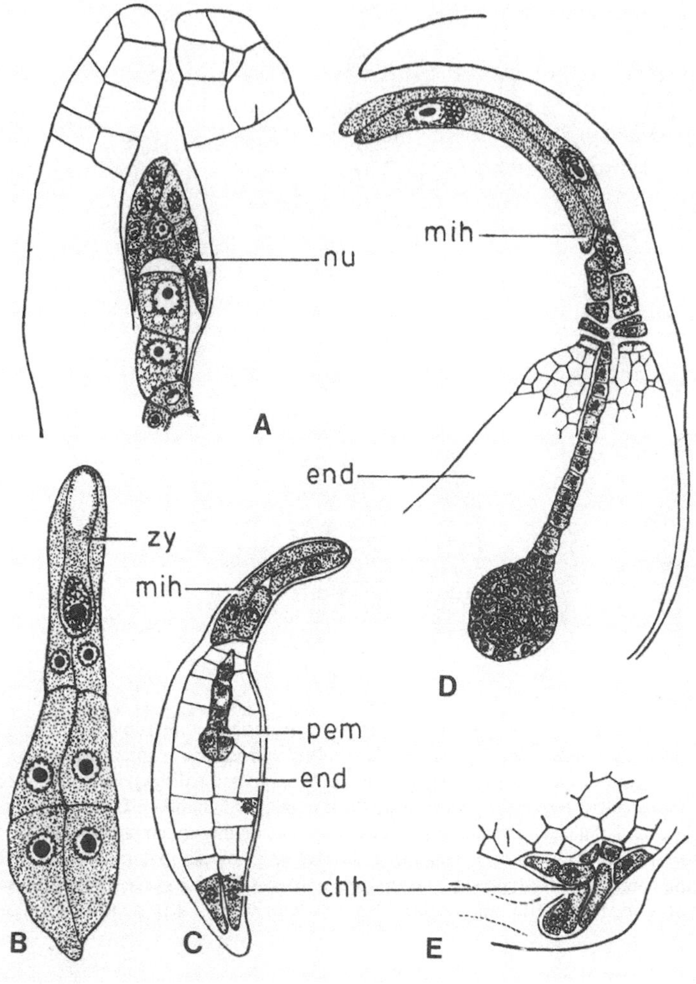

Fig. 48.9.4A–E. *Callitriche* sp. **A** Longisection of upper part of ovule to show single integument, nucellar (*nu*) epidermis, and two healthy megaspores capped by a binucleate dyad cell. **B** Twelve-celled endosperm, only six cells are seen. The terminal tiers produce the micropylar and chalazal haustorium, and the middle tier the endosperm proper. **C** Later stage, proembryo (*pem*), endosperm proper (*end*), and micropylar (*mih*) and chalazal (*chh*) haustorium. **D** Portion of embryo sac to show proembryo and upper part of endosperm (*end*); the micropylar haustorium (*mih*) extends through the micropyle up to the funicle. **E** Lower part of embryo sac to show eight-celled chalazal haustorium. (After Jørgensen 1923)

The seeds are minute with fleshy endosperm containing oily food reserve, and straight embryo. The seed-coat is membraneous, possibly the outer epidermis of the integument persists (see Corner 1976, Hutchinson 1973).

H Baillon (1858), B Clarke (1865), EJH Corner (1976), GL Davis (1966), J Hutchinson (1973), CA Jørgensen (1923, 1925), P Maheshwari (1950, 1964), F Pax and C Hoffman (1931), CT Philbrick (1984), MA Rau (1962), K Schnarf (1931), R Souèges (1952a)

48.10 Labiatae (Lamiaceae)

The anther wall comprises the epidermis (sometimes with prominent glands), fibrous endothecium, an ephemeral middle layer, and secretory tapetum of binucleate cells (Fig. 48.10.1A–F). The microspore mother cells undergo meiotic divisions, cytokinesis is simultaneous, and the tetrads are tetrahedral, isobilateral and decussate.

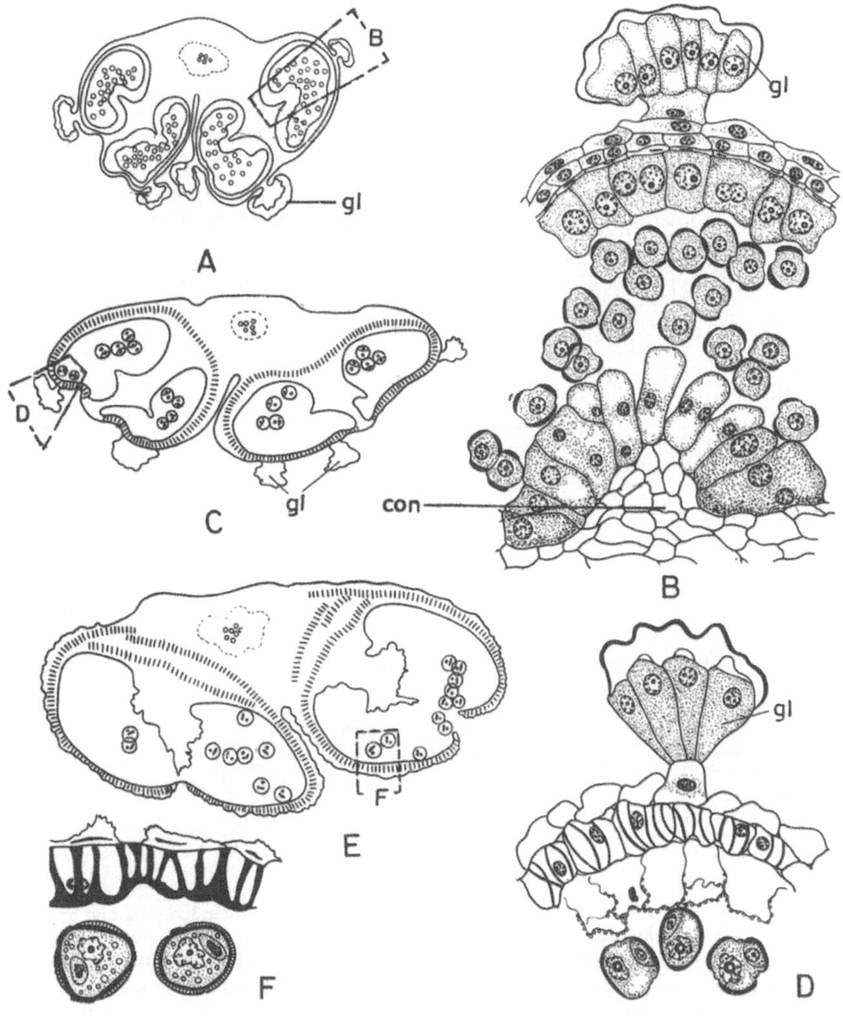

Fig. 48.10.1A–F. *Stachys trinervis.* **A, C, E** Cross-sections of anthers, outline diagrams for **B, D, F**. The *striations* in **C** and **E** represent endothecial fibrous thickenings. **B** *Inset* from **A**, secretory tapetum of binucleate cells which are more prominent on the connective-side (*con*), and uninucleate pollen grains. Note the anther wall in **B, D, F**, and peltate epidermal gland (*gl*) in **A–D**. **D** *Inset* from **C**, pollen grains two-celled; endothecium fibrous in **D** and **F**. **F** Later stage. (See Batygina and Yakovlev 1987)

The pollen grains are three- to six-colpate, and two- or three-celled at the dispersal stage.

The ovule is anatropous, unitegmic, and tenuinucellar. There is a well differentiated endothelium generally restricted to the middle region of the embryo sac, except in *Salvia coccinea* and *S. splendens* (Jaitly 1971). It develops at the tetrad stage in *Orthosiphon stamineus* (S.N. Murthy 1947). A funicular obturator, endothelium, and hypostase are reported in *Nepeta hindostana* (Dwivedi and Joshi 1981).

The hypodermal archesporial cell functions as megaspore mother cell, undergoes meiotic divisions, and the chalazal megaspore of linear tetrad develops into a Polygonum type of embryo sac. The egg is large and synergids hooked, polar nuclei fuse before fertilization, and antipodals are ephemeral. The latter persist in *Ajuga reptans* (Crété 1942).

The development of endosperm is of the Cellular type (Fig. 48.10.2A–F). The division of the primary endosperm nucleus is accompanied by a transverse wall separating a micropylar and a chalazal chamber (Fig. 48.10.2A).

According to further orientation of walls and the behaviour of daughter cells, the development of the endosperm is classified into four types (Schnarf 1931):

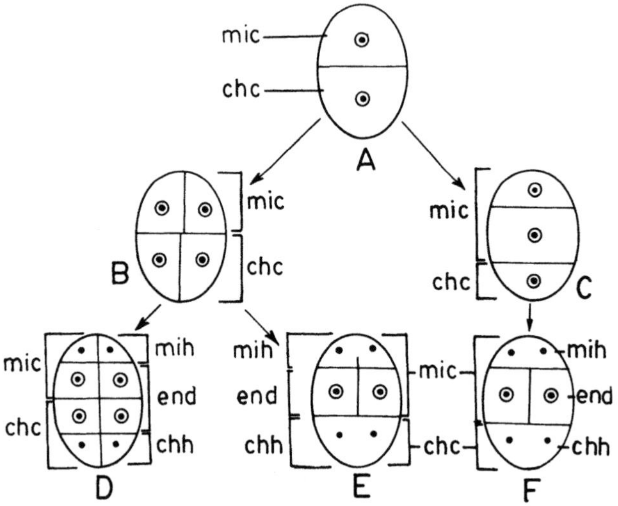

Fig. 48.10.2A–F. Schematic representation of development of endosperm in Labiatae and Verbenaceae (*black dots* denote the nuclei of haustorium). **A** Primary endosperm nucleus divides transversely to form micropylar (*mic*) and chalazal (*chc*) chamber. **B, C** The next division may be vertical in each chamber (**B**), or transverse only in micropylar chamber (**C**). **D Scutellaria type**: Transverse division in each chamber forms four tiers. The terminal tiers function as micropylar (*mih*) and chalazal (*chh*) haustorium, the central two tiers form initials of the endosperm (*end*) proper. **E Brunella type**: Incomplete vertical division in micropylar chamber is followed by a transverse division to form two-nucleate micropylar haustorium, and two initials of the endosperm proper. The chalazal haustorium is binucleate. **F Stachys type**: The terminal cells of three-celled endosperm (cf. **C**) develop into binucleate micropylar and chalazal haustorium, the central cell forms the endosperm proper. (Courtesy Dr M.A. Rau, Mysore Univ.)

1) Stachys type (Fig. 48.10.2C, F). The nuclear division in the micropylar chamber is followed by a transverse wall, resulting in a row of three superposed cells. The micropylar and chalazal cell develop into haustoria, and the middle cell gives rise to the endosperm proper in *Leucas aspera* (Murthy 1941b) and *Lavandula spica* (Fig. 48.10.3). Generally, the micropylar haustorium is four-nucleate. There are five nuclei in *Leucas biflora*, eight in *L. cephalotus*, and 12 in *L. aspera* (Jaitly 1968). In *Ocimum kilimandscharicum* and *Plectranthus incanus* (Santha Kumari 1976), it is four- to six-nucleate. The chalazal haustorium is binucleate in *Craniotome versicolor* (Dwivedi and Joshi 1980), *Hyptis suaveolens*, *Leucas* spp., *Ocimum sanctum* (Jaitly 1968), *Nepeta hindostana* (Dwivedi

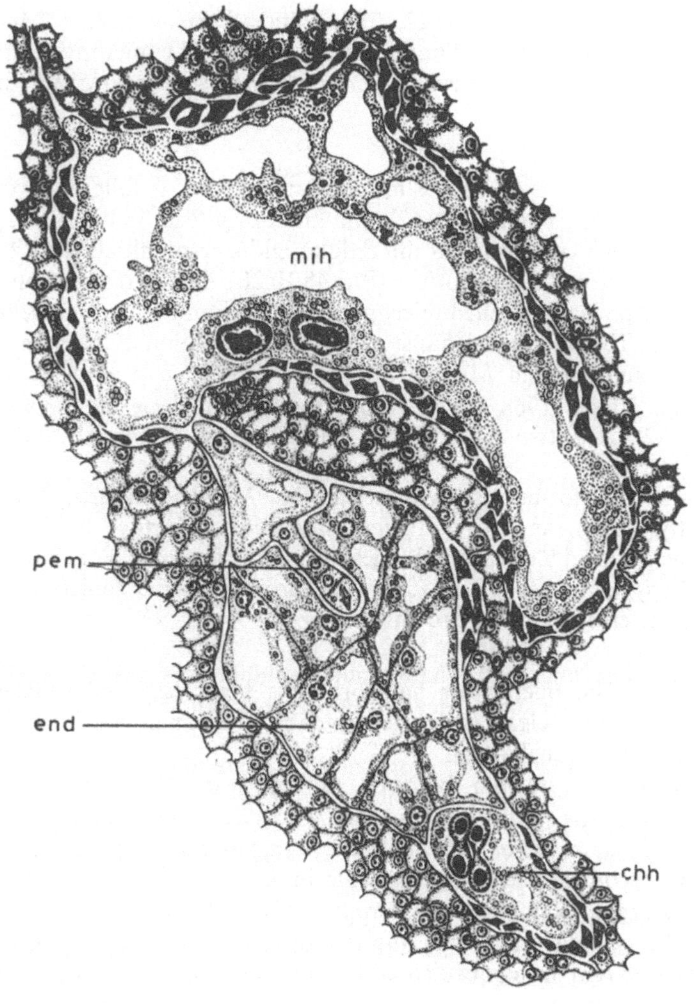

Fig. 48.10.3. *Lavandula spica*. Endosperm with binucleate micropylar (*mih*) and four(?)-nucleate chalazal (*chh*) haustorium, endosperm proper (*end*), and proembryo (*pem*). Note the starch grains. (See Batygina and Yakovlev 1987)

and Joshi 1981), *O. basilicum* (Crété 1942), and *Plectranthus rugosus* (B.M. Joshi and Dwivedi 1980b). In *O. kilimandscharicum* and *P. incanus* (Santha Kumari 1976), the haustorium develops sucker-like branches which invade the adjoining tissue.

2) Scutellaria type (Fig. 48.10.2B, D). The division of nuclei in the micropylar and chalazal chamber is followed by vertical walls. Divisions in transverse plane occur in all the four cells and four tiers of two cells each are produced. The middle two tiers produce the endosperm proper. The terminal tiers develop into haustoria. The micropylar haustorium is two- to four-nucleate, whereas the chalazal is binucleate: *Anisomeles indica* (Jaitly 1966), *Scutellaria altissima*, *S. galericula* (Schnarf 1917, 1931) and *S. minor* (Crété 1942), *S. violacea* and *Westringia rigida* (Santha Kumari 1976). The micropylar and chalazal haustorium is four-celled each in *Craniotome versicolor* (Dwivedi and Joshi 1980).

3) Brunella type. The division in the micropylar chamber is followed by a vertical wall which ends blindly towards the micropyle (Fig. 48.10.2B, E). A transverse wall in each cell separates the cellular endosperm initials from the partially bicelled micropylar haustorium (Fig. 48.10.2B). Due to dissolution of the blind wall, the partially bicelled micropylar haustorium becomes unicellular (Fig. 48.10.2E). The micropylar haustorium shows variation in the number of nuclei. It is binucleate in *Mentha arvensis*, *Pogostemon plectranthoides* and *Salvia* spp. (Jaitly 1968, 1971), four- to eight-nucleate in *Lamium* spp. (Crété 1942), and two-celled, each binucleate, in *P. plectranthoides* (Jaitly 1968).

The chalazal haustorium is two-nucleate in *Mentha arvensis*, *Salvia coccinea* (Jaitly 1968) and *Orthosiphon stamineus* (Murthy 1947). In *O. stamineus* the chalazal haustorium bulges out on the funicular side, but the thick-walled hypostase checks its downward growth in the funiculus. It collapses after the early globular proembryo.

4) Galeopsis type. The micropylar haustorium is binucleate; the chalazal haustorium remains uninucleate and degenerates early. It can be considered as a modification of the Brunella type.

The micropylar haustorium becomes multinucleate (eight nuclei) in *Lamium album*, *Leonurus cardiaca* (six nuclei), *Marrubium pannonicum* (two to six nuclei) and *Prasium majus* (16 nuclei). The cellular endosperm in the chalazal region becomes densely cytoplasmic in *Ajuga reptans*, *Amethystea coerulea*, *Perilla arguta* and *Teucrium scorodina* (Crété 1942).

In *Anisomeles indica* (Jaitly 1966), the micropylar haustorium lies in the broad micropyle. It remains connected with the cellular endosperm by richly cytoplasmic cells referred to as the "isthmus". The micropylar haustorium collapses after the late globular proembryo.

Development of the embryo corresponds to the Onagrad type (Crété 1942, 1963b, Jaitly et al. 1968a, c).

The seed is exotestal and exalbuminous (*Hyptis, Mentha, Ocimum, Salvia*), or albuminous (*Leucas, Pogostemon*), with a straight embryo. The endosperm is oily. The outer epidermal and hypodermal cells of the testa develop cellulosic spinules on the inner tangential walls. The spinules become thickened with lignin. The mesophyll and inner epidermis become crushed. The lignified cells of the outer epidermis and the hypodermis of the pericarp give protection to the seed. The rest of the wall layers of the pericarp also get crushed (Jaitly 1966).

In *Salvia* (Jaitly 1971), the testa is represented by two or three ruptured layers of cells. The radial walls of the outer epidermis are thick. As the testa is papery, the protective function is taken over by the pericarp (endocarp). The seeds are mucilaginous or non-mucilaginous, the mucilage is produced in the outer epidermis in *Brunella vulgaris* (Dwivedi 1984).

EP Bushnell (1936), EM Carlson and BC Stuart (1936), P Crété (1942, 1963b), NK Dwivedi (1984, 1985), NK Dwivedi and BM Joshi (1980, 1981, 1986), JK Ganguly (1948), MR Gurudev and B Gowda (1988), SC Jaitly (1966, 1968, 1971), SC Jaitly, JL Guignard and JC Mestre (1968a, c), BM Joshi and NK Dwivedi (1980b), S Junell (1937), SN Murthy (1940, 1941b, 1942, 1946a, b, 1947), M Rejdali (1990), D Santha Kumari (1976), K Schnarf (1917, 1931a), R Souèges (1951c), V Verechshagina (1987)

48.11 Nolanaceae

The anther is tetrasporangiate; placentoids develop in the anther loculi. The anther wall comprises an epidermis, fibrous endothecium, two ephemeral middle layers, and the secretory tapetum in *Nolana paradoxa* (di Fulvio 1969). Simultaneous cytokinesis accompanies meiotic divisions in microspore mother cells, and the tetrads are tetrahedral and decussate.

The trizonocolpate, oblate-spheroidal pollen grains have sexine thicker than nexine. They are two-celled at the dispersal stage.

The ovule is ana- to campylotropous, unitegmic, and tenuinucellate in *N. paradoxa*. The endothelium develops at the eight-nucleate stage of the embryo sac. The single archesporial cell functions as megaspore mother cell, undergoes meiotic divisions, forms a linear tetrad, and the chalazal megaspore develops into a Polygonum type of embryo sac. The synergids are hooked, the polar nuclei fuse before fertilization, and the antipodals persist for a short period after fertilization.

The development of the endosperm is of the Cellular type.

The embryogeny conforms to the Solanad type in *Nolana atriplicifolia* and *N. prostata* (S. Datta 1933). The proembryo is six-celled and filamentous. The apical cell gives rise to the cotyledons and the stem tip; the adjacent cell produces the hypocotyl and radicle, and the one next to it becomes the initial for the middle region of root cap. The suspensor is long and filamentous, and its uppermost cell is "haustorial".

According to Crété (1957b), in *N. prostata* the embryogeny conforms to Period I, Series C, Megarchetype V (Chenopodiad type). Di Fulvio (1969) mentions Polemonium variation, a transitional form between the Chenopodiad and Solanad type, in *N. paradoxa*.

The seeds are albuminous with a curved or spiral embryo (Hutchinson 1973). The outer epidermis of the integument persists as sclerified seed-coat.

P Crété (1957b), S Datta (1933), TE di Fulvio (1969), J Hutchinson (1973), W Rosén (1947)

48.12 Solanaceae

The anther wall comprises the epidermis, endothecium, two or three ephemeral middle layers, and glandular tapetum. The endothecium, outer middle layer and a few cells of the connective region develop fibrous thickenings, and tapetal cells (in the connective region) elongate in *Withania somnifera* (Mohan Ram and Kamini 1964). The endothecium is multilayered in *Nicotiana glutinosa* and *N. tabacum* (Jose and Singh 1968), whereas in *Solanum nigrum* (Saxena and Singh 1969) the fibrous thickenings are restricted only to the tip region. The tapetum in the connective region differentiates much earlier than on the other side in *Lycopersicon esculentum* (Orel and Zhakova 1976). In *Datura stramonium* (O' Neal 1920), the tapetal cells show three conditions: the majority of them are plasmodial, some are glandular, and some remain unchanged. Simultaneous cytokinesis accompanies meiosis in microspore mother cells, and tetrahedral, isobilateral and decussate tetrads are produced.

Placentoids develop in the anther loculi in *Capsicum annuum* var. *acuminatum* (Dharmadhaj and Prakash 1978).

The pollen grains are three- to five- or six-colpate, or colporate, and two-celled at the dispersal stage. Polyspory with micronuclei, and compound pollen grains occur in *Solanum nigrum* (Saxena and Singh 1969). The abnormal pollen grains show two to eight nuclei, but do not organize as pollen embryo sacs in *Capsicum* spp. (Dharmadhaj and Prakash 1978).

In nine species of *Capsicum*, Raghuvanshi and Singh (1984) report high male sterility. The failure of formation of fertile pollen is due to the malfunctioning of the tapetum, and erratic meiotic divisions in the pollen mother cells. The middle layers may disorganize, and the tapetum may degenerate precociously or become hypertrophied and highly vacuolate with irregular outline, the nuclei may enlarge, and excessive radial elongation of tapetal cells may occur. Because of erratic meiotic divisions, two- to four-nucleate microspores are formed. The nuclei finally degenerate and pollen grains become sterile.

The ovule is hemianatropous, anatropous, campylotropous or amphitropous, unitegmic and tenuinucellar. The nucellus persists for a long time in *S. phureja* (Dnyansagar and Cooper 1960).

A hypostase and an epistase are reported in *S. torvum* (Mohan and Singh 1969). An endothelium generally differentiates at the two-nucleate embryo sac stage, but is absent in *N. tabacum* (Raghavan and Srinivasan 1941a). It is prominent in *Capsicum*.

The archesporium is uni- to multicellular. A single cell functions as megaspore mother cell, undergoes meiotic divisions, produces a linear or T-shaped tetrad, and the chalazal megaspore develops into a Polygonum type of gametophyte. The embryo sac develops a chalazal caecum in *Solanum torvum*. The synergids are beaked. The antipodals are ephemeral.

Allium type of development of the embryo sac, rarely Polygonum type, occurs in *Capsicum frutescens* var. *tabasco*, *C. nigrum*, and *C. pendulum*. Both types occur in *C. annuum* var. *acuminatum* (Dharmadhaj and Prakash 1978). The frequency of the monosporic type is much higher in summer, of the bi- and tetrasporic in winter. Mohan (1970) reports Polygonum, Allium and Endymion types in *Solanum khasianum*. Monosporic gametophyte with a tendency towards bisporic type occurs in *Withania*. Bisporic type is reported in *Nicotiana glauca*, *N. rustica* × *N. glauca* (see S.C. Maheshwari 1955), and Adoxa type in *S. tuberosum* (Young 1922) and *S. tuberosum* var. jyoti gola (Govil 1980).

Multiple embryo sacs occur in *Solanum quineense* (Ivanova and Orel 1963). A case of reversed polarity in the mature embryo sac is reported in *Lycium europaeum* (Jain 1956).

In *Solanum* spp. (Lee and Cooper 1958), fertilization occurs between 24 to 48 h after pollination. The primary endosperm nucleus divides 48 h after pollination. The division of zygote occurs when the endosperm is approximately 40-celled.

All the three types of development of the endosperm are reported in Solanaceae. Nuclear type occurs in *Hyoscyamus orientalis*, *Salpiglossis picta*, *Scopolia atropoides* and *Schizanthus pinnatus* (see Mohan Ram and Kamini 1964), and Helobial in *Hyoscyamus niger* (Svensson 1926). Wunderlich (1959) described an "Intermediate" type in *H. niger*, but Swamy and Parmeswaran (1963) reject its Helobial affinity. The Cellular type is more common. In *Withania somnifera* (Mohan Ram and Kamini 1964), division of the primary endosperm nucleus is followed by a transverse wall. Further divisions may be longitudinal, transverse or oblique; the endosperm grows obliquely to the proembryo. The endosperm cells become large, and probably have polyploid nuclei.

In *Solanum phureja* (Dnyansagar and Cooper 1960), the endosperm is of the Cellular type. A multicellular chalazal haustorium develops 8 days after pollination, the cells have dense cytoplasm, are in direct contact with the chalazal cells, and establish connections with the vascular strand.

The embryogeny conforms to the Solanad type in *Sclerophylax caducifructus* (di Fulvio 1961, see also Prasad and Singh 1978). In *Capsicum annuum* it is of the Onagrad type (see Davis 1966).

Zygotic, synergid and integumentary polyembryony occurs in *Solanum torvum* (Mohan and Singh 1969). The development of reduced embryos in *Nicotiana* spp. (Belling and Blakeslee 1924), and reduced pseudogamy in *S. nigrum* (Jørgensen 1928) are noteworthy features.

Cooper (1943) observed the development of a four-celled haploid-triploid twin in a cross *Nicotiana glutinosa* (♀n = 12) × *N. tabacum* (♂n = 24). The larger triploid embryo formed as a result of fertilization of the egg cell, while the smaller developed parthenogenetically from a synergid.

Morgan and Rappleye (1954) observed haploid-haploid, haploid-diploid, diploid-diploid twins, and a diploid-diploid-diploid triplet in the reciprocal crosses of two varieties of *Capsicum frutescens*, Goliath and Hungarian yellow-wax (see Table 3.12.1).

The crosses *S. pinnatisectum* × *S. chacoense*, *S. pinnatisectum* × *S. kesselbrennari* and *S. pinnatisectum* × *S. yabari* exhibit failure of seed development (Lee and Cooper 1958).

In the cross *S. pinnatisectum* × *S. chacoense*, 10 days after pollination, the proembryo is club-shaped, and the endosperm is more or less disorganized. The proembryo collapses 18 days after pollination. The endothelium develops finger-like outgrowths which project into the embryo sac. Its further development has not been described.

In the cross *S. pinnatisectum* × *S. kesselbrennari*, the endosperm is two-celled 72 h after pollination; by this time syngamy has occurred. At 10 days after pollination, the proembryo and endosperm disintegrate. The endothelial cells proliferate 5 days after pollination, are highly vacuolate and elongated, and project and fill the central cavity of the embryo sac.

Fifteen days after pollination, in the cross *S. pinnatisectum* × *S. yabari*, the endosperm is ten-celled and the embryo disintegrates. The endothelial cells show proliferation.

The albuminous seed contains a straight or curved embryo, with flat or folded cotyledons. In *Withania somnifera* (Mohan Ram and Kamini 1964), the endosperm stores oil and starch as food reserve, in *S. phureja* (Dnyansagar and Cooper 1960) and *Petunia nyctaginiflora* (Cooper 1946) starch, and in *Datura* proteins. In *S. phureja* the testa is multiplicative. The outer surface of the outer epidermal cells of the seed-coat is cutinized. The epidermal cells elongate radially, develop thickenings on the inner tangential walls, and the innermost part of the radial walls. The outer walls of the epidermis break down, while the thickened radial walls persist and appear as minute hairs. In *Lycopersicon esculentum* the outer walls of the outer epidermal cells are thin, and later become mucilaginous; the inner and radial walls are lignified. In *Nicotiana* the epidermal cells are flattened, whereas in *Cestrum* they are polygonal. The outer hypodermis and the inner one, or two layers, are lignified in *Cestrum* and *Withania*. The mesophyll becomes crushed. The inner epidermis may collapse (*Cestrum*), or comprise small, rectangular thick-walled cells (*S. phureja*).

PJ Alemany (1985), P Ariz, N Iftikhar and NR Khan (1972), TJ Arnason (1943), MG Avanzi (1950), J Belling and AF Blakeslee (1924), C Bernard (1949), PN Bhaduri (1932, 1935, 1936), SW Brown (1949), AE Clarke (1940), DC Cooper (1931b, 1943, 1946), KVO Dahlgren (1923), GL Davis (1966), P Dharmadhaj and N Prakash (1978), TE di Fulvio (1961), VR Dnyansagar and DC Cooper (1960), M Fagherazzi-Abgrall and J Meyer (1963), CM Govil (1980), SR Grabovskaya (1968), JGTH Hermsen (1980), Hu Shi-yi, Le Le-gong and Zhu Cheng (1985), KV Ivanova and LI Orel (1963), TC Jain (1956), CA Jørgensen (1928), JS Jose and SP Singh (1968), H Kaur and RP Singh (1970), KV Krishnamurthy and KA Rao (1958), M Krüger (1932), R Lamm (1937), JH Lee and DC Cooper (1958), A Levan (1937), SC Maheshwari (1955), MV Magtang (1936), K Mohan (1966, 1968, 1970), K Mohan and B Singh (1969), HY Mohan Ram and I Kamini (1964), DT Morgan Jr and RD Rappleye (1954), A Müntzing (1937), A Nanetti (1912), F Novak and J Betlach (1970), AR Olson (1988), CE O'Neal (1920), LI Orel and MA Zhakova (1976), E Pacini and G Sarfatti (1978), T Prasad and D Singh (1978), TS Raghavan and AR Srinivasan (1941a), RK Raghuvanshi and D

Singh (1981, 1984), MS Ramanna (1979), OL Rees-Leonard (1935), S Satina, J Rapport and AF Blakeslee (1950), VK Sawhney and SK Bhadula (1988), T Saxena and D Singh (1969), R Souèges (1936f), J Stephen (1980), HG Svensson (1926), MS Swaminathan (1954), BGL Swamy and N Parmeswaran (1963), J Vallade (1970), TM Varghese (1970), IK Vasil (1962), JL van Went (1970), R von Wettstein (1895a), RI Walker (1955), EJ Williams (1955), R Wunderlich (1959), WJ Young (1922)

48.13 Duckeodendraceae

The embryology has not been studied.

48.14 Buddleiaceae

The anthers are bilocular, and the wall comprises the epidermis, fibrous endothecium, single middle layer, and secretory tapetum of binucleate cells in *Nicodemia diversifolia*. The microspore mother cells undergo meiotic divisions, cytokinesis is simultaneous, and the tetrads tetrahedral and isobilateral (Maheswari Devi and Lakshminarayana 1980).

The three- or four-colporate pollen grains are shed at the two-celled stage.

The ovule is anatropous, unitegmic, and tenuinucellar. An endothelium usually differentiates at the fertilized embryo sac stage in *Buddleia variabilis* (Crété 1942). A thick-walled lignified hypostase develops in *B. asiatica* (Bendre 1975). The archesporial cell functions as megaspore mother cell. Occasionally, two archesporial cells occur in *N. diversifolia*. The megaspore mother cell undergoes meiotic divisions, forms a linear tetrad, and the chalazal megaspore develops into a Polygonum type of embryo sac. The polar nuclei lie in the centre of the gametophyte. The antipodals are ephemeral (Bendre 1975).

Cellular type of endosperm occurs in *Buddleia asiatica* (Bendre 1975), *B. curvifolia* (Dop 1913), *B. lindbergiana* (K.V.O. Dahlgren 1922), *B. variabilis* (Crété 1942) and *Nicodemia diversifolia* (Maheswari Devi and Lakshminarayana 1980).

In *N. diversifolia* the division of the primary endosperm nucleus is followed by a transverse wall. Both the cells produce the endosperm proper. In *B. asiatica* only the upper cell gives rise to the endosperm proper. The lower cell divides vertically, and the daughter cells elongate and form a non-aggressive chalazal haustorium.

In *B. variabilis* (Crété 1942), the primary endosperm nucleus divides by a transverse wall separating a micropylar chamber from the chalazal chamber. Then the micropylar chamber probably divides by a transverse wall. Of the three tiers, the uppermost develops into a micropylar haustorium, the middle one into the endosperm proper, and the lowest tier gives rise to a chalazal haustorium (Crété 1942).

Two vertical divisions at right angle to each other produce the four-celled micropylar haustorium. All the four cells elongate beyond the micropyle and, while two of them extend in the same direction, the two peripheral cells extend laterally. These cells branch and ramify into the integument. The nuclei are highly hypertrophied and one of them is 10 μm in diameter. The branches grow deep into the integument, and some of them come into contact with the endothelium and raphe. The nuclei attain a diameter of 20 μm. The growth of the micropylar haustorium is intercellular. In *B. curvifolia* (Dop 1913) also the micropylar haustorium is aggressive.

A micropylar haustorium is absent in *B. asiatica* (Bendre 1975), and *N. diversifolia* (Maheswari Devi and Lakshminarayana 1980). This needs reinvestigation.

In *B. variabilis* the two-celled chalazal haustorium, by another vertical division, becomes four-celled. It is non-aggressive in *B. asiatica* and *B. variabilis.* It does not differentiate in *N. diversifolia* (Maheswari Devi and Lakshminarayana 1980).

The development of the embryo is of the Onagrad type (Bendre 1975).

The seed is albuminous with scanty endosperm; the embryo is small with leafy cotyledons. The seed-coat is multiplicative in *N. diversifolia.* During the post-fertilization period it becomes 12–15 layers thick. Later it becomes crushed. The thin testa of crushed cells is surrounded by 15–18 layers of pericarp of loosely arranged empty cells.

AM Bendre (1975), P Crété (1942), KVO Dahlgren (1922), P Dop (1913), H Maheswari Devi and K Lakshminarayana (1980), R Souèges (1940b)

48.15 Scrophulariaceae

The anther wall comprises the epidermis, fibrous endothecium, one or two ephemeral middle layers, and secretory tapetum of bi- or multinucleate cells. The anther wall is massive in *Chelone glabra* (Arekal 1963a), and the endothecium consists of two to six layers. The tapetum is of dual origin in *Alectra thompsoni* (Vijayaraghavan and Ratnaparkhi 1972); the cells of the connective adjoining the inner side of sporogenous tissue also contribute to its formation. (The dual origin of tapetum is not a novel feature, and is of common occurrence in angiosperms.) The tapetal cells derived from the parietal layer are isodiametric, while those from the connective are radially elongated. Kapoor et al. (1978), who studied *Celsia coromandeliana*, consider this variation as polymorphism of tapetal cells. This variation is known in several other taxa also. Nuclear fusions occur in the tapetal cells so that they become polyploid. At certain places the tapetum is multiseriate in *Anticharis linearis* (M.C. Joshi and Varghese 1963). Simultaneous cytokinesis follows meiotic divisions in microspore mother cells, and the tetrads are tetrahedral, isobilateral and decussate. Polyspory occurs in *Veronica anagallis* (B. Tiagi and Varghese 1970).

The family is multipalynous. In the majority of members the pollen grains are tricolporate, and two-celled at the dispersal stage.

The ovule is ana-, hemiana- or campylotropous, unitegmic, and tenuinucellate. In *Melampyrum lineare* (Arekal 1963b), the ovules exhibit heterotrophy; the micropyles of the upper two ovules are directed towards the septum, while the lower two have their micropyles facing the ovary wall.

The inner epidermis of the integument differentiates as endothelium at the tetrad stage, and it surrounds the lower three-fourths of the embryo sac. The endothelium is biseriate, and rarely forms a nodule into the endosperm in *Chelone glabra* (Arekal 1963a). In *M. lineare* (Arekal 1963b) and *M. pratense* (D. Raju and Arekal 1977), it disintegrates early, and the hypodermal layer around the micropylar region functions as endothelium.

The archesporium is one- or two-celled. A single cell functions as megaspore mother cell, undergoes meiotic divisions, and the chalazal megaspore of a linear tetrad produces a Polygonum type of embryo sac. The antipodals are ephemeral. They are persistent in *Mimulus ringens* (Arekal 1965). Of the three superposed antipodals, the uppermost is large and binucleate. The antipodal cells undergo secondary multiplication in *Dopatrium* (B. Tiagi and Varghese 1970) and *Orthocarpus luteus* (Arekal 1963b). The embryo sac becomes extramicropylar in *Euphrasia arctica* (Arekal 1963b), *Pedicularis silvatica* (Berg 1954), *Torenia fournieri* (Guilford and Fisk 1952) and *Vandellia hirsuta* (Krishna Iyengar 1940b). In *Melampyrum pratense* (D. Raju and Arekal 1977), the mature embryo sac develops two to four narrow micropylar tubular extensions, whereas its chalazal end becomes bulbous. Twin embryo sacs are reported in *Anticharis linearis* (M.C. Joshi and Varghese 1962).

In *Torenia fournieri* (Tiwari 1982), at an early stage, the embryo sac becomes extramicropylar and comes to lie near the placenta in the carpellary cavity. It does not derive food from the ovular cells. The wall of the exposed micropylar part of the embryo sac is impermeable to the flow of metabolites across it. It does not seem likely that the micropylar portion of the embryo sac can derive nutrients from the mucilaginous fluid of the carpellary cavity. Perhaps, the endothelium may provide the nutrients to the embryo sac laterally, through some cuticular interruptions, as shown in *Jasione* (Erdelská 1975).

Polygonum as well as Allium type of embryo sac is reported in *Alectra thompsoni* (Vijayaraghavan and Ratnaparkhi 1972) and *Scrophularia himalensis* (Natesh and Bhandari 1975). Arekal and D. Raju (1964) observed only the Allium type in *Linaria ramosissima*.

Syngamy and triple fusion occur simultaneously.

The development of the endosperm is of the Cellular type (Fig. 48.15.1A–J). The division of the primary endosperm nucleus is followed by a transverse wall separating a micropylar chamber from the chalazal chamber (Fig. 48.15.2A). A transverse division occurs in the micropylar chamber (Fig. 48.15.2B–D). The central cell may divide by a vertical wall (Fig. 48.15.2C) or by a series of transverse divisions (Fig. 48.15.2E). Instead of the transverse, a vertical division may occur in the micropylar chamber (Fig. 48.15.2F). Further stages are shown in Fig. 48.15.2G–I. The micropylar chamber contributes to the formation of initials of endosperm proper, and micropylar haustorium.

The formation of a linear row of three cells in the young endosperm, due to a transverse division in the micropylar chamber in *Bonnaya tenuifolia*, Gratiolae, *Ilysanthes hyssopioides*, *Vandellia hirsuta* and *V. scabra* (Krishna

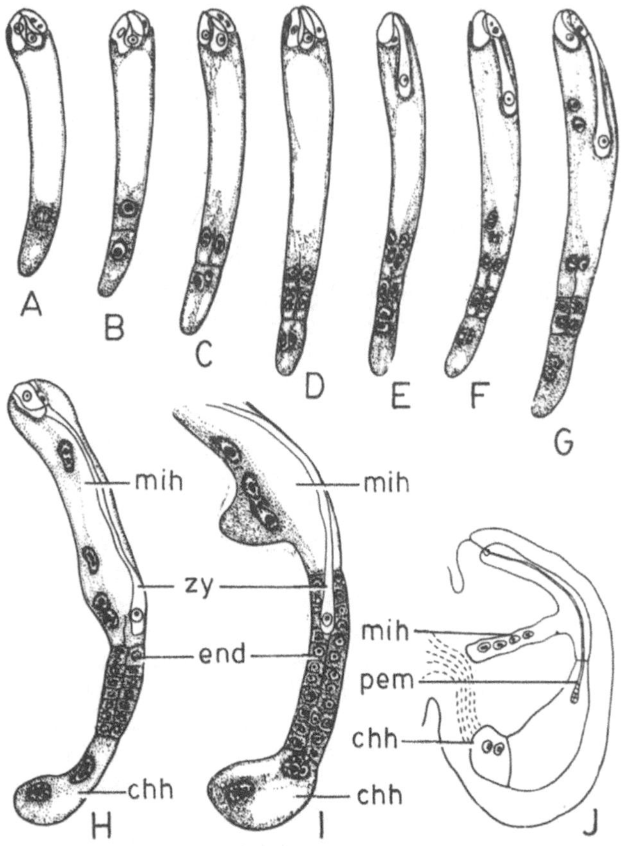

Fig. 48.15.1A–J. *Tozzia alpina.* **A** Fertilized embryo sac, primary endosperm nucleus in division. **B** Two-celled endosperm. **C** Embryo sac with zygote, incomplete vertical division in micropylar chamber, and two nucleate chalazal chamber. **D** Zygote, two-nucleate micropylar haustorium, endosperm proper initials (middle tier), and binucleate chalazal haustorium. **E, F** Elongated zygote, four-nucleate micropylar haustorium, endosperm proper initials, and binucleate chalazal haustorium. **G, H** Tubular zygote, four-nucleate micropylar haustorium (*mih*), endosperm proper (*end*), two-nucleate chalazal haustorium (*chh*). Note one persistent synergid in **A–H. I** Later stage of endosperm, four-nucleate micropylar haustorium (*mih*), and six-nucleate chalazal haustorium. **J** Longisection of ovule at young proembryo (*pem*) embedded in endosperm proper. The four-nucleate micropylar haustorium (*mih*) has a lateral extension towards the funicule, and the chalazal haustorium (*chh*) is two-nucleate. (After Nikiticheva and Tereokhyn 1973)

Iyengar 1940b) and *Mimulus tigrinus* (Guilford and Fisk 1952) has been considered doubtful by D. Raju (1976). However, such a doubt has been eliminated as D. Raju and Arekal (1978) demonstrated a transverse (Fig. 48.15.2B), or occasionally vertical division (Fig. 48.15.2F) in the micropylar chamber in *Linaria ramosissima.*

A vertical division may or may not occur in the chalazal chamber; the vertical division in the chalazal chamber may be incomplete as in *Chelone glabra* (Arekal 1963a), *Collinsia bicolor* (Crété 1958b) and *Tetranema*

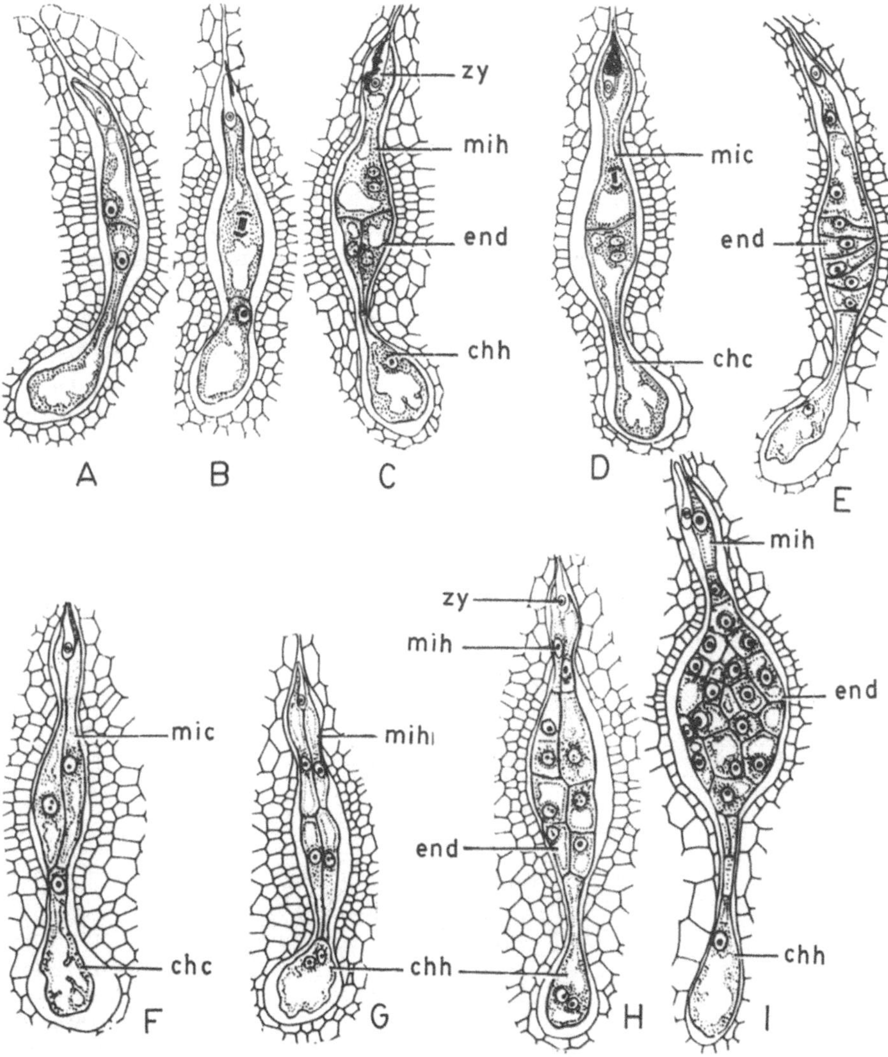

Fig. 48.15.2A–I. *Linaria ramosissima*, development of endosperm. **A** Zygote, and two-celled endosperm – micropylar and chalazal chamber. **B** Nucleus of micropylar chamber in division. **C** Embryo sac with zygote (*zy*), binucleate micropylar (*mih*) and uninucleate chalazal (*chh*) haustorium, and initials of the endosperm (*end*) proper. **D** Nuclei of micropylar (*mic*) and chalazal (*chc*) chamber in division. **E** Uninucleate micropylar and uninucleate chalazal haustorium, and endosperm (*end*) proper (a row of six cells). **F** Vertical division in micropylar chamber, chalazal chamber uninucleate. **G** Bicelled micropylar haustorium, two-celled endosperm proper, and two-nucleate bulbous chalazal haustorium. **H, I** Advanced stages; note uninucleate micropylar and uninucleate chalazal haustorium in **I**. The micropylar haustorium is non-aggressive. The zygote has not divided in **A–I**. (After D. Raju and Arekal 1978)

maxicanum (Crété 1954a). The chalazal chamber mostly develops into a haustorium.

The micropylar haustorium is aggressive, and its lateral branches ramify in the integument in *Euphrasia*, *Melampyrum* (Arekal 1963b, D. Raju and Arekal 1977), *Pedicularis* (Fig. 48.15.3A, B) (Berg 1954) and *Striga* (B. Tiagi 1956). In *Alectorolophus hirsutus* (Schmid 1906), *Orthocarpus luteus* (Arekal 1963b) and *Tozzia alpina* (Fig. 48.15.1I, J), the micropylar haustorium extends in the direction of the funiculus. It is branched in *Alonsoa*, bulbous in *Isoplexis tenuifolia* (Krishna Iyengar 1937, 1939a) and U-shaped in *Orthocarpus luteus* (Arekal 1963b).

The development of the embryo sac is tetrasporic, biphasic and bipolar, modified Drusa type in *Melampyrum angustissimum*, *M. arvense*, *M. barbatum*, *M. cristatum*, *M. nemorosum*, *M. pratense*, *M. sylvaticum* (Greilhuber 1973). The arrangement of the nuclei in a coenomegaspore is 1 + 3. The first mitosis is followed by the formation of a normal egg apparatus, upper polar nucleus, lower polar nucleus, and three antipodals. The polar nuclei do not fuse and only the micropylar polar nucleus is fertilized. The endosperm in *Melampyrum* spp. is, therefore, diploid (2n = 18). A similar condition is reported in *M. lineare* (Arekal 1963b) and *M. pratense* (D. Raju and Arekal 1977).

In *M. lineare* (Arekal 1963b), the embryo sac is tetrasporic, seven-nucleate. The hypodermal archesporial cell (Fig. 48.15.4A) enlarges, and functions as

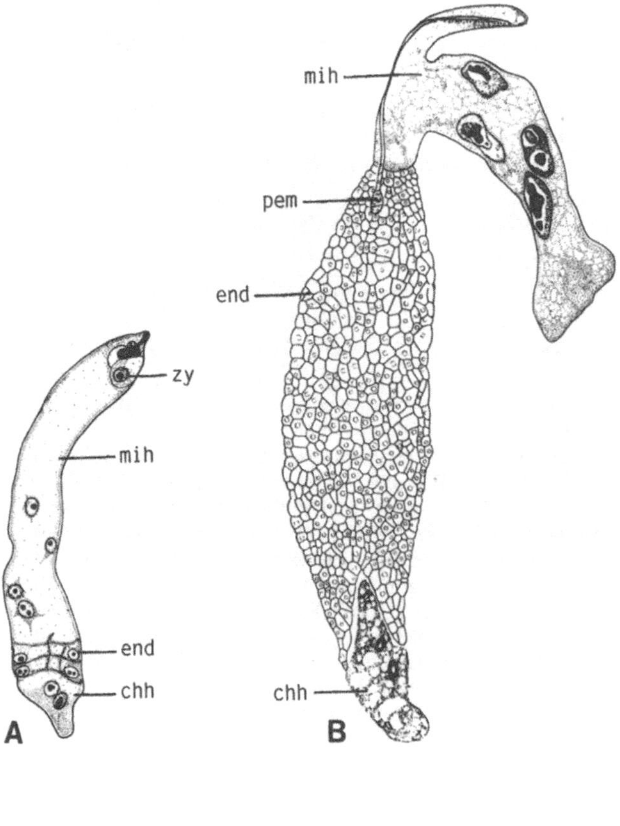

Fig. 48.15.3A, B. ***Pedicularis sylvatica.*** **A** Embryo sac with zygote (*zy*), four-nucleate micropylar haustorium (*mih*) with incomplete vertical wall (shown only above the endosperm proper), two tiers of endosperm proper (*end*), binucleate chalazal haustorium (*chh*), and remnants of one synergid and pollen tube. **B** Later stage, the branched micropylar haustorium contains four hypertrophied nuclei; *pem* proembryo. (After Berg 1954)

megaspore mother cell (Fig. 48.15.4B, C). Its nucleus undergoes meiosis I and II (Fig. 48.15.4D–F). The four nuclei of the coenomegaspore show a large micropylar and three small chalazal nuclei (Fig. 48.15.4F–H). The micropylar nucleus undergoes two mitotic divisions, and the four nuclei organize into the egg apparatus and upper polar nucleus (Fig. 48.15.4I–M). Of the three chalazal nuclei, one functions as the lower polar nucleus and comes to lie close to the upper polar nucleus (Fig. 48.15.4M). The other two nuclei degenerate. During maturation, the embryo sac develops eight lateral extensions in the micropylar region (Fig. 48.15.4N, O, Q).

Fertilization is porogamous. One male gamete comes to lie near the egg cell, and the other close to the upper polar nucleus (Fig. 48.15.4N). Syngamy takes place normally, but the secondary nucleus is not formed since the lower polar nucleus degenerates. The sperm nucleus fuses with the upper polar nucleus only, and the endosperm is diploid (Arekal 1963b).

This "endosperm" nucleus divides transversely to form a small micropylar and a large chalazal chamber. An incomplete vertical division occurs in the micropylar chamber, followed by a transverse wall; the upper tier develops into a four-nucleate micropylar haustorium.

The micropylar tubular part of the embryo sac becomes a constituent of the micropylar haustorium. The central cells are the initials of the endosperm proper, while the chalazal chamber develops into a massive binucleate haustorium (Fig. 48.15.4O).

The micropylar haustorium (Fig. 48.15.4N–S) produces eight branches (Fig. 48.15.4Q), and one of these contains four hypertrophied nuclei (Fig. 48.15.4S). Another conspicuous branch extends and ramifies in the funicle (Fig. 48.15.4P, R, T). This is quite prominent in a young seed at the early globular proembryo.

The endosperm proper is distinguishable into three distinct regions: small densely cytoplasmic cells in the micropylar region, narrow, short, constricted middle region of regularly arranged small cells, and sparsely protoplasmic cells in the chalazal region. The chalazal haustorium is massive (Fig. 48.15.4T).

In *Gerardia pedicularia* (Arekal 1964), the primary endosperm nucleus divides transversely (Fig. 48.15.5A), and the micropylar and chalazal chambers are formed (Fig. 48.15.5B). The division of the nucleus of the micropylar chamber is followed by an incomplete longitudinal wall (Fig. 48.15.5C). Then a transverse division separates the micropylar haustorium from the initials of the endosperm proper (Fig. 48.15.5D). The two cells of the micropylar haustorium are separated by an incomplete wall (Fig. 48.15.5D, E). The haustorium undergoes a nuclear division and becomes four-nucleate (Fig. 48.15.5F). It extends as a tubular process towards the raphe, grows along the conducting strand, and reaches almost up to the base of the chalaza (Fig. 48.15.5G). The nuclei migrate to the tubular region, become hypertrophied, and show irregular outlines. The wall between the two cells seems to disorganize before the tubular extension of the haustorium.

The chalazal chamber develops into a two- or four-nucleate prominent haustorium (Fig. 48.15.5F–H), which also extends as far as the conducting strand (Fig. 48.15.5G).

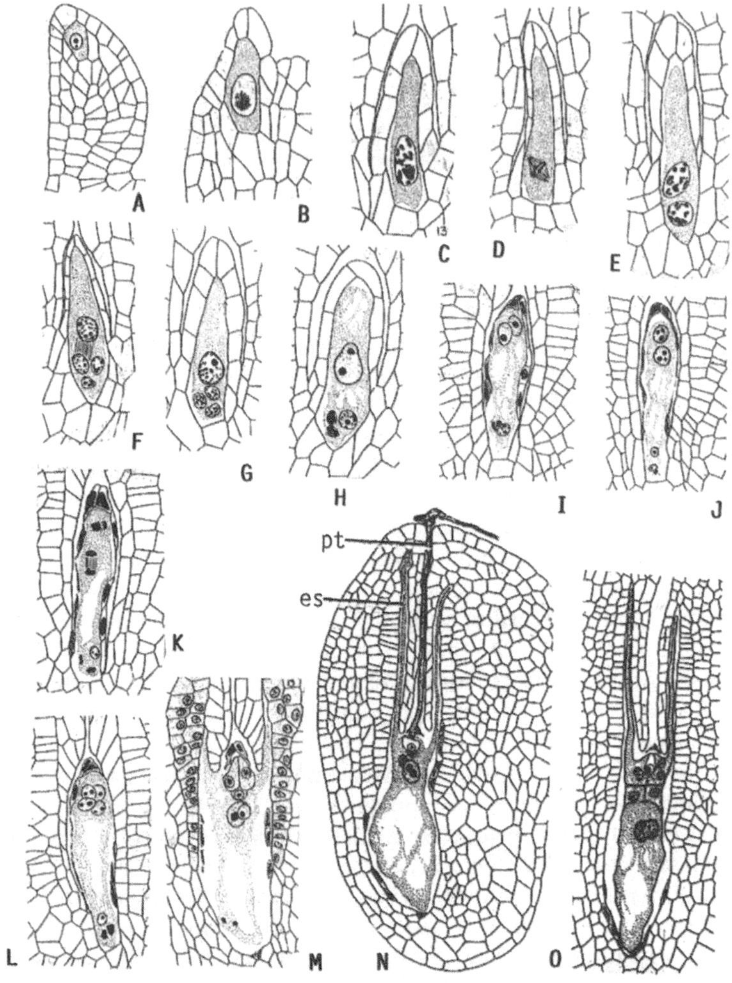

Fig. 48.15.4A–T. *Melampyrum lineare* (**Q–S** from dissections). **A** Archesporial cell. **B** Megaspore mother cell. **C** Prophase. **D** Metaphase I. **E** Dyad nuclei at chalazal end. **F** Meiosis II, coenomegaspore with four (1 + 3) nuclei. **G** One large (1 + 3) and three small nuclei. **H** 1 + 3 arrangement of nuclei. **I, J** Two daughter nuclei formed by division of larger nucleus at micropylar pole, two (or three ?) chalazal nuclei show signs of degeneration. **K** Both micropylar nuclei in division, of the three nuclei at chalazal pole two have degenerated. **L** Later stage. **M** Mature embryo sac with egg apparatus, two polar nuclei, and two degenerated chalazal nuclei. **N** Embryo sac at fertilization with one sperm cell attached to egg cell, and the other to secondary nucleus. The pollen tube (*pt*) and extensions (*es*) of embryo sac are quite conspicuous. **O** Embryo sac with zygote, endosperm proper initials, four-nucleate micropylar and two-nucleate massive chalazal haustorium.

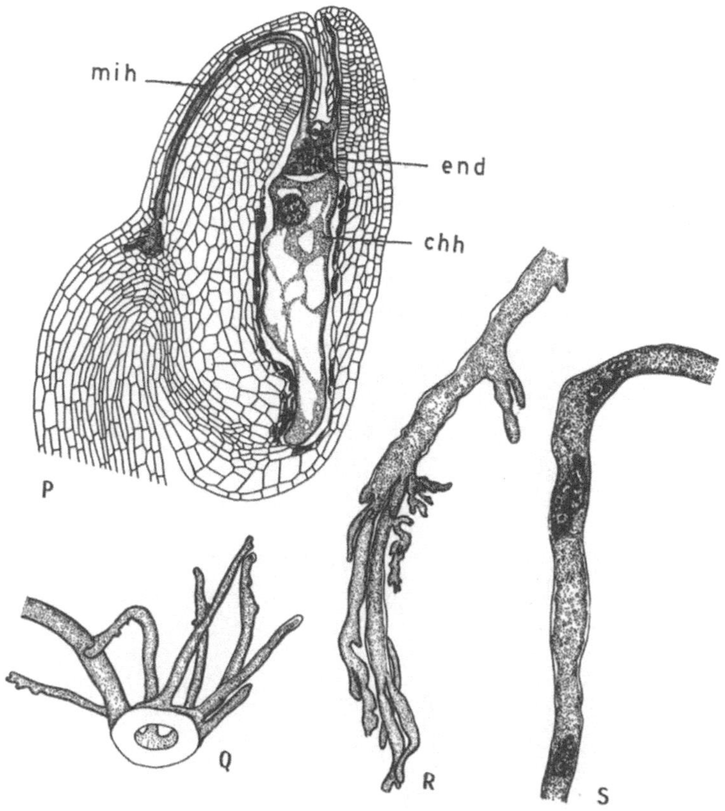

Fig. 48.15.4P–S. P Longisection of young seed with zygote, four-nucleate micropylar haustorium (*mih*), endosperm (*end*), and binucleate chalazal haustorium (*chh* – only one nucleus is seen). One of the branches of micropylar haustorium extends into the integument and abuts on the funicular vascular bundle. **Q** Basal part of micropylar haustorium with eight tubular extensions. **R** Irregular hypha-like branching of a broad haustorial tube in the funiculus. **S** Portion of another micropylar haustorial tube containing four nuclei.

A noteworthy feature is that both the micropylar and the chalazal haustorium apparently remain healthy during the development of seed (Fig. 48.15.5I).

The unequal enlargement of the endothelium causes ruminated endosperm (Fig. 48.15.5J, K).

In *Ellisiophyllum pinnatum* var. *reptans* (Yamazaki 1957b), the division of the primary endosperm nucleus is followed by a transverse wall producing a micropylar and a chalazal chamber. A vertical division occurs in each cell. The cells of the uppermost tier develop into a two-celled micropylar haustorium. A transverse division occurs in each daughter cell of the chalazal chamber. All the four cells contribute to the endosperm proper, and the chalazal haustorium does not differentiate.

The micropylar haustorium comprises four uninucleate cells as in *Agathelpis dubia* (D. Raju 1973), *Capraria biflora* (Hakki 1975), *Deinostema* sp.

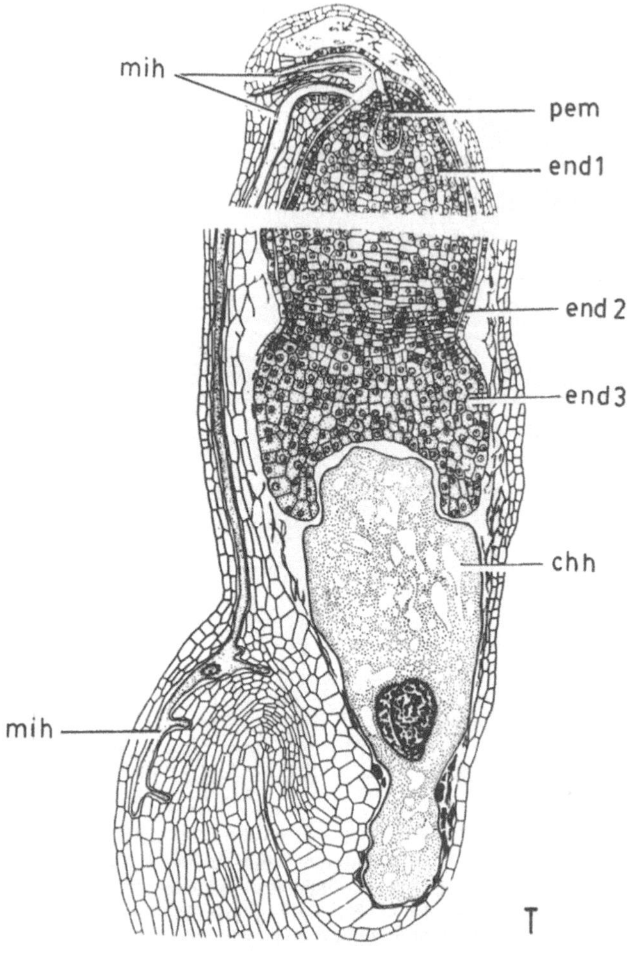

Fig. 48.15.4T. **T** Longisection of seed at globular proembryo, micropylar and chalazal region. Cellular endosperm is distinguishable into three zones: densely cytoplasmic cells (*end 1*), a narrow constricted region of small cells (*end 2*), and chalazal region of large somewhat sparsely cytoplasmic cells (*end 3*). One branch of the micropylar haustorium extends through the integument as far as the funiculus. The massive chalazal haustorium is binucleate. (After Arekal 1963b)

(Yamazaki 1953), *Hebenstreitia integrefolia* (Kapil and Masand 1964), *Lindernia hyssopioides* (Arekal et al. 1970), and *Scoparia dulcis* (Arekal et al. 1971); two juxtaposed uninucleate cells as in *Linaria ramosissima* (Fig. 48.15.2G), *Mimulus moschatus* (Crété 1964b), and *Parentucellia latifolia* (Greilhuber 1973). It consists of two binucleate cells as in *Anticharis linearis* (M.C. Joshi and Varghese 1963) and *Mimulus ringens* (Arekal 1965); is tetranucleate and one-celled as in *Pedicularis silvatica* (Fig. 48.15.3B), *Veronica agrestis* (Varghese 1963) and *V. serpyllifolia* (Arekal 1966). A four- to six-nucleate one-celled condition occurs in *Rehmannia angulata* (Krishna Iyengar 1942b), and eight- to 16-nucleate in *Melampyrum pratense* (D. Raju and Arekal 1977). Rarely, the micropylar haustorium is uninucleate and one-celled as in *Linaria ramosissima* (Fig. 48.15.2I).

The micropylar haustorium is multicellular in *Chelone glabra* (Arekal 1963a) and *Gratiola officinalis* (Glišić 1933). In *Chelone* the behaviour of the nuclei is interesting. "The nuclei in the older cells of the haustorium enlarge further, often acquiring irregular shape, and bud off small nuclei. In such cases the cells become multinucleate with one large parent nucleus and two or more smaller derivatives. The cells of the micropylar haustorium, whether four or more, never fuse but degenerate as such (Arekal 1963b)." This aspect requires further attention.

The micropylar haustorium is absent in *Chaenorrhinum minus* (Arekal 1963c) and *Linaria genistaefolia* (Persidsky 1934).

The chalazal haustorium is composed of:

1) Four uninucleate cells as in *Capraria biflora* (Hakki 1975), *Celsia coromandeliana* (Kapoor et al. 1978), *Mimulus moschatus* (Crété 1964b) and *Verbascum thapsus* (Krishna Iyengar 1942a).
2) Two juxtaposed uninucleate cells as in *Agathelpis dubia* (D. Raju 1973), *Linaria bipartita* (S.P. Bhatnagar and Kallarackal 1980) and *Lindernia hyssopioides* (Arekal et al. 1970).
3) Two cells separated by an incomplete wall as in *Chelone glabra* (Arekal 1963a).
4) Unicellular, binucleate as in *Buchnera hispida*, *Chaenorhinum minus* (Arekal 1963c), *Collinsia* spp. (Schrock and Palser 1967), *Melampyrum arvense*, *M. nemorosum* (B. Tiagi 1965b), *Orthocarpus luteus* (Arekal 1963b), *Rhinanthus major* and *R. serotinus* (B. Tiagi 1972), *Sopubia trifida* (Arekal and Raju 1967), *Striga densiflora* (Nagendran et al. 1980) and *Veronica serpyllifolia* (Arekal 1966).
5) Uninucleate, one-celled as in *Parentucellia latifolia* (Greilhuber 1973).

The chalazal haustorium is aggressive in *Euphrasia arctica* (Arekal 1963b) and *Melampyrum* spp. (Arekal 1963b, D. Raju and Arekal 1977, B. Tiagi 1965b). In *Euphrasia*, in its active state, the haustorium is densely protoplasmic and develops a narrow tubular process which extends through the nucellus and comes in direct contact with the conducting strand. Finally, it branches into two and, rarely, one of the branches extends further and becomes extra chalazal. In *Melampyrum lineare* (Arekal 1963b), the haustorium enlarges a great deal, and comes into contact with the vascular strand of the young seed. The massive haustorium is broad in the upper part, and narrow in the lower (Fig. 48.15.4P, T). The two nuclei also enlarge. A similar haustorium is also reported in *M. arvense* and *M. nemorosum* (B. Tiagi 1965b), and *M. pratense* (D. Raju and Arekal 1977).

The chalazal haustorium is absent in *Ellisiophyllum pinnatum* var. *reptans* (Yamazaki 1957b), and *Nemesia floribunda* and *N. melissaefolia* (Crété 1950a, b).

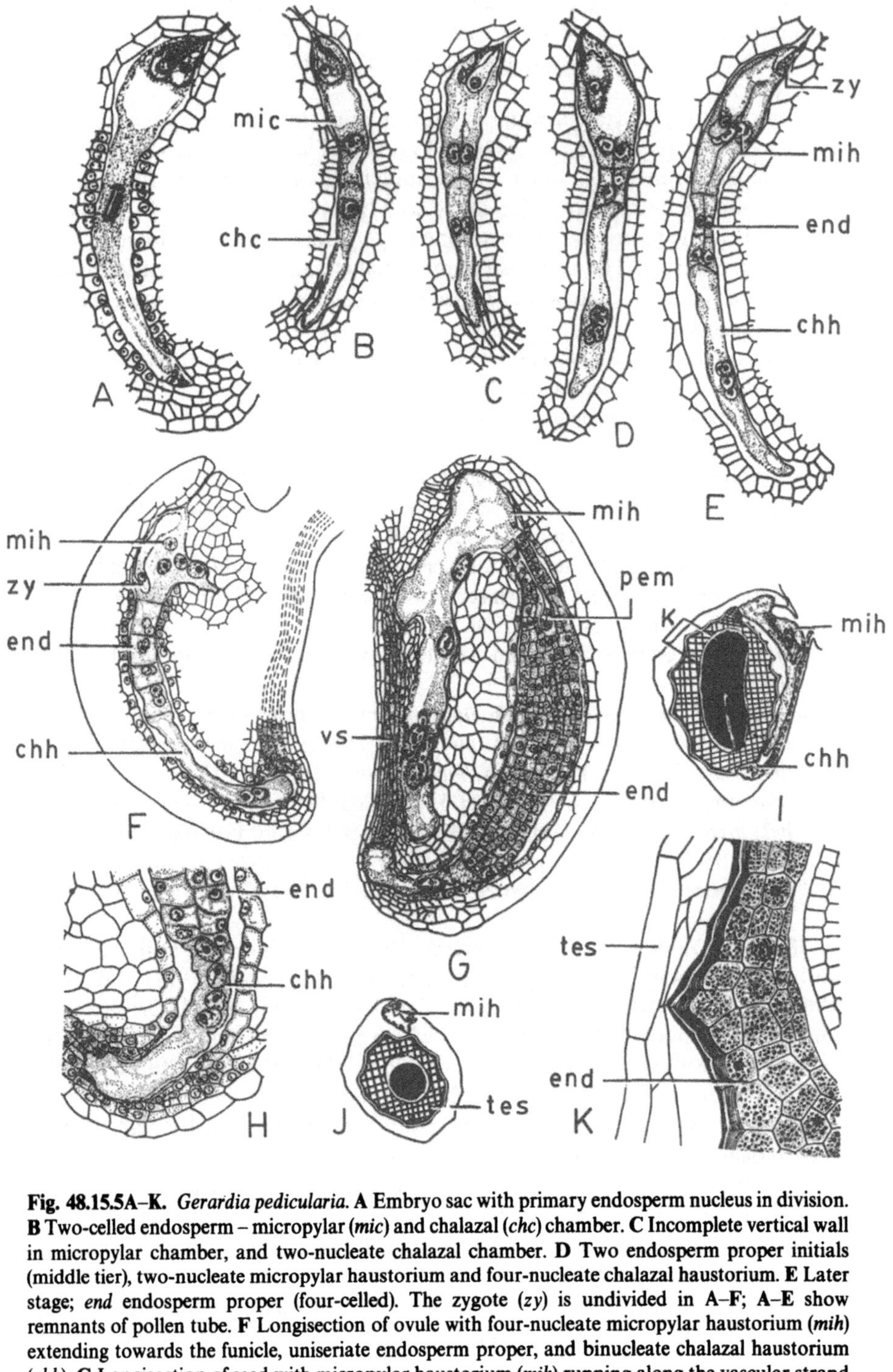

Fig. 48.15.5A–K. *Gerardia pedicularia.* **A** Embryo sac with primary endosperm nucleus in division. **B** Two-celled endosperm – micropylar (*mic*) and chalazal (*chc*) chamber. **C** Incomplete vertical wall in micropylar chamber, and two-nucleate chalazal chamber. **D** Two endosperm proper initials (middle tier), two-nucleate micropylar haustorium and four-nucleate chalazal haustorium. **E** Later stage; *end* endosperm proper (four-celled). The zygote (*zy*) is undivided in **A–F**; **A–E** show remnants of pollen tube. **F** Longisection of ovule with four-nucleate micropylar haustorium (*mih*) extending towards the funicle, uniseriate endosperm proper, and binucleate chalazal haustorium (*chh*). **G** Longisection of seed with micropylar haustorium (*mih*) running along the vascular strand

The four-nucleate micropylar and binucleate chalazal haustorium is weakly developed in *Centranthera hispida* (Krishna Iyengar 1942c). In addition, one-celled secondary haustoria develop from the peripheral cells of the cellular endosperm, extend into the integumentary tissue, and grow intercellularly. The adjacent haustoria may sometimes unite, resulting in a binucleate condition. In *Alectra thompsoni* (Vijayaraghavan and Ratnaparkhi 1972), the secondary haustoria beneath the micropylar region are small, uninucleate, tubular and unbranched.

According to Schmid (1906), Lundqvist (1915) and Krishna Iyengar (1942c), in some species of *Pedicularis*, the endosperm shows a primary three-celled condition (micropylar chamber, central cell, chalazal chamber) following the division of the primary endosperm nucleus. Berg (1954) states that a vertical wall is laid down in the micropylar chamber, and the development is of the Brunella type (Fig. 48.15.3A). The micropylar haustorium is aggressive, branched, and four-nucleate (Fig. 48.15.3B). The chalazal haustorium is also well developed with two hypertrophied nuclei (Fig. 48.15.3B). In *P. palustris* (Steffen 1956), the nuclei of the micropylar haustorium exhibit ploidy up to 192–384 ×, of the endosperm proper up to 6–12 ×, and of the chalazal haustorium up to 96 ×.

The development of the endosperm in *Rhinanthus major* and *R. serotinus* (B. Tiagi 1972) is very unique, and differs from the other members of the Scrophulariaceae. It also conforms to the Brunella type. The micropylar haustorium is bicelled at the base, unicellular and multinucleate in the upper region, invades the nodular placenta, and ramifies in it especially in the vicinity of the provascular strand. The chalazal haustorium is tubular, two to four-nucleate, and aggressive. It invades the chalazal tissue and extends up to the base of the seed. The development and structure of other species of *Rhinanthus* may reveal further interesting features, and require a careful study.

After the inactivation of haustoria, the cells of the endosperm proper adjoining the micropylar and the chalazal haustorium transform into cambial strips (B. Tiagi 1972). These strips produce additional tissue referred to as "micropylar haustorial endosperm", and "chalazal haustorial endosperm". The cells of the micropylar haustorial endosperm, adjoining the micropylar haustorium, are large with dense cytoplasm and conspicuous nuclei, as compared to the vacuolate cells in the lower region. The chalazal haustorial endosperm also shows a similar zonation. The staining reactions confirm the haustorial function of both the micropylar and the chalazal haustorial endosperm.

The micropylar haustorial endosperm pushes the micropylar haustorium proper towards the micropyle. After depleting the contents of the placenta, the

(*vs*), and chalazal haustorium abutting at the vascular strand; *pem* proembryo. H Lower portion of embryo sac showing four-nucleate chalazal haustorium. I Longisection of mature seed to show ruminate endosperm (*cross-hatched*), and remnants of micropylar (*mih*) and chalazal (*chh*) haustorium. J Transection of seed; *tes* testa. K Portion of seed-coat at *K* marked in I. The testa (*tes*) is thin-walled, inner walls of endothelial cells are thick, and ruminate endosperm (*end*) is followed by a portion of embryo. (After Arekal 1964)

micropylar haustorium becomes inactive and inconspicuous, but the remnants persist in the shrivelled placenta.

The growth of the chalazal haustorial endosperm is peripheral, and it comes to lie in the chalazal cavity. The remnants of the degenerated chalazal haustorium also lie in the centre of the cavity.

In later stages the micropylar and chalazal haustorial endosperm become tanniniferous and collapse. These cells do not store any food reserve.

The cambial strips also contribute to the endosperm proper which stores reserve food.

A similar report of formation of a cambial strip from the chalazal cellular endosperm concerns *Melampyrum nemorosum* (B. Tiagi 1965b). The cambial strip gives rise to the endosperm tissue which fills the chalazal cavity, first producing the cells towards the periphery, and then centripetally.

Considering the criteria of reduction in the number of haustorial cells and nuclei as an advanced feature, Banerji (1961) proposed a scheme of classification of the development of endosperm in Scrophulariaceae. This has been considerably improved by B. Tiagi and Varghese (1970), and they distinguish six groups, 16 series, and 31 types.

Following the arrangement suggested by Crété (1951) and D. Raju (1976), Dr M.A. Rau has drawn up the present scheme (Fig. 48.15.6A–K_4) which consists of four series B, B_1, B_2 and B_3, and 15 types of development. In all cases the micropylar chamber of the two-celled endosperm (Fig. 48.15.6A) gives rise to the initials of the micropylar haustorium, and cellular endosperm. The chalazal chamber invariably gives rise to chalazal haustorium. There is a progressive reduction in the formation of partition walls between the initials of haustorial cells.

Vertical division occurs in the micropylar chamber in B, B_1 and B_2, while in B_3 the division is transverse. A vertical division also occurs in the chalazal chamber in B and the haustorium becomes two-celled, whereas it is binucleate in B_1 and uninucleate in B_2 and B_3.

A vertical division occurs in each cell in the micropylar chamber in C, D, E and F (Fig. 48.15.6). It is followed by a transverse wall so that four initials of the endosperm proper are separated from the four-celled micropylar haustorium in the Verbascum type (C_1), Chaenostoma type (D_1), Linaria type (E_1) and

→

Fig. 48.15.6A–K. Development of endosperm in Scrophulariaceae. The variations in development are indicated by *arrows*. *Black dots* indicate nuclei of haustorium. **A** Two-celled endosperm, lower cell functions as uninucleate chalazal haustorium. **B, B_1, B_2, B_3** Series derived from **A**. **B** Vertical division in each cell. **B_1** Vertical division in micropylar cell, chalazal haustorium binucleate. **B_2** Vertical division in micropylar cell, chalazal haustorium uninucleate. **B_3** Transverse division in micropylar chamber, chalazal haustorium uninucleate. **C-K_4** Types of endosperm derived from series **B, B_1, B_2** and **B_3**. **B, C, C_1** Verbascum type. **B, D, D_1** Chaenostoma type. **B, E, E_1** Linaria type. **B, F, F_1** Limosella type. **B, G** Mimulus/Microcarpea type. **B, G, G_1** and **B, G, G_2** Veronica type. **B, H, H_2** Mimulus/Lindernia/Ilysanthes type. **B, H, H_3** Veronica type. **B, H, H_1, H_4** Gratiola I type. **B_1, I, I_1** Mazus/Bonnaya/Alecterolophus type. **B_2, J, J_1** Limosella type. **B_2, J, J_2** Torenia/Bacopa type. **B_2, J, J_3** Stemodia type. **B_2, J, K_1, K_2** Gratiola II type. **B_2, J, K_3, K_4** Gratiola III type. **B_3, K, K_1, K_2** Gratiola II type. **B_3, K, K_3, K_4** Gratiola III type. (Adapted after Crété 1951, and D. Raju 1976; Courtesy Dr M.A. Rau, Mysore)

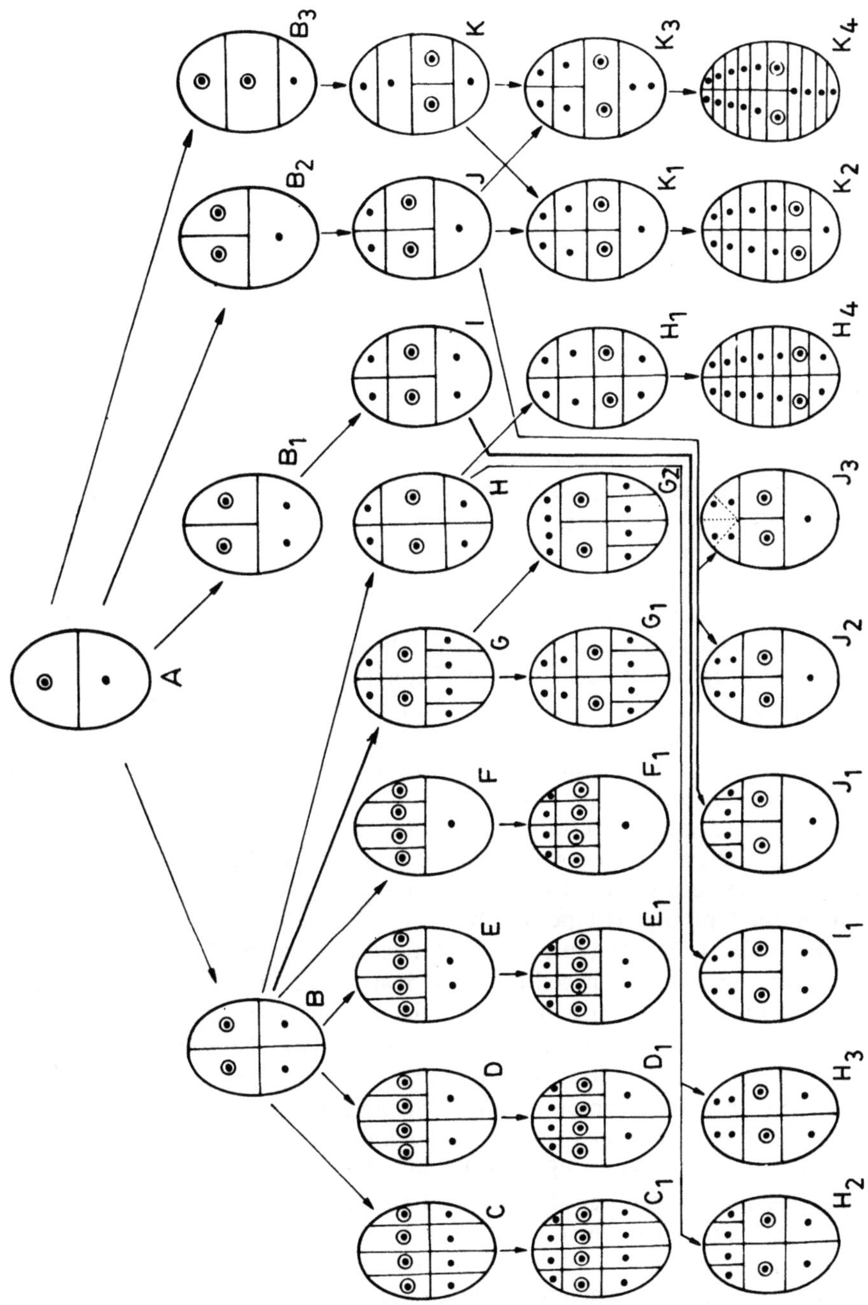
A
B
B1
B2
B3
C
C1
D
D1
E
E1
F
F1
G
G1
G2
H
H1
H2
H3
H4
I
I1
J
J1
J2
J3
K
K1
K2
K3
K4

the Limosella type (F_1). The chalazal haustorium is four-celled in *Verbascum* (C_1), two-celled in *Chaenostoma* (D_1), binucleate in *Linaria* (E_1) and uninucleate in *Limosella* (F_1).

In the Mimulus/Microcarpea type, the chalazal haustorium is four-celled. The Veronica type (G_1, G_2) can be derived from the Mimulus type. The micropylar haustorium becomes four-celled, or the walls are transitory to form a four-nucleate micropylar haustorium. The chalazal haustorium is four-celled, in either case.

The micropylar haustorium becomes four-celled in H_2, i.e. in the Mimulus, Lindernia and Ilysanthes type, or the walls may be transitory in H_3 and the micropylar haustorium becomes two-celled, each cell binucleate as in the Veronica IV type. There are two initials for the endosperm proper, and the chalazal haustorium is two-celled.

Repeated transverse divisions occur in the micropylar chamber in B, H and H_1, and the micropylar haustorium becomes ten-celled in the Gratiola I type (H_4). The two initials give rise to the endosperm proper, and the chalazal haustorium remains two-celled.

A transverse division in each cell of the micropylar chamber in B_1 separates the initials of the endosperm proper from the two-celled micropylar haustorium (I), each cell becomes binucleate. The chalazal haustorium is binucleate in the Mazus Bonnaya/Alecterolophus type (B_1, I, I_1).

In the Limosella type the micropylar haustorium becomes four-celled, whereas the chalazal haustorium is uninucleate (B_2, J, J_1). The walls in the micropylar haustorium are transitory and the micropylar haustorium becomes two-celled, each cell binucleate in the Torenia and Bacopa type (B_2, J, J_2), and one-celled, four-nucleate in the Stemodia type (B_3, J, J_3). The chalazal haustorium is one-celled, uninucleate.

Three consecutive transverse divisions in the micropylar chamber of J produce eight-celled micropylar haustorium. The chalazal haustorium remains uninucleate in the Gratiola II type (K_1, K_2). Four consecutive transverse divisions occur in the micropylar chamber and three in the chalazal chamber of J, to form ten-celled micropylar haustorium and four-celled chalazal haustorium in the Gratiola III type (K_3, K_4).

Gratiola II and III types can be derived from series B_3. The micropylar haustorium becomes four-celled by two divisions at right angles to each other (K, K_3). Further transverse divisions in the micropylar hasutorium form a ten-celled structure. Transverse divisions also occur in the chalazal chamber and it becomes four-celled in four tiers in the Gratiola III type.

The embryogeny conforms to the Onagrad type in *Celsia coromandeliana* (Kapoor et al. 1978), *Euphrasia arctica* (Arekal 1963b), *Mimulus ringens* (Arekal 1965), *Nemesia strumosa* (Taneja 1969b), *Pedicularis sylvatica* (Berg 1954), *Scoparia dulcis* (Arekal et al. 1971), and *Striga orobanchoides* (B. Tiagi 1956).

In *Ellisiophyllum pinnatum* var. *reptans* (Yamazaki 1957b), the suspensor comprises 40–50 cells and becomes twisted. The cells in the basal region become vesicular and vacuolate, and act as a haustorium.

Twin proembryos have been observed in *Anticharis linearis* (M.C. Joshi and Varghese 1963). In one case both the proembryos were at the octant stage, while in another one was at quadrant and the other at octant stage. The synergids degenerate and, therefore, the additional proembryo may be of zygotic origin.

The minute seed is albuminous with a straight or slightly curved embryo. In *Scrophularia himalensis* (Bhandari et al. 1976), the seed-coat comprises outer epidermis and endothelium; the mesophyll becomes crushed. The outer tangential wall of the outer epidermis collapses; the inner tangential and radial walls become thickened with insoluble polysaccharides and serve as mechanical support to the seed. The thickened radial walls project and form seven or eight longitudinal ridges on the seed surface. The cells of the endothelium are vacuolate, and their inner tangential walls are thickly cutinized. In *Antirrhinum* (Corner 1976), the outer epidermis comprises lignified cells; inner tangential and radial walls become thick and project in groups to form ridges. The inner epidermis does not persist.

In *Anticharis linearis* (M.C. Joshi and Varghese 1963), the endospermic seed has a multiplicative seed-coat. The testa, except for the endothelium, forms the seed-coat which exhibits ridges and furrows; thick-walled cells are confined to the ridges. The outer epidermis is thin-walled, whereas the hypodermis is thick-walled followed by a single layer of large thin-walled cells. Then there are three or four layers of elongated, spindle-shaped, compressed cells with dark contents.

In *Ellisiophyllum pinnatum* var. *reptans* (Yamazaki 1957b), the epidermal cells of the testa enlarge, elongate, and contain mucilage. The mesophyll and inner epidermis become crushed and persist as a membrane.

In *Celsia coromandeliana* (Kapoor et al. 1978, Kapoor and Vijayaraghavan 1978), the seed-coat comprises the tenuous epidermis, degenerated middle layers, and the endothelium. The outer tangential walls of the endothelium are thin-walled, whereas the inner tangential walls are extremely thick and occupy almost two-thirds of the lumen of the cells. The endosperm adjoining the endothelium is also thick-walled.

The inner epidermis of the testa is thick-walled in *Angelonia, Collinsia, Gerardia, Torenia* and *Vandellia* (Corner 1976). The inner epidermal cells undergo unequal radial and tangential elongations which cause rumination of the Verbascum type (Periasamy 1962a). In *Verbascum thapsus* (Krishna Iyengar 1942a), the alternate cells of the endothelial layer cause rumination, whereas in *Celsia* the cells causing the ruminations are distributed randomly.

In *Veronica agrestis* (Varghese 1963), the integument is multiplicative, but the mesophyll and endothelium collapse. The epidermal cells become spinous and lose their contents. In mature seed they cover the endosperm. The function of the seed-coat is served by the persistent endosperm cells.

A spongy eliasome, derived from the micropylar haustorium, develops in *Deinostema* (Yamazaki 1953) and *Pedicularis sylvatica* (Berg 1954).

In *Alectra parasitica* var. *chitrakutensis* (S. Chandra et al. 1980), the integument (except the epidermis of the testa) is consumed. The two outermost layers of the pericarp are thick-walled and give protection.

Elisens and Tomb (1983, see also Armstrong 1985), using LM and SEM, studied the seeds of 39 species of Antirrhineae. On the basis of surface ornamentation, they recognize six distinct morphological categories, and one miscellaneous category. Most generic categories are characterized by a single morphological type; *Maurandya* shows several distinctive seed types: (a) Cristate – *Antirrhinum* spp., *Galvezia* spp., (b) Tetracostata – *Maurandya coccinea*, (c) Foveolata – *M. petrophila*, (d) Tumid Tuberculata/Cristate – Most spp. of *Maurandya*, (e) Circumalate – (i) wings unilayered, radially expanded epidermal cells with a reticulum of lignified thickenings on radial walls, e.g. *Epixiphium wislizeni*, *Maurandya* spp., *Rhodochiton astrosanguineum*, (ii) wing multilayered, upper and lower epidermal cells surround a central hypodermal layer, e.g. *Linaria vulgaris*, (f) Medusiform – *Antirrhinum cyathiferum*, *Mohavea* spp., (g) Miscellaneous – *Kickxia*, *Linaria*.

NG Afanaséva (1968), GD Arekal (1963a, b, c, 1964, 1965, 1966), GD Arekal and D Raju (1964, 1967), GD Arekal and SN Ramaswamy (1974), GD Arekal, S Girijamma and SN Ramaswamy (1970), GD Arekal, S Rajeshwari and SN Ramaswamy (1971), CL Argue (1985, 1986), JE Armstrong (1985), I Banerji (1961), RY Berg (1954), CA Berger, ER Witkus and TC Joseph (1951), NN Bhandari and S Natesh (1985), NN Bhandari, F Bouman and S Natesh (1976), NN Bhandari, SL Tandon and S Jai (1969), SP Bhatnagar and J Kallarackal (1979, 1980), SP Bhatnagar and NS Singh (1964), DH Campbell (1930), S Chandra and RP Singh (1980), S Chandra, RP Singh and AK Pandey (1980), MT Cook (1924a), EJH Corner (1976), P Crété (1948b, 1949d, 1950a, b, 1951, 1953b, 1954a, b, 1958b, 1964b), WJ Elisens and AS Tomb (1983), O Erdelská (1975), A Gentry (1980), LM Glišic (1933, 1937), J Greilhuber (1973, 1974), VB Guilford and EL Fisk (1952), MI Hakki (1975, 1977), D Hartl (1957, 1960), MC Joshi and TM Varghese (1962, 1963), S Junell (1962), J Kallarackal and SP Bhatnagar (1978, 1980), RN Kapil and P Masand (1964), T Kapoor and MR Vijayaraghavan (1975, 1978), T Kapoor, NK Parulekar and MR Vijayaraghavan (1975), T Kapoor, MR Vijayaraghavan and NK Parulekar (1978), CV Krishna Iyengar (1937, 1939a, b, 1940a, b, c, 1941, 1942a, b, c), G Lundqvist (1915), N Magin, R Classen and G Claudia (1989), P Maheshwari and HS Navalakha (1941), JC Mestre and JL Guignard (1969), RM Michell (1915), V Millsaps (1936), JV Monachino (1949), CR Nagendran, TM Ramakrishna and BM Narayana (1980), S Natesh and NN Bhandari (1975, 1984), ZI Nikiticheva (1971), ZI Nikiticheva and ES Teryokhin (1973), N Pal (1959), K Periasamy (1962a), D Persidsky (1934), M Pogliani (1964), TS Raghavan and VK Srinivasan (1941a, b), D Raju (1973, 1976), D Raju and GD Arekal (1977, 1978), KM Safeeulla and HC Govindu (1950), A Sahai and KR Shivanna (1985a, b), FM Schertz (1919), E Schmid (1906), GF Schrock and BF Palser (1967), VK Sharma (1962), ARP Sinha (1985), KV Srinath (1934, 1940), AR Srinivasan (1946), VK Srinivasan (1940), K Steffen (1955, 1956), HG Svensson (1928), S Taneja (1969a, b, 1970), B Tiagi (1956, 1965b, 1966, 1972), B Tiagi and TM Varghese (1970), SC Tiwari (1982, 1983), C Tsan-Lang and LR Heckard (1973), R van Hulst, A Thériault and B Shiplay (1986), TM Varghese (1963, 1966), MR Vijayaraghavan and S Ratnaparkhi (1972), R von Wettstein (1895b), T Yamazaki (1953, 1954, 1957a, b).

48.16 Globulariaceae

The embryology has not been studied in sufficient detail (see Rosén 1940a).

The ovule is anatropous, unitegmic, and tenuinucellar. The archesporial cell functions directly as megaspore mother cell, undergoes meiotic divisions, and a linear tetrad is formed. The chalazal megaspore develops into a Polygonum type of embryo sac. The polar nuclei fuse before fertilization. The antipodals are ephemeral. An endothelium differentiates.

The development of the endosperm is of the Cellular type. The division of the primary endosperm nucleus results in a small chalazal and a large micropylar chamber. Two consecutive vertical divisions in the micropylar chamber are followed by transverse divisions in each cell. The four micropylar cells develop into haustoria. The separating walls degenerate so that the haustorium becomes unicellular and coenocytic in *Globularia vulgaris* (Rosén 1940a). The middle tier develops into the endosperm proper. The chalazal chamber forms a binucleate haustorium. Both the haustoria are aggressive and penetrate the ovary wall. According to Rosén (1940a), the development corresponds to the Brunella type (see Schnarf 1929).

Somatic apospory occurs in *Globularia vulgaris*.

The embryogeny corresponds to the Onagrad type. The seed has fleshy endosperm with a straight embryo in the middle (Hutchinson 1973).

P Crété (1943), J Hutchinson (1973), W Rosén (1940a), K Schnarf (1929)

48.17 Bignoniaceae

The anther is usually tetrasporangiate (Hutchinson 1973). Davis (1966) mentions bisporangiate anthers in *Dolichandrone* and *Parmentiera*, and each anther in these taxa is subdivided by a tapetal ingrowth. The anther wall consists of a cutinized epidermis, multilayered endothecium with fibrous thickenings, four middle layers, and a secretory tapetum of two- or four-nucleate cells in *Oroxylum indicum* (Ghatak 1956). While Davis (1966) reports the absence of endothecial thickenings in *Bignonia*, Nanda and Gupta (1971) point out that they are present in *Pyrostegia venusta* (= *Bignonia venusta*). The thickenings are of α-cellulose, and appear at the pollen grain stage, in restricted areas, in the four microsporangia. They are present throughout the length of the anther. It therefore appears that there are longitudinal bands of fibrous endothecial thickenings. This aspect requires further study.

The absence of endothecium is also mentioned in *Stereospermum chelonoides* and *Tabebuia rosea* (Mehra and Kulkarni 1985); the tapetal cells are multinucleate. In *S. chelonoides* the tapetal cells at the connective region become radially elongated with vacuoles towards the connective.

The microspore mother cells undergo meiotic divisions, cytokinesis is simultaneous, and the tetrads tetrahedral, rarely isobilateral and decussate. Degeneration of microspore mother cells, tetrads, and pollen grains is frequent in *Tecoma stans*.

Mature pollen grains are two-celled in *Dolichandrone falcata*, *Heterophragma adenophyllum* and *Millingtonia hortensis*, but three-celled in *Tabebuia rosea* and *Stereospermum chelonoides* (Mehra and Kulkarni 1985).

The ovule is anatropous, unitegmic, and tenuinucellar. The persistent hypostase is lignified. The hypodermal archesporium is generally unicellular, but two-celled in *Kigelia pinnata* and *Tecoma stans* (Govindu 1950). A single cell functions as megaspore mother cell, undergoes meiotic divisions, and produces a linear or T-shaped tetrad. Double tetrads have been observed in *Bignonia megapatomica* (Swamy 1941a). The chalazal megaspore develops into a Polygonum type of embryo sac. The synergids are beaked; polar nuclei fuse before fertilization. The antipodals are ephemeral, or may persist (Govindu 1950). The endothelium encloses the embryo sac except in the chalazal region.

The development of the endosperm is of the Cellular type. The division of the primary endosperm nucleus separates a large micropylar from a small chalazal chamber. The two crosswise vertical divisions in the micropylar chamber are followed by a transverse division. The upper four micropylar cells develop into a four-celled haustorium. However, in *Bignonia megapatomica* (Swamy 1941a), the vertical walls separating the four nuclei do not develop towards the extreme micropylar tier. The micropylar chamber develops into a four-nucleate haustorium, the cells in the middle region produce the endosperm proper, and two vertical divisions in the chalazal chamber give rise to a four-celled haustorium.

The micropylar haustorium is two-celled in *Dolichandrone falcata*, *Millingtonia hortensis*, and *Stereospermum chelonoides*. In *Heterophragma adenophyllum* and *Tabebuia rosea*, the two cells of the micropylar tier appear to enlarge and elongate to form narrow, polynucleate, vacuolate cells of the micropylar haustorium (Mehra and Kulkarni 1985).

The development described above is referred to the Catalpa type (Mauritzon 1935c), and also occurs in *Catalpa kaempferi* (Crété 1942), *Jacaranda mimosaefolia* (Mauritzon 1935c, Govindu 1950), *Parmentiera cerifera*, *Pithecoctenium clemantideum* (Mauritzon 1935c), *Tecoma stans* (Govindu 1950).

The chalazal haustorium is two-celled in *Dolichandrone falcata*, *Heterophragma adenophyllum*, *Stereospermum chelonoides*, and *Tabebuia rosea*, whereas it is four-celled in *Millingtonia hortensis*. Both the micropylar and the chalazal haustorium remain active up to the globular proembryo (Mehra and Kulkarni 1985). In a fully ripe seed the micropylar haustorium disintegrates leaving a hollow space which acts as an air-sac and gives buoyancy to the seed. There is no such report in any other angiosperm, and the above finding needs to be confirmed.

The chalazal haustorium is four-celled in *C. kaempferi* and *Tecoma*, and two-celled in *Parmentiera*. It is highly aggressive in *Catalpa*, digests the hypostase, and its branches ramify in the integument. The cellular endosperm proper can be clearly differentiated into two zones in *Catalpa* (Crété 1942). The micropylar part comprises fairly large cells, and the chalazal part comparatively small cells. In *Catalpa*, *Tecoma* and *Parmentiera*, the micropylar region does not develop into a haustorium.

Catalpa type development has also been observed in *Dolichandrone falcata*, *Heterophragma adenophyllum*, *Millingtonia hortensis*, *Stereospermum chelonoides* and *T. rosea* (Mehra and Kulkarni 1985).

The second mode of development is the Incarvillea type: *Incarvillea compacta*, *I. delavayi* and *I. grandiflora* (Mauritzon 1935c). The pattern of cell divisions is more or less similar to that in the Catalpa type. However, in the Incarvillea type, both micropylar and chalazal haustoria are one-celled and the micropylar haustorium is vesicular.

In *Incarvillea compacta* (Mauritzon 1935c), a transverse wall following the division of primary endosperm nucleus gives rise to a large micropylar and a small chalazal chamber. The micropylar chamber undergoes a transverse division resulting in a primary three-celled endosperm (Fig. 48.17.1A). The middle cell divides transversely and then longitudinally and the lower region of the cellular endosperm is delimited (Fig. 48.17.1B). The terminal cell undergoes a transverse division; the upper cell develops into an uninucleate, vesicular micropylar haustorium (Fig. 48.17.1C–G); the lower cell, by crosswise vertical

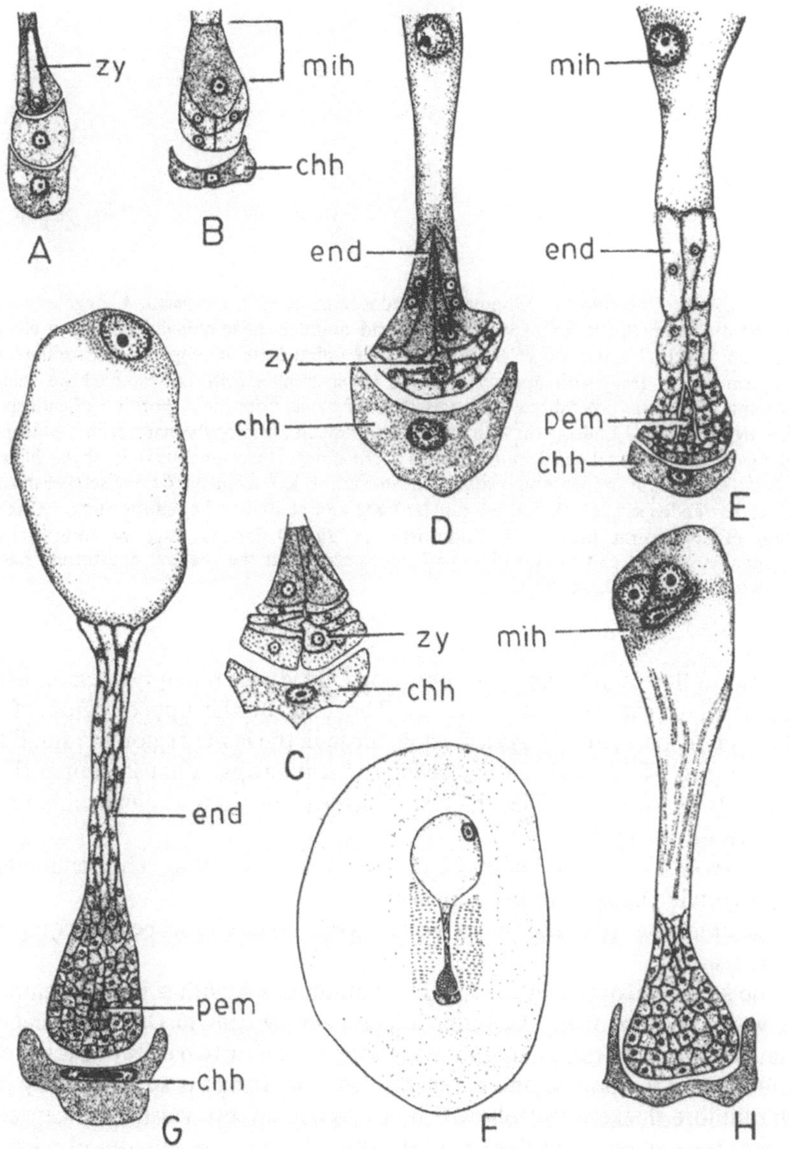

Fig. 48.17.1A–H

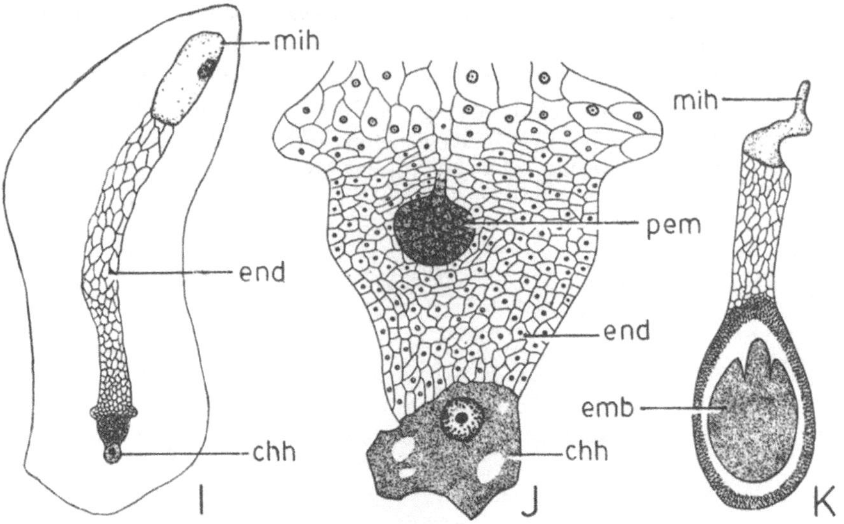

Fig. 48.17.1A–K. *Incarvillea*, development of endosperm. **A–G** *I. compacta*. **A** Zygote (*zy*) and three-celled endosperm; **B** Endosperm proper, and uninucleate micropylar (*mih*) and chalazal haustorium (*chh*); **C** Later stage; **D, E** Portion of uninucleate micropylar haustorium (*mih*), endosperm proper (*end*) with upper larger and lower smaller cells, and uninucleate chalazal haustorium (*chh*); *pem* proembryo; **F** Longisection of ovule, note the disposition of endosperm, outline figure for **G**; **G** Endosperm with uninucleate vesicular micropylar haustorium, endosperm proper (*end*), and chalazal haustorium (*chh*). **H** *I. grandiflora*, three-nucleate micropylar haustorium (*mih*), endosperm proper, and chalazal haustorium. **I–K** *I. delavayi*. **I** Longisection of ovule with cellular endosperm (*end*), and micropylar (*mih*) and chalazal (*chh*) haustorium; **J** Chalazal portion of endosperm from **I**. **K** Endosperm at dicotyledonous stage of embryo (*emb*), the remnants of micropylar haustorium (*mih*) still persist, but the chalazal haustorium has degenerated. (After Mauritzon 1935c)

divisions, followed by transverse partitions, produces the upper region of the endosperm proper (Fig. 48.17.1E–G). The cells in the upper region of the cellular endosperm are elongated, while those in the lower region are small and compact (Fig. 48.17.1E–I). In *I. grandiflora* the micropylar haustorium is three-nucleate (Fig. 48.17.1H). The chalazal chamber functions as chalazal haustorium (Fig. 48.17.1D–J).

Figure 48.17.1K represents the dicotyledonous embryo. The remnants of the micropylar haustorium still persist.

The embryogeny conforms to the Onagrad type. The suspensor comprises five or six cells.

The seed is exotestal, winged, and exalbuminous with a straight embryo. The wing develops from the outer integumentary cells in *Oroxylum indicum* (Ghatak 1956). The thin, membraneous wing is one or two cells thick; the cells are tangentially elongate and form a fringe. The fringe is supported by cells with cellulose thickenings, followed by multilayered, brownish, rib-like, radiating strands composed of thick-walled cells. The cells are pitted and sclerotic, and protect the embryo.

The seeds are of two types – hypogeal and epigeal (Gentry 1980). Epigeal seeds have flat, foliaceous cotyledons, usually deeply cordate at the base with a short radicle occupying the sinus. The hypogeal type has thick, fleshy, oval or rounded cotyledons which remain within the seed-coat. The mature seeds are exalbuminous.

The seed is albuminous in *Dolichandrone*, *Heterophragma*, *Millingtonia*, *Stereospermum* and *Tabebuia* (Mehra and Kulkarni 1985).

The seeds in the Bignoniaceae are usually considered exalbuminous because the endosperm, if present, is usually represented by a few layers of collapsed cells without any food reserve. The findings of Mehra and Kulkarni, therefore, need confirmation.

JE Armstrong (1985), P Crété (1942, 1952c), GL Davis (1966), BM Duggar (1899), A Gentry (1980), J Ghatak (1956), HC Govindu (1950), J Hutchinson (1973), J Mauritzon (1935c), KR Mehra and AR Kulkarni (1985), JV Monachino (1949), K Nanda and SC Gupta (1971), LW Sharp (1911), R Souèges (1939a, 1940a), BGL Swamy (1941a)

48.18 Henriqueziaceae

The embryology has not been studied.

48.19 Acanthaceae

The anther wall comprises the epidermis, fibrous endothecium, an ephemeral middle layer, and secretory tapetum with two- to four-nucleate cells. The endothecial thickenings are absent in *Barleria prionitis*, *Justicia betonica*, *Elytraria acaulis* (B.M. Johri and Singh 1959), *Thunbergia alata* (Wadhi 1962) and *T. mysorensis* (Khaleel and Boraiah 1973). The fibrous thickenings extend to the middle layers in *Nelsonia campestris* (Mohan Ram and Masand 1963). A protrusion of anther wall on the inner side into the loculus (placentoid) occurs in *Acanthus ilicifolius* (Rangaswamy 1941a), *Justicia betonica* (C.R. Narayanan 1956), *Thunbergia alata* (Wadhi 1962), and *T. mysorensis* (Khaleel and Boraiah 1973). The primary sporogenous tissue may develop directly into microsporocytes as in *T. mysorensis*, or may undergo secondary divisions as in *Nelsonia campestris*. Simultaneous cytokinesis follows meiotic division in microspore mother cells, and the tetrads are tetrahedral, isobilateral and decussate. Polyspory is frequent in *T. mysorensis* (Khaleel and Boraiah 1973), and as many as 22 microspores may be formed in a single polyad. The formation of micronuclei due to irregularities in meiosis leads to polyspory.

The family is multipalynous, and the pollen grains are polymorphic. These are colpate, colporate, or porate (Bhoj Raj 1961). In *Thunbergia*, they are acolpate with a smooth, spiral exine. They are two-celled at the dispersal stage.

The ovule is anatropous, hemianatropous, or campylotropous, unitegmic, and tenuinucellar. The integument is massive with a narrow micropyle. Its inner epidermis develops into the endothelium in *Elytraria acaulis* (B.M. Johri and Singh 1959), and *Nelsonia campestris* (Mohan Ram and Masand 1963). The obturator is funicular.

The jaculator is well developed except in two tribes of Acanthoideae – Aphelandrae and Acantheae. It is weakly developed in Nelsonioideae (Lindau 1895). However, B.M. Johri and Singh (1959), Bhoj Raj (1961), and Bremekamp (1965) negate its presence. It is absent in Mendencioideae and Thunbergioideae.

The archesporium is uni- to multicellular. A single cell functions as megaspore mother cell, and undergoes meiotic divisions, producing a linear tetrad. In *Thunbergia mysorensis* (Khaleel and Boraiah 1973), irregularities in meiotic divisions result in the formation of several micronuclei. The micronuclei may develop walls around them, and in a "tetrad" five or six megaspores may be formed. The chalazal megaspore develops into a Polygonum type of embryo sac.

The embryo sac extends into the micropyle, and the egg apparatus is organized in this region in *Dipteracanthus patulus* (P. Maheshwari and Negi 1955) and *Elytraria acaulis* (B.M. Johri and Singh 1959). In *Justicia betonica* (S.P. Bhatnagar and Puri 1970) and *J. simplex* (Mohan Ram and Sehgal 1958), the tip of the embryo sac extends into the schizogenous cavity (which differentiates in the funicular region) but the egg apparatus is left behind in situ. In *Nelsonia campestris* (Mohan Ram and Masand 1963), during the formation and maturation of the embryo sac, the tip with the nuclei extends into the funicle where the egg apparatus is organized.

At the four-nucleate stage, the embryo sac develops a chalazal caecum in *Acanthus ilicifolius* (Phatak and Ambegaokar 1957), *Andrographis* (Mohan Ram 1960b, Mohan Ram and Masand 1962), *Dipteracanthus patulus* (P. Maheshwari and Negi 1955), *Lepidagathis trinervis* (Ambegaokar 1959), *Thunbergia alata* (Wadhi 1962) and *Phlebophyllum kunthianum* (S.P. Bhatnagar and Kallarackal 1973). In *Thunbergia* the polar nuclei fuse in the caecum. The mature embryo sac contains starch grains. The synergids are beaked. In *Blepharis maderaspatensis* (Phatak and Ambegaokar 1963), sometimes the synergids are egg-like and one of them persists up to the quadrant proembryo. Occasionally, one synergid also persists in *Elytraria*, *Haplanthus* and *Justicia simplex*. The polar nuclei fuse before fertilization, but in *H. tentaculatus* they remain appressed to each other until one of them fuses with the male gamete (Phatak and Ambegaokar 1961a).

The antipodals are mostly ephemeral. However, they persist up to the early stages of endosperm development in *Aphelandra aurantica* (Hartman 1923), *Barleria cristata* (Mohan Ram 1962), *B. grandicalyx*, *B. spinisepata*, *B. taitensis*, *B. ventricosa* and *B. volkensii* (Karlström 1974b), *Elytraria acaulis* (B.M. Johri and Singh 1959) and *Nelsonia campestris* (Mohan Ram and Masand 1963). The antipodals are binucleate in *Elytraria*. They undergo secondary multiplication and there are eight to 14 cells in *Barleria prionitis*, four to five in *Blepharis maderaspatensis*, four in *Chamaeranthemum beyrichii*, and five in *Pseudoeranthemum tuberculatum* (Karlström 1974b, 1973). In *Blepharis* the antipodals simulate the egg cell.

The development of the endosperm is of the Cellular type. The primary endosperm nucleus moves to the chalazal region. A transverse wall follows its division separating a small chalazal from a large micropylar chamber. A transverse division in the micropylar chamber results in the formation of a row of three superposed cells – chalazal, central and micropylar. The growth of the endosperm is asymmetric. The micropylar and the chalazal chamber develop into haustoria. The central chamber gives rise to the endosperm proper. The development in the central chamber may be ab initio cellular or free-nuclear. In the case of interpolation of a free-nuclear phase in the central chamber, wall formation occurs sooner or later and the endosperm becomes completely cellular, or partially cellular, leaving the free-nucleate "basal apparatus". According to Karlström (1972, 1973, 1974a, b), wall formation does not occur following nuclear division in the micropylar chamber. When the two daughter nuclei divide, the spindles come to lie close together, the cell plates lie rectilinearly, and the wall separates a binucleate micropylar haustorium from the binucleate micropylar chamber. Such a mode of development has not been observed in any other member of Acanthaceae, or Sympetalae, and requires reinvestigation.

Mohan Ram and Wadhi (1964) recognized four developmental types, depending on the behaviour of the central chamber of three linearly placed endosperm cells (Fig. 48.19.1A–N).

1) Thunbergia type (Fig. 49.19.1A–F). The chalazal chamber does not develop into a haustorium, but produces the endosperm proper (48.19.1A–D). A good example is *Blepharis maderaspatensis* (Fig. 48.19.1A, B, E, F) (Phatak and Ambegaokar 1963), where a transverse wall follows the division of the primary endosperm nucleus, resulting in a micropylar and a chalazal chamber (Fig. 48.19.1A, B). The micropylar chamber develops into a binucleate haustorium; the chalazal chamber undergoes repeated divisions to produce the endosperm proper (Fig. 48.19.1E, F). The antipodals persist up to the globular proembryo. In *Thunbergia* (Mauritzon 1934f, Mohan Ram and Wadhi 1964), the micropylar haustorium is multinucleate and it branches in advanced stages (Fig. 48.19.2A–J). The cells of the endosperm proper abutting the micropylar haustorium transform into secondary haustoria (Fig. 48.19.2C–J). Due to partial dissolution of walls, a continuity is established between the micropylar and the secondary haustorium (Fig. 48.19.2I). The endosperm is consumed by the developing embryo, and only a few layers persist in the mature seed.

Khaleel and Boraiah (1973) describe an unusual type of development in *Thunbergia mysorensis*. The vertical division in the chalazal chamber is followed by a transverse division in each cell forming two tiers of two cells each. The micropylar chamber does not undergo any division and, finally, the endosperm collapses. This aspect needs further attention.

2) Acanthus type (Figs. 48.19.1H, 48.19.3A). The micropylar and the chalazal chamber develop into haustoria. The development of the endosperm in the central chamber is ab initio cellular (Fig. 48.19.1H). The central cell alone gives

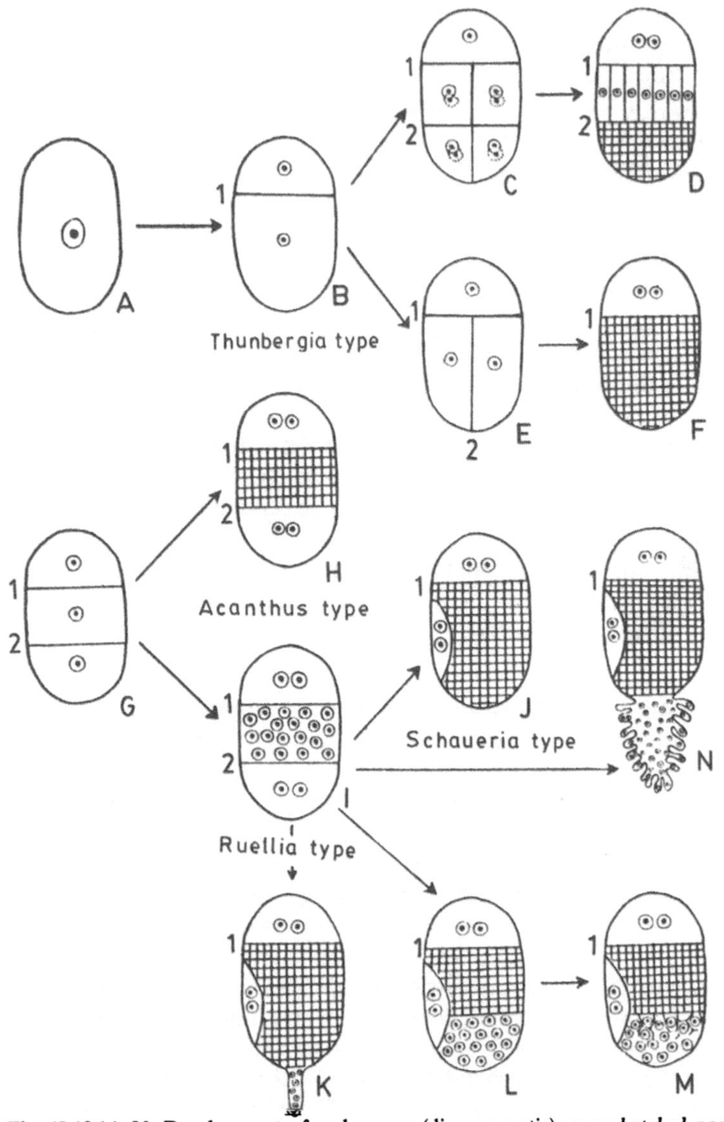

Fig. 48.19.1A–N. Development of endosperm (diagrammatic), *cross-hatched* portion represents cellular endosperm proper; embryo not shown. **A–G** Thunbergia type. **A** Primary endosperm nucleus; **B** Two-celled endosperm, micropylar and chalazal chamber. **C, D** *Thunbergia.* **C** Uninucleate micropylar haustorium, and two-tiered chalazal chamber; **D** Binucleate micropylar haustorium. The upper tier of the chalazal chamber forms secondary haustoria, the lower tier cellular endosperm. **E, F** *Blepharis.* **E** One-celled micropylar haustorium, and vertical division in chalazal chamber; **F** Two-nucleate micropylar haustorium, and cellular endosperm. **G** Micropylar, central, and chalazal chamber. **H** Acanthus type – cellular endosperm with binucleate micropylar and binucleate chalazal haustorium. **I** Ruellia type – free-nuclear endosperm in central chamber, and two-nucleate micropylar and two-nucleate chalazal haustorium. **J** Schaueria type – cellular endosperm in central chamber, binucleate micropylar, and binucleate lateral chalazal haustorium. **K–N** Ruellia type. **K** *Justicia,* cellular endosperm with a small tubular basal apparatus, two-nucleate micropylar and two-nucleate lateral chalazal haustorium; **L** *Ruellia squarrosa,* cellular

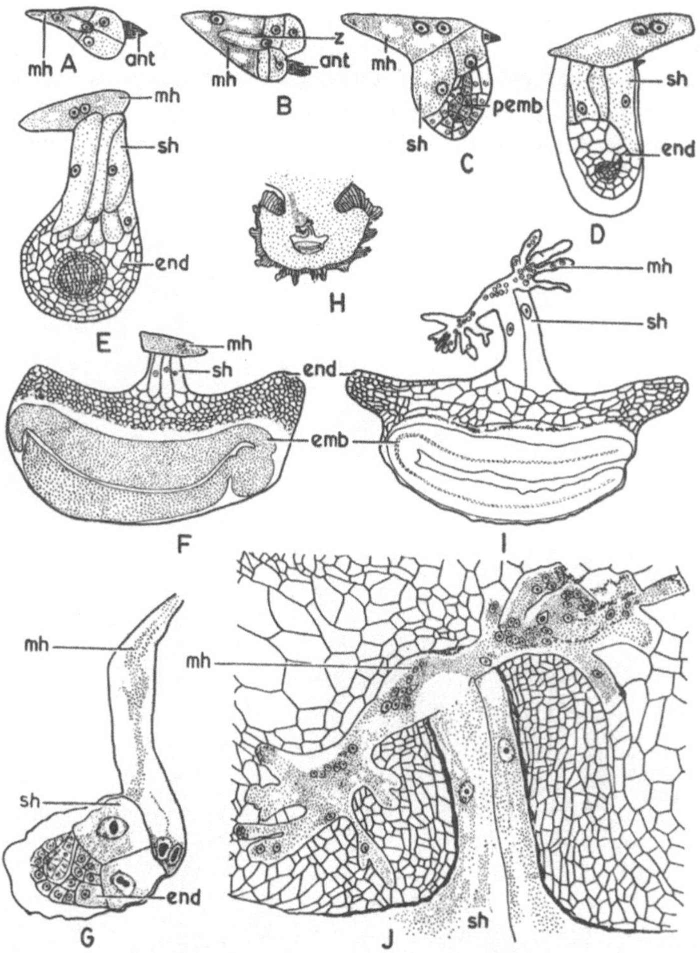

Fig. 48.19.2A–F. *Thunbergia reticulata*, **A** Embryo sac with zygote, two-celled endosperm, uninucleate micropylar haustorium (*mh*), and remnants of antipodals. Chalazal haustorium is absent; **B** Advanced stage; *z* zygote; *ant* antipodals; **C**, **D** Endosperm cells adjacent to binucleate micropylar haustorium (*mh*) develop as secondary haustoria (*sh*); *pem* proembryo; *end* endosperm proper; **E** Endosperm at globular proembryo, secondary haustoria and binucleate micropylar haustorium; **F** Endosperm at mature dicotyledonous (*emb*) embryo. (After Mauritzon 1934f). **G–J** *T. alata*. **G** Endosperm at globular proembryo with secondary (*sh*) haustoria and binucleate micropylar haustorium; **H** Longisection of seed, outline for **I**; **I** Dicotyledonous embryo (*emb*) with endosperm proper (*end*), secondary haustoria (*sh*), and much-branched micropylar haustorium (*mh*). **J** Portion of endosperm proper, branched multinucleate micropylar haustorium, and secondary haustoria. (After Mohan Ram and Wadhi 1964)

endosperm with coenocytic basal apparatus, two-nucleate micropylar and two-nucleate lateral chalazal haustorium; **M** *R. pulcherrima*, cellular endosperm with partially cellular basal apparatus, two-nucleate micropylar and two-nucleate lateral chalazal haustorium; **N** *Asystasia*, cellular endosperm with a conspicuous coenocytic haustorial basal apparatus, binucleate micropylar and binucleate lateral chalazal haustorium. (After Mohan Ram and Wadhi 1964) rise to the endosperm proper as in *Acanthus ilicifolius* (Fig. 48.19.3A), *Andrographis echioides* (Mohan Ram and Masand 1962), *Barleria prionitis* (Fig. 48.19.3B), *Crossandra undulifolia* (Mauritzon 1934f), *Elytraria acaulis* (B.M. Johri and Singh 1959), *Haplanthus tentaculatus* (Phatak and Ambegaokar 1956, 1957, 1961a), *Lepidagathis trinervis* (Fig. 48.19.3C) and *Nelsonia campestris* (Mohan Ram and Masand 1963). The endosperm proper, produced

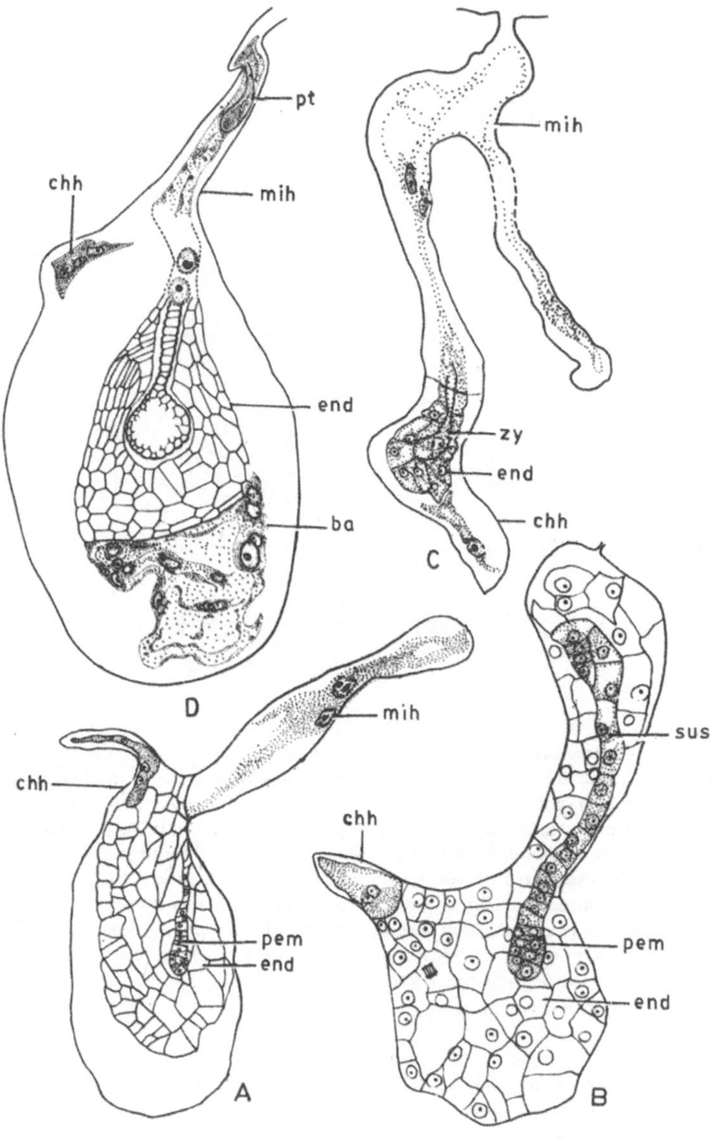

Fig. 48.19.3A–D

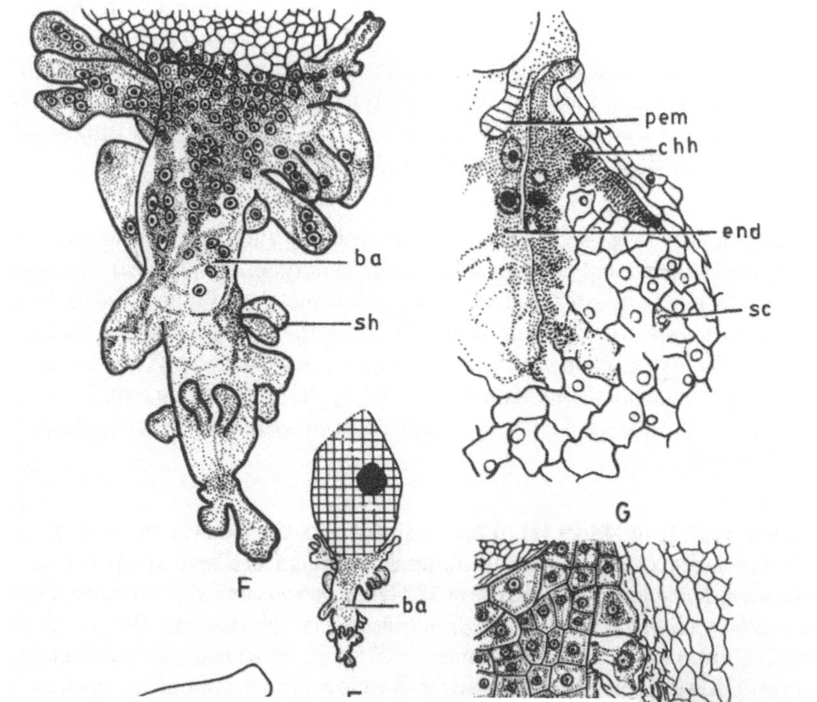

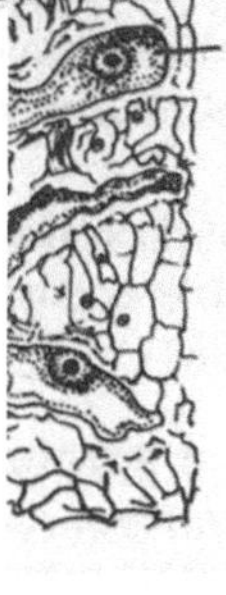

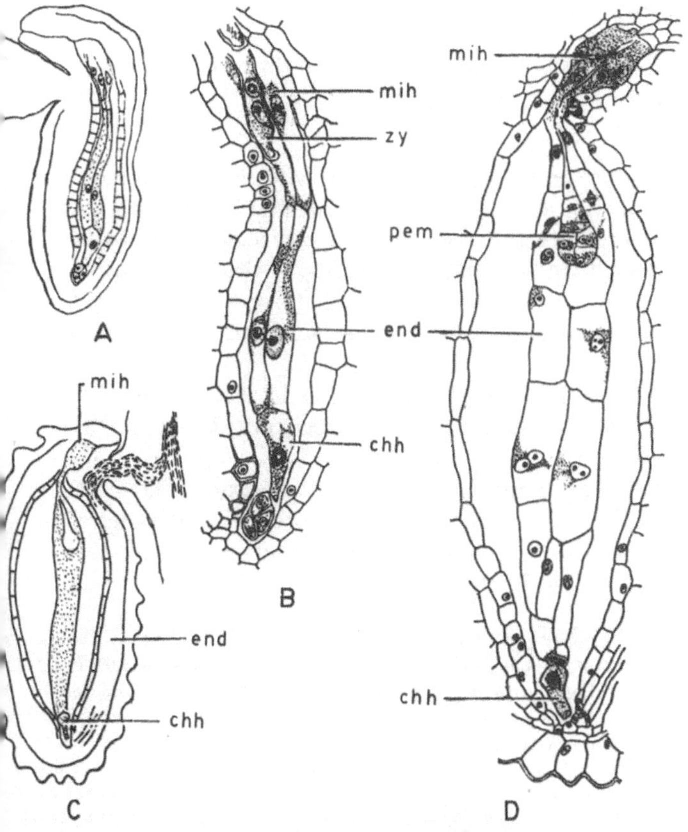

endosperm proper (*end*), ıaustorium. (After Phatak ar proembryo (*pem*) with :halazal haustorium (*chh*). ryo sac with zygote (*zy*), : micropylar haustorium. ılar proembryo, remnants ıs (*ba*), and four-nucleate *sia gangetica*. **E** Dissected rous finger-like processes basal apparatus from **E.** :ly proembryo (*pem*), free-al haustorium (*chh*) with '61b). **H, I** *Andrographis* and micropylar (*mih*) and (*sh*). (After Mohan Ram by the central chamber, occupies the micropylar haustorium in *Barleria prionitis*. In *Andrographis* (Fig. 48.19.3H, I), *Chamaeranthemum* and *Haplanthus*, the peripheral cells of the endosperm proper develop into secondary haustoria which persist up to the globular proembryo. The endosperm becomes ruminate due to the uneven growth of the peripheral endosperm cells in *Andrographis*, *Elytraria*, *Haplanthus* and *Nelsonia*.

3) **Schaueria type** (Fig. 48.19.1I, J). The micropylar and the chalazal chamber develop into haustoria. The central chamber undergoes free-nuclear divisions, wall formation sets in later and it becomes completely cellular (Fig. 48.19.1J) as in *Barleria cristata* (Mohan Ram 1962), *Barleria* spp., *Crabbea velutina* (Karlström 1974a), *Chamaeranthemum beyrichii* (Karlström 1974b), *Ecbolium innaeanum* var. *rotundifolium* (Mohan Ram 1958, 1962), *Jacobinia* and *Schaueria* (Mauritzon 1934f), and *Lankesteria* spp., *Pseudoeranthemum* and *Ruspolia* (Karlström 1973).

4) **Ruellia type** (Fig. 48.19.1I). This is similar to the Schaueria type. Cytokinesis takes place at the 16- or 32-nucleate stage of the endosperm as in *Eranthemum leuconeuron* (Mauritzon 1934f), *E. nervosum* (Mohan Ram 1958), *E. pulchellum* (*E. nervosum*) and *E. purphrascens* (Karlström 1973), or the maximum number of free nuclei may be 256 as in *Ruellia decaisniana* and *R. solitaria* (Mauritzon 1934f). However, wall formation remains incomplete so that the endosperm can be distinguished into the upper cellular region and the lower free-nuclear portion (Fig. 48.19.3D), the basal apparatus (Mauritzon 1934f).

The shape and size, as well as the number of nuclei in the basal apparatus, vary in different genera, and have been summarized by Mohan Ram and Wadhi (1964):

a) Basal apparatus narrow and weakly developed with numerous nuclei as in *Beloperone*, few nuclei as in *Adhatoda*, *Dicliptera* and *Justicia* (Fig. 48.19.1K), or nuclei absent as in *Peristrophe* and *Lepidagathis prostrata*.

b) Basal apparatus broad and persists in the free-nuclear stage as in *Asteracantha*, *Dipteracanthus* (Fig. 48.19.3D), *Eranthemum*, *Goldfussia*, *Hemigraphis*, *Ruellia decaisniana*, *R. rosea*, *R. speciosa*, and *R. squarrosa* (Fig. 48.19.1L).

c) Basal apparatus is broad and partially cellular as in *Blechum*, *Brillantaisia*, *Dyschoriste humilis* (Cosa 1978), *Ruellia pulcherrima* (Fig. 48.19.1M), *R. solitaria* and *R. tuberosa*.

d) Basal apparatus produces finger-like haustorial processes. These are enucleate in the beginning, but later become multinucleate due to the migration of nuclei. Such processes ramify into the integument as in *Asystasia gigentica* (Figs. 48.19.1N, 48.19.3E, F).

The micropylar haustorium is binucleate in *Acanthus* (Fig. 48.19.3A), *Andrographis*, *Asystasia*, *Barleria*, *Crabbea*, *Crossandra*, *Eranthemum*, *Goldfussia*, *Haplanthus*, *Hemigraphis latebrosa*, *Jacobinia*, *Justicia*, *Lepidagathis*,

Nelsonia, Peristrophe, Phlebophyllum, Pseudoeranthemum, Ruellia, Ruspolia and *Sericocalyx scaber*. It is two- or four-nucleate in *Adhatoda, Asystasia, Beloperone, Dicliptera, Ecbolium, Peristrophe* and *Schaueria*, and four-nucleate in *Thunbergia*. It is two-celled in *Elytraria acaulis*, and becomes multicellular in *Blepharis maderaspatensis*. Perhaps the cells of the central chamber project into the micropylar haustorium in *Blepharis* and *Barleria prionitis* (Mohan Ram and Wadhi 1964). The micropylar haustorium becomes aggressive, penetrates the funicle, and reaches as far as the vascular strand in *Hemigraphis primulaefolia, Thunbergia alata* and *Sericocalyx scaber*.

The chalazal haustorium is four-nucleate in *Asteracantha, Brillantaisia, Dicliptera, Elytraria* and *Hemigraphis*; two- or four-nucleate in *Acanthus* (Fig. 48.19.3A), *Dipteracanthus, Hemigraphis, Justicia betonica, Phlebophyllum, Ruellia* spp. and *Sericocalyx scaber* (Fig. 48.19.3G); one- to four-nucleate in *Justicia simplex*, and two-nucleate in *Andrographis, Dipteracanthus, Eranthemum, Elytraria* and *Nelsonia*. A chalazal haustorium is absent in *Blepharis maderaspatensis* and *Thunbergia alata*. At the globular proembryo, following the degeneration of the chalazal haustorium, in *Acanthus, Barleria, Elytraria, Hemigraphis, Justicia* and *Sericocalyx*, the adjoining cells of the endosperm proper take a denser stain.

The embryogeny is of the Onagrad type in *Elytraria acaulis* (B.M. Johri and Singh 1959). It appears to be of the Solanad type in *Adhatoda vasica* (Mohan Ram 1960c), *Dyschoriste humilis* (Cosa 1978), and *Justicia betonica* (S.P. Bhatnagar and Puri 1970). It conforms to Period II, Series A, Megarchetype VI (Caryophyllad type) in *Acanthus ilicifolius* (Phatak and Ambegaokar 1957), *Barleria prionitis* (Phatak and Ambegaokar 1956), *Haplanthus tentaculatus* (Phatak and Ambegaokar 1961a), *Hemigraphis latebrosa* and *Sericocalyx scaber* (Phatak and Ambegaokar 1961b). In *Blepharis maderaspatensis* (Phatak and Ambegaokar 1963), and occasionally in *H. latebrosa*, the embryogeny conforms to Period III, Series A, Megarchetype VI.

In *Nelsonia campestris* (Mohan Ram and Masand 1963), where the egg apparatus lies in the schizogenous cavity in the placenta, the tip of embryo comes to lie in the cellular endosperm due to elongation of the suspensor. In *Barleria ventricosa*, Karlström (1974a) observed a wall after the nuclear division in the zygote. However, he (Karlström 1972, 1973, 1974b) states later that a wall is not formed following the division of the zygote, and a binucleate proembryo elongates, and walls develop only when these two nuclei divide. These observations are highly suspect and a further investigation is called for.

Suspensor polyembryony is of frequent occurrence in Acanthaceae, and additional proembryos develop due to the proliferation of suspensor cells in *Acanthus, Barleria, Dipteracanthus, Haplanthus* and *Sericocalyx*.

The exotestal seed is exalbuminous with a straight embryo. The testa is crushed except for the outer epidermis, and one or two hypodermal layers. In several genera the epidermal cells develop into unicellular hairs. In *Acanthus ilicifolius*, the testa has two or three inner layers of thick-walled pitted cells (Corner 1976). The remnants of the seed-coat persist in *Andrographis* spp. (Mohan Ram 1960b, Mohan Ram and Masand 1962), *Elytraria acaulis* (B.M. Johri and Singh 1959), and *Haplanthus tentaculatus* (Phatak and Ambegaokar

1961a). The pericarp gives protection to the seed. The seed is characterized by a jaculator which develops from the funicle except in *Elytraria acaulis.*

KB Ambegaokar (1959), JM and CC Baskin (1971), S Bhaduri (1944), SP Bhatnagar and J Kallarackal (1973), SP Bhatnagar and S Puri (1970), CEB Bremekamp (1953, 1965), RK Brummitt, IK Ferguson and MM Poole (1980), EJH Corner (1976), MT Cosa (1978), R Gigante (1929), Y Gutterman and A Witztum (1979), A Hartmann (1923), BM Johri and H Singh (1959), G Karsten (1891), PC Karlström (1972, 1973, 1974a, b), TF Khaleel and G Boraiah (1973), G Lindau (1895), P Maheshwari and V Negi (1955), P Masand (1963a), J Mauritzon (1934f), HY Mohan Ram (1958, 1959, 1960a, b, c, 1962), HY Mohan Ram and P Masand (1962, 1963), HY Mohan Ram and P Sehgal (1958), HY Mohan Ram and M Wadhi (1964, 1965), DP Mullan (1936), CR Narayanan (1951, 1956), VG Phatak and KB Ambegaokar (1955, 1956, 1957, 1961a, b, 1963), Bhoj Raj (1961), K Rangaswamy (1941a), BG Sahi (1966), TS Sareen and RR Bhanwara (1978), M Wadhi (1962, 1970), WE Wall and RW Long (1965)

48.20 Pedaliaceae

The anther wall comprises the epidermis, fibrous endothecium, two or three middle layers, and secretory tapetum of binucleate cells in *Sesamum indicum* (S.P. Singh 1960). The endothecium is biseriate with poorly developed thickenings in *S. lacinatum* (Kulkarni and Pataskar 1973–74). Refractive droplets have been noted on degenerated tapetal cells. The nuclei of tapetal cells fuse and become polyploid in *Pedalium murex* (S.P. Singh 1963). Simultaneous cytokinesis follows meiotic divisions in microspore mother cells, and the tetrads are tetrahedral, isobilateral, decussate and T-shaped.

The pollen grains are six- (*Pedalium murex*) or ten-colpate (*Sesamum indicum*), and dispersal occurs at the three-celled stage.

The ovule is anatropous, unitegmic and tenuinucellate. The hypostase is thick-walled; an endothelium differentiates at the functional megaspore stage. The archesporium is one- or two-celled. A single cell functions directly as megaspore mother cell, undergoes meiotic divisions, produces a linear or T-shaped tetrad, and the chalazal megaspore develops into a Polygonum type of embryo sac. The polar nuclei fuse before or at the time of fertilization. The antipodals are ephemeral.

Development of the endosperm is of the Cellular type. The division of the primary endosperm nucleus separates a large micropylar chamber from a small chalazal chamber. A vertical division occurs in the micropylar chamber; the chalazal chamber becomes binucleate. A transverse and then a longitudinal division take place in each cell of the micropylar chamber. The upper tier gives rise to the micropylar haustorium, and the lower tier to the endosperm proper. The cellular endosperm can be differentiated into three zones: uppermost and lowest zone of richly cytoplasmic cells which show a higher rate of division as compared to the cells of the central zone. The chalazal chamber develops into a four-celled haustorium.

The micropylar haustorium is multicellular in *Pedalium* and *Sesamum*, and four-celled in *Trapella* (S.P. Singh 1960, 1963, 1970a). The chalazal haustorium is four-celled in *Sesamum* (S.P. Singh 1960), two-celled, with each cell coenocytic, in *Pedalium*, and of two elongated cells in *Trapella* (S.P. Singh 1960).

However, according to Srinivasan (1942), the micropylar haustorium is absent in *Pedalium* and *Sesamum*, the chalazal haustorium is four-nucleate and single-celled in *Pedalium*. It is two-celled, with each cell binucleate, in *Sesamum*. The findings of Srinivasan seem to be erroneous.

The embryogeny conforms to the Onagrad type. The mature embryo is spatulate.

The exotestal seed is albuminous with a straight embryo and prominent cotyledons. In *Sesamum indicum* (Vaughan 1970), the cotyledons exhibit a single layer of palisade cells on the inner surface, next to the epidermis. The rest of the cotyledonary cells are isodiametric and contain oil drops and aleurone grains. The endosperm contains 50% oil and 20–25% proteins.

The testa is ridged in *Sesamum*. The cells of the outer epidermis show thickenings on their radial walls. The epidermis contains cap-shaped ca-oxalate crystals. The two hypodermal layers are thin-walled with flattened cells. They also contain ca-oxalate crystals. Beneath the hypodermal layers lies a thick layer of cuticle which separates the testa from the endosperm.

The epidermis of testa, in *Pedalium murex* (S.P. Singh 1963), exhibits thick outer walls with pointed edges. The inner epidermis and most of the adjoining cells become crushed due to the enlarging endosperm and embryo. In *Harpagophytum* the epidermal cells show strong annular thickenings on the radial walls, and fine thickenings on the tangential walls. The mesophyll and endothelium degenerate (Corner 1976).

EJH Corner (1976), P Crété (1951), J Hanawa (1953), AR Kulkarni and PG Pataskar (1973–74), J Mauritzon (1936d), SP Singh (1960, 1963, 1970a), AR Srinivasan (1942), T Vaughan (1970)

48.21 Martyniaceae

The anther wall comprises the epidermis, fibrous endothecium, an ephemeral middle layer, and glandular tapetum of binucleate cells. Refractive droplets occur on the degenerated tapetum. The microspore mother cells undergo meiotic divisions, cytokinesis is simultaneous, and tetrads tetrahedral, isobilateral and decussate.

The pollen grains are two-celled at the dispersal stage. They are spheroidal and polyrugoidate with 15 rugae in *Martynia diandra* (S.P. Singh 1970b). Anderson (1922) reported a three-celled condition in *Proboscidea* (*Martynia*) *louisianica* (see Schnarf 1931).

The ovule is anatropous, unitegmic and tenuinucellar. The archesporium is two- or three-celled. A single cell functions as megaspore mother cell, undergoes meiotic divisions, produces a linear tetrad, and the chalazal megaspore develops into a Polygonum type of embryo sac. The polar nuclei fuse before fertilization. The antipodals are ephemeral.

One of the synergids persists up to 48 h after fertilization in *Proboscidea louisianica* (Mogensen 1978a). The pollen tube opens by a terminal pore.

Development of the endosperm is of the Cellular type. The division of the primary endosperm nucleus separates a large micropylar from a small chalazal chamber. A vertical division is followed by a transverse division in the micropylar chamber resulting in two central cells and a two-celled micropylar haustorium. The latter is non-aggressive, and becomes multicellular in *Martynia diandra*. The central cells produce the endosperm proper which can be differentiated into two distinct zones: micropylar with large cells, and chalazal with small cells (see Crété 1942). The chalazal chamber develops into a four-celled haustorium; each cell becomes coenocytic in *Martynia* (S.P. Singh 1964).

The embryogeny corresponds to the Onagrad type-Catalpa variation. The embryo is dicotyledonous with foliaceous cotyledons. The latter have two palisade layers in the hypodermis.

The seed is thinly albuminous, or exalbuminous, with a straight embryo. In *Martynia diandra* (S.P. Singh 1970b), the mature embryo is of the foliar type. The cotyledons are dorsiventral comprising the upper epidermis, two layers of palisade cells, parenchyma, and the lower epidermis. The large thin-walled epidermal cells of the testa form a subgelatinous pellicle. The mesophyll is crushed. The inner epidermis consists of small, tubular cells. In *Proboscidea* (Corner 1976), the epidermal cells are sclerotic, with thick inner and radial walls.

F Anderson (1922), EJH Corner (1976), P Crété (1942, 1951), LB Kajale and BA Razi (1954), E Martyni (1939), HL Mogensen (1978a, b, c), K Schnarf (1931), SP Singh (1964, 1970b)

48.22 Gesneriaceae

The anther wall comprises the epidermis, fibrous endothecium, a single ephemeral middle layer, and secretory tapetum of binucleate cells in *Platystemma violoides* (Adatia et al. 1971). The tapetal cells become two- to five-nucleate, and their nuclei fuse to form one or two polyploid masses in *Didymocarpus tomentosa* (Thathachar 1942b). The microspore mother cells undergo meiotic divisions, cytokinesis is simultaneous, and tetrads tetrahedral and decussate.

The triporate pollen grains are two-celled at the dispersal stage.

The ovule is anatropous, unitegmic and tenuinucellar. The endothelium is restricted to the chalazal region of the embryo sac. The hypostase is feebly developed in *Platystemma*. A hypodermal archesporial cell functions as megaspore mother cell, undergoes meiotic divisions, produces a linear tetrad, and the chalazal megaspore develops into a Polygonum type of embryo sac. The embryo sac is extramicropylar in *Roettlera* sp. (Glišić 1934). The synergids have long tapering tips. The polar nuclei fuse before fertilization; the antipodals are ephemeral. They persist in *Ramondia pyrenaica* (Crété 1942) through the first few divisions of the endosperm.

The development of endosperm is of the Cellular type. Figure 48.22.1A–K is a schematic representation of the variations met with in the development of endosperm in Gesneriaceae.

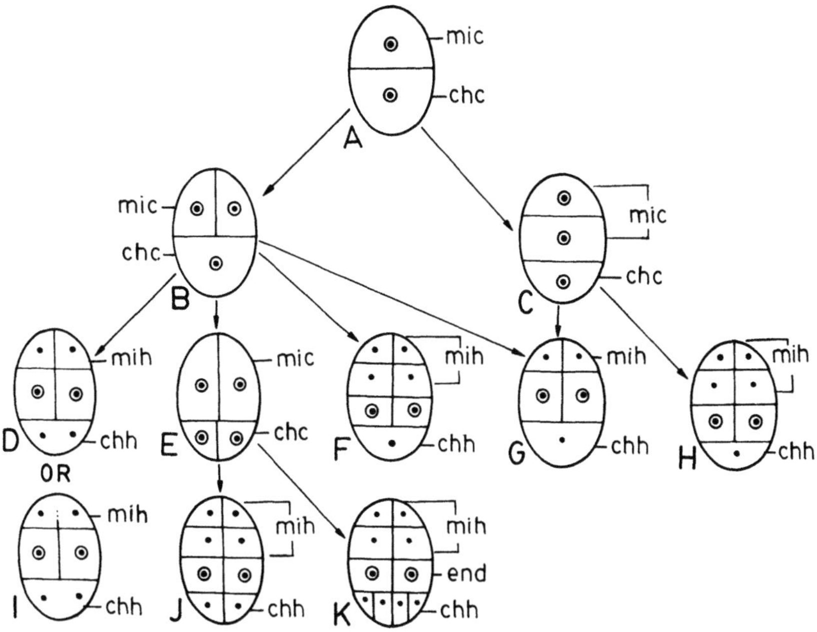

Fig. 48.22.1A–K. Development of endosperm in Gesneriaceae. (After Crété 1951). *Black dots without circles* indicate the nuclei of haustorium, *black dots with circles* (**D–K**) indicate initials of the endosperm proper. **A** Two-celled endosperm; *mic* micropylar chamber; *chc* chalazal chamber. **B, C** Vertical or transverse division in *mic*. **D** Transverse division in *mic*. Two-celled micropylar haustorium (*mih*), two initials of endosperm proper, and two-nucleate (*chh*) chalazal haustorium (*Corytholoma*). **I** Vertical wall in *mih* incomplete; *mih* binucleate (*Roettlera*). **E** Vertical division in *chc*. **J** *mih* four-celled, *chh* two-celled. Two initials of endosperm proper (*Haberlea*). **K** Four-celled *mih* and four-celled *chh*, two initials of endosperm proper (*Prohaberlea*). **F** Four-celled *mih*, uninucleate *chh*, and two initials of endosperm proper (*Ramondia pyrenaica*). **G** Two-celled *mih*, uninucleate *chh*, two initials of endosperm proper (*Klugia*). **H** Four-celled *mih*, uninucleate *chh*, and two initials of endosperm proper (*Ramondia nathaliae*). (Courtesy Dr M.A. Rau, Mysore)

In *Klugia zeylanica* (Schnarf 1921), *Ramondia nathaliae* and *R. serbica* (Glišić 1934, see Crété 1942) a linear row of three cells is produced due to another transverse division in the two-celled endosperm. The terminal cells develop into haustoria, while the central cell produces the endosperm proper.

A vertical division, followed by a transverse division in micropylar chamber separates the initials of the micropylar haustorium from the initials of the endosperm proper. Such a development occurs in *Alloplectus sanguineus* (Crété 1955c), *Corytholoma cyclophyllum* (Laurent 1923), *Chirita lavandulaca* (Crété 1949e), *Didymocarpus tomentosa* (Thathachar 1942b), *Epithema carnosum* (Padmanabhan 1961a), *Haberlea rhodopensis* (Glišić 1928), *Klugia notoniana* (Govindappa 1961b), *Platystemma violoides* (Adatia et al. 1971), *Ramondia pyrenaica* (Fig. 48.22.2A, B) (Crété 1942), *Rhyncoglossum obliqum* (Thathachar 1943) and *Roettlera* sp. (Glišić 1934).

The micropylar haustorium comprises two juxtaposed, uninucleate cells in *Corytholoma* and *Klugia*, whereas it becomes binucleate in *Roettlera* since the

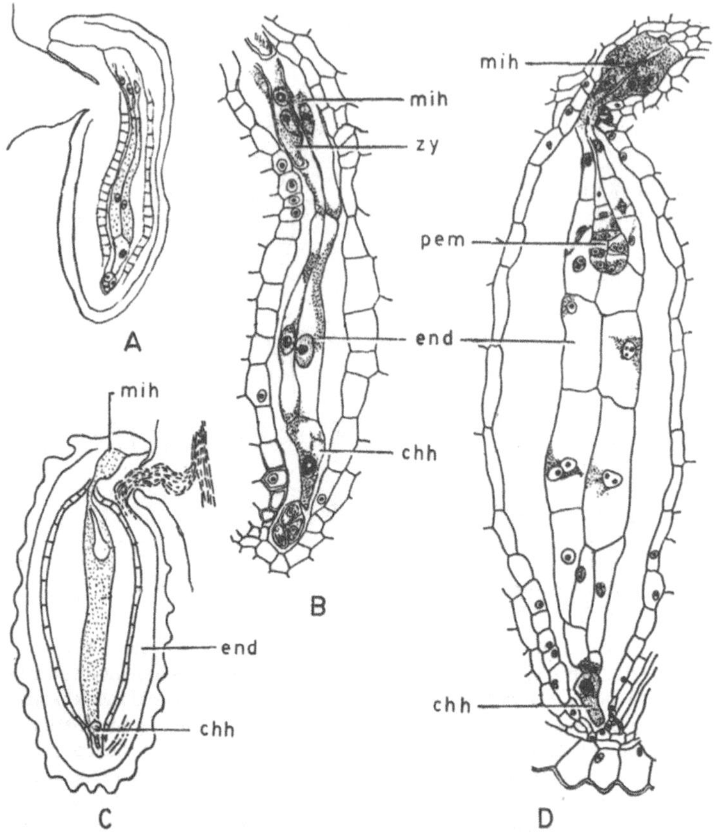

Fig. 48.22.2A–D. *Ramondia pyrenaica.* **A** Longisection of ovule, micropylar (*mih*) and chalazal (*chh*) haustorium, outline diagram for **B**. **B** Embryo sac with zygote (*zy*), two-celled micropylar haustorium (*mih*), endosperm proper initials (*end*), uninucleate chalazal haustorium (*chh*), and persistent antipodal cells. **C** Outline for **D**. **D** Embryo sac with young proembryo (*pem*), four-celled micropylar haustorium (*mih*), endosperm proper (*end*), and uninucleate chalazal haustorium (*chh*). (After Crété 1942)

vertical wall remains incomplete. The micropylar haustorium consists of two tiers of two cells each (four-celled) as in *Haberlea, Prohaberlea, Ramondia pyrenaica* (Fig. 48.22.2C, D) and *R. nathaliae.* It persists up to the dicotyledonous stage of embryo in *R. pyrenaica.* It is extramicropylar in *Didymocarpus tomentosa* (Thathachar 1942b).

In all the types cellular endosperm develops from the daughter cells of the micropylar chamber; the chalazal chamber transforms into a haustorium. The chalazal haustorium is uni- (*Klugia, Ramondia*) or binucleate (*Corytholoma, Roettlera*), or two-celled (*Haberlea*). In *Prohaberlea* a vertical division occurs in each cell resulting in a four-celled condition.

The chalazal haustorium is uninucleate in *Klugia zeylanica* (Schnarf 1921), *Ramondia nathaliae, R. serbica* (Glišic 1934) and *Rhyncoglossum obliqum*

(Thathachar 1943). In *K. notoniana* (Govindappa 1961b), the two nuclei of the binucleate haustorium fuse, and the uninucleate haustorium is highly aggressive (Fig. 48.22.3A–C). It grows laterally and upward, consuming the subepidermal cells of the integument. The terminal cell in the chalazal region of the endosperm does not develop into a haustorium in an interspecific hybrid (*S. achimenoeflorus* × *S. kewensis*) of *Streptocarpus* (Crété 1942). Probably this is due to the absence of conducting tissue in the funicular region, which needs confirmation. The micropylar haustorium is favourably placed for conducting nutrients to the endosperm.

The embryogeny conforms to the Onagrad type. Padmanabhan (1967) interprets the embryogeny of *Epithema carnosum* on the basis of five phases according to the morphogenetic concept.

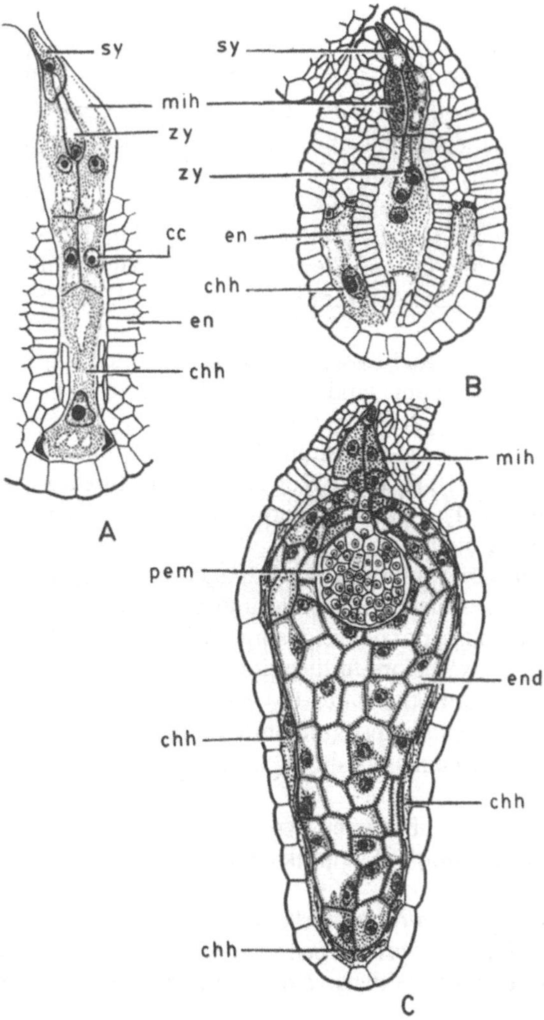

Fig. 48.22.3A–C. *Klugia notoniana.* **A** Embryo sac with persistent synergid, zygote (*zy*), micropylar haustorium (*mih*), central chamber (*cc*) which will produce endosperm proper, and uninucleate chalazal haustorium (*chh*); *en* endothelium. **B** Longisection of ovule; lateral branches of aggressive chalazal haustorium (*chh*) invade the integument. **C** Later stage, the proembryo (*pem*) is embedded in cellular endosperm (*end*), laterally the branches of chalazal haustorium reach almost up to the micropylar haustorium and consume the integumentary tissue except the epidermis. (After Govindappa 1961b)

The seed is exotestal with a small, straight embryo. The outer epidermis of the integument is variously thickened, the mesophyll becomes crushed, and the inner epidermis may persist but is unspecialized.

RD Adatia, YB Sharma and MR Vijayaraghavan (1971), GD Arekal (1961b), P Crété (1942, 1949e, 1955c), LM Glišić (1928, 1934), DA Govindappa (1961b), V Laurent (1923), D Padmanabhan (1961a, 1966, 1967), K Schnarf (1921), BGL Swamy and D Padmanabhan (1961b), T Thathachar (1942b, 1943)

48.23 Columelliaceae

The embryology of Columelliaceae has not been investigated.

48.24 Orobanchaceae

The anther wall comprises the epidermis, endothecium, two to four middle layers, and glandular tapetum of multinucleate cells. The tangential walls of the epidermis are cutinized. In *Aeginetia indica* (B. Tiagi 1952b), the endothecium and sometimes even the epidermis and the outer middle layer show well developed spiral thickenings. The endothecium is well developed and prominent in *Boschniakia tuberosa* (B. Tiagi 1963). The tapetum is biseriate in *Christisonia subacaulis* (D. Raju and Chamaiah 1972). The nuclei of the tapetal cells undergo irregular divisions and fusions. Simultaneous cytokinesis follows meiosis in microspore mother cells, and the tetrads are tetrahedral, isobilateral and decussate.

The pollen grains are tricolpate, and two-celled at the dispersal stage.

The ovule is anatropous, unitegmic and tenuinucellar with a narrow micropyle. The endothelium develops at the four- or eight-nucleate stage of the embryo sac and surrounds it. The endothelium differentiates only in the chalazal region in *Aeginetia indica* (B. Tiagi 1952b). It is two- to three-layered in *Conopholis* (B. Tiagi 1965c). The outer epidermis of the integument is elongated at right angles to the ovule.

The archesporium is uni- to multicellular. A single cell functions as megaspore mother cell, undergoes meiotic divisions and produces a linear or T-shaped tetrad. The chalazal megaspore develops into a Polygonum type of embryo sac. The polar nuclei fuse before fertilization. The remnants of the synergids and antipodals persist until the 14- to 16-celled stage of endosperm. Twin embryo sacs are common. The unfertilized embryo sacs become extra-micropylar and form a bulbous pendant lobe in the ovary in *Boschniakia himalaica* (B. Tiagi 1963).

Double fertilization takes place. Both the synergids may remain intact during fertilization (which is doubtful). The pollen tube is persistent (Crété 1942, B. Tiagi 1951b).

Development of the endosperm is of the Cellular type (Fig. 48.24.1A–P). The micropylar chamber produces two central cells, and a micropylar haustorium (Fig. 48.24.1A–G). The two central cells give rise to the endosperm proper

(Fig. 48.24.1H, I, K–M) and the "isthmus" connecting the micropylar haustorium (B. Tiagi 1970). The outer tangential walls of peripheral endosperm cells are highly cutinized and form a protective rim (Fig. 48.24.1K, M, O, P) (B. Tiagi 1970).

The two cells of the micropylar haustorium are aggressive in *Orobanche aegyptica* and *O. cernua* (B. Tiagi 1951b), and *O. eryangii* and *O. teucrii* (Crété 1942). The branches of the micropylar haustorium penetrate the middle region of integument and funicle, and absorb the integumentary tissue. In the micropylar region, the fine endings of the haustoria extend on to the surface of the ovule in *Aeginetia indica* (B. Tiagi 1952b). A similar condition has been reported in *O. hederae* (Glišić 1929, see Crété, 1942). Crété (1942) could not confirm it in this species, since he did not examine the older stages of seed development.

The micropylar haustorium has two, uninucleate cells in *Aeginetia* (Fig. 48.24.1L, M) (B. Tiagi 1952b), *Boschniakia* (B. Tiagi 1963), *Christisonia subacaulis* (D. Raju and Chamaiah 1972), and *Orobanche* spp. (Glišić 1929, Srivastava 1939, Crété 1942, B. Tiagi and Sankhla 1963). It has two binucleate cells in *Cistanche tinctoria* (Kadry 1952), *C. tubulosa* (B. Tiagi 1952a), *Conopholis americana* (B. Tiagi 1965c) and *Orobanche cernua* (B. Tiagi 1951b).

The chalazal haustorium is one-celled with a single nucleus in *Aeginetia* (Fig. 48.24.1I, J), *Christisonia subacaulis* and *O. eryngii*; one-celled but binucleate in *O. lucorum* (B. Tiagi and Sankhla 1963) and *O. teucrii* (Crété 1942), or two-celled, each cell binucleate, in *O. aegyptica* (B. Tiagi 1951b). During later stages, the nuclei of the endothelial cells, around the uninucleate chalazal haustorium, become hypertrophied (Fig. 48.24.1J).

Early embryogeny conforms to the Onagrad type in *Boschniakia himalaica* (B. Tiagi 1963), and Caryophyllad type in *Aeginetia indica* (B. Tiagi 1952b). The suspensor degenerates. The mature embryo is globose, and does not differentiate into the usual organs. It comprises about 45 cells in *Orobanche minor* (Kuijt 1969).

The minute seed is albuminous with a globular organless embryo. In *Aeginetia indica* (B. Tiagi 1952b), only the outer epidermis of the testa persists, the other layers collapse. The cells of the outer epidermis enlarge, and develop spiral thickenings on the inner tangential walls which give a reticulate appearance (Fig. 48.24.1M, O, P). In *O. cernua* (B. Tiagi 1951b), the outer epidermis of testa develops pitted thickenings on the inner tangential and radial walls. The mesophyll is absorbed. The inner tangential walls of the peripheral endosperm cells become highly cutinized. In *O. hederae* (Kuijt 1969), the seed-coat remains single-layered (outer epidermis of testa) with sclerified thickenings and pits on radial and inner tangential walls. The outer tangential walls collapse and the seed-coat is represented by a layer of "sclerified polygonal cups tightly fitting together".

In *Conopholis americana* (B. Tiagi 1965c), the cells of the outer epidermis and two or three hypodermal layers of testa are lignified. The endothelial cells also become lignified and form a thick coat. The remnants of micropylar haustorium and pollen tube persist in the mature seed.

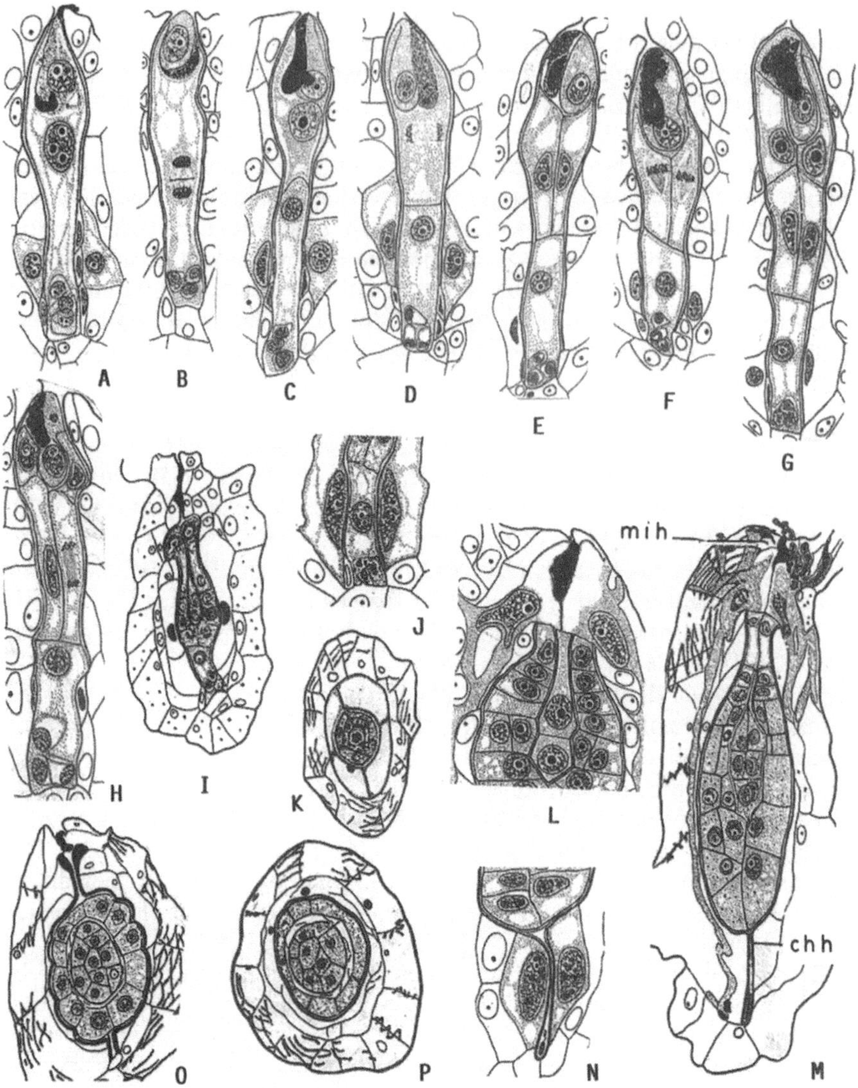

Fig. 48.24.1A–P. *Aeginetia indica*, development of endosperm. **A** Embryo sac with zygote, remnants of synergids, primary endosperm nucleus, and antipodal cells. Persistent antipodals are also seen in **B–J**, and remnants of pollen tube in **A–I**, **L**, **M** and **O**. **B** Primary endosperm nucleus in division. **C** Two-celled endosperm. **D** Embryo sac with zygote, micropylar chamber (nucleus at anaphase), and uninucleate chalazal chamber. **E** Vertical division in micropylar chamber. The uninucleate chalazal chamber develops into haustorium. **F**, **G** Transverse division in each of the two cells of micropylar chamber. **H**, **I** Embryo sac with zygote, two-celled micropylar haustorium, endosperm proper, and uninucleate chalazal haustorium. Endothelial cells are prominent in **I**. **J** Chalazal region of embryo sac to show uninucleate chalazal haustorium, and hypertrophied nuclei of endothelium. **K** Transection of young seed to show embryo and endosperm. The enlarged endothelial cells are vacuolate. **L** Micropylar portion of embryo sac with two-celled proembryo, endosperm proper, and two-celled micropylar haustorium with branches which have invaded the

JD Cassera (1935), P Crété (1942), WW Finn and T Rudenko (1930), M Glišić (1929), JB Juliano (1935b), AER Kadry (1952, 1955), JC Kuijt (1969), S and JD Olsen (1980), D Raju and HN Chamaiah (1972), GD Srivastava (1939), ES Teryokhin (1980, 1988), B Tiagi (1951b, 1952a, b, 1963, 1965c, 1970), B Tiagi and NS Sankhla (1963)

48.25 Lentibulariaceae

The anther is one- (*Pinguicula* and *Polypompholyx*) or two-locular (Hutchinson 1973).

The anther wall (Fig. 48.25.1K, L) comprises the epidermis, prominent fibrous endothecium, a middle layer, and glandular tapetum of binucleate cells. Simultaneous cytokinesis follows meiosis in the microspore mother cells, and the tetrads are tetrahedral, isobilateral and decussate (Fig. 48.25.1A–C).

The pollen grains are three- to 28-colporate with reticulate or smooth exine, and three-celled at the dispersal stage (Fig. 48.25.1D–K, M). Giant pollens, compound pollen grains, pollen with supernumerary nuclei, polyspermy, and double gametophytes have been observed in *Utricularia flexuosa* (Fig. 48.25.2A–K).

The ovule is anatropous, unitegmic and tenuinucellar. Occasionally, it is orthotropous or anacampylotropous and lacks vascular tissue in the funicle. The presence of nutritive tissue in the chalazal region of the ovule, and adjacent to the placenta, is noteworthy (Fig. 48.25.3). In *U. flexuosa* the micropyle is not organized, and the tip of the embryo sac remains exposed. Generally, the endothelium differentiates but it is absent in *U. arcuata* (Farooq 1965a) and *U. striatula* (Farooq 1966). Farooq's observations deserve confirmation.

A hypodermal archesporial cell functions as megaspore mother cell, undergoes meiotic divisions, and produces a linear, T-shaped or isobilateral tetrad. The chalazal megaspore develops into a Polygonum type of embryo sac. The synergids are hooked in *U. reticulata*. The polar nuclei fuse before fertilization. The antipodals are small; occasionally one may enlarge in *U. flexuosa*. The hypostase becomes lignified.

The behaviour of the embryo sac in relation to the nutritive tissue has been described in *Utricularia* by R. Khan (1954, 1970; see also Shivaramaiah 1964a). The embryo sac may remain confined to the ovule, or may grow out even during earlier stages of development and come to lie in contact with the placenta. The embryo sac derives nutrition from the placental and chalazal nutritive tissue, and through the endothelium (Fig. 48.25.4).

Several abnormal features have been observed in *U. flexuosa* (R. Khan 1954): (1) embryo sac with three polar nuclei, and one synergid, (2) 11-nucleate

integument. **M** Longisection of immature seed with proembryo and endosperm. The branches of bicelled micropylar haustorium (*mih*) ramify in the integumentary epidermis. The chalazal haustorium (*chh*) has collapsed. **N** Lower portion of embryo sac with remnants of chalazal haustorium. **O** Longisection of mature seed with spirally thickened cells of integumentary epidermis (also seen in **K**, **M** and **P**), cutinized inner wall of integument, one layer of endosperm, and the organless embryo. The micropylar and chalazal haustorium have degenerated. **P** Transection of mature seed. (After B. Tiagi 1952b)

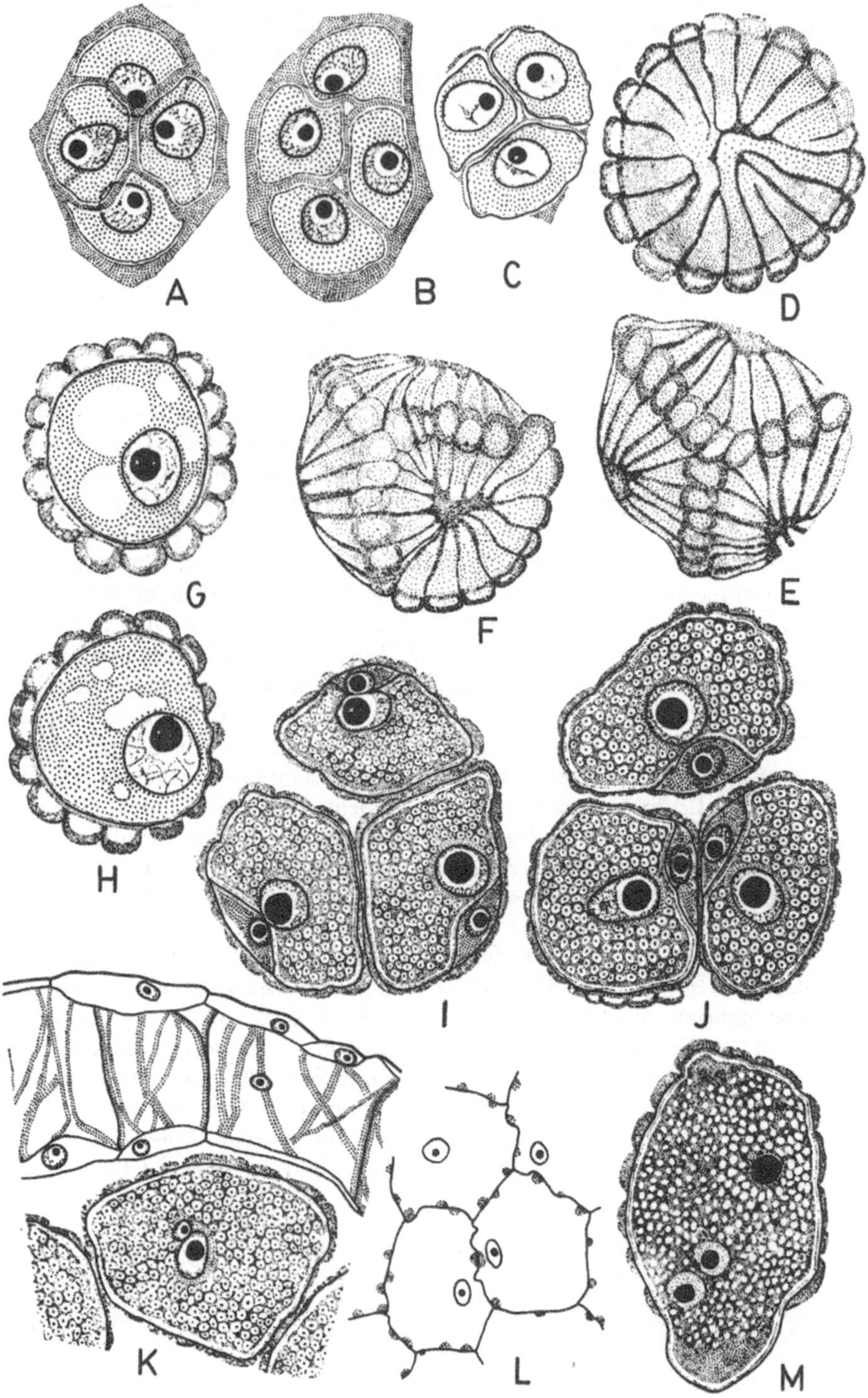
A
B
C
D
G
F
E
H
I
J
K
L
M

embryo sac with five nuclei in the position of polars, (3) four-nucleate embryo sac with reversed polarity, (4) twin embryos sacs, and (5) three to four embryo sacs in one ovule.

The degeneration of the embryo sac is quite frequent in *U. flexuosa* (R. Khan 1963), and *U. coerulea* var. *filicaulis* (Siddiqui and Farooq 1965). Formation of a multinucleate nodule by endothelial cells with its protrusion into the degenerated embryo sac is reported in *U. flexuosa* (R. Khan 1963) and *U. stellaris* var. *inflexa* (Farooq 1964a).

The pollen tube meets the embryo sac in the ovarian cavity. This mode of fertilization is described as exogamous (R. Khan 1970). Double fertilization takes place (Fig. 48.25.5A, B).

The development of the endosperm is of the Cellular type and conforms to the Scutellaria type in *U. reticulata* (Kausik and Raju 1955), *U. stellaris* (Farooq and Siddiqui 1964), *U. stellaris* var. *inflexa*, *U. arcuata*, *U. uliginosa*, *U. striatula* (Farooq 1964a, b, 1965a, b, 1966), *U. bifida* (Siddiqui 1975), *U. graminifolia* (M. Begum 1969), and *U. dichotoma* (Siddiqui 1978). The two cells at the micropylar end of the eight-celled endosperm develop into the micropylar haustorium, and the two cells at the chalazal end into the chalazal haustorium; the remaining four central cells contribute to the formation of the endosperm proper (Fig. 48.25.6). The haustoria are generally aggressive, the micropylar more aggressive than the chalazal. Their nuclei are hypertrophied.

The development of haustoria is totally absent in *Pinguicula alpina* (see R. Khan 1970). *P. vulgaris* shows a two-celled chalazal haustorium, but the micropylar haustorium is absent or feebly developed (Stolt 1936, Crété 1956c). A micropylar haustorium has been observed in *U. coerulea* var. *filicaulis* (Siddiqui 1975) but the chalazal haustorium is absent. Siddiqui (1975) reports frequent free-nuclear divisions in the cells of young endosperm; this aspect needs confirmation.

Numerous variations in the development of endosperm are reported (R. Khan 1970): (1) isobilateral, decussate or T-shaped arrangement at the four-celled stage in *U. flexuosa*, (2) seven-celled endosperm forming a linear row of superposed cells in *U. flexuosa*, (3) lateral growth of the micropylar haustorium in ovarian cavity, approaching the adjoining ovule, (4) free-nuclear divisions in the primary chalazal chamber in *U. arcuata*, *U. coerulea* var. *filicaulis* and *U. scandens*, (5) three-nucleate chalazal haustorium, and (6) direct development of the primary chalazal chamber into chalazal haustorium and its extension into the funicle; shows nuclear divisions in *U. stellaris* var. *inflexa*.

The zygote grows into a tube-like structure. The apical end of the tube with the nucleus elongates, and reaches the centre of the endosperm in *Utricularia flexuosa* (R. Khan 1954). A transverse division occurs in the zygotic tube. The four-celled proembryo is T-shaped. The two apical cells divide transversely to

◀

Fig. 48.25.1A–M. *Utricularia flexuosa.* **A–C** Microspore tetrads. **D–F** Pollen grains, exine pattern; **D** polar view. **G, H** Uninucleate pollen grains. **I, J** Two-celled pollen grains, generative cell cut off on the outer side (**I**) and inner side (**J**); the configuration is that of tetrads. **K** Part of anther wall (the tapetum has collapsed) with bicelled pollen grain. **L** Endothecial cells with thickenings. **M** Three-celled pollen grain. Note starch in **I–K**. (After Khan 1954)

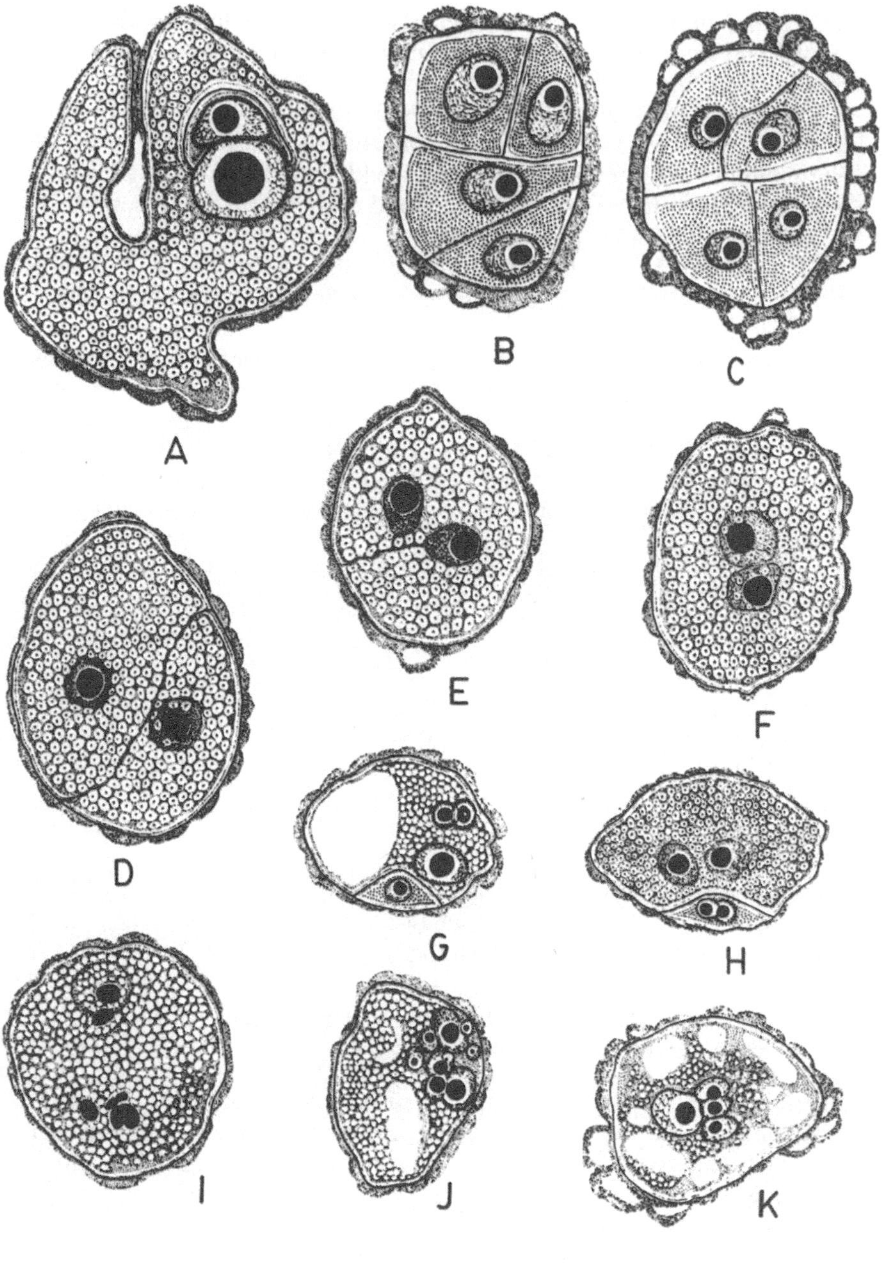

Fig. 48.25.2A–K. *Utricularia flexuosa.* **A** Abnormal two-celled "giant" pollen grain. **B, C** Tetrads with a common exine. This may be due to precocious formation of exine before the separation of microspores. **D–F** Pollen grains, each with two identical nuclei; note a separating wall in **D**, incomplete wall in **E**, and absence of wall in **F**. Such a condition may be due to incomplete cytokinesis, or failure of cytokinesis. **G–K** Pollen grains with supernumerary nuclei; **G** Two smaller nuclei in vegetative cell; **H** Two identical nuclei in the vegetative as well as in generative cell. This may be considered as double gametophyte; **I** Two groups of dissimilar nuclei; **J** Four pairs of

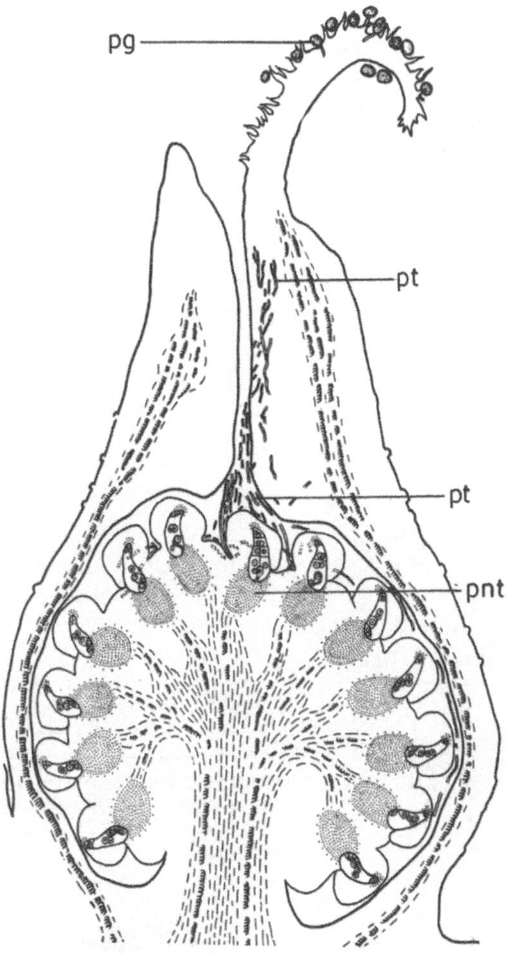

Fig. 48.25.3. *Utricularia flexuosa*. Longisection of gynoecium to show placental nutritive tissue (*pnt*), fertilized embryo sacs, pollen grains (*pg*) on stigma, and pollen tubes (*pt*) in style and ovary. (After Khan 1954)

form a quadrant in two tiers, *l* and *l'*. Vertical divisions occur in tier *l*, further divisions are irregular.

In *Utricularia bifida* (Basheer and Prakasa Rao 1984), the zygote divides at the ten-celled stage of the endosperm. The embryogeny conforms to Period I, Megarchetype III, Series C_2 (Chenopodiad variation of Johansen).

Early embryogeny conforms to the Onagrad type-Capsella variation in *Pinguicula leptoceras* (Crété 1956b) or Ruta variation in *U. coerulea* var. *filicaulis* (Siddiqui 1975) and *U. dichotoma* (Siddiqui 1978). In exceptional cases

nuclei of three different sizes; **K** One large and three small nuclei. The additional nuclei in **I–K** can result from (a) more than one microspore present in a common exine, and (b) failure of wall formation after the division of nucleus of uninucleate pollen grain, and supernumerary division of one or more nuclei. Starch is conspicuous in **A, D–K**. (After Khan 1954)

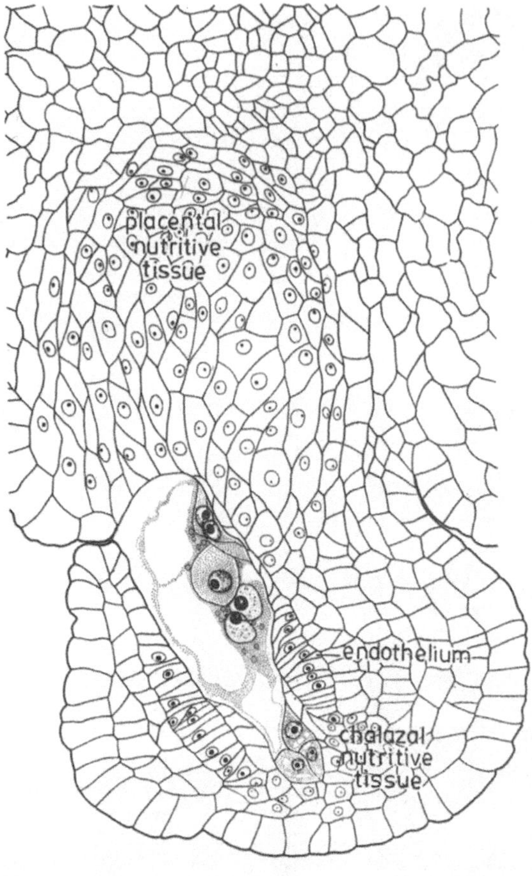

Fig. 48.25.4. *Utricularia flexuosa.* Longisection of part of ovary to show chalazal nutritive tissue, endothelium, and embryo sac with its tip extending into placental nutritive tissue, cf. Fig. 48.25.3. (After Khan 1954)

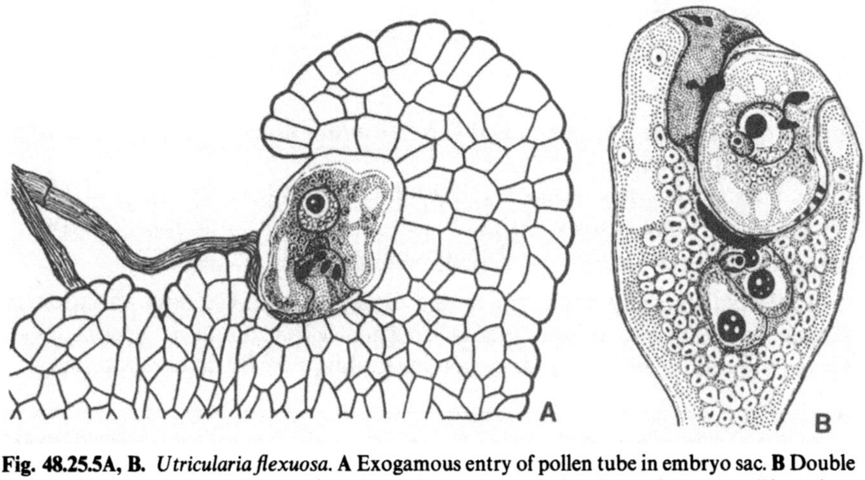

Fig. 48.25.5A, B. *Utricularia flexuosa.* **A** Exogamous entry of pollen tube in embryo sac. **B** Double fertilization; note the remnants of pollen tube and starch in the embryo sac. There is a supernumerary nucleus lying on one polar which may probably be of a synergid. (After Khan 1954)

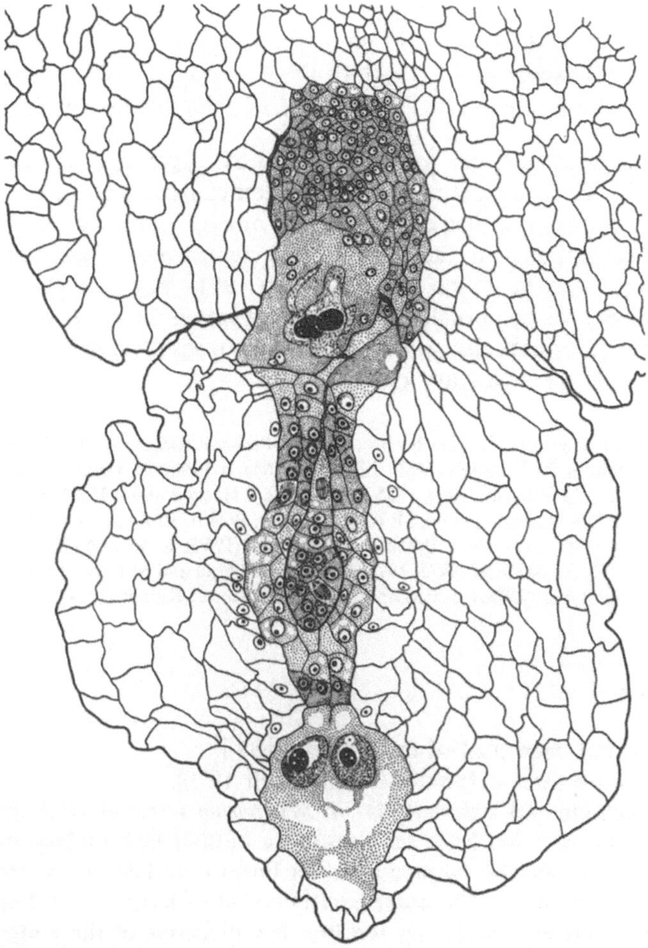

Fig. 48.25.6. *Utricularia flexuosa.* Longisection of ovule to show the endosperm and proembryo. The micropylar haustorium has digested some cells of the placental nutritive tissue (the smaller nuclei in the haustorium are those of the placental nutritive tissue). The chalazal haustorium extends up to the epidermis. (After Khan 1954)

the Piperad type-Scabiosa variation has been observed in *U. dichotoma.* Siddiqui (1975) reports occasional occurrence of the Chenopodiad or Solanad type. The Solanad type also occurs in *U. smithiana* (Sunder Rajan and Jeevan Kumar 1974) and *U. stricticaulis* (Shivaramaiah 1967). The Catalpa variation of the Onagrad type has been noted in *U. arcuata, U. uliginosa* (Farooq 1965a, b), *U. striatula* (Farooq 1966) and *U. scandens* (Farooq and Bilquis 1966), and the Chenopodiad type-Polemonium variation in *Polypompholyx* (Johansen 1950).

Cleavage polyembryony occurs in *U. arcuata* (Farooq 1965a).

The mature embryo shows only the differentiation of the epidermis, and the meristematic apical region. In *Pinguicula* the mature embryo is fully differentiated with root primordium and cotyledons (Stolt 1936), in *Genlisea* only the cotyledons differentiate, and in *Polypompholyx* there is no differentiation of root or cotyledons.

The small, albuminous seed has an undifferentiated (organless), straight embryo in *Utricularia*. The endosperm is oily. The outer epidermis of the testa forms the seed-coat. The thick-walled epidermal cells develop numerous short, pointed spine-like projections on the surface. The micropylar haustorium persists in the mature seed in *U. flexuosa* (R. Khan 1954). The size, shape, pattern of arrangement of epidermal cells, the thickenings of their walls, projections from the epidermal cells, and wings form the basis of classification of seeds (Abraham and Subramanyam 1965).

V Abraham and K Subramanyam (1965), MD Basheer and PS Prakasa Rao (1984), M Begum (1965, 1969), P Crété (1956b, c), M Farooq (1958b, 1964a, b, 1965a, b, 1966), M Farooq and S Bilquis (1966), M Farooq and SA Siddiqui (1964, 1965, 1966, 1967), B Haccius and E Hartl-Baude (1956), J Hutchinson (1973), DA Johansen (1950), SB Kausik (1935a, 1938d), SB Kausik and MVS Raju (1955, 1956), R Khan (1953, 1954, 1963, 1970), K Kopezynska (1964), K Schnarf (1931), G Shivaramaiah (1964a, b, 1967), SA Siddiqui (1975, 1978, 1979), SA Siddiqui and M Farooq (1965), KAH Stolt (1936), S Sunder Rajan (1974a), S Sunder Rajan and D Jeevan Kumar (1974)

48.26 Myoporaceae

The embryology has not been studied in sufficient detail.

The pollen grains are three-colporate (P.K.K. Nair 1970).

The ovule is unitegmic and tenuinucellar in *Myoporum serratum* (Billings 1901). The archesporial cell functions as megaspore mother cell, undergoes meiotic divisions, and the chalazal megaspore of the linear tetrad develops into a Polygonum type of embryo sac. The polar nuclei fuse before fertilization. The antipodals are small, and persist during the first few divisions of the endosperm. The endothelium encloses the lower portion of the embryo sac.

The development of the endosperm is of the Cellular type. The terminal cells develop into the micropylar and the chalazal haustorium. The micropylar haustorium is multicellular, four-celled in *Oftia africana* (de Vos 1947), while the chalazal is four-celled.

The embryogeny conforms to the Onagrad type. The suspensor is long, slender and uniseriate.

FH Billings (1901), GL Davis (1966), S Junell (1934), PKK Nair (1970), K Schnarf (1931), MP de Vos (1947)

48.27 Phrymaceae

The embryology has not been fully investigated.

The pollen is tricolpate.

The ovule is hemianatropous (Whipple 1972), unitegmic, and tenuinucellate in *Phryma leptostachya* (D.C. Cooper 1941). A hypostase, and endothelium differentiate at the eight-nucleate stage of the embryo sac.

The archesporial cell functions as megaspore mother cell, undergoes meiotic divisions and produces a linear tetrad. The chalazal megaspore develops into a Polygonum type of embryo sac. The polar nuclei fuse before fertilization. Both the synergids persist during the earlier stages of proembryos. The antipodals persist up to the first few divisions of the endosperm.

The division of the primary endosperm nucleus is transverse in *P. leptostachya* (D.C. Cooper 1941), resulting in micropylar and chalazal chamber. Two crosswise vertical divisions occur in each chamber so that two tiers, each of four cells, are formed. The micropylar tier, by further divisions, gives rise to the endosperm proper. The four chalazal cells develop into haustoria. They elongate, the nuclei enlarge, and the cytoplasm becomes denser. Each haustorium extends in the funiculus, and forms a direct connection between the conducting tissue and the cellular endosperm. Finally, at their chalazal end the haustoria grow up to the conducting strand, and their tip becomes bulbous.

The embryogeny conforms to the Solanad type. The suspensor is 10–20 cells long.

The albuminous seed has an oblong embryo with broad and folded cotyledons. The endosperm is two-layered. The cells of the integument collapse and form the membraneous testa. The pericarp is also membraneous (Hutchinson 1973).

DC Cooper (1941), J Hutchinson (1973), HL Whipple (1972)

48.28 Taxonomic Considerations

Tubiflorae is a large order of herbaceous plants with gamopetalous corolla, epipetalous stamens, and unitegmic ovules. It includes 26 families distributed over six suborders: Convolvulineae, Boragininineae, Verbenineae, Solanineae, Myoporineae, and Phyrmineae (Melchior 1964). Bentham and Hooker (1965b) place them under Bicarpellatae in three cohorts: Polemoniales, Personales, Lamiales; Hutchinson (1973) distributes them over eight orders.

According to Cocucci (1983; see also Prósperi and Cocucci 1979, di Fulvio 1979), the callose in pollen tubes can be correlated with other embryological characters. Using fluorescence microscopy on fresh and dry specimens of several members of the Tubiflorae to detect callose in pollen tubes, and correlating it with floral organization and endosperm haustoria, the families can be segregated into three groups.

1) Flower actinomorphic, stamens five, callose in pollen tubes present, endosperm haustoria absent: Polemoniales – Polemoniaceae (Sect. 48.2), Convolvulaceae (48.4), Solanaceae (48.12), Nolanaceae (48.11) and Boraginaceae (48.6).

2) Flower actinomorphic, stamens five, callose in pollen tube present or absent, endosperm haustoria present: Hydrophyllales – Hydrophyllaceae (48.5) and Heliotropaceae (48.6).
3) Flower zygomorphic, stamens two or four, callose in pollen tubes absent, endosperm haustoria present: Scrophulariales – Acanthaceae (48.19), Verbenaceae (48.8), Lamiaceae (48.10), Scrophulariaceae (48.15), Bignoniaceae (48.17), Orobanchaceae (48.24), Globulariaceae (48.16) and Lentibulariaceae (48.25).

We do not agree with this categorization since other embryological characters have not been considered to support the above grouping.

Polemoniaceae. The suborder Convolvulineae comprises three families: Polemoniaceae, Fouquieriaceae and Convolvulaceae. Bentham and Hooker, Takhtajan (1980) and Cronquist (1981) place Polemoniaceae close to Convolvulaceae in the Polemoniales. However, Hutchinson (1973) separates them into two orders: Polemoniaceae with Cuscutaceae and Hydrophyllaceae in Polemoniales, and Convolvulaceae with Solanaceae and Nolanaceae in Solanales.

Hallier (1905) and Hutchinson (1973) relate Polemoniaceae to Geraniales, Wettstein (1935) and Rendle (1971) to sympetalous relatives of Rosales, whereas Bessey (1915) and Dawson (1936) conclude that Polemoniaceae are related to Caryophyllaceae and Geraniaceae.

The general floral plan and embryology of Polemoniaceae is closer to Convolvulaceae: both have unitegmic, tenuinucellate ovule, Polygonum type of embryo sac, and Nuclear type of endosperm.

Sunder Rao (1940a) compares the embryological features of Polemoniaceae, Geraniaceae and Caryophyllaceae. Polemoniaceae show close alliance with Geraniaceae: amoeboid anther tapetum (*Erodium* and *Geranium*), anatropous ovule with Polygonum type of embryo sac, Nuclear type of endosperm, and straight embryo. The bitegmic, crassinucellar, campylotropous ovule, Caryophyllad type of embryogeny, and Centrospermalean seed with perisperm of Caryophyllaceae have no parallel in Polemoniaceae. According to Kapil et al. (1968), Polemoniaceae have originated from Geraniaceae and show several characteristic features: two-celled pollen grains, hemianatropous, tenuinucellate ovule with prominent endothelium and hypostase, feebly developed nucellus, absence of parietal cells, and characteristic testa. They emphasize that the origin of Polemoniaceae is from Geraniaceae, as also pointed out by Hallier (1905) and Hutchinson (1973).

Convolvulaceae. Convolvulaceae differ from Polemoniaceae: integumentary vascular strand reaching up to the micropyle, absence of endothelium and hypostase, occasional presence of parietal cell (H. Kaur 1969), Allium type of embryo sac in *Cuscuta reflexa* (B.M. Johri and Tiagi 1952), folded, curved or coiled embryo, and mechanical layers of palisade-like cells in the seed-coat.

Wettstein (1935) and Rendle (1959) separate Convolvulaceae from Tubiflorae, and raise them to an ordinal rank, Convolvulales, on the basis of a large embryo with folded cotyledons. According to Corner (1976), "seed-coat with multiplicative outer hypodermis forming a hypodermal palisade of one to four cells thick" is unique in Tubiflorae. Bessey (1915) associates them with Polemoniales, and Hutchinson (1973) and R. Dahlgren (1980b) with Solanales.

S. Datta (1933) negates the relationship of Convolvulaceae with Nolanaceae, due to the Caryophyllad type of embryogeny and the massive development of suspensor and embryo in Convolvulaceae. In Nolanaceae the ovules are anacampylotropous with an endothelium. The endosperm is of the Cellular type; Nuclear, Cellular and Helobial types occur in Solanaceae. The embryogeny in the latter is of the Solanad type, and the structure of seed also differs. In Convolvulaceae, beneath the subepidermis a palisade tissue of lignified cells occurs, whereas it is exotestal in Nolanaceae.

The genus *Cuscuta* comprises parasitic species. It shows the occurrence of polyspory, placental obturator, occasional presence of a parietal cell, Polygonum and Allium type of embryo sac, persistent synergid, unique type of embryogeny, coenocytic vesicular suspensor, and a coiled hypocotyl. The hypodermal single layer of testa forms the lignified palisade tissue. B.M. Johri and Tiagi (1952) and Govil (1980) support the separation of the genus *Cuscuta* from Convolvulaceae, and the erection of the family Cuscutaceae.

The phytochemical data of Cuscutaceae differ from those of Convolvulaceae. Tropane alkaloids are present in some members of the Convolvulaceae, but are absent in Cuscutaceae (see R. Dahlgren 1977b). Their position as a distinct family has been recognized by R. Dahlgren (1980b), Takhtajan (1980) and Cronquist (1981).

Fouquieriaceae. Bessey (1897) associates Fouquieriaceae with Ebenales, Wettstein (1935) places them in Parietales, while Hutchinson (1973) groups them with Tamaricales. Cronquist (1981) recognizes their isolated position and retains them in Violales, whereas Thorne (1968) places them in Solanales. R. Dahlgren (1980b) erected a separate order, Fouquieriales, close to Ericales in the superorder Corniflorae.

Fouquieriaceae have bitegmic, tenuinucellate ovule with endothelium, L-shaped Polygonum type of embryo sac (R. Khan 1943a), Cellular type of endosperm with chalazal haustorium, and Asterad type of embryogeny (Govil 1970b). These features have not been observed either in Convolvulaceae or Polemoniaceae, and the position of Fouquieriaceae in the suborder Convolvulineae is doubtful. In contrast to the other two families of Convolvulineae, Polemoniaceae and Convolvulaceae, Fouquieriaceae contain iridoids and ellagitannins (R. Dahlgren et al. 1976).

The anatropous, bitegmic, tenuinucellar ovule with endothelium, and the micropyle formed by the endostome in Fouquieriaceae are comparable to those in Ebenaceae. However, in Ebenales the Cellular type of endosperm occurs only in Styracaceae and Ebenaceae, where the embryogeny is of the Solanad type.

The Parietales have crassinucellar ovules without endothelium, Nuclear type of endosperm, and Onagrad type of embryogeny. They do not show any relation to Fouquieriaceae. The Polygonum type of embryo sac (Mauritzon

1936d, R. Khan 1943a) in Fouquieriaceae also rules out their association with Tamaricales, where the development is tetrasporic.

R. Dahlgren et al. (1976) compare the vegetative and floral morphology, anatomy, cytology, palynology and chemistry of Fouquieriaceae with those of Tamaricales, Ericales, Cornales and Solanales (Polemoniales), and point out that most of the affinities are directed towards Ericales and Cornales, and the chemical data support the segregation of Fouquieriaceae from Solanales. In contrast to the unitegmic ovule of most of the iridoid-containing families, *Fouquieria* shows a bitegmic condition. However, the apotropous, pendulous, tenuinucellate ovule with a long micropyle approaches that in Ericaceae. Similarly, the Cellular type of endosperm with at least a haustorium at the chalazal end shows affinity with Ericales and Cornales, in contrast to the Solanales and Tamaricales (R. Dahlgren 1975a).

R. Dahlgren et al. (1976) and R. Dahlgren (1977b) support the erection of a separate order, Fouquieriales, close to Ericales and Cornales. Thorne (1977), however, retains Fouquieriaceae in the order Solanales, as suborder Fouquierineae. He regards them as allied to the primitive genera of Polemoniaceae, in habit and floral structure. The distribution pattern of the population in Fouquieriaceae is more like in the members of Polemoniaceae and Solanaceae than in Ericaceae. An array of similar chemical compounds – except iridoids (peculiar to Fouquieriaceae) – are present in Fouquieriaceae, Polemoniaceae, Solanaceae and Convolvulaceae (see Thorne 1977). Thorne (1977), therefore, rejects the transfer of Fouquieriaceae to Ericaceae.

The second suborder Boraginineae comprises Hydrophyllaceae, Boraginaceae and Lennoaceae. Iridoid compounds and ellagitannins are absent in all the three families.

Hydrophyllaceae. Bessey (1915) and Hutchinson (1973) assign Hydrophyllaceae to Polemoniales, due to helicoid cyme-like inflorescence. The embryological features common to Polemoniaceae and Hydrophyllaceae are: anatropous, unitegmic, tenuinucellate ovule with endothelium, Polygonum type of embryo sac, albuminous seed with a straight embryo, and the outer epidermis of integument alone functions as seed-coat. The chief difference between the two lies in the development of the endosperm. It is of the Nuclear type in Polemoniaceae, whereas Hydrophyllaceae show Nuclear, Cellular and Ce–Nu–Ce type (Svensson 1925). The cellular endosperm develops haustoria in *Nemophila* spp., *Phacelia divericata* and *P. viscida*. The Ce–Nu–Ce type in *P. parryi* (Svensson 1925) is rather peculiar, and is considered as "of reduced ontogeny" (Schnarf 1929). Such a situation has also been observed in Boraginaceae (*Echium* and *Lycopsis*). Moreover, the Solanad type of embryogeny also occurs in Boraginaceae.

Swamy (1974a) discusses the Ce–Nu–Ce condition of the endosperm in Hydrophyllaceae and Boraginaceae, and concludes that its ontogeny is towards Cellular rather than Nuclear type. Thus, the position assigned to the family Hydrophyllaceae near Boraginaceae is appropriate.

Di Fulvio (1989a) points out that *Nama jamaicense* and *Phacelia congesta* (Phacelieae) show similar embryological characters. The development of the endosperm in *Nama* and *Hydrolea* is very different and di Fulvio (1989b) recommends the removal of *Hydrolea* from Hydrophyllaceae, give the latter family rank, and place them in the order Scrophulariales near the Acanthaceae. The assignment of *Hydrolea* should be re-examined.

Boraginaceae. Hutchinson (1973) erects a separate order Boraginales for the Boraginaceae. Boraginaceae show several diverse features: orthotropous, anatropous, hemianatropous or campylotropous, tenui- or crassinucellar ovule, Polygonum or Allium type of embryo sac, Nuclear, Cellular and Ce–Nu–Ce type of endosperm, and Chenopodiad, Asterad, Solanad and Onagrad type of embryogeny, occurrence of apospory, and seeds with or without endosperm with straight or curved embryo. The development of a modified Helobial type (Schnarf 1931, P. Maheshwari 1950) has been ruled out by Swamy and Parmeswaran (1963). In spite of the diversity of characters, Boraginaceae are well placed in Tubiflorae (Khaleel 1974).

Lennoaceae. Hutchinson (1973) groups Lennoaceae with Ericales, due to morphological similarities. Engler and Gilg (1924) remove them to the order Tubiflorae, near Hydrophyllaceae and Boraginaceae. Süssenguth (1927) and Copeland (1935) support this assignment on the basis of morphological and anatomical studies. The floral plan, as well as their chemistry, do not support their position in Ericales. Likewise, the short, solid style, fibrous endothecium, separate pollen grains, and multilayered seed-coat segregate Lennoaceae from Ericales (see P. Maheshwari 1950). Due to small, exotegmic and unitegmic seed, Corner (1976) associates Lennoaceae with Ericales. The affinities of Lennoaceae to the order Tubiflorae are exhibited by the presence of anatropous, unitegmic, tenuinucellar ovule with endothelium, Polygonum type of embryo sac, Cellular type of endosperm, and Caryophyllad type of embryogeny.

Callitrichaceae. Callitrichaceae, Verbenaceae and Labiatae comprise the third suborder. P. Maheshwari (1950) and Rau (1962) discuss the systematic position of Callitrichaceae. Bentham and Hooker (1965b) and Hutchinson (1973) relate Callitrichaceae to Haloragidaceae. Baillon (1858) includes them in Euphorbiaceae, and Clarke (1865) in Caryophyllales. Pax and Hoffman (1931) accept the family status of Callitrichaceae, and place them close to Euphorbiaceae. The findings of Jørgensen (1923, 1925) and Souèges (1952a), such as a tenuinucellate ovule with a single massive integument, and Cellular type of endosperm with well-developed haustoria (see P. Maheshwari 1950), bring them closer to Tubiflorae, related to Verbenaceae and Labiatae. The mode of development of the endosperm in *Callitriche* is identical with that in *Verbena*, and of the embryo (Souèges 1952a) with that in *Verbena*, or *Mentha* (Period I, Megarchetype IV, Group I, in *Callitriche verna* – see Rau 1962). The embryological features of Callitrichaceae are comparable to Verbenaceae and Labiatae (Rau 1962): anatropous, tenuinucellar, unitegmic ovule with endothelium, and Polygonum type of embryo sac. Endosperm of the Brunella and Stachys type is common to both families. The embryogeny is of the Onagrad type. The presence of iridoids in all the three families is also worthy of attention (R. Dahlgren 1977b).

Verbenaceae. Most taxonomists include Verbenaceae and Labiatae in Tubiflorae; Bessey (1915) places them in his Lamiales. Hutchinson (1973) assigns Verbenaceae to Verbenales, and Labiatae to Lamiales. Verbenaceae show close affinities with Labiatae in major embryological features. Even the early development of endosperm conforms to the Brunella and the Stachys type of Labiatae.

Avicennia. Moldenke (1946) raises the genus *Avicennia* to a family level. Endlicher, Eichler and others support the family status on the basis of wood anatomy, articulate branches, free-central placentation, and pendent orthotropous ovules. On palynotaxonomic basis, M.R. Saxena (1981) also proposes family rank for the genus *Avicennia*. The orthotropous condition of the ovule, and free-central placentation are incorrectly described in *Avicennia* (Padmanabhan 1970b). The development of the endosperm is similar to that in other members of the Verbenaceae. The exceptions are the aggressiveness of the micropylar haustorium, the behaviour of the central cellular endosperm (merely as a means of transportation of nutrients instead of storage), extra-ovular growth of proembryo due to unorganized micropyle, and viviparous germination of seed due to halophytic habitat. As these exceptions are due to adaptation to a different habitat, Padmanabhan (1970b) rejects the erection of a separate family, Avicenniaceae, on embryological grounds, and confirms its inclusion in Verbenaceae.

Nyctanthes. De Jussieu (1789) assigned *Nyctanthes* to Oleaceae. Bentham and Hooker (1965b), Eichler (1878), Engler and Prantl (1897) and Hutchinson (1973) accept this assignment. Airy Shaw (1951) proposes the removal of *Nyctanthes* from Oleaceae to Verbenaceae; Melchior (1964) also assigns it to Verbenaceae (see also Sect. 46.3).

Kapil and Vani (1966) point out several similarities between *Nyctanthes* and Oleaceae (Table 48.28.1): (1) contorted corolla, (2) two stamens with short filaments, and carpelloid anthers, (3) spheroidal tricolpate pollen, (4) one-celled, rarely two-celled female archesporium, (5) mono- or bisporic embryo sac with ephemeral antipodals, twin embryo sacs, (6) Cellular endosperm with initially a row of three cells; absence of micropylar and chalazal haustorium, (7) embryogeny conforms to Period II, Megarchetype III (Caryophyllad type), (8) vascularized seed-coat, and (9) capsular fruit (Table 48.28.1).

In contrast, in Verbenaceae the corolla is quincuncial, there are four anthers with long filaments, and carpelloid stamens are absent. The pollen grains are two- to five-colpate. The female archesporium is single-celled. The Cellular type of endosperm develops haustoria. The embryogeny refers to Period I, Megarchetype IV. The seed-coat is unvascularized, and the fruit is drupaceous (Table 48.28.1).

An evaluation of the similarities between Oleaceae and *Nyctanthes*, and differences with Verbenaceae, by Kapil and Vani (1966), confirms that *Nyctanthes* should be transferred from Verbenaceae to Oleaceae. Maheswari Devi

(1975) accepts the assignment to Oleaceae, and further suggests that *Nyctanthes* should be included in a new subfamily, Nyctanthioideae, in Oleaceae (Table 48.28.1).

On the basis of vegetative anatomy, floral morphology, characters of integument and testa, palynology, chromosome number and phytochemistry, *Nyctanthes* is a valid member of the Oleaceae.

Labiatae. Labiatae has been treated as a highly evolved herbaceous family. Embryologically, it is allied to Verbenaceae. The development of the Cellular type of endosperm with terminal haustoria has been a subject of critical study. The transverse division of the primary endosperm nucleus separates a micropylar and a chalazal chamber. The sequence of further divisions and mode of orientation of walls in these chambers, lead to the formation of four types: Scutellaria, Brunella, Stachys and Galeopsis (see Crété 1942). The Galeopsis type is merely a modification of the Brunella type. The haustoria are well developed, and functional till a young stage of proembryo. Zygomorphic flowers, adapted to entomophily, also indicate its advanced condition.

The fourth suborder, Solanineae, comprises 15 families. Engler and Prantl (1931) place all the families in the order Tubiflorae. These are treated here according to their taxonomic placement by Melchior (1964).

Nolanaceae. R. Dahlgren (1980b) treats *Nolana* as a genus in Solanaceae. Bessey (1915) places it, together with Solanaceae, in Polemoniales. Wettstein (1935), Rendle (1971) and Melchior (1964) accept the family status, and include Nolanaceae in Tubiflorae. Hutchinson (1973) also accepts the family status of Nolanaceae, but places them in Solanales. The development of pollen, ovule, embryo sac and embryo is similar to that in Solanaceae (di Fulvio 1969). The development of the endosperm is of the Cellular type, and haustoria are absent. The mericarpous, nutlet-like fruit (Hutchinson 1973) segregates it from Solanaceae.

Solanaceae. Solanaceae are associated with Convolvulaceae in Polemoniales (Bentham and Hooker 1965b), and in Solanales (Hutchinson 1973, Cronquist 1981). Apart from the characters of the order, Solanaceae exhibit polyspory, occasional development of Allium, Endymion and Adoxa types of embryo sacs, Nuclear, Cellular and Ce–Nu–Ce type (*Hyoscyamus niger* – Svensson 1926) of endosperm, apomixis in some species, and polyembryony. The development of the endosperm in *Hyoscyamus niger* (Svensson 1926) is treated as of the Cellular type. The variation in the ontogeny of the endosperm is due to some alterations in the genetic conditions, i.e. "genetic instability" (Swamy 1974a). Solanaceae and Convolvulaceae both have intraxylary phloem, chemically allied alkaloids (Frohne and Jensen 1973), and capacity to store excess starch (Wagenitz 1977a).

The alliance of Solanaceae is always sought with Scrophulariales (see Takhtajan 1980). However, Scrophulariaceae invariably show Cellular type of endosperm and Onagrad type of embryogeny. Perhaps, both families (Solanaceae and Scrophulariaceae) originated from a common stock, and Scrophulariaceae are more advanced (B. Tiagi and Varghese 1970).

Table 48.28.1. Morphology and embryology of *Nyctanthes*, Oleaceae and Verbenaceae (see Kapil and Vani 1966)

	Nyctanthes	Oleaceae	Verbenaceae
Corolla	Contorted	Contorted	Quincuncial
Stamens	2, filaments short, carpelloid anthers present	2, filaments short, carpelloid anthers common	4, didynamous, filaments long, carpelloid anthers absent
Pollen	Spheroidal, tricolpate, tenuinexinous, sexine reticulate with piloid excrescences	Spheroidal, tricolpate, tenuinexinous, sexine reticulate with piloid excrescences	Euripalynous, peroblate-prolate, 2–5-colpate or tricolporate, sexine and nexine of equal thickness, sexine reticulate without piloid excrescences
Nucellar archesporium	Single-celled, rarely 2-celled	Single-celled, rarely 2- or 3-celled	Single-celled
Embryo sac	Mono- (Polygonum type 50%) or bisporic (Allium 35%, Endymion 15%), antipodal cells small and ephemeral, twin embryo sacs occur	Mono- (Polygonum type) or bisporic (*Olea*–Allium type), antipodal cells small and ephemeral, occasionally twin embryo sacs occur	Monosporic (Polygonum type), antipodal cells large and persistent, twin embryo sacs not reported
Endosperm	Cellular type, first two divisions transverse, chalazal haustorium absent	Cellular type, first two divisions transverse, chalazal haustorium absent; Nuclear in *Jasminum*	Cellular type, first division transverse, second longitudinal or transverse, chalazal haustorium common
Embryogeny	Caryophyllad type	Caryophyllad and Solanad, polyembryony occurs	Onagrad
Seed-coat	Vascularized	Vascularized	Non-vascularized
Fruit	Capsule	Capsule, drupe, berry or samara	Mostly drupaceous

According to R. Dahlgren (1980b), Solanaceae and Scrophulariaceae belong to two different superorders; Solanaceae to Solaniflorae, a non-iridoid group, and Scrophulariaceae to Lamiiflorae, an iridoid group.

Bentham and Hooker (1965b) place the rest of the families of the suborder Solanineae in his order Personales. Hutchinson (1973) separates Bignoniaceae and Pedaliaceae and includes them in the order Bignoniales, while he assigns all the other families to his order Personales.

Buddleiaceae. According to Punt and Leenhouts (1967): "Palynology reveals a close relationship to the Scrophulariaceae and hardly any to the Loganiaceae". Wagenitz (1977b) and Takhtajan (1980) also state that Buddleiaceae is closer to Scrophulariaceae than to Loganiaceae.

Buddleiaceae and Scrophulariaceae (Sects. 48.14 and 48.15) have several common features: secretory anther tapetum, two-celled pollen grains, anatropous, unitegmic, tenuinucellar ovule with endothelium and hypostase, Polygonum type of embryo sac with ephemeral antipodals, Cellular type of endosperm with a micropylar and a chalazal haustorium, Onagrad type of embryogeny, and albuminous seed. According to Jensen et al. (1975), *Buddleia* contains Group I iridoids as in Scrophulariales; absent in Gentianales.

Considering the above data, there is no doubt that the affinities of Buddleiaceae are with Scrophulariaceae. The position assigned to it near Scrophulariaceae by Melchior (1964) is appropriate.

Scrophulariaceae. The embryological features of the family Scrophulariaceae are in conformity with those of the order. The Cellular type of endosperm with haustoria has been studied extensively. On the basis of the sequence and plane of division of endosperm cells, differentiation of micropylar and chalazal haustoria, and the number and organization of haustoria and their subsequent development, Banerji (1961), B. Tiagi and Varghese (1970) and Rau (pers. comm. 1987) classify the development into several types (Fig. 48.15.6A–K_4). Banerji (1961) considers reduction in the number of haustorial cells, and their nuclei, as an advanced feature.

Banerji (1961) and B. Tiagi and Varghese (1970) treat the formation of a linear row of three cells in the young endosperm as a normal feature. D. Raju and Arekal (1978) observed the three-celled condition of the endosperm in *Linaria ramosissima*, and consider it as a possible link between Scrophulariaceae and Acanthaceae.

The tribe Rhinanthoideae (members parasitic) is treated as related to Acanthaceae. The embryological data do not support this alliance (Sect. 48.19). Acanthaceae do not have any members which are parasitic. The behaviour of the central cellular endosperm, after the inactivation of the micropylar and the chalazal haustoria in *Rhinanthus major* (B. Tiagi 1972), is noteworthy; the development of a cambium-like strip in the cellular endosperm is unique.

The tribe Selagineae has been assigned family rank, Selaginaceae, by Bentham and Hooker (1965b) and Hutchinson (1973) and placed in Lamiales. Wettstein (1935) and Lawrence (1951) treat it as a tribe of Antirrhinoideae, while Melchior (1964) raises it to a subfamily Selagınoideae of Scrophulariaceae. The common embryological features are: secretory anther tapetum, two-celled pollen, anatropous, unitegmic, tenuinucellar ovule with endothelium, Polygonum type of embryo sac, Cellular type of endosperm with haustoria, Onagrad type of embryogeny, and fleshy seed with a large straight embryo. Kapil and Masand (1964) and D. Raju (1973), on the basis of embryology, include it as a tribe in Scrophulariaceae. Taneja (1970) considers it as an offshoot of Scrophulariaceae from Gratioleae.

While dealing with the embryology of *Striga orobanchoides*, B. Tiagi (1956, 1963) points out that its embryological features are so identical with those of Orobanchaceae, that the latter can be derived from Scrophulariaceae (Sect. 48.15).

Globbulariaceae. Bentham and Hooker (1965b) merge Globbulariaceae with the family Selaginaceae, Hallier includes them in Scrophulariaceae, while Engler, Bessey, Rendle, Wettstein and Hutchinson treat them as a distinct family. Wettstein derives them from Scrophulariaceae (see Lawrence 1951). Hutchinson (1973) groups them in Lamiales with Myoporaceae, Selaginaceae and Labiatae.

The anatropous, unitegmic, tenuinucellar ovule with endothelium, Polygonum type of embryo sac, Cellular type of endosperm with haustoria, and Onagrad type of embryogeny in Globbulariaceae are comparable to similar features in other members of Tubiflorae. The development of endosperm in the micropylar chamber is similar to that in Selaginaceae. However, due to the disintegration of intervening walls, the four-celled micropylar haustorium becomes coenocytic. The chalazal haustorium is one-celled. This family appears to be derived from Scrophulariaceae. In fact, R. Dahlgren (1980b) includes Orobanchaceae, Selaginaceae and Globbulariaceae in Scrophulariales.

Bignoniaceae. Hallier (1912) derives Bignoniaceae from Scrophulariaceae. Wettstein (1935) recognizes its advanced status, and allies it with Scrophulariaceae and Gesneriaceae. Hutchinson (1973) gives Bignoniaceae ordinal rank, Bignoniales, which also include the families Cobaeaceae, Pedaliaceae and Martyniaceae. He recognizes Pedaliaceae and Martyniaceae as advanced members in Bignoniales, due to their herbaceous habit, and greatly modified fruits. Pedaliaceae have four-loculed ovary with axile placentation, Martyniaceae have parietal placentation, whereas the members of Bignoniaceae exhibit both types.

Bignoniaceae and Scrophulariaceae. The delimitation of Bignoniaceae and Scrophulariaceae has long been a taxonomic problem. *Paulownia, Schlegelia, Gibsoniothamnus* and *Synapsis* (not studied) are considered as "intermediate" or "borderline" genera. Armstrong (1985) discusses the delimitation of Bignoniaceae and Scrophulariaceae, and concludes that *Paulownia, Schlegelia,* and *Gibsoniothamnus* are related to Scrophulariaceae, and they cannot be considered of intermediate status between Bignoniaceae and Scrophulariaceae.

This is especially so because in all definitive characters – gynoecial anatomy, placental morphology, presence of endosperm, embryo morphology, seed-wing morphology – these genera have the characters of Scrophulariaceae (see also Campbell 1930).

Paulownia. The stylar transmission tissue separates into two channels and enters the top of each locule. The membraneous circumalate seed-wing is composed of radially elongated epidermal cells with a reticulum of lignified thickenings on the radial walls. The seed-wings of *Paulownia* are very similar to those described for *Lophospermum* (= *Rhodochiton*), *Epixiphium*, and *Maurandya*. The embryo, seed-wing, and cotyledon morphology and venation, along with the presence of endosperm in the seeds of *Paulownia*, are also consistent with placement in Scrophulariaceae (Armstrong 1985).

Schlegelia. *Schlegelia* does not conform to the floral anatomy of the rest of Bignoniaceae, and is consistent with scrophulariaceous floral structure. The seeds of *Schlegelia* are angular, unwinged, endospermous, and contain columnar embryos. While larger than typical scrophulariaceous seeds, they are nonetheless very similar, as pointed out by Monachino (1949). The Schlegelieae are considered an anomalous tribe within Bignoniaceae (Gentry 1980), but relationships within Scrophulariaceae are equally uncertain. The placement of *Schlegelia* in Scrophulariaceae must await further studies (Armstrong 1985).

Martyniaceae. Embryological evidence indicates a close alliance between Bignoniaceae and Martyniaceae. The pollen grains of *Martynia* are similar to those in Bignoniaceae (Erdtman 1952). Crété (1942) compares the development of endosperm in Bignoniaceae and Martyniaceae. In both families the central cellular endosperm can be differentiated into two distinct zones: micropylar region with large cells, and chalazal with a compact tissue of small cells. The chalazal haustorium is four-nucleate. The micropylar haustorium differentiates after the chalazal one (Crété 1942). Moreover, the embryogeny in both families is of the Onagrad type-Catalpa variation. However, the exalbuminous, winged seed of Bignoniaceae has no match in Martyniaceae. In fact, Martyniaceae have been associated with Bignoniaceae (Hallier 1912, Hutchinson 1973). Bentham and Hooker (1965b) merge them with Pedaliaceae, while Bessey, Rendle and Wettstein recognize their independent family status.

Pedaliaceae. Palynologically, Pedaliaceae is closer to Bignoniaceae and Scrophulariaceae. The pollen grains of *Pedalium* and *Sesamum* are allied to those of Bignoniaceae, and those of *Trapella* to Scrophulariaceae. The belated development of micropylar haustorium occurs in all the three families – Bignoniaceae, Martyniaceae and Pedaliaceae (Crété 1942, S.P. Singh 1970b).

The pollen grains of *Martynia* are pantoporate and polyrugoidate with a thick reticulate exine, and two-celled at the dispersal stage; they are polycolpate and three-celled in *Pedalium* (S.P. Singh 1960). The nature of the endosperm haustoria in Pedaliaceae is more or less similar to that in Scrophulariaceae (Crété 1951). The marked difference between *Martynia* and *Pedalium* lies in the structure of the cotyledons (S.P. Singh 1970a, b). They are dorsiventral with upper and lower epidermis, and hypodermal palisade parenchyma (S.P. Singh 1970a, b). The palisade tissue is single-layered in Pedaliaceae, and two-layered in Martyniaceae. Also, the spines on the fruits of *Pedalium* are morphologically identical with the hooks of *Martynia* (S.P. Singh 1963, 1964); only their number varies.

Embryologically, Martyniaceae can be segregated from Pedaliaceae on the basis of the structure of pollen, placentation, endosperm, embryo and fruit (S.P. Singh 1970a, b).

S.P. Singh (1960, 1970a) favours Erdtman's suggestion of separating the genus *Trapella* from Pedaliaceae, and erects a new family, Trapellaceae. Palynological data, and inferior condition of the ovary support this segregation. The morphology of fruit in *Trapella* and *Pedalium* (S.P. Singh 1960, 1963) is also different. According to S.P. Singh (1963), the spines in *Pedalium* are associated with the lateral bundles of carpels, while in *Trapella* the protuberances correspond to the sides of the gamosepalous calyx. In addition, S.P. Singh (1970a) cites features, such as ventral raphe, short ovular vascular supply ending near the chalaza, and thin-walled epidermis of tangentially elongated cells forming the seed-coat in *Trapella*, which differ from *Pedalium* and *Sesamum*. Embryologically also, *Trapella* differs from *Martynia* and *Pedalium*: persistence of antipodals, simultaneous differentiation of micropylar and chalazal haustorium, and a four-celled micropylar haustorium and chalazal haustorium of two elongated cells. Trapellaceae can no doubt be segregated from Pedaliaceae.

Acanthaceae. Acanthaceae are a natural taxon. The major embryological features are identical with the order Tubiflorae.

The family is multipalynous; the acolpate pollen grains, with smooth exine in *Thunbergia alata*, are rare in Tubiflorae. The jaculator (retinaculum) is a characteristic feature, but its occurrence in the tribe Nelsonioideae is controversial (Bhoj Raj 1961, Bremekamp 1965). The jaculator is non-functional in *Elytraria acaulis* (B.M. Johri and Singh 1959), and *Nelsonia campestris* (Mohan Ram and Masand 1963). The formation of micropylar and chalazal caeca of embryo sacs is quite common. The development of the endosperm is of the Cellular type with the formation of a linear row of three cells in the majority of the members.

The asymmetric development of central cellular endosperm, formation of the "basal apparatus" in the central chamber, finger-like haustorial processes of the basal apparatus in *Asystasia*, well organized endosperm haustoria, enchroachment of the cells of the central cellular endosperm on the micropylar haustorium in *Barleria prionitis*, absence of chalazal haustorium in *Blepharis* and *Thunbergia* (Mohan Ram and Wadhi 1964), development of secondary endosperm haustoria in *Andrographis*, *Haplanthus* and *Thunbergia*, ruminate endosperm in *Elytraria*, *Haplanthus* and *Nelsonia*, and absence of seed-coat in mature seed in *Andrographis*, *Elytraria* and *Haplanthus* are some of the noteworthy features of Acanthaceae.

On the basis of endosperm formation in the central chamber, Mohan Ram and Wadhi (1964) classify the development of the endosperm into four types – Thunbergia, Acanthus, Schaueria and Ruellia. However, variations have been observed in different species of the same genus, as in *Barleria* and *Lepidagathis*. The development in *B. prionitis* and *L. trinervis* (Ambegaokar 1959) conforms to the Acanthus type, in *B. cristata* and *L. prostata* (Mohan Ram 1962), to the Ruellia type.

The Ce–Nu–Ce sequence of development of the endosperm in the central chamber has been interpreted as "reverted" Helobial type (K.V.O. Dahlgren 1934a), as a variant of the ab initio cellular endosperm (P. Maheshwari 1950), or as an "intermediate" type (Wunderlich 1959). Swamy (1974a), however, considers this phase to be determined by genetic constitution in this family.

Bremekamp (1953) raises the tribes Acanthoideae, Mendoncioideae and Thunbergioideae to family rank. The tribe Nelsonioideae has been proposed for transfer to Rhinantheae under Scrophulariaceae, which has been rejected by B.M. Johri and Singh (1959), Mohan Ram and Masand (1963), and P. Maheshwari (1964). The presence of jaculator in *Elytraria* (B.M. Johri and Singh 1959) and *Nelsonia* (Mohan Ram and Masand 1963), and asymmetric development of cellular endosperm in both genera are acanthaceous features lacking in Rhinantheae. Mohan Ram and Masand (1963) add that features like alternate leaves, parietal placentation, endothelium, funicular obturator, and albuminous seed, shared by *Elytraria* and *Nelsonia*, suggest their assignment, according to Lindau (1895), as a subfamily under Acanthaceae. Wadhi (1970) compares the sequence and plane of the first few divisions in the endosperm of *Elytraria* and *Nelsonia* with those in *Rhinanthus* and *Pedicularis* of Rhinantheae, and negates their relationship. The embryogeny in Acanthaceae is of the Onagrad (B.M. Johri and Singh 1959), Solanad (Mohan Ram 1962), and Caryophyllad (Ambegaokar 1959) type, whereas it is of the Onagrad and Solanad type in Scrophulariaceae.

Gesneriaceae. Gesneriaceae are closely allied to Scrophulariaceae, Bignoniaceae, and probably to Orobanchaceae (see Lawrence 1951). Hutchinson (1973) groups them with Scrophulariaceae, Acanthaceae, Orobanchaceae and Lentibulariaceae in Personales. The unilocular ovary with parietal placentation, loculicidal type of fruit, and minute seeds with abundant endosperm separate them from Scrophulariaceae and Bignoniaceae, and their non-parasitic habit differs from Orobanchaceae.

The embryological features of Gesneriaceae (Sect. 48.22) are in conformity with the order Tubiflorae. The first few divisions of the Cellular type of endosperm are either according to the Brunella type in some spp., or lead to the formation of a linear row of three cells. The aggressive chalazal haustorium in *Klugia* is noteworthy (Fig. 48.22.3A–C), whereas a chalazal haustorium is absent in *Streptocarpus* (Crété 1942).

Orobanchaceae. Some taxonomists derive the family Orobanchaceae from Gesneriaceae, others from Scrophulariaceae. An unilocular ovary with parietal placentation occurs in Gesneriaceae, Orobanchaceae, and some members of Scrophulariaceae (Varghese 1963, see also B. Tiagi and Varghese 1970). Parasitic habit is common to the members of Gerardeae and Rhinanthoideae (Scrophulariaceae). Floral morphology also supports the alliance of Orobanchaceae with Scrophulariaceae (Cassera 1935). Glišić (1929) suggests the merging of Orobanchaceae with Scrophulariaceae, following the similarity in endosperm development. B. Tiagi (1970) draws attention to remarkable similarities in the embryological features of Orobanchaceae and *Striga orobanchoides*, and concludes that Orobanchaceae can be derived through forms like *S. orobanchoides* from Scrophulariaceae.

Even though the type of placentation differs between Orobanchaceae (parietal) and *S. orobanchoides* (axile), both exhibit similarity in several features: (1) two-celled tricolpate pollen grain, (2) curvature of the functional (chalazal) megaspore and, consequently, of the female gametophyte, towards the funiculus, (3) Brunella type of endosperm, (4) extremely aggressive, two-celled (each cell binucleate) micropylar haustorium, and a binucleate chalazal haustorium, (5) Onagrad type of embryogeny, (6) minute, albuminous seed with undifferentiated or feebly differentiated embryo, and reticulately thickened testa, and (7) persistence of remnants of pollen tube, and endosperm haustoria in seed (*Orobanche cernua* – B. Tiagi 1951b).

Lentibulariaceae. Lentibulariaceae are morphologically allied to Scrophulariaceae. On the basis of embryology, R. Khan (1970) suggests their derivation from the Scrophulariaceae. The characteristic features of Lentibulariaceae are: insectivorous nature, unilocular ovary with free-central placentation, three-celled pollen grains, extra-ovular extension of embryo sac, "exogamy", Scutellaria type of early development of endosperm, aggressive micropylar and chalazal haustoria, and organless (undifferentiated) embryo in *Utricularia* spp. The embryo is fully differentiated only in *Pinguicula* (Stolt 1936).

Myoporaceae. The fifth suborder Myoporineae has a single family, Myoporaceae. Hutchinson (1973) includes it in Lamiales with Globbulariaceae, Selaginaceae and Labiatae. The embryological features, such as secretory anther tapetum, Polygonum type of embryo sac, Cellular endosperm with haustoria, and Onagrad type of embryogeny, are in conformity with the order. A detailed study would be helpful to determine its phyletic position.

Phrymaceae. The last suborder of Tubiflorae, Phrymineae comprises a single family Phrymaceae. Bentham and Hooker (1965b), Hallier (1912) and Cronquist (1981) treat the latter as a genus in Verbenaceae. Bessey (1915) and Wettstein (1935) accept it as a separate family. Hutchinson (1973) includes Phyrmaceae in his Verbenales. The solitary, erect ovule, persistent antipodals, absence of micropylar haustorium, and convolute cotyledons distinguish it from Lamiales.

49 Plantaginales

Plantaginaceae is the only family in the order Plantaginales.

49.1 General Features

Plantaginales are characterized by: (1) glandular anther tapetum, (2) simultaneous cytokinesis in pollen mother cells, (3) pantoporate, two-celled pollen grains, (4) anatropous, unitegmic, tenuinucellate ovule with endothelium, and single archesporial cell, (5) Polygonum type of embryo sac, (6) Cellular endosperm with four-celled micropylar, and binucleate chalazal haustorium, (7) Onagrad type of embryogeny, and (8) albuminous, winged seed with a characteristic mucilagenous testa.

49.2 Plantaginaceae

The anther wall comprises the epidermis, fibrous endothecium, an ephemeral middle layer, and glandular tapetum in *Plantago major* (R.C. Misra 1970). Simultaneous cytokinesis follows meiosis in microspore mother cells, and the tetrads are tetrahedral, isobilateral and decussate.

The pollen grains are pantoporate with smooth exine, and are two-celled at the dispersal stage. Large pollen grains are reported in *P. lagopus* and *P. major* (R.C. Misra 1970).

The ovule is anatropous, unitegmic, and tenuinucellate. An endothelium differentiates in the lower half of the embryo sac. The archesporial cell functions as megaspore mother cell, undergoes meiotic divisions, produces a linear tetrad, and the chalazal megaspore develops into a Polygonum type of embryo sac in *Plantago lanceolata* (G.O. Cooper 1942a). The synergids are beaked. The polar nuclei fuse before fertilization. The antipodals persist up to fertilization.

The development of endosperm is of the Cellular type. The division of the primary endosperm nucleus separates a large chalazal from a small micropylar chamber. The micropylar chamber becomes four-celled by two vertical divisions at right angle to each other. A transverse division in each of these cells results in eight cells, four cells in each tier. Each cell of the upper tier develops into a micropylar haustorium (Fig. 49.2.1A, B). The lower tier of four cells gives rise to the endosperm proper. The chalazal chamber develops into a binucleate haustorium (Fig. 49.2.1A, C, D).

The haustoria are aggressive and the nuclei become hypertrophied in *Plantago lanceolata* (Crété 1942). The four micropylar haustoria remain separated by the ovular tissue (Fig. 49.2.1B), and in *P. lanceolata* they exhibit palmate branching at their tip (Fig. 49.2.1C, D). They are unbranched in *P. coronopsis*, *P. major* and *P. pumila* (R.C. Misra 1970). The micropylar

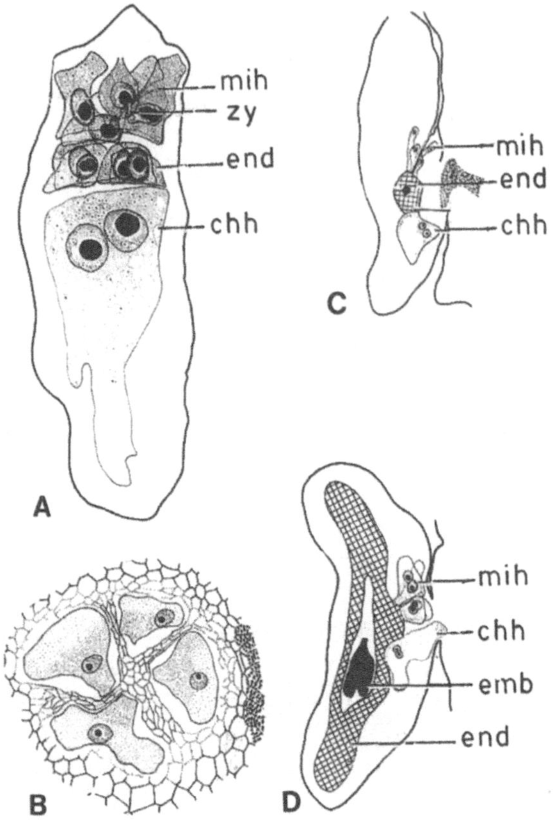

Fig. 49.2.1A–D. Development of the endosperm. **A–C** *Plantago major*; **A** Embryo sac with zygote (*zy*), four-celled micropylar haustoria (*mih*), endosperm proper (*end*), and binucleate chalazal haustorium (*chh*); **B** Transection of ovule to show four micropylar haustoria; **C** Longisection of ovule at globular proembryo (*black*). **D** *P. lanceolata*, longisection of ovule at dicotyledonous embryo (*emb*), the haustoria are very close to the placenta. (After R.C. Misra 1970)

haustoria consume the integumentary and placental tissue but, finally, this activity ceases as the placental cells abutting the haustoria develop thickenings.

The binucleate chalazal haustorium is less aggressive. Due to the asymmetric development of the endosperm tissue, the chalazal haustorium also comes to lie in the vicinity of placenta, adjacent to the micropylar haustoria (Fig. 49.2.1D).

The chalazal haustorium generally produces a lateral outgrowth towards the placenta (Fig. 49.2.1C, D). In *P. lagopus*, the haustorium is rather massive and barrel-shaped. Its lower portion is swollen (termed "bulla" by the author) and contains two hypertrophied nuclei. From the lower part, two enucleate branches grow laterally up to the anterior end of the ovule encircling the main body of the endosperm. The apices of both the branches become considerably broad and lobed and remain appressed to each other. The micropylar and chalazal haustoria remain healthy up to the dicotyledonous stage of embryo. The development of the endosperm in *Plantago* needs further investigation.

The embryogeny conforms to the Onagrad type. A synergid embryo (without fertilization) is formed in *P. lanceolata* (Souèges 1926f).

The seed is albuminous, exotestal, and winged with a large spatulate embryo. The seed-coat consists of outer and inner epidermis of the integument. The middle layers collapse. The outer epidermal cells are large, thin-walled, and contain mucilage. The cells of the inner epidermis are small, thin-walled, and contain pigment.

GO Cooper (1942a), P Crété (1942), D Czapska-Dziekanowska (1966), RC Misra (1964a, b, 1965, 1970), W Rosén (1940b), R Souèges (1923e, 1926f)

49.3 Taxonomic Considerations

Plantaginaceae. Bentham and Hooker (1965b) consider the family Plantaginaceae as anomalous. Hallier (1912) assigns it to Tubiflorae, allied to Scrophulariaceae. Bessey (1915) places it in Primulales and derives its ancestry from Plumbaginaceae, while Hutchinson (1973) places it in Plantaginales. Following Melchior (1964), Cronquist (1981) erects the order Plantaginales, whereas R. Dahlgren (1980b) places it in the order Scrophulariales.

Bitegmic ovules, multicellular female archesporium, Nuclear type of endosperm, Caryophyllad type of embryogeny, and dwarf seeds of Primulaceae differ from those of Plantaginaceae.

Plantaginaceae do not show any affinity with Plumbaginaceae. Plumbago or Plumbagella type of embryo sac, Nuclear type of endosperm, and Solanad type of embryogeny in Plumbaginaceae do not occur in Plantaginaceae. Their relation to Plumbaginaceae (see Bessey 1915) and their advancement over Primulales (see Hutchinson 1973) can be ruled out.

R.C. Misra (1970) indicates a close alliance of Plantaginaceae with Scrophulariaceae. Both families have similar type of anther, pollen, ovule, embryo sac, and seed structure. The cellular endosperm has aggressive micropylar and chalazal haustoria. The haustoria lie close to the hilum, and persist in mature seed, as in Scrophulariaceae and Labiatae. The embryogeny in both families corresponds to the Onagrad type.

There is no doubt that the families Plantaginaceae and Scrophulariaceae are closely related. (The difference lies in the floral structure and mode of pollination.) Also, the order Plantaginales is related to Tubiflorae through Scrophulariaceae.

50 Dipsacales

The order Dipsacales comprises Caprifoliaceae, Adoxaceae, Valerianaceae and Dipsacaceae.

50.1 General Features

Dipsacales are characterized by: (1) fibrous endothecium, and amoeboid anther tapetum; both secretory and amoeboid in Caprifoliaceae, (2) three-celled pollen, (3) inferior ovary with anatropous, unitegmic, tenuinucellar ovules, and a single-celled archesporium; crassinucellar in *Viburnum* (Caprifoliaceae), (4) Polygonum type of embryo sac with well-developed endothelium, except in Adoxaceae; Adoxa type in Adoxaceae, and Allium and Adoxa (*Sambucus*) types in Caprifoliaceae, (5) Cellular type of endosperm; ruminate in *Viburnum* (Caprifoliaceae) and *Morina* (Dipsacaceae), (6) Asterad type of embryogeny; Piperad in *Scabiosa*, and Solanad in *Morina* (Dipsacaceae), and (7) albuminous seed with a straight embryo; exalbuminous in Valerianaceae.

50.2 Caprifoliaceae

The anther wall comprises the epidermis, fibrous endothecium, an ephemeral middle layer, and the tapetum. The tapetum is of the secretory type in *Sambucus* and *Viburnum*, and amoeboid in *Lonicera* and *Symphoricarpus*. Amoeboid tapetum also occurs in *Kolkwitzia amabilis*, *Leycesteria formosa*, *Triosteum erythrocarpum* and *T. pinnatifidum* (von Weberling and Hildenbrand 1982). Periplasmodium is formed at the uninucleate pollen. The microspore mother cells undergo meiotic divisions, and the tetrads are tetrahedral.

The three-colporate (endocolpium indistinct) pollen grains are three-celled at the dispersal stage.

In *Diervilla lonicera* (Thompson 1985), the ovule development is limited by insufficient pollination early in the blooming period, and during extensive rainy periods. The number of ovules per flower increases through the blooming period.

The ovule is anatropous, unitegmic, and tenuinucellar with an integumentary vascular bundle; crassi- to tenuinucellar in *Viburnum* (Suneson 1933). A single archesporial cell functions as megaspore mother cell (except in *Viburnum*), undergoes meiotic divisions, and forms a linear tetrad. The chalazal megaspore develops into a Polygonum type of embryo sac in *Viburnum* (Moissel 1941). Allium type of development is reported in *Lonicera myrtilloides* (Feng 1934), *Viburnum acerifolium* and *V. lanata* (Suneson 1933), and Adoxa type in *Sambucus nigra* (Dolcher 1951); tetrasporic development also occurs in the sterile ovules of *Viburnum* (Moissel 1941). The antipodals are ephemeral. An endothelium differentiates.

Tokć (1981) investigated the embryology of *Sambucus racemosa* from four localities. In *Sambucus* sp. from Chelm and Sikornik there are sporadic disturbances during meiotic divisions of the megaspore mother cell: (1) formation of a vertical wall after meiosis I, (2) inhibition of wall formation and the two nuclei of a dyad may fuse, (3) formation of three-nucleate embryo sac at the time of fertilization, (4) formation of micronuclei, and (5) retardation of the division of nuclei and delay in differentiation of the chalazal pole.

In *S. racemosa* (Tokć 1981), the development of the embryo sac is unstable. Of the embryo sacs examined, 58% were of the Adoxa type; 38% exhibited irregular meiosis, or irregular mitotic divisions, in a coenomegaspore. The number of nuclei in abnormal embryo sacs varied from four to nine; their arrangement also varied. There may be two, three or four polar nuclei. In addition, micronuclei may develop in the central cell.

The polar nuclei fuse before fertilization; triple fusion is frequent. During fertilization the pollen tubes enter laterally through the integumentary cells in *S. racemosa* (Tokć 1981). Mesogamy and chalazogamy have also been observed. Fertilization occurs even in extremely modified embryo sacs.

The development of the endosperm is of the Cellular type. In *Sambucus*, the division of the primary endosperm nucleus is vertical or oblique; subsequent divisions are in various planes. Rarely, the development of the endosperm is autonomous.

The embryogeny probably corresponds to the Asterad type. In *Sambucus* polyembryony occurs in 2.8% of ovules. The additional embryo at the micropylar end shows the same number of cells as the zygotic one. One of them may grow parallel to the long axis of the embryo sac, the other perpendicular to the first, extending into the micropyle. From the appearance of the nuclei, one appeared to be diploid and the other haploid which may have developed from an unfertilized synergid. Of the embryo sacs 4.6% develop due to somatic nucellar apospory (Tokć 1981).

The development of the embryo sac, endosperm, embryo, polyembryony, apospory and apogamy in *Sambucus* requires detailed reinvestigation.

The albuminous seed contains a small and linear embryo (Hutchinson 1973). In *Viburnum foetens* the endosperm is subruminate due to the folds of the endocarp (Periasamy 1962a). The testa is composed of two or three layers. The cells of the outer epidermis are flat, with thick outer walls in *Sambucus*, or radially elongated in *Viburnum*. The radial and inner walls are pitted and lignified in *Lonicera*. In *Symphoricarpus* the epidermal cells are extremely thick-walled and the two hypodermal layers are lignified (Corner 1976).

M Charzynska and E Lewandowska (1990), EJH Corner (1976), P Crété (1953c), KVO Dahlgren (1927b), T Dolcher (1951), YA Feng (1934), N Fukuoka (1972), GR Hildebrand and DE Fairbrothers (1970b), F Höck (1892), J Hutchinson (1973), E Moissel (1941), K Periasamy (1962a), F von Schwerin (1920), R Souèges (1945d, 1948, 1959b), S Suneson (1933), JD Thompson (1985), E Tokć (1981), F von Weberling and M Hildenbrand (1982)

50.3 Adoxaceae

The anther wall comprises the epidermis, fibrous endothecium, an ephemeral middle layer, and amoeboid tapetum with two-nucleate cells. The microspore mother cells undergo meiotic divisions, cytokinesis is simultaneous, and the tetrads are tetrahedral.

The pollen grains are three colporate (endocolpium indistinct), and three-celled at the dispersal stage.

The ovule is anatropous, unitegmic, and tenuinucellar. The single archesporial cell functions as megaspore mother cell, undergoes meiotic divisions, and forms a four-nucleate coenomegaspore. These nuclei undergo one mitotic division, and the eight nuclei organize into a three-celled egg apparatus, two polar nuclei, and three antipodal cells (Adoxa type) in *Adoxa moschatellina* (Fagerlind 1938c). The polar nuclei fuse at the time of triple fusion. One of the antipodals is larger than the others.

The development of the endosperm is of the Cellular type in *A. moschatellina* (Lagerberg 1909). The first two vertical walls, following the division of the primary endosperm nucleus, result in the formation of four cylindrical cells. Then a transverse division occurs in each cell followed by another transverse division. Further divisions are irregular.

The embryogeny conforms to the Asterad type; synergid polyembryony occurs in *A. moschatellina* (Czapik 1976).

The seed is albuminous with a minute embryo. The endosperm stores oil. The testa is single-layered, and comprises thin-walled cells.

R Czapik (1976), F Fagerlind (1938c), B Jönsson (1879–1880), OP Kamelena (1980), T Lagerberg (1909), G Wagenitz (1977b)

50.4 Valerianaceae

The anther wall comprises the epidermis, fibrous endothecium, an ephemeral middle layer, and amoeboid tapetum. Simultaneous cytokinesis follows meiotic divisions of microspore mother cells in *Valeriana supina* (Asplund 1920).

The three-colporate pollen grains are three-celled at the dispersal stage.

The ovule is anatropous, unitegmic, and tenuinucellar. The single archesporial cell develops into the megaspore mother cell, undergoes meiotic divisions, produces a linear tetrad, and the chalazal megaspore develops into a Polygonum type of embryo sac. The polar nuclei fuse before fertilization. The antipodals are persistent. An endothelium differentiates.

In rudimentary ovules of *Valeriana* (Asplund 1920), the development of the embryo sac is tetrasporic, as the walls between the megaspore nuclei fail to develop.

The development of the endosperm is of the Cellular type. In *Kentranthus macrosiphon* (Asplund 1920), the division of the primary endosperm nucleus is vertical.

The embryogeny corresponds to the Asterad type.

The exalbuminous seed contains a straight embryo. The testa is single-layered, and unspecialized.

E Asplund (1920), OP Kamelena (1980, 1983), R Souèges (1923a, 1941a)

50.5 Dipsacaceae

Kamelena (1980) gives a detailed account of the embryology of *Dipsacus, Cephalaria, Pterocephalus, Succisa, Scabiosa, Knautia* and *Morina.*

The development of the anther is shown in Figures 50.5.1A–E and 50.5.2A–F. The wall comprises the epidermis, fibrous endothecium, an ephemeral middle layer, and secretory tapetum of large multinucleate vacuolate cells. In *Morina* there are two middle layers and the outer one develops fibrous thickenings (Kamelena 1980).

The tapetal cells extend into the locule (pseudo-amoeboid) but there is no evidence of periplasmodium formation (cf. Fig. 50.5.2A–F). However, in *Morina lehmanniana* (Kamelena 1980), the tapetal cells break down and form a periplasmodium. A periplasmodium is also reported in *Cephalaria ambrosioides*, *Dipsacus leschenaultii* and *Scabiosa caucasia* (Razi and Subramanium 1952). According to Vijayaraghavan and Sarveshwari (1968), the tapetum is secretory in *M. longifolia*. In view of the conflicting reports, this aspect should be reinvestigated.

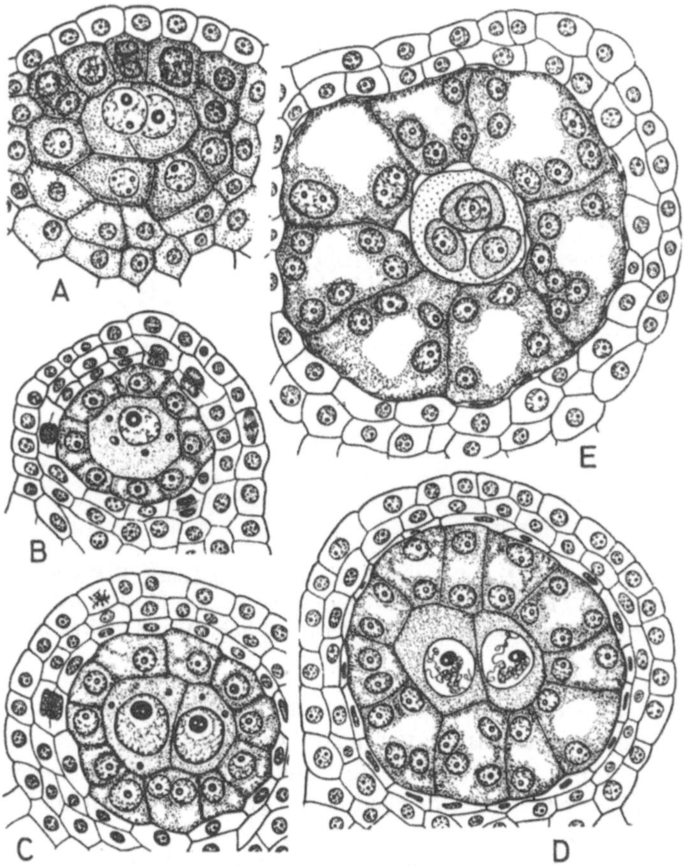

Fig. 50.5.1A–E. Transections of anthers. **A** *Knautia arvensis.* **B** *Scabiosa lucida.* **C, D** *S. bipinnata.* **E** *S. lucida.* **A–E** Secretory tapetum, cells uni- and binucleate (**A–C**), multinucleate and vacuolate (**D**) and unusually large (**E**); microspore mother cells (**A–D**) and tetrad (**E**). (After Kamelena 1980)

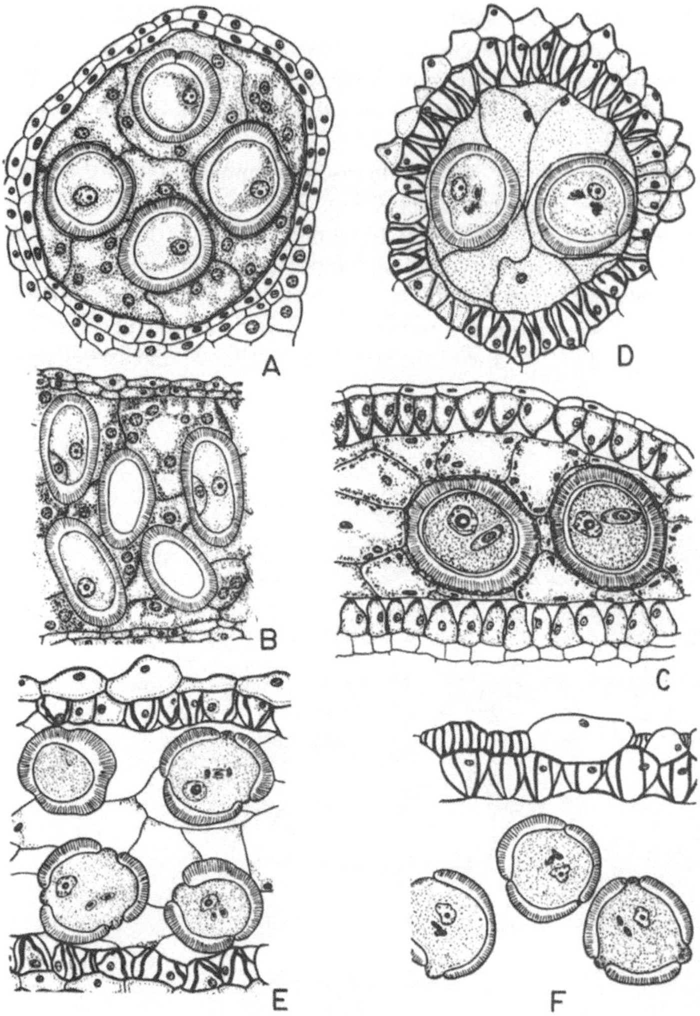

Fig. 50.5.2A–F. **A, D** *Scabiosa lucida.* **B** *S. gorica.* **C** *Pterocephalus afghanicus.* **E, F** *S. bipinnata.* **A–F** Secretory tapetum (**A–E**), fibrous endothecium (**D–F**), uninucleate (**A**), bi- (**B, C**) and tricelled (**D–F**) pollen grains. Note fibrous thickenings in epidermal cells in **F**. (After Kamelena 1980)

The reduction divisions in pollen mother cells are followed by simultaneous cytokinesis, and the tetrads are tetrahedral (Fig. 50.5.1E).

The pollen grains are tricolpate and invariably three-celled (Fig. 50.5.3 A–F). The report of two-celled pollen grains in *Morina longifolia* appears to be erroneous.

Due to absence and/or irregular cytokinesis, a number of abnormal pollen grains are produced (Fig. 50.5.4A–F).

The ovule is anatropous, unitegmic and tenuinucellar. The integument is massive with a very long and narrow micropyle, and the funicular vascular

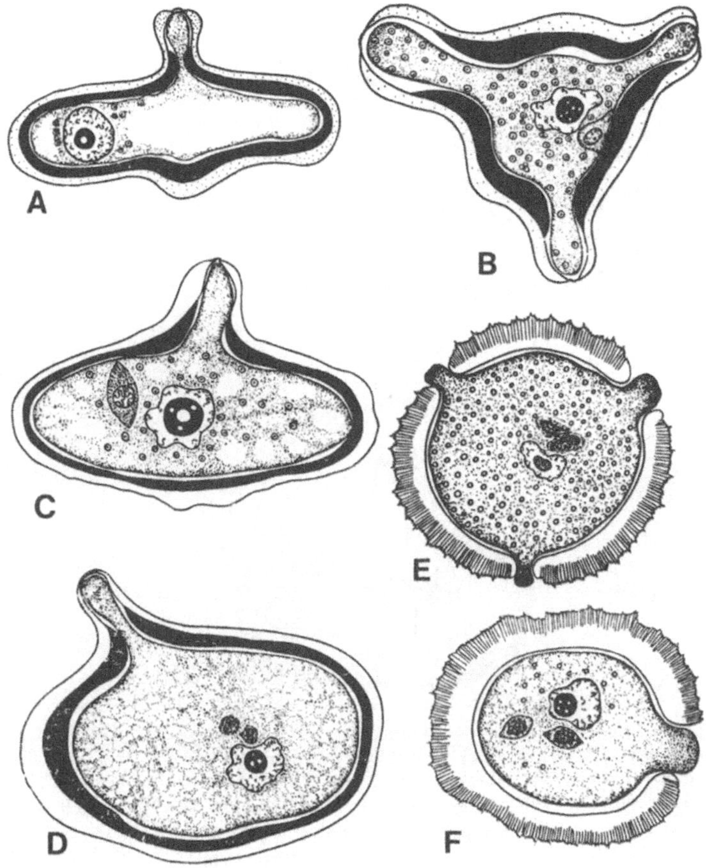

Fig. 50.5.3A–F. A–D *Morina lehmanniana.* **E, F** *Scabiosa bipinnata.* **A** Uninucleate, **B, C** two- and **D–F** three-celled pollen grains. (After Kamelena 1980)

supply traverses up to the tip of the micropyle (Fig. 50.5.5A–E). Vieth (1958) also observed the extension of the vascular strand in several taxa.

In *Morina longifolia* (Vijayaraghavan and Sarveshwari 1968) and *M. kokanica* (Kamelena 1980), the integumentary vascular strand is absent. In the latter species, another significant difference is that the micropyle is short and the embryo sac is situated very close to it.

A prominent thick-walled hypostase, and endothelium (Fig. 50.5.7C–G) differentiate.

The hypodermal archesporial cell functions as megaspore mother cell, undergoes meiotic divisions, produces a linear tetrad, and the chalazal megaspore develops into a Polygonum type of embryo sac. The embryo sac has the usual organization, the polar nuclei fuse before fertilization, and the three antipodal cells show much variation (Figs. 50.5.6A–F, 50.5.7A–G). They are

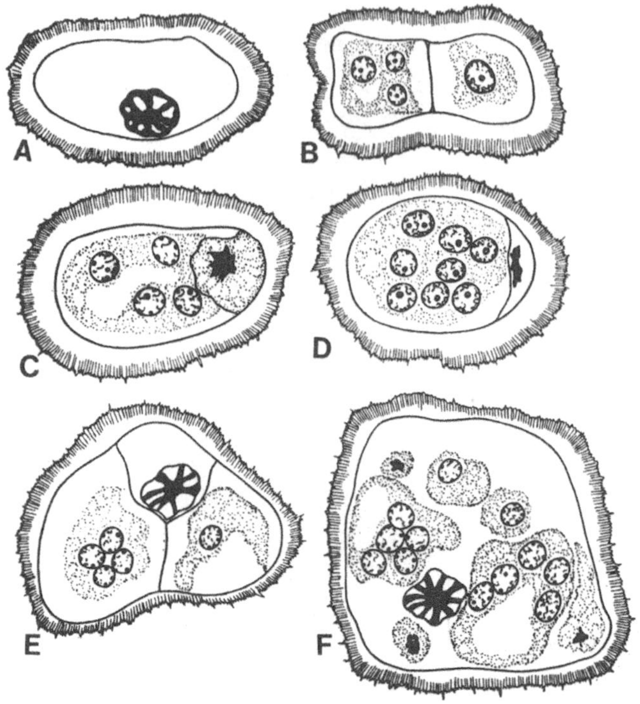

Fig. 50.5.4A–F. *Scabiosa rotata.* Abnormal pollen grains formed due to irregular cytokinesis, **B–F** have a common exine. **A** Uninucleate condition, meiosis in pollen mother cell and cytokinesis did not occur (?). **B** "Dyad", uni- and three-nucleate cells. **C, D** "Dyad", one cell with "degenerated" nucleus, the other cell has four (**C**) and eight (**D**) nuclei. **E** Triad (?), one "dyad" cell with "degenerated" nucleus, and uni- and four-nucleate "microspores". **F** "Polyspore", two uninucleate small "microspores", two large five-nucleate "microspores", three small "microspores" with "degenerated" nucleus, and one large "degenerated" nucleus. The multinucleate condition in a "dyad" cell in **B–D** and of "microspores" in **E** and **F** is due to supernumerary divisions. (After Kamelena 1980)

two- to four-nucleate, and nuclear fusion leads to polyploid condition (Figs. 50.5.6C–F, 50.5.7A, B). A very unusual feature is the "internal" division/s in the antipodal cells (Figs. 50.5.6F, 50.5.7C). In *Morina kokanica* and *M. lehmanniana* (Fig. 50.5.7D–G), each antipodal becomes two- to multicelled and simulates a "proembryo". Most so-called antipodal embryos in several taxa probably belong to this category of multicelled antipodals and should be critically re-examined.

Twin embryo sacs – four-nucleate and organized, and both organized – have been observed in *Scabiosa bipinnata* (Kamelena 1980).

The development of endosperm is of the Cellular type. In most members, the division of the primary endosperm nucleus is followed by an incomplete vertical wall (Figs. 50.5.8A, B, B′, 50.5.9A, B). If the next nuclear division is not followed by a wall, both the cells (open at lower end) become binucleate (Figs. 50.5.8C, 50.5.9B, C). If the walls are laid down, four uninucleate cells

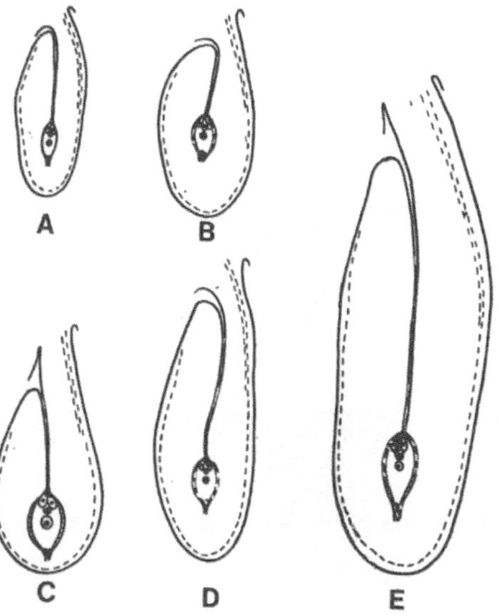

Fig. 50.5.5A–E. **A** *Succisa pratensis.* **B** *Scabiosa lucida,* **C** *Knautia montana.* **D** *Dipsacus laciniatus.* **E** *Cephalaria gigantea.* Longisection of ovules at mature embryo sac, the micropyle is conspicuously long so that the gametophyte comes to lie in the chalazal region. (After Kamelena 1980)

(open at one end) are formed (Figs. 50.5.8C′, 50.5.9D). Likewise, after the third division, four binucleate cells (open at one end) or eight uninucleate cells (lower four cells open at one end) are formed (Figs. 50.5.8D, D′, 50.5.9D). Finally, however, the entire endosperm becomes cellular. Due to fusion of nuclei in *Cephalaria gigantea* (Kamelena 1980), large polyploid multinucleolate nuclei are formed.

Morina kokanica and *M. lehmanniana* show an entirely different mode of development (Fig. 50.5.8E–I). The division of primary endosperm nucleus is followed by a transverse wall in the equatorial plane. After the next division, an oblique wall is laid down in the upper cell, and a transverse one in the lower cell. The third division gives rise to eight cells, with the difference that the third cell from below may have an oblique or a vertical wall (Kamelena 1980).

Vijayaraghavan and Sarveshwari (1968) report ruminate endosperm with a sinuous outline in *M. longifolia,* and Kamelena (1980) in *M. kokanica* (Fig. 50.5.11D).

The development of the embryo has been studied in *Scabiosa columbaria* and *S. succisa* (Souèges 1937d, 1957), *S. lucida* and *S. ochroleuca* (Frey 1976), and *Cephalaria gigantea* (Fig. 50.5.10A–L). The nuclear division in the zygote is followed by an oblique vertical wall and the development corresponds to the Piperad (?) type. The embryogenesis conforms to the following plan:

$$\text{ca} = \tfrac{1}{2}\,\text{pco} + \tfrac{1}{2}\,\text{pvt} + \tfrac{1}{2}\,\text{phy}$$
$$\text{cb} = \tfrac{1}{2}\,\text{pco} + \tfrac{1}{2}\,\text{pvt} + \tfrac{1}{2}\,\text{phy} + \text{icc} + \text{iec} + \text{co} + \text{s}$$

In *Morina longifolia* (Vijayaraghavan and Sarveshwari 1968), the division in the zygote is invariably transverse and the development of the embryo

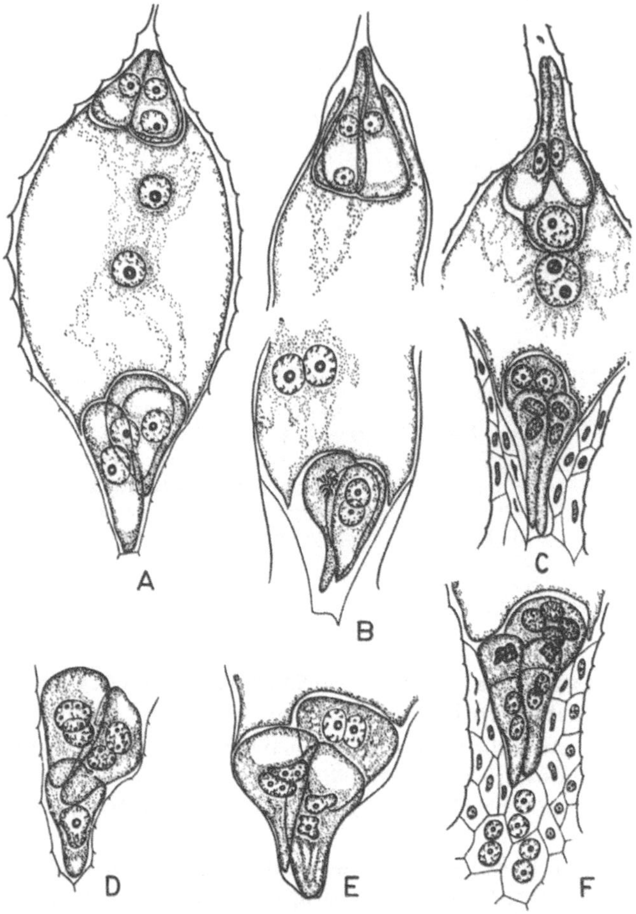

Fig. 50.5.6A–F. **A, B.** *Scabiosa rotata.* **C** *S. olivierii.* **D** *S. lucida.* **E, F** *S. bipinnata.* **A–C** Mature embryo sac, upper and lower part in **B** and **C**, antipodals uninucleate in **A** and **B**, one nucleus in division in **B**, antipodals two-nucleate in **C**. **D–F** Antipodals, one uni- and two binucleate in **D**, all three binucleate in **E**, two bicelled (upper cell binucleate) in **F**. **E** shows fused (?) nuclei in two cells, and **F** shows four nuclei in one antipodal. (After Kamelena 1980)

corresponds to the Solanad type. In *M. kokanica* (Kamelena 1980) also the zygote divides transversely and embryogenesis conforms to the following plan:

$$ca = pco = pvt$$
$$cb = phy = icc + iec + co + s$$

The two- to four-seriate suspensor is conspicuous. In the mature embryo the cotyledons are foliaceous.

The albuminous seed (Fig. 50.5.11A–D) has a large chlorophyllous embryo (except in *Morina*). The ruminate endosperm is oily. The seed-coat is composed of the outer epidermis and hypodermis, and rest of the tissue becomes crushed.

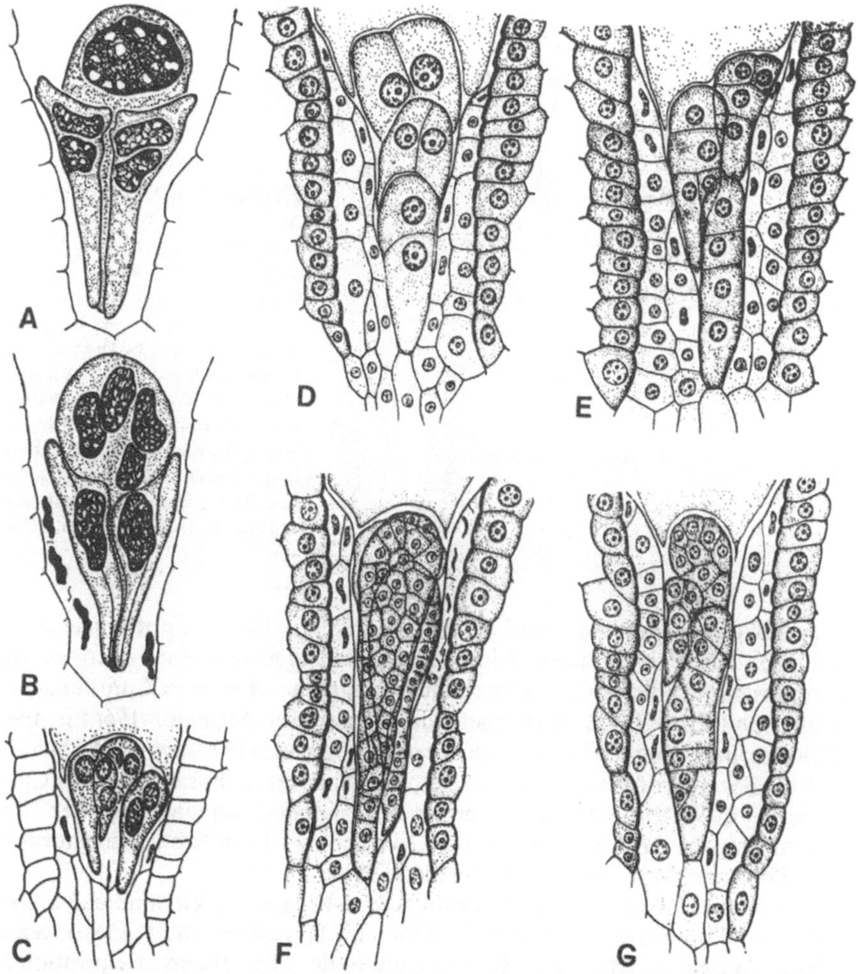

Fig. 50.5.7A, B. *Cephalaria gigantea*, antipodals with polyploid nuclei. **C** *Pterocephalus plumosis*, two antipodals binucleate, one bicelled and lower cell binucleate. **D**, **E** *M. kokanica*, **F**, **G** *M. lehmanniana*, antipodals with two cells each (**D**), four cells each (**E**), and multicelled (**F**, **G**). **D–G** Simulate proembryos. (After Kamelena 1980)

K Carniel (1963), EJH Corner (1976), L Frey (1976), OP Kamelena (1980, 1983), VA Poddubnaya-Arnoldi (1933), BA Razi and K Subramanyam (1952), R Souèges (1937d, 1957, 1962, 1963a,b), MJ Vieth (1958), MR Vijayaraghavan and Sarveshwari (1968)

50.6 Taxonomic Considerations

Kamelena (1980) compares the embryology of Dipsacaceae, Valerianaceae, Adoxaceae, Caprifoliaceae, Calyceraceae, Compositae, Rubiaceae and Acanthaceae (Table 50.6.1).

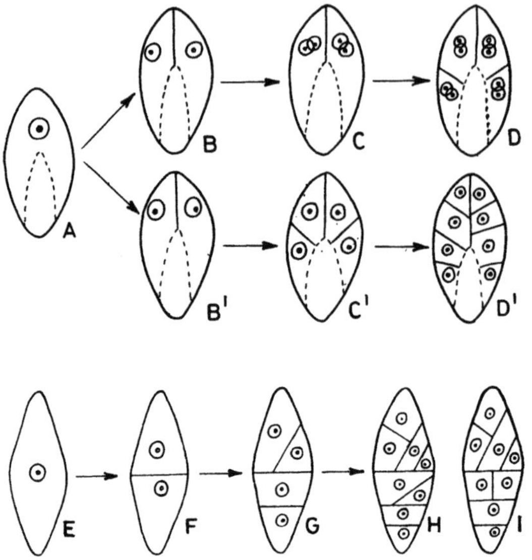

Fig. 50.5.8A–I. Development of the endosperm in Dipsacaceae. **A, B, B′** *Succisa pratensis*, **C** *Dipsacus dipsacoides*, *C′ Scabiosa rotata*, **D** *Cephalaria gigantea*, **D′** *C. brevipalea*, **E–I** *Morina*. **A, E** Primary endosperm nucleus. **B–D** Division of nucleus followed by an incomplete wall, next division without wall formation, and third division followed by wall formation. **B′–D′** Every division followed by a wall. **E–I** Two-, four- and eight-celled endosperm. (After Kamelena 1980)

Caprifoliaceae. Bentham and Hooker (1965b) place Caprifoliaceae in Rubiales, and Valerianaceae, Dipsacaceae and Calyceraceae in Asterales with Compositae (Table 50.6.1). They include the genus *Adoxa* in Caprifoliaceae. Engler and Diels (1936) treat the order Dipsacales (Melchior 1964) in their Rubiales, including the family Rubiaceae. Hutchinson (1973) groups Rubiaceae in Rubiales, Caprifoliaceae in Araliales, Valerianaceae, Dipsacaceae and Calyceraceae in Valerianales, and derives Valerianales from Saxifragales. The relationship of Rubiaceae is discussed in Sect. 47.9, and the segregation of Caprifoliaceae from Rubiales is justified (Table 50.6.1).

Cronquist (1981) includes Caprifoliaceae, Adoxaceae, Valerianaceae, Dipsacaceae (Dipsacales of Melchior 1964) in Dipsacales, and Calyceraceae (Campanulatae of Melchior 1964) in Calycerales. The systematic position of Calyceraceae is treated along with Campanulatae (Sect. 51.10). Cronquist (1981) recognizes Caprifoliaceae as a primitive, and Adoxaceae as a specialized family. On serological evidence (Hildebrand and Fairbrothers 1970a, b), Caprifoliaceae are related to Cornaceae through *Sambucus*. Caprifoliaceae (including *Sambucus*) and Cornaceae have secretory anther tapetum, tricolporate pollen, anatropous, unitegmic ovules with Polygonum type of embryo sac. The development of the Adoxa type in *Sambucus* occurs in only 58% of ovules, the rest show variations and, according to Tokć (1981), the development of the embryo sac is not stable in *Sambucus*. In Cornaceae, monosporic as well as tetrasporic type occurs. The development of the endosperm is of the Cellular type, and embryogeny of the Asterad type.

Sambucus is morphologically and anatomically an isolated member of the Caprifoliaceae (Fukuoka 1972, see also Wagenitz 1977b). It has been placed in Caprifoliaceae in most systems of classification; R. Dahlgren (1980b) gives it an alternate position in the iridoid-containing order Cornales (a transfer from

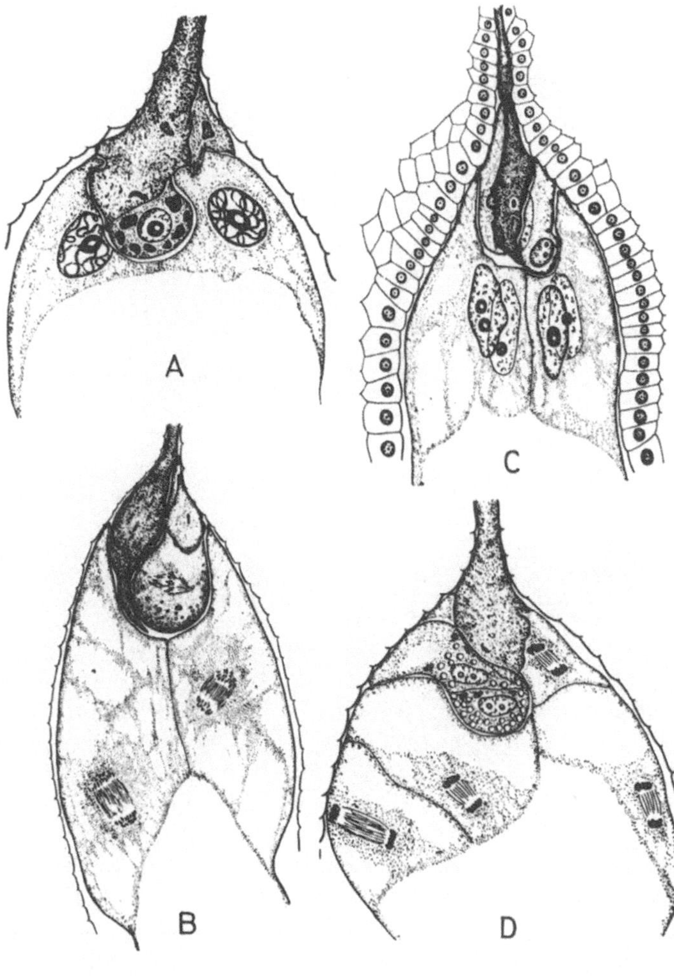

Fig. 50.5.9A–D. A, D *Scabiosa rotata.* **B** *Succisa pratensis.* **C** *Dipsacus dipsacoides.* **A** Zygote and two endosperm nuclei. Note remnants of synergids in **A–C**, and of pollen tube in **A–D**. **B** Zygote nucleus in division, and longitudinal incomplete wall separating two endosperm cells, both nuclei in division. **C** Zygote still undivided, both endosperm cells binucleate. **D** Bicelled proembryo, four-celled endosperm and each nucleus in division. (After Kamelena 1980)

Asteridae to polypetalous families). Wagenitz (1977b) favours its retention in Asteridae.

The available embryological data – tendency in the reduction of nucellar tissue, anatropous, unitegmic ovule with endothelium, Cellular endosperm, and Asterad type of embryogeny – are inadequate to justify the position of Caprifoliaceae in the Araliales (Hutchinson 1973). Further detailed investigations are necessary to decide the relationships.

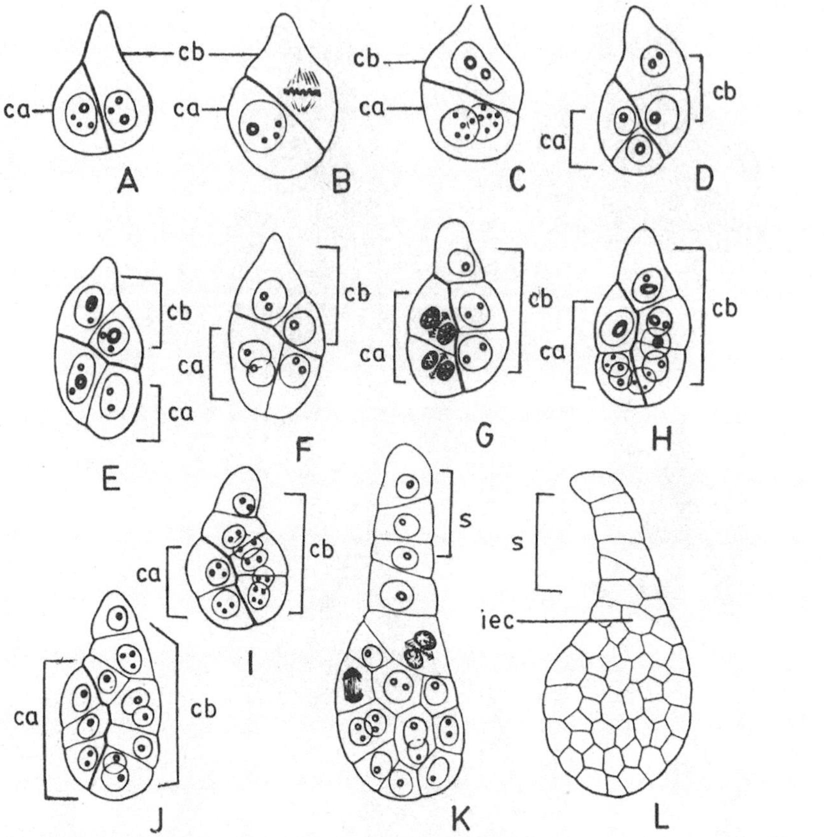

Fig. 50.5.10A–L. *Cephalaria gigantea.* Embryogeny of Piperad type. *s* suspensor (After Kamelena 1980)

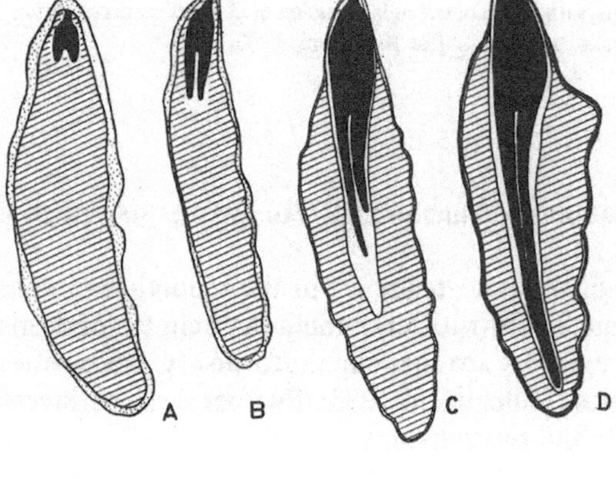

Fig. 50.5.11A–D. *Morina kokanica.* Development of seed, embryo *black*, endosperm obliquely hatched, seed-coat *dotted* (*bold line* in **D**). The endosperm is ruminate in **B–D**, and the cotyledons are much longer than the hypocotyl in **D**. (After Kamelena 1980)

Adoxaceae. The genus *Adoxa*, or the monogeneric family Adoxaceae, is placed in Saxifragaceae (Gundersen 1950), in Saxifragales (Hutchinson 1973), Cornales (R. Dahlgren 1980b), Rubiales (Wettstein 1935, Engler and Diels 1936), and Dipsacales (Wagenitz 1959, 1977b, Melchior 1964, Cronquist 1968, Thorne 1968, Takhtajan 1980).

P. Maheshwari (1964) evaluates the systematic position of Adoxaceae. There is a similarity of floral structure between Adoxaceae and Caprifoliaceae, and tetrasporic embryo sac occurs in *Adoxa* and *Sambucus* (Moissel 1941). The embryogeny in Caprifoliaceae (Souèges 1959b) also supports their alliance. Therefore, P. Maheshwari (1964) does not approve the assignment of Adoxaceae to Saxifragales, and favours its retention, as a separate family, near Caprifoliaceae.

In addition, features such as amoeboid anther tapetum in *Adoxa*, *Lonicera* and *Symphoricarpus* (Caprifoliaceae), tricolpate, three-celled pollen grains with indistinct endocolpium (P.K.K. Nair 1970), anatropous, unitegmic, tenuinucellate ovule, and Cellular type of endosperm indicate the close relationship between Adoxaceae and Caprifoliaceae.

Adoxaceae is appropriately placed in the Dipsacales (Table 50.6.1), and its transfer to the Cornales, along with the Sambucaceae (as proposed by R. Dahlgren 1977b, 1980b), needs further study.

Valerianaceae and Dipsacaceae. The families Valerianaceae and Dipsacaceae are closely allied. Both have amoeboid anther tapetum, three-celled pollen grains, and similar ovule, embryo sac and endosperm. Valerianaceae show the Asterad type of embryogeny as in Caprifoliaceae and Adoxaceae, whereas it is of the Piperad or Solanad type in Dipsacaceae. The seeds in Valerianaceae are exalbuminous; albuminous in other members of Dipsacales (Table 50.6.1).

Dipsacaceae also show an integumentary vascular strand, and endosperm with sinuous outline (*Morina*). Ruminate endosperm occurs in Caprifoliaceae (*Viburnum*).

Morina. *Morina* differs from the rest of the members of Dipsacaceae in several important respects (Table 50.6.1): (a) two persistent middle layers in anther wall, (b) outer layer develops fibrous thickenings, (c) tapetal cells two-nucleate, (d) pollen grains are three-armed, (e) the ovular integument is unvascularized, (f) the micropyle is very short, (g) the embryo sac lies close to the micropyle, (h) the three antipodals undergo internal divisions and each becomes multicelled, (i) the division of primary endosperm nucleus is followed by a transverse (complete) wall, (j) the endosperm is ruminate, (k) the zygote divides by a transverse wall, (l) Solanad type of embryogeny, and (m) non-chlorophyllous embryo.

Most taxonomist include *Morina* in Dipsacaceae. Takhtajan (1980) has erected an independent family, Morinaceae, which is fully justified on the embryological characters enumerated above (see also Vijayaraghavan and Sarveshwari 1968, Kamelena 1980).

Table 50.6.1. Embryology of Dipsacaceae and "related" families (Adapted from Vijayaraghavan and Sarweshwari 1968, Kamelena 1980, see also Batygina and Yakovlev 1987)

	Dipsacaceae	*Morina*	Valerianaceae	Adoxaceae
Anther	Wall 4-layered, tapetum secretory, cells multinucleate and pseudo-amoeboid, periplasmodium reported in some taxa, cytokinesis simultaneous, tetrads tetrahedral, isobilateral and decussate, pollen 3-celled	Wall 5-layered, tapetum amoeboid, cells binucleate, cytokinesis simultaneous, tetrads tetrahedral, pollen 3-celled	Wall 4-layered, tapetum amoeboid, cells multinucleate, cytokinesis simultaneous, tetrads tetrahedral, pollen 3-celled	Wall 4-layered, tapetum amoeboid, cells 2-nucleate, cytokinesis simultaneous, tetrads tetrahedral, pollen 3-celled
Ovule	Anatropous, unitegmic, tenuinucellate, integument vascularized, micropyle long and narrow, endothelium differentiates	Anatropous, unitegmic, tenuinucellate, integument unvascularized, chalaza massive, endothelium differentiates	Anatropous, unitegmic, tenuinucellate, integument vascularized, endothelium differentiates	Anatropous, unitegmic, tenuinucellate, endothelium differentiates
Embryo sac	Polygonum type, lies close to chalazal region, synergids ephemeral, antipodals 3, each 2–4-nucleate, become highly polyploid	Polygonum, lies close to micropyle, antipodal cells 3, each multicelled with uninucleate daughter cells, polyploidy absent, simulate proembryos	Polygonum (rarely tetrasporic in *Valeriana*), antipodals 3, each cell 2–4-nucleate, persistent	Adoxa, antipodals 3, one larger than others

Caprifoliaceae	Calyceraceae	Compositae	Rubiaceae	Acanthaceae
Wall 4-layered, tapetum secretory in some taxa, amoeboid in others, cells multinucleate, cytokinesis simultaneous, tetrads tetrahedral and decussate, pollen 3-celled	Wall 4-layered, tapetum secretory, pollen 2-celled	Wall 4-layered, tapetum secretory in some taxa, amoeboid in others, cells 2- to multinucleate, cytokinesis simultaneous as well as successive, tetrads of all types, polyspory occurs, pollen usually 3-celled	Wall 5-layered, tapetum secretory, cells uninucleate, cytokinesis simultaneous, tetrads tetrahedral, isobilateral and decussate, pollen 2- or 3-celled; pollen buds reported in some taxa	Wall 4-layered, tapetum secretory, cells 2–4-nucleate, endothecium lacks thickenings in some taxa, cytokinesis simultaneous, tetrads tetrahedral, isobilateral and decussate, polyspory frequent, pollen grains polymorphic, 2-celled
Anatropous, unitegmic, tenuinucellate, crassi- to tenuinucellate in *Viburnum*, integument vascularized, endothelium differentiates	Anatropous, unitegmic, tenuinucellate, endothelium differentiates	Anatropous, unitegmic, tenuinucellar, integument vascularized, endothelium differentiates and shows much variation	Anatropous, unitegmic, tenuinucellate, reduction of nucellus, endothelium differentiates, aril and obturator reported	Ana-, hemiana- or campylotropous, unitegmic, tenuinucellar, endothelium differentiates, funicular obturator reported, jaculator well-developed in some taxa
Polygonum, Allium and Adoxa type, antipodals 3, ephemeral, many irregularities occur; nucellar aposporous embryo sacs develop in *Sambucus*	Polygonum, antipodals ephemeral	Polygonum, synergid and antipodal haustoria reported in some taxa, supernumerary and multinucleate antipodals occur in some taxa, Allium, Adoxa, Drusa, Peperomia, Fritillaria and Chrysanthemum type of embryo sacs also develop; apomixis common	Polygonum, the lowest antipodal forms a conspicuous tubular extension; Allium and Peperomia type in some taxa	Polygonum, sometimes extends into the micropyle, a chalazal caecum develops, antipodals usually ephemeral, sometimes 2-nucleate, and up to 14

Table 50.6.1. (*Contd.*)

	Dipsacaceae	*Morina*	Valerianaceae	Adoxaceae
Endosperm	Cellular type, division of primary endosperm nucleus longitudinal but wall incomplete towards chalaza, entire endosperm becomes cellular, rumination absent	Cellular, division of primary endosperm nucleus transverse, entire endosperm becomes cellular, ruminate (false rumination in *Morina kokanica*)	Cellular, in *Kentranthus macrosiphon* division of primary endosperm nucleus vertical	Cellular, primary endosperm nucleus divides vertically, next division also vertical forming 4 cylindrical cells, next two divisions transverse and later divisions irregular
Embryo	Piperad type – *scabiosa* variation (*Cephalaria*), chlorophyllous	Solanad (*Morina*), non-chlorophyllous	Asterad type, chlorophyllous	Asterad, synergid polyembryony occurs
Seed	Exotestal, albuminous, stores oil, cotyledons long, epi- and hypodermis of integument produce testa	Exotestal, albuminous, cotyledons long and tapering	Exalbuminous, embryo straight, testa single-layered and unspecialized	Albuminous, stores oil, embryo minute, testa single-layered and thin-walled

51 Campanulatae

Campanulatae comprise eight families: Campanulaceae, Sphenocleaceae, Pentaphragmataceae, Goodeniaceae, Brunoniaceae, Stylidiaceae, Calyceraceae and Compositae.

Caprifoliaceae	Calyceraceae	Compositae	Rubiaceae	Acanthaceae
Cellular, division of primary endosperm nucleus vertical or oblique, later divisions irregular, ruminate in *Viburnum*	Cellular, primary endosperm nucleus divides by a transverse wall, then another transverse wall forming a row of 4 cells, later divisions longitudinal and irregular	Cellular mostly, Nuclear type in some taxa, nucleate and enucleate cytoplasmic nodules reported; 2–3 layers persist and envelop the embryo, cells contain proteins and fatty oils	Cellular mostly, Nuclear type in some taxa, enucleate cytoplasmic nodules reported; ruminate	Cellular, micropylar and chalazal haustoria develop, endosperm development shows much variation
Asterad type, polyembryony occurs in *Sambucus*, nonchlorophyllous	Solanad, well-differentiated	Asterad, polyembryony occurs in *Tridax*, integumentary in *Melampodium* and *Carthamus*	Solanad, suspensor uniseriate, nonhaustorial, more often of multiseriate vesicular cells, haustorial, in *Asperula* – some basal cells bulge out into the endosperm	Onagrad, Solanad, Caryophyllad type, suspensor polyembryony frequent
Albuminous, embryo small and linear, testa composed of 2 or 3 layers	Albuminous, embryo straight	Exotestal, exalbuminous or only a few layers of endosperm persist, embryo straight	Exotestal, strophiolate, vascular strand extends up to the micropyle in pachychalazal seed, albuminous, embryo small to large; seed often covered with placental outgrowths	Exotestal, exalbuminous, embryo straight, some seeds develop epidermal hairs, a jaculator develops from the funicle, pericarp gives protection to the seed

51.1 General Features

Companulatae have the following embryological features: (1) fibrous endothecium, and secretory anther tapetum; amoeboid in Compositae, (2) two- or three-celled pollen grain, (3) anatropous, unitegmic, tenuinucellate ovule with endothelium, (4) Polygonum type of embryo sac; Polygonum, Allium, Peperomia, Drusa, Fritillaria and Adoxa type in Compositae, (5) Cellular type of endosperm; both Cellular and Nuclear type in Compositae, (6) endosperm haustoria in Campanulaceae, Sphenocleaceae, Pentaphragmataceae and Stylidiaceae, and (7) Solanad type of embryogeny in Campanulaceae, Pentaphragmataceae, Goodeniaceae and Stylidiaceae; Onagrad in Sphenocleaceae; Asterad in Calyceraceae and Compositae.

51.2 Campanulaceae

The anther wall comprises the epidermis, fibrous endothecium, single ephemeral middle layer, and secretory tapetum of binucleate cells in *Isotoma longiflora* (Kausik and Subramanyam 1945b). The microspore mother cells undergo meiotic divisions, cytokinesis is simultaneous, and the tetrads are tetrahedral and isobilateral.

The pollen grains are three- or four- to ten-colpate, colporoidate or colporate with smooth or spinuliferous exine in Campanulioideae, three- or four-colporate in Cyphioideae, and two-, three- or four-colporoidate with a smooth exine in Lobelioideae (Erdtman 1952). They are two-celled at the dispersal stage in *Downingia* (Kaplan 1969a), and three-celled in *Isotoma longiflora* (Kausik and Subramanyam 1945b) and *Wahlenbergia bicolor* (Want 1963). In *Lobelia trialata* (Kausik and Subramanyam 1945a) and *L. pyramidalis* (Subramanyam 1949), the pollen grains occasionally germinate in situ.

The ovule is anatropous, unitegmic, and tenuinucellate. The archesporium is uni- or multicellular (*Campanula carpathica* – Rosén 1932). A single cell functions directly as megaspore mother cell, undergoes meiotic divisions, forms a linear, T-shaped or oblique tetrad, and the chalazal megaspore develops into a Polygonum type of embryo sac. The synergids are hooked, polar nuclei fuse before fertilization, and antipodals are ephemeral. Starch grains occur in the mature embryo sac. In *Lobelia nicotianaefolia* (*L. heyniana*) (Kausik 1938e) and *L. pyramidalis* (Subramanyam 1949), the embryo sac produces a small chalazal caecum beyond the antipodals. The antipodals remain in situ. Erdelská (1975) describes the presence of a cuticle which covers the gametophyte in *Jasione montana*. An endothelium differentiates at the two-nucleate embryo sac in *Isotoma longiflora*.

The development of the endosperm is of the Cellular type (Fig. 51.2.1A–D), and Subramanyam (1970b) has given a detailed account. The development in Campanulaceae is further classified, according to the sequence of walls following early divisions in the endosperm, and the number of cells of the haustorium, into four types: Codonopsis, Phyteuma, Isotoma and Scutellaria. Rosén (1949) recognized only three basic types: Codonopsis, Phyteuma and Sphenoclea. The Sphenoclea type occurs in the family Sphenocleaceae (Sect. 51.3). The developmental sequence of all these types is represented in Figure 51.2.2A–F_1.

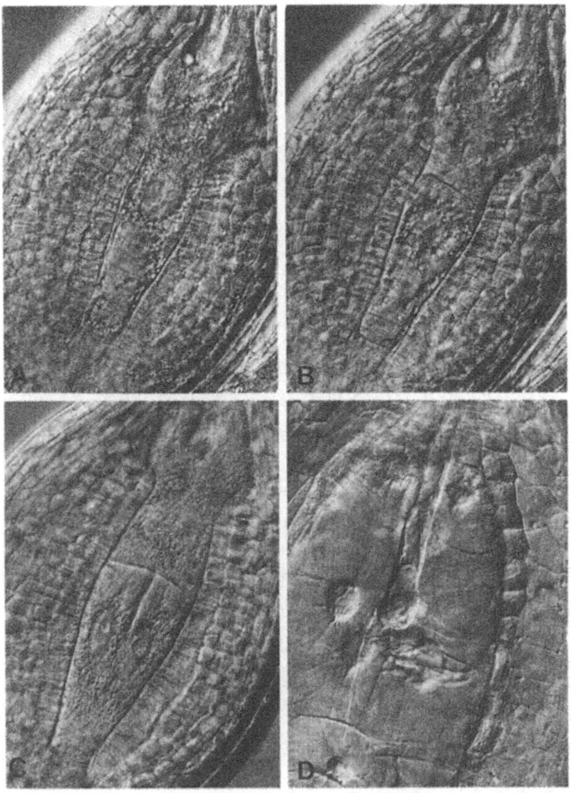

Fig. 51.2.1A–D. *Jasione montana*; **A** Embryo sac with primary endosperm nucleus; **B** Two-celled endosperm; **C** Four-celled endosperm; **D** Micropylar region of embryo sac with zygote and endosperm. Note the endothelium in **A–D**. All observations based on living material. **A–C** cine film E 1759/1971, **D** cine film E 1759/1971 — Encyclopaedia Cinematographica, IWF, Göttingen. (Courtesy Prof. Dr. O. Erdelská, Bratislava, Czechoslovakia)

1) Codonopsis type. The division of the primary endosperm nucleus (Fig. 51.2.2A) forms a micropylar and a chalazal chamber. Vertical divisions occur in both chambers (first in the micropylar and then in the chalazal) forming a four-celled endosperm (Fig. 51.2.2C). Transverse divisions occur in each cell of the micropylar chamber (Fig. 51.2.2D). The upper two cells develop into a micropylar haustorium, whereas the lower two contribute to the endosperm proper (Fig. 51.2.2D). The vertical wall which separates the two micropylar cells ends blindly in *Downingia bacigalupii* and *D. pulchella* (Kaplan 1969b). The cells of the chalazal chamber undergo one or two successive transverse divisions (Fig. 51.2.2E_1). The lower tier of two cells develops into a

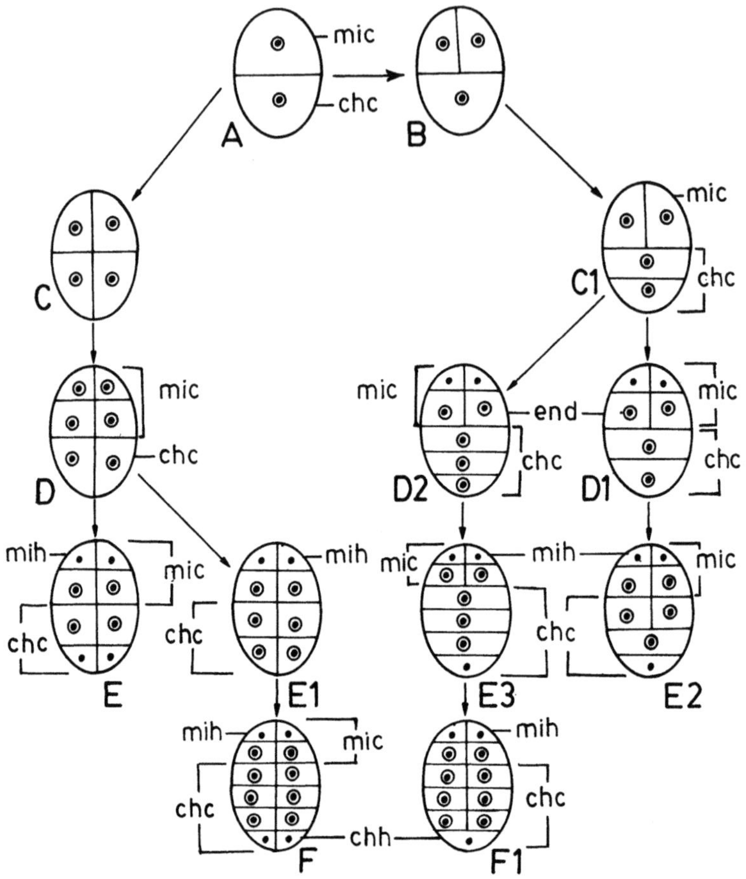

Fig. 51.2.2A–F. Development of the endosperm in Campanulaceae, Lobeliaceae, Sphenocleaceae and Stylidiaceae. *Black dots without circles* indicate the nuclei of haustorium, the tiers with proper nuclei (*black dots with circles*) produce the endosperm proper. Codonopsis type – **A, C, D, E1, F; A** Two-celled endosperm; *chc* chalazal chamber; *mic* micropylar chamber; **C** Vertical division in each chamber. **D** Further transverse division in *mic*. **E1** Uppermost tier of micropylar chamber develops into two-celled haustorium. Transverse division in *chc*. **F** Further transverse division in *chc*. The terminal tier at the chalazal end develops into two-celled haustorium. Four-tiers of initials of the endosperm proper. Sphenoclea type – **A, C, D, E** Development up to **D** as in Codonopsis type; **E** Transverse division in *chc*. Terminal tier develops into two-celled chalazal haustorium, and two tiers of initials of endosperm proper (Sect. 51.3). Isotoma type – **A, B, C1, D2, E3, F1. C1** Transverse division in *chc*; **D2** Transverse division in *mic* deliniates two-celled micropylar haustorium, another transverse division in *chc*; **E3** Another transverse division in *chc* separates a uninucleate chalazal haustorium, and initials (four tiers) of the endosperm proper; **F1** Vertical divisions in the derivatives of *chc* except the terminal cell, i.e. chalazal haustorium. Phyteuma type – **A–C1, D1, E2** Development similar to Isotoma type. **A, B, C1, D1** Transverse division in *mic*, the terminal tier forms two-celled micropylar haustorium; **E2** *mih* two-celled, transverse division in chalazal cell separates uninucleate chalazal haustorium, three tiers of endosperm proper initials. (Courtesy Dr M.A. Rau, Mysore)

chalazal haustorium, the other cells contribute to the formation of the cellular endosperm (Fig. 51.2.2F), e.g. *Cephalostigma schimperi* (Kausik and Subramanyam 1947a), *Downingia bacigalupii* and *D. pulchella* (Fig. 51.2.3A–J), and *Wahlenbergia bicolor* (Want 1963).

2) Phyteuma type. The division of the primary endosperm nucleus forms a micropylar and a chalazal chamber (Fig. 51.2.2A). Vertical divisions occur in

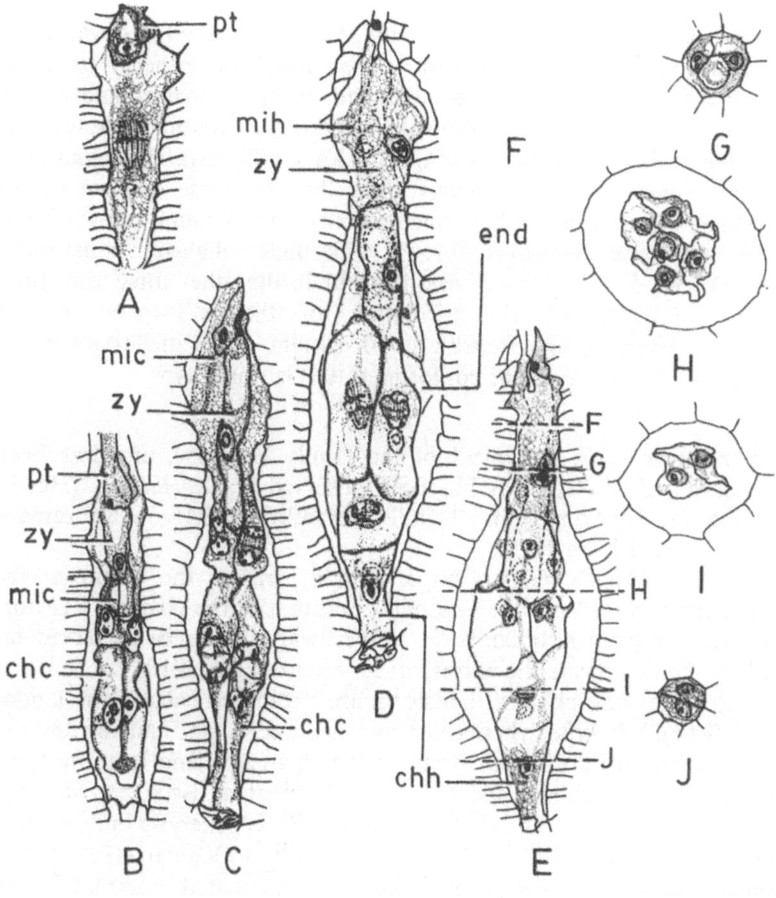

Fig. 51.2.3A–D. *Downingia pulchella.* **A** Embryo sac with zygote, primary endosperm nucleus in metaphase; **B** Zygote (*zy*), and two-celled micropylar (*mic*) and two-celled chalazal (*chc*) chamber. Remnants of pollen tube (*pt*) persist in **A–E**; **C** Transverse divisions in micropylar (*mic*) and chalazal chamber (*chc*); **D** Later stage, zygote (outline shown with *broken line* in **D, E**), endosperm proper (*end*), bicelled micropylar (*mih*) and bicelled chalazal haustorium. **E–J** *D. bacigalupii.* **E** Cellular endosperm, two-celled micropylar and two-celled chalazal haustorium; **F–J** Transections at levels marked in **E**; **F, G** Portion of suspensor encircled by micropylar haustorium; **H** Zygote encircled by four cells of endosperm proper; **I** Two cells of endosperm proper; **J** Two-celled chalazal haustorium. (After Kaplan 1969b)

the micropylar, and transverse in the chalazal chamber (Fig. 51.2.2B, C_1). The two cells of the micropylar chamber undergo transverse division. The upper tier develops into a two-celled micropylar haustorium, while the lower contributes to the formation of the endosperm proper (Fig. 51.2.2D_1). A vertical division occurs in the upper cell of the chalazal chamber, and the lower undergoes transverse division. The lowermost single cell develops into the uninucleate chalazal haustorium, e.g. *Isotoma longiflora* (Kausik and Subramanyam 1945b), *Laurentia tenella* (Rosén 1932), *Lobelia pyramidalis* and *L. cardinalis* (Subramanyam 1949, 1951c) (Fig. 51.2.2E_2).

3) Isotoma type. Kausik and Subramanyam (1947b) describe this deviation of the Phyteuma type in 50% of ovules of *Isotoma longiflora*, either in the same ovary or in different ovaries. The development up to four-celled endosperm (Fig. 51.2.2C_1), and further development in micropylar chamber, are exactly similar to that in the Phyteuma type (Fig. 51.2.2A, C_1, D_1, E_2). The deviation is that the two superposed cells, formed after the transverse division of the chalazal chamber, undergo further transverse divisions forming a row of four cells. The lowermost develops into a unicellular chalazal haustorium (Fig. 51.2.2D_2, E_3, F_1). Vertical divisions occur in the other three, and their derivatives contribute to the formation of the endosperm proper (Fig. 51.2.2F_1). Even though Beltran (1970) assigns the Scutellaria type to *Isotoma petrae*, the development conforms to the Isotoma type.

4) Scutellaria type. This is typical of the family Labiatae, and has been described earlier (Sect. 48.10). It has been observed in *Lobelia* spp. (Hewitt 1939, Kausik and Subramanyam 1945a, P. Maheshwari 1944a, b, Subramanyam 1949, 1951b, Beltran 1970).

In *Lobelia cardinalis*, G.O. Cooper (1942b) reports the origin of the micropylar haustorium from the synergids, and that of the chalazal haustorium from two or three antipodals. P. Maheshwari (1944a, b) corrected his observations, and confirms the endospermal origin of haustoria.

The micropylar and chalazal haustoria are two-celled each in the Codonopsis and Scutellaria type, whereas a two-celled micropylar haustorium and a one-celled chalazal haustorium occur in the Phyteuma and Isotoma type. The micropylar haustorium is located in a rich nutritive tissue, and remains functional for a long time. The chalazal haustorium is generally bulbous and small, not very aggressive in Lobelioideae. According to Kaplan (1969b): "the cells appear devoid of cytoplasm and are permeated completely by the meshwork of fibrous strands". ...... "Both the chalazal and micropylar haustoria remain as prominent components of the dormant seed of *Downingia*".

The embryogeny conforms to the Solanad type. The suspensor is long and filamentous. Polyembryony is reported in *Lobelia syphilitica* (Crété 1939), *Isotoma longiflora* (*Laurentia longiflora*) and *Wahlenbergia gracilis* (*W. marginata*) (Subramanyam 1947). It is due to the proliferation of suspensor cells in *Isotoma* and *Lobelia*.

The minute seed is albuminous with a small straight embryo and copious endosperm. The endosperm stores oil. It is starchy in *Cephalostigma* and *Isotoma*. The outer tangential wall of the outer epidermis of the testa is thickly cuticularized, the radial walls are lignified and pitted. The mesophyll collapses. The cells of the endothelium acquire thickenings, and accumulate ergastic substances (Kaplan 1969b). The chalazal and the micropylar haustorium appear as darkly staining structures with prominent nuclei, in the dormant seed (Kausik and Subramanyam 1945b, Kaplan 1969b).

IC Beltran (1970), G Bentham (1877), C Berger and O Erdelská (1973), GO Cooper (1942b), P Crété (1938, 1939, 1956f), R Dahlgren and A-m Lu (1985), O Erdelská (1969, 1975), O Erdelská and A Klasová (1978), G Erdtman (1952), WC Hewitt (1939), DR Kaplan (1969a, b), SB Kausik (1935b, 1938e), SB Kausik and K Subramanyam (1945a, b, 1946a, b, 1947a, b), P Maheshwari (1944a, b), W Rosén (1932, 1949), LD Safijowska (1935a, b), R Souèges (1936c, 1938a), K Subramanyam (1947, 1948b, 1949, 1951b, c, f, 1953, 1970b), MR Telesca (1962), G Want (1963)

51.3 Sphenocleaceae

The anther wall comprises the epidermis, fibrous endothecium, a single middle layer, and secretory tapetum of binucleate cells in *Sphenoclea zeylanica* (Kausik and Subramanyam 1946c). The microspore mother cells undergo meiotic divisions, cytokinesis takes place by furrowing, and the tetrads are tetrahedral, rarely isobilateral.

The tricolporate pollen grains are three-celled at the dispersal stage, in *S. zeylanica* (Kausik and Subramanyam 1946c).

The ovule is anatropous, unitegmic, and tenuinucellate. The single hypodermal archesporial cell functions as megaspore mother cell, undergoes meiotic divisions, forms a linear tetrad, and the chalazal megaspore develops into a Polygonum type of embryo sac. The synergids are elongated with pointed apices, polar nuclei fuse before fertilization, and the antipodals persist up to fertilization. An endothelium differentiates.

The development of the endosperm is of the Cellular type, and conforms to the "Modified" Scutellaria type. After the formation of the eight-celled endosperm according to the Scutellaria type, vertical walls are laid down in each cell so as to form four superposed tiers, each of four cells. The cells in the central region produce the endosperm proper. The terminal tiers develop into the micropylar and the chalazal haustorium, and each comprises four large uninucleate cells. Both the haustoria are aggressive, lie in close vicinity to the nutritive tissue, and remain functional for a fairly long time (Kausik and Subramanyam 1946c).

The embryogeny conforms to the Onagrad type (Swamy and Padmanabhan 1961c). The suspensor is short.

The dwarf seed contains a straight dicotyledonous embryo, and one or two layers of endosperm. The seed-coat comprises two to three cell layers. The inner walls of the epidermal cells are thickened with radial spinelike outgrowths, the outer walls are thin. The remnants of the micropylar and the chalazal haustorium persist in dormant seed (Subramanyam 1970c).

HK Airy Shaw (1948), SB Kausik and K Subramanyam (1946c), W Rosén (1949), K Subramanyam (1950a, 1970c), BGL Swamy and K Padmanabhan (1961c)

51.4 Pentaphragmataceae

The anther wall comprises the epidermis, fibrous endothecium, two middle layers, and secretory tapetum of binucleate cells in *Pentaphragma horsfieldii* (Kapil and Vijayaraghavan 1962, 1965). The microspore mother cells undergo meiotic divisions, cytokinesis is simultaneous, and tetrads tetrahedral and decussate.

The tricolporate pollen grains are two-celled at the dispersal stage.

The ovule is anatropous, unitegmic, and tenuinucellar. The archesporial cell functions as megaspore mother cell, undergoes meiotic divisions, produces a linear or T-shaped tetrad, and the chalazal megaspore develops into a Polygonum type of embryo sac. At the four-nucleate stage, the embryo sac extends beyond the micropyle. The synergids are hooked, polar nuclei fuse before fertilization, and antipodals degenerate after fertilization. The endothelium differentiates at the eight-nucleate embryo sac, and encircles its lower half.

The development of the endosperm is of the Cellular type. The division of the primary endosperm nucleus separates a small micropylar and a large chalazal chamber. The former develops into a prominent micropylar haustorium with a single hypertrophied nucleus. The chalazal chamber undergoes a few vertical divisions followed by transverse divisions producing the cellular endosperm proper. A chalazal haustorium is absent (Kapil and Vijayaraghavan 1965).

The embryogeny conforms to the Solanad type.

The albuminous seed contains a dicotyledonous straight embryo (Hutchinson 1973). The seed-coat consists of a single layer of cells. Lignified, pitted thickenings develop along their inner tangential and radial walls (Subramanyam 1970d).

HK Airy Shaw (1941, 1954), A Dunbar (1978), J Hutchinson (1973), RN Kapil and MR Vijayaraghavan (1962, 1965), SB Kausik and K Subramanyam (1946c), K Subramanyam (1950a, 1970c, d), BGL Swamy and D Padmanabhan (1961c)

51.5 Goodeniaceae

The anther wall comprises the epidermis, fibrous endothecium, one or two ephemeral middle layers, and secretory tapetum of multinucleate cells. The nuclei of the tapetal cells fuse in *Scaevola frutescens* (Vijayaraghavan and Malik 1972). Calcium oxalate crystals occur in the endothecial cells and middle layers. The microspore mother cells undergo meiotic divisions, cytokinesis is simultaneous, and tetrads tetrahedral and isobilateral.

The triporate pollen grains are two-celled at the dispersal stage. In *Leschenaultia* (Martin and Peacock 1959), however, they remain attached in tetrads.

The ovule is anatropous, unitegmic, and tenuinucellar (Fig. 51.5.1A, B). A hypostase is prominent in *Scaevola frutescens* (Vijayaraghavan and Malik

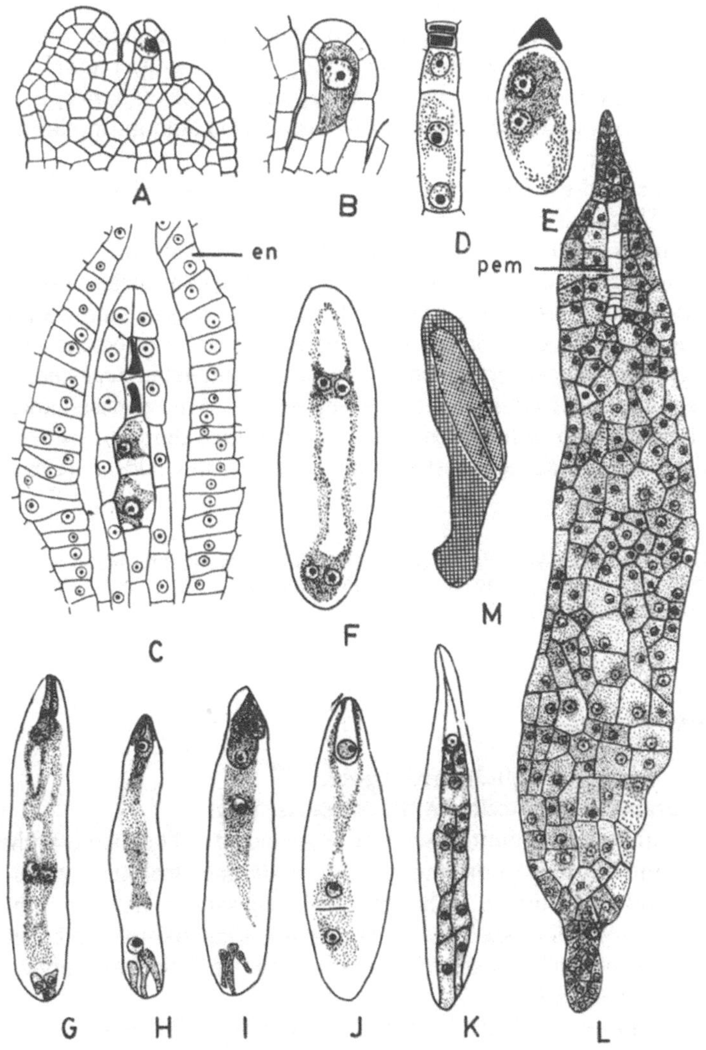

Fig. 51.5.1A–M. *Scaevola frutescens.* **A, B** Longisection of nucellus to show hypodermal archesporial and megaspore mother cell. **C** Linear tetrad with two degenerated megaspores. The endothelium (*en*) has differentiated. **D–F** Two- and four-nucleate embryo sac. **G** Mature embryo sac. **H, I** Embryo sac with zygote, degenerated synergids, antipodals, and primary endosperm nucleus. **J** Zygote and two-celled endosperm. **K** Zygote and multicelled endosperm. **L** Proembryo (*pem*), and advanced stage of endosperm. **M** Endosperm (*hatched*) at dicotyledonous embryo (*dotted*). (After Vijayaraghavan and Malik 1972)

1972). The endothelium is well-differentiated (Fig. 51.5.1C), but it does not enclose the extremities of the embryo sac in *Scaevola*. The vascular strand extends beyond the chalaza, to some extent, in *Anthotium*, *Dampiera*, *Leschenaultia*; there is a double strand in *Dampiera*. The single bundle extends up to the micropyle in *Calogyne*, *Catosperma*, *Diaspasis*, *Goodenia*, *Neogoodenia*, *Pentaptilon*, *Scaevola*, *Selliera*, *Velleia* and *Verreauxia* (Corner 1976).

The archesporium comprises three to four cells in *Leschenaultia biloba* (Rosén 1946). They function as megaspore mother cells, undergo meiotic divisions, and produce linear tetrads. Only a single tetrad is functional, and the chalazal megaspore develops into a Polygonum type of embryo sac (Fig. 51.5.1C–G). The polar nuclei fuse before fertilization, and antipodals are ephemeral (Fig. 51.5.1H, I).

The development of the endosperm is of the Cellular type. The haustoria are absent (Fig. 51.5.1J–M).

The development of the embryo conforms to the Solanad type.

The albuminous seed contains a straight embryo. The endosperm is oily; it may be copious or reduced to a single layer. The testa contains calcium oxalate crystals in *Scaevola*. The cells of the outer epidermis are thin-walled with reticulate thickenings. The few hypodermal layers are lignified in *Scaevola*; the other tissues collapse. In *Goodenia* the outer epidermis alone persists, and is a palisade of lignified cells with pitted walls. In *Anthotium* the radial walls of the epidermal cells are thickened. A strophiole is present in *Goodenia* (Corner 1976).

P Brough (1927), RC Carolin (1966), EJH Corner (1976), SB Kausik (1939b), RB Knox and E Friederich (1974), PG Martin and WJ Peacock (1959), W Rosén (1937, 1946, 1949), MR Vijayaraghavan and U Malik (1972)

51.6 Brunoniaceae

The embryology has not been fully investigated.

The pollen grains are two-celled at the dispersal stage.

The ovule is anatropous, unitegmic and tenuinucellar. The single archesporial cell functions as megaspore mother cell, undergoes meiotic divisions, and a linear tetrad is formed. The chalazal megaspore develops into a Polygonum type of embryo sac. The synergids are elongated and enter into the micropyle, polar nuclei fuse before fertilization, and antipodals are ephemeral. An endothelium differentiates.

The development of endosperm is of the Cellular type. The haustoria are absent (Rosén 1946).

A multinucleate suspensor haustorium develops in *Brunonia australis* (Rosén 1946).

The seed is exalbuminous; the embryo is straight with ovate cotyledons (Hutchinson 1973). The seed epidermis is composed of thin-walled more or less compressed cells (Corner 1976). As the pericarp collapses, the hardened calyx gives protection.

EJH Corner (1976), GL Davis (1966), J Hutchinson (1973), W Rosén (1946), JC Willis (1973)

51.7 Stylidiaceae

The anther wall comprises the epidermis, fibrous endothecium, an ephemeral middle layer, and secretory tapetum of binucleate cells in *Stylidium graminifolium* (Subramanyam 1951e). The epidermal cells enlarge to form a conspicuous stomium. The microspore mother cells undergo meiotic divisions, cytokinesis is simultaneous, and tetrahedral tetrads are produced.

The pollen grains are three- or four-colporate with spinuliferous exine. In *Donatia* they are three- or four-colporoidate. Occasionally, the pollen grains germinate in situ in *Levenhookia dubia*, *Stylidium graminifolium* and *S. tenellum* (Subramanyam 1950b, c, 1951e).

The ovule is anatropous, unitegmic, and tenuinucellar. A single archesporial cell functions as megaspore mother cell, undergoes meiotic divisions, and produces a linear or T-shaped tetrad. Occasionally, two archesporial cells develop into a twin dyad and tetrad. Only the chalazal megaspore develops into a Polygonum type of embryo sac. The endothelium (Fig. 51.7.1A–D) is strongly developed around the central, bulbous region of the embryo sac. The synergids are fairly long with tapering ends and are hooked, polar nuclei fuse before fertilization, and antipodals are ephemeral in *Forstera tenella* (W.R. and M.N. Philipson 1973). The mature embryo sac contains starch, is narrow at both ends, and in *Stylidium graminifolium* the chalazal end develops a small caecum. The chalazal region shows a "cytoplasmic" hypostase in *Donatia novae-zelandiae* (W.R. and M.N. Philipson 1973).

The development of the endosperm is of the Cellular type, and conforms to the Scutellaria type (Subramanyam 1970e). A transverse wall following the division of the primary endosperm nucleus results in a micropylar and a chalazal chamber in *Forstera tenella* (Fig. 51.7.1A). A vertical division occurs first in the micropylar (Fig. 51.7.1B) and then in the chalazal chamber (Fig. 51.7.1C). A transverse division occurs in each tier giving rise to an eight-celled endosperm. The terminal tiers develop into haustoria (Fig. 51.7.1D).

The aggressive two-celled micropylar haustorium invades the integument, and extends into the micropyle (Fig. 51.7.1E). The haustorial cells develop irregular lobes and processes which ramify between the integumentary cells (Fig. 51.7.1E–G).

The chalazal haustorium is also two-celled, and invades the chalazal end of the ovule. It remains much smaller than the micropylar haustorium (Fig. 51.7.1D, E).

In *Stylidium* the micropylar haustorium develops lateral processes which invade the integument, and persist up to late stages of seed development. The aggressive chalazal haustorium also develops a number of processes which ramify in the hypostase. Then they grow laterally, and protrude into the integumentary tissue. In *Levenhookia* these processes extend up to the epidermis of the chalazal region of young seed.

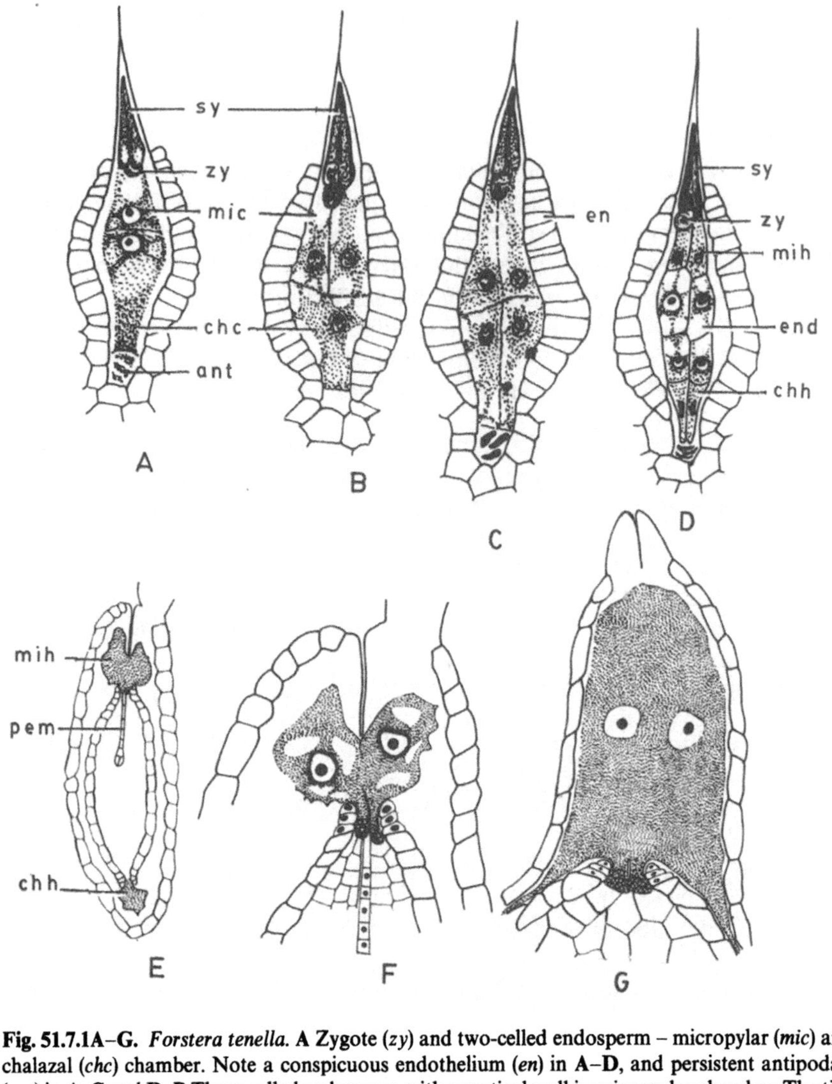

Fig. 51.7.1A–G. *Forstera tenella.* **A** Zygote (*zy*) and two-celled endosperm – micropylar (*mic*) and chalazal (*chc*) chamber. Note a conspicuous endothelium (*en*) in **A–D**, and persistent antipodals (*ant*) in **A, C** and **D. B** Three-celled endosperm with a vertical wall in micropylar chamber. The tips of synergids (*sy*) extend into the micropyle in **A–D. C** Endosperm with vertical walls in both chambers. **D** Zygote (*zy*), endosperm proper (*end*), two-celled micropylar (*mih*) and two-celled chalazal (*chh*) haustorium. **E** Longisection of ovule showing proembryo (*pem*), micropylar (*mih*) and chalazal (*chh*) haustorium (*stippled*); cellular endosperm omitted. **F** Micropylar haustorium from **E. G** Later stage. (After W.R. and M.N. Philipson 1973)

The embryogeny conforms to the Solanad type. The suspensor is filamentous with elongated cells in *Forstera tenella*, and disc-shaped in *Donatia*. The proembryo is linear in *Forstera* (Fig. 51.7.1E) and ovate in *Donatia*.

The small albuminous seed contains a small embryo and fleshy endosperm (Hutchinson 1973).

The outer epidermal cells of the integument and endothelium contain darkly staining material. Generally, only the outer epidermis persists as seed-coat and is sclerosed.

B Chandler (1911), P Crété (1951), J Hutchinson (1973), WR and MN Philipson (1973), W Rosén (1935), K Subramanyam (1950b, c, 1951d, e, f, 1970e)

51.8 Calyceraceae

The embryology has not been fully investigated.

The anther wall shows fibrous endothecium, and glandular tapetum in *Acicarpha tribuloides* (Juel 1915, see Schnarf 1931).

The pollen grains are two-celled at the dispersal stage.

The ovule is anatropous, unitegmic, and tenuinucellar. The archesporial cell functions as megaspore mother cell, undergoes meiotic divisions, and forms a linear tetrad. The chalazal megaspore develops into a Polygonum type of embryo sac. The polar nuclei fuse before fertilization, and antipodals are ephemeral.

The development of endosperm is of the Cellular type in *Acicarpha tribuloides* (Juel 1915) and *Calycera crenata* (K.V.O. Dahlgren 1923).

The seed is albuminous with a straight embryo and copious endosperm (Hutchinson 1973).

KVO Dahlgren (1915, 1923), GL Davis (1964 c), J Hutchinson (1973), HO Juel (1915), K Schnarf (1931)

51.9 Compositae

The anther wall comprises the epidermis, fibrous endothecium, an ephemeral middle layer, and tapetum of binucleate cells. The fibrillar thickenings of endothecium are absent in *Emilia sonchifolia* (Sunder Rajan 1968) and *Gerbera jamesonii* (Maheswari Devi 1957). The tapetal cells become bi-, tetra- or 12-nucleate in *Erigeron bonariensis* (H.P. Sharma and Murty 1978). Due to fusion, the nuclei become polyploid in *Flaveria repanda* (S. Misra 1957), niger and safflower (Maheswari Devi and Pullaiah 1976), and *Siegesbeckia orientalis* (S. Misra 1965). The polyploid nuclei are variously shaped in *Tithonia rotundifolia* (Pullaiah 1978a). The multinucleate tapetal cells (protoplasts) develop finger-like processes into the locule of the anther in *Emilia flammea* (Prakasa Rao et al. 1979). In *Carthamus tinctorius* (Chatterji and Jayaramu 1981), the tapetal nuclei exhibit endopolyploidy up to $28\times(2n = 24)$.

The tapetum is glandular in *Ainsliaea aptera* (Kapil and Sethi 1962a), *Emilia flammea* (Prakasa Rao et al. 1979), *Platycarpha carlinoides* (Ahlstrand 1979c), and *Sonchus arvensis* and *S. asper* (Kaul et al. 1975). True periplasmodium occurs in *Achillea squarrosa* (Pullaiah 1983b), *Cyathocline purpurea* (P.K. Deshpande and Kothare 1976), *Conyza stricta*, *Erigeron bonariensis*,

Felicia bergeriana (H.P. Sharma and Murty 1978), *Gundelia tournefortii* (Ahlstrand 1979c), *Macraea larcifolia* (Eliasson 1972), *Myriactis wightii* (L. Ram 1982) and *Sonchus deraceus* (Pullaiah 1982).

While studying the embryology of *Calotis lappulacea*, Davis (1968a) mentions an unusual behaviour of some tapetal cells, not reported so far in any other angiosperm. At the microspore tetrad formation, one or two tapetal cells enlarge and become vacuolate. They detach from the anther wall, enlarge considerably, and resemble developing embryo sacs. In one anther (in material collected from Cobar), an elongated tapetal cell simulated an embryo sac – complete egg apparatus, two antipodal cells (one binucleate), and an elongated three-nucleate polar cell. The rest of the pollen sac contained only plasmodial tapetal cells in one- or two-nucleate condition. In the corresponding pollen sac of the same anther, only two such embryo sac-like structures were observed along with well-developed periplasmodium and one-nucleate pollen grains. To us it appears that the resemblance to the embryo sac is very remote. Each cell (so-called three-celled egg apparatus, polar cell with an enlarged cell lying close to it, and two isolated antipodal cells) is an independent enlarged tapetal cell which has failed to form periplasmodium. What Davis refers to as the enlarged tapetal cell is essentially the protoplast.

Davis (1968a) refers to Ram's report of the origin of embryo sac-like structures, in *Leptomeria billardieri*, from microspore mother cells (actually sporogenous cells). She considers that the condition in *Leptomeria* may be the same as in *Calotis*. The investigations on *Leptomeria* were conducted under the supervision of one of us (BMJ), and the microscopic preparations were also examined by Professor J. Heslop-Harrison. We are convinced that in *Leptomeria* the embryo sac-like structures were indeed formed from sporogenous cells. This is an obligate apomict, and pollen grains do not ever develop in anthers.

The archesporium is a single vertical row of hypodermal cells. The microspore mother cells undergo meiotic divisions followed by both simultaneous and successive cytokinesis in *Conyza stricta* (H.P. Sharma and Murty 1978) and *Vernonia cinerea* (Dakshini and Dadlani 1978). The tetrads are tetrahedral, isobilateral, decussate and linear. One, two or three microspores of a tetrad reach maturity. Polyspory is reported in *Carthamus tinctorius* (Maheswari Devi and Pullaiah 1977) and *Podolepis jaceoides* (Davis 1961a). Davis (1964a) also noted compound four-nucleate pollen grains, due to failure of cytokinesis, in *Brachycome ciliaris*.

In *Tridax procumbens* (Banerji 1940b), large, multinucleate microspore mother cells were observed in the microsporangium of some anthers. As many as eight nuclei were present at interphase or metaphase. According to Banerji, their origin might be due to the fusion of two or more microspore mother cells.

The pollen grains are tricolpate with spinuous exine, and three-celled at the dispersal stage (Fig. 51.9.1A–F); two-celled in *Lecocarpus decaisne* (Eliasson 1971). Pollen polymorphism is frequent in *Melampodium divaricatum* (Maheswari Devi and Pullaiah 1976). There are four different sizes of pollen grains: smaller (12.9 μm) and medium size (15.8 μm) are non-functional, have thin exine and abort; larger pollen grains (21.6 μm) are functional and three-celled at the dispersal stage. Occasionally, the pollen grains are of still larger size,

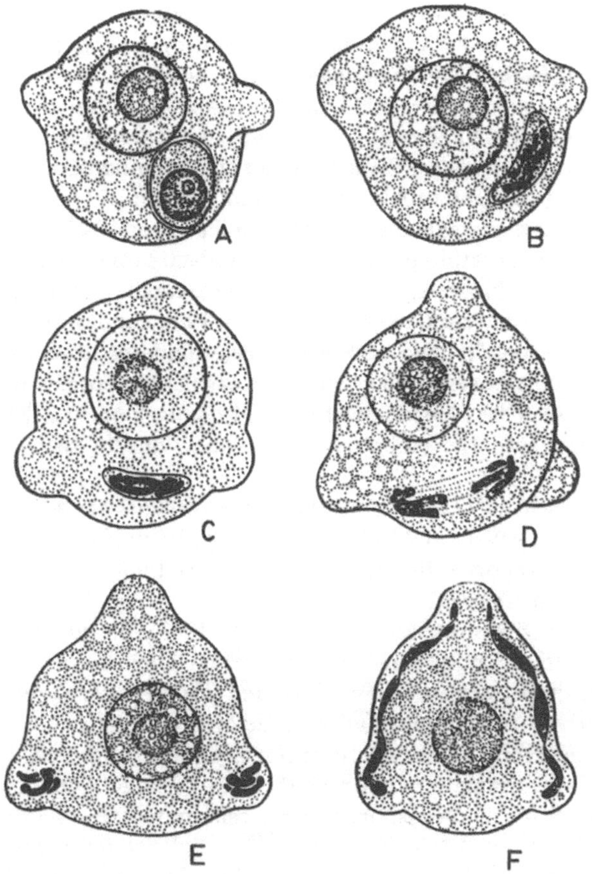

Fig. 51.9.1A–F. *Crepis tectorum.* **A** Two-celled pollen grain. **B–D** Generative cell in prophase (**B**, **C**) and telophase (**D**). **E**, **F** Three-celled pollen grains. (After Gerassimova-Navashina 1957, see Poddubnaya-Arnoldi 1982)

then become binucleate and abort. Pollen degeneration is a common feature in several members of Compositae.

The ovule is anatropous, unitegmic, and tenuinucellar. The funicular vascular strand passes into the integument in *Ainsliaea aptera* (Kapil and Sethi 1962a) and *Siegesbeckia orientalis* (S. Misra 1965). In *Tridax procumbens* (P. Maheshwari and Roy 1952), it reaches up to the tip of the micropyle. Integumentary vascular traces are also reported in *Cosmos sulphureus, Helianthus annuus, H. debilis, Xanthium strumarium* and *Zinnia pauciflora* (Maheswari Devi 1963). In *X. strumarium* the funicular vascular strand branches and rebranches, and ramifies all over the integument.

The endothelium differentiates at the tetrad stage in *Erigeron bonariensis* and *Felicia bergeriana* (H.P. Sharma and Murty 1978). The endothelial cells become binucleate (P.K. Deshpande 1962a, b, 1964a, b), and the nuclei fuse and attain polyploidy. The endothelium becomes biseriate in *Glossocardia bosvallea* (P.K. Deshpande 1962a) and *Zinnia angustifolia* (Pullaiah 1981), and multiseriate in *Berkheya, Gazania* (Ahlstrand 1979b), *Lagascea mollis* (Pullaiah 1981), *Siegesbekia* (S. Misra 1965), and *Tithonia rotundifolia* (Pullaiah 1978a).

Persistent endothelium in the mature seed has been observed in *Bidens biternata* and *Volutarella ramosa* (P.K. Deshpande 1964a, b), *Caesulia axillaris* (P.K. Deshpande 1960a), *Flaveria repanda* (P.K. Deshpande 1962b), *Galinsoga parviflora* and *Lagascea mollis* (Pullaiah 1981), *Glossocardia bosvallea* (P.K. Deshpande 1962a), and *Launaea nudicaulis* (Chikkannaiah and Hiremath 1982). In *Tridax procumbens*, Padmanabhan (1962b) noted that the endothelium persists only in the micropylar and chalazal region.

In *Melampodium divaricatum* (Maheswari Devi and Pullaiah 1976), the behaviour of the endothelium is rather peculiar. The endothelial cells divide in all planes resulting in a multilayered massive structure which eventually destroys the embryo sac, as in sunflower (Savchenko 1960) and *Flaveria repanda* (S. Misra 1964). Secondly, the endothelial cells undergo divisions and develop into adventive embryos which lack suspensor. Sometimes, the sequence of division of these cells is identical with that of zygotic embryos. The adventive embryos project into the embryo sac. At times, one of them is functional.

According to S. Misra (1965, 1972), the endothelium persists only as a layer of cuticle, and not as a cell-layer, in *Siegesbeckia orientalis* and *Verbesina encelioides*. Pullaiah (1979b) reported that it persists in the seed only as a non-cellular pellicle in *Blumea malabarica*.

In *Launaea nudicaulis* (Chikkannaiah and Hiremath 1982), the nutritive activity of the uni-, or biseriate endothelium is noteworthy. The cells around the endothelium differ in shape, size and stainability, and form a periendothelial zone. Its nutritive activity is at its peak at the time of fertilization.

The archesporium is generally unicellular, occasionally two- to eight-celled in *Launaea nudicaulis* (Chikkannaiah and Hiremath 1982). A single hypodermal archesporial cell functions as megaspore mother cell, undergoes meiotic divisions, and forms a linear or T-shaped tetrad. The chalazal megaspore develops into a Polygonum type of embryo sac (Fig. 51.9.2A). Twin tetrads have been observed in *Tridax procumbens* (P. Maheshwari and Roy 1952), and twin embryo sacs in *Guizotia abyssinica* (S. Chopra 1969).

The synergids are hooked, and project into the micropyle. In *Cotula australis* (Davis 1961c, 1962a), and occasionally in *Platycarpha carlinoides* (Ahlstrand 1979c), one of them becomes haustorial. The synergid haustorium is well developed, and faces the funiculus in *Ursinia* spp. (Ahlstrand 1978). In *Elephantopus scaber* (Pullaiah 1979a), the synergids as well as the antipodals (Fig. 51.9.2B–G) show haustorial development. One of the synergids persists up to the two-celled proembryo in *Tithonia* (Pullaiah 1978a) and *Eclipta prostrata* (P.K. Deshpande 1961). The polar nuclei fuse before fertilization.

The antipodals exhibit a good deal of variation. They are elongated and egg-like (see Fig. 51.9.4A–C) in *Rudbeckia bicolor* (P. Maheshwari and Srinivasan 1944). There may be two or three antipodals; when only two, one is binucleate as in *Siegesbeckia.* They persist and undergo secondary multiplication in several species, up to 51 in *Podolepis arachnoides* (Davis 1961b). The multinucleate condition is also very frequent. They may become one- to six-nucleate as in *Bidens pilosa, Coreopsis tinctoria* and *Cosmos sulphureus* (Maheswari Devi 1963), or 18- or 19-nucleate as in *Aster* (Oppermann 1904). The nuclei

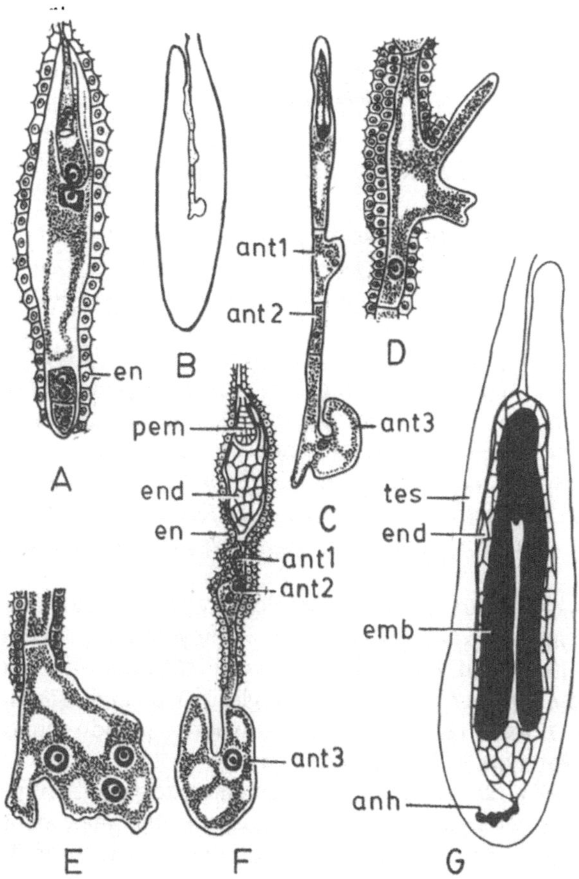

Fig. 51.9.2A–G. **A** *Vernonia elaeagnifolia*, mature embryo sac. *en* endothelium. **B–G** *Elephantopus scaber*. **B** Longisection of ovule, outline diagram for **C**; **C** Embryo sac with haustorial antipodals (*ant 1*, *ant 3*), the uppermost antipodal cell is binucleate; **D, E** Haustorial antipodal, three-nucleate condition in **E**; **F** Embryo sac at globular proembryo with cellular endosperm, and haustorial antipodals (*ant 1*, *ant 2*, *ant 3*); **G** Longisection of seed. Note the remnants of haustorial antipodal (*anh*); *tes* testa; *end* endosperm; *emb* embryo. (After Pullaiah 1979a)

may fuse and become polyploid as in *Tithonia rotundifolia* (Pullaiah 1978a). Antipodal haustoria have been reported in *Grindelia squarrosa* and *Prionopsis ciliata* (Howe 1926, 1975), and *Tridax procumbens* (Rogers 1969).

In *Chrysothamnus* spp. (Anderson 1970), there are two antipodal cells (one binucleate), or they become multinucleate (two to ten nuclei), or their number increases up to 16. A strong lateral antipodal haustorium develops in *C. viscidiflorus* ssp. *humilis*, and a weak one in *C. linifolius*. Frequently, the chalazal antipodal cell elongates. In *C. viscidiflorus* ssp. *lanceolatus* (Anderson 1970), the largest chalazal antipodal is over 200 μm. Soon after fertilization, the antipodals become inactive.

A large, empty space encircles the embryo sac in the region of antipodals in *Gazania* (Ahlstrand 1979b).

Bisporic development of the embryo sac occurs in *Ammobium alatum* (Avanzi 1948, Davis 1962b), *Chrysanthemum anethifolium*, *C. frutescens*, *C. jacobaeifolium*, *C. webbi* (Harling 1951a), *Erigeron* spp. (Holmgren 1919, Carano 1921, Chiarugi 1927a, b, Bergman 1942, Harling 1951b), and *Tridax procumbens* (Hjelmqvist 1951).

In *Tridax trilobata* (Hjelmqvist 1951), the failure of wall formation after the first meiotic division of the megaspore mother cell results in a binucleate cell. The second meiotic division is followed by walls, and a binucleate cell with two megaspore nuclei is separated in the middle. The other two megaspores lie on either side. The central binucleate cell develops into an eight-nucleate embryo sac. This mode of development is of an unusual bisporic type and needs to be confirmed.

Both Polygonum and Allium type of development have been observed in *Tridax trilobata* (Kapil and Sethi 1962b), *Sanvitalia procumbens* (Hjelmqvist and Holmberg 1961) and *Vernonia cinerascens* (B. Tiagi and Taimani 1960).

Bi- and tetrasporic development occurs in *Erigeron alpinus* (Chiarugi 1927b) and *E. frigidus* (Harling 1951b). *Othake texanum* (Asplund 1978) exhibits a rare combination of mono- and tetrasporic type in fertile ovules. In *Anthemis tinctoria* (Bijok et al. 1977), the development is tetrasporic.

Mono-, bi- and tetrasporic types occur in *Erigeron* spp. (Fig. 51.9.3). Monosporic (Polygonum) and tetrasporic types are reported in *Anacyclus radiatus* (Pogliani 1966) and *Chrysanthemum leucanthemum* (Martin and Smith 1955).

Tetrasporic development occurs in several species: Pyrethrum parthenifolium type in *Chrysanthemum maximum* (Ervandyan 1975), Fritillaria type in *Ratibida tagetes* (Howe 1964) and *Rudbeckia bicolor* (P. Maheshwari and Srinivasan 1944), and Drusa type in *C. viscosum* (Battaglia 1951b) and *Minuria denticulata* (Davis 1963). Borgen (1971) reports tetrasporic, pseudo-monosporic, triphasic and bipolar type in *C. ferulaceum*.

Capineri (1966) describes a rare mode of development of tetrasporic type. In *Anthemis montana*, in a tetranucleate coenomegaspore, the micropylar megaspore nucleus undergoes three successive mitotic divisions to produce eight nuclei, while the other three nuclei persist undivided at the chalazal end. The latter may fuse, or remain separate, or only one, two, or three, or all the megaspore nuclei divide, so that a whole series of mature gametophytes with a different number of nuclei is produced.

It would be very profitable to re-examine embryo sac development in such Compositae as show more than one type of development, and determine the percentage of each type.

The entry of the pollen tube is porogamous, and double fertilization takes place. The remnants of the pollen tube persist until the globular proembryo. Sometimes, two pollen tubes enter the embryo sac in *Siegesbeckia*, but one fails to burst (S. Misra 1965). The sperm cells lose the cytoplasmic sheath in one of the synergids in merigold (*Calendula*) (Belyaeva 1975). In a rare case, polyspermy has been observed in *Melampodium*.

The Cellular type of development of the endosperm generally occurs in the family, as in *Arctotis fastuosa* (Ahlstrand 1979a), *Adenostemma lavenia*, *A. rugosum* (Pullaiah 1979a), *Blainvillea rhomboidea* (Sunder Rajan 1972), *Berkheya* and *Gazania* (Ahlstrand 1979b), *Blumea laciniata* (Banerji 1942b), *B. membranacea* and *Laggera pterodonta* (Pullaiah 1979b), *Caesulia axillaris* (P.K. Deshpande 1962c), *Cotula australis* (Davis 1962a), *Eclipta erecta* (Bhargava

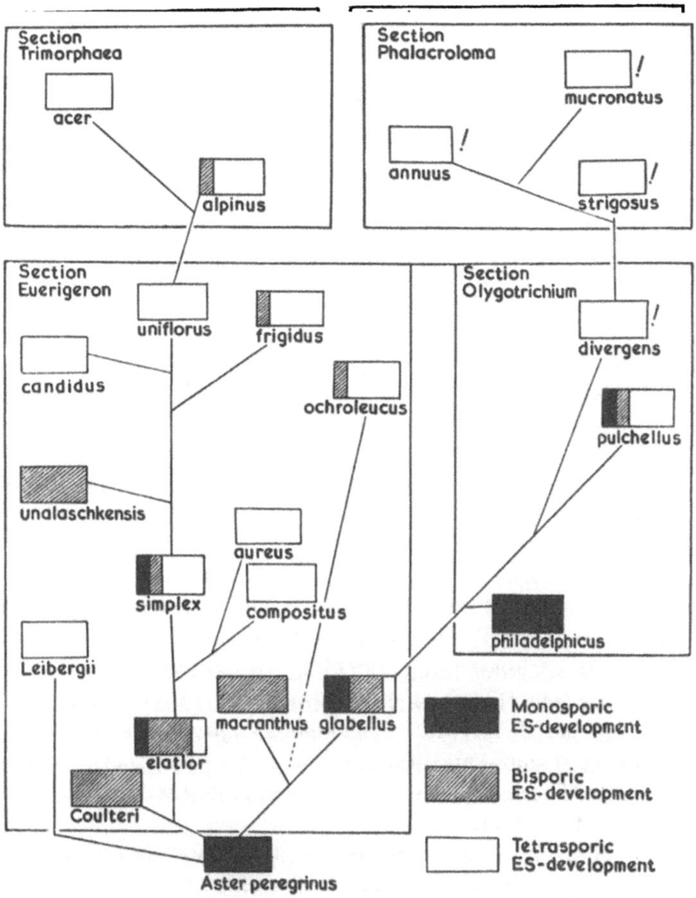

Fig. 51.9.3. *Erigeron*. Occurrence of different types of embryo sacs. More than one type in the same species has been marked by *divided fields*, the *broadest region* represents the dominant type of development. The *exclamation mark* (!) indicates agamospermous species. The genus *Erigeron* has probably been derived from *Aster peregrinus* with a monosporic embryo sac. In *Erigeron* there is a tendency toward the suppression of wall formation during meiosis so that higher up in the above "genealogical tree" the embryo sac becomes predominantly tetrasporic. (After Harling 1951b)

1935b), *Felicia bergeriana* (H.P. Sharma and Murty 1978), *Podolepis jaceoides* (Davis 1961b), and *Vernonia elaeagnifolia* and *V. divergens* (Pullaiah 1979a).

The Nuclear type of endosperm develops in *Ainsliaea aptera* (Kapil and Sethi 1962a), *Bidens biternata* (P.K. Deshpande 1964b), *Blumea malabarica* (Pullaiah 1979b), *Conyza stricta*, *Erigeron bonariensis* (H.P. Sharma and Murty 1978), *Elephantopus scaber* (Pullaiah 1979a), *Glossocardia bosvallea* (P.K. Deshpande 1962a), *Grindelia squarrosa* (Howe 1926), *Launaea nudicaulis* (Chikkannaiah and Hiremath 1982), *Melampodium divaricatum* (Maheswari Devi and Pullaiah 1976), *Taraxacum kok-saghyz* (Warmke 1943) and *Tithonia rotundifolia* (Pullaiah 1978a). Nucleate as well as enucleate cytoplasmic nodules occur in *Carthamus tinctorius* (Maheswari Devi and Pullaiah 1977).

Finally, the endosperm becomes cellular. The two peripheral layers persist as jacket layers in *Pterotheca falconeri* (Gill and Iqbal 1961).

The cells of endosperm tissue in the seeds (achenes) of *Helianthus annuus* and *Lactuca sativa* show a differential behaviour during germination (Psaras 1985). The endosperm cells of *L. sativa* have thick walls and dense cytoplasm which contains numerous protein bodies. During germination, before radicle emergence, micropylar cells of the endosperm tissue undergo drastic protoplast alterations. The walls of endosperm cells degrade only after the emergence of the radicle. The micropylar cells remain metabolically active long after radicle emergence, while the rest of the tissue is almost completely degenerated, suggesting that the endosperm cells of the micropylar area may have a role other than being a main reserve site like the rest of endosperm.

The presence of endosperm in the achene of Compositae has been known since 1849. In all tribes, one or three layers of endosperm envelop the embryo. It consists of "living cells" rich in proteins and fatty oils; there is no starch. The completely closed and firm cover of the endosperm provides an additional protection for the embryos (as in many Liliiflorae) which is only mechanically disrupted during germination (Grau and Hopf 1985).

The embryogeny conforms to the Asterad type-Senecio variation in *Achillea squarrosa* (Pullaiah 1983b), *Myriactis wightii* (L. Ram 1982) and *Sonchus oleraceous* (Pullaiah 1982). The suspensor is filamentous with three to eight cells. In *Eclipta alba* (Sunder Rajan 1974c), it is one-celled.

In *Chrysanthemum makinoi* (Tanaka and Watanabe 1972), the growth of the embryo is normal, but in the hybrid *C. makinoi* × *C. japonense* the rate of endosperm development is slow and ultimately the endosperm, endothelium and embryo degenerate. The cause for the failure of development of endosperm and embryo must be within them.

Polyembryony occurs in *Tridax procumbens* (P. Maheshwari and Roy 1952). Maheswari Devi and Pullaiah (1976, 1977) describe adventive embryos from integumentary tapetum in *Melampodium divericatum* (Fig. 3.12.1I, J) and *Carthamus tinctorius*. Antipodal embryos are reported in *Aster* (Oppermann 1904) and *Solidago canadensis* (Pullaiah 1978b). This needs confirmation.

The seed is exotestal with a large straight embryo. The testa comprises two to three layers (Maheswari Devi and Ranjalkar 1980). The cells of the outer epidermal layer develop annular bands on the radial walls. The seed-coat is reported to be absent in *Vernonia* (B. Tiagi and Taimani 1963) which is doubted by Pullaiah (1979a). In *V. divergens* and *V. elaeagnifolia* (Pullaiah 1979a), the thick-walled outer epidermis of the integument persists as seed-coat. According to Harris (1935), P. Maheshwari and Roy (1952), Kapil and Sethi (1962b), Padmanabhan (1962b), and Pullaiah (1978b), the peripheral layer of endosperm differentiates as a jacket layer and persists in the seed. It becomes "glandular" in *Tithonia rotundifolia* (Pullaiah 1978a). In *Bidens biternata*, *Glossocardia bosvallea* and *Tridax procumbens*, the endosperm is completely consumed, but the endothelium persists in the mature seed (P.K. Deshpande 1962a, d, 1964b). This aspect needs further study.

In the tribe Senecioneae, the seed-coat consists of a single cell layer, and a thin layer of endosperm surrounds the embryo which shows resin ducts; resin ducts have not been reported in other tribes of the Compositae (Pandey and Singh 1982b).

Apomixis is quite frequent in Compositae (Battaglia 1963, Davis 1967). According to Davis (1967), three types of apomixis occur in the ovule – aneuspory and gonial apospory in the megaspore mother cell, and somatic apospory in the nucellus – leading to the formation of non-reduced embryo sacs.

Aneuspory in Compositae is exhibited in three forms: Taraxacum type, Ixeris type, and Rudbeckia Type I, II and III (Battaglia 1963). In each type a restitution nucleus is formed during the first reduction (heterotypic) division in the megaspore mother cell.

1) Taraxacum type. The second reduction division (homotypic) in the megaspore mother cell is followed by wall formation, and a dyad of unreduced nuclei is formed. The chalazal dyad develops into an eight-nucleate embryo sac. Battaglia (1963) enumerates several instances in which this type has been reported. Engelskjon (1967) observed it in *Taraxacum* from Bear Island, and Chennaveeraiah and Patil (1972) in two triploid species, *Blumea eriantha* (male sterile) and *B. oxyodonta.*

V. Kaul and Singh (1980) investigated the embryology of nine polyploid apomictic species of *Taraxacum.* Meiotic division I of the megaspore mother cell is irregular in *T. duplidens.* During metaphase the chromosomes remain scattered along the spindle, and form a restitution nucleus which gives rise to an unreduced dyad. The chalazal unreduced dyad develops into an eight-nucleate embryo sac. The development of endosperm and embryo is apomictic. The unreduced egg develops parthenogenetically. The synergids and antipodals persist until the proembryo reaches the globular stage in the apomictic embryo sacs. In approximately 5% of ovules of the apomictic species, the primary endosperm nucleus fails to divide. However, in such ovules the proembryo grows only up to the globular stage.

The only exception is the diploid *T. wallichii* (V. Kaul and Singh 1980) which is eugamous, and where both syngamy and triple fusion take place.

V. Kaul and Singh (1982) report that in *Taraxacum* spp. the sporogenous cells, microspore mother cells, and male gametophyte show normal as well as abnormal behaviour. In *T. duplidens* the microspore mother cells degenerate and there is total male sterility. Considerable meiotic disturbances occur in *T. harbhajansinghii* spp. *pahalgamense*, *T. polyodon*, *T. tenebristylum* and *T. vulpinum* f. *pallidocarpum.* In *T. wallichii*, the only diploid and sexual species, meiosis is fairly regular and more than 90% of normal tetrahedral and isobilateral tetrads are formed.

In triploid *T. harbhajansinghii* spp. *pahalgamense* and tetraploid *T. tenebristylum*, the lagging of univalents occurs frequently, during anaphase I. The two to four lagging univalents form micronuclei which lead to polyspory with different chromosome numbers in the microspores. At anaphase I and II, due to unequal distribution of chromosomes, nuclei of unequal size are produced.

In *T. harbhajansinghii* spp. *pahalgamense*, in 50% to 70% of pollen mother cells the chromosome pairing is poor or lacking, and the univalents are scattered along the spindle. The chromosomes fail to separate, forming a restitution nucleus. The restitution nucleus may also divide, forming two nuclei. Occasionally, during this division, unequal distribution of chromosome occurs. The dyad (binucleate) rounds up and secretes a wall. The nuclei of the dyad may fuse to form a monad in *T. tenebristylum*. The meiotic irregularities result in the production of sporads with a varied number of microspores.

Both regular and irregular meiosis result in the formation of regular dyads, triads, tetrads, pentads and hexads, etc. The resulting spores are of reduced, unreduced, as well as hyperunreduced type.

According to Poddubnaya-Arnoldi et al. (1935), the meiotic disturbances are attributed to thermal shocks. Malecka (1961) observed the meiotic disturbances in the apical part of the anther of *T. pieninicum*, irrespective of the weather. She regards them to be due to some physiological factor such as a weak supply of nutrients. V. Kaul and Singh (1982) consider that they are due to the inherent quality of species and controlled by genetic factors.

Due to high meiotic irregularities, the pollen grains are of different size. The diploid species (*T. wallichii*) show a limited variability, whereas polyploid forms show much variation.

In *T. serotinum*, Malecka (1964) observed giant pollen grains with two, three or four nuclei. This is due to the inhibition of cytokinesis.

Cooper and Brink (1949) studied the correlation of the development of embryo and endosperm in *Taraxacum officinale*, an autonomous apomict. In *T. kok-saghyz*, a species used as control, there is a correlation between the development of embryo and endosperm initiated by the process of fertilization. In an apomict, embryo and endosperm grow irrespective of each other as the growth is not triggered by double fertilization. The apomictic embryo may be well developed, while the endosperm remains one-celled, or the endosperm may be multicellular prior to the division of the egg. According to these investigators, the nutritive function of the endosperm is superfluous as there is extensive prestorage of food in the ovule, and the growth of the embryo can proceed even though the endosperm remains one-celled.

T. crassiceps (Malecka 1982) is an obligate apomict. Melecka observed several irregularities in the meiotic divisions of microspore mother cells in the anther. Meiosis II produces dyads instead of tetrads, or tetrads of unequal size. If meiosis II is omitted, only the dyads develop.

The unreduced embryo sac conforms to the Taraxacum type. The endosperm develops autonomously from the central nucleus, in closed buds. It may develop even from unfused polar nuclei. The endosperm is of the Cellular type. In closed flowers, the development of the embryo is parthenogenetic. The antipodals remain healthy up to the globular proembryo.

In *T. alatum* (Malecka 1982), the restitution nucleus in the megaspore mother cell is very irregular in shape. The egg develops parthenogenetically, and the endosperm autonomously. The initiation of endosperm and embryo development occurs prior to the expansion of florets.

In the anthers of *T. duplidens* (Malecka 1982), dyads, tetrads with unequal microspores, and polyads are common. The tapetal cells may become 16-nucleate, or polyploid nuclei are formed due to mitotic disturbances. However, mostly the tapetal cells and microspores degenerate. The unreduced embryo sac of the Taraxacum type develops from the chalazal megaspore. There is no possibility of fertilization as the ligule is rolled and covers the anthers and the stigma. The egg develops parthenogenetically, and the endosperm autonomously.

2) Ixeris type. The formation of a restitution nucleus in the megaspore mother cell is similar to that in the Taraxacum type. The chief difference is that cytokinesis does not follow the second meiotic division and a binucleate cell, a "gynospore", is formed. By two further successive mitotic divisions, an eight-nucleate embryo sac of normal organization is produced. Parthenogenesis occurs in the unreduced egg as in *Erigeron karwinskianus* var. *mucronatus* (Battaglia 1950). However, the secondary nucleus does not divide. Ixeris type occurs in *E. annuus* (Fagerlind 1947c), *Rudbeckia laciniata* (Battaglia 1945, 1946b), *R. speciosa* (Battaglia 1945, 1946b), *R. deamii* and *R. triloba* (Battaglia 1952, 1955).

Irregular meiotic division in the megaspore mother cell (with restitution nucleus) leads to the formation of univalents. They become scattered on the meridional plane during the metaphase. This results in the formation of reduced megaspores. Reductional aneuspory is reported in *E. karwinskianus* var. *mucronatus* (Battaglia 1950, Fagerlind 1947c), *E. annuus* (Fagerlind 1947c) and *R. laciniata* (Fagerlind 1946).

3a) Rudbeckia type I. The development is similar to that of the Ixeris type, but some of the chromosomes are left out in the cytoplasm during the first meiotic division of the megaspore mother cell. They form a micronucleus which undergoes two mitotic divisions and four micronuclei are produced at the micropylar end of the embryo sac. These organize into a miniature egg apparatus and polar nucleus. The remaining eight nuclei organize into egg-like antipodals. Sometimes, the last mitotic division is suppressed resulting in a four-nucleate condition. This has been reported in *R. laciniata* (Battaglia 1946c, 1947b).

3b) Rudbeckia type II. The development of the embryo sac up to the formation of eight nuclei is similar to the Ixeris type. Sometimes, the last mitotic division is suppressed giving rise to a four-nucleate embryo sac without any micronuclei or egg apparatus. The four nuclei organize into a pro-endospermic cell and three antipodals; two of the antipodals are egg-like. Such a behaviour has been observed in *R. laciniata* and *R. speciosa* (Battaglia 1946a, b). Occasionally, one more mitotic division may take place so that besides the central cell there are seven antipodals, five egg-like.

3c) Rudbeckia type III. In this type a birestitutional nucleus is formed as in *R. speciosa*, *R. laciniata* (Battaglia 1946a, b), and *Erigeron karwinskianus* var.

mucronatus (Battaglia 1950). The embryo sac contains a tetraploid set of chromosomes in the egg apparatus, octoploid secondary nucleus, and a tetraploid set in the three antipodals.

Gonial apospory. The megaspore mother cell becomes vacuolate and functions directly as a uninucleate embryo sac, without undergoing any meiotic division. By three successive mitotic divisions, it develops into a normally organized eight-nucleate embryo sac. In *Brachycome ciliaris* (Davis 1964a, 1967), the egg develops parthenogenetically even before the opening of the flower. In one instance the polar nuclei behaved like the first two endosperm nuclei. In such cases microsporogenesis also shows irregularities. Gonial apospory has been observed in *Arnica* spp., *Bellis perennis* (Murty and Sharma 1977), *Eupatorium glandulosum*, *Hieracium* spp., *Parthenium argentatum* and *P. incanum* (see Davis 1967). Urbanska-Worytkiewicz (1974) reported an unreduced female gametophyte in *Antennaria* sp., and development of seed either by pseudogamy or autonomous apomixis. Gonial apospory is associated with true polyembryony.

Somatic apospory. The most common occurrence is in Compositae. The nucellar cells in the chalazal region enlarge, become multinucleate, behave as embryo sacs, and polarity is established. In *Carthamus tinctorius* (Maheswari Devi and Pullaiah 1977), in addition to nucellar cells, one to three cells of the integument also function as initials of the aposporous embryo sacs. Meiotic divisions in the megaspore mother cell and formation of an aposporous embryo sac are independent phenomena. Very often, the aposporous embryo sacs may develop up to the eight-nucleate stage, and then degenerate. At times, concurrent development in three or four aposporous embryo sacs leads to false polyembryony.

Davis (1964b) cites a good example of somatic apospory in *Minuria integerrima*, a somatic apomict. In this species irregularities occur during microsporogenesis so that the male gametes fail to develop. During megasporogenesis degeneration occurs at the tetrad stage. The space thus created is occupied by an aposporous embryo sac which grows to maturity. Occasionally, two to three aposporous embryo sacs may grow in that region. The parthenogenetic development of the unreduced egg follows the Asterad type of embryogeny, as in normal species. Davis (1967) has quoted numerous examples: *Artemisia tridentata* (Howe 1926), *Aster capensis* (K.V.O. Dahlgren 1920), *A. novae-angliae* (Chamberlain 1895, Palm 1914, Carano 1921), *A. undulatus* (Oppermann 1904), *Bellis perennis* (Carano 1921), *Bidens tripartitus* (K.V.O. Dahlgren 1920), *Calotis lappulacea* (Davis 1968a), *Erigeron elatius* (Bergman 1942), *E. glabellus* (Carano 1921), *Gerbera jamesonii* (Maheswari Devi 1957), *Solidago serotina* (Palm 1914) and *Tridax trilobata* (Hjelmqvist 1951).

Hemigamy (Section 3.13). *Rudbeckia laciniata* and *R. triloba* (Fig. 51.9.4A–G) are pseudogamous (hemigamic) apomicts. One male gamete enters the egg cell but does not fuse with the nucleus. The male gamete and the egg nucleus

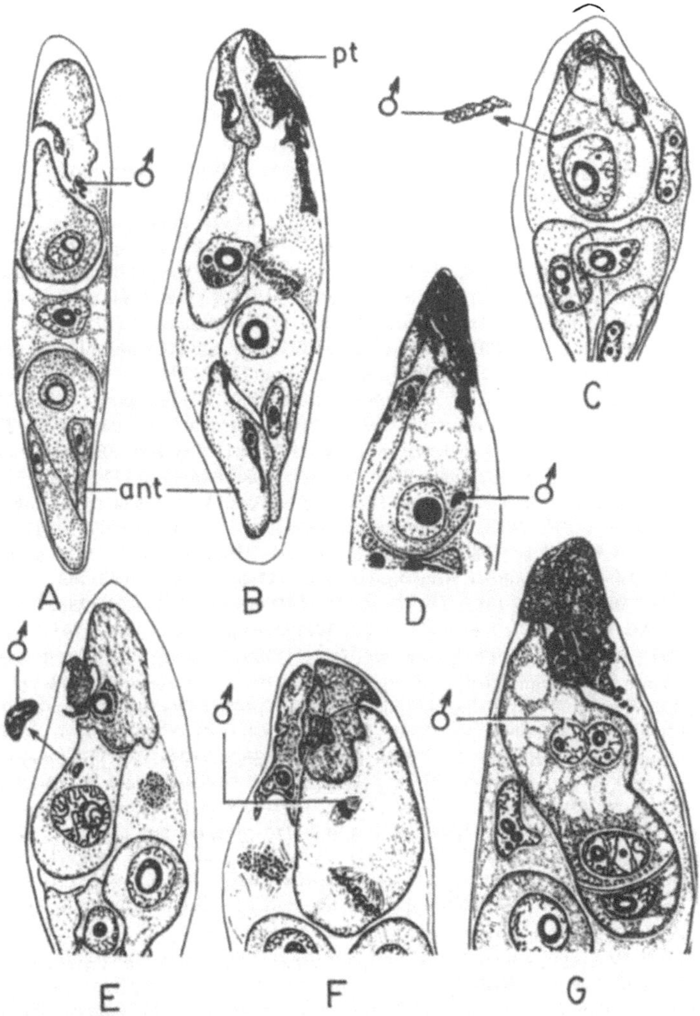

Fig. 51.9.4A–G. *Rudbeckia laciniata*, hemigamy. **A** Embryo sac shows a sperm lying outside the egg cell, primary endosperm nucleus, and antipodal cells (*ant*). **B** Syngamy, primary endosperm nucleus in metaphase *pt* pollen tube. **C** Sperm in egg cell, enlarged view of sperm also shown in **C** and **E. D** Rounding off of sperm in egg cell. **E** Sperm and egg nuclei in prophase; primary endosperm nucleus in metaphase. **F** Synchronous division (metaphase) of sperm and egg nucleus, and primary endosperm nucleus. **G** Proembryo with two large nuclei (egg nucleus derivatives) and two small nuclei (sperm derivatives). (After Solntzeva 1978)

divide at the same time so that a parthenogenetic two-celled proembryo shows two smaller nuclei (derivatives of the male gamete) in the "suspensor" (Fig. 51.9.4F, G).

K Afzelius (1936), L Ahlstrand (1978, 1979a, b, c, d, 1985), Anantaswamy Rau (1951g), LC Anderson (1970), E Annen (1945), B Anzalone (1949), I Asplund (1978), HG Avanzi (1948), C

Banchetti (1961), I Banerji (1940a, b, 1942b), I Banerji and S Pal (1959), E Battaglia (1940, 1945, 1946a, b, c, d, 1947a, b, c, d, e, 1948, 1950, 1951b, 1952, 1955, 1963), NS Belyaeva (1975), C Berger and O Erdelská (1973), B Bergman (1942, 1944), R Beruti (1961), HR Bhargava (1935b), K Bijok (1960), K Bijok, T Pawlak and B Krenska (1977), L Borgen (1971, 1972), R Capineri (1966), E Carano (1921), J Carapetian and EA Rupert (1989), G Cesca (1961), CJ Chamberlain (1895), B Chandler (1911), AK Chatterji and M Jayaramu (1981), MS Chennaveeraiah and RM Patil (1972), A Chiarugi (1927a, b), PS Chikkannaiah and BS Hiremath (1982), S Chopra (1969), MA Cichan and BF Palser (1982), DC Cooper and RA Brink (1949), KVO Dahlgren (1920, 1924), KMM Dakshini and N Dadlani (1978), GL Davis (1961a, b, c, 1962a, b, 1963, 1964a, b, c, 1966, 1967, 1968a), PK Deshpande (1960a, b, 1961, 1962a, b, c, d, 1964a, b, 1970), PK Deshpande and MP Kothare (1976), L Desole (1951), M Dittrich (1977), KJ Dormer (1962), S Dutt (1939), U Eliasson (1971, 1972), K Engell (1970), T Englskjon (1967), SG Ervandyan (1975), K Esau (1944), F Fagerlind (1941b, 1946, 1947b, c, e), S Ferri (1961), G Gentcheff (1937), H Gerassimova (1933, 1936), H Gerassimova-Navashina (1947a, b, c), RB Ghosh (1962b), RB Ghosh and AG Ghosal (1971), BS Gill and M Iqbal (1961), J Grau and H Hopf (1985), Å Gustafsson (1932, 1933), G Harling (1950, 1951a, b, 1960, 1962), CJ Harris (1935), J Heslop-Harrison (1969), H Hjelmqvist (1951), H Hjelmqvist and V Holmberg (1961), I Holmgren (1919), HT Horner Jr and C Pearson (1978), TD Howe (1926, 1959, 1964, 1969, 1970, 1975), R Izmailow (1984), HO Juel (1898, 1900), RN Kapil and SB Sethi (1962a, b), V Kaul and D Singh (1980, 1982), V Kaul, ASR Dathan and D Singh (1975), PSH Khan (1970), M Kościńska (1980), H Kuroiwa (1989), M Lankosz-Mróz (1984a, b), B Longly (1984), H Maheswari Devi (1957, 1963), H Maheswari Devi and T Pullaiah (1976, 1977), H Maheswari Devi and MM Ranjalkar (1980), P Maheshwari and A Haque (1949), P Maheshwari and SK Roy (1952), P Maheshwari and AR Srinivasan (1944), J Malecká (1961, 1964, 1982, 1990), DR Marshall and AHD Brown (1974, 1981), DR Marshall and BS Brown (1979), RW Martin and FH Smith (1955), GO Martinoli (1967), E Maugini (1962), WD Merrell (1900), S Misra (1957, 1964, 1965, 1972), JN Mitra (1947b), PK Mukherjee and SG Desai (1990), YS Murty and HP Sharma (1977), S Murugesan and KV Krishnamurthy (1984), K Musial (1990), W Newcomb (1973a, b), A Nygren (1950), M Oppermann (1904), E Pacini and CJ Keijzer (1989), D Padmanabhan (1962b, 1968), B Palm (1914), AK Pandey (1980), AK Pandey and RP Singh (1976, 1980, 1982a, b, 1983a, b), AK Pandey, S Chopra and RP Singh (1983, 1986a, b), WR Philipson (1953), VA Poddubnaya-Arnoldi, N Stechina and A Scsonovetz (1935), M Pogliani (1966), RA Popham (1938), PS Prakasa Rao, KT Sundri and LL Narayana (1979), G Psaras (1984, 1985), T Pullaiah (1978a, b, 1979a, b, 1981, 1982, 1983a, b, 1984), G Rajashekhar (1980, 1981), L Ram (1982, 1986), V Rangaswamy and T Pullaiah (1986), H Reese (1989), G Renzoni-Cela (1963, 1972), AJ Richards (1973), S Rogers (1969), E Rozenblum, S Maldonado and CE Waisman (1988), MI Savchenko (1960), CB Sehgal (1966), HP Sharma and YS Murty (1977, 1978), RP Singh and AK Pandey (1984), BB Smith and LK Johnson (1980), S Sunder Rajan (1968, 1972, 1974b, c), P Swarajya Lakshmi and T Pullaiah (1988, 1989), R Tanaka and K Watanabe (1972), B Tiagi and S Taimani (1960, 1963), BS Trivedi and R Tripathi (neé Trivedi) (1985), K Urbanska-Worytkiewicz (1974), PB Vaidya (1961b), J Venkateswarlu (1941a, b, 1959), J Venkateswarlu and H Maheswari Devi (1955a, b), S Venugopalan (1966), G Wagenitz (1977a, b), HE Warmke (1943), HGR Yurukova and PD Ncharova (1978)

51.10 Taxonomic Consideration

Campanulaceae. Bentham and Hooker (1965b) place Calyceraceae and Compositae in Asterales, and the other families in Campanulales. Engler and Prantl (1931) and Engler and Diels (1936) place them in Campanulatae. Hutchinson (1973) classifies Campanulaceae and Lobeliaceae in Campanulales; Goodeniaceae, Brunoniaceae and Stylidiaceae in Goodeniales; Compositae in Asterales. Melchior (1964) divides the family Campanulaceae into three subfamilies: Campanulioideae, Cyphioideae and Lobelioideae. (There are no embryological data on Cyphioideae.) Hutchinson (1973) recognizes the subfamily

Lobelioideae as a separate family, Lobeliaceae, allied to Campanulaceae. Subramanyam (1970b) evaluates the embryological features of the three subfamilies, and shows their close affinities in morphological and embryological features.

However, three- or four- to ten-colpate, colporoidate or colporate pollen, two- to five-carpellary syncarpous inferior or superior ovary, absence of embryo sac haustoria and Codonopsis or Phyteuma type of endosperm development in the Campanulioideae differ from the two- to three- or four-colporoidate pollen with smooth exine, inferior, bicarpellary, bilocular, or rarely even unilocular ovary, embryo sac with a caecum at the chalazal end, and the Scutellaria type of endosperm in the Lobelioideae. The endosperm haustoria in Lobelioideae are not so aggressive as in Campanulioideae (Rosén 1932). Subramanyam (1970b) considers that the floral characters and embryological features adequately justify the separation of these two subfamilies. The erection of a separate family Lobeliaceae is also justified (Table 51.10.2).

Sphenocleaceae. Engler and Prantl (1931) include the genus *Sphenoclea* in the family Campanulaceae. Airy Shaw (1948) raises *Sphenoclea* to family rank and proposes that the Sphenocleaceae be placed between Phytolaccaceae and Primulaceae. Hutchinson (1973) contradicts Airy Shaw's opinion as, according to the former, "the resemblance to *Phytolacca* is entirely superficial and due to parallel evolution of the inflorescence". He prefers the assignment of *Sphenoclea* in Campanulaceae. Kausik and Subramanyam (1946c) and Subramanyam (1950a, 1970c) investigated the embryology of *Sphenoclea* and discuss its phylogenetic relationships.

Sphenocleaceae and Campanulaceae have anatropous, unitegmic ovules with ephemeral nucellus, Polygonum type of embryo sac with endothelium, and Cellular type of endosperm with well-developed haustoria. However, the centrospermalean characters of the Phytolaccaceae, such as anacampylotropous, bitegmic ovules with well-developed nucellus which persists as perisperm, absence of endothelium, and Nuclear endosperm without haustoria, are quite contrasting.

The Scutellaria type of endosperm with four-celled micropylar and chalazal haustorium, Onagrad type of embryogeny, and exalbuminous seeds in *Sphenoclea* differ from the members of Campanulaceae. Subramanyam (1950a) justifies its assignment to a separate family, Sphenocleaceae, and places it adjacent to Campanulaceae.

B.M. Johri (1963b) and P. Maheshwari (1964) compare the embryological features of Sphenocleaceae with Campanulaceae, Phytolaccaceae and Primulaceae, and conclude that Sphenocleaceae do not show any affinity with Phytolaccaceae or Primulaceae, but resemble Campanulaceae.

Pentaphragmataceae. Bentham and Hooker (1965b) and Engler and Prantl (1931) place the genus *Pentaphragma* in Campanulaceae. B.M. Johri (1963b) evaluates its systematic position in relation to embryology. The resemblance of its inflorescence to that of Boraginaceae, Convolvulaceae (*Jacquemontia*) and Loasaceae (*Kissenia*) has been pointed out by Hallier (1912). Airy Shaw (1941,

1954) removes *Pentaphragma* to a monogeneric family, Pentaphragmataceae, on the basis of habitat and morphology. Kapil and Vijayaraghavan (1962) investigated its embryology and compare it with that of Boraginaceae and Campanulaceae (Table 51.10.1, for Campanulaceae see Table 51.10.2). *Pentaphragma* shows a close alliance with Campanulaceae in the type of anther tapetum, ovary, ovule with endothelium, embryo sac, endosperm, and embryo development. Due to significant differences, they negate any relationship with the Boraginaceae (Table 51.10.1). Kapil (1962) and Kapil and Vijayaraghavan (1962) maintain that *Pentaphragma* is correctly placed in the family Campanulaceae. However, its distinctive features, such as (1) extra-ovular embryo sac, (2) single-celled micropylar haustorium, (3) absence of chalazal haustorium, and (4) a short suspensor, suggest its assignment to a separate subfamily, Pentaphragmoideae, in the Campanulaceae. Hutchinson (1973) assigns *Pentaphragma* to the Campanulaceae.

Subramanyam (1970c) seeks the ancestry of Sphenocleaceae and Pentaphragmataceae in the Campanulaceae, and that of Goodeniaceae, Brunoniaceae and Stylidiaceae in the Lobelioideae. Hutchinson (1973) also groups the latter three families together in Goodeniales.

Goodeniaceae. Vijayaraghavan and Malik (1972) discuss the embryological affinities of Goodeniaceae with those of Brunoniaceae, Campanulaceae, Lobeliaceae and Stylidiaceae (Table 51.10.2). Goodeniaceae differ from the latter three families in having (1) calcium oxalate crystals in endothecium, middle layer of anther wall, and seed-coat, (2) tricolporate pollen grains, occasionally liberated in tetrads in some species, (3) occasional occurrence of integumentary vascular bundle, (4) presence of hypostase, (5) absence of starch in mature embryo sac, (6) Cellular endosperm without haustoria, (7) absence of polyembryony, and (8) uni- or multilayered seed-coat. They suggest that the family Goodeniaceae should be segregated to a separate order, Goodeniales, and placed close to Campanulales. In our opinion, there is no need to erect a new order, and Goodeniaceae is correctly assigned to Campanulales.

Brunoniaceae. Brunoniaceae show anatropous, unitegmic, tenuinucellate ovule, and Polygonum type of embryo sac with three ephemeral antipodal cells as in Campanulaceae (Table 51.10.2). It appears to be more advanced than Goodeniaceae, due to the presence of a single ovule in a reduced ovary, suspensor haustorium, and exalbuminous seed (Rosén 1946). A detailed study of the embryology would be very useful in discussing the relationships of Brunoniaceae.

Stylidiaceae. Rosén (1935, 1937) suggests the origin of Stylidiaceae from lobeliaceous forms. Stylidiaceae show affinities to Lobeliaceae (Table 51.10.2) in floral structure and embryological features: (1) fibrillar endothecium, and secretory (binucleate cells) anther tapetum, (2) three-celled pollen grains with a precocious tendency to germinate in situ, (3) Polygonum type of embryo sac with conspicuous endothelium, and a chalazal caecum, (4) Scutellaria type of

Table 51.10.1. Embryology of Boraginaceae (Sect. 48.6) and Pentaphragmataceae (51.4). (Adapted after Kapil 1962, Kapil and Vijayaraghavan 1962, B.M. Johri 1963b). For Campanulaceae (51.2) see Table 51.10.2.

	Boraginaceae	Pentaphragmataceae (*Pentaphragma*)
Anther	Endothecium fibrous, tapetum secretory, cells multinucleate, cytokinesis simultaneous	Endothecium fibrous, tapetum secretory, cells binucleate, cytokinesis simultaneous
Tetrads	Tetrahedral, isobilateral, decussate, polyspory may occur	Tetrahedral, decussate
Pollen	Multipalynous, 2- or 3-celled	Tricolporate, 2-celled
Ovule	Ortho-, hemiana-, anatropous or anacampylotropous, unitegmic, tenui-, crassi- or pseudo-crassinucellar, endothelium differentiates, hypostase develops; integument may or may not be vascularized	Anatropous, unitegmic, tenuinucellar, endothelium around the lower half of embryo sac
Tetrad	Linear, T-shaped	Linear, T-shaped
Embryo sac	Polygonum type, antipodals degenerate before fertilization, twin embryo sacs occur; bisporic embryo sac may develop in some taxon	Polygonum, 4-nucleate embryo sac extends beyond the micropyle, antipodals degenerate after fertilization
Endosperm	Nuclear or Cellular type, when Cellular the micropylar haustorium is 2- or 4-celled, chalazal haustorium 4-celled or uni- or binucleate	Cellular, only uninucleate micropylar haustorium develops
Embryogeny	Onagrad, Asterad, Solanad or Chenopodiad type	Solanad
Seed-coat	Exotestal, or unspecialized, seed remains enclosed in nutlet, or protected by stony pericarp in some taxa	Exotestal
Seed	Albuminous, embryo straight or curved	Albuminous, embryo straight
Apospory	Embryo sacs develop from chalazal nucellar cells, mostly remain arrested	Absent

Table 51.10.2. Embryology of Campanulaceae (Sect. 51.2), Lobeliaceae, Goodeniaceae (51.5), Brunoniaceae (51.6) and Stylidiaceae (51.7). (See Vijayaraghavan and Malik 1972, Hutchinson 1973, Corner 1976, Batygina and Yakovlev 1987)

	Campanulaceae	Lobeliaceae	Goodeniaceae	Brunoniaceae	Stylidiaceae
Endothecium	Fibrous, ca-oxalate crystals absent	Fibrous, crystals absent	Fibrous, crystals present	—	Fibrous, crystals absent
Middle layer	1, ephemeral, crystals absent	1, ephemeral, crystals absent	1 or 2, ephemeral, crystals present	—	1, ephemeral, crystals absent
Tapetum secretory	Cells binucleate, polyploid	Binucleate, polyploid	Multinucleate, polyploid	—	Binucleate, polyploid
Cytokinesis	Simultaneous	—	Simultaneous	—	Simultaneous
Tetrad	Tetrahedral, isobilateral	Tetrahedral	Tetrahedral, isobilateral	—	Tetrahedral
Pollen	Multipalynous, 2- or 3-celled	Multipalynous, 2-celled	Triporate, 2-celled remain attached in tetrads in *Leschenaultia*	2-celled	3-celled
Ovule	Anatropous, unitegmic, tenuinucellar, endothelium conspicuous, integumentary vascular bundle absent	Anatropous, unitegmic, tenuinucellar, endothelium conspicuous, integumentary vascular bundle absent	Anatropous, unitegmic, tenuinucellar, endothelium conspicuous, integumentary vascular bundle present or absent	Anatropous, unitegmic, tenuinucellar, integumentary vascular bundle absent	Anatropous, unitegmic, tenuinucellar, endothelium conspicuous, integumentary vascular bundle absent
Hypostase	Absent	Absent	Present, cells densely cytoplasmic	Absent	Absent

Tetrad	Linear, T-shaped	—	Tetrahedral, isobilateral	—	Tetrahedral
Embryo sac	Polygonum type, antipodals 3, uninucleate, ephemeral, starch present	Polygonum, antipodals 3, uninucleate, ephemeral, chalazal caecum develops, starch present	Polygonum, antipodals 3, uninucleate, ephemeral, starch absent	Polygonum, antipodals 3, uninucleate, ephemeral	Polygonum, antipodals 3, uninucleate, persist after fertilization, starch present
Endosperm	Cellular type, micropylar and chalazal haustoria present	Cellular, micropylar and chalazal haustoria present	Cellular, haustoria absent	Cellular, haustoria absent	Cellular, micropylar and chalazal haustoria present
Embryogeny	Solanad type, suspensor long and filamentous	Solanad	Solanad	Multinucleate suspensor haustorium develops	Solanad
Polyembryony	Present	Present	Absent	—	Absent
Testa	1–4 layers, only exotesta persists; endothelial cells become thick-walled	1–4 layers, only exotesta persists	7–14 layers, exo- and endotesta persist	—	1–4 layers, only exotesta persists
Seed	Minute, albuminous, small straight embryo	Small, albuminous, small straight embryo	Albuminous or exalbuminous, straight embryo	Exalbuminous, embryo straight	Small, albuminous, embryo minute

endosperm with prominent, aggressive haustoria, (5) Solanad type of embryogeny, and (6) outer epidermis of the integument forms the seed-coat (Subramanyam 1950b, c, 1970e, W.R. and M.N. Philipson 1973). These embryological features of the family are comparable to those of Lobilioideae of Campanulaceae.

Stylidiaceae also show some features similar to those in Compositae: (1) two-loculed ovary which gradually becomes one-chambered (Subramanyam 1951d), (2) elongated synergids, (3) rather persistent antipodals, (4) tendency of arrested growth of the chalazal haustorium, and (5) presence of inulin crystals.

Thorne (1968) places Stylidiaceae in the suborder Saxifragineae in Rosales. Jensen et al. (1975) object to the placement of Stylidiaceae in or next to Campanulales, or Asterales, as Stylidiaceae have glandular hairs but lack laticiferous ducts and differ from Campanulaceae. In addition, *Stylidium* contains Group I iridoids, tannins, and leucoanthocyanins (Hegnauer 1973). R. Dahlgren (1977b) proposes the position of Stylidiaceae (including Donatiaceae) in his Order Cornales. W.R. and M.N. Philipson (1973) investigated the embryology of *Forstera* and emphasize that Stylidiaceae are correctly placed close to Campanulaceae (Table 51.10.2).

Donatia. The systematic position of the genus *Donatia* is controversial, because of its polypetalous corolla, absence of connation between the stamens, and free styles. It has been placed in Saxifragaceae (Engler 1891), in Saxifragales as Donatiaceae (Hutchinson 1973), in Rosales (Thorne 1968), and in Cornales (R. Dahlgren 1975b, see Wagenitz 1977b).

Chandler (1911) and Skottsberg (1915) regard *Donatia* as distinct from Stylidiaceae, and raise it to a monogeneric family, Donatiaceae. *Donatia* has been shifted to Asteridae and placed in Campanulales-Stylidiaceae (Wagenitz in Melchior 1964, Cronquist 1968), or Campanulales-Donatiaceae (see Wagenitz 1977a, b, Takhtajan 1980).

W.R. and M.N. Philipson (1973) investigated the embryology of *Donatia novae-zelandiae*. Except for the nature of the proembryo, other features, such as anatropous, unitegmic, tenuinucellate ovule with endothelium, hypodermal female archesporium, Polygonum type of embryo sac with long tapering synergids, and development of endosperm with haustoria, are identical with those of *Forstera* sp. (W.R. and M.N. Philipson 1973) and other members of the Stylidiaceae (Rosén 1935, Subramanyam 1950a, b, c). The shape of the proembryo of *Donatia* is ovate, linear in *Forstera* and *Stylidium*. W.R. and M.N. Philipson conclude that the systematic affinities of *Donatia* are with Stylidiaceae, but the erection of the family Donatiaceae is justified.

R. Dahlgren (1977b) groups the family Stylidiaceae with the iridoid Corniflorae and not with the non-iridoid Campanuliflorae. Embryological data do not support its assignment on a chemical basis.

Calyceraceae. Morphologically, Calyceraceae are closely allied to Compositae (Lawrence 1951). Bentham and Hooker (1965b) group them in the Asterales. Thorne (1968) includes Calyceraceae in Dipsacales (see Table 50.6.1). Takhtajan (1980) and Cronquist (1981) place Calyceraceae in a separate order,

Calycerales, as it shows (1) alternate leaves, (2) absence of glandular hairs, (3) Asteraceae-like inflorescence, and (4) two-celled pollen grains (three-celled in Dipsacales). Jensen et al. (1975) rule out its alliance with Campanulales or Asterales as their members show the presence of simple sicologanin, and a simple sico-iridoid, as in Dipsacales.

The family Calyceraceae shares the following characters with the order Campanulatae: (1) fibrous endothecium, (2) secretory anther tapetum, (3) two-celled pollen grains, (4) anatropous, unitegmic, tenuinucellate ovule, (5) Polygonum type of embryo sac with ephemeral antipodals, and (6) Cellular type of endosperm.

Compositae. Compositae are the largest family of angiosperms. In recent systems of classification it is designated as Asteraceae (see Takhtajan 1980). Gundersen (1950) places Compositae with Campanulaceae, Dipsacaceae and Calyceraceae in the Asterales. Cronquist (1981) erects the order Asterales for the family Compositae (Asteraceae), and traces its ancestry in the Rubiales-Dipsacales complex (see Table 50.6.1). Thorne (1968) and R. Dahlgren (1977b) place Asteraceae in the superorder Asteriflorae, order Asterales. Robyns (1972) erects the order Synandrae (Asterales).

In Compositae secretory anther tapetum occurs only in *Ainsliaea*, *Epaltes*, *Palicuria* and *Sonchus*, otherwise it is amoeboid. The antipodals show variations in form and number and are persistent; the development of the endosperm corresponds to the Cellular as well as to the Nuclear type. Takhtajan (1980) points out certain differences between Compositae and Calyceraceae: (1) basal attachment of ovule, (2) bifid style, (3) three-celled pollen grains, (4) absence of endosperm in the seed (which is doubtful), (5) presence of external glands and secretory cavities, (6) absence of iridoid compounds, and (7) presence of inulin and acetylenes. Thus, Compositae do not show much similarity with Calyceraceae (Table 50.6.1). The embryogeny and seed anatomy of Calyceraceae are still to be investigated.

The floral morphology and inflorescence of Calyceraceae resemble those of Compositae (Cronquist 1981). Pollen grains of these two families are also similar. However, the available embryological data on Calyceraceae do not show much similarity.

Poddubnaya-Arnoldi (1933) draws attention to a close similarity in the development of male gametophytes in Dipsacaceae, Umbelliferae, Compositae, and Campanulaceae.

Rosén (1949) derives Compositae from Lobeliaceae and Campanulaceae (see Table 51.10.2). However, Crété (1951) rejects this proposal on the basis of the development of the endosperm and embryo. The early development of the endosperm is of the Cellular type (regular divisions) in Campanulaceae, whereas in Compositae both the Cellular (early divisions irregular) and the Nuclear type occur. The embryogeny also differs from that of Campanulaceae and Lobeliaceae.

The chemical data favour the alliance between Campanulales and Asterales (see Wagenitz 1977a, b). The presence of polyacetylenes (Bohlman et al. 1973) and storage of inulin occurs in both the orders. Phytochemically, Asteriflorae are also similar to Araliiflorae (R. Dahlgren 1977a). Both superorders show the presence of polyacetylenes and sesquiterpene lactones, and the absence of iridoids and tannins (see Hegnauer 1971). R. Dahlgren (1977b) associates these data with some other common characters between Asteriflorae and Araliiflorae: epigyny (in combination with a tendency for reduction or specialization of calyx segments), unitegmic ovule, Nuclear type of endosperm, and absence of endosperm haustoria.

Embryologically, Compositae show many similarities with the members of the Dipsacales (Table 50.6.1): (1) amoeboid anther tapetum, (2) anatropous, unitegmic, tenuinucellar ovule with endothelium, (3) Polygonum type of embryo sac, and (4) Cellular type of endosperm. The development of the embryo in *Centranthus angustifolia*, *C. ruber* and *Valerio alitoris* of Valerianaceae, and *Scabiosa succisa* of Dipsacaceae, conforms to the Asterad type (see P.K. Deshpande 1970), as in Compositae. However, phytochemical data in Compositae and Dipsacales differ.

Ahlstrand (1979d) studied 18 species of the tribe Arctotideae s.l., and 31 species of the tribe Calenduleae. He discusses the relationships and classification up to the generic and subgeneric level on the basis of time of formation of periplasmodium during microsporogenesis, occurrence of one or many-celled female archesporium (*Gibbaria*), nature of the endothelium (radially stretched in *Dimorphotheca* or multilayered in *Berkheya*), type of embryo sac, presence (*Ursinia*) or absence of synergid haustoria, number of antipodal cells and their haustoria, presence of a space outside the embryo sac in the chalazal region (*Gazania*), Cellular (*Chrysanthemoides*), or Nuclear (*Osteospermum*) type of endosperm, and the formation of aposporous embryo sacs.

Variations in the nature of the anther tapetum, different types of embryo sac even in the same genus such as *Erigeron* (Harling 1951b) and *Minuria* (Davis 1963, 1964b), the presence of synergid and antipodal haustoria in some members, Nuclear as well as Cellular type of endosperm, apomictic tendency in several members, and the complexity of their behaviour indicate the highly evolved position of the Compositae amongst the sympetalous families.

III Monocotyledons

52 Helobiae

Helobiae comprise four suborders: Alismatineae, Hydrocharitineae, Scheuchzeriineae and Potamogetonineae. The Alismatineae include the families Alismataceae and Butomaceae; Hydrocharitineae – Hydrocharitaceae; Scheuchzeriineae – Scheuchzeriaceae, Potamogetonineae – Aponogetonaceae, Juncaginaceae, Potamogetonaceae, Zannichelliaceae and Najadaceae.

52.1 General Features

The characteristic features of Helobiae are: (1) aquatic habit, (2) plasmodial (amoeboid) anther tapetum, (3) successive cytokinesis in pollen mother cells, (4) three-celled pollen grains, two-celled in Juncaginaceae, (5) bitegmic ovule with inner integument forming the micropyle, crassinucellate except in Alismataceae and Butomaceae (crassinucellar in *Butomus*), (6) Polygonum type of embryo sac, Allium type in Alismataceae, Butomaceae (except *Butomus*) and Zannichelliaceae, (7) mostly Helobial type of endosperm, Nuclear in Juncaginaceae, Najadaceae, and some members of Alismataceae and Potamogetonaceae, (8) Caryophyllad type of embryogeny with a vesicular cell at the upper end of the suspensor, (9) origin of shoot apex and cotyledon from the same terminal tier, and (10) exalbuminous seeds.

52.2 Alismataceae

The anther wall comprises the epidermis, fibrous endothecium, an ephemeral middle layer, and tapetum with uninucleate cells. The tapetum forms periplasmodium as the microspores separate from one another in *Limnophyton obtusifolium*, *Sagittaria sagittifolia* and *S. guayanensis* (B.M. Johri 1935a, b, d). The microspore mother cells undergo meiotic divisions, cytokinesis is successive, and the tetrads isobilateral.

The pollen grains are pantoporate (P.K.K. Nair 1970, Chanda et al. 1988) with spiny exine, and three-celled at the dispersal stage. The male gametes are spherical or sickle-shaped (sickle-shaped gametes appear spherical in cross-section).

The ovule is anatropous, bitegmic and tenuinucellar with a narrow micropyle formed by the inner integument. The nucellar epidermis divides periclinally in *S. guayanensis* (B.M. Johri 1935d). The hypodermal archesporium is one- or two-celled. A single cell functions directly as megaspore mother cell. It undergoes the first meiotic division, and dyads are produced. The upper one degenerates, the chalazal one functions and develops into a four-nucleate embryo sac. The two micropylar nuclei divide once and the daughter nuclei organize into the egg apparatus and the upper polar nucleus. The two chalazal nuclei may not divide further. The upper one functions as lower polar nucleus, while the lower represents a single antipodal. The nucleus at the antipodal end undergoes fragmentation (this is doubtful and mitosis seems to have been missed) in *Limnophyton* and *Sagittaria.* The development of gametophyte is of the Allium type and the embryo sac is generally six-nucleate, sometimes seven- or eight-nucleate if one or both of the chalazal nuclei divide once or twice. In *Echinodorus ranunculoides* (K.V.O. Dahlgren 1928), the chalazal nucleus of the two-nucleate embryo sac fails to divide so that it remains five-nucleate. One of the synergids persists during post-fertilization development. The polar nuclei fuse before or at the time of fertilization, and the secondary nucleus lies in the lower part of the embryo sac.

Generally, syngamy and triple fusion occur simultaneously. Occasionally, more than one pollen tube enters an embryo sac. The additional male gametes lie either in the vicinity of egg, or may get attached to the endosperm nuclei (one each to two daughter nuclei of the primary endosperm nucleus) in *Sagittaria graminea* (B.M. Johri 1936a).

Development of the endosperm is of the Nuclear, or Helobial type. Nuclear type occurs in *Alisma plantago*, *Damasonium alisma*, *Elisma natans* (K.V.O. Dahlgren 1928), *Alisma plantago-aquatica* (B.M. Johri 1936a), and *Machaerocarpus* (P. Maheshwari and Singh 1943). In the Helobial type the division of the primary endosperm nucleus leads to the formation of a large micropylar and a small chalazal chamber. Free-nuclear divisions occur in the micropylar chamber. Wall formation occurs only around the periphery of the embryo sac in the micropylar chamber in *Limnophyton obtusifolium* (B.M. Johri 1935a), whereas in *S. guayanensis* and *S. latifolia* (B.M. Johri 1935d) it remains free-nuclear throughout. The nucleus of the chalazal chamber divides once or twice but the daughter nuclei soon degenerate. Helobial type of endosperm also occurs in *Echinodorus macrophyllus*, *S. sagittifolia* (K.V.O. Dahlgren 1934b), and *S. graminea* (B.M. Johri 1936a).

The developmental sequence of the embryo has been studied by Hanstein (1870) in *Alisma plantago*, Schaffner (1897) in *Sagittaria variabilis*, Cook (1907) in *S. lancifolia*, Souèges (1917a, b, 1931a) in *S. sagittifolia*, and B.M. Johri (1935a, d) in *Limnophyton obtusifolium* and *S. guayanensis*. Swamy (1980) critically examined and reinterpreted the embryogeny in *S. sagittifolia*, on the hypothesis that the origin of the epicotyledonary and cotylary initials is from the terminal cell of the filamentous proembryo (Fig. 52.2.1A–S).

The terminal cell *ca* of the two-celled proembryo (Fig. 52.2.1A, B) divides transversely to form a three-celled proembryo, *c*, *d*, *cb* (Fig. 52.2.1C). The apical cell *c* undergoes two vertical divisions (Fig. 52.2.1D–F). By a transverse division the derivatives of *c* form two superposed tiers *l* and *l'* (Fig. 52.2.1G). Periclinal divisions in these tiers demarcate a protodermal layer (Fig. 52.2.1H, I). The basal cell *cb* divides to produce the tiers *m* and *ci* (Fig. 52.2.1D).

A quarter sector of *l'*, exhibits quiescent growth and forms the epicotylary initial (nucleated cell), whereas the remaining portion derived from *c* contributes to the formation of the single cotyledon (Fig. 52.2.1J–L).

The transverse sections (Fig. 52.2.1J_1–J_3, K_1–K_3, L_2, L_3) of the embryos (Fig. 52.2.1J–L), at three levels, clearly demarcate the quiescent centre (nucleated cell/s) in the quarter-sector of *l'*. The nucleated cell may divide once or twice (Fig. 52.2.1K, L), whereas the other sectors, along with the derivatives of *l*, undergo periclinal and anticlinal divisions so that the embryo grows in length (Fig. 52.2.1M–S). A paracotyledonary section of the embryo, passing through the quiescent centre, shows four epicotyledonary cells (Fig. 52.2.1Q); the one passing through cotyledonary plane shows massive growth (Fig. 52.2.1Q'). Due to vigorous growth of the cotyledonary portion, as

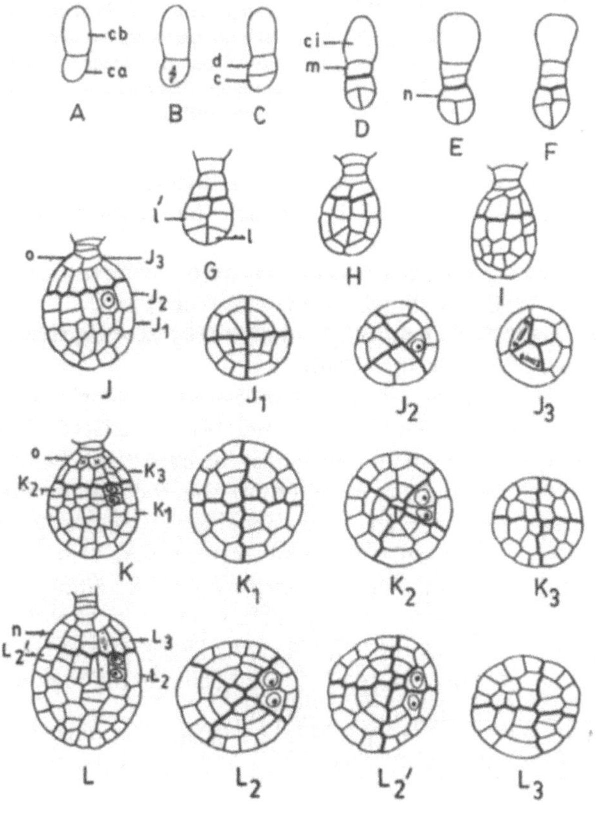

Fig. 52.2.1A–L. (*Contd.*)

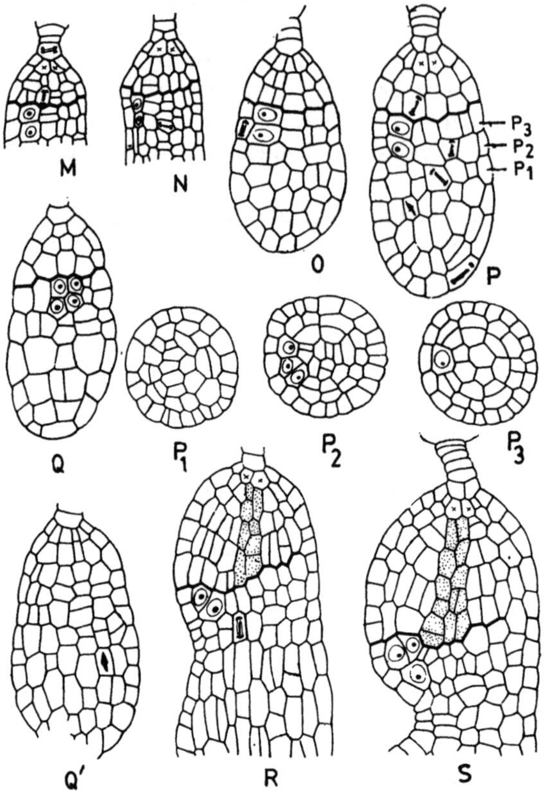

Fig. 52.2.1A–S. *Sagittaria sagittifolia*, embryogeny. **A, B** Two-celled proembryo. **C** Three-celled proembryo. **D, E** Vertical division in apical cell. **F** Quadrant. **G** Later stage. **H, I** Demarcation of protoderm in globular proembryo. **J, K, L** Epicotyl initials (nucleated cell/s) in the quarter sector of *l′*. **J_1, J_2, J_3** Transections of **J**; **K_1, K_2, K_3** transections of **K**; **L_2, $L_{2'}$, L_3** transections of **L**, at levels *l*, *l′* and hypocotyl region. The initials of tap root are marked *x* in **K**. **M–P** Advanced stages, note the two epicotyl initials (with nuclei), initials of tap root (marked *x*), and initiation of notch and divisions in the cotyledonary part in **P**. **P_1, P_2, P_3** Transections at three levels of **P**. **Q, Q′** Longisections of same embryo in paracotylary plane; **Q** through epicotylary and **Q′** through cotyledonary sections. **R, S** Older embryos showing epicotylary (nucleated cells), and cotyledonary relationships with axis of embryo; *stippled* cells denote locus of differentiation of procambium. (After Swamy 1980)

compared to the epicotylary one, the embryo shows a distinct notch at the level of epicotyl (when the longisection is through the median plane) (Fig. 52.2.1R, S).

Thus, " . . . the cotyledonary and epicotyledonary sectors differentiate in adjacent loci and these are strictly terminal on the embryonal axis". . . . " . . . the epicotylary sector is organized by one-fourth volume of the working sphere of the embryonic shoot apex" (see Swamy 1980).

The lowermost derivative of *cb*, the cell *m*, undergoes vertical divisions and delimits the epicotylary portion (marked by heavy lines) (Fig. 52.2.1E–L). This layer, by further divisions in both the planes, produces the hypocotyl. The tier *o*, above it, cuts off protodermal initials, the axial group from the initials of the quiescent centre of the tap root (marked "x" in longisections of embryos) (Fig. 52.2.1P, R, S). The rest of the cells are richly cytoplasmic and show meristematic activity which is responsible for elongation of the embryo. A central group of cells, below the quiescent centre of the tap root, differentiates into promeristem of the stele (stippled cells in Fig. 52.2.1R, S), and abruptly ends below the epicotylary/cotyledonary region which delimits the region of shoot apex from the hypocotylary axis of the embryo.

Hanstein (1870) and Schaffner (1897) allocated the development of cotyledon to the terminal cell of the three-celled proembryo, and the shoot apex to the middle cell. Souèges (1917a, b, 1931a) also confirmed that the terminal cell of the four-celled proembryo gives rise to cotyledon, and its subjacent (middle) cell forms the initial cell of the shoot apex. However, Swamy (1980) demonstrated, through transections of embryos at different levels, that "the loci of initiation of the epicotyl and cotyledons constituting the shoot apex are adjacent to one another and both in turn are terminal on the radicular axis".

Polyembryony is reported in *S. graminea* (B.M. Johri 1936a), and the two two-celled additional proembryos originate from fertilized synergids.

The exalbuminous curved seed has a horseshoe-shaped embryo.

S Chanda, S Nilsson and S Blackmore (1988), MT Cook (1907), KVO Dahlgren (1928, 1934b), J Hanstein (1870), BM Johri (1933a, b, 1934a, b, 1935a, b, d, 1936a, b, 1970c), LV Kudryashov and EI Savich (1963), P Maheshwari and B Singh (1943), PKK Nair (1970), SK Narasimhamurthy (1933a, b, 1935b), M Pichon (1946), JH Schaffner (1896, 1897, 1908), CK Shah and SN Pandey (1978), R Souèges (1917a, b, 1931a), MY Stant (1964), BGL Swamy (1980), H von Guttenberg and C Jakuszeit (1957)

52.3 Butomaceae

The anther wall comprises the epidermis, fibrous endothecium, an ephemeral middle layer, and tapetum of uninucleate cells (Fig. 52.3.1A) (B.M. Johri 1936b). Periplasmodium is formed as the microspores start to separate from the tetrad and persists until the pollen grains attain maturity (Fig. 52.3.1B–E). Successive cytokinesis follows meiosis in the microspore mother cells, and the tetrads are isobilateral, tetrahedral, linear and T-shaped.

The pollen grains (P.K.K. Nair 1970) are monocolpate in *Butomus*, and pantoporate in *Butomopsis*, and three-celled at the dispersal stage (Fig. 52.3.2A). The male gametes are spindle-shaped (Fig. 52.3.2A–E).

The ovule is anatropous and bitegmic with the inner integument forming the micropyle. The nucellar cells elongate along the long axis of the ovule (Roper 1952). The hypodermal archesporium is unicellular, multicellular in *Butomus umbellatus* (Holmgren 1913). A parietal cell may be formed but, generally, the archesporial cell functions directly as megaspore mother cell. It undergoes the first meiotic division, the upper smaller dyad collapses, and the larger one develops into a two-nucleate embryo sac with the chalazal nucleus smaller, which usually degenerates. Rarely, the chalazal nucleus divides in *Hydrocleis nymphoides* and *Limnocharis emarginata* (B.M. Johri 1938a, b). The micropylar nucleus of the two-nucleate embryo sac undergoes two mitotic divisions, and the four nuclei organize into the egg apparatus and upper polar nucleus. The degenerated chalazal nucleus represents the antipodal. The lower polar nucleus is never formed. The development of the embryo sac is of the Allium type, and the mature embryo sac is five-nucleate.

In *Butomus umbellatus* (Holmgren 1913, Roper 1952) the chalazal megaspore of linear or T-shaped tetrad develops into an eight-nucleate, Polygonum type of embryo sac. The polar nuclei fuse in the middle of the embryo sac.

Twin nucelli, enclosed in a common integument, have been observed in *Hydrocleis nymphoides* (B.M. Johri 1938a).

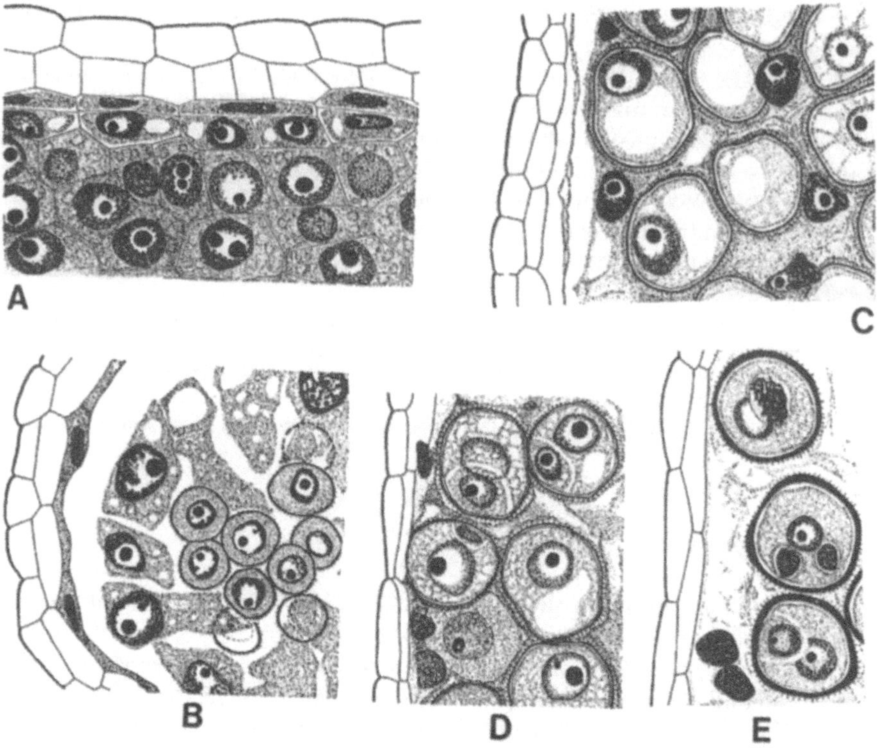

Fig. 52.3.1A–E. *Butomopsis lanceolata.* **A** Part of anther lobe to show microspore mother cells, tapetum, middle layer, endothecium, and epidermis. **B** Later stage, beginning of periplasmodium formation around microspores; the middle layer has degenerated. **C, D** Uni- and bi-nucleate pollen grains embedded in tapetal periplasmodium. **E** Traces of periplasmodium, generative cell in division in uppermost pollen grain, central pollen grain three-celled, lowermost two-celled. The tapetal nuclei are quite prominent in **C–E**. (After B.M. Johri 1936b)

Syngamy occurs normally, and the upper polar nucleus is fertilized (Fig. 52.3.2F, G), except in *Butomus umbellatus* where triple fusion occurs.

Occasional presence of mature pollen grains in the hollow stylar canal, and rarely of a germinated pollen grain attached to the ovule, is reported in *Butomopsis lanceolata* (B.M. Johri 1936b). The occurrence of pollen grains, and the possible mechanism of their entry into the carpel are regarded as gymnospermous features (Sahni and Johri 1936). Sahni assumed that even though *Butomopsis* is a confirmed angiosperm, it has retained a primitive character of "open stylar canal" (there are some other angiosperms also with a hollow stylar canal) which is considered as a "relic of gymnosperms". B.M. Johri and Ambegaokar (1984) have recently discussed this issue.

The development of the endosperm is of the Helobial type (Hall 1902, Holmgren 1913, B.M. Johri 1963b, P. Maheshwari 1963a). The nucleus of the chalazal chamber remains undivided and degenerates. It is extremely large

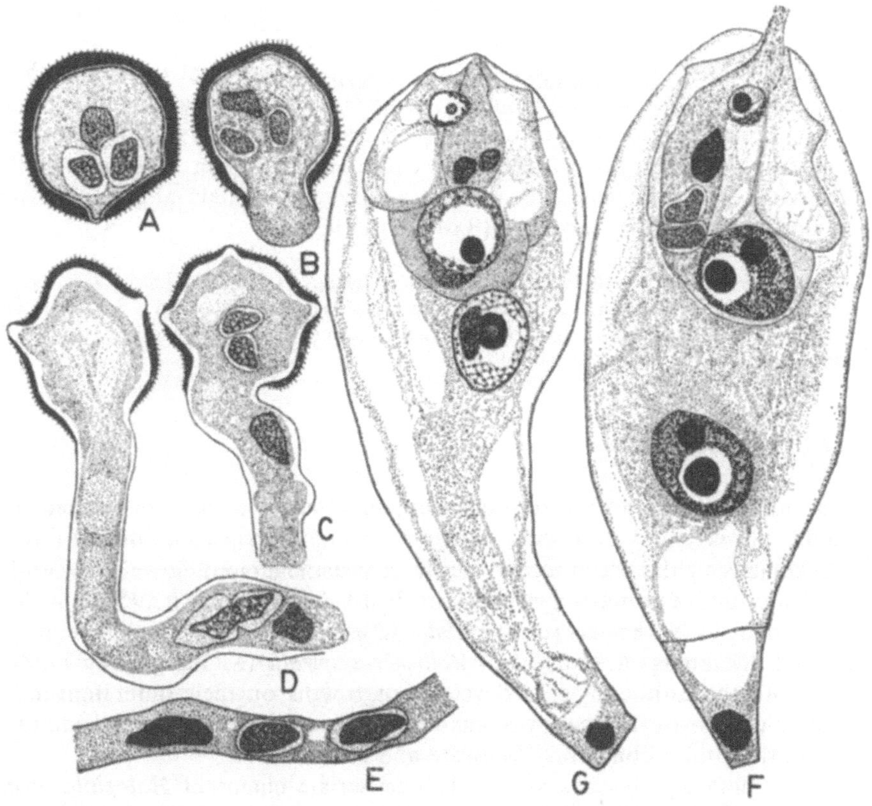

Fig. 52.3.2A–G. *Butomopsis lanceolata.* **A** Mature three-celled pollen grain. **B–D** Germinated pollen grains. **E** Part of pollen tube with male gametes and vegetative nucleus. **F, G** Double fertilization. (After B.M. Johri 1936b)

in *Butomus.* Only in *Butomopsis* it may divide once. Free-nuclear divisions occur in the large micropylar chamber, giving rise to 30–40 endosperm nuclei (Roper 1952).

The embryogeny conforms to the Caryophyllad type. In *Butomopsis lanceolata* (B.M. Johri 1936b) the zygote undergoes a transverse division, and apical (*ca*) and basal (*cb*) cells are produced. The derivatives of *ca* form the cylindrical, notched embryonal mass and a suspensor of two to four cells. *cb* does not divide, persists as uppermost cell of the suspensor, and becomes vesicular and haustorial.

According to Hall (1902), in *Limnocharis emarginata* (syn. *flava*) the basal cell *cb* of the two-celled proembryo normally does not divide and becomes vesicular. But in one case it divided repeatedly to form an embryogenic mass from which several embryos budded off (polyembryony). B.M. Johri and Bhatnagar (1955b) reinvestigated the embryogeny and observed that the basal cell *cb* may divide transversely to form two cells, but in no case does it give rise to an embryogenic mass. According to these authors, there is no polyembryony in this species as well as in other members of Butomaceae. The material studied by Hall (1902) may have been wrongly identified.

The seed is exalbuminous and exotestal with a straight or a horseshoe-shaped embryo. Both the integuments contribute to the formation of seed-coat. The outer wall of the outer epidermis of the testa and the inner wall of the inner epidermis of the tegmen are heavily cutinized. The outer epidermis of the tegmen degenerates. The cells of testa and tegmen contain abundant starch grains in *Butomopsis lanceolata* (B.M. Johri 1936b).

S Carter (1960), EL Core (1941), JG Hall (1902), J Holmgren (1913), BM Johri (1935e, 1936b, 1938a, b, 1963b, 1970c), BM Johri and SP Bhatnagar (1955b), BM Johri and KB Ambegaokar (1984), P Maheshwari (1963a), PKK Nair (1970), RB Roper (1952), B Sahni and BM Johri (1936), MY Stant (1967), Y Sunder Rao (1953)

52.4 Hydrocharitaceae

The anther wall comprises the epidermis, endothecium, one or two ephemeral middle layers, and amoeboid tapetum forming periplasmodium. Fibrous thickenings in the endothecium occur in chasmogamous flowers of *Ottelia ovalifolia*, and *Xystrolobos yunnanensis* (Ernst-Schwarzenbach 1956), but they are absent in *Nechamandra alternifolia* (Padhye and Rao 1963). The endothecial thickenings are absent in *Vallisneria spiralis* (K. Rangaswamy 1934) also, but the exothecial cells develop outgrowths on their outer tangential walls. Successive cytokinesis occurs in microspore mother cells, and the tetrads are tetrahedral, isobilateral, decussate and linear.

The pollen grains are arranged in uniseriate chains in *Halophila ovata* (Kausik and Rao 1942). The nucleus of the microspore divides and a large vegetative and a small generative cell are formed. The spindle-shaped generative cell divides followed by a transitory cell-plate. Finally, two sperm cells are produced (Fig. 52.4.1A–L).

The monocolpate pollen grains are three-celled at the dispersal stage, except in *Blyxa octandra* (Lakshmanan 1961) and *Ottelia alismoides* (Islam 1950) where they are two-celled. The generative cell is exceptionally long, and occupies the entire width of the pollen grain. The presence of pollen grains inside the ovary wall is reported in *O. alismoides* (Islam 1950).

Pettitt (1980) studied reproduction in three sea-grasses. In *Enhalus acoroides* and *Thalassia hemiprichii*, the pollen grains are spherical (Fig. 52.4.2A) with ornamental exine, while in *Halophila stiphlacea* they are reniform and the exine is absent. They occur within transparent monoliform tubes (Fig. 52.4.2B). The stigma in these sea-grasses is papillate and "dry" with a continuous proteinaceous pellicle subtended by the cuticle. The flowering time of *Thalassia* corresponds with the spring tides and the pollen grains are released as "a mass suspended in thecal slime". The slime contains monosaccharides as mannose. The investigator suggests that the pollen-stigma reaction of sea-grasses is comparable to that of the land plants.

The ovules are ana- or orthotropous, bitegmic and crassinucellar. The inner integument forms the micropyle. In *Blyxa oryzetorum* (Govindappa and

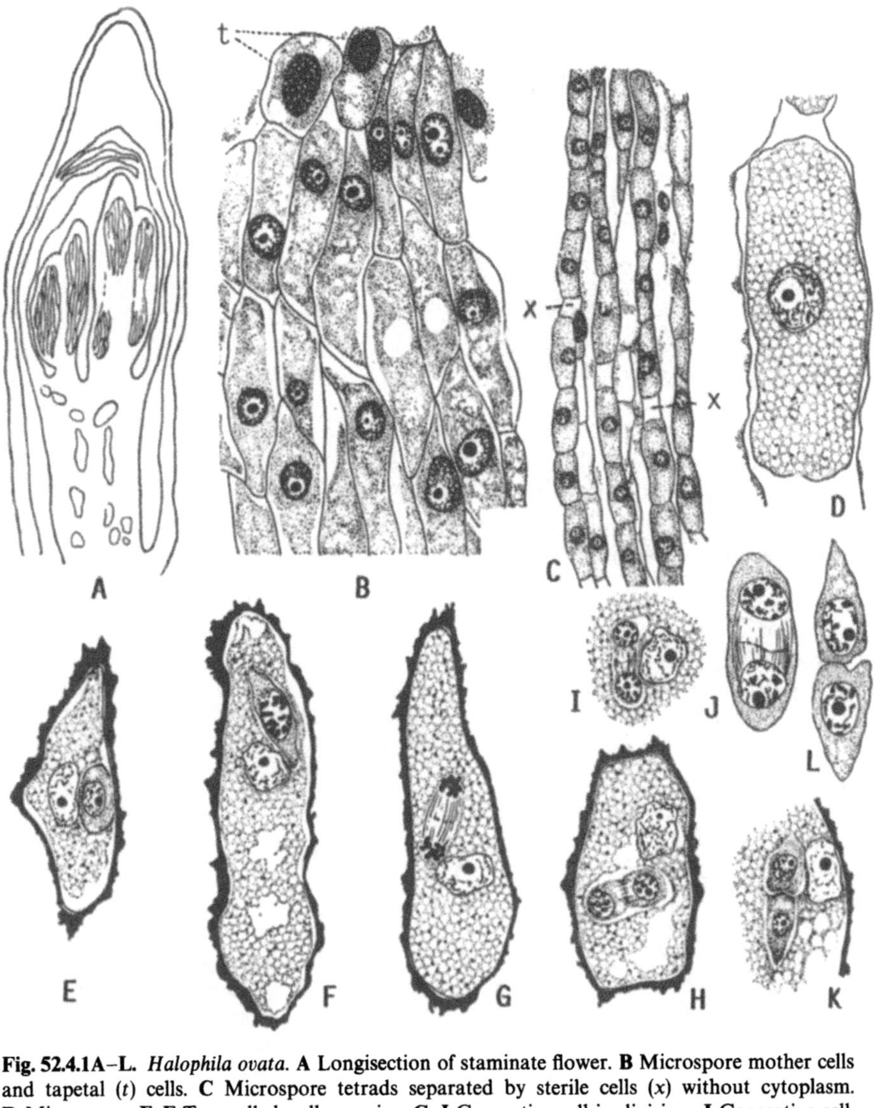

Fig. 52.4.1A–L. *Halophila ovata*. **A** Longisection of staminate flower. **B** Microspore mother cells and tapetal (*t*) cells. **C** Microspore tetrads separated by sterile cells (*x*) without cytoplasm. **D** Microspore. **E, F** Two-celled pollen grains. **G–I** Generative cell in division. **J** Generative cell. **K** Vegetative nucleus and male gametes. **L** Male gametes. (After Kausik and Rao 1942, see P. Maheshwari 1950)

Naidu 1956), the cells of nucellar epidermis divide periclinally. The archesporium is unicellular, except in *Hydrilla verticillata* (P. Maheshwari 1933b) and *Ottelia* (Narasimhamurthy 1935a). The archesporial cell cuts off a parietal cell, except in *Stratiotes aloides* (Baude 1956). The parietal tissue is two to four layers thick. The megaspore mother cell undergoes meiotic divisions, and the chalazal megaspore of a linear, T-shaped or oblique tetrad develops into a Polygonum type of embryo sac. The synergids are large, polar nuclei fuse

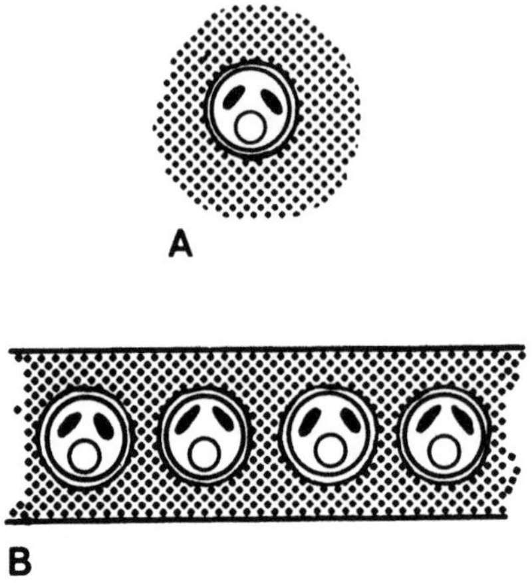

Fig. 52.4.2A, B. Two different adaptations of sea-grass pollen. **A** Spherical grain of *Thalassia* or *Enhalus*, embedded in a droplet of slime. **B** Ellipsoidal grains of *Halophila* held within a mucilagenous tube. (See Knox 1984)

before fertilization, and secondary nucleus lies in the chalazal region of the embryo sac. The three antipodals are small and persistent (Fig. 52.4.3A). The chalazal end of the embryo sac forms a pouch in *Blyxa oryzetorum* (Govindappa and Naidu 1956).

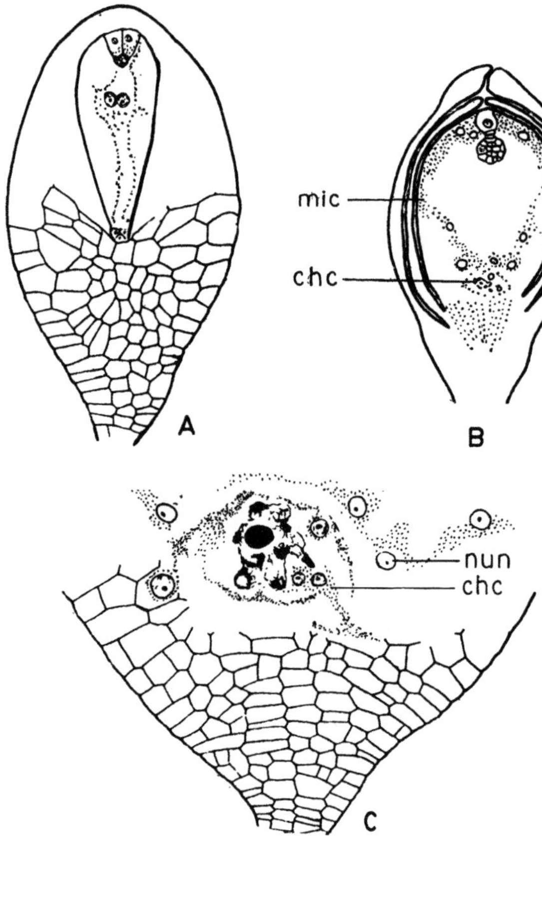

Fig. 52.4.3A–C. *Halophila ovata.* **A** Longisection of nucellus at organized embryo sac. **B** Longisection of ovule at globular proembryo, the endosperm is of the Helobial type; *mic* micropylar chamber; *chc* chalazal chamber. **C** Uninucleate chalazal endosperm chamber from **B**. Some of the nucellar cells loose their walls and the nuclei (*nun*) migrate into the chalazal region of embryo sac. The smaller nuclei in the chalazal chamber are from the broken down nucellar cells. There is meristematic activity in the chalazal tissue. (After Lakshmanan 1963a)

The entry of the pollen tube in *Hydrilla verticillata* and *Ottelia alismoides* (P. Maheshwari and Johri 1950b) is porogamous. The remnants of the pollen tube persist in the seed as a darkly staining structure. There is no indication of any functional activity and, therefore, it cannot be considered to have any haustorial role.

The primary endosperm nucleus divides before or after the division of zygote or, in some members, both divide simultaneously.

The development of the endosperm is of the Helobial type (Fig. 52.4.3B). The chalazal chamber, or the basal apparatus, becomes haustorial and its nucleus becomes hypertrophied in *O. alismoides* (Islam 1950). Free-nuclear divisions occur in the micropylar chamber. During later stages, the lower free-nuclear region of the micropylar chamber overgrows and pushes deep into the chalazal nucellar tissue. Wall formation in the endosperm commences from the micropylar end, and a peripheral layer of uninucleate cells is produced. A central sap-filled cavity occurs in *Enhalus acoroides* (Kausik 1940c). In *Blyxa octandra* (Lakshmanan 1961), the lateral position of the chalazal chamber is due to perichalazal growth. In *Ottelia alismoides* (Shah and Apparao 1982), free-nuclear divisions occur both in the micropylar and the chalazal chamber. The nucleus of the chalazal chamber divides twice so that four nuclei are formed. These nuclei fuse and produce a single polyploid mass.

A noteworthy feature is reported in *Halophila ovata* (Lakshmanan 1963a). The nucellar cells, just below the chalazal end of the embryo sac, become meristematic and add to the nucellar tissue. As the nucleus of the chalazal chamber degenerates, the cell walls between the embryo sac and adjoining tissue break down so that the nuclei of nucellar cells, along with the cytoplasm, migrate into the chalazal chamber (Fig. 52.4.3B–C).

The embryogeny conforms to the Caryophyllad type-Sagittaria variation in *Vallisneria spiralis* (K. Rangaswamy 1934). The shoot apex has been considered to be lateral and the cotyledon terminal in *Hydrocharis morsus-ranae* (Souèges 1959a).

In *Ottelia alismoides* (Swamy 1963) the division of zygote results in the formation of the apical cell *ca* and basal cell *cb*. *ca*, by further three transverse divisions, give rise to *cc*, *m*, *n*, *n'*. A vertical division occurs in the hemispherical cell *cc* and the initials of the cotyledon and epicotyl differentiate. Haccius (1952) and Swamy (1963) working on *Ottelia alismoides*, Swamy and Lakshmanan (1962a) on *Halophila ovata* (Fig. 52.4.4A–L) and Lakshmanan (1965b) on *Hydrilla verticillata*, confirm that the epicotyl (vegetative shoot apex) and the cotyledon originate from the same terminal tier *cc*. The derivatives of *m*, *n*, and *n'* give rise to hypocotyl. The cell *cb* does not divide.

In *Halophila ovata* (Swamy and Lakshmanan 1962a), the derivatives of *m* and *n* enlarge, and maintain the horizontal arrangement of cotyledonary and epicotyledonary loci.

According to Swamy (1963), "... the concept of quadrant or octant stages is not applicable to the embryogenesis in *Ottelia*. The hemispherical terminal cell itself has been looked upon as the extremely protracted shoot apex and the cotyledon as the first product of its activity". The procambium of cotyledon originates independently of the pleurome tissue in the hypocotyl.

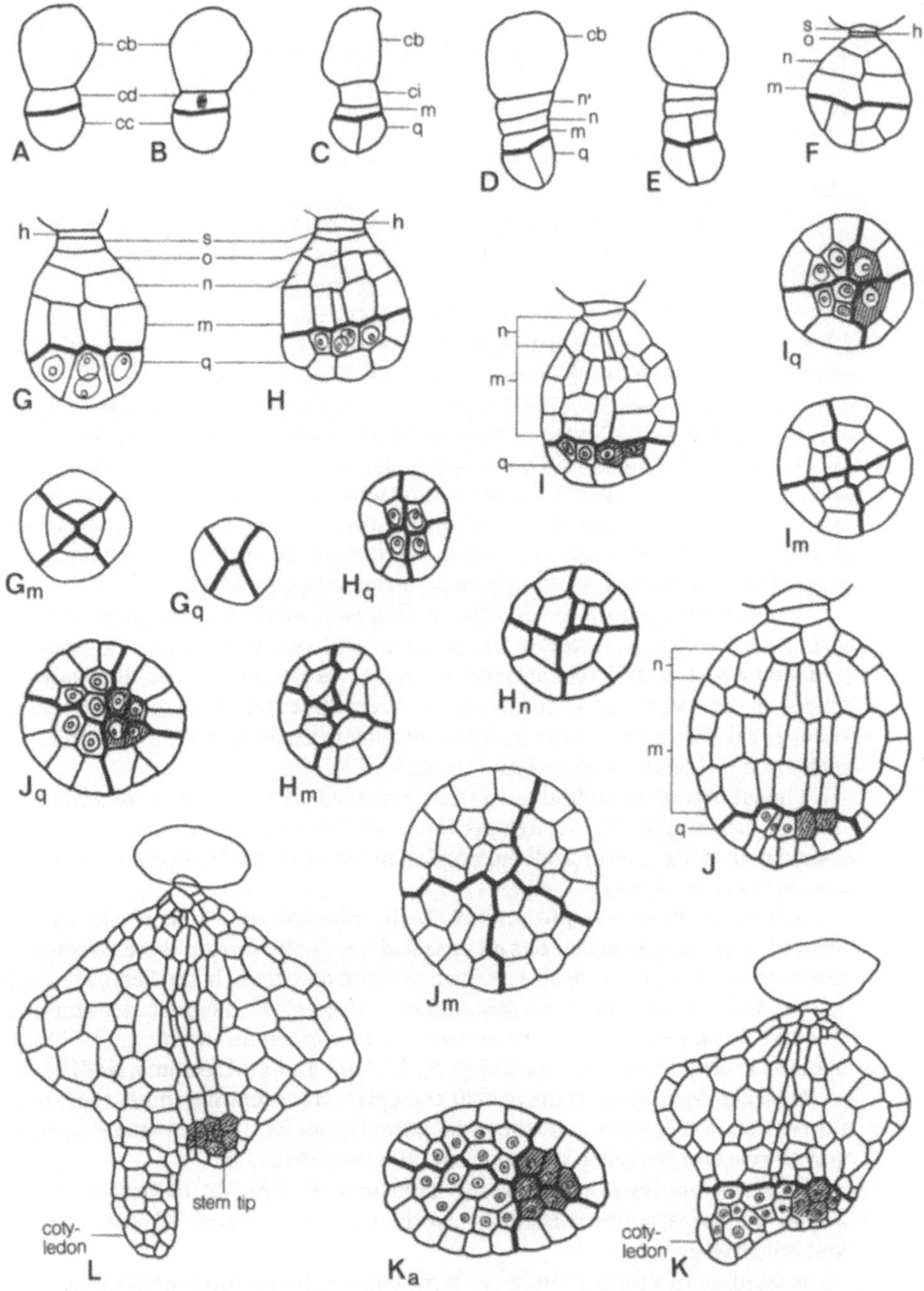

Fig. 52.4.4A–L. *Halophila ovata*, stages in the development of embryo. In monocotyledons half the derivatives of apical cell exhibit retarded growth (*shaded region* in **I–L**) and give rise to the stem tip, whereas the other derivatives grow rapidly by cell divisions and form the cotyledon. Thus, the shoot tip is not lateral but terminal in origin. The cross-sections are appropriately marked. (After Swamy and Lakshmanan 1962a)

Due to overgrowth of the cotyledon and its excessive enlargement, and the divisions in tiers *m* and *n*, the epicotyl is pushed to one side, in *Ottelia*.

The exalbuminous seed has a straight embryo with an inconspicuous plumule and thick radicle.

The structure of seed is the basis for generic classification. Both integuments take part in the formation of seed-coat. The outer epidermis of the testa is hairy in *Ottelia*, and spiny in *Blyxa*. The mesophyll develops thickenings along the middle lamella in *Blyxa*. The radial walls of the outer epidermis develop banded thickenings in *Halophila ovata* (Lakshmanan 1963a) and *Vallisneria spiralis* (Lakshmanan 1970a).

In *Hydrilla* the outer tangential wall of the outer epidermis and part of the radial walls develop thickenings, whereas in *Nechamandra* the inner tangential and part of the radial walls develop lamellar thickenings. The outer wall shows banded thickenings. In *Blyxa*, *Enhalus*, *Halophila* and *Hydrilla*, the tegmen does not show any change. In *Nechamandra*, *Ottelia* and *Vallisneria*, the inner tangential wall of the tegmen alone persists. It is smooth in *Nechamandra*, and shows minute protuberances in *Ottelia* and *Vallisneria* (Lakshmanan 1970a).

E Baude (1956), CDK Cook (1982), PA Cox and PB Tomlinson (1988), M Ernst-Schwarzenbach (1945, 1956), DA Govindappa and TRB Naidu (1956), B Haccius (1952), R Indra and KV Krishnamurthy (1984, 1985), AS Islam (1950), SB Kausik (1939c, 1940c, 1941b), SB Kausik and PVK Rao (1942), KK Lakshmanan (1961, 1963a, b, 1965b, 1970a), P Maheshwari (1933b, 1943), P Maheshwari and BM Johri (1950b), SK Narasimhamurthy (1935a), MD Padhye (1962b), MD Padhye and H Rao (1963), JM Pettitt (1980), JM Pettitt and AC Jermy (1975), JM Pettitt, SC Ducker and RB Knox (1981), K Rangaswamy (1934, 1941b), CK Shah and BJ Apparao (1982), R Souèges (1959a), N Svedelius (1904), BGL Swamy (1963), BGL Swamy and KK Lakshmanan (1962a), SW Witmer (1937), RB Wylie (1904)

52.5 Scheuchzeriaceae

The anther wall comprises the epidermis, endothecium, two ephemeral middle layers, and amoeboid tapetum of uninucleate cells, in *Scheuchzeria palustris* (Stenar 1935). Successive cytokinesis follow meiosis in microspore mother cells.

The pollen grains are inaperturate (P.K.K. Nair 1970) and three-celled at the dispersal stage. They are connected in pairs, and remain enclosed in a continuous reticulate exine.

The ovule is anatropous, bitegmic and crassinucellate. The inner integument forms the micropyle. The hypodermal archesporial cell cuts off a parietal cell, the parietal tissue is three to four cells thick. The megaspore mother cell undergoes meiotic divisions, and the chalazal megaspore of the linear tetrad develops into a Polygonum type of embryo sac. The polar nuclei fuse before fertilization, and the secondary nucleus lies in the chalazal region. The antipodal nuclei are ephemeral.

The development of the endosperm is of the Helobial type. Free-nuclear divisions occur in the micropylar chamber, and wall formation is initiated at the micropylar end. The uninucleate chalazal chamber develops into a large, highly vacuolate cell. The endosperm is consumed during the maturation of seed.

The embryogeny probably corresponds to the Caryophyllad type. The large embryo has a conspicuous suspensor cell with hypertrophied nucleus in *S. palustris* (Stenar 1935).

The seed is exalbuminous, the embryo has a rounded cotyledon and small plumule; the testa is smooth.

JS Agrawal (1952), KK Lakshmanan (1970b), PKK Nair (1970), H Stenar (1935)

52.6 Aponogetonaceae

Bouman (1985) has made a detailed study of 26 species and a hybrid of *Aponogeton.*

The anther is tetrasporangiate and the development of wall is of the Monocotyledonous type. It comprises the epidermis, fibrous endothecium, an ephemeral middle layer and tapetum of uninucleate cells. The tapetum forms a periplasmodium at tetrad stage. The microspore mother cells undergo meiotic divisions. Simultaneous cytokinesis (needs confirmation, see Bouman 1985) occurs in *Aponogeton crispus, A. monostachyon* (Sâné 1939) and *A. ulvaceus* (Stenar 1925b), and successive in *A. abyssinicus*; both types in *A. distachys* (see Schnarf 1931, Davis 1966). The tetrads are generally tetrahedral, sometimes decussate and isobilateral.

The pollen grains are monocolpate and trichotomocolpate (P.K.K. Nair 1970), and three-celled at the dispersal stage. Gardner (1976) reports two-celled pollen grains in *A. distachys*, which appears doubtful.

The ovule is ana- or hemianatropous, bitegmic and crassinucellar. The inner integument forms the micropyle. The integuments are partially or completely fused in *Aponogeton crispus* (Sâné 1939), *A. quadrangularis* (Afzelius 1920), and many other species (Bouman 1985). The unitegmy has arisen due to integumentary fusion as in *A. distachys.* The nucellar epidermal cells undergo periclinal divisions and add to the parietal tissue.

The archesporial cell cuts off a parietal cell which produces two to three layers. The megaspore mother cell undergoes meiotic divisions, and the chalazal megaspore of the linear tetrad develops into a Polygonum type of embryo sac. The polar nuclei fuse before fertilization, and secondary nucleus lies near the antipodal cells. The three antipodals persist up to 16-celled proembryo in *A. monostachyon* (Sâné 1939).

The development of the endosperm is of the Helobial type. The nucleus of the chalazal chamber rarely divides once in *A. monostachyon* (Sâné 1939). Free-nuclear divisions occur in the micropylar chamber; the thin layer of the endosperm is absorbed during maturity of the seed.

The embryogeny corresponds to the Sagittaria variation of Caryophyllad type (Johansen 1950). Judging from Sâné's (1939) figures, the embryogeny in *A. monostachyon* appears to be of the Onagrad type. The basal cell of the two-celled proembryo does not divide, enlarges into a bladder-like structure, contains a hypertrophied nucleus, and persists until a late stage of the embryo.

The zygote divides transversely and the apical cell gives rise to the entire mature embryo. The apical cell divides by two longitudinally cruciform walls, and becomes transformed into a quadrant. By one further division in each an octant is formed (Shah 1970, Yamashita 1976a).

According to Shah (1974), in *A. natans* the apical cell divides either by a longitudinal or a transverse wall, and ultimately an octant is formed. He suggests that the embryogeny is anomalous and can be assigned to "Aponogetonad" type. In the absence of a suspensor, the embryo is directly attached to the vesicular basal cell. The mature embryo shows the cotyledon, shoot apex, hypocotyl and root, which are all derived from the apical cell. The mature embryo is straight, chlorophyllous and rich in starch.

In *A. crispus* (Sâné 1939), the embryo shows rudiments of three or four foliage leaves, and initials of adventitious root.

According to Swamy (1979, see also Lakshmanan 1972), in Aponogetonales the cotyledon and the epicotyl share equal volumes of the shoot apex.

The exalbuminous seed contains a straight embryo. The cotyledon is cylindrical, and the plumule lies in a notch at its base on one side. The testa shows faint longitudinal ridges on the outer surface.

Twins and triplets are reported in *A. natans* (Shah 1970, 1974); the additional proembryos may arise due to cleavage of the zygotic embryonal mass.

The development of the seed-coat is interesting and deserves special attention. According to Bouman (1985), the seeds are of three types: (a) double-coated seeds from bitegmic ovules, (b) single-coated seeds from ovules with one or two fused integument(s), and (c) single-coated seeds from bitegmic ovules.

In the first case the exotesta forms the seed wings or ridges. The lower part of the anticlinal walls of the exotesta develop into band-like annular thickenings of pectic compounds (*A. natans*) or suberin (*A. dioecus*). The middle layers of cells remain parenchymatous. The endotestal cells remain small and are unspecialized. The tegmen is two-layered; the outer layer collapses at an early stage, whereas the inner has dark contents, probably mainly tanniniferous. In *A. ulvaceus*, the testa disintegrates except for the thickened anticlinal walls of the exotesta which form a reticulum around the central part. The tegmen is protective. The embryo is achlorophyllous.

In the second type, the testa is not protective and only the outer layer may develop faint cell wall thickenings. The middle layers are parenchymatous with conspicuous intercellular spaces. During development the nucellus and inner layers of the seed-coat are compressed and resorbed. The seeds are smooth with an oily surface. The single seed-coat is transparent, and the embryo is chlorophyllous: *A. appendiculatus*, *A. bernierianus*, *A. capuronii*, *A. crispus*, *A. distachys*, *A. jacobsenii*, *A. ranunculiflorus*, *A. rehmanni*, *A. rigidifolius* and *A. undulatus*.

The third type is intermediate between the first two groups: *A. boivinianus*, *A. bullosus*, *A. cordatus*, *A. decaryi*, *A. elongatum*, *A. junceus*, *A. longiplumulosus*, *A. loriae* and *A. madagascariensis*. The outer integument is three- to five-layered, the inner two-layered; both the integuments are unspecialized. Only in *A. decaryi* has the exotesta a relatively large number of small cells with thickened cell walls. The seeds are smooth without wings or ridges. In *A. bullosus*, however, the outer testal layer shows longitudinal zones of small cells with underlying intercellular spaces, and alternating zones of enlarged cells. There is a slight tanniniferous deposit in the cells of the inner layer of the tegmen. A noteworthy feature in this type of seed is the presence of relatively large intercellular spaces in the exostomal part of the seed-coat.

The wings and ridges are supposed to be concerned in hydrochoric dispersal. The seeds float on the surface of water due to air-filled intercellular spaces of the testa. The seed-coat swells, becomes mucilaginous, and ruptures at the micropyle.

The group of single-coated seeds is heterogeneous and the unitegmic single-coated condition is derived.

K Afzelius (1920), F Bouman (1985), GL Davis (1966), RO Gardner (1976), DA Johansen (1950), KK Lakshmanan (1972), PKK Nair (1970), YK Sâné (1939), K Schnarf (1931), CK Shah (1970, 1974), M Sharma (1967), H Stenar (1925b), BGL Swamy (1979), T Yamashita (1976a)

52.7 Juncaginaceae

The anther wall comprises the epidermis, fibrous endothecium, one or two ephemeral middle layers, and tapetum of uninucleate cells which produce a periplasmodium. The microspore mother cells undergo meiotic divisions, cytokinesis is successive, and the tetrads are isobilateral.

The pollen grains are inaperturate, two-celled at the dispersal stage (Gardner 1976).

Each carpel has a solitary ovule. The ovule is anatropous, bitegmic and crassinucellar. The micropyle is formed by the inner integument. The archesporium is multicellular in *Lilaea subulata* (J.S. Agarwal 1952). The archesporial cell cuts off a parietal cell, the parietal tissue is two or three layers thick. The megaspore mother cell undergoes meiotic divisions, and the chalazal megaspore of the linear tetrad develops into a Polygonum type of embryo sac. The polar nuclei do not fuse until fertilization. The upper cell of the three antipodals is large, persistent and undergoes secondary multiplication in *L. subulata* (Lakshmanan 1970c) and *Triglochin maritimum* (Hill 1900).

The development of the endosperm is of the Nuclear type. The free nuclei lie peripherally in the cytoplasm enclosing a large vacuole. Wall formation is initiated (around the embryo) at a very late stage of embryo development (J.S. Agarwal 1952).

The embryogeny conforms to the Caryophyllad type-Sagittaria variation, in *T. maritimum* (Souèges 1943a). According to J.S. Agarwal (1952), the division of the zygote is transverse producing the cells *ca* and *cb*; *cb* does not divide but becomes vesicular. *ca* undergoes repeated divisions resulting in a globular mass which continues to elongate until it fills the entire embryo sac. The mature embryo has a single cotyledon, lateral stem tip, and terminal radicle. Lakshmanan (1970c) has emphasized that the origin of epicotyl and cotyledon is terminal.

The seed is exalbuminous, and contains a straight embryo. The outer epidermis of the testa is thinly cuticularized, the inner one is crushed. The inner epidermis of the tegmen has a thick cuticle, the outer one is crushed (J.S. Agarwal 1952).

JS Agarwal (1952), DH Campbell (1898), RO Gardner (1976), TG Hill (1900), KK Lakshmanan (1970c, 1972), R Souèges (1943a), HD Wülff (1939)

52.8 Potamogetonaceae

The anther wall comprises the persistent tanniniferous epidermis, endothecium devoid of fibrous thickenings, two persistent middle layers, and two- to five-layered tapetum derived from the peripheral sporogenous tissue in *Syringodium isoetifolium* (Lakshmanan and Rajeshwari 1979). B.L. Gupta (1934) mentions the formation of fibrous thickenings in the endothecium at the three-celled pollen grain stage in *Potamogeton crispus*. The epidermis and the fibrous endothecium persist in the mature anther.

A periplasmodium develops when the microspore mother cells undergo meiotic divisions, and persists until a late stage. Cytokinesis is successive, and the walls are laid in a plane parallel to the axis of the cell so that the tetrads are linearly arranged in *Syringodium* (Lakshmanan and Rajeshwari 1979).

The pollen grains are inaperturate and tricolpate (P.K.K. Nair 1970). They are filiform in *Phyllospadix* and *Zostera*. The exine is apparently absent, and the intine is thin and membraneous in *Syringodium* (Lakshmanan and Rajeshwari 1979) and *Zostera* (Lakshmanan 1970d). The filamentous pollen grains of *Syringodium* lie in a bundle in the pollen sac. They are three-celled at the dispersal stage.

Pettitt et al. (1984) point out that in *Z. marina* the pollen grain is bicelled (Fig. 52.8.1A–I).

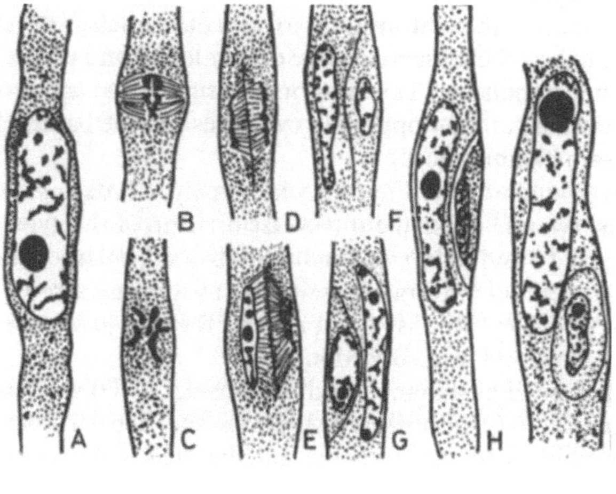

Fig. 52.8.1A–I. *Zostera marina*, **A** Uninucleate pollen grain. **B–E** Nucleus in division. **F–I** Two-celled pollen grains. Only a portion of the filiform pollen grain is shown in each figure. (After Rosenberg 1901)

The ovule is ana-, ortho- or campylotropous, bitegmic and crassinucellate. The inner integument forms the micropyle, and the cells at the tip are large and tanniniferous in *Syringodium*.

In *Potamogeton natans* (Takaso and Bouman 1984) the ovular primordium is trizonate. It consists of two tunica layers enveloping the corpus portion. The ovule is initially orthotropous, but during megasporogenesis it bends so that the micropyle is directed downward (Fig. 52.8.2A, B). At the pollination stage it becomes almost ortho-campylotropous. The curvature of the ovule is due to the active periclinal cell divisions on the inside of the adaxial wall of carpel. The inner and outer integuments are two-layered each, and the nucellus is large (Fig. 52.8.2B–D).

The archesporial cell cuts off a parietal cell, and the parietal tissue is massive. The cells of the nucellar epidermis also undergo periclinal divisions and add to the parietal tissue. The megaspore mother cell undergoes meiotic divisions, and a linear or T-shaped tetrad is produced. The chalazal megaspore develops into a Polygonum type of embryo sac. The polar nuclei fuse before fertilization, and the secondary nucleus lies near the antipodal cells. The antipodals are ephemeral.

In *Zostera marina* (K.V.O. Dahlgren 1939b) the egg is fertilized and develops into a young embryo. The secondary nucleus is not fertilized, and does not divide.

Nuclear type of endosperm occurs in *Ruppia maritima* (Graves 1908), *Syringodium isoetifolium* (Lakshmanan and Rajeshwari 1979) and *Zostera marina* (K.V.O. Dahlgren 1939b). Helobial type has been observed in *Potamogeton lucens* (Cook 1908) and *P. natans* (Holferty 1901). In Nuclear type, the endosperm remains free-nuclear throughout. Further studies on endosperm development would be very useful.

Vijayaraghavan and Kapoor (1985b) report that in *P. nodosus* the development of endosperm is of the Helobial type. Free-nuclear divisions occur in the micropylar chamber, but the chalazal chamber remains uninucleate.

The two-celled proembryo comprises the apical and basal cell. The latter does not divide any further, but enlarges and becomes vesicular with a hypertrophied polyploid nucleus.

A feature of interest is that, at the mature embryo stage, the nucleus of the basal cell as well as of the chalazal chamber become deeply lobed and extrude nucleolar bodies into the cytoplasm. The authors presume that nucleo-cytoplasm interaction "may help in the stoppage of cytokinesis in the basal cell and the chalazal endosperm chamber (cell)".

The embryogeny corresponds to the Caryophyllad type. In *Potamogeton indicus* (Swamy and Parmeswaran 1962), the transverse division of the zygote produces the apical cell *ca* and basal cell *cb*. *cb* remains undivided, enlarges and has a haustorial role. *ca* undergoes two transverse divisions forming a row of three cells – *cc*, *m* and *ci*. A vertical wall is laid in *cc*, and the two juxtaposed cells become the initials of epicotyl and cotyledon.

In *Syringodium isoetifolium* (Lakshmanan and Rajeshwari 1979) the suspensor is a single enlarged cell (*cb*) with dense cytoplasm and hypertrophied nucleus. The hypocotyl region expands radially, and its cells enlarge and

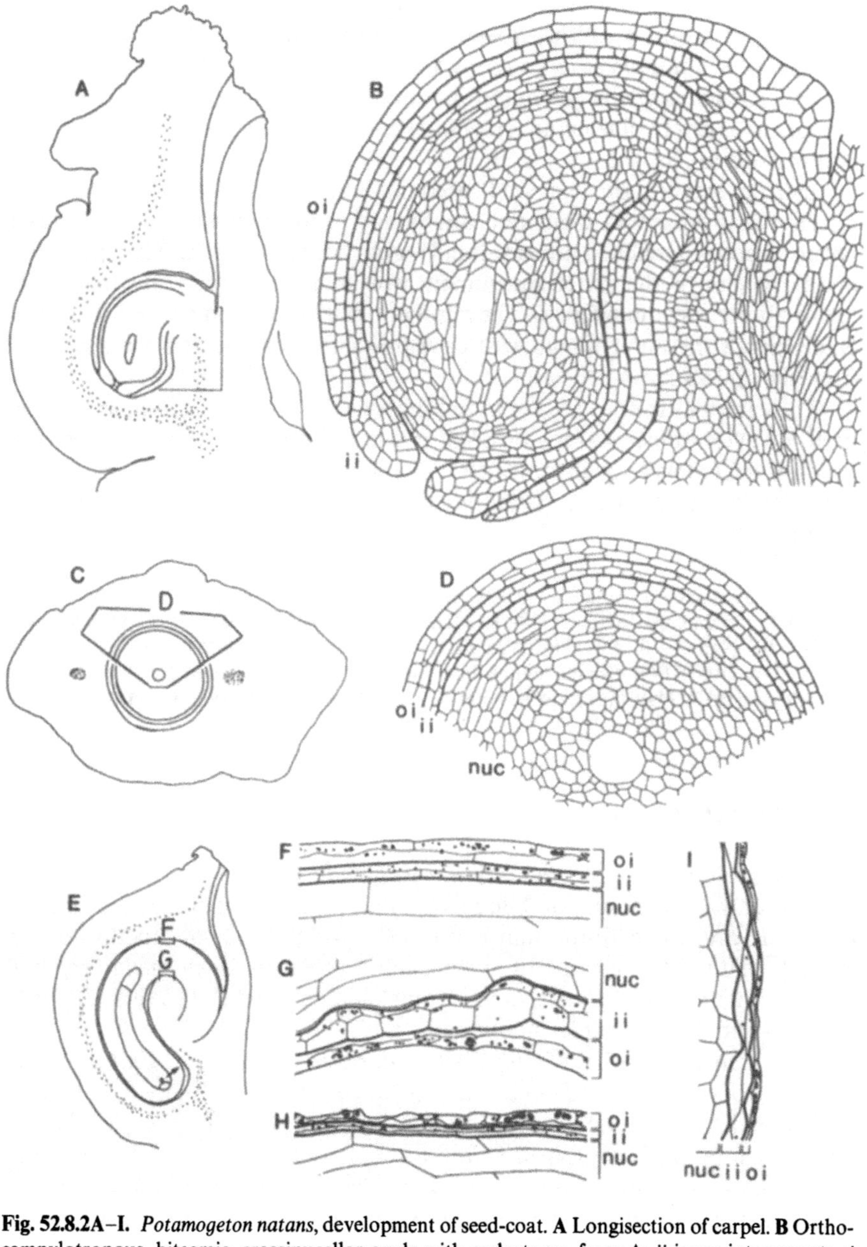

Fig. 52.8.2A–I. *Potamogeton natans*, development of seed-coat. **A** Longisection of carpel. **B** Ortho-campylotropous, bitegmic, crassinucellar ovule with endostome, from **A**; *ii* inner integument; *oi* outer integument. **C** Cross-section of carpel. **D** *Inset D* from **C**; *nuc* nucellus. **E** Longisection of immature fruit with curved seed. **F**, **G** Seed-coat marked *F* and *G* in **E**. Note the cuticle (*dark lines*) separating the testa (*oi*), tegmen (*ii*) and nucellus (*nuc*). **H** Later stage (same portion as *F*). **I** Transection of immature seed-coat at the level of *arrow* in **E**. (After Takaso and Bouman 1984)

accumulate starch grains. The mature embryo is distinguishable into two regions – the pear-shaped radially expanded storage region, and the lateral organogenic region. The latter comprises an elongated cotyledon, epicotyl with two leaf primordia, a raised hypocotyl, and an obliquely placed root.

The exalbuminous seed contains a curved embryo with a single, massive cotyledon enclosing the plumule. The outer epidermis of the seed-coat with two or three layers of testa acquires thickenings and lignification. The tegmen degenerates in *Zostera*.

During maturation the seed becomes curved. The testa, tegmen and nucellus are separated from one another by cuticle (Fig. 52.8.2E–I). These layers become stretched longitudinally and contain starch. The inner and middle layers of the outer integument are compressed and absorbed. The tegmen becomes tanniniferous. The nucellar tissue is also compressed. The outer epidermis of the outer integument develops a thin cuticle and remains loose from the endocarp.

The campylotropous seed contains embryo up to twice as long as the seed itself, or even longer; campylotrophy is a way to accommodate longer embryos. Long embryos have longer roots, hypocotyls and/or cotyledons, and are supposed to have better possibilities in seedling establishment (Takaso and Bouman 1984).

MT Cook (1908), KVO Dahlgren (1939b), AH Graves (1908), BL Gupta (1934), GM Holferty (1901), KK Lakshmanan (1968, 1970d), KK Lakshmanan and M Rajeshwari (1979), S Murbeck (1902), PKK Nair (1970), JM Pettitt, CA McConchie, SC Ducker and RB Knox (1980, 1984), O Rosenberg (1901), BGL Swamy and N Parmeswaran (1962), T Takaso and F Bouman (1984), ARA Taylor (1957), MR Vijayaraghavan and T Kapoor (1985b), KM Wiegand (1899)

52.9 Zannichelliaceae

The embryology has not been fully investigated.

The anther is mono- or bilocular. The epidermis of the anther wall is persistent, and the endothecium is fibrous. A single layer of tapetum occurs in each of the four groups of sporogenous tissue in *Zannichellia palustris* (Venkatesh 1952). *Halodule* also has a single file of tapetal cells (Fig. 52.9.1A). The tapetum forms periplasmodium.

The premeiotic anther of *Amphibolis antarctica* (Pettitt et al. 1984) contains microsporocytes and sterile cells. The latter begin to break down as meiosis begins, and the product of degeneration later supplements the periplasmodium. Each tetrad is contained within a vesicle bound by a membrane which disappears as wall initiation starts. After release from the tetrad, the microspores increase in length and rapidly become filiform. The binucleate pollen then becomes immersed in the tapetal residue.

The microspore mother cells are longitudinally elongate and arranged end-to-end in *Diplanthera uninervis* (Lakshmanan 1968). After meiosis I and II (Fig. 52.9.1B, C) the microspores in a tetrad are linearly arranged in *Diplanthera* and *Halodule*.

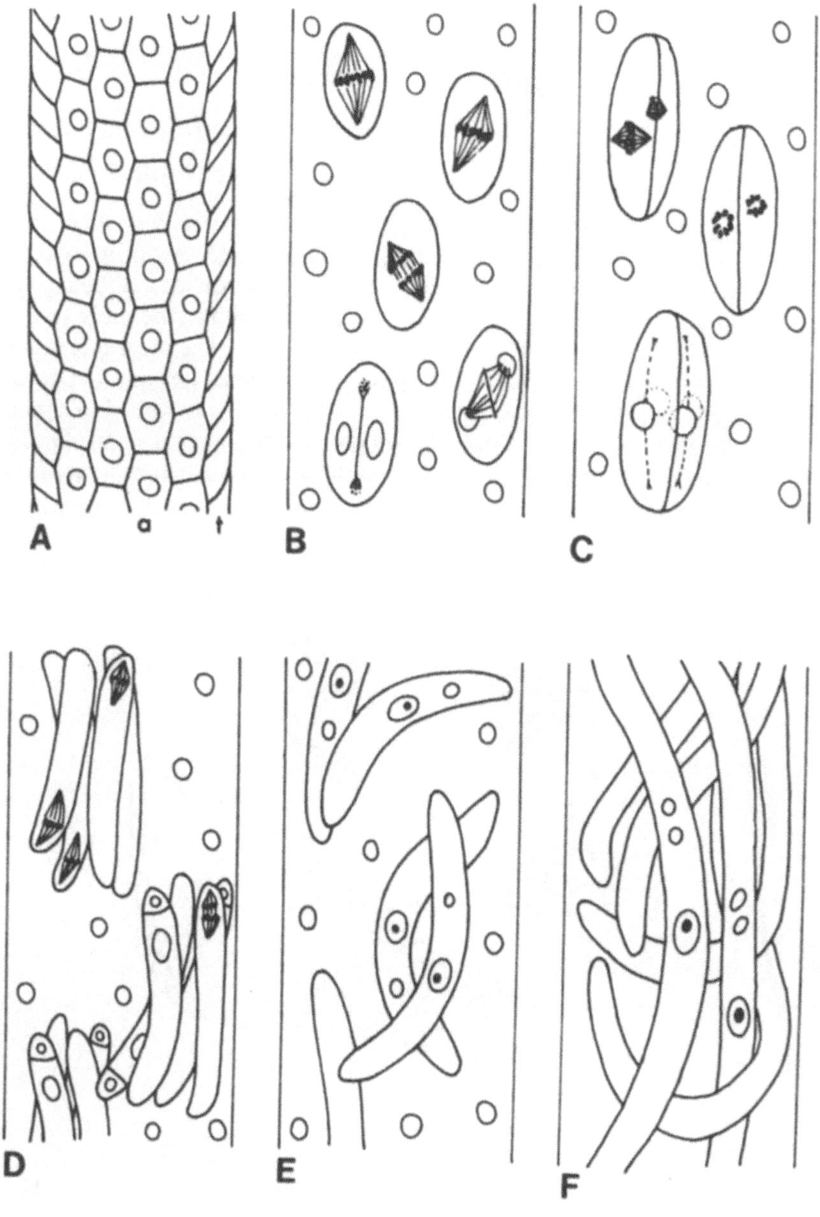

Fig. 52.9.1A–F. Filiform pollen of the sea-grass *Halodule*. **A** Archesporial cells (*a*) are surrounded by a single file of tapetal cells (*t*). **B** Meiosis I in pollen mother cells surrounded by plasmodial tapetum (nuclei indicated by *open circles* in **B** and **C**). **C** Meiosis II, young tetrads surrounded by plasmodial tapetum. **D** Young spore period, first mitotic division in pollen grains and early two-celled stage. **E** Bicelled grains. **F** Tricelled filiform grains in mature anther. (After Yamashita 1976b)

The pollen grain is thread-like, devoid of exine, and contains starch grains. It is three-celled (Venkatesh 1952), or two-celled (Lakshmanan 1970e), at the dispersal stage.

The immature pollen grains are two-celled which become three-celled at dispersal (Figs. 52.9.1D–F).

In *Cymodocea nodosa* the female flower has two carpels with bibrachiate styles to capture the filiform pollen grains (Fig. 52.9.2A–D). A three-celled pollen grain of *Amphibolis* is shown in Fig. 52.9.2E.

The ovule is anatropous, bitegmic and crassinucellate with inner integument forming the micropyle. The archesporial cell produces a primary parietal cell and megaspore mother cell. The parietal cell divides anticlinally, or periclinally; two to four parietal cells are present above the embryo sac in *Z. palustris* (Vijayaraghavan and Vidya Kumari 1974). The megaspore mother cell undergoes meiotic divisions, the upper dyad degenerates and the development of the embryo sac is of the Allium type in *Z. palustris*. Lakshmanan (1970e), however, describes Polygonum type of development in the family. A further study of the development of the embryo sac is necessary to confirm or

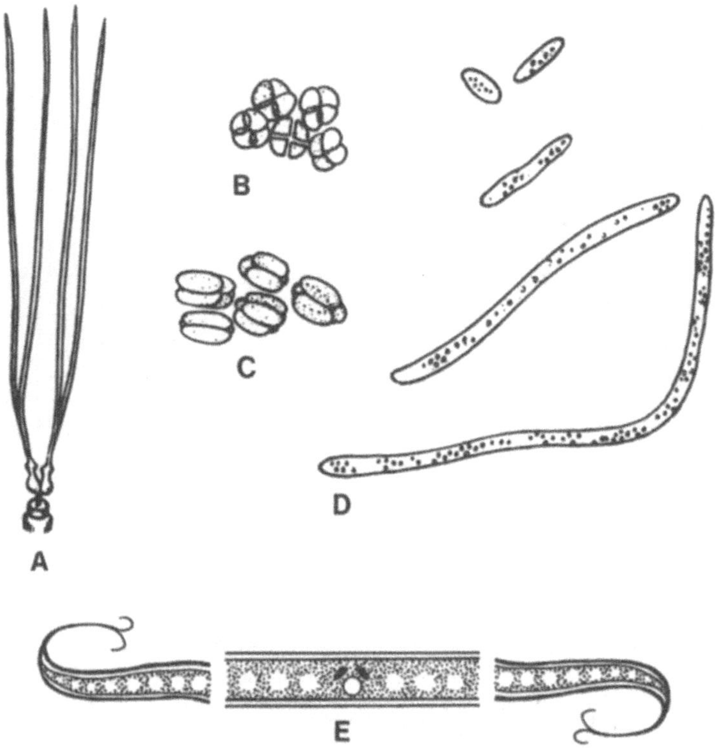

Fig. 52.9.2A–E. A–D *Cymodocea nodosa.* **A** Female flower, two carpels with bibrachiate styles suitable for capturing filiform pollen grains. **B, C** Pollen tetrads. **D** Developmental stages (five) of filiform pollen grain. (After Emberger 1960). **E** *Amphibolis*, pollen grain. (After Knox 1984)

reject the contradictory reports. The polar nuclei fuse before fertilization near the antipodals, which are ephemeral. Occasionally, secondary multiplication of antipodals has been observed (Lakshmanan 1970e).

The development of the endosperm is of the Helobial type in *Zannichellia palustris* (Lakshmanan 1965a, Vijayaraghavan and Vidya Kumari 1974). The chalazal chamber is richly cytoplasmic with a large nucleus. The basal apparatus persists up to the monocotyledonous stage of embryo. Free-nuclear divisions occur in the micropylar chamber, and cytokinesis takes place as the organs differentiate in the embryo. Nuclear type of endosperm occurs in *Cymodocea isoetifolia* and *Diplanthera uninervis* (Lakshmanan 1968, 1970d).

The embryogeny conforms to the Caryophyllad type. The cell *ca* of the two-celled proembryo, by transverse division, produces *cc* and *cd*. The tier *cc* gives rise to epicotyl and cotyledon while *cd*, by further transverse divisions, gives rise to *m*, *n*, *o* and *p*. The derivatives of *cc* produce the cotyledon and epicotyl. The basal cell *cb* does not divide, functions as suspensor, and becomes hypertrophied with a polyploid nucleus. It persists in the mature seed (Vijayaraghavan and Kapoor 1985a). The mature embryo is coiled (Vijayaraghavan and Vidya Kumari 1974).

The integuments degenerate in *Z. palustris*. The hypodermis of the pericarp develops ridges and furrows, and forms the immediate covering of the embryo.

SC Ducker and RB Knox (1976), SC Ducker, JM Pettitt and RB Knox (1978), KK Lakshmanan (1965a, 1968, 1970d, e), JM Pettitt (1976), V Posluszny and R Sattler (1976), R Souèges (1954), CS Venkatesh (1952), MR Vijayaraghavan and T Kapoor (1985a), MR Vijayaraghavan and A Vidya Kumari (1974), T Yamashita (1976b)

52.10 Najadaceae

The anther is unilocular (Hutchinson 1973). In *Najas graminea* and *N. lacerata* (Lakshmanan 1970f) it is quadrilocular. The epidermis of the anther wall persists and the endothecium lacks fibrous thickenings in *N. graminea* and *N. lacerata* (Swamy and Lakshmanan 1962b). There is no definite site for dehiscence of the anther. The tapetum has dual origin (Swamy and Lakshmanan 1962b), the outer part originates from the derivatives of primary archesporium, and the inner part from the ground tissue (Lakshmanan 1970f). It comprises uninucleate cells, and forms periplasmodium. The microspore mother cells undergo meiotic divisions resulting in tetrahedral, isobilateral, rarely linear and T-shaped tetrads. Cytokinesis is successive.

The formation of vegetative and generative cells occurs when the microspores are still attached in tetrads in *Najas palustris* (Venkatesh 1956b), and *N. graminea* and *N. lacerata* (Swamy and Lakshmanan 1962b).

The pollen grain is non-aperculate and ellipsoidal with a thin intine. The exine as well as any pore or furrow are absent (Kuprianova 1954). It is three-celled at the dispersal stage.

The ovule is anatropous, bitegmic and crassinucellate. The inner integument forms the micropyle. Secretory papillae develop at the base of the funicle, and style (Swamy and Lakshmanan 1962b). The archesporial cell cuts off a parietal cell, which does not divide further. The megaspore mother cell undergoes meiotic divisions, and the chalazal megaspore of a linear tetrad develops into a Polygonum type of embryo sac. The polar nuclei fuse before fertilization; the antipodals are persistent.

After fertilization the primary endosperm nucleus, in *Najas flexilis* and *N. marina* (Kapoor and Vijayaraghavan 1986), moves towards the chalazal end of the embryo sac. Its division leads to the formation of a large micropylar and a small chalazal chamber. Free-nuclear divisions take place in the micropylar chamber, and cell formation occurs very late. The nucleus of the chalazal chamber remains undivided and becomes polyploid. (Rarely, the chalazal chamber becomes binucleate). Thus, the development of the endosperm conforms to the Helobial type.

The zygote divides transversely to form the apical cell *ca* and basal cell *cb* (Fig. 52.10.1A). *ca*, by a further transverse division, produces *c* and *d*. The cell *c*, by two crosswise vertical divisions, forms *q* (Fig. 52.10.1B–E). Periclinal divisions occur in *q* and an axial group of four cells is separated from the protoderm initials (Fig. 52.10.1F–H). One of the axial cells (a quarter-sector) remains quiescent and gives rise to epicotyl, while the other three grow at a faster rate and form the cotyledon (Fig. 52.10.1I–L) in *Najas graminea* and *N. lacerata* (Swamy and Lakshmanan 1962b).

The tier *d*, by further transverse divisions, gives rise to *m*, *n*, *o*, *p*, and *s* (not shown). The tiers *m* and *n* undergo a vertical and then a transverse division, and the derivatives form the hypocotyl, *o* divides transversely and the daughter cells contribute to the hypocotyl. The central cells of tiers *q* to *n* become densely (*stippled*) cytoplasmic (Fig. 52.10.1J–L). *p* divides vertically, but does not contribute to the completion of the root cap and remains as such in the mature embryo. The tier *s* produces two- or three-celled suspensor.

The cell *cb* does not divide further, but becomes cyst-like and haustorial. It has a hypertrophied nucleus and a central vacuole.

In *Najas marina*, Vijayaraghavan and Kapoor (1985a) report intraseminal leaves in the mature embryo (Fig. 52.10.2). This is quite common in embryos of several monocots.

Polyembryony occurs due to the fertilization of the egg and a synergid (by two male gametes) in *Najas major* (Guignard 1901). In such cases the endosperm fails to develop. This plant needs to be re-examined.

The middle layer of the outer integument degenerates. The inner tangential wall of the outer epidermis of the testa and the outer tangential wall of the inner epidermis are juxtaposed. These are highly thickened and minutely pitted. The tegmen is represented by the external tangential wall of the outer epidermis alone, and the rest of it degenerates. The outer surface of this wall is studded with minute particles.

DH Campbell (1897), L Guignard (1901), J Hutchinson (1973), T Kapoor and MR Vijayaraghavan (1986), LA Kuprianova (1954), KK Lakshmanan (1970f), BGL Swamy and KK Lakshmanan (1962b), CS Venkatesh (1956b), MR Vijayaraghavan and T Kapoor (1985a), R Wunderlich (1954, 1959)

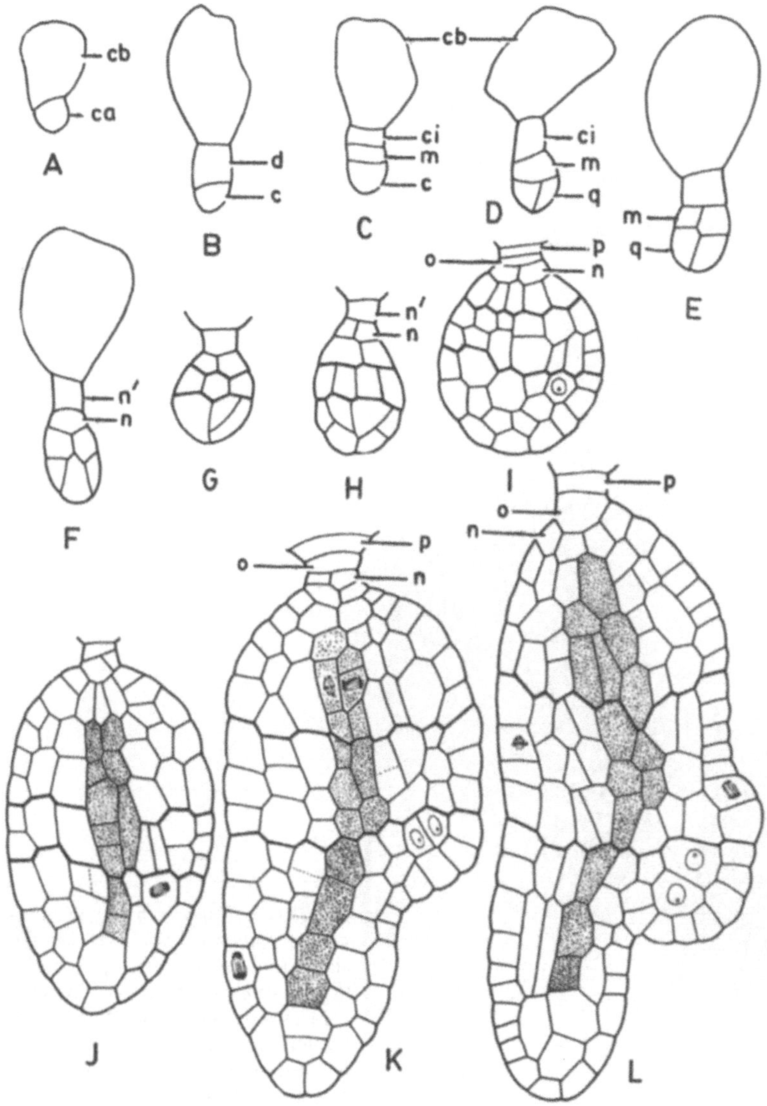

Fig. 52.10.1A–L. *Najas lacerata*, development of embryo. **A** Two-celled proembryo. **B–E** Formation of quadrant. **F–H** Periclinal divisions to delimit protoderm. **I, J** Initiation of epicotyl and cotyledon. **K, L** Rapid development of the cotyledonary zone, the development of the epicotyl is quiescent; both are derived from *q*. The central cells of tiers *q* to *n* become densely cytoplasmic (*stippled*). (After Lakshmanan 1972)

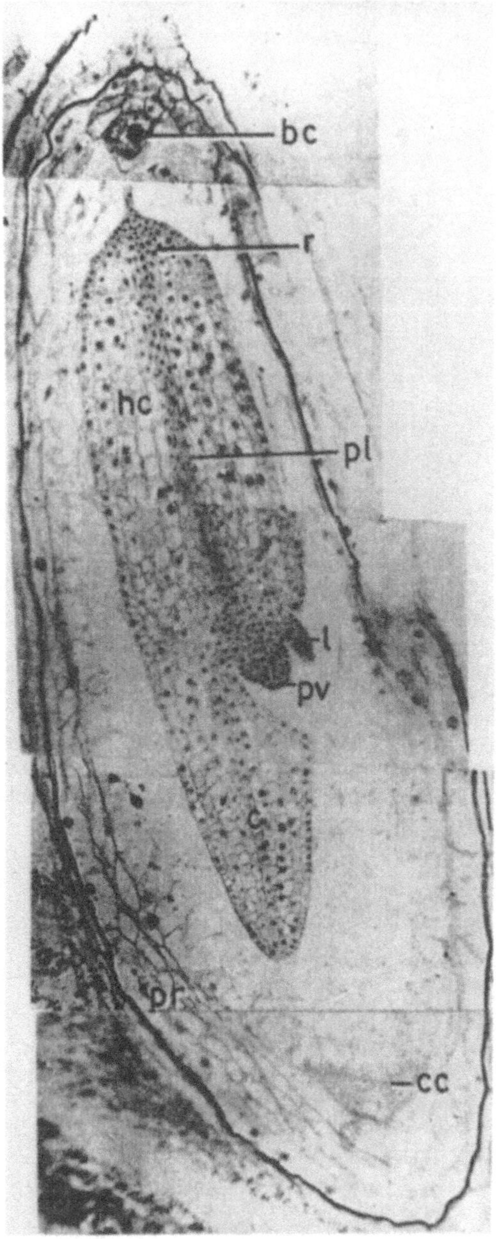

Fig. 52.10.2. *Najas marina*, longisection of portions of seed. The embryo shows root primordium (*r*), hypocotyl (*hc*), procambial strand (*pl*), intraseminal leaves (*l*), shoot apex (*pv*), and cotyledon (*c*). Note the basal cell (*bc*), part of suspensor, chalazal endosperm chamber (*cc*), and persistent nucellus, *pr* perisperm. (After Vijayaraghavan and Kapoor 1985a)

52.11 Taxonomic Considerations

Alismataceae and Butomaceae. Alismataceae and Butomaceae comprise the suborder Alismatineae (Melchior 1964). The taxonomists have variously assigned the genera in these families.

Hooker (1894) assigns *Butomus* and *Butomopsis* along with *Alisma*, *Limnophyton*, *Sagittaria* and *Wisneria* to Alismataceae. Markgraf (1936) and Engler and Diels (1936) erect two suborders, Alismatineae (with family Alismataceae) and Butomineae (with families Butomaceae and Hydrocharitaceae), under the Helobiae. Rendle (1959) retains the family Alismataceae but splits it into two subfamilies, Butomoideae and Alismoideae. Hutchinson (1973) raises these families to orders, Butomales and Alismatales. Takhtajan (1980) recognizes the order Alismatales with four families: Butomaceae (*Butomus*), Limnocharitaceae [*Limnocharis*, *Hydrocleis* and *Tenagocharis* (*Butomopsis*)], Alismataceae and Hydrocharitaceae. Cronquist (1981) recognizes only the first three families. Butomaceae usually include *Butomus*, *Butomopsis* (*Tenagocharis*), *Hydrocleis* and *Limnocharis*.

The families Alismataceae and Butomaceae, except *Butomus*, show close affinities in the development of anther, ovule, embryo sac, Helobial endosperm, and Caryophyllad type of embryogeny.

The position of *Butomus* is controversial. Melchior (1964) includes *Butomus* in the subfamily Butomoideae, and *Hydrocleis*, *Limnocharis* and *Tenagocharis* (*Butomopsis*) in Limnocharitoideae. B.M. Johri (1963b, 1970c) and P. Maheshwari and Kapil (1966) evaluate its systematic position.

The pollen morphology of *Butomus* (monocolpate, reticulate) differs from that of the other genera (pantoporate, spinulate) in Butomaceae (M. Sharma 1967). The formation of a parietal cell and development of Polygonum type of embryo sac occurs in *Butomus* alone. On the other hand, *Butomopsis*, *Hydrocleis* and *Limnocharis*, as in other alismataceous genera, show absence of parietal cell, and Allium type of embryo sac. *Butomus* also lacks laticiferous ducts, and the occurrence of sessile linear leaves, persistent petals, anatropous ovules, and seeds with straight embryo is in contrast to the other members of Butomaceae and Alismataceae (Pichon 1946). Anatomically, there is a close relationship between Alismataceae and Butomaceae but *Butomus* differs from both (Stant 1964, 1967). Cytological data confirm the transfer of *Butomopsis*, *Hydrocleis*, *Limnocharis* and *Ostenia* to Alismataceae, and retain only *Butomus* in Butomaceae (Sunder Rao 1953). B.M. Johri (1970c) supports this view.

Hydrocharitaceae. The suborder Hydrocharitineae is represented by a single family Hydrocharitaceae. Bentham and Hooker (1965c) place Hydrocharitaceae along with Burmanniaceae and Orchidaceae in the series Microspermae. Engler and Prantl (1889), and Fritsch (1932) include them in Butomineae of Helobiae. Hutchinson (1973) places Butomaceae and Hydrocharitaceae in Butomales. Cronquist (1968, 1981) erects a separate order Hydrocharitales for Hydrocharitaceae, and derives it from an ancestor similar to Butomaceae. R. Dahlgren (1977a, 1980b) includes Butomaceae, Hydrocharitaceae and Aponogetonaceae in Hydrocharitales.

The aquatic herbs of Hydrocharitaceae with regular flowers, hydrophillous pollination, and indehiscent berry-like fruits differ from those of Burmanniaceae and Orchidaceae (Lakshmanan 1970a). The unisexual flower with an inferior ovary and characteristic placentation indicates alliance to Butomaceae (Hutchinson 1973). The embryological features show greater similarity with the order Helobiae than Microspermae. Hydrocharitaceae are closer to Butomaceae, particularly the genus *Butomus* (Islam 1950, B.M. Johri 1963a), and Aponogetonaceae (monocolpate pollen, Polygonum type of embryo sac, Helobial endosperm, and Caryophyllad type of embryogeny). The assignment of Hydrocharitaceae in Helobiae, near Butomaceae, in a separate suborder is justified.

In Hydrocharitaceae the presence of a meristematic zone of nucellar cells at the base of the embryo sac, and migration of nuclei and cytoplasm of the nucellar cells into the basal apparatus of *Halophila ovata* (Lakshmanan 1963a) is worth mentioning.

Juncaginaceae and Scheuchzeriaceae. Bentham and Hooker (1965c), Engler and Prantl (1889) and Wettstein (1935) place the genera *Lilaea*, *Triglochin* and *Scheuchzeria* together in one family Juncaginaceae. Hutchinson (1973) raises the three genera into three independent families and assigns *Scheuchzeria* to Scheuchzeriaceae in Alismatales; *Lilaea* to Lilaeaceae, and *Triglochin* to Juncaginaceae in Juncaginales. Hutchinson also suggests a separate name, Heterostylaceae, for Lilaeaceae. B.M. Johri (1963b) and Lakshmanan (1970c) compare the embryological features of the three genera. *Lilaea* and *Triglochin* show close similarities, whereas *Scheuchzeria* differs in having three-celled pollen grains, two or more ovules in each carpel, and Helobial type of endosperm. In addition, *Lilaea* and *Triglochin* show secondary multiplication of antipodals.

The evidence of floral morphology and anatomy (Uhl 1947; see Lakshmanan 1970b, c) indicates that the three genera are closely related, but *Scheuchzeria* is closer to *Aponogeton*. The embryological data favour this evidence. The features of similarities are: (1) amoeboid anther tapetum, (2) successive cytokinesis in pollen mother cells, (3) anatropous, bitegmic, crassinucellate ovule, (4) micropyle formed by the inner integument, (5) formation of parietal cell, (6) Polygonum type of embryo sac, (7) Helobial type of endosperm, and (8) exalbuminous seed. It is, therefore, appropriate to separate *Scheuchzeria* from the other two genera, as done by Melchior (1964), and to keep it in a separate suborder near the family Aponogetonaceae. The other two genera, *Lilaea* and *Triglochin*, have been placed in Juncaginaceae. However, J.S. Agrawal (1952) favours inclusion of all the three genera in Scheuchzeriaceae.

The families Aponogetonaceae, Juncaginaceae, Potamogetonaceae, Zannichelliaceae and Najadaceae comprise the suborder Potamogetonineae. All the families have aquatic plants with spadix type of inflorescence. While the bracts are absent, the petals are bract-like (Hutchinson 1973).

Aponogetonaceae. Aponogetonaceae are a unigeneric family of floating perennials with embryological features in conformity with the order Helobiae. It resembles Scheuchzeriaceae more than any other family of Potamogetonineae. The occasional occurrence of secretory tapetum, monocolpate and trichotomocolpate pollen grains, persistent antipodals, and absence of wall formation in the micropylar chamber of the endosperm segregate it from *Scheuchzeria*.

According to Takhtajan (1980), Aponogetonaceae are closely related to Alismataceae. The embryological features of similarity are: (1) amoeboid anther tapetum, (2) anatropous, bitegmic ovules, (3) periclinal divisions in nucellar epidermis, and (4) Helobial type of endosperm.

Potamogetonaceae. Morphologically and embryologically Potamogetonaceae are a heterogeneous group. Some of its members were previously assigned to Najadaceae (Bentham and Hooker 1965c), and Zosteraceae (Benson 1957). Hutchinson (1967, 1973) distributes them over four families: *Zostera* to Zosteraceae, *Potamogeton* to Potamogetonaceae, *Ruppia* to Ruppiaceae, and *Cymodocea*, *Diplanthera* and *Zannichellia* to Zannichelliaceae. Cronquist (1981) groups all the families in the order Najadales, and recognizes the family status of Ruppiaceae and Zosteraceae. Takhtajan (1980) suggests the transfer of the genus *Cymodocea* to the family Cymodoceaceae, near Najadaceae. R. Dahlgren (1977a) places them (except Aponogetonaceae) in Zosterales, and erects the families Zosteraceae and Cymodoceaceae. The data based on floral morphology and anatomy (Uhl 1947; see Lakshmanan 1970b, c) support Hutchinson's classification.

Lakshmanan (1970d) describes the embryological features of the family Potamogetonaceae.

The pollen grains are filiform in *Zostera* and *Diplanthera*, and there is no exine in *Syringodium* and *Zostera*. The ovule is orthotropous in *Zostera*, and anatropous in *Diplanthera* and *Syringodium*. The parietal cell is absent in *Zostera*. The three members show similarity in a Nuclear type of endosperm which never becomes cellular. *Potamogeton* and *Ruppia* exhibit reticulate exine in the pollen grains. It is beaded in the former, and delicate and thin in the latter. Both have campylotropous ovules and Helobial type of endosperm; Nuclear type occurs in *Potamogeton natans* (Holferty 1901). The development of the embryo is essentially similar in all the members. Lakshmanan (1970d) favours the erection of families Zosteraceae and Ruppiaceae. On the basis of embryological and taxonomic characters, he recognizes the heterogeneity of Potamogetonaceae.

Takhtajan (1980) draws attention to the relationship between Potamogetonaceae and Juncaginaceae. Both show (1) amoeboid anther tapetum, (2) successive cytokinesis of pollen mother cells, (3) anatropous, bitegmic, crassinucellate ovule, (4) Polygonum type of embryo sac, (5) Nuclear type of endosperm, and (6) Caryophyllad type of embryogeny. The position assigned to Potamogetonaceae (Melchior 1964) is appropriate.

Zannichelliaceae and Najadaceae. Lawrence (1951) includes *Zannichellia* in Potamogetonaceae, and D.H. Campbell (1897) in Najadaceae. Hooker (1894), Willis (1973), Hutchinson (1973), R. Dahlgren (1975a), Takhtajan (1980) and Cronquist (1981) assign it to a separate family, Zannichelliaceae. *Zannichellia* appears to be a connecting link between Potamogetonaceae and Najadaceae.

Zannichellia is a perennial aquatic herb with a pistil of one to nine carpels and pendulous anatropous ovule, and shows Allium type of embryo sac and Helobial type of endosperm. On the other hand, *Najas* is a fresh-water annual with unicarpellate, unilocular pistil with a single basal or erect anatropous ovule, and has Polygonum type of embryo sac and Nuclear endosperm. The seed lacks seed-coat in Zannichelliaceae, whereas testa forms the covering in Najadaceae.

The submerged flowers of *Najas* and *Zannichellia* do not have fibrous thickenings in the endothecium, and exine in pollen grains. In *Najas* even pores and furrows are absent. Both have amoeboid type of anther tapetum, successive cytokinesis in pollen mother cells, anatropous, bitegmic and crassinucellar ovule, and Caryophyllad type of embryogeny. Both the families have been kept together in Najadales by Hutchinson (1973). But other embryological differences (type of embryo sac, endosperm, and seed-coat) are equally prominent. Najadaceae exhibit extreme reduction in floral organs (single stamen, uniovule ovary). R. Dahlgren (1980b) includes Najadaceae and Zannichelliaceae in Zosterales, superorder Alismatiflorae.

53 Triuridales

53.1 General Features

Triuridales is a small monotypic order. The family Triuridaceae comprises saprophytic herbs with unisexual flowers. The female flowers have an apocarpous pistil. Each carpel contains a single basal ovule. The fruits are of an aggregate type bearing endospermic seeds with a small, globular and undifferentiated embryo.

(1) Successive cytokinesis in pollen mother cells, (2) monocolpate, three-celled pollen grains, (3) anatropous, bitegmic, tenuinucellar ovule, (4) *Polygonum* (probably also *Fritillaria*) type of embryo sac, and (5) Nuclear type of endosperm are important embryological characters.

53.2 Triuridaceae

Rübsamen-Weustenfeld (1991) has made a monographic study of the morphology, anatomy, cytology, embryology, palynology, and taxonomic relationships of the Triuridaceae – Tribe Sciaphileae: *Seychellaria, Andruris, Sciaphila* and *Soridium*; Tribe Triurideae: *Peltophyllum* and *Triuris*.

The anthers are two-, three- or four-locular. The wall is of the monocotyledonous type and shows persistent epidermis, fibrous endothecium with U-shaped thickenings, and a single ephemeral middle layer. The tapetal cells are uni-, bi- or three-nucleate, plasmodial in many Sciaphileae, secretory in Triurideae and some Sciaphileae, with intermediate forms in Sciaphileae. The microspore mother cells undergo meiotic divisions, cytokinesis is successive and the tetrads are tetrahedral and isobilateral.

The mature pollen grains are three-celled, inaperturate or monocolpate, and the exine is finely to coarsely granulose, verrucose, rarely finely reticulate, or more or less psilate (Wirz 1910, Rübsamen-Weustenfeld 1991).

The ovule is anatropous, bitegmic and tenuinucellar. Both the integuments are two-layered. The inner one forms the micropyle; the inner layer differentiates as endothelium and later becomes crushed. The hypodermal archesporial cell functions directly as the megaspore mother cell which undergoes meiotic divisions, and the chalazal megaspore of a linear or T-shaped tetrad develops into a *Polygonum* type of embryo sac. It is probably of the *Fritillaria* type in *Triuris* (Rübsamen-Weustenfeld 1991). The polar nuclei fuse before fertilization. The antipodal nuclei/cells are ephemeral.

Double fertilization has been observed in *Soridium spruceanum* (Rübsamen-Weustenfeld 1991). Development of the endosperm is of the Nuclear type. The chalazal region shows dense cytoplasm and aggregation of polyploid nuclei. The food reserve consists of protein and starch in Sciaphileae, but exclusively starch in Triurideae (Rübsamen-Weustenfeld 1991).

The zygote divides transversely into the apical and basal cell. The basal cell again divides transversely, and divisions in the derivatives of the basal cell may be transverse, longitudinal, or oblique. The suspensor remains short, but in *Sciaphila* it may comprise 40–50 cells. The apical cell forms a quadrant and an octant followed by demarcation of the protoderm. The mature embryo is small, undifferentiated and globose (Rübsamen-Weustenfeld 1991).

The endospermic seeds are compact, elliptical to elongated. The outer integument alone forms the testa, and the cells of the subepidermal layer elongate unequally, and produce ribs and furrows. In later stages, the cells show characteristic thickenings. The cuticle adjacent to the endosperm is strongly thickened.

Parthenocarpy has been noticed in *Sciaphila* spp. (Poulsen 1905; Wirz 1910).

I Ohga and Y Sinotô (1924, 1932), VA Poulsen (1886, 1905), T Rübsamen-Weustenfeld (1991), H Wirz (1910)

53.3 Taxonomic Considerations

Triuridaceae. Bentham and Hooker (1965c) place Triuridaceae in Apocarpae along with Alismataceae and Najadaceae. Engler and Prantl (1931) and Hutchinson (1973) treat them as the only member of the order Triuridales. They are considered as an advanced but reduced form of Helobiae, and derived from it. Thorne (1977) traces the possibility of a common ancestry of Liliiflorae and Triuridales, on a morphological basis. R. Dahlgren (1980b) assigns Triuridales to the superorder Triuridiflorae next to Alismatiflorae. Cronquist (1981) includes Petrosaviaceae and Triuridaceae in Triuridales in the subclass Alismatidae.

Rübsamen-Weustenfeld (1991) discusses the relationships of Triuridaceae with Burmanniales–Burmanniaceae; Liliiflorae–Petrosaviaceae and Melanthiaceae; Alismatiflorae–Alismatales: Alismataceae, and Najadales: Scheuchzeriaceae. The chief resemblances and differences in their embryology are: (a) anther four-locular, two-, three- or four-locular in Triuridaceae; (b) wall four-layered, sometimes five-layered in Burmanniaceae and five- to six-layered in Melanthiaceae; (c) endothecium fibrous, except in Burmanniaceae; (d) tapetal cells one- to four-nucleate; secretory in Burmanniaceae, Petrosaviaceae and Melanthiaceae, secretory/plasmodial in Triuridaceae, plasmodial in Alismataceae and Scheuchzeriaceae; (e) cytokinesis successive, rarely simultaneous (?) in Petrosaviaceae, successive/simultaneous in Melianthaceae; (f) pollen in monads; monads, dyads and tetrads in Burmanniaceae, dyads in Scheuchzeriaceae; three-celled, two-/three-celled in Burmanniaceae, two-celled in Petrosaviaceae and Melanthiaceae; (g) ovule anatropous, ortho-, hemiana- or anatropous in Melanthiaceae; bitegmic; tenuinucellate, tenui-/crassinucellate in Melanthiaceae, crassinucellate in Scheuchzeriaceae; (h) embryo sac of the *Polygonum* type, *Allium* type in Alismataceae, *Polygonum* and *Allium* type in Burmanniaceae, *Polygonum* and *Fritillaria* (?) type in Triuridaceae; (i) endosperm Nuclear type in Petrosaviaceae and Triuridaceae, Helobial in Burmanniaceae, Melanthiaceae and Scheuchzeriaceae, Helobial and Nuclear in Alismataceae; reserve food protein and starch in Burmanniaceae, protein in Petrosaviaceae, Melanthiaceae and Triuridaceae, starch in Alismataceae and Najadaceae; (j) embryo Onagrad or Asterad type in Burmanniaceae, and Caryophyllad in Alismataceae and Scheuchzeriaceae, small and undifferentiated in Burmanniaceae, Petrosaviaceae and Triuridaceae, small and differentiated in Melanthiaceae, large and differentiated in Alismataceae and Scheuchzeriaceae, suspensor reported in Burmanniaceae, Petrosaviaceae, Triuridaceae, Alismataceae and Scheuchzeriaceae; (k) testa formed by both integuments, only outer integument in Triuridaceae, and outer layer of outer integument in Scheuchzeriaceae.

The above summary of embryological characters shows several resemblances between the Triuridaceae and Burmanniaceae (Burmanniales), but they (Triuridiflorae, Triuridales) resemble the Alismatiflorae (Alismatales, Helobiae) and Liliiflorae (Liliales) much more. Therefore, their placement between the latter two superorders/orders is fully justified (Melchior 1964, Rübsamen-Weustenfeld 1991).

54 Liliiflorae

The large order Liliiflorae comprises 17 families, distributed over 5 suborders: Liliineae, Pontederiineae, Iridineae, Burmanniineae and Philydrineae. Liliineae includes Liliaceae, Xanthorrhoeaceae, Stemonaceae, Agavaceae, Haemodoraceae, Cynastraceae, Amaryllidaceae, Hypoxidaceae, Velloziaceae, Taccaceae and Dioscoreaceae; Pontedereineae–Pontederiaceae;

Iridineae–Iridaceae and Geosiridaceae; Burmanniineae–Burmanniaceae and Corsiaceae; Philydrineae–Philydraceae.

54.1 General Features

Liliiflorae exhibit variations in embryological features. However, the common features are: (1) secretory anther tapetum, amoeboid in Pontederiaceae, both types in Haemodoraceae and Hypoxidaceae, (2) successive cytokinesis in pollen mother cells, only the simultaneous type in Cynastraceae, Taccaceae, Dioscoreaceae and Iridaceae, both types in Liliaceae, Agavaceae and Haemodoraceae, (3) two-celled pollen grains, two- as well as three-celled in Liliaceae and Burmanniaceae, pollen grains in permanent tetrads in Velloziaceae and Philydraceae, (4) anatropous, bitegmic ovule with micropyle formed by the inner integument, crassinucellar ovules common, only tenuinucellar in Velloziaceae and Burmanniaceae, (5) Polygonum type of embryo sac, Allium, Endymion, Drusa, Adoxa and Fritillaria in Liliaceae, Drusa in Dioscoreaceae, and Allium in Hypoxidaceae and Burmanniaceae, (6) Helobial type of endosperm, only the Nuclear type in Stemonaceae, Cynastraceae, Taccaceae, Dioscoreaceae and Iridaceae; mature endosperm with fat, oil, protein, and hemicellulose on cell walls, and (7) Asterad type of embryogeny, Onagrad, Caryophyllad or probably Chenopodiad type in Liliaceae, Caryophyllad in Iridaceae, Onagrad in Phylidraceae, and undifferentiated embryo in Burmanniaceae.

54.2 Liliaceae

The anther wall comprises the epidermis, fibrous endothecium, two ephemeral middle layers, and secretory tapetum with binucleate cells in *Lilium neilgherrense* (Subramanyam and Govindu 1949). The tapetal cells are two-, four-, six-nucleate, and the nuclei fuse to become polyploid in *Dipcadi saxorum* (Chennaveeraiah and Mahabale 1962). The microspore mother cells undergo meiotic divisions, and cytokinesis is successive. The tetrads are isobilateral, T-shaped and decussate in *Asparagus* (Venkateswarlu and Raju 1958), tetrahedral in *Dianella nemorosa* (M.V.S. Raju 1957) and linear, T-shaped or ⊥-shaped in *Dipcadi*. Simultaneous cytokinesis has been reported in *Asphodelus*, *Asphodeline*, *Bulbine*, *Eremurus*, *Kniphofia* and *Tofieldia*.

The family is multipalynous. Pollen grains are trichotomocolpate, one-, two- or more-colpate (colpus distal or zonal), or porate and spiraperturate (P.K.K. Nair 1970). They are three-celled at the dispersal stage.

The pollen grains are two-celled in *Urginea indica*, *Convallaria majalis* and *Eriospermum* (Lu 1985). In *Tradescantia virginica* (Poddubnaya-Arnoldi 1976), the generative cell is horseshoe-shaped. The nucleus of the vegetative cell becomes amoeboid in *Erythronium albidum*. The generative cell divides, and male gametes are formed in the pollen tube.

In *Paris axialis* (Liang 1985), normal pollen grains are 77.1%, and abnormal 22.9%. More than 70% of abnormal pollen grains show two equal nuclei (after the division of nucleus of the uninucleate pollen grain). During the first mitotic division the nuclear spindle is at right angles to pollen wall, cytokinesis is asymmetric and a complete wall is formed. In the abnormal pollen grains the mitotic spindle is parallel to the wall of the pollen grain, cytokinesis is symmetric and, sometimes, a partial wall is formed.

The development of "pollen embryo sacs" is a unique phenomenon (see Johri and Ambegaokar 1984). Němec (1898), for the first time, discovered pollen embryo sac-like structures in the anthers of *Hyacinthus orientalis*. The pollen grains are of two types: small and large. The small pollen is devoid of starch and may become binucleate, or two-celled, or germinate into an uninucleate pollen embryo sac. The large pollen grain contains abundant food reserve. It may become binucleate, or the vegetative nucleus enlarges and forms an uninucleate, binucleate, four-nucleate, or six-nucleate pollen embryo sac. The pollen embryo sacs exhibit polarity. All the four nuclei divide synchronously. In most cases, the remnants of exine persist. Němec mentions that one of the components of a tetrad may develop into a pollen embryo sac, and the other three may remain attached to it.

Stow (1930) induced pollen embryo sacs in *Hyacinthus orientalis* by subjecting the bulbs to a temperature of 17°–20°C during meiosis of the pollen mother cells. The pollen embryo sacs are produced only in normal anthers. Stow (1934) placed the pollen embryo sac on agar medium along with pollen grains. The pollen grains germinated and the pollen tube coiled around the pollen embryo sac. He emphasizes that under abnormal conditions "necro-hormones" are produced (from the dead pollen grains) which weaken the male potency so that the female potency becomes stronger.

According to Naithani (1937), the temperature treatment given to the immature bulbs also resulted in the production of pollen embryo sacs, in *H. orientalis* var. yellow hammer.

Attention is also invited to pollen embryo sacs in *Heuchera micrantha* (Vijayaraghavan and Ratnaparkhi 1977; Sect. 26.7).

The ovule is anatropous, bitegmic, and tenuinucellar or crassinucellar. An obturator develops in *Aletris aurea* (Browne 1961) and *Chlorophytum tuberosum* and formation of postament is reported in *Allium* sp. (Mirkamilov 1975).

In *Ornithogalum caudatum* (Tilton and Horner 1980), an obturator develops from the proliferation of tissue at the base of the funiculus, and from the tip of each carpellary margin. *Ornithogalum*, thus, has a placental-funicular type of obturator. It completely covers the placenta, forming a pad of tissue which extends axially almost to the entire length of the locule, and horizontally all the way across the two rows of ovules. The base of each funiculus becomes completely surrounded by obturator, and the entire central column of tissue in the ovary provides a surface (coated with an exudate) conducive to pollen tube growth.

In *Eriospermum* (Lu 1985), the bitegmic crassinucellate ovule undergoes extensive growth in the chalazal region. The placental obturator and hypostase are well developed.

The inner integument forms the micropyle in *Blandfordia nobilis* (di Fulvio and Cave 1964) and *Allium textile* (Khaleel and Mitchell 1982). As the tip of the inner integument swells, an operculum develops in *Albuca transvalensis* (Eunus 1950a). Endothelium has been observed in *Androcymbium* sp. (Cave 1967) at eight-nucleate embryo sac. The nucellar epidermal cells undergo periclinal divisions as in *Dianella nemorosa* (M.V.S. Raju 1957).

During megagametogenesis in *Androcymbium longipes* (Cave 1967), the apical region of nucellus develops a cap of thin-walled cells above the embryo sac, and the lateral nucellar cells disintegrate. The cap is four- to six-layered in *Odontostamum hartwegii* (Cave 1952). The nucellar cap persists in the seed in *Blandfordia nobilis* (di Fulvio and Cave 1964). The degeneration of nucellus starts at the first division of the functional megaspore. In *Schoenolirion album* (Cave 1974), at the chalazal end the nucellus becomes massive during the post-fertilization period.

In *Ornithogalum caudatum* (Tilton 1980a), in the micropylar region, the nucellar epidermis becomes modified to form a nucellar cap. The cells become palisade-like with lignified walls and have dense RNA-rich cytoplasm with a large distal vacuole. At anthesis the nucellar cap and the inner integument (facing the micropylar region) secrete an exudate in the micropyle and seals the exostomium. The exudate of the nucellar cap and the hymen provide the surface for the deposition of synergid-synthesized chemotropic materials. The chemosynthesized substances are localized in the micropylar region, and they do not flood the entire locule.

A funicular obturator develops at the eight-nucleate embryo sac in *Polygonatum cirrhifolium* (Mohana Rao and Kaur 1979b).

The hypodermal archesporial cell either cuts off a parietal cell (Scilloideae) or functions directly (Lilioideae) as megaspore mother cell. Both types occur in *Amianthium muscae-toxicum, Hemerocallis fulva, Ophiopogon wallichianus, Polygonatum communatum* and *Smilacina* spp. (see Davis 1966). The parietal tissue is two- or three-layered in *Dipcadi saxorum* (Chennaveeraiah and Mahabale 1962). The megaspore mother cell undergoes meiotic divisions, and a linear or T-shaped tetrad is produced.

During meiosis of the megaspore mother cell, in *Gasteria verrucosa* (Willemse and Kapil 1981), unequal dyad cells are produced. The next division is non-synchronous, forming a triad. In tetrads the cell wall between the micropylar megaspores is irregular and, sometimes, incomplete. Usually, the micropylar megaspore degenerates first. Rarely, the second megaspore from below may function or degenerate. The degeneration of megaspores is related to deficient nutrient supply and loss of turgidity of the megaspore, and physical pressure of the functional megaspore.

Simultaneous development of megaspores up to the binucleate stage is reported in *Androcymbium* spp. (Cave 1967), *Disporum* spp. (Sugihara et al. 1969), *Gloriosa* sp. (Afzelius 1918) and *Iphigenia indica* (Sulbha 1954b). The chalazal megaspore is functional, and develops into a Polygonum type of embryo sac. The synergids are hooked in some members. Giant synergids show endomitosis in the Fritillaria type of embryo sac of *Gagea chlorantha* and *G. tenuifolia* (Gvaladze 1974), during post-fertilization stages. The polar nuclei fuse before fertilization, and the secondary nucleus lies near the antipodals. The antipodals are large, conspicuous, persistent, and mostly arranged in a T-shaped manner. They become two- or four-nucleate in *Amianthium muscaetoxicum* (Eunus 1951a) and *Iphigenia indica* (Sulbha 1954a). Their secondary multiplication has been observed in *Iphigenia, Smilacina* and *Veratrum.*

The embryo sac produces a chalazal caecum in *Aletris aurea* (Browne 1961), *Asparagus* spp. (Venkateswarlu and Raju 1958) and *Dipcadi* spp. (Mahabale and Chennaveeraiah 1962). Aposporous embryo sacs occur in *Ornithogalum gussonei* (Zabinska 1972), and twin and triple embryo sacs in *Amianthium muscaetoxicum* (Eunus 1951a). The development of twin tetrads, twin embryo sacs, their fusion into a compound structure, and fertilization of both egg apparatus by two pollen tubes are reported in *Maianthemum canadense* (Swamy 1949b).

In *Ornithogalum caudatum* (Tilton 1980b), a hypostase differentiates during the meiotic-mitotic interphase. The differentiation proceeds to the micropylar half of the embryo sac. Around the hypostase, a large quantity of starch accumulates in the chalazal region of the nucellus. The location of the hypostase between the embryo sac and the chalazal nutritive tissue suggests its probable function in translocation of nutrients into the megagametophyte.

The development of the embryo sac is of the Allium type in *Allium textile* (Khaleel and Mitchell 1982). Allium type is also reported in *A. cernuum* (Murphy 1946), *Convallaria majalis, Polygonatum odoratum* (Björnstad 1970) and *Veratrum lobelianum* (Sokolowska-Kulczycka 1973). In *A. aflatunense* (Mirkamilov 1969) the mature embryo sac is five-nucleate with the egg apparatus and two polar nuclei. Endymion type is reported in *Smilacina amplexicaulis* (Björnstad 1970).

In *Trillium camschatcense* (Naumova 1978) also, the development of the embryo sac is of the Allium type. However, several variations occur due to: (1) deviations in the number of nuclear divisions at the chalazal end, (2) delay in the number of nuclear divisions at the micropylar and the chalazal end, and (3) disturbed polarity of the nuclear distribution.

In the first case the division in the chalazal nucleus of a binucleate embryo sac is suppressed; the single chalazal nucleus in the mature embryo sac behaves as the lower polar nucleus. More often, the delay in the division of the chalazal nucleus occurs at the four-nucleate stage; either one or both the nuclei are capable of division, or neither may divide. The antipodal cells are not organized.

The second type of deviation is rare in angiosperms. One of the micropylar nuclei of the four-nucleate embryo sac fails to divide. The remaining three nuclei differentiate into two cells of the egg apparatus and upper polar nucleus, the two cells are either the synergids, or the egg cell and one synergid. The antipodal cells are absent as nuclear divisions are suppressed at the chalazal end. According to Naumova (1978), "both the egg cell and the synergids remain capable of division, a feature leading to the production of two embryos: diploid sexual embryo and haploid synergid one". The development of the female gametophyte in *T. camschatcense* needs to be reinvestigated.

Stenar (1953) described modified *Allium* type in *Polygonatum latifolium*, *P. multiflorum* and *P. odoratum*. Meiosis I in the megaspore mother cell results in the formation of a dyad. After meiosis II the walls between the newly formed nuclei breakdown. Thus, each dyad becomes binucleate; the chalazal dyad develops into an eight-nucleate embryo sac. Development of the Oenothera type is rare and occurs in *Clintonia borealis* (W.R. Smith 1911).

Drusa type has been observed in *Maianthemum canadense* (Swamy 1949b), *M. bifolium*, *Polygonatum multiflorum*, *Smilacina stellata* (Björnstad 1970, see also Eunus 1950b), *S. trifolia* (Stenar 1953), and *Tulipa aitchisonii* (Koul and Wafai 1982) and *T. clusiana* (Wafai and Koul 1984). In *M. canadense* (Swamy 1949b), in addition to the Drusa type, there seems to be a tendency for the Fritillaria type of development. During the division following the 2 + 6-nucleate stage, the metaphase spindles at the chalaza fuse in pairs so that six binucleolate nuclei are produced. The mature embryo sac is ten-nucleate.

Haque (1951) describes both the *Adoxa* and the *Fritillaria* type of embryo sac in *Erythronium americanum*.

Fritillaria type of embryo sac was first recognized by Sargant (1896) in *Lilium martagon*. She noticed fusion of the three chalazal spindles of the primary 1 + 3 four-nucleate embryo sac. Consequently, the secondary four-nucleate embryo sac had two haploid micropylar and two triploid chalazal nuclei.

For the first time, the sequence of development of the Fritillaria type was described by Bambacioni (1928a) in *Fritillaria persica*. Fritillaria type occurs in *F. liliacea* (J.S. Agrawal 1950), *F. pudica* (Eunus 1951b), *Gagea chlorantha* and *G. tenuifolia* (Gvaladze 1974), *G. stipitata* (Koul et al. 1976), *Lilium henryi* (D.C.Cooper 1935b), *L. neilgherrense* (Subramanyam and Govindu 1949), *Maianthemum canadense* (Swamy 1949b), *Tulipa affinis*, *T. anadroma*, *T. butkovii*, *T. tschimganica* and *T. vedenski* (Pechenitsyn 1972a). The three chalazal megaspore nuclei of the coenomegaspore degenerate, and the upper nucleus develops into the embryo sac in *Clintonia andrewsiana* and *C. uniflora* (Bjönrstad 1970). The development is "pseudomonosporic". A similar type is described in *Clintonia* sp. (Pahuja and Kumar 1970). In addition to the daughter nuclei of the micropylar megaspore nucleus, abortive nuclei at the chalazal end have been observed. A deviation of the Fritillaria type occurs in *Lilium regale* (Krusheva and Belcheva 1976). The four megaspore nuclei in a coenomegaspore lie in 1 + 3 position. A mitotic division in each nucleus results into two haploid nuclei at the micropylar pole and two triploid nuclei at the chalazal pole (2 + 2). During second mitosis one of the haploid nuclei (at micropylar pole) degenerates. Thus, the embryo sac contains two haploid nuclei at the micropylar pole, and these differentiate into an egg cell and the upper polar nucleus. Of the four triploid nuclei, three differentiate as antipodals, and the fourth as lower polar (triploid) nucleus. The synergids are not formed. A reinvestigation is suggested.

A noteworthy feature of the family is the formation of different types of embryo sac in the same taxon. S.C. Maheshwari (1955), in his review on bisporic embryo sacs, quotes several examples: *Allium* spp., *Canavallaria* sp., *Leucocoryne*, *Nothoscordum*, *Polygonatum* spp., *Scilla* spp., *Smilacina* and *Trillium* spp.

Double fertilization of the post-mitotic type occurs in the Fritillaria type of embryo sacs of *Tulipa* spp. (Pechenitsyn 1972b). But as the lower polar nucleus (triploid) does not take part in the formation of the secondary nucleus, the primary endosperm nucleus is diploid.

Massive development of nucellus at the chalazal end, after fertilization and before the division of zygote, occurs in *Chlorogalum*, *Schoenolirion* (Cave 1974) and *Stypandra glauca* (Cave 1975).

The development of the endosperm is of the Helobial as well as the Nuclear type. Helobial type occurs in *Asphodelus tenuifolius* (Eunus 1952), *Eucomis*, *Heloniopsis*, *Muscari*, *Ornithogalum*, *Puschkinia*, *Veltheimia* and *Veratrum* (Cave 1953), *Albuca transvalensis* (Eunus 1950a), *Blandfordia nobilis* (di Fulvio and Cave 1964), *Chlorogalum pomeridianum* and *Schoenolirion album* (Cave 1974), *Stypandra glauca* (Cave 1975), *Dipcadi* (Chennaverriah and Mahabale 1962), *Ornithogalum caudatum* (Chiappini 1962) and *Scilla indica* (Govindappa 1956b). The chalazal chamber is usually two- to four-nucleate, but in *Blandfordia nobilis* and *Dipcadi* it becomes multinucleate. Free-nuclear divisions occur in the large micropylar chamber; cytokinesis follows and the tissue becomes cellular.

The chalazal chamber of the Helobial endosperm in *Trillium undulatum* (Swamy 1948e) contains 25–30 nuclei, at the same time the micropylar chamber contains thrice this number. Wall formation occurs centripetally in both chambers. The cells are large and vacuolate in the micropylar chamber, but small with dense cytoplasm in the chalazal chamber.

Nuclear type of endosperm is reported in *Androcymbium albanense* (Cave 1967), *Eriospermum* (Lu 1985), *Iphigenia indica* (Sulbha 1954a),*Canavallaria majallis*, *Clintonia andrewsiana*, *Disporum hookeri*, *Maianthemum bifolium*, *Polygonatum multiflorum*, *P. verticillatum*, *Smilacina amplexicaulis*, *S. stellata* (Björnstad 1970), and *Gagea stipitata* (Koul et al. 1976).

The embryogeny is variable. It conforms to the Onagrad, Asterad or Caryophyllad type. In *Eriospermum* (Lu 1985), the embryogeny is of the Solanad type-Nicotiana variation (Fig. 54.2.1A–I). In *Tulipa kolpakowskiana* (Haccius and Hausner 1972), it belongs to Period V or VI of Souèges's system of classification (Fig. 3.11.19A–N; see Sect. 2.6.2). According to Swamy (1979), in Liliales one half of the shoot apex develops into the epicotyl, while the other develops into a cotyledon. In *Scilla indica* (Naidu 1984) the cells of the embryo contain raphides.

In *Erythronium americanum* (Jeffrey 1895), the zygote divides and the daughter cells produce an irregular mass of cells. This mass grows and its lower end cleaves into two to four embryonal buds. Cleavage polyembryony has also been reported in *Tulipa clusiana* (Haccius and Hausner 1972) and two- or four-celled outgrowths arise from the basal cell complex. None of these embryonal buds reach maturity.

In *Lilium elegans*, *L. henryi*, *L. martagon* and *L. regale* (D.C. Cooper 1943), concurrently with the development of a zygotic embryo, a haploid four-celled proembryo develops from the synergid. It does not grow any further and collapses.

In *Allium odorum*, Haberlandt (1923, 1925) reported the development of embryos from synergids, antipodals, and cells of the inner integument. This aspect needs re-investigation. Irregularities occur in the meiotic divisions of the megaspore mother cell (aneuspory) in *A. odorum* (Håkansson and Levan 1947), and the functional chalazal dyad is either unreduced (2n) or its two nuclei show 4n chromosomes (see Battaglia 1963).

Smilacina racemosa (Gorham 1953) reproduces only apomictically; abscission of flowers and degeneration of ovules is quite common. The pollen grains are two-celled and the pollen tubes burst in the stylar canal. The sub-hypodermal archesporial cell develops directly into megaspore mother cell. The meiotic divisions result in the formation of an upper dyad, and two chalazal megaspores. The chalazal megaspores degenerate and the upper dyad develops into an eight-nucleate embryo sac. The egg apparatus also shows signs of degeneration, and one to four nucellar embryos develop. The development of the endosperm is autonomous.

The albuminous seed contains proteins and oil. The embryo is straight or curved (Hutchinson 1973). The seed is arillate in *Asphodelus fistulosus* (Stenar 1928). The testa alone forms the seed-coat, the tegmen collapses. The outer epidermal cells of the testa are large and have dark contents (Eunus 1952).

In *Eriospermum* (Lu 1985) the seed contains a large embryo. There is no endosperm but the perisperm is very conspicuous (Fig. 54.2.1H, I). The outer epidermal cells of the testa form trichomes which develop into long hairs. The cells of the tegmen contain tannin, and their walls become cuticularized.

Figure 54.2.2A–D shows the seed and seed surface of some members of Liliaceae. The elaisome is very prominent in *Chionodoxa sardensis*.

K Afzelius (1918), JS Agarwal (1950), LE Anderson (1939), RE Angold (1968), GD Arekal (1955), V Bambacioni (1928a, b, c), V Bambacioni-Mezzetti (1931, 1940, 1943, 1947), E Battaglia (1958a, b, c, 1959, 1963), E Battaglia and E Feeley (1959), JM Bellows and R Bamford (1941), RY Berg (1959), RY Berg and JR Maze (1978), NN Bhandari and A Mal (1989), NN Bhandari and A Sachdeva (1983), NN Bhandari, M Bhargava and T Geier (1981), R Bianchi (1946), IN Björnstad (1970), FD Boesewinkel (1990), A Bresinsky (1963), RD Brock (1955), ET Browne Jr (1961), L Buchner (1948), F Buxbaum (1937a, b), SP Capoor (1937a), C Cappelletti (1931), MS Cave (1939, 1941, 1942, 1952, 1953, 1967, 1974, 1975), MS Chennaveeraiah and TS Mahabale (1952, 1962), M Chiappini (1962), DC Cooper (1934, 1935b, 1936, 1939, 1943), P Crété (1952d, e, 1954c), M Gi Chung and SB Jones Jr (1989), GL Davis (1966), BLT de Silva (1945), TE di Fulvio and MS Cave (1964), BSM Dutt (1975), OJ Eigsti (1941), A Ernst (1901, 1902), AM Eunus (1950a, b, 1951a, b, 1952), F Fagerlind (1941c), FF Flint and DA Johansen (1958), L Geitler (1941, 1955), H Gerassimova-Navashina and TB Batygina (1958), H Gerassimova-Navashina and SN Korobova (1959), Z Golaszewaka (1934), AL Gorham (1953), DA Govindappa (1956b), DA Govindappa and A Sheriff (1951), P Guérin (1930, 1931, 1937), GE Gvaladze (1961, 1974), G Haberlandt (1922, 1923, 1925), B Haccius (1954), B Haccius and G Hausner (1972), Å Håkansson (1953), Å Håkansson and A Levan (1947), A Haque (1951), G Hasitschka-Jenschke (1958), M Heatley (1916), J Heslop-Harrison (1968a, b, c, 1971, 1981), G Hoare (1934), CA Hoffman (1933), HT Horner Jr and HJ Arnott (1966), TD Howe (1940), C Hruby (1934, 1938), H Huber (1969), J Hutchinson (1973), EC Jeffrey (1895), BM Johri and KB Ambegaokar (1984), MM Johri (1962), HA Jones and SL Emsweller (1936), AC Joshi (1937b, 1939c, 1940), TF Khaleel and BB Mitchell (1982), MH Kirch (1936), K Kostriukova (1939), AK Koul and BA Wafai (1982), AK Koul, BA Wafai and AK Wakhlu (1976), RM Krusheva and RG

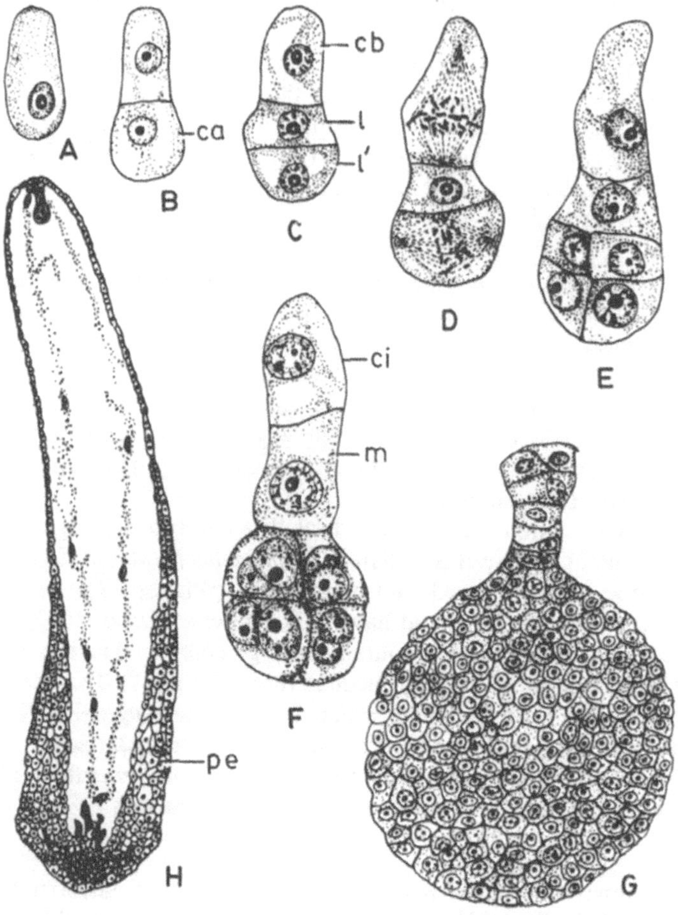

Fig. 54.2.1A–H. (*Contd.*)

Belcheva (1976), N Lakshmi (1979), JE Lazarte and BF Palser (1979), A Levan (1936), Liang-Han-Xing (1985), HF Linskens (1958), A-m Lu (1985), TS Mahabale and MS Chennaveeraiah (1962), TS Mahabale and AC Inamdar (1975), P Maheshwari (1933a, 1934b), P Maheshwari and UB Singh (1930), SC Maheshwari (1955), G Mazzolani (1978), F McAllister (1914), MA Mirkamilov (1969, 1975), J Modilewski (1925), PR Mohana Rao and A Kaur (1979b), J Moret, H Couderc, A Bari and Y Delarue (1990), JB Murphy (1946), TRB Naidu (1984), PKK Nair (1970), SP Naithani (1937), AN Namboodiri and M Bhasker (1983), K Nanda and SC Gupta (1974), TN Naumova (1978), SG Nawaschin (1898), B Němec (1898, 1931), H Ogura (1964, 1966), K Oikawa (1940), T Oksala and E Therman (1977), E Pacini and GG Franchi (1983), AN Pahuja and V Kumar (1970), UP Pechenitsyn (1972a, b), VA Poddubnaya-Arnoldi (1976), TR Porter (1936), MVS Raju (1957), SA Reznickova and MTM Willemse (1980), B Rodkiewicz (1961), WE Roever (1935), ID Romanov (1936, 1938, 1939, 1957, 1959), O Rosenberg (1946), JK Santos (1937), E Sargant (1896), K Schnarf and R Wunderlich (1939), PN Schürhoff (1922), FH Smith (1942, 1943, 1955), WR Smith (1911), ID Sokolav (1971), A Sokolowska-Kulczycka (1973, 1985), RC Spangler (1925), H Stenar (1928, 1931, 1932, 1933, 1934, 1941, 1949, 1950, 1951a, 1952, 1953), I Stow (1930, 1934), K Subramanyam and HC Govindu (1949), Y Sugihara, H Yoshimura and N Umezu (1969), K Sulbha (1954a, b), Y Sunder Rao (1940b), BGL Swamy (1948e, 1949b, 1979), M Takahashi (1987), VR Tilton (1980a, b), VR

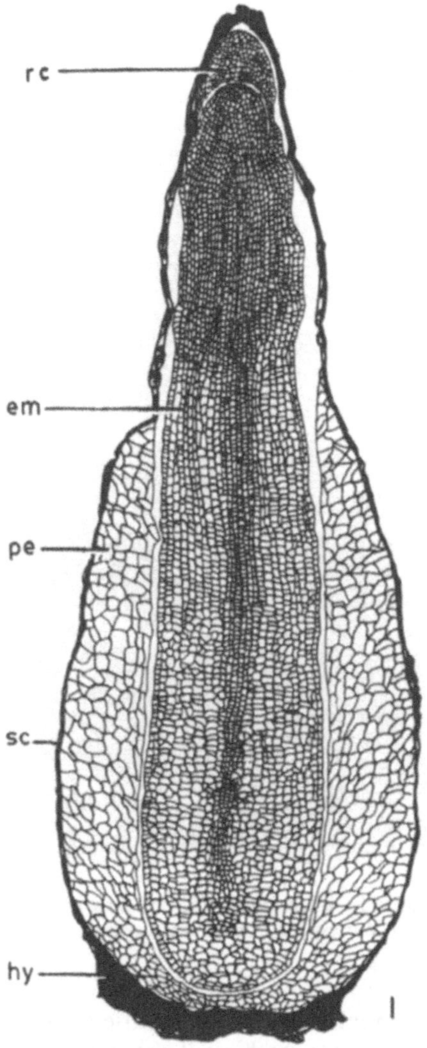

Fig. 54.2.1A–I. **A–C** *Eriospermum zeyheri*; **A** Zygote; **B** Two-celled proembryo; **C** Three-celled stage; *ca* has divided into tiers *l* and *l'*. **D–H** *E. luteorulerum*; **D** Later stage, apical cell divides longitudinally, basal one transversely; **E** Proembryo, quadrant; **F** Octant stage, *cb* has divided into *m* and *ci*; **G** Globular proembryo; **H** Embryo sac at quadrant proembryo and nuclear endosperm; *pe* perisperm. **I** *E. capense*, longisection of seed; *rc* root cap; *em* embryo; *sc* seed-coat; *hy* hypostase. (After Lu 1985)

Tilton and HT Horner Jr (1980), M Treub and J Mellink (1880), B Varitchak (1940), J Venkateswarlu and CSK Raju (1958), MR Vijayaraghavan and S Ratnaparkhi (1977), H von Guttenberg and H Heydel (1957), H von Guttenberg and C Jakuszeit (1957), BA Wafai and AK Koul (1984), E Webber (1929), KM Wiegand (1900), MTM Willemse and J Bednara (1979), MTM Willemse and RN Kapil (1981), R Wunderlich (1936, 1937), D Zabinska (1972), C Zhou, K Orndorrff, RD Allen and AE DeMaggio (1986)

54.3 Xanthorrhoeaceae

The embryology has not been fully investigated.

The ovule is anatropous, bitegmic, and crassinucellar with the inner integument forming the micropyle. The cells of the nucellar epidermis divide

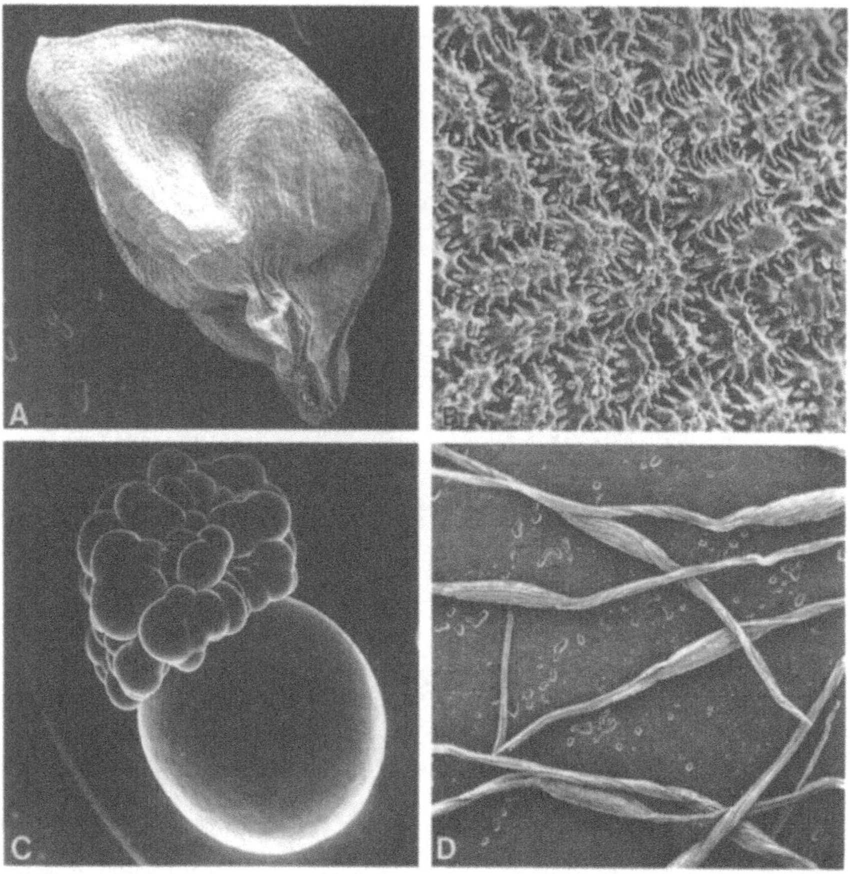

Fig. 54.2.2A–D. (SEM) **A, B** *Allium porum*, seed and seed surface. **C** *Chionodoxa sardensis*, seed with elaiosome. **D** *Narthecium ossifragum* seeds. (Courtesy Dr F. Bouman, Amsterdam)

and form a nucellar cap. The archesporial cell cuts off a parietal cell; the parietal tissue is well developed. The megaspore mother cell undergoes meiotic divisions. The chalazal megaspore of a linear tetrad forms a Polygonum type of embryo sac in *Lomandra rigida*. The polar nuclei fuse before fertilization, and the antipodals are large. The mature embryo sac is located at the apex of the nucellar column.

The albuminous seed has a straight embryo.

GL Davis (1966), K Schnarf and R Wunderlich (1939)

54.4 Stemonaceae

The embryology has not been fully investigated.

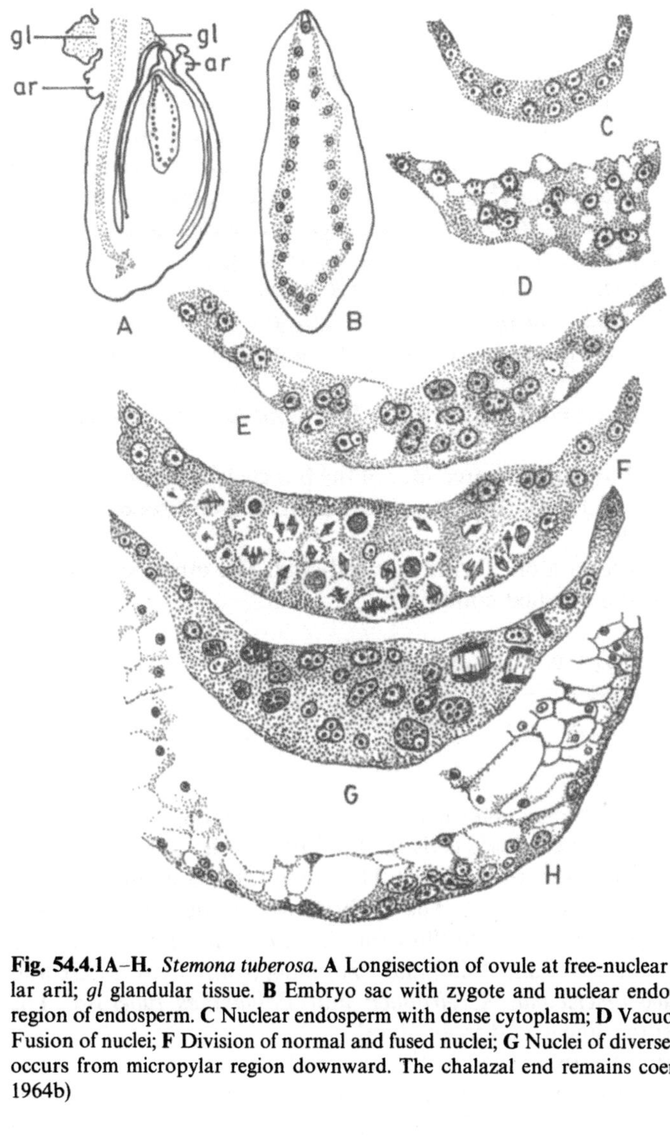

Fig. 54.4.1A–H. *Stemona tuberosa.* **A** Longisection of ovule at free-nuclear endosperm; *ar* funicular aril; *gl* glandular tissue. **B** Embryo sac with zygote and nuclear endosperm. **C–H** Chalazal region of endosperm. **C** Nuclear endosperm with dense cytoplasm; **D** Vacuolation in cytoplasm; **E** Fusion of nuclei; **F** Division of normal and fused nuclei; **G** Nuclei of diverse ploidy; **H** Cytokinesis occurs from micropylar region downward. The chalazal end remains coenocytic. (After Swamy 1964b)

Meiotic divisions in the microspore mother cells are successive in *Stemona* sp. (Palm 1920).

The ovule is anatropous, bitegmic, and crassinucellar in *S. tuberosa* (Swamy 1964b). Swamy describes it as an intermediate form between crassinucellar and tenuinucellar type. The micropylar part of the inner integument is dilated and forms the micropyle (Fig. 54.4.1A). The archesporial cell cuts off a parietal cell. The latter undergoes one more division, or remains undivided. The megaspore mother cell undergoes meiotic divisions, and the chalazal megaspore of a linear tetrad develops into a Polygonum type of embryo sac. The synergids are hooked, polar nuclei fuse before fertilization, and antipodals are ephemeral.

The development of the endosperm is of the Nuclear type (Fig. 54.4.1B), in *Stemona tuberosa* (Swamy 1964b). The nuclei lie around the central vacuole, and divisions are synchronous up to the 128-nucleate stage. Then the nuclei aggregate at the chalazal end (Fig. 54.4.1C), vacuoles appear between the nuclei (Fig. 54.4.1D) and, as the vacuoles enlarge, the nuclei form groups of four or five (Fig. 54.4.1E, F). The nuclei fuse and divide again (Fig. 54.4.1G) so that the daughter nuclei are polyploid (Fig. 54.4.1G). Cytokinesis commences from the micropylar end, and proceeds to the chalaza. At first the chalazal region remains coenocytic (Fig. 54.4.1H) but, ultimately, wall formation occurs.

The seed contains copious endosperm. The embryo is small and axile (Hutchinson 1973).

An aril is initiated towards the free-face of the funicle which spreads to the antiraphe side along the rim of the outer integument. It is composed of short, cylindrical, hollow lobes.

The tanniniferous testa alone forms the seed-coat. The inner epidermis of the tegmen persists in a crushed condition.

J Hutchinson (1973), B Palm (1920), BGL Swamy (1964b)

54.5 Agavaceae

The anther wall comprises the epidermis, fibrous endothecium, a middle layer, and secretory tapetum of binucleate cells in *Aloe ciliaris* (Govindappa 1955a). The endothecium is multilayered in *Agave*. The tapetal cells are multinucleate in *Polyanthes*, and the nuclei fuse. The microspore mother cells undergo meiotic divisions, cytokinesis is simultaneous in *Aloe ciliaris* (Govindappa 1955a), *Doryanthes excelsa* (Palm 1920) and *Phormium tenax* (Cave 1955), and successive in *Agave americana*, *A. virginica* and *Polyanthes tuberosa*. The tetrads are tetrahedral, linear and isobilateral.

The pollen grains are two-celled at the dispersal stage.

The ovule is anatropous, bitegmic, and crassinucellar with the inner integument forming the micropyle. The presence of a third integument around the base of the ovule is described in *Aloe ciliaris* (Govindappa 1955a) and *A. africana* (McNaughton and Robertson 1974). The cells of the nucellar epidermis undergo periclinal divisions. In *Agave* carbohydrate, protein and lipid accumulate in the hypostase (Tilton 1980b).

The archesporial cell cuts off a parietal cell and the parietal tissue is one- to two-layered. In *Phormium* it functions directly as megaspore mother cell. The latter undergoes meiotic divisions, and the chalazal megaspore of a linear or T-shaped tetrad develops into a Polygonum type of embryo sac. The synergids are hooked, the polar nuclei fuse before fertilization, and the secondary nucleus lies in the chalazal part of the embryo sac in *Agave parryi* (Mogensen 1970b).

At the chalazal end the embryo sac develops a narrow tube which may have a haustorial role, in *Yucca filamentosa* (H.S. Reed 1903).

The development of endosperm is of the Helobial type in *Yucca filamentosa* (Wunderlich 1938) and *Beschorneria californica* (Wunderlich 1950). Both the chambers contain a large number of nuclei. Centripetal wall formation occurs in the micropylar chamber. The endosperm is ruminate in *Yucca* spp.

The embryogeny conforms to the Caryophyllad type in *Yucca glauca* (see Johansen 1950).

The seed is albuminous, and the fleshy endosperm surrounds the small embryo. Arnott (1961) compared the seed characters, such as weight, volume, density, colour, surface texture, and shape of 22 species of *Yucca*. In *Yucca* (Horner and Arnott 1966), the seed is perispermous. The cells of the perisperm have thick pitted walls, and they contain lipoids. The seed-coat consists of an outer pigmented epidermis and 10–25 layers of parenchyma.

HJ Arnott (1959a, b, 1961, 1962), NV Biradar (1968, 1978), G Catalano (1930, 1931), MS Cave (1953, 1955), F Gioelli (1930, 1933), DA Govindappa (1955a), HT Horner Jr and AJ Arnott (1966), DH Johansen (1950), AC Joshi (1937b), AC Joshi and JV Pantulu (1941), N Lakshmi (1979), JE McNaughton and BE Robertson (1974), HL Mogensen (1970b), B Palm (1920), HS Reed (1903), L Regen (1941), H Stenar (1942), BGL Swamy and KV Krishnamurthy (1970), VR Tilton (1980b), GM Watkins (1937), FT Wolf (1940), R Wunderlich (1938, 1950)

54.6 Haemodoraceae

The embryology has not been fully investigated.

The anther tapetum is secretory in *Cyanella capensis* and *Lanaria plumosa* (de Vos 1950, 1963), and amoeboid in *Anigozanthos* and *Xiphidium* (Stenar 1927, 1938b), and *Wachendorfia paniculata* (Fig. 54.6.1A). The microspore mother cells undergo meiotic divisions, and the tetrads are tetrahedral and isobilateral. Cytokinesis is successive, except in *Cyanella* and *Lanaria* where it is simultaneous.

The monocolpate pollen grains are two-celled at the dispersal stage (Fig. 54.6.1B).

The ovule is hemianatropous or orthotropous (*Anigozanthos*), bitegmic and crassinucellar. The micropyle is formed either by the inner integument (*Dilatris*, *Wachendorfia*), or both integuments. The hypodermal archesporial cell cuts off a parietal cell, and the parietal tissue is well developed. The cells of the nucellar epidermis divide to produce two or three layers. The megaspore mother cell undergoes meiotic divisions (Fig. 54.6.1C) producing a linear tetrad. The chalazal megaspore develops into a Polygonum type of embryo sac. The polar nuclei fuse before fertilization, and the antipodals are ephemeral. The embryo sac develops a lateral caecum in *Cyanella*.

The development of the endosperm is of the Helobial or Nuclear type. Helobial type occurs in *Dilatris*, *Wachendorfia* (Fig. 54.6.1D, E) and *Lanaria* (de Vos 1956, 1963). The chalazal chamber is four-celled and haustorial in the first two members; Swamy and Parmeswaran (1963) suggest reinvestigation. It is multinucleate in *Lanaria*. Nuclear type of development occurs in *Cyanella*

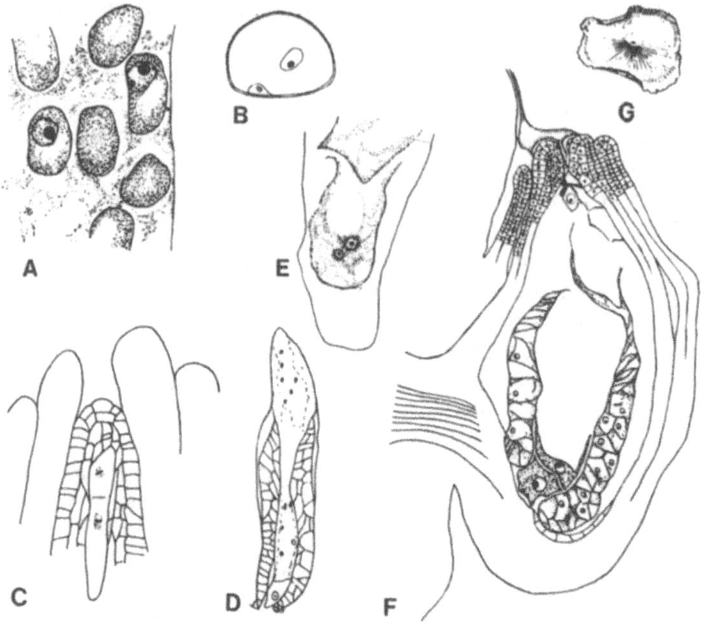

Fig. 54.6.1A–G. A–F *Wachendorfia paniculata*. **A** Amoeboid tapetum at uninucleate pollen; **B** Two-celled pollen grain; **C** Nuclei of micropylar and chalazal dyad in division; **D** Nuclear endosperm in micropylar chamber and uninucleate chalazal chamber; **E** Two-nucleate chalazal chamber; **F** Longisection of hemianatropous, bitegmic ovule. Embryo sac shows zygote, cellular endosperm in micropylar chamber, and two-celled chalazal chamber (?). **G** *Haemodorum planifolium*, seed. (**A, B, F** After Dellert 1933, **C–E** after de Vos 1956, **G** after van Steenis 1954)

capensis (de Vos 1950). Wall formation is initiated from the micropylar region at ca 100 free-nucleate stage of the endosperm, and proceeds to the chalaza (Fig. 54.6.1F). The chalazal region remains free-nuclear, and forms a haustorium.

The embryogeny conforms to the Asterad type. Nucellar polyembryony occurs in *Cyanella capensis* (de Vos 1950).

The seed (Fig. 54.6.1G) contains abundant endosperm, and a small embryo. The seed-coat is smooth, black and hard in *Lanaria plumosa* (de Vos 1963). The testa forms the seed-coat. The outer epidermis of the testa comprises palisade-like, thick-walled cells, the thickenings are of cellulose. The mesophyll is thin-walled and contains starch. The inner epidermis comprises small, thin-walled cells. The tegmen is two-layered and thin-walled, and may not persist. The seed-coat has three cuticular thickenings – one on the outer epidermis of the testa, the other between the testa and tegmen, and the third close to the tegmen.

G Bentham (1883), MS Cave (1952), R Dellert (1933), MP de Vos, (1950, 1956, 1961, 1963), BSM Dutt (1970a), F Pax (1930), MG Simpson (1988), H Steinecke and U Hamman (1989), H Stenar (1927, 1938b), BGL Swamy and N Parmeswaran (1963)

54.7 Cynastraceae

The embryology has not been fully studied.

The anther wall comprises several wall layers. The tapetum is secretory with multinucleate cells in *Cyanastrum cordifolium* (Nietsch 1941). Nuclear fusions occur in tapetal cells. Simultaneous cytokinesis accompanies meiotic divisions in microspore mother cells. The tetrads are isobilateral and tetrahedral.

The pollen grains are mono- and trichotomo-colpate, and two-celled at the dispersal stage.

The ovule is anatropous, bitegmic, and crassinucellate with the inner integument forming the micropyle. The funicular vascular strand traverses the outer integument. A single hypodermal archesporial cell cuts off a parietal cell, and the parietal tissue is two or three layers thick. The megaspore mother cell undergoes meiotic divisions. The chalazal megaspore of a linear tetrad develops into a Polygonum type of embryo sac in *C. cordifolium* (Nietsch 1941). The antipodals are ephemeral.

The obturator is placental.

The development of the endosperm is of the Nuclear type (B.S.M. Dutt 1970b). In *Cyanastrum johnstoni* (Fries 1919) the endosperm degenerates following the formation of a few nuclei. The seed is exalbuminous.

Only one seed matures in an ovary, and during its development the endosperm and nucellus are absorbed. The chalazal part of the ovule divides actively, and forms chalazosperm (Fig. 3.15.6E) (Fries 1919). It gets filled with fat and starch, and acts as a nutritive tissue after the absorption of the endosperm.

BSM Dutt (1970b), TCE Fries (1919), H Nietsch (1941)

54.8 Amaryllidaceae

The anther wall comprises the epidermis with stomata, fibrous endothecium, four middle layers, and secretory tapetum of binucleate cells in *Pancratium longiflorum* (B.S.M. Dutt 1970c). In *Furcraea gigantea* (Panchaksharappa and Rangel 1966), there are five to six wall layers and all of them, except the tapetum, develop fibrous thickenings. Übisch granules have been noted on the inner walls of tapetal cells in *Crinum defixum* (B.S.M. Dutt 1962). Amoeboid tapetum is reported in *Galanthus nivalis* (Pankow 1958). The microspore mother cells undergo meiotic divisions, and the tetrads are tetrahedral, isobilateral and decussate. Cytokinesis is successive.

The pollen grains are monosulcate, two-sulcate in *Crinum*. They are two-celled at the dispersal stage. In *Hippeastrum rutilum* (B.S.M. Dutt 1970d) they germinate in situ.

The ovules are anatropous, bitegmic and crassinucellar; the inner integument forms the micropyle. It is tenuinucellate in *Zephyranthes citrina*

(B.S.M. Dutt 1958). The ovule lacks the integument in *Crinum giganteum* (Toilliez-Genoud 1965). It is unitegmic and pseudocrassinucellar with a massive integument in *C. defixum* (B.S.M. Dutt 1962). The feebly developed nucellus of one to three cells overlies the primary archesporial cell. The nucellar epidermal cell/s divide periclinally, but soon get crushed. The single archesporial cell cuts off a parietal cell in *Furcraea gigantea* (Panchaksharappa and Rangel 1966), *Pancratium longiflorum* (B.S.M. Dutt 1970c) and *P. maritimum* (B.S.M. Dutt 1964). It develops directly into the megaspore mother cell in *Cooperia drummondii* (Church 1916), *Crinum defixum* (B.S.M. Dutt 1962), *C. latifolium* (B.S.M. Dutt 1959) and *Zephyranthes citrina* (B.S.M. Dutt 1958).

Stenar (1925b) reported Allium type of embryo sac in *Crinum latifolium*, and Brown (1951) mentions Adoxa type in *Zephyranthes texana*. B.S.M. Dutt (1959) reinvestigated the development in *C. latifolium* and stated (B.S.M. Dutt 1970d) that the earlier accounts of Allium and Adoxa type are erroneous, and the development of the embryo sac conforms to the Polygonum type. Swamy (1946f) reported, in *C. asiaticum*, a case of reversed polarity of the embryo sac with egg apparatus at the chalazal end and antipodals at the micropylar end. The development in *Z. texana* is apomictic. The synergids are hooked, polar nuclei fuse before fertilization, and the antipodals are large and persistent. Secondary multiplication of antipodals occurs in *Nothoscordum*.

Development of the endosperm is of the Helobial type in *Brodiaea, Clivia, Haemanthus, Hippeastrum, Ixiolirion, Tulbaghia, Vallota, Zephyranthes* (Stenar 1949, 1951a, b, see also Swamy and Parmeswaran 1963), *Haemanthus multiflorus, Hippeastrum rutilum, Pancratium longiflorum, Zephyranthes citrina* (B.S.M. Dutt 1958), *Ixiolirion* (Stenar 1925b) and *Nothoscordum* (Stenar 1932, Håkansson 1953). The chalazal chamber becomes 16-nucleate in *Z. citrina* (B.S.M. Dutt 1958) and 25- to 30-nucleate in *Z. andersonii* (Swamy 1946c). Free-nuclear divisions are followed by cytokinesis in the micropylar chamber.

The development is of the Nuclear type in *Crinum asiaticum* (Schlimbach 1924), *C. defixum* and *C. latifolium* (Tomita 1931). Later, the tissue becomes cellular. In *C. defixum*, the cellular endosperm consumes the ovular cells and abuts on the pericarp. It remains exposed in the ovary wall. The peripheral cells of the endosperm behave as phellogen and develop cork on the outside. Another peculiar feature is the presence of chloroplasts in the peripheral cells lying below the phellogen. The endosperm is starchy.

The embryogeny conforms to the Asterad type-Muscari variation in *Pancratium maritimum* and *P. longiflorum* (B.S.M. Dutt 1970c). In *Zephyranthes andersonii* (see Lakshmanan 1972), the origin of the shoot apex is strictly terminal, and the epicotyl and cotyledon occupy equal areas in the terminal tier. Synergid polyembryony has been observed in *Crinum defixum*.

The seed contains a small embryo and fleshy endosperm.

Zephyranthes microsiphon (Solntzeva and Vorsobina 1972) exhibits hemigamy. The male gametes enter the egg cell and synergid. The sperm nucleus comes in contact with the egg nucleus; the synergids remain healthy. The egg nucleus enters prophase, but divisions of egg nucleus and sperm nucleus do not occur simultaneously. By the time the proembryo becomes two-celled, the sperm nucleus is in telophase. In a four- or six-celled proembryo, the upper cells show small male nuclei, and the lower cells show long female nuclei. There is abnormal mitosis of male and female nuclei, and formation of restitution nuclei.

In the ovule of *Crinum latifolium* (Tomita 1931) the nucellus is ephemeral, and the integument is said to be absent. In *C. giganteum*, Toilliez-Genoud (1965) described ategmic ovule and overgrown seed. P. Maheshwari (1950) and Eames (1961) stress the need for further study. B.S.M. Dutt (1962) reported unitegmic condition of the ovule with ephemeral nucellus. During its development the endosperm consumes the ovular tissue (integument), ruptures the pericarp and becomes completely exposed. Thus, the seed is "naked" lacking testa.

WV Brown (1951), MB Church (1916), GE Coe (1954), F D'Amato (1948–49), BSM Dutt (1957a, b, 1958, 1959, 1962, 1964, 1970c, d), AJ Eames (1961), ML Eckles (1940), ME Farrell (1914), JL Guignard and JC Mestre (1969b), Å Håkansson (1953), G Howell and N Prakash (1990), K Kostriukova and G Benetskaia (1939), KK Lakshmanan (1972), P Maheshwari (1950), AW Meerow and B Dehgan (1988), W Newcomb (1978), MG Panchaksharappa and JF Rangel (1966), H Pankow (1958), H Schlimbach (1924), MP Solntzeva and LI Vorsobina (1972), R Souèges (1931b, 1952b), H Stenar (1925b, 1932, 1949, 1951a, b), N Suita (1937), BGL Swamy (1946c, f), BGL Swamy and N Parmeswaran (1963), SL Tandon and BM Kapoor (1961, 1962a, b), J Toilliez-Genoud (1965), K Tomita (1931), E Werker and A Fahn (1975), MR Whitehead and CA Brown (1940), R Wunderlich (1936), C Zhou, K Orndorff, RD Allen and AF DeMaggio (1986)

54.9 Hypoxidaceae

The anther wall comprises the epidermis, fibrous endothecium, two ephemeral middle layers, and secretory tapetum of multinucleate cells in *Campynema lineare* (B.S.M. Dutt 1970e). The tapetal cells are uninucleate in *Hypoxis aurea* (Arekal 1967). In *Curculigo crassifolia* (B.S.M. Dutt 1970e) the outermost middle layer develops fibrous thickenings, and multinucleate cells of the tapetum form periplasmodium. The microspore mother cells undergo meiotic divisions, cytokinesis is successive, and the tetrads are isobilateral and tetrahedral.

The pollen grains are monocolpate and two-celled at the dispersal stage.

The ovule is anatropous in *Campynema* and *Curculigo crassifolia*, hemi-anatropous in *Pauridia minida*, and camylotropous in *Empodium plicatum*. The micropyle is formed by the inner integument in *Campynema lineare*, and by both integuments in *Curculigo*. It is bitegmic and tenuinucellar in *Curculigo*, *Empodium* and *Hypoxis*, and crassinucellate in *Campynema*. The single archesporial cell cuts off a parietal cell in *Campynema lineare*; otherwise it functions directly as megaspore mother cell (Fig. 54.9.1A). The parietal cell divides periclinally and anticlinally. The megaspore mother cell undergoes meiotic divisions, and the chalazal megaspore of a linear or T-shaped tetrad develops into a Polygonum type of embryo sac (Fig. 54.9.1B–G). The polar nuclei fuse before fertilization. The antipodals are persistent in *Curculigo crassifolia*, *Hypoxis aurea* and *Spiloxene* (Fig. 54.9.1H–M), and they undergo secondary multiplication in *Ianthe*.

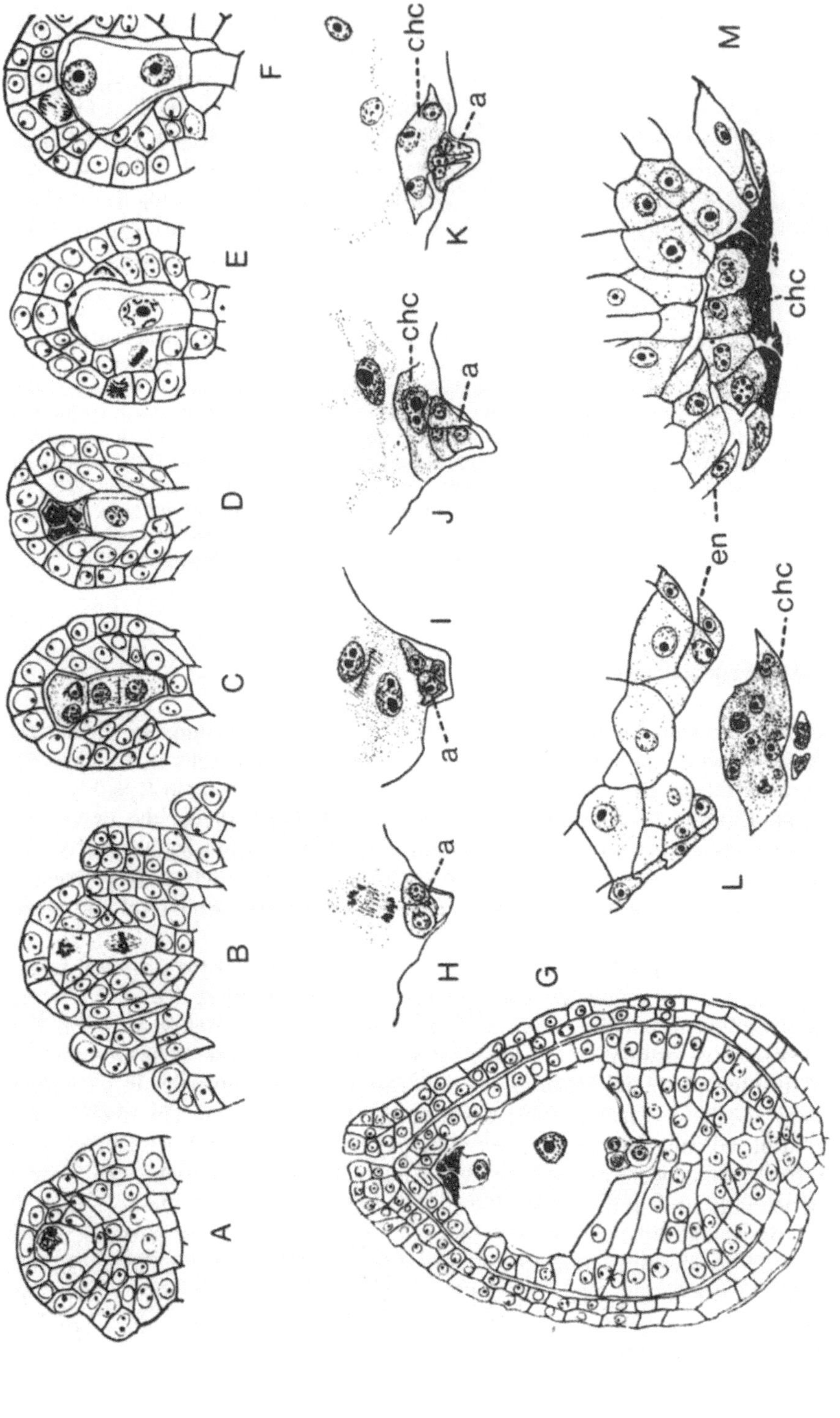
A
B
C
D
E
F
G
H
I
J
K
L
M
a
chc
en

The development of the endosperm is of the Helobial type in *Curculigo crassifolia* (B.S.M. Dutt 1970e), *Forbesia plicata* (*Empodium plicatum*) (de Vos 1949) and *Hypoxis aurea* (Fig. 54.9.2A–D) (Arekal 1967). Free-nuclear divisions occur in both the micropylar and the chalazal chamber (Fig. 54.9.1H–K); wall formation occurs in both chambers in *Curculigo crassifolia*. In *H. aurea*, the chalazal chamber degenerates in a coenocytic condition. Nuclear type occurs in *Pauridia* (de Vos 1949), and wall formation is initiated at the quadrant proembryo. Both types occur in *Ianthe* (de Vos 1948), Nuclear in *I. alba*, *I. aquatica* and *I. minuta*, and Helobial in *I. schlechteri*. Further studies on *Ianthe* are desirable.

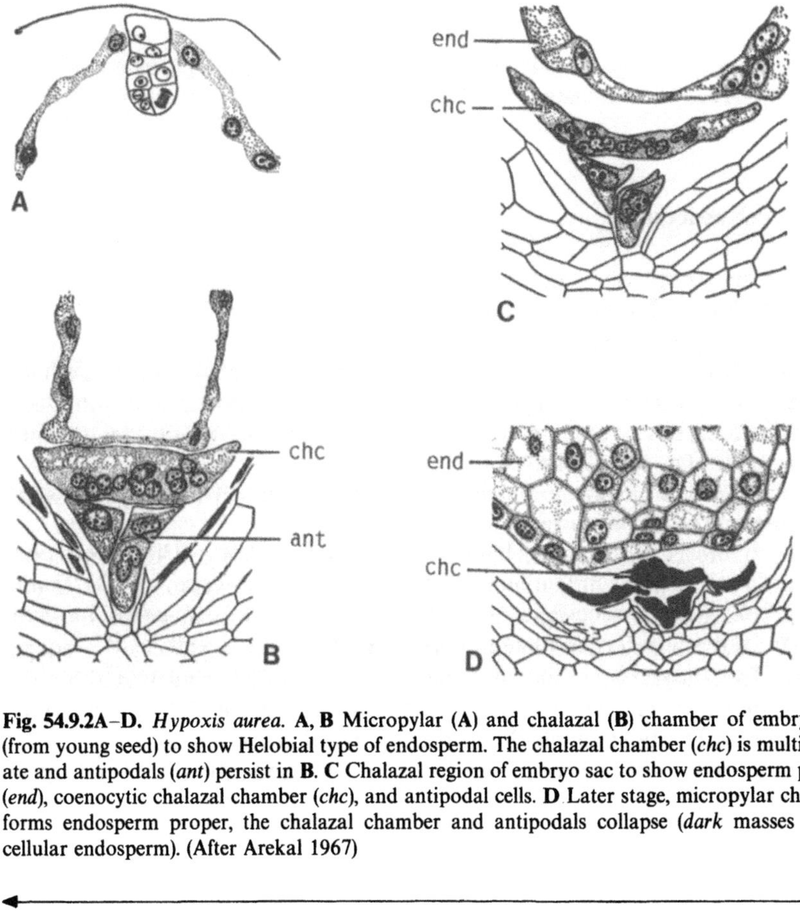

Fig. 54.9.2A–D. *Hypoxis aurea.* **A, B** Micropylar (**A**) and chalazal (**B**) chamber of embryo sac (from young seed) to show Helobial type of endosperm. The chalazal chamber (*chc*) is multinucleate and antipodals (*ant*) persist in **B**. **C** Chalazal region of embryo sac to show endosperm proper (*end*), coenocytic chalazal chamber (*chc*), and antipodal cells. **D** Later stage, micropylar chamber forms endosperm proper, the chalazal chamber and antipodals collapse (*dark* masses below cellular endosperm). (After Arekal 1967)

Fig. 54.9.1A–M. A–G *Spiloxene aquatica.* **A** Young ovule with hypodermal megaspore mother cell; **B** Dyads in division; **C, D** T-shaped tetrad; **E** Functional megaspore; **F** Binucleate embryo sac; **G** Mature embryo sac with polars fused. **H–M** *S. schlechteri.* **H–J** Division of primary endosperm nucleus with a wall between micropylar and chalazal chamber (*chc*); *a* antipodals; **K** Nuclear divisions in chalazal (*chc*) and micropylar chamber; **L** Cellular endosperm (*en*) in micropylar chamber, and free nuclei in chalazal chamber; **M** Part of endosperm proper (*en*), the antipodals and chalazal chamber degenerate. (After de Vos 1948)

The embryogeny conforms to Period I, Megarchetype I (Asterad type) in *Hypoxis aurea* (Govindappa and Shamakumari 1957). While the mature embryo is well differentiated in *Hypoxis*, an undifferentiated condition in the nearly mature seed is reported in *Curculigo crassifolia* (B.S.M. Dutt 1970e). This aspect requires further study.

The seed contains a small embryo surrounded by copious endosperm. The small seed is marked with a distinct lateral hilum. The outer epidermal cells of the testa develop lamellate thickenings, and their lumen becomes narrow. A very heavy black (cutin) deposition occurs on the outer tangential walls. The other cell layers remain thin-walled. The outer epidermis of the tegmen gets crushed, the inner one persists. A single layer of nucellus persists in *Hypoxis aurea* (Arekal 1967).

GD Arekal (1967), R Dahlgren and A-m Lu (1985), MP de Vos (1948, 1949), BSM Dutt (1970e), DA Govindappa and K Shamakumari (1957), MG Simpson (1989), R Wilsnack and JL Warren (1967)

54.10 Velloziaceae

The family comprises taxa with xerophytic habitat, and dichotomously branched stems. The placentae are swollen.

The embryology has not been fully investigated.

The anther wall comprises four to five layers in *Vellozia elegans* (Stenar 1925b). The middle layers degenerate; the tapetum is secretory with uninucleate cells. Successive cytokinesis accompanies meiotic divisions in microspore mother cells. The tetrads are isobilateral.

The pollen grains are monocolpate and two-celled at the dispersal stage. They remain united in tetrads in *V. compacta*.

The ovule is anatropous, bitegmic and tenuinucellate with a micropyle formed by the inner integument. An endothelium develops in *V. elegans* (Stenar 1925b). The archesporial cell functions directly as megaspore mother cell. The latter undergoes meiotic divisions forming a linear or T-shaped tetrad. The megaspores show a tendency to develop concurrently up to binucleate stage. The chalazal megaspore develops into a Polygonum type of embryo sac. The polar nuclei lie in the middle of embryo sac.

The albuminous seed contains a small embryo, and rather hard endosperm (Hutchinson 1973).

BSM Dutt (1970f), J Hutchinson (1973), H Stenar (1925b)

54.11 Taccaceae

The embryology has not been fully studied.

The anther wall comprises persistent epidermis, radially elongate endothecium, two ephemeral middle layers, and secretory tapetum of binucleate cells in *Schizocapsa* and *Tacca*. Simultaneous cytokinesis accompanies meiotic divisions in the microspore mother cells. The tetrads are tetrahedral.

Pollen grains are monocolpate and two-celled at the dispersal stage in *S. plantaginea* (see Schnarf 1931).

The ovule is anatropous, bitegmic, and crassinucellate with the micropyle formed by the inner integument. The archesporial cell cuts off a parietal cell which produces a parietal layer. The megaspore mother cell undergoes meiotic divisions. The chalazal megaspore of a linear or ⊥-shaped tetrad develops into a Polygonum type of embryo sac. The synergids degenerate even before the entry of the pollen tube (Paetow 1931). The polar nuclei fuse before fertilization in the chalazal region of the embryo sac; the three antipodals are ephemeral. Apospory has been observed in *S. plantaginea* (see Schnarf 1931).

Development of the endosperm is of the Nuclear type. The nuclei with cytoplasm aggregate in the chalazal region of the embryo sac. Wall formation takes place simultaneously in the entire embryo sac in *Tacca* (Paetow 1931), and the endosperm becomes cellular.

The seed contains copious endosperm and a minute embryo (Hutchinson 1973).

Å Håkansson (1921), H Huber (1969), J Hutchinson (1973), W Paetow (1931), B Palm (1920), K Schnarf (1931), HG Solms-Laubach (1978), K Süssenguth (1921)

54.12 Dioscoreaceae

The anther wall comprises the epidermis, fibrous endothecium, an ephemeral middle layer, and glandular tapetum of binucleate cells in *Dioscorea oppositifolia* and *Trichopus zeylanicus* (Nagaraja Rao 1953b, 1955c). The microspore mother cells undergo meiotic divisions. Cytokinesis is simultaneous (see Davis 1966) or successive (Nagaraja Rao 1955c). The tetrads are tetrahedral, isobilateral, linear and decussate.

The pollen grains are one-, two- or three-colpate or pantoporate, and two-celled at the dispersal stage. In *Trichopus*, they occasionally germinate in situ (Nagaraja Rao 1955c).

The ovule is anatropous, bitegmic and crassinucellar. Both integuments form the micropyle. A placental obturator develops in *Trichopus*. A hypostase differentiates in *D. oppositifolia* and persists in mature seed. The archesporial cell cuts off a parietal cell, the parietal tissue is several layers thick. The megaspore mother cell undergoes meiotic divisions, and the chalazal megaspore develops into a Polygonum type of embryo sac. The polar nuclei fuse prior to fertilization in the chalazal region, and antipodals persist up to fertilization. The chalazal end of the embryo sac elongates (after fertilization) in *Trichopus zeylanicus* (Nagaraja Rao 1955c).

Occasionally, Drusa type of development occurs in *Dioscorea izuensis* and *D. septemlaba* (Takeuchi 1971); Polygonum type is reported in *D. nipponica* and *D. tokoro* (Takeuchi and Chugai 1968), and *D. japonica* and *D. quinqueloba* (Takeuchi 1972).

The development of the endosperm is of the Nuclear type. Wall formation is initiated at the micropylar end, and proceeds to the chalaza, at the globular proembryo. The cellular endosperm grows asymmetrically in *T. zeylanicus* (Nagaraja Rao 1955c). The food reserve is oil, and there is a heavy deposition of hemicellulose in the cells of the outer three or four layers of the endosperm.

The embryogeny conforms to the Asterad type. The division of zygote is transverse, rarely oblique in *D. oppositifolia* (Nagaraja Rao 1953b). A vertical division occurs in *ca* and *cb*. There is another vertical division in each cell, and *ca* forms a quadrant. Further divisions are irregular, and the mature embryo is cylindrical with a cotyledon, stem tip, root tip, and root cap. The cotyledon is terminal, and the shoot apex appears to be lateral in *T. zeylanicus* (Nagaraja Rao 1955c).

The synergid develops into a proembryo in *Dioscorea composita* (Thirumaran and Lakshmanan 1982), and *T. zeylanicus* (Nagaraja Rao 1954). The ploidy-level was not ascertained.

The albuminous seed contains a small embryo, and is often winged (Hutchinson 1973) in the chalazal region. The seed-coat comprises two-layered testa and two-layered tegmen, remnants of nucellus persist, and the endosperm is thick-walled (Nagaraja Rao 1955c).

GL Davis (1966), JL Guignard (1963), J Hutchinson (1973), A Nagaraja Rao (1951, 1953b, 1954, 1955c), PM Smith (1916), Y Takeuchi (1971, 1972), Y Takeuchi and K Chugai (1968), K Thirumaran and KK Lakshmanan (1982)

54.13 Pontederiaceae

The anther is tetrasporangiate. The endothecial layer develops fibrous thickenings in *Eichhornia crassipes* (W.R. Smith 1898). The anther tapetum has binucleate cells and forms plasmodium. The microspore mother cells undergo meiotic divisions, cytokinesis is successive in *Eichhornia, Heteranthera, Monochoria* and *Pontederia*, and the tetrads are isobilateral and decussate.

The pollen grains are two- or three-colpate and two-celled at the dispersal stage.

The ovule is anatropous, bitegmic and crassinucellar with the micropyle formed by both integuments. The hypodermal archesporial cell cuts off a parietal cell. The latter produces a layer of parietal tissue. The megaspore mother cell undergoes meiotic divisions. The chalazal megaspore of a linear tetrad develops into a Polygonum type of embryo sac. The nucellus is massive, and the embryo sac is located at the micropylar tip. The polar nuclei fuse before fertilization. The antipodals are ephemeral in *Monochoria hastaefolia* (Banerji and Haldar 1942).

The development of endosperm is of the Helobial type, and free-nuclear divisions occur in the micropylar and the chalazal chamber. The endosperm becomes cellular in both the chambers in *Pontederia cordata* (Coker 1907).

In *Monochoria* spp. the micropylar chamber is multinucleate, and the chalazal four-nucleate. From the micropylar chamber two lateral four-nucleate haustoria develop in *M. korsakowii* (Ono 1928) and *M. vaginalis* (Ono 1928, Juliano 1931). They grow towards the chalaza and enclose a nucellar column. As the micropylar chamber increases in size by consuming the nucellus, the two haustoria merge with the micropylar chamber. The chalazal chamber remains four-nucleate. Centripetal wall formation occurs and the micropylar chamber becomes cellular. Further study is suggested.

In *Eichhornia crassipes* (Swamy 1966), the zygote divides transversely, forming the apical cell *ca* and basal cell *cb*. By a transverse division *ca* produces *c* and *d*; *d* divides further and gives rise to *m* and *ci*. *cb* remains undivided, and has a haustorial role. A vertical division in *c* is followed by another vertical division, at right angles to the previous wall, to form *q*. The tier *m* also undergoes a vertical division. Each of the four cells of *q* divides periclinally so that four inner cells are separated from the peripheral cells. The outer layer forms the protoderm initials. Each of the inner terminal cells divides transversely to form two tiers, *l* and *l'*. Periclinal divisions take place in both the tiers so as to separate an axial group of four cells which is surrounded by a circlet of eight cells.

Further growth takes place at a differential rate, and is asymmetrical. The divisions in one particular quarter-sector of the tier *l* are delayed, and this quarter-sector gives rise to epicotyl. The other three quarter-sectors form the initials of the cotyledon. These divide in diverse planes and grow at a rapid rate, whereas the epicotylary locus lags behind. Thus, the cotyledonary and epicotylary loci are derived from the same terminal tier. The procambium of cotyledon develops independently of the hypocotyl. The tier *m* and the derivatives of *ci* contribute to the formation of the hypocotyl and root.

The seed is longitudinally ribbed, and contains copious endosperm and a large cylindrical embryo.

JM Anderson and SCH Barrett (1986), I Banerji and HC Gangulee (1937), I Banerji and S Halder (1942), WE Coker (1907), JB Juliano (1931), T Ono (1928), K Schnarf (1931), WR Smith (1898), BGL Swamy (1966)

54.14 Iridaceae

The anther wall comprises the epidermis, fibrous endothecium, one or two ephemeral middle layers, and glandular tapetum of uni- or binucleate cells in *Sisyrinchium californicum* and *S. striatum* (Lakshmanan and Philip 1971). The hypertrophied condition of the tapetal cells and the formation of persistent bands in *Belamcanda chinensis* (S.C. Gupta and Nanda 1973) inhibit the formation of fibrous thickenings in the endothecium, and cause sterility in pollen grains. The microspore mother cells undergo meiotic divisions, and tetrahedral, ⊥-shaped, isobilateral and linear tetrads are produced. Cytokinesis is simultaneous.

In *Crocus speciosus* (Rudall et al. 1984), meiosis in microspore mother cells occurs when the corm is still underground.

The pollen grains are one- or two-colpate, spiraperturate or pantoporate (*Crocus* spp.). They are two-celled at dispersal.

The ovule is anatropous, bitegmic, tenuinucellar in *Eleutherine chinensis* (Venkateswarlu et al. 1980). The inner integument forms the micropyle (Steyn 1973a). A small funicular obturator and hypostase develop in *Romulea rosea* var. *reflexa* (Steyn 1973a) and *C. speciosus*. The archesporial cell cuts off a parietal cell, except in *E. plicata* (Venkateswarlu et al. 1980). The parietal cell degenerates in *Sisyrinchium* spp. (Lakshmanan and Philip 1971). The megaspore mother cell undergoes meiotic divisions, and the chalazal megaspore of a linear tetrad develops into a Polygonum type of embryo sac. The synergids are hooked, polar nuclei fuse prior to fertilization, and antipodal nuclei do not organize into cells in *Sisyrinchium* spp.

In *C. speciosus*, the tips of synergids extend into the micropylar canal. The antipodals are large and persistent, and sometimes there are eight cells in an embryo sac.

Development of the endosperm is of the Nuclear type. Wall formation occurs at about the 16-nucleate stage in *S. californicum*. The endosperm becomes horny in the ripe seed, and stores proteins and oil in *Romulea rosea* var. *reflexa* (Steyn 1973b). It does not persist in *S. striatum*.

The embryogeny conforms to the Asterad type-Muscari variation in *Eleutherine plicata* (Venkateswarlu et al. 1980). In *R. rosea* var. *reflexa* (Steyn 1973b) and *Sisyrinchium* spp., it corresponds to the Caryophyllad type.

The apical cell *ca* of the two-celled proembryo develops into the embryo proper, and *cb* becomes haustorial, except in *S. californicum* (Lakshmanan and Philip 1971) where it becomes multicellular and bulbous. *ca* undergoes transverse divisions to form a row of five to seven cells. Two vertical walls are laid in the terminal tier to produce the quadrant *q*. One of the sectors of the quadrant is rather quiescent and gives rise to epicotyl, while the other three exhibit rapid growth and produce the cotyledon (Lakshmanan and Philip 1971). The growth in the subterminal tier of the epicotylary region is slow as compared to that in the cotylary region.

The albuminous seed contains a small embryo. The testa is three- to four-layered in *Sisyrinchium striatum* (Lakshmanan and Philip 1971), and is filled with starch. The cells are elongated along the longitudinal axis. The tegmen is two-layered. The cells of the outer epidermis are isodiametric, whereas the cells of the inner epidermis are longitudinally elongated.

A Blumenthal, HK Lerner, F Wesker and A Poljakoff-Mayber (1986), G Chichiricco (1989), AE Clarke, JA Considine, R Ward and RB Knox (1977), P Crété (1957c), L Diels (1930a), SC Gupta and K Nanda (1973), I Haeckel (1930), U Hamann (1961), M Karagozova (1963), RB Knox (1971), KK Lakshmanan and VJ Philip (1971), PC Pande and V Singh (1981), HP Riley (1942), PJ Rudall, SJ Owens and AY Kenton (1984), E Steyn (1973a, b), J Venkateswarlu, P Sarojini Devi and A Nirmala (1980), K Yasui and N Sawada (1940)

54.15 Geosiridaceae

The embryology has not been studied.

54.16 Burmanniaceae

The anther wall comprises the epidermis, endothecium, a middle layer, and secretory tapetum of binucleate cells. Meiotic divisions occur in the microspore mother cells, and cytokinesis is successive.

The pollen grains are monocolpate and two- or three-celled in *Burmannia pusilla* (Arekal and Ramaswamy 1973) and *Thismia americana* (Pfeiffer 1918). They are two-celled in *B. candida* (Ernst and Bernard 1912a). The pollen grains germinate in situ in cleistogamous flowers. Further studies on pollen are necessary.

The ovule is anatropous, bitegmic, and tenuinucellar with a long funicle.

The archesporial cell functions directly as megaspore mother cell. At times, the archesporium is two-celled in *B. coelestis* (Ernst and Bernard 1912b). The megaspore mother cell undergoes meiotic divisions, and the chalazal megaspore of a linear tetrad develops into a Polygonum type of embryo sac. In *B. candida* (Ernst and Bernard 1912a), the micropylar dyad degenerates (Fig. 54.16.1A, B) whereas the chalazal develops into an Allium type of embryo sac.

The polar nuclei fuse before fertilization. The secondary nucleus lies near the antipodals in *B. coelestis*, whereas it lies near the egg apparatus in *B. championii* (Ernst and Bernard 1912a) and *Thismia americana* (Pfeiffer 1918). The antipodals are usually persistent, but in *T. americana* the nuclei do not organize into cells (Fig. 54.16.1C) and degenerate.

B. coelestis (Ramaswamy 1970) exhibits diplospory of the Antennaria alpina type (unreduced gonial apospory). The megaspore mother cell undergoes a long resting period, divides mitotically and produces an eight-nucleate embryo sac (Fig. 54.16.1D–G).

The entry of the pollen tube into the embryo sac is through the micropyle (Fig. 54.16.1H), and double fertilization takes place as usual.

The development of the endosperm is of the Helobial type. The smaller chalazal chamber is binucleate (Fig. 54.16.1I). It becomes four- to six-nucleate in *B. pusilla* (Ramaswamy 1970). Free-nuclear divisions occur in the micropylar chamber, and later wall formation sets in (Fig. 54.16.1J, K). The number of endosperm cells in the mature seed is six to eight in *B. championii* (Ernst and Bernard 1912a), and 20 in *B. coelestis* (Ramaswamy 1970). The occurrence of Cellular type of endosperm in *Thismia leutzelburgi*, as described by Goebel and Süssenguth (1924), is doubtful.

The development of embryo ceases at a very early stage in *B. pusilla* (Arekal and Ramaswamy 1973). In a ripe seed, the embryo is three-celled in *B. candida* and *Gymnosiphon tenellus*, four-celled in *Dictyostega orobanchoides* and *Apteria setacea*, six-celled in *B. coelestis* and *B. pusilla*, and ten-celled in *B. capitata*. However, it is well developed in *Thismia*.

Synergid polyembryony (Fig. 54.16.1K) has been observed in diplosporous *Burmannia coelestis* (Ramaswamy 1970).

The small mature seed contains an organless (undifferentiated) embryo, and scanty endosperm.

The testa mainly forms the seed-coat. The tegmen is membraneous.

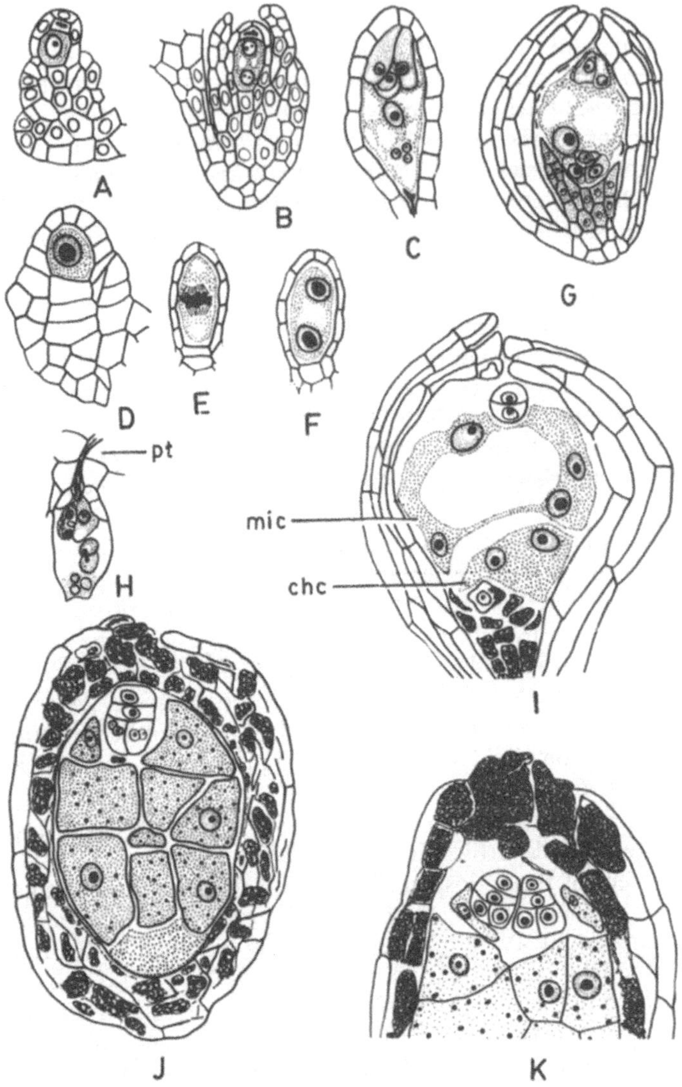

Fig. 54.16.1A–K. A, B *Burmannia candida*. **A** Longisection of ovule shows dyad cells, the upper has degenerated; **B** Binucleate bisporic embryo sac. (After Ernst and Bernard 1912a). **C** *Thismia americana*, mature embryo sac with free antipodal nuclei. (After Pfeiffer 1918). **D–J** *B. coelestis*; **D** Longisection of ovule with megaspore mother cell; **E** Megaspore mother cell in division; **F** Binucleate embryo sac; **G** Organized embryo sac; **D–G** show diplospory of Antennaria alpina type. (After Ramaswamy 1967). **H** *B. candida*, mature embryo sac after the release of male gametes from pollen tube. (After Ernst and Bernard 1912a). **I–K** *B. coeleostis*; **I** Longisection of ovule, the endosperm is of the Helobial type; *mic* micropylar and *chc* chalazal chamber; **J** Longisection of mature seed, the embryo is organless, and tegmen is crushed; **K** Micropylar region of seed (l. s.), one proembryo is zygotic and the other has probably originated from a synergid. (After Ramaswamy 1970)

GD Arekal and SN Ramaswamy (1973), A Ernst and Ch Bernard (1912a, b), K Goebel and K Süssenguth (1924), NE Pfeiffer (1918), SN Ramaswamy (1970)

54.17 Corsiaceae

The embryology has not been investigated.

54.18 Philydraceae

The anther wall comprises persistent epidermis, fibrous endothecium, two to four ephemeral middle layers, and irregularly biseriate secretory tapetum of binucleate cells in *Philydrum lanuginosum* (Kapil and Walia 1965). Successive cytokinesis follows meiotic divisions of microspore mother cells. The tetrads are tetrahedral and isobilateral (Fig. 54.18.1A, B).

The generative cell is usually cut off towards the proximal pole forming two-celled pollen grains (Fig. 54.18.1C–F). Occasionally, in a tetrad, one, two or three microspores degenerate (Fig. 54.18.1C, D). The monosulcate pollen grains have reticulate exine (Fig. 54.18.1G). They usually germinate in situ, and many pollen tubes enter the stigma of the same flower (Kapil and Walia 1965).

The ovule is anatropous, bitegmic, and crassinucellar with the inner integument forming the micropyle. The archesporial cell cuts off a parietal cell which divides once or twice. The megaspore mother cell undergoes meiotic divisions resulting in a linear or T-shaped tetrad (Hamann 1966b). The chalazal megaspore develops into a Polygonum type of embryo sac. The synergids are hooked, and the polar nuclei fuse in the chalazal region of the embryo sac. The antipodals are ephemeral.

The development of the endosperm is of the Helobial type (Kapil and Walia 1965). Free-nuclear divisions occur in the micropylar chamber; wall formation is initiated in the chalazal region and proceeds upward. The entire endosperm becomes cellular at the globular proembryo, and contains crystals. In the chalazal chamber the development of the endosperm follows two pathways: either it remains uninucleate and then degenerates, or its nucleus divides precociously, earlier than that of the micropylar chamber, and the chalazal chamber becomes two- to six-celled. It remains conspicuous up to the two-celled proembryo.

The embryogeny conforms to the Onagrad type-Juncus variation and Philydrum variation in *Philydrum lanuginosum* (Kapil and Walia 1965). The mature embryo is straight.

The seed is caruncúlate with a conspicuous chalazal cap which develops from the outer integument during the post-fertilization period (Fig. 54.18.2A). Both the integuments take part in the formation of the seed-coat (Fig. 54.18.2B–D). The outer epidermal cells have tangentially thickened walls. The inner epidermal cells are thickly cuticularized and contain tannin. The cells of the outer epidermis of the tegmen are strongly thick-walled with a small lumen. Its inner epidermis is tanniniferous and cuticularized.

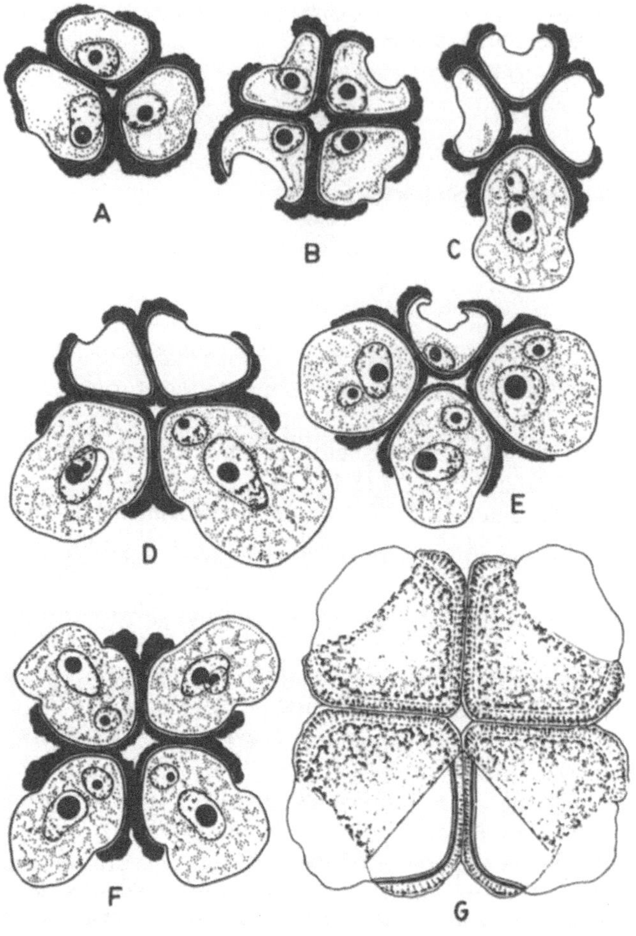

Fig. 54.18.1A–G. *Philydrum lanuginosum,* male gametophyte. **A, B** Uninucleate pollen grains united in permanent tetrads. **C–F** One, two, three or four binucleate pollen grains in permanent tetrads. **G** Palynogram, exine reticulate. (After Kapil and Walia 1965)

U Hamann (1961, 1963, 1966a, b), RN Kapil and K Walia (1965), R Pilger (1930)

54.19 Taxonomic Considerations

Bentham and Hooker (1965c) assign Burmanniaceae to Microspermae; Roxburghiaceae, Liliaceae, Pontederiaceae, Philydraceae and Cynastraceae to Farinosae (Commelinales of Melchior 1964); and along with Juncaceae include the remaining families in Liliiflorae: Xanthorrhoeaceae, Stemonaceae, Agavaceae, Haemodoraceae, Amaryllidaceae, Hypoxidaceae, Velloziaceae, Taccaceae, Dioscoreaceae, Iridaceae, Geosiridaceae, Corsiaceae, Thurniaceae and Bromeliaceae. Hutchinson (1973) distributes these families over seven orders.

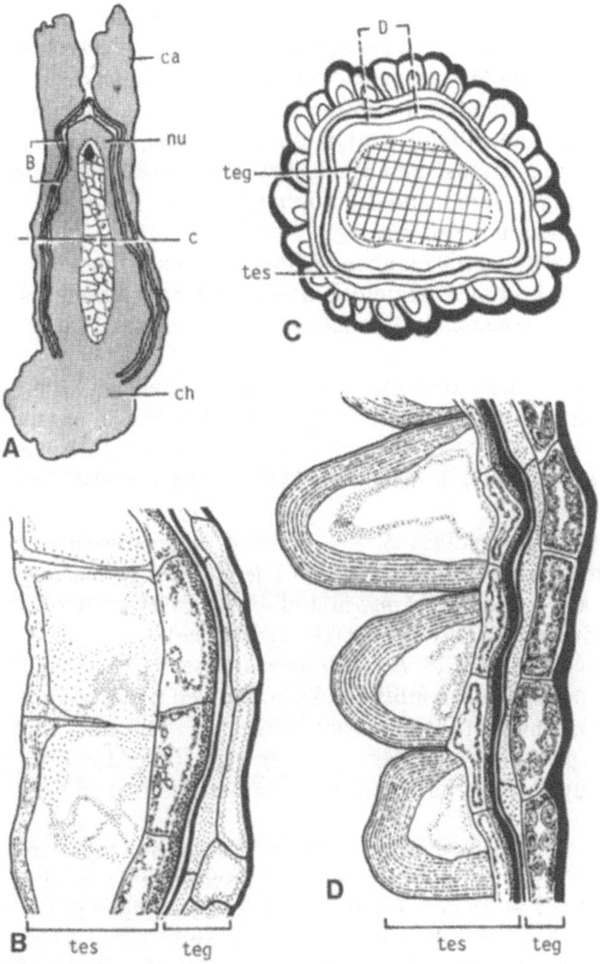

Fig. 54.18.2A–D. *Philydrum lanuginosum.* **A** Longisection of caruncular seed at globular proembryo; *ca* caruncle; *nu* nucellus; *ch* chalaza. **B** *Inset B* marked in **A**, note the deposition of thickenings on the outer epidermis of testa (*tes*) and tannin and cuticle on the inner epidermis. *teg* tegmen. **C** Transection (along level *C* marked in **A**) of mature seed; endosperm *cross-hatched.* **D** *Inset D* marked in **C**, note lamellate thickenings on exotesta, and tannin in endotegmen. (After Kapil and Walia 1965)

The suborder Liliineae comprises 11 families with less advanced features. Various views have been put forward about the origin of this suborder, such as from Ranales (*Cabomba*), from *Scheuchzeria* (*Helobieae*), or from Triuridales (with saprophytic habit and apocarpous condition of pistil) through *Petrosavia* (Liliaceae).

Liliaceae. Liliaceae are a large family showing diversity in habit, and marked variability in cytological and anatomical features. The family is multipalynous, and the floral structure is adapted to cross-pollination. Hutchinson (1973) defines it, within closer limits, by recognizing independent status of the families Tecophilaeaceae, Trilliaceae, Smilacaceae, Ruscaceae, Philesiaceae, Xanthorrhoeaceae and Agavaceae, and assigns the tribes Agapantheae, Allieae and Gilliesieae to Amaryllidaceae.

Liliaceae also exhibit variability in embryological features such as successive or simultaneous cytokinesis in pollen mother cells, two- or three-celled pollen grains, tenui- or crassinucellar ovules (at times both conditions coexist in one and the same species), extreme variation in the formation of different types of embryo sac (Polygonum, Allium, Endymion, Drusa and Fritillaria), and Helobial or Nuclear type of endosperm. The embryogeny exhibits further variations, such as Onagrad, Asterad and Caryophyllad type. The seed is albuminous, but the endosperm is never mealy.

Liliaceae are considered as the basic monocotyledonous stock, from whose ancestral stocks have evolved today's petaloid families, such as Amaryllidaceae, Iridaceae, Palmaceae, Araceae and Orchidaceae (see Lawrence 1951).

Eriospermum. *Eriospermum* has been treated as a subtribe Eriospermae within the tribe Asphodeleae (Bentham and Hooker 1883, Engler 1888), as a family "Eriospermeae" by Endlicher (1836–1840) and Eriospermaceae by Huber (1969) and R. Dahlgren et al. (1985). Melchior (1964) grouped *Eriospermum* with Bowieeae (Eryosperminae).

Lu (1985) has studied the embryology of *E. capense*, *E. luteorulerum* and *E. zeyheri*. He discusses the systematic position with reference to *Asphodelus*, *Lilium* and *Bowiea* (Liliaceae), *Cyanella*, *Lanaria* and *Walleria* (Haemodoraceae), *Cyanastrum* (Cynastraceae), and *Hypoxis* (Hypoxidaceae).

The main embryological features of *Eriospermum* are: secretory anther tapetum, successive cytokinesis in pollen mother cells, two-celled pollen grains, anatropous, bitegmic, crassinucellate ovule with placental obturator and hypostase, Polygonum type of embryo sac, Nuclear type of endosperm, Solanad type of embryogeny, and exalbuminous perispermous hairy seed with a large embryo.

With Liliaceae (Sect. 54.2) the common features are: secretory tapetum, successive type of cytokinesis, anatropous, bitegmic, crassinucellate ovule with obturator and hypostase, and Nuclear type of endosperm. In Liliaceae the pollen grains are three-celled, the female gametophyte is mono-, bi- or tetrasporic, endosperm may be Nuclear or Helobial, embryogeny is variable and conforms to Onagrad, Asterad or Caryophyllad type. The seed is albuminous, never perispermous.

Some embryological features of Haemadoraceae (Sect. 54.6) common to *Eriospermum* are: successive cytokinesis, two-celled pollen grains, bitegmic, crassinucellate ovule, sometimes Nuclear type of endosperm. But hemiana- to orthotropous ovule, sometimes Helobial type of endosperm, Asterad type of embryogeny, and albuminous seed of Haemadoraceae are in marked contrast.

Hypoxidaceae (Sect. 54.9) have secretory tapetum, successive cytokinesis, two-celled pollen grain, sometimes crassinucellate ovule, Polygonum type of embryo sac akin to *Eriospermum*, but ana- to campylotropous ovule, Helobial endosperm, Asterad type of embryogeny, and small embryo in an albuminous seed are significant dissimilarities.

The resemblances between *Eriospermum* and Cynastraceae (Sect. 54.7) are: secretory tapetum, two-celled pollen grains, anatropous, bitegmic and crassinucellate ovule with placental obturator, Polygonum type of embryo sac, and Nuclear endosperm. Chalazosperm in Cynastraceae is in marked contrast. The embryogeny of Cynastraceae has not been studied, but the seed is not perispermous as in *Eriospermum.*

On the basis of embryology, *Eriospermum* is closer to Liliaceae than to Haemadoraceae, Hypoxidaceae and Cynastraceae. The assignment to the subfamily Eriosperminae (Bowiceae) of Liliaceae is fully justified. Therefore, Lu's (1985) suggestion that *Eriospermum* should be raised to an independent family, Eriospermaceae, is not acceptable.

Xanthorrhoeaceae. Xanthorrhoeaceae were previously treated as a subfamily of Liliaceae. Hutchinson (1973) removes them to (his) Agavales. Embryologically, the family is insufficiently investigated to justify its segregation from Liliaceae.

Stemonaceae. Hutchinson (1973) places the family Stemonaceae with Dioscoreaceae in his Dioscoreales. The family is unipalynous and has monocolpate pollen grains, while Dioscoreaceae are multipalynous and are characterized by supernumerary colpi and pantoporate condition. Swamy (1964b) describes the embryology of *Stemona tuberosa*: anatropous, bitegmic, crassinucellate ovule with Polygonum type of embryo sac, ephemeral antipodals, and Nuclear type of endosperm as in Dioscoreaceae and Liliaceae. The family belongs to Liliineae. Further investigations on its anther, pollen and embryo development would be fruitful in discussing the affinities.

R. Dahlgren (1977a) erects two distinct orders, Dioscoreales and Stemonales; both contain abundant steroid saponins. Later, he (1980b) includes Dioscoreaceae and Stemonaceae in Dioscoreales.

Agavaceae. Hutchinson (1973) elevates the tribe Agavoideae of Amaryllidaceae to Agavaceae, and adds to it Yuccae, Nolineae and Dracaeneae from Liliaceae. He includes Xanthorrhoeaceae and Agavaceae in his Agavales. The presence of multilayered endothecium in *Agave* is notable. The family shows successive cytokinesis in pollen mother cells, two-celled pollen grains, anatropous, bitegmic, crassinucellar ovule, Polygonum type of embryo sac, Nuclear (*Agave*) or Helobial (*Yucca*) type of endosperm, and Asterad (*Yucca*) type of embryogeny as in Amaryllidaceae and Liliaceae. Ruminate endosperm in *Yucca* spp. resembles that of palmaceous families. Hutchinson assigns it an intermediate position between Liliaceae and Palmae (woody and xerophytic habit).

Haemodoraceae. Haemodoraceae is a unipalynous (monocolpate) family with rarely trichotomocolpate pollen grains. R. Dahlgren (1980b) places Haemodoraceae in Haemodorales, Pontederiaceae in Pontederiales and Philydraceae in the order Philydrales. These three families have Helobial endosperm, and the mature endosperm is starchy. The affinities of Haemodoraceae with Philydraceae are discussed later. Haemodoraceae are described as "appear to be crassinucellate Hypoxideae with superior or inferior ovary, the ovules being usually ortho- or hemianatropous". Embryologically, both families,

Haemodoraceae and Hypoxidaceae, resemble each other in such features as monosulcate pollen grains, secretory or amoeboid tapetum, hemiana- or orthotropous bitegmic ovule, Polygonum type of embryo sac, Helobial endosperm, and Asterad type of embryogeny.

On the basis of the embryology of *Anigozanthos*, *Conostylis* and *Haemodorum*, and some observations on *Dilatris*, *Tribonanthes*, *Wachendorfia* and *Xiphidium*, Steinecke and Hamann (1989) conclude that the family Haemodoraceae is embryologically homogeneous. There is no justification to raise the subfamilies Haemodoroideae and Conostylidoideae to family rank. These authors also confirm the position of Haemodoraceae near Pontederiaceae and Philydraceae.

The genus *Cyanella* (Conanthereae) shows operculate pollen grains, secretory anther tapetum, simultaneous cytokinesis in pollen mother cells, chalazal (lateral) embryo sac caecum, and Nuclear type of endosperm. B.S.M. Dutt (1970a) suggests its exclusion from Haemodoraceae and Amaryllidaceae, and supports Hutchinson's view to raise Conanthereae to a separate family Tecophilaeaceae. We ågree with this assignment.

According to de Vos (1963), the genus *Lanaria* is allied to *Odontostomum* (Cave 1952) of Tecophilaeaceae rather than to Haemodoraceae and Conostylae (Hypoxidaceae).

Cynastraceae. Cynastraceae is a monogeneric family with monosulcate, or occasionally trichotomosulcate (*Cyanastrum hostifolium*–Erdtman 1952) pollen grains. The genus *Cyanastrum* has been placed in Haemodoraceae, in Pontederiaceae as a subfamily Cynastroideae, or in a separate family Cynastraceae. Hutchinson (1973) allies it with *Cyanella* and places *Cyanastrum* in Tecophilaeaceae. Cronquist (1981) recognizes its family status.

Cave (1952) compares the embryological features of *Cyanastrum* and *Odontostomum*, and B.S.M. Dutt (1970b) of *Cyanastrum* and *Cyanella*. These genera resemble one another in simultaneous cytokinesis in pollen mother cells, secretory anther tapetum, formation of parietal tissue in ovule, increase in nucellar tissue below the micropyle during megagametogenesis, Polygonum type of embryo sac, and Nuclear type of endosperm. Cynastraceae are characterized by the presence of chalazosperm (a storage tissue containing starch and fat) in the absence of endosperm in the mature seed. Melchior (1964) adopts Nietsch's (1941) view of placing the family Cynastraceae under Liliiflorae in the neighbourhood of Dioscoreaceae and Taccaceae.

Amaryllidaceae. According to Hutchinson (1973, p. 790): ". . . . . Amaryllidaceae may have either a superior or an inferior ovary; they have nearly always six stamens, and the flowers are umbellate and subtended by an involucre of two or more bracts, or rarely the flowers are umbellate with a reduction to one bract or even one flower".

While discussing the affinity and origin of Amaryllidaceae, Hutchinson (1973) considers that instead of superior (Liliaceae) and inferior (Amaryllidaceae) ovary, the umbellate type of inflorescence and involucre of bracts are of greater taxonomic importance as these characters give a more natural grouping. On this basis, he removed the tribes Agapantheae, Gilliesieae and Allieae from Liliaceae, and included them in Amaryllidaceae. He (Hutchinson 1973, p. 666) further added that the genus *Butomus* has umbel type of inflorescence surrounded by an involucre of bracts. This perhaps is ". . . . the beginning of a line of development which may have culminated in the Amaryllidaceae, the intermediate stages, in which the ovary became syncarpous and inferior, have disappeared".

Embryologically, Amaryllidaceae have uniform characters and show close alliance with Liliaceae. Secretory anther tapetum, successive cytokinesis in pollen mother cells, anatropous bitegmic ovule, Polygonum type of embryo sac, and Helobial and Nuclear type of endosperm occur in Liliaceae as well as in Amaryllidaceae. But, Liliaceae show much variation in the formation of different types of embryo sac. Occasionally, there is delayed embryogeny as in *Tulipa*.

Amaryllidaceae also show several characters comparable to *Butomus*. The successive cytokinesis in pollen mother cells, monocolpate or monosulcate pollen grains, anatropous bitegmic ovule, Polygonum type of embryo sac, and Helobial type of endosperm are common to both. Therefore, Hutchinson's conclusion that a line of development originates from *Butomus* and culminates in Amaryllidaceae is appropriate.

Hypoxidaceae. The alliance of Hypoxidaceae with Haemadoraceae is beyond doubt. The occurrence of unilocular ovary with parietal placentation (de Vos 1949, B.S.M. Dutt 1970e) in *Empodium plicatum* (*Forbesia plicata*) and *Curculigo crassifolia* indicates that Hypoxidaceae also have unilocular ovary with parietal placentation as in Taccaceae and Philydraceae.

B.S.M. Dutt (1970e) draws attention to the position of *Campynema* and *Campynemanthe* of the subfamily Campynematoideae. In contrast to features such as amoeboid tapetum, tenuinucellate ovule with micropyle formed by both integuments in Hypoxidaceae, *Campynema* has secretory tapetum, a parietal cell in the ovule, and the micropyle is formed by the inner integument. The secretory anther tapetum, and monosulcate two-celled pollen grains show its alliance with Amaryllidaceae. B.S.M. Dutt (1970e) also discusses the occurrence of the extrose nature of anthers, and three free styles in Campynematoideae. These features are absent in Hypoxidaceae of Melchior (1964). Therefore, B.S.M. Dutt (1970e) suggests the placement of *Campynema* and *Campynemanthe* in a separate subfamily Campynematoideae within Amaryllidaceae.

Velloziaceae. Velloziaceae is a small xerophytic family. It shows secretory anther tapetum, successive cytokinesis in pollen mother cells, two-celled pollen, anatropous, bitegmic, tenuinucellate ovule, and Polygonum type of embryo sac as in Amaryllidaceae. But the family is unipalynous, and in *Vellozia campacta* the pollen grains remain in permanent tetrads. An endothelium develops in *V. elegans* (Stenar 1925b). Morphologically, the family differs from Amaryllidaceae. *Vellozia* exhibits a dichotomously branched woody stem bearing persistent leaf bases and tufts of narrow pointed leaves at the apices of branches. The solitary terminal flowers have branched stamens, and inferior trilocular ovaries with thick placentae. The family is considered as a close relative of Bromeliaceae. Their embryology is also comparable. The knowledge about the development of the endosperm and embryo may be helpful in deciding the phyletic position.

Taccaceae. Similar to Velloziaceae, Taccaceae also resemble Amaryllidaceae in the development of pollen (except simultaneous cytokinesis in pollen mother cells), nature of ovule, embryo sac, and endosperm. Taccaceae are the only monocotyledonous family which shows "Dicotyledonous" type (Davis 1966) of development of the anther wall. Huber (1969) mentions their remote affinity with Haemodoraceae, due to peculiar pigmentation of the seed-coat. R. Dahlgren (1980b) includes Taccaceae in the order Dioscoreales, superorder Liliiflorae.

Dioscoreaceae. Dioscoreaceae are derived from Liliaceae. The unisexual flowers with inferior ovary suggest their advanced features. They show probable phylogeny with dicotyledons in the reticulate venation of leaves, and absence of sheath at the leaf-base. Their alliance with Stemonaceae has already been discussed earlier. The secretory anther tapetum, simultaneous cytokinesis in pollen mother cells, two-celled pollen, similar nature of the ovule, Polygonum type of embryo sac (occasionally Drusa type), Nuclear type of endosperm, and Asterad type of embryogeny distinctly show their alliance with Liliaceae. The presence of chelidonic acid supports this alliance.

The suborder Pontederiineae comprises the family Pontederiaceae. Bentham and Hooker (1965c) associated the family with Liliaceae, while Engler and Prantl (1931) included it in Farinosae. Pontederiaceae are supposed to be a link between Commelinales (Farinosae) and Liliiflorae; their habit and structure of inflorescence with spathe-like bracts are similar to that of aroids.

Pontederiaceae. Pontederiaceae are closely allied to Liliaceae in most of the embryological features, except amoeboid anther tapetum and mealy endosperm which do not occur in Liliaceae. On the other hand, they resemble Farinosae, particularly Commelinaceae, in the zygomorphic nature of flower, reduction in androecium, amoeboid anther tapetum, successive cytokinesis in pollen mother cells, Polygonum type of embryo sac, mealy endosperm, Asterad type of embryogeny, and a small embryo. But the orthotropous ovule, and Nuclear type of endosperm of Commelinaceae do not occur in Pontederiaceae.

Hamann (1966a) relates Philydraceae to Pontederiaceae. Both the families have Helobial type of endosperm which stores starch in mature seed. But, the permanent pollen tetrads, their germination in situ, Onagrad type of embryogeny, and caruneulate seed of Philydraceae differ from those of Pontederiaceae.

Iridaceae, and Geosiridaceae form the suborder Iridineae. The embryology of the latter family has not been investigated.

Iridaceae. Iridaceae are herbaceous plants with advanced features such as zygomorphic epigynous flowers adapted to entomophily. Bessey (1915), Diels (1930a) and Wettstein (1935) derive Iridaceae from Amaryllidaceae, while Hutchinson (1967) segregates them from all other families of Liliiflorae and erects an independent order, Iridales. The general structure of flower resembles Liliaceae, while its multipalynous condition (one- or two-colpate, spiraperturate or pantoporate pollen-*Crocus* spp.) and anatomy indicate relationship with Liliaceae and Amaryllidaceae.

Embryologically also Iridaceae are allied to Amaryllidaceae except for a few features such as secretory anther tapetum, endosperm haustorium, Caryophyllad type of embryogeny, small embryo, and arillate seeds. The secretory anther tapetum and Caryophyllad type of embryogeny are met with in Liliaceae. Iridaceae have presumably originated from Liliflorean stock.

Burmanniaceae. The suborder Burmanniineae consists of two families: Burmanniaceae and Corsiaceae. Bentham and Hooker (1965c) and Engler and Prantl (1931) place Burmanniaceae with Orchidaceae in Microspermae, and treat Corsieae as a tribe in Burmanniaceae. Wettstein (1935), Melchior (1964), Hutchinson (1973), R. Dahlgren (1980b) and Cronquist (1981) recognize Corsieae as a family Corsiaceae. The embryology of Corsiaceae has not been investigated.

Burmanniaceae are treated either as an independent order, or allied to the members of Liliiflorae (Lawrence 1951). Hutchinson (1973) and R. Dahlgren (1980b) elevate the family to ordinal rank, as Burmanniales. Of all the families of Liliiflorae, embryologically Burmanniaceae show a close relationship to Hypoxidaceae. Both have secretory anther tapetum, successive cytokinesis in pollen mother cells, two-celled monosulcate or monocolpate pollen grains, bitegmic tenuinucellate ovules, Polygonum and Allium type of embryo sac, Helobial type of endosperm with coenocytic chalazal chamber (*Hypoxis*), and an organless (undifferentiated) embryo.

However, the affinity of Burmanniaceae to Orchidaceae cannot be ruled out. They resemble one another in saprophytic habit, have flowers with inferior tricarpellate ovary, single pollen grains (Dyandrae), bitegmic tenuinucellate ovule, belated tendency of differentiation in ovules, mono-, bi- and diplosporic (Monandrae) embryo sac, and minute seeds with undifferentiated embryo. Even though there are several major differences, the Englerian view that "Orchidaceae are derived from Burmanniaceae" finds some support in their embryology. Cronquist (1981) has also placed Burmanniaceae in Orchidales. According to Hutchinson (1973), these two families (Burmanniaceae and Orchidaceae) may have a common origin, and then they evolved on parallel lines.

In contrast to the embryological features of Orchidaceae, Burmanniaceae have Helobial endosperm, suspensor of hardly two or three cells, and albuminous seed.

Philydraceae. Philydrineae, the last suborder of Liliiflorae, are represented by a single family Philydraceae. Bentham and Hooker (1965c) place them in

Coronarieae, Wettstein (1935) in Liliiflorae, Engler and Prantl (1931) in Farinosae, Deyl (1955) in Orchidales, and Hutchinson (1973) in Haemodorales.

While discussing the phyletic position of Philydraceae, Kapil and Walia (1965) compare the embryological features with those of Farinosae – Commelinaceae, Liliiflorae, Juncaceae and Orchidales – Orchidaceae. They reject the possibility of including Philydraceae in Farinosae, Liliiflorae and Orchidales. According to them, the genus *Philydrum* has zygomorphic flowers with a spathe-like bract, a single stamen, glandular tapetum, permanent union of pollen grains in tetrads, Helobial type of endosperm with precocious activity of the chalazal chamber, Juncus and Philydrum variations of the Onagrad type of embryogeny, and presence of caruncle and a chalazal cap.

Philydraceae show some similarities with Haemodoraceae: secretory anther tapetum, successive cytokinesis in pollen mother cells, bitegmic crassinucellate ovule with inner integument forming the micropyle, Polygonum type of embryo sac with ephemeral antipodals, and Helobial type of endosperm. But in Philydraceae, the presence of a spathe-like bract, single stamen, and tetrads of pollen grains mark their individuality. Kapil and Walia (1965) do not support the inclusion of Philydraceae in Haemodorales, and point out that Philydraceae should be elevated to an ordinal rank, Philydrales, and placed near the Haemodorales.

55 Juncales

The order Juncales comprises two families: Juncaceae and Thurniaceae.

55.1 General Features

Juncales exhibit the following characters: (1) secretory anther tapetum, (2) simultaneous cytokinesis, (3) pollen grains in permanent tetrads, three-celled at shedding stage, (4) anatropous, bitegmic, crassinucellate ovule with inner integument forming the micropyle, (5) Polygonum type of embryo sac with ephemeral antipodals, (6) Helobial endosperm, (7) embryogeny of Onagrad type-Juncus variation; the major part of the shoot apex is involved in the development of cotyledon, and (8) albuminous endotestal seed.

55.2 Juncaceae

The anther wall comprises the epidermis, fibrous endothecium, an ephemeral middle layer, and secretory tapetum of uninucleate cells in *Juncus bufonius* (Shah 1963). The endothecium is irregularly biseriate in *J. effusus* and *J. prismatocarpus* (Zaman 1950). The middle layer persists in *Oxychloe andina* and develops fibrous thickenings; the tapetum becomes irregularly biseriate.

Simultaneous cytokinesis follows meiotic divisions of microspore mother cells. The tetrads are tetrahedral and isobilateral (Fig. 55.2.1A–D).

The pollen grains remain permanently attached in tetrads up to their germination on the stigma. The generative cell is cut off towards the inner angle of each microspore in *Juncus* (Fig. 55.2.1B, C) (Zaman 1950, Shah 1963). Rarely, it is cut off towards the outer side (Fig. 55.2.1D). The pollen grains are three-celled at the dispersal stage (Fig. 55.2.1E).

The ovule is anatropous, bitegmic and crassinucellate. The inner integument forms the micropyle. The archesporium is two- or three-celled. A parietal cell is cut off, the parietal tissue is three or four layers thick in *J. bufonius*. The chalazal megaspore of a linear tetrad develops into a Polygonum type of embryo sac. The synergids are large, polar nuclei fuse before fertilization, and antipodals are small and ephemeral.

The development of the endosperm is of the Helobial type (Brenner 1922). The chalazal chamber remains uninucleate in *J. bufonius* (Shah 1963), but becomes four- to eight-nucleate in *J. effusus* and *J. prismatocarpus* (Zaman 1950). Free-nuclear divisions occur in the micropylar chamber, later walls are formed. The cells store starch and oil.

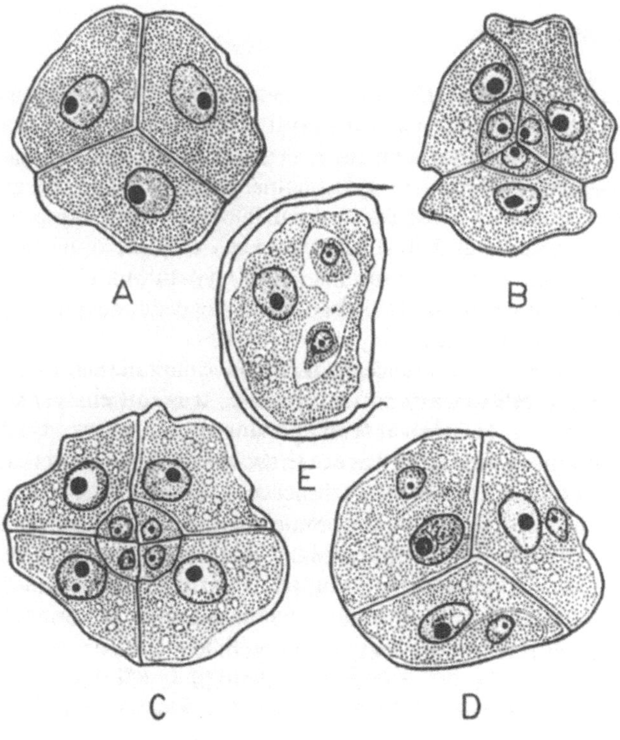

Fig. 55.2.1A–E. *Juncus bufonius.* **A** Single-celled pollen grains in a tetrahedral tetrad. **B** Two-celled condition. **C** Two-celled pollen grains in an isobilateral tetrad. The generative cell is cut off towards the centre in **B** and **C**. **D** Two-celled pollen grains in a tetrahedral tetrad, generative cell cut off towards the periphery. **E** Three-celled mature pollen from a tetrad. (After Shah 1963)

The embryogeny corresponds to the Onagrad type-Juncus variation. The dermatogen initials are cut off immediately after the quadrant stage (Shah 1963). According to Swamy (1979), a larger volume of quadrant contributes to the formation of the cotyledon, while a smaller portion gives rise to shoot apex.

The seed contains a small straight embryo enclosed by the starchy endosperm. The outer epidermis of the testa becomes stretched and distorted, and the inner is wavy and thinly cuticularized. The inner epidermis of the tegmen develops conspicuous thickenings on the tangential walls, and contains tannin-like substances.

W Brenner (1922), LN Eluterius (1984), M Juguet (1969a), KK Lakshmanan (1972), CK Shah (1963, 1967), P Souèges (1923d), BGL Swamy (1979), B Zaman (1950)

55.3 Thurniaceae

The embryology has not been investigated.

55.4 Taxonomic Considerations

Juncaceae. Bentham and Hooker (1965c) place Juncaceae in Calycinae with Flagellariaceae and Cyperaceae. Engler and Prantl (1931) include them in the suborder Juncinae of Liliiflorae, and treat them as allied to Liliaceae, either as ancestral to these or as derived from them. Hutchinson (1973) assigns them to an independent order, Juncales, and places them in Glumiferae along with Centrolepidaceae, Thurniaceae and Restionaceae. He treats them as an advanced order over his Liliales, and as a forerunner of Cyperales and Graminales. R. Dahlgren (1980b) and Cronquist (1981) accept the order Juncales with the families Juncaceae and Thurniaceae.

Juncaceae show alliance with Liliaceae. Both have simultaneous cytokinesis in pollen mother cells, anatropous, bitegmic, crassinucellate ovule, Polygonum type of embryo sac, Helobial type of endosperm, Onagrad type of embryogeny, and albuminous seeds. But Liliaceae show variations in all major embryological features which are absent in Juncaceae. Moreover, the endosperm is starchy in Juncales, which is a Commelinean feature. Anatomical evidence (presence of vessels in the entire plant body, and stomates with two subsidiary cells) also negates their association with liliaceous families (Cronquist 1968). In addition, the occurrence of pollen grains in permanent tetrads segregates Juncaceae from Liliaceae, Palmaceae, Restionaceae, Centrolepidaceae and Gramineae. The latter four families exhibit Nuclear type of endosperm; Palmae and Gramineae have successive cytokinesis in pollen mother cells, and Asterad type of embryogeny. Krupko (1962) suggests their (Juncaceae) closeness to Restionaceae due to (1) strongly four-layered anther wall, (2) similarity in the opening of anther, and (3) initially anatropous condition of ovule (orthotropous in Restionaceae and Centrolepidaceae).

These differences are significant. The similarity lies with Centrolepidaceae in the nature of the anther wall, and development of the embryo.

Of all the families, Juncaceae show close affinities with Cyperaceae. The glandular tapetum, simultaneous cytokinesis in pollen mother cells, Polygonum type of embryo sac, and embryogeny of Onagrad type-Juncus variation are common to both. In Cyperaceae the pollen grain is a pseudomonad; in Juncaceae they are in tetrads. The only exception is the development of the endosperm. In both families embryogeny reflects evolved character and strengthens close affinities (Juguet 1969a). The seed-coat develops from both the integuments, and is free from the pericarp (Shah 1967).

As Juncales exhibit pluriovular condition of ovary, Helobial type of endosperm, and starch as storage product in the endosperm, their assignment near Commelinales is appropriate.

56 Bromeliales

56.1 General Features

Bromeliaceae represent the order Bromeliales. The family exhibits: (1) secretory anther tapetum, (2) simultaneous divisions in pollen mother cells, (3) one- or two-aperturate, two-celled pollen grains, (4) anatropous, bitegmic, crassinucellate ovule, (5) Polygonum type of embryo sac, occasional proliferation of antipodals, (6) Helobial type of endosperm, (7) Asterad type of embryogeny, and (8) albuminous (starchy endosperm) seed with a small laterally placed embryo.

56.2 Bromeliaceae

The anther wall comprises the epidermis, fibrous endothecium, two ephemeral middle layers, and secretory tapetum of binucleate cells in pineapple (Wee and Nagaraja Rao 1974). The microspore mother cells undergo meiotic divisions producing linear and isobilateral tetrads. Cytokinesis is successive.

The pollen grains are one- or two-aperturate (colpate – distal, zonal or global), and two-celled at the dispersal stage.

The ovule is anatropous, bitegmic and crassinucellate. The inner integument forms the micropyle. Almost one-third of the ovules of pineapple are orthotropous and unitegmic or bitegmic.

The archesporial cell cuts off a parietal cell; the parietal tissue is well developed. The cells of the nucellar epidermis also divide and add to the parietal tissue. The megaspore mother cell undergoes meiotic divisions, and produces a linear tetrad. The chalazal megaspore develops into a Polygonum type of embryo sac. The synergids are hooked, polar nuclei fuse before fertilization, and antipodals undergo secondary multiplication in *Lindmania penduriflora* (Lakshmanan 1967). The secondary nucleus lies near the antipodals.

Development of the endosperm is of the Helobial type. The primary endosperm nucleus divides 6 h after fertilization in *Ananas comosus* (Nagaraja Rao and Wee 1979), and a large micropylar and a small chalazal chamber is formed. Free-nuclear divisions occur in the micropylar chamber. The nucleus of the chalazal chamber also divides followed by cytokinesis, and a triangular tissue is produced. Centripetal cytokinesis occurs in the micropylar chamber on day 24 after fertilization, and cellularization is completed during the next 4 to 6 days. Ultimately, the endosperm tissue in both the chambers merges; the cells in the micropylar region are larger than in the chalazal region.

In orthotropous ovules fertilization occurs rarely, but the formation of free-nuclear endosperm has been observed in the absence of fertilization (Nagaraja Rao and Wee 1979). The aspect needs further investigation.

In *Lindmania penduriflora* (Lakshmanan 1967) the nucleus of the chalazal chamber may not divide so that it remains uninucleate, or it may divide so that it becomes binucleate, or the division is followed by a vertical wall and it becomes two-celled. Free-nuclear divisions occur in the micropylar chamber, but later it becomes cellular.

The embryogeny conforms to the Asterad type (Wee and Nagaraja Rao 1974). The zygote divides transversely to form the apical cell *ca* and basal cell *cb*. *ca* divides vertically and *cb* transversely to form a four-celled proembryo. Another vertical division occurs in *ca*, at right angles to the first wall, and a quadrant is formed. The derivatives of the lower cell of *cb* also contribute to the proembryo. The mature embryo is elongated with a distinct cotyledon.

The seed is mainly exotegmic with a small embryo and mealy endosperm. The outer epidermis of the testa develops thickenings on the radial and inner walls; the outer epidermis of the tegmen develops into a sclerenchymatous layer. The rest of the cell layers degenerate.

KK Lakshmanan (1967), A Nagaraja Rao and YC Wee (1979), LB Smith (1934), YC Wee and A Nagaraja Rao (1974)

56.3 Taxonomic Considerations

Bromeliaceae. Bentham and Hooker (1965c) include Bromeliaceae in Epigynae along with Scitamineae, Haemadoraceae, Iridaceae, Amaryllidaceae, Taccaceae and Dioscoreaceae. Engler and Prantl (1931), after considering the common features, such as mealy endosperm and generally superior ovary, place Bromeliaceae in Farinosae (mostly Commelinales of Melchior 1964). According to L.B. Smith (1934), this family has a strong affinity with Rapateaceae (Commelinales), and is the most primitive member in Farinosae. Hutchinson (1973), however, raises it to ordinal rank as Bromeliales, and seeks from it the origin of Zingiberales (Scitamineae). Cronquist (1981) also places it in a distinct order, Bromeliales. R. Dahlgren (1980b) includes Bromeliales in the superorder Liliiflorae.

Erdtman (1952) reports the similarity of pollen grains in Bromeliaceae and Scitamineae. Bromeliaceae and Zingiberaceae show several similar features and appear to be closely allied: anatropous bitegmic ovule, Polygonum type of embryo sac, Helobial type of endosperm, and Asterad type of embryogeny. The rest of the members of Scitamineae have Nuclear type of endosperm. Zingiberaceae differ since they have amoeboid anther tapetum, a nucellar pad, epistase, hypostase, arillate seed with perisperm, collar, lid and chalazal tissue.

Similarly, Bromeliaceae show some alliance with Haemadoraceae and Amaryllidaceae. However, Iridaceae, Taccaceae and Dioscoreaceae stand apart in the Nuclear type of endosperm.

Phytochemically, due to the presence of steroidal saponins Bromeliaceae resemble the members of superorder Liliiflorae (R. Dahlgren 1980b), but chelidonic acid is absent in Bromeliaceae.

The embryology of Rapateaceae (Sect. 57.5) and Bromeliaceae (Sect. 56.2), and their affinities are discussed under Section 57.10.

The order Commelinales is characterized by amoeboid anther tapetum, orthotropous ovules, and Nuclear type of endosperm. These features are absent in Bromeliales. The ovule with the outer integument of two cell layers, Polygonum type of embryo sac, starchy endosperm, Asterad type of embryogeny, and lateral position of embryo are common to both. P.K.K. Nair (1970) cites the similarity between pollen grains of Bromeliaceae and Commelinaceae. Both the orders also show the accumulation of silica.

Thus, Melchior's (1964) assignment of Bromeliaceae to an independent order, Bromeliales, near the order Commelinales, is justified.

57 Commelinales

The order Commelinales comprises eight families distributed over four suborders. The suborder Commelinineae consists of Commelinaceae, Mayacaceae, Xyridaceae and Rapateaceae; Eriocaulineae of Eriocaulaceae, Restionineae of Restionaceae and Centrolepidaceae, and Flagellariineae of Flagellariaceae.

57.1 General Features

Commelinales exhibit: (1) secretory anther tapetum; amoeboid in Commelinaceae, (2) successive cytokinesis in pollen mother cells, (3) orthotropous bitegmic ovules with Polygonum type of embryo sac; Allium type in Xyridaceae and Flagellariaceae; Polygonum, Allium and Adoxa in Commelinaceae, (4) Nuclear type of endosperm, and (5) Asterad type of embryogeny in Commelinineae and Eriocaulineae; Onagrad in Centrolepidaceae. The starchy endosperm and the lateral position of the embryo (except in Commelinaceae) in seed are taxonomic features for the classification of the order.

57.2 Commelinaceae

The anther wall comprises the persistent epidermis, fibrous endothecium, an ephemeral middle layer, and amoeboid tapetum (Chikkannaiah 1962). In *Cyanotis axillaris, C. cuculata* and *Streptolirion volubile* (D. Rao 1968b), the endothecial thickenings are band-like. During the meiosis of microspore mother cells, a periplasmodium is formed and is completely absorbed by the time the pollen grains mature. Meiotic divisions are followed by successive cytokinesis which results in the formation of isobilateral, tetrahedral and decussate tetrads.

The pollen grains are monosulcate and two-celled at the dispersal stage in *Pollia sorzogenensis* (Lakshmanan 1962); three-celled in *Floscopa scandens* (Chikkannaiah 1964–65a).

The ovule is ortho- (*Pollia*), ortho-campylo- (*Cyanotis*), or hemianatropous (*Aneilema*), bitegmic and tenui- or crassinucellar. The micropyle is formed only by the inner integument in *Tinantia fugax*, and by both integuments in *Cyanotis axillaris*. The nucellus is two or three layers thick at the micropylar end in *Commelina subulata* (Chikkannaiah 1963). In *Aneilema paniculatum, Commelina subulata*, and *Floscopa scandens*, the tip of the nucellus is exposed (Chikkannaiah 1962). The nucellar cells in the exposed region enlarge and acquire thickenings.

Cartonema spicatum (Grootjen 1983a) has a trilocular ovary with two ovules in each cell. The placental region is dizonate, i.e. the corpus layer divides periclinally and the single tunica layer divides anticlinally. The second subdermal tunica layer originates from the corpus. Thus, the primordium is trizonate. The hypodermal archesporial cell cuts off a parietal cell, the parietal tissue is one-layered.

The inner integument is initiated as a complete ring within the dermal layer of the ovular primordium. It is two layers thick, three layers around the micropyle. The outer integument is initiated later than the inner one, and is dermal in origin. It forms an asymmetrical ring as the development is much less on the raphal side. The outer integument is five layers thick and longer than the inner integument.

In the anatropous bitegmic ovule, the funicle is broad. A vascular strand differentiates in the raphe. The micropyle is formed by both the integuments. The nucellus is urceolate in longitudinal view, and contains starch.

A conspicuous micropylar collar develops as a result of activity of the integuments (Figs. 57.2.1A–F). The chalazal region of the ovule expands.

The ovule is tenuinucellar in *Amischophacellus cuculata, Cyanotis tuberosa* (Chikkannaiah and Hemaraddi 1979), *Aneilema paniculatum* (Chikkannaiah 1969), and *C. cuculata* (Kulkarni and Patil 1969).

The archesporial cell cuts off a parietal cell except in *A. paniculatum, C. axillaris* and *C. cuculata* (D. Rao 1968b). The parietal cell degenerates, but in *Pollia sorzogonensis* (Lakshmanan 1962) it produces two or three layers of parietal tissue. The megaspore mother cell undergoes meiotic divisions, and a linear or T-shaped tetrad is produced. The chalazal megaspore develops into a Polygonum type of embryo sac (Fig. 57.2.1A–C). The polar nuclei fuse before

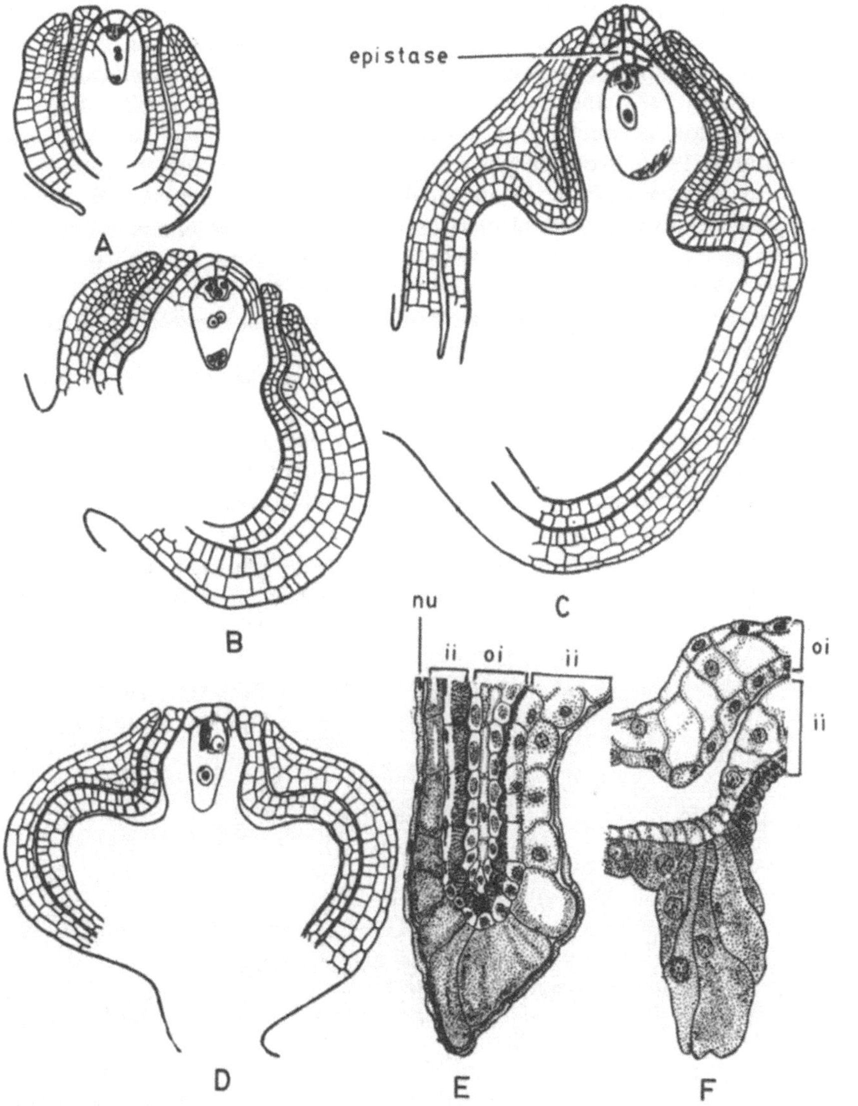

Fig. 57.2.1A–F. A *Floscopa scandens*, **B** *Commelina subulata*, **C** *Tinantia fugax*, **D, E** *Aneilema paniculatum*, **F** *Cyanotis axillaris*. **A–C** Longisections of ovule at mature embryo sac and **D** at fertilization to show the formation of collar. **E, F** Portion of ovules in the region of the collar; *nu* nucellus; *ii* inner integument; *oi* outer integument. (After Chikkannaiah 1962)

fertilization; the antipodals are ephemeral except in *Tinantia fugax* (Chikkannaiah 1964–65b) where they enlarge and persist. An epistase and a hypostase differentiate.

Allium type of development is reported in *Commelina stricta* and *Rhoeo discolor* (Tschermak-Woess 1947); the Adoxa type in *Commelina angustifolia* (McCollum 1939) needs a reinvestigation.

The development of the endosperm is of the Nuclear type. Enucleate cytoplasmic vesicles develop in *Floscopa scandens* (Chikkannaiah 1964–65a). The chalazal region develops into a haustorium in *Amischophacellus axillaris* (Chikkannaiah 1973) and *Cyanotis axillaris* (Chikkannaiah 1962); otherwise the aggregation of nuclei is a common feature. In *Pollia sorzogonensis* (Lakshmanan 1962), free endosperm nuclei aggregate in the upper region. Wall formation commences from the micropylar to the chalazal region, and cellular endosperm exhibits peripheral meristematic activity. The endospermal cells may become binucleate; the growth of the endosperm is centripetal. In *Cyanotis tuberosa* (Chikkannaiah and Hemaraddi 1979), the micropylar part does not become cellular, and perhaps serves as a haustorium. Further investigations are necessary to decide this issue.

The development of the embryo in *Commelina forskalaei* is shown in Figure 57.2.2A–R.

The embryogeny conforms to the Asterad type-Commelina variation in *Aneilema gigentem, A. paniculatum* (Chikkannaiah 1971), *Commelina communis, Rhoeo discolor* (Souèges 1958a,b), *C. subulata, Murdannia simplex, Floscopa scandens* and *Tinantia fugax* (Chikkannaiah 1963, 1964, 1964–65a,b). The derivatives of *ca* give rise to cotyledon. The tier *m*, daughter cell of *cb*, produces the shoot apex and upper part of the hypocotyl, while *ci* gives rise to the remaining part of the hypocotyl, root, and root cap. However, in *Commelina benghalensis* (see Lakshmanan 1972) and *Murdania semiters* (Philip and Gouri 1975), cotyledonary and epicotyledonary loci develop from the terminal cell *ca* of the two-celled proembryo, and are adjacent to one another. Thus, the shoot apex originates terminally.

In *Cartonema spicatum* (Grootjen 1983a), the exo- and endostome are zigzag. The tanniniferous endotegmen becomes crushed and shows stratification. The exotegmen stretches radially, especially at the micropylar end. The innermost layer of the outer integument forms a thin cuticle and lies firmly adpressed against the inner integument. The radial walls become thickened by the deposition of silica. The mesophyll degenerates. The outermost layer of the outer integument is thin-walled and silica gets deposited.

The operculum forms as a subconical elevation on the seed. It appears as a black ring due to the hyaline nature of the exotegmen. The operculum remains attached to the seed on one side; it does not get completely detached.

The seed has a small, poorly differentiated embryo. The thin-walled large endosperm cells contain starch. The seed breaks away from the broad funicle along the endotesta and the silica layer is exposed. The cells bordering the plane of rupture are discontinuous so that the endotegmen ruptures at a different site than the exotegmen.

The micropylar collar and operculum develop in *Stanfieldiella imperforata*. The collar is initiated in the middle layers close to the tip of the outer integument followed by complementary changes in the inner integument and nucellus. The operculum develops from the distal part of the integument, together with the epistase. In the ripe seed the outer integument, except its silicified inner layer, gets detached leaving the micropylar collar around the operculum.

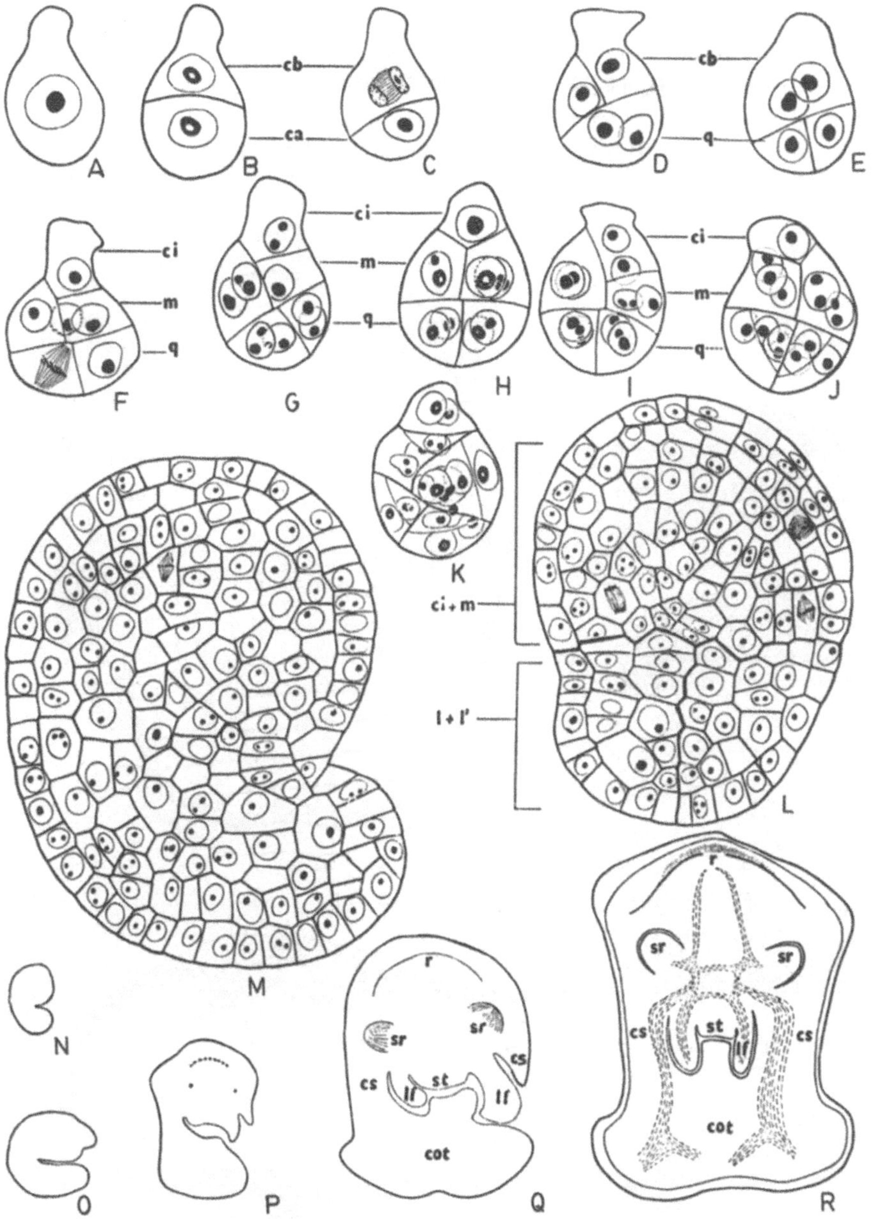

Fig. 57.2.2A–R. *Commelina forskalaei.* **A** Zygote. **B, C** Two-celled proembryo. **D, E** Four-celled proembryo, longitudinal division of *cb* in **D**. **F** Six-celled proembryo. **G, H** Eight-celled proembryo, the quadrant develops from the terminal tier. **I–K** Later stages of proembryo. The cells of the quadrant undergo periclinal divisions to delimit dermatogen (**J, K**). **L, M** Differentiation of notch (on the side of embryo) and cotyledon. **N–Q** Outline diagrams to show progressive development of embryo; *cot* cotyledon; *cs* cotyledonary sheath; *lf* leaf; *r* root; *sr* secondary root; *st* stem tip. **R** Longisection of mature embryo, note the vascular supply. (After S.C. Maheshwari and Baldev 1958)

The endosperm is starchy; below the seed-coat lie scattered peripheral endosperm cells which contain fatty substances.

The development of the testa in *Commelina subulata* and *Floscopa scandens* is represented in Fig. 57.2.3A–F.

The exotestal, exotegmic seed is albuminous, with a small and marginal cylindrical embryo. Both the integuments form the seed-coat. The cells of

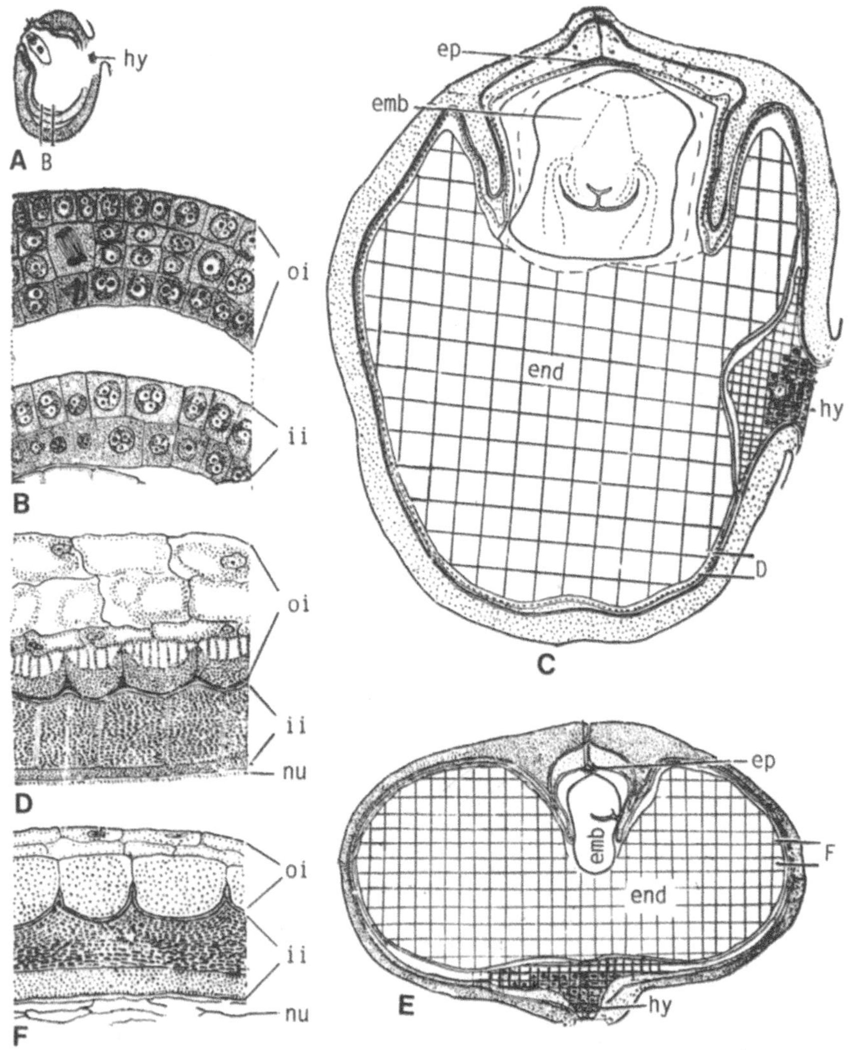

Fig. 57.2.3A–F. A–D *Commelina subulata*, **E, F** *Floscopa scandens*. **A** Longisection of young seed; *hy* hypostase; **B** Portions of outer (*oi*) and inner (*ii*) integument and nucellus (*nu*) from *inset B* marked in **A**; **C, E** Longisections of mature seeds, *ep* epistase; *emb* embryo; *end* endosperm. **D** *Inset D* marked in **C**. Note the endotestal thickenings in **D**. The outer layer of inner integument also develops thickenings. **F** *Inset F* marked in **E**. The endotesta shows thickenings on the inner walls of cells. The cells of inner integument are filled with refractive granules. (After Chikkannaiah 1962)

the outer epidermis of the testa enlarge and in the collar region they form a small cap.

JW Beatty and AV Betty (1953), JPM Brenan (1966), G Brückner (1926, 1930), PS Chikkannaiah (1961, 1962, 1963, 1964, 1964–1965a, b, 1969, 1970, 1971, 1973), PS Chikkannaiah and B Hemaraddi (1979), CJ Grootjen (1983a), CJ Grootjen and F Bouman (1981a), AR Kulkarni and VD Patil (1969), KK Lakshmanan (1962, 1972, 1977), RE Lee (1961), SC Maheshwari and B Baldev (1958), RL McCollum (1939), RH Mepham and GR Lane (1969), KL Murthy (1934, 1938), K Nanda and SC Gupta (1977), SJ Owens and FM Kimmins (1981), M Parks (1935), VJ Phillip and U Gouri (1975), D Rao (1968b). BB Simpson, JL Neff and G Dieninger (1986), R Souèges (1958a, b), E Tshermak-Woess (1947), K Yasui and N Suita (1939), T Yamashita (1983)

57.3 Mayacaceae

The flowers are partly cleistogamous. The anther is tetrasporangiate with poricidal dehiscence. The anther wall comprises an epidermis, endothecium, two to four middle layers, and secretory tapetum with uninucleate cells in *Mayaca fluviatilis* (Venturelli and Bouman 1986). The development of the anther wall is of the Basic type (Davis 1966). After the uninucleate stage of pollen grains, the inner tangential walls of the epidermis become reticulately thickened and form the exothecium. In mature anther the inner walls of the exothecium also develop thickenings. The endothecial cells lack fibrous thickenings. The microspore mother cells undergo meiotic divisions, cytokinesis is successive, and the tetrads isobilateral and decussate. The nucleus of the spherical microspore divides to form a generative and a vegetative cell.

The pollen grain is monosulcate with a finely reticulate exine. It is two-celled at the dispersal stage; germination occurs in situ. The sperm cells are fusiform with condensed nucleus and surrounded by a sheath of cytoplasm. The pollen tube passes through the tip of style and grows ectotrophically along the stylar transmitting tissue.

The orthotropous, bitegmic, tenuinucellate ovules are arranged in two or three rows on parietal placentae. The micropyle is formed by both integuments. The endostome is narrow, while the exostome forms a broad channel. The hypostase is well developed. The nucellar cells radiate from the epidermis to the embryo sac. The hypodermal archesporial cell develops directly into a megaspore mother cell which undergoes meiotic divisions and forms a T-shaped tetrad (Fig. 57.3.1A–C). The chalazal megaspore develops into a Polygonum type of embryo sac (Fig. 57.3.1D–G). The polars fuse before fertilization. The antipodals are enclosed in a pouch formed by the hypostase (Fig. 57.3.1G). They are ephemeral.

The primary endosperm nucleus remains in the middle of the embryo sac; the development of the endosperm is of the Nuclear type (Fig. 57.3.2A, B). Cellularization starts from the micropylar end at the young globular proembryo, and proceeds downward (Fig. 57.3.2C). During cellularization, the chalazal chamber remains distinct. The nuclei undergo divisions producing irregularly shaped daughter nuclei. Later, this chamber also becomes cellular,

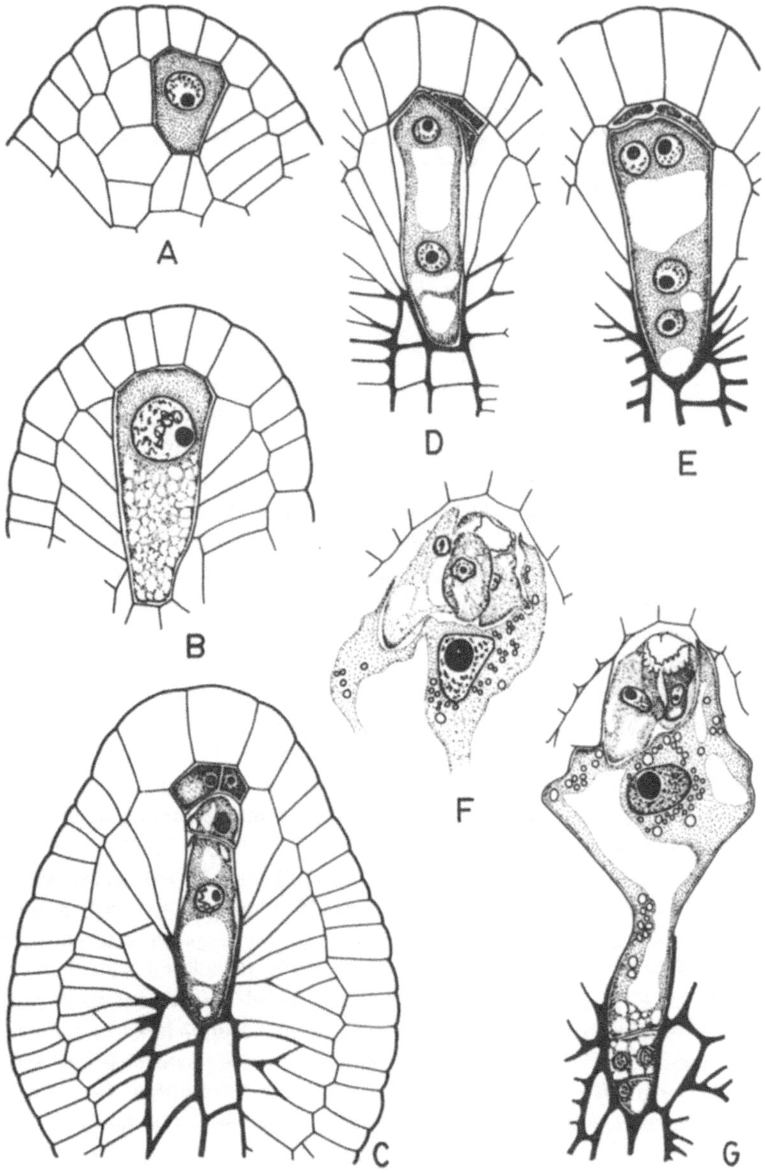

Fig. 57.3.1A–G. *Mayaca fluviatilis.* **A, B** Megaspore mother cell. **C** T-shaped tetrad. **D, E** Two- and four-nucleate embryo sac. **F** Egg apparatus and secondary nucleus. **G** Mature embryo sac. (After Venturelli and Bouman 1986)

does not store starch grains, and degenerates. The peripheral cells of endosperm are uninucleate and form the aleurone layer. The rest of the endosperm comprises multinucleate cells and stores reserve food.

The embryogeny corresponds to the Onagrad type. The mature embryo assumes a depressed-ovoid shape, and remains small and organless (undifferentiated). The cells contain amyloplasts, and may form small intercellular spaces (Fig. 57.3.3A–L).

The non-operculate albuminous seed contains starchy endosperm and broad embryo. The testa and tegmen are two-layered. The hypostase is very conspicuous (Figs. 57.3.2A–C, F). Mature seeds float with the help of air-filled exotesta. The cells of the outer layer of the testa elongate radially and accumulate starch grains (Fig. 57.3.2D, E). The lateral and inner tangential walls are strongly thickened and pitted. The cell walls are lignified. The inner layer of the testa is unspecialized. The tegmen shows differentiation only in the micropylar region (Fig. 57.3.2C). The inner layer of the testa and the outer and inner layer of the tegmen undergo radial elongation to close the exo- and endostome. The inner layer of the tegmen is tanniniferous and the micropylar cells do not have distinct wall thickenings. Seed lids are always rigid due to the presence of sclerotic tissue to protect the underlying embryo. They get lifted during germination.

EJH Corner (1976), GL Davis (1966), J Hutchinson (1973), DWM Stevenson (1983), M Venturelli and F Bouman (1986), MS Zavada (1983)

57.4 Xyridaceae

The anther wall comprises persistent epidermis, an endothecium, a single ephemeral middle layer, and secretory tapetum in *Xyris pauciflora* (Govindappa 1955b). The endothecium lacks fibrous thickenings, and the tapetal cells occasionally become binucleate. The meiotic divisions of microspore mother cells are accompanied by successive cytokinesis. The tetrads are isobilateral.

The anther wall in *Abolboda grandis* (Tiemann 1985) is of the Monocotyledonous type. The endothecium shows fibrous thickenings, and the tapetum is amoeboid with binucleate cells. Cytokinesis is successive.

The four microspores of a tetrad divide simultaneously; the generative cell is cut off towards the proximal pole in *Xyris pauciflora* (Govindappa 1955b). The pollen grains are monocolpate (pantoperculate with inaperturate form) and two-celled at the dispersal stage in *X. pauciflora*, and three-celled in *X. indica*. In *A. grandis* the pollen grains are inaperturate and two-celled with a large generative cell.

The ovule is orthotropous, bitegmic and tenuinucellar. Both the integuments form micropyle. The archesporial cell functions as megaspore mother cell, and undergoes meiotic division. In *X. pauciflora* (Govindappa 1955b) and *X. schoenoides* (Ramaswamy and Raju 1982), the chalazal dyad develops into an Allium type of embryo sac. Polygonum type of development is reported in *X. indica* (Weinzieher 1914). The synergids are hooked, polar nuclei fuse before

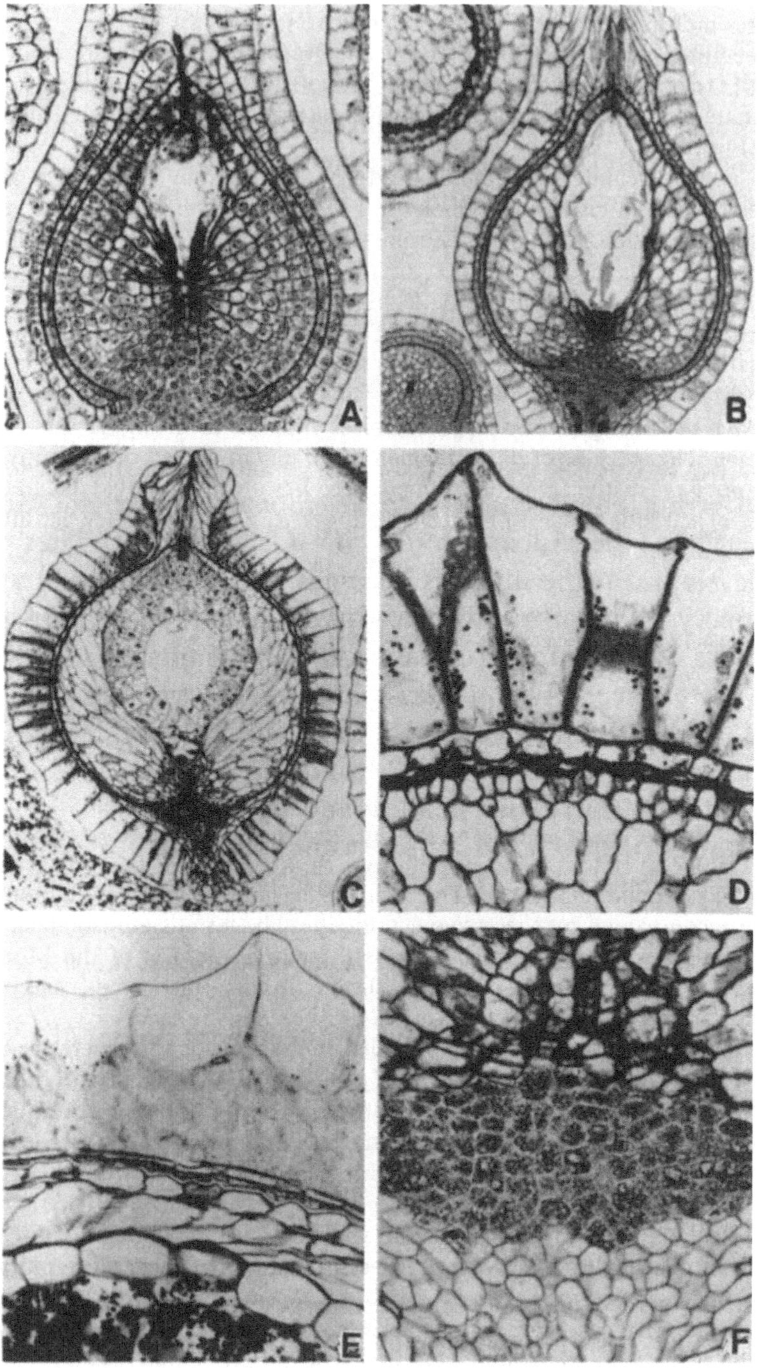
A
B
C
D
E
F

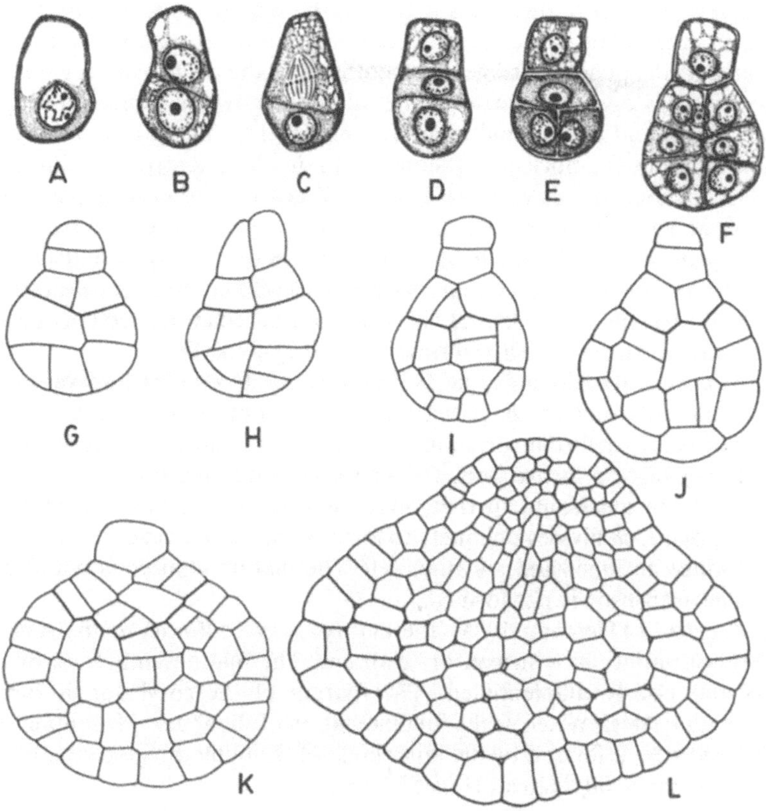

Fig. 57.3.3A–L. *Mayaca fluviatilis.* **A** Zygote. **B–E** Stages in the formation of proembryo, the apical cell has divided longitudinally (**E**). **F, G** Globular proembryo, the derivatives of the apical cell have undergone transverse division. **H–K** Later stages of globular proembryo. **L** Organless embryo from mature seed. (After Venturelli and Bouman 1986)

fertilization, and antipodals do not organize into cells in *X. pauciflora* and *X. schoenoides.*

In *A. grandis* the ovule is anatropous, bitegmic, tenuinucellar. The nucellar epidermis elongates radially and laterally and divides (irregularly) periclinally. The archesporial cell develops into megaspore mother cell, and the chalazal megaspore of a linear tetrad develops into a Polygonum type of embryo sac. The antipodals are ephemeral.

Fig. 57.3.2A–F. *Mayaca fluviatilis.* **A** Ovule at fertilization of embryo sac, the nucellar epidermis remains one-layered at the micropylar region and is three or four cells thick laterally and basally. **B** Young seed, shortly after fertilization, with free-nuclear endosperm. The endo- and exostome have become closed. **C** Immature seed with conspicuous outer integument, nucellus, cellular endosperm, and globular proembryo. **D** Seed-coat and nucellus (longisection) corresponding to **C**. **E** Transection of mature seed-coat to show thickened exotestal cells, and cuticle between nucellus and inner integument. **F** Chalazal region of seed with tanniniferous hypostase. (After Venturelli and Bouman 1986)

The development of the endosperm is of the Nuclear type; free nuclei and cytoplasm accumulate in the chalazal region of the embryo sac. Cell formation is initiated at the quadrant stage of proembryo, in the micropylar region. The endosperm cells are uninucleate in the micropylar region, binucleate in the central region, and a large multinucleate cell develops in the chalazal region. As growth proceeds, nuclear divisions occur in the micropylar uninucleate cells. The coenocytic cells undergo centripetal growth and occupy the central cavity of the embryo sac; nuclear fusions form irregular polyploid nuclei.

In *A. grandis*, the development of the endosperm is of the Helobial type (Fig. 57.4.1A). The micropylar chamber becomes cellular and the cells contain starch and proteins (Fig. 57.4.1B, C). The chalazal chamber remains coenocytic, and shows transfer cell wall-ingrowths (Fig. 57.4.1D, E).

In *X. pauciflora* (Govindappa 1955b), the zygote divides transversely to produce the cells *ca* and *cb*. A vertical division occurs in *ca*, and the two juxtaposed cells, by further division, give rise to a quadrant *q*. By periclinal division of *q*, eight cells are formed. Then tangential division occurs in each axial cell of the quadrant. Further divisions take place in longitudinal and transverse plane. *cb* divides obliquely to form *m* and *ci*. *m* divides in such a way as to produce an inner and an outer cell. The mature embryo is undifferentiated. The suspensor is not formed.

In *A. grandis* (Tiemann 1985), the embryogeny conforms to the Asterad type-Penaea variation; a suspensor is absent. The mature embryo is broad, conical and poorly differentiated. The margin of the cotyledon is turned upward with a lateral notch as the first indication of shoot apex. It contains oil.

The seed-coat is formed by the inner layer of the outer integument and two layers of the inner integument (Fig. 57.4.1F).

DA Govindappa (1953, 1955b), SN Ramaswamy and MVS Raju (1982), A Tiemann (1985), S Weinzieher (1914)

57.5 Rapateaceae

Tiemann (1985) has investigated the embryology of several taxa: *Cephalostemon*, *Maschalocephalus*, *Potarophytum*, *Rapatea*, *Saxofridericia*, *Spathanthus* and *Stegolepis*.

The anther wall comprises an epidermis, one to three fibrous endothecial layers, one to four middle layers, and secretory tapetum with binucleate cells. The development is of the Monocotyledonous type (Davis 1966). The microspore mother cells undergo meiotic divisions, and cytokinesis is simultaneous. The tetrads are tetrahedral and decussate.

The anther opens by a subapical longitudinal split, except in *Spathanthus* where dehiscence occurs by an apical pore.

The pollen grains are unisulcate or two-zonisulcate, and two-celled at the shedding stage.

The pistil is tricarpellary, syncarpous, with two ovules in each cell. The ovule is anatropous, bitegmic, and crassinucellate with a funicular obturator and chalazal cap. The micropyle is organized by both the integuments. The

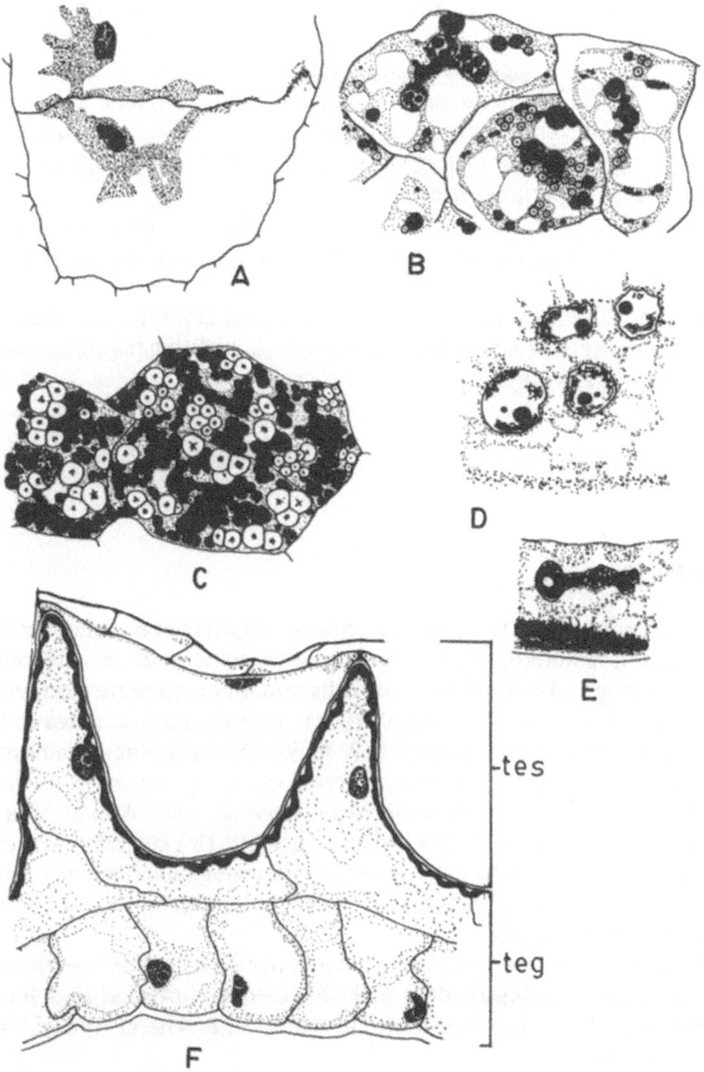

Fig. 57.4.1A–F. *Abolboda grandis.* **A** Lower part of embryo sac to show two-celled endosperm. **B** Micropylar region of endosperm with multinucleate cells containing reserve food. **C** Endosperm cells from micropylar region of mature seed, with abundant reserve food – starch and proteins. **D** Free-nuclear endosperm in chalazal region. **E** Chalazal wall of endosperm shows transfer cell wall-ingrowths. **F** Cross-section of seed-coat of nearly mature seed, testa (*tes*) thin-walled and two-layered tegmen (*teg*). (After Tiemann 1985)

outer integument consists of three to ten cell layers, while the inner is two-layered.

The single hypodermal archesporial cell cuts off a parietal cell. The nucellar epidermis is radially elongated and partially periclinally divided in *Maschalocephalus*, *Patarophytum*, *Saxofridericia* and *Stegolepis*. The megaspore mother cell undergoes meiotic divisions and the chalazal megaspore of a linear or T-shaped tetrad, or triad, develops into an embryo sac of the Polygonum type. Usually, there are three ephemeral antipodals but as many as 40 small antipodals may develop in *Maschalocephalus* and *Potarophytum*.

Development of the endosperm is of the Nuclear type, the nuclei aggregate in the chalazal region, and wall formation is delayed. The endosperm stores starch and proteins in peripheral layers.

The development of the embryo is of the Asterad type – Penaea variation; suspensor is absent. The mature embryo is lenticular and undifferentiated (Fig. 57.5.1A–H).

The albuminous seed contains a broad, unilateral embryo. The seed-coat comprises the innermost layer of the testa with U-shaped thickenings, and two layers of tanniniferous tegmen.

GL Davis (1966), A Engler (1988), Y Kimura (1956), A Tiemann (1985)

57.6 Eriocaulaceae

The anther wall comprises the epidermis, fibrous endothecium, an ephemeral middle layer, and glandular tapetum with binucleate cells in *Eriocaulon quinquangulare* (Begum 1968). The tapetal cells remain uninucleate throughout in *E. cinereum* (C.M. and D.S. Patel 1964), *E. hookerianum* (Arekal and Ramaswamy 1980), and *E. septangulare* (W.R. Smith 1910). The development of the anther wall follows the Monocotyledonous type (Davis 1966) in *E. hookerianum* and *E. xeranthemum* (Ramaswamy and Arekal 1982b), Dicotyledonous according to C.M. and D.S. Patel (1964). The archesporium is two-rowed in *E. cinereum*, and single-rowed in *E. hookerianum*. The microspore mother cells undergo meiotic divisions resulting in isobilateral and decussate tetrads. Cytokinesis is successive.

The pollen grains are spiraperturate. They are two-celled at the dispersal stage in *E. trilobum*, and three-celled in *E. hookerianum* (Arekal and Ramaswamy 1980) and *E. xeranthemum* (Ramaswamy and Arekal 1982b). This aspect needs further study.

The ovule is orthotropous, bitegmic, and tenuinucellar, with the inner integument forming the micropyle (Fig. 57.6.1A–E); in *E. trilobum* both integuments take part. The integuments are two layers thick. The hypodermal archesporial cell functions as megaspore mother cell. The latter undergoes meiotic divisions and linear, obliquely linear, or T-shaped tetrad is produced. The chalazal megaspore develops into a Polygonum type of embryo sac. The polar nuclei fuse before fertilization. The formation of endothelium is reported (Begum 1966, 1968) but, according to Arekal and Ramaswamy (1980), it differs from that of dicotyledons in histology and function. A further study is necessary.

The antipodals are linearly arranged (Begum 1966, Swamy and Krishnamurthy 1970). In *E. hookerianum* (Arekal and Ramaswamy 1980) and

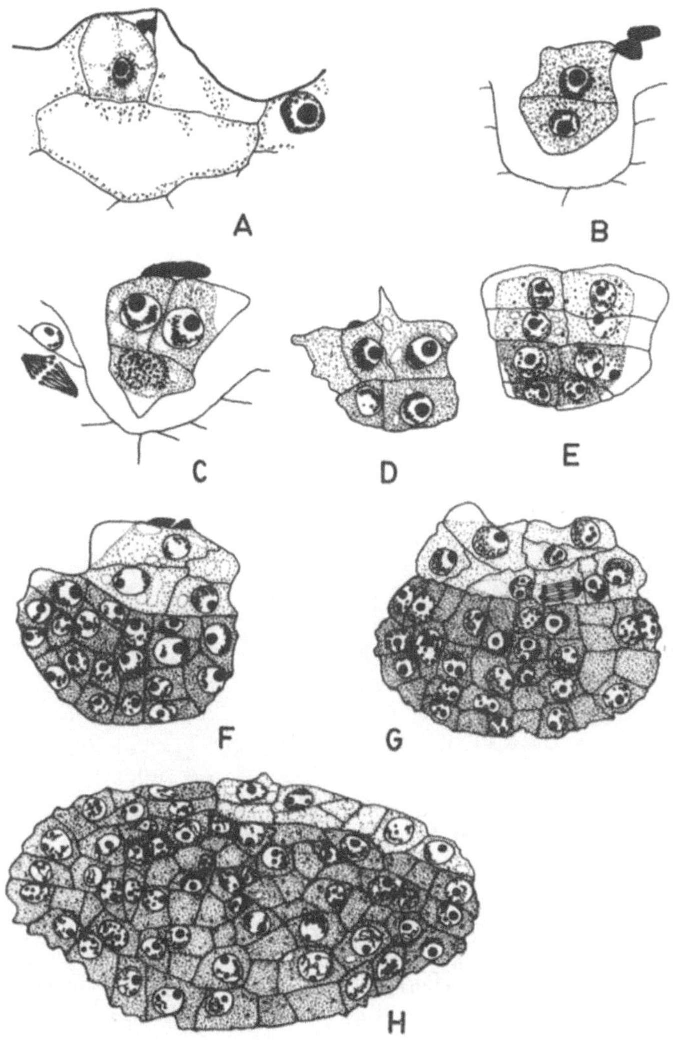

Fig. 57.5.1A–H. *Spathanthus bicolor*, embryogeny. **A** Micropylar portion of embryo sac with zygote and an endosperm cell. **B** Two-celled proembryo. **C** Three-celled proembryo, one endospermal nucleus at metaphase. The remnants of pollen tube persist in **A–C**. **D** Proembryo with vertical division in apical and basal cell. **E** Four-tiered proembryo. Derivatives of basal cell are *lightly stippled* in **E–H**. **F** Young monosymmetric embryo. **G** Later stage. **H** Lens-shaped organless (undifferentiated) embryo. (After Tiemann 1985)

E. xeranthemum (Ramaswamy and Arekal 1982b), the walls of the antipodal cells disorganize, their protoplasts fuse, and form an antipodal "cyst" (Fig. 57.6.1A-D). Before the formation of secondary nucleus, the antipodal cyst grows upward. Following the enlargement of the central cell, after the formation of the secondary nucleus, the antipodal cyst comes to lie in the chalazal region of the embryo sac in *E. hookerianum* (Arekal and Ramaswamy 1980).

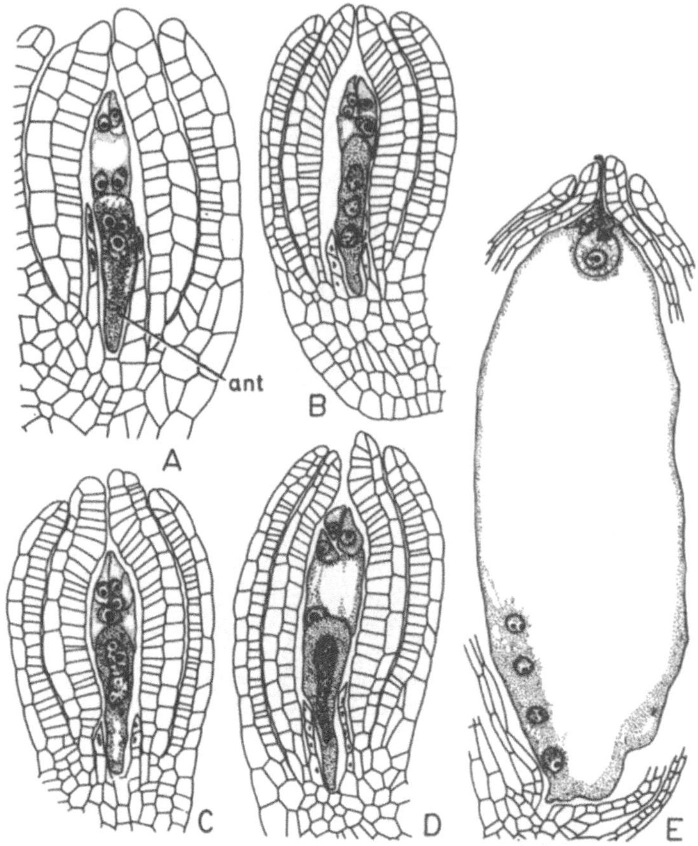

Fig. 57.6.1A–E. *Eriocaulon hookerianum*. **A** Longisection of ovule at mature embryo sac, one synergid is not shown. The antipodal cell (*ant*) is trinucleate. **B** Later stage, secondary nucleus lies close to egg apparatus. **C, D** The uninucleate condition of cyst is due to fusion of three antipodal nuclei. **E** Longisection of embryo sac with zygote and four-nucleate endosperm, note remnants of pollen tube and synergids, the antipodal cyst has collapsed. (After Arekal and Ramaswamy 1980)

The development of the endosperm is of the Nuclear type (Fig. 57.6.1E). Wall formation is initiated at the chalazal end, and proceeds upward (Arekal and Ramaswamy 1980, Ramaswamy and Arekal 1982b). The cells contain starch and oil globules as food reserve.

In *E. xeranthemum* (Ramaswamy and Arekal 1982b), the zygote divides in a transverse plane to form two hemispherical cells of equal volume, *ca* and *cb*. Vertical divisions occur in each cell giving rise to a quadrant. Each cell divides periclinally to form a peripheral layer surrounding axial cells. The peripheral cells divide anticlinally and produce a protodermal layer. The octants are not formed. Internal differentiation occurs and the growth activity is concentrated much more in the derivatives of apical cell *ca*, generally confined to the lower half of the hemisphere. In the derivatives of the apical tier, one sector (of the hemisphere) grows more rapidly and can be distinguished as the cotylary zone. In the other sector the growth is quiescent and it is distinguishable as the epicotylary locus. Thus, both cotylary and epicotylary zones are of terminal origin. In the ripe seed, the epicotylary zone consists of a group of two to three cells.

The contribution of basal cell *cb*, of the two-celled proembryo, is only a few layers, which are not histologically differentiated. The suspensor and true radicle are not formed. The multicelled embryo in the mature seed is bell-shaped.

The embryogeny conforms to the Asterad type in *E. quinquangulare* (Begum 1968). The zygote divides transversely to form *ca* and *cb*, both the cells divide vertically. The divisions in the derivatives of *ca* cut off a protodermal layer. The division in one quartet of the proembryo is quiescent (epicotyledonary locus), while in other sectors rapid divisions occur (cotyledonary locus). The mature embryo is undifferentiated. Due to the rapid development in other sectors, the embryo appears dome-shaped and becomes broad and disc-shaped with recurved margins. Ramaswamy et al. (1981) wrongly interpret the recurved portion as a part of the endosperm. The epicotylary sector remains only three- or four-celled at the dispersal of seed.

Both the integuments contribute to the formation of seed-coat. The inner layer of the testa and the outer layer of the tegmen are lignified along their outer tangential walls. The endotegmen contains dense contents. The cells of the inner layer of pericarp develop band-like thickenings on the inner tangential wall. These assist in the dehiscence of fruit (Arekal and Ramaswamy 1980, Ramaswamy et al. 1981).

The germination of an undifferentiated embryo is of much interest. In *E. robusta-brownianum* (Ramaswamy et al. 1981), the narrower pole of the bell-shaped embryo is adpressed to the micropyle. The epidermal cells of the cotyledonary sector elongate in the direction of the endosperm, while the subepidermal cells become meristematic and undergo a series of transverse divisions so that the epicotylary locus gets separated from the apex of the cotyledon. With continued linear growth of the cotyledon, the embryo pushes out of the micropyle and the cotyledon remains inside the seed functioning as an absorbing and conducting organ. As the radicular end of the embryo pushes out of the seed-coat, the primordium of the first leaf appears close to the shoot apex opposite to the cotyledon. The protodermal cells opposite to the shoot apex develop rhizoid-like hairs. At the same time, a group of richly cytoplasmic cells forms primary root initials opposite to the shoot apex. As the second leaf emerges, the shoot apex can be recognized as a small group of cells. The cotyledon remains inside the seed in contact with the endosperm and serves as an absorptive organ.

During the establishment of the seedling of *E. robusta-brownianum*, the lateral strands of the procambium of the first leaf appear. Below it two to four tiers of epidermal cells develop into two- to three-celled uniseriate hairy outgrowths. The primary root lies away from the shoot-apex and the intervening space is occupied by parenchyma. The procambium differentiates and connects the primary root and the first node. One of the branch passes to the cotyledon. Adventive root primordium arises between cotyledon and the first leaf. The primary root pushes out of the radicular tissue. Now the seedling grows on the substratum.

GD Arekal and SN Ramaswamy (1980), M Begum (1966, 1968), GL Davis (1966), WR Monteiro-Scanavacca and SC Mazzoni (1978), CM and DS Patel (1964), SN Ramaswamy and GD Arekal (1981a,b,c, 1982a,b), SN Ramaswamy and MVS Raju (1982), SN Ramaswamy, GD Arekal and MVS Raju (1983), SN Ramaswamy, BGL Swamy and GD Arekal (1981), WR Smith (1910), BGL Swamy and KV Krishnamurthy (1970)

57.7 Restionaceae

The embryology of 18 genera and 36 species has been investigated by Kircher (1986).

The anther is usually monothecous and bisporangiate (bithecous and tetrasporangiate in Anarthrieae). The wall comprises the epidermis, fibrous endothecium, ephemeral middle layer, and mostly glandular tapetum with binucleate cells in *Hypodiscus aristatus* (Krupko 1962). Sometimes, the tapetal cells may be three-, or four-, or just uninucleate (Kircher 1986). Successive cytokinesis follows meiotic divisions in the microspore mother cells. The tetrads are usually isobilateral, rarely decussate and T-shaped.

The monocolpate pollen grains are three-celled in *Hypodiscus.* Krupko (1962) mentions that three-celled condition is rare in *Chondropetalum hookerianum.* Kircher points out that even four-nucleate (fourth nucleus component of vegetative cell) pollen grains develop in *Leptocarpus similis.* Chanda (1966) studied the pollen morphology of 118 species of 18 genera.

The ovule is anatropous, bitegmic (mostly two-layered integuments), and crassinucellar; it becomes orthotropous secondarily. Both the integuments form the micropyle. In *Restio dodii* (Borwein et al. 1949), the integuments do not close up till the tetrad stage. In a few taxa – *L. similis and Willdenowia lucaeana* – the micropyle is formed only by the inner integument. An obturator is reported in *L. paniculatus* and *Restio calystachyme.* A hypostase is common in *R. callustachyus* (Kircher 1986).

According to Kircher, the nucellar organization shows much variation. In *L. similis,* the epidermis remains undivided and the cells elongate in the micropylar region (*L. paniculatus* has smaller cells). In *Elegia,* the epidermis undergoes periclinal divisions, and the derivatives surround the tip of the embryo sac. *Restio* and *Willdenowia* also show divisions in nucellar epidermis. During post-fertilization development, only a few nucellar cells persist in *L. similis.* In *Hypodiscus,* the massive nucellar tissue protects the embryo.

The archesporial cell functions as megaspore mother cell, and undergoes meiotic divisions. The chalazal megaspore of a linear tetrad develops into a Polygonum type of embryo sac. Twin embryo sacs develop in *L. similis.* In *Elegia,* the embryo sac elongates over the antipodals deep into the nucellus so that the antipodals come to lie in a lateral position (Borwein et al. 1949). In *E. capensis* (Kircher 1986), the synergids simulate the egg, and the egg shows a vacuole above as well as below the nucleus. The polar nuclei fuse before fertilization, and antipodal cells persist up to fertilization. They undergo secondary multiplication (six to 14) in *E. racemosa* and *Restio dodii* (Borwein et al. 1949, see also Kircher 1986).

Occasionally, syngamy is delayed and occurs at 16-nucleate stage of the endosperm; this requires confirmation.

The development of the endosperm is of the Nuclear type in *Thamnochortus fruticosus* (Mlodzianowski 1964). At first the endosperm nuclei divide synchronously, then in a wave from the micropylar to the chalazal end, and later on irregularly (Kircher 1986). Wall formation occurs and the endosperm becomes cellular. At maturity an aleurone layer can be distinguished; rest of the endosperm is starchy.

In *L. similis*, there may be even 150 endosperm nuclei while the zygote is still undivided (Kircher 1986, see also Krupko 1962, 1963, Mlodzianowski 1964). It divides by an oblique wall into the cells *ca* and *cb*. The latter produces a short suspensor which later collapses. *ca* gives rise to the quadrant and octant in which periclinal divisions in the outer cells cut off the protodermal layer. The mature embryo is not fully differentiated. It is lens-shaped with a lateral plumule in *Elegia asperifolia* (Kircher 1986). The development of the embryo has not been studied in sufficient detail.

After fertilization the outer integument shrivels, and the seed-coat is derived from the inner layer of the inner integument. In the nut-fruit species, *Hypodiscus argenteus* (Kircher 1986), the testa is reduced to a thin layer and the thick-walled pericarp takes over the protective function. In contrast, in the capsule-fruit wind-disseminated *Chondropetalum* sp., the testa persists. Its anticlinal walls form a characteristic pattern on the seed surface. In *C. mucronatum*, the testa is longitudinally ribbed and winged. In *Leptocarpus similis*, the anticlinal walls project so that the net-like sculpture of the testa becomes apparent.

The pendulous seed contains a small laterally placed embryo, and abundant endosperm (Hutchinson 1973).

B Borwein, ML Coetsee and S Krupko (1949), S Chanda (1966), Ch Gilg-Benedict (1930), J Hutchinson (1973), P Kircher (1986), S Krupko (1962, 1963), F Mlodzianowski (1964), PI Rudall and HP Linder (1988)

57.8 Centrolepidaceae

The wall of bisporangiate (unilocular) anther comprises the epidermis, fibrous endothecium, single middle layer, and secretory tapetum with binucleate cells in *Centrolepis fascicularis*. The microspore mother cells undergo meiotic divisions, cytokinesis is successive, and the tetrads are isobilateral, linear or T-shaped.

The pollen grains are monocolpate, and three-celled at the dispersal stage.

The ovule is orthotropous, bitegmic and tenuinucellar. The nucellar epidermis shows marked radial elongation of cells. The archesporial cell functions as megaspore mother cell. The latter undergoes meiotic divisions resulting in a linear tetrad. The chalazal megaspore develops into a Polygonum type of embryo sac. The polar nuclei fuse before fertilization. The antipodals persist up to fertilization, they undergo secondary multiplication in *C. aristata* and *C. strigosa* (Hamann 1962).

Development of the endosperm is of the Nuclear type, and it becomes cellular at the late globular proembryo.

The embryogeny conforms to the Onagrad type-Juncus variation. In *C. fascicularis* (N. Prakash 1969e), the cell *ca* of the two-celled proembryo undergoes a vertical division, and produces two juxtaposed cells which develop into the quadrant *q*. The latter develops into an octant, and by further anticlinal and periclinal divisions into a globular proembryo. By a transverse division *cb* produces *m* and *ci*. *m*, by a vertical division, forms the initials of root cortex and root cap. The embryo is conical, small and unorganized at the shedding stage of seed.

The seed is albuminous with a small marginal embryo, and contains compressed nucellar epidermis. The outer epidermis of the testa comprises small, thick-walled cells. Cuticular thickenings occur at three sites: on the epidermis or testa, between the two integuments, and between the inner integument and the nucellar epidermis. The inner epidermis of the testa and outer epidermis of the tegmen disorganize. The cells of the inner epidermis of the tegmen are large and contain tannin.

S Chanda (1966), U Hamann (1962, 1975, 1976), B Palm (1920), N Prakash (1969e)

57.9 Flagellariaceae

Development of the endosperm, embryo and seed-coat has not been investigated.

The anther wall comprises the epidermis, fibrous endothecium, two ephemeral middle layers, and secretory tapetum with binucleate cells. The microspore mother cells undergo meiotic divisions, cytokinesis is successive in *Flagellaria indica* (Palm 1920), and the tetrads are isobilateral and linear. According to Shetty and Subramanyam (1964), the divisions in the pollen mother cells are simultaneous in *F. indica*, and occasionally giant pollen grains are formed.

The pollen grains are mono-ulcerate, and two-celled at the dispersal stage.

The ovule is pendulous, orthotropous, bitegmic and crassinucellar. The inner integument grows faster and forms the micropyle. The hypodermal archesporium is multicellular. A single cell divides to form a parietal cell which does not undergo further divisions, but elongates radially. The cells of the nucellar epidermis divide periclinally. The megaspore mother cell undergoes heterotypic division, and the chalazal dyad develops into an Allium type of embryo sac in *F. indica* (Subramanyam and Narayana 1972). The synergids are flask-shaped. The upper polar nucleus is larger than the lower. They fuse before fertilization, and the secondary nucleus lies near the egg apparatus. The antipodals are ephemeral.

The seed contains a small embryo and copious endosperm (Hutchinson 1973).

S Chanda (1966), J Hutchinson (1973), B Palm (1920), PI Rudall and HP Linder (1988), BV Shetty and K Subramanyam (1964), K Subramanyam and HS Narayana (1972)

57.10 Taxonomic Considerations

Bentham and Hooker (1965c) place the families of the suborder Commelinineae (Melchior 1964) in the series Coronarieae, and Flagellariaceae with Juncaceae in Calycinae; Eriocaulaceae, Restionaceae and Centrolepidaceae are placed in Glumaceae. Engler and Prantl (1931) include all the families of Commelinales in Farinosae. Hutchinson (1973) distributes them in three orders: Commelinales, Xyridales and Eriocaulales in Division I – Calyciferae; and Centrolepidaceae, Restionaceae in Juncales along with Juncaceae in Division III – Glumiflorae.

On the basis of comparative embryology, Hamann (1975) discusses the systematic position of various families included in the Farinosae: Restionaceae, Centrolepidaceae, Xyridaceae, Eriocaulaceae, Bromeliaceae, Commelinaceae, Pontederiaceae, Cynastraceae and Phylidraceae. The anther tapetum is mostly secretory, plasmodial in Commelinaceae, secretory or plasmodial in Pontederiaceae; successive cytokinesis in microspore mother cells, simultaneous in Cynastraceae; pollen is shed at the two-celled stage, two- or three-celled in Restionaceae and Xyridaceae, rarely three-celled in Commelinaceae, three-celled in Eriocaulaceae; the ovule is atropous in Restionaceae, Centrolepidaceae, Eriocaulaceae, anatropous in Bromeliaceae, Pontederiaceae, Cynastraceae and Philydraceae, atropous or anatropous in Xyridaceae, atropous, ana- or hemianatropous in Commelinaceae; the ovules are invariably bitegmic; the embryo sac is mostly of the Polygonum type, Polygonum or Allium type in Xyridaceae, rarely Allium type in Commelinaceae; the endosperm is of the Nuclear type in Restionaceae, Centrolepidaceae, Xyridaceae, and Helobial type in Bromeliaceae, Commelinaceae, Pontederiaceae and Philydraceae.

The suborder Commelinineae comprises four families: Commelinaceae, Mayacaceae, Xyridaceae and Rapateaceae. Cronquist (1981) groups these four families in the order Commelinales.

Commelinaceae and Xyridaceae. These families are closely related, and show persistent anther epidermis, successive cytokinesis in pollen mother cells, similar type of pollen grains, ovule, and development of embryo sac, endosperm (usually Nuclear type; Helobial in *Abolboda* of Xyridaceae) and embryo. Both have albuminous seeds. But the amoeboid anther tapetum, occasionally crassinucellar ovule, and occasional tetrasporic type of embryo sac in Commelinaceae are absent in Xyridaceae. In addition, Xyridaceae segregate out due to absence of fibrous thickenings in endothecium, secretory anther tapetum, and unorganized antipodal cells.

Cartonema. *Cartonema spicatum* (Grootjen 1983a) has an isolated position in the Commelinaceae, due to non-succulent habit, spicate inflorescence, and aberrant shape of seed (Brückner 1926, 1930). The seed is operculate as in Commelinaceae, but the micropylar collar is absent. Anatomical evidence segregates it from Commelinaceae, and this is supported by Cronquist (1981) and R. Dahlgren and Clifford (1982).

The placental region is initially dizonate, later becomes trizonate; this aspect has not been investigated in Commelinaceae. The absence of micropylar collar, an annular intrusion of the integument into the nucellus around the embryo sac, and anatropous ovule are its characteristic features.

The similarities with Commelinaceae are: two placentae in each ovarian locule, but the ovules lie in a single row, bitegmic, crassinucellate ovule, both integuments are of dermal origin, micropyle formed by both the integuments, Polygonum type of embryo sac, Nuclear type of endosperm which later becomes cellular, and operculate seed (Grootjen 1983a).

Grootjen points out that, on the basis of some anatomical features, the erection of an independent family, Cartonemataceae, is justified (see also Hutchinson 1973). This, however, is not supported by subsequent anatomical studies. The embryological features also do not warrant raising *Cartonema* to a family status, and this taxon should be included in the Commelinaceae.

Mayacaceae. Mayacaceae are assigned to the suborder Commelinineae. They show persistent anther epidermis, absence of fibrous thickenings in endothecium, successive cytokinesis in pollen mother cells, orthotropous, bitegmic and tenuinucellate ovule, Polygonum type of embryo sac, Nuclear type of endosperm, Onagrad type of embryogeny, and a broad organless embryo. An operculum is absent. The radial elongation of the nucellar cells is akin to that of Cyclanthaceae. In Commelinaceae, the endothecium is fibrous, tapetum amoeboid, ovule tenui- or crassinucellate, embryo sac of bi- or tetrasporic type, embryogeny of Asterad type, and the embryo is well differentiated and cylindrical. The seed is operculate and has a micropylar collar.

Mayacaceae resemble Commelinaceae only in the Polygonum type of embryo sac and Nuclear type of endosperm. Mayacaceae resemble more to Xyridaceae: Both show absence of fibrous thickenings in endothecium, orthotropous, bitegmic and tenuinucellate ovule, Polygonum type of embryo sac in some species, Nuclear type of endosperm, and a broad organless embryo. Thus, Mayacaceae can be placed in Commelinales near Xyridaceae.

Rapateaceae. Melchior (1964) places Rapateaceae in the order Commelinales, suborder Commelinineae, along with Commelinaceae, Mayacaceae and Xyridaceae. Now Rapateaceae are either placed along with Xyridaceae in Xyridales (Hutchinson 1973), or together with Eriocaulaceae in Eriocaulales (R. Dahlgren and Clifford 1982). According to Cronquist (1981), Rapateaceae are associated with Commelinaceae, Mayacaceae and Xyridaceae in the order Commelinales. Kimura (1956) proposes the order Xyridales which includes Xyridaceae, Rapateaceae and Mayacaceae. Rapateaceae and Xyridaceae are invariably placed in the same order.

Tiemann (1985) compares the embryological features of Rapateaceae (Sect. 57.5) with those of Commelinineae – Commelinaceae (Sect. 57.2), Mayacaceae (Sect. 57.3), Xyridaceae (Sect. 57.4); Eriocaulinae – Eriocaulaceae (Sect. 57.6); Bromeliales – Bromeliaceae (Sect. 56.2); Juncales – Juncaceae (Sect. 55.2), Thurniaceae (Sect. 55.3); Cyperales – Cyperaceae (Sect. 63.2).

Rapateaceae have secretory tapetum with binucleate cells, simultaneous cytokinesis in pollen mother cells, and monosulcate two-celled pollen grains. The ovule is anatropous, bitegmic, crassinucellar with exo- and endostome. The nucellar epidermis is radially elongated but the micropylar collar and operculum are absent. The embryo sac is of the Polygonum type. Supernumerary antipodals have been observed. The development of the endosperm is of the Nuclear type, with nuclei aggregating in the chalazal region. The reserve food is starch and proteins. The embryogeny is of the Asterad type-Penaea variation. Embryo in a mature seed is undifferentiated.

In Commelinaceae the tapetum is amoeboid, cytokinesis is successive and the pollen grains are two-celled. The bitegmic ovule is orthotropous to anatropous, crassinucellate with endostome, and micropylar collar. Sometimes, the nucellar apex is free. The embryo sac is either of the Polygonum or the Allium type. There are three small antipodals. The development of the endosperm is of the Nuclear type. Reserve food is starch and proteins. The embryogeny is of the Asterad type-Penaea variation. The embryo is well differentiated in the mature albuminous seed.

In view of the foregoing resemblances and differences, there is every justification to retain Rapateaceae, as a separate family, in Commelinales.

Mayacaceae and Rapateaceae have some common features: tetrasporangiate anthers with poricidal dehiscence, binucleate pollen grains, Polygonum type of embryo sac, and Nuclear type of endosperm. They also differ in several features: cytokinesis in pollen mother cells, structure of ovule, and type of embryogeny. However, both families should be included in Commelinales.

The embryological features of Rapateaceae are more akin to Xyridaceae. The development of anther wall is of the Monocotyledonous type. The endothecium is without fibrous thickenings in *Xyris*. The tapetum is of amoeboid type and cytokinesis of successive type in *Abolboda*. The pollen grains are inaperturate. The embryo sac is of the Allium type with antipodal nuclei (not cells) in *Xyris*. It is of the Polygonum type in *Abolboda* with three antipodals; the number goes up to 40 in Rapateaceae. The endosperm is of the Nuclear type in both families (Helobial in *Abolboda*). The embryogeny is of the Asterad type and in the mature seed the embryo remains undifferentiated.

Rapateaceae and Xyridaceae show more embryological differences than resemblances. Within the Commeliniflorae, Xyridaceae show maximum similarities with the embryology of Eriocaulaceae (Tiemann 1985).

Rapateaceae can also be compared to Eriocaulaceae. Both the families have secretory anther tapetum, Polygonum type of embryo sac, Nuclear type of endosperm, Asterad type of embryogeny, and undifferentiated embryo. But, simultaneous cytokinesis, two-celled pollen grains, anatropous crassinucellate ovule, and supernumerary antipodals in Rapateaceae are not common to

Eriocaulaceae. The antipodal cyst in *Eriocaulon* is also unique. The pollen grains are three- to four-aperturate.

Rapateaceae should, therefore, be grouped with Xyridaceae near Eriocaulaceae, in Commelinales.

Rapateaceae and Bromeliaceae have bi- or tri-seriate secretory anther tapetum with binucleate cells, anatropous crassinucellate ovule, Polygonum type of embryo sac with ephemeral antipodals, and Asterad type of embryogeny. They differ in cytokinesis in pollen mother cells, formation of micropyle, type of endosperm, embryogeny, and structure of mature embryo. Rapateaceae are, indeed, distant from Bromeliaceae which have been placed in a separate order, Bromeliales (Melchior 1964).

Rapateaceae and Thurniaceae have a many-flower-head inflorescence on long stalk. The occurrence of tetrahedral pollen tetrads, anatropous crassinucellate ovule, and albuminous seed are common features in both. The embryology of Thurniaceae has not been studied in sufficient detail and, therefore, any further discussion is not possible.

Rapateaceae show several common features with Juncaceae and Cyperaceae (Tiemann 1985): fibrous endothecium, secretory tapetum, simultaneous cytokinesis in pollen mother cells, anatropous crassinucellate ovule with funicular obturator, Polygonum type of embryo sac with small antipodals, Nuclear type of endosperm (Helobial type in Juncaceae), and albuminous seed. However, the embryogeny and the mature embryo differ markedly. Therefore, the placement of Rapateaceae in Commelinales rather than Cyperales and Juncales is justified.

The suborder Commelinineae comprises Commelinaceae, Mayacaceae, Xyridaceae and Rapateaceae. The common features in these families are: successive cytokinesis in pollen mother cells (except Rapateaceae), Polygonum type of embryo sac, Nuclear type of endosperm (except *Abolboda*, Xyridaceae), Asterad type of embryogeny (except Mayacaceae), broad embryo in the mature seed, and starchy endosperm. The independent status of each family is fully justified.

Eriocaulaceae. Eriocaulaceae represent the suborder Eriocaulineae. It differs from the rest of the suborders in the secretory anther tapetum, spiraperturate pollen grains, unisexual flowers with a single pendulous, orthotropous and tenuinucellate ovule per carpel, conspicuous antipodal cyst, chalazal position of the primary endosperm nucleus which produces Nuclear type of endosperm, and an "undifferentiated" embryo in mature seed. The segregation to a separate order, Eriocaulales, is justified. Arekal and Ramaswamy (1980) also favour its elevation to ordinal rank.

Restionaceae and Centrolepidaceae. Restionaceae and Centrolepidaceae constitute the suborder Restionineae. N. Prakash (1969e, see also Kircher 1986) discusses several common features: (1) bisporangiate anthers with persistent epidermis, (2) secretory anther tapetum, (3) orthotropous, bitegmic, tenuinucellar ovules with both integuments forming the micropyle, (4) radial elongation of apical cells of the nucellar epidermis, (5) Polygonum type of embryo sac with starch grains, (6) fusion of polar nuclei before fertilization, (7) Nuclear type of endosperm, and (8) peripheral position of the embryo. But the ovule is initially anatropous in Restionaceae, and becomes orthotropous secondarily. The embryogeny in Restionaceae is somewhat comparable to that of Centrolepidaceae. The embryogeny in Centrolepidaceae (Onagrad type-Juncus variation) is similar to that of Juncaceae. However, both families negate any close affinity with Juncaceae (Sect. 55.4).

Hydatella. *Hydatella* is included in the tribe Trithurieae of Centrolepidaceae, order Commelinales (Melchior 1964). Hutchinson (1973) also includes it in Centrolepidaceae. Hamann (1976) raises it to an independent family, Hydatellaceae. Takhtajan (1980) and Cronquist (1981) accept the family rank, but in an independent order Hydatellales. According to Cronquist, the operculate swelling of the seed-coat suggests its possible relationship with Commelinales.

R. Dahlgren (1980b) refers *Hydatella* to Hydatellales, superorder Liliiflorae, and points out that its phylogenetic position is a riddle. He emphasizes that it is difficult to put *Hydatella* in any of the superorders and states: "The cellular endosperm formation agrees with the Ariflorae, the perisperm-rich seeds with the Zingiberiflorae, the graminoid habit and vessel-containing stem with the Commeliniflorae, the anomocytic stomata and monocolpate pollen grains perhaps best with the Liliiflorae."

Hamann (1976) studied the embryology of *Hydatella inconspicua* and discusses whether this taxon should be included in Centrolepidaceae, or deserves an independent family rank. *Hydatella* differs from *Centrolepis* in several embryological features: (1) *Hydatella* shows tetrasporangiate, and *Centrolepis* bisporangiate anther, (2) monosulcate pollen, ulcerate in *Centrolepis*, (3) pendulous, anatropous, bitegmic and tenuinucellar ovule with micropyle directed upward; the development of the embryo sac has not been studied. In *Centrolepis*, the pendulous ovule is orthotropous, bitegmic and tenuinucellar with micropyle directed downward. The development of the embryo sac is of the Polygonum type, (4) the development of the endosperm is of the Cellular type, Nuclear type in *Centrolepis*, (5) endosperm restricted to only a few cells in *Hydatella*, starchy food reserve in copious perisperm; endosperm starchy in *Centrolepis*, perisperm absent, (6) seed exotestal in *Hydatella*, inner integument forms an operculum. The seed of *Centrolepis* lacks an operculum, the exotesta comprises small, thick-walled cells. The inner epidermis of the inner integument persists and contains tannin, and (7) the minute embryo of *Hydatella* is lens-shaped, peripheral and incompletely organized, whereas in *Centrolepis* the embryo is small, conical and totally unorganized.

"The Centrolepidaceae s. str. agree in their embryology (and also in some other characters) best with Restionaceae and Poaceae (Hamann 1962, 1975), whereas the Hydatellaceae differ in this respect (embryology) from all graminoid families."

The embryological features of *Hydatella* differ from those of *Centrolepis* and their placement in two different tribes, Trithurieae (*Hydatella*) and Centrolepideae (*Centrolepis*) of Centrolepidaceae is appropriate.

Flagellariaceae. The suborder Flagellariineae includes a single family, Flagellariaceae. The pollen grains (mono-porate) are similar to those of Restionineae. Flagellariaceae have been associated with Commelinaceae by Bentham and Hooker. Both have common features: two-celled pollen grains, orthotropous, bitegmic ovules, and Allium type of embryo sac with ephemeral antipodals. The family Flagellariaceae has starchy endosperm, and laterally placed embryo in seed. It is correctly placed near the suborder Restionineae.

Kircher (1986) mentions that the Restionaceae are especially close to Gramineae and Centrolepidaceae and, together with Flagellariaceae, form a natural order Poales (R. Dahlgren 1980b). He also discusses the relationships to Cyperaceae, Juncaceae, Xyridaceae and Eriocaulaceae (see earlier under Rapateaceae).

Gramineae and Restionaceae show some similarities: three-celled pollen grains, micropyle is formed from the inner integument, multiplication of antipodal cells, and the inner layer of the inner integument forms the seed-coat.

In spite of the similarities with Gramineae, there is no doubt that Restionaceae are much more closely related to Centrolepidaceae and their placement in suborder Restioninae, order Commelinales, is fully justified.

58 Graminales

The family Gramineae, also designated Poaceae, represents the order Graminales.

58.1 General Features

The embryological features of Graminales are: (1) secretory anther tapetum, (2) successive cytokinesis in pollen mother cells, (3) uniporate-operculate, three-celled pollen grains, (4) ana- or campylotropous, tenui- or crassinucellar, or pseudo-crassinucellar, bitegmic ovule, (5) Polygonum type of embryo sac with hypertrophied, persistent antipodals, secondary multiplication of antipodal cells and their polyploid condition, (6) Nuclear type of endosperm, (7) Asterad type of embryogeny, (8) albuminous seed with starchy endosperm, and peripheral embryo, (9) fruit a caryopsis, and (10) occurrence of apomixis.

58.2 Gramineae

The anther wall comprises the epidermis, fibrous endothecium, one or two ephemeral middle layers, and secretory tapetum of binucleate cells. The tapetum is uninucleate in *Eleusine compressa* (Mahalingappa 1977) and *Pennisetum typhoideum* (Narayanaswami 1953). Successive cytokinesis accompanies meiotic divisions in microspore mother cells. The tetrads are isobilateral and tetrahedral. Frequently, linear, decussate and T-shaped tetrads occur in *Eleusine coracana* (Narayanaswami 1955c) and *Echinochloa colonum* (Bhanwra and Choda 1986).

The pollen grains are uniporate-operculate and three-celled at the dispersal stage. Diploid gametes due to endoduplication have been observed in *E. coracana* (Narayanaswami 1952).

Lakshmanan and Jayalakshmi (1980) report simultaneous cytokinesis and formation of four types of tetrads – isobilateral, linear, tetrahedral and decussate – in *Spinifex littoreus*. Moreover, the pollen grains are two-celled at the shedding stage.

In most florets of rye, within a floret, meiosis in microspore and megaspore mother cells occurs simultaneously. When the functional chalazal megaspore is still without vacuoles, the pollen grain has a single excentric nucleus. The functional megaspore becomes vacuolate and two-, four-, eight-nucleate and organized embryo sac is formed; the pollen grain is three-celled (Xiang-Yuan and de Mason 1984).

The ovule is ana-, hemiana-, or campylotropous, bitegmic, tenui-, crassi-, or pseudo-crassinucellar. The outer integument invests the ovule except in the micropylar region in Pooideae. In Panicoideae its growth is arrested (N. Chandra 1963a, Choda et al. 1982). In *Melocanna bambusoides* (Yakovlev 1970), the integuments are represented by small primordia, while in *Dendrocalamus hamiltonii* (Hari Gopal and Ram 1981) the integuments remain short and do not cover the micropyle. In *Echinochloa colonum* and *E. crusgalli* (Bhanwra and Choda 1986), both the integuments and the nucellus are said to degenerate in the mature caryopsis.

The nucellar tissue is well developed in Panicoideae, and the nucellar epidermal cells divide to produce six to seven layers. Such a growth is absent in Pooideae (N. Chandra 1963a). The nucellar epidermal cells are conspicuous, and become multinucleate in *Pennisetum ramosum* (Narayan 1962).

The nucellar cells below the megaspore mother cell in *Triticum* differentiate as an axial row of tabular cells. They are in close contact with the double connective bundle and probably have a role as conductive tissue.

An epistase is conspicuous in *Themeda australis* (Woodland 1964).

The ovule is described as pseudo-crassinucellar with an archesporium of two or three cells in *Apluda aristata*, *Ischaemum rugosum* and *Panicum repens* (Seshavatharam and Satyamurthi 1976). In these species a single archesporial cell cuts off a parietal cell which gets obliterated. According to Davis (1966), such an ovule is described as crassinucellar. In most of the species the parietal cell is absent, and the archesporial cell functions as megaspore mother cell. The latter undergoes meiotic divisions resulting in linear or T-shaped tetrad. The chalazal megaspore develops into a Polygonum type of embryo sac. Rarely, the egg cell develops a haustorium in the micropylar region in the apomictic species *Cortaderia jubata* (M.N. Philipson 1978a). The synergids are hooked; the polar nuclei fuse before fertilization.

The synergids in the genus *Cortaderia* (Connor 1974, M.N. Philipson 1977) show a massive haustorial development. Their tips elongate and grow along and out of the micropyle between the ovule and the ovary wall. There are two extensions in *C. selloana* (M.N. Philipson 1977), whereas in the apomictic species *C. jubata* (M.N. Philipson 1978b) the nucellar embryo sac contains a single synergid which becomes haustorial. The remnants of the other synergid are visible under electron microscope. Ultrastructural studies reveal that the transfer cell wall-ingrowths in the haustorium develop strongly, and the endoplasmic reticulum system parallel to its long axis is dense. Thus, these synergids are highly adapted for absorption and transfer of nutrients to the embryo sac (M.N. Philipson 1977). It is suggested that the haustoria be re-examined in cleared wholemounts and cross-sections.

M.N. Philipson and Connor (1984) describe a similar synergid haustorium in eight arundinoid genera: *Chionochloa*, *Cortaderia* (see also M.N. Philipson 1977, 1978b), *Danthonia*, *Erythranthera*, *Lamprothyrsus*, *Pyrrhanthera*, *Rytidosperma* (Fig. 58.2.1) and *Sieglingia*. The genera *Arundo* and *Phragmites* lack the haustorium. To the best of our knowledge, synergid haustorium is not known in any other family of the monocots.

In Panicoideae, the antipodals lie in the chalazal region, whereas in Pooideae they lie laterally (N. Chandra 1963a). The enlargement, persistence and secondary multiplication of antipodals is a very common feature (Mehlenbacher 1970, Maze and Bohm 1974, B.M. Johri and Ambegaokar 1975, Mahalingappa 1977, Hari Gopal and M. Ram 1981, Choda et al. 1982). The antipodals become multinucleate. Their nuclei become multinucleolate, and polyploid due to endomitosis. In *Sasa paniculata*, their number goes up to 300 (see Davis 1966, see also Anton and Cocucci 1984).

In *Elytrophorous spicata* (Satyamurthi 1985), the antipodal cells organize precociously and are left in a lateral position due to extension of the embryo sac after fertilization.

In *Phleum boehmeri* (Joachimiak 1981), the number of antipodals in an embryo sac varies considerably as well as their ploidy-level. Thirty-eight per cent of mature embryo sacs had three antipodals each, but the number increases to ten at the cellular stage of endosperm.

The nucleus of the antipodal cell may be spherical, dumb-bell-shaped or irregular.

The restitution nucleus (after anaphase) is constricted in the central region. They undergo further inhibited divisions.

The polyploid nuclei often show regular shape, and rhythmical growth in volume. With ploidy-level the volume increases in geometrical proportion: n = 70–80 μm^3, 2n = 151–173 μm^3, 4n = 315–376 μm^3, 8n = 681–777 μm^3, 16n = 1121–1452 μm^3, 32n = 2073–3076 μm^3, 64n = 4820–5060 μm^3, 128n = 10685–12075 μm^3, and 256n = 20093–10420 μm^3. The nuclei with irregular shape have many nucleoli.

The highly polyploid antipodal cell may become multinucleate. In one case, out of the four nuclei two were 16n, one 32n, and one 256n. They degenerate at the multicellular stage of the embryo. The degeneration is marked by vacuolation of cytoplasm, collapse of cell walls, and agglutination of chromatin, or disintegration of nuclei.

Antipodal haustoria are reported in *Agrostis vulgaris*, *Alopecurus pratensis*, *Festuca pratensis*, *Holcus lanatus* (Terzijski and Kuristou 1973). In *Eleusine*

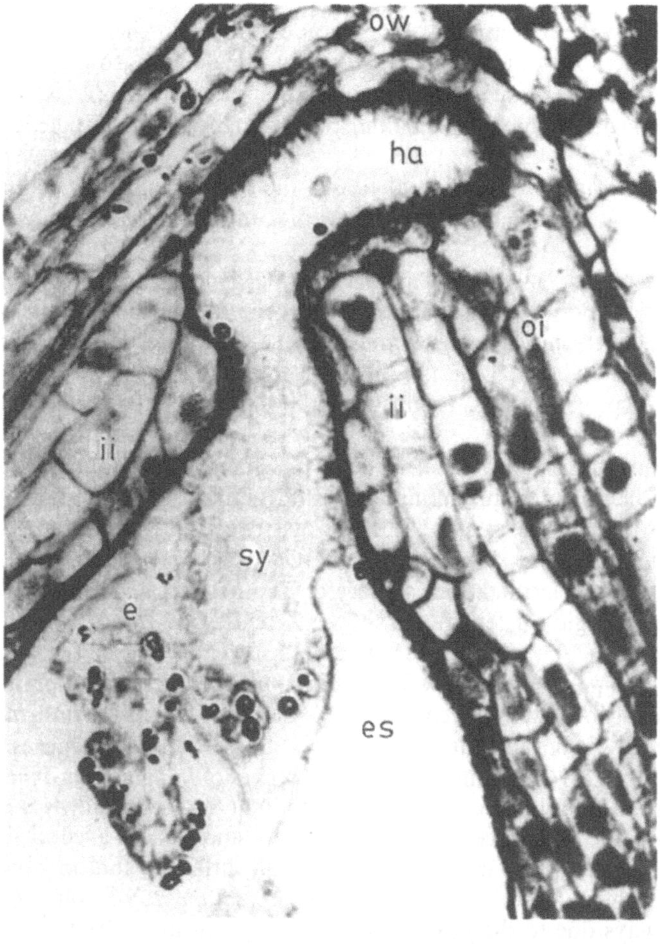

Fig. 58.2.1. *Rytidosperma setifolium*, light micrograph of micropylar end of ovary. A synergid (*sy*) with a haustorial extension (*ha*) which projects through the micropyle between the inner (*ii*) and outer (*oi*) integument of the ovule, and lies against the ovary wall (*ow*). The inner wall of the haustorium has transfer cell wall-ingrowths. Note the portion of egg cell (*e*) beside the synergid; *es* embryo sac. (After M.N. Philipson and H.E. Connor 1984)

(Narayanaswami 1955c, Mahalingappa 1977), the antipodals are large and richly cytoplasmic.

The nutritional role of antipodals is evident, and in rye and *Triticale* they degenerate earlier in shrivelled seeds than in "full" seeds (Kaltsikes 1973). However, in T6437, a highly shrivelled variety of triticales, they are of giant-size and persist for a fairly long time (B.M. Johri and Ambegaokar 1975). A similar feature is reported in an octoploid species of *Hierochloe odorata* (Weimarck 1967a).

Multiple embryo sacs develop in *Echinochloa frumentosa* (Narayanaswami 1955b). Extra-ovular embryo sacs occur in *Digitaria valida* and *Fingerhuthia africana* (Emery and Brown 1957). In *D. valida*, the upper part of the embryo sac containing the egg apparatus protrudes through the micropyle; in *F. africana*, the embryo sac invades the ovarian locule and growth continues till the formation of endosperm and embryo. Bisporic embryo sac is reported in *Panicum repens* (Seshavatharam and Satyamurthi 1976); this needs confirmation.

In *Phleum boehmeria* (Joachimiak 1981), the mature embryo sac and the endosperm grow asymmetrically so that the antipodals occupy a lateral position.

The archesporial cell develops into the megaspore mother cell in the triploid variety of rice (Sapre 1968), while it degenerates in the haploid variety (Sapre 1970a). The multicaryoptic variety of rice shows sterility of seeds. The stigma develops abnormal fleshy, hyperplastic outgrowths which inhibit germination of pollen grains, or obstruct the entry of pollen tube into the embryo sac.

The entire stigma is receptive in *Triticum* (S. Chandra and Bhatnagar 1974).

In barley, the pollen grains germinate within a minute of pollination. As the pollen tube enters the embryo sac, it discharges two male gametes, vegetative nucleus and starch in the degenerated synergid (Cass and Jensen 1970). The sperms migrate, and fertilization is effected. The degenerated vegetative nucleus and the synergid nucleus represent the X-bodies. In *Cortaderia selloana* (M.N. Philipson 1977), the pollen tube enters the extended end of the synergid, penetrates it, and grows within the haustorium until it reaches that part of the synergid which is located in the embryo sac. This aspect requires confirmation.

It is generally believed that pollination in grasses is anemophilous. Pant et al. (1982a, b) observed that in the umbellate spikes of *Dichanthium annulatum* pollination is entomophilous. The insects with light body-weight (*Apis indica*) can remain on the inflorescence for a long time and collect a good deal of pollen grains on their hind legs, or on the ventral surface of thorax, abdomen and wings. The heavy insects, such as *Megapis dorsata*, perch on the inflorescence which sways due to the jerk and the pollen grains are collected momentarily. Certain insects, as *Apis indica*, which visit the inflorescence during different seasons, are specific or "faithfuls". The others show a mixture of pollen grains of nearby growing plants. In *Dichanthium*, the pollen grains remain compactly adhered to the abdominal brush of *Nomia andrenina*, with the help of strands of regurgitated honey. When the insects alight on the feathery stigmas, pollination is effected. These insects visit the plants for pollen and nectar.

The relationship in the sexual development of cross-pollinated rye (Xiang-Yuang and de Mason 1984) is similar to that of self-pollinated wheat (*Triticum aestivum* L.).

The development of the endosperm is of the Nuclear type. Cytoplasmic vesicles with five to eight nuclei are of sporadic occurrence in *Triticum durum* (Deshpande and Raju 1979). Wall formation is initiated from the micropylar end, and proceeds to the chalazal end at the globular proembryo. The peripheral layer of the cellular endosperm differentiates into secondary cambium, and contributes additional tissue to the endosperm. Ultimately, it develops into an aleurone layer. There may be two aleurone layers in *Eleusine* (Mahalingappa 1977). The secondary cambium does not differentiate in *Echinochloa frumentosa* (Narayanaswami 1955b). The mature endosperm becomes horny and stores abundant starch. Liquid endosperm is reported in *Limnodea arkansana* (Brown 1955), and *Helictotrichon hookeri, Koeleria cristata, Sphenopholis filiformis* and *Trisetum bifidum* (Dore 1956). Probably, this endosperm is colloidal in nature and includes starch grains and oil droplets. The grain remains within the spikelet till germination, and is thus protected during the period of dormancy (Brown 1955).

In maize (McClintock 1977), following the division of the primary endosperm nucleus, one daughter nucleus moves to the micropylar pole and the other to the chalazal pole of the embryo sac. Both the nuclei divide in a plane at right angles to the previous one. Each of these four nuclei undergoes synchronous nuclear divisions. Cytokinesis leads to four distinct zones of cellular endosperm tissue which remain recognizable in the mature caryopsis.

In *Zea mays* (Cooper 1951), the hybrid kernels collapse, between a cross of diploid and tetraploid ($2n \times 4n$) strains. Fertilization occurs between 18 and 24 h after pollination. The endosperm becomes abnormally meristematic and encloses the embryo. The central portion of the endosperm fails to develop, and the adjoining tissue forms lobes which protrude into the central cavity. The embryo disintegrates. In a reciprocal cross ($4n \times 2n$), both endosperm and embryo develop normally up to 6 days after pollination. Starch deposition occurs in the endosperm up to 16 days of pollination, and then the growth of endosperm ceases. The embryo differentiates, but ultimately degenerates.

The embryogeny corresponds to the Asterad type. According to the Souèges system, the embryogeny in Triticale-1 can be classified to Period I, Series C_2, Megarchetype II (B.M. Johri and Ambegaokar 1977), and in *Eragrostis ciliensis* (Kulkarni and Dnyansagar 1984) to Period I, Series A, Megarchetype II (Asterad type).
In *Triticum aestivum* "Sonara" 64 (S.P. Bhatnagar and Chandra 1975), the embryogeny conforms to the Graminad type (Fig. 58.2.2A–T).

Reeder (1957) investigated about 300 species from 150 genera of Gramineae and classified the mature embryos into six categories on the basis of vasculature, presence or absence of epiblast, development of the lower part of scutellum, and morphology and anatomy of embryonic leaves.

The caryopsis is albuminous and contains a small laterally placed embryo and starchy endosperm. The peripheral layer of the endosperm differentiates as aleurone layer. The testa degenerates soon after pollination. The inner epidermis of the tegmen persists as radially elongated cells; the pericarp gives protection.

In *Bambusa arundinacea* (Philip and Haccius 1976), in the two-celled proembryo (Fig. 58.2.3A) transverse and vertical divisions occur in three different ways: (1) the apical cell *ca* divides vertically and the basal cell *cb* obliquely, (2) the zygote divides transversely followed by a transverse division in *cb* (Fig. 58.2.3B) and vertical division in *ca*, and (3) the zygote divides transversely followed by further transverse divisions so that a four- to five-celled filamentous proembryo (Fig. 58.2.3C) is formed. Later, crosswise vertical divisions occur in the terminal cell to form a quadrant. Further growth occurs in vertical and horizontal planes (Fig. 58.2.3D–G). The differentiation of *l* and *l′* becomes obscure, and the differentiation of protoderm is delayed. The growth in the terminal tier is asymmetric, and on one side appears a depression where the shoot apex differentiates in due course. A major portion of the proembryonic apex grows at a faster rate and develops into a cotyledon. It overgrows the shoot apex with cells of smaller size and lesser vacuolation (Philip 1972), and the shoot apex appears to be lateral.

The cells of the globular proembryo, located directly below the suspensor, function as initials of the root. They divide periclinally; periclinal divisions also occur in the protodermal cells leading to the differentiation of the rhizodermis

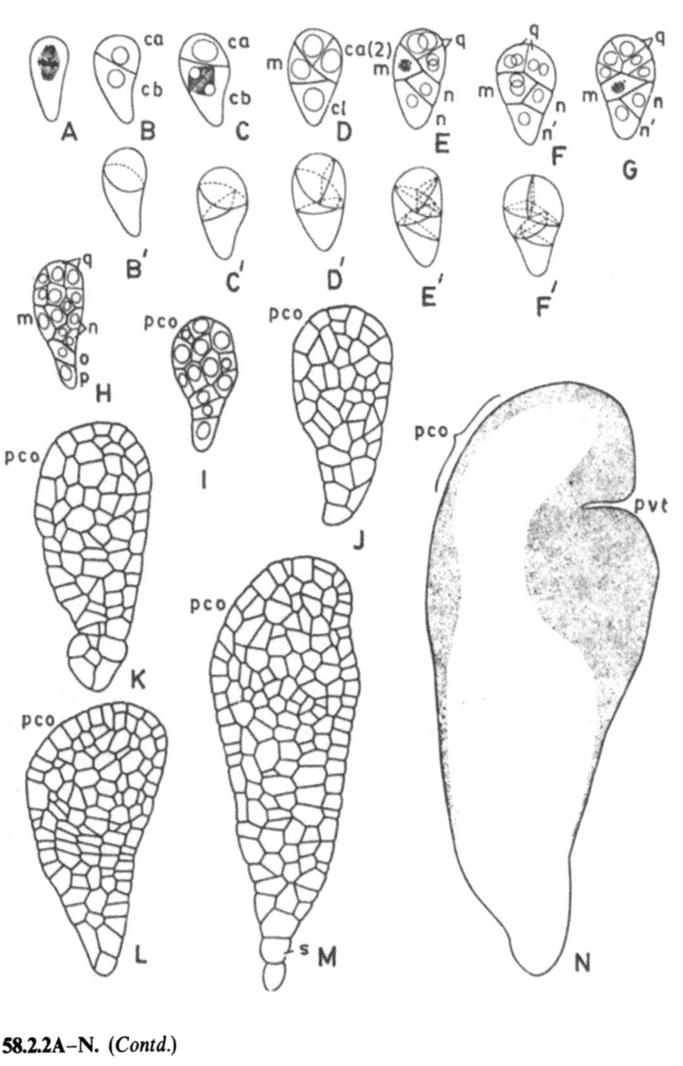

Fig. 58.2.2A–N. (*Contd.*)

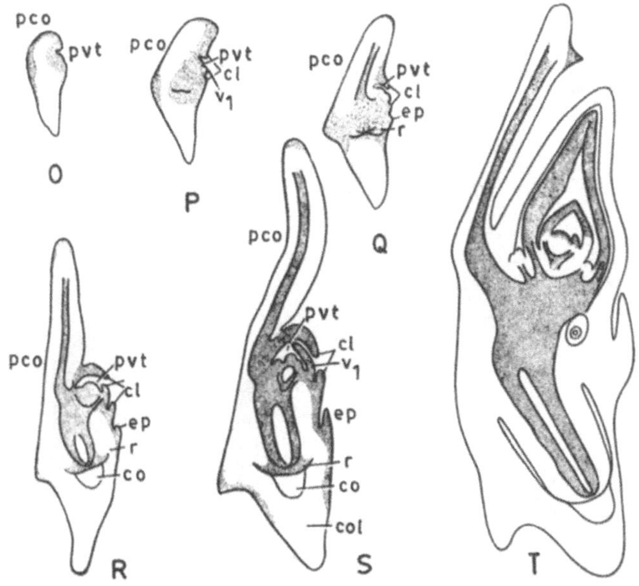

Fig. 58.2.2A–T. *Triticum aestivum*, sequential development of embryo from zygote (nucleus in metaphase) to mature embryo. **B′–F′** Stereoscopic representation of **B–F** (schematic); *cl* coleoptile; *co* root cap; *col* coleorhiza; *ep* epiblast; *pco* shield initial; *pvt* shoot apex; *r* root cap; *s* suspensor; *v* first leaf of plumule. (After Batygina 1969)

(Fig. 58.2.3H–J). The central region differentiates a columella of six to eight layers (Fig. 58.2.3K, L). The formation of radicle ceases before histogen differentiation, and the functional root originates endogenously (Fig. 58.2.3M). According to Philip and Haccius (1976), the coleorrhiza is the rudiment of an aborted primary root and encloses the seminal root as in other members of Gramineae.

The caryopsis of *Arundinaria alpina* and *Bambusa arundinacea* has a linear hilum, strongly curved small embryo, epiblast, and an internode (between the point of divergence of the scutellum); coleoptile is absent – all festucoid characters. The panicoid character is the presence of a distinct cleft between scutellum and coleorhiza (Fig. 58.2.3N, O). An axillary bud develops between the coleoptile and the first plumular leaf. In the species of *Bambusa*, the number of vascular bundles in the coleoptile varies. There are nine or ten bundles in *Arundinaria tecta*, two each in *Dendrocalamus strictus* and *Guadua aculeata* (Reeder 1957), five in *Streptochaeta spicata*, and four in *A. alpina* and *B. arundinacea*. The margins of coleoptile overlap in *Streptochaeta*, but in *Bambusa* they meet end to end.

The rate of growth in the cotyledonary and hypocotyledonary region varies from the globular proembryo up to the maturation of the embryo. In the initial stages, i.e. at the early globular stage, the average length correlation of terminal and basal cells is 2:4 (1:2), at the late globular stage it becomes 3:4.6, and when the notch (or the depression) occurs in the embryo, it is 5:5 (1:1). In the

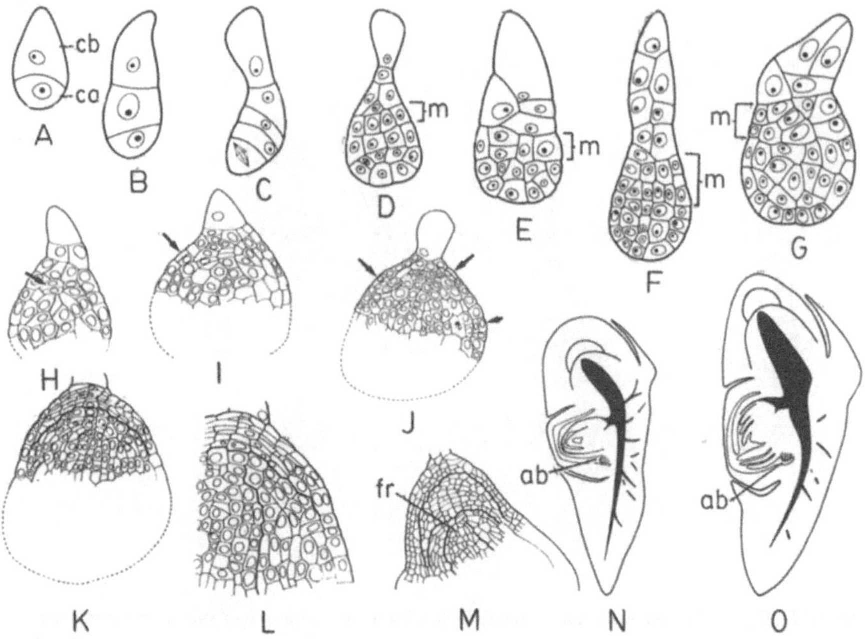

Fig. 58.2.3A–O. **A–M, O** *Bambusa arundinacea*, **N** *Arundinaria alpina*. **A, B** Two- and three-celled proembryo. **C** Five-celled linear proembryo with vertical division in terminal cell. **D** Globular proembryo. The tier *m*, derivative of *cb*, is well differentiated. **E** Globular proembryo with suspensor, the distinction between *l* and *l'* is not conspicuous. **F, G** Globular proembryos with root initials, and suspensor: **H** Periclinal divisions (*arrow*) in the cell below suspensor. **I, J** Periclinal divisions (*arrows*) in protodermal cells, initiation of rhizodermis. **K, L** Portions of older proembryo show six to eight cell layers of columella. **M** Portion of embryo, the origin of functional root (*fr*) is endogenous. **N, O** Median longisections of embryos. There is a distinct cleft between scutellum and coleorhiza, and there is no internode between the point of divergence of scutellar trace and that of coleoptile. The epiblast is ill-defined. The young leaves enclose shoot apex and an axillary bud (*ab*) between the coleoptile and primordium of first leaf. The root is covered by root-cap and protected by coleorhiza. (After Philip and Haccius 1976)

mature embryo the cotyledonary portion is quite large and the ratio is 10:4 (Fig. 58.2.3N, O).

Philip and Haccius (1976) compare the embryo of *Bambusa* with that of *Setaria* (Gramineae), both at the same stage of development. In *Bambusa*, the cotyledon is fairly large as compared to the hypocotyl, while in *Setaria* (see Guignard 1961a, b) the cotyledon is small and the hypocotyl forms the main bulk of the embryo. Thus, the cotyledon–hypocotyl ratio is higher in *Bambusa* than in *Setaria*. This feature signifies a higher state of evolution in *Bambusa*.

Hari Gopal and Mohan Ram (1985) examined the mature embryos of seven species belonging to five genera of Indian bamboos: *Bambusa tulda*, *Dendrocalamus hamiltonii*, *D. longispathus*, *Melocalamus maclellandii*, *M. compactiflorus*, *Ochlandra travancorica* and *Pseudostachyum polymorphum*. They observed that the basic pattern of the embryo is more or less similar to that described by Philip and Haccius (1976). The features unique to fleshy-fruited bamboos are: presence of a massive scutellum, juxtaposition of plumule and radicle, and occurrence of a bud in the axil of coleoptile.

In *Pappophorum subbulbosum* (Rost et al. 1984), the caryopsis-coat consists of: (a) an outer epidermal layer of pericarp, a collapsed layer of middle pericarp, and a discontinuous epidermal layer of inner pericarp, (b) the collapsed integuments form a dense layer, and (c) the nucellus also collapses and forms a discontinuous layer. The placental pad, at the ventral surface of caryopsis, opposite to scutellum and coleorrhiza, consists of 15–20 collapsed layers including the pigment strand and the placental vascular supply. The inner epidermis of the pericarp does not extend across the pad. The aleurone layer is a continuous uniseriate layer around the entire caryopsis except the placental pad. The outer epidermis of the pericarp is a continuous layer of transfer cells across the pad. These cells contain membraneous inclusions suggesting that they have a role during germination.

In contrast to the normal fruit of Gramineae, *Melocanna bambusoides* has a berry-like large fruit with non-endospermic seed, and a well-developed embryo with large scutellum.

The phenomena of male sterility, somatic nucellar apospory, diplosporous formation of four-nucleate unreduced embryo sacs, pseudogamous fertilization and non-pseudogamy, autonomous development of endosperm, and precocious division of egg nucleus are of common occurrence in Gramineae, and mostly associated with polyploidy. However, in several members, both sexual and apomictic embryo sacs develop simultaneously, and multiple embryo sacs are a common feature.

Mostly, the megaspore mother cell degenerates and nucellar cells in the chalazal region behave as aposporous embryo sac initials (Hieracium type) as in *Bothriochloa pertusa* and *Dichanthium annulatum* (Reddy and D'Cruz 1969a).

In hexaploid and octoploid varieties of *Hierochloë odorata* (Weimarck 1967a) and *H. montica* (Weimarck 1967b, 1970), uninucleate aposporous embryo sacs develop from the thecal wall in the anthers; this needs confirmation.

At times, as many as eleven aposporous embryo sac initials have been observed in the chalazal region of the nucellus in *H. australis* (Weimarck 1967a). Aposporous embryo sac initials also differentiate in the integument (Weimarck 1970).

The tendency of apospory is very conspicuous in *Pennisetum hohenackeri*, *P. orientale*, *P. pedicellatum*, *P. ramosum*, *P. setaceum* and *P. villosum* (Narayan 1962). The aposporous embryo sac initials differentiate in the nucellus, chalaza, funiculus, integuments, and even in the pericarp.

In *P. ciliare* (Snyder et al. 1955), eight-nucleate reduced Polygonum type, and four-nucleate Oenothera type of development have been observed. In addition, due to nucellar apospory, unreduced four-nucleate embryo sacs (Panicum type) are also formed. In *P. orientale*, *P. setaceum* and *P. villosum* (Narayan 1962), such functional embryo sacs are present in most of the apomicts of Panicoideae (Brown and Emery 1958). They are monopolar; all the four nuclei lie in the micropylar region and differentiate into the normal egg apparatus and a single polar nucleus (Brown and Emery 1957, P.K. Gupta 1968, 1969–70, Reddy and D'Cruz 1969a, b).

Pollination is necessary for the initiation of endosperm which normally precedes the division of the unreduced unfertilized egg. The precocious division of the egg nucleus appears to be a common feature in Festucoideae.

In *Paspalum secans* (Snyder 1957), *P. cromyorhizon*, *P. dilatatum*, *P. exaltatum*, *P. lividum*, *P. mandiocanum*, *P. quenoarum*, and *P. paspaloides* (Srivastava 1982), from the nucellar cells one to three unreduced embryo sacs develop up to the eight-nucleate stage.

Srivastava (1982) maintains that if mutation occurs in an obligate apomict, the segregation of characters is not possible due to absence of sexuality. The vigorous and well-developed genotypes survive, but due to sexual sterility the race cannot be continued. On the contrary, it builds up a large population of genetically similar individuals which can rapidly colonize the newly available habitats.

Young et al. (1979) report apospory in *Cenchrus ciliaris* (see Nogler 1984).

C. glaucus (Shanthamma 1982) is an obligate apomict. Microsporogenesis is irregular, and pollen viability is 49%. This species shows total absence of mature sexual female gametophyte. The degenerated dyads and tetrads indicate that meiosis occurs. Aposporous embryo sacs develop not only in the nucellus, but also in the integuments, and the broad base of the ovule where it is attached to the pericarp wall. These are four-nucleate, three nuclei form the egg apparatus and the fourth functions as upper polar nucleus. The development of the endosperm is pseudogamous (fertilization of upper polar nucleus), and of the embryo autonomous.

Twin seedlings have been observed in 4% of seeds. Triple seedlings are absent.

In the *Sorghum* variety R473 (N.G.P. Rao et al. 1978), the female archesporial cell directly develops into megaspore mother cell which degenerates. Therefore, sexual embryo sacs are very rare. The embryo sac develops from a nucellar cell, and Hieracium type of apospory occurs. The pollen tubes grow to a very short length in the stigma and do not reach up to the ovary. Due to the stimulus of pollination, endosperm and embryo develop autonomously. The apomixis in R 473 is associated with self-incompatibility. Like the two parents, the F_1 progeny exhibits somatic apospory associated with self-incompatibility.

Panicum maximum (Brown and Emery 1957, 1958) is 28% apomictic. The megaspores are not produced, but the megasporocyte develops directly into a four-nucleate embryo sac of Panicum type. This may be associated with diplospory.

Most of the non-pseudogamous grasses exhibit gonial apospory (diplospory) as in *Calamagrostis chalybaea*, *C. lapponica*, *C. purpurea* (Nygren 1946, 1954, 1967), *Nardus stricta* (Rychlewski 1961), and *Poa nervosa* (Grun 1955). *C. stricta* subsp. *inexpansa* (Greene 1984) includes apomictic variants that produce megagametophytes by diplospory. The single archesporial cell divides mitotically to produce an embryo sac appearing identical to those formed by sexual reproduction. The embryo from the egg cell, and endosperm from the secondary nucleus (formed by the fusion of polars), develop autonomously. Some apomictic individuals occasionally produce pollen grains, and they may have a potential for reproducing sexually.

Chao (1980) reported diplospory in *Paspalum conjugatum*. The chromosomes do not pair but form a restitution nucleus during meiosis I. At the end of meiosis II, in many ovules, only two unreduced "megaspores" are formed. The embryo sac is diplosporous and develops from the chalazal megaspore. Due to irregularities during megasporogenesis, seed-set is only 45.5%. The development of the egg is autonomous, and starts about 2 days prior to anthesis. However, pollination is necessary for the development of the endosperm. Triple fusion results in the development of an endosperm with 120 chromosomes in each cell.

Tripsacum dactyloides (Farquharson 1955) is a pseudogamous apomict. The endosperm fails to develop, but the apomictic embryos grow up to 30 cells. The ovaries continue to enlarge and probably pollination and development of the embryo stimulate the growth of ovaries.

In a tetraploid collection, twin megaspore mother cells, twin tetrads, twin embryo sacs, and twin embryos were observed. The unfertilized synergid develops into a haploid embryo; 50% of caryopses are polyembryonate.

Bothriochloa ischaemum and *Themeda triandra* (Brown and Emery 1957) are aposporous apomicts with probably pseudogamous development of embryos. The apomictic mechanism is essentially similar in both species. It involves: (1) degeneration of the megaspore mother cell, (2) initiation of one to several aposporous embryo sac initials in the region immediately adjacent to the degenerated sporogenous cell, or from the cells in the hypodermal layer of the nucellus near the micropyle, or in the region of ovular attachment (*Themeda*), and (3) two nuclear divisions in each sac. The organized four-nucleate embryo sac contains the usual egg apparatus, and a single polar nucleus. The development of the embryo is not dependent on prior development of the endosperm.

Reduced sexual embryo sacs may occasionally develop in these strains.

In *Eremopogon foveolatus* (Bhanwra and Choda 1981), micro- and megasporogenesis and gametogenesis are normal. The mature pollen is three-celled, and the embryo sac is seven-celled. In addition, one to six aposporous four-nucleate embryo sacs develop in the nucellus. One of these reaches maturity and contains the usual egg apparatus and one polar nucleus, or rarely one egg cell, one synergid, and two polar nuclei. Triple fusion occurs, but the egg develops parthenogenetically. This species is pseudogamous, and fertilization of the polar/secondary nucleus is essential for the development of the endosperm.

Bhanwra et al. (1982) observed similar four-nucleate unreduced embryo sacs in *Capillipedium parviflorum*; megasporogenesis occurs but the megaspores degenerate. The egg develops parthenogenetically, whereas triple fusion is essential for the development of the endosperm.

Choda et al. (1982) report 90% pollen fertility in *Cymbopogon nardus* and 70% in *C. martinii*. Development of the embryo sac is of the Polygonum type.

The pollen grains germinate on the stigmatic hairs, but the tubes fail to reach the embryo sac, and eventually the ovule shrivels. The development of the endosperm may be autonomous, but the egg does not develop. In *C. parkeri*, the seed-set is 8%.

In *Sorghum* varieties R473 and 302 (N.G.P. Rao et al. 1978), and the F_1 progeny of reciprocal crosses of these two varieties (R.N. Reddy et al. 1979), microsporogenesis is normal, the pollen grains are three-celled at the dispersal stage, and the fertility is 90%.

In anatropous, bitegmic, pseudocrassinucellar ovule the archesporial or megaspore mother cell soon degenerates, and the asexual one or two unreduced nucellar embryo sacs develop. The apospory is of the Hieracium type (Battaglia 1963). The pollen grains germinate on the stigma, but the pollen tubes do not enter the ovule (or the embryo sac). Both the synergids remain healthy up to four- to five-celled proembryo indicating that fertilization does not occur. Only the stimulus of pollination is necessary to induce autonomous development of egg and endosperm. The embryo (2n), endosperm (4n), and maternal tissue (2n) do not cause any imbalance, and the seed-set is normal (N.G.P. Rao et al. 1978).

Somatic apospory in R473 and F_1 progeny of reciprocal crosses between R 473 and 302 is associated with self-incompatibility.

In the absence of androecium, *Cortaderia jubata* (M.N. Philipson 1978b) does not produce pollen grains. The development of the megaspore mother cell ceases and it degenerates. Two to three aposporous embryo sacs develop in the nucellus, and are well developed by the time the inflorescence emerges. This species exhibits non-pseudogamy (Fig. 3.13.2D).

In *C. selloana*, the eight-nucleate embryo sac is of the Polygonum type, and in *C. jubata* the four-nucleate aposporous embryo sac is of the Panicum type (Costas-Lippmann 1979). The diploid egg develops parthenogenetically. Costas-Lippmann states: "The advantages of agamospermy to weedy *C. jubata* are the reproduction of the same genotype in every generation, the economy of energy and material due to lack of an androecium and its products, and the development of all the ovules into seeds."

Fertilization of aposporous embryo sacs is described in *Dichanthium annulatum* (Reddy and D'Cruz 1969a). In four-nucleate embryo sacs the polar nucleus is always fertilized by the male gamete, and the other male gamete either enters the egg cell or remains outside (pseudogamy). In eight-nucleate unreduced embryo sacs both the polar nuclei get independently fertilized by the male gamete. Frequently, syngamy is eliminated due to: (1) failure of the sperm nucleus to enter the egg, (2) degeneration of the male nucleus inside the egg cell, (3) non-availability of the male gamete for fertilizing the egg cell in the eight-nucleate embryo sac where both the polar nuclei are independently fertilized, and (4) development of nucellar embryos. Reports of pseudogamy are not rare (see Choda and Bhanwra 1977, Grun 1955, Nygren 1954, Rychlewski 1961, Shanthamma 1982).

Muniyamma (1977) reports that *Brachiaria setigera* (2n = 36) is a facultative apomict. In this species polyembryony is due to formation of embryos from egg, synergids and endosperm. Six aposporous embryo sacs from 675 post-fertilized ovules contained embryos which had originated from endosperm. Meristematic activity in localized areas of the mature endosperm produce globular or spindle-shaped structures (Fig. 58.2.4A–E). Enlargement and organization of these structures led to viable embryos. The sequence of events beginning with a single cell and followed by a complete organized

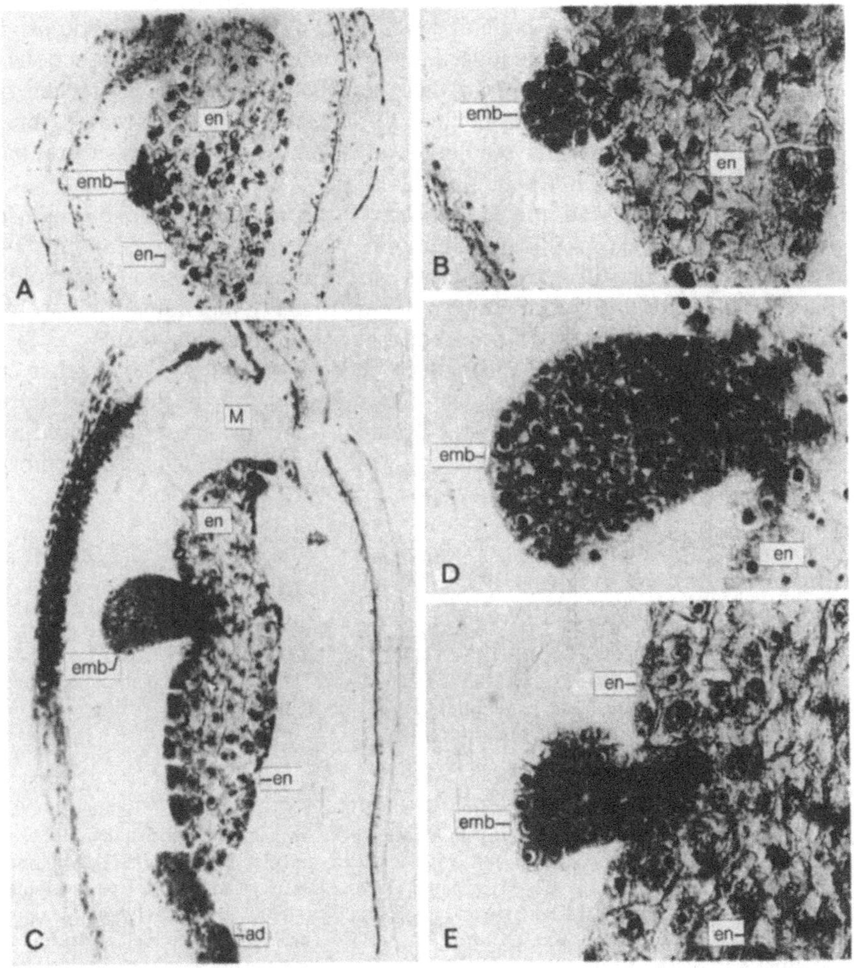

Fig. 58.2.4A–E. *Brachiaria setigera.* **A** Longisection of ovule, early initiation of "embryo" (*emb*) from endosperm (*en*). **B** "Embryo" from **A**. **C** Longisection of ovule shows a large embryo (*emb*) which originated from the endosperm (*en*). An adventitious embryo (*ad*) at the chalazal end which originated from an aposporous embryo sac; *M* Micropyle. **D** Endospermal embryo (*emb*) from **C**. **E** Embryo (*emb*) from endosperm (*en*) with a suspensor-like attachment. (After Muniyamma 1977)

embryo was not observed. Cytological studies of root tips of the twin and triplet seedlings showed combinations of diploid-diploid, diploid-triploid, and diploid-diploid-triploid. Only two triploids were obtained from 200 seeds germinated.

The globular and spindle-shaped structures arising from mature endosperm appear (to us) like tumors. There is no semblance at all to any stage of embryogenesis. That they rarely germinate and produce triploid seedlings needs a critical confirmation. In this connection, reference is invited to the so-called embryos (Fig. 24.3.1A–F′) originating from the endosperm in *Isomeris* (Billings 1937). After a very careful reinvestigation, P. Maheshwari and Sachar (1954a) demonstrated that the embryo is quite normal and of zygotic origin. The endosperm nodules were indeed a wrong interpretation of transections of the highly coiled suspensor. So far there is no confirmed report of an embryo originating from the endosperm (in vivo). Therefore, Muniyamma's report raises many doubts. Also, the seedlings and chromosome plates to substantiate her observations have not been illustrated.

Reddy (1977) traced the evolutionary trend in apomictic mechanism. According to him, diplospory, commonly associated with the apomicts of Festucoideae, is primitive to apospory. He emphasizes the significant role of the four-nucleate unreduced embryo sac as it easily maintains the chromosomal balance between the embryo, endosperm, and maternal tissue (2:3:2). In Andropogoneae, the sexual mechanism ends at the megaspore mother cell stage (MMC degenerates), and the occurrence of four-nucleate unreduced embryo sacs is predominant. In the apomicts of Panicoideae the unreduced eight-nucleate embryo sacs occur along with unreduced four-nucleate embryo sacs. This denotes the intermediate position of the Panicoideae.

KB Ambegaokar and BM Johri (1977), AM Anton and E Cocucci (1984), CA Aulbach-Smith and JM Herr Jr (1984), GS Avery Jr (1930), KN Bahadur, D Ram and HB Naithani (1976), UC Banerjee and ES Barghoorn (1971a), ME Barkworth (1982), E Battaglia (1963), TB Batygina (1968, 1969, 1971), TB Batygina, ES Teryokhin, GK Alimova and MS Yakovlev (1963), JR Beaudry (1951), NN Bhandari and R Khosla (1982), RK Bhanwra (1988), RK Bhanwra and SP Choda (1981, 1986), RK Bhanwra, SP Choda and S Kumar (1982), SP Bhatnagar and S Chandra (1975, 1976), SP Bhatnagar and BM Johri (1972), FH Billings (1937), JW Boyes and WP Thompson (1937), RA Brink and DC Cooper (1944, 1947), K Brix (1974), WV Brown (1955, 1960), WV Brown and WHP Emery (1957, 1958), GW Burton (1982), PA Buss Jr and NR Lersten (1972), DD Cass and GC Fabi (1988), DD Cass and WA Jensen (1970), DD Cass and I Karas (1975), RP Celarier and JR Harlan (1957), N Chandra (1963a, b, 1976), S Chandra and SP Bhatnagar (1974), C-Y Chao (1974, 1980), S Chao (1975), PC Cheng, RI Greyson and DB Walden (1986), PS Chikkannaiah and MS Mahalingappa (1975), SP Choda and RK Bhanwra (1977), SP Choda, H Mitter and RK Bhanwra (1982), HE Connor (1974), DC Cooper (1951), M Costas-Lippman (1979), MP Cummins (1929), R Dahlgren and FN Rasmussen (1983), GL Davis (1966), JMJ de Wet and JR Harlan (1970), JMJ de Wet, DE Brink and CE Cohen (1983), JMJ de Wet, JR Harlan and DE Brink (1982), PK Deshpande and PSG Raju (1979, 1985), AG Diboll and DA Larson (1966), WG Dore (1956), M Dujardin and WW Hanna (1983a, b), WHP Emery (1957), WHP Emery and WV Brown (1957), K Engell (1989), O Erdelská (1966), LI Farquharson (1955), BA Fineran, DJC Wild and M Ingerfeld (1982), CW Greene (1984), P Grun (1955), J-L Guignard (1961a, b), J-L Guignard and JC Mestre (1970), PK Gupta (1968, 1969–70), Å Håkansson (1943, 1946), WW Hanna, KF Schertz and EC Bashaw (1970), B Hari Gopal and HY Mohan Ram (1985), B Hari Gopal and M Ram (1981), MP Hayward and MAP Manthriratna (1972), R Izmailow (1984), A Joachimiak (1981), GW Johnston (1953), BM Johri and KB Ambegaokar (1975, 1977), JB Juliano and MJ Aldama (1937), PJ Kaltsikes (1973), PJ Kaltsikes and DG Roupakias (1975), C Kempanna, A Seetharam and G Shivashankar (1971), S Khattra and G Singh (1989), S Khosla (1946), MV Klyuchareva (1983), S-N Korobova (1971), AR Kulkarni and VR Dnyansagar (1984), S Kumari (1960), KK Lakshmanan (1972), KK Lakshmanan and R Jayalakshmi (1980), M Luxová (1967), MS Mahalingappa (1977, 1978), P Maheshwari and RC Sachar (1954a), DJ Mares, K Norstog and BA Stone (1975), J Maze and LR Bohm (1973, 1974, 1977), J Maze and SCh Lin (1975), J Maze, LR Bohm and LE Mehlenbecher Jr (1970), B McClintock (1977), CC McDonald (1927), LE Mehlenbacher Jr (1970), MH Mengeha and AT Guard (1966), J Merry (1941), HL Mogensen (1982, 1984), HY Mohan Ram and B Hari Gopal (1981), JW Morrison (1955), RD Moskova and MS Yakovlev (1974), A Mujeeb-Kazi (1981), M Muniyamma (1977), Y Nagato (1981), KN Narayan (1955, 1962), PS Narayana (1986), S Narayanaswami (1952, 1953, 1954, 1955a, b, c, 1956), ER Nielsen (1946), GA Nogler (1984), K Norstog (1963), A Nygren (1946, 1954, 1967), DD Pant, DD Nautiyal and SK Chaturvedi (1982a, b), VJ Philip (1972), VJ Philip and B Haccius (1976), MN Philipson (1977, 1978a, b), MN Philipson and HE Connor (1984), VA Poddubnaya-Arnoldi and MA Makhaline (1969), Y Pomeranz and DB Bechtel (1978), V Raghavan (1988), NGP Rao, LL Narayana and RN Reddy (1978), PN Rao and DS Narayana (1980), PS Reddy (1977), PS Reddy and R D'Cruz (1969a, b), RN Reddy, LL Narayana and NGP Rao (1979), JR Reeder (1953, 1957, 1962, 1982), BL Robertson (1976b), SO Rogers and BS Squatrano (1983), WM Ross and JA Wilson (1969), TL Rost, PI de Artucio and EB Risley (1984), P Rudall and S Dransfield (1989), J Rychlewski (1961), AB Sapre (1968, 1970a, b), TVC Satyamurthi (1985), MI Savchenko (1960, 1973), YH Savidan (1983), YH Savidan and J Pernès (1982), AE Shadowsky (1926), CK Shah (1975), C Shanthamma (1982), C Shanthamma and KN Narayan (1977), V Seshavatharam and TVC Satyamurthi (1976), ANR Sindhe, BGL Swamy and DA Govindappa (1980), LA Snyder (1957), LA Snyder, AR Hernandez and HE Warmke (1955), R Souèges (1924a), AK Srivastava (1982), A Szkukalek, M Hausksecht and O Erdelská (1989), D Terzijski and MA Khristou (1973), WP Thompson and D Johnson (1945), HJ Tillich (1977), FW Tinney (1940), AG Untawale, PK Deshpande and KB Sharma (1969), AAM van Lammeren (1981), S Veerasamy (1982), J Venkateswarlu and I Devi (1964), LD Vijendra Das and SR Rangaswamy (1974), S Wakakuwa (1934), HE Warmke (1954), G Weimarck (1967a, b, 1970), PS Woodland (1964), Xi Xiang-Yuan and Cui de Cai (1983), Xi Xiang-Yuan and DA de Mason (1984), MS Yakovlev (1950, 1970), MS Yakovlev and OP Kamelena (1973), BA Yeung, RT Sherwood and EC Bashaw (1979), Sh Yu and Ch Y Chao (1979)

58.3 Taxonomic Considerations

Gramineae. Bentham and Hooker (1965c) associate Gramineae with Eriocaulaceae, Centrolepidaceae, Restionaceae and Cyperaceae in the series Glumaceae. Engler and Prantl (1931) and Engler and Diels (1936) assign Cyperaceae and Gramineae to Glumiflorae. Hutchinson (1973) segregates grasses from sedges, and assigns Gramineae to the order Graminales. He considers Graminales to be advanced over Cyperales, and derives both Gramineae and Cyperaceae from liliaceous ancestors through the Juncaceae complex. Gramineae and Cyperaceae both have spikelet as the basic floral unit. But this does not indicate any phyletic relationship, and they have evolved along two divergent lines. The study of floral morphology and morphology of inflorescence also confirms that these two families are not closely related (see Lawrence 1951).

Cronquist (1981) places Cyperaceae and Gramineae (Poaceae) in the order Cyperales, assuming that they have diverged along different lines from a common ancestor, the Commelinales. R. Dahlgren (1975a, 1977a, 1980b) includes Poaceae in the order Poales, superorder Commelinanae, Commeliniflorae, along with Restionaceae and Flagellariaceae. According to him, features such as vessels with entire perforations, presence of silica particles, and seeds with starchy endosperm segregate the superorder Commelinanae from the superorder Lilianae. On the basis of lateral position of embryo to the endosperm, Huber (1977) segregates Gramineae – along with Centrolepidaceae, Eriocaulaceae, Flagellariaceae, Rapateaceae and Restionaceae – from Cyperaceae. According to Thorne (1977): "The poaceous seed with marginal embryo lateral to the abundant horny and mealy, starchy endosperm in itself epitomizes the position of the Poaceae both as a member of the Commelinales and of the independent suborder Poineae".

The embryological features of Cyperaceae, such as simultaneous cytokinesis in pollen mother cells, formation of pseudomonads, ovule with air chambers on the funicular side, ephemeral antipodals, Onagrad type of embryogeny, and embryo embedded in the endosperm, are in contrast to those in Gramineae.

Embryologically, both families differ widely, and the position assigned to them in separate orders by Melchior (1964) and Hutchinson (1967, 1973) is appropriate.

59 Principes

59.1 General Features

Palmae represent the order Principes. These are woody plants with compound spadices and spatheceous bracts, superior tricarpellate ovary, and endospermic seeds.

Some of the embryological features are: (1) secretory anther tapetum, (2) two-celled pollen grains, (3) bitegmic, crassinucellate ovule with well-developed integumentary vascular bundles, (4) Polygonum and Allium type of embryo sac, (5) Nuclear type of endosperm, ruminations at maturity, and (6) Asterad and Onagrad type of embryogeny.

59.2 Palmae

The anther wall consists of the epidermis, fibrous endothecium, an ephemeral middle layer, and secretory tapetum of binucleate cells. The microspore mother cells undergo meiotic divisions. Cytokinesis is successive in *Nipa fruticans* (Radermacher 1924) and *Phoenix sylvestris* (Mahabale and Biradar 1968), and simultaneous in *Caryota urens* (Shirke and Mahabale 1972) and *Jubaeopsis caffra* (Robertson 1976a). Venkata Rao (1959b) mentions simultaneous type in 11 species of Palmae. The tetrads are isobilateral and tetrahedral.

The family is multipalynous. The pollen grains are two-celled at the dispersal stage.

The ovule is ana- (*Caryota, Cocos*), hemiana- (*Areca*) or orthotropous (*Actinophloeus, Bactris* and *Howea*), bitegmic and crassinucellate (Venkata

Rao 1959b). Both the integuments form the micropyle, except in *Sabal* sp. and *Washingtonia* (Venkata Rao 1959a) where it is formed by the inner integument. In *Hyphaene indica* (Mahabale and Chennaveeraiah 1957), the outer integument is massive and a ring of 16–18 vascular traces extend to two-thirds its length. An endothelium differentiates in *Licuala peltata, Pritchardia pacifica, Trachycarpus* sp., *Washingtonia* sp. (Venkata Rao 1959a), *Cocos plumosa, C. schizophylla* (Venkata Rao 1959b), and *Livistona chinensis* (Kulkarni and Mahabale 1974). (The differentiation of the endothelium needs to be further examined.) A well-developed funicular obturator develops in *L. rotundifolia*, and in *Elaeis guineensis* (Kajale and Ranade 1953) there is a conspicuous tanniniferous nucellar beak. The development of the chalaza is very prominent in Sabaleae (Venkata Rao 1959a). A hypostase differentiates during post-fertilization stages.

The archesporial cell cuts off a parietal cell. The parietal tissue is six layers thick in *Areca catechu* and *Cocos nucifera*, otherwise it is one or two layers thick. The megaspore mother cell undergoes meiotic divisions. Kajale and Ranade (1952) report linear, isobilateral, T- and ⊥-shaped tetrads in *E. guineensis*. The chalazal megaspore develops into a Polygonum type of embryo sac. The synergids are beaked. They are egg-like in *Caryota urens* (Shirke and Mahabale 1972) and *E. guineensis* (Kajale and Ranade 1953). They show lateral haustoria in *Elaeis*. The polar nuclei fuse before fertilization. The antipodals are ephemeral. The antipodal nuclei remain unorganized in *Hyphaene indica* (Mahabale and Chennaveeraiah 1957). The antipodal cells are large and sac-like in *Areca catechu, Actinophloeus macarthurii* and *Howea belmoreana*, multinucleate in *Chrysalidocarpus lutescens* (Venkata Rao 1959b), and persistent in *Areca catechu* (Swamy 1942b). Twin embryo sacs have been observed in *Elaeis*.

Allium type of embryo sac is reported in *Borassus flabellifer, Chamaedorea latifolia, Hyphaene indica* and *Nipa fruticans*. The development in *C. latifolia* (Jönsson 1879–80) and *N. fruticans* needs reinvestigation (see S.C. Maheshwari 1955). Reports on tetrasporic type of embryo sac are unreliable and need confirmation.

The development of the endosperm is of the Nuclear type. The initiation of walls takes place after the formation of numerous free nuclei, and at about the octant proembryo.

The endosperm in *Cocos nucifera* shows many special features. In early stages the endosperm is in the form of a liquid syncytium, i.e. a sap or "milk" with free polyploid nuclei (M. Dutt 1953, M. Datta 1955), and fine cytoplasmic particles. Later, vesicles appear around the free nuclei and they coalesce (Cutter et al. 1955). A large vacuole appears in the centre, and a gelatinous mass of these nuclei and cytoplasm surround it. Then cell walls are laid down in the gelatinous mass producing a meristematic ring around the central cavity. The meristematic ring produces the "coconut meat".

Voigt (1888) reports rumination in *Actinophloeus ambiguus* and *Caryota furfurcata*. This has been confirmed by Periasamy (1962a). Ruminate endosperm also occurs in *Areca catechu, A. concinna, Pinanga kuhlii* and others. The ruminations (Myristica type) are due to the ingrowths of outer layers of the inner integument (Periasamy 1962a). In *Licuala grandis* (Venkata Rao 1959a), the chalaza develops warty protuberances which cause ruminations in the endosperm. In *Bentinckia condapanna* (Padmanabhan and Regupathy 1981, 1983), it is the fruit wall which causes ruminations; the pericarp is closely appressed to the seed-coat. The growth centres are located in the fruit-wall which develop ingrowths and affect the seed-coat-endosperm outline. The undulations of the wall of the embryo sac appear even before the endosperm becomes cellular.

Ruminated seeds are also reported in *Arikuryroba schizophylla* and *Rhyticocos amara* (Reddy and Kulkarni 1985).

The embryogeny corresponds to the Asterad type-Geum variation. It is of Asterad type-Geum and Polygonum variations in *Phoenix robusta*, and Onagrad type in *Cocos coronata*, *C. plumosa*, and *C. schizophylla*. Onagrad type has also been observed in *Areca catechu* (Venkata Rao 1955b).

In *Cocos nucifera* (Haccius and Philip 1979), the zygote divides obliquely or transversely. *ca* divides more or less vertically and *cb* transversely to form a T-shaped tetrad; *cb* may divide vertically. A transverse division separates a single cell from the hemispherical daughter cell of *ca*, designated epiphysis-like cell (Fig. 59.2.1A). This is richly cytoplasmic, and divides repeatedly to produce the entire embryo proper (Fig. 59.2.1B–I). The other daughter cells of *ca*, along with those of *cb*, form a multicellular suspensor (Fig. 59.2.1B–I). Its cells are vacuolate and contain starch grains. Peripheral meristematic activity in the derivatives of the epiphysis-like cell lead to the differentiation of shoot tip and cotyledon.

Haccius and Philip emphasize that at the globular proembryo a conspicuous shoot tip differentiates in the exact terminal position, and the cotyledon in lateral position (Fig. 59.2.2A). At the same time, the cotyledonary primordium develops as a crescent-shaped mound expanding around the stem tip (Fig. 59.2.2A), and forms a cotyledonary sheath. A broad disc-shaped apical meristem in the cotyledonary region forms a mound or a "cylindro-conical body" which pushes the sheathing base on one side so as to enclose the shoot tip.

Haccius and Philip (1979) refute the formation of two cotyledonary primordia which is indeed due to an erroneous interpretation of sections cut in an incorrect plane (cf. Fig. 59.2.2A–D).

Guignard (1984) and Robertson (1976a, b) wrongly interpret the lateral origin of the shoot tip in the embryos of *Chamaerops humilis* and *Jubaeopsis*, possibly due to the smaller size of the embryo. These taxa need to be reinvestigated.

The reports of two cotyledons in certain embryos of Palmae (see Mahabale and Biradar 1968, Biradar 1968, Biradar and Mahabale 1968, Kulkarni and Mahabale 1974) are erroneous conclusions due to examining merely the longisections of young embryos (see Table 59.2.1).

Haccius and Philip refer to "delayed embryo differentiation" as an embryogenic type in which the entire basal cell *cb* as well as some derivatives of the terminal cell *ca* do not contribute to the formation of the embryo proper. Development of the embryo shows a number of highly primitive characters.

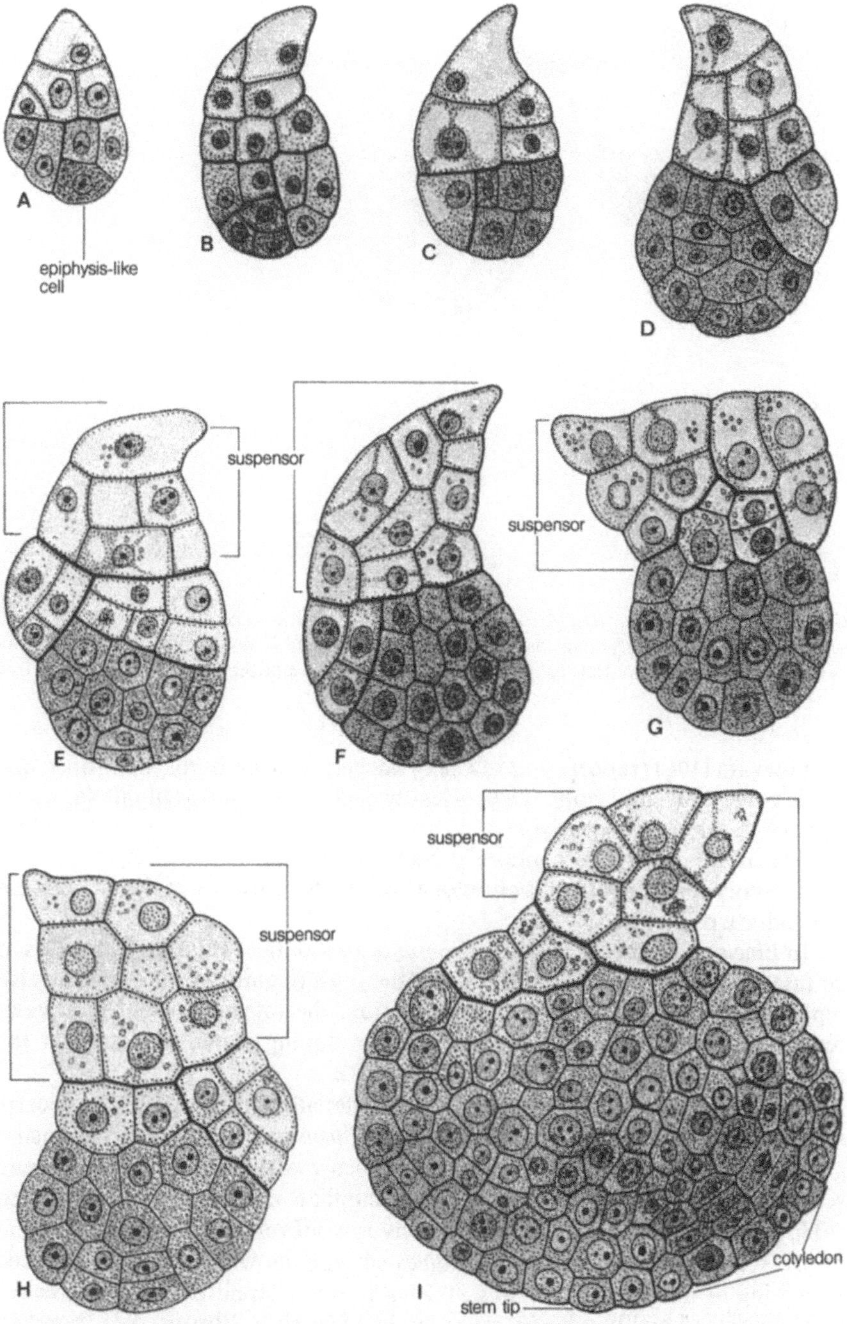

Fig. 59.2.1A–I. *Cocos nucifera*, stages in embryogeny. **A–C** Late quadrants. **D–H** Early globular proembryo. **I** Organization of stem tip and cotyledon. (After Haccius and Philip 1979)

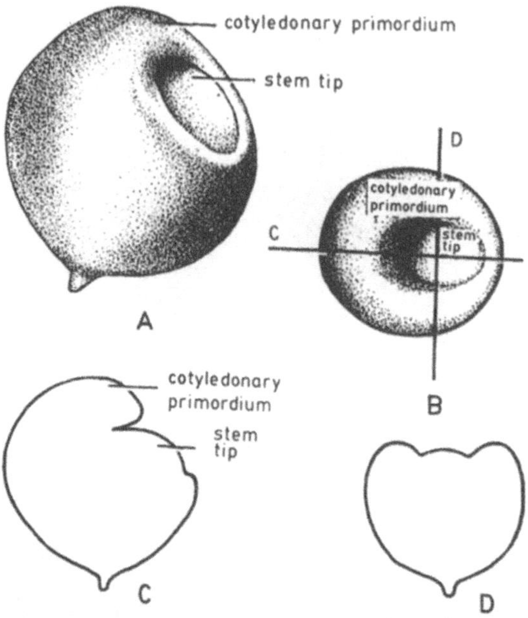

Fig. 59.2.2A–D. *Cocos nucifera.* **A** Excised young embryo with cotyledonary primordium which surrounds the stem tip. **B** Young embryo indicates planes *C* and *D* along which longisections can be cut to obtain configurations as in **C** and **D**. (After Haccius and Philip 1979)

Guevara (1961) reports that in *Cocos nucifera* twin embryos are formed due to cleavage polyembryony. Two or more cells of the pluricellular late proembryo develop into embryos.

The albuminous seed contains a small embryo.

The cocosoid palm fruit is characterized by the presence of three "eyes" on the endocarp.

In *Elaeis guineensis* (*E. laeioid*), the major part of the ovular body appears to be fused with the septal tissues even at the mature embryo sac stage, as also reported in *E. oleifera.* Due to this adnation, the integuments of the ovules remain fused to the inner surface of the developing endocarp except at the micropylar region (Reddy and Kulkarni 1985).

In a mature palm fruit the pericarp is differentiated into a tough epicarp, middle massive fibrous mesocarp (spongy in *Cocos nucifera*, fleshy and mucilaginous in other species), and inner stony endocarp with pits or eyes (Reddy and Kulkarni 1985). The epidermis comprises papillate cells, the hypodermal and subhypodermal mesocarp has one to many rows of vertically running fibre and fibrovascular strands, except in *Aiphanes caryotifolia* where they are replaced by a wing of obliquely traversing stegmatiferous sclereid patches. The endocarp consists of highly sclerotic cells and includes some fibrovascular bundles.

NV Biradar (1968), NV Biradar and TS Mahabale (1968), KN Chandra Sekhar and DA de Mason (1988), VM Cutter Jr and B Freeman (1954, 1955), VM Cutter Jr, KS Witson and B Freeman

Table 59.2.1. Embryogenesis in palms, *Cocos* (see Haccius and Philip 1979)

Taxa investigated	Contribution of *ca* and *cb* to embryo (*em*) and suspensor (*s*)	Origin of stem tip	Number of cotyledons	References
Areca *Pritchardia* *Actinophloeus* *Howea*	ca — *em* cb — *s* Onagrad type	Terminal from an epiphysis initial	1	C.V. Rao (1955, 1956, 1959a, b)
Phoenix *Livistona*	Whole of *ca* + upper part of *cb* — *em* lower part of *cb* — *s*	Terminal from an epiphysis initial	2	Biradar (1968) Biradar et al. (1968) Mahabale et al. (1968) Kulkarni et al. (1974)
Chamaerops	Asterad type	Lateral[a]	1	Guignard (1961)
Jubaeopsis		Lateral[a]	1	Robertson (1976)
Cocos	One cell of *ca* — *em* rest of *ca* + whole of *cb* — *s*	Terminal, but not from an epiphysis initial	1	Haccius and Philip (1979)

[a] see text.

(1955), NK Das (1965), M Datta (1955), DA de Mason, KN Chandra Sekhar and M Harris (1989), RA de Poerck (1950), M Dutt (1953), GG Gassner (1941), VF Guevara (1961), JL Guignard (1984), B Haccius and VJ Philip (1979), B Jönsson (1879–80), LB Kajale and SG Ranade (1952, 1953), KM Kulkarni and TS Mahabale (1974), TS Mahabale (1966), TS Mahabale and NV Biradar (1968), TS Mahabale and MS Chennaveeraiah (1957), SC Maheshwari (1955), D Padmanabhan and D Regupathy (1981, 1983), K Periasamy (1962a), E Quisumbing and JB Juliano (1927), A Radermacher (1924), GN Reddy and AR Kulkarni (1985, 1988), BL Robertson (1976a, b), EM Selvaratnam (1952), NS Shirke and TS Mahabale (1972), K Stephen (1974), BGL Swamy (1942b), C Venkata Rao (1955b, 1959a, b), J Venkateswarlu and PN Rao (1975a), S Veerasamy (1982), A Voigt (1888)

59.3 Taxonomic Considerations

Palmae. Palmae are included in the series Calycinae (calycine bracts) by Bentham and Hooker (1965c), and in the order Principes by Engler and Prantl (1931). Hallier (1912) places Spathiflorae (Palmae), Synantheae (Calycanthaceae), and Arales (Araceae and Lemnaceae), together with Pandanales, in Spadiciflorae. Rendle (1959) relates Palmae to Araceae under Spadiciflorae, and Wettstein (1935) to Pandanaceae and Cyclanthaceae. Hutchinson (1973) places them in the order Palmales near Agavales, and seeks their ancestry from Liliaceae. R. Dahlgren (1977a, 1980b) groups the orders Arecales (Arecaceae, i.e., Palmae), Pandanales (Pandanaceae) and Cyclanthales (Cyclanthaceae) in the superorder Arecanae Areciflorae. Cronquist (1981) accepts the order Arecales for the family Palmae (Arecaceae).

The family is multipalynous, and P.K.K. Nair (1970) holds the view that palynologically "Araceae, Palmae, Cyclanthaceae and Lemnaceae form one order, the 'Spadiciflorae' as defined by Hallier, but excluding the families of Pandanales. Within the Spathiflorae, the Palmae, Araceae (and possibly Lemnaceae) evolved along independent lines, and the Cyclanthaceae culminates the evolutionary line of the Palmae".

The relationship of Palmae with Cyclanthaceae is discussed in Section 60.3.

Palmae show a close alliance with Liliaceae in secretory anther tapetum, successive or simultaneous cytokinesis in pollen mother cells, two-celled pollen grains, anatropous, bitegmic and crassinucellar ovule, Polygonum and Allium type of embryo sacs, Nuclear type of endosperm, Asterad type of embryogeny, and seed with ruminate endosperm which stores hemicellulose, aleurone grains, fats and oils (starch is absent). Hutchinson's view about their origin is acceptable.

60 Synanthae

Cyclanthaceae, the family of palm-like perrenial herbs, represent the order Synanthae.

60.1 General Features

Cyclanthaceae show (1) secretory anther tapetum, (2) successive cytokinesis in pollen mother cells, (3) monocolpate, two-celled pollen grains, (4) anatropous, bitegmic, tenui- or pseudo-crassinucellate ovule, (5) Polygonum type of embryo sac, and (6) Helobial endosperm.

60.2 Cyclanthaceae

The embryology has not been fully investigated.

The anther wall comprises the epidermis, fibrous endothecium, an ephemeral middle layer, and secretory tapetum of two- to four-nucleate cells. In *Asplundia* spp. and *Evodianthus funifer* (Harling 1958), the epidermis is ephemeral, and the endothecium acts as "pseudo-exothecium". Successive cytokinesis accompanies meiotic divisions of microspore mother cells. The tetrads are isobilateral and decussate.

The pollen grains are monoporate, monosulcate and ulceroidate. They are two-celled at the dispersal stage.

The ovule is anatropous, bitegmic and pseudocrassinucellate (Fig. 60.2.1A), or tenuinucellate. The nucellar epidermal cells divide periclinally (Fig. 60.2.1C–F), except in *Cyclanthus bipartitus*. The lateral nucellar cells undergo radial elongation (Fig. 60.2.1A–D). The elongated cells become more prominent during the development of the female gametophyte (Fig. 60.2.1C, D), and have been designated "ray cells". Usually, the nucellus degenerates during the expansion of the embryo sac, but in *Asplundia polymera* subsp. *reitzii* (Harling 1958) the nuclei of the ray cells divide once or twice and then fuse to form irregularly shaped syncaryons. Due to these syncaryons, the mature embryo sac appears as an "hour-glass" (Fig. 60.2.1E, F).

The hypodermal archesporial cell (Fig. 60.2.1B) functions as megaspore mother cell, undergoes meiotic divisions to produce a linear, T-shaped, or obliquely T-shaped tetrad in *Asplundia nonoensis* and *Evodianthus funifer* (Harling 1958). The chalazal megaspore (Fig. 60.2.1G) develops into a Polygonum type of embryo sac (Fig. 60.2.1H, I). The three antipodals are ephemeral, and the polar nuclei lie in the chalazal region.

The development of the endosperm is of the Helobial type. The wall separating the large micropylar and the small chalazal chamber may be concave (Fig. 60.2.1J), transverse (Fig. 60.2.1K), or lateral (Fig. 60.2.1L). Free-nuclear divisions in the micropylar chamber are followed by wall formation. The mature endosperm contains proteins and fats as food reserve; starch is present in *Dicranopygium*. The chalazal chamber becomes four-nucleate; there are 12 nuclei in *Carludovica drudei* (Harling 1958). However, it does not become cellular and degenerates.

The endotestal, albuminous seed contains a small cylindrical embryo and fatty endosperm. The testa is five or six layers thick. The cells, except those of the inner epidermis, become mucilaginous and contain raphides. In *Asplundia*

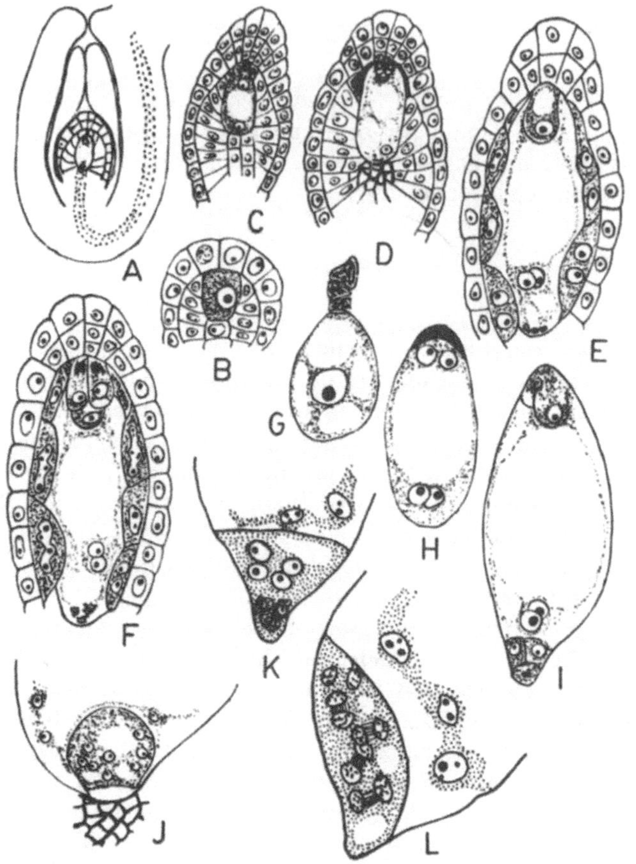

Fig. 60.2.1A–L. **A–D, G–J** *Carludovica palmata*, **E, F** *Asplundina polymera* subsp. *reitzii*, **K, L** *Dicranopygium* aff. *Coma-Pyrrhae*. **A** Anatropous, bitegmic, pseudocrassinucellar ovule. **B** Archesporial cell. **C, D** Longisections of nucellus, the lateral cells show a characteristic radiating pattern. There is a thick-walled hypostase in **D** and **J**. **E, F** Organized mature embryo sac. The innermost nucellar cells show syncaryons. **G** Linear tetrad with functional chalazal megaspore. **H** Four-nucleate embryo sac. **I** Eight-nucleate condition. **J** Lower region of embryo sac shows a part of micropylar endosperm chamber and free-nuclear chalazal chamber (Helobial type). **K** The chalazal chamber is four-nucleate, and the antipodals still persist. **L** The chalazal chamber is laterally situated, and several nuclei are in division. (After Harling 1958)

spp., the entire testa becomes mucilaginous. The inner epidermal cells elongate radially and form a thick-walled palisade layer. The two-layered thin-walled tegmen is supported on either side by a thick cuticle.

GL Davis (1966), G Harling (1958), K Schnarf (1931)

60.3 Taxonomic Considerations

Cyclanthaceae. Bentham and Hooker (1965c) place Cyclanthaceae in Nudiflorae, along with Araceae, Lemnaceae, Pandanaceae and Typhaceae. Engler and Prantl (1931) treat them in a separate order, Synanthae, between palms and aroids. Hutchinson (1973) recognizes them as a lone representative of the order Cyclanthales. He derives the family from Palmae as it is characterized by the much-reduced unisexual flowers, and the inflorescence is in a deciduous spathe. Palynological data support Hutchinson's view about its origin (P.K.K. Nair 1970).

Cyclanthaceae have close affinities with Palmae in secretory anther tapetum, successive cytokinesis in pollen mother cells, two-celled pollen grains, anatropous, bitegmic ovule with both integuments forming the micropyle, Polygonum type of embryo sac, and albuminous seed. However, tenuinucellate ovule and Helobial type of endosperm in Cyclanthaceae segregate them from Palmae.

The family Cyclanthaceae cannot be associated with aroids, nor with Pandanales. Amoeboid anther tapetum, occasional occurrence of pollen tetrads, and operculum in Araceae, and reduced type of anther wall, three-celled pollen grains, operculum, orthotropous ovule (*Wolffia*), and Allium type of embryo sac in Lemnaceae, and Cellular type of endosperm in both families are not met with in Cyclanthaceae. In addition, Nuclear type of endosperm occurs in Pandanaceae, and amoeboid anther tapetum and three-celled pollen grains in Sparganiaceae and Typhaceae.

Cyclanthaceae cannot be grouped with the Nudiflorae of Bentham and Hooker either, but their position near Palmae, as recognized by Melchior (1964), Hutchinson (1973) and R. Dahlgren (1977a, 1980b) is appropriate.

61 Spathiflorae

The order Spathiflorae comprises two families: Araceae and Lemnaceae.

61.1 General Features

Spathiflorae exhibit: (1) amoeboid anther tapetum, (2) successive cytokinesis in pollen mother cells, (3) three-celled pollen grains (occasionally two-celled in Araceae), (4) ortho- or anatropous bitegmic ovule, (5) nucellar cap, (6) Polygonum and Allium type of embryo sac, (7) Cellular type of endosperm with basal apparatus, (8) probably Caryophyllad type of embryogeny in Araceae, Asterad in Lemnaceae, and (9) operculate seed.

61.2 Araceae

The disposition of anthers is shown in Figure 61.2.1A–K. The anther wall (Fig. 61.2.1L) shows persistent epidermis, fibrous endothecium (see also French 1985a, b, 1986a), three middle layers, and tapetum in *Synandrospadix vermitoxicus* (Cocucci 1966).

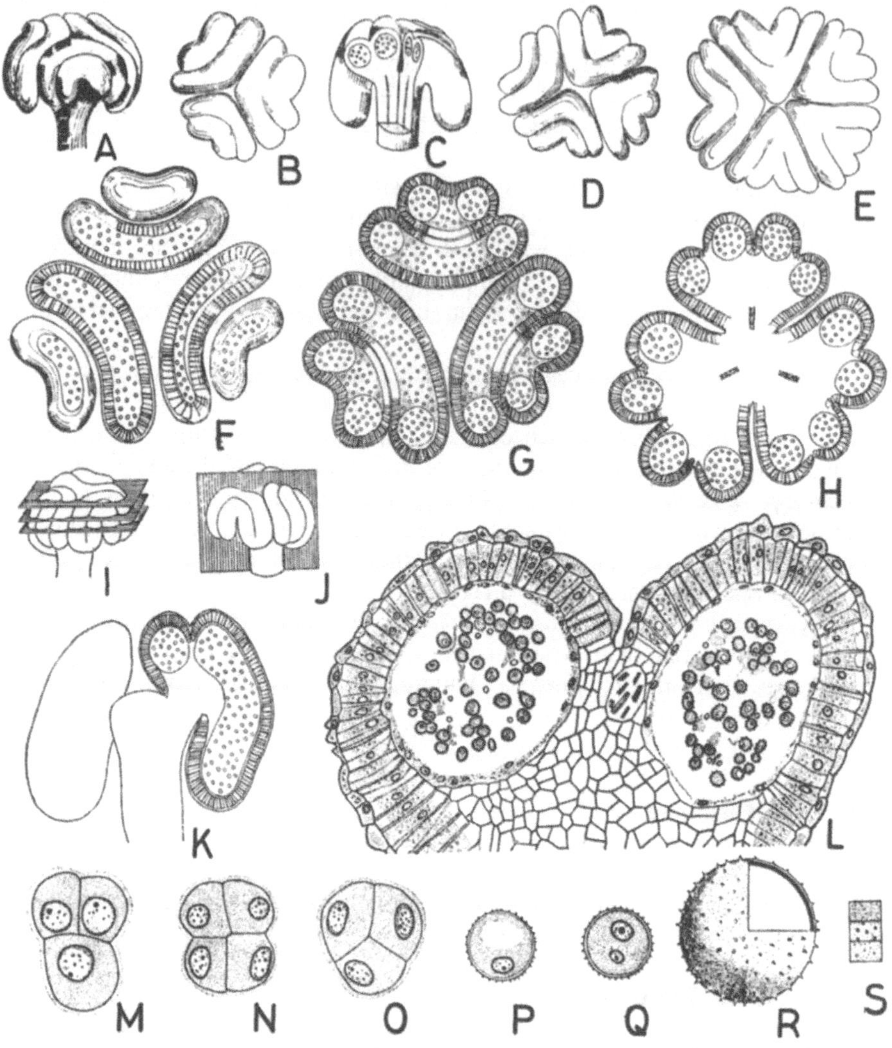

Fig. 61.2.1A–S. *Arisaema wallichianum*. **A, B** Male flower, lateral (**A**) and dorsal (**B**) view. **C** Three-dimensional sectional view of male flower (diagrammatic). **D, E** Top view of male flowers bearing four (**D**) and five (**E**) anthers. **F–I** Transections (**F–H**) of male flower cut at levels shown in **I**. **J, K** Plane of sectioning (**J**) for **K**. **L** Transection of anther. The middle layers have collapsed, remnants of periplasmodium still persist. **M–O** Tetrads. **P, Q** Microspore and two-celled pollen grain. **R, S** Palynogram and Lo-diagram of pollen grain. (After S.C. Maheshwari and Khanna 1956)

The tapetum is amoeboid and irregularly biseriate in *Theriophonum minutum* (Parmeswaran 1959, see also Pacini and Juniper 1983). The tapetal cells are uninucleate. The number of middle layers varies in different members. Successive cytokinesis follows meiotic divisions in microspore mother cells. The tetrads (Fig. 61.2.1M–O) are isobilateral, tetrahedral and decussate.

Polyspory occurs in *Colocasia* (Jose and Vijaya Bai 1978) and *Synandrospadix vermitoxicus* (Cocucci 1966).

The pollen grains are one-, three- or four-sulcate, three-porate and non-aperturate. They are dimorphic in *Colocasia antiquorum*. Sometimes, they remain united in tetrads in *Caladium* and *Xanthosoma* (Kapil 1967). The pollen grains (Fig. 61.2.1P–S) are two-celled at the dispersal stage in *Arisaema*, *Dieffenbachia*, *Spathyema*, *Symplocarpus*, *Synandrospadix* and *Theriophonum*; three-celled in *Aglaonema*, *Arum*, *Peltandra* and *Zantedeschia* (Kapil 1967).

Galil (1990) has studied pollen germination in *Arisarum vulgare*. Germination occurs in two phases: the aeration phase, and the germination proper phase. Aeration is a crucial prerequisite for germination. In fact, in the absence of aeration, the pollen scattered on the inner wall of the chamber of inflorescence fails to germinate. Aeration is made possible by the decrease in relative humidity within the chamber at noon, while the increase in humidity at night promotes germination proper.

The ovular vasculature in Araceae, 97 genera and 280 species, has been studied by French (1986b). The ovule is ortho-, hemi- or anatropous, bitegmic and pseudocrassinucellar. It is tenuinucellate in *Pistia*, *Synandrospadix* and *Theriophonum*. The micropyle is organized by the inner integument in *Arisaema*. An air-space develops in the micropylar region under the outer integument between exostome and endostome in *Synandrospadix vermitoxicus* (Cocucci 1966). The nucellar epidermis, by periclinal divisions, produces a nucellar cap (Fig. 61.2.2A–C, E–G). In *Theriophonum minutum* (Parmeswaran 1959), the nucellar cap develops at the four-nucleate embryo sac; its remnants persist up to the late zygote stage. The nucellar cap is absent in *Synandrospadix*. An endothelium differentiates at the eight-nucleate embryo sac in *Arisaema wallichianum* (S.C. Maheshwari and Khanna 1956) and *Lagenandra ovata* (Panicker 1965). A hypostase develops in *Acorus* and *Theriophonum*, and postament in *Typhonium*.

The archesporial cell functions as megaspore mother cell. The latter undergoes meiotic divisions, and the chalazal megaspore develops into a Polygonum type of embryo sac. The polar nuclei fuse before fertilization, and the secondary nucleus lies near the egg apparatus in *Theriophonum*. The antipodals are ephemeral; persistent in *Theriophonum*. They enlarge in *Arum orientale* and *Symplocarpus foetidus*, and multiply in *Aglaonema*, *Lysichiton* and *Xanthosoma*. Twin embryo sacs are reported in *Arisaema triphyllum* (Piekett 1913).

The development of the endosperm is of the Cellular type. The division of the primary endosperm nucleus (rarely it fails to divide – Fig. 61.2.2E) is followed by a wall separating a micropylar chamber from the chalazal chamber. The micropylar chamber develops into cellular endosperm (Fig. 61.2.2A–D, F, G). The chalazal chamber remains uninucleate (sometimes binucleate in later stages – Fig. 61.2.2G), and becomes haustorial with a hypertrophied nucleus. The highest ploidy level in *Arum maculatum* (Erbrich 1965) is 24,576 ×. The chalazal chamber is aggressive in *Arisaema* and *Pistia*.

Helobial type of endosperm occurs in *Ariopsis peltata* (Gowda 1980). In the micropylar chamber, after the formation of eight nuclei, cytokinesis begins

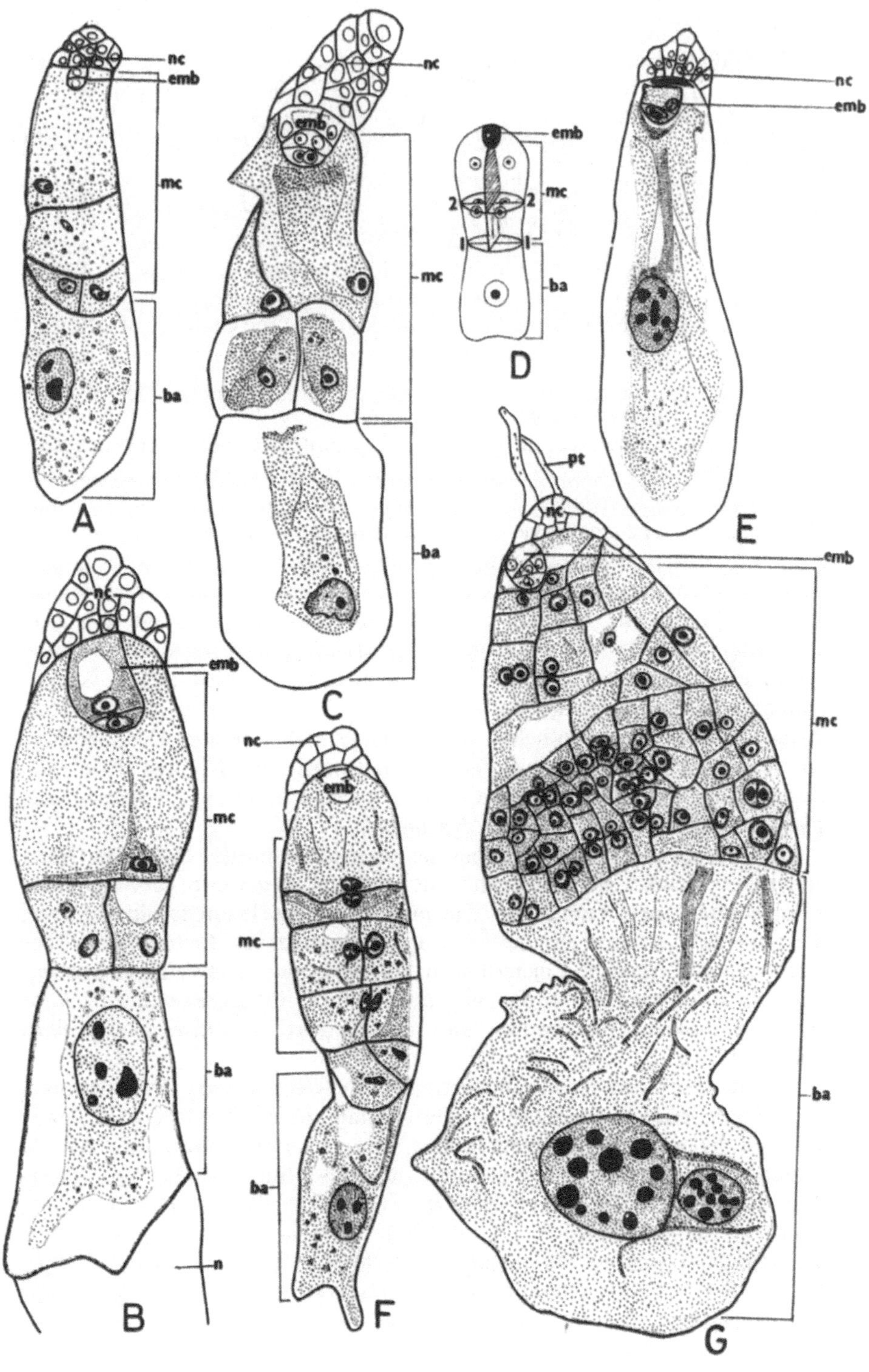
nc
emb
mc
ba
A
B
C
D
E
F
G
pt
n
2
1

close to the chalazal chamber, the latter remains uninucleate. This taxon needs to be reinvestigated.

The zygote undergoes a transverse division. The basal cell *cb* does not divide but elongates and forms a suspensor. The terminal cell *ca* undergoes repeated divisions and forms an embryonal mass in *Theriophonum minutum* (Parmeswaran 1959). In *Pistia stratiotes* (Haccius and Lakshmanan 1966), the epicotyl and cotyledon develop side by side from the embryonal apex. Polyembryony is reported in *Spathiphyllum*.

The seed is operculate (Buell 1935) with a cylindrical embryo in *Theriophonum*. The testa is multiplicative, parenchymatous and contains raphides. Some of its cells, and the cells of inner epidermis of the tegmen are tanniniferous.

I Banerji (1934, 1937, 1947), M Bedalov and M Hesse (1989), MF Buell (1935, 1938), AE Cocucci (1966), P Crété (1957c), P Erbrich (1965), JC French (1985a, b, 1986a, b), J Galil (1990), B Goldberg (1941), MTG Gowda (1980), B Haccius and KK Lakshmanan (1966), JS Jose and K Vijaya Bai (1978, 1986), RN Kapil (1967), AR Kulkarni, D Dosi and VM Manoj (1990), SC Maheshwari (1956b), SC Maheshwari and PP Khanna (1956), SJ Mayo (1989), E Pacini and BE Juniper (1983), TKB Panicker (1965), N Parmeswaran (1959), FL Piekett (1913), CO Rosendahl (1909), AE Shadowsky (1931), H von Guttenberg, A Tober, W Dretke and M Ruickhold (1960)

61.3 Lemnaceae

The location of the flowers on the frond is shown in Figure 61.3.1A–F, S).

The anther wall comprises the epidermis, fibrous endothecium, and amoeboid tapetum with uninucleate cells in *Lemna paucicostata* (Fig. 61.3.1G–O, S) and *Wolffia microscopica* (S.C. Maheshwari 1954). According to S.C. Maheshwari (1954), and S.C. Maheshwari and Kapil (1963a), the middle layer is not formed in *Wolffia microscopica* and *Lemna paucicostata*, (Fig. 61.3.1K–M). Davis (1966) describes the development of the anther wall as of the "Reduced" type. Lodkina (1977) reports a middle layer in *Lemna gibba*, *L. trisulca*, *L. paucicostata* and *Wolffia microscopica*. In the latter two species, the "middle" cells and the tapetal cells occur in the same layer, as the alternate cell differentiates into a tapetal cell. This requires confirmation. The tapetum forms true periplasmodium (Fig. 61.3.1L, N) at the time of meiotic divisions in the microspore mother cells. Cytokinesis is successive, and the tetrads (Fig. 61.3.1L, P, Q) are isobilateral, tetrahedral and decussate.

Fig. 61.2.2A–G. *Arisaema wallichianum* (all diagrams from dissections). **A** Embryo sac with two-celled proembryo (*emb*), and four-celled micropylar chamber (*mc*) and one-celled basal (*ba*) apparatus (= chalazal chamber). The nucellar cap (*nc*) is quite conspicuous in **A–C**, **E–G**. **B** Two-celled proembryo. The micropylar chamber is four-celled, the upper cell is undivided; basal apparatus with hypertrophied nucleus; *n* nucellar column. **C** Young globular proembryo, six-celled micropylar chamber, and uninucleate chalazal haustorium (*ba*). **D** Schematic representation for the sequence of wall formation in **C**. **E** Abnormal embryo sac with two-celled proembryo, and undivided primary endosperm nucleus. **F, G** Later stages of endosperm development. In **G** the basal apparatus (haustorium) is binucleate. The remnants of pollen tube (*pt*) are also seen. (After S.C. Maheshwari and Khanna 1956)

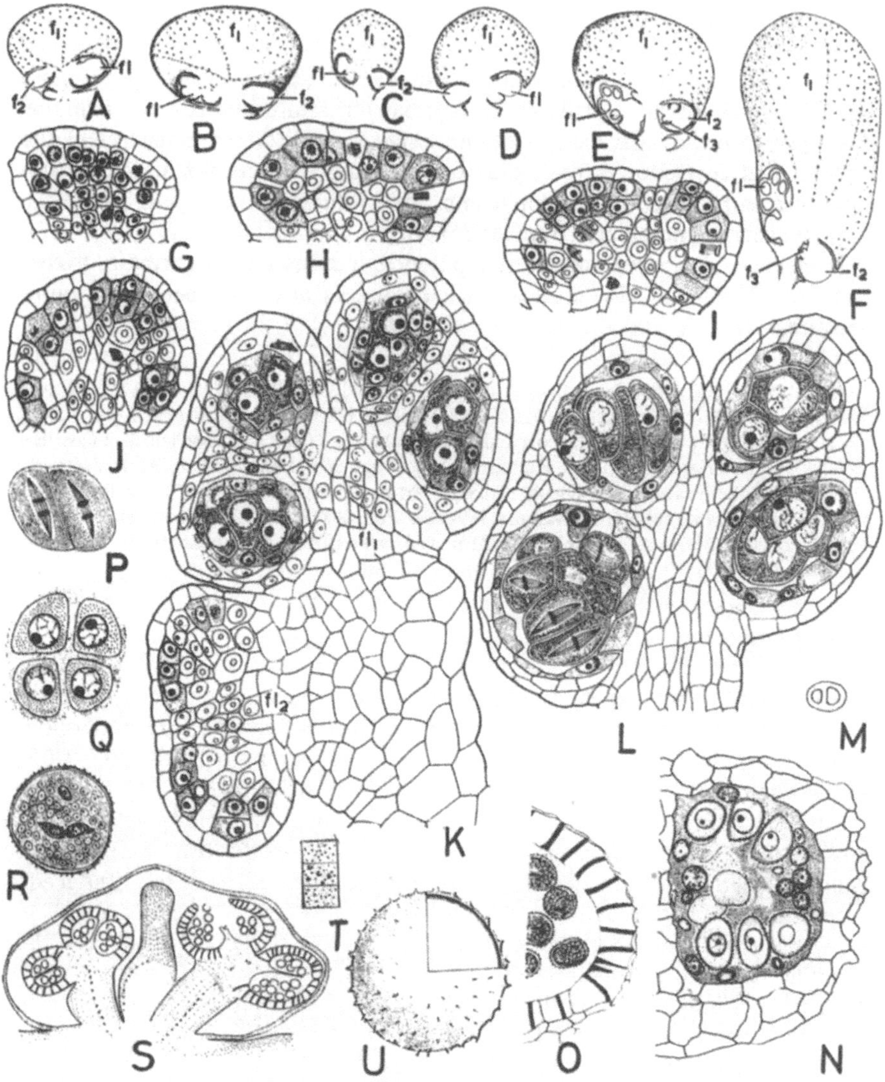

Fig. 61.3.1A–U. *Lemna paucicostata.* **A–F** Outline diagrams of longisections of mother fronds at various stages of development; the respective anthers are shown in **G–L**; f_1, f_2, f_3 mother and daughter fronds; *fl* flower. **G** Longisection of young anther. **H** Multicelled archesporium. **I** Delimitation of primary parietal layer. **J** Four sporogenous groups. **K** Pair of stamens (fl_1, fl_2), filaments fused; in the larger anther note the endothecium and tapetum (middle layer is not formed). **L** Meiosis I in microspore mother cells. **M** Transection of anther, two locules (outline diagram). **N** One lobe from **M**; microspores and tapetal plasmodium. **O** Portion of mature anther shows flattened epidermis and fibrous endothecium. **P** Microspore mother cell, meiosis II; **Q** Isobilateral tetrad. **R** Three-celled pollen grain. **S** Vertical section of inflorescence at dehiscence of anther (diagrammatic). The carpel is in the middle of the two anthers. **T, U** Lo-diagram and palynogram of mature pollen grain (diagrammatic). (After S.C. Maheshwari and Kapil 1963a)

The pollen grains are three-celled at the dispersal stage in *L. paucicostata* (Fig. 61.3.1R) and *W. microscopica* (S.C. Maheshwari 1954). They are monoporate with spinulose exine (Fig. 61.3.1T, U), and contain starch in *Lemna* and *Wolffia*.

The ovule is orthotropous in *Wolffia* (S.C. Maheshwari 1954), and anatropous in *Spirodela* (S.C. and N. Maheshwari 1963). It becomes hemianatropous in *L. paucicostata*. It is bitegmic and crassinucellate with the micropyle formed by the inner integument (Fig. 61.3.2A–I). Both the integuments form the micropyle in *S. polyrrhiza* (S.C. and N. Maheshwari 1963), the exostome is widely away from the endostome. The inner integument develops an operculum (Fig. 61.3.4O) at the micropylar end. A conspicuous nucellar cap (Figs. 61.3.2I, 61.3.4A–E) and hypostase also develop.

The archesporial cell cuts off a parietal cell which produces three to four cells (by anticlinal divisions) in *Spirodela*; in *Lemna* and *Wolffia* the parietal tissue is two to three layers thick. The megaspore mother cell undergoes meiosis I and produces a dyad.

Allium type of embryo sac occurs in *Wolffia arrhiza* (B.L. Gupta 1935), *W. microscopica* (S.C. Maheshwari 1954), and *Lemna* (Fig. 61.3.2A–I). The lower dyad undergoes meiosis II and the chalazal megaspore develops into a Polygonum type of embryo sac in *Spirodela* (Fig. 61.3.3B–F). The polar nuclei fuse before fertilization. The antipodals are ephemeral.

The development of the endosperm is of the Cellular type (Figs. 61.3.3G, H, 61.3.4B–E). The division of the primary endosperm nucleus separates a micropylar chamber from the chalazal chamber in *Spirodela*. The former develops into cellular endosperm, while the latter forms a slightly aggressive chalazal haustorium with a single hypertrophied nucleus (Fig. 61.3.3G, H). Rarely, the chalazal chamber gets divided in *Lemna minor* (S.C. Maheshwari and Kapil 1963b). In *L. paucicostata* and *Wolffia microscopica* also the chalazal chamber becomes cellular (Fig. 61.3.4C–E) and forms a caecum (Fig. 61.3.4O).

In *Lemna paucicostata* (Fig. 61.3.4D–N), the zygote divides when the endosperm is four-celled. The terminal cell (*ca*) undergoes two longitudinal divisions, at right angles to each other, to form a quadrant. A transverse division produces an octant in two tiers. These tiers divide periclinally and anticlinally resulting in a symmetrical apical hemisphere with central cells enclosed by an epidermis (Fig. 61.3.5A, *ca*).

The basal cell (*cb*) divides transversely (daughter cells *m* and *ci*) followed by longitudinal division so that each tier becomes four-celled. The cells of the tier *ci* divide transversely to form tiers *n* and *n′*. The tier *m* also increases in length. Later growth is mostly in the derivatives of *m* and *ca*, while there is not much addition to the basal portion (cf. Fig. 61.3.5A–F). A small protuberance differentiates from *m* and the division of cells form a mound, the primordium of a frond (Fig. 61.3.5D–F). Subsequent growth shifts this young frond from a lateral to a basal position in the proembryo. It becomes completely vertical in a mature embryo (Figs. 61.3.4O, 61.3.5H, I). Multiplication of the derivatives of *ca* gives rise to a club-shaped cotyledon. The embryo continues to elongate and extends up to the base of the endosperm (Fig. 61.3.4O). There is no distinction

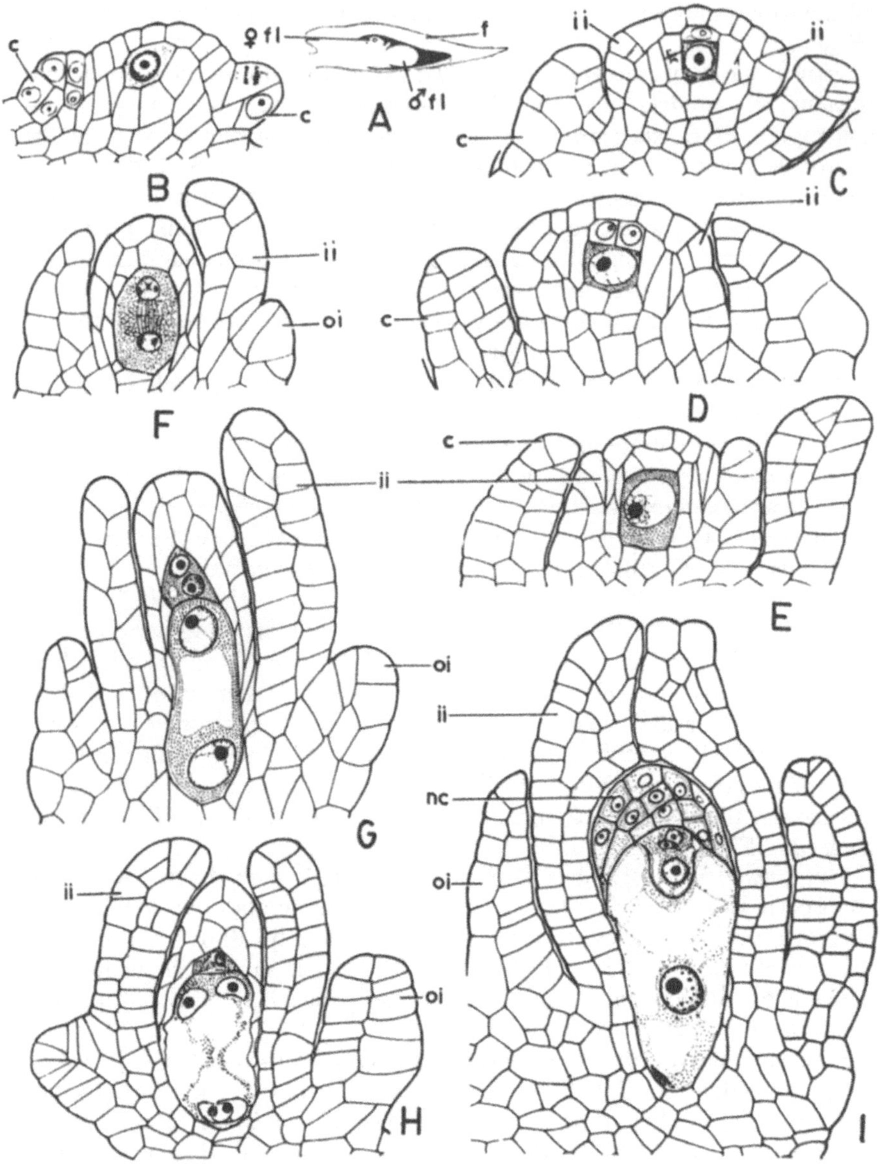

Fig. 61.3.2A–I. *Lemna paucicostata.* **A** young frond (v.s.) bearing an inflorescence; *f* frond; *fl* flower. **B** Female archesporial cell, note the carpellary primordium (*c*). **C–E** Megaspore mother cell, and parietal cell; *ii* inner integument. **F** Megaspore mother cell, meiosis I; *oi* outer integument. **G** Two-nucleate embryo sac and binucleate upper dyad cell. **H** Four-nucleate embryo sac. **I** Mature embryo sac, polars fused, and antipodals have degenerated; *nc* nucellar cap. (After S.C. Maheshwari and Kapil 1963a)

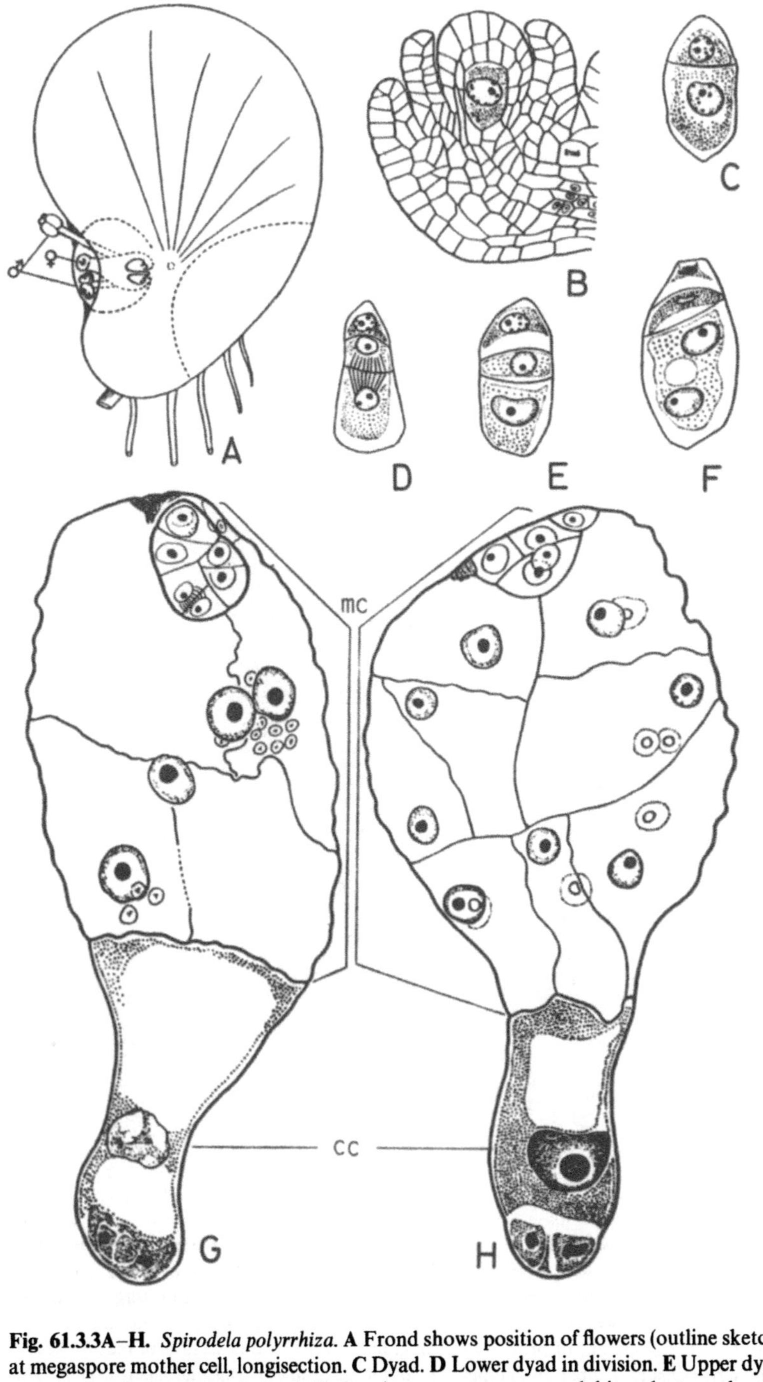

Fig. 61.3.3A–H. *Spirodela polyrrhiza.* **A** Frond shows position of flowers (outline sketch). **B** Ovule at megaspore mother cell, longisection. **C** Dyad. **D** Lower dyad in division. **E** Upper dyad, and two megaspores from chalazal dyad. **F** Dyad, one megaspore and bi-nucleate embryo sac. **G, H** Endosperm with micropylar (*mc*) and chalazal chamber (*cc*). Note degenerated antipodals. (After S.C. Maheshwari 1958)

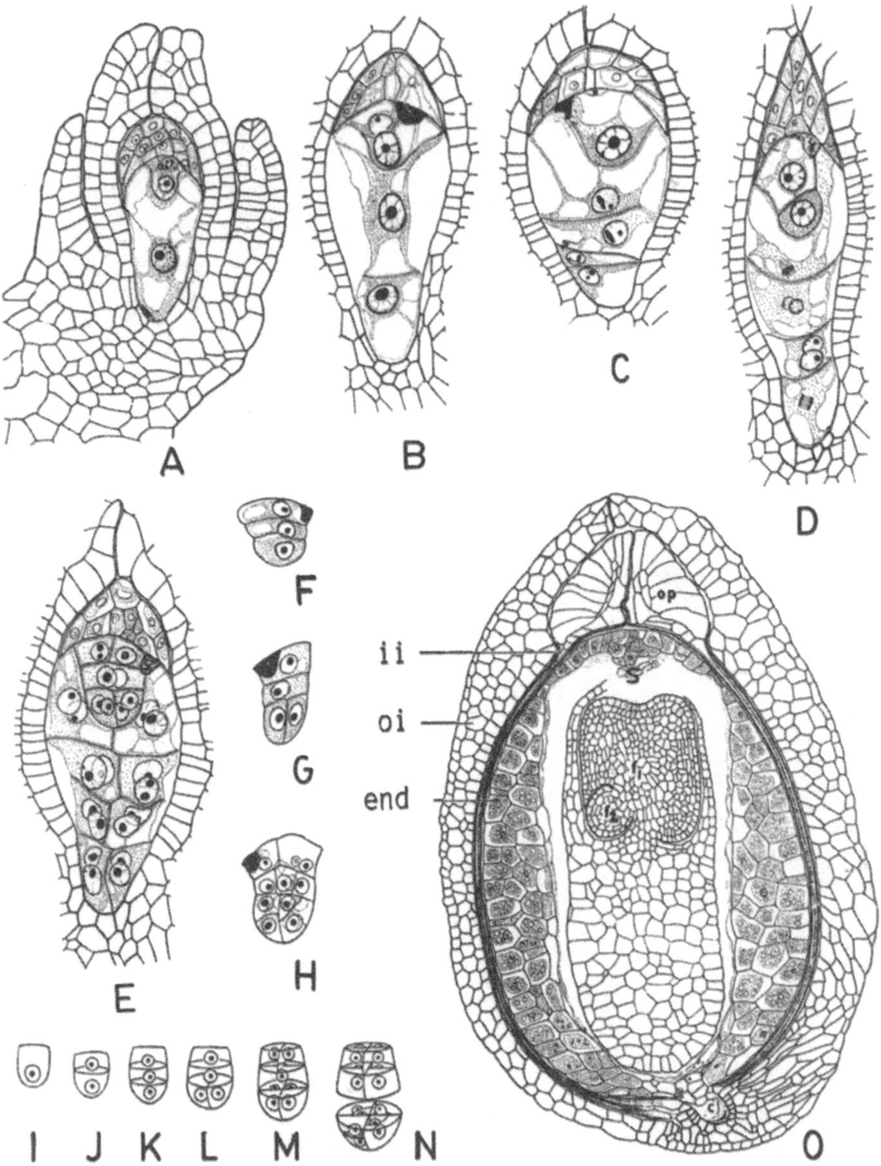

Fig. 61.3.4A–O. *Lemna paucicostata.* **A** Longisection of ovule at mature embryo sac, note nucellar cap in **A–E**. **B** Embryo sac with one synergid, zygote and two-celled endosperm; remnants of pollen tube *black*. **C** Four-celled endosperm, zygote still undivided. **D** Later stage, proembryo bicelled. **E** Endosperm 16-celled, and proembryo eight-celled. **F–H** Three-, four- and 16-celled proembryo. **I–N** Development of proembryo (diagrammatic). **O** Longisection of mature seed; *op* operculum; *s* suspensor; *ii* inner integument; *oi* outer integument; *end* endosperm; f_1, f_2 frond; *c* caecum (chalazal endosperm haustorium). (After S.C. Maheshwari and R.N. Kapil 1963b)

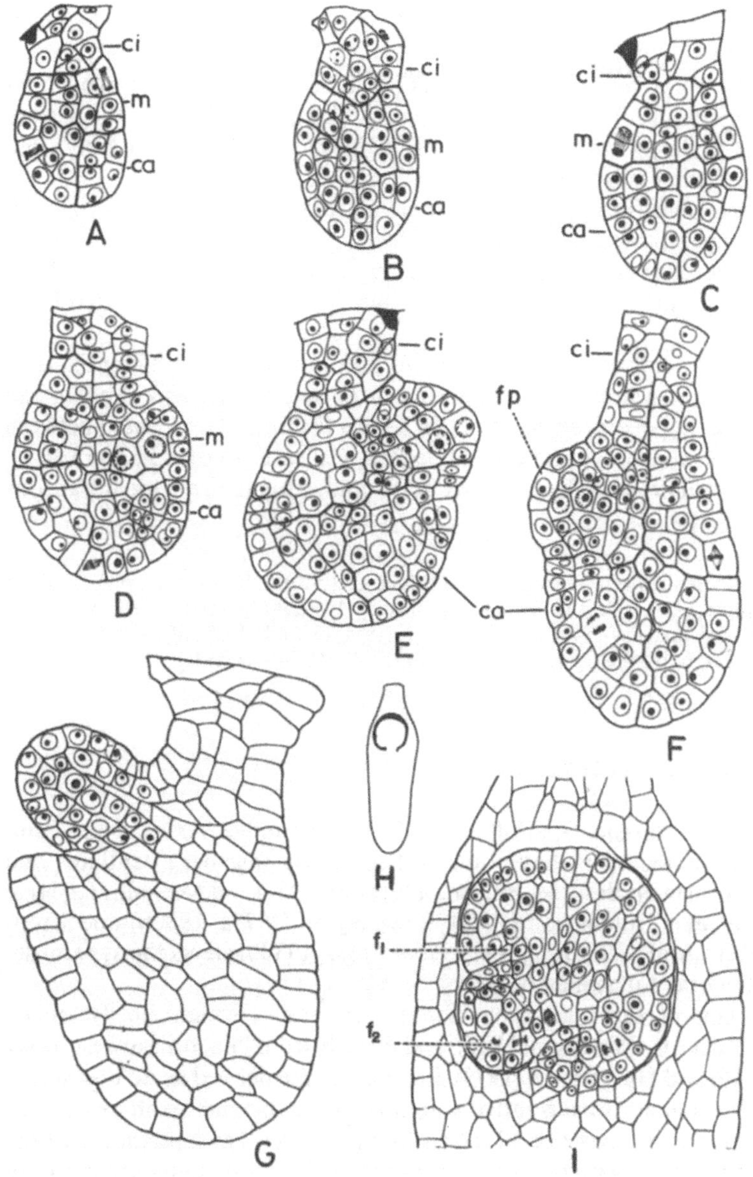

Fig. 61.3.5A–I. *Lemna paucicostata*. **A–D** Growth of proembryo is mainly due to divisions in the derivatives of *ca* and *m*. **E, F** Frond primordium (*fp*) formed as a result of divisions in the daughter cells of *m*. **G** Later stage. **H** Longisection of a nearly-mature embryo cut parallel to the surface of the frond, outline diagram for **I**. **I** Frond (f_1) and daughter frond (f_2). (After S.C. Maheshwari and Kapil 1963b)

between the hypocotyl and suspensor, and the authors designate this portion as "hypocotyl-suspensor" (four cells broad and 10 – 15 cells in length).

The epidermal cells below the frond primordium form a cotyledonary sheath which surrounds the first frond (Fig. 61.3.5G) connected to the main body of the embryo by a short pedicel. The frond develops from the hypocotyledonary region and not from the cotyledon. The primordium of the second frond arises at the junction of the pedicel and laminar portion of the frond, and develops like the first frond (Figs. 61.3.4O, 61.3.5I).

In *Lemna gibba* (Haccius and Lakshmanan 1966), the cotyledon originates from the upper part of the embryo, whereas the epicotyl, even though it appears to develop laterally, arises partly from the derivatives of the cell *ca*. The lateral position of shoot apex is due to accelerated growth of the cotyledon.

The mature embryo is cylindrical. A root primordium differentiates in *Lemna minor* and *Spirodela*, but is absent in *L. paucicostata* and *Wolffia*, and the suspensor–hypocotyl axis is short. In *Lemna* the fronds are located between the suspensor–hypocotyl and the cotyledonary sheath (Fig. 61.3.6A–L).

The operculate seed has a thick testa and cuticularized membraneous tegmen.

FA Bernard, JM Bernard and P Denny (1990), FH Blodgett (1923), OW Caldwell (1899), GL Davis (1966), BL Gupta (1935), B Haccius and KK Lakshmanan (1966), B Jönsson (1879–80), A Lawalrée (1945, 1952), MM Lodkina (1977), SC Maheshwari (1954, 1956a, b, 1958), SC Maheshwari and RN Kapil (1963a, b, 1964), SC and N Maheshwari (1963), HL Mason (1938)

61.4 Taxonomic Considerations

Araceae and Lemnaceae. Spathiflorae include two families, Araceae and Lemnaceae. The systematists have always kept them together, Bentham and Hooker (1965c) in Nudiflorae, Engler and Prantl (1931) and Melchior (1964) in Spathiflorae, and Hutchinson (1973), R. Dahlgren (1977a, 1980b) and Cronquist (1981) in Arales. R. Dahlgren (1975a, 1977a, 1980b) erects the superorder Aranae, Ariflorae for Arales.

The origin of the herbaceous family Araceae is sought in Helobiae-Liliiflorae stocks by Wettstein (1935), and directly from liliaceous ancestors by Bessey (1915) and Hutchinson (1973). Lemnaceae are derived as an offshoot of Araceae. The studies on vegetative and floral anatomy and cytology confirm this derivation (see Lawrence 1951). Eichler (1878) assigns Lemnaceae to the order Helobiales. Lawalrée (1945, 1952) accepts this view, and supports it on the basis of morphology and embryology of *Lemna*.

The embryology of Araceae has been investigated by S.C. Maheshwari and Khanna (1956), and, of Lemnaceae by S.C. Maheshwari (1954, 1956a, b, 1958), S.C. Maheshwari and Kapil (1963a, b, 1964), and S.C. and N. Maheshwari (1963). The close alliance of these two families is supported by the following features (S.C. Maheshwari 1956b): (1) successive cytokinesis in pollen mother cells, (2) true periplasmodium, (3) three-celled pollen (sometimes two-celled

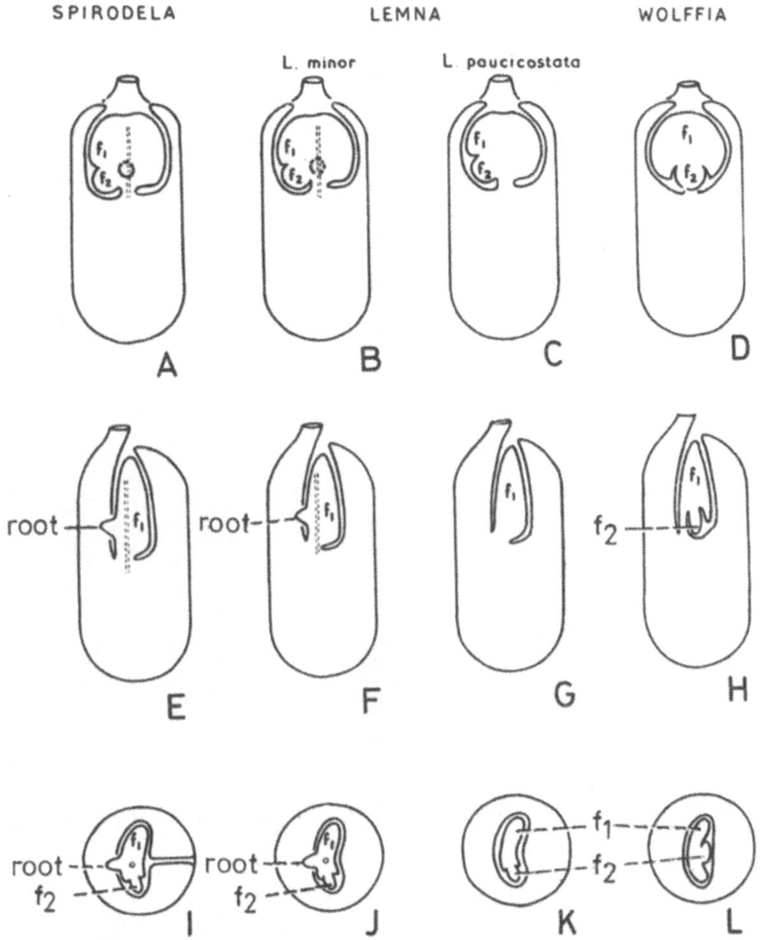

Fig. 61.3.6A–L. Embryo in Lemnaceae. **A–D** Longisections parallel to the surface of frond; **E–H** At right angles to the surface of frond; **I–L** Transections at the level of frond. **A, E, I** *Spirodela* shows a well-developed root primordium, lateral daughter frond f_2, and a vascular strand in the first frond f_1; **B, F, J** *Lemna minor* is similar to *Spirodela*; **C, G, K** *L. paucicostata* lacks root primordium, and provascular strand; **D, H, L** *Wolffia*, the root primordium and vascular strand are absent, the daughter frond (f_2) is placed symmetrically in the median line. (After S.C. Maheshwari and Kapil 1963b)

in aroids), (4) periclinal divisions in nucellar epidermis, (5) nucellar cap, (6) Cellular type of endosperm, (7) endosperm caecum at the chalazal end, and (8) formation of an operculum.

To determine the systematic position of the Lemnaceae, S.C. Maheshwari (1956b) compares the embryological features of Helobiales, Araceae and Lemnaceae (Table 61.4.1). In Helobiales the endosperm is of the Helobial or Nuclear type without a chalazal caecum, and the basal cell of the two-celled

Table 61.4.1. Embryology of Helobiales (Sect. 52), Araceae (61.2) and Lemnaceae (61.3). (Adapted after S.C. Maheshwari 1956b)

	Helobiales	Araceae	Lemnaceae
Anther	Wall 4-layered, endothecium usually fibrous, thickenings absent in some taxa, tapetum amoeboid, cells uninucleate	Wall 6-layered, endothecium fibrous, tapetum irregularly biseriate, amoeboid, cells uninucleate	Wall 3-layered, endothecium fibrous, middle layer absent, tapetum amoeboid, cells uninucleate
Cytokinesis	Successive	Successive	Successive
Tetrads	Tetrahedral, isobilateral, decussate, linear	Tetrahedral, isobilateral, decussate, polyspory occurs	Tetrahedral, isobilateral, decussate
Pollen	Monocolpate, pentaporate, usually 3-celled, 2-celled in *Blyxa* and *Ottelia*; arranged in uniseriate chains in *Halophila*	1-, 3- or 4-sulcate, 3-porate, nonaperturate, 2- or 3-celled, sometimes dimorphic; may remain attached in pairs	Monoporate, exine spinulose, 3-celled
Ovule	Ana- or orthotropous, bitegmic, crassi- or tenuinucellar, bitegmic, micropyle formed by inner integument	Vascularized, ortho-, ana- or hemianatropous, bi- or unitegmic, pseudocrassinucellar or tenuinucellar, micropyle organized by inner integument (when bitegmic); nucellar cap, endothelium and hypostase differentiate	Ortho-, ana- or hemianatropous, bitegmic, crassinucellate, micropyle formed by inner or both integuments (micropyle zigzag), operculum develops from inner integument, nucellar cap and hypostase also develop

Embryo sac	Polygonum type, Allium in Alismataceae, Butomaceae (except *Butomus*) and Zannichelliaceae, much variation in the behaviour of chalazal nuclei of 2- and 4-nucleate embryo sac, antipodals ephemeral	Polygonum, antipodals persistent or ephemeral	Allium in *Lemna* and *Wolffia*, Polygonum type in *Spirodela*, antipodals ephemeral
Endosperm	Mostly Helobial type, the nucleus of chalazal chamber may or may not divide; Nuclear type in some taxa	Cellular, chalazal chamber haustorial with a highly polyploid nucleus; Helobial type in *Ariopsis*, micropylar chamber becomes cellular and chalazal chamber remains uninucleate	Cellular, chalazal haustorium in *Spirodela*, chalazal chamber cellular in *Lemna* and *Wolffia* and forms a caecum
Embryo	Caryophyllad (?) type, *cb* may or may not divide, becomes vesicular with an hypertrophied nucleus, suspensor usually small, the epicotyl and cotyledon develop from the same terminal tier, due to overgrowth of the cotyledon the epicotyl is pushed to one side, polyembryony occurs	Caryophyllad (?), *cb* elongates and forms a suspensor, *ca* forms an embryonal mass, epicotyl and cotyledon develop side by side, polyembryony occurs	Asterad (?), basal cell inconspicuous, suspensor short and stocky, *ca* and *cb* both contribute to the embryo, mature embryo lacks radicle, terminal cotyledon cylindrical, plumular apex more or less completely enclosed by hypocotyl and cotyledonary sheath
Seed	Exalbuminous, embryo straight or horseshoe-shaped, exotestal	Operculate, exalbuminous or albuminous, embryo curved or cylindrical, testa multiplicative and parenchymatous	Operculate, exalbuminous or albuminous, embryo straight, thick testa, cuticularized membraneous tegmen

proembryo becomes hypertrophied. A nucellar cap and operculum are also absent. The duckweeds are closely related to the aroids, and Lawalrée's view that they belong to the Helobiales is erroneous.

S.C. Maheshwari and Kapil (1963b), and S.C. and N. Maheshwari (1963) suggest that the evolution of *Lemna* is from Araceae: *Pistia*-like ancestors → *Spirodela* → *Lemna minor* → *L. paucicostata* → other Lemnaceae such as *Wolffia*. The megasporogenesis, and development of endosperm in *Spirodela* indicate a possible link between the aroids and duckweeds (S.C. Maheshwari 1958). The development of the embryo sac is of the monosporic type in Araceae and *Spirodela polyrrhiza*, while in Lemnaceae it is bisporic. In Araceae, however, triads are formed due to suppression of meiosis II in the micropylar dyad.

The chalazal endosperm chamber in *Arisaema*, *Pistia* and *Spirodela* remains undivided, and shows hypertrophied nucleus indicating their alliance. In the first two taxa, the chalazal chamber forms a uninucleate haustorium. A division of the chalazal chamber rarely occurs in *L. minor*, but in *L. paucicostata* and *Wolffia microscopica* the chalazal chamber becomes eight-celled. Thus, both chambers contribute to the endosperm proper.

B.M. Johri (1963b) and P. Maheshwari and Kapil (1966) also discuss the interrelationships of the two families. According to them, not only the mode of development of the cellular endosperm, but also the mode of embryo development provides sufficient evidence of relationship. Both the families have cylindrical, terminal cotyledon, and a vertically directed plumule apex ("frond-apex" in Lemnaceae) which is completely enclosed by the hypocotyl and cotyledonary sheath except for a small pore near the hypocotyl (S.C. Maheshwari and Kapil 1963b).

62 Pandanales

Pandanales comprise three families: Pandanaceae, Sparganiaceae and Typhaceae.

62.1 General Features

Pandanales are characterized by: (1) amoeboid anther tapetum, (2) uniporate or ulcerate pollen grains, two-celled at the dispersal stage, (3) anatropous, bitegmic, crassinucellate ovule, (4) Polygonum type of embryo sac, (5) Nuclear type of endosperm in Pandanaceae, and Helobial in Sparganiaceae and Typhaceae, (6) Onagrad type of embryogeny in Sparganiaceae, and Asterad in Typhaceae, and (7) albuminous seed with a minute embryo.

62.2 Pandanaceae

The embryology has not been fully investigated.

The anther wall comprises the epidermis, fibrous endothecium, one or two ephemeral middle layers, and amoeboid tapetum in *Pandanus parvus* (Cheah and Stone 1975). Davis (1966) mentions secretory tapetum in Pandanaceae, which requires confirmation. The meiotic divisions in microspore mother cells result in the formation of decussate and isobilateral tetrads.

The pollen grain is spinulose, and has a single circular germinal pore. It is two-celled at the dispersal stage.

The ovule is anatropous, bitegmic and crassinucellate. The inner integument forms the micropyle, both integuments are concerned in *Freycinetia funicularis* (Strömberg 1956). The cells of the nucellar epidermis are papillate and may divide periclinally producing a few layers. The hypostase appears to act as a barrier to prevent complete breakdown of nucellar cells in the chalazal region in *P. parvus*. The archesporial cell cuts off a parietal cell, and the parietal tissue is three to four layers thick. In *P. parvus*, the parietal cell as well as its derivatives may rarely divide anticlinally. The megaspore mother cell undergoes meiotic divisions resulting in a linear or T-shaped tetrad, and the chalazal megaspore develops into a Polygonum type of embryo sac. The polar nuclei fuse to form the secondary nucleus, and the three antipodals are small in *Freycinetia*. Three archesporials, triple tetrads, and twin embryo sacs are reported in *F. funicularis*.

In *Pandanus* (Fagerlind 1940c, Cheah and Stone 1975), the nucellar nuclei in the chalazal region have the same appearance as those of the embryo sac. As the developing embryo sac encroaches upon the chalazal nucellar tissue, the intervening walls are ruptured. The somatic nuclei of the nucellar tissue migrate into the embryo sac even at the four-nucleate stage (Fagerlind 1940c). At maturity, the embryo sac contains the egg apparatus, secondary nucleus, primary antipodal cells, and numerous supernumerary nuclei organized as secondary antipodal cells. Nearly 200 secondarily added antipodals have been observed (see Strömberg 1956). In *P. parvus* (Cheah and Stone 1975), they remain in a free state and may fuse with each other, or with the secondary (fused) nuclei to form polyploid masses. Cheah and Stone suggest (without any justification) the possibility that they may contribute to the occurrence of parthenocarpy, and agamospermy.

The perforations in the wall of the embryo sac, in the chalazal region as in *Pandanus*, are totally absent in *F. funicularis* (Strömberg 1956). Therefore, there is no migration of somatic nuclei into the embryo sac.

The development of the endosperm is of the Nuclear type, and it becomes centripetally cellular. The cellular endosperm is fleshy and oily.

The seed is albuminous, and contains a minute organless (undifferentiated) embryo (Hutchinson 1973).

CH Cheah and BC Stone (1975), GL Davis (1966), Th Eckardt (1964), F Fagerlind (1940c), I Harda (1947), J Hutchinson (1973), P Maheshwari (1950), B Strömberg (1956)

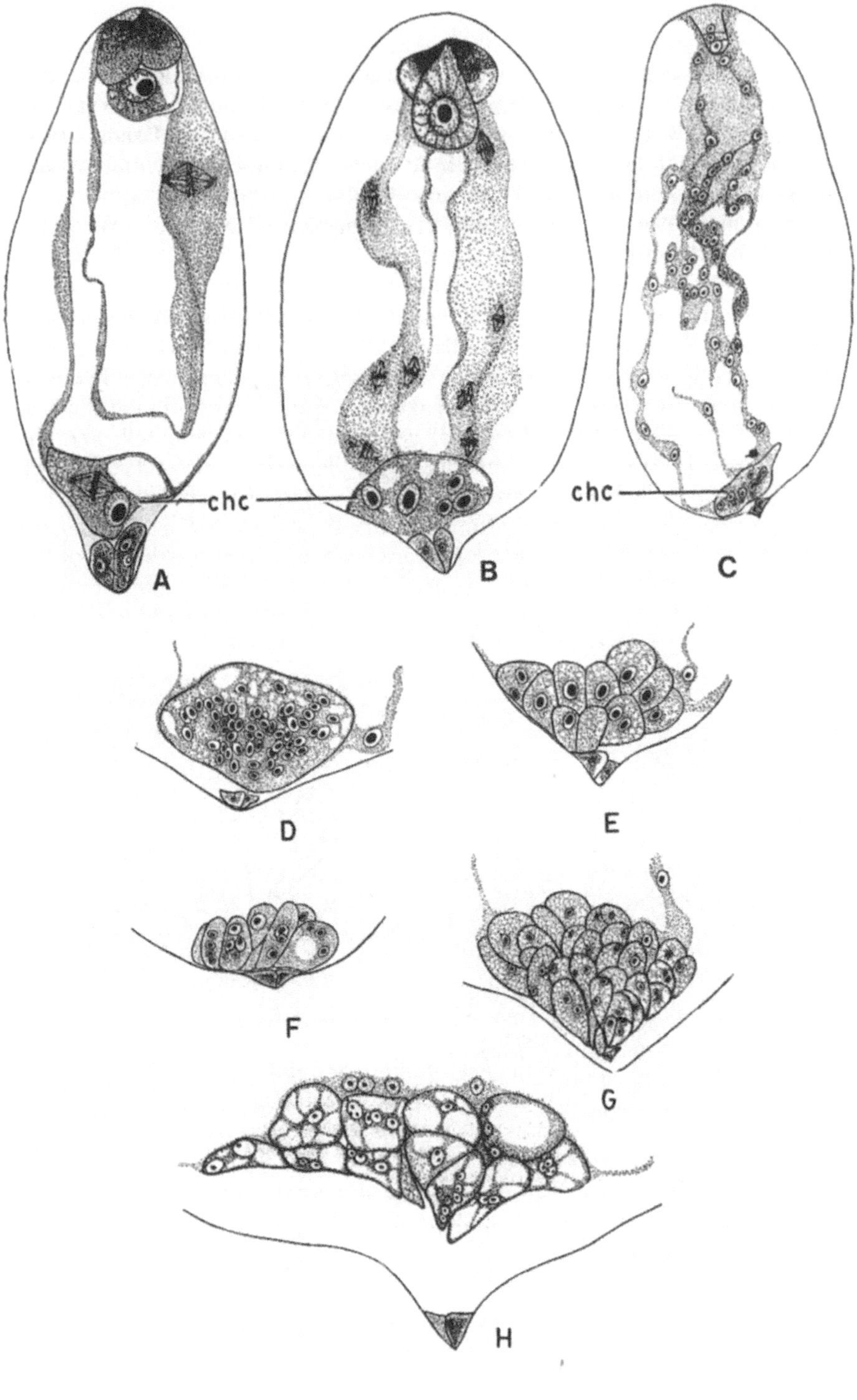
chc
chc
A
B
C
D
E
F
G
H

62.3 Sparganiaceae

The anther wall consists of the epidermis, fibrous endothecium, a middle layer, and an amoeboid tapetum. The tapetum may be irregularly two-layered. Initially, the tapetal cells are uninucleate, but asynchronous divisions of the nuclei take place, and within a single cell seven nuclei of different sizes have been observed. The microspore mother cells undergo meiotic divisions, cytokinesis is successive, and the tetrads are isobilateral, decussate, linear, tetrahedral, or oblique-isobilateral in *Sparganium erectum* (Asplund 1973). In the post-meiotic phase of microspore mother cells, at the separation of microspores, the walls of the tapetal cells disintegrate and a periplasmodium develops.

The pollen grains are spheroidal, ulcerate with a thick sexine, and thick reticulate nexine. The ulcus is faintly delimited in *Sparganium* (Erdtman 1952, see also Asplund 1973). They are two-celled at the dispersal stage.

The ovule is anatropous, bitegmic and crassinucellate with the micropyle formed by both integuments. The sub-epidermal archesporial cell cuts off a parietal cell and the parietal tissue is four to five, or eight, cells thick. Meiotic divisions in the megaspore mother cell result in an unequal dyad, and then a linear tetrad. The chalazal megaspore gives rise to a Polygonum type of embryo sac. The polar nuclei fuse before fertilization, and the secondary nucleus lies near the egg cell. The antipodals persist (Fig. 62.3.1A–H) up to a very late stage of endosperm development (U. Müller-Doblies 1969a, Asplund 1973). There is no secondary increase in the number of antipodals, as previously recorded by Campbell (1899) in *Sparganium eurycarpum*, *S. greenii* and *S. simplex*. Rarely, twin embryo sacs occur in *S. erectum*.

The development of the endosperm is of the Helobial type. The primary endosperm nucleus lies near the antipodals and its division results in a large micropylar and a small chalazal chamber (Fig. 62.3.1A). The divisions in the micropylar chamber are synchronous and free-nuclear (Fig. 63.3.1B, C), in the chalazal chamber also the divisions are free-nuclear (Fig. 62.3.1A–D).

Wall formation in the micropylar chamber does not always precede that in the chalazal chamber (Asplund 1973). The endosperm cells become multinucleate.

The development in the chalazal chamber follows two different courses: (1) Due to free-nuclear divisions a coenocyte with as many as 59 nuclei may be formed (Fig. 62.3.1D). Later, the nuclei disintegrate. (2) Wall formation occurs after the second nuclear division, further divisions are followed by wall

Fig. 62.3.1A–H. Development of endosperm. **A** *Sparganium minimum*, **B, F** *S. glomeratum*, **C–E** *S. simplex*, **G** *S. augustifolium*, **H** *S. erectum*. **A** Embryo sac with zygote and Helobial type of endosperm. The nucleus of micropylar chamber is in division, chalazal chamber (*chc*) binucleate with one nucleus in division; the nuclear division in chalazal chamber precedes that in micropylar chamber. Note antipodals in **A** and **B**. **B** Synchronous divisions in micropylar chamber, chalazal chamber five-nucleate. **C** Three-celled proembryo, multinucleate micropylar chamber, coenocytic chalazal chamber. Remnants of antipodals persist in **C–H**. **D** Large chalazal chamber with ca. 59 nuclei. **E–H** Cellular chalazal endosperm, cells uni-, bi- and multinucleate. (After Asplund 1973) formation, and a hemispherical tissue of 300 to 400 cells develops (Fig. 62.3.1E–H). The cells are vacuolate and multinucleate. U. Müller-Doblies' (1969a) report of two-celled condition in the chalazal chamber seems to be erroneous. In *S. simplex*, the chalazal chamber may be coenocytic (Fig. 62.3.1C, D) or multicelled (Fig. 62.3.1E).

The embryogeny conforms to the Onagrad type (see Davis 1966). The differentiation of epicotyl and the cotyledonary initial occurs at the onset of two vertical divisions in the hemispherical cell *ca* in *S. simplex*. They arise from the same terminal tier *ca*. One of the sectors of quadrant *q* is involved in the organization of the epicotyl, and the other three give rise to the cotyledon (Swamy 1966).

The seed has a thin testa. The embryo is embedded in the centre of the mealy endosperm.

I Asplund (1968, 1973, 1978), DH Campbell (1899), GL Davis (1966), E Erdtman (1952), I Harda (1947), KK Lakshmanan (1972), A and D Löve (1948), U Müller-Doblies (1969a), R Souèges (1924c), BGL Swamy (1966)

62.4 Typhaceae

The anther wall comprises the epidermis, fibrous endothecium, an ephemeral middle layer, and tapetum. In *Typha latifolia* (Asplund 1972), the tapetum forms a periplasmodium at the time of development of microspore tetrads; also reported earlier by Clausen (1927) in the same plant. However, Davis (1966) mentions a glandular type with binucleate cells, which needs confirmation. The microspore mother cells undergo meiotic divisions producing 80% of isobilateral tetrads, the rest are decussate, tetrahedral, linear and T-shaped. Cytokinesis is successive.

The pollen grains with reticulate exine are two-celled at the dispersal stage.

The ovule is anatropous, bitegmic and crassinucellar with micropyle formed by the inner integument. The hypodermal archesporial cell cuts off a parietal cell. The megaspore mother cell undergoes meiotic divisions, and the chalazal megaspore of a linear tetrad develops into a Polygonum type of embryo sac. The polar nuclei fuse before fertilization. The antipodals are ephemeral and do not multiply.

The development of the endosperm is reported to be of the Nuclear type (see Schnarf 1931, Davis 1966) in *Typha latifolia*. However, Asplund's (1972) reinvestigation shows it to be of the Helobial type. Asynchronous divisions occur in the micropylar as well as the chalazal chamber. The divisions in the micropylar chamber are initially free-nuclear, but later multinucleate cells are formed. The development in the chalazal chamber follows two pathways (Asplund 1972):

(1) nuclear divisions are followed by wall formation,
(2) free-nuclear divisions occur giving rise to a coenocyte.

The embryogeny corresponds to the Asterad type (see Davis 1966). A larger volume of shoot-apex takes part in the formation of the cotyledon.

The seed contains a narrow and cylindrical embryo. It has a striate testa, and mealy endosperm.

I Asplund (1968, 1972, 1973, 1978), P Clausen (1927), GL Davis (1966), I Harda (1947), DW Lee and DE Fairbrothers (1972), D Müller-Doblies (1969b), MV Roscoe (1927), N Sahasi and MIkuse (1974), K Schnarf (1931), H von Güttenberg, A Tober, W Dretke and M Ruickholdt (1960)

62.5 Taxonomic Considerations

The three families of the order Pandanales – Pandanaceae, Sparganiaceae and Typhaceae – are usually placed together, in Nudiflorae by Bentham and Hooker (1963c), in Spadiciflorae by Hallier (1912), in Alismatales by Bessey (1915), and in Pandanales by Engler and Prantl (1931). Hutchinson (1973) assigns Typhaceae and Sparganiaceae to Typhales, and Pandanaceae to Pandanales, and considers the latter as an advanced family due to the coalescence of fruits into a syncarp.

Pandanaceae. Pandanaceae are terrestrial plants with woody palm-like habit, and stout aerial prop roots. Apart from the embryological features of the order, the genus *Pandanus* is characterized by the presence of (1) stigmatic channel, (2) glandular epithelial hairs on the funicle, placenta, ovary wall, and obturator, (3) dense nucellar cells at the chalazal end of the embryo sac, (4) hypostase, and (5) migration of somatic nucellar nuclei into the embryo sac (Cheah and Stone 1975).

The increase in the number of "antipodals" (up to 200) in *Pandanus* (Fagerlind 1940b, Strömberg 1956) is due to the migration of nucellar nuclei in the embryo sac. Cheah and Stone (1975) make no mention of the organization of these nuclei (nucellar) in the embryo sac into cells (secondary antipodal cells) in *P. parvus*. No such migration of nuclei has been observed in the other genus *Freycynetia*.

The genera *Freycynetia* (Strömberg 1956) and *Pandanus* (Fagerlind 1940b, Cheah and Stone 1975), however, resemble one another closely in the construction of integuments, papillose nucellar epidermis, hypostase, and anatropous, bitegmic, crassinucellar ovule with Polygonum type of embryo sac. The albuminous seed contains a minute embryo. The typical fruit clearly distinguishes the family from the other two families (Sparganiaceae and Typhaceae).

Sparganiaceae and Typhaceae. The Sparganiaceae and Typhaceae are very closely allied. Fagerlind (1940b), von Güttenberg et al. (1960), D. and U. Müller-Doblies (1969b), Asplund (1973) and Hutchinson (1973) suggest their removal from the order Pandanales. This view is supported by morphological, palynological, cytological, embryological, and serological data (Lee and Fairbrothers 1972, see Asplund 1973).

Asplund (1968, 1973) describes the embryology of *Sparganium* spp. and compares it with that of *Typha latifolia* (Asplund 1972). The development of the anther is similar in both families. The amoeboid anther tapetum with multinucleate cells, asynchronous divisions of pollen mother cells, several types of microspore tetrads, and spheroidal ulcerate two-celled pollen grains occur in both. These pollen grains differ very much from the spheroidal, spinuliferous pollen grains of *Pandanus*, and ellipsoidal, psilate pollen grains with sulcoid apertures of *Freycinetia*. *Typha* (Roscoe 1927, Clausen 1927, Harda 1947) and *Sparganium* (A. and D. Löve 1948) have the same chromosome number ($n = 15$).

The anatropous, bitegmic, crassinucellate ovule with Polygonum type of embryo sac with three antipodal cells, and Helobial type of endosperm are common to both families. Moreover, in the members investigated so far, the development of the endosperm in the chalazal chamber follows two different patterns in both the families (Asplund 1972, 1973). Asynchronous nuclear divisions occur in both; the difference is only in the number of free-nuclei in the coenocytic cell, or in the cells of the hemispherical mass.

Sparganiaceae and Typhaceae differ significantly from Pandanaceae in morphology, palynology, cytology, and embryology. The papillose nucellar epidermis, hypostase, migration of nucellar nuclei in the embryo sac in *Pandanus*, and Nuclear type of endosperm segregate Pandanaceae from the other two families.

After critically evaluating the morphological, palynological, cytological and embryological data within Pandanales, Asplund (1973) emphasizes that Sparganiaceae and Typhaceae are to be removed from the order Pandanales. These two families together form the order Typhales with a single family Typhaceae, and two sub-families, Typhoideae and Sparganioideae. We do not support the subfamily level; the family status is justified, under the order Typhales.

63 Cyperales

63.1 General Features

Cyperaceae represent the order Cyperales, and exhibit the following features: (1) secretory anther tapetum, (2) simultaneous cytokinesis in pollen mother cells, (3) three-celled, monocolpate pollen grain as pseudomonad, (4) anatropous, bitegmic, crassinucellar ovule with inner integument forming the micropyle, (5) cutinized hypostase, (6) Polygonum type of embryo sac with ephemeral antipodals, (7) Nuclear type of endosperm, (8) Onagrad type of embryogeny, and (9) albuminous seed.

63.2 Cyperaceae

The anther wall comprises persistent tanniniferous epidermis, fibrous endothecium, one or two middle layers, and glandular tapetum with uninucleate cells in *Cyperus alopecuroides* (Nagaraj and Nijalingappa 1973). The endothecial cells have spiral lignified thickenings (Makde 1981). The tapetal cells may be binucleate as in *Cyperus iria* (Padhye 1971a), *C. tegum* (Padhye and Moharir 1958), *Kyllinga brevifolia* (Padhye 1971b) and *K. triceps* (Padhye 1960), or multinucleate. The microspore mother cell undergoes meiotic divisions and forms four nuclei of which three occupy wallward position, and the fourth lies in the centre (Fig. 63.2.1A–D). Cytokinesis is very much delayed and is of the "simultaneous" type. The four nuclei of the tetrad are distinctly separated by walls (Fig. 63.2.1E). The three effete microspores degenerate in situ, while the larger one develops into a pollen grain. The non-functional spore nuclei may divide (Fig. 62.2.1F) (P. Khanna 1965b), but they finally degenerate.

The pollen grain is a pseudomonad (Fig. 63.2.1G) showing three degenerated microspores. It is monocolpate with an ulceroid aperture on the side opposite to the three degenerated nuclei. They are three-celled at the dispersal stage (Fig. 63.2.1H). Two-celled condition is reported in *Fimbrystylis dichotoma* (M.N. Gupta 1962) which requires confirmation. Occasionally, double pollen grains, or all the four nuclei in healthy condition, have been observed (Shah 1962, see also Nijalingappa 1976). It would have been useful if the percentage of normal and abnormal pollen grains, in the same pollen sac, had been determined.

The ovule is anatropous, bitegmic and crassinucellar with the inner integument forming the micropyle. Air chambers occur on the funicular side in *Cyperus alternifolius* (Untawale and Makde 1971). The hypostase is cutinized. The archesporial cell cuts off a parietal cell, and the parietal tissue is four layers

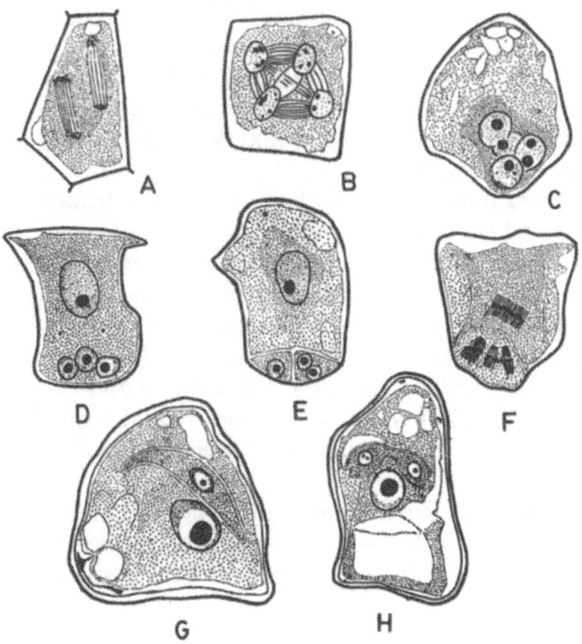

Fig. 63.2.1A–H. *Carex wallichiana*, development of pollen. **A, B** Meiosis II in microspore mother cell. **C** Microspore mother cell with four identical nuclei. **D** One functional (central) and three non-functional (basal) nuclei. **E** Delimitation of non-functional nuclei. **F** All the four nuclei in metaphase. **G** Two-celled pollen grain. **H** Three-celled condition (pseudomonad). (After Shah 1962)

thick. The archesporium is three-celled in *Pycreus nitens*. The megaspore mother cell undergoes meiotic divisions resulting in a linear or T-shaped tetrad. The chalazal megaspore develops into a Polygonum type of embryo sac. The synergids are hooked, and polar nuclei fuse before fertilization. The antipodals are ephemeral. A funicular obturator develops in *Cyperus* and *Kyllinga* (P. Khanna 1963, 1965b, Padhye 1960, 1968, 1971a, b, Makde and Bhuskute 1984, 1987).

The development of the endosperm is of the Nuclear type (Fig. 63.2.2A–D), the first few divisions are synchronous. Centripetal wall formation (Fig. 63.2.2E, F) commences after the formation of about 120 nuclei in *Lipocarpha sphacelata* (Nagaraj et al. 1976), but after only eight nuclei in *Diplacrum caricinum* (Nijalingappa and Devaki 1978). The entire endosperm becomes cellular in *Cyperus hyalinus* (Fig. 63.2.2G) which is unusual. In *Scleria foliosa* (Nijalingappa and Devaki 1979, Nijalingappa 1986), the binucleate endosperm develops a tubular micropylar extension which behaves as an enucleate haustorium. Centripetal wall formation occurs in the upper one-third portion of the embryo sac, whereas the chalazal coenocytic portion elongates up to the base of nucellus and forms a haustorium. Two- to three-nucleate nodules develop in the coenocytic portion. The central core of the endosperm disintegrates, and a cavity is produced which persists in the seed.

Cyperus pseudokyllingioides shows a micropylar tubular "caecum" at the free-nucleate stage of the endosperm (Fig. 63.2.3A–C). It grows towards the ovarian tissue invading the placenta, and eventually reaches as far as the outer epidermis of the pericarp. Along with the rest of the endosperm, this extension becomes cellular. The abnormal micropylar and chalazal endosperm caeca, as reported by the above investigators in 5% of the ovules, need to be further examined.

The embryogeny conforms to the Onagrad type-Juncus variation (Padhye and Makde 1982, Makde 1983, Makde and Bhuskute 1987). P. Khanna (1965b) reports Asterad type in *Scirpus mucronatus*. On reinvestigation, Nijalingappa and Tejavathi (1977) confirm it to be of the Onagrad type-Juncus variation. Similar embryogeny is also reported in *Cyperus alternifolius*, *Mariscus paniceus*, *Pycreus pumilus* and *Remirea maritima* – Period I, Series A2, Megarchetype IV (Padhye and Makde 1982).

In *Diplacrum caricinum* (Nijalingappa and Devaki 1978), the zygote divides transversely to form the cells *ca* and *cb*. Two crosswise vertical divisions occur in *ca* to produce *q*, whereas *cb* divides transversely and gives rise to *m* and *ci*. In the formation of an octant the protoderm cells are separated from axial cells by periclinal divisions. One-half of terminal tier is "quiescent" and grows at a slower rate, and ultimately gives rise to the epicotylary region. The other half – the initial of the cotyledonary region – grows at a faster rate and ensheaths the epicotyledonary region. The middle cell *m* gives rise to hypocotyl and the radicle; the cell *ci* gives rise to a four- or five-celled suspensor which degenerates.

The development of the embryo in *Cyperus* is shown in Figure 63.2.4A–L.

During the maturation of embryo (Fig. 63.2.4H, I), due to unequal growth the opening of the cotyledonary sheath shifts from lateral to basal position,

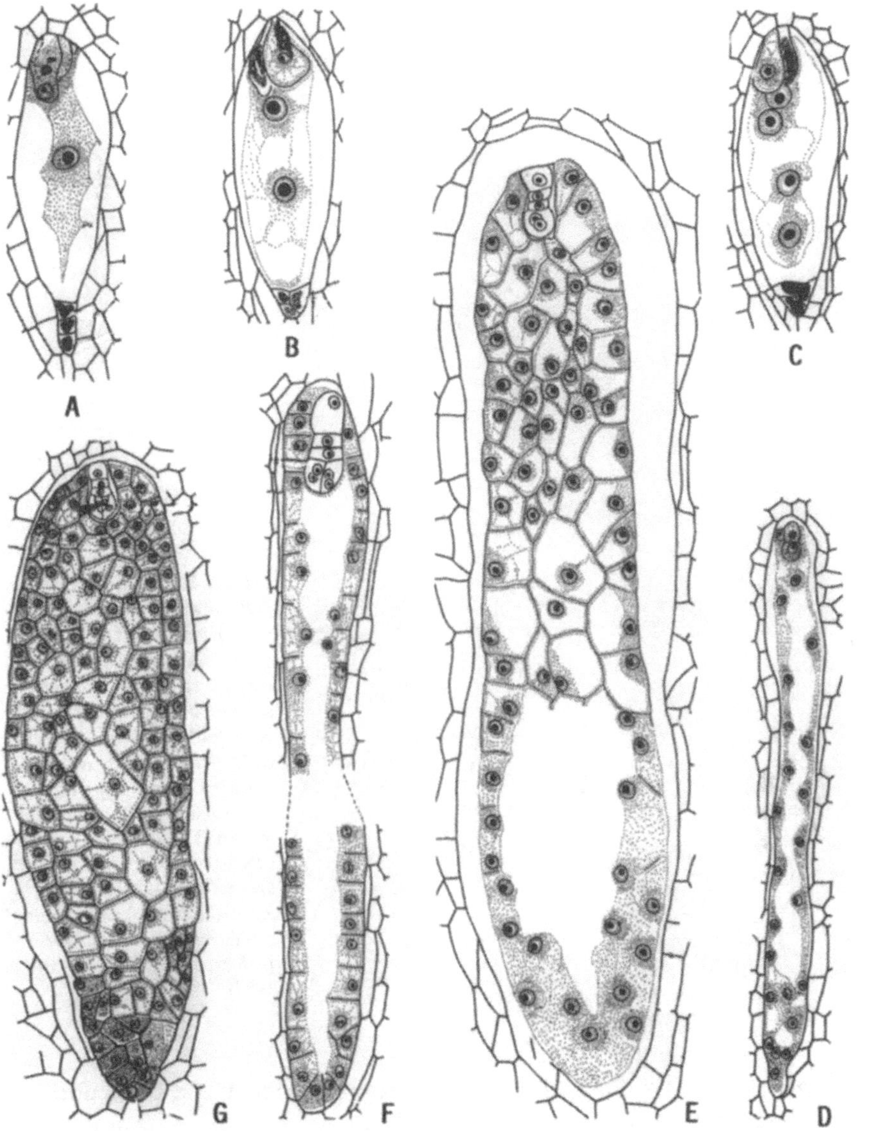

Fig. 63.2.2A–G. Development of endosperm. **A** *Cyperus pachyrrhizus*, **B–E**, **G** *C. hyalinus*, **F** *C. pseudokyllingioides*. **A** Fertilized embryo sac. **B–D** Development of free-nuclear endosperm, two-celled proembryo in **D**, note remnants of pollen tube in **B** and **C**, and antipodals in **A–C**. **E** Wall formation in endosperm from micropylar region downward, lower part free-nuclear; proembryo in **E–G**. **F** Centripetal wall formation in endosperm. **G** Cellular endosperm. (After Tejavathi 1987)

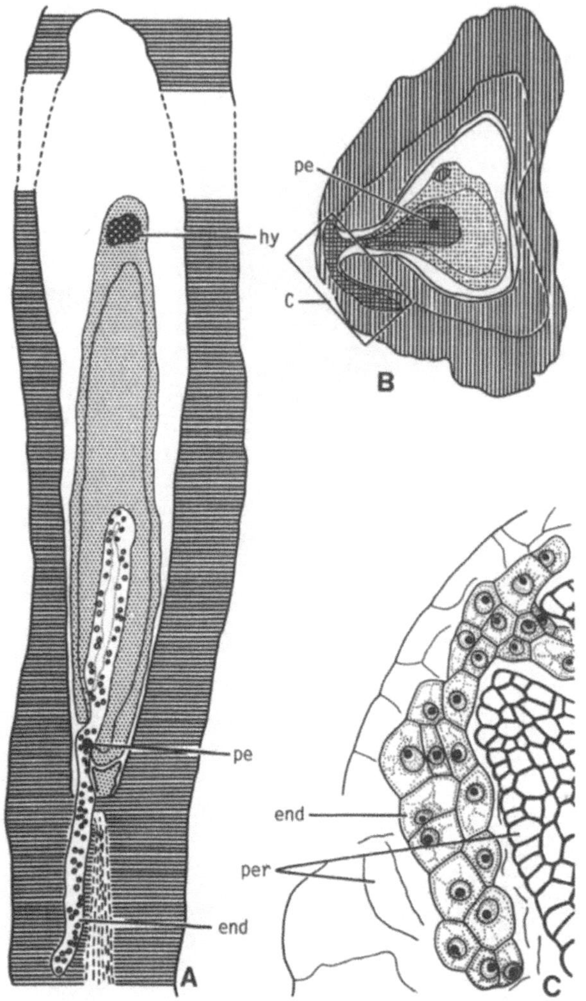

Fig. 63. 2. 3 A–C. *Cyperus pseudokyllingioides*, abnormal development of endosperm. **A** Longisection of ovary (*pericarp hatched*) shows extraovular micropylar extension of free-nuclear endosperm (*end*) into the pericarp. Ovule *stippled*; *hy* hypostase; *pe* proembryo (*black*). **B** Transection of ovary, extension of the endosperm (*cross-hatched*) in pericarp (*hatched*) has become cellular. **C** *Inset C* in **B**; *per* pericarp. (After Tejavathi 1987)

while the radicle is shifted from basal to lateral position; the cotyledon is terminal (see also Makde and Bhuskute 1987).

Shah (1965) classifies the mature embryo into four types, according to the disposition of organs: (1) Carex type – The plumule is lateral and radicle basal. (2) Cyperus type – The plumule is basal and radicle lateral. (3) Bulbostylis type – Both plumule and radicle are basal. (4) Shoenus type – Both plumule and radicle are lateral. In all these cases, the cotyledon is terminal.

Both the integuments form the seed-coat (Fig. 63.2.5A–I). The outer epidermis of the testa and the inner epidermis of the tegmen develop thickenings on their inner tangential walls and contain a tannin-like amorphous mass. The inner epidermis of the testa and the outer epidermis of the tegmen degenerate. The pericarp becomes hard due to the thickened outer and inner epidermal

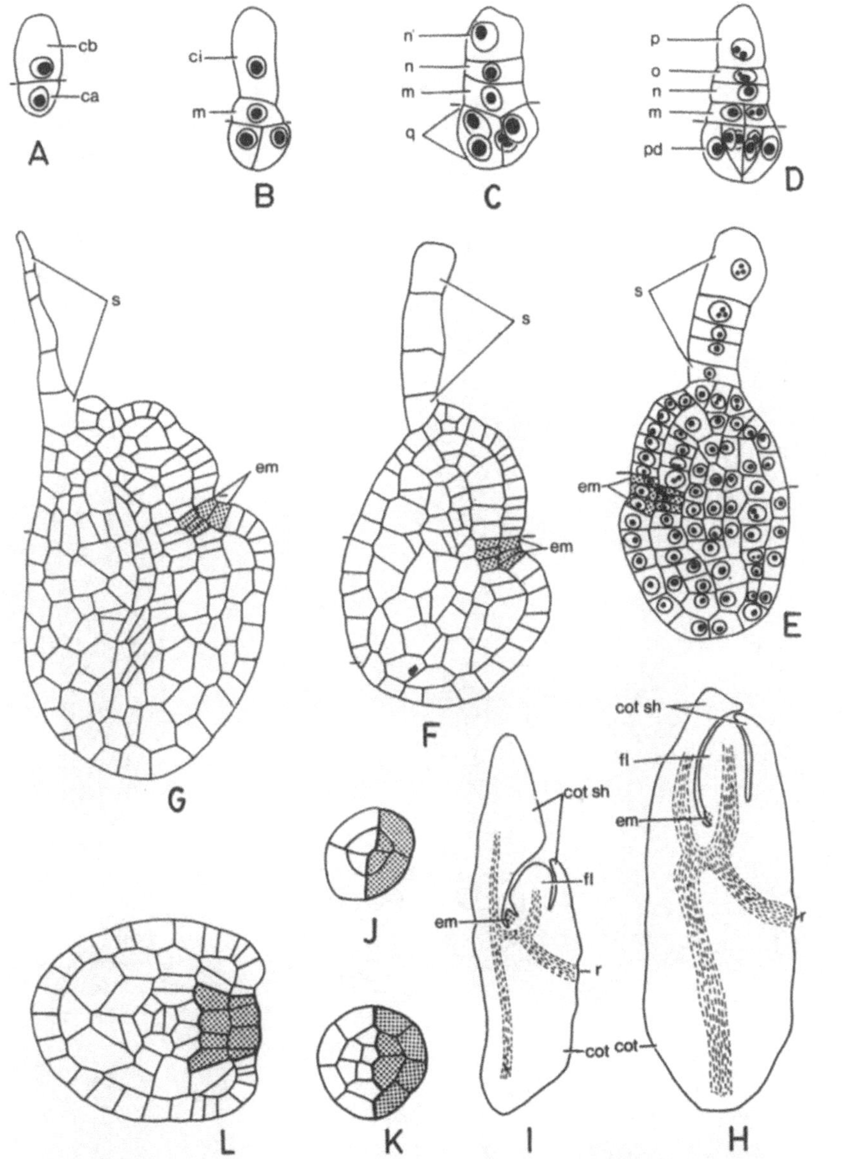

Fig. 63.2.4A–L. Development of embryo. **A–C, F, G, L** *Cyperus laevigatus,* **D, E, I–K** *C. pseudokyllingioides,* **H** *Ascopholis gamblei.* **A** Bi-celled proembryo. **B.** Proembryonal tetrad. **C** Quadrant (*q*). **D** Demarcation of protoderm initials (*pd*). **E** Globular proembryo, epicotyledonary meristem (*em*) *stippled*; *s* suspensor. **F, G** Later stages. **H, I** Mature embryo, longisection; *cot sh* cotyledonary sheath; *fl* first leaf; *cot* cotyledon; *r* radicle. **J–L** Transection of proembryos; epicotyledonary sector *stippled*, cotyledonary sector *unstippled*. (After Tejavathi 1987)

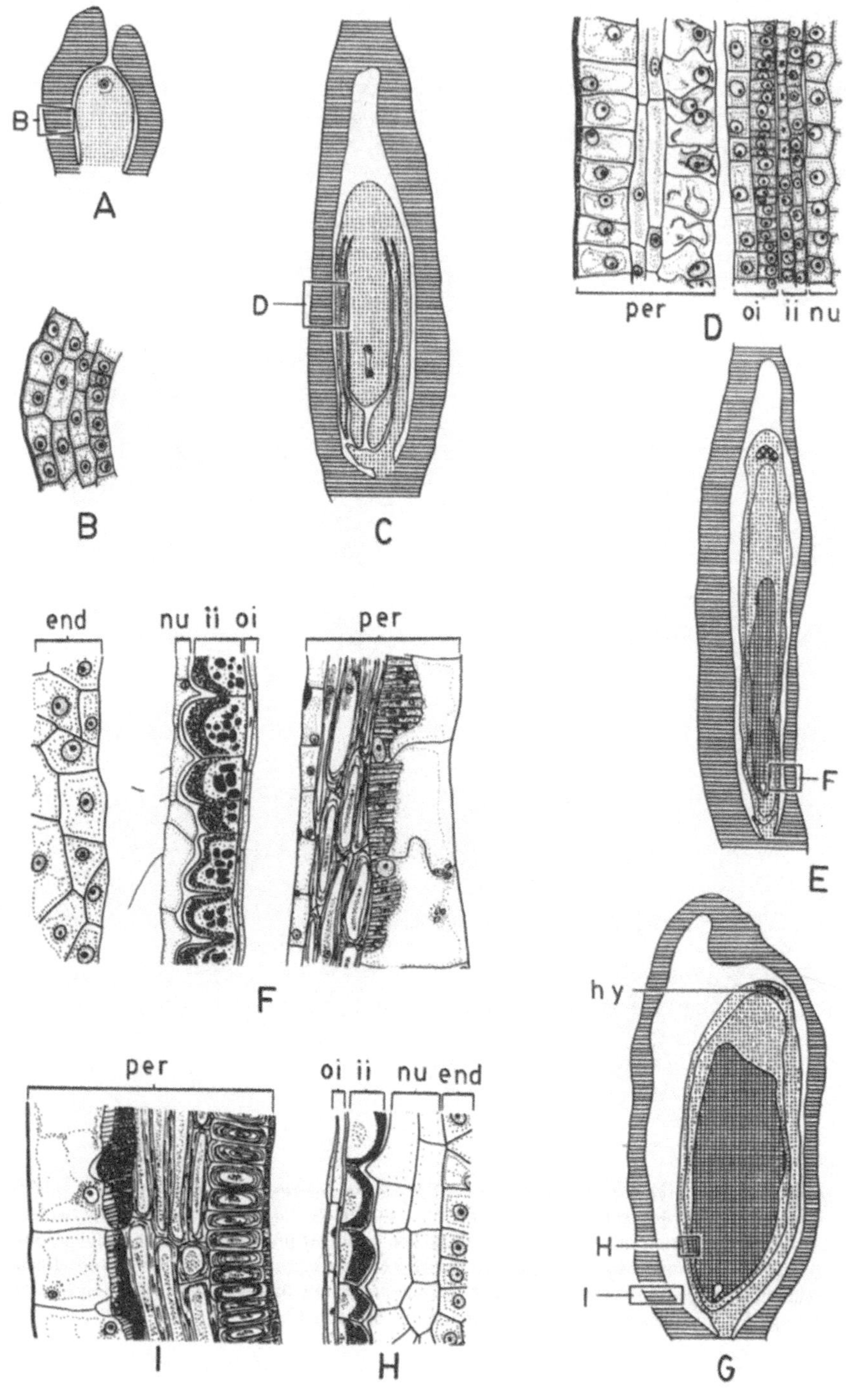
B
A
D
C
per
D
oi
ii
nu
B
end
nu
ii
oi
per
F
E
F
hy
per
oi
ii
nu
end
H
I
I
H
G

layers, and fibrous middle layers (Tejavathi 1987). The testa and pericarp remain free (Tiwari 1968).

PK Deshpande and AG Untawale (1972), VR Dnyansagar and DK Tiwari (1956), A Dunbar (1973), MN Gupta (1962), M Juguet (1966b, 1969b), P Khanna (1963, 1965b), X Kostrioukoff (1930), KH Makde (1981, 1982, 1983), KH Makde and SM Bhuskute (1984, 1987), YS Murty and V Kumar (1967), M Nagaraj and BHM Nijalingappa (1973), M Nagaraj, BHM Nijalingappa and DH Tejavathi (1976), BHM Nijalingappa (1976, 1977, 1986), BHM Nijalingappa and N Devaki (1978, 1979), BHM Nijalingappa and M Nagaraj (1980), BHM Nijalingappa and DH Tejavathi (1977, 1983), MD Padhye (1959, 1960, 1968, 1971a, b), MD Padhye and KH Makde (1982), MD Padhye and SK Moharir (1958), MD Padhye, SN Chaube and AN Iyer (1970), CM Patel and CK Shah (1962), M Schneider (1932), CK Shah (1962, 1964, 1965, 1967, 1968, 1972), CK Shah and N Neelakandan (1971), N Tanaka (1950), DH Tejavathi (1987), DK Tiwari (1968, 1978), AG Untawale and KH Makde (1971, 1984), P van der Veken (1965)

63.3 Taxonomic Considerations

Cyperaceae. Cyperaceae and Gramineae are grouped together in Glumaceae by Bentham and Hooker (1963c), and in Glumiflorae by Engler and Prantl (1931), and Engler and Diels (1936). Hutchinson (1973) assigns these families to two distinct orders, Cyperaceae to Cyperales and Gramineae to Graminales, and considers the latter family as more advanced. He derives both the families from liliaceous ancestors through Juncaceae.

The embryological features of Cyperaceae, in relation to Juncaceae, have already been discussed (Sect. 55.4). The pollen grains in permanent tetrads, and Helobial type of endosperm, segregate Juncaceae from Cyperaceae.

Cyperaceae, even though always aligned with Gramineae, differ from them in several embryological features: simultaneous cytokinesis in pollen mother cells, anatropous ovule, and ephemeral antipodals in contrast with the successive cytokinesis in pollen mother cells, campylotropous, tenui-, crassi- or pseudocrassinucellar ovule, and hypertrophied persistent antipodals which often proliferate in the Gramineae. The formation of pseudomonad in Cyperaceae has no parallel in angiosperms except in Epacridaceae. The peripheral cells of the cellular endosperm never exhibit meristematic activity, and the aleurone layer is absent. The embryogeny in Cyperaceae and Gramineae also differs. The pericarp and seed-coat lie separately in Cyperaceae, in contrast to their position in the caryopsis of Gramineae.

Embryological data support Melchior's classification, and an independent order Cyperales is justified.

Fig. 63.2.5A–I. Development of testa and pericarp. **A, B** *Cyperus laevigatus*, **C, D** *Ascopholis gambeli*, **E, F** *C. pseudokyllingioides*, **G–I** *C. hyalinus*. **A, C, E, G** Longisections of ovary; pericarp *hatched*, ovule stippled. **A** Ovule at archesporial cell. **B** *Inset B* from **A**. **C** Ovule at mature embryo sac. **D** *Inset D* from **C**. *per* pericarp; *oi, ii* outer and inner integument; *nu* nucellus. **E, G** Ovule at globular proembryo, endosperm *cross-hatched*; *hy* hypostase. **F** *Inset F* in **E**; *end* endosperm. **H, I** *Insets H* and *I* in **G**. (After Tejavathi 1987)

64 Scitamineae

The order Scitamineae forms a natural group of related families: Musaceae, Zingiberaceae, Cannaceae, Marantaceae and Lowiaceae.

64.1 General Features

The chief embryological characters are: (1) secretory as well as amoeboid anther tapetum in Zingiberaceae, amoeboid in Cannaceae, (2) successive cytokinesis in pollen mother cells, (3) inaperturate pollen grains, (4) anatropous, bitegmic crassinucellar ovule, (5) radial elongation of nucellar epidermal cells, or formation of a "nucellar pad", (6) Polygonum type of embryo sac with ephemeral antipodals, secondary multiplication of antipodals in Marantaceae (*Calathea*), (7) Nuclear type of endosperm, Helobial in Zingiberaceae, and (8) perispermous, albuminous seed.

64.2 Musaceae

The embryology has not been studied in sufficient detail.

Meiotic divisions in the microspore mother cells are followed by successive cytokinesis.

The pollen grains are three-celled at the dispersal stage. They are inaperturate, and sometimes uniporate.

The ovule is anatropous, bitegmic and crassinucellar. The micropyle is formed by both the integuments. The nucellar epidermal cells are radially elongated in *Strelitzia reginae* (Mauritzon 1936e). The archesporial cell cuts off a parietal cell and the parietal tissue is two or three layers thick. The megaspore mother cell undergoes meiotic divisions, the tetrads are linear, T-shaped or ⊥-shaped, the chalazal megaspore develops into a Polygonum type of embryo sac, the polar nuclei fuse at the time of fertilization, and the antipodals are ephemeral. In *Musa* var. IR 53, Dodds (1945) reports a "compound" embryo sac with two egg apparatus and four polar nuclei; another contained two egg apparatus, four polar nuclei, and seven nuclei at the chalazal end.

The development of the endosperm is of the Nuclear type; cytoplasmic nucleate nodules are formed in *M. errans* (Juliano and Alcala 1933). Later, the tissue becomes cellular.

The seed contains a straight embryo and copious endosperm; sterility is very high. An aril develops from the funicle and exostome in *Strelitzia*. A functional operculum has a well-developed sclerotic layer (see Grootjen and Bouman 1981b). The testa is thick.

J Bouharmont (1963), GL Davis (1966), KS Dodds (1945), CJ Grootjen and F Bouman (1981b), JB Juliano and PE Alcala (1933), E Kronestedt and P-A Bystedt (1981), EC Kronestedt-Robards and

JR Rowley (1989), J Mauritzon (1936e), MW McGahan (1961), K Schnarf (1931), K Shepherd (1954), MW Simmonds (1961)

64.3 Zingiberaceae

The anther wall comprises the epidermis, fibrous endothecium, four or five middle layers, and secretory tapetum of binucleate cells in *Costus speciosus* (Panchaksharappa 1962a). Fibrous thickenings also develop in the outer middle layers as well as in the epidermis. The tapetum is amoeboid in *Amomum dealbatum* (Sachar and Arora 1963) and *Nicolaia atropurpuria* (Boehm 1931), and forms a periplasmodium during dyad and tetrad stage in *Amomum*. Successive cytokinesis follows meiotic divisions in microspore mother cells. The tetrads are isobilateral and decussate; linear or T-shaped tetrads also occur in *Alpinia* (T.S. Raghavan and Venkatasubban 1941b). Extranuclear bodies are present in the sporogenous tissue, dyads, tetrads and anther wall in *Costus, Curcuma, Elettaria* and *Hitchenia* (Panchaksharappa 1962a, b, 1966). Raphides occur in the sporogenous tissue of *Elettaria*.

The pollen grains are non-aperturate. They are uniporate in *Zingiber*, and show a single furrow in *Costus*. The exine is variously sculptured. It is thick and spinescent in *Costus*, and thin in other members. The pollen grains are two-celled at the dispersal stage.

The ovule is anatropous (Fig. 64.3.1A), rarely orthotropous (*Hitchenia*), bitegmic and crassinucellar. Both the integuments are dermally initiated in *Costus cuspidatus* (Grootjen and Bouman 1981b). The inner integument forms the micropyle. The cells of the nucellar epidermis, below the micropyle, are richly cytoplasmic and elongate to form a characteristic "nucellar pad" in *Costus* (Figs. 64.3.1A, 64.3.2A) and *Hedychium*. A nucellar cap is produced in *Hitchenia* and *Zingiber* (Panchaksharappa 1962b, 1966). An epistase occurs in some members. A hypostase develops in *Costus igneus* (Mauritzon 1936e) at the functional megaspore stage. A postament has been observed in *Costus, Elettaria* and *Zingiber* (Panchaksharappa 1962b).

The archesporial cell cuts off a parietal cell which divides once or twice. It divides only anticlinally to form a single layer in *Amomum* (Sachar and Arora 1963) and *Costus* (Panchaksharappa 1962a, b). The megaspore mother cell undergoes meiotic divisions, the tetrad is either linear or T-shaped, and the chalazal megaspore develops into a Polygonum type of embryo sac. The synergids are hooked in *Costus* (Panchaksharappa 1962a) and *Hedychium acuminatum* (Sachar and Arora 1963). The polar nuclei fuse before fertilization. The antipodals are ephemeral, they do not organize into cells in *Elettaria*. Nucellar nuclei migrate into the embryo sac through the perforated intervening walls in *Costus* and *Hedychium* (see P. Maheshwari 1950). Sachar and Arora (1963) also report migration of nucellar nuclei into the embryo sac, and twin nucelli and twin embryo sacs in a common integument.

The pollen grains germinate on the stigma, except in *Elettaria* where they enter the stylar canal. The structure of stigma and style at pollination is shown in Figure 64.3.1B–G.

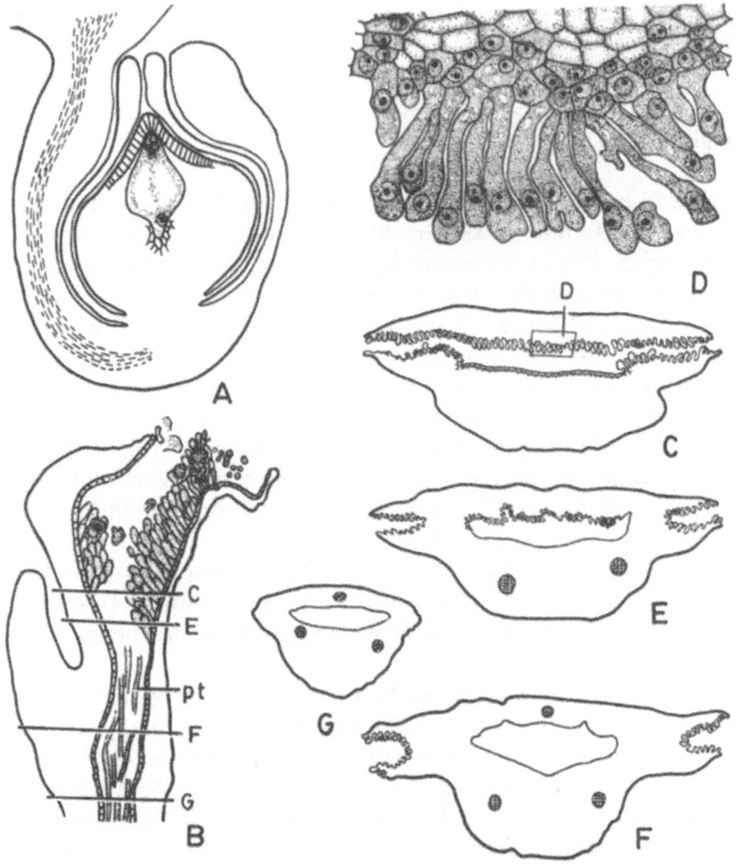

Fig. 64.3.1A–G. *Costus speciosus.* **A** Longisection of ovule at mature embryo sac, with prominent epistase and hypostase. **B** Longisection of glandular, bilipped stigma with bits of pollen tubes (*pt*) in stylar canal. **C** Transection of style at level *C* marked in **B**. **D** *Inset D* from **C**, glandular tissue lining the "mouth" of stylar canal. **E–G** Transections of style at levels marked in **B**, pollen tubes not shown in **F**, **G**. (After Panchaksharappa 1962a)

The development of the endosperm is of the Helobial type (Fig. 64.3.2B, C). The micropylar chamber undergoes free-nuclear divisions and wall formation occurs at the globular proembryo. The behaviour of the chalazal chamber is variable. Usually, it becomes binucleate and then degenerates. In *Costus*, it forms a multinucleate basal apparatus and the nuclei become hypertrophied (Fig. 64.3.2D–H). In *Hedychium*, the basal apparatus may remain uninucleate, or become binucleate, or a cell wall may separate a lower uninucleate cell from the upper cell which becomes multinucleate. It remains healthy up to the globular proembryo (Sachar and Arora 1963). Autonomous development of endosperm occurs in *Hedychium* (Harling 1949); the egg apparatus degenerates. The endosperm stores starch but in *Elettaria* it also contains calcium

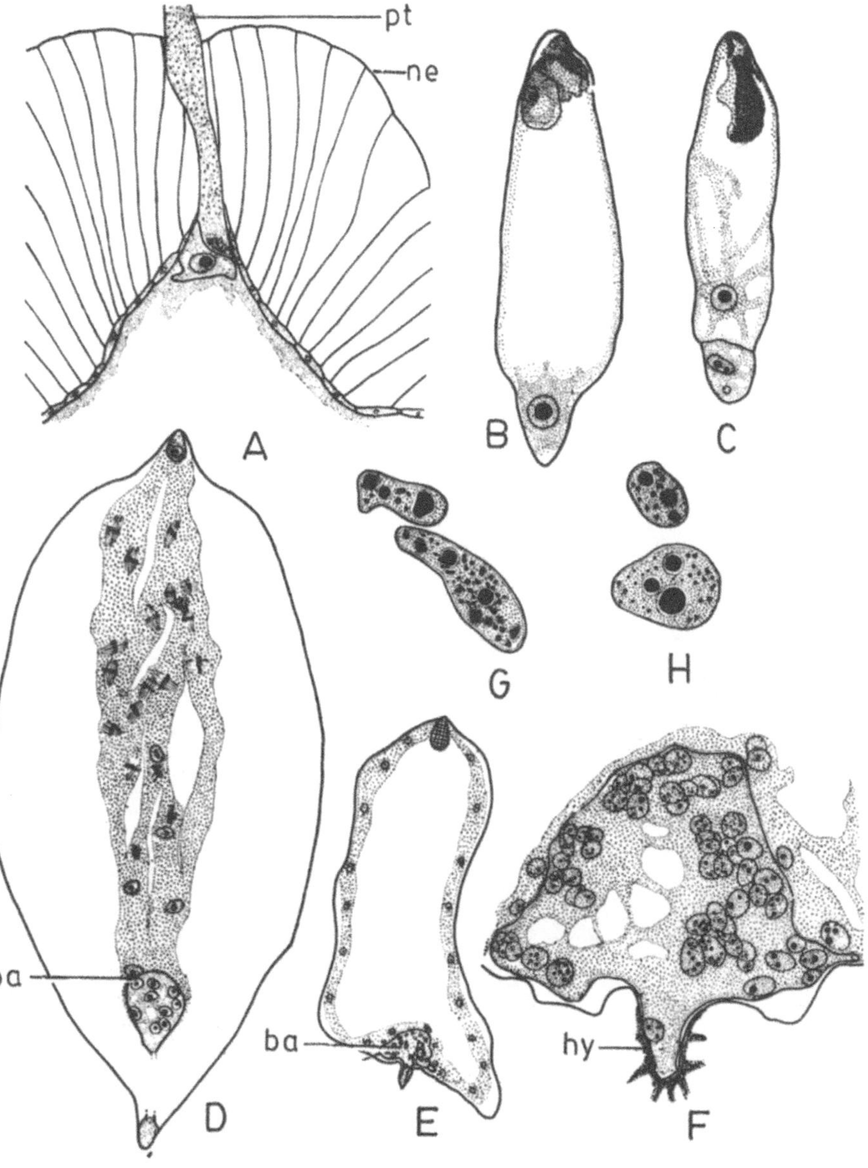

Fig. 64.3.2A–H. *Costus speciosus.* **A** Longisection of upper part of nucellus shows zygote, remnants of pollen tube (*pt*), and radially elongate cells of nucellar epidermis (*ne*). **B** Embryo sac with zygote and primary endosperm nucleus. **C** Division of primary endosperm nucleus followed by wall formation. **D** Zygote, nuclear endosperm, and coenocytic basal apparatus (*ba*). The development of the endosperm is of the Helobial type. **E** Globular proembryo, free-nuclear endosperm, and multinucleate basal apparatus. **F** Basal apparatus from **E**; *hy* hypostase. **G, H** Hypertrophied nuclei from basal apparatus. (After Panchaksharappa 1962a)

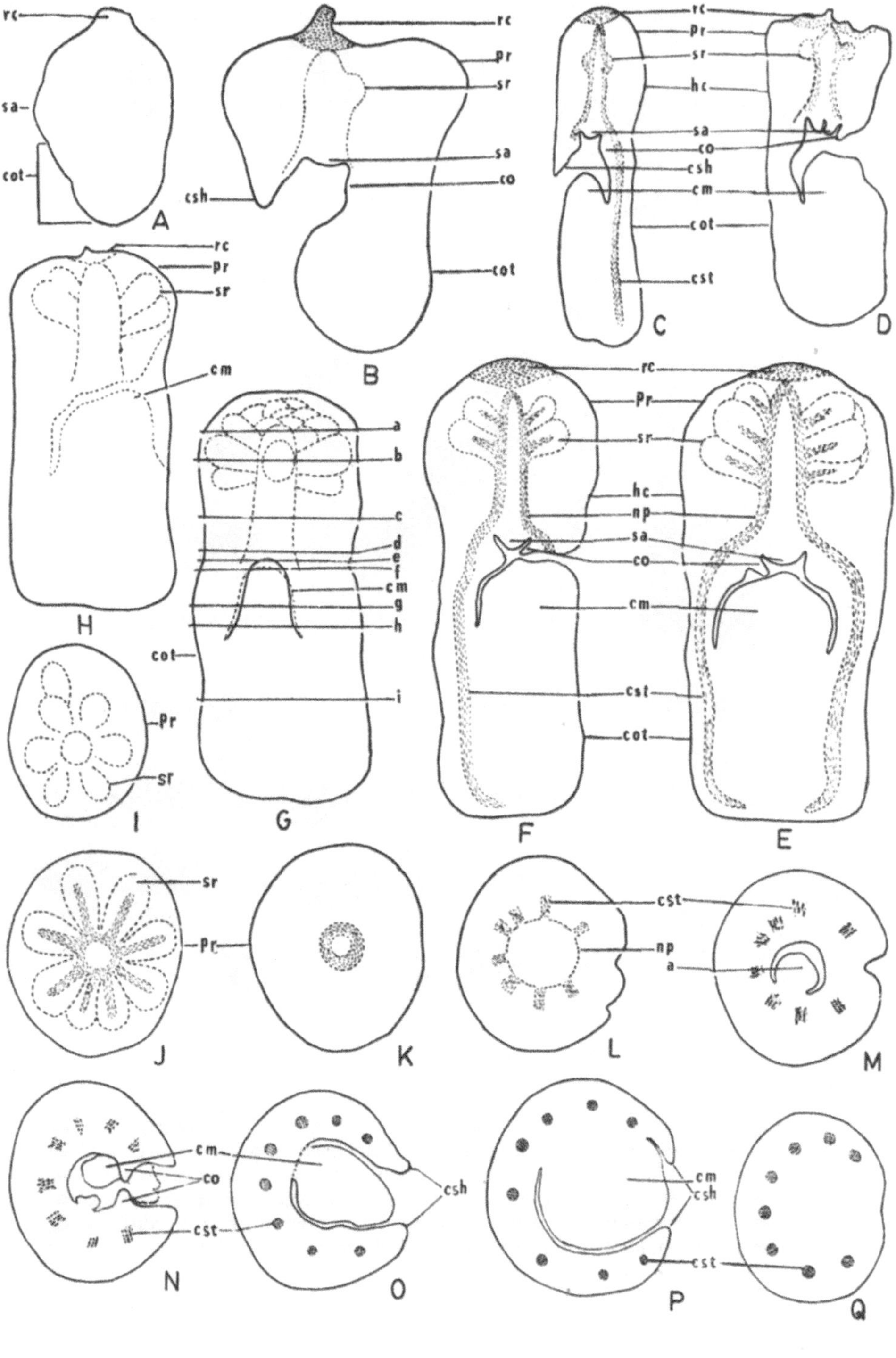
rc
sa
cot
pr
sr
csh
co
hc
cm
cst
np
a
b
c
d
e
f
g
h
i
A
B
C
D
E
F
G
H
I
J
K
L
M
N
O
P
Q

oxalate crystals (Berger 1958), and there is only oil in *Costus* (Panchaksharappa 1962a).

The embryogeny conforms to the Asterad type. A transverse division in the zygote is followed by a vertical division in each cell, *ca* and *cb*, in *Elettaria cardamomum* (Philip 1968). Both the cells contribute to the formation of embryo proper; the suspensor is absent. With the initiation of a vertical wall in *ca* two adjacent cells are produced, one of these gives rise to epicotyl and the other to cotyledon. In *Costus* (Panchaksharappa 1962a), the embryogeny conforms to Caryophyllad type; the development of the embryo up to maturity is shown in Figure 64.3.3A–Q. Synergid polyembryony occurs in *Hedychium acuminatum* (Sachar and Arora 1963).

The endotestal seed with mealy endosperm and perisperm is characterized by such structures as aril, collar, lid, and chalazal tissue (Fig. 64.3.4A–H). The aril is an outgrowth of the funicle and tip of the outer integument. It grows around the seed, and forms a sheath or a veil-like covering in *Elettaria* (Panchaksharappa 1962b, 1966) and *Hedychium*. It is sweet-smelling and fringed in *Zingiber*, and lobed in *Hitchenia*. A caruncle develops in *Costus*. A collar develops as a sclerotic ring from the outer integument due to the meristematic activity of its cells around the micropylar end of the embryo sac, and is protective in function. From the outer integument develops a lid in *Costus* (Panchaksharappa 1962a), *Amomum* and *Elettaria* (Panchaksharappa 1962b, 1966, Berger 1958). This lid is a patch of sclerenchymatous tissue around the micropyle (away from the inner epidermis) in *Costus*; in *Amomum* and *Elettaria* it is represented by the thickened inner epidermis. A multilayered chalazal tissue develops at the base of the nucellus (Panchaksharappa 1970). It contains dense cytoplasm and oil bodies in *Costus*.

In *Costus cuspidatus* (Grootjen and Bouman 1981b), the operculum is functional and consists of epistase (remnants of nucellar pad), compressed sclerotic layers of endostome, and the parenchymatous part of exostome. The parenchymatous part partially originates from the outer integument and from the raphe and has a dual origin – dermal and sub-dermal.

The outer epidermis of the testa comprises thick-walled, elongated cells. The mesophyll cells are slightly thickened with pitted walls. The inner epidermis is a mechanical layer; a dark-coloured granular substance is deposited along the inner tangential and radial walls and in the lumen. Silica is also deposited (Grootjen and Bouman 1981b). In *Aframomum*, *Amomum* and *Elettaria*, the inner epidermis contains only silica (Berger 1958). The tegmen is papillose and multiplicative along the endostome, otherwise its cells become stretched.

Fig. 64.3.3A–Q. *Costus speciosus*, development of embryo. **A, B** Longisection of younger embryos; *rc* root cap; *sa* shoot apex; *cot* cotyledon; *pr* primary root; *sr* secondary root; *co* coleoptile; *csh* cotyledonary sheath. **C, D** Nearly mature embryos; *hc* hypocotyl; *cm* cotyledonary mound; *cst* cotyledonary strand. **E, F** Mature embryos along tangential (**E**) and median (**F**) planes; *np* nodal plate. **G, H** Front and side views of mature embryo (wholemounts). **I–Q** Transections of a mature embryo at levels *a, b, c, d, e, f, g, h* and *i* marked in **G**. (After Panchaksharappa 1962a)

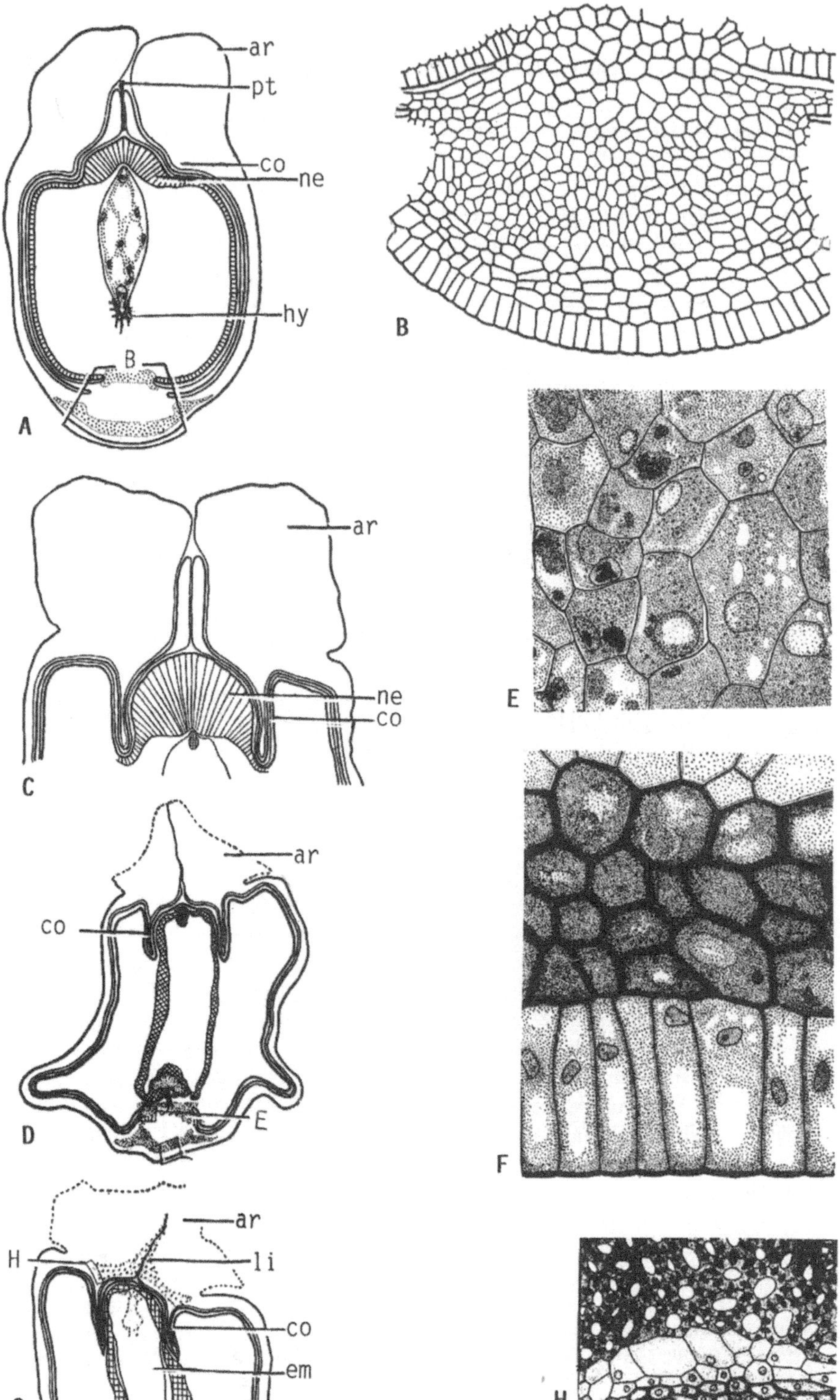
ar
pt
co
ne
hy
B
A
B
ar
ne
co
C
E
co
ar
E
D
F
ar
H
li
co
em
G
H

I Banerji (1940c), I Banerji and V Venkateswarlu (1935), F Berger (1958), L Boehm (1931), CJ Grootjen and F Bouman (1981b), G Harling (1949), MK Jose (1989), MAP Madge (1934), P Maheshwari (1950), JK Mangaly and J Nayar (1990), JK Mangaly and K Swarupanandan (1977), J Mauritzon (1936e), MG Panchaksharappa (1962a, b, 1966, 1970), VJ Philip (1968), TS Raghavan and KR Venkatasubban (1941b), RC Sachar and U Arora (1963), PB Tomlinson (1956)

64.4 Cannaceae

The embryology has not been fully investigated.

The solitary anther is monolocular. The anther tapetum is amoeboid, and several layers thick in *Canna indica* (Kracauer 1930). Meiosis in the microspore mother cells is followed by simultaneous cytokinesis (Kracauer 1930); Täckholm and Söderberg (1918) report successive type. This aspect may be reinvestigated.

According to Tiwari and Gunning (1986), in *Canna indica* × *C.* sp. hybrid the anthers are bilocular. The anther wall comprises the epidermis, multilayered fibrous endothecium, two to four middle layers, and four to five tapetal layers. The inner outline of the tapetum is highly irregular and columns of tapetal cells extend deep into the spaces between the sporogenous cells; they never completely traverse the locular cavity.

Prior to meiosis, the tapetal cells become two- to six-nucleate. During early meiosis I the tapetal cell walls break down and release the protoplasts into the locule. The protoplasts do not fuse to form a plasmodium, although they may become amoeboid in outline.

The middle layers persist until the binucleate stage of the pollen grains. In mature anther there is no trace of the middle layers.

The microspore mother cells undergo meiotic divisions followed by successive cytokinesis. The tetrads are isobilateral. Tapetal protoplasts surround the developing microspores.

The pollen mitosis results in a vegetative cell, and a generative cell anchored to the wall of pollen grain.

The pollen grains are inaperturate and three-celled at shedding (see Davis 1966).

The ovule is anatropous, bitegmic, and crassinucellar. The inner integument forms the micropyle. The cells of the nucellar epidermis divide, and add to the parietal tissue. A hypostase differentiates. The archesporial cell cuts off a parietal cell which produces a single parietal layer. The development of the embryo sac is of the Polygonum type. The polar nuclei fuse before fertilization. The antipodals are ephemeral.

Fig. 64.3.4A–H. *Costus speciosus.* **A** Longisection of ovule at zygote and nuclear endosperm. Note aril (*ar*), remnants of pollen tube (*pt*), collar (*co*), nucellar epidermis (*ne*) and hypostase (*hy*). **B** *Inset B* from A, chalazal tissue. **C** Micropylar region of young seed shows further growth of aril, collar and nucellar epidermis. **D** Longisection of immature seed with well-developed chalazal tissue. **E, F** *Insets E* and *F* from **D**. **G** Upper-half of nearly mature seed shows aril, lid (*li*), collar, and embryo (*em*). **H** *Inset H* from *G* shows pitted sclerenchymatous cells and bordering thin-walled parenchymatous cells of lid. (After Panchaksharappa 1962a)

The development of the endosperm is of the Nuclear type. The tissue later becomes cellular.

The arillate seed contains a straight embryo, and very hard endosperm.

GL Davis (1966), CJ Grootjen and F Bauman (1988), P Kracauer (1930), J Mauritzon (1936e), K Schnarf (1931), U Scheer and WW Franke (1972), G Täckholm and E Söderberg (1918), SC Tiwari and BES Gunning (1986), PB Tomlinson (1962)

64.5 Marantaceae

The embryology has not been studied in sufficient detail.

The meiotic divisions of microspore mother cells are followed by successive cytokinesis. The pollen grains are inaperturate.

The ovule is campylotropous, bitegmic and crassinucellate; the inner integument forms the micropyle. Both the integuments take part in *Phrynium capitatum* (Venkateswarlu 1937b). The archesporial cell cuts off a parietal cell which divides anticlinally. The cells of the nucellar epidermis elongate in the region of the micropyle. The megaspore mother cell undergoes meiotic divisions, and the chalazal megaspore of a linear tetrad develops into a Polygonum type of embryo sac. In *Calathea picturata* (Grootjen 1983b) and *Stromanthe lutea* (Mauritzon 1936e), the embryo sac is located at the tip of nucellus.

The polar nuclei fuse before fertilization. The antipodals are ephemeral in *Thalia*, and persist in *Phrynium.* They become binucleate, and undergo secondary multiplication in *Calathea.* There are five antipodals (two binucleate) in *C. litzei*, and seven (five binucleate) in *C. variegata* (Mauritzon 1936e).

The development of the endosperm is of the Nuclear type.

The embryogeny corresponds to the Chenopodiad type (see Davis 1966).

The outer integument is six to seven layers thick and multiplicative. The inner integument is two-layered. The outer and inner epidermis of the testa become palisade-like. During maturation of seed, the outer epidermis of the testa becomes radially stretched and almost collapses. The mesophyll contains tannin. The endotesta is sclerotic and contains silica particles. The outer epidermis of the tegmen also collapses, and the inner becomes sclerenchymatous. The seed is perispermous and arillate.

The seeds of Marantaceae are characterized by a perisperm channel, aril, micropylar collar, and an operculum (Fig. 64.5.1C–E).

In *Calathea picturata* (Grootjen 1983b), a special feature is the formation of a perisperm channel. The mature ovule becomes campylotropous and the chalaza intrudes into the nucellus. The chalazal ingrowth is parenchymatous and encircled by a sheath of tanniniferous cells (Fig. 64.5.1A–E). The central and peripheral vascular bundles are not well differentiated. A resinous cavity develops in the chalazal parenchyma, vascular elements also differentiate, and a perisperm channel arises by the degeneration of parenchyma and vascular elements. During maturation of seed, the chalazal intrusion in the perisperm is cut off by the silicification of tissue at the narrow basal end. The intrusive tissue degenerates at the distal side leaving some remnants at the inner side of the tannin-sheath, and thus a closed off perisperm channel exists (Grootjen 1983b).

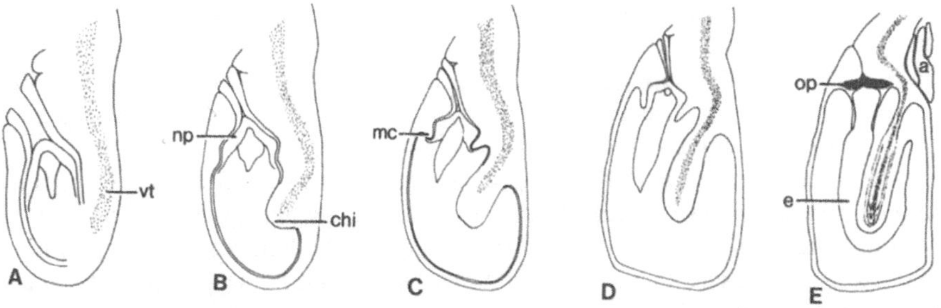

Fig. 64.5.1A–E. *Calathea picturata*, longisections of ovules (**A, B**), immature seeds (**C, D**), and a nearly mature seed (**E**) show the intrusion of chalazal tissue (*chi*) into the nucellus, and campylotropous curvature; *a* aril; *e* embryo; *mc* micropylar collar; *np* nucellar pad; *op* sclerotic part of operculum; *vt* vascular tissue. (After Grootjen 1983b)

The aril develops from the epidermis of the outer integument along with a part of raphe. In *Calathea picturata*, the meristematic activity occurs at two sides of the raphe and a two-lobed aril is produced (Fig. 64.5.1E). It is white and gorged with starch; in mature seed it contains only oil. Finally, it unfolds, becomes turgid, and due to pressure the fruit dehisces (along predetermined lines of rupture) releasing the seeds.

The middle layers of the outer integument give rise to a micropylar collar which grows in the form of a ring around the exostome (Fig. 64.5.1B–D). The cells of the micropylar collar divide repeatedly and intrude into the outer integument in the distal part of the nucellus. By silicification of the outer layer (facing the perisperm), the collar forms a sclerotic ring around the embryo.

The innermost layer of the testa gives rise to the operculum (Fig. 64.5.1E). It originates distally and consists mainly of silicified sclereids. The sclerotic part of the operculum fuses with the endotesta. It gives protection to the interior of seed by silicified sclerotic tissue surrounded by a tanniniferous layer.

The arillate seed contains a folded embryo and copious endosperm (Fig. 64.5.1E). The operculum is functional and has a thick sclerotic layer (Grootjen and Bouman 1981b).

GL Davis (1966), CJ Grootjen (1983b), CJ Grootjen and F Bouman (1981b), J Mauritzon (1936e), J Venkateswarlu (1937b)

64.6 Lowiaceae

The embryology has not been investigated.

64.7 Taxonomic Considerations

Scitamineae are placed just before the Orchidales. The former are a natural taxon of closely related highly evolved families. Habitat, leaf and floral structure, inaperturate pollen grains, mode of pollination (cross), and perispermous seeds indicate the highly evolved features. Scitamineae are always derived from Liliiflorean stock, as a result of elaboration of floral parts and gradual suppression in fertile organs. Musaceae are considered as phylogenetically primitive, while Marantaceae as highly evolved.

Bentham and Hooker (1965c) treat Scitamineae as a natural order in the series Epigynae, whereas Engler and Prantl (1931) recognize them as an order with four families. Hutchinson (1973) includes all the families in Zingiberales, and treats Lowiaceae as a separate family. He considers the group as parallel to Orchidaceae in evolution.

Takhtajan (1980) refers Zingiberales to Liliidae, Cronquist (1981) to Commelinidae, and R. Dahlgren (1980b) to superorder Zingiberiflorae with a single order Zingiberales.

Scitamineae and Liliales both have chelidonic acid, and vessels are restricted to the roots alone. The bitegmic crassinucellate ovule, Polygonum type of embryo sac, and Nuclear as well as Helobial (Zingiberaceae) type of endosperm are common features. However, the embryogeny differs. Likewise, the perisperm (absent in Liliales) and starchy endosperm markedly differ from that of Liliales.

Scitamineae have flowers with well-differentiated calyx and corolla adapted to insect pollination; anatropous (rarely orthotropous in Zingiberaceae), bitegmic, crassinucellar ovule (mostly tenuinucellar in Commelinales), Polygonum type of embryo sac, Nuclear type of endosperm (Helobial in Zingiberaceae), Asterad type of embryogeny, and starchy endosperm as well as perisperm (absent in Commelinales). The seed-coat is variously modified. Even though, to some extent, Scitamineae resemble Commelinales, they form a distinct close unit.

Musaceae. Musaceae have five fertile stamens (six in *Ravenala*); only one fertile stamen is present in Zingiberaceae and Cannaceae, and Marantaceae have a single fertile stamen with one anther lobe suppressed. In the highly evolved family Marantaceae, the ovary is trilocular with one ovule in each cell, or unilocular with only one ovule, due to the suppression of the other two carpels.

Cannaceae and Marantaceae. Embryologically, the families Cannaceae and Marantaceae are very closely related. The nature of pollen, ovule, and development of embryo sac and endosperm are exactly similar. Even the parietal layer is single in both families. Musaceae, which is considered as primitive, also show alliance with these two families in three-celled pollen grains, anatropous, bitegmic, crassinucellate ovule, Polygonum type of embryo sac, and Nuclear type of endosperm.

Zingiberaceae. To some extent, Zingiberaceae stand apart from the other three families. The anther tapetum may be secretory or amoeboid, pollen grains are two-celled, ovule may be ana- or orthotropous and may show nucellar pad, hypostase, epistase, or postament, the endosperm is of the Helobial type, and seed may show aril, collar, lid, and chalazal tissue.

On morphological basis, Zingiberaceae has two subfamilies: Costoideae and Zingiberoideae. The study of vegetative anatomy and evidences from other fields led Tomlinson (1962) to raise Costoideae to a family rank. Only the genus *Costus* is investigated.

Panchaksharappa (1962a, b) states that embryologically Costoideae is distinct from Zingiberoideae, and have unilayered sporogenous cells, fibrous thickenings in anther wall, coenocytic and persistent basal apparatus, Caryophyllad type of embryogeny, mature embryo of a different nature, bulbous aril, micropylar lid, and well-developed chalazal tissue. He, however, recognizes an underlying basic similarity between the two subfamilies: single anther with equal number of wall layers, secretory anther tapetum, isobilateral tetrads, two-celled pollen grains at the dispersal stage, presence of hypostase and epistase in the ovule, Polygonum type of embryo sac, Helobial endosperm, and perisperm, aril and collar in the seed. He is not in favour of raising the subfamily Costoideae to a family level, unless it is corroborated by the study of other genera belonging to this subfamily.

65 Microspermae

Microspermae are represented by a single family, Orchidaceae.

65.1 General Features

Orchidaceae show the following characters: (1) secretory anther tapetum, (2) pollen grains either single (Diandrae) or in tetrads, in pollinia or massulae, (3) two-celled single as well as compound pollen grains, (4) anatropous, bitegmic, tenuinucellate ovule, (5) belated differentiation in ovule, (6) mono-, bi- and tetrasporic type of embryo sac, generally bisporic with well-developed antipodals in Diandrae, (7) long interval between pollination and fertilization, double fertilization (Diandrae), failure of triple fusion common, (8) endosperm generally absent, Nuclear type and short-lived when present, (9) embryo (organless) undifferentiated, chlorophyllous and with well-developed haustorial suspensor in Monandrae, (10) seed-coat with tannin, lignified in Monandrae, (11) polyembryony, and (12) apomixis (common in Monandrae).

65.2 Orchidaceae

The anther wall comprises persistent epidermis, fibrous endothecium, one or two ephemeral middle layers, and secretory tapetum (Sood and Mohana Rao 1986). The fibrous thickenings in endothecium are strongly developed in epiphytic forms, develop only in the basal region of pollinium in *Arundina graminifolia* (Nagaraja Rao 1967). The tapetal cells are uninucleate in *Zeuxine sulcata* (Swamy 1946e), binucleate in *Spathoglottis plicata* (N. Prakash and Lee-Lee 1973).

The behaviour of the epidermis of the anther wall, similar to tapetal cells, in *Zeuxine longilabris* (Karanth et al. 1979) is unique. When the middle layer and tapetum break down, the endothecial cells enlarge radially, and the epidermal cells elongate tangentially. The epidermal cells are densely cytoplasmic, and become multinucleate during the formation of the male gametophyte. As the endothecial cells develop band-like thickenings, the epidermal cells degenerate, and in the mature anther form a thin membrane identical with the tapetum. The multinucleate densely cytoplasmic epidermal cells show nuclear fusions, and their degeneration indicates a nutritional function. They are protective in the early stages of development of the anther wall.

Simultaneous cytokinesis follows meiotic divisions in the microspore mother cells. The tetrads are isobilateral, tetrahedral, linear and T-shaped in *Spathoglottis plicata.*

The pollen grains are single (Fig. 65.2.1A) in diandrous forms, and some monandrous forms (*Cephalanthera, Vanilla*). They remain attached in tetrads (Fig. 65.2.1C–K) as in *Aa. achalensis* (Cocucci 1964), *Bromheadia finlaysoniana* (Jeyanayaghy and Rao 1966), *Calanthe* and *Polystachya flavescens*, or form massulae (Fig. 65.2.1L, M) as in *Epipogium roseum* (Arekal and Karanth 1981), *Liparis viridiflora* (Mohana Rao and K.M. Rao 1983), *Platanthera* and *Peristylus spiralis* (Swamy 1949c), or pollinia as in Monandrae.

The mitosis in a tetrad (*Spiranthes australis* – P. Maheshwari and Narayanaswami 1952), or a massula (*P. spiralis* – Swamy 1949c), occurs synchronously. The pollen grains are non-aperturate, or one- or two-sulcate, and two-celled at the dispersal stage.

Figure 65.2.1A–M shows single pollen grains, tetrads, and massulae, and germination stages. The pollen grains in a tetrad germinate simultaneously (Fig. 65.2.1H–K), whereas in a massula the peripheral pollen grains germinate prior to the central ones (Fig. 65.2.1M).

Schill and Wolter (1986) point out that elastoviscin is a sticky substance, not resistant to acetolysis. It is of tapetal origin, is a special form of pollenkit, and has been shown to be a feature of all subfamilies of the Orchidaceae. Elastoviscin strands connect the tetrads forming a pollinium, and glues massulae together.

The ovule is anatropous, bitegmic and tenuinucellate with inner integument forming the micropyle (Sood and Mohana Rao 1986). The micropyle is unorganized in *Epipogium roseum* (Arekal and Karanth 1981). The report of unitegmic ovule in this taxon appears to be erroneous.

The ovule is undifferentiated at the time of pollination in Diandrae. The archesporium is uni- to multicellular, and a single cell functions as megaspore mother cell. The megaspore mother cell undergoes meiotic divisions, and a linear or T-shaped tetrad, often a triad, is formed. The chalazal megaspore develops into a Polygonum type of embryo sac in Monandrae. The polar nuclei fuse before fertilization as in *Polystachya flavescens* and *Spathoglottis plicata*, or at the time of fertilization. In *Spiranthes australis* (P. Maheshwari and Narayanaswami 1950), the polar nuclei degenerate. The antipodals rarely

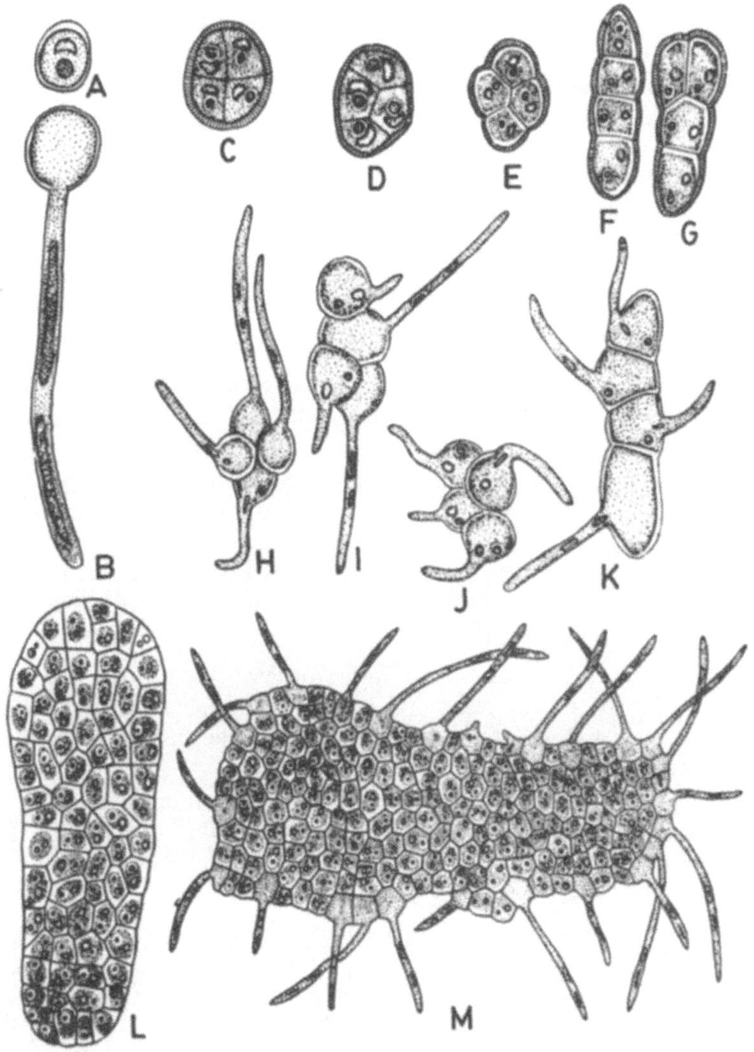

Fig. 65.2.1A, B. *Cypripedium insigne*. **A** Two-celled pollen grain; **B** Elongated sperm cells in pollen tube. **C–K** *Calanthe veitochii*. **C–G** Two-celled pollen grains in tetrads; **H–K** Germinated pollen grains in tetrads. **L, M** *Platanthera bifolia*. **L** Massula; **M** Massula with germinated pollen grains. (After Poddubnaya-Arnoldi 1976)

organize into cells as in *Spathoglottis*. In *Tricyrtis latifolia* (Ogura 1964), they become hypertrophied with large nuclei.

Allium type of development is the rule in Diandrae. The antipodals are well developed and undergo secondary multiplication in *Cypripedium* spp. (Poddubnaya-Arnoldi 1967). Four-, five- or six-nucleate mature embryo sacs occur very often in both Monandrae and Diandrae. This is due to the degeneration, or suppression of mitotic division, or only one division of the chalazal nucleus of the two-nucleate embryo sac.

Abe (1971d) classifies the monosporic female gametophyte into five types: G 1, G 2, G 3, G 3′ and G 4.

G 1 type. The mature embryo sac is eight-nucleate, and the antipodal nuclei may or may not organize into cells.

G 2 type. In a four-nucleate gametophyte, one of the chalazal nuclei does not divide, so that the mature embryo sac is seven-nucleate. The antipodal nuclei may or may not organize into cells.

G 3 type. G 3 type is subdivided into G 3-a, G 3-b, G 3-c, G 3-d and G 3-d′. Variations occur due to suppression of division of the chalazal nuclei, and the embryo sac is six-nucleate. In all cases, the four nuclei at the micropylar pole differentiate into the egg apparatus and upper polar nucleus. Of the two chalazal nuclei, one usually behaves as the lower polar, the other organizes as antipodal cell (G 3-a). The lower chalazal nucleus may not organize as an antipodal, and degenerates before fertilization (G 3-b). Both the chalazal nuclei (of the six-nucleate embryo sac) degenerate before fertilization, and the mature embryo sac is four-nucleate (G 3-c), e.g. *Gastrodia elata* (Abe 1976).

In a six-nucleate embryo sac of *Calanthe torifera* (Abe 1971d), the two chalazal nuclei move to the centre (G 3-d); in *Cirrhopetalum fimbriatum* (Ekanthappa and Arekal 1977a), the two chalazal nuclei fuse in the centre (G 3-d′), and the embryo sac is five-nucleate.

In *Gastrodia alba* (Abe 1976), there is extreme reduction in the number of nuclei. The two chalazal nuclei (of the four-nucleate embryo sac) fuse. The diploid nucleus is ephemeral and degenerates prior to fertilization. The mature embryo sac is four-nucleate.

G 3′ type. In *Polystachya flavescens* (Ekanthappa and Arekal 1977b), the nuclei of the four-nucleate embryo sac undergo mitotic division. The two chalazal spindles fuse forming two diploid nuclei. One of these functions as a polar nucleus and the other as antipodal.

G 4 type. The chalazal nucleus of the two-nucleate embryo sac fails to divide. The mature embryo sac is five-nucleate, four at the micropylar and one at the chalazal pole.

The phenomenon of 'strike' has been observed at the micropylar pole of the embryo sac in *Paphiopedilum* spp. (Francini 1931) – a rare feature. One of the micropylar nuclei and two chalazal nuclei of the four-nucleate embryo sac fail to divide, so that the mature embryo sac has only five nuclei, three at the micropylar and two at the chalazal end. Abe (1976) suggests a reinvestigation.

We do not think that it is at all useful to establish subtypes on the basis of variations of nuclear behaviour at the two- and four-nucleate stages of the embryo sac. It would be very rewarding to study the percentage of variation of embryo sacs in different taxa.

In Monandrae, occasionally bi- and tetrasporic embryo sacs occur. Abe (1971b) reports Polygonum, Allium and Adoxa type in *Oreorchis patens*, and Polygonum and Adoxa type in *Bletilla striata* (Abe 1971a). Occurrence of twin nucelli and multiple embryo sacs (sporadically) leading to polyembryony is reported in *Bletilla*.

The diploid (2n = 40) species – *Nigritella lithopolitanica* and *N. nigra* – are sexual and embryos develop from fertilized egg cells. The tetraploid (2n = 80) species – *N. archiducisjoannis*, *N. miniata* and *N. stiriaca* – are apomictic and produce nucellar embryos (Teppner and Klein 1985).

In *Dendrobium speciosum* (Calder and Slater 1985), the stigmatic column is short and broad, and bears an anther containing a single, four-lobed pollinium without viscidium, stipe, rostellum or caudicles. The cup-shaped stigma faces downward. The stigmatic cup contains a wet jelly-like viscid material or mucilage. The pollinium is placed in the stigmatic cup at pollination.

Under SEM, the stigmatic cup appears to contain a large number of elongated detached cells suspended in a matrix. These cells are thick-walled and have vacuolate cytoplasm containing amyloplasts. They are not attached to the tissue of stigma or column and are oriented parallel to the axis of the column. These cells are spindle-shaped with longitudinally sculptured wall. Towards the stigmatic surface they are densely cytoplasmic. The stigmatic cells give a positive non-specific esterase reaction which indicates physiological receptivity. These cells also give a positive reaction to fluorescein diacetate indicating that the plasma membrane is intact. The matrix shows high levels of acidic polysaccharides with low levels of lipids and proteins. The authors consider that such a stigma is more highly evolved than the wet papillate stigma.

The relative volume of this mucilage appears to vary according to the age of flower, relative humidity, and soil environment.

The stigmatic cup narrows towards the proximal end and extends as a continuum into the narrow stylar canal. The canal is filled with the material similar to that in the stigma, and leads into the ovary which, at pollination, contains a large number of developing ovules.

It is suggested that the stigmatic material, consisting of matrix and detached cells, has evolved in response to the special needs of both the pollination process and the support and nourishment of the large number of pollen tubes.

The term "pseudocopulation" (Stoutamire 1983) is appropriate where insects transfer pollen during attempted mating with the flower, as described in *Ophrys* and *Cryptostylis*. The term is less appropriate for most *Caladenia* spp. because the wasp's activity is not immediately and clearly copulatory. The author prefers to use the phrase "sexual attraction" until immediate copulatory behaviour can be more consistently demonstrated.

Caladenia spp. are pollinated after sexually attracting male triphiid wasps of the subfamily Thynninae. The mechanism is well developed in other Australian orchid genera *Chiloglottis* and *Drakaea*, as well as in *Spiculaea*, *Paracaleana* and *Arthrochilus* in which the wasps attempt copulation with the flowers.

Many species of *Caladenia* of Western Australia attract more wasps (Triphiidae, subfamily Thynninae) by visual and pheromonal mechanisms. Floral colours are predominantly combinations of green, cream and brown. Labella bear dark-maroon tips or dark calli on the upper surface which are visual wasp equivalents.

Glandular areas on perianth tips are sources of sexual attractants in some species. Flower form and position are adjusted to the flight behaviour of the wasps. Each *Caladenia* sp. appears to attract and be pollinated by only one wasp species at the orchid site although there is limited evidence of geographic pollinator replacement in some species.

Thickened sepal tips bearing ovoid unicellular trichomes occur in most of the sexually attracting *Caladenia* sp. When sepals appear unspecialized, as in sexually attracting *C. cairnsiana*, the unicellular trichomes appear on the thickened apical labellum margin and wasps orient first to this margin.

In *Epidendrum scutella*, Cocucci and Jensen (1971) studied the time, size and stage correlation in the development of flower bud, micro- and megasporogenesis, male and female gametophyte, and pollination and fertilization (Fig. 65.2.2A–D).

During the first month the bud undergoes torsion, microsporogenesis is completed, and isobilateral tetrads are formed. The primordia of ovules also

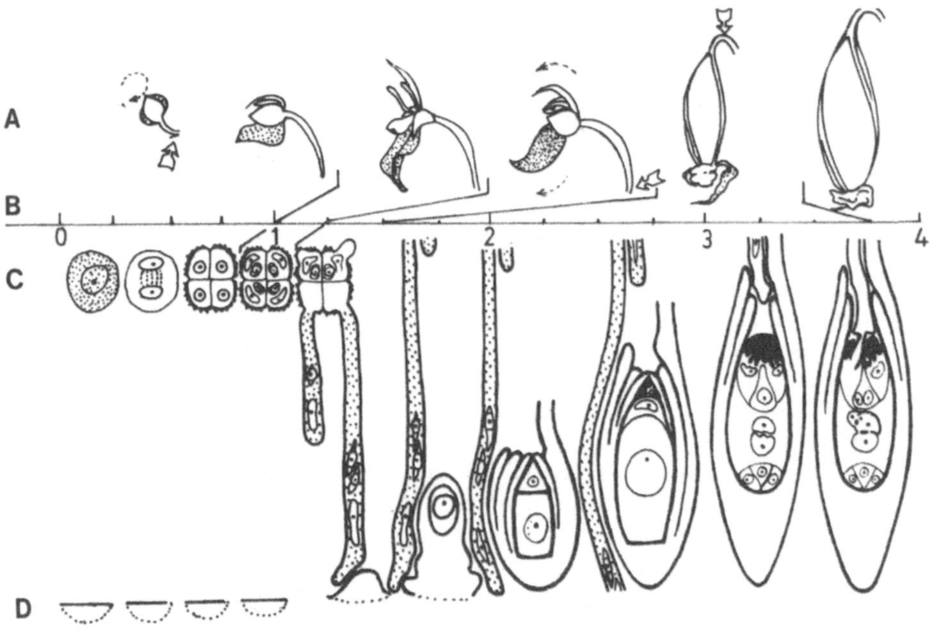

Fig. 65.2.2A–D. *Epidendrum scutella.* **A** Development of bud into mature fruit. Labellum *stippled.* Direction of flower movements *long arrows*; torsion of flexion point *short arrows.* **B** Time scale in months divided into weeks. **C** Development of male and female gametophyte, pollination and fertilization. **D** Development of ovule starting with primordium. Note corresponding duration of development and respective stage of bud/flower/fruit. (After Cocucci and Jensen 1971)

differentiate. In the first week of the second month the pollen grains (in a pollinium) become two-celled. In the second week (second month) the bud develops into a flower, and the labellum projects and becomes the landing site for pollinators. The pollinium germinates and the pollen tubes enter the ovary containing ovular primordia. In the third and fourth week (second month), the labellum becomes receptive. The female archesporial cell differentiates in the ovules. The bitegmic and anatropous ovule shows a dyad in the first and second week of the third month. Megasporogenesis proceeds up to the end of the third month and the labellum remains receptive. In the first week of the fourth month, the female gametophyte matures, stigma becomes receptive, and pollination takes place. The labellum shrivels and the ovary grows rapidly. In the next two weeks, syngamy occurs and the second male gamete comes in contact with the two polar nuclei.

A very long period elapses between pollination and fertilization; it is two-and-a-half months in Diandrae. Double fertilization is normal in Diandrae, whereas usually only syngamy occurs in Monandrae. Savina (1972) describes that in *Ophrys insectifera* the second male gamete remains in the disrupted synergid, and does not enter the central cell to facilitate triple fusion. In species like *Achalensis schlechter* (Cocucci 1964), even though double fertilization is normal, the primary endosperm nucleus degenerates.

Tohda (1967) studied pollination and fertilization in a Japanese saprophytic orchid, *Hetaeria shikokiana*, in relation to the time factor. This species is self-pollinated. On the fourth day of pollination, the pollen tube reaches the placenta and ovules begin to develop. The development of the embryo sac is of the Polygonum type. On the eighth day after fertilization, the primary endosperm nucleus degenerates; the zygote remains in resting condition. In *Cirrhopetalum fimbriatum* (Ekanthappa and Arekal 1977a), the primary endosperm nucleus persists up to the quadrant proembryo and then degenerates.

In *Spiranthes sinensis* (Terasaka et al. 1979), the generative cell does not divide and behaves as a sperm cell. It fuses with the egg cell and single fertilization is effected. As the other sperm is not formed, triple fusion does not take place. This aspect needs careful reinvestigation. In *Listera ovata* and *Ophrys insectifera* (Savina 1978), syngamy alone occurs and the second sperm does not fuse with the secondary nucleus.

Germination of pollen grains, fertilization, development of endosperm and proembryo, mature embryo, and zygotic polyembryony are shown in Figures 65.2.3 and 65.2.4.

In *Cymbidium bicolor*, the germination of pollen grains in a tetrad is simultaneous (Fig. 65.2.3A). The pollen tube enters through the micropyle and destroys one of the synergids in *Cypripedium spectabile*. Syngamy and triple fusion occur at the same time (Fig. 65.2.3B). The proembryo develops up to the globular stage.

The development of the endosperm, when present, is of the Nuclear type and four to six nuclei are present in *C. insigne* (Fig. 65.2.3C). In *Ophrys insectifera*, twin embryo sacs which develop from twin megaspores contain free endosperm nuclei as well as proembryo (Fig. 65.2.3D). Free-nuclear endosperm occurs in *Cypripedium* (Fig. 65.2.4A–E). Six nuclei are

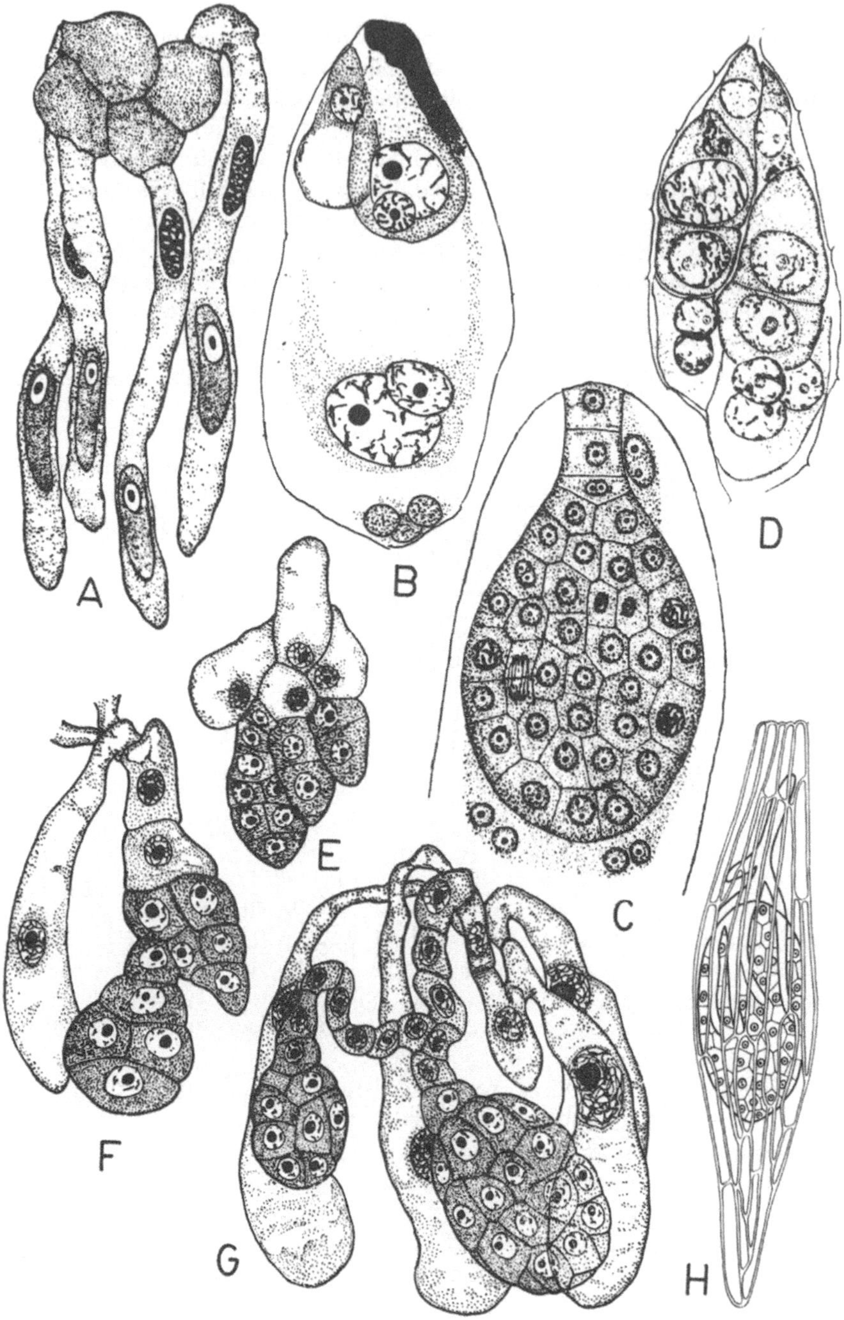

Fig. 65.2.3A. *Cymbidium bicolor*, germination of pollen grains in a tetrad. (After Swamy 1941b). **B** *Cypripedium spectabile*, double fertilization. (After Swamy 1945a). **C** *C. insigne*, micropylar portion of embryo sac shows embryo and five endosperm nuclei. (After Poddubnaya-Arnoldi 1967). **D** *Ophrys insectifera*, twin embryo sacs (from twin megaspores); each has a two-celled proembryo and a few endosperm nuclei. (After Savina 1971). **E–G** *Eulophia epidendraea* (wholemounts); **E**

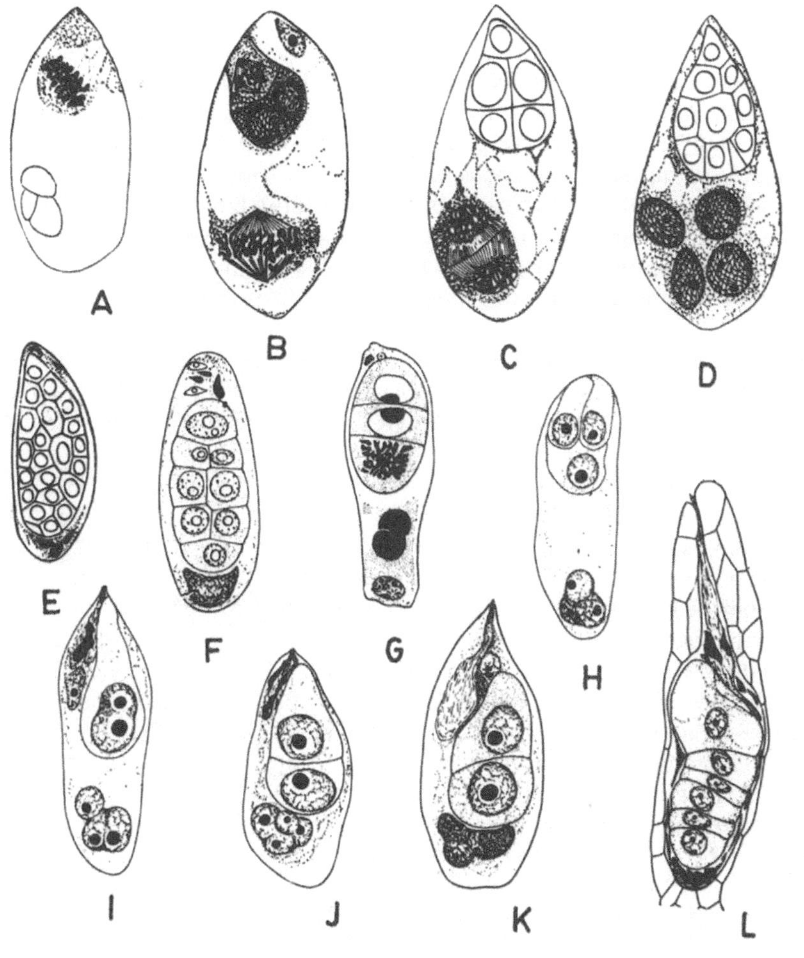

Fig. 65.2.4A–E. *Cypripedium.* **A** Embryo sac with zygote in division; **B** Three-celled proembryo, and primary endosperm nucleus in division; **C** Later stage; **D** Proembryo and four endosperm nuclei; **E** Older proembryo, endosperm nuclei have degenerated. (After Pace 1909). **F** *Orchis maculata,* proembryo and degenerated primary endosperm nucleus. (After Hagerup 1944). **G** *Spiranthes australis,* three-celled proembryo and degenerated polar nuclei. The chalazal nucleus represents the antipodal. (After P. Maheshwari and Narayanaswami 1952). **H–J** *Corallorhiza maculata.* **H** Egg apparatus, and polar nuclei and antipodal lying in a clump; **I** Double fertilization; **J** Two-celled proembryo, primary endosperm nucleus and antipodals. **K, L** *Bletia shepherdi.* **K** Two-celled proembryo and degenerated primary endosperm nucleus and antipodals; **L** Later stage. (After Sharp 1912)

Proliferation of zygote and origin of three independent proembryos; **F** Additional embryo derived by "budding" of zygotic embryo; **G** Twin embryos due to proliferation of suspensor; the suspensor cells are balloon-shaped. (After Swamy 1943b). **H** *Dendrobium nobile,* mature seed; seed-coat transparent, embryo organless (undifferentiated) with a branched suspensor. (After Poddubnaya-Arnoldi 1976) formed in *Arundina graminifolia* (Nagaraja Rao 1967) and four in *Lecanorchis japonica* (Hiroshi 1971a). According to Chugai (1971), the maximum number of endosperm nuclei observed is 12 in *Vanilla* and 16 in *Galeola septentrionalis*. In *Orchis maculata*, the embryo sac contains the degenerated primary endosperm nucleus and the proembryo (Fig. 65.2.4F). As both the polar nuclei degenerate in *Spiranthes australis*, the endosperm does not develop (Fig. 65.2.4G–J). In *Corallorhiza maculata*, the remnants of pollen tubes indicate that fertilization does occur and embryos are formed. In *Bletia shepherdi* (Fig. 65.2.4K, L), whereas the primary endosperm nucleus degenerates, the nuclei at the chalazal end fuse and form the endosperm. This aspect needs further investigation.

When the primary endosperm nucleus degenerates, there is no possibility of development of endosperm as in *Habenaria edgeworthii*, *H. elisabethae* and *H. galeandra* (Sood 1986) and *Microstylis wallichii* (Sood and Mohana Rao 1986).

The suspensor cells show much variation. They are balloon-shaped in *Eulophia epidendraea*. Polyembryony occurs due to proliferation/budding of the zygote, or by splitting of a single proembryo (Fig. 65.2.3E–G).

The mature seed has a transparent seed-coat, and undifferentiated embryo with branched suspensor in *Dendrobium nobile* (Fig. 65.2.3H).

Parthenogenetic development of egg is reported in *Spiranthes australis* (P. Maheshwari and Narayanaswami 1950), and the investigators discuss several possibilities: (1) the egg divides even though the pollen tube does not discharge the gametes, or the discharged gametes do not effect fertilization, (2) the undischarged pollen tube lies in the vicinity of the proembryo, (3) the synergids remain intact, (4) the parthenogenetic egg is haploid ($n = \pm 15$) while the fertilized egg is diploid ($n = \pm 30$), and (5) during the divisional stages of the proembryo, haploid condition has been observed.

Swamy (1949d) classifies orchid embryos into three types: Group A (Asterad type), Group B (Onagrad type), and Group C (Cymbidium type). In Cymbidium type, the divisions in terminal and basal cell of the two-celled proembryo are irregular.

The embryogeny corresponds to the Asterad type in *Epidendrum prismatocarpum* (Swamy 1948d), *Liparis viridifolia* (Mohana Rao and K.M. Rao 1983) and *Neottia listeroides* (Sood 1984). Onagrad type occurs in *Herminium angustifolium* (Mohana Rao and Sood 1979b), *Habenaria densa* (Mohana Rao and Sood 1979c), *H. edgeworthii*, *H. elisabethae* and *H. galeandra* (Sood 1986), and *Microstylis wallichii* (Sood and Mohana Rao 1986). In *Cheirostylis flabellata* (Swamy 1979) and *Zeuxine longilabris* (Karanth et al. 1981), equal volume of shoot apex contributes to the formation of cotyledonary and epicotyledonary regions.

The mature embryo is undifferentiated. The embryo is chlorophyllous with a well-developed haustorial suspensor in Monandrae. The haustorial suspensor branches in *Arundina graminifolia* (Nagaraja Rao 1967), *Bromheadia finlaysoniana* (Jeyanayaghy and A.N. Rao 1966), *Cymbidium bicolor* (Swamy 1942a) and *Eulophia epidendraea* (Swamy 1943b). The suspensor is unicellular in *Spathoglottis plicata* (N. Prakash and Lee-Lee 1973) and vestigial in

Lecanorchis (Hiroshi 1971a). In *Hetaeria shikokima* (Hiroshi 1967), the unicellular suspensor develops into a long tube which projects through the micropyle, becomes twisted, and enters the tissue of the placenta. The much-branched haustorial suspensor persists in ripe seed in *Cymbidium, Dendrobium, Phalaenopsis* and *Stanhopea* (Poddubnaya-Arnoldi 1967). In *Neottia nidus avis* (at germination), mycorrhiza has been observed in the embryo.

Cleavage polyembryony occurs in *Cymbidium, Eulophia, Gastrodia, Gymnodia* and *Spiranthes*. Polyembryony due to proliferation of the zygote, budding or splitting of the embryo occurs in *Cymbidium bicolor* (Swamy 1942a), *E. epidendraea* (Fig. 65.2.3E, F) and *Zeuxine sulcata* (Swamy 1946e). In *Eulophia*, proliferation of suspensor may also lead to polyembryony (Fig. 65.2.3G). Synergid polyembryony is reported in *Calanthe, Cymbidium, Goodyera* and *Phalaenopsis* (Poddubnaya-Arnoldi 1967). Nucellar polyembryony occurs in *Zeuxine sulcata* (Swamy 1946e). In *Lecanorchis japonica* (Hiroshi 1971a), twin embryos develop due to fertilization of two embryo sacs within the same ovule. In *Spiranthes cernua*, adventive proembryos develop from the inner integument. Their formation does not depend on fertilization of the egg cell. Some of the integumentary cells enlarge, become richly cytoplasmic, divide and produce two to six supernumerary proembryos in the micropylar region.

In *Epidendrum nocturnum* (Silva Stort et al. 1985), the flowers are cleistogamous and self-pollinated. The lateral portions of the rostellum are better developed than the middle portion, to facilitate contact of pollen with the stigma. Ripe seeds have one to four embryos, or none. Single embryos are of sexual origin and occur more frequently. Supernumerary embryos are of asexual origin, and result from natural pollination (31%) than from artificial pollination (9%).

Additional embryos arise by polyembryony and apomixis. In the first case (a) the zygote divides irregularly, (b) the suspensor cells of the sexual embryo branch, or (c) the proembryo itself develops branches which form "buds" and produce several embryos (see Fig. 65.2.4E–G).

Multiple embryos are also formed by apomixis in two ways. The supernumerary embryos originate from the cells of the embryo sac following mitotic divisions. Secondly, the embryos develop from integumentary cells.

The seed is exalbuminous with (organless) undifferentiated embryo.

On the basis of seed size, shape of cells of the testa, thickness of cells, and pattern of thickenings on the wall, Clifford and Smith (1969) classify 49 species of the tribes Epidendreae and Neottieae. In Epidendreae the thin-walled cells vary in shape; in Neottieae they are of uniform shape and show bead-like thickenings on the walls.

K Abe (1967, 1968, 1971a, b, c, d, e, 1976, 1977), K Afzelius (1916, 1954, 1959), GB Amici (1847), GD Arekal and KA Karanth (1981), S Arora and RN Kapil (1989), P Balogh (1982), HN Barber (1942), N Bernard (1909), WH Brown and LW Sharp (1911), P Burns-Balogh (1983), DM Calder and AT Slater (1985), MC Carlson (1945), K Chugai (1971), HT Clifford and WK Smith (1969), AE Cocucci (1964, 1973b), AE Cocucci and WA Jensen (1969a, b, 1971), Ch Dannenbaum, M Walter and R Schill (1989), KG Ekanthappa and GD Arekal (1977a, b), E Francini (1931, 1945), O Hagerup (1944, 1945, 1947), J Heslop-Harrison (1968d), T Hiroshi (1967, 1971a, b), S Jeyanayaghy and AN Rao (1966), KA Karanth, PK Bhat and DA Arekal (1979), KA Karanth, BGL Swamy and GD Arekal (1981), S Krupko, F Israelstam and B Martinovic (1954), H Kurzweil (1990), SK Law and EC Yeung (1989), P Maheshwari and S Narayanaswami (1950, 1952), J Modilewski (1918), PR Mohana Rao and A Kaur (1979a, b), PR Mohana Rao and KM Rao (1983, 1984), PR Mohana Rao and SK Sood (1979a, b, c, 1986, 1987), A Nagaraja Rao (1967), DH Niimoto and Y Sagawa (1961, 1962), H Ogura (1964), O Olsson (1967), L Pace (1907, 1909), MD Pastrana and JK Santos (1931), R Peakall (1989), VA Poddubnaya-Arnoldi (1960, 1967), N Prakash and A Lee-Lee (1973), MN Prosina (1930), VP Razmologov (1958), J Sagawa and DH Nimoto (1961), G-I Savina (1971, 1972, 1978), R Schill and M Wolter (1986), KN Seshagiriah (1932a, b, 1941), LW Sharp (1912), MN Silva Stort, S Dos and EA Pavanelli (1985), SK Sood (1983, 1984, 1985, 1986, 1989), SK Sood and PR Mohana Rao (1986), KR Sporne (1974), H Stenar (1937b), WP Stoutamire (1983), D Swaminathan (1967), BGL Swamy (1941b, 1942a, 1943a, b, 1944c, 1945a, 1946b, c, d, e, 1947, 1948c, d, e, 1949c, d, 1979), H Teppner and E Klein (1985), O Terasaka, T Niitsu and R Tanaka (1979), ES Teryokhin and OP Kamelena (1969), H Tohda (1967, 1974), Y Veyret (1981a, b), SP Vij, M Sharma and N Shekhar (1982), SP Vij and M Sharma (1987), MR Vijayaraghavan and AK Shukla (1980), M Wolter, C Seuffert and R Schill (1988), S Yasugi (1983)

65.3 Taxonomic Considerations

Orchidaceae. Bentham and Hooker (1963c) and Engler and Prantl (1931) place Orchidaceae in Microspermae along with Burmanniaceae. Melchior (1964) includes only Orchidaceae in his Microspermae, and places this order at the highest level in angiosperms. Hutchinson (1973) elevates the family to the rank of an order, Orchidales. According to Engler, Orchidaceae have been derived from Burmanniaceae. Their relationships with Burmanniaceae have been discussed earlier (Sect. 54.19).

Cronquist (1968, 1981) includes Orchidaceae and Burmanniaceae in the order Orchidales, and derives Orchidales from Liliales. Orchidales show distinct advanced evolutionary features such as highly specialized floral structure adapted to insect pollination (these vary with different pollinators), inferior ovary, loss of endosperm, and production of a large quantity of tiny seeds for effective distribution. Orchidales and Liliales both contain steroid saponins (see Huber 1977). R. Dahlgren (1977a, 1980b) places Orchidales next to Burmanniales, in the superorder Lilianae, Liliiflorae.

Orchidaceae is a diverse family with adaptive characters. Their members are terrestrial, and have evolved towards epiphytism and saprophytism. The family has bee-flowers. The mode of pollination observed in the family needs a very high mutual adaptation between insect and flower. This represents an advanced state (Sporne 1974).

The Orchidaceae exhibit several interesting features among the angiosperms: formation of massulae and pollinia, belated differentiation of ovules, reduction of female gametophyte, irregularities in double fertilization leading to suppression of triple fusion, absence of endosperm in Monandrae, presence of suspensor haustorium, and different modes of polyembryony.

Poddubnaya-Arnoldi (1967) enumerates several embryological features which separate the two subfamilies, Diandrae and Monandrae. The simple single pollen grains, Allium type of embryo sac with secondary multiplication of antipodal cells, double fertilization, Nuclear type of short-lived endosperm, embryo without chlorophyll, feebly developed suspensor, tannin deposition in seed-coat, absence of polyembryony, and apomixis in Diandrae are in marked contrast with those in Monandrae. In Monandrae the pollen grains are compound, united in massulae or pollinia, development of the embryo sac is mono-, bi- as well as tetrasporic, and the antipodals may not organize into cells. There is single fertilization with total absence of endosperm formation, embryo is chlorophyllous with a well-developed haustorial suspensor, seed-coat is transparent and lignified, polyembryony and apomixis are common. Diandrae is primitive, while Monandrae show advanced features and greater deviation from the common angiosperms. According to Poddubnaya-Arnoldi (1967), both the subfamilies, Diandrae and Monandrae, have originated independently from a common ancestor.

66 Concluding Remarks

We have limited the 'Concluding Remarks' to (a) primitive and advanced embryological characters, (b) embryological features in Magnoliiflorae s.l. (dicots) and certain Liliiflorae (monocots), (c) distribution and phylogeny of anther tapetum, (d) affinities of the megagametophyte of Poaceae, (e) archegonium in gymnosperms, and the angiosperm embryo sac, and (f) prospects.

Primitive and Advanced Embryological Characters

The primitive embryological features usually comprise: a thick anther wall, widespread endothecial thickenings, several middle layers, secretory (glandular) tapetum (Sect. 3.1); successive cytokinesis in pollen mother cells after meiosis (3.2); two-celled pollen grains (3.3); anatropous, bitegmic, crassinucellate ovule, micropyle formed by both integuments, massive parietal tissue including nucellar cap formed by divisions in apical cells of the nucellar epidermis and their derivatives (3.4); monosporic *Polygonum* type of embryo sac (3.6); Nuclear type of endosperm, sometimes ruminate (3.10); typical dicotyledonous embryo with a conspicuous suspensor (3.11); seed albuminous, sometimes perispermous; seed coat thick, formed by both testa and tegmen (3.14).

The advanced embryological characters include: a thin anther wall, fibrous thickenings limited mostly to endothecium, ephemeral middle layer, amoeboid (plasmodial, invasive, intrusive) tapetum (Sect. 3.1); simultaneous cytokinesis by furrowing in pollen mother cells after meiosis (3.2); three-celled pollen grains (3.3); anatropous, unitegmic, tenuinucellate ovule, micropyle narrow and short, the inner epidermis of integument differentiates as endothelium (3.4); mono-, bi- and tetrasporic embryo sac, synergids sometimes absent (*Plumbago*), antipodal nuclei may or may not organize into cells, ephemeral

(3.5–3.7); endosperm of the Cellular type, chalazal, micropylar or both micropylar and chalazal haustoria develop (3.10), sometimes the haustoria invade the placenta and even pedicel; embryo with a few-celled suspensor, sometimes suspensor haustoria conspicuous (3.11); seed exalbuminous, perisperm usually absent (3.14); seed-coat exotestal.

These embryological characters cannot be considered as *strictly* primitive or advanced, since many primitive features have been observed in advanced families and vice versa. A few examples will illustrate this point.

1. Secretory and amoeboid (Figs. 3.1.2A–I, 52.3.1A–E) tapetum, two- and three-celled pollen grains (Fig. 3.3.1A–C), successive and simultaneous cytokinesis in post-meiosis microspore mother cells are randomly distributed in Polypetalae, Sympetalae and Monocots.

2. Single pollen grains, and pollen grains in permanent dyads (Fig. 28.2.1A), tetrads (Figs. 18.5.1A, C, D; 54.18.1A–G), pollinia (Fig. 47.7.1A–D), and massulae (Fig. 65.2.1L, M) occur in primitive as well as advanced taxa. All these conditions are known in the Orchidaceae (Fig. 65.2.1C–M), a highly evolved monocot family.

3. Since an anatropous ovule occurs in Magnoliiflorae, it is considered a primitive character. In fact, the simplest ovule is orthotropous which undergoes curvature to give rise to anatropous, hemianatropous, campylotropous, and sometimes a circinotropous condition (Fig. 3.4.2A–H). Seeking homology with the gymnosperms, orthotropous ovules should indeed be considered primitive.

4. The development and organization of the female gametophyte is highly variable in Polypetalae, Sympetalae and Monocots. Further consideration becomes even more debatable since mono- and bisporic conditions may occur in the same species – *Tellima grandiflora* (Ratnaparkhi 1972); mono-, bi- and tetrasporic embryo sacs develop in *Erigeron* (Fig. 51.9.3); several types of tetrasporic embryo sacs – Adoxa, Drusa, Fritillaria and Chrysanthemum cinerariaefolium – occur in the same genus, e.g. *Tamarix* (Sect. 37.15).

5. The Nuclear type of endosperm is characteristic of Polypetalae, Cellular type of Sympetalae, and Helobial type of Monocots (Fig. 3.10.1A–O). All three types have been observed in all three groups. Some families – Trapaceae (Sect. 39.3), Podostemaceae (28.2) – lack endosperm all together. These families have, therefore, developed alternative strategies for the nourishment of the embryo, e.g. pseudo-embryo sac (Fig. 28.2.1D), and haustorial suspensor (Fig. 65.2.3E–H).

6. The embryo is typically dicotyledonous in Polypetalae and Sympetalae, and has only a single cotyledon in Monocots. In a dicotyledonous taxon, *Trapa* (Fig. 39.3.1Q), one cotyledon is completely suppressed.

7. What about the organless (undifferentiated) embryo as in Lentibulariaceae (Sect. 48.25), Burmanniaceae (Fig. 54.16.1J), Mayacaceae (Fig. 57.3.3L), Xyridaceae (Sect. 57.4), Rapateaceae (Fig. 57.5.1H), Eriocaulaceae (Sect. 57.6),

Centrolepidaceae (Sect. 57.8), Orchidaceae (Fig. 65.2.3H), and others? We have no knowledge as to why some embryos have remained arrested. Their germination should be studied to throw light on their differentiation.

Embryological Features in Magnoliiflorae s.l. and Certain Liliiflorae

R. Dahlgren and Clifford (1981) compare the embryological features of Magnoliiflorae – especially Annonaceae (Sect. 18.6), Aristolochiaceae (21.2), Lactoridaceae (20.5), Chloranthaceae (20.4) and Myristicaceae (18.8), and the Liliiflorae – Dioscoreales [Dioscoreaceae (54.12), Trichopodaceae (see 54.12), Taccaceae (54.11), Trilliaceae (see 54.2)] and Asparagales [Smilacaceae (see 54.2), Philesiaceae (see 24.2), Convallariaceae (see 24.2) and a few more families]. Some similarities are:

1. Anther wall of the dicotyledonous type in *Tacca*, the only case known in monocots.

2. Tapetum of the secretory type in all Magnoliiflorae, as in all Dioscoreales and most Asparagales.

3. Post-meiotic cytokinesis in pollen mother cells of the simultaneous type in most dicots; successive type is widely distributed in monocots. Both types are known in *Aristolochia*, *Annona*, *Cananga* (Davis 1966) and certain other Magnoliiflorae. Simultaneous type is known in some species of *Dioscorea* and *Tacca* (monocots).

4. Monoaperturate (often sulcate) pollen is common in Monocots and some Magnoliiflorae, although inaperturate, biaperturate types occur as well.

5. Bicelled pollen grains are reported in practically all Magnoliiflorae and nearly all Liliiflorae; three-celled pollen grains occur in both groups (see Brewbaker 1967).

6. The ovules are usually anatropous, bitegmic and crassinucellate in the Magnoliiflorae; in the Dioscoreales and closely related families of Asparagales, the early stages of development are similar although exceptions occur. The ovules are orthotropous in *Stemona*; *Trichopus* lacks a parietal cell. The Polygonum type of embryo sac is common in all these groups; Allium type is known in the Trilliaceae and Orchidaceae.

7. The Magnoliiflorae have a Cellular type of endosperm development. The Nuclear type is widespread in Dioscoreales and many berry-fruited Asparagales.

8. In some families, as in Saxifragaceae (Sect. 26.7), all three types of endosperm are reported.

In some families, the initial development of endosperm is of the Cellular type, but free-nuclear divisions occur in each cell. Finally, the endosperm becomes completely cellular. This is described as the Cellular-Nuclear-Cellular type (Fig. 47.2.2A–F) and is rare.

9. Ruminate endosperm (Figs. 3.10.3A–K, 18.3.1G, H), common in Annonaceae, Aristolochiaceae, Canellaceae, Degeneriaceae, Eupomatiaceae and Myristicaceae (Magnoliiflorae), is also known in *Avetra* and *Trichopus* (Dioscoreales).

10. The embryo is small in relation to the endosperm in most Magnoliiflorae and some Liliiflorae (*Paris, Trillium, Madiola*), and fairly small in most Dioscoreales. In both, monocots and dicots, the stem tip is invariably terminal and the cotyledon/s lateral. The interpretation of a lateral stem tip in the monocot embryo has been unequivocally disproved (see Swamy and Padmanabhan 1962, Swamy and Krishnamurthy 1977, Natesh and Rau 1984).

11. The seeds have a copious starchy endosperm; starch is occasionally present in the Dioscoreales (*Croomia*, sometimes *Paris* and *Trillium*) and related Asparagales (*Ripogonum, Streptopus*). In the Magnoliiflorae, starchy endosperm is reported in several members of Annonaceae, Aristolochiaceae, Myristicaceae and others.

12. Polyembryony and apomixis are common to mono- and dicots.

The above account clearly indicates that the Magnoliiflorae show a number of monocot characters, and Dioscoreales and berry-fruited Asparagales (Liliiflorae) show a number of dicot features; both superorders also have some common characters. It is quite obvious that in many cases the dicots and monocots are not very distinctly demarcated. However, the similarities *may* not indicate any relationship.

The assignment of the Nymphaeaceae to the monocots has been repeatedly questioned, and it has been emphasized again and again that Nymphaeaceae are a dicotyledonous group with a distinct dicot embryo (Figs. 19.7.2A–M, 19.7.3, 19.7.5A–C). In spite of the fact that R. Dahlgren and Clifford (1981) accept that in the Nymphaeaceae the leaves are net-veined, and the embryo has two distinct cotyledons, they still assign this family to the monocots. This controversial issue is discussed in Section 19.7.

The taxonomic alliances of many genera of disputed assignment, families and orders are discussed under 'Taxonomic Considerations', with respect to each order (see Contents).

Distribution and Phylogeny of Anther Tapetum

Pacini et al. (1985) discuss the form, function and possible phylogeny of tapetum in the Embryophyta. The tapetum is of considerable physiological significance because all nutritional material entering the sporogenous cells/microspores/pollen grains originates or passes through it. Between the 'secretory' (glandular) and 'amoeboid' (intrusive, plasmodial) types, an intermediate 'pseudoamoeboid' type is also recognized; the tapetal protoplasts extend into the pollen sac but do not fuse to form a true periplasmodium as in Dipsacaceae (Fig. 50.5.2A–F; see also Sect. 53.2).

According to Pacini et al., in 207 angiosperm families the type of tapetum is a family character. The amoeboid type occurs mostly in monocots. Both types of tapetum have been reported in at least 12 (either primitive or advanced) families: Caesalpiniaceae, Caprifoliaceae, Chenopodiaceae, Droseraceae, Euphorbiaceae, Gentianaceae, Haemodoraceae, Helleboraceae, Lauraceae, Solanaceae, Vitaceae, Winteraceae, and Zingiberaceae (the only monocot family).

Sporne (1973) states that the secretory tapetum is primitive, and the amoeboid type advanced. The type of tapetum depends on locule diameter, pollen shape or dry/wet habitats. There is no unequivocal proof that the secretory type is primitive, but it appears to be so because of the widespread occurrence in gymnosperms and angiosperms.

Affinities of the Megagametophyte of Poaceae

Anton and Cocucci (1984) attempt to establish the affinities of the megagametophyte of the Poaceae (Gramineae) with those of other monocot families.

The Polygonum type of embryo sac with supernumerary antipodal cells (6–300) has been reported in the Poaceae. In *Echinocloa frumentacea*, there are three cells, each two- or three-nucleate, in *Panicum miliare*, each of the three antipodal cells is uninucleate (Narayanaswami 1955a, c).

On the basis of the number and position of antipodals, and the number of nuclei in each cell, Shadowsky (1926) recognizes eight variants of the Polygonum type of embryo sac in Poaceae. Anton and Cocucci point out that in view of the common features in the variants, and more than one variant occurring in the same genus, the rank of a variant or subtype is not at all convincing.

Proliferation of antipodals – both in dicots and monocots – has arisen independently, in unrelated families even in monocots: Amaryllidaceae, Araceae, Hypoxidaceae, Juncaginaceae, Liliaceae, Marantaceae, Restionaceae and Zannichelliaceae.

Clayton (1981) considers the possibility of origin of Poaceae from Flagellariaceae (see also R. Dahlgren 1981, 1983a, b, R. Dahlgren and Clifford 1982). However, *Flagellaria indica* shows a bisporic Allium type of embryo sac, and three ephemeral antipodals (Subramanyam and Narayana 1972). Therefore, the origin of Poaceae – monosporic Polygonum type of embryo sac with supernumerary persistent antipodals – from the Flagellariaceae is doubtful.

According to Cronquist (1981) and R. Dahlgren and Clifford (1982), other monocot families which may be considered as ancestors of Poaceae are Joinvilleaceae, Centrolepidaceae and Restionaceae – all members (along with Flagellariaceae) of the order Restionales. Of these, Restionaceae is very closely related to Poaceae; Anton and Cocucci (1984) confirm this alliance.

Archegonium in Gymnosperms and the Angiosperm Embryo Sac

Cocucci (1973a) and Favre-Duchartre (1984) again raise the controversy of homology of a four-celled archegonium with the egg apparatus and the upper polar nucleus in angiosperms. The closest example is the Oenothera type of monosporic embryo sac (Fig. 39.13.2D–H); the lower polar and antipodal nuclei are not formed. Another good example is *Butomopsis lanceolata* (Fig. 52.3.2F, G; embryo sac at double fertilization); the chalazal nucleus of the two-nucleate bisporic embryo sac degenerates. The upper nucleus divides twice and a three-celled egg apparatus and the upper polar nucleus are produced.

An eight-nucleate gametophyte is equated to two archegonia, and a 16-nucleate Penaea type of tetrasporic embryo sac to four archegonia. In *Oenothera*, *Polygonum* and *Allium*, the two synergids are sister cells, and so are the egg and the upper polar nucleus. In the Penaea type (Fig. 30.9.1G–K), only the micropylar quartet functions as in *Oenothera*, *Allium* and *Butomopsis*. The homology is carried further by equating the three antipodals and the lower polar nucleus (in *Polygonum*, *Allium* and *Adoxa*) to a second archegonium. The two lateral groups and one chalazal group in *Penaea* are equated to three archegonia. The above homology is completely unconvincing since all archegonia in gymnosperms organize in monosporic gametophytes. Moreover, the ventral canal cell always lies above the egg and cannot be equated to a polar nucleus; even though sometimes the ventral canal nucleus is fertilized as in *Ephedra* (Khan 1943).

In spite of the homologies discussed above, the embryo sac in angiosperms cannot be derived from the archegonium-bearing female gametophyte of gymnosperms. Cocucci (1973a) and Favre-Duchartre (1977, 1984) have not put forward any new facts in support of their hypothesis (see also P. Maheshwari 1950). Favre-Duchartre (1977) discusses a series of interpretations seeking homology and therotical and derived comparisons from algae, pteridophytes and gymnosperms in support of the 'archegonial theory' (see also Battaglia 1982, 1989). We do not agree with this view on the origin of the angiosperm embryo sac.

Battaglia (1989) discusses the evolution of the embryo sac of angiosperms and devotes considerable attention to the female gametophyte of *Welwitschia* and *Gnetum*, the concept of the primitive embryo sac, and variations in organization and structure of mono-, bi- and tetrasporic embryo sacs in angiosperms. However, the origin and evolution of the female gametophytes still elude us.

The angiosperm embryo sac is undoubtedly a reduced structure in comparison to the condition in gymnosperms. It should, however, be emphasized that considerable reduction and elaboration have taken place in the angiosperm embryo sac (Fig. 3.6.1).

Prospects

The families which have not yet been investigated or have not been fully investigated should receive immediate attention.

Among the families not yet investigated and those on which we have meagre information are:

Dicotyledons: Archichlamydeae – Balanopaceae, Didymelaceae, Rhoipteleaceae, Dipentodontaceae, Medusandraceae, Gyrostemonaceae, Achatocarpaceae, Dysphaniaceae, Trimeniaceae, Amborellaceae, Gomortegaceae, Tetracentraceae, Sargentodoxaceae, Lactoridaceae, Eucryphiaceae, Medusagynaceae, Strasburgeriaceae, Caryocaraceae, Quiinaceae, Ancistrocladaceae, Bataceae, Cephalotaceae, Brunelliaceae, Davidsoniaceae, Bruniaceae, Neuradaceae, Picrodendraceae, Akaniaceae, Bretschneideraceae, Julianiaceae, Pandaceae, Cardiopteridaceae, Sarcolaenaceae, Dichapetalaceae, Peridiscaceae, Scyphostegiaceae, Achariaceae, Sphaerosepataceae, Dialypetalanthaceae, Davidiaceae.

Dicotyledons: Sympetalae – Sarcospermataceae, Lissocarpaceae, Hoplestigmataceae, Desfontainiaceae, Lennoaceae, Duckeodendraceae.

Monocotyledons: Xanthorrhoeaceae, Taccaceae, Geosiridaceae, Corsiaceae, Thurniaceae, Lowiaceae, Cynastraceae.

When we have detailed embryological information on the above families, the inter- and intrafamilial and inter- and intraordinal relationships can be more meaningfully discussed.

These relationships can be further strengthened by using comparative data from (a) floral morphology and anatomy, (b) vegetative anatomy, (c) cytology, (d) palynology, and (e) chemotaxonomy.

In fact, it is now time that separate monographs on different orders, based on all the above evidence, be prepared. Every effort should be made to fill the gaps in our knowledge. Separate monographs should also be prepared on (a) major crops, including the breeding/reproductive behaviour of parents and reciprocal crosses, (b) parasites and semi-parasites (see Kuijt 1955), saprophytes, and (c) insectivorous plants. There are very few such monographs available at present.

The percentage of variations encountered during various developmental phases should be studied to determine the dominant condition.

Another important aspect which needs attention is the use of cross- and longisections as well as cleared and dissected whole mounts to trace the structure and disposition of extensions of the embryo sac, synergid, endosperm and suspensor haustoria. It is also necessary to follow the development of the embryo from longisections in various planes and transections of the various tiers. The seed-coat structure should be studied from paradermal, tran- and longisections, and macerations.

The application of histochemical techniques, and electron (TEM/SEM), fluorescence, interference (Nomaraski effect), phase contrast and polarizing microscopy has already resulted in significant advances in our knowledge.

For critical embryological investigations, a method par excellence is to study frozen sections of living material (see Coe 1954). The study of living material is also possible from enzymatically isolated male gametes from pollen tubes; the central cell of the embryo sac, synergids, egg cell; embryo sac, synergid and antipodal haustoria; endosperm, and endosperm haustoria; embryo and suspensor haustoria (see Zhou and Yang 1985; Zhou et al. 1986; Huang and Russell 1989).

The above procedures will add a dynamic approach to the study of reproductive biology, and will provide answers to many controversial and unsolved questions.

67 References

Abe K (1967) Contributions to the embryology of the family Orchidaceae 1. Development of the embryo sac in *Cymbidium goeringii* Reichb. f. Sci Rep Tôhoku Univ Ser 4 Biol 33: 79–82

Abe K (1968) Contributions to the embryology of the Orchidaceae 2. Development of the embryo sac in *Pogonia japonica*. Sci Rep Tôhoku Univ Ser 4 Biol 34: 59–65.

Abe K (1971a) Contributions to the embryology of the family Orchidaceae 4. Development of the embryo sac in *Bletilla striata*. Sci Rep Tôhoku Univ Ser 4 Biol 35: 213–218

Abe K (1971b) Contributions to the embryology of the family Orchidaceae 5. Development of embryo sac in *Oreorchis patens*. Sci Rep Tôhoku Univ Ser 4 Biol 35: 219–224

Abe K (1971c) Contributions to the embryology of the family Orchidaceae 6. Development of the embryo sac in 15 species of orchids. Sci Rep Tôhoku Univ Ser 4 Biol 36: 135–178

Abe K (1971d) Contributions to the embryology of the family Orchidaceae 7. A comparative study of the orchid embryo sac. Sci Rep Tôhoku Univ Ser 4 Biol 36: 179–201

Abe K (1976) A reinvestigation of the development of the embryo sac in *Gastrodia elata* Blume (Orchidaceae). Ann Bot (London) 40: 99–102

Abe K (1977) Development of embryo sac in *Amitostigma kinoshitae* (Makino) Schltr. (Orchidaceae). Ann Bot (London) 41: 897–899

Abraham V, Subramanyam K (1965) Exomorphic characters of seed of *Utricularia*. Proc Indian Acad Sci B 62: 97–102

Adatia RD (1946) The occurrence of entire pollen grains inside the carpels of some Annonaceae. J Univ Bombay B 14: 47–68

Adatia RD, Gavde SG (1962) Embryology of Celastraceae. In: Symp Plant embryology. Council Sci Ind Res, New Delhi, pp 1–11

Adatia RD, Shah GL (1952) A contribution to the life history of *Coriandrum sativum* Linn. J Univ Bombay B 20: 34–46

Adatia RD, Mulay BN, Hingorani GR (1950) A contribution to the embryology of *Vitis trifolia* Linn. 1. J Univ Bombay B 19: 1–10

Adatia RD, Mulay BN, Hingorani GR (1953) A contribution to the embryology of *Vitis trifolia* L. 2. J Univ Bombay B 21: 51–60

Adatia RD, Sharma YB, Vijayaraghavan MR (1971) Studies in Gesneriaceae 1. Morphology and embryology of *Platystemma violoides* Wall. Bot Not 125: 25–38

Afanaséva NG (1968) A comparative embryological study of female gametophyte of the genus *Veronica* L. (in Italian). SB Kratk Soobshtch Kazan Univ Bot Pochvoved 1: 68–72

Afzelius K (1916) Zur Embryosackentwicklung der Orchideen. Svensk Bot Tidskr 10: 183–227

Afzelius K (1918) Zur Entwicklungsgeschichte der Gattung *Gloriosa*. Acta Hortic Bergiani 6: 3–12

Afzelius K (1920) Einige Beobachtungen über die Samenentwicklung der Aponogetonaceae. Svensk Bot Tidskr 14: 168–175
Afzelius K (1936) Apomixis in der Gattung *Arnica*. Svensk Bot Tidskr 30: 572–679
Afzelius K (1954) Embryo sac development in *Epipogium aphyllum*. Svensk Bot Tidskr 48: 513–520
Afzelius K (1959) Apomixis and polyembryony in *Zygopetalum mackayi*. Acta Hortic Bergiani 19: 7–13
Agardh JG (1858) Theoria systematis plantarum. Lund (Sweden)
Agarwal Saroj (1961a) Synergid haustoria of *Quinchamalium chilense* Lam. Nature (London) 192: 1313–1314
Agarwal Saroj (1961b) The embryology of *Strombosia* Blume. Phytomorphology 11: 269–272
Agarwal Saroj (1962) Embryology of *Quinchamalium chilense* Lam. In: Symp Plant embryology. Council Sci Ind Res, New Delhi, pp 162–169
Agarwal Saroj (1963a) Morphological and embryological studies in the family Olacaceae 1. *Olax* L. Phytomorphology 13: 185–196
Agarwal Saroj (1963b) Morphological and embryological studies in the family Olacaceae 2. *Strombosia* Blume. Phytomorphology 13: 348–356
Agarwal Shikha (1984) Seed structure in some *Indigophera* species. J Indian Bot Soc 63: 11–19
Agharkar SP, Banerji I (1930) The development of the embryo sac in *Carica papaya*. J Sci Calcutta Univ 10: 1–10
Agoram S, Krishnamurthy KV (1980) Further contributions to the embryology of *Antigonon leptopus* H.K. & A. In: Periasamy K (ed) Symp Histochemistry, developmental and structural anatomy of angiosperms. P&B, Tiruchirapalli (India), pp 1–8
Agrawal JS (1950) The embryo sac of *Fritillaria liliacea*. Proc Natl Inst Sci India 16: 87–92
Agrawal JS (1952) The embryology of *Lilaea subulata* H.B.K. with a discussion on its systematic position. Phytomorphology 2: 15–29
Agrawal JS (1954) Female gametophyte, endosperm and embryo of *Scurrula* Linn. In: Proc 41 Indian Sci Congr, Hyderabad, pt 3: 137
Agrawal JS, Singh SP (1957) Nucellar polyembryony in *Momordica charantia* Linn. Sci Cult 22: 630–631
Ahlstrand L (1978) Embryology of *Ursinia* (Compositae). Bot Not 131: 487–496
Ahlstrand L (1979a) Embryology of Arctoteae (Compositae). Bot Not 132: 109–116
Ahlstrand L (1979b) Embryology of Arctotideae-Gorteriinae (Compositae). Bot Not 132: 371–376
Ahlstrand L (1979c) Embryology of Arctotideae-Gundeliinae (Compositae). Bot Not 132: 377–380
Ahlstrand L (1979d) Embryology of Arctotideae and Calenduleae (Compositae). PhD Thesis, Univ Göteborg (Sweden)
Ahlstrand L (1985) Embryology of Calenduleae (Compositae). Nord J Bot 5: 79–97
Ahuja BS (1964) Floral anatomy and gametophytes of *Hibiscus micranthus* Linn. Proc Natl Acad Sci India B 34: 73–83
Ahuja YR, Bhaduri PN (1956) The embryology of *Brassica campestris* L. var. *toria* Duth. & Full. Phytomorphology 6: 63–67
Airy Shaw HK (1941) Additions to the flora of Borneo and other Malay Islands: The Pentaphragmataceae of the Oxford University Expedition to Sarawak, 1932. Kew Bull 3: 233–236
Airy Shaw HK (1948) Sphenocleaceae. Flora Malesiana 4: 27–28
Airy Shaw HK (1951) Note on the taxonomic position of *Nyctanthes* L. and *Dimetra* Kerr. Kew Bull 2: 271–272
Airy Shaw HK (1954) Pentaphragmataceae. In: Van Steenis CGGJ (ed) Flora Malesiana 4: 517–528
Airy Shaw HK (1973) JC Willis: a dictionary of flowering plants and ferns, edn 8. Univ Press, Cambridge (England)
Al-Attar AAA (1977) Occurrence of non-nucleate cytoplasmic vesicles during development of embryo and endosperm in *Torilis nodosa* and *Pseudorlaya pumila*. Proc Indian Acad Sci B 85: 364–368
Al-Attar AAA (1979) Morphological and embryological investigation of *Daucus muricatus* (L.) L. (Caucalideae, Umbelliferae). Proc Indian Acad Sci B 88: 435–443
Alemany PJ (1985) Flor, Esporogenesis Y gametogenesis de *Nierembergia hippomanica* (Solanaceae). Bol Soc Arg Bot 24: 49–69
Alpi A, Tognoni F, D'Amato F (1975) Growth regulator levels in embryo and suspensor of *Phaseolus coccineus* at two stages of development. Planta 127: 153–165

Ambegaokar KB (1959) Embryological studies on the family Acanthaceae. PhD Thesis, Univ Baroda (India)
Ambegaokar KB (1976) Seed anatomy of some legumes. Indian J Agric Sci 46: 8–15
Ambegaokar KB, Johri BM (1977) Seed development in Triticale-1. Phytomorphology 27: 190–198
Amici GB (1824) Observations microscopiques sur diverses espèces de plantes. Ann Sci Nat Bot 2: 41–70, 211–248
Amici GB (1830) Note sur le mode d'action du pollen sur le stigmate. Extrait d'une lattre d'Amici à Mirbel. Ann Sci Nat Bot 21: 329–332
Amici GB (1844) Quatrième réunion des naturalistes italiens, Padua 1843. Flora 1: 359
Amici GB (1847) Sur la fécondation des Orchidées. Ann Sci Nat Bot 718: 193–205
Anantaswamy Rau M (= Rau MA) (1940a) An embryological study of *Suriana maritima* Linn. Proc Indian Acad Sci B 11: 100–106
Anantaswamy Rau M (1940b) Studies in the Apocynaceae. J Indian Bot Soc 19: 33–44
Anantaswamy Rau M (1942) Development of the embryo sac and embryo in *Streblus asper* Lour. J Mysore Univ B 2: 109–114
Anantaswamy Rau M (1950a) Integumentary vascular tissue in *Cassia tora* Linn. Curr Sci 19: 186–187
Anantaswamy Rau M (1950b) The suspensor haustoria of some species of *Crotalaria* Linn. Ann Bot (London) 14: 557–562
Anantaswamy Rau M (1950c) Development of the embryo in *Trigonella foenum-graecum* L. J Indian Bot Soc 29: 210–213
Anantaswamy Rau M (1950d) Endosperm in *Cassia tora* Linn. Nature (London) 163: 157
Anantaswamy Rau M (1951a) The endosperm in some species of *Cassia* L. Svensk Bot Tidskr 45: 516–522
Anantaswamy Rau M (1951b) The mechanism of nutrition in the developing seed of *Vigna catjang* Endl. New Phytol 50: 121–123
Anantaswamy Rau M (1951c) The endosperm in *Crotalaria*. Curr Sci 20: 73–74
Anantaswamy Rau M (1951d) Development of embryo in some members of the Papilionaceae. Phytomorphology 1: 80–86
Anantaswamy Rau M (1951e) The endosperm in some of the Papilionaceae. Phytomorphology 1: 153–158
Anantaswamy Rau M (1951f) Development of the embryo in *Aeschynomene indica* Linn. New Phytol 50: 124–126
Anantaswamy Rau M (1951g) The endosperm in *Rothia trifoliata* Pers. Ann Bot (London) 15: 175–177
Anantaswamy Rau M (1953) Some observations on the endosperm in Papilionaceae. Phytomorphology 3: 209–222
Anantaswamy Rau M (1954) The development of the embryo of *Cyamopsis*, *Desmodium* and *Lespedeza* with a discussion of the position of the Papilionaceae in the system of embryogenic classification. Phytomorphology 4: 418–430
Anantaswamy Rau M (1955) Embryological studies in the Leguminosae. J Mysore Univ B 14: 63–75
Anderson A (1931) Studien über der Embryologie der Familien Celastraceae, Oleaceae und Apocynaceae. Acta Univ Lund 27: 1–112
Anderson F (1922) The development of the flower and embryology of *Martynia louisiana*. Bull Torrey Bot Club 49: 141–155
Anderson JM, Barrett SCH (1986) Pollen tube growth in tristylous *Pontederia cordata* (Pontederiaceae). Can J Bot 64: 2602–2607
Anderson LC (1970) Embryology of Chrysanthemums (Astereae-Compositae). Madrono 20: 337–342
Anderson LE (1939) Cytoplasmic inclusions in the male gametes of *Lilium*. Am J Bot 26: 761–766
Andrews FM (1895) Development of the embryo sac of *Jeffersonia diphylla*. Bot Gaz 20: 423–424
Andrews FM (1910) The development of the embryo sac of *Hybanthus concolor*. Bull Torrey Bot Club 37: 477–478
Angold RE (1968) The formation of the generative cell in the pollen grain of *Endymion nonscriptus*. J Cell Sci 3: 573–578

Ankermann F (1927) Die Phylogenie der Monocotyledonen. Bot Arch 19: 51–60

Annen E (1945) Die Embryosack- und Pollenentwicklung bei einigen polyploiden Garten-Astern im Vergleich mit der wildwachsenden *Aster amellus*. Ber Schweiz Bot Ges 55: 81

Anton AM, Cocucci AE (1984) The grass megagametophyte and its possible phylogenetic implications. Plant Syst Evol 146: 117–121

Anzalone B (1949) Osservazioni citoembryologiche in *Taraxacum megalorrhizon*. Ann Bot (Rome) 23: 31–41

Arekal GD (= Govindappa DA) (1955) Development of the gametophytes in *Aloe ciliaris* Haw. J Indian Bot Soc 34: 146–150

Arekal GD (1961a) Contribution to the embryology of *Hoppea dichotoma*. Can J Bot 39: 1001–1006

Arekal GD (1961b) Embryology of *Klugia notoniana*. Bot Gaz 123: 144–150

Arekal GD (1963a) Contribution to the embryology of *Chelone glabra* L. Phytomorphology 13: 376–388

Arekal GD (1963b) Embryological studies in Canadian representatives of the tribe Rhinantheae, Scrophulariaceae. Can J Bot 41: 267–302

Arekal GD (1963c) Contribution to the embryology of *Chaenorrhinum minus* (L.) Lange. Proc Indian Acad Sci B 58: 375–385

Arekal GD (1964) Contribution to the embryology of *Gerardia pedicularia* L. (Scrophulariaceae). J Indian Bot Soc 43: 409–423

Arekal GD (1965) Embryology of *Mimulus ringens*. Bot Gaz 126: 58–66

Arekal GD (1966) Embryology of *Veronica serpyllifolia* L. Proc Indian Acad Sci B 64: 241–253

Arekal GD (1967) Contribution to the embryology of *Hypoxis aurea* Lour. J Indian Bot Soc 46: 193–198

Arekal GD, Karanth KA (1981) The embryology of *Epipogium roseum*. Plant Syst Evol 138: 1–8

Arekal GD, Nagendran CR (1975a) Is there a Podostemum type of embryo sac in genus *Farmeria?* Caryologia 28: 229–235

Arekal GD, Nagendran CR (1975b) Embryo sac of *Hydrobryopsis sessilis* (Podostemaceae): Origin, organization and significance. Bot Not 128: 332–338

Arekal GD, Nagendran CR (1976) A new type of embryo sac organization in angiosperms. Curr Sci 45: 717–719

Arekal GD, Nagendran CR (1977a) Female gametophytes of *Zeylanidium* (Podostemaceae): A clarification. Phytomorphology 27: 123–129

Arekal GD, Nagendran CR (1977b) The female gametophyte in two Indian genera of Tristichoideae (Podostemaceae): a reinvestigation. Proc Indian Acad Sci B 86: 287–294

Arekal GD, Raju D (1964) The female gametophyte of *Linaria ramosissima* Wall. Curr Sci 33: 591–592

Arekal GD, Raju D (1967) Endosperm in the genus *Sopubia* Ham. Curr Sci 36: 22–23

Arekal GD, Ramaswamy SN (1973) Embryology of *Burmannia pusilla* (Wall. ex Miers) THW. and its taxonomic status. Beitr Biol Pflanzen 49: 35–46

Arekal GD, Ramaswamy SN (1974) The endosperm organization in *Microcarpaea* R. Br. (Scrophulariaceae). Curr Sci 43: 87–88

Arekal GD, Ramaswamy SN (1980) Embryology of *Eriocaulon hookerianum* Stapf. and the systematic position of Eriocaulaceae. Bot Not 133: 295–309

Arekal GD, Shivamurthy GR (1976) 'Seed' germination in *Balanophora abbreviata*. Phytomorphology 26: 135–138

Arekal GD, Shivamurthy GR (1978) Female gametophyte in two species of *Balanophora*: a reinvestigation. Phytomorphology 28: 7–13

Arekal GD, Girijamma S, Ramaswamy SN (1970) Contribution to the embryology of *Lindernia hyssopioides* (L.) Haines. Proc Indian Acad Sci B 72: 221–235

Arekal GD, Rajeshwari S, Ramaswamy SN (1971) Contribution to the embryology of *Scoparia dulcis* L. Bot Not 124: 237–248

Argue CL (1985) Pollen morphology in the genera *Mouttea* and *Melosperma* (Scrophulariaceae). Am J Bot 72: 1248–1255

Argue CL (1986) Some taxonomic implications of pollen and seed morphology in *Mimulus hymenophyllus* and *M. jungermannioides* and comparisons with other putative members of the *M. moschatus* alliance (Scrophulariaceae). Can J Bot 64: 1331–1337

Ariz P, Iftikhar N, Khan NR (1972) The megasporogenesis and the development of embryo sac in *Withania somnifera*. Pak J Sci Ind Res 15: 196–198

Armbruster WS (1984) The role of resin in angiosperm pollination: Ecological and chemical considerations. Am J Bot 71: 1149–1160

Armstrong JE (1985) The delimitation of Bignoniaceae and Scrophulariaceae based on floral anatomy, and the placement of problem genera. Am J Bot 72: 755–756

Arnason TJ (1943) Female sterility in potatoes. Can J Res C21: 41–56

Arnott HJ (1959a) Vivipary in *Cordyline australis*. Madrono 15: 71–73

Arnott HJ (1959b) The cotyledon of *Yucca*. In: 9th Int Bot Congr Montreal 2: 9–10

Arnott HJ (1961) The morphology and anatomy of *Yucca* seeds. Am J Bot 48: 531

Arnott HJ (1962) The seed germination and seedling of *Yucca*. Univ Cal Publ Bot 35: 1–44

Arora N (1953) The embryology of *Zizyphus rotundifolia* Lamk. Phytomorphology 3: 88–98

Arora S, Kapil RN (1989) Comparative study of pollinia of two species of *Pholidota* Lindl. Phytomorphology 39: 343–352

Artschwager E (1927) Development of flowers and seed in the sugar beet. J Agric Res (USA) 34: 1–25

Artschwager E (1947) Pollen degeneration in male-sterile sugar beets, with special reference to the tapetal plasmodium. J Agric Res (USA) 75: 191–197

Artschwager E, Starrett RC (1933) The time factor in fertilization and embryo development in the sugar beet. J Agric Res (USA) 47: 823–843

Arunalakshmi V (1989) Structure and development of seed-coat in *Cleome*. J Indian Bot Soc 68: 116–121

Asana JJ, Adatia RD (1932) Microsporogenesis in *Luffa aegyptica*. J Indian Bot Soc 11: 181–187

Asana JJ, Adatia RD (1947) Contributions to the embryology of the Annonaceae 1. *Artabotrys odoratissimus* R. Br. J Univ Bombay B 16: 7–21

Ashley T (1975a) Fine structure of early endosperm development in *Hibiscus*. Caryologia 28: 63–71

Ashley T (1975b) Alterations in the fine structure of developing endosperm of *Hibicus* hybrids. Caryologia 28: 73–80

Ashley T (1975c) Zygote shrinkage and subsequent development in some *Hibiscus* hybrids. Planta 108: 303–317

Asker S (1970) Apomixis and sexuality in the *Potentilla argentea* complex 2. Crosses within the species. Hereditas 66: 189–204

Asker S (1977) Pseudogamy, hybridization, and evolution in *Potentilla*, Hereditas 87: 179–184

Asplund E (1920) Studien über die Entwicklungsgeschichte der Blüten einiger Valerianaceen. Kon Svensk Vet Akad Handl 61: 1–66

Asplund I (1968) Embryological studies in the genera *Sparganium* and *Typha*: A preliminary report. Svensk Bot Tidskr 62: 410–412

Asplund I (1972) Embryological studies in genus *Typha*. Svensk Bot Tidskr 66: 1–17

Asplund I (1973) Embryological studies in the genus *Sparganium*. Svensk Bot Tidskr 67: 177–200

Asplund I (1978) Embryological studies in Typhaceae, Sparganiaceae and Compositae. PhD Thesis, Univ Göteborg (Sweden)

Atabekova AI, Ling C-H (1962) The embryology of lupine. In: Lupin (Lupine). Nauka Moscow, pp 97–113 (in Russian)

Audran JC, Batcho M (1981) Microsporogenesis and pollen grains in *Silene dioica* (L.) Cl. and alterations in its anthers parasitised by *Ustilago violaceae* (Pers.) Rouss. (Ustilaginales). Acta Soc Bot Pol 50: 29–32

Aulbach-Smith CA, Herr Jr JM (1984) Development of the ovule and female gametophyte in *Eustachys petraea* and *E. glauca* (Poaceae). Am J Bot 71: 427–438

Avanzi HG (1948) Osservazioni cito-embryilogiche su *Ammobium alatum* R. Br. (Asteraceae). Caryologia 1: 83–91

Avanzi HG (1950) Endomitosi e mitosi a diplocromosomi nello sviluppo delle cellule del tappeto di *Solanum tuberosum* L. Caryologia 2: 205–222

Avery GS Jr (1930) Comparative anatomy and morphology of embryos and seedlings of maize, oats and wheat. Bot Gaz 89: 1–39

Babis H (1966) Embryo development in *Delphinium cardiopetalum* D.C. Genet Pol 7: 137–151
Babis H (1973) General description of a mature ovule and embryo sac, early stages of development of the endosperm and of embryo development in some species of *Delphinium* L. genus. Genet Pol 14: 29–36
Babis H (1974) Morphological classification of tetrads occurring in *Delphinium* L. genus: Variation in their structure. Genet Pol 15: 7–12
Babis H (1976a) Early stage of embryo development in *Delphinium belladonna* (Kelw) Bergmans, and philogenesis of this species. Genet Pol 17: 325–331
Babis H (1976b) Development of the embryo in some species of the genus *Delphinium* L. Acta Soc Bot Pol 45: 135–142
Babis H (1978) Early stages of embryo development in *Delphinium paniculatum* Host. Acta Soc Bot Pol 47: 359–363
Babis H (1979a) Diversity of embryo development in *Delphinium* L. genus 1. The zygote divides transversely. Embryogenesis in *Delphinium staphisagria* L. and *D. elatum* L. Genet Pol 20: 103–116
Babis H (1979b) Diversity of embryo development in *Delphinium* L. genus 2. The zygote divides obliquely. Embryogenesis in *Delphinium tricolor* Bernh. Genet Pol 20: 155–166
Babis H (1979c) Diversity of embryo development in *Delphinium* L. genus 3. The zygote divides obliquely. Early stages of embryogenesis in *Delphinium cashmirianum* Royle, *D. grandiflorum* L., *D. nudicaule* Torr. et Gray and *D. zalil* Aitch. Genet Pol 20: 167–178
Bacchi O (1943) Cytological observations in *Citrus* 2. Megasporogenesis, fertilization, endosperm and polyembryony. Bot Gaz 105: 221–225
Baciu E (1971) Contributii la stadium microsporogenesci la unele specii din subfamilia Prunoideae. Lucr Sci Inst Agron N Balescu 14: 241–245
Baciu E (1972) Contributions to the study of macrosporogenesis and embryo sac formation in some Prunoideae. Lucr Sci Inst Agron N Balescu 15: 251–256
Bahadur KN, Ram D, Naithani HB (1976) Proliferation in Indian grasses. Acta Bot Indica 4: 151–153
Bailey IW, Smith AC (1942) Degeneriaceae: A new family of flowering plants from Fiji. J Arnold Arbor 23: 356–365
Bailey IW, Swamy BGL (1948) *Amborella trichopoda* Baill: A new morphological type of vesselless dicotyledon. J Arnold Arbor 29: 245–253
Bailey IW, Swamy BGL (1949) The morphology and relationships of *Austrobaileya*. J Arnold Arbor 30: 211–226
Bailey IW, Nast CG, Smith AC (1943) The family Himatandraceae. J Arnold Arbor 24: 190–206
Baillon H (1858) Recherches sur l'organogenie du *Callitriche* et sur ses rapports naturels. Bull Soc Bot Fr 5: 337–341
Baillon H (1871) Nouvelles notes sur les Hamamélidacées. Adansonia 10: 120–137
Baillon H (1874) Histoire des plantes, Vol 5. Paris
Baillon H (1892) Histoire des plantes, Vol 11. Paris
Bajpai PN (1965) Studies on the vegetative growth and development of male and female gametophytes in aonla, *Emblica officinalis* Gaertn. Agra Univ J Res Sci 14: 167–186
Baker HG, Baker I (1979) Starch in angiosperm pollen grains and its evolutionary significance. Am J Bot 66: 591–600
Bakshi TS (1952) Floral morphology and embryology of *Psilostachys serica* Hook. f. Phytomorphology 2: 151–161
Bakshi TS (1954) The antipodals in the family Amaranthaceae. Curr Sci 23: 128–129
Bakshi TS (1959) Ecology and morphology of *Pterospora andromeda*. Bot Gaz 120: 203–217
Balasubramanium G, Sunderaraj DD (1957) Vivipary in *Benincasa cerifera* (ash gourd). Sci Cult 22: 509–510
Balasubramanium R (1931) Parthenogenesis in cotton. Madras Agric J 19: 509–516
Balasubramanian V (1980) Development of gametophytes in *Rouwolfia beddomei* Hook. Curr Sci 49: 559–560
Balasubramanian V (1982) Studies in Apocynaceae 1. Embryology of *Aganosma cymosa* G. Don. Swamy Bot Club Newslett 1: 24–31
Balasubramanian V (1983) Embryology of *Anodendron paniculatum* A. DC. Curr Sci 83: 883–885

Balasubramanian V, Padmanabhan D (1985) Contributions to the embryology of *Canscora sessiliflora* Roem. J Swamy Bot Club 2: 27–32
Balfour E (1957) The development of the vascular systems in *Macropiper excelsa* Forst. 1. The embryo and seedling. Phytomorphology 7: 354–364
Ballantine AJ (1909) A preliminary note on the embryo sac of *Protea lepidocarpon* R. Br. Ann Bot (London) 23: 161–162
Balogh P (1982) Generic redefinition in subtribe Spiranthinae (Orchidaceae). Am J Bot 69: 1119–1132
Bambacioni V (1928a) Ricerche sulla ecologia e sulla embriologia di *Fritillaria persica*. Ann Bot (Rome) 18: 7–37
Bambacioni V (1928b) Come avvienne in *Fritillaria persica* lo sviluppo del gametofito femminile e l'aumento dei cromosomi nella regione chalazale. CR Acad Lincei (Rome) 6: 544–546
Bambacioni V (1928c) Contributo alla embriologia di *Lilium candidum* CR Acad Lincei (Rome) 8: 612–618
Bambacioni-Mezzetti V (1931) Sullo sviluppo dell'embrione in *Tulipa gesneriana*. Ann Bot (Rome) 19: 145–155
Bambacioni-Mezzetti V (1935) Ricerche morfologiche sulle Lauraceae. Lo sviluppo dell'ovulo e dei sacchi pollinici nel *Laurus nobilis* L. Ann Bot (Rome) 21: 1–19
Bambacioni-Mezzetti V (1938) Ricerche morfologiche sulle Lauraceae. Lo sviluppo dell'ovulo e dei sacchi pollinica nel *Laurus nobilis*. Ann Bot (Rome) 24: 186–204
Bambacioni-Mezzetti V (1940) Le modificazioni dell'apice nucellare di *Tulipa gesneriana* in seguito all'impollinazioni. Nuovo G Bot Ital 47: 421–423
Bambacioni-Mezzetti V (1943) Sulla constituzione del gametofitd femminile adulto e sull'origine dell'embrione in *Tulipa gesneriana* L. Ann Bot (Rome) 22: 153–172
Bambacioni-Mezzetti V (1947) Sull'embriologia della *Fritillaria messanensis* e sull'opportunita di distinguere una serie *Fritillaria* nel tipo *Euphorbia dulcis*. Ann Bot (Rome) 23: 1–23
Banchetti C (1961) The development of the female gametophyte of *Senecio leucanthemifolia* var. *pinnatifidus*. Caryologia 14: 303–311
Banerji I (1932) The development of the embryo sac and fertilization in jute. J Indian Bot Soc 11: 228–240
Banerji I (1933) The development of the flower and pollen in jute. Indian J Agric Sci 3: 116–126
Banerji I (1934) Sterility of the female gametophyte of *Colocasia antiquorum*. Curr Sci 2: 432
Banerji I (1937) Sterility in *Colocasia antiquorum* Schott. J Indian Bot Soc 16: 159–164
Banerji I (1940a) A contribution to the life history of *Tridax procumbens* Linn. J Bombay Nat Hist Soc 42: 89–99
Banerji I (1940b) A contribution to the life history of *Tridax procumbens*. Proc Indian Acad Sci B 11: 325–336
Banerji I (1940c) A contribution to the life history of *Costus speciosus* Smith. J Indian Bot Soc 19: 181–196
Banerji I (1942a) The development of the female gametophyte and floss in *Bombax malabaricum* DC. Proc Indian Acad Sci B 16: 205–211
Banerji I (1942b) A contribution to the life history of *Blumea lacinata* L. J Indian Bot Soc 21: 295–307
Banerji I (1947) Life history of *Typhonium trilobatum* Schott. Proc Natl Inst Sci India B 13: 207–230
Banerji I (1949) A contribution to the life history of *Acalypha fallax* Muell. Bull Bot Soc Bengal 3: 29–32
Banerji I (1954) Morphological and cytological studies on *Citrus grandis* Osbeck. Phytomorphology 4: 390–396
Banerji I (1961) The endosperm in Scrophulariaceae. J Indian Bot Soc 40: 1–11
Banerji I, Chaudhuri KL (1944) A contribution to the life history of *Litchi chinensis* Sonn. Proc Indian Acad Sci B 19: 19–37
Banerji I, Dutt MK (1944) The development of the female gametophyte in some members of the Euphorbiaceae. Proc Indian Acad Sci B 20: 51–60
Banerji I, Gangulee HC (1937) Spermatogenesis in *Eiçhhornia crassipes* Solms. J Indian Bot Soc 16: 289–295

Banerji I, Hakkim A (1954) A contribution to the life-history of *Artocarpus lakoocha* Roxb. Proc Indian Acad Sci B 39: 128–132
Banerji I, Haldar S (1942) A contribution to the morphology and cytology of *Monocharia hastaefolia* Presl. Proc Indian Acad Sci B 16: 91–106
Banerji I, Pal S (1959) A contribution to the life history of *Synedrella nodiflora* Gaertn. Bot J Linn Soc 55: 810–817
Banerji I, Sen S (1954) A contribution to the cytology and embryology of *Hylocereus undatus* (Haw.) Br. & R. Bull Bot Soc Bengal 8: 18–23
Banerji I, Venkateswarlu V (1935) A preliminary note on the development of the female gametophyte in *Costus speciosus*. Curr Sci 4: 414
Banerjee N, Sharma AK (1988) Cytomixis in microsporocytes of *Rauwolfia serpentina* Benth. Curr Sci 57: 267–268
Banerjee UC, Barghoorn ES (1971a) The tapetal membranes in grasses and Übisch body control of mature exine pattern. In: Heslop-Harrison J (ed) Pollen development and physiology. Butterworth London, pp 126–127
Banerjee UC, Barghoorn ES (1971b) How is pollen grain and spore exine pattern controlled (Abstr)? Am J Bot 58: 457
Barber HN (1942) The pollen grain division in the Orchidaceae. J Genet 43: 97–103
Barkworth ME (1982) Embryological characters and taxonomy of the Stipeae (Gramineae). Taxon 31: 233–243
Barlow BA (1983) Biogeography of Loranthaceae and Viscaceae. In: Calder M, Bernhardt P (eds) The biology of mistletoes. Academic Press, New York London. Sydney, pp 19–46
Bartels F (1956) Zur Entwicklung der Keimpflanze von *Epilobium hirsutum* 1. Die ablaufenden Zellteilungen bis zum beginnenden "herzförmingen Embryo". Flora 144: 105–120
Barthlott W (1981) Epidermal and seed surface characters of plants: Systematic applicability and some evolutionary aspects. Norw J Bot 1: 345–355
Barthlott W, Voit G (1979) Seed-coat morphology and taxonomy of Cactaceae: A scanning electron microscopic survey. Plant Syst Evol 132: 205–230
Bartling Fr Th (1830) Ordines naturales plantarum eorumque characters et affinitates adjecta generum Enumeratione. Göttingen (FRG)
Basheer MD, Prakasa Rao PS (1984) The genesis of embryo in *Utricularia bifida* Linn.: a reinvestigation. Curr Sci 53: 502–504
Baskin JM, Baskin CC (1971) Polyembryony in *Ruellia humilis* Nutt. Bull Torrey Bot Club 98: 109–110
Battaglia E (1940) Embriologia di *Ligularia kaempferi*. Nuovo G Bot Ital 47: 271–286
Battaglia E (1941) Contributo all'embriologia delle Tamaricaceae. Nuovo G Bot Ital 48: 575–612
Battaglia E (1943) Alcune osservazione sullo sviluppo del gametofito femmineo della *Myricaria germanica*. Nuovo G Bot Ital 49: 464–467
Battaglia E (1945) Fenomeni citologici nuovi nella embriogenesi (semigamia) e nella microsporogenesi (dopio nucleo di restituzione) di *Rudbeckia laciniata* L. Nuovo G Bot Ital 52: 34–38
Battaglia E (1946a) Ricerche cariologiche e embriologiche sul genere *Rudbeckia* 1–5: Il gametofito femminile e maschile di *R. bicolor*, *R. hirta*, *R. hirta* var. *meine freude*, *R. amplexicaule* e *R. purpurea* (*Echinacea purpurea*). Nuovo G Bot Ital 53: 1–26
Battaglia E (1946b) Ricerche cariologiche e embriologiche sul genere *Rudbeckia* (Asteraceae) 6. Apomissia in *Rudbeckia speciosa* Wender. Nuovo G Bot Ital 53: 27–69
Battaglia E (1946c) Ricerche cariologiche e embriologiche sul genere *Rudbeckia* (Asteraceae) 7. Apomissa in *Rudbeckia laciniata* L. ed. amfimissia nella sua varieta a fiori doppi. Nuovo G Bot Ital 53: 437–482
Battaglia E (1946d) Ricerche cariologiche e embriologiche sul genere *Rudbeckia* 8. Semigamia in *R. laciniata*. Nuovo G Bot Ital 53: 483–511
Battaglia E (1947a) Apomissia in *Hieracium ramosum*. Atti Soc Tosc Sci Nat Pisa B 54: 50–59
Battaglia E (1947b) Ricerche cariologiche e embriologiche sul genere *Rudbeckia* (Asteraceae) 9. Le anomalie del gametofito femminile cellularizzato di *Rudbeckia laciniata* L. Nuovo G Bot Ital 54: 377–405
Battaglia E (1947c) Ricerche cariologiche e embryologiche sul genere *Rudbeckia* 10. Le anomalie della meiosi durante la microsporogenesi di *Rudbeckia laciniata* L. con particolare riguardo alla formazione del nucleo di restituzione. Nuovo G Bot Ital 54: 406–431

Battaglia E (1947d) Ricerche cariologiche e embriologiche sul genere *Rudbeckia* 11. Semigamia in *R. speciosa.* Nuovo G Bot Ital 54: 531–559

Battaglia E (1947e) Ricerche cariologiche e embriologiche sul genere *Rudbeckia* 12. Gametofito femminile e maschile di *R. flava.* Nuovo G Bot Ital 54: 560–567

Battaglia E (1948) Ricerche sulla parameiosi restituzionale nel genere *Taraxacum.* Caryologia 1: 1–47

Battaglia E (1950) L'alterazione della meiosi nella riproduzione apomittica di *Erigeron karuinskianus* D.C. var. *mucronatus* D.C. (Asteraceae). Caryologia 2: 165–204

Battaglia E (1951a) The male and female gametophytes of angiosperms: An interpretation. Phytomorphology 1: 1–30

Battaglia E (1951b) Development of tetrasporic embryo sac of *Chrysanthemum viscosum.* Bot Gaz 112: 490–494

Battaglia E (1952) Nuovi reperti di apomissia e di anfimissia nel genere *Rudbeckia* (Compositae). Atti Soc Tosc Sci Nat Pisa B 59: 203–209

Battaglia E (1955) Unusual cytological features in the apomictic *Rudbeckia sullivantii* Boynton et Beadle. Caryologia 8: 1–32

Battaglia E (1958a) L'abolizione del tipo embryiologico *Scilla* e la creazione dei nuovi tipo *Endymion* e *Allium.* Caryologia 11: 247–252

Battaglia E (1958b) Osservazioni critiche sul tipo di sviluppo del gametofito femminile attribuito a *Scilla hyacinthoides.* Caryologia 11: 253–258

Battaglia E (1958c) Ricerche embriologiche in *Scilla autumnalis* e *Scilla obtusifolia.* Nota preliminare. Nuovo G Bot Ital 65: 397–398

Battaglia E (1959) New types of embryo sac development in *Scilla autumnalis.* In: 9th Int Bot Congr Montreal 2: 23–24

Battaglia E (1963) Apomixis. In: Maheshwari P (ed) Recent advances in the embryology of angiosperms. Int Soc Plant Morphol, Univ Delhi, pp 221–264

Battaglia E (1971) The embryo sac of Podostemaceae: an interpretation. Caryologia 24: 403–420

Battaglia E (1981) Embryological questions 3. Semigamy, hemigamy, and gynandroembryony. Ann Bot (Rome) 39: 173–175

Battaglia E (1982) Embryological questions: 4. Gynogonium versus archegonium and the generalization of the prefixes andro- and gyno- in plant reproduction. Appendix: Bischoff TG (1835) "De Hepaticis . . . " Ann Bot (Rome) 40: 1–45

Battaglia E (1988) Embryological questions 13. Can the atypical 4-nucleate embryo sacs of *Tamarix* assigned to the Plumbagella type (haploid egg) be reinterpreted as new Tamarix types (diploid egg)? Atti Soc Tosc Sci Nat Mern Anno Ser B 95: 69–82

Battaglia E (1989) The evolution of the female gametophyte of angiosperms: an interpretative key (embryological questions 14). Ann Bot (Rome) 47: 7–144

Battaglia E, Feeley E (1959) The embryo sac of *Scilla pratensis.* Caryologia 11: 407–414

Batygina TB (1968) Embryogenesis in the genus *Triticum* L. (as related to the problems of monocotyledony and remote hybridization in Gramineae). Bot Zh 53: 480–490 (in Russian)

Batygina TB (1969) On the possibility of separation of a new type of embryogenesis in angiospermae. Rev Cytol Biol Veg 32: 335–341

Batygina TB (1971) Some peculiar features of macrosporogenesis and development of the female gametophyte in wheat. Ann Univ ARERS 9: 46–50

Batygina TB, Yakovlev MS (eds) (1985) Comparative embryology of flowering plants, vol 3. Brunelliaceae-Tremandraceae. Nauka, Leningrad (in Russian)

Batygina TB, Yakovlev MS (eds) (1987) Comparative embryology of flowering plants, vol 4. Davidiaceae-Asteraceae. Nauka, Leningrad (in Russian)

Batygina TB, Kolesova GE, Vasiljeva VE (1983) Embryology of Nymphaeales 3. Embryogenesis of *Nelumbo nucifera.* Bot Zh 68: 311–325

Batygina TB, Teryokhin ES, Alimova GK, Yakovlev MS (1963) Genesis of male sporangia in families Gramineae and Ericaceae. Bot Zh 48: 1108–1120 (in Russian)

Baude E (1956) Die Embryoentwicklung von *Stratiotes aloides* L. Planta 46: 640–671

Bawa SB (1969a) Embryological studies on the Haloragidaceae 3. *Myriophyllum intermedium* D.C. Beitr Biol Pflanzen 45: 447–464

Bawa SB (1969b) Embryological studies on the Haloragidaceae 2. *Laurenbergia brevipes* Schindi, and a discussion on systematic considerations. Proc Natl Inst Sci India B 35: 275–290

Bawa SB (1970a) Theaceae. In: Symp Comparative embryology of angiosperms. Indian Natl Sci Acad Bull 41: 75–77
Bawa SB (1970b) Haloragaceae. In: Symp Comparative embryology of angiosperms. Indian Natl Sci Acad Bull 41: 226–229
Beardsell DV, William EG, Knox RB (1989) The structure and histochemistry of the nectary and anther secretory tissue of the flowers of *Thryptomene calycina* (Lindl) Stapf (Myrtaceae). Aust J Bot 37: 65–79
Beattie AJ, Turnbull C, Knox RB, Williams EG (1984) Ant inhibition of pollen function: A possible reason why ant pollination is rare. Am J Bot 71: 421–426
Beatty JW, Betty AV (1953) Duration of the stages in microspore development and in the first microspore division of *Tradescantia paludosa*. Am J Bot 40: 593–596
Beaudry JR (1951) Seed development following the mating *Elymus virginalis* L. × *Agropyron repens* (L.) Beauv. Genetics 36: 109–133
Bedalov M, Hesse M (1989) On some pollen types within *Arum* (Araceae). Plant Syst Evol 166: 41–44
Begum M (1965) Studies on the embryology of *Utricularia graminifolia* Vahl. Curr Sci 34: 355–356
Begum M (1966) Sporogenesis and the development of gametophytes in *Eriocaulon quinquangulare* Linn. Curr Sci 35: 262–263
Begum M (1968) Embryological studies in *Eriocaulon quinquangulare* Linn. Proc Indian Acad Sci B 67: 148–156
Begum M (1969) A note on the development of endosperm and embryo in *Utricularia graminifolia* Vahl. Curr Sci 38: 445–446
Behl HM, Tiagi B (1976) Seed development in Fabaceae: Phaseoleae and Vicieae (Abstr). In: Embryology of crop plants. Indo-Sov Symp. Indian Natl Sci Acad, New Delhi; and Univ Delhi, p 1
Behl HM, Tiagi B (1977a) Developmental anatomy of seed and fruit in *Cicer arietinum* Linn. J Indian Bot Soc 56: 313–321
Behl HM, Tiagi B (1977b) Development of seed in *Rhynchosia*, *Cajanus* and *Atylosia* (subtribe Cajaneae). Proc Indian Acad Sci B 86: 303–310
Behl HM, Tiagi B (1978) The haustorial embryonal suspensor and endosperm in Fabaceae (Phaseoleae and Vicieae). In: Malik CP, Srivastava AK, Bhattacharya NC, Singh R (eds) Physiology of sexual reproduction in flowering plants. Kalyani, Ludhiana, pp 527–531
Bell HP (1957) The development of the blueberry seed. Can J Bot 35: 139–153
Belling J, Blakeslee HF (1924) The distribution of chromosomes in tetraploid Daturas. Am Nat 58: 60–61
Bellows JM, Bamford R (1941) Megagametophyte development in a triploid tulip. Bot Gaz 104: 699–711
Beltran IC (1970) Embryology of *Isotoma petraea*. Aust J Bot 18: 213–221
Belyaeva NS (1975) Subsidiary sperm cells in embryo sacs of Angiospermae and their significance in studying the process of double fertilization. Izv Akad Nauk Turkm SSR Ser Biol Nauk 1: 13–19 (in Russian)
Belyaeva NS (1977) On interrelation of male gametes and female cells in double fertilization (Abstr). In: Embryology of crop plants. Indo-Sov Symp. USSR Acad Sci, Moscow, and Indian Natl Sci Acad, New Delhi, at Komarov Bot Inst, Leningrad, p 43
Belyaeva NS (1981) Relation of the male gametes with embryo sac cells. The hypothesis of double fertilization. Acta Soc Bot Pol 50: 173–176
Belyaeva LY (1968) Development of male and female gametophytes of *Erysimum cheiranthoides* L. Ukr Bot Zh 25: 54–59 (in Russian)
Benavente RS, Skorupska H, Palmer RG, Shoemaker RC (1989) Embryo sac development in the cv KS male sterile, female sterile line of soybean (*Glycine max*). Am J Bot 76: 1759–1768
Bendre AM (1975) Studies in the family Loganiaceae 2. Embryology of *Buddleia* and *Strychnos*. J Indian Bot Soc 54: 272–279
Benikova NL (1952) Polyembryony in beans (*Phaseolus*). Bull Soc Nat Moscow 57: 65–66
Benson L (1957) Plant classification. Heath, Boston
Benson L (1962) Plant taxonomy: principles and methods. The Ronald Press Company, New York
Benson M (1894) Contribution to the embryology of the Amentiferae 1. Trans Linn Soc London 3: 409–424

Bentham G (1862a) Frankeniaceae. In: Bentham G, Hooker JD (eds) Genera plantarum, vol 1, pt 1. Reeve, London, pp 140–141
Bentham G (1862b) Linaceae. In: Bentham G, Hooker JD (eds) Genera plantarum, vol 1, pt 1. Reeve, London, pp 241–246
Bentham G (1876a) Oleaceae. In: Bentham G, Hooker JD (eds) Genera plantarum, vol 2, pt 2. Reeve, London, pp 672–679
Bentham G (1876b) Salvadoraceae. In: Bentham G, Hooker JD (eds) Genera plantarum, vol 2, pt 2. Reeve, London, pp 680–681
Bentham G (1877) Note on the gamopetalous orders belonging to the campanulaceous and oleaceous groups. Bot J Linn Soc 15: 1–16
Bentham G (1880) Urticaceae. In: Bentham G, Hooker JD (eds) Genera plantarum, vol 3, pt 1. Reeve, London, pp 341–395
Bentham G (1883) Haemodoraceae. In: Bentham G, Hooker JD (eds) Genera plantarum, vol 3, pt 2. Reeve, London, pp 671–681
Bentham G, Hooker JD (1965a, original edn 1862–1867) Genera plantarum, vol 1. Kramer, Weinheim (FRG)
Bentham G, Hooker JD (1965b; 1873–1876) Genera plantarum, vol 2. Kramer, Weinheim (FRG)
Bentham G, Hooker JD (1965c; 1880–1883) Genera plantarum, vol 3. Kramer, Weinheim (FRG)
Bensel CR, Palser BF (1975) Floral anatomy in the Saxifragaceae sensu lato 4. Baueroideae and Conclusions. Am J Bot 62: 688–694
Berezenko NP (1971) Embryology of *Rubus idaeus* L. Ukr Bot Zh 28: 57–60 (in Russian)
Berg CC (1977) Urticales: their differentiation and systematic position. In: Kubitzki H (ed) Flowering plants. Plant Syst Evol Suppl 1. Springer, Wien, pp 349–374
Berg RY (1954) Development and dispersal of the seed of *Pedicularis silvatica*. Nytt Mag Bot 2: 1–60
Berg RY (1958) Seed dispersal, morphology and phylogeny of *Trillium*. Skr Nors Vid Akad Oslo 1: 1–36
Berg RY (1959) Seed dispersal, morphology and taxonomic position of *Scoliopus*, Liliaceae. Skr Norsk Vidensk Akad Oslo I. Mat Naturvidensk Kl 4
Berg RY (1967) Megagametogenesis and seed development in *Dendromecon rigida* (Papaveraceae). Phytomorphology 17: 223–233
Berg RY (1969) Adaptation and evolution in *Dicentra* (Fumariaceae), with special reference to seed, fruit, and dispersal mechanism. Nytt Mag Bot 16: 49–75
Berg RY (1985) Gynoecium and development of embryo sac, endosperm and seed in *Pholistoma* (Hydrophyllaceae) relative to taxonomy. Am J Bot 72: 1775–1778
Berg RY, Maze JR (1978) Development of ovule, embryo sac, and endosperm in *Brodiaea* (Liliales). Norw J Bot 25: 1–7
Berger CA, Erdelská O (1973) Ultrastructural aspects of the embryo sac of *Jasione montana* L.: Cell walls. Caryologia 25: 109–120
Berger CA, Witkus ER, Joseph TC (1951) Tapetal cell and meiotic divisions in *Antirrhinum majus* L. and *Linaria vulgaris* Hill. Caryologia 4: 110–114
Berger F (1958) Zur Samenanatomie der Zingiberazeen: Gattungen *Elettaria*, *Amomum* und *Aframomum*. Sci Pharm 26: 224–258
Bergfeld R, Kühnl T, Schopfer P (1980) Formation of protein storage bodies during embryogenesis in cotyledons of *Sinapis alba* L. Planta 148: 146–156
Bergman B (1942) Zur Embryologie der Gattung *Erigeron*. Svensk Bot Tidskr 36: 429–443
Bergman B (1944) A contribution to the knowledge of the embryo sac mother cell and its development in two apomicts. Svensk Bot Tidskr 38: 249–259
Bernard FA, Bernard JM, Denny P (1990) Flower structure, anatomy and life history of *Wolffia australiana* (Benth.) den Hartog & van der Plas. Bull Torrey Bot Club 117: 18–26
Bernard C (1949) Microsporogenesis, macrosporogenesis and development of the macrogametophyte and seeds of *Duboisia leichhardtii* and *D. myoporoides*. Aust J Sci Res B 2: 241–248
Bernard N (1909) L'évolution dans la symbiose. Les Orchidées et leur champignons commensaux. Ann Sci Nat Ser Bot 9: 1–196
Bernardo FA, Ramirez DA (1959) Cytology of Philippine plants 3. *Lansium domesticum* Correa. Philipp Agric 43: 375–377

Bernardo FA, Jessena Jr CC, Ramirez DA (1961) Parthenocarpy and apomixis in *Lansium domesticum* Correa. Philipp Agric 44: 421–451

Beruti R (1961) Lo sviluppo del gametofito femininile di *Hypochoeris aetnensis*. Caryologia 14: 319–325

Bessey CE (1897) Phylogeny and taxonomy of angiosperms. Bot Gaz 24: 145–178

Bessey CE (1915) Phylogenetic taxonomy of flowering plants. Ann Missouri Bot Garden 2: 101–155

Bevilacqua LR, Fossati F, Dondero G (1987) 'Callose' in the impermeable seed-coat of *Sesbania punicea*. Ann Bot (London) 59: 335–341

Bevilacqua LR, Profumo P, Gastaldo P, Barella P (1988) Cytochemical study on the dormancy-imposing endosperm of *Cercis siliquastrum*. Ann Bot (London) 61: 561–565

Bhaduri PN (1932) The development of ovule and embryo sac in *Solanum melongena* L. J Indian Bot Soc 11: 202–224

Bhaduri PN (1935) Studies on the female gametophyte in Solanaceae. J Indian Bot Soc 14: 133–149

Bhaduri PN (1936) Studies on the embryogeny of the Solanaceae 1. Bot Gaz 98: 283–295

Bhaduri S (1944) A contribution to the morphology of pollen grains of Acanthaceae and its bearing on taxonomy. J Dep Sci Calcutta Univ 1: 25–38

Bhagawathi PR (1938) Microsporogenesis in *Drimys*. Presidency Coll Bot Mag (Madras) 5: 22–25

Bhalla PL, Malik CP (1982) Localization and activity of some glycosidases during early embryogenesis in *Tropaeolum majus* L. J Indian Bot Soc 61: 91–94

Bhalla PL, Singh MB, Malik CP (1979) Physiology of sexual reproduction 6. Embryogenesis in *Tropaeolum majus* L. enzyme changes. Acta Bot Indica 7: 72–86

Bhalla PL, Singh HB, Malik CP (1982) Post-fertilization developmental time table for *Tropaeolum majus*. Acta Bot Indica 10: 201–205

Bhandari NN (1962) Studies in the family Ranunculaceae 3. Development of female gametophyte in *Adonis annua* L. Phytomorphology 12: 70–74

Bhandari NN (1963a) Embryology of *Pseudowintera colorata*: a vesselless dicotyledon. Phytomorphology 13: 302–316

Bhandari NN (1963b) Studies in the family Ranunculaceae 5. The female gametophyte of *Adonis aestivalis* L.: a reinvestigation. Phytomorphology 13: 317–320

Bhandari NN (1965) Studies in the family Ranunculaceae 8. Variations in the development of the embryo sac of *Anemone vitifolia* Buch.-Ham. ex DC. Phytomorphology 15: 285–291

Bhandari NN (1966) Studies in the family Ranunculaceae 9. Embryology of *Adonis* Dill. ex Linn. Phytomorphology 16: 578–587

Bhandari NN (1968) Studies in the family Ranunculaceae 10. Embryology of *Anemone* L. Phytomorphology 18: 487–497

Bhandari NN (1970a) Degeneriaceae. In: Symp Comparative embryology of angiosperms. Indian Natl Sci Acad Bull 41: 36–37

Bhandari NN (1970b) Schisandraceae. In: Symp Comparative embryology of angiosperms. Indian Natl Sci Acad Bull 41: 42–44

Bhandari NN (1971) Embryology of the Magnoliales and comments on their relationships. J Arnold Arbor 52: 1–39, 285–304

Bhandari NN (1984) The microsporangium. In: Johri BM (ed) Embryology of angiosperms. Springer, Berlin Heidelberg New York, pp 53–122

Bhandari NN, Asnani S (1968) Studies in the family Ranunculaceae 11. Morphology and embryology of *Ceratocephalus falcatus* Pers. Beitr Biol Pflanzen 45: 271–290

Bhandari NN, Bhargava M (1983) Wall ingrowths in the antipodal cells in *Papaver somniferum* L. Beitr Biol Pflanzen 58: 53–58

Bhandari NN, Chaube RK (1969) Studies in the Viscaceae: Morphology and embryology of *Korthalsella opuntia* Merr. (Abstr). In: Johri BM, Kapil RN, Rashid A (eds) Sem Morphology, anatomy and embryology of land plants, Univ Delhi, pp 59–60

Bhandari NN, Chitralekha P (1989) Cellularization of the female gametophyte in *Ranunculus sceleratus*. Can J Bot 67: 1325–1330

Bhandari NN, Indira K (1969) Studies in Viscaceae 4. Embryology of *Eubrachion* (Hook. et Arn.) Engl. Bot Not 122: 183–203

Bhandari NN, Kapil RN (1964) Studies in the family Ranunculaceae 7. Two types of embryo sacs in *Trollius* Linn. Beitr Biol Pflanzen 40: 113–120

Bhandari NN, Khosla R (1982) Development and histochemistry of anther in *Triticale* cv. Tri-1 1. Some new aspects in early ontogeny. Phytomorphology 32: 18–27

Bhandari NN, Kishori R (1973) Development of tapetal membrane and Übisch granules in *Nigella damascena*: A histochemical approach. Beitr Biol Pflanzen 49: 59–72

Bhandari NN, Mal A (1989) Wall ingrowth or trabeculae in the free-nuclear embryo sac of *Scilla sibirica* Sâto. Phytomorphology 39: 21–28

Bhandari NN, Nanda K (1968a) Studies in the Viscaceae 1. Morphology and embryology of the Indian dwarf mistletoe: *Arceuthobium minutissimum* Hook. f. Phytomorphology 18: 435–450

Bhandari NN, Nanda K (1968b) Studies in the Viscaceae 2. A reinvestigation of the female gametophyte of *Arceuthobium doughlasii* Engelm. Am J Bot 55: 1028–1031

Bhandari NN, Natesh S (1985) Development of endosperm in *Scrophularia himalensis* with a discussion on the variation in the endosperm of the tribe Scrophularieae (Scrophulariaceae). Plant Syst Evol 148: 177–184

Bhandari NN, Sachdeva A (1983) Some aspects of organization and histochemistry of the embryo sac of *Scilla sibirica* Sâto. Protoplasma 116: 170–178

Bhandari NN, Venkataramani R (1968) Embryology of *Drimys winteri*. J Arnold Arbor 49: 509–525

Bhandari NN, Vohra SCA (1983) Embryology and affinities of Viscaceae. In: Calder M, Bernhardt P (eds) The biology of mistletoes. Academic Press, New York, London, Sydney, pp 69–86

Bhandari NN, Bhargava M, Geier T (1981) A persisting cellulosic wall of microspore mother cells during microsporogenesis in *Allium tuberosum* Rottl. and *Cyclamen persicum* Mill. Ann Bot (London) 48: 425–431

Bhandari NN, Bhargava M, Sachdeva A (1985) The mature ovule of *Papaver somniferum*. A histochemical study. Phytomorphology 35: 111–119

Bhandari NN, Bouman F, Natesh S (1976) Ovule ontogeny and seed-coat structure of *Scrophularia himalensis* Royle. Bot Jahrb 95: 535–548

Bhandari NN, Kishori R, Natesh S (1976) Ontogeny, cytology and histochemistry of anther tapetum in relation to pollen development in *Nigella damascena*. Phytomorphology 26: 46–59

Bhandari NN, Tandon SL, Jai S (1969) Some observations on the cytology and cytomixis in *Canavalia* DC. Cytologia 34: 22–28

Bhanwra RK (1988) Embryology in relation to systematics of Gramineae. Ann Bot (London) 62: 215–233

Bhanwra RK, Choda SP (1981) Apomixis in *Eremopogon foveolatus* (Gramineae). Nord J Bot 1: 97–101

Bhanwra RK, Choda SP (1986) Comparative embryology of *Echinochloa colonum* and *E. crusgalli* (Poaceae). Proc Indian Acad Sci (Plant Sci) 96: 71–78

Bhanwra RK, Choda SP, Kumar S (1982) Comparative embryology of some grasses. Proc Indian Natl Sci Acad B 48: 152–162

Bharathi M, Murty VR (1984) Comparative embryology of wild and cultivated species of *Arachis*. Phytomorphology 34: 48–56

Bharathi M, Kirti PB, Murty VR, Rao NGP (1982) Occurrence of atypical embryo sac-like structures in *Arachis glabrata* Benth. Curr Sci 51: 612–613

Bhargava HR (1932) Contribution to the morphology of *Boerhaavia repanda*. J Indian Bot Soc 11: 303–326

Bhargava HR (1934) Contribution to the morphology of *Mollugo nudicaulis* Lamk. Proc Indian Acad Sci B 1: 271–278

Bhargava HR (1935a) The life history of *Trianthema monogyna* Linn. Proc Indian Acad Sci B 2: 49–58

Bhargava HR (1935b) Contribution to the morphology of *Eclipta erecta* Linn. Proc Indian Acad Sci B 1: 325–339

Bhargava HR (1936) The life history of *Chenopodium album* Linn. Proc Indian Acad Sci B 4: 179–200

Bhargava HR, Sawhney RN (1958) Morphological studies in the Polygonaceae l. Contribution to the life history of *Rumex dentatus*. J Univ Saugar (India) B 7: 21–28

Bhargava YR (1980) Microsporogenesis and male gametophyte in *Jasminum pubescens* Willd. Curr Sci 49: 911–912

Bhasin RK (1971) Studies in Apocynaceae 2. A contribution to the embryology of *Catharanthus pusillus* (Murr.) G. Don. Botanique (Nagpur, India) 2: 111–122

Bhatnagar AK, Garg M (1977) Affinities of *Daphniphyllum*: Palynological approach. Phytomorphology 27: 92–97

Bhatnagar AK, Kapil RN (1979) Ontogeny and taxonomic significance of anther in *Bischofia javanica*. Phytomorphology 29: 298–306

Bhatnagar SP (1959) Some observations on the post-fertilization development of the embryo sac of *Santalum*. Phytomorphology 9: 87–91

Bhatnagar SP (1960) Morphological and embryological studies in the family Santalaceae 4. *Mida salicifolia* A. Cunn. Phytomorphology 10: 198–207

Bhatnagar SP (1965a) Studies in angiospermic parasites 2. *Santalum album*: The sandalwood tree. Natl Bot Garden Lucknow (India) Bull 112: 1–90

Bhatnagar SP (1965b) Some observations on the embryology of *Holboellia latifolia* Wall. Curr Sci 34: 28–29

Bhatnagar SP (1966) Microsporogenesis and male gametophyte in *Santalum* L. Proc Natl Inst Sci India B 32: 47–52

Bhatnagar SP (1970) Santalaceae. In: Symp Comparative embryology of angiosperms. Indian Natl Sci Acad Bull 41: 15–18

Bhatnagar SP, Agarwal S (1961) Morphological and embryological studies in the family Santalaceae 6. *Thesium* L. Phytomorphology 11: 273–282

Bhatnagar SP, Chandra S (1968) Endosperm in *Psittacanthus*. Curr Sci 37: 704–706

Bhatnagar SP, Chandra S (1975) Reproductive biology of *Triticum* 3. Unfertilized ovule and embryo sac, fertilization; post-fertilization changes in embryo sac and transformation of pistil into caryopsis in relation to time. Phytomorphology 25: 471–477

Bhatnagar SP, Chandra S (1976) Reproductive biology of *Triticum* 5. Post-pollination development of nucellus, integuments and pericarp in relation to time. Phytomorphology 26: 139–143

Bhatnagar SP, Garg R (1984) Endothecial thickenings in the anther wall of *Phthirusa adunca* (Loranthaceae). Plant Syst Evol 146: 265–267

Bhatnagar SP, Gupta M (1969) Structure and ontogeny of stomata in *Mimusops elengi*. J Indian Bot Soc 48: 362–367

Bhatnagar SP, Gupta M (1970) Morphology and embryology of *Mimusops elengi* L. Bot Not 123: 455–473

Bhatnagar SP, Johri BM (1972) Development of angiosperm seeds. In: Kozlowski TT (ed) Seed biology, vol I. Academic Press, New York London, pp 77–149

Bhatnagar SP, Johri BM (1983) Embryology of Loranthaceae. In: Calder M, Bernhardt P (eds) The biology of mistletoes. Academic Press, New York London Sydney, pp 47–67

Bhatnagar SP, Joshi PC (1965) Morphological and embryological studies in the family Santalaceae 7. *Exocarpus bidwillii* Hook. f. Proc Natl Inst Sci India B 31: 34–44

Bhatnagar SP, Kallarackal J (1973) Morphology and embryology of a pleistesial species (*Phlebophyllum kunthianum*). Phytomorphology 23: 205–211

Bhatnagar SP, Kallarackal J (1979) Development of ovule in *Linaria bipartita*. Phytomorphology 29: 1–6

Bhatnagar SP, Kallarackal J (1980) Cytochemical studies on the endosperm of *Linaria bipartita* (Vent.) Willd. with a note on the role of endosperm haustoria. Cytologia 45: 247–256

Bhatnagar SP, Kaur I (1984) Structure of style in *Brassica campestris* var. yellow sarson. Curr Sci 53: 391

Bhatnagar SP, Puri S (1970) Morphology and embryology of *Justicia betonica* Linn. Öst Bot Z 118: 55–71

Bhatnagar SP, Sabharwal G (1966) Female gametophyte and endosperm of *Iodina rhombifolia* Hook. & Arn. Phytomorphology 16: 588–591

Bhatnagar SP, Sabharwal G (1968) Morphology and embryology of *Iodina rhombifolia* Hook. & Arn. Beitr Biol Pflanzen 45: 465–479

Bhatnagar SP, Sawhney V (1981) Endosperm: Its morphology, ultrastructure and histochemistry. Int Rev Cytol 73: 55–98

Bhatnagar SP, Singh NS (1964) Development and structure of angiosperm seed 2. *Antirrhinum majus*. Natl Bot Garden Lucknow (India) Bull 92: 1–10

Bhattacharya SS, Mitra JN (1952) The development of female gametophyte in *Triumfetta rhomboidea* Jacq. Sci Cult 17: 343–344
Bhojwani SS, Bhatnagar SP (1986) The embryology of angiosperms. Vikas, New Delhi
Bianchi R (1946) Untersuchungen über die Fortpflanzungsverhältnisse von *Gager fistulosa* und *Lloydia serotina*. Ber Schweiz Bot Ges 56: 523–582
Biddle JA (1979) Anther and pollen development in garden pea and cultivated lentil. Can J Bot 57: 1883–1900
Biddle JA, Christophel DC (1978) Intergynoecial development in Tremandraceae. Phytomorphology 28: 411–418
Bijok K (1960) Studies in the karyological differentiation of the anther tapetum in *Chrysanthemum subcorymbosum*, *C. segetum* and *C. corymbosum*. Acta Biol Cracov Ser Bot 3: 15–24
Bijok K, Pawlak T, Krenska B (1977) Embryological studies on *Anthemis tinctoria* L. Acta Soc Bot Pol 46: 547–558
Billings FH (1901) Beiträge zur Kenntnis der Samenentwicklung. Flora 88: 253–318
Billings FH (1934) Male gametophyte of *Atriplex hymenelytra*. Bot Gaz 95: 477–484
Billings FH (1937) Some new features in the reproductive cytology of angiosperms illustrated by *Isomeris arborea*. New Phytol 36: 301–326
Biradar NV (1968) Studies on Palms: Embryology of *Phoenix pusilla* Gaertn., *P. acaulis* Buch. and *P. reclinata* Jacq. Proc Indian Acad Sci B 67: 165–173
Biradar NV (1978) Studies in monocotyledons: Micro-, megasporogenesis and development of female gametophyte in *Yucca gloriosa-superba* L. Biovigyanam (Pune, India) 4: 33–42
Biradar NV, Mahabale TS (1968) Studies on palms: embryology of *Phoenix robusta* Hook. Proc Indian Acad Sci B 68: 1–8
Biswas I (1957) Embryological studies in *Daemia extensa* R. Br. J Indian Bot Soc 36: 207–222
Björnstad IN (1970) Comparative embryology of Asparagoideae: Polygonateae, Liliaceae. Nytt Mag Bot 17: 169–207
Blackmore S, Meconchie CA, Knox RB (1987) Phylogenetic analysis of the male ontogenetic program in aquatic and terrestrial monocotyledons. Cladistics 3: 333–347
Bley F (1925) Zur Embryologie von *Laurenbergia javanica* (Micq.) Schindler. Inaug Diss, Univ Zürich
Bliss MC (1912) Contribution to the life history of *Viola*. Ann Bot (London) 26: 155–163
Blodgett FH (1923) The embryo of *Lemna*. Am J Bot 10: 336–342
Blumenthal A, Lerner HK, Wesker F, Poljakoff-Mayber A (1986) Germination preventing mechanisms in *Iris* seeds. Ann Bot (London) 58: 551–561
Bobde PL (1963) Male and female gametophytes of *Operculina turpenthum* (L.) Silva Manso. Bull Bot Soc Coll Sci (Nagpur, India) 4: 33–38
Böcher TW (1947) Cytological studies of *Arabis holboellii*. Hereditas 33: 573
Böcher TW (1951) Cytological and embryological studies in the amphiapomict *Arabis holboellii* Complex. KL Dansk Vid Selsk Biol 6: 1–59
Bocquet G (1959) The campylotropous ovule. Phytomorphology 9: 222–227
Bocquet G, Bersier JD (1960) La valeur systématique de l'ovule: développement teratalogiques. Arch Sci (Geneva) 13: 475–496
Boehm K (1931) Embryologische Untersuchungen an Zingiberaceen. Planta 14: 411–440
Boesewinkel FD (1977) Development of ovule and testa in Rutaceae 1: *Ruta*, *Zanthoxylum* and *Skimmia*. Acta Bot Neerl 26: 193–211
Boesewinkel FD (1978) Development of ovule and testa in Rutaceae 3. Some representatives of the Aurantioideae. Acta Bot Neerl 27: 341–354
Boesewinkel FD (1980a) Development of ovule and testa of *Linum usitatissimum* L. Acta Bot Neerl 29: 17–32
Boesewinkel FD (1980b) Development of ovule and seed-coat of *Dichapetalum mombuttense* Engl., with notes on other species. Acta Bot Neerl 29: 103–115
Boesewinkel FD (1980c) Development of ovule and seed-coat of *Erythroxylum coca* Lamk. Acta Bot Neerl 29: 231–241
Boesewinkel FD (1981) Development of the seed of *Trichilia grandiflora* Oliv. (Meliaceae). Acta Bot Neerl 30: 459–464
Boesewinkel FD (1984a) A comparative SEM study of the seed coats of recent and 900–1100 years old subfossil linseed. Ber Dtsch Bot Ges 97: 443–450

Boesewinkel FD (1984b) Development of ovule and seed-coat in *Cneorum tricoccon* L. (Cneoraceae). Acta Bot Neerl 33: 61–70

Boesewinkel FD (1984c) Ovule and seed structure in Datiscaceae. Acta Bot Neerl 33: 419–429

Boesewinkel FD (1985a) The ovule and seed of *Humiria balsamifera* (Aubl.) St. Hil. Acta Bot Neerl 34: 183–191

Boesewinkel FD (1985b) Development of ovule and seed-coat in *Averrhoa* (Oxalidaceae), with notes on some related genera. Acta Bot Neerl 34: 413–424

Boesewinkel FD (1987) Ovules and seeds of Trigoniaceae. Acta Bot Neerl 36: 81–91

Boesewinkel FD (1988) The seed structure and taxonomic relationships of *Hypseocharis* Remy. Acta Bot Neerl 37: 111–120

Boesewinkel FD (1989) Ovule and seed development in Droseraceae. Acta Bot Neerl 38: 295–311

Boesewinkel FD (1990) Ovule and seed development of *Tovaria pendula* Ruiz et Pavon. Bot Jahrb 111: 389–401

Boesewinkel FD, Been W (1979) Development of ovule and testa of *Geranium pratense* L. and some other representatives of the Geraniaceae. Acta Bot Neerl 28: 335–348

Boesewinkel FD, Bouman F (1980) Development of ovule and seed-coat of *Dichapetalum mombuttense* Engl. with notes on other species. Acta Bot Neerl 29: 103–115

Boesewinkel FD, Bouman F (1984) The seed: Structure. In: Johri BM (ed) Embryology of angiosperms. Springer, Berlin Heidelberg New York, pp 567–610

Boesewinkel FD, de Lange A (1983) Development of ovule and seed in *Begonia squamulosa* Hook. f. Acta Bot Neerl 32: 417–425

Boesewinkel FD, Geenen J (1980) Development of ovule and seed-coat of *Erythroxylum coca* Lamk. Acta Bot Neerl 29: 231–241

Boesewinkel FD, Venturelli M (1987) Ovule and seed structure in Vochysiaceae. Bot Jahrb 108: 547–566

Bohdanowicz J, Turala-Szybowska K (1985) Ultrastructure of endopolyploid antipodals in *Aconitum vulparia* Rchb. 1. Antipodals in the mature embryo sac. Protoplasma 127: 163–170

Bohdanowicz J, Turala-Szybowska K (1987) Ultrastructure of endopolyploid antipodals in *Aconitum vulparia* Rchb. 2. Antipodals in the period of free-nuclear endosperm. Protoplasma 140: 13–21

Bohlmann F, Burkhardt T, Zdero C (1973) Naturally occurring acetylenes. Academic Press, New York London

Bolenbach A (1928) Microsporogenesis in *Tropaeolum majus* with special reference to the cleavage process in tetrad formation. Bull Torrey Bot Club 55: 105–112

Böös G (1924) Neue embryologische Studien über *Alchemilla arvensis* (L.) Scop. Bot Not 1924: 209–250

Bor J (1978) A note on anatropy versus orthotropy. Phytomorphology 28: 219–224

Bor J, Bouman F (1974) Development of ovule and integuments in *Euphorbia milii* and *Codiaeum variegatum*. Phytomorphology 24: 280–296

Bor J, Kapil RN (1975) *Euphorbia geniculata*: Ovule to seed. Acta Bot Neerl 24: 257–258

Bor J, Kapil RN (1976) Anatropy and ontogeny of the bitegmic ovule in *Chrozophora* A.H.L. Jussieu (Euphorbiaceae). Acta Bot Neerl 25: 385–400

Borgen L (1971) A new type of tetrasporic embryo sac in *Chrysanthemum ferulaceum* (Webb) Sunding. Norw J Bot 18: 169–175

Borgen L (1972) Embryology and achene morphology of endemic canarian species of *Chrysanthemum* sub-genus *Argyranthemum* (Asteraceae). Norw J Bot 19: 149–170

Borthwick RA (1931) Development of the macrogametophyte and embryo of *Daucus carota*. Bot Gaz 92: 23–44

Borwein B, Coetsee ML, Krupko S (1949) Development of the embryo sac of *Restio dodii* and *Elegia racemosa*. J S Afr Bot 15: 1–11

Bose PK, Banerji I (1933) The development of the female gametophyte, and chromosome number of *Argemone mexicana* Linn. Curr Sci 2: 94

Bouharmont J (1963) Evolution de l'ovule fécondé chez *Musa acuminata* subsp. *burmannica*. Cellule 63: 259–279

Bouman F (1971a) Integumentary studies in the Polycarpicae 1. Lactoridaceae. Acta Bot Neerl 20: 565–569

Bouman F (1971b) The application of integumentary studies to taxonomic and phylogenetic problems. Ber Dtsch Bot Ges 84: 169–722
Bouman F (1974) Developmental studies of the ovule, integuments and seed in some angiosperms. Los, Naarden, pp 1–180
Bouman F (1975) Integument initiation and testa development in some Cruciferae. Bot J Linn Soc 70: 213–229
Bouman F (1978) Integumentary studies in the Polycarpicae 5. *Nigella damascena* L. Acta Bot Neerl 27: 175–182
Bouman F (1984) The ovule. In: Johri BM (ed) Embryology of angiosperms. Springer, Berlin Heidelberg New York, pp 123–157
Bouman F (1985) Embryology. In: Van Bruggen HWE (ed) Monograph of the genus *Aponogeton* (Aponogetonaceae). Bibl Bot 137: 4–9
Bouman F, Boesewinkel FD (1969) On a case of polyembryony in *Pterocarya fraxinifolia* (Juglandaceae) and polyembryony in general. Acta Bot Neerl 18: 50–57
Bouman F, Calis JIM (1977) Integumentary shifting. A third way to unitegmy. Ber Dtsch Bot Ges 90: 15–28
Bouman F, de Lange A (1983) Structure, micromorphology of *Begonia* seeds. Begonian May–June 1983: 70–78, 91
Bouman F, Louis A (1989) Seed structure in *Voyria primuloides* Baker (Gentianaceae): taxonomic and ecological implications. In: Paré J, Bugnicourt M (eds) Some aspects and actual orientations in plant embryology. Dedicatory volume to Prof A. Lebègue. Univ Press Picardie, Amiens (Fr), pp 261–270
Bouman F, Meijer W (1987) Seeds of Rafflesiaceae. In: 14th Int Bot Congr, Berlin, Abstr, p 221
Bouman F, Schrier S (1979) Ovule ontogeny and seed-coat development in *Gentiana* with a discussion on the evolutionary origin of the single integument. Acta Bot Neerl 28: 467–478
Boursnell JG (1950) The symbiotic seed-borne fungus in the Cistaceae. Ann Bot (London) 14: 217–243
Boyes JW (1939a) Embryo sac development in *Plumbagella*. Proc Natl Acad Sci USA 25: 141–145
Boyes JW (1939b) Development of the embryo sac of *Plumbagella micrantha*. Am J Bot 26: 539–547
Boyes JW, Battaglia E (1951a) The tetrasporic embryo sacs of *Plumbago coccinea*, *P. scandens*, and *Ceratostigma willmottianum*. Bot Gaz 112: 485–489
Boyes JW, Battaglia E (1951b) Embryo sac development in the Plumbaginaceae. Caryologia 3: 305–310
Boyes JW, Thompson WP (1937) The development of the endosperm and embryo in reciprocal interspecific crosses in cereals. J Genet 34: 203–217
Brandbyge J (1989) The diversity of micromorphological features in the genus *Coccoloba* (Polygonaceae). Nord J Bot 10: 25–44
Bregman R, Bouman F (1983) Seed germination in Cactaceae. Bot J Linn Soc 86: 357–374
Bremekamp CEB (1953) The delimitation of the Acanthaceae. Proc Konigl Nederl Akad Wet Amsterdam Ser C 56: 533–546
Bremekamp CEB (1965) Delimitation and subdivision of the Acanthaceae. Bull Bot Serv India (Calcutta) 7: 21–30
Bremer G (1915) Reliquiae Treubianae 2. The development of the ovule and embryo sac of *Pittosporum ramiflorum* and *Pittosporum trimorense*. Ann Jardin Bot Buitenz 14: 161–164
Brenan JPM (1966) The classification of Commelinaceae. Bot J Linn Soc 59: 349–370
Brenner W (1922) Zur Kenntnis der Blütenentwicklung einiger Juncaceen. Acta Soc Sci Fennicae 50: 1–37
Bresinsky A (1963) Bau, Entwicklungsgeschichte und Inhaltsstoffe der Elaiosomen. Bibl Bot 126: 1–54
Bretzler E (1924) Beiträge zur Kenntnis der Gattung *Platanus*. Bot Arch 7: 388–417
Brewbaker JL (1967) The distribution and phylogenetic significance of binucleate and trinucleate pollen grains in the angiosperms. Am J Bot 54: 1069–1083
Briggs CL (1989) Early post-fertilization development in *Swainsona greyana* (Fabaceae): relationship of the persisting synergid and pollen tube to the developing embryo. In: Paré J, Bugnicourt M (eds) Some aspects and actual orientations in plant embryology. Dedicatory volume to Prof. A Lebègue. Univ Press Picardie, Amiens (Fr), pp 60–75

Brink RA, Cooper DC (1936) The mechanism of pollination in alfalfa, *Medicago sativa*. Am J Bot 23: 678–683

Brink RA, Cooper DC (1944) The antipodals in relation to abnormal endosperm behaviour in *Hordeum jubatum* × *Secale cereale* hybrid seeds. Genetics 29: 391–406

Brink RA, Cooper DC (1947) The endosperm in seed development. Bot Rev 13: 423–541

Brisson JD, Peterson RL (1975) SEM of fractured plant material embedded in glycol methacrylate. Microsc Soc Can 2: 64–65

Brix K (1974) Sexual reproduction in *Eragrostis curvula* (Schrad.) Nees. Z Pflanzenzücht 71: 25–32

Brizicky GK (1964) Polyembryony in *Euonymus* (Celastraceae). J Arnold Arbor 45: 251–259

Brock RD (1955) Chromosome balance and endosperm failure in *Hyacinthus*. Heredity 9: 199–222

Brofferia J (1930) Osservazioni sullo sviluppo della Calycanthaceae. Ann Bot (Rome) 18: 387–394

Brogniart A (1827) Mémoire sur la génération et le développement de l'embryon dans les végétaux phanérogamiques. Ann Sci Natl Bot 12: 14–52, 145–172, 225–298

Brough P (1923) Preliminary note on the embryo sac of *Styphelia longifolia*. Proc Linn Soc NSW 48: 674–680

Brough P (1924) Studies in the Epacridaceae 1. The life history of *Styphelia longifolia*. Proc Linn Soc NSW 49: 162–178

Brough P (1927) Studies in Goodeniaceae 1. The life history of *Dampiera stricta* R. Br. Proc Linn Soc NSW 52: 471–498

Brough P (1933) The life history of *Grevillea robusta*. Proc Linn Soc NSW 58: 33–73

Brouland M (1975) Recherches sur la structure du sac embryonnaire et sur les phenomènes de fécondation chez la *Clematis recta* L. Bot Ser 57 Fasc 1: 5–14

Brouland M, Guerven C (1978) La fécondation et le développement de l'albumen chez l'*Helleborus foetidus*. Rev Gen Bot 85: 353–370

Brouland M, Guerven C, Le Cog C (1976) Etude cytologique qualitative et quantitative de la réproduction chez le *Delphinium ajacis* L. Rev Cytol Biol Veg 39: 121–138

Brouwer J (1924) Studies in Platanaceae. Rec Trav Bot Neerl 21: 269–382

Brown MM (1917) The development of the embryo sac and the embryo in *Phaseolus vulgaris*. Bull Torrey Bot Club 44: 535–544

Brown RC, Mogensen HL (1972) Late ovule and early embryo development in *Quercus gambelii*. Am J Bot 59: 311–316

Brown SW (1949) Endomitosis in the tapetum of tomato. Am J Bot 36: 703–716

Brown WH, Sharp LW (1911) The embryo sac of *Epipactis*. Bot Gaz 52: 439–452

Brown WV (1951) Apomixis in *Zephyranthes texana* Herb. Am J Bot 38: 697–702

Brown WV (1955) A species of grass with liquid endosperm. Bull Torrey Bot Club 82: 284–285

Brown WV (1960) The morphology of the grass embryo. Phytomorphology 10: 215–223

Brown WV, Emery WHP (1957) Apomixis in Gramineae, tribe Andropogoneae: *Themada triandra* and *Bothriochloa ischaemum*. Bot Gaz 118: 246–253

Brown WV, Emery WHP (1958) Apomixis in the Gramineae: Panicoideae. Am J Bot 45: 253–263

Browne ET Jr (1961) Morphological studies in *Aletris* 1. Development of the ovule, megaspores and megagametophyte of *A. aurea* and their connection with systematics of the genus. Am J Bot 48: 143–147

Bruck DK, Walker DB (1985) Cell determination during embryogenesis in *Citrus jambhiri* 1. Ontogeny of the epidermis. Bot Gaz 146: 188–195

Brückner G (1926) Beträge zur Anatomie, Morphologie und Systematik der Commelinaceae. Bot Jahrb 61 Beibl 137: 1–70

Brückner G (1930) Commelinaceae. In: Engler A, Prantl K (eds) Die natürlichen Pflanzenfamilien. 2nd edn. 15a: 159–181

Brummitt RK, Ferguson IK, Poole MM (1980) A unique and extraordinary pollen tube in the genus *Crossandra* (Acanthaceae). Pollen Spores 22: 11–16

Brundell DI (1975) Flower development of the Chinese gooseberry (*Actinidia chinensis* Planch.) 1. Development of the flowering shoot. NZ J Bot 13: 473–483

Bruni A, Tosi B (1978) Distribuzione, morphologie e composizione chimica dei laticiferi embrionali di *Euphorbia marginata*. Nuovo G Bot Ital 112: 312–313

Bruun, Olesen P (1989) A structural investigation of the ovule in sugar beet, *Beta vulgaris*: The micropyle and nucellus. Nord J Bot 9: 81–87

Buchner L (1948) Vergleichende embryologische Studien an Scilloideae. Öst Bot Z 95: 428–450

Buell KM (1952a) Developmental morphology in *Dianthus* 1. Structure of the pistil and seed development. Am J Bot 39: 194–210

Buell KM (1952b) Developmental morphology in *Dianthus* 2. Starch accumulation in ovule. Am J Bot 39: 458–467

Buell MF (1935) Seed and seedling of *Acorus calamus*. Bot Gaz 96: 758–765

Buell MF (1938) Embryology of *Acorus calamus*. Bot Gaz 99: 556–568

Bugnicourt M (1969) Polyembryonie chez l'*Hypericum tetrapterum* Fries. (= *Hypericum acutum* Moench.). Bull Soc Bot Fr 116: 495–504

Bugnicourt M (1971a) Embryogénie des Hypericacées. Développement de l'embryon chez l'*Hypericum tetrapterum* Fries. Bull Soc Bot Fr 118: 319–334

Bugnicourt M (1971b) Bourgeonnement du suspenseur chez l'*Hypericum tetrapterum* Fries. Bull Soc Bot Fr 118: 335–340

Burns-Balogh P (1983) A theory on the evolution of the exine in Orchidaceae. Am J Bot 70: 1304–1312

Burs Jr PA, Lersten NR (1972) Crystal in tapetal cells of the Leguminosae. Bot J Linn Soc 65: 81–85

Burton GW (1982) Effect of environment on apomixis in bahiagrass. Crop Sci 22: 109–111

Bushnell EP (1936) Development of the macrogametophyte in certain Labiatae. Bot Gaz 98: 190–197

Buss Jr PA, Lersten NR (1975) Survey of tapetal nuclear number as a taxonomic character in Leguminosae. Bot Gaz 136: 388–395

Buss Jr PA, Galen FD, Lersten NR (1969) Pollen and tapetum development in *Desmodium glutinosum* and *D. illinoense* (Papilionoideae, Leguminosae). Am J Bot 56: 1203–1208

Buxbaum F (1937a) Die Entwicklungslinien der Lilioideae 1. Die Wurmbeoidea. Bot Arch 38: 213–293

Buxbaum F (1937b) Die Entwicklungslinien der Lilioideae 2. Die systematische Stellung der Gattung *Gagea*. Bot Arch 38: 305–398

Buxbaum F (1944) Untersuchungen zur Morphologie der Kakteenblüte 1. Das Gynoecium. Bot Arch 45: 190–247

Buxbaum F (1955) Morphology of cacti, sect 3. Fruits and seeds. Abbey Garden, Pasadena (USA)

Buxbaum F (1958) The phylogenetic division of the subfamily Cereoideae, Cactaceae. Madrono 14: 177–126

Caccavari MA (1986) Estudio de las caracters. Del Polen en las *Mimosa* Lepidotae. Pollen Spores 28: 29–42

Calder DM (1983) Mistletoes in focus: An introduction. In: Calder M, Bernhardt P (eds) The biology of mistletoes. Academic Press, New York London Sydney, pp 1–18

Calder DM, Slater AT (1985) The stigma of *Dendrobium speciosum* Sm. (Orchidaceae): a new stigma type comprising detached cells within a mucilaginous matrix. Ann Bot (London) 55: 297–307

Caldwell OW (1899) On the life history of *Lemna minor*. Bot Gaz 27: 37–80

Camerarius RJ (1694) De sexu plantarum epistola. Tübingen

Cameron BG, Prakash N (1990) Occurrence of giant antipodals in the female gametophytes of Australian Bossiaeae, Indigofereae and Mirbelieae (Leguminosae). Aust J Bot 38: 395–401

Camp WH, Hubbard MM (1963) Vascular supply and structure of the ovule and aril in peony and of the aril in nutmeg. Am J Bot 50: 174–178

Campbell DH (1897) A morphological study of *Najas* and *Zannichellia*. Proc Cal Acad Sci 1: 1–71

Campbell DH (1898) The development of the flower and embryo in *Lilaea subulata*. H.B.K. Ann Bot (London) 12: 1–28

Campbell DH (1899) Notes on the structure of the embryo sac in *Sparganium* and *Lysichiton*. Bot Gaz 27: 153–166

Campbell DH (1930) The relationships of *Paulownia*. Bull Torrey Bot Club 57: 47–50

Campbell SC, Greene CW, Neubauer BF, Higgins JM (1985) Apomixis in *Amelanchier laevis*, shadbush (Rosaceae, Maloideae). Am J Bot 72: 1397–1403

Canright JE (1952) The comparative morphology and relationships of the Magnoliaceae 1. Trends of specialization in the stamen. Am J Bot 39: 484–497

Canright JE (1953) The comparative morphology and relationships of the Magnoliaceae 2. Significance of pollen. Phytomorphology 3: 355–365
Canright JE (1962) Comparative morphology of pollen of Annonaceae. Pollen Spores 4: 338–339
Canright JE (1963) Contribution of pollen morphology to the phylogeny of some Ranalian families. Grana Palynol 4: 64–72
Capineri R (1966) Ricerche cariologiche ed embriologiche su *Anthemis mixta* L. (*Ormentis mixta* L. Dum.). Nuovo G Bot Ital 73: 94–95
Capoor SP (1937a) Contribution to the morphology of some Indian Liliaceae 2. The gametophytes of *Urginea indica*. Beih Bot Zentralbl 56A: 156–170
Capoor SP (1937b) The life history of *Holoptelea integrifolia* Planch. Beih Bot Zentralbl 57A: 233–249
Cappelletti C (1931) Sull'azione dei prodotti del ricambio di miceli microrizogeni sulle piante ospiti. Recerche fisiologiche e morfologiche. Ann Bot (Rome) 19: 1–62
Capuron R (1962) Revision des Rhopalocarpacees. Adansonia 11: 228–267
Carano E (1921) Nuove ricerche sulla embriologia delle Asteraceae. Ann Bot (Rome) 15: 97–195
Carapetian J, Rupert EA (1989) Ovule and female gametophyte development in fertile and sterile safflower plants (*Carthamus tinctorius* L.). Aust J Bot 37: 519–528
Carey G (1934) Further investigations on the embryogeny of viviparous seeds. Proc Linn Soc NSW 59: 392–410
Carey G, Fraser L (1932) The embryology and seedling development of *Aegiceros majus*. Proc Linn Soc NSW 57: 341–360
Carlson EM, Stuart BC (1936) Development of spores and gametophytes in certain New World species of *Salvia*. New Phytol 35: 68–91
Carlson MC (1945) Megasporogenesis and development of the embryo sac of *Cypripedium parviflorum*. Bot Gaz 107: 107–114
Carlquist S (1964) Morphology and relationships of Lactoridaceae. Aliso 5: 421–435
Carlquist S (1977) A revision of Grubbiaceae. J S Afr Bot 43: 115–128
Carlquist S (1978) Vegetative anatomy and systematics of Grubbiaceae. Bot Not 131: 117–126
Carlquist S, Debuhr L (1970) Wood anatomy of Penaeaceae (Myrtales). Comparative, phylogenetic and ecological implications. Bot J Linn Soc 75: 211–228
Carniel K (1963) Contributions to the developmental history of the sporogenous tissues and of the tapetum in some Dipsacaceae. Öst Bot Z 110: 547–555
Carniel K (1967) Über die Embryobildung in der Gattung *Paeonia*. Öst Bot Z 114: 4–19
Carolin RC (1966) Seeds and fruits of the Goodeniaceae. Proc Linn Soc NSW 91: 58–83
Carolin RC (1987) A review of the family Portulacaceae. Aust J Bot 35: 383–412
Carter S (1960) Butomaceae. In: Milne-Redhead E, Polhill RM (eds) Flora of tropical East Africa. Crown Agents for Overseas Govt. & Administrations, London
Cass DD (1972) Occurrence and development of a filiform apparatus in the egg of *Plumbago capensis*. Am J Bot 59: 279–283
Cass DD, Fabi GC (1988) Structure and properties of sperm cells isolated from the pollen of *Zea mays*. Can J Bot 66: 819–825
Cass DD, Jensen WA (1970) Fertilization in barley. Am J Bot 57: 62–70
Cass DD, Karas I (1975) Development of sperm cells in barley. Can J Bot 53: 1051–1062
Cassera JD (1935) Origin and development of the female gametophyte, endosperm and embryo in *Orobanche uniflora*. Bull Torrey Bot Club 62: 455–466
Catalano G (1930) Contributo alla conoscenza delle cause della sterilita in *Agave* e *Fourcroya*. Lav R Ist Bot Palermo 1: 1–60
Catalano G (1931) I rapporti genetici e la fruttificazione nelle *Agave* del gruppo Salamiane. Lav R Ist Bot Palermo 2: 22–45
Cavara FC (1899) Studi sul the Ricerche intorno allo sviluppo del frutto della *Thea chinensis*, cultivata nel orto botanico di Pavia. Atti Ist Bot Univ Lab Crittogam Pavia 5: 265–326
Cave MS (1939) Macrosporogenesis in *Leucocoryne ixioides*. Cytologia 9: 407–411
Cave MS (1941) Megasporogenesis and embryo development in *Calochortus*. Am J Bot 28: 390–394
Cave MS (1942) Development of the macrogametophyte of *Miersia chilensis*. Bot Gaz 104: 185–187

Cave MS (1952) Sporogenesis and gametogenesis in *Odontostamum hartwegii* Torr. Phytomorphology 2: 210–214
Cave MS (1953) Cytology and embryology in the delimitation of genera. In: Plant genera: their nature and definition. Chron Bot 14: 140–153
Cave MS (1955) Sporogenesis and the female gametophyte of *Phormium tenax*. Phytomorphology 15: 247–253
Cave MS (1967) The megagametophyte of *Androcymbium*. Phytomorphology 17: 233–239
Cave MS (1974) Female gametophytes of *Chlorogalam* and *Schoenolirion* (*Hastingsia*). Phytomorphology 24: 56–60
Cave MS (1975) Embryological studies in *Stypandra glauca* (Liliaceae). Phytomorphology 25: 95–99
Cave MS, Arnott HJ, Cook SA (1961) Embryogeny in the California peonies with reference to their taxonomic position. Am J Bot 48: 397–404
Cebrat J (1967) Preliminary embryological investigation on the cause of lowered fertility of lucerene (*Medicago media* Pers.). Acta Agrobot 20: 71–86 (in Russian)
Celarier RP, Harlan JR (1957) Apomixis in *Bothriochloa*, *Dichanthium* and *Capillipedium*. Phytomorphology 7: 93–102
Cesca G (1961) Ricerche embryologische su *Rudbeckia missouriensis*. Caryologia 14: 129–139
Cesca G (1969) Cytological and embryological studies in the genus *Euphorbia*: *Euphorbia epithymoides* L. Proc Natl Inst Sci India B 35: 139–152
Chadefaud M (1974) Sur la formule florale de la capucine (*Tropaeolum majus* L.). Bull Soc Bot Fr 121: 347–357
Chakravarthy RS (1935) Polyembryony in *Murraya koenigii* Spreng. Curr Sci 3: 361–362
Chakravarthy RS (1936) Nucellar polyembryony in the Rutaceae. Curr Sci 5: 202–203
Chakravorti AK (1947) The development of the female gametophyte and seed of *Coccinia indica* W. et A. J Indian Bot Soc 26: 95–104
Chamberlain CJ (1895) Contributions to the embryology of *Aster novae-anglica*. Bot Gaz 20: 205–212
Chamberlain CJ (1897) Contribution to the life history of *Salix*. Bot Gaz 28: 147–179
Chan SCK, Hillson CJ (1968) Developmental morphology of the megasporangium and embryogeny in *Ipomoea reptans* Poir. and related species. Bot Gaz 129: 150–156
Chan SCK, Hillson CJ (1971) Developmental morphology of the microsporangium in *Ipomoea reptans* Poir. Bot Gaz 132: 224–229
Chance GD, Bacon D (1984) Systematic implications of seed-coat morphology in *Naona* (Hydrophyllaceae). Am J Bot 71: 829–842
Chanda S (1966) On the pollen morphology of the Centrolepidaceae, Restionaceae and Flagellariaceae, with special reference to taxonomy. Grana Palynol 6: 355–415
Chanda S, Ghosh K, Nilsson S (1979) On the polarity and tetrad arrangements in some mono- and diaperturate angiosperm pollen grains. Grana 18: 21–31
Chanda S, Nilsson S, Blackmore S (1988) Phylogenetic trends in Alismatales with reference to pollen grains. Grana 27: 257–272
Chandler B (1911) Note on *Donatia novae-zelandiae* Hook. f. Notes R Bot Garden Edinburgh 6: 43–48
Chandra C, Shankar G (1964) Polyembryonic studies in certain indigenous species of *Citrus*. Sci Cult 31: 373–374
Chandra N (1963a) Some ovule characters in the systematics of Gramineae. Curr Sci 32: 277–278
Chandra N (1963b) Morphological studies in the Gramineae 4. Embryology of *Eleusine indica* Gaertn. and *Dactyloctenium aegypticum* (Desf.) Beauv. Proc Indian Acad Sci B 58: 117–127
Chandra N (1976) Embryology of some species of *Eragrostis*. Acta Bot Indica 4: 36–43
Chandra S (Sudhir), Bhatnagar SP (1974) Reproductive biology of *Triticum* 2. Pollen germination, pollen tube growth and its entry into the ovule. Phytomorphology 24: 210–217
Chandra S, Bhatnagar SP (1975a) Pollination and nucellus in *Abelmoschus esculentus*. Biol Plant (Prague) 17: 154–155
Chandra S, Bhatnagar SP (1975b) Seed development of *Abelmoschus esculentus* (Malvaceae). Plant Syst Evol 123: 255–262
Chandra S, Bhatnagar SP (1976) Reproductive biology of *Abelmoschus esculentus* 4. Ovule, fertilization, endosperm and embryo development. J Indian Bot Soc 55: 1–13

Chandra S (Saroj), Singh RP (1980) Embryological studies in *Antirrhinum orontium* var. *indicum* and var. *parviflorum*. J Indian Bot Soc 59 (Suppl): 64
Chandra S (Suman), Singh RP, Pandey AK (1980) Embryological studies in *Alectra parasitica* A. Rich. var. *chitrakutensis* Rau (Scrophulariaceae). Flora 169: 111–120
Chandravadana P (1963) A note on the embryology of *Erythrina indica* Lam. (Leguminosae). Curr Sci 32: 229–230
Chandravadana P (1965) A note on the embryology of *Gliricidia sepium* (Jacq.) Steud. Curr Sci 34: 185–186
Chandra Sekhar KN, de Mason A (1988) Quantitative ultrastructure and protein composition of date palm (*Phoenix dactylifera*) seeds: A comparative study of endosperm vs embryo. Am J Bot 75: 323–329
Chao C-Y (1974) Megasporogenesis and megagametogenesis in *Paspalum commersonii* and *P. longifolium* at two polyploid levels. Bot Not 127: 267–275
Chao C-Y (1980) Autonomous development of embryo in *Paspalum conjugatum* Berg. Bot Not 133: 215–222
Chao S (1975) A study on embryogeny of maize by total dissection of the embryo sac and embryo. Acta Bot Sin 17: 328–329
Chapman M (1933) The ovule and embryo sac of *Saxifraga verginiensis*. Am J Bot 20: 151–158
Charzynska M (1973) Translocation of the generative cell towards the centre of the pollen grain due to changes in the vacuolar system in *Ranunculus repens* L. Acta Soc Bot Pol 42: 295–300
Charzynska M, Lewandowska E (1990) Generative cell division and sperm cell association in the pollen grain of *Sambucus nigra*. Ann Bot (London) 65: 685–690
Chatterji AK, Jayaramu M (1981) Tapetal endoploidy in safflower (*Carthamus tinctorius* L.) cultivars. Curr Sci 50: 141–142
Chaudhari HK (1978) Use of semigamy in the production of cotton hybrids. Bull Torrey Bot Club 105: 98–103
Chaudhuri KL (1940) A note on the morphology and chromosome number of *Litchi chinensis*. Curr Sci 9: 416
Chaudhury IP, Narayana HS (1978) Studies on the development and structure of seed in the Caryophyllaceae 1. *Spergula arvensis* L. J Indian Bot Soc 57: 46–52
Chaudhury MR (1961) On the structure of embryo sac in 4n *Corchorus capsularis* L. (Strain D 54). Indian Agric 5: 29–31
Chauhan SVS (1970) Micro- and megasporogenesis in *Luffa echinata* Roxb. Agra Univ J Res Sci 19: 37–42
Chauhan SVS (1979) Development of endothecium in relation to tapetal behaviour in some male sterile plants. Phytomorphology 29: 245–251
Chauhan TS (1979) Morphological studies in Apocynaceae 2. Embryology of *Plumeria*. J Indian Bot Soc 58: 363–368
Cheah CH, Stone BC (1975) Embryo sac and microsporangium development in *Pandanus* (Pandanaceae). Phytomorphology 25: 228–238
Cheesman EE (1927) Fertilization and embryogeny in *Theobroma cacao*. Ann Bot (London) 41: 107–126
Chen C, Gibson PB (1971) Seed development following the mating of *Trifolium repens* × *T. uniflorum*. Crop Sci 11: 667–672
Cheng PC, Greyson RI, Walden DB (1986) The anther cuticle of *Zea mays*. Can J Bot 64: 2088–2097
Chennaveeraiah MS, Mahabale TS (1952) A note on sporogenesis in *Dipcadi serotinum* (L.) Medic. Can J Bot 37: 345–352
Chennaveeraiah MS, Mahabale TS (1962) Morphological and embryological studies in *Dipcadi*. In: Symp Plant embryology. Council Sci Ind Res, New Delhi, pp 12–22
Chennaveeraiah MS, Patil RM (1972) Apomixis in *Blumea*. Phytomorphology 22: 71–76
Chennaveeraiah MS, Sivakumar PM (1983) Pollen bud formation and its role in *Ophiorrhiza* spp. Ann Bot (London) 51: 449–452
Chiappini M (1962) Osservazioni carioembriologiche in *Ornithogalum* di *Sardegna* 1. Sporogenesi e sviluppo del gametofito *Ornithogalum caudatum* Air. Nuovo G Bot Ital 69: 91–102
Chiarugi A (1925) Embriologia delle Cistaceae. Nuovo G Bot Ital 32: 223–314
Chiarugi A (1927a) Ricerche sulla embriologia delle Asteraceae. Nuovo G Bot Ital 34: 717–777

Chiarugi A (1927b) L'evoluzione della cellule del tappeto e la formazione del periplasmodio in alcune Asteraceae. Nuovo G Bot Ital 34: 783–828
Chiarugi A (1930) Parthenocarpia in *Zizyphus sativa.* Nuovo G Bot Ital 37: 287–312
Chiarugi A, Francini E (1930) Apomissia in *Ochna serrulata.* Nuovo G Bot Ital 37: 1–250
Chichiricco G (1989) Embryology of *Crocus thomassi* (Iridaceae). Plant Syst Evol 165: 39–47
Chifflot JBJ (1902) Contributions à l'étude de la classe de Nymphéinées. Ann Univ Lyon N Ser 1 Sci Med 40: 1–294
Chikkannaiah PS (1954) The embryo sac of *Ochna squarrosa* Linn. Curr Sci 23: 232
Chikkannaiah PS (1961) The embryo sac of *Rhoeo discolor* Hance: A reinvestigation. Karnatak Univ J Sci (India) 6: 57–60
Chikkannaiah PS (1962) Morphological and embryological studies in the Commelinaceae. In: Plant embryology. Symp Council Sci Ind Res, New Delhi, pp 23–36
Chikkannaiah PS (1963) Embryology of some members of the family Commelinaceae: *Commelina subulata* Roth. Phytomorphology 13: 174–184
Chikkannaiah PS (1964) An embryological study of *Murdannia simplex* (Vahl) Brenan. J Indian Bot Soc 43: 238–248
Chikkannaiah PS (1964–1965a) A contribution to the life history of *Floscopa scandens* Lour. Karnatak Univ J Sci (India) 9: 110–122
Chikkannaiah PS (1964–1965b) Studies in *Tradescantia erecta* (= *Tinantia fugax* Scheidw). Karnatak Univ J Sci (India) 9: 123–137
Chikkannaiah PS (1969) Floral morphology, sporogenesis and gametogenesis in *Aneilema paniculatum* Wall. Karnatak Univ J Sci (India) 14: 130–141
Chikkannaiah PS (1970) Sporogenesis and gametogenesis in *Amischophacellus axillaris* (L.) Rao & Kam. (= *Cyanotis axillaris* Roem & Sch.). Karnatak Univ J Sci (India) 15: 100–109
Chikkannaiah PS (1971) Post-fertilization studies in *Murdannia semiteres* (= *Aneilema paniculatum* Wall.). Proc Indian Acad Sci B 73: 96–104
Chikkannaiah PS (1973) Development of endosperm, embryo and seed-coat in *Amischophacellus axillaris* L. Proc Indian Acad Sci B 77: 166–175
Chikkannaiah PS, Hemaraddi B (1979) A note on the megasporogenesis and endosperm of some Commelinaceae. Curr Sci 48: 363
Chikkannaiah PS, Hiremath BS (1982) Embryological studies in *Launaea nudicaulis* Hook. Proc Indian Acad Sci Plant Sci 91: 519–527
Chikkannaiah PS, Mahalingappa MS (1975) Antipodals in some members of Gramineae. Curr Sci 44: 22–23
Choda SP, Bhanwra RK (1977) The mode of reproduction in *Bothriochloa odorata* (Lisboa) A. Camus and *Paspalum disticum* Linn. Proc Indian Natl Sci Acad Sec B 43: 175–183
Choda SP, Mitter H, Bhanwra RK (1982) Embryological studies in three species of *Cymbopogon* Spreng (Poaceae). Proc Indian Acad Sci Plant Sci 91: 55–60
Chodat R (1916) Hydnoraceae. Bull Soc Bot Geneva 8: 186–201
Chodat R, Bernard Ch (1900) Sur le sac embryonnaire d'*Helosis guyanensis.* J Bot Mort 14: 72–79
Cholakhyan DP (1965) Cytological and embryological data from investigations of some representatives of the Cucurbitaceae in the conditions of Ararat plain, Armenian SSR. VCH ZAP Ereuan Univ 97: 3–9
Chopra RN (1953) The endosperm of some Cucurbitaceae. Curr Sci 22: 383–384
Chopra RN (1955) Some observations on endosperm haustoria in the Cucurbitaceae. Phytomorphology 5: 219–230
Chopra RN (1957) The mode of embryo sac development in *Opuntia aurantiaca* Lindl.: a reinvestigation. Phytomorphology 7: 403–406
Chopra RN, Agarwal S (1958) Some further observations on the endosperm haustoria in the Cucurbitaceae. Phytomorphology 8: 194–201
Chopra RN, Agarwal S (1960) The female gametophyte of *Benincasa cerifera* Savi. Bot Not 113: 192–205
Chopra RN, Basu B (1965) Female gametophyte and endosperm of some members of the Cucurbitaceae. Phytomorphology 15: 217–223
Chopra RN, Kaur H (1965a) Embryology of *Bixa orellana* Linn. Phytomorphology 15: 211–214

Chopra RN, Kaur H (1965b) Some aspects of the embryology of *Cornus*. Phytomorphology 15: 353–359
Chopra RN, Kaur H (1969) Pollination and fertilization in some *Ficus* spp. Beitr Biol Pflanzen 45: 441–446
Chopra RN, Mukkada AJ (1966) Gametogenesis and pseudo-embryo sac in *Indotristicha ramosissima* (Wight) van Royen. Phytomorphology 16: 182–188
Chopra RN, Sachar RC (1963) Endosperm. In: Maheshwari P (ed) Recent advances in the embryology of angiosperms. Int Soc Plant Morphol, Univ Delhi, pp 135–170
Chopra RN, Seth PN (1977) Some aspects of endosperm development in Cucurbitaceae. Phytomorphology 27: 112–115
Chopra S (1969) Development of female gametophyte in niger (*Guizotia abyssinica* Cass). Curr Sci 38: 392–393
Chopra S (1970) Development of female gametophyte in *Sapium sebiferum* Roxb. Curr Sci 39: 17–18
Chopra S, Singh RP (1969) Structure and development of seed in *Phyllanthus urinaria* L. J Indian Bot Soc 48: 212–216
Chopra S, Agarwal S, Singh RP (1984) Seed morphology in some *Phaseolus* species. Geophytology (Lucknow, India) 14: 114–117
Chou YL (1952) Floral morphology of three species of *Gaultheria*. Bot Gaz 114: 198–221
Choudhury JK, Mitra JN (1953) Abnormal tricotyledonous embryo and morphological structure of normal fruit and seed of *Cinnamomum camphora*. Sci Cult 19: 159–160
Chowdhury KA, Buth GM (1970) Seed-coat structure and anatomy of Indian pulses. In: Robson NKB, Cutler DF, Gregory M (eds) New research in plant anatomy. Academic Press, London New York, pp 169–179
Christensen PB (1986) Pollen morphological studies in the Malvaceae. Grana 25: 95–117
Chugai K (1971) Embryological studies of *Galeola septentrionalis* Reichb. fil (Orchidaceae). Sci Rep Tôhoku Univ Ser Biol 35: 253–258
Chung M Gi, Jones Jr SB (1989) Pollen morphology of *Hosta* Tratt (Funkiaceae) and related genera. Bull Torrey Bot Club 116: 31–44
Church MB (1916) The development of embryo sac and embryo of *Cooperia drummondii*. Bull Torrey Bot Club 43: 397–405
Cichan MA, Palser BF (1982) Development of normal and seedless achenes in *Cichorium intybus* (Compositae). Am J Bot 69: 885–895
Cionini PG, Bennici A, Alpi A, D'Amato F (1976) Suspensor, gibberellin and in vitro development of *Phaseolus coccineus* embryos. Planta 131: 115–117
Clarke AE (1940) Fertilization and early embryo development in the potato. Am Potato J 17: 20–25
Clarke B (1865) On the structure and affinities of Callitrichaceae. J Bot 3
Clarke AE, Considine JA, Ward R, Knox RB (1977) Mechanism of pollination in *Gladiolus*: roles of the stigma and pollen tube guide. Ann Bot (London) 41: 15–20
Claussen P (1927) Über das Verhalten des Antherentapetums bei einigen Monokotylen und Ranales. Bot Arch 18: 1–27
Clayton WD (1981) Evolution and distribution of grasses. Ann Missouri Bot Garden 68: 5–14
Clifford HT, Smith WK (1969) Seed morphology and classification of Orchidaceae. Phytomorphology 19: 133–139
Cocucci AE (1961) Embriologia de *Trianthema argentina*. Kurtziana 1: 105–122
Cocucci AE (1964) The life history of *Aa achalensis* Schlecht. (Orchidaceae). Phytomorphology 14: 588–597
Cocucci AE (1966) Embryology of *Synandrospadix vermitoxicus* (Araceae). Kurtziana 3: 157–181
Cocucci AE (1969) El proceso sexual en Angiospermas. Kurtziana 5: 407–423
Cocucci AE (1973a) Some suggestions on the evolution of gametophytes of higher plants. Phytomorphology 23: 109–124
Cocucci AE (1973b) Orchid embryology: the membrane systems and the pollen tube growth. Caryologia 25: 201–206
Cocucci AE (1975) Estudios en el género Prosopanche (Hydnoraceae) 2. Organización de la flor. Kurtziana 8: 7–15

Cocucci AE (1976) Estudios en el género *Prosopanche* (Hydnoraceae) 3. Embriología. Kurtziana 9: 19–39
Cocucci AE (1983) New evidence from embryology in angiosperm classification. Nord J Bot 3: 67–73
Cocucci AE, Jensen WA (1969a) Orchid embryology: pollen tetrads of *Epidendrum scutella* in the anther and on the stigma. Planta 84: 215–229
Cocucci AE, Jensen WA (1969b) Orchid embryology: megagametophyte of *Epidendrum scutella* following fertilization. Am J Bot 56: 629–640
Cocucci AE, Jensen WA (1971) Orchid embryology: germinating male gametophyte of *Epidendrum scutella*. Kurtziana 6: 25–39
Cocucci AE, Venturelli M (1982) El óvulo y el genecio en Loranthaceae. Bol Soc Argentina Bot 21: 131–141
Cocucci AE, Venturelli M (1984) Sexualidad en especies con vacuoma oosférico de polaridad invertida (*Struthanthus*, Loranthaceae). Kurtziana 17: 9–24
Coe GE (1953) Cytology of reproduction in *Cooperia peduncrelata*. Am J Bot 40: 335–343
Coe GE (1954) Distribution of carbon^{14} in ovules of *Zephyranthes drummondi*. Bot Gaz 115: 342–346
Cohen LI (1963) Studies on the ontogeny of the dwarf mistletoes, *Arceuthobium* 1. Embryogeny and histogenesis. Am J Bot 50: 400–407
Coker WE (1907) The development of the seed in the Pontederiaceae. Bot Gaz 44: 293–301
Connor HE (1974) Breeding systems in *Cortaderia* (Gramineae). Evolution 27: 663–678
Conrad AH (1900) The contribution to the life history of *Quercus*. Bot Gaz 29: 408–418
Conrad HS (1905) The waterlilies: a monograph of the genus *Nymphaea*. Carnegie Inst Washington 4: 1–278
Cook CDK (1982) Pollination mechanisms in the Hydrocharitaceae. In: Symoens JJ, Hooper SS, Compere P (eds) Studies on aquatic vascular plants. R Bot Soc Belg, Brussels, pp 1–15
Cook MT (1903) The development of the embryo sac and embryo of *Claytonia virginiana*. Ohio Nat 3: 349–353
Cook MT (1906) The embryogeny of some Cuban Nymphaeaceae. Bot Gaz 42: 376–392
Cook MT (1907) The embryology of *Sagittaria lancifolia* L. Ohio Nat 7: 97–101
Cook MT (1908) The development of the embryo sac and embryo of *Potamogeton luceus*. Bull Torrey Bot Club 35: 209–218
Cook MT (1909) Notes on the embryo sac of *Passiflora adenophylla*. Bull Torrey Bot Club 36: 373–374
Cook MT (1924a) Development of the seed of *Linaria vulgaris*. Bot Gaz 77: 225–227
Cook MT (1924b) Development of seed of *Crotalaria sagittalis*. Bot Gaz 77: 440–445
Cooper DC (1931a) Microsporogenesis in *Bougainvillaea glabra*. Am J Bot 18: 337–355
Cooper DC (1931b) Macrosporogenesis and the development of the macrogametophyte of *Lycopersicon esculentum*. Am J Bot 18: 739–748
Cooper DC (1933) Nuclear divisions in the tapetal cells of certain angiosperms. Am J Bot 20: 358–364
Cooper DC (1934) Development of the embryo sac of *Lilium henryi*. Proc Natl Acad Sci USA 20: 163–166
Cooper DC (1935a) Microsporogenesis and the development of the male gametes in *Portulaca oleracea*. Am J Bot 22: 453–479
Cooper DC (1935b) Macrosporogenesis and development of embryo sac of *Lilium henryi*. Bot Gaz 97: 346–355
Cooper DC (1936) Development of the male gametes in *Lilium*. Bot Gaz 98: 169–177
Cooper DC (1938) Embryology of *Pisum sativum*. Bot Gaz 100: 123–132
Cooper DC (1939) Development of megagametophyte in *Erythronium albidum*. Bot Gaz 101: 862–867
Cooper DC (1940) Macrosporogenesis and embryology of *Portulaca oleracea*. Am J Bot 27: 326–330
Cooper DC (1941) Macrosporogenesis and the development of the seed of *Phryma leptostachya*. Am J Bot 28: 755–761
Cooper DC (1943) Haploid-diploid twin embryos in *Lilium* and *Nicotiana*. Am J Bot 30: 408–413
Cooper DC (1946) Double fertilization in *Petunia*. Am J Bot 33: 54–57

Cooper DC (1949) Flower and seed development in *Oxybaphus nyctagineus*. Am J Bot 36: 348–355
Cooper DC (1951) Caryopsis development following matings between diploid and tetraploid strain of *Zea mays*. Am J Bot 38: 702–708
Cooper DC, Brink RA (1949) The endosperm-embryo relationship in an autonomous apomict, *Taraxacum officinale*. Bot Gaz 111: 139–153
Cooper GO (1935) Cytological studies in the Chenopodiaceae 1. Microsporogenesis and pollen development. Bot Gaz 97: 169–178
Cooper GO (1942a) Development of the ovule and formation of seed in *Plantago lanceolata* L. Am J Bot 29: 577–581
Cooper GO (1942b) Microsporogenesis and development of seed in *Lobelia cardinalis*. Bot Gaz 104: 72–81
Copeland HF (1933) The development of seeds in certain Ericales. Am J Bot 20: 513–517
Copeland HF (1935) The structure of the flower of *Pholisma arenarium*. Am J Bot 22: 366–383
Copeland HF (1938) The *Styrax* of Northern California and the relationships of the Styracaceae. Am J Bot 25: 771–780
Copeland HF (1941) Further studies on Monotropoideae. Madrono 6: 97–119
Copeland HF (1943) A study, anatomical and taxonomic of the genera of Rhododendroideae. Am Midl Nat 30: 533–625
Copeland HF (1953) Observations on the Cyrillaceae particularly on the reproductive structures of the North American species. Phytomorphology 3: 405–411
Copeland HF (1954) Observations on certain Epacridaceae. Am J Bot 41: 215–222
Copeland HF (1955) The reproductive structures of *Pistacia chinensis* (Anacardiaceae). Phytomorphology 5: 440–449
Copeland HF (1959) The reproductive structures of *Schinus molle* (Anacardiaceae). Madrono 15: 14–25
Copeland HF (1960) The reproductive structures of *Fraxinus velutina*. Madrono 15: 161–172
Copeland HF (1961) Observations on the reproductive structures of *Anacardium occidentale*. Phytomorphology 11: 315–325
Copeland HF (1963) Structural notes on hollies (*Ilex aquifolium* and *I. cornuta*, family Aquifoliaceae). Phytomorphology 13: 455–464
Copeland HF (1966) Morphology and embryology of *Euomymus japonica*. Phytomorphology 16: 326–334
Core EL (1941) *Butomus umbellatus* in America. Ohio J Sci 79–85
Corner EJH (1949) The annonaceous seed and its four integuments. New Phytol 48: 332–364
Corner EJH (1951) The leguminous seed. Phytomorphology 1: 117–150
Corner EJH (1976) The seeds of dicotyledons, vols 1, 2. Univ Press, Cambridge (England)
Corner EJH (1981) Angiosperm classification and phylogeny: A criticism. Bot J Linn Soc 82: 81–87
Corpal R, Pevez-Gargia F, Pita JM (1989) Seed morphology and histology in four species of *Cistus* L. (Cistaceae). Phytomorphology 39: 75–80
Correa JP (1958) Morphological and embryological studies in Loranthaceae-Viscoideae. PhD Thesis, Univ Delhi
Corti R (1948) Sul dioicismo di *Idesia polycarpa*. Nuovo G Bot Ital 55: 446–495
Cortini C (1958) Sviluppo del gametofito femminile in *Malpighia fulcata* Ker-Gawl (Malpighiaceae). Caryologia 11: 42–56
Cosa MT (1978) Embryology of *Dyschoriste humilis* (Acanthaceae) (in Spanish). Bot Acad Nacl Cienc (Cordoba) 52: 253–266
Costa TJ, Mackenzie KAD (1990) Ovule and embryo sac development in *Malus punila* L. cv Cox's orange Pippin, from dormancy to blossom. Ann Bot (London) 66: 443–450
Costas-Lippmann M (1979) Embryogeny of *Cortaderia selloana* and *C. jubata* (Gramineae). Bot Gaz 140: 393–397
Coulter JM (1898) Contributions to the life history of *Ranunculus*. Bot Gaz 25: 73–78
Coulter JM, Chamberlain CJ (1903) Morphology of angiosperms. Univ Press, Chicago
Cox PA, Knox RB (1986) Pollination postulates and two-dimensional pollination. In: Williams EG, Knox RB, Irvine D (eds) Pollination '86. School Bot, Univ Melbourne, pp 48–57
Cox PA, Tomlinson PB (1988) Pollination ecology of a sea-grass, *Thalassia testudinum* (Hydrocharitaceae) in St. Croix. Am J Bot 75: 958–965

Coy GV (1928) Morphology of *Sassafras* in relation to the phylogeny of angiosperms. Bot Gaz 80: 149–171

Crane MB (1940a) Reproductive versatility in *Rubus* 1. Morphology and inheritance. J Genet 40: 109–118

Crane MB (1940b) The origin of new forms in *Rubus* 2. The loganberry *R. loganobaccus* Bayley. J Genet 40: 129–140

Crang RE, Millay MA (1972) Microscopical studies of *Lychnis alba* pollen walls during germination. Grana 12: 87–92

Creasy WD (1961) Development of the megaspore, megagametophyte and embryo of *Arenaria fenlleri* var. *fendleri*. Castanea 26: 140–146

Creech JL (1955) An embryological study in *Rhododendron*, subgenus *Anthodendron* Endl. Bot Gaz 116: 234–243

Crété P (1938) Embryogénie des Lobeliacées. Développement de l'embryon chez le *Lobelia syphilitica*. C R Acad Sci Paris 207: 177–179

Crété P (1939) La polyembryonie chez le *Lobelia syphilitica* L. Bull Soc Bot Fr 85: 580–583

Crété P (1942) Recherches histologiques et physiologiques sur l'embryologie des labiatiflores. Contribution à l'étude des formations haustoriales. D Sci These, Univ Paris

Crété P (1943) Recherches histologiques et physiologiques sur l'embryologie des labiatiflores. Embryologie du *Globularia vulgaris*. Bull Soc Bot Fr 90: 36–39

Crété P (1944) Polyembryonie chez l'*Actinidia chinensis* Planch. Bull Soc Bot Fr 91: 89–92

Crété P (1946a) Développement d l'albumen chez le *Loasa lateritia* Gill. C R Acad Sci Paris 222: 509–511

Crété P (1946b) Embryogénie des Loasacées. Développement de l'embryon chez le *Loasa lateritia* Gill. C R Acad Sci Paris 222: 920–921

Crété P (1946c) Embryogénie des Crassulacées. Développement de l'embryon chez le *Cotyledon umbilicus* L. C R Acad Sci Paris 222: 1311–1313

Crété P (1946d) Embryogénie des Hydrophyllacées. Développement de l'embryon chez le *Phacelia tanacetifolia*. C R Acad Sci Paris 223: 459–460

Crété P (1948a) Embryogénie des Papilionacées. Développement de l'embryon chez le *Lens esculenta* Moench (*Ervum lens* L., *Orbus lens* Stokes, *Vicia lens* Coss et G de St P). C R Acad Sci Paris 226: 590–592

Crété P (1948b) Recherches embryologiques chez les Scrofulariacées. Développement de l'albumen et de l'embryon chez le *Chaenostoma foetidum* (Jacq.) Benth. Bull Soc Bot Fr 95: 142–146

Crété P (1949a) Un cas de polyembryonie chez une Gentianacées l'*Erythraea centaurium* Pers. Bull Soc Bot Fr 96: 113–115

Crété P (1949b) Embryogénie des Gentianacées. Développement de l'embryon chez le *Gentiana asclepiadea* L. C R Acad Sci Paris 228: 768–770

Crété P (1949c) Embryogénie des Gentianacées. Développement de l'embryon chez l'*Erythraea centaurium* Pers. C R Acad Sci Paris 228: 1448–1449

Crété P (1949d) Recherches embryologiques chez les Scrofulariacées. Développement de l'albumen chez le *Lyperia violacea* Benth. Bull Soc Bot Fr 96: 186–188

Crété P (1949e) Recherches embryologiques sur les Gesneracées. Développement de l'albumen et de l'embryon chez le *Chirita lavandulacea*. Bull Soc Bot Fr 96: 234–235

Crété P (1950a) Embryologie des Scrofulariacées. Développement de l'albumen chez les *Nemesia*. C R Acad Sci Paris 231: 711–713

Crété P (1950b) Embryologie des Scrofulariacées. L'albumen et l'embryon chez les *Nemesia*. Bull Soc Bot Fr 97: 196–197

Crété P (1951) Repartition et intérêt phylogénétique des albumens à formations haustoriales chez les Angiospermes et plus particulièrement chez Gamopétales. Ann Sci Nat 12: 131–191

Crété P (1952a) Contribution à l'étude embryologique des Datiscacées. L'origine du sac embryonnaire et la transformation de l'ovule en graine chez le *Datisca cannabina*. Bull Soc Bot Fr 99: 152–156

Crété P (1952b) Embryogénie des Datiscacées. Développement de l'embryon chez le *Datisca cannabina* L. C R Acad Sci Paris 234: 1082–1084

Crété P (1952c) Embryologie des Bignoniacées. Les premiers stades du développement de l'albumen chez l'*Amphicome arguta*. Bull Soc Bot Fr 99: 315–318

Crété P (1952d) Embryogénie des Liliacées. Développement de l'embryon chez le *Scilla autumnalis*. C R Acad Sci Paris 235: 313–316

Crété P (1952e) Embryogénie des Liliacées. Développement de l'embryon chez le *Gasteria verrucosa*. C R Acad Sci Paris 235: 1421–1423

Crété P (1953a) Embryogénie des Boragacées. Développement de l'embryon chez l'*Eritrichium strictum*. C R Acad Sci Paris 236: 224–226

Crété P (1953b) Embryologie des Scrofulariacées. Développement de l'albumen et de l'embryon chez le *Digitalis purpurea* L. Phytomorphology 3: 168–172

Crété P (1953c) Embryogénie des Caprifoliacées. Développement de l'embryon chez le *Leycesteria formosa* Wall. C R Acad Sci Paris 237: 1432–1434

Crété P (1954a) Le développement de l'albumen et de l'embryon chez le *Tetranema mexicanum* Benth. Bull Soc Bot Fr 101: 232–235

Crété P (1954b) Embryologie de l'*Erinus alpinus* L. (Scrofulariacées). Les relations entre les genres *Erinus* L. et *Digitalis* L. Phytomorphology 4: 325–328

Crété P (1954c) Embryogénie des Liliacées. Développement de l'embryon chez l'*Endymion nutans*. C R Acad Sci Paris 238: 280–282

Crété P (1955a) Embryogénie des Résedacées. Développement de l'embryon chez le *Reseda glauca*. C R Acad Sci Paris 240: 645–647

Crété P (1955b) Embryogénie des Boragacées. Développement de l'embryon chez le *Cynoglossum officinale*. C R Acad Sci Paris 241: 660–662

Crété P (1955c) L'origine du sac embryonnaire et le développement de l'albumen chez l'*Alloplectus sanguineus*. Bull Soc Bot Fr 102: 205–208

Crété P (1955d) Embryogénie des Gentianacées. Développement de l'embryon chez le *Chlora perfoliata* L. C R Acad Sci Paris 241: 1825–1828

Crété P (1956a) A propos de l'embryologie de l'*Argemone mexicana*. Phytomorphology 6: 144–148

Crété P (1956b) Ményanthacées. Développement de l'embryon chez le *Limnanthemum nymphoides* Hoflg. et Link. C R Acad Sci Paris 242: 3110–3113

Crété P (1956c) Embryologie des Lentibulariacées. Développement de l'albumen chez le *Pinguicula*. C R Acad Sci Paris 242: 1346–1349

Crété P (1956d) Embryogénie végétale: Lentibulariacées. Développement de l'embryon chez le *Pinguicula leptoceras* Rchb. C R Acad Sci Paris 242: 1063–1066

Crété P (1956e) Embryogénie des Rubiacées. Développement de l'embryon chez le *Diodia dasycephala*. C R Acad Sci Paris 243: 861–863

Crété P (1956f) Contribution à l'étude de l'albumen et de l'embryon chez les Campanulacées et les Lobeliacées. Bull Soc Bot Fr 103: 446–454

Crété P (1956g) A propos de l'embryogénie du *Sonerila wallichii*. Bull Soc Bot Fr 103: 599–603

Crété P (1957a) Embryogénie des Melastomacées. Développement de l'embryon chez le *Clidemia hirta*. C R Acad Sci Paris 244: 374–376

Crété P (1957b) Embryogénie des Nolanacées. Développement de l'embryon chez le *Nolana prostrata*. C R Acad Sci Paris 245: 853–855

Crété P (1957c) Embryogénie des Saururacées. Développement de l'embryon chez l'*Houttuynia cordata* Thunb. C R Acad Sci Paris 245: 1331–1334

Crété P (1958a) Embryogénie des Cucurbitacées. Développement de l'albumen et de l'embryon chez le *Sicyos angulata* L. C R Acad Sci Paris 246: 456–459

Crété P (1958b) Développement de l'albumen et l'embryon chez le *Collinsia bicolor* Benth. Phytomorphology 8: 302–305

Crété P (1958c) Papilionacées. Développement de l'albumen et de l'embryon chez le *Cantharospermum barbatum* Koorders (*Atylosia barbata* Barker). C R Acad Sci Paris 246: 821–823

Crété P (1959) Embryogénie des Gentianacées. Développement de l'embryon chez le *Gentiana punctata* L. C R Acad Sci Paris 248: 3594–3596

Crété P (1960) Embryogénie des Cucurbitacées. Developpement de l'embryon chez l'*Ecballium elaterium* A Rich. C R Acad Sci Paris 251: 968–970

Crété P (1962) Embryogénie des Cucurbitacées. Developpement de l'embryon chez le *Cyclanthera explodens* Naud. C R Acad Sci Paris 254: 3411–3412

Crété P (1963a) Embryo. In: Maheshwari P (ed) Recent advances in the embryology of angiosperms. Int Soc Plant Morphol, Univ Delhi, pp 171–220

Crété P (1963b) Développement de l'albumen et de l'embryon chez le *Bystropogon origanifolius*. Phytomorphology 13: 363–367
Crété P (1964a) L'embryogénie et son rôle dans les essais de classification phylogénétique. Phytomorphology 14: 70–78
Crété P (1964b) L'albumen et l'embryon du *Mimulus moschatus* Doul. (Scrofulariacées). Phytomorphology 14: 468–472
Cronquist A (1957) Outline of a new system of families and orders of Dicotyledons. Bull Jardin Bot Etat Bruxelles 27: 13–40
Cronquist A (1965) The status of general system of classification of flowering plants. Ann Missouri Bot Garden 52: 281–303
Cronquist A (1968) The evolution and classification of flowering plants. Mifflin, Boston
Cronquist A (1981) An integrated system of classification of flowering plants. Columbia Univ Press, New York
Cronquist A (1983) Some realignments in the dicotyledons. Nord J Bot 3: 75–83
Croziat L (1941) On the systematic position of *Daphniphyllum* and its allies. Lingnan Sci J 20: 79–104
Cruden RW (1977) Pollen ovule ratio: a conservative indicator of breeding system in flowering plants. Evolution 31: 32–46
Cruden RW, Jensen WA (1979) Viscin threads, pollination efficiency and low pollen-ovule ratios. Am J Bot 66: 876–879
Cruden RW, Lyon DL (1985) Correlations among stigma depth, style length, and the pollen grain size. Do they reflect function or phylogeny? Bot Gaz 146: 143–149
Cummins MP (1929) Development of integument and germination of the seed of *Eleusine indica*. Bull Torrey Bot Club 56: 155–162
Cutter VM Jr, Freeman B (1954) The development of syncytial endosperm of *Cocos nucifera*. Nature (London) 173: 827–828
Cutter VM Jr, Freeman B (1955) Nuclear aberrations in the syncytial endosperm of *Cocos nucifera*. J Elisha Mitchell Sci Soc 71: 49–58
Cutter VM Jr, Witson KS, Freeman B (1955) Nuclear behaviour and cell formation in the developing endosperm of *Cocos nucifera*. Am J Bot 42: 109–115
Czapik R (1962) Embryological studies in the genus *Potentilla* L. 2. *Potentilla arenaria* (Rosaceae). Acta Biol Cracov Ser Bot 5: 29–42
Czapik R (1974) Embryology of five species of the *Arabis hirsuta* complex. Acta Biol Cracov Ser Bot 17: 13–25
Czapik R (1976) Variability in the embryological cycle of *Adoxa moschatellina* L. Acta Biol Cracov Ser Bot 19: 1–13
Czapik R (1981) Embryology of *Rubus saxatilis* L. Acta Biol Cracov Ser Bot 23: 7–13
Czapik R (1985) Apomictic embryo sacs in diploid *Waldsteinia geoides* Willd. (Rosaceae). Acta Biol Cracov Ser Bot 27: 29–37
Czapska-Dziekanowska D (1966) Studies in the mode of reproduction and the differentiation of the endosperm of *Plantago atrata* var. *capitata* (in Russian). Acta Biol Cracov Ser Bot 8: 101–112

Dahl AE (1989) Taxonomic and morphological studies in *Hypecoum* Sect. Hypecoum (Papaveraceae). Plant Syst Evol 163: 227–280
Dahl OA, Rowley JR (1965) Pollen of *Degeneria vitiensis*. J Arnold Arbor 46: 308–323
Dahlgren G (1989) The updated angiosperm classification. Bot J Linn Soc 100: 197–203
Dahlgren KVO (1915) Über die Embryologie von *Acicarpha tribuloides*. Svensk Bot Tidskr 9: 184–191
Dahlgren KVO (1916) Zytologische und embryologische Studien über die Reihen Primulales and Plumbaginales. Kon Svensk Vet Akad Handl 56: 1–80
Dahlgren KVO (1920) Zur Embryologie der Kampositen mit besonderer Berücksichtigung der Endospermbildung. Z Bot 12: 481–516
Dahlgren KVO (1922) Die Embryologie der Loganiazeen: Gattung *Spigelia*. Svensk Bot Tidskr 16: 77–87
Dahlgren KVO (1923) Notes on the ab initio cellular endosperm. Bot Not 1923: 1–24

Dahlgren KVO (1924) Studien über die Endospermbildung der Kompositen. Svensk Bot Tidskr 18: 177–203
Dahlgren KVO (1927a) Über das Vorkommen von Stärke in den Embryosäcken der Angiospermen. Ber Dtsch Bot Ges 45: 374–384
Dahlgren KVO (1927b) Morphologie des Nuzellus mit besonderer Berücksichtigung der deckzellosen Typen. Jahrb Wiss Bot 67: 347–426
Dahlgren KVO (1928) Die embryologie einiger Alismatazeen. Svensk Bot Tidskr 22: 1–17
Dahlgren KVO (1934a) Die Embryologie von *Impatiens roylei*. Svensk Bot Tidskr 28: 103–125
Dahlgren KVO (1934b) Die Embryosackentwicklung von *Echinodorus macrophyllus* und *Sagittaria sagittifolia*. Planta 21: 602–612
Dahlgren KVO (1937) Die Entwicklung des Embryosackes bei *Plumbago zeylanica*. Bot Not 1937: 487–498
Dahlgren KVO (1939a) Sur la présence d'amidon dans le sac embryonnaire chez les Angiospermes. Bot Not 1939: 221–231
Dahlgren KVO (1939b) Endosperm und Embryobildung bei *Zostera marina*. Bot Not 1939: 607–615
Dahlgren R (1975a) A system of classification of the angiosperms to be used to demonstrate the distribution of characters. Bot Not 128: 119–147
Dahlgren R (1975b) Current topics: The distribution of characters within an angiosperm system 1. Some embryological characters. Bot Not 128: 181–197
Dahlgren R (1977a) A commentary on a diagrammatic presentation of the angiosperms in relation to the distribution of character states. In: Kubitzki H (ed) Flowering plants: evolution and classification of higher categories. Plant Syst Evol Suppl 1: 253–283
Dahlgren R (1977b) A note on the taxonomy of the Sympetalae and related groups. Cairo Univ Herb 7, 8: 83–102
Dahlgren R (1980a) The taxonomic significance of chlorophyllous embryos in angiosperm seeds. Bot Not 133: 337–341
Dahlgren R (1980b) A revised system of classification of the angiosperms. Bot J Linn Soc 80: 91–124
Dahlgren R (1981) Angiosperm classification and phylogeny: A rectifying comment. Bot J Linn Soc 82: 89–92
Dahlgren R (1983a) General aspects of angiosperm evolution and macrosystematics. Nord J Bot 3: 119–149
Dahlgren R (1983b) The importance of modern serological research for angiosperm classification. In: Jensen U, Fairbrothers DE (eds) Proteins and nucleic acids in plant systematics. Springer, Berlin Heidelberg New York, pp 371–394
Dahlgren R, Clifford HT (1981) Some conclusions from a comparative study of the monocotyledons and related dicotyledonous orders. Ber Dtsch Bot Ges 94: 203–227
Dahlgren R, Clifford HT (1982) The monocotyledons: a comparative study. Academic Press, New York London
Dahlgren R, Lu A-m (1985) *Campynemanthe* (Campynemaceae): morphology, microsporogenesis, early ovule ontogeny and relationships. Nord J Bot 5: 321–330
Dahlgren R, Rasmussen FN (1983) Monocotyledon evolution: Characters and phylogenetic estimation. In: Hecht MK, Wallace B, Prance GT (eds). Evol Biol 16: 255–395
Dahlgren R, Thorne RF (1984) The order Myrtales: circumscription, variation and relationships. Ann Missouri Bot Garden 71: 633–699
Dahlgren R, Clifford HT, Yeo PF (1985) The families of the monocotyledons: structure, evolution and taxonomy. Springer, Berlin Heidelberg New York
Dahlgren R, Jensen SR, Nielsen BJ (1976) Iridoid compounds in Fouquieriaceae and notes on its possible affinities. Bot Not 129: 207–212
Dakshini KMM, Dadlani N (1978) Unusual diversity in microsporogenesis in *Vernonia cinerea*. Bot Mag (Tokyo) 91: 131–135
D'Amato F (1939) Ricerche embryologiche e caryologiche sul genere *Euphorbia*. Nuovo G Bot Ital 46: 470–509
D'Amato F (1940) Embriologia di *Ulmus campestris*. Nuovo G Bot Ital 47: 247–263
D'Amato F (1943) Nuovo contributo alla embriologia delle Plumbaginaceae. Nuovo G Bot Ital 50: 79–99

D'Amato F (1946a) Nuove ricerche embryologiche e caryologiche sul genere *Euphorbia*. Nuovo G Bot Ital 53: 405–436

D'Amato F (1946b) Osservazioni cito-embriologiche su *Cornus mas* con particolare riguardo alla sterilita di un Biotipo triploide. Nuovo G Bot Ital 55: 170–210

D'Amato F (1948–1949) Risultati di una analisi cario-embriologica in una popolazione di *Nothoscordum fragrans*. Caryologia 1: 194–200

D'Amato F (1949) Triploidia e apomissia in *Statice oleaefolia* Scop. var. *confusa* Goodr. Caryologia 2: 71–84

D'Amato F (1977) Nuclear cytology in relation to development. Univ Press, Cambridge (England)

D'Amato F (1984) Role of polyploidy in reproductive organs and tissues. In: Johri BM (ed) Embryology of angiosperms. Springer, Berlin Heidelberg New York, pp 519–566

Dang Van Liem (1962) Recherches sur l'embryogénie des Tricoques. Thèse, Univ Paris

Dannenbaum Ch, Walter M, Schill R (1989) Stigma morphology of the orchids. Bot Jahrb 110: 441–459

Danser BH (1933) A new system for the genera of Loranthaceae-Loranthoideae, with a nomenclator for the Old World species of this subfamily. Verh Kon Akad Wet 29: 1–128

Darlington CD, Wylie AP (1955) Chromosome atlas of flowering plants. George Allen & Unwin Ltd, London

Darrow GM, Waldo GF (1933) Pseudogamy in blackberry crosses. J Hered 24: 313–315

Das BC (1961) Polyembryony in mulberry. Curr Sci 30: 203

Das LDV, Ramkrishnan V (1974) Ovule sterility in *Bougainvillea* (Nyctaginaceae). Sci Cult 40: 81–82

Das NK (1965) Polyembryony in areca nuts (*Areca catechu* L.). Sci Cult 32: 469

Das VSR, Rao KN (1966) Chemotaxonomical investigation of *Nyctanthes*. Naturwissenschaften 53: 439

Dastur RH (1922) Notes on the development of the ovule, embryo sac and embryo of *Hydnora africana* Thunb. Trans Soc S Afr 10: 27–31

Dathan ASR (1974a) Embryology of *Melothria leucocarpa* (Bl.) Cogn. Sci Cult 40: 119–121

Dathan ASR (1974b) Embryology and seed development of *Dactyliandra welwitschii* Hook. f. Proc Indian Acad Sci B 80: 207–215

Dathan ASR, Singh D (1970) Female gametophyte and seed of *Carica candamarensis* Hook. f. Plant Sci (Lucknow, India) 2: 52–60

Dathan ASR, Singh D (1971a) Embryology and seed development in *Bergia* L. J Indian Bot Soc 50: 362–370

Dathan ASR, Singh D (1971b) Morphology and embryology of *Marah macrocarpa* Greene. Proc Indian Acad Sci B 73: 241–250

Dathan ASR, Singh D (1972a) Development of embryo sac and seed in *Bixa* L. and *Cochlospermum* Kunth. J Indian Bot Soc 51: 254–266

Dathan ASR, Singh D (1972b) Female gametophyte and endosperm of *Hydnocarpus laurifolia* (Dennst.) Sleumer. Curr Sci 41: 647–649

Dathan ASR, Singh D (1973a) Structure and development of ovule and seed in *Flacourtia indica* (Burn. F) Merrill. Proc Indian Natl Sci Acad B 39: 172–179

Dathan ASR, Singh D (1973b) Development and structure of seed in *Tacsonia* Juss. and *Passiflora* L. Proc Indian Acad Sci B 77: 5–18

Dathan ASR, Singh D (1973c) Structure and development of seed-coat in *Viola* spp. J Indian Bot Soc 52: 119–126

Dathan ASR, Singh D (1976) Embryological studies on papaya and mountain papaya (Abstr). In: Embryology of crop plants. Indo-Sov Symp. Indian Natl Sci Acad, New Delhi, and Univ Delhi, pp 8–9

Dathan ASR, Singh D (1979) Structure and development of female gametophyte and seed in *Hydnocarpus laurifolia* (Dennst.) Sleumer. J Indian Bot Soc 59: 256–263

Datta AK, Biswas AK (1984) Cytomixis and a trisomic in *Nigella sativa* L. Cytologia 49: 437–445

Datta M (1955) The occurrence and division of free nuclei in the endospermal milk of some Palmae. Trans Bose Res Inst (Calcutta) 19: 117–125

Datta PC, Chowdhury A (1981) Anatomy of pregermination to postgermination embryos of *Hibiscus cannabinus*. Flora 171: 266–279
Datta RM (1934) On the development of the embryo sac and pollen grain in *Cassia tora* Linn. J Indian Bot Soc 13: 277–299
Datta RM (1941) Tricotyledony in *Morus multicaulis* Perr. Curr Sci 10: 440
Datta RM (1954) Tricotyledony in *Corchorus capsularis* Linn. Sci Cult 20: 240–241
Datta RM (1955) Male gametophyte in cultivated jute. Curr Sci 24: 104
Datta RM, Mitra JN (1947) The systematic position of the family Moringaceae based on the study of *Moringa pterygosperma* Gaert. J Bombay Nat Hist Soc 47: 355–357
Datta S (1933) Embryology and cytological studies in *Nolana arbiplicifolia* and *Nolana prostrata*. J Indian Bot Soc 12: 131–152
David E (1938) Embryologische Untersuchungen an Myoporaceen, Salvadoraceen, Sapindaceen und Hippocrateaceen. Planta 28: 680–703
Davis GL (1961a) The life history of *Podolepis jaceoides* (Sims) Voss 1. Microsporogenesis and male gametogenesis. Phytomorphology 11: 86–97
Davis GL (1961b) The life history of *Podolepis jaceoides* (Sims) Voss 2. Megasporogenesis, female gametophyte and embryogeny. Phytomorphology 11: 206–219
Davis GL (1961c) The occurrence of synergid haustoria in *Cotula australis* (Less) Hook. f. Aust J Sci 24: 296
Davis GL (1962a) Embryological studies in the Compositae 1. Sporogenesis, gametogenesis and embryogeny of *Cotula australis* (Less.) Hook. f. Aust J Bot 10: 1–12
Davis GL (1962b) Embryological studies in the Compositae 2. Sporogenesis, gametogenesis and embryogeny in *Ammobium alatum* R. Br. Aust J Bot 10: 65–75
Davis GL (1963) Embryological studies in Compositae 3. Sporogenesis, gametogenesis and early embryogeny in *Minuria denticulata*. Proc Linn Soc NSW 88: 35–40
Davis GL (1964a) Embryological studies in Compositae 4. Sporogenesis, gametogenesis and embryogeny in *Brachycome ciliaris* (Labill.) Less. Aust J Bot 12: 142–156
Davis GL (1964b) The embryology of *Minuria integerrima* (DC) Benth. a somatic apomict. Phytomorphology 14: 230–239
Davis GL (1964c) Development of the female gametophyte of *Minuria cunninghamii*. Aust J Bot 12: 152–156
Davis GL (1966) Systematic embryology of the angiosperms. John Wiley & Sons, New York
Davis GL (1967) Apomixis in the Compositae. Phytomorphology 17: 270–277
Davis GL (1968a) Apomixis and abnormal anther development in *Calotis lappulacea* Benth. (Compositae). Aust J Bot 16: 1–17
Davis GL (1968b) Floral morphology and the development of gametophytes in *Eucalyptus melliodora* A. Cunn. Aust J Bot 16: 19–35
Davis GL (1969) Floral morphology and the development of the gametophytes in *Eucalyptus stellulata* Sieb. Aust J Bot 17: 177–190
Davis GL (1972) Apomixis. In: Murty YS, Johri BM, Mohan Ram HY, Varghese TM (eds) Advances in plant morphology. Prof. V. Puri Comm. Vol. Sarita Prakashan, Meerut (India), pp 126–132
Dawson ML (1936) The floral morphology of the Polemoniaceae. Am J Bot 23: 501–511
De Boer R, Bouman F (1972) Integumentary studies in the Polycarpicae. Acta Bot Neerl 21: 617–629
De Buhr LE (1975) Phylogenetic relationships of the Sarraceniaceae. Taxon 24: 297–306
De Candolle A (1857) Note sur la famille. Bibliotheque Univ Geneva
De Jongh Ph (1938) On the symbiosis of *Ardisia crispa* (Thunb.) A. DC. Kon Ned Akad Wet Verh (Tweede Sec) 37: 1–74
De Jussieu AL (1789) Genera plantarum. Paris
De Lange A, Bouman F (1986) Micromorphology of the seeds in *Begonia* section *Solananthera* A. DC. Acta Bot Neerl 35: 489–495
De Mason DA, Chandra Sekhar KN, Harris M (1989) Endosperm development in the date palm (*Phoenix dactylifera*) (Arecaceae). Am J Bot 76: 1255–1265
De M, Sharma AK (1983) Cytomixis in pollen mother cells of an apomictic ornamental *Ervatamia divaricata* (Linn.) Alston. Cytologia 48: 201–207

De Poerck RA (1950) Contribution à l'étude de Palmier à huile Africain, *Elaeis guineennsis.* Oleagineux 5: 623–628
De Vos MP (1947) Die Ontwikkeling van die Saadknop en saad by die Myoporaceae en die Systematiese Posisie van *Oftia.* S Afr J Sci 43: 171–187
De Vos MP (1948) The development of the ovule and seed in the Hypoxideae 1. *Ianthe* Salisb. J S Afr Bot 14: 159–169
De Vos MP (1949) The development of the ovule and the seed in the Hypoxideae 2. The genera *Pauridia* Harv. and *Forbesia* Ecklon. J S Afr Bot 15: 13–22
De Vos MP (1950) Die Ontwikkeling van die Saadknop en Saad by *Cyanella capensis.* S Afr J Sci 46: 220–226
De Vos MP (1956) Studies on the embryology and relationships of the South African genera of the Haemodoraceae: *Dilatris* and *Wachendorfia.* J S Afr Bot 22: 41–63
De Vos MP (1961) On the embryology and relationships of the South African genera of the Haemodoraceae. Rec Adv Bot 2: 694–698
De Vos MP (1963) Studies on the embryology and relationships of South African genera of the Haemodoraceae: *Lanaria* Ait. J S Afr Bot 29: 79–90
De Wet JMJ, Harlan JR (1970) Apomixis, polyploidy and speciation in *Dichanthium.* Evolution 24: 270–277
De Wet JMJ, Brink DE, Cohen CE (1983) Systematics of *Tripsacum,* Section Fasciculata (Gramineae). Am J Bot 70: 1139–1146
De Wet JMJ, Harlan JR, Brink DE (1982) Systematics of *Tripsacum dactyloides.* Am J Bot 69: 1251–1257
Dell B (1980) Structure and function of the strophiolar plug in seeds of *Albizia lophantha.* Am J Bot 67: 556–563
Dellert R (1933) Zur systematischen Stellung von *Wachendorfia.* Öst Bot Z 82: 335
Den Hartog C (1970) *Ondinea*: a new genus of Nymphaeaceae. Blumea 18: 413–417
Dermen H (1936) Aposporic parthenogenesis in a triploid apple, *Malus hupehensis.* J Arnold Arbor 17: 90–105
Dermen H (1949) Ploidy in the Hibernal apple and in some *Malus* species. J Hered 40: 162–164
Derschau MV (1918) Über disperme Befruchtung der Antipoden von *Nigella arvensis.* Ber Dtsch Bot Ges 36: 260–262
Desai S (1961) Studies on apospory in *Skimmia.* Phyton (Argentina) 16: 147–152
Desai S (1962a) Cytology and embryology of the Rutaceae. Phytomorphology 12: 178–184
Desai S (1962b) Polyembryony in *Xanthoxylum* Mill. Phytomorphology 12: 184–190
Deshpande PK (1960a) Morphology of endosperm in *Caesulia axillaris* Roxb. Curr Sci 29: 56
Deshpande PK (1960b) Studies in the family Compositae. Male and female gametophytes of *Parthenium hysterophorus.* J Biol Sci (Bombay) 2: 26–29
Deshpande PK (1961) Fertilization and development of endosperm, embryo and seed-coat in *Eclipta prostrata* Linn. Bull Bot Soc Coll Sci (Nagpur, India) 2: 1–8
Deshpande PK (1962a) Studies in the life history of *Glossocardia bosvallea* (Linn. f.) DC. Bull Bot Soc Coll Sci (Nagpur, India) 3: 84–91
Deshpande PK (1962b) Fertilization and development of endosperm, embryo and fruit in *Flaveria repanda.* J Indian Bot Soc 41: 505–509
Deshpande PK (1962c) Contribution to the embryology of *Caesulia axillaris* Roxb. J Indian Bot Soc 41: 540–550
Deshpande PK (1962d) A reinvestigation of the endosperm in *Tridax procumbens* L. Curr Sci 31: 113–114
Deshpande PK (1964a) A contribution to the life history of *Volutarella ramosa* Roxb. (= *Volutarella divericata* Benth & Hook. f.). J Indian Bot Soc 43: 141–148
Deshpande PK (1964b) A contribution to the embryology of *Bidens biternata* (Lour.) Mer. & Sherff. J Indian Bot Soc 43: 149–157
Deshpande PK (1970) Compositae. In: Symp Comparative embryology of angiosperms. Indian Natl Sci Acad Bull 41: 325–333
Deshpande PK, Bhasin RK (1974) Embryological studies in *Phaseolus aconitifolius* Jacq. Obs. Bot Gaz 135: 104–113
Deshpande PK, Bhasin RK (1976) A contribution to the life history of *Zornia diphylla* Pers. J Indian Bot Soc 55: 115–124

Deshpande PK, Gomkale KD (1982) Embryological studies in *Prosopis juliflora* (SW.) DC. J Indian Bot Soc 61: 39–46
Deshpande BD, Joneja P (1962) Studies in Asclepiadaceae 1. Morphology and embryology of *Leptadenia pyrotechnica* Deue. Phyton (Argentina) 19: 73–84
Deshpande PK, Kothare MP (1976) Embryology of *Cyathocline purpurea* Don. J Indian Bot Soc 55: 205–212
Deshpande PK, Raju PSG (1979) Development of caryopsis and localization of DNA and proteins in *Triticum* species. Phytomorphology 29: 100–111
Deshpande PK, Raju PSG (1985) Meiotic behaviour and microspore tetrad patterns in *Triticum* species. In: Govil CM, Kumar V (eds) Trends in plant research. (Prof. Y.S. Murty Comm Vol). Bishen Singh & Mahendra Pal Singh, Dehra Dun, pp 188–193
Deshpande PK, Thakkar SJ (1982) Contribution to the embryology of *Cassia mimosoides* L. J Indian Bot Soc 61: 9–18
Deshpande PK, Untawale AG (1971) Development of seed and fruit in *Indigofera enneaphylla* L. Bot Gaz 132: 96–102
Deshpande PK, Untawale AG (1972) Embryology of *Pycreus puncticulatus* Proc Natl Acad Sci India B 41: 303–310
Deshpande PK, Bhuskute SM, Makde KH (1984) A case of diplospory in *Luffà cylindrica* Linn. Curr Sci 53: 820–822
De Silva BLT (1945) A contribution to the life history of *Gloriosa superba*. Ceylon J Sci A 12: 155–159
Desole L (1951) First contribution to a knowledge of the embryological development of the genus *Centaurea*. Nuovo G Bot Ital 58: 162–168
Dessiatoff N (1911) Zur Entwicklung des Embryosackes von *Euphorbia virgata*. Ber Dtsch Bot Ges 29: 33–39
Dexheimer J (1970) Recherches cytophysiologiques sur les grains de pollen. Rev Cytol Biol Veg Bot 33: 169–234
Deyl M (1955) The evolution of the plants and the taxonomy of the monocotyledons. Acta Mus Nat Prague 11B: 1–143
Dhakre JS (1968) Morphological studies in the order Umbelliflorae: Umbelliferae 2. *Anethum graveolens* Linn. J Indian Bot Soc 47: 13–21
Dhar U, Vijayaraghavan MR (1979) Ontogeny, structure and breakdown of protein bodies in *Linum usitatissimum* Linn. Ann Bot (London) 43: 107–111
Dharamadhaj P, Prakash N (1978) Development of the anther and ovule in *Capsicum* L. Aust J Bot 26: 433–439
Dhawan V, Bhojwani SS (1985) Polyembryony in *Leucaena leucocephala*. Phytomorphology 35: 147–149
Di Fulvio TE (1961) El género *Sclerophylax* (Solanaceae). Kurtziana 1: 9–103
Di Fulvio TE (1965) Sobre la Embriología de *Cortesia cuneifolia* (Boraginaceae) 1. Anatomía floral esporogénesis y gametogénesis. Kurtziana 2: 7–25
Di Fulvio TE (1966) Embriología de *Plagiobothrys tenellus* (Boraginaceae). Kurtziana 3: 183–199
Di Fulvio TE (1969) Embryology of *Nolana paradoxa* (Nolanaceae). Kurtziana 5: 39–54
Di Fulvio TE (1978) Sobre la vasculatura floral, embriología y cromosomas de *Ixorhea tschudiana* (Heliotropiaceae). Kurtziana 11: 75–105
Di Fulvio TE (1979) El endosperma y el embrion en el sistema de Tubiflorae, con especial referencia a Boraginaceae e Hydrophyllaceae. Kurtziana 12–13: 101–112
Di Fulvio TE (1980) Sobre el polen de *Ixorhea tschudiana* (Heliotropiaceae). Bol Soc Argentina Bot 19: 83–90
Di Fulvio TE (1981) La embriología en la sistemática de Angiospermas. Kurtziana 14: 21–39
Di Fulvio TE (1987) La endospermogénesis en Hydrophylleae (Hydrophyllaceae) con relación a la taxonomía. Kurtziana 19: 13–34
Di Fulvio TE (1989a) Embriología de *Nama jamaicense* (Phacelieae, Hydrophyllaceae). Kurtziana 20: 9–31
Di Fulvio TE (1989b) Observaciones embriológicas en especies Argentinas de *Hydrolea* (Hydrophyllaceae) con especial referencia a la endospermogénesis. Kurtziana 20: 33–64

Di Fulvio TE, Cave MS (1964) Embryology of *Blandfordia nobilis* Smith (Liliaceae), with special reference to its taxonomic position. Phytomorphology 14: 487–499
Diboll AG, Larson DA (1966) An electron microscopic study of the mature megagametophyte in *Zea mays*. Am J Bot 53: 391–402
Dickison WC (1972) Observations of the floral morphology of some species of *Saurauia*, *Actinidia* and *Clematoclethra*. J Elisha Mitchell Sci Soc 88: 43–54
Dickison WC (1975a) Floral morphology and anatomy of *Bauera*. Phytomorphology 25: 69–76
Dickison WC (1975b) Studies on the floral anatomy of the Cunoniaceae. Am J Bot 62: 433–447
Dickison WC, Swetrer EM (1970) The morphology and relationships of *Barbeya oleoides*. Am J Bot 57: 468–476
Dickinson HG (1982) The development of pollen. Rev Cytol Biol Veg Bot 5: 5–19
Dickinson HG, Lawson J (1975) The growth of the pollen tube wall in *Oenothera organensis*. J Cell Sci 18: 519–532
Diels L (1930a) Iridaceae. In: Engler A, Prantl K (eds) Die natürlichen Pflanzenfamilien 15a: 463–505
Diels L (1930b) Roridulaceae. In: Engler A, Prantl K (eds) Die natürlichen Pflanzenfamilien 18a: 346–348
Diels L (1936) Engler's Syllabus der Pflanzenfamilien. Bornträger, Berlin
Ding H (1967) *Sarawakodendron*, a new genus of Celastraceae. Blumea 15: 139–143
Dittrich M (1977) Cynareae: Systematic review. In: Heywood VH, Harborne JB, Turner BL (eds) The biology and chemistry of Compositae, vol 2. Academic Press, New York London, pp 999–1055
Dixit SN (1954) Occurrence of tracheids in the mamelon of *Amyema* van Tiegh. Sci Cult 20: 39–41
Dixit SN (1958a) Morphological and embryological studies in the family Loranthaceae 4. *Amyema* van Tiegh. Phytomorphology 8: 346–364
Dixit SN (1958b) Morphological and embryological studies in the family Loranthaceae 5. *Lepeostegeres gemmiflorus* (Bl.) Bl. Phytomorphology 8: 365–376
Dixit SN (1961) Morphological and embryological studies in the family Loranthaceae 8. *Tolypanthus* Bl. Phytomorphology 11: 335–345
Dnyansagar VR (1949) Embryological studies in the Leguminosae 1. A contribution to the embryology of *Leucaena glauca* Benth. J Indian Bot Soc 28: 95–107
Dnyansagar VR (1951a) Embryological studies in the Leguminosae 2. A contribution to the embryology of *Mimosa hamata*. J Indian Bot Soc 30: 100–107
Dnyansagar VR (1951b) Embryological studies in Leguminosae 3. A contribution to the embryology of *Pithecolobium saman* Benth. (syn. *Enterolobium saman* Prain). Proc Indian Acad Sci B 34: 188–198
Dnyansagar VR (1952) Embryological studies in the Leguminosae 4. A contribution to the embryology of *Neptunia triquerta* Benth. Proc Indian Acad Sci B 36: 1–11
Dnyansagar VR (1954a) Behaviour of tapetal cells during microsporogenesis of *Adenanthera pavonina* Linn. Curr Sci 23: 131
Dnyansagar VR (1954b) Embryological studies in the Leguminosae 6. Inflorescence, sporogenesis, and gametophytes of *Dichrostachys cinerea* W. & A. and *Parkia biglandulosa* W. & A. Lloydia 17: 263–274
Dnyansagar VR (1954c) Embryological studies in the Leguminosae. 7. Endosperm and embryo development in *Neptunia triquetra* Benth. and *Prosopis spicigera* Linn. J Indian Bot Soc 33: 247–253
Dnyansagar VR (1954d) Embryological studies in the Leguminosae 9. Development of the endosperm and embryo in *Dichrostachys cinerea* W. & A. and *Parkia biglandulosa* W. & A. J Indian Bot Soc 33: 423–432
Dnyansagar VR (1954e) Embryological studies in the Leguminosae 10. Supplementary observations on the development of the endosperm and embryo in *Leucaena glauca* Benth. and *Mimosa hamata* Willd. J Indian Bot Soc 33: 433–442
Dnyansagar VR (1955) Embryological studies in the Leguminosae 11. Embryological features and formula and taxonomy of the Mimosaceae. J Indian Bot Soc 34: 362–374
Dnyansagar VR (1955–56) Embryological studies in Leguminosae 12. Status of the Leguminosae. Saugar Univ J Sci (Sagar, India) 1: 39–53

Dnyansagar VR (1957) Embryological studies in the Leguminosae 5. *Prosopis spicigera* and *Desmanthus virgatus*. Bot Gaz 118: 180–186

Dnyansagar VR (1958) Embryological studies in the Leguminosae 8. *Acacia auriculaeformis* A. Cunn., *Adenanthera pavonina* Linn., *Calliandra hematocephala* Hassk. and *Calliandra grandiflora* Benth. Lloydia 21: 1–25

Dnyansagar VR (1970) Leguminosae. In: Symp Comparative embryology of angiosperms. Indian Natl Sci Acad Bull 41: 93–103

Dnyansagar VR, Cooper DC (1960) Development of the seed of *Solanum phureja*. Am J Bot 47: 176–186

Dnyansagar VR, Malkhede SR (1963) Development of the seed of *Trianthema portulacastrum*. Proc Indian Acad Sci B 57: 343–355

Dnyansagar VR, Sudhakaran IV (1977) Seed development in diploid and auto-tetraploid of *Vinca rosea* syn. *Catharanthus roseus* (*Lochnera rosea*). Proc Indian Natl Sci Acad B 43: 133–143

Dnyansagar VR, Tiwari DK (1956) Sporogenesis and gametophytes of *Fimbristylis quinquangularis* Kunth. Bull Bot Soc Univ Saugar (Sagar, India) 8: 3–6

Dnyansagar VR, Tiwari HN (1966) Development of seed of *Euphorbia rothiana* Spreng. J Biol Sci (Bombay) 9: 1–7

Dodds KS (1945) Genetical and cytological studies of *Musa* 6. The development of female cells of certain edible diploids. J Genet 46: 161–179

Doida Y (1960) Developmental studies in the genus *Polygonum* 1. Microsporogenesis of *Polygonum persicaria*. Bot Mag (Tokyo) 73: 278–282

Dolcher T (1947) Ricerche embryologiche sulla famiglia delle Rhamnaceae. Nuovo G Bot Ital 54: 648–673

Dolcher T (1951) La meiosi e lo sviluppo del gametofito nell' ovulo di *Sambucus nigra* L. (Caprifoliaceae). Caryologia 4: 58–76

Don D (1827) On the affinities of the Empetraceae: a natural group of plants. Edinburgh New Philos Mag 2: 59–63

Donati G (1913) Ricerche embriologiche sulle Euphorbiaceae. Ann Bot (Rome) 11: 395–399

Dop P (1903) Recherches sur la structure et le développement de la fleur des Asclepiadées. These, Univ Toulouse (Fr)

Dop P (1913) Recherches sur la développement et la nutrition du sac embryonnaire et de l'endosperme des *Buddleija*. Bull Soc Bot Fr 68: 9–16, 45–50, 92–98

Dorasami LS, Gopinath DM (1945) An embryological study of *Linum mysorense* Hyn. Proc Indian Acad Sci B 22: 6–9

Dore WG (1956) Some grass genera with liquid endosperm. Bull Torrey Bot Club 83: 335–337

Dormer KJ (1962) The fibrous layer in the anthers of Compositae. New Phytol 61: 150–153

Dowrick GJ (1961) Biology of reproduction in *Rubus*. Nature (London) 191: 680–682

Doyle JA (1973) The monocotyledons. Their evolution and comparative biology 5. Fossil evidence on early evolutions of the monocotyledons. Q Rev Biol 48: 399–413

Drexler U, Hakki MI (1979) Embryologische und morphologische Untersuchungen an Pflanzen aus Westindien 2. Zur Embryologie von *Eustoma exaltatum* (Gentianaceae), mit einer Bemerkung zum Phänomen der "instant pollen tubes". Willdenowia 9: 131–147

Drinnan AN, Ladiges PY (1989) Operculum development in *Eucalyptus cloeziana* and *Eucalyptus* informal subg. *Monocalyptus* (Myrtaceae). Plant Syst Evol 166: 183–196

Drude O (1889) Ericaceae. In: Engler A, Prantl K (eds) Die natürlichen Pflanzenfamilien 4: 15–65

Drugg WS (1962) Pollen morphology of the Lennoaceae. Am J Bot 49: 1027–1032

Ducamp L (1902) Recherches sur l'embryogénie des Araliacées. Ann Sci Nat Bot 15: 311–402

Ducker SC, Knox RB (1976) Submarine pollination in sea grasses. Nature (London) 263: 705–706

Ducker SC, Pettitt JM, Knox RB (1978) Biology of Australian seagrasses: Pollen development and submarine pollination in *Amphibolis antarctica* and *Thalossodendron ciliatum* (Cymodoceaceae). Aust J Bot 26: 265–285

Dudgeon W (1918) Morphology of *Rumex crispus*. Bot Gaz 66: 393–421

Duggar BM (1899) On the development of the pollen grain and the embryo sac in *Bignonia venusta*. Bull Torrey Bot Club 26: 89–105

Dujardin M, Hanna WW (1983a) Meiotic and reproductive behaviour of facultative apomictic BC_1 offspring derived from *Pennisetum americanum* – *P. orientale* interspecific hybrids. Crop Sci 23: 156–160

Dujardin M, Hanna WW (1983b) Apomictic and sexual pearl millet × *Pennisetum squamulatum* hybrids. J Hered 74: 277–279
Dulberger R (1989) The apertural wall in pollen of *Linum grandiflorum*. Ann Bot (London) 63: 421–432
Dulcy JE, Lakshmanan KK (1981) Ruminate endosperm in *Dillenia suffruticosa* (Griff.) Martelli. Curr Sci 50: 965–966
Dulcy JE, Lakshmanan KK (1984) Further contributions to the embryology of *Dillenia suffruticosa* (Griff.) Martelli. J Indian Bot Soc 63: 353–359
Dumortier BC (1829) Analyse des familles des plantes. Tournay (Fr)
Dunbar A (1973) Pollen development in the *Eleocharis palustris* group (Cyperaceae) 1. Ultrastructure and ontogeny. Bot Not 126: 197–254
Dunbar A (1978) Pollen morphology and taxonomic position of the genus *Pentaphragma* Wall. (Pentaphragmaceae). Grana 17: 141–147
Duncan EJ (1970) Ovule and embryo ontogenesis in *Bombacopsis glabra* (Pasg.) A. Robyns. Ann Bot (London) 34: 677–683
Duncan EJ, Todd AW (1972) Structure of the mature embryo of *Theobroma cacao* L. Ann Bot (London) 36: 939–945
Dute RR, Peterson CM, Kushing AE (1989) Ultrastructural changes of egg apparatus associated with fertilization and proembryo development of Soybean, *Glycine max* (Fabaceae). Ann Bot (London) 64: 123–135
Dutt BSM (1957a) Morphology of the ovule of *Crinum defixum* Ker. Curr Sci 26: 22–24
Dutt BSM (1957b) Ovule and embryo sac of *Crinum asiaticum* L. Sci Cult 22: 688–691
Dutt BSM (1958) Morphology of the ovule of *Zephyranthes citrina* Baker. Curr Sci 27: 501–502
Dutt BSM (1959) Ovule and embryo sac of *Crinum latifolium* L.: A reinvestigation. Curr Sci 28: 293–294
Dutt BSM (1962) A contribution to the life history of *Crinum defixum* Ker. In: Symp Plant embryology. Council Sci Ind Res, New Delhi, pp 37–48
Dutt BSM (1964) Ovule and embryo sac of *Pancratium maritimum* L.: A reinvestigation. Curr Sci 33: 150–151
Dutt BSM (1970a) Haemodoraceae. In: Symp Comparative embryology of angiosperms. Indian Natl Sci Acad Bull 41: 358–361
Dutt BSM (1970b) Cynastraceae. In: Symp Comparative embryology of angiosperms. Indian Natl Sci Acad Bull 41: 362–364
Dutt BSM (1970c) Embryology of *Pancratium longiflorum*. Phytomorphology 20: 1–5
Dutt BSM (1970d) Amaryllidaceae. In: Symp Comparative embryology of angiosperms. Indian Natl Sci Acad Bull 41: 365–367
Dutt BSM (1970e) Hypoxidaceae. In: Symp Comparative embryology of angiosperms. Indian Natl Sci Acad Bull 41: 368–372
Dutt BSM (1970f) Velloziaceae. In: Symp Comparative embryology of angiosperms. Indian Natl Sci Acad Bull 41: 373–374
Dutt BSM (1975) Embryo sac development in *Scilla peruviana* L. Curr Sci 44: 91–92
Dutt BSM (1978a) Anther in *Moringa concanensis* Nimmo. Curr Sci 47: 589
Dutt BSM (1978b) Embryogeny in *Gyrocarpus americanus* Jacq. Curr Sci 47: 636
Dutt BSM (1978c) Embryo development in *Moringa concanensis* Nimmo. Curr Sci 47: 693
Dutt BSM (1979a) Fruit and seed of *Gyrocarpus americanus* Jacq. Curr Sci 48: 594–596
Dutt BSM (1979b) Ovule and seed of *Moringa concanensis* Nimmo. Curr Sci 48: 652–654
Dutt BSM, Prakasa Rao PS (1977) Pollen grains in the stylar canal of *Moringa concanensis* Nimmo. Curr Sci 46: 535–536
Dutt M (1953) Dividing nuclei in coconut milk. Nature (London) 171: 799
Dutt MK (1943) Development of the microspores and the nuclear behaviour in tapetal cells of *Putranjiva roxburghii*. Sci Cult 8: 309–310
Dutt S (1939) A note on the development of the female gametophyte in some Indian Compositae. Curr Sci 8: 471–472
Dwivedi NK (1984) Embryology of *Brunella vulgaris* L. J Indian Bot Soc 63: 381–386
Dwivedi NK (1985) Development of embryo, seed-coat and pericarp in *Craniotome versicolor* Reichb. Acta Bot Indica 13: 185–188

Dwivedi NK (1990) Sporogenesis and gametogenesis in paper mulberry (*Broussonetia papyrifera* Vent). Acta Bot Indica 18: 99–100
Dwivedi NK, Joshi BM (1980) Embryological studies in Lamiaceae 7. Development of endosperm in *Craniotome versicolor* Reichb. Curr Sci 49: 274–275
Dwivedi NK, Joshi BM (1981) Studies in the Lamiaceae 10. A note on the sporogenesis and gametogenesis in *Nepeta hindostana* (Roth.) Haines. Curr Sci 50: 647–649
Dwivedi NK, Joshi BM (1986) Dual origin and dimorphism of anther tapetum in *Melissa parviflora* Benth. Acta Bot Indica 14: 57–60
Dzevaltovsky AK (1963) Development of the female gametophyte in *Benincasa hispida* (*Benincasa cerifera*). In: Questions of physiology, cytoembryology and flora of the Ukraine. Akad Nauk Ukr RSR Kiev, pp 60–72 (in Russian)

Eames AJ (1951) Floral anatomy as an aid in generic limitation. Chron Bot 14: 126–132
Eames AJ (1961) Morphology of the angiosperms. McGraw-Hill, New York
Eckardt Th (1957) Zur systematischen Stellung von *Eucommia ulmoides*. Ber Dtsch Bot Ges 69: 487–498
Eckardt Th (1963) Some observations on the morphology and embryology of *Eucommia ulmoides*. In: Sadasivan TS (ed) J Indian Bot Soc (Maheshwari Comm Vol) 42A: 27–34
Eckardt Th (1964) Pandanales. In: Engler A (ed) Syllabus der Pflanzenfamilien 2: 598–602
Eckles ML (1940) Megasporogenesis and development of the megagametophyte in *Nothascordum bivalve*. Bot Gaz 102: 806–809
Edlin HL (1935) A critical revision of certain taxonomic groups of the Malvales 2. New Phytol 34: 122–143
Edwards J (1920) Flower and seed of *Hedyosmum nutans*. Bot Gaz 70: 409–429
Egley GH, Paul RN (1981) Morphological observations on the early imbibition of water by *Sida spinosa* (Malvaceae) seed. Am J Bot 68: 1056–1065
Ehdaie M, Russell SD (1984) Megagametophyte development of *Nandina domestica* and its taxonomic implications. Phytomorphology 34: 221–225
Ehrendorfer F (1973) Adaptive significance of major taxonomic characters and morphological trends in angiosperms. In: Heywood VH (ed) Taxonomy and ecology. Academic Press, London New York, pp 317–327
Eichler AW (1875) Blüthendiagramme. Leipzig Berlin
Eichler AW (1878) Blüthendiagramme 2: 128–129
Eigner J (1973) Zur Stempel- und Fruchtentwicklung ausgewählter Brassicaceae (Cruciferae) unter neueren Gesichtspunkten der Blütenmorphologie und der Systematik. Beitr Biol Pflanzen 49: 359–427
Eigsti OJ (1941) The occurrence of a pollen tube with four sperms and two tube nuclei in *Polygonatum*. Proc Oklahoma Acad Sci 21: 134–136
Einset J (1945) The spontaneous origin of polyploid apples. Proc Am Soc Hortic Sci 51: 61–63
Einset J (1948) The occurrence of spontaneous triploids and tetraploids in apples. Proc Am Soc Hortic Sci 51: 61–63
Einset J (1951) Apomixis in American polyploid black-berries. Am J Bot 38: 768–772
Ekambaram T, Panje RR (1935) Contributions to our knowledge of *Balanophora* 2. Life history of *B. dioica*. Proc Indian Acad Sci B 1: 522–543
Ekanthappa KG, Arekal GD (1977a) A contribution to the embryology of *Cirrhopetalum fimbriatum* Lindl. Proc Indian Acad Sci B 86: 211–216
Ekanthappa KG, Arekal GD (1977b) Ontogeny and organisation of female gametophyte in *Polystachya flavescens* (Bl.) JJ Smith: A reinvestigation. Curr Sci 46: 684–686
Ekdahl I (1941) Die Entwicklung von Embryosack und Embryo bei *Ulmus glabra*. Svensk Bot Tidskr 35: 143–156
Eliasson U (1971) Studies in Galapagos plants 10. The genus *Lecocarpus* Decaisne. Svensk Bot Tidskr 65: 245–277
Eliasson U (1972) Studies in Galapagos plants 11. Embryology of *Macraea larcifolia* Hook. f. (Compositae). Svensk Bot Tidskr 66: 43–47
Elisens WJ, Tomb AS (1983) Seed morphology in new world Antirrhineae (Scrophulariaceae): Systematic and phylogenetic implication. Plant Syst Evol 142: 23–47

Elizabeth JD, Lakshmanan KK (1984) Further contributions to the embryology of *Dillenia suffruticosa* (Griff.) Martelli. J Indian Bot Soc 63: 353–359
Ellerström S, Zagorcheva L (1977) Sterility and apomictic embryo sac formation in *Raphanobrassica.* Hereditas 87: 107–120
Eluterius LN (1984) Female gametophyte development in *Juncus roemerianus.* Bull Torrey Bot Club 111: 445–450
Emberger L (1960) Les végétaux vasculaires. In: Chaudefaud M, Emberger L (eds) Traité de botanique, vol 2. Masson, Paris
Emery WHP (1957) A study of reproduction in *Setaria macrostachya* and its relatives in the southwestern United States and northern Mexico. Bull Torrey Bot Club 84: 106–121
Emery WHP, Brown WV (1957) Extra-ovular development of embryos in two grass species. Bull Torrey Bot Club 84: 361–365
Endlicher SL (1836–1840) Genera plantarum. Secundum ordines naturales disposita Vindobonae
Endress PK (1969a) Gesichtspunkte zur systematischen Stellung der Eupteleaceen (Magnoliales). Untersuchungen über Bau und Entwicklung der generativen Region bei *Euptelea polyandra* Sieb. et Zucc. Ber Schweiz Bot Ges 79: 229–278
Endress PK (1969b) Eupteleaceae – Gesichtpunkte zur systematischen Stellung der *Euptelea*-(Magnoliales). Mitt Bot Mus Univ Zürich 241: 229–278
Endress PK (1971) Bau der weiblichen Blüten von *Hedyosmum maxicanum* Cordemoy (Chloranthaceae). Bot Jahrb 91: 39–60
Endress PK (1972) Zur vergleichenden Entwicklungsmorphologie, Embryologie und Systematik bei Laurales. Bot Jahrb 92: 331–428
Endress PK (1977a) Evolutionary trends in the Hamamelidales-Fagales group. In: Kubitzki H (ed) Flowering plants. Plant Syst Evol Suppl 1. Springer, Wien, pp 321–347
Endress PK (1977b) Über Blütenbau und Verwandtshaft der Eupomatiaceae und Himatandraceae (Magnoliales). Ber Dtsch Bot Ges 90: 83–103
Endress PK (1979) Noncarpellary pollination and "hyperstigma" in an angiosperm (*Tambourissa religiosa*, Monimiaceae). Experientia 35: 45
Endress PK (1980a) The reproductive structures and systematic position of the Austrobaileyaceae. Bot Jahrb 101: 393–433
Endress PK (1980b) Floral structure and relationships of *Hortonia* (Monimiaceae). Plant Syst Evol 133: 199–221
Endress PK (1980c) Ontogeny, function and evolution of extreme floral construction in Monimiaceae. Plant Syst Evol 134: 79–120
Endress PK (1987a) The early evolution of the angiosperm flower. Trends Ecol Evol 2: 300–304
Endress PK (1987b) The Chloranthaceae: reproductive structures and phylogenetic position. Bot Jahrb 109: 153–226
Endress PK, Honegger R (1980) The pollen of the Austrobaileyaceae and its phylogenetic significance. Grana 19: 177–182
Endress PK, Hufford LD (1989) The diversity of stamen structures and dehiscence patterns among Magnoliidae. Bot J Linn Soc 100: 45–85
Endress PK, Sampson FB (1983) Floral structure and relationships of the Trimeniaceae (Laurales). J Arnold Arbor 64: 447–473
Endress PK, Vosser P (1975) Zur Androeciumanlage und Antherenentwicklung bei *Caloncoba echinata* (Flacourtiaceae). Plant Syst Evol 123: 241–253
Endriss W (1902) Monographie von *Pilostyles ingae.* Flora 91: 209–236
Engell K (1970) Embryological investigations in *Arnica alpina* from Greenland. Bot Tidsskr 65: 225–244
Engell K (1973) A preliminary morphological, cytological and embryological investigation in *Polygonum viviparum.* Bot Tidsskr 67: 305–316
Engell K (1979) Morphology and embryology of Scybalioideae (Balanophoraceae) 1. *Corynaea crassa* Hook. f. var. *sprucei* (Eichl.) B. Hansen. Bot Tidsskr 73: 155–166
Engell K (1989) Embryology of barley: time course and analysis of controlled fertilization and early embryo formation based on serial sections. Nord J Bot 9: 265–280
Engleman EM (1960) Ovule and seed development in certain cacti. Am J Bot 47: 460–467

Engler A (1876) Beiträge zur Kenntnis der Antherenbildung der Metaspermen. Jahrb Wiss Bot 10: 275–316

Engler A (1888) Rapateaceae. In: Engler A, Prantl K (eds) Die natürlichen Pflanzenfamilien. 2(4): 28–31

Engler A (1891) Saxifragaceae. In: Engler A, Prantl K (eds) Die natürlichen Pflanzenfamilien. 3(29): 41–93

Engler A (1894) Saxifragaceae. In: Engler A, Prantl K (eds) Die natürlichen Pflanzenfamilien. 3(1): 41–93

Engler A (1907) Syllabus der Pflanzenfamilien. Berlin

Engler A, Diels L (1936) Syllabus der Pflanzenfamilien. Bornträger, Berlin

Engler A, Gilg E (1924) Syllabus der Pflanzenfamilien. Engelmann, Leipzig

Engler A, Melchior K (1925) Medusagynaceae. In: Engler A, Prantl K (eds) Die natürlichen Pflanzenfamilien 21: 50–52

Engler A, Prantl K (eds) (1894) Die natürlichen Pflanzenfamilien. Leipzig

Engler A, Prantl K (eds) (1897) Die natürlichen Pflanzenfamilien. Leipzig

Engler A, Prantl K (eds) (1931) Die natürlichen Pflanzenfamilien. Leipzig

Englskjon T (1967) Cytoembryological studies in Arctic-Alpine *Taraxacum* I. The Bear Island *Taraxacum.* Nytt Mag Bot 14: 125–137

Erbrich P (1965) Über Endopolyploidie und Kernstrukturen in Endospermhaustorien. Öst Bot Z 112: 197–262

Erdelská O (1966) Influence of fertilization on the development of antipodal nuclei in barley. Biologia (Bratislava) 21: 857–864

Erdelská O (1967) Type of endosperm of the species *Linum austriacum* L. Biologia (Bratislava) 22: 172–177

Erdelská O (1969) Les premières phases du développement de l'albumen de *Jasione montana* L. Résultats d'une étude microcinématographique. Rev Cytol Biol Veg 32: 397–400

Erdelská O (1975) Pre-fertilization development of ovule of *Jasione montana.* Phytomorphology 25: 76–80

Erdelská O, Klasova A (1978) La region micropylaire du sac embryonnaire de *Jasione montana* L. avant et après la fécondation. Bull Soc Bot Fr 125: 249–252

Erdtman G (1945) Pollen morphology and plant taxonomy 5. On the occurrence of tetrads and dyads. Svensk Bot Tidskr 39: 286–297

Erdtman G (1952) Pollen morphology and plant taxonomy. Chronica Botanica, Waltham (Mass, USA)

Erdtman G (1969) Handbook of palynology. Hafner, New York

Ernst A (1901) Beiträge zur Kenntnis der Entwicklung des Embryosackes und des Embryos (Polyembryonie) von *Tulipa gesneriana* L. Flora 88: 37–77

Ernst A (1902) Chromosomenreduktion, Entwicklung des Embryosackes und Befruchtung bei *Paris quadrifolia* und *Trillium grandiflorum.* Flora 91: 1–46

Ernst A (1913) Zur Kenntnis von Parthenogenesis und Apogamie bei Angiospermen. Verh Schweiz Natl Jahresv Ges 96: 1–13

Ernst A (1914) Embryobildung bei *Balanophora.* Flora 106: 129–158

Ernst A (1918) Bastardierung als Ursache der Apogamie im Pflanzenreich. Fischer, Jena

Ernst A, Bernard Ch (1912a) Entwicklungsgeschichte des Embryosackes und des Embryos von *Burmannia candida* Engl. und *B. championii* Thw. Ann Jardin Bot Buitenz 10: 161–168

Ernst A, Bernard Ch (1912b) Entwicklungsgeschichte des Embryosackes des Embryos und des Endospermas von *Burmannia coelestis* Don. Ann Jardin Bot Buitenz 11: 234–257

Ernst A, Schmid E (1913) Über Blüte und Frucht von Rafflesiaceae. Ann Jardin Bot Buitenz 12: 1–58

Ernst-Schwarzenbach M (1945) Zur Blütenbiologie einiger Hydrocharitaceen. Ber Schweiz Bot Ges 55: 33–69

Ernst-Schwarzenbach M (1956) Kleistogamie und Antherenbau in der Hydrocharitaceen-Gattung *Ottelia.* Phytomorphology 6: 296–311

Ervandyan SG (1975) Study of development of megagametophyte of *Chrysanthemum maximum* R. Biol Zh 28: 81–85

Esau K (1944) Apomixis in guayule. Proc Natl Acad Sci USA 30: 352–355
Etheridge AL, Herr Jr JM (1968) The development of the ovule and megagametophyte in *Rhexia mariana*. Can J Bot 46: 133–139
Eunus AM (1950a) Contribution to the embryology of the Liliaceae 1. Development of the embryo sac and endosperm of *Albuca transvalensis* Moss-Verdoorn. J Indian Bot Soc 29: 68–78
Eunus AM (1950b) Contributions to the embryology of the Liliaceae 4. Gametophytes of *Smilacina stellata*. New Phytol 49: 269–273
Eunus AM (1951a) Contributions to the embryology of the Liliaceae 5. Life history of *Amianthium muscae-toxicum* Walt. Phytomorphology 1: 73–79
Eunus AM (1951b) Development of the embryo sac and fertilization in *Frittilaria pudica* Spreng. Pakistan J Sci Res 3: 106–113
Eunus AM (1952) Contributions to the embryology of Liliaceae 3. Embryogeny and development of the seed of *Asphodelus tenuifolius* Cave. Lloydia 15: 149–156
Eyde RH (1964) Inferior ovary and generic affinities of *Garrya*. Am J Bot 51: 1083–1092
Eyde RH (1966) The Nyssaceae in the southeastern United States. J Arnold Arbor 47: 117–125
Eyde RH (1967) The peculiar gynoecial vasculature of Cornaceae and its systematic significance. Phytomorphology 17: 172–182

Faegri K, van der Pijl L (1971) The principles of pollination ecology. Pergamon, Oxford
Fagerlind F (1936a) Die Embryologie von *Putoria*. Svensk Bot Tidskr 30: 362–372
Fagerlind F (1936b) Embryologische Beobachtungen über die Gattung *Phyllis*. Bot Not 1936: 577–584
Fagerlind F (1937) Embryologische, Zytologische und bestäubungsexperimentelle Studien in der Familie Rubiaceae nebst Bemerkungen über einige Polyploiditätsprobleme. Acta Hortic Bergiani 11: 195–470
Fagerlind F (1938a) Bau und Entwicklung der floralen Organe von *Helosis cayennensis*. Svensk Bot Tidskr 32: 139–159
Fagerlind F (1938b) Embryosack von *Plumbagella* und *Plumbago*. Ark Bot 29B: 1–8
Fagerlind F (1938c) Wo kommen Tetrasporiche durch drei Teilungsschritte vollentwickelter Embryosäcke unter den Angiospermen vor? Bot Not 1938: 461–498
Fagerlind F (1939a) Die Entwicklung des Embryosackes bei *Peperomia pellucida*. Ark Bot 29A: 1–15
Fagerlind F (1939b) Kritische und revidierende Untersuchungen über das Vorkommen des Adoxa (Lilium)-Typus. Acta Hortic Bergiani 13: 1–49
Fagerlind F (1939c) Drei Beispiele des Fritillaria-Typus. Svensk Bot Tidskr 33: 188–204
Fagerlind F (1940a) Die Terminologie der Apomixis-Prozesse. Hereditas 26: 1–22
Fagerlind F (1940b) Zytologie und Gametophytenbildung in der Gattung *Wikstroemia*. Hereditas 26: 23–50
Fagerlind F (1940c) Stempelbau und Embryosackentwicklung bei einigen Pandanazeen. Ann Jardin Bot Buitenz 49: 55–78
Fagerlind F (1941a) Der Bau der Samenanlage und des Makrogametophyten bei *Quisqualis indica*. Bot Not 1941: 217–222
Fagerlind F (1941b) Die Embryosackentwicklung bei *Tanacetum vulgare* und einigen *Chrysanthemum*-Arten. Svensk Bot Tidskr 35: 157–176
Fagerlind F (1941c) Die Embryosackbildung bei *Camassia*. Svensk Bot Tidskr 35: 258–260
Fagerlind F (1944a) Der tetrasporische Angiospermen: Embryosack und dessen Bedeutung für das Verhältnis der Entwicklungsmechanik und Phylogenie des Embryosacks. Ark Bot 31A: 1–71
Fagerlind F (1944b) Die Samenbildung und die Zytologie bei agamospermischen und sexuellen Arten von *Elastostema* und einigen nahestehenden Gattungen nebst Beleuchtung einiger damit zusammenhängender Probleme. Kgl Svensk Vet Akad Handl 21: 1–130
Fagerlind F (1945a) Blüte und Blütenstand der Gattung *Balanophora*. Bot Not 1945: 330–350
Fagerlind F (1945b) Bildung und Entwicklung des Embryosacks bei sexuellen und agamospermischen *Balanophora*-Arten. Svensk Bot Tidskr 39: 65–82
Fagerlind F (1945c) Bau der floralen Organe bei der Gattung *Langsdorffia*. Svensk Bot Tidskr 39: 197–210
Fagerlind F (1945d) Bau des Gynöceums, des Samenanlage und des Embryosackes bei einigen Repräsentanten der Familie Icacinaceae. Svensk Bot Tidskr 39: 346–364

Fagerlind F (1946) Sporogenesis, Embryosackentwicklung und pseudogame Samenbildung bei *Rudbeckia laciniata* L. Acta Hortic Bergiani 14: 39–90
Fagerlind F (1947a) Gynöceummorphologische und embryologische Studien in der Familie Olacaceae. Bot Not 100: 207–230
Fagerlind F (1947b) Die Restitutions- und Kontraktionskerne der *Hieracium*-Mikrosporogenese. Svensk Bot Tidskr 41: 247–263
Fagerlind F (1947c) Macrogametophyte formation in two agamospermous *Erigeron* species. Acta Hortic Bergiani 14: 221–247
Fagerlind F (1947d) Die systematische Stellung der Familie Grubbiaceae. Svensk Bot Tidskr 41: 315–320
Fagerlind F (1947e) Macrosporogenesis und Embryosackbildung bei agamospermischen *Taraxacum*-Biotypen. Svensk Bot Tidskr 41: 365–390
Fagerlind F (1948) Beitrage zur Kenntnis der Gynöceummorphologie und Phytogenie der Santalales-Familien. Svensk Bot Tidskr 42: 195–229
Fagerlind F (1959) Development and structure of the flower and gametophytes in the genus *Exocarpus*. Svensk Bot Tidskr 53: 147–282
Fagherazzi-Abgrall M, Meyer J (1963) Développement de l'albumen avec différenciation d'un tissu chalazien d'aspect haustorial chez deux *Schizanthus*. Phytomorphology 13: 22–29
Fallen ME (1985) The gynoecial development and systematic position of *Allamanda* (Apocynaceae). Am J Bot 72: 572–579
Farley HM, Hutchinson AH (1941) Seed development in *Medicago* (Alfalfa) hybrids 1. The normal ovule. Can J Res C19: 421–437
Farooq M (1952) The embryology of *Borreria hispida* K. Schum. (= *Spermacoce hispida* Linn.) (Rubiaceae): A reinvestigation. Curr Sci 21: 252–253
Farooq M (1953) Endosperm and seed structure of *Oldenlandia corymbosa*. Curr Sci 22: 280–282
Farooq M (1958a) The embryology of *Oldenlandia corymbosa* Linn. J Indian Bot Soc 37: 358–364
Farooq M (1958b) The development of embryo in *Utricularia stellaris* Linn. f. var. *inflexa*. Sci Cult 23: 479–480
Farooq M (1959) The embryology of *Borreria hispida* K. Schum. (= *Spermacoce hispida* Linn.) (Rubiaceae): A reinvestigation. J Indian Bot Soc 38: 280–287
Farooq M (1960) The embryology of *Galium asperifolium* Wall. J Indian Bot Soc 39: 171–175
Farooq M (1964a) Studies in the Lentibulariaceae 1. The embryology of *Utricularia stellaris* Linn. f. var. *inflexa* Clarke 1. Flower organogeny, ovary, megasporogenesis and female gametophyte. Proc Natl Inst Sci India B 30: 263–279
Farooq M (1964b) Studies in the Lentibulariaceae 1. The embryology of *Utricularia stellaris* Linn. f. var. *inflexa* Clarke 2. Microsporangium, male gametophyte, fertilization, endosperm, embryo and seed. Proc Natl Inst Sci India B 30: 280–299
Farooq M (1965a) Studies in the Lentibulariaceae 2. The embryology of *Utricularia arcuata* Wr. J Indian Bot Soc 44: 326–346
Farooq M (1965b) Studies in the Lentibulariaceae 3. The embryology of *Utricularia uliginosa* Vahl. Phytomorphology 15: 122–131
Farooq M (1966) Studies in the Lentibulariaceae 4. The embryology of *Utricularia striatula* Sm. J Indian Bot Soc 45: 1–13
Farooq M, Bilquis S (1966) Studies in the Lentibulariaceae 8. The life history of *Utricularia scandens* Benj. Beitr Biol Pflanzen 42: 363–371
Farooq M, Inamuddin M (1969) The embryology of *Oldenlandia nudicaulis*. J Indian Bot Soc 48: 160–173
Farooq M, Siddiqui SA (1964) The embryology of *Utricularia stellaris*. Sci Cult 30: 394–395
Farooq M, Siddiqui SA (1965) Abnormal ovules and embryo sacs in *Utricularia vulgaris* var. *americana* A. Gray. Naturwissenschaften 52: 90
Farooq M, Siddiqui SA (1966) Studies in the Lentibulariaceae 5. The development of endosperm in *Utricularia vulgaris* var. *americana* A. Gray.: a reinvestigation. New Phytol 65: 50–53
Farooq M, Siddiqui SA (1967) Studies in the Lentibulariaceae 6. The embryology of *Utricularia stellaris* Linn. J Indian Bot Soc 46: 31–44
Farrell ME (1914) The ovary and embryo of *Cyrtanthus sanguineus*. Bot Gaz 57: 428–436
Farquharson LI (1955) Apomixis and polyembryony in *Tripsacum dactyloides*. Am J Bot 42: 737–743

Favre-Duchartre M (1975) About the marginal situation of the angiospermous egg cell. Phytomorphology 25: 66–68

Favre-Duchartre M (1977) Eight interpretations of embryo sac. Phytomorphology 27: 407–418

Favre-Duchartre M (1984) Homologies and phylogeny. In: Johri BM (ed) Embryology of angiosperms. Springer, Berlin Heidelberg New York, pp 697–737

Fedortschuk W (1931) Embryologische Untersuchung von *Cuscuta monogyna* Vahl. und *C. epithymum* L. Planta 14: 94–111

Feng YA (1934) Recherches cytologiques sur la caryocinèse, la spermatogénèse et la fécundation chez les Caprifoliacées, en particulier sur la présence de centrosomes présidant à la caryocinèse dans les *Lonicera*. Botaniste 26: 3–84

Fernandez MC, Rodriguez-Garcia MI (1989) Developmental changes in the aperture during pollen grain ontogeny in *Olea europaea* L. New Phytol 111: 717–723

Ferri S (1961) Ricerche sullo sviluppo embriologico di *Erigeron speciosus*. Nuovo G Bot Ital 68: 269–284

Feuer S (1981) Pollen morphology and relationships of the Myzodendraceae (Santalales). Nord J Bot 1: 731–734

Feuer S, Niezgoda CJ, Nevling LI (1985) Ultrastructure of *Parkia* polyads (Mimosoideae, Leguminosae). Am J Bot 72: 1871–1890

Fineran BA, Wild DJC, Ingerfeld M (1982) Initial wall formation in the endosperm of wheat, *Triticum aestivum*: a re-evaluation. Can J Bot 60: 1776–1795

Finn WW (1925) Male cells in angiosperms: spermatogenesis and fertilization in *Asclepias cornuti*. Bot Gaz 80: 1–25

Finn WW, Rudenko T (1930) Spermatogenesis und Befruchtung bei einigen Orobanchaceae. Bull Jardin Bot Kieff 11: 69–82

Fischer JB (1980) The vegetative and reproductive structure of papaya (*Carica papaya*). Lloydia 1: 191–207

Fish RK (1970) Megagametogenesis in *Clematis* and its taxonomic and phylogenetic implications. Phytomorphology 20: 317–327

Fisher GC (1914) Seed development in the genus *Peperomia*. Bull Torrey Bot Club 41: 137–156

Flint FF (1957a) Megasporogenesis and megagametogenesis in *Forthergilla garden* Murr. and *F. major* Lodd. Trans Am Microsc Soc 76: 307–311

Flint FF (1957b) Megasporogenesis and megagametogenesis in *Hamamelis virginiana*. Virginia J Sci 8: 185–189

Flint FF (1959) Development of the megagametophyte in *Liquidambar styraciflua*. Madrono 15: 25–29

Flint FF, Johansen DA (1958) Nucleocytoplasmic relationships in the Fritillaria type of megagametogenesis. Am J Bot 45: 464–473

Folsom MW, Cass DD (1989) Embryo sac development in soybean: ultrastructure of megasporogenesis and early megagametogenesis. Can J Bot 67: 2841–2849

Foster LT (1943) Morphological and cytological studies in *Carica papaya*. Bot Gaz 105: 116–126

Fotidar AN (1939) An example of a naked ovule in *Galphimia gracilis*. J Indian Bot Soc 18: 59–63

Fougère-Rifot M (1975) L'édification de l'appareil filiforme et l'évolution cytoplasmique des synergides du sac embryonnaire d'*Aquilegia vulgaris*. CR Acad Sci Paris 280: 2445–2447

Fougère-Rifot M (1979) Les synergides du sac embryonnaire d'*Eschscholtzia californica* (Papavéracée): etude ultrastructurale et cytochimique de l'appareil filiforme. Rev Cytol Biol Veg Bot 2: 199–211

Fougère-Rifot M (1981) Intercellular relations and wall structures in mature embryo sac of four species of angiosperms. Acta Soc Bot Pol 50: 131–138

Francini E (1931) Ricerche embriologiche cariologiche sul genere *Cypripedium* s.l. Nuovo G Bot Ital 38: 155–212

Francini E (1945) Ibridazioni interspecifica nel genere *Papiopedilum*. Nuovo G Bot Ital 52: 21–29

Franchi GG, Pacini E (1980) Wall projections in the vegetative cell of *Parietaria officinalis* L. pollen. Protoplasma 104: 67–74

Frean ML (1979) Cytoplasmic changes during microsporogenesis in *Euphorbia obesa* Hook. f. Proc Electron Microsc Soc S Afr 9: 87–88

Freed HJ, Grant WF (1976) Polytene chromosomes in the suspensor cells of *Lotus* (Fabaceae). Caryologia 29: 387–390

French JC (1985a) Patterns of endothecial wall thickenings in Araceae: Subfamilies Pothoideae and Monsteroideae. Am J Bot 72: 472–486

French JC (1985b) Patterns of endothecial wall thickenings in Araceae: Subfamilies Calloideae, Lasioideae and Philodendroideae. Bot Gaz 146: 521–533

French JC (1986a) Patterns of endothecial wall thickenings in Araceae: subfamilies Colocasioideae, Aroideae and Pistioideae. Bot Gaz 147: 166–179

French JC (1986b) Ovular vasculature in Araceae. Bot Gaz 147: 478–495

Frey L (1976) Embryological studies on *Scabiosa lucida* Vill. and *S. ochroleuca* L. in Poland. Acta Biol Cracov Ser Bot 19: 173–179

Fries TCE (1919) Der Samenbau bei *Cyanastrum*. Svensk Bot Tidskr 13: 295–304

Frisendahl A (1912) Cytologische und entwicklungsgeschichtliche Studien über *Myricaria germanica*. Kon Svensk Vet Akad Handl 48: 1–62

Frisendahl A (1927) Über die Entwicklung chasmogamer und kleistogamer Blüten bei der Gattung *Elatine*. Acta Hortic Götheborg 3: 99–142

Fritsh K (1932) Die systematische Gruppierung der Monokotyledon. Ber Dtsch Bot Ges 50: 162–184

Frohne D, Jensen U (1973) Systematik des Pflanzenreichs unter besonderer Berücksichtigung chemischer Merkmale und pflanzlicher Drogen. Fischer, Stuttgart

Frost HB (1938) Nucellar embryony and juvenile characters in clonal varieties of *Citrus*. J Hered 29: 423–432

Frye TC (1901) Development of the pollen in some Asclepiadaceae. Bot Gaz 32: 325–331

Frye TC (1903) The embryo sac of *Casuarina stricta*. Bot Gaz 36: 101–112

Frye TC, Blodgett EB (1905) A contribution to the life history of *Apocynum androsaemifolium*. Bot Gaz 40: 49–53

Fryxell PA (1968) A redefinition of the tribe Gossypieae. Bot Gaz 129: 296–308

Fryxell PA (1969) The genus *Hampea* (Malvaceae). Brittonia 21: 359–396

Fuchs A (1936) Untersuchungen über den männlichen Gametophyten von *Eleagnus angustifolia*. Öst Bot Z 85: 1–15

Fuchs A (1938) Beiträge zur Embryologie der Thymelaeaceae. Öst Bot Z 87: 1–41

Fukuoka N (1972) Taxonomic study of the Caprifoliaceae. Mem Fac Sci Kyoto Univ Ser Biol 6: 15–58

Furusato K (1951) Studies on polyembryony in *Citrus*. Jpn J Genet 26: 223–224

Furusato K (1953) Studies on polyembryony in *Citrus*. Annu Rep Natl Inst Genet Jpn 4: 56

Furusato K, Ohta Y, Ishibashi K (1957) Studies on polyembryony in *Citrus*. Rep Kihara Inst Biol Res 8: 40–48

Gachechiladze MI (1974a) Development of the Oenothera type embryo sac (in Russian). Soobshch Akad Nauk Gruz SSR 76: 433–436

Gachechiladze MI (1974b) Deviations in the embryo sac of Oenothera type (in Russian). Soobshch Akad Nauk Gruz SSR 76: 437–440

Gaekwad LK, Phatak VG (1952) Studies in Mimosae: The female gametophyte in *Desmanthus virgatus*. J MS Univ Baroda 1: 63–71

Gagnepain F (1910) Un genre méconnu: classification des *Cissus* et *Cayratia*. Not Syst 1: 339–362

Gagnepain F, Boureau E (1947) Nouvelles considerations systématiques à propos du *Sarcopus aberrans* Gagnepain. Bull Soc Bot Fr 94: 182–185

Galil J (1977) Fig biology. Endeavour NS 1: 52–56

Galil J (1990) Pollen germination in *Arisarum vulgare* Targ.-Tozz. Flora 184: 51–61

Galil J, Eisikowitch D (1968a) On the pollination ecology of *Ficus religiosa* in Israel. Phytomorphology 18: 356–363

Galil J, Eisikowitch D (1968b) On the pollination ecology of *Ficus sycomorus* in East Africa. Ecology 49: 259–269

Galil J, Eisikowitch D (1969) Further studies on the pollination ecology of *Ficus sycomorus* L. (Hymenoptera, Chalcidoidea, Agaonidae). Tijdsch Ent 112: 1–13

Galil J, Snitzer-Pasternak Y (1970) Pollination in *Ficus religiosa* L. as connected with the structure and mode of action of the pollen pockets of *Blastophaga quadraticeps* Mayr. New Phytol 69: 775–784

Galimberti Z (1963) Embriologia di *Codiaeum veriegatum* (L.) Blume var. *Pictum* (Lodd.) Müll. Nuovo G Bot Ital 70: 21–32
Ganapathy PM (1956) Female gametophyte and endosperm of *Ophiorrhiza mungos*. J Madras Univ B 26: 589–592
Ganapathy PS (1967a) Pyrolaceae (Abstr). In: Johri BM, Kapil RN, Bhatnagar SP, Bhandari NN, Vijayaraghavan MR (eds) Sem Comparative embryology of angiosperms. Univ Delhi, and Natl Inst Sci India, New Delhi, p 83
Ganapathy PS (1967b) Empetraceae (Abstr). In: Johri BM, Kapil RN, Bhatnagar SP, Bhandari NN, Vijayaraghavan MR (eds) Sem Comparative embryology of angiosperms. Univ Delhi, and Natl Inst Sci India, New Delhi, p 85
Ganapathy PS (1970) Clethraceae. In: Symp Comparative embryology of angiosperms. Indian Natl Sci Acad Bull 41: 239–240
Ganapathy PS, Palser BF (1964) Studies of floral morphology in the Ericales 7. Embryology in the Phyllodoceae. Bot Gaz 125: 280–297
Ganesan AP, Shah SS, Swaminathan MS (1957) Cause of the failure of seed-setting in the cross *Corchorus olitorius* × *C. capsularis*. Curr Sci 26: 292–293
Ganguly JK (1948) Studies in the embryology of *Anisomeles indica* O. Kze. and *Leonurus sibiricus* Linn. Proc Natl Inst Sci India B 14: 181–212
Garagaty-Feissly C (1969–1970) Sur les modifications de l'ultrastructure des cellules tapétales du genre *Daphne* au cours du développement des grains de pollen. Ber Schweiz Bot Ges 79: 221–228
Garcia V (1962a) Embryological studies in the Loasaceae with special reference to the endosperm haustoria. In: Symp Plant embryology. Council Sci Ind Res, New Delhi, pp 157–161
Garcia V (1962b) Embryological studies in the Loasaceae. Development of the endosperm in *Blumenbachia hieronymi* Urb. Phytomorphology 12: 307–312
Gardner EJ (1978) History of biology, 3rd edn. Wiley Eastern, New Delhi
Gardner IC, Miller IM, Scott A (1981) The fine structure of the leaf nodules of *Ardisia cripsa* (Thunb.) A. DC. (Myrsinaceae). Bot J Linn Soc 83: 93–102
Gardner RO (1974) Trinucleate pollen in *Beilschmiedia* Nees (Lauraceae). NZ J Bot 12: 243–244
Gardner RO (1976) Binucleate pollen in *Triglochin* L. NZ J Bot 14: 115–116
Garg M (1980) Pollen morphology and systematic position of *Coriaria*. Phytomorphology 30: 5–10
Garudamma GK (1956) Studies in Meliaceae 1. Development of the embryo in *Azadirachta indica* A. Juss. J Indian Bot Soc 35: 222–226
Garudamma GK (1957) Studies in Meliaceae 2. Gametogenesis in *Melia azadirachta* Linn. J Indian Bot Soc 36: 227–231
Gassner GG (1941) Über Bau der männlichen Blüten und Pollenentwicklung einiger Palmen der Unterfamilie der Ceroxylinae. Beih Bot Zentralbl 61A: 237
Gates RR (1911) Pollen formation in *Oenothera gigas*. Ann Bot (London) 25: 909–940
Gauba E, Pryor LD (1958) Seed-coat anatomy and taxonomy in *Eucalyptus* 1. Proc Linn Soc NSW. 83: 20–32
Gauba E, Pryor LD (1961) Seed-coat anatomy and taxonomy in *Eucalyptus* 3. Proc Linn Soc NSW. 86: 96–111
Gaur RD (1979) Pollen grains in Malvaceae (Abstr.). J Indian Bot Soc 58 (Suppl): 54
Gavade SG (1963) A note on post-fertilization in *Cipadessa baccifera*. J Biol Sci (Bombay) 6: 11–13
Geitler L (1941) Embryosäcke aus Pollenkernen bei *Ornithogalum*. Ber Dtsch Bot Ges 59: 419–423
Gietler L (1955) Riesenkerne im Endosperm von *Allium ursinum*. Öst Bot Z 102: 460–475
Gentcheft G (1937) Zygotogische und embryologische Studien über einige *Hieracium*-Arten. Planta 27: 165–195
Gentry A (1980) Bignoniaceae 1. Crescentieae and Tourrettieae. Flora Neotropica. New York Bot Garden Monogr 25: 1–130
George GP, George RA, Herr Jr JM (1979) A comparative study of ovule and megagametophyte development in field-grown and green-house-grown plants of *Glycine max* and *Phaseolus aureus* (Papilionaceae). Am J Bot 66: 1033–1043

George K, Geethamma S (1984) Cytological and other evidences for the taxonomic position of *Nyctanthes arbor-tristis* L. Curr Sci 53: 439–441
Gerassimova H (1933) Fertilization in *Crepis capillaris*. Cellule 42: 103–148
Gerassimova H (1936) Experimentell erhaltene haploide Pflanzen von *Crepis tectorum* L. Planta 25: 696–702
Gerassimova-Navashina H (1947a) Morphological observations on cytoplasm of the male gametophyte in *Crepis*. Dokl Acad Nauk SSSR 56: 415–418
Gerassimova-Navashina H (1947b) On the development and structure of sperms in *Crepis*. Dokl Akad Nauk SSSR 56: 643–646
Gerassimova-Navashina (1947c) On the behaviour of sperms within the pollen tube in *Crepis*. Dokl Akad Nauk SSSR 57: 285–288
Gerassimova-Navashina H (1957) On some cytological principles underlying double fertilization. Phytomorphology 7: 150–167
Gerassimova-Navashina H (1960) A contribution to the cytology of fertilization in flowering plants. Nucleus 3: 111–120
Gerassimova-Navashina H (1961) Fertilization and events leading up to fertilization, and their bearing on the origin of angiosperms. Phytomorphology 11: 139–146
Gerassimova-Navashina H (1982) Process of double fertilization in angiosperms and mitotic cycle of the cell. Phytomorphology 32: 222–233
Gerassimova-Navashina H, Batygina TB (1958) Fertilization in *Scilla sibirica*. Bot Zh 43: 959–988 (in Russian)
Gerassimova-Navashina H, Korobova SN (1959) The role of synergids in the process of fertilization. Bjull Mosk Obshch Isp Prirody Otd Biol 64: 69–76 (in Russian)
Gerlach-Cryse D (1970) Experimentelle Auslösung von Semigamie bei *Arabidopsis thaliana* (L.) Heyn. Biol Zentralbl 89: 435–456
Ghatak J (1956) A contribution to the life history of *Oroxylum indicum* Vent. Proc Indian Acad Sci B 43: 72–87
Ghosh M (= Ram M) (1954) The embryology and systematic position of *Trapa bispinosa* Roxb. Curr Sci 23: 24–26
Ghosh RB (1962a) A note on the nucellar polyembryony in *Aphanamixis polystachya* (Wall.) Parker. Curr Sci 31: 165
Ghosh RB (1962b) A contribution to the life history of *Wedelia calendulacea* Less. J Indian Bot Soc 41: 197–206
Ghosh RB (1964) Supernumerary archesporial cells in *Ochna atropurpurea* DC. Curr Sci 33: 116–117
Ghosh RB (1966a) Studies in the family Meliaceae 1. Development of the female gametophyte of *Aphanamixis polystachya* (Wall.) Parker [= *Amoora rohituka* (W.A.)]. Beitr Biol Pflanzen 42: 133–138
Ghosh RB (1966b) Studies in the family Meliaceae 2. The development of the gametophytes in *Walsura piscidia* Roxb. Beitr Biol Pflanzen 42: 373–380
Ghosh RB (1969) Embryological studies in family Rutaceae 1. The development of embryo in *Ravenia spectabilis* (= *Lemonia spectabilis* Lindl.). Bull Bot Soc Bengal 23: 149–151
Ghosh RB (1970) Studies on the embryology of Burseraceae: Embryogeny of *Garuga pinnata* Roxb. (Abstr). Broteria (Lisboa) 39: 17–23
Ghosh RB (1975) Studies on embryogenesis of the Indian taxa of Rutaceae. A review. Ann Estacion Exp Aula 13: 201–207
Ghosh RB (1983a) Studies on *Triphasia trifoliata* (Linn.) DC. 1. Microsporogenesis and male gametophyte. Swamy Bot Club Newslett 2: 8–13
Ghosh RB (1983b) Studies on *Triphasia trifoliata* (Linn.) DC. 2. Megasporogenesis, female gametophyte and endosperm. Swamy Bot Club Newslett 2: 163–167
Ghosh RB, Ghosal AG (1971) Studies on the embryogeny of *Vicoa indica* DC. Bull Bot Soc Bengal 25: 77–78
Ghosh RB, Shahi R (1957) A case of polyembryony in *Shorea robusta* Gaertn. Sci Cult 23: 254–256
Ghosh S, Shivanna KR (1984) Interspecific incompatibility in *Linum*. Phytomorphology 34: 128–135

Gibbs RD (1974) Chemotaxonomy of flowering plants, vols 1–4. McGill-Queen's University Press, Montreal & London
Gigante R (1929) Embriologia dell *Acanthus mollis* L. Nuovo G Bot Ital 36: 5–32
Gilg E (1896) Vitaceae (Ampelidaceae). In: Engler A, Prantl K (eds) Die natürlichen Pflanzenfamilien 3(5): 427–456
Gilg-Benedict Ch (1930) Restionaceae, Centrolepidaceae. In: Engler A, Prantl K (eds) Die natürlichen Pflanzenfamilien 2nd edn, 15a
Gill BS, Iqbal M (1961) Embryological studies in *Pterotheca falconeri* Hook. f. Curr Sci 50: 781–782
Gilissen LJW (1976) The role of the style as a sense-organ in relation to wilting of the flower. Planta 131: 201–202
Gioelli F (1930) Ricerche sullo sviluppo del gametofito femmineo del polline nel genero *Aloe*. Lav R Ist Bot Palermo 1: 61–84
Gioelli F (1933) Polimorfismo sessuale e sviluppo del gametofito femmineo e del polline in *Nolina longifolia*. Lav R Ist Bot Palermo 4: 17
Glišic LM (1928) Development of the female gametophyte and endosperm in *Haberlea rhodopensis* Friv. Bull Inst Jardin Bot Univ Beograd 1: 1–13
Glišic LM (1929) Über die Endosperm- und Haustoriendbildung bei *Orobanche hederae* Duby. und *Orobanche gracilis* Sm. (Zugleich ein Beitrag zur Phylogenie der Orobanchaceae). Bull Inst Jardin Bot Univ Beograd 2: 106–141
Glišic LM (1933) Entwicklungsgeschichte von *Gratiola officinale*. Bull Inst Jardin Bot Univ Beograd 2: 20–56
Glišic LM (1934) Zur Kenntnis der Samenentwicklung der Gesneriaceen. Über die Endosperm- und Haustorienbildung von *Roettlera*. Bull Inst Jardin Bot Univ Beograd 3: 94–111
Glišic LM (1937) Ein Versuch der Verwertung der Endospermmerkmale für typologische und phylogenetische Zwecke innerhalb der Scrophulariaceen. Bull Inst Jardin Bot Univ Beograd 4: 42–73
Godwin H (1968) The origin of exine. New Phytol 67: 667–676
Goebel K, Süssenguth K (1924) Beiträge zur Kenntnis der südamerikanischen Burmanniaceen. Flora 117: 55–90
Golaszewska Z (1934) Die Entwicklung des Embryosackes bei *Aspidistra elatior*. Acta Soc Bot Pol 11: 399–407
Goldberg B (1941) Life history of *Peltandra virginica*. Bot Gaz 102: 641–662
Goldblatt P, Bernhardt P, Manning J (1989) Notes on the pollination mechanisms of *Moraea inclinata* (Loasaceae). Plant Syst Evol 163: 201–210
Gomkale KD (1980) Studies in the embryology of *Bauhinia purpurea* Linn. (Abstr.). In: Proc 67th Indian Sci Congr, Calcutta, pt 3: 36
Gomkale KD, Deshpande PK (1979) Embryological studies in *Bauhinia acuminata* Linn. (Abstr). J Indian Bot Soc 58 (Suppl): 54
Good R (1957) Features of evolution in the flowering plants. Longmans Green, London
Gopinath DM (1942) A preliminary note on the sterility in flax (*Linum usitatissimum*). Curr Sci 11: 286–287
Gopinath DM (1943) Cases of antipodal polyembryony in *Alangium lamarkii* Thw. Curr Sci 12: 329–330
Gopinath DM (1944) Gametogenesis and embryogeny in a few members of the Araliaceae. Proc Indian Acad Sci B 20: 175–186
Gopinath DM (1945) A contribution to the embryology of *Alangium lamarkii* Thw. with a discussion on systematic position of the family Alangiaceae. Proc Indian Acad Sci B 22: 225–231
Gopinath DM (1946) Development of the female gametophyte in *Flacourtia cataphracta* and *F. ramontchili* (Abstr.). In: Proc 33rd Indian Sci Congr, Bangalore, pt 3: 109
Gopinath PM, Chennaveeraiah MS (1961) A contribution to the embryology of *Ophiorrhiza harrisiana* Heyne. Karnatak Univ J Sci 6: 61–70
Gore UR (1932) Development of the female gametophyte in cotton. Am J Bot 19: 295–807
Gorham AL (1953) The question of fertilization in *Smilacina racemosa*. Phytomorphology 3: 44–50
Gori C (1957) Sull'embriologia e citologia di alcune specie del genere *Reseda*. Caryologia 10: 391–401

Gori P (1977) Wall ingrowths in the embryo sac of *Euphorbia helioscopia*. Isr J Bot 26: 202–208

Gottsberger G (1989a) Beetle pollination and flowering rhythm of *Annona* spp. (Annonaceae) in Brazil. Plant Syst Evol 167: 165–188

Gottsberger G (1989b) Comments on flower evolution and beetle pollination in the genera *Annona* and *Rollinia* (Annonaceae). Plant Syst Evol 167: 189–194

Gottsberger G, Silberbauer-Gottsberger I, Ehrendorfer F (1980) Reproductive biology in the primitive relict angiosperm *Drimys brasiliensis* (Winteraceae). Plant Syst Evol 135: 11–39

Govil CM (1970a) Polemoniaceae. In: Symp Comparative embryology of angiosperms. Indian Natl Sci Acad Bull 41: 241–243

Govil CM (1970b) Fouquieriaceae. In: Symp Comparative embryology of angiosperms. Indian Natl Sci Acad Bull 41: 244–245

Govil CM (1970c) Convolvulaceae. In: Symp Comparative embryology of angiosperms. Indian Natl Sci Acad Bull 41: 246–249

Govil CM (1971) Morphological studies in the family Convolvulaceae 1. Development and structure of the seed-coat. J Indian Bot Soc 50: 32–38

Govil CM (1978) Structure and development of seed in *Averrhoa carambola* (Abstr). J Indian Bot Soc 57 (Suppl): 43

Govil CM (1980) Embryo sac development in *Solanum tuberosum* var. *Jyoti Gola*. Acta Bot Indica 8: 263–264

Govil CM, Kaur Maninder (1989) Studies in ovule and seed development in *Averrhoa carambola* L. Acta Bot Indica 17: 101–107

Govil CM, Saxena NP (1976) Anatomy and embryology of *Weinmannia fraxinea* Sm. (Cunoniaceae). J Indian Bot Soc 55: 219–226

Govindappa DA (1953) The embryo sac of *Xyris pauciflora* Willd. Curr Sci 22: 386

Govindappa DA (1955a) Development of the gametophytes in *Aloe ciliaris* Haw. J Indian Bot Soc 34: 146–150

Govindappa DA (1955b) Embryological studies in *Xyris pauciflora* Willd. Proc Indian Acad Sci B 42: 47–57

Govindappa DA (1956a) Morphological study of *Hypericum japonicum* Humb. J Mysore Univ B 15: 69–79

Govindappa DA (1956b) The development of endosperm in *Scilla indica* Baker. Curr Sci 25: 198–199

Govindappa DA (1961) Embryology of *Klugia notoniana*. Bot Gaz 123: 144–150

Govindappa DA, Boriah G (1956) The development of female gametophyte in *Biophytum sensitivum*. (L.) DC. Curr Sci 25: 403–404

Govindappa DA, Naidu TRB (1956) The embryo sac and endosperm of *Blyxa oryzetorum* Hook. f. J Indian Bot Soc 35: 417–422

Govindappa DA, Shamakumari K (1957) Development of the embryo in *Hypoxis aurea* Lour. J Indian Bot Soc 36: 324–327

Govindappa DA, Sheriff A (1951) Contribution to the life history of *Scilla indica* Baker. J Mysore Univ B 12: 15–21

Govindu HC (1950) Studies in the embryology of some members of the Bignoniaceae. Proc Indian Acad Sci B 32: 164–178

Gowda MTG (1980) Endosperm in *Ariopsis peltata* Nimmo (Araceae). Curr Sci 49: 601–602

Goyal S (1959) Floral morphology and embryology of *Cissus quadrangularis*. MSc Thesis, Univ Rajasthan, Jaipur, India

Grabovskaya SR (1968) The development of multicellular archesporia in the ovules of species of the genus *Capsicum* Tourn. (in Russian). Trud Pritkl Bot Genet Selek 40: 65–73

Graef PE (1958) The ovule and embryo sac of *Galax aphylla*. Virginia J Sci 9: 319–322

Grant I, Bevers-dorf WD, Peterson RL (1986) A comparative light and electron microscopic study of microspore and tapetal development in male fertile and cytoplasmic male sterile oilseed rape (*Brassica napus*). Can J Bot 64: 1055–1068

Grant V, Wilken DH (1986) Taxonomy of the *Ipomopsis aggregata* group (Polemoniaceae). Bot Gaz 147: 359–371

Grau J, Hopf H (1985) Das Endosperm der Compositae. Bot Jahrb 107: 251–268

Graves AH (1908) The morphology of *Ruppia maritima*. Trans Conn Acad Sci New Haven 14: 59–170

Gray A (1858) Manual of the botany of the Northern United States. New York
Green MW (1978) A nomarski interference study of megasporogenesis and megagametogenesis in *Smelowskia calycina* (Cruciferae). Am J Bot 65: 353–358
Greenday A, French TC (1988) Endothecial thickenings in anthers of porate monocotyledons. Am J Bot 75: 22–25
Greene CW (1984) Sexual and apomictic reproduction in *Calamagrostis* (Gramineae) from Eastern North America. Am J Bot 71: 285–293
Greenwood JS, Bewley JD (1982) Seed development in *Ricinus communis* (castor bean) 1. Descriptive morphology. Can J Bot 60: 1751–1760
Gregory WC (1941) Phylogenetic and cytological studies in the Ranunculaceae. Trans Am Philos Soc NS 31: 443–521
Greilhuber J (1973) Über die Entwicklung des Embryosacks von *Melampyrum* und *Parentucellia latifolia* (Scrophulariaceae, Pedicularieae). Öst Bot Z 121: 81–97
Greilhuber J (1974) Uncommon karyological features in the tapetum of some Pedicularieae (Scrophulariaceae). Caryologia 27: 169–182
Grew N (1682) The anatomy of plants. London
Griffith W (1843) On the ovulum of *Santalum, Osyris, Loranthus* and *Viscum*. Trans Linn Soc London Bot 19: 171–214
Grimm J (1912) Entwicklungsgeschichtliche Untersuchungen an *Rhus* und *Coriaria*. Flora 104: 309–334
Grootjen CJ (1983a) Development of ovule and seed in *Cartonema spicatum* R. Br. (Cartonemataceae). Aust J Bot 31: 297–305
Grootjen CJ (1983b) Development of ovule and seed in Marantaceae. Acta Bot Neerl 32: 69–86
Grootjen CJ, Bouman F (1981a) Development of ovule and seed in *Stanfieldiella imperforata* (Commelinaceae). Acta Bot Neerl 30: 265–275
Grootjen CJ, Bouman F (1981b) Development of the ovule and seed in *Costus cuspidatus* (N. E. Br.) Maas (Zingiberaceae), with special reference to the formation of operculum. Bot J Linn Soc 83: 27–39
Grootjen CJ, Bouman F (1988) Seed structure in Cannaceae: taxonomic and ecological implications. Ann Bot (London) 61: 363–371
Grover H (1969) Development of the female gametophyte in some *Ficus* species. J Indian Bot Soc 48: 334–338
Grover H (née Kaur), Chopra RN (1971) Observations on oviposition, nutrition and emergence of some fig insects. J Indian Bot Soc (Golden Jubilee Vol) 50A: 107–115
Grun P (1955) Cytogenic studies in *Poa* 3. Variation within *Poa nervosa*, an obligate apomict. Am J Bot 42: 778–784
Grundwag M (1976) Embryology and fruit development in four species of *Pistacia* L. (Anacardiaceae). Bot J Linn Soc 73: 355–370
Grundwag M, Fahn A (1969) The relation of embryology to the low seed-set in *Pistacia vera* (Anacardiaceae). Phytomorphology 19: 225–235
Guard A (1943) The development of the seed of *Liriodendron tulipifera* L. Proc Indian Acad Sci B 53: 75–77
Guérin P (1901) Développement de la graine et en particulier du tégument séminal de quelques Sapindacees. J Bot (Paris) 15: 336–362
Guérin P (1930) Le développement de l'oeuf et la polyembryonie chez l'*Erythronium dens-canis*. CR Acad Sci Paris 191: 1369–1372
Guérin P (1931) L'ovule et la graine de l'*Erythronium dens-canis*. Botaniste 23: 185–195
Guérin P (1937) L'ovule et la graine des *Erythronium* et des *Calochortus*. Ann Sci Nat Bot 19: 255–265
Guevara VF (1961) The occurrence of twin embryos in *Cocos nucifera*. Philipp Agric 44: 520–522
Guignard L (1881) Recherches d'embryogénie végétale comparée. l. Memoire: legumineuses. Ann Sci Nat Bot Ser 6 12: 5–166
Guignard L (1901) La double fécondation dans le *Najas major*. J Bot (Paris) 15: 205–213
Guignard L (1917a) L'ovule chez les Apocynacées et Asclepiadacées. Mem Acad Sci Inst Fr 55: 1–34

Guignard L (1917b) Sur le développement et la structure de l'ovule chez les Apocynacées et les Asclepiadacées. CR Acad Sci Paris 165: 981–987
Guignard JL (1961a) Recherches sur l'embryogénie des Graminées; rapports des Graminées avec les autres Monocotylédones. Ann Sci Nat Bot Ser 12 2: 491–610
Guignard JL (1961b) Embryogeny of the Gramineae. Development of embryo of *Setaria verticillata* Beauv. Bull Soc Bot Fr 108: 212–217
Guignard JL (1963) Embryogénie des Dioscoreacées. Développement de l'embryon chez le *Tamus communis*. CR Acad Sci Paris 256: 3172–3175
Guignard JL (1966) Embryogénie des Polygonacées. Développement de l'embryon chez le *Rumex aquaticus* L. Bull Soc Bot Fr 113: 224–226
Guignard JL (1975a) Du cotylédon des monocotylédons. Phytomorphology 25: 193–200
Guignard JL (1975b) Embryogénie et classification embryogénique. Bull Soc Bot Fr 122: 281–294
Guignard JL (1984) The development of cotyledon and shoot apex in monocotyledons. Can J Bot 62: 1316–1318
Guignard JL, Ly-Thi-Ba (1989) L'embryogenie des Helobiales (The embryology of Helobiae). In: Paré J, Bugnicourt M (eds) Some aspects and actual orientations in plant embryology. Dedicatory volume to Prof A Lebègue. Univ Press Picardie, Amiens (Fr), pp 223–232
Guignard JL, Mestre JC (1966) Sur le développement d'embryons a partir des antipodes chez l'*Ulmus campensis* L. Bull Soc Bot Fr 113: 227–228
Guignard JL, Mestre JC (1967) Embryogénie des Caryophyllacées. Développement de l'embryon chez le *Cerastium pumilum* Curt. CR Acad Sci Paris 265: 111–114
Guignard JL, Mestre JC (1969a) L'origine du cotylédon et du cône végétatif chez les monocotylédons. Bull Soc Bot Fr 116: 207–214
Guignard JL, Mestre JC (1969b) Embryogénie des Amaryllidacées: l'origine du cotylédon chez l'*Agapanthus umbellatus* L'Hérit. CR Acad Sci Paris 268: 3068–3070
Guignard JL, Mestre JC (1970) L'embryon des Graminées. Phytomorphology 20: 190–197
Guilford VB, Fisk EL (1952) Megasporogenesis and seed development in *Mimulus tigrinus* and *Torenia fournieri*. Bull Torrey Bot Club 79: 6–24
Gulati N, Mathur S (1977) Embryology and taxonomy of *Filicium decipiens*. Phytomorphology 27: 261–266
Gulati M, Ravishankar S (1980) Studies in Symplocaceae l. Microsporangium, microsporogenesis and male gametophyte. Phytomorphology 30: 189–195
Gulyayev VA (1961) Notes on fertilization in *Citrullus vulgaris* Schrad. Byull Vses Inst Rasteniev 9: 27–30 (in Russian)
Gunasekera SP, Sultanbawa NU (1973) Mangostin from the barks of *Hydnocarpus*. Phytochemistry 12: 232
Gundersen A (1927) The Frankeniaceae as a link in the classification of dicotyledons. Torreia 27: 65–71
Gundersen A (1943) Flower forms and groups of dicotyledons. Bull Torrey Bot Club 70: 511–517
Gundersen N (1950) Families of dicotyledons. Chron Bot, Waltham (Mass, USA)
Gunning BES, Pate JS (1969) Transfer cells – Plant cells with wall ingrowth, specialized in relation to short distance transport of solutes: Their occurrence, structure, and development. Protoplasma 68: 107–133
Gunning BES, Pate JS (1974) Transfer cells. In: Robards AW (ed) Dynamic aspects of plant ultrastructures. McGraw-Hill, New York, pp 441–480
Gupta BL (1934) Contribution to the life history of *Potamogeton crispus* L. J Indian Bot Soc 13: 51–65
Gupta BL (1935) Studies in the development of pollen grain and embryo sac of *Wolffia arrhiza*. Curr Sci 4: 104–105
Gupta MN (1962) Morphological studies in Cyperaceae l. Development of the ovule and the gametophytes in *Fimbristylis dichotoma*. Agra Univ J Res Sci 11: 59–67
Gupta PK (1968) Observations on degree of apospory in three members of Andropogonae. Curr Sci 37: 295–296
Gupta PK (1969–70) Apomixis in *Bothriochloa pertusa* (L.) A. Camus. Port Acta Biol Ser A 11: 279–287
Gupta SC (1962a) Development of exembryonate seeds in *Foeniculum vulgare* Mill. In: Symp Plant embryology. Council Sci Ind Res, New Delhi, pp 188–191

Gupta SC (1962b) Occurrence of exembryonate seeds in Umbelliferae. Curr Sci 31: 203–205
Gupta SC (1964) The embryology of *Coriandrum sativum* L. and *Foeniculum vulgare* Mill. Phytomorphology 14: 530–547
Gupta SC (1970) Umbelliferae. In: Symp Comparative embryology of angiosperms. Indian Natl Sci Acad Bull 41: 233–238
Gupta SC, Ahluwalia R (1977) The carpel of *Nelumbo nucifera*. Phytomorphology 27: 274–282
Gupta SC, Ahluwalia R (1979) The anther and ovule of *Nelumbo nucifera*: A reinvestigation. J Indian Bot Soc 58: 177–182
Gupta SC, Ahuja R (1967) Is *Nelumbo* a monocot? Naturwissenschaften 52: 498
Gupta SC, Gupta M (1964) Embryological observations on *Bupleurum tenue* Buch.-Ham. ex D. Don. Beitr Biol Pflanzen 40: 301–323
Gupta SC, Nanda K (1972) Occurrence and histochemistry of the anther tapetal membrane. Grana 12: 99–104
Gupta SC, Nanda K (1973) Fibrous endothecium, tapetum and pollen development in *Belamcanda chinensis* DC. Bot Gaz 134: 125–129
Gupta SC, Rajeswari VM (1977) Nucellar tracheids in ovules of *Luffa cylindrica*. Ann Bot (London) 41: 1005–1007
Gupta SC, Paliwal GS, Gupta M (1965) The development of stomata in vegetative and reproductive organs of *Bupleurum tenue* Buch.-Ham. ex D. Don. Ann Bot (London) 116: 546–554
Gupta V, Lamba LC (1985) Structure and development of seed in *Eschscholzia californica* Cham. Acta Bot Indica 13: 240–245
Gupta V, Vijayaraghavan MR (1985) Translator apparatus in *Leptadenia sporteum* (Asclepiadaceae). Curr Sci 54: 395–396
Gürke M (1896) Melianthaceae. In: Engler A, Prantl K (eds) Die natürlichen Pflanzenfamilien. 3(5): 374–383
Gürke M (1897) Symplocaceae. In: Engler A, Prantl K (eds) Die natürlichen Pflanzenfamilien. 4(1): 165–172
Gurudev MR, Gowda B (1988) Development of male gametophyte and origin of tapetum in *Scutellaria discolor* Wall. ex Bentham. Curr Sci 57: 148–150
Gustafsson Å (1930) Kastrierungen und Pseudogamie bei *Rubus*. Bot Not 1930: 477–494
Gustafsson Å (1932) Zytologische und experimentelle Studien in der Gattung *Taraxacum*. Hereditas 16: 41–62
Gustafsson Å (1933) Über die Teilung der Embryosackmutterzelle bei *Taraxacum*. Bot Not 1933: 531–534
Gustafsson Å (1946) Apomixis in the higher plants 1. The mechanism of apomixis. Acta Univ Lund 42: 1–66
Gustafsson Å (1947a) Apomixis in the higher plants 2. The causal aspects of apomixis. Acta Univ Lund 43: 71–178
Gustafsson Å (1947b) Apomixis in higher plants 3. Biotype and species formation. Acta Univ Lund 44: 183–370
Guth CJ, Weller SG (1986) Pollination, fertilization and ovule abortion in *Oxalis magnifica*. Am J Bot 73: 246–253
Gutterman Y, Witztum A (1979) The movement of integumentary hairs in *Blepharis ciliaris* (L.) Burtt. Bot Gaz 138: 29–34
Guzowska L (1964) Reinvestigation of embryo sac development, fertilization and early embryogeny in *Cystinus hypocistis*. Acta Soc Bot Pol 33: 157–166
Gvaladze GE (1961) The embryology of the genus *Allium*. Soobsh Akad Nauk Gruz SSR 26: 193–200 (in Russian)
Gvaladze GE (1974) Synergids of some *Gagea* species. Soobsh Akad Nauk Gruz SSR 74: 421–423 (in Russian)

Haberlandt G (1922) Die Vorstufen und Ursachen der Adventivembryonie. Sitzungsber Preuss Akad Wiss Berlin 25: 386–406
Haberlandt G (1923) Zur Embryologie von *Allium odorum*. Ber Dtsch Bot Ges 41: 174–179
Haberlandt G (1925) Zur Embryologie und Cytologie von *Allium odorum*. Ber Dtsch Bot Ges 43: 559–564

Haccius B (1952) Die Embryoentwicklung bei *Ottelia alismoides* und das Problem des terminalen Monokotylen-Keimblatts. Planta 40: 443–460
Haccius B (1954) Embryologische und histogenetische Studien an "Monokotylen Dikotylen" 1. *Claytonia virginica* L. Öst Bot Z 101: 285–303
Haccius B (1963) Restitution in acidity-damaged plant embryos: Regeneration or regulation? Phytomorphology 13: 107–115
Haccius B (1971) Zur derzeitigen Situation der Angiospermen-Embryologie. Bot Jahrb 91: 309–379
Haccius B, Bhandari NN (1975) Delayed histogen differentiation as a common primitive character in all types of non-zygotic embryos. Phytomorphology 25: 91–94
Haccius B, Hartl-Baude E (1956) Embryologische und histogenetische Studien an "Monokotylen Dikotylen" 2. *Pinguicula vulgaris* und *Pinguicula alpina*. Öst Bot Z 103: 567–587
Haccius B, Hausner G (1972) Frühembryogenese und Polyembryonie: Problem in der Gattung *Tulipa* L. Beitr Biol Pflanzen 48: 207–228
Haccius B, Lakshmanan KK (1966) Vergleichende Untersuchung der Entwicklung von Kotyledon und Sprosscheitel bei *Pistia stratiotes* und *Lemna gibba*. Ein Beitrag zum Problem der sogenannten terminalen Blattorgane. Beitr Biol Pflanzen 42: 425–443
Haccius B, Philip VJ (1979) Embryo development in *Cocos nucifera* L.: A critical contribution to a general understanding of palm embryogenesis. Plant Syst Evol 132: 91–106
Haeckel I (1930) Über Iridaceen. Flora 125: 1–82
Håfliger E (1943) Zytologisch-embryologische Untersuchungen pseudogamer Ranunkeln des *Auricomus*-Gruppe. Ber Schweiz Bot Ges 53: 317–382
Hagemann R (1981) Unequal plastid distribution during the development of male gametophyte of angiosperms. Acta Soc Bot Pol 50: 321–327
Hagerup O (1927) *Empetrum hermaphroditum* (Lange) Hager. A new tetraploid bisexual species. Dansk Bot Ark 5: 1–17
Hagerup O (1928) Morphological and cytological studies of the Bicornes. Dansk Bot Ark 6: 1–26
Hagerup O (1944) On fertilization, polyploidy and haploidy in *Orchis maculatus*. Dansk Bot Ark 11: 1–26
Hagerup O (1945) Facultative parthenogenesis and haploidy in *Epipactis latifolia*. Kon Dansk Vidensk Selsk Biol Medd 19: 1–13
Hagerup O (1947) The spontaneous formation of haploid, polyploid and aneuploid embryos in some orchids. Kon Dansk Vidensk Selsk Biol Medd 20: 1–22
Håkansson Å (1921) Beiträge zur Entwicklungsgeschichte der Taccaceen. Bot Not 1921: 189–268, 257–268
Håkansson Å (1923) Studien über die Entwicklungsgeschichte der Umbelliferen. Acta Univ Lund 18: 1–120
Håkansson Å (1927) Der sechzehnkernige Embryosack von *Azorella trifurcata* (Gaertn.) Hook. Ber Dtsch Bot Ges 45: 654–664
Håkansson Å (1943) Die Entwicklung des Embryosackes und die Befruchtung bei *Poa alpina*. Hereditas 29: 25–61
Håkansson Å (1946) Some observations on the seed development in two strains of *Triticale*. Acta Agric Suecl 1: 377–384
Håkansson Å (1952) Seed development in *Bowlesia tenera*. Bot Not 1952: 33–45
Håkansson Å (1953) Die Samenbildung bei *Nothoscordum fragrans*. Bot Not 1953: 129–139
Håkansson Å (1954) Endosperm formation in *Salix*. Bot Not 1954: 326–332
Håkansson Å (1955) Endosperm formation in *Myrica gale* L. Bot Not 108: 6–16
Håkansson Å (1956) Seed development of *Brassica oleracea* and *B. rapa* after certain reciprocal pollinations. Hereditas 42: 373–396
Håkansson Å (1957) Notes on endosperm formation in *Betula*. Bot Not 109: 201–204
Håkansson Å, Levan A (1947) Endoduplicational meiosis in *Allium odorum*. Hereditas 43: 179–200
Hakki MI (1971) Blütenmorphologische und embryologische Untersuchungen an *Chenopodium capitatum* und *Chenopodium foliosum* sowie weiteren Chenopodiaceae. Bot Jahrb 92: 178–330
Hakki MI (1973) Blüntenmorphologische und embryologische Untersuchungen an Chenopodiaceae: *Monolepis trifida* Schrad. Bot Jahrb 93: 481–497
Hakki MI (1974) Embryologische und morphologische Beobachtungen an *Succowia balearica* (L.) Medik. (Brassicaceae). Bot Jahrb 94: 360–382

Hakki MI (1975) Embryologische und morphologische Untersuchungen an Pflanzen aus Westindien 1. Zur Embryologie von *Capraria biflora* L. (Scrophulariaceae). Bot Jahrb 96: 125–153

Hakki MI (1977) Embryologische und morphologische Untersuchungen an Pflanzen aus Südafrika 1. Über die Embryologie, Morphologie und systematische Zugehörigkeit von *Dermatobotrys saundersii* Bolus. Bot Jahrb 98: 93–119

Hakki MI (1985) Studies on West Indian plants 3. On floral morphology, anatomy and relationship of *Picrodendron baccatum* (L.) Krug & Urban (Euphorbiaceae). Bot Jahrb 107: 379–394

Hall JG (1902) The embryological study of *Limnocharis emarginata*. Bot Gaz 33: 214–219

Hall JB (1971) New Podostemaceae from Ghana with notes on related species. Kew Bull 26: 125–136

Hallier H (1901) Über die Verwandschaftsverhältnisse der Tubifloren und Ebenalten, den polyphyletischen Ursprung der Sympetalen und Apetalen. Naturwissenschaften 16: 1–112

Hallier H (1903) Über den Umfang, die Gliederung und die Verwandtschaft der Familie der Hamamelidaceen. Beih Bot Zentralbl 14: 247–260

Hallier H (1905) Provisional scheme of the natural system of flowering plants. New Phytol 4: 151–162

Hallier H (1908) Über *Juliania*, etc. Neue Beiträge zur Stammesgeschichte der Dicotyledonen. Bot Zentralbl 23: 81–265

Hallier H (1912) L'origine et le système phylétique das Angiospermes exposés à l'aide de leur arbre généalogique. Arch Néerl Sci Exact Nat Ser 3B 1: 146–234

Halward TM, Stalker HT (1987) Comparison of embryo development in wild and cultivated *Arachis* species. Ann Bot (London) 59: 9–14

Hamann U (1961) Merkmalabestand und Verwandtschaftsbeziehungen der Farinosae. Ein Beitrag zum System der Monokotyledonen. Willdenowia 2: 639–768

Hamann U (1962) Beitrag zur Embryologie der Centrolepidaceae mit Bemerkungen über den Bau der Blüten und Blütenstände und die systematische Stellung der Familie. Ber Dtsch Bot Ges 75: 153–171

Hamann U (1963) Die Embryologie von *Philydrum lanuginosum* (Monocotyledoneae-Philydraceae). Ber Dtsch Bot Ges 76: 203–208

Hamann U (1966a) Embryologische, morphologische, anatomische und systematische Untersuchungen an Philydraceen. Willdenowia 4: 1–178

Hamann U (1966b) Nochmals zur Embryologie von *Philydrum lanuginosum*. Beitr Biol Pflanzen 42: 151–159

Hamann U (1975) Neue Untersuchungen zur Embryologie und Systematik der Centrolepidaceae. Bot Jahrb 96: 154–191

Hamann U (1976) Hydatellaceae – a new family of Monocotyledoneae. NZ J Bot 14: 193–196

Hamann U (1977) Über Konvergenzen bei embryologischen Merkmalen der Angiospermen. Ber Dtsch Bot Ges 90: 369–384

Hammond BL (1937) Development of *Podostemon ceratophyllum*. Bull Torrey Bot Club 64: 17–36

Hanawa J (1953) The embryo sac formation and the embryology of *Sesamum indicum* in 2 × and 4 × strains. Bot Mag (Tokyo) 66: 98–102

Hanna WW, Schertz KF, Bashaw EC (1970) Apospory in *Sorghum bicolor* (L.) Moench. Science 170: 338–339

Hanstein J (1870) Die Entwicklungsgeschichte des Keimes der Monokotylen und Dikotylen. Bot Abhandl Bonn 1: 1–112

Hanumantha Rao B, Prakasa Rao PS (1977) Anther in *Capparis decidua* (Forsk.) Pax. Curr Sci 46: 683–684

Hanumantha Rao B, Prakasa Rao PS (1980) Female gametophyte in *Anchusa officinalis* L.: a reinvestigation. Curr Sci 49: 77–79

Hanumantha Rao B, Prakasa Rao PS (1984) Sporogenesis and genesis of gametophytes in *Cordia sebestena* L. Acta Bot Indica 12: 69–76

Haque SM (1911) The morphological study of *Diospyros virginiana*. Bot Gaz 52: 34–44

Haque A (1951) The embryo sac of *Erythronium americanum*. Bot Gaz 112: 495–500

Harda I (1947) Chromosome numbers in *Pandanus*, *Sparganium* and *Typha*. Cytologia 14: 214–218

Hari Gopal B, Ram M (1981) Floral morphology, development of sporangia and sporogenesis in *Dendrocalamus hamiltonii*. Proc Indian Natl Sci Acad B 47: 519–526

Hari Gopal B, Mohan Ram HY (1985) Systematic significance of mature embryo of bamboos. Plant Syst Evol 148: 239–246
Harland SC (1936) Haploids in polyembryonic seeds of Sea Island cotton. J Hered 27: 229–231
Harling G (1949) Zur Embryologie der Gattung *Hedychium*. Svensk Bot Tidskr 43: 357–364
Harling G (1950) Embryological studies in the Compositae 1. Anthemideae-Anthemidinae. Acta Hortic Bergiani 15: 135–168
Harling G (1951a) Embryological studies in the Compositae 2. Anthemideae-Chrysantheminae. Acta Hortic Bergiani 16: 1–56
Harling G (1951b) Embryological studies in the Compositae 3. Astereae. Acta Hortic Bergiani 16: 73–120
Harling G (1958) Monograph of the Cyclanthaceae. Acta Hortic Bergiani 18: 1–428
Harling G (1960) Further embryological and taxonomical studies in *Anthemis* and some related genera. Svensk Bot Tidskr 54: 571–590
Harling G (1962) On some Compositae endemic to the Galapagos Islands. Acta Hortic Bergiani 20: 63–120
Harms H (1897a) Die Gattungen der Cornaceen. Ber Dtsch Bot Ges 15: 21–29
Harms H (1897b) Über die Stellung der Gattung *Tetracentron* und die Familie der Trochodendraceen. Ber Dtsch Bot Ges 15: 350–360
Harms H (1918) Zur Kenntniss der Gattung *Cercidiphyllum*. Mitt Dtsch Dendrol Ges 26: 71–87
Harms H (1925) Achariaceae. In: Engler A, Prantl K (eds) Die natürlichen Pflanzenfamilien 21: 507–510
Harms H (1930) Eucommiaceae. In: Engler A, Prantl K (eds) Die natürlichen Pflanzenfamilien 18a: 348–351
Harms H (1935) Reihen Santalales, Aristolochiales, Balanophorales. Geschichtliche Entwicklung der Ansichten über die Umgrenzung der Reihen und ihre Zusammensetzung. In: Engler A, Prantl K (eds) Die natürlichen Pflanzenfamilien 16b. Rafflesiaceae: 243–281, Hydnoraceae: 282–294, Balanophoraceae 296–339
Harris CJ (1935) The development of flower and seed in *Galinsoga ciliata*. Univ Pittsburg Bull 32: 131–137
Hartl D (1957) Die Stellung von *Lindenbergia* im System der Scrophulariaceen. Beitr Biol Pflanzen 33: 265–277
Hartl D (1960) Das alveolierte Endosperm bei Scrophulariaceen, seine Enstehung, Anatomie und taxonomische Bedeutung. Beitr Biol Pflanzen 35: 95–110
Hartman A (1923) Zur Entwicklungsgeschichte und Biologie der Acanthaceen. Flora 116: 216–258
Harvey CF, Fraser LG (1988) Floral biology of two species of *Actinidia* (Actinidiaceae) 2. Early embryology. Bot Gaz 149: 37–44
Hashmi S, Siddiqui SA (1974) Development of endosperm, embryo and seed in *Paederia scandens* Lour. Geobios (Jodhpur, India) 1: 183–184
Hasitschka-Jenschke G (1956) Bildung von Chromosomenbünden nach Art der Speicheldrüsenchromosomen, spiralisierte Ruhekernchromosomen und andere Struktureigentümlichkeiten in den endopolyploiden Riesenkernen der Antipoden von *Papaver rhoeas*. Chromosoma 9: 87–113
Hasitschka-Jenschke G (1958) Zur Karyologie der Samenanlage dreier Allium-Arten. Öst Bot Z 105: 71–82
Hasitschka-Jenschke G (1959) Bermerkenswerte Kernstrukturen in Endosperm und im Suspensor zweier Helobiae. Öst Bot Z 106: 301–314
Hasitschka-Jenschke G (1962) Notizen über endopolyploide Kerne im Bereich der Samenanlage von Angiospermen. Öst Bot Z 109: 125–137
Haskell DA, Postlethwait SN (1971) Structure and histogenesis of the embryo of *Acer sachharinum* 1. Embryo sac and proembryo. Am J Bot 58: 595–603
Haupt AW (1934) Ovule and embryo sac of *Plumbago capensis*. Bot Gaz 95: 649–659
Hayashi Y (1960) On the microsporogenesis and pollen morphology in the family Magnoliaceae. Sci Rep Tôhoku Univ Ser Biol 26: 45–52
Hayashi Y (1963a) The embryology of the family Magnoliaceae sensu lato 1. Megasporogenesis, female gametophyte and embryogeny of *Illicium anisatum* L. Sci Rep Tôhoku Univ Ser Biol 29: 27–33
Hayashi Y (1963b) The embryology of the family Magnoliaceae sensu lato 2. Megasporogenesis,

female gametophyte and embryogeny of *Schisandra repandra* Rodlkofer and *Kadsura japonica* Dunal. Sci Rep Tôhoku Univ Ser Biol 29: 403–411
Hayashi Y (1964) Megasporogenesis, female gametophyte and embryogeny of *Magnolia liliflora* and *Michelia fuscata*. Sci Rep Tôhoku Univ Ser Biol 30: 89–98
Hayashi Y (1965) The comparative embryology of the Magnoliaceae (s.l.) in relation to the systematic consideration of the family. Sci Rep Tôhoku Univ Ser Biol 31: 29–44
Hayashi Y (1966) The embryology of the family Magnoliaceae s.l. 4. Microsporogenesis and development of the male gametophyte of *Michelia figo* Sprengel. Sci Rep Tôhoku Univ Ser Biol 32: 111–118
Hayashi Y (1984) Embryology of *Magnolia salicifolia* Maxim. (Magnoliaceae). Jpn J Bot 59: 298–307
Hayward MP, Manthriratna MAP (1972) Pollen development and variation in the genus *Lolium* 1. Pollen size and tapetal relationships in male fertiles and male steriles. Z Pflanzenzücht 67: 131–144
Heatley M (1916) A study of the life history of *Trillium cernuum*. Bot Gaz 61: 425–429
Hegelmaier F (1880) Über aus mehrkernigen Zellen aufgebaute Dicotyledonen-Keimträger. Bot Z 38: 407–506, 513–522
Hegnauer R (1963, 1964, 1969, 1973) Chemotaxonomie der Pflanzen, vols 2, 3, 5, 6. Birkhäuser, Basel Stuttgart
Hegnauer R (1971) Chemical patterns and relationships of Umbelliferae. In: Heywood VH (ed) The biology and chemistry of the Umbelliferae. Academic Press, New York London, pp 267–277
Heilborn O (1928) Taxonomical and embryological notes on *Carica*. Acta Hortic Bergiani 9: 105–108
Heilborn O (1931) Studies on the taxonomy, geographical distribution and embryology of the genus *Siparuna*. Svensk Bot Tidskr 25: 202–228
Heimsch C (1942) The comparative anatomy of the secondary xylem in the Gruinales and Terebinthales of Wettstein with reference to taxonomic grouping. Lilloa 8: 83–198
Henrickson J (1973) Pollen morphology of the Fouquieriaceae. Aliso 6: 137–160
Herms WB (1907) Contributions to the life history of *Asimina triloba*. Ohio Nat 8: 211–216
Hermsen JGTH (1980) Breeding for apomixis in potato: Pursuing an utopian scheme? Euphytica 29: 595–607
Herr JM Jr (1959a) Notes on the structures and origin of the ovule in the genus *Ilex*. J Elisha Mitchell Sci Soc 75: 67
Herr JM Jr (1959b) The development of the ovule and the megagametophyte in the genus *Ilex*. J Elisha Mitchell Sci Soc 75: 107–128
Herr JM Jr (1959c) Embryological evidence for the relationship of Aquifoliaceae to Celastraceae. Virginia J Sci 10: 259
Herr JM Jr (1961) Endosperm development and associated ovule modifications in the genus *Ilex*. J Elisha Mitchell Sci Soc 77: 26–32
Herr JM Jr (1972) An extended investigation on the megagametophyte in *Oxalis corniculata* L. In: Murty YS, Johri BM, Mohan Ram HY, Varghese TM (eds) Advances in plant morphology (Prof. V. Puri Comm. Vol.). Sarita Prakashan, Meerut (India), pp 92–101
Herr JM Jr (1984) Embryology and taxonomy. In: Johri BM (ed) Embryology of angiosperms. Springer, Berlin Heidelberg New York, pp 647–696
Herr JM Jr, Maryhruce D (1968) Development of the ovule and megagametophyte in *Oxalis corniculata* L. Phytomorphology 18: 43–53
Herscovitch JC, Martin ARH (1989) Pollen–pistil interactions in *Grevillea banksii*. The pollen grain, stigma, transmitting tissue and in vitro pollinations. Grana 28: 69–82
Herscovitch JC, Martin ARH (1990) Pollen–pistil interactions in *Grevillea banksii* 2: Pollen tube ultrastructure and interactions and results of field experiments. Grana 29: 5–17
Heslop-Harrison J (1963) An ultrastructural study of the pollen wall ontogeny in *Silene pendula*. Grana Palynol 4: 7–24
Heslop-Harrison J (1968a) Wall development within the microspore tetrad of *Lilium longiflorum*. Can J Bot 46: 1185–1191
Heslop-Harrison J (1968b) Tapetal origin of pollen-coat substances in *Lilium*. New Phytol 67: 779–786
Heslop-Harrison J (1968c) Pollen-wall development. Science 161: 230–237

Heslop-Harrison J (1968d) Synchronous pollen mitosis and the formation of the generative cell in massulate orchids. J Cell Sci 3: 457–466
Heslop-Harrison J (1969) An acetolysis-resistant membrane investing tapetum and sporogenous tissue in the anthers of certain Compositae. Can J Bot 47: 541–542
Heslop-Harrison J (ed) (1971) Pollen development and physiology. Butterworth, London
Heslop-Harrison J (1972) Sexuality in angiosperms. In: Steward FC (ed) Plant physiology, vol 6C. Academic Press, New York London, pp 133–289
Heslop-Harrison Y (1981) Stigma characteristics and angiosperm taxonomy. Norw J Bot 1: 401–420
Heslop-Harrison Y (1990) Stigma form and surface in relation to self-incompatibility in the Onagraceae. Nord J Bot 10: 1–19
Hesse M (1978) Entwicklungeschichte und Ultrastruktur von Pollenkitt und Exine von beinahe verwandten entomophilen und anemophilen Angiospermensippen: Ranunculaceae, Hamamelidaceae, Platanaceae und Fagaceae. Plant Syst Evol 130: 13–42
Hesse M (1981) Pollen kitt and viscin threads: their roles in pollination. Grana 20: 145–152
Hewitt WC (1939) Seed development in *Lobelia amoena.* J Elisha Mitchell Sci Soc 55: 63–82
Hichs DJ, Wyatt R, Meagher TR (1985) Reproductive biology of distylous partridge berry *Mitchella repens.* Am J Bot 72: 1503–1514
Hildebrand GR, Fairbrothers DE (1970a) Serological investigation of the systematic position of the Caprifoliaceae 1. Correspondence with selected Rubiaceae and Cornaceae. Am J Bot 57: 810–815
Hildebrand GR, Fairbrothers DE (1970b) Phytoserological systematic survey of the Caprifoliaceae. Brittonia 22: 125–133
Hilger HH (1989) Flower and fruit development in the macaronesian endemic *Ceballosia fruticosa* (syn. *Messerschmidia fruticosa* – Boraginaceae, Heliotropioideae). Plant Syst Evol 166: 119–130
Hill TG (1900) The structure and development of *Triglochin maritimum.* Ann Bot (London) 14: 83–107
Hils MS, Dickison WC, Lucansky TW, Stern WL (1988) Comparative anatomy and systematics of woody Saxifragaceae: *Tetracarpaea.* Am J Bot 75: 1687–1700
Himmelbaur W (1909) Eine blütenmorphologie Studie über *Datisca canonabina.* Sitzungsber Akad Wiss Wien 118: 92–113
Hindmarsh GJ (1964) Gametophyte development in *Trifolium pratense* L. Aust J Bot 12: 1–14
Hindmarsh GJ (1965) An embryological study of five species of *Bassia.* Proc Linn Soc NSW 90: 274–289
Hiroshi T (1967) An embryological study of *Hetaeria shikokiana*, a saprophytic orchid in Japan. Sci Rep Tôhoku Univ Ser Biol 33: 83–95
Hiroshi T (1971a) Development of the embryo of *Lecanorchis japonica.* Sci Rep Tôhoku Univ Ser Biol 35: 245–251
Hiroshi T (1971b) Development of the embryo of *Orchis aristata.* Sci Rep Tôhoku Univ Ser Biol 35: 239–243
Hiroshi T (1973) Embryo sac development in *Trautvetteria japonica* Sieb. et Zucc. (Ranunculaceae). Sci Rep Tôhoku Univ Ser Biol 36: 255–259
Hiroshi T (1974) Embryology of *Cimicifuga acerina* Sieb. et Zucc., *C. tanaka.* Sci Rep Tôhoku Univ Ser Biol 37: 55–61
Hiroshi T, Toshiko A, Osamu I (1975) Embryological studies on *Hepatica nobilis* Schreber var. *japonica* Nacai (Ranunculaceae). J Coll Arts Sci Chiba Univ Ser B 8: 93–102
Hjelmqvist H (1948) Studies on the floral morphology and phylogeny of the Amentiferae. Bot Not Suppl 1948: 147–166
Hjelmqvist H (1951) The embryo sac development of *Tridax trilobata.* Bot Not 104: 180–187
Hjelmqvist H (1953) The embryo sac development of *Quercus robur* L. Phytomorphology 3: 377–384
Hjelmqvist H (1957a) Some notes on the endosperm and embryo development in Fagales and related orders. Bot Not 110: 173–175
Hjelmqvist H (1957b) The apomictic development in *Malus sieboldii.* Bot Not 110: 455–467
Hjelmqvist H (1959) Studien über Embryologie und Variabilität bei einigen Aphanes-Arten. Bot Not 112: 17–64

Hjelmqvist H (1962) The embryo sac development of some *Cotoneaster* species. Bot Not 115: 208–236
Hjelmqvist H (1964) Variations in embryo sac development. Phytomorphology 14: 186–196
Hjelmqvist H (1967) The origin of the Plumbagella type. Phytomorphology 17: 211–214
Hjelmqvist H, Grazi F (1964) Studies on variation in embryo sac development. Bot Not 117: 141–166
Hjelmqvist H, Grazi F (1965) Studies on variation in embryo sac development. Bot Not 118: 329–360
Hjelmqvist H, Holmberg V (1961) The development of the embryo sac in *Sanvitallia procumbens*. Bot Not 114: 353–360
Hoare G (1934) Gametogenesis and fertilization in *Scilla nonscripta*. Cellule 42: 269–292
Höck F (1892) Zur systematischen Stellung von *Sambucus*. Bot Zentralbl 51: 233–234
Hoffman CA (1933) Developmental morphology of *Allium cepa*. Bot Gaz 95: 279–299
Hofmeister W (1847) Untersuchungen des Vorgangs bei der Befruchtung der Oenotheren. Bot Z 5: 785–792
Hofmeister W (1848) Über die Entwicklung des Pollens. Bot Z 6: 425–434, 649–658, 670–674
Hofmeister W (1849) Die Entstehung des Embryo der Phanerogamen. Leipzig
Hofmeister W (1859) Neue Beiträge zur Kenntnis der Embryobildung der Phanerogamen I. Dikotyledonen mit ursprünglich einzelligem, nur durch Zellteilung wachsendem Endosperm. Abh Königl Sächs Ges Wiss 1859: 535–672
Hofmeister W (1861) Neue Beiträge zur kenntnis der Embryobildung der Phanerogamen II. Monokotyledonen. Abh Königl Sächs Ges Wiss 7: 629–760
Holferty GM (1901) Ovule and embryo of *Potamogeton natans*. Bot Gaz 31: 339–346
Holmgren J (1913) Zur Entwicklungsgeschichte von *Butomus umbellatus* L. Svensk Bot Tidskr 7: 58–77
Holmgren J (1919) Zytologische Studien über die Fortpflanzung bei den Gattungen *Erigeron* und *Eupatorium*. Kon Svensk Vet Akad Handl 59: 1–118
Hooker JD (1846) Flora antarctica, pt 2. Loranthaceae, pp 289–302
Hooker JD (1856) On the structure and affinities of Balanophoraceae. Trans Linn Soc London 22: 1–68
Hooker JD (1862) Burseraceae. In: Bentham G, Hooker JD (eds) Genera plantarum, vol 1. pt 1. Reeve, London, pp 321–327
Hooker JD (1885) The flora of British India, London
Hooker JD (1894) The flora of British India, vol 6. London
Hopping ME (1976) Structure and development of fruit and seeds in Chinese gooseberry (*Actinidia chinensis* Planch.). NZ J Bot 14: 63–68
Hopping ME, Jerram EM (1979) Pollination of kiwifruit (*Actinidia chinensis* Planch.): Stigma-style structure and pollen tube growth. NZ J Bot 17: 233–240
Horn CL (1940) Existence of only one variety of cultivated mangosteen explained by asexually formed 'seed'. Science 92: 237–238
Horne AS (1909) The structure and affinities of *Davidia involucrata* Baill. Trans Linn Soc London 7: 303–326
Horner Jr HT, Arnott HJ (1966) Histochemical and ultrastructural study of pre- and post-germinated *Yucca* seeds. Bot Gaz 127: 48–84
Horner Jr HT, Lersten NR (1971) Microsporogenesis in *Citrus limon* (Rutaceae). Am J Bot 58: 72–79
Horner Jr HT, Pearson C (1978) Pollen wall and aperture development in *Helianthus annuus* (Compositae: Heliantheae). Am J Bot 65: 293–309
Hossaert M, Valévo M (1988) Effect of ovule position in the pod on patterns of seed formation in two species of *Lathyrus* (Leguminosae, Papilionoideae). Am J Bot 75: 1714–1731
Hotchkiss AT (1958) Pollen and pollination in the Eupomatiaceae. Proc Linn Soc NSW 83: 86–91
Houk WG (1938) Endosperm and perisperm of coffee with notes on the morphology of the ovule and seed development. Am J Bot 25: 56–61
Howe TD (1926) Development of embryo sac in *Grindelia squarrosa*. Bot Gaz 81: 280–296
Howe TD (1940) Development of the embryo sac in *Trillium grandiflorum*. Am J Bot 51: 678
Howe TD (1959) Recent studies of the female gametophyte in the Compositae, especially in the tribe Astereae. In: 9th Int Bot Congr Montreal 2: 171–172
Howe TD (1964) Development of the embryo sac in *Ratibida tagetes* (Abstr). Am J Bot 51: 678

Howe TD (1969) The female gametophyte of *Machaeranthera pattersonii* (*Aster patersonii*) and *M. tanacetifolia*. Am J Bot 56: 641–645
Howe TD (1970) The female gametophyte in *Grindelia arizonica* and *G. stricta* ssp. *blakei* (Abstr). Am J Bot 57: 744
Howe TD (1975) The female gametophyte of three species of *Grindelia* and of *Prionopsis ciliata* (Compositae). Am J Bot 62: 273–279
Howell G, Prakash N (1990) Embryology and reproductive ecology of the darling lily, *Crinum flaccidum* Herbert. Aust J Bot 38: 433–444
Hrubý C (1934) A contribution to the cytology and embryology of *Erythronium dens-canis*. Bull Int Acad Sci Boheme 35: 124–132
Hrubý C (1938) Embryo sac development in *Erythronium dens-canis*. Chron Bot 4: 20–21
Hu CY, Ferreira AG (1989) In vitro embryology of *Ilex*. In: Paré J, Bugnicourt M (eds) Some aspects and actual orientations in plant embryology. Dedicatory volume to Prof A. Lebègue. Univ Press Picardie, Amiens (Fr), pp 76–91
Huang BQ, Russell SD (1989) Isolation of fixed and viable eggs, central cells, and embryo sacs from ovules of *Plumbago zeylanica*. Plant Physiol 90: 9–12
Hufford LD, Endress PK (1989) The diversity of anther structures and dehiscence patterns among Hamamelididae. Bot J Linn Soc 99: 301–346
Hu Shi-yi, Li Le-gong, Zhu Cheng (1985) Isolation of viable embryo sacs and their protoplasts of *Nicotiana tabacum* (in Chinese). Acta Bot Sin 27: 337–344
Hu Shiu-Ying (1960) A revision of the genus *Clethra* in China. J Arnold Arbor 41: 164–190
Huber H (1969) Die Samenmerkmale und Verwandtschaftsverhältnisse der Lilliifloren. Mitt Bot Staat München 8: 219–538
Huber H (1977) The treatment of monocotyledons in evolutionary system of classification. In: Kubitzki H (ed) Flowering plants. Plant Syst Evol Suppl 1. Springer, Wien, pp 285–298
Hulbary RL, Nagaraja Rao A (1959) Flower development and gametogenesis in *Oenothera laciniata* Hill. Proc Iowa Acad Sci 66: 91–97
Hurusawa I (1954) Eine nochmalige Durchsicht des herkömmlichen Systems der Euphorbiaceen im weiteren Sinne. J Fac Sci Tôhoku Univ Bot 21: 209–340
Hutchinson J (1926) The families of flowering plants. Dicotyledon, vol 1. Univ Press, Oxford (England)
Hutchinson J (1967) The genera of flowering plants, vols 1, 2. Univ Press, Oxford (England)
Hutchinson J (1973) The families of flowering plants, edn 3. Clarendon Press, Oxford (England)
Huynh K-L (1976) L'arrangement du pollen du genre *Schisandra* (Schisandraceae) et sa signification phylogénique chez les Angiospermes. Beitr Biol Pflanzen 52: 227–253

Iglesias L, Lima H, Simon JP (1974) Isoenzyme identification of zygotic and nucellar seedlings in *Citrus*. J Hered 29: 417–419
Il'ina GM (1962) Embryological study of mustard *Brassica juncea* (L.) Czern. (in Russian). Vest Mosk Univ Biol Ser 6 Pochvovedeniya 1: 34–35
Indra R, Krishnamurthy KV (1984) Ovary wall obturator in *Ottelia alismoides*. Curr Sci 53: 442–443
Indra R, Krishnamurthy KV (1985) On the persistent pollen tube of *Ottelia alismoides* (L.) Pers. Phyton (Argentina) 45: 41–44
Ishikawa M (1918) Studies in the embryo sac and fertilization in *Oenothera*. Ann Bot (London) 32: 279–317
Islam AS (1950) A contribution to the life history of *Ottelia alismoides* Pers. J Indian Bot Soc 29: 79–91
Ivanova KV, Orel LI (1963) Some biological peculiarities and an embryological characterization of *Solanum guineense* L. Trud Prikl Bot Genet Selek 35: 146–150 (in Russian)
Ivanov MA, Sharenkova EA (1970) The cytoembryology of lilac (*Syringa billosa* Vahl.). Iza Biol Geogr Nauch Inst Irkutsk Univ 23: 19–31 (in Russian)
Iwamasa M, Isamu V, Masao N (1970) Location of zygotic embryo in polyembryonic *Citrus* seed (in Japanese). Bull Hortic Res Stn Ser B (Okitsu) 10: 7–16
Iyengar GS (1937) Life history of *Santalum album*. J Indian Bot Soc 16: 175–195
Iyer RD (1962) Embryological studies in relation to interspecific hybridization in jute. In: Symp Plant embryology. Council Sci Ind Res, New Delhi, pp 49–54

Iyer RD, Sulbha K, Swaminathan MS (1961) Fertilization and seed development in crosses between *Corchorus olitorius* and *C. capsularis*. Indian J Genet Plant Breed 21: 191–200
Izmailow R (1965) Megasporogenesis in the apomictic species *Ranunculus cassubicus* L. Acta Biol Cracov Ser Bot 8: 183–195
Izmailow R (1973) Cyto-embryological studies in experimental hybrids of the apomictic species *Ranunculus cassubicus* L. Acta Biol Cracov Ser Bot 16: 99–120
Izmailow R (1984) Cyto-embryological studies on *Alchemilla* L. (series Calycinae Buser) 1. Microsporangium and microsporogenesis. Acta Biol Cracov Ser Bot 26: 1–17

Jacobsen JV (1984) The seed: germination. In: Johri BM (ed) Embryology of angiosperms. Springer, Berlin Heidelberg New York, pp 611–646
Jäger-Zürn I (1965) Zur Frage der systematischen Stellung der Hydrostachyaceae auf Grund ihrer Embryologie, Blüten- und Infloreszenzmorphologie. Öst Bot Z 112: 621–639
Jäger-Zürn I (1966) Infloreszenz- und blütenmorphologische sowie embryologische Untersuchungen an *Myrothamnus* Welw. Beitr Biol Pflanzen 42: 241–271
Jäger-Zürn I (1967) Embryologische Untersuchungen an vier Podostemaceen. Öst Bot Z 114: 20–45
Jain TC (1956) The gametophytes of *Lycium europaeum* Linn. J Indian Bot Soc 35: 181–188
Jaitly SC (1966) Development of seeds and fruits in *Anisomeles indica* (Linn) O. Kze. Phytomorphology 16: 430–436
Jaitly SC (1968) Le développement de l'albumen chez les genres *Salvia*, *Hyptis*, *Ocimum*, *Pogostemon*, *Mentha* et *Leucas*. Bull Soc Bot Fr 115: 373–378
Jaitly SC (1971) Studies in Labiatae l. Structure and development of seed and fruit in two species of *Salvia*. J Indian Bot Soc 50: 182–183
Jaitly SC, Guignard JL, Mestre JC (1968a) Embryogénie des Labiées. Développement de l'embryon chez le *Mentha aquatica* L. Bull Soc Bot Fr 115: 369–372
Jaitly SC, Guignard JL, Mestre JC (1968b) Embryogénie des Verbenacées. Développement de l'embryon chez le *Caryopteris mastacanthus* Schau. CR Acad Sci Paris 267: 59–61
Jaitly SC, Guignard JL, Mestre JC (1968c) Embryogénie des Labiées: le développement de l'embryon chez le *Scutellaria columnae* Host. CR Acad Sci Paris 267: 1717–1719
Jalan S (1963) Studies in the family Ranunculaceae 4. The embryology of *Actaea spicata* Linn. Phytomorphology 13: 338–347
Jalan S, Bhandari NN (1963) Extracarpellary ovules in some Ranales. Curr Sci 32: 230–232
Jalan S, Kapil RN (1964) Pollen grains of *Schisandra* Michaux. Grana Palynol 5: 216–221
Jalouzet MF (1975) Aspects ultrastructuraux du sac embryonnaire d'Oenothera lamarckiana. CR Acad Sci Paris 281: 1305–1308
Jaranowski J (1961) Haploid-diploid twin embryos in *Melilotus*. Genet Pol 2: 129–137
Jaranowski J (1962) Fertilization and embryo development in the genus *Lupinus* 1. Seed development in cases of autogamy (in Polish). Genet Pol 3: 209–246
Jarrett FM (1976) The syncarp of *Artocarpus*: A unique biological phenomenon. Garden Bull (Singapore) 29: 35–39
Jassem B (1980) Origin and reproduction of higher polyploids within the Corollinae section of the genus *Beta*. Genet Pol 21: 17–27
Jay M (1968) Distribution des flavonoides chez les Hamamélidacées et familles affines. Taxon 17: 136–147
Jay M (1969) Contribution biochimique à la connaissance taxonomique et phylogénétique des saxifragacées et familles affines. 1. Thèse, Univ Lyon (Fr)
Jayalakshmi R, Lakshmanan KK (1990) Embryogeny in the wild species of *Arachis* L. Acta Bot Indica 18: 231–234
Jeffrey EC (1895) Polyembryony in *Erythronium americanum*. Ann Bot (London) 9: 537–541
Jensen WA, Fisher DB (1968) Cotton embryogenesis: The entrance and discharge of the pollen tube into the embryo sac. Planta 78: 158–183
Jensen SR, Nielsen BJ, Dahlgren R (1975) Iridoid compounds, their occurrence and systematic importance in the angiosperms. Bot Not 128: 148–180
Jensen WA, Schulz P, Ashton ME (1977) An ultrastructural study of early endosperm development and synergid changes in unfertilized ovules of cotton. Planta 133: 179–189

Jensen WA, Ashton ME, Beasley CA (1983) Pollen tube-embryo sac interaction in cotton. In: Mulcahy D, Ottavinao F (eds) Pollen: biology and implications for plant breeding. Elsevier, New York, pp 57–72
Jernstedt JA, Clark C (1979) Stomata on the fruits and seeds of *Eschscholtzia* (Papaveraceae). Am J Bot 66: 586–590
Jeyanayaghy S, Rao AN (1966) Flower and seed development in *Bromheadia finlaysoniana*. Bull Torrey Bot Club 93: 97–103
Jha SS (1985) Seed structure in some *Dalbergia* sp. J Indian Bot Soc 64: 269–274
Jha SS, Pandey AK (1989) Seed-coat structure in *Melilotus* (Fabaceae). Phytomorphology 39: 221–230
Joachimiak A (1981) Antipodals in *Phleum boehmeri* Wib.: Development and structure. Acta Biol Cracov Ser Bot 23: 25–35
Johansen DA (1928a) The hypostase: Its presence and functions in the ovule of Onagraceae. Proc Natl Acad Sci USA 14: 710–713
Johansen DA (1928b) The hypostase and seed sterility in the Onagraceae. Madrono 1: 165–167
Johansen DA (1930a) Studies on the morphology of the Onagraceae 2. Embryonal manifestation in *Clarkia elegans*. Bot Gaz 90: 75–91
Johansen DA (1930b) Studies on the morphology of the Onagraceae 4. *Stenosiphon linifolium*. Bull Torrey Bot Club 57: 315–326
Johansen DA (1931a) Studies on the morphology of Onagraceae 3. *Taraxia ovata*. Ann Bot (London) 45: 111–124
Johansen DA (1931b) Studies on the morphology of the Onagraceae 5. *Zauschneria latifolia*, typical of a genus characterised by irregular embryology. Ann NY Acad Sci 33: 1–26
Johansen DA (1933) Studies on the morphology of the Onagraceae 7. *Gayophytum ramosissimum*. Bull Torrey Bot Club 60: 1–8
Johansen DA (1934) Studies on the morphology of Onagraceae 8. *Circaea pacifica*. Am J Bot 21: 508–510
Johansen DA (1936) Morphology and embryology of *Fouquieria*. Am J Bot 23: 95–99
Johansen DA (1945) A critical survey of the present status of plant embryology. Bot Rev 11: 87–107
Johansen DA (1950) Plant embryology. Chronica Botanica, Waltham (Mass, USA)
Johansson JT (1987) Pollen morphology of the tribe Morindeae (Rubiaceae). Grana 26: 134–150
Johnson AM (1936) Polyembryony in *Eugenia hookeri*. Am J Bot 23: 83–85
Johnson DS (1900) On the development of *Saururus cernuus*. Bull Torrey Bot Club 27: 365–372
Johnson DS (1914) Studies in the development of Piperaceae 2. The structure and seed development of *Peperomia hispidula*. Am J Bot 1: 323–339, 357–397
Johnson LAS, Briggs BG (1984) Myrtales and Myrtaceae: A phylogenetic analysis. Ann Missouri Bot Garden 71: 700–756
Johnson T (1889) The nursing of the embryo and some other points in *Myzodendron punctatum* Banks et Sol. Ann Bot (London) 3: 179–206
Johnston GW (1953) Observation on twin embryo sacs in *Sorghum vulgare*. Phytomorphology 3: 313–315
Johnston GW (1959) Further evidence of some so-called abnormalities in the development of the male gametophyte of angiosperms. Phytomorphology 9: 130–133
Johnston GW (1961) Microsporogenesis in *Exochorda racemosa*. Phytomorphology 11: 41–45
Johri BM (1933a) Contributions to the morphology of *Limnophyton obtusifolium* Miq. Curr Sci 2: 12–13
Johri BM (1933b) The life history of *Limnophyton obtusifolium* Miq. Curr Sci 2: 99–100
Johri BM (1934a) A note on the life history of *Sagittaria guayanensis* H.B.K. Curr Sci 2: 428–429
Johri BM (1934b) A note on the life history of *Sagittaria sagittifolia* L. Curr Sci 3: 17–18
Johri BM (1935a) Studies in the family Alismataceae 1. *Limnophyton obtusifolium* Miq. J Indian Bot Soc 14: 49–66
Johri BM (1935b) Studies in the family Alismataceae 2. *Sagittaria sagittifolia* L. Proc Indian Acad Sci B1: 340–348
Johri BM (1935c) The gametophytes of *Berberis nepalensis* Spreng. Proc Indian Acad Sci B 1: 640–649

Johri BM (1935d) Studies in the family Alismataceae 3. *Sagittaria guayanensis* H.B.K. and *S. latifolia* Willd. Proc Indian Acad Sci B 2: 33–48
Johri BM (1935e) Life history of *Butomopsis lanceolata* Kunth. Nature (London) 136: 338
Johri BM (1936a) Studies in the family Alismataceae 4. *Alisma plantago* L., *A. plantago-aquatica* L. and *Sagittaria graminea* Mich. Proc Indian Acad Sci B 4: 128–138
Johri BM (1936b) The life history of *Butomopsis lanceolata* Kunth. Proc Indian Acad Sci B 4: 139–162
Johri BM (1938a) The embryo sac of *Hydrocleis nymphoides* Buchen. Beih Bot Zentralbl 58A: 165–172
Johri BM (1938b) The embryo sac of *Limnocharis emarginata* L. New Phytol 37: 279–285
Johri BM (1951) Endosperm and embryo development in *Cuscuta reflexa* Roxb. Curr Sci 20: 189–191
Johri BM (1956) Polyploid pollen grains in *Helixanthera ligustrina*. In: Proc 43rd Indian Sci Congr Agra 3: 258
Johri BM (1962a) The female gametophyte of the Santalales. In: Symp Plant embryology Council Sci Ind Res, New Delhi, pp 192–198
Johri BM (1962b) Nutrition of the embryo sac. In: Maheshwari P, Johri BM, Vasil IK (eds) Proc Summer school on botany, Darjeeling (India), pp 106–118
Johri BM (1963a) Female gametophyte. In: Maheshwari P (ed) Recent advances in the embryology of angiosperms. Int Soc Plant Morphol, Univ Delhi, pp 69–104
Johri BM (1963b) Embryology and taxonomy. In: Maheshwari P (ed) Recent advances in the embryology of angiosperms. Int Soc Plant Morphol, Univ Delhi, pp 395–444
Johri BM (1967) Angiosperm embryology and taxonomy. In: Symp Newer trends in taxonomy. Natl Inst Sci India Bull 34: 263–268
Johri BM (1970a) Symp Comparative embryology of angiosperms. Indian Natl Sci Acad Bull 41: 1–385
Johri BM (1970b) Limnanthaceae. In: Symp Comparative embryology of angiosperms. Indian Natl Sci Acad Bull 41: 110–113
Johri BM (1970c) Alismaceae and Butomaceae. In: Symp Comparative embryology of angiosperms. Indian Natl Sci Acad Bull 41: 334–335
Johri BM (1987a) The ovary-ovule complex in Loranthaceae. 14th Int Bot Congr, Berlin, p 219
Johri BM (1987b) Embryology of *Cuscuta* L. (Cuscutaceae). In: Weber H Chr, Forstreuter W (eds) Parasitic flowering plants. Proc 4th Int Symp, Marburg (Germany), pp 445–448
Johri BM (1987c) Reproductive biology of mistletoes (Loranthaceae and Viscaceae). In: Weber H Chr, Forstreuter W (eds) Parasitic flowering plants. Proc 4th Int Symp, Marburg (Germany), pp 449–456
Johri BM (1989) The fig and the fig wasp. In: Paré J, Bugnicourt M (eds) Some aspects and actual orientations in plant embryology. Dedicatory volume to Professor A Lebègue. Univ Picardie, Amiens (Fr), pp 144–157
Johri BM (1992) Haustorial role of pollen tubes. Ann Bot (London) (in press)
Johri BM, Agrawal JS (1954) Morphology and embryology of *Helicanthes elastica* (Desr.) Dans. Curr Sci 23: 96–98
Johri BM, Agarwal S (1965) Morphological and embryological studies in the family Santalaceae 8. *Quinchamalium chilense* Lam. Phytomorphology 15: 360–372
Johri BM, Ahuja MR (1956) Development of endosperm and nucellar polyembryony in *Aegle marmelos* Correa. Curr Sci 25: 162–164
Johri BM, Ahuja MR (1957) A contribution to the floral morphology and embryology of *Aegle marmelos* Correa. Phytomorphology 7: 10–24
Johri BM, Ambegaokar KB (1975) The antipodals in *Triticale*. Phytomorphology 25: 112–117
Johri BM, Ambegaokar KB (1977) Seed development in Triticale-1. Phytomorphology 27: 190–197
Johri BM, Ambegaokar KB (1984) Embryology: then and now. In: Johri BM (ed) Embryology of angiosperms. Springer, Berlin Heidelberg, New York, pp 1–52
Johri BM, Bhatnagar SP (1955a) A contribution to the morphology and life history of *Aristolochia*. Phytomorphology 5: 123–157
Johri BM, Bhatnagar SP (1955b) Does polyembryony occur in *Limnocharis emarginata* L. (syn. *L. flava* Buchem.)? Sci Cult 21: 276

Johri BM, Bhatnagar SP (1957) Intra-carpellary pollen grains in angiosperms. Phytomorphology 7: 292–296
Johri BM, Bhatnagar SP (1960) Embryology and taxonomy of the Santalales 1. Proc Natl Inst Sci India B 26: 199–220
Johri BM, Bhatnagar SP (1969) Endosperm in the Santalales. Rev Cytol Biol Veg 32: 353–369
Johri BM, Bhatnagar SP (1970) Bibliography: embryology of angiosperms (1936–1969). Cent Adv Stud Bot, Univ Delhi, pp 1–104
Johri BM, Bhatnagar SP (1972) Loranthaceae. Botanical monograph 8. Council Sci Ind Res, New Delhi, pp 1–155
Johri BM, Bhatnagar SP (1973) Some histochemical and ultrastructural aspects of the female gametophyte and fertilization in angiosperms. Caryologia 25 (Suppl): 9–25
Johri BM, Bhojwani SS (1965) Growth responses of mature endosperm in cultures. Nature (London) 208: 1345–1347
Johri BM, Garg S (1959) Development of endosperm haustoria in some Leguminosae. Phytomorphology 9: 34–46
Johri BM, Kak D (1954) The embryology of *Tamarix* Linn. Phytomorphology 4: 230–247
Johri BM, Kapil RN (1953) Contribution to the morphology and life history of *Acalypha indica* L. Phytomorphology 3: 137–151
Johri BM, Konar RN (1955) A contribution to the morphology and embryology of *Ficus religiosa* Linn. Curr Sci 24: 382–385
Johri BM, Konar RN (1956) The floral morphology and embryology of *Ficus religiosa* Linn. Phytomorphology 6: 97–111
Johri BM, Maheshwari P (1951) The embryo sac of *Floerkea proserpinacoides* Willd. Curr Sci 20: 44–46
Johri BM, Nand S (1934) The development of the male and female gametophytes in *Cuscuta reflexa* Roxb. Proc Indian Acad Sci B 1: 283–290
Johri BM, Prakash S (1965) Morphological and embryological studies in the family Loranthaceae 11. *Tapinostemma acaciae* (Zucc.) van Tiegh. Phytomorphology 15: 150–158
Johri BM, Pundir NS (1976) Embryological studies in cultivated cottons (Abstr). In: Embryology of crop plants. Indo-Sov Symp. Indian Natl Sci Acad, New Delhi, and Cent Adv Stud Bot, Univ Delhi, pp 16–17
Johri BM, Raj B (1965) Embryo sac development in *Moquiniella rubra*. Nature (London) 205: 415–416
Johri BM, Raj B (1969) Morphological and embryological studies in the family Loranthaceae 12. *Moquiniella rubra* (Spreng. f.) Balle. Öst Bot Z 116: 475–485
Johri BM, Roy Chowdhury C (1957) A contribution to the embryology of *Citrulus colocynthis* and *Melothria maderaspatana*. New Phytol 56: 51–60
Johri BM, Shivanna KR (1977) Physiology of 2- and 3-celled pollen. Phytomorphology 27: 98–106
Johri BM, Singh H (1959) The morphology, embryology and systematic position of *Elytraria acaulis* (Linn. f.) Lindau. Bot Not 112: 227–251
Johri BM, Tiagi B (1952) Floral morphology and seed formation in *Cuscuta reflexa* Roxb. Phytomorphology 2: 162–180
Johri BM, Vasil IK (1956) The embryology of *Ehretia laevis* Roxb. Phytomorphology 6: 134–143
Johri BM, Agrawal JS, Garg S (1957) Morphological and embryological studies in the family Loranthaceae 1. *Helicanthes elastica* (Desr.) Dans. Phytomorphology 7: 336–354
Johri MM (1962) Intracarpellary pollen grains in *Fritillaria* and *Lilium*. Curr Sci 31: 225–256
Jones HA, Emsweller SL (1936) The development of the flower and megagametophyte of *Allium cepa*. Hilgardia 10: 415–428
Jones BL, Gordon CC (1965) Embryology and development of endosperm haustorium of *Arceuthobium doughlasii*. Am J Bot 52: 127–132
Jones RL (1974) The structure of the lettuce endosperm. Planta 121: 133–147
Jönsson B (1879/80) Om embryosäckens utveckling hos Angiosperma. Acta Univ Lund 16: 1–86
Jorapur SM, Noronha HD (1963) Contributions to the embryology of Combretaceae. Karnatak Univ J Sci 8: 59–66
Jordaan PG (1946) Die Saadknop en Embriologie van *Brabejum stellatifolium*. J S Afr Bot 12: 15–26
Jørgensen CA (1923) Studies on Callitrichaceae. Bot Tidskr 38: 81–126

Jørgensen CA (1925) Zur Frage der systematischen Stellung der Callitrichaceen. Jahrb Wiss Bot 64: 440–442

Jørgensen CA (1928) The experimental formation of heteroploid plants in the genus *Solanum*. J Genet 19: 133–211

Jose JS (1963) The structure and development of seeds in Convolvulaceae: *Ipomoea* species. Agra Univ J Res Sci 12: 247–260

Jose JS, Singh SP (1968) Gametophyte development and embryogeny in the Genus *Nicotiana*. J Indian Bot Soc 47: 117–128

Jose JS, Vijaya Bai K (1978) Supernumerary gametes in the pollen tube of *Colocasia*. Curr Sci 47: 281–282

Jose JS, Vijaya Bai K (1986) Seed-set and polyembryony in *Amorphophallus campanulatus* Blume. J Indian Bot Soc 65: 178–184

Jose MK (1989) Morphology of the ovule and ovular appendages in south Indian Zingiberaceae. Acta Bot Indica 17: 82–87

Joshi AC (1933) Megaspore formation and embryo sac of *Argemone mexicana* Linn. J Indian Bot Soc 12: 83–91

Joshi AC (1936a) A contribution to the embryology and cytology of *Rivina humilis* Linn. J Indian Bot Soc 15: 91–103

Joshi AC (1936b) A note on the antipodals of *Digera arvensis* Forsk. Curr Sci 4: 741–742

Joshi AC (1937a) Contributions to the embryology of the Menispermaceae 1. *Cocculus villosus* DC. Proc Indian Acad Sci B 5: 57–63

Joshi AC (1937b) Megasporogenesis in *Aloe vera* Linn. J Indian Bot Soc 15: 297–300

Joshi AC (1938a) Parthenocarpy in *Dodonaea viscosa* Facq. J Indian Bot Soc 17: 97–99

Joshi AC (1938b) A note on the morphology of the gynoecium, ovule and embryo sac of *Psoralea corylifolia* L. J Indian Bot Soc 17: 169–172

Joshi AC (1938c) A note on the morphology of the ovule of Rubiaceae, with special reference to *Cinchona* and coffee. Curr Sci 7: 236–237

Joshi AC (1938d) Systematic distribution of the Fritillaria type of embryo sac and the mono- or polyphyletic origin of angiosperms. Chron Bot 4: 507–508

Joshi AC (1939a) Morphology of *Tinospora cordifolia* with some observations on the origin of the single integument, nature of synergidae and affinities of the Menispermaceae. Am J Bot 26: 433–439

Joshi AC (1939b) Embryological evidence for relationships of the Lythraceae and related families. Curr Sci 8: 112–113

Joshi AC (1939c) Structure and development of the embryo sac of *Iphigenia indica* Kunth. Proc Natl Inst Sci India B 5: 289–305

Joshi AC (1940) Development of the embryo sac of *Gagea fascicularis* Salisb. Bull Torrey Bot Club 67: 155–158

Joshi AC (1943) The anatomy of the male flower of *Myristica fragrans* with special reference to the origin of the trimerous perianth. J Indian Bot Soc, MOP Iyengar Comm Vol, pp 91–95

Joshi AC (1944) Structure and development of the ovule and embryo sac of *Piper longum* L. Proc Natl Inst Sci India 10: 105–112

Joshi AC (1946) A note on the development of pollen of *Myristica fragrans* van Houtten and the affinities of the family Myristicaceae. J Indian Bot Soc 25: 139–143

Joshi AC, Kajale LB (1936) A note on the structure and development of the embryo sac, ovule and fruit of *Tamarix dioica* Roxb. Ann Bot (London) 50: 421–426

Joshi AC, Kajale LB (1937) Fertilization and seed development in Amaranthaceae. Proc Indian Acad Sci B 5: 91–100

Joshi AC, Pantulu JV (1941) A morphological and cytological study of *Polianthes tuberosa*. J Indian Bot Soc 20: 37–71

Joshi AC, Rama Rao BV (1934) Fruit and seed development in *Tinospora cordifolia* without fertilization and embryo formation inside. Curr Sci 3: 62–63

Joshi AC, Rama Rao BV (1935) A study of microsporogenesis in two Menispermaceae. Cellule 44: 219–234

Joshi AC, Rama Rao BV (1936) The embryology of *Gisekia pharnaceoides* Linn. Proc Indian Acad Sci B 3: 71–92

Joshi AC, Venkateswarlu J (1935a) Embryological studies in the Lythraceae 1. *Lawsonia inermis* Linn. Proc Indian Acad Sci B 2: 481–493
Joshi AC, Venkateswarlu J (1935b) Embryological studies in the Lythraceae 2. *Lagerstroemia* Linn. Proc Indian Acad Sci B 2: 523–534
Joshi AC, Venkateswarlu J (1935c) A case of reversed polarity in the embryo sac. Ann Bot (London) 49: 841–843
Joshi AC, Venkateswarlu J (1936a) Embryological studies in Lythraceae 3. Proc Indian Acad Sci B 3: 377–400
Joshi AC, Venkateswarlu J (1936b) Structure and development of the synergids in *Ammania baccifera* Linn. A correction. New Phytol 35: 131
Joshi BM, Dwivedi NK (1980a) Studies on Indian desert plants 5. Floral anatomy and embryology of *Heliotropium supinum* L. Trans Isd Ucds 5: 32–38
Joshi BM, Dwivedi NK (1980b) Studies in the Lamiaceae 3. Embryology of *Plectranthus rugosus* Wall. J Indian Bot Soc 59: 127–132
Joshi MC, Varghese TM (1962) Presence of twin embryos and twin ovules in *Anticharis linearis* Hochst.: a member of Scrophulariaceae. Sci Cult 28: 489
Joshi MC, Varghese TM (1963) A contribution to the life history of *Anticharis linearis* Hochst. Proc Indian Acad Sci B 57: 164–177
Joshi PC (Prakash Chand) (1936) Contribution to the life history of *Stellaria media* L. Proc Indian Acad Sci B 3: 8–22
Joshi PC (Puran Chand) (1960) Morphological and embryological studies in the family Santalaceae 5. *Osyris wightiana* Wall. Phytomorphology 10: 239–248
Joshi PC, Wadhwani (née Ramchandani) AM, Johri BM (1967) Morphological and embryological studies of *Gossypium* L. Proc Natl Inst Sci India B 33: 37–93
Juel HO (1898) Parthenogenesis bei *Antennaria alpina* (L.). Beih Bot Zentralbl 74: 369–372
Juel HO (1900) Vergleichende Untersuchungen über typische und parthenogenetische Fortpflanzung bei der Gattung *Antennaria*. Kon Svensk Vet Akad Handl 33: 1–59
Juel HO (1903) Zur Entwicklungsgeschichte des Samens von *Cynomorium*. Beih Bot Zentralbl 13: 194–202
Juel HO (1910) *Cynomorium* und *Hippuris* Sv. Svensk Bot Tidskr 4: 151–159
Juel HO (1911) Studien über die Entwicklungsgeschichte von *Hippuris vulgaris*. Nov Acta Reg Soc Sci Upsala 2: 1–26
Juel HO (1915) Untersuchungen über die Auflösung der Tapetenzellen in den Pollensäcken der Angiospermen. Jahrb Wiss Bot 56: 337–364
Juguet M (1966a) Polyembryonie chez le *Geum rivale* L. Bull Soc Bot Fr 113: 133–137
Juguet M (1966b) Polyembryonie chez le *Carex arenaria* L. Bull Soc Bot Fr 113: 448–452
Juguet M (1969a) Embryogénie des Juncacées: développement de l'embryon chez le *Luzula pilosa* (L.) Willd. CR Acad Sci Paris 268: 3036–3039
Juguet M (1969b) Développement de l'embryon chez l'*Eriophorum angustifolium* Roth. (Cyperacees). Bull Soc Bot Fr 116: 137–143
Juguet M (1989) Not all the monocotyledons are as monocotyledonous as it is believed. In: Paré J, Bugnicourt M (eds) Some aspects and actual orientations in plant embryology. Dedicatory volume to Prof A. Lebègue. Univ Press Picardie, Amiens (Fr), pp 206–222
Juliano JB (1931) Morphological study of the flower of *Monochoria vaginalls* (Burn. f.) Pressl. Philipp Agric 20: 177–186
Juliano JB (1934) Studies on the morphology of Meliaceae 1. *Sandoricum koetjape*. Philipp Agric 23: 11–48
Juliano JB (1935a) Morphological contributions to the genus *Annona* L. Philipp Agric 24: 528–541
Juliano JB (1935b) Anatomy and morphology of the Bunga, *Aeginetia indica* L. Philipp J Sci 56: 405–451
Juliano JB (1937) Embryos of carabao mango, *Mangifera indica* L. Philipp J Agric 25: 749–760
Juliano JB, Alcala PE (1933) Floral morphology of *Musa errans* (Blanco) Teodoro. var. botoan Teodoro. Philipp Agric 22: 91–126
Juliano JB, Aldama MJ (1937) Morphology of *Oryza sativa* L. Philipp Agric 26: 1–134
Juncosa AM (1982) Developmental morphology of the embryo and seedling of *Rhizophora mangle* L. (Rhizophoraceae). Am J Bot 69: 1599–1611

Juncosa AM (1984a) Embryogenesis and seedling development in *Cassipourea elliptica* (Sw.) Poir. (Rhizophoraceae). Am J Bot 71: 170–179

Juncosa AM (1984b) Embryogenesis and developmental morphology of the seedling in *Bruguiera exaristata* Ding Hou. (Rhizophoraceae). Am J Bot 71: 180–191

Junell S (1930) Die Entwicklungsgeschichte von *Circaeaster agrestis*. Svensk Bot Tidskr 25: 238–270

Junell S (1934) Zur Gynaceummorphologie und Systematik der Verbenaceen und Labiateen nebst Bemerkungen über ihre Samenentwicklung. Symp Bot Upsala 4: 1–219

Junell S (1937) Die Samenentwicklung bei einigen Labiateen. Svensk Bot Tidskr 31: 67–110

Junell S (1938) Über den Fruchtknotenbau der Borraginazeen mit pseudomonomeren Gynäzeen. Svensk Bot Tidskr 32: 261–273

Junell S (1962) Embryology of *Hebenstreitia, Dischisma, Sutera* and *Zaluzianskya* (Scrophulariaceae). Acta Hortic Göteborg 25: 91–101

Kadry AER (1946) Embryology of *Cardiospermum halicacabum* L. Svensk Bot Tidskr 40: 111–126

Kadry AER (1952) The development of microsporangia and pollen grains in *Cistanche tinctoria*. Bot Not 1952: 46–57

Kadry AER (1955) The development of endosperm and embryo in *Cistanche tinctoria* (Forsk.) G. Beck. Bot Not 1955: 231–243

Kaeiser M, Boyce SG (1962) Embryology of *Liriodendron tulipifera* L. Phytomorphology 12: 103–109

Kainradl E (1927) Beiträge zur Biologie von *Hydrolea spinosa* L. mit besonderer Berücksichtigung von Fruchtwand und Samenentwicklung. Sitzungsber Akad Wiener Akad Math Nat Kl Abt 1 136: 167–194

Kajale LB (1935) The female gametophyte of *Alternanthera sessilis* R. Br. Proc Indian Acad Sci B 2: 476–480

Kajale LB (1936) Embryo development in *Boerhaavia diffusa*. Curr Sci 4: 743

Kajale LB (1937a) A case of polyembryony in the Nyctaginaceae. Curr Sci 5: 429

Kajale LB (1937b) Embryology of *Achyranthes aspera* Linn. Proc Indian Acad Sci B 5: 195–205

Kajale LB (1938) Embryo and seed development in the Nyctaginaceae 1. Studies in the genus *Boerhaavia*. J Indian Bot Soc 17: 243–254

Kajale LB (1939) A contribution to the life history of *Bergia ammanioides* Roxb. J Indian Bot Soc 18: 157–167

Kajale LB (1940) A contribution to the embryology of the Amaranthaceae. Proc Natl Inst Sci India B 6: 597–625

Kajale LB (1942) A contribution to the embryology of the genus *Portulaca*. J Indian Bot Soc 21: 1–19

Kajale LB (1944a) A contribution to the embryology of the Phytolaccaceae 1. Studies in the genus *Phytolacca*. J Patna Univ 1: 10–21

Kajale LB (1944b) A contribution to the life history of *Zizyphus jujuba* Lamk. Proc Natl Inst Sci India B 10: 387–391

Kajale LB (1954a) A contribution to the embryology of Phytolaccaceae 2. Fertilization and development of embryo, seed and fruit in *Rivina humilis* Linn. and *Phytolacca dioica* Linn. J Indian Bot Soc 33: 206–225

Kajale LB (1954b) The antipodals in the family Amaranthaceae. Curr Sci 23: 165–166

Kajale LB (1954c) Fertilization and the development of embryo and seed in *Euphorbia hirta* Linn. Proc Natl Inst Sci India B 20: 353–360

Kajale LB, Murthy KSN (1954) The embryo sac of *Acalypha ciliata*. J Indian Bot Soc 33: 417–422

Kajale LB, Ranade SG (1952) Occurrence of four kinds of megaspore tetrads in *Elaeis guineensis* Jacq. Curr Sci 21: 170

Kajale LB, Ranade SG (1953) The embryo sac of *Elaeis guineensis* Jacq.: A reinvestigation. J Indian Bot Soc 32: 101–107

Kajale LB, Rao GV (1943) Pollen and embryo sac of two Euphorbiaceae. J Indian Bot Soc 22: 229–236

Kajale LB, Razi BA (1954) Contribution to the embryology of *Martynia diandra* Clox. J Univ Poona 6: 28–33

Kallarackal J, Bhatnagar SP (1978) Integument initiation and seed-coat development in *Linaria bipartita* (Vent.) Willd. Bot Jahrb 4: 579–585
Kallarackal J, Bhatnagar SP (1980) Cytochemical studies on the developing female gametophyte of *Linaria bipartita*. Acta Bot Indica 8: 11–22
Kaltsikes PJ (1973) Early seed development in hexaploid triticale. Can J Bot 51: 2291–2300
Kaltsikes PJ, Roupakias DG (1975) Endosperm abnormalities in *Triticum-Secale* combinations 2. Addition and substitution lines. Can J Bot 53: 2068–2076
Kamelena OP (1980) Comparative embryology of Dipsacaceae and Morinaceae families. Nauka, Leningrad (in Russian)
Kamelena OP (1983) Basic results of the comparative embryological investigations of Dipsacaceae and Morinaceae. In: Erdelská O, Ciamporová M, Luxova A, Pretóvá A, Tupý J (eds) Fertilization and embryogenesis in ovulated plants. VEDA, Bratislava (Czech), pp 343–346
Kandasamy MK, Vivekanandan M (1984) A comparative study on the structural features of the stigma and style in three members of Leguminosae. J Swamy Bot Club 1: 79–88
Kanta K (1961) Embryolessness in *Piper nigrum*. Phytomorphology 11: 304–306
Kanta K (1962) Morphology and embryology of *Piper nigrum*. Phytomorphology 12: 207–221
Kapil RN (1956a) Development of embryo sac and endosperm in *Crozophora rottleri* A. Juss. Bot Gaz 117: 242–247
Kapil RN (1956b) A further contribution to the morphology and life history of *Crozophora* Neck. Phytomorphology 6: 278–288
Kapil RN (1960) Embryology of *Acalypha* Linn. Phytomorphology 10: 174–184
Kapil RN (1961) Some embryological aspects of *Euphorbia dulcis* L. Phytomorphology 11: 24–36
Kapil RN (1962) Some recent examples of the value of embryology in relation to taxonomy. Bull Bot Surv India 4: 57–66
Kapil RN (1967) Araceae. In: Johri BM, Kapil RN, Bhatnagar SP, Bhandari NN, Vijayaraghavan MR (eds) Comparative embryology of angiosperms. Univ Press, Delhi, pp 126, 127
Kapil RN (1970a) Crossosomataceae. In: Symp Comparative embryology of angiosperms. Indian Natl Sci Acad Bull 41: 63–68
Kapil RN (1970b) Podostemaceae. In: Symp Comparative embryology of angiosperms. Indian Natl Sci Acad Bull 41: 104–109
Kapil RN (1974) Contemporary spectacle in angiosperm embryology. Acta Bot Indica 2: 79–106
Kapil RN, Ahluwalia K (1963) Embryology of *Peganum harmala* Linn. Phytomorphology 13: 127–140
Kapil RN, Bawa SB (1968) Embryological studies on the Haloragidaceae 1. *Haloragis colensoi* Skottisb. Bot Not 121: 11–28
Kapil RN, Bhandari NN (1964) Morphology of *Magnolia* Linn. Proc Natl Inst Sci India B 30: 245–262
Kapil RN, Bhatnagar AK (1972) Endosperm in Euphorbiaceae: A critical appraisal. In: Murty YS, Johri BM, Mohan Ram HY, Varghese TM (eds) Advances in plant morphology (Prof. V. Puri Comm. Vol). Sarita Prakashan, Meerut (India), pp 376–393
Kapil RN, Bhatnagar AK (1975) A fresh look at the process of double fertilization in angiosperms. Phytomorphology 25: 334–369
Kapil RN, Bhatnagar AK (1980) Embryology in relation to angiosperm taxonomy: In: Nair PKK (ed) Glimpses in plant research, vol 5. Vikas, New Delhi, pp 272–295
Kapil RN, Bhatnagar AK (1981) Ultrastructure and biology of female gametophyte in flowering plants. Int Rev Cytol 70: 291–337
Kapil RN, Bhatnagar AK (1983) Central cell: pre- and post-fertilization parameters. Phytomorphology 33: 157–167
Kapil RN, Jalan S (1962) Studies in the Family Ranunculaceae 1. The embryology of *Caltha palustris* L. In: Plant embryology Symp. Council Sci Ind Res, New Delhi, pp 205–214
Kapil RN, Jalan S (1964) *Schisandra michaux*: Its embryology and systematic position. Bot Not 117: 285–305
Kapil RN, Kaul U (1972) Embryologically little-known taxon: *Parrotiopsis jacquemontiana*. Phytomorphology 22: 334–345
Kapil RN, Maheshwari R (1964) Embryology of *Helianthemum vulgare* Gaertn. Phytomorphology 14: 547–557

Kapil RN, Masand P (1964) Embryology of *Hebernstretia integrefolia* Linn. Proc Natl Inst Sci India B 30: 99–113

Kapil RN, Mohana Rao PR (1966a) Embryology and systematic position of *Theligonum* Linn. Proc Natl Inst Sci India B 32: 218–232

Kapil RN, Mohana Rao PR (1966b) Studies of the Garryaceae 2. Embryology and systematic position of *Garrya* Douglas ex Lindley. Phytomorphology 16: 564–578

Kapil RN, Nanchahal A (1989) Consequences of restrained pollination on structure and behaviour of antipodal cells in *Argemone mexicana* (Papaveraceae). In: Paré J, Bugnicourt M (eds) Some aspects and actual orientations in plant embryology. Dedicatory volume to Prof A Lebègue. Univ Press Pieardie, Amiens (Fr), pp 251–260

Kapil RN, Prakash N (1966) Co-existence of mono-, bi- and tetrasporic embryo sacs in *Delosperma cooperi* Hook. f. (Aizoaceae). Beitr Biol Pflanzen 42: 381–392

Kapil RN, Prakash N (1969) Embryology of *Cereus jamacaru* and *Ferocactus wislizeni* and comments on the systematic position of the Cactaceae. Bot Not 122: 409–426

Kapil RN, Sethi SB (1962a) Gametogenesis and seed development in *Ainsliaea aptera* DC. Phytomorphology 12: 222–234

Kapil RN, Sethi SB (1962b) Development of seed in *Tridax trilobata* Hemsl. Phytomorphology 12: 235–239

Kapil RN, Sethi SB (1963) Development of male and female gametophytes in *Camellia sinensis* (L.) O. Kuntze. Proc Natl Inst Sci India B 29: 567–574

Kapil RN, Tiwari SC (1978) The integumentary tapetum. Bot Rev 44: 457–490

Kapil RN, Vani RS (1963) Embryology and systematic position of *Crossosoma californicum* Nutt. Curr Sci 32: 493–495

Kapil RN, Vani RS (1966) *Nyctanthes arbor-tristis* Linn: Embryology and relationships. Phytomorphology 16: 553–563

Kapil RN, Vasil IK (1963) Ovule. In: Maheshwari P (ed) Recent advances in the embryology of angiosperms. Int Soc Plant Morphol, Univ Delhi, pp 41–68

Kapil RN, Vijayaraghavan MR (1962) Embryology and systematic position of *Pentaphragma horsfieldii* (Miq.) Airy Shaw. Curr Sci 31: 270–272

Kapil RN, Vijayaraghavan MR (1965) Embryology of *Pentaphragma horsfieldii* (Miq.) Airy Shaw with a discussion on the systematic position of the genus. Phytomorphology 15: 93–102

Kapil RN, Walia K (1965) The embryology of *Philydrum lanuginosum* Banks ex Gaertn. and the systematic position of the Philydraceae. Beitr Biol Pflanzen 41: 381–404

Kapil RN, Bor J, Bouman F (1980) Seed appendages in angiosperms 1. Bot Jahrb 101: 555–573

Kapil RN, Rustagi PN, Venkataraman R (1968) A contribution to the embryology of Polemoniaceae. Phytomorphology 18: 403–412

Kaplan DR (1969a) Sporogenesis and gametogenesis in *Downingia* (Campanulaceae: Lobelioideae). Bull Torrey Bot Club 96: 418–434

Kaplan DR (1969b) Seed development in *Downingia*. Phytomorphology 19: 253–278

Kaplan K (1976) Zur Embryologie und Systematik der Gattung *Saxifraga*. Bot Jahrb 97: 61–71

Kaplan K, Strohschneider M (1984) Mikromorphologische Untersuchungen an Samenoberflächen in der Gattung *Chrysosplenium* (Saxifragaceae). Bot Jahrb 104: 469–482

Kapoor T, Vijayaraghavan MR (1975) Ontogeny of ruminate endosperm in *Celsia coromandeliana*. Beitr Biol Pflanzen 51: 359–365

Kapoor T, Vijayaraghavan MR (1978) Seed structure in *Celsia coromandeliana* Vahl. In: Malik CP, Srivastava AK, Bhattacharya NC, Singh R (eds) Physiology of sexual reproduction in flowering plants. Kalyani, New Delhi, pp 245–247

Kapoor T, Vijayaraghavan MR (1986) The endosperm in *Najas*: A reassessment. Phytomorphology 36: 37–42

Kapoor T, Parulekar NK, Vijayaraghavan MR (1975) Contribution to the embryology of *Celsia coromandeliana* Vahl. with a discussion on its affinities with *Verbascum thapsus* L. Bot Not 128: 438–449

Kapoor T, Vijayaraghavan MR, Parulekar NK (1978) Ontogeny, structure and differentiation of anther tapetum in *Celsia coromandeliana*. Phyton (Argentina) 18: 209–216

Kappert H (1933) Erbliche Polyembryonie bei *Linum usitatissimum*. Biol Zentralbl 53: 276–307

Kapuskar AT (1964) Studies in the embryology of Leguminosae 2. Embryo development in *Aeschynomene aspera* Linn. Proc Indian Acad Sci B 60: 86–94

Karagozova M (1963) Embryological studies on *Iris pseudocorus*. Bulg Akad Nauk Bot Inst Izv 11: 111–123
Karanth KA, Bhat PK, Arekal DA (1979) Tapetum-like anther epidermis in *Zeuxine longilabris* (Lindl.) Benth. ex HK (Orchidaceae). Curr Sci 48: 542–553
Karanth KA, Swamy BGL, Arekal GD (1981) Embryogenesis in sexual and asexual species of *Zeuxine*. Proc Indian Acad Sci B 90: 1–8
Karlström P-O (1972) Embryological studies in Acanthaceae 1. The genus *Ruellia* L. Svensk Bot Tidskr 66: 303–313
Karlström P-O (1973) Embryological studies in the Acanthaceae 2. The genera *Pseuderanthemum* Radlk., *Ruspolia* Lindau, *Eranthemum* L. and *Lankesteria* Lindl. Svensk Bot Tidskr 67: 257–280
Karlström P-O (1974a) Embryological studies in the Acanthaceae 3. The genera *Barleria* L. and *Crabbea* Harv. Svensk Bot Tidskr 68: 121–135
Karlström P-O (1974b) Embryological studies in Acanthaceae 4. The genera *Asystasia* B1. and *Chamaeranthemum* Nees. Svensk Bot Tidskr 68: 315–328
Karsten G (1891) Über die Mangroven-Vegetation im malayischen Archipel. Eine morphologische-biologische Studie. Bibl Bot (Stuttgart) 22: 11–18, 31–41
Karsten G (1902) Über die Entwicklung der weiblichen Blüten bei einigen Juglandaceen. Flora 90: 316–333
Kashyap G (1955) Studies in the family Vitaceae 1. Floral morphology of *Vitis trifolia* Linn. Agra Univ J Res Sci 4: 777–784
Kashyap G (1958) Studies in the family Vitaceae 3. Floral morphology of *Vitis latifolia* Roxb., *Vitis himalayana* Brandis, and *Vitis trifolia* Linn. J Indian Bot Soc 37: 240–248
Kaul U (1969) Endosperm in *Parrotiopsis jacquemontiana*. Phytomorphology 19: 197–199
Kaul U, Kapil RN (1974) *Exbucklandia populnea*: From flower to fruit. Phytomorphology 24: 217–228
Kaul V, Dathan ASR, Singh D (1975) Embryological studies on the genus *Sonchus* L. J Indian Bot Soc 54: 238–245
Kaul V, Rouse JL, Williams EG (1986) Early events in the embryo sac after intraspecific and interspecific pollinations in *Rhododendron kawakamii* and *R. retusum*. Can J Bot 64: 282–291
Kaul V, Singh D (1980) Embryological studies on Indian *Taraxacum* Weber: Embryo sac, endosperm and embryo. Giorn Bot Ital 114: 277–294
Kaul V, Singh D (1982) Cytoembryological studies on Indian *Taraxacum* Weber 1. Microsporangium, microsporogenesis and microgametogenesis, Cytologia 47: 125–135
Kaur A, Jong K, Sands VE, Soepadmo E (1986) Cytoembryology of some Malaysian dipterocarps, with some evidence of apomixis. Bot J Linn Soc 92: 75–88
Kaur D (1970) Turneraceae. In: Symp Comparative embryology of angiosperms. Indian Natl Sci Acad Bull 41: 194–198
Kaur H (Harbans) (1969) Structure and development of seed in *Ipomoea obscura* Ker-Gawl. J Indian Bot Soc 48: 346–351
Kaur H (1981a) Development of female gametophyte in *Ipomoea hispida* Parodl. Curr Sci 50: 912–913
Kaur H (1981b) Embryological studies in *Jacquemontia tamnifolia* Grisev. (Abstr). J Indian Bot Soc 60 (Suppl): 56–57
Kaur H, Singh RP (1970) Structure and development of seeds in three *Ipomaea* species. J Indian Bot Soc 49: 168–174
Kaur H, Singh RP (1981) Development of male and female gametophytes in some Convolvulaceae. New Bot 8: 91–98
Kaur H (Harjinder) (1969) Embryological investigations on *Bixa orellana* Linn. Proc Natl Inst Sci India B 35: 487–506
Kaur H (1970a) Moraceae. In: Symp Comparative embryology of angiosperms. Indian Natl Sci Acad Bull 41: 1–3
Kaur H (1970b) Bixaceae. In: Symp Comparative embryology of angiosperms. Indian Natl Sci Acad Bull 41: 205–207
Kaur H (1970c) Cornaceae. In: Symp Comparative embryology of angiosperms. Indian Natl Sci Acad Bull 41: 230–232
Kausik SB (1935a) The life history of *Utricularia coerulea* L. Curr Sci 3: 357–359

Kausik SB (1935b) The life history of *Lobelia trigona* Roxb. with special reference to the embryo sac. Proc Indian Acad Sci B 2: 410–418
Kausik SB (1938a) Studies in the Proteaceae 1. Cytology and floral morphology of *Grevillea robusta* Cunn. Ann Bot (London) 2: 899–910
Kausik SB (1938b) Studies in the Proteaceae 2. Floral anatomy and morphology of *Macadamia ternifolia* F. Muell. Proc Indian Acad Sci B 8: 45–63
Kausik SB (1938c) The endosperm in *Grevillea robusta*. Curr Sci 6: 332–333
Kausik SB (1938d) Pollen development and seed formation in *Utricularia coerulea* L. Beih Bot Zentralbl 58A: 365–378
Kausik SB (1938e) Gametogenesis and embryogeny in *Lobelia nicotianaefolia* Heyne. J Indian Bot Soc 17: 269–278
Kausik SB (1939a) Studies in the Proteaceae 3. Embryology of *Grevillea banksii* R. Br. Ann Bot (London) 3: 815–824
Kausik SB (1939b) A cytological study of *Scaevola lobelia* Linn. Proc Indian Acad Sci B 9: 39–48
Kausik SB (1939c) Pollination and its influence on the behaviour of pistillate flower in *Vallisnaria spiralis*. Am J Bot 26: 207–211
Kausik SB (1940a) Studies in the Proteaceae 4. Structure and development of the ovule of *Hakea saligna* Knight. Ann Bot (London) 4: 73–80
Kausik SB (1940b) Structure and development of the ovule and embryo sac of *Lasiosiphon eriocephalus* Decne. Proc Natl Inst Sci India B 6: 117–132
Kausik SB (1940c) A contribution to the embryology of *Enalus acoroides* (L. fil.) Steud. Proc Indian Acad Sci B 11: 83–99
Kausik SB (1941a) Development of the vermiform appendage in *Grevillea robusta* Cunn. Proc Indian Acad Sci B 14: 137–140
Kausik SB (1941b) Structure and development of the staminate flower and male gametophyte of *Enalus acoroides* (L. fil.) Steud. Proc Indian Acad Sci B 14: 1–16
Kausik SB (1942) Studies in the Proteaceae 7. The endosperm of *Grevillea robusta* Cunn. with special reference to the structure and development of the vermiform appendage. Proc Indian Acad Sci B 16: 121–140
Kausik SB, Raju MVS (1955) A contribution to the floral morphology and embryology of *Utricularia reticulata* Smith. Proc Indian Acad Sci B 41: 155–166
Kausik SB, Raju MVS (1956) Variations in the development of proembryo in *Utricularia coerulea* L. Curr Sci 25: 296–297
Kausik SB, Rao PVK (1942) The male gametophyte of *Halophila ovata*. J Mysore Univ B 3: 43–49
Kausik SB, Subramanyam K (1945a) A contribution to the embryology of *Lobelia trialata* Buch.-Ham. J Indian Bot Soc 24: 175–181
Kausik SB, Subramanyam K (1945b) An embryological study of *Isotoma longiflora* Presl. Proc Indian Acad Sci B 21: 269–278
Kausik SB, Subramanyam K (1946a) Development of endosperm in *Lobelia nicotianaefolia* Heyne. Curr Sci 15: 78–79
Kausik SB, Subramanyam K (1946b) A case of polyembryony in *Isotoma longiflora* Presl. Curr Sci 15: 257–258
Kausik SB, Subramanyam K (1946c) A contribution to the life-history of *Sphenoclea zeylanica* Gaertn. Proc Indian Acad Sci B 23: 274–278
Kausik SB, Subramanyam K (1947a) Embryology of *Cephalostigma schimperi* Hochst. Bot Gaz 109: 85–90
Kausik SB, Subramanyam K (1947b) Embryogeny of *Isotoma longiflora* Presl. Proc Indian Acad Sci B 46: 164–167
Kavaljian LG (1952) The floral morphology of *Clethra alnifolia* with some notes on *C. acuminata* and *C. arborea*. Bot Gaz 113: 392–413
Kavathekar KY, Pillai A (1981) Studies on the developmental anatomy of Ranales 5. Late embryogeny and histogenesis in *Delphinium*, *Nigella* and *Ranunculus*. J Indian Bot Soc 60: 58–64
Keefe JM, Moseley Jr MF (1978) Wood anatomy and phylogeny of *Paeonia* section Moutan. J Arnold Arbor 59: 274–297
Kelkar SS (1954) Occurrence of twin embryo sacs in *Rhus mysorensis* Heyne. J Univ Poona 6: 36–37

Kelkar SS (1958a) Embryology of *Rhus mysorensis* Heyne. J Indian Bot Soc 37: 114–122
Kelkar SS (1958b) A contribution to the embryology of *Lannea coromandelica* (Houtt.) Merr. J Univ Bombay 26: 152–159
Kelkar SS (1960) The female gametophyte in *Euphorbia tirucalli* Linn. Sci Cult 25: 633–635
Kelkar SS (1961) The development of endosperm and embryo in *Lannea coromandelica* (Houtt) Merr. J Univ Bombay 29: 1–5
Kelkar SS (1974) Abnormal embryo sac formation in *Nothopegia dalzellii* Gamble var. *angustifolia* Gamble. Curr Sci 42: 759–760
Kempanna C, Seetharam A, Shivashankar G (1971) Multinucleate pollen mother cells in triticales. Curr Sci 40: 301–302
Keng H (1967) Observations on *Ancistrocladus tectorius*. Garden Bull Singapore 22: 113–121
Kennedy MJ, Prakash N (1981) A morphological and embryological study of *Callicoma serratifolia* Andr. (Cunoniaceae). Aust J Bot 29: 721–731
Kennell JC, Horner HT (1985) Megasporogenesis and megagametogenesis in soybean *Glycine max*. Am J Bot 72: 1553–1564
Kenyan FMG (1928) A morphological and cytological study of *Ipomoea trifida*. Bull Torrey Bot Club 55: 499–512
Kenrick J, Knox RB (1979) Pollen development and cytochemistry in some Australian species of *Acacia*. Aust J Bot 27: 413–427
Kenrick J, Knox RB (1982) Function of the polyad in reproduction of *Acacia* (Leguminosae, Mimosoideae). Ann Bot (London) 50: 721–727
Khaleel TF (1974) Embryology of *Cynoglossum denticulatum* DC. Bot Not 127: 193–210
Khaleel TF (1975) Embryology of *Cordia*. Bot Gaz 136: 380–387
Khaleel TF (1977) Embryology of *Trichodesma* (Boraginaceae). Bot Not 130: 441–452
Khaleel TF (1978) Embryology of *Heliotropium scabrum* and *H. strigosum* (Boraginaceae). Plant Syst Evol 129: 45–62
Khaleel TF (1982) Embryology of *Cordia sebestena* (Boraginaceae). Plant Syst Evol 139: 303–311
Khaleel TF, Boraiah G (1973) Cytoembryology of *Thunbergia mysorensis* T. Anders. Proc Indian Natl Sci Acad B 39: 157–171
Khaleel TF, Mitchell BB (1982) Cytoembryology of *Allium textile* Nels. and Macbr. Am J Bot 69: 950–956
Khan AM (1973) Development of ovule and female gametophyte in *Talinum patens* Willd. Curr Sci 42: 364–365
Khan AM (1975) Structure and development of seed in *Talinum patens* Willd. Bangladesh J Bot 4: 169–172
Khan PSH (1970) Structure and development of seed-coat and fruit-wall in *Catharanthus pusillus* (Murr.) G. Don. Proc Natl Acad Sci India B 40: 21–24
Khan R (1942) A contribution to the embryology of *Jussieua repens* Linn. J Indian Bot Soc 21: 267–282
Khan R (1943a) The ovule and embryo sac of *Fouquieria*. Proc Natl Inst Sci India B 9: 253–256
Khan R (1943b) Contributions to the morphology of *Ephedra foliata* Boiss 2. Fertilization and embryogeny. Proc Natl Acad Sci India 13: 357–375
Khan R (1953) Haustorial behaviour of the chalazal end of embryo sac and reversion of polarity in *Utricularia flexuosa* Vahl. Curr Sci 22: 179–180
Khan R (1954) A contribution to the embryology of *Utricularia flexuosa* Vahl. Phytomorphology 4: 80–117
Khan R (1963) The behaviour of the integumentary tapetum in the ovules containing degenerating gametophytes in *Utricularia flexuosa* Vahl. Proc Natl Acad Sci India B 33: 651–655
Khan R (1970) Lentibulariaceae. In: Symp Comparative embryology of angiosperms. Indian Natl Sci Acad Bull 41: 290–297
Khanna P (1963) Male gametophyte and obturator in *Cyperus rotundus* L. and *C. triceps* Endl. (*Kyllinga triceps* Rottb.). Curr Sci 32: 133–134
Khanna P (1964a) Morphological and embryological studies in Nymphaeaceae 1. *Euryale ferox* Salisb. Proc Indian Acad Sci B 59: 237–243
Khanna P (1964b) Embryology of *Trichodesma amplexicaule* Roth. Bull Torrey Bot Club 91: 105–114
Khanna P (1964c) Embryology of *Mertensia*. J Indian Bot Soc 43: 192–202

Khanna P (1965a) Morphological and embryological studies in Nymphaeaceae 2. *Brasenia schreberi* Gmel. and *Nelumbo nucifera* Gaertn. Aust J Bot 13: 379–387

Khanna P (1965b) A contribution to the embryology of *Cyperus rotundus* L., *Scripus mucrinatus* L. and *Kyllinga melanospora* Nees. Can J Bot 43: 1539–1547

Khanna P (1967) Morphological and embryological studies in Nymphaeaceae 3. *Victoria cruziana* D'Orb. and *Nymphaea stellata* Willd. Bot Mag (Tokyo) 80: 305–312

Khattra S (neé Sharma), Singh G (1989) Histochemical studies of anther development in male sterile *Pennisetum typhoides*. Acta Bot Indica 17: 159–162

Khosla S (1946) Developmental morphology in some Indian millets. Proc Indian Acad Sci B 24: 207–224

Khushalani I (1963a) Floral morphology and embryology of *Melianthus major* L. Phyton (Argentina) 10: 145–156

Khushalani I (1963b) Floral morphology and embryology of *Acer oblongum*. Phyton (Argentina) 10: 275–284

Kimura Y (1956) Système et phylogénie des Monocotylédones. Not Syst 15: 137–159

King JR (1938) Morphological development of fruit of olive. Hilgardia 11: 437–458

King JR (1947) Development of ovule and megagametophyte in pomegranate. Bot Gaz 108: 394–398

Kirch MH (1936) Some abnormalities in the development in the embryo sac of *Lilium longiflorum*. Bull Torrey Bot Club 63: 383–396

Kircher P (1986) Untersuchungen zur Blüte- und Infloreszenzmorphologie, Embryologie und Systematik der Restionaceen im Vergleich mit Gramineen and verwandten Familien. Diss Bot, Ruhr-Univ, Bochum 94: 1–219

Kirsten E (1973) A preliminary morphological, cytological and embryological investigation in *Polygonum viviparum*. Bot Tidsskr 67: 305–316

Klebs G (1881–1885) Beiträge zur Morphologie und Biologie der Keimung. In: Untersuchungen aus dem Botanischen Institut zu Tübingen 1: 536–635

Kloos A, Bouman F (1980) Case studies in aril development: *Passiflora tuberosa* L. and *Turnera ulmifolia* L. Beitr Biol Pflanzen 55: 49–66

Klyuchareva MV (1983) Extrusion of nuclear material in proembryos of gramineous plants. Dokl Bot Sci Proc Acad Sci USSR 268–270: 19–21

Knox RB (1971) Pollen wall proteins: Localization, enzymic and antigenic activity during development in *Gladiolus* (Iridaceae). J Cell Sci 9: 209–237

Knox RB (1982) Pollination systems. In: Williams EG, Knox RB, Gilbert JH, Bernhardt P (eds) Pollination 1982. Plant Cell Biol Res Cent, School Bot Univ Melbourne, pp 7–14

Knox RB (1984) The pollen grain. In: Johri BM (ed) Embryology of angiosperms. Springer, Berlin Heidelberg New York, pp 197–271

Knox RB, Ducker SC (1976) Submarine pollination in seagrasses. Helob Newslett 1: 25–26

Knox RB, Friederich E (1974) Tetrad pollen grain development and sterility in *Leschenaultia formosa*. New Phytol 73: 251–258

Knox RB, Williams EG, Dumas C (1986) Pollen, pistil, and reproductive function in crop plants. In: Janick J (ed) Plant breeding reviews, vol 4. AVI, Wesport (Conn), pp 9–79

Kobuski CE (1948) Studies in the Theaceae 17. A review of the genus *Bonnetia*. J Arnold Arbor 29: 393–413

Kölreuter JC (1761–1766) Vorläufige Nachricht von einigen das Geschlecht der Pflanzen betreffenden Versuchen und Beobachtungen.

Kooiman P (1974) Iridoid glycosides in the Loasaceae and the taxonomic position of the family. Acta Bot Neerl 19: 329–340

Kopezynska K (1964) Embryo sac development in *Pinguicula vulgaris*. Acta Soc Bot Pol 33: 141–156

Kordyum EL (1957) On the modes of reproduction of endosperm nuclei in *Nigella sativa*. Ukr Bot Zh 14: 40–46 (in Russian)

Kordyum EL (1959) Comparative embryological investigations of the family Ranunculaceae DC. Ukr Bot Zh 16: 32–43 (in Russian)

Korobova S-N (1971) On the development of the "specific" cells during micro- and macrosporangium formation in *Zea mays*. Ann Univ ARERS 9: 58–61

Korobova SN (1974) On the behaviour of sperms in the process of fertilization of higher plants. In: Linskens HF (ed) Fertilization in higher plants. North-Holland, Amsterdam, pp 261–270
Kościńska M (1980) Embryo and endosperm in *Rudbeckia bicolor*. Acta Biol Cracov Ser Bot 22: 71–79
Kostriukova K (1939) Observations in vivo on the formation of the male sex cells in *Lilium martagon* (in Russian). Dokl Akad Nauk SSSR 22: 444–447
Kostriukova K, Benetskaia G (1939) Spermatogenesis in *Narcissus poeticus*. Observation in vivo. Bot Zh 24: 209–220 (in Russian)
Kostrikova LN (1972) Embryology of *Amorpha glabra* Desf. Vestn Mosk Univ Ser 6 Biol Pochvoved 27: 112–114 (in Russian)
Kostrioukoff X (1930) Cellules males dans le *Scripus lacustris*. Bull Jardin Bot Kieff 11: 10–20
Koul AK, Wafai BA (1982) Analysis of breeding systems in the genus *Tulipa* 1. Sporogenesis, gametogenesis and embryogeny in diploid *T. aitchisonii*. Phytomorphology 32: 287–305
Koul AK, Wafai BA, Wakhlu AK (1976) Studies on the genus *Gagea* 3. Sporogenesis, gametogenesis, early embryogeny and endosperm development in hexaploid *Gagea stipitata*. Phytomorphology 26: 255–262
Kracauer P (1930) Die Haploidgeneration von *Canna indica*. Diss, Univ Berlin
Kratzer J (1918) Die verwandtschaftlichen Beziehungen der Cucurbitaceen auf Grund ihrer Samenentwicklung mit spezieller Berücksichtigung der Caricaceen, Passifloraceen, Aristolochiaceen und Loasaceen. Flora 110: 275–343
Krishna Iyengar CV (1937) Development of the embryo sac and endosperm haustoria in some members of Scrophularineae 1. *Sopubia delfinifolia* G. Don. and *Alonsoa* sp. J Indian Bot Soc 16: 99–109
Krishna Iyengar CV (1939a) Development of embryo sac and endosperm haustoria in some members of Scrophularineae 2. *Isoplexis canariensis* Lindl. and *Celsia coromandeliana* Vahl. J Indian Bot Soc 18: 13–20
Krishna Iyengar CV (1939b) Development of embryo sac and endosperm haustoria in some members of Scrophularineae 3. *Limnophila heterophylla* Benth. and *Stemodia viscosa* Roxb. J Indian Bot Soc 18: 35–42
Krishna Iyengar CV (1940a) Development of embryo sac and endosperm haustoria in some members of Scrophularineae 5. *Ilysanthes hyssopioides* Benth. and *Bonnaya tenuifolia* Spreng. J Indian Bot Soc 19: 5–17
Krishna Iyengar CV (1940b) Development of embryo sac and endosperm haustoria in some members of Scrophularineae 4. *Vandellia hirsuta* Ham. and *V. scabra* Benth. J Indian Bot Soc 19: 179–189
Krishna Iyengar CV (1940c) Structure and development of seed in *Sopubia trifida* Ham. J Indian Bot Soc 19: 251–261
Krishna Iyengar CV (1941) Development of embryo sac and endosperm haustoria in *Torenia cordifolia* Roxb. and *T. hirsuta* Benth. Proc Natl Inst Sci India B 7: 61–71
Krishna Iyengar CV (1942a) Development of embryo sac and endosperm haustoria in *Tetranema mexicana* Benth. and *Verbascum thapsus* Linn. Proc Natl Inst Sci India B 8: 59–69
Krishna Iyengar CV (1942b) Development of embryo sac and endosperm haustoria in *Rehmannia angulata* Hemsl. J Indian Bot Soc 21: 51–57
Krishna Iyengar CV (1942c) Development of seed and its nutritional mechanism in Scrophulariaceae 1. *Rhamphicarpa longiflora* Benth., *Centranthera hispida* R. Br., and *Pedicularis zeylanica* Benth. Proc Natl Inst Sci India B 8: 249–261
Krishnamurthy KV (1988) Endosperm controls symmetry changes in the developing embryos of angiosperms. Proc Indian Acad Sci Plant Sci 98: 257–259
Krishnamurthy KV, Rao KA (1958) Abnormal pollen tube development in a *Nicotiana* hybrid. Curr Sci 27: 397–398
Kroh M, Helsper JPFG (1974) Transmitting tissue and pollen tube growth. In: Linskens HF (ed) Fertilization in higher plants. North-Holland, Amsterdam, pp 167–178
Kroh M, Munting A (1967) Pollen germination and pollen tube growth in *Diplotaxis tenuifolia* after cross-pollination. Acta Bot Neerl 16: 182–187
Kronestedt E, Bystedt P-A (1981) Thread-like formations in the anthers of *Strelitzia reginae*. Nord J Bot 1: 523–529

Kronestedt-Robards EC, Rowley JR (1989) Pollen grain development and tapetal changes in *Strelitzia reginae*. Am J Bot 76: 856–870

Krüger M (1932) Vergleichende entwicklungsgeschichtliche Untersuchungen an den Fruchtknoten und Früchten zweier *Solanum*-Chimären and ihrer Elternarten. Planta 17: 372–436

Krupko S (1962) Embryological and cytological investigations in *Hypodiscus aristatus* Nees (Restionaceae). J S Afr Bot 28: 21–44

Krupko S (1963) Macrosporogenesis and embryo sac development in *Chondropetalum hookerianum*. Acta Soc Bot Pol 32: 171–190

Krupko S (1973) Megasporogenesis and development of the embryo sac in Palestine variety of *Trifolium alexandrinum* L. Acta Soc Bot Pol 42: 617–625

Krupko S, Israelstam F, Martinovic B (1954) Embryo sac development and chromosome number in *Vanilla roscheri* from Inhaca Island. S Afr J Sci 51: 115–117

Krusheva RM, Belcheva RG (1976) Aspects of megagametogenesis in *Lilium regale* Wils. Tsitol Genet 10: 45–48 (in Russian)

Kubitzki K (1969) Monographie der Hernandiaceen. Bot Jahrb 89: 78–148

Kudlicka K, Kuras M, Rodkiewicz B (1981) Embryo sac development in *Stellaria media*. Acta Soc Bot Pol 50: 143–146

Kudryashov LV, Savich EI (1963) Some data on the embryology of *Alisma plantago-aquatica* L. Byul Moskow Obsh Isp Prirod Otdes Biol 68: 50–63 (in Russian)

Kühl R (1933) Vergleichende entwicklungsgeschichtliche Untersuchungen an der Insektivore *Nepenthes*. Beih Bot Zentralbl 51: 311–334

Kuijt J (1955) Dwarf mistletoes. Bot Rev 21: 569–626

Kuijt J (1967) The genus *Ixocactus* (Loranthaceae s.s.): description of its first species. Brittonia 19: 62–67

Kuijt J (1969) The biology of parasitic flowering plants. Univ Cal Press, Berkeley

Kuijt J, Weberling F (1972) The flower of *Phthirusa pyrifolia* (Loranthaceae). Ber Dtsch Bot Ges 85: 467–480

Kulkarni AR, Dnyansagar VR (1984) Embryology of *Eragrostis ciliensis* (All.) Link. J Indian Bot Soc 63: 206–213

Kulkarni AR, Patil VD (1969) On the embryology of *Cyanotis cuculata* Kunth. Curr Sci 38: 220–221

Kulkarni KM, Mahabale TS (1974) Studies on palms: embryology of *Livistonia chinensis* R. Br. Proc Indian Acad Sci B 80: 1–12

Kulkarni AR, Pataskar PG (1973–74) Gametogenesis in *Sesamum laciniatum* Kle in ex Willd. J Biol Sci (Bombay) 16/17: 86–91

Kulkarni AR, Dosi D, Manoj VM (1990) Fruit and seed structure in Araceae. Proc Indian Acad Sci B 100: 61–70

Kumar A (1976) Studies in Geraniales 7. The embryology of *Erodium stephanianum* Willd. Acta Bot Indica 4: 105–110

Kumar P, Singh D (1990) Development and structure of seed-coat in *Lavatera* L. and *Anoda* Cav. J Indian Bot Soc 69: 121–124

Kumar LSS (1942) False polyembryony in viviparous *Rhizophora mucronata*. Curr Sci 11: 242

Kumar S (1978) Embryological studies on Indian species of *Morus* 1. Microsporogenesis. Curr Sci 47: 807–808

Kumar S, Jain DK (1977) Microsporogenesis in *Gilia capitata* Sims. Curr Sci 46: 749–750

Kumari S (1960) Cytogenetic investigations in Paniceae: Occurrence of apospory in a diploid species of *Panicum-Panicum antidotale* Retz. Curr Sci 29: 191

Kumari VL, Narayana LL (1980) A contribution to the embryology of Oxalidaceae. Jpn J Bot 55: 171–179

Kumazawa M (1935) The structure and affinities of *Paeonia*. Bot Mag (Tokyo) 49: 306–315

Kumazawa M (1972) Structure and affinities of *Glaucidium* and its allied genera. Bot Mag (Tokyo) 44: 479–490

Kuprianova LA (1954) Sur la phylogénie des monocotylédons. Essais Bot 1: 120–127

Kuran H (1972) Callose localization in the walls of megasporocytes and megaspores in the course of development of monosporic embryo sacs. Acta Soc Bot Pol 41: 519–534

Kuras M (1978) Activation of embryo during rape (*Brassica napus* L.) seed germination. Acta Soc Bot Pol 47: 65–82

Kuriachen PM, Dave YS (1989) Structural studies in the fruits of Oleaceae with discussion on the systematic position of *Nyctanthes* L. Phytomorphology 39: 51–60
Kuroiwa H (1989) Ultrastructural examination of embryogenesis in *Crepis capillaris* (L.) Wallr. 1. The synergid before and after pollination. Bot Mag (Tokyo) 102: 9–24
Kurzweil H (1990) Floral morphology and ontogeny in Orchidaceae subtribe Disinae. Bot J Linn Soc 102: 61–83
Kuta E (1990) Biosystematic studies on the genus *Viola* L. Sect. *Plagiostigma* Godr. 2. Embryological analysis of *V. epipsila* Ledeb, *V. palustris* L. and their hybrids from Poland. Acta Biol Cracov Ser Bot 31: 45–62
Kuwada Y (1928) An occurrence of restitution nuclei in the formation of the embryo sacs in *Balanophora japonica* Mak. Bot Mag (Tokyo) 42: 117–129

Labrecque M, Barabe D (1984) Developpement du fruit de *Prunus serotina* (Rosaceae). Can J Bot 62: 195–206
Lagerberg T (1909) Studien über die Entwicklungsgeschichte und systematische Stellung von *Adoxa moschatellina* L. Kon Svenska Vet Akad Handl 44: 1–86
Lakshmanan KK (1961) Embryological studies in the Hydrocharitaceae 1. *Blyxa octandra* Planch. J Madras Univ B 31: 133–142
Lakshmanan KK (1962) Embryological features of *Pollia sorzogenesis* Endl. Curr Sci 31: 208
Lakshmanan KK (1963a) Embryological studies in the Hydrocharitaceae 2. *Halophila ovata* Goudich. J Indian Bot Soc 42: 15–18
Lakshmanan KK (1963b) Embryological studies in the Hydrocharitaceae 3. *Nechamandra alternifolia.* Phyton (Argentina) 20: 49–58
Lakshmanan KK (1965a) Note on the endosperm formation in *Zannichellia palustris* L. Phyton (Argentina) 22: 13–14
Lakshmanan KK (1965b) Embryological studies in Hydrocharitaceae 4. Post-fertilization development in *Hydrilla verticillata* Royle. Phyton (Argentina) 22: 45–50
Lakshmanan KK (1967) Embryological studies in the Bromeliaceae 1. *Lindmania penduliflora* (C.H. Wright) Stapf. Proc Indian Acad Sci B 65: 49–55
Lakshmanan KK (1968) Preliminary note on the embryology of *Diplanthera uninervis* Aschers. Curr Sci 37: 534–535
Lakshmanan KK (1970a) Hydrocharitaceae. In: Symp Comparative embryology of angiosperms. Indian Natl Sci Acad Bull 41: 336–341
Lakshmanan KK (1970b) Scheuchzeriaceae. In: Symp Comparative embryology of angiosperms. Indian Natl Sci Acad Bull 41: 342–343
Lakshmanan KK (1970c) Juncaginaceae. In: Symp Comparative embryology of angiosperms. Indian Natl Sci Acad Bull 41: 344–347
Lakshmanan KK (1970d) Potamogetonaceae. In: Symp Comparative embryology of angiosperms. Indian Natl Sci Acad Bull 41: 348–351
Lakshmanan KK (1970e) Zannichelliaceae. In: Symp Comparative embryology of angiosperms. Indian Natl Sci Acad Bull 41: 352–353
Lakshmanan KK (1970f) Najadaceae. In: Symp Comparative embryology of angiosperms. Indian Natl Sci Acad Bull 41: 354–357
Lakshmanan KK (1972) Monocot embryo. In: Varghese TM, Grover RK (eds) Vistas in plant sciences, vol 2. Int Bio-Sci Publ, Hissar (India), pp 61–110
Lakshmanan KK (1977) Studies on the development of *Commelina benghalensis* L. 1. Zygote to globular proembryo. Proc Indian Acad Sci B 86: 167–174
Lakshmanan KK, Ambegaokar KB (1984) Polyembryony. In: Johri BM (ed) Embryology of angiosperms. Springer, Berlin Heidelberg New York, pp 445–474
Lakshmanan KK, Dulcy EJ (1982) Tapetum in Chenopodiaceae. Beitr Biol Pflanzen 52: 469–480
Lakshmanan KK, Jayalakshmi R (1980) Preliminary observations on the embryology of *Spinifex littoreus* Merr. Curr Sci 49: 325–326
Lakshmanan KK, Philip VJ (1971) A contribution to the embryology of Iridaceae. Proc Indian Acad Sci B 73: 110–116
Lakshmanan KK, Rajeshwari M (1979) Sea-grasses of Krusadai Island in the Gulf of Mannar, India 2. *Syringodium isoetifolium* (Aschers) Dandy. Indian J Bot 2: 87–95

Lakshmi N (1979) A contribution to the embryology of *Sansevieria zeylanica* Willd. Curr Sci 48: 410–412

Lakshminarayana K (1978a) Embryology of *Jasminum pubescens* Willd. In: Proc 65th Indian Sci Congr, Ahmedabad, p 85 (Abstr)

Lakshminarayana K (1978b) Male and female gametophytes of *Sarcostemma brunonianum* W. & H. (Abstr.) In: Proc 65th Indian Sci Congr, Ahmedabad, pt 3: 79

Lakshminarayana K (1988) Male and female gametophytes of *Rauwolfia sumatrana* Jack. Curr Sci 59: 85–86

Lakshminarayana K (1989) Embryological studies in a few taxa of Asclepiadaceae. J Indian Bot Soc 68: 29–34

Lakshminarayana K, Maheswari Devi H (1985) Embryological studies in Gentianaceae. Proc Indian Acad Sci Plant Sci 95: 213–219

Lakshminarayana K, Suresh Babu M (1988) Developmental morphology of the micro- and megasporangium in *Cryptolepis buchananii* R. & S. Curr Sci 57: 852–854

Lam HJ (1948) Classification and the new morphology. Acta Biotheor 8: 108–154

Lamba LC (1974) Megasporogenesis and female gametophyte in *Rauwolfia serpentina* (L.) Benth. ex Kurz. Acta Bot Indica 2: 154–156

Lamba LC (1975) Structure and development of seed in *Brassica nigra* Koch. J Indian Bot Soc 54: 225–233

Lamba LC, Arora BB (1981a) Anatomical and morphological studies on field-ripe seeds of *Eruca sativa* Mill. Acta Bot Indica 9: 88–93

Lamba LC, Arora BB (1981b) Stomata on the pericarp of *Lepidium sativum* L. and *Raphanus sativus* var. *caudatus* L. J Indian Bot Soc 60: 69–72

Lamm R (1937) A contribution to the embryology of potato. Svensk Bot Tidskr 31: 217–220

Lampton RK (1957) Floral morphology in *Asimina triloba* 1. Development of ovule and embryo sac. Bull Torrey Bot Club 84: 151–156

Landes M (1946) Seed development in *Acalypha rhomboidea* and some other Euphorbiaceae. Am J Bot 33: 562–568

Lang FX (1901) Untersuchungen über Morphologie, Anatomie und Samenentwicklung von *Polypompholyx* und *Biblis gigantea*. Flora 88: 149–206

Langdon LM (1934) Embryogeny of *Carya* and *Juglans*: a comparative study. Bot Gaz 96: 93–117

Langlet O (1928) Einige Beobachtungen über die Zytologie der Berberidazeen. Svensk Bot Tidskr 22: 169–184

Lankosz-Mróz M (1984a) Embryological studies on two subspecies of *Tripleurospermum maritimum* s.l. Acta Biol Cracov Ser Bot 26: 19–32

Lankosz-Mróz M (1984b) Embryological studies on experimental hybrids in *Tripleurospermum* s.l. Acta Biol Cracov Ser Bot 26: 55–73

Lantis U (1912) Development of the microsporangia and microspores of *Abutilon theophrasti*. Bot Gaz 54: 330–335

Larston DA, Lewis CW (1962) Pollen wall development in *Parkinsonia aculeata*. Grana Palynol 3: 21–27

Lattoo CS (1974) Morphology and embryology of *Holarrhena antidysenterica* Wall. Bot Gaz 135: 173–180

Laurent V (1923) Zur Entwicklungsgeschichte von *Corytholoma cyclophyllum* Dus. n. sp. ined. Svensk Bot Tidskr 17: 164–174

Law SK, Yeung EC (1989) Embryology of *Calypso bulbosa*. 1. Ovule development. Am J Bot 76: 1668–1674

Lawalrée A (1945) La position systématique des Lemnaceae at leur classification. Bull Soc R Bot Belg 71: 27–38

Lawalrée A (1952) L'embryologie des Lemnaceae. Observations sur *Lemna minor*. Cellule 54: 305–326

Lawrence GHM (1951) Taxonomy of vascular plants. Macmillan, New York

Lazarte JE, Palser BF (1979) Morphology, vascular anatomy and embryology of pistillate and staminate flowers of *Asparagus officinalis*. Am J Bot 66: 753–764

Leandri J (1937) Sur l'aire et la position systématique du genre malgache *Didymeles* Thouars. Ann Sci Nat Bot 19: 309–317

Lebègue A (1948a) Embryogénie des Oenotheracées. Développement de l'embryon chez l'*Epilobium montanum*. Bull Soc Bot Fr 95: 173–178

Lebègue A (1948b) Embryologie des Crucifères. Polyembryonie chez l'*Arabis lyalli* S. Bull Soc Bot Fr 95: 250–252

Lebègue A (1948c) Embryogénie des Oenotheracées. Développement de l'embryon chez l'*Epilobium spicatum*. CR Acad Sci Paris 226: 429

Lebègue A (1949) Embryologie des Saxifragacées, Polyembryonie chez le *Bergenia delavayi* Engl. Bull Soc Bot Fr 96: 38–39

Lebègue A (1953) Embryogénie des Parnassiacées. Développement de l'embryon chez le *Parnessia palustris* L. CR Acad Sci Paris 236: 1693–1965

Lebègue A (1989) Reflections about an embryogenic system. In: Paré J, Bugnicourt M (eds) Some aspects and actual orientations in plant embryology. Dedicatory volume to Prof A. Lebègue. Univ Press Picardie, Amiens (Fr), pp 45–59

Le Cog C (1967) La mégasporogenèse chez le *Saxifraga cochlearis*, Rev Cytol Biol Veg 30: 207–313

Le Cog C, Brounland M, Guerven C (1978) Le problème de la régulation nucléaire au cours de l'ontogenèse de l'albumen, à la lumière des résultats obtenus chez les Renonculacées. Bull Soc Bot Fr 125: 261–265

Lecompte H (1896) Sur la formation du pollen chez Annonacées. Bull Mus Hist Nat Paris 2: 152–153

Lee JH, Cooper DC (1958) Seed development following hybridization between diploid *Solanum* species from Mexico, Central and South America. Am J Bot 45: 104–110

Lee DW, Fairbrothers DE (1972) Taxonomic placement of the Typhales within the monocotyledons: Preliminary serological investigation. Taxon 21: 39–44

Lee RE (1961) Pollen dimorphism in *Tripodandra grandiflora*. Baileya 9: 53–56

Leeman A (1927) Contribution à l'étude de l'*Asarum europaeum* avec une étude particulière sur le développement des cellules sécrétius. Bull Soc Bot Gen Paris 19: 92–173

Leins P, Bonnery-Brachten R (1977) Entwicklungsgeschichtliche Untersuchungen an Blüten von *Datisca cannabina*. Beitr Biol Pflanzen 53: 142–155

Leliveld JA (1935) Cytological studies in the genus *Ulmus* 2. The embryo sac and seed development in the common Dutch elm. Rec Trav Bot Neerl 32: 543–573

Lemesle R (1929) Embryologie des Elatinacées. Développement de l'embryon chez l'*Elatine alsinastrum*. CR Acad Sci Paris 18: 1569

Lemesle R (1946) Contribution à l'étude morphologique et phylogénétique des Eupteleacées, Cercidiphyllacées, Eucommiacées. Ann Sci Nat Bot 11 Ser 7: 41–52

Lemesle R (1948) Position phylogénétique de l'*Hydrastis canadensis* L. et du *Crossosoma californicum* Nutt., d'après les particularités histologiques du xyleme. CR Acad Sci Paris 227: 221–223

Lemesle R (1955) Contribution à l'étude de quelques familles de dicotylédones considérées comme primitives. Phytomorphology 5: 11–45

Leroy JF (1955) Étude sur les Juglandaceae. Mem Mus Hist Nat Belg 6: 1–246

Lersten NR (1971) A review of septate microsporangia in vascular plants. Iowa State J Sci 45: 487–497

Lersten NR (1982) Tracheid bar and vestured pits in legume seeds (Leguminosae, Papilionoideae). Am J Bot 69: 89–107

Lersten NR (1983) Suspensors in Leguminosae. Bot Rev 49: 233–257

Levan A (1936) Zytologische Studien an *Allium schoenoprasum*. Hereditas 22: 353–394

Levan A (1937) Eine erbliche Anomalie der Samenanlage bei *Petunia*. Bot Not 1937: 35–55

Lewis JF (1905) Notes on the development of *Phytolacca decandra*. John Hopkins Univ Circ 178: 38–42

Li HL (1952) A taxonomic review of the genus *Actinidia*. J Arnold Arbor 33: 1–61

Li HL (1955) Classification and phylogeny of Nymphaeaceae and allied families. Am Midl Nat 54: 33–41

Liang-Han-Xing (1985) The formation of pollen dimorphism in *Paris axialis* (in Chinese). Acta Bot Sin 27: 465–468
Lijefors A (1953) Studies on propagation, embryology and pollination in *Sorbus*. Acta Hortic Bergiani 16: 277–329
Lindau G (1895) Acanthaceae. In: Engler A, Prantl K (eds) Die natürlichen Pflanzenfamilien 4, 3B: 274–354
Lindsey AA (1938) Anatomical evidence for the Menyanthaceae. Am J Bot 25: 480–485
Linskens HF (1958) Physiologische Untersuchungen zur Reifeteilung, 2. Mitteilung. Über die Änderung des Nukleinsäuregehaltes während der Pollenmeiose und Pollenentwicklung von *Lilium henryi*. Acta Bot Neerl 7: 61–68
Linskens HF (1973) Activation of the ovary. Caryologia 25: 27–41
List A, Steward FC (1965) The nucellus, embryo sac, endosperm, and embryo of *Aesculus* and their interdependence during growth. Ann Bot (London) 29: 1–15
Livingston M (1950) Microsporogenesis in *Sarracenia minor* Walt. Q J Fla Acad Sci 13: 61–71
Lloyd FE (1902) The comparative embryology of the Rubiaceae. Mem Torrey Bot Club 8: 1–112
Locke JF (1936) Microsporogenesis and cytokinesis in *Asimina triloba*. Bot Gaz 98: 159–168
Lodkina MM (1977) A special pattern for cell differentiation in tapetal and in middle layer of pollen sacs in some aspects of Najadaceae and Lemnaceae (Abstr). In: Embryology of crop plants. Indo-Sov Symp, USSR Acad Sci, Moscow, and Indian Natl Sci Acad, New Delhi, at Komarov Bot Inst, Leningrad (USSR), p 48
Loesener T (1942) Celastraceae. In: Engler A, Prantl K (eds) Die natürlichen Pflanzenfamilien 20b: 87–197
Long RW (1965) Aberrant microgametogenesis and pollen tube growth in *Polycarpaea nebulosa*. Bull Torrey Bot Club 92: 46–50
Longevialle M (1970) Embryogénie des Caryophyllacées. Développement de l'embryon chez *Silene laeta* (Ait) A. Br., étude d'embryons en coupe et d'embryons extraits. Bull Soc Bot Fr 117: 367–371
Longevialle M (1971) Embryogénie des Caryophyllacées. Développement de l'embryon chez *Honkenya peploides* Ehrh. Etude d'embryons extraits. CR Acad Sci Paris 273: 2067–2070
Longly B (1984) Tétrades irrégulières, développements apoméiotiques et aposporie dans les ovules de *Cichorium intybus* L. (Chicorée de Bruxelles). Can J Bot 62: 635–646
Longly B, Waterkeyn L (1979) Etude de la cytokinèse 3. Les cloisonnements simultanes et successifs des microsporocytes. Cellule 73: 66–80
Longo B (1901) La mesogamia nella commune zucca (*Cucurbita pepo* L.). CR Accad Lincei (Rome) 10: 168–172
Longo B (1903) La nutrizione dell'embrione delle *Cucurbita operata* per mezzo del tubetto pollinico. Ann Bot (Rome) 1: 71–74
Longo B (1908) La poliembrionia nello *Xanthoxylum bungei* Planch. senza fecondazione. Bull Soc Bot Ital 1908: 113–115
Löve A, Löve D (1948) Chromosome numbers of northern plant species. Reykjavik
Lu A-m (1985) Embryology and possible relationships of *Eriospermum* (Eriospermaceae). Nord J Bot 5: 229–240
Lubliner OK (1925) Recherches sur le développement de l'ovule et de la graine dans le genre *Podophyllum*. Bull Int Acad Pol Cracov Sci Ser B 1925: 379–402
Lundqvist G (1915) Die Embryosackentwicklung von *Pedicularis sceptrum-carolinum*. Z Bot 7: 545–559
Luxemburg A (1927) Recherches cytologiques sur les grains de pollen chez les Malvacées. Bull Int Acad Pol Cracov Sci Ser B 1927: 363–396
Luxova M (1967) Fertilization of barley (*Hordeum distichum* L.). Biol Plant (Prague) 9: 301–307
Luza JG, Polito VS (1988) Microsporogenesis and anther differentiation in *Juglans regia* L.: a developmental basis for heterodichogamy in walnut. Bot Gaz 149: 30–36
Lyon FM (1898) A contribution to the life history of *Euphorbia corollata*. Bot Gaz 25: 418–426
Lyshade OB (1984) Seed structure and germination in *Cuscuta pedicellata* with some notes on *C. compestris*. Nord J Bot 4: 669–674
Ly-Thi-Ba (1960) Embryogénie des Rubiacées. Développement de l'embryon chez le *Galium aparine*. CR Acad Sci Paris 251: 1812–1814

Ly-Thi-Ba (1961) Embryogénie des Rubiacées. Développement de l'embryon chez le *Spermacoce tenuior*. CR Acad Sci Paris 252: 927–929
Ly-Thi-Ba (1962) Embryogénie des Calycanthacées. Développement de l'embryon chez le *Chimonanthus fragrans* Lindl. CR Acad Sci Paris 254: 1323–1325
Ly-Thi-Ba (1971) Embryologie végétale. Embryogénie des Renonculacées. Développement de l'embryon chez le *Cimicifuga europaea* Schipcz. CR Acad Sci Paris 272: 1248–1251
Ly-Thi-Ba (1974) Embryogénie comparée et phylogénie des Helleborées. Rev Gen Bot Veg 81: 151–191
Ly-Thi-Ba (1976) Embryogénie des Berberidacées. Développement de l'embryon chez le *Mahonia aquifolium* Nutt. CR Acad Sci Paris 282: 445–447
Ly-Thi-Ba, Guignard JL, Mestre JC (1970) Embryogénie des Magnoliacées. Développement de l'embryon chez *Magnolia grandiflora* L. CR Acad Sci Paris 271: 1272–1275

Macdonald AD, Mothersill DH (1987) Shoot development in *Betula papyrifera* 6. Development of the reproductive structures. Can J Bot 65: 466–475
Mackiewicz T (1973) Gametogenesis, embryo sac development and pollen grain morphology in *Brassica oleracea* vâr. *caritata* L. × *B. oleraceae* var. *acephala* DC. hybrid as compared with the parental forms. Genet Pol 14: 1–10
Madge M (1929) Spermatogenesis and fertilization in the cleistogamous flowers of *Viola odorata* var. *praecox* Gregory. Ann Bot (London) 43: 545–577
Madge M (1934) Nuclear migrations in *Hedychium*. Proc Linn Soc London 146: 108–109
Madhulata (1956) A contribution to the life history of *Begonia picta* Smith. (Abstr). In: Proc 43rd Indian Sci Congr, Agra, pt 3: 232
Magin N, Classen R, Clandia G (1989) The morphology of false anthers in *Craterostigma plantagineum* and *Torenia polygonoides* (Scrophulariaceae). Can J Bot 67: 1931–1937
Magnus W (1913) Die atypische Embryosackentwicklung der Podostemaceen. Flora 105: 275–336
Magtang MV (1936) Floral biology and morphology of the egg plant. Philipp Agric 25: 30–53
Mahabale TS (1966) The embryo in Palmae. In: Mahabale TS (ed) Proc Autumn School Bot, Univ Poona (India), pp 163–185
Mahabale TS, Biradar NV (1968) Studies in palms: embryology of *Phoenix sylvestris* Roxb. Proc Indian Acad Sci B 67: 77–96
Mahabale TS, Chennaveeraiah MS (1957) Studies on *Hyphaene indica* 1. Morphology. Phytomorphology 7: 184–194
Mahabale TS, Chennaveeraiah MS (1962) Cyto-embryology of *Dipcadi montanum* Baker. Proc Natl Inst Sci India B 27: 153–171
Mahabale TS, Inamdar AC (1975) Morphology and embryology of *Asparagus virgatus* Baker. Proc Indian Acad Sci B 81: 162–169
Mahabale TS, Solanky IN (1953a) Studies in the Chenopodiaceae 1. Embryology of *Suaeda fruticosa* Forsk. J Univ Bombay 21: 81–92
Mahabale TS, Solanky IN (1953b) Studies in the Chenopodiaceae 3. Embryology of *Kochia scoparia* Schrad. J Univ Bombay 22: 18–25
Mahabale TS, Solanky IN (1954a) Studies in the Chenopodiaceae 2. Embryology of *Arthrocnemum indicum* Moq. Proc Indian Acad Sci B 39: 212–222
Mahabale TS, Solanky IN (1954b) Studies in the Chenopodiaceae 4. Embryology of *Chenopodium ambrosioides* Linn. J Univ Bombay 22: 31–42
Mahabale TS, Solanky IN (1954c) Studies in the Chenopodiaceae 5. Embryology of *Chenopodium murale* Linn. J Univ Bombay 23: 25–37
Mahalingappa MS (1975) Anther and male gametophyte development in *Turnera ulmifolia* Linn. (var. *angustifolia* Willd.). Curr Sci 44: 640–641
Mahalingappa MS (1977) Gametophytes of *Eleusine compressa*. Phytomorphology 27: 231–239
Mahalingappa MS (1978) Microsporogenesis and male gametophyte in *Cynodon dactylon*. Curr Sci 47: 594–595
Maheshwari JK (1954) Floral morphology and embryology of *Lippia nodiflora* Rich. Phytomorphology 4: 217–230
Maheshwari P (1929) Contribution to the morphology of *Boerhaavia diffusa* L. J Indian Bot Soc 8: 219–234

Maheshwari P (1931) Contribution to the morphology of *Albizzia lebbek*. J Indian Bot Soc 10: 241–264
Maheshwari P (1933a) The development of the endosperm in *Asphodelus tenuifolius* Cav. Curr Sci 2: 13
Maheshwari P (1933b) A note on the life history of *Hydrilla verticillata* Press. Curr Sci 2: 13–14
Maheshwari P (1934a) The Indian mango. Curr Sci 3: 97–98
Maheshwari P (1934b) Contributions to the morphology of some Indian Liliaceae 1. The gametophytes of *Ophiopogon wallichianus* Hook. f. Proc Indian Acad Sci B 1: 197–204
Maheshwari P (1937) A critical review of the types of embryo sacs in angiosperms. New Phytol 36: 359–417
Maheshwari P (1941) Recent work on the types of embryo sacs in angiosperms. J Indian Bot Soc 20: 229–261
Maheshwari P (1942) The embryo sac of *Euphorbia heterophylla* L.: A reinvestigation. Proc Indian Acad Sci B 15: 158–166
Maheshwari P (1943) The mode of endosperm formation in *Vallisneria*, *Ottelia* and *Limnocharis*. Proc Natl Acad Sci India B 13: 260–263
Maheshwari P (1944a) The origin of the haustoria in the ovule of *Lobelia*. Curr Sci 13: 186–187
Maheshwari P (1944b) The origin of the haustoria in the ovule of *Lobelia*. J Indian Bot Soc 23: 79–81
Maheshwari P (1944c) The seed structure of *Ipomoea*: a criticism. Sci Cult 9: 557
Maheshwari P (1945) On the presence of an obturator in the ovule of *Phyllanthus niruri* L. Curr Sci 14: 132–133
Maheshwari P (1946) The Fritillaria type of embryo sac: a critical review. J Indian Bot Soc (MOP Iyengar Comm Vol): 101–109
Maheshwari P (1947) Tetranucleate embryo sacs in angiosperms. Lloydia 10: 1–18
Maheshwari P (1948) The angiosperm embryo sac. Bot Rev 14: 1–56
Maheshwari P (1950) An introduction to the embryology of angiosperms. McGraw-Hill, New York
Maheshwari P (1956) The embryology of angiosperms: a retrospect and prospect. Curr Sci 25: 106–110
Maheshwari P (ed) (1963a) Recent advances in the embryology of angiosperms. Int Soc Plant Morphol, Univ Delhi
Maheshwari P (1963b) History and present status of plant embryology. In: Maheshwari P (ed) Recent advances in the embryology of angiosperms. Int Soc Plant Morphol, Univ Delhi, pp 1–14
Maheshwari P (1964) Embryology in relation to taxonomy. In: Turrill WB (ed) Vistas in botany, vol 4. Pergamon, Oxford, pp 55–97
Maheshwari P, Chopra RN (1954) Polyembryony in *Opuntia dillenii* Haw. Curr Sci 23: 130–131
Maheshwari P, Chopra RN (1955) The structure and development of the ovule and seed of *Opuntia dillenii* Haw. Phytomorphology 5: 112–122
Maheshwari P, Chowdhry OR (1937) A note on the development of embryo sac in *Phyllanthus niruri* Linn. Curr Sci 5: 535–536
Maheshwari P, Gangulee H (1942) The development of embryo sac in *Heckeria umbellata* Kunth. J Indian Bot Soc 21: 245–248
Maheshwari P, Gupta BL (1934) The development of female gametophyte of *Ludwigia parviflora* Roxb. and *Jussieua repens* Linn. Curr Sci 3: 107–108
Maheshwari P, Haque A (1949) The embryo sac of *Chrysanthemum parthenium* L. (Bernh.). New Phytol 48: 255–258
Maheshwari P, Johri BM (1940) A note on the embryo sac of *Acalypha indica* L. Curr Sci 9: 322–323
Maheshwari P, Johri BM (1941) The embryo sac of *Acalypha indica* L. Beih Bot Zentralbl 61A: 125–136
Maheshwari P, Johri BM (1950a) Development of embryo sac, embryo and endosperm in *Helixanthera ligustrina* (Wall.) Dans. Nature (London) 165: 978–979
Maheshwari P, Johri BM (1950b) The occurrence of persistent pollen tubes in *Hydrilla*, *Ottelia* and *Boerhaavia*, together with a discussion of the possible significance of this phenomenon in the life history of angiosperms. J Indian Bot Soc 29: 47–51

Maheshwari P, Johri BM (1956) The morphology and embryology of *Floerkea proserpinacoides* Willd. with a discussion on the systematic position of the family Limnanthaceae. Bot Mag (Tokyo - Ogura Comm Vol) 69: 410–423

Maheshwari P, Kapil RN (1963) Fifty years of science in India: progress of botany. Indian Sci Congr Assoc, Calcutta, pp 82–106

Maheshwari P, Kapil RN (1966) Some Indian contributions to the embryology of angiosperms. Phytomorphology 16: 239–291

Maheshwari P, Khan R (1953) Development of the embryo sac, endosperm and embryo in *Isomeris arborea*: a reinvestigation. Phytomorphology 3: 446–459

Maheshwari P, Narayanaswami S (1950) Parthenogenetic development of the egg in *Spiranthes australis* Lindl. Curr Sci 19: 249–250

Maheshwari P, Narayanaswami S (1952) Embryological studies in *Spiranthes australis* Lindl. Bot J Linn Soc 53: 474–486

Maheshwari P, Navalakha HS (1941) A note on the embryology of *Scoparia dulcis* Linn. and *Angelonia grandiflora* C. Morr. Curr Sci 10: 297–298

Maheshwari P, Negi V (1955) The embryology of *Dipteracanthus patulus* (Jacq.) Nees. Phytomorphology 5: 456–472

Maheshwari P, Rangaswamy NS (1960) The embryology of cotton. Indian Cent Cotton Committ Monogr 1: 118–136

Maheshwari P, Roy SK (1951) The embryo sac of *Salix*. Phytomorphology 1: 70–72

Maheshwari P, Roy SK (1952) The embryo sac and embryo of *Tridax procumbens* L. Phytomorphology 2: 245–252

Maheshwari P, Sachar RC (1954a) A reinvestigation of the embryology of *Isomeris arborea* Nutt. Curr Sci 23: 61–63

Maheshwari P, Sachar RC (1954b) The endosperm nodules of *Capsella bursa-pastoris* Moench. Sci Cult 19: 413–414

Maheshwari P, Sachar RC (1955) So-called endosperm embryos in *Isomeris arborea* Nutt. Nature (London) 176: 470–471

Maheshwari P, Sachar RC (1963) Polyembryony. In: Maheshwari P (ed) Recent advances in the embryology of angiosperms. Int Soc Plant Morphol, Univ Delhi, pp 265–296

Maheshwari P, Singh B (Bahadur) (1952) Embryology of *Macrosolen cochinchinensis*. Bot Gaz 114: 20–32

Maheshwari P, Singh B (Balwant) (1943) Studies in the family Alismataceae 5. The embryology of *Machaerocarpus californicus* (Torr.) Small. Proc Natl Inst Sci India B 9: 311–322

Maheshwari P, Singh UB (1930) The development of female gametophyte of *Asphodelus tenuifolius*. J Indian Bot Soc 9: 31–39

Maheshwari P, Srinivasan AR (1944) A contribution to the embryology of *Rudbeckia bicolor* Nutt. New Phytol 43: 135–142

Maheshwari P, Johri BM, Dixit SN (1957) The floral morphology and embryology of the Loranthoideae. J Madras Univ Sec B 27: 121–136

Maheshwari SC (1954) The embryology of *Wolffia*. Phytomorphology 4: 355–365

Maheshwari SC (1955) The occurrence of bisporic embryo sacs in angiosperms: a critical review. Phytomorphology 5: 67–99

Maheshwari SC (1956a) Endosperm and seed of *Wolffia*. Nature (London) 178: 925–926

Maheshwari SC (1956b) The endosperm and embryo of *Lemna* and systematic position of the Lemnaceae. Phytomorphology 6: 51–55

Maheshwari SC (1958) *Spirodela polyrrhiza*: The link between the aroids and the duckweeds. Nature (London) 181: 1745–1746

Maheshwari SC, Baldev B (1958) A contribution to the morphology and embryology of *Commelina forskalaei* Wall. Phytomorphology 8: 277–298

Maheshwari SC, Kapil RN (1963a) Morphological and embryological studies on the Lemnaceae 1. The floral structure and gametophytes of *Lemna paucicostata*. Am J Bot 50: 677–686

Maheshwari SC, Kapil RN (1963b) Morphological and embryological studies on the Lemnaceae 2. The endosperm and embryo of *Lemna paucicostata*. Am J Bot 50: 907–914

Maheshwari SC, Kapil RN (1964) Morphological and embryological studies on the Lemnaceae 3. Seed and seedling of *Lemna paucicostata*. J Indian Bot Soc 43: 270–277

Maheshwari SC, Khanna PP (1956) The embryology of *Arisaema wallichianum* Hook. f. and the systematic position of the Araceae. Phytomorphology 6: 379–388

Maheshwari SC, Maheshwari N (1963) The female gametophyte, endosperm and embryo of *Spirodela polyrrhiza*. Beitr Biol Pflanzen 39: 179–188

Maheswari Devi H (1957) Embryological studies in Compositae 3. *Gerbera jamesoni* Bolus. Proc Indian Acad Sci B 46: 68–74

Maheswari Devi H (1962) Embryological studies in Gentianaceae: Gentianoideae and Menyanthoideae. Proc Indian Acad Sci B 56: 195–216

Maheswari Devi H (1963) Embryological studies in Compositae 4. Helianthae. Proc Indian Acad Sci B 58: 274–290

Maheswari Devi H (1964) Embryological studies in Asclepiadaceae. Proc Indian Acad Sci B 60: 53–65

Maheswari Devi H (1970) Embryology of *Rauvolfia serpentina* and *Catharanthus roseus*. Curr Sci 39: 376–377

Maheswari Devi H (1971) Embryology of Apocynaceae 1. Plumiereae. J Indian Bot Soc 50: 74–85

Maheswari Devi H (1972) Salvadoraceae: A study of its embryology and systematics. J Indian Bot Soc 51: 56–62

Maheswari Devi H (1975) Embryology of jasminums and its bearing on the composition of Oleaceae. Acta Bot Indica 3: 52–61

Maheswari Devi H, Raju PVSNGK (1990) An embryological approach to the taxonomical status of *Hedyotis* Linn. Proc Indian Acad Sci B 100: 51–60

Maheswari Devi H, Lakshminarayana K (1977a) Embryological studies in Gentianaceae. J Indian Bot Soc 56: 182–188

Maheswari Devi H, Lakshminarayana K (1977b) Embryology of *Oxystelma esculentum*. Phytomorphology 27: 59–67

Maheswari Devi H, Lakshminarayana K (1978) Life history of *Caralluma adseendens*. Acta Bot Indica 6: 177–184

Maheswari Devi H, Lakshminarayana K (1980) Embryological studies in Loganiaceae. J Indian Bot Soc 59: 370–375

Maheswari Devi H, Manorama K (1984) Embryology of two species of *Polygonum*. J Indian Bot Soc 63: 75–82

Maheswari Devi H, Manorama K (1985) Embryological studies in Polygonaceae. In: Govil CM, Kumar V (eds) Trends in plant research (Prof Y.S. Murty Comm Vol). Bishen Singh & Mahendra Pal Singh, Dehra Dun, pp 206–213

Maheswari Devi H, Naidu KC (1978) Embryological studies in the family Begoniaceae. J Indian Bot Soc 57 (Suppl): 36

Maheswari Devi H, Naidu KC (1982) Embryo development in two species of *Begonia*. Curr Sci 51: 616–617

Maheswari Devi H, Naidu KC (1983) Embryological studies in Cucurbitaceae 1. Melothricae. Jpn J Bot 58: 366–375

Maheswari Devi H, Naidu KC (1984a) Life history of *Bryonopsis lanciniosa*. Acta Bot Indica 12: 89–92

Maheswari Devi H, Naidu KC (1984b) Embryology of three species of *Melothria*: Cucurbitaceae 2. J Indian Bot Soc 63: 306–312

Maheswari Devi H, Pullaiah T (1975) Life history of *Basella rubra* Linn. and taxonomic status of the family Basellaceae. J Indian Bot Soc 54: 154–166

Maheswari Devi H, Pullaiah T (1976) Embryological investigations in the Melampodinae 1. *Melampodium divaricatum*. Phytomorphology 26: 77–86

Maheswari Devi H, Pullaiah T (1977) Embryological abnormalities in *Carthamus tinctorius* Linn. Acta Bot Indica 5: 8–15

Maheswari Devi H, Krishnam Raju PVSNG (1980) Embryology of two species of *Dentella* (*Dentella repens* and *Dentella serpyllifolia*). Proc Indian Acad Sci Plant Sci 89: 213–218

Maheswari Devi H, Rama Rao MS (1976) Embryology of *Suaeda maritima* (Linn.) Dumort. Curr Sci 45: 383–384

Maheswari Devi H, Rama Rao MS (1982) Embryological studies in the tribe Salicornieae: *Salicornia brachiata* and *S. fruticosa*. Acta Bot Indica 10: 261–266

Maheswari Devi H, Ranjalkar MM (1980) Developmental and structural anatomy of seeds and fruits in Heliantheae (Asteraceae). In: Periasamy K (ed) Symp Histochemistry, developmental and structural anatomy of angiosperms. Autonomous PG Centre, Univ Madras, Tiruchirapalli, pp 59–71

Maheswari Devi H, Santa Ram A (1985) Embryology of Apiaceae: *Pimpinella*. Phytomorphology 35: 183–188

Mahony KL (1935) Morphological and cytological studies on *Fagopyrum esculentum* 1. Am J Bot 22: 460–475

Maiti G, Banerji ML (1976) Exomorphic seed structure of the Himalayan species of *Swertia* Linn. (Gentianaceae). Proc Indian Acad Sci B 84: 231–237

Makde KH (1971) Gametophytes and fertilization of *Indigofera pulchella* Roxb. J Indian Bot Soc 50: 308–311

Makde KH (1981) Endothecium in Cyperaceae. Curr Sci 50: 98–99

Makde KH (1982) Pollen development in Cyperaceae. J Indian Bot Soc 61: 242–249

Makde KH (1983) Embryology of *Lipocarpha* with a brief discussion of its systematic position. J Indian Bot Soc 62: 36–43

Makde KH, Bhuskute SM (1984) Trends of specialization in endosperm of the Cyperaceae. Curr Sci 53: 504–506

Makde KH, Bhuskute MS (1987) Embryology of *Kyllinga monocephala* (Cyperaceae) and its systematic position. Plant Syst Evol 156: 143–150

Maldonado S (1989) Embryological studies in *Spigelia humboldtiana* (Loganiaceae). Phytomorphology 39: 299–310

Malecka J (1961) Studies in the mode of reproduction of the diploid endemic species *Taraxacum pieninicum*. Acta Biol Cracov Ser Bot 4: 25–42

Malecka J (1964) Multinucleate pollen grains in the genus *Taraxacum serotinum*. Acta Biol Cracov Ser Bot 7: 107–116

Malecka J (1981) Cytological processes during the differentiation of the elements of embryo sac in *Cerinthe minor* L. Acta Biol Cracov Ser Bot 23: 55–67

Malecka J (1982) Further embryological studies in the genus *Taraxacum* L. Acta Biol Cracov Ser Bot 24: 143–157

Malecka J (1990) Studies on the genus *Solidago* L. 4. Cytoembryology of *Solidago canadensis* L. var. *scabra*. Acta Biol Cracov Ser Bot 31: 85–96

Malik CP, Vermani S (1975) Histochemical characteristics of suspensor during embryo development in *Brassica campestris* var. *sarson* (Abstr). In: All-India Symp Form, structure and function in plants. Sardar Patel Univ, Vallabh Vidyanagar (Gujarat, India), pp 16–17

Mandryk Vyu (1972) Embryological study of *Ribes nigrum* L. (Development of the female generative sphere, fertilization and development of the endosperm and embryo). Bot Zh 57: 234–244 (in Russian)

Mandryk Vyu, Andrusyak MA (1969) Cytoembryological study of microsporogenesis in gooseberry (*Grossularia reclinata* Mill.) and alpine currant (*Ribes alpinum* L.) Bot Zh 54: 1388–1396 (in Russian)

Mandryk Vyu, Petrus YY (1971) Differentiation of the ovule, development of female gametophyte and embryogenesis in *Rubus idaeus* L. and *Rubus caesius* L. (in Russian). Ukr Bot Zh 28: 469–476

Maneval WE (1914) The development of *Magnolia* and *Leriodendron*, including a discussion of the primitiveness of the Magnoliaceae. Bot Gaz 57: 1–31

Mangaly JK, Nayar J (1990) Palynology of South Indian Zingiberaceae. Bot J Linn Soc 103: 351–366

Mangaly JK, Swarrupanandan K (1977) Some aspects of the morphology of the ovule and seed of *Costus malortieanus* (Zingiberaceae). Proc Indian Acad Sci B 86: 175–79

Manshard E (1936) Embryologische Untersuchungen an *Styrax obassia*. Planta 25: 364–383

Mares DJ, Norstog K, Stone BA (1975) Early stages in development of wheat endosperm. Aust J Bot 23: 311–326

Markgraf Fr (1936) Blütenbau und Verwandtschaft bei den einfachsten Helobiae. Ber Dtsch Bot Ges 54: 191–229

Marloth R (1903) Some recent observations on the biology of *Roridula*. Ann Bot (London) 17: 151–157

Marloth R (1925) The Flora of South Africa, vol II. London.

Marshall DR, Brown AHD (1974) Estimation of the level of apomixis in plant populations. Heredity 32: 321–333
Marshall DR, Brown AHD (1979) Maintenance of genetic variation in apomictic plant populations. Heredity 42: 159–172
Marshall DR, Brown AHD (1981) The evolution of apomixis. Heredity 47: 1–15
Martin AC (1946) The comparative internal morphology of seeds. Am Midl Nat 36: 513–660
Martin ARH, Herscovitch JC (1989) Fertilizing pollen tubes parasitize ovary wall in *Grevillea*. Ann Bot (London) 64: 163–166
Martin JN (1914) Comparative morphology of some Leguminosae. Bot Gaz 58: 154–167
Martin PG, Peacock WJ (1959) Pollen tetrad patterns in *Leschenaultia*. Proc Linn Soc NSW 87: 388–396
Martin RW, Smith FH (1955) Megagametophyte development in *Chrysanthemum leucanthemum* L. Bot Gaz 116: 243–249
Martinoli GO (1967) Position and significance of the Pyrethrum cinerariaefolium (Chrysanthemum cinerariaefolium) type among the tetrasporic types of female gametophyte. Phytomorphology 17: 207–210
Martinson VA (1972a) Polyembryony in *Theobroma cacao* L. Ann Bot (London) 36: 947–951
Martinson VA (1972b) Embryological studies on hybridization between *Theobroma cacao* and *Theobroma grandiflora*. Can J Bot 50: 2117–2124
Martyni E (1939) Reserche embryologiche sulle Martyniaceae. Nuovo G Bot Ital 46: 197–226
Masand P (1963a) Embryology of *Nelsonia campestris*. Phytomorphology 13: 82–91
Masand P (1963b) Embryology of *Zygophyllum fabago* Linn. Phytomorphology 13: 293–302
Masand P (1970) Zygophyllaceae. In: Symp Comparative embryology of angiosperms. Indian Natl Sci Acad Bull 41: 123–126
Masand P, Kapil RN (1966) Nutrition of the embryo sac and embryo: A morphological approach. Phytomorphology 16: 158–175
Mason CT (1950) Development of the embryo sac in the genus *Limnanthes*. Am J Bot 38: 17–22
Mason HL (1938) The flowering of *Wolffiella lingulata*. Madrono 4: 241–251
Mathew CJ (1978) Development of male and female gametophytes in *Camellia sasanqua*. Phytomorphology 28: 262–269
Mathew CJ (1980) Embryological studies in Hamamelidaceae: Development of female gametophyte and embryogeny in *Hamamelis virginiana*. Phytomorphology 30: 172–179
Mathew CJ, Chaphekar M (1977) Development of female gametophyte and embryogeny in *Stachyurus chinensis*. Phytomorphology 27: 68–78
Mathew PM (1958) Studies on the Menispermaceae. Proc Indian Acad Sci B 47: 274–286
Mathur G, Mohan Ram HY (1978) Significance of petal colour in thrips-pollinated *Lantana camara* L. Ann Bot (London) 42: 1473–1476
Mathur G, Mohan Ram HY (1986) Floral biology and pollination of *Lantana camara*. Phytomorphology 36: 79–100
Mathur KL (1934) A note on the presence of parietal cells in the nucellus of *Convolvulus arvensis*. Curr Sci 3: 160–161
Mathur KL (1940) A note on the development of the embryo sac of *Vogelia indica* Lamk. Curr Sci 9: 180–182
Mathur KL, Khan R (1941) The development of the embryo sac in *Vogelia indica* Lamk. Proc Indian Acad Sci B 13: 360–368
Mathur N (1956) The embryology of *Limnanthes*. Phytomorphology 6: 41–51
Mathur SL (1968) Development of the male and female gametophytes of *Calycanthus fertilis* Walt. Proc Natl Inst Sci India B 34: 323–329
Mathur S (1977) Embryology of *Melianthus major* L. 1. Microsporangium and pollen. Curr Sci 46: 820–821
Mathur S (1980) Embryology of *Allophylus cobbe* Blume. Microsporogenesis, pollen and male gametophyte. Proc 67 Indian Sci Congr, Calcutta, pt 3, p 38
Mathur S, Gulati N (1979) Embryology of *Melianthus major* L. 2. Megasporangium and female gametophyte. Proc 66th Indian Sci Congr, Hyderabad, pt 3: 69–70
Matthiessen A (1962) A contribution to the embryology of *Paeonia*. Acta Hortic Bergiani 20: 57–61

Maugini E (1953) Ricerche cito-embriologiche su *Piper medium* var. *ceanothifolium*. Caryologia 5: 282–287

Maugini E (1962) Flower morphology, embryology and embryogenesis in *Leontopodium alpinum* Cass var. *typicum* Fiori et Paeoletti. (in Italian). Nuovo G Bot Ital 69: 1–18

Mauritzon J (1930) Beitrag zur Embryologie der Crassulaceen. Bot Not 1930: 233–250

Mauritzon J (1933a) Studien über die Embryologie der Familien Crassulaceae und Saxifragaceae. Diss, Univ Lund

Mauritzon J (1933b) Über die systematische Stellung der Familien Hydrostachyaceae und Podostemonaceae. Bot Not 1933: 172–180

Mauritzon J (1934a) Ein Beitrag zur Embryologie der Phytolaccaceen und Cactaceen. Bot Not 1934: 111–135

Mauritzon J (1934b) Die Embryologie einiger Capparidaceen sowie von *Tovaria pendula*. Ark Bot 26A: 1–14

Mauritzon J (1934c) Zur Embryologie der Elaeocarpaceae. Ark Bot 26A: 1–8

Mauritzon J (1934d) Zur Embryologie einiger Gruinales. Svensk Bot Tidskr 28: 84–102

Mauritzon J (1934e) Zur Embryologie einiger Lythraceen. Acta Hortic Göteborg 9: 1–21

Mauritzon J (1934f) Die Endosperm- und Embryoentwicklung einiger Acanthaceen. Acta Univ Lund 30: 1–41

Mauritzon J (1935a) Über die Embryologie der Familie Rutaceae. Svensk Bot Tidskr 29: 319–347

Mauritzon J (1935b) Embryologische Angaben über Theophrastaceen. Ark Bot 28B: 1–4

Mauritzon J (1935c) Etwas über die Embryologie der Bignoniaceen. Bot Not 1935: 60–77

Mauritzon J (1936a) Zur Embryologie einiger Parietales-Familien. Svensk Bot Tidskr 30: 79–113

Mauritzon J (1936b) Zur Embryologie und systematischen Abgrenzung der Reihen Terebinthales und Celastrales. Bot Not 1936: 161–212

Mauritzon J (1936c) Embryologische Angaben über Stackhousiaceae, Hippocrateaceae und Icacinaceae. Svensk Bot Tidskr 30: 541–550

Mauritzon J (1936d) Die Endospermentwicklung von *Sesamum indicum* und *orientale*. Ark Bot 28B: 1–6

Mauritzon J (1936e) Samenbau und Embryologie einiger Scitamineen. Acta Univ Lund 30: 1–31

Mauritzon J (1939a) Über die Embryologie von *Marcgravia*. Bot Not 1939: 249–255

Mauritzon J (1939b) Contribution to the embryology of the orders Rosales and Myrtales. Acta Univ Lund 35: 1–121

Maury G (1980) Intérêt systématique et phylogénique des caractères juvéniles (germination, embryon, mûr, plantules) chez les Dipterocarpacees. Bull Soc Bot Fr 126: 13–21

Maximowicz CJ (1872) Diagnoses breves plantarum novarum Japoniae et Mandshriae. Bull Acad Imp Sci St Petersburg 17: 142–179

Mayer SJ, Snow AA, Stanton ML (1986) Fertilization dynamics and parental effects upon fruit development in *Raphanus raphanistrum*: Consequences for seed size variation. Am J Bot 73: 500–511

Mayo SJ (1989) Observations of gynoecial structure in *Philodendron* (Araceae). Bot J Linn Soc 100: 139–172

Maze J, Bohm LR (1973) Comparative embryology of *Stipa elmerea* (Gramineae). Can J Bot 51: 235–247

Maze J, Bohm LR (1974) Embryology of *Agrostis interrupta* (Gramineae). Can J Bot 52: 365–379

Maze J, Bohm LR (1977) Embryology of *Festuca microstachys*. Can J Bot 55: 1768–1782

Maze J, Lin SCh (1975) A study of the mature megagametophyte of *Stipa elmerea*. Can J Bot 53: 2958–2977

Maze J, Bohm LR, Mehlenbecher Jr LE (1970) Embryo sac and early ovule development in *Oryzopsis miliacea* and *Stipa tortilis*. Can J Bot 48: 27–41

Mazzolani G (1978) Embriologia vegetale: ricordo della prof. ssa Valeria Bambacioni-Mezzetti. Ann Bot (Rome) 37: 65–102

McAllister F (1914) The development of the embryo sac in Convallariaceae. Bot Gaz 58: 137–153

McClintock B (1977) The clonal basis of development. In: Subtelny S, Sussex IM (eds) Symp Soc Dev Biol 36: 217–246

McCollum RL (1939) The development of the embryo sac and seed of *Commelina angustifolia* Michx. Bull Torrey Bot Club 66: 539–545

McCoy RW (1949) On the embryology of *Swertia carolinensis*. Bull Torrey Bot Club 76: 430–439
McDonald CC (1927) The study of seed development in three species of *Erigeron* with special reference to apogamy. Bull Torrey Bot Club 54: 479–496
McGahan MW (1961) Studies on the seed of Banana 1. Anatomy of the seed and embryo of *Musa balbisiana*. Am J Bot 48: 230–238
McGlone MS (1978) Pollen structure of the New Zealand members of the Styphelieae (Epacridaceae). NZ J Bot 16: 91–101
McKay JW (1947) Embryology of pecan. J Agric Res 74: 263–283
McNaughton JE, Robertson BE (1974) Megasporogenesis and megagametogenesis in *Aloe africana* Mill. J S Afr Bot 40: 75–79
Meerow AW, Dehgan B (1988) Pollen morphology of the Euchariдeae (Amaryllidaceae). Am J Bot 75: 1857–1870
Meeuse ADJ (1975) Floral evolution in the Hamamelididae 3. Hamamelidales and associated groups including Urticales, and final conclusions. Acta Bot Neerl 24: 181–191
Mehlenbacher Jr LE (1970) Floret development, embryology and systematic position of *Oryzopsis hendersoni* (Gramineae). Can J Bot 48: 1741–1758
Mehra KR, Kulkarni AR (1985) Embryological studies in Bignoniaceae. Phytomorphology 35: 239–252
Meier KI (1960) The embryology of *Nuphar luteum* Sm. Byu Moskow Obsh Isp Prirod Otdes Biol 65: 48–59 (in Russian)
Melchior H (1964) A. Engler's Syllabus der Pflanzenfamilien, vol 2. Bornträger, Berlin
Melikian AP (1973) Seed-coat types of Hamamelidaceae and allied families in relation to their systematics. Bot Zh 58: 350–359 (in Russian)
Mendes AJT (1941) Cytological observations in *Coffea* 6. Embryo and endosperm development in *Coffea arabica*. Am J Bot 28: 784–789
Mendoza R (1973) Embryogenesis in *Amaranthus spinosus* L. and *Amaranthus viridis* L. Philipp J Sci 103: 243–268
Mengeha MH, Guard AT (1966) Development of the embryo sac and embryo of Teff, *Eragrostis* Tef. Can J Bot 44: 1071–1075
Mepham RH, Lane GR (1969) Formation and development of the tapetal periplasmodium in *Tradescantia bracteata*. Protoplasma 68: 175–192
Mercy ST, Kakkar SN, Varghese TM (1974) Embryology of *Cicer arietinum* and *C. soongaricum*. Bull Torrey Bot Club 101: 26–30
Merrell WD (1900) A contribution to the life history of *Silphium*. Bot Gaz 29: 99–133
Merry J (1941) Studies on the embryo of *Hordeum sativum* 1. The development of the embryo. Bull Torrey Bot Club 68: 585–598
Messeri A (1928) Embriologia di *Grevillea macrostachya*. Nuovo G Bot Ital 34: 1037–1042
Messeri A (1952) Alcuni dati sulla embriologia ed embriogenesi di *Olea europea* L. Nuovo G Bot Ital 57: 149–169
Mestre JC (1964) Embryogénie des Amentacées. Développement de l'embryon chez l'*Alnus glutinosa*. CR Acad Sci Paris 258: 5949–5951
Mestre JC (1967) La signification phylogénétique de l'embryogénie. Rev Gen Bot Veg 74: 273–322
Mestre JC, Guignard JL (1969) Embryogénie des Scrofulariacées. Le développement de l'embryon chez le *Zaluzianskya capensis* Walp. CR Acad Sci Paris 268: 3170–3172
Mestre JC, Guignard JL (1970) Embryogénie des Cornacées. Développement de l'embryon chez l'*Aucuba japonica* Thunb. CR Acad Sci Paris 271: 1358–1360
Mestre JC, Guignard JL (1971) Embryogénie des Umbellifères. Développement de l'embryon chez le *Daucus carota* L. CR Acad Sci Paris 272: 1383–1386
Metcalfe CR (1936) An interpretation of the morphology of the single cotyledon of *Ranunculus ficaria*, based on embryology and seedling anatomy. Ann Bot (London) 50: 103–120
Metcalfe CR (1961) The anatomical approach to systematics. Recent Adv Bot, Univ Toronto Press 2: 146–150
Metcalfe CR, Chalk L (1950) Anatomy of the dicotyledons, vols 1, 2. Clarendon, Oxford
Meyer S (1938) Studies in the family Apocynaceae. J Dep Sci Cal Univ 1: 131–158

Mianny N (1972) Embryologie des Crucifères. Polyembryonie chez l'*Arabis hirsuta* Scop. et l'*Arabis halleri* L. Bull Soc Bot Fr 118: 341–344

Michell RM (1915) The embryo sac and embryo of *Striga lutea.* Bot Gaz 59: 124–135

Miers J (1872) On the Hippocrateaceae of South America. Trans Linn Soc London 28: 328–432

Mihov A, Zagorcheva L (1966) A case of apogamy in *Cucumis sativus* L. CR Acad Bulg Sci 19: 527–529

Mikatadze TA (1975) Megasporogenesis and development of the female gametophyte in some forms of *Thea sinensis* L. Sobsch Akd Nauk Gruz SSR 78: 677–680 (in Russian)

Miller HA, Wetmore RH (1945) Studies in the developmental anatomy of *Phlox drummondii* 1. The embryo. Am J Bot 32: 588–599

Millsaps V (1936) The structure and development of the seeds of *Paulownia tomentosa.* J Elisha Mitchell Sci Soc 52: 56–75

Millsaps V (1940) Structure and development of the seed of *Cynoglossum amabile* Stapf & Drumm. J Elisha Mitchell Sci Soc 56: 140–164

Mirick S, Guinn J (1981) Some observations on the reproductive biology of *Gaultheria procumbens* (Ericaceae). Am J Bot 68: 1298–1305

Mirkamilov MA (1969) Development of female gametophyte in *Allium aflatunense* B. Fedtsch. Uzb Biol Zh 13: 33–36 (in Russian)

Mirkamilov MA (1975) Structure of the ovule in species of the genus *Allium* (Section Mollium Don.). Uzb Biol Zh 19: 75–76 (in Russian)

Misra KC (1937) The antipodals of Verbenaceae. Curr Sci 6: 98–99

Misra KC (1939) A contribution to the embryology of the Verbenaceae. Proc Indian Acad Sci B 9: 49–56

Misra RC (1962) Contribution to the embryology of *Arabidopsis thalianum* Gay & Monn. Agra Univ J Res Sci 11: 191–199

Misra RC (1964a) Ovule in *Plantago.* Curr Sci 33: 438–439

Misra RC (1964b) Development and structure of angiosperm seed 3. *Plantago* L. Natl Bot Garden Lucknow (India) Bull 105: 1–14

Misra RC (1965) Development of female gametophyte and endosperm in *Plantago coronopus* L. Agra Univ J Res Sci 14: 9–13

Misra RC (1966) Morphological studies on *Sisymbrium irio* Linn. J Indian Bot Soc 45: 14–23

Misra RC (1970) Plantaginaceae. In: Symp Comparative embryology of angiosperms. Indian Natl Sci Acad Bull 41: 298–305

Misra RC, Sahu RC (1970) Embryology and seed structure in green grain (*Phaseolus aureus* Roxb.). Indian J Agric Sci 40: 216–222

Misra S (1957) Floral morphology of the family Compositae 1. The flower and the gametophytes of *Flaveria repanda* Lag. J Indian Bot Soc 36: 503–512

Misra S (1964) Floral morphology of the family Compositae 2. Development of the seed and fruit in *Flaveria repanda.* Bot Mag (Tokyo) 77: 290–296

Misra S (1965) Floral morphology of the family Compositae 3. Embryology of *Siegesbeckia orientalis* L. Aust J Bot 13: 1–10

Misra S (1972) Floral morphology of the family Compositae 5. The seed-coat and pericarp in *Verbesina encelioides* (Cav.) Benth. & Hook. f. ex A. Gray. J Indian Bot Soc 51: 332–341

Mitchell JP (1975) Megasporogenesis and microsporogenesis in *Vicia faba.* Can J Bot 53: 2804–2812

Mitra JN (1947a) A contribution to the life history of *Hydrolea zeylanica* Val. J Indian Bot Soc 26: 51–61

Mitra JN (1947b) A contribution to the embryology of some Compositae. J Indian Bot Soc 26: 105–123

Mitra JN, Datta RM (1947) A note on the development of the embryo sac and endosperm in *Alangium begonifolium* Roxb. (= *Marlea begoniaefolia* Roxb.). Sci Cult 12: 451–452

Mitra JN, Datta RM (1949) A note on the development of embryo sac and endosperm in *Alangium chinensis* (Lour.) Harms. Bull Bot Soc Bengal 3: 11–15

Miyaji Y (1927) Über die somatischen Chromosome einiger Renunculaceen. Bot Mag (Tokyo) 41: 568–569

Mlodzianowski F (1964) The structure and the later stages of development in the embryo sac of *Thamnochortus fruticosus* Berg. (Restionaceae). Bull Soc Amis Sci Leh (Poznan) D4: 3–11
Modilewski J (1908a) Zur Embryobildung von *Gunnera chilensis*. Ber Dtsch Bot Ges 26: 550–556
Modilewski J (1908b) Zur Samenentwicklung einiger Urticifloren. Flora 98: 423–470
Modilewski J (1909a) Zur Embryobildung von *Euphorbia procera*. Ber Dtsch Bot Ges 27: 21–26
Modilewski J (1909b) Zur Embryobildung einiger Onagraceen. Ber Dtsch Bot Ges 27: 287–292
Modilewski J (1910) Weitere Beiträge zur Embryobildung einiger Euphorbiaceen. Ber Dtsch Bot Ges 28: 413–418
Modilewski J (1911) Über die anormale Embryosackbildung bei *Euphorbia palustris* L. und anderen Euphorbiaceen. Ber Dtsch Bot Ges 29: 430–436
Modilewski J (1918) Cytological and embryological studies of *Neottia nidus-avis* (L.) Rich. Verh Kiev Ges Nat 26: 1–55 (in Russian)
Modilewski J (1925) Zur Kenntnis der Polyembryonie von *Allium odorum* L. Bull Jardin Bot Kiev 2: 9–19
Mogensen HL (1970a) Syncotyly in *Quercus arizonica*. Am J Bot 57: 1207–1210
Mogensen HL (1970b) Megagametophyte development in *Agave parryi*. Phytomorphology 20: 16–22
Mogensen HL (1978a) Synergids of *Proboscidea louisianica* (Martyniaceae) before fertilization. Phytomorphology 28: 114–122
Mogensen HL (1978b) Pollen tube-synergid interactions in *Proboscidea louisianica* (Martyniaceae). Am J Bot 65: 953–964
Mogensen HL (1978c) The effect of the pollen tube on synergid ultrastructure before and after penetration in *Proboscidea louisianica* (Mill.) Thall. Soc Bot Fr Acta Bot 1–2: 237–241
Mogensen HL (1982) Double fertilization in barley and the cytological exploration for haploid embryo formation, embryoless caryopsis and ovule abortion. Carlsberg Res Commun 47: 313–354
Mogensen HL (1984) Quantitative observations on the pattern of synergid degeneration in barley. Am J Bot 71: 1448–1451
Mohan K (1966) The ovule and embryo sac development in *Browallia demissa* Linn. Proc Indian Acad Sci B 64: 26–31
Mohan K (1968) Morphological studies in Solanaceae 2. Morphology, development, and structure of seed of *Browallia demissa* Linn. Proc Natl Inst Sci India B 34: 142–148
Mohan K (1970) Morphological studies in Solanaceae 6. The development of ovule and embryo sac in *Solanum khasianum* Clarke. Agra Univ J Res Sci 19: 39–44
Mohan K, Singh B (1969) Morphological studies in Solanaceae 3. Structure and development of seed of *Solanum torvum* Sw. J Indian Bot Soc 48: 338–345
Mohan Ram HY (1958) Post-fertilization studies in the ovules of some Acanthaceae. PhD Thesis, Univ Delhi
Mohan Ram HY (1959) Endosperm in *Eranthemum nervosum*. Curr Sci 28: 169–170
Mohan Ram HY (1960a) Post-fertilization studies in the ovule of *Ruellia tuberosa* Linn. Lloydia 23: 21–27
Mohan Ram HY (1960b) The development of the seed in *Andrographis serpyllifolia*. Am J Bot 47: 215–219
Mohan Ram HY (1960c) Seed development in *Adhatoda vasica* Nees. Can J Bot 39: 207–214
Mohan Ram HY (1962) Post-fertilization development of the ovule in *Barleria cristata* Linn. J Indian Bot Soc 41: 288–296
Mohan Ram HY, Hari Gopal B (1981) Some observations on the flowering of bamboos in Mizoram. Curr Sci 50: 708–710
Mohan Ram HY, Kamini I (1964) Embryology and fruit development in *Withania somnifera* Dunal. Phytomorphology 14: 574–587
Mohan Ram HY, Masand P (1962) Endosperm and seed development in *Andrographis echioides* Nees. Curr Sci 31: 7–8
Mohan Ram HY, Masand P (1963) The embryology of *Nelsonia campestris* R. Br. Phytomorphology 13: 82–91
Mohan Ram HY, Nath R (1964) The morphology and embryology of *Cannabis sativa* Linn. Phytomorphology 14: 414–429

Mohan Ram HY, Sehgal P (1958) The life history of *Justicia simplex* Don. Phytomorphology 8: 124–136
Mohan Ram HY, Wadhi M (1964) Endosperm in Acanthaceae. Phytomorphology 14: 388–413
Mohan Ram HY, Wadhi M (1965) Embryology and delimitation of the Acanthaceae. Phytomorphology 15: 201–205
Mohana Rao PR (1963) Suspensor polyembryony in *Garrya veatchii* Kell. Curr Sci 32: 468–469
Mohana Rao PR (1972a) Embryology of *Nyssa sylvatica* and systematic consideration of the family Nyssaceae. Phytomorphology 22: 8–21
Mohana Rao PR (1972b) Morphology and embryology of *Tieghemopanax sambucifolius* with comments on the affinities of the family Araliaceae. Phytomorphology 22: 75–87
Mohana Rao PR (1974a) Seed anatomy in some Hamamelidaceae and phylogeny. Phytomorphology 24: 113–139
Mohana Rao PR (1974b) Nutmeg seed: Its morphology and developmental anatomy. Phytomorphology 24: 262–273
Mohana Rao PR (1975a) Seed anatomy in *Michelia champaca* Linn. with comments on the primitiveness of the magnoliaceous seed. Phytomorphology 25: 81–90
Mohana Rao PR (1975b) Seed anatomy of *Artabotrys odoratissimus* with a discussion on chalaza, integumentary bundles, and ruminate endosperm. Phytomorphology 25: 215–227
Mohana Rao PR (1976) Seed and fruit anatomy of *Pterospermum acerifolium* (Sterculiaceae). Phytomorphology 26: 363–369
Mohana Rao PR (1979) Seed and fruit anatomy in *Polyalthia longifolia* (Annonaceae). Bot Jahrb 101: 83–89
Mohana Rao PR (1980) Seed and fruit anatomy of *Cassytha filiformis* with comments on its systematic position. Isr J Bot 28: 44–54
Mohana Rao PR (1981) Seed and fruit anatomy of *Trochodendron aralioides*. Phytomorphology 31: 18–23
Mohana Rao PR (1982) Seed and fruit anatomy in *Asimina triloba*, with a discussion on the affinities of Annonaceae. Bot Jahrb 103: 47–57
Mohana Rao PR (1983) Seed and fruit anatomy in *Eupomatia laurina* with a discussion on the affinities of Eupomatiaceae. Flora 173: 311–319
Mohana Rao PR (1986) Seed and fruit anatomy in *Gyrocarpus americanus* with a discussion on the affinities of Hernandiaceae. Isr J Bot 35: 133–152
Mohana Rao PR, Kaur A (1979a) Embryology and systematic position of *Ophiopogon intermedius*. Proc Indian Natl Sci Acad B 45: 115–187
Mohana Rao PR, Kaur A (1979b) Sporogenesis and gametophytes of *Polygonatum cirrhifolium*. Phytomorphology 29: 93–97
Mohana Rao PR, Rao KM (1983) Embryology of *Liparis viridiflora*. Acta Bot Indica 11: 228–234
Mohana Rao PR, Rao KM (1984) Embryology of *Habenaria pectinata*. Phytomorphology 34: 237–242
Mohana Rao PR, Sood SK (1979a) Life history of *Satyrium nepalense* (Orchidaceae). Norw J Bot 26: 285–295
Mohana Rao PR, Sood SK (1979b) Development of the embryo in *Herminium angustifolium* Benth. Ann Bot (London) 43: 661–663
Mohana Rao PR, Sood SK (1979c) Embryology of *Habenaria densa* (Orchidaceae). Bot Not 132: 145–148
Mohana Rao PR, Sood SK (1986) Embryology of *Cephalanthera ensifolia* (Orchidaceae). Acta Bot Indica 14: 38–44
Mohana Rao PR, Sood SK (1987) Embryology of *Oreorchis foliosa*. Phytomorphology 37: 1–8
Mohrbutter C (1936) Embryologische Studien an Loganiaceen. Planta 26: 64–80
Moissel E (1941) Vergleichende embryologische Studien über die Familie Caprifoliaceae. Öst Bot Z 90: 153–212
Moldenke HN (1946) A brief historical survey of the Verbenaceae and related families. Plant Life 2: 13–98
Monachino JV (1949) A note on *Schlegelia* and *Deimatocalyx*. Phytologia 3: 102–105
Money L, Bailey IW, Swamy BGL (1950) The morphology and the relationships of the Monimiaceae. J Arnold Arbor 31: 372–404

Monteiro-Scanavacca WR, Mazzoni SC (1978) Embryological studies in *Leiothrix fluitans* (Mark.) Ruhl. (Eriocaulaceae). Rev Res Bot 1: 59–64

Moore RJ (1946) Investigations on rubber-bearing plants 3. Development of normal and aborting seeds in *Asclepias syriaca* L. Can J Res C24: 55–65

Moore RJ (1948) Cytotaxonomic studies in Loganiaceae 2. Embryology of *Polypremum procumbens.* Am J Bot 35: 403–410

Moret J, Couderc H, Bari A, Delarue Y (1990) Micromorphology of seeds of *Ornithogalum* (Hyacinthaceae) in North Africa. Nord J Bot 9: 461–468

Morgan Jr DT, Rappleye RD (1954) A cytogenetic study of origin of multiple seedlings of *Capsicum frutescens.* Am J Bot 41: 576–585

Morley T (1953) The genus *Mouriri* (Melastomataceae): a sectional revision based on anatomy and morphology. Univ Cal Publ Bot 26: 223–311

Morrison JW (1955) Fertilization and post-fertilization development in wheat. Can J Bot 33: 168–176

Morrow LA (1979) Studies in the reproductive biology of leafy spurge (*Euphorbia esula* L.). Ned Sci 27: 106–109

Morse WC (1907) Contribution to the life history of *Cosmus florida.* Ohio Nat 8: 197–204

Mortier J (1969) Embryogénie des Malvacées. Développement de l'embryon chez l'*Althae officinalis* L. Bull Soc Bot Fr 116: 505–509

Moscov IV (1964) On the development of the embryo in several species of *Paeonia* (in Russian). Bot Zh 49: 887–894

Moseley Jr MF (1948) Comparative anatomy and phylogeny of the Casuarinaceae. Bot Gaz 110: 232–280

Moseley Jr MF, Beeks RM (1955) Studies of the Garryaceae 1. The comparative morphology and phylogeny. Phytomorphology 5: 314–346

Moskova RD, Yakovlev MS (1974) Apomixis of *Bothriochloa ischaemum* L. Flowering, pollination, fertilization. In: Linskens HF (ed) Fertilization in higher plants. North-Holland, Amsterdam, pp 341–349

Mu Xi-jin (1983) Ovule, female and male gametophyte and fertilization of *Kingdonia uniflora* Balfour F. et W.W. Smith. Acta Bot Sin 25: 497–504

Mu Xi-jin (1984) Early development of the endosperm in *Kingdonia.* Acta Bot Sin 26: 668–671

Mu Xi-jin, Wang Fu (1985a) The early development of embryo and endosperm of *Paeonia lactiflora.* Acta Bot Sin 27: 7–12

Mu Xi-jin, Wang Fu (1985b) Fertilization and histochemical investigation of *Pulsatilla chinensis* (Bung) Regel. Acta Bot Sin 27: 225–232

Mujeeb-Kazi A (1981) Apomictic progeny derived from intergeneric *Hordeum-Triticum* hybrids. J Hered 72: 284–285

Mukherjee PK (1957) Studies in the embryology of *Euphorbia hypericifolia* Linn. Bull Bot Soc Univ Saugar (India) 9: 7–18

Mukherjee PK (1958) The female gametophyte of *Acalypha malabarica* Muell., with a brief discussion on the Penaea type of embryo sac. J Indian Bot Soc 37: 504–508

Mukherjee PK (1961a) Embryology of two Euphorbiaceae. Proc Indian Acad Sci B 53: 217–229

Mukherjee PK (1961b) Embryology of *Polygala.* Bull Bot Soc Coll Sci Nagpur (India) 2: 58–70

Mukherjee PK (1961c) Study of nature of embryo, seed-coat and fruit wall in *Euphorbia cristata.* J Biol Sci (Bombay) 4: 1–5

Mukherjee PK (1962) Gametophytes of three Euphorbiaceae. Bull Bot Soc Coll Sci Nagpur (India) 3: 10–14

Mukherjee PK (1964) Further contribution to the embryology of the genus *Acalypha* Linn. Proc Natl Acad Sci India B 34: 129–141

Mukherjee PK (1965) Contribution to the embryology of *Euphorbia peltata* Roxb. Proc Natl Acad Sci India B 35: 327–337

Mukherjee PK (1968) Development of endosperm and embryo in *Euphorbia heterophylla* L. Proc Natl Acad Sci India B 38: 39–42

Mukherjee PK (1972) Embryological studies in *Polygonum barbarum* L. J Indian Bot Soc 51: 106–112

Mukherjee PK, Desai SG (1990) Reduced and unreduced embryo sacs in *Solidago purpurea* Linn. J Indian Bot Soc 69: 75–78
Mukherjee PK, Padhye MD (1964) Contribution to the embryology of the genus *Phyllanthus*. Proc Natl Acad Sci India B 34: 117–128
Mukherjee PK, Santhianathan KN (1985) Reproductive biology in *Phyllanthus debilis* Ham. In: Govil CM, Kumar V (eds) Trends in plant research (Prof. Y.S. Murty Comm Vol). Bishen Singh & Mahendra Pal Singh, Dehra Dun, pp 200–205
Mukherjee PK, Thakur OS (1959) Studies on the embryology of Leguminosae: *Caesalpinia crista* Linn. J Biol Sci (Bombay) 2: 84–98
Mukherjee PK, Venugopalan P (1957) Male and female gametophytes of *Ionidium suffruticosum* Ging. J Indian Bot Soc 36: 513–518
Mukherjee SK (1952) Abnormal microsporocytes in *Caesalpina pulcherrima* Sw. Curr Sci 21: 290
Mukkada AJ (1962) Some observations on the embryology of *Dicraea stylosa* Wight. In: Symp Plant embryology. Council Sci Ind Res, New Delhi, pp 139–145
Mukkada AJ (1964) An addition to the bisporic embryo sac types: The Dicraea type. New Phytol 63: 289–292
Mukkada AJ (1969) Some aspects of the morphology, embryology and biology of *Terniola zeylanica* (Gardner) Tulsane. New Phytol 68: 1145–1158
Mukkada AJ, Chopra RN (1973) Post-fertilization development in *Indotristicha ramosissima* (Wight) van Royen. New Phytol 72: 639–646
Mulay BN, Deshpande BD, Tolani U (1965) Studies in Asclepiadaceae 2. Floral morphology and gametogenesis in certain members of Asclepiadaceae. J Indian Bot Soc 44: 95–104
Mulay BN, Nair NC, Sastry MSRK (1953) Contributions to the embryology of Vitaceae 1. Male and female gametophyte of *Vitis pedata*. Proc Rajasthan Acad Sci 4: 17–28
Mulcahy GB, Mulcahy DL (1985) Ovarian influence on pollen tube growth as indicated by the semivivo technique. Am J Bot 72: 1078–1080
Mullan DP (1936) On the seed structure and germination of *Acanthus ilicifolius*. J Indian Bot Soc 15: 143–147
Muller J (1975) Note on the pollen morphology of Crypteroniaceae sensu lato. Blumea 22: 275–294
Muller-Doblies U (1969a) Uber der Blütenstände und Blüten sowie zur Embryologie von *Sparaganium*. Bot Jahrb 89: 359–450
Müller-Doblies U (1969b) Über die Verwandtschaft von *Typha* und *Sparaganium* im Infloreszenz- und Blütenbau. Bot Jahrb 89: 451–462
Muller J, Leenhouts PW (1976) A general survey of pollen types in Sapindaceae in relation to taxonomy. Linn Soc Symp Ser 1: 407–455
Muniyamma M (1977) Triploid embryos from endosperm in vivo. Ann Bot (London) 41: 1077–1079
Munshi M, Farooq M (1969) The embryology of *Stigmaphyllon periplocaefolium* A. Juss. (Abstr). In: Johri BM, Kapil RN, Rashid A (eds) Sem Morphology, anatomy and embryology of land plants, Univ Delhi, pp 73–74
Müntzing A (1937) Polyploidy from twin seedlings. Cytologia (Fujii Jubil Vol): 211–227
Murbeck S (1901) Parthenogenetische Embryobildung in der Gattung *Alchemilla*. Acta Univ Lund 36: 1–45
Murbeck S (1902) Über die Embryologie von *Ruppia rostellata* Koch. Kon Svensk Vet Akad Handl 36: 1–21
Murdy WH, Carter MEB (1987) Regulation of the timing of pollen germination by the pistil in *Talinum mengesii* (Portulacaceae). Am J Bot 74: 1888–1892
Murgai P (1959) The development of the embryo in *Paeonia*: A reinvestigation. Phytomorphology 9: 275–277
Murgai P (1962) Embryology of *Paeonia* together with a discussion on its systematic position. In: Symp Plant embryology. Council Sci Ind Res, New Delhi, pp 215–223
Murgia M, Wilms HJ, Cresti M, Cesca G (1986) Ultrastructure of pollen development in *Euphorbia dulcis* L. 1. Diploid plants. Acta Bot Neerl 35: 405–424
Murphy JB (1946) Megasporogenesis and development of the embryo sac of *Allium cernuum*. Bot Gaz 108: 129–136

Murray BG (1990) Heterostyly and pollen tube interactions in *Lugulia gratissima* (Rubiaceae). Ann Bot (London) 65: 691–698

Murray MG (1985) Figs (*Ficus* spp.) and fig wasps (Chalcidoidea, Agaonidae): Hypotheses for ancient symbiosis. Bot J Linn Soc 26: 69–81

Murray DR (1987) Nutritive role of seed coats in developing legume seeds. Am J Bot 74: 1122–1137

Murthy KL (1934) Gametogenesis and embryogeny in some Commelinaceae. Curr Sci 3: 258–259

Murthy KL (1938) Gametogenesis and embryogeny in some Commelinaceae. J Indian Bot Soc 17: 101–116

Murthy SN (1940) Studies in the Labiatae 1. Embryology of *Ocimum sanctum* Linn., *O. canum* Sims. and *O. basilicum* Benth. J Mysore Univ B 1: 97–104

Murthy SN (1941a) Morphological studies in Sapotaceae 1. Embryology of *Bassia latifolia* Roxb. and related genera. J Mysore Univ B 2: 67–80

Murthy SN (1941b) The development of endosperm in *Leucas aspera* Spreng. Curr Sci 10: 257–259

Murthy SN (1942) Endosperm formation in *Anisomeles* sp. Curr Sci 11: 284–285

Murthy SN (1946a) Studies in the Labiatae 2. Contribution to the morphology of *Ocimum adscendens*. J Univ Bombay 14: 37–46

Murthy SN (1946b) Studies in the Labiatae 5. Contribution to the morphology of *Anisomeles malabarica* R. Br. and *A. indica* O. Kze. J Univ Bombay 15: 1–14

Murthy SN (1947) Studies in Labiatae 4. Contribution to the morphology of *Orthosiphon stamineus* Benth. J Indian Bot Soc 26: 87–94

Murty UR, Rao NGP, Kirit PB, Bharathi M (1981) Fertilization in groundnut, *Arachis hypogaea* 1. Oleagineux 36: 73–76

Murty YS (1959a) Studies in the order Piperales 3. A contribution to the study of floral morphology of some species of *Peperomia*. J Indian Bot Soc 38: 120–139

Murty YS (1959b) Studies in the order Piperales 5. A contribution to the study of floral morphology of some species of *Piper*. Proc Indian Acad Sci B 49: 52–65

Murty YS (1960) Studies in the order Piperales 8. A contribution to the morphology of *Houttuynia cordata* Thunb. Phytomorphology 10: 329–341

Murty YS, Chauhan TS (1966) Morphological studies in Apocynaceae 1. *Lochnera pusilla* K. Schum. Agra Univ J Res Sci 15: 147–162

Murty YS, Kumar V (1967) Development of the female gametophyte and embryo in *Fimbristylis diphalla* Vahl. Proc Indian Acad Sci B 65: 185–191

Murty YS, Sharma HP (1977) Embryological studies in Compositae 2. *Bellis perennis* Linn. A casual apomict. In: Dnyansagar VR, Deshpande PK, Padhye MD (eds) Recent trends and contacts between cytogenetics, embryology and morphology. Today and Tomorrow, New Delhi, pp 251–260

Murugesan S, Krishnamurthy KV (1984) Occurrence of multinucleate microspores in Suf-2 variety of sunflower. J Swamy Bot Club 1: 99–100

Musial K (1990) Studies on the genus *Solidago* L. 3. Embryology of *Solidago canadensis* L. var *canadensis*. Acta Biol Cracov Ser Bot 31: 75–84

Musselman LJ (1984) Some parasitic angiosperms of Sudan: Hydnoraceae, Orobanchaceae, and *Cuscuta* (Convolvulaceae). Notes R Bot Garden Edinburgh 42: 21–38

Myers Jr O (1964) Megasporogenesis, megagametophyte and endosperm development in *Hypericum virginicum* (Abstr). Am J Bot 51: 664

Nagamasu H (1989a) Pollen morphology and relationship of *Symplocos tinctoria* (L.f.) L'Hev (Symplocaceae). Bot Gaz 150: 314–318

Nagamasu H (1989b) Pollen morphology of Japanese *Symplocos* (Symplocaceae). Bot Mag (Tokyo) 102: 149–164

Nagaraj M (1952) Floral morphology of *Populus deltoides* and *P. tremuloides*. Bot Gaz 114: 222–243

Nagaraj M (1954a) Contributions to the floral morphology of *Terminalia chebula* Retz. Curr Sci 23: 299–300

Nagaraj M (1954b) Megasporogenesis and male gametophyte development in *Combretum ovalifolium* Roxb. Curr Sci 23: 370

Nagaraj M (1954c) Contributions to the floral morphology of *Terminalia catappa* L. Curr Sci 23: 408
Nagaraj M (1955a) Some aspects of floral morphology in *Combretum ovalifolium* Roxb. J Mysore Univ B 14: 77–85
Nagaraj M (1955b) Floral morphology of *Terminalia belerica* Roxb. Curr Sci 24: 89–90
Nagaraj M, Nijalingappa BHM (1967) Embryological studies in *Myriophyllum intermedium* DC. Proc Indian Acad Sci B 65: 210–220
Nagaraj M, Nijalingappa BHM (1973) Embryological studies in *Cyperus alopecuroides* Rottb. Proc Indian Acad Sci B 77: 252–263
Nagaraj M, Nijalingappa BHM (1974) Embryological studies in *Laurenbergia hirsuta*. Bot Gaz 135: 19–28
Nagaraj M, Nijalingappa BHM, Tejavathi DH (1976) A note on the embryology of *Lipocarpa sphacelata* Kunth. Curr Sci 45: 67–69
Nagaraj M, Tasneem F (1967) Embryological studies in *Rotula aquatica* Lour. Proc Indian Acad Sci B 66: 106–116
Nagaraj M, Tasneem F (1968) A note on the sporogenesis and gametogenesis in *Adelocaryum*. Curr Sci 37: 265–267
Nagaraj M, Tasneem F (1971) Studies on the structure and development of pericarp and seed-coat in *Rotula aquatica* Lour. Proc Indian Acad Sci B 74: 314–318
Nagaraja Rao A (= Rao AN) (1951) A note on the embryology of *Dioscorea oppositifolia* L. Curr Sci 21: 162–163
Nagaraja Rao A (1953a) Inverted polarity in the embryo sac of *Saurauia napaulensis* DC. Curr Sci 22: 282
Nagaraja Rao A (1953b) Embryology of *Dioscorea oppositifolia* L. Phytomorphology 3: 121–126
Nagaraja Rao A (1953c) Embryology of *Shorea talura* Roxb. Phytomorphology 3: 476–484
Nagaraja Rao A (1954) A case of polyembryony in *Trichopus zeylanicus* Gaertn. Sci Cult 20: 143–144
Nagaraja Rao (1955a) Gametogenesis in *Dillenia pentagyna* Roxb. Curr Sci 24: 62
Nagaraja Rao A (1955b) Development of gametophytes in *Hopea wightiana* Wall. J Mysore Univ B 15: 7–15
Nagaraja Rao A (1955c) Embryology of *Trichopus zeylanicus* Gaertn. J Indian Bot Soc 34: 213–221
Nagaraja Rao A (1955d) A contribution to the embryology of *Vateria indica* Linn. Proc Natl Inst Sci India B 21: 247–256
Nagaraja Rao A (1956) Life history of *Shorea robusta* Gaertn. Curr Sci 25: 128–129
Nagaraja Rao A (1957a) A contribution to the embryology of Dilleniaceae. Proc Iowa Acad Sci 64: 172–176
Nagaraja Rao A (1957b) The embryology of *Hypericum patulum* Thunb. and *H. mysorense* Heyne. Phytomorphology 7: 36–45
Nagaraja Rao A (1959) Flower development and gametogenesis in *Oenothera laciniata*. Proc Iowa Acad Sci 66: 91–97
Nagaraja Rao A (1961) Fibrous thickenings in the anther epidermis of *Wormia burdigei*. Curr Sci 30: 426
Nagaraja Rao A (1962) Floral anatomy and gametogenesis in *Hopea racophloea* Dyer. J Indian Bot Soc 41: 557–562
Nagaraja Rao A (1964a) Notes on the embryology of *Hevea brasiliensis* Muell. Curr Sci 33: 739–740
Nagaraja Rao A (1964b) An embryological study of *Salomonia cantoniensis* Lour. New Phytol 63: 281–288
Nagaraja Rao A (1967) Flower and seed development in *Arundina graminifolia*. Phytomorphology 17: 291–300
Nagaraja Rao A, Shamanna S (1963) Rudimentary aril in *Sanguinaria canadensis*. Can J Bot 41: 1529–1530
Nagaraja Rao A, Wee YC (1979) Embryology of the pineapple, *Ananas comosus* (L.) Merr. New Phytol 83: 485–497
Nagato Y (1981) Embryological studies on the evolutionary path in Asian rice. Bot Gaz 142: 274–278

Nagendran CR (1974) Is the embryo sac of Podostemaceae bisporic? Curr Sci 43: 259–260
Nagendran CR, Arekal GD (1976) Embryo sac of *Griffithella hookeriana*: A reinvestigation. Phytomorphology 26: 359–362
Nagendran CR, Arekal GD (1979) A new species of *Polypleurum* (Podostemaceae) from India. Bot Not 132: 49–50
Nagendran CR, Anand VV, Arekal GD (1980) The embryo sac of *Podostemum subulatus* (Podostemaceae). Plant Syst Evol 134: 121–125
Nagendran CR, Arekal GD, Subramanyam K (1977) Embryo sac studies in three species of *Polypleurum* (Podostemaceae). Plant Syst Evol 128: 215–226
Nagendran CR, Ramakrishna TM, Narayana BM (1980) Development of embryo sac and endosperm haustoria in *Striga densiflora* Benth. (Scrophulariaceae). Indian J Bot (Hyderabad, India) 3: 51–54
Nagendran CR, Subramanyam K, Arekal GD (1976) Development of the female gametophyte in *Hydrobryum griffithii* (Podostemaceae). Ann Bot (London) 40: 511–513
Nägl W (1962) Über Endopolyploidie, Restitutionskernbildung und Kernstrukturen im Suspensor von Angiospermen und einer Gymnosperme. Öst Bot Z 109: 431–494
Nägl W (1974) The *Phaseolus* suspensor and its polytene chromosomes. Z Pflanzenphysiol 73: 1–44
Nägl W, Kuhner S (1976) Early embryogenesis in *Tropaeolum majus* L.: Diversification of plastids. Planta 133: 15–19
Naidu TRB (1984) Raphide crystals in the embryos of *Scilla indica* Baker (Liliaceae). Curr Sci 53: 500–502
Nair NC (1958) Studies on Meliaceae 3. Floral morphology and embryology of *Sandoricum indicum* Cav. Phyton (Argentina) 10: 145–151
Nair NC (1959a) Studies on Meliaceae 1. Floral morphology and embryology of *Naregamia alata* W. & A. J Indian Bot Soc 38: 353–366
Nair NC (1959b) Studies on Meliaceae 2. Floral morphology and embryology of *Melia azedarach* Linn.: A reinvestigation. J Indian Bot Soc 38: 367–378
Nair NC (1965) Development of the ovule in a species of *Cissus*. Nature (London) 208: 402
Nair NC (1968) Contribution to the floral morphology and embryology of two species of *Leea* with a discussion on the taxonomic position of the genus. J Indian Bot Soc 47: 193–205
Nair NC (1970a) Meliaceae. In: Symp Comparative embryology of angiosperms. Indian Natl Sci Acad Bull 41: 151–155
Nair NC (1970b) Rhamnaceae. In: Symp Comparative embryology of angiosperms. Indian Natl Sci Acad Bull 41: 168–173
Nair NC (1970c) Vitaceae. In: Symp Comparative embryology of angiosperms. Indian Natl Sci Acad Bull 41: 174–179
Nair NC (1970d) Leeaceae. In: Symp Comparative embryology of angiosperms. Indian Natl Sci Acad Bull 41: 180–184
Nair NC (1972) Floral morphology and embryology of *Myristica malabarica* Lamk. with a discussion on certain aspects of the systematics of Myristicaceae. In: Murty YS, Johri BM, Mohan Ram HY, Varghese TM (eds) Advances in plant morphology (Prof. V. Puri Comm. Vol.). Sarita Prakashan, Meerut (India), pp 264–277
Nair NC, Abraham V (1963) A contribution to the morphology and embryology of *Micrococca mercurialis* Benth. J Indian Bot Soc 42: 583–593
Nair NC, Abraham V (1980) Floral morphology and embryology of *Tragia involucrata* L. var. *angustifolia* Hook. f. In: Periasamy K (ed) Symp Histochemistry, developmental and structural anatomy of angiosperms. Autonomous PG Centre, Univ Madras, Tiruchirapalli (India), pp 95–101
Nair NC, Bajaj YPS (1966) Floral morphology and embryology of *Cyphostemma setosum* (Wall) Alston, and a discussion on the taxonomic position of the genus *Cyphostemma* (Planch.) Alston. J Indian Bot Soc 45: 103–115
Nair NC, Gupta I (1961) A contribution to the floral morphology and embryology of *Fagonia cretica* Linn. J Indian Bot Soc 40: 635–640
Nair NC, Jain RK (1956) Floral morphology and embryology of *Balanites roxburghii* Planch. Lloydia 19: 269–279

Nair NC, Joseph TC (1957) Floral morphology and embryology of *Samadera indica*. Bot Gaz 119: 104–115
Nair NC, Joseph TC (1960) Morphology and embryology of *Cardiospermum halicacabum* Linn. J Indian Bot Soc 39: 176–194
Nair NC, Kahate S (1961) Floral morphology and embryology of *Parkinsonia aculeata* L. Phyton (Argentina) 17: 77–90
Nair NC, Kanta K (1961) Studies in Meliaceae 4. Floral morphology and embryology of *Azadirachta indica* A. Juss.: a reinvestigation. J Indian Bot Soc 40: 382–396
Nair NC, Maitreyi N (1962) Morphology and embryology of *Sebastiana chamaelea* Muell. Bot Gaz 124: 58–68
Nair NC, Mani KV (1960) Organography and floral anatomy of some species of Vitaceae. Phytomorphology 10: 138–144
Nair NC, Nambisan PN (1957) Contribution to the floral morphology and embryology of *Leea sambucina* Willd. Bot Not 110: 160–172
Nair NC, Narayanan KR (1961) Studies on Aristolochiaceae 2. Contribution to the embryology of *Bragantia wallichii*. Lloydia 24: 199–203
Nair NC, Parasuraman V (1954) Gametophytes of *Vitis pallida* W. & A. Curr Sci 23: 163–164
Nair NC, Parasuraman V (1962) Contribution to the embryology of *Vitis pallida* W. & A. Phyton (Argentina) 18: 157–164
Nair NC, Sukumaran NP (1960) Floral morphology and embryology of *Brucea amarissima*. Bot Gaz 121: 175–185
Nair NC, Suri PL (1957) Contribution to the embryology of *Vitis latifolia* Roxb. (syn. *Cissus latifolia* Vahl.). Sci Cult 22: 387–389
Nair PKK (1970) Pollen morphology of angiosperms. Vikas, Delhi
Naithani SP (1937) Chromosome studies in *Hyacinthus orientalis* L. 2. Reversal of sexual state in the anthers of *H. orientalis* L. var. yellow Hammer. Ann Bot (London) 1: 369–377
Namboodiri AN, Bhaskar M (1983) Pollen wall diffusates of *Chlorophytum* and their role in pollen germination. Phytomorphology 33: 31–36
Nanda K, Gupta SC (1971) Endothecium in *Pyrostegia venusta*. Curr Sci 40: 470–471
Nanda K, Gupta SC (1974) Mal-functioning tapetum and callose wall behaviour in *Allium cepa* microsporangia. Beitr Biol Pflanzen 50: 465–472
Nanda K, Gupta SC (1977) Development of tapetal periplasmodium in *Rhoeo spathacea*. Phytomorphology 27: 308–313
Nanetti A (1912) Sulle probabili cause della parthenocarpia del *Solanum muricatum*. Nuovo G Bot Ital 19: 91–111
Narang AK, Chandra N (1977) Structure and development of seed-coat in *Crotalaria* and *Tephrosia* species. Proc Indian Acad Sci B 85: 283–291
Narang N (1953) The life history of *Stackhousia linariaefolia* A. Cunn. with a discussion on its systematic position. Phytomorphology 3: 485–493
Narasimhamurthy SK (1933a) Studies on the life history of *Limnophyton obtusifolium*. Curr Sci 2: 53–54
Narasimhamurthy SK (1933b) Cytological and morphological studies in *Limnophyton obtusifolium* Miq. J Mysore Univ B 7: 1–32
Narasimhamurthy SK (1935a) The life history of *Ottelia alismoides* Pers. Proc Indian Acad Sci B 2: 59–66
Narasimhamurthy SK (1935b) A reply to the postscript of the paper on studies in the family Alismaceae 1. *Limnophyton obtusifolium* by BM Johri. J Indian Bot Soc 14: 351
Narayan KN (1955) Cytogenetic studies of apomixis in *Pennisetum*: *Pennisetum clandestinum* Hochst. Proc Indian Acad Sci B 41: 196–208
Narayan KN (1962) Apomixis in some species of *Pennisetum* and in *Panicum antidotale*. In: Symp Plant embryology. Council Sci Ind Res, New Delhi, pp 55–61
Narayana HS (1962a) Seed structure in the Aizoaceae. In: Maheshwari P, Johri BM, Vasil IK (eds) Proc Summer School Bot, Darjeeling. Min Sci Res Cult Aff, New Delhi, pp 220–230
Narayana HS (1962b) Post-fertilization study of *Moringa oleifera* Lamk.: A reinvestigation. Phytomorphology 12: 65–69
Narayana HS (1962c) Studies in Capparidaceae 1. The embryology of *Capparis decidua* (Forsk.) Pax. Phytomorphology 12: 167–177

Narayana HS (1965) Studies in Capparidaceae 2. Floral morphology and embryology of *Cadaba indica* Lamk. and *Crataeva nurvala* Buch-Ham. Phytomorphology 15: 158–175
Narayana HS (1970a) Capparaceae. In: Symp Comparative embryology of angiosperms. Indian Natl Sci Acad Bull 41: 78–80
Narayana HS (1970b) Moringaceae. In: Symp Comparative embryology of angiosperms. Indian Natl Sci Acad Bull 41: 81–83
Narayana HS, Arora PK (1963) The embryology of *Monsonia senegalensis* Guill et Perr. Am Midl Nat 70: 309–318
Narayana HS, Jain K (1962) A contribution to the embryology of *Limeum indicum*. Lloydia 25: 100–108
Narayana HS, Lodha BC (1963) Embryology of *Orygia decumbens* Forsk. (Aizoaceae). Phytomorphology 13: 54–59
Narayana HS, Lodha BC (1972) Embryology of *Glinus lotoides* Loefl. Proc Indian Acad Sci B 74: 77–85
Narayana HS, Prakasa Rao CG (1963) Floral morphology and embryology of *Seetzenia orientalis* Decne. Phytomorphology 13: 197–205
Narayana HS, Sharma MK (1976) Embryology of fruit crops of Rosaceae (Abstr.). In: Embryology of crop plants. Indo-Sov Symp, Indian Natl Sci Acad, New Delhi, and Univ Delhi, pp 28–29
Narayana HS, Sharma MK (1979) Studies on Rosaceae 1. The ovule and megagametogenesis in *Rubus recemosus* Roxb. and *R. canadensis* Linn. In: Bir SS (ed) Recent researches in plant sciences. Kalyani, New Delhi, pp 141–145
Narayana LL (1957) Embryology of two Simaroubaceae. Curr Sci 26: 323–324
Narayana LL (1958) Floral anatomy and embryology of *Cipadessa baccifera* Miq. J Indian Bot Soc 37: 147–154
Narayana LL (1959) Microsporogenesis and female gametophyte in *Boswellia serrata* Roxb. Curr Sci 28: 77–78
Narayana LL (1960a) Embryology of Staphyleaceae. Curr Sci 29: 403–404
Narayana LL (1960b) Studies in Erythroxylaceae 1. Proc Indian Acad Sci B 51: 270–275
Narayana LL (1960c) Studies in Burseraceae 2. J Indian Bot Soc 39: 402–409
Narayana LL (1962) Development of embryo in *Biophytum intermedium* Wight and *Oxalis pubescens* H.B.K. J Indian Bot Soc 41: 156–159
Narayana LL (1963a) A note on the embryology of a few Rutaceae. Curr Sci 32: 516–517
Narayana LL (1963b) Contributions to the embryology of Balsaminaceae 1. J Indian Bot Soc 42: 102–109
Narayana LL (1964a) A contribution to the floral anatomy and embryology of Linaceae. J Indian Bot Soc 43: 343–357
Narayana LL (1964b) Embryology of a few species of *Erythroxylum*. Curr Sci 33: 441–442
Narayana LL (1965) Contributions to the embryology of Balsaminaceae 2. Jpn J Bot 40: 104–116
Narayana LL (1970a) Oxalidaceae. In: Symp Comparative embryology of angiosperms. Indian Natl Sci Acad Bull 41: 114–116
Narayana LL (1970b) Geraniaceae. In: Symp Comparative embryology of angiosperms. Indian Natl Sci Acad Bull 41: 117–120
Narayana LL (1970c) Tropaeolaceae. In: Symp Comparative embryology of angiosperms. Indian Natl Sci Acad Bull 41: 121–122
Narayana LL (1970d) Linaceae. In: Symp Comparative embryology of angiosperms. Indian Natl Sci Acad Bull 41: 127–132
Narayana LL (1970e) Erythroxylaceae. In: Symp Comparative embryology of angiosperms. Indian Natl Sci Acad Bull 41: 133–135
Narayana LL (1970f) Balsaminaceae. In: Symp Comparative embryology of angiosperms. Indian Natl Sci Acad Bull 41: 158–162
Narayana LL (1975) A contribution to the floral anatomy and embryology of Ochnaceae. Jpn J Bot 50: 329–336
Narayana LL, Rao D (1978) Systematic position of Humiriaceae, Linaceae and Erythroxylaceae in the light of their comparative floral morphology and embryology. A discussion. J Indian Bot Soc 57: 258–266
Narayana LL, Sayeeduddin M (1959) A study of the gametophytes in *Impatiens leschenaultii* Wall. J Indian Bot Soc 38: 391–397

Narayana LL, Sundri KT (1977) Embryology of Pittosporaceae 1. Jpn J Bot 52: 204–209
Narayana LL, Sundri KT (1978) Embryology of Pittosporaceae 3. Proc Indian Acad Sci B 87: 205–214
Narayana LL, Sundri KT (1979) Embryology of Pittosporaceae 2. Curr Sci 48: 892–893
Narayana LL, Sundri KT (1981) Embryology of Pittosporaceae 4. Indian J Bot (Hyderabad, India) 4: 123–131
Narayana LL, Sundri KT (1983) Embryology of Pittosporaceae 5. J Indian Bot Soc 62: 159–165
Narayana LL, Sundri KT (1984) Embryology of Pittosporaceae 6. J Swamy Bot Club 1: 71–78
Narayana PS (1986) Reproductive behaviour of VZM 2B: A male sterile *Sorghum* line. J Indian Bot Soc 65: 131–135
Narayana PS, Narayana LL (1986) The embryology of Stegnospermataceae, with a discussion on its status, affinities and systematic position. Plant Syst Evol 154: 137–145
Narayana R (1954) Contribution to the embryology of *Dendrophthoe* Mart. Phytomorphology 4: 173–179
Narayana R (1955) Morphological and embryological studies in the family Loranthaceae-Loranthoideae. PhD Thesis, Univ Delhi
Narayana R (1956) Morphological and embryological studies in the Loranthaceae: Loranthoideae 3. *Dendrophthoe neelgherrensis* (W. & A.) Van Tiegh. J Mysore Univ B 16: 185–205
Narayana R (1958a) Morphological and embryological studies in the family Loranthaceae 2. *Lysiana exocarpi* (Behr.) Van Tieghem. Phytomorphology 8: 146–168
Narayana R (1958b) Morphological and embryological studies in the family Loranthaceae 3. *Nuytsia floribunda* (Labill.) R. Br. Phytomorphology 8: 306–323
Narayanan CR (1951) Somatic chromosomes of the Acanthaceae. J Madras Univ B 21: 220–231
Narayanan CR (1956) Studies in the Acanthaceae: Microsporogenesis in *Justicia betonica, Barleria prionitis* and *Ruellia tuberosa*. J Madras Univ B26: 189–207
Narayanan CR (1968) Polyembryony in *Streblus taxoides* (Heyne) Kurz. J Indian Bot Soc 47: 354–358
Narayanaswami S (1952) Microsporogenesis and male gametophyte of *Eleusine coracana* Gaertn. Curr Sci 21: 19–21
Narayanaswami S (1953) The structure and development of the caryopsis in some Indian millets 1. *Pennisetum typhoideum* Rich. Phytomorphology 3: 98–112
Narayanaswami S (1954) The structure and development of the caryopsis in some Indian millets 2. *Paspalum scrobiculatum* L. Bull Torrey Bot Club 81: 288–299
Narayanaswami S (1955a) The structure and development of the caryopsis in some millets 4. *Echinochloa frumentacea* Link. Phytomorphology 5: 161–170
Narayanaswami S (1955b) The structure and development of the caryopsis in some Indian millets 5. *Eleusine coracana* Gaertn. Pap Mich Acad Sci 11: 33–46
Narayanaswami S (1955c) The structure and development of the caryopsis in some Indian millets 3. *Panicum miliare* Lamk. and *P. miliaceum* Linn. Lloydia 18: 61–73
Narayanaswami S (1956) Structure and development of the caryopsis in some Indian millets 6. *Setaria italica*. Bot Gaz 118: 112–122
Narayanaswami S, Roy SK (1960a) Embryology of the genus *Psidium*. J Indian Bot Soc 39: 35–45
Narayanaswami S, Roy SK (1960b) Embryo sac development and polyembryony in *Syzygium cumini* (Linn.) Skeels. Bot Not 113: 273–284
Narayanaswami S, Sawhney S (1959) Microsporogenesis and embryo sac development in *Casearia tomentosa* Roxb. Phyton (Argentina) 13: 133–144
Narasimhachar SG (1938) Megasporogenesis and female gametophyte of *Malpigia punicifolia* Linn. Curr Sci 6: 507–509
Narasimhachar SG (1948) A contribution to the embryology of *Acacia farnesiana* L. (Willd.). Proc Indian Acad Sci B 28: 144–149
Narasimhachar SG (1949) A contribution to the embryology of *Drosera burmanni* Vahl. Proc Indian Acad Sci B 29: 98–104
Narasimhachar SG (1951a) An embryological study of *Mimosa pudica* Linn. Proc Indian Acad Sci B 33: 192–198
Narasimhachar SG (1951b) A contribution to the embryology of *Drosera indica*. Proc Indian Acad Sci B 33: 290–295

Narmatha Bai V, Lakshmanan KK (1982) Occurrence of nucellar polyembryony in *Toddalia asiatica* Lamk. Curr Sci 51: 374

Narmatha Bai V, Lakshmanan KK (1983) Embryological studies in Rubiaceae 4. *Mussaenda hirsutissima* Hutch. Acta Bot Indica 11: 224–227

Narmatha Bai V, Lakshmanan KK (1984) Embryological studies on Rubiaceae 3. *Oldenlandia verticillaris* O. Kze. J Indian Bot Soc 63: 20–24

Nast CG (1941) The embryology and seedling morphology of *Juglans regia*. Lilloa 6: 163–205

Nast CG, Bailey IW (1945) Morphology and relationships of *Trochodendron* and *Tetracentron* 2. Inflorescence, flowers and fruits. J Arnold Arbor 26: 267–276

Nast CG, Bailey IW (1946) Morphology of *Euptelea* and comparison with *Trochodendron*. J Arnold Arbor 27: 186–192

Natesh S, Bhandari NN (1975) Two types of embryo sac development in *Scrophularia himalensis* Royle. Ann Bot (London) 39: 729–732

Natesh S, Bhandari NN (1984) Post-fertilization development in *Scrophularia himalensis*. Phytomorphology 34: 208–220

Natesh S, Rau MA (1984) The embryo. In: Johri BM (ed) Embryology of angiosperms. Springer, Berlin Heidelberg New York, pp 377–443

Naumova TN (1970a) Development of sexual and apomictic embryos of *Euonymus europea* (in Russian). Bot Zh 55: 638–648

Naumova TN (1970b) Polyembryony in *Euonymus macroptera* Rupr. and *E. planipes* (Kochne) Kochne. Bot Zh 55: 9 (in Russian)

Naumova TN (1978) Peculiarity in macrogametogenesis and post-fertilization development of *Trillium camschatcense* Kev-Gawl. Bull Soc Bot Fr 125: 183–186

Naumova TN (1979) Ovule, macrosporogenesis, micro- and macrogametogenesis in *Sarcococca humilis* Hort. (Buxaceae). Bot Zh 64: 635–645

Naumova TN (1980) Nucellar polyembryony in *Sarcococca* (Buxaceae). Bot Zh 65: 230–240

Nautiyal DD, Gupta SC (1985) Ovule ontogeny in *Chenopodium album* Linn. and *Kochia scoparia* Schrad. In: Govil CM, Kumar V (eds) Trends in plant research (Prof. Y.S. Murty Comm. Vol.). Bishen Singh & Mahendra Pal Singh, Dehra Dun, pp 225–232

Nawaschin SG (1894) Über die gemeine Birke, *Betula alba* und die morphologische Deutung der Chalazogamie. Mem Acad Imp Sci St Petersburg (7) 52 (12)

Nawaschin SG (1895a) Ein neues Beispiel der Chalazogamie. Bot Zentralbl 63: 353–357

Nawaschin SG (1895b) Neue Ergebnisse über die Embryologie der Hasel, *Corylus avellana*. Bot Zentralbl 63: 104–106

Nawaschin SG (1898) Resultate einer Revision der Befruchtungsvorgänge bei *Lilium martagon* und *Fritillaria tenella*. Bull Acad Imp Sci St Petersburg 9: 377–382

Nawaschin SG, Finn WW (1913) Zur Entwicklungsgeschichte der Chalazogamen *Juglans regia* und *Juglans nigra*. Mem Acad Imp Sci St Petersburg 31: 1–59

Neeman G, Galil J (1978) Seed-set in the "male syconia" of the common fig *Ficus carica* L. (Caprificus). New Phytol 81: 375–380

Němec B (1898) Über den Pollen der petaloiden Antheren von *Hyacinthus orientalis* L. Bull Int Acad Sci Boh 5: 17–23

Němec B (1931) Fecundation in *Gagea lutea*. Preslia 10: 104–110

Neogodi G (1937) Gametogenesis cariolog del l'*Erodium corsicum* Lem. Nuovo G Bot Ital 55: 667–672

Netolitzky F (1926) Anatomie der Angiospermen-Samen. In: Linsbauer K (ed) Handbuch der Pflanzenanatomie, vol X 4. Bornträger, Berlin

Neubauer BF (1971) The development of the achene of *Polygonum pensylvanicum*: Embryo, endosperm and pericarp. Am J Bot 58: 655–664

Neumann M (1935) Die Entwicklung des pollens der Samenanlagen und des Embryosackes von *Pereskia amapola* var. *argentina*. Öst Bot Z 84: 1–30

Newcomb W (1973a) The development of the embryo sac of the sunflower *Helianthus annuus* before fertilization. Can J Bot 51: 863–878

Newcomb W (1973b) The development of the embryo sac of sunflower *Helianthus annuus* after fertilization. Can J Bot 51: 879–890

Newcomb W (1978) The development of cells in the coenocytic endosperm of the African blood lily *Haemanthus katherinae*. Can J Bot 56: 483–501

Newcomb W, Fowke LC (1973) The fine structure of the change from the free-nuclear to cellular condition in the endosperm of chickweed *Stellaria media*. Bot Gaz 134: 236–241

Newcomb W, Fowke LC (1974) *Stellaria media* embryology: the development and ultrastructure of the suspensor. Can J Bot 52: 60

Newman IV (1934) Studies in the Australian Acacias 4. The life history of *Acacia baileyana* F.V.M. 2. Gametophytes, fertilization, seed production, germination and general conclusion. Proc Linn Soc NSW 59: 277–313

Nichols M (1908) The development of the pollen of *Sarracenia*. Bot Gaz 45: 31–37

Nicoloff T (1911) L'ovule et le sac embryonnaire de Platanacées. C R Acad Sci Paris 153: 287–290

Niedenzu F (1891) Hamamelidaceae. In: Engler A, Prantl K (eds) Die natürlichen Pflanzenfamilien III 2a: 115–130

Nielsen ER (1946) The origin of multiple macrogametophytes in *Poa pratensis*. Bot Gaz 108: 41–50

Nietsch H (1941) Zur systematischen Stellung von *Cyanastrum*. Öst Bot Z 90: 31–52

Niimoto DH, Sagawa Y (1961) Ovule development in *Dendrobium*. Am Orchid Soc Bull: 813–819

Niimoto DH, Sagawa Y (1962) Ovule development in *Phalaenopsis*. Caryologia 15: 89–97

Nijalingappa BHM (1975) Embryological studies in *Haloragis micrantha*. Phytomorphology 25: 146–155

Nijalingappa BHM (1976) Sporogenesis and gametogenesis in some Cyperaceae. Proc Indian Acad Sci Plant Sci 83A: 66–72

Nijalingappa BHM (1977) Post-pollination studies in some Cyperaceae. Proc Indian Acad Sci B 85: 391–398

Nijalingappa BHM (1986) Embryology of *Scleria foliosa* (Cyperaceae). Plant Syst Evol 152: 219–230

Nijalingappa BHM, Devaki N (1978) Embryological studies in *Diplacrum caricinum* R. Br. Beitr Biol Pflanzen 54: 217–225

Nijalingappa BHM, Devaki N (1979) Endosperm of *Scleria foliosa* Hochstetter ex. A. Richard. Curr Sci 48: 451–452

Nijalingappa BHM, Nagaraj M (1980) Glimpses of embryological studies in Cyperaceae. In: Nagaraj M, Malik CP (eds) Current trends in botanical research. Kalyani, New Delhi, pp 51–58

Nijalingappa BHM, Tejavathi DH (1977) Embryo development in *Scirpus mucronatus* L. Curr Sci 46: 84–85

Nijalingappa BHM, Tejavathi DH (1983) Occurrence of endosperm haustorium in *Rhynchospora corymbosa* (L.) Britton (Cyperaceae). Curr Sci 52: 262–263

Nikam ER (1976) Embryological studies in *Lagenaria siceraria* (Molina) Standley. Curr Sci 45: 250–253

Nikiticheva ZI (1971) Embryogenesis in certain parasitic species from the families Scrophulariaceae and Orobanchaceae. Bot Zh 56: 93–105 (in Russian)

Nikiticheva ZI (1981) Embryological features of some Piperales. Acta Soc Bot Pol 50: 329–332

Nikiticheva ZI, Teryokhin ES (1973) Embryology of *Tozzia alpina* L. (Scrophulariaceae). Bot Zh 58: 5

Niklas KJ, Buchman SL (1985) Aerodynamics of wind pollination in *Simmondsia chinensis* (Link) Schneider. Am J Bot 72: 530–539

Nirula RL (1945) Embryological and cytological studies in the family Asclepiadaceae 4. On the development of embryo sac and endosperm in *Daemia extensa* Br. Proc Indian Acad Sci B 21: 181–185

Nirula RL, Richharia RH (1945) Embryological and cytological studies in the family Asclepiadaceae 3. The development and arrangement of microspores in *Hemidesmus indicus* R. Br. Proc Indian Acad Sci B 21: 178–180

Nogler GA (1984) Gametophytic apomixis. In: Johri BM (ed) Embryology of angiosperms. Springer, Berlin Heidelberg New York, pp 475–518

Noguchi Y (1928) Cytological studies on a case of pseudogamy in the genus *Brassica*. Proc Imp Acad Jpn 4: 617–619

Noll W (1935) Embryonalentwicklung von *Biophytum tendroides* DC. Planta 24: 609–648

Nordberg G (1967) Embryological studies in *Sanguisorba minor* Complex (Rosaceae). Bot Not 120: 109–119

Norman EM, Clayton D (1986) Reproductive biology of two Florida pawpaws: *Asimina obovata* and *A. pygmaea* (Annonaceae). Bull Torrey Bot Club 113: 16–22

Norstog K (1963) Apomixis and polyembryony in *Hierchlöe odorata*. Am J Bot 50: 815–821

Norton G, Haris JF (1975) Compositional changes in developing rape seed (*Brassica napus* L.). Planta 123: 163–174

Novak F, Betlach J (1970) Development and karyology of the tapetal layer and anther in sweet pepper (*Capsicum annuum* L.). Biol Plant (Prague) 12: 275–280

Nowicke JW, Skvarla JJ (1981) Pollen morphology and phylogenetic relationships of the Berberidaceae. No. 50. Smithsonian Inst Press, Washington DC

Nowicke JW, Skvarla JJ (1984) Pollen morphology and the relationships of *Simmondsia chinensis* to the order Euphorbiales. Am J Bot 71: 210–215

Nygren A (1946) The genesis of some Scandinavian species of *Calamagrostis*. Hereditas 32: 131–262

Nygren A (1950) A cytological and embryological study of *Antennaria porsildii*. Hereditas 36: 483–486

Nygren A (1954) Apomixis in the angiosperms 2. Bot Rev 20: 577–649

Nygren A (1967) Apomixis in angiosperms. In: Ruhland W (ed) Encycl Plant Physiol 18: 551–596

Ogebuadze NV (1977) Autecology and embryology of *Carya illinoinensis* (Abstr). In: Embryology of crop plants. Indo-Sov Symp, USSR Acad Sci, Moscow, and Indian Natl Sci Acad, New Delhi, at Komarov Bot Inst, Leningrad, p 27

Ogura H (1964) On the embryo sac of two species of *Tricyrtis*. Sci Rep Tôhoku Univ Ser Biol 30: 219–222

Ogura H (1966) On the embryo sac of *Tricyrtis macranthopsis* Masam (Liliaceae). Sci Rep Tôhoku Univ Ser Biol 32: 31–34

Ohga I, Sinotô Y (1924) Cytological studies on *Sciaphila japonica* 1. Bot Mag (Tokyo) 38: 202–207

Ohga I, Sinotô Y (1932) Cytological studies on *Sciaphila japonica* 2. On pollen and embryo sac development. Bot Mag (Tokyo) 46: 311–315

Oikawa K (1940) The embryo sac of *Erythronium japonicum*. Bot Mag (Tokyo) 54: 366–369

Ojeaga OO, Samyaolu MO (1970) Ovule formation and embryo development in persisting and abortive fruits of the cowpea, *Vigna unguiculata* (L.) Walp. Niger J Sci 4: 31–40

Okamoto M (1989) A comparative study of the ontogenic development of the cupules in *Castanea* and *Lithocarpus* (Fagaceae). Plant Syst Evol 168: 7–78

Oksala T, Therman E (1977) Endomitosis in tapetal cells of *Eremurus* (Liliaceae). Am J Bot 64: 866–872

Olden EJ (1947) En pentaploid äpple planta. Sver Pomol For Arsskr 1946: 76–79

Olden EJ (1953) Sexual and apomictic reproduction in *Malus sieboldii* Rehd. Bot Not 106: 105–128

Oliver D (1894) *Rhynchocalyx lawsonioides* Oliv. In: Hooker WJ (ed) Icon Plant Ser 4, vol 1, pt 1, 24: 23–48

Olson AR (1980) Seed morphology of *Monotropa uniflora* L. (Ericaceae). Am J Bot 67: 968–974

Olson AR (1988) Post-pollinational placental development of a diploid *Solanum tuberosum*. Can J Bot 66: 1813–1817

Olsen S, Olsen JD (1980) The seed of *Boschniaka hookeri* (Orobanchaceae). Bot Tidsskr 75: 159–172

Olsson O (1967) Embryological studies in the Orchidaceae. The genus *Hetaeria*. Svensk Bot Tidskr 61: 31–42

Öltmann O (1971) Pollenmorphologisch-systematische Untersuchungen innerhalb der Geraniales. Diss Bot Kiel 11: 1–163

O'Neal CE (1920) Microsporogenesis in *Datura stramonium*. Bull Torrey Bot Club 47: 231–241

O'Neal CE (1923) A study of embryo sac development and accompanying phenomena in *Oenothera rubrinervis*. Bull Torrey Bot Club 50: 133–146

Ono T (1928) Embryologische Studien an einigen Pontederiaceen. Sci Rep Tôhoku Univ Ser Biol 3: 407–415

Oppermann M (1904) A contribution to the life history of *Aster*. Bot Gaz 37: 353–362
Orel LI, Zhakova MA (1976) Differentiation of anther walls and tapetum dimorphism in the tomato *Lycopersicon esculentum* Mill. (Solanaceae). Bot Zh 61: 1720–1729 (in Russian)
Orel LI, Ogorodnikova VF, Golubeva EA, Ephendueva KK (1977) Ultrastructure and ontogenesis of tapetal film in some cultivated plants. In: Embryology of crop plants. Indo-Sov Symp, USSR Acad Sci, Moscow, and Indian Natl Sci Acad, New Delhi, at Komarov Bot Inst, Leningrad, p 49
Orr MY (1921) Observations on the structure of the seed in the Capparidaceae and Resedaceae. Notes R Bot Garden Edinburgh 12: 259–260
Orr MY (1923) Polyembryony in *Sarcococca ruscifolia*. Notes R Bot Garden Edinburgh 14: 21–23
Osawa I (1913) On the development of the pollen grain and embryo sac of *Daphne* with special reference to the sterility of *Daphne oxora*. J Cell Agric Univ Tokyo 4: 237–264
Osterwalder A (1898) Beiträge zur Embryologie von *Aconitum napellus* L. Flora 85: 254–292
Owens SJ, Kimmins FM (1981) Stigma morphology in Commelinaceae. Ann Bot (London) 47: 771–783

Pace L (1907) Fertilization in *Cypripedium*. Bot Gaz 44: 353–374
Pace L (1909) The gametophytes of *Calopogon*. Bot Gaz 48: 126–137
Pace L (1912) *Parnassia* and some allied genera. Bot Gaz 54: 306–329
Pacini E, Franchi GG (1982) Germination of pollen inside anthers in some non-cleistogamous species. Caryologia 35: 205–215
Pacini E, Franchi GG (1983) Pollen grain development in *Smilax aspera* L. and possible functions of the loculus. In: Mulcahy GB, Ottaviana E (eds) Pollen: biology and implications for plant breeding. Elsevier, Amsterdam, pp 183–189
Pacini E, Juniper BE (1983) The ultrastructure of the formation and development of the amoeboid tapetum in *Arum italicum* Miller. Protoplasma 117: 116–129
Pacini E, Keijzer CJ (1989) Ontogeny and intruding non-periplasmodial tapetum in the wild chicory, *Cichorium intybus*. Plant Syst Evol 167: 149–164
Pacini E, Sarfatti G (1978) The reproductive calendar of *Lycopersicum peruvianum* Mill. Bull Soc Bot Fr 125: 295–299
Pacini E, Cresti M, Sarfatti G (1972) Incorporation of integumentary nuclei in *Eranthis hiemalis* endosperm and their disaggregation by the endoplasmic reticulum. J Submicrosc Cytol 4: 19–31
Pacini E, Franchi GG, Hesse M (1985) The tapetum: Its form, function, and possible phylogeny in embryophyta. Plant Syst Evol 149: 155–185
Padhye MD (1959) Microsporogenesis and male gametophyte of *Cyperus compressus* Linn. J Biol Sci (Bombay) 2: 116–117
Padhye MD (1960) A contribution to the life history of *Kyllinga triceps*. Bull Bot Soc Coll Sci Nagpur (India) 1: 1–15
Padhye MD (1962a) Solanad type of embryo development in *Gomphrena celosioides* Mart.: a member of the Amaranthaceae. J Indian Bot Soc 41: 52–63
Padhye MD (1962b) Fertilization and seed development in *Nechamandra alternifolia* (Roxb.) Thw. Bull Bot Soc Coll Sci Nagpur (India) 3: 92–100
Padhye MD (1963) Two types of embryo development in *Passiflora foetida* Linn. Curr Sci 32: 372–374
Padhye MD (1968) Male and female gametophytes of *Eleocharis geniculata* Roem et Schutt. Proc Indian Acad Sci B 67: 54–60
Padhye MD (1971a) Studies in Cyperaceae 1. Embryology of *Cyperus iria* L. Proc Indian Natl Sci Acad B 37: 1–10
Padhye MD (1971b) Studies in Cyperaceae 3. Life history of *Kyllinga brevifolia* Rottb. with brief discussion on systematic position of *Kyllinga*. Bot Gaz 132: 172–179
Padhye MD, Deshpande BG (1960) The male and female gametophytes of *Passiflora foetida* Linn. Proc Indian Acad Sci B 52: 124–130
Padhye MD, Makde KH (1982) Embryogeny in some Cyperaceae. Plant Syst Evol 139: 279–287
Padhye MD, Moharir SK (1958) Studies in embryology of *Cyperus tagetum*. Proc Indian Acad Sci B 48: 89–95

Padhye MD, Rao H (1963) Contribution to the embryology of *Nechamandra alternifolia* (Roxb.) Thw. Bull Bot Soc Coll Sci Nagpur (India) 4: 39–44
Padhye MD, Chaube SN, Iyer AN (1970) Studies in Cyperaceae 7. Gametophytes and fertilization in two members of Cyperaceae. J Indian Bot Soc 49: 86–92
Padmanabhan D (1959) The development of endosperm in *Lantana camara* L. Proc Indian Acad Sci B 49: 420–427
Padmanabhan D (1960a) A contribution to the embryology of *Michelia champaca*. J Madras Univ B 30: 155–165
Padmanabhan D (1960b) The embryology of *Avicennia officinalis* 1. Floral morphology and gametophytes. Proc Indian Acad Sci B 52: 131–145
Padmanabhan D (1961a) A contribution to the embryology of *Epithema carnosum*. J Madras Univ B 31: 37–46
Padmanabhan D (1961b) Embryogenesis in *Dorstenia indica*. J Madras Univ B 31: 47–51
Padmanabhan D (1961c) A contribution to the embryology of *Gomphandra polymorpha*. Proc Natl Inst Sci India B 27: 389–398
Padmanabhan D (1962a) The embryology of *Avicennia officinalis* 3. The embryo. J Madras Univ B 32: 13–19
Padmanabhan D (1962b) A reinvestigation of the endosperm and endothelium in *Tridax procumbens* L. Phytomorphology 12: 356–361
Padmanabhan D (1964) The embryology of *Avicennia officinalis* 2. Endosperm. Phytomorphology 14: 442–451
Padmanabhan D (1966) The octant proembryo of *Epithema carnosum* Benth. Proc Indian Acad Sci B 64: 283–292
Padmanabhan D (1967) Embryogenesis in *Epithema carnosum*. J Indian Bot Soc 46: 90–107
Padmanabhan D (1968) Development from zygote to seedling in *Tridax procumbens* Linn. J Indian Bot Soc 47: 94–112
Padmanabhan D (1970a) Nymphaeaceae. In: Symp Comparative embryology of angiosperms. Indian Natl Sci Acad Bull 41: 59–62
Padmanabhan D (1970b) Verbenaceae. In: Symp Comparative embryology of angiosperms. Indian Natl Sci Acad Bull 41: 250–254
Padmanabhan D, Ramji MV (1966) Developmental studies on *Cabomba caroliniana* Gray. 2. Floral anatomy and microsporogenesis. Proc Indian Acad Sci B 64: 216–223
Padmanabhan D, Regupathy D (1981) Studies on *Bentinckia condapanna* 1. The fruit and the seed. Principes 25: 172–177
Padmanabhan D, Regupathy D (1983) Is the endosperm of *Bentinckia* ruminate? J Indian Bot Soc 62: 215–219
Paetow W (1931) Embryologische Untersuchungen an Taccaceen, Meliaceen und Dilleniaceen. Planta 14: 441–470
Pahuja AN, Kumar V (1970) Embryo sac development, cytology and systematic position of *Clintonia*. Phytomorphology 20: 97–102
Pal A (1968) Structure and development of seed of *Celosia cristata* Linn. J Indian Bot Soc 47: 381–387
Pal A (1974) Structure and development of seed in *Croton oblongifolius*. Acta Bot Indica 2: 147–150
Pal N (1951a) On the embryology of *Terminalia catapa* Linn. Sci Cult 17: 178–179
Pal N (1951b) Studies in the embryology of some Verbenaceae. J Indian Bot Soc 30: 59–74
Pal N (1951c) Studies in the embryology of the Verbenaceae. J Indian Bot Soc 30: 297–315
Pal N (1951d) A note on the embryology of *Stellaria media* L. Sci Cult 16: 374
Pal N (1952) A contribution to the life histories of *Stellaria media* Linn. and *Polycarpon loeflingiae* Benth. & Hook. Proc Natl Inst Sci India B 18: 363–378
Pal N (1959) Floral organogeny and development of male and female gametophytes in two species of Scrophulariaceae. J Asiat Soc Bengal 1: 245–249
Pal N (1960) Development of the seed of *Millettia ovalifolia*. Bot Gaz 122: 130–137
Pal PK (1963) Comparative studies in four species of *Heliotropium* L. Proc Natl Inst Sci India B 29: 1–36
Pal S (1974) Microsporogenesis in *Stellaria aquatica* Scop. Curr Sci 43: 192

Pal S (1975) Development of embryo in *Stellaria glauca* Withering (Caryophyllaceae). Curr Sci 44: 363–364

Pal S, Murty YS (1972) Structure and development of female gametophyte and embryo in *Stellaria aquatica* Scop. J Indian Bot Soc 51: 232–234

Pal S, Murty YS (1974) A contribution to the embryology of *Stellaria paniculata* Edgew. J Indian Bot Soc 53: 288–292

Paliwal RL (1951) Polarity of male cells in pollen grains of some Umbelliferae. Curr Sci 20: 17

Paliwal RL (1956) Morphological and embryological studies in some Santalaceae. Agra Univ J Res Sci 5: 191–284

Palm B (1914) Zur Embryologie der Gattungen *Aster* und *Solidago*. Acta Hortic Bergiani 5: 1–18

Palm B (1920) Preliminary notes on pollen development in tropical monocotyledons. Svensk Bot Tidskr 14: 261–266

Palm B (1922) Das Endosperm von *Hypericum*. Svensk Bot Tidskr 16: 60–68

Palser BF (1951) Early endosperm development and embryogeny in *Cassiope hypnoides*. Trans Ill State Acad Sci 44: 51–57

Palser BF (1952) Studies of floral morphology in the Ericales 2. Megasporogenesis and megagametophyte development in Andromedeae. Bot Gaz 114: 33–52

Palser BF (1961) Some aspects of embryology in the Ericales. In: Recent advances in botany, vol. 1. Univ Press, Toronto, pp 685–689

Palser BF (1963) Studies on the floral morphology in the Ericales 6. The Diapensiaceae. Bot Gaz 124: 200–219

Palser BF (1975) The bases of angiosperms phylogeny: Embryology. Ann Missouri Bot Garden 62: 621–646

Palser BF, Philipson WR, Philipson MN (1971) Embryology of *Rhododendron* 1. Introduction and ovule, megagametophyte and early endosperm development in *R. yunnanense*. J Indian Bot Soc (Golden Jubl Vol) 50A: 172–188

Palser BF, Philipson WR, Philipson MN (1985) The ovary, ovule and megagametophyte in *Rhododendron*. Notes R Bot Garden Edinburgh 43: 133–160

Palser BF, Rouse JL, Williams EC (1987) Daily progression of megagametophyte, pollen tube and early post-fertilization development in *Rhododendron* from anthesis to three weeks after pollination (Abstr). In: 14th Int Bot Congr, Berlin, p 220

Palser BF, Philipson WR, Philipson MN (1989) Development of ovule, megagametophyte and early endosperm in representative species of *Rhododendron* L. (Ericaceae). Bot J Linn Soc 101: 363–393

Palser BF, Rouse JL, Williams EG (1989) Coordinated time tables for megagametophyte development and pollen tube growth in *Rhododendron nuttallii* from anthesis to early post-fertilization. Am J Bot 76: 1167–1202

Panchaksharappa MG (1962a) Embryological studies in the family Zingiberaceae 1. *Costus speciosus* Smith. Phytomorphology 12: 418–430

Panchaksharappa MG (1962b) Embryological studies in some members of the Zingiberaceae. In: Symp Plant embryology. Council Sci Ind Res, New Delhi, pp 224–238

Panchaksharappa MG (1966) Embryological studies in some members of Zingiberaceae 2. *Elettaria cardamomum*, *Hitchenia caulina* and *Zingiber macrostachyum*. Phytomorphology 16: 412–417

Panchaksharappa MG (1970) Zingiberaceae. In: Symp Comparative embryology of angiosperms. Indian Natl Sci Acad Bull 41: 380–385

Panchaksharappa MG, Rangel JF (1966) Floral morphology, sporogenesis and gametogenesis in *Furcraea gigantea* Vent. Karnatak Univ J Sci 11: 161–173

Pandey AK (1980) Structure and behaviour of endothelium in some Compositae. Acta Bot Indica 8: 50–56

Pandey AK, Singh RP (1976) Development of female gametophyte in *Hymentherum tenuifolium* Cass. Curr Sci 45: 29–30

Pandey AK, Singh RP (1980) Development and structure of seeds and fruits in tribe Vernonieae – some *Vernonia* and *Elephantopus* species. Flora 169: 443–452

Pandey AK, Singh RP (1982a) Development and structure of seeds and fruits in Compositae: *Coreopsis* species. J Indian Bot Soc 61: 417–425

Pandey AK, Singh RP (1982b) Development and structure of seeds and fruits in the Compositae: tribe Senecioneae. Bot Jahrb 103: 413–422
Pandey AK, Singh RP (1983a) Development and structure of seeds and fruits in Compositae: tribe Eupatorieae. J Indian Bot Soc 62: 276–281
Pandey AK, Singh RP (1983b) Development and structure of seeds and fruits in Compositae: tribe Arctotideae. Flora 174: 323–328
Pandey AK, Chopra S, Singh RP (1983) Development and structure of seeds and fruits in Compositae: tribe Inuleae. Proc Indian Acad Sci Plant Sci 92: 467–471
Pandey AK, Chopra S, Singh RP (1986a) Development and structure of seeds and fruits in Compositae: *Cosmos* species. J Indian Bot Soc 65: 362–368
Pandey AK, Chopra S, Singh RP (1986b) Development and structure of seeds and fruits in Compositae – tribe Helenieae (sensu Bentham). J Indian Bot Soc 65: 416–421
Pandey AK, Jha SS, Jha A (1990) Development and structure of seed in some Dalbergieae (Papilionoideae, Fabaceae). J Indian Bot Soc 69: 107–114
Pandey BP, Saxena NP (1977) Structure of gametophyte in *Lithophragma affinis* Gray. Acta Bot Indica 5: 207–209
Pandey KK (1955) Seed development in diploid, tetraploid and diploid-tetraploid crosses of *Trifolium pratense*. Indian J Genet Pl Breed 15: 25–35
Pande PC, Singh V (1981) A contribution to the embryology of the Iridaceae. J Indian Bot Soc 60: 160–167
Panicker TKB (1965) Female gametophyte and endosperm in *Lagenandra ovata* Thw. Curr Sci 34: 614–615
Pankow H (1958) Über den Pollenkitt bei *Galanthus nivalis*. Flora 146: 240–253
Pant DD, Nautiyal DD, Chaturvedi SK (1982a) Pollination ecology of some Indian asclepiads. Phytomorphology 32: 302–313
Pant DD, Nautiyal DD, Chaturvedi SK (1982b) Insect pollination in some Indian Glumiflorae. Beitr Biol Pflanzen 57: 229–236
Pantulu JV (1942) A contribution to the life history of *Desmodium gangeticum* DC. J Indian Bot Soc 21: 137–144
Pantulu JV (1945) Studies in the Caesalpiniaceae 1. A contribution to the embryology of the genus *Cassia*. J Indian Bot Soc 24: 10–24
Pantulu JV (1951) Studies in the Caesalpiniaceae 2. Development of the endosperm and embryo in *Cassia occidentalis* L. J Indian Bot Soc 30: 95–99
Parmeswaran N (1959) A contribution to the embryology of *Theriophonum minutum* Engl. Proc Indian Acad Sci B 50: 15–25
Parmeswaran N (1961) Ruminate endosperm in the Canellaceae. Curr Sci 30: 344–345
Parmeswaran N (1962) Floral morphology and embryology in some taxa of the Canellaceae. Proc Indian Acad Sci B 55: 167–182
Pardi P (1933) Contributo all'embryologia delle Asclepiadaceae. Nuovo G Bot Ital 40: 560–569
Parks M (1935) Embryo sac development and cleistogamy in *Commelinantia pringlei*. Bull Torrey Bot Club 62: 91–104
Parulekar NK (1970) Annonaceae. In: Symp Comparative embryology of angiosperms. Indian Natl Sci Acad Bull 41: 38–41
Parvathi A, Narayana LL, Dutt BSM (1978) Embryology of *Cuscuta chinensis* Lamk. Natl Acad Sci Lett (Allahabad, India) 1: 433–435
Pastrana MD, Santos JK (1931) A contribution to the life history of *Dendrobium anosmum*. Natl Appl Sci Bull Philipp Univ 1: 133–144
Patankar TB (1956) Further contribution to the embryology of *Drosera burmanni* Vahl. Proc Indian Acad Sci B 43: 161–171
Pate JS, Gunning BES (1972) Transfer cells. A Rev Plant Physiol 23: 173–196
Patel CM, Patel DS (1964) The morphological and embryological studies in *Eriocaulon cinereum* R. Br. Vidya (Gujarat Univ, India) 7: 58–70
Patel CM, Shah CK (1962) The life history of *Cyperus niveus* Retz. Vidya (Gujarat Univ, India) 5: 148–158
Patel NK (1960) Some preliminary observations on the floral structures of the Oleaceae. Curr Sci 29: 59

Patel TK, Cook CDK (1972) Embryological studies on *Ranunculus tripartitus* DC. Bot Jahrb 91: 438–461
Patel V, Skvarla JJ, Raven PH (1983) Half pseudocolpi: A unique feature of *Olinia* (Oliniaceae) pollen. Am J Bot 70: 469–473
Paterson BR (1961) Studies of floral morphology in the Epacridaceae. Bot Gaz 122: 259–279
Pattee HE, Mohapatra SC (1987) Anatomical changes during ontogeny of the peanut (*Arachis hypogaea* L.) fruit: Mature megagametophyte through heart-shaped embryo. Bot Gaz 148: 156–164
Patwardhan U (1953) Megasporogenesis in *Daemia extensa* R. Br. Sci Cult 19: 157–158
Paul AK (1937) Development of ovule and embryo sac of *Tamarindus indica* Linn. J Indian Bot Soc 16: 151–157
Pax F (1889) Aizoaceae. In: Engler A, Prantl K (eds) Die natürlichen Pflanzenfamilien 3(1B): 33–51
Pax F (1891) Lauraceae. In: Engler A, Prantl K (eds) Die natürlichen Pflanzenfamilien 3: 106–126
Pax F (1896a) Euphorbiaceae. In: Engler A, Prantl K (eds) Die natürlichen Pflanzenfamilien 3(5): 1–119
Pax F (1896b) Hippocastanaceae. In: Engler A, Prantl K (eds) Die natürlichen Pflanzenfamilien 3(5): 273–276
Pax F (1930) Haemodoraceae. In: Engler A, Prantl K (eds) Die natürlichen Pflanzenfamilien 15a: 386–391
Pax F, Hoffman C (1931) Callitrichaceae. In: Engler A, Prantl K (eds) Die natürlichen Pflanzenfamilien 19c: 236–240
Peakall R (1989) The unique pollination of *Leporella fimbriata* (Orchidaceae): pollination by pseudocopulating male ants (*Myrmecia urens*, Formicidae). Plant Syst Evol 167: 137–148
Pearson NL (1948) Observations on seed-hair growth in *Asclepias syriaca* L. Am J Bot 35: 27–36
Pechan PM (1988) Ovule fertilization and seed number per pod determination in oil seed rape (*Brassica napus*). Ann Bot (London) 61: 201–207
Pechenitsyn UP (1972a) Development of the Fritillaria type embryo sac in some Asiatic species of *Tulipa*. Bot Zh 57: 221–229 (in Russian)
Pechenitsyn UP (1972b) Double fertilization in species of *Tulipa* with Fritillaria type embryo sac. Bot Zh 57: 465–469 (in Russian)
Periasamy K (1961) Studies on seeds with ruminate endosperm 1. Morphology of ruminating tissue in *Myristica fragrans*. J Madras Univ B 31: 53–58
Periasamy K (1962a) The ruminate endosperm: development and types of rumination. In: Symp Plant embryology. Council Sci Ind Res, New Delhi, pp 62–74
Periasamy K (1962b) Studies on seeds with ruminate endosperm 2. Development of the rumination in Vitaceae. Proc Indian Acad Sci B 56: 13–26
Periasamy K (1963) Studies on seeds with ruminate endosperm 3. Development of rumination in certain members of Apocynaceae. Proc Indian Acad Sci B 58: 325–332
Periasamy K (1964a) Studies on seeds with ruminate endosperm 4. Development of rumination in *Coccoloba uvifera*. J Indian Bot Soc 43: 543–548
Periasamy K (1964b) Studies on seeds with ruminate endosperm 5. Seed development and rumination in two genera of Rubiaceae. Proc Indian Acad Sci B 60: 351–360
Periasamy K (1966) Studies on seeds with ruminate endosperm 6. Rumination in the Araliaceae, Aristolochiaceae, Caprifoliaceae and Ebenaceae. Proc Indian Acad Sci B 64: 127–134
Periasamy K (1977) A new approach to the classification of angiosperm embryos. Proc Indian Acad Sci B 86: 1–13
Periasamy K, Indira C (1986) The carpel of *Moringa*. Ann Bot (London) 58: 897–901
Periasamy K, Kandasamy MK (1981) Development of the anther of *Annona squamosa* L. Ann Bot (London) 48: 885–893
Periasamy K, Parmeswaran N (1962) Extra-ovular outgrowths in Rubiaceae. Curr Sci 31: 300–301
Periasamy K, Parmeswaran N (1965) A contribution to the floral morphology and embryology of *Tarenna asiatica*. Beitr Biol Pflanzen 41: 123–138
Periasamy K, Sivaramakrishnan R (1980) Anther development in *Aegiceras corniculatum* Blanco. In: Periasamy K (ed) Symp Histochemistry, developmental and structural anatomy of angiosperms. Autonomous PG Centre, Univ Madras, Tiruchirapalli (India), pp 134–139
Periasamy K, Swamy BGL (1959) Studies in the Annonaceae 1. Microsporogenesis in *Cananga odorata* and *Miliusa wightiana*. Phytomorphology 9: 251–263

Periasamy K, Swamy BGL (1961) Studies in the Annonaceae 2. The development of ovule and seed in *Cananga odorata* and *Miliusa wightiana*. J Indian Bot Soc 40: 206–216
Periasamy K, Swamy BGL (1964) Is the microsporangium of angiosperms wall-less? Curr Sci 33: 735–738
Periasamy K, Swamy BGL (1966) Morphology of the anther tapetum in angiosperms. Curr Sci 35: 427–430
Periasamy K, Thangavel S (1988) Anther development in *Xylopia nigricans*. Proc Indian Acad Sci Plant Sci 98: 251–255
Periasamy K, Vaidyanathan R (1980) Development of rumination in *Borreria hispida* K. Sch. In: Periasamy K (ed) Symp Histochemistry, developmental and structural anatomy of angiosperms. Autonomous PG Centre, Univ Madras, Tiruchirapalli (India), pp 140–144
Perkins JR, Gilg E (1911) Monimiaceae. In: Engler A (ed) Das Pflanzenreich 4(4 101): 1–67
Perrot E, Guérin P (1903) Les *Didierea* de Madagascar. J Bot Paris 17: 233–251
Persidsky D (1934) On the development of the endosperm and haustorium in *Linaria genistraefolia* (in Russian). Bull Jardin Bot Kieff 17: 11–18
Peter J (1920) Zur Entwicklungsgeschichte einiger Calycanthaceen. Beitr Biol Pflanzen 14: 59–84
Petrov DF (1939) On the occurrence of facultative pseudogamy in a triploid variety of raspberries "Immer tragende", *Rubus idaeus*. Dokl Akad Nauk SSSR 22: 352–353 (in Russian)
Pettitt JM (1976) Pollen wall and stigma surface in the marine angiosperms *Thalassia* and *Thalassodendron*. Micron 7: 21–32
Pettitt JM (1980) Reproduction in seagrasses. Nature of pollen and receptive surface of the stigma in the Hydrocharitaceae. Ann Bot (London) 45: 257–271
Pettitt JM, Ducker SC, Knox RB (1981) Submarine pollination in seagrasses. Sci Am 244: 135–143
Pettitt JM, Jermy AC (1975) Pollen in hydrophilous angiosperms. Micron 5: 377–405
Pettitt JM, McConchie CA, Ducker SC, Knox RB (1980) Unique adaptations for submarine pollination in seagrasses. Nature (London) 286: 487–489
Pettitt JM, McConchie CA, Ducker SC, Knox RB (1984) Reproduction in sea grasses: pollen wall morphogenesis in *Amphibolis antarctica* and wall structure in filiform grains. Nord J Bot 4: 199–216
Pfeiffer A (1891) Die Arillargebilde der Pflanzensamen. Bot Jahrb 13: 492–540
Pfeiffer NE (1918) The sporangia of *Thismia americana*. Bot Gaz 66: 354–363
Pfeiffer WM (1912) The morphology of *Leitneria floridata*. Bot Gaz 53: 189–203
Phatak VG, Ambegaokar KB (1955) Embryological studies in Acanthaceae 1. The female gametophyte. J MS Univ Baroda (India) 4: 87–95
Phatak VG, Ambegaokar KB (1956) Embryological studies in Acanthaceae 2. Endosperm and embryo development in *Barleria prionitis* L. J MS Univ Baroda (India) 5: 73–87
Phatak VG, Ambegaokar KB (1957) Embryological studies in Acanthaceae 3. Endosperm and embryo development in *Acanthus ilicifolius* Linn. J MS Univ Baroda (India) 6: 1–10
Phatak VG, Ambegaokar KB (1961a) Embryological studies in Acanthaceae 4. Development of embryo sac and seed formation in *Haplanthus tentaculatus* Nees. J Indian Bot Soc 40: 525–534
Phatak VG, Ambegaokar KB (1961b) Embryological studies in Acanthaceae 6. Development of embryo sac and endosperm in *Hemigraphis latebrosa* var. *heyneana* Bremek. and *Sericocalyx scaber* (Nees) Bremek. J Univ Poona Sci (India) 22: 139–145
Phatak VG, Ambegaokar KB (1963) Embryological studies in Acanthaceae 5. Development of embryo sac and endosperm in *Blepharis maderaspatensis* (Linn). Roth. Proc Indian Acad Sci B 57: 88–95
Phatak VG, Savant SD (1955) Embryological studies in Euphorbiaceae. The female gametophyte in *Jatropha podagrica* Hook. J MS Univ Baroda (India) 4: 113–118
Philbrick CT (1984) Pollen tube growth within vegetative tissues of *Callitriche* (Callitrichaceae). Am J Bot 71: 882–886
Philip VJ (1968) The origin of cotyledon and epicotyl in *Elettaria cardamomum* Maton. Aust J Bot 16: 427–431
Philip VJ (1972) On embryogenesis in *Bambusa arundinacea* Willd. Curr Sci 41: 153
Phillip VJ, Gouri V (1975) Development of the cotyledon and epicotyl in *Murdania semiters* (Dalz) Santapau (*Aneilema paniculata* Wall). Cellule 71: 31–38
Philip VJ, Haccius B (1976) Embryogenesis in *Bambusa arundinacea* Willd. and structure of the mature embryo. Beitr Biol Pflanzen 52: 83–100

Philipson MN (1977) Haustorial synergids in *Cortaderia* (Gramineae). NZ J Bot 15: 777–778

Philipson MN (1978a) Nucellar degeneration in *Cortaderia* (Gramineae). Protoplasma 95: 361–369

Philipson MN (1978b) Apomixis in *Cortaderia jubata* (Gramineae). NZ J Bot 16: 45–59

Philipson MN, Connor HE (1984) Haustorial synergids in Danthonioid grasses. Bot Gaz 145: 78–82

Philipson WR (1953) The relationship of the Compositae particularly as illustrated by the morphology of the inflorescence in the Rubiales and the Campanulatae. Phytomorphology 3: 391–404

Philipson WR (1974) Ovular morphology and the major classification of dicotyledons. Bot J Linn Soc 68: 89–108

Philipson WR (1975) Evolutionary lines within the dicotyledons. NZ J Bot 13: 73–91

Philipson WR (1977) Ovular morphology and the classification of dicotyledons. In: Kubitzki H (ed) Flowering plants. Plant Syst Evol Suppl 1. Springer, Wien, pp 123–140

Philipson WR, Philipson MN (1973) A comparison of the embryology of *Forstera* L. and *Donatia* J.R. & G. Forst. NZ J Bot 11: 449–460

Phouphas C (1952) Développement du proembryon chez le *Gynandropsis pentaphylla* DC. (Capparidacées). Bull Soc Bot Fr 99: 18–20

Picci V (1964) Ricerche embriologiche sul genere *Thalictrum* 1. Embriologia di *Thalictrum angustifolium* L. var. *flavum* L. Giorn Bot Ital 71: 258–267

Picci V (1969) Ricerche embriologiche sul genere *Thalictrum* 2. Embriologia è cariologia di *Thalictrum* L. e di *Thalictrum minus* L. subsp. *minus*. Giorn Bot Ital 103: 475–483

Pichon M (1946) Sur les Alismatacées et les Butomacées. Not Syst Paris 12: 170–183

Piekett FL (1913) The development of the embryo sac of *Arisaema triphyllum*. Bull Torrey Bot Club 40: 229–235

Pieringer AP, Edwards GJ (1967) Identification of nucellar and zygotic *Citrus* seedlings by infra-red spectroscopy. Proc Am Soc Hortic Sci 86: 226–234

Pigni AM, Corsi G, Cappelletti EM (1986) Fruit morpho-anatomical aspects and secretory structures in three related *Athamanta* sp. (Umbelliferae). Bot Jahrb 106: 211–220

Pigott CD, Warr SJ (1989) Pollination, fertilization and fruit development in Sycamore (*Acer pseudoplatanus*). New Phytol 111: 99–104

Pigott EM (1915) Notes on *Nothopanax arboreum*, with some reference to the development of the gametophytes. Trans Proc NZ Inst 47: 599–612

Pigott EM (1927) Observations on *Corynocarpus laevigata* Forst, the Karaka. Trans Proc NZ Inst 58: 57–71

Pijl L van der (1955) Sarcotesta, aril, pulpa and the evolution of the angiosperm fruit 1,2. Proc Koln Ned Akad Wet, Amsterdam C 58: 154–161, 307–312

Pilger R (1915) Celastraceae. In: Engler A, Prantl K (eds) Die natürlichen Pflanzenfamilien 4 (2–4): 198–202

Pilger R (1925) Caryocaraceae. In: Engler A, Prantl K (eds) Die natürlichen Pflanzenfamilien 21: 90–92

Pilger R (1930) Philydraceae. In: Engler A, Prantl K (eds) Die natürlichen Pflanzenfamilien 15a: 190–191

Pilger R (1935) Santalaceae. In: Engler A, Prantl K (eds) Die natürlichen Pflanzenfamilien 16b: 52–91

Pillai SK, Ramasita K, Ishwar D (1974) Embryogeny, histogenesis, apical meristems, seed-coat and seedling anatomy of *Albizzia lebbek* (Linn.) Benth. New Bot (India) 1: 23–33

Pimienta E, Polito VS (1983) Embryo sac development in almond [*Prunus dulcis* (Mill.) D.A. Webb] as affected by cross-, self-, and non-pollination. Ann Bot (London) 51: 469–479

Pisek A (1924) Antherenentwicklung und meiotische Teilung bei der Wacholdermistel [*Arceuthobium oxycedri* (DC) M.B.]; Anthrenbau und Chromosomenzahlen von *Loranthus europaeus* Jacq. SB Akad Wiss Wien 133: 1–15

Planchon JE (1845a) Sur affinités des Genres *Henslowia* Wall. (*Crypteronia*?) Blume, *Quillamum* Blanco, *Raleighia* Gdn. et *Alzatea* Ruiz et Pav. London J Bot 4: 474–478

Planchon JE (1845b) Développement et caractères des Vreis et des faux arilles. Ann Sci Nat Ser Bot 3: 275–312

Planchon JE (1848) Sur la familie des Droseracées. Ann Sci Nat Ser Bot 9: 79–99

Planchon JE (1887) Monographie des Ampelidacées vries. Candolle Monogr Phanerogam 5: 305–648

Poddubnaya-Arnoldi VA (1933) Spermazellen in die Familie der Dipsacaceae. Planta 21: 380–386

Poddubnaya-Arnoldi VA (1960) Study of fertilization in the living material of some angiosperms. Phytomorphology 10: 185–198

Poddubnaya-Arnoldi VA (1964) General embryology of angiosperms. Nauka, Moscow (in Russian)

Poddubnaya-Arnoldi VA (1967) Comparative embryology of the Orchidaceae. Phytomorphology (Prof. P. Maheshwari Comm. Vol.) 17: 312–320

Poddubnaya-Arnoldi VA (ed) (1976) Cytoembryology of angiosperms: Principles and perspectives. Nauka, Moscow (in Russian)

Poddubnaya-Arnoldi VA (1982) Characteristics of Angiospermae families according to cytoembryological characters. Nauka, Moscow (in Russian)

Poddubnaya-Arnoldi VA, Makhaline MA (1969) Microsporogenèse et spermiogenèse chez certains hybrides éloignés de la famille des Gramineae. Rev Cytol Biol Veg 32: 59–66

Poddubnaya-Arnoldi VA, Stechina N, Scsonovetz A (1935) Der Charakter und die Ursachen der Sterilität bei *Scorzonera kak-saghyz*. Beih Bot Zentralbl 53: 309–339

Pogan E, Wcislo H (1981) Studies in *Ranunculus ficaria* L. 4. Cytoembryological studies. Acta Biol Cracov Ser Bot 23: 37–54

Pogliani M (1964) Ricerche ecologiche e cita-embriologiche di *Antirrhinum latifolium* Mill. Giorn Bot Ital 71: 558

Pogliani M (1966) Ricerche embriologiche in *Anacyclus radiatus* Lois. Ann Bot (Rome) 28: 725–729

Polunina NN (1957) Comparative studies of embryology and flowering in some varieties of *Eucalyptus*. Dokl Akad Nauk SSSR 115: 819–821 (in Russian)

Polunina NN (1958a) Concerning the biology of flowering and embryology of *Callistemon lanceolatus*. Bot Zh 43: 1169–1178 (in Russian)

Polunina NN (1958b) On the flowering biology and embryology of *Eugenia myrtifolia* Sims. Izv Glavn Bot Sad SSSR 30: 59–65 (in Russian)

Polunina NN (1959) The embryology of *Eucalyptus*. Byul Glavn Bot Sad Akad Nauk SSSR 16: 69–79 (in Russian)

Polunina NN (1963) A comparative embryological study of some members of the Myrtaceae. Byul Glavn Bot Sad Akad Nauk SSSR 49: 82–90 (in Russian)

Pomeranz Y, Bechtel DB (1978) Structure of cereal grains as related to end-use properties. In: Hultin HO, Milner M (eds) Post-harvest biology and biochemistry. Food and nutrition press, Westport (Conne, USA)

Poole AL (1952) The development of *Nothofagus* seed. Trans R Soc NZ 80: 207–212

Popham RA (1938) A contribution to the life history of *Galinsoga ciliata*. Bot Gaz 99: 543–555

Popova LK, Kolesnikov (1972) Intertissue interaction in ovules of apomictic forms of *Potentilla*. Izv Akad Nauk Mold SSR Ser Biol Khim Nauk 1: 88–89 (in Russian)

Porter TR (1936) Development of the megagametophyte and embryo of *Allium mutabile*. Bot Gaz 98: 317–327

Posluszny V, Sattler R (1976) Floral development of *Zannichellia palustris*. Can J Bot 54: 651–662

Poulsen VA (1886) Bidrag till Triuridaceernes naturhistorie. Vidensk Medd Foren Kjöbenhavn 1884–86: 161–179

Poulsen VA (1905) *Sciaphila nana* Bl. Vidensk Medd Foren Kjöbenhavn (see Bot Zentralbl 49, 1905) 1884–1886

Prakasa Rao PS (1963) Some embryological observations of *Guiera senegalensis* Lam. Curr Sci 32: 30–31

Prakasa Rao PS, Rajeswari Devi KR (1974) Some embryological features of *Euphorbia vermiculata* Raf. Curr Sci 43: 257–259

Prakasa Rao PS, Rajeswari Devi KR (1975) Embryo development in *Euphorbia marginata* Pursh. Jpn J Bot 50: 59–64

Prakasa Rao PS, Sundri KT, Narayana LL (1979) Some embryological features of *Emilia flammea* Cass. Curr Sci 48: 129–131

Prakash N (1967a) Aizoaceae: a study of its embryology and systematics. Bot Not 120: 305–323

Prakash N (1967b) Life history of *Tetragonia tetragonioides* (Pall.) O. Kuntze. Aust J Bot 15: 413–424
Prakash N (1967c) Gametogenesis and seed development in *Hereroa hesperantha* (Dtr.) Dtr. et Schwant (Aizoaceae). Aust J Bot 15: 425–435
Prakash N (1969a) The embryology of *Kunzea capitata* Reichb. Aust J Bot 17: 97–106
Prakash N (1969b) Some aspects of the life history of *Callistemon citrinus* (Curt.) Skeels. Aust J Bot 17: 107–117
Prakash N (1969c) Reproductive development in two species of *Darwinia* Rudge (Myrtaceae). Aust J Bot 17: 215–217
Prakash N (1969d) A contribution to the life history of *Angophora floribunda* (Sm) Sweet (Myrtaceae). Aust J Bot 17: 457–469
Prakash N (1969e) The floral development and embryology of *Centrolepis fascicularis*. Phytomorphology 19: 285–291
Prakash N (1973) A study of embryology of *Melaleuca*. Malays J Sci 2: 81–85
Prakash N, Chan YY (1976) Embryology of *Glycine max*. Phytomorphology 26: 302–308
Prakash N, Herr JM Jr (1979) Embryological study in *Glottidium vesicarium* through the use of clearing technique. Phytomorphology 29: 71–77
Prakash N, McAlister EJ (1977) An embryological study of *Bauera capitata* with comments on the systematic position of *Bauera*. Aust J Bot 22: 615–622
Prakash N, Lau YY (1976) Morphology of *Ploiarium alternifolium* and the taxonomic position of *Ploiarium*. Bot Not 129: 279–285
Prakash N, Lee-Lee A (1973) Life history of common Malaysian orchid *Spathoglottis plicata*. Phytomorphology 23: 9–17
Prakash N, Bohm LR, Maze J (1979) Developmental anatomy of the achene in *Cannabis*. Can J Bot 57: 1243–1251
Prakash N, Foreman DB, Griffith SJ (1984) Gametogenesis in *Galbulimima belgraveana* (Himatandraceae). Aust J Bot 32: 605–612
Prakash N, Lim AL, Manurung R (1977) Embryology of duku and langsat varieties of *Lansium domesticum*. Phytomorphology 27: 50–58
Prakash S (1960a) The endosperm of *Arachis hypogaea* Linn. Phytomorphology 10: 60–64
Prakash S (1960b) Morphological and embryological studies in the family Loranthaceae 6. *Peraxilla tetrapetala* (Linn. f.) Van Tiegh. Phytomorphology 10: 224–234
Prakash S (1961) Morphological and embryological studies in the family Loranthaceae 7. *Atkinsonia ligustrina* (Cunnigh.) F.V. Muell. Phytomorphology 11: 325–334
Prakash S (1963) Morphological and embryological studies in the family Loranthaceae 10. *Barathranthus axanthus* (Korth.) Miq. Phytomorphology 13: 97–103
Pramanik A, Thothathri K (1986) On the status of *Desmodiastrum* Prain (Fabaceae). J Indian Bot Soc 65: 373–379
Prance GT (1972) Chrysobalanaceae. In: Flora neotropica. Monogr 9. Hafner, New York
Prasad K (1969) Occurrence of cytoplasmic nodules in the endosperm of *Farsetia hamiltonii* Royle (Cruciferae) (Abstr.) Proc 56th Indian Sci Congr, Bombay, pt 3: 389
Prasad K (1974) Studies in the Cruciferae Gametophytes, structure and development of seed in *Eruca sativa* Mill. J Indian Bot Soc 53: 24–33
Prasad K (1977a) Seed-coat structure and development in *Lepidium sativum* L. and *Tulaspi perfoliatum* L. Bot Jahrb 97: 508–515
Prasad K (1977b) The development and structure of basal body in the ovule and seed of certain species of Cruciferae. Bot Jahrb 98: 266–272
Prasad K (1979) Morphology and histochemistry of the nucellus and endosperm in certain species of Cruciferae. Bot Jahrb 100: 536–541
Prasad T, Singh D (1978) Gametophytes and seed development in *Nicandra physaloides* Gaertn. J Indian Bot Soc 57: 76–83
Preil W, Keyser D (1976) Pollenkeimung und Samenansatz bei einigen "gymnospermen" Begonien. Naturwissenschaften 29: 360–361
Prichard EC (1955) Morphological studies in Rhamnaceae. J Elisha Mitchell Sci Soc 71: 82–106
Priyadarshan PM, Ramachandran K (1984) Cytology and exceptional pollen development in *Mussaenda* Linn. Cytologia 49: 407–413

Prosina MN (1930) Über die vom Cypripedium-Typus abweichende Embryosackentwicklung von *Cypripedium guttatum*. Planta 12: 532–544

Prósperi CH, Cocucci AE (1979) Importancia taxonómica de la calosa de los tubos polínicos en Tubiflorae. Kurtziana 12–13: 75–81

Psaras G (1984) On the structure of lettuce (*Lactuca sativa* L.) endosperm during germination. Ann Bot (London) 54: 187–194

Psaras G (1985) The structure of sunflower (*Helianthus annuus*) endosperm during germination. Phytomorphology 35: 141–146

Ptak K (1990) Cytoembryological investigations of the polish representatives of the genus *Crataegus* L. 2. Embryology of triploid species. Acta Biol Cracov Ser Bot 31: 97–113

Pullaiah T (1978a) Embryology of *Tithonia*. Phytomorphology 28: 437–444

Pullaiah T (1978b) Studies in the embryology of Compositae 3. The tribe Astereae. Bot Mag (Tokyo) 91: 197–205

Pullaiah T (1979a) Embryology of *Adenostemma*, *Elephantopus* and *Vernonia* (Compositae). Bot Not 132: 51–56

Pullaiah T (1979b) Studies in the embryology of Compositae 4. The tribe Inuleae. Am J Bot 66: 1119–1127

Pullaiah T (1981) Studies in the embryology of Heliantheae (Compositae). Plant Syst Evol 137: 203–214

Pullaiah T (1982) Embryology of *Sonchus oleraceus* Linn. Curr Sci 51: 708–710

Pullaiah T (1983a) Studies in the embryology of Senecioneae (Compositae). Plant Syst Evol 142: 61–70

Pullaiah T (1983b) Embryology of *Achillea squarrosa* Ait. (Anthemideae, Compositae). J Indian Bot Soc 62: 44–47

Pullaiah T (1984) Embryology of Compositae. Int Biosci Monogr (New Delhi) 13: 1–192

Pundir NS (1972) Experimental embryology of *Gossypium arboreum* L. and *G. hirsutum* L. and their reciprocal crosses. Bot Gaz 133: 7–26

Punt W (1975) Pollen morphology of the Dichapetalaceae with special reference to evolutionary trends and mutual relationships of pollen types. Rev Palaeobot Palynol 9: 1–97

Punt W, Leenhouts PW (1967) Pollen morphology and taxonomy in the Loganiaceae. Grana Palynol 7: 469–516

Purewal SS, Randhawa GS (1947) Studies in *Hibiscus esculentus* (lady's finger) 1. Chromosome and pollination studies. Indian J Agric Sci 17: 129–136

Puri V (1934) A note on the embryo sac and embryo of *Moringa oleifera* Lamk. Proc Indian Acad Sci B 1: 279–282

Puri V (1939a) Studies in the order Parietales 1. A contribution to the morphology of *Tamarix chinensis* Lour. Beih Bot Zentralbl 59A: 335–349

Puri V (1939b) Studies in the order Parietales 2. A contribution to the morphology of *Garcinia livingstonii* T. Anders. Proc Indian Acad Sci B 9: 74–86

Puri V (1941) The life history of *Moringa oleifera* Lamk. J Indian Bot Soc 20: 263–284

Puri V (1970) The angiosperm ovule. Presidential address, Sect Bot. In: Proc 57th Indian Sci Congr, Kharagpur, pt 6: 1–36

Puri V, Singh B (1935) Studies in the family Amaranthaceae 1. The life history of *Digera arvensis* Forsk. Proc Indian Acad Sci B 1: 893–908

Purohit KM, Panigrahi G (1983) A revision of the genus *Geum* (Rosaceae) in India. J Indian Bot Soc 62: 59–67

Pyykkö M (1968) Embryological and anatomical studies on Finnish species of the Pyrolaceae. Ann Bot (Fenn) 5: 153–165

Quibell CH (1941) Floral anatomy and morphology of *Anemopsis californica*. Bot Gaz 102: 749–758

Quisumbing E, Juliano JB (1927) Development of ovule and embryo sac of *Cocos nucifera*. Bot Gaz 84: 279–293

Radcliffe-Smith A (1987) Segregate families from the Euphorbiaceae. Bot J Lin Soc 94: 47–66

Radlkofer L (1879) Über *Cupania* und damit verwandte Pflanzen. Sitzungsber Kön Bayr Akad Wiss Math Phys Kl, pp 457–468

Radermacher A (1924) Die Gametophyten von *Nipa fruticans* und *Actinophloeus macarthuri*, sowie ein Versuch, die Systematik der Angiospermen durch die haploide Generation zu ergänzen. Ann Jardin Bot Buitenz 35: 1–54
Radford AE, Dickison WC, Massey JR, Bell CR (eds) (1974) Vascular plant systematics. Harper & Row, New York
Raghava Rao KV (1940) Gametogenesis and embryogeny in five species of the Convolvulaceae. J Indian Bot Soc 19: 53–69
Raghavan TS (1937) Studies in the Capparidaceae 1. The life history of *Cleome chelidonii* Linn. fil. Bot J Linn Soc 51: 43–72
Raghavan TS, Rangaswami K (1941) Studies in Rubiaceae 1. Development of female gametophyte and embryo formation in *Dentella repens* Forst. and *Oldenlandia* Koch., and some cyto-taxonomical considerations. J Indian Bot Soc 20: 341–356
Raghavan TS, Srinivasan AR (1941a) Cytogenetical studies in *Nicotiana* 2. Morphological features of *Nicotiana glutinosa* and the hybrid between *Nicotiana glutinosa* and *N. tabacum*. Proc Indian Acad Sci B 14: 35–46
Raghavan TS, Srinivasan AR (1941b) Studies in Rubiaceae 2. *Spermacoce hispida* Linn. and *Guettarda speciosa* Linn., and some cytomorphological considerations. Proc Indian Acad Sci B 14: 412–426
Raghavan TS, Srinivasan AR (1941c) Cytomorphological features of *Portulaca tuberosa* Roxb. Proc Indian Acad Sci B 14: 472–488
Raghavan TS, Srinivasan VK (1940) A contribution to the life history of *Bergia capensis*. J Indian Bot Soc 19: 283–291
Raghavan TS, Srinivasan VK (1941a) Morphological and cytological studies in the Scrophulariaceae 3. A contribution to the life history of *Ilysanthes parviflora* Benth. Proc Indian Acad Sci B 13: 24–32
Raghavan TS, Srinivasan VK (1941b) Morphological and cytological studies in Scrophulariaceae 4. The development of the embryo sac and endosperm in *Scoparia dulcis* Linn. Proc Indian Acad Sci B 13: 229–234
Raghavan TS, Srinivasan VK (1942) A contribution to the life history of *Vahlia viscosa* Roxb. and *Vahlia oldenlandioides* Roxb. Proc Indian Acad Sci B 15: 83–105
Raghavan TS, Venkatasubban KR (1941a) Studies in Capparidaceae 7. The floral morphology of *Crataeva religiosa* Forst. Proc Indian Acad Sci B 13: 235–243
Raghavan TS, Venkatasubban KR (1941b) A contribution to the morphology and cytology of *Alpinia calcarata* Rosc. with special reference to the theory of zingiberous flowering. Proc Indian Acad Sci B 13: 325–344
Raghavan TS, Venkatasubban KR, Wülff HD (1939) Division of the generative cell in *Impatiens balsamina*. Cytologia 9: 389–392
Raghavan V (1986a) Embryogenesis in angiosperms: The changing scene. In: Raghavan V (ed) Embryogenesis in angiosperms. A developmental and experimental study. Univ Press, Cambridge (England), pp 1–12
Raghavan V (1986b) Developmental embryogenesis. In: Raghavan V (ed) Embryogenesis in angiosperms. A developmental and experimental study. Univ Press, Cambridge (England), pp 13–45
Raghavan V (1988) Anther and pollen development in rice (*Oryza* sativa). Am J Bot 75: 183–196
Raghuvanshi RK, Singh D (1981) Development and structure of resorption and structure of resorption tissue in *Capsicum* L. Curr Sci 50: 30–31
Raghuvanshi RK, Singh D (1984) Embryological manifestations of male sterility in *Capsicum* L. Acta Bot Indica 12: 213–215
Raj B (Baldev) (1964) Female gametophyte of *Buckleya lanceolata* Sieb. et Zucc. Curr Sci 33: 348–349
Raj B (1969) Morphological and embryological studies in the family Santalaceae – *Buckleya lanceolata* Sieb. et Zucc. Beitr Biol Pflanzen 46: 193–209
Raj B (1970a) Morphological and embryological studies in the family Loranthaceae 13. *Amylotheca dictyophleba* Van Tiegh. Öst Bot Z 118: 417–430
Raj B (1970b) Loranthaceae. In: Symp Comparative embryology of angiosperms. Indian Natl Sci Acad Bull 41: 22–28
Raj B (Bhoj) (1961) Pollen morphological studies in Acanthaceae. Grana Palynol 3: 1–108

Rajashekhar G (1980) The development of embryo sac, endosperm and the structure of fruit wall in *Blumea mollis* (D. Don) Merr. Curr Sci 49: 156–158
Rajashekhar G (1981) Ontogeny of the ovule, seed-coat and fruit wall of *Dicoma tomentosa* Cass. (Asteraceae). Proc Indian Acad Sci Plant Sci 90: 177–182
Rajeswari Devi KR, Prakasa Rao PS (1975) Embryo development in *Euphorbia peplus* L. Curr Sci 44: 57–59
Raju D (1973) Embryological studies in the tribe Selaginae: *Agathelpis dubia*. Phytomorphology 23: 194–205
Raju D (1976) Gametogenesis, endosperm development and embryogeny in *Microcarpaea muscosa* (Scrophulariaceae, Gratiolae). Plant Syst Evol 126: 209–220
Raju D, Arekal GD (1977) Contribution to the embryology of *Melampyrum pratensis* L. Proc Indian Acad Sci B 86: 23–31
Raju D, Arekal GD (1978) Endosperm and embryo in *Linaria ramosissima*. Phytomorphology 28: 88–95
Raju D, Chamaiah HN (1972) Contribution to the embryology of *Christisonia subacaulis* Gardn. (Orobanchaceae). Proc Indian Acad Sci B 74: 129–139
Raju MVS (1952a) Embryology of Sabiaceae. Curr Sci 21: 107–108
Raju MVS (1952b) Embryology of Passifloraceae. Curr Sci 21: 288–289
Raju MVS (1952c) Embryology of *Anagallis pumila* Swartz. Proc Indian Acad Sci B 36: 34–42
Raju MVS (1956a) Development of embryo and seed-coat in *Turnera ulmifolia* L. var. *angustifolia* Willd. Bot Not 109: 308–312
Raju MVS (1956b) Embryology of the Passifloraceae 1. Gametogenesis and seed development of *Passiflora calcarata* Mast. J Indian Bot Soc 35: 126–137
Raju MVS (1957) Some aspects of the embryology of *Dianella nemorosa* Lamk. J Indian Bot Soc 36: 223–226
Raju MVS (1958) Seed development and fruit dehiscence in *Ionidium suffruticosum* Ging. Phytomorphology 8: 218–224
Raju MVS (1961) Morphology and anatomy of the Saururaceae 1. Floral anatomy and embryology. Ann Missouri Bot Garden 48: 107–124
Raju MVS, Nagaraja Rao A (1953) The development of the male and female gametophytes in *Mallotus albus* Muell. J Mysore Univ B 13: 5–8
Ram L (1982) Embryology of *Myriactis wightii*. Phytomorphology 32: 234–240
Ram L (1986) Embryology of *Pluchea tomentosa* DC. J Indian Bot Soc 65: 47–52
Ram M (= Ghosh M) (1956) Floral morphology and embryology of *Trapa bispinosa* Roxb. with a discussion on the systematic position of the genus. Phytomorphology 6: 312–323
Ram M (1957) Morphological and embryological studies in the family Santalaceae 1. *Comandra umbellata* (L.) Nutt. Phytomorphology 7: 24–35
Ram M (1959a) Morphological and embryological studies in the family Santalaceae 2. *Exocarpus* with a discussion on its systematic position. Phytomorphology 9: 4–19
Ram M (1959b) Morphological and embryological studies in the family Santalaceae 3. *Leptomeria* R. Br. Phytomorphology 9: 20–33
Ram M (1959c) Occurrence of embryo sac-like structures in the microsporangia of *Leptomeria billardierii* R. Br. Nature (London) 184: 914–915
Ram M (1960) Occurrence of endosperm haustorium in *Cannabis sativa* L. Ann Bot (London) 24: 79–82
Ram M (1970a) Olacaceae. In: Symp Comparative embryology of angiosperms. Indian Natl Sci Acad Bull 41: 4–8
Ram M (1970b) Opiliaceae. In: Symp Comparative embryology of angiosperms. Indian Natl Sci Acad Bull 41: 9–12
Ram M (1970c) Grubbiaceae. In: Symp Comparative embryology of angiosperms. Indian Natl Sci Acad Bull 41: 13–14
Ram M (1970d) Myzodendraceae. In: Symp Comparative embryology of angiosperms. Indian Natl Sci Acad Bull 41: 19–21
Ramadevi D, Narayana LL (1990) Morphology of the flower and fruit of *Hydrocera triflora* Wight and Arn. emend Venkat. and Dutt: an elucidation. Proc Indian Acad Sci B 100: 43–50
Ramamurthi BA (1958) Life history of *Heliotropium curassavicum*. J Vikram Univ (Ujjain, India) 2: 134–137

Raman SS (1954) Gametogenesis and fertilization of *Stephegyne parviflora* Korth. Agra Univ J Res Sci 3: 343–348
Ramanna MS (1979) A re-examination of the mechanisms of 2n gamete formation in potato and its implications for breeding. Euphytica 28: 537–561
Ramaswamy N, Bir B (1984) Pollen flow in dimorphic *Turnera subulata* (Turneraceae). New Phytol 98: 205–209
Ramaswamy SN (1967) A contribution to the embryology of *Burmannia coelestis* Dm. In: Proc 54th Indian Sci Congr, Hyderabad, pt 3: 331
Ramaswamy SN (1970) Burmanniaceae. In: Symp Comparative embryology of angiosperms. Indian Natl Sci Acad Bull 41: 375–379
Ramaswamy SN, Arekal GD (1981a) Embryology of *Eriocaulon setaceum* L. (Eriocaulaceae). Plant Syst Evol 138: 175–188
Ramaswamy SN, Arekal GD (1981b) Is there a proliferation of antipodal cells in *Leiothrix* Ruhland (Eriocaulaceae)? Curr Sci 50: 382–383
Ramaswamy SN, Arekal GD (1981c) The development of male and female gametophytes in two Indian species of *Eriocaulon* L. Proc Indian Natl Sci Acad B 47: 682–693
Ramaswamy SN, Arekal GD (1982a) On the embryogeny of three taxa of Paepalanthoideae (Eriocaulaceae). Ann Bot (London) 49: 99–102
Ramaswamy SN, Arekal GD (1982b) Embryology of *Eriocaulon xeranthemum* Mart. Acta Bot Neerl 31: 41–54
Ramaswamy SN, Raju MVS (1982) The embryo sac of *Xyris schoenoides* Mart. (Xyridaceae). Bull Torrey Bot Club 109: 325–329
Ramaswamy SN, Arekal GD, Raju MVS (1983) Developmental anatomy of seed-coat and pericarp in two species of *Eriocaulon* L. (Eriocaulaceae). Bull Torrey Bot Club 110: 287–291
Ramaswamy SN, Swamy BGL, Arekal GD (1981) From zygote to seedling in *Eriocaulon robusta-brownianum* Ruhl. Beitr Biol Pflanzen 55: 179–188
Ramchandani S, Joshi PC, Pundir NS (1966) Seed development in *Gossypium* Linn. Indian Cotton J 20: 97–106
Ramji MV (1961) Ontogenetic studies in *Sarcandra irvingbaileyi* Swamy 1. The structure of the shoot apex and ontogeny of leaf. Proc Indian Acad Sci B 53: 20–35
Ramji MV (1975) Histology of growth with regard to embryos and apical meristems in some angiosperms 1. Embryogeny of *Stellaria media*. Phytomorphology 25: 131–145
Ramji MV, Padmanabhan D (1965) Developmental studies on *Cabomba caroliniana* Gray 1. Ovule and carpel. Proc Indian Acad Sci B 62: 215–223
Rangaswamy K (1934) Contribution to the life history of *Vallisneria spiralis*. J Indian Bot Soc 13: 129–148
Rangaswamy K (1941a) Cytomorphological studies in *Asteracantha longifolia* (*Hydrophilla spinosa* T. Hud.). Proc Indian Acad Sci B 14: 149–165
Rangaswamy K (1941b) A morphological study of the flower of *Blyxa echinosperma*. J Indian Bot Soc 20: 123–134
Rangaswamy NS, Nandakumar L (1985) Correlative studies on seed-coat structure, chemical composition and impermeability in the legume *Rhynchosia minima*. Bot Gaz 146: 501–509
Rangaswamy V, Pullaiah T (1986) Studies in the embryology of *Senecio candicans* DC. (Compositae). J Indian Bot Soc 65: 509–512
Rao AVN (1967) Embryological studies in *Cleome monophylla* Linn. Proc Indian Acad Sci B 65: 249–256
Rao BA, Mukherjee PK (1974) A note on the embryology of *Polygonum glabrum* Willd. Curr Sci 42: 250–251
Rao D (1966) A contribution to the floral anatomy and embryology of Fumariaceae, Linaceae and Erythroxylaceae. Ph D Thesis, Osmania Univ, Hyderabad (India)
Rao D (1968a) A contribution to the embryology of Erythroxylaceae. Proc Natl Inst Sci India B 38: 53–65
Rao D (1968b) Contributions to the embryology of Commelinaceae 1. Proc Natl Inst Sci India B 38: 81–89
Rao D (1970a) Simaroubaceae. In: Symp Comparative embryology of angiosperms. Indian Natl Sci Acad Bull 41: 142–147

Rao D (1970b) Burseraceae. In: Symp Comparative embryology of angiosperms. Indian Natl Sci Acad Bull 41: 148–156
Rao D, Narayana LL (1965) Embryology of Linaceae. Curr Sci 34: 92–93
Rao LN (1932) Studies in the life history of *Balanophora indica*. Curr Sci 1: 134–135
Rao LN (1942) Studies in the Santalaceae. Ann Bot (London) 6: 151–175
Rao NGP, Narayana LL, Reddy RN (1978) Apomixis and its utilization in grain *Sorghum* 1. Embryology of two apomictic parents. Caryologia 31: 427–433
Rao PN (1962) A note on the embryology of *Micrococca mercurialis* Benth. Curr Sci 31: 426–427
Rao PN (1970) Euphorbiaceae. In: Symp Comparative embryology of angiosperms. Indian Natl Sci Acad Bull 41: 136–141
Rao PN (1976) Embryology of *Croton klotzschianus*. J Indian Bot Soc 55: 84–89
Rao PN, Narayana DS (1980) Occurrence and identification of semigamy in *Coix aquatica*. J Hered 71: 117–120
Rao PN, Sarveswara Rao D (1976) Embryology of Cassava (*Manihot*). Proc Indian Natl Sci Acad B 42: 111–121
Rao PS, Hanumantha Rao B (1975) Anther development in *Cleome tenella* Linn. Curr Sci 44: 438–440
Rao PS, Hanumantha Rao B (1976) Embryo development in *Cleome aspera* L. Jpn J Bot 51: 110–117
Rao PS, Hanumantha Rao B (1977) Anther in *Capparis decidua* (Forsk.) Pax. Curr Sci 46: 683–684
Rao PS, Prakasa DK, Rajeswari Devi KR (1975) Embryo development in *Euphorbia marginata* Pursh. Jpn J Bot 50: 59–64
Rao PS, Prakasa DK, Rajeswari Devi KR (1977) A note on the occurrence of intracarpellary pollen sacs in *Securinega virosa*. Jpn J Bot 52: 60–62
Rao VS (1936a) Studies on Capparidaceae 1. The embryo sac of *Maerua arenaria* Forsk. J Indian Bot Soc 15: 71–75
Rao VS (1936b) A contribution to the morphology of *Antigonon leptopus* Hook. & Arn. J Indian Bot Soc 15: 105–114
Rao VS (1936c) Studies on Capparidaceae 2. The embryology of *Gynandropsis pentaphylla*. J Indian Bot Soc 15: 335–344
Rao VS (1938) Studies on Capparidaceae 3. Genus *Capparis*. J Indian Bot Soc 17: 69–80
Rao VS (1944) Development of the embryo sac in the Convolvulaceae. J Indian Bot Soc 23: 164–169
Rao VS, Dahlgren R (1969) The floral anatomy and relationships of Oliniaceae. Bot Not 122: 160–171
Rathore RKS, Singh RP (1968) Embryological studies in *Brassica campestris* L. var. Yellow Sarson Prain. J Indian Bot Soc 47: 341–349
Ratnaparkhi S (1972) Co-existence of monosporic and bisporic embryo sacs in *Tellima grandiflora* Pursh. Phyton (Argentina) 15: 39–42
Rau MA (1962) Review of recent work on the embryogeny of some families and genera of disputed systematic position. In: Symp Plant embryology. Council Sci Ind Res, New Delhi, pp 75–80
Rau MA (1970) Saxifragaceae. In: Symp Comparative embryology of angiosperms. Indian Natl Sci Acad Bull 41: 90–92
Rau MA, Sharma VK (1967) Early embryogeny in *Saxifraga* Linn. (Panchanan Maheshwari Mem. Vol.). Phytomorphology 17: 252–255
Rau MA, Sharma VK (1970a) Coriariaceae. In: Symp Comparative embryology of angiosperms. Indian Natl Sci Acad Bull 41: 156–157
Rau MA, Sharma VK (1970b) Eleagnaceae. In: Symp Comparative embryology of angiosperms. Indian Natl Sci Acad Bull 41: 185–187
Rauh W, Jäger-Zürn I (1966) Zur Kenntnis der Hydrostachyaceae, 1. Teil: Blütenmorphologische und embryologische Untersuchungen an Hydrostachyaceen unter besonderer Berücksichtigung ihrer systematischen Stellung. Sitzung der Heidelberg Akad Wiss Math Naturwiss Kl 1: 1–117
Rauh W, Schölch HF (1965) Weitere Untersuchungen an Didiereaceen, 2. Teil: Infloreszenz-,

blütenmorphologische und embryologische Untersuchungen mit Ausblick auf die systematische Stellung der Didiereaceen. Sitzung der Heidelberg Akad Wiss Math Naturwiss Kl 3–4: 217–443

Rauh W, Barthlott W, Ehler N (1975) Morphologie und Funktion der Testa staubförmiger Flugsamen. Bot Jahrb 96: 353–374

Raven PH (1975) The bases of angiosperm phylogeny: Cytology. Ann Missouri Bot Garden 62: 724–764

Ravishankar S, Gulati N (1980) Embryology of *Symplocos cochinchinensis* Moore. (Abstr). In: Proc 67th Indian Sci Congr, Calcutta, pt 3: 42

Razi BA (1949) Embryological studies of two members of the Podostemaceae. Bot Gaz 111: 211–218

Razi BA (1955) Some aspects of the embryology of *Zeylanidium olivaceum* (Tul.) Engl. and *Lawia zeylanica* Tul. Bull Bot Soc Bengal 9: 36–41

Razi BA (1966) Some observations on embryology of the Podostemaceae. In: Mahabale TS (ed) Proc Autumn School Bot Mahabaleshwar, Univ Poona (India), pp 303–315

Razi BA, Subramanyam K (1952) Embryology of the Dipsacaceae 1. Glands, the male and female gametophytes. Proc Indian Acad Sci B 36: 249–257

Razmologov VP (1958) Embryologic investigations of the orchid, *Phalaenopsis schilleriana*. Byul Glavn Bot Sad Akad Nauk 32: 67–72 (in Russian)

Reddi EUB, Reddi CS (1985) Wind and insect pollination in a monoecious and a dioecious species of Euphorbiaceae. Proc Indian Natl Sci Acad B 51: 468–482

Reddy GN, Kulkarni AR (1985) Contribution to the anatomy of palm fruits – Cocosoid palms. Proc Indian Acad Sci Plant Sci 95: 153–165

Reddy GN, Kulkarni AR (1988) Cases of polyembryony in Cocosoid palms. Curr Sci 57: 733–734

Reddy KB, Murthy S, Narayana R (1966) Male and female gametophytes in *Vaccinium leschenaultii* W. Curr Sci 55: 130–131

Reddy PS (1977) Evolution of apomictic mechanisms in Gramineae: A concept. Phytomorphology 27: 45–49

Reddy PS, D'Cruz R (1969a) Mechanism of apomixis in *Dichanthium annulatum* (Forsk.) Stapf. Bot Gaz 130: 71–79

Reddy PS, D'Cruz R (1969b) Development of caryopsis in *Dichanthium armatum* (Hook. f.) Blatt et McCann. J Indian Bot Soc 48: 351–359

Reddy RN, Narayana LL, Rao NGP (1979) Apomixis and its utilization in grain sorghum 2. Embryology of F_1 progeny of reciprocal crosses between R473 and 302. Proc Indian Acad Sci B 88: 455–461

Reed CF (1955) The comparative morphology of the Olacaceae, Opiliaceae and Octoknemataceae. Mem Soc Broteriana 10: 29–79

Reed HL (1924) Anatomy, embryology and ecology of *Arachis hypogaea*. Bot Gaz 78: 289–310

Reed HS (1903) The development of the macrosporangium of *Yucca filamentosa*. Bot Gaz 35: 209–214

Reeder JR (1953) The embryo of *Streptochaeta* and its bearing on the homology of the coleoptile. Am J Bot 40: 77–80

Reeder JR (1957) The embryo in grass systematics. Am J Bot 44: 756–768

Reeder JR (1962) The bambusoid embryo: A reappraisal. Am J Bot 49: 639–641

Reeder JR (1982) Systematics of the tribe Orcutlineae (Gramineae) and the description of a new segregate genus, *Tuctoria*. Am J Bot 69: 1082–1095

Reese H (1989) Entwicklung von Perikarp und Testa bei Calenduleae und Arctotideae (Asteraceae) ein Beitrag zur systematik. Bot Jahrb 110: 325–419

Rees-Leonard OL (1935) Macrosporogenesis and development of the macrogametophyte of *Solanum tuberosum*. Bot Gaz 96: 734–750

Reeves JH (1957) Megasporogenesis and megagametogenesis in *Polygala polygana*. Virginia Coop Wild Life Res Release 57: 1–3, and Virg J Sci 8: 303

Reeves RG (1930a) Nuclear and cytoplasmic division in the microsporocytes of alfalfa. Am J Bot 17: 29–40

Reeves RG (1930b) Development of ovule and embryo sac of alfalfa. Am J Bot 17: 239–246
Regen L (1941) The development of the embryo sac in *Agave virginica*. Bull Torrey Bot Club 68: 229–236
Rejdali M (1990) Seed morphology and taxonomy of the North African species of *Sideritis* (Lamiaceae). Bot J Linn Soc 103: 317–324
Rembert DH Jr (1966) Megasporogenesis in *Laburnum anagyroides* Medik. A case of bisporic development in Leguminosae. Trans Ky Acad Sci 27: 47–50
Rembert DH Jr (1969a) Comparative megasporogenesis in Caesalpiniaceae. Bot Gaz 130: 47–52
Rembert DH Jr (1969b) Comparative megasporogenesis in Papilionaceae. Am J Bot 56: 584–591
Rembert DH Jr (1971) Phylogenetic significance of megaspore tetrad patterns in Leguminales. Phytomorphology 21: 1–9
Rembert DH Jr (1977) Ovule ontogeny, megasporogenesis, and early gametogenesis in *Trifolium repens* (Papilionaceae). Am J Bot 64: 483–488
Rendle AB (1952) The classification of flowering plants, vol 2. Dicotyledons. Univ Press, Cambridge (England)
Rendle AB (1959) The classification of flowering plants, vol 1. Gymnosperms and monocotyledons. Univ Press, Cambridge (England)
Rendle AB (1971) The classification of flowering plants, vol 2. Dicotyledons. Univ Press, Cambridge (England)
Renzoni-Cela G (1963) Ricerche cito-embriologiche e distribuzione geografica di *Cirsium casabonae* Lam. et D.C. (Compositae). Nuovo G Bot Ital 70: 493–504
Renzoni-Cela G (1969) *Matthiola tricuspidata* R. Br. (Cruciferae), analisi cariologica ed embriologica. Nuovo G Bot Ital 103: 531–545
Renzoni-Cela G (1972) Studies on the genus *Centaurea* L. (Asteraceae) 1. Embryology of *Centaurea cineraria* L. var. *veneris* Sommier. Nuovo G Bot Ital 104: 457–468 (in Italian)
Reynolds JD (1968) Morphological studies in Diapensiaceae 1. *Pyxidanthera brevifolia* Wells. Bull Torrey Bot Club 95: 653–656
Reznickova SA, Willemse MTM (1980) Formation of pollen in the anther of *Lilium* 2. The function of the surrounding tissue in the formation of pollen and pollen wall. Acta Bot Neerl 29: 141–156
Ricardi SM (1967) Revisión taxonómica de las Malesherbiáceas. Gayana Bot 16: 1–139
Richards AJ (1973) The origin of *Taraxacum* agamospecies. Bot J Linn Soc 66: 189–211
Richards AJ, Ibrahim HB (1982) The breeding system in *Primula veris* L.2. Pollen tube growth and seed-set. New Phytol 90: 305–314
Ridlle LC (1905) Development of the embryo sac and embryo of *Staphylea trifoliata*. Ohio Nat 5: 320–325
Rifot M (1973) Evolution structurale du pole chalazien du sac embryonnaire d'*Aquilegia vulgaris* en liaison avec son activité trophique. CR Acad Sci Paris 277: 1313–1316
Riley HP (1942) Development of the embryo sac of *Iris fulva* and *I. hexagona* var. *giganticoerula*. Trans Am Microsc Soc 61: 328–335
Riley NP (1963) Families of flowering plants of South Africa. Kentucky, USA
Ritter F (1980) Kakteen in Südamerika, vol 4., Ritter Selbstverlag, Spangenberg [see Bregman R, Bouman F (1983) Bot J Linn Soc 86: 357–374]
Ritzerow J (1908) Über Bau und Befruchtung kleistogamer Blüten. Flora 98: 163–212
Robertson BL (1976a) Embryology of *Jubaeopsis caffra* Becc. 2. Megasporangium, megasporogenesis and megagametogenesis. J S Afr Bot 42: 173–184
Robertson BL (1976b) Embryology of *Jubaeopsis caffra* Becc. 3. Endosperm and embryogeny. J S Afr Bot 42: 285–299
Robeva-Davidova PN, Fileva-Mikhajlova MA (1979) Etude embryologique de *Thalictrum minus* L. 1. Etude du gamétophyte femelle. Fitologia 11: 15–25
Robyns W (1972) Outline of a new system of orders and families of Sympetalae. Bull Jardin Bot Nat Belg 42: 363–372
Rocén T (1926) Några drag ur caryophyllaceernas embryologi. Svensk Bot Tidskr 20: 97–98
Rocén T (1927) Zur Embryologie der Centrospermen. Diss, Univ Uppasala

Rodionova GB (1972) Development of the embryo and integuments in *Heliophila amplexicaulis* L. Bot Zh 57: 1153–1161 (in Russian)
Rodionova GB (1983) The place of Cruciferae and their relative families in the system according to the embryology data. In: Erdelská O (ed) Fertilization and embryogenesis in ovulated plants. VEDA, Bratislava (Czech), pp 347–350
Rodkiewicz B (1961) Phases of embryo sac development and pollen grain development in *Lilium candidum* (in Polish). Acta Soc Bot Pol 30: 749–754
Rodkiewicz B (1967) Walls with callose in the megaspores and hypostase of ovules of *Antirrhinum majus* observed in a fluorescence microscope. Bull Acad Pol Sci 15: 493–495
Rodkiewicz B (1970) Callose in cell walls during megasporogenesis in angiosperms. Planta 93: 39–47
Rodkiewicz B, Bednara J, Pora H (1971) Alternative localization of the active megaspore in tetrads in *Oenothera muricata*. Bull Acad Pol Sci 19: 691–694
Rodriguez RL (1971) The relationship of the Umbellales. In: Heywood VH (ed) The biology and chemistry of the Umbelliferae. Academic Press, New York London, pp 63–91
Roever WE (1935) Nuclear behaviour in the tapetum of *Hosta caerulea*, with special reference to the divisions. Bull Torrey Bot Club 62: 345–358
Rogers S (1969) Some observations on the reproduction of *Tridax procumbens* L. 2. Development of embryo sac and embryo. Cytologia 34: 196–201
Rogers SO, Squatrano RS (1983) Morphological staging of wheat caryopsis development. Am J Bot 70: 308–311
Rohweber O (1970) Cetrospermen-Studien 4. Morphologie und Anatomie der Blüten, Früchte und Samen bei Alsinoideen und Poronychioideen s.l. Bot Jahrb 90: 201–271
Romanov ID (1936) Die Embryosackentwicklung in der Gattung *Gagea*. Planta 25: 438–458
Romanov ID (1938) Eine neue Form des Embryosackes von Adoxa-Typus bei *Tulipa tetraphylla* und *T. ostrovskina*. Dokl Akad Nauk SSSR Bot 19: 113–115
Romanov ID (1939) Zwei neue Embryosackformen in der Gattung *Tulipa*. Dokl Akad Nauk SSSR Bot 22: 140–142
Romanov ID (1957) Embryo sac in the genus *Tulipa*. Dokl Akad Nauk SSSR Bot 115: 1025–1027 (in Russian)
Romanov ID (1959) The embryo sac and pollen morphology in *Tulipa*. 9th Int Bot Congr (Montreal) 2: 331–332
Roper RB (1952) The embryo sac of *Butomus umbellatus* L. Phytomorphology 2: 61–74
Roscoe MV (1927) Cytological studies in the genus *Typha*. Bot Gaz 84: 392–406
Rosen W (1932) Zur Embryologie der Campanulaceen und Lobeliaceen. Acta Hortic Gothob 7: 31–42
Rosen W (1935) Beiträge zur Embryologie der Stylidiaceen. Bot Not 1935: 273–278
Rosen W (1937) Beiträge zur Kenntnis der Embryologie der Goodeniaceen. Acta Hortic Gothob 12: 1–10
Rosen W (1940a) Notes on the embryology of *Globularia vulgaris*. Bot Not 1940: 252–261
Rosen W (1940b) Seed development of *Litorella uniflora*. Acta Hortic Gothob 14: 177–184
Rosen W (1946) Further notes on the embryology of the Goodeniaceae. Acta Hortic Gothob 16: 235–249
Rosen W (1947) The female gametophyte in *Nolana* and endosperm development in Tubiflorae. Bot Not 1947: 372–382
Rosen W (1949) Endosperm development in Campanulaceae and closely related families. Bot Not 1949: 137–147
Rosenberg O (1901) Über die Embryologie von *Zostera marina*. Kon Svensk Vatensk Akad Handl 27: 1–26
Rosenberg O (1909) Cytologische und morphologische Studien an *Drosera longifolia* × *rotundifolia*. Kon Svensk Vet Akad Handl 43: 1–64
Rosenberg O (1946) The influence of low temperature on the development of the embryo sac mother cell in *Lilium longiflorum*. Hereditas 32: 65–92
Rosendahl CO (1909) Embryo sac development and embryology of *Symplocarpus foetidus*. Minn Bot Stud 4: 1–9

Rosenthal K (1931) Daphniphyllaceae. In: Engler A, Prantl K (eds) Die natürlichen Pflanzenfamilien 19C: 233–235
Ross WM, Wilson JA (1969) Polyembryony in *Sorghum*. Crop Sci 9: 842–843
Rossignol L, Rossignol M, Haicour R (1987) A systematic revision of *Phyllanthus* subsection *Urinaria* (Euphorbiaceae). Am J Bot 74: 1853–1862
Rost TL, De Artucio PI, Risley EB (1984) Transfer cells in the placental pad and caryopsis coat of *Pappophorum subbulbosum* Arech. (Poaceae) Am J Bot 71: 948–957
Rost TL, Simper AD, Schell P, Allen S (1977) Anatomy of Jojoba (*Simmondsia chinensis*) seed and the utilization of liquid wax during germination. J Econ Bot 31: 146–148
Roth I, Lindorf H (1974a) The morphological interpretation of the seed of the Rubiaceae and especially that of coffee. Acta Bot Venezuela 9: 141–147 (in Spanish)
Roth I, Lindorf H (1974b) Desarrollo Y anatomía del fruta y de la semilla de *Myristica fragrans* van Houtt. Acta Bot Venezuela 9: 149–176
Rousi A (1956) Cytotaxonomy and reproduction in the apomictic *Ranunculus auricomus* complex. Ann Bot Soc Vanamo 29: 1–64
Roy B (1933) Studies in the development of the female gametophyte in some leguminous crop plants of India. Indian J Agric Sci 1933: 1098–1107
Roy SK (1953) Embryology of *Eugenia jambos* L. Curr Sci 22: 249–250
Roy SK (1955) Embryology of *Eugenia bracteata* Roxb. Curr Sci 24: 348–349
Roy SK (1960) Embryology of *Eugenia malaccensis* Lam. Curr Sci 29: 189
Roy SK (1962a) Embryology of *Eugenia myrtifolia* Roxb. Sci Cult 28: 376–378
Roy SK (1962b) A contribution to the embryology of *Myrtus communis* L. Proc Natl Inst Sci India B 32: 305–311
Roy SK, Sahai R (1962) The embryo sac and embryo of *Syzygium caryophyllifolium* DC. J Indian Bot Soc 41: 45–51
Rozanova MA (1932) Versuch einer analytischen Monographie conspecies *Ranunculus auricomus* Korsh. Trav Inst Sci Nat Peterhoff 8: 19–148
Rozenblum E, Maldonado S, Waisman CE (1988) Apomixis in *Eupatorium tanacetifolium* (Compositae). Am J Bot 75: 311–322
Rudall PJ, Dransfield S (1989) Fruit structure and development in *Dinochloa* and *Ochlandra* (Gramineae, Bambusoideae). Ann Bot (London) 63: 29–38
Rudall PJ, Linder HP (1988) Megagametophyte and nucellus in Restionaceae and Flagellariaceae. Am J Bot 75: 7777–7786
Rudall PJ, Owens SJ, Kenton AY (1984) Embryology and breeding systems in *Crocus* (Iridaceae). A study in causes of chromosome variation. Plant Syst Evol 148: 119–134
Russell A (1919) A comparative study of *Floerkea proserpinacoides* and allies. Contrib Bot Lab Univ Penn 4: 401–418
Russell SD (1980a) Gamete delivery and syngamy in *Plumbago zeylanica*. Bot Soc Am Misc Publ Ser 158: 98–99
Russell SD (1980b) Participation of male cytoplasm during gamete fusion in an angiosperm, *Plumbago zeylanica*. Science 210: 200–201
Russell SD (1982) Fertilization in *Plumbago zeylanica*: Entry and discharge of the pollen tube in the embryo sac. Can J Bot 60: 2219–2230
Russell SD (1983) Fertilization in *Plumbago zeylanica*. Gametic fusion and fate of the male cytoplasm. Am J Bot 70: 416–434
Russell SD (1984) Ultrastructure of the sperm of *Plumbago zeylanica* 2. Quantitative cytology and three-dimensional organization. Planta 162: 385–391
Russell SD (1989) Preferential fertilization in the synergid-lacking angiosperm *Plumbago zeylanica*. Phytomorphology 39: 1–20
Russell SD, Cass DD (1978) Post-fertilization development in *Plumbago zeylanica* (Abstr). Bot Soc Am Misc Publ Ser 156: 41
Russell SD, Cass DD (1981) Ultrastructure of fertilization in *Plumbago zeylanica*. Acta Soc Bot Pol 50: 185–190
Russell SD, Cass DD (1983) Unequal distribution of plastids and mitochondria during sperm cell formation in *Plumbago zeylanica*. In: Mulcahy DL, Ottaviano E (eds) Pollen: biology and implications for plant breeding. Elsevier, Amsterdam, pp 135–140

Russell SD, Cass DD (1988) Fertilization in *Plumbagella micrantha*. Am J Bot 75: 778–781
Rutgers FL (1923) Embryo sac and embryo of *Moringa oleifera* Lamk. Ann Jardin Bot Buitenz 33: 1–66
Rutishauser A (1935) Entwicklungsgeschichtliche und zytologische Untersuchungen an *Korthalsella dacrydii*. Ber Schweiz Bot Ges 44: 389–436
Rutishauser A (1937) Blütenmorphologische und embryologische Untersuchungen an den Viscoideen *Korthalsella opuntia* Merr. und *Ginalloa linearis* Dans. Ber Schweiz Bot Ges 47: 5–28
Rutishauser A (1943) Über die Entwicklungsgeschichte pseudogamer Potentillen. Arch Julius Klaus Stift 18: 687–691
Rutishauser A (1954) Entwicklungserregung der Eizelle bei pseudogamen Arten der Gattung *Ranunculus*. Bull Schweiz Akad Med Wiss 10: 491–512
Ruys JD (1925) Contribution à l'histoire du développement des Mélastomatacées. Enumération des plantes phanérogames angiospermes examinées au point de vue de la Karyologie. Ann Jardin Bot Buitenz 34: 1–123
Rychlewski J (1961) Cytoembryological studies in the apomictic species *Nardus stricta* L. Acta Biol Cracov Ser Bot 4: 1–23

Sabet YS (1931) Development of the embryo sac in *Calotropis procera* with special reference to the endosperm formation. Ann Bot (London) 45: 503–518
Sachar RC (1953) A reinvestigation of the embryology of *Argemone mexicana*. Curr Sci 22: 381–383
Sachar RC (1955) The embryology of *Argemone mexicana* L.: A reinvestigation. Phytomorphology 5: 200–218
Sachar RC (1956a) The embryogeny of *Argemone mexicana* L.: A criticism. Phytomorphology 6: 148–151
Sachar RC (1956b) The embryology of *Isomeris*: A reinvestigation. Phytomorphology 6: 346–363
Sachar RC, Arora U (1963) Some embryological aspects of *Amomum dealbatum* and *Hedychium accuminatum*. Bot Gaz 124: 353–360
Sachar RC, Chopra RN (1957) A study of endosperm and embryo in *Mangifera indica*. Indian J Agric Sci 27: 219–228
Sachar RC, Mohan Ram HY (1958) The embryology of *Eschscholtzia californica*. Phytomorphology 8: 114–124
Sachar RC, Murgai P (1958) Embryology of *Aerva tomentosa* Forsk. Curr Sci 27: 105–107
Sachar RC, Murgai P (1959) Diplosporous parthenogenesis in *Aerva tomentosa* Forsk. Phytomorphology 9: 119–129
Schlimbach H (1924) Beiträge zur Kenntnis der Samenanlagen und Samen der Amaryllidaceen mit Berücksichtigung der Wassergehaltes der Samen. Flora 117: 41–54
Safeeulla KM, Govindu HC (1950) Development of the female gametophyte and endosperm in *Bacopa hamiltoniana*. Lloydia 13: 179–182
Safijowska LD (1935a) Spermatogenesis bei Campanulaceae. Bull Sci Univ Kyew 1: 265–278
Safijowska LD (1935b) Do embriologii *Adenophora liliiflora*. J Inst Bot Acad Sci Ukr: 85–98
Safijowska LD (1960) Male gametophyte in *Enkianthus*. Bot Gaz 121: 185–190
Sagawa J, Nimoto DH (1961) Ovule development in *Dendrobium phalaenopsis*. ASB Bull 8: 33
Sahai A, Shivanna KR (1985a) Seed germination and seedling growth in *Sopubia delphinifolia*. A hemi-root-parasite: Germination requirements. Ann Bot (London) 55: 775–783
Sahai A, Shivanna KR (1985b) Seed germination and seedling growth in *Sopubia delphinifolia*. A hemi-root-parasite: requirements for seedling growth and role of cotyledons. Ann Bot (London) 55: 785–791
Sahasi S, Ikuse M (1974) Morphological studies on some transitional series in the pollen tetrads of *Typha latifolia*. Jpn J Bot 49: 54–62
Sahi BG (1966) Embryo sac and endosperm in *Andrographis paniculata* Nees. Curr Sci 35: 368–369
Sahni B, Johri BM (1936) Pollen grains in the stylar canal and in the ovary of an angiosperm. Curr Sci 4: 587–589
Saksena HB (1954) Floral morphology and embryology of *Fumaria parviflora* Lamk. Phytomorphology 4: 409–417

Salgare SA (1974a) On the megagametophyte of *Sesbania aculeata* 1. Curr Sci 43: 24–25
Salgare SA (1974b) On the megagametophyte of *Sesbania aegyptica* Poir 1. J Biol Sci (Bombay) 16/17: 108–110
Samal KK (1936) The development of embryo sac and embryo in *Crotalaria juncea* L. J Indian Bot Soc 15: 19–31
Sampathkumar R, Ayyangar KR (1982) Seed morpho-anatomy of Convolvulaceae. J Indian Bot Soc 62: 346–354
Samant K (1973) Embryo development in *Jatropha gossypifolia* Linn. Maharashtra Vigyan Mandir Patrika Poona (India) 8: 47–50
Sampson FB (1963) The floral morphology of *Pseudowintera*: the New Zealand member of vesselless Winteraceae. Phytomorphology 13: 402–423
Sampson FB (1969a) Studies on the Monimiaceae 1. Floral morphology and gametophyte development of *Hedycarya arborea* J.R. et G. Frost (subfamily Monimioideae). Aust J Bot 17: 403–424
Sampson FB (1969b) Cytokinesis with pollen mother cells of angiosperms with emphasis on *Laurelia novae-zelandiae* (Monimiaceae). Cytologia 34: 627–634
Sampson FB (1969c) Studies on the Monimiaceae 3. Gametophyte development of *Laurelia novae-zeylandiae* A. Cunn. (subfamily Antherospermoideae). Aust J Bot 17: 425–439
Sampson FB (1970) Unusual features of cytokinesis in meiosis of pollen mother cells of *Pseudowintera traversii* (Buchen) Dandy (Winteraceae). Beitr Biol Pflanzen 47: 71–77
Sampson FB (1974) A new pollen type in the Winteraceae. Grana 14: 11–15
Sampson FB (1975) Aperture orientation in *Laurelia* pollen (Antherospermataceae syn. subfamily Antherospermoideae of Monimiaceae). Grana 15: 153–157
Sampson FB (1977) Pollen tetrads of *Hedycarya arborea* J.R. et G. Forst (Monimiaceae). Grana 16: 61–73
Sampson FB, Endress PK (1984) Pollen morphology in the Trimeniaceae. Grana 23: 129–137
Samuels JA (1912) Etudes sur le développement du sac embryonnaire et sur la fécondation du *Gunnera macrophylla* Bl. Arch Zellforsch 8: 52–120
Samuelsson G (1913) Studien Über die Entwicklungsgeschichte einiger Bicornes-Typen. Ein Beitrag zur Kenntnis der systematischen Stellung der Diapensiaceen und Empetraceen. Svensk Bot Tidskr 7: 97–188
Samuelsson G (1914) Über die Pollenentwicklung von *Annona* und *Aristolochia* und ihre systematische Bedeutung. Svensk Bot Tidskr 8: 181–189
Sanchez ST (1938) Embryo sac development in *Euphorbia heterophylla* L. Bull Univ Philipp Nat Appl Sci 6: 59–75
Sàné YK (1939) A contribution to the embryology of the Aponogetonaceae. J Indian Bot Soc 18: 79–92
Sangduen N, Kreitner GL, Sorensen EL (1983a) Light and electron microscopy of embryo development in perennial and annual *Medicago* species. Can J Bot 61: 837–849
Sangduen N, Kreitner GL, Sorensen EL (1983b) Light and electron microscopy of embryo development in an annual × perennial *Medicago* sp. cross. Can J Bot 61: 1241–1257
Sankara Rao K (1972) Studies in Myrsinaceae 1. A contribution to the embryology of *Maesa dubia* Wall. Proc Indian Acad Sci B 75: 160–166
Sankara Rao K (1975) Gametophytes and endosperm development in *Swertia corymbosa* Wt. Acta Bot Indica 3: 165–166
Sankara Rao K (1978) Gametophytes in *Hoppea dichotoma* Willd. Curr Sci 47: 786
Sankara Rao K, Chinnappa CC (1983) Studies in Gentianaceae: Microsporangium and pollen. Can J Bot 61: 324–336
Sankara Rao K, Nagaraj M (1982) Studies in Gentianaceae: Embryology of *Swertia minor* (Gentianinae). Can J Bot 60: 141–151
Sankara Rao L, Shivaramaiah G (1977) The embryology of *Ximenia americana* L. Proc Indian Acad Sci B 87: 23–27
Santha Kumari D (1976) Endosperm development in some Labiatae (Lamiaceae). J Indian Bot Soc 58: 148–161
Santos JK (1937) Macrosporogenesis of *Lilium philippinense*. Cytologia (Fujii Jubl Vol.) 41: 822–835
Sapre AB (1968) Embryological studies in a triploid plant of rice. J Univ Poona (India) 34: 85–89

Sapre AB (1970a) Embryological studies in haploid rice. Sci Cult 36: 225–226
Sapre AB (1970b) Embryological studies in relation to the sterility in the F_1 between *Oryza sativa* and *Oryza alta*. Sci Cult 36: 667
Sarada D, Pullaiah T (1985) Embryology of *Luffa tuberosa*. Phytomorphology 35: 47–52
Sareen TS, Bhanwra RR (1978) Embryological studies in *Dicliptera*. Bull Bot Soc Bengal 32: 1–8
Sargant E (1896) The formation of the sexual nuclei in *Lilium martagon*. Ann Bot (London) 10: 445–472
Sargant E (1900) Recent work on the results of fertilization in angiosperms. Ann Bot (London) 22: 121–186
Sarweswara Rao D (1978) Embryological studies in *Jatropha heterophylla* Heyne. In: Proc 65th Indian Sci Congr, Ahmedabad, pt 3: 84
Sassen MMA (1974) The stylar transmitting tissue. Acta Bot Neerl 23: 99–108
Sastri RLN (1952) Studies in Lauraceae 1. Floral anatomy of *Cinnamomum iners* and *Cassytha filiformis*. J Indian Bot Soc 31: 240–246
Sastri RLN (1954a) Embryological studies in Menispermaceae 1. *Tiliacora racemosa* Coleb. Proc Natl Inst Sci India B 20: 494–502
Sastri RLN (1954b) Development of embryo of *Cocculus villosus* DC. Curr Sci 23: 187–188
Sastri RLN (1955a) Development of embryo of *Polyalthia longifolia* Hook. f. & Thomas. Curr Sci 24: 51
Sastri RLN (1955b) Structure and development of nutmeg seed. Curr Sci 24: 172–173
Sastri RLN (1955c) Embryology of *Ceratophyllum demersum* L. (Abstr). In: Proc 42nd Indian Sci Congr, Baroda, pt 3: 226
Sastri RLN (1956) Embryo sac haustoria in *Cassytha filiformis* Linn. Curr Sci 25: 401–402
Sastri RLN (1957) On the division of pollen mother cells in some Annonaceae. Sci Cult 22: 633–634
Sastri RLN (1958a) Studies in Lauraceae 2. Embryology of *Cinnamomum* and *Litsea*. J Indian Bot Soc 37: 266–278
Sastri RLN (1958b) Floral morphology and embryology of some Dilleniaceae. Bot Not 111: 495–511
Sastri RLN (1962) Studies in Lauraceae 3. Embryology of *Cassytha*. Bot Gaz 123: 197–206
Sastri RLN (1963) Studies in the Lauraceae 4. Comparative embryology and phylogeny. Ann Bot (London) 27: 425–433
Sastri RLN (1964) Embryological studies in the Menispermaceae 2. Embryo and seed development. Bull Torrey Bot Club 91: 79–85
Sastri RLN (1969a) Floral morphology, embryology and relationships of Berberidaceae. Aust J Bot 17: 69–79
Sastri RLN (1969b) Comparative morphology and phylogeny of the Ranales. Biol Rev 44: 291–319
Satina S, Rapport J, Blackslee AF (1950) Ovular tumors connected with incompatible crosses in *Datura*. Am J Bot 37: 576–586
Sâto Y (1971) Embryological study in *Aucuba japonica* Thunb. with special reference to the unusual development of embryo sac. Sci Rep Tôhoku Univ Ser Biol 35: 201–206
Sâto Y (1972) Development of the embryo sac of *Daphniphyllum macropodum* var. *humile* (Maxim.) Rosenth. Sci Rep Tôhoku Univ Ser Biol 36: 129–133
Sâto Y (1976a) Embryological studies of some cornaceous taxa. Sci Rep Tôhoku Univ Ser Biol 37: 117–130
Sâto Y (1976b) Embryological studies on *Stachyurus praecox* and its variety. Sci Rep Tôhoku Univ Ser Biol 37: 131–138
Satyamurthi TVC (1985) Embryology of *Elytrophorus spicata*. Phytomorphology 35: 11–15
Satyanarayana P, Narayana LL (1985) Embryology of Stegnospermaceae. Curr Sci 54: 882–884
Savchenko MI (1960) Anomalies in the structure of angiosperm ovules. Dokl Akad Nauk SSSR 130: 15–17 (in Russian)
Savchenko MI (1973) Ovule morphology in angiosperms. Nauka, Leningrad
Savidan YH (1983) Embryological analysis of facultative apomixis in *Panicum maximum*. Crop Sci 22: 467–469
Savidan YH, Pernès J (1982) Diploid-tetraploid-dihaploid cycles and the evolution of *Panicum maximum* Jacq. Evolution 36: 596–600

Savina GI (1971) Development of the embryo sac in *Ophrys insectifera*. Ann Univ ARERS 9: 62–65
Savina GI (1972) Fertilization process in *Ophrys insectifera* L. Bot Zh 57: 382–392 (in Russian)
Savina GI (1978) Certain peculiarities in the embryology of Orchids. Proc Indian Natl Sci Acad B 44: 141–145
Sawhney VK, Bhadula SK (1988) Microsporogenesis in the normal and male sterile stamenless-2-mutant of tomato (*Lycopersicon esculentum*). Can J Bot 66: 2013–2021
Saxena MR (1975) Pollen morphology of the Nyctanthoideae (Verbenaceae). J Indian Bot Soc 54: 71–74
Saxena MR (1981) Contribution to the palynotaxonomy of Avicenniaceae Endl. J Indian Bot Soc 60: 28–32
Saxena NP (1963) Twin embryo sacs in *Saxifraga diversifolia* Wall. Sci Cult 29: 614–615
Saxena NP (1964a) Studies in the family Saxifragaceae 1. A contribution to the morphology and embryology of *Saxifraga diversifolia* Wall. Proc Indian Acad Sci B 60: 38–51
Saxena NP (1964b) Studies in the family Saxifragaceae 2. Development of ovule and megagametophyte in *Parnassia nubicola* Wall. Proc Indian Acad Sci B 60: 196–202
Saxena NP (1969a) Studies in the family Saxifragaceae 3. A contribution to the morphology and embryology of *Astilbe rivularis* Ham. J Indian Bot Soc 48: 368–374
Saxena NP (1969b) Studies in the family Saxifragaceae 4. A contribution to the embryology of *Bergenia ciliata* (Royle) Raizada. Proc Indian Acad Sci B 70: 104–110
Saxena NP (1970) Studies in the family Saxifragaceae 7. Floral morphology and development of gametophytes in *Ribes* L. Proc Indian Acad Sci B 71: 213–220
Saxena NP (1971) Studies in the family Saxifragaceae 6. Structure and development of gametophytes in *Deutzia corymbosa* R. Br. and *Dichora febrifuga* Lour. Ber Schweiz Bot Ges 81: 91–96
Saxena NP (1973) Studies in the family Saxifragaceae 8. Floral anatomy, seed development and systematics of *Bistella adans*. J Indian Bot Soc 52: 244–251
Saxena NP (1976) Studies in the family Saxifragaceae 10. Floral morphology and systematic position of *Parnassia*. J Indian Bot Soc 55: 282–288
Saxena T, Singh D (1969) Embryology and seed development of tetraploid form of *Solanum nigrum* Linn. J Indian Bot Soc 48: 148–157
Saxton WT (1910) The ovule of the Bruniaceae. Trans R Soc S Afr 2: 27–31
Schaeppi H, Steindl F (1942) Blütenmorphologische und embryologische Untersuchungen an Loranthoideen. J Naturforsch Ges Zürich 87: 301–372
Schaeppi H, Steindl F (1945) Blütenmorphologische und embryologische Untersuchungen an einigen Viscoideen. J Naturforsch Ges Zürich 90: 1–46
Schaffner JH (1896) The embryo sac of *Alisma plantago*. Bot Gaz 21: 123–132
Schaffner JH (1897) The life history of *Sagittaria variabilis*. Bot Gaz 23: 252–273
Schaffner JH (1901) A contribution to the life history and cytology of *Erythronium*. Bot Gaz 31: 369–387
Schaffner JH (1908) On the origin of polar conjugation in the angiosperms. Ohio Nat 8: 255–258
Scheer U, Franke WW (1972) Annulate lamellae in plant cells: formation during microsporogenesis and pollen development in *Canna generalis* Bailey. Planta 107: 131
Schertz FM (1919) Early development of floral organs and embryonic structures of *Scrophularia marylandica*. Bot Gaz 68: 441–450
Schill R, Wolter M (1986) On the presence of elastoviscin in all subfamilies of the orchidaceae and the homology to pollenkitt. N J Bot 6: 321–324
Schleiden MJ (1837) Einige Blicke auf die Entwicklungsgeschichte des vegetablischen Organismus bei den Phanerogamen. Arch Bwl Naturgesch III 1: 289–320
Schlimbach H (1924) Beiträge zur Kenntnis der Samenanlagen und Samen der Amaryllidaceen mit Berücksichtigung des Wassergehaltes der Samen. Flora 117: 41–54
Schmid E (1906) Beiträge zur Entwicklungsgeschichte der Scrophulariaceen. Beih Bot Zentralbl 20: 175–299
Schmid R (1964) Die systematische Stellung der Dioncophyllaceen. Bot Jahrb 83: 1–56
Schmid R (1978a) Reproductive anatomy of *Actinidia chinensis* (Actinidiaceae). Bot Jahrb 100: 149–195
Schmid R (1978b) Actinidiaceae, Davidiaceae and Paracryphiaceae: Systematic considerations. Bot Jahrb 100: 196–204

Schmid R (1980) Comparative anatomy and morphology of *Psiloxylon* and *Heteropyxis*, and the subfamilial and tribal classification of Myrtaceae. Taxon 89: 559–595
Schmid R (1982) Descriptors used to indicate abundance and frequency in ecology and systematics. Taxon 91: 89–94
Schmid W (1925) Morphologische, anatomische und entwicklungsgeschichtliche Untersuchungen an *Mesembryanthemum pseudotruncatellum*. J Naturforsch Ges Zürich 70: 1–96
Schnarf K (1914) Beiträge zur Kenntnis der Samenentwicklung einiger europäischer *Hypericum*-Arten. SB Akad Wiss Wien 123: 1
Schnarf K (1917) Beiträge zur Kenntnis der Samenentwicklung der Labiaten. Denkschr Akad Wiss Wien 94: 211–275
Schnarf K (1921) Kleine Beiträge zur Kenntnis der Entwicklungsgeschichte der Angiospermen 2. *Klugia zeylanica* (R. Brown) Gardn. Öst Bot Z 70: 255–261
Schnarf K (1929) Embryologie der Angiospermen. Bornträger, Berlin
Schnarf K (1931) Vergleichende Embryologie der Angiospermen. Bornträger, Berlin
Schnarf K (1933) Die Bedeutung des embryologischen Forschung für das natürliche System der Pflanzen. Biol Gen 10: 271–288
Schnarf K (1936) Contemporary understanding of embryo sac development among angiosperms. Bot Rev 2: 565–585
Schnarf K (1939) Variation im Bau des Pollenkornes der Angiospermen. Tab Biol Period 27: 72–89
Schnarf K, Wunderlich R (1939) Zur vergleichenden Embryologie der Liliaceae-Asphodeloideae. Flora 133: 297–327
Schnepf E, Nägl W (1970) Über einige Strukturbesonderheiten der Suspensorzellen von *Phaseolus vulgaris*. Protoplasma 69: 133–143
Schnegg H (1902) Beiträge zur Kenntniss der Gattung *Gunnera*. Flora 90: 161–208
Schneider EL (1978) Morphological studies of the Nymphaeaceae 9. The seed of *Barclaya longifolia* Wall. Bot Gaz 139: 223–230
Schneider EL (1983) Gross morphology and floral biology of *Ondinea purpurea* den Hartog. Aust J Bot 31: 371–382
Schneider EL, Buchanan J (1980) Morphological studies of the Nymphaeaceae 11. The floral biology of *Nelumbo pentapetala*. Am J Bot 67: 182–193
Schneider EL, Ford EG (1978) Morphological studies of the Nymphaeaceae 10. The seed of *Ondinea purpurea* Den Hartog. Bull Torrey Bot Club 105: 192–200
Schneider EL, Nichols DM (1984) Floral biology of *Argemone aurantiaca* (Papaveraceae). Bull Torrey Bot Club 111: 1–7
Schneider M (1932) Untersuchungen über die Embryobildung und Entwicklung der Cyperaceen. Beih Bot Zentralbl 49: 649–674
Schölch HF (1963) Die systematische Stellung der Didiereaceen im Lichte neuer Untersuchungen über ihren Blütenbereich. Ber Dtsch Bot Ges 76: 49–55
Scholz H (1964) Geraniales. In: Melchior H (ed) A Englers Syllabus der Pflanzenfamilien, vol 2. Bornträger, Berlin, pp 246–261
Schrock GF, Palser BF (1967) Floral development, anatomy and embryology of *Collinsia heterophylla* with some notes on ten other species of *Collinsia* and on *Tonella tenella*. Bot Gaz 128: 83–104
Schroeder CA (1952) Floral development, sporogenesis and embryology in the avocado, *Persea americana*. Bot Gaz 113: 270–278
Schulle H (1933) Zur Entwicklungsgeschichte von *Thesium montanum*. Flora 27: 140–184
Schulz P, Jensen WA (1969) *Capsella* embryogenesis: the suspensor and the basal cell. Protoplasma 67: 139–163
Schulz P, Jensen WA (1986) Prefertilization ovule development in *Capsella*. The dyad, tetrad, developing megaspore and two-nucleate gametophyte. Can J Bot 64: 875–884
Schürhoff PN (1921) Die Entwicklungsgeschichte von *Ilex aquifolium*. Ber Dtsch Bot Ges 39: 377–379
Schürhoff PN (1922) Zur Polyembryonie von *Allium ordorum*. Ber Dtsch Bot Ges 40: 374–381
Schürhoff PN (1923) Zur Apogamie von *Calycanthus*. Flora 116: 73–84
Schürhoff PN (1924) Zytologische Untersuchungen in der Reihe der Geraniales. Jahrb Wiss Bot 63: 707–759

Schürhoff PN (1926) Die Zytologie der Blütenpflanzen. Stuttgart

Schwarzenbach AM (1972) Aleuronvakuolen und Sphärosomen im Endosperm von *Ricinus communis* während der Samenreifung und Keimung. Ber Schweiz Bot Ges 81: 70–90

Seaton S (1908) The development of the embryo sac of *Nymphaea advena*. Bull Torrey Bot Club 35: 283–289

Secor DL, Russell SD (1988) Megagametophyte organization in a polyembryonic line of *Linum usitatissimum*. Am J Bot 75: 114–122

Sedgley M (1976) Control by the embryo sac over pollen tube growth in the style of the avocado (*Persea americana* Mill.). New phytol 77: 149–152

Sedgley M (1979) Light study of pollen tube growth, fertilization and early embryo and endosperm development in the avocado varieties Fuerte and Hass. Ann Bot (London) 44: 353–359

Sedgley M (1982) Floral anatomy and pollen tube growth in the Quandong [*Santalum acuminatum* (R. Br.) A. DC.]. Aust J Bot 30: 601–609

Sedgley M (1989) Ovule and seed development in *Eucalyptus woodwardii* Maiden (*Symphyomyrtus*). Bot Gaz 150: 271–280

Sedgley M, Blesing MA (1985) Polysaccharide secretion by the watermelon stigma. Ann Bot (London) 55: 269–273

Sedgley M, Hand FC, Smith RM, Griffin AR (1989) Pollen tube growth and early seed development in *Eucalyptus regnans* F. Muell (Myrtaceae) in relation to ovule structure and preferential outcrossing. Aust J Bot 37: 397–411

Seefeldner G (1912) Die Polyembryonie bei *Cynanchun vincetoxicum* L. (Pers.) Sitzungsber Acad Wiss Wien 121: 274–296

Seetharam YN (1989) Diversity of androecium and pollen grains in the genus *Garcinia* L. and its bearing on geographical distribution and evolution. Proc Indian Acad Sci B 99: 107–116

Sehgal CB (1965) The embryology of *Cuminum cyminum* L. and *Trachyspermum ammi* (L.) Sprague (*Carum copticum* Clarke). Proc Natl Inst Sci India B 31: 175–201

Sehgal CB (1966) Morphological and embryological studies on *Erigeron bonariensis* L. Beitr Biol Pflanzen 42: 161–183

Sehgal CB, Gandhi V (1985) Studies on the endosperm of *Cajanus cajan* with special emphasis on the function. Phytomorphology 35: 201–206

Selvaratnan EM (1952) Embryo of the coconut. Nature (London) 169: 714–715

Semple KS (1974) Pollination in Piperaceae. Ann Missouri Bot Garden 61: 868–871

Sen PK, Malik PC (1940) The embryo of the Indian mangoes, *Mangifera indica*. Indian J Agric Sci 10: 750–760

Serrato VG, Modenesi P, Roti-Michelozzi G, Bevilacqua L (1986) Structural and histochemical characters of the *Prosopis tamarugo* Phil. seed-coat, in relation to its hardness. Acta Bot Neerl 35: 475–487

Seshagiriah KN (1932a) Development of the female gametophyte and embryo in *Spiranthes australis*. Curr Sci 1: 102

Seshagiriah KN (1932b) Pollen sterility in *Zeuxine sulcata*. Curr Sci 3: 205–206

Seshagiriah KN (1941) Morphological studies in Orchidaceae 1. *Zeuxine sulcata*. J Indian Bot Soc 20: 357–365

Seshavatharam V (1969) Development of embryos in Papilionatae. In: Johri BM, Kapil RN, Rashid A (eds) Sem Morphology, anatomy and embryology of land plants. Univ Delhi, Delhi, pp 78–80

Seshavatharam V (1970) Onagraceae. In: Symp Comparative embryology of angiosperms. Indian Natl Sci Acad Bull 41: 220–225

Seshavatharam V (1978) A contribution to the embryology of the genus *Phaseolus* Linn. In: Proc 65th Indian Sci Congr, Ahmedabad, pt 3: 87

Seshavatharam V (1982) A contribution to the embryology of the genus *Tephrosia* Pers. J Indian Bot Soc 61: 377–386

Seshavatharam V, Satyamurthi TVC (1976) Embryology of some forage grasses. In: Embryology of crop plants (Abstr). Indo-Sov Symp Indian Natl Sci Acad, New Delhi, and Univ Delhi, p 39

Seth N, Vijayaraghavan MR (1990) Differentiation of seed-coat in *Sesbania speciosa*. Proc Indian Acad Sci Plant Sci 100: 301–309

Sethi ML (1930) Microsporogenesis in *Cassia didymobotrya.* J Indian Bot Soc 9: 126–139
Sethi SB (1965) Structure and development of seed in *Camellia sinensis* (L.) O. Kuntze. Proc Natl Inst Sci India B 31: 24–33
Shadowsky AE (1926) Der antipodiale Apparat bie Gramineen. Flora 120: 344–370
Shadowsky AE (1931) Einige Angaben über die Embryologie von *Pistia stratioites.* Ber Dtsch Bot Ges 49: 350–356
Shah CK (1962) Pollen development in some members of the Cyperaceae. In: Symp Plant embryology. Council Sci Ind Res, New Delhi, pp 81–93
Shah CK (1963) The life history of *Juncus bufonius* Linn. J Indian Bot Soc 42: 238–251
Shah CK (1964) Embryo development in *Eleocharis palustris.* Nat Can 35: 41–49
Shah CK (1965) Embryology in some Cyperaceae. Phytomorphology 15: 1–9
Shah CK (1967) A taxonomic evaluation on the families Cyperaceae and Juncaceae. In: Symp Newer trends in taxonomy. Natl Inst Sci India Bull 34: 248–256
Shah CK (1968) Development of pericarp and seed-coat in the Cyperaceae. Nat Can 95: 39–48
Shah CK (1970) Embryogenesis in Monocotyledons 3. *Aponogeton natans* (L.) Engl. et Krause. Proc Natl Acad Sci India Biol Sci 40: 1–5
Shah CK (1972) Morphology and embryology of the family Cyperaceae, an appraisal. In: Murty YS, Johri BM, Mohan Ram HY, Varghese TM (eds) Advances in plant morphology (Prof. V. Puri Comm. Vol.). Sarita Prakashan, Meerut (India), pp 102–112
Shah CK (1974) Some peculiar structures in the embryogeny of *Aponogeton natans* (L.) Engl. et Krause. Geobios (Jodhpur, India) 1: 104–107
Shah CK (1975) The morphology of the mesocotyl. In: Mohan Ram HY, Shah JJ, Shah CK (eds) Form, structure and function in plants (Prof. B.M. Johri Comm. Vol.). Sarita Prakashan, Meerut (India), pp 368–379
Shah CK, Apparao BJ (1982) Endosperm in *Ottelia alismoides* – a quantitative analysis. Acta Bot Indica 10: 85–87
Shah CK, Neelakandan N (1971) Embryogeny in some Cyperaceae 2. Beitr Biol Pflanzen 47: 215–227
Shah CK, Pandey SN (1978) Histochemical studies during embryogenesis in *Limnophyton obtusifolium.* Phytomorphology 28: 31–42
Shah GL (1953) On the megaspore arrangement in *Peucedanum grande* Clarke. Curr Sci 22: 311–312
Shamanna S (1954) A contribution to the embryology of *Olax wightiana* Wall. Proc Indian Acad Sci B 39: 249–256
Shamanna S (1955) A contribution to the embryology of *Opilia amentacea* Roxb. Curr Sci 24: 165–167
Shamanna S (1961) A contribution to the embryology of *Strombosia ceylanica* Gardn. Proc Indian Acad Sci B 54: 12–16
Shanthamma C (1982) Apomixis in *Cenchrus glaucus* Mudaliar et Sundaraj. Proc Indian Acad Sci Plant Sci 91: 25–35
Shanthamma C, Narayan KN (1977) Formation of nucellar embryos with total absence of embryo sacs in two species of Gramineae. Ann Bot (London) 41: 469–470
Sharma G (1956) Studies in the family Euphorbiaceae 1. The gametophytes of *Chrozophora rottleri* A. Juss. J Indian Bot Soc 35: 189–193
Sharma HP, Murty YS (1977) Embryological studies in the Compositae 3. Tribe Astereae, *Brachycome iberidifolia* Benth. A casual apomict. J Indian Bot Soc 56: 321–328
Sharma HP, Murty YS (1978) Embryological studies in the Compositae, Asteraceae 2. Proc Indian Acad Sci B 87: 149–156
Sharma KD (1955) Cytoembryological studies in some Indian Euphorbiaceae. Sci Cult 21: 270–271
Sharma M (1964) Morphological and anatomical investigations on *Artocarpus* Forst. 4. The fruit. Proc Indian Acad Sci B 60: 380–393
Sharma M (1967) Pollen morphology of Indian monocotyledons. J Palynol Spec Vol, pp 1–97
Sharma M, Narayana HS (1971) Development of male and female gametophytes in *Prinsepia utilis* Royle. Curr Sci 40: 7–8

Sharma VK (1962) The embryology of *Lagotis glauca*. Curr Sci 31: 342–344
Sharma VK (1966) Embryology of *Elaeagnus conferta* Roxb. Curr Sci 35: 185–188
Sharma VK (1968a) Floral morphology, anatomy and embryology of *Coriaria nepalensis* Wall. with a discussion on the interrelationships of the family Coriariaceae. Phytomorphology 18: 143–153
Sharma VK (1968b) Morphology, floral anatomy and embryology of *Parnassia nubicola* Wall. Phytomorphology 18: 193–204
Sharma YML (1938) A note on gametogenesis in a few members of Sterculiaceae. Curr Sci 7: 284–285
Sharma YML (1939) Gametogenesis and embryogeny of *Tamarix ericoides* Rottl. Ann Bot (London) 3: 861–870
Sharma YML (1940) Evidence for the Fritillaria type of embryo sac in *Tamarix ericoides* Rottl. Curr Sci 9: 26
Sharp LW (1911) The embryo sac of *Pyrostegia*. Bot Gaz 52: 218–225
Sharp LW (1912) The orchid embryo sac. Bot Gaz 54: 372–385
Shattuck CH (1905) A morphological study of *Ulmus americana*. Bot Gaz 40: 209–223
Shaw E, Gibbs RD (1961) Comparative chemistry and the relationships of the Hamamelidaceae. Nature (London) 190: 463–464
Shealy HE Jr, Herr JM Jr (1973) Carpelloid stamens in *Rubus trivialis* Michx. Bot Gaz 134: 77–87
Sheela R, Narayana LL (1966) Embryology of Pittosporaceae. Curr Sci 35: 74–75
Shen YE (1954) Phylogeny and wood anatomy of *Nandina*. Taiwania 5: 85–92
Shepherd K (1954) Seed fertility of the Gros Michel banana in Jamaica. J Hortic Sci 29: 1–11
Shetty BV, Subramanyam K (1964) Cytology of *Flagellaria indica* Linn. Curr Sci 42: 279–280
Shirke NS, Mahabale TS (1972) Studies on Palms: Embryology of *Caryota urens* L. In: Murty YS, Johri BM, Mohan Ram HY, Varghese TM (eds) Advances in plant morphology (Prof. V. Puri Comm. Vol.). Sarita Prakashan, Meerut (India), pp 218–232
Shivamurthy GR, Arekal GD (1980) Ontogeny, structure and morphology of the female flower of *Balanophora*. In: Periasamy K (ed) Symp Histochemistry, developmental and structural anatomy of angiosperms. PG Centre Univ Madras, Tiruchirapalli (India), pp 245–251
Shivanna KR (1978) Pollen-stigma interaction: Recognition, acceptance and rejection. In: Symp Basic sciences and agriculture. Indian Natl Sci Acad Bull 55: 53–61
Shivanna KR (1982) Pollen-pistil interaction and control of fertilization. In: Johri BM (ed) Experimental embryology of vascular plants. Springer, Berlin Heidelberg New York, pp 131–174
Shivanna KR, Johri BM (1985) The angiosperm pollen: Structure and function. Wiley Eastern, New Delhi
Shivanna KR, Johri BM, Sastri DC (1979) Development and physiology of pollen. Adv Pollen Spore Res 3: 1–117
Shivaramaiah G (1964a) A contribution to the embryology of *Utricularia exoleta* R. Br. Curr Sci 33: 501–503
Shivaramaiah G (1964b) A contribution to the embryology of *Utricularia wallichiana* Wt. Curr Sci 33: 657–658
Shivaramaiah G (1967) Observations on the floral morphology and embryology of *Utricularia stricticaulis* Stapf. Proc Indian Acad Sci B 65: 56–62
Shivaramaiah G (1971) A contribution to the embryology of *Anotis wightiana* Wall. Curr Sci 40: 606–607
Shivaramaiah G, Dutt RC (1964) A contribution to the embryology of *Rondeletia amoena* Hemsl. Curr Sci 33: 280–281
Shivaramaiah G, Ganapathy PS (1961) A contribution to the embryology of *Knoxia corymbosa* Willd. Curr Sci 30: 190–191
Shivaramaiah G, Sunder Rajan S (1973) A contribution to the embryology of *Oldenlandia umbellata* Linn. Proc Indian Acad Sci B 77: 19–24
Shreve F (1906) The development and anatomy of *Sarracenia purpurea*. Bot Gaz 42: 107–126
Shukla RD (1954) Gametophytes in *Balsamodendron mukul* Hook. Curr Sci 23: 333
Shukla TC (1954) Behaviour of tapetal cells during microsporogenesis of *Lens esculenta* Moench. Curr Sci 23: 365

Shutts CF (1960) Wood anatomy of Hernandiaceae and Gyrocarpaceae. Trop Woods 113: 85–123

Siddiqui SA (1968) The microsporangium and the male gametophyte in *Malpighia coccigera* Linn. Beitr Biol Pflanzen 44: 361–364

Siddiqui SA (1975) Studies in the Lentibulariaceae 7. The development of endosperm and embryo in *Utricularia coerulea* var. *filicaulis* Clarke. Bot Not 128: 432–437

Siddiqui SA (1978) Studies in the Lentibulariaceae 9. Pollination, fertilization, endosperm, embryo and seed in *Utricularia dichotoma* Labill. Bot Jahrb 100: 237–245

Siddiqui SA (1979) Studies in the Lentibulariaceae 11. The development of endosperm and embryo in *Utricularia cornuta* Mixch. Proc Indian Acad Sci Plant Sci 88: 213–218

Siddiqui SA, Farooq M (1965) Degeneration in the ovules of *Utricularia coerulea* L. var. *filicaulis* Clarke. Bull Torrey Bot Club 92: 245–249

Siddiqui SA, Siddiqui SB (1968a) Studies in Rubiaceae 1. A contribution to the embryology of *Oldenlandia dichotoma* Hook. f. Beitr Biol Pflanzen 44: 343–351

Siddiqui SA, Siddiqui SB (1968b) A contribution to the embryology of *Borreria stricta* Linn. Beitr Biol Pflanzen 44: 353–360

Siddiqui SA, Siddiqui SB (1979) The development of gametophytes in *Hymenodictyon excelsum* Wall. Sci Environ 1: 61–64

Silva Stort MN, Dos S, Pavanelli EA (1985) Formation of multiple or adventive embryos in *Epidendrum nocturnum* Jacq (Orchidaceae). Ann Bot (London) 55: 331–336

Simmonds NW (1961) Megasporogenesis and female fertility in three edible triploid bananas. J Genet 57: 269–278

Simoncioli C (1974) Ultrastructural characteristics of *Diplotaxis erucoides* (L.) DC. suspensor. Giorn Bot Ital 108: 175–189

Simpson BB, Skvarla JJ (1981) Pollen morphology and ultrastructure of *Krameria* (Krameriaceae): utility in questions of intrafamilial and interfamilial classification. Am J Bot 68: 277–294

Simpson BB, Neff JL, Dieninger G (1986) Reproductive biology of *Tinantia anomala* (Commelinaceae). Bull Torrey Bot Club 113: 149–158

Simpson MG (1988) Embryological development of *Lachnanthes coroliniana* (Haemodoraceae). Am J Bot 75: 1394–1408

Simpson MG (1989) Pollen wall development of *Xiphidium coeruleum* (Haemodoraceae) and its systematic implications. Ann Bot (London) 64: 257–270

Sindhe ANR, Swamy BGL, Govindappa DA (1980) Synergid embryo in *Pennisetum squamulatum*. Curr Sci 49: 914–915

Singh AP, Mogensen HL (1975) Fine structure of the zygote and early embryo in *Quercus gambelii*. Am J Bot 62: 105–115

Singh AP, Mogensen HL (1976) Fine structure of early endosperm in *Quercus gambelii*. Cytologia 41: 345–361

Singh B (Bahadur) (1934) The seeds of *Tinospora cordifolia*. Curr Sci 3: 265

Singh B (1936) The life history of *Ranunculus sceleratus* Linn. Proc Indian Acad Sci B 4: 75–91

Singh B (1952a) A contribution to the floral morphology and embryology of *Dendrophthoe falcata* (L. fil) Ettings. Bot J Linn Soc 53: 449–472

Singh B (1952b) Studies on the structure and development of seeds of Cucurbitaceae 1. Seed of *Echinocystis wrightii* Cogn. Phytomorphology 2: 201–209

Singh B (1953) Studies on the structure and development of seeds of Cucurbitaceae. Phytomorphology 3: 224–239

Singh B (1959) Studies in the family Malpighiaceae 1. The morphology of *Thryallis glauca* Kuntze. Hortic Adv (Saharanpur, India) 3: 1–19

Singh B (1961a) Studies in the family Malpighiaceae 2. Morphology of *Malpighia glabra* Linn. Hortic Adv (Saharanpur, India) 5: 83–95

Singh B (1961b) Studies in the family Malpighiaceae 3. Development and structure of the seed and fruit of *Malpighia glabra*. Hortic Adv (Saharanpur, India) 5: 145–155

Singh B (1964) Development and structure of angiosperm seed 1. Review of the Indian work. Natl Bot Garden (Lucknow, India) Bull 89: 1–115

Singh B (1967) The structure and development of *Abelmoschus moschatus* Medic. seed. Phytomorphology (Panchanan Maheshwari Mem. Vol.) 17: 282–290

Singh B, Shivapuri TN (1935) The gametophytes of *Neptunia oleracea* Lour. Proc Indian Acad Sci B 1: 423–434

Singh D (1954) Floral morphology and embryology of *Hedera nepalensis* K. Koch. Agra Univ J Res Sci 3: 289–299

Singh D (1955a) Embryological studies in *Cucumis melo* L. var *pubescens* Willd. J Indian Bot Soc 36: 72–78

Singh D (1955b) Cytoplasmic nodules in the endosperm of some Cucurbitaceae. Nature (London) 176: 607–608

Singh D (1960) Studies on endosperm and development of seeds of *Carica papaya* L. Hortic Adv (Saharanpur, India) 4: 89–96

Singh D (1961a) Studies on endosperm and development of seeds in Cucurbitaceae and some related families. Agra Univ J Res Sci 10: 117–123

Singh D (1961b) Development of embryo in the Cucurbitaceae. J Indian Bot Soc 40: 620–623

Singh D (1962) The structure and development of ovule and seed of *Passiflora foetida* Linn. Agra Univ J Res Sci 11: 99–111

Singh D (1963) Structure and development of ovule and seed of *Viola tricolor* L. and *Ionidium suffruticosum* Ging. J Indian Bot Soc 42: 448–462

Singh D (1964a) A further contribution to the endosperm of the Cucurbitaceae. Proc Indian Acad Sci B 60: 399–413

Singh D (1964b) Cytoplasmic nodules in the endosperm of angiosperms. Bull Torrey Bot Club 91: 86–94

Singh D (1965a) Ovule and seed of *Sechium edule* Sw.: A reinvestigation. Curr Sci 34: 696–697

Singh D (1965b) Ovule and seed of *Dicoelospermum* C.B. Clarke, together with a note on its systematic position. J Indian Bot Soc 44: 183–190

Singh D (1966) Development and behaviour of pollen grains in *Viola odorata* Linn. Palynol Bull 3, 4: 23–29

Singh D (1967) Structure and development of seed-coat of the Cucurbitaceae 1. Seeds of *Biswarea* Cogn., *Edgaria* Clarke and *Herpetospermium* Hook. f. Proc Indian Acad Sci B 65: 267–274

Singh D (1970a) Violaceae. In: Symp Comparative embryology of angiosperms. Indian Natl Sci Acad Bull 41: 188–193

Singh D (1970b) Passifloraceae. In: Symp Comparative embryology of angiosperms. Indian Natl Sci Acad Bull 41: 199–204

Singh D (1970c) Caricaceae. In: Symp Comparative embryology of angiosperms. Indian Natl Sci Acad Bull 41: 208–211

Singh D (1970d) Cucurbitaceae. In: Symp Comparative embryology of angiosperms. Indian Natl Sci Acad Bull 41: 212–219

Singh D, Dathan ASR (1972a) Structure and development of the seed-coat in Cucurbitaceae 6. Seeds of *Cucurbita*. Phytomorphology 22: 29–45

Singh D, Dathan ASR (1972b) Structure and development of the seed-coat in Cucurbitaceae 7. Seeds of *Marah* Kell. Bull Torrey Bot Club 99: 239–242

Singh D, Dathan ASR (1974) Structure and development of the seed-coat in Cucurbitaceae 9. Seeds of *Corallocarpus*, *Kedrostis* and *Ibervillea*. Bull Torrey Bot Club 101: 78–82

Singh D, Gupta JS (1956) Cytoplasmic nodules in the endosperm of *Coriandrum sativum* L. Sci Cult 22: 343–344

Singh D, Gupta S (1963) Some interesting features in the male gametophyte of *Viola odorata* Linn. Curr Sci 32: 183–184

Singh D, Gupta S (1968) The seeds of the Violaceae and Resedaceae: A comparison. J Indian Bot Soc 46: 248–256

Singh D, Taneja CP (1985) Ovule in Centrospermae with remarks on types and interrelationship of ovule in angiosperms. In: Govil CM, Kumar V (eds) Trends in plant research (Prof. Y.S. Murty Comm. Vol.). Bishen Singh & Mahendra Pal Singh, Dehra Dun, pp 214–224

Singh RP (1954) Structure and development of seeds in Euphorbiaceae: *Ricinus communis* L. Phytomorphology 4: 118–123

Singh RP (1956) Development of endosperm and embryo in *Phyllanthus niruri* L. Agra Univ J Res Sci 5: 163–167

Singh RP (1962) Forms of ovules in Euphorbiaceae. In: Symp Plant embryology. Council Sci Ind Res, New Delhi, pp 124–128

Singh RP (1965) Structure and development of seeds in *Codiaeum variegatum* Blume. J Indian Bot Soc 44: 205–210
Singh RP (1968) Structure and development of seeds in Euphorbiaceae: *Melanthesa rhamnoides* Wt. Beitr Biol Pflanzen 45: 127–133
Singh RP (1969) Structure and development of seeds in *Euphorbia helioscopia* L. Bot Mag (Tokyo) 82: 287–293
Singh RP (1970a) Structure and development of seeds in *Putranjiva roxburghii* Wall. J Indian Bot Soc 49: 99–105
Singh RP (1970b) Structure and development of seeds in Euphorbiaceae: *Jatropha* species. Beitr Biol Pflanzen 47: 79–90
Singh RP (1985) Anatomical patterns of seed-coat in some dicotyledonous families. In: Govil CM, Kumar V (eds) Trends in plant research (Prof. Y.S. Murty Comm. Vol.). Bishen Singh & Mahendra Pal Singh, Dehra Dun, pp 233–242
Singh RP, Chauhan MPS (1984) Seed structure and systematic position of *Hampea nutricia* (Malvaceae). Plant Syst Evol 147: 55–61
Singh RP, Chopra S (1970) Structure and development of seeds in *Croton bonplandianum* Baill. Phytomorphology 20: 83–87
Singh RP, Jain JL (1965) Development of female gametophyte in *Euphorbia pilosa* L. Curr Sci 34: 611–612
Singh RP, Khan AM (1974) Development of ovule and female gametophyte in *Eomecon chionantha* Hanc. Curr Sci 43: 695–697
Singh RP, Pandey AK (1984) Development and structure of seeds and fruits in Compositae-Cynareae. Phytomorphology 34: 1–10
Singh SP (1954) Morphological studies in *Morus* spp. Agra Univ J Res Sci 3: 323–341
Singh SP (1959) Structure and development of seeds in *Euphorbia geniculata* Orteg. J Indian Bot Soc 38: 103–108
Singh SP (1960) Morphological studies in some members of the family Pedaliaceae 1. *Sesamum indicum* DC. Phytomorphology 10: 65–82
Singh SP (1963) Morphological studies in some members of the family Pedaliaceae 2. *Pedalium murex* Linn. Agra Univ J Res Sci 12: 143–161
Singh SP (1964) Morphological studies in some members of the family Pedaliaceae 3. *Martynia diandra* Glox. Agra Univ J Res Sci 13: 107–127
Singh SP (1970a) Pedaliaceae. In: Symp Comparative embryology of angiosperms. Indian Natl Sci Acad Bull 41: 273–277
Singh SP (1970b) Martyniaceae. In: Symp Comparative embryology of angiosperms. Indian Natl Sci Acad Bull 41: 278–281
Singh UP (1959) Embryogeny in loquat, *Eriobotrya japonica*. Curr Sci 28: 162–163
Sinha ARP (1985) Cytomixis during microsporogenesis in natural populations of some plant species (Scrophulariaceae). Cytologia 50: 341–346
Siva Rao BS (1952) Persistent pollen tube in *Antigonon leptopus* Hook. et Arn. Curr Sci 21: 106–107
Siva Rao BS (1975) Embryo development in five species of *Mollugo*. Curr Sci 44: 712–713
Sivaramkrishna D (1978) Size relationships of apical cell and basal cell in two-celled embryos in angiosperms. Can J Bot 56: 1434–1439
Skottsberg C (1913) Morphologische und embryologische Studien über die Myzodendraceae. Kon Svensk Vet Akad Handl 51: 1–34
Skottsberg C (1915) Notes on the relations between the floras of Subantarctic America and New Zealand. Plant World 18: 129–142
Small BB (1933) Manual of Southeastern flora. Univ North Carolina Press, Chapel Hill (USA)
Small JK (1914) Monotropaceae. N Am Flora 29: 11–18
Smart C (1952) The life history of *Tupeia* Cham. et Sche. Trans R Soc NZ 79: 459–466
Smith AC (1972) An appraisal of the orders and families of primitive extant angiosperms. J Indian Bot Soc (Golden Jubl Vol) 50A: 215–226
Smith BB (1973a) The use of a new clearing technique for the study of early ovule development, megasporogenesis and megagametogenesis in five species of *Cornus* L. Am J Bot 60: 322–338
Smith BB (1973b) Quantification in morphological studies with reference to the megagametophyte of angiosperms. ASB Bull 20: 82

Smith BB (1975) A quantitative analysis of the megagametophyte of five species of *Cornus* L. Am J Bot 62: 387–394
Smith BB, Herr JM Jr (1971) Ovule development, megagametogenesis and early embryogeny in *Ammania coccinea* Roth. J Elisha Mitchell Sci Soc 87: 192–199
Smith BB, Johnson LK (1980) Early ovule development, megasporogenesis and megagametogenesis in *Solidago graminifolia* var. *nuttallii* and *Solidago canadensis* var. *canadensis* (Asterales, Asteraceae; Tubiflorae, Astereae). Am J Bot 67: 612–618
Smith BW (1956) *Arachis hypogaea*: Normal megasporogenesis and syngamy with occasional single fertilization. Am J Bot 43: 81–89
Smith DL (1971) Nuclear changes in the cotyledons of *Pisum arvense* L. during germination. Ann Bot (London) 35: 511–521
Smith FG (1968) "Dyads" in the Proteaceae. Grana Palynol 8: 86–87
Smith FH (1942) Development of the gametophytes and fertilization in *Camassia*. Am J Bot 29: 657–663
Smith FH (1943) Megagametophyte of *Clintonia*. Bot Gaz 105: 263–267
Smith FH (1955) Megagametophyte development in five species of *Erythronium*. Am J Bot 42: 213–224
Smith FH, Smith EC (1942) Floral anatomy of the Santalaceae and some related forms. Ore St Monogr Bot 5: 1–93
Smith GL (1963) Studies in *Potentilla* 1. Embryological investigations into the mechanism of agamospermy in British *P. tabernaemontani* Aschers. New Phytol 62: 264–282
Smith LB (1934) Geographical evidence on the lines of evolution in the Bromeliaceae. Bot Jahrb 66: 446–468
Smith LJ, Scarisbrick DH (1990) Reproductive development in oilseed rape (*Brassica napus* cv Bienvenu). Ann Bot (London) 65: 205–212
Smith PM (1916) The development of the embryo and seedling of *Dioscorea villosa*. Bull Torrey Bot Club 43: 545–558
Smith WR (1898) A contribution to the life history of the Pontederiaceae. Bot Gaz 25: 324–337
Smith WR (1910) The floral development and embryogeny of *Eriocaulon septangulare*. Bot Gaz 49: 281–289
Smith WR (1911) The tetranucleate embryo sac of *Clintonia*. Bot Gaz 52: 209–217
Smith-White S (1948) A survey of chromosome numbers in the Epacridaceae. Proc Linn Soc NSW 73: 37–56
Smith-White S (1955a) Chromosome numbers and pollen types in the Epacridaceae. Aust J Bot 3: 48–67
Smith-White S (1955b) The life history and genetic system of *Leucopogon juniperinus*. Heredity 9: 79–91
Smith-White S (1959) Pollen development patterns in the Epacridaceae: A problem in cytoplasm-nucleus interaction. Proc Linn Soc NSW 48: 8–35
Snigirevskaya NS (1964) Information on the morphology and systematics of the genus *Nelumbo adans*. In: Shishkina BK (ed) Flora on systematics of the higher plants, vol 13. Nauka, Moscow, pp 104–172 (in Russian)
Snyder LA (1957) Apomixis in *Paspalum secans*. Am J Bot 44: 318–324
Snyder LA, Hernandez AR, Warmke HE (1955) The mechanism of apomixis in *Pennisetum ciliare*. Bot Gaz 116: 209–221
Sobajima YM, Ishida M, Inaba A (1975) Studies on the fruit development of Japanese persimmon (*Diospyros kaki* L.) 11. Seed abortion in Hiratanenashi. J Jpn Soc Hortic Sci 44: 1–6
Soerodikoesomo W (1970) Development of the embryo sac in some cauliflorous plants. Ann Bogor 5: 93–98
Sokolav ID (1971) Early stage of embryo development in *Allium porrum* L. Tsitol Genet 5: 555–556
Sokolowska-Kulczycka A (1973) Development of bisporic embryo sacs in *Veratrum lobelianum* Bernh. Acta Biol Cracov Ser Bot 16: 85–98
Sokolowska-Kulczycka A (1985) The influence of temperature on the frequency of developmental types of ES in *Tofieldia calyculata* (L.) Whlb. Acta Biol Cracov Ser Bot 27: 1–7
Solereder H (1908) Systematic anatomy of the dicotyledons, vols 1, 2. (English translation by L.A. Boodle, F.E. Fritsch; revised by D.H. Scott). Clarendon Press, Oxford

Solms-Laubach HG (1978) Über monokotyle Embryonen mit scheitelartigem Vegetationspunkt. Bot Z 36: 65–74, 81–93
Solntzeva MP (1969) Principles of embryological classification of apomixis in angiosperms. Rev Cytol Biol Veg 32: 371–377
Solntzeva MP (1974) Disturbances in the process of fertilization in angiosperms under hemigamy. In: Linskens HF (ed) Fertilization in higher plants. North-Holland, Amsterdam, pp 311–324
Solntzeva MP (1978) Apomixis and hemigamy as one of its forms. Proc Indian Natl Sci Acad B 44: 78–90
Solntzeva MP, Ogebuadze NV (1974) Embryology of the pecan, *Carya illinoensis* (Wageheim) K. Koch (Juglandaceae). Bot Zh 59: 1794–1804 (in Russian)
Solntzeva MP, Vorsobina LI (1972) Semigamy in *Zephyranthes macrosiphon* (in Russian). Dokl Akad Nauk SSSR 206: 1006–1109
Sood SK (1983) Development of the embryo in *Oreorchis foliosa* Lindl. Curr Sci 52: 890–891
Sood SK (1984) An embryological study of *Neottia listeroides*. A saprophytic orchid in India. J Plant Anat Morphol 1: 69–75
Sood SK (1985) Gametophytes, integuments initiation and embryogeny in *Microstylis cylindrostachya* (Orchidaceae, Epidendreae). Proc Indian Acad Sci Plant Sci 95: 379–387
Sood SK (1986) Gametogenesis, integuments initiation and embryogeny in three species of *Habenaria* (Orchidaceae, Orchideae). Proc Indian Acad Sci Plant Sci 96: 487–494
Sood SK (1989) Embryology and systematic position of *Liparis* (Orchidaceae). Plant Syst Evol 166: 1–10
Sood SK, Mohana Rao PR (1986) Gametophytes, embryogeny and pericarp of *Microstylis wallichii* Lindl. (Orchidaceae). Bot Mag (Tokyo) 99: 351–359
Souèges R (1914) Recherches sur l'embryogénie des Renonculacées. Conclusions générales. Bull Soc Bot Fr 61: 27–32
Souèges R (1917a) Embryogénie des Alismacées. Développement du proembryon chez le *Sagittaria sagittifolia* L. C R Acad Sci Paris 165: 715–717
Souèges R (1917b) Embryogénie des Alismacées. Différentiation du cône végétatif de la tige chez le *Sagittaria sagittifolia* L. C R Acad Sci Paris 165: 1014–1017
Souèges R (1919) Recherches sur l'embryogénie des Polygonacées. Bull Soc Bot Fr 66: 168–199; and 67: 1–11, 75–85
Souèges R (1920a) Développement de l'embryon chez le *Chenopodium bonus-henricus*. Bull Soc Bot Fr 67: 233–257
Souèges R (1920b) Embryogénie des Oenotheracées. Développement de l'embryon chez l'*Oenothera biennis*. C R Acad Sci Paris 170: 946–948
Souèges R (1920c) Embryogénie des Urticacées. Développement de l'embryon chez l'*Urtica pilulifera*. C R Acad Sci Paris 171: 1009–1011
Souèges R (1921) Développement de l'embryon chez l'*Urtica pilulifera* L. Bull Soc Bot Fr 68: 172–188, 280–294
Souèges R (1922a) Recherches embryogéniques sur l'*Hippuris vulgaris*. C R Acad Sci Paris 175: 529–532
Souèges R (1922b) Embryogénie des Malvacées. Développement de l'embryon chez le *Malva rotundifolia* L. C R Acad Sci Paris 175: 1435–1436
Souèges R (1923a) Embryogénie des Valerianacées. Développement de l'embryon chez le *Valerianella olitoria*. C R Acad Sci Paris 176: 1081–1083
Souèges R (1923b) Embryogénie des Geraniacées. Développement de l'embryon chez l'*Erodium cicutarium*. C R Acad Sci Paris 176: 1565–1567
Souèges R (1923c) Embryogénie des Geraniacées. Développement de l'embryon chez le *Geranium molle*. C R Acad Sci Paris 177: 556–559
Souèges R (1923d) Embryogénie des Joncacées. Développement de l'embryon chez le *Luzula forsteri* DC. C R Acad Sci Paris 177: 705–708
Souèges R (1923e) Embryogénie des Plantaginacées. Développement de l'embryon chez le *Plantago lanceolata* L. C R Acad Sci Paris 177: 964–967
Souèges R (1923f) Embryogénie des Salicacées. Développement de l'embryon chez le *Salix triandra*. C R Acad Sci Paris 177: 1234–1237
Souèges R (1924a) Embryogénie des Graminées. Développement de l'embryon chez le *Poa annua* L. C R Acad Sci Paris 178: 860–862

Souèges R (1924b) Embryogénie des Rubiacées. Développement de l'embryon chez le *Sherardia arvensis*. C R Acad Sci Paris 178: 1919–1921

Souèges R (1924c) Embryogénie des Typhacées. Développement de l'embryon chez le *Sparganium simplex*. C R Acad Sci Paris 179: 198–200

Souèges R (1924d) Développement de l'embryon chez le *Linum catharticum* Mill. Bull Soc Bot Fr 71: 925–938

Souèges R (1925a) Embryogénie des Hypericacées. Développement de l'embryon chez l'*Hypericum perforatum*. C R Acad Sci Paris 180: 949–951

Souèges R (1925b) Embryogénie des Lythracées. Développement chez le *Lythrum salicaria*. C R Acad Sci Paris 180: 1417–1418

Souèges R (1925c) Embryogénie der Crassulacées. Développement de l'embryon chez le *Sedum acre* L. C R Acad Sci Paris 181: 521–522

Souèges R (1925d) Développement de l'embryon chez le *Sherardia arvensis* L. Bull Soc Bot Fr 72: 546–565

Souèges R (1926a) Embryogénie des Ombelliféres. Développement de l'embryon chez le *Carum carvi* L. C R Acad Sci Paris 182: 339–341

Souèges R (1926b) Embryogénie des Papaveracées. Développement du proembryon chez le *Papaver rhoeas* L. C R Acad Sci Paris 183: 902–904

Souèges R (1926c) Embryogénie de Papaveracées. Les derniers stades du développement de l'embryo chez le *Papaver rhoeas* L. C R Acad Sci Paris 183: 1119–1120

Souèges R (1926d) Développement de l'embryon chez le *Ruta graveolens*. Bull Soc Bot Fr 73: 248–260

Souèges R (1926e) Un mode d'avortement du sac embryonnaire chez le *Papaver rhoeas*. Bull Soc Bot Fr 73: 636–640

Souèges R (1926f) Polyembryonie chez le *Plantago lanceolata* L. Bull Soc Bot Fr 73: 955–957

Souèges R (1927) Développement de l'embryon chez le *Sedum acre* L. Bull Soc Bot Fr 74: 234–251

Souèges R (1928) Développement de l'embryon chez le *Papaver rhoeas*. Bull Soc Bot Fr 75: 452–469

Souèges R (1931a) L'embryon chez le *Sagittaria sagittifolia*. Le cône végétatif de la tige et l'extremité radiculaire chez le monocotyledones. Ann Sci Nat Bot 13: 353–402

Souèges R (1931b) Recherches sur l'embryogénie des Liliacées. *Anthericum ranosum* et *Allium ursinum*. Bull Soc Bot Fr 78: 662–681

Souèges R (1935a) Embryogénie des Oenotheracées. Les principaux termes du développement de l'embryon chez le *Ludwigia palustris* Elliott. C R Acad Sci Paris 200: 1626–1628

Souèges R (1935b) Embryogénie des Verbenacées. Développement de l'embryon chez le *Verbena officinalis* L. C R Acad Sci Paris 201: 234–236

Souèges R (1936a) Embryogénie des Saxifragacées. Développement de l'embryon chez le *Saxifraga granulata*. C R Acad Sci Paris 202: 240–242

Souèges R (1936b) Embryogénie des Droseracées. Développement de l'embryon chez le *Drosera rotundifolia*. C R Acad Sci Paris 202: 1457–1459

Souèges R (1936c) Embryogénie des Campanulacées. Développement de l'embryon chez le *Campanula patula* L. C R Acad Sci Paris 202: 2009–2011

Souèges R (1936d) Modifications au tableau récapitulatif des lois du développement chez *Sedum acre* L. Le type embryonomique de cette espèce chez les autres Crassulacées. Bull Soc Bot Fr 83: 13–18

Souèges R (1936e) Les relations embryogéniques des Crassulacées, Saxifragacées et Hypericacées. Bull Soc Bot Fr 83: 317–329

Souèges R (1936f) Développement de l'embryon chez le *Schizanthus* et les *Petunias*. Bull Soc Bot Fr 83: 570–577

Souèges R (1937a) Développement de l'embryon chez le *Radiola linoides* Roth. Bull Soc Bot Fr 84: 279–306

Souèges R (1937b) Développement de l'embryon chez l'*Helianthemum guttatum*. Bull Soc Bot Fr 84: 400–407

Souèges R (1937c) Embryogénie des Primulacées. Développement de l'embryon chez le *Samolus valerandi* L. C R Acad Sci Paris 204: 145–147

Souèges R (1937d) Embryogénie des Dipsacacées. Développement de l'embryon chez le *Scabiosa succisa* L. C R Acad Sci Paris 204: 292–294
Souèges R (1937e) Embryogénie des Violacées. Développement de l'embryon chez le *Viola tricolor* L. C R Acad Sci Paris 205: 169–171
Souèges R (1937f) Embryogénie des Convolvulacées. Développement de l'embryon chez le *Convolvulus arvensis*. C R Acad Sci Paris 205: 813–815
Souèges R (1938a) Embryogénie des Campanulacées. Développement de l'embryon chez le *Jasione montana* L. C R Acad Sci Paris 206: 278–280
Souèges R (1938b) Embryogénie des Portulacacées. Développement de l'embryon chez le *Portulaca oleracea*. C R Acad Sci Paris 206: 768–770
Souèges R (1938c) Embryogénie des Nyctaginacées. Développement de l'embryon chez l'*Oxybaphus viscosus*. C R Acad Sci Paris 206: 1830–1832
Souèges R (1939a) Embryogénie des Begoniacées. Développement de l'embryon chez le *Begonia semperflorens* Link et Otto. C R Acad Sci Paris 208: 534–536
Souèges R (1939b) Embryogénie des Polemoniacées. Développement de l'embryon chez le *Polemonium caeruleum*. C R Acad Sci Paris 208: 1338–1340
Souèges R (1939c) Embryogénie des Rhamnacées. Développement de l'embryon chez le *Ceanothus azureus*. C R Acad Sci Paris 208: 1673–1675
Souèges R (1939d) Embryogénie des Oxalidacées. Développement de l'embryon chez le *Oxalis corniculata* L. C R Acad Sci Paris 209: 698–700
Souèges R (1939e) Embryogénie et classification 2. Essai d'un système embryogénique. Partie générale. Paris
Souèges R (1940a) Embryogénie des Bignoniacées. Développement de l'embryon chez le *Catalpa kaempferi* Sieb. et Zucc. C R Acad Sci Paris 210: 116–118
Souèges R (1940b) Embryogénie des Loganiacées. Développement de l'embryon chez le *Buddleia variabilis*. C R Acad Sci Paris 211: 139–140
Souèges R (1940c) Embryogénie des Haloragacées. Développement de l'embryon chez le *Myriophyllum alterniflorum* DC. C R Acad Sci Paris 211: 185–187
Souèges R (1941a) Embryogénie des Valerianacées. Développement de l'embryon chez le *Centranthus*. C R Acad Sci Paris 212: 718–720
Souèges R (1941b) Embryogénie des Tiliacées. Développement de l'embryon chez le *Tilia platyphyllos*. C R Acad Sci Paris 212: 998–1000
Souèges R (1941c) Embryogénie des Rhamnacées. Développement de l'embryon chez le *Rhamnus frangula* L. C R Acad Sci Paris 213: 39–41
Souèges R (1941d) Embryogénie des Polygalacées. Développement de l'embryon chez le *Polygala vulgaris* L. C R Acad Sci Paris 213: 446–448
Souèges R (1941e) Embryogénie des Fumariacées. L'origine du corps de l'embryon chez le *Fumaria officinalis* L. C R Acad Sci Paris 213: 528–530
Souèges R (1941f) Embryogénie des Fumariacées. La différentiation des regions fondamentales du corps chez le *Fumaria officinalis* L. C R Acad Sci Paris 213: 699–702
Souèges R (1942a) Embryogénie des Thyméleacées. Développement de l'embryon chez le *Daphne mezereum*. C R Acad Sci Paris 214: 569–571
Souèges R (1942b) Embryogénie des Oleacées. Développement de l'embryon chez le *Ligustrum ovalifolium* Hassk. C R Acad Sci Paris 215: 328–330
Souèges R (1942c) Embryogénie des Polémoniacées. Développement de l'embryon chez le *Gilia tricolor* Benth. C R Acad Sci Paris 215: 543–545
Souèges R (1943a) Embryogénie des Scheuchzériacées. Développement de l'embryon chez le *Triglochin maritimum* L. C R Acad Sci Paris 216: 746–748
Souèges R (1943b) Embryogénie des Actinidiacées. Développement de l'embryon chez l'*Actinidia chinensis* Planch. C R Acad Sci Paris 217: 430–432
Souèges R (1943c) Embryogénie des Gentianacées. Développement de l'embryon chez le *Menyanthes trifoliata* L. C R Acad Sci Paris 217: 488–490
Souèges R (1943d) Embryogénie des Boragacées. Développement de l'embryon chez l'*Heliotropium peruvianum*. C R Acad Sci Paris 217: 551–553
Souèges R (1945a) Embryogénie des Balsaminacées. Développement de l'embryon chez l'*Impatiens bulfourii* Hook. C R Acad Sci Paris 220: 837–840

Souèges R (1945b) Embryogénie des Polemoniacées. Développement de l'embryon chez le *Polemonium pauciflorum*. C R Acad Sci Paris 220: 897–900
Souèges R (1945c) Embryogénie des Portulacacées. Développement de l'embryon chez le *Claytonia perfoliata*. C R Acad Sci Paris 221: 111–113
Souèges R (1945d) Embryogénie des Caprifoliacées. Développement de l'embryon chez le *Lonicera biflora*. C R Acad Sci Paris 221: 480–482
Souèges R (1948) Embryogénie des Caprifoliacées. Développement de l'embryon chez le *Symphoricarpos occidentalis* Hook. C R Acad Sci Paris 227: 1066–1068
Souèges R (1949a) Embryogénie des Papaveracées. Développement de l'embryon chez l'*Eschscholtzia californica* Cham. C R Acad Sci Paris 229: 485–487
Souèges R (1949b) Embryogénie des Papaveracées. Développement de l'embryon chez l'*Argemone mexicana* L. C R Acad Sci Paris 229: 573–576
Souèges R (1949c) Embryogénie des Papilionacées. Développement de l'embryon chez le *Glycine soja* Sieb. et Zucc. *Soja hispida* Moench. C R Acad Sci Paris 229: 1183–1185
Souèges R (1951a) Embryogénie des Papilionacées. L'embryon et le suspenseur chez l'*Anthyllis vulneraria* L. C R Acad Sci Paris 232: 581–584
Souèges R (1951b) Embryogénie des Rubiacées. Développement de l'embryon chez le *Richardsonia pilosa* H.B. et K. (*R. brasiliensis* Hayme). C R Acad Sci Paris 233: 5–8
Souèges R (1951c) Développement de l'embryon chez l'*Utrica pillulifera*. Bull Soc Bot Fr 13: 189–193
Souèges R (1952a) L'albumen et l'embryon chez le *Callitriche vernalis* Kuetz. (*C. verna* L.). C R Acad Sci Paris 235: 453–456
Souèges R (1952b) Embryogénie des Amaryllidacées. Développement de l'embryo chez le *Pancratium maritimum*. C R Acad Sci Paris 234: 1300
Souèges R (1953) Embryogénie des Peganacées. Développement de l'embryon chez le *Pegannum harmala* L. C R Acad Sci Paris 236: 2186–2188
Souèges R (1954) L'origine due cône végétatif de la tige et la question de la "terminalité" du cotyledon des monocotylédones. Ann Sci Nat Bot 15: 1–20
Souèges R (1955) Embryogénie des Cneoracées. Développement de l'embryon chez le *Cneorum tricoccum* L. C R Acad Sci Paris 241: 1240–1243
Souèges R (1956) Essai d'embryogénie comparée dans les limites des Hédysarées. Ann Sci Nat Bot 17: 325–351
Souèges R (1957) Embryogénie des Dipsacacées. Développement de l'embryon chez le *Scabiosa columbaria* L. C R Acad Sci Paris 245: 465–468
Souèges R (1958a) Embryogénie des Commelinacées. Développement de l'embryon chez le *Commelina communis* L. C R Acad Sci Paris 246: 2082–2086
Souèges R (1958b) Embryogénie des Commelinacées. Développement de l'embryon chez le *Rhoeo discolor* Hance. C R Acad Sci Paris 246: 2436–2440
Souèges R (1958c) Embryogénie des Boragacées. Développement de l'embryon chez l'*Omphalodes linifera* Moench. C R Acad Sci Paris 247: 249–253
Souèges R (1959a) Embryogénie des Hydrocharitacées. Développement de l'embryon chez l'*Hydrocharis morsus-ranae* L. C R Acad Sci Paris 248: 45–59
Souèges R (1959b) Embryogénie des Caprifoliacées. Développement de l'embryon chez le *Sambucus nigra*. C R Acad Sci Paris 248: 1072–1078
Souèges R (1962) Embryogénie des Dipsacacées. Développement de l'embryon chez le *Dipsacus sylvestris*. C R Acad Sci Paris 256: 2268–2273
Souèges R (1963a) Embryogénie des Dipsacacées. Développement de l'embryon chez le *Cephalaria tatarica*. C R Acad Sci Paris 256: 45–48
Souèges R (1963b) Embryogénie des Dipsacacées. Développement de l'embryon chez le *Kanautia arvensis*. C R Acad Sci Paris 256: 1190–1194
Souèges R, Crété P (1952) Les acquisitions les plus récentes de l'embryogénie des Angiospermes (1947–1951). Ann Biol 28: 9–45
Spangler RC (1925) Female gametophyte of *Trillium sessile*. Bot Gaz 79: 217–221
Spooner DM (1984) Reproductive features of *Dentaria laciniata* and *D. diphylla* (Cruciferae), and the implications in the taxonomy of the eastern North American *Dentaria* complex. Am J Bot 71: 999–1005

Sporne KR (1969) The ovule as an indicator of evolutionary status in angiosperms. New Phytol 68: 555–566
Sporne KR (1973) A note on the evolutionary status of tapetal types in dicotyledons. New Phytol 72: 1173–1174
Sporne KR (1974) The morphology of angiosperms. Hutchinson, London
Sprecher A (1919) Etude sur la semence et la germination de *Garcinia mangostana*. Rev Gen Bot 31: 513–531, 609–633
Sridhar, Singh D (1986) Development of anther and male gametophyte in Cucurbitaceae. J Indian Bot Soc 65: 487–493
Srinath KV (1934) Life history of *Herpestis monniera* H.B. & K. Curr Sci 3: 257–258
Srinath KV (1940) Morphological studies in the genera *Calceolaria* and *Herpestis*. Proc Linn Soc London 2: 152–174
Srinivasachar D (1940a) Morphological studies in the family Anacardiaceae. J Mysore Univ B 1: 83–91
Srinivasachar D (1940b) Embryological studies of some members of Rhamnaceae. Proc Indian Acad Sci B 11: 107–115
Srinivasachar D (1942) A contribution to the embryology of the genus *Polygala*. Proc Natl Acad Sci India B 12: 1–6
Srinivasan AR (1941) Cytomorphological features of *Limnanthemum cristatum* Griseb. and *Enicostemma littorale* Blume. Proc Indian Acad Sci B 14: 529–542
Srinivasan AR (1942) Contribution to the morphology of *Pedalium murex* Linn. and *Sesamum indicum* DC. Proc Indian Acad Sci B 16: 155–164
Srinivasan AR (1946) Morphological and cytological studies in Scrophulariaceae 5. *Striga euphrasioides* Benth. Proc Indian Acad Sci B 24: 21–33
Srinivasan VK (1940) Morphological and cytological studies in Scrophulariaceae 2. Floral morphology and embryology of *Angelonia grandiflora* C. Morr. and related genera. J Indian Bot Soc 19: 197–222
Srivastava AK (1982) Apomixis in *Paspalum paspaloides*. Acta Bot Indica 10: 111–113
Srivastava BL, Tandon S (1985) Adventive embryony in *Morus* species (Mulberry). Phytomorphology 35: 181–182
Srivastava GD (1939) Contribution to the morphology of *Orobanche aegyptiaca*. Proc Natl Acad Sci India B 9: 58–68
Srivastava GN (1964) Male and female gametophytes and development of the seeds in *Bursera delpechiana* Poiss. J Indian Bot Soc 47: 53–59
Srivastava GS, Chandra V (1958) Tricotyly in *Rauwolfia serpentina* Benth. Sci Cult 24: 87–88
Srivastava RK, Agarwal GP (1953) Development of female gametophyte and endosperm in *Chrozophora rottleri*. Bot Gaz 114: 348–350
Stafleu FA (1952) A monograph of Vochysiaceae 2. Acta Bot Neerl 1: 222–242
Stairs GR (1964) Microsporogenesis and embryogenesis in *Quercus*. Bot Gaz 125: 115–121
Stanley RG, Linskens HF (1974) Pollen: biology, biochemistry and management. Springer, Berlin Heidelberg New York
Stant MY (1952) Anatomical evidence for including *Nyctanthes* and *Dimetra* in the Verbenaceae. Kew Bull 1952: 273–276
Stant MY (1964) Anatomy of the Alismataceae. Bot J Linn Soc 59: 1–42
Stant MY (1967) Anatomy of the Butomaceae. Bot J Linn Soc 60: 31–60
Stauffer HU (1957) Zur Stellung der Gattung *Okoubaka* Santales – Studien 1. Ber Schweiz Bot Ges 67: 422–427
Stead AD, Roberts IN, Dickinson HG (1979) Pollen-pistil interaction in *Brassica oleracea*. Planta 146: 211–216
Stebbins GL (1974a) Flowering plants. Origin and dispersal. Oliver & Boyd, Edinburgh
Stebbins GL (1974b) Flowering plants evolution above the species level. Arnold, London
Steffen K (1951) Zur Kenntnis des Befruchtungsvorganges bei *Impatiens glandulifera* Lindl. Cytologische Studien am Embryosack der Balsamineen. Planta 39: 175–244
Steffen K (1955) Kern- und Nucleolenwachstum bei endomitotischer Polyploidisierung (Ein Beitrag zur karyologischen Anatomie von *Pedicularis palustris* L.). Planta 45: 379–394
Steffen K (1956) Endomitosen im Endosperm von *Pedicularis palustris* L. Planta 47: 625–652

Steffen K (1963a) Male gametophyte. In: Maheshwari P (ed) Recent advances in the embryology of angiosperms. Int Soc Plant Morphol, Univ Delhi, pp 15–40
Steffen K (1963b) Fertilization. In: Maheshwari P (ed) Recent advances in the embryology of angiosperms. Int Soc Plant Morphol, Univ Delhi, pp 105–134
Steinbauer GP (1943) Frequency of polyembryony in *Fraxinus* seeds. Bot Gaz 105: 85
Steindl F (1945) Beitrag zur Pollen- und Embryobildung bei *Cynomorium coccineum* L. Arch Julius Klaus Stift 20: 342–355
Steinecke H, Hamman U (1989) Embryologisch-systematische Untersuchungen an Haemodoraceen. Bot Jahrb 111: 247–262
Stenar H (1925a) Embryologische und zytologische Studien über *Limnanthes douglassi* R. Br. Svensk Bot Tidskr 19: 133–152
Stenar H (1925b) Embryologische Studien, 1: Zur Embryologie einiger Columniferen. 2: Die Embryologie der Amaryllidaceen. Diss, Univ Uppsala
Stenar H (1927) Zur Entwicklungsgeschichte der Gattung *Anigosanthus*. Bot Not 1927: 104–114
Stenar H (1928) Zur Embryologie der Asphodeline-Gruppe. Ein Beitrag zur systematischen Stellung der Gattungen *Bulbine* und *Paradisia*. Svensk Bot Tidskr 22: 145–159
Stenar H (1931) Die Art der Pollenbildung bei *Narthecium ossifragum*. Bot Not 1931: 51–54
Stenar H (1932) Studien über die Entwicklungsgeschichte von *Nothoscordum fragrans* und *N. striatum* Kunth. Svensk Bot Tidskr 26: 25–44
Stenar H (1933) Zur Embryologie der *Agapanthus* Gruppe. Bot Not 1933: 520–530
Stenar H (1934) Embryologiche und zytologische Beobachtungen über *Maianthemum biflorum* und *Smilacina stellata*. Ark Bot 26A: 1–20
Stenar H (1935) Embryologische Beobachtungen über *Scheuchzeria palustris* L. Bot Not 1935: 78–86
Stenar H (1937a) Zur Embryosackentwicklung einiger Malpighiazeen. Bot Not 1937: 110–118
Stenar H (1937b) Om *Achroanthes monophyllos*, dess geografiska utbredning och embryologi. Heim Tidsk Jämten 1937–1938: 177–221
Stenar H (1938a) Das Endosperm bei *Hypericum acutum* Moench. Bot Not 1938: 515–527
Stenar H (1938b) Die systematische Stellung der Gattung *Xiphidium*. Svensk Bot Tidskr 32: 274
Stenar H (1941) Über die Entwicklung des Embryosackes bei *Convallaria majalis* L. Bot Not 94: 123–128
Stenar H (1942) Zur Embryologie der *Dracaena*-Gruppe. Heim Tidsk Fornvard 8: 183–195
Stenar H (1949) Zur Kenntnis der Embryologie und der Rapidenzellen bei *Bowiea volubilis* und anderen Liliazeen. Acta Hortic Bergiani 15: 45–63
Stenar H (1950) Studien über das Endosperm bei *Galtonia candicans* und anderen Scilloideen. Acta Hortic Bergiani 15: 169–184
Stenar H (1951a) Zur Embryologie von *Haemanthus katharinae*, nebst Erörterungen über das helobiale Endosperm in den Amaryllidaceae und Liliaceae. Acta Hortic Bergiani 16: 57–72
Stenar H (1951b) Zur Embryologie der Gattung *Phaedranassa* nebst einer Übersicht über den Endospermtyp bei den Amaryllidaceae. Bot Not 1951: 209–228
Stenar H (1952) Notes on the embryology and anatomy of *Luzuriaga latifolia*. Acta Hortic Bergiani 16: 219–232
Stenar H (1953) The embryo sac type in *Smilacina*, *Polygonatum* and *Theropogon* (Liliaceae). Phytomorphology 3: 326–338
Stephen J (1980) Mechanism of endopolyploidization in *Datura* endosperm. Cytologia 45: 657–661
Stephen K (1974) Cytological investigations on the endosperm of *Borassus flabellifer* Linn. Cytologia 39: 195–207
Stephens EL (1908) A preliminary note on the embryo sac in certain Penaeaceae. Ann Bot (London) 22: 329–330
Stephens EL (1909) The embryo sac and embryo of certain Penaeaceae. Ann Bot (London) 23: 363–378
Stephens EL (1910) The embryo sac and embryo of *Geissoloma marginata*. New Phytol 8: 345–348
Stern K (1917) Beiträge zur Kenntnis der Nepenthaceen. Flora 109: 213–282
Stevans NE (1919) The development of the endosperm in *Vaccinium corymbosum*. Bull Torrey Bot Club 46: 465–468
Stevens PF (1971) A classification of Ericaceae. Subfamilies and tribes. Bot J Linn Soc 64: 1–53

Stevenson DWM (1983) Systematic implications of the floral morphology of the Mayacaceae. Am J Bot 70 (Suppl): 32
Steyn E (1973a) The embryology of *Romulea rosea* Eckl. var *reflexa* Beg. 1. The structure, development and fertilization of the ovule. J S Afr Bot 39: 113–121
Steyn E (1973b) The embryology of *Romulea rosea* Eckl. var *reflexa* Beg. 2. Development of the endosperm, embryo and seed-coat. J S Afr Bot 39: 235–243
Steyn E (1975) Embryogeny of *Melianthus major* L. J S Afr Bot 41: 199–205
Stolt KAH (1921) Zur Embryologie der Gentianaceen. Kon Svensk Vet Akad Handl 61: 1–56
Stolt KAH (1928) Die Embryologie von *Myriophyllum alterniflorum* D.C. Svensk Bot Tidskr 22: 305–319
Stolt KAH (1936) Beiträge zur Embryologie der Lentibulariaceen. Svensk Bot Tidskr 30: 690–696
Stone DE, Broome CR (1971) Pollen ultrastructure evidence for relationship of the Juglandaceae and the Rhoipteleaceae. Pollen Spores 13: 5–14
Stoudt HN (1960) Sporogenesis in *Magnolia kobus* 1. Microsporogenesis. Proc Penn Acad Sci 34: 29–33
Stoutamire WP (1983) Wasp-pollinated species of *Caladenia* (Orchidaceae) in South-western Australia. Aust J Bot 31: 383–394
Stover E (1932) Life history of *Nymphoides peltatum*. Bot Gaz 93: 474–483
Stow I (1930) Experimental studies on the formation of the embryo sac-like giant pollen grains in the anther of *Hyacinthus orientalis*. Cytologia 1: 417–439
Stow I (1934) On the female tendencies of the embryo sac-like giant pollen grains of *Hyacinthus orientalis*. Cytologia 5: 88–108
Strasburger E (1879) Die Angiospermen und die Gymnospermen. Fischer, Jena
Strasburger E (1884) Neuere Untersuchungen über den Befruchtungsvorgang bei den Phanerogamen als Grundlage für eine Theorie der Zeugung. Fischer, Jena
Strasburger E (1902) Ein Beitrag zur Kenntnis von *Ceratophyllum submersum* und phylogenetische Erörterungen. Bot Jahrb 37: 477–526
Strasburger E (1905) Die Samenanlagen von *Drimys winteri* und die Endospermbildung bei Angiospermen. Flora 95: 215–231
Strasburger E (1910) Sexuelle und apogame Fortpflanzung beinden Urticaceen. Bot Jahrb 47: 245–288
Strömberg B (1956) The embryo sac development of the genus *Freycinetia*. Svensk Bot Tidskr 50: 129–134
Stushnoff C, Palser BF (1969) Embryology of five *Vaccinium* taxa including diploid, tetraploid and hexaploid species or cultivars. Phytomorphology 19: 312–331
Subba Rao AM (1937) A note on the development of the female gametophytes of some Malpighiaceae and polyembryony in *Hiptage madablota*. Curr Sci 6: 280–282
Subba Rao AM (1939) The ovule and embryo sac development of some Malpighiaceae. Curr Sci 8: 79–81
Subba Rao AM (1940a) Cytology and embryogeny in *Artocarpus integrifolia* Linn. J Mysore Univ B 1: 63–73
Subba Rao AM (1940b) Studies in Malpighiaceae 1. Embryo sac development and embryology in the genera *Hiptage*, *Banisteria* and *Stigmatophyllum*. J Indian Bot Soc 18: 145–156
Subba Rao AM (1941) Studies in Malpighiaceae 2. Structure and development of the ovules and embryo sacs of *Malpighia coccifera* Linn. and *Tristellateia australis* Linn. Proc Natl Inst Sci India B 7: 393–404
Subramanian D (1983) Seed morphological studies in *Phaseolus*, *Vigna* and *Macroptilium*. J Indian Bot Soc 62: 77–83
Subramanyam K (1942) Gametogenesis and embryogeny in a few members of the Melastamaceae. J Indian Bot Soc 21: 69–85
Subramanyam K (1944) A contribution to the life history of *Sonerila wallichii* Benn. Proc Indian Acad Sci B 19: 115–120
Subramanyam K (1946) Further contribution to the embryology of the genus *Osbeckia* Linn. J Mysore Univ B 7: 1–12
Subramanyam K (1947) A case of polyembryony in *Wahlenbergia gracilis* Schrad. Curr Sci 16: 250
Subramanyam K (1948a) An embryological study of *Melastoma malabathricum* L. J Indian Bot Soc 29: 11–19

Subramanyam K (1948b) A contribution to the embryology of *Wahlenbergia gracilis* Schard. Proc Natl Inst Sci India B 14: 359–366
Subramanyam K (1949) An embryological study of *Lobelia pyramidalis* Wall., with special reference to the mechanism of nutrition of the embryo sac in the family Lobeliaceae. New Phytol 48: 365–373
Subramanyam K (1950a) A contribution to the knowledge of the systematic position of the Sphenocleaceae. Proc Indian Acad Sci B 31: 60–65
Subramanyam K (1950b) An embryological study of *Levenhookia dubia* Sond in Lehm. Proc Natl Inst Sci India B 16: 245–253
Subramanyam K (1950c) Development of embryo sac and endosperm in *Stylidium tenellum* Swartz. Curr Sci 19: 294
Subramanyam K (1951a) Embryology of *Oxyspora paniculata* DC. Phytomorphology 1: 205–212
Subramanyam K (1951b) Flower structure and seed development in *Isotoma fluviatilis* F.V.M. Proc Natl Inst Sci India B 17: 275–285
Subramanyam K (1951c) The origin and nature of haustoria in *Lobelia cardinalis* L. Bot Gaz 112: 319–322
Subramanyam K (1951d) On the probable origin of the unilocular ovary of the Compositae from the Stylidiaceae. Proc Indian Acad Sci B 33: 327–330
Subramanyam K (1951e) A morphological study of *Stylidium graminifolium*. Lloydia 14: 65–81
Subramanyam K (1951f) Interrelationships of Campanulatae. J Mysore Univ B 12: 5–13
Subramanyam K (1953) The nutritional mechanism of embryo sac and embryo in the families Campanulaceae, Lobeliaceae and Stylidiaceae. J Mysore Univ B 13: 1–4
Subramanyam K (1960) Nutritional mechanism of the seed. 1 Nutritional mechanism of embryo sac. 2 Nutritional mechanism of the embryo. J Madras Univ B 30: 29–56
Subramanyam K (1962) Embryology in relation to systematic botany with particular reference to the Crassulaceae. In: Symp Plant embryology. Council Sci Ind Res, New Delhi, pp 94–112
Subramanyam K (1963) Embryology of *Sedum ternatum* Mich. J Indian Bot Soc (Prof. P. Maheshwari Comm. Vol.) 42A: 259–275
Subramanyam K (1967) Some aspects on the embryology of *Sedum chrysanthum* (Boissier) Rayomond-Hamlet with a discussion on its systematic position. Phytomorphology (Panchanan Maheshwari Memm. Vol.) 17: 240–247
Subramanyam K (1970a) Crassulaceae: In: Symp Comparative embryology of angiosperms. Indian Natl Sci Acad Bull 41: 84–89
Subramanyam K (1970b) Campanulaceae. In: Symp Comparative embryology of angiosperms. Indian Natl Sci Acad Bull 41: 306–312
Subramanyam K (1970c) Sphenocleaceae. In: Symp Comparative embryology of angiosperms. Indian Natl Sci Acad Bull 41: 313–316
Subramanyam K (1970d) Pentaphragmataceae. In: Symp Comparative embryology of angiosperms. Indian Natl Sci Acad Bull 41: 317–320
Subramanyam K (1970e) Stylidiaceae. In: Symp Comparative embryology of angiosperms. Indian Natl Sci Acad Bull 41: 321–324
Subramanyam K, Govindu HC (1948) The development of female gametophyte in *Oenothera odorata*. J Indian Bot Soc 27: 25–30
Subramanyam K, Govindu HC (1949) The development of the male and female gametophyte in *Lilium neilgherrense* Wight. J Indian Bot Soc 28: 36–41
Subramanyam K, Narayana HS (1972) Some aspects of the floral morphology and embryology of *Flagellaria indica* Linn. In: Murty YS, Johri BM, Mohan Ram HY, Varghese TM (eds) Advances in plant morphology (Prof. V. Puri Comm. Vol.). Sarita Prakashan, Meerut (India), pp 211–217
Subramanyam K, Narayana LL (1968) Floral anatomy and embryology of *Primula floribunda* Wall. Phytomorphology 18: 105–113
Subramanyam K, Narayana LL (1976) A contribution to the floral anatomy and embryology in certain members of Primulaceae. J Indian Bot Soc 55: 274–282
Subramanyam K, Narayana LL, Sundri KT (1985) A contribution to the embryology of *Nepenthes khasiana* Hook. f. J Indian Bot Soc 64: 156–160

Sud KC (1984a) Microsporogenesis and male gametophyte in *Beaumontia grandiflora* Wall. Acta Bot Indica 12: 210–212

Sud KC (1984b) A contribution to the embryology of *Trachelospermum fragrans* Hook. f. (Apocynaceae). Proc Indian Acad Sci Plant Sci 93: 495–501

Sugihara Y, Yoshimura H, Umezu N (1969) Embryo sac development in *Disporum*. Phytomorphology 19: 112–114

Suita N (1937) Studies on the male gametophyte in angiosperms 2. Differentiation and behaviour of the vegetative and generative elements in the pollen grains of *Crinum*. Cytologia (Fuji Jubl Vol): 920–933

Sulbha K (1954a) The embryology of *Iphigenia indica* Kunth. Phytomorphology 4: 180–191

Sulbha K (1954b) Embryo sac of *Scilla*. Curr Sci 23: 89–90

Sulbha K (1957) Embryology of *Brassica juncea* Czern. & Cess. J Indian Bot Soc 36: 292–301

Sumner MJ, van Caeseele L (1989) The ultrastructure and cytochemistry of the egg apparatus of *Brassica campestris*. Can J Bot 67: 177–190

Sun AC, Wang FH (1984) Megasporogenesis and microsporogenesis in *Pulsatilla chinense*. Phytomorphology 34: 232–237

Sunder Rajan S (1968) Embryological studies in Compositae. Floral morphology, sporogenesis and gametogenesis in *Emilia sonchifolia* (*Linn.*) D.C. Curr Sci 37: 26–28

Sunder Rajan S (1972) Embryological studies in Compositae 3. A contribution to the embryology of *Blainvillea rhomboidea* Cass. (*B. latifolia* DC.). Proc Indian Acad Sci B 74: 167–176

Sunder Rajan S (1974a) Embryological studies in Lentibulariaceae. Proc Indian Acad Sci B 80: 18–25

Sunder Rajan S (1974b) Embryological studies in Compositae 4. A contribution to the life history of *Spilanthes acmilla* Murr. (*S. calva* Wt. Ic.). Proc Indian Acad Sci B 79: 267–282

Sunder Rajan S (1974c) Embryological studies in Compositae 5. On the embryology of *Eclipta alba*. Curr Sci 43: 56–57

Sunder Rajan S (1976) Micropyle formation in the ovule as an indicator of primitiveness in angiosperms. Proc Indian Acad Sci B 84: 124–127

Sunder Rajan S, Jeevan Kumar D (1974) Embryological studies in Lentibulariaceae 1. Floral morphology and embryology of *Utricularia smithiana* Wt. Ic. Proc Indian Acad Sci B 80: 18–25

Sunder Rao Y (1940a) Male and female gametophytes of *Polemonium coeruleum* Linn. Proc Natl Inst Sci India B 6: 695–704

Sunder Rao Y (1940b) Structure and development of the embryo sac of *Drimiopsis kirki* Baker and *Allium govanianum* Wall. J Indian Bot Soc 19: 273–282

Sunder Rao Y (1953) Karyo-systematic studies in Helobiales 1. Butomaceae. Proc Natl Inst Sci India B 19: 563–581

Sundri KT, Narayana LL (1977) A note on the embryology of *Marianthus bignonaceus* F. Muell. Curr Sci 46: 282–283

Sundri KT, Radhakrishnan M, Narayana LL (1982) Chemotaxonomy of *Ceratophyllum*. Acta Bot Indica 10: 304–305

Suneson S (1933) Zur Embryologie der Gattung *Viburnum*. Bot Not 1933: 181–194

Suri RK, Deshpande PK (1981) Development of seed in *Cassia absus* and *C. auriculata*. Phytomorphology 31: 40–47

Süssenguth K (1921) Bemerkungen zur meiotischen und somatischen Kernteilung bei einigen Monocotyledon. Flora 119: 313–328

Süssenguth K (1927) Über die Gattung *Lennoa*. Flora 122: 264–305

Süssenguth K (1953) Vitaceae. In: Engler A, Prantl K (eds) Die natürlichen Pflanzenfamilien.

Svedelius N (1904) On the life history of *Enalus acoroides*. Ann Roy Bot Garden Paradeniya 2: 267–297

Svensson HG (1922) Embryologien hos *Lycopsis arvensis*. Svensk Bot Tidskr 16: 137–138

Svensson HG (1925) Zur Embryologie der Hydrophyllaceen, Boraginaceen und Heliotropiaceen mit besonderer Rücksicht auf die Endospermbildung. Uppsala Univ Arssk 2: 1–176

Svensson HG (1926) Zytologisch-embryologische Solanaceenstudien 1. Über die Samenentwicklung von *Hyoscyamus niger* L. Svensk Bot Tidskr 20: 420–434

Svensson HG (1928) Zur Entwicklungsgeschichte der Blüten und Samen von *Limosella aquatica*. Svensk Bot Tidskr 22: 462–476
Swaminathan D (1967) Embryological studies in *Polystachya flavescens*. Curr Sci 36: 384–385
Swaminathan MS (1954) Microsporogenesis in some commercial potato varieties. J Hered 45: 265–272
Swamy BGL (1941a) Contribution to the life history of *Bignonia megaspatomica*. J Indian Bot Soc 20: 299–305
Swamy BGL (1941b) The development of the male gamete in *Cymbidium bicolor* Lindl. Proc Indian Acad Sci B 14: 454–460
Swamy BGL (1942a) Female gametophyte and embryogeny in *Cymbidium bicolor* Lindl. Proc Indian Acad Sci B 15: 194–201
Swamy BGL (1942b) Embryological studies in Palmae. A preliminary note on the megasporogenesis in *Areca catechu*. Curr Sci 11: 109
Swamy BGL (1943a) Embryology of Orchidaceae. Curr Sci 12: 13–17
Swamy BGL (1943b) Gametogenesis and embryogeny of *Eulophia epidendraea*. Proc Natl Inst Sci India B 9: 59–65
Swamy BGL (1944a) A preliminary note on the embryology of *Casuarina equisetifolia* Forst. Proc Indian Acad Sci B 20: 187–191
Swamy BGL (1944b) A reinvestigation of the embryo sac of *Piper betel*. Proc Natl Acad Sci India B 14: 109–113
Swamy BGL (1944c) The embryo sac and the embryo of *Satyrium nepalense* D. Don. J Indian Bot Soc 23: 66–70
Swamy BGL (1945a) Embryo sac and fertilization in *Cypripedium spectabile*. Bot Gaz 107: 291–295
Swamy BGL (1945b) The embryo sac of *Heckeria subpeltata* Kunth. J Indian Bot Soc 24: 1–3
Swamy BGL (1946a) Endosperm in *Hypericum mysorense* Heyne. Ann Bot (London) 9: 165–169
Swamy BGL (1946b) Some notes on the embryo of *Cymbidium bicolor* Lindl. Curr Sci 15: 139–140
Swamy BGL (1946c) Development of endosperm in *Zephyranthes andersonii* Baker. Proc Natl Inst Sci India B 12: 187–190
Swamy BGL (1946d) Embryology of *Habenaria*. Proc Natl Inst Sci India B 12: 413–426
Swamy BGL (1946e) The embryology of *Zeuxine sulcata* Lindl. New Phytol 45: 132–136
Swamy BGL (1946f) Inverted polarity of the embryo sac of angiosperms and its relation to the archegonium theory. Ann Bot (London) 9: 171–183
Swamy BGL (1947) On the life history of *Vanilla planifolia*. Bot Gaz 108: 449–456
Swamy BGL (1948a) A contribution to the life history of *Casuarina*. Proc Am Acad Arts Sci 77: 1–32
Swamy BGL (1948b) A contribution to the embryology of the Marcgraviaceae. Am J Bot 35: 628–633
Swamy BGL (1948c) The embryology of *Epidendrum prismatocarpum*. Bull Torrey Bot Club 75: 245–249
Swamy BGL (1948d) Agamospermy in *Spiranthes cernua*. Lloydia 11: 149–162
Swamy BGL (1948e) On the post-fertilization development of *Trillium undulatum*. Cellule 52: 7–14
Swamy BGL (1949a) Further contribution to the morphology of the Degeneriaceae. J Arnold Arbor 30: 10–38
Swamy BGL (1949b) A reinvestigation of the embryo sac of *Maianthemum canadense*. Bull Torrey Bot Club 76: 17–23
Swamy BGL (1949c) Embryological studies in the Orchidaceae 1. Gametophytes. Am Midl Nat 41: 184–201
Swamy BGL (1949d) Embryological studies in the Orchidaceae 2. Embryogeny. Am Midl Nat 41: 202–232
Swamy BGL (1952) Some aspects in the embryology of *Zygogynum bailloni* Van Tiegh. Proc Natl Inst Sci India B 18: 390–406
Swamy BGL (1953a) On the floral structure of *Scyphostegia*. Proc Natl Inst Sci India B 19: 127–142
Swamy BGL (1953b) *Sarcandra irvingbaileyi*: a new species of vesselless dicotyledon from South India. Proc Natl Inst Sci India B 19: 301–306
Swamy BGL (1953c) Some observations on the embryology of *Decaisnea insignis* Hook et Thoms. Proc Natl Inst Sci India B 19: 307–310

Swamy BGL (1953d) Morphology and relationships of the Chloranthaceae. J Arnold Arbor 34: 375–408
Swamy BGL (1960) Contributions to the embryology of *Cansjera rheedii*. Phytomorphology 10: 397–409
Swamy BGL (1962) The embryo of monocotyledons: A working hypothesis from a new approach. In: Symp Plant embryology. Council Sci Ind Res, New Delhi, pp 113–123
Swamy BGL (1963) The origin of cotyledon and epicotyl in *Ottelia alismoides*. Beitr Biol Pflanzen 39: 1–16
Swamy BGL (1964a) Macrogametophytic ontogeny in *Schisandra chinensis*. J Indian Bot Soc 43: 391–396
Swamy BGL (1964b) Observations on the floral morphology and embryology of *Stemona tuberosa* Lour. Phytomorphology 14: 458–468
Swamy BGL (1966) Origin and organisation of the embryonic shoot apex in *Eichhornia crassipes*. Bull Torrey Bot Club 93: 20–34
Swamy BGL (1974a) A critique on the "intermediate" type of endosperm in angiosperms. Phytomorphology 24: 35–41
Swamy BGL (1974b) On the presumed ancestory of angiosperm embryo sac. Phytomorphology 24: 102–106
Swamy BGL (1979) Embryogenesis in *Cheirostylis flabellata*. Phytomorphology 29: 199–203
Swamy BGL (1980) Embryogenesis in *Sagittaria sagittifolia*. Phytomorphology 30: 204–212
Swamy BGL, Bailey IW (1949) The morphology and relationships of *Cercidiphyllum*. J Arnold Arbor 30: 197–210
Swamy BGL, Balakrishna BP (1946) Female gametophyte of *Acalypha tricolor*. J Indian Bot Soc 25: 67–69
Swamy BGL, Dayanand Rao J (1963) The endosperm of *Opilia amentacea* Roxb. Phytomorphology 13: 423–428
Swamy BGL, Ganapathy PM (1957a) On endosperm in dicotyledons. Bot Gaz 119: 47–50
Swamy BGL, Ganapathy PM (1957b) A new type of endosperm haustorium in *Nothapodytes foetida*. Phytomorphology 7: 331–336
Swamy BGL, Krishnamurthy KV (1970) On the so-called endothelium in the monocotyledons. Phytomorphology 20: 262–269
Swamy BGL, Krishnamurthy KV (1973) The Helobial endosperm: A decennial review. Phytomorphology 23: 74–79
Swamy BGL, Krishnamurthy KV (1975) Embryo sac ontogenies in angiosperms: An elucidation. Phytomorphology 25: 12–18
Swamy BGL, Krishnamurthy KV (1977) Certain conceptual aspects of meristems 2. Epiphysis and shoot apex. Phytomorphology 27: 1–8
Swamy BGL, Krishnamurthy KV (1980) From flower to fruit. McGraw-Hill, New Delhi
Swamy BGL, Lakshmanan KK (1962a) The origin of epicotylary meristem and cotyledon in *Halophila ovata* Goudich. Ann Bot (London) 26: 243–249
Swamy BGL, Lakshmanan KK (1962b) Contributions to the embryology of the Najadaceae. J Indian Bot Soc 42: 247–266
Swamy BGL, Padmanabhan D (1961a) Notulae embryologicae 1. The functions of endosperm in *Aviccinia officinalis*. Curr Sci 30: 424–425
Swamy BGL, Padmanabhan D (1961b) The quadrant proembryo of *Epithema carnosum*. Proc Indian Acad Sci B 53: 166–172
Swamy BGL, Padmanabhan D (1961c) Embryogenesis in *Sphenoclea zeylanica*. Proc Indian Acad Sci B 54: 169–187
Swamy BGL, Padmanabhan D (1962) A reconnaissance of angiosperm embryogenesis. J Indian Bot Soc 41: 422–439
Swamy BGL, Parmeswaran N (1960) A contribution to the embryology of *Begonia crenata*. J Indian Bot Soc 39: 140–148
Swamy BGL, Parmeswaran N (1962) On the origin of cotyledon and epicotyl in *Potamogeton indicus*. Öst Bot Z 109: 344–349
Swamy BGL, Parmeswaran N (1963) The helobial endosperm. Biol Rev 38: 1–50
Swamy BGL, Periasamy K (1955) Contribution to the embryology of *Acrotrema arnottianum*. Phytomorphology 5: 301–314

Swanson SD, Sohmer SH (1976) The biology of *Podophyllum peltatum* L. (Berberidaceae), the may apple 2. The transfer of pollen and success of sexual reproduction. Bull Torrey Bot Club 103: 223–226
Swarajya Lakshmi P, Pullaiah T (1988) Embryology of *Pulicaria*. Phytomorphology 38: 141–145
Swarajya Lakshmi P, Pullaiah T (1989) A contribution to the embryology of *Youngia japonica*. Phytomorphology 39: 149–156
Swarupanandan K (1986) Late embryogenesis and morphology of mature embryos in three species of Dipterocarpaceae. Can J Bot 64: 2582–2587
Sylla Th, Albers F (1989) Samenentwicklung und samenmorphologic Krautiger und sukkutenter Asclepiadaceae. Bot Jahrb 110: 479–491
Szkukalek A, Hausksecht M, Erdelská O (1989) Some aspects of the developmental correlation of the maize embryo and endosperm. In: Paré J, Bugnicourt M (eds) Some aspects and actual orientations in plant embryology. Dedicatory volume to Prof A. Lebègue. Univ Press Picardie, Amiens (Fr), pp 245–250
Szujkó-Lacza J (1976) Fertilization and fusion of free endosperm nuclei in *Pimpinella anisum* L. Acta Biol Acad Sci Hung 27: 65–70
Szujkó-Lacza J (1978) Hypostase, embryonic sac and endosperm in *Anethum graveolens* L., and in various families. Acta Biol Acad Sci Hung 29: 255–271
Szwabowicz A (1976) Microsporogenesis of *Betula oycoviensis* Bess. and its progeny. Acta Soc Bot Pol 45: 169–184

Täckholm G (1914) Zur Kenntnis der Embryosackentwicklung von *Lopezia coronata*. Svensk Bot Tidskr 8: 223–234
Täckholm G (1915) Beobachtungen über die Samenentwicklung einiger Onagraceen. Svensk Bot Tidskr 9: 294–361
Täckholm G, Söderberg E (1918) Neue Beispiele der simultanen und successiven Wandbildung in den Pollen-Mutterzellen. Svensk Bot Tidskr 12: 189–201
Tainter FH (1968) The embryology of *Arceuthobium pusilum*. Can J Bot 46: 1473–1476
Takahashi H (1988) Ontogenetic development of pollen tetrads of *Drosera capensis* L. Bot Gaz 149: 275–282
Takahashi M (1986) Microsporogenesis in a parthenogenetic species *Hottuynia cordata* Thunb. (Saururaceae). Bot Gaz 147: 47–54
Takahashi M (1987) Development of omniaperturate pollen in *Trillium kamtschaticum* (Liliaceae). Am J Bot 74: 1842–1852
Takahashi M (1989) Pattern determination of the exine in *Caesalpinia japonica* (Leguminosae, Caesalpinioideae). Am J Bot 76: 1615–1626
Takahashi M, Kouchi J (1988) Ontogenetic development of spinous exine in *Hibiscus syriacus* (Malvaceae). Am J Bot 75: 1549–1558
Takao S (1966) A study of development of the embryo sac in *Impatiens balsamina*. Bot Mag (Tokyo) 79: 437–446
Takao S (1968) A study in the development of embryo sac in *Impatiens textori*. Bot Mag (Tokyo) 81: 310–317
Takao S (1976) Embryogenical study of *Impatiens textori*. Phytomorphology 26: 200–210
Takaso T (1987) Ovule ontogeny and morphology in Ulmaceae (Abstr). In: 14th Int Bot Congr, Berlin, p 220
Takaso T, Bouman F (1984) Ovule ontogeny and seed development in *Potamogeton natans* L. (Potamogetonaceae), with a note on the campylotropous ovule. Acta Bot Neerl 33: 519–533
Takaso T, Tobe H (1990) Seed-coat morphology and evolution in Celtidaceae and Ulmaceae (Urticales). Bot Mag (Tokyo) 103: 25–41
Takeda H (1936) On the coma or hairy tufts on the seed of willow. Bot Mag (Tokyo) 50: 283–289
Takeuchi Y (1971) Embryological development in *Dioscorea izuensis* Akahori. Sci Rep Tôhoku Univ Ser Biol 35: 225–231
Takeuchi Y (1972) Embryo sac development in *Dioscorea japonica* Thunb. and *D. quinqueloba* Thunb. Acta Phytotaxon Geobot 25: 57–60
Takeuchi Y, Chugai K (1968) On the embryo sac formation of *Dioscorea nipponica* Makino and *D. tokoro* Makino. Sci Rep Tôhoku Univ Ser Biol 34: 137–140

Takhtajan A (1959) Die Evolution der Angiospermen. Fischer, Jena
Takhtajan A (1966) System and phylogeny of flowering plants. Nauka, Moscow
Takhtajan A (1969) Flowering plants: origin and dispersal (translated from Russian by C. Jeffrey). Oliver & Boyd, Edinburgh
Takhtajan A (1973) Evolution und Ausbreitung der Blütenpflanzen. Fischer, Stuttgart
Takhtajan AL (1976) Neoteny and the origin of flowering plants. In: Beck CB (ed) Origin and early evolution of angiosperms. Columbia Univ Press, New York, pp 207–219
Takhtajan AL (1980) Outline of the classification of flowering plants (Magnoliophyta). Bot Rev 46: 225–359
Takhtajan A (ed) (1985) Anatomia seminum comparativa, vol 1. Liliopsida seu monocotyledons. Nauka, Leningrad (in Russian)
Takhtajan A (1987) Systema Magnoliophytorum. Nauka, Leningrad
Takhtajan A (ed) (1988) Anatomia seminum comparativa, vol 2. Dicotyledons: Magnoliidae, Ranunculidae. Nauka, Leningrad (in Russian)
Tamura M (1972) Morphology and phyletic relationship of the Glaucidiaceae. Bot Mag (Tokyo) 85: 29–41
Tamura M, Mizumoto I (1972) Stages of embryo development in ripe seeds or achenes of the Ranunculaceae. Jpn J Bot 47: 225–237
Tamura M, Mizumoto I (1974) The cotyledon and growing point in the monocotyledonous embryos of *Shibateranthis pinnatifida* and *Anemone flaccida*. Jpn J Bot 49: 123–128
Tan K (1980) Studies in the Thymelaeaceae 1. Germination, seedling, fruit and seeds. Notes R Bot Garden Edinburgh 38: 149–164
Tanaka N (1950) Gametogenesis and fertilization in the genus *Carex*. Coord Comm Res Genet 1: 133–137
Tanaka R, Watanabe K (1972) Embryological studies in *Chrysanthemum makinoi* and its hybrid crossed with hexaploid cv. *japonense*. J Sci Hiroshima Univ B 14: 75–84
Tandon SL, Kapoor BM (1961) Amitosis in endosperm of *Zephyranthes ajax* Sprenger. Curr Sci 30: 72–73
Tandon SL, Kapoor BM (1962a) Contributions to the cytology of endosperm in some angiosperms 1. *Zephyranthes ajax*. Caryologia 15: 21–41
Tandon SL, Kapoor BM (1962b) Nuclear variation in the endosperm of *Zephyranthes ajax* and its causes. Sci Cult 28: 189–190
Tandon SR, Herr Jr JM (1971) Embryological features of taxonomic significance in genus *Nyssa*. Can J Bot 49: 505–519
Taneja S (1969a) Embryology, seed and fruit structure of *Hebenstreitia comosa* Hochst. with discussion on the systematic position of the family Selaginaceae. J Indian Bot Soc 48: 300–310
Taneja S (1969b) Embryogeny in *Nemesia strumosa* Benth. Curr Sci 38: 522
Taneja S (1970) Embryology, structure and development of seed and fruit of *Calceolaria mexicana* Benth. Plant Sci (Lucknow, India) 2: 61–66
Tang SH (1962) Sporogenesis and gametophyte development in *Eucommia ulmoides* (in Chinese). Acta Bot Sin 10: 29–34
Tang Y (1943) Systematic anatomy of the woods of the Hamamelidaceae. Bull Fan Mem Inst Biol 1: 8–63
Tasneem F (= Khaleel TF) (1966) Sporogenesis and the development of gametophytes in *Cordia alba* L. Curr Sci 35: 73–74
Tasneem F (1967) Embryological studies in *Trichodesma zeylanicum*. Curr Sci 36: 53
Tasneem F, Kusuma KP (1970) Embryological studies in Capparidaceae 1. Sporogenesis and the development of gametophytes in *Cleome viscosa* L. Proc Indian Acad Sci B 72: 207–215
Tateishi S (1927) On the development of the embryo sac and fertilization of *Acalypha australis*. Bot Mag (Tokyo) 41: 477–485
Taylor ARA (1957) Studies of the development of *Zostera marina* 1. The embryo and seed. Can J Bot 35: 477–497
Tejavathi DH (1987) Seed development in some members of Cyperaceae. Beitr Biol Pflanzen 62: 43–55
Telesca MR (1962) Contributo alla embriologia di *Phyteuma limonifolium* S. et S. della Dott. Nuovo G Bot Ital 69: 372–378

Teppner H, Klein E (1985) Karyologie und Forstpflanzungsmodus von *Nigritella* (Orchidaceae-Orchideae), inkl. *N. archiducisjoannis* spec. nov. und Zweier Neukombinationen. Phyton (Austria) 25: 147–176

Terasaka O, Niitsu T, Tanaka R (1979) Single fertilization in *Spiranthes sinensis*. Bot Mag (Tokyo) 92: 59–67

Teryokhin ES (1962) The development of embryos in certain Pyroleae: Monotropoideae. Bot Zh 47: 1811–1816 (in Russian)

Teryokhin ES (1963) Development of the ovule and female gametophyte in Pyroleae and Monotropeae. Bot Zh 48: 406–414 (in Russian)

Teryokhin ES (1980) Embryology of Orobanchaceae 3. *Mannagettaea hummelii*. Bot Zh 64: 1390–1400 (in Russian)

Teryokhin ES (1988) Reproductive biology of weed broomrapes. Nauka, Leningrad

Teryokhin ES, Kamelina OP (1969) The endosperm of Orchidaceae: on the question of reduction. Bot Zh 54: 657–666 (in Russian)

Teryokhin ES, Kotov VA (1988) Embryology of *Cuscuta japonica*. Bot Zh 73: 222–230

Teryokhin ES, Yakovlev MS (1967) Embryology of Balanophoraceae (on the problem of homologies of the "flower" of *Balanophora*). Bot Zh 52: 745–748 (in Russian)

Terzijski D (1971) Development of the endosperm and embryo in *Trigonella gladiata* Stov. Genet Sci 4: 343–351

Terzijski D, Dumanova A (1974) Cytoembryological study on the ontogenesis of the endosperm in some species of the genus *Lathyrus* L. (Papilionaceae). Genet Sci 7: 367–379

Terzijski D, Kuristou MA (1973) Cytoembryologic study of *Festuca protensis* Huds., *Alopecurus pratensis* L., *Holcus lanatus* L. and *Agrostis vulgaris*. Genet Sci 6: 45–57

Thakur DR, Bajwa BS (1971) Extent of polyembryony in some species and varieties of *Citrus*. Indian J Hortic 28: 25–28

Thathachar T (1940) Development of the embryo sac and formation of haustoria in *Lantana indica* Roxb. and *Stachytarpheta indica* Vahl. J Indian Bot Soc 19: 45–52

Thathachar T (1942a) Studies in Oxalidaceae: *Biophytum sensitivum* DC., *Averrhoa carambola* L. and *Averrhoa billimbi* L. J Indian Bot Soc 21: 21–31

Thathachar T (1942b) Studies in Gesneriaceae. Gametogenesis and embryogeny in *Didymocarpus tomentosa* Wt. J Indian Bot Soc 21: 185–193

Thathachar T (1943) Development of seed in *Rhyncoglossum obliquum* BL. J Indian Bot Soc 22: 151–157

Thathachar T (1952) Morphological studies in the Euphorbiaceae 1. *Acalypha lanceolata* Willd. Phytomorphology 2: 197–201

Thathachar T (1953) Morphological studies in the Euphorbiaceae 2. *Mallotus philippinensis* M. Arg. Proc Natl Inst Sci India B 19: 469–474

Thathachar T, Nagaraja Rao A (1956) A contribution to the embryology of *Agrostistachys meeboldii* Pax et Hofman. Sci Cult 21: 470–472

Thien LB (1974) Floral biology of *Magnolia*. Am J Bot 61: 1037–1045

Thirumalachar MJ, Razi BA (1941) Megasporogenesis and endosperm formation in *Eriodendron anfractuosum* DC. Proc Indian Acad Sci B 14: 461–465

Thirumaran K, Lakshmanan KK (1981) The development of endosperm in *Priva cordifolia* (Linn. f.) Druce. Curr Sci 50: 775–776

Thirumaran K, Lakshmanan KK (1982) Synergid embryo in *Dioscorea composita* Hemsl. Curr Sci 57: 428–429

Thirumaran K, Lakshmanan KK (1984) Embryological studies on *Priva cordifolia*. Acta Bot Indica 12: 103–106

Thomas A, Lugardon B (1972) Sur la structure fine des tétrades de deux Annonacées (*Asteranthe asterias* et *Hexalobus monopetalus*). C R Acad Sci Paris 275: 1749–1752

Thomas JL (1960) A monographic study of the Cyrillaceae. Univ Press, Cambridge (England)

Thomas JL (1961) The genera of the Cyrillaceae and Clethraceae of the south-eastern United States. J Arnold Arbor 42: 96–106

Thomson JD (1985) Pollination and seed-set in *Diervilla lonicera* (Caprifoliaceae): Temporal patterns of flower and ovule development. Am J Bot 72: 737–740

Thompson WP, Johnson D (1945) The cause of incompatibility between barley and rye. Can J Res 23C: 1–5

Thorne RF (1963) Some problems and guiding principles of angiosperm phylogeny. Am Nat 97: 287–305
Thorne RF (1968) Synopsis of a putatively phylogenetic classification of the flowering plants. Aliso 6: 57–66
Thorne RF (1973) The "Amentiferae" or Hamamelidae as an artificial group: A summary statement. Brittonia 25: 395–404
Thorne RF (1976) A phylogenetic classification of the angiospermae. Evol Biol 9: 35–106
Thorne RF (1977) Some realignments in the angiospermae. In: Kubitzki H (ed) Flowering plants. Plant Syst Evol, Suppl 1. Springer, Wien, pp 299–319
Thorne RF (1983) Proposed new realignments in angiosperms. N J Bot 3: 85–117
Tiagi B (1951a) A contribution to the morphology and embryology of *Cuscuta hyalina* Roth. and *C. planiflora* Tenore. Phytomorphology 1: 9–21
Tiagi B (1951b) Studies in the family Orobanchaceae 3. A contribution to the embryology of *Orobanche cernua* Loeffl. and *O. aegyptiaca* Pers. Phytomorphology 1: 158–169
Tiagi B (1952a) Studies in the family Orobanchaceae 1. A contribution to the embryology of *Cistanche tubulosa* Wight. Lloydia 15: 129–148
Tiagi B (1952b) Studies in the family Orobanchaceae 2. A contribution to the embryology of *Aeginetia indica* Linn. Bull Torrey Bot Club 79: 63–68
Tiagi B (1956) A contribution to the embryology of *Striga orobanchoides* Benth. and *Striga euphrasioides* Benth. Bull Torrey Bot Club 83: 154–171
Tiagi B (1963) Studies in the family Orobanchaceae 5. Embryology of *Boschniakia himalaica* Hook. and *B. tuberosa* (Hook.) Jepson, with remarks on the evolution of the family. Bot Not 1963: 81–93
Tiagi B (1965a) Development of the embryo sac and embryology of *Cuscuta lupuliformis*. Curr Sci 34: 671–672
Tiagi B (1965b) Development of the seed and fruit in *Melampyrum nemorosum* L. and *M. arvense* L. Can J Bot 43: 1511–1521
Tiagi B (1965c) Studies in the family Orobanchaceae 4. Development of the seed in *Conopholis americana* (L. fil.) Wallr. Acta Bot Hung 11: 253–261
Tiagi B (1966) Development of the seed and fruit in *Rhinanthus major* and *R. serotinus*. Am J Bot 53: 645–651
Tiagi B (1970) Orobanchaceae. In: Symp Comparative embryology of angiosperms. Indian Natl Sci Acad Bull 41: 282–289
Tiagi B (1972) A new kind of endosperm in *Rhinanthus major* Ehr. and *R. serotinus* (Schoenl) Schinz and Thell. In: Murty YS, Johri BM, Mohan Ram HY, Varghese TM (eds) Advances in plant morphology (Prof. V. Puri Comm. Vol.). Sarita Prakashan, Meerut (India), pp 370–375
Tiagi B, Gupta PK (1963) Floral morphology and embryology of *Evolvulus alsinoides* Linn. Proc Rajasthan Acad Sci (Jaipur) 10: 51–58
Tiagi B, Sankhla NS (1963) A contribution to the embryology of *Orobanche lucorum*. Bot Mag (Tokyo) 76: 81–88
Tiagi B, Taimani S (1960) Embryo sac development in *Vernonia cinerascens* Schult. and seed development in *V. cinerea* Lees. Curr Sci 29: 406
Tiagi B, Taimani S (1963) Floral morphology and embryology of *Vernonia cinerascens* and *V. cinerea*. Agra Univ J Res Sci 12: 123–137
Tiagi B, Varghese TM (1970) Scrophulariaceae. In: Symp Comparative embryology of angiosperms. Indian Natl Sci Acad Bull 41: 259–266
Tiagi YD (1954) Studies in the floral morphology of *Opuntia dillenii* Haworth 1. Development of ovule and gametophytes. Bot Not 4: 343–346
Tiagi YD (1956) Polyembryony in *Mammillaria tenuis* DC. Bull Bot Soc Univ Saugar (Sagar, India) 8: 25–27
Tiagi YD (1957) Studies in floral morphology 3. A contribution to the floral morphology of *Mammillaria tenuis* DC. Madhya Bharati J Univ Saugar (Sagar, India) B 6: 7–31
Tiagi YD (1959) Embryological and anatomical studies of several species of the family Cactaceae in connection with the position of the family in the systems of angiosperms. In: Proc All Un Congr Plant morphogenesis, Moscow, pp 320–321
Tiagi YD (1961) Studies in the floral morphology 7. The development of anther and ovule of

Mammillaria carnea Zucc. with a discussion on the systematic position of the family Cactaceae. Madhya Bharati J Univ Saugar (Sagar, India) B 10: 12–24

Tiagi YD (1963) Vascular anatomy of the flower of certain species of the Calycanthaceae. Proc Indian Acad Sci B 58: 224–234

Tiagi YD (1970a) Cactaceae. In: Symp Comparative embryology of angiosperms. Indian Natl Sci Acad Bull 41: 29–35

Tiagi YD (1970b) Paeoniaceae. In: Symp Comparative embryology of angiosperms. Indian Natl Sci Acad Bull 41: 53–58

Tiemann A (1985) Untersuchungen zur Embryologie, Blütenmorphologie und Systematik der Rapateaceen und der Xyridaceen-Gattung *Abolboda* (Monocotyledoneae). Diss Bot, Ruhr-Univ Bochum 82: 1–201

Tillich HJ (1977) Vergleichende morphologische Untersuchungen zur Identität der Gramineen-Primärwurzel. Flora 166: 415–421

Tilquin JP, Brouwer DE, Mathieu A, Calier M (1983) Haustorial pollen tubes in *Fuchsia boliviana*. Ann Bot (London) 52: 425–428

Tilton VR (1980a) The nucellar epidermis and micropyle of *Ornithogalum caudatum* (Liliaceae) with a review of these structures in other taxa. Can J Bot 58: 1872–1884

Tilton VR (1980b) Hypostase development in *Ornithogalum caudatum* (Liliaceae), and notes on other types of modifications in the chalaza of angiosperm ovules. Can J Bot 58: 2059–2066

Tilton VR, Horner Jr HT (1980) Stigma, style and obturator of *Ornithogalum caudatum* (Liliaceae) and their function in the reproductive process. Am J Bot 67: 1113–1131.

Tilton VR, Wilcox LW, Palmer RC (1984) Post-fertilization and labyrinthe formation and function in the central cell of soybean, *Glycine max* (L.) Merr. (Leguminosae). Bot Gaz 145: 334–379

Tinney FW (1940) Cytology of parthenogenesis in *Poa pratensis*. J Agric Res 60: 351–360

Tippo O (1938) Comparative anatomy of the Moraceae and their presumed allies. Bot Gaz 100: 1–99

Tippo O (1940) The comparative anatomy of the secondary xylem and the phylogeny of the Eucommiaceae. Am J Bot 27: 832–838

Tischler G (1917) Über die Entwicklung und phylogenetische Bedeutung des Embryosackes von *Lythrum salicaria*. Ber Dtsch Bot Ges 35: 233–245

Tiwari DK (1968) Embryological studies in the Cyperaceae 2. Endosperm and embryo development in *Fimbristylis quinquangularis* Kunth. J Indian Bot Soc 47: 264–271

Tiwari DK (1978) Taxonomic status of Cyperaceae on embryological grounds. In: Dnyansagar VR, Deshpande PK, Padhye MD (eds) Recent trends and contacts between cytogenetics, embryology and morphology. Today and Tomorrow, New Delhi, pp 437–444

Tiwari SC (1982) Callose in the walls of mature embryo sac of *Torenia fournieri*. Protoplasma 110: 1–4

Tiwari SC (1983) The hypostase in *Torenia fournieri* Lindl: A histochemical study of the cell walls. Ann Bot (London) 51: 17–26

Tiwari SC, Bouman F, Kapil RN (1977) Ovule ontogeny in *Tropaeolum majus*. Phytomorphology 27: 350–358

Tiwari SC, Gunning BES (1986) Development of tapetum and microspores in *Canna* L.: an example of an invasive but non-syncytial tapetum. Ann Bot (London) 57: 557–563

Tiwary NK (1926) On the occurrence of polyembryony in the genus *Eugenia*. J Indian Bot Soc 5: 124–136

Tiwary NK (1934) Fruit and seed development of *Tinospora cordifolia*. Curr Sci 3: 262–264

Tiwary NK (1936) Megasporogenesis and embryo development in *Cleome viscosa* Linn. Sci Cult 1: 476–477

Tobe H (1973) Embryo sac development in *Trautvetteria japonica* Sieb. et Zucc. (Ranunculaceae). Sci Rep Tôhoku Univ Ser Biol 36: 255–259

Tobe H (1974a) The embryology of *Cimicifuga acerina* (Sieb. et Zucc.) C. Tanaka (Ranunculaceae). Sci Rep Tôhoku Univ Ser Biol 37: 55–61

Tobe H (1974b) The embryology of *Anemonopsis macrophylla* Sieb. et Zucc. Sci Rep Tôhoku Univ Ser Biol 37: 63–69

Tobe H (1981) Embryological studies in *Glaucidium palmatum* Sieb. et Zucc. with a discussion on the taxonomy of the genus. Bot Mag (Tokyo) 94: 207–224

Tobe H (1987) Seed ontogeny and morphology in Ulmaceae (Abstr). In: 14th Int Bot Congr, Berlin, p 220

Tobe H (1989) The embryology of angiosperms: its broad application to the systematic and evolutionary study. Bot Mag (Tokyo) 102: 351–367

Tobe H, Keating RC (1985) The morphology and anatomy of *Hydrastis* (Ranunculales). Systematic re-evaluation of the genus. Bot Mag (Tokyo) 98: 291–316

Tobe H, Peng C-I (1990) The embryology and taxonomic relationships of *Bretschneidera* (Bretschneideraceae). Bot J Linn Soc 103: 139–152

Tobe H, Raven PH (1983a) An embryological analysis of the Myrtales: Its definition and characteristics. Ann Missouri Bot Garden 70: 71–94

Tobe H, Raven PH (1983b) The embryology of *Axinandra zeylanica* (Crypteroniaceae) and the relationships of the genus. Bot Gaz 144: 426–432

Tobe H, Raven PH (1984a) The number of cells in the pollen of Melastomataceae (Myrtales). Bot Mag (Tokyo) 97: 131–136

Tobe H, Raven PH (1984b) An embryological contribution to systematics of the Chrysobalanaceae 1. Tribe Chrysobalaneae. Bot Mag (Tokyo) 97: 397–411

Tobe H, Raven PH (1984c) The embryology and relationships of Oliniaceae. Plant Syst Evol 146: 105–116

Tobe H, Raven PH (1984d) The embryology and relationships of *Rhynchocalyx* Oliv. (Rhynchocalycaceae). Ann Missouri Bot Garden 71: 836–843

Tobe H, Raven PH (1984e) The embryology and relationships of *Alzatea* Ruiz & Pav. (Alzateaceae, Myrtales). Ann Missouri Bot Garden 71: 844–852

Tobe H, Raven PH (1984f) The embryology and relationships of Penaeaceae (Myrtales). Plant Syst Evol 146: 181–195

Tobe H, Raven PH (1985) The histogenesis and evolution of integuments in Onagraceae. Ann Missouri Bot Garden 72: 451–468

Tobe H, Raven PH (1986a) Evolution of polysporangiate anthers in Onagraceae. Am J Bot 73: 475–488

Tobe H, Raven PH (1986b) A comparative study of the embryology of *Ludwigia* (Onagraceae): characteristics, variation, and relationships. Ann Missouri Bot Garden 73: 768–787

Tobe H, Raven PH (1987a) The embryology and relationships of *Crypteronia* (Crypteroniaceae). Bot Gaz 148: 96–102

Tobe H, Raven PH (1987b) The embryology and relationships of *Dactylocladus* (Crypteroniaceae) and a discussion of the family. Bot Gaz 148: 103–111

Tobe H, Raven PH (1987c) Embryology and systematic position of *Heteropyxis* (Myrtales). Am J Bot 74: 197–208

Tobe H, Raven PH (1988a) Additional notes on the embryology of *Polygonanthus* (Anisophyllaceae) and relationships of the family. Ann Missouri Bot Garden 75: 1425–1428

Tobe H, Raven PH (1988b) Seed morphology and anatomy of Rhizophoraceae: inter- and intrafamilial relationships. Ann Missouri Bot Garden 75: 1319–1342

Tobe H, Adachi T, Yoshida O (1975) Embryological studies on *Hepatica nobilis* Schreber var. *Japonica* Nakai (Ranunculaceae). J Coll Arts Sci Chiba Univ Ser B 8: 93–102

Tohda H (1967) An embryological study of *Hetaeria shikokiana*, a saprophytic orchid in Japan. Sci Rep Tôhoku Univ Ser Biol 33: 83–95

Tohda H (1971) Embryological study on *Idesia polycarpa*. Sci Rep Tôhoku Univ Ser Biol 35: 231–237

Tohda H (1974) Development of the embryo of *Pogonia* (Orchidaceae). Sci Rep Tôhoku Univ Ser Biol 37: 89–93

Toilliez-Genoud J (1965) L'ovule et la graine de *Crinum giganteum* Andr. (Amaryllidacees). Adansonia 5: 525–530

Tokć E (1981) Embryology of *Sambucus racemosa* L. Acta Biol Cracov Ser Bot 22: 173–188

Tomer E, Gazit S (1979) Early stages in avocado (*Persea americana* Mill.) fruit development: anatomical aspects. Bot Gaz 140: 304–309

Tomer E, Gottreich M (1975) Observations on the fertilization process in avocado with fluorescent light. Euphytica 24: 531–535

Tomer E, Gottreich M (1978) Abnormalities in avocado (*Persea americana* Mill.) ovule development. Bot Gaz 139: 81–86

Tomita K (1931) Über die Entwicklung des nackten Embryos von *Crinum latifolium*. Sci Rep Tôhoku Univ Ser Biol 6: 163–169
Tomlinson PB (1956) Studies in the systematic anatomy of the Zingiberaceae. Bot J Linn Soc 55: 547–592
Tomlinson PB (1962) Phylogeny of the Scitamineae: Morphological and anatomical considerations. Evolution 16: 192–213
Tong KY (1930) Studien über die Familie der Hamamelidaceae, mit besonderer Berücksichtigung der Systematik und Entwicklungsgeschichte von *Corylopsis*. Bull Dep Biol Coll Sci Sun Yatsen Univ Canton 2: 1–72
Topham PB (1970) The histology of seed development in diploid and tetraploid raspberries (*Rubus idaeus* L.). Ann Bot (London) 34: 123–125
Traub HP (1939) Polyembryony in *Myricaria cauliflora*. Bot Gaz 101: 233–234
Trela-Sawicka Z (1974) Embryological studies in *Anemone ranunculoides* L. from Poland. Acta Biol Cracov Ser Bot 17: 1–11
Trela-Sawicka Z (1977) Further embryological studies in the genus *Anemone* L. from Poland. Acta Biol Cracov Ser Bot 19: 83–92
Trela-Sawicka Z (1978) Embryological studies in *Trapa natans* L. Acta Biol Cracov Ser Bot 21: 101–108
Treub M (1883) Observations sur les Loranthacées 4. *Loranthus pentandrus* L. Ann Jardin Bot Buitenz 3: 184–189
Treub M (1891) Sur les Casuarinées et leur place dans le système naturel. Ann Jardin Bot Buitenz 10: 145–231
Treub M (1898) L'organe fémelle et l'apogamie du *Balanophora elongata* Bl. Ann Jardin Bot Buitenz 15: 1–25
Treub M, Mellink J (1880) Notice sur le développement du sac embryonnaire dans quelques angiospermes. Arch Neerl Sci 15: 452–457
Trivedi BS, Tripathi R (1985) Mechanism of pollen attachment on the stigma of *Zinnia elegans* Jacq. J Indian Bot Soc 64: 287–289
Tsai AH, Harney PM, Peterson RL (1973) Megasporogenesis and megagametogenesis in *Pelargonium hortorum*. Can J Bot 51: 607–617
Tsan-Lang C, Heckard LR (1973) Seed-coat morphology in *Cordylanthus* (Scrophulariaceae) and its taxonomic significance. Am J Bot 59: 258–265
Tschermak-Woess E (1947) Cytologische und embryologische Untersuchungen an *Rhoeo discolor*. Öst Bot Z 94: 128–135
Tschermak-Woess E (1956) Notizen über die Riesenkerne und "Riesenchromosomen" in den Antipoden von *Aconitum*. Chromosoma 8: 114–134
Turala K (1969) Endopolyploidy and structure of the nuclei in the endosperm of the Cucurbitaceae. Öst Bot Z 116: 95–111
Turala K (1971) Disturbed mitotic division and endomitose during the differentiation of the endosperm in *Ecballium elaterium* (Cucurbitaceae). Acta Biol Cracov Ser Bot 14: 27–35
Turala-Szybowska K (1974) Karyological anatomy of the endosperm of *Bryonia dioica* and *Cucurbita pepo*. Acta Biol Cracov Ser Bot 17: 75–90
Turala-Szybowska K (1975) Ultrastructure of the endosperm of *Echinocystis lobata* (Cucurbitaceae) at the cellular stage. Acta Biol Cracov Ser Bot 18: 27–39
Turala-Szybowska K (1981) Cytoembryological studies in self-incompatible populations of *Ranunculus pedicellatus* (Dumort.) Bab. from Poland. Acta Biol Cracov Ser Bot 21: 9–21
Turala-Szybowska K (1984) Inhibited prophases in the differentiation of the anthers' tapetum in *Ranunculus fluitans* Lam. Acta Biol Cracov Ser Bot 26: 33–42
Turala-Szybowska K, Wolanska J (1990) Two mechanisms of polyploidization of the antipodals in *Aquilegia vulgaris* L. Acta Biol Cracov Ser Bot 31: 63–74
Turkel HS, Rebuck AL, Grove AR Jr (1955) Floral morphology of white oak. Penn State Univ Bull 593
Tybirk K (1989) Flowering, pollination and seed production of *Acacia nelotica*. Nord J Bot 9: 375–381
Tykarska T (1976) Rape embryogenesis 1. The proembryo development. Acta Soc Bot Pol 45: 3–16
Tykarska T (1979) Rape embryogenesis 2. Development of embryo proper. Acta Soc Bot Pol 48: 391–421

Ugemuge NR (1982) Embryological studies in *Acacia leucophloea* Willd. J Indian Bot Soc 61: 278–285
Ugemuge NR, Padhye MD (1986) Contribution to the life history of *Mucuna prurita* Hook. J Indian Bot Soc 65: 74–80
Ugemuge NR, Padhye MD (1989) Development of seed in *Psoralea corylifolia*. J Indian Bot Soc 68: 100–106
Uhl NW (1947) Studies in the floral morphology and anatomy of certain members of the Helobiae. Ph D Thesis, Cornell Univ, Ithaca (USA)
Umiker O (1920) Entwicklungsgeschichtlich-zytologische Untersuchungen an *Helosis guayannensis*. Diss, Univ Zürich
Ünal M (1978) Cytological and embryological studies on *Anemone pavonina* Lam. Istanbul Univ Fen Fak Med B 43: 185–208
Untawale AG, Bhasin RK (1973) On endothecial thickenings in some monocotyledonous families. Curr Sci 42: 398–400
Untawale AG, Deshpande PK (1968) Studies in the Leguminosae 1. Male and female gametophytes of *Indigofera enneaphylla* Linn. Proc Natl Acad Sci India B 38: 16–24
Untawale AG, Deshpande PK, Sharma KB (1969) Studies in Gramineae 1. Male and female gametophytes of *Eragrostis unioloides* (Retz.) Nees ex Stends. J Indian Bot Soc 48: 386–392
Untawale AG, Makde KH (1971) Gametophytes of *Cyperus alternifolius* Linn. Curr Sci 40: 327–328
Untawale AG, Makde KH (1984) A contribution to the embryology of *Scirpus royelei* Nees. J Indian Bot Soc 63: 298–305
Urbanska-Worytkiewicz K (1974) L'agamospermie, système de reproduction important dans la spéciation des Angiospermes. Bull Soc Bot Fr 121: 329–346

Vaidya PB (1961a) Occurrence of twin embryo sacs in *Capparis aphylla* Roth. Curr Sci 30: 187–188
Vaidya PB (1961b) Male and female gametophytes of *Acanthospermum hispidum* DC. J Biol Sci (Bombay) 4: 11–15
Vallade J (1970) Développement embryonnaire chez un *Petunia hybrida* Hort. C R Acad Sci Paris 270: 1893–1896
Van Caeseele L, Mills JT, Sumner M, Gillespie R (1981) Cytology of mucilage production in the seed-coat of candle canola (*Brassica campestris*). Can J Bot 59: 292–300
Van der Heijden, Bouman F (1988) Studies in Annonaceae: seed anatomy of the *Annona* group. Bot Jahrb 110: 117–135
Van der Pijl L (1934) Über die Polyembryonie bei *Eugenia*. Rec Trav Bot Neerl 31: 113–187
Van der Pijl L (1955) Sarcotesta, aril, pulpa and the evolution of the angiosperm fruit 1, 2. Proc Kon Ned Akad Wetensch Amsterdam C 58: 154–161, 307–312
Van der Veken P (1965) Contribution à l'embryographie systématique des Cyperacéae, Cyperoidéae. Bull Jardin Bot Etat Bruxelles 35: 285–384
Van Heel WA (1966) Morphology of the androecium in Malvales. Blumea 13: 177–394
Van Heel WA (1967) Anatomical and ontogenetic investigations on the morphology of the flowers and the fruit of *Scyphostegia borneensis* Stapf. (Scyphostegiaceae). Blumea 15: 107–125
Van Heel WA (1970a) Some unusual tropical labyrinth seeds. Proc Kon Ned Akad Wetensch Amsterdam C 73: 298–301
Van Heel WA (1970b) Distally-lobed integuments in some angiosperms ovules. Blumea 18: 67–70
Van Heel WA (1971a) The labyrinth seed of *Hernandia peltata* Meissn. in DC. Proc Kon Ned Akad Wetensch Amsterdam C 74: 46–51
Van Heel WA (1971b) Note on some more tropical labyrinth seeds. Blumea 19: 109–111
Van Heel WA (1971c) The distally-lobed inner integument of *Hernandia peltata* Meissn. DC. (Hernandiaceae). Blumea 19: 147–148
Van Heel WA (1973) Flowers and fruits in Flacourtiaceae 1. *Scaphocalyx spathacea* Ridl. Blumea 21: 259–279
Van Heel WA (1974) Flowers and fruits in Flacourtiaceae 2. The seeds of *Pangium edule* Reinw. Blumea 22: 15–19
Van Heel WA (1976) Distally-lobed integuments in *Exochorda*, *Juglans*, *Leontice* and *Bongardia*. Phytomorphology 26: 1–4

Van Heel WA (1977) Flowers and fruits in Flacourtiaceae 3. Some Oncobeae. Blumea 23: 349–369

Van Heel WA (1979) Flowers and fruits in Flacourtiaceae 4. *Hydnocarpus* spp., *Kiggelaria africana* L., *Casearia* spp., *Berberidopsis corallina* Hook. f. Blumea 25: 513–529

Van Heel WA (1982) Note on the structure of developing seeds of *Knema* and *Horsfieldia* (Myristicaceae). Blumea 28: 53–60

Van Hulst R, Theriault A, Shiplay B (1986) The systematic position of the genus *Rhinanthus* (Scrophulariaceae) in North America. Can J Bot 64: 1443–1449

Van Lammeren AAM (1981) Early events during embryogenesis in *Zea mays* L. Acta Soc Bot Pol 50: 289–290

Van Steenis CGGJ (1954) Haemodoraceae. Flora Malesiana, Ser 1. Spermatophyta 5: 111–113

Van Tieghem Ph (1895) Sur la structure et les affinités des *Tupeia, Ginalloa, Phoradendron,* et *Dendrophthora,* de la famille des Loranthacées. Bull Soc Bot Fr 30: 643–652

Van Tieghem Ph (1896) Sur les Phanérogames à ovules sans nucelle, formant le groupe des Innucellées ou Santalinées. Bull Soc Bot Fr 43: 543–577

Van Tieghem Ph (1897) Sur les Caractères et les affinités des Grubbiacées. J Bot (Paris) 11: 127–238

Van Tieghem Ph (1899) Sur les genres *Actinidie* et *Sauraule* considérés comme types d'une famille nouvelle, les Actinidiacées. J Bot (Paris) 13: 170–173

Van Tieghem Ph (1901) L'hypostase, sa structure et son rôle constantes, sa position et sa forme variables. Bull Mus Natl Hist Nat 7: 412–418

Van Vliet GJCM (1975) Wood anatomy of Crypteroniaceae sensu lato. J Microsc 104: 65–82

Van Went JL (1970) The ultrastructure of the synergids of *Petunia.* Acta Bot Neerl 19: 121–132

Van Went JL (1989) Integumentary tapetum in *Impatiens wallesiana* ovules. In: Paré J, Bugnicourt M (eds) Some aspects and actual orientations in plant embryology. Dedicatory volume to Prof A. Lebègue. Univ Press Picardie, Amiens (Fr), pp 271–279

Van Went JL, Willemse MTM (1984) Fertilization. In: Johri BM (ed) Embryology of angiosperms. Springer, Berlin Heidelberg New York, pp 273–317

Vani-Hardev (1972) Systematic embryology of *Roridula gorgonias* Planch. Beitr Biol Pflanzen 48: 339–351

Varghese TM (1963) Studies in the family Scrophulariaceae 1. A contribution to the embryology of *Veronica agrestis* L. Proc Indian Acad Sci B 58: 333–347

Varghese TM (1966) Morphological studies in Scrophulariaceae. Agra Univ J Res Sci 16: 153–157

Varghese TM (1970) Solanaceae. In: Symp Comparative embryology of angiosperms. Indian Natl Sci Acad Bull 41: 255–258

Varghese TM, Chowdhury JB (1972) Development of microsporangium and pollen in angiosperms. In: Varghese TM, Grover RK (eds) Vistas in plant sciences, vol 2. Int Bio-Sci Publ, Hissar (India), pp 159–183

Varitchak B (1940) La développement du sac embryonnaire et le nombre de chromosomes chez la plante *Narthecium scardicum.* Bull Acad Sci (Beograd) Ser B 6: 97–105

Vasil IK (1960) Studies on pollen germination of certain Cucurbitaceae. Am J Bot 47: 239–247

Vasil IK (1962) Formation of male gametes in the pollen tubes of some crop plants. In: Symp Plant embryology. Council Sci Ind Res, New Delhi, pp 254–260

Vasil IK, Johri MM (1964) The style, stigma and pollen tube 1. Phytomorphology 14: 352–369

Vaughan JG (1968) Seed anatomy and taxonomy. Proc Linn Soc London 179: 251–255

Vaughan JG (1970) The structure and utilization of oil seeds. Chapman & Hall, London

Vaughan JG, Rest JA (1969) A note on the testa structure of *Panda pierre, Gallaria* Zoll. et Mor. and *Microdesmis* Hook. f. (Pandaceae). Kew Bull 23: 215–218

Vaughan JG, Whitehouse JM (1971) Seed structure and taxonomy of Cruciferae. Bot J Linn Soc 64: 383–409

Veerasamy S (1982) Polyembryony and twin seedlings in *Borassus flabellifer* L. (Palmae). Bot J Linn Soc 85: 147–152

Veillet-Bartoszewska M (1957a) Embryogénie des Ericacées. Développement de l'embryon chez le *Rhododendron ferrugineum.* C R Acad Sci Paris 244: 1952–1954

Veillet-Bartoszewska M (1957b) Embryogénie des Primulacées. Développement de l'embryon chez le *Primula auricula* L. C R Acad Sci Paris 245: 2363–2366

Veillet-Bartoszewska M (1957c) La polyembryonie chez le *Primula auricula* L. Bull Soc Bot Fr 104: 473–475

Veillet-Bartoszewska M (1958a) Embryogénie des Ericacées. Développement de l'embryon chez le *Vaccinium myrtillus*. C R Acad Sci Paris 246: 824–826
Veillet-Bartoszewska M (1958b) Embryogénie des Plumbagacées. Développement de l'embryon chez le *Plumbago europaea*. C R Acad Sci Paris 247: 2178–2181
Veillet-Bartoszewska M (1959a) Embryogénie des Ericacées. Développement de l'embryon chez le *Gaultheria shalon* Pursh. C R Acad Sci Paris 248: 720–722
Veillet-Bartoszewska M (1959b) Embryogénie des Empetracées. Développement du proembryon chez le *Corema album* (L.) Don. C R Acad Sci Paris 249: 857–859
Veillet-Bartozewska M (1960a) Embryogénie des Ericacées. Développement de l'embryon chez l'*Erica tetralix* L. C R Acad Sci Paris 250: 1711–1714
Veillet-Bartoszewska M (1960b) Embryogénie des Ericacées. Développement de l'embryon chez le *Ledum palustre* L. C R Acad Sci Paris 251: 777–779
Veillet-Bartoszewska M (1960c) Embryogénie des Styracacées. Développement de l'embryon chez le *Styrax officinalis* L. C R Acad Sci Paris 250: 905–907
Veillet-Bartoszewska M (1960d) Embryogénie des Clethracées. Développement de l'embryon chez le *Clethra alnifolia* L. C R Acad Sci Paris 251: 2572–2574
Veillet-Bartoszewska M (1961a) Embryogénie des Ericacées. Développement de l'embryon chez le *Calluna vulgaris* Salisb. C R Acad Sci Paris 252: 1192–1194
Veillet-Bartoszewska M (1961b) Embryogénie des Epacridacées. Développement de l'embryon chez le *Dracophyllum secundum*. C R Acad Sci Paris 253: 1000–1002
Venkata Rao C (1948) Development of the embryo sac in some Sterculiaceae. Curr Sci 17: 304–305
Venkata Rao C (1949) Contributions to the embryology of Sterculiaceae 1. J Indian Bot Soc 28: 180–197
Venkata Rao C (1950) Contributions to the embryology of Sterculiaceae 2. *Waltheria indica* Linn. J Indian Bot Soc 29: 163–176
Venkata Rao C (1951a) Life history of *Mutingia calabura* L. Curr Sci 20: 47–48
Venkata Rao C (1951b) Contributions to the embryology of Sterculiaceae 3. *Melochia corcorifolia* L. J Indian Bot Soc 30: 122–131
Venkata Rao C (1952a) The embryology of *Mutingia calabura* L. J Indian Bot Soc 31: 87–101
Venkata Rao C (1952b) Contributions to the embryology of Sterculiaceae 4. Development of the gametophytes in *Pterospermum suberifolium* Lam. J Indian Bot Soc 31: 251–260
Venkata Rao C (1953a) Life history of *Ochroma lagopus* Sw. Curr Sci 22: 18–19
Venkata Rao C (1953b) Floral anatomy and embryology of two species of *Elaeocarpus*. J Indian Bot Soc 32: 21–33
Venkata Rao C (1953c) Contributions to the embryology of Sterculiaceae 5. J Indian Bot Soc 32: 208–238
Venkata Rao C (1954a) A contribution to the embryology of Bombacaceae. Proc Indian Acad Sci B 39: 51–75
Venkata Rao C (1954b) Embryological studies in Malvaceae 1. Development of gametophytes. Proc Natl Inst Sci India B 20: 127–150
Venkata Rao C (1955a) Embryological studies in Malvaceae 2. Fertilization and seed development. Proc Natl Inst Sci India B 21: 53–67
Venkata Rao C (1955b) Embryo development in arecanut. Nature (London) 175: 432–433
Venkata Rao C (1959a) Contributions to the embryology of Palmae 1. Sabaleae. Proc Natl Inst Sci India B 25: 143–168
Venkata Rao C (1959b) Contributions to the embryology of Palmae 2. Ceroxylinae. J Indian Bot Soc 38: 46–75
Venkata Rao C (1960) Studies in the Proteaceae 1. Tribe Persoonieae. Proc Natl Inst Sci India B 26: 300–337
Venkata Rao C (1961a) Studies in the Proteaceae 2. Tribes Placospermeae and Conospermeae. Proc Natl Inst Sci India B 27: 126–151
Venkata Rao C (1961b) Pollen types in the Epacridaceae. J Indian Bot Soc 40: 409–423
Venkata Rao C (1962) Morphology and embryology of *Lomatia* R. Br. with a discussion on its probable origin. In: Symp Plant embryology. Council Sci Ind Res, New Delhi, pp 261–273
Venkata Rao C (1963) Studies in the Proteaceae 3. Tribe Oriteae. Proc Natl Inst Sci India B 29: 489–510

Venkata Rao C (1964) Pollen grains of *Grevillea vestita* Meissn. Curr Sci 33: 722–724
Venkata Rao C (1965) Studies in the Proteaceae 6. Tribe Franklandiae. J Indian Bot Soc 44: 479–494
Venkata Rao C (1966) Studies in the Proteaceae 8. Floral morphology, floral anatomy and embryology of *Grevillea* R. Br. Proc Natl Inst Sci India B 33: 162–199
Venkata Rao C (1967) Evolution of the endosperm in Proteaceae. New Phytol 66: 755–768
Venkata Rao C (1968) Studies in the Proteaceae 11. Morphology, floral anatomy and embryology of *Lambertia* Sm. with a discussion on the taxonomic position of the genus. Proc Indian Acad Sci B 68: 11–24
Venkata Rao C (1969a) Studies in the Proteaceae 10. Morphology and embryology of *Hakea* Schrad. J Indian Bot Soc 48: 310–321
Venkata Rao C (1969b) Studies in the Proteaceae 12. Floral morphology, floral anatomy and embryology of *Aulax* Berg. with a discussion on the taxonomic position of the genus. J Indian Bot Soc 48: 238–251
Venkata Rao C (1969c) Studies in the Proteaceae 13. Tribe Grevilleae. Proc Natl Inst Sci India B 35: 471–486
Venkata Rao C (1970) Studies in Proteaceae 14. Tribe Macadamieae. Proc Indian Natl Sci Acad B 36: 345–363
Venkata Rao C, Rama Rao S (1954) Embryology of *Cryptostegia grandiflora* R. Br. and *Caralluma attenuata* Wt. J Indian Bot Soc 33: 453–472
Venkata Rao C, Sambasiva Rao KV (1952) A contribution to the embryology of *Triumfetta rhomboidea* Jacq. and *Corchorus acutangulus* L. J Indian Bot Soc 31: 56–68
Venkataramani KS (1950) An instance of polyembryony in tea. Planters Chron 45: 180–181
Venkatasubban K (1950) Studies in the Droseraceae 2. A contribution to the embryology of three species of *Drosera*. Proc Indian Acad Sci B 32: 23–38
Venkatesh CS (1952) Anther and pollen grains of *Zannichellia palustris* L. Curr Sci 21: 225–226
Venkatesh CS (1956a) The form, structure and special modes of dehiscence in anthers of *Cassia* 1. subgenus *fistula*. Phytomorphology 6: 168–176
Venkatesh CS (1956b) Structure and dehiscence of the anther in *Najas*. Bot Not 109: 75–82
Venkateswarlu J (1936a) A preliminary note on the embryology of *Duabanga sonneratioides*. Curr Sci 4: 742–743
Venkateswarlu J (1936b) A preliminary note on the development of embryo sac of *Averrhoa carambola* Linn. Curr Sci 5: 23
Venkateswarlu J (1936c) Some observations on the ovule and embryo sac of *Sonneratia apetala*. Curr Sci 5: 201
Venkateswarlu J (1937a) A contribution to the embryology of Sonneratiaceae. Proc Indian Acad Sci B 58: 206–223
Venkateswarlu J (1937b) A note on the development of the embryo sac in *Phrynium capitatum* W. J Indian Bot Soc 16: 95–98
Venkateswarlu J (1937c) Structure and development of the embryo sac of *Pemphis acidula* Forst. J Indian Bot Soc 16: 259–262
Venkateswarlu J (1941a) Vascular supply in the ovules of some Compositae. Curr Sci 10: 367–368
Venkateswarlu J (1941b) Vascular supply in the ovules of some Compositae 1. *Launaea pinnatifida*. Proc Indian Acad Sci B 38: 38–46
Venkateswarlu J (1945) Embryological studies in the Thymelaeaceae 1. *Thymelaea arvensis* Lamk. J Indian Bot Soc 24: 45–66
Venkateswarlu J (1946) A case of polyembryony in *Daphne cannabina* Wall. Curr Sci 15: 169
Venkateswarlu J (1947a) Embryological studies in Thymelaeaceae 2. *Daphne cannabina* Wall. and *Wikstroemia canescens* Meissn. J Indian Bot Soc 26: 13–39
Venkateswarlu J (1947b) A contribution to the embryology of *Pisonia aculeata* Linn. J Indian Bot Soc 26: 183–194
Venkateswarlu J (1947c) Development of the embryo of *Thymelaea arvensis* Lamk. Bot Gaz 108: 581–586
Venkateswarlu J (1947d) A case of polyembryony in *Poivrea coccinea* DC. Curr Sci 16: 95
Venkateswarlu J (1948) A contribution to the embryology of *Pisonia aculeata* Linn. J Indian Bot Soc 26: 183–194
Venkateswarlu J (1952a) Embryological studies in Lecythidaceae 1. J Indian Bot Soc 31: 103–116

Venkateswarlu J (1952b) Contributions to the embryology of Combretaceae 1. *Poivrea coccinea* DC. Phytomorphology 2: 231–240
Venkateswarlu J (1952c) Reversed polarity in the embryo sac of *Napoleona imperialis* Pal. B. Curr Sci 21: 167
Venkateswarlu J (1959) A note on the structure and development of the ovule and the embryo sac in two species of *Launaea*. Curr Sci 8: 556–557
Venkateswarlu J, Atchutarammurthi B (1955a) Life history of *Coldenia procumbens* Linn. Curr Sci 24: 26–27
Venkateswarlu J, Atchutarammurthi B (1955b) Embryological studies in Boraginaceae 1. *Coldenia procumbens* Linn. J Indian Bot Soc 34: 235–247
Venkateswarlu J, Devi I (1964) Embryology of some Indian grasses. Curr Sci 33: 104–106
Venkateswarlu J, Lakshminarayan L (1957) A contribution to the embryology of *Hydrocera triflora* W. & A. Phytomorphology 7: 194–203
Venkateswarlu J, Maheswari Devi H (1955a) Embryological studies in the Compositae 1. *Launaea pinnatifida* Cass. Proc Indian Acad Sci B 41: 38–46
Venkateswarlu J, Maheswari Devi H (1955b) Embryological studies in Compositae 2. Helenieae. Proc Natl Inst Sci India B 21: 149–161
Venkateswarlu J, Maheswari Devi H (1973) An embryological approach to the taxonomic status of Vacciniaceae. Proc Indian Acad Sci B 78: 282–291
Venkateswarlu J, Prakasa Rao PS (1972) Embryological studies in some Combretaceae. Bot Not 125: 161–179
Venkateswarlu J, Rajeswara Rao G (1958) A contribution to the life history of *Rubia cordifolia* Linn. J Indian Bot Soc 37: 442–454
Venkateswarlu J, Raju CSK (1958) Male and female gametophytes in four species of *Asparagus*. J Indian Bot Soc 37: 290–299
Venkateswarlu J, Rao PN (1963) Endosperm in Euphorbiaceae and occurrence of endosperm haustoria in two species of *Croton* Linn. Curr Sci 32: 514–516
Venkateswarlu J, Rao PN (1967) Embryogeny in the Euphorbiaceae with special reference to the embryo development in *Euphorbia rothiana* Spr. Proc 64th Indian Sci Congr, Hyderabad, pt 3: 322–323
Venkateswarlu J, Rao PN (1975a) Apomixis in *Coix aquatica* Roxb. Ann Bot (London) 39: 1131–1136
Venkateswarlu J, Rao PN (1975b) A contribution to the embryology of tribe Hippomaneae of the Euphorbiaceae. J Indian Bot Soc 54: 98–103
Venkateswarlu J, Rao PN, Rao SD (1974) Occurrence of stylar obturator in two Euphorbiaceae. Curr Sci 43: 128–129
Venkateswarlu J, Sarojini Devi P, Nirmala A (1980) Embryological studies in *Eleutherine plicata* Herb. and *Belamcanda chinensis* Lem. Proc Indian Acad Sci B 89: 361–367
Venkateswarlu J, Seshavatharam V (1971) A contribution to the embryology of *Abrus precatorius* Linn. with remarks on its systematic position. J Indian Bot Soc 50: 332–337
Ventura M (1930) Osservazioni sulla embriologia di *Daphniphyllum macropodum*. Ann Bot (Rome) 18: 395–401
Ventura M (1933) Sviluppo del gametofito femminile di *Euphorbia mauritanica*. Ann Bot (Rome) 20: 267–273
Ventura M (1934) Sulla poliembryonia di *Mallotus japonicus*. Ann Bot (Rome) 20: 568–578
Ventura M (1938) Osservazioni embriologiche su *Arechavaletaia uruguayensis* Speg. Ann Bot (Rome) 21: 527–533
Venturelli M (1980) Desenvolvimento anatômico do haustorio primário de *Struthanthus vulgaris* Mart. (Anatomical development of the primary haustorium of *Struthanthus vulgaris* Mart.). Bol Bot Univ Sao Paulo 8: 47–64
Venturelli M (1981a) Embriologia de *Struthanthus vulgaris* (Loranthaceae-Loranthoideae). Kurtziana 14: 73–100
Venturelli M (1981b) Estudos sobre *Struthanthus vulgaris* Mart.: Anatomia do fruto e semente e aspectos de germinacão, crecimento e de senvolvimento. Rev Bras Bot 4: 131–147
Venturelli M (1983) Estudos embriológicos em Loranthaceae: Gênero *Tripodanthus*. Kurtziana 16: 71–90

Venturelli M (1984a) Morfologia e anatomia floral de *Struthanthus vulgaris* Mart. (Loranthaceae). Rev Bras Bot 7: 121–128

Venturelli M (1984b) Estudos sobre *Struthanthus vulgaris* Mart.: Aspectos anatômicos de raiz adventicia, caule with folha. Rev Bras Bot 7: 79–89

Venturelli M (1984c) Estudos embriológicos em Loranthaceae: *Struthanthus flexicaulis* Mart. Rev Bras Bot 7: 107–119

Venturelli M, Bouman F (1986) Embryology and seed development in *Mayaca fluviatilis* (Mayacaceae). Acta Bot Neerl 35: 497–516

Venugopalan S (1966) Embryology of *Haplopappus gracilis* (Nutt) Gray. Univ Kans Sci Bull 46: 487–493

Verechshagina V (1987) Embryology of some gynodioecious Lamiaceae (Abstr). In: 14th Int Bot Congr, Berlin, p 220

Verkerke W (1984) Ovule and seed of *Xanthophyllum* (Polygalaceae). Blumea 29: 409–421

Verkerke W (1985) Ovules and seeds of the Polygalaceae. J Arnold Arbor 66: 353–394

Verkerke W (1986a) Anatomy of *Ficus ottoniifolia* (Moraceae) syconia and its role in the fig-fig wasp symbiosis. Proc Kon Ned Akad Wetensch Ser C 89: 443–469

Verkerke W (1986b) Ovule ontogeny and seed-coat development in *Krameria loefling* (Krameriaceae). Beitr Biol Pflanzen 60: 341–351

Verkerke W (1987a) Ovule dimorphism in *Ficus asperifolia* Miquel. Acta Bot Neerl 36: 121–124

Verkerke W (1987b) Syconium morphology and its relation to fig wasp pollination in *Ficus* (Abstr). In: 14th Int Bot Congr, Berlin, p 224

Verkerke W (1990) Fig anatomy and reproductive biology of African *Ficus* species (Moraceae). Proc 12th Plenary Meeting, AETFAT, Hamburg, pp 427–431

Verkerke W, Bouman F (1980) Ovule ontogeny and its relation to seed-coat structure in some species of *Polygala* (Polygalaceae). Bot Gaz 141: 227–282

Verma RC, Sarkar A, Das BC (1986) Cytoplasmic channels and chromatin migration in mulberry. Cytologia 51: 731–736

Vestal PA (1937) The significance of comparative anatomy in establishing the relationship of the Hypericaceae to the Guttiferae and their allies. Philipp J Sci 64: 199–256

Veyret Y (1981a) L'ovule au cours de la vie des fleurs non pollinisées chez les Orchideae et les Neottieae (Orchidaceae). Bull Mus Natl Hist Nat Paris Ser 4 Sec B Adansonia 2: 179–184

Veyret Y (1981b) Quelques aspects du pistil et de son devenir chez quelques Sobraliinae (Orchidaceae) de Guyane. Bull Mus Natl Hist Nat Paris Ser 4 Sec B Adansonia 3: 75–83

Vieth MJ (1958) Sur le faisceau tégumentaire chez quelques Dipsacacées. Rev Gen Bot 65: 667–673

Vignon-Fétré J (1966) Polyembryonie chez le *Sedum spectabile* Boreau. Bull Soc Bot Fr 113: 137–140

Vij SP, Sharma M (1987) Embryological studies in Orchidaceae 5. *Epipactis* Adams. Phytomorphology 37: 81–86

Vij SP, Sharma M, Shekhar N (1982) Embryological studies in Orchidaceae 2. *Zeuxine strateumatica* complex. Phytomorphology 32: 257–264

Vijaya MD (1986) Embryological notes on *Ruta graveolens* L. J Swamy Bot Club 3: 63–65

Vijayaraghavan MR (1962) Studies in family Ranunculaceae 2. The female gametophyte of *Clematis gauriana* Roxb. Phytomorphology 12: 45–49

Vijayaraghavan MR (1964) Morphology and embryology of a vesselless dicotyledon – *Sarcandra irvingbaileyi* Swamy, and systematic position of Chloranthaceae. Phytomorphology 14: 429–441

Vijayaraghavan MR (1965) Morphology and embryology of *Actinidia polygama* Franch. & Sav. and systematic position of the family Actinidiaceae. Phytomorphology 15: 224–235

Vijayaraghavan MR (1969) Studies in the family Cyrillaceae 1. Development of male and female gametophytes in *Cliftonia monophylla* (Lam.) Britton ex Sarg. Bull Torrey Bot Club 96: 484–489

Vijayaraghavan MR (1970a) Ranunculaceae. In: Symp Comparative embryology of angiosperms. Indian Natl Sci Acad Bull 41: 45–52

Vijayaraghavan MR (1970b) Actinidiaceae. In: Symp Comparative embryology of angiosperms. Indian Natl Sci Acad Bull 41: 69–74

Vijayaraghavan MR (1970c) Cyrillaceae. In: Symp Comparative embryology of angiosperms. Indian Natl Sci Acad Bull 41: 163–167

Vijayaraghavan MR, Bhandari NN (1970) Studies in the family Ranunculaceae: Embryology of *Thalictrum japonicum* Bl. Flora 159: 450–458

Vijayaraghavan MR, Bhat V (1983) Synergids before and after fertilization. Phytomorphology 33: 74–84

Vijayaraghavan MR, Dhar U (1975) *Kadsura heterodita* – Microsporangium and pollen. J Arnold Arbor 56: 176–182

Vijayaraghavan MR, Dhar U (1976) *Scytopetalum tieghemii* – embryologically unexplored taxon and affinities of the family Scytopetalaceae. Phytomorphology 26: 16–22

Vijayaraghavan MR, Dhar U (1977) Structure of the carpel with special reference to lysigenous cavities in *Kadsura roxburghiana*. Bot Jahrb 98: 273–277

Vijayaraghavan MR, Dhar U (1978) Embryology of *Cyrilla* and *Cliftonia* (Cyrillaceae). Bot Not 131: 127–138

Vijayaraghavan MR, Kapoor T (1985a) Embryogenesis in *Najas marina* L. Structural and histochemical approach. Aquat Bot 22: 45–60

Vijayaraghavan MR, Kapoor T (1985b) Nucleocytoplasmic interaction in the embryo-basal cell and in the chalazal-endosperm chamber cell in *Potamogeton nodosus*. Cytologia 50: 333–340

Vijayaraghavan MR, Kaur D (1966) Morphology and embryology of *Turnera ulmifolia* L. and affinities of the family Turneraceae. Phytomorphology 16: 539–553

Vijayaraghavan MR, Malik U (1972) Morphology and embryology of *Scaevola frutescens* K. and affinities of the family Goodeniaceae. Bot Not 125: 241–254

Vijayaraghavan MR, Marwah KN (1968) Studies in the family Ranunculaceae. Microsporangium, microsporogenesis and Übisch granules in *Nigella damascena* Linn. Phyton (Austria) 13: 203–209

Vijayaraghavan MR, Marwah KN (1969) Studies in the family Ranunculaceae. Morphology and embryology of *Nigella damascena*. Phytomorphology 19: 147–153

Vijayaraghavan MR, Padmanabhan U (1969) Morphology and embryology of *Centaurium ramosissimum* Druce and affinities of the family Gentianaceae. Beitr Biol Pflanzen 46: 15–37

Vijayaraghavan MR, Prabhakar K (1984) The endosperm. In: Johri BM (ed) Embryology of angiosperms. Springer, Berlin Heidelberg New York, pp 319–376

Vijayaraghavan MR, Ratnaparkhi S (1972) Some aspects of embryology of *Alectra thompsoni*. Phytomorphology 22: 1–8

Vijayaraghavan MR, Ratnaparkhi S (1975) Megasporogenesis and female gametophyte in Saxifragaceae. In: Mohan Ram HY, Shah JJ, Shah CK (eds) Form, structure and function in plants (Prof. B.M. Johri Comm. Vol.). Sarita Prakashan, Meerut (India), pp 215–222

Vijayaraghavan MR, Ratnaparkhi S (1977) Pollen embryo sacs in *Heuchera micrantha* Dougl. Caryologia 30: 105–119

Vijayaraghavan MR, Ratnaparkhi S (1979) Histological dynamics of anther tapetum in *Heuchera micrantha*. Proc Indian Acad Sci Plant Sci 88: 309–316

Vijayaraghavan MR, Sarveshwari GS (1968) Embryology and systematic position of *Morina longifolia* Wall. Bot Not 121: 383–402

Vijayaraghavan MR, Shukla AK (1976) The nature of covering around the aggregate of microspores in *Pergularia daemia* (Forsk.) Mc & Blat. Ann Bot (London) 40: 417–421

Vijayaraghavan MR, Shukla AK (1977) Absence of callose around the microspore tetrad and poorly-developed exine in *Pergularia daemia*. Ann Bot (London) 41: 923–926

Vijayaraghavan MR, Shukla AK (1978a) Role of persistent tapetal membrane in the germination of pollinium in *Pergularia daemia*. In: 4th Int Palynology Conf (Lucknow, India) 1: 347–348

Vijayaraghavan MR, Shukla AK (1978b) Role of callose during microsporogenesis. J Cytol Genet 13: 129–131

Vijayaraghavan MR, Shukla AK (1980) Viscin threads in *Zeuxine strateumatica* (Orchidaceae). Grana 19: 173–175

Vijayaraghavan MR, Vidya Kumari A (1974) Embryology and systematic position of *Zannichellia palustris* L. J Indian Bot Soc 53: 292–302

Vijayaraghavan MR, Misra G, Saxena P (1988) Nutrient transfer during embryonic development of angiosperm plants. Sci Prog Oxford 72: 467–480

Vijayaraghavan MR, Sudesh, Gautam N (1990) Ontogeny and histochemistry of the zygotic transfer cell in *Acrotrema arnottianum*. Cytologia 55: 321–326

Vijayalakshmi A, Lakshmanan KK (1985) Embryology of *Ligustrum lucidum* Ait. J Swamy Bot Club 2: 137–138

Vijendra Das LD, Rangaswamy SR (1974) Embryological studies in the sterile interspecific hybrid of rice between *O. sativa* (autotetraploid) and *O. perennis* (diploid). Sci Cult 40: 253–254

Villamil P, Palser B (1980) Studies of floral morphology in the Ericales 9. Organography, vascular anatomy and megagametophyte in three species of Gaultherieae. Phytomorphology 30: 250–265

Vink W (1970) The Winteraceae of the Old World 1. *Pseudowintera* and *Drimys* – Morphology and taxonomy. Blumea 18: 225–354

Vink W (1977) The Winteraceae of the Old World 2. *Zygogynum* – Morphology and taxonomy. Blumea 23: 219–250

Virdi BV, Eaton GW (1969) Embryo sac and early embryo development in the salmon berry (*Rubus spectabilis*). Can J Bot 47: 1891–1893

Virkki N (1962) Meiosis and development of the embryo sac in *Gunnera insignis* (Oerstei) D.C. (Halorhagaceae). J Agric Univ Puerto Rico 46: 254–268

Viswanathan K, Lakshmanan KK (1984) New techniques to determine pollen viability of *Calotropis gigantea* R. Br. Curr Sci 53: 661

Voigt A (1888) Untersuchungen über Bau und Entwicklung von Samen mit ruminiertem Endosperm aus den Familien der Palmen, Myristicaceen und Anonaceen. Ann Jardin Bot Buitenz 7: 150–190

Von Guttenberg H, Heydel H (1957) Vergleichende Studien über die Entwicklung von Primär-, Seiten- und sprossburtigen Wurzeln bei einigen Liliaceen. Bot Stud Jena 7: 40–90

Von Guttenberg H, Jakuszeit C (1957) Die Entwicklung des Embryos und der Primärwurzel von *Galtonia candidans* nebst Untersuchungen über die Differenzierung des Wurzelvegetationspunktes von Stud *Alisma plantago*. Bot Stud Jena 7: 91–126

Von Guttenberg H, Tober A, Dretke W, Ruickholdt M (1960) Embryologische Studien an Monocotyledonen 3. Die Embryogenese von *Triglochin maritimum* L., *Arum maculatum* L. und *Typha latifolia* L. Flora 149: 243–281

Von Schwerin F (1920) Revisio generis *Sambucus*. Mitt Dtsch Dendrol Ges 29: 194–231

Von Teichman I (1988) Notes on the ontogeny and structure of the seed-coat of *Sclerocarya birrea* (Richard) Hochst. subsp. *caffra* (Sonder) Kokwaro (Anacardiaceae). Bot J Linn Soc 98: 153–158

Von Teichman I, Robbertse PJ (1986a) Development and structure of the drupe in *Sclerocarya birrea* (Richard) Hochst. subsp. *caffra* Kokwaro (Anacardiaceae) with special reference to the pericarp and the opercula. Bot J Linn Soc 92: 303–322

Von Teichman I, Robbertse PJ (1986b) Development and structure of the pericarp and seed of *Rhus lancea* L. fil. (Anacardiaceae), with taxonomic notes. Bot J Linn Soc 93: 291–306

Von Teichman I, Van Wyk AE (1988) The ontogeny and structure of the pericarp and seed-coat of *Harpephyllum caffrum* Bernh. ex Krauiss (Anacardiaceae). Bot J Linn Soc 98: 159–176

Von Weberling F, Hildenbrand M (1982) Zur Tapetumentwicklung bei *Triosteum* L., *Leycesteria* Wall. und *Kolkwitzia* Graebn. (Caprifoliaceae). Beitr Biol Pflanzen 57: 481–486

Von Wettstein R (1895a) Solanaceae. In: Engler A, Prantl K (eds) Die natürlichen Pflanzenfamilien 4(3b): 4–38

Von Wettstein R (1895b) Scrophulariaceae. In: Engler A, Prantl K (eds) Die natürlichen Pflanzenfamilien 4(3b): 39–107

Wadhi M (1962) Morphology and embryology of *Thunbergia alata* Boj. ex Sims. M Sc. Thesis, Univ Delhi

Wadhi M (1970) Acanthaceae. In: Symp Comparative embryology of angiosperms. Indian Natl Sci Acad Bull 41: 267–272

Wafai BA, Koul AK (1984) Impact of polyploidy on the gametophytes, ovule and seeds in some Himalayan tulips. Phytomorphology 34: 64–69

Wagenitz G (1959) Die systematische Stellung der Rubiaceae. Ein Beitrag zum System der Sympetalen. Bot Jahrb 79: 17–35

Wagenitz G (1977a) Systematics and phylogeny of the Compositae (Asteraceae). Plant Syst Evol 125: 29–46

Wagenitz G (1977b) New aspects of the systematics of Asteridae. In: Kubitzki H (ed) Flowering plants. Plant Syst Evol Suppl 1: 375–395

Wakakuwa S (1934) Embryological studies on the different seed development in reciprocal interspecific crosses of wheat. Jpn J Bot 7: 151–185

Wakana A, Uemoto S (1988) Adventive embryogenesis in *Citrus* (Rutaceae) 2. Post-fertilization development. Am J Bot 75: 1033–1047

Walia K, Kapil RN (1965) Embryology of *Frankenia* Linn. with some comments on the systematic position of the Frankeniaceae. Bot Not 118: 412–429

Walker JW (1971) Unique type of angiosperm pollen from family Annonaceae. Science 172: 565–567

Walker JW (1974a) Evolution of exine structure in the pollen of primitive angiosperms. Am J Bot 61: 891–902

Walker JW (1974b) Aperture evolution in the pollen of primitive angiosperms. Am J Bot 61: 1112–1137

Walker JW (1976) Comparative pollen morphology and phylogeny of the Ranalean complex. In: Beck CB (ed) The origin and early evolution of angiosperms. Columbia Univ Press, New York, pp 241–299

Walker JW, Doyle JA (1975) The bases of angiosperm phylogeny: Palynology. Ann Missouri Bot Garden 62: 664–723

Walker JW, Skvarla JJ (1975) Primitively columellaless pollen: A new concept in the evolutionary morphology of angiosperms. Science 18: 445–447

Walker RI (1938) Macrosporogenesis and embryo development in *Ulmus fulva*. Bot Gaz 99: 592–598

Walker RI (1947) Megasporogenesis and embryo development in *Tropaeolum majus*. Bull Torrey Bot Club 74: 240–249

Walker RI (1950) Megasporogenesis and development of megagametophyte in *Ulmus*. Am J Bot 37: 47–52

Walker RI (1955) Cytological and embryological studies in *Solanum* section *tuberarium*. Bull Torrey Bot Club 82: 87–101

Wall WE, Long RW (1965) Megasporogenesis and embryo sac development in *Ruellia caroliniensis*. Bull Torrey Bot Club 92: 372–377

Wallace GD (1975) Interrelationships of the sub-families of the Ericaceae and derivation of the Monotropoideae. Bot Not 128: 286–298

Walters JL (1962) Megasporogenesis and gametophyte selection in *Paeonia californica*. Am J Bot 49: 787–794

Wangerin W (1910a) Nyssaceae. In: Engler A (ed) Das Pflanzenreich 41(4) 220a: 1–20

Wangerin W (1910b) Cornaceae. In: Engler A (ed) Das Pflanzenreich 41(3) 229: 1–110

Wangerin W (1910c) Garryaceae. In: Engler A (ed) Das Pflanzenreich 41(4) 56a: 1–18

Wannan BS, Quinn CJ (1990) Pericarp structure and generic affinities in the Anacardiaceae. Bot J Linn Soc 102: 225–252

Wannan BS, Waterhouse JT, Quinn CJ (1987) A taxonomic reassessment of *Blepharocarya* F. Muell. Bot J Linn Soc 95: 61–72

Want G (1963) Sporogenesis, gametogenesis and embryogeny of *Wahlenbergia bicolor* N. Lothian. Aust J Bot 11: 152–167

Warburg EF (1938) Taxonomy and interrelationships in the Geraniales in the light of their cytology. New Phytol 37: 130–159, 189–220

Wardlaw CW (1955) Embryogenesis in plants. Methuen, London

Wardrop AB (1983) The opening mechanism of follicles of some species of *Banksia*. Aust J Bot 31: 485–500

Warmke HE (1943) Macrosporogenesis, fertilization and early embryology of *Taraxacum kok-saghyz*. Bull Torrey Bot Club 70: 164–173

Warmke HE (1954) Apomixis in *Panicum maximum*. Am J Bot 41: 5–11

Warmke HE, Overman MA (1972) Tapetal behaviour in fertile and sterile anthers. J Hered 63: 227–233

Watanabe K (1936) Morphologisch-biologische Studien über die Gattung *Mitrastemon* 4. Jpn J Bot 12: 848–858

Watkins GM (1937) Embryo sac development in *Yucca rupicola*. Am J Bot 24: 481–484
Watson L (1965) The taxonomic significance of certain anatomical variations among Ericaceae. J Linn Soc Bot 59: 111–125
Webb CJ (1984) Humming bird pollination of *Malvaviscus arboreus* in Costa Rica. NZ J Bot 22: 575–581
Webber E (1929) Entwicklungsgeschichtliche Untersuchungen über die Gattung *Allium*. Bot Arch 25: 1–44
Webber JM (1940) Polyembryony. Bot Rev 6: 575–598
Webber HJ, Batchelor LD (1943) The citrus industry, vol I. UCLA Press, Berkeley (USA)
Wee YC, Nagaraja Rao A (1974) Gametophytes and seed development in pineapple. Curr Sci 43: 171–173
Weimarck G (1967a) Apomixis and sexuality in *Hierochloë australis* and in Swedish *H. odorata* on different polyploid levels. Bot Not 120: 209–235
Weimarck G (1967b) Apomixis in *Hierochloë monticola* (Gramineae). Bot Not 120: 448–455
Weimarck G (1970) Apomixis and sexuality in *Hierochloë* (Gramineae) from Finland and Greenland and in *H. monticola* from Greenland. Bot Not 123: 495–504
Weinstein AJ (1926) Cytological studies on *Phaseolus vulgaris*. Am J Bot 13: 248–263
Weinzieher S (1914) Beiträge zur Entwicklungsgeschichte von *Xyris indica*. Flora 106: 393–432
Went FAFC (1909) The development of the ovule, embryo sac and egg in Podostemaceae. Rec Trav Bot Neerl 5: 1–16
Went FAFC (1910) Untersuchungen über Podostemaceen 1. Verh Kon Akad Wetensch Amsterdam Sec 2 16: 1–88
Went FAFC (1912) Untersuchungen über Podostemaceen 2. Verh Kon Akad Wetensch Amsterdam Sec 2 17: 5–18
Went FAFC (1926) Untersuchungen über Podostemaceen 3. Verh Kon Akad Wetensch Amsterdam Sec 2 25: 3–58
Werker E, Fahn A (1975) Seed anatomy of *Pancratium* species from three different habitats. Bot Gaz 136: 396–403
Werner E (1915) Zur Ökologie atypischer Samenanlagen. Beih Bot Zentralbl 33: 1–11
West G (1930) Cleistogamy in *Viola riviniana*, with special reference to the cytological aspects. Ann Bot (London) 44: 87–109
Wettstein R (1911) Handbuch der systematischen Botanik. Deuticke, Leipzig
Wettstein R (1933, 1935) Handbuch der systematischen Botanik, 4th edn. Deuticke, Leipzig
Wheeler R, Lye KA (1975) Seedlings of Nymphaeaceae. Bot J Linn Soc 70: 255–266
Whelan EDP (1974) Discontinuities in the callose wall intermeiocyte connections, and cytomixis in angiosperm meiocytes. Can J Bot 52: 1219–1224
Whipple HL (1972) Structure and systematics of *Phryma leptostachya* L. J Elisha Mitchell Sci Soc 88: 1–17
White DWR, Williams E (1976) Early seed development after crossing of *Trifolium semiipilosum* and *T. repens*. NZ J Bot 14: 161–168
White J (1990) Pollen development in *Actinidia deliciosa* var. *deliciosa*: histochemistry of the microspore mother cell walls. Ann Bot (London) 65: 231–240
Whitehead DR (1965) Pollen morphology in the Juglandaceae. J Arnold Arbor 46: 369–410
Whitehead MR, Brown CA (1940) The seed of the spider lily, *Hymenocallis occidentalis*. Am J Bot 27: 199–203
Wiegand KM (1899) The development of the microsporangium and microspores in *Convallaria* and *Potamogeton*. Bot Gaz 28: 328–368
Wiegand KM (1900) The development of the embryo sac in some monocotyledonous plants. Bot Gaz 30: 25–47
Wiger J (1935) Embryological studies on the families Buxaceae, Meliaceae, Simaroubaceae and Burseraceae. Diss, Univ Lund
Wiggins IL (1959) Development of the ovule and megagametophyte in *Saxifraga hieracifolia*. Am J Bot 46: 692–697
Williams EG, Knox RB, Irvine D (eds) (1986) Pollination '86. School Bot, Univ Melbourne
Williams EG, O'Neill P, Hough T (1982) The pollen tube as an experimental system. In: Williams EG, Knox RB, Gilbert JH, Bernhardt P (eds) Pollination 1982. School Bot, Univ Melbourne, pp 15–26

Williams EG, Kaul V, Rouse JL, Palser BF (1986) Overgrowth of pollen tubes in embryo sacs of *Rhododendron* following interspecific pollinations. Aust J Bot 34: 413–423

Williams EJ (1955) Seed failure in the chippewa variety of *Solanum tuberosum*. Bot Gaz 117: 10–15

Williams ME (1932) The development of the embryo of *Kochia scoparia*. Bull Torrey Bot Club 59: 391–398

Willemse MTM (1981) Polarity during megasporogenesis and megagametogenesis. Phytomorphology 31: 124–134

Willemse MTM, Bednara J (1979) Polarity during megasporogenesis in *Gasteria verrucosa*. Phytomorphology 29: 156–165

Willemse MTM, Kapil RN (1981) On some anomalies during megasporogenesis in *Gasteria verrucosa* (Liliaceae). Acta Bot Neerl 30: 439–447

Willemse MTM, Van Went JL (1984) The female gametophyte. In: Johri BM (ed) Embryology of angiosperms. Springer, Berlin Heidelberg New York, pp 159–196

Willis JC (1973) A dictionary of the flowering plants and ferns. 8 edn (revised by Airy Shaw HK). Univ Press, Cambridge (England)

Wilms HJ (1980a) Ultrastructure of the stigma and style of spinach in relation to pollen germination and pollen tube growth. Acta Bot Neerl 29: 33–47

Wilms HJ (1980b) Development and composition of the spinach ovule. Acta Bot Neerl 29: 243–260

Wilms HJ (1981a) Pollen tube penetration and fertilization in spinach. Acta Bot Neerl 30: 101–122

Wilms HJ (1981b) Pollen tube growth through nucellus into embryo sac of *Spinacia*: An ultrastructural investigation. Acta Soc Bot Pol 50: 191–193

Wilsnack R, Warren JL (1967) Cytological observations on *Hypoxis* 3. Embryo development in *Hypoxis rooperii* and *H. filiformis*. J S Afr Bot 33: 133–140

Wilson TK (1966) Comparative morphology of the Canellaceae 4. Floral morphology and conclusions. Am J Bot 53: 336–343

Winkler H (1906) Botanische Untersuchungen aus Buitenzorg 2. Über Parthenogenesis bei *Wikstroemia indica*. Ann Jardin Bot Buitenz 5: 208–276

Winkler H (1908) Über Parthenogenesis und Apogamie im Pflanzenreich. Prog Rei Bot (Jena) 2: 293–454

Winter DM (1960) The development of the seed of *Abutilon theophrasti* 1. Ovule and embryo. Am J Bot 47: 8–14

Wirz H (1910) Beiträge zur Entwicklungsgeschichte von *Sciaphila* spec. und von *Epirrhizanthes elongata* Bl. Flora 101: 395–446

Witmer SW (1937) Morphology and cytology of *Vallisneria spiralis* L. Am Midl Nat 18: 309–333

Wodehouse RP (1936) Evolution of pollen grains. Bot Rev 2: 67–84

Wojciechowska B (1971) Embryogenesis in *Ornithopus* sp. Genet Pol 12: 61–75

Wolf FT (1940) Macrosporogenesis and development of the embryo sac in *Yucca aloifolia*. Bull Torrey Bot Club 67: 755–761

Wolter M, Seuffert C, Schill R (1988) The ontogeny of pollinia and elastoviscin in the anther of *Doritis pulcherrima* (Orchidaceae). Nord J Bot 8: 77–88

Woodburn WL (1911) Development of the embryo sac and endosperm in some seedless persimmons. Bull Torrey Bot Club 38: 379–383

Woodcock EF (1929) Seed studies in Nyctaginaceae. Papers Mich Acad Sci Arts Lett 9: 495–502

Woodcock CLF, Bell PR (1968) Features of the ultrastructure of the female gametophyte of *Myosurus minimus*. J Ultrastruct Res 22: 546–563

Woodland PS (1964) The floral morphology and embryology of *Themeda australis* (R. Br.) Stapf. Aust J Bot 12: 157–172

Woodroof JG (1924) The development of pecan buds and the quantitative production of pollen. Georgia Agric Exp Stn Bull 144: 134–161

Woodroof JG, Woodroof NC (1927) The development of the pecan nut (*Hicoria pecan*) from the flower to the maturity. J Agric Res 34: 1049–1063

Woodroof NC (1928) Development of the embryo sac and young embryo in *Hicoria pecan*. Am J Bot 15: 416–421

Woodson RE (1941) The north American Asclepiadaceae 1. Perspective of the genera. Ann Missouri Bot Garden 28: 193–244

Woodworth RH (1929) Cytological studies in the Betulaceae 2. *Corylus* and *Alnus*. Bot Gaz 88: 383–399

Woodworth RH (1930a) Cytological studies in the Betulaceae 3. Parthenogenesis and polyembryony in *Alnus rugosa*. Bot Gaz 89: 402–409

Woodworth RH (1930b) Cytological studies in the Betulaceae 4. *Betula, Carpinus, Ostrya, Ostryopsis*. Bot Gaz 90: 108–115

Worsdell WC (1908) The affinities of *Paeonia*. J Bot (London) 46: 114–116

Wu HK (1960) Embryogenesis in tea plant. Bot Bull Acad Sin Taipei 1: 165–168

Wulff HD (1939) Die Entwicklung der Pollenkörner von *Triglochin palustris* und die verschiendenen Typen der Pollenkornentwicklung des Angiospermen. Bot Jahrb 88: 141–168

Wulff HD, Raghavan TS (1937) Beobachtungen an Pollenschlauchkulturen von der Hydrophyllaceae *Nemophila insignis*. Planta 27: 466–473

Wunderlich R (1936) Vergleichende Untersuchungen von Pollenkörnern einiger Liliaceen und Amaryllidaceen. Öst Bot Z 85: 32–55

Wunderlich R (1937) Zur vergleichenden Embryologie der Liliaceae-Scilloideae. Flora 132: 48–90

Wunderlich R (1938) Ein künstlich bestäubter Fruchtknoten von *Yucca filamentosa*. Öst Bot Z 87: 109–113

Wunderlich R (1950) Die Agavaceae Hutchinsons im Lichte ihrer Embryologie, ihres Gynözeum-, Staubblatt- und Blattbaues. Öst Bot Z 97: 437–502

Wunderlich R (1954) Über das Antherentapetum mit besonderer Berücksichtigung seiner Kernzahl. Öst Bot Z 101: 1–63

Wunderlich R (1959) Zur Frage der Phylogenie der Endospermtypen bei den Angiospermen. Öst Bot Z 106: 203–293

Wunderlich R (1966) Zur Deutung der eigenartigen Embryoentwicklung von *Paeonia*. Öst Bot Z 113: 395–407

Wunderlich R (1967) Some remarks on the taxonomic significance of the seed-coat. Phytomorphology 17: 301–311

Wunderlich R (1971) Die systematische Stellung von *Theligonum*. Öst Bot Z 119: 329–394

Wyatt RL (1955) An embryological study of four species of *Asarum*. J Elisha Mitchell Sci Soc 71: 64–82

Wylie RB (1904) The morphology of *Elodea canadensis*. Bot Gaz 37: 1–22

Xiang-Yuan Xi, de-Cai Cui (1983) Relationship between pollen and embryo sac development in wheat, *Triticu.n aestivum* L. Bot Gaz 144: 191–200

Xiang-Yuan Xi, de Mason DA (1984) Relationship between male and female gametophyte development in rye. Am J Bot 71: 1067–1079

Yakovlev MS (1950) The structure of endosperm and embryo in cereals as a systematic feature. Izv Akad Nauk Arm SSR Bot 1: 121–218 (in Russian)

Yakovlev MS (1951) On the similarity of embryogenesis of angiosperms and gymnosperms Trans Bot Inst Akad Nauk 2: 356–363 (in Russian)

Yakovlev MS (1959) On a new type of embryogenesis in *Paeonia*. In: 9th Int Bot Congr, Seattle (USA) 2: 437–438

Yakovlev MS (1961) Further studies of the new type of embryogenesis in angiosperms. Bot Zh 46: 1402–1421

Yakovlev MS (1965) The embryology of the genus *Paeonia*. In: Yakovlev MS (ed) Flower morphology and reproductive processes of angiosperms. Nauka, Leningrad

Yakovlev MS (1967) Polyembryony in higher plants and principles of its classification. Phytomorphology (Panchanan Maheshwari Mem. Vol.) 17: 278–282

Yakovlev MS (1969) Embryogenesis and some problems of phylogenesis. Rev Cytol Biol Veg 32: 325–330

Yakovlev MS (1970) On the embryology of *Melocanna bambusoides* Turn. Ann Bogr 5: 109–115

Yakovlev MS (1974) Gametogenesis, embryo sac and pollen grains (Contribution to the origin in Angiospermae). In: Linskens HF (ed) Fertilization in higher plants. North-Holland, Amsterdam, pp 253–260

Yakovlev MS (ed) (1981) Comparative embryology of flowering plants, vol 1. Winteraceae-Juglandaceae. Nauka, Leningrad (in Russian)

Yakovlev MS (ed) (1983) Comparative embryology of flowering plants, vol 2. Phytolaccaceae-Thymelaeaceae. Nauka, Leningrad (in Russian)
Yakovlev MS, Alimova GK (1976) Embryogenesis in *Arabidopsis thaliana* (L.) Heynh (Cruciferae). Bot Zh 6: 12–24 (in Russian)
Yakovlev MS, Kamelina OP (1973) Development of the embryo of *Melocanna bambusoides* Trin. Bot Zh 58: 248–263
Yakovlev MS, Yoffe MD (1957) On some peculiar features in the embryogeny of *Paeonia* L. Phytomorphology 7: 74–82
Yakovlev MS, Yoffe MD (1961) Further studies of the new type of embryogenesis in angiosperms. Bot Zh 46: 1402–1421
Yakovlev MS, Zhukova GY (1975) On the presence of chlorophyll in embryos of angiosperm seeds. In: Mohan Ram HY, Shah JJ, Shah CK (eds) Form, structure and function in plants (Prof. B.M. Johri Comm. Vol.). Sarita Prakashan, Meerut (India), pp 223–229
Yakovlev MS, Zhukova GY (1980) Chlorophyll in embryos of angiosperm seeds: A review. Bot Not 133: 323–336
Yamada H, Yoshida O (1979) Embryological study of the family Ulmaceae 1. Embryology of *Zelkova serrata* Makino. J Coll Art Sci Chiba Univ (Nat Sci) 12: 27–43
Yamashita T (1976a) Über die Embryo- und Wurzelentwicklung bei *Aponogeton madagascariensis* (Mirbel) van Bruggen. J Fac Sci Univ Tokyo 12: 37–64
Yamashita T (1976b) Über die Pollenbildung bei *Halodule pinifolia* und *H. uninervis*. Beitr Biol Pflanzen 52: 217–226
Yamashita T (1983) Embryo- und Wurzelentwicklung bei *Commelina communis* mit besonderer Berücksichtigung der Coleorrhiza. Proc Int Symp Gumpenstein (A-8952 Irdning), pp 21–32
Yamazaki T (1953) On the floral structure, seed development, and affinities of *Deinostema*: a new genus of Scrophulariaceae 2. Bot Mag (Tokyo) 66: 141–149
Yamazaki T (1954) Notes on *Lindernia*, *Vandellia*, *Torenia* and their allied genera in eastern Asia 2. Bot Mag (Tokyo) 68: 14–23
Yamazaki T (1957a) Taxonomic and phylogenetic studies of Scrophulariaceae: Veronicae with special reference to *Veronica* and *Veronicastrum* in eastern Asia. J Fac Sci Univ Tokyo 7: 91–162
Yamazaki T (1957b) Seed formation of *Ellisiophyllum pinnatum* var. *reptans*. Bot Mag (Tokyo) 70: 162–168
Yamazaki T (1963) Embryology of *Mitrasacme alsinoides* var. *indica*. Sci Rep Tôhoku Univ Ser Biol 29: 201–205
Yamazaki T (1966) The embryology of *Shortia uniflora* with a brief review of the systematic position of the Dipensiaceae. Jpn J Bot 41: 245–251
Yamazaki T (1970a) Embryological studies in Ebenales 1. Styracaceae. Jpn J Bot 45: 267–273
Yamazaki T (1970b) Embryological studies in Ebenales 2. Symplocaceae. Jpn J Bot 45: 353–358
Yamazaki T (1971) Embryological studies in Ebenales 3. Sapotaceae. Jpn J Bot 46: 161–166
Yamazaki T (1972) Embryological studies in Ebenales 4. Ebenaceae. Jpn J Bot 47: 20–28
Yamazaki T (1974) A system of Gamopetalae based on the embryology. J Fac Sci Univ Tokyo 11: 263–281
Yamazaki T (1975) Floral anatomy and embryology of *Tripetaleia paniculata* and *T. bracteata* (Ericaceae). Bot Mag (Tokyo) 88: 267–279
Yamazaki T (1982) Recognized types in early development of the embryo and phylogeny in the dicotyledons. Acta Phytotax Geobot 33: 240–249 (in Japanese)
Yasugi S (1983) Ovule and embryo development in *Doritis pulcherrima* (Orchidaceae). Am J Bot 70: 555–560
Yasui K (1937) Karyological studies in *Magnolia*, with special reference to the cytokinesis in pollen mother cells. Bot Mag (Tokyo) 51: 539–546
Yasui K, Suita N (1939) A note on the refractive granules in the microspore mother cell and the microspores of *Tradescantia*. Bot Mag (Tokyo) 53: 521–524
Yasui K, Sawada N (1940) On the spore and embryo sac formation with special reference to the sterility of *Iris japonica*. Bot Mag (Tokyo) 54: 96–102 (in Japanese)
Yen TK (1950) Structure and development of the flower and the fruit of *Myrica rubra*. Peking Nat Hist Bull 19: 2–20

Yeung BA, Sherwood RT, Bashaw EC (1979) Cleared-pistil and thick-sectioning techniques for detecting aposporous apomixis in grasses. Can J Bot 57: 1668–1672

Yeung EC (1980) Embryogeny of *Phaseolus*: The role of the suspensor. Z Pflanzenphysiol 96: 17–28

Yeung EC, Cavey MJ (1988) Cellular endosperm formation in *Phaseolus vulgaris* 1. Light and scanning electron microscopy. Can J Bot 66: 1209–1216

Yeung EC, Clutter ME (1978) Embryogeny of *Phaseolus coccineus*: growth and microanatomy. Protoplasma 94: 19–40

Yoffe MD (1952) The presence of chlorophyll in the endosperm of cruciferous plants. Dokl Akad Nauk SSSR Ser Biol 82: 473–476

Yoffe MD (1957) Embryo and endosperm development in wheat, horse-bean, and radish. Trans Bot Inst Acad Nauk SSSR Sec VII 4: 211–269 (in Russian)

Yoffe MD (1962) On the embryology of *Trochodendron aralioides* Sieb. et Zucc.: Development of pollen and embryo sac. Acta Inst Bot Acad Sci URSS Morphol Anat 5: 250–259

Yoffe MD (1965) On the embryology of *Trochodendron aralioides*: Embryo and endosperm development. In: Yakovlev MS (ed) Flower morphology and reproductive processes of angiosperms. Nauka, Leningrad, pp 177–185

Yoffe MD (1969) Fertilization and formation of the primary endosperm nucleus in *Anemone nemorosa* L. Rev Cytol Biol Veg 32: 319–323

York HH (1913) The origin and development of the embryo sac and embryo of *Dendrophthora opuntioides* and *D. gracile*. Bot Gaz 56: 89–111, 200–216

Yoshida O (1957) Embryologische Studien über die Ordnung Piperales 1. Embryologie von *Chloranthus japonicus* Sieb. J Coll Art Sci Chiba Univ (Nat Sci) 2: 172–178

Yoshida O (1959) Embryologische Studien über die Ordnung Piperales 2. Embryologie von *Chloranthus serratus*. J Coll Art Sci Chiba Univ (Nat Sci) 2: 295–303

Yoshida O (1960) Embryologische Studien über die Ordnung Piperales 3. Embryologie von *Sarcandra glabra*. J Coll Art Sci Chiba Univ (Nat Sci) 3: 55–60

Yoshida O (1961) Embryologische Studien über die Ordnung Piperales 5. Embryologie von *Saururus loureiri* Decne. J Coll Art Sci Chiba Univ (Nat Sci) 3: 311–316

Yoshida O (1962) Embryologische Studien über *Schisandra chinensis*. J Coll Art Sci Chiba Univ (Nat Sci) 4: 459–462

Yoshida O, Kawaguchi H (1970) Embryology of *Korthalsella japonica* (Thunb.) Engler. J Coll Art Sci Chiba Univ (Nat Sci) 4: 37–47

Yoshida O, Michikawa A (1973) Embryological studies of the genus *Akebia* Decne. J Coll Art Sci Chiba Univ (Nat Sci) 6: 25–37

Yoshida O, Nakajima Y (1978) Embryological study on *Stauntonia hexaphylla* Decne. J Coll Art Sci Chiba Univ (Nat Sci) 11: 45–57

Yoshida O, Yamada H (1973) Embryological studies of *Cocculus trilobus* DC. J Coll Art Sci Chiba Univ (Nat Sci) 6: 39–57

Young WJ (1922) Potato ovules with twin embryo sacs. Am J Bot 9: 213–214

Yu Sh, Chao Ch Y (1979) Histochemical studies of ovary tissues during the embryo sac development in *Paspalum longifolium* Roxb. Caryologia 32: 147–160

Yurukova HGR, Ncharova PD (1978) Female gametophyte development of two species from the genus *Leontodon* L. Fitologia 9: 52–69 (in Bulgarian)

Zabinska D (1972) Embryo sac and endosperm development in some representatives of the group *Ornithogalum gussonei* from Poland. Acta Biol Cracov Ser Bot 15: 177–190

Zaman B (1950) The embryology of *Juncus prismatocarpus* Br. and *J. effusus* Linn. Proc Indian Acad Sci B 31: 223–234

Zhalatov G, Dzevaltovskii AK (1975) Cytoembryology of the West Indian gherkin (*Cucumis anguria* L.). Uzb Biol Zh 19: 41–43 (in Russian)

Zavada MS (1983) Comparative morphology of monocot pollen and evolutionary trends of apertures and wall structures. Bot Rev 49: 331–379

Zhou C, Yang HY (1982) Enzymatic isolation of embryo sacs in angiosperms: isolation and microscopical observation on fixed materials. Acta Bot Sin 24: 403–407

Zhou C, Yang HY (1985) Observations on enzymatically isolated, living and fixed embryo sacs in several angiosperm species. Planta 165: 225–231

Zhou C, Orndorff K, Allen RD, De Maggio AE (1986) Direct observations on generative cells isolated from pollen grains of *Haemanthus katherinae* Baker. Plant Cell Rep 5: 306–309

Zhukova GYA, Sokolovskaya RB (1977) Ultrastructure of antipodals of *Aconitum napellus* L. (Ranunculaceae) embryo sac before fertilization. Bot Zh 62: 1600–1611

Ziegler A (1925) Beiträge zur Kenntnis des Androeciums und der Samenentwicklung einiger Melastomaceen. Bot Arch Mez 9: 398–467

Zweifel R (1939) Cytologisch-embryologische Untersuchungen *Balanophora àbbreviata* Blume und *B. indica* Wall. Vierteljahresschr Naturf Ges Zürich 84: 245–306

Plant Index

GPSR Compliance

The European Union's (EU) General Product Safety Regulation (GPSR) is a set of rules that requires consumer products to be safe and our obligations to ensure this.

If you have any concerns about our products, you can contact us on ProductSafety@springernature.com

In case Publisher is established outside the EU, the EU authorized representative is:

Springer Nature Customer Service Center GmbH
Europaplatz 3
69115 Heidelberg, Germany

Batch number: 10755036

Printed by Printforce, the Netherlands